The Elements

Name	Symbol	Atomic Number	Atomic Mass*
Actinium	Ac	89	(227)
Aluminum	Al	13	26.98
Americium	Am	95	(243)
Antimony	Sb	51	121.8
Argon	Ar	18	39.95
Arsenic	As	33	74.92
Astatine	At	85	(210)
Barium	Ba	56	137.3
Berkelium	Bk	97	(247)
Beryllium	Be	4	9.012
Bismuth	Bi	83	209.0
Bohrium	Bh	107	(267)
Boron	B	5	10.81
Bromine	Br	35	79.90
Cadmium	Cd	48	112.4
Calcium	Ca	20	40.08
Californium	Cf	98	(249)
Carbon	C	6	12.01
Cerium	Ce	58	140.1
Cesium	Cs	55	132.9
Chlorine	Cl	17	35.45
Chromium	Cr	24	52.00
Cobalt	Co	27	58.93
Copper	Cu	29	63.55
Curium	Cm	96	(247)
Darmstadtium	Ds	110	(281)
Dubnium	Db	105	(262)
Dysprosium	Dy	66	162.5
Einsteinium	Es	99	(254)
Erbium	Er	68	167.3
Europium	Eu	63	152.0
Fermium	Fm	100	(253)
Fluorine	F	9	19.00
Francium	Fr	87	(223)
Gadolinium	Gd	64	157.3
Gallium	Ga	31	69.72
Germanium	Ge	32	72.61
Gold	Au	79	197.0
Hafnium	Hf	72	178.5
Hassium	Hs	108	(277)
Helium	He	2	4.003
Holmium	Ho	67	164.9
Hydrogen	H	1	1.008
Indium	In	49	114.8
Iodine	I	53	126.9
Iridium	Ir	77	192.2
Iron	Fe	26	55.85
Krypton	Kr	36	83.80
Lanthanum	La	57	138.9
Lawrencium	Lr	103	(257)
Lead	Pb	82	207.2
Lithium	Li	3	6.941
Lutetium	Lu	71	175.0
Magnesium	Mg	12	24.31
Manganese	Mn	25	54.94
Meitnerium	Mt	109	(268)
Mendelevium	Md	101	(256)

Name	Symbol	Atomic Number	Atomic Mass*
Mercury	Hg	80	200.6
Molybdenum	Mo	42	95.94
Neodymium	Nd	60	144.2
Neon	Ne	10	20.18
Neptunium	Np	93	(244)
Nickel	Ni	28	58.70
Niobium	Nb	41	92.91
Nitrogen	N	7	14.01
Nobelium	No	102	(253)
Osmium	Os	76	190.2
Oxygen	O	8	16.00
Palladium	Pd	46	106.4
Phosphorus	P	15	30.97
Platinum	Pt	78	195.1
Plutonium	Pu	94	(242)
Polonium	Po	84	(209)
Potassium	K	19	39.10
Praseodymium	Pr	59	140.9
Promethium	Pm	61	(145)
Protactinium	Pa	91	(231)
Radium	Ra	88	(226)
Radon	Rn	86	(222)
Rhenium	Re	75	186.2
Rhodium	Rh	45	102.9
Roentgenium	Rg	111	(272)
Rubidium	Rb	37	85.47
Ruthenium	Ru	44	101.1
Rutherfordium	Rf	104	(263)
Samarium	Sm	62	150.4
Scandium	Sc	21	44.96
Seaborgium	Sg	106	(266)
Selenium	Se	34	78.96
Silicon	Si	14	28.09
Silver	Ag	47	107.9
Sodium	Na	11	22.99
Strontium	Sr	38	87.62
Sulfur	S	16	32.07
Tantalum	Ta	73	180.9
Technetium	Tc	43	(98)
Tellurium	Te	52	127.6
Terbium	Tb	65	158.9
Thallium	Tl	81	204.4
Thorium	Th	90	232.0
Thulium	Tm	69	168.9
Tin	Sn	50	118.7
Titanium	Ti	22	47.88
Tungsten	W	74	183.9
Uranium	U	92	238.0
Vanadium	V	23	50.94
Xenon	Xe	54	131.3
Ytterbium	Yb	70	173.0
Yttrium	Y	39	88.91
Zinc	Zn	30	65.41
Zirconium	Zr	40	91.22
		112**	(285)
		113	(284)
		114	(289)
		115	(288)
		116	(292)

*All atomic masses are given to four significant figures. Values in parentheses represent the mass number of the most stable isotope.

**The names and symbols for elements 112 through 116 have not been chosen.

FIFTH EDITION

CHEMISTRY

The Molecular Nature of Matter and Change

Martin S. Silberberg

Annotations by John Pollard, Arizona State University

McGraw-Hill
Higher Education

Boston   Burr Ridge, IL   Dubuque, IA   New York   San Francisco   St. Louis
Bangkok   Bogotá   Caracas   Kuala Lumpur   Lisbon   London   Madrid   Mexico City
Milan   Montreal   New Delhi   Santiago   Seoul   Singapore   Sydney   Taipei   Toronto

McGraw-Hill
Higher Education

CHEMISTRY: THE MOLECULAR NATURE OF MATTER AND CHANGE, FIFTH EDITION

Published by McGraw-Hill, a business unit of The McGraw-Hill Companies, Inc., 1221 Avenue of the Americas, New York, NY 10020. Copyright © 2009 by The McGraw-Hill Companies, Inc. All rights reserved. Previous editions © 2006, 2003, 2000, and 1996. No part of this publication may be reproduced or distributed in any form or by any means, or stored in a database or retrieval system, without the prior written consent of The McGraw-Hill Companies, Inc., including, but not limited to, in any network or other electronic storage or transmission, or broadcast for distance learning.

Some ancillaries, including electronic and print components, may not be available to customers outside the United States.

This book is printed on acid-free paper.

1 2 3 4 5 6 7 8 9 0 DOW/DOW 0 9 8

ISBN 978–0–07–304859–8
MHID 0–07–304859–3

ISBN 978–0–07–304862–8 (Annotated Instructor's Edition)
MHID 0–07–304862–3

Publisher: *Thomas D. Timp*
Senior Sponsoring Editor: *Tamara Good-Hodge*
Senior Developmental Editor: *Donna Nemmers*
Freelance Developmental Services: *Karen Pluemer*
Senior Marketing Manager: *Todd Turner*
Lead Project Manager: *Peggy J. Selle*
Lead Production Supervisor: *Sandy Ludovissy*
Lead Media Project Manager: *Stacy A. Patch*
Senior Designer: *David W. Hash*
USE Cover Image: *Michael Goodman; based on research from the laboratory of Charles M. Lieber,
    Department of Chemistry and Chemical Biology, Harvard University*
Illustrations: *Federico/Goodman Studios; Daniel Silberberg*
Page Layout/Special Features Designer: *Ruth Melnick*
Senior Photo Research Coordinator: *Lori Hancock*
Photo Research: *Jerry Marshall*/www.pictureresearching.com
Supplement Producer: *Mary Jane Lampe*
Compositor: *Aptara*
Typeface: *10.5/12 Times Roman*
Printer: *R. R. Donnelley Willard, OH*

The credits section for this book begins on page C-1 and is considered an extension of the copyright page.

Library of Congress Cataloging-in-Publication Data

Silberberg, Martin S. (Martin Stuart), 1945–
  Chemistry : the molecular nature of matter and change / Martin S. Silberberg. — 5th ed.
    p. cm.
  Includes index.
    ISBN 978–0–07–304859–8 — ISBN 0–07–304859–3 (hard copy : alk. paper) 1. Chemistry. I. Title.
QD33.2.S55 2009
540—dc22
                                                                    2007039578

To Ruth and Daniel,
for all their love and confidence—and patience

BRIEF CONTENTS

Preface xvi

Acknowledgments xxv

1   Keys to the Study of Chemistry 2

2   The Components of Matter 40

3   Stoichiometry of Formulas and Equations 89

4   Three Major Classes of Chemical Reactions 140

5   Gases and the Kinetic-Molecular Theory 186

6   Thermochemistry: Energy Flow and Chemical Change 235

7   Quantum Theory and Atomic Structure 268

8   Electron Configuration and Chemical Periodicity 302

9   Models of Chemical Bonding 340

10   The Shapes of Molecules 377

11   Theories of Covalent Bonding 410

12   Intermolecular Forces: Liquids, Solids, and Phase Changes 436

13   The Properties of Mixtures: Solutions and Colloids 500

Interchapter: A Perspective on the Properties of the Elements 553

14   Periodic Patterns in the Main-Group Elements 564

15   Organic Compounds and the Atomic Properties of Carbon 628

16   Kinetics: Rates and Mechanisms of Chemical Reactions 684

17   Equilibrium: The Extent of Chemical Reactions 737

18   Acid-Base Equilibria 782

19   Ionic Equilibria in Aqueous Systems 831

20   Thermodynamics: Entropy, Free Energy, and the Direction of Chemical Reactions 880

21   Electrochemistry: Chemical Change and Electrical Work 922

22   The Elements in Nature and Industry 980

23   The Transition Elements and Their Coordination Compounds 1022

24   Nuclear Reactions and Their Applications 1064

Appendix A Common Mathematical Operations in Chemistry A-1

Appendix B Standard Thermodynamic Values for Selected Substances A-5

Appendix C Equilibrium Constants for Selected Substances A-8

Appendix D Standard Electrode (Half-Cell) Potentials A-14

Appendix E Answers to Selected Problems A-15

Glossary G-1

Credits C-1

Index I-1

DETAILED CONTENTS

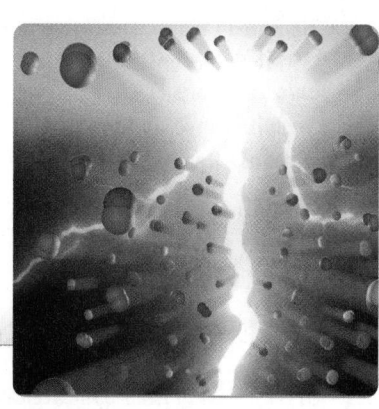

1 CHAPTER

Keys to the Study of Chemistry 2

1.1 Some Fundamental Definitions 4
1.2 Chemical Arts and the Origins of Modern Chemistry 10
1.3 The Scientific Approach: Developing a Model 12
1.4 Chemical Problem Solving 14

1.5 Measurement in Scientific Study 18
1.6 Uncertainty in Measurement: Significant Figures 27
CHAPTER PERSPECTIVE 32

Chemical Connections to Interdisciplinary Science: Chemistry Problem Solving in the Real World 33
CHAPTER REVIEW GUIDE 34
PROBLEMS 35

2 CHAPTER

The Components of Matter 40

2.1 Elements, Compounds, and Mixtures: An Atomic Overview 41
2.2 The Observations That Led to an Atomic View of Matter 44
2.3 Dalton's Atomic Theory 47
2.4 The Observations That Led to the Nuclear Atom Model 48
2.5 The Atomic Theory Today 52

Tools of the Laboratory: Mass Spectrometry 55
2.6 Elements: A First Look at the Periodic Table 57
2.7 Compounds: Introduction to Bonding 60
2.8 Compounds: Formulas, Names, and Masses 64
GALLERY: Picturing Molecules 74

2.9 Mixtures: Classification and Separation 75
Tools of the Laboratory: Basic Separation Techniques 76
CHAPTER PERSPECTIVE 78
CHAPTER REVIEW GUIDE 79
PROBLEMS 81

3 CHAPTER

Stoichiometry of Formulas and Equations 89

3.1 The Mole 90
3.2 Determining the Formula of an Unknown Compound 98
3.3 Writing and Balancing Chemical Equations 104

3.4 Calculating Amounts of Reactant and Product 109
3.5 Fundamentals of Solution Stoichiometry 121

CHAPTER PERSPECTIVE 127
CHAPTER REVIEW GUIDE 128
PROBLEMS 131

4 CHAPTER

Three Major Classes of Chemical Reactions 140

4.1 The Role of Water as a Solvent 141
4.2 Writing Equations for Aqueous Ionic Reactions 145
4.3 Precipitation Reactions 146
4.4 Acid-Base Reactions 150

4.5 Oxidation-Reduction (Redox) Reactions 158
4.6 Elements in Redox Reactions 166
4.7 Reaction Reversibility and the Equilibrium State 173

CHAPTER PERSPECTIVE 175
CHAPTER REVIEW GUIDE 176
PROBLEMS 178

5 CHAPTER

Gases and the Kinetic-Molecular Theory 186

5.1 An Overview of the Physical States of Matter 187
5.2 Gas Pressure and Its Measurement 189
5.3 The Gas Laws and Their Experimental Foundations 193
5.4 Further Applications of the Ideal Gas Law 203

5.5 The Ideal Gas Law and Reaction Stoichiometry 208
5.6 The Kinetic-Molecular Theory: A Model for Gas Behavior 210
■ *Chemical Connections to Planetary Science:* Structure and Composition of Earth's Atmosphere 218

5.7 Real Gases: Deviations from Ideal Behavior 221
CHAPTER PERSPECTIVE 224
CHAPTER REVIEW GUIDE 224
PROBLEMS 227

6 CHAPTER

Thermochemistry: Energy Flow and Chemical Change 235

6.1 Forms of Energy and Their Interconversion 236
6.2 Enthalpy: Heats of Reaction and Chemical Change 243
6.3 Calorimetry: Laboratory Measurement of Heats of Reaction 246

6.4 Stoichiometry of Thermochemical Equations 249
6.5 Hess's Law of Heat Summation 251
6.6 Standard Heats of Reaction ($\Delta H^{\circ}_{\text{rxn}}$) 253

■ *Chemical Connections to Environmental Science:* The Future of Energy Use 256
CHAPTER PERSPECTIVE 259
CHAPTER REVIEW GUIDE 260
PROBLEMS 261

7 CHAPTER

Quantum Theory and Atomic Structure 268

7.1 The Nature of Light 269
7.2 Atomic Spectra 276
■ *Tools of the Laboratory:* Spectrophotometry in Chemical Analysis 281

7.3 The Wave-Particle Duality of Matter and Energy 283
7.4 The Quantum-Mechanical Model of the Atom 287

CHAPTER PERSPECTIVE 295
CHAPTER REVIEW GUIDE 295
PROBLEMS 297

8 CHAPTER

Electron Configuration and Chemical Periodicity 302

8.1 Development of the Periodic Table 303
8.2 Characteristics of Many-Electron Atoms 304
8.3 The Quantum-Mechanical Model and the Periodic Table 308

8.4 Trends in Three Key Atomic Properties 317
8.5 Atomic Structure and Chemical Reactivity 325

CHAPTER PERSPECTIVE 334
CHAPTER REVIEW GUIDE 334
PROBLEMS 336

9 CHAPTER

Models of Chemical Bonding 340

9.1 Atomic Properties and Chemical Bonds 341
9.2 The Ionic Bonding Model 344
9.3 The Covalent Bonding Model 351
9.4 Bond Energy and Chemical Change 356

■ Tools of the Laboratory: Infrared Spectroscopy 357
9.5 Between the Extremes: Electronegativity and Bond Polarity 363

9.6 An Introduction to Metallic Bonding 369
CHAPTER PERSPECTIVE 371
CHAPTER REVIEW GUIDE 371
PROBLEMS 373

10 CHAPTER

The Shapes of Molecules 377

10.1 Depicting Molecules and Ions with Lewis Structures 378
10.2 Valence-Shell Electron-Pair Repulsion (VSEPR) Theory and Molecular Shape 388
GALLERY: Molecular Beauty: Odd Shapes with Useful Functions 398

10.3 Molecular Shape and Molecular Polarity 399
CHAPTER PERSPECTIVE 401

■ Chemical Connections to Sensory Physiology: Molecular Shape, Biological Receptors, and the Sense of Smell 402

CHAPTER REVIEW GUIDE 404
PROBLEMS 405

11 CHAPTER

Theories of Covalent Bonding 410

11.1 Valence Bond (VB) Theory and Orbital Hybridization 411
11.2 The Mode of Orbital Overlap and the Types of Covalent Bonds 418

11.3 Molecular Orbital (MO) Theory and Electron Delocalization 422

CHAPTER PERSPECTIVE 430
CHAPTER REVIEW GUIDE 431
PROBLEMS 432

12 CHAPTER

Intermolecular Forces: Liquids, Solids, and Phase Changes 436

12.1 An Overview of Physical States and Phase Changes 437
12.2 Quantitative Aspects of Phase Changes 440
12.3 Types of Intermolecular Forces 450
12.4 Properties of the Liquid State 457
GALLERY: Properties of a Liquid 459

12.5 The Uniqueness of Water 460
12.6 The Solid State: Structure, Properties, and Bonding 463
■ *Tools of the Laboratory:* X-Ray Diffraction Analysis and Scanning Tunneling Microscopy 468

12.7 Advanced Materials 476
CHAPTER PERSPECTIVE 491
CHAPTER REVIEW GUIDE 492
PROBLEMS 493

13 CHAPTER

The Properties of Mixtures: Solutions and Colloids 500

13.1 Types of Solutions: Intermolecular Forces and Solubility 502
13.2 Intermolecular Forces and Biological Macromolecules 507
13.3 Why Substances Dissolve: Understanding the Solution Process 514
13.4 Solubility as an Equilibrium Process 519

13.5 Quantitative Ways of Expressing Concentration 522
13.6 Colligative Properties of Solutions 527
GALLERY: Colligative Properties in Industry and Biology 533
13.7 The Structure and Properties of Colloids 539

■ *Chemical Connections to Environmental Engineering:* Solutions and Colloids in Water Purification 541
CHAPTER PERSPECTIVE 543
CHAPTER REVIEW GUIDE 543
PROBLEMS 546

Interchapter

A Perspective on the Properties of the Elements 553

Topic 1 The Key Atomic Properties 554
Topic 2 Characteristics of Chemical Bonding 556
Topic 3 Metallic Behavior 558

Topic 4 Acid-Base Behavior of the Element Oxides 559
Topic 5 Redox Behavior of the Elements 560

Topic 6 Physical States and Phase Changes 562

14 CHAPTER

Periodic Patterns in the Main-Group Elements 564

14.1 Hydrogen, the Simplest Atom 565
14.2 Trends Across the Periodic Table: The Period 2 Elements 567
14.3 Group 1A(1): The Alkali Metals 570
14.4 Group 2A(2): The Alkaline Earth Metals 574
14.5 Group 3A(13): The Boron Family 578

14.6 Group 4A(14): The Carbon Family 584
GALLERY: Silicate Minerals and Silicone Polymers 592
14.7 Group 5A(15): The Nitrogen Family 595
14.8 Group 6A(16): The Oxygen Family 603

14.9 Group 7A(17): The Halogens 610
14.10 Group 8A(18): The Noble Gases 617
CHAPTER PERSPECTIVE 619
CHAPTER REVIEW GUIDE 619
PROBLEMS 620

15 CHAPTER

Organic Compounds and the Atomic Properties of Carbon 628

15.1 The Special Nature of Carbon and the Characteristics of Organic Molecules 629

15.2 The Structures and Classes of Hydrocarbons 632

■ *Chemical Connections to Sensory Physiology:* Geometric Isomers and the Chemistry of Vision 642

■ *Tools of the Laboratory:* Nuclear Magnetic Resonance (NMR) Spectroscopy 646

15.3 Some Important Classes of Organic Reactions 646

15.4 Properties and Reactivities of Common Functional Groups 650

15.5 The Monomer-Polymer Theme I: Synthetic Macromolecules 662

15.6 The Monomer-Polymer Theme II: Biological Macromolecules 665

CHAPTER PERSPECTIVE 673

■ *Chemical Connections to Genetics and Forensics:* DNA Sequencing and Fingerprinting 674

CHAPTER REVIEW GUIDE 676

PROBLEMS 678

16 CHAPTER

Kinetics: Rates and Mechanisms of Chemical Reactions 684

16.1 Factors That Influence Reaction Rate 686

16.2 Expressing the Reaction Rate 687

16.3 The Rate Law and Its Components 691

■ *Tools of the Laboratory:* Measuring Reaction Rates 692

16.4 Integrated Rate Laws: Concentration Changes over Time 699

16.5 The Effect of Temperature on Reaction Rate 705

16.6 Explaining the Effects of Concentration and Temperature 708

16.7 Reaction Mechanisms: Steps in the Overall Reaction 714

16.8 Catalysis: Speeding Up a Chemical Reaction 720

■ *Chemical Connections to Enzymology:* Kinetics and Function of Biological Catalysts 723

■ *Chemical Connections to Atmospheric Science:* Depletion of the Earth's Ozone Layer 725

CHAPTER PERSPECTIVE 726

CHAPTER REVIEW GUIDE 726

PROBLEMS 728

17 CHAPTER

Equilibrium: The Extent of Chemical Reactions 737

17.1 The Equilibrium State and the Equilibrium Constant 738

17.2 The Reaction Quotient and the Equilibrium Constant 741

17.3 Expressing Equilibria with Pressure Terms: Relation Between K_c and K_p 748

17.4 Reaction Direction: Comparing Q and K 749

17.5 How to Solve Equilibrium Problems 752

17.6 Reaction Conditions and the Equilibrium State: Le Châtelier's Principle 761

■ *Chemical Connections to Cellular Metabolism:* Design and Control of a Metabolic Pathway 770

■ *Chemical Connections to Industrial Production:* The Haber Process for the Synthesis of Ammonia 771

CHAPTER PERSPECTIVE 772

CHAPTER REVIEW GUIDE 773

PROBLEMS 775

18 CHAPTER

Acid-Base Equilibria 782

18.1 Acids and Bases in Water 784
18.2 Autoionization of Water and the pH Scale 789
18.3 Proton Transfer and the Brønsted-Lowry Acid-Base Definition 793
18.4 Solving Problems Involving Weak-Acid Equilibria 798

18.5 Weak Bases and Their Relation to Weak Acids 805
18.6 Molecular Properties and Acid Strength 810
18.7 Acid-Base Properties of Salt Solutions 812
18.8 Generalizing the Brønsted-Lowry Concept: The Leveling Effect 816

18.9 Electron-Pair Donation and the Lewis Acid-Base Definition 817
CHAPTER PERSPECTIVE 821
CHAPTER REVIEW GUIDE 821
PROBLEMS 823

19 CHAPTER

Ionic Equilibria in Aqueous Systems 831

19.1 Equilibria of Acid-Base Buffer Systems 832
19.2 Acid-Base Titration Curves 841
19.3 Equilibria of Slightly Soluble Ionic Compounds 851

■Chemical Connections to Geology:
Creation of a Limestone Cave 859

19.4 Equilibria Involving Complex Ions 862

■Chemical Connections to Environmental Science:
The Acid-Rain Problem 863

CHAPTER PERSPECTIVE 870
CHAPTER REVIEW GUIDE 870
PROBLEMS 872

20 CHAPTER

Thermodynamics: Entropy, Free Energy, and the Direction of Chemical Reactions 880

20.1 The Second Law of Thermodynamics: Predicting Spontaneous Change 881
20.2 Calculating the Change in Entropy of a Reaction 893

■Chemical Connections to Biology:
Do Living Things Obey the Laws of Thermodynamics? 897

20.3 Entropy, Free Energy, and Work 899

■Chemical Connections to Biological Energetics: The Universal Role of ATP 908

20.4 Free Energy, Equilibrium, and Reaction Direction 909
CHAPTER PERSPECTIVE 914
CHAPTER REVIEW GUIDE 914
PROBLEMS 916

21 CHAPTER

Electrochemistry: Chemical Change and Electrical Work 922

21.1 Redox Reactions and Electrochemical Cells 923
21.2 Voltaic Cells: Using Spontaneous Reactions to Generate Electrical Energy 929
21.3 Cell Potential: Output of a Voltaic Cell 934
21.4 Free Energy and Electrical Work 943

21.5 Electrochemical Processes in Batteries 952
21.6 Corrosion: A Case of Environmental Electrochemistry 956
21.7 Electrolytic Cells: Using Electrical Energy to Drive Nonspontaneous Reactions 959

■Chemical Connections to Biological Energetics: Cellular Electrochemistry and the Production of ATP 967
CHAPTER PERSPECTIVE 969
CHAPTER REVIEW GUIDE 969
PROBLEMS 972

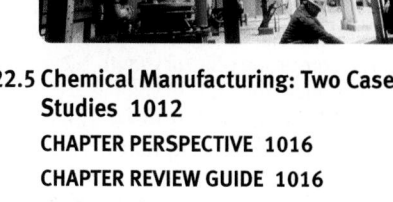

22 CHAPTER

The Elements in Nature and Industry 980

22.1 How the Elements Occur in Nature 981

22.2 The Cycling of Elements Through the Environment 986

22.3 Metallurgy: Extracting a Metal from Its Ore 993

22.4 Tapping the Crust: Isolation and Uses of Selected Elements 998

22.5 Chemical Manufacturing: Two Case Studies 1012

CHAPTER PERSPECTIVE 1016

CHAPTER REVIEW GUIDE 1016

PROBLEMS 1017

23 CHAPTER

The Transition Elements and Their Coordination Compounds 1022

23.1 Properties of the Transition Elements 1024

23.2 The Inner Transition Elements 1030

23.3 Highlights of Selected Transition Metals 1032

23.4 Coordination Compounds 1037

23.5 Theoretical Basis for the Bonding and Properties of Complexes 1046

CHAPTER PERSPECTIVE 1054

Chemical Connections to Nutritional Science: Transition Metals as Essential Dietary Trace Elements 1055

CHAPTER REVIEW GUIDE 1057

PROBLEMS 1058

24 CHAPTER

Nuclear Reactions and Their Applications 1064

24.1 Radioactive Decay and Nuclear Stability 1066

24.2 The Kinetics of Radioactive Decay 1074

Tools of the Laboratory: Counters for the Detection of Radioactive Emissions 1075

24.3 Nuclear Transmutation: Induced Changes in Nuclei 1080

24.4 The Effects of Nuclear Radiation on Matter 1082

24.5 Applications of Radioisotopes 1087

24.6 The Interconversion of Mass and Energy 1090

24.7 Applications of Fission and Fusion 1094

Chemical Connections to Cosmology: Origin of the Elements in the Stars 1099

CHAPTER PERSPECTIVE AND EPILOG 1101

CHAPTER REVIEW GUIDE 1101

PROBLEMS 1103

Appendix A Common Mathematical Operations in Chemistry A-1

Appendix B Standard Thermodynamic Values for Selected Substances A-5

Appendix C Equilibrium Constants for Selected Substances A-8

Appendix D Standard Electrode (Half-Cell) Potentials A-14

Appendix E Answers to Selected Problems A-15

Glossary G-1

Credits C-1

Index I-1

LIST OF SAMPLE PROBLEMS (Molecular-scene problems are shown in color.)

Chapter 1
1.1 Visualizing Change on the Atomic Scale 5
1.2 Distinguishing Between Physical and Chemical Change 7
1.3 Converting Units of Length 17
1.4 Converting Units of Volume 21
1.5 Converting Units of Mass 22
1.6 Calculating Density from Mass and Volume 24
1.7 Converting Units of Temperature 26
1.8 Determining the Number of Significant Figures 28
1.9 Significant Figures and Rounding 31

Chapter 2
2.1 Distinguishing Elements, Compounds, and Mixtures at the Atomic Scale 43
2.2 Calculating the Mass of an Element in a Compound 45
2.3 Visualizing the Mass Laws 48
2.4 Determining the Number of Subatomic Particles in the Isotopes of an Element 54
2.5 Calculating the Atomic Mass of an Element 56
2.6 Predicting the Ion an Element Forms 62
2.7 Naming Binary Ionic Compounds 66
2.8 Determining Formulas of Binary Ionic Compounds 67
2.9 Determining Names and Formulas of Ionic Compounds of Elements That Form More Than One Ion 68
2.10 Determining Names and Formulas of Ionic Compounds Containing Polyatomic Ions 69
2.11 Recognizing Incorrect Names and Formulas of Ionic Compounds 69
2.12 Determining Names and Formulas of Anions and Acids 70
2.13 Determining Names and Formulas of Binary Covalent Compounds 71
2.14 Recognizing Incorrect Names and Formulas of Binary Covalent Compounds 71
2.15 Calculating the Molecular Mass of a Compound 72
2.16 Using Molecular Depictions to Determine Formula, Name, and Mass 73

Chapter 3
3.1 Calculating the Mass in a Given Number of Moles of an Element 94
3.2 Calculating Number of Atoms in a Given Mass of an Element 95
3.3 Calculating the Moles and Number of Formula Units in a Given Mass of a Compound 96
3.4 Calculating Mass Percents and Masses of Elements in a Sample of a Compound 97

3.5 Determining an Empirical Formula from Masses of Elements 99
3.6 Determining a Molecular Formula from Elemental Analysis and Molar Mass 100
3.7 Determining a Molecular Formula from Combustion Analysis 101
3.8 Balancing Chemical Equations 107
3.9 Balancing an Equation from a Molecular Depiction 108
3.10 Calculating Amounts of Reactants and Products 111
3.11 Writing an Overall Equation for a Reaction Sequence 113
3.12 Using Molecular Depictions to Solve a Limiting-Reactant Problem 115
3.13 Calculating Amounts of Reactant and Product in a Limiting-Reactant Problem 117
3.14 Calculating Percent Yield 119
3.15 Calculating the Molarity of a Solution 121
3.16 Calculating Mass of Solute in a Given Volume of Solution 122
3.17 Preparing a Dilute Solution from a Concentrated Solution 124
3.18 Visualizing Changes in Concentration 125
3.19 Calculating Amounts of Reactants and Products for a Reaction in Solution 125
3.20 Solving Limiting-Reactant Problems for Reactions in Solution 126

Chapter 4
4.1 Determining Moles of Ions in Aqueous Ionic Solutions 143
4.2 Predicting Whether a Precipitation Reaction Occurs; Writing Ionic Equations 148
4.3 Using Molecular Depictions to Understand a Precipitation Reaction 149
4.4 Determining the Molarity of H^+ (or OH^-) Ions 152
4.5 Writing Ionic Equations for Acid-Base Reactions 153
4.6 Finding the Concentration of Acid from an Acid-Base Titration 154
4.7 Determining the Oxidation Number of an Element 160
4.8 Identifying Redox Reactions 161
4.9 Recognizing Oxidizing and Reducing Agents 162
4.10 Balancing Redox Equations by the Oxidation Number Method 163
4.11 Finding a Concentration by a Redox Titration 165
4.12 Identifying the Type of Redox Reaction 172

Chapter 5
5.1 Converting Units of Pressure 192
5.2 Applying the Volume-Pressure Relationship 199
5.3 Applying the Pressure-Temperature Relationship 200
5.4 Applying the Volume-Amount Relationship 200
5.5 Solving for an Unknown Gas Variable at Fixed Conditions 201
5.6 Using Gas Laws to Determine a Balanced Equation 202
5.7 Calculating Gas Density 203
5.8 Finding the Molar Mass of a Volatile Liquid 205
5.9 Applying Dalton's Law of Partial Pressures 206
5.10 Calculating the Amount of Gas Collected over Water 207
5.11 Using Gas Variables to Find Amounts of Reactants or Products 209
5.12 Using the Ideal Gas Law in a Limiting-Reactant Problem 209
5.13 Applying Graham's Law of Effusion 216

Chapter 6
6.1 Determining the Change in Internal Energy of a System 241
6.2 Drawing Enthalpy Diagrams and Determining the Sign of ΔH 245
6.3 Finding Quantity of Heat from Specific Heat Capacity 247
6.4 Determining the Heat of a Reaction 247
6.5 Calculating the Heat of a Combustion Reaction 249
6.6 Using the Heat of Reaction (ΔH_{rxn}) to Find Amounts 250
6.7 Using Hess's Law to Calculate an Unknown H 252
6.8 Writing Formation Equations 254
6.9 Calculating the Heat of Reaction from Heats of Formation 255

Chapter 7
7.1 Interconverting Wavelength and Frequency 271
7.2 Calculating the Energy of Radiation from Its Wavelength 275
7.3 Determining ΔE and λ of an Electron Transition 280
7.4 Calculating the de Broglie Wavelength of an Electron 284
7.5 Applying the Uncertainty Principle 286
7.6 Determining Quantum Numbers for an Energy Level 290
7.7 Determining Sublevel Names and Orbital Quantum Numbers 291
7.8 Identifying Incorrect Quantum Numbers 291

Chapter 8

8.1 Determining Quantum Numbers from Orbital Diagrams 310
8.2 Determining Electron Configurations 316
8.3 Ranking Elements by Atomic Size 320
8.4 Ranking Elements by First Ionization Energy 323
8.5 Identifying an Element from Successive Ionization Energies 324
8.6 Writing Electron Configurations of Main-Group Ions 329
8.7 Writing Electron Configurations and Predicting Magnetic Behavior of Transition Metal Ions 331
8.8 Ranking Ions by Size 333

Chapter 9

9.1 Depicting Ion Formation 345
9.2 Comparing Bond Length and Bond Strength 354
9.3 Using Bond Energies to Calculate ΔH°_{rxn} 361
9.4 Determining Bond Polarity from EN Values 366

Chapter 10

10.1 Writing Lewis Structures for Molecules with One Central Atom 380
10.2 Writing Lewis Structures for Molecules with More than One Central Atom 380
10.3 Writing Lewis Structures for Molecules with Multiple Bonds 381
10.4 Writing Resonance Structures 383
10.5 Writing Lewis Structures for Octet-Rule Exceptions 387
10.6 Predicting Molecular Shapes with Two, Three, or Four Electron Groups 395
10.7 Predicting Molecular Shapes with Five or Six Electron Groups 396
10.8 Predicting Molecular Shapes with More Than One Central Atom 397
10.9 Predicting the Polarity of Molecules 400

Chapter 11

11.1 Postulating Hybrid Orbitals in a Molecule 417
11.2 Describing the Types of Bonds in Molecules 421
11.3 Predicting Stability of Species Using MO Diagrams 424
11.4 Using MO Theory to Explain Bond Properties 428

Chapter 12

12.1 Finding the Heat of a Phase Change Depicted by Molecular Scenes 443
12.2 Using the Clausius-Clapeyron Equation 446
12.3 Drawing Hydrogen Bonds Between Molecules of a Substance 453
12.4 Predicting the Types of Intermolecular Force 456
12.5 Determining Atomic Radius from Crystal Structure 469

Chapter 13

13.1 Predicting Relative Solubilities of Substances 505
13.2 Using Henry's Law to Calculate Gas Solubility 522
13.3 Calculating Molality 523
13.4 Expressing Concentrations in Parts by Mass, Parts by Volume, and Mole Fraction 525
13.5 Converting Concentration Terms 526
13.6 Using Raoult's Law to Find Vapor Pressure Lowering 529
13.7 Determining the Boiling Point Elevation and Freezing Point Depression of a Solution 531
13.8 Determining Molar Mass from Osmotic Pressure 535
13.9 Finding Colligative Properties from Molecular Scenes 538

Chapter 15

15.1 Drawing Hydrocarbons 634
15.2 Naming Alkanes, Alkenes, and Alkynes 643
15.3 Recognizing the Type of Organic Reaction 648
15.4 Predicting the Reactions of Alcohols, Alkyl Halides, and Amines 654
15.5 Predicting the Steps in a Reaction Sequence 657
15.6 Predicting Reactions of the Carboxylic Acid Family 660
15.7 Recognizing Functional Groups 662

Chapter 16

16.1 Expressing Rate in Terms of Changes in Concentration with Time 690
16.2 Determining Reaction Order from Rate Laws 695
16.3 Determining Reaction Orders from Initial Rate Data 697
16.4 Determining Reaction Orders from a Series of Molecular Scenes 698
16.5 Determining the Reactant Concentration at a Given Time 700
16.6 Using Molecular Scenes to Determine Half-Life 703
16.7 Determining the Half-Life of a First-Order Reaction 704
16.8 Determining the Energy of Activation 706
16.9 Drawing Reaction Energy Diagrams and Transition States 713
16.10 Determining Molecularity and Rate Laws for Elementary Steps 716

Chapter 17

17.1 Writing the Reaction Quotient from the Balanced Equation 743
17.2 Writing the Reaction Quotient and Finding K for an Overall Reaction 744
17.3 Finding the Equilibrium Constant for an Equation Multiplied by a Common Factor 746
17.4 Converting Between K_c and K_p 749
17.5 Using Molecular Scenes to Determine Reaction Direction 750
17.6 Comparing Q and K to Determine Reaction Direction 751
17.7 Calculating K_c from Concentration Data 754
17.8 Determining Equilibrium Concentrations from K_c 754
17.9 Determining Equilibrium Concentrations from Initial Concentrations and K_c 755
17.10 Calculating Equilibrium Concentrations with a Simplifying Assumption 758
17.11 Predicting Reaction Direction and Calculating Equilibrium Concentrations 759
17.12 Predicting the Effect of a Change in Concentration on the Equilibrium Position 763
17.13 Predicting the Effect of a Change in Volume (Pressure) on the Equilibrium Position 766
17.14 Predicting the Effect of a Change in Temperature on the Equilibrium Position 767
17.15 Determining Equilibrium Parameters from Molecular Scenes 769

Chapter 18

18.1 Classifying Acid and Base Strength from the Chemical Formula 788
18.2 Calculating $[H_3O^+]$ or $[OH^-]$ in Aqueous Solution 790
18.3 Calculating $[H_3O^+]$, pH, $[OH^-]$, and pOH 792
18.4 Identifying Conjugate Acid-Base Pairs 795
18.5 Predicting the Net Direction of an Acid-Base Reaction 796
18.6 Using Molecular Scenes to Predict the Net Direction of an Acid-Base Reaction 798
18.7 Finding K_a of a Weak Acid from the Solution 800
18.8 Determining Concentrations from K_a and Initial [HA] 801
18.9 Using Molecular Scenes to Determine the Extent of HA Dissociation 802
18.10 Calculating Equilibrium Concentrations for a Polyprotic Acid 804
18.11 Determining pH from K_b and Initial [B] 807
18.12 Determining the pH of a Solution of A^- 809
18.13 Predicting Relative Acidity of Salt Solutions 814
18.14 Predicting the Relative Acidity of Salt Solution from K_a and K_b of the Ions 815
18.15 Identifying Lewis Acids and Bases 819

Chapter 19
19.1 Calculating the Effect of Added H_3O^+ or OH^- on Buffer pH 835
19.2 Using Molecular Scenes to Examine Buffers 839
19.3 Preparing a Buffer 840
19.4 Finding the pH During a Weak Acid–Strong Base Titration 847
19.5 Writing Ion-Product Expressions for Slightly Soluble Ionic Compounds 853
19.6 Determining K_{sp} from Solubility 854
19.7 Determining Solubility from K_{sp} 855
19.8 Calculating the Effect of a Common Ion on Solubility 856
19.9 Predicting the Effect on Solubility of Adding Strong Acid 857
19.10 Predicting Whether a Precipitate Will Form 858
19.11 Using Molecular Scenes to Predict Whether a Precipitate Will Form 860
19.12 Separating Ions by Selective Precipitation 861
19.13 Calculating the Concentration of a Complex Ion 866
19.14 Calculating the Effect of Complex-Ion Formation on Solubility 867

Chapter 20
20.1 Predicting Relative Entropy Values 892
20.2 Calculating the Standard Entropy of Reaction, ΔS°_{rxn} 894
20.3 Determining Reaction Spontaneity 896
20.4 Calculating ΔG°_{rxn} from Enthalpy and Entropy Values 901
20.5 Calculating ΔG°_{rxn} from ΔH°_f Values 902
20.6 Using Molecular Scenes to Determine the Signs of ΔH, ΔS, and ΔG 905
20.7 Determining the Effect of Temperature on ΔG 906
20.8 Using Molecular Scenes to Find ΔG for a Reaction at Nonstandard Conditions 910
20.9 Calculationg ΔG at Nonstandard Conditions 913

Chapter 21
21.1 Balancing Redox Reactions by the Half-Reaction Method 927
21.2 Describing a Voltaic Cell with a Diagram and Notation 932
21.3 Calculating an Unknown $E^\circ_{half-cell}$ from E°_{cell} 937
21.4 Writing Spontaneous Redox Reactions and Ranking Oxidizing and Reducing Agents by Strength 940
21.5 Calculating K and ΔG° from E°_{cell} 945
21.6 Using the Nernst Equation to Calculate E_{cell} 947
21.7 Calculating the Potential of a Concentration Cell 950
21.8 Predicting the Electrolysis Products of a Molten Salt Mixture 962
21.9 Predicting the Electrolysis Products of Aqueous Salt Solutions 964
21.10 Applying the Relationship Among Current, Time, and Amount of Substance 966

Chapter 23
23.1 Writing Electron Configurations of Transition Metal Atoms and Ions 1025
23.2 Finding the Number of Unpaired Electrons 1031
23.3 Writing Names and Formulas of Coordination Compounds 1041
23.4 Determining the Type of Stereoisomerism 1046
23.5 Ranking Crystal Field Splitting Energies for Complex Ions of a Given Metal 1051
23.6 Identifying Complex Ions as High Spin or Low Spin 1053

Chapter 24
24.1 Writing Equations for Nuclear Reactions 1070
24.2 Predicting Nuclear Stability 1072
24.3 Predicting the Mode of Nuclear Decay 1073
24.4 Finding the Number of Radioactive Nuclei 1077
24.5 Applying Radiocarbon Dating 1079
24.6 Calculating the Binding Energy per Nucleon 1092

SPECIAL FEATURES

Chemical Connections
To Interdisciplinary Science: Chemistry Problem Solving in the Real World 33
To Planetary Science: Structure and Composition of Earth's Atmosphere 218
To Environmental Science: The Future of Energy Use 256
To Sensory Physiology: Molecular Shape, Biological Receptors, and the Sense of Smell 402
To Environmental Engineering: Solutions and Colloids in Water Purification 541
To Sensory Physiology: Geometric Isomers and the Chemistry of Vision 642
To Genetics and Forensics: DNA Sequencing and Fingerprinting 674
To Enzymology: Kinetics and Function of Biological Catalysts 723
To Atmospheric Science: Depletion of the Earth's Ozone Layer 725
To Cellular Metabolism: Design and Control of a Metabolic Pathway 770

To Industrial Production: The Haber Process for the Synthesis of Ammonia 771
To Geology: Creation of a Limestone Cave 859
To Environmental Science: The Acid-Rain Problem 863
To Biology: Do Living Things Obey the Laws of Thermodynamics? 897
To Biological Energetics: The Universal Role of ATP 908
To Biological Energetics: Cellular Electrochemistry and the Production of ATP 967
To Nutritional Science: Transition Metals as Essential Dietary Trace Elements 1055
To Cosmology: Origin of the Elements in the Stars 1099

Tools of the Laboratory
Mass Spectrometry 55
Basic Separation Techniques 76
Spectrophotometry in Chemical Analysis 281
Infrared Spectroscopy 357

X-Ray Diffraction Analysis and Scanning Tunneling Microscopy 468
Nuclear Magnetic Resonance (NMR) Spectroscopy 646
Measuring Reaction Rates 692
Counters for the Detection of Radioactive Emissions 1075

Galleries
Picturing Molecules 74
Molecular Beauty: Odd Shapes with Useful Functions 398
Properties of a Liquid 459
Colligative Properties in Industry and Biology 533
Silicate Minerals and Silicone Polymers 592

ABOUT the AUTHOR

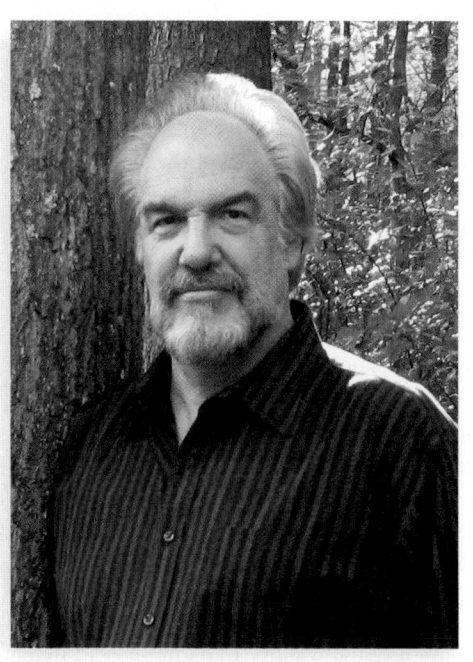

Martin S. Silberberg received a B.S. in Chemistry from the City University of New York and a Ph.D. in Chemistry from the University of Oklahoma. He then accepted a position as research associate in analytical biochemistry at the Albert Einstein College of Medicine in New York City, where he developed methods to study neurotransmitter metabolism in Parkinson's disease and other neurological disorders. Following six years as research associate, Dr. Silberberg joined the faculty of Simon's Rock College of Bard, a liberal arts college known for its excellence in teaching small classes of highly motivated students. As head of the Natural Sciences Major and Director of Premedical Studies, he taught courses in general chemistry, organic chemistry, biochemistry, and liberal-arts chemistry. The small class size and close student contact afforded him insights into how students learn chemistry, where they have difficulties, and what strategies can help them succeed. Dr. Silberberg then decided to apply these insights in a broader context and established a textbook writing, editing, and consulting company. Before writing his own texts, he worked as a consulting and developmental editor on chemistry, biochemistry, and physics texts for several other major college publishers. He resides with his wife and son in the Pioneer Valley near Amherst, Massachusetts, where he enjoys the rich cultural and academic life of the area and relaxes by cooking and hiking.

PREFACE

CHEMISTRY AT THE CORE

Some years ago, a question occasionally heard was "Why study chemistry?"— but no longer. At the core of the natural sciences, chemistry is crucial to an understanding of molecular biology, genetics, pharmacology, ecology, atmospheric science, nuclear studies, materials science, and numerous other fields. Because chemistry is so central to understanding these fields, it is a core requirement for an increasing number of academic majors. Some major societal issues also have chemical principles at their core, including climate change, energy options, materials recycling, diet, nutrition, exercise, and traditional vs. alternative medicine. Clearly, the study of chemistry as an integral part of our world is essential.

To respond to numerous modern challenges, chemistry is evolving in new directions to design "greener" plastics and fuels, monitor atmosphere and oceans to model global warming, determine our genetic makeup to treat disease, and synthesize nanomaterials with revolutionary properties, among many others.

Nevertheless, as the applications change, the basic concepts of chemistry still form the essence of the course. The mass laws and the mole concept still apply to the amounts of substances involved in a reaction. Atomic properties, and the periodic trends and types of bonding derived from them, still determine molecular structure, which in turn still governs the forces between molecules and the resulting physical behavior of substances and mixtures. And the central concepts of kinetics, equilibrium, and thermodynamics still account for the dynamic aspects of chemical change.

The challenge for a modern text surveying this enormous field is to present the core concepts of chemistry clearly *and* show how they apply to current practice. The fifth edition of *Chemistry: The Molecular Nature of Matter and Change* has evolved in important ways to meet this challenge.

STILL SETTING THE STANDARD

Since its first edition, *Chemistry: The Molecular Nature of Matter and Change* has set—and continues to raise—the standard for general chemistry texts. While the content has been repeatedly updated to reflect chemistry's new ideas and changing impact in the world, the mechanisms of the text—the teaching approaches that are so admired and emulated—have remained the same. Three hallmarks continue to make this text a market leader:

- **Visualizing Chemical Models—Macroscopic to Molecular**

Chemistry deals with observable changes caused by unobservable atomic-scale events, which means a size gap of mind-boggling proportions must be understood. One of the text's goals is consonant with that of so many instructors: to help the student visualize chemical events on the molecular scale. Thus, concepts are explained first at the macroscopic level and then from a molecular point of view, with groundbreaking illustrations always placed next to the discussion to bring the point home for today's visually oriented students.

- **Thinking Logically to Solve Problems**

The problem-solving approach, based on the four-step method widely accepted by experts in science education, is introduced in Chapter 1 and employed *consistently* throughout the text. It encourages students to first plan a logical approach to a problem, and only then proceed to solve it mathematically. Each problem includes a check, which fosters the habit of assessing the reasonableness and magnitude of the answer. Finally, for practice and reinforcement, a similar follow-up problem is provided immediately, for which an abbreviated solution, not merely a numerical answer, is given at the end of the chapter. In this edition, solving problems and visualizing models have been integrated in a large number of molecular-scene problems in both worked examples and homework sets.

- **Applying Ideas to the Real World**

An understanding of modern chemistry influences attitudes about melting glaciers and global food supplies, while also explaining the spring in a running shoe and the display of a laptop screen. Today's students may enter one of the emerging chemistry-related, hybrid fields—biomaterials science or planetary geochemistry, for example—and the text that introduces them to chemistry should point out the relevance of chemical concepts to such career directions. Chemical Connections, Tools of the Laboratory, Galleries, and margin notes are up-to-date pedagogic features that complement content of this application-rich text.

EMBRACING CHANGE: HOW WE EVALUATED YOUR NEEDS

Just as the applications of chemistry change, so do your needs in the classroom. Martin Silberberg and McGraw-Hill listened—and responded. They invited instructors like you from across the nation—with varying teaching styles, class sizes, and student backgrounds—to provide feedback through reviews, focus groups, and class testing. Many of the suggestions were incorporated into this revision, and they helped mold the new edition of *Chemistry: The Molecular Nature of Matter and Change*, resulting in new topic coverage, succinct and logical presentation, and expanded treatment in key areas.

WHAT'S NEW IN THE FIFTH EDITION?

Enhancements in Pedagogy

A new edition always brings an opportunity to enhance the text's pedagogy. As in earlier editions, writing has been clarified wherever readers felt ideas could flow more smoothly. Updates made to rapidly changing areas of chemistry always tie the application to fundamental principles. Each chapter ends with a Chapter Review Guide, which offers ways to review the chapter content through Learning Objectives, Key Terms, Key Equations and Relationships, Highlighted Figures and Tables, and, most importantly, Brief Solutions to Follow-Up Problems, which effectively doubles the number of worked problems. But, by far the greatest pedagogical change is the addition of many new worked sample problems and end-of-chapter problems that use simple molecular scenes to teach quantitative concepts.

Molecular-Scene Sample Problems

Many texts include molecular-scene problems in their end-of-chapter sets, but none makes the attempt to explain how to reason toward a solution. It seemed most productive to help students solve these end-of-chapter problems by working out similar ones within the chapter, just as the text does with other types of worked problems. In the previous (4th) edition, in addition to the inclusion of more molecular-scene problems in the end-of-chapter sets, 5 worked-out, molecular-scene sample problems were introduced, and they used the same multistep problem-solving approach as in other sample problems. Responses from students and teachers alike were very positive, so 17 new molecular-scene sample problems and an equal number of follow-up problems have been included in this edition. Together with the original 5, they make a total of 44 such problems, providing a rich source for learning how to understand quantitative concepts via simple chemical scenes.

End-of-Chapter Problems

In each edition, a special effort is made to create new problems that address pedagogical needs and real applications. In the 5th edition, in addition to the quantitative revision of hundreds of end-of-chapter problems, over 135 completely new ones have been added. Of these, 88 are molecular-scene problems, which, together with the 52 already present from the 4th edition, offer abundant practice for the skills learned in the molecular-scene sample and follow-up problems. The remaining new problems incorporate realistic, up-to-date scenarios in biological, organic, environmental, or engineering/industrial applications and are at the challenging level.

Content Changes to Individual Chapters

After four successful editions, *Chemistry: The Molecular Nature of Matter and Change* has reached a level of matu-rity in that much of the mainstream content works well for teacher and student. But everyone's course is unique, so the content is presented with many section and subsection breaks so that it can, in most cases, be rearranged with minimal loss of continuity. Thus, for example, redox balancing can be covered in Chapter 4, in Chapter 21, or, as done in the text, both in Chapter 4 (oxidation-number method) *and* Chapter 21 (half-reaction method, in preparation for electrochemistry). Likewise, several chapters can be taught in a different order. For instance, gases (Chapter 5) can be covered in sequence to explore the mathematical modeling of physical behavior or, with no loss of continuity, just before liquids and solids (Chapter 12) to show the effects of intermolecular forces on the three states of matter. In fact, based on user feedback, many of you already move chapters and sections around, for example, covering descriptive chemistry (Chapter 14) and organic chemistry (Chapter 15) in the more traditional placement at the end of the course. The topic sequence is flexible, and you should feel comfortable making these, or any of numerous other changes, to suit your course.

In the 5th edition, small content changes have been made to many chapters, but a few sections, and even one whole chapter, have been revised considerably. Among the most important changes to this edition are the following:

- Chapter 3 now introduces reaction tables in the discussion of limiting reactants to show the changes in amounts in a stoichiometry problem, just as similar tables are used later to show changes in amounts in an equilibrium problem.
- Chapter 5 includes an updated discussion of how gas behavior relates to Earth's atmosphere.
- Chapter 6 provides updated coverage of how thermochemical ideas relate to the future of energy sources, with expanded coverage of climate change.
- Chapter 12 contains an updated discussion of the relation between the solid state and nanotechnology.
- Chapter 15 includes new material on the role of H-bonding in DNA profiling for forensic chemistry.
- Chapter 16 offers an updated discussion of the catalytic basis of ozone depletion in the stratosphere.
- Chapter 19 covers quantitative analysis by selective precipitation in an earlier section and eliminates the outdated discussion of ion-group qualitative analysis.
- Chapter 20 has been revised further to clarify the discussion of entropy, with several new pieces of art to illustrate key ideas.
- Chapter 24 has been thoroughly revised to more accurately reflect modern ideas in nuclear chemistry.
- Appendices of equilibrium constants for weak acids and bases now include structures of the species.

A VISUAL TOUR THROUGH THE FEATURES OF THE TEXT

Many pedagogical tools are interwoven throughout the chapters to guide students on their learning journey.

Chapter Openers

Each chapter begins with a thought-provoking *opener figure and legend* that relate to the main topic of the chapter. The chapter opening page also contains the *Chapter Outline* that shows the sequence of topics and subtopics, and the final paragraph of the introduction, called *In This Chapter,* ties the main topics to the outline. In the margin next to the introduction, *Concepts and Skills to Review* refers to key material from earlier chapters that you should understand before you start reading the current one.

Problem Solving

A worked-out *sample problem* appears whenever an important new concept or skill is introduced, and the problem-solving approach helps you think through all problems logically and systematically. The stepwise approach, based on the universally accepted four-step approach of plan, solve, check, and practice, is used consistently for every sample problem in the text. These steps are as follows:

- **Plan:** analyzes the problem so that you can use what is known to find what is unknown. This step develops the habit of thinking through the solution before performing calculations. Most quantitative problems are accompanied in the margin by a *roadmap*, a flow diagram that leads you visually through the planned steps for each specific problem.
- **Solution:** presents the calculation steps *in the same order* as they are discussed in the plan and shown in the roadmap.
- **Check:** fosters the habit of going over your work with a rough calculation to make sure the answer is both chemically and mathematically reasonable—a great way to avoid careless errors. In many cases, following the check is a *Comment* that provides an additional insight, alternative approach, or common mistake to avoid.
- **Follow-up Problem:** presents a similar problem to provide immediate practice, with an abbreviated multistep solution appearing at the end of the chapter.

In this edition, in addition to sample problems involving only calculations, a large number of **molecular-scene sample problems** utilize depictions of chemical species to solve quantitative problems.

Applications

Tools of the Laboratory essays describe the key instruments and techniques that chemists use in modern practice to obtain the data that underlie their theories.

Chemical Connections essays show the interdisciplinary nature of chemistry by applying chemical principles directly to related scientific fields, including physiology, geology, biochemistry, engineering, and environmental science.

Gallery features show how common and unusual substances and processes relate to chemical principles. You'll learn how a towel dries you, why bubbles in a drink are round, why contact-lens rinse must have a certain concentration, and many other intriguing facts about everyday applications.

Margin notes are brief, lively explanations that apply ideas presented in the text. You'll learn how water controls the temperature of your body and our planet, how crime labs track illegal drugs, how gas behavior affects lung function, how fat-free chips and decaf coffee are made, in addition to handy tips for memorizing relationships, and much more.

Illustrated Summaries of Facts and Concepts

The multipage *Interchapter* is a Perspective on the Properties of the Elements that reviews major concepts from Chapters 7–13, covering atomic and bonding properties and their resulting effects on element behavior.

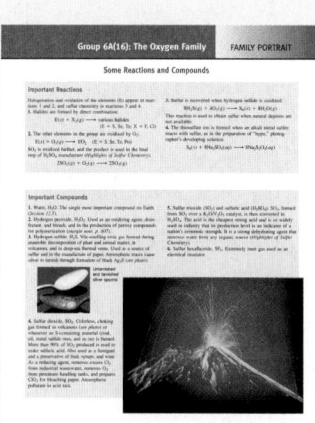

Family Portraits (within Chapter 14) display the atomic and physical properties of each main group of elements and present their representative chemical reactions and some important compounds.

Three-Level Illustrations

A hallmark of this text, the three-level illustrations help you connect the macroscopic and molecular levels of reality with the symbolic level in the form of a chemical equation.

Figure 12.47 The random-coil shape of mer chain. Note the random coiling of the carbon atoms (*black*). Sections of several chains (*red*, *green*, and *yellow*) are entangl this chain, kept near one another by dis forces. In reality, entangling chains fill ar shown here. The radius of gyration (R_g) rep the average distance from the center of mas coiled molecule to its outer edge.

Accurate, Cutting-Edge Molecular Models

Author and illustrator worked side by side to create ground-breaking visual representations.

Page Layout

Author and pager collaborated on page layout to ensure that all figures, tables, margin notes, and sample problems are as close as possible to their related text.

Section Summaries and Chapter Perspective

Concise summary paragraphs conclude each section, immediately restating the major ideas just covered. Each chapter ends with a brief perspective that places its topics in the context of previous and upcoming chapters.

8.5 Atomic Structure and Chemical Reactivity   327

Figure 8.23 The change in metallic behavior in Group 5A(15) and Period 3. Moving down from nitrogen to bismuth shows an *increase* in metallic behavior (and thus a decrease in ionization energy). Moving left to right from sodium to chlorine shows a *decrease* in metallic behavior (and thus a general increase in ionization energy).

Some metals and many metalloids form oxides that are **amphoteric:** they can act as acids *or* as bases in water.

Figure 8.24 classifies the acid-base behavior of some common oxides, focusing once again on the elements in Group 5A(15) and Period 3. Note that *as the elements become more metallic down a group, their oxides become more basic.* In Group 5A, dinitrogen pentaoxide, N_2O_5, forms nitric acid:

$$N_2O_5(s) + H_2O(l) \longrightarrow 2HNO_3(aq)$$

Tetraphosphorus decaoxide, P_4O_{10}, forms the weaker acid H_3PO_4:

$$P_4O_{10}(s) + 6H_2O(l) \longrightarrow 4H_3PO_4(aq)$$

The oxide of the metalloid arsenic is weakly acidic, whereas that of the metalloid antimony is weakly basic. Bismuth, the most metallic of the group, forms a basic oxide that is insoluble in water but that forms a salt and water with acid:

$$Bi_2O_3(s) + 6HNO_3(aq) \longrightarrow 2Bi(NO_3)_3(aq) + 3H_2O(l)$$

Note that *as the elements become less metallic across a period, their oxides become more acidic.* In Period 3, sodium and magnesium form the strongly basic oxides Na_2O and MgO. Metallic aluminum forms amphoteric aluminum oxide (Al_2O_3), which reacts with acid or with base:

$$Al_2O_3(s) + 6HCl(aq) \longrightarrow 2AlCl_3(aq) + 3H_2O(l)$$
$$Al_2O_3(s) + 2NaOH(aq) + 3H_2O(l) \longrightarrow 2NaAl(OH)_4(aq)$$

Silicon dioxide is weakly acidic, forming a salt and water with base:

$$SiO_2(s) + 2NaOH(aq) \longrightarrow Na_2SiO_3(aq) + H_2O(l)$$

The common oxides of phosphorus, sulfur, and chlorine form acids of increasing strength: H_3PO_4, H_2SO_4, and $HClO_4$.

Figure 8.24 The trend in acid-base behavior of element oxides. The trend in acid-base behavior for some common oxides of Group 5A(15) and Period 3 elements is shown as a gradation in color (*red* = acidic; *blue* = basic). Note that the metals form basic oxides and the nonmetals form acidic oxides. Aluminum forms an oxide (*purple*) that can act as an acid or as a base. Thus, as atomic size increases, ionization energy decreases, and oxide basicity increases.

Section Summary

A stepwise process converts a molecular formula into a Lewis structure, a *two-dimensional* representation of a molecule (or ion) that shows the placement of atoms and distribution of valence electrons among bonding and lone pairs. • When two or more Lewis structures can be drawn for the same relative placement of atoms, the actual structure is a hybrid of those resonance forms. • Formal charges are often useful for determining the most important contributor to the hybrid. • Electron-deficient molecules (central Be or B) and odd-electron species (free radicals) have less than an octet around the central atom but often attain an octet in reactions. • In a molecule (or ion) with a central atom from Period 3 or higher, the atom can hold more than eight electrons because it is larger and uses *d* orbitals to expand its valence shell.

Chapter Review Guide

A rich assortment of study aids ends each chapter to help you review its content.

- **Learning Objectives** are listed, with section and/or sample problem numbers, to focus you on key concepts and skills.
- **Key Terms** that are boldfaced within the chapter are listed here by section (with page numbers) and defined again in the end-of-book Glossary.
- **Key Equations and Relationships** are screened and numbered within the chapter and listed here with page numbers.
- **Highlighted Figures and Tables** are listed with page numbers so that you can review their essential content.
- **Brief Solutions to Follow-up Problems** double the number of worked problems by offering multistep calculations at the end of the chapter, rather than just numerical answers at the back of the book.

End-of-Chapter Problems

An exceptionally large number of problems end each chapter. These include three types of problems keyed by chapter section followed by a number of comprehensive problems:

- **Concept Review Questions** test your qualitative understanding of key ideas.
- **Skill-Building Exercises** are presented in pairs that cover a similar idea, with one of each pair answered in the back of the book. These exercises begin with simple questions and increase in difficulty, gradually eliminating your need for multistep directions.
- **Problems in Context** apply the skills learned in the Skill-Building Exercises to interesting scenarios, including examples from industry, medicine, and the environment.
    - **Comprehensive Problems,** based on realistic applications, are more challenging and rely on concepts and skills from any section of the current chapter or from previous chapters.

Moreover, in this edition, 140 **molecular-scene problems** are included.

SUPPLEMENTS FOR THE INSTRUCTOR

Annotated Instructor's Edition

The Annotated Instructor's Edition, with annotations prepared by John Pollard of Arizona State University, is your guide to all pertinent information for assigning homework and for integrating media and extra content into your lecture. The general difficulty of every problem in an end-of-chapter set is indicated, and, where appropriate, the relevant field of chemistry or related science is given. Application icons are located throughout the text to denote:

🧬 **Biological Applications**

⚙ **Engineering Applications**

🌎 **Environmental/Green Chemistry Applications**

🧪 **Organic Chemistry Applications**

🧫 **Class Demonstrations**

📙 **Journal and Literature References** to related scholarly publications. A number of journal articles are available online and links to them can be found within the instructor's tools on the ARIS site for the text.

 Each figure, photo, or table that is available within the **Presentation Center** is noted for easy integration into your lecture presentation.

 Transparency/Presentation Center icons alert you to figures that are available as overhead transparencies as well as in digital format.

♻ **Animations** related to specific chapter content are available within the ARIS site for the text.

ARIS 🤸 ARIS™

The unique Assessment, Review, and Instruction System, known as ARIS and accessed at aris.mhhe.com, is an electronic homework and course management system designed to have greater flexibility, power, and ease of use than any other system. Whether you are looking for a preplanned course or one you can customize to fit your course needs, ARIS is your solution. In addition to having access to all student digital learning objects, ARIS allows instructors to:

Build Assignments

- Choose from prebuilt assignments or create custom assignments by importing content or editing an existing prebuilt assignment.
- Include quiz questions, animations, and videos—anything found on the ARIS website—in your assignments.

- Create announcements and utilize full-course or individual student communication tools.
- Assign questions that use the problem-solving strategy presented within the text, allowing students to carry over the structured learning process from the text into their homework assignments.
- Assign algorithmic questions, providing students with multiple chances to practice and gain skill at problem-solving involving the same concept.

Track Student Progress

- Assignments are automatically graded.
- Gradebook functionality allows full course management including:
  - dropping the lowest grades
  - weighting grades or manually adjusting grades
  - exporting your gradebook to Excel, WebCT, or BlackBoard
  - manipulating data to track student progress through multiple reports

Offer More Flexibility

- **Share Course Materials with Colleagues**—create and share course materials and assignments with colleagues through just a few clicks of the mouse, allowing multiple-section courses with many instructors to be continually in synch.
- **Integration with Blackboard or WebCT**—once a student is registered in the course, all of the student's activity within ARIS is automatically recorded and available to the instructor through a fully integrated gradebook that can be downloaded to Excel, WebCT, or Blackboard.

To access ARIS, instructors may request a registration code from their McGraw-Hill sales representative.

Presentation Center

Accessed from the instructor's side of the ARIS website, **Presentation Center** is an online digital library containing photos, artwork, animations, and other media types that can be used to create customized lectures, visually enhanced tests and quizzes, compelling course websites, or attractive printed support materials. All assets are copyrighted by McGraw-Hill Higher Education, but can be used by instructors for classroom purposes. The visual resources in this collection include the following:

- **Art** Full-color digital files of all illustrations in the book can be readily incorporated into lecture presentations, exams, or custom-made classroom materials.
- **Photos** The photo collection contains digital files of photographs from the text, which can be reproduced for multiple classroom uses.
- **Tables** Every table that appears in the text has been saved in electronic form for use in classroom presentations and/or quizzes.

- **Animations** Numerous full-color animations illustrating important processes are also provided. Harness the visual impact of concepts in motion by importing these files into classroom presentations or online course materials.

Also residing on the ARIS website are PowerPoint resources:

- **PowerPoint Lecture Outlines** Updated by Angela Cannon, ready-made presentations that combine art and lecture notes are provided for each chapter of the text.
- **PowerPoint Slides** For instructors who prefer to create their lectures from scratch, all illustrations, photos, and tables are pre-inserted into blank PowerPoint slides, arranged by chapter.

The Presentation Center library includes thousands of assets from many McGraw-Hill titles. This ever-growing resource gives instructors the power to utilize assets specific to an adopted textbook as well as content from all other books in the library.

Presentation Center's dynamic search engine allows you to explore by discipline, course, textbook chapter, asset type, or keyword. Simply browse, select, and download the files you need to build engaging course materials. To access ARIS, request registration information from your McGraw-Hill sales representative.

Computerized Test Bank Online

Prepared by Walter Orchard, a comprehensive bank of test questions is provided within a computerized test bank powered by McGraw-Hill's flexible electronic testing program EZ Test Online (www.eztestonline.com). EZ Test Online allows you to create paper and online tests or quizzes anywhere, at any time, without installing the testing software. With EZ Test Online, instructors can select questions from multiple McGraw-Hill test banks or author their own, and then either print the test for distribution on paper or administer it online.

Test Creation

- Author/edit questions online using 14 different question type templates
- Create printed tests or deliver online to get instant scoring and feedback
- Create questions pools to offer multiple versions online—great for practice
- Export your tests for use in WebCT, Blackboard, Page-Out and Apple's iQuiz
- Take advantage of compatibility with EZ Test Desktop tests you've already created
- Share tests easily with colleagues, adjuncts, and TAs

Online Test Management

- Set availability dates and time limits for your quiz or test
- Control how your test will be presented
- Assign points by question or question type with drop-down menu

- Provide immediate feedback to students or delay feedback until all finish the test
- Create practice tests online to enable student mastery of concepts
- Upload class roster to enable student self-registration

Online Scoring and Reporting

- Use automated scoring for most of EZ Test's numerous question types
- Use manual scoring for essay and other open response questions
- Use manual re-scoring and feedback if desired
- Export EZ Test's gradebook to your gradebook easily
- View basic statistical reports

Support and Help

- User's Guide and built-in page-specific help
- Flash tutorials for getting started on the support site
- Support Website, including Online Training and Registration at www.mhhe.com/eztest
- Product specialist available at 1-800-331-5094

Instructor's Solutions Manual

This supplement, prepared by Patricia Amateis of Virginia Tech, contains complete, worked-out solutions for *all* the end-of-chapter problems in the text. It can be found within the Instructors Resources for this text on the ARIS website.

Student Response System

Wireless technology brings interactivity into the classroom or lecture hall. Instructors and students receive immediate feedback through wireless response pads that are easy to use and engage students. This system can be used by instructors to:

- Take attendance
- Administer quizzes and tests
- Create a lecture with intermittent questions
- Manage lectures and student comprehension through the use of the gradebook
- Integrate interactivity into PowerPoint presentations

Content Delivery Flexibility

Chemistry: The Molecular Nature of Matter and Change by Martin Silberberg is available in other formats in addition to the traditional textbook to give instructors and students more choices when deciding on the format of their chemistry text. Choices include:

Color Custom by Chapter

For even more flexibility, Silberberg's *Chemistry: The Molecular Nature of Matter and Change* is available in a full-color, custom version that allows instructors to pick the chapters they want included. Students pay for only what the instructor chooses.

Electronic Books

If you or your students are ready for an alternative version of the traditional textbook, McGraw-Hill offers two options to bring you innovative and inexpensive electronic textbooks.

CourseSmart provides a new way for faculty to find and review Ebooks. It's also a great option for students who are interested in accessing their course materials digitally and saving up to 50% of the cost of a print textbook. Students are able to reduce their impact on the environment as well as gain access to powerful web tools for learning including full text search, notes and highlighting, and email tools for sharing notes between classmates.

McGraw-Hill and VitalSource have partnered to provide a media-enhanced Ebook. In addition to a powerful suite of built-in tools that allow detailed searching, highlighting, note taking, and student-to-student or instructor-to-student note sharing, the media-rich Ebook for Silberberg's *Chemistry: The Molecular Nature of Matter and Change* integrates relevant animations and videos into the textbook content for a true multimedia learning experience. By purchasing Ebooks from McGraw-Hill & VitalSource, students can save as much as 50% on selected titles delivered on the most advanced Ebook platform available, Vital-Source Bookshelf.

Contact your McGraw-Hill sales representative to discuss Ebook packaging options.

Cooperative Chemistry Laboratory Manual

Prepared by Melanie Cooper of Clemson University, this innovative guide features open-ended problems designed to simulate experience in a research lab. Working in groups, students investigate one problem over a period of several weeks, so they might complete three or four projects during the semester, rather than one preprogrammed experiment per class. The emphasis is on experimental design, analysis, problem solving, and communication.

SUPPLEMENTS FOR THE STUDENT

Student Study Guide

This valuable ancillary, prepared by Libby Bent Weberg, is designed to help students recognize their learning style; understand how to read, classify, and create a problem-solving list; and practice problem-solving skills. For each section of a chapter, Dr. Weberg provides study objectives and a summary of the corresponding text. Following the summary are sample problems with detailed solutions. Each chapter has true-false questions and a self-test, with all answers provided at the end of the chapter.

Student Solutions Manual

This supplement, prepared by Patricia Amateis of Virginia Tech, contains detailed solutions and explanations for all even-numbered problems in the main text.

Intelligent Tutors

Powered by Quantum Tutors, Intelligent Tutors is an Internet-based, artificial intelligence tutoring software that supports McGraw-Hill's chemistry textbooks. It provides real-time personal tutoring help for struggling and advanced students with step-by-step feedback and detailed instruction based on the student's own work. Immediate answers are provided to the student over the Internet, day or night.

Animations and Media Player/MPEG Content

A number of animations are available for download to your media player through the ARIS website. Also, audio summaries of each chapter are available for media player download.

ARIS

Assessment, Review, and Instruction System, also known as ARIS, is an electronic homework and course management system designed to have greater flexibility, power, and ease of use than any other system. Students will benefit from independent study tools such as quizzes, animations, and key term flashcards and also will be able to complete homework assignments electronically if instructors desire. Visit the ARIS site at aris.mhhe.com.

ACKNOWLEDGMENTS

As in previous editions, a small army of extremely talented people helped me prepare this one. Once again, I am indebted to Dorothy B. Kurland, who meticulously reviewed the accuracy of every chapter and new homework problem. Many other academic and research chemists contributed their expertise to improve the content of specific chapters: Aric Dutelle, University of Wisconsin, provided information for the essay on DNA Fingerprinting (Chapter 15); Sarina Ergas, University of Massachusetts, updated the essays on the Future of Energy Use (Chapter 6) and Solutions and Colloids in Water Purification (Chapter 13); Jon Kurland updated the essays on Depletion of Earth's Ozone Layer (Chapter 16) and the Acid Rain Problem (Chapter 19), reviewed the revised discussion of entropy (Chapter 20), and suggested more realistic scenarios for many of the Comprehensive problems; Frank Lambert, Emeritus of Occidental College, graciously served as a consultant on the entropy coverage (Chapter 20); Mike Lipschutz, Emeritus of Purdue University, updated the discussions on the Effects of Nuclear Radiation and Applications of Radioisotopes (Chapter 24) and consulted on the revision of that chapter. William McHarris, Michigan State University, provided a comprehensive review of the revised coverage of nuclear reactions and applications (Chapter 24); and Jason Telford, University of Iowa, updated the discussion on nanotechnology (Chapter 12).

Other professors devoted their efforts to creating superb special features or supplements; Patricia Amateis, Virginia Tech, prepared the Appendix E solutions as well as the Instructor's and Student Solutions Manuals; Angela Cannon updated the PowerPoint Lecture Outlines and prepared a supplement covering Beer's Law (which can be found on the ARIS site at aris.mhhe.com; Hon-kie Ng, Florida State University, wrote CAPA Problems to accompany the text; S. Walter Orchard, Emeritus of Tacoma Community College, prepared the Test Bank; John Pollard, University of Arizona, researched and coordinated the annotations for the Annotated Instructor's Edition; and Elizabeth Bent Weberg prepared the Student Study Guide.

And, finally, special thanks go to several others for help with key parts of the text; Charles M. Lieber, Department of Chemistry and Chemical Biology, Harvard University, for consulting on the cover image, which depicts a key aspect of his research; Sue Nurrenbern, Purdue University, for insightful reviewing of the new molecular-scene sample problems; and Jon Kurland, S. Walter Orchard, and Jason Overby, College of Charleston, for writing many excellent new homework problems.

I am grateful for the support of the Board of Advisors, a select group of chemical educators dedicated to helping make this text the optimum teaching tool:

Bill Donovan,
  The University of Akron
Greg Gellene,
  Texas Tech University
Mike Lipschutz,
  Purdue University

MaryKay Orgill,
  University of Nevada, Las Vegas
Jessica Orvis,
  Georgia Southern University
John Pollard,
  University of Arizona

Dawn Rickey,
  Colorado State University
Thomas Schleich,
  University of California, Santa Cruz
Jason Telford,
  University of Iowa

Included with this group of professors are all those who participated in focus groups, reviewed content, and class-tested the previous edition:

Edwin H. Abbott,
  Montana State University
Rosa Alvarez Bolainez,
  East Carolina University
Russell S. Andrews, Jr.,
  University of South Alabama
Gabriele Backes,
  Portland Community College
David Ball,
  California State University, Chico
Chris Bauer,
  University of New Hampshire
Terri Beam,
  Mt. San Antonio College

Debbie Beard,
  Mississippi State University
Adriana Bishop,
  Spokane Falls Community College
Dan Black,
  Snow College
Jo Blackburn,
  Richland College
Gary L. Blackmer,
  Western Michigan University
Bob Blake,
  Texas Tech University
Barry Boatwright,
  Gadsden State Community College

Mary J. Bojan,
  The Pennsylvania State University
Donald C. Bowman,
  Central Virginia Community College
Stephen Bradforth,
  University of Southern California
Mark Braiman,
  Syracuse University
Bryan Breyfogle,
  Missouri State University
Kenneth G. Brown,
  Old Dominion University
Donna M. Budzynski,
  San Diego Mesa College

Acknowledgments

Brian Buffin,
Western Michigan University
Lillian Bird,
University of Puerto Rico, Rio Piedras
William Burns,
Arkansas State University
Stephen Cabaniss,
University of New Mexico
Sebastian G. Canagaratna,
Ohio Northern University
Frank Carey,
Wharton County Junior College
Jon W. Carnahan,
Northern Illinois University
Charles Carraher,
Florida Atlantic University
Tara S. Carpenter,
University of Maryland, Baltimore County
Christopher M. Cheatum,
University of Iowa
William Cleaver,
University of Vermont
James E. Collins,
East Carolina University
Leon L. Combs,
Kennesaw State University
Kevin D. Crawford,
The Citadel
Karen Pressprich Creager,
Clemson University
Patrick L. Daubenmire,
Loyola University, Chicago
Steven R. Davis,
University of Mississippi
David O. De Haan,
University of San Diego
Roger de la Rosa,
St. Louis University
John C. Deàk,
University of Scranton
Timothy O. Deschaines,
University of New Hampshire
Anthony L. Diaz,
Central Washington University
John DiVincenzo,
Middle Tennessee State University
William Donovan,
University of Akron
Ronald J. Duchovic,
Indiana University-Purdue University Fort Wayne
David C. Easter,
Texas State University, San Marcos
Francisco J. Echegaray,
University of Puerto Rico, Rio Piedras
Lourdes E. Echegoyen,
Clemson University
Rebecca A. Eikey,
College of the Canyons
Brian Enderle,
University of California, Davis
Deborah Berkshire Exton,
University of Oregon
Matthew P. Fasnacht,
Southeast Missouri State University

Luis C. Fernández-Torres,
University of Puerto Rico, Cayey
Joanna Fischer,
Spokane Falls Community College
Walter A. Flomer,
Northwestern State University
Daniel Freedman,
SUNY, New Paltz
Mark Freilich,
University of Memphis
Herb Fynewever,
Western Michigan University, Kalamazoo
John I. Gelder,
Oklahoma State University
Leanna Giancarlo,
University of Mary Washington
Paul Gilletti,
Mesa Community College
Sharon Fetzer Gislason,
University of Illinois, Chicago
Ken Goldsby,
Florida State University
John A. Goodwin,
Coastal Carolina University
Donna L. Gosnell,
Valdosta State University
Pierre Y. Goueth,
Santa Monica College
Derek E. Gragson,
California Polytechnic State University
David L. Greene,
Rhode Island College
Steven M. Gunther,
Albuquerque Technical Vocational Institute
Ram Gurumurthy,
San Diego City College
John Hagen,
California Polytechnic State University
Ewan J. M. Hamilton,
The Ohio State University of Lima
Kathleen A. Harter,
Community College of Philadelphia
Cynthia Harwood,
University of Illinois, Chicago
C. Alton Hassell,
Baylor University
Michael Hecht,
Princeton University
Monte L. Helm,
Fort Lewis College
Michael R. Hempstead,
York University
Rick Holz,
Utah State University, Logan
Robert P. Houser,
University of Oklahoma, Norman
James Hovick,
University of North Carolina, Charlotte
Wendy Innis-Whitehouse,
University of Texas Pan American
T.G. Jackson,
University of South Alabama
Milton D. Johnston, Jr.,
University of South Florida

Michael Jones,
Texas Tech University
Jeffrey J. Keaffaber,
University of Florida
Catherine A. Keenan,
Chaffey College
Farooq A. Khan,
University of West Georgia
Charles C. Kirkpatrick,
Saint Louis University
Phillip E. Klebba,
University of Oklahoma
Bert Knesel,
Midlands Technical College
Deborah Koeck,
Texas State University-San Marcos
Bette Kreuz,
The University of Michigan, Dearborn
Julie Ellefson Kuehn,
William Rainey Harper College
Cynthia M. Lamberty,
Nicholls State University
David Laude,
University of Texas at Austin
Joan Lebsack,
Fullerton College
Neocles Leontis,
Bowling Green State University
Hong-Chang Liang,
San Diego State University
Bernard A. Liburd,
Grand Rapids Community College
Pippa Lock,
McMaster University
Madhu Mahalingam,
University of the Sciences in Philadelphia
Diana Mason,
University of North Texas
Maryann McDermott-Jones,
University of Maryland
Wm. C. McHarris,
Michigan State University
Lauren E. H. McMills,
Ohio University
Robert Milofsky,
Fort Lewis College
Tracy Morkin,
Emory University
James Murphy,
Santa Monica College
Kathy Nabona,
Austin Community College
David F. Nachman,
Mesa Community College
Richard L. Nafshun,
Oregon State University
Chip Nataro,
Lafayette College
Christian Nelson,
Chemeketa Community College
Anne Marie Nickel,
Milwaukee School of Engineering
Daphne Norton,
Emory University

MaryKay Orgill,
University of Nevada, Las Vegas
Jason Overby,
College of Charleston
G. S. Owens,
University of Utah
Stephen J. Paddison,
University of Alabama, Huntsville
Maria K. Paukstelis,
Kansas State University
Eugenia Paulus,
North Hennepin Community College
Ralf M. Peetz,
City University of New York, Staten Island
Joanna Petridou,
Spokane Falls Community College
Giuseppe Petrucci,
University of Vermont
Cortland Pierpont,
University of Colorado
Patricia A. Pleban,
Old Dominion University
Amy Pollack,
Michigan State University, East Lansing
Marie K. Pomije,
Minnesota State University, Mankato
Victoria G. Prevatt,
Tulsa Community College, Southeast Campus
Jeffrey R. Pribyl,
Minnesota State University, Mankato
Jeffrey J. Rack,
Ohio University
Daniel Raftery,
Purdue University
David Rainville,
University of Wisconsin, River Falls
Jerry Reed-Mundell,
Cleveland State University
Jimmy Reeves,
University of North Carolina, Wilmington
Philip J. Reid,
University of Washington

Theresa M. Reineke,
University of Cincinnati
Kimberly A. Rickert,
California University of Pennsylvania
Tom Ridgway,
University of Cincinnati
Lydia Martinez Rivera,
University of Texas, San Antonio
Jimmy Rogers,
University of Texas, Arlington
Gillian E. A. Rudd,
Northwestern State University
Kresimir Rupnik,
Louisiana State University
Gregory T. Rushton,
Kennesaw State University
Jerry L. Sarquis,
Miami University, Oxford Ohio
Reva A. Savkar,
Northern Virginia Community College
Thomas Schleich,
University of California, Santa Cruz
Kerri Scott,
University of Mississippi
Raymond Scott,
University of Mary Washington
Brett K. Simpson,
Coastal Carolina University
John D. Sink,
Southern Polytechnic State University
Roger D. Sommer,
DePaul University
Ram Subramaniam,
Santa Clara University
Jerry P. Suits,
University of Northern Colorado
Brian K. Taylor,
The University of Texas, Tyler
John D. Thoemke,
Minnesota State University, Mankato
Richard E. Thompson,
LSU, Shreveport
Joe Thrasher,
University of Alabama

Paul J. Toscano,
University of Albany, SUNY
Joe Toto,
Mesa Community College, San Diego
Christopher L. Truitt,
Texas Tech University
Ellen Verdel,
University of South Florida
Edward A. Walters,
University of New Mexico
Lihua Wang,
Kettering University
Rachel E. Ward,
East Carolina University
Steve Watkins,
Louisiana State University
Thomas Webb,
Auburn University
Wayne Wesolowski,
University of Arizona
Daniel J. Williams,
Kennesaw State University
Lou Anne Williams,
Hinds Community College
Donald R. Wirz,
University of California, Riverside
Peter R. Witt,
Midlands Technical College
Gary L. Wood,
Valdosta State University
Gene G. Wubbels,
University of Nebraska, Kearney
Warren Yeakel,
Henry Ford Community College
David E. Young,
Baylor University
James A. Zimmerman,
Missouri State University
Susan Moyer Zirpoli,
Slippery Rock University
Lisa A. Zuraw,
The Citadel

Many exceptional publishing professionals deserve my deepest gratitude. First, my wonderful team at McGraw-Hill Higher Education, always there with expertise, warmth, and support, consists of Publisher Thomas Timp, Senior Sponsoring Editor Tami Hodge, Senior Development Editor Donna Nemmers, Lead Project Manager Peggy Selle, Senior Designer David Hash, and Senior Marketing Manager Todd Turner.

Second, a group of superb freelancers contributed their talents. Once again, Jane Hoover did a masterful copyediting job, and, once again, Katie Aiken followed with expert proofreading. Michael Goodman created beautiful in-text molecular art and the striking cover. And, with friendship and skill, Karen Pluemer coordinated the complex flow of reviews, chapters, and art to keep me close to the schedule.

And finally, I am indebted to my son Daniel and wife Ruth for their love and confidence. Daniel, an accomplished artist at 19, drafted the initial design for the cover and contributed several key pieces of text art. As in past editions, Ruth was indispensible to the completion of this one—laying out the pages of text, art, and tables, collaborating on style and design, checking copyediting and page proofs, and helping author and publisher maintain the highest standards of quality and consistency.

CHEMISTRY

The Molecular Nature of Matter and Change

1

In the Eye of the Storm Lightning supplies the energy for a myriad of explosive chemical reactions. Right near the bolt, at temperatures hotter than the Sun's surface, molecules of nitrogen (blue) and oxygen (red) break into atoms. A few meters away, the chaos continues as these atoms collide with others and with water molecules to form various species, including nitrogen oxides and ozone. Kilometers farther away, the final products eventually rain down as natural fertilizers. Peering at reality on the molecular scale gives us an astonishing point of view, and this chapter opens the door for you to enter that world.

Keys to the Study of Chemistry

1.1 Some Fundamental Definitions
Properties of Matter
States of Matter
Central Theme in Chemistry
Importance of Energy

1.2 Chemical Arts and the Origins of Modern Chemistry
Prechemical Traditions
Impact of Lavoisier

1.3 The Scientific Approach: Developing a Model

1.4 Chemical Problem Solving
Units and Conversion Factors
Solving Chemistry Problems

1.5 Measurement in Scientific Study
Features of SI Units
SI Units in Chemistry

1.6 Uncertainty in Measurement: Significant Figures
Determining Significant Digits
Significant Figures in Calculations
Precision and Accuracy

Today, as always, the science of chemistry, together with the other sciences that depend on it, stands at the forefront of discovery. Developing "greener" energy sources to power society and using our newfound knowledge of the human genome to cure disease are but two of the areas that will occupy researchers in the chemical, biological, and engineering sciences for much of the 21st century. Addressing these and countless other challenges and opportunities depends on an understanding of the concepts you will learn in this course.

The impact of chemistry on your personal, everyday life is mind-boggling. Consider what the beginning of a typical day might look like from a chemical point of view. Molecules align in the liquid crystal display of your clock and electrons flow to create a noise. A cascade of neuronal activators arouses your brain, and you throw off a thermal insulator of manufactured polymer and jump in the shower to emulsify fatty substances on your skin and hair with purified water and formulated detergents. Then, you adorn yourself in an array of processed chemicals—pleasant-smelling pigmented materials suspended in cosmetic gels, dyed polymeric fibers, synthetic footwear, and metal-alloyed jewelry. Breakfast is a bowl of nutrient-enriched, spoilage-retarded cereal and milk, a piece of fertilizer-grown, pesticide-treated fruit, and a cup of a hot aqueous solution of stimulating alkaloid. After abrading your teeth with artificially flavored, dental-hardening agents in a colloidal dispersion, you're ready to leave, so you grab your laptop (an electronic device containing ultrathin, microetched semiconductor layers powered by a series of voltaic cells), collect some books (processed cellulose and plastic, electronically printed with light- and oxygen-resistant inks), hop in your hydrocarbon-fueled, metal-vinyl-ceramic vehicle, electrically ignite a synchronized series of controlled gaseous explosions, and you're off to class!

The influence of chemistry extends to the natural environment as well. The air, water, and land and the organisms that thrive there form a remarkably complex system of chemical interactions. While modern chemical products have enhanced the quality of our lives, their manufacture and use also pose increasing dangers, such as toxic wastes, acid rain, global warming, and ozone depletion. If our careless disregard of chemical principles has led to some of these problems, our careful adherence to the same principles is helping to solve them.

Perhaps the significance of chemistry is most profound when you contemplate the chemical nature of biology. Molecular events taking place within you right now allow your eyes to scan this page and your brain cells to translate fluxes of electric charge into thoughts. The most vital biological questions—How did life arise and evolve? How does an organism reproduce, grow, and age? What is the essence of health and disease?—ultimately have chemical answers.

This course comes with a bonus—the development of two mental skills you can apply to many fields. The first, common to all science courses, is the ability to solve problems systematically. The second is specific to chemistry, for as you comprehend its ideas, your mind's eye will learn to see a hidden level of the universe, one filled with incredibly minute particles moving at fantastic speeds, colliding billions of times a second, and interacting in ways that determine how all the matter inside and outside of you behaves. This first chapter holds the keys to help you enter this new world.

IN THIS CHAPTER... We begin with fundamental definitions and concepts of matter and energy and their changes. Then, a brief discussion of chemistry's historical origins leads to an overview of how scientists build models to study nature. We consider chemical problem solving, including unit conversion, modern systems of measurement—focusing on mass, length, volume, density, and temperature—and the manipulation of numbers in calculations. A final essay examines how modern chemists work with other scientists for society's benefit.

Concepts & Skills to Review
before you study this chapter

• exponential (scientific) notation (Appendix A)

Literature Keller, Sarah L.; Smith, Andri L. "Advice for New Faculty Teaching Undergraduate Science," *J. Chem. Educ.* **2006,** *83,* 401.

Literature Short, Barbara J.; Carver, Jeffrey S.; Hunter, William J. F.; Young, John R. "Moments in Constructivism: How Does Accepting Failures Allow Us to Examine Our Teaching?" *Chem. Educator* 6 **2001,** *5,* 277–287.

Literature Toby, Sidney; Plano, Richard J. "Testing, Testing: Good Teaching Is Difficult; So Is Meaningful Testing," *J. Chem. Educ.* **2004,** *81,* 180.

Literature Beall, Herbert. "The Ubiquitous Metaphors of Chemistry Teaching," *J. Chem. Educ.* **1999,** *76,* 366.

Literature Long, Steve. "The Science Teacher: Summer 2003," *J. Chem. Educ.* **2003,** *80,* 987.

Literature DePierro, Ed; Garafalo, Fred; Toomey, Richard T. "Using a Socratic Dialog to Help Students Construct Fundamental Concepts," *J. Chem. Educ.* **2003,** *80,* 1408.

Literature King, Jessica R.; Hunter, William J. F.; Szczepura, Lisa F. "Design, Implementation, and Evaluation of Novel Discussion Exercises for First-Year Chemistry Students," *Chem. Educator* 7 **2002,** *6,* 329–333.

1.1 SOME FUNDAMENTAL DEFINITIONS

The science of chemistry deals with the makeup of the entire physical universe. A good place to begin our discussion is with the definition of a few central ideas, some of which may already be familiar to you. **Chemistry** is *the study of matter and its properties, the changes that matter undergoes, and the energy associated with those changes.*

The Properties of Matter

Matter is the "stuff" of the universe: air, glass, planets, students—*anything that has mass and volume.* (In Section 1.5, we discuss the meanings of mass and volume in terms of how they are measured.) Chemists are particularly interested in the **composition** of matter, that is, *the types and amounts of simpler substances that make it up.* A *substance* is a type of matter that has a defined, fixed composition.

We learn about matter by observing its **properties,** *the characteristics that give each substance its unique identity.* To identify a person, we observe such properties as height, weight, hair and eye color, fingerprints, and, now, even DNA pattern, until we arrive at a unique identification. To identify a substance, chemists observe two types of properties, physical and chemical, which are closely related to two types of change that matter undergoes. **Physical properties** are those that a substance shows *by itself, without changing into or interacting with another substance.* Some physical properties are color, melting point, electrical conductivity, and density.

A **physical change** occurs when a substance *alters its physical form, **not** its composition.* Thus, a physical change results in different physical properties. For example, when ice melts, several physical properties change, such as hardness, density, and ability to flow. But the composition of the sample does *not* change: it is still water. The photograph in Figure 1.1A shows what this change looks like in everyday life. In your imagination, try to see the magnified view that appears

Animation: The Three States of Matter
ARIS Presentation Center
Chapter 1

Transparency/ARIS Presentation Center,
Photo Library

Solid water

Liquid water

Hydrogen gas

Oxygen gas

A Physical change:
Solid form of water becomes liquid form; composition does *not* change because particles are the same.

B Chemical change:
Electric current decomposes water into different substances (hydrogen and oxygen); composition *does* change because particles are different.

Figure 1.1 The distinction between physical and chemical change.

in the "blow-up" circles. Here we see the particles that make up the sample; note that the same particles appear in solid and liquid water.

Physical change (same substance before and after):

$$\text{Water (solid form)} \longrightarrow \text{water (liquid form)}$$

On the other hand, **chemical properties** are those that a substance shows *as it changes into or interacts with another substance (or substances).* Some examples of chemical properties are flammability, corrosiveness, and reactivity with acids. A **chemical change,** also called a **chemical reaction,** occurs when *a substance (or substances) is converted into a different substance (or substances).*

Figure 1.1B shows the chemical change (reaction) that occurs when you pass an electric current through water: the water decomposes (breaks down) into two other substances, hydrogen and oxygen, each with physical and chemical properties different from each other *and* from water. The sample *has* changed its composition: it is no longer water, as you can see from the different particles in the magnified views.

Chemical change (different substances before and after):

$$\text{Water} \xrightarrow{\text{electric current}} \text{hydrogen gas + oxygen gas}$$

Let's work through a sample problem that helps you visualize this important distinction between physical and chemical change.

SAMPLE PROBLEM 1.1 Visualizing Change on the Atomic Scale

PROBLEM The scenes below represent an atomic-scale view of a sample of matter, A, undergoing two different changes, left to B and right to C:

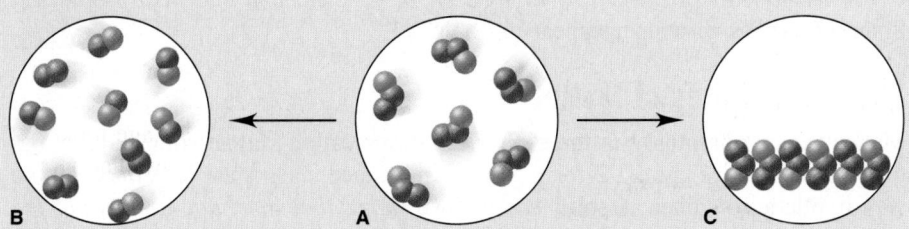

Decide whether each depiction shows a physical or chemical change.

PLAN Given depictions of the changes, we have to determine whether each represents a physical or a chemical change. The number and color of the little spheres that make up each particle tell its "composition." Samples with particles of the *same* composition but in a different form depict a *physical* change, and particles of a *different* composition depict a *chemical* change.

SOLUTION In A, each particle consists of one blue and two red spheres. The particles in A change into two types in B, one made of red and blue spheres and the other made of two red spheres; therefore, they have undergone a chemical change to form different particles in B. The particles in C are the same as those in A, though they are closer together and aligned; therefore, the conversion from A to C represents a physical change.

FOLLOW-UP PROBLEM 1.1 Is the following change chemical or physical?

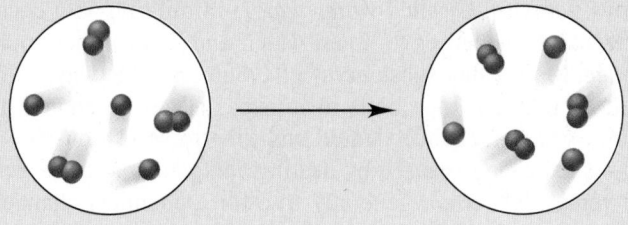

ARIS Presentation Center,
Worked Examples Library

ARIS Presentation Center,
Art Library

Demo Saint-Antonin, François. "Visual Experiments Supporting Four Basic Concepts in Chemistry," *J. Chem. Educ.* **2003,** *80,* 288.

Table 1.1	Some Characteristic Properties of Copper		
Physical Properties		**Chemical Properties**	

Reddish brown, metallic luster

Easily shaped into sheets (malleable) and wires (ductile)

Slowly forms a blue-green carbonate in moist air

Good conductor of heat and electricity

Can be melted and mixed with zinc to form brass

Reacts with nitric acid (photo) or sulfuric acid

Slowly forms deep-blue solution in aqueous ammonia

Density = 8.95 g/cm³
Melting point = 1083°C
Boiling point = 2570°C

● **The Incredible Range of Physical Change** Scientists often study physical change in remarkable settings, using instruments that allow observation far beyond the confines of the laboratory. Instruments aboard the *Voyager* and *Galileo* spacecrafts and the Hubble Space Telescope have measured temperatures on Jupiter's moon Io (shown here) that are hot enough to maintain lakes of molten sulfur and cold enough to create vast snowfields of sulfur dioxide and polar caps swathed in hydrogen sulfide frost. (On Earth, sulfur dioxide is one of the gases released from volcanoes and coal-fired power plants, and hydrogen sulfide occurs in swamp gas.)

A substance is identified by its own set of physical and chemical properties. Some properties of copper appear in Table 1.1.

The Three States of Matter

Matter occurs commonly in three physical forms called **states:** solid, liquid, and gas. As shown in Figure 1.2 for a general substance, each state is defined by the way it fills a container. A **solid** *has a fixed shape that does not conform to the container shape.* Solids are *not* defined by rigidity or hardness: solid iron is rigid, but solid lead is flexible and solid wax is soft. A **liquid** *conforms to the container shape but fills the container only to the extent of the liquid's volume;* thus, a liquid forms a *surface.* A **gas** *conforms to the container shape also, but it fills the entire container,* and thus, does *not* form a surface. Now, look at the views within the blow-up circles of the figure. The particles in the solid lie next to each other in a regular, three-dimensional array with a definite pattern. Particles in the liquid also lie close together but move randomly around one another. Particles in the gas have great distances between them as they move randomly throughout the container.

Depending on the temperature and pressure of the surroundings, many substances can exist in each of the three physical states and undergo changes in state as well. As the temperature increases, solid water melts to liquid water, which boils to gaseous water (also called *water vapor*). Similarly, with decreasing temperature, water vapor condenses to liquid water, and with further cooling, the liquid freezes to ice. Many other substances behave in the same way: solid iron melts to liquid (molten) iron and then boils to iron gas at a high enough temperature. Cooling the iron gas changes it to liquid and then to solid iron. ●

Thus, a physical change caused by heating can generally be reversed by cooling, and vice versa. This is *not* generally true for a chemical change. For example, heating iron in moist air causes a chemical reaction that yields the brown, crumbly substance known as rust. Cooling does not reverse this change; rather, another chemical change (or series of them) is required.

Figure 1.2 **The physical states of matter.** The magnified (blow-up) views show the atomic-scale arrangement of the particles in the three states of matter.

Transparency/ARIS Presentation Center, Art Library

Solid
Particles close together and organized

Liquid
Particles close together but disorganized

Gas
Particles far apart and disorganized

Literature Wilcox, C. Jayne. "An Inquiry-Based Classroom Activity on States of Matter," *Chem. Educator* 9 **2004,** *5,* 270–271.

To summarize the key distinctions:

- A physical change leads to a different form of the same substance (same composition), whereas a chemical change leads to a different substance (different composition).
- A physical change caused by a temperature change can generally be reversed by the opposite temperature change, but this is not generally true of a chemical change.

The following sample problem provides some familiar examples of these types of changes.

ARIS Presentation Center, Worked Examples Library

SAMPLE PROBLEM 1.2 Distinguishing Between Physical and Chemical Change

PROBLEM Decide whether each of the following processes is primarily a physical or a chemical change, and explain briefly:
(a) Frost forms as the temperature drops on a humid winter night.
(b) A cornstalk grows from a seed that is watered and fertilized.
(c) A match ignites to form ash and a mixture of gases.
(d) Perspiration evaporates when you relax after jogging.
(e) A silver fork tarnishes slowly in air.
PLAN The basic question we ask to decide whether a change is chemical or physical is, "Does the substance change composition or just change form?"
SOLUTION **(a)** Frost forming is a physical change: the drop in temperature changes water vapor (gaseous water) in humid air to ice crystals (solid water).
(b) A seed growing involves chemical change: the seed uses water, substances from air, fertilizer, and soil, and energy from sunlight to make complex changes in composition.
(c) The match burning is a chemical change: the combustible substances in the matchhead are converted into other substances.
(d) Perspiration evaporating is a physical change: the water in sweat changes its form, from liquid to gas, but not its composition.
(e) Tarnishing is a chemical change: silver changes to silver sulfide by reacting with sulfur-containing substances in the air.

Literature Yezierski, Ellen J.; Birk, James P. "Misconceptions about the Particulate Nature of Matter: Using Animations to Close the Gender Gap," *J. Chem. Educ.* **2006,** *83,* 954.

FOLLOW-UP PROBLEM 1.2 Decide whether each of the following processes is primarily a physical or a chemical change, and explain briefly:
(a) Purple iodine vapor appears when solid iodine is warmed.
(b) Gasoline fumes are ignited by a spark in an automobile engine's cylinder.
(c) A scab forms over an open cut.

Methane and oxygen form carbon dioxide and water in the flame of a lab burner.

ARIS Presentation Center, Art Library

The Central Theme in Chemistry

Understanding the properties of a substance and the changes it undergoes leads to the central theme in chemistry: *macroscopic* properties and behavior, those we can see, are the results of *submicroscopic* properties and behavior that we cannot see. The distinction between chemical and physical change is defined by composition, which we study macroscopically. But it ultimately depends on the makeup of substances at the atomic scale, as the magnified views of Figure 1.1 show. Similarly, the defining properties of the three states of matter are macroscopic, but they arise from the submicroscopic behavior shown in the magnified views of Figure 1.2. Picturing a chemical event on the molecular scale, even one as common as the flame of a laboratory burner (see margin), helps clarify what is taking place. What is happening when water boils or copper melts? What events occur in the invisible world of minute particles that cause a seed to grow, a neon light to glow, or a nail to rust? Throughout the text, we return to this central idea: we study *observable* changes in matter to understand their *unobservable* causes.

The Importance of Energy in the Study of Matter

In general, physical and chemical changes are accompanied by energy changes. **Energy** is often defined as *the ability to do work.* Essentially, all work involves moving something. Work is done when your arm lifts a book, when an engine moves a car's wheels, or when a falling rock moves the ground as it lands. The object doing the work (arm, engine, rock) transfers some of the energy it possesses to the object on which the work is done (book, wheels, ground).

The total energy an object possesses is the sum of its potential energy and its kinetic energy. **Potential energy** is the *energy due to the **position** of the object.* **Kinetic energy** is the *energy due to the **motion** of the object.* Let's examine four systems that illustrate the relationship between these two forms of energy: (1) a weight raised above the ground, (2) two balls attached by a spring, (3) two electrically charged particles, and (4) a fuel and its waste products. A key concept illustrated by all four cases is that *energy is conserved: it may be converted from one form to the other, but it is not destroyed.*

Suppose you lift a weight off the ground, as in Figure 1.3A. The energy you use to move the weight against the gravitational attraction of Earth increases the weight's potential energy (energy due to its position). When the weight is dropped, this additional potential energy is converted to kinetic energy (energy due to motion). Some of this kinetic energy is transferred to the ground as the weight does work, such as driving a stake or simply moving dirt and pebbles. As you can see, the added potential energy does not disappear, but is converted to kinetic energy.

In nature, situations of lower energy are typically favored over those of higher energy: because the weight has less potential energy (and thus less total energy) at rest on the ground than held in the air, it will fall when released. Therefore, the situation with the weight elevated and higher in potential energy is *less stable,* and the situation after the weight has fallen and is lower in potential energy is *more stable.*

Next, consider the two balls attached by a relaxed spring depicted in Figure 1.3B. When you pull the balls apart, the energy you exert to stretch the spring increases the system's potential energy. This change in potential energy is converted

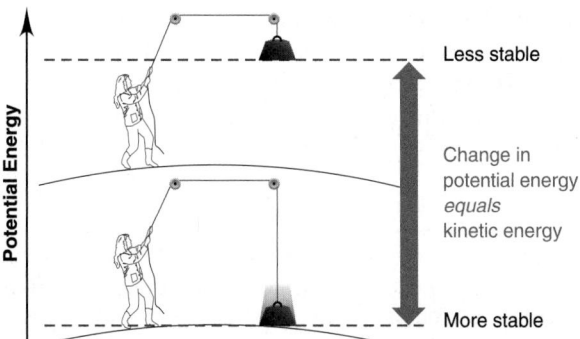

A **A gravitational system.** The potential energy gained when a weight is lifted is converted to kinetic energy as the weight falls.

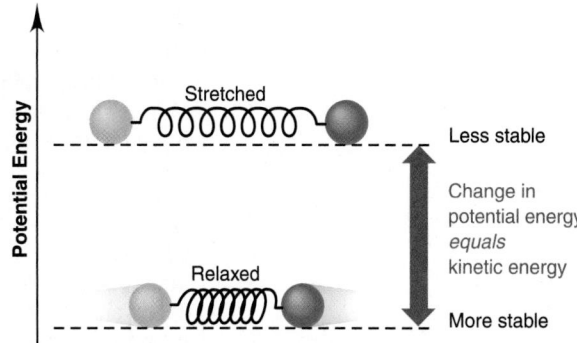

B **A system of two balls attached by a spring.** The potential energy gained when the spring is stretched is converted to the kinetic energy of the moving balls when it is released.

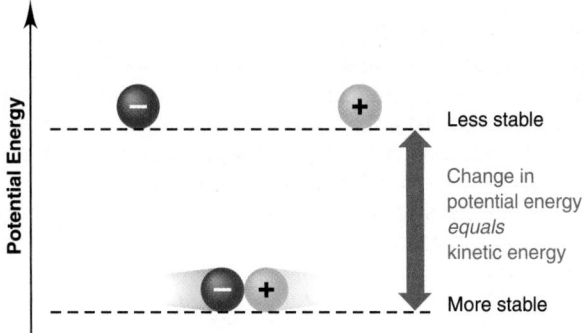

C **A system of oppositely charged particles.** The potential energy gained when the charges are separated is converted to kinetic energy as the attraction pulls them together.

D **A system of fuel and exhaust.** A fuel is higher in chemical potential energy than the exhaust. As the fuel burns, some of its potential energy is converted to the kinetic energy of the moving car.

Figure 1.3 **Potential energy is converted to kinetic energy.** In all four parts of the figure, the dashed horizontal lines indicate the potential energy of the system in each situation.

Transparency/ARIS Presentation Center, Art Library

ARIS Presentation Center, Active Art Library

to kinetic energy when you release the balls and they move closer together. The system of balls and spring is less stable (has more potential energy) when the spring is stretched than when it is relaxed.

There are no springs in a chemical substance, of course, but the following situation is similar in terms of energy. Much of the matter in the universe is composed of positively and negatively charged particles. A well-known behavior of charged particles (similar to the behavior of the poles of magnets) results from interactions known as *electrostatic forces: opposite charges attract each other, and like charges repel each other.* When work is done to separate a positive particle from a negative one, the potential energy of the particles increases. As Figure 1.3C shows, that increase in potential energy is converted to kinetic energy when the particles move together again. Also, when two positive (or two negative) particles are pushed toward each other, their potential energy increases, and when they are allowed to move apart, that increase in potential energy is changed into kinetic energy. Like the weight above the ground and the balls connected by a spring, charged particles move naturally toward a position of lower energy, which is more stable.

The chemical potential energy of a substance results from the relative positions and the attractions and repulsions among all its particles. Some substances are richer in this chemical potential energy than others. Fuels and foods, for example, contain more potential energy than the waste products they form. Figure 1.3D shows that when gasoline burns in a car engine, substances with higher chemical potential energy (gasoline and air) form substances with lower potential energy (exhaust gases). This difference in potential energy is converted into the kinetic energy that moves the car, heats the passenger compartment, makes the lights shine, and so forth. Similarly, the difference in potential energy between the food and air we take in and the waste products we excrete is used to move, grow, keep warm, study chemistry, and so on. Note again the essential point: *energy is neither created nor destroyed—it is always conserved as it is converted from one form to the other.*

Section Summary

Chemists study the composition and properties of matter and how they change. • Each substance has a unique set of physical properties (attributes of the substance itself) and chemical properties (attributes of the substance as it interacts with or changes to other substances). • Changes in matter can be physical (different form of the same substance) or chemical (different substance). • Matter exists in three physical states—solid, liquid, and gas. The observable features that distinguish these states reflect the arrangement of a substance's particles. • A change in physical state brought about by heating may be reversed by cooling. A chemical change can be reversed only by other chemical changes. Macroscopic changes result from submicroscopic changes. • Changes in matter are accompanied by changes in energy. • An object's potential energy is due to its position; an object's kinetic energy is due to its motion. Energy used to lift a weight, stretch a spring, or separate opposite charges increases the system's potential energy, which is converted to kinetic energy as the system returns to its original condition. • Chemical potential energy arises from the positions and interactions of the particles in a substance. Higher energy substances are less stable than lower energy substances. When a less stable substance is converted into a more stable substance, some potential energy is converted into kinetic energy, which can do work.

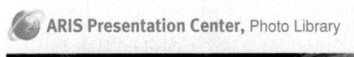
ARIS Presentation Center, Photo Library

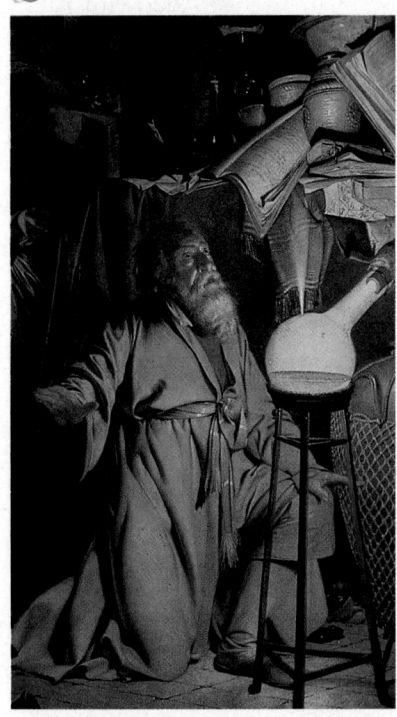

Figure 1.4 An alchemist at work. The apparatus shown here is engaged in distillation, a process still commonly used to separate substances. This is a portion of a painting by the Englishman Joseph Wright. Some think it portrays the German alchemist Hennig Brand in his laboratory, lit by the glow of phosphorus, which he discovered in 1669.

1.2 CHEMICAL ARTS AND THE ORIGINS OF MODERN CHEMISTRY

Chemistry has a rich, colorful history. Even some concepts and discoveries that led temporarily along confusing paths have contributed to the heritage of chemistry. This brief overview of early breakthroughs and false directions provides some insight into how modern chemistry arose and how science progresses.

Prechemical Traditions

Chemistry has its origin in a prescientific past that incorporated three overlapping traditions: alchemy, medicine, and technology.

The Alchemical Tradition The occult study of nature practiced in the 1st century AD by Greeks living in northern Egypt later became known by the Arabic name **alchemy.** Its practice spread through the Near East and into Europe, where it dominated Western thinking about matter for more than 1500 years! Alchemists were influenced by the Greek idea that matter naturally strives toward perfection, and they searched for ways to change less valued substances into precious ones. What started as a search for spiritual properties in matter evolved over a thousand years into an obsession with potions to bestow eternal youth and elixirs to transmute "baser" metals, such as lead, into "purer" ones, such as gold.

Alchemy's legacy to chemistry is mixed at best. The confusion arising from alchemists' use of different names for the same substance and from their belief that matter could be altered magically was very difficult to eliminate. Nevertheless, through centuries of laboratory inquiry, alchemists invented the chemical methods of distillation, percolation, and extraction, and they devised apparatus that today's chemists use routinely (Figure 1.4). Most important, alchemists encouraged the widespread acceptance of observation and experimentation, which replaced the Greek approach of studying nature solely through reason.

The Medical Tradition Alchemists greatly influenced medical practice in medieval Europe. Since the 13th century, distillates and extracts of roots, herbs, and other plant matter have been used as sources of medicines. Paracelsus (1493–1541), an active alchemist and important physician of the time, considered the body to be a chemical system whose balance of substances could be restored by medical treatment. His followers introduced mineral drugs into 17th-century pharmacy. Although many of these drugs were useless and some harmful, later practitioners employed other mineral prescriptions with increasing success. Thus began the indispensable alliance between medicine and chemistry that thrives today.

The Technological Tradition For thousands of years, people have developed technological skills to carry out changes in matter. Pottery making, dyeing, and especially metallurgy (begun about 7000 years ago) contributed greatly to experience with the properties of materials. During the Middle Ages and the Renaissance, such technology flourished. Books describing how to purify, assay, and coin silver and gold and how to use balances, furnaces, and crucibles were published and regularly updated. Other writings discussed making glass, gunpowder, and other materials. Some even introduced quantitative measurement, which had been lacking in alchemical writings. Many creations of these early artisans are still unsurpassed today, and we marvel at them in the great centers of world art. Nevertheless, even though the artisans' working knowledge of substances was expert, their approach to understanding matter shows little interest in exploring *why* a substance changes or *how to predict* its behavior.

The Phlogiston Fiasco and the Impact of Lavoisier

Chemical investigation in the modern sense—inquiry into the causes of changes in matter—began in the late 17th century but was hampered by an incorrect theory of **combustion,** the process of burning.

At the time, most scientists embraced the **phlogiston theory,** which held sway for nearly 100 years. The theory proposed that combustible materials contain varying amounts of an undetectable substance called *phlogiston,* which is released when the material burns. Highly combustible materials like charcoal contain a lot of phlogiston and thus release a lot when they burn, whereas slightly combustible materials like metals contain very little and thus release very little.

However, the theory could not answer some key questions from its critics: "Why is air needed for combustion, and why does charcoal stop burning in a closed vessel?" The theory's supporters responded that air "attracted" the phlogiston out of the charcoal, and that burning in a vessel stops when the air is "saturated" with phlogiston. When a metal burns, it forms its *calx,* which weighs more than the metal, so critics asked, "How can the *loss* of phlogiston cause a *gain* in mass?" Supporters proposed that phlogiston had negative mass! These responses seem ridiculous now, but they point out that the pursuit of science, like any other endeavor, is subject to human failings; even today, it is easier to dismiss conflicting evidence than to give up an established idea.

Into this chaos of "explanations" entered the young French chemist Antoine Lavoisier (1743–1794), who demonstrated the true nature of combustion. ● In a series of careful measurements, Lavoisier heated mercury calx, decomposing it into mercury and a gas, whose combined masses equaled the starting mass of calx. The reverse experiment—heating mercury with the gas—re-formed the mercury calx, and again, *the total mass remained constant.* Lavoisier proposed that when a metal forms its calx, it does not lose phlogiston but rather combines with this gas, which must be a component of air. To test this idea, Lavoisier heated mercury in a measured volume of air to form mercury calx and noted that only four-fifths of the air volume remained. He placed a burning candle in the remaining air, and it went out, showing that the gas that had combined with the mercury

🏛 **Literature** Laing, Michael. "Lavoisier Preempted Gay-Lussac by 20 Years!" *J. Chem. Educ.* **1998,** *75,* 177.

🏛 **Literature** Giunta, Carmen J. "Using History to Teach Scientific Method: The Role of Errors," *J. Chem. Educ.* **2001,** *78,* 623.

🌐 **ARIS Presentation Center,** Photo Library

● **Scientific Thinker Extraordinaire** Lavoisier's fame would be widespread, even if he had never performed a chemical experiment. A short list of his other contributions: He improved the production of French gunpowder, which became a key factor in the success of the American Revolution. He established on his farm a scientific balance between cattle, pasture, and cultivated acreage to optimize crop yield. He developed public assistance programs for widows and orphans. He quantified the relation of fiscal policy to agricultural production. He proposed a system of free public education and of societies to foster science, politics, and the arts. He sat on the committee that unified weights and measures in the new metric system. His research into combustion clarified the essence of respiration and metabolism. To support these pursuits, he joined a firm that collected taxes for the king, and only this role was remembered during the French Revolution. Despite his contributions to French society, the father of modern chemistry was guillotined at the age of 50.

● **A Great Chemist Yet Strict Phlogistonist** Despite the phlogiston theory, chemists made key discoveries during the years it held sway. Many were made by the English clergyman Joseph Priestley (1733–1804), who systematically studied the physical and chemical properties of many gases (inventing "soda water," carbon dioxide dissolved in water, along the way). The gas obtained by heating mercury calx was of special interest to him. In 1775, he wrote to his friend Benjamin Franklin: "Hitherto only two mice and myself have had the privilege of breathing it." Priestley also demonstrated that the gas supports combustion, but he drew the wrong conclusion about it. He called the gas "dephlogisticated air," air devoid of phlogiston, and thus ready to attract it from a burning substance. Priestley's contributions make him one of the great chemists of all time. He was a liberal thinker, favoring freedom of conscience and supporting both the French and American Revolutions, positions that caused severe personal problems throughout his later life. But, scientifically, he remained a conservative, believing strictly in phlogiston and refusing to accept the new theory of combustion.

was necessary for combustion. Lavoisier named the gas *oxygen* and called metal calxes *metal oxides.*

Lavoisier's new theory of combustion made sense of the earlier confusion. A combustible substance such as charcoal stops burning in a closed vessel once it combines with all the available oxygen, and a metal oxide weighs more than the metal because it contains the added mass of oxygen. This theory triumphed because it relied on *quantitative, reproducible measurements,* not on the strange properties of undetectable substances. Because this approach is at the heart of science, many propose that the *science* of chemistry began with Lavoisier. ●

Section Summary

Alchemy, medicine, and technology established processes that have been important to chemists since the 17th century. These prescientific traditions placed little emphasis on objective experimentation, focusing instead on practical experience or mystical explanations. • The phlogiston theory dominated thinking about combustion for almost 100 years, but in the mid-1770s, Lavoisier showed that oxygen, a component of air, is required for combustion and combines with a substance as it burns.

1.3   THE SCIENTIFIC APPROACH: DEVELOPING A MODEL

The principles of chemistry have been modified over time and are *still* evolving. At the dawn of human experience, our ancestors survived through knowledge acquired by *trial and error:* which types of stone were hard enough to shape others, which plants were edible and which were poisonous, and so forth. Today, the science of chemistry, with its powerful *quantitative theories,* helps us understand the essential nature of materials to make better use of them and create new ones: specialized drugs, advanced composites, synthetic polymers, and countless other new materials (Figure 1.5).

Figure 1.5 Modern materials in a variety of applications. A, High-tension polymers in synthetic hip joints. **B,** Specialized polymers in clothing and sports gear. **C,** Medicinal agents in pills. **D,** Liquid crystals and semiconductors in electronic devices.

A   C   D

Figure 1.6 The scientific approach to understanding nature. Note that hypotheses and models are mental pictures that are changed to match observations and experimental results, *not* the other way around.

Transparency/ARIS Presentation Center, Art Library

Is there something special about the way scientists think? If we could break down a "typical" modern scientist's thought processes, we could organize them into an approach called the **scientific method.** This approach is not a stepwise checklist, but rather a flexible process of creative thinking and testing aimed at objective, verifiable discoveries about how nature works. It is very important to realize that there is no typical scientist and no single method, and that luck can and often has played a key role in scientific discovery. In general terms, the scientific approach includes the following parts (Figure 1.6):

1. ***Observations.*** These are the facts that our ideas must explain. Observation is basic to scientific thinking. The most useful observations are quantitative because they can be compared and allow trends to be seen. Pieces of quantitative information are **data.** When the same observation is made by many investigators in many situations with no clear exceptions, it is summarized, often in mathematical terms, and called a **natural law.** The observation that mass remains constant during chemical change—made by Lavoisier and numerous experimenters since—is known as the law of mass conservation (discussed in Chapter 2).

2. ***Hypothesis.*** Whether derived from actual observation or from a "spark of intuition," a hypothesis is a proposal made to explain an observation. A sound hypothesis need not be correct, but it must be *testable.* Thus, a hypothesis is often the reason for performing an experiment. If the hypothesis is inconsistent with the experimental results, it must be revised or discarded.

3. ***Experiment.*** An experiment is a clear set of procedural steps that tests a hypothesis. Experimentation is the connection between our hypotheses about nature and nature itself. Often, hypothesis leads to experiment, which leads to revised hypothesis, and so forth. Hypotheses can be altered, but the results of an experiment cannot.

An experiment typically contains at least two **variables,** quantities that can have more than a single value. A well-designed experiment is **controlled** in that it measures the effect of one variable on another while keeping all others constant. For experimental results to be accepted, they must be *reproducible,* not only by the person who designed the experiment, but also by others. Both skill and creativity play a part in experimental design.

4. ***Model.*** Formulating conceptual models, or **theories,** *based on experiments* is what distinguishes scientific thinking from speculation. As hypotheses are revised according to experimental results, a model gradually emerges that describes how the observed phenomenon occurs. A model is not an exact representation of nature, but rather a simplified version of nature that can be used to make *predictions* about related phenomena. Further investigation refines a model by testing its predictions and altering it to account for new facts.

Literature Robinson, William R. "Chemistry Problem-Solving: Symbol, Macro, Micro, and Process Aspects," *J. Chem. Educ.* **2003**, *80*, 978.

Literature Holme, Thomas. "Divergence of Faculty Perceptions of General Chemistry and Problem Solving Skills," *J. Chem. Educ.* **2001**, *78*, 1578.

Literature DeLorenzo, Ronald. "Expanded Dimensional Analysis: A Blending of English and Math," *J. Chem. Educ.* **1994**, *71*, 789.

Literature Williams, Kathryn R. "Don't Forget the Units!" *J. Chem. Educ.* **1999**, *76*, 313.

Lavoisier's overthrow of the phlogiston theory demonstrates the scientific approach. *Observations* of burning and smelting led some to *hypothesize* that combustion involved the loss of phlogiston. *Experiments* by others showing that air is required for burning and that a metal gains mass during combustion led Lavoisier to propose a new *hypothesis,* which he tested repeatedly with quantitative *experiments.* Accumulating evidence supported his developing *model (theory)* that combustion involves combination with a component of air (oxygen). Innumerable *predictions* based on this theory have supported its validity. A sound theory remains useful even when minor exceptions appear. An unsound one, such as the phlogiston theory, eventually crumbles under the weight of contrary evidence and absurd refinements.

Section Summary

The scientific method is not a rigid sequence of steps, but rather a dynamic process designed to explain and predict real phenomena. • Observations (sometimes expressed as natural laws) lead to hypotheses about how or why something occurs. • Hypotheses are tested in controlled experiments and adjusted if necessary. • If all the data collected support a hypothesis, a model (theory) can be developed to explain the observations. A good model is useful in predicting related phenomena but must be refined if conflicting data appear. ●

1.4   CHEMICAL PROBLEM SOLVING

In many ways, learning chemistry is learning how to solve chemistry problems, not only those in exams or homework, but also more complex ones in professional life and society. (The Chemical Connections essay at the end of this chapter provides an example.) This textbook was designed to help strengthen your problem-solving skills. Almost every chapter contains sample problems that apply newly introduced ideas and skills and are worked out in detail. In this section, we discuss the problem-solving approach. Most problems include calculations, so let's first go over some important ideas about measured quantities.

Units and Conversion Factors in Calculations

All measured quantities consist of a number *and* a unit; a person's height is "6 feet," not "6." Ratios of quantities have ratios of units, such as miles/hour. (We discuss the most important units in chemistry in Section 1.5.) To minimize errors, try to make a habit of including units in all calculations.

The arithmetic operations used with measured quantities are the same as those used with pure numbers; in other words, units can be multiplied, divided, and canceled:

• A carpet measuring 3 feet (ft) by 4 ft has an area of

$$\text{Area} = 3 \text{ ft} \times 4 \text{ ft} = (3 \times 4)(\text{ft} \times \text{ft}) = 12 \text{ ft}^2$$

• A car traveling 350 miles (mi) in 7 hours (h) has a speed of

$$\text{Speed} = \frac{350 \text{ mi}}{7 \text{ h}} = \frac{50 \text{ mi}}{1 \text{ h}} \text{ (often written 50 mi·h}^{-1})$$

• In 3 hours, the car travels a distance of

$$\text{Distance} = 3 \text{ h} \times \frac{50 \text{ mi}}{1 \text{ h}} = 150 \text{ mi}$$

Conversion factors are ratios used to express a measured quantity in different units. Suppose we want to know the distance of that 150-mile car trip in feet. To convert the distance between miles and feet, we use equivalent quantities to construct the desired conversion factor. The equivalent quantities in this case are 1 mile and the number of feet in 1 mile:

$$1 \text{ mi} = 5280 \text{ ft}$$

We can construct two conversion factors from this equivalency. Dividing both sides by 5280 ft gives one conversion factor (shown in blue):

$$\frac{1 \text{ mi}}{5280 \text{ ft}} = \frac{5280 \text{ ft}}{5280 \text{ ft}} = 1$$

And, dividing both sides by 1 mi gives the other conversion factor (the inverse):

$$\frac{1 \text{ mi}}{1 \text{ mi}} = \frac{5280 \text{ ft}}{1 \text{ mi}} = 1$$

It's very important to see that, since the numerator and denominator of a conversion factor are equal, multiplying by a conversion factor is the same as multiplying by 1. Therefore, *even though the number and unit of the quantity change, the size of the quantity remains the same.*

In our example, we want to convert the distance in miles to the equivalent distance in feet. Therefore, we choose the conversion factor with units of feet in the numerator, because it cancels units of miles and gives units of feet:

$$\text{Distance (ft)} = 150 \text{ mi} \times \frac{5280 \text{ ft}}{1 \text{ mi}} = 792{,}000 \text{ ft}$$

$$\text{mi} \implies \text{ft}$$

Choosing the correct conversion factor is made much easier if you think through the calculation to decide whether the answer expressed in the new units should have a larger or smaller number. In the previous case, we know that a foot is *smaller* than a mile, so the distance in feet should have a *larger* number (792,000) than the distance in miles (150). The conversion factor has the larger number (5280) in the numerator, so it gave a larger number in the answer. The main goal is that *the chosen conversion factor cancels all units except those required for the answer.* Set up the calculation so that the unit you are converting *from* (beginning unit) is in the *opposite position in the conversion factor* (numerator or denominator). It will then cancel and leave the unit you are converting *to* (final unit):

$$\text{beginning unit} \times \frac{\text{final unit}}{\text{beginning unit}} = \text{final unit} \qquad \text{as in} \qquad \text{mi} \times \frac{\text{ft}}{\text{mi}} = \text{ft}$$

Or, in cases that involve units raised to a power,

$$(\text{beginning unit} \times \text{beginning unit}) \times \frac{\text{final unit}^2}{\text{beginning unit}^2} = \text{final unit}^2$$

$$\text{as in} \qquad (\text{ft} \times \text{ft}) \times \frac{\text{mi}^2}{\text{ft}^2} = \text{mi}^2$$

Or, in cases that involve a ratio of units,

$$\frac{\text{beginning unit}}{\text{final unit}_1} \times \frac{\text{final unit}_2}{\text{beginning unit}} = \frac{\text{final unit}_2}{\text{final unit}_1} \qquad \text{as in} \qquad \frac{\text{mi}}{\text{h}} \times \frac{\text{ft}}{\text{mi}} = \frac{\text{ft}}{\text{h}}$$

We use the same procedure to convert between systems of units, for example, between the English (or American) unit system and the International System (a revised metric system discussed fully in Section 1.5). Suppose we know the height of Angel Falls in Venezuela (Figure 1.7) to be 3212 ft, and we find its height in miles as

$$\text{Height (mi)} = 3212 \text{ ft} \times \frac{1 \text{ mi}}{5280 \text{ ft}} = 0.6083 \text{ mi}$$

$$\text{ft} \implies \text{mi}$$

Now, we want its height in kilometers (km). The equivalent quantities are

$$1.609 \text{ km} = 1 \text{ mi}$$

Figure 1.7 Angel Falls. The world's tallest waterfall is 3212 ft high.

Because we are converting from miles to kilometers, we use the conversion factor with kilometers in the numerator in order to cancel miles:

$$\text{Height (km)} = 0.6083 \; \cancel{\text{mi}} \times \frac{1.609 \text{ km}}{1 \; \cancel{\text{mi}}} = 0.9788 \text{ km}$$

$$\text{mi} \implies \text{km}$$

Notice that, since kilometers are *smaller* than miles, this conversion factor gave us a *larger* number (0.9788 is larger than 0.6083).

If we want the height of Angel Falls in meters (m), we use the equivalent quantities 1 km = 1000 m to construct the conversion factor:

$$\text{Height (m)} = 0.9788 \; \cancel{\text{km}} \times \frac{1000 \text{ m}}{1 \; \cancel{\text{km}}} = 978.8 \text{ m}$$

$$\text{km} \implies \text{m}$$

In longer calculations, we often string together several conversion steps:

$$\text{Height (m)} = 3212 \text{ ft} \times \frac{1 \; \cancel{\text{mi}}}{5280 \; \cancel{\text{ft}}} \times \frac{1.609 \; \cancel{\text{km}}}{1 \; \cancel{\text{mi}}} \times \frac{1000 \text{ m}}{1 \; \cancel{\text{km}}} = 978.8 \text{ m}$$

$$\text{ft} \implies \text{mi} \implies \text{km} \implies \text{m}$$

The use of conversion factors in calculations is known by various names, such as the factor-label method or **dimensional analysis** (because units represent physical dimensions). We use this method in quantitative problems throughout the text.

A Systematic Approach to Solving Chemistry Problems

The approach we use in this text provides a systematic way to work through a problem. It emphasizes reasoning, not memorizing, and is based on a very simple idea: plan how to solve the problem *before* you go on to solve it, and then check your answer. Try to develop a similar approach on homework and exams. In general, the sample problems consist of several parts:

1. **Problem.** This part states all the information you need to solve the problem (usually framed in some interesting context).

2. **Plan.** The overall solution is broken up into two parts, *plan* and *solution,* to make a point: *think* about how to solve the problem *before* juggling numbers. There is often more than one way to solve a problem, and the plan shown in a given problem is just one possibility; develop a plan that suits you best. The plan will
   • Clarify the known and unknown. (What information do you have, and what are you trying to find?)
   • Suggest the steps from known to unknown. (What ideas, conversions, or equations are needed to solve the problem?)
   • Present a "roadmap" of the solution for many problems in early chapters (and some in later ones). The roadmap is a visual summary of the planned steps. Each step is shown by an arrow labeled with information about the conversion factor or operation needed.

3. **Solution.** In this part, the steps appear in the same order as in the plan.

4. **Check.** In most cases, a quick check is provided to see if the results make sense: Are the units correct? Does the answer seem to be the right size? Did the change occur in the expected direction? Is it reasonable chemically? We often do a rough calculation to see if the answer is "in the same ballpark" as the calculated result, just to make sure we didn't make a large error. *Always* check your answers, especially in a multipart problem, where an error in an early step can affect all later steps. Here's a typical "ballpark" calculation in everyday life. You are at the music store and buy three CDs at $14.97 each. With a 5% sales tax, the bill comes to $47.16. In your mind, you quickly check that 3 times approximately $15 is $45, and, given the sales tax, the cost should be a bit more. So, the amount of the bill *is* in the right ballpark.

Literature McCalla, Joanne. "Problem Solving with Pathways," *J. Chem. Educ.* **2003,** *80,* 92.

Literature Cohen, Jeff; Kennedy-Justice, Meghan; Pai, Sunny; Torres, Carmen; Toomey, Rick; DePierro, Ed; Garafalo, Fred. "Encouraging Meaningful Quantitative Problem Solving," *J. Chem. Educ.* **2000,** *77,* 1166.

Literature Akers, Brian N. "Appalachian Trail Problems," *J. Chem. Educ.* **1998,** *75,* 1571.

5. **Comment.** This part is included occasionally to provide additional information, such as an application, an alternative approach, a common mistake to avoid, or an overview.

6. **Follow-up Problem.** This part consists of a problem statement only and provides practice by applying the same ideas as the sample problem. Try to solve it *before* you look at the brief worked-out solution at the end of the chapter.

Of course, you can't learn to solve chemistry problems, any more than you can learn to swim, by reading about an approach. Practice is the key to mastery. Here are a few suggestions that can help:

- Follow along in the sample problem with pencil, paper, and calculator.
- Do the follow-up problem as soon as you finish studying the sample problem. Check your answer against the solution at the end of the chapter.
- Read the sample problem and text explanations again if you have trouble.
- Utilize the practice quizzes in the Online Learning Center for this text at www.mhhe.com/silberberg. For each chapter, two interactive quizzes provide conceptual and problem-solving practice and offer feedback on areas where you may need additional review.
- Work on as many of the problems at the end of the chapter as you can. They review and extend the concepts and skills in the text. Answers are given in the back of the book for problems with a colored number, but try to solve them yourself first. Let's apply this approach in a unit-conversion problem.

SAMPLE PROBLEM 1.3 Converting Units of Length

PROBLEM To wire your stereo equipment, you need 325 centimeters (cm) of speaker wire that sells for $0.15/ft. What is the price of the wire?

PLAN We know the length of wire in centimeters and the cost in dollars per foot ($/ft). We can find the unknown price of the wire by converting the length from centimeters to inches (in) and from inches to feet. Then the cost (1 ft = $0.15) gives us the equivalent quantities to construct the factor that converts feet of wire to price in dollars. The roadmap starts with the known and moves through the calculation steps to the unknown.

SOLUTION Converting the known length from centimeters to inches: The equivalent quantities alongside the roadmap arrow are the ones needed to construct the conversion factor. We choose 1 in/2.54 cm, rather than the inverse, because it gives an answer in inches:

$$\text{Length (in)} = \text{length (cm)} \times \text{conversion factor} = 325 \text{ cm} \times \frac{1 \text{ in}}{2.54 \text{ cm}} = 128 \text{ in}$$

Converting the length from inches to feet:

$$\text{Length (ft)} = \text{length (in)} \times \text{conversion factor} = 128 \text{ in} \times \frac{1 \text{ ft}}{12 \text{ in}} = 10.7 \text{ ft}$$

Converting the length in feet to price in dollars:

$$\text{Price (\$)} = \text{length (ft)} \times \text{conversion factor} = 10.7 \text{ ft} \times \frac{\$0.15}{1 \text{ ft}} = \$1.60$$

Length (cm) of wire

2.54 cm = 1 in

Length (in) of wire

12 in = 1 ft

Length (ft) of wire

1 ft = $0.15

Price ($) of wire

CHECK The units are correct for each step. The conversion factors make sense in terms of the relative unit sizes: the number of inches is *smaller* than the number of centimeters (an inch is *larger* than a centimeter), and the number of feet is *smaller* than the number of inches. The total price seems reasonable: a little more than 10 ft of wire at $0.15/ft should cost a little more than $1.50.

COMMENT 1. We could also have strung the three steps together:

$$\text{Price (\$)} = 325 \text{ cm} \times \frac{1 \text{ in}}{2.54 \text{ cm}} \times \frac{1 \text{ ft}}{12 \text{ in}} \times \frac{\$0.15}{1 \text{ ft}} = \$1.60$$

2. There are usually alternative sequences in unit-conversion problems. Here, for example, we would get the same answer if we first converted the cost of wire from $/ft to $/cm and kept the wire length in cm. Try it yourself.

FOLLOW-UP PROBLEM 1.3 A furniture factory needs 31.5 ft^2 of fabric to upholster one chair. Its Dutch supplier sends the fabric in bolts of exactly 200 m^2. What is the maximum number of chairs that can be upholstered by 3 bolts of fabric (1 m = 3.281 ft)?

Section Summary

A measured quantity consists of a number and a unit. A conversion factor is used to express a quantity in different units and is constructed as a ratio of equivalent quantities. • The problem-solving approach used in this text usually has four parts: (1) devise a plan for the solution, (2) put the plan into effect in the calculations, (3) check to see if the answer makes sense, and (4) practice with similar problems.

1.5 MEASUREMENT IN SCIENTIFIC STUDY

Almost everything we own—clothes, house, food, vehicle—is manufactured with measured parts, sold in measured amounts, and paid for with measured currency. Measurement is so commonplace that it's easy to take for granted, but it has a history characterized by the search for *exact, invariable standards.* ●

Our current system of measurement began in 1790, when the newly formed National Assembly of France, of which Lavoisier was a member, set up a committee to establish consistent unit standards. This effort led to the development of the *metric system.* In 1960, another international committee met in France to establish the International System of Units, a revised metric system now accepted by scientists throughout the world. The units of this system are called **SI units,** from the French **S**ystème **I**nternational d'Unités.

General Features of SI Units

As Table 1.2 shows, the SI system is based on a set of seven **fundamental units,** or **base units,** each of which is identified with a physical quantity.

Table 1.2 SI Base Units		
Physical Quantity (Dimension)	**Unit Name**	**Unit Abbreviation**
Mass	kilogram	kg
Length	meter	m
Time	second	s
Temperature	kelvin	K
Electric current	ampere	A
Amount of substance	mole	mol
Luminous intensity	candela	cd

All other units, called **derived units,** are combinations of these seven base units. For example, the derived unit for speed, meters per second (m/s), is the base unit for length (m) divided by the base unit for time (s). (Derived units that occur as a ratio of two or more base units can be used as conversion factors.) For quantities that are much smaller or much larger than the base unit, we use decimal prefixes and exponential (scientific) notation. Table 1.3 shows the most important prefixes. (If you need a review of exponential notation, read Appendix A.) Because these prefixes are based on powers of 10, SI units are easier to use in calculations than are English units such as pounds and inches.

Some Important SI Units in Chemistry

Some of the SI units we use early in the text are for quantities of length, volume, mass, density, temperature, and time. (Units for other quantities are presented in later chapters, as they are used.) Table 1.4 shows some useful SI quantities for length, volume, and mass, along with their equivalents in the English system.

Length The SI base unit of length is the **meter (m).** The standard meter is now based on two quantities, the speed of light in a vacuum and the second. ● A meter

● **How Many Barleycorns from His Majesty's Nose to His Thumb?** Systems of measurement began thousands of years ago as trade, building, and land surveying spread throughout the civilized world. For most of that time, however, measurement was based on inexact physical standards. For example, an inch was the length of three barleycorns (seeds) placed end to end; a yard was the distance from the tip of King Edgar's nose to the tip of his thumb with his arm outstretched; and an acre was the area tilled by one man working with a pair of oxen in a day.

Literature Freeman, Robert D. "SI for Chemists: Persistent Problems, Solid Solutions," *J. Chem. Educ.* **2003,** *80,* 16.

ARIS Presentation Center, Tables Library

Demo Earley, Clarke W. "A Simple Demonstration for Introducing the Metric System to Introductory Chemistry Classes," *J. Chem. Educ.* **1999,** *76,* 1215.

● **How Long Is a Meter?** The history of the meter exemplifies the ongoing drive to define units based on unchanging standards. The French scientists who set up the metric system defined the meter as 1/10,000,000 of the distance from the equator (through Paris!) to the North Pole. The meter was later redefined as the distance between two fine lines engraved on a corrosion-resistant metal bar kept at the International Bureau of Weights and Measures in France. Fear that the bar would be damaged by war led to the adoption of an exact, unchanging, universally available atomic standard: 1,650,763.73 wavelengths of orange-red light from electrically excited krypton atoms. The current standard is even more reliable: 1 meter is the distance light travels in a vacuum in 1/299,792,458 second.

Table 1.3 Common Decimal Prefixes Used with SI Units

Prefix*	Prefix Symbol	Word	Conventional Notation	Exponential Notation
tera	T	trillion	1,000,000,000,000	1×10^{12}
giga	G	billion	1,000,000,000	1×10^{9}
mega	M	million	1,000,000	1×10^{6}
kilo	k	thousand	1,000	1×10^{3}
hecto	h	hundred	100	1×10^{2}
deka	da	ten	10	1×10^{1}
—	—	one	1	1×10^{0}
deci	d	tenth	0.1	1×10^{-1}
centi	c	hundredth	0.01	1×10^{-2}
milli	m	thousandth	0.001	1×10^{-3}
micro	μ	millionth	0.000001	1×10^{-6}
nano	n	billionth	0.000000001	1×10^{-9}
pico	p	trillionth	0.000000000001	1×10^{-12}
femto	f	quadrillionth	0.000000000000001	1×10^{-15}

*The prefixes most frequently used by chemists appear in bold type.

Table 1.4 Common SI-English Equivalent Quantities

Quantity	SI	SI Equivalents	English Equivalents	English to SI Equivalent
Length	1 kilometer (km)	1000 (10^3) meters	0.6214 mile (mi)	1 mile = 1.609 km
	1 meter (m)	100 (10^2) centimeters	1.094 yards (yd)	1 yard = 0.9144 m
		1000 millimeters (mm)	39.37 inches (in)	1 foot (ft) = 0.3048 m
	1 centimeter (cm)	0.01 (10^{-2}) meter	0.3937 inch	1 inch = 2.54 cm (exactly)
Volume	1 cubic meter (m^3)	1,000,000 (10^6) cubic centimeters	35.31 cubic feet (ft^3)	1 cubic foot = 0.02832 m^3
	1 cubic decimeter (dm^3)	1000 cubic centimeters	0.2642 gallon (gal)	1 gallon = 3.785 dm^3
			1.057 quarts (qt)	1 quart = 0.9464 dm^3
				1 quart = 946.4 cm^3
	1 cubic centimeter (cm^3)	0.001 dm^3	0.03381 fluid ounce	1 fluid ounce = 29.57 cm^3
Mass	1 kilogram (kg)	1000 grams	2.205 pounds (lb)	1 pound = 0.4536 kg
	1 gram (g)	1000 milligrams (mg)	0.03527 ounce (oz)	1 ounce = 28.35 g

is a little longer than a yard (1 m = 1.094 yd); a centimeter (10^{-2} m) is about two-fifths of an inch (1 cm = 0.3937 in; 1 in = 2.54 cm). Biological cells are often measured in micrometers (1 μm = 10^{-6} m). On the atomic-size scale, nanometers and picometers are used (1 nm = 10^{-9} m; 1 pm = 10^{-12} m). Many proteins have diameters of about 2 nm; atomic diameters are about 200 pm (0.2 nm). An older unit still in use is the angstrom (1 Å = 10^{-10} m = 0.1 nm = 100 pm).

Volume Any sample of matter has a certain **volume (V),** the amount of space that the sample occupies. The SI unit of volume is the **cubic meter (m^3).** In chemistry, the most important volume units are non-SI units, the **liter (L)** and the **milliliter (mL)** (note the uppercase L). Physicians and other medical practitioners measure body fluids in cubic decimeters (dm^3), which is equivalent to liters:

$$1\ L = 1\ dm^3 = 10^{-3}\ m^3$$

As the prefix *milli-* indicates, 1 mL is $\frac{1}{1000}$ of a liter, and it is equal to exactly 1 cubic centimeter (cm^3):

$$1\ mL = 1\ cm^3 = 10^{-3}\ dm^3 = 10^{-3}\ L = 10^{-6}\ m^3$$

Figure 1.8 **Some volume relationships in SI.** The cube on the left is 1 dm^3. Each edge is 1 dm long and is divided into ten 1-cm segments. One of those segments forms an edge of the middle cube, which is 1 cm^3, and is divided into ten 1-mm segments. Each one of those segments forms an edge of the right cube, which is 1 mm^3.

Some volume equivalents:
$1 \ m^3 \ = 1000 \ dm^3$
$1 \ dm^3 = 1000 \ cm^3$
$\quad\quad = 1 \ L = 1000 \ mL$
$1 \ cm^3 = 1000 \ mm^3$
$\quad\quad = 1 \ mL = 1000 \ \mu L$
$1 \ mm^3 = 1 \ \mu L$

1 dm

1 dm

1 dm

1 cm

1 mm

1 cm

1 cm

A liter is slightly larger than a quart (qt) (1 L = 1.057 qt; 1 qt = 946.4 mL); 1 fluid ounce ($\frac{1}{32}$ of a quart) equals 29.57 mL (29.57 cm^3).

Figure 1.8 is a life-size depiction of the two 1000-fold decreases in volume from the cubic decimeter to the cubic millimeter. The edge of a cubic meter would be about 2.5 times the width of this textbook when open.

Figure 1.9 shows some of the types of laboratory glassware designed to contain liquids or measure their volumes. Many come in sizes from a few milliliters to a few liters. Erlenmeyer flasks and beakers are used to contain liquids. Graduated cylinders, pipets, and burets are used to measure and transfer liquids. Volumetric flasks and many pipets have a fixed volume indicated by a mark on the

Figure 1.9 **Common laboratory volumetric glassware. A,** From left to right are two graduated cylinders, a pipet being emptied into a beaker, a buret delivering liquid to an Erlenmeyer flask, and two volumetric flasks. **Inset,** In contact with glass, this liquid forms a concave meniscus (curved surface). **B,** Automatic pipets deliver a given volume of liquid.

A

B

neck. In quantitative work, liquid solutions are prepared in volumetric flasks, measured in cylinders, pipets, and burets, and then transferred to beakers or flasks for further chemical operations. Automatic pipets transfer a given volume of liquid accurately and quickly.

SAMPLE PROBLEM 1.4 Converting Units of Volume

ARIS Presentation Center, Worked Examples Library

PROBLEM The volume of an irregularly shaped solid can be determined from the volume of water it displaces. A graduated cylinder contains 19.9 mL of water. When a small piece of galena, an ore of lead, is added, it sinks and the volume increases to 24.5 mL. What is the volume of the piece of galena in cm^3 and in L?

PLAN We have to find the volume of the galena from the change in volume of the cylinder contents. The volume of galena in mL is the difference in the known volumes before and after adding it. The units mL and cm^3 represent identical volumes, so the volume of the galena in mL equals the volume in cm^3. We construct a conversion factor to convert the volume from mL to L. The calculation steps are shown in the roadmap.

SOLUTION Finding the volume of galena:

$$\text{Volume (mL)} = \text{volume after} - \text{volume before} = 24.5 \text{ mL} - 19.9 \text{ mL} = 4.6 \text{ mL}$$

Converting the volume from mL to cm^3:

$$\text{Volume (cm}^3\text{)} = 4.6 \text{ mL} \times \frac{1 \text{ cm}^3}{1 \text{ mL}} = 4.6 \text{ cm}^3$$

Converting the volume from mL to L:

$$\text{Volume (L)} = 4.6 \text{ mL} \times \frac{10^{-3} \text{ L}}{1 \text{ mL}} = 4.6 \times 10^{-3} \text{ L}$$

CHECK The units and magnitudes of the answers seem correct. It makes sense that the volume expressed in mL would have a number 1000 times larger than the volume expressed in L, because a milliliter is $\frac{1}{1000}$ of a liter.

FOLLOW-UP PROBLEM 1.4 Within a cell, proteins are synthesized on particles called ribosomes. Assuming ribosomes are generally spherical, what is the volume (in dm^3 and μL) of a ribosome whose average diameter is 21.4 nm (V of a sphere $= \frac{4}{3}\pi r^3$)?

Mass The **mass** of an object refers to the quantity of matter it contains. The SI unit of mass is the **kilogram (kg),** the only base unit whose standard is a physical object—a platinum-iridium cylinder kept in France. It is also the only base unit whose name has a prefix. (In contrast to the practice with other base units, however, we attach prefixes to the word "gram," as in "microgram," rather than to the word "kilogram"; thus, we never say "microkilogram.")

The terms *mass* and *weight* have distinct meanings. Since a given object's quantity of matter cannot change, its *mass is constant.* Its **weight,** on the other hand, depends on its mass *and* the strength of the local gravitational field pulling on it. Because the strength of this field varies with height above Earth's surface, the object's weight also varies. For instance, you actually weigh slightly less on a high mountaintop than at sea level.

Does this mean that if you weighed an object on a laboratory balance in Miami (sea level) and in Denver (about 1.7 km above sea level), you would obtain different results? Fortunately not, because such balances are designed to measure mass rather than weight. (We are actually "massing" an object when we weigh it on a balance, but we rarely use that term.) Mechanical balances compare the object's unknown mass with known masses built into the balance, so the local gravitational field pulls equally on them. Electronic (analytical) balances determine mass by generating an electric field that counteracts the local gravitational field. The magnitude of the current needed to restore the pan to its zero position is then displayed as the object's mass. Therefore, an electronic balance must be readjusted with standard masses when it is moved to a different location.

SAMPLE PROBLEM 1.5 Converting Units of Mass

PROBLEM International computer communications are often carried by optical fibers in cables laid along the ocean floor. If one strand of optical fiber weighs 1.19×10^{-3} lb/m, what is the mass (in kg) of a cable made of six strands of optical fiber, each long enough to link New York and Paris (8.84×10^3 km)?

PLAN We have to find the mass of cable (in kg) from the given mass/length of fiber, number of fibers/cable, and the length (distance from New York to Paris). One way to do this (as shown in the roadmap) is to first find the mass of one fiber and then find the mass of cable. We convert the length of one fiber from km to m and then find its mass (in lb) by using the lb/m factor. We multiply the fiber mass by six to get the cable mass, and finally we convert lb to kg.

SOLUTION Converting the fiber length from km to m:

$$\text{Length (m) of fiber} = 8.84 \times 10^3 \text{ km} \times \frac{10^3 \text{ m}}{1 \text{ km}} = 8.84 \times 10^6 \text{ m}$$

Converting the length of one fiber to mass (lb):

$$\text{Mass (lb) of fiber} = 8.84 \times 10^6 \text{ m} \times \frac{1.19 \times 10^{-3} \text{ lb}}{1 \text{ m}} = 1.05 \times 10^4 \text{ lb}$$

Finding the mass of the cable (lb):

$$\text{Mass (lb) of cable} = \frac{1.05 \times 10^4 \text{ lb}}{1 \text{ fiber}} \times \frac{6 \text{ fibers}}{1 \text{ cable}} = 6.30 \times 10^4 \text{ lb/cable}$$

Converting the mass of cable from lb to kg:

$$\text{Mass (kg) of cable} = \frac{6.30 \times 10^4 \text{ lb}}{1 \text{ cable}} \times \frac{1 \text{ kg}}{2.205 \text{ lb}} = 2.86 \times 10^4 \text{ kg/cable}$$

CHECK The units are correct. Let's think through the relative sizes of the answers to see if they make sense: The number of m should be 10^3 larger than the number of km. If 1 m of fiber weighs about 10^{-3} lb, about 10^7 m should weigh about 10^4 lb. The cable mass should be six times as much, or about 6×10^4 lb. Since 1 lb is about $\frac{1}{2}$ kg, the number of kg should be about half the number of lb.

COMMENT Actually, the pound (lb) is the English unit of *weight,* not *mass.* The English unit of mass, called the *slug,* is rarely used.

FOLLOW-UP PROBLEM 1.5

An intravenous bag delivers a nutrient solution to a hospital patient at a rate of 1.5 drops per second. If a drop weighs 65 mg on average, how many kilograms of solution are delivered in 8.0 h?

Figure 1.10 shows the ranges of some common lengths, volumes, and masses.

Density The **density** (*d*) of an object is its mass divided by its volume:

$$\text{Density} = \frac{\text{mass}}{\text{volume}} \tag{1.1}$$

Whenever needed, you can isolate mathematically each of the component variables by treating density as a conversion factor:

$$\text{Mass} = \text{volume} \times \text{density} = \text{volume} \times \frac{\text{mass}}{\text{volume}}$$

or,

$$\text{Volume} = \text{mass} \times \frac{1}{\text{density}} = \text{mass} \times \frac{\text{volume}}{\text{mass}}$$

Because volume may change with temperature, density may change also. But, under given conditions of temperature and pressure, *density is a characteristic physical property of a substance* and has a specific value. Mass and volume are examples of **extensive properties,** those dependent on the amount of substance. Density, on the other hand, is an **intensive property,** one that is independent of the amount of substance. For example, the mass of a gallon of water is four times the mass of a quart of water, but its volume is also four times greater; therefore, the

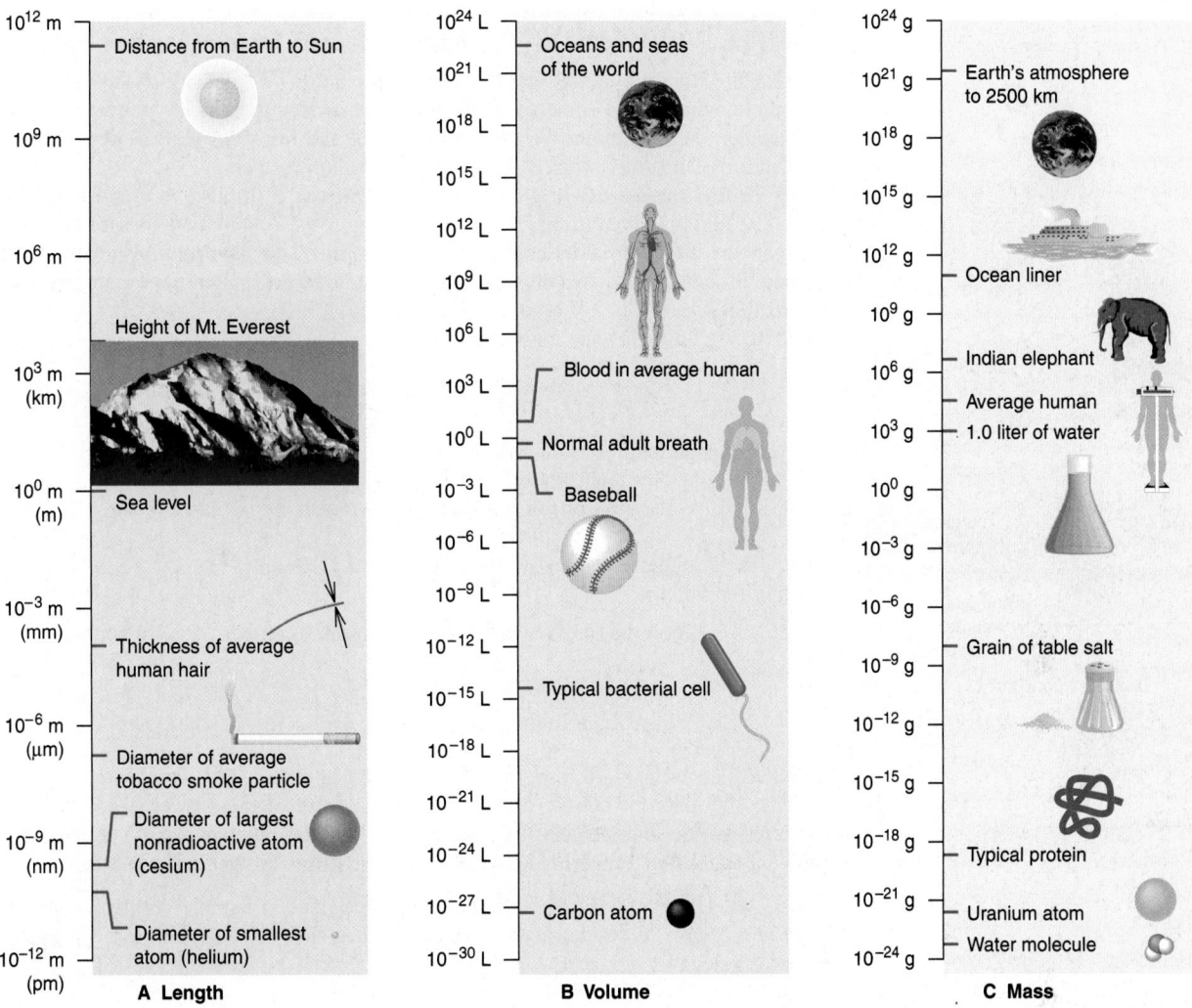

Figure 1.10 Some interesting quantities of length (A), volume (B), and mass (C). Note that the scales are exponential.

Transparency/ARIS Presentation Center, Art Library

density of water, the *ratio* of its mass to its volume, is constant at a particular temperature and pressure, regardless of the sample size.

The SI unit of density is the kilogram per cubic meter (kg/m^3), but in chemistry, density is typically given in units of g/L (g/dm^3) or g/mL (g/cm^3). For example, the density of liquid water at ordinary pressure and room temperature (20°C) is 1.0 g/mL. The densities of some common substances are given in Table 1.5. As you might expect from the magnified views of the physical states (see Figure 1.2), the densities of gases are much lower than those of liquids or solids.

ARIS Presentation Center, Tables Library

Table 1.5	**Densities of Some Common Substances***	
Substance	**Physical State**	**Density (g/cm³)**
Hydrogen	gas	0.0000899
Oxygen	gas	0.00133
Grain alcohol	liquid	0.789
Water	liquid	0.998
Table salt	solid	2.16
Aluminum	solid	2.70
Lead	solid	11.3
Gold	solid	19.3

*At room temperature (20°C) and normal atmospheric pressure (1 atm).

SAMPLE PROBLEM 1.6   Calculating Density from Mass and Volume

PROBLEM Lithium is a soft, gray solid that has the lowest density of any metal. It is an essential component of some advanced batteries, such as the one in your laptop. If a small rectangular slab of lithium weighs 1.49×10^3 mg and has sides that measure 20.9 mm by 11.1 mm by 11.9 mm, what is the density of lithium in g/cm^3?

PLAN To find the density in g/cm^3, we need the mass of lithium in g and the volume in cm^3. The mass is given in mg, so we convert mg to g. Volume data are not given, but we can convert the given side lengths from mm to cm, and then multiply them to find the volume in cm^3. Finally, we divide mass by volume to get density. The steps are shown in the roadmap.

SOLUTION Converting the mass from mg to g:

$$\text{Mass (g) of lithium} = 1.49 \times 10^3 \text{ mg} \left(\frac{10^{-3} \text{ g}}{1 \text{ mg}} \right) = 1.49 \text{ g}$$

Converting side lengths from mm to cm:

$$\text{Length (cm) of one side} = 20.9 \text{ mm} \times \frac{1 \text{ cm}}{10 \text{ mm}} = 2.09 \text{ cm}$$

Similarly, the other side lengths are 1.11 cm and 1.19 cm.
Finding the volume:

$$\text{Volume (cm}^3) = 2.09 \text{ cm} \times 1.11 \text{ cm} \times 1.19 \text{ cm} = 2.76 \text{ cm}^3$$

Calculating the density:

$$\text{Density of lithium} = \frac{\text{mass}}{\text{volume}} = \frac{1.49 \text{ g}}{2.76 \text{ cm}^3} = \boxed{0.540 \text{ g/cm}^3}$$

CHECK Since 1 cm = 10 mm, the number of cm in each length should be $\frac{1}{10}$ the number of mm. The units for density are correct, and the size of the answer (~0.5 g/cm^3) seems correct since the number of g (1.49) is about half the number of cm^3 (2.76). Since the problem states that lithium has a very low density, this answer makes sense.

FOLLOW-UP PROBLEM 1.6 The piece of galena in Sample Problem 1.4 has a volume of 4.6 cm^3. If the density of galena is 7.5 g/cm^3, what is the mass (in kg) of that piece of galena?

Temperature There is a common misunderstanding about heat and temperature. **Temperature** (*T*) is a measure of how hot or cold a substance is *relative to another substance*. **Heat** is the energy that flows between objects that are at different temperatures. Temperature is related to the *direction* of that energy flow: when two objects at different temperatures touch, energy flows from the one with the higher temperature to the one with the lower temperature until their temperatures are equal. When you hold an ice cube, its "cold" seems to flow *into* your hand; actually, heat flows *from* your hand into the ice. (In Chapter 6, we will see how heat is measured and how it is related to chemical and physical change.) Energy is an *extensive* property (as is volume), but temperature is an *intensive* property (as is density): a vat of boiling water has more energy than a cup of boiling water, but the temperatures of the two water samples are the same.

In the laboratory, the most common means for measuring temperature is the **thermometer,** a device that contains a fluid that expands when it is heated. When the thermometer's fluid-filled bulb is immersed in a substance hotter than itself, heat flows from the substance through the glass and into the fluid, which expands and rises in the thermometer tube. If a substance is colder than the thermometer, heat flows outward from the fluid, which contracts and falls within the tube.

The three temperature scales most important for us to consider are the Celsius (°C, formerly called centigrade), the Kelvin (K), and the Fahrenheit (°F) scales. The SI base unit of temperature is the **kelvin (K);** note that the kelvin has no degree sign (°). Figure 1.11 shows some interesting temperatures in this scale. The Kelvin scale, also known as the *absolute scale,* is preferred in all scientific work, although the Celsius scale is used frequently. In the United States, the Fahrenheit scale is still used for weather reporting, body temperature, and other everyday purposes. *The three scales differ in the size of the unit and/or the temperature of the zero point.* Figure 1.12 shows the freezing and boiling points of water in the three scales.

The **Celsius scale,** devised in the 18th century by the Swedish astronomer Anders Celsius, is based on changes in the physical state of water: 0°C is set at water's freezing point, and 100°C is set at its boiling point (at normal atmospheric pressure). The **Kelvin (absolute) scale** was devised by the English physicist William Thomson, known as Lord Kelvin, in 1854 during his experiments on the expansion and contraction of gases. *The Kelvin scale uses the same size degree unit as the Celsius scale—$\frac{1}{100}$ of the difference between the freezing and boiling points of water—but it differs in zero point.* The zero point in the Kelvin scale, 0 K, is called *absolute zero* and equals -273.15°C. In the Kelvin scale, *all temperatures have positive values.* Water freezes at $+273.15$ K (0°C) and boils at $+373.15$ K (100°C).

- 6×10^3: Surface of the Sun (interior $\approx 10^7$ K)
- 3683: Highest melting point of a metal element (tungsten)
- 1337: Melting point of gold
- 600: Melting point of lead
- 373: Boiling point of H_2O
- 370: Day on Moon
- 273: Melting point of H_2O
- 140: Jupiter cloud top
- 120: Night on Moon
- 90: Boiling point of oxygen
- 27: Boiling point of neon
- 0 K: Absolute zero (lowest attained temperature $\approx 10^{-9}$ K)

Figure 1.11 Some interesting temperatures.

Transparency/ARIS Presentation Center, Art Library

Figure 1.12 The freezing point and the boiling point of water in the Celsius, Kelvin (absolute), and Fahrenheit temperature scales. As you can see, this range consists of 100 degrees on the Celsius and Kelvin scales, but 180 degrees on the Fahrenheit scale. At the bottom of the figure, a portion of each of the three thermometer scales is expanded to show the sizes of the units. A Celsius degree (°C; *left*) and a kelvin (K; *center*) are the same size, and each is $\frac{9}{5}$ the size of a Fahrenheit degree (°F; *right*).

Literature Mossotti, Victor G. "A Simplified Algorithm for Temperature Conversions," *J. Chem. Educ.* **2003,** *80,* 1380.

Literature Mitschele, Jonathan. "An Interactive Graphical Approach to Temperature Conversions," *J. Chem. Educ.* **2002,** *79,* 1235.

ARIS Presentation Center, Worked Examples Library

We can convert between the Celsius and Kelvin scales by remembering the difference in zero points: since $0°C = 273.15$ K,

$$T \text{ (in K)} = T \text{ (in °C)} + 273.15 \qquad (1.2)$$

Solving Equation 1.2 for T (in °C) gives

$$T \text{ (in °C)} = T \text{ (in K)} - 273.15 \qquad (1.3)$$

The Fahrenheit scale differs from the other scales in its zero point *and* in the size of its unit. Water freezes at 32°F and boils at 212°F. Therefore, 180 Fahrenheit degrees (212°F − 32°F) represents the same temperature change as 100 Celsius degrees (or 100 kelvins). Because 100 Celsius degrees equal 180 Fahrenheit degrees,

$$1 \text{ Celsius degree} = \tfrac{180}{100} \text{ Fahrenheit degrees} = \tfrac{9}{5} \text{ Fahrenheit degrees}$$

To convert a temperature in °C to °F, first change the degree size and then adjust the zero point:

$$T \text{ (in °F)} = \tfrac{9}{5} T \text{ (in °C)} + 32 \qquad (1.4)$$

To convert a temperature in °F to °C, do the two steps in the opposite order; that is, first adjust the zero point and then change the degree size. In other words, solve Equation 1.4 for T (in °C):

$$T \text{ (in °C)} = [T \text{ (in °F)} - 32]\tfrac{5}{9} \qquad (1.5)$$

(The only temperature with the same numerical value in the Celsius and Fahrenheit scales is −40°; that is, −40°F = −40°C.)

SAMPLE PROBLEM 1.7 Converting Units of Temperature

PROBLEM A child has a body temperature of 38.7°C.
(a) If normal body temperature is 98.6°F, does the child have a fever?
(b) What is the child's temperature in kelvins?
PLAN (a) To find out if the child has a fever, we convert from °C to °F (Equation 1.4) and see whether 38.7°C is higher than 98.6°F.
(b) We use Equation 1.2 to convert the temperature in °C to K.
SOLUTION (a) Converting the temperature from °C to °F:

$T \text{ (in °F)} = \tfrac{9}{5} T \text{ (in °C)} + 32 = \tfrac{9}{5}(38.7°C) + 32 = 101.7°F$; yes, the child has a fever.

(b) Converting the temperature from °C to K:

$T \text{ (in K)} = T \text{ (in °C)} + 273.15 = 38.7°C + 273.15 = 311.8$ K

CHECK (a) From everyday experience, you know that 101.7°F is a reasonable temperature for someone with a fever.
(b) We know that a Celsius degree and a kelvin are the same size. Therefore, we can check the math by approximating the Celsius value as 40°C and adding 273: 40 + 273 = 313, which is close to our calculation, so there is no large error.

FOLLOW-UP PROBLEM 1.7 Mercury melts at 234 K, lower than any other pure metal. What is its melting point in °C and °F?

Time The SI base unit of time is the **second (s).** Although time was once measured by the day and year, it is now based on an atomic standard: microwave radiation absorbed by cesium atoms (Figure 1.13). In the laboratory, we study the speed (or rate) of a reaction by measuring the time it takes a fixed amount of substance to undergo a chemical change. The range of reaction rates is enormous: a fast reaction may be over in less than a nanosecond (10^{-9} s), whereas slow ones, such as rusting or aging, take years. Chemists now use lasers to study changes that occur in a few picoseconds (10^{-12} s) or femtoseconds (10^{-15} s).

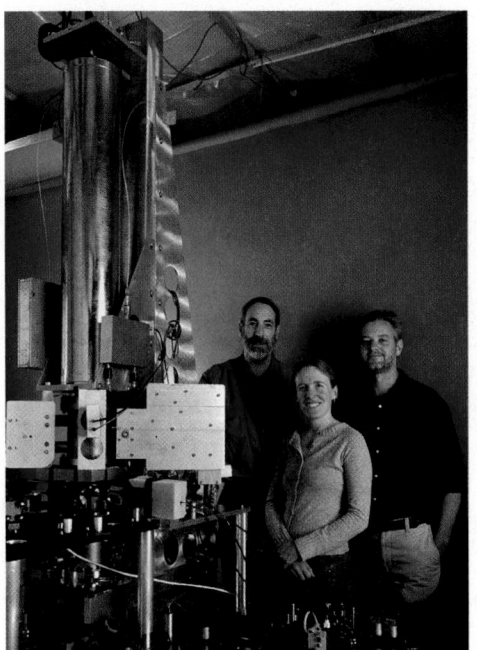

Figure 1.13 The cesium atomic clock.
The accuracy of the best pendulum clock is to within 3 seconds per year and that of the best quartz clock is 1000 times greater. The most recent version of the atomic clock, NIST-F1, developed by the Physics Laboratory of the National Institute of Standards and Technology, is over 6000 times more accurate still, to within 1 second in 20 million years! Rather than using the oscillations of a pendulum, the atomic clock measures the oscillations of microwave radiation absorbed by gaseous cesium atoms: 1 second is defined as 9,192,631,770 of these oscillations. This new clock cools the cesium atoms with infrared lasers to around 10^{-6} K, which allows much longer observation times of the atoms, and thus much greater accuracy.

ARIS Presentation Center, Photo Library

Section Summary

SI units consist of seven base units and numerous derived units. • Exponential notation and prefixes based on powers of 10 are used to express very small and very large numbers. • The SI base unit of length is the meter (m). Length units on the atomic scale are the nanometer (nm) and picometer (pm). Volume units are derived from length units; the most important volume units in chemistry are the cubic meter (m^3) and the liter (L). • The mass of an object, a measure of the quantity of matter present in it, is constant. The SI unit of mass is the kilogram (kg). The weight of an object varies with the gravitational field influencing it. Density (d) is the ratio of mass to volume of a substance and is one of its characteristic physical properties. • Temperature (T) is a measure of the relative hotness of an object. Heat is energy that flows from an object at higher temperature to one at lower temperature. Temperature scales differ in the size of the degree unit and/or the zero point. In chemistry, temperature is measured in kelvins (K) or degrees Celsius (°C). • Extensive properties, such as mass, volume, and energy, depend on the amount of a substance. Intensive properties, such as density and temperature, are independent of amount. •

> ● **The Central Importance of Measurement in Science** It's important to keep in mind *why* scientists measure things: "When you can measure what you are speaking about, and express it in numbers, you know something about it; but when you cannot measure it, . . . your knowledge is of a meager and unsatisfactory kind; it may be the beginning of knowledge, but you have scarcely, in your thoughts, advanced to the stage of science" (William Thomson, Lord Kelvin, 1824–1907).

Literature Crute, Thomas D. "Teaching Significant Figures Using Age Conversions," *J. Chem. Educ.* **2005,** *82,* 1507.

Literature Pacer, Richard A. "How Can an Instructor Best Introduce the Topic of Significant Figures to Students Unfamiliar with the Concept?" *J. Chem. Educ.* **2000,** *77,* 1435.

1.6 UNCERTAINTY IN MEASUREMENT: SIGNIFICANT FIGURES

We can never measure a quantity exactly, because measuring devices are made to limited specifications and we use our imperfect senses and skills to read them. Therefore, every measurement includes some **uncertainty.**

The measuring device we choose in a given situation depends on how much uncertainty we are willing to accept. When you buy potatoes, a supermarket scale that measures in 0.1-kg increments is perfectly acceptable; it tells you that the mass is, for example, 2.0 ± 0.1 kg. The term "± 0.1 kg" expresses the uncertainty in the measurement: the potatoes weigh between 1.9 and 2.1 kg. For a large-scale reaction, a chemist uses a lab balance that measures in 0.001-kg increments in order to obtain 2.036 ± 0.001 kg of a chemical, that is, between 2.035 and 2.037 kg. The greater number of digits in the mass of the chemical indicates that we know its mass with *more certainty* than we know the mass of the potatoes.

A 32.33°C   **B** 32.3°C

Figure 1.14 The number of significant figures in a measurement depends on the measuring device. A, Two thermometers measuring the same temperature are shown with expanded views. The thermometer on the left is graduated in 0.1°C and reads 32.33°C; the one on the right is graduated in 1°C and reads 32.3°C. Therefore, a reading with more significant figures (more certainty) can be made with the thermometer on the left. **B,** This modern electronic thermometer measures the resistance through a fine platinum wire in the probe to determine temperatures to the nearest microkelvin (10^{-6} K).

 ARIS Presentation Center, Art Library

Literature Abel, Kenton B.; Hemmerlin, William M. "Significant Figures," *J. Chem. Educ.* **1990,** *67,* 213.

We *always estimate the rightmost digit* when reading a measuring device. The uncertainty can be expressed with the ± sign, but generally we drop the sign and *assume an uncertainty of one unit in the rightmost digit.* The digits we record in a measurement, both the certain and the uncertain ones, are called **significant figures.** There are four significant figures in 2.036 kg and two in 2.0 kg. *The greater the number of significant figures in a measurement, the greater is the certainty.* Figure 1.14 shows this point for two thermometers.

Determining Which Digits Are Significant

When you take measurements or use them in calculations, you must know the number of digits that are significant. In general, *all digits are significant, except zeros that are not measured but are used only to position the decimal point.* Here is a simple procedure that applies this general point:

1. Make sure that the measured quantity has a decimal point.
2. Start at the left, move right until you reach the first nonzero digit.
3. Count that digit and every digit to its right as significant.

A complication may arise with zeros that *end* a number. Zeros that end a number and lie either after or before the decimal point *are* significant; thus, 1.030 mL has four significant figures, and 5300. L has four significant figures also. If there is no decimal point, as in 5300 L, we assume that the zeros are *not* significant; exponential notation is needed to show which of the zeros, if any, were measured and therefore are significant. Thus, 5.300×10^3 L has four significant figures, 5.30×10^3 L has three, and 5.3×10^3 L has only two. A terminal decimal point is used to clarify the number of significant figures; thus, 500 mL has one significant figure, but 5.00×10^2 mL, 500. mL, and 0.500 L have three.

ARIS Presentation Center, Worked Examples Library

SAMPLE PROBLEM 1.8 Determining the Number of Significant Figures

PROBLEM For each of the following quantities, underline the zeros that are significant figures (sf), and determine the number of significant figures in each quantity. For (**d**) to (**f**), express each in exponential notation first.
(a) 0.0030 L   **(b)** 0.1044 g   **(c)** 53,069 mL
(d) 0.00004715 m   **(e)** 57,600. s   **(f)** 0.0000007160 cm^3
PLAN We determine the number of significant figures by counting digits, as just presented, paying particular attention to the position of zeros in relation to the decimal point.

SOLUTION (a) 0.0030 L has 2 sf

(b) 0.1044 g has 4 sf

(c) 53,069 mL has 5 sf

(d) 0.00004715 m, or 4.715×10^{-5} m, has 4 sf

(e) 57,600. s, or 5.7600×10^4 s, has 5 sf

(f) 0.0000007160 cm^3, or 7.160×10^{-7} cm^3, has 4 sf

CHECK Be sure that every zero counted as significant comes after nonzero digit(s) in the number.

FOLLOW-UP PROBLEM 1.8 For each of the following quantities, underline the zeros that are significant figures and determine the number of significant figures (sf) in each quantity. For **(d)** to **(f)**, express each in exponential notation first.

(a) 31.070 mg   **(b)** 0.06060 g   **(c)** 850.°C

(d) 200.0 mL   **(e)** 0.0000039 m   **(f)** 0.000401 L

Significant Figures in Calculations

Measurements often contain differing numbers of significant figures. In a calculation, we keep track of the number of significant figures in each quantity so that we don't claim more significant figures (more certainty) in the answer than in the original data. If we have too many significant figures, we **round off** the answer to obtain the proper number of them.

The general rule for rounding is that *the least certain measurement sets the limit on certainty for the entire calculation and determines the number of significant figures in the final answer.* Suppose you want to find the density of a new ceramic. You measure the mass of a piece on a precise laboratory balance and obtain 3.8056 g; you measure its volume as 2.5 mL by displacement of water in a graduated cylinder. The mass has five significant figures, but the volume has only two. Should you report the density as 3.8056 g/2.5 mL = 1.5222 g/mL or as 1.5 g/mL? The answer with five significant figures implies more certainty than the answer with two. But you didn't measure the volume to five significant figures, so you can't possibly know the density with that much certainty. Therefore, you report the answer as 1.5 g/mL.

Significant Figures and Arithmetic Operations The following two rules tell how many significant figures to show based on the arithmetic operation:

1. *For multiplication and division.* The answer contains the same number of *significant figures* as in the measurement with the fewest significant figures. Suppose you want to find the volume of a sheet of a new graphite composite. The length (9.2 cm) and width (6.8 cm) are obtained with a meterstick and the thickness (0.3744 cm) with a set of fine calipers. The volume calculation is

$$\text{Volume (cm}^3) = 9.2 \text{ cm} \times 6.8 \text{ cm} \times 0.3744 \text{ cm} = 23 \text{ cm}^3$$

The calculator shows 23.4225 cm^3, but you should report the answer as 23 cm^3, with two significant figures, because the length and width measurements determine the overall certainty, and they contain only two significant figures.

2. *For addition and subtraction.* The answer has the same number of *decimal places* as there are in the measurement with the fewest decimal places. Suppose you measure 83.5 mL of water in a graduated cylinder and add 23.28 mL of protein solution from a buret. The total volume is

$$\text{Volume (mL)} = 83.5 \text{ mL} + 23.28 \text{ mL} = 106.8 \text{ mL}$$

Here the calculator shows 106.78 mL, but you report the volume as 106.8 mL, with one decimal place, because the measurement with fewer decimal places (83.5 mL) has one decimal place. (Appendix A covers significant figures in calculations involving logarithms, which will be used later in the text.)

Demo Kirksey, H. Graden. "Significant Figures: A Classroom Demonstration," *J. Chem. Educ.* **1992,** *69,* 497.

Rules for Rounding Off In most calculations, you need to round off the answer to obtain the proper number of significant figures or decimal places. Notice that in calculating the volume of the graphite composite above, we removed the extra digits, but in calculating the total protein solution volume, we removed the extra digit and increased the last digit by one. Here are rules for rounding off:

1. If the digit removed is *more than 5*, the preceding number is increased by 1: 5.379 rounds to 5.38 if three significant figures are retained and to 5.4 if two significant figures are retained.
2. If the digit removed is *less than 5*, the preceding number is unchanged: 0.2413 rounds to 0.241 if three significant figures are retained and to 0.24 if two significant figures are retained.
3. If the digit removed *is 5*, the preceding number is increased by 1 if it is odd and remains unchanged if it is even: 17.75 rounds to 17.8, but 17.65 rounds to 17.6. If the 5 is followed only by zeros, rule 3 is followed; if the 5 is followed by nonzeros, rule 1 is followed: 17.6500 rounds to 17.6, but 17.6513 rounds to 17.7.
4. *Always carry one or two additional significant figures through a multistep calculation and round off the final answer **only**.* Don't be concerned if you string together a calculation to check a sample or follow-up problem and find that your answer differs in the last decimal place from the one in the book. To show you the correct number of significant figures in text calculations, *we round off intermediate steps,* and this process may sometimes change the last digit.

Demo Zipp, Arden P. "A Simple but Effective Demonstration for Illustrating Significant Figure Rules When Making Measurements and Doing Calculations," *J. Chem. Educ.* **1992**, *69*, 291.

ARIS Presentation Center, Photo Library

Figure 1.15 Significant figures and measuring devices. The mass (6.8605 g) measured with an analytical balance *(top)* has more significant figures than the volume (68.2 mL) measured with a graduated cylinder *(bottom)*.

Significant Figures and Electronic Calculators A calculator usually gives answers with too many significant figures. For example, if your calculator displays ten digits and you divide 15.6 by 9.1, it will show 1.714285714. Obviously, most of these digits are not significant; the answer should be rounded off to 1.7 so that it has two significant figures, the same as in 9.1. A good way to prove to yourself that the additional digits are not significant is to perform two calculations, including the uncertainty in the last digits, to obtain the highest and lowest possible answers. For $(15.6 \pm 0.1)/(9.1 \pm 0.1)$,

$$\text{The highest answer is } \frac{15.7}{9.0} = 1.744444\ldots$$

$$\text{The lowest answer is } \frac{15.5}{9.2} = 1.684782\ldots$$

No matter how many digits the calculator displays, the values differ in the first decimal place, so the answer has two significant figures and should be reported as 1.7. Many calculators have a FIX button that allows you to set the number of digits displayed.

Significant Figures and Choice of Measuring Device The measuring device you choose determines the number of significant figures you can obtain. Suppose you are doing an experiment that requires mixing a liquid with a solid. You weigh the solid on the analytical balance and obtain a value with five significant figures. It would make sense to measure the liquid with a buret or pipet, which measures volumes to more significant figures than a graduated cylinder. If you chose the cylinder, you would have to round off more digits in the calculations, so the certainty in the mass value would be wasted (Figure 1.15). With experience, you'll choose a measuring device based on the number of significant figures you need in the final answer.

Exact Numbers Some numbers are called **exact numbers** because they have no uncertainty associated with them. Some exact numbers are part of a unit definition: there are 60 minutes in 1 hour, 1000 micrograms in 1 milligram, and

2.54 centimeters in 1 inch. Other exact numbers result from actually counting individual items: there are exactly 3 quarters in my hand, 26 letters in the English alphabet, and so forth. Because they have no uncertainty, *exact numbers do not limit the number of significant figures in the answer.* Put another way, exact numbers have as many significant figures as a calculation requires.

SAMPLE PROBLEM 1.9   Significant Figures and Rounding

PROBLEM Perform the following calculations and round the answers to the correct number of significant figures:

(a) $\dfrac{16.3521 \text{ cm}^2 - 1.448 \text{ cm}^2}{7.085 \text{ cm}}$

(b) $\dfrac{(4.80 \times 10^4 \text{ mg})\left(\dfrac{1 \text{ g}}{1000 \text{ mg}}\right)}{11.55 \text{ cm}^3}$

PLAN We use the rules just presented in the text. In (a), we subtract before we divide. In (b), we note that the unit conversion involves an exact number.

SOLUTION (a) $\dfrac{16.3521 \text{ cm}^2 - 1.448 \text{ cm}^2}{7.085 \text{ cm}} = \dfrac{14.904 \text{ cm}^2}{7.085 \text{ cm}} = 2.104 \text{ cm}$

(b) $\dfrac{(4.80 \times 10^4 \text{ mg})\left(\dfrac{1 \text{ g}}{1000 \text{ mg}}\right)}{11.55 \text{ cm}^3} = \dfrac{48.0 \text{ g}}{11.55 \text{ cm}^3} = 4.16 \text{ g/cm}^3$

CHECK Note that in (a) we lose a decimal place in the numerator, and in (b) we retain 3 sf in the answer because there are 3 sf in 4.80. Rounding to the nearest whole number is always a good way to check: (a) $(16 - 1)/7 \approx 2$; (b) $(5 \times 10^4 / 1 \times 10^3)/12 \approx 4$.

FOLLOW-UP PROBLEM 1.9 Perform the following calculation and round the answer to the correct number of significant figures: $\dfrac{25.65 \text{ mL} + 37.4 \text{ mL}}{73.55 \text{ s}\left(\dfrac{1 \text{ min}}{60 \text{ s}}\right)}$

ARIS Presentation Center, Worked Examples Library

Literature Treptow, Richard S. "Precision and Accuracy in Measurements: A Tale of Four Graduated Cylinders," *J. Chem. Educ.* **1998**, *75*, 992.

Precision, Accuracy, and Instrument Calibration

Precision and accuracy are two aspects of certainty. We often use these terms interchangeably in everyday speech, but in scientific measurements they have distinct meanings. **Precision,** or *reproducibility,* refers to how close the measurements in a series are to each other. **Accuracy** refers to how close a measurement is to the actual value.

Precision and accuracy are linked with two common types of error:

1. **Systematic error** produces values that are *either* all higher or all lower than the actual value. Such error is part of the experimental system, often caused by a faulty measuring device or by a consistent mistake in taking a reading.
2. **Random error,** in the absence of systematic error, produces values that are higher *and* lower than the actual value. Random error *always* occurs, but its size depends on the measurer's skill and the instrument's precision.

Precise measurements have low random error, that is, small deviations from the average. *Accurate measurements have low systematic error and, generally, low random error as well.* In some cases, when many measurements are taken that have a high random error, the *average* may still be accurate.

Suppose each of four students measures 25.0 mL of water in a pre-weighed graduated cylinder and then weighs the water *plus* cylinder on a balance. If the density of water is 1.00 g/mL at the temperature of the experiment, the actual mass of 25.0 mL of water is 25.0 g. Each student performs the operation four

Figure 1.16 Precision and accuracy in a laboratory calibration. Each graph represents four measurements made with a graduated cylinder that is being calibrated (see text for details).

Literature Guare, Charles J. "Error, Precision, and Uncertainty," *J. Chem. Educ.* **1991**, *68*, 649.

times, subtracts the mass of the empty cylinder, and obtains one of the four graphs shown in Figure 1.16. In graphs A and B, the random error is small; that is, the precision is high (the weighings are reproducible). In A, however, the accuracy is high as well (all the values are close to 25.0 g), whereas in B the accuracy is low (there is a systematic error). In graphs C and D, there is a large random error; that is, the precision is low. Large random error is often called large *scatter.* Note, however, that in D there is also a systematic error (all the values are high), whereas in C the average of the values is close to the actual value.

Systematic error can be avoided, or at least taken into account, through **calibration** of the measuring device, that is, by comparing it with a known standard. The systematic error in graph B, for example, might be caused by a poorly manufactured cylinder that reads "25.0" when it actually contains about 27 mL. If you detect such an error by means of a calibration procedure, you could adjust all volumes measured with that cylinder. Instrument calibration is an essential part of careful measurement.

Section Summary

The final digit of a measurement is always estimated. Thus, all measurements have a limit to their certainty, which is expressed by the number of significant figures. • The certainty of a calculated result depends on the certainty of the data, so the answer has as many significant figures as in the least certain measurement. Excess digits are rounded off in the final answer. The choice of laboratory device depends on the certainty needed. Exact numbers have as many significant figures as the calculation requires. • Precision (how close values are to each other) and accuracy (how close values are to the actual value) are two aspects of certainty. • Systematic errors give values that are either all higher or all lower than the actual value. Random errors result in some values that are higher and some that are lower than the actual value. Precise measurements have low random error; accurate measurements have low systematic error and often low random error. The size of random errors depends on the skill of the measurer and the precision of the instrument. A systematic error, however, is often caused by faulty equipment and can be compensated for by calibration.

Chapter Perspective

This chapter has provided key ideas to use repeatedly in your study of chemistry: descriptions of some essential concepts; insight into how scientists think; the units of modern measurement and the mathematical skills to apply them; and a systematic approach to solving problems. You can begin using these keys in the next chapter, where we discuss the components of matter and their classification and trace the winding path of scientific discovery that led to our current model of atomic structure.

Chemical Connections *to Interdisciplinary Science*

Chemistry Problem Solving in the Real World

Learning chemistry is essential to many fields, including medicine, engineering, and environmental science. It is also essential to an understanding of complex science-related issues, such as the recycling of plastics, the reduction of urban smog, and the application of genetic cloning—to mention just three of many.

Any major scientific discipline such as chemistry consists of several subdisciplines that form connections with other sciences to spawn new fields. Traditionally, chemistry has five main branches—organic, inorganic, analytical, physical, and biological chemistry—but these long ago formed interconnections, such as physical organic and bioinorganic chemistry. Solving the problems of today requires further connections, such as ecological chemistry, materials science, atmospheric chemistry, and molecular genetics. The more complex the system under study is, the greater the need for interdisciplinary scientific thinking.

Environmental issues are especially complex, and one of the most intractable is the acid rain problem. Let's see how it is being approached by chemists interacting with scientists in related fields. Acid rain results in part from burning high-sulfur coal, a fuel used throughout much of North America and Europe. As the coal burns, the gaseous products, including an oxide of its sulfur impurities called sulfur dioxide, are carried away by prevailing winds. In contact with oxygen and rain, sulfur dioxide undergoes chemical changes, yielding acid rain. (We discuss the reactions in later chapters.) In the northeastern United States and adjacent parts of Canada, acid rain has killed fish, injured forests and crops, and released harmful substances into the soil. Acid rain has severely damaged many forests and lakes in Germany, Sweden, Norway, and several countries in central and eastern Europe. And acidic precipitation has now been confirmed at both Poles!

Chemists and other scientists are currently working together to solve this problem (Figure B1.1). As geochemists search for low-sulfur coal deposits, their engineering colleagues design better ways of removing sulfur dioxide from smokestack gases. Atmospheric chemists and meteorologists track changes through the affected regions, develop computer models to predict the changes, and coordinate their findings with environmental chemists at ground stations. Ecological chemists, microbiologists, and aquatic biologists monitor the effects of acid rain on microbes, insects, birds, and fish. Agricultural chemists and agronomists study ways to protect crop yields. Biochemists and genetic engineers develop new, more acid-resistant crop species. Soil chemists measure changes in mineral content, sharing their data with forestry scientists to save valuable timber and recreational woodlands. Organic chemists and chemical engineers convert coal to cleaner fuels. Working in tandem with this intense experimental activity are scientifically trained economic and policy experts who provide business and government leaders with the information to make decisions and foster "greener" approaches to energy use. With all this input, interdisciplinary understanding of the acid rain problem has increased enormously and certainly will continue to do so.

These professions are just a few of those involved in studying a single chemistry-related issue. Chemical principles apply to many other specialties, from medicine and pharmacology to art restoration and forensics, from genetics and space research to archaeology and oceanography. Chemistry problem solving has far-reaching relevance to many aspects of your daily life and your future career as well.

Figure B1.1 The central role of chemistry in solving real-world problems. Researchers in many chemical specialties join with those in other sciences to investigate complex modern issues such as acid rain. **A,** Atmospheric chemists study the location and concentration of air pollutants with balloons that carry monitoring equipment aloft. **B,** Ecologists sample lake, pond, and river water and observe wildlife to learn the effects of acidic precipitation on aquatic environments. **C,** The sulfur dioxide in power plant emissions will be reduced by devices that remove it from smokestack gases.

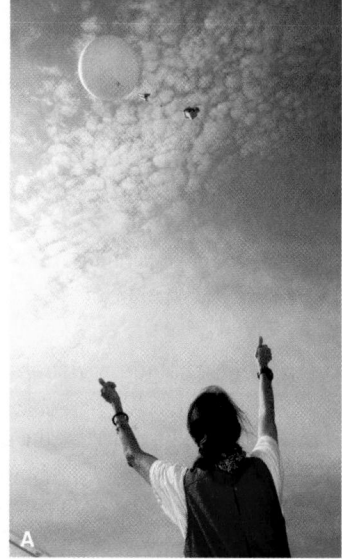

ARIS Presentation Center, Photo Library

CHAPTER REVIEW GUIDE

The following sections provide many aids to help you study this chapter. (Numbers in parentheses refer to pages, unless noted otherwise.)

Learning Objectives These are concepts and skills you should know after studying this chapter.

Relevant section and/or sample problem (SP) numbers appear in parentheses.

Understand These Concepts

1. The distinction between physical and chemical properties and changes (1.1; SPs 1.1, 1.2)
2. The defining features of the states of matter (1.1)
3. The nature of potential and kinetic energy and their interconversion (1.1)
4. The process of approaching a phenomenon scientifically and the distinctions between observation, hypothesis, experiment, and model (1.3)
5. The common units of length, volume, mass, and temperature and their numerical prefixes (1.5)

6. The distinctions between mass and weight, heat and temperature, and intensive and extensive properties (1.5)
7. The meaning of uncertainty in measurements and the use of significant figures and rounding (1.6)
8. The distinctions between accuracy and precision and between systematic and random error (1.6)

Master These Skills

1. Using conversion factors in calculations (1.4; SPs 1.3–1.5)
2. Finding density from mass and volume (SP 1.6)
3. Converting among the Kelvin, Celsius, and Fahrenheit scales (SP 1.7)
4. Determining the number of significant figures (SP 1.8) and rounding to the correct number of digits (SP 1.9)

Key Terms These important terms appear in boldface in the chapter and are defined again in the Glossary.

Section 1.1
chemistry (4)
matter (4)
composition (4)
property (4)
physical property (4)
physical change (4)
chemical property (5)
chemical change (chemical reaction) (5)
state of matter (6)
solid (6)
liquid (6)
gas (6)
energy (8)
potential energy (8)
kinetic energy (8)

Section 1.2
alchemy (10)
combustion (11)
phlogiston theory (11)
Section 1.3
scientific method (13)
observation (13)
data (13)
natural law (13)
hypothesis (13)
experiment (13)
variable (13)
controlled experiment (13)
model (theory) (13)
Section 1.4
conversion factor (14)
dimensional analysis (16)

Section 1.5
SI unit (18)
base (fundamental) unit (18)
derived unit (18)
meter (m) (18)
volume (V) (19)
cubic meter (m^3) (19)
liter (L) (19)
milliliter (mL) (19)
mass (21)
kilogram (kg) (21)
weight (21)
density (d) (22)
extensive property (22)
intensive property (22)
temperature (T) (24)
heat (24)

thermometer (24)
kelvin (K) (25)
Celsius scale (25)
Kelvin (absolute) scale (25)
second (s) (26)
Section 1.6
uncertainty (27)
significant figures (28)
round off (29)
exact number (30)
precision (31)
accuracy (31)
systematic error (31)
random error (31)
calibration (32)

Key Equations and Relationships Numbered and screened concepts are listed for you to refer to or memorize.

1.1 Calculating density from mass and volume (22):

$$\text{Density} = \frac{\text{mass}}{\text{volume}}$$

1.2 Converting temperature from °C to K (26):

$$T \text{ (in K)} = T \text{ (in °C)} + 273.15$$

1.3 Converting temperature from K to °C (26):

$$T \text{ (in °C)} = T \text{ (in K)} - 273.15$$

1.4 Converting temperature from °C to °F (26):

$$T \text{ (in °F)} = \tfrac{9}{5}T \text{ (in °C)} + 32$$

1.5 Converting temperature from °F to °C (26):

$$T \text{ (in °C)} = [T \text{ (in °F)} - 32]\tfrac{5}{9}$$

Highlighted Figures and Tables These figures (F) and tables (T) provide a visual review of key ideas.

Entries in **bold** contain frequently used data.

F1.1 The distinction between physical and chemical change (4)
F1.2 The physical states of matter (7)
F1.3 Potential energy and kinetic energy (9)
F1.6 The scientific approach (13)

T1.2 SI base units (18)
T1.3 Decimal prefixes used with SI units (19)
T1.4 SI-English equivalent quantities (19)
F1.8 Some volume relationships in SI (20)

Brief Solutions to FOLLOW-UP PROBLEMS Compare your solutions to these calculation steps and answers.

1.1 Chemical. The red-and-blue and separate red particles on the left become paired red and separate blue particles on the right.

1.2 (a) Physical. Solid iodine changes to gaseous iodine.
(b) Chemical. Gasoline burns in air to form different substances.
(c) Chemical. In contact with air, torn skin and blood react to form different substances.

1.3 No. of chairs

$$= 3\ \cancel{bolts} \times \frac{200\ \cancel{m^2}}{1\ \cancel{bolt}} \times \frac{3.281\ \cancel{ft}}{1\ \cancel{m}} \times \frac{3.281\ \cancel{ft}}{1\ \cancel{m}} \times \frac{1\ chair}{31.5\ \cancel{ft^2}}$$

$$= 205\ chairs$$

1.4 Radius of ribosome (dm) $= \dfrac{21.4\ \cancel{nm}}{2} \times \dfrac{1\ dm}{10^8\ \cancel{nm}}$

$$= 1.07 \times 10^{-7}\ dm$$

Volume of ribosome $(dm^3) = \frac{4}{3}\pi r^3 = \frac{4}{3}(3.14)(1.07 \times 10^{-7}\ dm)^3$

$$= 5.13 \times 10^{-21}\ dm^3$$

Volume of ribosome $(\mu L) = (5.13 \times 10^{-21}\ \cancel{dm^3})\left(\dfrac{1\ \cancel{L}}{1\ \cancel{dm^3}}\right)\left(\dfrac{10^6\ \mu L}{1\ \cancel{L}}\right)$

$$= 5.13 \times 10^{-15}\ \mu L$$

1.5 Mass (kg) of solution $= 8.0\ \cancel{h} \times \dfrac{60\ \cancel{min}}{1\ \cancel{h}} \times \dfrac{60\ \cancel{s}}{1\ \cancel{min}} \times \dfrac{1.5\ \cancel{drops}}{1\ \cancel{s}}$

$$\times \dfrac{65\ \cancel{mg}}{1\ \cancel{drop}} \times \dfrac{1\ \cancel{g}}{10^3\ \cancel{mg}} \times \dfrac{1\ kg}{10^3\ \cancel{g}}$$

$$= 2.8\ kg$$

1.6 Mass (kg) of sample $= 4.6\ \cancel{cm^3} \times \dfrac{7.5\ \cancel{g}}{1\ \cancel{cm^3}} \times \dfrac{1\ kg}{10^3\ \cancel{g}}$

$$= 0.034\ kg$$

1.7 T (in °C) $= 234\ K - 273.15 = -39°C$
T (in °F) $= \frac{9}{5}(-39°C) + 32 = -38°F$
Answer contains two significant figures (see Section 1.6).

1.8 (a) 31.0$\underline{7}$0 mg, 5 sf   (b) 0.060$\underline{6}$0 g, 4 sf
(c) 85$\underline{0}$.°C, 3 sf   (d) 2.$\underline{000}$×10² mL, 4 sf
(e) 3.9×10⁻⁶ m, 2 sf   (f) 4.$\underline{0}$1×10⁻⁴ L, 3 sf

1.9 $\dfrac{25.65\ mL + 37.4\ mL}{73.55\ s\left(\dfrac{1\ min}{60\ s}\right)} = 51.4\ mL/min$

PROBLEMS

Problems with colored numbers are answered in Appendix E and worked in detail in the Student Solutions Manual. Problem sections match those in the text and provide the numbers of relevant sample problems. Most offer Concept Review Questions, Skill-Building Exercises (grouped in pairs covering the same concept), and Problems in Context. The Comprehensive Problems are based on material from any section.

Some Fundamental Definitions
(Sample Problems 1.1 and 1.2)

● Concept Review Question
• **1.1.** Scenes A–D represent atomic-scale views of different samples of substances:

(a) Under one set of conditions, the substances in A and B mix and the result is depicted in C. Does this represent a chemical or a physical change?
(b) Under a second set of conditions, the same substances mix and the result is depicted in D. Does this represent a chemical or a physical change?

(c) Under a third set of conditions, the sample depicted in C changes to that in D. Does this represent a chemical or a physical change?
(d) When the change in part (c) occurs, does the sample have different chemical properties? Physical properties?

● Skill-Building Exercises (grouped in similar pairs)
• **1.2** Describe solids, liquids, and gases in terms of how they fill a container. Use your descriptions to identify the physical state (at room temperature) of the following: (a) helium in a toy balloon; (b) mercury in a thermometer; (c) soup in a bowl.
• **1.3** Use your descriptions in the previous problem to identify the physical state (at room temperature) of the following: (a) the air in your room; (b) tablets in a bottle of vitamins; (c) sugar in a packet.

• **1.4** Define *physical property* and *chemical property*. Identify each type of property in the following statements:
(a) Yellow-green chlorine gas attacks silvery sodium metal to form white crystals of sodium chloride (table salt).
(b) A magnet separates a mixture of black iron shavings and white sand.
• **1.5** Define *physical change* and *chemical change*. State which type of change occurs in each of the following statements:
(a) Passing an electric current through molten magnesium chloride yields molten magnesium and gaseous chlorine.
(b) The iron in discarded automobiles slowly forms reddish brown, crumbly rust.

• **1.6** Which of the following is a chemical change? Explain your reasoning: (a) boiling canned soup; (b) toasting a slice of bread; (c) chopping a log; (d) burning a log.
• **1.7** Which of the following changes can be reversed by changing the temperature: (a) dew condensing on a leaf; (b) an egg turning hard when it is boiled; (c) ice cream melting; (d) a spoonful of batter cooking on a hot griddle?

1.8 For each pair, which has higher potential energy?
  (a) The fuel in your car or the gaseous products in its exhaust
  (b) Wood in a fire or the ashes after the wood burns
1.9 For each pair, which has higher kinetic energy?
  (a) A sled resting at the top of a hill or a sled sliding down the hill
  (b) Water above a dam or water falling over the dam

Chemical Arts and the Origins of Modern Chemistry
● Concept Review Questions
1.10 The alchemical, medical, and technological traditions were precursors to chemistry. State a contribution that each made to the development of the science of chemistry.
1.11 How did the phlogiston theory explain combustion?
1.12 One important observation that supporters of the phlogiston theory had trouble explaining was that the calx of a metal weighs more than the metal itself. Why was that observation important? How did the phlogistonists respond?
1.13 Lavoisier developed a new theory of combustion that overturned the phlogiston theory. What measurements were central to his theory, and what key discovery did he make?

The Scientific Approach: Developing a Model
● Concept Review Questions
1.14 How are the key elements of scientific thinking used in the following scenario? While making toast, you notice it fails to pop out of the toaster. Thinking the spring mechanism is stuck, you notice that the bread is unchanged. Assuming you forgot to plug in the toaster, you check and find it is plugged in. When you take the toaster into the dining room and plug it into a different outlet, you find the toaster works. Returning to the kitchen, you turn on the switch for the overhead light and nothing happens.
1.15 Why is a quantitative observation more useful than a non-quantitative one? Which of the following are quantitative?
(a) The Sun rises in the east. (b) A person weighs one-sixth as much on the Moon as on Earth. (c) Ice floats on water. (d) A hand pump cannot draw water from a well more than 34 ft deep.
1.16 Describe the essential features of a well-designed experiment.
1.17 Describe the essential features of a scientific model.

Chemical Problem Solving
(Sample Problem 1.3)
● Concept Review Question
1.18 When you convert feet to inches, how do you decide which portion of the conversion factor should be in the numerator and which in the denominator?

● Skill-Building Exercises *(grouped in similar pairs)*
1.19 Write the conversion factor(s) for
  (a) in^2 to m^2 (b) km^2 to cm^2
  (c) mi/h to m/s   (d) lb/ft^3 to g/cm^3
1.20 Write the conversion factor(s) for
  (a) cm/min to in/s   (b) m^3 to in^3
  (c) m/s^2 to km/h^2 (d) gal/h to L/min

Measurement in Scientific Study
(Sample Problems 1.4 to 1.7)
● Concept Review Questions
1.21 Describe the difference between intensive and extensive properties. Which of the following properties are intensive: (a) mass; (b) density; (c) volume; (d) melting point?

1.22 Explain the difference between mass and weight. Why is your weight on the Moon one-sixth that on Earth?
1.23 For each of the following cases, state whether the density of the object increases, decreases, or remains the same:
  (a) A sample of chlorine gas is compressed.
  (b) A lead weight is carried up a high mountain.
  (c) A sample of water is frozen.
  (d) An iron bar is cooled.
  (e) A diamond is submerged in water.
1.24 Explain the difference between heat and temperature. Does 1 L of water at 65°F have more, less, or the same quantity of energy as 1 L of water at 65°C?
1.25 A one-step conversion is sufficient to convert a temperature in the Celsius scale into the Kelvin scale, but not into the Fahrenheit scale. Explain.

● Skill-Building Exercises *(grouped in similar pairs)*
1.26 The average radius of a molecule of lysozyme, an enzyme in tears, is 1430 pm. What is its radius in nanometers (nm)?
1.27 The radius of a barium atom is 2.22×10^{-10} m. What is its radius in angstroms (Å)?

1.28 What is the length in inches (in) of a 100.-m soccer field?
1.29 The center on your basketball team is 6 ft 10 in tall. How tall is the player in millimeters (mm)?

1.30 A small hole in the wing of a space shuttle requires a 20.7-cm^2 patch. (a) What is the patch's area in square kilometers (km^2)? (b) If the patching material costs NASA $3.25/in^2$, what is the cost of the patch?
1.31 The area of a telescope lens is 7903 mm^2. (a) What is the area in square feet (ft^2)? (b) If it takes a technician 45 s to polish 135 mm^2, how long does it take her to polish the entire lens?

1.32 Express your body weight in kilograms (kg).
1.33 There are 2.60×10^{15} short tons of oxygen in the atmosphere (1 short ton = 2000 lb). How many metric tons of oxygen are present (1 metric ton = 1000 kg)?

1.34 The average density of Earth is 5.52 g/cm^3. What is its density in (a) kg/m^3; (b) lb/ft^3?
1.35 The speed of light in a vacuum is 2.998×10^8 m/s. What is its speed in (a) km/h; (b) mi/min?

1.36 The volume of a certain bacterial cell is 2.56 μm^3. (a) What is its volume in cubic millimeters (mm^3)? (b) What is the volume of 10^5 cells in liters (L)?
1.37 (a) How many cubic meters of milk are in 1 qt (946.4 mL)? (b) How many liters of milk are in 835 gal (1 gal = 4 qt)?

1.38 An empty vial weighs 55.32 g. (a) If the vial weighs 185.56 g when filled with liquid mercury ($d = 13.53$ g/cm^3), what is its volume? (b) How much would the vial weigh if it were filled with water ($d = 0.997$ g/cm^3 at 25°C)?
1.39 An empty Erlenmeyer flask weighs 241.3 g. When filled with water ($d = 1.00$ g/cm^3), the flask and its contents weigh 489.1 g. (a) What is the flask's volume? (b) How much does the flask weigh when filled with chloroform ($d = 1.48$ g/cm^3)?

1.40 A small cube of aluminum measures 15.6 mm on a side and weighs 10.25 g. What is the density of aluminum in g/cm^3?
1.41 A steel ball-bearing with a circumference of 32.5 mm weighs 4.20 g. What is the density of the steel in g/cm^3 (V of a sphere $= \frac{4}{3}\pi r^3$; circumference of a circle $= 2\pi r$)?

- **1.42** Perform the following conversions:
  (a) 68°F (a pleasant spring day) to °C and K
  (b) −164°C (the boiling point of methane, the main component of natural gas) to K and °F
  (c) 0 K (absolute zero, theoretically the coldest possible temperature) to °C and °F
- **1.43** Perform the following conversions:
  (a) 106°F (the body temperature of many birds) to K and °C
  (b) 3410°C (the melting point of tungsten, the highest for any metallic element) to K and °F
  (c) 6.1×10^3 K (the surface temperature of the Sun) to °F and °C

● Problems in Context

- **1.44** A 25.0-g sample of each of three unknown metals is added to 25.0 mL of water in graduated cylinders A, B, and C, and the final volumes are depicted in the circles below. Given their densities, identify the metal in each cylinder: zinc (7.14 g/mL), iron (7.87 g/mL), nickel (8.91 g/mL).

- **1.45** The distance between two adjacent peaks on a wave is called the *wavelength*.
  (a) The wavelength of a beam of ultraviolet light is 247 nanometers (nm). What is its wavelength in meters?
  (b) The wavelength of a beam of red light is 6760 pm. What is its wavelength in angstroms (Å)?
- **1.46** Each of the beakers depicted below contains two liquids that do not dissolve in each other. Three of the liquids are designated A, B, and C, and water is designated W.

  (a) Which of the liquids is(are) more dense than water and which less dense?
  (b) If the densities of W, C, and A are 1.0 g/mL, 0.88 g/mL, and 1.4 g/mL, respectively, which of the following densities is possible for liquid B: 0.79 g/mL, 0.86 g/mL, 0.94 g/mL, 1.2 g/mL?
- **1.47** A cylindrical tube 9.5 cm high and 0.85 cm in diameter is used to collect blood samples. How many cubic decimeters (dm^3) of blood can it hold (V of a cylinder $= \pi r^2 h$)?
- **1.48** Copper can be drawn into thin wires. How many meters of 34-gauge wire (diameter $= 6.304 \times 10^{-3}$ in) can be produced from the copper in 5.01 lb of covellite, an ore of copper that is 66% copper by mass? (*Hint:* Treat the wire as a cylinder: V of cylinder $= \pi r^2 h$; d of copper $= 8.95$ g/cm^3.)

Uncertainty in Measurement: Significant Figures
(Sample Problems 1.8 and 1.9)

● Concept Review Questions

- **1.49** What is an exact number? How are exact numbers treated differently from other numbers in a calculation?
- **1.50** Which procedure(s) decrease(s) the random error of a measurement: (1) taking the average of more measurements; (2) calibrating the instrument; (3) taking fewer measurements? Explain.
- **1.51** A newspaper reported that the attendance at Slippery Rock's home football game was 16,532. (a) How many significant figures does this number contain? (b) Was the actual number of people counted? (c) After Slippery Rock's next home game, the newspaper reported an attendance of 15,000. If you assume that this number contains two significant figures, how many people could actually have been at the game?

● Skill-Building Exercises (grouped in similar pairs)

- **1.52** Underline the significant zeros in the following numbers:
  (a) 0.41   (b) 0.041   (c) 0.0410   (d) 4.0100×10^4
- **1.53** Underline the significant zeros in the following numbers:
  (a) 5.08   (b) 508   (c) 5.080×10^3 (d) 0.05080
- **1.54** Round off each number to the indicated number of significant figures (sf): (a) 0.0003554 (to 2 sf); (b) 35.8348 (to 4 sf); (c) 22.4555 (to 3 sf).
- **1.55** Round off each number to the indicated number of significant figures (sf): (a) 231.554 (to 4 sf); (b) 0.00845 (to 2 sf); (c) 144,000 (to 2 sf).
- **1.56** Round off each number in the following calculation to one fewer significant figure, and find the answer:
  $$\frac{19 \times 155 \times 8.3}{3.2 \times 2.9 \times 4.7}$$
- **1.57** Round off each number in the following calculation to one fewer significant figure, and find the answer:
  $$\frac{10.8 \times 6.18 \times 2.381}{24.3 \times 1.8 \times 19.5}$$
- **1.58** Carry out the following calculations, making sure that your answer has the correct number of significant figures:
  (a) $\dfrac{2.795 \text{ m} \times 3.10 \text{ m}}{6.48 \text{ m}}$
  (b) $V = \frac{4}{3}\pi r^3$, where $r = 17.282$ mm
  (c) $1.110 \text{ cm} + 17.3 \text{ cm} + 108.2 \text{ cm} + 316 \text{ cm}$
- **1.59** Carry out the following calculations, making sure that your answer has the correct number of significant figures:
  (a) $\dfrac{2.420 \text{ g} + 15.6 \text{ g}}{4.8 \text{ g}}$ (b) $\dfrac{7.87 \text{ mL}}{16.1 \text{ mL} - 8.44 \text{ mL}}$
  (c) $V = \pi r^2 h$, where $r = 6.23$ cm and $h = 4.630$ cm
- **1.60** Write the following numbers in scientific notation:
  (a) 131,000.0   (b) 0.00047   (c) 210,006   (d) 2160.5
- **1.61** Write the following numbers in scientific notation:
  (a) 282.0   (b) 0.0380   (c) 4270.8   (d) 58,200.9
- **1.62** Write the following numbers in standard notation. Use a terminal decimal point when needed:
  (a) 5.55×10^3 (b) 1.0070×10^4 (c) 8.85×10^{-7} (d) 3.004×10^{-3}
- **1.63** Write the following numbers in standard notation. Use a terminal decimal point when needed:
  (a) 6.500×10^3 (b) 3.46×10^{-5} (c) 7.5×10^2 (d) 1.8856×10^2

1.64 Convert the following into correct scientific notation:
(a) 802.5×10^2 (b) 1009.8×10^{-6} (c) 0.077×10^{-9}

1.65 Convert the following into correct scientific notation:
(a) 14.3×10^1 (b) 851×10^{-2} (c) 7500×10^{-3}

1.66 Carry out each of the following calculations, paying special attention to significant figures, rounding, and units (J = joule, the SI unit of energy; mol = mole, the SI unit for amount of substance):

(a) $\dfrac{(6.626 \times 10^{-34}\ \text{J·s})(2.9979 \times 10^8\ \text{m/s})}{489 \times 10^{-9}\ \text{m}}$

(b) $\dfrac{(6.022 \times 10^{23}\ \text{molecules/mol})(1.23 \times 10^2\ \text{g})}{46.07\ \text{g/mol}}$

(c) $(6.022 \times 10^{23}\ \text{atoms/mol})(2.18 \times 10^{-18}\ \text{J/atom})\left(\dfrac{1}{2^2} - \dfrac{1}{3^2}\right)$,

where the numbers 2 and 3 in the last term are exact.

1.67 Carry out each of the following calculations, paying special attention to significant figures, rounding, and units:

(a) $\dfrac{4.32 \times 10^7\ \text{g}}{\frac{4}{3}(3.1416)(1.95 \times 10^2\ \text{cm})^3}$ (The term $\frac{4}{3}$ is exact.)

(b) $\dfrac{(1.84 \times 10^2\ \text{g})(44.7\ \text{m/s})^2}{2}$ (The term 2 is exact.)

(c) $\dfrac{(1.07 \times 10^{-4}\ \text{mol/L})^2(3.8 \times 10^{-3}\ \text{mol/L})}{(8.35 \times 10^{-5}\ \text{mol/L})(1.48 \times 10^{-2}\ \text{mol/L})^3}$

1.68 Which statements include exact numbers?
(a) Angel Falls is 3212 ft high.
(b) There are eight known planets in the Solar System.
(c) There are 453.59 g in 1 lb.
(d) There are 1000 mm in 1 m.

1.69 Which of the following include exact numbers?
(a) The speed of light in a vacuum is a physical constant; to six significant figures, it is 2.99792×10^8 m/s.
(b) The density of mercury at 25°C is 13.53 g/mL.
(c) There are 3600 s in 1 h.
(d) In 2003, the United States had 50 states.

● **Problems in Context**

1.70 How long is the metal strip shown below? Be sure to answer with the correct number of significant figures.

1.71 These organic solvents are used to clean compact discs:

Solvent	Density (g/mL) at 20°C
Chloroform	1.492
Diethyl ether	0.714
Ethanol	0.789
Isopropanol	0.785
Toluene	0.867

(a) If a 15.00-mL sample of CD cleaner weighs 11.775 g at 20°C, which solvent is most likely to be present?
(b) The chemist analyzing the cleaner calibrates her equipment and finds that the pipet is accurate to ±0.02 mL, and the balance is accurate to ±0.003 g. Is this equipment precise enough to distinguish between ethanol and isopropanol?

1.72 A laboratory instructor gives a sample of amino-acid powder to each of four students, I, II, III, and IV, and they weigh the samples. The true value is 8.72 g. Their results for three trials are
I: 8.72 g, 8.74 g, 8.70 g   II: 8.56 g, 8.77 g, 8.83 g
III: 8.50 g, 8.48 g, 8.51 g   IV: 8.41 g, 8.72 g, 8.55 g
(a) Calculate the average mass from each set of data, and tell which set is the most accurate.
(b) Precision is a measure of the average of the deviations of each piece of data from the average value. Which set of data is the most precise? Is this set also the most accurate?
(c) Which set of data is both the most accurate and most precise?
(d) Which set of data is both the least accurate and least precise?

1.73 The following dartboards illustrate the types of errors often seen in measurements. The bull's-eye represents the actual value, and the darts represent the data.

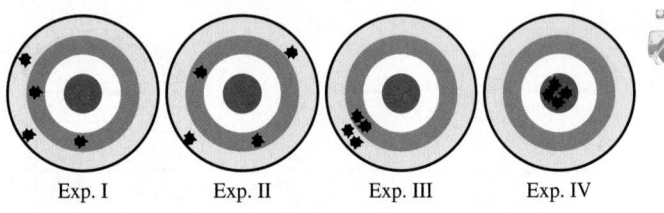

Exp. I   Exp. II   Exp. III   Exp. IV

(a) Which experiments yield the same average result?
(b) Which experiment(s) display(s) high precision?
(c) Which experiment(s) display(s) high accuracy?
(d) Which experiment(s) show(s) a systematic error?

Comprehensive Problems

1.74 Two blank potential energy diagrams appear below. Beneath each diagram are objects to place in the diagram. Draw the objects on the dashed lines to indicate higher or lower potential energy and label each case as more or less stable:

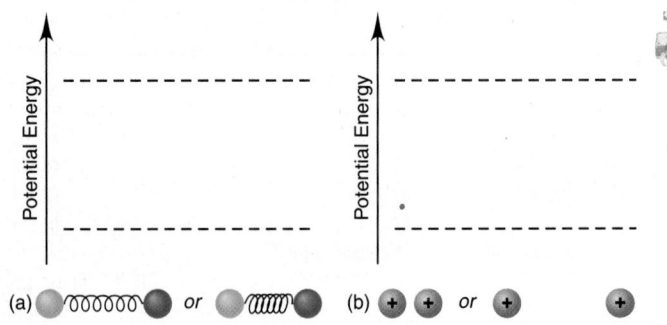

(a) Two balls attached to a relaxed *or* a compressed spring.
(b) Two positive charges near *or* apart from each other.

1.75 The scenes below illustrate two different mixtures. When mixture A at 273 K is heated to 473 K, mixture B results.

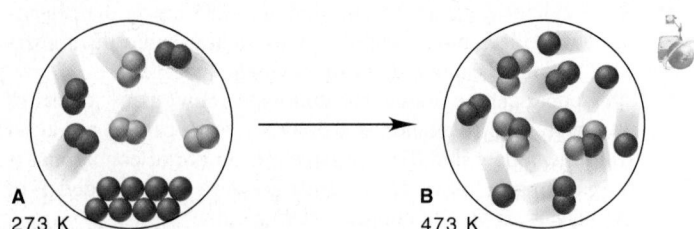

A
273 K

B
473 K

(a) How many different chemical changes occur?

(b) How many different physical changes occur?

•• **1.76** Bromine is used to prepare the pesticide methyl bromide and flame retardants for plastic electronic housings. It is recovered from seawater, underground brines, and the Dead Sea. The average concentrations of bromine in seawater (d = 1.024 g/mL) and the Dead Sea (d = 1.22 g/mL) are 0.065 g/L and 0.50 g/L, respectively. What is the mass ratio of bromine in the Dead Sea to that in seawater?

•• **1.77** An Olympic-size pool is 50.0 m long and 25.0 m wide. (a) How many gallons of water (d = 1.0 g/mL) are needed to fill the pool to an average depth of 4.8 ft? (b) What is the mass (in kg) of water in the pool?

••• **1.78** At room temperature (20°C) and pressure, the density of air is 1.189 g/L. An object will float in air if its density is less than that of air. In a buoyancy experiment with a new plastic, a chemist creates a rigid, thin-walled ball that weighs 0.12 g and has a volume of 560 cm^3.

(a) Will the ball float if it is evacuated?

(b) Will it float if filled with carbon dioxide (d = 1.830 g/L)?

(c) Will it float if filled with hydrogen (d = 0.0899 g/L)?

(d) Will it float if filled with oxygen (d = 1.330 g/L)?

(e) Will it float if filled with nitrogen (d = 1.165 g/L)?

(f) For any case that will float, how much weight must be added to make the ball sink?

•• **1.79** Asbestos is a fibrous silicate mineral with remarkably high tensile strength. But it is no longer used because airborne asbestos particles can cause lung cancer. Grunerite, a type of asbestos, has a tensile strength of 3.5×10^2 kg/mm^2 (thus, a strand of grunerite with a 1-mm^2 cross-sectional area can hold up to 3.5×10^2 kg). The tensile strengths of aluminum and Steel No. 5137 are 2.5×10^4 lb/in^2 and 5.0×10^4 lb/in^2, respectively. Calculate the cross-sectional area (in mm^2) of wires of aluminum and of Steel No. 5137 that have the same tensile strength as a fiber of grunerite with a cross-sectional area of 1.0 μm^2.

•• **1.80** Drugs called COX-2 inhibitors (e.g., Vioxx, Bextra, and Celebrex) were thought to relieve the pain and inflammation of osteoarthritis without the stomach bleeding and ulcers nonsteroidal anti-inflammatory drugs (NSAIDs) cause. In a 12-month trial, Vioxx caused fewer gastrointestinal side effects than the NSAID ibuprofen. However, a study of the recurrence of colon polyps after three years of Vioxx found an increased risk for heart attack and stroke beginning after 18 months of treatment. As a result, Vioxx was withdrawn from the market, and an FDA panel concluded that COX-2 inhibitors as a class have increased cardiovascular risk that varies by drug and dose. The FDA then caused the withdrawal of Bextra and required a warning label on Celebrex. Based on this information, list (a) an observation, (b) a hypothesis, (c) an experiment, and (d) a theory.

•• **1.81** Earth's oceans have an average depth of 3800 m, a total area of 3.63×10^8 km^2, and an average concentration of dissolved gold of 5.8×10^{-9} g/L. (a) How many grams of gold are in the oceans? (b) How many m^3 of gold are in the oceans? (c) Assuming the price of gold is $370.00/troy oz, what is the value of gold in the oceans (1 troy oz = 31.1 g; d of gold = 19.3 g/cm^3)?

••• **1.82** Brass is an alloy of copper and zinc. Varying the mass percentages of the two metals produces brasses with different properties. A brass called *yellow zinc* has high ductility and strength and is 34%–37% zinc by mass. (a) Find the mass range (in g) of

copper in 185 g of yellow zinc. (b) What is the mass range (in g) of zinc in a sample of yellow zinc that contains 46.5 g of copper?

•• **1.83** Liquid nitrogen is obtained from liquefied air and is used industrially to prepare frozen foods. It boils at 77.36 K. (a) What is this temperature in °C? (b) What is this temperature in °F? (c) At the boiling point, the density of the liquid is 809 g/L and that of the gas is 4.566 g/L. How many liters of liquid nitrogen are produced when 895.0 L of nitrogen gas is liquefied at 77.36 K?

•• **1.84** The speed of sound varies according to the material. Sound travels at 5.4×10^3 cm/s through rubber and at 1.97×10^4 ft/s through granite. Calculate each of these speeds in m/s.

•• **1.85** If a raindrop weighs 0.52 mg on average and 5.1×10^5 raindrops fall on a lawn every minute, what mass (in kg) of rain falls on the lawn in 1.5 h?

•• **1.86** A jogger runs at an average speed of 5.9 mi/h. (a) How fast is she running in m/s? (b) How many kilometers does she run in 98 min? (c) If she starts a run at 11:15 am, what time is it after she covers 4.75×10^4 ft?

•• **1.87** Scenes A and B depict changes in matter at the atomic scale:

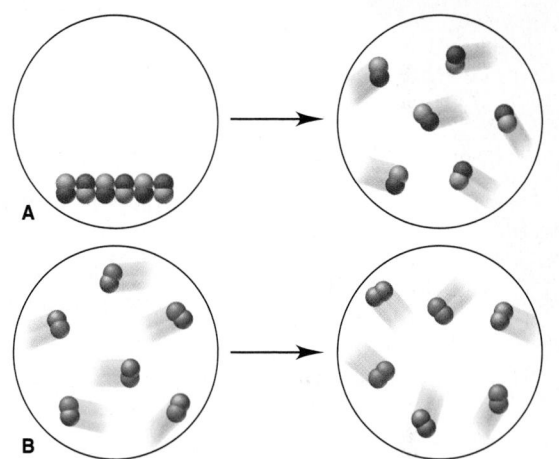

(a) Which show(s) a physical change?

(b) Which show(s) a chemical change?

(c) Which result(s) in different physical properties?

(d) Which result(s) in different chemical properties?

(e) Which result(s) in a change in state?

•• **1.88** Nutritional tables give the potassium content of a standard apple (about 3 apples/lb) as 159 mg. How many grams of potassium are in 3.25 kg of apples?

•• **1.89** Describe, in general terms, the changes in potential energy and kinetic energy as an automobile (a) starts moving, (b) climbs a hill, (c) descends a hill, and (d) comes to a stop.

••• **1.90** If a temperature scale were based on the freezing point (5.5°C) and boiling point (80.1°C) of benzene and the temperature difference between these points was divided into 50 units (called °X), what would be the freezing and boiling points of water in °X? (See Figure 1.12, p. 25.)

••• **1.91** Earth's surface area is 5.10×10^8 km^2, and its crust has a mean thickness of 35 km and mean density of 2.8 g/cm^3. The two most abundant elements in the crust are oxygen (4.55×10^5 g/metric ton, t) and silicon (2.72×10^5 g/t), and the two rarest nonradioactive elements are ruthenium and rhodium, each with an abundance of 1×10^{-4} g/t. What is the total mass of each of these elements in Earth's crust (1 t = 1000 kg)?

2

Great Art on the Molecular Scale Any sample of matter has components, which in turn have smaller components, and so on down to the subatomic scale. This magnificent ancient Roman statue of a discus thrower is composed largely of marble, which consists of calcium and carbonate ions attracted to each other in a regular array. Each carbonate ion consists of carbon *(black)* and oxygen *(red)* atoms bonded to each other. In this chapter, you'll learn what matter is made of.

The Components of Matter

2.1 Elements, Compounds, and Mixtures: An Atomic Overview

2.2 The Observations That Led to an Atomic View of Matter
Mass Conservation
Definite Composition
Multiple Proportions

2.3 Dalton's Atomic Theory
Postulates of the Theory
Explanation of Mass Laws

2.4 The Observations That Led to the Nuclear Atom Model
Discovery of the Electron
Discovery of the Nucleus

2.5 The Atomic Theory Today
Structure of the Atom
Atomic Number, Mass Number, and Atomic Symbol
Isotopes and Atomic Masses
Reassessing the Atomic Theory

2.6 Elements: A First Look at the Periodic Table

2.7 Compounds: Introduction to Bonding
Formation of Ionic Compounds
Formation of Covalent Compounds
Elements of Life

2.8 Compounds: Formulas, Names, and Masses
Types of Chemical Formulas
Learning Names and Formulas
Ionic Compounds
Binary Covalent Compounds
Organic Compounds
Molecular Masses

2.9 Mixtures: Classification and Separation

It may seem surprising, but questioning what things are made of is as common today as it was among the philosophers of ancient Greece, even though we approach the question very differently. They believed everything was made of one or, at most, a few elemental substances (elements). Some believed the elemental substance was water because rivers and oceans extend everywhere. Others thought it was air, which was "thinned" into fire or "thickened" into clouds, rain, and rock. Still others believed there were four elements—fire, air, water, and earth—whose properties accounted for hotness, wetness, sweetness, and all other characteristics of things.

Democritus (c. 460–370 BC), the father of atomism, took a different approach. He focused on the ultimate components of *all* substances, and his reasoning went something like this: if you cut a piece of, say, copper smaller and smaller, you must eventually reach a particle of copper so small that it can no longer be cut. Therefore, matter is ultimately composed of indivisible particles, with nothing between them but empty space. He called the particles *atoms* (Greek *atomos*, "uncuttable") and proclaimed: "According to convention, there is a sweet and a bitter, a hot and a cold, and according to convention, there is order. In truth, there are atoms and a void." However, Aristotle (384–322 BC), who elaborated the idea of four elements, held that it was impossible for "nothing" to exist, and his influence was so great that the concept of atoms was suppressed for 2000 years.

Finally, in the 17th century, the great English scientist Robert Boyle argued that an element is composed of "simple Bodies, not made of any other Bodies, of which all mixed Bodies are compounded, and into which they are ultimately resolved," a description that is remarkably similar to today's idea of an element, in which the "simple Bodies" are atoms. Boyle's hypothesis began the wonderful process of discovery, debate, and rediscovery that marks scientific inquiry, as exemplified by Lavoisier's work. Further studies in the 18th century gave rise to laws concerning the relative masses of substances that react with each other. Then, at the beginning of the 19th century, John Dalton proposed an atomic model that explained these mass laws and soon led to rapid progress in chemistry. By that century's close, however, further observation exposed the need to revise Dalton's model. A burst of creativity in the early 20th century gave rise to a picture of the atom with a complex internal structure, which led to our current model.

IN THIS CHAPTER . . . We compare the properties and composition of the three types of matter—elements, compounds, and mixtures—on the macroscopic and atomic scales. We examine the mass laws and Dalton's theory to explain them and then cover key experiments that led to our current model of the atom. Atomic structure is described, and then we see how elements are organized and classified in the periodic table. We discuss the two ways elements combine to form compounds, and learn how to derive compound names, formulas, and masses. Finally, we see how mixtures are classified and separated.

Concepts & Skills to Review
before you study this chapter

- physical and chemical change (Section 1.1)
- states of matter (Section 1.1)
- attraction and repulsion between charged particles (Section 1.1)
- meaning of a scientific model (Section 1.3)
- SI units and conversion factors (Section 1.5)
- significant figures in calculations (Section 1.6)

2.1 ELEMENTS, COMPOUNDS, AND MIXTURES: AN ATOMIC OVERVIEW

Matter can be classified into three types based on its composition—elements, compounds, and mixtures. An **element** is the simplest type of matter with unique physical and chemical properties. *An element consists of only one kind of atom.* Therefore, it cannot be broken down into a simpler type of matter by any physical or chemical methods. An element is one kind of **substance,** matter whose composition is fixed. Each element has a name, such as silicon, oxygen, or copper. A sample of silicon contains only silicon atoms. A key point to remember is that the *macroscopic* properties of a piece of silicon, such as color, density, and combustibility, are different from those of a piece of copper because silicon atoms are different from copper atoms; in other words, *each element is unique because the properties of its atoms are unique.*

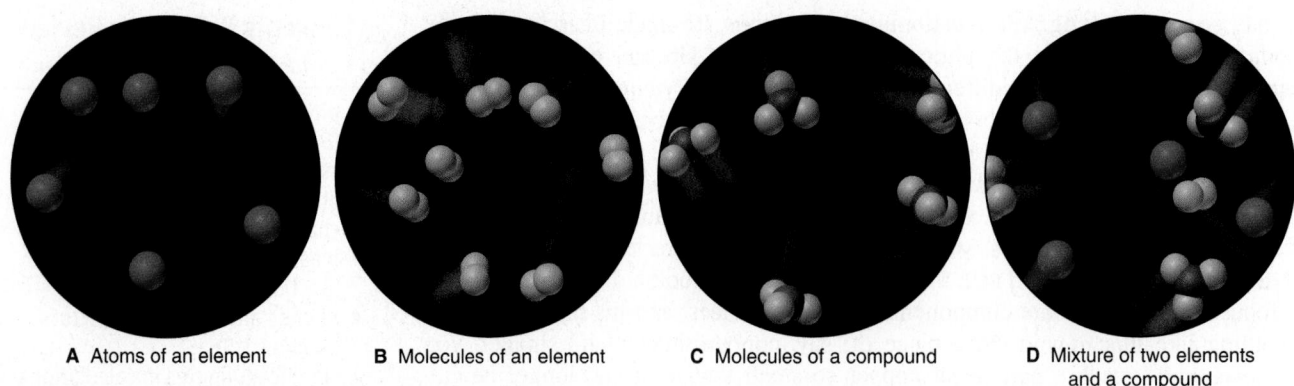

A Atoms of an element   **B** Molecules of an element   **C** Molecules of a compound   **D** Mixture of two elements and a compound

Figure 2.1 **Elements, compounds, and mixtures on the atomic scale. A,** Most elements consist of a large collection of identical atoms. **B,** Some elements occur as molecules. **C,** A molecule of a compound consists of characteristic numbers of atoms of two or more elements chemically bound together. **D,** A mixture contains the individual units of two or more elements and/or compounds that are physically intermingled. The samples shown here are gases, but elements, compounds, and mixtures occur as liquids and solids also.

Transparency/ARIS Presentation Center, Active Art Library

Most elements exist in nature as populations of atoms. Figure 2.1A shows atoms of a gaseous element such as neon. However, several elements occur naturally as molecules: a **molecule** is an independent structure consisting of two or more atoms chemically bound together (Figure 2.1B). Elemental oxygen, for example, occurs in air as *diatomic* (two-atom) molecules.

A **compound** is a type of matter composed of *two or more different elements that are chemically bound together.* Be sure you understand that the elements in a compound are not just mixed together; rather, their atoms have joined chemically (Figure 2.1C). Ammonia, water, and carbon dioxide are some common compounds. One defining feature of a compound is that *the elements are present in fixed parts by mass* (fixed mass ratio). Because of this fixed composition, *a compound is also considered a substance.* Any sample of the compound has the same fixed parts by mass because each of its molecules consists of *fixed numbers* of atoms of the component elements. For example, any sample of ammonia is 14 parts nitrogen by mass plus 3 parts hydrogen by mass. Since 1 nitrogen atom has 14 times the mass of 1 hydrogen atom, a molecule of ammonia must consist of 1 nitrogen atom for every 3 hydrogen atoms:

Ammonia is 14 parts N and 3 parts H by mass.
1 N atom has 14 times the mass of 1 H atom.
Therefore, ammonia has 1 N atom for every 3 H atoms.

Another defining feature of a compound is that *its properties are different from those of its component elements.* Table 2.1 shows a striking example. Soft, silvery sodium metal and yellow-green, poisonous chlorine gas have very different properties from the compound they form—white, crystalline sodium chloride, or common table salt! Unlike an element, a compound *can* be broken down into simpler substances—its component elements. For example, an electric current breaks down molten sodium chloride into metallic sodium and chlorine gas. Note that this breakdown is a *chemical change,* not a physical one.

Figure 2.1D depicts a **mixture,** a group of two or more substances (elements and/or compounds) that are physically intermingled. In contrast to a compound,

 ARIS Presentation Center, Tables Library

Table 2.1   **Some Properties of Sodium, Chlorine, and Sodium Chloride**

Property	Sodium	+	Chlorine	⟶	Sodium Chloride	
Melting point	97.8°C		−101°C		801°C	
Boiling point	881.4°C		−34°C		1413°C	
Color	Silvery		Yellow-green		Colorless (white)	
Density	0.97 g/cm^3		0.0032 g/cm^3		2.16 g/cm^3	
Behavior in water	Reacts		Dissolves slightly		Dissolves freely	

*the components of a mixture **can** vary in their parts by mass.* Because its composition is not fixed, a mixture is *not* a substance. A mixture of the two compounds sodium chloride and water, for example, can have many different parts by mass of salt to water. Because the components are physically mixed, not chemically combined, a mixture at the atomic scale is merely a group of the individual units that make up its component elements and/or compounds. Therefore, *a mixture retains many of the properties of its components.* Saltwater, for instance, is colorless like water and tastes salty like sodium chloride. Unlike compounds, mixtures can be separated into their components by *physical changes;* chemical changes are not needed. For example, the water in saltwater can be boiled off, a physical process that leaves behind the solid sodium chloride. The following sample problem will help differentiate these types of matter.

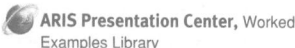

ARIS Presentation Center, Worked Examples Library

SAMPLE PROBLEM 2.1 **Distinguishing Elements, Compounds, and Mixtures at the Atomic Scale**

PROBLEM The scenes below represent an atomic-scale view of three samples of matter:

(a) **(b)** **(c)**

Describe each sample as an element, compound, or mixture.

PLAN From depictions of the samples, we have to determine the type of matter by examining the component particles. If a sample contains only one type of particle, it is either an element or a compound; if it contains more than one type, it is a mixture. Particles of an element have only one kind of atom (one color), and particles of a compound have two or more kinds of atoms.

SOLUTION **(a)** This sample is a mixture: there are three different types of particles, two types contain only one kind of atom, either green or purple, so they are elements, and the third type contains two red atoms for every one yellow, so it is a compound. **(b)** This sample is an element: it consists of only blue atoms, **(c)** This sample is a compound: it consists of molecules that each have two black and six blue atoms.

FOLLOW-UP PROBLEM 2.1 Describe the following reaction in terms of elements, compounds, and mixtures.

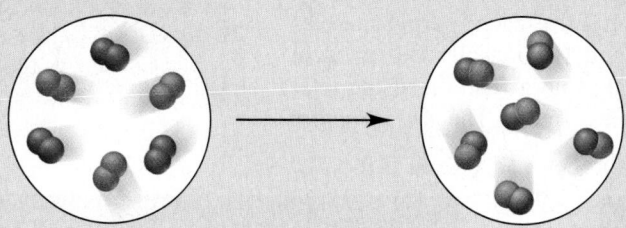

Section Summary

All matter exists as either elements, compounds, or mixtures. • Elements and compounds are referred to as *substances* because their compositions are fixed. An element consists of only one type of atom. A compound contains two or more elements in chemical combination and exhibits different properties from its component elements. The elements of a compound occur in fixed parts by mass because each unit of the compound has fixed numbers of each type of atom. • A mixture consists of two or more substances mixed together, not chemically combined. The components retain their individual properties and can be present in any proportion.

Literature Clark, Roy W. "Why Do Chemists Weigh Things in Grams Instead of Gamus?" *Chem. Educator* 7 **2002**, *4*, 192–194.

2.2 THE OBSERVATIONS THAT LED TO AN ATOMIC VIEW OF MATTER

Any model of the composition of matter had to explain two extremely important chemical observations that were well established by the end of the 18th century: the *law of mass conservation* and the *law of definite (or constant) composition.* As you'll see, John Dalton's atomic theory explained these laws and another observation now known as the *law of multiple proportions.*

Mass Conservation

The most fundamental chemical observation of the 18th century was the **law of mass conservation:** *the total mass of substances does not change during a chemical reaction.* The *number* of substances may change and, by definition, their properties must, but the *total amount* of matter remains constant. Lavoisier had first stated this law on the basis of his combustion experiments. Figure 2.2 illustrates mass conservation in a reaction that occurs in water.

Even in a complex biochemical change, such as the metabolism of the sugar glucose, which involves many reactions, mass is conserved:

$$180 \text{ g glucose} + 192 \text{ g oxygen gas} \longrightarrow 264 \text{ g carbon dioxide} + 108 \text{ g water}$$
$$372 \text{ g material before change} \longrightarrow 372 \text{ g material after change}$$

Mass conservation means that, based on all chemical experience, *matter cannot be created or destroyed.* (As you'll see later, however, mass *does* change in nuclear reactions.) ●

Definite Composition

Another fundamental chemical observation is summarized as the **law of definite (or constant) composition:** *no matter what its source, a particular compound is composed of the same elements in the same parts (fractions) by mass.* The **fraction by mass (mass fraction)** is that part of the compound's mass that each element contributes. It is obtained by dividing the mass of each element by the total mass of compound. The **percent by mass (mass percent, mass %)** is the fraction by mass expressed as a percentage.

Let's see what these ideas mean in terms of a box of marbles *(right).* The box contains three types of marbles: yellow marbles weigh 1.0 g each, purple marbles 2.0 g each, and red marbles 3.0 g each. Each type makes up a fraction of the total mass of marbles, 16.0 g. The *mass fraction*

● **Immeasurable Changes in Mass** Based on the work of Albert Einstein (1879–1955), we now know that some mass *does* change into energy during a chemical reaction. But the amount is too small to measure, even by the best modern balance. For example, when 100 g of carbon burns in oxygen, carbon dioxide is formed, and only 0.000000036 g $(3.6 \times 10^{-8} \text{ g})$ of mass is converted to energy. The energy yields of chemical reactions are relatively so small that, for all practical purposes, mass *is* conserved. As you'll see later, however, energy changes in nuclear reactions are so large that mass changes are measured easily.

ARIS Presentation Center, Art Library

ARIS Presentation Center, Photo Library

Figure 2.2 The law of mass conservation: mass remains constant during a chemical reaction. The total mass of lead nitrate solution and sodium chromate solution before they react **(A)** is the same as the total mass after they have reacted **(B)** to form lead chromate (yellow solid) and sodium nitrate solution.

of the yellow marbles is their number times their mass divided by the total mass: $(3 \times 1.0 \text{ g})/16.0 \text{ g} = 0.19$. The *mass percent* (parts per 100 parts) of the yellow marbles is $0.19 \times 100 = 19\%$ by mass. The purple marbles have a mass fraction of 0.25 and are 25% of the total by mass, and the red marbles have a mass fraction of 0.56 and are 56% by mass. Similarly, in a compound, each element has a *fixed* mass fraction (and mass percent).

Consider calcium carbonate, the major compound in marble. It is composed of three elements—calcium, carbon, and oxygen—and each is present in a fixed fraction (or percent) by mass. The following results are obtained for the elemental mass composition of 20.0 g of calcium carbonate (for example, 8.0 g of calcium/20.0 g = 0.40 parts of calcium):

Analysis by Mass (grams/20.0 g)	Mass Fraction (parts/1.00 part)	Percent by Mass (parts/100 parts)
8.0 g calcium	0.40 calcium	40% calcium
2.4 g carbon	0.12 carbon	12% carbon
9.6 g oxygen	0.48 oxygen	48% oxygen
20.0 g	1.00 part by mass	100% by mass

As you can see, the sum of the mass fractions (or mass percents) equals 1.00 part (or 100%) by mass. The law of definite composition tells us that pure samples of calcium carbonate, no matter where they come from, always contain these elements in the same percents by mass (Figure 2.3).

Because a given element always constitutes the same mass fraction of a given compound, we can use that mass fraction to find the actual mass of the element in any sample of the compound:

$$\text{Mass of element} = \text{mass of compound} \times \frac{\text{part by mass of element}}{\text{one part by mass of compound}}$$

Or, more simply, because mass analysis tells us the parts by mass, we can use that ratio directly with *any* mass unit and skip the need to find the mass fraction first:

$$\text{Mass of element in sample}$$
$$= \text{mass of compound in sample} \times \frac{\text{mass of element in compound}}{\text{mass of compound}} \quad (2.1)$$

CALCIUM CARBONATE
40 mass % calcium
12 mass % carbon
48 mass % oxygen

Figure 2.3 The law of definite composition. Calcium carbonate is found naturally in many forms, including marble *(top)*, coral *(bottom)*, chalk, and seashells. The mass percents of its component elements do not change regardless of the compound's source.

ARIS Presentation Center, Photo Library

Literature DeMeo, Stephen. "Revisiting Molar Mass, Atomic Mass, and Mass Number: Organizing, Integrating, and Sequencing Fundamental Chemical Concepts," *J. Chem. Educ.* **2006,** *83,* 617.

ARIS Presentation Center, Worked Examples Library

SAMPLE PROBLEM 2.2 Calculating the Mass of an Element in a Compound

PROBLEM Pitchblende is the most commercially important compound of uranium. Analysis shows that 84.2 g of pitchblende contains 71.4 g of uranium, with oxygen as the only other element. How many grams of uranium can be obtained from 102 kg of pitchblende?

PLAN We have to find the mass of uranium in a known mass of pitchblende, given the mass of uranium in a different mass of pitchblende. The mass ratio of uranium/pitchblende is the same for any sample of pitchblende. Therefore, as shown by Equation 2.1, we multiply the mass (in kg) of pitchblende by the ratio of uranium to pitchblende that we construct from the mass analysis. This gives the mass (in kg) of uranium, and we just convert kilograms to grams.

SOLUTION Finding the mass (kg) of uranium in 102 kg of pitchblende:

$$\text{Mass (kg) of uranium} = \text{mass (kg) of pitchblende} \times \frac{\text{mass (kg) of uranium in pitchblende}}{\text{mass (kg) of pitchblende}}$$

$$\text{Mass (kg) of uranium} = 102 \, \cancel{\text{kg pitchblende}} \times \frac{71.4 \text{ kg uranium}}{84.2 \, \cancel{\text{kg pitchblende}}} = 86.5 \text{ kg uranium}$$

Converting the mass of uranium from kg to g:

$$\text{Mass (g) of uranium} = 86.5 \, \cancel{\text{kg}} \text{ uranium} \times \frac{1000 \text{ g}}{1 \, \cancel{\text{kg}}} = 8.65 \times 10^4 \text{ g uranium}$$

Mass (kg) of pitchblende

multiply by mass ratio of uranium to pitchblende from analysis

Mass (kg) of uranium

1 kg = 1000 g

Mass (g) of uranium

CHECK The analysis showed that most of the mass of pitchblende is due to uranium, so the large mass of uranium makes sense. Rounding off to check the math gives:

$$\sim\!100 \text{ kg pitchblende} \times \frac{70}{85} = 82 \text{ kg uranium}$$

FOLLOW-UP PROBLEM 2.2 How many metric tons (t) of oxygen are combined in a sample of pitchblende that contains 2.3 t of uranium? (*Hint:* Remember that oxygen is the only other element present.)

Demo Shakhashiri, Bassam Z. "Reaction of Iron and Sulfur," *Chemical Demonstrations, A Handbook for Teachers of Chemistry*, Vol. 1 (The University of Wisconsin Press, Madison, 1985): 55.

Multiple Proportions

Dalton described a phenomenon that occurs when two elements form more than one compound. His observation is now called the **law of multiple proportions:** *if elements A and B react to form two compounds, the different masses of B that combine with a fixed mass of A can be expressed as a ratio of small whole numbers.* Consider two compounds that form from carbon and oxygen; for now, let's call them carbon oxides I and II. They have very different properties. For example, measured at the same temperature and pressure, the density of carbon oxide I is 1.25 g/L, whereas that of II is 1.98 g/L. Moreover, I is poisonous and flammable, but II is not. Analysis shows that their compositions by mass are

Carbon oxide I: 57.1 mass % oxygen and 42.9 mass % carbon
Carbon oxide II: 72.7 mass % oxygen and 27.3 mass % carbon

To see the phenomenon of multiple proportions, we use the mass percents of oxygen and of carbon in each compound to find the masses of these elements in a given mass, for example, 100 g, of each compound. Then we divide the mass of oxygen by the mass of carbon in each compound to obtain the mass of oxygen that combines with a fixed mass of carbon:

	Carbon Oxide I	Carbon Oxide II
g oxygen/100 g compound	57.1	72.7
g carbon/100 g compound	42.9	27.3
g oxygen/g carbon	$\dfrac{57.1}{42.9} = 1.33$	$\dfrac{72.7}{27.3} = 2.66$

If we then divide the grams of oxygen per gram of carbon in II by that in I, we obtain a ratio of small whole numbers:

$$\frac{2.66 \text{ g oxygen/g carbon in II}}{1.33 \text{ g oxygen/g carbon in I}} = \frac{2}{1}$$

The law of multiple proportions tells us that in two compounds of the same elements, the mass fraction of one element relative to the other element changes in *increments based on ratios of small whole numbers.* In this case, the ratio is 2:1—for a given mass of carbon, II contains *2 times* as much oxygen as I, not 1.583 times, 1.716 times, or any other intermediate amount. As you'll see next, Dalton's theory allows us to explain the composition of carbon oxides I and II on the atomic scale.

Section Summary

Three fundamental observations are known as the mass laws. The law of mass conservation states that the total mass remains constant during a chemical reaction. • The law of definite composition states that any sample of a given compound has the same elements present in the same parts by mass. • The law of multiple proportions states that, in different compounds of the same elements, the masses of one element that combine with a fixed mass of the other can be expressed as a ratio of small whole numbers.

2.3 DALTON'S ATOMIC THEORY

With 200 years of hindsight, it may be easy to see how the mass laws could be explained by an atomic model—matter existing in indestructible units, each with a particular mass—but it was a major breakthrough in 1808 when John Dalton (1766–1844) presented his atomic theory of matter in *A New System of Chemical Philosophy.* ●

Postulates of the Atomic Theory

Dalton expressed his theory in a series of postulates. Like most great thinkers, Dalton incorporated the ideas of others into his own to create the new theory. As we go through the postulates, which are presented here in modern terms, let's see which were original and which came from others. (Later, we can examine the key differences between Dalton's postulates and our present understanding.)

1. All matter consists of **atoms,** tiny indivisible particles of an element that cannot be created or destroyed. (Derives from the "eternal, indestructible atoms" proposed by Democritus more than 2000 years earlier and conforms to mass conservation as stated by Lavoisier.)
2. Atoms of one element *cannot* be converted into atoms of another element. In chemical reactions, the atoms of the original substances recombine to form different substances. (Rejects the alchemical belief in the magical transmutation of elements.)
3. Atoms of an element are identical in mass and other properties and are different from atoms of any other element. (Contains Dalton's major new ideas: unique mass and properties for all the atoms of a given element.)
4. Compounds result from the chemical combination of a specific ratio of atoms of different elements. (Follows directly from the law of definite composition.)

How the Theory Explains the Mass Laws

Let's see how Dalton's postulates explain the mass laws:

- *Mass conservation.* Atoms cannot be created or destroyed (postulate 1) or converted into other types of atoms (postulate 2). Since each type of atom has a fixed mass (postulate 3), a chemical reaction, in which atoms are just combined differently with each other, cannot possibly result in a mass change.
- *Definite composition.* A compound is a combination of a *specific* ratio of different atoms (postulate 4), each of which has a particular mass (postulate 3). Thus, each element in a compound constitutes a fixed fraction of the total mass.
- *Multiple proportions.* Atoms of an element have the same mass (postulate 3) and are indivisible (postulate 1). The masses of element B that combine with a fixed mass of element A give a small, whole-number ratio because different numbers of B atoms combine with each A atom in different compounds.

The *simplest* arrangement consistent with the mass data for carbon oxides I and II in our earlier example is that one atom of oxygen combines with one atom of carbon in compound I (carbon monoxide) and that two atoms of oxygen combine with one atom of carbon in compound II (carbon dioxide):

Carbon oxide I
(carbon monoxide)

Carbon oxide II
(carbon dioxide)

Let's work through a sample problem that reviews the mass laws.

● **Dalton's Revival of Atomism** Although John Dalton, the son of a poor weaver, had no formal education, he established one of the most powerful concepts in science. Dalton began teaching science at 12 years of age, and later studied color blindness, a personal affliction still known as *daltonism.* In 1787, he began his life's work in meteorology, recording daily weather data until his death 57 years later. His studies on humidity and dew point led to a key discovery about the behavior of gases (Section 5.4) and eventually to his atomic theory. In 1803, he stated, "I am nearly persuaded that [the mixing of gases and their solubility in water] depends upon the mass and number of the ultimate particles. . . . An enquiry into the relative masses of [these] particles of bodies is a subject . . . I have lately been prosecuting . . . with remarkable success." The atomic theory was published 5 years later.

ARIS Presentation Center, Photo Library

ARIS Presentation Center, Art Library

SAMPLE PROBLEM 2.3 Visualizing the Mass Laws

PROBLEM The scenes below represent an atomic-scale view of a chemical reaction:

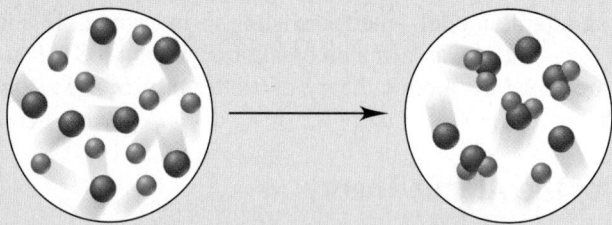

Which of the mass laws—mass conservation, definite composition, or multiple proportions—is (are) illustrated?

PLAN From the depictions, we note the number, color, and combinations of atoms (spheres) to see which mass laws pertain. If the numbers of each atom are the same before and after the reaction, the total mass did not change (mass conservation). If a compound forms that always has the same atom ratio, the elements are present in fixed parts by mass (definite composition). When the same elements form different compounds and the ratio of the atoms of one element that combine with one atom of the other element is a small whole number, the ratio of their masses is a small whole number as well (multiple proportions).

SOLUTION There are seven purple and nine green atoms in each circle, so mass is conserved. The compound formed has one purple and two green atoms, so it has definite composition. Only one compound forms, so the law of multiple proportions does not pertain.

FOLLOW-UP PROBLEM 2.3 Which sample(s) best display(s) the fact that compounds of bromine (orange) and fluorine (yellow) exhibit the law of multiple proportions? Explain.

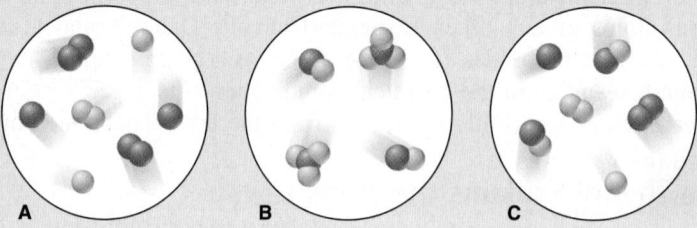

Section Summary

Dalton's atomic theory explained the mass laws by proposing that all matter consists of indivisible, unchangeable atoms of fixed, unique mass. • Mass is conserved during a reaction because atoms form new combinations. • Each compound has a fixed mass fraction of each of its elements because it is composed of a fixed number of each type of atom. • Different compounds of the same elements exhibit multiple proportions because they each consist of whole atoms.

2.4 THE OBSERVATIONS THAT LED TO THE NUCLEAR ATOM MODEL

After publication of the atomic theory, investigators tried to determine the masses of atoms from the mass fractions of elements in compounds. Because an individual atom is so small, the mass of the atoms of one element was determined *relative* to the mass of the atoms of another element, based on a mass standard. Dalton's model was crucial because it originated the idea that masses of reacting elements could be explained in terms of atoms. However, the model did not explain why atoms bond as they do: for example, why do two, and not three, hydrogen atoms bond with one oxygen atom in water? Also, Dalton's "billiard ball" atom did not account for the charged particles observed in later experiments. Clearly, a more complex atomic model was needed. ●

Basic research into the nature of electricity eventually led to the discovery of *electrons,* negatively charged particles that are part of all atoms. Soon thereafter,

● **Atoms? Humbug!** Rarely does a major new concept receive unanimous acceptance. Despite the atomic theory's impact, several major scientists denied the existence of atoms for another century. In 1877, Adolf Kolbe, an eminent organic chemist, said, "[Dalton's atoms are] . . . no more than stupid hallucinations . . . mere table-tapping and supernatural explanations." The influential physicist Ernst Mach believed that scientists should look at facts, not hypothetical entities such as atoms. It was not until 1908 that the famous chemist and outspoken opponent of atomism Wilhelm Ostwald wrote, "I am now convinced [by recent] experimental evidence of the discrete or grained nature of matter, which the atomic hypothesis sought in vain for hundreds and thousands of years." He was referring to the discovery of the electron.

other experiments revealed that the atom has a *nucleus*—a tiny, central core of mass and positive charge. In this section, we examine some key experiments that led to our current model of the atom.

Discovery of the Electron and Its Properties

Nineteenth-century investigators of electricity knew that matter and electric charge were somehow related. When amber is rubbed with fur, or glass with silk, positive and negative charges form—the same charges that make your hair crackle and cling to your comb on a dry day. They also knew that an electric current could decompose certain compounds into their elements.

What they did not know, however, was what an electric current itself might consist of. Some investigators tried passing current from a high-voltage source through nearly evacuated glass tubes fitted with metal electrodes that were sealed in place and connected to an external source of electricity. When the power was turned on, a "ray" could be seen striking the phosphor-coated end of the tube and emitting a glowing spot of light. The rays were called **cathode rays** because they originated at the negative electrode (cathode) and moved to the positive electrode (anode). Cathode rays typically travel in a straight line, but in a magnetic field the path is bent, indicating that the particles are charged, and in an electric field the path bends toward the positive plate. The ray is identical no matter what metal is used as the cathode (Figure 2.4). It was concluded that cathode rays consist of negatively charged particles found in all matter. The rays appear when these particles collide with the few remaining gas molecules in the evacuated tube. ● Cathode ray particles were later named *electrons*.

● **The Familiar Glow of Colliding Particles** The electric and magnetic properties of charged particles that collide with gas particles or hit a phosphor-coated screen have familiar applications. A "neon" sign glows because electrons collide with the gas particles in the tube, causing them to give off light. An aurora display occurs when Earth's magnetic field bends streams of charged particles coming from the Sun, which then collide with gases in the atmosphere. In a television tube or computer monitor, the cathode ray passes back and forth over the coated screen, creating a pattern that the eye sees as a picture.

ARIS Presentation Center, Photo Library

Transparency/ARIS Presentation Center, Art Library

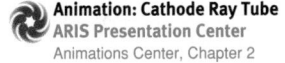
Animation: Cathode Ray Tube
ARIS Presentation Center
Animations Center, Chapter 2

Figure 2.4 Experiments to determine the properties of cathode rays. A cathode ray forms when high voltage is applied to a partially evacuated tube. The ray passes through a hole in the anode and hits the coated end of the tube to produce a glow.

OBSERVATION	CONCLUSION
1. Ray bends in magnetic field	Consists of charged particles
2. Ray bends toward positive plate in electric field	Consists of negative particles
3. Ray is identical for any cathode	Particles found in all matter

In 1897, the British physicist J. J. Thomson (1856–1940) used magnetic and electric fields to measure the ratio of the cathode ray particle's mass to its charge. By comparing this value with the mass/charge ratio for the lightest charged particle in solution, Thomson estimated that the cathode ray particle weighed less than $\frac{1}{1000}$ as much as hydrogen, the lightest atom! He was shocked because this implied that, contrary to Dalton's atomic theory, *atoms are divisible into even smaller particles.* Thomson concluded, "We have in the cathode rays matter in a new state, . . . in which the subdivision of matter is carried much further . . . ; this matter being the substance from which the chemical elements are built up." Fellow scientists reacted with disbelief, and some even thought he was joking.

In 1909, the American physicist Robert Millikan (1868–1953) measured the *charge* of the electron. He did so by observing the movement of tiny droplets of the "highest grade clock oil" in an apparatus that contained electrically charged

Figure 2.5 Millikan's oil-drop experiment for measuring an electron's charge. The motion of a given oil droplet depends on the variation in electric field and the total charge on the droplet, which depends in turn on the number of attached electrons. Millikan reasoned that the total charge must be some whole-number multiple of the charge of the electron.

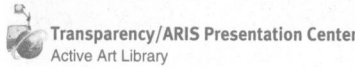
Transparency/ARIS Presentation Center, Active Art Library

Literature Pearson, Earl F. "Millikan: Good to the Last (Oil) Drop," *J. Chem. Educ.* **2006,** *83,* 1312A.

Literature Eckey, Doris. "A Millikan Oil Drop Analogy," *J. Chem. Educ.* **1996,** *73,* 237.

Animation: Millikan Oil Drop ARIS Presentation Center Chapter 2

Literature Pearson, Earl F. "Magnet and BB Analogy for Millikan's Oil-Drop Experiment," *J. Chem. Educ.* **2006,** *83,* 1313.

❷ Oil droplets fall through hole in positively charged plate
❶ Fine mist of oil sprayed into apparatus
❸ X-rays knock electrons from surrounding air, which stick to droplets
❹ Electrically charged plates influence droplet's motion
X-ray source
❺ Observer times droplet's motion and controls electric field

plates and an x-ray source (Figure 2.5). Here is a description of the basis of the experiment: X-rays knocked electrons from gas molecules in the air, and as an oil droplet fell through a hole in the positive (upper) plate, the electrons stuck to the drop, giving it a negative charge. With the electric field off, Millikan measured the mass of the droplet from its rate of fall. By turning on the field and varying its strength, he could make the drop fall more slowly, rise, or pause suspended. From these data, Millikan calculated the total charge of the droplet.

After studying many droplets, Millikan calculated that the various charges of the droplets were always some *whole-number multiple of a minimum charge*. He reasoned that different oil droplets picked up different numbers of electrons, so this minimum charge must be that of the electron itself. The value that he calculated a century ago was within 1% of the modern value of the electron's charge, -1.602×10^{-19} C (C stands for *coulomb,* the SI unit of charge). Using the electron's mass/charge ratio from work by Thomson and others and this value for the electron's charge, let's calculate the electron's *extremely* small mass the way Millikan did:

$$\text{Mass of electron} = \frac{\text{mass}}{\text{charge}} \times \text{charge} = \left(-5.686\times10^{-12}\,\frac{\text{kg}}{\text{C}}\right)(-1.602\times10^{-19}\,\text{C})$$
$$= 9.109\times10^{-31}\,\text{kg} = 9.109\times10^{-28}\,\text{g}$$

Discovery of the Atomic Nucleus

Clearly, the properties of the electron posed problems about the inner structure of atoms. If everyday matter is electrically neutral, the atoms that make it up must be neutral also. But if atoms contain negatively charged electrons, what positive charges balance them? And if an electron has such an incredibly tiny mass, what accounts for an atom's much larger mass? To address these issues, Thomson proposed a model of a spherical atom composed of diffuse, positively charged matter, in which electrons were embedded like "raisins in a plum pudding."

Near the turn of the 20th century, French scientists discovered radioactivity, the emission of particles and/or radiation from atoms of certain elements. Just a few years later, in 1910, the New Zealand–born physicist Ernest Rutherford (1871–1937) used one type of radioactive particle in a series of experiments that solved this dilemma of atomic structure.

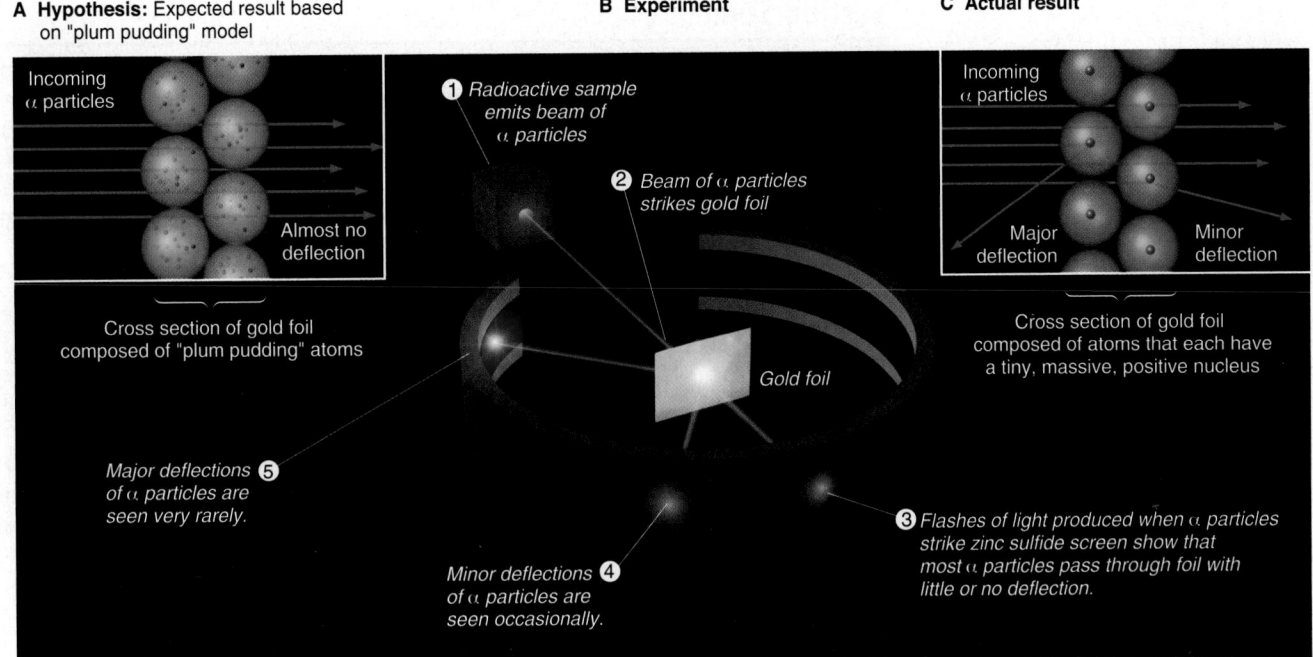

A Hypothesis: Expected result based on "plum pudding" model

B Experiment

C Actual result

Incoming α particles

Almost no deflection

Cross section of gold foil composed of "plum pudding" atoms

Major deflections ❺ of α particles are seen very rarely.

❶ *Radioactive sample emits beam of α particles*

❷ *Beam of α particles strikes gold foil*

Gold foil

Minor deflections ❹ of α particles are seen occasionally.

❸ *Flashes of light produced when α particles strike zinc sulfide screen show that most α particles pass through foil with little or no deflection.*

Incoming α particles

Major deflection   Minor deflection

Cross section of gold foil composed of atoms that each have a tiny, massive, positive nucleus

Figure 2.6 Rutherford's α-scattering experiment and discovery of the atomic nucleus. A, HYPOTHESIS: Atoms consist of electrons embedded in diffuse, positively charged matter, so the speeding α particles should pass through the gold foil with, at most, minor deflections. **B,** EXPERIMENT: α Particles emit a flash of light when they pass through the gold atoms and hit a phosphor-coated screen. **C,** RESULTS: Occasional minor deflections and very infrequent major deflections are seen. This means very high mass and positive charge are concentrated in a small region within the atom, the nucleus.

 ARIS Presentation Center, Active Art Library

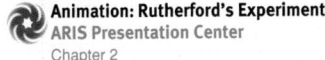 **Animation: Rutherford's Experiment**
ARIS Presentation Center
Chapter 2

Figure 2.6 is a three-part representation of Rutherford's experiment. Tiny, dense, positively charged alpha (α) particles emitted from radium were aimed, like minute projectiles, at thin gold foil. The figure illustrates (A) the "plum pudding" hypothesis, (B) the apparatus used to measure the deflection (scattering) of the α particles from the light flashes created when the particles struck a circular, coated screen, and (C) the actual result.

With Thomson's model in mind, Rutherford expected only minor, if any, deflections of the α particles because they should act as tiny, dense, positively charged "bullets" and go right through the gold atoms. According to the model, one of the embedded electrons could not deflect an α particle any more than a Ping-Pong ball could deflect a speeding baseball. Initial results confirmed this, but soon the unexpected happened. As Rutherford recalled: "Then I remember two or three days later Geiger [one of his coworkers] coming to me in great excitement and saying, 'We have been able to get some of the α particles coming backwards . . .' It was quite the most incredible event that has ever happened to me in my life. It was almost as incredible as if you fired a 15-inch shell at a piece of tissue paper and it came back and hit you."

The data showed that very few α particles were deflected at all, and that only 1 in 20,000 was deflected by more than 90° ("coming backwards"). It seemed that these few α particles were being repelled by something small, dense, and positive within the gold atoms. From the mass, charge, and velocity of the α particles, the frequency of these large-angle deflections, and the properties of electrons, Rutherford calculated that *an atom is mostly space occupied by electrons,* but in the center of that space is a tiny region, which he called the **nucleus,** that contains *all the positive charge and essentially all the mass of the atom.* He proposed that positive particles lay within the nucleus and called them *protons,* and then he calculated the magnitude of the nuclear charge with remarkable accuracy. Rutherford's model explained the charged nature of matter, but it could not account for all the atom's mass. After more than 20 years, this issue was resolved when, in 1932, James Chadwick discovered the *neutron,* an uncharged dense particle that also resides in the nucleus.

Section Summary

Several major discoveries at the turn of the 20th century led to our current model of atomic structure. • Cathode rays were shown to consist of negative particles (electrons) that exist in all matter. J. J. Thomson measured their mass/charge ratio and concluded that they are much smaller and lighter than atoms. • Robert Millikan determined the charge of the electron, which he combined with other data to calculate its mass. • Ernest Rutherford proposed that atoms consist of a tiny, massive, positive nucleus surrounded by electrons.

2.5   THE ATOMIC THEORY TODAY

For over 200 years, scientists have known that all matter consists of atoms, and they have learned astonishing things about them. Dalton's tiny indivisible particles have given way to atoms with "fuzzy," indistinct boundaries and an elaborate internal architecture of subatomic particles. In this section, we examine our current model and begin to see how the properties of subatomic particles affect the properties of atoms. Then we'll see how Dalton's theory stands up today.

Structure of the Atom

An *atom* is an electrically neutral, spherical entity composed of a positively charged central nucleus surrounded by one or more negatively charged electrons (Figure 2.7). The electrons move rapidly within the available atomic volume, held there by the attraction of the nucleus. The nucleus is incredibly dense: it contributes 99.97% of the atom's mass but occupies only about 1 quadrillionth of its volume. (A nucleus the size of a period on this page would weigh about 100 tons, as much as 50 cars!) An atom's diameter ($\sim 10^{-10}$ m) is about 100,000 times the diameter of its nucleus ($\sim 10^{-15}$ m).

An *atomic nucleus consists of protons and neutrons* (the only exception is the simplest hydrogen nucleus, which is a single proton). The **proton (p^+)** has a positive charge, and the **neutron (n^0)** has no charge; thus, the positive charge of the nucleus results from its protons. The *magnitude* of charge possessed by a proton is equal to that of an **electron (e^-)**, but the *signs* of the charges are opposite. *An atom is neutral because the number of protons in the nucleus equals the number of electrons surrounding the nucleus.* Some properties of these three subatomic particles are listed in Table 2.2.

📖 **Literature** Peake, Barrie M. "The Discovery of the Electron, Proton, and Neutron," *J. Chem. Educ.* **1989,** *66,* 738.

Figure 2.7   **General features of the atom. A,** A "cloud" of rapidly moving, negatively charged electrons occupies virtually all the atomic volume and surrounds the tiny, central nucleus. **B,** The nucleus contains virtually all the mass of the atom and consists of positively charged protons and uncharged neutrons. If the nucleus were actually the size in the figure (~1 cm across), the atom would be about 1000 m (1 km) across.

🎞 **Animation: Alpha, Beta, and Gamma Rays**
ARIS Presentation Center
Chapter 2

 ARIS Presentation Center, Active Art Library

Table 2.2   Properties of the Three Key Subatomic Particles

Name (Symbol)	Charge		Mass		Location in Atom
	Relative	Absolute (C)*	Relative (amu)†	Absolute (g)	
Proton (p⁺)	1+	$+1.60218\times10^{-19}$	1.00727	1.67262×10^{-24}	Nucleus
Neutron (n⁰)	0	0	1.00866	1.67493×10^{-24}	Nucleus
Electron (e⁻)	1−	-1.60218×10^{-19}	0.00054858	9.10939×10^{-28}	Outside nucleus

*The coulomb (C) is the SI unit of charge.
†The atomic mass unit (amu) equals 1.66054×10^{-24} g; discussed later in this section.

Transparency/ARIS Presentation Center, Tables Library

Atomic Number, Mass Number, and Atomic Symbol

The **atomic number (Z)** of an element equals the number of protons in the nucleus of each of its atoms. *All atoms of a particular element have the same atomic number, and each element has a different atomic number from that of any other element.* All carbon atoms (Z = 6) have 6 protons, all oxygen atoms (Z = 8) have 8 protons, and all uranium atoms (Z = 92) have 92 protons. There are currently 116 known elements, of which 90 occur in nature; the remaining 26 have been synthesized by nuclear scientists.

The total number of protons and neutrons in the nucleus of an atom is its **mass number (A).** Each proton and each neutron contributes one unit to the mass number. Thus, a carbon atom with 6 protons and 6 neutrons in its nucleus has a mass number of 12, and a uranium atom with 92 protons and 146 neutrons in its nucleus has a mass number of 238.

The nuclear mass number and charge are often written with the **atomic symbol** (or *element symbol*). Every element has a symbol based on its English, Latin, or Greek name, such as C for carbon, O for oxygen, S for sulfur, and Na for sodium (Latin *natrium*). The atomic number (Z) is written as a left *sub*script and the mass number (A) as a left *super*script to the symbol, so element X would be $^{A}_{Z}X$. Since the mass number is the sum of protons and neutrons, the number of neutrons (N) equals the mass number minus the atomic number:

Number of neutrons = mass number − atomic number,   or $N = A - Z$ **(2.2)**

Thus, a chlorine atom, which is symbolized as $^{35}_{17}Cl$, has A = 35, Z = 17, and N = 35 − 17 = 18. Each element has its own atomic number, so we know the atomic number from the symbol. For example, every carbon atom has 6 protons. Therefore, instead of writing $^{12}_{6}C$ for carbon with mass number 12, we can write ^{12}C (spoken "carbon twelve"), with Z = 6 understood. Another way to write this atom is carbon-12.

Isotopes and Atomic Masses of the Elements

All atoms of an element are identical in atomic number but not in mass number. **Isotopes** of an element are atoms that have *different numbers of neutrons* and therefore different mass numbers. For example, all carbon atoms (Z = 6) have 6 protons and 6 electrons, but only 98.89% of naturally occurring carbon atoms have 6 neutrons in the nucleus (A = 12). A small percentage (1.11%) have 7 neutrons in the nucleus (A = 13), and even fewer (less than 0.01%) have 8 (A = 14). These are carbon's three naturally occurring isotopes—^{12}C, ^{13}C, and ^{14}C. Five other carbon isotopes—^{9}C, ^{10}C, ^{11}C, ^{15}C, and ^{16}C—have been created in the laboratory. Figure 2.8 depicts the atomic number, mass number, and symbol for four atoms, two of which are isotopes of the same element.

A key point is that the chemical properties of an element are primarily determined by the number of electrons, so *all isotopes of an element have nearly identical chemical behavior,* even though they have different masses.

ARIS Presentation Center, Active Art Library

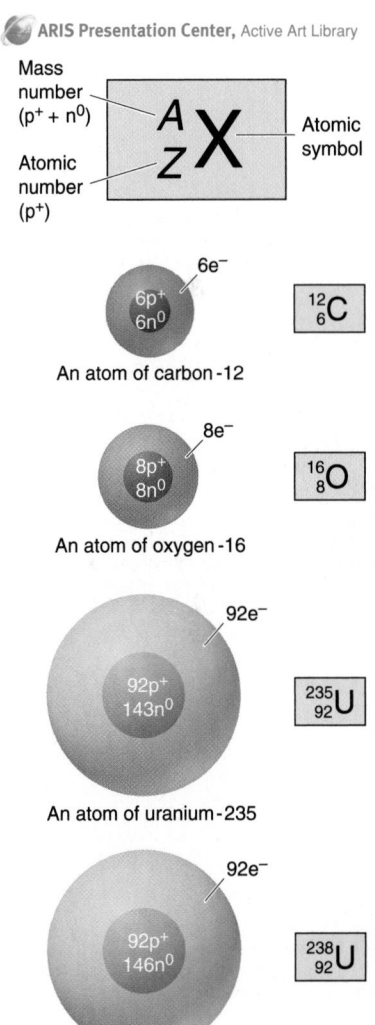

Figure 2.8 Depicting the atom. Atoms of carbon-12, oxygen-16, uranium-235, and uranium-238 are shown (nuclei not drawn to scale) with their symbolic representations. The sum of the number of protons (Z) and the number of neutrons (N) equals the mass number (A). An atom is neutral, so the number of protons in the nucleus equals the number of electrons around the nucleus. The two uranium atoms are isotopes of the element.

SAMPLE PROBLEM 2.4 Determining the Number of Subatomic Particles in the Isotopes of an Element

PROBLEM Silicon (Si) is essential to the computer industry as a major component of semiconductor chips. It has three naturally occurring isotopes: ^{28}Si, ^{29}Si, and ^{30}Si. Determine the numbers of protons, neutrons, and electrons in each silicon isotope.

PLAN The mass number (A) of each of the three isotopes is given, so we know the sum of protons and neutrons. From the elements list on the text's inside front cover, we find the atomic number (Z, number of protons), which equals the number of electrons. We obtain the number of neutrons from Equation 2.2.

SOLUTION From the elements list, the atomic number of silicon is 14. Therefore,

$$^{28}\text{Si has } 14p^+, 14e^-, \text{ and } 14n^0 \ (28 - 14)$$
$$^{29}\text{Si has } 14p^+, 14e^-, \text{ and } 15n^0 \ (29 - 14)$$
$$^{30}\text{Si has } 14p^+, 14e^-, \text{ and } 16n^0 \ (30 - 14)$$

FOLLOW-UP PROBLEM 2.4 How many protons, neutrons, and electrons are in (a) $^{11}_{5}$Q? (b) $^{41}_{20}$R? (c) $^{131}_{53}$X? What element symbols do Q, R, and X represent?

Literature Blauch, David N.; Schuh, Merlyn D.; Carroll, Felix A. "Determination of the Natural Abundances of Krypton and Xenon Isotopes Using Mass Spectrometry: A Demonstration of Isotopes and the Basis of Atomic Mass," *J. Chem. Educ.* **2002**, *79*, 584.

The mass of an atom is measured *relative* to the mass of an atomic standard. The modern atomic mass standard is the carbon-12 atom. Its mass is defined as *exactly* 12 atomic mass units. Thus, the **atomic mass unit (amu)** is $\frac{1}{12}$ the mass of a carbon-12 atom. Based on this standard, the ^{1}H atom has a mass of 1.008 amu; in other words, a ^{12}C atom has almost 12 times the mass of an ^{1}H atom. We will continue to use the term *atomic mass unit* in the text, although the name of the unit has been changed to the **dalton (Da);** thus, one ^{12}C atom has a mass of 12 daltons (12 Da, or 12 amu). The atomic mass unit, which is a unit of relative mass, has an absolute mass of 1.66054×10^{-24} g.

The isotopic makeup of an element is determined by **mass spectrometry,** a method for measuring the relative masses and abundances of atomic-scale particles very precisely (see the Tools of the Laboratory essay). For example, using a mass spectrometer, we measure the mass ratio of ^{28}Si to ^{12}C as

$$\frac{\text{Mass of } ^{28}\text{Si atom}}{\text{Mass of } ^{12}\text{C standard}} = 2.331411$$

From this mass ratio, we find the **isotopic mass** of the ^{28}Si atom, the mass of the isotope relative to the mass of the standard carbon-12 isotope:

$$\text{Isotopic mass of } ^{28}\text{Si} = \text{measured mass ratio} \times \text{mass of } ^{12}\text{C}$$
$$= 2.331411 \times 12 \text{ amu} = 27.97693 \text{ amu}$$

Along with the isotopic mass, the mass spectrometer gives the relative abundance (fraction) of each isotope in a sample of the element. For example, the percent abundance of ^{28}Si is 92.23%. Such measurements provide data for obtaining the **atomic mass** (also called *atomic weight*) of an element, the *average* of the masses of its naturally occurring isotopes weighted according to their abundances.

Literature Last, Arthur M.; Webb, Michael J. "Using Monetary Analogies to Teach Average Atomic Mass," *J. Chem. Educ.* **1993**, *70*, 234.

Literature Marshall, James L.; Marshall, Virginia R. "Rediscovery of the Elements: Ytterby Gruva (Ytterby Mine)," *J. Chem. Educ.* **2001**, *78*, 1343.

Literature Spindel, William; Ishida, Takanobu. "Isotope Separation," *J. Chem. Educ.* **1991**, *68*, 312.

Each naturally occurring isotope of an element contributes a certain portion to the atomic mass. For instance, as just noted, 92.23% of Si atoms are ^{28}Si. Using this percent abundance as a fraction and multiplying by the isotopic mass of ^{28}Si gives the portion of the atomic mass of Si contributed by ^{28}Si:

$$\text{Portion of Si atomic mass from } ^{28}\text{Si} = 27.97693 \text{ amu} \times 0.9223 = 25.8031 \text{ amu}$$
(retaining two additional significant figures)

Similar calculations give the portions contributed by ^{29}Si (28.976495 amu $\times$ 0.0467 = 1.3532 amu) and by ^{30}Si (29.973770 amu $\times$ 0.0310 = 0.9292 amu), and adding the three portions together (rounding to two decimal places at the end) gives the atomic mass of silicon:

$$\text{Atomic mass of Si} = 25.8031 \text{ amu} + 1.3532 \text{ amu} + 0.9292 \text{ amu}$$
$$= 28.0855 \text{ amu} = 28.09 \text{ amu}$$

Mass Spectrometry

Mass spectrometry, the most powerful technique for measuring the mass and abundance of charged particles, emerged from electric and magnetic deflection studies on particles formed in cathode ray experiments. When a high-energy electron collides with an atom of neon-20, for example, one of the atom's electrons is knocked away and the resulting particle has one positive charge, Ne^+ (Figure B2.1). Thus, its mass/charge ratio (*m/e*) equals the mass divided by $1+$. The *m/e* values are measured to identify the masses of different isotopes of an element.

Figure B2.2, parts A–C, depicts the core of one type of mass spectrometer and the data it provides. The sample is introduced and vaporized (if liquid or solid), then bombarded by high-energy electrons to form positively charged particles. These are attracted toward a series of negatively charged plates with slits in them, and some particles pass through into an evacuated tube exposed to a magnetic field. As the particles zoom through this region, they are deflected (their paths are bent) according to their *m/e:* the lightest particles are deflected most and the heaviest particles least. At the end of the magnetic region, the particles strike a detector, which records their relative positions and abundances. For very precise work, such as determining isotopic masses and abundances, the instrument is calibrated with a substance of known amount and mass.

① High-energy electron collides with neon atom in gas sample

② Neon electron is knocked away by impact

③ Positively charged neon particle is produced that has 10p⁺ and 10n⁰ in nucleus but only 9e⁻

$m/e = 19.992435$

Figure B2.1 Formation of a positively charged neon (Ne) particle.

Mass spectrometry is also used in structural chemistry and separations science to measure the mass of virtually any atom, molecule, or molecular fragment. The technique is employed by biochemists determining protein structures (Figure B2.2, part D), materials scientists examining catalyst surfaces, forensic chemists analyzing criminal evidence, pharmaceutical chemists designing new drugs, industrial chemists investigating petroleum components, and many others. In fact, John B. Fenn and Koichi Tanaka shared part of the 2002 Nobel Prize in chemistry for developing methods to study proteins by mass spectrometry.

Figure B2.2 The mass spectrometer and its data.
A, Charged particles are separated on the basis of their *m/e* values. Ne is the sample here. **B,** The data show the abundances of three Ne isotopes. **C,** The percent abundance of each isotope. **D,** The mass spectrum of a protein molecule. Each peak represents a molecular fragment.

Transparency/ARIS Presentation Center, Art Library

② If necessary, heater vaporizes sample

③ Electron beam knocks electrons from atoms (see Figure B2.1)

① Sample enters chamber

Electron source

④ Electric field accelerates particles toward magnetic region

Magnet

⑤ Magnetic field separates particles according to their mass/charge ratio

Detector

Lightest particles in sample

Charged particle beam

Heaviest particles in sample

$^{20}Ne^+$ $^{21}Ne^+$ $^{22}Ne^+$

A

B Mass/charge   19 20 21 22
Abundance of Ne^+ particles

C Mass/charge   20 21 22
Percent abundance
$^{20}Ne^+$ (90.5%)
$^{21}Ne^+$ (0.3%)
$^{22}Ne^+$ (9.2%)

D
MYOGLOBIN
MW 16951.47
21⁺
15⁺

Note that atomic mass is an average value, and averages must be interpreted carefully. Although the average number of children in an American family in 1985 was 2.4, no family actually had 2.4 children; similarly, no individual silicon atom has a mass of 28.09 amu. But for most laboratory purposes, we consider a sample of silicon to consist of atoms with this average mass.

ARIS Presentation Center, Worked Examples Library

SAMPLE PROBLEM 2.5 Calculating the Atomic Mass of an Element

PROBLEM Silver (Ag; $Z = 47$) has 46 known isotopes, but only two occur naturally, ^{107}Ag and ^{109}Ag. Given the following mass spectrometric data, calculate the atomic mass of Ag:

Isotope	Mass (amu)	Abundance (%)
^{107}Ag	106.90509	51.84
^{109}Ag	108.90476	48.16

PLAN From the mass and abundance of the two Ag isotopes, we have to find the atomic mass of Ag (weighted average of the isotopic masses). We multiply each isotopic mass by its fractional abundance to find the portion of the atomic mass contributed by each isotope. The sum of the isotopic portions is the atomic mass.

SOLUTION Finding the portion of the atomic mass from each isotope:

Portion of atomic mass from ^{107}Ag: = isotopic mass × fractional abundance
$$= 106.90509 \text{ amu} \times 0.5184 = 55.42 \text{ amu}$$
Portion of atomic mass from ^{109}Ag: = 108.90476 amu × 0.4816 = 52.45 amu

Finding the atomic mass of silver:

$$\text{Atomic mass of Ag} = 55.42 \text{ amu} + 52.45 \text{ amu} = \boxed{107.87 \text{ amu}}$$

CHECK The individual portions seem right: $\sim$100 amu × 0.50 = 50 amu. The portions should be almost the same because the two isotopic abundances are almost the same. We rounded each portion to four significant figures because that is the number of significant figures in the abundance values. This is the correct atomic mass (to two decimal places), as shown in the list of elements (*inside front cover*).

FOLLOW-UP PROBLEM 2.5 Boron (B; $Z = 5$) has two naturally occurring isotopes. Find the percent abundances of ^{10}B and ^{11}B given the atomic mass of B = 10.81 amu, the isotopic mass of ^{10}B = 10.0129 amu, and the isotopic mass of ^{11}B = 11.0093 amu. (*Hint:* The sum of the fractional abundances is 1. If x = abundance of ^{10}B, then $1 - x$ = abundance of ^{11}B.)

Mass (g) of each isotope

multiply by fractional abundance of each isotope

Portion of atomic mass from each isotope

add isotopic portions

Atomic mass

Literature Reeves, Perry C.; Pamplin, Kim L. "A Strategy for Incorporating Hands-On GC-MS into the General Chemistry Lecture and Laboratory Courses," *J. Chem. Educ.* 2001, *78*, 368.

Demo Ellis, Arthur B.; Adler, Edward A.; Juergens, Frederick H. "Dramatizing Isotopes: Deuterated Ice Cubes Sink," *J. Chem. Educ.* 1990, *67*, 159.

A Modern Reassessment of the Atomic Theory

We began discussing the atomic basis of matter with Dalton's model, which proved inaccurate in several respects. What happens to a model whose postulates are found by later experiment to be incorrect? No model can predict every possible future observation, but a powerful model evolves and remains useful. Let's reexamine the atomic theory in light of what we know now:

1. *All matter is composed of atoms.* We now know that atoms *are* divisible and composed of smaller, subatomic particles (electrons, protons, and neutrons), but the atom is still the smallest body that *retains the unique identity* of an element.

2. *Atoms of one element cannot be converted into atoms of another element in a chemical reaction.* We now know that, in *nuclear* reactions, atoms of one element often change into atoms of another, but this *never* happens in a *chemical* reaction. ●

3. *All atoms of an element have the same number of protons and electrons, which determines the chemical behavior of the element.* We now know that isotopes of an element differ in the number of neutrons, and thus in mass number, but a sample of the element is treated as though its atoms have an *average* mass.

4. *Compounds are formed by the chemical combination of two or more elements in specific ratios.* We now know that a few compounds can have slight variations in their atom ratios, but this postulate remains essentially unchanged.

● **The Heresy of Radioactive "Transmutation"** In 1902, Rutherford performed a series of experiments with radioactive elements that shocked the scientific world. When a radioactive atom of thorium ($Z = 90$) emits an α particle ($Z = 2$), it becomes an atom of radium ($Z = 88$), which then emits another α particle and becomes an atom of radon ($Z = 86$). He proposed that when an atom emits an α particle, it turns into a different atom—one element changes into another! Many viewed this conclusion as a return to alchemy, and, as with Thomson's discovery that atoms contain smaller particles, Rutherford's findings fell on disbelieving ears.

Even today, our picture of the atom is being revised. Although we are confident about the distribution of electrons within the atom (Chapters 7 and 8), the interactions among protons and neutrons within the nucleus are still on the frontier of discovery (Chapter 24).

Section Summary

An atom has a central nucleus, which contains positively charged protons and uncharged neutrons and is surrounded by negatively charged electrons. An atom is neutral because the number of electrons equals the number of protons. • An atom is represented by the notation $^{A}_{Z}X$, in which Z is the atomic number (number of protons), A the mass number (sum of protons and neutrons), and X the atomic symbol. • An element occurs naturally as a mixture of isotopes, atoms with the same number of protons but different numbers of neutrons. Each isotope has a mass relative to the ^{12}C mass standard. • The atomic mass of an element is the average of its isotopic masses weighted according to their natural abundances and is determined by modern instruments, especially the mass spectrometer.

2.6 ELEMENTS: A FIRST LOOK AT THE PERIODIC TABLE

At the end of the 18th century, Lavoisier compiled a list of the 23 elements known at that time; by 1870, 65 were known; by 1925, 88; today, there are 116 and still counting! These elements combine to form millions of compounds, so we clearly need some way to organize what we know about their behavior. By the mid-19th century, enormous amounts of information concerning reactions, properties, and atomic masses of the elements had been accumulated. Several researchers noted recurring, or *periodic,* patterns of behavior and proposed schemes to organize the elements according to some fundamental property.

In 1871, the Russian chemist Dmitri Mendeleev (1836–1907) published the most successful of these organizing schemes as a table of the elements listed by increasing atomic mass and arranged so that elements with similar chemical properties fell in the same column. The modern **periodic table of the elements,** based on Mendeleev's earlier version (but arranged by atomic number, not mass), is one of the great classifying schemes in science and is now an indispensable tool to chemists. Throughout your study of chemistry, the periodic table will guide you through an otherwise dizzying amount of chemical and physical behavior.

Organization of the Periodic Table A modern version of the periodic table appears in Figure 2.9 on the next page and inside the front cover. It is formatted as follows:

1. Each element has a box that contains its atomic number, atomic symbol, and atomic mass. The boxes lie in order of *increasing atomic number* (number of protons) as you move from left to right.
2. The boxes are arranged into a grid of **periods** (horizontal rows) and **groups** (vertical columns). Each period has a number from 1 to 7. Each group has a number from 1 to 8 *and* either the letter A or B. A new system, with group numbers from 1 to 18 but no letters, appears in parentheses under the number-letter designations. (Most chemists still use the number-letter system, so the text retains it, but shows the new numbering system in parentheses.)
3. The eight A groups (two on the left and six on the right) contain the *main-group,* or *representative,* elements. The ten B groups, located between Groups 2A(2) and 3A(13), contain the *transition elements.* Two horizontal series of *inner transition elements,* the lanthanides and the actinides, fit *between* the elements in Group 3B(3) and Group 4B(4) and are usually placed below the main body of the table.

Literature Talanquer, Vicente. "Recreating a Periodic Table: A Tool for Developing Pedagogical Content Knowledge," *Chem. Educator* 10 **2005,** *2,* 95–99.

Literature Dopke, Nancy Carter; Treichel, Paul M., Jr.; Vestling, Martha M. "Significant Figures, the Periodic Table, and Mass Spectrometry: The Challenge of Large Biomolecules," *J. Chem. Educ.* **2000,** *77,* 1065.

Literature Dutch, Steven I. "Periodic Tables of Elemental Abundance," *J. Chem. Educ.* **1999,** *76,* 356.

Literature He, Fu-cheng; Li, Xiang-yuan. "The Periodic Building of the Elements: Can a Periodic Table Be Transformed into a Stereo One?" *J. Chem. Educ.* **1997,** *74,* 792.

Figure 2.9 The modern periodic table. The table consists of element boxes arranged by *increasing* atomic number into groups (vertical columns) and periods (horizontal rows). Each box contains the atomic number, atomic symbol, and atomic mass. (A mass in parentheses is the mass number of the most stable isotope of that element.) The periods are numbered 1 to 7. The groups (sometimes called *families*) have a number-letter designation and a new group number in parentheses. The A groups are the main-group elements; the B groups are the transition elements. Two series of inner transition elements are placed below the main body of the table but actually fit between the elements indicated. Metals lie below and to the left of the thick "staircase" line [top of 3A(13) to bottom of 6A(16) in Period 6] and include main-group metals (*purple-blue*), transition elements (*blue*), and inner transition elements (*gray-blue*). Nonmetals (*yellow*) lie to the right of the line. Metalloids (*green*) lie along the line. We discuss the placement of hydrogen in Chapter 14. As of mid-2007, elements 112–116 had not been named.

ARIS Presentation Center, Active Art Library

Literature Jensen, William B. "Why Helium Ends in "-ium"," *J. Chem. Educ.* **2004,** *81,* 944.

At this point in the text, the clearest distinction among the elements is their classification as metals, nonmetals, or metalloids. The "staircase" line that runs from the top of Group 3A(13) to the bottom of Group 6A(16) in Period 6 is a dividing line for this classification. The **metals** (three shades of blue) appear in the large lower-left portion of the table. About three-quarters of the elements are metals, including many main-group elements and all the transition and inner transition elements. They are generally shiny solids at room temperature (mercury is the only liquid) that conduct heat and electricity well and can be tooled into sheets (malleable) and wires (ductile). The **nonmetals** (yellow) appear in the small upper-right portion of the table. They are generally gases or dull, brittle solids at room temperature (bromine is the only liquid) and conduct heat and electricity poorly. Along the staircase line lie the **metalloids** (green; also called **semimetals**), elements that have properties between those of metals and nonmetals. Several

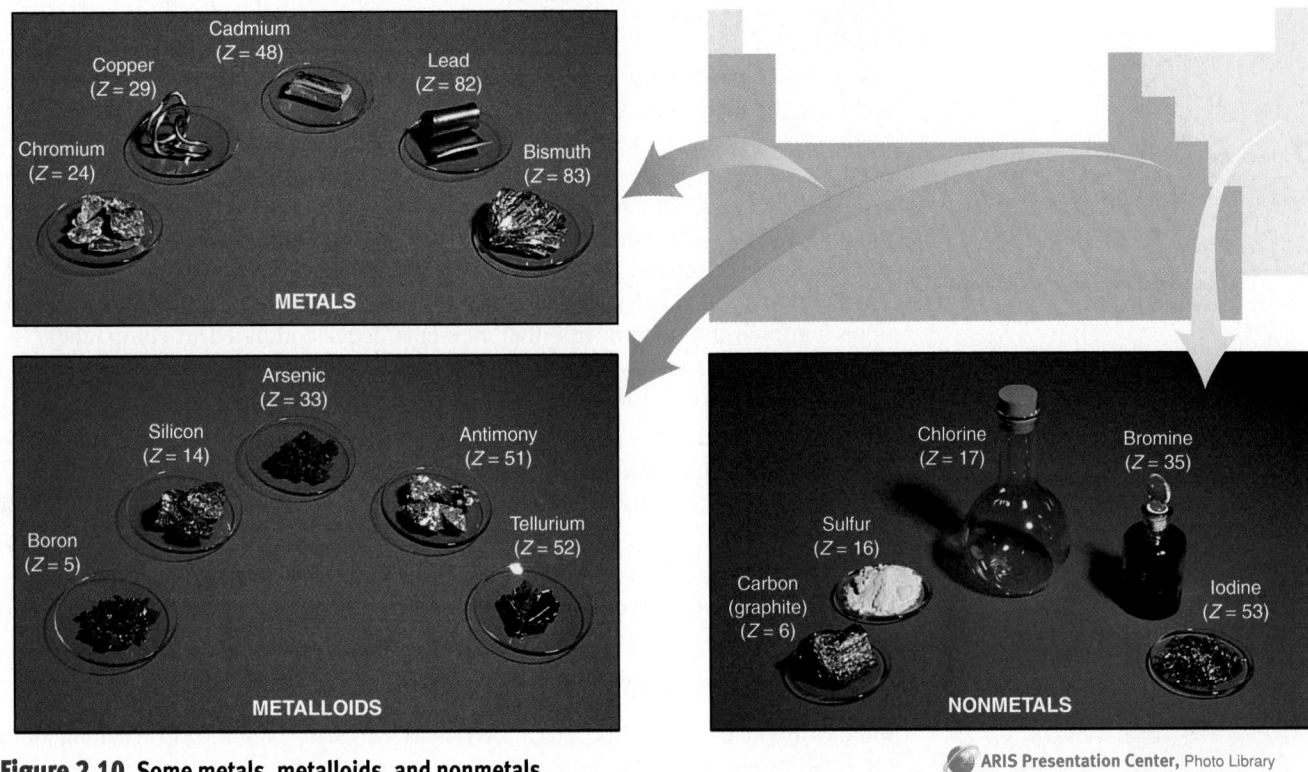

Figure 2.10 Some metals, metalloids, and nonmetals.

ARIS Presentation Center, Photo Library

metalloids, such as silicon (Si) and germanium (Ge), play major roles in modern electronics. Figure 2.10 shows examples of these three classes of elements.

Two of the major branches of chemistry have traditionally been defined by the elements that each studies. *Organic chemistry* studies the compounds of carbon, specifically those that contain hydrogen and often oxygen, nitrogen, and a few other elements. This branch is concerned with fuels, drugs, dyes, polymers, and the like. *Inorganic chemistry,* on the other hand, focuses mainly on the compounds of all the other elements. It is concerned with catalysts, electronic materials, metal alloys, mineral salts, and the like. With the explosive growth in biomedical and materials research, the line between these branches has all but disappeared.

It is important to learn some of the group (family) names. Group 1A(1), except for hydrogen, consists of the *alkali metals,* and Group 2A(2) consists of the *alkaline earth metals.* Both groups consist of highly reactive elements. The *halogens,* Group 7A(17), are highly reactive nonmetals, whereas the *noble gases,* Group 8A(18), are relatively unreactive nonmetals. Other main groups [3A(13) to 6A(16)] are often named for the first element in the group; for example, Group 6A is the *oxygen family.*

A key point that we return to many times is that, in general, *elements in a group have **similar** chemical properties and elements in a period have **different** chemical properties.* We begin applying the organizing power of the periodic table in the next section, where we discuss how elements combine to form compounds.

Literature Nelson, P. G. "Important Elements," *J. Chem. Educ.* **1991,** *68,* 732.

Literature Ringnes, Vivi. "Origin of the Names of Chemical Elements," *J. Chem. Educ.* **1989,** *66,* 731.

Literature Cronyn, Marshall W. "The Proper Place for Hydrogen in the Periodic Table," *J. Chem. Educ.* **2003,** *80,* 947.

Section Summary

In the periodic table, the elements are arranged by atomic number into horizontal periods and vertical groups. • Because of the periodic recurrence of certain key properties, elements within a group have similar behavior, whereas elements in a period have dissimilar behavior. • Nonmetals appear in the upper-right portion of the table, metalloids lie along a staircase line, and metals fill the rest of the table.

2.7 | COMPOUNDS: INTRODUCTION TO BONDING

The overwhelming majority of elements occur in chemical combination with other elements. In fact, only a few elements occur free in nature. The noble gases—helium (He), neon (Ne), argon (Ar), krypton (Kr), xenon (Xe), and radon (Rn)—occur in air as separate atoms. In addition to occurring in compounds, oxygen (O), nitrogen (N), and sulfur (S) occur in the most common elemental form as the molecules O_2, N_2, and S_8, and carbon (C) occurs in vast, nearly pure deposits of coal. Some of the metals, such as copper (Cu), silver (Ag), gold (Au), and platinum (Pt), may also occur uncombined with other elements. But these few exceptions reinforce the general rule that elements occur combined in compounds.

It is the electrons of the atoms of interacting elements that are involved in compound formation. Elements combine in two general ways:

1. *Transferring electrons* from the atoms of one element to those of another to form **ionic compounds**
2. *Sharing electrons* between atoms of different elements to form **covalent compounds**

These processes generate **chemical bonds,** the forces that hold the atoms of elements together in a compound. We'll introduce compound formation next and have much more to say about it in later chapters.

The Formation of Ionic Compounds

Ionic compounds are composed of **ions,** charged particles that form when an atom (or small group of atoms) gains or loses one or more electrons. The simplest type of ionic compound is a **binary ionic compound,** one composed of just two elements. It typically forms *when a metal reacts with a nonmetal.* Each metal atom loses a certain number of its electrons and becomes a **cation,** a positively charged ion. The nonmetal atoms gain the electrons lost by the metal atoms and become **anions,** negatively charged ions. In effect, the metal atoms *transfer electrons* to the nonmetal atoms. The resulting cations and anions attract each other through electrostatic forces and form the ionic compound. *All binary ionic compounds are solids.* A cation or anion derived from a single atom is called a **monatomic ion;** we'll discuss polyatomic ions, those derived from a small group of atoms, later.

The formation of the binary ionic compound sodium chloride, common table salt, is depicted in Figure 2.11, from the elements through the atomic-scale

Literature Chimeno, Joseph S.; Wulfsberg, Gary P.; Sanger, Michael J.; Melton, Tammy J. "The Rainbow Wheel and Rainbow Matrix: Two Effective Tools for Learning Ionic Nomenclature," *J. Chem. Educ.* **2006,** *8,* 651.

Animation: Formation of an Ionic Compound
ARIS Presentation Center
Chapter 2

ARIS Presentation Center, Photo Library

ARIS Presentation Center, Active Art Library

A The elements (lab view)

Chlorine gas

Sodium metal

B The elements (atomic view)

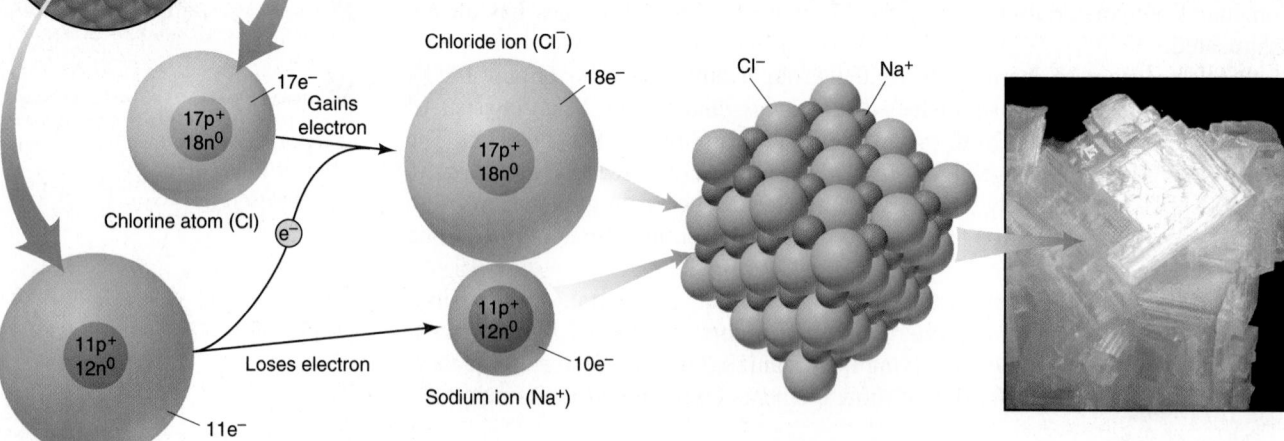

C Electron transfer

D The compound (atomic view): Na^+ and Cl^- in the crystal

E The compound (lab view): sodium chloride crystal

Figure 2.11 The formation of an ionic compound. A, The two elements as seen in the laboratory. **B,** The elements on the atomic scale. **C,** The neutral sodium atom loses one electron to become a sodium cation (Na^+), and the chlorine atom gains one electron to become a chloride anion (Cl^-). (Note that when atoms lose electrons, they become ions that are smaller, and when they gain electrons, they become ions that are larger.) **D,** Na^+ and Cl^- ions attract each other and form a regular three-dimensional array. **E,** This arrangement of the ions is reflected in the structure of crystalline NaCl, which occurs naturally as the mineral halite, hence the name *halogens* for the Group 7A(17) elements.

electron transfer to the compound. In the electron transfer, a sodium atom, which is neutral because it has the same number of protons as electrons, *loses* one electron and forms a sodium cation, Na^+. (The charge on the ion is written as a *right superscript*.) A chlorine atom *gains* the electron and becomes a chloride anion, Cl^-. (The name change from the nonmetal atom to the ion is discussed in the next section.) Even the tiniest visible grain of table salt contains an enormous number of sodium and chloride ions. The oppositely charged ions (Na^+ and Cl^-) attract each other, and the similarly charged ions (Na^+ and Na^+, or Cl^- and Cl^-) repel each other. The resulting solid aggregation is a regular array of alternating Na^+ and Cl^- ions that extends in all three dimensions.

The strength of the ionic bonding depends to a great extent on the net strength of these attractions and repulsions and is described by *Coulomb's law,* which can be expressed as follows: *the energy of attraction (or repulsion) between two particles is directly proportional to the product of the charges and inversely proportional to the distance between them.*

$$\text{Energy} \propto \frac{\text{charge 1} \times \text{charge 2}}{\text{distance}}$$

In other words, ions with higher charges attract (or repel) each other more strongly than ions with lower charges. Likewise, smaller ions attract (or repel) each other more strongly than larger ions, because their charges are closer together. These effects are summarized in Figure 2.12.

Ionic compounds are neutral; that is, they possess no net charge. For this to occur, they must contain equal numbers of positive and negative *charges*—not necessarily equal numbers of positive and negative *ions*. Because Na^+ and Cl^- each bear a unit charge (1+ or 1−), equal numbers of these ions are present in sodium chloride; but in sodium oxide, for example, there are two Na^+ ions present to balance the 2− charge of each oxide ion, O^{2-}.

Can we predict the number of electrons a given atom will lose or gain when it forms an ion? In the formation of sodium chloride, for example, why does each sodium atom give up only 1 of its 11 electrons? Why doesn't each chlorine atom gain two electrons, instead of just one? For A-group elements, the periodic table provides an answer. We generally find that metals lose electrons and nonmetals gain electrons to *form ions with the same number of electrons as in an atom of the nearest noble gas* [Group 8A(18)]. Noble gases have a stability (low reactivity) that is related to their number (and arrangement) of electrons. A sodium atom ($11e^-$) can attain the stability of neon ($10e^-$), the nearest noble gas, by losing one electron. Similarly, by gaining one electron, a chlorine atom ($17e^-$) attains the stability of argon ($18e^-$), its nearest noble gas. Thus, when an element located near a noble gas forms a monatomic ion, *it gains or loses enough electrons to attain the same number as that noble gas.* Specifically, the elements that are in Group 1A(1) lose one electron, those in Group 2A(2) lose two, and aluminum in Group 3A(13) loses three; the elements in Group 7A(17) gain one electron, oxygen and sulfur in Group 6A(16) gain two, and nitrogen in Group 5A(15) gains three.

With the periodic table printed on a two-dimensional surface, as in Figure 2.9, it is easy to get the false impression that the elements in Group 7A(17) are "closer" to the noble gases than the elements in Group 1A(1). Actually, both groups are only one electron away from having the same number of electrons as the noble gases. To make this point, Figure 2.13 shows a modified periodic table of monatomic ions that is cut and rejoined as a cylinder. Now you can see that fluorine (F; $Z = 9$) has one electron fewer than the noble gas neon (Ne; $Z = 10$) and sodium (Na; $Z = 11$) has one electron more; thus, they form the F^- and Na^+ ions. Similarly, oxygen (O; $Z = 8$) gains two electrons and magnesium (Mg; $Z = 12$) loses two to form the O^{2-} and Mg^{2+} ions and attain the same number of electrons as neon.

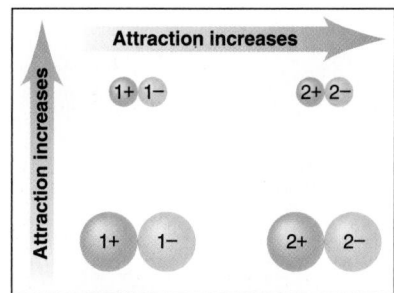

Figure 2.12 Factors that influence the strength of ionic bonding. For ions of a given size, strength of attraction *(arrows)* increases with higher ionic charge *(left to right)*. For ions of a given charge, strength of attraction increases with smaller ionic size *(bottom to top).*

Figure 2.13 The relationship between ions formed and the nearest noble gas. This periodic table was redrawn to show the positions of other nonmetals *(yellow)* and metals *(blue)* relative to the noble gases and to show the ions these elements form. The ionic charge equals the number of electrons lost (+) or gained (−) to attain the same number of electrons as the nearest noble gas. Species in the same row have the same number of electrons. For example, H^-, He, and Li^+ all have two electrons. [Note that H is shown here in Group 7A(17).]

A No interaction

B Attraction begins

C Covalent bond

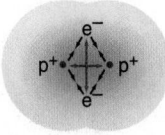

D Interaction of forces

Figure 2.14 Formation of a covalent bond between two H atoms. A, The distance is too great for the atoms to affect each other. **B,** As the distance decreases, the nucleus of each atom begins to attract the electron of the other. **C,** The covalent bond forms when the two nuclei mutually attract the pair of electrons at some optimum distance. **D,** The H_2 molecule is more stable than the separate atoms because the attractive forces *(black arrows)* between each nucleus and the two electrons are greater than the repulsive forces *(red arrows)* between the electrons and between the nuclei.

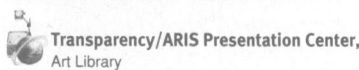

Transparency/ARIS Presentation Center, Art Library

ARIS Presentation Center, Art Library

SAMPLE PROBLEM 2.6 Predicting the Ion an Element Forms

PROBLEM What monatomic ions do the following elements form?
(a) Iodine ($Z = 53$)   **(b)** Calcium ($Z = 20$)   **(c)** Aluminum ($Z = 13$)
PLAN We use the given Z value to find the element in the periodic table and see where its group lies relative to the noble gases. Elements in Groups 1A, 2A, and 3A *lose* electrons to attain the same number as the nearest noble gas and become positive ions; those in Groups 5A, 6A, and 7A *gain* electrons and become negative ions.
SOLUTION **(a)** I^- Iodine ($_{53}I$) is a nonmetal in Group 7A(17), one of the halogens. Like any member of this group, it gains 1 electron to have the same number as the nearest Group 8A(18) member, in this case $_{54}Xe$.
(b) Ca^{2+} Calcium ($_{20}Ca$) is a member of Group 2A(2), the alkaline earth metals. Like any Group 2A member, it loses 2 electrons to attain the same number as the nearest noble gas, in this case, $_{18}Ar$.
(c) Al^{3+} Aluminum ($_{13}Al$) is a metal in the boron family [Group 3A(13)] and thus loses 3 electrons to attain the same number as its nearest noble gas, $_{10}Ne$.

FOLLOW-UP PROBLEM 2.6 What monatomic ion does each of the following elements form: **(a)** $_{16}S$; **(b)** $_{37}Rb$; **(c)** $_{56}Ba$?

The Formation of Covalent Compounds

Covalent compounds form when elements share electrons, which usually occurs between nonmetals. Even though relatively few nonmetals exist, they interact in many combinations to form a very large number of covalent compounds.

The simplest case of electron sharing occurs not in a compound but between two hydrogen atoms (H; $Z = 1$). Imagine two separated H atoms approaching each other, as in Figure 2.14. As they get closer, the nucleus of each atom attracts the electron of the other atom more and more strongly, and the separated atoms begin to interpenetrate each other. At some optimum distance between the nuclei, the two atoms form a **covalent bond,** a pair of electrons mutually attracted by the two nuclei. The result is a hydrogen molecule, in which each electron no longer "belongs" to a particular H atom: the two electrons are *shared* by the two nuclei. Repulsions between the nuclei and between the electrons also occur, but the net attraction is greater than the net repulsion. (We discuss the properties of covalent bonds in great detail in Chapter 9.)

A sample of hydrogen gas consists of these diatomic molecules (H_2)—pairs of atoms that are chemically bound and behave as an independent unit—*not* separate H atoms. Other nonmetals that exist as diatomic molecules at room temperature are nitrogen (N_2), oxygen (O_2), and the halogens [fluorine (F_2), chlorine (Cl_2), bromine (Br_2), and iodine (I_2)]. Phosphorus exists as tetratomic molecules (P_4), and sulfur and selenium as octatomic molecules (S_8 and Se_8) (Figure 2.15). At room temperature, covalent substances may be gases, liquids, or solids.

Figure 2.15 Elements that occur as molecules.

Diatomic molecules
Tetratomic molecules
Octatomic molecules

Atoms of different elements share electrons to form the molecules of a covalent compound. A sample of hydrogen fluoride, for example, consists of molecules in which one H atom forms a covalent bond with one F atom; water consists of molecules in which one O atom forms covalent bonds with two H atoms:

Hydrogen fluoride, HF   Water, H$_2$O

(As you'll see in Chapter 9, covalent bonding provides another way for atoms to attain the same number of electrons as the nearest noble gas.)

Distinguishing the Entities in Covalent and Ionic Substances

There is a key distinction between the chemical entities in covalent substances and ionic substances. *Most covalent substances consist of molecules.* A cup of water, for example, consists of individual water molecules lying near each other. In contrast, under ordinary conditions, *no molecules exist in a sample of an ionic compound.* A piece of sodium chloride, for example, is a continuous array of oppositely charged sodium and chloride ions, *not* a collection of individual "sodium chloride molecules."

Another key distinction exists between the nature of the particles attracting each other. Covalent bonding involves the mutual attraction between two (positively charged) nuclei and the two (negatively charged) electrons that reside between them. Ionic bonding involves the mutual attraction between positive and negative ions.

Polyatomic Ions: Covalent Bonds Within Ions

Many ionic compounds contain **polyatomic ions,** which consist of two or more atoms bonded *covalently* and have a net positive or negative charge. For example, the ionic compound calcium carbonate is an array of polyatomic carbonate anions and monatomic calcium cations attracted to each other. The carbonate ion consists of a carbon atom covalently bonded to three oxygen atoms, and two additional electrons give the ion its 2− charge (Figure 2.16). In many reactions, a polyatomic ion stays together as a unit.

The Elements of Life

About one-quarter of all the elements have known roles in organisms. As you can see in Figure 2.17, metals, nonmetals, and metalloids are among these essential elements. But, except for some diatomic oxygen and nitrogen molecules inhaled into the lungs, none of the elements in organisms occurs in pure form; rather, they appear in compounds or as ions in solution.

Figure 2.16 A polyatomic ion. Calcium carbonate is a three-dimensional array of monatomic calcium cations *(purple spheres)* and polyatomic carbonate anions. As the bottom structure shows, each carbonate ion consists of four covalently bonded atoms.

ARIS Presentation Center, Photo Library

Literature Marshall, James L. "A Living Periodic Table," *J. Chem. Educ.* **2000,** *77,* 979.

Biological application

Figure 2.17 A biological periodic table. The building-block elements and major minerals are required by all organisms. Most organisms, including humans, require the trace elements as well. Many other elements (not shown) are found in organisms but have no known role.

ARIS Presentation Center, Art Library

The elements of life are often classified by the amount present in organisms. The four nonmetals carbon (C), oxygen (O), hydrogen (H), and nitrogen (N) are the *building-block elements* because they make up the major portion of biological molecules. Over 99% of the atoms in organisms are C, O, H, and N; in humans, they account for over 96% by mass of body weight. The nonmetals O and H make up the water in organisms, of course, and together with C occur in all four major classes of biological molecules—carbohydrates, fats, proteins, and nucleic acids. All proteins and nucleic acids also contain N.

The seven *major minerals* (or *macronutrients*) range from around 2% by mass for calcium (Ca) to around 0.14% by mass for chlorine (Cl). The alkali metals sodium and potassium and the halogen chlorine are dissolved in cell fluids as the ions Na^+, K^+, and Cl^-. The alkaline earth metals magnesium and calcium occur as Mg^{2+} and Ca^{2+}, most often bound to proteins or, in the case of calcium, in bones and teeth. Sulfur (S) occurs mostly in proteins, but phosphorus (P) also occurs in nucleic acids, many fats, and sugars, and as part of a polyatomic ion in bone and cell fluids.

The *trace elements* (or *micronutrients*) are present in much lower amounts, with iron (Fe) the most abundant at only 0.005% by mass. Most of them are associated with protein functions. We will look more closely at the trace elements in Chapter 23.

Section Summary

Although a few elements occur uncombined in nature, the great majority exist in compounds. • Ionic compounds form when a metal *transfers* electrons to a nonmetal, and the resulting positive and negative ions attract each other to form a three-dimensional array. In many cases, metal atoms lose and nonmetal atoms gain enough electrons to attain the same number of electrons as in atoms of the nearest noble gas. • Covalent compounds form when elements, usually nonmetals, *share* electrons. Each covalent bond is an electron pair mutually attracted by two atomic nuclei. • Monatomic ions are derived from single atoms. Polyatomic ions consist of two or more covalently bonded atoms that have a net positive or negative charge due to a deficit or excess of electrons. • The elements in organisms are found as ions or bonded in large biomolecules. Four building-block elements (C, O, H, N) form these compounds, seven other elements (major minerals, or macronutrients) are also common, and many others (trace elements, or micronutrients) occur in tiny amounts and play specific roles.

Literature Wirtz, Michael C.; Kaufmann, Joan; Hawley, Gary. "Nomenclature Made Practical: Student Discovery of the Nomenclature Rules," *J. Chem. Educ.* **2006,** *83,* 595.

Literature Chimeno, Joseph. "How to Make Learning Chemical Nomenclature Fun, Exciting, and Palatable," *J. Chem. Educ.* **2000,** *77,* 144.

Literature Lind, Gerhard. "Teaching Inorganic Nomenclature: A Systematic Approach," *J. Chem. Educ.* **1992,** *69,* 613.

2.8 COMPOUNDS: FORMULAS, NAMES, AND MASSES

Names and formulas of compounds form the vocabulary of the chemical language. In this discussion, you'll learn the names and formulas of ionic and simple covalent compounds and how to calculate the mass of a unit of a compound from its formula.

Types of Chemical Formulas

In a **chemical formula,** element symbols and numerical subscripts show the type and number of each atom present in the smallest unit of the substance. There are several types of chemical formulas for a compound:

1. The **empirical formula** shows the *relative* number of atoms of each element in the compound. It is the simplest type of formula and is derived from the masses of the component elements. For example, in hydrogen peroxide, there is 1 part by mass of hydrogen for every 16 parts by mass of oxygen. Therefore, the empirical formula of hydrogen peroxide is HO: one H atom for every O atom.

2. The **molecular formula** shows the *actual* number of atoms of each element in a molecule of the compound. The molecular formula of hydrogen peroxide is H_2O_2; there are two H atoms and two O atoms in each molecule.
3. A **structural formula** shows the number of atoms and *the bonds between them;* that is, the relative placement and connections of atoms in the molecule. The structural formula of hydrogen peroxide is H—O—O—H; each H is bonded to an O, and the O's are bonded to each other.

Some Advice about Learning Names and Formulas

Perhaps in the future, systematic names for compounds will be used by everyone. However, many reference books, chemical supply catalogs, and practicing chemists still use many common (trivial) names, so you should learn them as well.

Here are some points to note about ion formulas:

- Members of a periodic table group have the same ionic charge; for example, Li, Na, and K are all in Group 1A and all have a 1+ charge.
- For A-group cations, ion charge = group number: for example, Na^+ is in Group 1A, Ba^{2+} in Group 2A. (Exceptions in Figure 2.18 are Sn^{2+} and Pb^{2+}.)
- For anions, ion charge = group number minus 8: for example, S is in Group 6A ($6 - 8 = -2$), so the ion is S^{2-}.

Here are some suggestions about how to learn names and formulas:

1. Memorize the A-group monatomic ions of Table 2.3 (all except Ag^+, Zn^{2+}, and Cd^{2+}) according to their positions in Figure 2.18. These ions have the same number of electrons as an atom of the nearest noble gas.
2. Consult Table 2.4 (page 67) and Figure 2.18 for some metals that form two different monatomic ions.
3. Divide the tables of names and charges into smaller batches, and learn a batch each day. Try flash cards, with the name on one side and the ion formula on the other. The most common ions are shown in **boldface** in Tables 2.3, 2.4, and 2.5, so you can focus on learning them first.

Table 2.3 Common Monatomic Ions*

Charge	Formula	Name
Cations		
1+	H^+	hydrogen
	Li^+	**lithium**
	Na^+	**sodium**
	K^+	**potassium**
	Cs^+	cesium
	Ag^+	**silver**
2+	**Mg^{2+}**	**magnesium**
	Ca^{2+}	**calcium**
	Sr^{2+}	strontium
	Ba^{2+}	**barium**
	Zn^{2+}	**zinc**
	Cd^{2+}	cadmium
3+	**Al^{3+}**	**aluminum**
Anions		
1−	H^-	hydride
	F^-	**fluoride**
	Cl^-	**chloride**
	Br^-	**bromide**
	I^-	**iodide**
2−	**O^{2-}**	**oxide**
	S^{2-}	**sulfide**
3−	N^{3-}	nitride

*Listed by charge; those in **boldface** are most common.

ARIS Presentation Center, Tables Library

Transparency/ARIS Presentation Center, Art Library

Figure 2.18 **Some common monatomic ions of the elements.** Main-group elements usually form a single monatomic ion. Note that members of a group have ions with the same charge. [Hydrogen is shown as both the cation H^+ in Group 1A(1) and the anion H^- in Group 7A(17).] Many transition elements form two different monatomic ions. (Although Hg_2^{2+} is a diatomic ion, it is included for comparison with Hg^{2+}.)

Names and Formulas of Ionic Compounds

All ionic compound names give the positive ion (cation) first and the negative ion (anion) second.

Compounds Formed from Monatomic Ions Let's first consider how to name binary ionic compounds, those composed of ions of two elements.

- *The name of the cation is the same as the name of the metal.* Many metal names end in *-ium*.
- *The name of the anion takes the root of the nonmetal name and adds the suffix -ide.*

For example, the anion formed from brom*ine* is named brom*ide* (brom+ide). Therefore, the compound formed from the metal calcium and the nonmetal bromine is named *calcium bromide*.

ARIS Presentation Center, Worked Examples Library

SAMPLE PROBLEM 2.7   Naming Binary Ionic Compounds

PROBLEM Name the ionic compound formed from the following pairs of elements:
(a) Magnesium and nitrogen   **(b)** Iodine and cadmium
(c) Strontium and fluorine   **(d)** Sulfur and cesium

PLAN The key to naming a binary ionic compound is to recognize which element is the metal and which is the nonmetal. When in doubt, check the periodic table. We place the cation name first, add the suffix *-ide* to the nonmetal root, and place the anion name last.

SOLUTION **(a)** Magnesium is the metal; *nitr-* is the nonmetal root: magnesium nitride

(b) Cadmium is the metal; *iod-* is the nonmetal root: cadmium iodide

(c) Strontium is the metal; *fluor-* is the nonmetal root: strontium fluoride (Note the spelling is flu*o*ride, not flo*u*ride.)

(d) Cesium is the metal; *sulf-* is the nonmetal root: cesium sulfide

FOLLOW-UP PROBLEM 2.7 For the following ionic compounds, give the name and periodic table group number of each of the elements present: **(a)** zinc oxide; **(b)** silver bromide; **(c)** lithium chloride; **(d)** aluminum sulfide.

Ionic compounds are arrays of oppositely charged ions rather than separate molecular units. Therefore, we write a formula for the **formula unit,** which gives the *relative* numbers of cations and anions in the compound. Thus, ionic compounds generally have only empirical formulas.* The compound has zero net charge, so the positive charges of the cations must balance the negative charges of the anions. For example, calcium bromide is composed of Ca^{2+} ions and Br^- ions; therefore, two Br^- balance each Ca^{2+}. The formula is $CaBr_2$, not Ca_2Br. In this and all other formulas,

- The subscript refers to the element *preceding* it.
- The *subscript 1 is understood* from the presence of the element symbol alone (that is, we do not write Ca_1Br_2).
- The charge (without the sign) of one ion becomes the subscript of the other:

$$Ca^{2+} \quad Br^{1-} \qquad gives \qquad Ca_1Br_2 \quad or \quad CaBr_2$$

Reduce the subscripts to the smallest whole numbers that retain the ratio of ions. Thus, for example, from the ions Ca^{2+} and O^{2-} we have Ca_2O_2, which we reduce to the formula CaO (but see the footnote).

---

*Compounds of the mercury(I) ion, such as Hg_2Cl_2, and peroxides of the alkali metals, such as Na_2O_2, are the only two common exceptions. Their empirical formulas are HgCl and NaO, respectively.

SAMPLE PROBLEM 2.8 Determining Formulas of Binary Ionic Compounds

PROBLEM Write empirical formulas for the compounds named in Sample Problem 2.7.

PLAN We write the empirical formula by finding the smallest number of each ion that gives the neutral compound. These numbers appear as *right subscripts* to the element symbol.

SOLUTION

(a) Mg^{2+} and N^{3-}; three Mg^{2+} ions (6+) balance two N^{3-} ions (6−): Mg_3N_2

(b) Cd^{2+} and I^-; one Cd^{2+} ion (2+) balances two I^- ions (2−): CdI_2

(c) Sr^{2+} and F^-; one Sr^{2+} ion (2+) balances two F^- ions (2−): SrF_2

(d) Cs^+ and S^{2-}; two Cs^+ ions (2+) balance one S^{2-} ion (2−): Cs_2S

COMMENT Note that ion charges do *not* appear in the compound formula. That is, for cadmium iodide, we do *not* write $Cd^{2+}I_2^-$.

FOLLOW-UP PROBLEM 2.8 Write the formulas of the compounds named in Follow-up Problem 2.7.

Compounds with Metals That Can Form More Than One Ion Many metals, particularly the transition elements (B groups), can form more than one ion, each with its own particular charge. Table 2.4 lists some examples, and the earlier Figure 2.18 shows their placement in the periodic table. Names of compounds containing these elements include a *Roman numeral within parentheses* immediately after the metal ion's name to indicate its ionic charge. For example, iron can form Fe^{2+} and Fe^{3+} ions. The two compounds that iron forms with chlorine are $FeCl_2$, named iron(II) chloride (spoken "iron two chloride"), and $FeCl_3$, named iron(III) chloride.

In common names, the Latin root of the metal is followed by either of two suffixes:

- The suffix *-ous* for the ion with the lower charge
- The suffix *-ic* for the ion with the higher charge

Thus, iron(II) chloride is also called ferr*ous* chloride and iron(III) chloride is fer-*ic* chloride. (You can easily remember this naming relationship because there is an *o* in *-ous* and *lower,* and an *i* in *-ic* and *higher.*)

Table 2.4 Some Metals That Form More Than One Monatomic Ion*

Element	Ion Formula	Systematic Name	Common (Trivial) Name
Chromium	Cr^{2+}	chromium(II)	chromous
	$\mathbf{Cr^{3+}}$	**chromium(III)**	chromic
Cobalt	Co^{2+}	cobalt(II)	
	Co^{3+}	cobalt(III)	
Copper	$\mathbf{Cu^+}$	**copper(I)**	cuprous
	$\mathbf{Cu^{2+}}$	**copper(II)**	cupric
Iron	$\mathbf{Fe^{2+}}$	**iron(II)**	ferrous
	$\mathbf{Fe^{3+}}$	**iron(III)**	ferric
Lead	$\mathbf{Pb^{2+}}$	**lead(II)**	
	Pb^{4+}	lead(IV)	
Mercury	Hg_2^{2+}	mercury(I)	mercurous
	$\mathbf{Hg^{2+}}$	**mercury(II)**	mercuric
Tin	$\mathbf{Sn^{2+}}$	**tin(II)**	stannous
	Sn^{4+}	tin(IV)	stannic

*Listed alphabetically by metal name; those in **boldface** are most common.

ARIS Presentation Center, Worked Examples Library

Table 2.5	Common Polyatomic Ions*
Formula	**Name**
Cations	
NH_4^+	**ammonium**
H_3O^+	**hydronium**
Anions	
CH_3COO^- (or $C_2H_3O_2^-$)	**acetate**
CN^-	cyanide
OH^-	**hydroxide**
ClO^-	hypochlorite
ClO_2^-	chlorite
ClO_3^-	**chlorate**
ClO_4^-	**perchlorate**
NO_2^-	nitrite
NO_3^-	**nitrate**
MnO_4^-	**permanganate**
CO_3^{2-}	**carbonate**
HCO_3^-	**hydrogen carbonate** (or **bicarbonate**)
CrO_4^{2-}	chromate
$Cr_2O_7^{2-}$	**dichromate**
O_2^{2-}	peroxide
PO_4^{3-}	**phosphate**
HPO_4^{2-}	hydrogen phosphate
$H_2PO_4^-$	dihydrogen phosphate
SO_3^{2-}	sulfite
SO_4^{2-}	**sulfate**
HSO_4^-	hydrogen sulfate (or bisulfate)

****Boldface** ions are most common.

ARIS Presentation Center, Tables Library

ARIS Presentation Center, Art Library

	Prefix	Root	Suffix
	per	*root*	ate
		root	ate
		root	ite
	hypo	*root*	ite

(No. of O atoms — increasing upward)

Figure 2.19 Naming oxoanions. Prefixes and suffixes indicate the number of O atoms in the anion.

Literature Marmorino, M. G. "Mnemonic for Oxoanion Formulas," *Chem. Educator* 10 **2005,** *6,* 415.

SAMPLE PROBLEM 2.9 Determining Names and Formulas of Ionic Compounds of Elements That Form More Than One Ion

PROBLEM Give the systematic names for the formulas or the formulas for the names of the following compounds: **(a)** tin(II) fluoride; **(b)** CrI_3; **(c)** ferric oxide; **(d)** CoS.

SOLUTION (a) Tin(II) is Sn^{2+}; fluoride is F^-. Two F^- ions balance one Sn^{2+} ion: tin(II) fluoride is SnF_2. (The common name is stannous fluoride.)

(b) The anion is I^-, iodide, and the formula shows three I^-. Therefore, the cation must be Cr^{3+}, chromium(III): CrI_3 is chromium(III) iodide. (The common name is chromic iodide.)

(c) Ferric is the common name for iron(III), Fe^{3+}; oxide ion is O^{2-}. To balance the ionic charges, the formula of ferric oxide is Fe_2O_3. [The systematic name is iron(III) oxide.]

(d) The anion is sulfide, S^{2-}, which requires that the cation be Co^{2+}. The name is cobalt(II) sulfide.

FOLLOW-UP PROBLEM 2.9 Give the systematic names for the formulas or the formulas for the names of the following compounds: **(a)** lead(IV) oxide; **(b)** Cu_2S; **(c)** $FeBr_2$; **(d)** mercuric chloride.

Compounds Formed from Polyatomic Ions

Ionic compounds in which one or both of the ions are polyatomic are very common. Table 2.5 gives the formulas and the names of some common polyatomic ions. Remember that *the polyatomic ion stays together as a charged unit.* The formula for potassium nitrate is KNO_3: each K^+ balances one NO_3^-. The formula for sodium carbonate is Na_2CO_3: two Na^+ balance one CO_3^{2-}. *When two or more of the same polyatomic ion are present in the formula unit, that ion appears in parentheses with the subscript written outside.* For example, calcium nitrate, which contains one Ca^{2+} and two NO_3^- ions, has the formula $Ca(NO_3)_2$. Parentheses and a subscript are *not* used unless *more than one* of the polyatomic ions is present; thus, sodium nitrate is $NaNO_3$, *not* $Na(NO_3)$.

Families of Oxoanions

As Table 2.5 shows, most polyatomic ions are **oxoanions,** those in which an element, usually a nonmetal, is bonded to one or more oxygen atoms. There are several families of two or four oxoanions that differ only in the number of oxygen atoms. The following simple naming conventions are used with these ions.

With two oxoanions in the family:

- The ion with *more* O atoms takes the nonmetal root and the suffix *-ate*.
- The ion with *fewer* O atoms takes the nonmetal root and the suffix *-ite*.

For example, SO_4^{2-} is the sulf*ate* ion, and SO_3^{2-} is the sulf*ite* ion; similarly, NO_3^- is nitr*ate*, and NO_2^- is nitr*ite*.

With four oxoanions in the family (usually a halogen bonded to O), as Figure 2.19 shows:

- The ion with *most* O atoms has the prefix *per-*, the nonmetal root, and the suffix *-ate*.
- The ion with *one fewer* O atom has just the root and the suffix *-ate*.
- The ion with *two fewer* O atoms has just the root and the suffix *-ite*.
- The ion with *least (three fewer)* O atoms has the prefix *hypo-*, the root, and the suffix *-ite*.

For example, for the four chlorine oxoanions,

ClO_4^- is *per*chlor*ate*, ClO_3^- is chlor*ate*, ClO_2^- is chlor*ite*, ClO^- is *hypo*chlor*ite*

Hydrated Ionic Compounds Ionic compounds called **hydrates** have a specific number of water molecules associated with each formula unit. In their formulas, this number is shown after a centered dot. It is indicated in the systematic name by a Greek numerical prefix before the word *hydrate*. Table 2.6 shows these prefixes. For example, Epsom salt has the formula $MgSO_4 \cdot 7H_2O$ and the name magnesium sulfate *hepta*hydrate. Similarly, the mineral gypsum has the formula $CaSO_4 \cdot 2H_2O$ and the name calcium sulfate *di*hydrate. The water molecules, referred to as "waters of hydration," are part of the hydrate's structure. Heating can remove some or all of them, leading to a different substance. For example, when heated strongly, blue copper(II) sulfate pentahydrate ($CuSO_4 \cdot 5H_2O$) is converted to white copper(II) sulfate ($CuSO_4$).

ARIS Presentation Center, Worked Examples Library

Table 2.6	Numerical Prefixes for Hydrates and Binary Covalent Compounds	
Number	**Prefix**
1	mono-
2	di-
3	tri-
4	tetra-
5	penta-
6	hexa-
7	hepta-
8	octa-
9	nona-
10	deca-

ARIS Presentation Center, Tables Library

SAMPLE PROBLEM 2.10 Determining Names and Formulas of Ionic Compounds Containing Polyatomic Ions

PROBLEM Give the systematic names for the formulas or the formulas for the names of the following compounds:
(a) $Fe(ClO_4)_2$ **(b)** Sodium sulfite   **(c)** $Ba(OH)_2 \cdot 8H_2O$
SOLUTION (a) ClO_4^- is perchlorate, which has a 1− charge, so the cation must be Fe^{2+}. The name is iron(II) perchlorate. (The common name is ferrous perchlorate.)
(b) Sodium is Na^+; sulfite is SO_3^{2-}. Therefore, two Na^+ ions balance one SO_3^{2-} ion. The formula is Na_2SO_3. **(c)** Ba^{2+} is barium; OH^- is hydroxide. There are eight (*octa-*) water molecules in each formula unit. The name is barium hydroxide octahydrate.

FOLLOW-UP PROBLEM 2.10 Give the systematic names for the formulas or the formulas for the names of the following compounds:
(a) Cupric nitrate trihydrate   **(b)** Zinc hydroxide   **(c)** LiCN

SAMPLE PROBLEM 2.11 Recognizing Incorrect Names and Formulas of Ionic Compounds

 ARIS Presentation Center, Worked Examples Library

PROBLEM Something is wrong with the second part of each statement. Provide the correct name or formula.
(a) $Ba(C_2H_3O_2)_2$ is called barium diacetate.
(b) Sodium sulfide has the formula $(Na)_2SO_3$.
(c) Iron(II) sulfate has the formula $Fe_2(SO_4)_3$.
(d) Cesium carbonate has the formula $Cs_2(CO_3)$.
SOLUTION (a) The charge of the Ba^{2+} ion *must* be balanced by *two* $C_2H_3O_2^-$ ions, so the prefix *di-* is unnecessary. For ionic compounds, we do not indicate the number of ions with numerical prefixes. The correct name is barium acetate.
(b) Two mistakes occur here. The sodium ion is monatomic, so it does *not* require parentheses. The sulfide ion is S^{2-}, *not* SO_3^{2-} (called "sulfite"). The correct formula is Na_2S.
(c) The Roman numeral refers to the charge of the ion, *not* the number of ions in the formula. Fe^{2+} is the cation, so it requires one SO_4^{2-} to balance its charge. The correct formula is $FeSO_4$.
(d) Parentheses are *not* required when only one polyatomic ion of a kind is present. The correct formula is Cs_2CO_3.

FOLLOW-UP PROBLEM 2.11 State why the second part of each statement is incorrect, and correct it:
(a) Ammonium phosphate is $(NH_3)_4PO_4$.
(b) Aluminum hydroxide is $AlOH_3$.
(c) $Mg(HCO_3)_2$ is manganese(II) carbonate.
(d) $Cr(NO_3)_3$ is chromic(III) nitride.
(e) $Ca(NO_2)_2$ is cadmium nitrate.

Acid Names from Anion Names Acids are an important group of hydrogen-containing compounds that have been used in chemical reactions since before alchemical times. In the laboratory, acids are typically used in water solution. When naming them and writing their formulas, we consider them as anions connected to the number of hydrogen ions (H^+) needed for charge neutrality. The two common types of acids are binary acids and oxoacids:

1. *Binary acid* solutions form when certain gaseous compounds dissolve in water. For example, when gaseous hydrogen chloride (HCl) dissolves in water, it forms a solution whose name consists of the following parts:

    Prefix *hydro-* + nonmetal *root* + suffix *-ic* + separate word *acid*
    hydro   +   chlor   +   ic   +   acid

    or *hydrochloric acid.* This naming pattern holds for many compounds in which hydrogen combines with an anion that has an *-ide* suffix.

2. *Oxoacid* names are similar to those of the oxoanions, except for two suffix changes:
    • *-ate* in the anion becomes *-ic* in the acid
    • *-ite* in the anion becomes *-ous* in the acid
    The oxoanion prefixes *hypo-* and *per-* are kept. Thus,

    BrO_4^- is *per*brom*ate*, and $HBrO_4$ is *per*brom*ic* acid
    IO_2^- is iod*ite*, and HIO_2 is iod*ous* acid

SAMPLE PROBLEM 2.12 Determining Names and Formulas of Anions and Acids

PROBLEM Name the following anions and give the names and formulas of the acids derived from them: **(a)** Br^-; **(b)** IO_3^-; **(c)** CN^-; **(d)** SO_4^{2-}; **(e)** NO_2^-.

SOLUTION **(a)** The anion is bromide; the acid is hydrobromic acid, HBr.

(b) The anion is iodate; the acid is iodic acid, HIO_3.

(c) The anion is cyanide; the acid is hydrocyanic acid, HCN.

(d) The anion is sulfate; the acid is sulfuric acid, H_2SO_4. (In this case, the suffix is added to the element name *sulfur,* not to the root, *sulf-*.)

(e) The anion is nitrite; the acid is nitrous acid, HNO_2.

COMMENT We added *two* H^+ ions to the sulfate ion to obtain sulfuric acid because SO_4^{2-} has a 2− charge.

FOLLOW-UP PROBLEM 2.12 Write the formula for the name or name for the formula of each acid: **(a)** chloric acid; **(b)** HF; **(c)** acetic acid; **(d)** sulfurous acid; **(e)** HBrO.

Names and Formulas of Binary Covalent Compounds

Binary covalent compounds are formed by the combination of two elements, usually nonmetals. Some are so familiar, such as ammonia (NH_3), methane (CH_4), and water (H_2O), we use their common names, but most are named in a systematic way:

1. The element with the lower group number in the periodic table is the first word in the name; the element with the higher group number is the second word. (*Exception:* When the compound contains oxygen and any of the halogens chlorine, bromine, and iodine, the halogen is named first.)

2. If both elements are in the same group, the one with the higher period number is named first.

3. The second element is named with its root and the suffix *-ide.*

4. Covalent compounds have Greek numerical prefixes (see Table 2.6) to indicate the number of atoms of each element in the compound. The first word has a prefix *only* when more than one atom of the element is present; the second word *usually* has a numerical prefix.

SAMPLE PROBLEM 2.13 Determining Names and Formulas of Binary Covalent Compounds

PROBLEM (a) What is the formula of carbon disulfide?
(b) What is the name of PCl_5?
(c) Give the name and formula of the compound whose molecules each consist of two N atoms and four O atoms.
SOLUTION (a) The prefix *di-* means "two." The formula is CS_2.
(b) P is the symbol for phosphorus; there are five chlorine atoms, which is indicated by the prefix *penta-*. The name is phosphorus pentachloride.
(c) Nitrogen (N) comes first in the name (lower group number). The compound is dinitrogen tetraoxide, N_2O_4.

FOLLOW-UP PROBLEM 2.13 Give the name or formula for (a) SO_3; (b) SiO_2; (c) dinitrogen monoxide; (d) selenium hexafluoride.

SAMPLE PROBLEM 2.14 Recognizing Incorrect Names and Formulas of Binary Covalent Compounds

PROBLEM Explain what is wrong with the name or formula in the second part of each statement and correct it: (a) SF_4 is monosulfur pentafluoride. (b) Dichlorine heptaoxide is Cl_2O_6. (c) N_2O_3 is dinitrotrioxide.
SOLUTION (a) There are two mistakes. *Mono-* is not needed if there is only one atom of the first element, and the prefix for four is *tetra-*, not *penta-*. The correct name is sulfur tetrafluoride.
(b) The prefix *hepta-* indicates seven, not six. The correct formula is Cl_2O_7.
(c) The full name of the first element is needed, and a space separates the two element names. The correct name is dinitrogen trioxide.

FOLLOW-UP PROBLEM 2.14 Explain what is wrong with the second part of each statement and correct it: (a) S_2Cl_2 is disulfurous dichloride. (b) Nitrogen monoxide is N_2O. (c) $BrCl_3$ is trichlorine bromide.

Table 2.7	The First 10 Straight-Chain Alkanes
Name (Formula)	**Model**
Methane (CH_4)	
Ethane (C_2H_6)	
Propane (C_3H_8)	
Butane (C_4H_{10})	
Pentane (C_5H_{12})	
Hexane (C_6H_{14})	
Heptane (C_7H_{16})	
Octane (C_8H_{18})	
Nonane (C_9H_{20})	
Decane ($C_{10}H_{22}$)	

An Introduction to Naming Organic Compounds

Organic compounds typically have complex structural formulas that consist of chains, branches, and/or rings of carbon atoms bonded to hydrogen atoms and, often, to atoms of oxygen, nitrogen, and a few other elements. At this point, we'll look at one or two basic principles for naming them. Much more on the rules of organic nomenclature appears in Chapter 15.

Hydrocarbons, the simplest type of organic compound, contain *only* carbon and hydrogen. *Alkanes* are the simplest type of hydrocarbon; many function as important fuels, such as methane, propane, butane, and the mixture of alkanes in gasoline. The simplest alkanes to name are the *straight-chain alkanes* because the carbon chains have no branches. Alkanes are named with a *root,* based on the number of C atoms in the chain, followed by the suffix *-ane.* Table 2.7 gives the names, molecular formulas, and space-filling models (discussed shortly) of the first 10 straight-chain alkanes. Note that the roots of the four smallest ones are new, but those for the larger ones are the same as the Greek prefixes shown in Table 2.6.

Alkanes (and other organic compounds) *with* branches have a *prefix* in the name as well. *The prefix names the length of the branch and numbers the carbon atom in the main chain that the branch is attached to.* A prefix consists of a root plus the ending *-yl.* Thus, for example, the compound with a one-carbon ("meth") branch attached to the second carbon of the main chain of butane is *2-methylbutane,* where "2-methyl" is the prefix (see margin).

Ball-and-stick model of 2-methylbutane

2-butanol

ethanamine

pentanoic acid

ARIS Presentation Center, Art Library

Organic chemistry application

Organic compounds other than alkanes have names derived from a particular region of the molecule, called the *functional group,* which consists of one or a few atoms bonded in a specific way. *The functional group determines how the compound reacts.* The *alcohol* functional group is a hydroxyl group, O—H, in place of one of the H atoms in the hydrocarbon; an *amine* has an amino group, —NH$_2$; a *carboxylic acid* has a carboxyl group, —COOH; and so forth. Each functional group has its own suffix. Thus, the compound with a hydroxyl group attached to the second carbon in butane is called *2-butanol* (see margin); the compound with an amino group bonded to a two-carbon chain is *ethanamine;* the compound with a carboxyl group bonded to a four-carbon chain is *pentanoic acid* (the C of the carboxyl group counts as one of the carbons). We'll examine how the different functional groups react in Chapter 15.

Molecular Masses from Chemical Formulas

In Section 2.5, we calculated the atomic mass of an element. Using the periodic table and the formula of a compound to see the number of atoms of each element, we calculate the **molecular mass** (also called *molecular weight*) of a formula unit of the compound as the sum of the atomic masses:

$$\text{Molecular mass} = \text{sum of atomic masses} \qquad (2.3)$$

The molecular mass of a water molecule (using atomic masses to four significant figures from the periodic table) is

$$\text{Molecular mass of } H_2O = (2 \times \text{atomic mass of H}) + (1 \times \text{atomic mass of O})$$
$$= (2 \times 1.008 \text{ amu}) + 16.00 \text{ amu} = 18.02 \text{ amu}$$

Ionic compounds are treated the same, but because they do not consist of molecules, we use the term **formula mass** for an ionic compound. To calculate its formula mass, *the number of atoms of each element inside the parentheses is multiplied by the subscript outside the parentheses.* For barium nitrate, Ba(NO$_3$)$_2$,

Formula mass of Ba(NO$_3$)$_2$
$$= (1 \times \text{atomic mass of Ba}) + (2 \times \text{atomic mass of N}) + (6 \times \text{atomic mass of O})$$
$$= 137.3 \text{ amu} + (2 \times 14.01 \text{ amu}) + (6 \times 16.00 \text{ amu}) = 261.3 \text{ amu}$$

Atomic, not ionic, masses are used because electron loss equals electron gain in the compound, so electron mass is balanced. In the next two problems, the name or molecular depiction is used to find a compound's formula and molecular mass.

ARIS Presentation Center, Worked Examples Library

SAMPLE PROBLEM 2.15 Calculating the Molecular Mass of a Compound

PROBLEM Using data in the periodic table, calculate the molecular (or formula) mass of:
(a) Tetraphosphorus trisulfide   **(b)** Ammonium nitrate
PLAN We first write the formula, then multiply the number of atoms (or ions) of each element by its atomic mass, and find the sum.
SOLUTION (a) The formula is P$_4$S$_3$.

$$\text{Molecular mass} = (4 \times \text{atomic mass of P}) + (3 \times \text{atomic mass of S})$$
$$= (4 \times 30.97 \text{ amu}) + (3 \times 32.07 \text{ amu}) = 220.09 \text{ amu}$$

(b) The formula is NH$_4$NO$_3$. We count the total number of N atoms even though they belong to different ions:

Formula mass
$$= (2 \times \text{atomic mass of N}) + (4 \times \text{atomic mass of H}) + (3 \times \text{atomic mass of O})$$
$$= (2 \times 14.01 \text{ amu}) + (4 \times 1.008 \text{ amu}) + (3 \times 16.00 \text{ amu}) = 80.05 \text{ amu}$$

CHECK You can often find large errors by rounding atomic masses to the nearest 5 and adding: **(a)** $(4 \times 30) + (3 \times 30) = 210 \approx 220.09$. The sum has two decimal places because the atomic masses have two. **(b)** $(2 \times 15) + 4 + (3 \times 15) = 79 \approx 80.05$.

FOLLOW-UP PROBLEM 2.15 What is the formula and molecular (or formula) mass of each of the following compounds: **(a)** hydrogen peroxide; **(b)** cesium chloride; **(c)** sulfuric acid; **(d)** potassium sulfate?

SAMPLE PROBLEM 2.16 **Using Molecular Depictions to Determine Formula, Name, and Mass**

PROBLEM Each circle contains a representation of a binary compound. Determine its formula, name, and molecular (formula) mass.

(a)

(b)

- sodium
- fluorine
- nitrogen

PLAN Each of the compounds contains only two elements, so to find the formula, we find the simplest whole-number ratio of one atom to the other. Then we determine the name and formula (see Sample Problems 2.7–2.9 and 2.13) and the mass (see Sample Problem 2.15).

SOLUTION **(a)** There is one brown (sodium) for each green (fluorine), so the formula is NaF. A metal and nonmetal form an ionic compound, in which the metal is named first: sodium fluoride.

$$\text{Formula mass} = (1 \times \text{atomic mass of Na}) + (1 \times \text{atomic mass of F})$$
$$= 22.99 \text{ amu} + 19.00 \text{ amu} = 41.99 \text{ amu}$$

(b) There are three green (fluorine) for each blue (nitrogen), so the formula is NF_3. Two nonmetals form a covalent compound. Nitrogen has a lower group number, so it is named first: nitrogen trifluoride.

$$\text{Molecular mass} = (1 \times \text{atomic mass of N}) + (3 \times \text{atomic mass of F})$$
$$= 14.01 \text{ amu} + (3 \times 19.00 \text{ amu}) = 71.01 \text{ amu}$$

CHECK **(a)** For binary ionic compounds, we predict ionic charges from the periodic table (see Figure 2.13). Na forms a 1+ ion, and F forms a 1− ion, so the charges balance with one Na^+ per F^-. Also, ionic compounds are solids, consistent with the picture. **(b)** Covalent compounds often occur as individual molecules, as in the picture. Rounding in (a) gives 25 + 20 = 45; in (b), we get 15 + (3 × 20) = 75, so there are no large errors.

FOLLOW-UP PROBLEM 2.16 Each circle contains a representation of a binary compound. Determine its name, formula, and molecular (formula) mass.

(a)

(b)

- sodium
- oxygen
- nitrogen

The Gallery on the next page shows some of the ways that chemists picture molecules and the enormous range of molecular sizes.

Section Summary

Chemical formulas describe the simplest atom ratio (empirical formula), actual atom number (molecular formula), and atom arrangement (structural formula) of one unit of a compound. • An ionic compound is named with cation first and anion second. For metals that can form more than one ion, the charge is shown with a Roman numeral. • Oxoanions have suffixes, and sometimes prefixes, attached to the root of the element name to indicate the number of oxygen atoms. • Names of hydrates give the number of associated water molecules with a numerical prefix. • Acid names are based on anion names. • Covalent compounds have as the first word of the name the element that is farther left or lower down in the periodic table, and prefixes show the numbers of each atom. • The molecular (or formula) mass of a compound is the sum of the atomic masses in the formula. • Molecules are three-dimensional objects that range in size from H_2 to biological and synthetic macromolecules.

Picturing Molecules

The most exciting thing about learning chemistry is training your mind to imagine a molecular world, one filled with tiny objects of various shapes. Molecules are depicted in a variety of useful ways, as shown at right for the water molecule:

All molecules are minute, with their relative sizes depending on composition. A water molecule is small because it consists of only three atoms. Many air pollutants, such as ozone, carbon monoxide, and nitrogen dioxide, also consist of small molecules.

Chemical formulas show only the relative numbers of atoms.

Electron-dot and *bond-line formulas* show a bond between atoms as either a pair of dots or a line.

Ball-and-stick models show atoms as spheres and bonds as sticks, with accurate angles and relative sizes, but distances are exaggerated.

Space-filling models are accurately scaled-up versions of molecules, but they do not show bonds.

Electron-density models show the ball-and-stick model within the space-filling shape and color the regions of high (*red*) and low (*blue*) electron charge.

H_2O

H:O:H

H–O–H

Ozone (O$_3$, 48.00 amu) contributes to smog; natural component of stratosphere that absorbs harmful solar radiation.

Carbon monoxide (CO, 28.01 amu), toxic component of car exhaust and cigarette smoke.

C≡O

Nitrogen dioxide (NO$_2$, 46.01 amu) forms from nitrogen monoxide and contributes to smog and acid rain.

Many household chemicals, such as butane, acetic acid, and aspirin, consist of somewhat larger molecules. The biologically essential molecule heme is larger still.

Butane (C$_4$H$_{10}$, 58.12 amu), fuel for cigarette lighters and camping stoves.

Acetic acid (CH$_3$COOH, 60.05 amu), component of vinegar.

Aspirin (C$_9$H$_8$O$_4$, 180.15 amu), most common pain reliever in the world.

Heme (C$_{34}$H$_{32}$FeN$_4$O$_4$, 616.49 amu), part of the blood protein hemoglobin, which carries oxygen through the body.

Very large molecules, called macromolecules, can be synthetic, like nylon, or natural, like DNA, and typically consist of thousands of atoms.

Nylon-66 (~15,000 amu), relatively small, synthetic macromolecule used to make textiles and tires.

Deoxyribonucleic acid (DNA, ~10,000,000 amu), cellular macromolecule that contains genetic information.

2.9 | MIXTURES: CLASSIFICATION AND SEPARATION

Although chemists pay a great deal of attention to pure substances, this form of matter almost never occurs around us. In the natural world, *matter usually occurs as mixtures.* A sample of clean air, for example, consists of many elements and compounds physically mixed together, including oxygen (O_2), nitrogen (N_2), carbon dioxide (CO_2), the six noble gases [Group 8A(18)], and water vapor (H_2O). The oceans are complex mixtures of dissolved ions and covalent substances, including Na^+, Mg^{2+}, Cl^-, SO_4^{2-}, O_2, CO_2, and of course H_2O. Rocks and soils are mixtures of numerous compounds—such as calcium carbonate ($CaCO_3$), silicon dioxide (SiO_2), aluminum oxide (Al_2O_3), and iron(III) oxide (Fe_2O_3)— perhaps a few elements (gold, silver, and carbon in the form of diamond), and petroleum and coal, which are complex mixtures themselves. Living things contain thousands of substances: carbohydrates, lipids, proteins, nucleic acids, and many simpler ionic and covalent compounds.

There are two broad classes of mixtures. A **heterogeneous mixture** has one or more visible boundaries between the components. Thus, its composition is *not* uniform. Many rocks are heterogeneous, showing individual grains and flecks of different minerals. In some cases, as in milk and blood, the boundaries can be seen only with a microscope. A **homogeneous mixture** has no visible boundaries because the components are mixed as individual atoms, ions, and molecules. Thus, its composition *is* uniform, unvarying from one region to another. A mixture of sugar dissolved in water is homogeneous, for example, because the sugar molecules and water molecules are uniformly intermingled on the molecular level. We have no way to tell visually whether an object is a substance (element or compound) or a homogeneous mixture.

A homogeneous mixture is also called a **solution.** Although we usually think of solutions as liquid, they can exist in all three physical states. For example, air is a gaseous solution of mostly oxygen and nitrogen molecules, and wax is a solid solution of several fatty substances. Solutions in water, called **aqueous solutions,** are especially important in chemistry and comprise a major portion of the environment and of all organisms.

Recall that mixtures differ fundamentally from compounds in three ways: (1) the proportions of the components can vary; (2) the individual properties of the components are observable; and (3) the components can be separated by physical means. In some cases, if we apply enough energy to the components of the mixture, they react with each other chemically and form a compound, after which their individual properties are no longer observable. Figure 2.20 shows such a case with a mixture of iron and sulfur.

In order to investigate the properties of substances, chemists have devised many procedures for separating a mixture into its component elements and compounds. Indeed, the laws and models of chemistry could never have been formulated without this ability. Many of Dalton's critics, who thought they had found compounds with varying composition, were unknowingly studying mixtures! The Tools of the Laboratory essay on the next two pages describes some of the more common laboratory separation methods.

A

B

Figure 2.20 The distinction between mixtures and compounds. A, A *mixture* of iron and sulfur can be separated with a magnet because only the iron is magnetic. The blow-up shows separate regions of the two elements. **B,** After strong heating, the *compound* iron(II) sulfide forms, which is no longer magnetic. The blow-up shows the structure of the compound, in which there are no separate regions of the elements.

ARIS Presentation Center, Art Library

Section Summary

Heterogeneous mixtures have visible boundaries between the components. • Homogeneous mixtures have no visible boundaries because mixing occurs at the molecular level. A solution is a homogeneous mixture and can occur in any physical state. • Components of mixtures (unlike those of compounds) can have variable proportions, can be separated physically, and retain their properties. • Common physical separation processes include filtration, crystallization, extraction, chromatography, and distillation.

Tools of the Laboratory

Basic Separation Techniques

Some of the most challenging and time-consuming laboratory procedures involve separating mixtures and purifying the components. Several common separation techniques are described here. All of these methods depend on the *physical properties* of the substances in the mixture; no chemical changes occur.

ARIS Presentation Center, Photo Library

Filtration separates the components of a mixture on the basis of *differences in particle size*. It is used most often to separate a liquid (smaller particles) from a solid (larger particles). Figure B2.3 shows simple filtration of a solid reaction product. In vacuum filtration, reduced pressure within the flask speeds the flow of the liquid through the filter. Filtration is a key step in the purification of the tap water you drink.

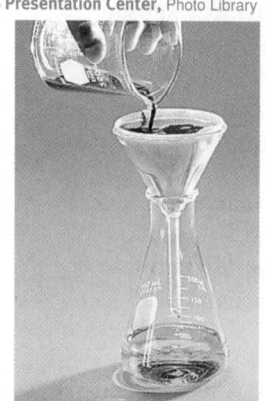

Figure B2.3 Filtration.

Crystallization is based on *differences in solubility*. The *solubility* of a substance is the amount that dissolves in a fixed volume of solvent at a given temperature. The result shown in Figure B2.4 applies the fact that many substances are more soluble in hot solvent than in cold. The purified compound crystallized as the solution was cooled. Key substances in computer chips and other electronic devices are purified by a type of crystallization.

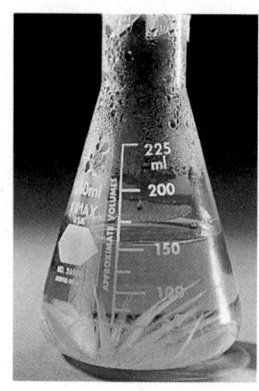

Figure B2.4 Crystallization.

Distillation separates components through *differences in volatility,* the tendency of a substance to become a gas. Ether, for example, is more volatile than water, which is much more volatile than sodium chloride. The simple distillation apparatus shown in Figure B2.5 is used to separate components with *large* differences in volatility, such as water from dissolved ionic compounds. As the mixture boils, the vapor is richer in the more volatile component, which is condensed and collected separately. Separating components with small volatility differences requires many vaporization-condensation steps (discussed in Chapter 13).

Extraction is also based on *differences in solubility*. In a typical procedure, a natural (often plant or animal) material is ground in a blender with a solvent that extracts (dissolves) soluble compound(s) embedded in insoluble material. This extract is separated further by the addition of a second solvent that does not dissolve in the first. After shaking in a separatory funnel, some components are extracted into the new solvent. Figure B2.6 shows the extraction of plant pigments from water into hexane, an organic solvent.

Thermometer

① Mixture is heated and volatile component vaporizes

② Vapors in contact with cool glass condense to form pure liquid distillate

Distilling flask

Water-cooled condenser

Water out to sink

Water in

③ Distillate collected in separate flask

Figure B2.5 Distillation. ARIS Presentation Center, Art Library

① Hexane shaken with water solution of plant material extracts some dissolved substances

Hexane layer

Water layer

② Upon standing, two layers separate

③ Stopcock is opened to drain bottom layer, and top layer is poured from separatory funnel

Figure B2.6 Extraction. ARIS Presentation Center, Art Library

Chromatography is a third technique based on *differences in solubility*. The mixture is dissolved in a gas or liquid called the *mobile phase*, and the components are separated as this phase moves over a solid (or viscous liquid) surface called the *stationary phase*. A component with low solubility in the stationary phase spends less time there, thus moving faster than a component that is highly soluble in that phase. Figure B2.7 depicts the separation of a mixture of pigments in ink.

Many types of chromatography are used to separate a wide variety of substances, from simple gases to biological macromolecules. In *gas-liquid chromatography (GLC),* the mobile phase is an inert gas, such as helium, that carries the previously vaporized components into a long tube that contains the stationary phase (Figure B2.8, part A). The components emerge separately and reach a detector to create a chromatogram. A typical chromatogram has numerous peaks of specific position and height, each of which represents the amount of a given component (Figure B2.8, part B).

The principle of *high-performance (high-pressure) liquid chromatography (HPLC)* is very similar. However, in this technique the mixture is not vaporized, so a more diverse group of components, which may include nonvolatile compounds, can be separated (Figure B2.9).

Solvent (mobile phase)

Solvent

Stationary phase packed in column

① Ink mixture is placed carefully on stationary phase

② Fresh solvent flows through the column

③ Components move through column at different rates

④ The component with the greatest preference for the mobile phase moves fastest

1 2 3 **Later time** 1 2 3

Collecting flasks

⑤ Separated components are collected as they emerge from column

Figure B2.7 Procedure for column chromatography.

He gas   He gas

Before interacting with the stationary phase

He gas   He gas

A   After interacting with the stationary phase

B   Time (minutes)

Figure B2.8 Principle of gas-liquid chromatography (GLC). A, The mobile phase *(purple arrow)* carries the sample mixture into a tube packed with the stationary phase *(gray outline on yellow spheres),* and each component dissolves in the stationary phase to a different extent.

A component *(red)* that dissolves less readily than another *(blue)* emerges from the tube sooner. **B,** A typical gas-liquid chromatogram of a complex mixture displays each component as a peak.

Figure B2.9 A high-performance liquid chromatograph.

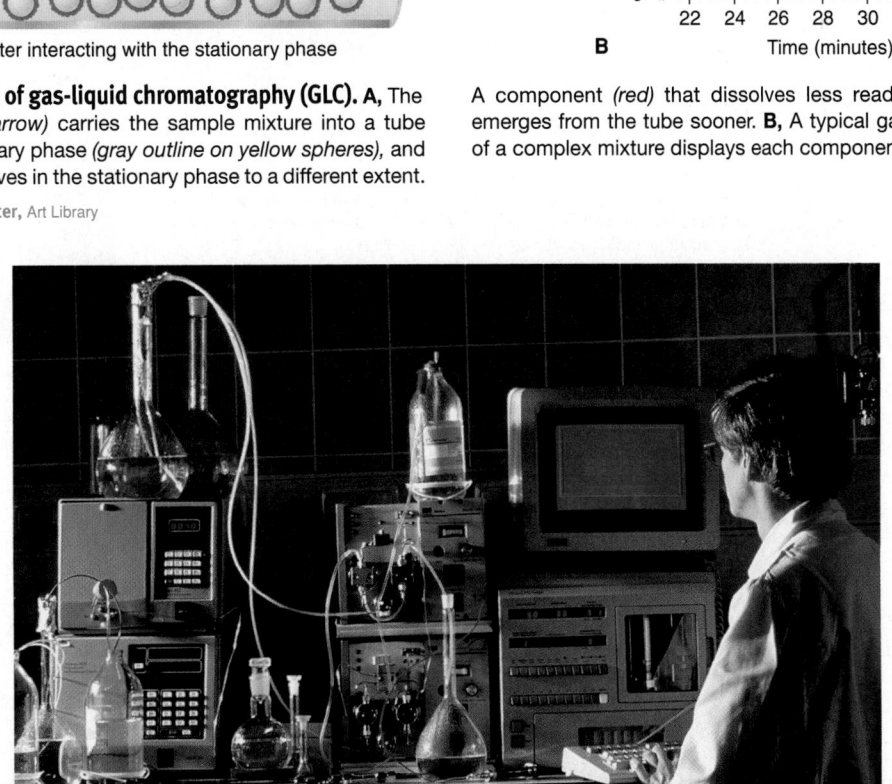

Chapter Perspective

An understanding of matter at the observable and atomic levels is the essence of chemistry. In this chapter, you have learned how matter is classified in terms of its composition and how it is named in words and formulas, which are major steps toward that understanding. Figure 2.21 provides a visual review of many key terms and ideas in this chapter. In Chapter 3, we explore one of the central quantitative ideas in chemistry: how the observable amount of a substance relates to the number of atoms, molecules, or ions that make it up.

Figure 2.21 The classification of matter from a chemical point of view. Mixtures are separated by physical changes into elements and compounds. Chemical changes are required to convert elements into compounds, and vice versa.

CHAPTER REVIEW GUIDE

The following sections provide many aids to help you study this chapter. (Numbers in parentheses refer to pages, unless noted otherwise.)

Learning Objectives These are concepts and skills you should know after studying this chapter.

Relevant section and/or sample problem (SP) numbers appear in parentheses.

Understand These Concepts

1. The defining characteristics of the three types of matter—element, compound, and mixture—on the macroscopic and atomic levels (2.1)
2. The significance of the three mass laws—mass conservation, definite composition, and multiple proportions (2.2)
3. The postulates of Dalton's atomic theory and how it explains the mass laws (2.3)
4. The major contribution of experiments by Thomson, Millikan, and Rutherford concerning atomic structure (2.4)
5. The structure of the atom, the main features of the subatomic particles, and the importance of isotopes (2.5)
6. The format of the periodic table and general location and characteristics of metals, metalloids, and nonmetals (2.6)
7. The essential features of ionic and covalent bonding and the distinction between them (2.7)
8. The types of mixtures and their properties (2.9)

Master These Skills

1. Distinguishing elements, compounds, and mixtures at the atomic scale (SP 2.1)
2. Using the mass ratio of element to compound to find the mass of an element in a compound (SP 2.2)
3. Visualizing the mass laws (SP 2.3)
4. Using atomic notation to express the subatomic makeup of an isotope (SP 2.4)
5. Calculating an atomic mass from isotopic composition (SP 2.5)
6. Predicting the monatomic ion formed from a main-group element (SP 2.6)
7. Naming and writing the formula of an ionic compound formed from the ions in Tables 2.3 to 2.5 (SP 2.7–2.12, 2.16)
8. Naming and writing the formula of an inorganic binary covalent compound (SP 2.13, 2.14, 2.16)
9. Calculating the molecular or formula mass of a compound (SP 2.15)

Key Terms These important terms appear in boldface in the chapter and are defined again in the Glossary.

Section 2.1
element (41)
substance (41)
molecule (42)
compound (42)
mixture (42)

Section 2.2
law of mass conservation (44)
law of definite (or constant) composition (44)
fraction by mass (mass fraction) (44)
percent by mass (mass percent, mass %) (44)
law of multiple proportions (46)

Section 2.3
atom (47)

Section 2.4
cathode ray (49)
nucleus (51)

Section 2.5
proton (p^+) (52)
neutron (n^0) (52)
electron (e^-) (52)
atomic number (Z) (53)
mass number (A) (53)
atomic symbol (53)
isotope (53)
atomic mass unit (amu) (54)
dalton (Da) (54)
mass spectrometry (54)
isotopic mass (54)
atomic mass (54)

Section 2.6
periodic table of the elements (57)

period (57)
group (57)
metal (58)
nonmetal (58)
metalloid (semimetal) (58)

Section 2.7
ionic compound (60)
covalent compound (60)
chemical bond (60)
ion (60)
binary ionic compound (60)
cation (60)
anion (60)
monatomic ion (60)
covalent bond (62)
polyatomic ion (63)

Section 2.8
chemical formula (64)
empirical formula (64)

molecular formula (65)
structural formula (65)
formula unit (66)
oxoanion (68)
hydrate (69)
binary covalent compound (70)
molecular mass (72)
formula mass (72)

Section 2.9
heterogeneous mixture (75)
homogeneous mixture (75)
solution (75)
aqueous solution (75)
filtration (76)
crystallization (76)
distillation (76)
volatility (76)
extraction (76)
chromatography (77)

Key Equations and Relationships Numbered and screened concepts are listed for you to refer to or memorize.

2.1 Finding the mass of an element in a given mass of compound (45):

Mass of element in sample

$$= \text{mass of compound in sample} \times \frac{\text{mass of element}}{\text{mass of compound}}$$

2.2 Calculating the number of neutrons in an atom (53):

Number of neutrons = mass number − atomic number

or
$$N = A - Z$$

2.3 Determining the molecular mass of a formula unit of a compound (72):

$$\text{Molecular mass} = \text{sum of atomic masses}$$

Highlighted Figures and Tables These figures (F) and tables (T) provide a visual review of key ideas.

Entries in **bold** contain frequently used data.

F2.1 Elements, compounds, and mixtures on atomic scale (42)
F2.7 General features of the atom (52)
T2.2 Properties of the three key subatomic particles (53)
F2.9 The modern periodic table (58)
F2.12 Factors that influence the strength of ionic bonding (61)
F2.13 The relationship of ions formed to the nearest noble gas (61)

F2.14 Formation of a covalent bond between two H atoms (62)
F2.18 Some common monatomic ions of the elements (65)
T2.3 Common monatomic ions (65)
T2.4 Some metals that form more than one monatomic ion (67)
T2.5 Common polyatomic ions (68)
T2.6 Numerical prefixes for hydrates and binary covalent compounds (69)
F2.21 Classification of matter from a chemical point of view (78)

Brief Solutions to FOLLOW-UP PROBLEMS Compare your solutions to these calculation steps and answers.

2.1 There are two types of particles reacting (left circle), one with two blue atoms and the other with two orange, so the depiction shows a mixture of two elements. In the product (right circle), all the particles have one blue atom and one orange; this is a compound.

2.2 Mass (t) of pitchblende

$$= 2.3 \text{ t uranium} \times \frac{84.2 \text{ t pitchblende}}{71.4 \text{ t uranium}} = 2.7 \text{ t pitchblende}$$

Mass (t) of oxygen

$$= 2.7 \text{ t pitchblende} \times \frac{(84.2 - 71.4 \text{ t oxygen})}{84.2 \text{ t pitchblende}} = 0.41 \text{ t oxygen}$$

2.3 Sample B. Two bromine-fluorine compounds appear. In one, there are three fluorine atoms for each bromine; in the other, there is one fluorine for each bromine. Therefore, in the two compounds, the ratio of fluorines combining with one bromine is 3/1.

2.4 (a) $5p^+, 6n^0, 5e^-$; Q = B
(b) $20p^+, 21n^0, 20e^-$; R = Ca
(c) $53p^+, 78n^0, 53e^-$; X = I

2.5 $10.0129x + [11.0093(1 - x)] = 10.81$; $0.9964x = 0.1993$; $x = 0.2000$ and $1 - x = 0.8000$; % abundance of $^{10}B = 20.00\%$; % abundance of $^{11}B = 80.00\%$

2.6 (a) S^{2-}; (b) Rb^+; (c) Ba^{2+}

2.7 (a) Zinc [Group 2B(12)] and oxygen [Group 6A(16)]
(b) Silver [Group 1B(11)] and bromine [Group 7A(17)]
(c) Lithium [Group 1A(1)] and chlorine [Group 7A(17)]
(d) Aluminum [Group 3A(13)] and sulfur [Group 6A(16)]

2.8 (a) ZnO; (b) AgBr; (c) LiCl; (d) Al_2S_3

2.9 (a) PbO_2; (b) copper(I) sulfide (cuprous sulfide); (c) iron(II) bromide (ferrous bromide); (d) $HgCl_2$

2.10 (a) $Cu(NO_3)_2 \cdot 3H_2O$; (b) $Zn(OH)_2$; (c) lithium cyanide

2.11 (a) $(NH_4)_3PO_4$; ammonium is NH_4^+ and phosphate is PO_4^{3-}.
(b) $Al(OH)_3$; parentheses are needed around the polyatomic ion OH^-.
(c) Magnesium hydrogen carbonate; Mg^{2+} is magnesium and can have only a 2+ charge, so it does not need (II); HCO_3^- is hydrogen carbonate (or bicarbonate).
(d) Chromium(III) nitrate; the -*ic* ending is not used with Roman numerals; NO_3^- is nit*rate*.
(e) Calcium nitrite; Ca^{2+} is calcium and NO_2^- is nit*rite*.

2.12 (a) $HClO_3$; (b) hydrofluoric acid; (c) CH_3COOH (or $HC_2H_3O_2$); (d) H_2SO_3; (e) hypobromous acid

2.13 (a) Sulfur trioxide; (b) silicon dioxide; (c) N_2O; (d) SeF_6

2.14 (a) Disulfur dichloride; the -*ous* suffix is not used.
(b) NO; the name indicates one nitrogen.
(c) Bromine trichloride; Br is in a higher period in Group 7A(17), so it is named first.

2.15 (a) H_2O_2, 34.02 amu; (b) CsCl, 168.4 amu; (c) H_2SO_4, 98.09 amu; (d) K_2SO_4, 174.27 amu

2.16 (a) Na_2O. This is an ionic compound, so the name is sodium oxide.

Formula mass
$$= (2 \times \text{atomic mass of Na}) + (1 \times \text{atomic mass of O})$$
$$= (2 \times 22.99 \text{ amu}) + 16.00 \text{ amu} = 61.98 \text{ amu}$$

(b) NO_2. This is a covalent compound, and N has the lower group number, so the name is nitrogen dioxide.

Molecular mass
$$= (1 \times \text{atomic mass of N}) + (2 \times \text{atomic mass of O})$$
$$= 14.01 \text{ amu} + (2 \times 16.00 \text{ amu}) = 46.01 \text{ amu}$$

PROBLEMS

Problems with colored numbers are answered in Appendix E and worked in detail in the Student Solutions Manual. Problem sections match those in the text and provide the numbers of relevant sample problems. Most offer Concept Review Questions, Skill-Building Exercises (grouped in pairs covering the same concept), and Problems in Context. The Comprehensive Problems are based on material from any section or previous chapter.

Elements, Compounds, and Mixtures: An Atomic Overview
(Sample Problem 2.1)

● Concept Review Questions

• **2.1** What is the key difference between an element and a compound?

• **2.2** List two differences between a compound and a mixture.

• **2.3** Which of the following are pure substances? Explain.
(a) Calcium chloride, used to melt ice on roads, consists of two elements, calcium and chlorine, in a fixed mass ratio.
(b) Sulfur consists of sulfur atoms combined into octatomic molecules.
(c) Baking powder, a leavening agent, contains 26% to 30% sodium hydrogen carbonate and 30% to 35% calcium dihydrogen phosphate by mass.
(d) Cytosine, a component of DNA, consists of H, C, N, and O atoms bonded in a specific arrangement.

• **2.4** Classify each substance in Problem 2.3 as an element, compound, or mixture, and explain your answers.

• **2.5** Explain the following statement: The smallest particles unique to an element may be atoms or molecules.

• **2.6** Explain the following statement: The smallest particles unique to a compound cannot be atoms.

• **2.7** Can the relative amounts of the components of a mixture vary? Can the relative amounts of the components of a compound vary? Explain.

● Problems in Context

• **2.8** The tap water found in many areas of the United States leaves white deposits when it evaporates. Is this tap water a mixture or a compound? Explain.

• **2.9** Each scene below represents a mixture. Describe each one in terms of the number of elements and/or compounds present.

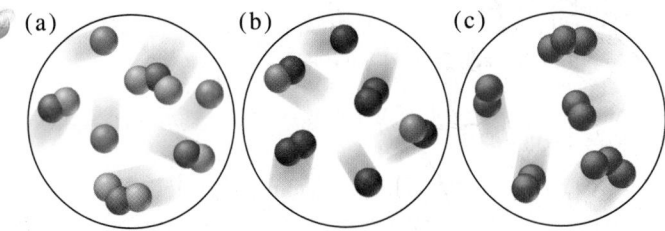
(a) (b) (c)

• **2.10** Samples of illicit "street" drugs often contain an inactive component, such as ascorbic acid (vitamin C). After obtaining a sample of cocaine, government chemists calculate the mass of vitamin C per gram of drug sample, and use it to track the drug's distribution. For example, if different samples of cocaine obtained on the streets of New York, Los Angeles, and Paris all contain 0.6384 g of vitamin C per gram of sample, they very likely come from a common source. Do these street samples consist of a compound, element, or mixture? Explain.

The Observations That Led to an Atomic View of Matter
(Sample Problem 2.2)

● Concept Review Questions

• **2.11** Why was it necessary for separation techniques and methods of chemical analysis to be developed before the laws of definite composition and multiple proportions could be formulated?

• **2.12** To which classes of matter—element, compound, and/or mixture—do the following apply: (a) law of mass conservation; (b) law of definite composition; (c) law of multiple proportions?

• **2.13** In our modern view of matter and energy, is the law of mass conservation still relevant to chemical reactions? Explain.

• **2.14** Identify the mass law that each of the following observations demonstrates, and explain your reasoning:
(a) A sample of potassium chloride from Chile contains the same percent by mass of potassium as one from Poland.
(b) A flashbulb contains magnesium and oxygen before use and magnesium oxide afterward, but its mass does not change.
(c) Arsenic and oxygen form one compound that is 65.2 mass % arsenic and another that is 75.8 mass % arsenic.

• **2.15** Which of the following scenes illustrate(s) the fact that compounds of chlorine (green) and oxygen (red) exhibit the law of multiple proportions? Name the compounds.

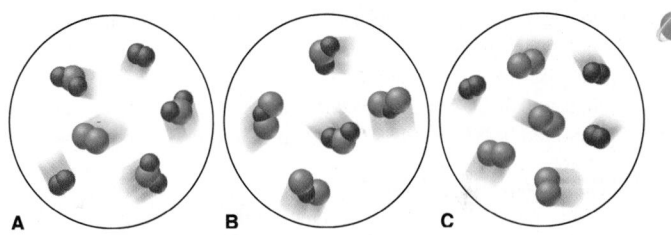
A B C

•• **2.16** (a) Does the percent by mass of each element in a compound depend on the amount of compound? Explain. (b) Does the mass of each element in a compound depend on the amount of compound? Explain.

• **2.17** Does the percent by mass of each element in a compound depend on the amount of that element used to make the compound? Explain.

● Skill-Building Exercises *(grouped in similar pairs)*

• **2.18** State the mass law(s) demonstrated by the following experimental results, and explain your reasoning:
Experiment 1: A student heats 1.00 g of a blue compound and obtains 0.64 g of a white compound and 0.36 g of a colorless gas.
Experiment 2: A second student heats 3.25 g of the same blue compound and obtains 2.08 g of a white compound and 1.17 g of a colorless gas.

• **2.19** State the mass law(s) demonstrated by the following experimental results, and explain your reasoning:
Experiment 1: A student heats 1.27 g of copper and 3.50 g of iodine to produce 3.81 g of a white compound; 0.96 g of iodine remains.
Experiment 2: A second student heats 2.55 g of copper and 3.50 g of iodine to form 5.25 g of a white compound, and 0.80 g of copper remains.

2.20 Fluorite, a mineral of calcium, is a compound of the metal with fluorine. Analysis shows that a 2.76-g sample of fluorite contains 1.42 g of calcium. Calculate the (a) mass of fluorine in the sample; (b) mass fractions of calcium and fluorine in fluorite; (c) mass percents of calcium and fluorine in fluorite.

2.21 Galena, a mineral of lead, is a compound of the metal with sulfur. Analysis shows that a 2.34-g sample of galena contains 2.03 g of lead. Calculate the (a) mass of sulfur in the sample; (b) mass fractions of lead and sulfur in galena; (c) mass percents of lead and sulfur in galena.

2.22 Magnesium oxide (MgO) forms when the metal burns in air. (a) If 1.25 g of MgO contains 0.754 g of Mg, what is the mass ratio of magnesium to oxide? (b) How many grams of Mg are in 534 g of MgO?

2.23 Zinc sulfide (ZnS) occurs in the zinc blende crystal structure. (a) If 2.54 g of ZnS contains 1.70 g of Zn, what is the mass ratio of zinc to sulfide? (b) How many kilograms of Zn are in 3.82 kg of ZnS?

2.24 A compound of copper and sulfur contains 88.39 g of metal and 44.61 g of nonmetal. How many grams of copper are in 5264 kg of compound? How many grams of sulfur?

2.25 A compound of iodine and cesium contains 63.94 g of metal and 61.06 g of nonmetal. How many grams of cesium are in 38.77 g of compound? How many grams of iodine?

2.26 Show, with calculations, how the following data illustrate the law of multiple proportions:
Compound 1: 47.5 mass % sulfur and 52.5 mass % chlorine
Compound 2: 31.1 mass % sulfur and 68.9 mass % chlorine

2.27 Show, with calculations, how the following data illustrate the law of multiple proportions:
Compound 1: 77.6 mass % xenon and 22.4 mass % fluorine
Compound 2: 63.3 mass % xenon and 36.7 mass % fluorine

● **Problems in Context**

2.28 Dolomite is a carbonate of magnesium and calcium. Analysis shows that 7.81 g of dolomite contains 1.70 g of Ca. Calculate the mass percent of Ca in dolomite. On the basis of the mass percent of Ca, and neglecting all other factors, which is the richer source of Ca, dolomite or fluorite (see Problem 2.20)?

2.29 The mass percent of sulfur in a sample of coal is a key factor in the environmental impact of the coal because the sulfur combines with oxygen when the coal is burned and the oxide can then be incorporated into acid rain. Which of the following coals would have the smallest environmental impact?

	Mass (g) of Sample	Mass (g) of Sulfur in Sample
Coal A	378	11.3
Coal B	495	19.0
Coal C	675	20.6

Dalton's Atomic Theory
(Sample Problem 2.3)

● **Concept Review Questions**

2.30 Which of Dalton's postulates about atoms are inconsistent with later observations? Do these inconsistencies mean that Dalton was wrong? Is Dalton's model still useful? Explain.

2.31 Use Dalton's theory to explain why potassium nitrate from India or Italy has the same mass percents of K, N, and O.

The Observations That Led to the Nuclear Atom Model

● **Concept Review Questions**

2.32 Thomson was able to determine the mass/charge ratio of the electron but not its mass. How did Millikan's experiment allow determination of the electron's mass?

2.33 The following charges on individual oil droplets were obtained during an experiment similar to Millikan's. Determine a charge for the electron (in C, coulombs), and explain your answer: -3.204×10^{-19} C; -4.806×10^{-19} C; -8.010×10^{-19} C; -1.442×10^{-18} C.

2.34 Describe Thomson's model of the atom. How might it account for the production of cathode rays?

2.35 When Rutherford's coworkers bombarded gold foil with α particles, they obtained results that overturned the existing (Thomson) model of the atom. Explain.

The Atomic Theory Today
(Sample Problems 2.4 and 2.5)

● **Concept Review Questions**

2.36 Define *atomic number* and *mass number*. Which can vary without changing the identity of the element?

2.37 Choose the correct answer. The difference between the mass number of an isotope and its atomic number is (a) directly related to the identity of the element; (b) the number of electrons; (c) the number of neutrons; (d) the number of isotopes.

2.38 Even though several elements have only one naturally occurring isotope and all atomic nuclei have whole numbers of protons and neutrons, no atomic mass is a whole number. Use the data from Table 2.2 to explain this fact.

● **Skill-Building Exercises** (grouped in similar pairs)

2.39 Argon has three naturally occurring isotopes, ^{36}Ar, ^{38}Ar, and ^{40}Ar. What is the mass number of each? How many protons, neutrons, and electrons are present in each?

2.40 Chlorine has two naturally occurring isotopes, ^{35}Cl and ^{37}Cl. What is the mass number of each isotope? How many protons, neutrons, and electrons are present in each?

2.41 Do both members of the following pairs have the same number of protons? Neutrons? Electrons?
(a) $^{16}_{8}$O and $^{17}_{8}$O   (b) $^{40}_{18}$Ar and $^{41}_{19}$K   (c) $^{60}_{27}$Co and $^{60}_{28}$Ni
Which pair(s) consist(s) of atoms with the same Z value? N value? A value?

2.42 Do both members of the following pairs have the same number of protons? Neutrons? Electrons?
(a) $^{3}_{1}$H and $^{3}_{2}$He   (b) $^{14}_{6}$C and $^{15}_{7}$N   (c) $^{19}_{9}$F and $^{18}_{9}$F
Which pair(s) consist(s) of atoms with the same Z value? N value? A value?

2.43 Write the $^{A}_{Z}$X notation for each atomic depiction:

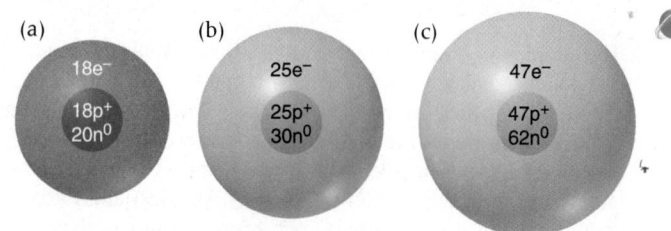

(a)   18e⁻   18p⁺   20n⁰
(b)   25e⁻   25p⁺   30n⁰
(c)   47e⁻   47p⁺   62n⁰

2.44 Write the $_Z^A X$ notation for each atomic depiction:

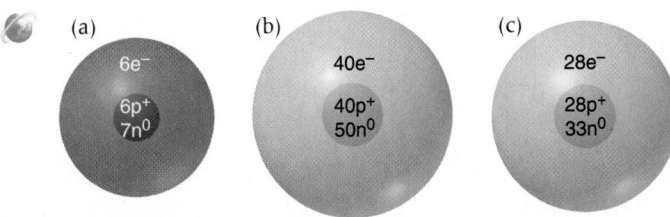

(a)
6e⁻
6p⁺
7n⁰

(b)
40e⁻
40p⁺
50n⁰

(c)
28e⁻
28p⁺
33n⁰

2.45 Draw atomic depictions similar to those in Problem 2.43 for (a) $_{22}^{48}Ti$; (b) $_{34}^{79}Se$; (c) $_5^{11}B$.

2.46 Draw atomic depictions similar to those in Problem 2.43 for (a) $_{82}^{207}Pb$; (b) $_4^9Be$; (c) $_{33}^{75}As$.

2.47 Gallium has two naturally occurring isotopes, ^{69}Ga (isotopic mass 68.9256 amu, abundance 60.11%) and ^{71}Ga (isotopic mass 70.9247 amu, abundance 39.89%). Calculate the atomic mass of gallium.

2.48 Magnesium has three naturally occurring isotopes, ^{24}Mg (isotopic mass 23.9850 amu, abundance 78.99%), ^{25}Mg (isotopic mass 24.9858 amu, abundance 10.00%), and ^{26}Mg (isotopic mass 25.9826 amu, abundance 11.01%). Calculate the atomic mass of magnesium.

2.49 Chlorine has two naturally occurring isotopes, ^{35}Cl (isotopic mass 34.9689 amu) and ^{37}Cl (isotopic mass 36.9659 amu). If chlorine has an atomic mass of 35.4527 amu, what is the percent abundance of each isotope?

2.50 Copper has two naturally occurring isotopes, ^{63}Cu (isotopic mass 62.9396 amu) and ^{65}Cu (isotopic mass 64.9278 amu). If copper has an atomic mass of 63.546 amu, what is the percent abundance of each isotope?

Elements: A First Look at the Periodic Table

● Concept Review Questions

2.51 How can iodine ($Z = 53$) have a higher atomic number yet a lower atomic mass than tellurium ($Z = 52$)?

2.52 Correct each of the following statements:
(a) In the modern periodic table, the elements are arranged in order of increasing atomic mass.
(b) Elements in a period have similar chemical properties.
(c) Elements can be classified as either metalloids or nonmetals.

2.53 What class of elements lies along the "staircase" line in the periodic table? How do their properties compare with those of metals and nonmetals?

2.54 What are some characteristic properties of elements to the left of the elements along the "staircase"? To the right?

2.55 The elements in Groups 1A(1) and 7A(17) are all quite reactive. What is a major difference between them?

● Skill-Building Exercises (grouped in similar pairs)

2.56 Give the name, atomic symbol, and group number of the element with the following Z value, and classify it as a metal, metalloid, or nonmetal:
(a) $Z = 32$ (b) $Z = 15$ (c) $Z = 2$ (d) $Z = 3$ (e) $Z = 42$

2.57 Give the name, atomic symbol, and group number of the element with the following Z value, and classify it as a metal, metalloid, or nonmetal:
(a) $Z = 33$ (b) $Z = 20$ (c) $Z = 35$ (d) $Z = 19$ (e) $Z = 13$

2.58 Fill in the blanks:
(a) The symbol and atomic number of the heaviest alkaline earth metal are _____ and _____.
(b) The symbol and atomic number of the lightest metalloid in Group 4A(14) are _____ and _____.
(c) Group 1B(11) consists of the *coinage metals*. The symbol and atomic mass of the coinage metal whose atoms have the fewest electrons are _____ and _____.
(d) The symbol and atomic mass of the halogen in Period 4 are _____ and _____.

2.59 Fill in the blanks:
(a) The symbol and atomic number of the heaviest nonradioactive noble gas are _____ and _____.
(b) The symbol and group number of the Period 5 transition element whose atoms have the fewest protons are _____ and _____.
(c) The elements in Group 6A(16) are sometimes called the *chalcogens*. The symbol and atomic number of the first metallic chalcogen are _____ and _____.
(d) The symbol and number of protons of the Period 4 alkali metal atom are _____ and _____.

Compounds: Introduction to Bonding
(Sample Problem 2.6)

● Concept Review Questions

2.60 Describe the type and nature of the bonding that occurs between reactive metals and nonmetals.

2.61 Describe the type and nature of the bonding that often occurs between two nonmetals.

2.62 How can ionic compounds be neutral if they consist of positive and negative ions?

2.63 Given that the ions in LiF and in MgO are of similar size, which compound has stronger ionic bonding? Use Coulomb's law in your explanation.

2.64 Are molecules present in a sample of BaF_2? Explain.

2.65 Are ions present in a sample of P_4O_6? Explain.

2.66 The monatomic ions of Groups 1A(1) and 7A(17) are all singly charged. In what major way do they differ? Why?

2.67 Describe the formation of solid magnesium chloride ($MgCl_2$) from large numbers of magnesium and chlorine atoms.

2.68 Describe the formation of solid potassium sulfide (K_2S) from large numbers of potassium and sulfur atoms.

2.69 Does potassium nitrate (KNO_3) incorporate ionic bonding, covalent bonding, or both? Explain.

● Skill-Building Exercises (grouped in similar pairs)

2.70 What monatomic ions do potassium ($Z = 19$) and iodine ($Z = 53$) form?

2.71 What monatomic ions do barium ($Z = 56$) and selenium ($Z = 34$) form?

2.72 For each ionic depiction, give the name of the parent atom, its mass number, and its group and period numbers:

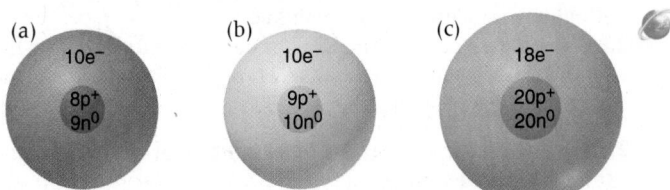

(a)
10e⁻
8p⁺
9n⁰

(b)
10e⁻
9p⁺
10n⁰

(c)
18e⁻
20p⁺
20n⁰

• **2.73** For each ionic depiction, give the name of the parent atom, its mass number, and its group and period numbers:

(a) 36e⁻ 35p⁺ 44n⁰   (b) 10e⁻ 7p⁺ 8n⁰   (c) 36e⁻ 37p⁺ 48n⁰

•• **2.74** An ionic compound forms when lithium ($Z = 3$) reacts with oxygen ($Z = 8$). If a sample of the compound contains 8.4×10^{21} lithium ions, how many oxide ions does it contain?

•• **2.75** An ionic compound forms when calcium ($Z = 20$) reacts with iodine ($Z = 53$). If a sample of the compound contains 7.4×10^{21} calcium ions, how many iodide ions does it contain?

• **2.76** The radii of the sodium and potassium ions are 102 pm and 138 pm, respectively. Which compound has stronger ionic attractions, sodium chloride or potassium chloride?

• **2.77** The radii of the lithium and magnesium ions are 76 pm and 72 pm, respectively. Which compound has stronger ionic attractions, lithium oxide or magnesium oxide?

Compounds: Formulas, Names, and Masses
(Sample Problems 2.7 to 2.16)

● Concept Review Questions

• **2.78** What is the difference between an empirical formula and a molecular formula? Can they ever be the same?

• **2.79** How is a structural formula similar to a molecular formula? How is it different?

•• **2.80** Consider a mixture of 10 billion O_2 molecules and 10 billion H_2 molecules. In what way is this mixture similar to a sample containing 10 billion hydrogen peroxide (H_2O_2) molecules? In what way is it different?

• **2.81** For what type(s) of compound do we use Roman numerals in the name?

• **2.82** For what type(s) of compound do we use Greek numerical prefixes in the name?

• **2.83** For what type of compound are we unable to write a molecular formula?

● Skill-Building Exercises *(grouped in similar pairs)*

• **2.84** Write an empirical formula for each of the following:
(a) Hydrazine, a rocket fuel, molecular formula N_2H_4
(b) Glucose, a sugar, molecular formula $C_6H_{12}O_6$

• **2.85** Write an empirical formula for each of the following:
(a) Ethylene glycol, car antifreeze, molecular formula $C_2H_6O_2$
(b) Peroxodisulfuric acid, a compound used to make bleaching agents, molecular formula $H_2S_2O_8$

•• **2.86** Give the name and formula of the compound formed from the following elements: (a) sodium and nitrogen; (b) oxygen and strontium; (c) aluminum and chlorine.

•• **2.87** Give the name and formula of the compound formed from the following elements: (a) cesium and bromine; (b) sulfur and barium; (c) calcium and fluorine.

•• **2.88** Give the name and formula of the compound formed from the following elements:
(a) $_{12}$L and $_9$M   (b) $_{30}$L and $_{16}$M   (c) $_{17}$L and $_{38}$M

•• **2.89** Give the name and formula of the compound formed from the following elements:
(a) $_{37}$Q and $_{35}$R   (b) $_8$Q and $_{13}$R   (c) $_{20}$Q and $_{53}$R

• **2.90** Give the systematic names for the formulas or the formulas for the names: (a) tin(IV) chloride; (b) FeBr₃; (c) cuprous bromide; (d) Mn_2O_3.

• **2.91** Give the systematic names for the formulas or the formulas for the names: (a) Na_2HPO_4; (b) potassium carbonate dihydrate; (c) $NaNO_2$; (d) ammonium perchlorate.

• **2.92** Give the systematic names for the formulas or the formulas for the names: (a) CoO; (b) mercury(I) chloride; (c) $Pb(C_2H_3O_2)_2 \cdot 3H_2O$; (d) chromic oxide.

• **2.93** Give the systematic names for the formulas or the formulas for the names: (a) $Sn(SO_3)_2$; (b) potassium dichromate; (c) $FeCO_3$; (d) copper(II) nitrate.

• **2.94** Correct each of the following formulas:
(a) Barium oxide is BaO_2.
(b) Iron(II) nitrate is $Fe(NO_3)_3$.
(c) Magnesium sulfide is $MnSO_3$.

• **2.95** Correct each of the following names:
(a) CuI is cobalt(II) iodide.
(b) $Fe(HSO_4)_3$ is iron(II) sulfate.
(c) $MgCr_2O_7$ is magnesium dichromium heptaoxide.

• **2.96** Give the name and formula for the acid derived from each of the following anions:
(a) hydrogen sulfate   (b) IO_3^- (c) cyanide   (d) HS^-

• **2.97** Give the name and formula for the acid derived from each of the following anions:
(a) perchlorate   (b) NO_3^- (c) bromite   (d) F^-

• **2.98** Many chemical names are similar at first glance. Give the formulas of the species in each set: (a) ammonium ion and ammonia; (b) magnesium sulfide, magnesium sulfite, and magnesium sulfate; (c) hydrochloric acid, chloric acid, and chlorous acid; (d) cuprous bromide and cupric bromide.

• **2.99** Give the formulas of the compounds in each set: (a) lead(II) oxide and lead(IV) oxide; (b) lithium nitride, lithium nitrite, and lithium nitrate; (c) strontium hydride and strontium hydroxide; (d) magnesium oxide and manganese(II) oxide.

• **2.100** Give the name and formula of the compound whose molecules consist of two sulfur atoms and four fluorine atoms.

• **2.101** Give the name and formula of the compound whose molecules consist of two chlorine atoms and one oxygen atom.

• **2.102** Correct the name to match the formula of the following compounds: (a) calcium(II) dichloride, $CaCl_2$; (b) copper(II) oxide, Cu_2O; (c) stannous tetrafluoride, SnF_4; (d) hydrogen chloride acid, HCl.

• **2.103** Correct the formula to match the name of the following compounds: (a) iron(III) oxide, Fe_3O_4; (b) chloric acid, HCl; (c) mercuric oxide, Hg_2O; (d) dichlorine heptaoxide, Cl_2O_6.

•• **2.104** Give the number of atoms of the specified element in a formula unit of each of the following compounds, and calculate the molecular (formula) mass:
(a) Oxygen in aluminum sulfate, $Al_2(SO_4)_3$
(b) Hydrogen in ammonium hydrogen phosphate, $(NH_4)_2HPO_4$
(c) Oxygen in the mineral azurite, $Cu_3(OH)_2(CO_3)_2$

2.105 Give the number of atoms of the specified element in a formula unit of each of the following compounds, and calculate the molecular (formula) mass:
(a) Hydrogen in ammonium benzoate, $C_6H_5COONH_4$
(b) Nitrogen in hydrazinium sulfate, $N_2H_6SO_4$
(c) Oxygen in the mineral leadhillite, $Pb_4SO_4(CO_3)_2(OH)_2$

2.106 Write the formula of each compound, and determine its molecular (formula) mass: (a) ammonium sulfate; (b) sodium dihydrogen phosphate; (c) potassium bicarbonate.

2.107 Write the formula of each compound, and determine its molecular (formula) mass: (a) sodium dichromate; (b) ammonium perchlorate; (c) magnesium nitrite trihydrate.

2.108 Calculate the molecular (formula) mass of each compound: (a) dinitrogen pentaoxide; (b) lead(II) nitrate; (c) calcium peroxide.

2.109 Calculate the molecular (formula) mass of each compound: (a) iron(II) acetate tetrahydrate; (b) sulfur tetrachloride; (c) potassium permanganate.

2.110 Give the formula, name, and molecular mass of the following molecules:

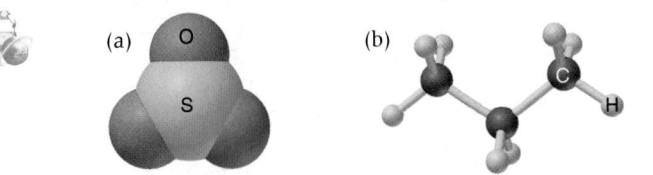

(a)   (b)

2.111 Give the formula, name, and molecular mass of the following molecules:

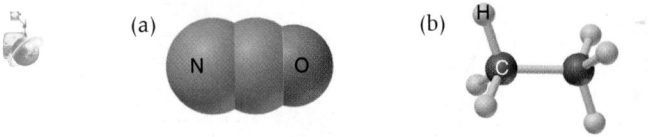

(a)   (b)

2.112 Give the name, empirical formula, and molecular mass of the molecule depicted in Figure P2.112.

2.113 Give the name, empirical formula, and molecular mass of the molecule depicted in Figure P2.113.

Figure P2.112   Figure P2.113

● **Problems in Context**

2.114 Before the use of systematic names, many compounds had common names. Give the systematic name for each of the following: (a) blue vitriol, $CuSO_4\cdot5H_2O$; (b) slaked lime, $Ca(OH)_2$; (c) oil of vitriol, H_2SO_4; (d) washing soda, Na_2CO_3; (e) muriatic acid, HCl; (f) Epsom salt, $MgSO_4\cdot7H_2O$; (g) chalk, $CaCO_3$; (h) dry ice, CO_2; (i) baking soda, $NaHCO_3$; (j) lye, NaOH.

2.115 Each circle contains a representation of a binary compound. Determine its name, formula, and molecular (formula) mass.

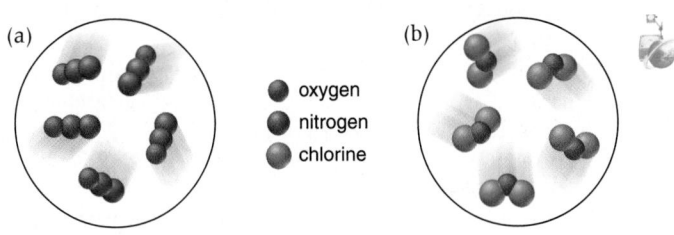

(a)   (b)

● oxygen
● nitrogen
● chlorine

Mixtures: Classification and Separation

● **Concept Review Questions**

2.116 In what main way is separating the components of a mixture different from separating the components of a compound?

2.117 What is the difference between a homogeneous and a heterogeneous mixture?

2.118 Is a solution a homogeneous or a heterogeneous mixture? Give an example of an aqueous solution.

● **Skill-Building Exercises** (grouped in similar pairs)

2.119 Classify each of the following as a compound, a homogeneous mixture, or a heterogeneous mixture: (a) distilled water; (b) gasoline; (c) beach sand; (d) wine; (e) air.

2.120 Classify each of the following as a compound, a homogeneous mixture, or a heterogeneous mixture: (a) orange juice; (b) vegetable soup; (c) cement; (d) calcium sulfate; (e) tea.

2.121 Name the technique(s) and briefly describe the procedure you would use to separate each of the following mixtures into two components: (a) table salt and pepper; (b) table sugar and sand; (c) drinking water contaminated with fuel oil; (d) vegetable oil and vinegar.

2.122 Name the technique(s) and briefly describe the procedure you would use to separate each of the following mixtures into two components: (a) crushed ice and crushed glass; (b) table sugar dissolved in ethanol; (c) iron and sulfur; (d) two pigments (chlorophyll a and chlorophyll b) from spinach leaves.

● **Problems in Context**

2.123 Which separation method is operating in each of the following procedures: (a) pouring a mixture of cooked pasta and boiling water into a colander; (b) removing colored impurities from raw sugar to make refined sugar; (c) preparing coffee by pouring hot water through ground coffee beans?

2.124 A quality-control laboratory analyzes a product mixture using gas-liquid chromatography. The separation of components is more than adequate, but the process takes too long. Suggest two ways, other than changing the stationary phase, to shorten the analysis time.

Comprehensive Problems

2.125 Helium is the lightest noble gas and the second most abundant element (after hydrogen) in the universe.
(a) The radius of a helium atom is 3.1×10^{-11} m; the radius of its nucleus is 2.5×10^{-15} m. What fraction of the spherical atomic volume is occupied by the nucleus (V of a sphere $= \frac{4}{3}\pi r^3$)?
(b) The mass of a helium-4 atom is 6.64648×10^{-24} g, and each of its two electrons has a mass of 9.10939×10^{-28} g. What fraction of this atom's mass is contributed by its nucleus?

•• **2.126** From the following ions and their radii (in pm), choose a pair that gives the strongest ionic bonding and a pair that gives the weakest: Mg^{2+} 72; K^+ 138; Rb^+ 152; Ba^{2+} 135; Cl^- 181; O^{2-} 140; I^- 220.

•• **2.127** Prior to 1961, the atomic mass standard was defined as $\frac{1}{16}$ of the mass of ^{16}O. Based on that standard: (a) What was the mass of carbon-12, given the modern atomic mass of oxygen is 15.9994 amu? (b) What was the mass of potassium-39, given its modern isotopic mass is 38.9637 amu?

• **2.128** Give the molecular mass of each compound depicted below, and provide a correct name for any that are named incorrectly.

•• **2.129** Transition metals, located in the center of the periodic table, have many essential uses as elements and form many important compounds as well. Calculate the molecular mass of the following transition metal compounds:
(a) $[Co(NH_3)_6]Cl_3$ (b) $[Pt(NH_3)_4BrCl]Cl_2$
(c) $K_4[V(CN)_6]$ (d) $[Ce(NH_3)_6][FeCl_4]_3$

••• **2.130** A rock is 5.0% by mass fayalite (Fe_2SiO_4), 7.0% by mass forsterite (Mg_2SiO_4), and the remainder silicon dioxide. What is the mass percent of each element in the rock?

•• **2.131** Polyatomic ions are named by patterns that apply to elements in a given group. Using the periodic table and Table 2.5, give the name of each of the following: (a) SeO_4^{2-}; (b) AsO_4^{3-}; (c) BrO_2^-; (d) $HSeO_4^-$; (e) TeO_3^{2-}.

•• **2.132** Ammonium dihydrogen phosphate, formed from the reaction of phosphoric acid with ammonia, is used as a crop fertilizer as well as a component of some fire extinguishers. (a) What are the mass percentages of N and P in the compound? (b) How much ammonia is incorporated into 100. g of compound?

•• **2.133** Nitrogen forms more oxides than any other element. The percents by mass of N in three different nitrogen oxides are (I) 46.69%; (II) 36.85%; (III) 25.94%. (a) Determine the empirical formula of each compound. (b) How many grams of oxygen per 1.00 g of nitrogen are in each compound?

••• **2.134** Boron has two naturally occurring isotopes, ^{10}B (19.9%) and ^{11}B (80.1%). Although the B_2 molecule does not exist naturally on Earth, it has been produced in the laboratory and been observed in stars. (a) How many different B_2 molecules are possible? (b) What are the masses and percent abundances of each?

•• **2.135** Dimercaprol ($HSCH_2CHSHCH_2OH$) was developed during World War I as an antidote to arsenic-based poison gas and is used today to treat heavy-metal poisoning. It binds the toxic element and carries it out of the body.
(a) If each molecule binds one arsenic (As) atom, how many atoms of As could be removed by 175 mg of dimercaprol?
(b) If one molecule binds one metal atom, calculate the mass % of each of the following metals in a metal-dimercaprol combination: mercury, thallium, chromium.

•• **2.136** Scenes A–I depict various types of matter on the atomic scale. Choose the correct scene(s) for each of the following:
(a) A mixture that fills its container
(b) A substance that cannot be broken down into simpler ones
(c) An element with a very high resistance to flow
(d) A homogeneous mixture
(e) An element that conforms to the walls of its container and displays a surface
(f) A gas consisting of diatomic particles
(g) A gas that can be broken down into simpler substances
(h) A substance with a 2/1 ratio of its component atoms
(i) Matter that can be separated into its component substances by physical means
(j) A heterogeneous mixture
(k) Matter that obeys the law of definite composition

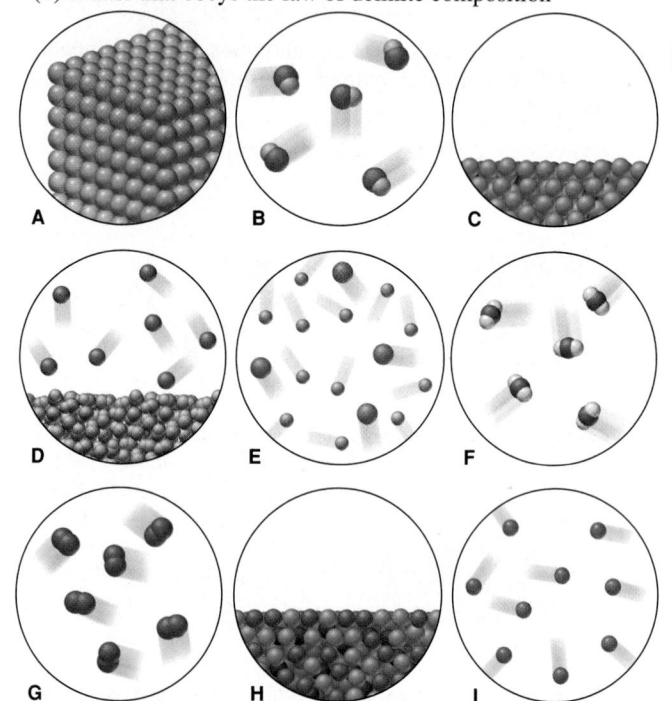

•• **2.137** The number of atoms in 1 dm^3 of aluminum is nearly the same as the number of atoms in 1 dm^3 of lead, but the densities of these metals are very different (see Table 1.5). Explain.

•• **2.138** You are working in the laboratory preparing sodium chloride. Consider the following results for three preparations of the compound:
Case 1: 39.34 g Na + 60.66 g Cl_2 $\longrightarrow$ 100.00 g NaCl
Case 2: 39.34 g Na + 70.00 g Cl_2 $\longrightarrow$
                   100.00 g NaCl + 9.34 g Cl_2
Case 3: 50.00 g Na + 50.00 g Cl_2 $\longrightarrow$
                   82.43 g NaCl + 17.57 g Na
Explain these results in terms of the laws of conservation of mass and definite composition.

•• **2.139** The seven most abundant ions in seawater make up more than 99% by mass of the dissolved compounds. They are listed in units of mg ion/kg seawater: chloride 18,980; sodium 10,560; sulfate 2650; magnesium 1270; calcium 400; potassium 380; hydrogen carbonate 140.
(a) What is the mass % of each ion in seawater?
(b) What percent of the total mass of ions is sodium ion?

(c) How does the total mass % of alkaline earth metal ions compare with the total mass % of alkali metal ions?

(d) Which makes up the larger mass fraction of dissolved components, anions or cations?

•• **2.140** The scenes below represent a mixture of two monatomic gases undergoing a reaction when heated. Which mass law(s) is (are) illustrated by this change?

••• **2.141** When barium (Ba) reacts with sulfur (S) to form barium sulfide (BaS), each Ba atom reacts with an S atom. If 2.50 cm³ of Ba reacts with 1.75 cm³ of S, are there enough Ba atoms to react with the S atoms (d of Ba = 3.51 g/cm³; d of S = 2.07 g/cm³)?

•• **2.142** Succinic acid *(below)* is an important metabolite in biological energy production. Give the molecular formula, empirical formula, and molecular mass of succinic acid, and calculate the mass percent of each element.

•• **2.143** Fluoride ion is poisonous in relatively low amounts: 0.2 g of F⁻ per 70 kg of body weight can cause death. Nevertheless, in order to prevent tooth decay, F⁻ ions are added to drinking water at a concentration of 1 mg of F⁻ ion per L of water. How many liters of fluoridated drinking water would a 70-kg person have to consume in one day to reach this toxic level? How many kilograms of sodium fluoride would be needed to treat a 8.50×10^7-gal reservoir?

•• **2.144** Which of the following models represent compounds having the same empirical formula? What is the molecular mass of this common empirical formula?

A   B   C   D   E

•• **2.145** Antimony has many uses, for example, in infrared devices and as part of an alloy in lead storage batteries. The element has two naturally occurring isotopes, one with mass 120.904 amu, the other with mass 122.904 amu. (a) Write the $_{Z}^{A}X$ notation for each isotope. (b) Use the atomic mass of antimony from the periodic table to calculate the natural abundance of each isotope.

•• **2.146** Dinitrogen monoxide (N₂O; nitrous oxide) is a greenhouse gas that enters the atmosphere principally from natural fertilizer breakdown. Some studies have shown that the isotope ratios of ¹⁵N to ¹⁴N and of ¹⁸O to ¹⁶O in N₂O depend on the source, which

can thus be determined by measuring the relative abundance of molecular masses in a sample of N₂O.

(a) What different molecular masses are possible for N₂O?

(b) The percent abundance of ¹⁴N is 99.6%, and that of ¹⁶O is 99.8%. Which molecular mass of N₂O is least common, and which is most common?

•• **2.147** Silver acetylide (AgC₂H) is a shock-sensitive explosive. The synthesis of an organic compound in the presence of silver salts leaves a residue whose mass spectrum shows an ion with *m/e* 132.9 but no other ions between *m/e* 130 and 135. Should the chemist be concerned that the residue may be explosive?

•• **2.148** Choose the box color(s) in the periodic table below that match(es) each of the following:

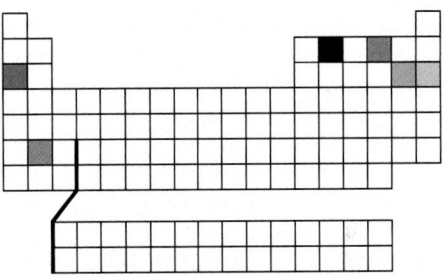

(a) Four elements that are nonmetals

(b) Two elements that are metals

(c) Three elements that are gases at room temperature

(d) Three elements that are solid at room temperature

(e) One pair of elements likely to form a covalent compound

(f) Another pair of elements likely to form a covalent compound

(g) One pair of elements likely to form an ionic compound with formula MX

(h) Another pair of elements likely to form an ionic compound with formula MX

(i) Two elements likely to form an ionic compound with formula M₂X

(j) Two elements likely to form an ionic compound with formula MX₂

(k) An element that forms no compounds

(l) A pair of elements whose compounds exhibit the law of multiple proportions

(m) Two elements that are building blocks in biomolecules

(n) Two elements that are macronutrients in organisms

•• **2.149** The two isotopes of potassium with significant abundance in nature are ³⁹K (isotopic mass 38.9637 amu, 93.258%) and ⁴¹K (isotopic mass 40.9618 amu, 6.730%). Fluorine has only one naturally occurring isotope, ¹⁹F (isotopic mass 18.9984 amu). Calculate the formula mass of potassium fluoride.

•• **2.150** Boron trifluoride is used as a catalyst in the synthesis of organic compounds. When this compound is analyzed by mass spectrometry (see Tools of the Laboratory, p. 55), several different 1+ ions form, including ions representing the whole molecule as well as molecular fragments formed by the loss of one, two, and three F atoms. Given that boron has two naturally occurring isotopes, ¹⁰B and ¹¹B, and fluorine has one, ¹⁹F, calculate the masses of all possible 1+ ions.

•• **2.151** Nitrogen monoxide (NO) is a bioactive molecule in blood. Low NO concentrations cause respiratory distress and the formation of blood clots. Doctors prescribe nitroglycerin,

$C_3H_5N_3O_9$, and isoamyl nitrate, $(CH_3)_2CHCH_2CH_2ONO_2$, to increase NO. If each compound releases one molecule of NO per atom of N, calculate the mass percent of NO in each medicine.

•••• **2.152** TNT (trinitrotoluene; *below*) is used as an explosive in construction. Calculate the mass of each element in 1.00 lb of TNT.

•••• **2.153** Carboxylic acids react with alcohols to form esters, which are found in all plants and animals. Some are responsible for the flavors and odors of fruits and flowers. What is the percent by mass of carbon in each of the following esters?

Name	Formula	Odor
Isoamyl isovalerate	$C_4H_9COOC_5H_{11}$	apple
Amyl butyrate	$C_3H_7COOC_5H_{11}$	apricot
Isoamyl acetate	$CH_3COOC_5H_{11}$	banana
Ethyl butyrate	$C_3H_7COOC_2H_5$	pineapple

•• **2.154** Nuclei differ in their stability, and some are so unstable that they undergo radioactive decay. The ratio of the number of neutrons to number of protons (N/Z) in a nucleus correlates with its stability. Calculate the N/Z ratio for (a) ^{144}Sm; (b) ^{56}Fe; (c) ^{20}Ne; (d) ^{107}Ag. (e) The radioactive isotope ^{238}U decays in a series of nuclear reactions that includes another uranium isotope, ^{234}U, and three lead isotopes, ^{214}Pb, ^{210}Pb, and ^{206}Pb. How many neutrons, protons, and electrons are in each of these five isotopes?

•• **2.155** The anticancer drug Platinol (Cisplatin), $Pt(NH_3)_2Cl_2$, reacts with the cancer cell's DNA and interferes with its growth. (a) What is the mass % of platinum (Pt) in Platinol? (b) If Pt costs \$19/g, how many grams of Platinol can be made for \$1.00 million (assume that the cost of Pt determines the cost of the drug)?

•• **2.156** Grignard reagents, which contain a C—Mg bond, have the general formula CH_3—$(CH_2)_x$—$MgBr$ and are essential in the synthesis of organic compounds. (a) Calculate the mass percent of Mg if $x = 0$. (b) Calculate the mass percent of Mg if $x = 5$. (c) Calculate the value of x if the mass percent of Mg is 16.5%.

•• **2.157** In a sample of any metal, spherical atoms pack closely together, but the space between them means that the density of the sample is less than that of the atoms themselves. Iridium (Ir) is one of the densest elements: 22.56 g/cm^3. The atomic mass of Ir is 192.22 amu, and the mass of the nucleus is 192.18 amu. Determine the density (in g/cm^3) of (a) an Ir atom and (b) an Ir nucleus. (c) How many Ir atoms placed in a row would extend 1.00 cm [radius of Ir atom = 1.36 Å; radius of Ir nucleus = 1.5 femtometers (fm); V of a sphere = $\frac{4}{3}\pi r^3$]?

•• **2.158** Which of the following steps in an overall process involve(s) a physical change and which involve(s) a chemical change?

Mass and Number. During any chemical reaction, such as this spectacular one between sodium and bromine to form sodium bromide, total mass is constant but individual masses change. As you'll see in this chapter, weighing provides a means for knowing not only the *mass* of each substance involved in a reaction but also the *number* of atoms, ions, or molecules undergoing the change.

Stoichiometry of Formulas and Equations

3.1 The Mole
Defining the Mole
Molar Mass
Mole-Mass-Number Conversions
Mass Percent

3.2 Determining the Formula of an Unknown Compound
Empirical Formulas
Molecular Formulas
Formulas and Structures

3.3 Writing and Balancing Chemical Equations

3.4 Calculating Amounts of Reactant and Product
Molar Ratios from Balanced Equations
Reaction Sequences
Limiting Reactants
Reaction Yields

3.5 Fundamentals of Solution Stoichiometry
Molarity
Solution Mole-Mass-Number Conversions
Preparation of Molar Solutions
Reactions in Solution

Concepts & Skills to Review
before you study this chapter

- isotopes and atomic mass (Section 2.5)
- names and formulas of compounds (Section 2.8)
- molecular mass of a compound (Section 2.8)
- empirical and molecular formulas (Section 2.8)
- mass laws in chemical reactions (Section 2.2)

 Literature Jensen, William B. "The Origin of the Mole Concept," *J. Chem. Educ.* **2004**, *81*, 1409.

Literature Gorin, George. "Mole, Mole per Liter, and Molar: A Primer on SI and Related Units for Chemistry Students," *J. Chem. Educ.* **2003**, *80*, 103.

Literature Yin, Mali; Ochs, Raymond S. "The Mole, the Periodic Table, and Quantum Numbers: An Introductory Trio," *J. Chem. Educ.* **2001**, *78*, 1345.

ARIS Presentation Center, Art Library

● **Imagine a Mole of . . .** A mole of any ordinary object is a staggering amount: a mole of periods (.) lined up side by side would equal the radius of our galaxy; a mole of marbles stacked tightly together would cover the United States 70 miles deep. However, atoms and molecules are not ordinary objects: a mole of water molecules (about 18 mL) can be swallowed in one gulp!

C hemistry is a practical science. Just imagine how useful it could be to determine the formula of a compound from the masses of its elements or to predict the amounts of substances consumed and produced in a reaction. Suppose you are a polymer chemist preparing a new plastic: how much of this new material will a given polymerization reaction yield? Or suppose you're a chemical engineer studying rocket engine thrust: what amount of exhaust gases will a test of this fuel mixture produce? Perhaps you are on a team of environmental chemists examining coal samples: what quantity of air pollutants will this sample release when burned? Or, maybe you're a biomedical researcher who has extracted a new cancer-preventing substance from a tropical plant: what is its formula, and what quantity of metabolic products will establish a safe dosage level? You can answer countless questions like these with a knowledge of **stoichiometry** (pronounced "stoy-key-AHM-uh-tree"; from the Greek *stoicheion*, "element or part," and *metron*, "measure"), the study of the quantitative aspects of chemical formulas and reactions.

IN THIS CHAPTER . . . We relate the mass of a substance to the number of chemical entities comprising it (atoms, ions, molecules, or formula units). We convert the results of mass analysis into a chemical formula and distinguish the types of chemical formulas and their relation to molecular structures. Reading, writing, and thinking in the language of chemical equations are applied to finding the amounts of reactants and products involved in a reaction. These methods are also applied to reactions that occur in solution.

3.1 THE MOLE

All the ideas and skills discussed in this chapter depend on an understanding of the *mole* concept, so let's begin there. In daily life, we typically measure things out by counting or by weighing, with the choice based on convenience. It is more convenient to weigh beans or rice than to count individual pieces, and it is more convenient to count eggs or pencils than to weigh them. To measure such things, we use mass units (a kilogram of rice) or counting units (a dozen pencils). Similarly, daily life in the laboratory involves measuring substances to prepare a solution or "run" a reaction. However, an obvious problem arises when we try to do this. The atoms, ions, molecules, or formula units are the entities that react with one another, so we would like to know the numbers of them that we mix together. But, how can we possibly count entities that are so small? To do this, chemists have devised a unit called the mole to *count chemical entities by weighing them.*

Defining the Mole

The **mole** (abbreviated **mol**) is the SI unit for amount of substance. It is defined as *the amount of a substance that contains the same number of entities as there are atoms in exactly 12 g of carbon-12.* This number is called **Avogadro's number,** in honor of the 19th-century Italian physicist Amedeo Avogadro, and as you can tell from the definition, it is enormous: ●

> One mole (1 mol) contains 6.022×10^{23} entities (to four significant figures) (3.1)

Thus,

1 mol of carbon-12	contains	6.022×10^{23} carbon-12 atoms
1 mol of H_2O	contains	6.022×10^{23} H_2O molecules
1 mol of NaCl	contains	6.022×10^{23} NaCl formula units

However, the mole is not just a counting unit like the dozen, which specifies only the *number* of objects. The definition of the mole specifies the *number* of objects in a fixed *mass* of substance. Therefore, *1 mole of a substance represents a fixed number of chemical entities **and** has a fixed mass.* To see why this is

Figure 3.1 Counting objects of fixed relative mass. A, If marbles had a fixed mass, we could count them by weighing them. Each red marble weighs 7 g, and each yellow marble weighs 4 g, so 84 g of red marbles and 48 g of yellow marbles each contains 12 marbles. Equal numbers of the two types of marbles always have a 7/4 mass ratio of red/yellow marbles. **B,** Because atoms of a substance have a fixed mass, we can weigh the substance to count the atoms; 55.85 g of Fe (left pan) and 32.07 g of S (right pan) each contains 6.022×10^{23} atoms (1 mol of atoms). Any two samples of Fe and S that contain equal numbers of atoms have a 55.85/32.07 mass ratio of Fe/S.

ARIS Presentation Center,
Photo Library

important, consider the marbles in Figure 3.1A, which we'll use as an analogy for atoms. Suppose you have large groups of red marbles and yellow marbles; each red marble weighs 7 g and each yellow marble weighs 4 g. Right away you know that there are 12 marbles in 84 g of red marbles or in 48 g of yellow marbles. Moreover, because one red marble weighs $\frac{7}{4}$ as much as one yellow marble, any given *number* of red and of yellow marbles always has this 7/4 *mass* ratio. By the same token, any given *mass* of red and of yellow marbles always has a 4/7 *number* ratio. For example, 280 g of red marbles contains 40 marbles, and 280 g of yellow marbles contains 70 marbles. As you can see, the fixed masses of the marbles allow you to count marbles by weighing them.

Atoms have fixed masses also, and the mole gives us a practical way to determine the number of atoms, molecules, or formula units in a sample by weighing it. Let's focus on elements first and recall a key point from Chapter 2: the atomic mass of an element (which appears on the periodic table) is the weighted average of the masses of its naturally occurring isotopes. For purposes of weighing, all atoms of an element are considered to have this atomic mass. That is, all iron (Fe) atoms have an atomic mass of 55.85 amu, all sulfur (S) atoms have an atomic mass of 32.07 amu, and so forth.

The central relationship between the mass of one atom and the mass of 1 mole of those atoms is that *the atomic mass of an element expressed in **amu** is numerically the same as the mass of 1 mole of atoms of the element expressed in **grams**.* You can see this from the definition of the mole, which referred to the number of atoms in "**12** g of carbon-**12**." Thus,

Demo Solomon, Sally; Hur, Chinhyu. "Measuring Avogadro's Number on the Overhead Projector," *J. Chem. Educ.* **1993**, *70*, 252.

Literature Svensson, Christer. "How Many Digits Should We Use in Formula or Molar Mass Calculations?" *J. Chem. Educ.* **2004**, *81*, 827.

1 Fe atom	has a mass of	55.85 amu	and	1 mol of Fe atoms	has a mass of	55.85 g
1 S atom	has a mass of	32.07 amu	and	1 mol of S atoms	has a mass of	32.07 g
1 O atom	has a mass of	16.00 amu	and	1 mol of O atoms	has a mass of	16.00 g
1 O_2 molecule	has a mass of	32.00 amu	and	1 mol of O_2 molecules	has a mass of	32.00 g

Moreover, because of their fixed atomic masses, we know that 55.85 g of Fe atoms and 32.07 g of S atoms each contains 6.022×10^{23} atoms. As with marbles of fixed mass, one Fe atom weighs $\frac{55.85}{32.07}$ as much as one S atom, and 1 mol of Fe atoms weighs $\frac{55.85}{32.07}$ as much as 1 mol of S atoms (Figure 3.1B).

A similar relationship holds for compounds: *the molecular mass (or formula mass) of a compound expressed in **amu** is numerically the same as the mass of 1 mole of the compound expressed in **grams**.* Thus, for example,

1 molecule of H_2O	has a mass of	18.02 amu	and	1 mol of H_2O (6.022×10^{23} molecules)	has a mass of	18.02 g
1 formula unit of NaCl	has a mass of	58.44 amu	and	1 mol of NaCl (6.022×10^{23} formula units)	has a mass of	58.44 g

Figure 3.2 One mole of some familiar substances. One mole of a substance is the amount that contains 6.022×10^{23} atoms, molecules, or formula units. From left to right: 1 mol (172.19 g) of writing chalk (calcium sulfate dihydrate), 1 mol (32.00 g) of gaseous O_2, 1 mol (63.55 g) of copper, and 1 mol (18.02 g) of liquid H_2O.

ARIS Presentation Center,
   Photo Library

ARIS Presentation Center,
   Tables Library

Literature de Sanabia, Josefina Arce. "Relative Atomic Mass and the Mole: A Concrete Analogy to Help Students Understand These Abstract Concepts," *J. Chem. Educ.* **1993**, *70*, 233.

To summarize the two key points about the usefulness of the mole concept:

- The mole maintains the *same mass relationship* between macroscopic samples as exists between individual chemical entities.
- The mole relates the *number* of chemical entities to the *mass* of a sample of those entities.

A grocer cannot obtain 1 dozen eggs by weighing them because eggs vary in mass. But a chemist *can* obtain 1 mol of copper atoms (6.022×10^{23} atoms) simply by weighing 63.55 g of copper. Figure 3.2 shows 1 mol of some familiar elements and compounds.

Molar Mass

The **molar mass** ($\mathcal{M}$) of a substance is the mass per mole of its entities (atoms, molecules, or formula units). Thus, molar mass has units of grams per mole (g/mol). Table 3.1 summarizes the meanings of mass units used in this text.

Table 3.1   Summary of Mass Terminology*

Term	Definition	Unit
Isotopic mass	Mass of an isotope of an element	amu
Atomic mass (also called atomic weight)	Average of the masses of the naturally occurring isotopes of an element weighted according to their abundance	amu
Molecular (or formula) mass (also called molecular weight)	Sum of the atomic masses of the atoms (or ions) in a molecule (or formula unit)	amu
Molar mass ($\mathcal{M}$) (also called gram-molecular weight)	Mass of 1 mole of chemical entities (atoms, ions, molecules, formula units)	g/mol

*All terms based on the ^{12}C standard: 1 atomic mass unit $= \frac{1}{12}$ mass of one ^{12}C atom.

The periodic table is indispensable for calculating the molar mass of a substance. Here's how the calculations are done:

1. **Elements.** You find the molar mass of an element simply by looking up its atomic mass in the periodic table and then noting whether the element occurs naturally as individual atoms or as molecules.
   - *Monatomic elements.* For elements that occur as individual atoms, the molar mass is the numerical value from the periodic table expressed in units of grams per mole.* Thus, the molar mass of neon is 20.18 g/mol, the molar mass of iron is 55.85 g/mol, and the molar mass of gold is 197.0 g/mol.
   - *Molecular elements.* For elements that occur as molecules, you must know the molecular formula to determine the molar mass. For example, oxygen exists normally in air as diatomic molecules, so the molar mass of O_2 molecules is twice that of O atoms:

   Molar mass ($\mathcal{M}$) of $O_2 = 2 \times \mathcal{M}$ of $O = 2 \times 16.00$ g/mol $= 32.00$ g/mol

   The most common form of sulfur exists as octatomic molecules, S_8:

   $\mathcal{M}$ of $S_8 = 8 \times \mathcal{M}$ of $S = 8 \times 32.07$ g/mol $= 256.6$ g/mol

---

*The mass value in the periodic table has no units because it is a *relative* atomic mass, given by the atomic mass (in amu) divided by 1 amu ($\frac{1}{12}$ mass of one ^{12}C atom in amu):

$$\text{Relative atomic mass} = \frac{\text{atomic mass (amu)}}{\frac{1}{12} \text{ mass of } ^{12}C \text{ (amu)}}$$

Therefore, you use the same number for the atomic mass (weighted average mass of one atom in amu) and the molar mass (mass of 1 mole of atoms in grams).

2. **Compounds.** *The molar mass of a compound is the sum of the molar masses of the atoms of the elements in the formula.* For example, the formula of sulfur dioxide (SO_2) tells us that 1 mol of SO_2 molecules contains 1 mol of S atoms and 2 mol of O atoms:

$$\mathcal{M} \text{ of } SO_2 = \mathcal{M} \text{ of } S + (2 \times \mathcal{M} \text{ of } O)$$
$$= 32.07 \text{ g/mol} + (2 \times 16.00 \text{ g/mol})$$
$$= 64.07 \text{ g/mol}$$

Similarly, for ionic compounds, such as potassium sulfide (K_2S), we have

$$\mathcal{M} \text{ of } K_2S = (2 \times \mathcal{M} \text{ of } K) + \mathcal{M} \text{ of } S$$
$$= (2 \times 39.10 \text{ g/mol}) + 32.07 \text{ g/mol}$$
$$= 110.27 \text{ g/mol}$$

A key point to note is that *the subscripts in a formula refer to individual atoms (or ions), as well as to moles of atoms (or ions).* Table 3.2 presents this idea for glucose, $C_6H_{12}O_6$ (see margin), the essential sugar in energy metabolism.

ARIS Presentation Center, Art Library

Table 3.2   Information Contained in the Chemical Formula of Glucose, $C_6H_{12}O_6$ ($\mathcal{M}$ = 180.16 g/mol)

	Carbon (C)	Hydrogen (H)	Oxygen (O)
Atoms/molecule of compound	6 atoms	12 atoms	6 atoms
Moles of atoms/mole of compound	6 mol of atoms	12 mol of atoms	6 mol of atoms
Atoms/mole of compound	$6(6.022 \times 10^{23})$ atoms	$12(6.022 \times 10^{23})$ atoms	$6(6.022 \times 10^{23})$ atoms
Mass/molecule of compound	6(12.01 amu) = 72.06 amu	12(1.008 amu) = 12.10 amu	6(16.00 amu) = 96.00 amu
Mass/mole of compound	72.06 g	12.10 g	96.00 g

Interconverting Moles, Mass, and Number of Chemical Entities

Transparency/ARIS Presentation Center, Tables Library

One of the reasons the mole is such a convenient unit for laboratory work is that it allows you to calculate the mass or number of entities of a substance in a sample if you know the amount or number of moles of the substance. Conversely, if you know the mass (or number of entities) of a substance, you can calculate the number of moles.

The molar mass, which expresses the equivalent relationship between 1 mole of a substance and its mass in grams, can be used as a conversion factor. We multiply by the molar mass of an element or compound ($\mathcal{M}$, in g/mol) to convert a given amount (in moles) to mass (in grams):

$$\text{Mass (g)} = \text{no. of moles} \times \frac{\text{no. of grams}}{1 \text{ mol}} \tag{3.2}$$

Or, we divide by the molar mass (multiply by $1/\mathcal{M}$) to convert a given mass (in grams) to amount (in moles):

$$\text{No. of moles} = \text{mass (g)} \times \frac{1 \text{ mol}}{\text{no. of grams}} \tag{3.3}$$

In a similar way, we use Avogadro's number, which expresses the equivalent relationship between 1 mole of a substance and the number of entities it contains, as a conversion factor. We multiply by Avogadro's number to convert amount of substance (in moles) to the number of entities (atoms, molecules, or formula units):

$$\text{No. of entities} = \text{no. of moles} \times \frac{6.022 \times 10^{23} \text{ entities}}{1 \text{ mol}} \tag{3.4}$$

Or, we divide by Avogadro's number to do the reverse:

$$\text{No. of moles} = \text{no. of entities} \times \frac{1 \text{ mol}}{6.022 \times 10^{23} \text{ entities}} \tag{3.5}$$

Literature Sarikaya, Mustafa. "The Application of an Activity Relating to the Determination of Avogadro's Number in a Class of First-Year Science Students," *Chem. Educator* 9 **2004**, *1*, 17–19.

Literature Bindel, Thomas H. "Crystal Models Made from Clear Plastic Boxes and Their Use in Determining Avogadro's Number," *J. Chem. Educ.* **2002**, *79*, 468.

Converting Moles of Elements For problems involving mass-mole-number relationships of elements, keep these points in mind:

- To convert between amount (mol) and mass (g), use the molar mass ($\mathcal{M}$ in g/mol).
- To convert between amount (mol) and number of entities, use Avogadro's number ($6.022{\times}10^{23}$ entities/mol). For elements that occur as molecules, use the molecular formula to find atoms/mol.
- Mass and number of entities relate directly to number of moles, *not* to each other. Therefore, to convert between number of entities and mass, *first convert to number of moles*. For example, to find the number of atoms in a given mass,

$$\text{No. of atoms} = \text{mass (g)} \times \frac{1 \text{ mol}}{\text{no. of grams}} \times \frac{6.022{\times}10^{23} \text{ atoms}}{1 \text{ mol}}$$

These relationships are summarized in Figure 3.3 and demonstrated in Sample Problems 3.1 and 3.2.

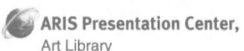

Figure 3.3 **Summary of the mass-mole-number relationships for elements.** The amount (mol) of an element is related to its mass (g) through the molar mass ($\mathcal{M}$ in g/mol) and to its number of atoms through Avogadro's number ($6.022{\times}10^{23}$ atoms/mol). For elements that occur as molecules, Avogadro's number gives *molecules* per mole.

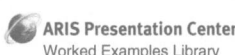

Amount (mol) of Ag
multiply by $\mathcal{M}$ of Ag (107.9 g/mol)
Mass (g) of Ag

SAMPLE PROBLEM 3.1   Calculating the Mass in a Given Number of Moles of an Element

PROBLEM Silver (Ag) is used in jewelry and tableware but no longer in U.S. coins. How many grams of Ag are in 0.0342 mol of Ag?

PLAN We know the number of moles of Ag (0.0342 mol) and have to find the mass (in g). To convert *moles* of Ag to *grams* of Ag, we multiply by the *molar mass* of Ag, which we find in the periodic table (see the roadmap).

SOLUTION Converting from moles of Ag to grams:

$$\text{Mass (g) of Ag} = 0.0342 \text{ mol Ag} \times \frac{107.9 \text{ g Ag}}{1 \text{ mol Ag}} = 3.69 \text{ g Ag}$$

CHECK We rounded the mass to three significant figures because the number of moles has three. The units are correct. About 0.03 mol × 100 g/mol gives 3 g; the small mass makes sense because 0.0342 is a small fraction of a mole.

FOLLOW-UP PROBLEM 3.1 Graphite is the crystalline form of carbon used in "lead" pencils. How many moles of carbon are in 315 mg of graphite?

SAMPLE PROBLEM 3.2 Calculating Number of Atoms in a Given Mass of an Element

PROBLEM Iron (Fe), the main component of steel, is the most important metal in industrial society. How many Fe atoms are in 95.8 g of Fe?

PLAN We know the grams of Fe (95.8 g) and need the number of Fe atoms. We cannot convert directly from grams to atoms, so we first convert to moles by dividing grams of Fe by its molar mass. (This is the reverse of the step in Sample Problem 3.1.) Then, we multiply number of moles by Avogadro's number to find number of atoms (see the roadmap).

SOLUTION Converting from grams of Fe to moles:

$$\text{Moles of Fe} = 95.8 \text{ g Fe} \times \frac{1 \text{ mol Fe}}{55.85 \text{ g Fe}} = 1.72 \text{ mol Fe}$$

Converting from moles of Fe to number of atoms:

$$\text{No. of Fe atoms} = 1.72 \text{ mol Fe} \times \frac{6.022 \times 10^{23} \text{ atoms Fe}}{1 \text{ mol Fe}}$$
$$= 10.4 \times 10^{23} \text{ atoms Fe} = 1.04 \times 10^{24} \text{ atoms Fe}$$

CHECK When we approximate the mass of Fe and the molar mass of Fe, we have $\sim 100 \text{ g}/(\sim 50 \text{ g/mol}) = 2$ mol. Therefore, the number of atoms should be about twice Avogadro's number: $2(6 \times 10^{23}) = 1.2 \times 10^{24}$.

FOLLOW-UP PROBLEM 3.2 Manganese (Mn) is a transition element essential for the growth of bones. What is the mass in grams of 3.22×10^{20} Mn atoms, the number found in 1 kg of bone?

Converting Moles of Compounds Solving mass-mole-number problems involving compounds requires a very similar approach to the one for elements. We need the chemical formula to find the molar mass and to determine the moles of a given element in the compound. These relationships are shown in Figure 3.4, and an example is worked through in Sample Problem 3.3.

Figure 3.4 Summary of the mass-mole-number relationships for compounds. Moles of a compound are related to grams of the compound through the molar mass ($\mathcal{M}$ in g/mol) and to the number of molecules (or formula units) through Avogadro's number (6.022×10^{23} molecules/mol). To find the number of molecules (or formula units) in a given mass, or vice versa, convert the information to moles first. With the chemical formula, you can calculate mass-mole-number information about each component element.

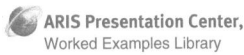

Mass (g) of $(NH_4)_2CO_3$

divide by $\mathcal{M}$ (g/mol)

Amount (mol) of $(NH_4)_2CO_3$

multiply by 6.022×10^{23}
formula units/mol

Number of $(NH_4)_2CO_3$ formula units

SAMPLE PROBLEM 3.3 Calculating the Moles and Number of Formula Units in a Given Mass of a Compound

PROBLEM Ammonium carbonate is a white solid that decomposes with warming. Among its many uses, it is a component of baking powder, fire extinguishers, and smelling salts. How many formula units are in 41.6 g of ammonium carbonate?

PLAN We know the mass of compound (41.6 g) and need to find the number of formula units. As we saw in Sample Problem 3.2, to convert grams to number of entities, we have to find number of moles first, so we must divide the grams by the molar mass ($\mathcal{M}$). For this, we need $\mathcal{M}$, so we determine the formula (see Table 2.5) and take the sum of the elements' molar masses. Once we have the number of moles, we multiply by Avogadro's number to find the number of formula units.

SOLUTION The formula is $(NH_4)_2CO_3$. Calculating molar mass:

$$\mathcal{M} = (2 \times \mathcal{M} \text{ of N}) + (8 \times \mathcal{M} \text{ of H}) + (1 \times \mathcal{M} \text{ of C}) + (3 \times \mathcal{M} \text{ of O})$$
$$= (2 \times 14.01 \text{ g/mol N}) + (8 \times 1.008 \text{ g/mol H}) + 12.01 \text{ g/mol C}$$
$$+ (3 \times 16.00 \text{ g/mol O})$$
$$= 96.09 \text{ g/mol } (NH_4)_2CO_3$$

Converting from grams to moles:

$$\text{Moles of } (NH_4)_2CO_3 = 41.6 \text{ g } (NH_4)_2CO_3 \times \frac{1 \text{ mol } (NH_4)_2CO_3}{96.09 \text{ g } (NH_4)_2CO_3}$$
$$= 0.433 \text{ mol } (NH_4)_2CO_3$$

Converting from moles to formula units:

$$\text{Formula units of } (NH_4)_2CO_3 = 0.433 \text{ mol } (NH_4)_2CO_3$$
$$\times \frac{6.022 \times 10^{23} \text{ formula units } (NH_4)_2CO_3}{1 \text{ mol } (NH_4)_2CO_3}$$
$$= 2.61 \times 10^{23} \text{ formula units } (NH_4)_2CO_3$$

CHECK The units are correct. Since the mass is less than half the molar mass ($\sim 42/96 < 0.5$), the number of formula units should be less than half Avogadro's number ($\sim 2.6 \times 10^{23}/6.0 \times 10^{23} < 0.5$).

COMMENT A *common mistake* is to forget the subscript 2 outside the parentheses in $(NH_4)_2CO_3$, which would give a much lower molar mass.

FOLLOW-UP PROBLEM 3.3 Tetraphosphorus decaoxide reacts with water to form phosphoric acid, a major industrial acid. In the laboratory, the oxide is used as a drying agent.
(a) What is the mass (in g) of 4.65×10^{22} molecules of tetraphosphorus decaoxide?
(b) How many P atoms are present in this sample?

Mass Percent from the Chemical Formula

Each element in a compound constitutes its own particular portion of the compound's mass. For an individual molecule (or formula unit), we use the molecular (or formula) mass and chemical formula to find the mass percent of any element X in the compound:

$$\text{Mass \% of element X} = \frac{\text{atoms of X in formula} \times \text{atomic mass of X(amu)}}{\text{molecular (or formula) mass of compound (amu)}} \times 100$$

Since the formula also tells the number of *moles* of each element in the compound, we use the molar mass to find the mass percent of each element on a mole basis:

$$\text{Mass \% of element X} = \frac{\text{moles of X in formula} \times \text{molar mass of X (g/mol)}}{\text{mass (g) of 1 mol of compound}} \times 100 \quad \textbf{(3.6)}$$

As always, the individual mass percents of the elements in the compound must add up to 100% (within rounding). As Sample Problem 3.4 demonstrates, an important practical use of mass percent is to determine the amount of an element in any size sample of a compound.

SAMPLE PROBLEM 3.4 Calculating Mass Percents and Masses of Elements in a Sample of a Compound

PROBLEM In mammals, lactose (milk sugar) is metabolized to glucose ($C_6H_{12}O_6$), the key nutrient for generating chemical potential energy.
(a) What is the mass percent of each element in glucose?
(b) How many grams of carbon are in 16.55 g of glucose?

(a) Determining the mass percent of each element

PLAN We know the relative numbers of moles of the elements in glucose from the formula (6 C, 12 H, 6 O). We multiply the number of moles of each element by its molar mass to find grams. Dividing each element's mass by the mass of 1 mol of glucose gives the mass fraction of each element, and multiplying each fraction by 100 gives the mass percent. The calculation steps for any element (X) are shown in the roadmap.

SOLUTION Calculating the mass of 1 mol of $C_6H_{12}O_6$:

$$\mathcal{M} = (6 \times \mathcal{M} \text{ of C}) + (12 \times \mathcal{M} \text{ of H}) + (6 \times \mathcal{M} \text{ of O})$$
$$= (6 \times 12.01 \text{ g/mol C}) + (12 \times 1.008 \text{ g/mol H}) + (6 \times 16.00 \text{ g/mol O})$$
$$= 180.16 \text{ g/mol } C_6H_{12}O_6$$

Converting moles of C to grams: There are 6 mol of C in 1 mol of glucose, so

$$\text{Mass (g) of C} = 6 \text{ mol C} \times \frac{12.01 \text{ g C}}{1 \text{ mol C}} = 72.06 \text{ g C}$$

Finding the mass fraction of C in glucose:

$$\text{Mass fraction of C} = \frac{\text{total mass C}}{\text{mass of 1 mol glucose}} = \frac{72.06 \text{ g C}}{180.16 \text{ g glucose}} = 0.4000$$

Finding the mass percent of C:

$$\text{Mass \% of C} = \text{mass fraction of C} \times 100 = 0.4000 \times 100 = 40.00 \text{ mass \% C}$$

Combining the steps for each of the other two elements in glucose:

$$\text{Mass \% of H} = \frac{\text{mol H} \times \mathcal{M} \text{ of H}}{\text{mass of 1 mol glucose}} \times 100 = \frac{12 \text{ mol H} \times \dfrac{1.008 \text{ g H}}{1 \text{ mol H}}}{180.16 \text{ g glucose}} \times 100$$
$$= 6.714 \text{ mass \% H}$$

$$\text{Mass \% of O} = \frac{\text{mol O} \times \mathcal{M} \text{ of O}}{\text{mass of 1 mol glucose}} \times 100 = \frac{6 \text{ mol O} \times \dfrac{16.00 \text{ g O}}{1 \text{ mol O}}}{180.16 \text{ g glucose}} \times 100$$
$$= 53.29 \text{ mass \% O}$$

CHECK The answers make sense: even though there are equal numbers of moles of O and C in the compound, the mass % of O is greater than the mass % of C because the molar mass of O is greater than the molar mass of C. The mass % of H is small because the molar mass of H is small. The total of the mass percents is 100.00%.

(b) Determining the mass (g) of carbon

PLAN To find the mass of C in the glucose sample, we multiply the mass of the sample by the mass fraction of C from part (a).

SOLUTION Finding the mass of C in a given mass of glucose (with units for mass fraction):

$$\text{Mass (g) of C} = \text{mass of glucose} \times \text{mass fraction of C} = 16.55 \text{ g glucose} \times \frac{0.4000 \text{ g C}}{1 \text{ g glucose}}$$
$$= 6.620 \text{ g C}$$

CHECK Rounding shows that the answer is "in the right ballpark": 16 g times less than 0.5 parts by mass should be less than 8 g.

COMMENT 1. A *more direct approach* to finding the mass of element in any mass of compound is similar to the approach we used in Sample Problem 2.2 (p. 45) and eliminates the need to calculate the mass fraction. Just multiply the given mass of compound by the ratio of the total mass of element to the mass of 1 mol of compound:

$$\text{Mass (g) of C} = 16.55 \text{ g glucose} \times \frac{72.06 \text{ g C}}{180.16 \text{ g glucose}} = 6.620 \text{ g C}$$

2. From here on, you should be able to determine the molar mass of a compound, so that calculation will no longer be shown.

FOLLOW-UP PROBLEM 3.4 Ammonium nitrate is a common fertilizer. Agronomists base the effectiveness of fertilizers on their nitrogen content.
(a) Calculate the mass percent of N in ammonium nitrate.
(b) How many grams of N are in 35.8 kg of ammonium nitrate?

Section Summary

A mole of substance is the amount that contains Avogadro's number (6.022×10^{23}) of chemical entities (atoms, molecules, or formula units). • The mass (in grams) of a mole has the same numerical value as the mass (in amu) of the entity. Thus, the mole allows us to count entities by weighing them. • Using the molar mass ($\mathcal{M}$, g/mol) of an element (or compound) and Avogadro's number as conversion factors, we can convert among amount (mol), mass (g), and number of entities. • The mass fraction of element X in a compound is used to find the mass of X in any amount of the compound.

3.2 DETERMINING THE FORMULA OF AN UNKNOWN COMPOUND

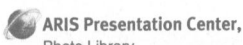 **Literature** Jensen, William B. "Empirical Formulas and the Solid State: A Proposal," *J. Chem. Educ.* **2004**, *81*, 1772.

Literature Thamburaj, P. K. "A Known-to-Unknown Approach to Teach about Empirical and Molecular Formulas," *J. Chem. Educ.* **2001**, *78*, 915.

ARIS Presentation Center, Photo Library

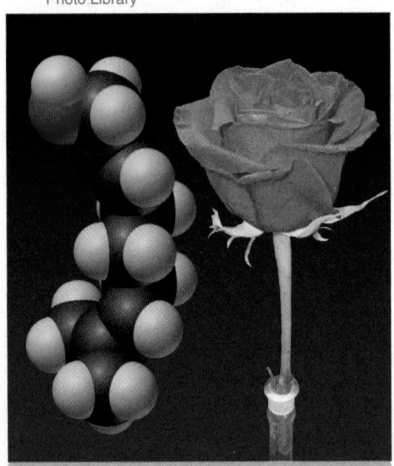

● **A Rose by Any Other Name ...**
Chemists studying natural substances obtained from animals and plants isolate compounds and determine their formulas. Geraniol ($C_{10}H_{18}O$), the main compound that gives a rose its odor, is used in many perfumes and cosmetics. Geraniol is also in citronella and lemongrass oils and is part of a larger compound in geranium leaves, from which its name is derived.

Organic chemistry application

In Sample Problem 3.4, we knew the formula and used it to find the mass percent (or mass fraction) of an element in a compound *and* the mass of the element in a given mass of the compound. In this section, we do the reverse: use the masses of elements in a compound to find its formula. We'll present the mass data in several ways and then look briefly at molecular structures.

Empirical Formulas

An analytical chemist investigating a compound decomposes it into simpler substances, finds the mass of each component element, converts these masses to numbers of moles, and then arithmetically converts the moles to whole-number (integer) subscripts. ● This procedure yields the empirical formula, the *simplest whole-number ratio* of moles of each element in the compound (see Section 2.8, p. 64). Let's see how to obtain the subscripts from the moles of each element.

Analysis of an unknown compound shows that the sample contains 0.21 mol of zinc, 0.14 mol of phosphorus, and 0.56 mol of oxygen. Because the subscripts in a formula represent individual atoms or moles of atoms, we write a preliminary formula that contains fractional subscripts: $Zn_{0.21}P_{0.14}O_{0.56}$. Next, we convert these fractional subscripts to whole numbers using one or two simple arithmetic steps (rounding when needed):

1. Divide each subscript by the smallest subscript:

$$Zn_{\frac{0.21}{0.14}}P_{\frac{0.14}{0.14}}O_{\frac{0.56}{0.14}} \longrightarrow Zn_{1.5}P_{1.0}O_{4.0}$$

This step alone often gives integer subscripts.

2. If any of the subscripts is still not an integer, multiply through by the *smallest integer* that will turn all subscripts into integers. Here, we multiply by 2, the smallest integer that will make 1.5 (the subscript for Zn) into an integer:

$$Zn_{(1.5 \times 2)}P_{(1.0 \times 2)}O_{(4.0 \times 2)} \longrightarrow Zn_{3.0}P_{2.0}O_{8.0}, \quad \text{or} \quad Zn_3P_2O_8$$

Notice that the *relative* number of moles has not changed because we multiplied *all* the subscripts by 2.

Always check that the subscripts are the smallest set of integers with the same ratio as the original numbers of moles; that is, 3/2/8 is *in the same ratio* as 0.21/0.14/0.56. A more conventional way to write this formula is $Zn_3(PO_4)_2$; the compound is zinc phosphate, a dental cement.

Sample Problems 3.5, 3.6, and 3.7 demonstrate how other types of compositional data are used to determine chemical formulas. In Sample Problem 3.5, the empirical formula is found from data given as grams of each element rather than as moles.

SAMPLE PROBLEM 3.5 Determining an Empirical Formula from Masses of Elements

ARIS Presentation Center, Worked Examples Library

PROBLEM Elemental analysis of a sample of an ionic compound showed 2.82 g of Na, 4.35 g of Cl, and 7.83 g of O. What is the empirical formula and name of the compound?
PLAN This problem is similar to the one on page 98, except that we are given element *masses*, so we must convert the masses into integer subscripts. We first divide each mass by the element's molar mass to find *number of moles*. Then we construct a preliminary formula and convert the numbers of moles to integers.
SOLUTION Finding moles of elements:

$$\text{Moles of Na} = 2.82 \text{ g Na} \times \frac{1 \text{ mol Na}}{22.99 \text{ g Na}} = 0.123 \text{ mol Na}$$

$$\text{Moles of Cl} = 4.35 \text{ g Cl} \times \frac{1 \text{ mol Cl}}{35.45 \text{ g Cl}} = 0.123 \text{ mol Cl}$$

$$\text{Moles of O} = 7.83 \text{ g O} \times \frac{1 \text{ mol O}}{16.00 \text{ g O}} = 0.489 \text{ mol O}$$

Constructing a preliminary formula: $Na_{0.123}Cl_{0.123}O_{0.489}$
Converting to integer subscripts (dividing all by the smallest subscript):

$$Na_{\frac{0.123}{0.123}} Cl_{\frac{0.123}{0.123}} O_{\frac{0.489}{0.123}} \longrightarrow Na_{1.00}Cl_{1.00}O_{3.98} \approx Na_1Cl_1O_4, \quad \text{or} \quad NaClO_4$$

We rounded the subscript of O from 3.98 to 4. The empirical formula is $NaClO_4$; the name is sodium perchlorate.

CHECK The moles seem correct because the masses of Na and Cl are slightly more than 0.1 of their molar masses. The mass of O is greatest and its molar mass is smallest, so it should have the greatest number of moles. The ratio of subscripts, 1/1/4, is the same as the ratio of moles, 0.123/0.123/0.489 (within rounding).

FOLLOW-UP PROBLEM 3.5 An unknown metal M reacts with sulfur to form a compound with the formula M_2S_3. If 3.12 g of M reacts with 2.88 g of S, what are the names of M and M_2S_3? (*Hint:* Determine the number of moles of S and use the formula to find the number of moles of M.)

Molecular Formulas

If we know the molar mass of a compound, we can use the empirical formula to obtain the molecular formula, the *actual* number of moles of each element in 1 mol of compound. In some cases, such as water (H_2O), ammonia (NH_3), and methane (CH_4), the empirical and molecular formulas are identical, but in many others the molecular formula is a *whole-number multiple* of the empirical formula. Hydrogen peroxide, for example, has the empirical formula HO and the molecular formula H_2O_2. Dividing the molar mass of H_2O_2 (34.02 g/mol) by the empirical formula mass (17.01 g/mol) gives the whole-number multiple:

$$\text{Whole-number multiple} = \frac{\text{molar mass (g/mol)}}{\text{empirical formula mass (g/mol)}}$$
$$= \frac{34.02 \text{ g/mol}}{17.01 \text{ g/mol}} = 2.000 = 2$$

Instead of giving compositional data in terms of masses of each element, analytical laboratories provide it as mass percents. From this, we determine the

Literature Gilbert, George L. "Percent Composition and Empirical Formula—A New View," *J. Chem. Educ.* **1998**, *75*, 851.

empirical formula by (1) assuming 100.0 g of compound, which allows us to express mass percent directly as mass, (2) converting the mass to number of moles, and (3) constructing the empirical formula. With the molar mass, we can also find the whole-number multiple and then the molecular formula.

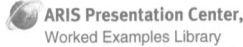

ARIS Presentation Center,
Worked Examples Library

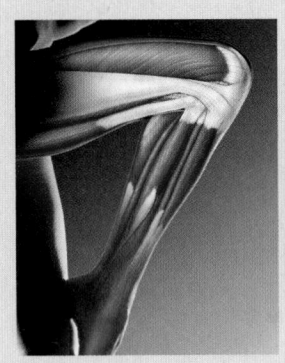

ARIS Presentation Center,
Photo Library

Biological application

SAMPLE PROBLEM 3.6 Determining a Molecular Formula from Elemental Analysis and Molar Mass

PROBLEM During excessive physical activity, lactic acid ($\mathcal{M}$ = 90.08 g/mol) forms in muscle tissue and is responsible for muscle soreness. Elemental analysis shows that this compound contains 40.0 mass % C, 6.71 mass % H, and 53.3 mass % O.
(a) Determine the empirical formula of lactic acid.
(b) Determine the molecular formula.

(a) Determining the empirical formula
PLAN We know the mass % of each element and must convert each to an integer subscript. Although the mass of lactic acid is not given, mass % is the same for any mass of compound, so we can assume 100.0 g of lactic acid and express each mass % directly as grams. Then, we convert grams to moles and construct the empirical formula as we did in Sample Problem 3.5.
SOLUTION Expressing mass % as grams, assuming 100.0 g of lactic acid:

$$\text{Mass (g) of C} = \frac{40.0 \text{ parts C by mass}}{100 \text{ parts by mass}} \times 100.0 \text{ g} = 40.0 \text{ g C}$$

Similarly, we have 6.71 g of H and 53.3 g of O.
Converting from grams of each element to moles:

$$\text{Moles of C} = \text{mass of C} \times \frac{1}{\mathcal{M} \text{ of C}} = 40.0 \text{ g C} \times \frac{1 \text{ mol C}}{12.01 \text{ g C}} = 3.33 \text{ mol C}$$

Similarly, we have 6.66 mol of H and 3.33 mol of O.
Constructing the preliminary formula: $C_{3.33}H_{6.66}O_{3.33}$
Converting to integer subscripts:

$$C_{\frac{3.33}{3.33}} H_{\frac{6.66}{3.33}} O_{\frac{3.33}{3.33}} \longrightarrow C_{1.00}H_{2.00}O_{1.00} \approx C_1H_2O_1, \text{ the empirical formula is } CH_2O$$

CHECK The numbers of moles seem correct: the masses of C and O are each slightly more than 3 times their molar masses (e.g., for C, 40 g/(12 g/mol) > 3 mol), and the mass of H is over 6 times its molar mass.

(b) Determining the molecular formula
PLAN The molecular formula subscripts are whole-number multiples of the empirical formula subscripts. To find this whole number, we divide the given molar mass (90.08 g/mol) by the empirical formula mass, which we find from the sum of the elements' molar masses. Then we multiply the whole number by each subscript in the empirical formula.
SOLUTION The empirical-formula molar mass is 30.03 g/mol. Finding the whole-number multiple:

$$\text{Whole-number multiple} = \frac{\mathcal{M} \text{ of lactic acid}}{\mathcal{M} \text{ of empirical formula}} = \frac{90.08 \text{ g/mol}}{30.03 \text{ g/mol}} = 3.000 = 3$$

Determining the molecular formula:

$$C_{(1\times3)}H_{(2\times3)}O_{(1\times3)} = C_3H_6O_3$$

CHECK The calculated molecular formula has the same ratio of moles of elements (3/6/3) as the empirical formula (1/2/1) and corresponds to the given molar mass:

$$\mathcal{M} \text{ of lactic acid} = (3 \times \mathcal{M} \text{ of C}) + (6 \times \mathcal{M} \text{ of H}) + (3 \times \mathcal{M} \text{ of O})$$
$$= (3 \times 12.01) + (6 \times 1.008) + (3 \times 16.00) = 90.08 \text{ g/mol}$$

FOLLOW-UP PROBLEM 3.6
One of the most widespread environmental carcinogens (cancer-causing agents) is benzo[a]pyrene ($\mathcal{M}$ = 252.30 g/mol). It is found in coal dust, in cigarette smoke, and even in charcoal-grilled meat. Analysis of this hydrocarbon shows 95.21 mass % C and 4.79 mass % H. What is the molecular formula of benzo[a]pyrene?

Figure 3.5 Combustion apparatus for determining formulas of organic compounds. A sample of compound that contains C and H (and perhaps other elements) is burned in a stream of O_2 gas inside a furnace. The CO_2 and H_2O formed are absorbed separately, while any other element oxides are carried through by the O_2 gas stream. H_2O is absorbed by $Mg(ClO_4)_2$; CO_2 is absorbed by NaOH on asbestos. The increases in mass of the absorbers are used to calculate the amounts (mol) of C and H in the sample.   ARIS Presentation Center, Art Library

Combustion Analysis of Organic Compounds Still another type of compositional data is obtained through **combustion analysis,** a method used to measure the amounts of carbon and hydrogen in a combustible organic compound. The unknown compound is burned in pure O_2 in an apparatus that consists of a combustion furnace and chambers containing compounds that absorb either H_2O or CO_2 (Figure 3.5). All the H in the unknown is converted to H_2O, which is absorbed in the first chamber, and all the C is converted to CO_2, which is absorbed in the second. By weighing the absorbers before and after combustion, we find the masses of CO_2 and H_2O and use them to calculate the masses of C and H in the compound, from which we find the empirical formula. As you've seen, many organic compounds also contain oxygen, nitrogen, or a halogen. As long as the third element doesn't interfere with the absorption of CO_2 and H_2O, we calculate its mass by subtracting the masses of C and H from the original mass of the compound.

Literature DeMeo, Stephen. "Making Assumptions Explicit: How the Law of Conservation of Matter Can Explain Empirical Formula Problems," *J. Chem. Educ.* **2001**, *78*, 1050.

SAMPLE PROBLEM 3.7 Determining a Molecular Formula from Combustion Analysis

ARIS Presentation Center, Worked Examples Library

PROBLEM Vitamin C ($\mathcal{M}$ = 176.12 g/mol) is a compound of C, H, and O found in many natural sources, especially citrus fruits. When a 1.000-g sample of vitamin C is placed in a combustion chamber and burned, the following data are obtained:

Mass of CO_2 absorber after combustion = 85.35 g

Mass of CO_2 absorber before combustion = 83.85 g

Mass of H_2O absorber after combustion = 37.96 g

Mass of H_2O absorber before combustion = 37.55 g

What is the molecular formula of vitamin C?

PLAN We find the masses of CO_2 and H_2O by subtracting the masses of the absorbers before the reaction from the masses after. From the mass of CO_2, we use the mass fraction of C in CO_2 to find the mass of C (see Comment in Sample Problem 3.4). Similarly, we find the mass of H from the mass of H_2O. The mass of vitamin C (1.000 g) minus the sum of the C and H masses gives the mass of O, the third element present. Then, we proceed as in Sample Problem 3.6: calculate numbers of moles using the elements' molar masses, construct the empirical formula, determine the whole-number multiple from the given molar mass, and construct the molecular formula.

SOLUTION Finding the masses of combustion products:

Mass (g) of CO_2 = mass of CO_2 absorber after − mass before

= 85.35 g − 83.85 g = 1.50 g CO_2

Mass (g) of H_2O = mass of H_2O absorber after − mass before

= 37.96 g − 37.55 g = 0.41 g H_2O

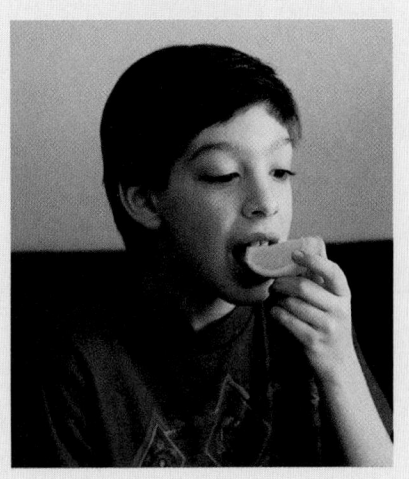

ARIS Presentation Center, Photo Library

Biological application

Calculating masses of C and H using their mass fractions:

$$\text{Mass of element} = \text{mass of compound} \times \frac{\text{mass of element in compound}}{\text{mass of 1 mol of compound}}$$

$$\text{Mass (g) of C} = \text{mass of } CO_2 \times \frac{1 \text{ mol C} \times \mathcal{M} \text{ of C}}{\text{mass of 1 mol } CO_2} = 1.50 \text{ g } CO_2 \times \frac{12.01 \text{ g C}}{44.01 \text{ g } CO_2}$$

$$= 0.409 \text{ g C}$$

$$\text{Mass (g) of H} = \text{mass of } H_2O \times \frac{2 \text{ mol H} \times \mathcal{M} \text{ of H}}{\text{mass of 1 mol } H_2O} = 0.41 \text{ g } H_2O \times \frac{2.016 \text{ g H}}{18.02 \text{ g } H_2O}$$

$$= 0.046 \text{ g H}$$

Calculating the mass of O:

$$\text{Mass (g) of O} = \text{mass of vitamin C sample} - (\text{mass of C} + \text{mass of H})$$

$$= 1.000 \text{ g} - (0.409 \text{ g} + 0.046 \text{ g}) = 0.545 \text{ g O}$$

Finding the amounts (mol) of elements: Dividing the mass in grams of each element by its molar mass gives 0.0341 mol of C, 0.046 mol of H, and 0.0341 mol of O.

Constructing the preliminary formula: $C_{0.0341}H_{0.046}O_{0.0341}$

Determining the empirical formula: Dividing through by the smallest subscript gives

$$C_{\frac{0.0341}{0.0341}} H_{\frac{0.046}{0.0341}} O_{\frac{0.0341}{0.0341}} = C_{1.00}H_{1.3}O_{1.00}$$

By trial and error, we find that 3 is the smallest integer that will make all subscripts approximately into integers:

$$C_{(1.00\times3)}H_{(1.3\times3)}O_{(1.00\times3)} = C_{3.00}H_{3.9}O_{3.00} \approx C_3H_4O_3$$

Determining the molecular formula:

$$\text{Whole-number multiple} = \frac{\mathcal{M} \text{ of vitamin C}}{\mathcal{M} \text{ of empirical formula}} = \frac{176.12 \text{ g/mol}}{88.06 \text{ g/mol}} = 2.000 = 2$$

$$C_{(3\times2)}H_{(4\times2)}O_{(3\times2)} = C_6H_8O_6$$

CHECK The element masses seem correct: carbon makes up slightly more than 0.25 of the mass of CO_2 (12 g/44 g > 0.25), as do the masses in the problem (0.409 g/1.50 g > 0.25). Hydrogen makes up slightly more than 0.10 of the mass of H_2O (2 g/18 g > 0.10), as do the masses in the problem (0.046 g/0.41 g > 0.10). The molecular formula has the same ratio of subscripts (6/8/6) as the empirical formula (3/4/3) and adds up to the given molar mass:

$$(6 \times \mathcal{M} \text{ of C}) + (8 \times \mathcal{M} \text{ of H}) + (6 \times \mathcal{M} \text{ of O}) = \mathcal{M} \text{ of vitamin C}$$

$$(6 \times 12.01) + (8 \times 1.008) + (6 \times 16.00) = 176.12 \text{ g/mol}$$

COMMENT In determining the subscript for H, if we string the calculation steps together, we obtain the subscript 4.0, rather than 3.9, and don't need to round:

$$\text{Subscript of H} = 0.41 \text{ g } H_2O \times \frac{2.016 \text{ g H}}{18.02 \text{ g } H_2O} \times \frac{1 \text{ mol H}}{1.008 \text{ g H}} \times \frac{1}{0.0341 \text{ mol}} \times 3 = 4.0$$

FOLLOW-UP PROBLEM 3.7 A dry-cleaning solvent ($\mathcal{M} = 146.99$ g/mol) that contains C, H, and Cl is suspected to be a cancer-causing agent. When a 0.250-g sample was studied by combustion analysis, 0.451 g of CO_2 and 0.0617 g of H_2O formed. Find the molecular formula.

Chemical Formulas and Molecular Structures

Let's take a short break from calculations to recall that *a formula represents a real three-dimensional object*. How much structural information is contained in each of the different types of chemical formulas?

1. *Different compounds with the same empirical formula.* The empirical formula tells the *relative* number of each type of atom, but it tells nothing about molecular structure. In fact, *different compounds can have the **same** empirical formula.* The oxides NO_2 and N_2O_4 are examples among inorganic compounds, but this phenomenon is especially common among organic compounds. While there is no stable hydrocarbon with the formula CH_2, compounds with the general

Table 3.3	Some Compounds with Empirical Formula CH_2O (Composition by Mass: 40.0% C, 6.71% H, 53.3% O)				
Name	**Molecular Formula**	**Whole-Number Multiple**	$\mathcal{M}$ **(g/mol)**	**Use or Function**	
Formaldehyde	CH_2O	1	30.03	Disinfectant; biological preservative	
Acetic acid	$C_2H_4O_2$	2	60.05	Acetate polymers; vinegar (5% solution)	
Lactic acid	$C_3H_6O_3$	3	90.08	Causes milk to sour; forms in muscle during exercise	
Erythrose	$C_4H_8O_4$	4	120.10	Forms during sugar metabolism	
Ribose	$C_5H_{10}O_5$	5	150.13	Component of many nucleic acids and vitamin B_2	
Glucose	$C_6H_{12}O_6$	6	180.16	Major nutrient for energy in cells	

CH_2O $C_2H_4O_2$ $C_3H_6O_3$ $C_4H_8O_4$ $C_5H_{10}O_5$ $C_6H_{12}O_6$

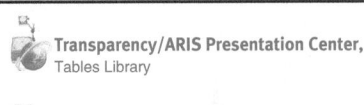

Transparency/ARIS Presentation Center, Tables Library

Organic chemistry application

formula C_nH_{2n} are well known; examples are ethylene (C_2H_4) and propylene (C_3H_6), starting materials for two very common plastics. Table 3.3 shows a few biologically important compounds with a given empirical formula.

2. *Isomers: different compounds with the same molecular formula.* A molecular formula tells the *actual* number of each type of atom, providing as much information as possible from mass analysis. Yet *different compounds can have the same molecular formula* because the atoms can bond to each other in different arrangements to give more than one *structural formula.* **Isomers** are compounds with the same molecular formula but different properties. The simplest type of isomerism, called *constitutional,* or *structural, isomerism,* occurs when the atoms link together in different arrangements. Table 3.4 shows two pairs of examples. The left pair is two compounds with the molecular formula C_4H_{10}, butane and 2-methylpropane. One has a four-C chain, and the other has a one-C branch attached to the second C of a three-C chain. Both are small alkanes, so their properties are similar, if not identical. The right pair of constitutional isomers share the molecular formula C_2H_6O but have very different properties because they are different types of compounds—one is an alcohol, and the other is an ether.

ARIS Presentation Center, Tables Library

Table 3.4	Two Pairs of Constitutional Isomers			
	C_4H_{10}		C_2H_6O	
Property	**Butane**	**2-Methylpropane**	**Ethanol**	**Dimethyl Ether**
$\mathcal{M}$ (g/mol)	58.12	58.12	46.07	46.07
Boiling point	$-0.5°C$	$-11.6°C$	$78.5°C$	$-25°C$
Density (at 20°C)	0.579 g/mL (gas)	0.549 g/mL	0.789 g/mL (liquid)	0.00195 g/mL (gas)
Structural formula				
Space-filling model				

Figure 3.6 Ampicillin. Of the many possible constitutional isomers with the formula $C_{16}H_{19}N_3O_4S$, only this particular arrangement of the atoms is the widely used antibiotic ampicillin.

ARIS Presentation Center, Art Library

Organic chemistry application

As the number and kinds of atoms increase, the number of isomers—that is, the number of structural formulas that can be written for a given molecular formula—also increases: C_2H_6O has two structural formulas, as you've seen, C_3H_8O three, and $C_4H_{10}O$, seven. Imagine how many there are for $C_{16}H_{19}N_3O_4S$! Of all the possible isomers with this formula, only one is the antibiotic ampicillin (Figure 3.6). Only by knowing a molecule's structure—the relative placement of atoms and the distances and angles separating them—can we begin to predict its behavior. (We'll discuss types of isomerism fully later in the text.)

Section Summary

From the masses of elements in an unknown compound, the relative amounts (in moles) are found and the empirical formula determined. • If the molar mass is known, the molecular formula can also be determined. • Methods such as combustion analysis provide data on the masses of elements in a compound, which are used to obtain the formula. • Because atoms can bond in different arrangements, more than one compound may have the same molecular formula (constitutional isomers).

3.3 WRITING AND BALANCING CHEMICAL EQUATIONS

Perhaps the most important reason for thinking in terms of moles is because it greatly clarifies the amounts of substances taking part in a reaction. Comparing masses doesn't tell the ratio of substances reacting but comparing numbers of moles does. It allows us to view substances as large populations of interacting particles rather than as grams of material. To clarify this idea, consider the formation of hydrogen fluoride gas from H_2 and F_2, a reaction that occurs explosively at room temperature. If we weigh the gases, we find that

$$2.016 \text{ g of } H_2 \text{ and } 38.00 \text{ g of } F_2 \text{ react to form } 40.02 \text{ g of HF}$$

This information tells us little except that mass is conserved. However, if we convert these masses (in grams) to amounts (in moles), we find that

$$1 \text{ mol of } H_2 \text{ and } 1 \text{ mol of } F_2 \text{ react to form } 2 \text{ mol of HF}$$

This information reveals that equal-size populations of H_2 and F_2 molecules combine to form twice as large a population of HF molecules. Dividing through by Avogadro's number shows us the chemical event that occurs between individual molecules:

$$1 \text{ } H_2 \text{ molecule and } 1 \text{ } F_2 \text{ molecule react to form } 2 \text{ HF molecules}$$

Figure 3.7 shows that when we express the reaction in terms of moles, *the macroscopic (molar) change corresponds to the submicroscopic (molecular) change.* As you'll see, a balanced chemical equation shows both changes.

A **chemical equation** is a statement in formulas that expresses the identities and quantities of the substances involved in a chemical or physical change. Equations are the "sentences" of chemistry, just as chemical formulas are the "words" and atomic symbols the "letters." The left side of an equation shows the amount of each substance present before the change, and the right side shows the amounts present afterward. *For an equation to depict these amounts accurately, it must be balanced; that is, the same number of each type of atom must appear on both sides of the equation.* This requirement follows directly from the mass laws and the atomic theory:

- In a chemical process, atoms cannot be created, destroyed, or changed, only rearranged into different combinations.
- A formula represents a fixed ratio of the elements in a compound, so a different ratio represents a different compound.

Literature Sanger, Michael J. "Evaluating Students' Conceptual Understanding of Balanced Equations and Stoichiometric Ratios Using a Particulate Drawing," *J. Chem. Educ.* **2005,** *82,* 131.

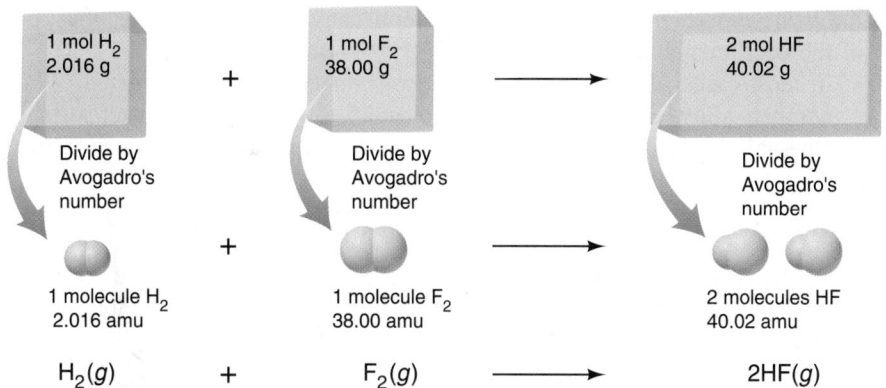

Figure 3.7 The formation of HF gas on the macroscopic and molecular levels. When 1 mol of H_2 (2.016 g) and 1 mol of F_2 (38.00 g) react, 2 mol of HF (40.02 g) forms. Dividing by Avogadro's number shows the change at the molecular level.

ARIS Presentation Center, Art Library

Literature Herndon, William C. "On Balancing Chemical Equations: Past and Present," *J. Chem. Educ.* **1997**, *74*, 1359.

Consider the chemical change that occurs in an old-fashioned photographic flashbulb, in many fireworks, and in a common lecture demo: a magnesium strip burns in oxygen gas to yield powdery magnesium oxide. (Light and heat are produced as well, we're only concerned with the substances involved.) Let's convert this chemical statement into a balanced equation through the following steps:

1. *Translating the statement.* We first translate the chemical statement into a "skeleton" equation: chemical formulas arranged in an equation format. All the substances that react during the change, called **reactants,** are placed to the left of a "yield" arrow, which points to all the substances produced, called **products:**

Literature Toth, Zoltan. "Balancing Chemical Equations by Inspection," *J. Chem. Educ.* **1997**, *74*, 1363.

$$\underbrace{\_Mg \quad + \quad \_O_2}_{\text{reactants}} \xrightarrow{\text{yield}} \underbrace{\_MgO}_{\text{product}}$$

magnesium and oxygen   yield   magnesium oxide

At the beginning of the balancing process, we put a blank in front of each substance to remind us that we have to account for its atoms.

2. *Balancing the atoms.* The next step involves shifting our attention back and forth from right to left in order to *match the number of each type of atom on each side.* At the end of this step, each blank will contain a **balancing (stoichiometric) coefficient,** a numerical multiplier of *all the atoms* in the formula that follows it. In general, balancing is easiest when we

Literature Olson, John A. "An Analysis of the Algebraic Method for Balancing Chemical Reactions," *J. Chem. Educ.* **1997**, *74*, 538.

• Start with the most complex substance, the one with the largest number of atoms or different types of atoms.
• End with the least complex substance, such as an element by itself.

In this case, MgO is the most complex, so we place a coefficient 1 *in front of* the compound:

$$\_Mg + \_O_2 \longrightarrow \underline{1}\,MgO$$

To balance the Mg in MgO on the right, we place a 1 in front of Mg on the left:

$$\underline{1}\,Mg + \_O_2 \longrightarrow \underline{1}\,MgO$$

The O atom on the right must be balanced by one O atom on the left. One-half an O_2 molecule provides one O atom:

$$\underline{1}\,Mg + \tfrac{1}{2}O_2 \longrightarrow \underline{1}\,MgO$$

In terms of number and type of atom, the equation is balanced.

3. *Adjusting the coefficients.* There are several conventions about the final form of the coefficients:

- In most cases, *the smallest whole-number coefficients are preferred.* Whole numbers allow entities such as O_2 molecules to be treated as intact particles. One-half of an O_2 molecule cannot exist, so we multiply the equation by 2:

$$2Mg + 1O_2 \longrightarrow 2MgO$$

- We used the coefficient 1 to remind us to balance each substance. In the final form, a coefficient of 1 is implied just by the presence of the formula of the substance, so we don't need to write it:

$$2Mg + O_2 \longrightarrow 2MgO$$

(This convention is similar to not writing a subscript 1 in a formula.)

4. *Checking.* After balancing and adjusting the coefficients, always check that the equation is balanced:

$$\text{Reactants (2 Mg, 2 O)} \longrightarrow \text{products (2 Mg, 2 O)}$$

5. *Specifying the states of matter.* The final equation also indicates the physical state of each substance or whether it is dissolved in water. The abbreviations that are used for these states are solid (*s*), liquid (*l*), gas (*g*), and aqueous solution (*aq*). From the original statement, we know that a Mg "strip" is solid, O_2 is a gas, and "powdery" MgO is also solid. The balanced equation, therefore, is

$$2Mg(s) + O_2(g) \longrightarrow 2MgO(s)$$

Of course, the key point to realize is, as was pointed out in Figure 3.7, *the balancing coefficients refer to both individual chemical entities and moles of chemical entities.* Thus, 2 mol of Mg and 1 mol of O_2 yield 2 mol of MgO. Figure 3.8 shows this reaction from three points of view—as you see it on the

Figure 3.8 A three-level view of the chemical reaction in a flashbulb. The photos present the macroscopic view that you see. Before the reaction occurs, a piece of magnesium ribbon will be added to a flask of oxygen (*left*). After the reaction, white, powdery magnesium oxide coats the flask's inner surface (*right*). The blow-up arrows lead to an atomic-scale view, a representation of the chemist's mental picture of the reaction. The darker colored spheres show the stoichiometry. By knowing the substances before and after a reaction, we can write a balanced equation (*bottom*), the chemist's symbolic shorthand for the change.

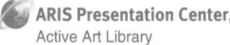
ARIS Presentation Center, Active Art Library

macroscopic level, as chemists (and you!) can imagine it on the atomic level (darker colored atoms represent the stoichiometry), and on the symbolic level of the chemical equation.

Keep in mind these other key points about the balancing process:

- A coefficient operates on *all the atoms in the formula* that follows it: 2MgO means 2 × (MgO), or 2 Mg atoms and 2 O atoms; $2Ca(NO_3)_2$ means 2 × $[Ca(NO_3)_2]$, or 2 Ca atoms, 4 N atoms, and 12 O atoms.
- In balancing an equation, *chemical formulas cannot be altered*. In step 2 of the example, we *cannot* balance the O atoms by changing MgO to MgO_2 because MgO_2 has a different elemental composition and thus is a different compound.
- We *cannot add other reactants or products* to balance the equation because this would represent a different reaction. For example, we *cannot* balance the O atoms by changing O_2 to O or by adding one O atom to the products, because the chemical statement does not say that the reaction involves O atoms.
- A balanced equation remains balanced even if you multiply all the coefficients by the same number. For example,

$$4Mg(s) + 2O_2(g) \longrightarrow 4MgO(s)$$

is also balanced: it is just the balanced equation we obtained above multiplied by 2. However, we balance an equation with the *smallest* whole-number coefficients.

SAMPLE PROBLEM 3.8 Balancing Chemical Equations

PROBLEM Within the cylinders of a car's engine, the hydrocarbon octane (C_8H_{18}), one of many components of gasoline, mixes with oxygen from the air and burns to form carbon dioxide and water vapor. Write a balanced equation for this reaction.

SOLUTION

1. *Translate* the statement into a skeleton equation (with coefficient blanks). Octane and oxygen are reactants; "oxygen from the air" implies molecular oxygen, O_2. Carbon dioxide and water vapor are products:

$$\_C_8H_{18} + \_O_2 \longrightarrow \_CO_2 + \_H_2O$$

2. *Balance the atoms.* We start with the most complex substance, C_8H_{18}, and balance O_2 last:

$$\underline{1}\,C_8H_{18} + \_O_2 \longrightarrow \_CO_2 + \_H_2O$$

The C atoms in C_8H_{18} end up in CO_2. Each CO_2 contains one C atom, so 8 molecules of CO_2 are needed to balance the 8 C atoms in each C_8H_{18}:

$$\underline{1}\,C_8H_{18} + \_O_2 \longrightarrow \underline{8}\,CO_2 + \_H_2O$$

The H atoms in C_8H_{18} end up in H_2O. The 18 H atoms in C_8H_{18} require the coefficient 9 in front of H_2O:

$$\underline{1}\,C_8H_{18} + \_O_2 \longrightarrow \underline{8}\,CO_2 + \underline{9}\,H_2O$$

There are 25 atoms of O on the right (16 in $8CO_2$ plus 9 in $9H_2O$), so we place the coefficient $\frac{25}{2}$ in front of O_2:

$$\underline{1}\,C_8H_{18} + \tfrac{25}{2}O_2 \longrightarrow \underline{8}\,CO_2 + \underline{9}\,H_2O$$

3. *Adjust the coefficients.* Multiply through by 2 to obtain whole numbers:

$$2C_8H_{18} + 25O_2 \longrightarrow 16CO_2 + 18H_2O$$

4. *Check* that the equation is balanced:

$$\text{Reactants (16 C, 36 H, 50 O)} \longrightarrow \text{products (16 C, 36 H, 50 O)}$$

5. *Specify* states of matter. C_8H_{18} is liquid; O_2, CO_2, and H_2O vapor are gases:

$$2C_8H_{18}(l) + 25O_2(g) \longrightarrow 16CO_2(g) + 18H_2O(g)$$

COMMENT This is an example of a combustion reaction. *Any* compound containing C and H that burns in an excess of air produces CO_2 and H_2O.

FOLLOW-UP PROBLEM 3.8 Write a balanced equation for each of the following chemical statements:
(a) A characteristic reaction of Group 1A(1) elements: chunks of sodium react violently with water to form hydrogen gas and sodium hydroxide solution.
(b) The destruction of marble statuary by acid rain: aqueous nitric acid reacts with calcium carbonate to form carbon dioxide, water, and aqueous calcium nitrate.
(c) Halogen compounds exchanging bonding partners: phosphorus trifluoride is prepared by the reaction of phosphorus trichloride and hydrogen fluoride; hydrogen chloride is the other product. The reaction involves gases only.
(d) Explosive decomposition of dynamite: liquid nitroglycerine ($C_3H_5N_3O_9$) explodes to produce a mixture of gases—carbon dioxide, water vapor, nitrogen, and oxygen.

Viewing an equation in a molecular scene is a great way to focus on the essence of the change—the rearrangement of the atoms from reactants to products. Here's a simple schematic representation of the combustion of octane:

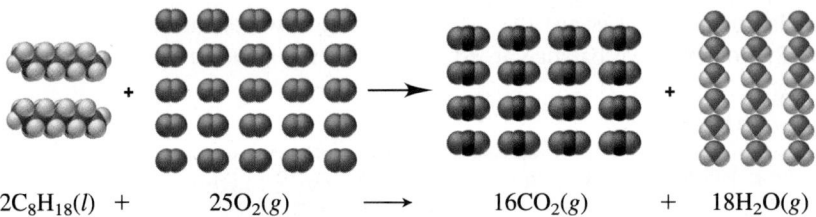

$$2C_8H_{18}(l) + 25O_2(g) \longrightarrow 16CO_2(g) + 18H_2O(g)$$

Now let's work through a sample problem to do the reverse—derive a balanced equation from a molecular scene.

SAMPLE PROBLEM 3.9 **Balancing an Equation from a Molecular Depiction**

PROBLEM The following molecular scene depicts an important reaction in nitrogen chemistry (nitrogen is blue; oxygen is red):

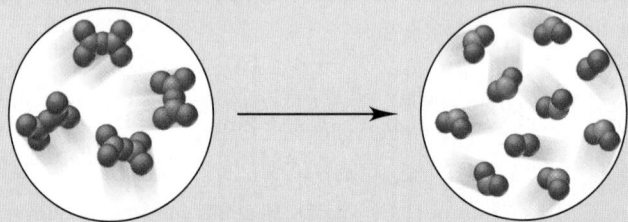

Write a balanced equation for this reaction.
PLAN To write a balanced equation from the depiction, we first have to determine the formulas of the molecules and obtain coefficients by counting the number of each molecule. Then, we arrange this information into the correct equation format, using the smallest whole-number coefficients and including states of matter.
SOLUTION The reactant circle shows only one type of molecule. It has two N and five O atoms, so the formula is N_2O_5; there are four of these molecules. The product circle shows two different molecules, one with one N and two O atoms, and the other with two O atoms; there are eight NO_2 and two O_2. Thus, we have:

$$4N_2O_5 \longrightarrow 8NO_2 + 2O_2$$

Writing the balanced equation with the smallest whole-number coefficients and all substances as gases:

$$2N_2O_5(g) \longrightarrow 4NO_2(g) + O_2(g)$$

CHECK Reactant (4 N, 10 O) $\longrightarrow$ products (4 N, 8 + 2 = 10 O)

FOLLOW-UP PROBLEM 3.9 Write a balanced equation for the important atmospheric reaction depicted below (carbon is black; oxygen is red):

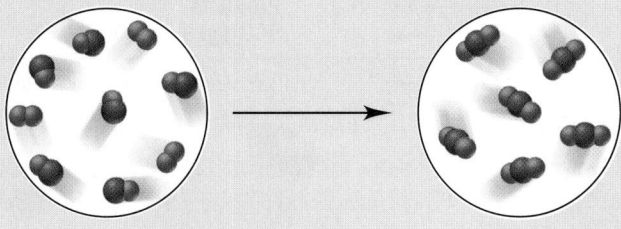

Section Summary

To conserve mass and maintain the fixed composition of compounds, a chemical equation must be balanced in terms of number and type of each atom. • A balanced equation has reactant formulas on the left of a yield arrow and product formulas on the right. • Balancing coefficients are integer multipliers for *all* the atoms in a formula and apply to the individual entity or to moles of entities.

3.4 CALCULATING AMOUNTS OF REACTANT AND PRODUCT

A balanced equation contains a wealth of quantitative information relating individual chemical entities, amounts of chemical entities, and masses of substances. It is essential for all calculations involving amounts of reactants and products: *if you know the number of moles of one substance, the balanced equation for the reaction tells you the number of moles of all the others.*

Stoichiometrically Equivalent Molar Ratios from the Balanced Equation

In a balanced equation, *the number of moles of one substance is stoichiometrically equivalent to the number of moles of any other substance.* The term *stoichiometrically equivalent* means that a definite amount of one substance is formed from, produces, or reacts with a definite amount of the other. These quantitative relationships are expressed as *stoichiometrically equivalent molar ratios* that we use as conversion factors to calculate these amounts. For example, consider the equation for the combustion of propane, a hydrocarbon fuel used in cooking and water heating:

$$C_3H_8(g) + 5O_2(g) \longrightarrow 3CO_2(g) + 4H_2O(g)$$

If we view the reaction quantitatively in terms of C_3H_8, we see that

> 1 mol of C_3H_8 reacts with 5 mol of O_2
> 1 mol of C_3H_8 produces 3 mol of CO_2
> 1 mol of C_3H_8 produces 4 mol of H_2O

Therefore, in this reaction,

> 1 mol of C_3H_8 is stoichiometrically equivalent to 5 mol of O_2
> 1 mol of C_3H_8 is stoichiometrically equivalent to 3 mol of CO_2
> 1 mol of C_3H_8 is stoichiometrically equivalent to 4 mol of H_2O

We chose to look at C_3H_8, but any two of the substances are stoichiometrically equivalent to each other. Thus,

> 3 mol of CO_2 is stoichiometrically equivalent to 4 mol of H_2O
> 5 mol of O_2 is stoichiometrically equivalent to 3 mol of CO_2

and so on. Table 3.5 on the next page presents various ways to view the quantitative information contained in this equation.

Literature Jensen, William B. "The Origin of Stoichiometry Problems," *J. Chem. Educ.* **2003**, *80*, 1248.

Literature Bird, Lillian. "Stoichiometric Calculations Using Equivalent Molar Expressions," *Chem. Educator* 11 **2006**, *6*, 380–382.

Literature Haim, Liliana; Cortón, Eduardo; Kocmur, Santiago; Galagovsky, Lydia. "Learning Stoichiometry with Hamburger Sandwiches," *J. Chem. Educ.* **2003**, *80*, 1021.

Literature Ault, Addison. "How to Say How Much: Amounts and Stoichiometry," *J. Chem. Educ.* **2001**, *78*, 1347.

Table 3.5   Information Contained in a Balanced Equation

Viewed in Terms of	Reactants $C_3H_8(g) + 5O_2(g)$	$\longrightarrow$	Products $3CO_2(g) + 4H_2O(g)$
Molecules	1 molecule C_3H_8 + 5 molecules O_2	$\longrightarrow$	3 molecules CO_2 + 4 molecules H_2O
Amount (mol)	1 mol C_3H_8 + 5 mol O_2	$\longrightarrow$	3 mol CO_2 + 4 mol H_2O
Mass (amu)	44.09 amu C_3H_8 + 160.00 amu O_2	$\longrightarrow$	132.03 amu CO_2 + 72.06 amu H_2O
Mass (g)	44.09 g C_3H_8 + 160.00 g O_2	$\longrightarrow$	132.03 g CO_2 + 72.06 g H_2O
Total mass (g)	204.09 g	$\longrightarrow$	204.09 g

Transparency/ARIS Presentation Center, Tables Library

Literature Sanger, Michael J. "Evaluating Students' Conceptual Understanding of Balanced Equations and Stoichiometric Ratios Using a Particulate Drawing," *J. Chem. Educ.* **2005,** *82,* 131.

Literature Arasasingham, Ramesh D.; Taagepera, Mare; Potter, Frank; Lonjers, Stacy. "Using Knowledge Space Theory to Assess Student Understanding of Stoichiometry," *J. Chem. Educ.* **2004,** *81,* 1517.

Here's a typical problem that shows how stoichiometric equivalence is used to create conversion factors: in the combustion of propane, how many moles of O_2 are consumed when 10.0 mol of H_2O are produced? To solve this problem, we have to find the molar ratio between O_2 and H_2O. From the balanced equation, we see that for every 5 mol of O_2 consumed, 4 mol of H_2O is formed:

5 mol of O_2 is stoichiometrically equivalent to 4 mol of H_2O

We can construct two conversion factors from this equivalence, depending on the quantity we want to find:

$$\frac{5 \text{ mol } O_2}{4 \text{ mol } H_2O} \quad \text{or} \quad \frac{4 \text{ mol } H_2O}{5 \text{ mol } O_2}$$

Since we want to find moles of O_2 and we know moles of H_2O, we choose "5 mol O_2/4 mol H_2O" to cancel "mol H_2O":

$$\text{Moles of } O_2 \text{ consumed} = 10.0 \text{ mol } H_2O \times \frac{5 \text{ mol } O_2}{4 \text{ mol } H_2O} = 12.5 \text{ mol } O_2$$

$$\underset{\substack{\text{molar ratio as} \\ \text{conversion factor}}}{\text{mol } H_2O \Longrightarrow \text{mol } O_2}$$

Obviously, we could not have solved this problem without the balanced equation. Here is a general approach for solving *any* stoichiometry problem that involves a chemical reaction:

1. Write a balanced equation for the reaction.
2. Convert the given mass (or number of entities) of the first substance to amount (mol).
3. Use the appropriate molar ratio from the balanced equation to calculate the amount (mol) of the second substance.
4. Convert the amount of the second substance to the desired mass (or number of entities).

Figure 3.9 summarizes the possible relationships, and multipart Sample Problem 3.10 applies three of them in an industrial setting.

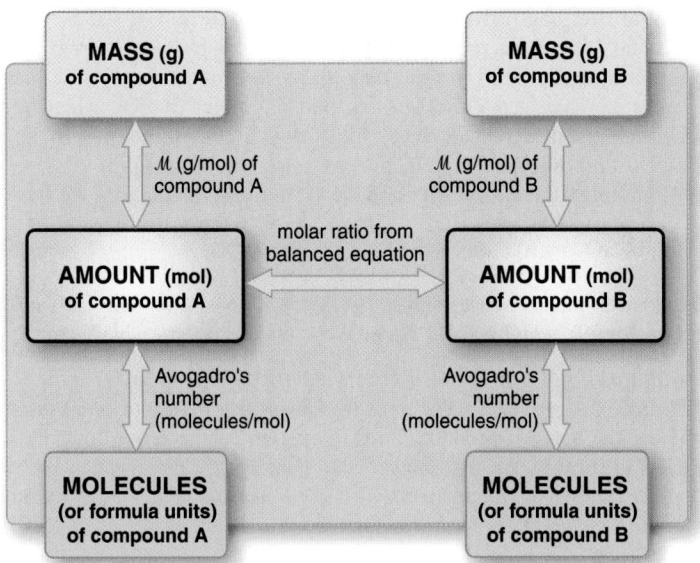

Figure 3.9 Summary of the mass-mole-number relationships in a chemical reaction. The amount of one substance in a reaction is related to that of any other. Quantities are expressed in terms of grams, moles, or number of entities (atoms, molecules, or formula units). Start at any box in the diagram (known) and move to any other box (unknown) by using the information on the arrows as conversion factors. As an example, if you know the mass (in g) of A and want to know the number of molecules of B, the path involves three calculation steps:

1. Grams of A to moles of A, using the molar mass ($\mathcal{M}$) of A
2. Moles of A to moles of B, using the molar ratio from the balanced equation
3. Moles of B to molecules of B, using Avogadro's number

Steps 1 and 3 refer to calculations discussed in Section 3.1 (see Figure 3.4).

ARIS Presentation Center, Art Library

ARIS Presentation Center, Worked Examples Library

SAMPLE PROBLEM 3.10 Calculating Amounts of Reactants and Products

PROBLEM In a lifetime, the average American uses 1750 lb (794 kg) of copper in coins, plumbing, and wiring. Copper is obtained from sulfide ores, such as chalcocite, or copper(I) sulfide, by a multistep process. After an initial grinding, the first step is to "roast" the ore (heat it strongly with oxygen gas) to form powdered copper(I) oxide and gaseous sulfur dioxide.

(a) How many moles of oxygen are required to roast 10.0 mol of copper(I) sulfide?

(b) How many grams of sulfur dioxide are formed when 10.0 mol of copper(I) sulfide is roasted?

(c) How many kilograms of oxygen are required to form 2.86 kg of copper(I) oxide?

(a) Determining the moles of O_2 needed to roast 10.0 mol of Cu_2S

PLAN We *always* write the balanced equation first. The formulas of the reactants are Cu_2S and O_2, and the formulas of the products are Cu_2O and SO_2, so we have

$$2Cu_2S(s) + 3O_2(g) \longrightarrow 2Cu_2O(s) + 2SO_2(g)$$

In this first part, we need just *one calculation step* to convert from amount (mol) of one substance to amount of another. We are given the *moles* of Cu_2S and need to find the *moles* of O_2. The balanced equation shows that 3 mol of O_2 is needed for every 2 mol of Cu_2S consumed, so the conversion factor is "3 mol O_2/2 mol Cu_2S" (see roadmap a).

SOLUTION Calculating number of moles of O_2:

$$\text{Moles of } O_2 = 10.0 \text{ mol } Cu_2S \times \frac{3 \text{ mol } O_2}{2 \text{ mol } Cu_2S} = 15.0 \text{ mol } O_2$$

CHECK The units are correct, and the answer is reasonable because this molar ratio of O_2 to Cu_2S (15/10) is equivalent to the ratio in the balanced equation (3/2).

COMMENT A *common mistake* is to use the incorrect conversion factor; the calculation would then be

$$\text{Moles of } O_2 = 10.0 \text{ mol } Cu_2S \times \frac{2 \text{ mol } Cu_2S}{3 \text{ mol } O_2} = \frac{6.67 \text{ mol}^2 Cu_2S}{1 \text{ mol } O_2}$$

Such strange units should signal that you made an error in setting up the conversion factor. In addition, the answer, 6.67, is *less* than 10.0, whereas the balanced equation shows that *more* moles of O_2 than of Cu_2S are needed. Be sure to think through the calculation when setting up the conversion factor and canceling units.

Amount (mol) of Cu_2S

molar ratio

Amount (mol) of O_2

(a)

(b)

(c)

(b) Determining the mass (g) of SO_2 formed from 10.0 mol of Cu_2S

PLAN The second part of the problem requires *two steps* to convert from amount of one substance to mass of another. Here we need the *grams* of product (SO_2) that form from the given *moles* of reactant (Cu_2S). We first find the moles of SO_2 using the molar ratio from the balanced equation (2 mol SO_2/2 mol Cu_2S) and then multiply by its molar mass (64.07 g/mol) to find grams of SO_2. The steps appear in roadmap b.

SOLUTION Combining the two conversion steps into one calculation, we have

$$\text{Mass (g) of } SO_2 = 10.0 \text{ mol } Cu_2S \times \frac{2 \text{ mol } SO_2}{2 \text{ mol } Cu_2S} \times \frac{64.07 \text{ g } SO_2}{1 \text{ mol } SO_2} = 641 \text{ g } SO_2$$

CHECK The answer makes sense, since the molar ratio shows that 10.0 mol of SO_2 is formed and each mole weighs about 64 g. We rounded to three significant figures.

(c) Determining the mass (kg) of O_2 that yields 2.86 kg of Cu_2O

PLAN In the final part, we need *three steps* to convert from mass of one substance to mass of another. Here the mass of product (Cu_2O) is known, and we need the mass of reactant (O_2) that reacts to form it. We first convert the quantity of Cu_2O from *kilograms* to *moles* (in two steps, as shown in roadmap c). Then, we use the molar ratio (3 mol O_2/2 mol Cu_2O) to find the *moles* of O_2 required. Finally, we convert *moles* of O_2 to *kilograms* (in two steps).

SOLUTION Converting from kilograms of Cu_2O to moles of Cu_2O: Combining the mass unit conversion with the mass-to-mole conversion gives

$$\text{Moles of } Cu_2O = 2.86 \text{ kg } Cu_2O \times \frac{10^3 \text{ g}}{1 \text{ kg}} \times \frac{1 \text{ mol } Cu_2O}{143.10 \text{ g } Cu_2O} = 20.0 \text{ mol } Cu_2O$$

Converting from moles of Cu_2O to moles of O_2:

$$\text{Moles of } O_2 = 20.0 \text{ mol } Cu_2O \times \frac{3 \text{ mol } O_2}{2 \text{ mol } Cu_2O} = 30.0 \text{ mol } O_2$$

Converting from moles of O_2 to kilograms of O_2: Combining the mole-to-mass conversion with the mass unit conversion gives

$$\text{Mass (kg) of } O_2 = 30.0 \text{ mol } O_2 \times \frac{32.00 \text{ g } O_2}{1 \text{ mol } O_2} \times \frac{1 \text{ kg}}{10^3 \text{ g}} = 0.960 \text{ kg } O_2$$

CHECK The units are correct. Round off to check the math: for example, in the final step, $\sim$30 mol $\times$ 30 g/mol $\times$ 1 kg/10^3 g = 0.90 kg. The answer seems reasonable: even though the amount (mol) of O_2 is greater than the amount (mol) of Cu_2O, the mass of O_2 is less than the mass of Cu_2O because $\mathcal{M}$ of O_2 is less than $\mathcal{M}$ of Cu_2O.

COMMENT This problem highlights a key point for solving stoichiometry problems: *convert the information given into moles*. Then, use the appropriate molar ratio and any other conversion factors to complete the solution.

FOLLOW-UP PROBLEM 3.10 Thermite is a mixture of iron(III) oxide and aluminum powders that was once used to weld railroad tracks. It undergoes a spectacular reaction to yield solid aluminum oxide and molten iron.
(a) How many grams of iron form when 135 g of aluminum reacts?
(b) How many atoms of aluminum react for every 1.00 g of aluminum oxide formed?

Chemical Reactions That Occur in a Sequence

In many situations, a product of one reaction becomes a reactant for the next reaction in a sequence of reactions. For stoichiometric purposes, when the same substance forms in one reaction and is used up in the next, we eliminate this common substance in an **overall (net) equation:**

1. Write the sequence of balanced equations.
2. Adjust the equations arithmetically to cancel out the common substance.
3. Add the adjusted equations together to obtain the overall balanced equation.

Sample Problem 3.11 shows the approach by continuing with the copper recovery process that was begun in Sample Problem 3.10.

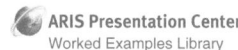

ARIS Presentation Center,
Worked Examples Library

SAMPLE PROBLEM 3.11 Writing an Overall Equation for a Reaction Sequence

PROBLEM Roasting is the first step in extracting copper from chalcocite (Sample Problem 3.10). In the next step, copper(I) oxide reacts with powdered carbon to yield copper metal and carbon monoxide gas. Write a balanced overall equation for the two-step sequence.

PLAN To obtain the overall equation, we write the individual equations in sequence, adjust coefficients to cancel the common substance (or substances), and add the equations together. In this case, only Cu_2O appears as a product in one equation and as a reactant in the other, so it is the common substance.

SOLUTION Writing the individual balanced equations:

$$2Cu_2S(s) + 3O_2(g) \longrightarrow 2Cu_2O(s) + 2SO_2(g) \quad \text{[equation 1; see Sample Problem 3.10(a)]}$$
$$Cu_2O(s) + C(s) \longrightarrow 2Cu(s) + CO(g) \quad \text{[equation 2]}$$

Adjusting the coefficients: Since 2 mol of Cu_2O is produced in equation 1 but only 1 mol of Cu_2O reacts in equation 2, we double *all* the coefficients in equation 2. Thus, the amount of Cu_2O formed in equation 1 is used up in equation 2:

$$2Cu_2S(s) + 3O_2(g) \longrightarrow 2Cu_2O(s) + 2SO_2(g) \quad \text{[equation 1]}$$
$$2Cu_2O(s) + 2C(s) \longrightarrow 4Cu(s) + 2CO(g) \quad \text{[equation 2, doubled]}$$

Adding the two equations and canceling the common substance: We keep the reactants of both equations on the left and the products of both equations on the right:

$$2Cu_2S(s) + 3O_2(g) + 2\cancel{Cu_2O(s)} + 2C(s) \longrightarrow 2\cancel{Cu_2O(s)} + 2SO_2(g) + 4Cu(s) + 2CO(g)$$

or, $\qquad 2Cu_2S(s) + 3O_2(g) + 2C(s) \longrightarrow 2SO_2(g) + 4Cu(s) + 2CO(g)$

CHECK Reactants (4 Cu, 2 S, 6 O, 2 C) $\longrightarrow$ products (4 Cu, 2 S, 6 O, 2 C)

COMMENT 1. Even though Cu_2O *does* participate in the chemical change, it is not involved in the reaction stoichiometry. An overall equation *may not* show which substances actually react; for example, $C(s)$ and $Cu_2S(s)$ do not interact directly here, even though both are shown as reactants.

2. The SO_2 formed in metal extraction contributes to acid rain (see the follow-up problem). To help control the problem, chemists have devised microbial and electrochemical methods to extract metals without roasting sulfide ores. Such methods are among many examples of *green chemistry;* we'll discuss another on page 120.

3. These reactions were shown to explain how to obtain an overall equation. The actual extraction of copper is more complex and will be discussed in Chapter 22.

FOLLOW-UP PROBLEM 3.11

The SO_2 produced in copper recovery reacts in air with oxygen and forms sulfur trioxide. This gas, in turn, reacts with water to form a sulfuric acid solution that falls as rain or snow. Write a balanced overall equation for this process.

Demo Antony, Erling; Mitchell, Lindsay; Nettenstrom, Lauren. "When A + B ≠ B + A," *J. Chem. Educ.* **2000**, *77*, 1180.

Demo Shakhashiri, Bassam Z. "Reaction Between Nitrogen(II) Oxide and Oxygen," *Chemical Demonstrations, A Handbook for Teachers of Chemistry,* Vol. 2 (The University of Wisconsin Press, Madison, 1985): 167–179.

Biological application

Reaction Sequences in Organisms Multistep reaction sequences called *metabolic pathways* occur throughout biological systems. In fact, in many cases, nearly every one of an organism's biomolecules is made within it from ingested nutrients. For example, in most cells, the chemical energy in glucose is released through a sequence of about 30 individual reactions. The product of each reaction step is the reactant of the next, so that all the common substances cancel. The overall equation is

$$C_6H_{12}O_6(aq) + 6O_2(g) \longrightarrow 6CO_2(g) + 6H_2O(l)$$

We eat food that contains glucose, inhale O_2, and excrete CO_2 and H_2O. In our cells, these reactants and products are many steps apart: O_2 never reacts *directly* with glucose, and CO_2 and H_2O are formed at various, often distant, steps along the sequence of reactions. Nevertheless, the molar ratios are the same as if the glucose burned in a combustion apparatus filled with pure O_2 and formed CO_2 and H_2O directly.

Literature Wood, Crystal; Breyfogle, Bryan. "Interactive Demonstrations for Mole Ratios and Limiting Reagents," *J. Chem. Educ.* **2006**, *83*, 741.

Animation: Limiting Reagent
ARIS Presentation Center
Chapter 3

Literature Toth, Zoltan. "Limiting Reactant: An Alternative Analogy," *J. Chem. Educ.* **1999**, *76*, 934.

Chemical Reactions That Involve a Limiting Reactant

In the problems we've considered up to now, the amount of *one* reactant was given, and we assumed there was enough of any other reactant for the first reactant to be completely used up. For example, to find the amount of SO_2 that forms when 100 g of Cu_2S reacts, we convert the grams of Cu_2S to moles and assume that the Cu_2S reacts with as much O_2 as needed. Because all the Cu_2S is used up, its initial amount determines, or limits, how much SO_2 can form. We call Cu_2S the **limiting reactant** (or *limiting reagent*) because the product stops forming once the Cu_2S is gone, no matter how much O_2 is present.

Suppose, however, that the amounts of both Cu_2S *and* O_2 are given in the problem, and we need to find out how much SO_2 forms. We first have to determine whether Cu_2S *or* O_2 is the limiting reactant (that is, which one is completely used up) because the amount of that reactant limits how much SO_2 can form. The other reactant is present *in excess,* and whatever amount of it is not used is left over.

To clarify the idea of limiting reactant, let's consider a much more appetizing situation. Suppose you have a job making sundaes in an ice cream parlor. Each sundae requires two scoops (12 oz) of ice cream, one cherry, and 50 mL of chocolate syrup:

$$2 \text{ scoops (12 oz)} + 1 \text{ cherry} + 50 \text{ mL syrup} \longrightarrow 1 \text{ sundae}$$

A mob of 25 ravenous school kids enters, and every one wants a vanilla sundae. Can you feed them all? You have 300 oz of vanilla ice cream, 30 cherries, and 1 L of syrup, so a quick calculation shows the number of sundaes you can make from each ingredient:

$$\text{Ice cream: No. of sundaes} = 300 \text{ oz} \times \frac{2 \text{ scoops}}{12 \text{ oz}} \times \frac{1 \text{ sundae}}{2 \text{ scoops}} = 25 \text{ sundaes}$$

$$\text{Cherries: No. of sundaes} = 30 \text{ cherries} \times \frac{1 \text{ sundae}}{1 \text{ cherry}} = 30 \text{ sundaes}$$

$$\text{Syrup: No. of sundaes} = 1000 \text{ mL syrup} \times \frac{1 \text{ sundae}}{50 \text{ mL syrup}} = 20 \text{ sundaes}$$

The syrup is the limiting "reactant" in this case because it limits the total amount of "product" (sundaes) that can "form": of the three ingredients, the syrup allows the *fewest* sundaes to be made. (Also see the example in Figure 3.10.) Some ice cream and cherries are still left "unreacted" when all the syrup has been used up, so they are present in excess:

300 oz (50 scoops) + 30 cherries + 1 L syrup $\longrightarrow$

20 sundaes + 60 oz (10 scoops) + 10 cherries

A good way to keep track of the quantities in a limiting-reactant problem is with a reaction table. It shows the initial amounts of reactants and products, the changes in their amounts due to the reaction, and their final amounts. For example, for the ice-cream sundae "reaction," the reaction table is

Quantity	2 scoops ice cream +	1 cherry +	50 mL syrup $\longrightarrow$	1 sundae
Initial	50 scoops	30 cherries	1000 mL syrup	0 sundaes
Change	−40 scoops	−20 cherries	−1000 mL syrup	+20 sundaes
Final	10 scoops	10 cherries	0 mL syrup	20 sundaes

At the top is the balanced equation, which provides column heads for the table. The first line of the table shows the initial amounts of reactants and products before the "reaction" starts. No "product" has yet formed, which is indicated in the table by "0 sundaes." The next line shows the changes in reactants and products as a result of the "reaction." Notice that since ice cream, cherries, and syrup were used to make the sundaes, the amounts of reactants decreased and their changes have a negative sign, while the amount of product increased so its change has a positive sign. We add the changes to the initial amounts to obtain the bottom line,

Figure 3.10 An ice cream sundae analogy for limiting reactants.
A, The "components" combine in specific amounts to form a sundae.
B, In this example, the number of sundaes possible is limited by the

amount of syrup, the limiting "reactant." Here, only two sundaes can be made. Four scoops of ice cream and four cherries remain "in excess." See the text for another situation involving these components.

Transparency/ARIS Presentation Center, Art Library

the final amounts after the reaction is over. Now we can see that some ice cream and cherries are in excess, and the syrup, the limiting reactant, is used up.

Now let's apply these ideas to solving chemical problems. In limiting-reactant problems, the amounts of two (or more) reactants are given, and we must first determine which is limiting. To do this, just as we did with the ice cream sundaes, we first note how much of each reactant *should* be present to completely use up the other, and then we compare it with the amount that is *actually* present. Simply put, the limiting reactant is the one there is not enough of; that is, *it is the reactant that limits the amount of the other reactant that can react, and thus the amount of product that can form.* In mathematical terms, *the limiting reactant is the one that yields the **lower** amount of product.*

We'll examine limiting reactants in the following two sample problems. Sample Problem 3.12 has two parts, and in both we have to identify the limiting reactant. In the first part, we'll look at a simple molecular view of a reaction and compare the number of molecules to find the limiting reactant; in the second part, we start with the amounts (mol) of two reactants and perform two calculations, each of which assumes an excess of one of the reactants, to see which reactant forms less product. Then, in Sample Problem 3.13, we go through a similar process but start with the masses of the two reactants. ●

ARIS Presentation Center, Worked Examples Library

● **Limiting "Reactants" in Everyday Life** Limiting-"reactant" situations arise in business all the time. A car assembly-plant manager must order more tires if there are 1500 car bodies and only 4000 tires, and a clothes manufacturer must cut more sleeves if there are 320 sleeves for 170 shirt bodies. You've probably faced such situations in daily life as well. A muffin recipe calls for 2 cups of flour and 1 cup of sugar, but you have 3 cups of flour and only $\frac{3}{4}$ cup of sugar. Clearly, the flour is in excess and the sugar limits the number of muffins you can make. Or, you're in charge of making cheeseburgers for a picnic, and you have 10 buns, 12 meat patties, and 15 slices of cheese. Here, the number of buns limits the cheeseburgers you can make. Or, there are 26 students and only 23 microscopes in a cell biology lab. You'll find that limiting-reactant situations are almost limitless.

SAMPLE PROBLEM 3.12 Using Molecular Depictions to Solve a Limiting-Reactant Problem

PROBLEM Nuclear engineers use chlorine trifluoride in the processing of uranium fuel for power plants. This extremely reactive substance is formed as a gas in special metal containers by the reaction of elemental chlorine and fluorine.
(a) Suppose the circle at right represents a container of the reactant mixture before the reaction occurs (with chlorine colored green). Name the limiting reactant, and draw the container contents after the reaction is complete.
(b) When the reaction is run again with 0.750 mol of Cl_2 and 3.00 mol of F_2, what mass of chlorine trifluoride will be prepared?

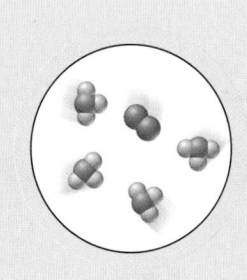

(a) Determining the limiting reactant and drawing the container contents

PLAN We first write the balanced equation. From its name, we know that chlorine trifluoride consists of one Cl atom bonded to three F atoms, ClF_3. Elemental chlorine and fluorine refer to the diatomic molecules Cl_2 and F_2. All the substances are gases. To find the limiting reactant, we compare the number of molecules we have of each reactant, with the number we need for the other to react completely. The limiting reactant limits the amount of the other reactant that can react and the amount of product that will form.

SOLUTION The balanced equation is

$$Cl_2(g) + 3F_2(g) \longrightarrow 2ClF_3(g)$$

The equation shows that two ClF_3 molecules are formed for every one Cl_2 molecule and three F_2 molecules that react. Before the reaction, there are three Cl_2 molecules (six Cl atoms). For all the Cl_2 to react, we need three times three, or nine, F_2 molecules (18 F atoms). But there are only six F_2 molecules (12 F atoms). Therefore, F_2 is the limiting reactant because it limits the amount of Cl_2 that can react, and thus the amount of ClF_3 that can form. After the reaction, as the circle at left depicts, all 12 F atoms and four of the six Cl atoms make four ClF_3 molecules, and one Cl_2 molecule remains in excess.

CHECK The equation is balanced: reactants (2 Cl, 6 F) $\longrightarrow$ products (2 Cl, 6 F), and, in the circles, the number of each type of atom before the reaction equals the number after the reaction. You can check the choice of limiting reactant by examining the reaction from the perspective of Cl_2: two Cl_2 molecules are enough to react with the six F_2 molecules in the container. But there are three Cl_2 molecules, so there is not enough F_2.

(b) Calculating the mass of ClF_3 formed

PLAN We first determine the limiting reactant by using the molar ratios from the balanced equation to convert the moles of each reactant to moles of ClF_3 formed, assuming an excess of the other reactant. Whichever reactant forms fewer moles of ClF_3 is the limiting reactant. Then we use the molar mass of ClF_3 to convert this lower number of moles to grams.

SOLUTION Determining the limiting reactant:
Finding moles of ClF_3 from moles of Cl_2 (assuming F_2 is in excess):

$$\text{Moles of } ClF_3 = 0.750 \text{ mol } Cl_2 \times \frac{2 \text{ mol } ClF_3}{1 \text{ mol } Cl_2} = 1.50 \text{ mol } ClF_3$$

Finding moles of ClF_3 from moles of F_2 (assuming Cl_2 is in excess):

$$\text{Moles of } ClF_3 = 3.00 \text{ mol } F_2 \times \frac{2 \text{ mol } ClF_3}{3 \text{ mol } F_2} = 2.00 \text{ mol } ClF_3$$

In this experiment, Cl_2 is limiting because it forms fewer moles of ClF_3.
Calculating grams of ClF_3 formed:

$$\text{Mass (g) of } ClF_3 = 1.50 \text{ mol } ClF_3 \times \frac{92.45 \text{ g } ClF_3}{1 \text{ mol } ClF_3} = 139 \text{ g } ClF_3$$

CHECK Let's check our reasoning that Cl_2 is the limiting reactant by assuming, for the moment, that F_2 is limiting. In that case, all 3.00 mol of F_2 would react to form 2.00 mol of ClF_3. Based on the balanced equation, however, that amount of product would require that 1.00 mol of Cl_2 reacted. But that is impossible because only 0.750 mol of Cl_2 is present.

COMMENT Note that a reactant can be limiting even though it is present in the greater amount. It is the *reactant molar ratio in the balanced equation* that is the determining factor. In both parts (a) and (b), F_2 is present in greater amount than Cl_2. However, in (a), the F_2/Cl_2 ratio is 6/3, or 2/1, which is less than the required molar ratio of 3/1, so F_2 is limiting; in (b), the F_2/Cl_2 ratio is 3.00/0.750, greater than the required 3/1, so F_2 is in excess. When we write a reaction table for part (b), this fact is revealed clearly:

Amount (mol)	$Cl_2(g)$ +	$3F_2(g)$ $\longrightarrow$	$2ClF_3(g)$
Initial	0.750	3.00	0
Change	−0.750	−2.25	+1.50
Final	0	0.75	1.50

FOLLOW-UP PROBLEM 3.12 B_2 (red spheres) reacts with AB as shown below:

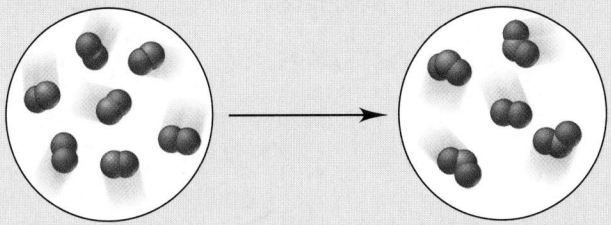

(a) Write a balanced equation for the reaction, and determine the limiting reactant.
(b) How many moles of product can form from the reaction of 1.5 mol of each reactant?

SAMPLE PROBLEM 3.13 Calculating Amounts of Reactant and Product in a Limiting-Reactant Problem

ARIS Presentation Center,
Worked Examples Library

PROBLEM A fuel mixture used in the early days of rocketry is composed of two liquids, hydrazine (N_2H_4) and dinitrogen tetraoxide (N_2O_4), which ignite on contact to form nitrogen gas and water vapor. How many grams of nitrogen gas form when 1.00×10^2 g of N_2H_4 and 2.00×10^2 g of N_2O_4 are mixed?

PLAN We first write the balanced equation. *Because the amounts of two reactants are given, we know this is a limiting-reactant problem.* To determine which reactant is limiting, we calculate the mass of N_2 formed from each reactant, *assuming an excess of the other*. We convert the grams of each reactant to moles and use the appropriate molar ratio to find the moles of N_2 each forms. Whichever yields *less* N_2 is the limiting reactant. Then, we convert this lower number of moles of N_2 to mass. The roadmap shows the steps.

SOLUTION Writing the balanced equation:

$$2N_2H_4(l) + N_2O_4(l) \longrightarrow 3N_2(g) + 4H_2O(g)$$

Finding the moles of N_2 from the moles of N_2H_4 (if N_2H_4 is limiting):

$$\text{Moles of } N_2H_4 = 1.00\times10^2 \text{ g } N_2H_4 \times \frac{1 \text{ mol } N_2H_4}{32.05 \text{ g } N_2H_4} = 3.12 \text{ mol } N_2H_4$$

$$\text{Moles of } N_2 = 3.12 \text{ mol } N_2H_4 \times \frac{3 \text{ mol } N_2}{2 \text{ mol } N_2H_4} = \textbf{4.68 mol } N_2$$

Finding the moles of N_2 from the moles of N_2O_4 (if N_2O_4 is limiting):

$$\text{Moles of } N_2O_4 = 2.00\times10^2 \text{ g } N_2O_4 \times \frac{1 \text{ mol } N_2O_4}{92.02 \text{ g } N_2O_4} = 2.17 \text{ mol } N_2O_4$$

$$\text{Moles of } N_2 = 2.17 \text{ mol } N_2O_4 \times \frac{3 \text{ mol } N_2}{1 \text{ mol } N_2O_4} = \textbf{6.51 mol } N_2$$

Thus, N_2H_4 is the limiting reactant because it yields fewer moles of N_2.
Converting from moles of N_2 to grams:

$$\text{Mass (g) of } N_2 = 4.68 \text{ mol } N_2 \times \frac{28.02 \text{ g } N_2}{1 \text{ mol } N_2} = 131 \text{ g } N_2$$

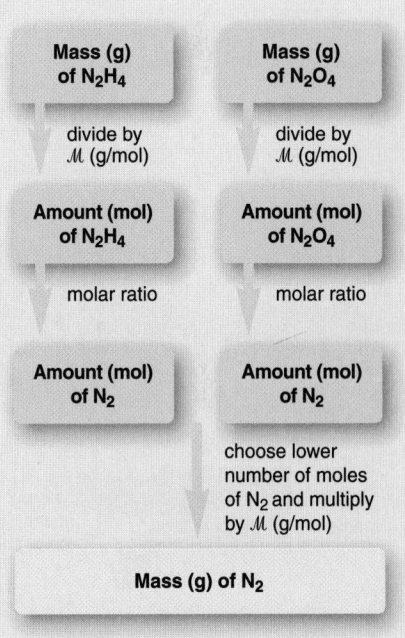

CHECK The mass of N_2O_4 is greater than that of N_2H_4, but there are fewer moles of N_2O_4 because its $\mathcal{M}$ is much higher. Round off to check the math: for N_2H_4, 100 g $N_2H_4 \times$ 1 mol/32 g $\approx$ 3 mol; ~3 mol $\times \frac{3}{2} \approx$ 4.5 mol N_2; ~4.5 mol $\times$ 30 g/mol $\approx$ 135 g N_2.

COMMENT 1. Here are two *common mistakes* in solving limiting-reactant problems:
• The limiting reactant is not the *reactant* present in fewer moles (2.17 mol of N_2O_4 vs. 3.12 mol of N_2H_4). Rather, it is the reactant that forms fewer moles of *product*.
• Similarly, the limiting reactant is not the *reactant* present in lower mass. Rather, it is the reactant that forms the lower mass of *product*.
2. Here is an *alternative approach* to finding the limiting reactant. Find the moles of each reactant that would be needed to react with the other reactant. Then see which amount actually given in the problem is sufficient. That substance is in excess, and the other

substance is limiting. For example, the balanced equation shows that 2 mol of N_2H_4 reacts with 1 mol of N_2O_4. The moles of N_2O_4 needed to react with the given moles of N_2H_4 are

$$\text{Moles of } N_2O_4 \text{ needed} = 3.12 \text{ mol } N_2H_4 \times \frac{1 \text{ mol } N_2O_4}{2 \text{ mol } N_2H_4} = 1.56 \text{ mol } N_2O_4$$

The moles of N_2H_4 needed to react with the given moles of N_2O_4 are

$$\text{Moles of } N_2H_4 \text{ needed} = 2.17 \text{ mol } N_2O_4 \times \frac{2 \text{ mol } N_2H_4}{1 \text{ mol } N_2O_4} = 4.34 \text{ mol } N_2H_4$$

We are given 2.17 mol of N_2O_4, which is *more* than the amount of N_2O_4 that is needed (1.56 mol) to react with the given amount of N_2H_4, and we are given 3.12 mol of N_2H_4, which is *less* than the amount of N_2H_4 needed (4.34 mol) to react with the given amount of N_2O_4. Therefore, N_2H_4 is limiting, and N_2O_4 is in excess. Once we determine this, we continue with the final calculation to find the amount of N_2.

3. Once again, a reaction table reveals the amounts of all the reactants and products before and after the reaction:

Amount (mol)	$2N_2H_4(g)$	+	$N_2O_4(g)$	$\longrightarrow$	$3N_2(g)$	+	$4H_2O(g)$
Initial	3.12		2.17		0		0
Change	−3.12		−1.56		+4.68		+6.24
Final	0		0.61		4.68		6.24

FOLLOW-UP PROBLEM 3.13 How many grams of solid aluminum sulfide can be prepared by the reaction of 10.0 g of aluminum and 15.0 g of sulfur? How much of the nonlimiting reactant is in excess?

Chemical Reactions in Practice: Theoretical, Actual, and Percent Yields

Up until now, we've been optimistic about the amount of product obtained from a reaction. We have assumed that 100% of the limiting reactant becomes product, that ideal separation and purification methods exist for isolating the product, and that we use perfect lab technique to collect all the product formed. In other words, we have assumed that we obtain the **theoretical yield,** the amount indicated by the stoichiometrically equivalent molar ratio in the balanced equation.

It's time to face reality. The theoretical yield is *never* obtained, for reasons that are largely uncontrollable. For one thing, although the major reaction predominates, many reactant mixtures also proceed through one or more **side reactions** that form smaller amounts of different products (Figure 3.11). In the rocket fuel reaction in Sample Problem 3.13, for example, the reactants might form some NO in the following side reaction:

$$N_2H_4(l) + 2N_2O_4(l) \longrightarrow 6NO(g) + 2H_2O(g)$$

This reaction decreases the amounts of reactants available for N_2 production (see Problem 3.122 at the end of the chapter). Even more important, as we'll discuss in Chapter 4, many reactions seem to stop before they are complete, which leaves some limiting reactant unused. But, even when a reaction does go completely to product, losses occur in virtually every step of a separation procedure (see Tools of the Laboratory, Section 2.9): a tiny amount of product clings to filter paper, some distillate evaporates, a small amount of extract remains in the separatory funnel, and so forth. With careful technique, you can minimize these losses but never eliminate them. The amount of product that you actually obtain is the **actual yield.** The **percent yield (% yield)** is the actual yield expressed as a percentage of the theoretical yield:

A + B
(reactants)

C
(main product)

D
(side product)

Figure 3.11 The effect of side reactions on yield. One reason the theoretical yield is never obtained is that other reactions lead some of the reactants along side paths to form undesired products.

ARIS Presentation Center,
Art Library

$$\% \text{ yield} = \frac{\text{actual yield}}{\text{theoretical yield}} \times 100 \qquad \textbf{(3.7)}$$

Because the actual yield *must* be less than the theoretical yield, the percent yield is *always* less than 100%. Theoretical and actual yields are expressed in units of amount (moles) or mass (grams).

ARIS Presentation Center,
Worked Examples Library

SAMPLE PROBLEM 3.14 Calculating Percent Yield

PROBLEM Silicon carbide (SiC) is an important ceramic material that is made by allowing sand (silicon dioxide, SiO_2) to react with powdered carbon at high temperature. Carbon monoxide is also formed. When 100.0 kg of sand is processed, 51.4 kg of SiC is recovered. What is the percent yield of SiC from this process?

PLAN We are given the actual yield of SiC (51.4 kg), so we need the theoretical yield to calculate the percent yield. After writing the balanced equation, we convert the given mass of SiO_2 (100.0 kg) to amount (mol). We use the molar ratio to find the amount of SiC formed and convert that amount to mass (kg) to obtain the theoretical yield [see Sample Problem 3.10(c)]. Then, we use Equation 3.7 to find the percent yield (see the roadmap).

SOLUTION Writing the balanced equation:

$$SiO_2(s) + 3C(s) \longrightarrow SiC(s) + 2CO(g)$$

Converting from kilograms of SiO_2 to moles:

$$\text{Moles of } SiO_2 = 100.0 \text{ kg } SiO_2 \times \frac{1000 \text{ g}}{1 \text{ kg}} \times \frac{1 \text{ mol } SiO_2}{60.09 \text{ g } SiO_2} = 1664 \text{ mol } SiO_2$$

Converting from moles of SiO_2 to moles of SiC: The molar ratio is 1 mol SiC/1 mol SiO_2, so

$$\text{Moles of } SiO_2 = \text{moles of SiC} = 1664 \text{ mol SiC}$$

Converting from moles of SiC to kilograms:

$$\text{Mass (kg) of SiC} = 1664 \text{ mol SiC} \times \frac{40.10 \text{ g SiC}}{1 \text{ mol SiC}} \times \frac{1 \text{ kg}}{1000 \text{ g}} = 66.73 \text{ kg SiC}$$

Calculating the percent yield:

$$\% \text{ yield of SiC} = \frac{\text{actual yield}}{\text{theoretical yield}} \times 100 = \frac{51.4 \text{ kg SiC}}{66.73 \text{ kg SiC}} \times 100 = 77.0\%$$

CHECK Rounding shows that the mass of SiC seems correct: ~1500 mol × 40 g/mol × 1 kg/1000 g = 60 kg. The molar ratio of SiC/SiO_2 is 1/1, and the $\mathcal{M}$ of SiC is about two-thirds ($\sim\frac{40}{60}$) the $\mathcal{M}$ of SiO_2, so 100 kg of SiO_2 should form about 66 kg of SiC.

FOLLOW-UP PROBLEM 3.14 Marble (calcium carbonate) reacts with hydrochloric acid solution to form calcium chloride solution, water, and carbon dioxide. What is the percent yield of carbon dioxide if 3.65 g of the gas is collected when 10.0 g of marble reacts?

Mass (kg) of SiO_2

1. multiply by 10^3
2. divide by $\mathcal{M}$ (g/mol)

Amount (mol) of SiO_2

molar ratio

Amount (mol) of SiC

1. multiply by $\mathcal{M}$ (g/mol)
2. divide by 10^3

Mass (kg) of SiC

Eq. 3.7

% Yield of SiC

Yields in Multistep Syntheses

In the multistep laboratory synthesis of a complex compound, *each step is expressed as a fraction of 1.00 and multiplied by the others to find the overall fractional yield and then by 100 to get the percent yield.* Even when the yield of each step is high, the final result can be surprisingly low. For example, suppose a six-step reaction sequence has a 90.0% yield for each step; that is, you are able to recover 90.0% of the theoretical yield of product in each step. Even so, overall recovery is only slightly more than 50%:

Overall % yield = (0.900 × 0.900 × 0.900 × 0.900 × 0.900 × 0.900) × 100 = 53.1%

Such multistep sequences are common in the laboratory synthesis of medicines, dyes, pesticides, and many other organic compounds. For example, the antidepressant Sertraline is prepared from a simple starting compound in six steps with yields of 80%, 80%, 50%, 100%, 48%, and 30%, respectively. Through a calculation like that above, we find the overall percent yield is only 4.6% (see margin). Because a typical synthesis begins with large amounts of inexpensive, simple reactants and ends with small amounts of expensive, complex products, the overall yield greatly influences the commercial potential of a product.

Organic chemistry application

Starting with 100 g of 1,2-dichlorobenzene

↓ 80%
↓ 80%
↓ 50%
↓ 100%
↓ 48%
↓ 30%

Sertraline

The yield of Sertraline is only 4.6 g

A Green Chemistry Perspective on Yield: Atom Economy Reactants are wasted by undesirable side reactions during each step in a synthesis, drastically lowering the overall yield. Moreover, many byproducts may be harmful. This fact is one of several concerns addressed by **green chemistry.** A major focus of the academic, industrial, and government chemists who work in this new field is to develop methods that reduce or prevent the release of harmful substances into the environment.

To fully evaluate alternative methods, several green chemistry principles are taken into account, including the quantity of energy needed and the nature of the solvents required. When these factors are similar, the *atom economy,* the proportion of reactant atoms that end up in the desired product, is a useful criterion for choosing the more efficient synthetic route. The efficiency of a synthesis is quantified in terms of the *percent atom economy:*

$$\% \text{ atom economy} = \frac{\text{no. of moles} \times \text{molar mass of desired product}}{\text{sum of (no. of moles} \times \text{molar mass) for all products}} \times 100$$

Consider two synthetic routes—one starting with benzene (C_6H_6), the other with butane (C_4H_{10})—for the production of maleic anhydride ($C_4H_2O_3$), a key industrial chemical used in the manufacture of polymers, dyes, medicines, pesticides, and other important products:

Route 1. $\quad 2C_6H_6(l) + 9O_2(g) \rightarrow \cdots \rightarrow 2C_4H_2O_3(l) + 4H_2O(l) + 4CO_2(g)$

Route 2. $\quad 2C_4H_{10}(g) + 7O_2(g) \rightarrow \cdots \rightarrow 2C_4H_2O_3(l) + 8H_2O(l)$

Let's compare the efficiency of these routes in terms of percent atom economy:

Route 1.

$$\% \text{ atom economy} = \frac{2 \times \mathcal{M} \text{ of } C_4H_2O_3}{(2 \times \mathcal{M} \text{ of } C_4H_2O_3) + (4 \times \mathcal{M} \text{ of } H_2O) + (4 \times \mathcal{M} \text{ of } CO_2)} \times 100$$

$$= \frac{2 \times 98.06 \text{ g}}{(2 \times 98.06 \text{ g}) + (4 \times 18.02 \text{ g}) + (4 \times 44.01 \text{ g})} \times 100$$

$$= 44.15\%$$

Route 2.

$$\% \text{ atom economy} = \frac{2 \times \mathcal{M} \text{ of } C_4H_2O_3}{(2 \times \mathcal{M} \text{ of } C_4H_2O_3) + (8 \times \mathcal{M} \text{ of } H_2O)} \times 100$$

$$= \frac{2 \times 98.06 \text{ g}}{(2 \times 98.06 \text{ g}) + (8 \times 18.02 \text{ g})} \times 100$$

$$= 57.63\%$$

Clearly, from the perspective of atom economy, route 2 is preferable because a larger percentage of reactant atoms end up in the desired product. It is also a "greener" approach than route 1 because it avoids the use of the toxic reactant benzene and does not produce CO_2, a gas that contributes to global warming.

Section Summary

The substances in a balanced equation are related to each other by stoichiometrically equivalent molar ratios, which are used as conversion factors to find the moles of one substance given the moles of another. • In limiting-reactant problems, the amounts of two (or more) reactants are given, and one of them limits the amount of product that forms. The limiting reactant is the one that forms the lower amount of product. • In practice, side reactions, incomplete reactions, and physical losses result in an actual yield of product that is less than the theoretical yield, the amount based on the molar ratio. The percent yield is the actual yield expressed as a percentage of the theoretical yield. In multistep reaction sequences, the overall yield is found by multiplying the percent yields for each step. • Atom economy, or the proportion of reactant atoms found in the product, is one criterion for choosing a "greener" reaction.

3.5 FUNDAMENTALS OF SOLUTION STOICHIOMETRY

In the popular media, you may have seen a chemist portrayed as a person in a white lab coat, surrounded by oddly shaped glassware, pouring one colored solution into another, which produces frothing bubbles and billowing fumes. Although most reactions in solution are not this dramatic and good technique usually requires safer mixing procedures, the image is true to the extent that aqueous solution chemistry is a central part of laboratory activity. Liquid solutions are more convenient to store and mix than solids or gases, and the amounts of substances in solution can be measured very precisely. Since many environmental reactions and almost all biochemical reactions occur in solution, an understanding of reactions in solution is extremely important in chemistry and related sciences.

We'll discuss solution chemistry at many places in the text, but here we focus on solution stoichiometry. Only one aspect of the stoichiometry of dissolved substances is different from what we've seen so far. We know the amounts of pure substances by converting their masses directly into moles. For dissolved substances, we must know the *concentration*—the number of moles present in a certain volume of solution—to find the volume that contains a given number of moles. Of the various ways to express concentration, the most important is *molarity,* so we discuss it here (and wait until Chapter 13 to discuss the other ways). Then, we see how to prepare a solution of a specific molarity and how to use solutions in stoichiometric calculations.

Expressing Concentration in Terms of Molarity

A typical solution consists of a smaller amount of one substance, the **solute,** dissolved in a larger amount of another substance, the **solvent.** When a solution forms, the solute's individual chemical entities become evenly dispersed throughout the available volume and surrounded by solvent molecules. The **concentration** of a solution is usually expressed as *the amount of solute dissolved in a given amount of solution.* Concentration is an *intensive* quantity (like density or temperature) and thus independent of the volume of solution: a 50-L tank of a given solution has the *same concentration* (solute amount/solution amount) as a 50-mL beaker of the solution. **Molarity (*M*)** expresses the concentration in units of *moles of solute per liter of solution:*

$$\text{Molarity} = \frac{\text{moles of solute}}{\text{liters of solution}} \quad \text{or} \quad M = \frac{\text{mol solute}}{\text{L soln}} \qquad (3.8)$$

ARIS Presentation Center, Worked Examples Library

SAMPLE PROBLEM 3.15 Calculating the Molarity of a Solution

PROBLEM Glycine (H_2NCH_2COOH) is the simplest amino acid. What is the molarity of an aqueous solution that contains 0.715 mol of glycine in 495 mL?

PLAN The molarity is the number of moles of solute in each liter of solution. We are given the number of moles (0.715 mol) and the volume (495 mL), so we divide moles by volume and convert the volume to liters to find the molarity (see the roadmap).

SOLUTION

$$\text{Molarity} = \frac{0.715 \text{ mol glycine}}{495 \text{ mL soln}} \times \frac{1000 \text{ mL}}{1 \text{ L}} = 1.44 \text{ } M \text{ glycine}$$

CHECK A quick look at the math shows about 0.7 mol of glycine in about 0.5 L of solution, so the concentration should be about 1.4 mol/L, or 1.4 *M*.

FOLLOW-UP PROBLEM 3.15 How many moles of KI are in 84 mL of 0.50 *M* KI?

Amount (mol) of glycine

divide by volume (mL)

Concentration (mol/mL) of glycine

10^3 mL = 1 L

Molarity (mol/L) of glycine

Mole-Mass-Number Conversions Involving Solutions

Molarity can be thought of as a conversion factor used to convert between volume of solution and amount (mol) of solute, from which we then find the mass or the number of entities of solute. Figure 3.12 shows this new stoichiometric relationship, and Sample Problem 3.16 applies it.

Figure 3.12 **Summary of mass-mole-number-volume relationships in solution.** The amount (in moles) of a compound in solution is related to the volume of solution in liters through the molarity (*M*) in moles per liter. The other relationships shown are identical to those in Figure 3.4, except that here they refer to the quantities *in solution*. As in previous cases, to find the quantity of substance expressed in one form or another, convert the given information to moles first.

ARIS Presentation Center,
Art Library

ARIS Presentation Center,
Worked Examples Library

Biological application

Animation: Making a Solution
ARIS Presentation Center
Chapter 3

SAMPLE PROBLEM 3.16   Calculating Mass of Solute in a Given Volume of Solution

PROBLEM A *buffered* solution maintains acidity as a reaction occurs. In living cells, phosphate ions play a key buffering role, so biochemists often study reactions in such solutions. How many grams of solute are in 1.75 L of 0.460 *M* sodium hydrogen phosphate?

PLAN We know the solution volume (1.75 L) and molarity (0.460 *M*), and we need the mass of solute. We use the known quantities to find the amount (mol) of solute and then convert moles to grams with the solute molar mass, as shown in the roadmap.

SOLUTION Calculating moles of solute in solution:

$$\text{Moles of Na}_2\text{HPO}_4 = 1.75 \text{ L soln} \times \frac{0.460 \text{ mol Na}_2\text{HPO}_4}{1 \text{ L soln}} = 0.805 \text{ mol Na}_2\text{HPO}_4$$

Converting from moles of solute to grams:

$$\text{Mass (g) Na}_2\text{HPO}_4 = 0.805 \text{ mol Na}_2\text{HPO}_4 \times \frac{141.96 \text{ g Na}_2\text{HPO}_4}{1 \text{ mol Na}_2\text{HPO}_4}$$
$$= 114 \text{ g Na}_2\text{HPO}_4$$

CHECK The answer seems to be correct: ~1.8 L of 0.5 mol/L contains 0.9 mol, and 150 g/mol × 0.9 mol = 135 g, which is close to 114 g of solute.

FOLLOW-UP PROBLEM 3.16 In biochemistry laboratories, solutions of sucrose (table sugar, $C_{12}H_{22}O_{11}$) are used in high-speed centrifuges to separate the parts of a biological cell. How many liters of 3.30 *M* sucrose contain 135 g of solute?

Preparing and Diluting Molar Solutions

Whenever you prepare a solution of specific molarity, remember that the volume term in the denominator of the molarity expression is the *solution* volume, **not** the *solvent* volume. The solution volume includes contributions from solute *and* solvent, so you cannot simply dissolve 1 mol of solute in 1 L of solvent and expect a 1 *M* solution. The solute would increase the solution volume above 1 L, resulting in a lower-than-expected concentration. The correct preparation of a solution containing a solid solute consists of four steps. Let's go through them to prepare 0.500 L of 0.350 *M* nickel(II) nitrate hexahydrate [Ni(NO₃)₂·6H₂O]:

1. *Weigh the solid needed.* Calculate the mass of solid needed by converting from liters to moles and from moles to grams:

$$\text{Mass (g) of solute} = 0.500 \text{ L soln} \times \frac{0.350 \text{ mol Ni(NO}_3)_2 \cdot 6\text{H}_2\text{O}}{1 \text{ L soln}}$$

$$\times \frac{290.82 \text{ g Ni(NO}_3)_2 \cdot 6\text{H}_2\text{O}}{1 \text{ mol Ni(NO}_3)_2 \cdot 6\text{H}_2\text{O}}$$

$$= 50.9 \text{ g Ni(NO}_3)_2 \cdot 6\text{H}_2\text{O}$$

2. *Carefully transfer the solid to a volumetric flask that contains about half the final volume of solvent.* Since we need 0.500 L of solution, we choose a 500-mL volumetric flask. Add about 250 mL of distilled water and then transfer the solid. Wash down any solid clinging to the neck with a small amount of solvent.
3. *Dissolve the solid thoroughly by swirling.* If some solute remains undissolved, the solution will be less concentrated than expected, so be sure the solute is dissolved. If necessary, wait for the solution to reach room temperature. (As we'll discuss in Chapter 13, the solution process is often accompanied by heating or cooling.)
4. *Add solvent until the solution reaches its final volume.* Add distilled water to bring the volume exactly to the line on the flask neck; cover and mix thoroughly again. Figure 3.13 shows the last three steps.

Figure 3.13 Laboratory preparation of molar solutions.

ARIS Presentation Center,
Photo Library

A concentrated solution (higher molarity) is converted to a dilute solution (lower molarity) by adding solvent to it. The solution volume increases while the number of moles of solute remains the same. Thus, a given volume of the final (dilute) solution contains fewer solute particles and has a lower concentration than the initial (concentrated) solution (Figure 3.14). If various lower concentrations of a solution are needed, it is common practice to prepare a more concentrated solution (called a *stock solution*), which is stored and diluted as needed.

Concentrated solution
More solute particles per
unit volume

Dilute solution
Fewer solute particles per
unit volume

Figure 3.14 Converting a concentrated solution to a dilute solution.
When a solution is diluted, only solvent is added. The solution volume increases while the total number of moles of solute remains the same. Therefore, as shown in the blow-up views, a unit volume of concentrated solution contains more solute particles than the same unit volume of dilute solution.

Transparency/ARIS Presentation Center,
Art Library

Volume (L) of dilute solution

multiply by *M* (mol/L)
of dilute solution

**Amount (mol) of NaCl
in dilute solution =
Amount (mol) of NaCl
in concentrated solution**

divide by *M* (mol/L)
of concentrated solution

Volume (L) of concentrated solution

SAMPLE PROBLEM 3.17 Preparing a Dilute Solution from a Concentrated Solution

PROBLEM Isotonic saline is a 0.15 *M* aqueous solution of NaCl that simulates the total concentration of ions found in many cellular fluids. Its uses range from a cleansing rinse for contact lenses to a washing medium for red blood cells. How would you prepare 0.80 L of isotonic saline from a 6.0 *M* stock solution?

PLAN To dilute a concentrated solution, we add only solvent, so the *moles of solute are the same in both solutions*. We know the volume (0.80 L) and molarity (0.15 *M*) of the dilute (dil) NaCl solution we need, so we find the moles of NaCl it contains and then find the volume of concentrated (conc; 6.0 *M*) NaCl solution that contains the same number of moles. Then, we add solvent *up to* the final volume (see the roadmap).

SOLUTION Finding moles of solute in dilute solution:

$$\text{Moles of NaCl in dil soln} = 0.80 \;\text{L soln} \times \frac{0.15 \text{ mol NaCl}}{1 \text{ L soln}}$$

$$= 0.12 \text{ mol NaCl}$$

Finding moles of solute in concentrated solution: Because we add only solvent to dilute the solution,

$$\text{Moles of NaCl in dil soln} = \text{moles of NaCl in conc soln}$$

$$= 0.12 \text{ mol NaCl}$$

Finding the volume of concentrated solution that contains 0.12 mol of NaCl:

$$\text{Volume (L) of conc NaCl soln} = 0.12 \;\text{mol NaCl} \times \frac{1 \text{ L soln}}{6.0 \text{ mol NaCl}}$$

$$= 0.020 \text{ L soln}$$

To prepare 0.80 L of dilute solution, place 0.020 L of 6.0 *M* NaCl in a 1.0-L cylinder, add distilled water (~780 mL) to the 0.80-L mark, and stir thoroughly.

CHECK The answer seems reasonable because a small volume of concentrated solution is used to prepare a large volume of dilute solution. Also, the ratio of volumes (0.020 L/0.80 L) is the same as the ratio of concentrations (0.15 *M*/6.0 *M*).

FOLLOW-UP PROBLEM 3.17
To prepare a fertilizer, an engineer dilutes a stock solution of sulfuric acid by adding 25.0 m³ of 7.50 *M* acid to enough water to make 500. m³. What is the mass (in g) of sulfuric acid per milliliter of the diluted solution?

A very useful way to solve dilution problems, and others involving a change in concentration, applies the following relationship:

$$M_{\text{dil}} \times V_{\text{dil}} = \text{number of moles} = M_{\text{conc}} \times V_{\text{conc}} \qquad (3.9)$$

where the *M* and *V* terms are the molarity and volume of the *dil*ute (subscript "dil") and *conc*entrated (subscript "conc") solutions. In Sample Problem 3.17, for example, we found the volume of concentrated solution. Solving Equation 3.9 for V_{conc} gives

$$V_{\text{conc}} = \frac{M_{\text{dil}} \times V_{\text{dil}}}{M_{\text{conc}}} = \frac{0.15 \;M \times 0.80 \text{ L}}{6.0 \;M}$$

$$= 0.020 \text{ L}$$

The method worked out in the solution to Sample Problem 3.17 is the same calculation broken into two parts to emphasize the thinking process:

$$V_{\text{conc}} = 0.80 \;\text{L} \times \frac{0.15 \;\text{mol NaCl}}{1 \;\text{L}} \times \frac{1 \text{ L}}{6.0 \;\text{mol NaCl}}$$

$$= 0.020 \text{ L}$$

In the upcoming sample problem, we'll use a variation of this relationship to visualize changes in concentration.

Literature Raviolo, Andrés. "An Analogic Model for Understanding the Preparation of Volumetric Solutions," *Chem. Educator* 9 **2004**, *4*, 211–215.

SAMPLE PROBLEM 3.18 Visualizing Changes in Concentration

PROBLEM The top circle at right represents a unit volume of a solution. Draw a circle representing a unit volume of the solution after each of these changes:
(a) For every 1 mL of solution, 1 mL of solvent is added.
(b) One third of the solution's total volume is boiled off.
PLAN Given the starting solution, we have to find the number of solute particles in a unit volume after each change. The number of particles per unit volume, N, is directly related to moles per unit volume, M, so we can use a relationship similar to Equation 3.9 to find the number of particles to show in each circle. In (a), the volume increases, so the final solution is more dilute—fewer particles per unit volume. In (b), some solvent is lost, so the final solution is more concentrated—more particles per unit volume.
SOLUTION **(a)** Finding the number of particles in the dilute solution, N_{dil}:

$$N_{dil} \times V_{dil} = N_{conc} \times V_{conc}$$

thus, $$N_{dil} = N_{conc} \times \frac{V_{conc}}{V_{dil}} = 8 \text{ particles} \times \frac{1 \text{ mL}}{2 \text{ mL}} = 4 \text{ particles}$$

(b) Finding the number of particles in the concentrated solution, N_{conc}:

$$N_{dil} \times V_{dil} = N_{conc} \times V_{conc}$$

thus, $$N_{conc} = N_{dil} \times \frac{V_{dil}}{V_{conc}} = 8 \text{ particles} \times \frac{1 \text{ mL}}{\frac{2}{3} \text{ mL}} = 12 \text{ particles}$$

CHECK In (a), the volume is doubled (from 1 mL to 2 mL), so the number of particles per unit volume should be half of the original; $\frac{1}{2}$ of 8 is 4. In (b), the volume is reduced to $\frac{2}{3}$ of the original, so the number of particles per unit volume should be $\frac{3}{2}$ of the original; $\frac{3}{2}$ of 8 is 12.
COMMENT In (b), we assumed that only solvent boils off. This is true with nonvolatile solutes, such as ionic compounds, but in Chapter 13, we'll encounter solutions in which both solvent *and* solute are volatile.

(a)

(a)

(b)

FOLLOW-UP PROBLEM 3.18
The circle labeled A represents a unit volume of a solution. Explain the changes that must be made to A to obtain the solutions depicted in B and C.

A   B   C

Stoichiometry of Chemical Reactions in Solution

Solving stoichiometry problems for reactions in solution requires the same approach as before, with the additional step of converting the volume of reactant or product to moles: (1) balance the equation, (2) find the number of moles of one substance, (3) relate it to the stoichiometrically equivalent number of moles of another substance, and (4) convert to the desired units.

SAMPLE PROBLEM 3.19 Calculating Amounts of Reactants and Products for a Reaction in Solution

 Biological application

PROBLEM Specialized cells in the stomach release HCl to aid digestion. If they release too much, the excess can be neutralized with an antacid to avoid discomfort. A common antacid contains magnesium hydroxide, $Mg(OH)_2$, which reacts with the acid to form water and magnesium chloride solution. As a government chemist testing commercial antacids, you use 0.10 M HCl to simulate the acid concentration in the stomach. How many liters of "stomach acid" react with a tablet containing 0.10 g of $Mg(OH)_2$?

PLAN We know the mass of $Mg(OH)_2$ (0.10 g) that reacts and the acid concentration (0.10 M), and we must find the acid volume. After writing the balanced equation, we convert the grams of $Mg(OH)_2$ to moles, use the molar ratio to find the moles of HCl that react with these moles of $Mg(OH)_2$, and then use the molarity of HCl to find the volume that contains this number of moles. The steps appear in the roadmap.

SOLUTION Writing the balanced equation:

$$Mg(OH)_2(s) + 2HCl(aq) \longrightarrow MgCl_2(aq) + 2H_2O(l)$$

Converting from grams of $Mg(OH)_2$ to moles:

$$\text{Moles of } Mg(OH)_2 = 0.10 \text{ g } Mg(OH)_2 \times \frac{1 \text{ mol } Mg(OH)_2}{58.33 \text{ g } Mg(OH)_2} = 1.7 \times 10^{-3} \text{ mol } Mg(OH)_2$$

Converting from moles of $Mg(OH)_2$ to moles of HCl:

$$\text{Moles of HCl} = 1.7 \times 10^{-3} \text{ mol } Mg(OH)_2 \times \frac{2 \text{ mol HCl}}{1 \text{ mol } Mg(OH)_2} = 3.4 \times 10^{-3} \text{ mol HCl}$$

Converting from moles of HCl to liters:

$$\text{Volume (L) of HCl} = 3.4 \times 10^{-3} \text{ mol HCl} \times \frac{1 \text{ L}}{0.10 \text{ mol HCl}} = 3.4 \times 10^{-2} \text{ L}$$

CHECK The size of the answer seems reasonable: a small volume of dilute acid (0.034 L of 0.10 M) reacts with a small amount of antacid (0.0017 mol).

COMMENT The reaction as written is an oversimplification; in reality, HCl and $MgCl_2$ exist as separated ions in solution (covered in great detail in Chapters 4 and 18).

FOLLOW-UP PROBLEM 3.19 Another active ingredient in some antacids is aluminum hydroxide. Which is more effective at neutralizing stomach acid, magnesium hydroxide or aluminum hydroxide? [*Hint:* Effectiveness refers to the amount of acid that reacts with a given mass of antacid. You already know the effectiveness of 0.10 g of $Mg(OH)_2$.]

In limiting-reactant problems for reactions in solution, we first determine which reactant is limiting and then determine the yield, as demonstrated in the next sample problem.

ARIS Presentation Center, Worked Examples Library

Environmental application

SAMPLE PROBLEM 3.20 Solving Limiting-Reactant Problems for Reactions in Solution

PROBLEM Mercury and its compounds have many uses, from fillings for teeth (as a mixture with silver, copper, and tin) to the industrial production of chlorine. Because of their toxicity, however, soluble mercury compounds, such as mercury(II) nitrate, must be removed from industrial wastewater. One removal method reacts the wastewater with sodium sulfide solution to produce solid mercury(II) sulfide and sodium nitrate solution. In a laboratory simulation, 0.050 L of 0.010 M mercury(II) nitrate reacts with 0.020 L of 0.10 M sodium sulfide. How many grams of mercury(II) sulfide form?

PLAN This is a limiting-reactant problem because *the amounts of two reactants are given*. After balancing the equation, we must determine the limiting reactant. The molarity (0.010 M) and volume (0.050 L) of the mercury(II) nitrate solution tell us the moles of one reactant, and the molarity (0.10 M) and volume (0.020 L) of the sodium sulfide solution tell us the moles of the other. Then, as in Sample Problem 3.12(b), we use the molar ratio to find the moles of product (HgS) that form from each reactant, *assuming the other reactant is in excess*. The limiting reactant is the one that forms fewer moles of HgS, which we convert to mass using the HgS molar mass. The roadmap shows the process.

SOLUTION Writing the balanced equation:

$$Hg(NO_3)_2(aq) + Na_2S(aq) \longrightarrow HgS(s) + 2NaNO_3(aq)$$

Finding moles of HgS assuming $Hg(NO_3)_2$ is limiting: Combining the steps gives

$$\text{Moles of HgS} = 0.050 \text{ L soln} \times \frac{0.010 \text{ mol } Hg(NO_3)_2}{1 \text{ L soln}} \times \frac{1 \text{ mol HgS}}{1 \text{ mol } Hg(NO_3)_2}$$
$$= 5.0 \times 10^{-4} \text{ mol HgS}$$

Finding moles of HgS assuming Na_2S is limiting: Combining the steps gives

$$\text{Moles of HgS} = 0.020 \text{ L soln} \times \frac{0.10 \text{ mol } Na_2S}{1 \text{ L soln}} \times \frac{1 \text{ mol HgS}}{1 \text{ mol } Na_2S}$$

$$= 2.0\times10^{-3} \text{ mol HgS}$$

$Hg(NO_3)_2$ is the limiting reactant because it forms fewer moles of HgS.
Converting the moles of HgS formed from $Hg(NO_3)_2$ to grams:

$$\text{Mass (g) of HgS} = 5.0\times10^{-4} \text{ mol HgS} \times \frac{232.7 \text{ g HgS}}{1 \text{ mol HgS}} = 0.12 \text{ g HgS}$$

CHECK As a check, let's use the alternative method for finding the limiting reactant (see Comment in Sample Problem 3.13, p. 117).
Finding moles of reactants available:

$$\text{Moles of } Hg(NO_3)_2 = 0.050 \text{ L soln} \times \frac{0.010 \text{ mol } Hg(NO_3)_2}{1 \text{ L soln}} = 5.0\times10^{-4} \text{ mol } Hg(NO_3)_2$$

$$\text{Moles of } Na_2S = 0.020 \text{ L soln} \times \frac{0.10 \text{ mol } Na_2S}{1 \text{ L soln}} = 2.0\times10^{-3} \text{ mol } Na_2S$$

The molar ratio of the reactants is 1 $Hg(NO_3)_2$/1 Na_2S. Therefore, $Hg(NO_3)_2$ is limiting because there are fewer moles of it than are needed to react with the available moles of Na_2S.
Finding grams of product from moles of limiting reactant and the molar ratio:

$$\text{Mass (g) of HgS} = 5.0\times10^{-4} \text{ mol } Hg(NO_3)_2 \times \frac{1 \text{ mol HgS}}{1 \text{ mol } Hg(NO_3)_2} \times \frac{232.7 \text{ g HgS}}{1 \text{ mol HgS}}$$

$$= 0.12 \text{ g HgS}$$

Let's use these amounts to prepare a reaction table:

Amount (mol)	$Hg(NO_3)_2(aq)$ +	$Na_2S(aq)$ ⟶	HgS(s) +	$2NaNO_3(aq)$
Initial	5.0×10^{-4}	2.0×10^{-3}	0	0
Change	-5.0×10^{-4}	-5.0×10^{-4}	$+5.0\times10^{-4}$	$+1.0\times10^{-3}$
Final	0	1.5×10^{-3}	5.0×10^{-4}	1.0×10^{-3}

Note the large excess of Na_2S that remains after the reaction.

FOLLOW-UP PROBLEM 3.20 Even though gasoline sold in the United States no longer contains lead, this metal persists in the environment as a poison. Despite their toxicity, many compounds of lead are still used to make pigments.
(a) What volume of 1.50 M lead(II) acetate contains 0.400 mol of Pb^{2+} ions?
(b) When this volume reacts with 125 mL of 3.40 M sodium chloride, how many grams of solid lead(II) chloride can form? (Sodium acetate solution also forms.)

Section Summary

When reactions occur in solution, reactant and product amounts are given in terms of concentration and volume. • Molarity is the number of moles of solute dissolved in one liter of solution. A concentrated solution (higher molarity) is converted to a dilute solution (lower molarity) by adding solvent. • Using molarity as a conversion factor, we apply the principles of stoichiometry to all aspects of reactions in solution.

Chapter Perspective

You apply the mole concept every time you weigh a substance, dissolve it, or think about how much of it will react. Figure 3.15 combines the individual stoichiometry summary diagrams into one overall review diagram. Use it for homework, to study for exams, or to obtain an overview of the various ways that the amounts involved in a reaction are interrelated. We apply stoichiometry next to some of the most important types of chemical reactions (Chapter 4), to systems of reacting gases (Chapter 5), and to the heat involved in a reaction (Chapter 6). These concepts and skills appear at many places later in the text as well.

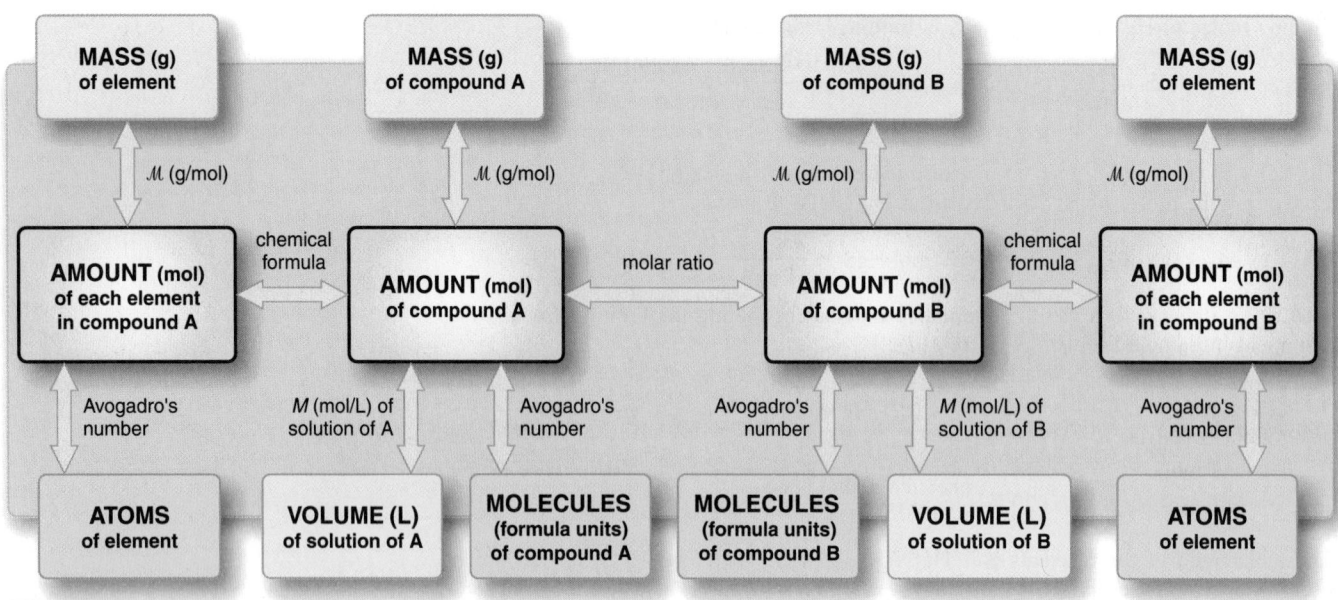

Figure 3.15 An overview of the key mass-mole-number stoichiometric relationships.

Transparency/ARIS Presentation Center,
Art Library

CHAPTER REVIEW GUIDE

The following sections provide many aids to help you study this chapter. (Numbers in parentheses refer to pages, unless noted otherwise.)

Learning Objectives These are concepts and skills you should know after studying this chapter.

Relevant section and/or sample problem (SP) numbers appear in parentheses.

Understand These Concepts

1. The meaning and usefulness of the mole (3.1)
2. The relation between molecular (or formula) mass and molar mass (3.1)
3. The relations among amount of substance (in moles), mass (in grams), and number of chemical entities (3.1)
4. The information in a chemical formula (3.1)
5. The procedure for finding the empirical and molecular formulas of a compound (3.2)
6. How more than one substance can have the same empirical formula and the same molecular formula (isomers) (3.2)
7. The importance of balancing equations for the quantitative study of chemical reactions (3.3)
8. The mole-mass-number information in a balanced equation (3.4)
9. The relation between amounts of reactants and products (3.4)
10. Why one reactant limits the yield of product (3.4)
11. The causes of lower-than-expected yields and the distinction between theoretical and actual yields (3.4)
12. The meanings of concentration and molarity (3.5)
13. The effect of dilution on the concentration of solute (3.5)
14. How reactions in solution differ from those of pure reactants (3.5)

Master These Skills

1. Calculating the molar mass of any substance (3.1; also SPs 3.3, 3.4)
2. Converting between amount of substance (in moles), mass (in grams), and number of chemical entities (SPs 3.1–3.3)
3. Using mass percent to find the mass of element in a given mass of compound (SP 3.4)
4. Determining empirical and molecular formulas of a compound from mass percent and molar mass of elements (SPs 3.5, 3.6)
5. Determining a molecular formula from combustion analysis (SP 3.7)
6. Converting a chemical statement or a molecular depiction into a balanced equation (SPs 3.8, 3.9)
7. Using stoichiometrically equivalent molar ratios to calculate amounts of reactants and products in reactions of pure and dissolved substances (SPs 3.10, 3.19)
8. Writing an overall equation from a series of equations (SP 3.11)
9. Solving limiting-reactant problems from molecular depictions and for reactions of pure and dissolved substances (SPs 3.12, 3.13, 3.20)
10. Calculating percent yield (SP 3.14)
11. Calculating molarity and the mass of solute in solution (SPs 3.15, 3.16)
12. Preparing a dilute solution from a concentrated one (SP 3.17)
13. Using molecular depictions to understand changes in volume (SP 3.18)

Key Terms These important terms appear in boldface in the chapter and are defined again in the Glossary.

stoichiometry (90)

Section 3.1

mole (mol) (90)

Avogadro's number (90)

molar mass ($\mathcal{M}$) (92)

Section 3.2

combustion analysis (101)

isomer (103)

Section 3.3

chemical equation (104)

reactant (105)

product (105)

balancing (stoichiometric)
coefficient (105)

Section 3.4

overall (net) equation (112)

limiting reactant (114)

theoretical yield (118)

side reaction (118)

actual yield (118)

percent yield (% yield) (118)

green chemistry (120)

Section 3.5

solute (121)

solvent (121)

concentration (121)

molarity (M) (121)

Key Equations and Relationships Numbered and screened concepts are listed for you to refer to or memorize.

3.1 Number of entities in one mole (90):

1 mol contains 6.022×10^{23} entities (to 4 sf)

3.2 Converting amount (mol) to mass using $\mathcal{M}$ (93):

$$\text{Mass (g)} = \text{no. of moles} \times \frac{\text{no. of grams}}{1 \text{ mol}}$$

3.3 Converting mass to amount (mol) using $1/\mathcal{M}$ (93):

$$\text{No. of moles} = \text{mass (g)} \times \frac{1 \text{ mol}}{\text{no. of grams}}$$

3.4 Converting amount (mol) to number of entities (93):

$$\text{No. of entities} = \text{no. of moles} \times \frac{6.022 \times 10^{23} \text{ entities}}{1 \text{ mol}}$$

3.5 Converting number of entities to amount (mol) (93):

$$\text{No. of moles} = \text{no. of entities} \times \frac{1 \text{ mol}}{6.022 \times 10^{23} \text{ entities}}$$

3.6 Calculating mass % (96):

Mass % of element X

$$= \frac{\text{moles of X in formula} \times \text{molar mass of X (g/mol)}}{\text{mass (g) of 1 mol of compound}} \times 100$$

3.7 Calculating percent yield (118):

$$\% \text{ yield} = \frac{\text{actual yield}}{\text{theoretical yield}} \times 100$$

3.8 Defining molarity (121):

$$\text{Molarity} = \frac{\text{moles of solute}}{\text{liters of solution}} \quad \text{or} \quad M = \frac{\text{mol solute}}{\text{L soln}}$$

3.9 Diluting a concentrated solution (124):

$$M_{\text{dil}} \times V_{\text{dil}} = \text{number of moles} = M_{\text{conc}} \times V_{\text{conc}}$$

Highlighted Figures and Tables These figures (F) and tables (T) provide a visual review of key ideas.

T3.1 Summary of mass terminology (92)

T3.2 Information contained in a formula (93)

F3.3 Mass-mole-number relationships for elements (94)

F3.4 Mass-mole-number relationships for compounds (95)

T3.5 Information contained in a balanced equation (110)

F3.9 Mass-mole-number relationships in a chemical reaction (111)

F3.12 Mass-mole-number-volume relationships in solution (122)

F3.15 Overview of mass-mole-number relationships (128)

Brief Solutions to FOLLOW-UP PROBLEMS Compare your solutions to these calculation steps and answers.

3.1 Moles of C = $315 \text{ mg C} \times \dfrac{1 \text{ g}}{10^3 \text{ mg}} \times \dfrac{1 \text{ mol C}}{12.01 \text{ g C}}$

$\qquad = 2.62 \times 10^{-2}$ mol C

3.2 Mass (g) of Mn = 3.22×10^{20} Mn atoms

$\qquad \times \dfrac{1 \text{ mol Mn}}{6.022 \times 10^{23} \text{ Mn atoms}} \times \dfrac{54.94 \text{ g Mn}}{1 \text{ mol Mn}}$

$\qquad = 2.94 \times 10^{-2}$ g Mn

3.3 (a) Mass (g) of P_4O_{10}

$\qquad = 4.65 \times 10^{22}$ molecules P_4O_{10}

$\qquad \times \dfrac{1 \text{ mol } P_4O_{10}}{6.022 \times 10^{23} \text{ molecules } P_4O_{10}} \times \dfrac{283.88 \text{ g } P_4O_{10}}{1 \text{ mol } P_4O_{10}}$

$\qquad = 21.9$ g P_4O_{10}

(b) No. of P atoms = 4.65×10^{22} molecules P_4O_{10}

$\qquad \times \dfrac{4 \text{ atoms P}}{1 \text{ molecule } P_4O_{10}}$

$\qquad = 1.86 \times 10^{23}$ P atoms

3.4 (a) Mass % of N = $\dfrac{2 \text{ mol N} \times \dfrac{14.01 \text{ g N}}{1 \text{ mol N}}}{80.05 \text{ g } NH_4NO_3} \times 100$

$\qquad = 35.00$ mass % N

(b) Mass (g) of N = $35.8 \text{ kg } NH_4NO_3 \times \dfrac{10^3 \text{ g}}{1 \text{ kg}} \times \dfrac{0.3500 \text{ g N}}{1 \text{ g } NH_4NO_3}$

$\qquad = 1.25 \times 10^4$ g N

3.5 Moles of S = $2.88 \text{ g S} \times \dfrac{1 \text{ mol S}}{32.07 \text{ g S}} = 0.0898$ mol S

Moles of M = $0.0898 \text{ mol S} \times \dfrac{2 \text{ mol M}}{3 \text{ mol S}} = 0.0599$ mol M

Molar mass of M = $\dfrac{3.12 \text{ g M}}{0.0599 \text{ mol M}} = 52.1$ g/mol

M is chromium, and M_2S_3 is chromium(III) sulfide.

3.6 Assuming 100.00 g of compound, we have 95.21 g of C and 4.79 g of H:

$$\text{Moles of C} = 95.21 \text{ g C} \times \frac{1 \text{ mol C}}{12.01 \text{ g C}}$$
$$= 7.928 \text{ mol C}$$

Also, 4.75 mol H

Preliminary formula = $C_{7.928}H_{4.75} \approx C_{1.67}H_{1.00}$

Empirical formula = C_5H_3

$$\text{Whole-number multiple} = \frac{252.30 \text{ g/mol}}{63.07 \text{ g/mol}} = 4$$

Molecular formula = $C_{20}H_{12}$

3.7 Mass (g) of C = $0.451 \text{ g CO}_2 \times \frac{12.01 \text{ g C}}{44.01 \text{ g CO}_2} = 0.123 \text{ g C}$

Also, 0.00690 g H

Mass (g) of Cl = 0.250 g − (0.123 g + 0.00690 g) = 0.120 g Cl

Moles of elements
$$= 0.0102 \text{ mol C}; \ 0.00685 \text{ mol H}; \ 0.00339 \text{ mol Cl}$$

Empirical formula = C_3H_2Cl; multiple = 2

Molecular formula = $C_6H_4Cl_2$

3.8 (a) $2Na(s) + 2H_2O(l) \longrightarrow H_2(g) + 2NaOH(aq)$

(b) $2HNO_3(aq) + CaCO_3(s) \longrightarrow$
$$H_2O(l) + CO_2(g) + Ca(NO_3)_2(aq)$$

(c) $PCl_3(g) + 3HF(g) \longrightarrow PF_3(g) + 3HCl(g)$

(d) $4C_3H_5N_3O_9(l) \longrightarrow$
$$12CO_2(g) + 10H_2O(g) + 6N_2(g) + O_2(g)$$

3.9 From the depiction, we have $6CO + 3O_2 \longrightarrow 6CO_2$

Or, $2CO(g) + O_2(g) \longrightarrow 2CO_2(g)$

3.10 $Fe_2O_3(s) + 2Al(s) \longrightarrow Al_2O_3(s) + 2Fe(l)$

(a) Mass (g) of Fe
$$= 135 \text{ g Al} \times \frac{1 \text{ mol Al}}{26.98 \text{ g Al}} \times \frac{2 \text{ mol Fe}}{2 \text{ mol Al}} \times \frac{55.85 \text{ g Fe}}{1 \text{ mol Fe}}$$
$$= 279 \text{ g Fe}$$

(b) No. of Al atoms = $1.00 \text{ g Al}_2O_3 \times \frac{1 \text{ mol Al}_2O_3}{101.96 \text{ g Al}_2O_3}$
$$\times \frac{2 \text{ mol Al}}{1 \text{ mol Al}_2O_3} \times \frac{6.022 \times 10^{23} \text{ Al atoms}}{1 \text{ mol Al}}$$
$$= 1.18 \times 10^{22} \text{ Al atoms}$$

3.11 $2SO_2(g) + O_2(g) \longrightarrow 2SO_3(g)$
$$\underline{2SO_3(g) + 2H_2O(l) \longrightarrow 2H_2SO_4(aq)}$$
$$2SO_2(g) + O_2(g) + 2H_2O(l) \longrightarrow 2H_2SO_4(aq)$$

3.12 (a) $2AB + B_2 \longrightarrow 2AB_2$

In the circles, the AB/B_2 ratio is 4/3, which is less than the 2/1 ratio in the equation. Thus, there is not enough AB, so it is the limiting reactant; note that one molecule of B_2 is in excess.

(b) Moles of $AB_2 = 1.5 \text{ mol AB} \times \frac{2 \text{ mol AB}_2}{2 \text{ mol AB}} = 1.5 \text{ mol AB}_2$

$$\text{Moles of } AB_2 = 1.5 \text{ mol B}_2 \times \frac{2 \text{ mol AB}_2}{1 \text{ mol B}_2} = 3.0 \text{ mol AB}_2$$

Therefore, 1.5 mol of AB_2 can form.

3.13 $2Al(s) + 3S(s) \longrightarrow Al_2S_3(s)$

Mass (g) of Al_2S_3 formed from 10.0 g of Al
$$= 10.0 \text{ g Al} \times \frac{1 \text{ mol Al}}{26.98 \text{ g Al}} \times \frac{1 \text{ mol Al}_2S_3}{2 \text{ mol Al}} \times \frac{150.17 \text{ g Al}_2S_3}{1 \text{ mol Al}_2S_3}$$
$$= 27.8 \text{ g Al}_2S_3$$

Similarly, mass (g) of Al_2S_3 formed from 15.0 g of S = 23.4 g Al_2S_3. Thus, S is the limiting reactant, and 23.4 g of Al_2S_3 forms.

Mass (g) of Al in excess
$$= \text{total mass of Al} - \text{mass of Al used}$$
$$= 10.0 \text{ g Al}$$
$$- \left(15.0 \text{ g S} \times \frac{1 \text{ mol S}}{32.07 \text{ g S}} \times \frac{2 \text{ mol Al}}{3 \text{ mol S}} \times \frac{26.98 \text{ g Al}}{1 \text{ mol Al}}\right)$$
$$= 1.6 \text{ g Al}$$

(We would obtain the same answer if sulfur were shown more correctly as S_8.)

3.14 $CaCO_3(s) + 2HCl(aq) \longrightarrow CaCl_2(aq) + H_2O(l) + CO_2(g)$

Theoretical yield (g) of CO_2
$$= 10.0 \text{ g CaCO}_3 \times \frac{1 \text{ mol CaCO}_3}{100.09 \text{ g CaCO}_3} \times \frac{1 \text{ mol CO}_2}{1 \text{ mol CaCO}_3}$$
$$\times \frac{44.01 \text{ g CO}_2}{1 \text{ mol CO}_2} = 4.40 \text{ g CO}_2$$

$$\% \text{ yield} = \frac{3.65 \text{ g CO}_2}{4.40 \text{ g CO}_2} \times 100 = 83.0\%$$

3.15 Moles of KI = $84 \text{ mL soln} \times \frac{1 \text{ L}}{10^3 \text{ mL}} \times \frac{0.50 \text{ mol KI}}{1 \text{ L soln}}$
$$= 0.042 \text{ mol KI}$$

3.16 Volume (L) of soln
$$= 135 \text{ g sucrose} \times \frac{1 \text{ mol sucrose}}{342.30 \text{ g sucrose}} \times \frac{1 \text{ L soln}}{3.30 \text{ mol sucrose}}$$
$$= 0.120 \text{ L soln}$$

3.17 M_{dil} of $H_2SO_4 = \frac{7.50 \ M \times 25.0 \text{ m}^3}{500. \text{ m}^3} = 0.375 \ M \ H_2SO_4$

Mass (g) of H_2SO_4/mL soln
$$= \frac{0.375 \text{ mol H}_2SO_4}{1 \text{ L soln}} \times \frac{1 \text{ L}}{10^3 \text{ mL}} \times \frac{98.09 \text{ g H}_2SO_4}{1 \text{ mol H}_2SO_4}$$
$$= 3.68 \times 10^{-2} \text{ g/mL soln}$$

3.18 To obtain B, the total volume of solution A was reduced by half:

$$V_{\text{conc}} = V_{\text{dil}} \times \frac{N_{\text{dil}}}{N_{\text{conc}}} = 1.0 \text{ mL} \times \frac{6 \text{ particles}}{12 \text{ particles}} = 0.50 \text{ mL}$$

To obtain C, $\frac{1}{2}$ of a volume of solvent was added for every volume of A:

$$V_{\text{dil}} = V_{\text{conc}} \times \frac{N_{\text{conc}}}{N_{\text{dil}}} = 1.0 \text{ mL} \times \frac{6 \text{ particles}}{4 \text{ particles}} = 1.5 \text{ mL}$$

3.19 $Al(OH)_3(s) + 3HCl(aq) \longrightarrow AlCl_3(aq) + 3H_2O(l)$

Volume (L) of HCl consumed
$$= 0.10 \text{ g Al(OH)}_3 \times \frac{1 \text{ mol Al(OH)}_3}{78.00 \text{ g Al(OH)}_3}$$
$$\times \frac{3 \text{ mol HCl}}{1 \text{ mol Al(OH)}_3} \times \frac{1 \text{ L soln}}{0.10 \text{ mol HCl}}$$
$$= 3.8 \times 10^{-2} \text{ L soln}$$

Therefore, $Al(OH)_3$ is more effective than $Mg(OH)_2$.

3.20 (a) Volume (L) of soln
$$= 0.400 \text{ mol Pb}^{2+}$$
$$\times \frac{1 \text{ mol Pb(C}_2H_3O_2)_2}{1 \text{ mol Pb}^{2+}} \times \frac{1 \text{ L soln}}{1.50 \text{ mol Pb(C}_2H_3O_2)_2}$$
$$= 0.267 \text{ L soln}$$

(b) $Pb(C_2H_3O_2)_2(aq) + 2NaCl(aq) \longrightarrow$
$$PbCl_2(s) + 2NaC_2H_3O_2(aq)$$

Mass (g) of $PbCl_2$ from $Pb(C_2H_3O_2)_2$ soln = 111 g $PbCl_2$

Mass (g) of $PbCl_2$ from NaCl soln = 59.1 g $PbCl_2$

Thus, NaCl is the limiting reactant, and 59.1 g of $PbCl_2$ can form.

PROBLEMS

Problems with **colored** numbers are answered in Appendix E and worked in detail in the Student Solutions Manual. Problem sections match those in the text and provide the numbers of relevant sample problems. Most offer Concept Review Questions, Skill-Building Exercises (grouped in pairs covering the same concept), and Problems in Context. The Comprehensive Problems are based on material from any section or previous chapter.

The Mole

(Sample Problems 3.1 to 3.4)

● Concept Review Questions

● **3.1** The atomic mass of Cl is 35.45 amu, and the atomic mass of Al is 26.98 amu. What are the masses in grams of 3 mol of Al atoms and of 2 mol of Cl atoms?

● **3.2** (a) How many moles of C atoms are in 1 mol of sucrose ($C_{12}H_{22}O_{11}$)?
(b) How many C atoms are in 2 mol of sucrose?

● **3.3** Why might the expression "1 mol of chlorine" be confusing? What change would remove any uncertainty? For what other elements might a similar confusion exist? Why?

● **3.4** How is the molecular mass of a compound the same as the molar mass, and how is it different?

● **3.5** What advantage is there to using a counting unit (the mole) in chemistry rather than a mass unit?

● **3.6** You need to calculate the number of P_4 molecules that can form from 2.5 g of $Ca_3(PO_4)_2$. Explain how you would proceed. (That is, write a solution "Plan," without actually doing any calculations.)

●● **3.7** Each of the following balances weighs the indicated numbers of atoms of two elements:

(a)   (b)
(c)   (d)

Which element—left, right, or neither,
(a) Has the higher molar mass?
(b) Has more atoms per gram?
(c) Has fewer atoms per gram?
(d) Has more atoms per mole?

● Skill-Building Exercises (grouped in similar pairs)

● **3.8** Calculate the molar mass of each of the following:
(a) $Sr(OH)_2$ (b) N_2O_3 (c) $NaClO_3$ (d) Cr_2O_3

● **3.9** Calculate the molar mass of each of the following:
(a) $(NH_4)_3PO_4$ (b) CH_2Cl_2 (c) $CuSO_4 \cdot 5H_2O$ (d) BrF_3

● **3.10** Calculate the molar mass of each of the following:
(a) SnO   (b) BaF_2 (c) $Al_2(SO_4)_3$ (d) $MnCl_2$

● **3.11** Calculate the molar mass of each of the following:
(a) N_2O_4 (b) C_4H_9OH (c) $MgSO_4 \cdot 7H_2O$ (d) $Ca(C_2H_3O_2)_2$

● **3.12** Calculate each of the following quantities:
(a) Mass in grams of 0.68 mol of $KMnO_4$
(b) Moles of O atoms in 8.18 g of $Ba(NO_3)_2$
(c) Number of O atoms in 7.3×10^{-3} g of $CaSO_4 \cdot 2H_2O$

● **3.13** Calculate each of the following quantities:
(a) Mass in kilograms of 4.6×10^{21} molecules of NO_2
(b) Moles of Cl atoms in 0.0615 g of $C_2H_4Cl_2$
(c) Number of H^- ions in 5.82 g of SrH_2

● **3.14** Calculate each of the following quantities:
(a) Mass in grams of 6.44×10^{-2} mol of $MnSO_4$
(b) Moles of compound in 15.8 kg of $Fe(ClO_4)_3$
(c) Number of N atoms in 92.6 mg of NH_4NO_2

● **3.15** Calculate each of the following quantities:
(a) Total number of ions in 38.1 g of SrF_2
(b) Mass in kilograms of 3.58 mol of $CuCl_2 \cdot 2H_2O$
(c) Mass in milligrams of 2.88×10^{22} formula units of $Bi(NO_3)_3 \cdot 5H_2O$

● **3.16** Calculate each of the following quantities:
(a) Mass in grams of 8.35 mol of copper(I) carbonate
(b) Mass in grams of 4.04×10^{20} molecules of dinitrogen pentaoxide
(c) Number of moles and formula units in 78.9 g of sodium perchlorate
(d) Number of sodium ions, perchlorate ions, Cl atoms, and O atoms in the mass of compound in part (c)

● **3.17** Calculate each of the following quantities:
(a) Mass in grams of 8.42 mol of chromium(III) sulfate decahydrate
(b) Mass in grams of 1.83×10^{24} molecules of dichlorine heptaoxide
(c) Number of moles and formula units in 6.2 g of lithium sulfate
(d) Number of lithium ions, sulfate ions, S atoms, and O atoms in the mass of compound in part (c)

● **3.18** Calculate each of the following:
(a) Mass % of H in ammonium bicarbonate
(b) Mass % of O in sodium dihydrogen phosphate heptahydrate

● **3.19** Calculate each of the following:
(a) Mass % of I in strontium periodate
(b) Mass % of Mn in potassium permanganate

● **3.20** Calculate each of the following:
(a) Mass fraction of C in cesium acetate
(b) Mass fraction of O in uranyl sulfate trihydrate (the uranyl ion is UO_2^{2+})

● **3.21** Calculate each of the following:
(a) Mass fraction of Cl in calcium chlorate
(b) Mass fraction of P in tetraphosphorus hexaoxide

● Problems in Context

● **3.22** Oxygen is required for the metabolic combustion of foods. Calculate the number of atoms in 38.0 g of oxygen gas, the amount absorbed from the lungs at rest in about 15 min.

● **3.23** Cisplatin (*right*), or Platinol, is used in the treatment of certain cancers. Calculate (a) the moles of compound in 285.3 g of cisplatin; (b) the number of hydrogen atoms in 0.98 mol of cisplatin.

Levels of Difficulty: ● Easy   ●● Medium   ●●● Challenging   Biological: 3.22, 3.23

• **3.24** Allyl sulfide *(below)* gives garlic its characteristic odor.

Calculate (a) the mass in grams of 2.63 mol of allyl sulfide; (b) the number of carbon atoms in 35.7 g of allyl sulfide.

• **3.25** Iron reacts slowly with oxygen and water to form a compound commonly called rust ($Fe_2O_3 \cdot 4H_2O$). For 45.2 kg of rust, calculate (a) the moles of compound; (b) the moles of Fe_2O_3; (c) the grams of iron.

• **3.26** Propane is widely used in liquid form as a fuel for barbecue grills and camp stoves. For 85.5 g of propane, calculate (a) the moles of compound; (b) the grams of carbon.

• **3.27** The effectiveness of a nitrogen fertilizer is determined mainly by its mass % N. Rank the following fertilizers, most effective first: potassium nitrate; ammonium nitrate; ammonium sulfate; urea, $CO(NH_2)_2$.

•• **3.28** The mineral galena is composed of lead(II) sulfide and has an average density of 7.46 g/cm^3. (a) How many moles of lead(II) sulfide are in 1.00 ft^3 of galena? (b) How many lead atoms are in 1.00 dm^3 of galena?

• **3.29** Hemoglobin, a protein in red blood cells, carries O_2 from the lungs to the body's cells. Iron (as ferrous ion, Fe^{2+}) makes up 0.33 mass % of hemoglobin. If the molar mass of hemoglobin is 6.8×10^4 g/mol, how many Fe^{2+} ions are in one molecule?

Determining the Formula of an Unknown Compound
(Sample Problems 3.5 to 3.7)

● Concept Review Questions

• **3.30** List three ways compositional data may be given in a problem that involves finding an empirical formula.

• **3.31** Which of the following sets of information allows you to obtain the molecular formula of a covalent compound? In each case that allows it, explain how you would proceed (write a solution "Plan").
(a) Number of moles of each type of atom in a given sample of the compound
(b) Mass % of each element and the total number of atoms in a molecule of the compound
(c) Mass % of each element and the number of atoms of one element in a molecule of the compound
(d) Empirical formula and mass % of each element in the compound
(e) Structural formula of the compound

• **3.32** Is $MgCl_2$ an empirical or a molecular formula for magnesium chloride? Explain.

● Skill-Building Exercises *(grouped in similar pairs)*

• **3.33** What is the empirical formula and empirical formula mass for each of the following compounds?
(a) C_2H_4 (b) $C_2H_6O_2$ (c) N_2O_5 (d) $Ba_3(PO_4)_2$ (e) Te_4I_{16}

• **3.34** What is the empirical formula and empirical formula mass for each of the following compounds?
(a) C_4H_8 (b) $C_3H_6O_3$ (c) P_4O_{10} (d) $Ga_2(SO_4)_3$ (e) Al_2Br_6

• **3.35** What is the molecular formula of each compound?
(a) Empirical formula CH_2 ($\mathcal{M} = 42.08$ g/mol)
(b) Empirical formula NH_2 ($\mathcal{M} = 32.05$ g/mol)

(c) Empirical formula NO_2 ($\mathcal{M} = 92.02$ g/mol)
(d) Empirical formula CHN ($\mathcal{M} = 135.14$ g/mol)

• **3.36** What is the molecular formula of each compound?
(a) Empirical formula CH ($\mathcal{M} = 78.11$ g/mol)
(b) Empirical formula $C_3H_6O_2$ ($\mathcal{M} = 74.08$ g/mol)
(c) Empirical formula HgCl ($\mathcal{M} = 472.1$ g/mol)
(d) Empirical formula $C_7H_4O_2$ ($\mathcal{M} = 240.20$ g/mol)

• **3.37** Find the empirical formula of the following compounds:
(a) 0.063 mol of chlorine atoms combined with 0.22 mol of oxygen atoms
(b) 2.45 g of silicon combined with 12.4 g of chlorine
(c) 27.3 mass % carbon and 72.7 mass % oxygen

• **3.38** Find the empirical formula of the following compounds:
(a) 0.039 mol of iron atoms combined with 0.052 mol of oxygen atoms
(b) 0.903 g of phosphorus combined with 6.99 g of bromine
(c) A hydrocarbon with 79.9 mass % carbon

• **3.39** An oxide of nitrogen contains 30.45 mass % N. (a) What is the empirical formula of the oxide? (b) If the molar mass is 90 ± 5 g/mol, what is the molecular formula?

• **3.40** A chloride of silicon contains 79.1 mass % Cl. (a) What is the empirical formula of the chloride? (b) If the molar mass is 269 g/mol, what is the molecular formula?

•• **3.41** A sample of 0.600 mol of a metal M reacts completely with excess fluorine to form 46.8 g of MF_2.
(a) How many moles of F are in the sample of MF_2 that forms?
(b) How many grams of M are in this sample of MF_2?
(c) What element is represented by the symbol M?

•• **3.42** A 0.370-mol sample of a metal oxide (M_2O_3) weighs 55.4 g.
(a) How many moles of O are in the sample?
(b) How many grams of M are in the sample?
(c) What element is represented by the symbol M?

● Problems in Context

• **3.43** Nicotine is a poisonous, addictive compound found in tobacco. A sample of nicotine contains 6.16 mmol of C, 8.56 mmol of H, and 1.23 mmol of N [1 mmol (1 millimole) = 10^{-3} mol]. What is the empirical formula?

•• **3.44** Cortisol ($\mathcal{M} = 362.47$ g/mol), one of the major steroid hormones, is a key factor in the synthesis of protein. Its profound effect on the reduction of inflammation explains its use in the treatment of rheumatoid arthritis. Cortisol is 69.6% C, 8.34% H, and 22.1% O by mass. What is its molecular formula?

• **3.45** Acetaminophen *(below)* is a popular nonaspirin, "over-the-counter" pain reliever. What is the mass % of each element in acetaminophen?

• **3.46** Menthol ($\mathcal{M} = 156.3$ g/mol), a strong-smelling substance used in cough drops, is a compound of carbon, hydrogen, and oxygen. When 0.1595 g of menthol was subjected to combustion analysis, it produced 0.449 g of CO_2 and 0.184 g of H_2O. What is menthol's molecular formula?

Biological: 3.29, 3.43, 3.44, 3.45; Organic: 3.24, 3.46

Writing and Balancing Chemical Equations
(Sample Problems 3.8 and 3.9)

● **Concept Review Questions**

• **3.47** What three types of information does a balanced chemical equation provide? How?

• **3.48** How does a balanced chemical equation apply the law of conservation of mass?

• **3.49** In the process of balancing the equation

$$Al + Cl_2 \longrightarrow AlCl_3$$

Student I writes: $Al + Cl_2 \longrightarrow AlCl_2$
Student II writes: $Al + Cl_2 + Cl \longrightarrow AlCl_3$
Student III writes: $2Al + 3Cl_2 \longrightarrow 2AlCl_3$
Is the approach of Student I valid? Student II? Student III? Explain.

•• **3.50** The scenes below represent a chemical reaction between elements A (red) and B (green):

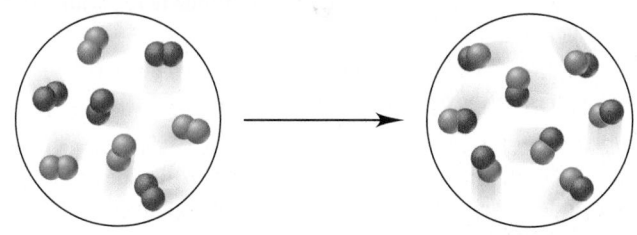

Which best represents the balanced equation for the reaction?
(a) $2A + 2B \longrightarrow A_2 + B_2$ (b) $A_2 + B_2 \longrightarrow 2AB$
(c) $B_2 + 2AB \longrightarrow 2B_2 + A_2$ (d) $4A_2 + 4B_2 \longrightarrow 8AB$

● **Skill-Building Exercises** (grouped in similar pairs)

• **3.51** Write balanced equations for each of the following by inserting the correct coefficients in the blanks:
(a) __$Cu(s)$ + __$S_8(s)$ ⟶ __$Cu_2S(s)$
(b) __$P_4O_{10}(s)$ + __$H_2O(l)$ ⟶ __$H_3PO_4(l)$
(c) __$B_2O_3(s)$ + __$NaOH(aq)$ ⟶
  __$Na_3BO_3(aq)$ + __$H_2O(l)$
(d) __$CH_3NH_2(g)$ + __$O_2(g)$ ⟶
  __$CO_2(g)$ + __$H_2O(g)$ + __$N_2(g)$

• **3.52** Write balanced equations for each of the following by inserting the correct coefficients in the blanks:
(a) __$Cu(NO_3)_2(aq)$ + __$KOH(aq)$ ⟶
  __$Cu(OH)_2(s)$ + __$KNO_3(aq)$
(b) __$BCl_3(g)$ + __$H_2O(l)$ ⟶ __$H_3BO_3(s)$ + __$HCl(g)$
(c) __$CaSiO_3(s)$ + __$HF(g)$ ⟶
  __$SiF_4(g)$ + __$CaF_2(s)$ + __$H_2O(l)$
(d) __$(CN)_2(g)$ + __$H_2O(l)$ ⟶ __$H_2C_2O_4(aq)$ + __$NH_3(g)$

• **3.53** Write balanced equations for each of the following by inserting the correct coefficients in the blanks:
(a) __$SO_2(g)$ + __$O_2(g)$ ⟶ __$SO_3(g)$
(b) __$Sc_2O_3(s)$ + __$H_2O(l)$ ⟶ __$Sc(OH)_3(s)$
(c) __$H_3PO_4(aq)$ + __$NaOH(aq)$ ⟶
  __$Na_2HPO_4(aq)$ + __$H_2O(l)$
(d) __$C_6H_{10}O_5(s)$ + __$O_2(g)$ ⟶ __$CO_2(g)$ + __$H_2O(g)$

• **3.54** Write balanced equations for each of the following by inserting the correct coefficients in the blanks:
(a) __$As_4S_6(s)$ + __$O_2(g)$ ⟶ __$As_4O_6(s)$ + __$SO_2(g)$
(b) __$Ca_3(PO_4)_2(s)$ + ___$SiO_2(s)$ + __$C(s)$ ⟶
  __$P_4(g)$ + __$CaSiO_3(l)$ + __$CO(g)$
(c) __$Fe(s)$ + __$H_2O(g)$ ⟶ __$Fe_3O_4(s)$ + __$H_2(g)$
(d) __$S_2Cl_2(l)$ + __$NH_3(g)$ ⟶
  __$S_4N_4(s)$ + __$S_8(s)$ + __$NH_4Cl(s)$

• **3.55** Convert the following into balanced equations:
(a) When gallium metal is heated in oxygen gas, it melts and forms solid gallium(III) oxide.
(b) Liquid hexane burns in oxygen gas to form carbon dioxide gas and water vapor.
(c) When solutions of calcium chloride and sodium phosphate are mixed, solid calcium phosphate forms and sodium chloride remains in solution.

• **3.56** Convert the following into balanced equations:
(a) When lead(II) nitrate solution is added to potassium iodide solution, solid lead(II) iodide forms and potassium nitrate solution remains.
(b) Liquid disilicon hexachloride reacts with water to form solid silicon dioxide, hydrogen chloride gas, and hydrogen gas.
(c) When nitrogen dioxide is bubbled into water, a solution of nitric acid forms and gaseous nitrogen monoxide is released.

● **Problems in Context**

•• **3.57** Loss of atmospheric ozone has led to an ozone "hole" over Antarctica. The process occurs in part by three consecutive reactions:
(1) Chlorine atoms react with ozone (O_3) to form chlorine monoxide and molecular oxygen.
(2) Chlorine monoxide forms $ClOOCl$.
(3) $ClOOCl$ absorbs sunlight and breaks into chlorine atoms and molecular oxygen.
(a) Write a balanced equation for each step.
(b) Write an overall balanced equation for the sequence.

Calculating Amounts of Reactant and Product
(Sample Problems 3.10 to 3.14)

● **Concept Review Questions**

• **3.58** What does the term *stoichiometrically equivalent molar ratio* mean, and how is it applied in solving problems?

•• **3.59** The circle below represents a mixture of A_2 and B_2 before they react to form AB_3.

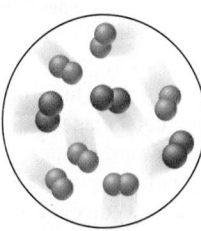

(a) What is the limiting reactant?
(b) How many molecules of product can form?

• **3.60** Percent yields are generally calculated from mass quantities. Would the result be the same if mole quantities were used instead? Why?

● **Skill-Building Exercises** (grouped in similar pairs)

• **3.61** Reactants A and B form product C. Write a detailed Plan to find the mass of C when 25 g of A reacts with excess B.

• **3.62** Reactants D and E form product F. Write a detailed Plan to find the mass of F when 27 g of D reacts with 31 g of E.

• **3.63** Chlorine gas can be made in the laboratory by the reaction of hydrochloric acid and manganese(IV) oxide:

$$4HCl(aq) + MnO_2(s) \longrightarrow MnCl_2(aq) + 2H_2O(g) + Cl_2(g)$$

When 1.82 mol of HCl reacts with excess MnO_2, (a) how many moles of Cl_2 form? (b) How many grams of Cl_2 form?

• **3.64** Bismuth oxide reacts with carbon to form bismuth metal:

$$Bi_2O_3(s) + 3C(s) \longrightarrow 2Bi(s) + 3CO(g)$$

When 283 g of Bi_2O_3 reacts with excess carbon, (a) how many moles of Bi_2O_3 react? (b) How many moles of Bi form?

• **3.65** Potassium nitrate decomposes on heating, producing potassium oxide and gaseous nitrogen and oxygen:

$$4KNO_3(s) \longrightarrow 2K_2O(s) + 2N_2(g) + 5O_2(g)$$

To produce 56.6 kg of oxygen, how many (a) moles of KNO_3 must be heated? (b) Grams of KNO_3 must be heated?

• **3.66** Chromium(III) oxide reacts with hydrogen sulfide (H_2S) gas to form chromium(III) sulfide and water:

$$Cr_2O_3(s) + 3H_2S(g) \longrightarrow Cr_2S_3(s) + 3H_2O(l)$$

To produce 421 g of Cr_2S_3, (a) how many moles of Cr_2O_3 are required? (b) How many grams of Cr_2O_3 are required?

• **3.67** Calculate the mass of each product formed when 43.82 g of diborane (B_2H_6) reacts with excess water:

$$B_2H_6(g) + H_2O(l) \longrightarrow H_3BO_3(s) + H_2(g) \text{ [unbalanced]}$$

• **3.68** Calculate the mass of each product formed when 174 g of silver sulfide reacts with excess hydrochloric acid:

$$Ag_2S(s) + HCl(aq) \longrightarrow AgCl(s) + H_2S(g) \text{ [unbalanced]}$$

•• **3.69** Elemental phosphorus occurs as tetratomic molecules, P_4. What mass of chlorine gas is needed to react completely with 455 g of phosphorus to form phosphorus pentachloride?

•• **3.70** Elemental sulfur occurs as octatomic molecules, S_8. What mass of fluorine gas is needed to react completely with 17.8 g of sulfur to form sulfur hexafluoride?

•• **3.71** Solid iodine trichloride is prepared in two steps: first, a reaction between solid iodine and gaseous chlorine to form solid iodine monochloride; then, treatment with more chlorine.
(a) Write a balanced equation for each step.
(b) Write a balanced equation for the overall reaction.
(c) How many grams of iodine are needed to prepare 2.45 kg of final product?

•• **3.72** Lead can be prepared from galena [lead(II) sulfide] by first roasting the galena in oxygen gas to form lead(II) oxide and sulfur dioxide. Heating the metal oxide with more galena forms the molten metal and more sulfur dioxide.
(a) Write a balanced equation for each step.
(b) Write an overall balanced equation for the process.
(c) How many metric tons of sulfur dioxide form for every metric ton of lead obtained?

• **3.73** Many metals react with oxygen gas to form the metal oxide. For example, calcium reacts as follows:

$$2Ca(s) + O_2(g) \longrightarrow 2CaO(s)$$

You wish to calculate the mass of calcium oxide that can be prepared from 4.20 g of Ca and 2.80 g of O_2.
(a) How many moles of CaO can be produced from the given mass of Ca?
(b) How many moles of CaO can be produced from the given mass of O_2?
(c) Which is the limiting reactant?
(d) How many grams of CaO can be produced?

• **3.74** Metal hydrides react with water to form hydrogen gas and the metal hydroxide. For example,

$$SrH_2(s) + 2H_2O(l) \longrightarrow Sr(OH)_2(s) + 2H_2(g)$$

You wish to calculate the mass of hydrogen gas that can be prepared from 5.70 g of SrH_2 and 4.75 g of H_2O.
(a) How many moles of H_2 can be produced from the given mass of SrH_2?
(b) How many moles of H_2 can be produced from the given mass of H_2O?
(c) Which is the limiting reactant?
(d) How many grams of H_2 can be produced?

•• **3.75** Calculate the maximum numbers of moles and grams of iodic acid (HIO_3) that can form when 635 g of iodine trichloride reacts with 118.5 g of water:

$$ICl_3 + H_2O \longrightarrow ICl + HIO_3 + HCl \text{ [unbalanced]}$$

What mass of the excess reactant remains?

•• **3.76** Calculate the maximum numbers of moles and grams of H_2S that can form when 158 g of aluminum sulfide reacts with 131 g of water:

$$Al_2S_3 + H_2O \longrightarrow Al(OH)_3 + H_2S \text{ [unbalanced]}$$

What mass of the excess reactant remains?

•• **3.77** When 0.100 mol of carbon is burned in a closed vessel with 8.00 g of oxygen, how many grams of carbon dioxide can form? Which reactant is in excess, and how many grams of it remain after the reaction?

•• **3.78** A mixture of 0.0375 g of hydrogen and 0.0185 mol of oxygen in a closed container is sparked to initiate a reaction. How many grams of water can form? Which reactant is in excess, and how many grams of it remain after the reaction?

•• **3.79** Aluminum nitrite and ammonium chloride react to form aluminum chloride, nitrogen, and water. What mass of each substance is present after 72.5 g of aluminum nitrite and 58.6 g of ammonium chloride react completely?

•• **3.80** Calcium nitrate and ammonium fluoride react to form calcium fluoride, dinitrogen monoxide, and water vapor. What mass of each substance is present after 16.8 g of calcium nitrate and 17.50 g of ammonium fluoride react completely?

• **3.81** Two successive reactions, A $\longrightarrow$ B and B $\longrightarrow$ C, have yields of 73% and 68%, respectively. What is the overall percent yield for conversion of A to C?

• **3.82** Two successive reactions, D $\longrightarrow$ E and E $\longrightarrow$ F, have yields of 48% and 73%, respectively. What is the overall percent yield for conversion of D to F?

•• **3.83** What is the percent yield of a reaction in which 45.5 g of tungsten(VI) oxide (WO_3) reacts with excess hydrogen gas to produce metallic tungsten and 9.60 mL of water ($d = 1.00$ g/mL)?

•• **3.84** What is the percent yield of a reaction in which 200. g of phosphorus trichloride reacts with excess water to form 128 g of HCl and aqueous phosphorous acid (H_3PO_3)?

•• **3.85** When 20.5 g of methane and 45.0 g of chlorine gas undergo a reaction that has a 75.0% yield, what mass of chloromethane (CH_3Cl) forms? Hydrogen chloride also forms.

•• **3.86** When 56.6 g of calcium and 30.5 g of nitrogen gas undergo a reaction that has a 93.0% yield, what mass of calcium nitride forms?

● **Problems in Context**

•• **3.87** Cyanogen, $(CN)_2$, has been observed in the atmosphere of Titan, Saturn's largest moon, and in the gases of interstellar nebulas. On Earth, it is used as a welding gas and a fumigant. In its reaction with fluorine gas, carbon tetrafluoride and nitrogen trifluoride gases are produced. What mass of carbon tetrafluoride forms when 60.0 g of each reactant is used?

•• **3.88** Gaseous dichlorine monoxide decomposes readily to chlorine and oxygen gases. (a) Which of the following circles best depicts the product mixture after the decomposition?

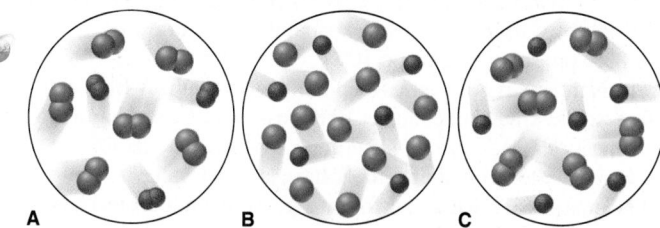

(b) Write the balanced equation for the decomposition.
(c) If each oxygen atom represents 0.050 mol, how many molecules of dichlorine monoxide were present before the decomposition?

•• **3.89** An intermediate step in the industrial production of nitric acid involves the reaction of ammonia with oxygen gas to form nitrogen monoxide and water. How many grams of nitrogen monoxide can form by the reaction of 485 g of ammonia with 792 g of oxygen?

•• **3.90** Butane gas is compressed and used as a liquid fuel in disposable cigarette lighters and lightweight camping stoves. Suppose a lighter contains 5.50 mL of butane ($d = 0.579$ g/mL).
(a) How many grams of oxygen are needed to burn the butane completely?
(b) How many moles of H_2O form when all the butane burns?
(c) How many total molecules of gas form when the butane burns completely?

•• **3.91** Sodium borohydride ($NaBH_4$) is used industrially in many organic syntheses. One way to prepare it is by reacting sodium hydride with gaseous diborane (B_2H_6). Assuming an 88.5% yield, how many grams of $NaBH_4$ can be prepared by reacting 7.98 g of sodium hydride and 8.16 g of diborane?

Fundamentals of Solution Stoichiometry
(Sample Problems 3.15 to 3.20)

● Concept Review Questions

• **3.92** Box A represents a unit volume of a solution. Choose from boxes B and C the one representing the same unit volume of solution that has (a) more solute added; (b) more solvent added; (c) higher molarity; (d) lower concentration.

• **3.93** A mathematical equation useful for dilution calculations is $M_{dil} \times V_{dil} = M_{conc} \times V_{conc}$. (a) What does each symbol mean, and why does the equation work? (b) Given the volume and molarity of a $CaCl_2$ solution, how do you determine the number of moles and the mass of solute?

•• **3.94** Six different aqueous solutions (with solvent molecules omitted for clarity) are represented in the beakers in the next column, and their total volumes are noted.

(a) Which solution has the highest molarity? (b) Which solutions have the same molarity? (c) If you mix solutions A and C, does the resulting solution have a higher, a lower, or the same molarity as solution B? (d) After 50. mL of water is added to solution D, is its molarity higher, lower, or the same as after 75 mL is added to solution F? (e) How much solvent must be evaporated from solution E for it to have the same molarity as solution A?

• **3.95** Are the following instructions for diluting a 10.0 M solution to a 1.00 M solution correct: "Take 100.0 mL of the 10.0 M solution and add 900.0 mL water"? Explain.

● Skill-Building Exercises (grouped in similar pairs)

• **3.96** Calculate each of the following quantities:
(a) Grams of solute in 185.8 mL of 0.267 M calcium acetate
(b) Molarity of 500. mL of solution containing 21.1 g of potassium iodide
(c) Moles of solute in 145.6 L of 0.850 M sodium cyanide

• **3.97** Calculate each of the following quantities:
(a) Volume in milliliters of 2.26 M potassium hydroxide that contains 8.42 g of solute
(b) Number of Cu^{2+} ions in 52 L of 2.3 M copper(II) chloride
(c) Molarity of 275 mL of solution containing 135 mmol of glucose

• **3.98** Calculate each of the following quantities:
(a) Grams of solute needed to make 475 mL of 5.62×10^{-2} M potassium sulfate
(b) Molarity of a solution that contains 7.25 mg of calcium chloride in each milliliter
(c) Number of Mg^{2+} ions in each milliliter of 0.184 M magnesium bromide

• **3.99** Calculate each of the following quantities:
(a) Molarity of the solution resulting from dissolving 46.0 g of silver nitrate in enough water to give a final volume of 335 mL
(b) Volume in liters of 0.385 M manganese(II) sulfate that contains 63.0 g of solute
(c) Volume in milliliters of 6.44×10^{-2} M adenosine triphosphate (ATP) that contains 1.68 mmol of ATP

•• **3.100** Calculate each of the following quantities:
(a) Molarity of a solution prepared by diluting 37.00 mL of 0.250 M potassium chloride to 150.00 mL
(b) Molarity of a solution prepared by diluting 25.71 mL of 0.0706 M ammonium sulfate to 500.00 mL
(c) Molarity of sodium ion in a solution made by mixing 3.58 mL of 0.348 M sodium chloride with 500. mL of 6.81×10^{-2} M sodium sulfate (assume volumes are additive)

3.101 Calculate each of the following quantities:
(a) Volume of 2.050 M copper(II) nitrate that must be diluted with water to prepare 750.0 mL of a 0.8543 M solution
(b) Volume of 1.63 M calcium chloride that must be diluted with water to prepare 350. mL of a 2.86×10^{-2} M chloride ion solution
(c) Final volume of a 0.0700 M solution prepared by diluting 18.0 mL of 0.155 M lithium carbonate with water

3.102 A sample of concentrated nitric acid has a density of 1.41 g/mL and contains 70.0% HNO_3 by mass.
(a) What mass of HNO_3 is present per liter of solution?
(b) What is the molarity of the solution?

3.103 Concentrated sulfuric acid (18.3 M) has a density of 1.84 g/mL.
(a) How many moles of H_2SO_4 are in each milliliter of solution?
(b) What is the mass % of H_2SO_4 in the solution?

3.104 How many milliliters of 0.383 M HCl are needed to react with 16.2 g of $CaCO_3$?
$$2HCl(aq) + CaCO_3(s) \longrightarrow CaCl_2(aq) + CO_2(g) + H_2O(l)$$

3.105 How many grams of NaH_2PO_4 are needed to react with 43.74 mL of 0.285 M NaOH?
$$NaH_2PO_4(s) + 2NaOH(aq) \longrightarrow Na_3PO_4(aq) + 2H_2O(l)$$

3.106 How many grams of solid barium sulfate form when 35.0 mL of 0.160 M barium chloride reacts with 58.0 mL of 0.065 M sodium sulfate? Aqueous sodium chloride forms also.

3.107 How many moles of excess reactant are present when 350. mL of 0.210 M sulfuric acid reacts with 0.500 L of 0.196 M sodium hydroxide to form water and aqueous sodium sulfate?

● **Problems in Context**

3.108 Ordinary household bleach is an aqueous solution of sodium hypochlorite. What is the molarity of a bleach solution that contains 20.5 g of sodium hypochlorite in a total volume of 375 mL?

3.109 Muriatic acid, an industrial grade of concentrated HCl, is used to clean masonry and cement. Its concentration is 11.7 M.
(a) Write instructions for diluting the concentrated acid to make 3.0 gallons of 3.5 M acid for routine use (1 gal = 4 qt; 1 qt = 0.946 L). (b) How many milliliters of the muriatic acid solution contain 9.66 g of HCl?

3.110 A sample of impure magnesium was analyzed by allowing it to react with excess HCl solution:
$$Mg(s) + 2HCl(aq) \longrightarrow MgCl_2(aq) + H_2(g)$$
After 1.32 g of the impure metal was treated with 0.100 L of 0.750 M HCl, 0.0125 mol of HCl remained. Assuming the impurities do not react, what is the mass % of Mg in the sample?

Comprehensive Problems

3.111 The mole is defined in terms of the carbon-12 atom. Use the definition to find (a) the mass in grams equal to 1 atomic mass unit; (b) the ratio of the gram to the atomic mass unit.

3.112 The study of sulfur-nitrogen compounds is an active area of chemical research, made more so by the discovery in the early 1980s of one such compound that conducts electricity like a metal. The first sulfur-nitrogen compound was prepared in 1835 and serves today as a reactant for preparing many of the others. Mass spectrometry of the compound shows a molar mass of 184.27 g/mol, and analysis shows it to contain 2.288 g of S for every 1.000 g of N. What is its molecular formula?

3.113 Hydroxyapatite, $Ca_5(PO_4)_3(OH)$, is the main mineral component of dental enamel, dentin, and bone, and thus has many medical uses. Coating it on metallic implants (such as titanium alloys and stainless steels) helps the body accept the implant. In the form of powder and beads, it is used to fill bone voids, which encourages natural bone to grow into the void. Hydroxyapatite is prepared by adding aqueous phosphoric acid to a dilute slurry of calcium hydroxide. (a) Write a balanced equation for this preparation. (b) What mass of hydroxyapatite could form from 100. g of 85% phosphoric acid and 100. g of calcium hydroxide?

3.114 Narceine is a narcotic in opium that crystallizes from solution as a hydrate that contains 10.8 mass % water and has a molar mass of 499.52 g/mol. Determine x in narceine·xH$_2$O.

3.115 Hydrogen-containing fuels have a "fuel value" based on their mass % H. Rank the following compounds from highest mass % H to lowest: ethane, propane, benzene, ethanol, cetyl palmitate (whale oil, $C_{32}H_{64}O_2$).

ethane   propane   benzene

ethanol

3.116 Serotonin ($\mathcal{M}$ = 176 g/mol) transmits nerve impulses between neurons. It contains 68.2 mass % C, 6.86 mass % H, 15.9 mass % N, and 9.08 mass % O. What is its molecular formula?

3.117 In 1961, scientists agreed that the atomic mass unit (amu) would be defined as $\frac{1}{12}$ the mass of an atom of ^{12}C. Before then, it was defined as $\frac{1}{16}$ the *average* mass of an atom of naturally occurring oxygen (a mixture of ^{16}O, ^{17}O, and ^{18}O). The current atomic mass of oxygen is 15.9994 amu. (a) Did Avogadro's number change after the definition of an amu changed and, if so, in what direction? (b) Did the definition of the mole change? (c) Did the mass of a mole of a substance change? (d) Before 1961, was Avogadro's number 6.02×10^{23} (when considered to three significant figures), as it is today?

3.118 Convert the following descriptions into balanced equations:
(a) In a gaseous reaction, hydrogen sulfide burns in oxygen to form sulfur dioxide and water vapor.
(b) When crystalline potassium chlorate is heated to just above its melting point, it reacts to form two different crystalline compounds, potassium chloride and potassium perchlorate.
(c) When hydrogen gas is passed over powdered iron(III) oxide, iron metal and water vapor form.
(d) The combustion of gaseous ethane in air forms carbon dioxide and water vapor.
(e) Iron(II) chloride is converted to iron(III) fluoride by treatment with chlorine trifluoride gas. Chlorine gas is also formed.

3.119 Isobutylene is a hydrocarbon used in the manufacture of synthetic rubber. When 0.847 g of isobutylene was analyzed by combustion analysis (see Figure 3.5), the gain in mass of the CO_2 absorber was 2.657 g and that of the H_2O absorber was 1.089 g. What is the empirical formula of isobutylene?

3.120 The multistep smelting of ferric oxide to form elemental iron occurs at high temperatures in a blast furnace. In the first step, ferric oxide reacts with carbon monoxide to form Fe_3O_4.

This substance reacts with more carbon monoxide to form iron(II) oxide, which reacts with still more carbon monoxide to form molten iron. Carbon dioxide is also produced in each step. (a) Write an overall balanced equation for the iron-smelting process. (b) How many grams of carbon monoxide are required to form 45.0 metric tons of iron from ferric oxide?

•• **3.121** One of the compounds used to increase the octane rating of gasoline is toluene *(right)*. Suppose 20.0 mL of toluene (d = 0.867 g/mL) is consumed when a sample of gasoline burns in air. (a) How many grams of oxygen are needed for complete combustion of the toluene? (b) How many total moles of gaseous products form? (c) How many molecules of water vapor form?

•• **3.122** During studies of the reaction in Sample Problem 3.13,

$$2N_2H_4(l) + N_2O_4(l) \longrightarrow 3N_2(g) + 4H_2O(g)$$

a chemical engineer measured a less-than-expected yield of N_2 and discovered that the following side reaction occurs:

$$N_2H_4(l) + 2N_2O_4(l) \longrightarrow 6NO(g) + 2H_2O(g)$$

In one experiment, 10.0 g of NO formed when 100.0 g of each reactant was used. What is the highest percent yield of N_2 that can be expected?

•• **3.123** A 0.652-g sample of a pure strontium halide reacts with excess sulfuric acid, and the solid strontium sulfate formed is separated, dried, and found to weigh 0.755 g. What is the formula of the original halide?

•• **3.124** The following circles represent a chemical reaction between AB_2 and B_2:

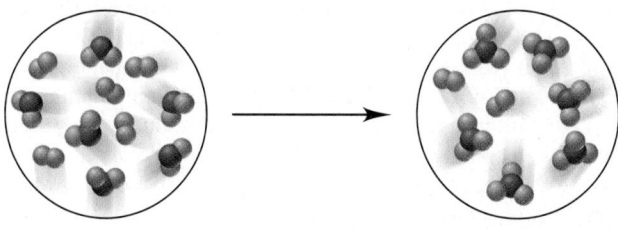

(a) Write a balanced equation for the reaction. (b) What is the limiting reactant? (c) How many moles of product can be made from 3.0 mol of B_2 and 5.0 mol of AB_2? (d) How many moles of excess reactant remain after the reaction in part (c)?

•• **3.125** Calculate each of the following quantities:
(a) Volume of 18.0 M sulfuric acid that must be added to water to prepare 2.00 L of a 0.429 M solution
(b) Molarity of the solution obtained by diluting 80.6 mL of 0.225 M ammonium chloride to 0.250 L
(c) Volume of water added to 0.130 L of 0.0372 M sodium hydroxide to obtain a 0.0100 M solution (assume the volumes are additive at these low concentrations)
(d) Mass of calcium nitrate in each milliliter of a solution prepared by diluting 64.0 mL of 0.745 M calcium nitrate to a final volume of 0.100 L

•• **3.126** A student weighs a sample of carbon on a balance that is accurate to ±0.001 g. (a) How many atoms are in 0.001 g of C? (b) The carbon is used in the following reaction:

$$Pb_3O_4(s) + C(s) \longrightarrow 3PbO(s) + CO(g)$$

What mass difference in the lead(II) oxide would be caused by an error in the carbon mass of 0.001 g?

•• **3.127** The zirconium oxalate $K_2Zr(C_2O_4)_3(H_2C_2O_4) \cdot H_2O$ was synthesized by mixing 1.68 g of $ZrOCl_2 \cdot 8H_2O$ with 5.20 g of $H_2C_2O_4 \cdot 2H_2O$ and an excess of aqueous KOH. After 2 months, 1.25 g of crystalline product was obtained, as well as aqueous KCl and water. Calculate the percent yield.

•• **3.128** Seawater is approximately 4.0% by mass dissolved ions. About 85% of the mass of the dissolved ions is from NaCl. (a) Find the mass % of NaCl in seawater. (b) Find the mass % of Na^+ ions and of Cl^- ions in seawater. (c) Find the molarity of NaCl in seawater at 15°C (d of seawater at 15°C = 1.025 g/mL).

•• **3.129** Is each of the following statements true or false? Correct any that are false:
(a) A mole of one substance has the same number of atoms as a mole of any other substance.
(b) The theoretical yield for a reaction is based on the balanced chemical equation.
(c) A limiting-reactant problem is presented when the quantity of available material is given in moles for one of the reactants.
(d) To prepare 1.00 L of 3.00 M NaCl, weigh 175.5 g of NaCl and dissolve it in 1.00 L of distilled water.
(e) The concentration of a solution is an intensive property, but the amount of solute in a solution is an extensive property.

•• **3.130** Box A represents one unit volume of solution A. Which box—B, C, or D—represents one unit volume after adding enough solvent to solution A to (a) triple its volume; (b) double its volume; (c) quadruple its volume?

A   **B**   **C**   **D**

•• **3.131** In each pair, choose the larger of the indicated quantities or state that the samples are equal:
(a) Entities: 0.4 mol of O_3 molecules or 0.4 mol of O atoms
(b) Grams: 0.4 mol of O_3 molecules or 0.4 mol of O atoms
(c) Moles: 4.0 g of N_2O_4 or 3.3 g of SO_2
(d) Grams: 0.6 mol of C_2H_4 or 0.6 mol of F_2
(e) Total ions: 2.3 mol of sodium chlorate or 2.2 mol of magnesium chloride
(f) Molecules: 1.0 g of H_2O or 1.0 g of H_2O_2
(g) Na^+ ions: 0.500 L of 0.500 M NaBr or 0.0146 kg of NaCl
(h) Mass: 6.02×10^{23} atoms of ^{235}U or 6.02×10^{23} atoms of ^{238}U

•• **3.132** For the reaction between solid tetraphosphorus trisulfide and oxygen gas to form solid tetraphosphorus decaoxide and sulfur dioxide gas, write a balanced equation. Show the equation (see Table 3.5) in terms of (a) molecules, (b) moles, and (c) grams.

•• **3.133** Hydrogen gas has been suggested as a clean fuel because it produces only water vapor when it burns. If the reaction has a 98.8% yield, what mass of hydrogen forms 105 kg of water?

•• **3.134** Assuming that the volumes are additive, what is the concentration of KBr in a solution prepared by mixing 0.200 L of 0.053 M KBr with 0.550 L of 0.078 M KBr?

•• **3.135** Solar winds composed of free protons, electrons, and α particles bombard Earth constantly, knocking gas molecules out of the atmosphere. In this way, Earth loses about 3.0 kg of matter per second. It is estimated that the atmosphere will be gone in about 50 billion years. Use this estimate to calculate (a) the mass (kg) of Earth's atmosphere and (b) the amount (mol) of nitrogen, which makes up 75.5 mass % of the atmosphere.

•• **3.136** Calculate each of the following quantities:
(a) Moles of compound in 0.588 g of ammonium bromide
(b) Number of potassium ions in 88.5 g of potassium nitrate
(c) Mass in grams of 5.85 mol of glycerol ($C_3H_8O_3$)
(d) Volume of 2.85 mol of chloroform ($CHCl_3$; d = 1.48 g/mL)
(e) Number of sodium ions in 2.11 mol of sodium carbonate
(f) Number of atoms in 25.0 μg of cadmium
(g) Number of atoms in 0.0015 mol of fluorine gas

•• **3.137** Elements X (green) and Y (purple) react according to the following equation: $X_2 + 3Y_2 \longrightarrow 2XY_3$. Which molecular scene represents the product of the reaction?

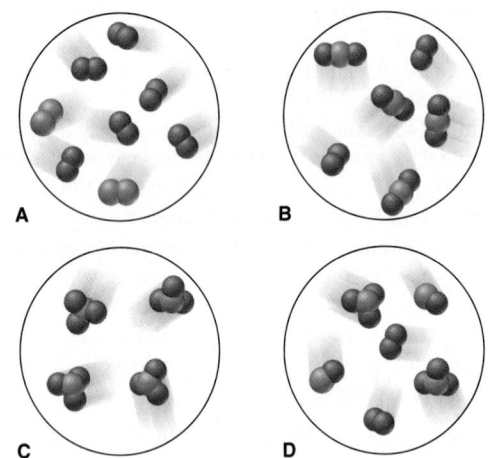

A   B

C   D

•• **3.138** Hydrocarbon mixtures are used as fuels. (a) How many grams of $CO_2(g)$ are produced by the combustion of 200. g of a mixture that is 25.0% CH_4 and 75.0% C_3H_8 by mass? (b) A 252-g gaseous mixture of CH_4 and C_3H_8 burns in excess O_2, and 748 g of CO_2 gas is collected. What is the mass % of CH_4 in the mixture?

•• **3.139** To 1.35 L of 0.325 M HCl, you add 3.57 L of a second HCl solution of unknown concentration. The resulting solution is 0.893 M HCl. Assuming the volumes are additive, calculate the molarity of the second HCl solution.

••• **3.140** Nitrogen (N), phosphorus (P), and potassium (K) are the main nutrients in plant fertilizers. By industry convention, the numbers on the label refer to the mass percents of N, P_2O_5, and K_2O, in that order. Calculate the N/P/K ratio of a 30/10/10 fertilizer in terms of moles of each element, and express it as $x/y/1.0$.

••• **3.141** What mass % of ammonium sulfate, ammonium hydrogen phosphate, and potassium chloride would you use to prepare 10/10/10 plant fertilizer (see Problem 3.140)?

•• **3.142** Methane and ethane are the two simplest hydrocarbons. What is the mass % C in a mixture that is 40.0% methane and 60.0% ethane by mass?

••• **3.143** Ferrocene, synthesized in 1951, was the first organic iron compound with Fe—C bonds. An understanding of the structure of ferrocene gave rise to new ideas about chemical bonding and led to the preparation of many useful compounds. In the combustion analysis of ferrocene, which contains only Fe, C, and H, a 0.9437-g sample produced 2.233 g of CO_2 and 0.457 g of H_2O. What is the empirical formula of ferrocene?

••• **3.144** When carbon-containing compounds are burned in a limited amount of air, some $CO(g)$ as well as $CO_2(g)$ is produced. A gaseous product mixture is 35.0 mass % CO and 65.0 mass % CO_2. What is the mass % C in the mixture?

•• **3.145** Write a balanced equation for the reaction depicted below:

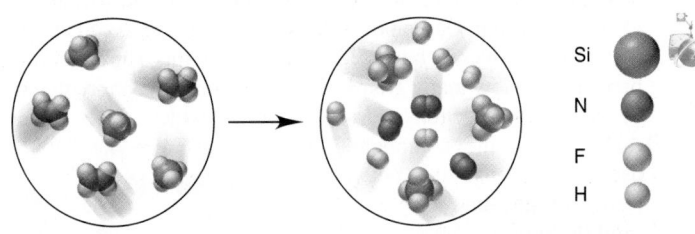

Si
N
F
H

If each reactant molecule represents 1.25×10^{-2} mol and the reaction yield is 87%, how many grams of Si-containing product form?

••• **3.146** Citric acid *(right)* is concentrated in citrus fruits and plays a central metabolic role in nearly every animal and plant cell. (a) What are the molar mass and formula of citric acid? (b) How many moles of citric acid are in 1.50 qt of lemon juice (d = 1.09 g/mL) that is 6.82% citric acid by mass?

••• **3.147** Various nitrogen oxides, as well as sulfur oxides, contribute to acidic rainfall through complex reaction sequences. Nitrogen and oxygen combine during high-temperature combustion of fuels in air to form nitrogen monoxide gas, which reacts with more oxygen to form nitrogen dioxide gas. In contact with water vapor, nitrogen dioxide forms aqueous nitric acid and more nitrogen monoxide. (a) Write balanced equations for these reactions. (b) Use the equations to write one overall balanced equation that does *not* include nitrogen monoxide and nitrogen dioxide. (c) How many metric tons (t) of nitric acid form when 1350 t of atmospheric nitrogen is consumed (1 t = 1000 kg)?

•• **3.148** Alum [$KAl(SO_4)_2 \cdot xH_2O$] is used in food preparation, dye fixation, and water purification. To prepare alum, aluminum is reacted with potassium hydroxide and the product with sulfuric acid. Upon cooling, alum crystallizes from the solution.
(a) A 0.5404-g sample of alum is heated to drive off the waters of hydration, and the resulting $KAl(SO_4)_2$ weighs 0.2941 g. Determine the value of x and the complete formula of alum.
(b) When 0.7500 g of aluminum is used, 8.500 g of alum forms. What is the percent yield?

•• **3.149** Nitrogen monoxide reacts with elemental oxygen to form nitrogen dioxide. The scene at right represents an initial mixture of reactants. If the reaction has a 66% yield, which of the scenes below (A, B, or C) best represents the final product mixture?

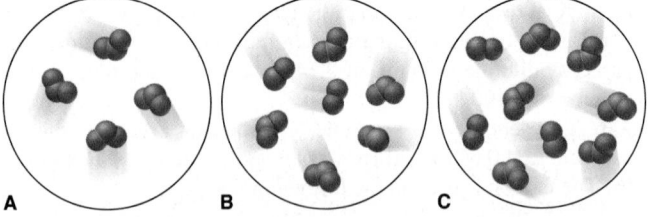

A   B   C

•• **3.150** When 1.5173 g of an organic iron compound containing Fe, C, H, and O was burned in O_2, 2.838 g of CO_2 and 0.8122 g of H_2O were produced. In a separate experiment to determine the mass % of iron, 0.3355 g of the compound yielded 0.0758 g of Fe_2O_3. What is the empirical formula of the compound?

3.151 Fluorine is so reactive that it forms compounds with materials inert to other treatments.
(a) When 0.327 g of platinum is heated in fluorine, 0.519 g of a dark red, volatile solid forms. What is its empirical formula?
(b) When 0.265 g of this red solid reacts with excess xenon gas, 0.378 g of an orange-yellow solid forms. What is the empirical formula of this compound, the first to contain a noble gas?
(c) Fluorides of xenon can be formed by direct reaction of the elements at high pressure and temperature. Under conditions that produce only the tetra- and hexafluorides, 1.85×10^{-4} mol of xenon reacted with 5.00×10^{-4} mol of fluorine, and 9.00×10^{-6} mol of xenon was found in excess. What are the mass percents of each xenon fluoride in the product mixture?

3.152 Hemoglobin is 6.0% heme ($C_{34}H_{32}FeN_4O_4$) by mass. To remove the heme, hemoglobin is treated with acetic acid and NaCl to form hemin ($C_{34}H_{32}N_4O_4FeCl$). At a crime scene, a blood sample contains 0.65 g of hemoglobin. (a) How many grams of heme are in the sample? (b) How many moles of heme? (c) How many grams of Fe? (d) How many grams of hemin could be formed for a forensic chemist to measure?

3.153 Manganese is a key component of extremely hard steel. The element occurs naturally in many oxides. A 542.3-g sample of a manganese oxide has an Mn/O ratio of 1.00/1.42 and consists of braunite (Mn_2O_3) and manganosite (MnO).
(a) What masses of braunite and manganosite are in the ore?
(b) What is the ratio Mn^{3+}/Mn^{2+} in the ore?

3.154 Sulfur dioxide is a major industrial gas used primarily for the production of sulfuric acid, but also as a bleach and food preservative. One way to produce it is by roasting iron pyrite (iron disulfide, FeS_2) in oxygen, which yields the gas and solid iron(III) oxide. What mass of each of the other three substances is involved in producing 1.00 kg of sulfur dioxide?

3.155 The human body excretes nitrogen in the form of urea, NH_2CONH_2. The key biochemical step in urea formation is the reaction of water with arginine to produce urea and ornithine:

Arginine   Water   Urea   Ornithine

(a) What is the mass percent of nitrogen in urea, arginine, and ornithine? (b) How many grams of nitrogen can be excreted as urea when 135.2 g of ornithine is produced?

3.156 Aspirin (acetylsalicylic acid, $C_9H_8O_4$) is made by reacting salicylic acid ($C_7H_6O_3$) with acetic anhydride [$(CH_3CO)_2O$]:

$$C_7H_6O_3(s) + (CH_3CO)_2O(l) \longrightarrow C_9H_8O_4(s) + CH_3COOH(l)$$

In one reaction, 3.077 g of salicylic acid and 5.50 mL of acetic anhydride react to form 3.281 g of aspirin. (a) Which is the limiting reactant (*d* of acetic anhydride = 1.080 g/mL)? (b) What is

the percent yield of this reaction? (c) What is the percent atom economy of this reaction?

3.157 The rocket fuel hydrazine (N_2H_4) is made by the three-step Raschig process, which has the following overall equation:

$$NaOCl(aq) + 2NH_3(aq) \longrightarrow N_2H_4(aq) + NaCl(aq) + H_2O(l)$$

What is the percent atom economy of this process?

3.158 Lead(II) chromate ($PbCrO_4$) is used as the yellow pigment in traffic lanes, but is banned from house paint because of the risk of lead poisoning. It is produced from chromite ($FeCr_2O_4$), an ore of chromium:

$$4FeCr_2O_4(s) + 8K_2CO_3(aq) + 7O_2(g) \longrightarrow$$
$$2Fe_2O_3(s) + 8K_2CrO_4(aq) + 8CO_2(g)$$

Lead(II) ion then replaces the K^+ ion. If a yellow paint is 0.511% $PbCrO_4$ by mass, how many grams of chromite are needed per kilogram of paint?

3.159 Ethanol (CH_3CH_2OH), the intoxicant in alcoholic beverages, is also used to make other organic compounds. In concentrated sulfuric acid, ethanol forms diethyl ether and water:

$$2CH_3CH_2OH(l) \longrightarrow CH_3CH_2OCH_2CH_3(l) + H_2O(g)$$

In a side reaction, some ethanol forms ethylene and water:

$$CH_3CH_2OH(l) \longrightarrow CH_2{=}CH_2(g) + H_2O(g)$$

(a) If 50.0 g of ethanol yields 35.9 g of diethyl ether, what is the percent yield of diethyl ether? (b) During the process, 45.0% of the ethanol that did not produce diethyl ether reacts by the side reaction. What mass of ethylene is produced?

3.160 When powdered zinc is heated with sulfur, a violent reaction occurs, and zinc sulfide forms:

$$Zn(s) + S_8(s) \longrightarrow ZnS(s) \text{ [unbalanced]}$$

Some of the reactants also combine with oxygen in air to form zinc oxide and sulfur dioxide. When 83.2 g of Zn reacts with 52.4 g of S_8, 104.4 g of ZnS forms. What is the percent yield of ZnS? (b) If all the remaining reactants combine with oxygen, how many grams of each of the two oxides form?

3.161 Cocaine ($C_{17}H_{21}O_4N$) is a natural substance found in coca leaves, which have been used for centuries as a local anesthetic and stimulant. Illegal cocaine arrives in the United States either as the pure compound or as the hydrochloride salt ($C_{17}H_{21}O_4NHCl$). At 25°C, the salt is very soluble in water (2.50 kg/L), but cocaine is much less so (1.70 g/L). (a) What is the maximum amount (in g) of the salt that can dissolve in 50.0 mL of water? (b) If the solution in part (a) is treated with NaOH, the salt is converted to cocaine. How much additional water (in L) is needed to dissolve it?

3.162 High-temperature superconducting oxides hold great promise in the utility, transportation, and computer industries.
(a) One superconductor is $La_{2-x}Sr_xCuO_4$. Calculate the molar masses of this oxide when $x = 0$, $x = 1$, and $x = 0.163$.
(b) Another common superconducting oxide is made by heating a mixture of barium carbonate, copper(II) oxide, and yttrium(III) oxide, followed by further heating in O_2:

$$4BaCO_3(s) + 6CuO(s) + Y_2O_3(s) \longrightarrow$$
$$2YBa_2Cu_3O_{6.5}(s) + 4CO_2(g)$$
$$2YBa_2Cu_3O_{6.5}(s) + \tfrac{1}{2}O_2(g) \longrightarrow 2YBa_2Cu_3O_7(s)$$

When equal masses of the three reactants are heated, which reactant is limiting?
(c) After the product in part (b) is removed, what is the mass percent of each reactant in the remaining solid mixture?

Biological: 3.152, 3.155, 3.161; Engineering: 3.162; Green chemistry: 3.157; Industrial: 3.153, 3.154; Organic: 3.156, 3.159

4

The Universality of Chemical Change A storm at sea becomes a natural "factory" where changes in the composition of matter take place. Many scientists believe that a similar event contributed to the origin of bio-molecules on the young Earth. In fact, we observe chemical change wherever we look. In this chapter, we investigate three major classes of reactions, giving special attention to those occurring in aqueous solution.

Three Major Classes of Chemical Reactions

4.1 The Role of Water as a Solvent
Polar Nature of Water
Ionic Compounds in Water
Covalent Compounds in Water

4.2 Writing Equations for Aqueous Ionic Reactions

4.3 Precipitation Reactions
The Key Event: Formation of a Solid
Predicting Whether a Precipitate Will Form

4.4 Acid-Base Reactions
The Key Event: Formation of Water
Acid-Base Titrations
Proton Transfer in Acid-Base Reactions

4.5 Oxidation-Reduction (Redox) Reactions
The Key Event: Movement of Electrons
Redox Terminology
Oxidation Numbers
Balancing Redox Equations
Redox Titrations

4.6 Elements in Redox Reactions

4.7 Reaction Reversibility and the Equilibrium State

R apid chemical changes occur among gas molecules as sunlight bathes the atmosphere or lightning rips through a stormy sky. Aqueous reactions go on unceasingly in the gigantic containers we know as oceans. And, in every cell of your body, thousands of reactions taking place right now allow you to function. Indeed, the amazing variety in nature is largely a consequence of the amazing variety of chemical reactions. Of the millions of chemical reactions occurring in and around you, we have examined only a tiny fraction so far, and it would be impossible to examine them all. Fortunately, it isn't necessary to catalog every reaction, because when we survey even a small percentage of reactions, especially those that occur often in aqueous solution, a few major patterns emerge.

IN THIS CHAPTER . . . We examine the underlying nature of three reaction processes that occur commonly in water. Because one of our main themes is aqueous reaction chemistry, we first investigate how the molecular structure of water influences its crucial role as a solvent in these reactions. We see how to use ionic equations to describe reactions. We then focus in turn on precipitation, acid-base, and oxidation-reduction reactions, examining why they occur and how to quantify them. We classify several important types of oxidation-reduction reactions that include elements as reactants or products. Finally, we take an introductory look at the reversible nature of all reactions.

Concepts & Skills to Review
before you study this chapter

- names and formulas of compounds (Section 2.8)
- nature of ionic and covalent bonding (Section 2.7)
- mole-mass-number conversions (Section 3.1)
- molarity and mole-volume conversions (Section 3.5)
- balancing chemical equations (Section 3.3)
- calculating amounts of reactants and products (Section 3.4)

4.1 THE ROLE OF WATER AS A SOLVENT

Our first step toward comprehending classes of aqueous reactions is to understand how water acts as a solvent. The role a solvent plays in a reaction depends on its chemical nature. Some solvents play a passive role, dispersing the dissolved substances into individual molecules but doing nothing further. Water plays a much more active role, interacting strongly with the substances and, in some cases, even reacting with them. To understand this active role, we'll examine the structure of water and how it interacts with ionic and covalent solutes.

The Polar Nature of Water

Of the many thousands of reactions that occur in the environment and in organisms, nearly all take place in water. Water's remarkable power as a solvent results from two features of its molecules: *the distribution of the bonding electrons* and *the overall shape.*

Recall from Section 2.7 that the electrons in a covalent bond are shared between the bonded atoms. In a covalent bond between identical atoms (as in H_2, Cl_2, O_2, etc.), the sharing is equal, so no imbalance of charge appears (Figure 4.1A). On the other hand, in covalent bonds between nonidentical atoms, the sharing is unequal: one atom attracts the electron pair more strongly than the other. For reasons discussed in Chapter 9, an O atom attracts electrons more strongly than an H atom. Therefore, in each O—H bond in water, the shared electrons spend more time closer to the O atom.

This unequal distribution of negative charge creates partially charged "poles" at the ends of each O—H bond (Figure 4.1B). The O end acts as a slightly negative pole (represented by the red shading and the $\delta-$), and the H end acts as a slightly positive pole (represented by the blue shading and the $\delta+$). Figure 4.1C indicates the bond's polarity with a *polar arrow* (the arrowhead points to the negative pole and the tail is crossed to make a "plus").

The H—O—H arrangement forms an angle, so the water molecule is bent. The combined effects of its bent shape and its polar bonds make water a **polar molecule:** the O portion of the molecule is the partially negative pole, and the region midway between the H atoms on the other end of the molecule is the partially positive pole (Figure 4.1D).

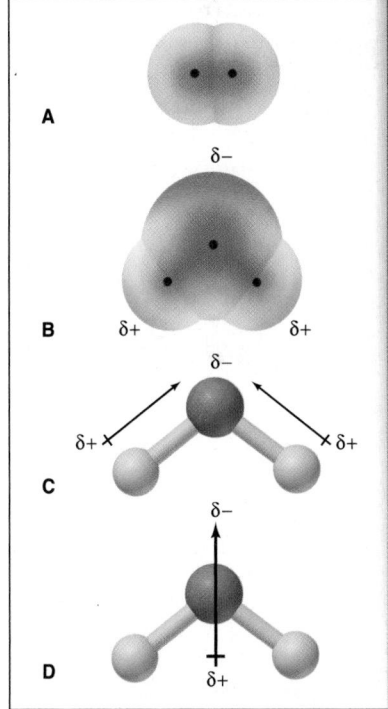

Figure 4.1 Electron distribution in molecules of H_2 and H_2O. **A,** In H_2, the identical nuclei attract the electrons equally. The central region of higher electron density *(red)* is balanced by the two outer regions of lower electron density *(blue)*. **B,** In H_2O, the O nucleus attracts the shared electrons more strongly than the H nucleus. **C,** In this ball-and-stick model, a polar arrow points to the negative end of each O—H bond. **D,** The two polar O—H bonds and the bent shape give rise to the polar H_2O molecule.

Transparency/ARIS Presentation Center, Art Library

Literature Johnson, J. L. H.; Yalkowsky, S. H. "A Three-Dimensional Model for Water," *J. Chem. Educ.* **2002,** *79,* 1088.

Ionic Compounds in Water

In an ionic solid, the oppositely charged ions are held next to each other by electrostatic attraction (see Figure 1.3C). *Water separates the ions by replacing that attraction with one between the water molecules and the ions.* Imagine a granule of an ionic compound surrounded by bent, polar water molecules. The negative ends of some water molecules are attracted to the cations, and the positive ends of others are attracted to the anions (Figure 4.2). Gradually, the attraction between each ion and the nearby water molecules outweighs the attraction of the ions for each other. By this process, the ions separate (dissociate) and become **solvated,** surrounded tightly by solvent molecules, as they move randomly in the solution. A similar scene occurs whenever an ionic compound dissolves in water.

Figure 4.2 The dissolution of an ionic compound. When an ionic compound dissolves in water, H_2O molecules separate, surround, and disperse the ions into the liquid. The negative ends of the H_2O molecules face the positive ions and the positive ends face the negative ions.

 Transparency/ARIS Presentation Center, Art Library

 Animation: Dissolution of an Ionic and a Covalent Compound
ARIS Presentation Center
Chapter 4

Although many ionic compounds dissolve in water, many others do not. In the latter cases, the electrostatic attraction among ions in the compound remains greater than the attraction between ions and water molecules, so the solid stays largely intact. Actually, these so-called insoluble substances *do* dissolve to a very small extent, usually several orders of magnitude less than so-called soluble substances. Compare, for example, the solubilities of NaCl (a "soluble" compound) and AgCl (an "insoluble" compound):

Solubility of NaCl in H_2O at 20°C = 365 g/L

Solubility of AgCl in H_2O at 20°C = 0.009 g/L

Actually, the process of dissolving is more complex than just a contest between the relative energies of attraction of the particles for each other and for the solvent. In Chapter 13, we'll see that it also involves the greater freedom of motion of the particles as they disperse randomly through the solution.

When an ionic compound dissolves, an important change occurs in the solution. Figure 4.3 shows this change with a simple apparatus that demonstrates *electrical conductivity,* the flow of electric current. When the electrodes are immersed in pure water or pushed into an ionic solid, such as potassium bromide (KBr), no current flows. In an aqueous KBr solution, however, a significant current flows, as shown by the brightly lit bulb. This current flow implies the *movement of charged particles:* when KBr dissolves in water, the K^+ and Br^- ions dissociate, become solvated, and move toward the electrode of opposite charge. A substance that conducts a current when dissolved in water is an **electrolyte.** Soluble ionic compounds are called *strong* electrolytes because they dissociate completely into ions and create a large current. We express the dissociation of KBr into solvated ions in water,

$$KBr(s) \xrightarrow{\text{H}_2\text{O}} K^+(aq) + Br^-(aq)$$

 Literature Kowalak, Albert. "When Is a Strong Electrolyte Strong?" *J. Chem. Educ.* **1988,** *65,* 607.

A Distilled water does not conduct a current

B Positive and negative ions fixed in a solid do not conduct a current

To (+) electrode

To (−) electrode

C In solution, positive and negative ions move and conduct a current

Figure 4.3 The electrical conductivity of ionic solutions. A, When electrodes connected to a power source are placed in distilled water, no current flows and the bulb is unlit. **B,** A solid ionic compound, such as KBr, conducts no current because the ions are bound tightly together. **C,** When KBr dissolves in H_2O, the ions separate and move through the solution toward the oppositely charged electrodes, thereby conducting a current.

The "H_2O" above the arrow indicates that water is required as the solvent but is not a reactant in the usual sense.

The formula of the compound tells the number of moles of different ions that result when the compound dissolves. Thus, 1 mol of KBr dissociates into 2 mol of ions—1 mol of K^+ and 1 mol of Br^-. The upcoming sample problem goes over this quantitative idea.

 ARIS Presentation Center, Worked Examples Library

 Transparency/ARIS Presentation Center, Art Library

SAMPLE PROBLEM 4.1 Determining Moles of Ions in Aqueous Ionic Solutions

PROBLEM How many moles of each ion are in each solution?
(a) 5.0 mol of ammonium sulfate dissolved in water
(b) 78.5 g of cesium bromide dissolved in water
(c) 7.42×10^{22} formula units of copper(II) nitrate dissolved in water
(d) 35 mL of 0.84 *M* zinc chloride
PLAN We write an equation that shows 1 mol of compound dissociating into ions. In part (a), we multiply the moles of ions by 5.0. In (b), we first convert grams to moles. In (c), we first convert formula units to moles. In (d), we first convert molarity and volume to moles.
SOLUTION (a) $(NH_4)_2SO_4(s) \xrightarrow{H_2O} 2NH_4^+(aq) + SO_4^{2-}(aq)$
Remember that, in general, *polyatomic ions remain as intact units in solution.*
Calculating moles of NH_4^+ ions:

$$\text{Moles of } NH_4^+ = 5.0 \text{ mol } (NH_4)_2SO_4 \times \frac{2 \text{ mol } NH_4^+}{1 \text{ mol } (NH_4)_2SO_4} = 10. \text{ mol } NH_4^+$$

5.0 mol of SO_4^{2-} is also present.

(b) $CsBr(s) \xrightarrow{H_2O} Cs^+(aq) + Br^-(aq)$

Converting from grams to moles:

$$\text{Moles of CsBr} = 78.5 \text{ g CsBr} \times \frac{1 \text{ mol CsBr}}{212.8 \text{ g CsBr}} = 0.369 \text{ mol CsBr}$$

Thus, 0.369 mol of Cs^+ and 0.369 mol of Br^- are present.

(c) $Cu(NO_3)_2(s) \xrightarrow{H_2O} Cu^{2+}(aq) + 2NO_3^-(aq)$

Converting from formula units to moles:

$$\text{Moles of } Cu(NO_3)_2 = 7.42 \times 10^{22} \text{ formula units } Cu(NO_3)_2$$

$$\times \frac{1 \text{ mol } Cu(NO_3)_2}{6.022 \times 10^{23} \text{ formula units } Cu(NO_3)_2}$$

$$= 0.123 \text{ mol } Cu(NO_3)_2$$

$$\text{Moles of } NO_3^- = 0.123 \text{ mol } Cu(NO_3)_2 \times \frac{2 \text{ mol } NO_3^-}{1 \text{ mol } Cu(NO_3)_2} = 0.246 \text{ mol } NO_3^-$$

0.123 mol of Cu^{2+} is also present.

(d) $ZnCl_2(aq) \longrightarrow Zn^{2+}(aq) + 2Cl^-(aq)$

Converting from liters to moles:

$$\text{Moles of } ZnCl_2 = 35 \text{ mL} \times \frac{1 \text{ L}}{10^3 \text{ mL}} \times \frac{0.84 \text{ mol } ZnCl_2}{1 \text{ L}} = 2.9 \times 10^{-2} \text{ mol } ZnCl_2$$

$$\text{Moles of } Cl^- = 2.9 \times 10^{-2} \text{ mol } ZnCl_2 \times \frac{2 \text{ mol } Cl^-}{1 \text{ mol } ZnCl_2} = 5.8 \times 10^{-2} \text{ mol } Cl^-$$

2.9×10^{-2} mol of Zn^{2+} is also present.

CHECK After you round off to check the math, see if the relative moles of ions are consistent with the formula. For instance, in (a), 10 mol NH_4^+/5.0 mol SO_4^{2-} = 2 NH_4^+/1 SO_4^{2-}, or $(NH_4)_2SO_4$. In (d), 0.029 mol Zn^{2+}/0.058 mol Cl^- = 1 Zn^{2+}/2 Cl^-, or $ZnCl_2$.

FOLLOW-UP PROBLEM 4.1 How many moles of each ion are in each solution?
(a) 2 mol of potassium perchlorate dissolved in water
(b) 354 g of magnesium acetate dissolved in water
(c) 1.88×10^{24} formula units of ammonium chromate dissolved in water
(d) 1.32 L of 0.55 M sodium bisulfate

Animation: Strong Electrolytes, Weak Electrolytes, and Nonelectrolytes
ARIS Presentation Center
Chapter 4

Demo Rizzo, Michelle M.; Halmi, Tracy A.; Jircitano, Alan J.; Kociolek, Martin G.; Magraw, Jerry A. "Revisiting the Electric Pickle Demonstration," *J. Chem. Educ.* **2005,** *82,* 545.

Demo Shakhashiri, Bassam Z. "Conductivity and Extent of Dissociation of Acids in Aqueous Solution," *Chemical Demonstrations, A Handbook for Teachers of Chemistry,* Vol. 2 (The University of Wisconsin Press, Madison, 1985), 140–145.

Covalent Compounds in Water

Water dissolves many covalent compounds also. Table sugar (sucrose, $C_{12}H_{22}O_{11}$), beverage (grain) alcohol (ethanol, CH_3CH_2OH), and automobile antifreeze (ethylene glycol, $HOCH_2CH_2OH$) are some familiar examples. All contain their own polar bonds, which interact with those of water. However, even though these substances dissolve, they do not dissociate into ions but remain as intact molecules. As a result, their aqueous solutions do not conduct an electric current, so these substances are called **nonelectrolytes.** (A small, but extremely important, group of H-containing covalent compounds interacts so strongly with water that their molecules *do* dissociate into ions. In aqueous solution, these substances are *acids,* as you'll see shortly.) Many other covalent substances, such as benzene (C_6H_6) and octane (C_8H_{18}), do not contain polar bonds, and these substances do not dissolve appreciably in water.

Section Summary

Water plays an active role in dissolving ionic compounds because it consists of polar molecules that are attracted to the ions. • When an ionic compound dissolves in water, the ions dissociate from each other and become solvated by water molecules. Because the ions are free to move, their solutions conduct electricity. • Water also dissolves many covalent substances with polar bonds, but the molecules remain intact so they do not conduct electricity.

4.2 WRITING EQUATIONS FOR AQUEOUS IONIC REACTIONS

Chemists use three types of equations to represent aqueous ionic reactions: molecular, total ionic, and net ionic equations. As you'll see, by balancing the atoms in the two types of ionic equations, we also balance the charges.

Let's examine a reaction to see what each of these equations shows. When solutions of silver nitrate and sodium chromate are mixed, the brick-red solid silver chromate (Ag_2CrO_4) forms. Figure 4.4 depicts three views of this reaction: the change you would see if you mixed these solutions in the lab, how you might imagine the change at the atomic level among the ions, and how you can symbolize the change with the three types of equations. (The ions that are reacting are shown in red type.)

The **molecular equation** (Figure 4.4, *top*) reveals the least about the species in solution and is actually somewhat misleading because *it shows all the reactants and products as if they were intact, undissociated compounds*:

$$2AgNO_3(aq) + Na_2CrO_4(aq) \longrightarrow Ag_2CrO_4(s) + 2NaNO_3(aq)$$

Only by examining the state-of-matter designations (*s*) and (*aq*) can you tell what change has occurred.

Molecular equation
$$2AgNO_3(aq) \quad + \quad Na_2CrO_4(aq) \quad \longrightarrow \quad Ag_2CrO_4(s) + 2NaNO_3(aq)$$
Silver nitrate · · · · · · · Sodium chromate · · · · · · · · Silver chromate · Sodium nitrate

Total ionic equation
$$2Ag^+(aq) + 2NO_3^-(aq) + 2Na^+(aq) + CrO_4^{2-}(aq) \longrightarrow Ag_2CrO_4(s) + 2Na^+(aq) + 2NO_3^-(aq)$$

Net ionic equation
$$2Ag^+(aq) \quad + \quad CrO_4^{2-}(aq) \quad \longrightarrow \quad Ag_2CrO_4(s)$$

Figure 4.4 An aqueous ionic reaction and its equations. When silver nitrate and sodium chromate solutions are mixed, a reaction occurs that forms solid silver chromate and a solution of sodium nitrate. The photos present the macroscopic view of the reaction, the view the chemist sees in the lab. The blow-up arrows lead to an atomic-scale view, a representation of the chemist's mental picture of the reactants and products. (The pale ions are spectator ions, present for electrical neutrality, but not involved in the reaction.) Three equations represent the reaction in symbols. (The ions that are reacting are shown in red type.) The *molecular equation* shows all substances intact. The *total ionic equation* shows all *soluble* substances as separate, solvated ions. The *net ionic equation* eliminates the spectator ions to show only the reacting species.

Transparency/ARIS Presentation Center, Photo Library

The **total ionic equation** (Figure 4.4, *middle*) is a much more accurate representation of the reaction because *it shows all the soluble ionic substances dissociated into ions.* Now the $Ag_2CrO_4(s)$ stands out as the only undissociated substance:

$$2Ag^+(aq) + 2NO_3^-(aq) + 2Na^+(aq) + CrO_4^{2-}(aq) \longrightarrow$$
$$Ag_2CrO_4(s) + 2Na^+(aq) + 2NO_3^-(aq)$$

Notice that charges balance: there are four positive and four negative charges on the left for a net zero charge, and there are two positive and two negative charges on the right for a net zero charge.

Note that $Na^+(aq)$ and $NO_3^-(aq)$ appear in the same form on both sides of the equation. They are called **spectator ions** because they are not involved in the actual chemical change. These ions are present as part of the reactants to balance the charge. That is, we can't add an Ag^+ ion without also adding an anion, in this case, NO_3^- ion.

The **net ionic equation** (Figure 4.4, *bottom*) is the most useful because *it eliminates the spectator ions and shows the actual chemical change taking place*:

$$2Ag^+(aq) + CrO_4^{2-}(aq) \longrightarrow Ag_2CrO_4(s)$$

The formation of solid silver chromate from silver ions and chromate ions *is* the only change. In fact, if we had originally mixed solutions of potassium chromate, $K_2CrO_4(aq)$, and silver acetate, $AgC_2H_3O_2(aq)$, instead of sodium chromate and silver nitrate, the same change would have occurred. Only the spectator ions would differ—$K^+(aq)$ and $C_2H_3O_2^-(aq)$ instead of $Na^+(aq)$ and $NO_3^-(aq)$.

Now, let's apply these types of equations to three important types of chemical reactions—precipitation, acid-base, and oxidation-reduction.

Section Summary

A molecular equation for an aqueous ionic reaction shows undissociated substances. • A total ionic equation shows all soluble ionic compounds as separate, solvated ions. Spectator ions appear unchanged on both sides of the equation. • The net ionic equation shows the actual chemical change by eliminating the spectator ions.

4.3 PRECIPITATION REACTIONS

Precipitation reactions are common in both nature and commerce. Many geological formations, including coral reefs, some gems and minerals, and deep-sea structures form, in part, through this type of chemical process. The chemical industry employs precipitation methods to produce several important inorganic compounds.

The Key Event: Formation of a Solid from Dissolved Ions

In **precipitation reactions,** two soluble ionic compounds react to form an insoluble product, a **precipitate.** The reaction you saw in Section 4.2 between silver nitrate and sodium chromate is an example. Precipitates form for the same reason that some ionic compounds do not dissolve: the electrostatic attraction between the ions outweighs the tendency of the ions to remain solvated and move randomly throughout the solution. When solutions of such ions are mixed, the ions collide and stay together, and the resulting substance "comes out of solution" as a solid, as shown in Figure 4.5 for calcium fluoride.

Thus, the key event in a precipitation reaction is *the formation of an insoluble product through the net removal of solvated ions from solution.* (As you'll see in Section 4.4, acid-base reactions have a similar result, but the product is water instead of an ionic compound.)

Literature Wruck, Betty J. "Reinforcing Net Ionic Equation Writing: Second Semester," *J. Chem. Educ.* **1996**, *73*, 149.

Animation: Precipitation Reactions ARIS Presentation Center Chapter 4

Demo Summerlin, L. R., Borgford, C. L., Ealy, J. B. "Name That Precipitate," *Chemical Demonstrations, A Sourcebook for Teachers*, Vol. 2 (American Chemical Society, Washington, DC, 1988), 121–123.

Figure 4.5 The precipitation of calcium fluoride. When an aqueous solution of NaF is added to a solution of $CaCl_2$, Ca^{2+} and F^- ions form particles of solid CaF_2.

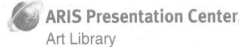
ARIS Presentation Center,
Art Library

Predicting Whether a Precipitate Will Form

If you mix aqueous solutions of two ionic compounds, can you predict if a precipitate will form? Consider this example. When solid sodium iodide and potassium nitrate are each dissolved in water, each solution consists of separated ions dispersed throughout the solution:

$$NaI(s) \xrightarrow{H_2O} Na^+(aq) + I^-(aq)$$
$$KNO_3(s) \xrightarrow{H_2O} K^+(aq) + NO_3^-(aq)$$

Let's follow three steps to predict whether a precipitate will form when solutions are mixed:

1. *Note the ions present in the reactants.* The reactant ions are

$$Na^+(aq) + I^-(aq) + K^+(aq) + NO_3^-(aq) \longrightarrow ?$$

2. *Consider the possible cation-anion combinations.* In addition to the two original ones, NaI and KNO_3, which you already know are soluble, the other possible cation-anion combinations are $NaNO_3$ and KI.

3. *Decide whether any of the combinations is insoluble.* A reaction does *not* occur when you mix these starting solutions because all the combinations—NaI, KNO_3, $NaNO_3$, and KI—are soluble. All the ions remain in solution. (You'll see shortly a set of rules for deciding if a product is soluble or not.)

Now, what happens if you substitute a solution of lead(II) nitrate, $Pb(NO_3)_2$, for the KNO_3? The reactant ions are Na^+, I^-, Pb^{2+}, and NO_3^-. In addition to the two soluble reactants, NaI and $Pb(NO_3)_2$, the other two possible cation-anion combinations are $NaNO_3$ and PbI_2. Lead(II) iodide is insoluble, so a reaction *does* occur as the Pb^{2+} and I^- ions are removed from solution (Figure 4.6):

$$2Na^+(aq) + 2I^-(aq) + Pb^{2+}(aq) + 2NO_3^-(aq) \longrightarrow 2Na^+(aq) + 2NO_3^-(aq) + PbI_2(s)$$

A close look (with color) at the molecular equation shows that *the ions are exchanging partners*:

$$2NaI(aq) + Pb(NO_3)_2(aq) \longrightarrow PbI_2(s) + 2NaNO_3(aq)$$

Figure 4.6 The reaction of $Pb(NO_3)_2$ and NaI. When aqueous solutions of these ionic compounds are mixed, the yellow solid PbI_2 forms.

ARIS Presentation Center, Photo Library

Such reactions are called double-displacement, exchange, or **metathesis** (pronounced *meh-TA-thuh-sis*) **reactions.** Several are important in industry, such as the preparation of silver bromide for the manufacture of black-and-white film:

$$AgNO_3(aq) + KBr(aq) \longrightarrow AgBr(s) + KNO_3(aq)$$

As stated earlier, there is no simple way to decide whether any given ion combination is soluble or not, so Table 4.1 provides a short list of solubility rules to memorize. They will allow you to predict the outcome of many precipitation reactions.

Table 4.1   Solubility Rules for Ionic Compounds in Water

Soluble Ionic Compounds	Insoluble Ionic Compounds
1. All common compounds of Group 1A(1) ions (Li^+, Na^+, K^+, etc.) and ammonium ion (NH_4^+) are soluble.	1. All common metal hydroxides are insoluble, *except* those of Group 1A(1) and the larger members of Group 2A(2) (beginning with Ca^{2+}).
2. All common nitrates (NO_3^-), acetates (CH_3COO^- or $C_2H_3O_2^-$), and most perchlorates (ClO_4^-) are soluble.	2. All common carbonates (CO_3^{2-}) and phosphates (PO_4^{3-}) are insoluble, *except* those of Group 1A(1) and NH_4^+.
3. All common chlorides (Cl^-), bromides (Br^-), and iodides (I^-) are soluble, *except* those of Ag^+, Pb^{2+}, Cu^+, and Hg_2^{2+}. All common fluorides (F^-) are soluble, *except* those of Pb^{2+} and Group 2A(2).	3. All common sulfides are insoluble *except* those of Group 1A(1), Group 2A(2), and NH_4^+.
4. All common sulfates (SO_4^{2-}) are soluble, *except* those of Ca^{2+}, Sr^{2+}, Ba^{2+}, Ag^+, and Pb^{2+}.	

 ARIS Presentation Center, Tables Library

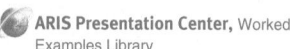 **ARIS Presentation Center,** Worked Examples Library

SAMPLE PROBLEM 4.2   Predicting Whether a Precipitation Reaction Occurs; Writing Ionic Equations

PROBLEM Predict whether a reaction occurs when each of the following pairs of solutions are mixed. If a reaction does occur, write balanced molecular, total ionic, and net ionic equations, and identify the spectator ions.
(a) Potassium fluoride(aq) + strontium nitrate(aq) $\longrightarrow$
(b) Ammonium perchlorate(aq) + sodium bromide(aq) $\longrightarrow$
PLAN For each pair of solutions, we note the ions present in the reactants, write the cation-anion combinations, and refer to Table 4.1 to see if any of them are insoluble. For the molecular equation, we predict the products. For the total ionic equation, we write the soluble compounds as separate ions. For the net ionic equation, we eliminate the spectator ions.
SOLUTION **(a)** In addition to the reactants, the two other ion combinations are strontium fluoride and potassium nitrate. Table 4.1 shows that strontium fluoride is insoluble, so a reaction *does* occur. Writing the molecular equation:

$$2KF(aq) + Sr(NO_3)_2(aq) \longrightarrow SrF_2(s) + 2KNO_3(aq)$$

Writing the total ionic equation:

$$2K^+(aq) + 2F^-(aq) + Sr^{2+}(aq) + 2NO_3^-(aq) \longrightarrow$$
$$SrF_2(s) + 2K^+(aq) + 2NO_3^-(aq)$$

Writing the net ionic equation:

$$Sr^{2+}(aq) + 2F^-(aq) \longrightarrow SrF_2(s)$$

The spectator ions are K^+ and NO_3^-.
(b) The other ion combinations are ammonium bromide and sodium perchlorate. Table 4.1 shows that all ammonium, sodium, and most perchlorate compounds are soluble, and all bromides are soluble except those of Ag^+, Pb^{2+}, Cu^+, and Hg_2^{2+}. Therefore, *no* reaction occurs. The compounds remain dissociated in solution as solvated ions.

FOLLOW-UP PROBLEM 4.2 Predict whether a reaction occurs, and write balanced total and net ionic equations:
(a) Iron(III) chloride(*aq*) + cesium phosphate(*aq*) $\longrightarrow$
(b) Sodium hydroxide(*aq*) + cadmium nitrate(*aq*) $\longrightarrow$
(c) Magnesium bromide(*aq*) + potassium acetate(*aq*) $\longrightarrow$
(d) Silver nitrate(*aq*) + barium chloride(*aq*) $\longrightarrow$

    In the next sample problem, we'll use molecular depictions to examine a precipitation reaction quantitatively.

SAMPLE PROBLEM 4.3   Using Molecular Depictions to Understand a Precipitation Reaction

ARIS Presentation Center, Worked Examples Library

PROBLEM Consider these molecular views of the reactant solutions in a precipitation reaction (with ions represented as simple spheres and solvent molecules omitted for clarity):

(a) Which compound is dissolved in solution A: KCl, Na_2SO_4, $MgBr_2$, or Ag_2SO_4?
(b) Which compound is dissolved in solution B: NH_4NO_3, $MgSO_4$, $Ba(NO_3)_2$, or CaF_2?
(c) Name the precipitate and the spectator ions that result when solutions A and B are mixed, and write balanced molecular, total ionic, and net ionic equations for the reaction.
(d) If each particle represents 0.010 mol of ions, what is the maximum mass of precipitate that can form (assuming complete reaction)?
PLAN (a) and **(b)** From the depictions of the solutions in the beakers, we note the charge and number of each kind of ion and use Table 4.1 to determine the likely compounds. **(c)** Once we know the possible ion combinations, Table 4.1 helps us determine which two make up the solid. The other two are spectator ions. **(d)** We use the formula of the solid in part (c) and count the number of each kind of ion, to see which ion is in excess, which means the amount of the other ion limits the amount of precipitate that forms. We multiply the number of limiting ion particles by 0.010 mol and then use the molar mass of the precipitate to find the mass in grams.
SOLUTION (a) In solution A, there are two 1+ particles for each 2− particle. Therefore, the dissolved compound cannot be KCl or $MgBr_2$. Of the remaining two choices, Ag_2SO_4 is insoluble, so it must be Na_2SO_4.
(b) In solution B, there are two 1− particles for each 2+ particle. Therefore, the dissolved compound cannot be NH_4NO_3 or $MgSO_4$. Of the remaining two choices, CaF_2 is insoluble, so it must be $Ba(NO_3)_2$.
(c) Of the remaining two ion combinations, the precipitate must be barium sulfate, and Na^+ and $NO_3{}^-$ are the spectator ions.

Molecular:	$Ba(NO_3)_2(aq) + Na_2SO_4(aq) \longrightarrow BaSO_4(s) + 2NaNO_3(aq)$
Total ionic:	$Ba^{2+}(aq) + 2NO_3{}^-(aq) + 2Na^+(aq) + SO_4{}^{2-}(aq) \longrightarrow$
	$\qquad\qquad\qquad\qquad BaSO_4(s) + 2NO_3{}^-(aq) + 2Na^+(aq)$
Net ionic:	$Ba^{2+}(aq) + SO_4{}^{2-}(aq) \longrightarrow BaSO_4(s)$

(d) The molar mass of $BaSO_4$ is 233.4 g/mol. There are four Ba^{2+} particles, so the maximum mass of $BaSO_4$ that can form is

$$\text{Mass (g) of } BaSO_4 = 4\ Ba^{2+} \text{ particles} \times \frac{0.010 \text{ mol } Ba^{2+} \text{ ions}}{1 \text{ particle}}$$

$$\times \frac{1 \text{ mol } BaSO_4}{1 \text{ mol } Ba^{2+} \text{ ions}} \times \frac{233.4 \text{ g } BaSO_4}{1 \text{ mol}}$$

$$= 9.3 \text{ g } BaSO_4$$

CHECK Let's use the method for finding the limiting reactant that was introduced in Sample Problems 3.12 and 3.13; that is, see which reactant gives fewer moles of product:

$$\text{Amount (mol) of BaSO}_4 = 4\ Ba^{2+}\ \text{particles} \times 0.010\ \text{mol Ba}^{2+}\ \text{ions/particle}$$
$$\times 1\ \text{mol of BaSO}_4/1\ \text{mol Ba}^{2+}\ \text{ions}$$
$$= 0.040\ \text{mol BaSO}_4$$

$$\text{Amount (mol) of BaSO}_4 = 5\ SO_4^{2-}\ \text{particles} \times 0.010\ \text{mol SO}_4^{2-}\ \text{ions/particle}$$
$$\times 1\ \text{mol of BaSO}_4/1\ \text{mol SO}_4^{2-}\ \text{ions}$$
$$= 0.050\ \text{mol BaSO}_4$$

Therefore, Ba^{2+} is the limiting reactant ion, and the mass, after rounding, is 0.040 mol $\times$ 230 g/mol = 9.2 g, close to our calculated answer.

FOLLOW-UP PROBLEM 4.3 Molecular views of the reactant solutions in a precipitation reaction are shown below (with ions represented as spheres and solvent omitted):

(a) Which compound is dissolved in beaker A: $Zn(NO_3)_2$, KCl, Na_2SO_4, or $PbCl_2$?
(b) Which compound is dissolved in beaker B: $(NH_4)_2SO_4$, $Cd(OH)_2$, $Ba(OH)_2$, or KNO_3?
(c) Name the precipitate and the spectator ions that result when the solutions in beakers A and B are mixed, and write balanced molecular, total ionic, and net ionic equations for the reaction.
(d) If each particle represents 0.050 mol of ions, what is the maximum mass of precipitate that can form (assuming complete reaction)?

Section Summary

Precipitation reactions involve formation of an insoluble ionic compound from two soluble ones. These reactions occur because electrostatic attractions among certain pairs of solvated ions are strong enough to cause their removal from solution. • Formation of a precipitate is predicted by noting whether any possible ion combinations are insoluble, based on a set of solubility rules.

4.4 ACID-BASE REACTIONS

Aqueous acid-base reactions involve water not only as solvent but also in the more active roles of reactant and product. These reactions occur in processes as diverse as the metabolic action of proteins, the industrial production of fertilizer, and some of the methods for revitalizing lakes damaged by acid rain.

Obviously, an **acid-base reaction** (also called a **neutralization reaction**) occurs when an acid reacts with a base, but the definitions of these terms and the scope of this reaction class have changed considerably over the years. For our purposes at this point, we'll use definitions that apply to chemicals you commonly encounter in the lab:

• An **acid** *is a substance that produces* H^+ *ions when dissolved in water.*

$$\text{HX} \xrightarrow{\text{H}_2\text{O}} H^+(aq) + X^-(aq)$$

• A **base** *is a substance that produces* OH^- *ions when dissolved in water.*

$$\text{MOH} \xrightarrow{\text{H}_2\text{O}} M^+(aq) + OH^-(aq)$$

(Other definitions of *acid* and *base* are presented later in this section and again in Chapter 18, along with a fuller meaning of *neutralization*.)

Acids and the Solvated Proton Acidic solutions arise from a special class of covalent molecules that do dissociate in water. In every case, these molecules contain a polar bond to hydrogen in which the atom bonded to H pulls more strongly on the shared electron pair. A good example is hydrogen chloride gas. The Cl end of the HCl molecule is partially negative, and the H end is partially positive. When HCl dissolves in water, the partially charged poles of H_2O molecules are attracted to the oppositely charged poles of HCl. The H—Cl bond breaks, with the H becoming the solvated cation $H^+(aq)$ and the Cl becoming the solvated anion $Cl^-(aq)$. Hydrogen bromide behaves similarly when it dissolves in water:

$$HBr(g) \xrightarrow{H_2O} H^+(aq) + Br^-(aq)$$

Water interacts strongly with many ions, but most strongly with the hydrogen cation, H^+, a very unusual species. The H atom is a proton surrounded by an electron, so the H^+ ion is just a proton. Because its full positive charge is concentrated in such a tiny volume, H^+ attracts the negative pole of surrounding water molecules so strongly that it actually forms a covalent bond to one of them. We usually show this interaction by writing the aqueous H^+ ion as H_3O^+ (hydronium ion). Thus, to show more accurately what takes place when $HBr(g)$ dissolves, we should write

$$HBr(g) + H_2O(l) \longrightarrow H_3O^+(aq) + Br^-(aq)$$

To make a point here about the interactions with water, let's write the hydronium ion as $(H_2O)H^+$. The hydronium ion associates with other water molecules to give species such as $H_5O_2^+$ [or $(H_2O)_2H^+$], $H_7O_3^+$ [or $(H_2O)_3H^+$], $H_9O_4^+$ [or $(H_2O)_4H^+$], and still larger aggregates; $H_7O_3^+$ is shown in Figure 4.7. These various species exist together, but we use $H^+(aq)$ as a general, simplified notation. Later in this chapter and much of the rest of the text, we show the solvated proton as $H_3O^+(aq)$ to emphasize water's role.

Water interacts covalently with many metal ions as well. For example, Fe^{3+} exists in water as $Fe(H_2O)_6^{3+}$, an Fe^{3+} ion bound to six H_2O molecules. Similarly, Zn^{2+} exists as $Zn(H_2O)_4^{2+}$ and Ni^{2+} as $Ni(H_2O)_6^{2+}$. We discuss these species fully in later chapters.

Acids and Bases as Electrolytes Acids and bases are categorized in terms of their "strength"—the degree to which they dissociate into ions in aqueous solution (Table 4.2). In water, *strong acids and strong bases dissociate completely* into ions. Therefore, like soluble ionic compounds, they are *strong* electrolytes and conduct a current well (Figure 4.8A). In contrast, *weak acids and weak bases dissociate into ions very little, and most of their molecules remain intact.* As a result, they conduct only a small current and are *weak* electrolytes (Figure 4.8B). Because a strong acid (or strong base) dissociates completely, we can determine the molarity of H^+ (or OH^-) in a given solution of one.

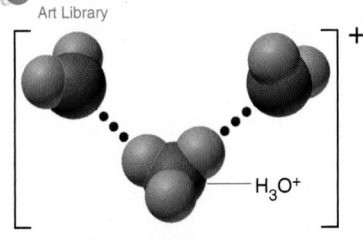

H_3O^+

Figure 4.7 The hydrated proton. The charge of the H^+ ion is highly concentrated because the ion is so small. In aqueous solution, it forms a covalent bond to a water molecule, yielding an H_3O^+ ion that associates tightly with other H_2O molecules. Here, the $H_7O_3^+$ ion is shown.

Table 4.2   Strong and Weak Acids and Bases

Acids

Strong

Hydrochloric acid, HCl
Hydrobromic acid, HBr
Hydriodic acid, HI
Nitric acid, HNO_3
Sulfuric acid, H_2SO_4
Perchloric acid, $HClO_4$

Weak

Hydrofluoric acid, HF
Phosphoric acid, H_3PO_4
Acetic acid, CH_3COOH
  (or $HC_2H_3O_2$)

Bases

Strong

Sodium hydroxide, NaOH
Potassium hydroxide, KOH
Calcium hydroxide, $Ca(OH)_2$
Strontium hydroxide, $Sr(OH)_2$
Barium hydroxide, $Ba(OH)_2$

Weak

Ammonia, NH_3

A

B

Figure 4.8 Acids and bases as electrolytes. A, Strong acids and bases are strong electrolytes, as indicated by the brightly lit bulb. **B,** Weak acids and bases are weak electrolytes.

SAMPLE PROBLEM 4.4 Determining the Molarity of H⁺ (or OH⁻) Ions

PROBLEM The strong acid nitric acid is a major chemical in the fertilizer and explosives industries. What is the molarity of $H^+(aq)$ in 1.4 M nitric acid?

PLAN We know the molarity of acid (1.4 M) and, from Table 4.2, we know that in aqueous solution, each molecule dissociates and the H becomes a solvated H^+ ion. Therefore, we use the formula to find the number of moles of $H^+(aq)$ present in 1 L of solution.

SOLUTION Nitrate ion is NO_3^-, so nitric acid is HNO_3. Thus, 1 mol of $H^+(aq)$ is released per mole of acid:

$$HNO_3(l) \xrightarrow{H_2O} H^+(aq) + NO_3^-(aq)$$

Therefore, 1.4 M HNO_3 contains 1.4 mol of $H^+(aq)$ per liter and is 1.4 M H^+ (aq).

FOLLOW-UP PROBLEM 4.4 How many moles of $OH^-(aq)$ are present in 451 mL of 1.20 M potassium hydroxide?

Strong and weak acids have one or more H atoms as part of their structure. Strong bases have either the OH^- or the O^{2-} ion as part of their structure. Soluble ionic oxides, such as K_2O, are strong bases because the oxide ion is not stable in water and reacts immediately to form hydroxide ion:

$$K_2O(s) + H_2O(l) \longrightarrow 2K^+(aq) + 2OH^-(aq)$$

Weak bases, such as ammonia, do not contain OH^- ions, but they produce them in a reaction with water that occurs to a small extent:

$$NH_3(g) + H_2O(l) \rightleftharpoons NH_4^+(aq) + OH^-(aq)$$

(Note the reaction arrow in the preceding equation. This type of arrow indicates that the reaction proceeds in both directions; we'll discuss this important idea further in Section 4.7.)

The Key Event: Formation of H₂O from H⁺ and OH⁻

Let's use the three types of equations (and color) to see what occurs in acid-base reactions. We begin with the molecular equation for the reaction between the strong acid HCl and the strong base $Ba(OH)_2$:

$$2HCl(aq) + Ba(OH)_2(aq) \longrightarrow BaCl_2(aq) + 2H_2O(l)$$

Because HCl and $Ba(OH)_2$ dissociate completely and H_2O remains undissociated, the total ionic equation is

$$2H^+(aq) + 2Cl^-(aq) + Ba^{2+}(aq) + 2OH^-(aq) \longrightarrow Ba^{2+}(aq) + 2Cl^-(aq) + 2H_2O(l)$$

In the net ionic equation, we eliminate the spectator ions $Ba^{2+}(aq)$ and $Cl^-(aq)$ and see the actual reaction:

$$2H^+(aq) + 2OH^-(aq) \longrightarrow 2H_2O(l)$$

or

$$H^+(aq) + OH^-(aq) \longrightarrow H_2O(l)$$

Thus, *the essential change in all aqueous reactions between a strong acid and a strong base is that an H^+ ion from the acid and an OH^- ion from the base form a water molecule.* In fact, only the spectator ions differ from one strong acid–strong base reaction to another.

Like precipitation reactions, acid-base reactions occur through *the electrostatic attraction of ions and their removal from solution* as the product. In this case, the ions are H^+ and OH^- and the product is H_2O, which consists almost entirely of undissociated molecules. (Actually, water molecules *do* dissociate, but *very* slightly. As you'll see in Chapter 18, this slight dissociation is very important, but the formation of water in a neutralization reaction nevertheless represents an enormous net removal of H^+ and OH^- ions.)

Evaporate the water from the above reaction mixture, and the ionic solid barium chloride remains. An ionic compound that results from the reaction of an acid and a base is called a **salt.** Thus, in a typical aqueous neutralization reaction, *the reactants are an acid and a base, and the products are a salt solution and water:*

$$HX(aq) + MOH(aq) \longrightarrow MX(aq) + H_2O(l)$$
$$\quad\text{acid} \qquad\quad \text{base} \qquad\qquad \text{salt} \qquad\quad \text{water}$$

The color shows that *the cation of the salt comes from the base and the anion comes from the acid.*

Notice that acid-base reactions, like precipitation reactions, are metathesis (double-displacement or exchange) reactions. The molecular equation for the reaction of aluminum hydroxide, the active ingredient in some antacid tablets, with HCl, the major component of stomach acid, shows this clearly:

$$3HCl(aq) + Al(OH)_3(s) \longrightarrow AlCl_3(aq) + 3H_2O(l)$$

Acid-base reactions occur frequently in industry, in the environment, and in the synthesis and breakdown of biological macromolecules. ●

SAMPLE PROBLEM 4.5 Writing Ionic Equations for Acid-Base Reactions

PROBLEM Write balanced molecular, total ionic, and net ionic equations for each of the following acid-base reactions and identify the spectator ions:
(a) Strontium hydroxide(*aq*) + perchloric acid(*aq*) ⟶
(b) Barium hydroxide(*aq*) + sulfuric acid(*aq*) ⟶
PLAN All are strong acids and bases (see Table 4.2), so the essential reaction is between H^+ and OH^-. The products are H_2O and a salt solution consisting of the spectator ions. Note that in (b), the salt ($BaSO_4$) is insoluble (see Table 4.1), so virtually all ions are removed from solution.
SOLUTION (a) Writing the molecular equation:

$$Sr(OH)_2(aq) + 2HClO_4(aq) \longrightarrow Sr(ClO_4)_2(aq) + 2H_2O(l)$$

Writing the total ionic equation:

$$Sr^{2+}(aq) + 2OH^-(aq) + 2H^+(aq) + 2ClO_4^-(aq) \longrightarrow$$
$$Sr^{2+}(aq) + 2ClO_4^-(aq) + 2H_2O(l)$$

Writing the net ionic equation:

$$2OH^-(aq) + 2H^+(aq) \longrightarrow 2H_2O(l) \text{ or } OH^-(aq) + H^+(aq) \longrightarrow H_2O(l)$$

$Sr^{2+}(aq)$ and $ClO_4^-(aq)$ are the spectator ions.
(b) Writing the molecular equation:

$$Ba(OH)_2(aq) + H_2SO_4(aq) \longrightarrow BaSO_4(s) + 2H_2O(l)$$

Writing the total ionic equation:

$$Ba^{2+}(aq) + 2OH^-(aq) + 2H^+(aq) + SO_4^{2-}(aq) \longrightarrow BaSO_4(s) + 2H_2O(l)$$

The net ionic equation is the same as the total ionic equation. This is a precipitation *and* a neutralization reaction. There are no spectator ions because all the ions are used to form the two products.

FOLLOW-UP PROBLEM 4.5 Write balanced molecular, total ionic, and net ionic equations for the reaction between aqueous solutions of calcium hydroxide and nitric acid.

● **Displacement Reactions Inside You**
The digestion of food proteins and the formation of an organism's own proteins form a continuous cycle of displacement reactions. A protein consists of hundreds or thousands of smaller molecules, called *amino acids,* linked in a long chain. When you eat proteins, your digestive processes use H_2O to displace one amino acid at a time. These are transported by the blood to your cells, where other metabolic processes link them together, displacing H_2O, to make your own proteins.

Acid-Base Titrations

Chemists study acid-base reactions quantitatively through titrations. In any **titration,** *one solution of known concentration is used to determine the concentration of another solution through a monitored reaction.*

Figure 4.9 An acid-base titration. A, In this procedure, a measured volume of the unknown acid solution is placed in a flask beneath a buret containing the known (standardized) base solution. A few drops of indicator are added to the flask; the indicator used here is phenolphthalein, which is colorless in acid and pink in base. After an initial buret reading, base (OH^- ions) is added slowly to the acid (H^+ ions). **B,** Near the end of the titration, the indicator momentarily changes to its base color but reverts to its acid color with swirling. **C,** When the end point is reached, a tiny excess of OH^- is present, shown by the permanent change in color of the indicator. The difference between the final buret reading and the initial buret reading gives the volume of base used.

ARIS Presentation Center,
Photo Library

In a typical acid-base titration, a *standardized* solution of base, one whose concentration is *known,* is added slowly to an acid solution of *unknown* concentration (Figure 4.9). A known volume of the acid solution is placed in a flask, and a few drops of indicator solution are added. An *acid-base indicator* is a substance whose color is different in acid than in base. (We examine indicators in Chapters 18 and 19.) The standardized solution of base is added slowly to the flask from a buret. As the titration is close to its end, indicator molecules near a drop of added base change color due to the temporary excess of OH^- ions there. As soon as the solution is swirled, however, the indicator's acidic color returns. The **equivalence point** in the titration occurs when *all the moles of H^+ ions present in the original volume of acid solution have reacted with an equivalent number of moles of OH^- ions added from the buret:*

Moles of H^+ (originally in flask) = moles of OH^- (added from buret)

The **end point** of the titration occurs when a tiny excess of OH^- ions changes the indicator permanently to its color in base. In calculations, we assume this tiny excess is insignificant, and therefore *the amount of base needed to reach the end point is the same as the amount needed to reach the equivalence point.*

ARIS Presentation Center,
Worked Examples Library

SAMPLE PROBLEM 4.6 Finding the Concentration of Acid from an Acid-Base Titration

PROBLEM You perform an acid-base titration to standardize an HCl solution by placing 50.00 mL of HCl in a flask with a few drops of indicator solution. You put 0.1524 *M* NaOH into the buret, and the initial reading is 0.55 mL. At the end point, the buret reading is 33.87 mL. What is the concentration of the HCl solution?
PLAN We must find the molarity of acid from the volume of acid (50.00 mL), the initial (0.55 mL) and final (33.87 mL) volumes of base, and the molarity of base (0.1524 *M*).

First, we balance the equation. We find the volume of base added from the difference in buret readings and use the base's molarity to calculate the amount (mol) of base added. Then, we use the molar ratio from the balanced equation to find the amount (mol) of acid originally present and divide by the acid's original volume to find the molarity.

SOLUTION Writing the balanced equation:

$$NaOH(aq) + HCl(aq) \longrightarrow NaCl(aq) + H_2O(l)$$

Finding volume (L) of NaOH solution added:

$$\text{Volume (L) of solution} = (33.87 \text{ mL soln} - 0.55 \text{ mL soln}) \times \frac{1 \text{ L}}{1000 \text{ mL}}$$

$$= 0.03332 \text{ L soln}$$

Finding amount (mol) of NaOH added:

$$\text{Moles of NaOH} = 0.03332 \text{ L soln} \times \frac{0.1524 \text{ mol NaOH}}{1 \text{ L soln}}$$

$$= 5.078 \times 10^{-3} \text{ mol NaOH}$$

Finding amount (mol) of HCl originally present: Since the molar ratio is 1/1,

$$\text{Moles of HCl} = 5.078 \times 10^{-3} \text{ mol NaOH} \times \frac{1 \text{ mol HCl}}{1 \text{ mol NaOH}} = 5.078 \times 10^{-3} \text{ mol HCl}$$

Calculating molarity of HCl:

$$\text{Molarity of HCl} = \frac{5.078 \times 10^{-3} \text{ mol HCl}}{50.00 \text{ mL}} \times \frac{1000 \text{ mL}}{1 \text{ L}}$$

$$= 0.1016 \, M \text{ HCl}$$

CHECK The answer makes sense: a larger volume of less concentrated acid neutralized a smaller volume of more concentrated base. Rounding shows that the moles of H^+ and OH^- are about equal: 50 mL $\times$ 0.1 M H^+ = 0.005 mol = 33 mL $\times$ 0.15 M OH^-.

FOLLOW-UP PROBLEM 4.6 What volume of 0.1292 M $Ba(OH)_2$ would neutralize 50.00 mL of the HCl solution standardized above in the sample problem?

Proton Transfer: A Closer Look at Acid-Base Reactions

We gain deeper insight into acid-base reactions if we look closely at the species in solution. Let's see what takes place when HCl gas dissolves in water. Polar water molecules pull apart each HCl molecule and the H^+ ion ends up bonded to a water molecule. In essence, HCl *transfers its proton* to H_2O:

$$\overset{\frown \text{ H}^+ \text{ transfer } \searrow}{HCl(g)} + H_2O(l) \longrightarrow H_3O^+(aq) + Cl^-(aq)$$

Thus, hydrochloric acid (an aqueous solution of HCl gas) actually consists of solvated H_3O^+ and Cl^- ions.

When sodium hydroxide solution is added, the H_3O^+ ion transfers a proton to the OH^- ion of the base (with the product water shown here as HOH):

$$[H_3O^+(aq) + Cl^-(aq)] + [Na^+(aq) + OH^-(aq)] \longrightarrow$$
$$H_2O(l) + Cl^-(aq) + Na^+(aq) + HOH(l)$$

Without the spectator ions, the transfer of a proton from H_3O^+ to OH^- is obvious:

$$\overset{\frown \text{ H}^+ \text{ transfer } \searrow}{H_3O^+(aq)} + OH^-(aq) \longrightarrow H_2O(l) + HOH(l) \quad [\text{or } 2H_2O(l)]$$

This net ionic equation is identical with the one we saw earlier (see p. 152),

$$H^+(aq) + OH^-(aq) \longrightarrow H_2O(l)$$

with the additional H_2O molecule coming from the H_3O^+. Thus, *an acid-base reaction is a proton-transfer process.* In this case, the Cl^- and Na^+ ions remain

Figure 4.10 **An aqueous strong acid–strong base reaction on the atomic scale.** When solutions of a strong acid (HX) and a strong base (MOH) are mixed, the H_3O^+ from the acid transfers a proton to the OH⁻ from the base to form an H_2O molecule. Evaporation of the water leaves the spectator ions, X⁻ and M⁺, as a solid ionic compound called a *salt*.

in solution, and if the water is evaporated, they crystallize as the salt NaCl. Figure 4.10 shows this process on the atomic level.

In the early 20th century, the chemists Johannes Brønsted and Thomas Lowry realized the proton-transfer nature of acid-base reactions. They defined *an acid as a molecule (or ion) that donates a proton, and a base as a molecule (or ion) that accepts a proton.* Therefore, in the aqueous reaction between strong acid and strong base, H_3O^+ *ion acts as the acid and* OH⁻ *ion acts as the base.* Because it ionizes completely, a given amount of strong acid (or strong base) creates an equivalent amount of H_3O^+ (or OH⁻) when it dissolves in water. (We discuss the Brønsted-Lowry concept thoroughly in Chapter 18.)

Gas-Forming Reactions Thinking of acid-base reactions as proton-transfer processes helps us understand another common type of aqueous ionic reactions, those that form a gaseous product. For example, when an ionic carbonate, such as K_2CO_3, is treated with an acid, such as HCl, one of the products is carbon dioxide. *Such reactions occur through the formation of a gas and water because both products remove reactant ions from solution:*

$$2HCl(aq) + K_2CO_3(aq) \longrightarrow 2KCl(aq) + [H_2CO_3(aq)]$$
$$[H_2CO_3(aq)] \longrightarrow H_2O(l) + CO_2(g)$$

The product H_2CO_3 is shown in square brackets to indicate that it is very unstable. It decomposes immediately into water and carbon dioxide. Combining these two equations gives the overall equation:

$$2HCl(aq) + K_2CO_2(aq) \longrightarrow 2KCl(aq) + H_2O(l) + CO_2(g)$$

When we show H_3O^+ ions from the HCl as the actual species in solution and write the net ionic equation, Cl^- and K^+ ions are eliminated. Note that each of the two H_3O^+ ions transfers a proton to the carbonate ion:

$$2H_3O^+(aq) + CO_3^{2-}(aq) \longrightarrow 2H_2O(l) + [H_2CO_3(aq)] \longrightarrow 3H_2O(l) + CO_2(g)$$

(2H⁺ transfer)

In essence, this is an acid-base reaction with carbonate ion accepting the protons and, thus, acting as the base. Several other polyatomic ions react similarly with an acid. In the formation of SO_2 from ionic sulfites, the net ionic equation is

$$2H_3O^+(aq) + SO_3^{2-}(aq) \longrightarrow 2H_2O(l) + [H_2SO_3(aq)] \longrightarrow 3H_2O(l) + SO_2(g)$$

(2H⁺ transfer)

Reactions of Weak Acids Ionic equations are written differently for the reactions of weak acids. When solutions of sodium hydroxide and acetic acid (CH_3COOH) are mixed, the molecular, total ionic, and net ionic equations are

Molecular equation:

$$CH_3COOH(aq) + NaOH(aq) \longrightarrow CH_3COONa(aq) + H_2O(l)$$

Total ionic equation:

$$CH_3COOH(aq) + Na^+(aq) + OH^-(aq) \longrightarrow CH_3COO^-(aq) + Na^+(aq) + H_2O(l)$$

Net ionic equation:

$$CH_3COOH(aq) + OH^-(aq) \longrightarrow CH_3COO^-(aq) + H_2O(l)$$

(H⁺ transfer)

Acetic acid dissociates very little because it is a weak acid (see Table 4.2). To show this, *it appears undissociated in both ionic equations.* Note that H_3O^+ does not appear; rather, the proton is transferred from CH_3COOH. Therefore, only $Na^+(aq)$ is a spectator ion; $CH_3COO^-(aq)$ is not. Figure 4.11 shows the gas-forming reaction between vinegar (an aqueous 5% solution of acetic acid) and baking soda (sodium hydrogen carbonate) solution.

Molecular equation
$$NaHCO_3(aq) + CH_3COOH(aq) \longrightarrow CH_3COONa(aq) + CO_2(g) + H_2O(l)$$

Total ionic equation
$$Na^+(aq) + HCO_3^-(aq) + CH_3COOH(aq) \longrightarrow$$
$$CH_3COO^-(aq) + Na^+(aq) + CO_2(g) + H_2O(l)$$

Net ionic equation
$$HCO_3^-(aq) + CH_3COOH(aq) \longrightarrow CH_3COO^-(aq) + CO_2(g) + H_2O(l)$$

Figure 4.11 An acid-base reaction that forms a gaseous product. Carbonates and hydrogen carbonates react with acids to form gaseous CO_2 and H_2O. Here, dilute acetic acid solution (vinegar) is added to sodium hydrogen carbonate (baking soda) solution, and bubbles of CO_2 gas form. (Note that the net ionic equation includes intact acetic acid because it does *not* dissociate into ions to an appreciable extent.)

ARIS Presentation Center, Photo Library

Section Summary

Acid-base (neutralization) reactions occur when an acid (an H^+-yielding substance) and a base (an OH^--yielding substance) react and the H^+ and OH^- ions form a water molecule. • Strong acids and bases dissociate completely in water; weak acids and bases dissociate slightly. • In a titration, a known concentration of one reactant is used to determine the concentration of the other. • An acid-base reaction can also be viewed as the transfer of a proton from an acid to a base. • An ionic gas-forming reaction is an acid-base reaction in which an acid transfers a proton to a polyatomic ion (carbonate or sulfite), forming a gas that leaves the reaction mixture. • Weak acids dissociate very little, so equations involving them show the acid as an intact molecule.

4.5 OXIDATION-REDUCTION (REDOX) REACTIONS

Redox reactions form the third and most important of the reaction classes we discuss here. They include the formation of a compound from its elements (and vice versa), all combustion reactions, the reactions that generate electricity in batteries, and the reactions that produce cellular energy. As some of these examples indicate, the redox process is so general that many redox reactions, including several we examine in this and the next section, do not occur in aqueous solution. In this section, we examine the process, learn some essential terminology, and see one way to balance redox equations and one way to apply them quantitatively.

The Key Event: Movement of Electrons Between Reactants

Demo Proksa, Miroslav. "The Sodium Rainbow," *J. Chem. Educ.* **1997**, *74*, 942.

In **oxidation-reduction** (or **redox**) **reactions,** the key chemical event is the *net movement of electrons from one reactant to the other*. This movement of electrons occurs *from the reactant (or atom in the reactant) with less attraction for electrons to the reactant (or atom) with more attraction for electrons.*

Such movement of electron charge occurs in the formation of both ionic and covalent compounds. As an example, let's reconsider the reaction shown in Figure 3.8 (p. 106), in which an ionic compound, MgO, forms from its elements:

$$2Mg(s) + O_2(g) \longrightarrow 2MgO(s)$$

Figure 4.12A shows that during the reaction, each Mg atom loses two electrons and each O atom gains them; that is, two electrons move from each Mg atom to each O atom. This change represents a ***transfer*** of electron charge away from

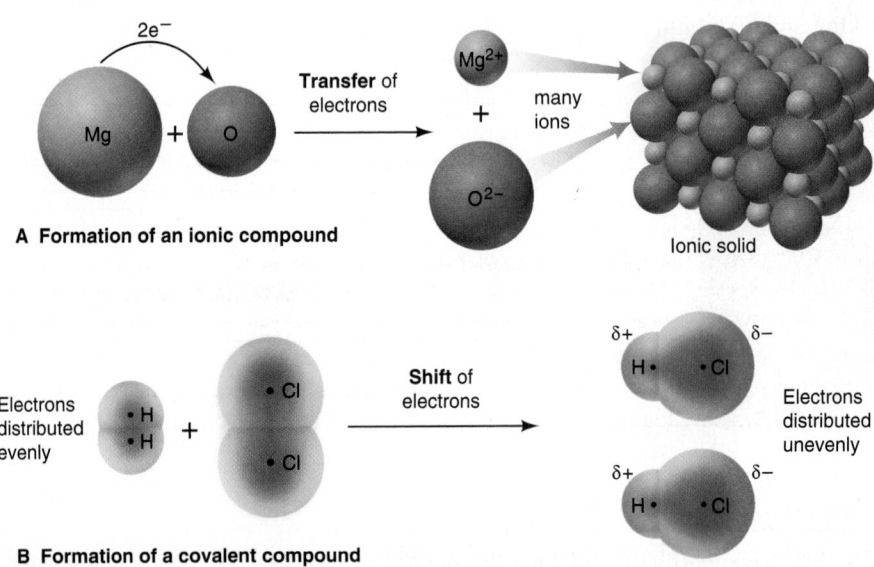

A Formation of an ionic compound

B Formation of a covalent compound

Figure 4.12 The redox process in compound formation. A, In forming the ionic compound MgO, each Mg atom transfers two electrons to each O atom. (Note that species become smaller when they lose electrons and larger when they gain electrons.) The resulting Mg^{2+} and O^{2-} ions aggregate with many others to form an ionic solid. **B,** In the reactants H_2 and Cl_2, the electron pairs are shared equally (indicated by even electron density shading). In the covalent product HCl, Cl attracts the shared electrons more strongly than H does. In effect, the H electron shifts toward Cl, as shown by higher electron density *(red)* near the Cl end of the molecule and lower electron density *(blue)* near the H end.

 Transparency/ARIS Presentation Center, Active Art Library

 Animation: Oxidation-Reduction Reactions ARIS Presentation Center Chapter 4

each Mg atom toward each O atom, resulting in the formation of Mg^{2+} and O^{2-} ions. The ions aggregate and form an ionic solid.

During the formation of a covalent compound from its elements, there is again a net movement of electrons, but it is more of a *shift* in electron charge than a full transfer. Thus, *ions do not form.* Consider the formation of HCl gas:

$$H_2(g) + Cl_2(g) \longrightarrow 2HCl(g)$$

To see the electron movement here, compare the electron charge distributions in the reactant bonds and in the product bonds. As Figure 4.12B shows, H_2 and Cl_2 molecules are each held together by covalent bonds in which the electrons are shared equally between the atoms (the tan shading is symmetrical). In the HCl molecule, the electrons are shared unequally because the Cl atom attracts them more strongly than the H atom does. Thus, in HCl, the H has less electron charge *(blue shading)* than it had in H_2, and the Cl has more charge *(red shading)* than it had in Cl_2. In other words, in the formation of HCl, there has been a relative **shift** *of electron charge* away from the H atom toward the Cl atom. This electron *shift* is not nearly as extreme as the electron *transfer* during MgO formation. In fact, in some reactions, the net movement of electrons may be very slight, but the reaction is still a redox process.

Some Essential Redox Terminology

Chemists use some important terminology to describe the movement of electrons in oxidation-reduction reactions. **Oxidation** is the *loss* of electrons, and **reduction** is the *gain* of electrons. (The original meaning of *reduction* comes from the process of reducing large amounts of metal ore to smaller amounts of metal, but you'll see shortly why we use the term "reduction" for the act of gaining.)

For example, during the formation of magnesium oxide, Mg undergoes oxidation (electron loss) and O_2 undergoes reduction (electron gain). The loss and gain are simultaneous, but we can imagine them occurring in separate steps:

Oxidation (electron loss by Mg): $\qquad Mg \longrightarrow Mg^{2+} + 2e^-$

Reduction (electron gain by O_2): $\frac{1}{2}O_2 + 2e^- \longrightarrow O^{2-}$

One reactant acts on the other. Thus, we say that *O_2 oxidizes Mg,* and that O_2 is the **oxidizing agent,** the species doing the oxidizing. Similarly, *Mg reduces O_2,* so Mg is the **reducing agent,** the species doing the reducing.

Note especially that O_2 takes the electrons that Mg loses or, put the other way around, Mg gives up the electrons that O_2 gains. This give-and-take of electrons means that *the oxidizing agent is reduced* because it takes the electrons (and thus gains them), and *the reducing agent is oxidized* because it gives up the electrons (and thus loses them). In the formation of HCl, Cl_2 oxidizes H_2 (H loses some electron charge and Cl gains it), which is the same as saying that H_2 reduces Cl_2. The reducing agent, H_2, is oxidized and the oxidizing agent, Cl_2, is reduced.

Using Oxidation Numbers to Monitor the Movement of Electron Charge

Chemists have devised a useful "bookkeeping" system to monitor which atom loses electron charge and which atom gains it. Each atom in a molecule (or ionic compound) is assigned an **oxidation number (O.N.),** or *oxidation state,* the charge the atom would have *if* electrons were not shared but were transferred completely. Based on this idea, the oxidation number for each element in a binary *ionic* compound equals the ionic charge. On the other hand, the oxidation number for each element in a *covalent* compound (or in a polyatomic ion) is not as obvious because the atoms don't have whole charges. In general, *oxidation numbers*

Literature ten Hoor, Marten J. "The Relative Explosive Power of Some Explosives," *J. Chem. Educ.* **2003,** *80,* 1397.

Demo Summerlin, L. R.; Ealy, J. B. "Oxidation States of Manganese: Mn^{7+}, Mn^{6+}, Mn^{4+} and Mn^{2+}," *Chemical Demonstrations, A Sourcebook for Teachers,* Vol. 1 (American Chemical Society, Washington, DC, 1988), 133–134.

Table 4.3   Rules for Assigning an Oxidation Number (O.N.)

General Rules

1. For an atom in its elemental form (Na, O_2, Cl_2, etc.):   O.N. = 0
2. For a monatomic ion: O.N. = ion charge
3. The sum of O.N. values for the atoms in a molecule or formula unit of a compound equals zero. The sum of O.N. values for the atoms in a polyatomic ion equals the ion's charge.

Rules for Specific Atoms or Periodic Table Groups

1. For Group 1A(1):   O.N. = +1 in all compounds
2. For Group 2A(2):   O.N. = +2 in all compounds
3. For hydrogen: O.N. = +1 in combination with nonmetals
    O.N. = −1 in combination with metals and boron
4. For fluorine: O.N. = −1 in all compounds
5. For oxygen: O.N. = −1 in peroxides
    O.N. = −2 in all other compounds (except with F)
6. For Group 7A(17):   O.N. = −1 in combination with metals, nonmetals (except O), and other halogens lower in the group

---

are determined by the set of rules in Table 4.3. (Oxidation numbers are assigned according to the relative attraction of an atom for electrons, so they are ultimately based on atomic properties, as you'll see in Chapters 8 and 9.) An O.N. has the sign *before* the number (e.g., +2), whereas an ionic charge has the sign *after* the number (e.g., 2+). Also, unlike a symbol with a 1+ ionic charge, an O.N. of +1 or −1 retains the numeral.

Figure 4.13 Highest and lowest oxidation numbers of reactive main-group elements. The A-group number shows the highest possible oxidation number (O.N.) for a main-group element. (Two important exceptions are O, which never has an O.N. of +6, and F, which never has an O.N. of +7.) For nonmetals *(yellow)* and metalloids *(green),* the A-group number minus 8 gives the lowest possible oxidation number.

SAMPLE PROBLEM 4.7   Determining the Oxidation Number of an Element

PROBLEM Determine the oxidation number (O.N.) of each element in these compounds:
(a) Zinc chloride   **(b)** Sulfur trioxide   **(c)** Nitric acid

PLAN We apply Table 4.3, noting the general rules that the O.N. values in a compound add up to zero, and the O.N. values in a polyatomic ion add up to the ion's charge.

SOLUTION **(a)** $ZnCl_2$. The sum of O.N.s for the monatomic ions in the compound must equal zero. The O.N. of the Zn^{2+} ion is +2. The O.N. of each $Cl^−$ ion is −1, for a total of −2. The sum of O.N.s is +2 + (−2), or 0.

(b) SO_3. The O.N. of each oxygen is −2, for a total of −6. The O.N.s must add up to zero, so the O.N. of S is +6.

(c) HNO_3. The O.N. of H is +1, so the O.N.s of the NO_3 group must add up to −1 to give zero for the compound. The O.N. of each O is −2 for a total of −6. Therefore, the O.N. of N is +5.

FOLLOW-UP PROBLEM 4.7 Determine the O.N. of each element in the following:
(a) Scandium oxide (Sc_2O_3)   **(b)** Gallium chloride ($GaCl_3$)
(c) Hydrogen phosphate ion   **(d)** Iodine trifluoride

You can find the highest and lowest oxidation numbers of most main-group elements from the periodic table, as Figure 4.13 shows:

- For most main-group elements, the A-group number (1A, 2A, and so on) is the *highest* oxidation number (always positive) of any element in the group. The exceptions are O and F (see Table 4.3).
- For main-group nonmetals and some metalloids, the A-group number minus 8 is the *lowest* oxidation number (always negative) of any element in the group.

For example, the highest oxidation number of S (Group 6A) is +6, as in SF_6, and the lowest is (6 − 8), or −2, as in FeS and other metal sulfides.

As Sample Problem 4.8 shows, a redox reaction can be defined as one in which *the oxidation numbers of the species change.* (In the rest of Section 4.5 and Section 4.6, blue type indicates oxidation, and red type indicates reduction.)

SAMPLE PROBLEM 4.8 Identifying Redox Reactions

ARIS Presentation Center, Worked Examples Library

PROBLEM Use oxidation numbers to decide which of the following are redox reactions:

(a) $CaO(s) + CO_2(g) \longrightarrow CaCO_3(s)$

(b) $4KNO_3(s) \xrightarrow{\Delta} 2K_2O(s) + 2N_2(g) + 5O_2(g)$

(c) $NaHSO_4(aq) + NaOH(aq) \longrightarrow Na_2SO_4(aq) + H_2O(l)$

PLAN To determine whether a reaction is an oxidation-reduction process, we use Table 4.3 to assign each atom an O.N. and see if it changes as the reactants become products.

SOLUTION

$$\overset{+2}{\underset{|}{Ca}}\overset{-2}{\underset{|}{O}}(s) + \overset{+4}{\underset{|}{C}}\overset{-2}{\underset{|}{O_2}}(g) \longrightarrow \overset{+2}{\underset{|}{Ca}}\overset{+4}{\underset{|}{C}}\overset{-2}{\underset{|}{O_3}}(s)$$

(a)

Because each atom in the product has the same O.N. that it had in the reactants, we conclude that this is *not* a redox reaction.

O.N. decreased: *reduction*

$$4\overset{+1}{\underset{|}{K}}\overset{+5}{\underset{|}{N}}\overset{-2}{\underset{|}{O_3}}(s) \xrightarrow{\Delta} 2\overset{+1}{\underset{|}{K_2}}\overset{-2}{\underset{|}{O}}(s) + 2\overset{0}{\underset{|}{N_2}}(g) + 5\overset{0}{\underset{|}{O_2}}(g)$$

(b)

O.N. increased: *oxidation*

In this case, the O.N. of N changes from +5 to 0, and the O.N. of O changes from −2 to 0, so this *is* a redox reaction.

$$\overset{+1}{\underset{|}{Na}}\overset{+1}{\underset{|}{H}}\overset{+6}{\underset{|}{S}}\overset{-2}{\underset{|}{O_4}}(aq) + \overset{+1}{\underset{|}{Na}}\overset{-2}{\underset{|}{O}}\overset{+1}{\underset{|}{H}}(aq) \longrightarrow \overset{+1}{\underset{|}{Na_2}}\overset{+6}{\underset{|}{S}}\overset{-2}{\underset{|}{O_4}}(aq) + \overset{+1}{\underset{|}{H_2}}\overset{-2}{\underset{|}{O}}(l)$$

(c)

The O.N. values do not change, so this is *not* a redox reaction.

COMMENT The reaction in part (c) is an acid-base reaction in which HSO_4^- transfers an H^+ to OH^- to form H_2O. In the net ionic equation for a strong acid–strong base reaction,

$$\overset{+1}{\underset{|}{H^+}}(aq) + \overset{-2}{\underset{|}{O}}\overset{+1}{\underset{|}{H^-}}(aq) \longrightarrow \overset{+1}{\underset{|}{H_2}}\overset{-2}{\underset{|}{O}}(l)$$

we see that the O.N. values remain the same on both sides of the equation. Therefore, *an acid-base reaction is* **not** *a redox reaction.*

FOLLOW-UP PROBLEM 4.8 Use oxidation numbers to decide which, if any, of the following equations represents a redox reaction:

(a) $NCl_3(l) + 3H_2O(l) \longrightarrow NH_3(aq) + 3HOCl(aq)$

(b) $AgNO_3(aq) + NH_4I(aq) \longrightarrow AgI(s) + NH_4NO_3(aq)$

(c) $2H_2S(g) + 3O_2(g) \longrightarrow 2SO_2(g) + 2H_2O(g)$

By assigning an oxidation number to each atom, we can see which species was oxidized and which reduced and, from that, which is the oxidizing agent and which the reducing agent:

Literature Calzaferri, Gion. "Oxidation Numbers," *J. Chem. Educ.* **1999**, *76*, 362.

• If a given atom has a higher (more positive or less negative) oxidation number in the product than it had in the reactant, the reactant species that contains the atom was oxidized (lost electrons) and is the reducing agent. Thus, *oxidation is represented by an increase in oxidation number.*

• If an atom has a lower (more negative or less positive) oxidation number in the product than it had in the reactant, the reactant species that contains the atom was reduced (gained electrons) and is the oxidizing agent. Thus, *the gain of electrons is represented by a decrease (a "reduction") in oxidation number.*

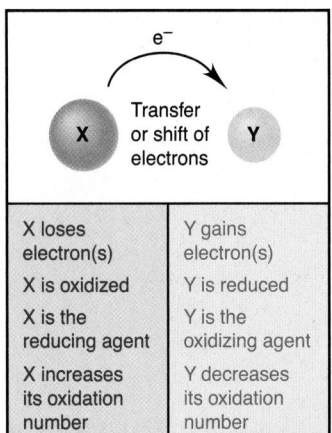

Figure 4.14 A summary of terminology for oxidation-reduction (redox) reactions.

Transparency/ARIS Presentation Center, Art Library

It is essential to realize that *the transferred electrons are never free, which means that the reducing agent loses electrons and the oxidizing agent gains them simultaneously.* In other words, a complete reaction *cannot* be just an "oxidation" *or* a "reduction"; it must be an "oxidation-reduction." Figure 4.14 summarizes redox terminology.

ARIS Presentation Center, Worked Examples Library

SAMPLE PROBLEM 4.9 Recognizing Oxidizing and Reducing Agents

PROBLEM Identify the oxidizing agent and reducing agent in each of the following:
(a) $2Al(s) + 3H_2SO_4(aq) \longrightarrow Al_2(SO_4)_3(aq) + 3H_2(g)$
(b) $PbO(s) + CO(g) \longrightarrow Pb(s) + CO_2(g)$
(c) $2H_2(g) + O_2(g) \longrightarrow 2H_2O(g)$

PLAN We first assign an oxidation number (O.N.) to each atom (or ion) based on the rules in Table 4.3. The reactant is the reducing agent if it contains an atom that is oxidized (O.N. increased from left side to right side of the equation). The reactant is the oxidizing agent if it contains an atom that is reduced (O.N. decreased).

SOLUTION **(a)** Assigning oxidation numbers:

$$\overset{0}{2Al(s)} + 3\overset{+1}{H_2}\overset{+6}{\underset{-2}{S}O_4}(aq) \longrightarrow \overset{+3}{Al_2}(\overset{+6}{\underset{-2}{S}O_4})_3(aq) + 3\overset{0}{H_2}(g)$$

The O.N. of Al increased from 0 to +3 (Al lost electrons), so Al was oxidized; Al is the reducing agent.

The O.N. of H decreased from +1 to 0 (H gained electrons), so H^+ was reduced; H_2SO_4 is the oxidizing agent.

(b) Assigning oxidation numbers:

$$\overset{+2}{Pb}\overset{-2}{O}(s) + \overset{+2}{C}\overset{-2}{O}(g) \longrightarrow \overset{0}{Pb}(s) + \overset{+4}{C}\overset{-2}{O_2}(g)$$

Pb decreased its O.N. from +2 to 0, so PbO was reduced; PbO is the oxidizing agent.

C increased its O.N. from +2 to +4, so CO was oxidized; CO is the reducing agent.
In general, when a substance (such as CO) becomes one with more O atoms (such as CO_2), it is oxidized; and when a substance (such as PbO) becomes one with fewer O atoms (such as Pb), it is reduced.

(c) Assigning oxidation numbers:

$$\overset{0}{2H_2}(g) + \overset{0}{O_2}(g) \longrightarrow \overset{+1-2}{2H_2O}(g)$$

O_2 was reduced (O.N. of O decreased from 0 to −2); O_2 is the oxidizing agent.

H_2 was oxidized (O.N. of H increased from 0 to +1); H_2 is the reducing agent.
Oxygen is always the oxidizing agent in a combustion reaction.

FOLLOW-UP PROBLEM 4.9 Identify each oxidizing agent and each reducing agent:
(a) $2Fe(s) + 3Cl_2(g) \longrightarrow 2FeCl_3(s)$
(b) $2C_2H_6(g) + 7O_2(g) \longrightarrow 4CO_2(g) + 6H_2O(g)$
(c) $5CO(g) + I_2O_5(s) \longrightarrow I_2(s) + 5CO_2(g)$

Balancing Redox Equations

We balance a redox reaction by making sure that *the number of electrons lost by the reducing agent equals the number of electrons gained by the oxidizing agent.*

Two methods used to balance redox equations are the *oxidation number* method and the *half-reaction* method. This section describes the oxidation number

Literature ten Hoor, Marten J. "Redox Balancing without Puzzling," *J. Chem. Educ.* **1997,** *74,* 1367.

method in detail; the half-reaction method is covered in Chapter 21. (If your professor chooses to cover that section here, it is completely transferable with no loss in continuity.)

The **oxidation number method** for balancing redox equations consists of five steps that use the changes in oxidation numbers to generate balancing coefficients. The first two steps are identical to those in Sample Problem 4.9:

Step 1. Assign oxidation numbers to all elements in the reaction.
Step 2. From the changes in oxidation numbers, identify the oxidized and reduced species.
Step 3. Compute the number of electrons lost in the oxidation and gained in the reduction from the oxidation number changes. (Draw tie-lines between these atoms to show the changes.)
Step 4. Multiply one or both of these numbers by appropriate factors to make the electrons lost equal the electrons gained, and use the factors as balancing coefficients.
Step 5. Complete the balancing by inspection, adding states of matter.

Literature Anderson, Wayne P. "Uncommon Oxidation Numbers of Nonmetals," *J. Chem. Educ.* **1998,** *75,* 187.

SAMPLE PROBLEM 4.10 Balancing Redox Equations by the Oxidation Number Method

PROBLEM Use the oxidation number method to balance the following equations:
(a) $Cu(s) + HNO_3(aq) \longrightarrow Cu(NO_3)_2(aq) + NO_2(g) + H_2O(l)$
(b) $PbS(s) + O_2(g) \longrightarrow PbO(s) + SO_2(g)$

(a) Reaction of copper and nitric acid
SOLUTION *Step 1.* Assign oxidation numbers to all elements:

$$\overset{0}{Cu} + \overset{+1\ +5\ -2}{HNO_3} \longrightarrow \overset{+2\ \ \ +5\ -2}{Cu(NO_3)_2} + \overset{+4\ -2}{NO_2} + \overset{+1-2}{H_2O}$$

Step 2. Identify oxidized and reduced species. The O.N. of Cu increased from 0 (in Cu metal) to +2 (in Cu^{2+}); Cu was oxidized. The O.N. of N decreased from +5 (in HNO_3) to +4 (in NO_2); HNO_3 was reduced. Note that some NO_3^- also acts as a spectator ion, appearing unchanged in the $Cu(NO_3)_2$; this is common in redox reactions.
Step 3. Compute e^- lost and e^- gained and draw tie-lines between the atoms. In the oxidation, $2e^-$ were lost from Cu. In the reduction, $1e^-$ was gained by N:

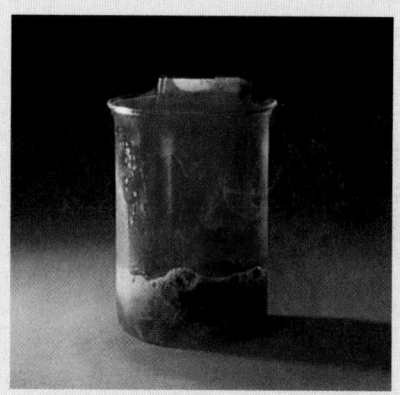

Copper in nitric acid

$$\underset{\text{gains 1e}^-}{\overset{\text{loses 2e}^-}{Cu + HNO_3 \longrightarrow Cu(NO_3)_2 + NO_2 + H_2O}}$$

ARIS Presentation Center, Photo Library

Step 4. Multiply by factors to make e^- lost equal e^- gained, and use the factors as coefficients. Cu lost $2e^-$, so the $1e^-$ gained by N should be multiplied by 2. We put the coefficient 2 before NO_2 and HNO_3:

$$Cu + 2HNO_3 \longrightarrow Cu(NO_3)_2 + 2NO_2 + H_2O$$

Step 5. Complete the balancing by inspection. Balancing N atoms requires a 4 in front of HNO_3 because two additional N atoms are in the NO_3^- ions in $Cu(NO_3)_2$:

$$Cu + 4HNO_3 \longrightarrow Cu(NO_3)_2 + 2NO_2 + H_2O$$

Then, balancing H atoms requires a 2 in front of H_2O, and we add states of matter:

$$Cu(s) + 4HNO_3(aq) \longrightarrow Cu(NO_3)_2(aq) + 2NO_2(g) + 2H_2O(l)$$

CHECK
Reactants (1 Cu, 4 H, 4 N, 12 O) $\longrightarrow$ products [1 Cu, 4 H, (2 + 2) N, (6 + 4 + 2) O]

ARIS Presentation Center, Worked Examples Library

COMMENT It is quite common that some portion of a species reacts and the rest doesn't. Note that, in this case, only 2 mol of NO_3^- is reduced to 2 mol of NO_2; the other 2 mol of NO_3^- remains as spectator ions.

(b) Reaction of lead(II) sulfide and oxygen
SOLUTION *Step 1.* Assign oxidation numbers:

$$\overset{+2}{P}\overset{-2}{b}\overset{}{S} + \overset{0}{O_2} \longrightarrow \overset{+2}{P}\overset{-2}{b}\overset{}{O} + \overset{+4}{S}\overset{-2}{O_2}$$

Step 2. Identify species that are oxidized and reduced. PbS was oxidized: the O.N. of S increased from -2 in PbS to $+4$ in SO_2. O_2 was reduced: the O.N. of O decreased from 0 in O_2 to -2 in PbO and in SO_2.
Step 3. Compute e^- lost and e^- gained and draw tie-lines. The S lost $6e^-$, and each O gained $2e^-$:

$$\overset{\text{loses } 6e^-}{PbS + O_2 \longrightarrow PbO + SO_2}$$
$$\underset{\text{gains } 4e^- \ (2e^- \text{ per O})}{}$$

Step 4. Multiply by factors to make e^- lost equal e^- gained. The S atom loses $6e^-$, and each O in O_2 gains $2e^-$, for a total gain of $4e^-$. Thus, placing the coefficient $\frac{3}{2}$ before O_2 gives 3 O atoms that each gain $2e^-$, for a total gain of $6e^-$:

$$PbS + \tfrac{3}{2}O_2 \longrightarrow PbO + SO_2$$

Step 5. Complete the balancing by inspection. The atoms are balanced, but all coefficients must be multiplied by 2 to obtain integers, and we add states of matter:

$$2PbS(s) + 3O_2(g) \longrightarrow 2PbO(s) + 2SO_2(g)$$

CHECK Reactants (2 Pb, 2 S, 6 O) $\longrightarrow$ products [2 Pb, 2 S, (2 + 4) O]

FOLLOW-UP PROBLEM 4.10 Use the oxidation number method to balance the following equation:

$$K_2Cr_2O_7(aq) + HI(aq) \longrightarrow KI(aq) + CrI_3(aq) + I_2(s) + H_2O(l)$$

Literature Tóth, Zoltán. "Double Redox Reactions," *J. Chem. Educ.* **1997,** *74,* 744.

Literature Cox, Amy L.; Cox, James R. "Determining Oxidation-Reduction on a Simple Number Line," *J. Chem. Educ.* **2002,** *79,* 965.

Redox Titrations

In an acid-base titration, a known concentration of base is used to find an unknown concentration of an acid (or vice versa). Similarly, in a redox titration, a known concentration of oxidizing agent is used to find an unknown concentration of reducing agent (or vice versa). This application of stoichiometry is used in a wide range of situations, including measuring the iron content in drinking water and the vitamin C content in fruits and vegetables.

The permanganate ion, MnO_4^-, is a common oxidizing agent in these titrations because it is strongly colored and, thus, also serves as an indicator. In Figure 4.15, MnO_4^- is used to oxidize the oxalate ion, $C_2O_4^{2-}$, to determine its concentration. As long as any $C_2O_4^{2-}$ is present, it reduces the deep purple MnO_4^- to the very faint pink (nearly colorless) Mn^{2+} ion (Figure 4.15, *left*). As soon as all the available $C_2O_4^{2-}$ has been oxidized, the next drop of MnO_4^- turns the solution light purple (Figure 4.15, *right*). This color change indicates the end point, the point at which the electrons lost by the oxidized species ($C_2O_4^{2-}$) equal the electrons gained by the reduced species (MnO_4^-). We then calculate the concentration of the $C_2O_4^{2-}$ solution from its known volume, the known volume and concentration of the MnO_4^- solution, and the balanced equation.

Preparing a sample for a redox titration sometimes requires several laboratory steps, as shown in Sample Problem 4.11 for the determination of the Ca^{2+} ion concentration of blood.

Net ionic equation:

$$\overset{+7}{2MnO_4^-}(aq) + \overset{+3}{5C_2O_4^{2-}}(aq) + 16H^+(aq) \longrightarrow \overset{+2}{2Mn^{2+}}(aq) + \overset{+4}{10CO_2}(g) + 8H_2O(l)$$

Figure 4.15 A redox titration. The oxidizing agent in the buret, $KMnO_4$, is strongly colored, so it also serves as the indicator. When it reacts with the reducing agent $C_2O_4^{2-}$ in the flask, its color changes from deep purple to very faint pink (nearly colorless) *(left)*. When all the $C_2O_4^{2-}$ is oxidized, the next drop of $KMnO_4$ remains unreacted and turns the solution light purple *(right)*, signaling the end point of the titration.

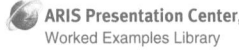
ARIS Presentation Center, Photo Library

SAMPLE PROBLEM 4.11 Finding a Concentration by a Redox Titration

PROBLEM Calcium ion (Ca^{2+}) is required for blood to clot and for many other cell processes. An abnormal Ca^{2+} concentration is indicative of disease. To measure the Ca^{2+} concentration, 1.00 mL of human blood is treated with $Na_2C_2O_4$ solution. The resulting CaC_2O_4 precipitate is filtered and dissolved in dilute H_2SO_4 to release $C_2O_4^{2-}$ into solution and allow it to be oxidized. This solution required 2.05 mL of 4.88×10^{-4} M $KMnO_4$ to reach the end point. The balanced equation is

$$2KMnO_4(aq) + 5CaC_2O_4(s) + 8H_2SO_4(aq) \longrightarrow$$
$$2MnSO_4(aq) + K_2SO_4(aq) + 5CaSO_4(s) + 10CO_2(g) + 8H_2O(l)$$

(a) Calculate the amount (mol) of Ca^{2+}.
(b) Calculate the Ca^{2+} ion concentration expressed in units of mg Ca^{2+}/100 mL blood.

(a) Calculating the moles of Ca^{2+}

PLAN Given the balanced equation, we have to find the amount (mol) of Ca^{2+} from the volume and concentration of $KMnO_4$ used to titrate it. All the Ca^{2+} ion in the 1.00-mL blood sample is precipitated and then dissolved in the H_2SO_4. We find the number of moles of $KMnO_4$ needed to reach the end point from the volume (2.05 mL) and molarity (4.88×10^{-4} M) and use the molar ratio to calculate the number of moles of CaC_2O_4 dissolved in the H_2SO_4. Then, from the chemical formula, we find moles of Ca^{2+} ions.
SOLUTION Converting from milliliters and molarity to moles of $KMnO_4$ to reach the end point:

$$\text{Moles of } KMnO_4 = 2.05 \text{ mL soln} \times \frac{1 \text{ L}}{1000 \text{ mL}} \times \frac{4.88\times10^{-4} \text{ mol } KMnO_4}{1 \text{ L soln}}$$
$$= 1.00\times10^{-6} \text{ mol } KMnO_4$$

Converting from moles of $KMnO_4$ to moles of CaC_2O_4 titrated:

$$\text{Moles of } CaC_2O_4 = 1.00\times10^{-6} \text{ mol } KMnO_4 \times \frac{5 \text{ mol } CaC_2O_4}{2 \text{ mol } KMnO_4} = 2.50\times10^{-6} \text{ mol } CaC_2O_4$$

Finding moles of Ca^{2+} present:

$$\text{Moles of } Ca^{2+} = 2.50\times10^{-6} \text{ mol } CaC_2O_4 \times \frac{1 \text{ mol } Ca^{2+}}{1 \text{ mol } CaC_2O_4}$$
$$= 2.50\times10^{-6} \text{ mol } Ca^{2+}$$

Volume (L) of $KMnO_4$ solution

multiply by M (mol/L)

Amount (mol) of $KMnO_4$

molar ratio

Amount (mol) of CaC_2O_4

ratio of elements in chemical formula

Amount (mol) of Ca^{2+}

CHECK A very small volume of dilute $KMnO_4$ is needed, so 10^{-6} mol of $KMnO_4$ seems reasonable. The molar ratio of 5/2 for CaC_2O_4 to $KMnO_4$ gives 2.5×10^{-6} mol of CaC_2O_4 and thus 2.5×10^{-6} mol of Ca^{2+}.

(b) Expressing the Ca^{2+} concentration as mg/100 mL blood
PLAN The amount in part (a) is the moles of Ca^{2+} ion present in 1.00 mL of blood. We multiply by 100 to obtain the moles of Ca^{2+} ion in 100 mL of blood and then use the atomic mass of Ca to convert that amount to grams and then milligrams.
SOLUTION Finding moles of Ca^{2+}/100 mL blood:

$$\text{Moles of } Ca^{2+}/100 \text{ mL blood} = \frac{2.50 \times 10^{-6} \text{ mol } Ca^{2+}}{1.00 \text{ mL blood}} \times 100$$
$$= 2.50 \times 10^{-4} \text{ mol } Ca^{2+}/100 \text{ mL blood}$$

Converting from moles of Ca^{2+} to milligrams:

$$\text{Mass (mg) } Ca^{2+}/100 \text{ mL blood} = \frac{2.50 \times 10^{-4} \text{ mol } Ca^{2+}}{100 \text{ mL blood}} \times \frac{40.08 \text{ g } Ca^{2+}}{1 \text{ mol } Ca^{2+}} \times \frac{1000 \text{ mg}}{1 \text{ g}}$$
$$= 10.0 \text{ mg } Ca^{2+}/100 \text{ mL blood}$$

CHECK The relative amounts of Ca^{2+} make sense. If there is 2.5×10^{-6} mol/mL blood, there is 2.5×10^{-4} mol/100 mL blood. A molar mass of about 40 g/mol for Ca^{2+} gives 100×10^{-4} g, or 10×10^{-3} g/100 mL blood. It is easy to make an order-of-magnitude (power of 10) error in this type of calculation, so be sure to include all units.
COMMENT 1. The normal range for the Ca^{2+} concentration in a human adult is 9.0 to 11.5 mg Ca^{2+}/100 mL blood, so our value seems reasonable.
2. When blood is donated, the receiving bag contains $Na_2C_2O_4$ solution, which precipitates the Ca^{2+} ion and, thus, prevents clotting.
3. A redox titration is analogous to an acid-base titration: in redox processes, electrons are lost and gained, whereas in acid-base processes, H^+ ions are lost and gained.

FOLLOW-UP PROBLEM 4.11 A 2.50-mL sample of low-fat milk was treated with sodium oxalate, and the precipitate was filtered and dissolved in H_2SO_4. This solution required 6.53 mL of 4.56×10^{-3} M $KMnO_4$ to reach the end point.
(a) Calculate the molarity of Ca^{2+} in the milk.
(b) What is the concentration of Ca^{2+} in g/L? Is this value consistent with the typical value for milk of about 1.2 g Ca^{2+}/L?

Section Summary

When one reactant has a greater attraction for electrons than another, there is a net movement of electron charge, and a redox reaction takes place. Electron gain (reduction) and electron loss (oxidation) occur simultaneously. • The redox process is tracked by assigning oxidation numbers to each atom in a reaction. The species that is oxidized (contains an atom that increases in oxidation number) is the reducing agent; the species that is reduced (contains an atom that decreases in oxidation number) is the oxidizing agent. • Redox reactions are balanced by keeping track of the changes in oxidation number. • A redox titration is used to determine the concentration of the oxidizing or the reducing agent from the known concentration of the other.

4.6 ELEMENTS IN REDOX REACTIONS

As we saw in Sample Problems 4.9 and 4.10, *whenever atoms appear in the form of a free element on one side of an equation and as part of a compound on the other, there must have been a change in oxidation state and the reaction is a redox process.* While there are many redox reactions that do *not* involve free elements, such as the one between MnO_4^- and $C_2O_4^{2-}$ that we saw in Section 4.5, we'll focus here on the many others that do. One way to classify these is by comparing the numbers of reactants and products. By that approach, we have three types:

- *Combination reactions:* two or more reactants form one product:

$$X + Y \longrightarrow Z$$

- *Decomposition reactions:* one reactant forms two or more products:

$$Z \longrightarrow X + Y$$

- *Displacement reactions:* the number of substances is the same but atoms (or ions) exchange places:

$$X + YZ \longrightarrow XZ + Y$$

Combining Two Elements Two elements may react to form binary ionic or covalent compounds. Here are some important examples:

1. *Metal and nonmetal form an ionic compound.* Figure 4.16 shows the reaction between an alkali metal and a halogen on the macroscopic and atomic scales. Note the change in oxidation numbers. As you can see, K is oxidized, so it is the reducing agent; Cl_2 is reduced, so it is the oxidizing agent.

Aluminum reacts with O_2, as does nearly every metal, to form an ionic oxide:

$$\overset{0}{4Al(s)} + \overset{0}{3O_2(g)} \longrightarrow \overset{+3-2}{2Al_2O_3(s)}$$

$$\overset{0}{2K(s)} + \overset{0}{Cl_2(g)} \longrightarrow \overset{+1-1}{2KCl(s)}$$
Potassium   Chlorine   Potassium chloride

Figure 4.16 Combining elements to form an ionic compound. When the metal potassium and the nonmetal chlorine react, they form the solid ionic compound potassium chloride. The photos *(top)* present the view the chemist sees in the laboratory. The blow-up arrows lead to an atomic-scale view *(middle)*; the stoichiometry is indicated by the more darkly colored spheres. The balanced redox equation is shown with oxidation numbers *(bottom)*.

2. *Two nonmetals form a covalent compound.* In one of thousands of examples, ammonia forms from nitrogen and hydrogen in a reaction that occurs in industry on an enormous scale:

$$\overset{0}{N_2}(g) + 3\overset{0}{H_2}(g) \longrightarrow 2\overset{-3}{\underset{}{N}}\overset{+1}{H_3}(g)$$

Halogens form many compounds with other nonmetals, as in the formation of phosphorus trichloride, a major reactant in the production of pesticides and other organic compounds:

$$\overset{0}{P_4}(s) + 6\overset{0}{Cl_2}(g) \longrightarrow 4\overset{+3}{\underset{}{P}}\overset{-1}{Cl_3}(l)$$

Nearly every nonmetal reacts with O_2 to form a covalent oxide, as when nitrogen monoxide forms from the nitrogen and oxygen in air at the very high temperatures created by lightning:

$$\overset{0}{N_2}(g) + \overset{0}{O_2}(g) \longrightarrow 2\overset{+2}{\underset{}{N}}\overset{-2}{O}(g)$$

Combining Compound and Element   Many binary covalent compounds react with nonmetals to form larger compounds. Many nonmetal oxides react with additional O_2 to form "higher" oxides (those with more O atoms in each molecule). For example,

$$2\overset{+2}{\underset{}{N}}\overset{-2}{O}(g) + \overset{0}{O_2}(g) \longrightarrow 2\overset{+4}{\underset{}{N}}\overset{-2}{O_2}(g)$$

Similarly, many nonmetal halides combine with additional halogen to form "higher" halides:

$$\overset{+3}{\underset{}{P}}\overset{-1}{Cl_3}(l) + \overset{0}{Cl_2}(g) \longrightarrow \overset{+5}{\underset{}{P}}\overset{-1}{Cl_5}(s)$$

Decomposing Compounds into Elements   A decomposition reaction occurs when a reactant absorbs enough energy for one or more of its bonds to break. The energy can take many forms—heat, electricity, light, mechanical, and so forth—but we'll focus in this discussion on heat and electricity. The products are either elements or elements and smaller compounds. Several common examples are:

1. *Thermal decomposition.* When the energy absorbed is heat, the reaction is called a thermal decomposition. (A Greek *delta*, Δ, shown above a reaction arrow indicates that significant heating, not just typical warming, is required for the reaction to proceed.) Many metal oxides, chlorates, and perchlorates release oxygen when strongly heated. The decomposition of mercury(II) oxide, used by Lavoisier and Priestley in their classic experiments, is shown on the macroscopic and atomic scales in Figure 4.17. Heating potassium chlorate is a modern method for forming small amounts of oxygen in the laboratory; the same reaction occurs in some explosives and fireworks:

$$2\overset{+1}{\underset{}{K}}\overset{+5}{\underset{}{Cl}}\overset{-2}{O_3}(s) \overset{\Delta}{\longrightarrow} 2\overset{+1}{\underset{}{K}}\overset{-1}{Cl}(s) + 3\overset{0}{O_2}(g)$$

Notice that, in these cases, the lone reactant is the oxidizing *and* the reducing agent. For example, in the case of HgO, the O^{2-} ion reduces the Hg^{2+} ion (which means that, at the same time, Hg^{2+} oxidizes O^{2-}).

Figure 4.17 Decomposing a compound to its elements. Heating solid mercury(II) oxide decomposes it to liquid mercury and gaseous oxygen: the macroscopic (laboratory) view *(top);* the atomic-scale view, with the more darkly colored spheres showing the stoichiometry *(middle);* and the balanced redox equation *(bottom).*

Transparency/ARIS Presentation Center, Art Library

2. *Electrolytic decomposition.* In the process of *electrolysis,* a compound absorbs electrical energy and decomposes into its elements. In the early 19th century, the observation of the electrolysis of water was crucial to the establishment of atomic masses:

$$\overset{+1-2}{2H_2O}(l) \xrightarrow{\text{electricity}} \overset{0}{2H_2}(g) + \overset{0}{O_2}(g)$$

Many active metals, such as sodium, magnesium, and calcium, are produced industrially by electrolysis of their molten halides:

$$\overset{+2-1}{MgCl_2}(l) \xrightarrow{\text{electricity}} \overset{0}{Mg}(l) + \overset{0}{Cl_2}(g)$$

(We'll examine the details of electrolysis in Chapter 21 and then its role in the industrial recovery of several elements in Chapter 22.)

Displacing One Element by Another; Activity Series As we said, displacement reactions have the same number of reactants as products. We mentioned double-displacement (metathesis) reactions in discussing precipitation and acid-base reactions. The other type, *single-displacement* reactions, are all oxidation-reduction processes. They occur when one atom displaces the ion of a different atom from solution. When the reaction involves metals, the atom reduces the ion; when it involves nonmetals (specifically halogens), the atom oxidizes the ion. Chemists rank various elements into activity series—one for metals and one for halogens—in order of their ability to displace one another.

Figure 4.18 An active metal displacing hydrogen from water. Lithium displaces hydrogen from water in a vigorous reaction that yields an aqueous solution of lithium hydroxide and hydrogen gas, as shown on the macroscopic scale *(top)*, at the atomic scale *(middle)*, and as a balanced equation *(bottom)*. (For clarity, the atomic-scale view of water has been greatly simplified, and only water molecules involved in the reaction are colored red and blue.)

ARIS Presentation Center, Active Art Library

$$\overset{0}{2Li(s)} + \overset{+1\;-2}{2H_2O(l)} \longrightarrow \overset{+1-2+1}{2LiOH(aq)} + \overset{0}{H_2(g)}$$
Lithium   Water   Lithium hydroxide   Hydrogen

1. *The activity series of the metals.* Metals can be ranked by their ability to displace H_2 from various sources or by their ability to displace one another from solution.

- *A metal displaces H_2 from water or acid.* The most reactive metals, such as those from Group 1A(1) and Ca, Sr, and Ba from Group 2A(2), displace H_2 from water, and they do so vigorously. Figure 4.18 shows this reaction for lithium. Heat is needed to speed the reaction of slightly less reactive metals, such as Al and Zn, so these metals displace H_2 from steam:

$$\overset{0}{2Al(s)} + \overset{+1-2}{6H_2O(g)} \longrightarrow \overset{+3-2\;\;+1}{2Al(OH)_3(s)} + \overset{0}{3H_2(g)}$$

  Still less reactive metals, such as nickel and tin, do not react with water but *do* react with acids. Because the concentration of H^+ is higher in acid solutions than in water, H_2 is displaced more easily (Figure 4.19). Here is the net ionic equation:

$$\overset{0}{Ni(s)} + \overset{+1}{2H^+(aq)} \longrightarrow \overset{+2}{Ni^{2+}(aq)} + \overset{0}{H_2(g)}$$

  Notice that in all such reactions, the metal is the reducing agent (O.N. of metal increases), and water or acid is the oxidizing agent (O.N. of H decreases). The least reactive metals, such as silver and gold, cannot displace H_2 from any source.

- *A metal displaces another metal ion from solution.* Direct comparisons of metal reactivity are clearest in these reactions. For example, zinc metal displaces copper(II) ion from (actually reduces Cu^{2+} in) copper(II) sulfate solution, as the total ionic equation shows:

$$\overset{+2}{Cu^{2+}(aq)} + \overset{+6\;-2}{SO_4^{2-}(aq)} + \overset{0}{Zn(s)} \longrightarrow \overset{0}{Cu(s)} + \overset{+2}{Zn^{2+}(aq)} + \overset{+6\;-2}{SO_4^{2-}(aq)}$$

Figure 4.19 The displacement of H_2 from acid by nickel.

ARIS Presentation Center, Photo Library

Copper wire

Silver nitrate solution

Copper wire coated with silver

Copper nitrate solution

Cu^{2+}

Ag^+

Ag^+

Ag atoms coating wire

$2e^-$

Cu atoms in wire

$$\underset{\text{2AgNO}_3(aq) + \text{Cu}(s)}{\overset{+1+5-2 \qquad\quad 0}{}} \longrightarrow \underset{\text{Cu(NO}_3)_2(aq) + \text{2Ag}(s)}{\overset{+2+5-2 \qquad\quad 0}{}}$$

Figure 4.20　Displacing one metal by another. More reactive metals displace less reactive metals from solution. In this reaction, Cu atoms each give up two electrons as they become Cu^{2+} ions and leave the wire. The electrons are transferred to two Ag^+ ions that become Ag atoms and deposit on the wire. With time, a coating of crystalline silver coats the wire. Thus, copper has displaced silver (reduced silver ion) from solution. The reaction is depicted as the laboratory view *(top)*, the atomic-scale view *(middle)*, and the balanced redox equation *(bottom)*.

Transparency/ARIS Presentation Center, Active Art Library

Figure 4.20 shows in atomic detail that copper metal can displace silver ion from solution: zinc is more reactive than copper, which is more reactive than silver.

The results of many such reactions between metals and water, aqueous acids, or metal-ion solutions form the basis of the **activity series of the metals.** In Figure 4.21 elements higher on the list are stronger reducing agents than elements lower down; that is, for those that are stable in water, elements higher on the list

Demo Shakhashiri, Bassam Z. "An Activity Series: Zinc, Copper, and Silver Half Cells," *Chemical Demonstrations, A Handbook for Teachers of Chemistry,* Vol. 4 (The University of Wisconsin Press, Madison, 1985), 101–106.

Strength as reducing agent

Li
K
Ba
Ca
Na

Can displace H_2 from water

$Ba(s) + 2H_2O(l) \longrightarrow Ba^{2+}(aq) + 2OH^-(aq) + H_2(g)$
(also see Figure 4.18)

Mg
Al
Mn
Zn
Cr
Fe
Cd

Can displace H_2 from steam

$Zn(s) + 2H_2O(g) \overset{\Delta}{\longrightarrow} Zn(OH)_2(s) + H_2(g)$

Co
Ni
Sn
Pb

Can displace H_2 from acid

$Sn(s) + 2H^+(aq) \longrightarrow Sn^{2+}(aq) + H_2(g)$
(also see Figure 4.19)

H_2

Cu
Hg
Ag
Au

Cannot displace H_2 from any source

$Ag(s) + 2H^+(aq) \longrightarrow$ no reaction

Figure 4.21　The activity series of the metals. This list of metals (and H_2) is arranged with the most active metal (strongest reducing agent) at the top and the least active metal (weakest reducing agent) at the bottom. The four metals below H_2 cannot displace it from any source. An example from each group appears to the right as a net ionic equation. (The ranking refers to behavior of ions in aqueous solution.)

Transparency/ARIS Presentation Center, Art Library

can reduce aqueous ions of elements lower down. The list also shows whether the metal can displace H_2 (reduce H^+) and, if so, from which source. Look at the metals in the equations we've just discussed. Note that Li, Al, and Ni lie above H_2, while Ag lies below it; also, Zn lies above Cu, which lies above Ag. The most reactive metals on the list are in Groups 1A(1) and 2A(2) of the periodic table, and the least reactive lie at the right of the transition elements in Groups 1B(11) and 2B(12).

2. *The activity series of the halogens.* Reactivity decreases down Group 7A(17), so we can arrange the halogens into their own activity series:

$$F_2 > Cl_2 > Br_2 > I_2$$

A halogen higher in the periodic table is a stronger oxidizing agent than one lower down. Thus, chlorine can oxidize bromide ions or iodide ions from solution, and bromine can oxidize iodide ions. Here, chlorine displaces bromine:

$$\overset{-1}{2Br^-}(aq) + \overset{0}{Cl_2}(aq) \longrightarrow \overset{0}{Br_2}(aq) + \overset{-1}{2Cl^-}(aq)$$

Combustion Reactions *Combustion* is the process of combining with oxygen, often with the release of heat and light, as in a flame. ● Combustion reactions do not fall neatly into classes based on the number of reactants and products, but *all are redox processes* because elemental oxygen is a reactant:

$$2CO(g) + O_2(g) \longrightarrow 2CO_2(g)$$

The combustion reactions that we commonly use to produce energy involve organic mixtures such as coal, gasoline, and natural gas as reactants. These mixtures consist of substances with many carbon-carbon and carbon-hydrogen bonds. During the reaction, these bonds break, and each C and H atom combines with oxygen. Therefore, the major products are CO_2 and H_2O. The combustion of the hydrocarbon butane, which is used in camp stoves, is typical:

$$2C_4H_{10}(g) + 13O_2(g) \longrightarrow 8CO_2(g) + 10H_2O(g)$$

Biological *respiration* is a multistep combustion process that occurs within our cells when we "burn" organic foodstuffs, such as glucose, for energy:

$$C_6H_{12}O_6(s) + 6O_2(g) \longrightarrow 6CO_2(g) + 6H_2O(g) + energy$$

● **Space-Age Combustion Without a Flame** Combustion reactions are used to generate large amounts of energy. In most applications, the fuel is burned and the energy is released as heat (e.g., in a furnace) or as a combination of work and heat (e.g., in a combustion engine). Aboard a space shuttle, *fuel cells* generate electrical energy from the flameless combustion of hydrogen gas. Here H_2 is the reducing agent, and O_2 is the oxidizing agent in a controlled reaction process that yields water—which the astronauts use for drinking. On Earth, fuel cells based on the reaction of either H_2 or methanol (CH_3OH) with O_2 are being developed for use in car engines.

Engineering application

ARIS Presentation Center, Worked Examples Library

Demo Shakhashiri, Bassam Z. "Reactions of Metals and Hydrochloric Acid," *Chemical Demonstrations, A Handbook for Teachers of Chemistry,* Vol. 1 (The University of Wisconsin Press, Madison, 1985), 25–26.

SAMPLE PROBLEM 4.12 Identifying the Type of Redox Reaction

PROBLEM Classify each of the following redox reactions as a combination, decomposition, or displacement reaction, write a balanced molecular equation for each, as well as total and net ionic equations for part (c), and identify the oxidizing and reducing agents:
(a) Magnesium(s) + nitrogen(g) ⟶ magnesium nitride(s)
(b) Hydrogen peroxide(l) ⟶ water + oxygen gas
(c) Aluminum(s) + lead(II) nitrate(aq) ⟶ aluminum nitrate(aq) + lead(s)
PLAN To decide on reaction type, recall that combination reactions produce fewer products than reactants, decomposition reactions produce more products, and displacement reactions have the same number of reactants and products. The oxidation number (O.N.) becomes more positive for the reducing agent and less positive for the oxidizing agent.
SOLUTION (a) Combination: two substances form one. This reaction occurs, along with formation of magnesium oxide, when magnesium burns in air:

$$\overset{0}{3Mg}(s) + \overset{0}{N_2}(g) \longrightarrow \overset{+2\;-3}{Mg_3N_2}(s)$$

Mg is the reducing agent; N_2 is the oxidizing agent.

(b) Decomposition: one substance forms two. This reaction occurs within every bottle of this common household antiseptic. Hydrogen peroxide is very unstable and breaks down from heat, light, or just shaking:

$$\overset{\overset{-1}{\overset{\mid}{+1-1}}}{\underset{2H_2O_2(l)}{\mid\ \mid}} \longrightarrow \overset{+1-2}{\underset{2H_2O(l)}{\mid\ \mid}} + \overset{0}{\underset{O_2(g)}{\mid}}$$

H_2O_2 is both the oxidizing *and* the reducing agent. The O.N. of O in peroxides is -1. It is shown in blue *and* red because it both increases to 0 in O_2 and decreases to -2 in H_2O.

(c) Displacement: two substances form two others. As Figure 4.21 shows, Al is more active than Pb and, thus, displaces it from aqueous solution:

$$\overset{0}{\underset{2Al(s)}{\mid}} + \overset{+2+5\ \overset{-2}{\mid}}{\underset{3Pb(NO_3)_2(aq)}{\mid\ \mid}} \longrightarrow \overset{+3+5\ \overset{-2}{\mid}}{\underset{2Al(NO_3)_3(aq)}{\mid\ \mid}} + \overset{0}{\underset{3Pb(s)}{\mid}}$$

Al is the reducing agent; $Pb(NO_3)_2$ is the oxidizing agent.

The total ionic equation is

$$2Al(s) + 3Pb^{2+}(aq) + 6NO_3^-(aq) \longrightarrow 2Al^{3+}(aq) + 6NO_3^-(aq) + 3Pb(s)$$

The net ionic equation is

$$2Al(s) + 3Pb^{2+}(aq) \longrightarrow 2Al^{3+}(aq) + 3Pb(s)$$

FOLLOW-UP PROBLEM 4.12 Classify each of the following redox reactions as a combination, decomposition, or displacement reaction, write a balanced molecular equation for each, as well as total and net ionic equations for parts (b) and (c), and identify the oxidizing and reducing agents:
(a) $S_8(s) + F_2(g) \longrightarrow SF_4(g)$ **(b)** $CsI(aq) + Cl_2(aq) \longrightarrow CsCl(aq) + I_2(aq)$
(c) $Ni(NO_3)_2(aq) + Cr(s) \longrightarrow Ni(s) + Cr(NO_3)_3(aq)$

Section Summary

A reaction that has the same atoms in elemental form and in a compound is a redox reaction. • In combination reactions, elements combine to form a compound, or a compound and an element combine. • Decomposition of compounds by absorption of heat or electricity forms elements or a compound and an element. • In displacement reactions, one element displaces another from solution. Activity series rank elements in order of reactivity. The activity series of the metals ranks metals by their ability to displace H_2 from water, steam, or acid or to displace one another from solution. • Combustion releases heat through reaction of a substance with O_2.

4.7 REACTION REVERSIBILITY AND THE EQUILIBRIUM STATE

So far, we have viewed reactions as occurring from "left to right," from reactants to products, and continuing until they are complete, that is, until the limiting reactant is used up. However, many reactions seem to stop before this happens. The reason is that two opposing reactions are taking place simultaneously. The forward (left-to-right) reaction has not stopped, but the reverse (right-to-left) reaction is occurring at the same rate. Therefore, *no further changes appear in the amounts of reactants or products*. At this point, the reaction mixture has reached **dynamic equilibrium.** On the macroscopic scale, the reaction is *static*, but it is *dynamic* on the molecular scale. In principle, all reactions are reversible and will eventually reach dynamic equilibrium as long as all products remain available for the reverse reaction.

Let's examine equilibrium with a particular set of substances. Calcium carbonate breaks down when heated to calcium oxide and carbon dioxide:

$$CaCO_3(s) \longrightarrow CaO(s) + CO_2(g) \ [\text{breakdown}]$$

It also forms when calcium oxide and carbon dioxide react:

$$CaO(s) + CO_2(g) \longrightarrow CaCO_3(s) \ [\text{formation}]$$

The formation is exactly the reverse of the breakdown. Suppose we place $CaCO_3$ in an *open* steel container and heat it to around 900°C, as shown in Figure 4.22A. The $CaCO_3$ starts breaking down to CaO and CO_2, and the CO_2 escapes from the open container. The reaction goes to completion because the reverse reaction (formation) can occur only if CO_2 is present.

In Figure 4.22B, we perform the same experiment in a *closed* container, so that the CO_2 remains in contact with the CaO. The breakdown (forward reaction) begins, but at first, when very little $CaCO_3$ has broken down, very little CO_2 and CaO are present; thus, the formation (reverse reaction) just barely begins. As the $CaCO_3$ continues to break down, the amounts of CO_2 and CaO increase. They react with each other more frequently, and the formation occurs a bit faster. As the amounts of CaO and CO_2 increase, the formation reaction gradually speeds up. Eventually, the reverse reaction (formation) happens just as fast as the forward reaction (breakdown), and the amounts of $CaCO_3$, CaO, and CO_2 no longer change: the system has reached equilibrium. We indicate this with a pair of arrows pointing in opposite directions:

$$CaCO_3(s) \rightleftharpoons CaO(s) + CO_2(g)$$

Bear in mind that equilibrium can be established only when *all the substances involved are kept in contact with each other.* The breakdown of $CaCO_3$ goes to completion in the open container because the CO_2 escapes.

Figure 4.22 The equilibrium state.
A, In an *open* steel reaction container, strong heating breaks down $CaCO_3$ completely because the product CO_2 escapes and is not present to react with the other product, CaO. **B,** When $CaCO_3$ breaks down in a *closed* container, the CO_2 *is* present to react with CaO and re-form $CaCO_3$ in a reaction that is the reverse of the breakdown. At a given temperature, no further change in the amounts of products and reactants means that the reaction has reached equilibrium.

Transparency/ARIS Presentation Center, Art Library

Reaction reversibility applies equally to the substances and reactions we discussed earlier. Weak acids and bases dissociate into ions only to a small extent because the dissociation quickly becomes balanced by reassociation. For example, when acetic acid dissolves in water, some of the CH_3COOH molecules transfer a proton to H_2O and form H_3O^+ and CH_3COO^- ions. As more of these ions form, they react with each other more often to re-form acetic acid and water:

$$CH_3COOH(aq) + H_2O(l) \rightleftharpoons H_3O^+(aq) + CH_3COO^-(aq)$$

In 0.1 M CH_3COOH at 25°C, only about 1.3% of the acid molecules are dissociated at any given moment. Different weak acids dissociate to different extents. For example, under the same conditions, propanoic acid (CH_3CH_2COOH) is only 1.1% dissociated, but hydrofluoric acid (HF) is 8.6% dissociated.

Similarly, the weak base ammonia reacts with water to form NH_4^+ and OH^- ions. As the ions interact, they re-form ammonia and water, and the rates of the reverse and forward reactions soon balance:

$$NH_3(aq) + H_2O(l) \rightleftharpoons NH_4^+(aq) + OH^-(aq)$$

Another weak base, methylamine (CH_3NH_2), reacts with water to a greater extent before reaching equilibrium, while still another, aniline ($C_6H_5NH_2$), reacts less.

Aqueous acid-base reactions that form a gas go to completion in an open container because the gas escapes. But, if the container were closed and the gas were present for the reverse reaction to take place, the reaction would reach equilibrium. Precipitation and other acid-base reactions seem to "go to completion," even with all the products present, because the ions are tied up either as an insoluble solid (precipitation) or as water molecules (acid-base). In truth, however, ionic precipitates and water do dissociate, but to an extremely small extent. Therefore, these reactions also reach equilibrium, but with almost all product formed.

Thus, some reactions proceed very little before they reach equilibrium, while others proceed almost completely, and still others reach equilibrium with a mixture of large amounts of both reactants *and* products. In Chapter 20, we'll examine the fundamental reason that different processes under the same conditions reach equilibrium with differing ratios of product concentrations to reactant concentrations. Many aspects of dynamic equilibrium are relevant to natural systems, from the cycling of water in the environment to the balance of lion and antelope on the plains of Africa to the nuclear processes occurring in stars. We examine equilibrium in chemical and physical systems in Chapters 12, 13, and 17 through 21.

Section Summary

Every reaction is reversible if all the substances are kept in contact with one another. As the amounts of products increase, the reactants begin to re-form. When the reverse reaction happens as rapidly as the forward reaction, the amounts of the substances no longer change, and the reaction mixture has reached dynamic equilibrium. • Weak acids and bases reach equilibrium in water with a very small proportion of their molecules dissociated. • A reaction "goes to completion" because a product is removed from the system (as a gas) or exists in a form that prevents it from reacting (precipitate or undissociated molecule).

Chapter Perspective

Classifying facts is the first step toward understanding them, and this chapter classified many important facts of reaction chemistry into three major processes—precipitation, acid-base, and oxidation-reduction—the first two of which occur most commonly in aqueous solution. We also examined the great influence that water has on reaction chemistry and introduced the ideas of reaction reversibility and dynamic equilibrium. All these topics appear again at many places in the text.

In the next chapter, our focus changes to the physical behavior of gases. You'll find that your growing appreciation of events on the molecular level is indispensable for understanding the nature of this physical state.

CHAPTER REVIEW GUIDE

The following sections provide many aids to help you study this chapter. (Numbers in parentheses refer to pages, unless noted otherwise.)

Learning Objectives
These are concepts and skills you should know after studying this chapter.

Relevant section and/or sample problem (SP) numbers appear in parentheses.

Understand These Concepts

1. Why water is a polar molecule and how it dissolves ionic compounds and dissociates them into ions (4.1)
2. The difference between the species present when ionic and covalent compounds dissolve in water and that between strong and weak electrolytes (4.1)
3. The use of ionic equations to specify the essential nature of an aqueous reaction (4.2)
4. The driving force for aqueous ionic reactions (4.3, 4.4, 4.5)
5. How to decide whether a precipitation reaction occurs (4.3)
6. The main distinction between strong and weak aqueous acids and bases (4.4)
7. The essential character of aqueous acid-base reactions as proton-transfer processes (4.4)
8. The importance of net movement of electrons in the redox process (4.5)
9. The relation between change in oxidation number and identity of oxidizing and reducing agents (4.5)
10. The presence of elements in some important types of redox reactions: combination, decomposition, displacement (4.6)

11. The balance between forward and reverse rates of a chemical reaction that leads to dynamic equilibrium; why some acids and bases are weak (4.7)

Master These Skills

1. Using the formula of a compound to find the number of moles of ions in solution (SP 4.1)
2. Predicting whether a precipitation reaction occurs (SP 4.2)
3. Using molecular depictions to understand a precipitation reaction (SP 4.3)
4. Determining the concentration of H^+ (or OH^-) ions in an aqueous acid solution (SP 4.4)
5. Writing ionic equations to describe precipitation and acid-base reactions (SPs 4.2, 4.3, 4.5)
6. Calculating an unknown concentration from an acid-base or redox titration (SPs 4.6, 4.11)
7. Determining the oxidation number of any element in a compound (SP 4.7)
8. Identifying redox reactions (SP 4.8)
9. Identifying the oxidizing and reducing agents in a redox reaction (SP 4.9)
10. Balancing redox equations (SP 4.10)
11. Identifying combination, decomposition, and displacement redox reactions (SP 4.12)

Key Terms
These important terms appear in boldface in the chapter and are defined again in the Glossary.

Section 4.1
polar molecule (141)
solvated (142)
electrolyte (142)
nonelectrolyte (144)

Section 4.2
molecular equation (145)
total ionic equation (146)
spectator ion (146)
net ionic equation (146)

Section 4.3
precipitation reaction (146)
precipitate (146)
metathesis reaction (148)

Section 4.4
acid-base reaction (150)
neutralization reaction (150)
acid (150)
base (150)
salt (153)

titration (153)
equivalence point (154)
end point (154)

Section 4.5
oxidation-reduction (redox) reaction (158)
oxidation (159)
reduction (159)
oxidizing agent (159)
reducing agent (159)

oxidation number (O.N.) (or oxidation state) (159)
oxidation number method (163)

Section 4.6
activity series of the metals (171)

Section 4.7
dynamic equilibrium (173)

Highlighted Figures and Tables
These figures (F) and tables (T) provide a visual review of key ideas.

Entries in **bold** contain frequently used data.

F4.1 Electron distribution in H_2 and H_2O (141)
F4.2 Dissolution of an ionic compound (142)
F4.4 An aqueous ionic reaction and its equations (145)
T4.1 Solubility rules for ionic compounds in water (148)
T4.2 Strong and weak acids and bases (151)
F4.10 An aqueous strong acid–strong base reaction on the atomic scale (156)

F4.12 The redox process in compound formation (158)
T4.3 Rules for assigning an oxidation number (160)
F4.13 Highest and lowest oxidation numbers of reactive main-group elements (160)
F4.14 A summary of terminology for redox reactions (162)
F4.21 The activity series of the metals (171)

Brief Solutions to FOLLOW-UP PROBLEMS Compare your solutions to these calculation steps and answers.

4.1 (a) $KClO_4(s) \xrightarrow{H_2O} K^+(aq) + ClO_4^-(aq)$;
2 mol of K^+ and 2 mol of ClO_4^-
(b) $Mg(C_2H_3O_2)_2(s) \xrightarrow{H_2O} Mg^{2+}(aq) + 2C_2H_3O_2^-(aq)$;
2.49 mol of Mg^{2+} and 4.97 mol of $C_2H_3O_2^-$
(c) $(NH_4)_2CrO_4(s) \xrightarrow{H_2O} 2NH_4^+(aq) + CrO_4^{2-}(aq)$;
6.24 mol of NH_4^+ and 3.12 mol of CrO_4^{2-}
(d) $NaHSO_4(s) \xrightarrow{H_2O} Na^+(aq) + HSO_4^-(aq)$;
0.73 mol of Na^+ and 0.73 mol of HSO_4^-
4.2 (a) $Fe^{3+}(aq) + 3Cl^-(aq) + 3Cs^+(aq) + PO_4^{3-}(aq) \longrightarrow$
$$FePO_4(s) + 3Cl^-(aq) + 3Cs^+(aq)$$
$Fe^{3+}(aq) + PO_4^{3-}(aq) \longrightarrow FePO_4(s)$
(b) $2Na^+(aq) + 2OH^-(aq) + Cd^{2+}(aq) + 2NO_3^-(aq) \longrightarrow$
$$2Na^+(aq) + 2NO_3^-(aq) + Cd(OH)_2(s)$$
$2OH^-(aq) + Cd^{2+}(aq) \longrightarrow Cd(OH)_2(s)$
(c) No reaction occurs
(d) $2Ag^+(aq) + 2NO_3^-(aq) + Ba^{2+}(aq) + 2Cl^-(aq) \longrightarrow$
$$2AgCl(s) + 2NO_3^-(aq) + Ba^{2+}(aq)$$
$Ag^+(aq) + Cl^-(aq) \longrightarrow AgCl(s)$
4.3 (a) Beaker A contains a solution of $Zn(NO_3)_2$.
(b) Beaker B contains a solution of $Ba(OH)_2$.
(c) The precipitate is zinc hydroxide, and the spectator ions are Ba^{2+} and NO_3^-.
Molecular: $Zn(NO_3)_2(aq) + Ba(OH)_2(aq) \longrightarrow$
$$Zn(OH)_2(s) + Ba(NO_3)_2(aq)$$
Total ionic:
$Zn^{2+}(aq) + 2NO_3^-(aq) + Ba^{2+}(aq) + 2OH^-(aq) \longrightarrow$
$$Zn(OH)_2(s) + Ba^{2+}(aq) + 2NO_3^-(aq)$$
Net ionic: $Zn^{2+}(aq) + 2OH^-(aq) \longrightarrow Zn(OH)_2(s)$
(d) The OH^- ion is limiting.
Mass (g) of $Zn(OH)_2$

$$= 6 \text{ OH}^- \text{ particles} \times \frac{0.050 \text{ mol OH}^- \text{ ions}}{1 \text{ OH}^- \text{ particle}}$$
$$\times \frac{1 \text{ mol Zn(OH)}_2}{2 \text{ mol OH}^- \text{ ions}} \times \frac{99.43 \text{ g Zn(OH)}_2}{1 \text{ mol Zn(OH)}_2}$$
$$= 15 \text{ g Zn(OH)}_2$$

4.4 Moles of $OH^- = 451 \text{ mL} \times \frac{1 \text{ L}}{10^3 \text{ mL}}$
$$\times \frac{1.20 \text{ mol KOH}}{1 \text{ L soln}} \times \frac{1 \text{ mol OH}^-}{1 \text{ mol KOH}}$$
$$= 0.541 \text{ mol OH}^-$$

4.5 $Ca(OH)_2(aq) + 2HNO_3(aq) \longrightarrow Ca(NO_3)_2(aq) + 2H_2O(l)$
$Ca^{2+}(aq) + 2OH^-(aq) + 2H^+(aq) + 2NO_3^-(aq) \longrightarrow$
$$Ca^{2+}(aq) + 2NO_3^-(aq) + 2H_2O(l)$$
$H^+(aq) + OH^-(aq) \longrightarrow H_2O(l)$
4.6 $Ba(OH)_2(aq) + 2HCl(aq) \longrightarrow BaCl_2(aq) + 2H_2O(l)$
Volume (L) of soln

$$= 50.00 \text{ mL HCl soln} \times \frac{1 \text{ L}}{10^3 \text{ mL}} \times \frac{0.1016 \text{ mol HCl}}{1 \text{ L soln}}$$
$$\times \frac{1 \text{ mol Ba(OH)}_2}{2 \text{ mol HCl}} \times \frac{1 \text{ L soln}}{0.1292 \text{ mol Ba(OH)}_2}$$
$$= 0.01966 \text{ L}$$

4.7 (a) O.N. of $Sc = +3$; O.N. of $O = -2$
(b) O.N. of $Ga = +3$; O.N. of $Cl = -1$
(c) O.N. of $H = +1$; O.N. of $P = +5$; O.N. of $O = -2$
(d) O.N. of $I = +3$; O.N. of $F = -1$

4.8

Redox; O.N. of N decreases, and O.N. of Cl increases.

Not redox; no changes in O.N. values.

Redox; O.N. of S increases, and O.N. of O decreases.
4.9 (a) Fe is the reducing agent; Cl_2 is the oxidizing agent.
(b) C_2H_6 is the reducing agent; O_2 is the oxidizing agent.
(c) CO is the reducing agent; I_2O_5 is the oxidizing agent.
4.10 $K_2Cr_2O_7(aq) + 14HI(aq) \longrightarrow$
$$2KI(aq) + 2CrI_3(aq) + 3I_2(s) + 7H_2O(l)$$
4.11 (a) Moles of $Ca^{2+} = 6.53 \text{ mL soln} \times \frac{1 \text{ L}}{10^3 \text{ mL}}$
$$\times \frac{4.56 \times 10^{-3} \text{ mol KMnO}_4}{1 \text{ L soln}}$$
$$\times \frac{5 \text{ mol CaC}_2O_4}{2 \text{ mol KMnO}_4} \times \frac{1 \text{ mol Ca}^{2+}}{1 \text{ mol CaC}_2O_4}$$
$$= 7.44 \times 10^{-5} \text{ mol Ca}^{2+}$$
Molarity of $Ca^{2+} = \frac{7.44 \times 10^{-2} \text{ mol Ca}^{2+}}{2.50 \text{ mL milk}} \times \frac{10^3 \text{ mL}}{1 \text{ L}}$
$$= 2.98 \times 10^{-2} \text{ M Ca}^{2+}$$
(b) Conc. of Ca^{2+} (g/L)
$$= \frac{2.98 \times 10^{-2} \text{ mol Ca}^{2+}}{1 \text{ L}} \times \frac{40.08 \text{ g Ca}^{2+}}{1 \text{ mol Ca}^{2+}}$$
$$= \frac{1.19 \text{ g Ca}^{2+}}{1 \text{ L}};$$
the value is consistent with the typical value.
4.12 (a) Combination:
$S_8(s) + 16F_2(g) \longrightarrow 8SF_4(g)$
S_8 is the reducing agent; F_2 is the oxidizing agent.
(b) Displacement:
$2CsI(aq) + Cl_2(aq) \longrightarrow 2CsCl(aq) + I_2(aq)$
$2Cs^+(aq) + 2I^-(aq) + Cl_2(aq) \longrightarrow 2Cs^+(aq) + 2Cl^-(aq) + I_2(aq)$
$2I^-(aq) + Cl_2(aq) \longrightarrow 2Cl^-(aq) + I_2(aq)$
Cl_2 is the oxidizing agent; CsI is the reducing agent.
(c) Displacement:
$3Ni(NO_3)_2(aq) + 2Cr(s) \longrightarrow 3Ni(s) + 2Cr(NO_3)_3(aq)$
$3Ni^{2+}(aq) + 6NO_3^-(aq) + 2Cr(s) \longrightarrow$
$$3Ni(s) + 2Cr^{3+}(aq) + 6NO_3^-(aq)$$
$3Ni^{2+}(aq) + 2Cr(s) \longrightarrow 3Ni(s) + 2Cr^{3+}(aq)$
Cr is the reducing agent; $Ni(NO_3)_2$ is the oxidizing agent.

PROBLEMS

Problems with colored numbers are answered in Appendix E and worked in detail in the Student Solutions Manual. Problem sections match those in the text and provide the numbers of relevant sample problems. Most offer Concept Review Questions, Skill-Building Exercises (grouped in pairs covering the same concept), and Problems in Context. The Comprehensive Problems are based on material from any section or previous chapter.

The Role of Water as a Solvent
(Sample Problem 4.1)

● Concept Review Questions

• **4.1** What two factors cause water to be polar?

• **4.2** What types of substances are most likely to be soluble in water?

• **4.3** What must be present in an aqueous solution for it to conduct an electric current? What general classes of compounds form solutions that conduct?

• **4.4** What occurs on the molecular level when an ionic compound dissolves in water?

• **4.5** Which of the following scenes best represents how the ions occur in an aqueous solution of: (a) $CaCl_2$; (b) Li_2SO_4; (c) NH_4Br?

• **4.6** Which of the following scenes best represents a volume from a solution of magnesium nitrate?

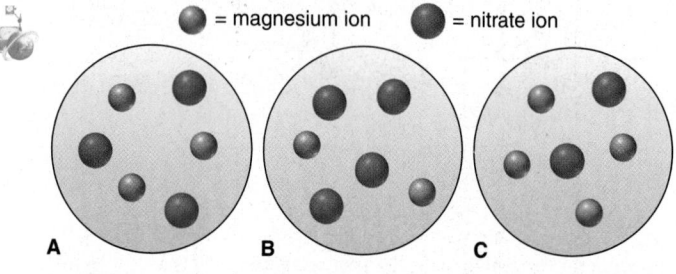

• **4.7** Why are some ionic compounds soluble in water and others are not?

• **4.8** Why are some covalent compounds soluble in water and others are not?

• **4.9** Some covalent compounds dissociate into ions when they dissolve in water. What atom do these compounds have in their structures? What type of aqueous solution do they form? Name three examples of such an aqueous solution.

● Skill-Building Exercises (grouped in similar pairs)

• **4.10** State whether each of the following substances is likely to be very soluble in water. Explain.
  (a) Benzene, C_6H_6 (b) Sodium hydroxide
  (c) Ethanol, CH_3CH_2OH (d) Potassium acetate

• **4.11** State whether each of the following substances is likely to be very soluble in water. Explain.

  (a) Lithium nitrate   (b) Glycine, H_2NCH_2COOH
  (c) Pentane   (d) Ethylene glycol, $HOCH_2CH_2OH$

• **4.12** State whether an aqueous solution of each of the following substances conducts an electric current. Explain your reasoning.
  (a) Cesium bromide   (b) Hydrogen iodide

• **4.13** State whether an aqueous solution of each of the following substances conducts an electric current. Explain your reasoning.
  (a) Potassium sulfate   (b) Sucrose, $C_{12}H_{22}O_{11}$

• **4.14** How many total moles of ions are released when each of the following samples dissolves completely in water?
  (a) 0.32 mol of NH_4Cl (b) 25.4 g of $Ba(OH)_2 \cdot 8H_2O$
  (c) 3.55×10^{19} formula units of LiCl

• **4.15** How many total moles of ions are released when each of the following samples dissolves completely in water?
  (a) 0.805 mol of Rb_2SO_4 (b) 3.85×10^{-3} g of $Ca(NO_3)_2$
  (c) 4.03×10^{19} formula units of $Sr(HCO_3)_2$

• **4.16** How many total moles of ions are released when each of the following samples dissolves completely in water?
  (a) 0.75 mol of K_3PO_4 (b) 6.88×10^{-3} g of $NiBr_2 \cdot 3H_2O$
  (c) 2.23×10^{22} formula units of $FeCl_3$

• **4.17** How many total moles of ions are released when each of the following samples dissolves completely in water?
  (a) 0.734 mol of Na_2HPO_4 (b) 3.86 g of $CuSO_4 \cdot 5H_2O$
  (c) 8.66×10^{20} formula units of $NiCl_2$

• **4.18** How many moles and numbers of ions of each type are present in the following aqueous solutions?
  (a) 130. mL of 0.45 M aluminum chloride
  (b) 9.80 mL of a solution containing 2.59 g lithium sulfate/L
  (c) 245 mL of a solution containing 3.68×10^{22} formula units of potassium bromide per liter

• **4.19** How many moles and numbers of ions of each type are present in the following aqueous solutions?
  (a) 88 mL of 1.75 M magnesium chloride
  (b) 321 mL of a solution containing 0.22 g aluminum sulfate/L
  (c) 1.65 L of a solution containing 8.83×10^{21} formula units of cesium nitrate per liter

• **4.20** How many moles of H^+ ions are present in the following aqueous solutions?
  (a) 1.40 L of 0.25 M perchloric acid
  (b) 6.8 mL of 0.92 M nitric acid
  (c) 2.6 L of 0.085 M hydrochloric acid

• **4.21** How many moles of H^+ ions are present in the following aqueous solutions?
  (a) 1.4 mL of 0.75 M hydrobromic acid
  (b) 2.47 mL of 1.98 M hydriodic acid
  (c) 395 mL of 0.270 M nitric acid

● Problems in Context

•• **4.22** To study a marine organism, a biologist prepares a 1.00-kg sample to simulate the ion concentrations in seawater. She mixes 26.5 g of NaCl, 2.40 g of $MgCl_2$, 3.35 g of $MgSO_4$, 1.20 g of $CaCl_2$, 1.05 g of KCl, 0.315 g of $NaHCO_3$, and 0.098 g of NaBr in distilled water. (a) If the density of the solution is 1.025 g/cm^3, what is the molarity of each ion? (b) What is the total molarity of alkali metal ions? (c) What is the total molarity of alkaline earth metal ions? (d) What is the total molarity of anions?

This figure is available as a transparency and is also on the Presentation Center, within the art library.

•• **4.23** Water "softeners" remove metal ions such as Ca^{2+} and Fe^{3+} by replacing them with enough Na^+ ions to maintain the same number of positive charges in the solution. If 1.0×10^3 L of "hard" water is 0.015 M Ca^{2+} and 0.0010 M Fe^{3+}, how many moles of Na^+ are needed to replace these ions?

Writing Equations for Aqueous Ionic Reactions

● Concept Review Questions

• **4.24** Which ions do not appear in a net ionic equation? Why?

• **4.25** Write two sets of equations (both molecular and total ionic) with different reactants that have the same net ionic equation as the following equation:

$$Ba(NO_3)_2(aq) + Na_2CO_3(aq) \longrightarrow BaCO_3(s) + 2NaNO_3(aq)$$

Precipitation Reactions
(Sample Problems 4.2 and 4.3)

● Concept Review Questions

• **4.26** Why do some pairs of ions precipitate and others do not?

• **4.27** Use Table 4.1 to determine which of the following combinations leads to a precipitation reaction. How can you identify the spectator ions in the reaction?
  (a) Calcium nitrate(aq) + sodium chloride(aq) $\longrightarrow$
  (b) Potassium chloride(aq) + lead(II) nitrate(aq) $\longrightarrow$

• **4.28** The beakers represent the aqueous reaction of $AgNO_3$ and NaCl. Silver ions are gray. What colors are used to represent NO_3^-, Na^+, and Cl^-? Write molecular, total ionic, and net ionic equations for the reaction.

● Skill-Building Exercises (grouped in similar pairs)

• **4.29** Complete the following precipitation reactions with balanced molecular, total ionic, and net ionic equations:
  (a) $Hg_2(NO_3)_2(aq) + KI(aq) \longrightarrow$
  (b) $FeSO_4(aq) + Sr(OH)_2(aq) \longrightarrow$

• **4.30** Complete the following precipitation reactions with balanced molecular, total ionic, and net ionic equations:
  (a) $CaCl_2(aq) + Cs_3PO_4(aq) \longrightarrow$
  (b) $Na_2S(aq) + ZnSO_4(aq) \longrightarrow$

• **4.31** When each of the following pairs of aqueous solutions is mixed, does a precipitation reaction occur? If so, write balanced molecular, total ionic, and net ionic equations:
  (a) Sodium nitrate + copper(II) sulfate
  (b) Ammonium bromide + silver nitrate

• **4.32** When each of the following pairs of aqueous solutions is mixed, does a precipitation reaction occur? If so, write balanced molecular, total ionic, and net ionic equations:
  (a) Potassium carbonate + barium hydroxide
  (b) Aluminum nitrate + sodium phosphate

• **4.33** When each of the following pairs of aqueous solutions is mixed, does a precipitation reaction occur? If so, write balanced molecular, total ionic, and net ionic equations.

(a) Potassium chloride + iron(III) nitrate
(b) Ammonium sulfate + barium chloride

• **4.34** When each of the following pairs of aqueous solutions is mixed, does a precipitation reaction occur? If so, write balanced molecular, total ionic, and net ionic equations:
  (a) Sodium sulfide + nickel(II) sulfate
  (b) Lead(II) nitrate + potassium bromide

• **4.35** If 38.5 mL of lead(II) nitrate solution reacts completely with excess sodium iodide solution to yield 0.628 g of precipitate, what is the molarity of lead(II) ion in the original solution?

• **4.36** If 25.0 mL of silver nitrate solution reacts with excess potassium chloride solution to yield 0.842 g of precipitate, what is the molarity of silver ion in the original solution?

• **4.37** With ions shown as spheres and solvent molecules omitted for clarity, the circle (right) illustrates the solid formed when a solution containing K^+, Mg^{2+}, Ag^+, or Pb^{2+} (blue) is mixed with one containing ClO_4^-, NO_3^-, or SO_4^{2-} (yellow). (a) Identify the solid. (b) Write a balanced net ionic equation for the reaction. (c) If each sphere represents 5.0×10^{-4} mol of ion, what mass of product forms?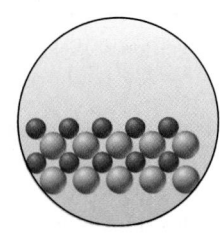

• **4.38** The precipitation reaction between 25.0 mL of a solution containing a cation (purple) and 35.0 mL of a solution containing an anion (green) is depicted below (with ions shown as spheres and solvent molecules omitted for clarity).

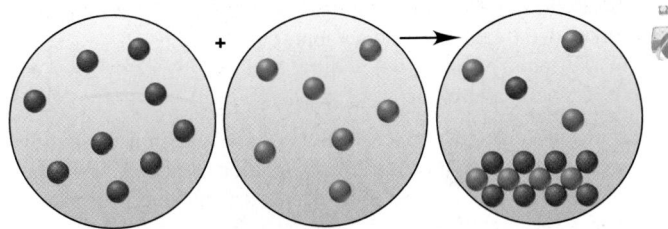

(a) Given the following choices of reactants, write balanced total ionic and net ionic equations that best represent the reaction:
  (1) $KNO_3(aq) + CuCl_2(aq) \longrightarrow$
  (2) $NaClO_4(aq) + CaCl_2(aq) \longrightarrow$
  (3) $Li_2SO_4(aq) + AgNO_3(aq) \longrightarrow$
  (4) $NH_4Br(aq) + Pb(CH_3COO)_2(aq) \longrightarrow$
(b) If each sphere represents 2.5×10^{-3} mol of ion, find the total number of ions present. (c) What is the mass of solid formed?

● Problems in Context

•• **4.39** The mass percent of Cl^- in a seawater sample is determined by titrating 25.00 mL of seawater with $AgNO_3$ solution, causing a precipitation reaction. An indicator is used to detect the end point, which occurs when free Ag^+ ion is present in solution after all the Cl^- has reacted. If 53.63 mL of 0.2970 M $AgNO_3$ is required to reach the end point, what is the mass percent of Cl^- in the seawater (d of seawater = 1.024 g/mL)?

•• **4.40** Aluminum sulfate, known as *cake alum*, has a wide range of uses, from dyeing leather and cloth to purifying sewage. In aqueous solution, it reacts with base to form a white precipitate. (a) Write balanced total and net ionic equations for its reaction with aqueous NaOH. (b) What mass of precipitate forms when 185.5 mL of 0.533 M NaOH is added to 627 mL of a solution that contains 15.8 g of aluminum sulfate per liter?

Acid-Base Reactions

(Sample Problems 4.4 to 4.6)

● Concept Review Questions

4.41 Is the total ionic equation the same as the net ionic equation when $Sr(OH)_2(aq)$ and $H_2SO_4(aq)$ react? Explain.

4.42 State a general equation for a neutralization reaction.

4.43 (a) Name three common strong acids. (b) Name three common strong bases. (c) What is a characteristic behavior of a strong acid or a strong base?

4.44 (a) Name three common weak acids. (b) Name one common weak base. (c) What is the major difference between a weak acid and a strong acid or between a weak base and a strong base, and what experiment would you perform to observe it?

4.45 Do either of the following reactions go to completion? If so, what factor(s) cause(s) each to do so?

(a) $MgSO_3(s) + 2HCl(aq) \longrightarrow$
$$MgCl_2(aq) + SO_2(g) + H_2O(l)$$

(b) $3Ba(OH)_2(aq) + 2H_3PO_4(aq) \longrightarrow$
$$Ba_3(PO_4)_2(s) + 6H_2O(l)$$

4.46 The net ionic equation for the aqueous neutralization reaction between acetic acid and sodium hydroxide is different from that for the reaction between hydrochloric acid and sodium hydroxide. Explain by writing balanced net ionic equations.

● Skill-Building Exercises (grouped in similar pairs)

4.47 Complete the following acid-base reactions with balanced molecular, total ionic, and net ionic equations:

(a) Potassium hydroxide(aq) + hydrobromic acid(aq) $\longrightarrow$

(b) Ammonia(aq) + hydrochloric acid(aq) $\longrightarrow$

4.48 Complete the following acid-base reactions with balanced molecular, total ionic, and net ionic equations:

(a) Cesium hydroxide(aq) + nitric acid(aq) $\longrightarrow$

(b) Calcium hydroxide(aq) + acetic acid(aq) $\longrightarrow$

4.49 Limestone (calcium carbonate) is insoluble in water but dissolves when a hydrochloric acid solution is added. Write balanced total ionic and net ionic equations, showing hydrochloric acid as it actually exists in water and the reaction as a proton-transfer process.

4.50 Zinc hydroxide is insoluble in water but dissolves when a nitric acid solution is added. Why? Write balanced total ionic and net ionic equations, showing nitric acid as it actually exists in water and the reaction as a proton-transfer process.

4.51 If 25.98 mL of a standard 0.1180 M KOH solution reacts with 52.50 mL of CH_3COOH solution, what is the molarity of the acid solution?

4.52 If 26.25 mL of a standard 0.1850 M NaOH solution is required to neutralize 25.00 mL of H_2SO_4, what is the molarity of the acid solution?

● Problems in Context

4.53 An auto mechanic spills 88 mL of 2.6 M H_2SO_4 solution from a rebuilt auto battery. How many milliliters of 1.6 M $NaHCO_3$ must be poured on the spill to react completely with the sulfuric acid?

4.54 Sodium hydroxide is used extensively in acid-base titrations because it is a strong, inexpensive base. A sodium hydroxide solution was standardized by titrating 25.00 mL of 0.1528 M standard hydrochloric acid. The initial buret reading of the sodium hydroxide was 2.24 mL, and the final reading was 39.21 mL. What was the molarity of the base solution?

4.55 An unknown amount of acid can often be determined by adding an excess of base and then "back-titrating" the excess. A 0.3471-g sample of a mixture of oxalic acid, which has two ionizable protons, and benzoic acid, which has one, is treated with 100.0 mL of 0.1000 M NaOH. The excess NaOH is titrated with 20.00 mL of 0.2000 M HCl. Find the mass % of benzoic acid.

4.56 One of the first steps in the enrichment of uranium for use in nuclear power plants involves a displacement reaction between UO_2 and aqueous HF:

$$UO_2(s) + HF(aq) \longrightarrow UF_4(s) + H_2O(l) \, [\text{unbalanced}]$$

How many liters of 2.40 M HF will react with 2.15 kg of UO_2?

4.57 A mixture of bases can sometimes be the active ingredient in antacid tablets. If 0.4826 g of a mixture of $Al(OH)_3$ and $Mg(OH)_2$ is neutralized with 17.30 mL of 1.000 M HNO_3, what is the mass % of $Al(OH)_3$ in the mixture?

Oxidation-Reduction (Redox) Reactions

(Sample Problems 4.7 to 4.11)

● Concept Review Questions

4.58 Describe how to determine the oxidation number of sulfur in (a) H_2S and (b) SO_3^{2-}.

4.59 Is the following a redox reaction? Explain.

$$NH_3(aq) + HCl(aq) \longrightarrow NH_4Cl(aq)$$

4.60 Explain why an oxidizing agent undergoes reduction.

4.61 Why must every redox reaction involve an oxidizing agent and a reducing agent?

4.62 In which of the following equations does sulfuric acid act as an oxidizing agent? In which does it act as an acid? Explain.

(a) $4H^+(aq) + SO_4^{2-}(aq) + 2NaI(s) \longrightarrow$
$$2Na^+(aq) + I_2(s) + SO_2(g) + 2H_2O(l)$$

(b) $BaF_2(s) + 2H^+(aq) + SO_4^{2-}(aq) \longrightarrow$
$$2HF(aq) + BaSO_4(s)$$

4.63 Identify the oxidizing agent and the reducing agent in the following reaction, and explain your answer:

$$8NH_3(g) + 6NO_2(g) \longrightarrow 7N_2(g) + 12H_2O(l)$$

● Skill-Building Exercises (grouped in similar pairs)

4.64 Give the oxidation number of carbon in the following:

(a) CF_2Cl_2 (b) $Na_2C_2O_4$ (c) HCO_3^- (d) C_2H_6

4.65 Give the oxidation number of bromine in the following:

(a) KBr (b) BrF_3 (c) $HBrO_3$ (d) CBr_4

4.66 Give the oxidation number of nitrogen in the following:

(a) NH_2OH (b) N_2F_4 (c) NH_4^+ (d) HNO_2

4.67 Give the oxidation number of sulfur in the following:

(a) $SOCl_2$ (b) H_2S_2 (c) H_2SO_3 (d) Na_2S

4.68 Give the oxidation number of arsenic in the following:

(a) AsH_3 (b) $H_2AsO_4^-$ (c) $AsCl_3$

4.69 Give the oxidation number of phosphorus in the following:

(a) $H_2P_2O_7^{2-}$ (b) PH_4^+ (c) PCl_5

4.70 Give the oxidation number of manganese in the following:

(a) MnO_4^{2-} (b) Mn_2O_3 (c) $KMnO_4$

4.71 Give the oxidation number of chromium in the following:

(a) CrO_3 (b) $Cr_2O_7^{2-}$ (c) $Cr_2(SO_4)_3$

4.72 Identify the oxidizing and reducing agents in the following:

(a) $5H_2C_2O_4(aq) + 2MnO_4^-(aq) + 6H^+(aq) \longrightarrow$
$$2Mn^{2+}(aq) + 10CO_2(g) + 8H_2O(l)$$

(b) $3Cu(s) + 8H^+(aq) + 2NO_3^-(aq) \longrightarrow$
$$3Cu^{2+}(aq) + 2NO(g) + 4H_2O(l)$$

• **4.73** Identify the oxidizing and reducing agents in the following:
(a) $Sn(s) + 2H^+(aq) \longrightarrow Sn^{2+}(aq) + H_2(g)$
(b) $2H^+(aq) + H_2O_2(aq) + 2Fe^{2+}(aq) \longrightarrow$
$$2Fe^{3+}(aq) + 2H_2O(l)$$

• **4.74** Identify the oxidizing and reducing agents in the following:
(a) $8H^+(aq) + 6Cl^-(aq) + Sn(s) + 4NO_3^-(aq) \longrightarrow$
$$SnCl_6^{2-}(aq) + 4NO_2(g) + 4H_2O(l)$$
(b) $2MnO_4^-(aq) + 10Cl^-(aq) + 16H^+(aq) \longrightarrow$
$$5Cl_2(g) + 2Mn^{2+}(aq) + 8H_2O(l)$$

• **4.75** Identify the oxidizing and reducing agents in the following:
(a) $8H^+(aq) + Cr_2O_7^{2-}(aq) + 3SO_3^{2-}(aq) \longrightarrow$
$$2Cr^{3+}(aq) + 3SO_4^{2-}(aq) + 4H_2O(l)$$
(b) $NO_3^-(aq) + 4Zn(s) + 7OH^-(aq) + 6H_2O(l) \longrightarrow$
$$4Zn(OH)_4^{2-}(aq) + NH_3(aq)$$

•• **4.76** Discuss each conclusion from a study of redox reactions:
(a) The sulfide ion functions only as a reducing agent.
(b) The sulfate ion functions only as an oxidizing agent.
(c) Sulfur dioxide functions as an oxidizing or a reducing agent.

•• **4.77** Discuss each conclusion from a study of redox reactions:
(a) The nitride ion functions only as a reducing agent.
(b) The nitrate ion functions only as an oxidizing agent.
(c) The nitrite ion functions as an oxidizing or a reducing agent.

•• **4.78** Use the oxidation number method to balance the following equations by placing coefficients in the blanks. Identify the reducing and oxidizing agents:
(a) __$HNO_3(aq)$ + __$K_2CrO_4(aq)$ + __$Fe(NO_3)_2(aq) \longrightarrow$
__$KNO_3(aq)$ + __$Fe(NO_3)_3(aq)$ + __$Cr(NO_3)_3(aq)$ + __$H_2O(l)$
(b) __$HNO_3(aq)$ + __$C_2H_6O(l)$ + __$K_2Cr_2O_7(aq) \longrightarrow$
__$KNO_3(aq)$ + __$C_2H_4O(l)$ + __$H_2O(l)$ + __$Cr(NO_3)_3(aq)$
(c) __$HCl(aq)$ + __$NH_4Cl(aq)$ + __$K_2Cr_2O_7(aq) \longrightarrow$
__$KCl(aq)$ + __$CrCl_3(aq)$ + __$N_2(g)$ + __$H_2O(l)$
(d) __$KClO_3(aq)$ + __$HBr(aq) \longrightarrow$
__$Br_2(l)$ + __$H_2O(l)$ + __$KCl(aq)$

•• **4.79** Use the oxidation number method to balance the following equations by placing coefficients in the blanks. Identify the reducing and oxidizing agents:
(a) __$HCl(aq)$ + __$FeCl_2(aq)$ + __$H_2O_2(aq) \longrightarrow$
__$FeCl_3(aq)$ + __$H_2O(l)$
(b) __$I_2(s)$ + __$Na_2S_2O_3(aq) \longrightarrow$
__$Na_2S_4O_6(aq)$ + __$NaI(aq)$
(c) __$HNO_3(aq)$ + __$KI(aq) \longrightarrow$
__$NO(g)$ + __$I_2(s)$ + __$H_2O(l)$ + __$KNO_3(aq)$
(d) __$PbO(s)$ + __$NH_3(aq) \longrightarrow$
__$N_2(g)$ + __$H_2O(l)$ + __$Pb(s)$

● **Problems in Context**

• **4.80** The active agent in many hair bleaches is hydrogen peroxide. The amount of H_2O_2 in 14.8 g of hair bleach was determined by titration with a standard potassium permanganate solution:
$$2MnO_4^-(aq) + 5H_2O_2(aq) + 6H^+(aq) \longrightarrow$$
$$5O_2(g) + 2Mn^{2+}(aq) + 8H_2O(l)$$
(a) How many moles of MnO_4^- were required for the titration if 43.2 mL of 0.105 M KMnO$_4$ was needed to reach the end point?
(b) How many moles of H_2O_2 were present in the 14.8-g sample of bleach?

(c) How many grams of H_2O_2 were in the sample?
(d) What is the mass percent of H_2O_2 in the sample?
(e) What is the reducing agent in the redox reaction?

• **4.81** A person's blood alcohol (C_2H_5OH) level can be determined by titrating a sample of blood plasma with a potassium dichromate solution. The balanced equation is
$$16H^+(aq) + 2Cr_2O_7^{2-}(aq) + C_2H_5OH(aq) \longrightarrow$$
$$4Cr^{3+}(aq) + 2CO_2(g) + 11H_2O(l)$$
If 35.46 mL of 0.05961 M $Cr_2O_7^{2-}$ is required to titrate 28.00 g of plasma, what is the mass percent of alcohol in the blood?

Elements in Redox Reactions
(Sample Problem 4.12)

● **Concept Review Questions**

• **4.82** Which type of redox reaction leads to the following?
(a) An increase in the number of substances
(b) A decrease in the number of substances
(c) No change in the number of substances

• **4.83** Why do decomposition reactions typically have compounds as reactants, whereas combination and displacement reactions have one or more elements?

• **4.84** Which of the three types of reactions discussed in Section 4.6 commonly produce one or more compounds?

• **4.85** Are all combustion reactions redox reactions? Explain.

•• **4.86** Give one example of a combination reaction that is a redox reaction and another that is not a redox reaction.

● **Skill-Building Exercises** (grouped in similar pairs)

• **4.87** Balance each of the following redox reactions and classify it as a combination, decomposition, or displacement reaction:
(a) $Ca(s) + H_2O(l) \longrightarrow Ca(OH)_2(aq) + H_2(g)$
(b) $NaNO_3(s) \longrightarrow NaNO_2(s) + O_2(g)$
(c) $C_2H_2(g) + H_2(g) \longrightarrow C_2H_6(g)$

• **4.88** Balance each of the following redox reactions and classify it as a combination, decomposition, or displacement reaction:
(a) $HI(g) \longrightarrow H_2(g) + I_2(g)$
(b) $Zn(s) + AgNO_3(aq) \longrightarrow Zn(NO_3)_2(aq) + Ag(s)$
(c) $NO(g) + O_2(g) \longrightarrow N_2O_4(l)$

• **4.89** Balance each of the following redox reactions and classify it as a combination, decomposition, or displacement reaction:
(a) $Sb(s) + Cl_2(g) \longrightarrow SbCl_3(s)$
(b) $AsH_3(g) \longrightarrow As(s) + H_2(g)$
(c) $Zn(s) + Fe(NO_3)_2(aq) \longrightarrow Zn(NO_3)_2(aq) + Fe(s)$

• **4.90** Balance each of the following redox reactions and classify it as a combination, decomposition, or displacement reaction:
(a) $Mg(s) + H_2O(g) \longrightarrow Mg(OH)_2(s) + H_2(g)$
(b) $Cr(NO_3)_3(aq) + Al(s) \longrightarrow Al(NO_3)_3(aq) + Cr(s)$
(c) $PF_3(g) + F_2(g) \longrightarrow PF_5(g)$

•• **4.91** Predict the product(s) and write a balanced equation for each of the following redox reactions:
(a) $Sr(s) + Br_2(l) \longrightarrow$
(b) $Ag_2O(s) \xrightarrow{\Delta}$
(c) $Mn(s) + Cu(NO_3)_2(aq) \longrightarrow$

•• **4.92** Predict the product(s) and write a balanced equation for each of the following redox reactions:
(a) $Mg(s) + HCl(aq) \longrightarrow$
(b) $LiCl(l) \xrightarrow{electricity}$
(c) $SnCl_2(aq) + Co(s) \longrightarrow$

•• **4.93** Predict the product(s) and write a balanced equation for each of the following redox reactions:
(a) $N_2(g) + H_2(g) \longrightarrow$
(b) $NaClO_3(s) \xrightarrow{\Delta}$
(c) $Ba(s) + H_2O(l) \longrightarrow$

•• **4.94** Predict the product(s) and write a balanced equation for each of the following redox reactions:
(a) $Fe(s) + HClO_4(aq) \longrightarrow$
(b) $S_8(s) + O_2(g) \longrightarrow$
(c) $BaCl_2(l) \xrightarrow{electricity}$

•• **4.95** Predict the product(s) and write a balanced equation for each of the following redox reactions:
(a) Cesium + iodine $\longrightarrow$
(b) Aluminum + aqueous manganese(II) sulfate $\longrightarrow$
(c) Sulfur dioxide + oxygen $\longrightarrow$
(d) Butane and oxygen $\longrightarrow$
(e) Write a balanced net ionic equation for (b).

•• **4.96** Predict the product(s) and write a balanced equation for each of the following redox reactions:
(a) Pentane (C_5H_{12}) + oxygen $\longrightarrow$
(b) Phosphorus trichloride + chlorine $\longrightarrow$
(c) Zinc + hydrobromic acid $\longrightarrow$
(d) Aqueous potassium iodide + bromine $\longrightarrow$
(e) Write a balanced net ionic equation for (d).

• **4.97** How many grams of O_2 can be prepared from the thermal decomposition of 4.27 kg of HgO? Name and calculate the mass (in kg) of the other product.

• **4.98** How many grams of chlorine gas can be produced from the electrolytic decomposition of 874 g of calcium chloride? Name and calculate the mass (in g) of the other product.

• **4.99** In a combination reaction, 1.62 g of lithium is mixed with 6.50 g of oxygen. (a) Which reactant is present in excess? (b) How many moles of product are formed? (c) After reaction, how many grams of each reactant and product are present?

• **4.100** In a combination reaction, 2.22 g of magnesium is heated with 3.75 g of nitrogen. (a) Which reactant is present in excess? (b) How many moles of product are formed? (c) After reaction, how many grams of each reactant and product are present?

•• **4.101** A mixture of $KClO_3$ and KCl with a mass of 0.950 g was heated to produce O_2. After heating, the mass of residue was 0.700 g. Assuming all the $KClO_3$ decomposed to KCl and O_2, calculate the mass percent of $KClO_3$ in the original mixture.

•• **4.102** A mixture of $CaCO_3$ and CaO weighing 0.693 g was heated to produce gaseous CO_2. After heating, the remaining solid weighed 0.508 g. Assuming all the $CaCO_3$ broke down to CaO and CO_2, calculate the mass percent of $CaCO_3$ in the original mixture.

● **Problems in Context**

•• **4.103** Before arc welding was developed, a displacement reaction involving aluminum and iron(III) oxide was commonly used to produce molten iron (the thermite process). This reaction was used, for example, to connect sections of iron railroad track. Calculate the mass of molten iron produced when 1.50 kg of aluminum reacts with 25.0 mol of iron(III) oxide.

•• **4.104** Iron reacts rapidly with chlorine gas to form a reddish brown, ionic compound (A), which contains iron in the higher of its two common oxidation states. Strong heating decomposes

compound A to compound B, another ionic compound, which contains iron in the lower of its two oxidation states. When compound A is formed by the reaction of 50.6 g of Fe and 83.8 g of Cl_2 and then heated, how much compound B forms?

Reaction Reversibility and the Equilibrium State

● **Concept Review Questions**

• **4.105** Why is the equilibrium state called "dynamic"?

• **4.106** In a decomposition reaction involving a gaseous product, what must be done for the reaction to reach equilibrium?

• **4.107** Describe what happens on the molecular level when acetic acid dissolves in water.

• **4.108** When either a mixture of NO and Br_2 or pure nitrosyl bromide (NOBr) is placed in a reaction vessel, the product mixture contains NO, Br_2, and NOBr. Explain.

● **Problems in Context**

•• **4.109** Ammonia is produced by the millions of tons annually for use as a fertilizer. It is commonly made from N_2 and H_2 by the Haber process. Because the reaction reaches equilibrium before going completely to product, the stoichiometric amount of ammonia is not obtained. At a particular temperature and pressure, 10.0 g of H_2 reacts with 20.0 g of N_2 to form ammonia. When equilibrium is reached, 15.0 g of NH_3 has formed. (a) Calculate the percent yield. (b) How many moles of N_2 and H_2 are present at equilibrium?

Comprehensive Problems

•• **4.110** Nutritional biochemists have known for decades that acidic foods cooked in cast-iron cookware can supply significant amounts of dietary iron (ferrous ion).
(a) Write a balanced net ionic equation, with oxidation numbers, that supports this fact.
(b) Measurements show an increase from 3.3 mg of iron to 49 mg of iron per $\frac{1}{2}$-cup (125-g) serving during the slow preparation of tomato sauce in a cast-iron pot. How many ferrous ions are present in a 26-oz (737-g) jar of the tomato sauce?

•• **4.111** Limestone ($CaCO_3$) is used to remove acidic pollutants from smokestack flue gases. It is heated to form lime (CaO), which reacts with sulfur dioxide to form calcium sulfite. Assuming a 70.% yield in the overall reaction, what mass of limestone is required to remove all the sulfur dioxide formed by the combustion of 8.5×10^4 kg of coal that is 0.33 mass % sulfur?

•• **4.112** The brewing industry uses yeast microorganisms to convert glucose to ethanol for wine and beer. The baking industry uses the carbon dioxide produced to make bread rise:
$$C_6H_{12}O_6(s) \xrightarrow{yeast} 2C_2H_5OH(l) + 2CO_2(g)$$
How many grams of ethanol can be produced from 100. g of glucose? What volume of CO_2 is produced? (Assume 1 mol of gas occupies 22.4 L at the conditions used.)

•• **4.113** A chemical engineer determines the mass percent of iron in an ore sample by converting the Fe to Fe^{2+} in acid and then titrating the Fe^{2+} with MnO_4^-. A 1.1081-g sample was dissolved in acid and then titrated with 39.32 mL of 0.03190 M $KMnO_4$. The balanced equation is
$$8H^+(aq) + 5Fe^{2+}(aq) + MnO_4^-(aq) \longrightarrow$$
$$5Fe^{3+}(aq) + Mn^{2+}(aq) + 4H_2O(l)$$
Calculate the mass percent of iron in the ore.

•• **4.114** Mixtures of $CaCl_2$ and $NaCl$ are used to melt ice on roads. A dissolved 1.9348-g sample of such a mixture was analyzed by using excess $Na_2C_2O_4$ to precipitate the Ca^{2+} as CaC_2O_4. The CaC_2O_4 was dissolved in sulfuric acid, and the resulting $H_2C_2O_4$ was titrated with 37.68 mL of 0.1019 M $KMnO_4$ solution. (a) Write the balanced net ionic equation for the precipitation reaction. (b) Write the balanced net ionic equation for the titration reaction. (See Sample Problem 4.11.) (c) What is the oxidizing agent? (d) What is the reducing agent? (e) Calculate the mass percent of $CaCl_2$ in the original sample.

•• **4.115** You are given solutions of HCl and NaOH and must determine their concentrations. You use 27.5 mL of NaOH to titrate 100. mL of HCl and 18.4 mL of NaOH to titrate 50.0 mL of 0.0782 M H_2SO_4. Find the unknown concentrations.

•• **4.116** The flask (right) represents the products of the titration of 25 mL of sulfuric acid with 25 mL of sodium hydroxide.
(a) Write balanced molecular, total ionic, and net ionic equations for the reaction.
(b) If each orange sphere represents 0.010 mol of sulfate ion, how many moles of acid and of base reacted?
(c) What are the molarities of the acid and the base?

•• **4.117** To find the mass percent of dolomite [$CaMg(CO_3)_2$] in a soil sample, a geochemist titrates 13.86 g of soil with 33.56 mL of 0.2516 M HCl. What is the mass percent of dolomite in the soil?

•• **4.118** On a lab exam, you have to find the concentrations of the monoprotic (one proton per molecule) acids HA and HB. You are given 43.5 mL of HA solution in one flask. A second flask contains 37.2 mL of HA, and you add enough HB solution to it to reach a final volume of 50.0 mL. You titrate the first HA solution with 87.3 mL of 0.0906 M NaOH and the mixture of HA and HB in the second flask with 96.4 mL of the NaOH solution. Calculate the molarity of the HA and HB solutions.

•• **4.119** Nitric acid, a major industrial and laboratory acid, is produced commercially by the multistep Ostwald process, which begins with the oxidation of ammonia:

Step 1. $4NH_3(g) + 5O_2(g) \longrightarrow 4NO(g) + 6H_2O(l)$

Step 2. $2NO(g) + O_2(g) \longrightarrow 2NO_2(g)$

Step 3. $3NO_2(g) + H_2O(l) \longrightarrow 2HNO_3(l) + NO(g)$

(a) What are the oxidizing and reducing agents in each step?
(b) Assuming 100% yield in each step, what mass (in kg) of ammonia must be used to produce 3.0×10^4 kg of HNO_3?

•• **4.120** For the following aqueous reactions, complete and balance the molecular equation and write a net ionic equation:
(a) Manganese(II) sulfide + hydrobromic acid
(b) Potassium carbonate + strontium nitrate
(c) Potassium nitrite + hydrochloric acid
(d) Calcium hydroxide + nitric acid
(e) Barium acetate + iron(II) sulfate
(f) Zinc carbonate + sulfuric acid
(g) Copper(II) nitrate + hydrosulfuric acid
(h) Magnesium hydroxide + chloric acid
(i) Potassium chloride + ammonium phosphate
(j) Barium hydroxide + hydrocyanic acid

•• **4.121** There are various methods for finding the composition of an alloy (a metal-like mixture). Show that calculating the mass % of Mg in a magnesium-aluminum alloy ($d = 2.40$ g/cm^3) gives the same answer (within rounding) using each of these methods: (a) a 0.263-g sample of alloy (d of Mg = 1.74 g/cm^3; d of Al = 2.70 g/cm^3); (b) an identical sample reacting with excess aqueous HCl forms 1.38×10^{-2} mol of H_2; (c) an identical sample reacting with excess O_2 forms 0.483 g of oxide.

•• **4.122** Use the oxidation number method to balance the following equations by placing coefficients in the blanks. Identify the reducing and oxidizing agents:
(a) __KOH(aq) + __H_2O_2(aq) + __Cr(OH)$_3$(s) $\longrightarrow$
        __K_2CrO_4(aq) + __H_2O(l)
(b) __MnO_4^-(aq) + __ClO_2^-(aq) + __H_2O(l) $\longrightarrow$
        __MnO_2(s) + __ClO_4^-(aq) + __OH^-(aq)
(c) __$KMnO_4$(aq) + __Na_2SO_3(aq) + __H_2O(l) $\longrightarrow$
        __MnO_2(s) + __Na_2SO_4(aq) + __KOH(aq)
(d) __CrO_4^{2-}(aq) + __$HSnO_2^-$(aq) + __H_2O(l) $\longrightarrow$
        __CrO_2^-(aq) + __$HSnO_3^-$(aq) + __OH^-(aq)
(e) __$KMnO_4$(aq) + __$NaNO_2$(aq) + __H_2O(l) $\longrightarrow$
        __MnO_2(s) + __$NaNO_3$(aq) + __KOH(aq)
(f) __I^-(aq) + __O_2(g) + __H_2O(l) $\longrightarrow$
        __I_2(s) + __OH^-(aq)

•• **4.123** In 1995, Mario Molina, Paul Crutzen, and F. Sherwood Rowland shared the Nobel Prize in chemistry for their work on atmospheric chemistry. One of several reaction sequences proposed for the role of chlorine in the decomposition of stratospheric ozone (we'll see another sequence in Chapter 16) is

(1) $Cl(g) + O_3(g) \longrightarrow ClO(g) + O_2(g)$

(2) $ClO(g) + ClO(g) \longrightarrow Cl_2O_2(g)$

(3) $Cl_2O_2(g) \xrightarrow{\text{light}} 2Cl(g) + O_2(g)$

Over the tropics, O atoms are more common in the stratosphere:

(4) $ClO(g) + O(g) \longrightarrow Cl(g) + O_2(g)$

(a) Which, if any, of these are oxidation-reduction reactions?
(b) Write an overall equation combining reactions 1–3.

•• **4.124** Sodium peroxide (Na_2O_2) is often used in self-contained breathing devices, such as those used in fire emergencies, because it reacts with exhaled CO_2 to form Na_2CO_3 and O_2. How many liters of respired air can react with 80.0 g of Na_2O_2 if each liter of respired air contains 0.0720 g of CO_2?

•• **4.125** A student forgets to weigh a mixture of sodium bromide dihydrate and magnesium bromide hexahydrate. Upon strong heating, the sample loses 252.1 mg of water. The mixture of anhydrous salts reacts with excess $AgNO_3$ solution to form 6.00×10^{-3} mol of solid AgBr. Find the mass % of each compound in the original mixture.

•• **4.126** Magnesium is used in lightweight alloys for airplane bodies and other structures. The metal is obtained from seawater in a process that includes precipitation, neutralization, evaporation, and electrolysis. How many kilograms of magnesium can be obtained from 1.00 km^3 of seawater if the initial Mg^{2+} concentration is 0.13% by mass (d of seawater = 1.04 g/mL)?

••• **4.127** A typical formulation for window glass is 75% SiO_2, 15% Na_2O, and 10.% CaO by mass. What masses of sand (SiO_2), sodium carbonate, and calcium carbonate must be combined to produce 1.00 kg of glass after carbon dioxide is driven off by thermal decomposition of the carbonates?

•• **4.128** Physicians who specialize in sports medicine routinely treat athletes and dancers. Ethyl chloride, a local anesthetic

Biological: 4.128; Engineering: 4.126; Environmental: 4.114, 4.123; Industrial: 4.119, 4.124

commonly used for simple injuries, is the product of the combination of ethylene with hydrogen chloride:

$$C_2H_4(g) + HCl(g) \longrightarrow C_2H_5Cl(g)$$

If 0.100 kg of C_2H_4 and 0.100 kg of HCl react:
(a) How many molecules of gas (reactants plus products) are present when the reaction is complete?
(b) How many moles of gas are present when half the product forms?

•• **4.129** The *salinity* of a solution is defined as the grams of total salts per kilogram of solution. An agricultural chemist uses a solution whose salinity is 35.0 g/kg to test the effect of irrigating farmland with high-salinity river water. The two solutes are NaCl and $MgSO_4$, and there are twice as many moles of NaCl as $MgSO_4$. What masses of NaCl and $MgSO_4$ are contained in 1.00 kg of the solution?

••• **4.130** Thyroxine ($C_{15}H_{11}I_4NO_4$) is a hormone synthesized by the thyroid gland and used to control many metabolic functions in the body. A physiologist determines the mass percent of thyroxine in a thyroid extract by igniting 0.4332 g of extract with sodium carbonate, which converts the iodine to iodide. The iodide is dissolved in water, and bromine and hydrochloric acid are added, which convert the iodide to iodate.
(a) How many moles of iodate form per mole of thyroxine?
(b) Excess bromine is boiled off and more iodide is added, which reacts as shown in the following *unbalanced* equation:

$$IO_3^-(aq) + H^+(aq) + I^-(aq) \longrightarrow I_2(aq) + H_2O(l)$$

How many moles of iodine are produced per mole of thyroxine? (*Hint:* Be sure to balance the charges as well as the atoms.) What are the oxidizing and reducing agents in the reaction?
(c) The iodine reacts completely with 17.23 mL of 0.1000 M thiosulfate as shown in the following *unbalanced* equation:

$$I_2(aq) + S_2O_3^{2-}(aq) \longrightarrow I^-(aq) + S_4O_6^{2-}(aq)$$

What is the mass percent of thyroxine in the thyroid extract?

•• **4.131** Over time, as their free fatty acid (FFA) content increases, edible fats and oils become rancid. To measure rancidity, the fat or oil is dissolved in ethanol, and any FFA present is titrated with KOH dissolved in ethanol. In a series of tests on olive oil, a stock solution of 0.050 M ethanolic KOH was prepared at 25°C, stored at 0°C, and then placed in a 100-mL buret to titrate any oleic acid [$CH_3(CH_2)_7CH{=}CH(CH_2)_7COOH$] present in the oil. Each of four 10.00-g samples of oil took several minutes to titrate: the first required 19.60 mL, the second 19.80 mL, and the third and fourth 20.00 mL of the ethanolic KOH.
(a) What is the apparent acidity of each sample, in terms of mass % of oleic acid? (*Note:* As the ethanolic KOH warms in the buret, its volume increases by a factor of 0.00104/°C.)
(b) Is the variation in acidity a random or systematic error? Explain.
(c) What is the actual acidity? How would you demonstrate this?

•• **4.132** Carbon dioxide is removed from the atmosphere of space capsules by reaction with a solid metal hydroxide. The products are water and the metal carbonate.
(a) Calculate the mass of CO_2 that can be removed by reaction with 3.50 kg of lithium hydroxide.
(b) How many grams of CO_2 can be removed by 1.00 g of each of the following: lithium hydroxide, magnesium hydroxide, and aluminum hydroxide?

•• **4.133** A chemist mixes solid AgCl, $CuCl_2$, and $MgCl_2$ in enough water to give a final volume of 50.0 mL.

(a) With ions shown as spheres and solvent molecules omitted for clarity, which of the following best represents the resulting mixture?

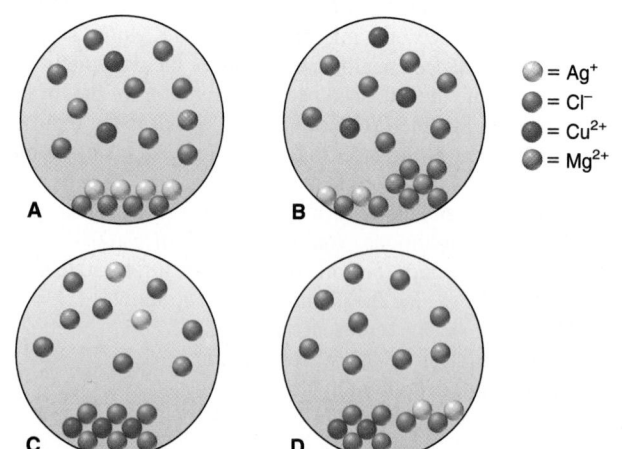

(b) If each sphere represents $5.0{\times}10^{-3}$ mol of ions, what is the total concentration of *dissolved* (separated) ions?
(c) What is the total mass of solid?

••• **4.134** Calcium dihydrogen phosphate, $Ca(H_2PO_4)_2$, and sodium hydrogen carbonate, $NaHCO_3$, are ingredients of baking powder that react with each other to produce CO_2, which causes dough or batter to rise:

$$Ca(H_2PO_4)_2(s) + NaHCO_3(s) \longrightarrow$$
$$CO_2(g) + H_2O(g) + CaHPO_4(s) + Na_2HPO_4(s)$$
$$\text{[unbalanced]}$$

If the baking powder contains 31% $NaHCO_3$ and 35% $Ca(H_2PO_4)_2$ by mass:
(a) How many moles of CO_2 are produced from 1.00 g of baking powder?
(b) If 1 mol of CO_2 occupies 37.0 L at 350°F (a typical baking temperature), what volume of CO_2 is produced from 1.00 g of baking powder?

•• **4.135** In a titration of HNO_3, you add a few drops of phenolphthalein indicator to 50.00 mL of acid in a flask. You quickly add 20.00 mL of 0.0502 M NaOH but overshoot the end point, and the solution turns deep pink. Instead of starting over, you add 30.00 mL of the acid, and the solution turns colorless. Then, it takes 3.22 mL of the NaOH to reach the end point. (a) What is the concentration of the HNO_3 solution? (b) How many moles of NaOH were in excess after the first addition?

••• **4.136** The active compound in Pepto-Bismol contains C, H, O, and Bi.
(a) When 0.22105 g of it was burned in excess O_2, 0.1422 g of bismuth(III) oxide, 0.1880 g of carbon dioxide, and 0.02750 g of water were formed. What is the empirical formula of this compound?
(b) Given a molar mass of 1086 g/mol, determine the molecular formula.
(c) Complete and balance the acid-base reaction between bismuth(III) hydroxide and salicylic acid ($HC_7H_5O_3$), which is used to form this compound.
(d) A dose of Pepto-Bismol contains 0.600 mg of the active ingredient. If the yield of the reaction in part (c) is 88.0%, what mass (in mg) of bismuth(III) hydroxide is required to prepare one dose?

Biological: 4.130, 4.131; Engineering: 4.132; Environmental: 4.129; Industrial: 4.134, 4.136

•• **4.137** Two aqueous solutions contain the ions indicated below.

250. mL + 250. mL

● = Na^+
● = CO_3^{2-}
● = Ca^{2+}
● = Cl^-

(a) Write balanced molecular, total ionic, and net ionic equations for the reaction that occurs when the solutions are mixed. (b) If each sphere represents 0.050 mol of ion, what mass (in g) of precipitate forms, assuming 100% reaction? (c) What is the concentration of each ion in solution after reaction?

••• **4.138** In 1997, at the United Nations Conference on Climate Change, the major industrial nations agreed to expand their research efforts to develop renewable sources of carbon-based fuels. For more than a decade, Brazil has been engaged in a program to replace gasoline with ethanol derived from the root crop manioc (cassava).
(a) Write separate balanced equations for the complete combustion of ethanol (C_2H_5OH) and of gasoline (represented by the formula C_8H_{18}).
(b) What mass of oxygen is required to burn completely 1.00 L of a mixture that is 90.0% gasoline ($d = 0.742$ g/mL) and 10.0% ethanol ($d = 0.789$ g/mL) by volume?
(c) If 1.00 mol of O_2 occupies 22.4 L, what volume of O_2 is needed to burn 1.00 L of the mixture?
(d) Air is 20.9% O_2 by volume. What volume of air is needed to burn 1.00 L of the mixture?

••• **4.139** In a car engine, gasoline (represented by C_8H_{18}) does not burn completely, and some CO, a toxic pollutant, forms along with CO_2 and H_2O. If 5.0% of the gasoline forms CO:
(a) What is the ratio of CO_2 to CO molecules in the exhaust?
(b) What is the mass ratio of CO_2 to CO?
(c) What percentage of the gasoline must form CO for the mass ratio of CO_2 to CO to be exactly 1/1?

••• **4.140** The amount of ascorbic acid (vitamin C; $C_6H_8O_6$) in tablets is determined by reaction with bromine and then titration of the hydrobromic acid with standard base:

$$C_6H_8O_6 + Br_2 \longrightarrow C_6H_6O_6 + 2HBr$$
$$HBr + NaOH \longrightarrow NaBr + H_2O$$

A certain tablet is advertised as containing 500 mg of vitamin C. One tablet was dissolved in water and reacted with Br_2. The solution was then titrated with 43.20 mL of 0.1350 *M* NaOH. Did the tablet contain the advertised quantity of vitamin C?

•• **4.141** In the process of *salting-in*, protein solubility in a dilute salt solution is increased by adding more salt. Because the protein solubility depends on the total ion concentration as well as the ion charge, salts yielding doubly charged ions are often more effective than those yielding singly charged ions. (a) How many grams of $MgCl_2$ must dissolve to equal the ion concentration of 12.4 g of NaCl? (b) How many grams of CaS must dissolve? (c) Which of the three salt solutions would dissolve the most protein?

•• **4.142** In the process of *pickling*, rust is removed from newly produced steel by washing the steel in hydrochloric acid:
(1) $6HCl(aq) + Fe_2O_3(s) \longrightarrow 2FeCl_3(aq) + 3H_2O(l)$
During the process, some iron is lost as well:
(2) $2HCl(aq) + Fe(s) \longrightarrow FeCl_2(aq) + H_2(g)$
(a) Which reaction, if either, is a redox process? (b) If reaction 2 did not occur and all the HCl were used, how many grams of Fe_2O_3 could be removed and $FeCl_3$ produced in a 2.50×10^3-L bath of 3.00 *M* HCl? (c) If reaction 1 did not occur and all the HCl were used, how many grams of Fe could be lost and $FeCl_2$ produced in a 2.50×10^3-L bath of 3.00 *M* HCl? (d) If 0.280 g of Fe is lost per gram of Fe_2O_3 removed, what is the mass ratio of $FeCl_2$ to $FeCl_3$?

•• **4.143** At liftoff, a space shuttle uses a solid mixture of ammonium perchlorate and aluminum powder to obtain great thrust from the volume change of solid to gas. In the presence of a catalyst, the mixture forms solid aluminum oxide and aluminum trichloride and gaseous water and nitrogen monoxide. (a) Write a balanced equation for the reaction, and identify the reducing and oxidizing agents. (b) How many total moles of gas (water vapor and nitrogen monoxide) are produced when 50.0 kg of ammonium perchlorate reacts with a stoichiometric amount of Al? (c) What is the volume change from this reaction? (*d* of $NH_4ClO_4 = 1.95$ g/cc, Al = 2.70 g/cc, $Al_2O_3 = 3.97$ g/cc, and $AlCl_3 = 2.44$ g/cc; assume 1 mol of gas occupies 22.4 L.)

••• **4.144** A *reaction cycle* for an element is a series of reactions beginning and ending with that element. In the following copper reaction cycle, copper has either a 0 or a +2 oxidation state. Write balanced molecular and net ionic equations for each step in the cycle.
(1) Copper metal reacts with aqueous bromine to produce a green-blue solution.
(2) Adding aqueous sodium hydroxide forms a blue precipitate.
(3) The precipitate is heated and turns black (water is released).
(4) The black solid dissolves in nitric acid to give a blue solution.
(5) Adding aqueous sodium phosphate forms a green precipitate.
(6) The precipitate forms a blue solution in sulfuric acid.
(7) Copper metal is recovered from the blue solution when zinc metal is added.

5

Rising on a Gas Law Heating the air inside this balloon causes it to expand and some of it to flow out, which lowers the mass and makes the balloon rise. In this chapter, we examine this and other behaviors of gases that influence everyday life.

Gases and the
Kinetic-Molecular Theory

5.1 An Overview of the Physical States of Matter

5.2 Gas Pressure and Its Measurement
Laboratory Devices
Units of Pressure

5.3 The Gas Laws and Their Experimental Foundations
Boyle's Law
Charles's Law
Avogadro's Law
Standard Conditions
The Ideal Gas Law
Solving Gas Law Problems

5.4 Further Applications of the Ideal Gas Law
Density of a Gas
Molar Mass of a Gas
Partial Pressure of a Gas

5.5 The Ideal Gas Law and Reaction Stoichiometry

5.6 The Kinetic-Molecular Theory: A Model for Gas Behavior
Explaining the Gas Laws
Effusion and Diffusion
Mean Free Path and Collision Frequency

5.7 Real Gases: Deviations from Ideal Behavior
Effects of Extreme Conditions
The van der Waals Equation

People have been observing the behavior of the states of matter throughout history; in fact, three of the four "elements" of the ancient Greek philosophers were air (gas), water (liquid), and earth (solid). Nevertheless, many questions remain. In Chapter 5 and its companion, Chapter 12, we examine these states and their interrelations. Here, we highlight the gaseous state, the one we understand best.

Gases are everywhere. Earth's atmosphere is a colorless, odorless mixture of nearly 20 elements and compounds that extends from the planet's surface upward more than 500 km and then merges with outer space. Some components—O_2, N_2, H_2O vapor, and CO_2—take part in complex redox reaction cycles throughout the environment, and you participate in those cycles with every breath you take. ● Gases also have essential roles in industry (Table 5.1).

Concepts & Skills to Review
before you study this chapter

- physical states of matter (Section 1.1)
- SI unit conversions (Section 1.5)
- mole-mass-number conversions (Section 3.1)

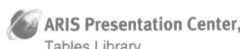 Literature Brown, Katherine E.; Micklos, Amy; Carver, Jeffrey S.; Hunter, William J. F. "A Teaching Plan for Introducing Gas Properties," *Chem. Educator 9* **2004**, *4*, 220–223.

 Literature Lin, Huann-shyang; Cheng, Hsiu-ju; Lawrenz, Frances. "The Assessment of Students' and Teachers' Understanding of Gas Laws," *J. Chem. Educ.* **2000**, *77*, 235.

ARIS Presentation Center, Tables Library

Table 5.1	Some Important Industrial Gases
Name (Formula)	**Origin; Use**
Methane (CH_4)	Natural deposits; domestic fuel
Ammonia (NH_3)	From $N_2 + H_2$; fertilizers, explosives
Chlorine (Cl_2)	Electrolysis of seawater; bleaching and disinfecting
Oxygen (O_2)	Liquefied air; steelmaking
Ethylene (C_2H_4)	High-temperature decomposition of natural gas; plastics

A key point in this chapter is that, while the *chemical* behavior of a gas depends on its composition, all gases have very similar *physical* behavior. For instance, although the particular gases differ, the same physical behaviors are at work in the operation of a car and in the baking of bread, in the thrust of a rocket engine and in the explosion of a kernel of popcorn. The process of breathing involves the same physical principles as the creation of thunder.

IN THIS CHAPTER . . . We first contrast gases with liquids and solids and then discuss gas pressure. We consider the mass laws, which describe observable gas behavior. We then examine the ideal gas law, which encompasses the other mass laws, and apply it to reaction stoichiometry. We explain the observable behavior of gases with the simple kinetic-molecular model. Next, we consider the properties of planetary atmospheres. Finally, we find that real gas behavior, especially under extreme conditions, requires refinements of the ideal gas law and the model.

Biological application

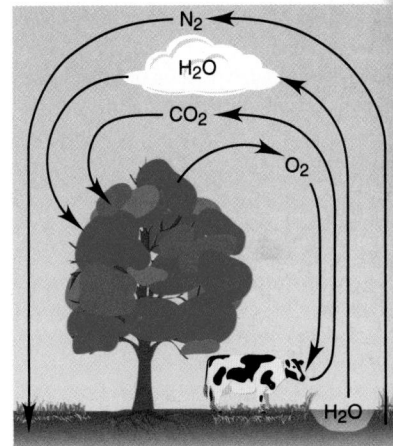

● **Atmosphere-Biosphere Redox Interconnections** The diverse organisms that make up the biosphere interact intimately with the gases of the atmosphere. Powered by solar energy, green plants reduce atmospheric CO_2 and incorporate the C atoms into their own substance. In the process, O atoms in H_2O are oxidized and released to the air as O_2. Certain microbes that live on plant roots reduce N_2 to NH_3 and form compounds that the plant uses to make its proteins. Other microbes that feed on dead plants (and animals) oxidize the proteins and release N_2 again. Animals eat plants and other animals, use O_2 to oxidize their food, and return CO_2 and H_2O to the air.

5.1 AN OVERVIEW OF THE PHYSICAL STATES OF MATTER

Under appropriate conditions of pressure and temperature, most substances can exist as a solid, a liquid, or a gas. In Chapter 1 we distinguished these physical states in terms of how each fills a container and began to develop a molecular view that explains this macroscopic behavior: a solid has a fixed shape regardless of the container shape because its particles are held rigidly in place; a liquid conforms to the container shape but has a definite volume and a surface because its particles are close together but free to move around each other; and a gas fills the container because its particles are far apart and moving randomly. Several other aspects of their behavior distinguish gases from liquids and solids:

1. *Gas volume changes greatly with pressure.* When a sample of gas is confined to a container of variable volume, such as a piston-cylinder assembly, an external force can compress the gas. Removing the external force allows the gas volume to increase again. In contrast, a liquid or solid resists significant changes in volume.

Literature DePierro, Ed; Garafalo, Fred. "Some Insights Regarding a Popular Introductory Gas Law Experiment," *J. Chem. Educ.* **2005,** *82,* 1194.

Literature David, Carl W. "An Alternative Problem-Solving Technique for Freshman Chemistry," *Chem. Educator* 10 **2005,** *2,* 102–106.

POW! P-s-s-s-t! POP! A jackhammer uses the force of rapidly expanding compressed air to break through rock and cement. When the nozzle on a can of spray paint is pressed, the pressurized propellant gases expand into the lower pressure of the surroundings and expel droplets of paint. The rapid expansion of heated gases results in such phenomena as the destruction caused by a bomb, the liftoff of a rocket, and the popping of kernels of corn.

2. *Gas volume changes greatly with temperature.* When a gas sample at constant pressure is heated, its volume increases; when it is cooled, its volume decreases. This volume change is 50 to 100 times greater for gases than for liquids or solids.
3. *Gases have relatively low viscosity.* Gases flow much more freely than liquids and solids. Low viscosity allows gases to be transported through pipes over long distances but also to leak rapidly out of small holes.
4. *Most gases have relatively low densities* under normal conditions. Gas density is usually tabulated in units of grams per *liter,* whereas liquid and solid densities are in grams per *milliliter,* about 1000 times as dense (see Table 1.5, p. 23). For example, at 20°C and normal atmospheric pressure, the density of $O_2(g)$ is 1.3 g/**L**, whereas the density of $H_2O(l)$ is 1.0 g/**mL** and that of NaCl(*s*) is 2.2 g/**mL.** When a gas is cooled, its density increases because its volume decreases: at 0°C, the density of $O_2(g)$ increases to 1.4 g/L.
5. *Gases are miscible. Miscible* substances mix with one another in any proportion to form a solution. Air, as we said, is a solution of nearly 20 gases. Two liquids, however, may or may not be miscible: water and ethanol are, but water and gasoline are not. Two solids generally do not form a solution at all unless they are mixed as molten liquids and then allowed to solidify.

Each of these observable properties offers a clue to the molecular properties of gases. For example, consider these density data. At 20°C and normal atmospheric pressure, gaseous N_2 has a density of 1.25 g/L. If cooled below −196°C, it condenses to liquid N_2 and its density becomes 0.808 g/mL. (Note the change in units.) The same amount of nitrogen occupies less than $\frac{1}{600}$ as much space! Further cooling to below −210°C yields solid N_2 ($d = 1.03$ g/mL), which is only slightly more dense than the liquid. These values show again that *the molecules are much farther apart in the gas than in either the liquid or the solid.* You can also see that a large amount of space between molecules is consistent with the miscibility, low viscosity, and compressibility of gases. ● Figure 5.1 compares macroscopic and atomic-scale views of the physical states of bromine.

Figure 5.1 The three states of matter.
Many pure substances, such as bromine (Br_2), can exist under appropriate conditions of pressure and temperature as a **A,** gas; **B,** liquid; or **C,** solid. The atomic-scale views show that molecules are much farther apart in a gas than in a liquid or solid.

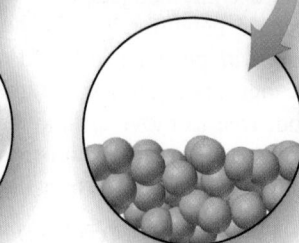

A Gas: Molecules are far apart, move freely, and fill the available space

B Liquid: Molecules are close together but move around one another

C Solid: Molecules are close together in a regular array and do not move around one another

Section Summary

The volume of a gas can be altered significantly by changing the applied external force or the temperature. The corresponding changes for liquids and solids are much smaller. • Gases flow more freely and have lower densities than liquids and solids, and they mix in any proportion to form solutions. • The reason for these differences in states is the greater distance between particles in a gas than in a liquid or a solid.

5.2 GAS PRESSURE AND ITS MEASUREMENT

Blowing up a balloon provides clear evidence that a gas exerts pressure on the walls of its container. **Pressure (P)** is defined as the force exerted per unit of surface area:

$$\text{Pressure} = \frac{\text{force}}{\text{area}}$$

Earth's gravitational attraction pulls the atmospheric gases toward its surface, where they exert a force on all objects. The force, or weight, of these gases creates a pressure of about 14.7 pounds per square inch (lb/in^2; psi) of surface. ●

As we'll discuss later, since the molecules in a gas are moving in every direction, the pressure of the atmosphere is exerted uniformly on the floor, walls, ceiling, and every object in a room. The pressure on the outside of your body is equalized by the pressure on the inside, so there is no net pressure on your body's outer surface. What would happen if this were not the case? As an analogy, consider the empty metal can attached to a vacuum pump in Figure 5.2. With the pump off, the can maintains its shape because the pressure on the outside is equal to the pressure on the inside. With the pump on, the internal pressure decreases greatly, and the ever-present external pressure easily crushes the can. A vacuum-filtration flask (and tubing), which you may have used in the lab, has thick walls that can withstand the external pressure when the flask is evacuated.

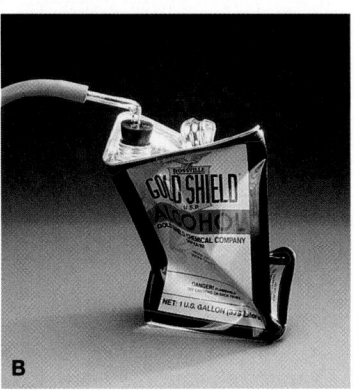

A   **B**

Laboratory Devices for Measuring Gas Pressure

The **barometer** is a common device used to measure atmospheric pressure. Invented in 1643 by Evangelista Torricelli, the barometer is still basically just a tube about 1 m long, closed at one end, filled with mercury, and inverted into a dish containing more mercury. When the tube is inverted, some of the mercury flows out into the dish, and a vacuum forms above the mercury remaining in the

● **The Meaning of Pressure in Daily Life**
Snowshoes allow you to walk on powdery snow without sinking because they distribute your weight over a much larger area than a boot does, thereby greatly decreasing the pressure per square inch. The area of a snowshoe is typically about 10 times as large as that of a boot sole, so the snowshoe exerts only about one-tenth as much pressure as the boot. The wide, padded paws of snow leopards accomplish this, too. For the same reason, high-heeled shoes exert much more pressure than flat shoes.

Demo Vitz, Ed. "Ammonia Can Crush," *J. Chem. Educ.* **1999**, *76*, 932.

Demo Shakhashiri, Bassam Z. "Collapsing Can," *Chemical Demonstrations, A Handbook for Teachers of Chemistry,* Vol. 2 (The University of Wisconsin Press, Madison, 1985), 14–19.

Figure 5.2 Effect of atmospheric pressure on objects at Earth's surface.
A, A metal can filled with air has equal pressure on the inside and outside.
B, When the air inside the can is removed, the atmospheric pressure crushes the can.

ARIS Presentation Center, Photo Library

Literature Sanger, Michael J.; Phelps, Amy J.; Fienhold, Jason. "Using a Computer Animation to Improve Students' Conceptual Understanding of a Can-Crushing Demonstration," *J. Chem. Educ.* **2000**, *77*, 1517.

 Demo Meyers, Robert D.; Yee, Gordon T. "A More Dramatic Container to Crush by Atmospheric Pressure," *J. Chem. Educ.* **1999**, *76*, 933.

Figure 5.3 A mercury barometer. (See text for explanation.)

tube, as shown in Figure 5.3. At sea level under ordinary atmospheric conditions, the outward flow of mercury stops when the surface of the mercury in the tube is about 760 mm above the surface of the mercury in the dish. It stops at 760 mm because at that point the column of mercury in the tube exerts the same pressure (weight/area) on the mercury surface in the dish as does the column of air that extends from the dish to the outer reaches of the atmosphere. The air pushing down keeps any more of the mercury in the tube from flowing out. Likewise, if you place an evacuated tube into a dish filled with mercury, the mercury rises about 760 mm into the tube because the atmosphere pushes the mercury up to that height.

Several centuries ago, people ascribed mysterious "suction" forces to a vacuum. We know now that a vacuum does not suck up mercury into the barometer tube any more than it sucks in the walls of the crushed can in Figure 5.2. Only matter—in this case, the atmospheric gases—can exert a force.

Notice that we did not specify the diameter of the barometer tube. If the mercury in a 1-cm diameter tube rises to a height of 760 mm, the mercury in a 2-cm diameter tube will rise to that height also. The *weight* of mercury is greater in the wider tube, but the area is larger also; thus the *pressure,* the *ratio* of weight to area, is the same.

Since the pressure of the mercury column is directly proportional to its height, a unit commonly used for pressure is mmHg, the height of the mercury (atomic symbol Hg) column in millimeters (mm). We discuss units of pressure shortly. At sea level and 0°C, normal atmospheric pressure is 760 mmHg; at the top of Mt. Everest (elevation 29,028 ft, or 8848 m), the atmospheric pressure is only about 270 mmHg. Thus, *pressure decreases with altitude:* the column of air above the sea is taller and weighs more than the column of air above Mt. Everest.

Laboratory barometers contain mercury because its high density allows the barometer to be a convenient size. For example, the pressure of the atmosphere would equal the pressure of a column of water about 10,300 mm, almost 34 ft, high. ● Note that, for a given pressure, the ratio of heights (h) of the liquid columns is inversely related to the ratio of the densities (d) of the liquids:

$$\frac{h_{H_2O}}{h_{Hg}} = \frac{d_{Hg}}{d_{H_2O}}$$

⊙⊙ **Engineering application**

● **The Mystery of the Suction Pump**
When you drink through a straw, you create lower pressure above the liquid, and the atmosphere pushes the liquid up. Similarly, a "suction" pump is a tube dipping into a water source, with a piston and handle that lower the air pressure above the water level. The pump can raise water from a well no deeper than 34 ft. This depth limit was a mystery until the great 17th-century Italian scientist Galileo showed that the atmosphere pushes the water up into the tube and that its pressure can support only a 34-ft column of water. Modern pumps that draw water from deeper sources use compressed air to increase the pressure exerted on the water.

Figure 5.4 Two types of manometer. A, A closed-end manometer with an evacuated flask attached has the mercury levels equal. **B,** A gas exerts pressure on the mercury in the arm closer to the flask. The difference in heights (Δh) equals the gas pressure. **C–E,** An open-end manometer is shown with gas pressure equal to atmospheric pressure, **(C),** gas pressure lower than atmospheric pressure **(D),** and gas pressure higher than atmospheric pressure **(E).**

Manometers are devices used to measure the pressure of a gas in an experiment. Figure 5.4 shows two types of manometer. Part A shows a *closed-end manometer,* a mercury-filled, curved tube, *closed* at one end and attached to a flask at the other. When the flask is evacuated, the mercury levels in the two arms of the tube are the same because no gas exerts pressure on either mercury surface. When a gas is in the flask (part B), it pushes down the mercury level in the near arm, so the level rises in the far arm. The *difference* in column heights (Δh) equals the gas pressure. Note that if we open the lower stopcock of the evacuated flask in part A, air rushes in, Δh equals atmospheric pressure, and the closed-end manometer becomes a barometer.

The *open-end manometer,* shown in parts C–E, also consists of a curved tube filled with mercury, but one end of the tube is *open* to the atmosphere and the other is connected to the gas sample. The atmosphere pushes on one mercury level and the gas pushes on the other. Since Δh equals the difference between two pressures, to calculate the gas pressure with an open-end manometer, we must measure the atmospheric pressure separately with a barometer.

Units of Pressure

Pressure results from a force exerted on an area. The SI unit of force is the newton (N): $1\ N = 1\ kg \cdot m/s^2$. The SI unit of pressure is the **pascal (Pa),** which equals a force of one newton exerted on an area of one square meter:

$$1\ Pa = 1\ N/m^2$$

A much larger unit is the **standard atmosphere (atm),** the average atmospheric pressure measured at sea level and 0°C. It is defined in terms of the pascal:

$$1\ atm = 101.325\ kilopascals\ (kPa) = 1.01325 \times 10^5\ Pa$$

Another common unit is the **millimeter of mercury (mmHg),** mentioned earlier, which is based on measurement with a barometer or manometer. In honor of Torricelli, this unit has been named the **torr:**

$$1\ torr = 1\ mmHg = \frac{1}{760}\ atm = \frac{101.325}{760}\ kPa = 133.322\ Pa$$

The *bar* is coming into more common use in chemistry:

$$1\ bar = 1 \times 10^2\ kPa = 1 \times 10^5\ Pa$$

Table 5.2   Common Units of Pressure

Unit	Atmospheric Pressure	Scientific Field
pascal (Pa); kilopascal (kPa)	1.01325×10^5 Pa; 101.325 kPa	SI unit; physics, chemistry
atmosphere (atm)	1 atm*	Chemistry
millimeters of mercury (mmHg)	760 mmHg*	Chemistry, medicine, biology
torr	760 torr*	Chemistry
pounds per square inch (lb/in² or psi)	14.7 lb/in^2	Engineering
bar	1.01325 bar	Meteorology, chemistry, physics

*This is an exact quantity; in calculations, we use as many significant figures as necessary.

ARIS Presentation Center,
Tables Library

Despite a gradual change to SI units, many chemists still express pressure in torrs and atmospheres, so they are used in this text, with reference to pascals and bars. Table 5.2 lists some important pressure units used in scientific fields.

ARIS Presentation Center,
Worked Examples Library

SAMPLE PROBLEM 5.1 Converting Units of Pressure

PROBLEM A geochemist heats a limestone ($CaCO_3$) sample and collects the CO_2 released in an evacuated flask attached to a closed-end manometer (see Figure 5.4B). After the system comes to room temperature, $\Delta h = 291.4$ mmHg. Calculate the CO_2 pressure in torrs, atmospheres, and kilopascals.

PLAN The CO_2 pressure is given in units of mmHg, so we construct conversion factors from Table 5.2 to find the pressure in the other units.

SOLUTION Converting from mmHg to torr:

$$P_{CO_2} \text{ (torr)} = 291.4 \text{ mmHg} \times \frac{1 \text{ torr}}{1 \text{ mmHg}} = 291.4 \text{ torr}$$

Converting from torr to atm:

$$P_{CO_2} \text{ (atm)} = 291.4 \text{ torr} \times \frac{1 \text{ atm}}{760 \text{ torr}} = 0.3834 \text{ atm}$$

Converting from atm to kPa:

$$P_{CO_2} \text{ (kPa)} = 0.3834 \text{ atm} \times \frac{101.325 \text{ kPa}}{1 \text{ atm}} = 38.85 \text{ kPa}$$

CHECK There are 760 torr in 1 atm, so ~300 torr should be <0.5 atm. There are ~100 kPa in 1 atm, so <0.5 atm should be <50 kPa.

COMMENT 1. In the conversion from torr to atm, we retained four significant figures because this unit conversion factor involves *exact* numbers; that is, 760 torr has as many significant figures as the calculation requires.

2. From here on, except in particularly complex situations, *the canceling of units in calculations is no longer shown.*

FOLLOW-UP PROBLEM 5.1 The CO_2 released from another mineral sample was collected in an evacuated flask connected to an open-end manometer (see Figure 5.4D). If the barometer reading is 753.6 mmHg and Δh is 174.0 mmHg, calculate P_{CO_2} in torrs, pascals, and lb/in^2.

Section Summary

Gases exert pressure (force/area) on all surfaces with which they make contact. • A barometer measures atmospheric pressure in terms of the height of the mercury column that the atmosphere can support (760 mmHg at sea level and 0°C). • Both closed-end and open-end manometers are used to measure the pressure of a gas sample. • Chemists measure pressure in units of atmospheres (atm), torr (equivalent to mmHg), and pascals (Pa, the SI unit).

5.3 THE GAS LAWS AND THEIR EXPERIMENTAL FOUNDATIONS

The physical behavior of a sample of gas can be described completely by four variables: pressure (*P*), volume (*V*), temperature (*T*), and amount (number of moles, *n*). The variables are interdependent: *any one of them can be determined by measuring the other three.* We know now that this quantitatively predictable behavior is a direct outcome of the structure of gases on the molecular level. Yet, it was discovered, for the most part, before Dalton's atomic theory was published!

Three key relationships exist among the four gas variables—Boyle's, Charles's, and Avogadro's laws. Each of these gas laws expresses *the effect of one variable on another, with the remaining two variables held constant.* Because gas volume is so easy to measure, the laws are expressed as the effect on gas volume of a change in the pressure, temperature, or amount of gas.

These three laws are special cases of an all-encompassing relationship among gas variables called the *ideal gas law.* This unifying observation quantitatively describes the state of a so-called **ideal gas,** one that exhibits simple linear relationships among volume, pressure, temperature, and amount. Although no ideal gas actually exists, most simple gases, such as N_2, O_2, H_2, and the noble gases, show nearly ideal behavior at ordinary temperatures and pressures. We discuss the ideal gas law after the three special cases.

The Relationship Between Volume and Pressure: Boyle's Law

Following Torricelli's invention of the barometer, the great 17th-century English chemist Robert Boyle performed a series of experiments that led him to conclude that at a given temperature, *the volume occupied by a gas is inversely related to its pressure.* Figure 5.5 shows Boyle's experiment and some typical data he might have collected. Boyle fashioned a J-shaped glass tube, sealed the shorter end, and poured mercury into the longer end, thereby trapping some air, the gas in the experiment. From the height of the trapped air column and the diameter of the tube, he calculated the air volume. The total pressure applied to the trapped air was the pressure of the atmosphere (measured with a barometer) plus that of the mercury column (part A). By adding mercury, Boyle increased the total pressure exerted on the air, and the air volume decreased (part B). With the temperature and amount of air constant, Boyle could directly measure the effect of the applied pressure on the volume of air.

Demo Shakhashiri, Bassam Z. "Effect of Pressure on the Size of a Balloon," *Chemical Demonstrations, A Handbook for Teachers of Chemistry,* Vol. 2 (The University of Wisconsin Press, Madison, 1985), 12–13.

Demo Broniec, Rick. "Boyle's Law and the Monster Marshmallow," *J. Chem. Educ.* **1982,** *59,* 974.

Demo Shakhashiri, Bassam Z. "Boyle's Law," *Chemical Demonstrations, A Handbook for Teachers of Chemistry,* Vol. 2 (The University of Wisconsin Press, Madison, 1985), 6–8.

Demo Shakhashiri, Bassam Z. "Boyle's Law and the Mass of a Textbook," *Chemical Demonstrations, A Handbook for Teachers of Chemistry,* Vol. 2 (The University of Wisconsin Press, Madison, 1985), 20–23.

Transparency/ARIS Presentation Center, Art Library

Figure 5.5 The relationship between the volume and pressure of a gas.
A, A small amount of air (the gas) is trapped in the short arm of a J tube; *n* and *T* are fixed. The total pressure on the gas (P_{total}) is the sum of the pressure due to the difference in heights of the mercury columns (Δh) plus the pressure of the atmosphere (P_{atm}). If $P_{atm} = 760$ torr, $P_{total} = 780$ torr. **B,** As mercury is added, the total pressure on the gas increases and its volume (*V*) decreases. Note that if P_{total} is doubled (to 1560 torr), *V* is halved (not drawn to scale). **C,** Some typical pressure-volume data from the experiment. **D,** A plot of *V* vs. P_{total} shows that *V* is inversely proportional to *P*. **E,** A plot of *V* vs. $1/P_{total}$ is a straight line whose slope is a constant characteristic of *any* gas that behaves ideally.

V (mL)	*P* (torr)			$\dfrac{1}{P_{total}}$	*PV* (torr•mL)
	Δh	$+ P_{atm}$	$= P_{total}$		
20.0	20.0	760	780	0.00128	1.56×10^4
15.0	278	760	1038	0.000963	1.56×10^4
10.0	800	760	1560	0.000641	1.56×10^4
5.0	2352	760	3112	0.000321	1.56×10^4

C

D

E

Note the following results, shown in Figure 5.5:

- The product of corresponding *P* and *V* values is a constant (part C, rightmost column).
- *V* is *inversely* proportional to *P* (part D).
- *V* is *directly* proportional to 1/*P* (part E) and generates a linear plot of *V* against 1/*P*. This *linear relationship between two gas variables* is a hallmark of ideal gas behavior.

The generalization of Boyle's observations is known as **Boyle's law:** *at constant temperature, the volume occupied by a fixed amount of gas is inversely proportional to the applied (external) pressure,* or

$$V \propto \frac{1}{P} \qquad [T \text{ and } n \text{ fixed}] \qquad \qquad \textbf{(5.1)}$$

This relationship can also be expressed as

$$PV = \text{constant} \qquad \text{or} \qquad V = \frac{\text{constant}}{P} \qquad [T \text{ and } n \text{ fixed}]$$

The constant is the same for the great majority of gases. Thus, tripling the external pressure reduces the volume to one-third its initial value; halving the external pressure doubles the volume; and so forth.

The wording of Boyle's law focuses on *external* pressure. In his experiment, however, adding more mercury caused the mercury level to rise until the pressure of the trapped air stopped the rise at some new level. At that point, the pressure exerted *on* the gas equaled the pressure exerted *by* the gas. In other words, by measuring the applied pressure, Boyle was also measuring the gas pressure. Thus, when gas volume doubles, gas pressure is halved. In general, if V_{gas} increases, P_{gas} decreases, and vice versa.

The Relationship Between Volume and Temperature: Charles's Law

One question raised by Boyle's work was why the pressure-volume relationship holds only at constant temperature. It was not until the early 19th century, through the separate work of French scientists J. A. C. Charles and J. L. Gay-Lussac, that the relationship between gas volume and temperature was clearly understood.

Let's examine this relationship by measuring the volume of a fixed amount of a gas under constant pressure but at different temperatures. A straight tube, closed at one end, traps a fixed amount of air under a small mercury plug. The tube is immersed in a water bath that can be warmed with a heater or cooled with ice. After each change of water temperature, we measure the length of the air column, which is proportional to its volume. The pressure exerted on the gas is constant because the mercury plug and the atmospheric pressure do not change (Figure 5.6A and B).

Some typical data are shown for different amounts and pressures of gas in Figure 5.6C. Again, note the linear relationships, but this time the variables are *directly* proportional: for a given amount of gas at a given pressure, *volume increases as temperature increases.* For example, the red line shows how the volume of 0.04 mol of gas at 1 atm pressure changes as the temperature changes. Extending (*extrapolating*) the line to lower temperatures (dashed portion) shows that the volume shrinks until the gas occupies a theoretical zero volume at −273.15°C (the intercept on the temperature axis). Similar plots for a different amount of gas (green) and a different gas pressure (blue) show lines with different slopes, but they all converge at this temperature.

Literature Goldwhite, Harold. "Gay-Lussac: Chemist Extraordinary," *J. Chem. Educ.* **1981,** *58,* 789.

Demo Shakhashiri, Bassam Z. "Thermal Expansion of Gases," *Chemical Demonstrations, A Handbook for Teachers of Chemistry,* Vol. 2 (The University of Wisconsin Press, Madison; 1985), 24–27.

Demo Shakhashiri, Bassam Z. "Charles' Law," *Chemical Demonstrations, A Handbook for Teachers of Chemistry,* Vol. 2 (The University of Wisconsin Press, Madison, 1985), 28–32.

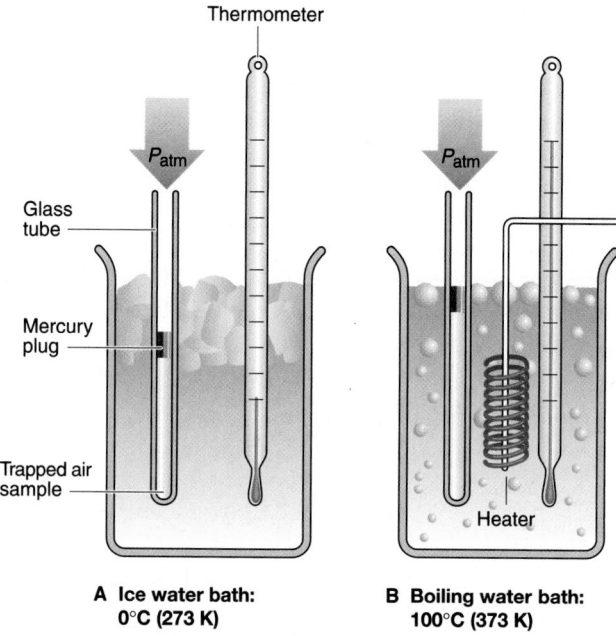

A Ice water bath:
0°C (273 K)

B Boiling water bath:
100°C (373 K)

C Temperature

Figure 5.6 **The relationship between the volume and temperature of a gas.** At constant *P*, the volume of a given amount of gas is *directly* proportional to the absolute temperature. A fixed amount of gas (air) is trapped under a small plug of mercury at a fixed pressure. **A,** The sample is in an ice water bath. **B,** The sample is in a boiling water bath. As the temperature increases, the volume of the gas increases. **C,** The three lines show the effect of amount (*n*) of gas (compare red and green) and pressure (*P*) of gas (compare red and blue). The dashed lines extrapolate the data to lower temperatures. For any amount of an ideal gas at any pressure, the volume is theoretically zero at −273.15°C (0 K).

Transparency/ARIS Presentation Center,
Art Library

A half-century after Charles's and Gay-Lussac's work, William Thomson (Lord Kelvin) used this linear relation between gas volume and temperature to devise the absolute temperature scale (Section 1.5). In this scale, absolute zero (0 K or −273.15°C) is the temperature at which an ideal gas would have zero volume. (Absolute zero has never been reached, but physicists have attained temperatures as low as 10^{-9} K.) Of course, no sample of matter can have zero volume, and every real gas condenses to a liquid at some temperature higher than 0 K. Nevertheless, *this linear dependence of volume on absolute temperature holds for most common gases over a wide temperature range.*

The modern statement of the volume-temperature relationship is known as **Charles's law:** *at constant pressure, the volume occupied by a fixed amount of gas is directly proportional to its absolute (Kelvin) temperature,* or

$$V \propto T \quad [P \text{ and } n \text{ fixed}] \quad (5.2)$$

This relationship can also be expressed as

$$\frac{V}{T} = \text{constant} \quad \text{or} \quad V = \text{constant} \times T \quad [P \text{ and } n \text{ fixed}]$$

If *T* increases, *V* increases, and vice versa. Once again, for any given *P* and *n*, the constant is the same for the great majority of gases.

The dependence of gas volume on *absolute* temperature means that you must *use the Kelvin scale in gas law calculations.* For instance, if the temperature changes from 200 K to 400 K, the volume of 1 mol of gas doubles. But, if the temperature changes from 200°C to 400°C, the volume increases by a factor of 1.42; that is, $\left(\dfrac{400°C + 273.15}{200°C + 273.15} \right) = \dfrac{673}{473} = 1.42$.

Demo Petty, John T. "Charles' Law of Gases: A Simple Experimental Demonstration," *J. Chem. Educ.* **1995,** *72,* 257.

Demo Haub, Elaine K. "Disappearing-Reappearing Rabbit Trick: A New Twist to an Old Liquid Nitrogen Demonstration," *J. Chem. Educ.* **2001,** *78,* 46.

Demo Adcock, Louis H. "The Egg in the Bottle Revisited: Air Pressure and Amontons' Law (Charles' Law)," *J. Chem. Educ.* **1998**, *75*, 1567.

Other Relationships Based on Boyle's and Charles's Laws Two other important relationships in gas behavior emerge from an understanding of Boyle's and Charles's laws:

1. *The pressure-temperature relationship.* Charles's law is expressed as the effect of a temperature change on gas *volume.* However, volume and pressure are interdependent, so the effect of temperature on volume is closely related to its effect on pressure (sometimes referred to as *Amontons's law*). Measure the pressure in your car's tires before and after a long drive, and you will find that it has increased. Heating due to friction between the tire and the road increases the air temperature inside the tire, but since the tire volume doesn't change appreciably, the air exerts more pressure. Thus, *at constant volume, the pressure exerted by a fixed amount of gas is directly proportional to the absolute temperature:*

$$P \propto T \qquad [V \text{ and } n \text{ fixed}] \tag{5.3}$$

or

$$\frac{P}{T} = \text{constant} \qquad \text{or} \qquad P = \text{constant} \times T$$

2. *The combined gas law.* A simple combination of Boyle's and Charles's laws gives the *combined gas law,* which applies to situations when two of the three variables (V, P, T) change and you must find the effect on the third:

$$V \propto \frac{T}{P} \qquad \text{or} \qquad V = \text{constant} \times \frac{T}{P} \qquad \text{or} \qquad \frac{PV}{T} = \text{constant}$$

The Relationship Between Volume and Amount: Avogadro's Law

Boyle's and Charles's laws both specify a fixed amount of gas. Let's see why. Figure 5.7 shows an experiment that involves two small test tubes, each fitted with a piston-cylinder assembly. We add 0.10 mol (4.4 g) of dry ice (frozen CO_2) to the first (tube A) and 0.20 mol (8.8 g) to the second (tube B). As the solid warms, it changes directly to gaseous CO_2, which expands into the cylinder and pushes up the piston. When all the solid has changed to gas and the temperature is constant, we find that cylinder A has half the volume of cylinder B. (We can neglect the volume of the tube because it is so much smaller than the volume of the cylinder.)

This experimental result shows that twice the amount (mol) of gas occupies twice the volume. Notice that, for both cylinders, the T of the gas equals room temperature and the P of the gas equals atmospheric pressure. Thus, *at fixed*

Figure 5.7 An experiment to study the relationship between the volume and amount of a gas. A, At a given external P and T, a given amount (n_1) of $CO_2(s)$ is put into the tube. When the CO_2 changes from solid to gas, it pushes up the piston until $P_{gas} = P_{atm}$, at which point it occupies a given volume of the cylinder. **B,** When twice the amount (n_2) of $CO_2(s)$ is used, twice the volume of the cylinder becomes occupied. Thus, at fixed P and T, the volume (V) of a gas is directly proportional to the amount of gas (n).

temperature and pressure, the volume occupied by a gas is directly proportional to the amount (mol) of gas:

$$V \propto n \qquad [P \text{ and } T \text{ fixed}] \qquad (5.4)$$

As *n* increases, *V* increases, and vice versa. This relationship is also expressed as

$$\frac{V}{n} = \text{constant} \qquad \text{or} \qquad V = \text{constant} \times n$$

The constant is the same for all gases at a given temperature and pressure. This relationship is another way of expressing **Avogadro's law,** which states that *at fixed temperature and pressure, equal volumes of **any** ideal gas contain equal numbers of particles (or moles).*

Many familiar phenomena are based on the relationships among volume, temperature, and amount of gas. For example, in a car engine, a reaction occurs in which fewer moles of gasoline and O_2 form more moles of CO_2 and H_2O vapor, which expand as a result of the released heat and push back the piston. Dynamite explodes because a solid decomposes rapidly to form hot gases. Dough rises in a warm room because yeast forms CO_2 bubbles in the dough, which expand during baking to give the bread a still larger volume. ●

Gas Behavior at Standard Conditions

To better understand the factors that influence gas behavior, chemists use a set of *standard conditions* called **standard temperature and pressure (STP):**

$$\text{STP:} \quad 0°C \ (273.15 \ K) \text{ and } 1 \ atm \ (760 \ torr) \qquad (5.5)$$

Under these conditions, the volume of 1 mol of an ideal gas is called the **standard molar volume:**

$$\text{Standard molar volume} = 22.4141 \ L \text{ or } 22.4 \ L \ [\text{to 3 sf}] \qquad (5.6)$$

Figure 5.8 compares the properties of three simple gases at STP.

$n = 1$ mol	$n = 1$ mol	$n = 1$ mol
$P = 1$ atm (760 torr)	$P = 1$ atm (760 torr)	$P = 1$ atm (760 torr)
$T = 0°C$ (273 K)	$T = 0°C$ (273 K)	$T = 0°C$ (273 K)
$V = 22.4$ L	$V = 22.4$ L	$V = 22.4$ L
Number of gas particles $= 6.022 \times 10^{23}$	Number of gas particles $= 6.022 \times 10^{23}$	Number of gas particles $= 6.022 \times 10^{23}$
Mass = 4.003 g	Mass = 28.02 g	Mass = 32.00 g
$d = 0.179$ g/L	$d = 1.25$ g/L	$d = 1.43$ g/L

Figure 5.8 Standard molar volume. One mole of an ideal gas occupies 22.4 L at STP (0°C and 1 atm). At STP, helium, nitrogen, oxygen, and most other simple gases behave ideally. Note that the *mass* of a gas, and thus its density (*d*), depends on its molar mass.

● **Breathing and the Gas Laws** Taking a deep breath is a combined application of the gas laws. When you inhale, muscles are coordinated such that your diaphragm moves down and your rib cage moves out. This movement increases the volume of the lungs, which decreases the pressure of the air inside them relative to that outside, so air rushes in (Boyle's). The greater amount of air stretches the elastic tissue of the lungs and expands the volume further (Avogadro's). The air also expands slightly as it warms to body temperature (Charles's). When you exhale, the diaphragm relaxes and moves up, the rib cage moves in, and the lung volume decreases. The inside air pressure becomes greater than the outside pressure, and air rushes out.

Biological application

Literature Pinkerton, K. David. "Sink or Swim: The Cartesian Diver," *J. Chem. Educ.* **2001,** *78,* 200A.

Transparency/ARIS Presentation Center, Art Library

Figure 5.9 **The volumes of 1 mol of an ideal gas and some familiar objects.** A basketball (7.5 L), 5-gal fish tank (18.9 L), 13-in television (21.6 L), and 22.4 L of He gas in a balloon.

ARIS Presentation Center, Photo Library

Literature Jensen, William B. "The Universal Gas Constant R," *J. Chem. Educ.* **2003,** *80,* 731.

Figure 5.9 compares the volumes of some familiar objects with the standard molar volume of an ideal gas.

The Ideal Gas Law

Each of the gas laws focuses on the effect that changes in one variable have on gas volume:

- Boyle's law focuses on pressure ($V \propto 1/P$).
- Charles's law focuses on temperature ($V \propto T$).
- Avogadro's law focuses on amount (mol) of gas ($V \propto n$).

We can combine these individual effects into one relationship, called the **ideal gas law** (or *ideal gas equation*):

$$V \propto \frac{nT}{P} \quad \text{or} \quad PV \propto nT \quad \text{or} \quad \frac{PV}{nT} = R$$

where R is a proportionality constant known as the **universal gas constant.** Rearranging gives the most common form of the ideal gas law:

$$PV = nRT \tag{5.7}$$

We can obtain a value of R by measuring the volume, temperature, and pressure of a given amount of gas and substituting the values into the ideal gas law. For example, using standard conditions for the gas variables, we have

$$R = \frac{PV}{nT} = \frac{1 \text{ atm} \times 22.4141 \text{ L}}{1 \text{ mol} \times 273.15 \text{ K}} = 0.082058 \frac{\text{atm·L}}{\text{mol·K}} = 0.0821 \frac{\text{atm·L}}{\text{mol·K}} \quad \text{[3 sf]} \tag{5.8}$$

This numerical value of R corresponds to the gas variables P, V, and T expressed *in these units*. R has a different numerical value when different units are used. For example, on p. 215, R has the value 8.314 J/mol·K (J stands for joule, the SI unit of energy).

Figure 5.10 makes a central point: the ideal gas law *becomes* one of the individual gas laws when two of the four variables are kept constant. When initial conditions (subscript $_1$) change to final conditions (subscript $_2$), we have

$$P_1V_1 = n_1RT_1 \quad \text{and} \quad P_2V_2 = n_2RT_2$$

Thus, $\quad \dfrac{P_1V_1}{n_1T_1} = R \quad \text{and} \quad \dfrac{P_2V_2}{n_2T_2} = R, \quad \text{so} \quad \dfrac{P_1V_1}{n_1T_1} = \dfrac{P_2V_2}{n_2T_2}$

Notice that if two of the variables remain constant, say P and T, then $P_1 = P_2$ and $T_1 = T_2$, and we obtain an expression for Avogadro's law:

$$\frac{P_1V_1}{n_1T_1} = \frac{P_2V_2}{n_2T_2} \quad \text{or} \quad \frac{V_1}{n_1} = \frac{V_2}{n_2}$$

We use rearrangements of the ideal gas law such as this one to solve gas law problems, as you'll see next. The point to remember is that there is no need to memorize the individual gas laws.

Figure 5.10 **Relationship between the ideal gas law and the individual gas laws.** Boyle's, Charles's, and Avogadro's laws are contained within the ideal gas law.

Transparency/ARIS Presentation Center, Art Library

Solving Gas Law Problems

Gas law problems are phrased in many ways, but they can usually be grouped into two main types:

Animation: Properties of Gases
ARIS Presentation Center
Chapter 5

1. *A change in one of the four variables causes a change in another, while the two remaining variables remain constant.* In this type, the ideal gas law reduces to one of the individual gas laws, and you solve for the new value of the variable. Units must be consistent, *T* must always be in kelvins, but *R* is not involved. Sample Problems 5.2 to 5.4 and 5.6 are of this type. [A variation on this type involves the combined gas law (p. 196) for simultaneous changes in two of the variables that cause a change in a third.]

2. *One variable is unknown, but the other three are known and no change occurs.* In this type, exemplified by Sample Problem 5.5, the ideal gas law is applied directly to find the unknown, and the units must conform to those in *R*.

These problems are far easier to solve if you follow a systematic approach:

- Summarize the information: identify the changing gas variables—knowns and unknown—and those held constant.
- Predict the direction of the change, and later check your answer against the prediction.
- Perform any necessary unit conversions.
- Rearrange the ideal gas law to obtain the appropriate relationship of gas variables, and solve for the unknown variable.

Sample Problems 5.2 to 5.6 apply the various gas behaviors.

SAMPLE PROBLEM 5.2 Applying the Volume-Pressure Relationship

ARIS Presentation Center,
Worked Examples Library

PROBLEM Boyle's apprentice finds that the air trapped in a J tube occupies 24.8 cm³ at 1.12 atm. By adding mercury to the tube, he increases the pressure on the trapped air to 2.64 atm. Assuming constant temperature, what is the new volume of air (in L)?

PLAN We must find the final volume (V_2) in liters, given the initial volume (V_1), initial pressure (P_1), and final pressure (P_2). The temperature and amount of gas are fixed. We convert the units of V_1 from cm³ to mL and then to L, rearrange the ideal gas law to the appropriate form, and solve for V_2. We can predict the direction of the change: since *P* increases, *V* will decrease; thus, $V_2 < V_1$. (Note that the roadmap has two parts.)

SOLUTION Summarizing the gas variables:

$P_1 = 1.12$ atm $\qquad\qquad P_2 = 2.64$ atm

$V_1 = 24.8$ cm³ (convert to L) $\qquad V_2 =$ unknown $\qquad$ *T* and *n* remain constant

Converting V_1 from cm³ to L:

$$V_1 = 24.8 \text{ cm}^3 \times \frac{1 \text{ mL}}{1 \text{ cm}^3} \times \frac{1 \text{ L}}{1000 \text{ mL}} = 0.0248 \text{ L}$$

Arranging the ideal gas law and solving for V_2: At fixed *n* and *T*, we have

$$\frac{P_1 V_1}{n_1 T_1} = \frac{P_2 V_2}{n_2 T_2} \qquad \text{or} \qquad P_1 V_1 = P_2 V_2$$

$$V_2 = V_1 \times \frac{P_1}{P_2} = 0.0248 \text{ L} \times \frac{1.12 \text{ atm}}{2.64 \text{ atm}} = 0.0105 \text{ L}$$

CHECK As we predicted, $V_2 < V_1$. Let's think about the relative values of *P* and *V* as we check the math. *P* more than doubled, so V_2 should be less than $\frac{1}{2}V_1$ (0.0105/0.0248 < $\frac{1}{2}$).

COMMENT Predicting the direction of the change provides another check on the problem setup: To make $V_2 < V_1$, we must multiply V_1 by a number *less than* 1. This means the ratio of pressures must be *less than* 1, so the larger pressure (P_2) must be in the denominator, or P_1/P_2.

FOLLOW-UP PROBLEM 5.2 A sample of argon gas occupies 105 mL at 0.871 atm. If the temperature remains constant, what is the volume (in L) at 26.3 kPa?

Literature Bare, William D.; Andrews, Lester. "A Demonstration of Ideal Gas Principles Using a Football," *J. Chem. Educ.* **1999**, *76*, 622.

ARIS Presentation Center,
Worked Examples Library

SAMPLE PROBLEM 5.3 Applying the Pressure-Temperature Relationship

PROBLEM A steel tank used for fuel delivery is fitted with a safety valve that opens if the internal pressure exceeds 1.00×10^3 torr. It is filled with methane at 23°C and 0.991 atm and placed in boiling water at exactly 100°C. Will the safety valve open?

PLAN The question "Will the safety valve open?" translates into "Is P_2 greater than 1.00×10^3 torr at T_2?" Thus, P_2 is the unknown, and T_1, T_2, and P_1 are given, with V (steel tank) and n fixed. We convert both T values to kelvins and P_1 to torrs in order to compare P_2 with the safety-limit pressure. We rearrange the ideal gas law to the appropriate form and solve for P_2. Since $T_2 > T_1$, we predict that $P_2 > P_1$.

SOLUTION Summary of gas variables:

$$P_1 = 0.991 \text{ atm (convert to torr)} \qquad P_2 = \text{unknown}$$
$$T_1 = 23°C \text{ (convert to K)} \qquad T_2 = 100°C \text{ (convert to K)}$$
$$V \text{ and } n \text{ remain constant}$$

Converting T from °C to K:

$$T_1 \text{ (K)} = 23°C + 273.15 = 296 \text{ K} \qquad T_2 \text{ (K)} = 100°C + 273.15 = 373 \text{ K}$$

Converting P from atm to torr:

$$P_1 \text{ (torr)} = 0.991 \text{ atm} \times \frac{760 \text{ torr}}{1 \text{ atm}} = 753 \text{ torr}$$

Arranging the ideal gas law and solving for P_2: At fixed n and V, we have

$$\frac{P_1 V_1}{n_1 T_1} = \frac{P_2 V_2}{n_2 T_2} \qquad \text{or} \qquad \frac{P_1}{T_1} = \frac{P_2}{T_2}$$

$$P_2 = P_1 \times \frac{T_2}{T_1} = 753 \text{ torr} \times \frac{373 \text{ K}}{296 \text{ K}} = 949 \text{ torr}$$

P_2 is less than 1.00×10^3 torr, so the valve will *not* open.

CHECK Our prediction is correct: because $T_2 > T_1$, we have $P_2 > P_1$. Thus, the temperature ratio should be >1 (T_2 in the numerator). The T ratio is about 1.25 (373/296), so the P ratio should also be about 1.25 (950/750 ≈ 1.25).

FOLLOW-UP PROBLEM 5.3
An engineer pumps air at 0°C into a newly designed piston-cylinder assembly. The volume measures 6.83 cm³. At what temperature (in K) will the volume be 9.75 cm³?

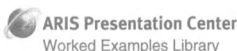
Engineering application

ARIS Presentation Center,
Worked Examples Library

SAMPLE PROBLEM 5.4 Applying the Volume-Amount Relationship

PROBLEM A scale model of a blimp rises when it is filled with helium to a volume of 55.0 dm³. When 1.10 mol of He is added to the blimp, the volume is 26.2 dm³. How many more grams of He must be added to make it rise? Assume constant T and P.

PLAN We are given the initial amount of helium (n_1), the initial volume of the blimp (V_1), and the volume needed for it to rise (V_2), and we need the additional mass of helium to make it rise. So we first need to find n_2. We rearrange the ideal gas law to the appropriate form, solve for n_2, subtract n_1 to find the additional amount ($n_{add'l}$), and then convert moles to grams. We predict that $n_2 > n_1$ because $V_2 > V_1$.

SOLUTION Summary of gas variables:

$$n_1 = 1.10 \text{ mol} \qquad n_2 = \text{unknown (find, and then subtract } n_1)$$
$$V_1 = 26.2 \text{ dm}^3 \qquad V_2 = 55.0 \text{ dm}^3$$
$$P \text{ and } T \text{ remain constant}$$

Arranging the ideal gas law and solving for n_2: At fixed P and T, we have

$$\frac{P_1 V_1}{n_1 T_1} = \frac{P_2 V_2}{n_2 T_2} \qquad \text{or} \qquad \frac{V_1}{n_1} = \frac{V_2}{n_2}$$

$$n_2 = n_1 \times \frac{V_2}{V_1} = 1.10 \text{ mol He} \times \frac{55.0 \text{ dm}^3}{26.2 \text{ dm}^3} = 2.31 \text{ mol He}$$

Engineering application

Finding the additional amount of He:

$$n_{add'l} = n_2 - n_1 = 2.31 \text{ mol He} - 1.10 \text{ mol He} = 1.21 \text{ mol He}$$

Converting moles of He to grams:

$$\text{Mass (g) of He} = 1.21 \text{ mol He} \times \frac{4.003 \text{ g He}}{1 \text{ mol He}}$$

$$= 4.84 \text{ g He}$$

CHECK Since V_2 is about twice V_1 ($55/26 \approx 2$), n_2 should be about twice n_1 ($2.3/1.1 \approx 2$). Since $n_2 > n_1$, we were right to multiply n_1 by a number >1 (that is, V_2/V_1). About $1.2 \text{ mol} \times 4 \text{ g/mol} \approx 4.8 \text{ g}$.

COMMENT 1. A different sequence of steps will give you the same answer: first find the additional volume ($V_{add'l} = V_2 - V_1$), and then solve directly for $n_{add'l}$. Try it for yourself.

2. You saw that Charles's law ($V \propto T$ at fixed P and n) translates into a similar relationship between P and T at fixed V and n. The follow-up problem demonstrates that Avogadro's law ($V \propto n$ at fixed P and T) translates into an analogous relationship at fixed V and T.

FOLLOW-UP PROBLEM 5.4 A rigid plastic container holds 35.0 g of ethylene gas (C_2H_4) at a pressure of 793 torr. What is the pressure if 5.0 g of ethylene is removed at constant temperature?

ARIS Presentation Center,
Worked Examples Library

Engineering application

SAMPLE PROBLEM 5.5 Solving for an Unknown Gas Variable at Fixed Conditions

PROBLEM A steel tank has a volume of 438 L and is filled with 0.885 kg of O_2. Calculate the pressure of O_2 at 21°C.

PLAN We are given V, T, and the mass of O_2, and we must find P. Since conditions are not changing, we apply the ideal gas law without rearranging it. We use the given V in liters, convert T to kelvins and mass of O_2 to moles, and solve for P.

SOLUTION Summary of gas variables:

$$V = 438 \text{ L} \qquad\qquad T = 21°C \text{ (convert to K)}$$
$$n = 0.885 \text{ kg } O_2 \text{ (convert to mol)} \qquad P = \text{unknown}$$

Converting T from °C to K:

$$T \text{ (K)} = 21°C + 273.15 = 294 \text{ K}$$

Converting from mass of O_2 to moles:

$$n = \text{mol of } O_2 = 0.885 \text{ kg } O_2 \times \frac{1000 \text{ g}}{1 \text{ kg}} \times \frac{1 \text{ mol } O_2}{32.00 \text{ g } O_2} = 27.7 \text{ mol } O_2$$

Solving for P (note the unit canceling here):

$$P = \frac{nRT}{V} = \frac{27.7 \text{ mol} \times 0.0821 \frac{\text{atm·L}}{\text{mol·K}} \times 294 \text{ K}}{438 \text{ L}}$$

$$= 1.53 \text{ atm}$$

CHECK The amount of O_2 seems correct: $\sim 900 \text{ g}/(30 \text{ g/mol}) = 30 \text{ mol}$. To check the approximate size of the final calculation, round off the values, including that for R:

$$P = \frac{30 \text{ mol } O_2 \times 0.1 \frac{\text{atm·L}}{\text{mol·K}} \times 300 \text{ K}}{450 \text{ L}} = 2 \text{ atm}$$

which is reasonably close to 1.53 atm.

FOLLOW-UP PROBLEM 5.5 The tank in the sample problem develops a slow leak that is discovered and sealed. The new measured pressure is 1.37 atm. How many grams of O_2 remain?

Finally, in a slightly different type of problem that depicts a simple laboratory scene, we apply the gas laws to determine the correct balanced equation for a process.

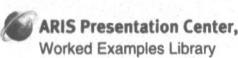

SAMPLE PROBLEM 5.6   Using Gas Laws to Determine a Balanced Equation

PROBLEM The piston-cylinders depicted below contain a gaseous reaction carried out at constant pressure. Before the reaction, the temperature is 150 K; when it is complete, the temperature is 300 K.

Before
150 K

After
300 K

Which of the following balanced equations describes the reaction?

(1) $A_2(g) + B_2(g) \longrightarrow 2AB(g)$ (2) $2AB(g) + B_2(g) \longrightarrow 2AB_2(g)$

(3) $A(g) + B_2(g) \longrightarrow AB_2(g)$ (4) $2AB_2(g) \longrightarrow A_2(g) + 2B_2(g)$

PLAN We are shown a depiction of a gaseous reaction and must choose the balanced equation. The problem says that P is constant, and the pictures show that T doubles and V stays the same. If n were also constant, the gas laws tell us that V should double when T doubles. Therefore, n cannot be constant, and the only way to maintain V with P constant and T doubling is for n to be halved. So we examine the four balanced equations and count the number of moles on each side to see in which equation n is halved.

SOLUTION In equation (1), n does not change, so doubling T would double V.

In equation (2), n decreases from 3 mol to 2 mol, so doubling T would increase V by one-third.

In equation (3), n decreases from 2 mol to 1 mol. Doubling T would exactly balance the decrease from halving n, so V would stay the same.

In equation (4), n increases, so doubling T would more than double V.

Equation (3) is correct:

$$A(g) + B_2(g) \longrightarrow AB_2(g)$$

FOLLOW-UP PROBLEM 5.6
The gaseous reaction in the piston-cylinders depicted below is carried out at constant pressure and an initial temperature of $-73°C$:

Before
$-73°C$

After
$?°C$

The *unbalanced* equation is $CD(g) \longrightarrow C_2(g) + D_2(g)$. What is the final temperature (in °C)?

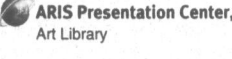

Section Summary

Four variables define the physical behavior of an ideal gas: volume (V), pressure (P), temperature (T), and amount (number of moles, n). • Most simple gases display nearly ideal behavior at ordinary temperatures and pressures. • Boyle's, Charles's, and Avogadro's laws relate volume to pressure, to temperature, and to amount of gas, respectively. • At STP (0°C and 1 atm), 1 mol of an ideal gas occupies 22.4 L. • The ideal gas law incorporates the individual gas laws into one equation: $PV = nRT$, where R is the universal gas constant.

Literature Ramette, Richard W. "Why Does **203** a Helium-Filled Balloon "Rise"? *J. Chem. Educ.* **2003**, *80*, 1149.

5.4 FURTHER APPLICATIONS OF THE IDEAL GAS LAW

The ideal gas law can be recast in additional ways to determine other properties of gases. In this section, we use it to find gas density, molar mass, and the partial pressure of each gas in a mixture.

The Density of a Gas

One mole of any gas occupies nearly the same volume at a given temperature and pressure, so differences in gas density ($d = m/V$) depend on differences in molar mass (see Figure 5.8). For example, at STP, 1 mol of O_2 occupies the same volume as 1 mol of N_2, but since each O_2 molecule has a greater mass than each N_2 molecule, O_2 is denser.

All gases are miscible when thoroughly mixed, but in the absence of mixing, a less dense gas will lie above a more dense one. ● There are many familiar examples of this phenomenon. Some types of fire extinguishers release CO_2 because it is denser than air and will sink onto the fire, preventing more O_2 from reaching the burning material. Enormous air masses of different densities and temperatures moving past each other around the globe give rise to much of our weather.

We can rearrange the ideal gas law to calculate the density of a gas from its molar mass. Recall that the number of moles (n) is the mass (m) divided by the molar mass ($\mathcal{M}$), $n = m/\mathcal{M}$. Substituting for n in the ideal gas law gives

$$PV = \frac{m}{\mathcal{M}}RT$$

Rearranging to isolate m/V gives

$$\frac{m}{V} = d = \frac{\mathcal{M} \times P}{RT} \tag{5.9}$$

Two important ideas are expressed by Equation 5.9:

- *The density of a gas is directly proportional to its molar mass* because a given amount of a heavier gas occupies the same volume as that amount of a lighter gas (Avogadro's law).
- *The density of a gas is inversely proportional to the temperature.* As the volume of a gas increases with temperature (Charles's law), the same mass occupies more space; thus, the density is lower.

Architectural designers and heating engineers apply the second idea when they place heating ducts near the floor of a room: the less dense warm air from the ducts rises and heats the room air. Safety experts recommend staying near the floor when escaping from a fire to avoid the hot, and therefore less dense, noxious gases. We use Equation 5.9 to find the density of a gas at any temperature and pressure near standard conditions.

● Gas Density and Human Disasters Many gases that are denser than air have been involved in natural and human-caused disasters. The dense gases in smog that blanket urban centers, such as Mexico City (see photo), contribute greatly to respiratory illness. In World War I, poisonous phosgene gas ($COCl_2$) was used against ground troops as they lay in trenches. In 1984, the unintentional release of methylisocyanate from a Union Carbide India Ltd. chemical plant in Bhopal, India, killed thousands of people as vapors spread from the outskirts into the city. In 1986 in Cameroon, CO_2 released naturally from Lake Nyos suffocated thousands as it flowed down valleys into villages. Some paleontologists suggest that a similar process in volcanic lakes may have contributed to dinosaur kills.

ARIS Presentation Center, Photo Library

Environmental application

Engineering application

ARIS Presentation Center, Worked Examples Library

Green chemistry application

SAMPLE PROBLEM 5.7 Calculating Gas Density

PROBLEM To apply a green chemistry approach, a chemical engineer uses waste CO_2 from a manufacturing process, instead of chlorofluorocarbons, as a "blowing agent" in the production of polystyrene containers. Find the density (in g/L) of CO_2 and the number of molecules per liter **(a)** at STP (0°C and 1 atm) and **(b)** at room conditions (20.°C and 1.00 atm).

PLAN We must find the density (d) and number of molecules of CO_2, given the two sets of P and T data. We find $\mathcal{M}$, convert T to kelvins, and calculate d with Equation 5.9. Then we convert the mass per liter to molecules per liter with Avogadro's number.

SOLUTION **(a)** Density and molecules per liter of CO_2 at STP. Summary of gas properties:

$$T = 0°C + 273.15 = 273 \text{ K} \qquad P = 1 \text{ atm} \qquad \mathcal{M} \text{ of } CO_2 = 44.01 \text{ g/mol}$$

Calculating density (note the unit canceling here):

$$d = \frac{\mathcal{M} \times P}{RT} = \frac{44.01 \text{ g/mol} \times 1.00 \text{ atm}}{0.0821 \dfrac{\text{atm·L}}{\text{mol·K}} \times 273 \text{ K}} = 1.96 \text{ g/L}$$

Converting from mass/L to molecules/L:

$$\text{Molecules } CO_2/L = \frac{1.96 \text{ g } CO_2}{1 \text{ L}} \times \frac{1 \text{ mol } CO_2}{44.01 \text{ g } CO_2} \times \frac{6.022 \times 10^{23} \text{ molecules } CO_2}{1 \text{ mol } CO_2}$$

$$= 2.68 \times 10^{22} \text{ molecules } CO_2/L$$

(b) Density and molecules of CO_2 per liter at room conditions. Summary of gas properties:

$$T = 20.°C + 273.15 = 293 \text{ K} \qquad P = 1.00 \text{ atm} \qquad \mathcal{M} \text{ of } CO_2 = 44.01 \text{ g/mol}$$

Calculating density:

$$d = \frac{\mathcal{M} \times P}{RT} = \frac{44.01 \text{ g/mol} \times 1.00 \text{ atm}}{0.0821 \dfrac{\text{atm·L}}{\text{mol·K}} \times 293 \text{ K}} = 1.83 \text{ g/L}$$

Converting from mass/L to molecules/L:

$$\text{Molecules } CO_2/L = \frac{1.83 \text{ g } CO_2}{1 \text{ L}} \times \frac{1 \text{ mol } CO_2}{44.01 \text{ g } CO_2} \times \frac{6.022 \times 10^{23} \text{ molecules } CO_2}{1 \text{ mol } CO_2}$$

$$= 2.50 \times 10^{22} \text{ molecules } CO_2/L$$

CHECK Round off to check the density values; for example, in (a), at STP:

$$\frac{50 \text{ g/mol} \times 1 \text{ atm}}{0.1 \dfrac{\text{atm·L}}{\text{mol·K}} \times 250 \text{ K}} = 2 \text{ g/L} \approx 1.96 \text{ g/L}$$

At the higher temperature in (b), the density should decrease, which can happen only if there are fewer molecules per liter, so the answer is reasonable.

COMMENT 1. An *alternative approach* for finding the density of most simple gases, but *at STP only,* is to divide the molar mass by the standard molar volume, 22.4 L:

$$d = \frac{\mathcal{M}}{V} = \frac{44.01 \text{ g/mol}}{22.4 \text{ L/mol}} = 1.96 \text{ g/L}$$

Once you know the density at one temperature (0°C), you can find it at any other temperature with the following relationship: $d_1/d_2 = T_2/T_1$.
2. Note that we have different numbers of significant figures for the pressure values. In (a), "1 atm" is part of the definition of STP, so it is an exact number. In (b), we specified "1.00 atm" to allow three significant figures in the answer.
3. Hot-air balloonists have always applied the change in density with temperature. ●

FOLLOW-UP PROBLEM 5.7 Compare the density of CO_2 at 0°C and 380 torr with its density at STP.

● **Up, Up, and Away!** When the gas in a hot-air balloon is heated, its volume increases and the balloon inflates. Further heating causes some of the gas to escape. By these means, the gas density decreases and the balloon rises. Two pioneering hot-air balloonists used their knowledge of gas behavior to excel at their hobby. Jacques Charles (of Charles's law) made one of the first balloon flights, in 1783. Twenty years later, Joseph Gay-Lussac (who studied the pressure-temperature relationship) set a solo altitude record that held for 50 years.

Demo Shakhashiri, Bassam Z. "Determination of the Molecular Mass of the Gas from a Butane Lighter," *Chemical Demonstrations, A Handbook for Teachers of Chemistry,* Vol. 2 (The University of Wisconsin Press, Madison, 1985), 48–50.

The Molar Mass of a Gas

Through another simple rearrangement of the ideal gas law, we can determine the molar mass of an unknown gas or volatile liquid (one that is easily vaporized):

$$n = \frac{m}{\mathcal{M}} = \frac{PV}{RT} \qquad \text{so} \qquad \mathcal{M} = \frac{mRT}{PV} \qquad \text{or} \qquad \mathcal{M} = \frac{dRT}{P} \qquad \textbf{(5.10)}$$

Notice that this equation is just a rearrangement of Equation 5.9.

The French chemist J. B. A. Dumas (1800–1884) pioneered an ingenious method for finding the molar mass of a volatile liquid. Figure 5.11 shows the apparatus. Place a small volume of the liquid in a preweighed flask of known volume. Close the flask with a stopper that contains a narrow tube and immerse it in a water bath whose fixed temperature exceeds the liquid's boiling point. As the liquid vaporizes, the gas fills the flask and some flows out the tube. When the liquid is gone, the pressure of the gas filling the flask equals the atmospheric pressure. Remove the flask from the water bath and cool it, and the gas condenses to a liquid. Reweigh the flask to obtain the mass of the liquid, *which equals the mass of gas* that remained in the flask.

By this procedure, you have directly measured all the variables needed to calculate the molar mass of the gas: the mass of gas (m) occupies the flask volume (V) at a pressure (P) equal to the barometric pressure and at the temperature (T) of the water bath.

ARIS Presentation Center, Worked Examples Library

SAMPLE PROBLEM 5.8 Finding the Molar Mass of a Volatile Liquid

PROBLEM An organic chemist isolates a colorless liquid from a petroleum sample. She uses the Dumas method and obtains the following data:

  Volume (V) of flask = 213 mL T = 100.0°C P = 754 torr

  Mass of flask + gas = 78.416 g   Mass of flask = 77.834 g

Calculate the molar mass of the liquid.

PLAN We are given V, T, P, and mass data and must find the molar mass ($\mathcal{M}$) of the liquid. We convert V to liters, T to kelvins, and P to atmospheres, find the mass of gas by subtracting the mass of the empty flask, and use Equation 5.10 to solve for $\mathcal{M}$.

SOLUTION Summary of gas variables:

$$m = 78.416 \text{ g} - 77.834 \text{ g} = 0.582 \text{ g} \qquad P \text{ (atm)} = 754 \text{ torr} \times \frac{1 \text{ atm}}{760 \text{ torr}} = 0.992 \text{ atm}$$

$$V \text{ (L)} = 213 \text{ mL} \times \frac{1 \text{ L}}{1000 \text{ mL}} = 0.213 \text{ L} \qquad T \text{ (K)} = 100.0°C + 273.15 = 373.2 \text{ K}$$

Solving for $\mathcal{M}$:

$$\mathcal{M} = \frac{mRT}{PV} = \frac{0.582 \text{ g} \times 0.0821 \frac{\text{atm·L}}{\text{mol·K}} \times 373.2 \text{ K}}{0.992 \text{ atm} \times 0.213 \text{ L}} = 84.4 \text{ g/mol}$$

CHECK Rounding to check the arithmetic, we have

$$\frac{0.6 \text{ g} \times 0.08 \frac{\text{atm·L}}{\text{mol·K}} \times 375 \text{ K}}{1 \text{ atm} \times 0.2 \text{ L}} = 90 \text{ g/mol} \qquad \text{which is close to 84.4 g/mol}$$

FOLLOW-UP PROBLEM 5.8 At 10.0°C and 102.5 kPa, the density of dry air is 1.26 g/L. What is the average "molar mass" of dry air at these conditions?

Figure 5.11 Determining the molar mass of an unknown volatile liquid. A small amount of unknown liquid is vaporized, and the gas fills the flask of known volume at the known temperature of the bath. Excess gas escapes through the capillary tube until $P_{\text{gas}} = P_{\text{atm}}$. When the flask is cooled, the gas condenses, the liquid is weighed, and the ideal gas law is used to calculate $\mathcal{M}$ (see text).

 Transparency/ARIS Presentation Center, Art Library

Demo Shakhashiri, Bassam Z. "Determination of the Molecular Mass of a Volatile Liquid," *Chemical Demonstrations, A Handbook for Teachers of Chemistry*, Vol. 2 (The University of Wisconsin Press, Madison, 1985), 51–54.

The Partial Pressure of a Gas in a Mixture of Gases

All of the behaviors we've discussed so far were observed from experiments with air, which is a complex mixture of gases. The ideal gas law holds for virtually any gas, whether pure or a mixture, at ordinary conditions for two reasons:

- Gases mix homogeneously (form a solution) in any proportions.
- Each gas in a mixture behaves as if it were the only gas present (assuming no chemical interactions).

Dalton's Law of Partial Pressures The second point above was discovered by John Dalton in his lifelong study of humidity. He observed that when water vapor is added to dry air, the total air pressure increases by an increment equal to the pressure of the water vapor:

$$P_{\text{humid air}} = P_{\text{dry air}} + P_{\text{added water vapor}}$$

In other words, each gas in the mixture exerts a **partial pressure,** a portion of the total pressure of the mixture, that is the same as the pressure it would exert

Demo Shakhashiri, Bassam Z. "Dalton's Law of Partial Pressures," *Chemical Demonstrations, A Handbook for Teachers of Chemistry*, Vol. 2 (The University of Wisconsin Press, Madison, 1985), 41–43.

by itself. This observation is formulated as **Dalton's law of partial pressures:** *in a mixture of unreacting gases, the total pressure is the sum of the partial pressures of the individual gases:*

$$P_{total} = P_1 + P_2 + P_3 + \cdots \tag{5.11}$$

As an example, suppose you have a tank of fixed volume that contains nitrogen gas at a certain pressure, and you introduce a sample of hydrogen gas into the tank. Each gas behaves independently, so we can write an ideal gas law expression for each:

$$P_{N_2} = \frac{n_{N_2}RT}{V} \quad \text{and} \quad P_{H_2} = \frac{n_{H_2}RT}{V}$$

Because each gas occupies the same total volume and is at the same temperature, the pressure of each gas depends only on its amount, *n*. Thus, the total pressure is

$$P_{total} = P_{N_2} + P_{H_2} = \frac{n_{N_2}RT}{V} + \frac{n_{H_2}RT}{V} = \frac{(n_{N_2} + n_{H_2})RT}{V} = \frac{n_{total}RT}{V}$$

where $n_{total} = n_{N_2} + n_{H_2}$.

Each component in a mixture contributes a fraction of the total number of moles in the mixture, which is the **mole fraction (X)** of that component. Multiplying *X* by 100 gives the mole percent. Keep in mind that the sum of the mole fractions of all components in any mixture must be 1, and the sum of the mole percents must be 100%. For N_2, the mole fraction is

$$X_{N_2} = \frac{n_{N_2}}{n_{total}} = \frac{n_{N_2}}{n_{N_2} + n_{H_2}}$$

Since the total pressure is due to the total number of moles, the partial pressure of gas A is the total pressure multiplied by the mole fraction of A, X_A:

$$P_A = X_A \times P_{total} \tag{5.12}$$

Equation 5.12 is a very important result. To see that it is valid for the mixture of N_2 and H_2, we recall that $X_{N_2} + X_{H_2} = 1$ and obtain

$$P_{total} = P_{N_2} + P_{H_2} = (X_{N_2} \times P_{total}) + (X_{H_2} \times P_{total}) = (X_{N_2} + X_{H_2})P_{total} = 1 \times P_{total}$$

ARIS Presentation Center,
Worked Examples Library

Biological application

Mole % of $^{18}O_2$

divide by 100

Mole fraction, $X_{^{18}O_2}$

multiply by P_{total}

Partial pressure, $P_{^{18}O_2}$

SAMPLE PROBLEM 5.9 Applying Dalton's Law of Partial Pressures

PROBLEM In a study of O_2 uptake by muscle at high altitude, a physiologist prepares an atmosphere consisting of 79 mole % N_2, 17 mole % $^{16}O_2$, and 4.0 mole % $^{18}O_2$. (The isotope ^{18}O will be measured to determine O_2 uptake.) The total pressure is 0.75 atm to simulate high altitude. Calculate the mole fraction and partial pressure of $^{18}O_2$ in the mixture.
PLAN We must find $X_{^{18}O_2}$ and $P_{^{18}O_2}$ from P_{total} (0.75 atm) and the mole % of $^{18}O_2$ (4.0). Dividing the mole % by 100 gives the mole fraction, $X_{^{18}O_2}$. Then, using Equation 5.12, we multiply $X_{^{18}O_2}$ by P_{total} to find $P_{^{18}O_2}$.
SOLUTION Calculating the mole fraction of $^{18}O_2$:

$$X_{^{18}O_2} = \frac{4.0 \text{ mol } \% \ ^{18}O_2}{100} = 0.040$$

Solving for the partial pressure of $^{18}O_2$:

$$P_{^{18}O_2} = X_{^{18}O_2} \times P_{total} = 0.040 \times 0.75 \text{ atm} = 0.030 \text{ atm}$$

CHECK $X_{^{18}O_2}$ is small because the mole % is small, so $P_{^{18}O_2}$ should be small also.
COMMENT At high altitudes, specialized brain cells that are sensitive to O_2 and CO_2 levels in the blood trigger an increase in rate and depth of breathing for several days, until a person becomes acclimated.

FOLLOW-UP PROBLEM 5.9 To prevent the presence of air, noble gases are placed over highly reactive chemicals to act as inert "blanketing" gases. A chemical engineer places a mixture of noble gases consisting of 5.50 g of He, 15.0 g of Ne, and 35.0 g of Kr in a piston-cylinder assembly at STP. Calculate the partial pressure of each gas.

① Water-insoluble gaseous product bubbles through water into collection vessel

② P_{gas} adds to vapor pressure of water (P_{H_2O}) to give P_{total}. As shown $P_{total} < P_{atm}$

③ P_{total} is made equal to P_{atm} by adjusting height of vessel until water level equals that in beaker

④ P_{total} equals P_{gas} plus P_{H_2O} at temperature of experiment. Therefore, $P_{gas} = P_{total} - P_{H_2O}$

Figure 5.12 Collecting a water-insoluble gaseous product and determining its pressure.

ARIS Presentation Center, Active Art Library

Collecting a Gas over Water The law of partial pressures is frequently used to determine the yield of a water-insoluble gas formed in a reaction. The gaseous product bubbles through water and is collected into an inverted container, as shown in Figure 5.12. The water vapor that mixes with the gas contributes a portion of the total pressure, called the *vapor pressure*, which depends only on the water temperature.

In order to determine the yield of gaseous product, we find the appropriate vapor pressure value from a list, such as the one in Table 5.3, and subtract it from the total gas pressure (corrected to barometric pressure) to get the partial pressure of the gaseous product. With V and T known, we can calculate the amount of product.

ARIS Presentation Center, Worked Examples Library

Table 5.3 Vapor Pressure of Water (P_{H_2O}) at Different T

T (°C)	P (torr)
0	4.6
5	6.5
10	9.2
12	10.5
14	12.0
16	13.6
18	15.5
20	17.5
22	19.8
24	22.4
26	25.2
28	28.3
30	31.8
35	42.2
40	55.3
45	71.9
50	92.5
55	118.0
60	149.4
65	187.5
70	233.7
75	289.1
80	355.1
85	433.6
90	525.8
95	633.9
100	760.0

ARIS Presentation Center, Tables Library

SAMPLE PROBLEM 5.10 Calculating the Amount of Gas Collected over Water

PROBLEM Acetylene (C_2H_2), an important fuel in welding, is produced in the laboratory when calcium carbide (CaC_2) reacts with water:

$$CaC_2(s) + 2H_2O(l) \longrightarrow C_2H_2(g) + Ca(OH)_2(aq)$$

For a sample of acetylene collected over water, total gas pressure (adjusted to barometric pressure) is 738 torr and the volume is 523 mL. At the temperature of the gas (23°C), the vapor pressure of water is 21 torr. How many grams of acetylene are collected?

PLAN In order to find the mass of C_2H_2, we first need to find the number of moles of C_2H_2, $n_{C_2H_2}$, which we can obtain from the ideal gas law by calculating $P_{C_2H_2}$. The barometer reading gives us P_{total}, which is the sum of $P_{C_2H_2}$ and P_{H_2O}, and we are given P_{H_2O}, so we subtract to find $P_{C_2H_2}$. We are also given V and T, so we convert to consistent units, and find $n_{C_2H_2}$ from the ideal gas law. Then we convert moles to grams using the molar mass from the formula.

SOLUTION Summary of gas variables:

$$P_{C_2H_2} \text{ (torr)} = P_{total} - P_{H_2O} = 738 \text{ torr} - 21 \text{ torr} = 717 \text{ torr}$$

$$P_{C_2H_2} \text{ (atm)} = 717 \text{ torr} \times \frac{1 \text{ atm}}{760 \text{ torr}} = 0.943 \text{ atm}$$

$$V \text{ (L)} = 523 \text{ mL} \times \frac{1 \text{ L}}{1000 \text{ mL}} = 0.523 \text{ L}$$

$$T \text{ (K)} = 23°C + 273.15 = 296 \text{ K}$$

$$n_{C_2H_2} = \text{unknown}$$

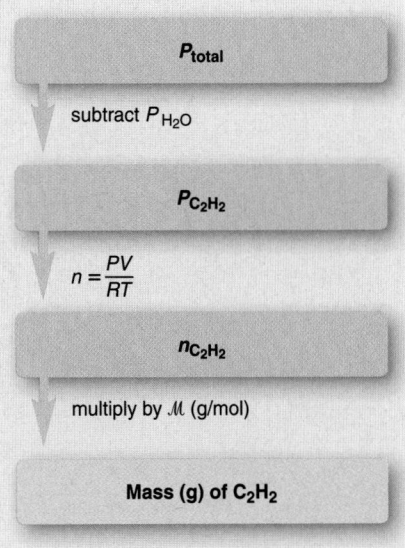

P_{total}

subtract P_{H_2O}

$P_{C_2H_2}$

$n = \dfrac{PV}{RT}$

$n_{C_2H_2}$

multiply by $\mathcal{M}$ (g/mol)

Mass (g) of C_2H_2

Solving for $n_{C_2H_2}$:

$$n_{C_2H_2} = \frac{PV}{RT} = \frac{0.943 \text{ atm} \times 0.523 \text{ L}}{0.0821 \frac{\text{atm·L}}{\text{mol·K}} \times 296 \text{ K}} = 0.0203 \text{ mol}$$

Converting $n_{C_2H_2}$ to mass:

$$\text{Mass(g) of } C_2H_2 = 0.0203 \text{ mol } C_2H_2 \times \frac{26.04 \text{ g } C_2H_2}{1 \text{ mol } C_2H_2}$$

$$= 0.529 \text{ g } C_2H_2$$

CHECK Rounding to one significant figure, a quick arithmetic check for n gives

$$n \approx \frac{1 \text{ atm} \times 0.5 \text{ L}}{0.08 \frac{\text{atm·L}}{\text{mol·K}} \times 300 \text{ K}} = 0.02 \text{ mol} \approx 0.0203 \text{ mol}$$

COMMENT The C_2^{2-} ion (called the *carbide*, or *acetylide, ion*) is an interesting anion. It is simply $^-C \equiv C^-$, which acts as a base in water, removing an H^+ ion from two H_2O molecules to form acetylene, $H—C \equiv C—H$.

FOLLOW-UP PROBLEM 5.10 A small piece of zinc reacts with dilute HCl to form H_2, which is collected over water at 16°C into a large flask. The total pressure is adjusted to barometric pressure (752 torr), and the volume is 1495 mL. Use Table 5.3 to help calculate the partial pressure and mass of H_2.

Section Summary

The ideal gas law can be rearranged to calculate the density and molar mass of a gas. • In a mixture of gases, each component contributes its own partial pressure to the total pressure (Dalton's law of partial pressures). The mole fraction of each component is the ratio of its partial pressure to the total pressure. • When a gas is in contact with water, the total pressure is the sum of the gas pressure and the vapor pressure of water at the given temperature.

5.5 THE IDEAL GAS LAW AND REACTION STOICHIOMETRY

In Chapters 3 and 4, we encountered many reactions that involved gases as reactants (e.g., combustion with O_2) or as products (e.g., acid treatment of a carbonate). From the balanced equation, we used stoichiometrically equivalent molar ratios to calculate the amounts (moles) of reactants and products and converted these quantities into masses, numbers of molecules, or solution volumes (see Figures 3.12, p. 122, and 3.15, p. 128). Figure 5.13 shows how you can expand your problem-solving repertoire by using the ideal gas law to convert between gas variables (P, T, and V) and amounts (moles) of gaseous reactants and products. In effect, you combine a gas law problem with a stoichiometry problem; it is more realistic to measure the volume, pressure, and temperature of a gas than its mass.

Figure 5.13 Summary of the stoichiometric relationships among the amount (mol, *n*) of gaseous reactant or product and the gas variables pressure (*P*), volume (*V*), and temperature (*T*).

SAMPLE PROBLEM 5.11 Using Gas Variables to Find Amounts of Reactants or Products

PROBLEM Copper dispersed in absorbent beds is used to react with oxygen impurities in the ethylene used for producing polyethylene. The beds are regenerated when hot H_2 reduces the metal oxide, forming the pure metal and H_2O. On a laboratory scale, what volume of H_2 at 765 torr and 225°C is needed to reduce 35.5 g of copper(II) oxide?

PLAN This is a stoichiometry *and* gas law problem. To find V_{H_2}, we first need n_{H_2}. We write and balance the equation. Next, we convert the given mass of CuO (35.5 g) to amount (mol) and use the molar ratio to find moles of H_2 needed (stoichiometry portion). Then, we use the ideal gas law to convert moles of H_2 to liters (gas law portion). A roadmap is shown, but you are familiar with all the steps.

SOLUTION Writing the balanced equation:

$$CuO(s) + H_2(g) \longrightarrow Cu(s) + H_2O(g)$$

Calculating n_{H_2}:

$$n_{H_2} = 35.5 \text{ g CuO} \times \frac{1 \text{ mol CuO}}{79.55 \text{ g CuO}} \times \frac{1 \text{ mol } H_2}{1 \text{ mol CuO}} = 0.446 \text{ mol } H_2$$

Summary of other gas variables:

$$V = \text{unknown} \qquad P \text{ (atm)} = 765 \text{ torr} \times \frac{1 \text{ atm}}{760 \text{ torr}} = 1.01 \text{ atm}$$

$$T \text{ (K)} = 225°C + 273.15 = 498 \text{ K}$$

Solving for V_{H_2}:

$$V = \frac{nRT}{P} = \frac{0.446 \text{ mol} \times 0.0821 \frac{\text{atm·L}}{\text{mol·K}} \times 498 \text{ K}}{1.01 \text{ atm}} = \boxed{18.1 \text{ L}}$$

CHECK One way to check the answer is to compare it with the molar volume of an ideal gas at STP (22.4 L at 273.15 K and 1 atm). One mole of H_2 at STP occupies about 22 L, so less than 0.5 mol occupies less than 11 L. T is less than twice 273 K, so V should be less than twice 11 L.

COMMENT The main point here is that the stoichiometry provides one gas variable (n), two more are given, and the ideal gas law is used to find the fourth.

FOLLOW-UP PROBLEM 5.11 Sulfuric acid reacts with sodium chloride to form aqueous sodium sulfate and hydrogen chloride gas. How many milliliters of gas form at STP when 0.117 kg of sodium chloride reacts with excess sulfuric acid?

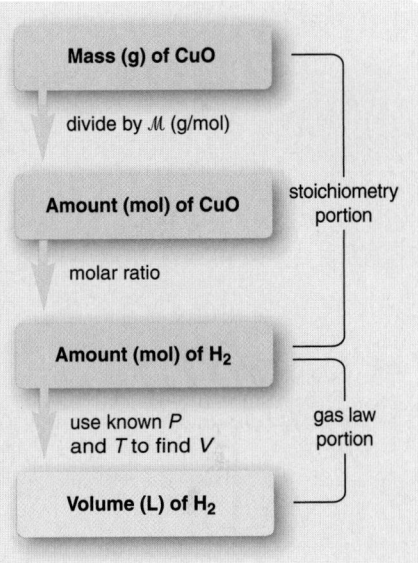

Mass (g) of CuO

divide by $\mathcal{M}$ (g/mol)

Amount (mol) of CuO

molar ratio

Amount (mol) of H_2

use known P and T to find V

Volume (L) of H_2

stoichiometry portion

gas law portion

SAMPLE PROBLEM 5.12 Using the Ideal Gas Law in a Limiting-Reactant Problem

PROBLEM The alkali metals [Group 1A(1)] react with the halogens [Group 7A(17)] to form ionic metal halides. What mass of potassium chloride forms when 5.25 L of chlorine gas at 0.950 atm and 293 K reacts with 17.0 g of potassium (see photo)?

PLAN The only difference between this and previous limiting-reactant problems (see Sample Problem 3.13, p. 117) is that here we use the ideal gas law to find the amount (n) of gaseous reactant from the known V, P, and T. We first write the balanced equation and then use it to find the limiting reactant and the amount of product.

SOLUTION Writing the balanced equation:

$$2K(s) + Cl_2(g) \longrightarrow 2KCl(s)$$

Summary of gas variables:

$$P = 0.950 \text{ atm} \qquad V = 5.25 \text{ L}$$
$$T = 293 \text{ K} \qquad n = \text{unknown}$$

Solving for n_{Cl_2}:

$$n_{Cl_2} = \frac{PV}{RT} = \frac{0.950 \text{ atm} \times 5.25 \text{ L}}{0.0821 \frac{\text{atm·L}}{\text{mol·K}} \times 293 \text{ K}} = 0.207 \text{ mol}$$

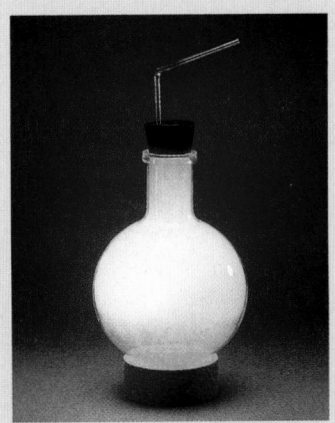

Chlorine gas reacting with potassium.

Converting from grams of potassium (K) to moles:

$$\text{Moles of K} = 17.0 \text{ g K} \times \frac{1 \text{ mol K}}{39.10 \text{ g K}} = 0.435 \text{ mol K}$$

Determining the limiting reactant: If Cl_2 is limiting,

$$\text{Moles of KCl} = 0.207 \text{ mol } Cl_2 \times \frac{2 \text{ mol KCl}}{1 \text{ mol } Cl_2} = 0.414 \text{ mol KCl}$$

If K is limiting,

$$\text{Moles of KCl} = 0.435 \text{ mol K} \times \frac{2 \text{ mol KCl}}{2 \text{ mol K}} = 0.435 \text{ mol KCl}$$

Cl_2 is the limiting reactant because it forms less KCl.
Converting from moles of KCl to grams:

$$\text{Mass (g) of KCl} = 0.414 \text{ mol KCl} \times \frac{74.55 \text{ g KCl}}{1 \text{ mol KCl}} = 30.9 \text{ g KCl}$$

CHECK The gas law calculation seems correct. At STP, 22 L of Cl_2 gas contains about 1 mol, so a 5-L volume would contain a bit less than 0.25 mol of Cl_2. Moreover, since P (in numerator) is slightly lower than STP and T (in denominator) is slightly higher than STP, these should lower the calculated n further below the ideal value. The mass of KCl seems correct: less than 0.5 mol of KCl gives $<0.5 \times \mathcal{M}$ (30.9 g $< 0.5 \times 75$ g).

FOLLOW-UP PROBLEM 5.12 Ammonia and hydrogen chloride gases react to form solid ammonium chloride. A 10.0-L reaction flask contains ammonia at 0.452 atm and 22°C, and 155 mL of hydrogen chloride gas at 7.50 atm and 271 K is introduced. After the reaction occurs and the temperature returns to 22°C, what is the pressure inside the flask? (Neglect the volume of the solid product.)

Section Summary

By converting the variables P, V, and T of gaseous reactants (or products) to amount (n, mol), we can solve stoichiometry problems for gaseous reactions.

5.6 THE KINETIC-MOLECULAR THEORY: A MODEL FOR GAS BEHAVIOR

So far we have discussed observations of macroscopic samples of gas: decreasing cylinder volume, increasing tank pressure, and so forth. This section presents the central model that explains macroscopic gas behavior at the level of individual particles: the **kinetic-molecular theory.** The theory draws conclusions through mathematical derivations, but here our discussion will be largely qualitative.

How the Kinetic-Molecular Theory Explains the Gas Laws

Developed by some of the great scientists of the 19th century, most notably James Clerk Maxwell and Ludwig Boltzmann, the kinetic-molecular theory was able to explain the gas laws that some of the great scientists of the century before had arrived at empirically.

Questions Concerning Gas Behavior To model gas behavior, we must rationalize certain questions at the molecular level:

1. *Origin of pressure.* Pressure is a measure of the force a gas exerts on a surface. How do individual gas particles create this force?
2. *Boyle's law ($V \propto 1/P$).* A change in gas pressure in one direction causes a change in gas volume in the other. What happens to the particles when external pressure compresses the gas volume? And why aren't liquids and solids compressible?

Literature Ball, David W. "Why Does Helium Have 92% of the Lifting Power of Hydrogen If It Has Twice the Density?" *J. Chem. Educ.* **1998**, *75*, 726.

Literature Hartwig, Dacio R.; Filho, Romeu C. Rocha. "Cinema, Flirts, Snakes, and Gases," *J. Chem. Educ.* **1982**, *59*, 295.
Paper presents analogies for use in teaching the kinetic molecular theory of gases.

3. *Dalton's law* $(P_{total} = P_1 + P_2 + P_3 + \cdots)$. The pressure of a gas mixture is the sum of the pressures of the individual gases. Why does each gas contribute to the total pressure in proportion to its mole fraction?

4. *Charles's law* $(V \propto T)$. A change in temperature is accompanied by a corresponding change in volume. What effect does higher temperature have on gas particles that increases the volume—or increases the pressure if volume is fixed? This question raises a more fundamental one: what does temperature measure on the molecular scale?

5. *Avogadro's law* $(V \propto n)$. Gas volume (or pressure) depends on the number of moles present, not on the nature of the particular gas. But shouldn't 1 mol of larger molecules occupy more space than 1 mol of smaller molecules? And why doesn't 1 mol of heavier molecules exert more pressure than 1 mol of lighter molecules?

Postulates of the Kinetic-Molecular Theory

The theory is based on three postulates (assumptions):

Postulate 1. *Particle volume.* A gas consists of a large collection of individual particles. The volume of an individual particle is *extremely* small compared with the volume of the container. In essence, the model pictures gas particles as points of mass with empty space between them.

Postulate 2. *Particle motion.* Gas particles are in constant, random, straight-line motion, except when they collide with the container walls or with each other.

Postulate 3. *Particle collisions.* Collisions are *elastic,* which means that, somewhat like minute billiard balls, the colliding molecules exchange energy but they do not lose any energy through friction. Thus, their total kinetic energy (E_k) is constant. Between collisions, the molecules do not influence each other by attractive or repulsive forces.

Picture the scene envisioned by the postulates: Countless particles, nearly points of mass, moving in every direction, smashing into the container walls and one another. Any given particle changes its speed with each collision, perhaps one instant standing nearly still from a head-on crash and the next instant zooming away from a smash on the side. Thus, *the particles have an average speed,* with most moving near the average speed, some moving faster, and some slower.

Figure 5.14 depicts the distribution of molecular speeds (u) for N_2 gas at three temperatures. The curves flatten and spread at higher temperatures. Note especially that the *most probable speed (the peak of each curve) increases as the temperature increases*. This increase occurs because the average kinetic energy of the molecules $(\overline{E_k}$; the overbar indicates the average value of a quantity), which incorporates the most probable speed, is proportional to the absolute temperature: $\overline{E_k} \propto T$, or $\overline{E_k} = c \times T$, where c is a constant that is the same for any gas. (We'll return to this equation shortly.) Thus, a major conclusion based on the distribution of speeds, which arises directly from postulate 3, is that *at a given temperature, all gases have the same average kinetic energy.*

A Molecular View of the Gas Laws

Let's continue visualizing the particles to see how the theory explains the macroscopic behavior of gases and answers the questions posed above:

1. *Origin of pressure.* When a moving object collides with a surface, it exerts a force. We conclude from postulate 2, which describes particle motion, that when a particle collides with the container wall, it too exerts a force. Many such collisions result in the observed pressure. The greater the number of molecules in a given container, the more frequently they collide with the walls, and the greater the pressure is.

Literature Hudson, Reggie L. "Toy Flying Saucers and Molecular Speeds," *J. Chem. Educ.* **1982,** *59,* 1025.

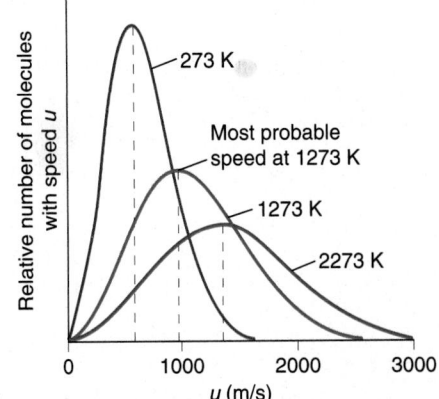

Figure 5.14 Distribution of molecular speeds at three temperatures. At a given temperature, a plot of the relative number of N_2 molecules vs. molecular speed (u) results in a skewed bell-shaped curve, with the most probable speed at the peak. Note that the curves spread at higher temperatures and the most probable speed is directly proportional to the temperature.

Transparency/ARIS Presentation Center, Art Library

2. *Boyle's law* ($V \propto 1/P$). Gas molecules are points of mass with empty space between them (postulate 1), so as the pressure exerted *on* the sample increases at constant temperature, the distance between molecules decreases, and the sample volume decreases. The pressure exerted *by* the gas increases simultaneously because in a smaller volume of gas, there are shorter distances between gas molecules and the walls and between the walls themselves; thus, collisions are more frequent (Figure 5.15). The fact that liquids and solids cannot be compressed means there is little, if any, free space between the molecules.

Figure 5.15 **A molecular description of Boyle's law.** At a given T, gas molecules collide with the walls across an average distance (d_1) and give rise to a pressure (P_{gas}) that equals the external pressure (P_{ext}). If P_{ext} increases, V decreases, and so the average distance between a molecule and the walls is shorter ($d_2 < d_1$). Molecules strike the walls more often, and P_{gas} increases until it again equals P_{ext}. Thus, V decreases when P increases.

Transparency/ARIS Presentation Center, Art Library

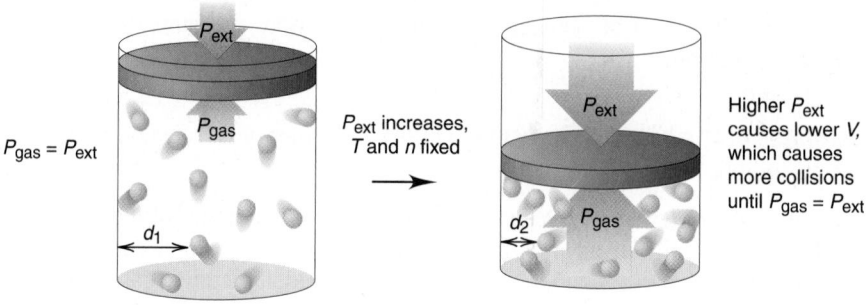

3. *Dalton's law of partial pressures* ($P_{total} = P_A + P_B$). Adding a given amount of gas A to a given amount of gas B causes an increase in the total number of molecules in proportion to the amount of A that is added. This increase causes a corresponding increase in the number of collisions per second with the walls (postulate 2), which causes a corresponding increase in the pressure (Figure 5.16). Thus, each gas exerts a fraction of the total pressure based on the fraction of molecules (or fraction of moles; that is, the mole fraction) of that gas in the mixture.

Transparency/ARIS Presentation Center, Art Library

Figure 5.16 **A molecular description of Dalton's law of partial pressures.** A piston-cylinder assembly containing 0.30 mol of gas A at 0.50 atm is connected to a tank of fixed volume containing 0.60 mol of gas B at 1.0 atm. When the piston is depressed at fixed temperature, gas A is forced into the tank of gas B and the gases mix. The new total pressure, 1.5 atm, equals the sum of the partial pressures, which is related to the new total amount of gas, 0.90 mol. Thus, each gas undergoes a fraction of the total collisions related to its fraction of the total number of molecules (moles), which is equal to its mole fraction.

4. *Charles's law* ($V \propto T$). As the temperature increases, the most probable molecular speed and the average kinetic energy increase (postulate 3). Thus, the molecules hit the walls more frequently *and* more energetically. A higher frequency of collisions causes higher internal pressure. As a result, the walls move outward, which increases the volume and restores the starting pressure (Figure 5.17).

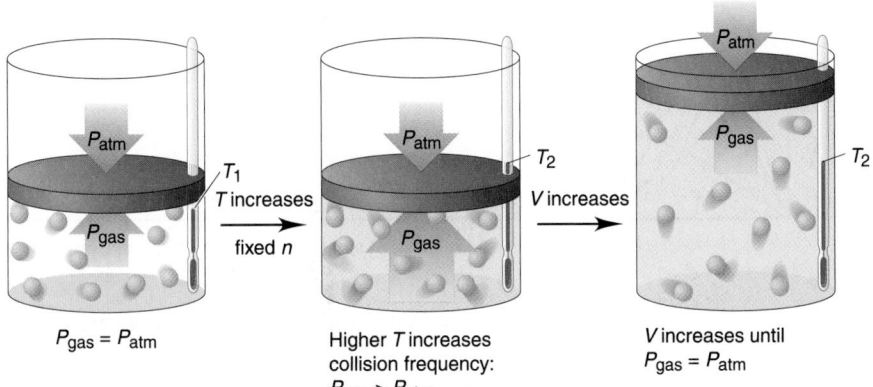

$P_{gas} = P_{atm}$ Higher T increases collision frequency: $P_{gas} > P_{atm}$ V increases until $P_{gas} = P_{atm}$

Figure 5.17 A molecular description of Charles's law. At a given temperature (T_1), $P_{gas} = P_{atm}$. When the gas is heated to T_2, the molecules move faster and collide with the walls more often, which increases P_{gas}. This increases V, and so the molecules collide less often until P_{gas} again equals P_{atm}. Thus, V increases when T increases.

Transparency/ARIS Presentation Center, Art Library

5. *Avogadro's law* ($V \propto n$). Adding more molecules to a container increases the total number of collisions with the walls and, therefore, the internal pressure. As a result, the volume expands until the number of collisions per unit of wall area is the same as it was before the addition (Figure 5.18).

$P_{gas} = P_{atm}$ More molecules increase collisions: $P_{gas} > P_{atm}$ V increases until $P_{gas} = P_{atm}$

Figure 5.18 A molecular description of Avogadro's law. At a given T, a certain amount (n) of gas gives rise to a pressure (P_{gas}) equal to P_{atm}. When more gas is added (n increases), collisions with the walls become more frequent, and P_{gas} increases. This leads to an increase in V until P_{gas} equals P_{atm} again. Thus, V increases when n increases.

Transparency/ARIS Presentation Center, Art Library

The Relationship Between Kinetic Energy and Temperature We still need to explain why equal numbers of molecules of two different gases, such as O_2 and H_2, occupy the same volume. Let's first see why heavier O_2 particles *do not* hit the container walls with more energy than lighter H_2 particles. To do so, we'll look more closely at the components of kinetic energy. From Chapter 1, the kinetic energy of an object is the energy associated with its motion. It is related to the object's mass *and* speed as follows:

$$E_k = \tfrac{1}{2}\,\text{mass} \times \text{speed}^2$$

This equation shows that *if a heavy object and a light object have the same kinetic energy, the heavy object must be moving more slowly.* As we said, postulate 3 leads to the conclusion that different gases at the same temperature have the same

Demo Shakhashiri, Bassam Z. "Relative Velocity of Sound Propagation: Musical Molecular Weights," *Chemical Demonstrations, A Handbook for Teachers of Chemistry*, Vol. 2 (The University of Wisconsin Press, Madison, 1985), 28–32.

Figure 5.19 **Relationship between molar mass and molecular speed.** At a given temperature, gases with lower molar masses (numbers in parentheses) have higher most probable speeds (peak of each curve).

Transparency/ARIS Presentation Center, Art Library

Literature Lieu, Van T.; Kalbus, Gene E. "A Simple Experiment for the Determination of Molecular Weights of Gases Lighter Than Air," *J. Chem. Educ.* **2002**, *79*, 473.

average kinetic energy. For this to be true, *molecules with a higher mass have, on average, a lower speed.* In other words, at the same temperature, O_2 molecules move more slowly than H_2 molecules. Figure 5.19 displays this fact for several gases.

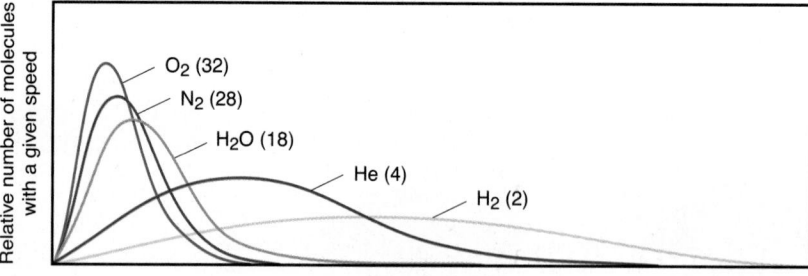

Molecular speed at a given *T*

Thus, with their greater speed, H_2 molecules collide with the walls more often than do O_2 molecules, but their lower mass means that each collision has less force. In keeping with Avogadro's law, at the same *T*, samples of H_2 and O_2 have the same pressure and, thus, the same volume because molecules hit the walls with the same average kinetic energy.

Now let's focus on a more fundamental idea—the relation between kinetic energy and temperature. Earlier we said that the average kinetic energy of a particle was equal to the absolute temperature multiplied by a constant, that is, $\overline{E_k} = c \times T$. Using definitions of velocity, momentum, force, and pressure, a derivation of this relationship gives the following equation:

$$\overline{E_k} = \tfrac{3}{2}\left(\frac{R}{N_A}\right)T$$

where *R* is the gas constant and N_A is Avogadro's number. This equation expresses the important point that *temperature is related to the average energy of molecular motion.* Note that it is not related to the *total* energy, which depends on the size of the sample, but to the *average* energy: as *T* increases, $\overline{E_k}$ increases. Thus, in the macroscopic world, the mercury rise we see in a thermometer when a beaker of water is heated over a flame is, in the molecular world, a sequence of kinetic energy transfers from higher energy particles in the flame to lower energy particles in the beaker glass, the water, the thermometer glass, and the mercury, such that each succeeding group of particles increases its average kinetic energy.

Finally, let's derive an expression for a special type of average molecular speed. From the general expression for kinetic energy of an object,

$$E_k = \tfrac{1}{2}\text{mass} \times \text{speed}^2$$

the average kinetic energy of each molecule in a large population is

$$\overline{E_k} = \tfrac{1}{2}m\overline{u^2}$$

where *m* is the molecular mass and $\overline{u^2}$ is the average of the squares of the molecular speeds. Setting this expression for average kinetic energy equal to the earlier one gives

$$\tfrac{1}{2}m\overline{u^2} = \tfrac{3}{2}\left(\frac{R}{N_A}\right)T$$

Multiplying through by Avogadro's number, N_A, gives the average kinetic energy for a mole of gas particles:

$$\tfrac{1}{2}N_A\,m\overline{u^2} = \tfrac{3}{2}RT$$

Avogadro's number times the molecular mass, $N_A \times m$, is just the molar mass, $\mathcal{M}$, and solving for $\overline{u^2}$, we have

$$\overline{u^2} = \frac{3RT}{\mathcal{M}}$$

The square root of $\overline{u^2}$ is the root-mean-square speed, or **rms speed (u_{rms}).** *A molecule moving at this speed has the average kinetic energy.** Taking the square root of both sides gives

$$u_{rms} = \sqrt{\frac{3RT}{\mathcal{M}}} \qquad (5.13)$$

where R is the gas constant, T is the absolute temperature, and $\mathcal{M}$ is the molar mass. (Because we want u in m/s and R includes the joule, which has units of $kg \cdot m^2/s^2$, we use the value 8.314 J/mol·K for R and express $\mathcal{M}$ in kg/mol.)

Thus, as an example, the root-mean-square speed of an average O_2 molecule ($\mathcal{M} = 3.200 \times 10^{-2}$ kg/mol) at room temperature (20°C, or 293 K) in the air you're breathing right now is

$$u_{rms} = \sqrt{\frac{3RT}{\mathcal{M}}} = \sqrt{\frac{3(8.314 \ \text{J/mol·K})(293 \ \text{K})}{3.200 \times 10^{-2} \ \text{kg/mol}}}$$

$$= \sqrt{\frac{3(8.314 \ \text{kg·m}^2/\text{s}^2/\text{mol·K})(293 \ \text{K})}{3.200 \times 10^{-2} \ \text{kg/mol}}}$$

$$= 478 \ \text{m/s}$$

Effusion and Diffusion

The movement of gases, either through one another or into regions of very low pressure, has many important applications.

The Process of Effusion

One of the early triumphs of the kinetic-molecular theory was an explanation of **effusion,** the process by which a gas escapes from its container through a tiny hole into an evacuated space. In 1846, Thomas Graham studied this process and concluded that the effusion rate was inversely proportional to the square root of the gas density. The effusion rate is the number of moles (or molecules) of gas effusing per unit time. Since density is directly proportional to molar mass, we state **Graham's law of effusion** as follows: *the rate of effusion of a gas is inversely proportional to the square root of its molar mass,*

$$\text{Rate of effusion} \propto \frac{1}{\sqrt{\mathcal{M}}}$$

Argon (Ar) is lighter than krypton (Kr), so it effuses faster, assuming equal pressures of the two gases. Thus, the ratio of the rates is

$$\frac{\text{Rate}_{Ar}}{\text{Rate}_{Kr}} = \frac{\sqrt{\mathcal{M}_{Kr}}}{\sqrt{\mathcal{M}_{Ar}}} \qquad \text{or, in general,} \qquad \frac{\text{Rate}_A}{\text{Rate}_B} = \frac{\sqrt{\mathcal{M}_B}}{\sqrt{\mathcal{M}_A}} = \sqrt{\frac{\mathcal{M}_B}{\mathcal{M}_A}} \qquad (5.14)$$

The kinetic-molecular theory explains that, at a given temperature and pressure, *the gas with the lower molar mass effuses faster because the most probable speed of its molecules is higher; therefore, more molecules escape per unit time.* ●

---

*The rms speed, u_{rms}, is proportional to, but slightly higher than, the average speed; for an ideal gas, $u_{rms} = 1.09 \times$ average speed.

Demo Shakhashiri, Bassam Z. "Graham's Law of Diffusion," and "Graham's Law of Effusion," *Chemical Demonstrations, A Handbook for Teachers of Chemistry,* Vol. 2 (The University of Wisconsin Press, Madison, 1985), 69–74.

●**Preparing Nuclear Fuel** One of the most important applications of Graham's law is the *enrichment* of nuclear reactor fuel: separating nonfissionable, more abundant ^{238}U from fissionable ^{235}U to increase the proportion of ^{235}U in the mixture. Since the two isotopes have identical chemical properties, they are separated by differences in the effusion rates of their gaseous compounds. Uranium ore is converted to gaseous UF_6 (a mixture of $^{238}UF_6$ and $^{235}UF_6$), which is pumped through a series of chambers with porous barriers. Because they move very slightly faster, molecules of $^{235}UF_6$ ($\mathcal{M} = 349.03$) effuse through each barrier 1.0043 times faster than do molecules of $^{238}UF_6$ ($\mathcal{M} = 352.04$). Many passes are made, each increasing the fraction of $^{235}UF_6$ until a mixture is obtained that contains enough $^{235}UF_6$. This isotope-enrichment process was developed during the latter years of World War II and produced enough ^{235}U for two of the world's first three atomic bombs.

 Engineering application

Literature Hawkes, Stephen J. "Graham's Law and Perpetuation of Error," *J. Chem. Educ.* **1997,** *74,* 1069.

SAMPLE PROBLEM 5.13 Applying Graham's Law of Effusion

PROBLEM A mixture of helium (He) and methane (CH_4) is placed in an effusion apparatus. Calculate the ratio of their effusion rates.
PLAN The effusion rate is inversely proportional to $\sqrt{\mathcal{M}}$, so we find the molar mass of each substance from the formula and take its square root. The inverse of the ratio of the square roots is the ratio of the effusion rates.
SOLUTION

$$\mathcal{M} \text{ of } CH_4 = 16.04 \text{ g/mol} \qquad \mathcal{M} \text{ of He} = 4.003 \text{ g/mol}$$

Calculating the ratio of the effusion rates:

$$\frac{\text{Rate}_{He}}{\text{Rate}_{CH_4}} = \sqrt{\frac{\mathcal{M}_{CH_4}}{\mathcal{M}_{He}}} = \sqrt{\frac{16.04 \text{ g/mol}}{4.003 \text{ g/mol}}} = \sqrt{4.007} = 2.002$$

CHECK A ratio >1 makes sense because the lighter He should effuse faster than the heavier CH_4. Because the molar mass of He is about one-fourth that of CH_4, He should effuse about twice as fast (the inverse of $\sqrt{\frac{1}{4}}$).

FOLLOW-UP PROBLEM 5.13 If it takes 1.25 min for 0.010 mol of He to effuse, how long will it take for the same amount of ethane (C_2H_6) to effuse?

Graham's law is also used to *determine the molar mass of an unknown gas.* By comparing the effusion rate of gas X with that of a known gas, such as He, we can solve for the molar mass of X:

$$\frac{\text{Rate}_X}{\text{Rate}_{He}} = \sqrt{\frac{\mathcal{M}_{He}}{\mathcal{M}_X}}$$

Squaring both sides and solving for the molar mass of X gives

$$\mathcal{M}_X = \mathcal{M}_{He} \times \left(\frac{\text{rate}_{He}}{\text{rate}_X}\right)^2$$

Demo Epp, Dianne N. "Overhead Projection of Graham's Law of Gaseous Diffusion," *J. Chem. Educ.* **1990**, *67*, 1061.

Demo Shakhashiri, Bassam Z. "Ratio of Diffusion Coefficients: The Ammonium Chloride Ring," *Chemical Demonstrations, A Handbook for Teachers of Chemistry,* Vol. 2 (The University of Wisconsin Press, Madison, 1985), 48–50.

The Process of Diffusion Closely related to effusion is the process of gaseous **diffusion,** the movement of one gas through another. Diffusion rates are also described generally by Graham's law:

$$\text{Rate of diffusion} \propto \frac{1}{\sqrt{\mathcal{M}}}$$

For two gases at equal pressures, such as NH_3 and HCl, moving through another gas or a mixture of gases, such as air, we find

$$\frac{\text{Rate}_{NH_3}}{\text{Rate}_{HCl}} = \sqrt{\frac{\mathcal{M}_{HCl}}{\mathcal{M}_{NH_3}}}$$

The reason for this dependence on molar mass is the same as for effusion rates: lighter molecules have higher molecular speeds than heavier molecules, so they move farther in a given amount of time.

But the presence of so many other gas particles is a major reason that diffusion rates are typically much lower than effusion rates. Put another way, if gas molecules move at hundreds of meters per second at ordinary temperatures (see Figure 5.14), why does it take a second or two after you open a bottle of perfume to smell the fragrance? Although convection plays an important role, a molecule moving by diffusion does not travel very far before it collides with a molecule in the air. As you can see from Figure 5.20, the path of each molecule is tortuous. Imagine how much quicker you can walk through an empty room compared with a room crowded with other moving people.

Diffusion also occurs in liquids (and even to a small extent in solids). However, because the distances between molecules are much shorter in a liquid than in a gas, collisions are much more frequent; thus, diffusion is *much* slower. Diffusion of a gas through a liquid is a vital process in biological systems. For

Figure 5.20 Diffusion of a gas particle through a space filled with other particles. In traversing a space, a gas molecule collides with many other molecules, which gives it a tortuous path. For clarity, the path of only one particle *(red dot)* is shown *(red lines)*.

example, it plays a key part in the movement of O_2 from lungs to blood. Many organisms have evolved elaborate ways to speed the diffusion of nutrients (for example, sugar and metal ions) through their cell membranes and to slow, or even stop, the diffusion of toxins.

Biological application

The Chaotic World of Gases: Mean Free Path and Collision Frequency

Refinements of the basic kinetic-molecular theory provide us with a view into the amazing, chaotic world of gas molecules. Imagine being able to visualize and follow an "average" N_2 molecule in a room at 20°C and 1 atm pressure—perhaps the room you are in now.

Distribution of Molecular Speeds Our molecule is hurtling through the room at an average speed of 0.47 km/s, or nearly 1100 mi/h (rms speed = 0.51 km/s); it is continually changing speed as it collides with other molecules. At any instant, this molecule may be traveling at 2500 mi/h or standing still as it collides head on, but these extreme speeds are *much* less likely than the most probable one (see Figure 5.19).

Mean Free Path From a molecule's diameter, we can use the kinetic-molecular theory to obtain the **mean free path,** the average distance the molecule travels between collisions at a given temperature and pressure. Our average N_2 molecule (3.7×10^{-10} m in diameter) travels 6.6×10^{-8} m before smashing into a fellow traveler, which means it travels about 180 molecular diameters between collisions. (An analogy in the macroscopic world would be an N_2 molecule the size of a billiard ball traveling about 30 ft before colliding with another.) Therefore, even though gas molecules are *not* points of mass, a gas sample *is* mostly empty space. Mean free path is a key factor in the rate of diffusion and the rate of heat flow through a gas.

Collision Frequency Divide the most probable speed (distance per second) by the mean free path (distance per collision) and you obtain the **collision frequency,** the average number of collisions per second that each molecule undergoes. As you can see, our average N_2 molecule experiences an enormous number of collisions every second: ●

$$\text{Collision frequency} = \frac{4.7 \times 10^2 \text{ m/s}}{6.6 \times 10^{-8} \text{ m/collision}} = 7.1 \times 10^9 \text{ collisions/s}$$

Distribution of speed (and kinetic energy) and collision frequency are essential ideas for understanding the speed of a chemical reaction, as you'll see in Chapter 16. As the upcoming Chemical Connections essay shows, the kinetic-molecular theory applies directly to the behavior of our planet's atmosphere.

●**Danger on Molecular Highways** To give you some idea of how astounding events are in the molecular world, we can express the collision frequency of a molecule in terms of a common experience in the macroscopic world: driving a compact car on the highway. Since a car is much larger than an N_2 molecule, to match the collision frequency of the molecule, you would have to travel at 2.8 billion mi/s (an impossibility, given that it is much faster than the speed of light) and would smash into another car every 700 yd!

Section Summary

The kinetic-molecular theory postulates that gas molecules take up a negligible portion of the gas volume, move in straight-line paths between elastic collisions, and have average kinetic energies proportional to the absolute temperature of the gas. • This theory explains the gas laws in terms of changes in distances between molecules and the container walls and changes in molecular speed. • Temperature is a measure of the average kinetic energy of molecules. • Effusion and diffusion rates are inversely proportional to the square root of the molar mass (Graham's law) because they are directly proportional to molecular speed. • Molecular motion is characterized by a temperature-dependent most probable speed within a range of speeds, a mean free path, and a collision frequency. • The atmosphere is a complex mixture of gases that exhibits variations in pressure, temperature, and composition with altitude.

Chemical Connections to Planetary Science

Structure and Composition of Earth's Atmosphere

An atmosphere is the envelope of gases that extends continuously from a planet's surface outward, eventually thinning to a point at which it is indistinguishable from interplanetary space. On Earth, complex changes in pressure, temperature, and composition occur within this mixture of gases, and the present atmosphere is very different from the one that existed during our planet's early history.

Variation in Pressure

Since gases are compressible (Boyle's law), the pressure of the atmosphere *decreases* smoothly with distance from the surface, with a more rapid decrease at lower altitudes (Figure B5.1). Although no specific boundary delineates the outermost fringe of the atmosphere, the density and composition at around 10,000 km from the surface are identical with those of outer space. About 99% of the atmosphere's mass lies within 30 km of the surface, and 75% lies within the lowest 11 km.

Variation in Temperature

Unlike the change in pressure, temperature does *not* decrease smoothly with altitude, and the atmosphere is usually classified into regions based on the direction of temperature change (Figure B5.1). In the *troposphere*, which includes the region from the surface to around 11 km, temperatures *drop* 7°C per kilometer to −55°C (218 K). All our weather occurs in the troposphere, and all but a few aircraft fly there. Temperatures then *rise* through the *stratosphere* from −55°C to about 7°C (280 K) at 50 km; we'll discuss the reason shortly.

In the *mesosphere*, temperatures *drop* smoothly again to −93°C (180 K) at around 80 km. Within the *thermosphere*, which extends to around 500 km, temperatures *rise* again, but vary between 700 and 2000 K, depending on the intensity of solar radiation and sunspot activity.

The *exosphere*, the outermost region, maintains these temperatures and merges with outer space.

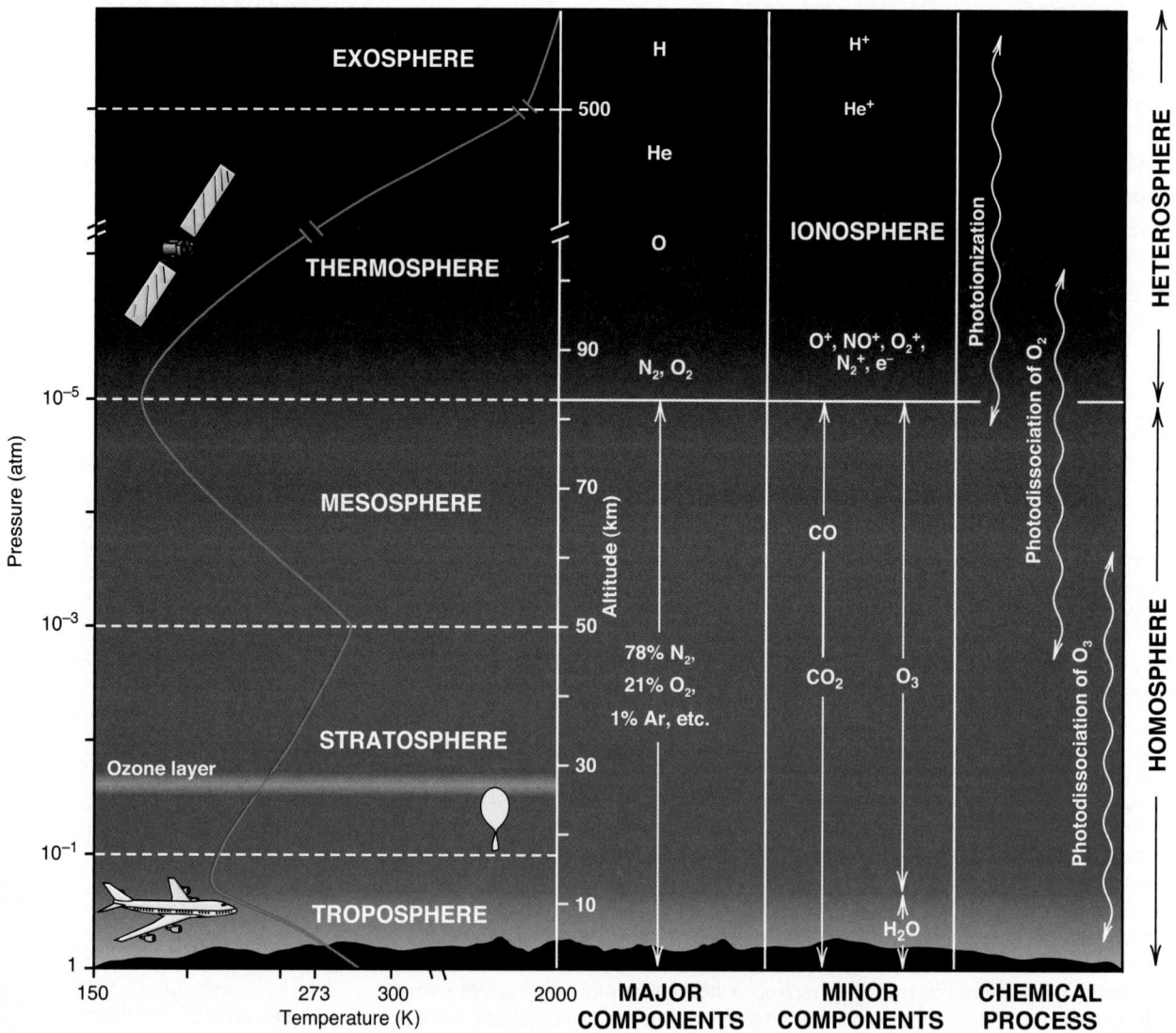

Figure B5.1 Variations in pressure, temperature, and composition of Earth's atmosphere.

Literature Heinhorst, Sabine; Cannon, Gordon. "*Nature:* Earth's Atmosphere and Beyond," *J. Chem. Educ.* **2003**, *80*, 1115.

What does it actually mean to have a temperature of 2000 K at 500 km (300 mi) above Earth's surface? Would a piece of iron (melting point ≈ 1700 K) glow red-hot and melt in the thermosphere? Our everyday use of the words "hot" and "cold" refers to measurements near the surface, where the density of the atmosphere is 10^6 times greater than in the thermosphere. At an altitude of 500 km, where collision frequency is extremely low, a thermometer, or any other object, experiences *very little transfer of kinetic energy*. Thus, the object does not become "hot" in the usual sense; in fact, it is very "cold." Recall, however, that absolute temperature is proportional to the average kinetic energy of the particles. The high-energy solar radiation reaching these outer regions is transferred to relatively few particles, so their average kinetic energy becomes extremely high, as indicated by the high temperature. For this reason, supersonic aircraft do not reach maximum speed until they reach maximum altitude, where the air is less dense, so that collisions with gas molecules are less frequent and the aircraft material becomes less hot.

Variation in Composition

In terms of chemical composition, the atmosphere is usually classified into two major regions, *homosphere* and *heterosphere*. Superimposing the regions defined by temperature on these shows that the homosphere includes the troposphere, stratosphere, and mesosphere, and the heterosphere includes the thermosphere and exosphere (Figure B5.1).

The Homosphere The homosphere has a relatively constant composition, containing, by volume, approximately 78% N_2, 21% O_2, and 1% a mixture of other gases (mostly argon). Under the conditions that occur in the homosphere, the atmospheric gases behave ideally, so volume percent is equal to mole percent (Avogadro's law), and the mole fraction of a component is directly related to its partial pressure (Dalton's law). Table B5.1 shows the components of a sample of clean, dry air at sea level.

The composition of the homosphere is uniform because of *convective mixing*. Air directly in contact with land is warmer than the air above it. The warmer air expands (Charles's law), its density decreases, and it rises through the cooler, denser air, thereby mixing the components. The cooler air sinks, becomes warmer by contact with the land, and the convection continues. The warm air currents that rise from the ground, called *thermals,* are used by soaring birds and glider pilots to stay aloft.

An important effect of convection is that the air above industrialized areas becomes cleaner as the rising air near the surface carries up ground-level pollutants, which are dispersed by winds. However, under certain weather and geographical conditions, a warm air mass remains stationary over a cool one. The resulting *temperature inversion* blocks normal convection, and harmful pollutants build up, causing severe health problems.

The Heterosphere The heterosphere has variable composition, consisting of regions dominated by a few atomic or molecular species. Convective heating does not reach these heights, so the gas particles become layered according to molar mass: nitrogen and oxygen molecules in the lower levels, oxygen atoms (O) in the next, then helium atoms (He), and free hydrogen atoms (H) in the highest level.

Embedded within the lower heterosphere is the ionosphere, containing ionic species such as O^+, NO^+, O_2^+, N_2^+, and free

Table B5.1	Composition of Clean, Dry Air at Sea Level
Component	**Mole Fraction**
Nitrogen (N_2)	0.78084
Oxygen (O_2)	0.20946
Argon (Ar)	0.00934
Carbon dioxide (CO_2)	0.00033
Neon (Ne)	1.818×10^{-5}
Helium (He)	5.24×10^{-6}
Methane (CH_4)	2×10^{-6}
Krypton (Kr)	1.14×10^{-6}
Hydrogen (H_2)	5×10^{-7}
Dinitrogen monoxide (N_2O)	5×10^{-7}
Carbon monoxide (CO)	1×10^{-7}
Xenon (Xe)	8×10^{-8}
Ozone (O_3)	2×10^{-8}
Ammonia (NH_3)	6×10^{-9}
Nitrogen dioxide (NO_2)	6×10^{-9}
Nitrogen monoxide (NO)	6×10^{-10}
Sulfur dioxide (SO_2)	2×10^{-10}
Hydrogen sulfide (H_2S)	2×10^{-10}

electrons (Figure B5.1). Ionospheric chemistry involves numerous light-induced bond-breaking (*photodissociation*) and light-induced electron-removing (*photoionization*) processes. One of the simpler ways that O atoms form, for instance, involves a four-step sequence that absorbs energy:

$$N_2 \longrightarrow N_2^+ + e^- \ [\text{photoionization}]$$
$$N_2^+ + e^- \longrightarrow N + N$$
$$N + O_2 \longrightarrow NO + O$$
$$N + NO \longrightarrow N_2 + O$$
$$\overline{O_2 \longrightarrow O + O \ [\text{overall photodissociation}]}$$

When the resulting high-energy O atoms collide with other neutral or ionic components, the average kinetic energy of thermospheric particles increases.

The Importance of Stratospheric Ozone Although most high-energy radiation is absorbed by the thermosphere, a small amount reaches the stratosphere and breaks O_2 into O atoms. The energetic O atoms collide with more O_2 to form ozone (O_3), another molecular form of oxygen:

$$O_2(g) \xrightarrow{\text{high-energy radiation}} 2O(g)$$
$$M + O(g) + O_2(g) \longrightarrow O_3(g) + M$$

where M is any particle that can carry away excess energy. This reaction releases heat, which is the reason stratospheric temperatures increase with altitude.

Stratospheric ozone is vital to life on Earth's surface because it absorbs a great proportion of solar ultraviolet (UV) radiation, which results in decomposition of the ozone:

$$O_3(g) \xrightarrow{\text{UV light}} O_2(g) + O(g)$$

UV radiation is extremely harmful because it is strong enough to break chemical bonds and, thus, interrupt normal biological processes. Without the presence of stratospheric ozone, much more of this radiation would reach the surface, resulting in increased mutation and cancer rates. The depletion of the ozone layer as a result of industrial gases is discussed in Chapter 16.

(continued) **219**

Chemical Connections *continued*

Earth's Primitive Atmosphere

The composition of the present atmosphere bears little resemblance to that covering the young Earth, but scientists disagree about what that primitive composition actually was: Did the carbon and nitrogen have low oxidation numbers, as in CH_4 (O.N. of C = −4) and NH_3 (O.N. of N = −3)? Or did these atoms have higher oxidation numbers, as in CO_2 (O.N. of C = +4) and N_2 (O.N. of N = 0)? One point generally accepted is that the primitive mixture did not contain free O_2.

Origin-of-life models propose that about 1 billion years after the earliest organisms appeared, blue-green algae evolved. These one-celled plants used solar energy to produce glucose by photosynthesis:

$$6CO_2(g) + 6H_2O(l) \xrightarrow{light} C_6H_{12}O_6(glucose) + 6O_2(g)$$

As a result of this reaction, the O_2 content of the atmosphere increased and the CO_2 content decreased. More O_2 allowed more oxidation to occur, which changed the geological and biological makeup of the early Earth. Iron(II) minerals changed to iron(III) minerals, sulfites changed to sulfates, and eventually organisms evolved that could use O_2 to oxidize other organisms to obtain energy. For these organisms to have survived exposure to the more energetic forms of solar radiation (particularly UV radiation), enough O_2 must have formed to create a protective ozone layer. Estimates indicate that the level of O_2 increased to the current level of about 20 mol % approximately 1.5 billion years ago.

A Survey of Planetary Atmospheres

Earth's combination of pressure and temperature and its oxygen-rich atmosphere and watery surface are unique in the Solar System. (Indeed, if similar conditions and composition were discovered on a planet circling any other star, excitement about the possibility of life there would be enormous.) Atmospheres on the Sun's other planets are strikingly different from Earth's. Some,

especially those on the outer planets, exist under conditions that cause extreme deviations from ideal gas behavior (Section 5.7). Based on current data from NASA spacecraft and Earth-based observations, Table B5.2 lists conditions and composition of the atmospheres on the planets within the Solar System and on some of their moons.

Biological application

Table B5.2	Planetary Atmospheres		
Planet (Satellite)	**Pressure* (atm)**	**Temperature[†] (K)**	**Composition (mol %)**
Mercury	$<10^{-12}$	~700 (day) ~100 (night)	He, H_2, O_2, Ar, Ne (Na and K from solar wind)
Venus	~90	~730	CO_2 (96), N_2 (3), He, SO_2, H_2O, Ar, Ne
Earth	1.0	avg. range 250–310	N_2 (78), O_2 (21), Ar (0.9), H_2O, CO_2, Ne, He, CH_4, Kr
(Moon)	$\sim2\times10^{-14}$	370 (day) 120 (night)	Ne, Ar, He
Mars	7×10^{-3}	300 (summer day) 140 (pole in winter) 218 average	CO_2 (95), N_2(3), Ar (1.6), O_2, H_2O, Ne, CO, Kr
Jupiter	$(\sim4\times10^6)$	(~140)	H_2 (89), He (11), CH_4, NH_3, C_2H_6, C_2H_2, PH_3
(Io)	$\sim10^{-10}$	~110	SO_2, S vapor
Saturn	$(\sim4\times10^6)$	(~130)	H_2 (93), He (7), CH_4, NH_3, H_2O, C_2H_6, PH_3
(Titan)	1.6	~94	N_2 (90), Ar (<6), CH_4 (3?), C_2H_6, C_2H_2, C_2H_4, HCN, H_2
Uranus	$(>10^6)$	(~60)	H_2 (83), He (15), CH_4 (2)
Neptune	$(>10^6)$	(~60)	H_2 (<90), He (~10), CH_4

*Values in parentheses refer to interior pressures.
[†]Values in parentheses refer to cloud-top temperatures.

ARIS Presentation Center, Tables Library

5.7 REAL GASES: DEVIATIONS FROM IDEAL BEHAVIOR

A fundamental principle of science is that simpler models are more useful than complex ones—as long as they explain the data. You can certainly appreciate the usefulness of the kinetic-molecular theory. With simple postulates, it explains the behavior of ideal gases (and simple real gases under ordinary conditions) in terms of particles acting like infinitesimal billiard balls, moving at speeds governed by the absolute temperature, and experiencing only perfectly elastic collisions.

In reality, however, you know that *molecules are not points of mass.* They have volumes determined by the sizes of their atoms and the lengths and directions of their bonds. You also know that atoms contain charged particles, and many bonds are polar, giving rise to *attractive and repulsive forces among molecules.* (In fact, such forces cause substances to undergo changes of state; we'll discuss these forces in great detail in Chapter 12.) Therefore, we expect these properties of real gases to cause deviations from ideal behavior under some conditions. We must alter the simple model and the ideal gas law to predict gas behavior at low temperatures and very high pressures.

Effects of Extreme Conditions on Gas Behavior

At ordinary conditions—relatively high temperatures and low pressures—most real gases exhibit nearly ideal behavior. Even at STP (0°C and 1 atm), however, real gases deviate *slightly* from ideal behavior. Table 5.4 shows the standard molar volumes of several gases to five significant figures. Note that they do not quite equal the ideal value. The phenomena that cause these slight deviations under standard conditions exert more influence as the temperature decreases toward the condensation point of the gas, the temperature at which it liquefies. As you can see, the largest deviations from ideal behavior in Table 5.4 are for Cl_2 and NH_3 because, at the standard temperature of 0°C, they are already close to their condensation points.

At pressures greater than 10 atm, we begin to see significant deviations from ideal behavior in many gases. Figure 5.21 shows a plot of PV/RT versus P_{ext} for 1 mol of several real gases and an ideal gas. For 1 mol of an *ideal* gas, the ratio PV/RT is equal to 1 at any pressure. The values on the horizontal axis are the external pressures at which the PV/RT ratios are calculated. The pressures range from normal (at 1 atm, $PV/RT = 1$) to very high (at ~1000 atm, $PV/RT \approx 1.6$ to 2.3).

Literature Braathen, Per Christian. "Determination of the Oxygen Content of Air," *J. Chem. Educ.* **2000**, *77*, 1410.

Literature Najdoski, Metodija; Petrusevski, Vladimir M. "A Novel Experiment for Fast and Simple Determination of the Oxygen Content in the Air," *J. Chem. Educ.* **2000**, *77*, 1447.

Literature Birk, James P.; Lawson, Anton E. "The Persistence of the Candle-and-Cylinder Misconception," *J. Chem. Educ.* **1999**, *76*, 914.

Table 5.4	Molar Volume of Some Common Gases at STP (0°C and 1 atm)	
Gas	**Molar Volume (L/mol)**	**Condensation Point (°C)**
He	22.435	−268.9
H_2	22.432	−252.8
Ne	22.422	−246.1
Ideal gas	**22.414**	—
Ar	22.397	−185.9
N_2	22.396	−195.8
O_2	22.390	−183.0
CO	22.388	−191.5
Cl_2	22.184	−34.0
NH_3	22.079	−33.4

Transparency/ARIS Presentation Center, Tables Library

Transparency/ARIS Presentation Center, Active Art Library

Figure 5.21 The behavior of several real gases with increasing external pressure. The horizontal line shows the behavior of 1 mol of ideal gas: $PV/RT = 1$ at all P_{ext}. At very high pressures, all real gases deviate significantly from such ideal behavior. Even at ordinary pressures, these deviations begin to appear (expanded portion).

The *PV/RT* curve shown in Figure 5.21 for 1 mol of methane (CH$_4$) is typical of that for most real gases: it decreases below the ideal value at moderately high pressures and then rises above it as pressure increases further. This shape arises from two overlapping effects of the two characteristics of real molecules just mentioned:

1. At moderately high pressure, values of *PV/RT* lower than ideal (less than 1) are due predominantly to *intermolecular attractions*.
2. At very high pressure, values of *PV/RT* greater than ideal (more than 1) are due predominantly to *molecular volume*.

Let's examine these effects on the molecular level:

1. *Intermolecular attractions.* Attractive forces between molecules are *much* weaker than the covalent bonding forces that hold a molecule together. Most intermolecular attractions are caused by slight imbalances in electron distributions and are important only over relatively short distances. At normal pressures, the spaces between the molecules of any real gas are so large that attractions are negligible and the gas behaves nearly ideally. As the pressure rises and the volume of the sample decreases, however, the average intermolecular distance becomes smaller and attractions have a greater effect.

Picture a molecule at these higher pressures (Figure 5.22). As it approaches the container wall, nearby molecules attract it, which lessens the force of its impact. *Repeated throughout the sample, this effect results in decreased gas pressure and, thus, a smaller numerator in the PV/RT ratio.* Lowering the temperature has the same effect because it slows the molecules, so attractive forces exert an influence for a longer time. At a low enough temperature, the attractions among molecules become overwhelming, and the gas condenses to a liquid.

Figure 5.22 The effect of intermolecular attractions on measured gas pressure. At ordinary pressures, the volume is large and gas molecules are too far apart to experience significant attractions. At moderately high external pressures, the volume decreases enough for the molecules to influence each other. As the close-up shows, a gas molecule approaching the container wall experiences intermolecular attractions from neighboring molecules that reduce the force of its impact. As a result, real gases exert *less* pressure than the ideal gas law predicts.

Transparency/ARIS Presentation Center, Art Library

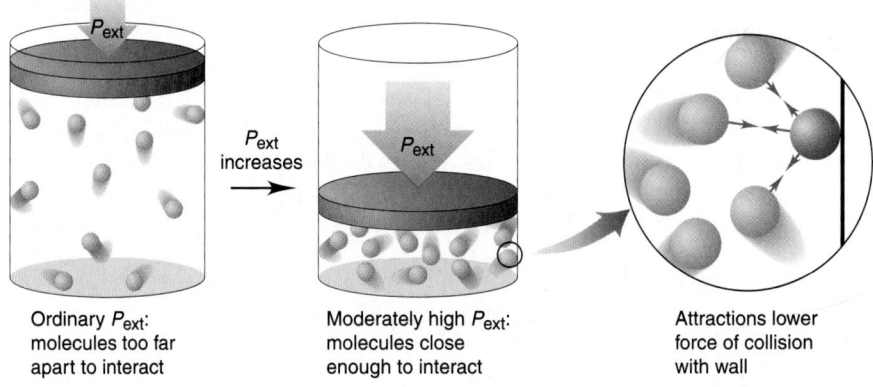

Ordinary P_{ext}: molecules too far apart to interact

Moderately high P_{ext}: molecules close enough to interact

Attractions lower force of collision with wall

2. *Molecular volume.* At normal pressures, the space between molecules of a real gas (free volume) is enormous compared with the volume of the molecules *themselves* (molecular volume), so the free volume is essentially equal to the container volume. As the applied pressure increases, however, and the free volume decreases, the molecular volume makes up a greater proportion of the container volume, which you can see in Figure 5.23. Thus, at very high pressures, the free volume becomes significantly *less* than the container volume. However, we continue to use the container volume as the *V* in the *PV/RT* ratio, so the ratio is artificially high. This makes the numerator artificially high. The molecular volume effect becomes more important as the pressure increases, eventually outweighing the effect of the intermolecular attractions and causing *PV/RT* to rise above the ideal value.

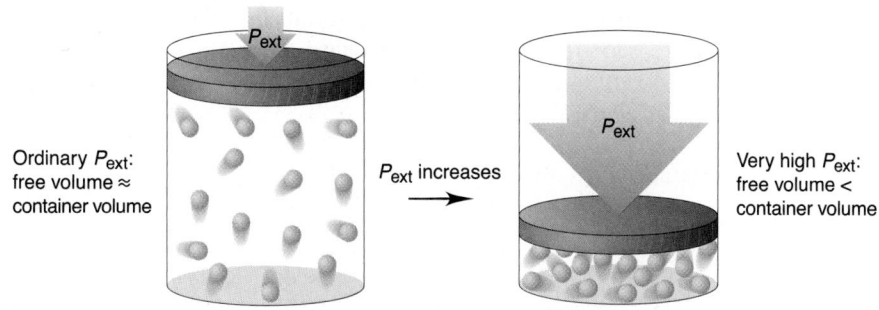

Figure 5.23 **The effect of molecular volume on measured gas volume.** At ordinary pressures, the volume *between* molecules of a real gas (free volume) is essentially equal to the container volume because the molecules occupy only a tiny fraction of the available space. At very high external pressures, however, the free volume is significantly *less* than the container volume because of the volume of the molecules themselves.

In Figure 5.21, the H_2 and He curves do not show the typical dip at moderate pressures. These gases consist of particles with such weak intermolecular attractions that the molecular volume effect predominates at all pressures.

The van der Waals Equation: The Ideal Gas Law Redesigned

To describe real gas behavior more accurately, we need to "redesign" the ideal gas equation to do two things:

1. Adjust the measured pressure *up* by adding a factor that accounts for intermolecular attractions, and
2. Adjust the measured volume *down* by subtracting a factor from the entire container volume that accounts for the molecular volume.

In 1873, Johannes van der Waals realized the limitations of the ideal gas law and proposed an equation that accounts for the behavior of real gases. The **van der Waals equation** for n moles of a real gas is

$$\left(P + \frac{n^2 a}{V^2}\right)(V - nb) = nRT \tag{5.15}$$

adjusts   adjusts
P up V down

where P is the measured pressure, V is the container volume, n and T have their usual meanings, and a and b are **van der Waals constants,** experimentally determined positive numbers specific for a given gas. Values of these constants for several gases are given in Table 5.5. The constant a relates to the number of electrons, which in turn relates to the complexity of a molecule and the strength of its intermolecular attractions. The constant b relates to molecular volume.

Consider this typical application of the van der Waals equation to calculate a gas variable. A 1.98-L vessel contains 215 g (4.89 mol) of dry ice. After standing at 26°C (299 K), the $CO_2(s)$ changes to $CO_2(g)$. The pressure is measured (P_{real}) and calculated by the ideal gas law (P_{IGL}) and, using the appropriate values of a and $b,$ by the van der Waals equation (P_{VDW}). The results are revealing:

$$P_{real} = 44.8 \text{ atm}$$
$$P_{IGL} = 60.6 \text{ atm}$$
$$P_{VDW} = 45.9 \text{ atm}$$

Comparing the real with each calculated value shows that P_{IGL} is 35.3% greater than P_{real}, but P_{VDW} is only 2.5% greater than P_{real}. At these conditions, CO_2 deviates so much from ideal behavior that the ideal gas law is not very useful.

Table 5.5	Van der Waals Constants for Some Common Gases	
Gas	$a\left(\dfrac{\text{atm·L}^2}{\text{mol}^2}\right)$	$b\left(\dfrac{\text{L}}{\text{mol}}\right)$
He	0.034	0.0237
Ne	0.211	0.0171
Ar	1.35	0.0322
Kr	2.32	0.0398
Xe	4.19	0.0511
H_2	0.244	0.0266
N_2	1.39	0.0391
O_2	1.36	0.0318
Cl_2	6.49	0.0562
CH_4	2.25	0.0428
CO	1.45	0.0395
CO_2	3.59	0.0427
NH_3	4.17	0.0371
H_2O	5.46	0.0305

Here is one final point to realize: According to kinetic-molecular theory, the constants *a* and *b* are zero for an ideal gas because the particles do not attract each other and have no volume. Yet, even for a real gas at ordinary pressures, the molecules are very far apart; thus,

- Attractive forces are miniscule, so $P + \dfrac{n^2a}{V^2} \approx P$

- The molecular volume is a miniscule fraction of the container volume, so $V - nb \approx V$

Therefore, *at ordinary conditions, the van der Waals equation becomes the ideal gas equation.*

Section Summary

At very high pressures or low temperatures, all gases deviate greatly from ideal behavior. • As pressure increases, most real gases exhibit first a lower and then a higher PV/RT ratio than the value of 1 for an ideal gas. • These deviations from ideality are due to attractions between molecules, which lower the pressure (and the PV/RT ratio), and to the larger fraction of the container volume occupied by the molecules, which increases the ratio. • By including parameters characteristic of each gas, the van der Waals equation corrects for deviations from ideal behavior.

Chapter Perspective

As with the atomic theory (Chapter 2), we have seen in this chapter how a simple molecular-scale model can explain macroscopic observations and how it often must be revised to predict a wider range of chemical behavior. Gas behavior is relatively easy to understand because gas structure is so randomized—very different from the structures of liquids and solids with their complex molecular interactions, as you'll see in Chapter 12. In Chapter 6, we return to chemical reactions, but from the standpoint of the heat involved in such changes. The meaning of kinetic energy, which we discussed in this chapter, bears directly on this central topic.

CHAPTER REVIEW GUIDE

The following sections provide many aids to help you study this chapter. (Numbers in parentheses refer to pages, unless noted otherwise.)

Learning Objectives These are concepts and skills you should know after studying this chapter.

Relevant section and/or sample problem (SP) numbers appear in parentheses.

Understand These Concepts

1. How gases differ in their macroscopic properties from liquids and solids (5.1)

2. The meaning of pressure and the operation of a barometer and a manometer (5.2)

3. The relations among gas variables expressed by Boyle's, Charles's, and Avogadro's laws (5.3)

4. How the individual gas laws are incorporated into the ideal gas law (5.3)

5. How the ideal gas law can be used to study gas density and molar mass (5.4)

6. The relation between the density and the temperature of a gas (5.4)

7. The meaning of Dalton's law and the relation between partial pressure and mole fraction of a gas; how Dalton's law applies to collecting a gas over water (5.4)

8. How the postulates of the kinetic-molecular theory are applied to explain the origin of pressure and the gas laws (5.6)

9. The relations among molecular speed, average kinetic energy, and temperature (5.6)

10. The meanings of *effusion* and *diffusion* and how their rates are related to molar mass (5.6)

11. The relations among mean free path, molecular speed, and collision frequency (5.6)

12. Why intermolecular attractions and molecular volume cause gases to deviate from ideal behavior at low temperatures and high pressures (5.7)

13. How the van der Waals equation corrects the ideal gas law for extreme conditions (5.7)

Master These Skills

1. Interconverting among the units of pressure (atm, mmHg, torr, pascal, psi) (SP 5.1)
2. Reducing the ideal gas law to the individual gas laws (SPs 5.2–5.5)
3. Applying gas laws to balancing a chemical equation (SP 5.6)
4. Rearranging the ideal gas law to calculate gas density (SP 5.7) and molar mass of a volatile liquid (SP 5.8)
5. Calculating the mole fraction and partial pressure of a gas (SP 5.9)
6. Using the vapor pressure of water to find the amount of a gas collected over water (SP 5.10)
7. Applying stoichiometry and gas laws to calculate amounts of reactants and products (SPs 5.11, 5.12)
8. Using Graham's law to solve problems of gaseous effusion (SP 5.13)

Key Terms These important terms appear in boldface in the chapter and are defined again in the Glossary.

Section 5.2
pressure (P) (189)
barometer (189)
manometer (191)
pascal (Pa) (191)
standard atmosphere (atm) (191)
millimeter of mercury (mmHg) (191)
torr (191)

Section 5.3
ideal gas (193)
Boyle's law (194)
Charles's law (195)
Avogadro's law (197)
standard temperature and pressure (STP) (197)
standard molar volume (197)
ideal gas law (198)
universal gas constant (R) (198)

Section 5.4
partial pressure (205)
Dalton's law of partial pressures (206)
mole fraction (X) (206)

Section 5.6
kinetic-molecular theory (210)
rms speed (u_{rms}) (215)
effusion (215)

Graham's law of effusion (215)
diffusion (216)
mean free path (217)
collision frequency (217)
atmosphere (218)

Section 5.7
van der Waals equation (223)
van der Waals constants (223)

Key Equations and Relationships Numbered and screened equations are listed for you to refer to or memorize.

5.1 Expressing the volume-pressure relationship (Boyle's law) (194):

$$V \propto \frac{1}{P} \quad \text{or} \quad PV = \text{constant} \quad [T \text{ and } n \text{ fixed}]$$

5.2 Expressing the volume-temperature relationship (Charles's law) (195):

$$V \propto T \quad \text{or} \quad \frac{V}{T} = \text{constant} \quad [P \text{ and } n \text{ fixed}]$$

5.3 Expressing the pressure-temperature relationship (Amontons's law) (196):

$$P \propto T \quad \text{or} \quad \frac{P}{T} = \text{constant} \quad [V \text{ and } n \text{ fixed}]$$

5.4 Expressing the volume-amount relationship (Avogadro's law) (197):

$$V \propto n \quad \text{or} \quad \frac{V}{n} = \text{constant} \quad [P \text{ and } T \text{ fixed}]$$

5.5 Defining standard temperature and pressure (197):
STP: 0°C (273.15 K) and 1 atm (760 torr)

5.6 Defining the volume of 1 mol of an ideal gas at STP (197):
Standard molar volume = 22.4141 L = 22.4 L [3 sf]

5.7 Relating volume to pressure, temperature, and amount (ideal gas law) (198):

$$PV = nRT \quad \text{and} \quad \frac{P_1 V_1}{n_1 T_1} = \frac{P_2 V_2}{n_2 T_2}$$

5.8 Calculating the value of R (198):

$$R = \frac{PV}{nT} = \frac{1 \text{ atm} \times 22.4141 \text{ L}}{1 \text{ mol} \times 273.15 \text{ K}}$$

$$= 0.082058 \frac{\text{atm} \cdot \text{L}}{\text{mol} \cdot \text{K}} = 0.0821 \frac{\text{atm} \cdot \text{L}}{\text{mol} \cdot \text{K}} \quad [3 \text{ sf}]$$

5.9 Rearranging the ideal gas law to find gas density (203):

$$PV = \frac{m}{\mathcal{M}} RT$$

so

$$\frac{m}{V} = d = \frac{\mathcal{M} \times P}{RT}$$

5.10 Rearranging the ideal gas law to find molar mass (204):

$$n = \frac{m}{\mathcal{M}} = \frac{PV}{RT}$$

so

$$\mathcal{M} = \frac{mRT}{PV} \quad \text{or} \quad \mathcal{M} = \frac{dRT}{P}$$

5.11 Relating the total pressure of a gas mixture to the partial pressures of the components (Dalton's law of partial pressures) (206):

$$P_{\text{total}} = P_1 + P_2 + P_3 + \cdots$$

5.12 Relating partial pressure to mole fraction (206):

$$P_A = X_A \times P_{\text{total}}$$

5.13 Defining rms speed as a function of molar mass and temperature (215):

$$u_{\text{rms}} = \sqrt{\frac{3RT}{\mathcal{M}}}$$

5.14 Applying Graham's law of effusion (215):

$$\frac{\text{Rate}_A}{\text{Rate}_B} = \frac{\sqrt{\mathcal{M}_B}}{\sqrt{\mathcal{M}_A}} = \sqrt{\frac{\mathcal{M}_B}{\mathcal{M}_A}}$$

5.15 Applying the van der Waals equation to find gas P and V under extreme conditions (223):

$$\left(P + \frac{n^2 a}{V^2} \right)(V - nb) = nRT$$

Highlighted Figures and Tables These figures (F) and tables (T) provide a visual review of key ideas.

Entries in **bold** contain frequently used data.

F5.1 The three states of matter (188)
T5.2 Common units of pressure (192)
F5.5 The relationship between the volume and pressure of a gas (193)
F5.6 The relationship between the volume and temperature of a gas (195)
F5.8 Standard molar volume (197)
F5.10 Relationship between the ideal gas law and the individual gas laws (198)

T5.3 Vapor pressure of water at different T (207)
F5.13 Stoichiometric relationships for gases (208)
F5.14 Distribution of molecular speeds at three T (211)
F5.15 Molecular description of Boyle's law (212)
F5.16 Molecular description of Dalton's law (212)
F5.17 Molecular description of Charles's law (213)
F5.18 Molecular description of Avogadro's law (213)
F5.19 Relation between molar mass and molecular speed (214)
F5.21 The behavior of several real gases with increasing external pressure (221)
T5.5 Van der Waals constants for some gases (223)

Brief Solutions to FOLLOW-UP PROBLEMS Compare your solutions to these calculation steps and answers.

5.1 P_{CO_2} (torr) = (753.6 mmHg − 174.0 mmHg) $\times \dfrac{1 \text{ torr}}{1 \text{ mmHg}}$

$\qquad$ = 579.6 torr

P_{CO_2} (Pa) = 579.6 torr $\times \dfrac{1 \text{ atm}}{760 \text{ torr}} \times \dfrac{1.01325 \times 10^5 \text{ Pa}}{1 \text{ atm}}$

$\qquad$ = 7.727×10^4 Pa

P_{CO_2} (lb/in^2) = 579.6 torr $\times \dfrac{1 \text{ atm}}{760 \text{ torr}} \times \dfrac{14.7 \text{ lb/in}^2}{1 \text{ atm}}$

$\qquad$ = 11.2 lb/in^2

5.2 P_2 (atm) = 26.3 kPa $\times \dfrac{1 \text{ atm}}{101.325 \text{ kPa}}$ = 0.260 atm

V_2 (L) = 105 mL $\times \dfrac{1 \text{ L}}{1000 \text{ mL}} \times \dfrac{0.871 \text{ atm}}{0.260 \text{ atm}}$ = 0.352 L

5.3 T_2 (K) = 273 K $\times \dfrac{9.75 \text{ cm}^3}{6.83 \text{ cm}^3}$ = 390. K

5.4 P_2 (torr) = 793 torr $\times \dfrac{35.0 \text{ g} - 5.0 \text{ g}}{35.0 \text{ g}}$ = 680. torr

(There is no need to convert mass to moles because the ratio of masses equals the ratio of moles.)

5.5 $n = \dfrac{PV}{RT} = \dfrac{1.37 \text{ atm} \times 438 \text{ L}}{0.0821 \dfrac{\text{atm·L}}{\text{mol·K}} \times 294 \text{ K}}$ = 24.9 mol O$_2$

Mass (g) of O$_2$ = 24.9 mol O$_2$ $\times \dfrac{32.00 \text{ g O}_2}{1 \text{ mol O}_2}$ = 7.97×10^2 g O$_2$

5.6 The balanced equation is 2CD(g) $\longrightarrow$ C$_2$(g) + D$_2$(g), so n does not change. Therefore, given constant P, the absolute temperature, T, must double: $T_1 = -73°\text{C} + 273.15 = 200$ K; so $T_2 = 400$ K, or 400 K − 273.15 = 127°C.

5.7 d (at 0°C and 380 torr) = $\dfrac{44.01 \text{ g/mol} \times \dfrac{380 \text{ torr}}{760 \text{ torr/atm}}}{0.0821 \dfrac{\text{atm·L}}{\text{mol·K}} \times 273 \text{ K}}$

$\qquad$ = 0.982 g/L

The density is lower at the smaller P because V is larger. In this case, d is lowered by one-half because P is one-half as much.

5.8 $\mathcal{M} = \dfrac{1.26 \text{ g} \times 0.0821 \dfrac{\text{atm·L}}{\text{mol·K}} \times 283.2 \text{ K}}{\dfrac{102.5 \text{ kPa}}{101.325 \text{ kPa/1 atm}} \times 1.00 \text{ L}}$ = 29.0 g/mol

5.9 $n_{\text{total}} = \left(5.50 \text{ g He} \times \dfrac{1 \text{ mol He}}{4.003 \text{ g He}}\right)$

$\qquad + \left(15.0 \text{ g Ne} \times \dfrac{1 \text{ mol Ne}}{20.18 \text{ g Ne}}\right)$

$\qquad + \left(35.0 \text{ g Kr} \times \dfrac{1 \text{ mol Kr}}{83.80 \text{ g Kr}}\right)$

$\qquad$ = 2.53 mol

$P_{\text{He}} = \left(\dfrac{5.50 \text{ g He} \times \dfrac{1 \text{ mol He}}{4.003 \text{ g He}}}{2.53 \text{ mol}}\right) \times 1 \text{ atm} = 0.543 \text{ atm}$

$P_{\text{Ne}} = 0.294$ atm $\qquad P_{\text{Kr}} = 0.165$ atm

5.10 $P_{H_2} = 752$ torr − 13.6 torr = 738 torr

Mass (g) of H$_2$ = $\left(\dfrac{\dfrac{738 \text{ torr}}{760 \text{ torr/atm}} \times 1.495 \text{ L}}{0.0821 \dfrac{\text{atm·L}}{\text{mol·K}} \times 289 \text{ K}}\right) \times \dfrac{2.016 \text{ g H}_2}{1 \text{ mol H}_2}$

$\qquad$ = 0.123 g H$_2$

5.11 H$_2$SO$_4$(aq) + 2NaCl(s) $\longrightarrow$ Na$_2$SO$_4$(aq) + 2HCl(g)

n_{HCl} = 0.117 kg NaCl $\times \dfrac{10^3 \text{ g}}{1 \text{ kg}} \times \dfrac{1 \text{ mol NaCl}}{58.44 \text{ g NaCl}} \times \dfrac{2 \text{ mol HCl}}{2 \text{ mol NaCl}}$

$\qquad$ = 2.00 mol HCl

At STP, V (mL) = 2.00 mol $\times \dfrac{22.4 \text{ L}}{1 \text{ mol}} \times \dfrac{10^3 \text{ mL}}{1 \text{ L}}$

$\qquad$ = 4.48×10^4 mL

5.12 NH$_3$(g) + HCl(g) $\longrightarrow$ NH$_4$Cl(s)

n_{NH_3} = 0.187 mol $\qquad n_{\text{HCl}}$ = 0.0522 mol

n_{NH_3} after reaction

$\qquad$ = 0.187 mol NH$_3$ − $\left(0.0522 \text{ mol HCl} \times \dfrac{1 \text{ mol NH}_3}{1 \text{ mol HCl}}\right)$

$\qquad$ = 0.135 mol NH$_3$

$P = \dfrac{0.135 \text{ mol} \times 0.0821 \dfrac{\text{atm·L}}{\text{mol·K}} \times 295 \text{ K}}{10.0 \text{ L}}$ = 0.327 atm

5.13 $\dfrac{\text{Rate of He}}{\text{Rate of C}_2\text{H}_6} = \sqrt{\dfrac{30.07 \text{ g/mol}}{4.003 \text{ g/mol}}}$ = 2.741

Time for C$_2$H$_6$ to effuse = 1.25 min $\times$ 2.741 = 3.43 min

PROBLEMS

Problems with colored numbers are answered in Appendix E and worked in detail in the Student Solutions Manual. Problem sections match those in the text and provide the numbers of relevant sample problems. Most offer Concept Review Questions, Skill-Building Exercises (grouped in pairs covering the same concept), and Problems in Context. The Comprehensive Problems are based on material from any section or previous chapter.

An Overview of the Physical States of Matter

● Concept Review Questions

- **5.1** How does a sample of gas differ in its behavior from a sample of liquid in each of the following situations?
  (a) The sample is transferred from one container to a larger one.
  (b) The sample is heated in an expandable container, but no change of state occurs.
  (c) The sample is placed in a cylinder with a piston, and an external force is applied.

- **5.2** Are the particles in a gas farther apart or closer together than the particles in a liquid? Use your answer to explain each of the following general observations:
  (a) Gases are more compressible than liquids.
  (b) Gases have lower viscosities than liquids.
  (c) After thorough stirring, all gas mixtures are solutions.
  (d) The density of a substance in the gas state is lower than in the liquid state.

Gas Pressure and Its Measurement
(Sample Problem 5.1)

● Concept Review Questions

- **5.3** How does a barometer work? Is the column of mercury in a barometer shorter when it is on a mountaintop or at sea level? Explain.

- **5.4** How can a unit of length such as millimeter of mercury (mmHg) be used as a unit of pressure, which has the dimensions of force per unit area?

- **5.5** In a closed-end manometer, the mercury level in the arm attached to the flask can never be higher than the mercury level in the other arm, whereas in an open-end manometer, it *can* be higher. Explain.

● Skill-Building Exercises *(grouped in similar pairs)*

- **5.6** On a cool, rainy day, the barometric pressure is 730 mmHg. Calculate the barometric pressure in centimeters of water (cmH_2O) (d of Hg = 13.5 g/mL; d of H_2O = 1.00 g/mL).

- **5.7** A long glass tube, sealed at one end, has an inner diameter of 10.0 mm. The tube is filled with water and inverted into a pail of water. If the atmospheric pressure is 755 mmHg, how high (in mmH_2O) is the column of water in the tube (d of Hg = 13.5 g/mL; d of H_2O = 1.00 g/mL)?

- **5.8** Convert the following:
  (a) 0.745 atm to mmHg   (b) 992 torr to bar
  (c) 365 kPa to atm   (d) 804 mmHg to kPa
- **5.9** Convert the following:
  (a) 76.8 cmHg to atm   (b) 27.5 atm to kPa
  (c) 6.50 atm to bar   (d) 0.937 kPa to torr

- **5.10** In Figure P5.10, what is the pressure of the gas in the flask (in atm) if the barometer reads 738.5 torr?
- **5.11** In Figure P5.11, what is the pressure of the gas in the flask (in kPa) if the barometer reads 765.2 mmHg?

Figure P5.10   **Figure P5.11**

- **5.12** If the sample flask in Figure P5.12 is open to the air, what is the atmospheric pressure (in atm)?
- **5.13** What is the pressure (in Pa) of the gas in the flask in Figure P5.13?

Figure P5.12   **Figure P5.13**

● Problems in Context

- **5.14** Convert each of the pressures described below to atm:
  (a) At the peak of Mt. Everest, atmospheric pressure is only 2.75×10^2 mmHg.
  (b) A cyclist fills her bike tires to 86 psi.
  (c) The surface of Venus has an atmospheric pressure of 9.15×10^6 Pa.
  (d) At 100 ft below sea level, a scuba diver experiences a pressure of 2.54×10^4 torr.

- **5.15** The gravitational force exerted by an object is given by $F = mg$, where F is the force in newtons, m is the mass in kilograms, and g is the acceleration due to gravity (9.81 m/s²).
  (a) Use the definition of the pascal to calculate the mass (in kg) of the atmosphere above 1 m² of ocean.
  (b) Osmium ($Z = 76$) is a transition metal in Group 8B(8) and has the highest density of any element (22.6 g/mL). If an osmium column is 1 m² in area, how high must it be for its pressure to equal atmospheric pressure? [Use the answer from part (a) in your calculation.]

Levels of Difficulty: ● Easy ●● Medium ●●● Challenging   This figure is available on the Presentation Center, within the art library.

The Gas Laws and Their Experimental Foundations
(Sample Problems 5.2 to 5.6)

● Concept Review Questions

5.16 A student states Boyle's law as follows: "The volume of a gas is inversely proportional to its pressure." How is this statement incomplete? Give a correct statement of Boyle's law.

5.17 In the following relationships, which quantities are variables and which are fixed: (a) Charles's law; (b) Avogadro's law; (c) Amontons's law?

5.18 Boyle's law relates gas volume to pressure, and Avogadro's law relates gas volume to number of moles. State a relationship between gas pressure and number of moles.

5.19 Each of the following processes caused the gas volume to double, as shown. For each process, state how the remaining gas variable changed or that it remained fixed:
(a) T doubles at fixed P.
(b) T and n are fixed.
(c) At fixed T, the reaction is $CD_2(g) \longrightarrow C(g) + D_2(g)$.
(d) At fixed P, the reaction is $A_2(g) + B_2(g) \longrightarrow 2AB(g)$.

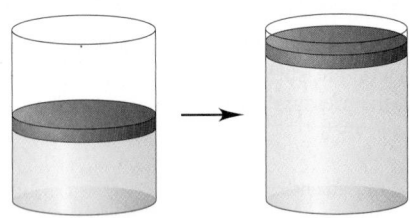

● Skill-Building Exercises *(grouped in similar pairs)*

5.20 What is the effect of the following on the volume of 1 mol of an ideal gas?
(a) The pressure is tripled (at constant T).
(b) The absolute temperature is increased by a factor of 3.0 (at constant P).
(c) Three more moles of the gas are added (at constant P and T).

5.21 What is the effect of the following on the volume of 1 mol of an ideal gas?
(a) The pressure is reduced by a factor of 4 (at constant T).
(b) The pressure changes from 760 torr to 202 kPa, and the temperature changes from 37°C to 155 K.
(c) The temperature changes from 305 K to 32°C, and the pressure changes from 2 atm to 101 kPa.

5.22 What is the effect of the following on the volume of 1 mol of an ideal gas?
(a) Temperature decreases from 800 K to 400 K (at constant P).
(b) Temperature increases from 250°C to 500°C (at constant P).
(c) Pressure increases from 2 atm to 6 atm (at constant T).

5.23 What is the effect of the following on the volume of 1 mol of an ideal gas?
(a) Half the gas escapes (at constant P and T).
(b) The initial pressure is 722 torr, and the final pressure is 0.950 atm; the initial temperature is 32°F, and the final temperature is 273 K.
(c) Both the pressure and temperature decrease to one-fourth of their initial values.

5.24 A sample of sulfur hexafluoride gas occupies 9.10 L at 198°C. Assuming that the pressure remains constant, what temperature (in °C) is needed to reduce the volume to 2.50 L?

5.25 A 93-L sample of dry air cools from 145°C to −22°C while the pressure is maintained at 2.85 atm. What is the final volume?

5.26 A sample of Freon-12 (CF_2Cl_2) occupies 25.5 L at 298 K and 153.3 kPa. Find its volume at STP.

5.27 A sample of carbon monoxide occupies 3.65 L at 298 K and 745 torr. Find its volume at −14°C and 367 torr.

5.28 A sample of chlorine gas is confined in a 5.0-L container at 328 torr and 37°C. How many moles of gas are in the sample?

5.29 If 1.47×10^{-3} mol of argon occupies a 75.0-mL container at 26°C, what is the pressure (in torr)?

5.30 You have 357 mL of chlorine trifluoride gas at 699 mmHg and 45°C. What is the mass (in g) of the sample?

5.31 A 75.0-g sample of dinitrogen monoxide is confined in a 3.1-L vessel. What is the pressure (in atm) at 115°C?

● Problems in Context

5.32 In preparation for a demonstration, your professor brings a 1.5-L bottle of sulfur dioxide into the lecture hall before class to allow the gas to reach room temperature. If the pressure gauge reads 85 psi and the temperature in the hall is 23°C, how many moles of sulfur dioxide are in the bottle? (*Hint:* The gauge reads zero when 14.7 psi of gas remains.)

5.33 A gas-filled weather balloon with a volume of 65.0 L is released at sea-level conditions of 745 torr and 25°C. The balloon can expand to a maximum volume of 835 L. When the balloon rises to an altitude at which the temperature is −5°C and the pressure is 0.066 atm, will it reach its maximum volume?

Further Applications of the Ideal Gas Law
(Sample Problems 5.7 to 5.10)

● Concept Review Questions

5.34 Why is moist air less dense than dry air?

5.35 To collect a beaker of H_2 gas by displacing the air already in the beaker, would you hold the beaker upright or inverted? Why? How would you hold the beaker to collect CO_2?

5.36 Why can we use a gas mixture, such as air, to study the general behavior of an ideal gas under ordinary conditions?

5.37 How does the partial pressure of gas A in a mixture compare to its mole fraction in the mixture? Explain.

5.38 The circle at right represents a portion of a mixture of four gases A (purple), B (brown), C (green), and D_2 (orange). (a) Which has the highest partial pressure? (b) Which has the lowest partial pressure? (c) If the total pressure is 0.75 atm, what is the partial pressure of D_2?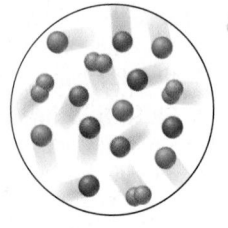

● Skill-Building Exercises *(grouped in similar pairs)*

5.39 What is the density of Xe gas at STP?

5.40 Find the density of Freon-11 ($CFCl_3$) at 120°C and 1.5 atm.

5.41 How many moles of gaseous arsine (AsH_3) occupy 0.0400 L at STP? What is the density of gaseous arsine?

5.42 The density of a noble gas is 2.71 g/L at 3.00 atm and 0°C. Identify the gas.

5.43 Calculate the molar mass of a gas at 388 torr and 45°C if 206 ng occupies 0.206 μL.

5.44 When an evacuated 63.8-mL glass bulb is filled with a gas at 22°C and 747 mmHg, the bulb gains 0.103 g in mass. Is the gas N_2, Ne, or Ar?

5.45 After 0.600 L of Ar at 1.20 atm and 227°C is mixed with 0.200 L of O_2 at 501 torr and 127°C in a 400-mL flask at 27°C, what is the pressure in the flask?

5.46 A 355-mL container holds 0.146 g of Ne and an unknown amount of Ar at 35°C and a total pressure of 626 mmHg. Calculate the moles of Ar present.

● **Problems in Context**

5.47 The air in a hot-air balloon at 744 torr is heated from 17°C to 60.0°C. Assuming that the moles of air and the pressure remain constant, what is the density of the air at each temperature? (The average molar mass of air is 28.8 g/mol.)

5.48 On a certain winter day in Utah, the average atmospheric pressure is 650. torr. What is the molar density (in mol/L) of the air if the temperature is −25°C?

5.49 A sample of a liquid hydrocarbon known to consist of molecules with five carbon atoms is vaporized in a 0.204-L flask by immersion in a water bath at 101°C. The barometric pressure is 767 torr, and the remaining gas weighs 0.482 g. What is the molecular formula of the hydrocarbon?

5.50 A sample of air contains 78.08% nitrogen, 20.94% oxygen, 0.05% carbon dioxide, and 0.93% argon, by volume. How many molecules of each gas are present in 1.00 L of the sample at 25°C and 1.00 atm?

5.51 An environmental chemist sampling industrial exhaust gases from a coal-burning plant collects a CO_2-SO_2-H_2O mixture in a 21-L steel tank until the pressure reaches 850. torr at 45°C.
(a) How many moles of gas are collected?
(b) If the SO_2 concentration in the mixture is 7.95×10^3 parts per million by volume (ppmv), what is its partial pressure? [*Hint:* ppmv = (volume of component/volume of mixture) $\times 10^6$.]

The Ideal Gas Law and Reaction Stoichiometry
(Sample Problems 5.11 and 5.12)

● **Skill-Building Exercises** (*grouped in similar pairs*)

5.52 How many grams of phosphorus react with 35.5 L of O_2 at STP to form tetraphosphorus decaoxide?
$$P_4(s) + 5O_2(g) \longrightarrow P_4O_{10}(s)$$

5.53 How many grams of potassium chlorate decompose to potassium chloride and 638 mL of O_2 at 128°C and 752 torr?
$$2KClO_3(s) \longrightarrow 2KCl(s) + 3O_2(g)$$

5.54 How many grams of phosphine (PH_3) can form when 37.5 g of phosphorus and 83.0 L of hydrogen gas react at STP?
$$P_4(s) + H_2(g) \longrightarrow PH_3(g) \quad [\text{unbalanced}]$$

5.55 When 35.6 L of ammonia and 40.5 L of oxygen gas at STP burn, nitrogen monoxide and water form. After the products return to STP, how many grams of nitrogen monoxide are present?
$$NH_3(g) + O_2(g) \longrightarrow NO(g) + H_2O(l) \quad [\text{unbalanced}]$$

5.56 Aluminum reacts with excess hydrochloric acid to form aqueous aluminum chloride and 35.8 mL of hydrogen gas over water at 27°C and 751 mmHg. How many grams of aluminum reacted?

5.57 How many liters of hydrogen gas are collected over water at 18°C and 725 mmHg when 0.84 g of lithium reacts with water? Aqueous lithium hydroxide also forms.

● **Problems in Context**

5.58 "Strike anywhere" matches contain the compound tetraphosphorus trisulfide, which burns to form tetraphosphorus decaoxide and sulfur dioxide gas. How many milliliters of sulfur dioxide, measured at 725 torr and 32°C, can be produced from burning 0.800 g of tetraphosphorus trisulfide?

5.59 Freon-12 (CF_2Cl_2), widely used as a refrigerant and aerosol propellant, is a dangerous air pollutant. In the troposphere, it traps heat 25 times as effectively as CO_2, and in the stratosphere, it participates in the breakdown of ozone. Freon-12 is prepared industrially by reaction of gaseous carbon tetrachloride with hydrogen fluoride. Hydrogen chloride gas also forms. How many grams of carbon tetrachloride are required for the production of 16.0 dm³ of Freon-12 at 27°C and 1.20 atm?

5.60 Xenon hexafluoride was one of the first noble gas compounds synthesized. The solid reacts rapidly with the silicon dioxide in glass or quartz containers to form liquid $XeOF_4$ and gaseous silicon tetrafluoride. What is the pressure in a 1.00-L container at 25°C after 2.00 g of xenon hexafluoride reacts? (Assume that silicon tetrafluoride is the only gas present and that it occupies the entire volume.)

5.61 The four sketches below represent cylinder-piston assemblies holding gases. The piston at far left holds a reactant about to undergo a reaction at constant T and P:

Which of the other three depictions best represents the products of the reaction?

5.62 Roasting galena [lead(II) sulfide] is an early step in the industrial isolation of lead. How many liters of sulfur dioxide, measured at STP, are produced by the reaction of 3.75 kg of galena with 228 L of oxygen gas at 220°C and 2.0 atm? Lead(II) oxide also forms.

5.63 In one of his most critical studies into the nature of combustion, Lavoisier heated mercury(II) oxide and isolated elemental mercury and oxygen gas. If 40.0 g of mercury(II) oxide is heated in a 502-mL vessel and 20.0% (by mass) decomposes, what is the pressure (in atm) of the oxygen that forms at 25.0°C? (Assume that the gas occupies the entire volume.)

The Kinetic-Molecular Theory: A Model for Gas Behavior
(Sample Problem 5.13)

● **Concept Review Questions**

5.64 Use the kinetic-molecular theory to explain the change in gas pressure that results from warming a sample of gas.

5.65 How does the kinetic-molecular theory explain why 1 mol of krypton and 1 mol of helium have the same volume at STP?

5.66 Is the rate of effusion of a gas higher than, lower than, or equal to its rate of diffusion? Explain. For two gases with molecules of approximately the same size, is the ratio of their effusion rates higher than, lower than, or equal to the ratio of their diffusion rates? Explain.

5.67 Consider two 1-L samples of gas: one is H_2 and the other is O_2. Both are at 1 atm and 25°C. How do the samples compare in terms of (a) mass, (b) density, (c) mean free path, (d) average molecular kinetic energy, (e) average molecular speed, and (f) time for a given fraction of molecules to effuse?

5.68 Three 5-L flasks, fixed with pressure gauges and small valves, each contain 4 g of gas at 273 K. Flask A contains H_2, flask B contains He, and flask C contains CH_4. Rank the flask contents in terms of (a) pressure, (b) average molecular kinetic energy, (c) diffusion rate after the valve is opened, (d) total kinetic energy of the molecules, (e) density, and (f) collision frequency.

● **Skill-Building Exercises** (grouped in similar pairs)

5.69 What is the ratio of effusion rates for the lightest gas, H_2, and the heaviest known gas, UF_6?

5.70 What is the ratio of effusion rates for O_2 and Kr?

5.71 The graph below shows the distribution of molecular speeds for argon and helium at the same temperature.

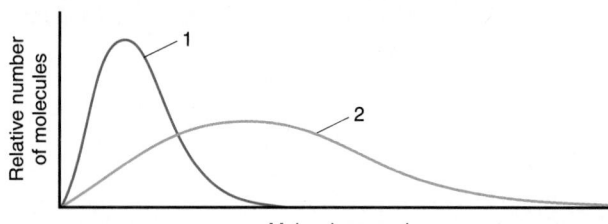

(a) Does curve 1 or 2 better represent the behavior of argon?
(b) Which curve represents the gas that effuses more slowly?
(c) Which curve more closely represents the behavior of fluorine gas? Explain.

5.72 The graph below shows the distribution of molecular speeds for a gas at two different temperatures.

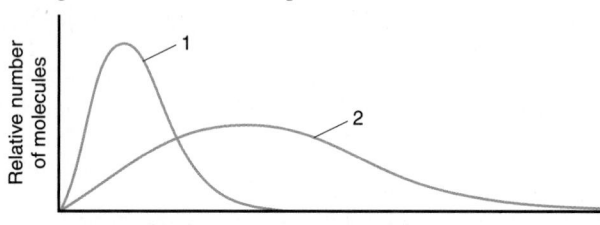

(a) Does curve 1 or 2 better represent the behavior of the gas at the lower temperature?
(b) Which curve represents the gas when it has a higher $\overline{E_k}$?
(c) Which curve is consistent with a higher diffusion rate?

5.73 At a given pressure and temperature, it takes 4.85 min for a 1.5-L sample of He to effuse through a membrane. How long does it take for 1.5 L of F_2 to effuse under the same conditions?

5.74 A sample of an unknown gas effuses in 11.1 min. An equal volume of H_2 in the same apparatus under the same conditions effuses in 2.42 min. What is the molar mass of the unknown gas?

● **Problems in Context**

5.75 White phosphorus melts and then vaporizes at high temperature. The gas effuses at a rate that is 0.404 times that of neon in the same apparatus under the same conditions. How many atoms are in a molecule of gaseous white phosphorus?

5.76 Helium (He) is the lightest noble gas component of air, and xenon (Xe) is the heaviest. [For this problem, use $R =$

8.314 J/(mol·K) and $\mathcal{M}$ in kg/mol.] (a) Find the rms speed of He in winter (0.°C) and in summer (30.°C). (b) Compare the rms speed of He with that of Xe at 30.°C. (c) Find the average kinetic energy per mole of He and of Xe at 30.°C. (d) Find the average kinetic energy per molecule of He at 30.°C.

5.77 A mixture of gaseous disulfur difluoride, dinitrogen tetrafluoride, and sulfur tetrafluoride is placed in an effusion apparatus. (a) Rank the gases in order of increasing effusion rate. (b) Find the ratio of effusion rates of disulfur difluoride and dinitrogen tetrafluoride. (c) If gas X is added, and it effuses at 0.935 times the rate of sulfur tetrafluoride, find the molar mass of X.

Real Gases: Deviations from Ideal Behavior

● **Skill-Building Exercises** (grouped in similar pairs)

5.78 Do intermolecular attractions cause negative or positive deviations from the PV/RT ratio of an ideal gas? Use Table 5.5 to rank Kr, CO_2, and N_2 in order of increasing magnitude of these deviations.

5.79 Does molecular size cause negative or positive deviations from the PV/RT ratio of an ideal gas? Use Table 5.5 to rank Cl_2, H_2, and O_2 in order of increasing magnitude of these deviations.

5.80 Does N_2 behave more ideally at 1 atm or at 500 atm? Explain.

5.81 Does SF_6 (boiling point = 16°C at 1 atm) behave more ideally at 150°C or at 20°C? Explain.

Comprehensive Problems

5.82 An "empty" gasoline can with dimensions 15.0 cm by 40.0 cm by 12.5 cm is attached to a vacuum pump and evacuated. If the atmospheric pressure is 14.7 lb/in^2, what is the total force (in pounds) on the outside of the can?

5.83 Hemoglobin is the protein that transports O_2 through the blood from the lungs to the rest of the body. In doing so, each molecule of hemoglobin combines with four molecules of O_2. If 1.00 g of hemoglobin combines with 1.53 mL of O_2 at 37°C and 743 torr, what is the molar mass of hemoglobin?

5.84 A baker uses sodium hydrogen carbonate (baking soda) as the leavening agent in a banana-nut quickbread. The baking soda decomposes according to two possible reactions:

(1) $2NaHCO_3(s) \longrightarrow Na_2CO_3(s) + H_2O(l) + CO_2(g)$
(2) $NaHCO_3(s) + H^+(aq) \longrightarrow H_2O(l) + CO_2(g) + Na^+(aq)$

Calculate the volume (in mL) of CO_2 that forms at 200.°C and 0.975 atm per gram of $NaHCO_3$ by each of the reaction processes.

5.85 A weather balloon containing 600. L of He is released near the equator at 1.01 atm and 305 K. It rises to a point where conditions are 0.489 atm and 218 K and eventually lands in the northern hemisphere under conditions of 1.01 atm and 250 K. If one-fourth of the helium leaked out during this journey, what is the volume (in L) of the balloon at landing?

5.86 Chlorine is produced from sodium chloride by the electrochemical chlor-alkali process. During the process, the chlorine is collected in a container that is isolated from the other products to prevent unwanted (and explosive) reactions. If a 15.50-L container holds 0.5950 kg of Cl_2 gas at 225°C, calculate:

(a) P_{IGL} (b) $P_{VDW}\left(\text{use } R = 0.08206 \dfrac{\text{atm·L}}{\text{mol·K}} \right)$

5.87 In a certain experiment, magnesium boride (Mg_3B_2) reacted with acid to form a mixture of four boron hydrides (B_xH_y), three as liquids (labeled I, II, and III) and one as a gas (IV).

(a) When a 0.1000-g sample of each liquid was transferred to an evacuated 750.0-mL container and volatilized at 70.00°C, sample I had a pressure of 0.05951 atm; sample II, 0.07045 atm; and sample III, 0.05767 atm. What is the molar mass of each liquid? (b) Boron is 85.63% by mass in sample I, 81.10% in II, and 82.98% in III. What is the molecular formula of each sample? (c) Sample IV was found to be 78.14% boron. Its rate of effusion was compared to that of sulfur dioxide and under identical conditions, 350.0 mL of sample IV effused in 12.00 min and 250.0 mL of sulfur dioxide effused in 13.04 min. What is the molecular formula of sample IV?

 5.88 Three equal volumes of gas mixtures, all at the same T, are depicted below (with gas A red, gas B green, and gas C blue):

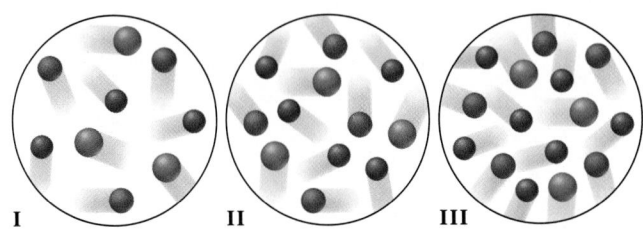

I   II   III

(a) Which sample, if any, has the highest partial pressure of A?
(b) Which sample, if any, has the lowest partial pressure of B?
(c) In which sample, if any, do the gas particles have the highest average kinetic energy?

5.89 Will the volume of a gas increase, decrease, or remain unchanged for each of the following sets of changes?
(a) The pressure is decreased from 2 atm to 1 atm, while the temperature is decreased from 200°C to 100°C.
(b) The pressure is increased from 1 atm to 3 atm, while the temperature is increased from 100°C to 300°C.
(c) The pressure is increased from 3 atm to 6 atm, while the temperature is increased from −73°C to 127°C.
(d) The pressure is increased from 0.2 atm to 0.4 atm, while the temperature is decreased from 300°C to 150°C.

5.90 When air is inhaled, it enters the alveoli of the lungs, and varying amounts of the component gases exchange with dissolved gases in the blood. The resulting alveolar gas mixture is quite different from the atmospheric mixture. The following table presents selected data on the composition and partial pressure of four gases in the atmosphere and in the alveoli:

| Gas | Atmosphere (sea level) | | Alveoli | |
	Mole %	Partial Pressure (torr)	Mole %	Partial Pressure (torr)
N_2	78.6	—	—	569
O_2	20.9	—	—	104
CO_2	00.04	—	—	40
H_2O	00.46	—	—	47

If the total pressure of each gas mixture is 1.00 atm, calculate:
(a) The partial pressure (in torr) of each gas in the atmosphere
(b) The mole % of each gas in the alveoli
(c) The number of O_2 molecules in 0.50 L of alveolar air (volume of an average breath of a person at rest) at 37°C

5.91 Radon (Rn) is the heaviest, and only radioactive, member of Group 8A(18) (noble gases). It is a product of the disintegration of heavier radioactive nuclei found in minute concentrations in many common rocks used for building and construction. In re-

cent years, health concerns about the cancers caused from inhaled residential radon have grown. If $1.0×10^{15}$ atoms of radium (Ra) produce an average of $1.373×10^4$ atoms of Rn per second, how many liters of Rn, measured at STP, are produced per day by 1.0 g of Ra?

5.92 At 1450. mmHg and 286 K, a skin diver exhales a 208-mL bubble of air that is 77% N_2, 17% O_2, and 6.0% CO_2 by volume. (a) How many milliliters would the volume of the bubble be if it were exhaled at the surface at 1 atm and 298 K? (b) How many moles of N_2 are in the bubble?

5.93 The mass of Earth's atmosphere is estimated as $5.14×10^{15}$ t (1 t = 1000 kg). (a) The average molar mass of air is 28.8 g/mol. How many moles of gas are in the atmosphere? (b) How many liters would the atmosphere occupy at 25°C and 1 atm? (c) If the surface area of Earth is $5.100×10^8$ km², how high should the atmosphere extend? Why does it actually extend much higher?

5.94 Nitrogen dioxide is used industrially to produce nitric acid, but it contributes to acid rain and photochemical smog. What volume of nitrogen dioxide is formed at 735 torr and 28.2°C by reacting 4.95 cm³ of copper ($d = 8.95$ g/cm³) with 230.0 mL of nitric acid ($d = 1.42$ g/cm³, 68.0% HNO_3 by mass)?

$$Cu(s) + 4HNO_3(aq) \longrightarrow$$
$$Cu(NO_3)_2(aq) + 2NO_2(g) + 2H_2O(l)$$

5.95 In the average adult male, the residual volume (RV) of the lungs, the volume of air remaining after a forced exhalation, is 1200 mL. (a) How many moles of air are present in the RV at 1.0 atm and 37°C? (b) How many molecules of gas are present under these conditions?

5.96 In a bromine-producing plant, how many liters of gaseous elemental bromine at 300°C and 0.855 atm are formed by the reaction of 275 g of sodium bromide and 175.6 g of sodium bromate in aqueous acid solution? (Assume no Br_2 dissolves.)

$$5NaBr(aq) + NaBrO_3(aq) + 3H_2SO_4(aq) \longrightarrow$$
$$3Br_2(g) + 3Na_2SO_4(aq) + 3H_2O(g)$$

5.97 In a collision of sufficient force, automobile air bags respond by electrically triggering the explosive decomposition of sodium azide (NaN_3) to its elements. A 50.0-g sample of sodium azide was decomposed, and the nitrogen gas generated was collected over water at 26°C. The total pressure was 745.5 mmHg. How many liters of dry N_2 were generated?

5.98 An anesthetic gas contains 64.81% carbon, 13.60% hydrogen, and 21.59% oxygen, by mass. If 2.00 L of the gas at 25°C and 0.420 atm weighs 2.57 g, what is the molecular formula of the anesthetic?

5.99 Aluminum chloride is easily vaporized above 180°C. The gas escapes through a pinhole 0.122 times as fast as helium at the same conditions of temperature and pressure in the same apparatus. What is the molecular formula of aluminum chloride gas?

5.100 Azodicarbonamide, $NH_2CON=NCONH_2$, is a blowing (foaming) agent for sponge rubber and expanded plastics. Its decomposition at 195°–202°C is given by

$$NH_2CON=NCONH_2(s) \longrightarrow$$
$$NH_3(g) + CO(g) + N_2(g) + HCNO(g)$$
$$NH_3(g) + HCNO(g) \longrightarrow \text{nonvolatile polymers}(s)$$

Calculate the volume (in mL) of gas, corrected to STP, in the final mixture from decomposition of 1.00 g of azodicarbonamide.

5.101 (a) What is the total volume of gaseous *products,* measured at 350°C and 735 torr, when an automobile engine burns 100. g of C_8H_{18} (a typical component of gasoline)?

(b) For part (a), the source of O_2 is air, which is about 78% N_2, 21% O_2, and 1.0% Ar by volume. Assuming all the O_2 reacts, but no N_2 or Ar does, what is the total volume of gaseous *exhaust*?

5.102 An atmospheric chemist studying the pollutant SO_2 places a mixture of SO_2 and O_2 in a 2.00-L container at 800. K and 1.90 atm. When the reaction occurs, gaseous SO_3 forms, and the pressure falls to 1.65 atm. How many moles of SO_3 form?

5.103 Liquid nitrogen trichloride is heated in a 2.50-L closed reaction vessel until it decomposes completely to gaseous elements. The resulting mixture exerts a pressure of 754 mmHg at 95°C. (a) What is the partial pressure of each gas in the container? (b) What is the mass of the original sample?

5.104 The thermal decomposition of ethylene occurs in many industrial contexts, for example, during ethylene transit in pipelines, formation of polyethylene, drying of the gas, and separating it from impurities. The decomposition reaction is

$$CH_2{=}CH_2(g) \longrightarrow CH_4(g) + C(graphite)$$

If the decomposition begins at 10°C and 50.0 atm with a gas density of 0.215 g/mL and the temperature increases by 950 K,
(a) What is the final pressure of the confined gas (ignore the volume of graphite and use the van der Waals equation)?
(b) How does the PV/RT value of CH_4 compare to that in Figure 5.21? Explain.

5.105 Ammonium nitrate, a common fertilizer, is used as an explosive in fireworks and by terrorists. It was the material used in the tragic explosion at the Oklahoma City federal building in 1995. How many liters of gas at 307°C and 1.00 atm are formed by the explosive decomposition of 15.0 kg of ammonium nitrate to nitrogen, oxygen, and water vapor?

5.106 An environmental engineer analyzes a sample of air contaminated with sulfur dioxide. To a 500.-mL sample at 700. torr and 38°C, she adds 20.00 mL of 0.01017 M aqueous iodine, which reacts as follows:

$$SO_2(g) + I_2(aq) + H_2O(l) \longrightarrow$$
$$HSO_4^-(aq) + I^-(aq) + H^+(aq) \quad [\text{unbalanced}]$$

Excess I_2 reacts with 11.37 mL of 0.0105 M sodium thiosulfate:

$$I_2(aq) + S_2O_3^{2-}(aq) \longrightarrow$$
$$I^-(aq) + S_4O_6^{2-}(aq) \quad [\text{unbalanced}]$$

What is the volume % of SO_2 in the air sample?

5.107 Canadian chemists have developed a modern variation of the 1899 Mond process for preparing extremely pure metallic nickel. A sample of impure nickel reacts with carbon monoxide at 50°C to form gaseous nickel carbonyl, $Ni(CO)_4$.
(a) How many grams of nickel can be converted to the carbonyl with 3.55 m^3 of CO at 100.7 kPa?
(b) The carbonyl is then decomposed at 21 atm and 155°C to pure (>99.95%) nickel. How many grams of nickel are obtained per cubic meter of the carbonyl?
(c) The released carbon monoxide is cooled and collected for reuse by passing it through water at 35°C. If the barometric pressure is 769 torr, what volume (in m^3) of CO is formed per cubic meter of carbonyl?

5.108 Analysis of a newly discovered gaseous silicon-fluorine compound shows that it contains 33.01 mass % silicon. At 27°C, 2.60 g of the compound exerts a pressure of 1.50 atm in a 0.250-L vessel. What is the molecular formula of the compound?

5.109 A gaseous organic compound containing only carbon, hydrogen, and nitrogen is burned in oxygen gas, and the volume of each reactant and product is measured under the same conditions of temperature and pressure. Reaction of four volumes of the compound produces four volumes of CO_2, two volumes of N_2, and ten volumes of water vapor. (a) What volume of O_2 was required? (b) What is the empirical formula of the compound?

5.110 A piece of dry ice (solid CO_2, $d = 0.900$ g/mL) weighing 10.0 g is placed in a 0.800-L bottle filled with air at 0.980 atm and 550.0°C. The bottle is capped, and the dry ice changes to gas. What is the final pressure inside the bottle?

5.111 Containers A, B, and C are attached by closed stopcocks of negligible volume.

If each particle shown in the picture represents 10^6 particles,
(a) How many blue particles and black particles are in B after the stopcocks are opened and the system reaches equilibrium?
(b) How many blue particles and black particles are in A after the stopcocks are opened and the system reaches equilibrium?
(c) If the pressure in C, P_C, is 750 torr before the stopcocks are opened, what is P_C afterward? (d) What is P_B afterward?

5.112 At the temperatures that exist in the thermosphere (see Figure B5.1), instruments would break down and astronauts would be killed. Yet satellites function in orbit there for many years and astronauts routinely repair equipment on space walks. Explain.

5.113 Combustible vapor-air mixtures are flammable over a limited range of concentrations. The minimum volume % of vapor that gives a combustible mixture is called the lower flammable limit (LFL). Generally, the LFL is about half the stoichiometric mixture, the concentration required for complete combustion of the vapor in air. (a) If oxygen is 20.9 vol % of air, estimate the LFL for *n*-hexane, C_6H_{14}. (b) What volume (in mL) of *n*-hexane ($d = 0.660$ g/cm³) is required to produce a flammable mixture of hexane in 1.000 m^3 of air at STP?

5.114 By what factor would a scuba diver's lungs expand if she ascended rapidly to the surface from a depth of 125 ft without inhaling or exhaling? If an expansion factor greater than 1.5 causes lung rupture, how far could she safely ascend from 125 ft without breathing? Assume constant temperature (d of seawater = 1.04 g/mL; d of Hg = 13.5 g/mL).

5.115 When 15.0 g of fluorite (CaF_2) reacts with excess sulfuric acid, hydrogen fluoride gas is collected at 744 torr and 25.5°C. Solid calcium sulfate is the other product. What gas temperature is required to store the gas in an 8.63-L container at 875 torr?

5.116 Dilute aqueous hydrogen peroxide is used as a bleaching agent and for disinfecting surfaces and small cuts. Its concentration is sometimes given as a certain number of "volumes hydrogen peroxide," which refers to the number of volumes of O_2 gas, measured at STP, that a given volume of hydrogen peroxide solution will release when it decomposes to O_2 and liquid H_2O. How many grams of hydrogen peroxide are in 0.100 L of "20 volumes hydrogen peroxide" solution?

•• **5.117** At a height of 300 km above Earth's surface, an astronaut finds that the atmospheric pressure is about 10^{-8} mmHg and the temperature is 500 K. How many molecules of gas are there per milliliter at this altitude?

•• **5.118** (a) What is the rms speed of O_2 at STP? (b) If the mean free path of O_2 molecules at STP is 6.33×10^{-8} m, what is their collision frequency? [Use $R = 8.314$ J/(mol·K) and $\mathcal{M}$ in kg/mol.]

•• **5.119** Acrylic acid (CH_2=CHCOOH) is used to prepare polymers, adhesives, and paints. The first step to make acrylic acid involves the vapor-phase oxidation of propylene (CH_2=CHCH$_3$) to acrolein (CH_2=CHCHO). This step is carried out at 330°C and 2.5 atm in a large bundle of tubes around which circulates a heat-transfer agent. The reactants spend an average of 1.8 s in the tubes, which have a void space of 100 ft^3. How many pounds of propylene must be added per hour in a mixture whose mole fractions are 0.07 propylene, 0.35 steam, and 0.58 air?

•• **5.120** Standard conditions are based on relevant environmental conditions. If normal average surface temperature and pressure on Venus are 730. K and 90 atm, respectively, what is the standard molar volume of an ideal gas on Venus?

•• **5.121** A barometer tube is 1.00×10^2 cm long and has a cross-sectional area of 1.20 cm^2. The height of the mercury column is 74.0 cm, and the temperature is 24°C. A small amount of N_2 is introduced into the evacuated space above the mercury, which causes the mercury level to drop to a height of 64.0 cm. How many grams of N_2 were introduced?

•• **5.122** What is the molar concentration of the cleaning solution formed when 10.0 L of ammonia gas at 33°C and 735 torr dissolves in enough water to give a final volume of 0.750 L?

•• **5.123** The Hawaiian volcano Kilauea emits an average of 1.5×10^3 m^3 of gas each day, when corrected to 298 K and 1.00 atm. The mixture contains gases that contribute to global warming and acid rain, and some are toxic. An atmospheric chemist analyzes a sample and finds the following mole fractions: 0.4896 CO_2, 0.0146 CO, 0.3710 H_2O, 0.1185 SO_2, 0.0003 S_2, 0.0047 H_2, 0.0008 HCl, and 0.0003 H_2S. How many metric tons (t) of each gas are emitted per year (1 t = 1000 kg)?

•• **5.124** To study a key fuel-cell reaction, a chemical engineer has 20.0-L tanks of H_2 and of O_2 and wants to use up both tanks to form 28.0 mol of water at 23.8°C. (a) Use the ideal gas law to find the pressure needed in each tank. (b) Use the van der Waals equation to find the pressure needed in each tank. (c) Compare the results from the two equations.

•• **5.125** For each of the following, which shows the greater deviation from ideal behavior at the same set of conditions? Explain.
(a) Argon or xenon   (b) Water vapor or neon
(c) Mercury vapor or radon (d) Water vapor or methane

•• **5.126** How many liters of gaseous hydrogen bromide at 29°C and 0.965 atm will a chemist need if she wishes to prepare 3.50 L of 1.20 M hydrobromic acid?

•• **5.127** A mixture consisting of 7.0 g of CO and 10.0 g of SO_2, two atmospheric pollutants, has a pressure of 0.33 atm when placed in a sealed container. What is the partial pressure of CO?

•• **5.128** Sulfur dioxide is used to make sulfuric acid. One method of producing it is by roasting mineral sulfides, for example,

$$FeS_2(s) + O_2(g) \longrightarrow SO_2(g) + Fe_2O_3(s) \quad \text{[unbalanced]}$$

A production error leads to the sulfide being placed in a 950-L vessel with insufficient oxygen. The partial pressure of O_2 is 0.64 atm, and the total pressure is initially 1.05 atm, with the balance N_2. The reaction is run until 85% of the O_2 is consumed, and the vessel is then cooled to its initial temperature. What is the total pressure and partial pressure of each gas in the vessel?

••• **5.129** A mixture of CO_2 and Kr weighs 35.0 g and exerts a pressure of 0.708 atm in its container. Since Kr is expensive, you wish to recover it from the mixture. After the CO_2 is completely removed by absorption with NaOH(s), the pressure in the container is 0.250 atm. How many grams of CO_2 were originally present? How many grams of Kr can you recover?

•• **5.130** When a car accelerates quickly, the passengers feel a force that presses them back into their seats, but a balloon filled with helium floats forward. Why?

•• **5.131** Gases such as CO are gradually oxidized in the atmosphere, not by O_2 but by the hydroxyl radical, ·OH, a hydroxide ion with one fewer electron. At night, the ·OH concentration is nearly zero but increases to 2.5×10^{12} molecules/m^3 in polluted air during the day. At daytime conditions of 1.00 atm and 22°C, what is the partial pressure and mole percent of ·OH in air?

••• **5.132** Aqueous sulfurous acid (H_2SO_3) was made by dissolving 0.200 L of sulfur dioxide gas at 19°C and 745 mmHg in water to yield 500.0 mL of solution. The acid solution required 10.0 mL of sodium hydroxide solution to reach the titration end point. What was the molarity of the sodium hydroxide solution?

•• **5.133** During World War II, a portable source of hydrogen gas was needed for weather balloons, and solid metal hydrides were the most convenient form. Many metal hydrides react with water to generate the metal hydroxide and hydrogen. Two candidates were lithium hydride and magnesium hydride. What volume of gas is formed from 1.00 lb of each hydride at 750. torr and 27°C?

•• **5.134** The lunar surface reaches 370 K at midday. The atmosphere consists of neon, argon, and helium at a total pressure of only 2×10^{-14} atm. Calculate the rms speed of each component in the lunar atmosphere. [Use $R = 8.314$ J/(mol·K) and $\mathcal{M}$ in kg/mol.]

••• **5.135** A person inhales air richer in O_2 and exhales air richer in CO_2 and water vapor. During each hour of sleep, a person exhales a total of about 300 L of this CO_2-enriched and H_2O-enriched air. (a) If the partial pressures of CO_2 and H_2O in exhaled air are each 30.0 torr at 37.0°C, calculate the masses of CO_2 and of H_2O exhaled in 1 h of sleep. (b) How many grams of body mass does the person lose in an 8-h sleep if all the CO_2 and H_2O exhaled come from the metabolism of glucose?

$$C_6H_{12}O_6(s) + 6O_2(g) \longrightarrow 6CO_2(g) + 6H_2O(g)$$

•• **5.136** Popcorn pops because the horny endosperm, a tough, elastic material, resists gas pressure within the heated kernel until it reaches explosive force. A 0.25-mL kernel has a water content of 1.6% by mass, and the water vapor reaches 170°C and 9.0 atm before the kernel ruptures. Assume the water vapor can occupy 75% of the kernel's volume. (a) What is the mass of the kernel? (b) How many milliliters would this amount of water vapor occupy at 25°C and 1.00 atm?

•• **5.137** Sulfur dioxide emissions from coal-based power plants are removed by *flue-gas desulfurization*. The flue gas passes through a scrubber, and a slurry of wet calcium carbonate reacts with it to form carbon dioxide and calcium sulfite. The calcium sulfite then reacts with oxygen to form calcium sulfate, which is sold as gypsum. (a) If the sulfur dioxide concentration is 1000 times higher than its mole fraction in clean dry air (2×10^{-10}), how much calcium sulfate (kg) can be made from scrubbing 4 GL of flue gas (1 GL = 1×10^9 L)? A state-of-the-art scrubber removes at least 95% of the sulfur dioxide. (b) If the mole fraction of oxygen in air is 0.209, what volume (L) of air at 1.00 atm and 25°C is needed to react with all the calcium sulfite?

••• **5.138** Many water treatment plants use chlorine gas to kill microorganisms before the water is released for residential use. A plant engineer has to maintain the chlorine pressure in a tank below the 85.0-atm rating and, to be safe, decides to fill the tank to 80.0% of this maximum pressure. (a) How many moles of Cl_2 gas can be kept in the 850.-L tank at 298 K if she uses the ideal gas law in the calculation? (b) What is the tank pressure if she uses the van der Waals equation for this amount of gas? (c) Did the engineer fill the tank to the desired pressure?

•• **5.139** What would you observe if you tilted a barometer 30° from the vertical? Explain.

•• **5.140** In A, the picture depicts a cylinder with 0.1 mol of a gas that behaves ideally. Choose the cylinder (B, C, or D) that correctly represents the volume of the gas after each of the following changes. If none of the cylinders is correct, specify "none."
(a) P is doubled at fixed n and T.
(b) T is reduced from 400 K to 200 K at fixed n and P.
(c) T is increased from 100°C to 200°C at fixed n and P.
(d) 0.1 mol of gas is added at fixed P and T.
(e) 0.1 mol of gas is added and P is doubled at fixed T.

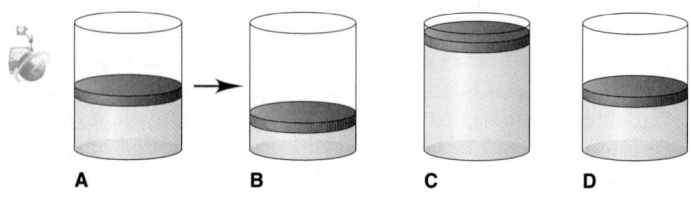

A B C D

•• **5.141** Ammonia is essential to so many industries that, on a molar basis, it is the most heavily produced substance in the world. Calculate P_{IGL} and P_{VDW} (in atm) of 51.1 g of ammonia in a 3.000-L container at 0°C and 400.°C, the industrial temperature. (From Table 5.5, for NH_3, $a = 4.17$ atm·L²/mol² and $b = 0.0371$ L/mol.)

•• **5.142** A 6.0-L flask contains a mixture of methane (CH_4), argon, and helium at 45°C and 1.75 atm. If the mole fractions of helium and argon are 0.25 and 0.35, respectively, how many molecules of methane are present?

•• **5.143** A large portion of metabolic energy arises from the biological combustion of glucose:

$$C_6H_{12}O_6(s) + 6O_2(g) \longrightarrow 6CO_2(g) + 6H_2O(g)$$

(a) If this reaction is carried out in an expandable container at 37°C and 780. torr, what volume of CO_2 is produced from 20.0 g of glucose and excess O_2? (b) If the reaction is carried out at the same conditions with the stoichiometric amount of O_2, what is the partial pressure of each gas when the reaction is 50% complete (10.0 g of glucose remains)?

•• **5.144** What is the average kinetic energy and rms speed of N_2 molecules at STP? Compare these values with those of H_2 molecules at STP. [Use $R = 8.314$ J/(mol·K) and $\mathcal{M}$ in kg/mol.]

•• **5.145** According to government standards, the *8-h threshold limit value* is 5000 ppmv for CO_2 and 0.1 ppmv for Br_2 (1 ppmv is 1 part by volume in 10^6 parts by volume). Exposure to either gas

for 8 h above these limits is unsafe. At STP, which of the following would be unsafe for 8 h of exposure?
(a) Air with a partial pressure of 0.2 torr of Br_2
(b) Air with a partial pressure of 0.2 torr of CO_2
(c) 1000 L of air containing 0.0004 g of Br_2 gas
(d) 1000 L of air containing 2.8×10^{22} molecules of CO_2

•• **5.146** One way to prevent emission of the pollutant NO from industrial plants is by a catalyzed reaction with NH_3:

$$4NH_3(g) + 4NO(g) + O_2(g) \xrightarrow{\text{catalyst}} 4N_2(g) + 6H_2O(g)$$

(a) If the NO has a partial pressure of 4.5×10^{-5} atm in the flue gas, how many liters of NH_3 are needed per liter of flue gas at 1.00 atm? (b) If the reaction takes place at 1.00 atm and 365°C, how many grams of NH_3 are needed per kL of flue gas?

••• **5.147** An equimolar mixture of Ne and Xe is accidentally placed in a container that has a tiny leak. After a short while, a very small proportion of the mixture has escaped. What is the mole fraction of Ne in the effusing gas?

••• **5.148** One way to utilize naturally occurring uranium (0.72% ^{235}U and 99.27% ^{238}U) as a nuclear fuel is to enrich it (increase its ^{235}U content) by allowing gaseous UF_6 to effuse through a porous membrane (see the margin note, p. 215). From the relative rates of effusion of $^{235}UF_6$ and $^{238}UF_6$, find the number of steps needed to produce uranium that is 3.0 mole % ^{235}U, the enriched fuel used in many nuclear reactors.

••• **5.149** A slight deviation from ideal behavior exists even at normal conditions. If it behaved ideally, 1 mol of CO would occupy 22.414 L and exert 1 atm pressure at 273.15 K. Calculate P_{VDW} for 1.000 mol of CO at 273.15 K. $\left(\text{Use } R = 0.08206 \dfrac{\text{atm·L}}{\text{mol·K}}\right)$

••• **5.150** In preparation for a combustion demonstration, a professor fills a balloon with equal molar amounts of H_2 and O_2, but the demonstration has to be postponed until the next day. During the night, both gases leak through pores in the balloon. If 35% of the H_2 leaks, what is the O_2/H_2 ratio in the balloon the next day?

••• **5.151** Phosphorus trichloride is important in the manufacture of insecticides, fuel additives, and flame retardants. Phosphorus has only one naturally occurring isotope, ^{31}P, whereas chlorine has two, ^{35}Cl (75%) and ^{37}Cl (25%). (a) What different molecular masses (amu) can be found for PCl_3? (b) Which is the most abundant? (c) What is the ratio of the effusion rates of the heaviest and the lightest PCl_3 molecules?

•• **5.152** A truck tire has a volume of 218 L and is filled with air to 35.0 psi at 295 K. After a drive, the air heats up to 318 K. (a) If the tire volume is constant, what is the pressure? (b) If the tire volume increases 2.0%, what is the pressure? (c) If the tire leaks 1.5 g of air per minute and the temperature is constant, how many minutes will it take for the tire to reach the original pressure of 35.0 psi ($\mathcal{M}$ of air = 28.8 g/mol)?

•• **5.153** Allotropes are different molecular forms of an element, such as dioxygen (O_2) and ozone (O_3). (a) What is the density of each oxygen allotrope at 0°C and 760 torr? (b) Calculate the ratio of densities, d_{O_3}/d_{O_2}, and explain the significance of this number.

••• **5.154** When gaseous F_2 and solid I_2 are heated to high temperatures, the I_2 sublimes and gaseous iodine heptafluoride forms. If 350. torr of F_2 and 2.50 g of solid I_2 are put into a 2.50-L container at 250. K and the container is heated to 550. K, what is the final pressure? What is the partial pressure of I_2 gas?

The Wonder of a Burning Match The chemical reaction that occurs when a match burns releases energy, as do many other processes. Still other processes absorb energy. In this chapter, you will begin to understand the importance of the changes in energy that occur every day and how we measure them.

Thermochemistry: Energy Flow and Chemical Change

6.1 Forms of Energy and Their Interconversion
System and Surroundings
Energy Flow to and from a System
Heat and Work
Energy Conservation
Units of Energy
State Functions

6.2 Enthalpy: Heats of Reaction and Chemical Change
Meaning of Enthalpy
Comparing ΔE and ΔH
Exothermic and Endothermic Processes
Types of Enthalpy Change

6.3 Calorimetry: Laboratory Measurement of Heats of Reaction
Specific Heat Capacity
Practice of Calorimetry

6.4 Stoichiometry of Thermochemical Equations

6.5 Hess's Law of Heat Summation

6.6 Standard Heats of Reaction (ΔH°_{rxn})
Formation Equations
Determining ΔH°_{rxn} from ΔH°_{f}

Concepts & Skills to Review
before you study this chapter

- energy and its interconversion (Section 1.1)
- distinction between heat and temperature (Section 1.5)
- nature of chemical bonding (Section 2.7)
- calculations of reaction stoichiometry (Section 3.4)
- properties of the gaseous state (Section 5.1)
- relation between kinetic energy and temperature (Section 5.6)

Literature Pétursson, Sigthór. "Three Forms of Energy," *J. Chem. Educ.* 2003, *80*, 776.

● **Wherever You Look, There Is a System**
In the example of the weight hitting the ground, if we define the falling weight as the system, the soil and pebbles that are moved and warmed are the surroundings. An astronomer may define a galaxy as the system and nearby galaxies as the surroundings. An ecologist studying African wildlife can define a zebra herd as the system and other animals, plants, and water supplies as the surroundings. A microbiologist may define a certain cell as the system and the extracellular solution as the surroundings. Thus, in general, it is the experiment and the experimenter that define the system and the surroundings.

Whenever matter changes composition, the energy content of the matter changes also. During a chemical change, as, for example, when a candle burns, the wax and oxygen reactants contain more energy than the gaseous CO_2 and H_2O products, and this difference in energy is *released* as heat and light. In contrast, some of the energy in a flash of lightning is *absorbed* when lower energy N_2 and O_2 in the air react to form higher energy NO. Energy content changes during a physical change, too. For example, energy is *absorbed* when snow melts and is *released* when water vapor condenses to rain.

As you probably already know, the production and utilization of energy have an enormous impact on society. Some of the largest industries manufacture products that release, absorb, or change the flow of energy. Common fuels—oil, wood, coal, and natural gas—release energy for heating and for powering combustion engines and steam turbines. Fertilizers enhance the ability of crops to absorb solar energy and convert it to the chemical energy of food, which our bodies then convert into other forms of energy. Metal products are often used to increase the flow of energy, while plastic, fiberglass, and ceramic materials serve as insulators that limit the flow of energy.

IN THIS CHAPTER . . . We investigate the heat, or *thermal energy,* associated with changes in matter. First, we examine some basic ideas of **thermodynamics,** the study of heat and its transformations. (The discussion begins here and is continued in Chapter 20.) Our focus here is on **thermochemistry,** the branch of thermodynamics that deals with the heat involved in chemical and physical change and especially with the concept of *enthalpy.* We describe how heat is measured in a calorimeter and how the quantity of heat released or absorbed is related to the amounts of substances involved in a reaction. You'll learn how to combine equations to obtain the heat change for another equation and see the importance of standardizing conditions in studying the heat change. The chapter ends with an overview of current and future energy sources and the conflicts between energy demand, environmental quality, and climate change.

6.1 FORMS OF ENERGY AND THEIR INTERCONVERSION

In Chapter 1, we discussed the fact that all energy is either potential or kinetic, and that these forms are convertible from one to the other. An object has potential energy by virtue of its position and kinetic energy by virtue of its motion. The potential energy of a weight raised above the ground is converted to kinetic energy as it falls (see Figure 1.3, p. 9). When the weight hits the ground, it transfers some of that kinetic energy to the soil and pebbles, causing them to move, and thereby doing *work*. In addition, some of the transferred kinetic energy appears as *heat*, as it slightly warms the soil and pebbles. Thus, the potential energy of the weight is converted to kinetic energy, which is transferred to the ground as work and heat.

Modern atomic theory allows us to consider other forms of energy—solar, electrical, nuclear, and chemical—as examples of potential and kinetic energy on the atomic and molecular scales. No matter what the details of the situation, *when energy is transferred from one object to another, it appears as work and/or heat*. In this section, we examine this idea in terms of the loss or gain of energy that takes place during a chemical or physical change.

The System and Its Surroundings

In order to observe and measure a change in energy, we must first define the **system**—the part of the universe that we are going to focus on. The moment that we define the system, everything else relevant to the change is defined as the **surroundings**. ●

ARIS Presentation Center,
Photo Library

Figure 6.1 shows a typical chemical system and its surroundings: the system is the contents of the flask; the flask itself, the other equipment, and perhaps the rest of the laboratory are the surroundings. In principle, the rest of the universe is the surroundings, but in practice, we need to consider only the portions of the universe relevant to the system. That is, it's not likely that a thunderstorm in central Asia or a methane blizzard on Neptune will affect the contents of the flask, but the temperature, pressure, and humidity of the lab might.

Energy Flow to and from a System

Each particle in a system has potential energy and kinetic energy, and the sum of these energies for all the particles in the system is the **internal energy, E,** of the system (some texts use the symbol U). When a chemical system, such as the contents of the flask in Figure 6.1, changes from reactants to products and the products return to the starting temperature, the internal energy has changed. To determine this change, ΔE, we measure the difference between the system's internal energy *after* the change (E_{final}) and *before* the change ($E_{initial}$):

$$\Delta E = E_{final} - E_{initial} = E_{products} - E_{reactants} \qquad (6.1)$$

where Δ (Greek *delta*) means "change (or difference) in." Note especially that Δ refers to the *final state of the system **minus** the initial state.*

Because the total energy must be conserved, *a change in the energy of the system is always accompanied by an **opposite** change in the energy of the surroundings.* We often represent this change with an *energy diagram* in which the final and initial states are horizontal lines on a vertical energy axis. The change in internal energy, ΔE, is the difference between the heights of the two lines. A system can change its internal energy in one of two ways:

1. By losing some energy *to* the surroundings, as shown in Figure 6.2A:

$$E_{final} < E_{initial} \qquad \Delta E < 0$$

2. By gaining some energy *from* the surroundings, as shown in Figure 6.2B:

$$E_{final} > E_{initial} \qquad \Delta E > 0$$

Note that the change in energy is always energy *transfer* from system to surroundings, or vice versa.

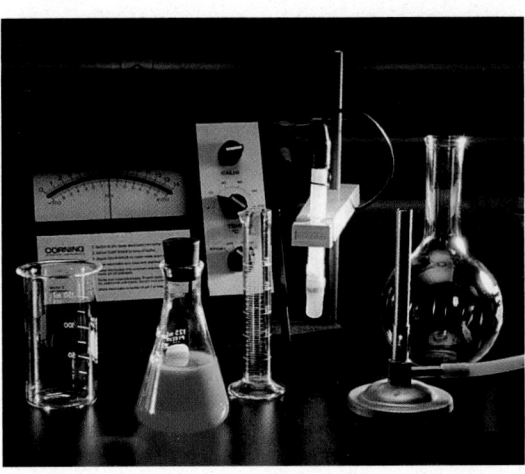

Figure 6.1 A chemical system and its surroundings. Once the contents of the flask (the orange solution) are defined as the system, the flask and the laboratory become defined as the surroundings.

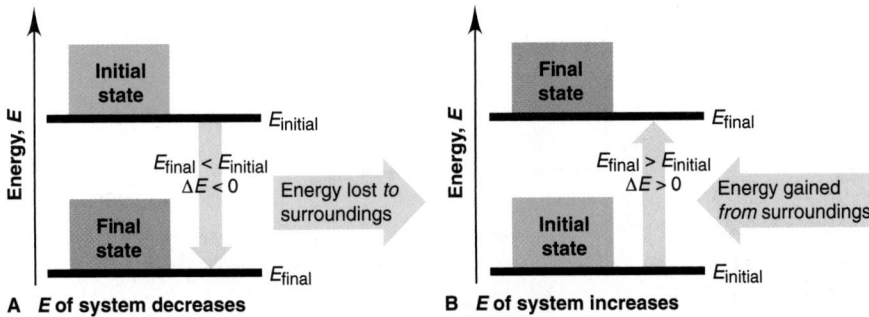

A *E* of system decreases   **B** *E* of system increases

Figure 6.2 Energy diagrams for the transfer of internal energy (E) between a system and its surroundings. A, When the internal energy of a system *decreases*, the change in energy (ΔE) is lost *to* the surroundings; therefore, ΔE of the system ($E_{final} - E_{initial}$) is negative. **B,** When the system's internal energy *increases*, ΔE is gained *from* the surroundings and is positive. Note that the vertical yellow arrow, which signifies the direction of the change in energy, *always* has its tail at the initial state and its head at the final state.

Heat and Work: Two Forms of Energy Transfer

Literature Wynn, Charles M. Sr. "Heat Flow vs. Cash Flow: A Banking Analogy," *J. Chem. Educ.* **1997,** *74,* 397.

Just as we saw when a weight hits the ground, energy transfer outward from the system or inward from the surroundings can appear in two forms, heat and work. **Heat** (or *thermal energy,* symbol *q*) is the energy transferred between a system and its surroundings as a result of a difference in their temperatures only. Energy in the form of heat is transferred from hot soup (system) to the bowl, air, and table (surroundings) because the surroundings have a lower temperature. All other forms of energy transfer (mechanical, electrical, and so on) involve some type of **work (*w*),** the energy transferred when an object is moved by a force. When you (system) kick a football (surroundings), energy is transferred as work to move the ball. When you inflate the ball, the inside air (system) exerts a force on the inner wall of the ball and nearby air (surroundings) and does work to move it outward.

The total change in a system's internal energy is the sum of the energy transferred as heat and/or work:

$$\Delta E = q + w \qquad\qquad (6.2)$$

The numerical values of *q* and *w* (and thus ΔE) can be either positive or negative, depending on the change the *system* undergoes. In other words, *we define the sign of the energy transfer from the system's perspective.* Energy coming *into* the system is *positive.* Energy going *out from* the system is *negative.* Of the innumerable changes possible in the system's internal energy, we'll examine the four simplest—two that involve only heat and two that involve only work.

Demo Hur, Chinhyu; Solomon, Sally; Wetzel, Christy. "Demonstrating Heat Changes on the Overhead Projector with a Projecting Thermometer," *J. Chem. Educ.* **1998,** *75,* 51.

Energy Transfer as Heat Only For a system that does no work but transfers energy only as heat (*q*), we know that $w = 0$. Therefore, from Equation 6.2, we have $\Delta E = q + 0 = q$. There are two possibilities:

1. *Heat flowing **out from** a system.* Suppose a sample of hot water is the system; then, the beaker containing it and the rest of the lab are the surroundings. The water transfers energy as heat to the surroundings until the temperature of the water equals that of the surroundings. The system's energy decreases as heat flows *out from* the system, so the final energy of the system is less than its initial energy. Heat was lost by the system, so *q is negative,* and therefore ΔE *is negative.* Figure 6.3A shows this situation.

2. *Heat flowing **into** a system.* If the system consists of ice water, it gains energy as heat from the surroundings until the temperature of the water equals that of the surroundings. In this case, energy is transferred *into* the system, so the final energy of the system is higher than its initial energy. Heat was gained by the system, so *q is positive,* and therefore ΔE *is positive* (Figure 6.3B). ●

● **Thermodynamics in the Kitchen**
The air in a refrigerator (surroundings) has a lower temperature than a newly added piece of food (system), so the food loses energy as heat to the refrigerator air, $q < 0$. The air in a hot oven (surroundings) has a higher temperature than a newly added piece of food (system), so the food gains energy as heat from the oven air, $q > 0$.

Figure 6.3 A system transferring energy as heat only. A, Hot water (the system, sys) transfers energy as heat (*q*) *to* the surroundings (surr) until $T_{sys} = T_{surr}$. Here $E_{initial} > E_{final}$ and $w = 0$, so $\Delta E < 0$ and the sign of *q* is negative. **B,** Ice water gains energy as heat (*q*) *from* the surroundings until $T_{sys} = T_{surr}$. Here $E_{initial} < E_{final}$ and $w = 0$, so $\Delta E > 0$ and the sign of *q* is positive.

Figure 6.4 A system losing energy as work only. The internal energy of the system decreases as the reactants form products because the $H_2(g)$ does work (w) *on* the surroundings by pushing back the piston. The reaction vessel is insulated, so $q = 0$. Here $E_{initial} > E_{final}$, so $\Delta E < 0$ and the sign of w is negative.

ARIS Presentation Center,
Art Library

Energy Transfer as Work Only

For a system that transfers energy only as work (w), $q = 0$; therefore, $\Delta E = 0 + w = w$. There are two possibilities:

1. *Work done **by** a system.* Consider the reaction between zinc and hydrochloric acid as it takes place in an insulated container attached to a piston-cylinder assembly. We define the system as the atoms that make up the substances. In the initial state, the system's internal energy is that of the atoms in the form of the reactants, metallic Zn and aqueous H^+ and Cl^- ions. In the final state, the system's internal energy is that of the same atoms in the form of the products, H_2 gas and aqueous Zn^{2+} and Cl^- ions:

$$Zn(s) + 2H^+(aq) + 2Cl^-(aq) \longrightarrow H_2(g) + Zn^{2+}(aq) + 2Cl^-(aq)$$

As the H_2 gas forms, some of the internal energy is used *by* the system to do work *on* the surroundings and push the piston outward. Energy is lost by the system as work, so *w is negative* and ΔE *is negative,* as you see in Figure 6.4. The H_2 gas is doing **pressure-volume work (PV work),** the type of work in which a volume changes against an external pressure. The work done here is not very useful because it simply pushes back the piston and outside air. But, if the system is a ton of burning coal and O_2, and the surroundings are a locomotive engine, much of the internal energy lost from the system does the work of moving a train.

2. *Work done **on** a system.* If we increase the external pressure on the piston in Figure 6.4, the system gains energy because work is done *on* the system *by* the surroundings: *w is positive,* so ΔE *is positive.*

Table 6.1 summarizes the sign conventions for q and w and their effect on the sign of ΔE.

Animation: Energy Flow
ARIS Presentation Center
Chapter 6

ARIS Presentation Center,
Tables Library

Table 6.1	**The Sign Conventions* for q, w, and ΔE**			
q	+	w	=	ΔE
+		+		+
+		−		Depends on *sizes* of q and w
−		+		Depends on *sizes* of q and w
−		−		−

*For q: + means system *gains* heat; − means system *loses* heat.
For w: + means work done *on* system; − means work done *by* system.

The Law of Energy Conservation

As you've seen, when a system gains energy, the surroundings lose it, and when a system loses energy, the surroundings gain it. Energy can be converted from one form to another as these transfers take place, but it cannot simply appear or disappear—it cannot be created or destroyed. The **law of conservation of energy** restates this basic observation as follows: *the total energy of the universe is constant.* This law is also known as the **first law of thermodynamics.**

Conservation of energy applies everywhere. As gasoline burns in a car engine, the released energy appears as an equivalent amount of heat and work. The heat warms the car parts, passenger compartment, and surrounding air. The work appears as mechanical energy to turn the car's wheels and belts. That energy is converted further into the electrical energy of the clock and radio, the radiant energy of the headlights, the chemical energy of the battery, the heat due to friction, and so forth. If you took the sum of all these energy forms, you would find that it equals the change in energy between the reactants and products as the gasoline is burned. Complex biological processes also obey energy conservation. Through photosynthesis, green plants convert radiant energy from the Sun into chemical energy, transforming low-energy CO_2 and H_2O into high-energy carbohydrates (such as wood) and O_2. When the wood is burned in air, those low-energy compounds form again, and the energy difference is released to the surroundings.

Thus, energy transfers between system and surroundings can be in the forms of heat and/or various types of work—mechanical, electrical, radiant, chemical—but *the energy of the system plus the energy of the surroundings remains constant: energy is conserved.* A mathematical expression of the law of conservation of energy (first law of thermodynamics) is

$$\Delta E_{universe} = \Delta E_{system} + \Delta E_{surroundings} = 0 \qquad \textbf{(6.3)}$$

This profound idea pertains to all systems, from a burning match to the movement of continents, from the inner workings of your heart to the formation of the Solar System.

Units of Energy

The SI unit of energy is the **joule (J),** a derived unit composed of three base units:

$$1 \text{ J} = 1 \text{ kg·m}^2/\text{s}^2$$

Both heat and work are expressed in joules. Let's see how these units arise in the case of work. The work (w) done on a mass is the force (F) times the distance (d) that the mass moves: $w = F \times d$. A *force* changes the velocity of (accelerates) a mass. Velocity has units of meters per second (m/s), so acceleration (a) has units of m/s^2. Force, therefore, has units of mass (m, in kilograms) times acceleration:

$$F = m \times a \quad \text{in units of} \quad \text{kg·m/s}^2$$

Therefore, $\quad w = F \times d \quad$ has units of $\quad (\text{kg·m/s}^2) \times \text{m} = \text{kg·m}^2/\text{s}^2 = \text{J}$

Potential energy, kinetic energy, and PV work are combinations of the same physical quantities and are also expressed in joules.

The **calorie (cal)** is an older unit that was defined originally as the quantity of energy needed to raise the temperature of 1 g of water by 1°C (from 14.5°C to 15.5°C). The calorie is now defined in terms of the joule:

$$1 \text{ cal} \equiv 4.184 \text{ J} \qquad \text{or} \qquad 1 \text{ J} = \frac{1}{4.184} \text{ cal} = 0.2390 \text{ cal}$$

Since the quantities of energy involved in chemical reactions are usually quite large, chemists use the kilojoule (kJ), or sometimes the kilocalorie (kcal):

$$1 \text{ kJ} = 1000 \text{ J} = 0.2390 \text{ kcal} = 239.0 \text{ cal}$$

The nutritional Calorie (note the capital C), the unit that diet tables use to show the energy available from food, is actually a kilocalorie. The *British thermal unit (Btu),* a unit in engineering that you may have seen used to indicate energy output of appliances, is the quantity of energy required to raise the temperature of 1 lb of water by 1°F and is equivalent to 1055 J. In general, the SI unit (J or kJ) is used throughout this text. Some interesting quantities of energy appear in Figure 6.5.

ARIS Presentation Center, Worked Examples Library

Figure 6.5 Some interesting quantities of energy. Note that the vertical scale is exponential.

Transparency/ARIS Presentation Center, Art Library

SAMPLE PROBLEM 6.1 Determining the Change in Internal Energy of a System

PROBLEM When gasoline burns in a car engine, the heat released causes the products CO_2 and H_2O to expand, which pushes the pistons outward. Excess heat is removed by the car's cooling system. If the expanding gases do 451 J of work on the pistons and the system loses 325 J to the surroundings as heat, calculate the change in energy (ΔE) in J, kJ, and kcal.

PLAN We must define system and surroundings, assign signs to q and w, and then calculate ΔE with Equation 6.2. The system is the reactants and products, and the surroundings are the pistons, the cooling system, and the rest of the car. Heat is released by the system, so q is negative. Work is done by the system to push the pistons outward, so w is also negative. We obtain the answer in J and then convert it to kJ and kcal.

SOLUTION Calculating ΔE (from Equation 6.2) in J:

$$q = -325 \text{ J}$$
$$w = -451 \text{ J}$$
$$\Delta E = q + w = -325 \text{ J} + (-451 \text{ J})$$
$$= -776 \text{ J}$$

Converting from J to kJ:

$$\Delta E = -776 \text{ J} \times \frac{1 \text{ kJ}}{1000 \text{ J}}$$
$$= -0.776 \text{ kJ}$$

Converting from kJ to kcal:

$$\Delta E = -0.776 \text{ kJ} \times \frac{1 \text{ kcal}}{4.184 \text{ kJ}}$$
$$= -0.185 \text{ kcal}$$

CHECK The answer is reasonable: combustion releases energy from the system, so $E_{\text{final}} < E_{\text{initial}}$ and ΔE should be negative. Rounding shows that, since 4 kJ ≈ 1 kcal, nearly 0.8 kJ should be nearly 0.2 kcal.

FOLLOW-UP PROBLEM 6.1 In a reaction, gaseous reactants form a liquid product. The heat absorbed by the surroundings is 26.0 kcal, and the work done on the system is 15.0 Btu. Calculate ΔE (in kJ).

State Functions and the Path Independence of the Energy Change

An important point to understand is that there is no particular sequence by which the internal energy (E) of a system must change. This is because E is a **state function,** a property dependent only on the *current* state of the system (its composition, volume, pressure, and temperature), *not* on the path the system took to reach that state; the current state depends only on the *difference* between the final and initial states.

In fact, the energy change of a system can occur by countless combinations of heat (q) and work (w). No matter what the combination, however, the same

Figure 6.6 **Two different paths for the energy change of a system.** The change in internal energy when a given amount of octane burns in air is the same no matter how the energy is transferred. On the left, the fuel is burned in an open can, and the energy is lost almost entirely as heat. On the right, it is burned in a car engine; thus, a portion of the energy is lost as work to move the car, and less is lost as heat.

Literature Czerwinski, Curtis J.; Cordes, Tanya J. "A Methane Balloon Inflation Chamber," *J. Chem. Educ.* **2005,** *82,* 248.

Demo Shakhashiri, Bassam Z. "Explosive Reaction of Hydrogen and Oxygen," *Chemical Demonstrations, A Handbook for Teachers of Chemistry,* Vol. 1 (The University of Wisconsin Press, Madison, 1985),106.

overall energy change occurs, because ΔE *does **not** depend on how the change takes place.* As an example, let's define a system in its initial state as 1 mol of octane (a component of gasoline) together with enough O_2 to burn it. In its final state, the system is the CO_2 and H_2O that form (a fractional coefficient is needed for O_2 because we specified 1 mol of octane):

$$C_8H_{18}(l) + \tfrac{25}{2}O_2(g) \longrightarrow 8CO_2(g) + 9H_2O(g)$$

initial state ($E_{initial}$)　　　　　final state (E_{final})

Energy is released to warm the surroundings and/or do work on them, so ΔE is negative. Two of the ways the change can occur are shown in Figure 6.6. If we burn the octane in an open container, ΔE appears almost completely as heat (with a small amount of work done to push back the atmosphere). If we burn it in a car engine, a much larger portion ($\sim$30%) of ΔE appears as work that moves the car, with the rest used to heat the car, exhaust gases, and surrounding air. If we burn the octane in a lawn mower or a plane, ΔE appears as other combinations of work and heat.

Thus, even though the separate quantities of work and heat available from the change *do* depend on how the change occurs, the change in internal energy (the *sum* of the heat and work) *does not.* In other words, for a given change, ΔE *(sum of q and w) is constant, even though q and w can vary.* Thus, heat and work are *not* state functions because their values *do* depend on the path the system takes in undergoing the energy change. ●

The pressure (P) of an ideal gas or the volume (V) of water in a beaker are other examples of state functions. This path independence means that *changes in state functions—ΔE, ΔP, and ΔV—depend only on their initial and final states.* (Note that symbols for state functions, such as E, P, and V, are capitalized.)

● **Your Personal Financial State Function** The *balance* in your checkbook is a state function of your personal financial system. You can open a new account with a birthday gift of $50, or you can open a new account with a deposit of a $100 paycheck and then write two $25 checks. The two paths to the balance are different, but the balance (current state) is the same.

Section Summary

Energy is transferred as heat (q) when the system and surroundings are at different temperatures; energy is transferred as work (w) when an object is moved by a force. • Heat or work gained by a system ($q > 0$; $w > 0$) increases its internal energy (E); heat or work lost by the system ($q < 0$; $w < 0$) decreases E. The total change in the system's internal energy is the sum of the heat and work: $\Delta E = q + w$. Heat and work are measured in joules (J). • Energy is always conserved: it changes from one form into another, moving into or out of the system, but the total quantity of energy in the universe (system *plus* surroundings) is constant. • Energy is a state function; therefore, the same ΔE can occur through any combination of q and w.

6.2 ENTHALPY: HEATS OF REACTION AND CHEMICAL CHANGE

Most physical and chemical changes occur at virtually constant atmospheric pressure—a reaction in an open flask, the freezing of a lake, a drug response in an organism. In this section, we define a thermodynamic variable that makes it much easier to measure energy changes at constant pressure.

The Meaning of Enthalpy

To determine ΔE, we must measure both heat and work. The two most important types of chemical work are electrical work, the work done by moving charged particles (Chapter 21), and PV work, the work done by an expanding gas. We find the quantity of PV work done by multiplying the external pressure (P) by the change in volume of the gas (ΔV, or $V_{\text{final}} - V_{\text{initial}}$). In an open flask (or a cylinder with a weightless, frictionless piston), a gas does work by pushing back the atmosphere. As Figure 6.7 shows, this work is done *on* the surroundings, so it has a negative sign because the system loses energy:

$$w = -P\Delta V \qquad (6.4)$$

For reactions at constant pressure, a thermodynamic variable called **enthalpy** (**H**) eliminates the need to consider PV work separately. The enthalpy of a system is defined as the internal energy *plus* the product of the pressure and volume:

$$H = E + PV$$

The **change in enthalpy (ΔH)** is the change in internal energy *plus* the product of the constant pressure and the change in volume:

$$\Delta H = \Delta E + P\Delta V \qquad (6.5)$$

Combining Equations 6.2 ($\Delta E = q + w$) and 6.4 leads to a key point about ΔH:

$$\Delta E = q + w = q + (-P\Delta V) = q - P\Delta V$$

At constant pressure, we denote q as q_P and solve for it:

$$q_P = \Delta E + P\Delta V$$

Notice the right side of this equation is identical to the right side of Equation 6.5:

$$q_P = \Delta E + P\Delta V = \Delta H \qquad (6.6)$$

Thus, *the change in enthalpy equals the heat gained or lost at constant pressure*. With most changes occurring at constant pressure, ΔH is more relevant than ΔE and easier to find: *to find ΔH, measure q_P.*

We discuss the laboratory method for measuring the heat involved in a chemical or physical change in Section 6.3.

Figure 6.7 Pressure-volume work. When the volume (V) of a system increases by an amount ΔV against an external pressure (P), the system pushes back, and thus does PV work *on* the surroundings ($w = -P\Delta V$).

ARIS Presentation Center, Art Library

Literature Canagaratna, Sebastian G. "A Visual Aid in Enthalpy Calculations," *J. Chem. Educ.* **2000**, *77*, 1178.

Comparing ΔE and ΔH

Knowing the *enthalpy* change of a system tells us a lot about its *energy* change as well. In fact, because many reactions involve little (if any) PV work, most (or all) of the energy change occurs as a transfer of heat. Here are three cases:

1. *Reactions that do not involve gases.* Gases do not appear in many reactions (precipitation, many acid-base, and many redox reactions). For example,

$$2KOH(aq) + H_2SO_4(aq) \longrightarrow K_2SO_4(aq) + 2H_2O(l)$$

Because liquids and solids undergo *very* small volume changes, $\Delta V \approx 0$; thus $P\Delta V \approx 0$ and $\Delta H \approx \Delta E$.

2. *Reactions in which the amount (mol) of gas does **not** change.* When the total amount of gaseous reactants equals the total amount of gaseous products, $\Delta V = 0$, so $P\Delta V = 0$ and $\Delta H = \Delta E$. For example,

$$N_2(g) + O_2(g) \longrightarrow 2NO(g)$$

3. *Reactions in which the amount (mol) of gas **does** change.* In these cases, $P\Delta V \neq 0$. However, q_P is usually so much larger than $P\Delta V$ that ΔH is very close to ΔE. For instance, in the combustion of H_2, 3 mol of gas yields 2 mol of gas:

$$2H_2(g) + O_2(g) \longrightarrow 2H_2O(g)$$

In this reaction, $\Delta H = -483.6$ kJ and $P\Delta V = -2.5$ kJ, so (from Equation 6.5), we get

$$\Delta E = \Delta H - P\Delta V = -483.6 \text{ kJ} - (-2.5 \text{ kJ}) = -481.1 \text{ kJ}$$

Obviously, most of ΔE occurs as heat transfer, so $\Delta H \approx \Delta E$.

The key point to realize from these three cases is that *for many reactions, ΔH equals, or is very close to, ΔE.*

Exothermic and Endothermic Processes

Because E, P, and V are state functions, H is also a state function, which means that ΔH depends only on the *difference* between H_{final} and $H_{initial}$. The enthalpy change of a reaction, also called the **heat of reaction, ΔH_{rxn},** *always refers to H_{final} **minus** $H_{initial}$:*

$$\Delta H = H_{final} - H_{initial} = H_{products} - H_{reactants}$$

Therefore, because $H_{products}$ can be either more or less than $H_{reactants}$, the sign of ΔH indicates whether heat is absorbed or released in the change. We determine the sign of ΔH by imagining the heat as a "reactant" or "product." When methane burns in air, for example, we know that heat is produced, so we show it as a product (on the right):

$$CH_4(g) + 2O_2(g) \longrightarrow CO_2(g) + 2H_2O(g) + heat$$

Because heat is released to the surroundings, the products (1 mol of CO_2 and 2 mol of H_2O) must have less enthalpy than the reactants (1 mol of CH_4 and 2 mol of O_2). Therefore, ΔH ($H_{final} - H_{initial}$) is negative, as the **enthalpy diagram** in Figure 6.8A shows. An **exothermic** ("heat out") **process** *releases* heat and results in a *decrease* in the enthalpy of the system:

$$\text{Exothermic:} \qquad H_{final} < H_{initial} \qquad \Delta H < 0$$

An **endothermic** ("heat in") **process** *absorbs* heat and results in an *increase* in the enthalpy of the system. When ice melts, for instance, heat flows *into* the ice from the surroundings, so we show the heat as a reactant (on the left):

$$heat + H_2O(s) \longrightarrow H_2O(l)$$

Because heat is absorbed, the enthalpy of the liquid water is higher than that of the solid water, as Figure 6.8B shows. Therefore, ΔH ($H_{water} - H_{ice}$) is positive:

$$\text{Endothermic:} \qquad H_{final} > H_{initial} \qquad \Delta H > 0$$

In general, the value of an enthalpy change refers to reactants and products at the same temperature.

Demo Shakhashiri, Bassam Z. "Evaporation As an Endothermic Process," *Chemical Demonstrations, A Handbook for Teachers of Chemistry,* Vol. 3 (The University of Wisconsin Press, Madison, 1985), 249–251.

Demo Shakhashiri, Bassam Z. "Endothermic Reactions of Hydrated Barium Hydroxide and Ammonium Salts," *Chemical Demonstrations, A Handbook for Teachers of Chemistry,* Vol. 1 (The University of Wisconsin Press, Madison, 1985), 10–12.

Demo Shakhashiri, Bassam Z. "Chemical Cold Pack," *Chemical Demonstrations, A Handbook for Teachers of Chemistry,* Vol. 1 (The University of Wisconsin Press, Madison, 1985), 8–9.

Figure 6.8 **Enthalpy diagrams for exothermic and endothermic processes.** **A,** Methane burns with a decrease in enthalpy because heat *leaves* the system. Therefore, $H_{final} < H_{initial}$, and the process is exothermic: $\Delta H < 0$. **B,** Ice melts with an increase in enthalpy because heat *enters* the system. Since $H_{final} > H_{initial}$, the process is endothermic: $\Delta H > 0$.

Transparency/ARIS Presentation Center, Art Library

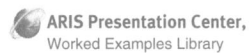

SAMPLE PROBLEM 6.2 Drawing Enthalpy Diagrams and Determining the Sign of ΔH

PROBLEM In each of the following cases, determine the sign of ΔH, state whether the reaction is exothermic or endothermic, and draw an enthalpy diagram:

(a) $H_2(g) + \frac{1}{2}O_2(g) \longrightarrow H_2O(l) + 285.8$ kJ

(b) 40.7 kJ $+ H_2O(l) \longrightarrow H_2O(g)$

PLAN From each equation, we see whether heat is a "product" (exothermic; $\Delta H < 0$) or a "reactant" (endothermic; $\Delta H > 0$). For exothermic reactions, reactants are above products on the enthalpy diagram; for endothermic reactions, reactants are below products. The ΔH arrow *always* points from reactants to products.

SOLUTION (a) Heat is a product (on the right), so $\Delta H < 0$ and the reaction is exothermic. The enthalpy diagram appears in the margin (*top*).

(b) Heat is a reactant (on the left), so $\Delta H > 0$ and the reaction is endothermic. The enthalpy diagram appears in the margin (*bottom*).

CHECK Substances that are on the same side of the equation as the heat have less enthalpy than substances on the other side, so make sure they are placed on the lower line of the diagram.

COMMENT ΔH values depend on conditions. In (b), for instance, $\Delta H = 40.7$ kJ at 1 atm and 100°C; at 1 atm and 25°C, $\Delta H = 44.0$ kJ.

FOLLOW-UP PROBLEM 6.2 When nitroglycerine decomposes, it causes a violent explosion and releases 5.72×10^3 kJ of heat per mole:

$$C_3H_5(NO_3)_3(l) \longrightarrow 3CO_2(g) + \tfrac{5}{2}H_2O(g) + \tfrac{1}{4}O_2(g) + \tfrac{3}{2}N_2(g)$$

Is the decomposition of nitroglycerine exothermic or endothermic? Draw an enthalpy diagram for it.

Some Important Types of Enthalpy Change

Some enthalpy changes are frequently studied and, therefore, they have special names:

- When 1 mol of a compound is produced from its elements, the heat of reaction is called the **heat of formation** (ΔH_f):

$$K(s) + \tfrac{1}{2}Br_2(l) \longrightarrow KBr(s) \quad \Delta H = \Delta H_f$$

- When 1 mol of a substance melts, the enthalpy change is called the **heat of fusion** (ΔH_{fus}):

$$NaCl(s) \longrightarrow NaCl(l) \quad \Delta H = \Delta H_{fus}$$

- When 1 mol of a substance vaporizes, the enthalpy change is called the **heat of vaporization** (ΔH_{vap}):

$$C_6H_6(l) \longrightarrow C_6H_6(g) \quad \Delta H = \Delta H_{vap}$$

We will encounter the heat of formation in discussions later in this chapter; other special enthalpy changes will be discussed in later chapters.

Section Summary

The change in enthalpy, ΔH, is equal to the heat lost or gained during a chemical or physical change that occurs at constant pressure, q_P. • In most cases, ΔH is equal, or very close, to ΔE. • A change that releases heat is exothermic ($\Delta H < 0$); a change that absorbs heat is endothermic ($\Delta H > 0$). • Some special enthalpy changes involve formation of a compound from its elements and physical changes from solid to liquid and liquid to gas.

6.3 CALORIMETRY: LABORATORY MEASUREMENT OF HEATS OF REACTION

Data about energy content and use are everywhere—the calories per serving of a slice of bread, the energy efficiency rating of a washing machine, the gas mileage of a new car, and so forth. How do we measure the heat released (or absorbed) by a change? To determine the energy content of a teaspoon of sugar, for example, you might think we can simply measure the enthalpies of the reactants (sucrose and O_2) and subtract them from the enthalpies of the products (CO_2 and H_2O). The problem is that the enthalpy (H) of a system in a given state cannot be measured because we have no starting point with which to compare it, no zero enthalpy. However, we *can* measure the *change* in enthalpy (ΔH) of a system. In this section, we'll see how ΔH values are determined.

To measure q_P, which is equal to ΔH, we construct "surroundings" that retain the heat, and we observe the temperature change. Then, we relate the quantity of heat released (or absorbed) to that temperature change through a physical property called the *specific heat capacity*.

Specific Heat Capacity

You know from everyday experience that the more you heat an object, the higher its temperature; that is, the quantity of heat (q) absorbed by an object is proportional to its temperature change:

$$q \propto \Delta T \quad \text{or} \quad q = \text{constant} \times \Delta T \quad \text{or} \quad \frac{q}{\Delta T} = \text{constant}$$

Every object has its own **heat capacity,** the quantity of heat required to change its temperature by 1 K. Heat capacity is the proportionality constant in the preceding equation:

$$\text{Heat capacity} = \frac{q}{\Delta T} \quad [\text{in units of J/K}]$$

A related property is **specific heat capacity (*c*),** the quantity of heat required to change the temperature of 1 *gram* of a substance by 1 K:*

$$\text{Specific heat capacity } (c) = \frac{q}{\text{mass} \times \Delta T} \quad [\text{in units of J/g·K}]$$

If we know c of the substance being heated (or cooled), we can measure its mass and temperature change and calculate the heat absorbed or released:

$$q = c \times \text{mass} \times \Delta T \qquad (6.7)$$

Notice that when an object gets hotter, ΔT (that is, $T_{\text{final}} - T_{\text{initial}}$) is positive. The object gains heat, so $q > 0$, as we expect. Similarly, when an object gets cooler, ΔT is negative; so $q < 0$ because heat is lost. Table 6.2 lists the specific heat capacities of some representative substances and materials.

Closely related to the specific heat capacity is the **molar heat capacity (*C*;** note capital letter), the quantity of heat required to change the temperature of 1 *mole* of a substance by 1 K:

$$\text{Molar heat capacity } (C) = \frac{q}{\text{moles} \times \Delta T} \quad [\text{in units of J/mol·K}]$$

The specific heat capacity of liquid water is 4.184 J/g·K, so

$$C \text{ of } H_2O(l) = 4.184 \frac{\text{J}}{\text{g·K}} \times \frac{18.02 \text{ g}}{1 \text{ mol}} = 75.40 \frac{\text{J}}{\text{mol·K}}$$

Literature Kimbrough, Doris R. "Heat Capacity, Body Temperature, and Hypothermia," *J. Chem. Educ.* 1998, *75*, 48.

Table 6.2 Specific Heat Capacities of Some Elements, Compounds, and Materials

Substance	Specific Heat Capacity (J/g·K)*
Elements	
Aluminum, Al	0.900
Graphite, C	0.711
Iron, Fe	0.450
Copper, Cu	0.387
Gold, Au	0.129
Compounds	
Water, $H_2O(l)$	4.184
Ethyl alcohol, $C_2H_5OH(l)$	2.46
Ethylene glycol, $(CH_2OH)_2(l)$	2.42
Carbon tetrachloride, $CCl_4(l)$	0.862
Solid materials	
Wood	1.76
Cement	0.88
Glass	0.84
Granite	0.79
Steel	0.45

*At 298 K (25°C).

*Some texts use the term *specific heat* in place of *specific heat capacity*. This usage is very common but somewhat incorrect. *Specific heat* is the ratio of the heat capacity of 1 g of a substance to the heat capacity of 1 g of H_2O and therefore has no units.

SAMPLE PROBLEM 6.3 Finding Quantity of Heat from Specific Heat Capacity

PROBLEM A layer of copper welded to the bottom of a skillet weighs 125 g. How much heat is needed to raise the temperature of the copper layer from 25°C to 300.°C? The specific heat capacity (c) of Cu is 0.387 J/g·K.

PLAN We know the mass and c of Cu and can find ΔT in °C, which equals ΔT in K. We use this ΔT and Equation 6.7 to solve for the heat.

SOLUTION Calculating ΔT and q:

$$\Delta T = T_{final} - T_{initial} = 300.°C - 25°C = 275°C = 275 \text{ K}$$

$$q = c \times \text{mass (g)} \times \Delta T = 0.387 \text{ J/g·K} \times 125 \text{ g} \times 275 \text{ K} = 1.33 \times 10^4 \text{ J}$$

CHECK Heat is absorbed by the copper bottom (system), so q is positive. Rounding shows that the arithmetic seems reasonable: $q \approx 0.4 \text{ J/g·K} \times 100 \text{ g} \times 300 \text{ K} = 1.2 \times 10^4 \text{ J}$.

FOLLOW-UP PROBLEM 6.3 Find the heat transferred (in kJ) when 5.50 L of ethylene glycol ($d = 1.11$ g/mL; see Table 6.2 for c) in a car radiator cools from 37.0°C to 25.0°C.

The Practice of Calorimetry

The **calorimeter** is a device used to measure the heat released (or absorbed) by a physical or chemical process. This apparatus is the "surroundings" that changes temperature when heat is transferred to or from the system. Two common types are the constant-pressure and constant-volume calorimeters.

Constant-Pressure Calorimetry

A "coffee-cup" calorimeter (Figure 6.9) is often used to measure the heat transferred (q_P) in processes open to the atmosphere. One common use is to find the specific heat capacity of a solid that does not react with or dissolve in water. The solid (system) is weighed, heated to some known temperature, and added to a sample of water (surroundings) of known temperature and mass in the calorimeter. With stirring, the final water temperature, which is also the final temperature of the solid, is measured.

The heat lost by the system ($-q_{sys}$, or $-q_{solid}$) is equal in magnitude but opposite in sign to the heat gained by the surroundings ($+q_{surr}$, or $+q_{H_2O}$):

$$-q_{solid} = q_{H_2O}$$

Substituting Equation 6.7 for each side of this equality gives

$$-(c_{solid} \times \text{mass}_{solid} \times \Delta T_{solid}) = c_{H_2O} \times \text{mass}_{H_2O} \times \Delta T_{H_2O}$$

All the quantities are known or measured except c_{solid}:

$$c_{solid} = -\frac{c_{H_2O} \times \text{mass}_{H_2O} \times \Delta T_{H_2O}}{\text{mass}_{solid} \times \Delta T_{solid}}$$

For example, suppose you heat a 25.64-g solid in a test tube to 100.00°C and carefully add it to 50.00 g of water in a coffee-cup calorimeter. The water temperature changes from 25.10°C to 28.49°C, and you want to find the specific heat capacity of the solid. Converting ΔT directly from °C to K, we know $\Delta T_{H_2O} = 3.39$ K (28.49°C − 25.10°C) and $\Delta T_{solid} = -71.51$ K (28.49°C − 100.00°C). Then, assuming all the heat lost by the solid is gained by the water, we have

$$c_{solid} = -\frac{c_{H_2O} \times \text{mass}_{H_2O} \times \Delta T_{H_2O}}{\text{mass}_{solid} \times \Delta T_{solid}} = -\frac{4.184 \text{ J/g·K} \times 50.00 \text{ g} \times 3.39 \text{ K}}{25.64 \text{ g} \times (-71.51 \text{ K})} = 0.387 \text{ J/g·K}$$

Follow-up Problem 6.4 applies this calculation, but Sample Problem 6.4 first shows how to find the heat of a reaction that takes place in the calorimeter.

Figure 6.9 Coffee-cup calorimeter. This apparatus is used to measure the heat at constant pressure (q_P).

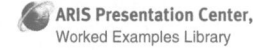

SAMPLE PROBLEM 6.4 Determining the Heat of a Reaction

PROBLEM You place 50.0 mL of 0.500 M NaOH in a coffee-cup calorimeter at 25.00°C and carefully add 25.0 mL of 0.500 M HCl, also at 25.00°C. After stirring, the final temperature is 27.21°C. Calculate q_{soln} (in J) and ΔH_{rxn} (in kJ/mol). (Assume the total volume is the sum of the individual volumes and that the final solution has the same density and specific heat capacity as water: $d = 1.00$ g/mL and $c = 4.184$ J/g·K.)

PLAN We first find the heat given off to the solution (q_{soln}) for the amounts given and then use the equation to find the heat per mole of reaction. We know the solution volumes (25.0 mL and 50.0 mL), so we can find their masses from the given density (1.00 g/mL). Multiplying their total mass by the change in T and the given c, we can find q_{soln}. Then, writing the balanced net ionic equation for the acid-base reaction, we use the volumes and the concentrations (0.500 M) to find moles of reactants (H^+ and OH^-) and, thus, product (H_2O). Dividing q_{soln} by the moles of water formed gives ΔH_{rxn} in kJ/mol.

SOLUTION Finding $mass_{soln}$ and ΔT_{soln}:

$$\text{Total mass (g) of solution} = (25.0 \text{ mL} + 50.0 \text{ mL}) \times 1.00 \text{ g/mL} = 75.0 \text{ g}$$

$$\Delta T = 27.21°C - 25.00°C = 2.21°C = 2.21 \text{ K}$$

Finding q_{soln}:

$$q_{soln} = c_{soln} \times mass_{soln} \times \Delta T_{soln} = (4.184 \text{ J/g·K})(75.0 \text{ g})(2.21 \text{ K}) = 693 \text{ J}$$

Writing the net ionic equation: $HCl(aq) + NaOH(aq) \longrightarrow H_2O(l) + NaCl(aq)$

$$H^+(aq) + OH^-(aq) \longrightarrow H_2O(l)$$

Finding moles of reactants and products:

$$\text{Moles of } H^+ = 0.500 \text{ mol/L} \times 0.0250 \text{ L} = 0.0125 \text{ mol } H^+$$

$$\text{Moles of } OH^- = 0.500 \text{ mol/L} \times 0.0500 \text{ L} = 0.0250 \text{ mol } OH^-$$

Therefore, H^+ is limiting, so 0.0125 mol of H_2O is formed.

Finding ΔH_{rxn}: Heat gained by the water was lost by the reaction; that is,

$$q_{soln} = -q_{rxn} = 693 \text{ J} \quad \text{so} \quad q_{rxn} = -693 \text{ J}$$

$$\Delta H_{rxn} \text{ (kJ/mol)} = \frac{q_{rxn}}{\text{mol } H_2O} \times \frac{1 \text{ kJ}}{1000 \text{ J}} = \frac{-693 \text{ J}}{0.0125 \text{ mol}} \times \frac{1 \text{ kJ}}{1000 \text{ J}} = -55.4 \text{ kJ/mol}$$

CHECK Rounding to check q_{soln} gives 4 J/g·K × 75 g × 2 K = 600 J. The volume of H^+ is half the volume of OH^-, so moles of H^+ determine moles of product. Taking the negative of q_{soln} to find ΔH_{rxn} gives −600 J/0.012 mol = -5×10^4 J/mol, or −50 kJ/mol.

FOLLOW-UP PROBLEM 6.4 In a purity check for industrial diamonds, a 10.25-carat (1 carat = 0.2000 g) diamond is heated to 74.21°C and immersed in 26.05 g of water in a constant-pressure calorimeter. The initial temperature of the water is 27.20°C. Calculate ΔT of the water and of the diamond ($c_{diamond} = 0.519$ J/g·K).

Constant-Volume Calorimetry

In the coffee-cup calorimeter, we assume all the heat is gained by the water, but some must be gained by the stirrer, thermometer, and so forth. For more precise work, as in constant-volume calorimetry, the *heat capacity of the entire calorimeter* must be known. One type of constant-volume apparatus is the *bomb calorimeter,* designed to measure very precisely the heat released in a combustion reaction. As Sample Problem 6.5 will show, this need for greater precision requires that we know (or determine) the heat capacity of the calorimeter.

Figure 6.10 depicts the preweighed combustible sample in a metal-walled chamber (the bomb), which is filled with oxygen gas and immersed in an insulated water bath fitted with motorized stirrer and thermometer. A heating coil connected to an electrical source ignites the sample, and the heat evolved raises the temperature of the bomb, water, and other calorimeter parts. Because we know the mass of the sample and the heat capacity of the entire calorimeter, we can use the measured ΔT to calculate the heat released.

Figure 6.10 A bomb calorimeter. This device (not drawn to scale) is used to measure the heat released in a combustion reaction at constant volume (q_V).

Labels: Electrical source; Motorized stirrer; Thermometer; System (combustible substance and compressed oxygen); O_2; Cutaway of steel bomb; Ignition coil; Cutaway of insulated jacket; Water bath; Heat being transferred

SAMPLE PROBLEM 6.5 Calculating the Heat of a Combustion Reaction

PROBLEM A manufacturer claims that its new dietetic dessert has "fewer than 10 Calories per serving." To test the claim, a chemist at the Department of Consumer Affairs places one serving in a bomb calorimeter and burns it in O_2 (heat capacity of the calorimeter = 8.151 kJ/K). The temperature increases 4.937°C. Is the manufacturer's claim correct?

PLAN When the dessert burns, the heat released is gained by the calorimeter:

$$-q_{sample} = q_{calorimeter}$$

To find the heat, we multiply the given heat capacity of the calorimeter (8.151 kJ/K) by ΔT (4.937°C).

SOLUTION Calculating the heat gained by the calorimeter:

$$q_{calorimeter} = \text{heat capacity} \times \Delta T = 8.151 \text{ kJ/K} \times 4.937 \text{ K} = 40.24 \text{ kJ}$$

Recall that 1 Calorie = 1 kcal = 4.184 kJ. Therefore, 10 Calories = 41.84 kJ, so the claim is correct.

CHECK A quick math check shows that the answer is reasonable: 8 kJ/K × 5 K = 40 kJ.

COMMENT Since the volume of the steel bomb is fixed, $\Delta V = 0$, and thus $P\Delta V = 0$. Thus, the energy change measured is the *heat at constant volume (q_V)*, which equals ΔE, not ΔH:

$$\Delta E = q + w = q_V + 0 = q_V$$

Recall from Section 6.2, however, that even though the number of moles of gas may change, ΔH is usually very close to ΔE. For example, ΔH is only 0.5% larger than ΔE for the combustion of H_2 and only 0.2% smaller for the combustion of octane.

FOLLOW-UP PROBLEM 6.5 A chemist burns 0.8650 g of graphite (a form of carbon) in a new bomb calorimeter, and CO_2 forms. If 393.5 kJ of heat is released per mole of graphite and T increases 2.613 K, what is the heat capacity of the bomb calorimeter?

Section Summary

We calculate ΔH of a process by measuring the heat at constant pressure (q_P). To do this, we determine ΔT of the surroundings and relate it to q_P through the mass of the substance and its specific heat capacity (c), the quantity of energy needed to raise the temperature of 1 g of the substance by 1 K. ● • Calorimeters measure the heat released from a system either at constant pressure ($q_P = \Delta H$) or at constant volume ($q_V = \Delta E$).

6.4 STOICHIOMETRY OF THERMOCHEMICAL EQUATIONS

A **thermochemical equation** is a balanced equation that includes the heat of reaction (ΔH_{rxn}). Keep in mind that the ΔH_{rxn} value shown refers to the *amounts (moles) of substances **and** their states of matter in that specific equation*. The enthalpy change of any process has two aspects:

1. *Sign.* The sign of ΔH depends on whether the reaction is exothermic (−) or endothermic (+). A forward reaction has the *opposite* sign of the reverse reaction.

   Decomposition of 2 mol of water to its elements (endothermic):

   $$2H_2O(l) \longrightarrow 2H_2(g) + O_2(g) \quad \Delta H_{rxn} = 572 \text{ kJ}$$

   Formation of 2 mol of water from its elements (exothermic):

   $$2H_2(g) + O_2(g) \longrightarrow 2H_2O(l) \quad \Delta H_{rxn} = -572 \text{ kJ}$$

2. *Magnitude.* The magnitude of ΔH is *proportional to the amount of substance* reacting.

   Formation of 1 mol of water from its elements (half the amount in the preceding equation):

   $$H_2(g) + \tfrac{1}{2}O_2(g) \longrightarrow H_2O(l) \quad \Delta H_{rxn} = -286 \text{ kJ}$$

● **Imagine an Earth Without Water** Liquid water has an unusually high specific heat capacity of nearly 4.2 J/g·K, about six times that of rock (~0.7 J/g·K). If Earth were devoid of oceans, the Sun's energy would heat a planet composed of rock. It would take only 0.7 J of energy to increase the temperature of each gram of rock by 1 K. Daytime temperatures would soar. The oceans also limit the temperature drop when the Sun sets, because the energy absorbed during the day is released at night. If Earth had a rocky surface, temperatures would be frigid every night.

Environmental application

Note that, in thermochemical equations, we often use fractional coefficients to specify the magnitude of ΔH_{rxn} for a *particular amount of substance*. Moreover, *in a particular reaction,* a certain amount of substance is thermochemically equivalent to a certain quantity of energy. In the reaction just shown,

286 kJ is thermochemically equivalent to 1 mol of $H_2(g)$

286 kJ is thermochemically equivalent to $\frac{1}{2}$ mol of $O_2(g)$

286 kJ is thermochemically equivalent to 1 mol of $H_2O(l)$

Just as we use stoichiometrically equivalent molar ratios to find amounts of substances, we use thermochemically equivalent quantities to find the heat of reaction for a given amount of substance. Also, just as we use molar mass (in g/mol of substance) to convert moles of a substance to grams, we use the heat of reaction (in kJ/mol of substance) to convert moles of a substance to an equivalent quantity of heat (in kJ). Figure 6.11 shows this new relationship, and the next sample problem applies it.

Figure 6.11 Summary of the relationship between amount (mol) of substance and the heat (kJ) transferred during a reaction.

 ARIS Presentation Center, Art Library

 ARIS Presentation Center, Worked Examples Library

SAMPLE PROBLEM 6.6 Using the Heat of Reaction (ΔH_{rxn}) to Find Amounts

PROBLEM The major source of aluminum in the world is bauxite (mostly aluminum oxide). Its thermal decomposition can be represented by

$$Al_2O_3(s) \xrightarrow{\Delta} 2Al(s) + \tfrac{3}{2}O_2(g) \quad \Delta H_{rxn} = 1676 \text{ kJ}$$

If aluminum is produced this way (see Comment), how many grams of aluminum can form when 1.000×10^3 kJ of heat is transferred?

PLAN From the balanced equation and the enthalpy change, we see that 2 mol of Al forms when 1676 kJ of heat is absorbed. With this equivalent quantity, we convert the given kJ transferred to moles formed and then convert moles to grams.

SOLUTION Combining steps to convert from heat transferred to mass of Al:

$$\text{Mass (g) of Al} = (1.000 \times 10^3 \text{ kJ}) \times \frac{2 \text{ mol Al}}{1676 \text{ kJ}} \times \frac{26.98 \text{ g Al}}{1 \text{ mol Al}} = 32.20 \text{ g Al}$$

CHECK The mass of aluminum seems correct: ~1700 kJ forms about 2 mol of Al (54 g), so 1000 kJ should form a bit more than half that amount (27 g).

COMMENT In practice, aluminum is not obtained by heating but by supplying electrical energy. Because ΔH is a state function, however, the total energy required for this chemical change is the same no matter how it occurs. (We examine the industrial method in Chapter 22.)

FOLLOW-UP PROBLEM 6.6 Organic hydrogenation reactions, in which H_2 and an "unsaturated" organic compound combine, are used in the food, fuel, and polymer industries. In the simplest case, ethene (C_2H_4) and H_2 form ethane (C_2H_6). If 137 kJ is given off per mole of C_2H_4 reacting, how much heat is released when 15.0 kg of C_2H_6 forms?

Section Summary

A thermochemical equation shows the balanced equation *and* its ΔH_{rxn}. The sign of ΔH for a forward reaction is opposite that for the reverse reaction. The magnitude of ΔH depends on the amount and physical state of the substance reacting and the ΔH per mole of substance. • We use the thermochemically equivalent amounts of substance and heat from the balanced equation as conversion factors to find the quantity of heat when a given amount of substance reacts.

Many reactions are difficult, even impossible, to carry out separately. A reaction may be part of a complex biochemical process; or it may take place only under extreme environmental conditions; or it may require a change in conditions while it is occurring. Even if we can't run a reaction in the lab, it is still possible to find its enthalpy change. One of the most powerful applications of the state-function property of enthalpy (H) allows us to find the ΔH of *any* reaction for which we can write an equation.

This application is based on **Hess's law of heat summation**: *the enthalpy change of an overall process is the sum of the enthalpy changes of its individual steps.* To use Hess's law, we imagine an overall reaction as the sum of a series of reaction steps, whether or not it really occurs that way. Each step is chosen because its ΔH is known. Because the overall ΔH depends only on the initial and final states, Hess's law says that we add together the known ΔH values for the steps to get the unknown ΔH of the overall reaction. Similarly, if we know the ΔH values for the overall reaction and all but one of the steps, we can find the unknown ΔH of that step.

Let's see how we apply Hess's law in the case of the oxidation of sulfur to sulfur trioxide, the central process in the industrial production of sulfuric acid and in the formation of acid rain. (To introduce the approach, we'll simplify the equations by using S as the formula for sulfur, rather than the more correct S_8.) When we burn S in an excess of O_2, sulfur dioxide (SO_2) forms, *not* sulfur trioxide (SO_3). Equation 1 shows this step and its ΔH. If we change conditions and then add more O_2, we can oxidize SO_2 to SO_3 (Equation 2). In other words, we cannot put S and O_2 in a calorimeter and find ΔH for the overall reaction of S to SO_3 (Equation 3). But, we *can* find it with Hess's law. The three equations are

$$\text{Equation 1:} \qquad S(s) + O_2(g) \longrightarrow SO_2(g) \qquad \Delta H_1 = -296.8 \text{ kJ}$$
$$\text{Equation 2:} \qquad 2SO_2(g) + O_2(g) \longrightarrow 2SO_3(g) \qquad \Delta H_2 = -198.4 \text{ kJ}$$
$$\text{Equation 3:} \qquad S(s) + \tfrac{3}{2}O_2(g) \longrightarrow SO_3(g) \qquad \Delta H_3 = ?$$

Hess's law tells us that if we manipulate Equations 1 and/or 2 so that they add up to Equation 3, then ΔH_3 is the sum of the manipulated ΔH values of Equations 1 and 2.

First, we identify Equation 3 as our "target" equation, the one whose ΔH we want to find, and we carefully note the number of moles of each reactant and product in it. We also note that ΔH_1 and ΔH_2 are the values for Equations 1 and 2 *as written*. Now we manipulate Equations 1 and/or 2 as follows to make them add up to Equation 3:

- Equations 1 and 3 contain the same amount of S, so we leave Equation 1 unchanged.
- Equation 2 has twice as much SO_3 as Equation 3, so we multiply it by $\tfrac{1}{2}$, being sure to halve ΔH_2 as well.
- With the targeted amounts of reactants and products now present, we add Equation 1 to the halved Equation 2 and cancel terms that appear on both sides:

$$\text{Equation 1:} \qquad S(s) + O_2(g) \longrightarrow SO_2(g) \qquad \Delta H_1 = -296.8 \text{ kJ}$$
$$\tfrac{1}{2}(\text{Equation 2}): \qquad SO_2(g) + \tfrac{1}{2}O_2(g) \longrightarrow SO_3(g) \qquad \tfrac{1}{2}(\Delta H_2) = -99.2 \text{ kJ}$$

Equation 3: $S(s) + O_2(g) + \cancel{SO_2(g)} + \tfrac{1}{2}O_2(g) \longrightarrow \cancel{SO_2(g)} + SO_3(g) \qquad \Delta H_3 = ?$

or, $\qquad\qquad\qquad\qquad S(s) + \tfrac{3}{2}O_2(g) \longrightarrow SO_3(g)$

Adding the ΔH values gives

$$\Delta H_3 = \Delta H_1 + \tfrac{1}{2}(\Delta H_2) = -296.8 \text{ kJ} + (-99.2 \text{ kJ}) = -396.0 \text{ kJ}$$

Once again, the key point is that H is a state function, so the overall ΔH depends on the difference between the initial and final enthalpies only. Hess's law tells us that the difference between the enthalpies of the reactants (1 mol of S and $\frac{3}{2}$ mol of O_2) and that of the product (1 mol of SO_3) is the same, whether S is oxidized directly to SO_3 (impossible) or through the formation of SO_2 (actual).

To summarize, calculating an unknown ΔH involves three steps:

1. Identify the target equation, the step whose ΔH is unknown, and note the number of moles of each reactant and product.
2. Manipulate the equations with known ΔH values so that the target numbers of moles of reactants and products are on the correct sides. Remember to:
   • Change the sign of ΔH when you reverse an equation.
   • Multiply numbers of moles and ΔH by the same factor.
3. Add the manipulated equations to obtain the target equation. All substances except those in the target equation must cancel. Add their ΔH values to obtain the unknown ΔH.

SAMPLE PROBLEM 6.7 Using Hess's Law to Calculate an Unknown ΔH

PROBLEM Two gaseous pollutants that form in auto exhaust are CO and NO. An environmental chemist is studying ways to convert them to less harmful gases through the following equation:

$$CO(g) + NO(g) \longrightarrow CO_2(g) + \tfrac{1}{2}N_2(g) \quad \Delta H = ?$$

Given the following information, calculate the unknown ΔH:

Equation A: $CO(g) + \tfrac{1}{2}O_2(g) \longrightarrow CO_2(g)$ $\Delta H_A = -283.0$ kJ
Equation B: $N_2(g) + O_2(g) \longrightarrow 2NO(g)$ $\Delta H_B = 180.6$ kJ

PLAN We note the numbers of moles of each substance in the target equation, manipulate Equations A and/or B *and* their ΔH values, and then add them together to obtain the target equation and the unknown ΔH.

SOLUTION Noting moles of substances in the target equation: There are 1 mol each of reactants CO and NO, 1 mol of product CO_2, and $\frac{1}{2}$ mol of product N_2.

Manipulating the given equations: Equation A has the same number of moles of CO and CO_2 as the target, so we leave it as written. Equation B has twice the needed amounts of N_2 and NO, and they are on the opposite sides from the target; therefore, we reverse Equation B, change the sign of ΔH_B, and multiply both by $\frac{1}{2}$:

$$\tfrac{1}{2}[2NO(g) \longrightarrow N_2(g) + O_2(g)] \quad \Delta H = -\tfrac{1}{2}(\Delta H_B) = -\tfrac{1}{2}(180.6 \text{ kJ})$$

or, $NO(g) \longrightarrow \tfrac{1}{2}N_2(g) + \tfrac{1}{2}O_2(g)$ $\Delta H = -90.3$ kJ

Adding the manipulated equations to obtain the target equation:

Equation A:	$CO(g) + \tfrac{1}{2}O_2(g) \longrightarrow CO_2(g)$	$\Delta H = -283.0$ kJ
$\tfrac{1}{2}$(Equation B reversed)	$NO(g) \longrightarrow \tfrac{1}{2}N_2(g) + \tfrac{1}{2}O_2(g)$	$\Delta H = -90.3$ kJ
Target:	$CO(g) + NO(g) \longrightarrow CO_2(g) + \tfrac{1}{2}N_2(g)$	$\Delta H = -373.3$ kJ

CHECK Obtaining the desired target equation is its own check. Be sure to remember to change the *sign* of ΔH for any equation you reverse.

FOLLOW-UP PROBLEM 6.7 Nitrogen oxides undergo many interesting reactions in the environment and in industry. Given the following information, calculate ΔH for the overall equation $2NO_2(g) + \tfrac{1}{2}O_2(g) \longrightarrow N_2O_5(s)$:

$$N_2O_5(s) \longrightarrow 2NO(g) + \tfrac{3}{2}O_2(g) \quad \Delta H = 223.7 \text{ kJ}$$
$$NO(g) + \tfrac{1}{2}O_2(g) \longrightarrow NO_2(g) \quad \Delta H = -57.1 \text{ kJ}$$

Section Summary

Because H is a state function, $\Delta H = H_{final} - H_{initial}$ and does not depend on how the reaction takes place. • Using Hess's law ($\Delta H_{total} = \Delta H_1 + \Delta H_2 + \cdots + \Delta H_n$), we can determine ΔH of any equation by manipulating the coefficients of other appropriate equations and their known ΔH values.

6.6 STANDARD HEATS OF REACTION (ΔH°_{rxn})

In this section, we see how Hess's law is used to determine the ΔH values of an enormous number of reactions. To begin we must take into account that thermodynamic variables, such as ΔH, vary somewhat with conditions. Therefore, to use heats of reaction, as well as other thermodynamic data that we will encounter in later chapters, chemists have established **standard states,** a set of specified conditions and concentrations:

- For a *gas,* the standard state is 1 atm* with the gas behaving ideally.
- For a substance in *aqueous solution,* the standard state is 1 *M* concentration.
- For a *pure substance* (element or compound), the standard state is usually the most stable form of the substance at 1 atm and the temperature of interest. In this text, that temperature is usually 25°C (298 K).†

We use the standard-state symbol (here shown typographically with a degree sign) to indicate these standard states. In other words, when the heat of reaction, ΔH_{rxn}, has been measured with all the reactants and products in their standard states, it is referred to as the **standard heat of reaction, ΔH°_{rxn}.**

Formation Equations and Their Standard Enthalpy Changes

In a **formation equation,** 1 mole of a compound forms from its elements. The **standard heat of formation (ΔH°_f)** is the enthalpy change for the formation equation when all the substances are in their standard states. For instance, the formation equation for methane (CH_4) is

$$C(graphite) + 2H_2(g) \longrightarrow CH_4(g) \quad \Delta H^\circ_f = -74.9 \text{ kJ}$$

Thus, the standard heat of formation of methane is -74.9 kJ/mol. Some other examples are

$$Na(s) + \tfrac{1}{2}Cl_2(g) \longrightarrow NaCl(s) \quad \Delta H^\circ_f = -411.1 \text{ kJ}$$
$$2C(graphite) + 3H_2(g) + \tfrac{1}{2}O_2(g) \longrightarrow C_2H_5OH(l) \quad \Delta H^\circ_f = -277.6 \text{ kJ}$$

Standard heats of formation have been tabulated for many substances. Table 6.3 shows ΔH°_f values for several, and a much more extensive table appears in Appendix B.

The values in Table 6.3 were selected to make two points:

1. *An element in its standard state is assigned a ΔH°_f value of zero.* For example, note that $\Delta H^\circ_f = 0$ for Na(s), but $\Delta H^\circ_f = 107.8$ kJ/mol for Na(g). These values mean that the gaseous state is *not* the most stable state of sodium at 1 atm and 298.15 K, and that heat is required to form Na(g). Note also that the standard state of chlorine is Cl_2 molecules, not Cl atoms. Several elements exist in different forms, only one of which is the standard state. Thus, the standard state of carbon is graphite, not diamond, so ΔH°_f of C(graphite) = 0. Similarly, the standard state of oxygen is dioxygen (O_2), not ozone (O_3), and the standard state of sulfur is S_8 in its rhombic crystal form, rather than its monoclinic form.

2. *Most compounds have a negative ΔH°_f.* That is, most compounds have exothermic formation reactions under standard conditions: *heat is given off when the compound forms.*

Table 6.3 Selected Standard Heats of Formation at 25°C (298 K)	
Formula	**ΔH°_f (kJ/mol)**
Calcium	
Ca(s)	0
CaO(s)	−635.1
CaCO₃(s)	−1206.9
Carbon	
C(graphite)	0
C(diamond)	1.9
CO(g)	−110.5
CO₂(g)	−393.5
CH₄(g)	−74.9
CH₃OH(l)	−238.6
HCN(g)	135
CS₂(l)	87.9
Chlorine	
Cl(g)	121.0
Cl₂(g)	0
HCl(g)	−92.3
Hydrogen	
H(g)	218.0
H₂(g)	0
Nitrogen	
N₂(g)	0
NH₃(g)	−45.9
NO(g)	90.3
Oxygen	
O₂(g)	0
O₃(g)	143
H₂O(g)	−241.8
H₂O(l)	−285.8
Silver	
Ag(s)	0
AgCl(s)	−127.0
Sodium	
Na(s)	0
Na(g)	107.8
NaCl(s)	−411.1
Sulfur	
S₈(rhombic)	0
S₈(monoclinic)	0.3
SO₂(g)	−296.8
SO₃(g)	−396.0

ARIS Presentation Center, Tables Library

---

*The definition of the standard state for gases has been changed to 1 bar, a slightly lower pressure than the 1 atm standard on which the data in this book are based (1 atm = 101.3 kPa = 1.013 bar). For most purposes, this makes *very* little difference in the standard enthalpy values.

†In the case of phosphorus, the most *common* form, white phosphorus (P_4), is chosen as the standard state, even though red phosphorus is more stable at 1 atm and 298 K.

SAMPLE PROBLEM 6.8 Writing Formation Equations

PROBLEM Write balanced equations for the formation of 1 mole of each of the following compounds from their elements in their standard states, and include ΔH_f°.
(a) Silver chloride, AgCl, a solid at standard conditions
(b) Calcium carbonate, $CaCO_3$, a solid at standard conditions
(c) Hydrogen cyanide, HCN, a gas at standard conditions
PLAN We write the elements as the reactants and 1 mol of the compound as the product, being sure all substances are in their standard states. Then, we balance the atoms and obtain the ΔH_f° values from Table 6.3 or Appendix B.
SOLUTION **(a)** $Ag(s) + \frac{1}{2}Cl_2(g) \longrightarrow AgCl(s) \quad \Delta H_f^\circ = -127.0 \text{ kJ}$

(b) $Ca(s) + C(\text{graphite}) + \frac{3}{2}O_2(g) \longrightarrow CaCO_3(s) \quad \Delta H_f^\circ = -1206.9 \text{ kJ}$

(c) $\frac{1}{2}H_2(g) + C(\text{graphite}) + \frac{1}{2}N_2(g) \longrightarrow HCN(g) \quad \Delta H_f^\circ = 135 \text{ kJ}$

FOLLOW-UP PROBLEM 6.8 Write balanced equations for the formation of 1 mol of **(a)** $CH_3OH(l)$, **(b)** $CaO(s)$, and **(c)** $CS_2(l)$ from their elements in their standard states. Include ΔH_f° for each reaction.

Determining ΔH_{rxn}° from ΔH_f° Values of Reactants and Products

By applying Hess's law, we can use ΔH_f° values to determine ΔH_{rxn}° for any reaction. All we have to do is view the reaction as an imaginary two-step process.

Step 1. Each reactant decomposes to its elements. This is the *reverse* of the formation reaction for each *reactant,* so each standard enthalpy change is $-\Delta H_f^\circ$.

Step 2. Each product forms from its elements. This step *is* the formation reaction for each *product,* so each standard enthalpy change is ΔH_f°.

According to Hess's law, we add the enthalpy changes for these steps to obtain the overall enthalpy change for the reaction (ΔH_{rxn}°). Figure 6.12 depicts the conceptual process. Suppose we want ΔH_{rxn}° for

$$TiCl_4(l) + 2H_2O(g) \longrightarrow TiO_2(s) + 4HCl(g)$$

We write this equation as though it were the sum of four individual equations, one for each compound. The first two of these equations show the decomposition of the reactants to their elements (*reverse* of their formation), and the second two show the formation of the products from their elements:

$$
\begin{array}{ll}
TiCl_4(l) \longrightarrow Ti(s) + 2Cl_2(g) & -\Delta H_f^\circ[TiCl_4(l)] \\
2H_2O(g) \longrightarrow 2H_2(g) + O_2(g) & -2\Delta H_f^\circ[H_2O(g)] \\
Ti(s) + O_2(g) \longrightarrow TiO_2(s) & \Delta H_f^\circ[TiO_2(s)] \\
2H_2(g) + 2Cl_2(g) \longrightarrow 4HCl(g) & 4\Delta H_f^\circ[HCl(g)]
\end{array}
$$

$$TiCl_4(l) + 2H_2O(g) + \cancel{Ti(s)} + \cancel{O_2(g)} + \cancel{2H_2(g)} + \cancel{2Cl_2(g)} \longrightarrow$$
$$\cancel{Ti(s)} + \cancel{2Cl_2(g)} + \cancel{2H_2(g)} + \cancel{O_2(g)} + TiO_2(s) + 4HCl(g)$$

or, $TiCl_4(l) + 2H_2O(g) \longrightarrow TiO_2(s) + 4HCl(g)$

Figure 6.12 The general process for determining ΔH_{rxn}° from ΔH_f° values. For any reaction, ΔH_{rxn}° can be considered as the sum of the enthalpy changes for the decomposition of reactants to their elements $[-\Sigma n\Delta H_{f(\text{reactants})}^\circ]$ and the formation of products from their elements $[+\Sigma m\Delta H_{f(\text{products})}^\circ]$. [The factors m and n are the amounts (mol) of the products and reactants and equal the coefficients in the balanced equation, and Σ is the symbol for "sum of."]

$$\Delta H_{rxn}^\circ = \Sigma m\Delta H_{f(\text{products})}^\circ - \Sigma n\Delta H_{f(\text{reactants})}^\circ$$

It's important to realize that when titanium(IV) chloride and water react, the reactants don't *actually* decompose to their elements, which then recombine to form the products. But that is the great usefulness of Hess's law and the state-function concept. Because ΔH°_{rxn} is the difference between two state functions, $H^\circ_{products}$ minus $H^\circ_{reactants}$, it doesn't matter *how* the change actually occurs. We simply add the individual enthalpy changes to find ΔH°_{rxn}:

$$\Delta H^\circ_{rxn} = \Delta H^\circ_f[TiO_2(s)] \;+ 4\Delta H^\circ_f[HCl(g)] \;+ \{-\Delta H^\circ_f[TiCl_4(l)]\} + \{-2\Delta H^\circ_f[H_2O(g)]\}$$
$$= \{\Delta H^\circ_f[TiO_2(s)] + 4\Delta H^\circ_f[HCl(g)]\} - \{\Delta H^\circ_f[TiCl_4(l)] \;+ 2\Delta H^\circ_f[H_2O(g)]\}$$

$\underbrace{\qquad\qquad}_{\text{Products}}$　　　　$\underbrace{\qquad\qquad}_{\text{Reactants}}$

Notice the pattern here. By generalizing it, we see that *the standard heat of reaction is the sum of the standard heats of formation of the **products** minus the sum of the standard heats of formation of the **reactants*** (Figure 6.12):

$$\Delta H^\circ_{rxn} = \Sigma m\Delta H^\circ_{f(products)} - \Sigma n\Delta H^\circ_{f(reactants)} \qquad (6.8)$$

where the symbol Σ means "sum of," and m and n are the amounts (mol) of the products and reactants indicated by the coefficients from the balanced equation.

SAMPLE PROBLEM 6.9　Calculating the Heat of Reaction from Heats of Formation

ARIS Presentation Center,
Worked Examples Library

Engineering application

PROBLEM Nitric acid, whose worldwide annual production is nearly 10 billion kilograms, is used to make many products, including fertilizers, dyes, and explosives. The first step in the production process is the oxidation of ammonia:

$$4NH_3(g) + 5O_2(g) \longrightarrow 4NO(g) + 6H_2O(g)$$

Calculate ΔH°_{rxn} from ΔH°_f values.

PLAN We use values from Table 6.3 (or Appendix B) and apply Equation 6.8 to find ΔH°_{rxn}.

SOLUTION Calculating ΔH°_{rxn}:

$$\Delta H^\circ_{rxn} = \Sigma m\Delta H^\circ_{f(products)} - \Sigma n\Delta H^\circ_{f(reactants)}$$
$$= \{4\Delta H^\circ_f[NO(g)] + 6\Delta H^\circ_f[H_2O(g)]\} - \{4\Delta H^\circ_f[NH_3(g)] + 5\Delta H^\circ_f[O_2(g)]\}$$
$$= (4\ mol)(90.3\ kJ/mol) + (6\ mol)(-241.8\ kJ/mol)$$
$$\qquad - [(4\ mol)(-45.9\ kJ/mol) + (5\ mol)(0\ kJ/mol)]$$
$$= 361\ kJ - 1451\ kJ + 184\ kJ - 0\ kJ$$
$$= -906\ kJ$$

CHECK One way to check is to write formation equations for the amounts of individual compounds in the correct direction and take their sum:

$4NH_3(g) \longrightarrow 2N_2(g) + 6H_2(g)$	$-4(-45.9\ kJ) =$	$184\ kJ$
$2N_2(g) + 2O_2(g) \longrightarrow 4NO(g)$	$4(90.3\ kJ) =$	$361\ kJ$
$6H_2(g) + 3O_2(g) \longrightarrow 6H_2O(g)$	$6(-241.8\ kJ) =$	$-1451\ kJ$
$4NH_3(g) + 5O_2(g) \longrightarrow 4NO(g) + 6H_2O(g)$		$-906\ kJ$

COMMENT In this problem, we know the individual ΔH°_f values and find the sum, ΔH°_{rxn}. In the follow-up problem, we know the sum and want to find an individual value.

FOLLOW-UP PROBLEM 6.9　Use the following information to find ΔH°_f of methanol [$CH_3OH(l)$]:

$$CH_3OH(l) + \tfrac{3}{2}O_2(g) \longrightarrow CO_2(g) + 2H_2O(g) \quad \Delta H^\circ_{rxn} = -638.5\ kJ$$
$$\Delta H^\circ_f \text{ of } CO_2(g) = -393.5\ kJ/mol$$
$$\Delta H^\circ_f \text{ of } H_2O(g) = -241.8\ kJ/mol$$

In the following Chemical Connections essay, we apply key ideas from this chapter to new approaches for energy utilization.

Chemical Connections *to Environmental Science*

The Future of Energy Use
Green chemistry application

Out of necessity, we must undertake a global rethinking of energy use. The dwindling supply of the world's fuels and the environmental impact of their polluting combustion products threaten human well-being and the survival of many other plant and animal species. For this reason, chemical aspects of energy production and utilization provide scientists, engineers, and political leaders with some of the greatest challenges of our time.

A changeover from the use of wood to coal and then petroleum took place over the past century. The **fossil fuels**—coal, petroleum, and natural gas—remain our major sources of energy today, but natural processes form them *much* more slowly than we consume them, so these resources are *nonrenewable:* once we use them up, they are gone. Experts agree that, in the United States, most existing large oil and gas fields have already been discovered and consumed. Thus, as progressively smaller fields are worked, greater effort and cost are needed to produce the same amount of fuel. Coal reserves are much more plentiful, so research seeks new and cleaner uses for coal. And novel ways to utilize important *renewable* fuels—wood, other *biomass* (plant and animal matter), and hydrogen gas—are the subject of intensive research as well.

In this discussion, we highlight more rational approaches to energy use that incorporate some aspects of *green chemistry,* the effort to devise processes that avoid the release of harmful products into the environment: converting coal to cleaner fuels, producing fuels from biomass, understanding the critical effect of carbon-based fuels on climate, developing a hydrogen-fueled economy, utilizing energy that does not involve combustion, and conserving the fuels we have.

Chemical Approaches to Cleaner Carbon-Based Fuels

Coal Current U.S. reserves of coal are enormous, but coal is a highly polluting fuel that produces SO_2 and releases particulates and mercury when it burns. Exposure to SO_2 and particulates causes an increased incidence of bronchitis, lung cancer, and other respiratory diseases. In the air, SO_2 is oxidized to H_2SO_4, which is the key contributor to acid rain (Chapter 19). When coal burns, trace amounts of Hg, a neurotoxin, are released to spread as Hg vapor and *bioaccumulate* in fish.

Two processes—desulfurization and gasification—are designed to reduce these noxious emissions. Flue-gas *desulfurization* devices (*scrubbers*) heat powdered limestone ($CaCO_3$) or spray lime-water slurries [$Ca(OH)_2$] to remove SO_2 from the gases produced during coal combustion:

$$CaCO_3(s) + SO_2(g) \xrightarrow{\Delta} CaSO_3(s) + CO_2(g)$$
$$2CaSO_3(s) + O_2(g) + 4H_2O(l) \longrightarrow 2CaSO_4 \cdot 2H_2O(s; gypsum)$$
$$2Ca(OH)_2(aq) + 2SO_2(g) + O_2(g) + 2H_2O(l) \longrightarrow$$
$$2CaSO_4 \cdot 2H_2O(s)$$

An innovative approach for disposing of the gypsum (nearly 1 ton per power-plant customer each year) has been to build a gypsum wallboard (drywall) plant adjacent to a coal-fired power plant. In 2002, 20% of drywall was made with synthetic gypsum.

In **coal gasification,** solid coal is converted to sulfur-free gaseous fuels for use in electricity generation, as well as valuable chemical feedstocks. Reaction of pulverized coal with limited oxygen and water at 800°–1500°C produces a gas stream consisting mainly of a 2/1 mixture of CO and H_2. Under these conditions, sulfur in the coal is reduced to H_2S, which is removed through an acid-base reaction with a nitrogen-containing compound such as ethanolamine ($HOCH_2CH_2NH_2$). The resulting salt is heated to release H_2S, which is converted to elemental sulfur, a valuable byproduct, by the Claus process (Chapter 22).

The principal reactions in coal gasification are exothermic oxidation of C in coal to CO and reaction of C with steam in the endothermic *water-gas reaction* (also known as the *steam-carbon reaction*) to form a mixture of CO and H_2 called *water gas*:

$$C(s) + \tfrac{1}{2}O_2(g) \longrightarrow CO(g) \qquad \Delta H° = -110 \text{ kJ}$$
$$C(s) + H_2O(g) \longrightarrow CO(g) + H_2(g) \quad \Delta H° = 131 \text{ kJ}$$

To produce synthesis gas (**syngas**), the H_2 content of water gas is enhanced by the *CO-shift (or water-gas shift) reaction*:

$$CO(g) + H_2O(g) \longrightarrow CO_2(g) + H_2(g) \quad \Delta H° = -41 \text{ kJ}$$

Water gas has a much lower fuel value than methane. For example, a mixture containing 0.5 mol of CO and 0.5 mol of H_2 releases about one-third as much energy as combustion of 1.0 mol of methane ($\Delta H° = -802$ kJ/mol):

$$\tfrac{1}{2}H_2(g) + \tfrac{1}{4}O_2(g) \longrightarrow \tfrac{1}{2}H_2O(g) \qquad \Delta H°_{rxn} = -121 \text{ kJ}$$
$$\underline{\tfrac{1}{2}CO(g) + \tfrac{1}{4}O_2(g) \longrightarrow \tfrac{1}{2}CO_2(g) \qquad \Delta H°_{rxn} = -142 \text{ kJ}}$$
$$\tfrac{1}{2}H_2(g) + \tfrac{1}{2}CO(g) + \tfrac{1}{2}O_2(g) \longrightarrow \tfrac{1}{2}H_2O(g) + \tfrac{1}{2}CO_2(g)$$
$$\Delta H° = -263 \text{ kJ}$$

To produce methane, syngas with a 3/1 ratio of H_2 to CO, from which the CO_2 has been removed, is employed:

$$CO(g) + 3H_2(g) \longrightarrow CH_4(g) + H_2O(g) \quad \Delta H° = -206 \text{ kJ}$$

Drying the product mixture gives **synthetic natural gas (SNG).** Thus, in three reaction steps, coal is converted to methane.

A newer method, called *integrated gasification combined cycle (IGCC),* produces electricity in two ways: syngas is used in a combustion turbine, and the hot product gases are then channeled to generate steam for a steam turbine. Another use of syngas receiving greater attention as petroleum supplies dwindle is the production of liquid hydrocarbon fuels by the *Fischer-Tropsch process*:

$$nCO(g) + (2n + 1)H_2(g) \longrightarrow C_nH_{2n+2}(l) + nH_2O(g)$$

Wood and Other Types of Biomass Nearly half the world's people rely on wood for energy. In principle, wood is renewable, but its worldwide use for fuel, lumber, and paper has resulted in widespread deforestation. On the other hand, vegetable and animal waste *is* renewable. **Biomass conversion** employs chemical and/or microbial methods to convert vegetable (sugarcane, corn, switchgrass) and tree waste into fuels such as ethanol (C_2H_5OH). In the United States, ethanol is mixed with gasoline to form *gasohol.* In Brazil, ethanol use is much more widespread, and cars with "flex-fuel" engines run on ethanol and/or gasoline; in fact, in many gas stations, pumps show "A" for alcohol or "G" for gasoline.

Methanogenesis produces methane by anaerobic (oxygen-free) biodegradation of plant and animal waste. Manure, vegetable waste, and municipal and industrial wastewater are used as feedstocks. China's biogas-generating facilities apply these techniques, and the United States funds similar ones to produce gaseous fuels from garbage and sewage. Methanogenesis has three clear benefits: methane fuel is produced, wastes do not end up as water pollutants, and the process residues are applied to improve soil. Anaerobic wastewater treatment is already common in Asia and South America.

Another use of biomass is the chemical conversion of vegetable oils (soybean, cottonseed, sunflower, canola, and even used cooking oil) into a mixture of methylated fatty acids called *biodiesel*, which substitutes for diesel fuel in nearly all applications. Compared to fossil-based diesel, biodiesel produces less CO, SO_2, and particulate matter. Even though the combustion of biodiesel produces CO_2, which harms the atmosphere as a greenhouse gas, this CO_2 has minimal impact on the net atmospheric load because it is part of a short-term loop that recycles the carbon taken up by plants grown as biodiesel feedstocks. An exciting new strategy is the direct conversion of biomass to electrical energy with *microbial fuel cells*. (We discuss fuel cells in Chapter 21.)

The Greenhouse Effect and Global Warming

All carbon-based fuels release CO_2 when burned, and it has become clear that our increased use of these fuels and release of CO_2 is changing Earth's climate. In pre-industrial times, the ability of CO_2 to absorb heat played a key temperature-regulating role in the atmosphere. Much of the sunlight that shines on Earth is absorbed by the land and oceans and converted to heat. Like the glass of a greenhouse, atmospheric CO_2 does not absorb visible light from the Sun, but it traps some of the heat radiating back from Earth's surface and, thus, helps warm the atmosphere. This process is called the natural *greenhouse effect* (Figure B6.1, *left*).

Over several billion years—due largely to the spread of plant life, which uses CO_2 in photosynthesis—the amount of CO_2 originally present in Earth's atmosphere decreased to 0.028% by volume. However, today, as a result of the human use of fossil fuels for the past 150 years, this amount has increased to slightly over 0.036%. Thus, although the same amount of solar energy passes

(continued)

Environmental/Green chemistry application

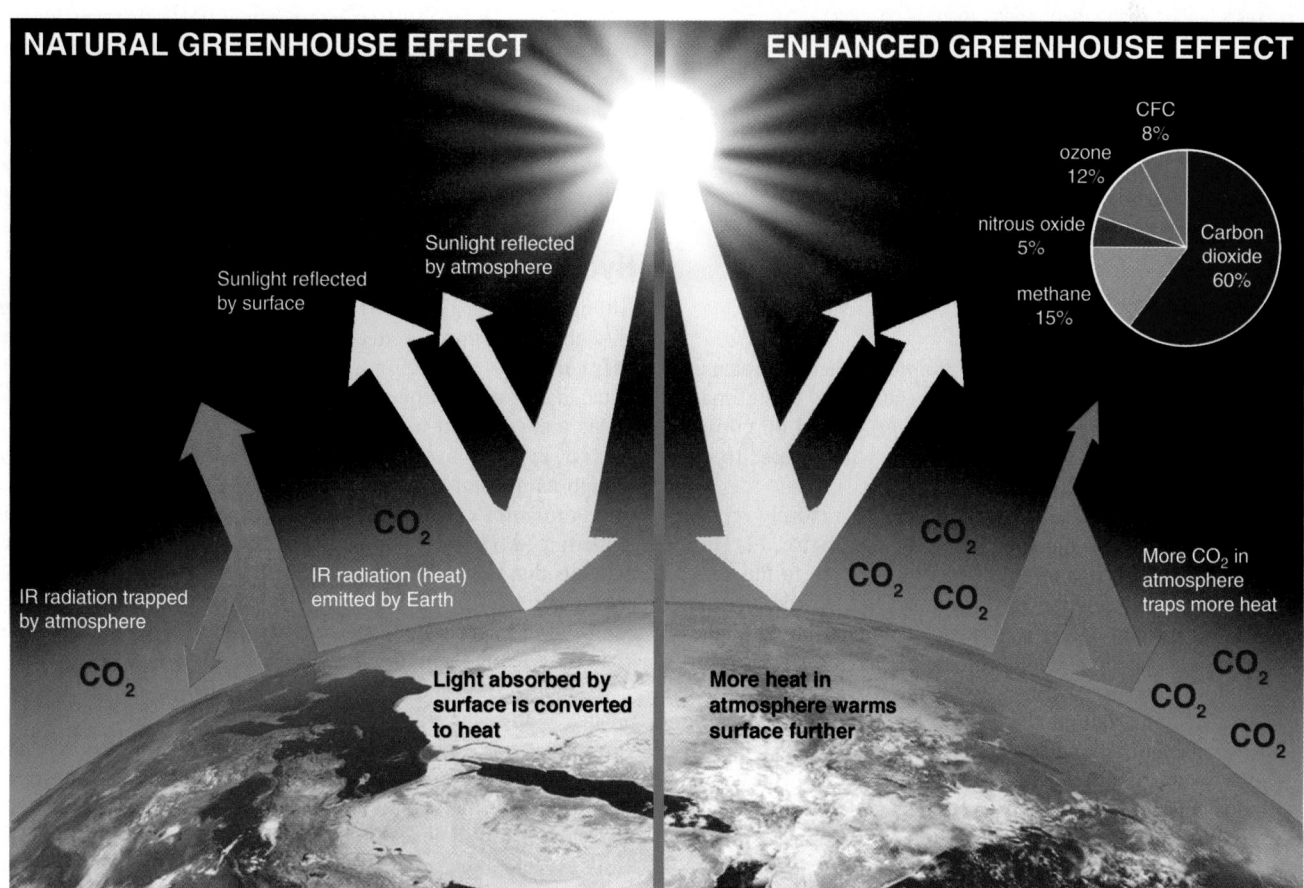

Figure B6.1 The trapping of heat by the atmosphere. Of the total sunlight reaching Earth, some is reflected by the atmosphere and some by the surface (especially snow, ice, and water). The remainder is absorbed by the surface and converted to IR radiation (heat). When this IR radiation is emitted by the surface, some is trapped by atmospheric components, especially CO_2. Without this *natural* greenhouse effect *(left)*, Earth's surface would have an average temperature of −18°C, far below water's freezing point, rather than its current average of 13°C. Since the early 19th century, and particularly in the past several decades, human activity has increased the amount of CO_2, along with several other greenhouse gases *(pie chart)*, and created an *enhanced* greenhouse effect *(right)*.

Chemical Connections *continued*

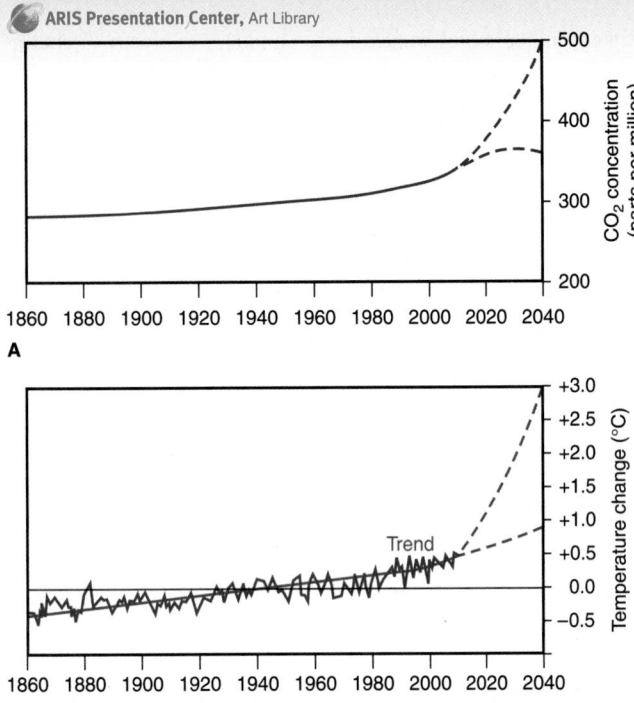

Figure B6.2 **The accumulating evidence for the greenhouse effect. A,** Since the mid-19th century, atmospheric CO_2 concentrations have increased. **B,** Coincident with this CO_2 increase, the average global temperature has risen about 0.6°C. (Zero equals the average global temperature from 1957 to 1970.) The projections in both graphs *(dashed lines)* are based on current fossil fuel consumption and deforestation either continuing *(upper line)* or being curtailed *(lower line).*

through the atmosphere, more is being trapped as heat to create an *enhanced* greenhouse effect, which is changing the climate through *global warming* (Figure B6.1, *right*). Based on current trends in fossil fuel use, CO_2 concentrations will increase to between 0.049% and 0.126% by 2100 (Figure B6.2).

If the projected increase in CO_2 occurs, two closely related questions arise: (1) How much will the temperature rise? (2) How will this temperature rise affect life on Earth? Due to the complexity of global circulation in the atmosphere and oceans, the best tools for gauging future changes are computer-based mathematical models that simulate the climate's behavior. Despite constantly improving models, answers to these two questions are difficult to obtain. Natural fluctuations in temperature must be taken into account, as well as cyclic changes in solar activity. Moreover, as the amount of CO_2 increases from fossil-fuel burning, so does the amount of particulate matter and SO_2, which may block sunlight and have a cooling effect. Water vapor also traps heat, and as temperatures rise, more water evaporates. The increased amounts of water vapor may thicken the cloud cover and lead to cooling.

Despite these opposing factors, the overwhelming majority of models predict a net warming of the atmosphere, and scientists are now documenting the predicted climate disruptions. The average temperature has increased by 0.6±0.2°C since the late 19th century and 0.2–0.3°C over the past 25 years. Globally, the 10

warmest years on record occurred in the last 15 years of the 20th century. Snow cover and glacier extent in the northern hemisphere and floating ice in the Arctic Ocean have decreased dramatically. Globally, sea level has risen 4–8 inches over the past century, and flooding and other extreme weather events have increased through much of the United States and Europe.

Ten years ago, the most accepted models predicted a temperature rise of 1.0–3.5°C. Today, the best predictions are more than 50% higher. Such increases would alter rainfall patterns and crop yields throughout the world and may increase sea level as much as 1 meter, thus flooding low-lying regions, such as the Netherlands, half of Florida, much of southern Asia, and many Pacific island nations. To make matters worse, as we burn fossil fuels that *release* CO_2, we cut down the forests that *absorb* it.

In addition to reducing fossil-fuel consumption, researchers are studying methods of increasing CO_2 *sequestration.* In its most common form, sequestration is performed by plants, and large-scale tree planting has been proposed. Another approach is to bury liquefied CO_2 underground or inject it into the deep oceans. The formations most suitable for underground storage may be those from which natural gas and petroleum are now extracted.

The Kyoto Protocol, created at the 1997 Conference on Climate Change in Japan, is an international treaty that sets legally binding targets to limit greenhouse gases. It recognizes the atmosphere as a shared resource that can be adversely affected by these emissions. The protocol was ratified by 189 countries, but the largest producer of CO_2, the United States, refused to do so. The 2005 Conference confirmed the human impact on climate change and proposed ways to address the dire effects.

Hydrogen

Although combustion of H_2 ($\Delta H^\circ_{rxn} = -242$ kJ/mol) produces only about one-third as much energy per mole as combustion of CH_4 ($\Delta H^\circ_{rxn} = -802$ kJ/mol), it yields nonpolluting water vapor. Thus, ideas for a hydrogen-based economy abound. In the United States today, almost all H_2 is produced by the *steam-reforming process,* in which hydrocarbons are treated with steam in a series of high-temperature, catalyzed reactions (Chapter 22).

Formation of H_2 from the decomposition of water is endothermic (ΔH°_{rxn} with liquid $H_2O = 286$ kJ/mol), and most *direct* methods that utilize electricity are still very costly. However, the energy derived from flowing water, wind, and geothermal sources can be used to generate this electricity. In fact, Iceland is making great progress in exploiting these resources.

Transporting and storing hydrogen fuel could be accomplished with technologies used currently by utility companies. However, the actual method by which the H_2 will power, say, the family car for a typical 300-mile driving range is still an open question. Nevertheless, the use of H_2 in fuel cells is a major area of electrochemical research (which we discuss in Chapter 21). Tests are underway to commercialize fuel-cell technology using the hydrogen that is a byproduct of the petrochemical and coal industries. In a related development, a major new government initiative, called FutureGen, will combine established and experimental technologies to create a zero-emission electric power plant within 10 years. This project will generate electricity through coal gasification and IGCC systems, produce H_2 for fuel cells, and sequester the waste CO_2.

Ultimately, the hydrogen needed for a sustainable economy may have to be derived from renewable sources, for example, by decomposing water with electricity from wind or sunlight. Newer, biological approaches employ certain algae that use sunlight to decompose water, and biomass gasification of peanut shells and other agricultural waste produces H_2, along with other useful chemicals and fuels.

Solar Energy and Nuclear Energy

Solar Energy The quantity of solar energy striking the United States annually is *600 times greater than the country's total energy needs*. Despite this abundance, it is difficult to collect and store. The Sun's output is enormous but not concentrated, so vast surface areas must be devoted to collecting it.

Storing the energy is necessary because intense sunlight is available over most regions for only 6 to 8 hours a day and is greatly reduced in rainy or cloudy weather. One storage approach uses ionic hydrates, such as $Na_2SO_4 \cdot 10H_2O$. When warmed by sunlight to over 32°C, the 3 mol of ions dissolve in the 10 mol of water in an endothermic process:

$$Na_2SO_4 \cdot 10H_2O(s) \xrightarrow{>32°C} Na_2SO_4(aq) \quad \Delta H° = 354 \text{ kJ}$$

When cooled below 32°C after sunset, the solution recrystallizes, releasing the absorbed energy for heating:

$$Na_2SO_4(aq) \xrightarrow{<32°C} Na_2SO_4 \cdot 10H_2O(s) \quad \Delta H° = -354 \text{ kJ}$$

Construction using passive solar heating can reduce heating and cooling costs by as much as 50%. Newer building designs incorporate features such as large, south-facing windows and materials that absorb and slowly release the Sun's heat.

Photovoltaic cells convert light directly into electricity, but the electrical energy produced is only about 10% of the radiant energy supplied. However, this output will increase as novel combinations of elements from Groups 3A(13) and 5A(15), such

as gallium arsenide (GaAs), show much higher yields. And mixing GaAs with other elements from these groups, such as indium, phosphorus, and aluminum, creates materials that respond to different wavelengths of sunlight.

Nuclear Energy Electricity from nuclear *fission* is used extensively, especially in France, despite major problems with the safe and permanent disposal of radioactive byproducts. Nuclear *fusion* avoids the problems of fission, but so far researchers have achieved fusion only with a net consumption of power. Practical application of fusion is still decades away. (We'll discuss these processes thoroughly in Chapter 24.)

Energy Conservation: More from Less

All systems on Earth, whether living organisms or industrial power plants, waste some energy, which in effect is a waste of fuel. In terms of the energy used to produce it, for example, every discarded aluminum can is equivalent to 0.25 L of gasoline. Energy conservation lowers production costs, extends our supply of fossil fuels, and reduces pollution.

One example of an energy-conserving device is a high-efficiency gas-burning furnace for domestic heating. Its design channels hot waste gases through the furnace to transfer more heat to the room and to an attached domestic hot water system. Moreover, the channeling path cools the gases to <100°C, so the water vapor condenses, thus releasing about 10% more heat:

$$\begin{array}{ll}
CH_4(g) + 2O_2(g) \longrightarrow CO_2(g) + 2H_2O(g) & \Delta H° = -802 \text{ kJ} \\
\underline{2H_2O(g) \longrightarrow 2H_2O(l)} & \underline{\Delta H° = -88 \text{ kJ}} \\
CH_4(g) + 2O_2(g) \longrightarrow CO_2(g) + 2H_2O(l) & \Delta H° = -890 \text{ kJ}
\end{array}$$

Engineers and chemists will continue to explore alternatives for energy production and use, but a more hopeful energy future ultimately depends on our wisdom in obtaining and conserving planetary resources.

Section Summary

Standard states are chosen conditions for substances. • When 1 mol of a compound forms from its elements with all substances in their standard states, the enthalpy change is $\Delta H_f°$. • Hess's law allows us to picture a reaction as the decomposition of reactants to their elements, followed by the formation of products from their elements. • We use tabulated $\Delta H_f°$ values to find $\Delta H_{rxn}°$ or use known $\Delta H_{rxn}°$ and $\Delta H_f°$ values to find an unknown $\Delta H_f°$. • Because of dwindling resources and environmental concerns, such as global warming, chemists are developing new energy alternatives, including coal and biomass conversion, hydrogen fuel, and noncombustible energy sources.

Chapter Perspective

Our investigation of energy in this chapter helps explain many chemical and physical phenomena: how gases do work, how reactions release or absorb heat, how the chemical energy in fuels and foods changes to other forms, and how we can use energy more wisely. We reexamine some of these ideas later in considering why physical and chemical changes occur (Chapter 20). In Chapter 7, you'll see how the energy released or absorbed by the atom gave rise to the revolutionary view of matter and energy that led to our current model of atomic energy states.

CHAPTER REVIEW GUIDE
The following sections provide many aids to help you study this chapter. (Numbers in parentheses refer to pages, unless noted otherwise.)

Learning Objectives These are concepts and skills you should know after studying this chapter.

Relevant section and/or sample problem (SP) numbers appear in parentheses.

Understand These Concepts

1. The distinction between a system and its surroundings (6.1)
2. How energy is transferred to or from a system as heat and/or work (6.1)
3. The relation among internal energy change, heat, and work (6.1)
4. The meaning of energy conservation (6.1)
5. The meaning of a state function and why ΔE is constant even though q and w vary (6.1)
6. The meaning of enthalpy and the relation between ΔE and ΔH (6.2)
7. The meaning of ΔH and the distinction between exothermic and endothermic reactions (6.2)
8. The relation between specific heat capacity and heat transferred (6.3)
9. How constant-pressure (coffee-cup) and constant-volume (bomb) calorimeters work (6.3)
10. The relation between ΔH°_{rxn} and amount of substance (6.4)
11. The importance of Hess's law and the manipulation of ΔH values (6.5)
12. The meaning of a formation equation and the standard heat of formation (6.6)
13. How a reaction can be viewed as the decomposition of reactants followed by the formation of products (6.6)

Master These Skills

1. Determining the change in energy in different units (SP 6.1)
2. Drawing enthalpy diagrams for chemical and physical changes (SP 6.2)
3. Solving problems involving specific heat capacity and heat of reaction (SPs 6.3–6.5)
4. Relating the heat of reaction and the amounts of substances changing (SP 6.6)
5. Using Hess's law to find an unknown ΔH (SP 6.7)
6. Writing formation equations and using ΔH°_f values to find ΔH°_{rxn} (SPs 6.8, 6.9)

Key Terms These important terms appear in boldface in the chapter and are defined again in the Glossary.

thermodynamics (236)
thermochemistry (236)
Section 6.1
system (236)
surroundings (236)
internal energy (E) (237)
heat (q) (238)
work (w) (238)
pressure-volume work (PV work) (239)
law of conservation of energy (first law of thermodynamics) (240)
joule (J) (240)

calorie (cal) (240)
state function (241)
Section 6.2
enthalpy (H) (243)
change in enthalpy (ΔH) (243)
heat of reaction (ΔH_{rxn}) (244)
enthalpy diagram (244)
exothermic process (244)
endothermic process (244)
heat of formation (ΔH_f) (245)
heat of fusion (ΔH_{fus}) (245)
heat of vaporization (ΔH_{vap}) (245)

Section 6.3
heat capacity (246)
specific heat capacity (c) (246)
molar heat capacity (C) (246)
calorimeter (247)
Section 6.4
thermochemical equation (249)
Section 6.5
Hess's law of heat summation (251)
Section 6.6
standard states (253)

standard heat of reaction (ΔH°_{rxn}) (253)
formation equation (253)
standard heat of formation (ΔH°_f) (253)
fossil fuel (256)
coal gasification (256)
syngas (256)
synthetic natural gas (SNG) (256)
biomass conversion (256)
methanogenesis (257)
photovoltaic cell (259)

Key Equations and Relationships Numbered and screened concepts are listed for you to refer to or memorize.

6.1 Defining the change in internal energy (237):
$$\Delta E = E_{final} - E_{initial} = E_{products} - E_{reactants}$$
6.2 Expressing the change in internal energy in terms of heat and work (238):
$$\Delta E = q + w$$
6.3 Stating the first law of thermodynamics (law of conservation of energy) (240):
$$\Delta E_{universe} = \Delta E_{system} + \Delta E_{surroundings} = 0$$
6.4 Determining the work due to a change in volume at constant pressure (PV work) (243):
$$w = -P\Delta V$$

6.5 Relating the enthalpy change to the internal energy change at constant pressure (243):
$$\Delta H = \Delta E + P\Delta V$$
6.6 Identifying the enthalpy change with the heat gained or lost at constant pressure (243):
$$q_P = \Delta E + P\Delta V = \Delta H$$
6.7 Calculating the heat absorbed or released when a substance undergoes a temperature change (246):
$$q = c \times mass \times \Delta T$$
6.8 Calculating the standard heat of reaction (255):
$$\Delta H^\circ_{rxn} = \Sigma m\Delta H^\circ_{f(products)} - \Sigma n\Delta H^\circ_{f(reactants)}$$

Highlighted Figures and Tables These figures (F) and tables (T) provide a visual review of key ideas.

F6.2 Energy diagrams for the transfer of internal energy (E) between a system and its surroundings (237)
T6.1 Sign conventions for q, w, and ΔE (239)
F6.8 Enthalpy diagrams for exothermic and endothermic processes (244)

F6.11 Summary of the relationship between amount (mol) of substance and heat (kJ) transferred during a reaction (250)
F6.12 The general process for determining ΔH_{rxn}° from ΔH_f° values (254)

Brief Solutions to FOLLOW-UP PROBLEMS Compare your solutions to these calculation steps and answers.

6.1 $\Delta E = q + w$

$$= \left(-26.0 \text{ kcal} \times \frac{4.184 \text{ kJ}}{1 \text{ kcal}}\right) + \left(+15.0 \text{ Btu} \times \frac{1.055 \text{ kJ}}{1 \text{ Btu}}\right)$$

$$= -93 \text{ kJ}$$

6.2 The reaction is exothermic.

$C_3H_5(NO_3)_3(l)$

Enthalpy, H

$\Delta H = -5.72 \times 10^3$ kJ

$3CO_2(g) + \frac{5}{2}H_2O(g) + \frac{1}{4}O_2(g) + \frac{3}{2}N_2(g)$

6.3 $\Delta T = 25.0°C - 37.0°C = -12.0°C = -12.0 \text{ K}$

Mass (g) $= 1.11 \text{ g/mL} \times \dfrac{1000 \text{ mL}}{1 \text{ L}} \times 5.50 \text{ L} = 6.10 \times 10^3 \text{ g}$

$q = c \times \text{mass} \times \Delta T$

$= (2.42 \text{ J/g·K})\left(\dfrac{1 \text{ kJ}}{1000 \text{ J}}\right)(6.10 \times 10^3 \text{ g})(-12.0 \text{ K})$

$= -177 \text{ kJ}$

6.4 $-q_{solid} = q_{water}$

$-[(0.519 \text{ J/g·K})(2.050 \text{ g})(x - 74.21)]$

$= [(4.184 \text{ J/g·K})(26.05 \text{ g})(x - 27.20)]$

$x = 27.65 \text{ K}$

$\Delta T_{diamond} = -46.56 \text{ K and } \Delta T_{water} = 0.45 \text{ K}$

6.5 $-q_{sample} = q_{calorimeter}$

$-(0.8650 \text{ g C})\left(\dfrac{1 \text{ mol C}}{12.01 \text{ g C}}\right)(-393.5 \text{ kJ/mol C}) = (2.613 \text{ K})x$

$x = 10.85 \text{ kJ/K}$

6.6 $C_2H_4(g) + H_2(g) \longrightarrow C_2H_6(g) + 137 \text{ kJ}$

Heat (kJ) $= 15.0 \text{ kg} \times \dfrac{1000 \text{ g}}{1 \text{ kg}} \times \dfrac{1 \text{ mol } C_2H_6}{30.07 \text{ g } C_2H_6} \times \dfrac{137 \text{ kJ}}{1 \text{ mol}}$

$= 6.83 \times 10^4 \text{ kJ}$

6.7

$2NO(g) + \frac{3}{2}O_2(g) \longrightarrow N_2O_5(s)$ $\Delta H = -223.7 \text{ kJ}$

$\underline{2NO_2(g) \longrightarrow 2NO(g) + O_2(g) \quad \Delta H = 114.2 \text{ kJ}}$

$\cancel{2NO(g)} + \frac{\cancel{3}}{\cancel{2}}\frac{3}{2}O_2(g) + 2NO_2(g) \longrightarrow$
$ N_2O_5(s) + \cancel{2NO(g)} + \cancel{O_2(g)}$

$2NO_2(g) + \frac{1}{2}O_2(g) \longrightarrow N_2O_5(s)$ $\Delta H = -109.5 \text{ kJ}$

6.8

(a) $C(graphite) + 2H_2(g) + \frac{1}{2}O_2(g) \longrightarrow CH_3OH(l)$
$ \Delta H_f^{\circ} = -238.6 \text{ kJ}$

(b) $Ca(s) + \frac{1}{2}O_2(g) \longrightarrow CaO(s) \quad \Delta H_f^{\circ} = -635.1 \text{ kJ}$

(c) $C(graphite) + \frac{1}{4}S_8(rhombic) \longrightarrow CS_2(l) \quad \Delta H_f^{\circ} = 87.9 \text{ kJ}$

6.9 ΔH_f° of $CH_3OH(l)$

$= -\Delta H_{rxn}^{\circ} + 2\Delta H_f^{\circ}[H_2O(g)] + \Delta H_f^{\circ}[CO_2(g)]$

$= 638.5 \text{ kJ} + (2 \text{ mol})(-241.8 \text{ kJ/mol})$

$ + (1 \text{ mol})(-393.5 \text{ kJ/mol})$

$= -238.6 \text{ kJ}$

PROBLEMS

Problems with colored numbers are answered in Appendix E and worked in detail in the Student Solutions Manual. Problem sections match those in the text and provide the numbers of relevant sample problems. Most offer Concept Review Questions, Skill-Building Exercises (grouped in pairs covering the same concept), and Problems in Context. The Comprehensive Problems are based on material from any section or previous chapter.

Forms of Energy and Their Interconversion
(Sample Problem 6.1)

● Concept Review Questions

• **6.1** Why do heat (q) and work (w) have positive values when entering a system and negative values when leaving?

• **6.2** If you feel warm after exercising, have you increased the internal energy of your body? Explain.

• **6.3** An *adiabatic* process is one that involves no heat transfer. What is the relationship between work and the change in internal energy in an adiabatic process?

• **6.4** State two ways that you increase the internal energy of your body and two ways that you decrease it.

• **6.5** Name a common device that is used to accomplish each energy change:
(a) Electrical energy to thermal energy
(b) Electrical energy to sound energy
(c) Electrical energy to light energy
(d) Mechanical energy to electrical energy
(e) Chemical energy to electrical energy

Levels of Difficulty: • Easy •• Medium ••• Challenging   Biological: 6.2, 6.4; Engineering: 6.3

6.6 In winter, an electric heater uses a certain amount of electrical energy to heat a room to 20°C. In summer, an air conditioner uses the same amount of electrical energy to cool the room to 20°C. Is the change in internal energy of the heater larger, smaller, or the same as that of the air conditioner? Explain.

6.7 Imagine lifting your textbook into the air and dropping it onto a desktop. Describe all the energy transformations (from one form to another) that occur, moving backward in time from a moment after impact.

● **Skill-Building Exercises** (grouped in similar pairs)

6.8 A system receives 425 J of heat and delivers 425 J of work to its surroundings. What is the change in internal energy of the system (in J)?

6.9 A system conducts 255 cal of heat to the surroundings while delivering 428 cal of work. What is the change in internal energy of the system (in cal)?

6.10 What is the change in internal energy (in J) of a system that releases 675 J of thermal energy to its surroundings and has 530 cal of work done on it?

6.11 What is the change in internal energy (in J) of a system that absorbs 0.615 kJ of heat from its surroundings and has 0.247 kcal of work done on it?

6.12 Complete combustion of 2.0 metric tons of coal (assuming pure carbon) to gaseous carbon dioxide releases 6.6×10^{10} J of heat. Convert this energy to (a) kilojoules; (b) kilocalories; (c) British thermal units.

6.13 Thermal decomposition of 5.0 metric tons of limestone to lime and carbon dioxide requires 9.0×10^6 kJ of heat. Convert this energy to (a) joules; (b) calories; (c) British thermal units.

● **Problems in Context**

6.14 The nutritional calorie (Calorie) is equivalent to 1 kcal. One pound of body fat is equivalent to about 4.1×10^3 Calories. Express this quantity of energy in joules and kilojoules.

6.15 If an athlete expends 1950 kJ/h, how long does she have to play to work off 1.0 lb of body fat? (See Problem 6.14.)

Enthalpy: Heats of Reaction and Chemical Change
(Sample Problem 6.2)

● **Concept Review Questions**

6.16 Why is the work done when a system expands against a constant external pressure assigned a negative sign?

6.17 Why is it usually more convenient to measure ΔH than ΔE?

6.18 "Hot packs" used by skiers, climbers, and others for warmth are based on the crystallization of sodium acetate from a highly concentrated solution. What is the sign of ΔH for this crystallization? Is the reaction exothermic or endothermic?

6.19 Classify the following processes as exothermic or endothermic: (a) freezing of water; (b) boiling of water; (c) digestion of food; (d) a person running; (e) a person growing; (f) wood being chopped; (g) heating with a furnace.

6.20 What are the two main components of the internal energy of a substance? On what are they based?

6.21 For each process, state whether ΔH is less than (more negative), equal to, or greater than ΔE of the system. Explain.
(a) An ideal gas is cooled at constant pressure.
(b) A mixture of gases undergoes an exothermic reaction in a container of fixed volume.

(c) A solid yields a mixture of gases in an exothermic reaction that takes place in a container of variable volume.

● **Skill-Building Exercises** (grouped in similar pairs)

6.22 Draw an enthalpy diagram for a general exothermic reaction; label the axis, reactants, products, and ΔH with its sign.

6.23 Draw an enthalpy diagram for a general endothermic reaction; label the axis, reactants, products, and ΔH with its sign.

6.24 Write a balanced equation and draw an approximate enthalpy diagram for each of the following: (a) the combustion of 1 mol of ethane in oxygen; (b) the freezing of liquid water.

6.25 Write a balanced equation and draw an approximate enthalpy diagram for each of the following: (a) the formation of 1 mol of sodium chloride from its elements (heat is released); (b) the vaporization of liquid benzene.

6.26 Write a balanced equation and draw an approximate enthalpy diagram for each of the following changes: (a) the combustion of 1 mol of liquid methanol (CH_3OH); (b) the formation of 1 mol of nitrogen dioxide from its elements (heat is absorbed).

6.27 Write a balanced equation and draw an approximate enthalpy diagram for each of the following changes: (a) the sublimation of dry ice [conversion of $CO_2(s)$ directly to $CO_2(g)$]; (b) the reaction of 1 mol of sulfur dioxide with oxygen.

6.28 The circles below represent a phase change occurring at constant temperature:

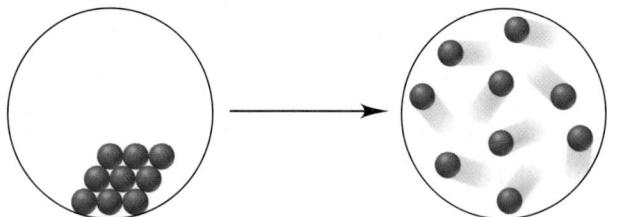

Is the value of each of the following positive (+), negative (−), or zero: (a) q_{sys}; (b) ΔE_{sys}; (c) ΔE_{univ}?

6.29 The piston-cylinder assemblies below represent a physical change occurring at constant pressure:

(a) Is w_{sys} +, −, or 0? (b) Is ΔH_{sys} +, −, or 0? (c) Can you determine whether ΔE_{surr} is +, −, or 0? Explain.

Calorimetry: Laboratory Measurement of Heats of Reaction
(Sample Problems 6.3 to 6.5)

● **Concept Review Questions**

6.30 Why can we measure only *changes* in enthalpy, not absolute enthalpy values?

6.31 What data do you need to determine the specific heat capacity of a substance?

6.32 Is the specific heat capacity of a substance an intensive or extensive property? Explain.

6.33 Distinguish between "specific heat capacity" and "heat capacity." Which parameter would you more likely use if you were calculating heat changes in (a) a chrome-plated, brass bathroom fixture; (b) a sample of high-purity copper wire; (c) a sample of pure water? Explain.

• **6.34** Both a coffee-cup calorimeter and a bomb calorimeter can be used to measure the heat involved in a reaction. Which measures ΔE and which measures ΔH? Explain.

● **Skill-Building Exercises** (grouped in similar pairs)

• **6.35** Find q when 22.0 g of water is heated from 25.°C to 100.°C.

• **6.36** Calculate q when 0.10 g of ice is cooled from 10.°C to -75°C ($c_{ice} = 2.087$ J/g·K).

• **6.37** A 295-g aluminum engine part at an initial temperature of 13.00°C absorbs 75.0 kJ of heat. What is the final temperature of the part (c of Al = 0.900 J/g·K)?

• **6.38** A 27.7-g sample of ethylene glycol, a car radiator coolant, loses 688 J of heat. What was the initial temperature of the ethylene glycol if the final temperature is 32.5°C (c of ethylene glycol = 2.42 J/g·K)?

• **6.39** Two iron bolts of equal mass—one at 100.°C, the other at 55°C—are placed in an insulated container. Assuming the heat capacity of the container is negligible, what is the final temperature inside the container (c of iron = 0.450 J/g·K)?

• **6.40** One piece of copper jewelry at 105°C has exactly twice the mass of another piece, which is at 45°C. Both pieces are placed inside a calorimeter whose heat capacity is negligible. What is the final temperature inside the calorimeter (c of copper = 0.387 J/g·K)?

• **6.41** When 155 mL of water at 26°C is mixed with 75 mL of water at 85°C, what is the final temperature? (Assume that no heat is lost to the surroundings; d of water is 1.00 g/mL.)

• **6.42** An unknown volume of water at 18.2°C is added to 24.4 mL of water at 35.0°C. If the final temperature is 23.5°C, what was the unknown volume? (Assume that no heat is lost to the surroundings; d of water is 1.00 g/mL.)

• **6.43** A 455-g piece of copper tubing is heated to 89.5°C and placed in an insulated vessel containing 159 g of water at 22.8°C. Assuming no loss of water and a heat capacity for the vessel of 10.0 J/K, what is the final temperature of the system (c of copper = 0.387 J/g·K)?

• **6.44** A 30.5-g sample of an alloy at 93.0°C is placed into 50.0 g of water at 22.0°C in an insulated coffee cup with a heat capacity of 9.2 J/K. If the final temperature of the system is 31.1°C, what is the specific heat capacity of the alloy?

● **Problems in Context**

•• **6.45** High-purity benzoic acid (C_6H_5COOH; ΔH_{rxn} for combustion = -3227 kJ/mol) is used as a standard for calibrating bomb calorimeters. A 1.221-g sample burns in a calorimeter (heat capacity = 1365 J/°C) that contains exactly 1.200 kg of water. What temperature change is observed?

•• **6.46** Two aircraft rivets, one of iron and the other of copper, are placed in a calorimeter that has an initial temperature of 20.°C. The data for the metals are as follows:

	Iron	Copper
Mass (g)	30.0	20.0
Initial T (°C)	0.0	100.0
c (J/g·K)	0.450	0.387

(a) Will heat flow from Fe to Cu or from Cu to Fe?

(b) What other information is needed to correct any measurements that would be made in an actual experiment?

(c) What is the maximum final temperature of the system (assuming the heat capacity of the calorimeter is negligible)?

• **6.47** A chemical engineer studying the properties of fuels placed 1.520 g of a hydrocarbon in the bomb of a calorimeter and filled it with O_2 gas (see Figure 6.10, p. 248). The bomb was immersed in 2.550 L of water and the reaction initiated. The water temperature rose from 20.00°C to 23.55°C. If the calorimeter (excluding the water) had a heat capacity of 403 J/K, what was the heat of reaction for combustion (q_V) per gram of the fuel?

• **6.48** When 25.0 mL of 0.500 M H_2SO_4 is added to 25.0 mL of 1.00 M KOH in a coffee-cup calorimeter at 23.50°C, the temperature rises to 30.17°C. Calculate ΔH of this reaction. (Assume that the total volume is the sum of the individual volumes and that the density and specific heat capacity of the solution are the same as for pure water.)

Stoichiometry of Thermochemical Equations
(Sample Problem 6.6)

● **Concept Review Questions**

• **6.49** Does a negative ΔH_{rxn} mean that the heat of reaction can be thought of as a reactant or as a product?

• **6.50** Would you expect $O_2(g) \longrightarrow 2O(g)$ to have a positive or a negative ΔH_{rxn}? Explain.

• **6.51** Is ΔH positive or negative when 1 mol of water vapor condenses to liquid water? Why? How does this value compare with that for the vaporization of 2 mol of liquid water to water vapor?

● **Skill-Building Exercises** (grouped in similar pairs)

• **6.52** Consider the following balanced thermochemical equation for a reaction sometimes used for H_2S production:

$$\tfrac{1}{8}S_8(s) + H_2(g) \longrightarrow H_2S(g) \quad \Delta H_{rxn} = -20.2 \text{ kJ}$$

(a) Is this an exothermic or endothermic reaction?
(b) What is ΔH_{rxn} for the reverse reaction?
(c) What is ΔH when 2.6 mol of S_8 reacts?
(d) What is ΔH when 25.0 g of S_8 reacts?

• **6.53** Consider the following balanced thermochemical equation for the decomposition of the mineral magnesite:

$$MgCO_3(s) \longrightarrow MgO(s) + CO_2(g) \quad \Delta H_{rxn} = 117.3 \text{ kJ}$$

(a) Is heat absorbed or released in the reaction?
(b) What is ΔH_{rxn} for the reverse reaction?
(c) What is ΔH when 5.35 mol of CO_2 reacts with excess MgO?
(d) What is ΔH when 35.5 g of CO_2 reacts with excess MgO?

• **6.54** When 1 mol of NO(g) forms from its elements, 90.29 kJ of heat is absorbed. (a) Write a balanced thermochemical equation for this reaction. (b) How much heat is involved when 3.50 g of NO decomposes to its elements?

• **6.55** When 1 mol of KBr(s) decomposes to its elements, 394 kJ of heat is absorbed. (a) Write a balanced thermochemical equation for this reaction. (b) How much heat is released when 10.0 kg of KBr forms from its elements?

● **Problems in Context**

• **6.56** Liquid hydrogen peroxide, an oxidizing agent in many rocket fuel mixtures, releases oxygen gas on decomposition:

$$2H_2O_2(l) \longrightarrow 2H_2O(l) + O_2(g) \quad \Delta H_{rxn} = -196.1 \text{ kJ}$$

How much heat is released when 652 kg of H_2O_2 decomposes?

• **6.57** Compounds of boron and hydrogen are remarkable for their unusual bonding (described in Section 14.5) and also for their reactivity. With the more reactive halogens, for example, diborane (B_2H_6) forms trihalides even at low temperatures:

$$B_2H_6(g) + 6Cl_2(g) \longrightarrow 2BCl_3(g) + 6HCl(g)$$
$$\Delta H_{rxn} = -755.4 \text{ kJ}$$

How much heat is released per kilogram of diborane that reacts?

6.58 Deterioration of buildings, bridges, and other structures through the rusting of iron costs millions of dollars every day. Although the actual process also requires water, a simplified equation (with rust shown as Fe_2O_3) is

$$4Fe(s) + 3O_2(g) \longrightarrow 2Fe_2O_3(s) \quad \Delta H_{rxn} = -1.65 \times 10^3 \text{ kJ}$$

(a) How much heat is evolved when 0.250 kg of iron rusts?
(b) How much rust forms when 4.85×10^3 kJ of heat is released?

6.59 A mercury mirror forms inside a test tube as a result of the thermal decomposition of mercury(II) oxide:

$$2HgO(s) \longrightarrow 2Hg(l) + O_2(g) \quad \Delta H_{rxn} = 181.6 \text{ kJ}$$

(a) How much heat is needed to decompose 555 g of the oxide?
(b) If 275 kJ of heat is absorbed, how many grams of Hg form?

6.60 Ethylene (C_2H_4) is the starting material for the preparation of polyethylene. Although typically made during the processing of petroleum, ethylene occurs naturally as a fruit-ripening hormone and as a component of natural gas. (a) The heat of reaction for the combustion of C_2H_4 is −1411 kJ/mol. Write a balanced thermochemical equation for the combustion of C_2H_4. (b) How many grams of C_2H_4 must burn to give 70.0 kJ of heat?

6.61 Sucrose ($C_{12}H_{22}O_{11}$, table sugar) is oxidized in the body by O_2 via a complex set of reactions that ultimately produces $CO_2(g)$ and $H_2O(g)$ and releases 5.64×10^3 kJ/mol sucrose.
(a) Write a balanced thermochemical equation for this reaction.
(b) How much heat is released per gram of sucrose oxidized?

Hess's Law of Heat Summation
(Sample Problem 6.7)

● Concept Review Questions

6.62 Express Hess's law in your own words.

6.63 What is the main use of Hess's law?

6.64 It is very difficult to burn carbon in a deficiency of O_2 and produce only CO; some CO_2 forms as well. However, carbon burns in excess O_2 to form only CO_2, and CO burns in excess O_2 to form only CO_2. Use the heats of the latter two reactions (from Appendix B) to calculate ΔH_{rxn} for the following reaction:

$$C(graphite) + \tfrac{1}{2} O_2(g) \longrightarrow CO(g)$$

● Skill-Building Exercises (grouped in similar pairs)

6.65 Calculate ΔH_{rxn} for

$$Ca(s) + \tfrac{1}{2}O_2(g) + CO_2(g) \longrightarrow CaCO_3(s)$$

given the following set of reactions:

$$Ca(s) + \tfrac{1}{2}O_2(g) \longrightarrow CaO(s) \qquad \Delta H = -635.1 \text{ kJ}$$
$$CaCO_3(s) \longrightarrow CaO(s) + CO_2(g) \quad \Delta H = 178.3 \text{ kJ}$$

6.66 Calculate ΔH_{rxn} for

$$2NOCl(g) \longrightarrow N_2(g) + O_2(g) + Cl_2(g)$$

given the following set of reactions:

$$\tfrac{1}{2}N_2(g) + \tfrac{1}{2}O_2(g) \longrightarrow NO(g) \qquad \Delta H = 90.3 \text{ kJ}$$
$$NO(g) + \tfrac{1}{2}Cl_2(g) \longrightarrow NOCl(g) \quad \Delta H = -38.6 \text{ kJ}$$

6.67 Write the balanced overall equation (equation 3) for the following process, calculate $\Delta H_{overall}$, and match the number of each equation with the letter of the appropriate arrow in Figure P6.67:

(1) $N_2(g) + O_2(g) \longrightarrow 2NO(g)$		$\Delta H = 180.6 \text{ kJ}$
(2) $2NO(g) + O_2(g) \longrightarrow 2NO_2(g)$		$\Delta H = -114.2 \text{ kJ}$
(3)		$\Delta H_{overall} = ?$

6.68 Write the balanced overall equation (equation 3) for the following process, calculate $\Delta H_{overall}$, and match the number of each equation with the letter of the appropriate arrow in Figure P6.68:

(1) $P_4(s) + 6Cl_2(g) \longrightarrow 4PCl_3(g)$		$\Delta H = -1148 \text{ kJ}$
(2) $4PCl_3(g) + 4Cl_2(g) \longrightarrow 4PCl_5(g)$		$\Delta H = -460 \text{ kJ}$
(3)		$\Delta H_{overall} = ?$

Figure P6.67   Figure P6.68

6.69 At a given set of conditions, 241.8 kJ is given off when 1 mol of $H_2O(g)$ forms from its elements. Under the same conditions, 285.8 kJ is given off when 1 mol of $H_2O(l)$ forms from its elements. Find the heat of vaporization of water at these conditions.

6.70 When 1 mol of $CS_2(l)$ forms from its elements at 1 atm and 25°C, 89.7 kJ is absorbed, and it takes 27.7 kJ to vaporize 1 mol of the liquid. How much heat is absorbed when 1 mol of $CS_2(g)$ forms from its elements at these conditions?

● Problems in Context

6.71 Diamond and graphite are two crystalline forms of carbon. At 1 atm and 25°C, diamond changes to graphite so slowly that the enthalpy change of the process must be obtained indirectly. Determine ΔH_{rxn} for

$$C(diamond) \longrightarrow C(graphite)$$

with equations from the following list:

(1) $C(diamond) + O_2(g) \longrightarrow CO_2(g)$	$\Delta H = -395.4 \text{ kJ}$
(2) $2CO_2(g) \longrightarrow 2CO(g) + O_2(g)$	$\Delta H = 566.0 \text{ kJ}$
(3) $C(graphite) + O_2(g) \longrightarrow CO_2(g)$	$\Delta H = -393.5 \text{ kJ}$
(4) $2CO(g) \longrightarrow C(graphite) + CO_2(g)$	$\Delta H = -172.5 \text{ kJ}$

Standard Heats of Reaction (ΔH_{rxn}°)
(Sample Problems 6.8 and 6.9)

● Concept Review Questions

6.72 What is the difference between the standard heat of formation and the standard heat of reaction?

6.73 How are ΔH_f° values used to calculate ΔH_{rxn}°?

6.74 Make any changes needed in each of the following equations to make ΔH_{rxn}° equal to ΔH_f° for the compound present:
(a) $Cl(g) + Na(s) \longrightarrow NaCl(s)$
(b) $H_2O(g) \longrightarrow 2H(g) + \tfrac{1}{2}O_2(g)$
(c) $\tfrac{1}{2}N_2(g) + \tfrac{3}{2}H_2(g) \longrightarrow NH_3(g)$

● Skill-Building Exercises (grouped in similar pairs)

6.75 Use Table 6.3 or Appendix B to write a balanced formation equation at standard conditions for each of the following compounds: (a) $CaCl_2$; (b) $NaHCO_3$; (c) CCl_4; (d) HNO_3.

• **6.76** Use Table 6.3 or Appendix B to write a balanced formation equation at standard conditions for each of the following compounds: (a) HI; (b) SiF_4; (c) O_3; (d) $Ca_3(PO_4)_2$.

• **6.77** Calculate ΔH°_{rxn} for each of the following:
(a) $2H_2S(g) + 3O_2(g) \longrightarrow 2SO_2(g) + 2H_2O(g)$
(b) $CH_4(g) + Cl_2(g) \longrightarrow CCl_4(l) + HCl(g)$ [unbalanced]

• **6.78** Calculate ΔH°_{rxn} for each of the following:
(a) $SiO_2(s) + 4HF(g) \longrightarrow SiF_4(g) + 2H_2O(l)$
(b) $C_2H_6(g) + O_2(g) \longrightarrow CO_2(g) + H_2O(g)$ [unbalanced]

• **6.79** Copper(I) oxide can be oxidized to copper(II) oxide:
$$Cu_2O(s) + \tfrac{1}{2}O_2(g) \longrightarrow 2CuO(s) \quad \Delta H^\circ_{rxn} = -146.0 \text{ kJ}$$
Given ΔH°_f of $Cu_2O(s) = -168.6$ kJ/mol, find ΔH°_f of $CuO(s)$.

• **6.80** Acetylene burns in air according to the following equation:
$$C_2H_2(g) + \tfrac{5}{2}O_2(g) \longrightarrow 2CO_2(g) + H_2O(g)$$
$$\Delta H^\circ_{rxn} = -1255.8 \text{ kJ}$$
Given ΔH°_f of $CO_2(g) = -393.5$ kJ/mol and ΔH°_f of $H_2O(g) = -241.8$ kJ/mol, find ΔH°_f of $C_2H_2(g)$.

● **Problems in Context**

• **6.81** The common lead-acid car battery produces a large burst of current, even at low temperatures, and is rechargeable. The reaction that occurs while recharging a "dead" battery is
$$2PbSO_4(s) + 2H_2O(l) \longrightarrow Pb(s) + PbO_2(s) + 2H_2SO_4(l)$$
(a) Use ΔH°_f values from Appendix B to calculate ΔH°_{rxn}.
(b) Use the following equations to check your answer to part (a):
(1) $Pb(s) + PbO_2(s) + 2SO_3(g) \longrightarrow 2PbSO_4(s)$
$$\Delta H^\circ = -768 \text{ kJ}$$
(2) $SO_3(g) + H_2O(l) \longrightarrow H_2SO_4(l)$ $\quad \Delta H^\circ = -132$ kJ

Comprehensive Problems

•• **6.82** Stearic acid ($C_{18}H_{36}O_2$) is a typical fatty acid, a molecule with a long hydrocarbon chain and an organic acid group (COOH) at the end. It is used to make cosmetics, ointments, soaps, and candles and is found in animal tissue as part of many saturated fats. In fact, when you eat meat, chances are that you are ingesting some fats that contain stearic acid.
(a) Write a balanced equation for the complete combustion of stearic acid to gaseous products.
(b) Calculate ΔH°_{rxn} for this combustion (ΔH°_f of $C_{18}H_{36}O_2 = -948$ kJ/mol).
(c) Calculate the heat (q) in kJ and kcal when 1.00 g of stearic acid is burned completely.
(d) Nutritional information for a candy bar states that one serving contains 11.0 g of fat and 100. Cal from fat (1 Cal = 1 kcal). Is this information consistent with your answer for (c)?

••• **6.83** Care must be taken when diluting sulfuric acid with water, because the dilution process is highly exothermic:
$$H_2SO_4(l) \xrightarrow{H_2O} H_2SO_4(aq) + heat$$
(a) Use Appendix B to find ΔH° for diluting 1.00 mol of $H_2SO_4(l)$ ($d = 1.83$ g/mL) to 1 L of 1.00 M $H_2SO_4(aq)$ ($d = 1.060$ g/mL).
(b) Suppose you carry out the dilution in a calorimeter. The initial T is 25.0°C, and the specific heat capacity of the final solution is 3.50 J/g·K. What is the final T?
(c) Use the ideas of density and heat capacity to explain why you should add acid to water rather than water to acid.

•• **6.84** A balloonist is preparing to make a trip in a helium-filled balloon. The trip begins in early morning when the temperature is 15°C. By midafternoon, the temperature has increased to 30.°C.

Assuming the pressure remains constant at 1.00 atm, for each mole of helium, calculate: (a) The initial and final volumes. (b) The change in internal energy, ΔE [*Hint:* Helium behaves like an ideal gas, so $E = \tfrac{3}{2}nRT$. Be sure the units of R are consistent with those of E]. (c) The work (w) done by the helium (in J). (d) The heat (q) transferred (in J). (e) ΔH for the process (in J). (f) Explain the relationship between the answers to (d) and (e).

•• **6.85** In winemaking, the sugars in grapes undergo *fermentation* by yeast to yield CH_3CH_2OH and CO_2. During cellular *respiration*, sugar and ethanol are "burned" to water vapor and CO_2.
(a) Using $C_6H_{12}O_6$ for sugar, calculate ΔH°_{rxn} of fermentation and of respiration (combustion).
(b) Write a combustion reaction for ethanol. Which has a higher ΔH°_{rxn} for combustion per mole of C, sugar or ethanol?

•• **6.86** Three of the reactions that occur when the paraffin of a candle (typical formula $C_{21}H_{44}$) burns are as follows:
(1) Complete combustion forms CO_2 and water vapor.
(2) Incomplete combustion forms CO and water vapor.
(3) Some wax is oxidized to elemental C (soot) and water vapor.
(a) Find ΔH°_{rxn} of each reaction (ΔH°_f of $C_{21}H_{44} = -476$ kJ/mol; use graphite for elemental carbon).
(b) Find q (in kJ) when a 254-g candle burns completely.
(c) Find q (in kJ) when 8.00% by mass of the candle burns incompletely and another 5.00% undergoes soot formation.

•• **6.87** Ethylene oxide (EO) is prepared by the vapor-phase oxidation of ethylene. Its main uses are in the preparation of the antifreeze ethylene glycol and in the production of poly(ethylene terephthalate), which is used to make beverage bottles and fibers. Pure EO vapor can decompose explosively:

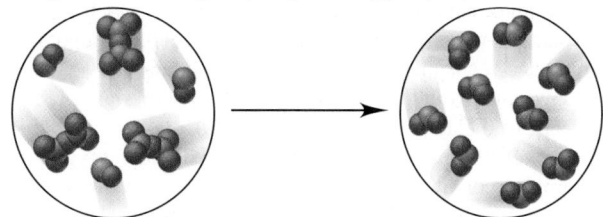

Liquid EO has $\Delta H^\circ_f = -77.4$ kJ/mol and $\Delta H^\circ_{vap} = 569.4$ J/g.
(a) Calculate ΔH°_{rxn} for the gas-phase reaction. (b) Due to external heating, the vapor decomposes at 10 bar and 93°C in a distillation column. What is the final temperature if the average specific heat capacity of the products is 2.5 J/g·°C?

•• **6.88** The following scenes represent a gaseous reaction between compounds of nitrogen (blue) and oxygen (red) at 298 K:

(a) Write a balanced equation and use Appendix B to calculate ΔH°_{rxn}. (b) If each molecule of product represents 1.50×10^{-2} mol, what quantity of heat (in J) is released or absorbed?

•• **6.89** Isooctane (C_8H_{18}; $d = 0.692$ g/mL) is used as the fuel in a test of a new automobile drive train.
(a) How much energy (in kJ) is released by complete combustion of the isooctane in a 20.4-gal fuel tank to gases ($\Delta H^\circ_{rxn} = -5.44 \times 10^3$ kJ/mol)?
(b) The energy delivered to the wheels at 65 mph is 5.5×10^4 kJ/h. Assuming all the energy is transferred to the wheels, what is the cruising range (in km) of the car on a full tank?
(c) If the actual range is 455 miles, explain your answer to (b).

6.90 Four 50.-g samples of different liquids are placed in beakers at $T_{initial}$ of 25.00°C. Each liquid is heated until 450. J is absorbed; T_{final} is shown on each beaker below. Rank the liquids in order of increasing specific heat capacity.

6.91 Oxidation of gaseous ClF by F_2 yields liquid ClF_3, an important fluorinating agent. Use the following thermochemical equations to calculate ΔH°_{rxn} for the production of ClF_3:

(1) $2ClF(g) + O_2(g) \longrightarrow Cl_2O(g) + OF_2(g)$ $\Delta H^\circ = 167.5$ kJ
(2) $2F_2(g) + O_2(g) \longrightarrow 2OF_2(g)$ $\Delta H^\circ = -43.5$ kJ
(3) $2ClF_3(l) + 2O_2(g) \longrightarrow Cl_2O(g) + 3OF_2(g)$
                                    $\Delta H^\circ = 394.1$ kJ

6.92 Silver bromide is used to coat ordinary black-and-white photographic film, while high-speed film uses silver iodide.
(a) When 50.0 mL of 5.0 g/L $AgNO_3$ is added to a coffee-cup calorimeter containing 50.0 mL of 5.0 g/L NaI, with both solutions at 25°C, what mass of AgI forms?
(b) Use Appendix B to find ΔH°_{rxn}.
(c) What is ΔT_{soln} (assume the volumes are additive and the solution has the density and specific heat capacity of water)?

6.93 The calorie (4.184 J) was originally defined as the quantity of energy required to raise the temperature of 1.00 g of liquid water 1.00°C. The British thermal unit (Btu) is defined as the quantity of energy required to raise the temperature of 1.00 lb of liquid water 1.00°F.
(a) How many joules are in 1.00 Btu (1 lb = 453.6 g; a change of 1.0°C = 1.8°F)?
(b) The *therm* is a unit of energy consumption and is defined as 100,000 Btu. How many joules are in 1.00 therm?
(c) How many moles of methane must be burned to give 1.00 therm of energy? (Assume water forms as a gas.)
(d) If natural gas costs $0.66 per therm, what is the cost per mole of methane? (Assume natural gas is pure methane.)
(e) How much would it cost to warm 318 gal of water in a hot tub from 15.0°C to 42.0°C (1 gal = 3.78 L)?

6.94 Whenever organic matter is decomposed under oxygen-free (anaerobic) conditions, methane is one of the products. Thus, enormous deposits of natural gas, which is almost entirely methane, serve as a major source of fuel for home and industry.
(a) Known sources of natural gas can produce 5600 EJ of energy (1 EJ = 10^{18} J). Current total global energy usage is 4.0×10^2 EJ per year. Find the mass (in kg) of known sources of natural gas (ΔH°_{rxn} for the combustion of $CH_4 = -802$ kJ/mol).
(b) For how many years could these sources supply the world's total energy needs?
(c) What volume (in ft^3) of natural gas, measured at STP, is required to heat 1.00 qt of water from 25.0°C to 100.0°C (d of H_2O = 1.00 g/mL; d of CH_4 at STP = 0.72 g/L)?
(d) The fission of 1 mol of uranium (about 4×10^{-4} ft^3) in a nuclear reactor produces 2×10^{13} J. What volume (in ft^3) of natural gas would produce the same amount of energy?

6.95 A reaction is carried out in a steel vessel within a chamber filled with argon gas. Shown below are molecular views of the argon adjacent to the surface of the reaction vessel before and after the reaction. Was the reaction exothermic or endothermic? Explain.

 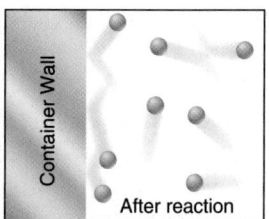

6.96 For each of the following events, the system is in *italics*. State whether heat, work, or both is (are) transferred, and specify the direction of each transfer.
(a) You pump air into an automobile *tire*.
(b) A *tree* rots in a forest.
(c) You strike a *match*.
(d) You cool juice in a *refrigerator*.
(e) You cook *food* on a kitchen range.

6.97 To make use of ionic hydrates for storing solar energy (see Chemical Connections, p. 259), you place 500.0 kg of sodium sulfate decahydrate on your house roof. Assuming complete reaction and 100% efficiency of heat transfer, how much heat (in kJ) is released to your house at night?

6.98 An aqueous waste stream that has a maximum concentration of 0.50 M H_2SO_4 (d = 1.030 g/mL at 25°C) will be neutralized by controlled addition of 40% caustic soda (NaOH; d = 1.430 g/L) before it goes to the process sewer and then to the chemical plant waste treatment facility. However, a safety review finds that the waste stream could meet a small stream of an immiscible organic compound, which could form a flammable vapor in air at 40.°C. The maximum temperature of the caustic soda and the waste stream is 31°C. Could the temperature increase due to the heat of neutralization cause the vapor to explode? Assume the specific heat capacity of each solution is 4.184 J/g·K.

6.99 Kerosene, a common space-heater fuel, is a mixture of hydrocarbons whose "average" formula is $C_{12}H_{26}$.
(a) Write a balanced equation, using the simplest whole-number coefficients, for the complete combustion of kerosene to gases.
(b) If $\Delta H^\circ_{rxn} = -1.50 \times 10^4$ kJ for the combustion equation as written in part (a), determine ΔH°_f of kerosene.
(c) Calculate the heat produced by combustion of 0.50 gal of kerosene (d of kerosene = 0.749 g/mL).
(d) How many gallons of kerosene must be burned for a kerosene furnace to produce 1250. Btu (1 Btu = 1.055 kJ)?

6.100 Coal gasification is a multistep process to convert coal into cleaner-burning fuels (see Chemical Connections, p. 256). In one step, a coal sample reacts with superheated steam:

$C(coal) + H_2O(g) \longrightarrow CO(g) + H_2(g)$ $\Delta H^\circ_{rxn} = 129.7$ kJ

(a) Combine this reaction with the following two to write an overall reaction for the production of methane:

$CO(g) + H_2O(g) \longrightarrow CO_2(g) + H_2(g)$ $\Delta H^\circ_{rxn} = -41$ kJ
$CO(g) + 3H_2(g) \longrightarrow CH_4(g) + H_2O(g)$ $\Delta H^\circ_{rxn} = -206$ kJ

(b) Calculate ΔH°_{rxn} for this overall change.

(c) Using the value in (b) and calculating ΔH°_{rxn} for the combustion of methane, find the total heat for gasifying 1.00 kg of coal and burning the methane formed (assume water forms as a gas and $\mathcal{M}$ of coal = 12.00 g/mol).

•• 6.101 Phosphorus pentachloride is used in the industrial preparation of organic phosphorus compounds. Equation 1 shows its preparation from PCl_3 and Cl_2:

(1) $PCl_3(l) + Cl_2(g) \longrightarrow PCl_5(s)$

Use equations 2 and 3 to calculate ΔH_{rxn} of equation 1:

(2) $P_4(s) + 6Cl_2(g) \longrightarrow 4PCl_3(l)$ $\Delta H = -1280$ kJ
(3) $P_4(s) + 10Cl_2(g) \longrightarrow 4PCl_5(s)$ $\Delta H = -1774$ kJ

•• 6.102 Consider the following hydrocarbon fuels:

(I) $CH_4(g)$ (II) $C_2H_4(g)$ (III) $C_2H_6(g)$

(a) Rank them in terms of heat released (a) per mole and (b) per gram. [Assume $H_2O(g)$ forms and combustion is complete.]

•• 6.103 A typical candy bar weighs about 2 oz (1.00 oz = 28.4 g).
(a) Assuming that a candy bar is 100% sugar and that 1.0 g of sugar is equivalent to about 4.0 Calories of energy, calculate the energy (in kJ) contained in a typical candy bar.
(b) Assuming that your mass is 58 kg and you convert chemical potential energy to work with 100% efficiency, how high would you have to climb to work off the energy in a candy bar? (Potential energy = mass × g × height, where g = 9.8 m/s².)
(c) Why is your actual conversion of potential energy to work less than 100% efficient?

•• 6.104 Silicon tetrachloride is produced annually on the multikiloton scale for making transistor-grade silicon. It can be made directly from the elements (reaction 1) or, more cheaply, by heating sand and graphite with chlorine gas (reaction 2). If water is present in reaction 2, some tetrachloride may be lost in an unwanted side reaction (reaction 3):

(1) $Si(s) + 2Cl_2(g) \longrightarrow SiCl_4(g)$
(2) $SiO_2(s) + 2C(\text{graphite}) + 2Cl_2(g) \longrightarrow SiCl_4(g) + 2CO(g)$
(3) $SiCl_4(g) + 2H_2O(g) \longrightarrow SiO_2(s) + 4HCl(g)$

$\Delta H^{\circ}_{rxn} = -139.5$ kJ

(a) Use reaction 3 to calculate the heats of reaction of reactions 1 and 2. (b) What is the heat of reaction for the new reaction that is the sum of reactions 2 and 3?

•• 6.105 Use the following information to find ΔH°_f of gaseous HCl:

$N_2(g) + 3H_2(g) \longrightarrow 2NH_3(g)$ $\Delta H^{\circ}_{rxn} = -91.8$ kJ
$N_2(g) + 4H_2(g) + Cl_2(g) \longrightarrow 2NH_4Cl(s)$ $\Delta H^{\circ}_{rxn} = -628.8$ kJ
$NH_3(g) + HCl(g) \longrightarrow NH_4Cl(s)$ $\Delta H^{\circ}_{rxn} = -176.2$ kJ

••• 6.106 You want to determine ΔH° for the reaction

$Zn(s) + 2HCl(aq) \longrightarrow ZnCl_2(aq) + H_2(g)$

(a) To do so, you first determine the heat capacity of a calorimeter using the following reaction, whose ΔH° is known:

$NaOH(aq) + HCl(aq) \longrightarrow NaCl(aq) + H_2O(l)$
$\Delta H^{\circ} = -57.32$ kJ

Calculate the heat capacity of the calorimeter from these data:
Amounts used: 50.0 mL of 2.00 M HCl and 50.0 mL of 2.00 M NaOH
Initial T of both solutions: 16.9°C
Maximum T recorded during reaction: 30.4°C

Density of resulting NaCl solution: 1.04 g/mL
c of 1.00 M $NaCl(aq)$ = 3.93 J/g·K

(b) Use the result from part (a) and the following data to determine ΔH°_{rxn} for the reaction between zinc and $HCl(aq)$:
Amounts used: 100.0 mL of 1.00 M HCl and 1.3078 g of Zn
Initial T of HCl solution and Zn: 16.8°C
Maximum T recorded during reaction: 24.1°C
Density of 1.0 M HCl solution = 1.015 g/mL
c of resulting $ZnCl_2(aq)$ = 3.95 J/g·K

(c) Given the values below, what is the error in your experiment?

ΔH°_f of $HCl(aq)$ = -1.652×10^2 kJ/mol
ΔH°_f of $ZnCl_2(aq)$ = -4.822×10^2 kJ/mol

•••• 6.107 One mole of nitrogen gas confined within a cylinder by a piston is heated from 0°C to 819°C at 1.00 atm.
(a) Calculate the work of expansion of the gas in joules (1 J = 9.87×10^{-3} atm·L). Assume all the energy is used to do work.
(b) What would be the temperature change if the gas were heated with the same amount of energy in a container of fixed volume? (Assume the specific heat capacity of N_2 is 1.00 J/g·K.)

•• 6.108 The chemistry of nitrogen oxides is very versatile. Given the following reactions and their standard enthalpy changes,

(1) $NO(g) + NO_2(g) \longrightarrow N_2O_3(g)$ $\Delta H^{\circ}_{rxn} = -39.8$ kJ
(2) $NO(g) + NO_2(g) + O_2(g) \longrightarrow N_2O_5(g)$ $\Delta H^{\circ}_{rxn} = -112.5$ kJ
(3) $2NO_2(g) \longrightarrow N_2O_4(g)$ $\Delta H^{\circ}_{rxn} = -57.2$ kJ
(4) $2NO(g) + O_2(g) \longrightarrow 2NO_2(g)$ $\Delta H^{\circ}_{rxn} = -114.2$ kJ
(5) $N_2O_5(s) \longrightarrow N_2O_5(g)$ $\Delta H^{\circ}_{subl} = 54.1$ kJ

calculate the heat of reaction for

$N_2O_3(g) + N_2O_5(s) \longrightarrow 2N_2O_4(g)$

••• 6.109 Electric generating plants transport large amounts of hot water through metal pipes, and oxygen dissolved in the water can cause a major corrosion problem. Hydrazine (N_2H_4) added to the water avoids the problem by reacting with the oxygen:

$N_2H_4(aq) + O_2(g) \longrightarrow N_2(g) + 2H_2O(l)$

About 4×10^7 kg of hydrazine is produced every year by reacting ammonia with sodium hypochlorite in the *Raschig process*:

$2NH_3(aq) + NaOCl(aq) \longrightarrow N_2H_4(aq) + NaCl(aq) + H_2O(l)$
$\Delta H^{\circ}_{rxn} = -151$ kJ

(a) If ΔH°_f of $NaOCl(aq)$ = -346 kJ/mol, find ΔH°_f of $N_2H_4(aq)$.
(b) What is the heat released when aqueous N_2H_4 is added to 5.00×10^3 L of plant water that is 2.50×10^{-4} M O_2?

••• 6.110 Liquid methanol (CH_3OH) can be used as an alternative fuel in pickup and SUV engines. An industrial method for preparing it uses the catalytic hydrogenation of carbon monoxide:

$CO(g) + 2H_2(g) \xrightarrow{\text{catalyst}} CH_3OH(l)$

How much heat (in kJ) is released when 15.0 L of CO at 85°C and 112 kPa reacts with 18.5 L of H_2 at 75°C and 744 torr?

••• 6.111 (a) How much heat is released when 25.0 g of methane burns in excess O_2 to form gaseous CO_2 and H_2O?
(b) Calculate the temperature of the product mixture if the methane and air are both at an initial temperature of 0.0°C. Assume a stoichiometric ratio of methane to oxygen from the air, with air being 21% O_2 by volume (c of CO_2 = 57.2 J/mol·K; c of $H_2O(g)$ = 36.0 J/mol·K; c of N_2 = 30.5 J/mol·K).

7

Atmospheric Excitement Charged particles (electrons and positive ions) in the solar wind collide with and excite atoms in the atmosphere, which then relax and emit the glorious light of an aurora. As you'll see in this chapter, TV screens and neon signs work by the same principle.

Quantum Theory and Atomic Structure

7.1 The Nature of Light
Wave Nature of Light
Particle Nature of Light

7.2 Atomic Spectra
Bohr Model of the Hydrogen Atom
Energy States of the Hydrogen Atom

7.3 The Wave-Particle Duality of Matter and Energy
Wave Nature of Electrons and Particle Nature of Photons
Heisenberg Uncertainty Principle

7.4 The Quantum-Mechanical Model of the Atom
The Atomic Orbital
Quantum Numbers
Shapes of Atomic Orbitals
The Special Case of the Hydrogen Atom

Over a few remarkable decades—from around 1890 to 1930—a revolution took place in how we view the makeup of the universe. ● But revolutions in science are not the violent upheavals of political overthrow. Rather, flaws appear in an established model as conflicting evidence mounts, a startling discovery or two widens the flaws into cracks, and the conceptual structure crumbles gradually from its inconsistencies. New insight, verified by experiment, then guides the building of a model more consistent with reality. So it was when Lavoisier's theory of combustion superseded the phlogiston model, when Dalton's atomic theory established the idea of individual units of matter, and when Rutherford's nuclear model substituted atoms with rich internal structure for "billiard balls" or "plum puddings." In this chapter, you will see this process unfold again with the development of modern atomic theory.

Almost as soon as Rutherford proposed his nuclear model, a major problem arose. A nucleus and an electron attract each other, so if they are to remain apart, the energy of the electron's motion (kinetic energy) must balance the energy of attraction (potential energy). However, the laws of classical physics had established that a negative particle moving in a curved path around a positive one *must* emit radiation and thus lose energy. If this requirement applied to atoms, why didn't the orbiting electron lose energy continuously and spiral into the nucleus? Clearly, if electrons behaved the way classical physics predicted, all atoms would have collapsed eons ago! The behavior of subatomic matter seemed to violate real-world experience and accepted principles.

The breakthroughs that soon followed Rutherford's model forced a complete rethinking of the classical picture of matter and energy. In the macroscopic world, the two are distinct. Matter occurs in chunks you can hold and weigh, and you can change the amount of matter in a sample piece by piece. In contrast, energy is "massless," and its quantity changes in a continuous manner. Matter moves in specific paths, whereas light and other types of energy travel in diffuse waves. As soon as 20th-century scientists probed the subatomic world, however, these clear distinctions between particulate matter and wavelike energy began to fade.

IN THIS CHAPTER . . . We discuss *quantum mechanics,* the theory that explains our current picture of atomic structure. We consider the wave properties of energy and then examine the theories and experiments that led to a quantized, or particulate, model of light. We see why the light emitted by excited hydrogen (H) atoms—the *atomic spectrum*—suggests an atom with distinct energy levels, and we look briefly at how atomic spectra are applied to chemical analysis. Wave-particle duality, which reveals two faces of matter and of energy, leads us to the current model of the H atom and the quantum numbers that identify the regions of space an electron occupies in an atom. In Chapter 8, we'll consider atoms that have more than one electron and relate electron number and distribution to chemical behavior.

7.1 THE NATURE OF LIGHT

Visible light is one type of **electromagnetic radiation** (also called *electromagnetic energy* or *radiant energy*). Other familiar types include x-rays, microwaves, and radio waves. All electromagnetic radiation consists of energy propagated by means of electric and magnetic fields that alternately increase and decrease in intensity as they move through space. This classical wave model distinguishes clearly between waves and particles; it is essential for understanding why rainbows form, how magnifying glasses work, why objects look distorted under water, and many other everyday observations. But, it cannot explain observations on the atomic scale because, as you'll shortly, in that unfamiliar realm, energy behaves as though it consists of particles!

● **Hooray for the Human Mind** The invention of the car, radio, and airplane fostered a feeling of unlimited human ability, and the discovery of x-rays, radioactivity, the electron, and the atomic nucleus led to the sense that the human mind would soon unravel all of nature's mysteries. Indeed, some people were convinced that few, if any, mysteries remained.

1895 Röntgen discovers x-rays.
1896 Becquerel discovers radioactivity.
1897 Thomson discovers the electron.
1898 Curie discovers radium.
1900 Freud proposes theory of the unconscious mind.
1900 Planck develops quantum theory.
1901 Marconi invents the radio.
1903 Wright brothers fly an airplane.
1905 Ford uses assembly line to build cars.
1905 Rutherford explains radioactivity.
1905 Einstein publishes relativity and photon theories.
1906 St. Denis develops modern dance.
1908 Matisse and Picasso develop modern art.
1909 Schoenberg and Berg develop modern music.
1911 Rutherford presents nuclear model.
1913 Bohr proposes atomic model.
1914 to 1918 World War I is fought.
1923 Compton demonstrates photon momentum.
1924 De Broglie publishes wave theory of matter.
1926 Schrödinger develops wave equation.
1927 Heisenberg presents uncertainty principle.
1932 Chadwick discovers the neutron.

The Wave Nature of Light

The wave properties of electromagnetic radiation are described by two interdependent variables, as Figure 7.1 shows:

- **Frequency** (ν, Greek *nu*) is the number of cycles the wave undergoes per second and is expressed in units of 1/second [s^{-1}; also called *hertz* (Hz)].
- **Wavelength** (λ, Greek *lambda*) is the distance between any point on a wave and the corresponding point on the next crest (or trough) of the wave, that is, the distance the wave travels during one cycle. Wavelength is expressed in meters and often, for very short wavelengths, in nanometers (nm, 10^{-9} m), picometers (pm, 10^{-12} m), or the non-SI unit angstroms (Å, 10^{-10} m).

The speed of the wave, the distance traveled per unit time (in units of meters per second), is the product of its frequency (cycles per second) and its wavelength (meters per cycle):

$$\text{Units for speed of wave:} \quad \frac{\text{cycles}}{\text{s}} \times \frac{\text{m}}{\text{cycle}} = \frac{\text{m}}{\text{s}}$$

In a vacuum, all types of electromagnetic radiation travel at 2.99792458×10^8 m/s (3.00×10^8 m/s to three significant figures), which is a physical constant called the **speed of light (*c*):**

$$c = \nu \times \lambda \qquad (7.1)$$

As Equation 7.1 shows, the product of ν and λ is a constant. Thus, the individual terms have a reciprocal relationship to each other: *radiation with a high frequency has a short wavelength, and vice versa.*

Another characteristic of a wave is its **amplitude,** the height of the crest (or depth of the trough) of each wave (Figure 7.2). The amplitude of an electromagnetic wave is a measure of the strength of its electric and magnetic fields. Thus, amplitude is related to the *intensity* of the radiation, which we perceive as brightness in the case of visible light. Light of a particular color—fire-engine red for instance—has a specific frequency and wavelength, but it can be dimmer (lower amplitude) or brighter (higher amplitude).

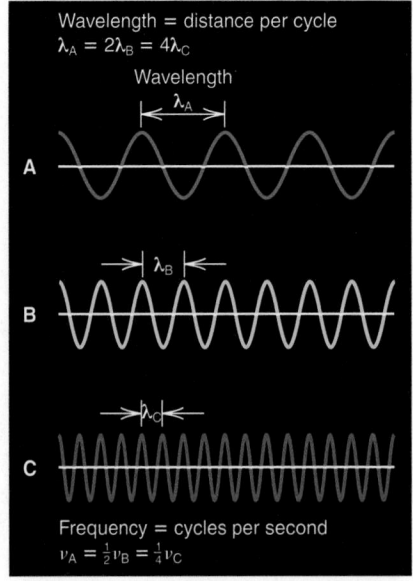

Figure 7.1 Frequency and wavelength. Three waves with different wavelengths (λ) and thus different frequencies (ν) are shown. Note that *as the wavelength decreases, the frequency increases, and vice versa.*

Figure 7.2 Amplitude (intensity) of a wave. Amplitude is represented by the height of the crest (or depth of the trough) of the wave. The two waves shown have the same wavelength (color) but different amplitudes and, therefore, different brightnesses (intensities).

Transparency/ARIS Presentation Center,
Art Library

Figure 7.3 **Regions of the electromagnetic spectrum.** The electromagnetic spectrum extends from the very short wavelengths (very high frequencies) of gamma rays through the very long wavelengths (very low frequencies) of radio waves. The relatively narrow visible region is expanded (and the scale made linear) to show the component colors.

The Electromagnetic Spectrum Visible light represents a small portion of the continuum of radiant energy known as the **electromagnetic spectrum** (Figure 7.3). *All the waves in the spectrum travel at the same speed through a vacuum but differ in frequency and, therefore, wavelength.* Some regions of the spectrum are utilized by particular devices; for example, the long-wavelength, low-frequency radiation is used by microwave ovens and radios. Note that each region meets the next. For instance, the **infrared (IR)** region meets the microwave region on one end and the visible region on the other.

We perceive different wavelengths (or frequencies) of *visible* light as different colors, from red ($\lambda \approx 750$ nm) to violet ($\lambda \approx 400$ nm). Light of a single wavelength is called *monochromatic* (Greek, "one color"), whereas light of many wavelengths is *polychromatic*. White light is polychromatic. The region adjacent to visible light on the short-wavelength end consists of **ultraviolet (UV)** radiation (also called *ultraviolet light*). Still shorter wavelengths (higher frequencies) make up the x-ray and gamma (γ) ray regions. Thus, a TV signal, the green light from a traffic signal, and a gamma ray emitted by a radioactive element all travel at the same speed but differ in their frequency (and wavelength). ●

● **Electromagnetic Emissions Everywhere** We are bathed in electromagnetic radiation from the Sun. Radiation from human activities bombards us as well: radio and TV signals; microwaves from traffic monitors and telephone relay stations; from lightbulbs, x-ray equipment, car motors, and so forth. Natural sources on Earth bombard us also: lightning, radioactivity, and even the glow of fireflies! And our knowledge of the distant universe comes from radiation entering our light, x-ray, and radio telescopes.

SAMPLE PROBLEM 7.1 Interconverting Wavelength and Frequency

PROBLEM A dental hygienist uses x-rays ($\lambda = 1.00$ Å) to take a series of dental radiographs while the patient listens to a radio station ($\lambda = 325$ cm) and looks out the window at the blue sky ($\lambda = 473$ nm). What is the frequency (in s^{-1}) of the electromagnetic radiation from each source? (Assume that the radiation travels at the speed of light, 3.00×10^8 m/s.)

PLAN We are given the wavelengths, so we use Equation 7.1 to find the frequencies. However, we must first convert the wavelengths to meters because c has units of m/s.

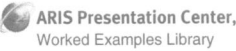
ARIS Presentation Center,
Worked Examples Library

ARIS Presentation Center,
Photo Library

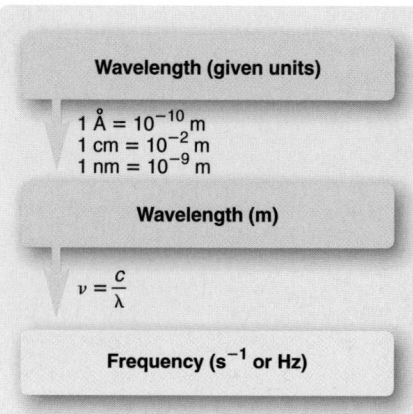

SOLUTION For the x-rays: Converting from angstroms to meters,

$$\lambda = 1.00 \text{ Å} \times \frac{10^{-10} \text{ m}}{1 \text{ Å}} = 1.00 \times 10^{-10} \text{ m}$$

Calculating the frequency:

$$\nu = \frac{c}{\lambda} = \frac{3.00 \times 10^8 \text{ m/s}}{1.00 \times 10^{-10} \text{ m}} = 3.00 \times 10^{18} \text{ s}^{-1}$$

For the radio signal: Combining steps to calculate the frequency,

$$\nu = \frac{c}{\lambda} = \frac{3.00 \times 10^8 \text{ m/s}}{325 \text{ cm} \times \dfrac{10^{-2} \text{ m}}{1 \text{ cm}}} = 9.23 \times 10^7 \text{ s}^{-1}$$

For the blue sky: Combining steps to calculate the frequency,

$$\nu = \frac{c}{\lambda} = \frac{3.00 \times 10^8 \text{ m/s}}{473 \text{ nm} \times \dfrac{10^{-9} \text{ m}}{1 \text{ nm}}} = 6.34 \times 10^{14} \text{ s}^{-1}$$

CHECK The orders of magnitude are correct for the regions of the electromagnetic spectrum (see Figure 7.3): x-rays (10^{19} to 10^{16} s^{-1}), radio waves (10^9 to 10^4 s^{-1}), and visible light (7.5×10^{14} to 4.0×10^{14} s^{-1}).

COMMENT The radio station here is broadcasting at 92.3×10^6 s^{-1}, or 92.3 million Hz (92.3 MHz), about midway in the FM range.

FOLLOW-UP PROBLEM 7.1 Some diamonds appear yellow because they contain nitrogen compounds that absorb purple light of frequency 7.23×10^{14} Hz. Calculate the wavelength (in nm and Å) of the absorbed light.

The Classical Distinction Between Energy and Matter In the everyday world, energy and matter behave very differently. Let's examine some important observations about light and see how its behavior contrasts with the behavior of particles. Light of a given wavelength travels at different speeds through different transparent media—vacuum, air, water, quartz, and so forth. Therefore, when a light wave passes from one medium into another, say, from air to water, the speed of the wave changes. Figure 7.4A shows the phenomenon known as **refraction.**

Figure 7.4 Different behaviors of waves and particles. A, A wave passing from air into water is *refracted* (bent at an angle). **B,** In contrast, a particle of matter (such as a pebble) entering a pond moves in a curved path, because gravity and the greater resistance (drag) of the water slow it down gradually. **C,** A wave is *diffracted* through a small opening, which gives rise to a circular wave on the other side. (The lines represent the crests of water waves as seen from above.) **D,** In contrast, when a collection of moving particles encounters a small opening, as when a handful of sand is thrown at a hole in a fence, some particles move through the opening and continue along their individual paths.

Transparency/ARIS Presentation Center, Art Library

If the wave strikes the boundary between air and water, at an angle other than 90°, the change in speed causes a change in direction, and the wave continues at a different angle. The new angle (angle of refraction) depends on the materials on either side of the boundary and the wavelength of the light. In the process of *dispersion,* white light separates (disperses) into its component colors, as when it passes through a prism, because each incoming wave is refracted at a slightly different angle. ●

In contrast, a particle, like a pebble, does not undergo refraction when passing from one medium to another. Figure 7.4B shows that if you throw a pebble through the air into a pond, its speed changes abruptly and then it continues to slow down gradually in a curved path.

When a wave strikes the edge of an object, it bends around it in a phenomenon called **diffraction.** If the wave passes through a slit about as wide as its wavelength, it bends around both edges of the slit and forms a semicircular wave on the other side of the opening, as shown in Figure 7.4C.

Once again, particles act very differently. Figure 7.4D shows that if you throw a collection of particles, like a handful of sand, at a small opening, some particles hit the edge, while others go through the opening and continue linearly in a narrower group.

If waves of light pass through two adjacent slits, the emerging circular waves interact with each other through the process of *interference.* If the crests of the waves coincide (*in phase*), they interfere *constructively* and the amplitudes add together. If the crests coincide with troughs (*out of phase*), they interfere *destructively* and the amplitudes cancel. The result is a diffraction pattern of brighter and darker regions (Figure 7.5). In contrast, particles passing through adjacent openings continue in straight paths, some colliding with each other and moving at different angles.

At the end of the 19th century, all everyday and laboratory experience seemed to confirm these classical distinctions between the wave nature of energy and the particle nature of matter.

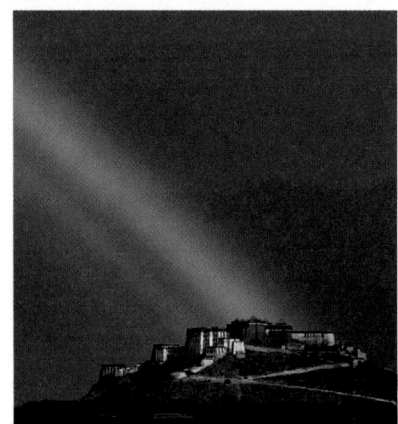

● **Rainbows and Diamonds** You can see a rainbow, like this one in Tibet, only when the Sun is at your back. Light entering the near surface of a water droplet is dispersed and reflected off the far surface. Because red light is bent least, it reaches your eye from droplets higher in the sky, whereas violet appears from lower droplets. The colors in a diamond's sparkle are due to its facets, which are cleaved at angles that disperse and reflect the incoming light, lengthening its path enough for the different wavelengths to separate.

A **B**

Figure 7.5 The diffraction pattern caused by waves passing through two adjacent slits. A, Constructive and destructive interference occurs as water waves viewed from above pass through two adjacent slits in a ripple tank. **B,** As light waves pass through two closely spaced slits, they also emerge as circular waves and interfere with each other. They create a diffraction (interference) pattern of bright and dark regions on a sheet of film. Bright regions appear where crests coincide and the amplitudes combine with each other (in phase); dark regions appear where crests meet troughs and the amplitudes cancel each other (out of phase).

 ARIS Presentation Center, Art Library

The Particle Nature of Light

Three phenomena involving matter and light confounded physicists at the turn of the 20th century: (1) blackbody radiation, (2) the photoelectric effect, and (3) atomic spectra. Explaining these phenomena required a radically new picture of energy. We discuss the first two here and the third in Section 7.2.

Blackbody Radiation and the Quantization of Energy

When a solid object is heated to about 1000 K, it begins to emit visible light, as you can see in the soft red glow of smoldering coal (Figure 7.6A). At about 1500 K, the light is brighter and more orange, like that from an electric heating coil (Figure 7.6B). At temperatures greater than 2000 K, the light is still brighter and whiter, as seen in the filament of a lightbulb (Figure 7.6C). These changes in intensity and wavelength of emitted light as an object is heated are characteristic of *blackbody radiation,* light given off by a hot *blackbody.** All attempts to account for these observed changes by applying classical electromagnetic theory failed.

In 1900, the German physicist Max Planck (1858–1947) developed a formula that fit the data perfectly; to find a physical explanation for his formula, however, Planck was forced to make a radical assumption, which eventually led to an entirely new view of energy. He proposed that the hot, glowing object could emit (or absorb) only *certain quantities* of energy:

$$E = nh\nu$$

where E is the energy of the radiation, ν is its frequency, n is a positive integer (1, 2, 3, and so on) called a **quantum number,** and h is a proportionality constant now known very precisely and called **Planck's constant.** With energy in joules (J) and frequency in s^{-1}, h has units of J·s:

$$h = 6.62606876 \times 10^{-34} \text{ J·s} = 6.626 \times 10^{-34} \text{ J·s} \quad \text{(4 sf)}$$

Later interpretations of Planck's proposal stated that the hot object's radiation is emitted by the atoms contained within it. If an atom can *emit* only certain quantities of energy, it follows that *the atom itself can have only certain quantities of energy.* Thus, the energy of an atom is *quantized:* it exists only in certain fixed quantities, rather than being continuous. Each change in the atom's energy results from the gain or loss of one or more "packets," definite amounts, of energy. Each energy packet is called a **quantum** ("fixed quantity"; plural, *quanta*), and its energy is equal to $h\nu$. Thus, *an atom changes its energy state by emitting (or absorbing) one or more quanta,* and the energy of the emitted (or absorbed) radiation is equal to the *difference in the atom's energy states:*

$$\Delta E_{\text{atom}} = E_{\text{emitted (or absorbed) radiation}} = \Delta n h\nu$$

Because the atom can change its energy only by integer multiples of $h\nu$, the smallest change occurs when an atom in a given energy state changes to an adjacent state, that is, when $\Delta n = 1$:

$$\Delta E = h\nu \quad \text{(7.2)}$$

The Photoelectric Effect and the Photon Theory of Light

Despite the idea that energy is quantized, Planck and other physicists continued to picture the emitted energy as traveling in waves. However, the wave model could not explain the **photoelectric effect,** the flow of current when monochromatic light of sufficient frequency shines on a metal plate (Figure 7.7). The existence of the current was not puzzling: it could be understood as arising when the light transfers energy to the electrons at the metal surface, which break free and are collected by the positive electrode. However, the photoelectric effect had certain confusing features, in particular, the *presence of a threshold frequency* and the *absence of a time lag:*

A Smoldering coal

B Electric heating element

C Lightbulb filament

Figure 7.6 Blackbody radiation. Familiar examples of the change in intensity and wavelength of light emitted by heated objects.

 ARIS Presentation Center,
Photo Library

---

*A blackbody is an idealized object that absorbs all the radiation incident on it. A hollow cube with a small hole in one wall approximates a blackbody.

1. *Presence of a threshold frequency.* Light shining on the metal must have a minimum *frequency*, or no current flows. (Different metals have different minimum frequencies.) The wave theory, however, associates the energy of the light with the *amplitude* (intensity) of the wave, not with its frequency (color). Thus, the wave theory predicts that an electron would break free when it absorbed enough energy from light of *any* color.

2. *Absence of a time lag.* Current flows the moment light of this minimum frequency shines on the metal, regardless of the light's intensity. The wave theory, however, predicts that in dim light there would be a time lag before the current flowed, because the electrons had to absorb enough energy to break free.

Carrying Planck's idea of quantized energy further, Einstein proposed that light itself is particulate, that is, quantized into small "bundles" of electromagnetic energy, which were later called **photons.** In terms of Planck's work, we can say that each atom changes its energy whenever it absorbs or emits one photon, one "particle" of light, whose energy is fixed by its *frequency:*

$$E_{photon} = h\nu = \Delta E_{atom}$$

Let's see how Einstein's photon theory explains the photoelectric effect:

1. *Explanation of the threshold frequency.* According to the photon theory, a beam of light consists of an enormous number of photons. Light intensity (brightness) is related to the number of photons striking the surface per unit time, but *not* to their energy. Therefore, a photon of a certain minimum *energy* must be absorbed for an electron to be freed. Since energy depends on frequency ($h\nu$), the theory predicts a threshold frequency.

2. *Explanation of the time lag.* An electron cannot "save up" energy from several photons below the minimum energy until it has enough to break free. Rather, one electron breaks free the moment it absorbs one photon of *enough* energy. The current is weaker in dim light than in bright light because fewer photons of enough energy are present, so fewer electrons break free per unit time. But some current flows the moment photons reach the metal plate. ●

Figure 7.7 Demonstration of the photoelectric effect. When monochromatic light of high enough frequency strikes the metal plate, electrons are freed from the plate and travel to the positive electrode, creating a current.

ARIS Presentation Center,
Active Art Library

● **Ping-Pong Photons** Consider this analogy for the fact that light of insufficient energy can't free an electron from the metal surface. If one Ping-Pong ball doesn't have enough energy to knock a book off a shelf, neither does a series of Ping-Pong balls, because the book can't save up the energy from the individual impacts. But one baseball traveling at the same speed does have enough energy.

ARIS Presentation Center,
Worked Examples Library

SAMPLE PROBLEM 7.2 Calculating the Energy of Radiation from Its Wavelength

PROBLEM A cook uses a microwave oven to heat a meal. The wavelength of the radiation is 1.20 cm. What is the energy of one photon of this microwave radiation?

PLAN We know λ in centimeters (1.20 cm) so we convert to meters, find the frequency with Equation 7.1, and then find the energy of one photon with Equation 7.2.

SOLUTION Combining steps to find the energy:

$$E = h\nu = \frac{hc}{\lambda} = \frac{(6.626 \times 10^{-34} \text{ J·s})(3.00 \times 10^8 \text{ m/s})}{(1.20 \text{ cm})\left(\dfrac{10^{-2} \text{ m}}{1 \text{ cm}}\right)} = 1.66 \times 10^{-23} \text{ J}$$

CHECK Checking the order of magnitude gives $\dfrac{10^{-33} \text{ J·s} \times 10^8 \text{ m/s}}{10^{-2} \text{ m}} = 10^{-23}$ J.

FOLLOW-UP PROBLEM 7.2 Calculate the energies of one photon of ultraviolet ($\lambda = 1 \times 10^{-8}$ m), visible ($\lambda = 5 \times 10^{-7}$ m), and infrared ($\lambda = 1 \times 10^{-4}$ m) light. What do the answers indicate about the relationship between the wavelength and energy of light?

Planck's quantum theory and Einstein's photon theory assigned properties to energy that, until then, had always been reserved for matter: fixed quantity and discrete particles. These properties have since proved essential to explaining the interactions of matter and energy at the atomic level. But how can a particulate model of energy be made to fit the facts of diffraction and refraction, phenomena explained only in terms of waves? As you'll see shortly, the photon model does not *replace* the wave model. Rather, we have to accept *both* to understand reality. Before we discuss this astonishing notion, however, let's see how the new idea of quantized energy led to a key understanding about atomic behavior.

Section Summary

Electromagnetic radiation travels in waves of specific wavelength (λ) and frequency (ν). All electromagnetic waves travel through a vacuum at the speed of light, c (3.00×10^8 m/s), which is equal to $\nu \times \lambda$. The intensity (brightness) of a light wave is related to its amplitude. • The electromagnetic spectrum ranges from very long radio waves to very short gamma rays and includes the visible region [750 nm (red) to 400 nm (violet)]. • Refraction and diffraction indicate that electromagnetic radiation is wavelike, but blackbody radiation and the photoelectric effect indicate that it is particle-like. • Light exists as photons (quanta) that have an energy proportional to the frequency. • According to quantum theory, an atom has only certain quantities of energy ($E = nh\nu$), which it can change only by absorbing or emitting a photon.

7.2   ATOMIC SPECTRA

The third key observation about matter and energy that late 19th-century physicists could not explain involved the light emitted when an element is vaporized and then thermally or electrically excited, as you see in a neon sign. Figure 7.8A shows the result when light from excited hydrogen atoms passes through a narrow slit and is then refracted by a prism. Note that this light does not create a *continuous spectrum,* or rainbow, as sunlight does. Instead, it creates a **line spectrum,** a series of fine lines of individual colors separated by colorless (black) spaces.* The wavelengths of these spectral lines are characteristic of the element producing them (Figure 7.8B).

---

*The appearance of the spectrum as a series of lines results from the construction of the apparatus. If the light passed through a small hole, rather than a narrow slit, the spectrum would appear as a circular field of dots rather than a horizontal series of lines. The key point is that *the spectrum is discrete, rather than continuous.*

Figure 7.8 The line spectra of several elements. A, A sample of gaseous H_2 is dissociated into atoms and excited by an electric discharge. The emitted light passes through a slit and a prism, which disperses the light into individual wavelengths. The line spectrum of atomic H is shown *(top).* **B,** The continuous spectrum of white light is compared with the line spectra of mercury and strontium. Note that each line spectrum is different from the others.

ARIS Presentation Center, Art Library

Animation: Atomic Line Spectra
ARIS Presentation Center
Chapter 7

ARIS Presentation Center, Art Library

Animation: Emission Spectra
ARIS Presentation Center
Chapter 7

Spectroscopists studying the spectrum of atomic hydrogen had identified several series of such lines in different regions of the electromagnetic spectrum. Figure 7.9 shows three of these series of lines. Equations of the following general form, called the *Rydberg equation,* were found to predict the position and wavelength of any line in a given series:

$$\frac{1}{\lambda} = R\left(\frac{1}{n_1^2} - \frac{1}{n_2^2}\right) \tag{7.3}$$

where λ is the wavelength of a spectral line, n_1 and n_2 are positive integers with $n_2 > n_1$, and R is the Rydberg constant (1.096776×10^7 m^{-1}). For the visible series of lines, $n_1 = 2$:

$$\frac{1}{\lambda} = R\left(\frac{1}{2^2} - \frac{1}{n_2^2}\right), \quad \text{with } n_2 = 3, 4, 5, \ldots$$

The Rydberg equation and the value of the constant are based on data rather than theory. No one knew *why* the spectral lines of hydrogen appear in this pattern. (Problems 7.23 and 7.24 are two of several that apply the Rydberg equation.)

The observed occurrence of line spectra did not correlate with classical theory for one major reason. As was mentioned in the chapter introduction, if an electron spiraled closer to the nucleus, it should emit radiation. Moreover, the frequency of the radiation should be related to the time of revolution. On the spiral path inward, that time should change smoothly, so the frequency of the radiation should change smoothly and create a continuous spectrum. Rutherford's nuclear model seemed totally at odds with atomic line spectra.

The Bohr Model of the Hydrogen Atom

Soon after the nuclear model was proposed, Niels Bohr (1885–1962), a young Danish physicist working in Rutherford's laboratory, suggested a model for the H atom that predicted the existence of line spectra. In his model, Bohr used Planck's and Einstein's ideas about quantized energy and proposed three postulates:

1. *The H atom has only certain allowable energy levels,* which Bohr called **stationary states.** Each of these states is associated with a fixed circular orbit of the electron around the nucleus.

2. *The atom does **not** radiate energy while in one of its stationary states.* That is, even though it violates the ideas of classical physics, the atom does not change energy while the electron moves *within* an orbit.

3. *The atom changes to another stationary state* (the electron moves to another orbit) *only by absorbing or emitting a photon whose energy equals the difference in energy between the two states:*

$$E_{\text{photon}} = E_{\text{state A}} - E_{\text{state B}} = h\nu$$

where the energy of state A is higher than that of state B. A spectral line results when a photon of specific energy (and thus specific frequency) is *emitted* as the electron moves from a higher energy state to a lower one. Therefore, Bohr's model explains that an atomic spectrum is not continuous because *the atom's energy has only certain discrete levels, or states.*

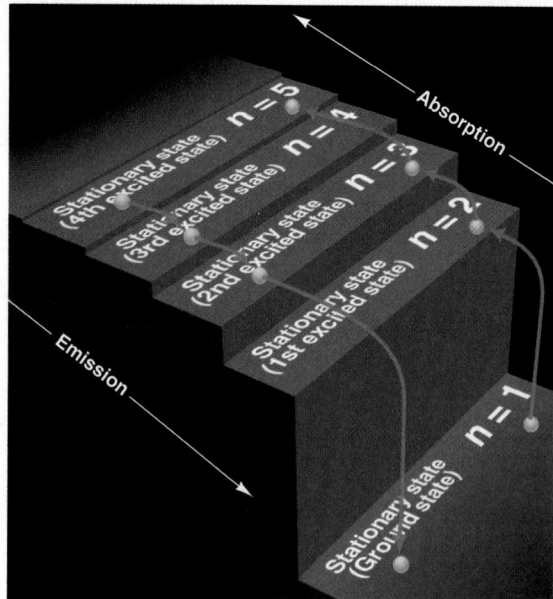

Figure 7.10 Quantum staircase. In this analogy for the energy levels of the hydrogen atom, an electron can absorb a photon and jump up to a higher "step" (stationary state) or emit a photon and jump down to a lower one. But the electron cannot lie between two steps. ARIS Presentation Center, Art Library

In Bohr's model, the quantum number n (1, 2, 3, . . .) is associated with the radius of an electron orbit, which is directly related to the electron's energy: *the lower the n value, the smaller the radius of the orbit, and the lower the energy level.* When the electron is in the first orbit ($n = 1$), the orbit closest to the nucleus, the H atom is in its lowest (first) energy level, called the **ground state.** If the H atom absorbs a photon whose energy equals the *difference* between the first and second energy levels, the electron moves to the second orbit ($n = 2$), the next orbit out from the nucleus. When the electron is in the second or any higher orbit, the atom is said to be in an **excited state.** If the H atom in the first excited state (the electron in the second orbit) emits a photon of that same energy, it returns to the ground state. Figure 7.10 shows a staircase analogy for this behavior.

Figure 7.11A shows how Bohr's model accounts for the three line spectra of hydrogen. When a sample of gaseous H atoms is excited, different atoms absorb different quantities of energy. Each atom has one electron, but so many atoms are present that all the energy levels (orbits) are populated by electrons. When the electrons drop from outer orbits to the $n = 3$ orbit (second excited state), the emitted photons create the *infrared* series of lines. The *visible* series arises when electrons drop to the $n = 2$ orbit (first excited state). Figure 7.11B shows that the *ultraviolet* series arises when electrons drop to the $n = 1$ orbit (ground state).

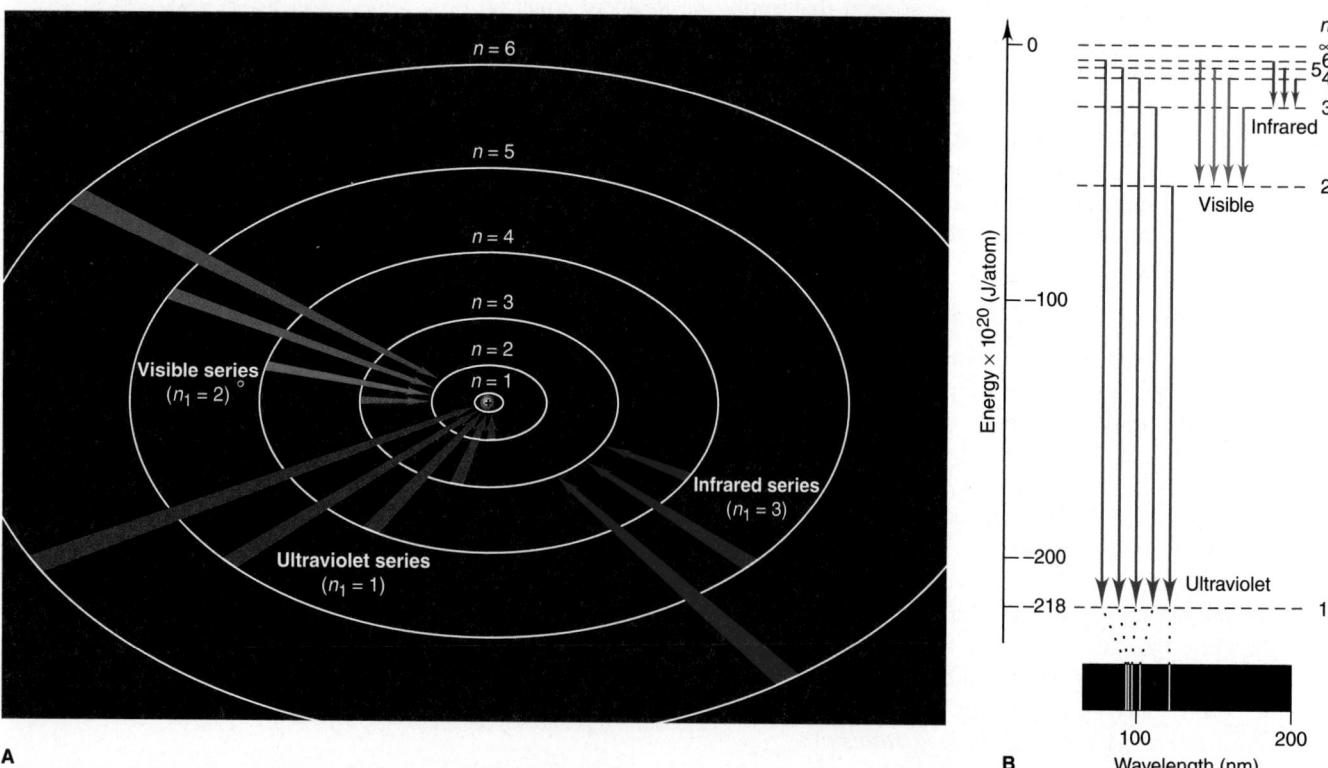

A

Figure 7.11 The Bohr explanation of three series of spectral lines. A, According to the Bohr model, when an electron drops from an outer orbit to an inner one, the atom emits a photon of specific energy that gives rise to a spectral line. In a given series, each electron drop, and thus each emission, has the same inner orbit, that is, the same value of n_1 in the Rydberg equation (see Equation 7.3). (The orbit radius is proportional to n^2. Only the first six orbits are shown.) **B,** An energy diagram shows how the ultraviolet series arises. Within each series,

B

the greater the *difference* in orbit radii, the greater the difference in energy levels (depicted as a downward arrow), and the higher the energy of the photon emitted. For example, in the ultraviolet series, in which $n_1 = 1$, a drop from $n = 5$ to $n = 1$ emits a photon with more energy (shorter λ, higher ν) than a drop from $n = 2$ to $n = 1$. [The axis shows negative values because $n = \infty$ (the electron completely separated from the nucleus) is *defined* as the atom with zero energy.]

Limitations of the Bohr Model Despite its great success in accounting for the spectral lines of the H atom, the Bohr model failed to predict the spectrum of any other atom, even that of helium, the next simplest element. In essence, the Bohr model is a one-electron model. It works beautifully for the H atom and for other one-electron species, such as He^+ ($Z = 2$), Li^{2+} ($Z = 3$), and Be^{3+} ($Z = 4$), which have either been created in the lab or seen in the spectra of stars. But it fails for atoms with more than one electron because in these systems, electron-electron repulsions and additional nucleus-electron attractions are present as well. Moreover, as you'll soon see, electron movement in an atom is not defined by fixed orbits. As a picture of the atom, the Bohr model is incorrect, but we still use the terms "ground state" and "excited state" and retain one of Bohr's central ideas in our current model: *the energy of an atom occurs in discrete levels.*

The Energy States of the Hydrogen Atom

A very useful result from Bohr's work is an equation for calculating the energy levels of an atom, which he derived from the classical principles of electrostatic attraction and circular motion:

$$E = -2.18 \times 10^{-18} \, J \left(\frac{Z^2}{n^2} \right)$$

where Z is the charge of the nucleus. For the H atom, $Z = 1$, so we have

$$E = -2.18 \times 10^{-18} \, J \left(\frac{1^2}{n^2} \right) = -2.18 \times 10^{-18} \, J \left(\frac{1}{n^2} \right)$$

Therefore, the energy of the ground state ($n = 1$) is

$$E = -2.18 \times 10^{-18} \, J \left(\frac{1}{1^2} \right) = -2.18 \times 10^{-18} \, J$$

Don't be confused by the negative sign for the energy values (see the axis in Figure 7.11B). It appears because we *define* the zero point of the atom's energy when *the electron is completely removed from the nucleus.* Thus, $E = 0$ when $n = \infty$, so $E < 0$ for any smaller n. As an analogy, consider a book resting on the floor. You can define the zero point of the book's potential energy in many ways. If you define zero when the book is on the floor, the energy is positive when the book is on a tabletop. But, if you define zero when the book is on a tabletop, the energy is negative when the book lies on the floor; the latter case is analogous to the energy of the H atom (Figure 7.12).

Since n is in the denominator of the energy equation, as the electron moves closer to the nucleus (n decreases), the atom becomes more stable (less energetic) and its energy becomes a *larger negative number.* As the electron moves away from the nucleus (n increases), the atom's energy increases (becomes a smaller negative number). This equation is easily adapted to find the energy difference between any two levels:

$$\Delta E = E_{final} - E_{initial} = -2.18 \times 10^{-18} \, J \left(\frac{1}{n_{final}^2} - \frac{1}{n_{initial}^2} \right) \quad (7.4)$$

Using Equation 7.4, we can predict the wavelengths of the spectral lines of the H atom. Note that if we combine Equation 7.4 with Planck's expression for the change in an atom's energy (Equation 7.2), we obtain the Rydberg equation (Equation 7.3):

$$\Delta E = h\nu = \frac{hc}{\lambda} = -2.18 \times 10^{-18} \, J \left(\frac{1}{n_{final}^2} - \frac{1}{n_{initial}^2} \right)$$

Therefore,

$$\frac{1}{\lambda} = -\frac{2.18 \times 10^{-18} \, J}{hc} \left(\frac{1}{n_{final}^2} - \frac{1}{n_{initial}^2} \right) = -\frac{2.18 \times 10^{-18} \, J}{(6.626 \times 10^{-34} \, J \cdot s)(3.00 \times 10^8 \, m/s)} \left(\frac{1}{n_{final}^2} - \frac{1}{n_{initial}^2} \right) = -1.10 \times 10^7 \, m^{-1} \left(\frac{1}{n_{final}^2} - \frac{1}{n_{initial}^2} \right)$$

where $n_{final} = n_2$, $n_{initial} = n_1$, and $1.10 \times 10^7 \, m^{-1}$ is the Rydberg constant $(1.096776 \times 10^7 \, m^{-1})$ to three significant figures. Thus, from the classical

Figure 7.12 A tabletop analogy for the H atom's energy.

ARIS Presentation Center, Art Library

relationships of charge and of motion combined with the idea that the H atom can have only certain values of energy, we obtain an equation from theory that leads directly to the empirical one. (Bohr's value for the Rydberg constant differed from the spectroscopists' value by only 0.05%!) In the following sample problem, we calculate the energy and wavelength of an electron jump in the H atom.

SAMPLE PROBLEM 7.3   Determining ΔE and λ of an Electron Transition

PROBLEM A hydrogen atom absorbs a photon of visible light (see Figure 7.11), and its electron enters the $n = 4$ energy level. Calculate **(a)** the change in energy of the atom and **(b)** the wavelength (in nm) of the photon.

PLAN (a) The H atom absorbs energy, so $E_{final} > E_{initial}$. We are given $n_{final} = 4$, and Figure 7.11 shows that $n_{initial} = 2$ because a visible photon is absorbed. We apply Equation 7.4 to find ΔE. **(b)** Once we know ΔE, we find the frequency with Equation 7.2 and the wavelength (in m) with Equation 7.1. Then we convert meters to nanometers.

SOLUTION (a) Substituting the values into Equation 7.4:

$$\Delta E = -2.18 \times 10^{-18}\, \text{J} \left(\frac{1}{n_{final}^2} - \frac{1}{n_{initial}^2} \right)$$

$$= -2.18 \times 10^{-18}\, \text{J} \left(\frac{1}{4^2} - \frac{1}{2^2} \right) = -2.18 \times 10^{-18}\, \text{J} \left(\frac{1}{16} - \frac{1}{4} \right)$$

$$= \boxed{4.09 \times 10^{-19}\, \text{J}}$$

(b) Combining Equations 7.2 and 7.1 and solving for λ:

$$\Delta E = h\nu = \frac{hc}{\lambda}$$

therefore, $\lambda = \dfrac{hc}{\Delta E} = \dfrac{(6.626 \times 10^{-34}\, \text{J·s})(3.00 \times 10^8\, \text{m/s})}{4.09 \times 10^{-19}\, \text{J}} = 4.86 \times 10^{-7}\, \text{m}$

Converting m to nm:

$$\lambda = 4.86 \times 10^{-7}\, \text{m} \times \frac{1\, \text{nm}}{10^{-9}\, \text{m}} = \boxed{486\, \text{nm}}$$

CHECK In (a), the energy change is positive, which is consistent with an absorption. In (b), the wavelength is consistent with a visible photon (400–750 nm); it is blue-green.

FOLLOW-UP PROBLEM 7.3 A hydrogen atom with its electron in the $n = 6$ energy level emits a photon of IR light. Calculate **(a)** the change in energy of the atom and **(b)** the wavelength (in Å) of the photon.

● What Are Stars and Planets Made Of?
In 1868, the French astronomer Pierre Janssen noted a bright yellow line in the solar emission spectrum and thought it was emitted by an element unique to the Sun, which he named helium (Greek *helios,* "sun"). In 1888, the British chemist William Ramsay saw the same line in the spectrum of the inert gas obtained by heating uranium-containing minerals. Analysis of starlight has shown many elements known on Earth, but helium is the only element discovered on a star. Recent analysis by the Hubble Space Telescope of light from a planet orbiting a star in the constellation Pegasus, 150 light-years from Earth, reveals a hydrogen-gas giant much like Jupiter, except so close to its sun that it appears to be losing 10,000 tons of its atmosphere every second!

We can also use Equation 7.4 to find the quantity of energy needed to completely remove the electron from an H atom. Let's find ΔE for the following change:

$$\text{H}(g) \longrightarrow \text{H}^+(g) + \text{e}^-$$

We substitute $n_{final} = \infty$ and $n_{initial} = 1$ and obtain

$$\Delta E = E_{final} - E_{initial} = -2.18 \times 10^{-18}\, \text{J} \left(\frac{1}{\infty^2} - \frac{1}{1^2} \right)$$

$$= -2.18 \times 10^{-18}\, \text{J}\, (0 - 1) = 2.18 \times 10^{-18}\, \text{J}$$

In this case, ΔE is positive because energy is *absorbed* to remove the electron from the vicinity of the nucleus. For 1 mol of H atoms,

$$\Delta E = \left(2.18 \times 10^{-18}\, \frac{\text{J}}{\text{atom}} \right) \left(6.022 \times 10^{23}\, \frac{\text{atoms}}{\text{mol}} \right) \left(\frac{1\, \text{kJ}}{10^3\, \text{J}} \right) = 1.31 \times 10^3\, \text{kJ/mol}$$

This is the *ionization energy* of the H atom, the quantity of energy required to form 1 mol of gaseous H^+ ions from 1 mol of gaseous H atoms. We return to this idea in Chapter 8.

Spectroscopic analysis of the H atom led to the Bohr model, the first step toward our current model of the atom. From its use by 19th-century chemists as a means of identifying elements and compounds, spectrometry has developed into a major tool of modern chemistry (see Tools of the Laboratory). ●

Tools of the Laboratory

Spectrophotometry in Chemical Analysis

The use of spectral data to identify and quantify substances is essential to modern chemical analysis. The terms *spectroscopy, spectrometry,* and **spectrophotometry** denote a large group of instrumental techniques that obtain spectra corresponding to a substance's atomic and molecular energy levels.

The two types of spectra most often obtained are emission and absorption spectra. An **emission spectrum,** such as the H atom line spectrum, is produced when atoms in an excited state *emit* photons characteristic of the element as they return to lower energy states. Some elements produce a very intense spectral line (or several closely spaced ones) that serves as a marker of their presence. Such an intense line is the basis of **flame tests,** rapid qualitative procedures performed by placing a granule of an ionic compound or a drop of its solution in a flame (Figure B7.1, A). Some of the colors of fireworks and flares are due to emissions from the same elements shown in the flame tests: crimson from strontium salts and blue-green from copper salts (Figure B7.1, B).

The characteristic colors of sodium-vapor and mercury-vapor streetlamps, seen in many towns and cities, are due to one or a few prominent lines in their emission spectra.

An **absorption spectrum** is produced when atoms *absorb* photons of certain wavelengths and become excited from lower to higher energy states. Therefore, the absorption spectrum of an element appears as dark lines against a bright background. When white light passes through sodium vapor, for example, it gives rise to a sodium absorption spectrum, on which dark lines appear at the same wavelengths as those for the yellow-orange lines in the sodium emission spectrum (Figure B7.2).

Instruments based on absorption spectra are much more common than those based on emission spectra, for several reasons. When a solid, liquid, or dense gas is excited, it *emits* so many lines that the spectrum is a continuum (recall the continuum of colors in sunlight). Absorption is also less destructive of fragile organic and biological molecules.

(continued)

Demo McKelvy, George M. "Flame Tests That Are Portable, Storable, and Easy to Use," *J. Chem. Educ.* **1998,** *75,* 55.

Demo Johnson, Kristin A., Schreiner, Rodney. "A Dramatic Flame Test Demonstration," *J. Chem. Educ.* **2001,** *78,* 640.

A

Figure B7.1 Flame tests and fireworks. A, In general, the color of the flame is created by a strong emission in the line spectrum of the element and therefore is often taken as preliminary evidence of the presence of the element in a sample. Shown here are the crimson of strontium and the blue-green of copper. **B,** The same emissions from compounds that contain these elements often appear in the brilliant displays of fireworks. **ARIS Presentation Center,** Photo Library

B

400 nm   Sodium emission spectrum   750 nm

400 nm   Sodium absorption spectrum   750 nm

Figure B7.2 Emission and absorption spectra of sodium atoms. The wavelengths of the bright emission lines correspond to those of the dark absorption lines because both are created by the same energy change: $\Delta E_{\text{emission}} = -\Delta E_{\text{absorption}}$. (Only the two most intense lines in the Na spectra are shown.)

ARIS Presentation Center, Art Library

Source *produces radiation in region of interest. Must be stable and reproducible. In most cases, the source emits many wavelengths.*

Lenses/slits/ collimators *narrow and align beam.*

Monochromator *(wavelength selector) disperses incoming radiation into continuum of component wavelengths that are scanned or individually selected.*

Sample *in compartment absorbs characteristic amount of each incoming wavelength.*

Detector *converts transmitted radiation into amplified electrical signal.*

Computer *converts signal into displayed data.*

Figure B7.3 The main components of a typical spectrometer.

ARIS Presentation Center,
Art Library

Despite differences that depend on the region of the electromagnetic spectrum used to irradiate the sample, all modern spectrometers have components that perform the same basic functions (Figure B7.3). (We discuss infrared spectroscopy and nuclear magnetic resonance spectroscopy in later chapters.)

Visible light is often used to study colored substances, which absorb only some of the wavelengths from white light. A leaf looks green, for example, because its chlorophyll absorbs red and blue wavelengths strongly and green weakly, so most of the green light is reflected. The absorption spectrum of chlorophyll *a* in ether solution appears in Figure B7.4.

The overall shape of the curve and the wavelengths of the major peaks are characteristic of chlorophyll *a,* so its spectrum serves as a means of identifying it from an unknown source. The curve varies in height because chlorophyll *a* absorbs incoming wavelengths to different extents. The absorptions appear as broad

bands, rather than as the distinct lines we saw earlier for individual gaseous atoms, because dissolved substances, as well as pure solids and liquids, absorb many more wavelengths. This broader absorbance is due to the greater numbers and types of energy levels within a molecule, among molecules, and between molecules and solvent.

In addition to identifying a substance, a spectrometer can be used to measure its concentration because *the absorbance, the amount of light of a given wavelength absorbed by a substance, is proportional to the number of molecules.* Suppose you want to determine the concentration of chlorophyll in an ether solution of leaf extract. You select a strongly absorbed wavelength from the chlorophyll spectrum (such as 663 nm in Figure B7.4, A), measure the absorbance of the leaf-extract solution, and compare it with the absorbances of a series of ether solutions with known chlorophyll concentrations.

663 nm

A   Wavelength (nm)

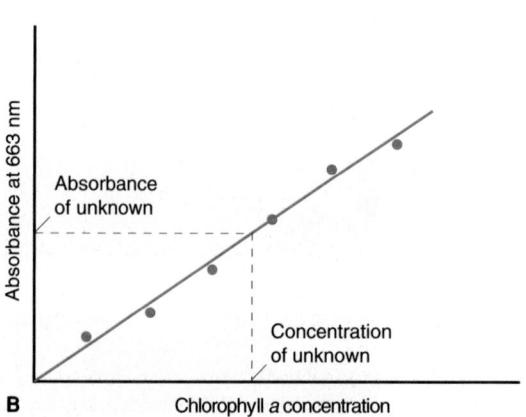

Absorbance
of unknown

Concentration
of unknown

B   Chlorophyll *a* concentration

Figure B7.4 Measuring chlorophyll *a* concentration in leaf extract.
Chlorophyll *a* is one of several leaf pigments. It absorbs red and blue wavelengths strongly but almost no green or yellow wavelengths. Thus, leaves containing large amounts of chlorophyll *a* appear green.

We can use the strong absorption at 663 nm in the spectrum **(A)** to quantify the amount of chlorophyll *a* present in a plant extract by comparing that absorbance to a series of known standards **(B)**.

ARIS Presentation Center, Art Library

Section Summary

To explain the line spectrum of atomic hydrogen, Bohr proposed that the atom's energy is quantized because electron motion is restricted to fixed orbits. The electron can move from one orbit to another only if the atom absorbs or emits a photon whose energy equals the difference in energy levels (orbits). Line spectra are produced because these energy changes correspond to photons of specific wavelengths. • Bohr's model predicted the spectrum of the H atom and other one-electron species, but not that of any other atom. Despite this, Bohr's idea that atoms have quantized energy levels is a cornerstone of our current atomic model. • Spectrophotometry is an instrumental technique in which emission and absorption spectra are used to identify and measure concentrations of substances.

● **"He'll Never Make a Success of Anything"** This comment, attributed to the principal of young Albert Einstein's primary school, remains a classic of misperception. Contrary to myth, the greatest physicist of the 20th century (some say of all time) was not a poor student but an independent one, preferring his own path to that prescribed by authority—a trait that gave him the intense focus characteristic of all his work. A friend recalls finding him in his small apartment, rocking his baby in its carriage with one hand while holding a pencil stub and scribbling on a pad with the other. At age 26, he was working on one of the four papers he published in 1905 that would revolutionize the way the universe is perceived and lead to his 1921 Nobel Prize.

ARIS Presentation Center,
Photo Library

7.3 THE WAVE-PARTICLE DUALITY OF MATTER AND ENERGY

The year 1905 was a busy one for Albert Einstein. In addition to presenting the photon theory of light and explaining the photoelectric effect, he found time to explain Brownian motion (Chapter 13), which helped establish the molecular view of matter, and to introduce a new branch of physics with his theory of relativity. One of its many startling revelations was that matter and energy are alternate forms of the same entity. This idea is embodied in his famous equation $E = mc^2$, which relates the quantity of energy equivalent to a given mass, and vice versa. Relativity theory does not depend on quantum theory, but together they have completely blurred the sharp divisions we normally perceive between matter (chunky and massive) and energy (diffuse and massless). ●

The early proponents of quantum theory demonstrated that *energy is particle-like*. Physicists who developed the theory turned this proposition upside down and showed that *matter is wavelike*. Strange as this idea may seem, it is the key to our modern atomic model.

The Wave Nature of Electrons and Particle Nature of Photons

Bohr's efforts were a perfect case of fitting theory to data: he *assumed* that an atom has only certain allowable energy levels in order to *explain* the observed line spectrum. However, his assumption had no basis in other physical theory. Then, in the early 1920s, a young French physics student named Louis de Broglie proposed a startling reason for fixed energy levels: *if energy is particle-like, perhaps matter is wavelike*. De Broglie had been thinking of other systems that display only certain allowed motions, such as the wave of a plucked guitar string. Figure 7.13 shows that, because the ends of the string are fixed, only certain vibrational frequencies (and wavelengths) are possible. De Broglie reasoned that *if electrons have wave-like motion* and are restricted to orbits of fixed radii, that would explain why they have only certain possible frequencies and energies.

Figure 7.13 Wave motion in restricted systems. A, In a musical analogy to electron waves, one half-wavelength (λ/2) is the "quantum" of the guitar string's vibration. The string length L is fixed, so the only allowed vibrations occur when L is a whole-number multiple (n) of λ/2. **B,** If an electron occupies a circular orbit, only whole numbers of wavelengths are allowed (n = 3 and n = 5 are shown). A wave with a fractional number of wavelengths (such as n = 3⅓) is "forbidden" and rapidly dies out through overlap of crests and troughs.

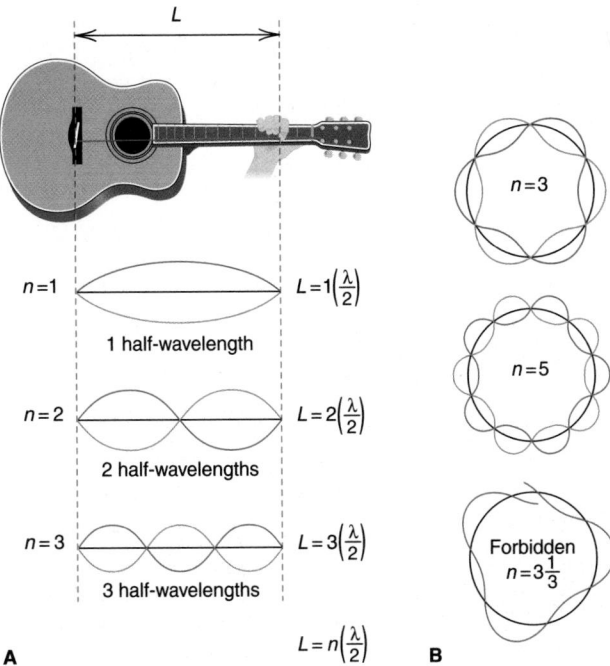

Transparency/ARIS Presentation Center,
Art Library

Combining the equation for mass-energy equivalence ($E = mc^2$) with that for the energy of a photon ($E = h\nu = hc/\lambda$), de Broglie derived an equation for the wavelength of any particle of mass m—whether planet, baseball, or electron—moving at speed u:

$$\lambda = \frac{h}{mu} \tag{7.5}$$

According to this equation for the **de Broglie wavelength,** *matter behaves as though it moves in a wave.* Note also that an object's wavelength is *inversely* proportional to its mass, so heavy objects such as planets and baseballs have wavelengths that are *many* orders of magnitude smaller than the object itself, as you can see in Table 7.1.

 Transparency/ARIS Presentation Center, Tables Library

Table 7.1 **The de Broglie Wavelengths of Several Objects**

Substance	Mass (g)	Speed (m/s)	λ (m)
Slow electron	9×10^{-28}	1.0	7×10^{-4}
Fast electron	9×10^{-28}	5.9×10^6	1×10^{-10}
Alpha particle	6.6×10^{-24}	1.5×10^7	7×10^{-15}
One-gram mass	1.0	0.01	7×10^{-29}
Baseball	142	25.0	2×10^{-34}
Earth	6.0×10^{27}	3.0×10^4	4×10^{-63}

 ARIS Presentation Center, Worked Examples Library

SAMPLE PROBLEM 7.4 Calculating the de Broglie Wavelength of an Electron

PROBLEM Find the de Broglie wavelength of an electron with a speed of 1.00×10^6 m/s (electron mass = 9.11×10^{-31} kg; $h = 6.626 \times 10^{-34}$ kg·m²/s).

PLAN We know the speed (1.00×10^6 m/s) and mass (9.11×10^{-31} kg) of the electron, so we substitute these into Equation 7.5 to find λ.

SOLUTION

$$\lambda = \frac{h}{mu} = \frac{6.626 \times 10^{-34} \text{ kg·m}^2/\text{s}}{(9.11 \times 10^{-31} \text{ kg})(1.00 \times 10^6 \text{ m/s})} = 7.27 \times 10^{-10} \text{ m}$$

CHECK The order of magnitude and units seem correct:

$$\lambda \approx \frac{10^{-33} \text{ kg·m}^2/\text{s}}{(10^{-30} \text{ kg})(10^6 \text{ m/s})} = 10^{-9} \text{ m}$$

COMMENT As you'll see in the upcoming discussion, such fast-moving electrons, with wavelengths in the range of atomic sizes, exhibit remarkable properties.

FOLLOW-UP PROBLEM 7.4 What is the speed of an electron that has a de Broglie wavelength of 100. nm?

If particles travel in waves, electrons should exhibit diffraction and interference (see Section 7.1). ● A fast-moving electron has a wavelength of about 10^{-10} m, so perhaps a beam of electrons would be diffracted by the spaces about this size between atoms in a crystal. Indeed, in 1927, C. Davisson and L. Germer guided a beam of electrons at a nickel crystal and obtained a diffraction pattern. Figure 7.14 shows the diffraction patterns obtained when either x-rays or electrons impinge on aluminum foil. Apparently, electrons—particles with mass and charge—create diffraction patterns, just as electromagnetic waves do! Even though electrons do not have orbits of fixed radius, as de Broglie thought, the energy levels of the atom *are* related to the wave nature of the electron.

 ARIS Presentation Center, Photo Library

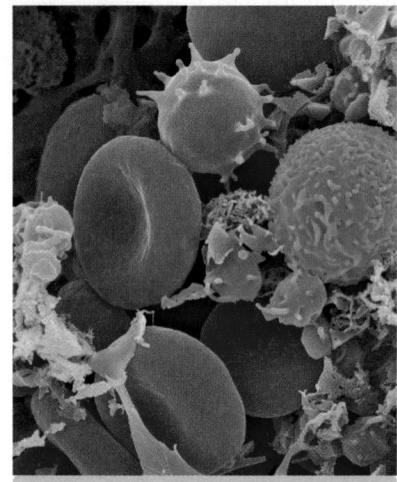

● **The Electron Microscope** In a transmission electron microscope, a beam is focused by a lens and passes through a thin section of the specimen to a second lens. The resulting image is then magnified by a third lens to a final image. The differences between this and a light microscope are that the "beam" consists of high-speed electrons and the "lenses" are electromagnetic fields, which can be adjusted to give up to 200,000-fold magnification and 0.5-nm resolution. In a scanning electron microscope, the electron beam scans the specimen, knocking electrons from it, which creates a current that varies with surface irregularities. The current generates an image that looks like the object's surface, as in this false-color micrograph of various types of blood cells ($\times 1200$). The great advantage of electron microscopes is that high-speed electrons have wavelengths much smaller than those of visible light and, thus, allow much higher image resolution.

If electrons have properties of energy, do photons have properties of matter? The de Broglie equation suggests that we can calculate the momentum (p), the product of mass and speed, for a photon of a given wavelength. Substituting the speed of light (c) for speed u in Equation 7.5 and solving for p gives

$$\lambda = \frac{h}{mc} = \frac{h}{p} \quad \text{and} \quad p = \frac{h}{\lambda}$$

Notice the inverse relationship between p and λ. This means that shorter wavelength (higher energy) photons have greater momentum. Thus, a decrease in a photon's momentum should appear as an increase in its wavelength. In 1923, Arthur Compton directed a beam of x-ray photons at a sample of graphite and observed that the wavelength of the reflected photons increased. This result means that the photons transferred some of their momentum to the electrons in the carbon atoms of the graphite, just as colliding billiard balls transfer momentum to one another. In this experiment, photons behaved as particles with momentum!

To scientists of the time, these results were very unsettling. Classical experiments had shown matter to be particle-like and energy to be wavelike, but these new studies showed that, on the atomic scale, every characteristic trait used to define the one now also defined the other. Figure 7.15 summarizes the conceptual and experimental breakthroughs that led to this juncture.

Figure 7.14 Comparing diffraction patterns of x-rays and electrons. A, X-ray diffraction pattern of aluminum. **B,** Electron diffraction pattern of aluminum. This behavior implies that both x-rays, which are electromagnetic radiation, and electrons, which are particles, travel in waves.

ARIS Presentation Center, Photo Library

Figure 7.15 Summary of the major observations and theories leading from classical theory to quantum theory. As often happens in science, an observation (experiment) stimulates the need for an explanation (theory), and/or a theoretical insight provides the impetus for an experimental test.

Transparency/ARIS Presentation Center, Art Library

The truth is that *both* matter and energy show *both* behaviors: each possesses both "faces." In some experiments, we observe one face; in other experiments, we observe the other face. The distinction between a particle and a wave is meaningful only in the macroscopic world, *not* in the atomic world. The distinction between matter and energy is in our minds and our limiting definitions, not inherent in nature. This dual character of matter and energy is known as the **wave-particle duality.**

The Heisenberg Uncertainty Principle

In the classical view of the world, a moving particle has a definite location at any instant, whereas a wave is spread out in space. If an electron has the properties of *both* a particle and a wave, what can we determine about its position in the atom? In 1927, the German physicist Werner Heisenberg postulated the **uncertainty principle,** which states that it is impossible to know the exact position *and* momentum (mass times speed) of a particle simultaneously. For a particle with constant mass *m*, the principle is expressed mathematically as

$$\Delta x \cdot m\Delta u \geq \frac{h}{4\pi} \tag{7.6}$$

where Δx is the uncertainty in position and Δu is the uncertainty in speed. The more accurately we know the position of the particle (smaller Δx), the less accurately we know its speed (larger Δu), and vice versa.

By knowing the position and speed of a pitched baseball and using the classical laws of motion, we can predict its trajectory and whether it will be a strike or a ball. For a baseball, Δx and Δu are insignificant because its mass is enormous compared with $h/4\pi$. Knowing the position and speed of an electron, and from them finding its trajectory, presents an entirely different situation, as Sample Problem 7.5 demonstrates.

Literature Ludwig, Oliver G. "On a Relation between the Heisenberg and deBroglie Principles," *J. Chem. Educ.* **1993,** *70,* 28.

Literature Bartell, Lawrence S. "Perspectives on the Uncertainty Principle and Quantum Reality," *J. Chem. Educ.* **1985,** *62,* 192.

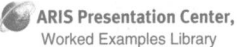

SAMPLE PROBLEM 7.5 Applying the Uncertainty Principle

PROBLEM An electron moving near an atomic nucleus has a speed of 6×10^6 m/s $\pm$ 1%. What is the uncertainty in its position (Δx)?

PLAN The uncertainty in the speed (Δu) is given as 1%, so we multiply *u* (6×10^6 m/s) by 0.01 to calculate the value of Δu, substitute it into Equation 7.6, and solve for the uncertainty in position (Δx).

SOLUTION Finding the uncertainty in speed, Δu:

$$\Delta u = 1\% \text{ of } u = 0.01(6\times10^6 \text{ m/s}) = 6\times10^4 \text{ m/s}$$

Calculating the uncertainty in position, Δx:

$$\Delta x \cdot m\Delta u \geq \frac{h}{4\pi}$$

Thus,
$$\Delta x \geq \frac{h}{4\pi m\Delta u} \geq \frac{6.626\times10^{-34} \text{ kg}\cdot\text{m}^2/\text{s}}{4\pi(9.11\times10^{-31} \text{ kg})(6\times10^4 \text{ m/s})} \geq 1\times10^{-9} \text{ m}$$

CHECK Be sure to round off and check the order of magnitude of the answer:

$$\Delta x \geq \frac{10^{-33} \text{ kg}\cdot\text{m}^2/\text{s}}{(10^1)(10^{-30} \text{ kg})(10^5 \text{ m/s})} = 10^{-9} \text{ m}$$

COMMENT The uncertainty in the electron's position is about 10 times greater than the diameter of the entire atom (10^{-10} m)! Therefore, we have no precise idea where in the atom the electron is located. In the follow-up problem, see if an umpire has any better idea where a baseball is located when calling balls and strikes.

FOLLOW-UP PROBLEM 7.5 How accurately can an umpire know the position of a baseball (mass = 0.142 kg) moving at 100.0 mi/h $\pm$ 1.00% (44.7 m/s $\pm$ 1.00%)?

As the results of Sample Problem 7.5 show, the uncertainty principle has profound implications for an atomic model. It means that *we cannot assign fixed paths for electrons,* such as the circular orbits of Bohr's model. As you'll see in the next section, the most we can ever hope to know is the *probability*—the odds—of finding an electron in a given region of space. However, we are not *sure* it is there any more than a gambler is sure of the next roll of the dice. ●

Section Summary

As a result of Planck's quantum theory and Einstein's relativity theory, we no longer view matter and energy as distinct entities. • The de Broglie wavelength refers to the idea that electrons (and all matter) have wavelike motion. Allowed atomic energy levels are related to allowed wavelengths of the electron's motion. • Electrons exhibit diffraction, as do waves of energy, and photons exhibit transfer of momentum, as do particles of mass. The wave-particle duality of matter and energy exists at all scales but is observable only on the atomic scale. • According to the uncertainty principle, we cannot know simultaneously the position and speed of an electron.

7.4 THE QUANTUM-MECHANICAL MODEL OF THE ATOM

Acceptance of the dual nature of matter and energy and of the uncertainty principle culminated in the field of **quantum mechanics,** which examines the wave nature of objects on the atomic scale. In 1926, Erwin Schrödinger derived an equation that is the basis for the quantum-mechanical model of the hydrogen atom. The model describes an atom that has certain allowed quantities of energy due to the allowed frequencies of an electron whose behavior is wavelike and whose exact location is impossible to know.

The Atomic Orbital and the Probable Location of the Electron

The electron's matter-wave occupies the three-dimensional space near the nucleus and experiences a continuous, but varying, influence from the nuclear charge. The **Schrödinger equation** is quite complex but is represented as

$$\mathcal{H}\psi = E\psi$$

where E is the energy of the atom. The symbol ψ (Greek *psi,* pronounced "sigh") is called a **wave function,** a mathematical description of the electron's matter-wave in terms of position in three dimensions. The symbol $\mathcal{H}$, called the Hamiltonian operator, represents a set of mathematical operations that, when carried out on a particular ψ, yields an allowed energy value.*

Each solution to the equation (that is, each energy state of the atom) is associated with a given wave function, also called an **atomic orbital.** It's important to keep in mind that an "orbital" in the quantum-mechanical model *bears no resemblance* to an "orbit" in the Bohr model: an *orbit* was, supposedly, an electron's path around the nucleus, whereas an *orbital* is a mathematical function with no direct physical meaning.

We cannot know precisely where the electron is at any moment, but we can describe where it *probably* is, that is, where it is most likely to be found, or where it spends most of its time. Although the wave function (atomic orbital) has no

---

*The complete form of the Schrödinger equation in terms of the three linear axes is

$$\left[-\frac{h^2}{8\pi^2 m_e} \left(\frac{d^2}{dx^2} + \frac{d^2}{dy^2} + \frac{d^2}{dz^2} \right) + V(x, y, z) \right] \psi(x, y, z) = E\psi(x, y, z)$$

where ψ is the wave function; m_e is the electron's mass; E is the total quantized energy of the atomic system; and V is the potential energy at point (x, y, z).

● **Uncertainty Is Unacceptable?** Although the uncertainty principle is universally accepted by physicists today, Einstein found some aspects of it difficult to accept, as reflected in his famous statement that "God does not play dice with the universe." Ernest Rutherford was also skeptical. When Niels Bohr, who had become a champion of the new physics, delivered a lecture in Rutherford's laboratory on the principle, Rutherford said, "You know, Bohr, your conclusions seem to me as uncertain as the premises on which they are built." Acceptance of radical ideas does not come easily, even among fellow geniuses.

Literature Leary, James J.; Kippeny, Tadd C. "A Framework for Presenting the Modern Atom," *J. Chem. Educ.* **1999,** *76,* 1217.

Literature Fortman, John J. "Pictorial Analogies VI: Radial and Angular Wave Function Plots," *J. Chem. Educ.* **1993**, *70*, 549.

direct physical meaning, the square of the wave function, ψ^2, is the *probability density,* a measure of the probability that the electron can be found within a particular tiny volume of the atom. (Whereas ψ can have positive or negative values, ψ^2 is always positive, which makes sense for a value that expresses a probability.) For a given energy level, we can depict this probability with an *electron probability density diagram,* or more simply, an **electron density diagram.** In Figure 7.16A, the value of ψ^2 for a given volume is represented pictorially by a certain density of dots: the greater the density of dots, the higher the probability of finding the electron within that volume.

Electron density diagrams are sometimes called **electron cloud** representations. If we *could* take a time-exposure photograph of the electron in wavelike motion around the nucleus, it would appear as a "cloud" of electron positions. The electron cloud is an *imaginary* picture of the electron changing its position rapidly over time; it does *not* mean that an electron is a diffuse cloud of charge. Note that *the electron probability density decreases with distance from the nucleus* along a line, *r*. The same concept is shown graphically in the plot of ψ^2 vs. *r* in Figure 7.16B. Note that due to the thickness of the printed line, the curve touches the axis; in reality, however, *the probability of the electron being far from the nucleus is very small, but not zero.*

The *total* probability of finding the electron at any distance *r* from the nucleus is also important. To find this, we mentally divide the volume around the nucleus into thin, concentric, spherical layers, like the layers of an onion (shown in cross section in Figure 7.16C), and ask in which *spherical layer* we are most likely to find the electron. This is the same as asking for the *sum of* ψ^2 *values* within each spherical layer. The steep falloff in probability density with distance (see Figure 7.16B) has an important effect. Near the nucleus, the volume of each layer increases faster than its probability density decreases. As a result, the *total* probability of finding the electron in the second layer is higher than in the first. Electron density drops off so quickly, however, that this effect soon diminishes with greater distance. Thus, even though the volume of each layer continues to increase, the total probability for a given layer gradually decreases. Because of

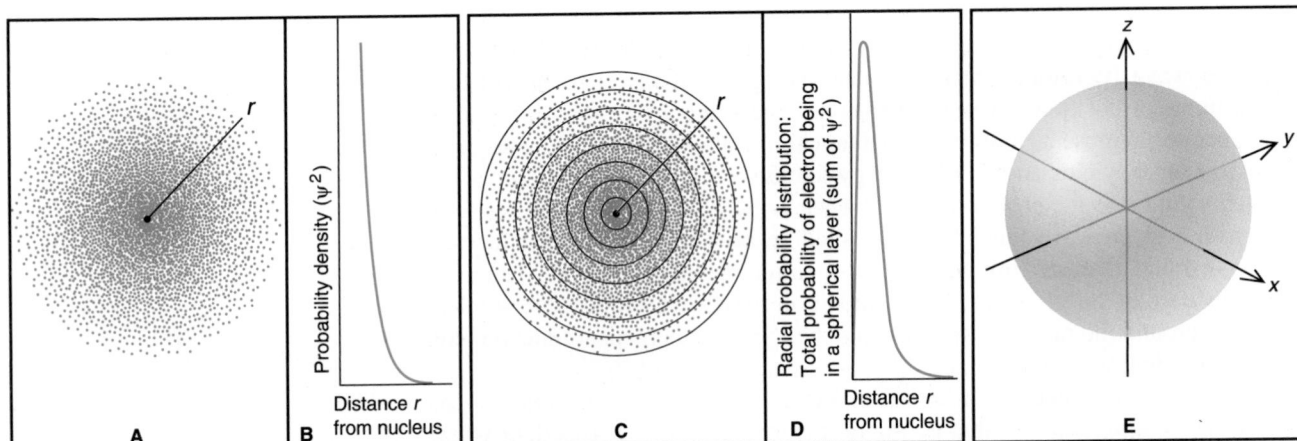

Figure 7.16 **Electron probability density in the ground-state H atom. A,** An electron density diagram shows a cross section of the H atom. The dots, each representing the probability of the electron being within a tiny volume, decrease along a line outward from the nucleus. **B,** A plot of the data from part A shows that the probability density (ψ^2) decreases with distance from the nucleus but does not reach zero (the thickness of the line makes it appear to do so). **C,** Dividing the atom's volume into thin, concentric, spherical layers (shown in cross section)

and counting the dots within each layer gives the total probability of finding the electron within that layer. **D,** A radial probability distribution plot shows total electron density in each spherical layer vs. *r*. Since electron density decreases more slowly than the volume of each concentric layer increases, the plot shows a peak. **E,** A 90% probability contour shows the ground state of the H atom (orbital of lowest energy) and represents the volume in which the electron spends 90% of its time.

these opposing effects of decreasing probability density and increasing layer volume, the total probability peaks in a layer some distance from the nucleus. Figure 7.16D shows this as a **radial probability distribution plot.** ●

The peak of the radial probability distribution for the ground-state H atom appears at the same distance from the nucleus (0.529 Å, or 5.29×10^{-11} m) as Bohr postulated for the closest orbit. Thus, at least for the ground state, the Schrödinger model predicts that the electron spends *most* of its time at the same distance that the Bohr model predicted it spent *all* of its time. The difference between "most" and "all" reflects the uncertainty of the electron's location in the Schrödinger model.

How far away from the nucleus can we find the electron? This is the same as asking "How large is the atom?" Recall from Figure 7.16B that the probability of finding the electron far from the nucleus is not zero. Therefore, we *cannot* assign a definite volume to an atom. However, we often visualize atoms with a 90% **probability contour,** such as in Figure 7.16E, which shows the volume within which the electron of the hydrogen atom spends 90% of its time.

Quantum Numbers of an Atomic Orbital

So far we have discussed the electron density for the *ground* state of the H atom. When the atom absorbs energy, it exists in an *excited* state, and the region of space occupied by the electron is described by a different atomic orbital (wave function). As you'll see, each atomic orbital has a distinctive radial probability distribution and 90% probability contour.

An atomic orbital is specified by three quantum numbers. One is related to the orbital's size, another to its shape, and the third to its orientation in space.* The quantum numbers have a hierarchical relationship: the size-related number limits the shape-related number, which limits the orientation-related number. Let's examine this hierarchy and then look at the shapes and orientations.

1. The **principal quantum number** (n) *is a positive integer* (1, 2, 3, and so forth). It indicates the relative *size* of the orbital and therefore the relative *distance from the nucleus* of the peak in the radial probability distribution plot. The principal quantum number specifies the *energy level* of the H atom: *the higher the n value, the higher the energy level.* When the electron occupies an orbital with $n = 1$, the H atom is in its ground state and has lower energy than when the electron occupies the $n = 2$ orbital (first excited state).

2. The **angular momentum quantum number** (l) *is an integer from 0 to* $n - 1$. It is related to the *shape* of the orbital and is sometimes called the *orbital-shape* (or *azimuthal*) *quantum number.* Note that the principal quantum number sets a limit on the values for the angular momentum quantum number; that is, n limits l. For an orbital with $n = 1$, l can have a value of only 0. For orbitals with $n = 2$, l can have a value of 0 or 1; for those with $n = 3$, l can be 0, 1, or 2; and so forth. Note that the number of possible l values equals the value of n.

3. The **magnetic quantum number** (m_l) *is an integer from* $-l$ *through* 0 *to* $+l$. It prescribes the *orientation* of the orbital in the space around the nucleus and is sometimes called the *orbital-orientation quantum number.* The possible values of an orbital's magnetic quantum number are set by its angular momentum quantum number; that is, l sets the possible values of m_l. An orbital with $l = 0$ can have only $m_l = 0$. However, an orbital with $l = 1$ can have any one of three m_l values, -1, 0, or $+1$; thus, there are three possible orbitals with $l = 1$, each with its own orientation. Note that the number of possible m_l values *equals* the number of orbitals, which is $2l + 1$ for a given l value.

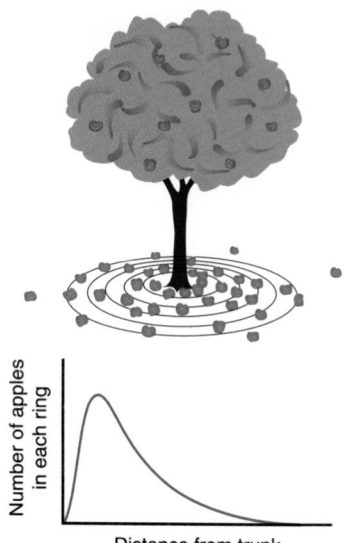

Number of apples in each ring

Distance from trunk

● **A Radial Probability Distribution of Apples** An analogy might clarify why the curve in the radial probability distribution plot peaks and then falls off. Picture fallen apples around the base of an apple tree: the density of apples is greatest near the trunk and decreases with distance. Divide the ground under the tree into foot-wide concentric rings and collect the apples within each ring. Apple density is greatest in the first ring, but the area of the second ring is larger, and so it contains a greater *total* number of apples. Farther out near the edge of the tree, rings have more area but lower apple "density," so the total number of apples decreases. A plot of "number of apples within a ring" vs. "distance of ring from trunk" shows a peak at some distance from the trunk, as in Figure 7.16D.

 ARIS Presentation Center, Art Library

---

*For ease in discussion, we refer to the size, shape, and orientation of an "atomic orbital," although we really mean the size, shape, and orientation of an "atomic orbital's radial probability distribution." This usage is common in both introductory and advanced texts.

Table 7.2   The Hierarchy of Quantum Numbers for Atomic Orbitals

Name, Symbol (Property)	Allowed Values	Quantum Numbers
Principal, n (size, energy)	Positive integer $(1, 2, 3, \ldots)$	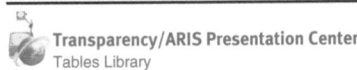
Angular momentum, l (shape)	0 to $n-1$	
Magnetic, m_l (orientation)	$-l, \ldots, 0, \ldots, +l$	

Table 7.2 summarizes the hierarchy among the three quantum numbers. (In Chapter 8, we'll discuss a fourth quantum number that relates to a property of the electron itself.) The total number of orbitals for a given n value is n^2.

SAMPLE PROBLEM 7.6   Determining Quantum Numbers for an Energy Level

PROBLEM What values of the angular momentum (l) and magnetic (m_l) quantum numbers are allowed for a principal quantum number (n) of 3? How many orbitals exist for $n = 3$?

PLAN We determine allowable quantum numbers with the rules from the text: l values are integers from 0 to $n - 1$, and m_l values are integers from $-l$ to 0 to $+l$. One m_l value is assigned to each orbital, so the number of m_l values gives the number of orbitals.

SOLUTION Determining l values: for $n = 3$, $l = 0, 1, 2$

Determining m_l for each l value:

$$\text{For } l = 0, \ m_l = 0$$
$$\text{For } l = 1, \ m_l = -1, 0, \ +1$$
$$\text{For } l = 2, \ m_l = -2, -1, 0, \ +1, \ +2$$

There are nine m_l values, so there are nine orbitals with $n = 3$.

CHECK Table 7.2 shows that we are correct. The total number of orbitals for a given n value is n^2, and for $n = 3$, $n^2 = 9$.

FOLLOW-UP PROBLEM 7.6 Specify the l and m_l values for $n = 4$.

The energy states and orbitals of the atom are described with specific terms and associated with one or more quantum numbers:

1. *Level.* The atom's energy **levels,** or *shells,* are given by the n value: the smaller the n value, the lower the energy level and the greater the probability of the electron being closer to the nucleus.

2. *Sublevel.* The atom's levels contain **sublevels,** or *subshells,* which designate the orbital shape. Each sublevel has a letter designation:

$l = 0$ is an s sublevel.
$l = 1$ is a p sublevel.
$l = 2$ is a d sublevel.
$l = 3$ is an f sublevel.

(The letters derive from the names of spectroscopic lines: *s*harp, *p*rincipal, *d*iffuse, and *f*undamental. Sublevels with l values greater than 3 are designated alphabetically: g sublevel, h sublevel, etc.) Sublevels are named by joining the n value and the letter designation. For example, the sublevel (subshell) with $n = 2$ and $l = 0$ is called the $2s$ sublevel.

3. *Orbital.* Each allowed combination of n, l, and m_l values specifies one of the atom's *orbitals*. Thus, the three quantum numbers that describe an orbital express its size (energy), shape, and spatial orientation. You can easily give the quantum numbers of the orbitals in any sublevel if you know the sublevel letter designation and the quantum number hierarchy. For example, the $2s$ sublevel has only one orbital, and its quantum numbers are $n = 2$, $l = 0$, and $m_l = 0$. The $3p$ sublevel has three orbitals: one with $n = 3$, $l = 1$, and $m_l = -1$; another with $n = 3$, $l = 1$, and $m_l = 0$; and a third with $n = 3$, $l = 1$, and $m_l = +1$.

SAMPLE PROBLEM 7.7 Determining Sublevel Names and Orbital Quantum Numbers

ARIS Presentation Center,
Worked Examples Library

PROBLEM Give the name, magnetic quantum numbers, and number of orbitals for each sublevel with the given quantum numbers:
(a) $n = 3$, $l = 2$ **(b)** $n = 2$, $l = 0$
(c) $n = 5$, $l = 1$ **(d)** $n = 4$, $l = 3$

PLAN To name the sublevel (subshell), we combine the n value and l letter designation. Since we know l, we can find the possible m_l values, whose total number equals the number of orbitals.

SOLUTION

n	l	Sublevel Name	Possible m_l Values	No. of Orbitals
(a) 3	2	$3d$	$-2, -1, 0, +1, +2$	5
(b) 2	0	$2s$	0	1
(c) 5	1	$5p$	$-1, 0, +1$	3
(d) 4	3	$4f$	$-3, -2, -1, 0, +1, +2, +3$	7

CHECK Check the number of orbitals in each sublevel using

$$\text{No. of orbitals} = \text{no. of } m_l \text{ values} = 2l + 1$$

FOLLOW-UP PROBLEM 7.7 What are the n, l, and possible m_l values for the $2p$ and $5f$ sublevels?

SAMPLE PROBLEM 7.8 Identifying Incorrect Quantum Numbers

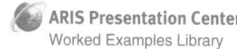
ARIS Presentation Center,
Worked Examples Library

PROBLEM What is wrong with each of the following quantum number designations and/or sublevel names?

n	l	m_l	Name
(a) 1	1	0	$1p$
(b) 4	3	+1	$4d$
(c) 3	1	-2	$3p$

SOLUTION (a) A sublevel with $n = 1$ can have only $l = 0$, not $l = 1$. The only possible sublevel name is $1s$.
(b) A sublevel with $l = 3$ is an f sublevel, not a d sublevel. The name should be $4f$.
(c) A sublevel with $l = 1$ can have only m_l of -1, 0, $+1$, not -2.

CHECK Check that l is always less than n, and m_l is always $\geq -l$ and $\leq +l$.

FOLLOW-UP PROBLEM 7.8 Supply the missing quantum numbers and sublevel names.

n	l	m_l	Name
(a) ?	?	0	$4p$
(b) 2	1	0	?
(c) 3	2	-2	?
(d) ?	?	?	$2s$

Shapes of Atomic Orbitals

Each sublevel of the H atom consists of a set of orbitals with characteristic shapes. As you'll see in Chapter 8, orbitals for the other atoms have similar shapes.

Literature Keeports, David. "How Is an Orbital Defined?" *Chem. Educator* 11 **2006,** *1,* 4-4.

The *s* Orbital   An orbital with *l* = 0 has a *spherical* shape with the nucleus at its center and is called an *s* **orbital.** The H atom's ground state, for example, has the electron in the 1*s* orbital, and *the electron probability density is highest at the nucleus.* Figure 7.17A shows this fact graphically *(top),* and an electron density *relief map (inset)* depicts this curve in three dimensions. The quarter-section of an electron cloud representation *(middle)* has the darkest shading at the nucleus. On the other hand, the radial probability distribution plot *(bottom),* which represents the probability of finding the electron (that is, where the electron spends

A 1*s* orbital   **B 2*s* orbital**   **C 3*s* orbital**

Figure 7.17 **The 1*s*, 2*s*, and 3*s* orbitals.** Information for each of the *s* orbitals is shown as a plot of probability density vs. distance *(top,* with the relief map, *inset,* showing the plot in three dimensions); as an electron cloud representation *(middle),* in which shading coincides with peaks in the plot above; and as a radial probability distribution *(bottom)* that shows where the electron spends its time. **A,** The 1*s* orbital. **B,** The 2*s* orbital. **C,** The 3*s* orbital. Nodes (regions of zero probability) appear in the 2*s* and 3*s* orbitals.

most of its time), is highest slightly out from the nucleus. Both plots fall off smoothly with distance.

The 2s orbital (Figure 7.17B) has two regions of higher electron density. The radial probability distribution (Figure 7.17B, *bottom*) of the more distant region is *higher* than that of the closer one because the sum of its ψ^2 is taken over a much larger volume. Between the two regions is a spherical **node,** a shell-like region where the probability drops to zero ($\psi^2 = 0$ at the node, analogous to zero amplitude of a wave). Because the 2s orbital is larger than the 1s, an electron in the 2s spends more time *farther* from the nucleus than when it occupies the 1s.

The 3s orbital, shown in Figure 7.17C, has three regions of high electron density and two nodes. Here again, the highest radial probability is at the greatest distance from the nucleus because the sum of all ψ^2 is taken over a larger volume. This pattern of more nodes and higher probability with distance continues for s orbitals of higher n value. An s orbital has a spherical shape, so it can have only one orientation and, thus, only one value for the magnetic quantum number: for any s orbital, $m_l = 0$.

The *p* Orbital An orbital with $l = 1$, called a *p* **orbital,** has two regions (lobes) of high probability, one on *either side* of the nucleus. Thus, as you can see in Figure 7.18, the *nucleus lies at the nodal plane* of this dumbbell-shaped orbital. The maximum value of l is $n - 1$, so only levels with $n = 2$ or higher can have a p orbital. Therefore, the lowest energy p orbital (the one closest to the nucleus) is the 2p. Keep in mind that *one p orbital consists of both lobes* and that the electron spends *equal* time in both. Similar to the pattern for s orbitals, a 3p orbital is larger than a 2p orbital, a 4p orbital is larger than a 3p orbital, and so forth.

Unlike an s orbital, each p orbital *does* have a specific orientation in space. The $l = 1$ value has three possible m_l values: -1, 0, and $+1$, which refer to three *mutually perpendicular p orbitals.* They are identical in size, shape, and energy, differing only in orientation. For convenience, we associate p orbitals with the x, y, and z axes (but there is no necessary relation between a spatial axis and a given m_l value): the p_x orbital lies along the x axis, the p_y along the y axis, and the p_z along the z axis.

Transparency/ARIS Presentation Center, Art Library

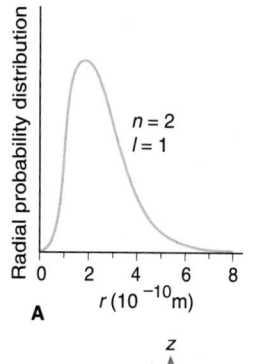

Figure 7.18 **The 2p orbitals. A,** A radial probability distribution plot of the 2p orbital shows a single peak. It lies at nearly the same distance from the nucleus as the larger peak in the 2s plot (shown in Figure 7.17B). **B,** A cross section shows an electron cloud representation of the 90% probability contour of the $2p_z$ orbital. An electron occupies both regions of a 2p orbital equally and spends 90% of its time within this volume. Note the nodal plane at the nucleus. **C,** An accurate representation of the $2p_z$ probability contour. The $2p_x$ and $2p_y$ orbitals lie along the x and y axes, respectively. **D,** The stylized depiction of the 2p probability contour used throughout the text. **E,** In an atom, the three 2p orbitals occupy mutually perpendicular regions of space, contributing to the atom's overall spherical shape.

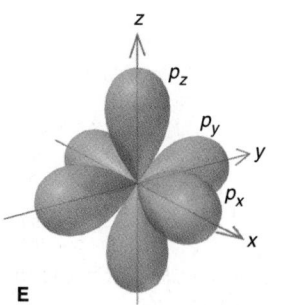

Figure 7.19 **The 3*d* orbitals. A,** A radial probability distribution plot. **B,** An electron cloud representation of the $3d_{yz}$ orbital in cross section. Note the mutually perpendicular nodal planes and the lobes lying *between* the axes. **C,** An accurate representation of the $3d_{yz}$ orbital probability contour. **D,** The stylized depiction of the $3d_{yz}$ orbital used throughout the text. **E,** The $3d_{xz}$ orbital. **F,** The $3d_{xy}$ orbital. **G,** The lobes of the $3d_{x^2-y^2}$ orbital lie *on* the x and y axes. **H,** The $3d_{z^2}$ orbital has two lobes and a central, donut-shaped region. **I,** A composite of the five 3*d* orbitals, which again contributes to an atom's overall spherical shape.

Transparency/ARIS Presentation Center, Art Library

The *d* Orbital An orbital with $l = 2$ is called a ***d* orbital.** There are five possible m_l values for the $l = 2$ value: -2, -1, 0, $+1$, and $+2$. Thus, a *d* orbital can have any one of five different orientations, as shown in Figure 7.19. Four of the five *d* orbitals have four lobes (a cloverleaf shape) prescribed by two mutually perpendicular nodal planes, with the nucleus lying at the junction of the lobes. Three of these orbitals lie in the mutually perpendicular *xy*, *xz*, and *yz* planes, with their lobes *between* the axes, and are called the d_{xy}, d_{xz}, and d_{yz} orbitals. A fourth, the $d_{x^2-y^2}$ orbital, also lies in the *xy* plane, but its lobes are directed *along* the axes. The fifth *d* orbital, the d_{z^2}, has a different shape: two major lobes lie along the *z* axis, and a donut-shaped region girdles the center. An electron associated with a given *d* orbital has equal probability of being in any of the orbital's lobes.

As we said for the *p* orbitals, an axis designation for a *d* orbital is *not* associated with a given m_l value. In keeping with the quantum number rules, a *d* orbital ($l = 2$) must have a principal quantum number of $n = 3$ or greater. The 4*d* orbitals extend farther from the nucleus than the 3*d* orbitals, and the 5*d* orbitals extend still farther.

Orbitals with Higher *l* Values Orbitals with $l = 3$ are *f* orbitals and must have a principal quantum number of at least $n = 4$. There are seven *f* orbitals ($2l + 1 = 7$), each with a complex, multilobed shape; Figure 7.20 shows one of them. Orbitals with $l = 4$ are *g* orbitals, but we will not discuss them further because they play no known role in chemical bonding.

Literature Ashkenazi Guy. "The Meaning of d-Orbital Labels," *J. Chem. Educ.* **2005,** 82, 323.

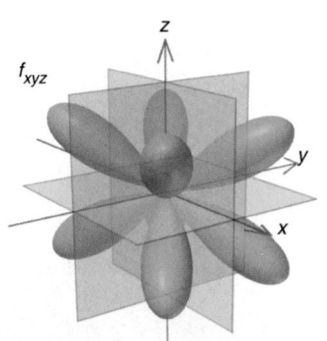

Figure 7.20 **One of the seven possible 4*f* orbitals.** The $4f_{xyz}$ orbital has eight lobes and three nodal planes. The other six 4*f* orbitals also have multilobed contours.

ARIS Presentation Center, Art Library

The Special Case of the Hydrogen Atom

The energy state of the H atom depends only on the principal quantum number, n. When an electron occupies an orbital with a higher *n* value, it occurs (on average) farther from the nucleus, so the atom is higher in energy. But *the H atom is a special case* because it has only one electron. As you'll see in Chapter 8, the energy states of all other atoms depend on *both* the *n* and *l* values of the occupied orbitals because of additional nucleus-electron attractions and electron-electron repulsions. In other words, *for the H atom only,* all four *n* = 2 orbitals (one 2*s* and three 2*p*) have the same energy, all nine *n* = 3 orbitals (one 3*s*, three 3*p*, and five 3*d*) have the same energy (Figure 7.21), and so forth.

Section Summary

The electron's wave function (ψ, atomic orbital) is a mathematical description of the electron's wavelike behavior in an atom. Each wave function is associated with one of the atom's allowed energy states. • The probability density of finding the electron at a particular location is represented by ψ^2. An electron density diagram and a radial probability distribution plot show how the electron occupies the space near the nucleus for a particular energy level. • Three features of an atomic orbital are described by quantum numbers: size (*n*), shape (*l*), and orientation (m_l). Orbitals with the same *n* and *l* values constitute a sublevel; sublevels with the same *n* value constitute an energy level. • A sublevel with *l* = 0 has a spherical (*s*) orbital; a sublevel with *l* = 1 has three, two-lobed (*p*) orbitals; and a sublevel with *l* = 2 has five multilobed (*d*) orbitals. • In the special case of the H atom, the energy levels depend on the *n* value only.

Chapter Perspective

In quantum physics, we see the everyday distinctions between matter and energy disappear and a new picture of the hydrogen atom come into focus. The very nature of the atom, however, makes this focus blurred. The sequel to these remarkable discoveries unfolds in Chapter 8, where we begin our discussion of how the periodic behavior of the elements emerges from the properties of the atom, which in turn arise from the electron occupancy of its orbitals.

Figure 7.21 The energy levels in the H atom. The H atom is the only atom in which the energy level depends *only* on the *n* value of the sublevels. For example, the 2*s* and the three 2*p* sublevels (shown as short lines) all have the same energy.

ARIS Presentation Center, Art Library

CHAPTER REVIEW GUIDE
The following sections provide many aids to help you study this chapter. (Numbers in parentheses refer to pages, unless noted otherwise.)

Learning Objectives These are concepts and skills you should know after studying this chapter.

Relevant section and/or sample problem (SP) numbers appear in parentheses.

Understand These Concepts

1. The wave characteristics of light (the interrelations of frequency, wavelength, and speed; the meaning of amplitude) and the general regions of the electromagnetic spectrum (7.1)
2. How particles and waves differ in terms of the phenomena of refraction, diffraction, and interference (7.1)
3. The quantization of energy and the fact that an atom changes its energy by emitting or absorbing quanta of radiation (7.1)
4. How the photon theory explains the photoelectric effect and the relation between photon absorbed and electron released (7.1)
5. How Bohr's theory explained the line spectra of the H atom; the importance of discrete atomic energy levels (7.2)
6. The wave-particle duality of matter and energy and the relevant theories and experiments that led to it (de Broglie wavelength, electron diffraction, photon momentum) (7.3)
7. The meaning of the uncertainty principle and how it limits our knowledge of electron properties (7.3)
8. The distinction between ψ (wave function, or atomic orbital) and ψ^2 (probability density) (7.4)

9. How electron density diagrams and radial probability distribution plots depict electron location within the atom (7.4)
10. The hierarchy of the quantum numbers that describe the size (*n*, energy), shape (*l*), and orientation (m_l) of an orbital (7.4)
11. The distinction between energy level (shell), sublevel (subshell), and orbital (7.4)
12. The shapes and nodes of *s*, *p*, and *d* orbitals (7.4)

Master These Skills

1. Interconverting wavelength and frequency (SP 7.1)
2. Calculating the energy of a photon from its wavelength (SP 7.2)
3. Finding the energy change and wavelength of the photon absorbed or emitted when an H atom's electron changes energy level (SP 7.3)
4. Applying de Broglie's equation to find the wavelength of an electron (SP 7.4)
5. Applying the uncertainty principle to see that the location and speed of a particle cannot be determined simultaneously (SP 7.5)
6. Determining quantum numbers and sublevel designations (SPs 7.6–7.8)

Key Terms These important terms appear in boldface in the chapter and are defined again in the Glossary.

Section 7.1
electromagnetic radiation (269)
frequency (ν) (270)
wavelength (λ) (270)
speed of light (c) (270)
amplitude (270)
electromagnetic spectrum (271)
infrared (IR) (271)
ultraviolet (UV) (271)
refraction (272)
diffraction (273)
quantum number (274)

Planck's constant (h) (274)
quantum (274)
photoelectric effect (274)
photon (275)

Section 7.2
line spectrum (276)
stationary state (277)
ground state (278)
excited state (278)
spectrophotometry (281)
emission spectrum (281)
flame test (281)
absorption spectrum (281)

Section 7.3
de Broglie wavelength (284)
wave-particle duality (286)
uncertainty principle (286)

Section 7.4
quantum mechanics (287)
Schrödinger equation (287)
wave function (atomic orbital) (287)
electron density diagram (288)
electron cloud (288)
radial probability distribution plot (289)

probability contour (289)
principal quantum number (n) (289)
angular momentum quantum number (l) (289)
magnetic quantum number (m_l) (289)
level (shell) (290)
sublevel (subshell) (290)
s orbital (292)
node (293)
p orbital (293)
d orbital (294)

Key Equations and Relationships Numbered and screened concepts are listed for you to refer to or memorize.

7.1 Relating the speed of light to its frequency and wavelength (270):

$$c = \nu \times \lambda$$

7.2 Determining the smallest change in an atom's energy (274):

$$\Delta E = h\nu$$

7.3 Calculating the wavelength of any line in the H atom spectrum (Rydberg equation) (277):

$$\frac{1}{\lambda} = R\left(\frac{1}{n_1^2} - \frac{1}{n_2^2}\right)$$

where n_1 and n_2 are positive integers and $n_2 > n_1$

7.4 Finding the difference between two energy levels in the H atom (279):

$$\Delta E = E_{\text{final}} - E_{\text{initial}} = -2.18 \times 10^{-18} \text{ J}\left(\frac{1}{n_{\text{final}}^2} - \frac{1}{n_{\text{initial}}^2}\right)$$

7.5 Calculating the wavelength of any moving particle (de Broglie wavelength) (284):

$$\lambda = \frac{h}{mu}$$

7.6 Finding the uncertainty in position or speed of a particle (Heisenberg uncertainty principle) (286):

$$\Delta x \cdot m\Delta u \geq \frac{h}{4\pi}$$

Highlighted Figures and Tables These figures (F) and tables (T) provide a visual review of key ideas.

F7.3 The electromagnetic spectrum (271)
F7.15 From classical to quantum theory (285)
F7.16 Electron probability density in the ground-state H atom (288)
T7.2 The hierarchy of quantum numbers (290)

F7.17 The $1s$, $2s$, and $3s$ orbitals (292)
F7.18 The $2p$ orbitals (293)
F7.19 The $3d$ orbitals (294)

Brief Solutions to FOLLOW-UP PROBLEMS Compare your solutions to these calculation steps and answers.

7.1 λ (nm) $= \dfrac{3.00\times10^8 \text{ m/s}}{7.23\times10^{14} \text{ s}^{-1}} \times \dfrac{10^9 \text{ nm}}{1 \text{ m}} = 415 \text{ nm}$

λ (Å) $= 415 \text{ nm} \times \dfrac{10 \text{ Å}}{1 \text{ nm}} = 4150 \text{ Å}$

7.2 UV: $E = hc/\lambda$

$= \dfrac{(6.626\times10^{-34} \text{ J·s})(3.00\times10^8 \text{ m/s})}{1\times10^{-8} \text{ m}} = 2\times10^{-17} \text{ J}$

Visible: $E = 4\times10^{-19}$ J; IR: $E = 2\times10^{-21}$ J
As λ increases, E decreases.

7.3 With $n_{\text{final}} = 3$ for an IR photon:

$\Delta E = -2.18\times10^{-18} \text{ J}\left(\dfrac{1}{n_{\text{final}}^2} - \dfrac{1}{n_{\text{initial}}^2}\right)$

$= -2.18\times10^{-18} \text{ J}\left(\dfrac{1}{3^2} - \dfrac{1}{6^2}\right)$

$= -2.18\times10^{-18} \text{ J}\left(\dfrac{1}{9} - \dfrac{1}{36}\right) = 1.82\times10^{-19} \text{ J}$

$\lambda = \dfrac{hc}{\Delta E} = \dfrac{(6.626\times10^{-34} \text{ J·s})(3.00\times10^8 \text{ m/s})}{1.82\times10^{-19} \text{ J}} \times \dfrac{1 \text{ Å}}{10^{-10} \text{ m}}$

$= 1.09\times10^4 \text{ Å}$

7.4 $u = \dfrac{h}{m\lambda} = \dfrac{6.626\times10^{-34} \text{ kg·m}^2/\text{s}}{(9.11\times10^{-31} \text{ kg})\left(100 \text{ nm} \times \dfrac{1 \text{ m}}{10^9 \text{ nm}}\right)}$

$= 7.27\times10^3 \text{ m/s}$

7.5 $\Delta x \geq \dfrac{6.626\times10^{-34} \text{ kg·m}^2/\text{s}}{4\pi(0.142 \text{ kg})(0.447 \text{ m/s})} \geq 8.31\times10^{-34} \text{ m}$

7.6 $n = 4$, so $l = 0, 1, 2, 3$. In addition to the m_l values in Sample Problem 7.6, we have those for $l = 3$:

$m_l = -3, -2, -1, 0, +1, +2, +3$

7.7 For $2p$: $n = 2$, $l = 1$, $m_l = -1, 0, +1$
For $5f$: $n = 5$, $l = 3$, $m_l = -3, -2, -1, 0, +1, +2, +3$

7.8 (a) $n = 4$, $l = 1$; (b) name is $2p$; (c) name is $3d$;
(d) $n = 2$, $l = 0$, $m_l = 0$

PROBLEMS

Problems with colored numbers are answered in Appendix E and worked in detail in the Student Solutions Manual. Problem sections match those in the text and provide the numbers of relevant sample problems. Most offer Concept Review Questions, Skill-Building Exercises (grouped in pairs covering the same concept), and Problems in Context. The Comprehensive Problems are based on material from any section or previous chapter.

The Nature of Light
(Sample Problems 7.1 and 7.2)

● Concept Review Questions

7.1 In what ways are microwave and ultraviolet radiation the same? In what ways are they different?

7.2 Consider the following types of electromagnetic radiation:
(1) Microwave   (2) Ultraviolet   (3) Radio waves
(4) Infrared   (5) X-ray   (6) Visible
(a) Arrange them in order of increasing wavelength.
(b) Arrange them in order of increasing frequency.
(c) Arrange them in order of increasing energy.

7.3 Define each of the following wave phenomena, and give an example of where each occurs: (a) refraction; (b) diffraction; (c) dispersion; (d) interference.

7.4 In the mid-17th century, Isaac Newton proposed that light existed as a stream of particles, and the wave-particle debate continued for over 250 years until Planck and Einstein presented their revolutionary ideas. Give two pieces of evidence for the wave model and two for the particle model.

7.5 Portions of electromagnetic waves A, B, and C are represented below (not drawn to scale):

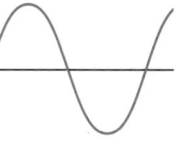

A   **B**   **C**

Rank them in order of (a) increasing frequency; (b) increasing energy; (c) increasing amplitude. (d) If wave B just barely fails to cause a current when shining on a metal, is wave A or C more likely to do so? (e) If wave B represents visible radiation, is wave A or C more likely to be IR radiation?

7.6 What new idea about light did Einstein use to explain the photoelectric effect? Why does the photoelectric effect exhibit a threshold frequency? Why does it *not* exhibit a time lag?

● Skill-Building Exercises (grouped in similar pairs)

7.7 An AM station broadcasts rock music at "950 on your radio dial." Units for AM frequencies are given in kilohertz (kHz). Find the wavelength of the station's radio waves in meters (m), nanometers (nm), and angstroms (Å).

7.8 An FM station broadcasts music at 93.5 MHz (megahertz, or 10^6 Hz). Find the wavelength (in m, nm, and Å) of these waves.

7.9 A radio wave has a frequency of 3.8×10^{10} Hz. What is the energy (in J) of one photon of this radiation?

7.10 An x-ray has a wavelength of 1.3 Å. Calculate the energy (in J) of one photon of this radiation.

7.11 Rank the following photons in terms of increasing energy: (a) blue ($\lambda = 453$ nm); (b) red ($\lambda = 660$ nm); (c) yellow ($\lambda = 595$ nm).

7.12 Rank the following photons in terms of decreasing energy: (a) IR ($\nu = 6.5 \times 10^{13}$ s^{-1}); (b) microwave ($\nu = 9.8 \times 10^{11}$ s^{-1}); (c) UV ($\nu = 8.0 \times 10^{15}$ s^{-1}).

● Problems in Context

7.13 Police often monitor traffic with "K-band" radar guns, which operate in the microwave region at 22.235 GHz (1 GHz = 10^9 Hz). Find the wavelength (in nm and Å) of this radiation.

7.14 Covalent bonds in a molecule absorb radiation in the IR region and vibrate at characteristic frequencies.
(a) The C—O bond absorbs radiation of wavelength 9.6 μm. What frequency (in s^{-1}) corresponds to that wavelength?
(b) The H—Cl bond has a frequency of vibration of 8.652×10^{13} Hz. What wavelength (in μm) corresponds to that frequency?

7.15 Cobalt-60 is a radioactive isotope used to treat cancers. A gamma ray emitted by this isotope has an energy of 1.33 MeV (million electron volts; 1 eV = 1.602×10^{-19} J). What is the frequency (in Hz) and the wavelength (in m) of this gamma ray?

7.16 (a) The first step in ozone formation in the upper atmosphere occurs when oxygen molecules absorb UV radiation of wavelengths ≤ 242 nm. Calculate the frequency and energy of the least energetic of these photons. (b) Ozone absorbs light having wavelengths of 2200 to 2900 Å, thus protecting organisms on Earth's surface from this high-energy UV radiation. What are the frequency and energy of the most energetic of these photons?

Atomic Spectra
(Sample Problem 7.3)

● Concept Review Questions

7.17 How is n_1 in the Rydberg equation (Equation 7.3) related to the quantum number n in the Bohr model of the atom?

7.18 How would a planetary (solar system) model of the atom differ from a key assumption in Bohr's model? What was the theoretical basis from which Bohr made this assumption?

7.19 Distinguish between an absorption spectrum and an emission spectrum. With which did Bohr work?

7.20 Which of these electron transitions correspond to absorption of energy and which to emission?
(a) $n = 2$ to $n = 4$ (b) $n = 3$ to $n = 1$
(c) $n = 5$ to $n = 2$ (d) $n = 3$ to $n = 4$

7.21 Why could the Bohr model not predict line spectra for atoms other than hydrogen?

7.22 The H atom and the Be^{3+} ion each have one electron. Does the Bohr model predict their spectra accurately? Would you expect their line spectra to be identical? Explain.

● Skill-Building Exercises (grouped in similar pairs)

7.23 Use the Rydberg equation (Equation 7.3) to calculate the wavelength (in nm) of the photon emitted when a hydrogen atom undergoes a transition from $n = 5$ to $n = 2$.

7.24 Use the Rydberg equation to calculate the wavelength (in Å) of the photon absorbed when a hydrogen atom undergoes a transition from $n = 1$ to $n = 3$.

7.25 What is the wavelength (in nm) of the least energetic spectral line in the infrared series of the H atom?

7.26 What is the wavelength (in nm) of the least energetic spectral line in the visible series of the H atom?

7.27 Calculate the energy difference (ΔE) for the transition in Problem 7.23 for 1 mol of H atoms.

7.28 Calculate the energy difference (ΔE) for the transition in Problem 7.24 for 1 mol of H atoms.

7.29 Arrange the following H atom electron transitions in order of *increasing* frequency of the photon absorbed or emitted:
(a) $n = 2$ to $n = 4$ (b) $n = 2$ to $n = 1$
(c) $n = 2$ to $n = 5$ (d) $n = 4$ to $n = 3$

7.30 Arrange the following H atom electron transitions in order of *decreasing* wavelength of the photon absorbed or emitted:
(a) $n = 2$ to $n = \infty$ (b) $n = 4$ to $n = 20$
(c) $n = 3$ to $n = 10$ (d) $n = 2$ to $n = 1$

7.31 The electron in a ground-state H atom absorbs a photon of wavelength 97.20 nm. To what energy level does it move?

7.32 An electron in the $n = 5$ level of an H atom emits a photon of wavelength 1281 nm. To what energy level does it move?

● Problems in Context

7.33 In addition to continuous radiation, fluorescent lamps emit sharp lines in the visible region from a mercury discharge within the tube. Much of this light has a wavelength of 436 nm. What is the energy (in J) of one photon of this light?

7.34 A Bohr-model representation of the H atom is shown below with several electron transitions depicted by arrows:

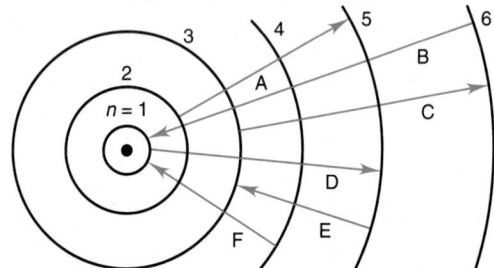

(a) Which transitions are absorptions and which are emissions?
(b) Rank the emissions in terms of increasing energy. (c) Rank the absorptions in terms of increasing wavelength of light emitted.

The Wave-Particle Duality of Matter and Energy
(Sample Problems 7.4 and 7.5)

● Concept Review Questions

7.35 In what sense is the wave motion of a guitar string analogous to the motion of an electron in an atom?

7.36 What experimental support did de Broglie's concept receive?

7.37 If particles have wavelike motion, why don't we observe that motion in the macroscopic world?

7.38 Why can't we overcome the uncertainty predicted by Heisenberg's principle by building more precise devices to reduce the error in measurements below the $h/4\pi$ limit?

● Skill-Building Exercises *(grouped in similar pairs)*

7.39 A 232-lb fullback runs the 40-yd dash at a speed of 19.8 ± 0.1 mi/h.
(a) What is his de Broglie wavelength (in meters)?
(b) What is the uncertainty in his position?

7.40 An alpha particle (mass = 6.6×10^{-24} g) emitted by radium travels at $3.4\times10^7 \pm 0.1\times10^7$ mi/h.
(a) What is its de Broglie wavelength (in meters)?
(b) What is the uncertainty in its position?

7.41 How fast must a 56.5-g tennis ball travel in order to have a de Broglie wavelength that is equal to that of a photon of green light (5400 Å)?

7.42 How fast must a 142-g baseball travel in order to have a de Broglie wavelength that is equal to that of an x-ray photon with $\lambda = 100.$ pm?

7.43 A sodium flame has a characteristic yellow color due to emissions of wavelength 589 nm. What is the mass equivalence of one photon of this wavelength (1 J = 1 kg·m²/s²)?

7.44 A lithium flame has a characteristic red color due to emissions of wavelength 671 nm. What is the mass equivalence of 1 mol of photons of this wavelength (1 J = 1 kg·m²/s²)?

The Quantum-Mechanical Model of the Atom
(Sample Problems 7.6 to 7.8)

● Concept Review Questions

7.45 What physical meaning is attributed to ψ^2, the square of the wave function?

7.46 What does "electron density in a particular tiny volume of space" mean?

7.47 Explain what it means for the peak in the radial probability distribution plot for the $n = 1$ level of a hydrogen atom to be at 0.529 Å. Is the probability of finding an electron at 0.529 Å from the nucleus greater for the 1s or the 2s orbital?

7.48 What feature of an orbital is related to each of the following quantum numbers?
(a) Principal quantum number (n)
(b) Angular momentum quantum number (l)
(c) Magnetic quantum number (m_l)

● Skill-Building Exercises *(grouped in similar pairs)*

7.49 How many orbitals in an atom can have each of the following designations: (a) 1s; (b) 4d; (c) 3p; (d) $n = 3$?

7.50 How many orbitals in an atom can have each of the following designations: (a) 5f; (b) 4p; (c) 5d; (d) $n = 2$?

7.51 Give all possible m_l values for orbitals that have each of the following: (a) $l = 2$; (b) $n = 1$; (c) $n = 4, l = 3$.

7.52 Give all possible m_l values for orbitals that have each of the following: (a) $l = 3$; (b) $n = 2$; (c) $n = 6, l = 1$.

7.53 Draw 90% probability contours (with axes) for each of the following orbitals: (a) s; (b) p_x.

7.54 Draw 90% probability contours (with axes) for each of the following orbitals: (a) p_z; (b) d_{xy}.

7.55 For each of the following, give the sublevel designation, the allowable m_l values, and the number of orbitals:
(a) $n = 4, l = 2$ (b) $n = 5, l = 1$ (c) $n = 6, l = 3$

7.56 For each of the following, give the sublevel designation, the allowable m_l values, and the number of orbitals:
(a) $n = 2, l = 0$ (b) $n = 3, l = 2$ (c) $n = 5, l = 1$

7.57 For each of the following sublevels, give the n and l values and the number of orbitals: (a) 5s; (b) 3p; (c) 4f.

7.58 For each of the following sublevels, give the n and l values and the number of orbitals: (a) 6g; (b) 4s; (c) 3d.

7.59 Are the following quantum number combinations allowed? If not, show two ways to correct them:
(a) $n = 2; l = 0; m_l = -1$ (b) $n = 4; l = 3; m_l = -1$
(c) $n = 3; l = 1; m_l = 0$ (d) $n = 5; l = 2; m_l = +3$

• **7.60** Are the following quantum number combinations allowed? If not, show two ways to correct them:
(a) $n = 1; l = 0; m_l = 0$ (b) $n = 2; l = 2; m_l = +1$
(c) $n = 7; l = 1; m_l = +2$ (d) $n = 3; l = 1; m_l = -2$

Comprehensive Problems

•• **7.61** The orange color of carrots and orange peel is due mostly to β-carotene, an organic compound insoluble in water but soluble in benzene and chloroform. Describe an experiment to determine the concentration of β-carotene in the oil from orange peel.

•• **7.62** The quantum-mechanical treatment of the hydrogen atom gives the energy, E, of the electron as a function of the principal quantum number, n:

$$E = -\frac{h^2}{8\pi^2 m_e a_0^2 n^2} \qquad (n = 1, 2, 3, \ldots)$$

where h is Planck's constant, m_e is the electron mass, and a_0 is 52.92×10^{-12} m.

(a) Write the expression in the form $E = -(\text{constant})\dfrac{1}{n^2}$, evaluate the constant (in J), and compare it with the corresponding expression from Bohr's theory.
(b) Use the expression to find ΔE between $n = 2$ and $n = 3$.
(c) Calculate the wavelength of the photon that corresponds to this energy change. Is this photon seen in the hydrogen spectrum obtained from experiment (see Figure 7.8, p. 276)?

•• **7.63** The photoelectric effect is illustrated in a plot of the kinetic energies of electrons ejected from the surface of potassium metal or silver metal at different frequencies of incident light.

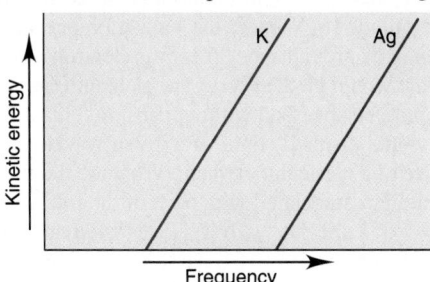

(a) Why don't the lines begin at the origin? (b) Why don't the lines begin at the same point? (c) From which metal will light of shorter wavelength eject an electron? (d) Why are the slopes equal?

•• **7.64** The human eye is a complex sensing device for visible light. The optic nerve needs a minimum of 2.0×10^{-17} J of energy to trigger a series of impulses that eventually reach the brain.
(a) How many photons of red light (700. nm) are needed?
(b) How many photons of blue light (475 nm)?

•• **7.65** One reason carbon monoxide (CO) is toxic is that it binds to the blood protein hemoglobin more strongly than oxygen does. The bond between hemoglobin and CO absorbs radiation of 1953 cm^{-1}. (The units are the reciprocal of the wavelength in centimeters.) Calculate the wavelength (in nm and Å) and the frequency (in Hz) of the absorbed radiation.

•• **7.66** A metal ion M^{n+} has a single electron. The highest energy line in its emission spectrum occurs at a frequency of 2.961×10^{16} Hz. Identify the ion.

•• **7.67** TV and radio stations transmit in specific frequency bands of the radio region of the electromagnetic spectrum.
(a) TV channels 2 to 13 (VHF) broadcast signals between the frequencies of 59.5 and 215.8 MHz, whereas FM radio stations broadcast signals with wavelengths between 2.78 and 3.41 m. Do these bands of signals overlap?
(b) AM radio signals have frequencies between 550 and 1600 kHz. Which has a broader transmission band, AM or FM?

•• **7.68** Compare the wavelengths of an electron (mass = 9.11×10^{-31} kg) and a proton (mass = 1.67×10^{-27} kg), each having (a) a speed of 3.4×10^6 m/s; (b) a kinetic energy of 2.7×10^{-15} J.

••• **7.69** Five lines in the H atom spectrum have wavelengths (in Å): (a) 1212.7; (b) 4340.5; (c) 4861.3; (d) 6562.8; (e) 10,938. Three lines result from transitions to $n_{\text{final}} = 2$ (visible series). The other two result from transitions in different series, one with $n_{\text{final}} = 1$ and the other with $n_{\text{final}} = 3$. Identify n_{initial} for each line.

•• **7.70** In his explanation of the threshold frequency in the photoelectric effect, Einstein reasoned that the absorbed photon must have a minimum energy to dislodge an electron from the metal surface. This energy is called the *work function* (φ) of that metal. What is the longest wavelength of radiation (in nm) that could cause the photoelectric effect in each of these metals: (a) calcium, φ = 4.60×10^{-19} J; (b) titanium, φ = 6.94×10^{-19} J; (c) sodium, φ = 4.41×10^{-19} J?

•• **7.71** You have three metal samples—A, B, and C—that are tantalum (Ta), barium (Ba), and tungsten (W), but you don't know which is which. Metal A emits electrons in response to visible light; metals B and C require UV light. (a) Identify metal A, and find the longest wavelength that removes an electron. (b) What range of wavelengths would distinguish B and C? [The work functions are Ta (6.81×10^{-19} J), Ba (4.30×10^{-19} J), and W (7.16×10^{-19} J); work function is explained in Problem 7.70.]

•• **7.72** Refractometry is an analytical method based on the difference between the speed of light as it travels through a substance (v) and its speed in a vacuum (c). In the procedure, light of known wavelength passes through a fixed thickness of the substance at a known temperature. The index of refraction equals c/v. Using yellow light ($\lambda = 589$ nm) at 20°C, for example, the index of refraction of water is 1.33 and that of diamond is 2.42. Calculate the speed of light in (a) water and (b) diamond.

•• **7.73** A laser (*l*ight *a*mplification by *s*timulated *e*mission of *r*adiation) provides a coherent (in-phase) nearly monochromatic source of high-intensity light. Lasers are used in eye surgery, CD/DVD players, basic research, etc. Some modern dye lasers can be "tuned" to emit a desired wavelength. Fill in the blanks in the following table of the properties of some common lasers:

Type	λ (nm)	v (s^{-1})	E (J)	Color
He-Ne	632.8	?	?	?
Ar	?	6.148×10^{14}	?	?
Ar-Kr	?	?	3.499×10^{-19}	?
Dye	663.7	?	?	?

•• **7.74** As space exploration increases, means of communication with humans and probes on other planets are being developed. (a) How much time (in s) does it take for a radio wave of frequency 8.93×10^7 s^{-1} to reach Mars, which is 8.1×10^7 km from Earth? (b) If it takes this radiation 1.2 s to reach the Moon, how far (in m) is the Moon from Earth?

•• **7.75** The following quantum number combinations are not allowed. Assuming the n and m_l values are correct, change the l value to create an allowable combination:
(a) $n = 3; l = 0; m_l = -1$ (b) $n = 3; l = 3; m_l = +1$
(c) $n = 7; l = 2; m_l = +3$ (d) $n = 4; l = 1; m_l = -2$

•••**7.76** A ground-state H atom absorbs a photon of wavelength 94.91 nm, and its electron attains a higher energy level. The atom then emits two photons: one of wavelength 1281 nm to reach an intermediate level, and a second to reach the ground state.
(a) What higher level did the electron reach?
(b) What intermediate level did the electron reach?
(c) What was the wavelength of the second photon emitted?

•••**7.77** Consider these ground-state ionization energies of one-electron species:

$$H = 1.31 \times 10^3 \text{ kJ/mol} \qquad He^+ = 5.24 \times 10^3 \text{ kJ/mol}$$
$$Li^{2+} = 1.18 \times 10^4 \text{ kJ/mol}$$

(a) Write a general expression for the ionization energy of any one-electron species. (b) Use your expression to calculate the ionization energy of B^{4+}. (c) What is the minimum wavelength required to remove the electron from the $n = 3$ level of He^+? (d) What is the minimum wavelength required to remove the electron from the $n = 2$ level of Be^{3+}?

••**7.78** Use the relative size of the $3s$ orbital represented below to answer the following questions about orbitals A–D.

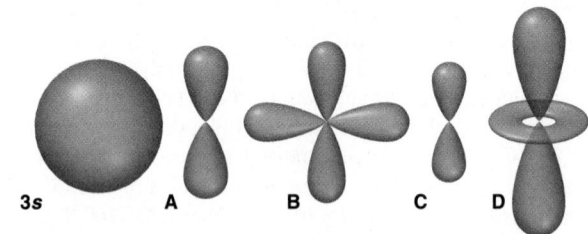

$3s$ A   B   C   D

(a) Which orbital has the highest value of n? (b) Which orbital(s) have a value of $l = 1$? $l = 2$? (c) How many other orbitals with the same value of n have the same shape as orbital B? Orbital C? (d) Which orbital has the highest energy? Lowest energy?

••**7.79** In the course of developing his model, Bohr arrived at the following formula for the radius of the electron's orbit: $r_n = n^2 h^2 \epsilon_0 / \pi m_e e^2$, where m_e is the electron mass, e is its charge, and ϵ_0 is a constant related to charge attraction in a vacuum. Given that $m_e = 9.109 \times 10^{-31}$ kg, $e = 1.602 \times 10^{-19}$ C, and $\epsilon_0 = 8.854 \times 10^{-12}$ C²/J·m, calculate the following:
(a) The radius of the 1st ($n = 1$) orbit in the H atom
(b) The radius of the 10th ($n = 10$) orbit in the H atom

•••**7.80** (a) Calculate the Bohr radius of an electron in the $n = 3$ orbit of a hydrogen atom. (See Problem 7.79.)
(b) What is the energy (in J) of the atom in part (a)?
(c) What is the energy of an Li^{2+} ion when its electron is in the $n = 3$ orbit?
(d) Why are the answers to parts (b) and (c) different?

••**7.81** Enormous numbers of microwave photons are needed to warm macroscopic samples of matter. A portion of soup containing 252 g of water is heated in a microwave oven from 20.°C to 98°C, with radiation of wavelength 1.55×10^{-2} m. How many photons are absorbed by the water in the soup?

•••**7.82** The quantum-mechanical treatment of the hydrogen atom gives an expression for the wave function, ψ, of the $1s$ orbital:

$$\psi = \frac{1}{\sqrt{\pi}} \left(\frac{1}{a_0} \right)^{3/2} e^{-r/a_0}$$

where r is the distance from the nucleus and a_0 is 52.92 pm. The electron probability density is the probability of finding the electron in a tiny volume at distance r from the nucleus and is proportional to ψ^2. The radial probability distribution is the total probability of finding the electron at all points at distance r from the nucleus and is proportional to $4\pi r^2 \psi^2$. Calculate the values (to three significant figures) of ψ, ψ^2, and $4\pi r^2 \psi^2$ to fill in the following table, and sketch plots of these quantities versus r. Compare the latter two plots with those in Figure 7.17A, p. 292:

r (pm)	ψ (pm$^{-3/2}$)	ψ^2 (pm^{-3})	$4\pi r^2 \psi^2$ (pm^{-1})
0			
50			
100			
200			

•••**7.83** Lines in one spectral series can overlap lines in another.
(a) Use the Rydberg equation to see if the range of wavelengths in the $n_1 = 1$ series overlaps the range in the $n_1 = 2$ series.
(b) Use the Rydberg equation to see if the range of wavelengths in the $n_1 = 3$ series overlaps the range in the $n_1 = 4$ series.
(c) How many lines in the $n_1 = 4$ series lie in the range of the $n_1 = 5$ series?
(d) What does this overlap imply about the hydrogen spectrum at longer wavelengths?

•••**7.84** The following values are the only allowable energy levels of a hypothetical one-electron atom:

$$E_6 = -2 \times 10^{-19} \text{ J} \qquad E_5 = -7 \times 10^{-19} \text{ J}$$
$$E_4 = -11 \times 10^{-19} \text{ J} \qquad E_3 = -15 \times 10^{-19} \text{ J}$$
$$E_2 = -17 \times 10^{-19} \text{ J} \qquad E_1 = -20 \times 10^{-19} \text{ J}$$

(a) If the electron were in the $n = 3$ level, what would be the highest frequency (and minimum wavelength) of radiation that could be emitted? (b) What is the ionization energy (in kJ/mol) of the atom in its ground state? (c) If the electron were in the $n = 4$ level, what would be the shortest wavelength (in nm) of radiation that could be absorbed without causing ionization?

••**7.85** In fireworks displays, light of a given wavelength indicates the presence of a particular element. What are the frequency and color of the light associated with each of the following?
(a) Li^+, $\lambda = 671$ nm   (b) Cs^+, $\lambda = 456$ nm
(c) Ca^{2+}, $\lambda = 649$ nm   (d) Na^+, $\lambda = 589$ nm

••**7.86** Photoelectron spectroscopy applies the principle of the photoelectric effect to study orbital energies of atoms and molecules. High-energy radiation (usually UV or x-ray) is absorbed by a sample and an electron is ejected. The orbital energy can be calculated from the known energy of the radiation and the measured energy of the electron lost. The following energy differences were determined for several electron transitions:

$$\Delta E_{2 \longrightarrow 1} = 4.098 \times 10^{-17} \text{ J} \qquad \Delta E_{3 \longrightarrow 1} = 4.854 \times 10^{-17} \text{ J}$$
$$\Delta E_{5 \longrightarrow 1} = 5.242 \times 10^{-17} \text{ J} \qquad \Delta E_{4 \longrightarrow 2} = 1.024 \times 10^{-17} \text{ J}$$

Calculate the energy change and the wavelength of a photon emitted in the following transitions:
(a) Level 3 $\longrightarrow$ 2   (b) Level 4 $\longrightarrow$ 1   (c) Level 5 $\longrightarrow$ 4

•••**7.87** Horticulturists know that, for many plants, leaf color depends on how brightly lit the growing area is: dark green leaves are associated with low light levels, and pale green with high levels. (a) Use the photon theory to explain this behavior. (b) What change in leaf composition might account for this behavior?

•••**7.88** In compliance with conservation of energy, Einstein explained that in the photoelectric effect, the energy of a photon ($h\nu$) absorbed by a metal is the sum of the work function (ϕ), the minimum energy needed to dislodge an electron from the metal's surface, and the kinetic energy (E_k) of the electron:

$hv = \phi + E_k$. When light of wavelength 358.1 nm falls on the surface of potassium metal, the speed (u) of the dislodged electron is 6.40×10^5 m/s. (a) What is E_k ($\frac{1}{2}mu^2$) of the dislodged electron? (b) What is ϕ (in J) of potassium?

•• **7.89** An electron microscope focuses electrons through magnetic lenses to observe objects at higher magnification than is possible with a light microscope. For any microscope, the smallest object that can be observed is one-half the wavelength of the radiation used. Thus, for example, the smallest object that can be observed with light of 400 nm is 2×10^{-7} m. (a) What is the smallest object observable with an electron microscope using electrons moving at 5.5×10^4 m/s? (b) At 3.0×10^7 m/s?

•• **7.90** In a typical fireworks device, the heat of the reaction between a strong oxidizing agent, such as $KClO_4$, and an organic compound excites certain salts, which emit specific colors. Strontium salts have an intense emission at 641 nm, and barium salts have one at 493 nm. (a) What colors do these emissions produce? (b) What is the energy (in kJ) of these emissions for 5.00 g each of the chloride salts of Sr and Ba? (Assume that all the heat released is converted to light emitted.)

••• **7.91** Atomic hydrogen produces well-known series of spectral lines in several regions of the electromagnetic spectrum. Each series fits the Rydberg equation with its own particular n_1 value. Calculate the value of n_1 (by trial and error if necessary) that would produce a series of lines in which:
(a) The *highest* energy line has a wavelength of 3282 nm.
(b) The *lowest* energy line has a wavelength of 7460 nm.

•• **7.92** Fish-liver oil is a good source of vitamin A, which is measured spectrophotometrically, after correcting for background, at a wavelength of 329 nm.
(a) Suggest a reason for using this wavelength.
(b) In what region of the spectrum does this wavelength lie?
(c) When 0.1232 g of fish-liver oil is dissolved in 500. mL of solvent, the absorbance is 0.724 units. When 1.67×10^{-3} g of vitamin A is dissolved in 250. mL of solvent, the absorbance is 1.018 units. Calculate the vitamin A concentration in the fish-liver oil.

•• **7.93** Many calculators use photocells to provide their energy. Find the maximum wavelength needed to remove an electron from silver ($\phi = 7.59\times10^{-19}$ J). Is silver a good choice for a photocell that uses visible light?

•• **7.94** The sodium salt of 2-quinizarinsulfonic acid forms a complex with Al^{3+} that absorbs strongly at 560 nm. (a) Use the data below to draw a plot of absorbance vs. concentration of a complex in solution and find the slope and y-intercept:

Concentration (M)	Absorbance (560 nm)
1.0×10^{-5}	0.131
1.5×10^{-5}	0.201
2.0×10^{-5}	0.265
2.5×10^{-5}	0.329
3.0×10^{-5}	0.396

(b) When 20.0 mL of this complex solution is diluted with water to 150. mL, its absorbance is 0.236. Find the concentrations of the diluted solution and of the original solution.

•• **7.95** In a game of "Clue," Ms. White is killed in the conservatory. You have a device in each room to help you find the murderer— a spectrometer that emits the entire visible spectrum to indicate who is in that room. For example, if someone wearing yellow is in a room, light at 580 nm is reflected. The suspects are Col. Mustard, Prof. Plum, Mr. Green, Ms. Peacock (blue), and Ms. Scarlet. At the time of the murder, the spectrometer in the dining room recorded a reflection at 520 nm, those in the lounge and study recorded reflections of lower frequencies, and the one in the library recorded a reflection of the shortest possible wavelength. Who killed Ms. White? Explain.

•• **7.96** Technetium (Tc; $Z = 43$) is a synthetic element used as a radioactive tracer in medical studies. A Tc atom emits a beta particle (electron) with a kinetic energy (E_k) of 4.71×10^{-15} J. What is the de Broglie wavelength of this electron ($E_k = \frac{1}{2}mv^2$)?

•• **7.97** Electric power is typically given in units of watts (1 W = 1 J/s). About 95% of the power output of an incandescent bulb is converted to heat and 5% to light. If 10% of that light shines on your chemistry text, how many photons per second shine on the book from a 75-W bulb? (Assume the photons have a wavelength of 550 nm.)

•• **7.98** The flame test for sodium is based on the intense yellow-orange emission at 589 nm; the test for potassium is based on its emission at 404 nm. When both elements are present, the Na^+ emission is so strong that the K^+ emission can't be seen, except by looking through a cobalt-glass filter. (a) What are the colors of these Na^+ and K^+ emissions? (b) What does the cobalt-glass filter do? (c) Why are the oxidizing agents in fireworks made of $KClO_4$ or $KClO_3$, rather than the corresponding sodium salts?

••• **7.99** The net change in the multistep biochemical process of photosynthesis is that CO_2 and H_2O form glucose ($C_6H_{12}O_6$) and O_2. Chlorophyll absorbs light in the 600 to 700 nm region. (a) Write a balanced thermochemical equation for formation of 1.00 mol of glucose. (b) What is the minimum number of photons with $\lambda = 680$. nm needed to prepare 1.00 mol of glucose?

••• **7.100** Only certain electron transitions are allowed from one energy level to another. In one-electron species, the change in the quantum number l of an allowed transition must be ±1. For example, a $3p$ electron can drop directly to a $2s$ orbital but not to a $2p$. Thus, in the UV series, where $n_{final} = 1$, allowed electron transitions can start in a p orbital ($l = 1$) of $n = 2$ or higher, not in an s ($l = 0$) or d ($l = 2$) orbital of $n = 2$ or higher. From what orbital do each of the allowed electron transitions start for the first four emission lines in the visible series ($n_{final} = 2$)?

•• **7.101** The discharge of phosphate compounds in detergents into the environment has led to serious imbalances in the natural life cycle of freshwater lakes. A chemist studying water pollution used a spectrophotometric method to measure total phosphate and obtained the following data for known standards:

Absorbance (880 nm)	Concentration (mol/L)
0	0.0
0.10	2.5×10^{-5}
0.16	3.2×10^{-5}
0.20	4.4×10^{-5}
0.25	5.6×10^{-5}
0.38	8.4×10^{-5}
0.48	10.5×10^{-5}
0.62	13.8×10^{-5}
0.76	17.0×10^{-5}
0.88	19.4×10^{-5}

(a) Draw a curve of absorbance vs. phosphate concentration.
(b) If a sample of lake water has an absorbance of 0.55, what is its phosphate concentration?

Biological: 7.92, 7.96, 7.99; Engineering: 7.93, 7.97; Environmental: 7.101

8

Patterns in Nature In the regularity of eclipses, the arrangement of seeds in the head of a sunflower, or, as shown here, the spiral of a *Nautilus* shell, nature exhibits periodic patterns. In this chapter, you'll see that the arrangement of electrons in atoms recurs periodically, too, which causes many properties of the elements to recur periodically and, thus, allows us to predict physical and chemical behavior.

Electron Configuration and Chemical Periodicity

8.1   Development of the Periodic Table

8.2   Characteristics of Many-Electron Atoms
The Electron-Spin Quantum Number
The Exclusion Principle
Electrostatic Effects and Energy-Level Splitting

8.3   The Quantum-Mechanical Model and the Periodic Table
Building Up Periods 1 and 2
Building Up Period 3
Electron Configurations Within Groups
Building Up Period 4
General Principles of Electron Configurations
Unusual Configurations: Transition and Inner Transition Elements

8.4   Trends in Three Key Atomic Properties
Trends in Atomic Size
Trends in Ionization Energy
Trends in Electron Affinity

8.5   Atomic Structure and Chemical Reactivity
Trends in Metallic Behavior
Properties of Monatomic Ions

I n Chapter 7, you saw the outpouring of scientific creativity by early 20th-century physicists that led to a new understanding of matter and energy, which in turn led to the quantum-mechanical model of the atom. But you can be sure that late 19th-century chemists were not sitting idly by, waiting for their colleagues in physics to develop that model. They were exploring the nature of electrolytes, establishing the kinetic-molecular theory, and developing chemical thermodynamics. The fields of organic chemistry and biochemistry were born, as were the fertilizer, explosives, glassmaking, soapmaking, bleaching, and dyestuff industries. And, for the first time, chemistry became a university subject in Europe and America. Superimposed on this activity was the accumulation of an enormous body of facts about the elements, which became organized into the periodic table.

The goal of this chapter is to show how the organization of the table, condensed from countless hours of laboratory work, was explained perfectly by the new quantum-mechanical atomic model. This model answers one of the central questions in chemistry: why do the elements behave as they do? Or, rephrasing the question to fit the main topic of this chapter: how does the **electron configuration** of an element—*the distribution of electrons within the orbitals of its atoms*—relate to its chemical and physical properties?

IN THIS CHAPTER... We first discuss the origin of the periodic table. Then we extend the quantum-mechanical model (Chapter 7) to many-electron atoms (those with more than one electron) to define a unique set of quantum numbers for each electron in the atoms of every element. Electrostatic effects lead to the order in which orbitals fill with electrons, and we'll see how that order correlates with the order of elements in the periodic table. We discuss how electron configuration and nuclear charge lead to periodic trends in atomic properties and how these trends account for the patterns of chemical reactivity. Finally, we apply these ideas to some properties of metals, nonmetals, and their ions.

Concepts & Skills to Review
before you study this chapter

- format of the periodic table (Section 2.6)
- characteristics of metals and nonmetals (Section 2.6)
- application of Coulomb's law to electrostatic attraction (Section 2.7)
- characteristics of acids and bases (Section 4.4)
- rules for assigning quantum numbers (Section 7.4)

8.1 DEVELOPMENT OF THE PERIODIC TABLE

An essential requirement for the amazing growth in theoretical and practical chemistry in the second half of the 19th century was the ability to organize the facts known about element behavior. The earliest organizing attempt was made by Johann Döbereiner, who placed groups of three elements with similar properties, such as calcium, strontium, and barium, into "triads." Later, John Newlands noted similarities between every eighth element (arranged by atomic mass), like the similarity between every eighth note in the musical scale, and placed elements into "octaves." As more elements were discovered, however, these early numerical schemes lost much of their validity.

In Chapter 2, you saw that the most successful organizing scheme was made by the Russian chemist Dmitri Mendeleev. In 1870, he arranged the 65 elements then known into a *periodic table* and summarized their behavior in the **periodic law:** when arranged by atomic mass, the elements exhibit a periodic recurrence of similar properties. ● It is a curious quirk of history that Mendeleev and the German chemist Julius Lothar Meyer arrived at virtually the same organization simultaneously, yet independently. Mendeleev focused on chemical properties and Meyer on physical properties. The greater credit has gone to Mendeleev because he was able to *predict* the properties of several as-yet-undiscovered elements, for which he had left blank spaces in his table. Table 8.1 (on the next page) compares the actual properties of germanium, which Mendeleev gave the provisional name "eka silicon" ("first under silicon"), with his predictions for it.

Today's periodic table, which appears on the inside front cover of the text, resembles Mendeleev's in most details, although it includes 51 elements that were unknown in 1870. The only substantive change is that the elements are now arranged in order of *atomic number* (number of protons) rather than atomic mass.

● **Mendeleev's Great Contribution** Born in a small Siberian town, Dmitri Ivanovich Mendeleev was the youngest of 17 children of the local schoolteacher. He showed early talent for mathematics and science, and his first research interests centered on the physical properties of gases and liquids. In developing his periodic table, he wrote a note card with the properties of each element and kept rearranging them until he realized that properties repeated when the elements were placed in order of increasing atomic mass.

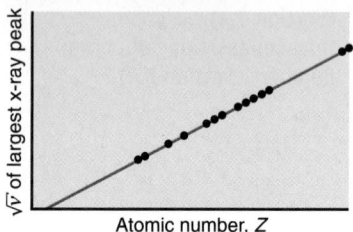

Atomic number, Z

● Moseley and Atomic Number When a metal is bombarded with high-energy electrons, an inner electron is knocked from the atom, an outer electron moves down to fill in the space, and x-rays are emitted. Bohr proposed that the x-ray spectrum of an element had wavelengths proportional to the nuclear charge. In 1913, Henry Moseley studied the x-ray spectra of a series of metals and correlated the largest peak in a metal's x-ray spectrum with its order in the periodic table (its *atomic number*). This correlation is known as *Moseley's law,* which may be expressed as $\sqrt{\nu} = K(Z - \sigma)$, where Z represents the atomic number, σ is 1 for electrons nearest the nucleus, ν is the frequency of the x-rays, and K is $5.0 \times 10^7 \ s^{-1}$. Moseley showed that the nuclear charge increased by 1 for each element (see the graph). Among other results, the findings confirmed the placement of Co ($Z = 27$) *before* Ni ($Z = 28$), despite cobalt's higher atomic mass, and also confirmed that the gap between Cl ($Z = 17$) and K ($Z = 19$) is the place for Ar ($Z = 18$). Tragically, Moseley died in 1915 at the age of 26 while serving as a pilot in the British army during World War I.

ARIS Presentation Center, Photo Library

Table 8.1	Mendeleev's Predicted Properties of Germanium ("eka Silicon") and Its Actual Properties	
Property	Predicted Properties of eka Silicon (E)	Actual Properties of Germanium (Ge)
Atomic mass	72 amu	72.61 amu
Appearance	Gray metal	Gray metal
Density	5.5 g/cm^3	5.32 g/cm^3
Molar volume	13 cm^3/mol	13.65 cm^3/mol
Specific heat capacity	0.31 J/g·K	0.32 J/g·K
Oxide formula	EO$_2$	GeO$_2$
Oxide density	4.7 g/cm^3	4.23 g/cm^3
Sulfide formula and solubility	ES$_2$; insoluble in H$_2$O; soluble in aqueous (NH$_4$)$_2$S	GeS$_2$; insoluble in H$_2$O; soluble in aqueous (NH$_4$)$_2$S
Chloride formula (boiling point)	ECl$_4$ ($<100°C$)	GeCl$_4$ ($84°C$)
Chloride density	1.9 g/cm^3	1.844 g/cm^3
Element preparation	Reduction of K$_2$EF$_6$ with sodium	Reduction of K$_2$GeF$_6$ with sodium

ARIS Presentation Center, Tables Library

This change was based on the work of the British physicist Henry G. J. Moseley, who found a direct dependence between an element's nuclear charge and its position in the periodic table. ●

Section Summary

The periodic law gave rise to the periodic table, which arranges the elements by atomic number into rows and columns that display recurring properties.

8.2 CHARACTERISTICS OF MANY-ELECTRON ATOMS

Like the Bohr model, the Schrödinger equation does not give *exact* solutions for many-electron atoms. However, unlike the Bohr model, the Schrödinger equation gives very good *approximate* solutions. These solutions show that the atomic orbitals of many-electron atoms resemble those of the H atom, which means we can use the same quantum numbers that we used for the H atom to describe the orbitals of other atoms. Nevertheless, the existence of more than one electron in an atom requires us to consider three features that were not relevant in the case of hydrogen: (1) the need for a fourth quantum number, (2) a limit on the number of electrons allowed in a given orbital, and (3) a more complex set of orbital energy levels. Let's examine these new features and then go on to determine the electron configuration for each element.

The Electron-Spin Quantum Number

Recall from Chapter 7 that the three quantum numbers n, l, and m_l describe the size (energy), shape, and orientation, respectively, of an atomic orbital. However, an additional quantum number is needed to describe a property of the electron itself, called *spin,* which is not a property of the orbital. Electron spin becomes important when more than one electron is present.

When a beam of H atoms passes through a nonuniform magnetic field, as shown in Figure 8.1, it splits into two beams that bend away from each other. The explanation of the split beam is that an electron generates a tiny magnetic field, as though it were a spinning charge. The single electron in each H atom can have one of two possible values of *spin,* each of which generates a tiny magnetic field. These two fields have opposing directions, so half of the electrons are

Figure 8.1 **Observing the effect of electron spin.** A nonuniform magnetic field, created by magnet faces with different shapes, splits a beam of hydrogen atoms in two. The split beam results from the two possible values of electron spin within each atom.

ARIS Presentation Center, Art Library

attracted into the large external magnetic field and the other half are repelled by it. As a result, the beam of H atoms splits.

Like its charge, spin is an intrinsic property of the electron, and the **spin quantum number** (m_s) has values of either $+\frac{1}{2}$ or $-\frac{1}{2}$. Thus, *each electron in an atom is described completely by a set of **four** quantum numbers: the first three describe its orbital, and the fourth describes its spin.* The quantum numbers are summarized in Table 8.2.

Transparency/ARIS Presentation Center, Tables Library

Table 8.2 **Summary of Quantum Numbers of Electrons in Atoms**

Name	Symbol	Permitted Values	Property
Principal	n	Positive integers (1, 2, 3, ...)	Orbital energy (size)
Angular momentum	l	Integers from 0 to $n-1$	Orbital shape (The l values 0, 1, 2, and 3 correspond to s, p, d, and f orbitals, respectively.)
Magnetic	m_l	Integers from $-l$ to 0 to $+l$	Orbital orientation
Spin	m_s	$+\frac{1}{2}$ or $-\frac{1}{2}$	Direction of e^- spin

Now we can write a set of four quantum numbers for any electron in the ground state of any atom. For example, the set of quantum numbers for the lone electron in hydrogen (H; $Z = 1$) is $n = 1$, $l = 0$, $m_l = 0$, and $m_s = +\frac{1}{2}$. (The spin quantum number for this electron could just as well have been $-\frac{1}{2}$, but by convention, we assign $+\frac{1}{2}$ for the first electron in an orbital.)

The Exclusion Principle

The element after hydrogen is helium (He; $Z = 2$), the first with atoms having more than one electron. The first electron in the He ground state has the same set of quantum numbers as the electron in the H atom, but the second He electron does not. Based on observations of the excited states of atoms, the Austrian physicist Wolfgang Pauli formulated the **exclusion principle**: *no two electrons in the same atom can have the same four quantum numbers.* That is, each electron must have a unique "identity" as expressed by its set of quantum numbers. Therefore, the second He electron occupies the same orbital as the first but has an opposite spin: $n = 1$, $l = 0$, $m_l = 0$, and $m_s = -\frac{1}{2}$. ●

Because the spin quantum number (m_s) can have only two values, the major consequence of the exclusion principle is that *an atomic orbital can hold a maximum of two electrons and they must have opposing spins.* We say that the 1s orbital in He is *filled* and that the electrons have *paired spins.* Thus, a beam of He atoms is not split in an experiment like that in Figure 8.1.

● **Baseball Quantum Numbers** The unique set of quantum numbers that describes an electron is analogous to the unique location of a box seat at a baseball game. The stadium (atom) is divided into section (n, level), box (l, sublevel), row (m_l, orbital), and seat (m_s, spin). Only one person (electron) can have this particular set of stadium "quantum numbers."

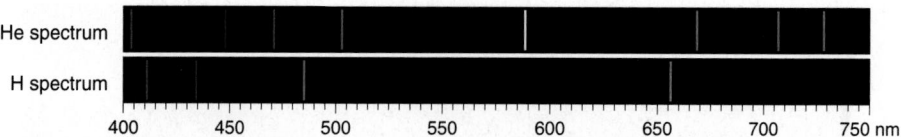

Figure 8.2 Spectral evidence of energy-level splitting in many-electron atoms. More spectral lines for He than for H indicates that an He atom has more available orbital energies, which is consistent with energy levels splitting into sublevels.

ARIS Presentation Center, Art Library

Transparency/ARIS Presentation Center, Active Art Library

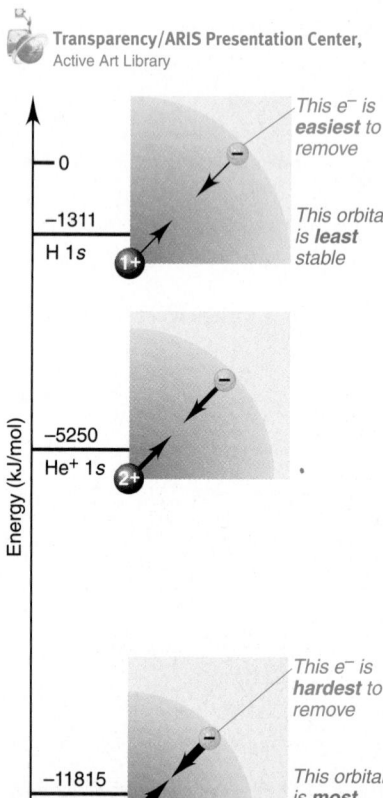

Figure 8.3 The effect of nuclear charge on orbital energy. Greater nuclear charge lowers orbital energy (more negative number), which makes the electron harder to remove. (Recall that an atom's energy is *defined* as negative.) In these one-electron species, the absolute value of the 1s orbital energy is related directly to Z^2: the value for He$^+$ ($Z = 2$) is about 4 times that for H ($Z = 1$), and the value for Li^{2+} ($Z = 3$) is about 9 times that for H.

Electrostatic Effects and Energy-Level Splitting

Electrostatic effects play a major role in determining the energy states of many-electron atoms. Recall that the energy state of the H atom is determined *only* by the *n* value of the occupied orbital. In other words, in the H atom, all sublevels of a given level, such as the 2s and 2p, have the same energy. The reason is that the only electrostatic interaction is the attraction between nucleus and electron. On the other hand, the energy states of many-electron atoms arise not only from nucleus-electron attractions, but also electron-electron repulsions. One major consequence of these additional interactions is *the splitting of energy levels into sublevels of differing energies: the energy of an orbital in a many-electron atom depends mostly on its n value (size) and to a lesser extent on its l value (shape).*

Thus, a more complex set of energy states exists for a many-electron atom than we saw in the H atom, and evidence for this appears in the more complex line spectra of many-electron atoms. Figure 8.2 shows that even though helium has only one more electron than hydrogen, it displays many more spectral lines, indicating more available orbital energies in its excited states.

Our first encounter with energy-level splitting in ground-state configurations occurs with lithium (Li; $Z = 3$). By definition, *the electrons of an atom in its ground state occupy the orbitals of lowest energy,* so the first two electrons in the ground state of Li fill its 1s orbital. Then, the third Li electron must go into the $n = 2$ level. But, this level has 2s *and* 2p sublevels: which has lower energy, that is, which does Li's third electron enter? As you'll see, the 2s is lower in energy than the 2p. The reasons for this energy difference are based on three factors—*nuclear charge, electron repulsions,* and *orbital shape* (more specifically, radial probability distribution). Their interplay leads to the phenomena of *shielding* and *penetration*, which occur in all the atoms in the periodic table—except hydrogen.

The Effect of Nuclear Charge (Z) on Orbital Energy Nuclear protons create an ever-present pull on the electrons. You know that higher charges attract each other more strongly than lower charges (Coulomb's law, Section 2.7). Therefore, *higher nuclear charge lowers orbital energy (stabilizes the system) by increasing nucleus-electron attractions.* We can see this effect clearly by comparing the 1s orbital energies of three species with one electron—the H atom ($Z = 1$), He$^+$ ion ($Z = 2$), and Li^{2+} ion ($Z = 3$). As Figure 8.3 shows, the H 1s orbital is the least stable (highest energy), so the least energy is needed to remove its electron; and the Li^{2+} 1s orbital is the most stable, so the most energy is needed to remove its electron.

Shielding: The Effect of Electron Repulsions on Orbital Energy In many-electron atoms, each electron "feels" not only the attraction to the nucleus but also the repulsion from other electrons. This repulsion counteracts the nuclear attraction somewhat, making each electron easier to remove by, in effect, helping to push it away. We speak of each electron "shielding" the other electrons somewhat from the nucleus. **Shielding** (also called *screening*) reduces the full nuclear charge to an **effective nuclear charge (Z_{eff}),** the nuclear charge an electron *actually experiences. This lower nuclear charge makes the electron easier to remove.* We see the effect of shielding by electrons in the *same* orbital when we compare the He atom and He$^+$ ion: both have a 2+ nuclear charge, but He has two electrons in the 1s orbital and He$^+$ has only one (Figure 8.4A). It takes less than half as much energy to remove an electron from He (2372 kJ/mol) than from He$^+$ (5250 kJ/mol) because the second electron in He repels the first, in effect shielding the first electron from the full nuclear charge (lowering Z_{eff}).

Much greater shielding is provided by inner electrons. Because they spend nearly all their time *between* the outer electrons and the nucleus, inner electrons shield outer electrons very effectively, in fact, *much more effectively* than do electrons in the same sublevel. We can see this by comparing two atomic systems with the same nucleus, one *with* inner electrons and the other *without*. The ground-state Li atom has two inner (1s) electrons and one outer (2s) electron, while the Li^{2+} ion has only one electron, which occupies the 2s orbital in the first excited state (Figure 8.4B). It takes about one-sixth as much energy to remove the 2s electron from the Li atom (520 kJ/mol) as from the Li^{2+} ion (2954 kJ/mol), because the inner electrons shield very effectively. *Shielding by inner electrons greatly lowers the Z_{eff} felt by outer electrons.*

Penetration: The Effect of Orbital Shape on Orbital Energy To return to the question of why the third electron occupies the 2s orbital in the Li ground state, rather than the 2p, we have to consider orbital shapes, that is, radial probability distributions (Figure 8.5). At first, we might expect that the electron would enter the 2p orbital (*orange curve*) because it is slightly closer to the nucleus, on average, than the major portion of the 2s orbital (*blue curve*). But note that a minor portion of the 2s radial probability distribution appears within the 1s region. As a result, an electron in the 2s orbital spends part of its time "penetrating" very close to the nucleus. Charges attract more strongly if they are near each other than far apart (Coulomb's law, Section 2.7). Therefore, **penetration** by the 2s electron increases its overall attraction to the nucleus relative to that for a 2p electron. At the same time, penetration into the 1s region decreases the shielding of the 2s electron by the 1s electrons. Evidence shows that, indeed, the 2s orbital of Li *is* lower in energy than the 2p orbital, because it takes more energy to remove a 2s electron (520 kJ/mol) than a 2p (341 kJ/mol).

In general, *penetration and the resulting effects on shielding cause an energy level to split into sublevels of differing energy.* The lower the l value of an orbital, the more its electrons penetrate, and so the greater their attraction to the nucleus. Therefore, *for a given n value, the lower the l value, the lower the sublevel energy:*

$$\text{Order of sublevel energies: } s < p < d < f \tag{8.1}$$

Thus, the 2s ($l = 0$) is lower in energy than the 2p ($l = 1$), the 3p ($l = 1$) is lower than the 3d ($l = 2$), and so forth.

Figure 8.5 Penetration and orbital energy. Radial probability distributions show that a 2s electron spends most of its time slightly farther from the nucleus than does a 2p electron but penetrates near the nucleus for a small part of the time. Penetration by the 2s electron increases its overall attraction to the nucleus; thus the 2s orbital is more stable (lower in energy) than the 2p.

Energy, *E*

Figure 8.6 **Order for filling energy sublevels with electrons.** In many-electron atoms, energy levels split into sublevels. The relative energies of the sublevels increase with the principal quantum number *n* (1 < 2 < 3, etc.) and the angular momentum quantum number *l* (*s* < *p* < *d* < *f*). As *n* increases, the energies become closer together. The penetration effect, together with this narrowing of energy differences, results in the overlap of some sublevels; for example, the 4*s* sublevel is slightly lower in energy than the 3*d*, so it is filled first. (Line color is by sublevel type; line lengths differ for ease in labeling.)

 Transparency/ARIS Presentation Center, Art Library

Literature Rioux, Frank. "Hund's Multiplicity Rule Revisited," *J. Chem. Educ.* **2007**, *84*, 358.

Literature Amoa, Kwesi. "Count the Boxes—A Simple Guide to Determining Electronic Configuration," *Chem. Educator* 11 **2006**, *1*, 1–3.

Figure 8.6 shows the general energy order of levels (*n* value) and how they are split into sublevels (*l* values) of differing energies. (Compare this with the H atom energy levels in Figure 7.21, p. 295.) Next, we use this energy order to construct a periodic table of ground-state atoms.

Section Summary

Identifying electrons in many-electron atoms requires four quantum numbers: three (*n*, *l*, m_l) describe the orbital, and a fourth (m_s) describes electron spin. • The Pauli exclusion principle requires each electron to have a unique set of four quantum numbers; therefore, an orbital can hold no more than two electrons, and their spins must be paired (opposite). • Electrostatic interactions determine orbital energies as follows:
1. Greater nuclear charge lowers orbital energy and makes electrons harder to remove.
2. Electron-electron repulsions raise orbital energy and make electrons easier to remove. Repulsions have the effect of *shielding* electrons from the full nuclear charge, reducing it to an effective nuclear charge, Z_{eff}. Inner electrons shield outer electrons most effectively.
3. Greater radial probability distribution near the nucleus (greater *penetration*) makes an electron harder to remove because it is attracted more strongly and shielded less effectively. As a result, an energy level (shell) is split into sublevels (subshells) with the energy order *s* < *p* < *d* < *f*.

8.3 THE QUANTUM-MECHANICAL MODEL AND THE PERIODIC TABLE

Quantum mechanics provides the theoretical foundation for the experimentally based periodic table. In this section, we fill the table with elements and determine their electron configurations—the distributions of electrons within their atoms' orbitals. Note especially the *recurring pattern in electron configurations, which is the basis for the recurring pattern in chemical behavior.*

Building Up Periods 1 and 2

A useful way to determine the electron configurations of the elements is to start at the beginning of the periodic table and add one electron per element to the *lowest energy orbital available.* (Of course, one proton and one or more neutrons are also added to the nucleus.) This approach is based on the **aufbau principle** (German *aufbauen,* "to build up"), and it results in *ground-state* electron configurations. Let's assign sets of quantum numbers to the electrons in the ground state of the first 10 elements, those in the first two periods (horizontal rows).

For the electron in H, as you've seen, the set of quantum numbers is

$$\text{H } (Z = 1): n = 1, l = 0, m_l = 0, m_s = +\tfrac{1}{2}$$

You also saw that the first electron in He has the same set as the electron in H, but the second He electron has opposing spin (exclusion principle):

$$\text{He } (Z = 2): n = 1, l = 0, m_l = 0, m_s = -\tfrac{1}{2}$$

(As we go through each element in this discussion, the quantum numbers that follow refer to the element's *last added* electron.)

Here are two common ways to designate the orbital and its electrons:

1. *The electron configuration.* This shorthand notation consists of the principal energy level (*n* value), the letter designation of the sublevel (*l* value), and the number of electrons (#) in the sublevel, written as a superscript: $nl^{\#}$. The electron configuration of H is $1s^1$ (spoken "one-ess-one"); that of He is $1s^2$ (spoken "one-ess-two," *not* "one-ess-squared"). This notation does *not* indicate electron spin but assumes you know that the two 1*s* electrons have paired (opposite) spins.

2. *The orbital diagram.* An **orbital diagram** consists of a box (or circle, or just a line) for each orbital in a given energy level, grouped by sublevel, with an

arrow indicating an electron *and* its spin. (Traditionally, ↑ is $+\frac{1}{2}$ and ↓ is $-\frac{1}{2}$, but these are arbitrary; it is necessary only to be consistent. Throughout the text, orbital occupancy is also indicated by color intensity: an orbital with no color is empty, pale color means half-filled, and full color means filled.) The electron configurations and orbital diagrams for the first two elements are

$$\text{H } (Z = 1) \;\; 1s^1 \quad \boxed{\uparrow} \qquad\qquad \text{He } (Z = 2) \;\; 1s^2 \quad \boxed{\uparrow\downarrow}$$
$$\qquad\qquad\qquad\qquad 1s \qquad\qquad\qquad\qquad\qquad\qquad\quad 1s$$

The exclusion principle tells us that an orbital can hold only two electrons, so the $1s$ orbital in He is filled, and the $n = 1$ level is also filled. The $n = 2$ level is filled next, beginning with the $2s$ orbital, the next lowest in energy. As we said earlier, the first two electrons in Li fill the $1s$ orbital, and the last added Li electron has quantum numbers $n = 2$, $l = 0$, $m_l = 0$, $m_s = +\frac{1}{2}$. The electron configuration for Li is $1s^2 2s^1$. Note that the orbital diagram shows all the orbitals for $n = 2$, whether or not they are occupied:

$$\xrightarrow{\qquad\qquad \textbf{Energy, } \textit{E} \qquad\qquad}$$

$$\text{Li } (Z = 3) \;\; 1s^2 2s^1 \quad \boxed{\uparrow\downarrow} \quad \boxed{\uparrow} \quad \boxed{||}$$
$$\qquad\qquad\qquad\qquad\qquad\quad 1s \qquad 2s \qquad\quad 2p$$

To save space on a page, orbital diagrams are often written horizontally, but note that *the energy of the sublevels increases from left to right*. Figure 8.7 emphasizes this point by arranging the orbital diagram of lithium vertically.

With the $2s$ orbital only half-filled in Li, the fourth electron of beryllium fills it with the electron's spin paired: $n = 2$, $l = 0$, $m_l = 0$, $m_s = -\frac{1}{2}$.

$$\text{Be } (Z = 4) \;\; 1s^2 2s^2 \quad \boxed{\uparrow\downarrow} \quad \boxed{\uparrow\downarrow} \quad \boxed{||}$$
$$\qquad\qquad\qquad\qquad\qquad\quad 1s \qquad 2s \qquad\quad 2p$$

The next lowest energy sublevel is the $2p$. A p sublevel has $l = 1$, so the m_l (orientation) values can be -1, 0, or $+1$. The three orbitals in the $2p$ sublevel have *equal energy* (same n and l values), which means that the fifth electron of boron can go into *any one of the $2p$ orbitals*. For convenience, let's label the boxes from left to right, -1, 0, $+1$. By convention, we place the electron in the $m_l = -1$ orbital: $n = 2$, $l = 1$, $m_l = -1$, $m_s = +\frac{1}{2}$.

$$\qquad\qquad\qquad\qquad\qquad\qquad\qquad\qquad\quad -1 \;\; 0 \;\; +1$$
$$\text{B } (Z = 5) \;\; 1s^2 2s^2 2p^1 \quad \boxed{\uparrow\downarrow} \quad \boxed{\uparrow\downarrow} \quad \boxed{\uparrow||}$$
$$\qquad\qquad\qquad\qquad\qquad\qquad 1s \qquad 2s \qquad\quad 2p$$

To minimize electron-electron repulsions, the last added (sixth) electron of carbon enters one of the *unoccupied* $2p$ orbitals; by convention, we place it in the $m_l = 0$ orbital. Experiment shows that the spin of this electron is *parallel* to (the same as) the spin of the other $2p$ electron: $n = 2$, $l = 1$, $m_l = 0$, $m_s = +\frac{1}{2}$.

$$\text{C } (Z = 6) \;\; 1s^2 2s^2 2p^2 \quad \boxed{\uparrow\downarrow} \quad \boxed{\uparrow\downarrow} \quad \boxed{\uparrow|\uparrow|}$$
$$\qquad\qquad\qquad\qquad\qquad\qquad 1s \qquad 2s \qquad\quad 2p$$

This placement of electrons for carbon exemplifies **Hund's rule:** *when orbitals of equal energy are available, the electron configuration of lowest energy has the maximum number of unpaired electrons with parallel spins.* Based on Hund's rule, nitrogen's seventh electron enters the last empty $2p$ orbital, with its spin parallel to the two other $2p$ electrons: $n = 2$, $l = 1$, $m_l = +1$, $m_s = +\frac{1}{2}$.

$$\text{N } (Z = 7) \;\; 1s^2 2s^2 2p^3 \quad \boxed{\uparrow\downarrow} \quad \boxed{\uparrow\downarrow} \quad \boxed{\uparrow|\uparrow|\uparrow}$$
$$\qquad\qquad\qquad\qquad\qquad\qquad 1s \qquad 2s \qquad\quad 2p$$

The eighth electron in oxygen must enter one of these three half-filled $2p$ orbitals and "pair up" with (have opposing spin to) the electron already present.

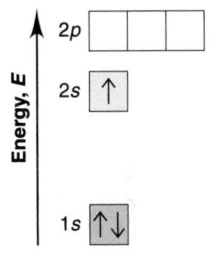

Figure 8.7 A vertical orbital diagram for the Li ground state. Although orbital diagrams are usually written horizontally, this vertical arrangement emphasizes that sublevel energy increases with increasing n and l.

ARIS Presentation Center, Art Library

Literature Garofalo, Anthony. "Housing Electrons: Relating Quantum Numbers, Energy Levels, and Electron Configuration," *J. Chem. Educ.* **1997**, *74*, 709.

Since the $2p$ orbitals all have the same energy, we proceed as before and place the electron in the orbital previously designated $m_l = -1$. The quantum numbers are $n = 2$, $l = 1$, $m_l = -1$, $m_s = -\frac{1}{2}$.

O ($Z = 8$) $1s^2 2s^2 2p^4$

$\begin{array}{ccccc} \boxed{\uparrow\downarrow} & \boxed{\uparrow\downarrow} & \boxed{\uparrow\downarrow\;\uparrow\;\uparrow} \\ 1s & 2s & 2p \end{array}$

Fluorine's ninth electron enters either of the two remaining half-filled $2p$ orbitals: $n = 2$, $l = 1$, $m_l = 0$, $m_s = -\frac{1}{2}$.

F ($Z = 9$) $1s^2 2s^2 2p^5$

$\begin{array}{ccccc} \boxed{\uparrow\downarrow} & \boxed{\uparrow\downarrow} & \boxed{\uparrow\downarrow\;\uparrow\downarrow\;\uparrow} \\ 1s & 2s & 2p \end{array}$

Only one unfilled orbital remains in the $2p$ sublevel, so the tenth electron of neon occupies it: $n = 2$, $l = 1$, $m_l = +1$, $m_s = -\frac{1}{2}$. With neon, the $n = 2$ level is filled.

Ne ($Z = 10$) $1s^2 2s^2 2p^6$

$\begin{array}{ccccc} \boxed{\uparrow\downarrow} & \boxed{\uparrow\downarrow} & \boxed{\uparrow\downarrow\;\uparrow\downarrow\;\uparrow\downarrow} \\ 1s & 2s & 2p \end{array}$

SAMPLE PROBLEM 8.1   Determining Quantum Numbers from Orbital Diagrams

PROBLEM Write a set of quantum numbers for the third electron and a set for the eighth electron of the F atom.

PLAN Referring to the orbital diagram, we count to the electron of interest and note its level (n), sublevel (l), orbital (m_l), and spin (m_s).

SOLUTION The third electron is in the $2s$ orbital. The upward arrow indicates a spin of $+\frac{1}{2}$:

$$n = 2, l = 0, m_l = 0, m_s = +\frac{1}{2}$$

The eighth electron is in the first $2p$ orbital, which is designated $m_l = -1$, and has a downward arrow:

$$n = 2, l = 1, m_l = -1, m_s = -\frac{1}{2}$$

FOLLOW-UP PROBLEM 8.1 Use the periodic table to identify the element with the electron configuration $1s^2 2s^2 2p^4$. Write its orbital diagram, and give the quantum numbers of its sixth electron.

Literature Fortman, John J. "Pictorial Analogies VII: Quantum Numbers and Orbitals," *J. Chem. Educ.* **1993**, *70*, 649.

With so much attention paid to these notations, it's easy to forget that atoms are real spherical objects and that the electrons occupy volumes with specific shapes and orientations. Figure 8.8 shows ground-state electron configurations and orbital contours for the first 10 elements arranged in periodic table format.

Even at this early stage of filling the table, we can make an important correlation between chemical behavior and electron configuration: elements in the same group have similar outer electron configurations. As an example, helium (He) and neon (Ne) in Group 8A(18) both have filled outer sublevels—$1s^2$ for helium and $2p^6$ for neon—and neither element forms compounds. As we'll see often, *filled outer sublevels make elements unreactive*.

Building Up Period 3

The Period 3 elements, sodium through argon, lie directly under the Period 2 elements, lithium through neon. The sublevels of the $n = 3$ level are filled in the order $3s$, $3p$, $3d$. Table 8.3 presents *partial* orbital diagrams ($3s$ and $3p$ sublevels only) and electron configurations for the eight elements in Period 3 (with *filled inner levels* in brackets and the sublevel to which the last electron is added in colored type). Note the *group similarities in outer electron configuration* with the elements in Period 2 (refer to Figure 8.9, p. 312).

Figure 8.8 Orbital occupancy for the first 10 elements, H through Ne. The first 10 elements are arranged in periodic table format with each box showing atomic number, atomic symbol, ground-state electron configuration, and a depiction of the atom based on the probability contours of its orbitals. Orbital occupancy is indicated with shading: lighter color for half-filled (one e⁻) orbitals, and darker color for filled (two e⁻) orbitals. For clarity, only the outer region of the 2s orbital is included.

1A(1)							8A(18)
Period 1 — 1 **H** $1s^1$							2 **He** $1s^2$
	2A(2)	3A(13)	4A(14)	5A(15)	6A(16)	7A(17)	
Period 2 — 3 **Li** $1s^2 2s^1$	4 **Be** $1s^2 2s^2$	5 **B** $1s^2 2s^2 2p^1$	6 **C** $1s^2 2s^2 2p^2$	7 **N** $1s^2 2s^2 2p^3$	8 **O** $1s^2 2s^2 2p^4$	9 **F** $1s^2 2s^2 2p^5$	10 **Ne** $1s^2 2s^2 2p^6$

In sodium (the second alkali metal) and magnesium (the second alkaline earth metal), electrons are added to the 3s sublevel, which contains the 3s orbital only, just as they filled the 2s sublevel in lithium and beryllium in Period 2. Then, just as for boron, carbon, and nitrogen in Period 2, the last electrons added to aluminum, silicon, and phosphorus in Period 3 half-fill the three 3p orbitals with spins parallel (Hund's rule). The last electrons added to sulfur, chlorine, and argon then successively enter the three half-filled 3p orbitals, thereby filling the 3p sublevel. With argon, the next noble gas after helium and neon, we arrive at the end of Period 3. (As you'll see shortly, the 3d orbitals are filled in Period 4.)

The rightmost column of Table 8.3 shows the *condensed electron configuration*. In this simplified notation, the electron configuration of the previous noble gas is shown by its element symbol in brackets, and it is followed by the electron configuration of the energy level being filled. The condensed electron configuration of sulfur, for example, is [Ne] $3s^2 3p^4$, where [Ne] stands for $1s^2 2s^2 2p^6$.

Table 8.3 Partial Orbital Diagrams and Electron Configurations* for the Elements in Period 3

Atomic Number	Element	Partial Orbital Diagram (3s and 3p Sublevels Only)	Full Electron Configuration	Condensed Electron Configuration
11	Na	3s ↑ 3p □ □ □	$[1s^2 2s^2 2p^6]\, 3s^1$	[Ne] $3s^1$
12	Mg	↑↓ □ □ □	$[1s^2 2s^2 2p^6]\, 3s^2$	[Ne] $3s^2$
13	Al	↑↓ ↑ □ □	$[1s^2 2s^2 2p^6]\, 3s^2 3p^1$	[Ne] $3s^2 3p^1$
14	Si	↑↓ ↑ ↑ □	$[1s^2 2s^2 2p^6]\, 3s^2 3p^2$	[Ne] $3s^2 3p^2$
15	P	↑↓ ↑ ↑ ↑	$[1s^2 2s^2 2p^6]\, 3s^2 3p^3$	[Ne] $3s^2 3p^3$
16	S	↑↓ ↑↓ ↑ ↑	$[1s^2 2s^2 2p^6]\, 3s^2 3p^4$	[Ne] $3s^2 3p^4$
17	Cl	↑↓ ↑↓ ↑↓ ↑	$[1s^2 2s^2 2p^6]\, 3s^2 3p^5$	[Ne] $3s^2 3p^5$
18	Ar	↑↓ ↑↓ ↑↓ ↑↓	$[1s^2 2s^2 2p^6]\, 3s^2 3p^6$	[Ne] $3s^2 3p^6$

*Colored type indicates the sublevel to which the last electron is added.

		1A (1)								8A (18)
	1	1 **H** $1s^1$	2A (2)	3A (13)	4A (14)	5A (15)	6A (16)	7A (17)	2 **He** $1s^2$
Period	2	3 **Li** [He] $2s^1$	4 **Be** [He] $2s^2$	5 **B** [He] $2s^2 2p^1$	6 **C** [He] $2s^2 2p^2$	7 **N** [He] $2s^2 2p^3$	8 **O** [He] $2s^2 2p^4$	9 **F** [He] $2s^2 2p^5$	10 **Ne** [He] $2s^2 2p^6$
	3	11 **Na** [Ne] $3s^1$	12 **Mg** [Ne] $3s^2$	13 **Al** [Ne] $3s^2 3p^1$	14 **Si** [Ne] $3s^2 3p^2$	15 **P** [Ne] $3s^2 3p^3$	16 **S** [Ne] $3s^2 3p^4$	17 **Cl** [Ne] $3s^2 3p^5$	18 **Ar** [Ne] $3s^2 3p^6$

Figure 8.9 Condensed ground-state electron configurations in the first three periods. The first 18 elements, H through Ar, are arranged in three periods containing two, eight, and eight elements. Each box shows the atomic number, atomic symbol, and condensed ground-state electron configuration. Note that elements in a group have similar outer electron configurations *(color)*.

A

B

Figure 8.10 Similar reactivities within a group. A, Potassium metal and water reacting. All alkali metals [Group 1A(1)] react vigorously with water and displace H_2. **B,** Chlorine and potassium metal reacting. All halogens [Group 7A(17)] react with metals to form ionic halides.

Electron Configurations Within Groups

One of the central points in all chemistry is that *similar outer electron configurations correlate with similar chemical behavior.* Figure 8.9 shows the condensed electron configurations of the first 18 elements. Note the similarities within each group. Here are some examples from just three groups:

- In Group 1A(1), lithium and sodium have the outer electron configuration ns^1 (where n is the quantum number of the outermost energy level), as do all the other alkali metals (K, Rb, Cs, Fr). All are highly reactive metals that form ionic compounds with nonmetals with formulas such as MCl, M_2O, and M_2S (where M represents the alkali metal), and all react vigorously with water to displace H_2 (Figure 8.10A).
- In Group 7A(17), fluorine and chlorine have the outer electron configuration ns^2np^5, as do the other halogens (Br, I, At). Little is known about rare, radioactive astatine (At), but all the others are reactive nonmetals that occur as diatomic molecules, X_2 (where X represents the halogen). All form ionic compounds with metals (KX, MgX_2) (Figure 8.10B), covalent compounds with hydrogen (HX) that yield acidic solutions in water, and covalent compounds with carbon (CX_4).
- In Group 8A(18), helium has the electron configuration ns^2 and all the others have the outer configuration ns^2np^6. Consistent with their *filled* energy levels, all group members are very unreactive monatomic gases.

To summarize the major connection between quantum mechanics and chemical periodicity: *orbitals are filled in order of increasing energy, which leads to outer electron configurations that recur periodically, which leads to chemical properties that recur periodically.*

The First *d*-Orbital Transition Series: Building Up Period 4

The 3*d* orbitals are filled in Period 4. Note, however, that *the 4s orbital is filled before the 3d.* This switch in filling order is due to the shielding and penetration effects that we discussed in Section 8.2. The radial probability distribution of the 3*d* orbital is greater outside the filled, inner $n = 1$ and $n = 2$ levels, so a 3*d* electron is shielded very effectively from the nuclear charge. In contrast, penetration by the 4*s* electron means that it spends a significant part of its time near the nucleus and feels a greater nuclear attraction. Thus, the 4*s* orbital is slightly *lower* in energy than the 3*d*, and so fills first. Similarly, the 5*s* orbital fills before the 4*d*, and the 6*s* fills before the 5*d*. In general, *the ns sublevel fills before the (n − 1)d sublevel.* As we proceed through the transition series, however, you'll see several exceptions to this pattern because the energies of the ns and $(n-1)d$ sublevels become extremely close at higher values of n.

Table 8.4 shows the partial orbital diagrams and ground-state electron configurations for the 18 elements in Period 4 (again with filled inner levels in brackets and the sublevel to which the last electron has been added in colored type). The first two elements of the period, potassium and calcium, are the next alkali and alkaline earth metals, respectively, and their electrons fill the $4s$ sublevel. The third element, scandium ($Z = 21$), is the first of the **transition elements,** those in which d orbitals are being filled. The last electron in scandium occupies any one of the five $3d$ orbitals because they are equal in energy. Scandium has the electron configuration [Ar] $4s^2 3d^1$.

The filling of $3d$ orbitals proceeds one at a time, as with p orbitals, except in two cases: chromium ($Z = 24$) and copper ($Z = 29$). Vanadium ($Z = 23$), the element before chromium, has three half-filled d orbitals ([Ar] $4s^2 3d^3$). Rather than having its last electron enter a fourth empty d orbital to give [Ar] $4s^2 3d^4$, chromium has one electron in the $4s$ sublevel and five in the $3d$ sublevel. Thus, both the $4s$ and the $3d$ sublevels are *half-filled* (see margin). In the next element, manganese ($Z = 25$), the $4s$ sublevel is filled again ([Ar] $4s^2 3d^5$).

Cr ($Z = 24$) [Ar] $4s^1 3d^5$

4s	3d	4p
↑	↑ ↑ ↑ ↑ ↑	☐ ☐ ☐

Table 8.4 Partial Orbital Diagrams and Electron Configurations* for the Elements in Period 4

Atomic Number	Element	Partial Orbital Diagram (4s, 3d, and 4p Sublevels Only) — 4s	3d	4p	Full Electron Configuration	Condensed Electron Configuration
19	K	↑	☐☐☐☐☐	☐☐☐	$[1s^2 2s^2 2p^6 3s^2 3p^6]\ 4s^1$	[Ar] $4s^1$
20	Ca	↑↓	☐☐☐☐☐	☐☐☐	$[1s^2 2s^2 2p^6 3s^2 3p^6]\ 4s^2$	[Ar] $4s^2$
21	Sc	↑↓	↑☐☐☐☐	☐☐☐	$[1s^2 2s^2 2p^6 3s^2 3p^6]\ 4s^2 3d^1$	[Ar] $4s^2 3d^1$
22	Ti	↑↓	↑↑☐☐☐	☐☐☐	$[1s^2 2s^2 2p^6 3s^2 3p^6]\ 4s^2 3d^2$	[Ar] $4s^2 3d^2$
23	V	↑↓	↑↑↑☐☐	☐☐☐	$[1s^2 2s^2 2p^6 3s^2 3p^6]\ 4s^2 3d^3$	[Ar] $4s^2 3d^3$
24	Cr	↑	↑↑↑↑↑	☐☐☐	$[1s^2 2s^2 2p^6 3s^2 3p^6]\ 4s^1 3d^5$	[Ar] $4s^1 3d^5$
25	Mn	↑↓	↑↑↑↑↑	☐☐☐	$[1s^2 2s^2 2p^6 3s^2 3p^6]\ 4s^2 3d^5$	[Ar] $4s^2 3d^5$
26	Fe	↑↓	↑↓↑↑↑↑	☐☐☐	$[1s^2 2s^2 2p^6 3s^2 3p^6]\ 4s^2 3d^6$	[Ar] $4s^2 3d^6$
27	Co	↑↓	↑↓↑↓↑↑↑	☐☐☐	$[1s^2 2s^2 2p^6 3s^2 3p^6]\ 4s^2 3d^7$	[Ar] $4s^2 3d^7$
28	Ni	↑↓	↑↓↑↓↑↓↑↑	☐☐☐	$[1s^2 2s^2 2p^6 3s^2 3p^6]\ 4s^2 3d^8$	[Ar] $4s^2 3d^8$
29	Cu	↑	↑↓↑↓↑↓↑↓↑↓	☐☐☐	$[1s^2 2s^2 2p^6 3s^2 3p^6]\ 4s^1 3d^{10}$	[Ar] $4s^1 3d^{10}$
30	Zn	↑↓	↑↓↑↓↑↓↑↓↑↓	☐☐☐	$[1s^2 2s^2 2p^6 3s^2 3p^6]\ 4s^2 3d^{10}$	[Ar] $4s^2 3d^{10}$
31	Ga	↑↓	↑↓↑↓↑↓↑↓↑↓	↑☐☐	$[1s^2 2s^2 2p^6 3s^2 3p^6]\ 4s^2 3d^{10} 4p^1$	[Ar] $4s^2 3d^{10} 4p^1$
32	Ge	↑↓	↑↓↑↓↑↓↑↓↑↓	↑↑☐	$[1s^2 2s^2 2p^6 3s^2 3p^6]\ 4s^2 3d^{10} 4p^2$	[Ar] $4s^2 3d^{10} 4p^2$
33	As	↑↓	↑↓↑↓↑↓↑↓↑↓	↑↑↑	$[1s^2 2s^2 2p^6 3s^2 3p^6]\ 4s^2 3d^{10} 4p^3$	[Ar] $4s^2 3d^{10} 4p^3$
34	Se	↑↓	↑↓↑↓↑↓↑↓↑↓	↑↓↑↑	$[1s^2 2s^2 2p^6 3s^2 3p^6]\ 4s^2 3d^{10} 4p^4$	[Ar] $4s^2 3d^{10} 4p^4$
35	Br	↑↓	↑↓↑↓↑↓↑↓↑↓	↑↓↑↓↑	$[1s^2 2s^2 2p^6 3s^2 3p^6]\ 4s^2 3d^{10} 4p^5$	[Ar] $4s^2 3d^{10} 4p^5$
36	Kr	↑↓	↑↓↑↓↑↓↑↓↑↓	↑↓↑↓↑↓	$[1s^2 2s^2 2p^6 3s^2 3p^6]\ 4s^2 3d^{10} 4p^6$	[Ar] $4s^2 3d^{10} 4p^6$

*Colored type indicates sublevel(s) whose occupancy changes when the last electron is added.

Transparency/ARIS Presentation Center, Tables Library

Cu $(Z = 29)$ [Ar] $4s^1 3d^{10}$

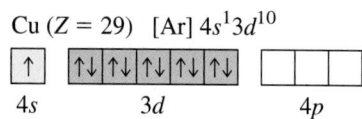

4s   3d   4p

The other anomalous filling pattern occurs with copper. Following nickel ([Ar] $4s^2 3d^8$), copper would be expected to have the [Ar] $4s^2 3d^9$ configuration. Instead, the $4s$ orbital of copper is half-filled (1 electron), and the $3d$ orbitals are *filled* with 10 electrons (see margin). Then, with zinc $(Z = 30)$, the $4s$ sublevel is filled ([Ar] $4s^2 3d^{10}$). The anomalous filling patterns in Cr and Cu lead us to conclude that *half-filled and filled sublevels are unexpectedly stable*. These are the first two cases of a pattern seen with many other elements.

In zinc, both the $4s$ and $3d$ sublevels are completely filled, and the first transition series ends. As Table 8.4 shows, the $4p$ sublevel is then filled by the next six elements. Period 4 ends with krypton, the next noble gas.

General Principles of Electron Configurations

There are 80 known elements beyond the 36 we have considered. Let's survey the ground-state electron configurations to highlight some key ideas.

Similar Outer Electron Configurations Within a Group
To repeat one of chemistry's central themes and the key to the usefulness of the periodic table, *elements in a group have similar chemical properties because they have similar outer electron configurations* (Figure 8.11). Among the main-group elements (A groups)— the *s*-block and *p*-block elements—outer electron configurations within a group

Transparency/ARIS Presentation Center, Art Library

Figure 8.11 A periodic table of partial ground-state electron configurations. These ground-state electron configurations show the electrons beyond the previous noble gas in the sublevel block being filled (excluding filled inner sublevels). For main-group elements, the group heading identifies the general outer configuration. Anomalous electron configurations occur often among the *d*-block and *f*-block elements, with the first two appearing for Cr $(Z = 24)$ and Cu $(Z = 29)$. Helium is colored as an *s*-block element but placed with the other members of Group 8A(18). Configurations for elements 112 through 116 have not yet been confirmed.

Period number: highest occupied energy level

Main-Group Elements (s block)

	1A (1)	2A (2)
	ns^1	ns^2
1	1 **H** $1s^1$	
2	3 **Li** $2s^1$	4 **Be** $2s^2$
3	11 **Na** $3s^1$	12 **Mg** $3s^2$
4	19 **K** $4s^1$	20 **Ca** $4s^2$
5	37 **Rb** $5s^1$	38 **Sr** $5s^2$
6	55 **Cs** $6s^1$	56 **Ba** $6s^2$
7	87 **Fr** $7s^1$	88 **Ra** $7s^2$

Main-Group Elements (p block)

	3A (13)	4A (14)	5A (15)	6A (16)	7A (17)	8A (18)
						$ns^2 np^6$
	$ns^2 np^1$	$ns^2 np^2$	$ns^2 np^3$	$ns^2 np^4$	$ns^2 np^5$	2 **He** $1s^2$
	5 **B** $2s^2 2p^1$	6 **C** $2s^2 2p^2$	7 **N** $2s^2 2p^3$	8 **O** $2s^2 2p^4$	9 **F** $2s^2 2p^5$	10 **Ne** $2s^2 2p^6$
	13 **Al** $3s^2 3p^1$	14 **Si** $3s^2 3p^2$	15 **P** $3s^2 3p^3$	16 **S** $3s^2 3p^4$	17 **Cl** $3s^2 3p^5$	18 **Ar** $3s^2 3p^6$
	31 **Ga** $4s^2 4p^1$	32 **Ge** $4s^2 4p^2$	33 **As** $4s^2 4p^3$	34 **Se** $4s^2 4p^4$	35 **Br** $4s^2 4p^5$	36 **Kr** $4s^2 4p^6$
	49 **In** $5s^2 5p^1$	50 **Sn** $5s^2 5p^2$	51 **Sb** $5s^2 5p^3$	52 **Te** $5s^2 5p^4$	53 **I** $5s^2 5p^5$	54 **Xe** $5s^2 5p^6$
	81 **Tl** $6s^2 6p^1$	82 **Pb** $6s^2 6p^2$	83 **Bi** $6s^2 6p^3$	84 **Po** $6s^2 6p^4$	85 **At** $6s^2 6p^5$	86 **Rn** $6s^2 6p^6$
	113 $7s^2 7p^1$	114 $7s^2 7p^2$	115 $7s^2 7p^3$	116 $7s^2 7p^4$		

Transition Elements (d block)

	3B (3)	4B (4)	5B (5)	6B (6)	7B (7)	8B (8)	8B (9)	8B (10)	1B (11)	2B (12)
4	21 **Sc** $4s^2 3d^1$	22 **Ti** $4s^2 3d^2$	23 **V** $4s^2 3d^3$	24 **Cr** $4s^1 3d^5$	25 **Mn** $4s^2 3d^5$	26 **Fe** $4s^2 3d^6$	27 **Co** $4s^2 3d^7$	28 **Ni** $4s^2 3d^8$	29 **Cu** $4s^1 3d^{10}$	30 **Zn** $4s^2 3d^{10}$
5	39 **Y** $5s^2 4d^1$	40 **Zr** $5s^2 4d^2$	41 **Nb** $5s^1 4d^4$	42 **Mo** $5s^1 4d^5$	43 **Tc** $5s^2 4d^5$	44 **Ru** $5s^1 4d^7$	45 **Rh** $5s^1 4d^8$	46 **Pd** $4d^{10}$	47 **Ag** $5s^1 4d^{10}$	48 **Cd** $5s^2 4d^{10}$
6	57 **La*** $6s^2 5d^1$	72 **Hf** $6s^2 5d^2$	73 **Ta** $6s^2 5d^3$	74 **W** $6s^2 5d^4$	75 **Re** $6s^2 5d^5$	76 **Os** $6s^2 5d^6$	77 **Ir** $6s^2 5d^7$	78 **Pt** $6s^1 5d^9$	79 **Au** $6s^1 5d^{10}$	80 **Hg** $6s^2 5d^{10}$
7	89 **Ac**** $7s^2 6d^1$	104 **Rf** $7s^2 6d^2$	105 **Db** $7s^2 6d^3$	106 **Sg** $7s^2 6d^4$	107 **Bh** $7s^2 6d^5$	108 **Hs** $7s^2 6d^6$	109 **Mt** $7s^2 6d^7$	110 **Ds** $7s^2 6d^8$	111 **Rg** $7s^2 6d^9$	112 $7s^2 6d^{10}$

Inner Transition Elements (f block)

		58 **Ce** $6s^2 4f^1 5d^1$	59 **Pr** $6s^2 4f^3$	60 **Nd** $6s^2 4f^4$	61 **Pm** $6s^2 4f^5$	62 **Sm** $6s^2 4f^6$	63 **Eu** $6s^2 4f^7$	64 **Gd** $6s^2 4f^7 5d^1$	65 **Tb** $6s^2 4f^9$	66 **Dy** $6s^2 4f^{10}$	67 **Ho** $6s^2 4f^{11}$	68 **Er** $6s^2 4f^{12}$	69 **Tm** $6s^2 4f^{13}$	70 **Yb** $6s^2 4f^{14}$	71 **Lu** $6s^2 4f^{14} 5d^1$
6	*Lanthanides														
7	**Actinides	90 **Th** $7s^2 6d^2$	91 **Pa** $7s^2 5f^2 6d^1$	92 **U** $7s^2 5f^3 6d^1$	93 **Np** $7s^2 5f^4 6d^1$	94 **Pu** $7s^2 5f^6$	95 **Am** $7s^2 5f^7$	96 **Cm** $7s^2 5f^7 6d^1$	97 **Bk** $7s^2 5f^9$	98 **Cf** $7s^2 5f^{10}$	99 **Es** $7s^2 5f^{11}$	100 **Fm** $7s^2 5f^{12}$	101 **Md** $7s^2 5f^{13}$	102 **No** $7s^2 5f^{14}$	103 **Lr** $7s^2 5f^{14} 6d^1$

are essentially identical, as shown by the group headings in Figure 8.11. Some variations in the transition elements (B groups, *d* block) and inner transition elements (*f* block) occur, as you'll see.

Orbital Filling Order When the elements are "built up" by filling their levels and sublevels in order of increasing energy, we obtain the actual sequence of elements in the periodic table. Thus, reading the table from left to right, as you read words on a page, gives the energy order of levels and sublevels, shown in Figure 8.12. The arrangement of the periodic table is the best way to learn the orbital filling order of the elements, but a useful memory aid is shown in Figure 8.13.

Categories of Electrons The elements have three categories of electrons:

1. **Inner (core) electrons** are those seen in the previous noble gas and any completed transition series. They fill all the *lower energy levels* of an atom.
2. **Outer electrons** are those in the *highest energy level* (highest *n* value). They spend most of their time farthest from the nucleus.
3. **Valence electrons** are those involved in forming compounds. *Among the main-group elements, the valence electrons are the outer electrons.* Among the transition elements, all the $(n - 1)d$ electrons are counted as valence electrons also, even though the elements Fe ($Z = 26$) through Zn ($Z = 30$) utilize only a few of their *d* electrons in bonding.

Group and Period Numbers Key information is embedded in the periodic table:

1. Among the main-group elements (A groups), *the group number equals the number of outer electrons* (those with the highest *n*): chlorine (Cl; Group **7A**) has 7 outer electrons, tellurium (Te; Group **6A**) has 6, and so forth.
2. *The period number is the n value of the highest energy level.* Thus, in Period 2, the $n = 2$ level has the highest energy; in Period 5, it is the $n = 5$ level.
3. The *n* value squared (n^2) gives the total number of *orbitals* in that energy level. Because an orbital can hold no more than two electrons (exclusion principle), $2n^2$ gives the maximum number of *electrons* (or elements) in the energy level. For example, for the $n = 3$ level, the number of orbitals is $n^2 = 9$: one 3*s*, three 3*p*, and five 3*d*. The number of electrons is $2n^2$, or 18: two 3*s* and six 3*p* electrons occur in the eight elements of Period 3, and ten 3*d* electrons are added in the ten transition elements of Period 4.

Unusual Configurations: Transition and Inner Transition Elements

Periods 4, 5, 6, and 7 incorporate the *d*-block transition elements. The general pattern, as you've seen, is that the $(n - 1)d$ orbitals are filled between the *ns* and *np* orbitals. Thus, Period 5 follows the same general pattern as Period 4. In Period 6, the 6*s* sublevel is filled in cesium (Cs) and barium (Ba), and then lanthanum

Figure 8.12 The relation between orbital filling and the periodic table. If we "read" the periods like the words on a page, the elements are arranged into sublevel blocks that occur in the order of increasing energy. This form of the periodic table shows the sublevel blocks. (The *f* blocks fit between the first and second elements of the *d* blocks in Periods 6 and 7.) *Inset:* A simple version of sublevel order.

Transparency/ARIS Presentation Center, Art Library

ARIS Presentation Center, Art Library

Figure 8.13 Aid to memorizing sublevel filling order. List the sublevels as shown, and read from 1*s*, following the direction of the arrows. Note that the
• *n* value is constant horizontally,
• *l* value is constant vertically, and
• *n* + *l* sum is constant diagonally.

(La; $Z = 57$), the first member of the $5d$ transition series, occurs. At this point, the first series of **inner transition elements,** those in which f orbitals are being filled, intervenes (Figure 8.12). The f orbitals have $l = 3$, so the possible m_l values are -3, -2, -1, 0, $+1$, $+2$, and $+3$; that is, there are seven f orbitals, for a total of 14 elements in *each* of the two inner transition series.

The Period 6 inner transition series fills the $4f$ orbitals and consists of the **lanthanides** (or **rare earths**), so called because they occur after and are similar to lanthanum. The other inner transition series holds the **actinides,** which fill the $5f$ orbitals that appear in Period 7 after actinium (Ac; $Z = 89$). In both series, the $(n - 2)f$ orbitals are filled, after which filling of the $(n - 1)d$ orbitals proceeds. Period 6 ends with the filling of the $6p$ orbitals as in other p-block elements. Period 7 is incomplete because only four elements with $7p$ electrons have been synthesized at this time.

Several irregularities in filling pattern occur in both the d and f blocks. Two already mentioned occur in chromium (Cr) and copper (Cu) in Period 4. Silver (Ag) and gold (Au), the two elements under Cu in Group 1B(11), follow copper's pattern. Molybdenum (Mo) follows the pattern of Cr in Group 6B(6), but tungsten (W) does not. Other anomalous configurations appear among the transition elements in Periods 5 and 6. Note, however, that even though minor variations from the expected configurations occur, the sum of ns electrons and $(n - 1)d$ electrons always equals the *new* group number. For instance, despite variations in the electron configurations in Group 6B(**6**)—Cr, Mo, W, and Sg—the sum of ns and $(n - 1)d$ electrons is 6; in Group 8B(**10**)—Ni, Pd, Pt, and Ds—the sum is 10.

Whenever our observations differ from our expectations, remember that the fact always takes precedence over the model; in other words, the electrons don't "care" what orbitals *we* think they should occupy. As the atomic orbitals in larger atoms fill with electrons, sublevel energies differ very little, which results in these variations from the expected pattern.

ARIS Presentation Center,
Worked Examples Library

Animation: Electron Configuration

ARIS Presentation Center
Chapter 8

SAMPLE PROBLEM 8.2 Determining Electron Configurations

PROBLEM Using the periodic table on the inside front cover of the text (not Figure 8.11 or Table 8.4), give the full and condensed electron configurations, partial orbital diagrams showing valence electrons, and number of inner electrons for the following elements:
(a) Potassium (K; $Z = 19$) **(b)** Molybdenum (Mo; $Z = 42$) **(c)** Lead (Pb; $Z = 82$)
PLAN The atomic number tells us the number of electrons, and the periodic table shows the order for filling sublevels. In the partial orbital diagrams, we include all electrons after those of the previous noble gas *except* those in *filled* inner sublevels. The number of inner electrons is the sum of those in the previous noble gas and in filled d and f sublevels.
SOLUTION (a) For K ($Z = 19$), the full electron configuration is $1s^2 2s^2 2p^6 3s^2 3p^6 4s^1$. The condensed configuration is $[\text{Ar}] 4s^1$.
The partial orbital diagram for valence electrons is

K is a main-group element in Group 1A(1) of Period 4, so there are 18 inner electrons.
(b) For Mo ($Z = 42$), we would expect the full electron configuration to be $1s^2 2s^2 2p^6 3s^2 3p^6 4s^2 3d^{10} 4p^6 5s^2 4d^4$. However, Mo lies under Cr in Group 6B(6) and exhibits the same variation in filling pattern in the ns and $(n - 1)d$ sublevels: $1s^2 2s^2 2p^6 3s^2 3p^6 4s^2 3d^{10} 4p^6 5s^1 4d^5$.
The condensed electron configuration is $[\text{Kr}] 5s^1 4d^5$.
The partial orbital diagram for valence electrons is

Mo is a transition element in Group 6B(6) of Period 5, so there are 36 inner electrons.

(c) For Pb ($Z = 82$), the full electron configuration is
$1s^2 2s^2 2p^6 3s^2 3p^6 4s^2 3d^{10} 4p^6 5s^2 4d^{10} 5p^6 6s^2 4f^{14} 5d^{10} 6p^2$.

The condensed electron configuration is [Xe] $6s^2 4f^{14} 5d^{10} 6p^2$.
The partial orbital diagram for valence electrons (no filled inner sublevels) is

↑↓	↑	↑
6s	6p	

Pb is a main-group element in Group 4A(14) of Period 6, so there are 54 (in Xe) + 14 (in $4f$ series) + 10 (in $5d$ series) = 78 inner electrons.
CHECK Be sure the sum of the superscripts (electrons) in the full electron configuration equals the atomic number, and that the number of *valence* electrons in the condensed configuration equals the number of electrons in the partial orbital diagram.

FOLLOW-UP PROBLEM 8.2 Without referring to Table 8.4 or Figure 8.11, give full and condensed electron configurations, partial orbital diagrams showing valence electrons, and the number of inner electrons for the following elements:
(a) Ni ($Z = 28$)   **(b)** Sr ($Z = 38$)   **(c)** Po ($Z = 84$)

Section Summary

In the aufbau method, one electron is added to an atom of each successive element in accord with Pauli's exclusion principle (no two electrons can have the same set of quantum numbers) and Hund's rule (orbitals of equal energy become half-filled, with electron spins parallel, before any pairing occurs). • The elements of a group have similar outer electron configurations and similar chemical behavior. • For the main-group elements, valence electrons (those involved in reactions) are in the outer (highest energy) level only. • For transition elements, $(n - 1)d$ electrons are also involved in reactions. In general, $(n - 1)d$ orbitals fill after ns and before np orbitals. • In Periods 6 and 7, $(n - 2)f$ orbitals fill between the first and second $(n - 1)d$ orbitals.

8.4 TRENDS IN THREE KEY ATOMIC PROPERTIES

All physical and chemical behavior of the elements is based ultimately on the electron configurations of their atoms. In this section, we focus on three properties of atoms that are directly influenced by electron configuration and, thus, effective nuclear charge: atomic size, ionization energy (the energy required to remove an electron from a gaseous atom), and electron affinity (the energy change involved in adding an electron to a gaseous atom). These properties are *periodic:* they generally increase and decrease in a recurring manner throughout the periodic table. As a result, their relative magnitudes can often be predicted, and they often exhibit consistent changes, or *trends,* within a group or period that correlate with element behavior.

Trends in Atomic Size

In Chapter 7, we noted that an electron in an atom can lie relatively far from the nucleus, so we commonly represent atoms as spheres in which the electrons spend 90% of their time. However, we often *define* atomic size in terms of how closely one atom lies next to another. In practice, we measure the distance between identical, adjacent atomic nuclei in a sample of an element and divide that distance in half. (The technique is discussed in Chapter 12.) Because atoms do not have hard surfaces, the size of an atom in a compound depends somewhat on the atoms near it. In other words, *atomic size varies slightly from substance to substance.*

Figure 8.14 shows two common definitions of atomic size. The **metallic radius** is one-half the distance between nuclei of adjacent atoms in a crystal of the element; we typically use this definition for metals. For elements commonly

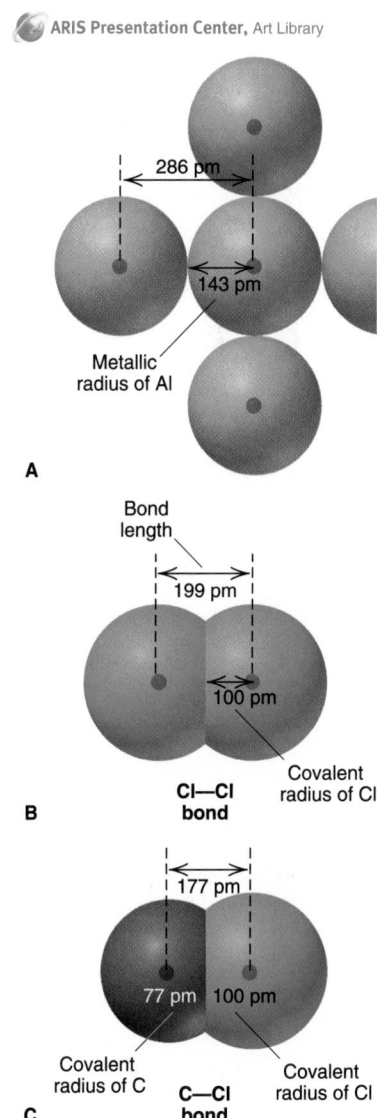

ARIS Presentation Center, Art Library

286 pm

143 pm

Metallic radius of Al

A

Bond length

199 pm

100 pm

Cl—Cl bond

Covalent radius of Cl

B

177 pm

77 pm   100 pm

Covalent radius of C

C—Cl bond

Covalent radius of Cl

C

Figure 8.14 Defining metallic and covalent radii. A, The metallic radius is one-half the distance between nuclei of adjacent atoms in a crystal of the element, as shown here for aluminum. **B,** The covalent radius is one-half the distance between bonded nuclei in a molecule of the element, as shown here for chlorine. In effect, it is one-half the bond length. **C,** In a covalent compound, the bond length and known covalent radii are used to determine other radii. Here the C—Cl bond length (177 pm) and the covalent radius of Cl (100 pm) are used to find a value for the covalent radius of C (177 pm − 100 pm = 77 pm).

occurring as molecules, mostly nonmetals, we define atomic size by the **covalent radius,** one-half the distance between nuclei of identical covalently bonded atoms.

Trends Among the Main-Group Elements Atomic size greatly influences other atomic properties and is critical to understanding element behavior. Figure 8.15 shows the atomic radii of the main-group elements and most of the transition elements. Among the main-group elements, note that atomic size varies within both a group and a period. These variations in atomic size are the result of two opposing influences:

1. *Changes in n.* As the principal quantum number (n) increases, the probability that the outer electrons will spend more time farther from the nucleus increases as well; thus, the atoms are larger.
2. *Changes in Z_{eff}.* As the effective nuclear charge (Z_{eff})—the positive charge "felt" by an electron—increases, outer electrons are pulled closer to the nucleus; thus, the atoms are smaller.

The net effect of these influences depends on shielding of the increasing nuclear charge by inner electrons:

1. *Down a group, n dominates.* As we move down a main group, each member has one more level of *inner* electrons that shield the *outer* electrons very effectively. Even though calculations show Z_{eff} on the outer electrons rising moderately for each element in the group, the atoms get larger as a result of the increasing n value. *Atomic radius generally **increases** in a group from top to bottom.*
2. *Across a period, Z_{eff} dominates.* As we move across a period of main-group elements, electrons are added to the same outer level, so the shielding by inner electrons does not change. Because outer electrons shield each other poorly, Z_{eff} on the outer electrons rises significantly, and so they are pulled closer to the nucleus. *Atomic radius generally **decreases** in a period from left to right.*

Trends Among the Transition Elements As Figure 8.15 shows, these trends hold well for the main-group elements but *not* as consistently for the transition elements. As we move from left to right, size shrinks through the first two or three transition elements because of the increasing nuclear charge. But, from then on, *the size remains relatively constant* because shielding by the inner *d* electrons counteracts the usual increase in Z_{eff}. For instance, vanadium (V; $Z = 23$), the third Period 4 transition metal, has the same atomic radius as zinc (Zn; $Z = 30$), the last Period 4 transition metal. This pattern of atomic size shrinking also appears in Periods 5 and 6 in the *d*-block transition series and in both series of inner transition elements. The *lack* of a vertical size increase from the Period 5 to 6 transition metal is especially obvious.

This shielding by *d* electrons causes a major size decrease from Group 2A(2) to Group 3A(13), the two main groups that flank the transition series. The size decrease in Periods 4, 5, and 6 (*with* a transition series) is much greater than in Period 3 (*without* a transition series). Because electrons in the *np* orbitals penetrate more than those in the $(n - 1)d$ orbitals, the first *np* electron [Group 3A(13)] "feels" a Z_{eff} that has been increased by the protons added to all the intervening transition elements. The greatest change in size occurs in Period 4, in which calcium (Ca; $Z = 20$) is nearly 50% larger than gallium (Ga; $Z = 31$). In fact, shielding by the *d* orbitals in the transition series causes such a major size contraction that gallium is slightly *smaller* than aluminum (Al; $Z = 13$), even though Ga is below Al in the same group!

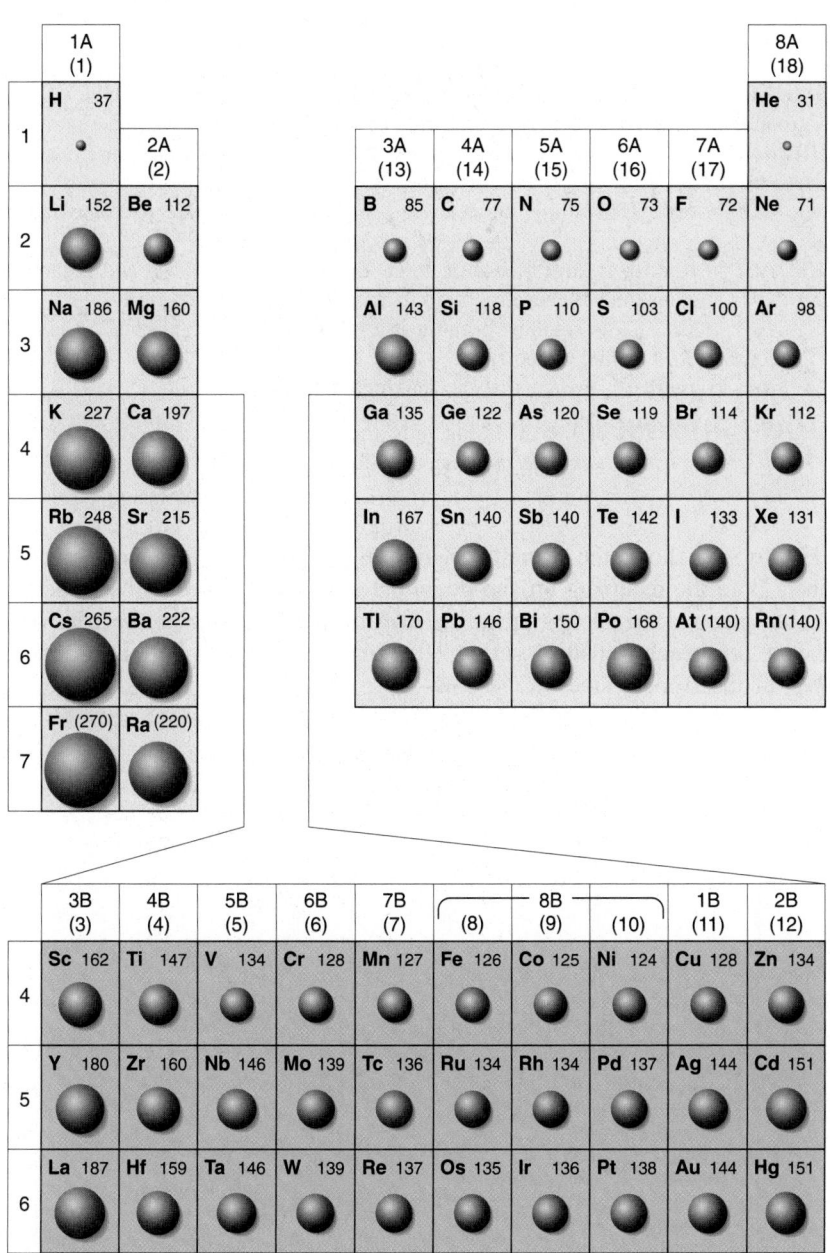

Figure 8.15 Atomic radii of the main-group and transition elements. Atomic radii (in picometers) are shown as half-spheres of proportional size for the main-group elements (*tan*) and the transition elements (*blue*). Among the main-group elements, atomic radius generally *increases* from top to bottom and *decreases* from left to right. The transition elements do not exhibit these trends so consistently. (Values in parentheses have only two significant figures; values for the noble gases are based on quantum-mechanical calculations.)

Transparency/ARIS Presentation Center,
Art Library

SAMPLE PROBLEM 8.3 Ranking Elements by Atomic Size

PROBLEM Using only the periodic table (not Figure 8.15), rank each set of main-group elements in order of *decreasing* atomic size:
(a) Ca, Mg, Sr   (b) K, Ga, Ca
(c) Br, Rb, Kr   (d) Sr, Ca, Rb
PLAN To rank the elements by atomic size, we find them in the periodic table. They are main-group elements, so size increases down a group and decreases across a period.
SOLUTION (a) Sr > Ca > Mg. These three elements are in Group 2A(2), and size decreases up the group.
(b) K > Ca > Ga. These three elements are in Period 4, and size decreases across a period.
(c) Rb > Br > Kr. Rb is largest because it has one more energy level and is farthest to the left. Kr is smaller than Br because Kr is farther to the right in Period 4.
(d) Rb > Sr > Ca. Ca is smallest because it has one fewer energy level. Sr is smaller than Rb because it is farther to the right.
CHECK From Figure 8.15, we see that the rankings are correct.

FOLLOW-UP PROBLEM 8.3 Using only the periodic table, rank the elements in each set in order of *increasing* size: (a) Se, Br, Cl; (b) I, Xe, Ba.

● **Packing 'Em In** The increasing nuclear charge shrinks the space in which each electron can move. For example, in Group 1A(1), the atomic radius of cesium (Cs; $Z = 55$) is only 1.7 times that of lithium (Li; $Z = 3$); so the volume of Cs is about five times that of Li, even though Cs has 18 times as many electrons. At the opposite ends of Period 2, neon (Ne; $Z = 10$) has about one-tenth the volume of Li ($Z = 3$), but Ne has three times as many electrons. Thus, whether size increases down a group or decreases across a period, the attraction caused by the larger number of protons in the nucleus greatly crowds the electrons.

Figure 8.16 shows the overall variation in atomic size with increasing atomic number. Note the recurring up-and-down pattern as size drops across a period to the noble gas and then leaps up to the alkali metal that begins the next period. Also note how each transition series, beginning with that in Period 4 (K to Kr), throws off the smooth size decrease. ●

Figure 8.16 **Periodicity of atomic radius.** A plot of atomic radius vs. atomic number for the elements in Periods 1 through 6 shows a periodic change: the radius generally decreases through a period to the noble gas [Group 8A(18); *purple*] and then increases suddenly to the next alkali metal [Group 1A(1); *brown*]. Deviation from the general decrease occurs among the transition elements.

Trends in Ionization Energy

The **ionization energy (IE)** is the energy (in kJ) required for the *complete removal* of 1 mol of electrons from 1 mol of gaseous atoms or ions. Pulling an electron away from a nucleus *requires* energy to overcome the attraction. Because energy flows *into* the system, the ionization energy is always positive (like ΔH of an endothermic reaction).

In Chapter 7, you saw that the ionization energy of the H atom is the energy difference between $n = 1$ and $n = \infty$, the point at which the electron is completely removed. Many-electron atoms can lose more than one electron. The first ionization energy (IE_1) removes an outermost electron (highest energy sublevel) from the gaseous atom:

$$\text{Atom}(g) \longrightarrow \text{ion}^+(g) + e^- \qquad \Delta E = IE_1 > 0 \qquad (8.2)$$

The second ionization energy (IE_2) removes a second electron. This electron is pulled away from a positively charged ion, so IE_2 is always larger than IE_1:

$$\text{Ion}^+(g) \longrightarrow \text{ion}^{2+}(g) + e^- \qquad \Delta E = IE_2 \ (\text{always} > IE_1)$$

The first ionization energy is a key factor in an element's chemical reactivity because, as you'll see, *atoms with a low IE_1 tend to form cations during reactions, whereas those with a high IE_1 (except the noble gases) often form anions.*

Variations in First Ionization Energy The elements exhibit a periodic pattern in first ionization energy, as shown in Figure 8.17. By comparing this figure with Figure 8.16, you can see a roughly *inverse* relationship between IE_1 and atomic size: *as size decreases, it takes more energy to remove an electron.* This inverse relationship appears throughout the groups and periods of the table.

Literature Lang, Peter F.; Smith, Barry C. "Ionization Energies of Atoms and Atomic Ions," *J. Chem. Educ.* **2003**, *80*, 938.

Literature Rioux, Frank; DeKock, Roger L. "The Crucial Role of Kinetic Energy in Interpreting Ionization Energies," *J. Chem. Educ.* **1998**, *75*, 537.

Figure 8.17 Periodicity of first ionization energy (IE_1). A plot of IE_1 vs. atomic number for the elements in Periods 1 through 6 shows a periodic pattern: the lowest values occur for the alkali metals (*brown*) and the highest for the noble gases (*purple*). This is the *inverse* of the trend in atomic size (see Figure 8.16).

Transparency/ARIS Presentation Center, Art Library

Let's examine the two trends and their exceptions.

1. *Down a group.* As we move *down* a main group, the orbital's *n* value increases and so does atomic size. As the distance from nucleus to outermost electron increases, the attraction between them lessens, which makes the electron easier to remove. Figure 8.18 shows that *ionization energy generally **decreases down** a group:* it is easier to remove an outer electron from an element in Period 6 than from an element in Period 2.

The only significant exception to this pattern occurs in Group 3A(13), right after the transition series, and is due to the effect of the series on atomic size: IE_1 decreases from boron (B) to aluminum (Al), but not for the rest of the group. Filling the *d* sublevels in Periods 4, 5, and 6 causes a greater-than-expected Z_{eff}, which holds the outer electrons more tightly in the larger Group 3A members.

2. *Across a period.* As we move left to right across a period, the orbital's *n* value stays the same, so Z_{eff} increases and atomic size decreases. As a result, the attraction between nucleus and outer electrons increases, which makes an electron harder to remove. *Ionization energy generally **increases across** a period:* it is easier to remove an outer electron from an alkali metal than from a noble gas.

There are several small "dips" in the otherwise smooth increase in ionization energy. These occur in Group 3A(13) for B and Al and in Group 6A(16) for O and S. The dips in Group 3A occur because these electrons are the first in the *np* sublevel. This sublevel is higher in energy than the *ns*, so the electron in it is pulled off more easily, leaving a stable, filled *ns* sublevel. The dips in Group 6A occur because the np^4 electron is the first to pair up with another *np* electron, and electron-electron repulsions raise the orbital energy. Removing the np^4 electron relieves the repulsions and leaves a stable, half-filled *np* sublevel; so the fourth *p* electron comes off more easily than the third one does.

Literature Lowe, John P. "Comments on Kinetic, Potential, and Ionization Energies," *J. Chem. Educ.* **2000,** *77,* 155.

Literature Cann, Peter. "Ionization Energies, Parallel Spins, and the Stability of Half-Filled Shells," *J. Chem. Educ.* **2000,** *77,* 1056.

Figure 8.18 First ionization energies of the main-group elements. Values for IE_1 (in kJ/mol) of the main-group elements are shown as posts of varying height. Note the general increase within a period and decrease within a group. Thus, the lowest value is at the bottom of Group 1A(1), and the highest is at the top of Group 8A(18).

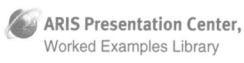
ARIS Presentation Center,
Worked Examples Library

SAMPLE PROBLEM 8.4 Ranking Elements by First Ionization Energy

PROBLEM Using the periodic table only, rank the elements in each of the following sets in order of *decreasing* IE_1:

(a) Kr, He, Ar   (b) Sb, Te, Sn

(c) K, Ca, Rb   (d) I, Xe, Cs

PLAN As in Sample Problem 8.3, we first find the elements in the periodic table and then apply the general trends of decreasing IE_1 down a group and increasing IE_1 across a period.

SOLUTION (a) He > Ar > Kr. These three are all in Group 8A(18), and IE_1 decreases down a group.

(b) Te > Sb > Sn. These three are all in Period 5, and IE_1 increases across a period.

(c) Ca > K > Rb. IE_1 of K is larger than IE_1 of Rb because K is higher in Group 1A(1). IE_1 of Ca is larger than IE_1 of K because Ca is farther to the right in Period 4.

(d) Xe > I > Cs. IE_1 of I is smaller than IE_1 of Xe because I is farther to the left. IE_1 of I is larger than IE_1 of Cs because I is farther to the right and in the previous period.

CHECK Because trends in IE_1 are generally the opposite of the trends in size, you can rank the elements by size and check that you obtain the reverse order.

FOLLOW-UP PROBLEM 8.4 Rank the elements in each of the following sets in order of increasing IE_1:

(a) Sb, Sn, I   (b) Sr, Ca, Ba

Variations in Successive Ionization Energies Successive ionization energies (IE_1, IE_2, and so on) of a given element increase because each electron is pulled away from an ion with a progressively higher positive charge. Note from Figure 8.19, however, that this increase is not smooth, but includes an enormous jump.

A more complete picture is presented in Table 8.5, which shows successive ionization energies for the elements in Period 2 and the first element in Period 3. Move horizontally through the values for a given element, and you reach a point that separates relatively low from relatively high IE values (shaded area to right of line). This jump appears *after* the outer (valence) electrons have been removed and, thus, reflects the much greater energy needed to remove an inner (core) electron. For example, follow the values for boron (B): IE_1 is lower than IE_2, which is lower than IE_3, which is *much* lower than IE_4. Thus, boron has three electrons in the highest energy level ($1s^2 2s^2 2p^1$). Because of the significantly greater energy needed to remove core electrons, they are *not* involved in chemical reactions.

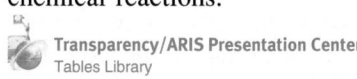
Transparency/ARIS Presentation Center,
Tables Library

ARIS Presentation Center, Art Library

Figure 8.19 The first three ionization energies of beryllium (in MJ/mol). Successive ionization energies always increase, but an exceptionally large increase occurs when the first core electron is removed. For Be, this occurs with the third electron (IE_3). (Also see Table 8.5.)

Table 8.5 Successive Ionization Energies of the Elements Lithium Through Sodium

Z	Element	Number of Valence Electrons	Ionization Energy (MJ/mol)*									
			IE_1	IE_2	IE_3	IE_4	IE_5	IE_6	IE_7	IE_8	IE_9	IE_{10}
3	Li	1	0.52	7.30	11.81							
4	Be	2	0.90	1.76	14.85	21.01		**CORE ELECTRONS**				
5	B	3	0.80	2.43	3.66	25.02	32.82					
6	C	4	1.09	2.35	4.62	6.22	37.83	47.28				
7	N	5	1.40	2.86	4.58	7.48	9.44	53.27	64.36			
8	O	6	1.31	3.39	5.30	7.47	10.98	13.33	71.33	84.08		
9	F	7	1.68	3.37	6.05	8.41	11.02	15.16	17.87	92.04	106.43	
10	Ne	8	2.08	3.95	6.12	9.37	12.18	15.24	20.00	23.07	115.38	131.43
11	Na	1	0.50	4.56	6.91	9.54	13.35	16.61	20.11	25.49	28.93	141.37

*MJ/mol, or megajoules per mole = 10^3 kJ/mol.

SAMPLE PROBLEM 8.5 Identifying an Element from Successive Ionization Energies

PROBLEM Name the Period 3 element with the following ionization energies (in kJ/mol), and write its electron configuration:

IE_1	IE_2	IE_3	IE_4	IE_5	IE_6
1012	1903	2910	4956	6278	22,230

PLAN We look for a large jump in the IE values, which occurs after all valence electrons have been removed. Then we refer to the periodic table to find the Period 3 element with this number of valence electrons and write its electron configuration.

SOLUTION The exceptionally large jump occurs after IE_5, indicating that the element has five valence electrons and, thus, is in Group 5A(15). This Period 3 element is phosphorus (P; $Z = 15$). Its electron configuration is $1s^2 2s^2 2p^6 3s^2 3p^3$.

FOLLOW-UP PROBLEM 8.5 Element Q is in Period 3 and has the following ionization energies (in kJ/mol):

IE_1	IE_2	IE_3	IE_4	IE_5	IE_6
577	1816	2744	11,576	14,829	18,375

Name element Q and write its electron configuration.

Trends in Electron Affinity

The **electron affinity (EA)** is the energy change (in kJ) accompanying the *addition* of 1 mol of electrons to 1 mol of gaseous atoms or ions. As with ionization energy, there is a first electron affinity, a second, and so forth. The *first electron affinity* (EA_1) refers to the formation of 1 mol of monovalent (1−) gaseous anions:

$$\text{Atom}(g) + e^- \longrightarrow \text{ion}^-(g) \qquad \Delta E = EA_1$$

In most cases, *energy is released when the first electron is added* because it is attracted to the atom's nuclear charge. Thus, EA_1 is usually negative (just as ΔH for an exothermic reaction is negative).* The second electron affinity (EA_2), on the other hand, is always positive because energy must be *absorbed* in order to overcome electrostatic repulsions and add another electron to a negative ion.

Factors other than Z_{eff} and atomic size affect electron affinities, so trends are not as regular as those for the previous two properties. For instance, we might expect electron affinities to decrease smoothly down a group (smaller negative number) because the nucleus is farther away from an electron being added. But, as Figure 8.20 shows, only Group 1A(1) exhibits this behavior. We might also expect a regular increase in electron affinities across a period (larger negative number) because size decreases and the increasing Z_{eff} should attract the electron being added more strongly. An overall left-to-right increase in magnitude is there, but we certainly cannot say that it is a regular increase. These exceptions arise from changes in sublevel energy and in electron-electron repulsion.

1A (1)							8A (18)
H −72.8	**2A** (2)	**3A** (13)	**4A** (14)	**5A** (15)	**6A** (16)	**7A** (17)	**He** (0.0)
Li −59.6	**Be** ≤0	**B** −26.7	**C** −122	**N** +7	**O** −141	**F** −328	**Ne** (+29)
Na −52.9	**Mg** ≤0	**Al** −42.5	**Si** −134	**P** −72.0	**S** −200	**Cl** −349	**Ar** (+35)
K −48.4	**Ca** −2.37	**Ga** −28.9	**Ge** −119	**As** −78.2	**Se** −195	**Br** −325	**Kr** (+39)
Rb −46.9	**Sr** −5.03	**In** −28.9	**Sn** −107	**Sb** −103	**Te** −190	**I** −295	**Xe** (+41)
Cs −45.5	**Ba** −13.95	**Tl** −19.3	**Pb** −35.1	**Bi** −91.3	**Po** −183	**At** −270	**Rn** (+41)

Figure 8.20 Electron affinities of the main-group elements. The electron affinities (in kJ/mol) of the main-group elements are shown. Negative values indicate that energy is released when the anion forms. Positive values, which occur in Group 8A(18), indicate that energy is absorbed to form the anion; in fact, these anions are unstable and the values are estimated.

*Tables of first electron affinity values often list them as positive if energy is *absorbed* to remove an electron from the anion. Keep this convention in mind when researching these values in reference texts. Electron affinities are difficult to measure, so values are frequently updated with more accurate data. Values for Group 2A(2) reflect recent changes.

Despite the irregularities, three key points emerge when we examine the relative values of ionization energy and electron affinity:

1. *Reactive nonmetals.* The elements in Groups 6A(16) and especially those in Group 7A(17) (halogens) have high ionization energies and highly negative (exothermic) electron affinities. These elements lose electrons with difficulty but attract them strongly. Therefore, *in their ionic compounds, they form negative ions.*

2. *Reactive metals.* The elements in Groups 1A(1) and 2A(2) have low ionization energies and slightly negative (exothermic) electron affinities. Both groups of elements lose electrons readily but attract them only weakly, if at all. Therefore, *in their ionic compounds, they form positive ions.*

3. *Noble gases.* The elements in Group 8A(18) have very high ionization energies and slightly positive (endothermic) electron affinities. Therefore, *these elements tend **not** to lose or gain electrons.* In fact, only the larger members of the group (Kr, Xe, Rn) form any compounds at all.

Section Summary

Trends in three atomic properties are summarized in Figure 8.21. Atomic size increases down a main group and decreases across a period. Across a transition series, size remains relatively constant. • First ionization energy (the energy required to remove the outermost electron from a mole of gaseous atoms) is inversely related to atomic size: IE_1 decreases down a main group and increases across a period. • An element's successive ionization energies show a very large increase when the first inner (core) electron is removed. • Electron affinity (the energy involved in adding an electron to a mole of gaseous atoms) shows many variations from expected trends. • Based on the relative sizes of IEs and EAs, in their ionic compounds, the Group 1A(1) and 2A(2) elements tend to form cations, and the Group 6A(16) and 7A(17) elements tend to form anions.

Figure 8.21 **Trends in three atomic properties.** Periodic trends are depicted as gradations in shading on miniature periodic tables, with arrows indicating the direction of general *increase* in a group or period. For electron affinity, Group 8A(18) is not shown, and the dashed arrows indicate the numerous exceptions to expected trends.

Transparency/ARIS Presentation Center, Art Library

8.5 ATOMIC STRUCTURE AND CHEMICAL REACTIVITY

Our main purpose for discussing atomic properties is, of course, to see how they affect element behavior. In this section, you'll see how the properties we just examined influence metallic behavior and determine the type of ion an element can form, as well as how electron configuration relates to magnetic properties.

Trends in Metallic Behavior

Metals are located in the left and lower three-quarters of the periodic table. They are typically shiny solids with moderate to high melting points, are good thermal and electrical conductors, can be drawn into wires and rolled into sheets, and tend to lose electrons to nonmetals. *Nonmetals* are located in the upper right quarter of the table. They are typically not shiny, have relatively low melting points, are poor thermal and electrical conductors, are mostly crumbly solids or gases, and tend to gain electrons from metals. *Metalloids* are located in the region between the other two classes and have properties between them as well. Thus, *metallic behavior decreases left to right and increases top to bottom* in the periodic table (Figure 8.22).

It's important to realize, however, that an element's properties may not fall neatly into our categories. For instance, the nonmetal carbon in the form of graphite is a good electrical conductor. Iodine, another nonmetal, is a shiny solid. Gallium and cesium are metals that melt at temperatures below body temperature, and mercury is a liquid at room temperature. And iron is quite brittle. Despite such exceptions, we can make several generalizations about metallic behavior.

Figure 8.22 **Trends in metallic behavior.** The gradation in metallic behavior among the elements is depicted as a gradation in shading from bottom left to top right, with arrows showing the direction of increase. (Hydrogen appears next to helium in this periodic table.)

Transparency/ARIS Presentation Center, Art Library

Relative Tendency to Lose Electrons Metals tend to lose electrons during chemical reactions because they have low ionization energies compared to nonmetals. The increase in metallic behavior down a group is most obvious in the physical and chemical behavior of the elements in Groups 3A(13) through 6A(16), which contain more than one class of element. For example, consider the elements in Group 5A(15), which appear vertically in Figure 8.23. Here, the change is so great that, with regard to monatomic ions, *elements at the top tend to form anions and those at the bottom tend to form cations.* Nitrogen (N) is a gaseous nonmetal, and phosphorus (P) is a solid nonmetal. Both occur occasionally as 3− anions in their compounds. Arsenic (As) and antimony (Sb) are metalloids, with Sb the more metallic of the two; neither forms ions readily. Bismuth (Bi), the largest member, is a typical metal, forming mostly ionic compounds in which it appears as a 3+ cation. Even in Group 2A(2), which consists entirely of metals, the tendency to form cations increases down the group. Thus, beryllium (Be) forms covalent compounds with nonmetals, whereas the compounds of barium (Ba) are ionic.

As we move across a period, it becomes more difficult to lose an electron (IE increases) and easier to gain one (EA becomes more negative). Therefore, with regard to monatomic ions, *elements at the left tend to form cations and those at the right tend to form anions.* The typical decrease in metallic behavior across a period is clear among the elements in Period 3, which appear horizontally in Figure 8.23. Sodium and magnesium are metals. Sodium is shiny when freshly cut under mineral oil, but it loses an electron so readily to O_2 that, if cut in air, its surface is coated immediately with a dull oxide. These metals exist naturally as Na^+ and Mg^{2+} ions in oceans, minerals, and organisms. Aluminum is metallic in its physical properties and forms the Al^{3+} ion in some compounds, but it bonds covalently in most others. Silicon (Si) is a shiny metalloid that does not occur as a monatomic ion. The most common form of phosphorus is a white, waxy nonmetal that, as noted above, forms the $P^{3−}$ ion in a few compounds. Sulfur is a crumbly yellow nonmetal that forms the sulfide ion ($S^{2−}$) in many compounds. Diatomic chlorine (Cl_2) is a yellow-green, gaseous nonmetal that attracts electrons avidly and exists in nature as the $Cl^−$ ion.

Acid-Base Behavior of the Element Oxides Metals are also distinguished from nonmetals by the acid-base behavior of their oxides in water:

- Most main-group metals *transfer* electrons to oxygen, so their *oxides are ionic. In water, these oxides act as bases,* producing $OH^−$ ions from $O^{2−}$ and reacting with acids. Calcium oxide is an example (turns indicator pink; see photo below, *left*).
- Nonmetals *share* electrons with oxygen, so *nonmetal oxides are covalent. In water, they act as acids,* producing H^+ ions and reacting with bases. Tetraphosphorus decaoxide is an example (turns indicator yellow; see photo below, *right*).

 Demo Shakhashiri, Bassam Z. "Acidic Properties of Nitrogen(IV) Oxide," *Chemical Demonstrations, A Handbook for Teachers of Chemistry,* Vol. 3 (The University of Wisconsin Press, Madison, 1985), 115.

ARIS Presentation Center,
Photo Library

CaO (basic) P_4O_{10} (acidic)

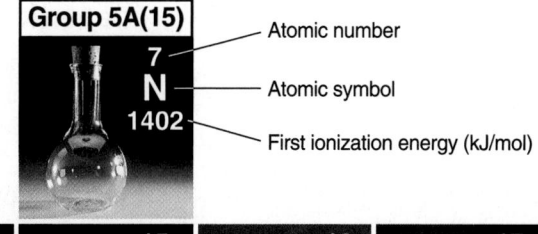

Group 5A(15)

7 N 1402 — Atomic number, Atomic symbol, First ionization energy (kJ/mol)

Figure 8.23 The change in metallic behavior in Group 5A(15) and Period 3.
Moving down from nitrogen to bismuth shows an *increase* in metallic behavior (and thus a decrease in ionization energy). Moving left to right from sodium to chlorine shows a *decrease* in metallic behavior (and thus a general increase in ionization energy).

Period 3

 11 Na 496
 12 Mg 738
 13 Al 577
 14 Si 786
 15 P 1012
 16 S 999
 17 Cl 1256

 33 As 947

51 Sb 834

 83 Bi 703

 Demo Shakhashiri, Bassam Z. "Amphoteric Properties of Hydroxides of Aluminum, Zinc, Chromium, and Lead," *Chemical Demonstrations, A Handbook for Teachers of Chemistry,* Vol. 3 (The University of Wisconsin Press, Madison, 1985), 128–135.

Some metals and many metalloids form oxides that are **amphoteric:** they can act as acids *or* as bases in water.

Figure 8.24 classifies the acid-base behavior of some common oxides, focusing once again on the elements in Group 5A(15) and Period 3. Note that *as the elements become more metallic down a group, their oxides become more basic.* In Group 5A, dinitrogen pentaoxide, N_2O_5, forms nitric acid:

$$N_2O_5(s) + H_2O(l) \longrightarrow 2HNO_3(aq)$$

Tetraphosphorus decaoxide, P_4O_{10}, forms the weaker acid H_3PO_4:

$$P_4O_{10}(s) + 6H_2O(l) \longrightarrow 4H_3PO_4(aq)$$

The oxide of the metalloid arsenic is weakly acidic, whereas that of the metalloid antimony is weakly basic. Bismuth, the most metallic of the group, forms a basic oxide that is insoluble in water but that forms a salt and water with acid:

$$Bi_2O_3(s) + 6HNO_3(aq) \longrightarrow 2Bi(NO_3)_3(aq) + 3H_2O(l)$$

Note that *as the elements become less metallic across a period, their oxides become more acidic.* In Period 3, sodium and magnesium form the strongly basic oxides Na_2O and MgO. Metallic aluminum forms amphoteric aluminum oxide (Al_2O_3), which reacts with acid or with base:

$$Al_2O_3(s) + 6HCl(aq) \longrightarrow 2AlCl_3(aq) + 3H_2O(l)$$

$$Al_2O_3(s) + 2NaOH(aq) + 3H_2O(l) \longrightarrow 2NaAl(OH)_4(aq)$$

Silicon dioxide is weakly acidic, forming a salt and water with base:

$$SiO_2(s) + 2NaOH(aq) \longrightarrow Na_2SiO_3(aq) + H_2O(l)$$

The common oxides of phosphorus, sulfur, and chlorine form acids of increasing strength: H_3PO_4, H_2SO_4, and $HClO_4$.

 Transparency/ARIS Presentation Center, Photo Library

Transparency/ARIS Presentation Center, Art Library

Figure 8.24 The trend in acid-base behavior of element oxides. The trend in acid-base behavior for some common oxides of Group 5A(15) and Period 3 elements is shown as a gradation in color (*red* = acidic; *blue* = basic). Note that the metals form basic oxides and the nonmetals form acidic oxides. Aluminum forms an oxide (*purple*) that can act as an acid or as a base. Thus, as atomic size increases, ionization energy decreases, and oxide basicity increases.

Properties of Monatomic Ions

So far we have focused on the reactants—the atoms—in the process of electron loss and gain. Now we focus on the products—the ions. We examine electron configurations, magnetic properties, and ionic radius relative to atomic radius.

Electron Configurations of Main-Group Ions In Chapter 2, you learned the symbols and charges of many monatomic ions. But *why* does an ion have a particular charge in its compounds? Why is a sodium ion Na^+ and not Na^{2+}, and why is a fluoride ion F^- and not F^{2-}? For elements at the left and right ends of the periodic table, the explanation concerns the very low reactivity of the noble gases. As we said earlier, because they have high IEs and positive (endothermic) EAs, the noble gases typically do not form ions but remain chemically stable with a *filled* outer energy level (ns^2np^6). *Elements in Groups 1A(1), 2A(2), 6A(16), and 7A(17) that readily form ions either lose or gain electrons to attain a filled outer level and thus a noble gas configuration.* Their ions are said to be **isoelectronic** (Greek *iso*, "same") with the nearest noble gas. Figure 8.25 shows this relationship (also see Figure 2.13, p. 61).

When an alkali metal atom [Group 1A(1)] loses its single valence electron, it becomes isoelectronic with the *previous* noble gas. The Na^+ ion, for example, is isoelectronic with neon (Ne):

$$Na\ (1s^22s^22p^63s^1) \longrightarrow Na^+\ (1s^22s^22p^6)\ [\text{isoelectronic with Ne } (1s^22s^22p^6)] + e^-$$

When a halogen atom [Group 7A(17)] adds a single electron to the five in its *np* sublevel, it becomes isoelectronic with the *next* noble gas. Bromide ion, for example, is isoelectronic with krypton (Kr):

$$Br\ ([Ar]\ 4s^23d^{10}4p^5) + e^- \longrightarrow Br^-\ ([Ar]\ 4s^23d^{10}4p^6)$$

$$[\text{isoelectronic with Kr } ([Ar]\ 4s^23d^{10}4p^6)]$$

The energy needed to remove the electrons from metals to attain the previous noble gas configuration is supplied during their exothermic reactions with nonmetals. Removing more than one electron from Na to form Na^{2+} or more than two from Mg to form Mg^{3+} means removing core electrons, which requires more energy than is available in a reaction. This is the reason that $NaCl_2$ and MgF_3 do *not* exist. Similarly, adding two electrons to F to form F^{2-} or three to O to form O^{3-} means placing the extra electron into the next energy level. With 18 electrons acting as inner electrons and shielding the nuclear charge very effectively, adding an electron to the negative ion, F^- or O^{2-}, requires too much energy. Thus, we never see Na_2F or Mg_3O_2.

The larger metals of Groups 3A(13), 4A(14), and 5A(15) form cations through a different process, because it would be energetically impossible for them to lose enough electrons to attain a noble gas configuration. For example, tin (Sn; $Z = 50$) would have to lose 14 electrons—two 5*p*, ten 4*d*, and two 5*s*—to be isoelectronic with krypton (Kr; $Z = 36$), the previous noble gas. Instead, tin loses far fewer electrons and attains two different stable configurations. In the tin(IV) ion (Sn^{4+}), the metal atom empties its outer energy level and attains the stability of empty 5*s* and 5*p* sublevels and a filled inner 4*d* sublevel. This $(n-1)d^{10}$ configuration is called a **pseudo–noble gas configuration:**

$$Sn\ ([Kr]\ 5s^24d^{10}5p^2) \longrightarrow Sn^{4+}\ ([Kr]\ 4d^{10}) + 4e^-$$

Alternatively, in the more common tin(II) ion (Sn^{2+}), the atom loses the two 5*p* electrons only and attains the stability of filled 5*s* and 4*d* sublevels:

$$Sn\ ([Kr]\ 5s^24d^{10}5p^2) \longrightarrow Sn^{2+}\ ([Kr]\ 5s^24d^{10}) + 2e^-$$

The retained ns^2 electrons are sometimes called an *inert pair* because they seem difficult to remove. Thallium, lead, and bismuth, the largest and most metallic

---

Figure 8.25 Main-group ions and the noble gas electron configurations.
Most of the elements that form monatomic ions that are isoelectronic with a noble gas lie in the four groups that flank Group 8A(18), two on either side.

 ARIS Presentation Center, Art Library

 Animation: Isoelectronic Series
ARIS Presentation Center
Chapter 8

members of Groups 3A(13) to 5A(15), commonly form ions that retain the ns^2 pair of electrons: Tl^+, Pb^{2+}, and Bi^{3+}.

Excessively high energy cost is also the reason that some elements do not form monatomic ions in any of their reactions. For instance, carbon would have to lose four electrons to form C^{4+} and attain the He configuration, or gain four to form C^{4-} and attain the Ne configuration, but neither ion forms. (Such multivalent ions *are* observed in the spectra of stars, however, where temperatures exceed 10^6 K.) As you'll see in Chapter 9, carbon and other atoms that do not form ions attain a filled shell by *sharing* electrons through covalent bonding.

SAMPLE PROBLEM 8.6 Writing Electron Configurations of Main-Group Ions

ARIS Presentation Center, Worked Examples Library

PROBLEM Using condensed electron configurations, write reactions for the formation of the common ions of the following elements:

(a) Iodine ($Z = 53$)   **(b)** Potassium ($Z = 19$)   **(c)** Indium ($Z = 49$)

PLAN We identify the element's position in the periodic table and recall two general points:

- Ions of elements in Groups 1A(1), 2A(2), 6A(16), and 7A(17) are isoelectronic with the nearest noble gas.
- Metals in Groups 3A(13) to 5A(15) lose the ns and np electrons or just the np electrons.

SOLUTION **(a)** Iodine is in Group 7A(17), so it gains one electron and is isoelectronic with xenon:

$$I\ ([Kr]\ 5s^2 4d^{10} 5p^5) + e^- \longrightarrow I^-\ ([Kr]\ 5s^2 4d^{10} 5p^6) \quad \text{(same as Xe)}$$

(b) Potassium is in Group 1A(1), so it loses one electron and is isoelectronic with argon:

$$K\ ([Ar]\ 4s^1) \longrightarrow K^+\ ([Ar]) + e^-$$

(c) Indium is in Group 3A(13), so it loses either three electrons to form In^{3+} (pseudo–noble gas configuration) or one to form In^+ (inert pair):

$$In\ ([Kr]\ 5s^2 4d^{10} 5p^1) \longrightarrow In^{3+}\ ([Kr]\ 4d^{10}) + 3e^-$$

$$In\ ([Kr]\ 5s^2 4d^{10} 5p^1) \longrightarrow In^+\ ([Kr]\ 5s^2 4d^{10}) + e^-$$

CHECK Be sure that the number of electrons in the ion's electron configuration, plus those gained or lost to form the ion, equals Z.

FOLLOW-UP PROBLEM 8.6 Using condensed electron configurations, write reactions showing the formation of the common ions of the following elements:

(a) Ba ($Z = 56$)   **(b)** O ($Z = 8$)   **(c)** Pb ($Z = 82$)

Electron Configurations of Transition Metal Ions In contrast to most main-group ions, *transition metal ions rarely attain a noble gas configuration,* and the reason, once again, is that energy costs are too high. The exceptions in Period 4 are scandium, which forms Sc^{3+}, and titanium, which occasionally forms Ti^{4+} in some compounds. The typical behavior of a transition element is to *form more than one cation by losing all of its ns and some of its $(n − 1)d$ electrons.* (We focus here on the Period 4 series, but these points hold for Periods 5 and 6 also.)

In the aufbau process of building up the ground-state atoms, Period 3 ends with the noble gas argon. At the beginning of Period 4, the radial probability distribution of the $4s$ orbital near the nucleus makes it more stable than the empty $3d$. Therefore, the first and second electrons added in the period enter the $4s$ in K and Ca. But, as soon as we reach the transition elements and the $3d$ orbitals begin to fill, the increasing nuclear charge attracts their electrons more and more strongly. Moreover, the added $3d$ electrons fill inner orbitals, so they are not very well shielded from the increasing nuclear charge by the $4s$ electrons. As a result,

Literature Matsumoto, Paul S. "Trends in Ionization Energy of Transition-Metal Elements," *J. Chem. Educ.* **2005**, *82*, 1660.

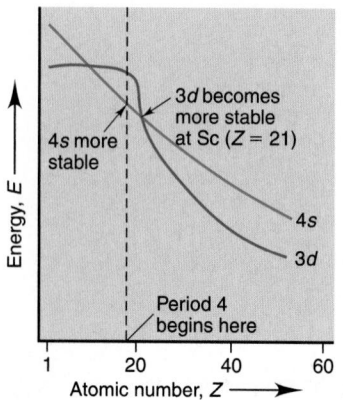

Figure 8.26 **The Period 4 crossover in sublevel energies.** The 3d orbitals are empty in elements at the beginning of Period 4. Because the 4s electron penetrates closer to the nucleus, the energy of the 4s orbital is lower in K and Ca; thus, the 4s fills before the 3d. But as the 3d orbitals fill, beginning with $Z = 21$, these inner electrons are attracted by the increasing nuclear charge, and they also shield the 4s electrons. As a result, there is an energy crossover, with the 3d sublevel becoming lower in energy than the 4s. For this reason, the 4s electrons are removed first when the transition metal ion forms. In other words,

- For a main-group metal ion, the highest n level of electrons is "last-in, first-out."
- For a transition metal ion, the highest n level of electrons is "first-in, first out."

ARIS Presentation Center, Art Library

the 3d orbital becomes *more stable* than the 4s. In effect, a *crossover in orbital energy* occurs as we enter the transition series (Figure 8.26). The effect on ion formation is critical: because the 3d orbitals are more stable, *the 4s electrons are lost before the 3d electrons to form the Period 4 transition metal ions.* Thus, the 4s electrons are added before the 3d to form the *atom* but lost before the 3d to form the *ion:* "first-in, first-out."

To summarize, *electrons with the highest n value are removed first.* Here are a few simple rules for forming the ion of any main-group or transition element:

- For main-group, s-block metals, remove all electrons with the highest *n* value.
- For main-group, p-block metals, remove *np* electrons before *ns* electrons.
- For transition (d-block) metals, remove *ns* electrons before $(n − 1)d$ electrons.
- For nonmetals, add electrons to the *p* orbitals of highest *n* value.

Magnetic Properties of Transition Metal Ions
If we can't see electrons in orbitals, how do we know that a particular electron configuration is correct? Although analysis of atomic spectra is the most important method for determining configuration, the magnetic properties of an element and its compounds can support or refute conclusions from spectra. Recall that electron spin generates a tiny magnetic field, which causes a beam of H atoms to split in an external magnetic field (see Figure 8.1). Only chemical species (atoms, ions, or molecules) with one or more *unpaired* electrons are affected by the external field. The species used in the original 1921 split-beam experiment was the silver atom:

Ag $(Z = 47)$ [Kr] $5s^1 4d^{10}$ [↑] [↑↓ ↑↓ ↑↓ ↑↓ ↑↓] [☐ ☐ ☐]
5s   4d   5p

Note the unpaired 5s electron. A beam of cadmium atoms, the element after silver, is not split because their 5s electrons are *paired* (Cd: [Kr] $5s^2 4d^{10}$).

A species with unpaired electrons exhibits **paramagnetism:** it is attracted by an external magnetic field. A species with all electrons paired exhibits **diamagnetism:** it is not attracted (and, in fact, is slightly repelled) by a magnetic field. Figure 8.27 shows how this magnetic behavior is studied. Many transition metals and their compounds are paramagnetic because their atoms and ions have unpaired electrons.

Let's see how studies of paramagnetism might be used to provide additional evidence for a proposed electron configuration. Spectral analysis of the titanium atom yields the configuration [Ar] $4s^2 3d^2$. Experiment shows that Ti metal is paramagnetic, which is consistent with the presence of unpaired electrons in its atoms. Spectral analysis of the Ti^{2+} ion yields the configuration [Ar] $3d^2$, indicating loss of the two 4s electrons. Once again, experiment supports these findings by showing

ARIS Presentation Center, Art Library

Figure 8.27 **Apparatus for measuring the magnetic behavior of a sample.** The substance is weighed on a very sensitive balance in the absence of an external magnetic field. **A,** If the substance is diamagnetic (has all *paired* electrons), its apparent mass is unaffected (or slightly reduced) when the magnetic field is "on." **B,** If the substance is paramagnetic (has *unpaired* electrons), its apparent mass increases when the field is "on" because the balance arm feels an additional force. This method is used to estimate the number of unpaired electrons in transition metal compounds.

A Electromagnet

B Electromagnet

that Ti^{2+} compounds are paramagnetic. If Ti had lost its two $3d$ electrons during ion formation, its compounds would be diamagnetic because the $4s$ electrons are paired. Thus, the [Ar] $3d^2$ configuration supports the conclusion that electrons of highest n value are lost first:

$$Ti\ ([Ar]\ 4s^2 3d^2) \longrightarrow Ti^{2+}\ ([Ar]\ 3d^2) + 2e^-$$

The partial orbital diagrams are

An increase in paramagnetism occurs when iron metal (Fe) forms Fe^{3+} compounds. This fact is consistent with Fe losing its $4s$ electrons and one of its paired $3d$ electrons:

$$Fe\ ([Ar]\ 4s^2 3d^6) \longrightarrow Fe^{3+}\ ([Ar]\ 3d^5) + 3e^-$$

Copper (Cu) is paramagnetic, but zinc (Zn) is diamagnetic, as are the Cu^+ and Zn^{2+} ions. These observations support the electron configurations proposed by spectral analysis. The two ions are isoelectronic:

$$Cu\ ([Ar]\ 4s^1 3d^{10}) \longrightarrow Cu^+\ ([Ar]\ 3d^{10}) + e^-$$
$$Zn\ ([Ar]\ 4s^2 3d^{10}) \longrightarrow Zn^{2+}\ ([Ar]\ 3d^{10}) + 2e^-$$

SAMPLE PROBLEM 8.7 Writing Electron Configurations and Predicting Magnetic Behavior of Transition Metal Ions

PROBLEM Use condensed electron configurations to write the reaction for the formation of each transition metal ion, and predict whether the ion is paramagnetic:

(a) Mn^{2+} ($Z = 25$)   **(b)** Cr^{3+} ($Z = 24$)   **(c)** Hg^{2+} ($Z = 80$)

PLAN We first write the condensed electron configuration of the atom, noting the irregularity for Cr in (b). Then we remove electrons, beginning with ns electrons, to attain the ion charge. If unpaired electrons are present, the ion is paramagnetic.

SOLUTION (a) $Mn\ ([Ar]\ 4s^2 3d^5) \longrightarrow Mn^{2+}\ ([Ar]\ 3d^5) + 2e^-$

There are five unpaired e^-, so Mn^{2+} is paramagnetic.

(b) $Cr\ ([Ar]\ 4s^1 3d^5) \longrightarrow Cr^{3+}\ ([Ar]\ 3d^3) + 3e^-$

There are three unpaired e^-, so Cr^{3+} is paramagnetic.

(c) $Hg\ ([Xe]\ 6s^2 4f^{14} 5d^{10}) \longrightarrow Hg^{2+}\ ([Xe]\ 4f^{14} 5d^{10}) + 2e^-$

There are no unpaired e^-, so Hg^{2+} is *not* paramagnetic (is diamagnetic).

CHECK We removed the ns electrons first, and the sum of the lost electrons and those in the electron configuration of the ion equals Z.

FOLLOW-UP PROBLEM 8.7 Write the condensed electron configuration of each transition metal ion, and predict whether it is paramagnetic:

(a) V^{3+} ($Z = 23$)   **(b)** Ni^{2+} ($Z = 28$)   **(c)** La^{3+} ($Z = 57$)

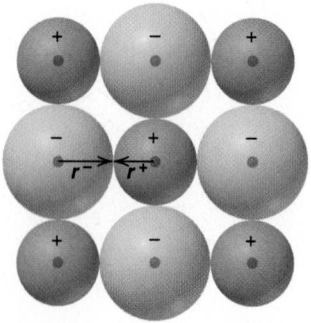

Figure 8.28 Depicting ionic radius. The cation radius (r^+) and the anion radius (r^-) each make up a portion of the total distance between the nuclei of adjacent ions in a crystalline ionic compound.

ARIS Presentation Center, Art Library

Ionic Size vs. Atomic Size The **ionic radius** is an estimate of the size of an ion in a crystalline ionic compound. You can picture it as one ion's portion of the distance between the nuclei of neighboring ions in the solid (Figure 8.28). From the relation between effective nuclear charge and atomic size, we can predict the size of an ion relative to its parent atom:

• *Cations are smaller than their parent atoms.* When a cation forms, electrons are *removed from* the outer level. The resulting decrease in electron repulsions allows the nuclear charge to pull the remaining electrons closer.
• *Anions are larger than their parent atoms.* When an anion forms, electrons are *added to* the outer level. The increase in repulsions causes the electrons to occupy more space.

Figure 8.29 shows the radii of some common main-group monatomic ions relative to their parent atoms. As you can see, *ionic size increases down a group* because the number of energy levels increases. Across a period, however, the pattern is more complex. Size decreases among the cations, then increases tremendously with the first of the anions, and finally decreases again among the anions.

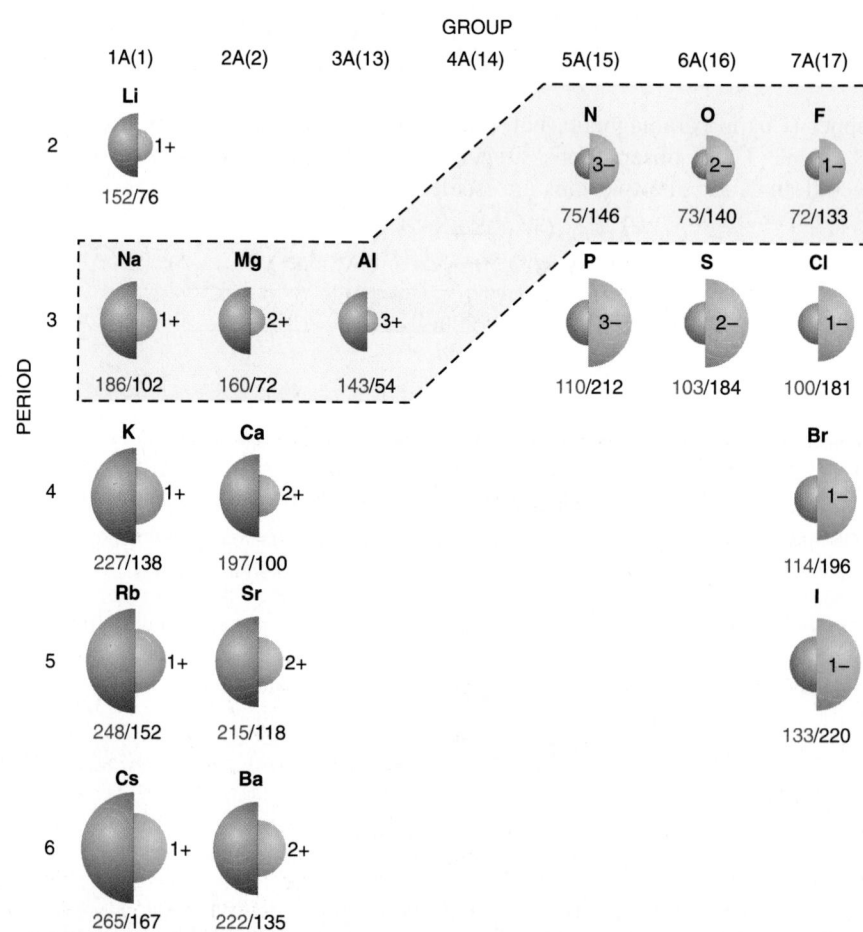

Figure 8.29 Ionic vs. atomic radii. The atomic radii (*colored half-spheres*) and ionic radii (*gray half-spheres*) of some main-group elements are arranged in periodic table format (with all radii values in picometers). Note that metal atoms (*blue*) form *smaller* positive ions, whereas nonmetal atoms (*red*) form *larger* negative ions. The dashed outline sets off ions of Period 2 nonmetals and Period 3 metals that are *isoelectronic* with neon. Note the size decrease from anions to cations.

Transparency/ARIS Presentation Center, Art Library

This pattern results from changes in effective nuclear charge and electron-electron repulsions. In Period 3 (Na through Cl), for example, increasing Z_{eff} from left to right makes Na^+ larger than Mg^{2+}, which in turn is larger than Al^{3+}. The great jump in size from cations to anions occurs because we are *adding* electrons rather than removing them, so repulsions increase sharply. For instance, P^{3-} has eight more electrons than Al^{3+}. Then, the ongoing rise in Z_{eff} makes P^{3-} larger than S^{2-}, which is larger than Cl^-. These factors lead to some striking effects even among ions with the same number of electrons. Look at the ions within the dashed outline in Figure 8.29, which are all isoelectronic with neon. Even though the cations form from elements in the next period, the anions are still much larger. The pattern is

$$3- \; > \; 2- \; > \; 1- \; > \; 1+ \; > \; 2+ \; > \; 3+$$

When an element forms more than one cation, *the greater the ionic charge, the smaller the ionic radius.* Consider Fe^{2+} and Fe^{3+}. The number of protons is the same, but Fe^{3+} has one fewer electron, so electron repulsions are reduced somewhat. As a result, Z_{eff} increases, which pulls all the electrons closer, so Fe^{3+} is smaller than Fe^{2+}.

To summarize the main points,

- Ionic size increases down a group.
- Ionic size decreases across a period but increases from cations to anions.
- Ionic size decreases with increasing positive (or decreasing negative) charge in an isoelectronic series.
- Ionic size decreases as charge increases for different cations of a given element.

SAMPLE PROBLEM 8.8　Ranking Ions by Size

PROBLEM Rank each set of ions in order of *decreasing* size, and explain your ranking:
(a) Ca^{2+}, Sr^{2+}, Mg^{2+}　　　**(b)** K^+, S^{2-}, Cl^-　　　**(c)** Au^+, Au^{3+}
PLAN We find the position of each element in the periodic table and apply the ideas presented in the text.
SOLUTION (a) Mg^{2+}, Ca^{2+}, and Sr^{2+} are all from Group 2A(2), so they decrease in size up the group: $Sr^{2+} > Ca^{2+} > Mg^{2+}$.
(b) The ions K^+, S^{2-}, and Cl^- are isoelectronic. S^{2-} has a lower Z_{eff} than Cl^-, so it is larger. K^+ is a cation, and has the highest Z_{eff}, so it is smallest: $S^{2-} > Cl^- > K^+$.
(c) Au^+ has a lower charge than Au^{3+}, so it is larger: $Au^+ > Au^{3+}$.

FOLLOW-UP PROBLEM 8.8 Rank the ions in each set in order of *increasing* size:
(a) Cl^-, Br^-, F^-　　　**(b)** Na^+, Mg^{2+}, F^-　　　**(c)** Cr^{2+}, Cr^{3+}

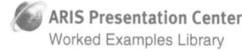

ARIS Presentation Center,
Worked Examples Library

Section Summary

Metallic behavior correlates with large atomic size and low ionization energy. Thus, metallic behavior increases down a group and decreases across a period. • Within the main groups, metal oxides are basic and nonmetal oxides acidic. Thus, oxides become more acidic across a period and more basic down a group. • Many main-group elements form ions that are isoelectronic with the nearest noble gas. Removing (or adding) more electrons than needed to attain the previous noble gas configuration requires a prohibitive amount of energy. • Metals in Groups 3A(13) to 5A(15) lose either their np electrons or both their ns and np electrons. • Transition metals lose ns electrons before $(n - 1)d$ electrons and commonly form more than one ion. • Many transition metals and their compounds are paramagnetic because their atoms (or ions) have unpaired electrons. • Cations are smaller and anions larger than their parent atoms. Ionic radius increases down a group. Across a period, cationic and anionic radii decrease, but a large increase occurs from cations to anions.

Chapter Perspective

This chapter is our springboard to understanding the chemistry of the elements. We have begun to see that recurring electron configurations lead to trends in atomic properties, which in turn lead to trends in chemical behavior. With this insight, we can go on to investigate how atoms bond (Chapter 9), how molecular shapes arise (Chapter 10), how molecular shapes and other properties can be explained using certain bonding models (Chapter 11), how the physical properties of liquids and solids emerge from atomic properties (Chapter 12), and how those properties influence the solution process (Chapter 13). We will briefly review these ideas and gain a perspective on where they lead (Interchapter) before we can survey elemental behavior in greater detail (Chapter 14) and see how it applies to the remarkable diversity of organic compounds (Chapter 15). Within this group of chapters, you will see chemical models come alive in chemical facts.

CHAPTER REVIEW GUIDE
The following sections provide many aids to help you study this chapter. (Numbers in parentheses refer to pages, unless noted otherwise.)

Learning Objectives These are concepts and skills you should know after studying this chapter.

Relevant section and/or sample problem (SP) numbers appear in parentheses.

Understand These Concepts

1. The meaning of the periodic law and the arrangement of elements by atomic number (8.1)
2. The reason for the spin quantum number and its two possible values (8.2)
3. How the exclusion principle applies to orbital filling (8.2)
4. The effects of nuclear charge, shielding, and penetration on the splitting of orbital energies; the meaning of effective nuclear charge (8.2)
5. How the order in the periodic table is based on the order of orbital energies (8.3)
6. How orbitals are filled in main-group and transition elements; the importance of Hund's rule (8.3)
7. How outer electron configuration within a group is related to chemical behavior (8.3)
8. The distinction among inner, outer, and valence electrons (8.3)
9. The meaning of atomic radius, ionization energy, and electron affinity (8.4)
10. How n value and effective nuclear charge give rise to the periodic trends of atomic size and ionization energy (8.4)
11. The importance of core electrons to the pattern of successive ionization energies (8.4)
12. How atomic properties are related to the tendency to form ions (8.4)
13. The general properties of metals and nonmetals (8.5)
14. How the vertical and horizontal trends in metallic behavior are related to ion formation and oxide acidity (8.5)
15. Why main-group ions are either isoelectronic with the nearest noble gas or have a pseudo–noble gas electron configuration (8.5)
16. Why transition elements lose ns electrons first (8.5)
17. The origin of paramagnetic and diamagnetic behavior (8.5)
18. The relation between ionic and atomic size and the trends in ionic size (8.5)

Master These Skills

1. Using orbital diagrams to determine the set of quantum numbers for any electron in an atom (SP 8.1)
2. Writing full and condensed electron configurations of an element (SP 8.2)
3. Using periodic trends to rank elements by atomic size and first ionization energy (SPs 8.3, 8.4)
4. Identifying an element from its successive ionization energies (SP 8.5)
5. Writing electron configurations of main-group and transition metal ions (SPs 8.6, 8.7)
6. Using periodic trends to rank ions by relative size (SP 8.8)

Key Terms These important terms appear in boldface in the chapter and are defined again in the Glossary.

electron configuration (303)

Section 8.1
periodic law (303)

Section 8.2
spin quantum number (m_s) (305)
exclusion principle (305)
shielding (306)

effective nuclear charge (Z_{eff}) (306)
penetration (307)

Section 8.3
aufbau principle (308)
orbital diagram (308)
Hund's rule (309)
transition elements (313)
inner (core) electrons (315)

outer electrons (315)
valence electrons (315)
inner transition elements (316)
lanthanides (rare earths) (316)
actinides (316)

Section 8.4
metallic radius (317)
covalent radius (318)
ionization energy (IE) (321)
electron affinity (EA) (324)

Section 8.5
amphoteric (327)
isoelectronic (328)
pseudo–noble gas configuration (328)
paramagnetism (330)
diamagnetism (330)
ionic radius (332)

Key Equations and Relationships
Numbered and screened concepts are listed for you to refer to or memorize.

8.1 Defining the energy order of sublevels in terms of the angular momentum quantum number (l value) (307):

Order of sublevel energies: $s < p < d < f$

8.2 Meaning of the first ionization energy (321):

$$\text{Atom}(g) \longrightarrow \text{ion}^+(g) + e^- \qquad \Delta E = \text{IE}_1 > 0$$

Highlighted Figures and Tables
These figures (F) and tables (T) provide a visual review of key ideas.

Entries in **bold** contain frequently used data.

T8.2 Summary of quantum numbers of electrons in atoms (305)
F8.3 The effect of nuclear charge on orbital energy (306)
F8.4 Shielding and orbital energy (307)
F8.5 Penetration and orbital energy (307)
F8.6 Order for filling energy sublevels with electrons (308)
F8.11 Periodic table of partial ground-state electron configurations (314)
F8.12 Relation between orbital filling and the periodic table (315)
F8.15 Atomic radii of the main-group and transition elements (319)

F8.16 Periodicity of atomic radius (320)
F8.17 Periodicity of first ionization energy (IE_1) (321)
F8.18 First ionization energies of the main-group elements (322)
F8.21 Trends in three atomic properties (325)
F8.22 Trends in metallic behavior (325)
F8.24 Trend in acid-base behavior of element oxides (327)
F8.25 Main-group ions and the noble gas electron configurations (328)
F8.26 The Period 4 crossover in sublevel energies (330)
F8.29 Ionic vs. atomic radii (332)

Brief Solutions to FOLLOW-UP PROBLEMS
Compare your solutions to these calculation steps and answers.

8.1 The element has eight electrons, so $Z = 8$: oxygen.
Sixth electron: $n = 2$, $l = 1$, $m_l = 0$, $m_s = +\frac{1}{2}$

↑↓	↑↓	↑↓	↑	↑
$1s$	$2s$		$2p$	

8.2 (a) For Ni, $1s^2 2s^2 2p^6 3s^2 3p^6 4s^2 3d^8$; [Ar] $4s^2 3d^8$

↑↓	↑↓	↑↓	↑↓	↑	↑			
$4s$		$3d$					$4p$	

Ni has 18 inner electrons.
(b) For Sr, $1s^2 2s^2 2p^6 3s^2 3p^6 4s^2 3d^{10} 4p^6 5s^2$; [Kr] $5s^2$

↑↓								
$5s$		$4d$					$5p$	

Sr has 36 inner electrons.
(c) For Po, $1s^2 2s^2 2p^6 3s^2 3p^6 4s^2 3d^{10} 4p^6 5s^2 4d^{10} 5p^6 6s^2 4f^{14} 5d^{10} 6p^4$; [Xe] $6s^2 4f^{14} 5d^{10} 6p^4$

↑↓	↑↓	↑	↑
$6s$		$6p$	

Po has 78 inner electrons.

8.3 (a) Cl < Br < Se; (b) Xe < I < Ba
8.4 (a) Sn < Sb < I; (b) Ba < Sr < Ca
8.5 Q is aluminum: $1s^2 2s^2 2p^6 3s^2 3p^1$
8.6 (a) Ba ([Xe] $6s^2$) $\longrightarrow$ Ba^{2+} ([Xe]) + 2e$^-$
(b) O ([He] $2s^2 2p^4$) + 2e$^-$ $\longrightarrow$ O^{2-} ([He] $2s^2 2p^6$) (same as Ne)
(c) Pb ([Xe] $6s^2 4f^{14} 5d^{10} 6p^2$) $\longrightarrow$ Pb^{2+} ([Xe] $6s^2 4f^{14} 5d^{10}$) + 2e$^-$
    Pb ([Xe] $6s^2 4f^{14} 5d^{10} 6p^2$) $\longrightarrow$ Pb^{4+} ([Xe] $4f^{14} 5d^{10}$) + 4e$^-$
8.7 (a) V^{3+}: [Ar] $3d^2$; paramagnetic
(b) Ni^{2+}: [Ar] $3d^8$; paramagnetic
(c) La^{3+}: [Xe]; not paramagnetic (diamagnetic)
8.8 (a) F$^-$ < Cl$^-$ < Br$^-$; (b) Mg^{2+} < Na$^+$ < F$^-$;
(c) Cr^{3+} < Cr^{2+}

PROBLEMS

Problems with **colored** numbers are answered in Appendix E and worked in detail in the Student Solutions Manual. Problem sections match those in the text and provide the numbers of relevant sample problems. Most offer Concept Review Questions, Skill-Building Exercises (grouped in pairs covering the same concept), and Problems in Context. The Comprehensive Problems are based on material from any section or previous chapter.

Development of the Periodic Table

● **Concept Review Questions**

• **8.1** What would be your reaction to a claim that a new element had been discovered, and it fit between tin (Sn) and antimony (Sb) in the periodic table?

• **8.2** Based on results of his study of atomic x-ray spectra, Moseley discovered a relationship that replaced atomic mass as the criterion for ordering the elements. By what criterion are the elements now ordered in the periodic table? Give an example of a sequence of element order that was confirmed by Moseley's findings.

● **Skill-Building Exercises** (grouped in similar pairs)

• **8.3** Before Mendeleev published his periodic table, Döbereiner grouped elements with similar properties into "triads," in which the unknown properties of one member could be predicted by averaging known values of the properties of the others. To test this idea, predict the values of the following quantities:
(a) The atomic mass of K from the atomic masses of Na and Rb
(b) The melting point of Br_2 from the melting points of Cl_2 ($-101.0°C$) and I_2 ($113.6°C$) (actual value = $-7.2°C$)

• **8.4** To test Döbereiner's idea (Problem 8.3), predict:
(a) The boiling point of HBr from the boiling points of HCl ($-84.9°C$) and HI ($-35.4°C$) (actual value = $-67.0°C$)
(b) The boiling point of AsH_3 from the boiling points of PH_3 ($-87.4°C$) and SbH_3 ($-17.1°C$) (actual value = $-55°C$)

Characteristics of Many-Electron Atoms

● **Concept Review Questions**

• **8.5** Summarize the rules for the allowable values of the four quantum numbers of an electron in an atom.

• **8.6** Which of the quantum numbers relate(s) to the electron only? Which relate(s) to the orbital?

• **8.7** State the exclusion principle. What does it imply about the number and spin of electrons in an atomic orbital?

• **8.8** What is the key distinction between sublevel energies in one-electron species, such as the H atom, and those in many-electron species, such as the C atom? What factors lead to this distinction? Would you expect the pattern of sublevel energies in Be^{3+} to be more like that in H or that in C? Explain.

• **8.9** Define *shielding* and *effective nuclear charge*. What is the connection between the two?

• **8.10** What is penetration? How is it related to shielding? Use the penetration effect to explain the difference in relative orbital energies of a $3p$ and a $3d$ electron in the same atom.

● **Skill-Building Exercises** (grouped in similar pairs)

• **8.11** How many electrons in an atom can have each of the following quantum number or sublevel designations?
(a) $n = 2, l = 1$ (b) $3d$ (c) $4s$

• **8.12** How many electrons in an atom can have each of the following quantum number or sublevel designations?
(a) $n = 2, l = 1, m_l = 0$ (b) $5p$ (c) $n = 4, l = 3$

• **8.13** How many electrons in an atom can have each of the following quantum number or sublevel designations?
(a) $4p$ (b) $n = 3, l = 1, m_l = +1$ (c) $n = 5, l = 3$

• **8.14** How many electrons in an atom can have each of the following quantum number or sublevel designations?
(a) $2s$ (b) $n = 3, l = 2$ (c) $6d$

The Quantum-Mechanical Model and the Periodic Table
(Sample Problems 8.1 and 8.2)

● **Concept Review Questions**

• **8.15** State the periodic law, and explain its relation to electron configuration. (Use Na and K in your explanation.)

• **8.16** State Hund's rule in your own words, and show its application in the orbital diagram of the nitrogen atom.

• **8.17** How does the aufbau principle, in connection with the periodic law, lead to the format of the periodic table?

• **8.18** For main-group elements, are outer electron configurations similar or different within a group? Within a period? Explain.

• **8.19** For which blocks of elements are outer electrons the same as valence electrons? For which are d electrons often included among valence electrons?

• **8.20** What is the electron capacity of the nth energy level? What is the capacity of the fourth energy level?

● **Skill-Building Exercises** (grouped in similar pairs)

• **8.21** Write a full set of quantum numbers for the following:
(a) The outermost electron in an Rb atom
(b) The electron gained when an S^- ion becomes an S^{2-} ion
(c) The electron lost when an Ag atom ionizes
(d) The electron gained when an F^- ion forms from an F atom

• **8.22** Write a full set of quantum numbers for the following:
(a) The outermost electron in an Li atom
(b) The electron gained when a Br atom becomes a Br^- ion
(c) The electron lost when a Cs atom ionizes
(d) The highest energy electron in the ground-state B atom

• **8.23** Write the full ground-state electron configuration for each:
(a) Rb (b) Ge (c) Ar

• **8.24** Write the full ground-state electron configuration for each:
(a) Br (b) Mg (c) Se

• **8.25** Write the full ground-state electron configuration for each:
(a) Cl (b) Si (c) Sr

• **8.26** Write the full ground-state electron configuration for each:
(a) S (b) Kr (c) Cs

• **8.27** Draw an orbital diagram showing valence electrons, and write the condensed ground-state electron configuration for each:
(a) Ti (b) Cl (c) V

• **8.28** Draw an orbital diagram showing valence electrons, and write the condensed ground-state electron configuration for each:
(a) Ba (b) Co (c) Ag

• **8.29** Draw an orbital diagram showing valence electrons, and write the condensed ground-state electron configuration for each:
(a) Mn (b) P (c) Fe

• **8.30** Draw an orbital diagram showing valence electrons, and write the condensed ground-state electron configuration for each:
(a) Ga   (b) Zn   (c) Sc

• **8.31** Draw the partial (valence-level) orbital diagram, and write the symbol, group number, and period number of the element:
(a) [He] $2s^22p^4$ (b) [Ne] $3s^23p^3$

• **8.32** Draw the partial (valence-level) orbital diagram, and write the symbol, group number, and period number of the element:
(a) [Kr] $5s^24d^{10}$ (b) [Ar] $4s^23d^8$

• **8.33** Draw the partial (valence-level) orbital diagram, and write the symbol, group number, and period number of the element:
(a) [Ne] $3s^23p^5$ (b) [Ar] $4s^23d^{10}4p^3$

• **8.34** Draw the partial (valence-level) orbital diagram, and write the symbol, group number, and period number of the element:
(a) [Ar] $4s^23d^5$ (b) [Kr] $5s^24d^2$

• **8.35** From each partial (valence-level) orbital diagram, write the ground-state electron configuration and group number:

(a)
   4s   3d   4p

(b)
  2s   2p

• **8.36** From each partial (valence-level) orbital diagram, write the ground-state electron configuration and group number:

(a)
  5s   4d   5p

(b)
  2s   2p

• **8.37** How many inner, outer, and valence electrons are present in an atom of each of the following elements?
(a) O   (b) Sn   (c) Ca   (d) Fe   (e) Se

• **8.38** How many inner, outer, and valence electrons are present in an atom of each of the following elements?
(a) Br   (b) Cs   (c) Cr   (d) Sr   (e) F

• **8.39** Identify each element below, and give the symbols of the other elements in its group:
(a) [He] $2s^22p^1$ (b) [Ne] $3s^23p^4$ (c) [Xe] $6s^25d^1$

• **8.40** Identify each element below, and give the symbols of the other elements in its group:
(a) [Ar] $4s^23d^{10}4p^4$ (b) [Xe] $6s^24f^{14}5d^2$ (c) [Ar] $4s^23d^5$

• **8.41** Identify each element below, and give the symbols of the other elements in its group:
(a) [He] $2s^22p^2$ (b) [Ar] $4s^23d^3$ (c) [Ne] $3s^23p^3$

• **8.42** Identify each element below, and give the symbols of the other elements in its group:
(a) [Ar] $4s^23d^{10}4p^2$ (b) [Ar] $4s^23d^7$ (c) [Kr] $5s^24d^5$

● **Problems in Context**

• **8.43** After an atom in its ground state absorbs energy, it exists in an excited state. Spectral lines are produced when the atom returns to its ground state. The yellow-orange line in the sodium spectrum, for example, is produced by the emission of energy when excited sodium atoms return to their ground state. Write the electron configuration and the orbital diagram of the first excited state of sodium. (*Hint:* The outermost electron is excited.)

• **8.44** One reason spectroscopists study excited states is to gain information about the energies of orbitals that are unoccupied in an atom's ground state. Each of the following electron configurations represents an atom in an excited state. Identify the element, and write its condensed ground-state configuration:
(a) $1s^22s^22p^63s^13p^1$ (b) $1s^22s^22p^63s^23p^44s^1$
(c) $1s^22s^22p^63s^23p^64s^23d^44p^1$ (d) $1s^22s^22p^53s^1$

Trends in Three Key Atomic Properties
(Sample Problems 8.3 to 8.5)

● **Concept Review Questions**

• **8.45** If the exact outer limit of an isolated atom cannot be measured, what criterion can we use to determine atomic radii? What is the difference between a covalent radius and a metallic radius?

•• **8.46** Given the following partial (valence-level) electron configurations, (a) identify each element, (b) rank the four elements in order of increasing atomic size, and (c) rank them in order of increasing ionization energy:

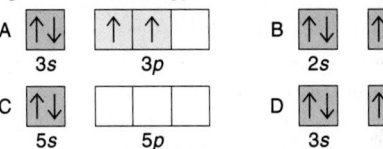

• **8.47** In what region of the periodic table will you find elements with relatively high IEs? With relatively low IEs?

•• **8.48** (a) Why do successive IEs of a given element always increase? (b) When the difference between successive IEs of a given element is exceptionally large (for example, between IE_1 and IE_2 of K), what do we learn about its electron configuration? (c) The bars represent the relative magnitudes of the first five ionization energies of an atom:

Identify the element and write its complete electron configuration, assuming it comes from (a) Period 2; (b) Period 3; (c) Period 4.

• **8.49** In a plot of IE_1 for the Period 3 elements (see Figure 8.17, p. 321), why do the values for elements in Groups 3A(13) and 6A(16) drop slightly below the generally increasing trend?

• **8.50** Which group in the periodic table has elements with high (endothermic) IE_1 and very negative (exothermic) first electron affinities (EA_1)? Give the charge on the ions these atoms form.

• **8.51** The EA_2 of an oxygen atom is positive, even though its EA_1 is negative. Why does this change of sign occur? Which other elements exhibit a positive EA_2? Explain.

• **8.52** How does *d*-electron shielding influence atomic size among the Period 4 transition elements?

● **Skill-Building Exercises** (*grouped in similar pairs*)

• **8.53** Arrange each set in order of *increasing* atomic size:
(a) Rb, K, Cs   (b) C, O, Be   (c) Cl, K, S   (d) Mg, K, Ca

• **8.54** Arrange each set in order of *decreasing* atomic size:
(a) Ge, Pb, Sn   (b) Sn, Te, Sr   (c) F, Ne, Na   (d) Be, Mg, Na

• **8.55** Arrange each set of atoms in order of *increasing* IE_1:
(a) Sr, Ca, Ba   (b) N, B, Ne   (c) Br, Rb, Se   (d) As, Sb, Sn

• **8.56** Arrange each set of atoms in order of *decreasing* IE_1:
  (a) Na, Li, K   (b) Be, F, C   (c) Cl, Ar, Na   (d) Cl, Br, Se

• **8.57** Write the full electron configuration of the Period 2 element with the following successive IEs (in kJ/mol):

  $IE_1 = 801$ $IE_2 = 2427$ $IE_3 = 3659$
  $IE_4 = 25,022$ $IE_5 = 32,822$

• **8.58** Write the full electron configuration of the Period 3 element with the following successive IEs (in kJ/mol):

  $IE_1 = 738$ $IE_2 = 1450$ $IE_3 = 7732$
  $IE_4 = 10,539$ $IE_5 = 13,628$

• **8.59** Which element in each of the following sets would you expect to have the *highest* IE_2?
  (a) Na, Mg, Al   (b) Na, K, Fe   (c) Sc, Be, Mg

• **8.60** Which element in each of the following sets would you expect to have the *lowest* IE_3?
  (a) Na, Mg, Al   (b) K, Ca, Sc   (c) Li, Al, B

Atomic Structure and Chemical Reactivity
(Sample Problems 8.6 to 8.8)

● **Concept Review Questions**

• **8.61** List three ways in which metals and nonmetals differ.

• **8.62** Summarize the trend in metallic character as a function of position in the periodic table. Is it the same as the trend in atomic size? Ionization energy?

• **8.63** Summarize the acid-base behavior of the main-group metal and nonmetal oxides in water. How does oxide acidity in water change down a group and across a period?

• **8.64** What ions are possible for the two largest stable elements in Group 4A(14)? How does each arise?

• **8.65** What is a pseudo–noble gas configuration? Give an example of one ion from Group 3A(13) that has it.

• **8.66** How are measurements of paramagnetism used to support electron configurations derived spectroscopically? Use Cu(I) and Cu(II) chlorides as examples.

• **8.67** The charges of a set of isoelectronic ions vary from 3+ to 3−. Place the ions in order of increasing size.

● **Skill-Building Exercises** (grouped in similar pairs)

• **8.68** Which element would you expect to be *more* metallic?
  (a) Ca or Rb   (b) Mg or Ra   (c) Br or I

• **8.69** Which element would you expect to be *more* metallic?
  (a) S or Cl   (b) In or Al   (c) As or Br

• **8.70** Which element would you expect to be *less* metallic?
  (a) Sb or As   (b) Si or P   (c) Be or Na

• **8.71** Which element would you expect to be *less* metallic?
  (a) Cs or Rn   (b) Sn or Te   (c) Se or Ge

• **8.72** Does the reaction of a main-group nonmetal oxide in water produce an acidic or a basic solution? Write a balanced equation for the reaction of a Group 6A(16) nonmetal oxide with water.

• **8.73** Does the reaction of a main-group metal oxide in water produce an acidic solution or a basic solution? Write a balanced equation for the reaction of a Group 2A(2) oxide with water.

• **8.74** Write the charge and full ground-state electron configuration of the monatomic ion most likely to be formed by each:
  (a) Cl   (b) Na   (c) Ca

• **8.75** Write the charge and full ground-state electron configuration of the monatomic ion most likely to be formed by each:
  (a) Rb   (b) N   (c) Br

• **8.76** Write the charge and full ground-state electron configuration of the monatomic ion most likely to be formed by each:
  (a) Al   (b) S   (c) Sr

• **8.77** Write the charge and full ground-state electron configuration of the monatomic ion most likely to be formed by each:
  (a) P   (b) Mg   (c) Se

• **8.78** How many unpaired electrons are present in the ground state of an atom from each of the following groups?
  (a) 2A(2)   (b) 5A(15)   (c) 8A(18)   (d) 3A(13)

• **8.79** How many unpaired electrons are present in the ground state of an atom from each of the following groups?
  (a) 4A(14)   (b) 7A(17)   (c) 1A(1)   (d) 6A(16)

• **8.80** Which of these are paramagnetic in their ground state?
  (a) Ga   (b) Si   (c) Be   (d) Te

• **8.81** Are compounds of these ground-state ions paramagnetic?
  (a) Ti^{2+} (b) Zn^{2+} (c) Ca^{2+} (d) Sn^{2+}

• **8.82** Write the condensed ground-state electron configurations of these transition metal ions, and state which are paramagnetic:
  (a) V^{3+} (b) Cd^{2+} (c) Co^{3+} (d) Ag^+

• **8.83** Write the condensed ground-state electron configurations of these transition metal ions, and state which are paramagnetic:
  (a) Mo^{3+} (b) Au^+ (c) Mn^{2+} (d) Hf^{2+}

• **8.84** Palladium (Pd; $Z = 46$) is diamagnetic. Draw partial orbital diagrams to show which of the following electron configurations is consistent with this fact:
  (a) $[Kr]\, 5s^2 4d^8$ (b) $[Kr]\, 4d^{10}$ (c) $[Kr]\, 5s^1 4d^9$

• **8.85** Niobium (Nb; $Z = 41$) has an anomalous ground-state electron configuration for a Group 5B(5) element: $[Kr]\, 5s^1 4d^4$. What is the expected electron configuration for elements in this group? Draw partial orbital diagrams to show how paramagnetic measurements could support niobium's actual configuration.

• **8.86** Rank the ions in each set in order of *increasing* size, and explain your ranking:
  (a) Li^+, K^+, Na^+ (b) Se^{2-}, Rb^+, Br^- (c) O^{2-}, F^-, N^{3-}

• **8.87** Rank the ions in each set in order of *decreasing* size, and explain your ranking:
  (a) Se^{2-}, S^{2-}, O^{2-} (b) Te^{2-}, Cs^+, I^- (c) Sr^{2+}, Ba^{2+}, Cs^+

Comprehensive Problems

•• **8.88** Some versions of the periodic table show hydrogen at the top of Group 1A(1) *and* at the top of Group 7A(17). What properties of hydrogen justify each of these placements?

•• **8.89** Name the element described in each of the following:
  (a) Smallest atomic radius in Group 6A(16)
  (b) Largest atomic radius in Period 6
  (c) Smallest metal in Period 3
  (d) Highest IE_1 in Group 4A(14)
  (e) Lowest IE_1 in Period 5
  (f) Most metallic in Group 5A(15)
  (g) Group 3A(13) element that forms the most basic oxide
  (h) Period 4 element with filled outer level
  (i) Condensed ground-state electron configuration of $[Ne]\, 3s^2 3p^2$
  (j) Condensed ground-state electron configuration of $[Kr]\, 5s^2 4d^6$
  (k) Forms 2+ ion with electron configuration $[Ar]\, 3d^3$
  (l) Period 5 element that forms 3+ ion with pseudo–noble gas configuration
  (m) Period 4 transition element that forms 3+ diamagnetic ion
  (n) Period 4 transition element that forms 2+ ion with a half-filled d sublevel

(o) Heaviest lanthanide
(p) Period 3 element whose 2− ion is isoelectronic with Ar
(q) Alkaline earth metal whose cation is isoelectronic with Kr
(r) Group 5A(15) metalloid with the most acidic oxide

8.90 Use electron configurations to account for the stability of the lanthanide ions Ce^{4+} and Eu^{2+}.

8.91 The NaCl crystal structure consists of alternating Na^+ and Cl^- ions lying next to each other in three dimensions. If the Na^+ radius is 56.4% of the Cl^- radius and the distance between Na^+ nuclei is 566 pm, what are the radii of the two ions?

8.92 When a nonmetal oxide reacts with water, it forms an oxoacid with the same nonmetal oxidation state. Give the name and formula of the oxide used to prepare each of these oxoacids: (a) hypochlorous acid; (b) chlorous acid; (c) chloric acid; (d) perchloric acid; (e) sulfuric acid; (f) sulfurous acid; (g) nitric acid; (h) nitrous acid; (i) carbonic acid; (j) phosphoric acid.

8.93 A fundamental relationship of electrostatics states that the energy required to separate opposite charges of magnitudes Q_1 and Q_2 that are the distance d apart is proportional to $\dfrac{Q_1 \times Q_2}{d}$.

Use this relationship and any other factors to explain the following observations: (a) the IE_2 of He ($Z = 2$) is *more* than twice the IE_1 of H ($Z = 1$); (b) the IE_1 of He is *less* than twice the IE_1 of H.

8.94 The energy difference between the $5d$ and $6s$ sublevels in gold accounts for its color. Assuming this energy difference is about 2.7 eV (electron volt; 1 eV $= 1.602 \times 10^{-19}$ J), explain why gold has a warm yellow color.

8.95 Write the formula and name of the compound formed from the following ionic interactions: (a) The 2+ ion and the 1− ion are both isoelectronic with the atoms of a chemically unreactive Period 4 element. (b) The 2+ ion and the 2− ion are both isoelectronic with the Period 3 noble gas. (c) The 2+ ion is the smallest with a filled d subshell; the anion forms from the smallest halogen. (d) The ions form from the largest and smallest ionizable atoms in Period 2.

8.96 The energy changes for many unusual reactions can be determined using Hess's law (Section 6.5).
(a) Calculate ΔE for the conversion of $F^-(g)$ into $F^+(g)$.
(b) Calculate ΔE for the conversion of $Na^+(g)$ into $Na^-(g)$.

8.97 Use Table 8.5, p. 323, to explain the major differences in the relative values of IE_1 through IE_5 for carbon and oxygen.

8.98 The hot glowing gases around the Sun, the *corona*, can reach millions of degrees Celsius, high enough to remove many electrons from gaseous atoms. Iron ions with charges as high as 14+ have been observed in the corona. Which ions from Fe^+ to Fe^{14+} are paramagnetic? Which would be most strongly attracted to a magnetic field?

8.99 There are some exceptions to the trends of first and successive ionization energies. For each of the following pairs, explain which ionization energy would be higher:
(a) IE_1 of Ga or IE_1 of Ge   (b) IE_2 of Ga or IE_2 of Ge
(c) IE_3 of Ga or IE_3 of Ge   (d) IE_4 of Ga or IE_4 of Ge

8.100 Use Figure 8.18, p. 322, to find: (a) the longest wavelength of electromagnetic (EM) radiation that can ionize an alkali metal atom; (b) the longest wavelength of EM radiation that can ionize an alkaline earth metal atom; (c) the elements, other than the alkali and alkaline earth metals, that could also be ionized by the radiation of part (b); (d) the region of the EM spectrum in which these photons are found.

8.101 Half of the first 18 elements have an odd number of electrons, and half have an even number. Show why atoms of these elements aren't half paramagnetic and half diamagnetic.

8.102 Rubidium and bromine atoms are depicted at right. (a) What monatomic ions do they form? (b) What electronic feature characterizes this pair of ions, and which noble gas are they related to? (c) Which pair best represents the relative ionic sizes?

A   **B**   **C**   **D**

8.103 Partial (valence-level) electron configurations for four different ions are shown below:

Identify the elements from which the ions are derived, and write the formula of the oxide each ion forms.

8.104 As a science major, you are assigned to the "Atomic Dorm" at college. The first floor has one suite with one bedroom that has a bunk bed for two students. The second floor has two suites, one like the one on the first floor, and the other with three bedrooms, each with a bunk bed for two students. The third floor has three suites, two like the ones on the second floor, and a third suite that has five bedrooms, each with a bunk bed for two students. Entering students choose room and bunk on a first-come, first-serve basis by criteria in the following order of importance:
(1) They want to be on the lowest available floor.
(2) They want to be in the smallest available dorm room.
(3) They want to be in a lower bunk if available.
(a) Which bunk does the 1st student choose? (b) How many students are in top bunks when the 17th student chooses? (c) Which bunk does the 21st student choose? (d) How many students are in bottom bunks when the 25th student chooses?

8.105 Data from the planet Zog for some main-group elements are shown below (Zoggian units are linearly related to Earth units but are not shown). Radio signals from Zog tell that balloonium is a monatomic gas with two positive nuclear charges. Use the data to deduce the names that Earthlings give to these elements:

Name	Atomic Radius	IE_1	EA_1
Balloonium	10	339	0
Inertium	24	297	+4.1
Allotropium	34	143	−28.6
Brinium	63	70.9	−7.6
Canium	47	101	−15.3
Fertilium	25	200	0
Liquidium	38	163	−46.4
Utilium	48	82.4	−6.1
Crimsonium	72	78.4	−2.9

9

ZAP! A single photon (yellow) causes H_2 to break up into its constituent electrons (blue) and protons (red). Using supercomputers, researchers have solved the Schrödinger equation for the light-induced breakage of a covalent bond and developed a model to study the interactions among nuclei and electrons. In this chapter, we examine three general models of chemical bonding.

Models of Chemical Bonding

9.1 Atomic Properties and Chemical Bonds
Three Types of Chemical Bonding
Lewis Electron-Dot Symbols

9.2 The Ionic Bonding Model
Importance of Lattice Energy
Periodic Trends in Lattice Energy
How the Model Explains the Properties of Ionic Compounds

9.3 The Covalent Bonding Model
Formation of a Covalent Bond
Bond Energy and Bond Length
How the Model Explains the Properties of Covalent Substances

9.4 Bond Energy and Chemical Change
Where Does ΔH°_{rxn} Come From?
Using Bond Energies to Calculate ΔH°_{rxn}
Bond Strengths in Fuels and Foods

9.5 Between the Extremes: Electronegativity and Bond Polarity
Electronegativity
Polar Covalent Bonds and Bond Polarity
Partial Ionic Character of Polar Covalent Bonds
Continuum of Bonding Across a Period

9.6 An Introduction to Metallic Bonding
The Electron-Sea Model
How the Model Explains the Properties of Metals

W hy do the substances around us behave as they do? That is, why is table salt (or any other ionic substance) a hard, brittle, high-melting solid that conducts a current only when molten or dissolved in water? Why is candle wax (along with most covalent substances) low melting, soft, and nonconducting, even though diamond and a few other exceptions are high melting and extremely hard? And why is copper (and most other metallic substances) shiny, malleable, and able to conduct a current whether molten or solid? The answers lie in the *type of bonding within the substance.* In Chapter 8, we examined the properties of individual atoms and ions. Yet, in virtually all the substances in and around you, these particles are bonded to one another. As you'll see, deeper insight comes as we discover how the properties of atoms influence the types of chemical bonds they form, because this is ultimately responsible for the behavior of substances.

IN THIS CHAPTER . . . We examine how atomic properties give rise to the three major types of bonding—ionic, covalent, and metallic—and how each model of bonding explains the properties of substances. For ionic bonding, we detail the steps and calculate the energy involved in the formation of an ionic solid from its elements. Covalent bonding occurs in the vast majority of compounds. We examine the formation and characteristics—order, energy, and length—of a covalent bond. We explore the range of bonding, from pure covalent to ionic, in terms of electronegativity and bond polarity and see that most bonds fall somewhere between these extremes. Finally, we consider a simple model that explains metallic bonding.

Concepts & Skills to Review
before you study this chapter

- characteristics of ionic and covalent bonding (Section 2.7)
- polar covalent bonds and the polarity of water (Section 4.1)
- Hess's law, ΔH°_{rxn}, and ΔH°_f (Sections 6.5 and 6.6)
- atomic and ionic electron configurations (Sections 8.3 and 8.5)
- trends in atomic properties and metallic behavior (Sections 8.4 and 8.5)

Literature Robinson, William R. "An Alternative Framework for Chemical Bonding," *J. Chem. Educ.* 1998, *75,* 1074.

9.1 ATOMIC PROPERTIES AND CHEMICAL BONDS

Before we examine the types of chemical bonding, we should ask why atoms bond at all. In general terms, *bonding lowers the potential energy between positive and negative particles,* whether those particles are oppositely charged ions or nuclei and the electrons between them. Just as the electron configuration and the strength of the nucleus-electron attraction(s) determine the properties of an atom, the type and strength of chemical bonds determine the properties of a substance.

The Three Types of Chemical Bonding

On the atomic level, we distinguish a metal from a nonmetal on the basis of several properties that correlate with position in the periodic table (Figure 9.1 and inside the front cover). Recall from Chapter 8 that, in general, there is a gradation

Transparency/ARIS Presentation Center, Art Library

Figure 9.1 A general comparison of metals and nonmetals. A, The positions of metals, nonmetals, and metalloids are shown within the periodic table. **B,** The relative magnitudes of some key atomic properties vary from left to right within a period and correlate with whether an element is metallic or nonmetallic.

PROPERTY	METAL ATOM	NONMETAL ATOM
Atomic size	Larger	Smaller
Z_{eff}	Lower	Higher
IE	Lower	Higher
EA	Less negative	More negative

Key:
- ▪ Metals
- ▫ Nonmetals
- ▪ Metalloids

Periodic table (A):

1A (1)												3A (13)	4A (14)	5A (15)	6A (16)	7A (17)	8A (18)
	2A (2)															H	He
Li	Be											B	C	N	O	F	Ne
Na	Mg	3B (3)	4B (4)	5B (5)	6B (6)	7B (7)	8B (8)	(9)	(10)	1B (11)	2B (12)	Al	Si	P	S	Cl	Ar
K	Ca	Sc	Ti	V	Cr	Mn	Fe	Co	Ni	Cu	Zn	Ga	Ge	As	Se	Br	Kr
Rb	Sr	Y	Zr	Nb	Mo	Tc	Ru	Rh	Pd	Ag	Cd	In	Sn	Sb	Te	I	Xe
Cs	Ba	La	Hf	Ta	W	Re	Os	Ir	Pt	Au	Hg	Tl	Pb	Bi	Po	At	Rn
Fr	Ra	Ac	Rf	Db	Sg	Bh	Hs	Mt	Ds	Rg	112	113	114	115	116		

Ce	Pr	Nd	Pm	Sm	Eu	Gd	Tb	Dy	Ho	Er	Tm	Yb	Lu
Th	Pa	U	Np	Pu	Am	Cm	Bk	Cf	Es	Fm	Md	No	Lr

A

B

Literature Pauling, Linus. "The Nature of the Chemical Bond-1992," *J. Chem. Educ.* **1992**, *69*, 519.

Literature Pauling, Linus. "The Origins of Bonding Concepts," *J. Chem. Educ.* **1985**, *62*, 362.

Literature Meek, Terry L.; Garner, Leah D. "Electronegativity and the Bond Triangle," *J. Chem. Educ.* **2005**, *82*, 325.

from more metal-like to more nonmetal-like behavior from left to right across a period and from bottom to top within most groups. Three types of bonding result from the three ways these two types of atoms can combine—metal with nonmetal, nonmetal with nonmetal, and metal with metal:

1. *Metal with nonmetal: electron transfer and ionic bonding* (Figure 9.2A). We typically observe **ionic bonding** between atoms with large differences in their tendencies to lose or gain electrons. Such differences occur between reactive metals [Groups 1A(1) and 2A(2)] and nonmetals [Group 7A(17) and the top of Group 6A(16)]. The metal atom (low IE) loses its one or two valence electrons, whereas the nonmetal atom (highly negative EA) gains the electron(s). *Electron transfer* from metal to nonmetal occurs, and each atom forms an ion with a noble gas electron configuration. The electrostatic attraction between these positive and negative ions draws them into the three-dimensional array of an ionic solid, whose chemical formula represents the cation-to-anion ratio (empirical formula).

2. *Nonmetal with nonmetal: electron sharing and covalent bonding* (Figure 9.2B). When two atoms have a small difference in their tendencies to lose or gain electrons, we observe *electron sharing* and **covalent bonding.** This type of bonding most commonly occurs between nonmetal atoms (although a pair of metal atoms can sometimes form a covalent bond). Each nonmetal atom holds onto its own electrons tightly (high IE) and tends to attract other electrons as well (highly negative EA). The attraction of each nucleus for the valence electrons of the other draws the atoms together. A shared electron pair is considered to be *localized* between the two atoms because it spends most of its time there, linking them in a covalent bond of a particular length and strength. In most cases, separate molecules form when covalent bonding occurs, and the chemical formula reflects the actual numbers of atoms in the molecule (molecular formula).

3. *Metal with metal: electron pooling and metallic bonding* (Figure 9.2C). In general, metal atoms are relatively large, and their few outer electrons are well

A Ionic bonding   **B Covalent bonding**   **C Metallic bonding**

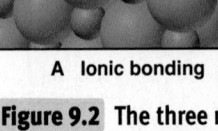

Figure 9.2 The three models of chemical bonding. A, In ionic bonding, metal atoms transfer electron(s) to nonmetal atoms, forming oppositely charged ions that attract each other to form a solid. **B,** In covalent bonding, two atoms share an electron pair localized between their nuclei (shown here as a bond line). Most covalent substances consist of individual molecules, each made from two or more atoms. **C,** In metallic bonding, many metal atoms pool their valence electrons to form a delocalized electron "sea" that holds the metal-ion cores together.

Transparency/ARIS Presentation Center, Art Library

shielded by filled inner levels. Thus, they lose outer electrons comparatively easily (low IE) but do not gain them very readily (slightly negative or positive EA). These properties lead large numbers of metal atoms to share their valence electrons, but in a way that differs from covalent bonding. In the simplest model of **metallic bonding,** all the metal atoms in a sample *pool* their valence electrons into an evenly distributed "sea" of electrons that "flows" between and around the metal-ion cores (nucleus plus inner electrons), attracting them and holding them together. Unlike the localized electrons in covalent bonding, electrons in metallic bonding are *delocalized,* moving freely throughout the piece of metal.

It's important to remember that in the world of real substances, there are exceptions to these idealized bonding models. You cannot always predict the bond type solely from the elements' positions in the periodic table. For instance, all binary ionic compounds contain a metal and a nonmetal, but all metals do not form binary ionic compounds with all nonmetals. As just one example, when the metal beryllium [Group 2A(2)] combines with the nonmetal chlorine [Group 7A(17)], the bonding fits the covalent model better than the ionic model. In other words, just as we see a gradation in metallic behavior within groups and periods, we also see a gradation in bonding from one type to another.

Literature Diemente, Damon. "The Gravity of the Situation," *J. Chem. Educ.* **1999,** *76,* 55.

Lewis Electron-Dot Symbols: Depicting Atoms in Chemical Bonding

Before turning to the individual models, let's discuss a method for depicting the valence electrons of interacting atoms. In the **Lewis electron-dot symbol** (named for the American chemist G. N. Lewis), the element symbol represents the nucleus *and* inner electrons, and the surrounding dots represent the valence electrons (Figure 9.3). Note that the pattern of dots is the same for elements within a group.

	1A(1)	2A(2)		3A(13)	4A(14)	5A(15)	6A(16)	7A(17)	8A(18)
	ns^1	ns^2		ns^2np^1	ns^2np^2	ns^2np^3	ns^2np^4	ns^2np^5	ns^2np^6
2	· Li	· Be ·		· B ·	· C·	· N·	:O·	:F:	:Ne:
3	· Na	·Mg·		· Al ·	· Si·	· P·	:S·	:Cl:	:Ar:

(left column label: Period)

Figure 9.3 **Lewis electron-dot symbols for elements in Periods 2 and 3.** The element symbol represents the nucleus and inner electrons, and the dots around it represent valence electrons, either paired or unpaired. The number of unpaired dots indicates the number of electrons a metal atom loses, or the number a nonmetal atom gains, or the number of covalent bonds a nonmetal atom usually forms.

 ARIS Presentation Center, Art Library

It's easy to write the Lewis symbol for any main-group element:

1. Note its A-group number (1A to 8A), which gives the number of valence electrons.
2. Place one dot at a time on the four sides (top, right, bottom, left) of the element symbol.
3. Keep adding dots, pairing the dots until all are used up.

The specific placement of dots is not important; that is, in addition to the one shown in Figure 9.3, the Lewis symbol for nitrogen can *also* be written as

$$\cdot\ddot{\text{N}}: \quad \text{or} \quad \cdot\dot{\text{N}}\cdot \quad \text{or} \quad :\dot{\text{N}}\cdot$$

The Lewis symbol provides information about an element's bonding behavior:

- For a metal, the *total* number of dots is the maximum number of electrons an atom loses to form a cation.
- For a nonmetal, the number of *unpaired* dots equals either the number of electrons an atom gains in becoming an anion or the number it shares in forming covalent bonds.

● **The Remarkable Insights of G. N. Lewis**
Many of the ideas in this text emerged from the mind of Gilbert Newton Lewis (1875–1946). As early as 1902, nearly a decade *before* Rutherford proposed the nuclear model of the atom, Lewis's notebooks show a scheme that involves the filling of outer electron "shells" to explain the way elements combine. His electron-dot symbols and associated structural formulas (discussed in Chapter 10) have become standards for representing bonding. Among his many other contributions is a more general model for the behavior of acids and bases (which we discuss in Chapter 18).

ARIS Presentation Center, Photo Library

To illustrate the last point, look at the Lewis symbol for carbon. Rather than one pair of dots and two unpaired dots, as its electron configuration ([He] $2s^22p^2$) would indicate, carbon has four unpaired dots because it forms four bonds. That is, in its compounds, carbon's four electrons are paired with four more electrons from its bonding partners for a total of eight electrons around carbon. (In Chapter 10, we'll see that larger nonmetals can form as many bonds as they have dots in the Lewis symbol.)

In his studies of bonding, Lewis generalized much of bonding behavior into the **octet rule:** *when atoms bond, they lose, gain, or share electrons to attain a filled outer level of eight (or two) electrons.* The octet rule holds for nearly all of the compounds of Period 2 elements and a large number of others as well. ●

Section Summary

Nearly all naturally occurring substances consist of atoms or ions bonded to others. Chemical bonding allows atoms to lower their energy. • Ionic bonding occurs when metal atoms transfer electrons to nonmetal atoms, and the resulting ions attract each other and form an ionic solid. • Covalent bonding most commonly occurs between nonmetal atoms and usually results in molecules. The bonded atoms share a pair of electrons, which remain localized between them. • Metallic bonding occurs when many metal atoms pool their valence electrons in a delocalized electron "sea" that holds all the atoms together. • The Lewis electron-dot symbol of an atom depicts the number of valence electrons for a main-group element. • In bonding, many atoms lose, gain, or share electrons to attain a filled outer level of eight (or two).

9.2 THE IONIC BONDING MODEL

The central idea of the ionic bonding model is the *transfer of electrons from metal atoms to nonmetal atoms to form ions that come together in a solid ionic compound.* For nearly every monatomic ion of a main-group element, the electron configuration has a filled outer level: either two or eight electrons, the same number as in the nearest noble gas (octet rule).

The transfer of an electron from a lithium atom to a fluorine atom is depicted in three ways in Figure 9.4. In each, Li loses its single outer electron and is left with a filled $n = 1$ level, while F gains a single electron to fill its $n = 2$ level. In this case, each atom is one electron away from its nearest noble gas—He for Li and Ne for F—so the number of electrons lost by each Li equals the number gained by each F. Therefore, equal numbers of Li^+ and F^- ions form, as the formula LiF indicates. That is, in ionic bonding, *the total number of electrons lost by the metal atoms equals the total number of electrons gained by the nonmetal atoms.*

Transparency/ARIS Presentation Center, Art Library

Figure 9.4 Three ways to represent the formation of Li$^+$ and F$^-$ through electron transfer. The electron being transferred is indicated in red.

SAMPLE PROBLEM 9.1 Depicting Ion Formation

PROBLEM Use partial orbital diagrams and Lewis symbols to depict the formation of Na^+ and O^{2-} ions from the atoms, and determine the formula of the compound the ions form.

PLAN First we draw the orbital diagrams and Lewis symbols for the Na and O atoms. To attain filled outer levels, Na loses one electron and O gains two. Thus, to make the number of electrons lost equal the number gained, two Na atoms are needed for each O atom.

SOLUTION

The formula is Na_2O.

FOLLOW-UP PROBLEM 9.1 Use condensed electron configurations and Lewis symbols to depict the formation of Mg^{2+} and Cl^- ions from the atoms, and write the formula of the ionic compound.

Energy Considerations in Ionic Bonding: The Importance of Lattice Energy

You may be surprised to learn that the electron-transfer process by itself actually *absorbs* energy! So why does it occur? As you'll learn now, the reason ionic substances exist at all is because of the enormous *release* of energy that occurs when the ions come together as a solid. Consider just the electron-transfer process for the formation of lithium fluoride, which involves two steps—a gaseous Li atom loses an electron, and a gaseous F atom gains it:

- The first ionization energy (IE_1) of Li is the energy change that occurs when 1 mol of gaseous Li atoms loses 1 mol of outer electrons:

$$Li(g) \longrightarrow Li^+(g) + e^- \qquad IE_1 = 520 \text{ kJ}$$

- The electron affinity (EA) of F is the energy change that occurs when 1 mol of gaseous F atoms gains 1 mol of electrons:

$$F(g) + e^- \longrightarrow F^-(g) \qquad EA = -328 \text{ kJ}$$

Note that the two-step electron-transfer process *by itself* requires energy:

$$Li(g) + F(g) \longrightarrow Li^+(g) + F^-(g) \qquad IE_1 + EA = 192 \text{ kJ}$$

The total energy needed for ion formation is even greater than this because metallic lithium and diatomic fluorine must first be converted to separate gaseous atoms, which also requires energy. Despite this, the standard heat of formation (ΔH_f°) of solid LiF is -617 kJ/mol; that is, 617 kJ is *released* when 1 mol of LiF(s) forms from its elements. The case of LiF is typical of many reactions between active metals and nonmetals: despite the endothermic electron transfer, ionic solids form readily, often vigorously. Figure 9.5 shows another example, the formation of NaBr.

Clearly, if the overall reaction of Li(s) and $F_2(g)$ to form LiF(s) releases energy, there must be some exothermic energy component large enough to overcome the endothermic steps. This component arises from the strong *attraction among many oppositely charged ions*. When 1 mol of $Li^+(g)$ and 1 mol of $F^-(g)$ form 1 mol of gaseous LiF molecules, a large quantity of heat is released:

$$Li^+(g) + F^-(g) \longrightarrow LiF(g) \qquad \Delta H^\circ = -755 \text{ kJ}$$

A

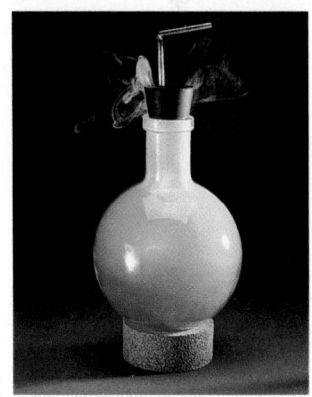

B

Figure 9.5 The reaction between sodium and bromine. A, Despite the endothermic electron-transfer process, all the Group 1A(1) metals react exothermically with any of the Group 7A(17) nonmetals to form solid alkali-metal halides. The reactants in the example shown are sodium (in beaker under mineral oil) and bromine (dark orange-brown substance in flask). **B,** The reaction is rapid and vigorous.

Literature Balaban, Alexandru T. "Chemistry Is As Easy As 2-3-4 (Types of Atoms—Chemical Bonds—Crystal Lattices)," *Chem. Educator* 9 **2004**, *6*, 356–358.

Literature Rakita, Philip E. "Dentifrice Fluoride," *J. Chem. Educ.* **2004**, *81*, 677.

As you know, under ordinary conditions, LiF does not consist of gaseous molecules. *Much more energy is released when the gaseous ions coalesce into a crystalline solid* because each ion attracts others of opposite charge:

$$Li^+(g) + F^-(g) \longrightarrow LiF(s) \qquad \Delta H° = -1050 \text{ kJ}$$

The negative of this enthalpy change is 1050 kJ, the lattice energy of LiF. The **lattice energy ($\Delta H°_{lattice}$)** is the enthalpy change that occurs when 1 mol of ionic solid separates into gaseous ions. It indicates the strength of ionic interactions, which influence melting point, hardness, solubility, and other properties.

Lattice energy plays the critical role in the formation of ionic compounds, but it cannot be measured directly. One way to determine lattice energy applies Hess's law, which states that the total enthalpy change of an overall reaction is the sum of the enthalpy changes of the individual reactions that make it up: $\Delta H_{total} = \Delta H_1 + \Delta H_2 + \cdots$ (see Section 6.5). Lattice energies are calculated by means of a **Born-Haber cycle,** a series of chosen steps from elements to ionic solid for which all the enthalpies* are known except the lattice energy.

Let's go through the Born-Haber cycle for lithium fluoride. Figure 9.6 shows two possible paths, either the direct combination reaction (black arrow) or the multistep path (orange arrows), one step of which involves the unknown lattice energy. Hess's law tells us both paths involve the same overall enthalpy change:

$$\Delta H°_f \text{ of LiF}(s) = \text{sum of } \Delta H° \text{ values for multistep path}$$

It's important to realize that we choose *hypothetical* steps whose enthalpy changes we can measure to depict the energy components of LiF formation, even though *these are **not** the actual steps that occur when lithium reacts with fluorine.*

Rather than working through a sample problem, let's examine the nature of the steps and describe how to calculate the lattice energy ($\Delta H°_{lattice}$) of LiF. (Problems 9.30 and 9.31 are among several that focus on Born-Haber cycles and the calculation of lattice energies.) We begin with the elements in their standard states, metallic lithium and gaseous diatomic fluorine. In the multistep process, the elements are converted to individual gaseous atoms (steps 1 and 2), gaseous ions form by electron transfers (steps 3 and 4), and the ions come together into a solid (step 5). We identify each $\Delta H°$ by its step number in Figure 9.6:

Step 1. Converting 1 mol of solid Li to separate gaseous Li atoms involves breaking metallic bonds, so this step requires energy:

$$Li(s) \longrightarrow Li(g) \qquad \Delta H°_{step\ 1} = 161 \text{ kJ}$$

(This process is called *atomization,* and the enthalpy change is $\Delta H°_{atom}$.)

Step 2. Converting an F_2 molecule to F atoms involves breaking a covalent bond, so it requires energy, which, as we discuss later, is called the *bond energy* (BE). We need 1 mol of F atoms for 1 mol of LiF, so we start with $\frac{1}{2}$ mol of F_2:

$$\frac{1}{2}F_2(g) \longrightarrow F(g) \quad \Delta H°_{step\ 2} = \frac{1}{2}(\text{BE of } F_2)$$
$$= \frac{1}{2}(159 \text{ kJ}) = 79.5 \text{ kJ}$$

Step 3. Removing the 2s electron from 1 mol of Li to form 1 mol of Li^+ requires energy:

$$Li(g) \longrightarrow Li^+(g) + e^- \qquad \Delta H°_{step\ 3} = IE_1 = 520 \text{ kJ}$$

Step 4. Adding an electron to each atom in 1 mol of F to form 1 mol of F^- releases energy:

$$F(g) + e^- \longrightarrow F^-(g) \qquad \Delta H°_{step\ 4} = EA = -328 \text{ kJ}$$

---

*Strictly speaking, ionization energy (IE) and electron affinity (EA) are internal energy changes (ΔE), not enthalpy changes (ΔH), but in these steps, $\Delta H = \Delta E$ because $\Delta V = 0$ (see Section 6.2).

Figure 9.6 **The Born-Haber cycle for lithium fluoride.** The formation of LiF(s) from its elements is shown happening either in one overall reaction *(black arrow)* or in five hypothetical steps, each with its own enthalpy change *(orange arrows).* The overall enthalpy change for the process (ΔH_f°) is known, as are $\Delta H_{step\ 1}^\circ$ through $\Delta H_{step\ 4}^\circ$. Therefore, $\Delta H_{step\ 5}^\circ$ ($-\Delta H_{lattice}^\circ$ of LiF) can be calculated to find the lattice energy. (ΔH_{atom}° is the heat of atomization; BE is the bond energy.)

Transparency/ARIS Presentation Center,
Active Art Library

Animation: Formation of an Ionic Compound
ARIS Presentation Center
Chapter 9

Step 5. Forming 1 mol of the crystalline ionic solid from the gaseous ions is the step whose enthalpy change (negative of the lattice energy) is unknown:

$$\text{Li}^+(g) + \text{F}^-(g) \longrightarrow \text{LiF}(s) \qquad \Delta H_{step\ 5}^\circ = -\Delta H_{lattice}^\circ \text{ of LiF} = ?$$

We know the enthalpy change of the formation reaction,

$$\text{Li}(s) + \tfrac{1}{2}\text{F}_2(g) \longrightarrow \text{LiF}(s) \qquad \Delta H_{overall}^\circ = \Delta H_f^\circ = -617 \text{ kJ}$$

Using Hess's law, we set the known ΔH_f° equal to the sum of the ΔH° values for the steps and calculate the lattice energy:

$$\Delta H_f^\circ = \Delta H_{step\ 1}^\circ + \Delta H_{step\ 2}^\circ + \Delta H_{step\ 3}^\circ + \Delta H_{step\ 4}^\circ + (-\Delta H_{lattice}^\circ \text{ of LiF})$$

Solving for $-\Delta H_{lattice}^\circ$ of LiF gives

$$
\begin{aligned}
-\Delta H_{lattice}^\circ \text{ of LiF} &= \Delta H_f^\circ - (\Delta H_{step\ 1}^\circ + \Delta H_{step\ 2}^\circ + \Delta H_{step\ 3}^\circ + \Delta H_{step\ 4}^\circ) \\
&= -617 \text{ kJ} - [\,161 \text{ kJ} + 79.5 \text{ kJ} + 520 \text{ kJ} + (-328 \text{ kJ})\,] \\
&= -1050 \text{ kJ}
\end{aligned}
$$

So, $\Delta H_{lattice}^\circ$ of LiF $= -(-1050 \text{ kJ}) = 1050 \text{ kJ}$

Note that *the magnitude of the lattice energy dominates the multistep process.*

The Born-Haber cycle reveals a central point: *ionic solids exist **only** because the lattice energy exceeds the energetically unfavorable electron transfer.* In other words, the energy *required* for elements to lose or gain electrons is *supplied* by the attraction among the ions formed: energy is expended to form the ions, but it is more than regained when they attract each other and form a solid.

Demo Shakhashiri, Bassam Z. "Thermite Reaction," *Chemical Demonstrations, A Handbook for Teachers of Chemistry,* Vol. 1 (The University of Wisconsin Press, Madison, 1985), 85–89.

Periodic Trends in Lattice Energy

Because the lattice energy is the result of electrostatic interactions among ions, we expect its magnitude to depend on several factors, including ionic size, ionic charge, and ionic arrangement in the solid. In Chapter 2, you were introduced to **Coulomb's law**, which tells us that the electrostatic energy between two charges (A and B) is directly proportional to the product of their magnitudes and inversely proportional to the distance between them:

$$\text{Electrostatic energy} \propto \frac{\text{charge A} \times \text{charge B}}{\text{distance}}$$

We can extend this relationship to lattice energy ($\Delta H^\circ_{\text{lattice}}$) because it is directly proportional to the electrostatic energy. In an ionic solid, cations and anions lie as close to each other as possible, so the distance between them is the distance between their centers, or the sum of their radii (see Figure 8.28, p. 332):

$$\text{Electrostatic energy} \propto \frac{\text{cation charge} \times \text{anion charge}}{\text{cation radius} + \text{anion radius}} \propto \Delta H^\circ_{\text{lattice}} \qquad \textbf{(9.1)}$$

This relationship helps us predict trends in lattice energy and explain the effects of ionic size and charge:

1. *Effect of ionic size.* As we move down a group in the periodic table, the ionic radius increases. Therefore, the electrostatic energy between cations and anions decreases because the interionic distance is greater; thus, the lattice energies of their compounds should decrease as well. This prediction is borne out by the alkali-metal halides (Figure 9.7): note the regular decrease in lattice energy down a group whether we hold the cation constant (LiF to LiI) or the anion constant (LiF to RbF).

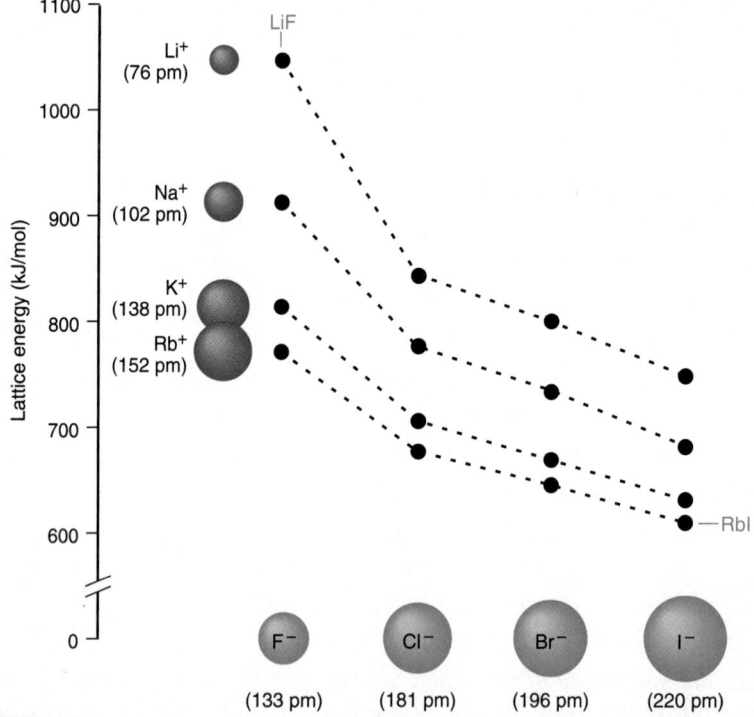

Figure 9.7 Trends in lattice energy. The lattice energies for many of the alkali-metal halides are shown. Each series of four points represents a given Group 1A(1) cation *(left side)* combining with each of the Group 7A(17) anions *(bottom)*. As ionic radii increase, the electrostatic attractions decrease, so the lattice energies of the compounds decrease as well. Thus, LiF (smallest ions shown) has the highest lattice energy, and RbI (largest ions) has the lowest. (Dotted lines connect data points; these are *not* continuous functions.)

2. *Effect of ionic charge.* When we compare lithium fluoride with magnesium oxide, we find cations of about equal radii ($Li^+ = 76$ pm and $Mg^{2+} = 72$ pm) and anions of about equal radii ($F^- = 133$ pm and $O^{2-} = 140$ pm). Thus, the only significant difference is the ionic charge: LiF contains the singly charged Li^+ and F^- ions, whereas MgO contains the doubly charged Mg^{2+} and O^{2-} ions. The difference in their lattice energies is striking:

$$\Delta H^\circ_{lattice} \text{ of LiF} = 1050 \text{ kJ/mol} \qquad \text{and} \qquad \Delta H^\circ_{lattice} \text{ of MgO} = 3923 \text{ kJ/mol}$$

This nearly fourfold increase in $\Delta H^\circ_{lattice}$ reflects the fourfold increase in the product of the charges (1×1 vs. 2×2) in the numerator of Equation 9.1.

We might ask why ionic solids like MgO, with its 2+ ions, even exist. After all, much more energy is needed to form 2+ ions than 1+ ions. Forming 1 mol of Mg^{2+} involves the sum of the first and second ionization energies for Mg:

$$Mg(g) \longrightarrow Mg^{2+}(g) + 2e^- \qquad \Delta H^\circ = IE_1 + IE_2 = 738 \text{ kJ} + 1450 \text{ kJ} = 2188 \text{ kJ}$$

Adding 1 mol of electrons to 1 mol of O atoms (first electron affinity, EA_1) is exothermic, but adding a second mole of electrons (second electron affinity, EA_2) is endothermic because an electron is being added to the negative O^- ion. The overall formation of 1 mol of O^{2-} ions is endothermic:

$$\begin{array}{lll}
O(g) + e^- \longrightarrow O^-(g) & \Delta H^\circ = EA_1 & = -141 \text{ kJ} \\
\underline{O^-(g) + e^- \longrightarrow O^{2-}(g)} & \underline{\Delta H^\circ = EA_2} & \underline{= \quad 878 \text{ kJ}} \\
O(g) + 2e^- \longrightarrow O^{2-}(g) & \Delta H^\circ = EA_1 + EA_2 = & 737 \text{ kJ}
\end{array}$$

In addition, there are the endothermic steps for converting 1 mol of Mg(s) to Mg(g) (148 kJ) and breaking $\frac{1}{2}$ mol of O_2 molecules into 1 mol of O atoms (249 kJ). Nevertheless, as a result of the ions' 2+ and 2− charges, solid MgO readily forms whenever Mg burns in air ($\Delta H^\circ_f = -601$ kJ/mol). Clearly, the enormous lattice energy ($\Delta H^\circ_{lattice}$ of MgO = 3923 kJ/mol) more than compensates for these endothermic steps.

How the Model Explains the Properties of Ionic Compounds

The first and most important job of any model is to explain the facts. By magnifying our view, we can see how the ionic bonding model accounts for the properties of ionic solids. You may have seen a piece of rock salt (NaCl). It is *hard* (does not dent), *rigid* (does not bend), and *brittle* (cracks without deforming). These properties are due to the powerful attractive forces that hold the ions *in specific positions* throughout the crystal. Moving the ions out of position requires overcoming these forces, so the sample resists denting and bending. If *enough* pressure is applied, ions of like charge are brought next to each other, and repulsive forces crack the sample suddenly (Figure 9.8).

Figure 9.8 Electrostatic forces and the reason ionic compounds crack. A, Ionic compounds are hard and will crack, rather than bend, when struck with enough force. **B,** The positive and negative ions in the crystal are arranged to maximize their attractions. When an external force is applied, like charges move near each other, and the repulsions crack the piece apart.

Figure 9.9 Electrical conductance and ion mobility. A, No current flows in the ionic solid because ions are immobile. **B,** In the molten compound, mobile ions flow toward the oppositely charged electrodes and carry a current. **C,** In an aqueous solution of the compound, mobile solvated ions carry a current.

 ARIS Presentation Center, Active Art Library

A Solid ionic compound   **B** Molten ionic compound   **C** Ionic compound dissolved in water

Table 9.1	Melting and Boiling Points of Some Ionic Compounds	
Compound	mp (°C)	bp (°C)
CsBr	636	1300
NaI	661	1304
$MgCl_2$	714	1412
KBr	734	1435
$CaCl_2$	782	>1600
NaCl	801	1413
LiF	845	1676
KF	858	1505
MgO	2852	3600

ARIS Presentation Center, Tables Library

ARIS Presentation Center, Art Library

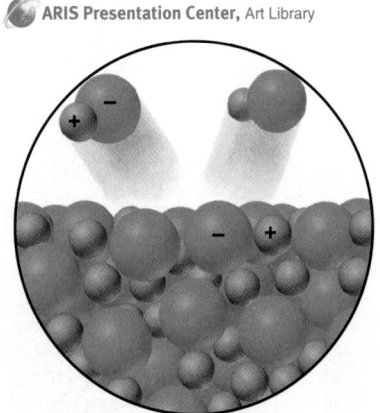

Figure 9.10 Vaporizing an ionic compound. Ionic compounds generally have very high boiling points because the ions must have high enough kinetic energies to break free from surrounding ions. In fact, ionic compounds usually vaporize as ion pairs.

Most ionic compounds *do not* conduct electricity in the solid state but *do* conduct it when melted or when dissolved in water. (Notable exceptions include so-called superionic conductors, such as AgI, and superconducting ceramics, which have remarkable conductivity in the solid state.) According to the ionic bonding model, the solid consists of immobilized ions. When it melts or dissolves, however, the ions are free to move and carry an electric current, as shown in Figure 9.9.

The model also explains that high temperatures are needed to melt and boil an ionic compound (Table 9.1) because freeing the ions from their positions (melting) requires large amounts of energy, and vaporizing them requires even more. In fact, the interionic attraction is so strong that, as Figure 9.10 shows, the vapor consists of **ion pairs,** gaseous ionic molecules rather than individual ions. But keep in mind that in their ordinary (solid) state, ionic compounds consist of arrays of alternating ions that extend in all directions, and *no separate molecules exist.*

Section Summary

In ionic bonding, a metal transfers electrons to a nonmetal, and the resulting ions attract each other strongly to form a solid. • Main-group elements often attain a filled outer level of electrons (either eight or two) by forming ions with the electron configuration of the nearest noble gas. • Ion formation by itself *requires* energy. However, the lattice energy, the energy *absorbed* when the solid separates into gaseous ions, is large and is the major reason ionic solids exist. The lattice energy, which depends on ionic size and charge, can be determined by applying Hess's law in a Born-Haber cycle. • The ionic bonding model pictures oppositely charged ions held in position by strong electrostatic attractions and explains why ionic solids crack rather than bend and why they conduct electric current only when melted or dissolved. Gaseous ion pairs form when an ionic compound vaporizes, which requires very high temperatures.

9.3 THE COVALENT BONDING MODEL

Look through any large reference source of chemical compounds, such as the *Handbook of Chemistry and Physics,* and you'll quickly find that the number of known covalent compounds dwarfs the number of known ionic compounds. Molecules held together by covalent bonds range from diatomic hydrogen to biological and synthetic macromolecules consisting of many hundreds or even thousands of atoms. We also find covalent bonds in many polyatomic ions. Without doubt, *sharing electrons is the principal way that atoms interact chemically.*

The Formation of a Covalent Bond

A sample of hydrogen gas consists of H_2 molecules. But why *do* the atoms bond to one another in pairs? Look at Figure 9.11 and imagine what happens to two isolated H atoms that approach each other from a distance (move *right* to *left* on the graph). When the atoms are far apart, each behaves as though the other were not present (point 1). As the distance between the nuclei decreases, each nucleus starts to attract the other atom's electron, which lowers the potential energy of the system. Attractions continue to draw the atoms closer, and the system becomes progressively lower in energy (point 2). As attractions increase, however, so do repulsions between the nuclei and between the electrons. At some internuclear distance, maximum attraction is achieved in the face of the increasing repulsion, and the system has its minimum energy (point 3, at the bottom of the energy "well"). Any shorter distance would increase repulsions and cause a rise in potential energy (point 4).* Thus a **covalent bond,** such as the one that holds the atoms together in the H_2 molecule, arises from the balance between nucleus-electron attractions and electron-electron

---

*In Chapter 11, we discuss valence bond theory, a quantum-mechanical model that explains covalent bonding in terms of the greater amplitude of overlapping atomic orbitals (wave functions) between two nuclei.

Transparency/ARIS Presentation Center, Art Library

Figure 9.11 Covalent bond formation in H_2. The potential energy of a system of two H atoms is plotted against the distance between the nuclei, with a depiction of the atomic systems above. At point 1, the atoms are too far apart to attract each other. At 2, each nucleus attracts the other atom's electron. At 3, the combination of nucleus-electron attractions and electron-electron and nucleus-nucleus repulsions gives the minimum energy of the system. The energy difference between points 1 and 3 is the H_2 bond energy (432 kJ/mol). It is released when the bond forms and must be absorbed to break the bond. The internuclear distance at point 3 is the H_2 bond length (74 pm). If the atoms move closer, as at point 4, repulsions increase the system's energy and force the atoms apart to point 3 again.

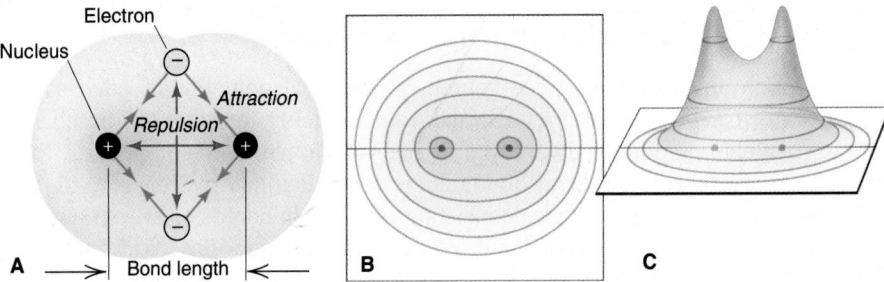

Figure 9.12 Distribution of electron density in H_2. A, At some optimum distance (bond length), attractions balance repulsions. Electron density *(blue shading)* is highest around and between the nuclei. **B,** This *contour map* represents a doubling of electron density with each contour line; the dots represent the nuclei. **C,** This *relief map* depicts the varying electron density of the contour map as peaks. The densest regions, by far, are around the nuclei (black dots on the "floor"), but the region between the nuclei—the bonding region—also has higher electron density.

ARIS Presentation Center, Art Library

Literature Shusterman, Gwendolyn P.; Shusterman, Alan J. "Teaching Chemistry with Electron Density Models," *J. Chem. Educ.* **1997,** *74,* 771.

Literature Matta, C.F.; Gillespie, R. "Understanding and Interpreting Molecular Electron Density Distributions," *J. Chem. Educ.* **2002,** *79,* 1141.

and nucleus-nucleus repulsions. Formation of a bond always results in *greater electron density between the nuclei.* Figure 9.12 depicts this fact in three ways: a cross-section of a space-filling model; an *electron density contour map,* with lines representing regular increments in electron density; and an *electron density relief map,* which portrays the contour map three-dimensionally as peaks of electron density.

Bonding Pairs and Lone Pairs In covalent bonding, as in ionic bonding, each atom achieves a full outer (valence) level of electrons, but this is accomplished by different means. *Each atom in a covalent bond "counts" the shared electrons as belonging entirely to itself.* Thus, the two electrons in the shared electron pair of H_2 simultaneously fill the outer level of *both* H atoms. The **shared pair,** or **bonding pair,** is represented by either a pair of dots or a line, H:H or H—H.

An outer-level electron pair that is *not* involved in bonding is called a **lone pair,** or **unshared pair.** The bonding pair in HF fills the outer level of the H atom *and,* together with three lone pairs, fills the outer level of the F atom as well:

bonding pair ⟨H ∶ F ∶⟩ lone pairs   or   H—F̈:

In F_2 the bonding pair and three lone pairs fill the outer level of *each* F atom:

:F̈ ∶ F̈:   or   :F̈—F̈:

(This text generally shows bonding pairs as lines and lone pairs as dots.)

Types of Bonds and Bond Order The **bond order** is the number of electron pairs being shared by any pair of bonded atoms. The covalent bond in H_2, HF, or F_2 is a **single bond,** one that consists of a single bonding pair of electrons. A *single bond has a bond order of* 1.

Single bonds are the most common type of bond, but many molecules (and ions) contain multiple bonds. Multiple bonds most frequently involve C, O, N, and/or S atoms. A **double bond** consists of two bonding electron pairs, four electrons shared between two atoms, so *the bond order is* 2. Ethylene (C_2H_4) is a simple hydrocarbon that contains a carbon-carbon double bond and four carbon-hydrogen single bonds:

structural formulas of ethylene   or   H₂C=CH₂

Each carbon "counts" the four electrons in the double bond and the four in its two single bonds to hydrogens to attain an octet. (In Chapter 10, we'll describe in detail how to draw a structural formula from a molecular formula.)

A **triple bond** consists of three bonding pairs; two atoms share six electrons, so *the bond order is* 3. In the N_2 molecule, the atoms are held together by a triple bond, and each N atom also has a lone pair:

:N ∶ N: or   :N≡N:

Six shared and two unshared electrons give *each* N atom an octet.

Animation: Formation of a Covalent Bond
ARIS Presentation Center
Chapter 9

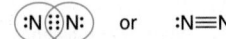

Properties of a Covalent Bond: Bond Energy and Bond Length

The strength of a covalent bond depends on the magnitude of the mutual attraction between bonded nuclei and shared electrons. The **bond energy (BE)** (also called *bond enthalpy* or *bond strength*) is the energy required to overcome this attraction and is defined as the standard enthalpy change for breaking the bond in 1 mol of gaseous molecules. *Bond breakage is an endothermic process, so the bond energy is always positive:*

$$A\text{—}B(g) \longrightarrow A(g) + B(g) \qquad \Delta H^\circ_{\text{bond breaking}} = BE_{A\text{—}B} \text{ (always} > 0)$$

Stated in another way, the bond energy is the difference in energy between the separated atoms and the bonded atoms (the potential energy difference between points 1 and 3 in Figure 9.11; the depth of the energy well). The energy absorbed to break the bond is released when the bond forms. *Bond formation is an exothermic process, so the sign of the enthalpy change is negative:*

$$A(g) + B(g) \longrightarrow A\text{—}B(g) \quad \Delta H^\circ_{\text{bond forming}} = -BE_{A\text{—}B} \text{ (always} < 0)$$

Because bond energies depend on characteristics of the bonded atoms—their electron configurations, nuclear charges, and atomic radii—each type of bond has its own bond energy. Energies of some common bonds are listed in Table 9.2, along with each bond's length, which we discuss next. *Stronger bonds are lower in energy (have a deeper energy well); weaker bonds are higher in energy (have a shallower energy well).* The energy of a given bond varies slightly from molecule to molecule and even within the same molecule, so each tabulated value is an *average* bond energy.

A covalent bond has a **bond length,** the distance between the nuclei of two bonded atoms. In Figure 9.11, bond length is shown as the distance between the

ARIS Presentation Center, Tables Library

Table 9.2	Average Bond Energies (kJ/mol) and Bond Lengths (pm)										
Bond	**Energy**	**Length**	**Bond**	**Energy**	**Length**	**Bond**	**Energy**	**Length**	**Bond**	**Energy**	**Length**
Single Bonds											
H—H	432	74	N—H	391	101	Si—H	323	148	S—H	347	134
H—F	565	92	N—N	160	146	Si—Si	226	234	S—S	266	204
H—Cl	427	127	N—P	209	177	Si—O	368	161	S—F	327	158
H—Br	363	141	N—O	201	144	Si—S	226	210	S—Cl	271	201
H—I	295	161	N—F	272	139	Si—F	565	156	S—Br	218	225
			N—Cl	200	191	Si—Cl	381	204	S—I	~170	234
C—H	413	109	N—Br	243	214	Si—Br	310	216			
C—C	347	154	N—I	159	222	Si—I	234	240	F—F	159	143
C—Si	301	186							F—Cl	193	166
C—N	305	147	O—H	467	96	P—H	320	142	F—Br	212	178
C—O	358	143	O—P	351	160	P—Si	213	227	F—I	263	187
C—P	264	187	O—O	204	148	P—P	200	221	Cl—Cl	243	199
C—S	259	181	O—S	265	151	P—F	490	156	Cl—Br	215	214
C—F	453	133	O—F	190	142	P—Cl	331	204	Cl—I	208	243
C—Cl	339	177	O—Cl	203	164	P—Br	272	222	Br—Br	193	228
C—Br	276	194	O—Br	234	172	P—I	184	243	Br—I	175	248
C—I	216	213	O—I	234	194				I—I	151	266
Multiple Bonds											
C=C	614	134	N=N	418	122	C≡C	839	121	N≡N	945	110
C=N	615	127	N=O	607	120	C≡N	891	115	N≡O	631	106
C=O	745	123	O_2	498	121	C≡O	1070	113			
	(799 in CO_2)										

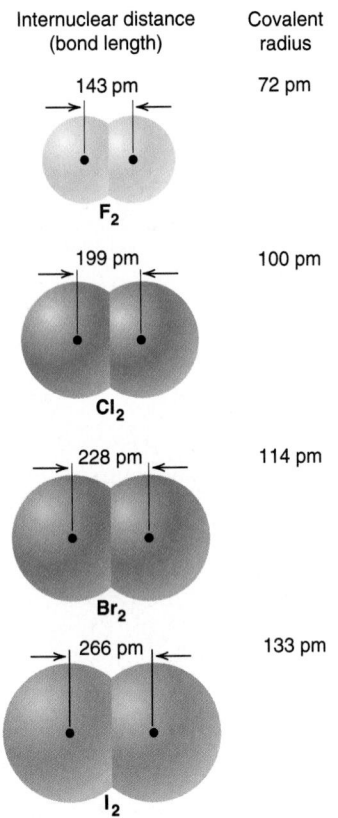

Internuclear distance (bond length)	Covalent radius

143 pm   72 pm

F₂

199 pm   100 pm

Cl₂

228 pm   114 pm

Br₂

266 pm   133 pm

I₂

Figure 9.13 Bond length and covalent radius. Within a series of similar molecules, such as the diatomic halogen molecules, bond length increases as covalent radius increases.

nuclei at the point of minimum energy, and Table 9.2 shows the lengths of some covalent bonds. Like the bond energies, these values are *average* bond lengths for the given bond in different substances. Bond length is related to the sum of the radii of the bonded atoms. In fact, most atomic radii are calculated from measured bond lengths (see Figure 8.14C, p. 317). Bond lengths for a series of similar bonds increase with atomic size, as shown in Figure 9.13 for the halogens.

A close relationship exists among bond order, bond length, and bond energy. Two nuclei are more strongly attracted to two shared electron pairs than to one: the atoms are drawn closer together *and* are more difficult to pull apart. Therefore, *for a given pair of atoms, a higher bond order results in a shorter bond length and a higher bond energy.* So, as Table 9.3 shows, for a given pair of atoms, *a shorter bond is a stronger bond.*

Table 9.3   The Relation of Bond Order, Bond Length, and Bond Energy

Bond	Bond Order	Average Bond Length (pm)	Average Bond Energy (kJ/mol)
C—O	1	143	358
C=O	2	123	745
C≡O	3	113	1070
C—C	1	154	347
C=C	2	134	614
C≡C	3	121	839
N—N	1	146	160
N=N	2	122	418
N≡N	3	110	945

In some cases, we can extend this relationship among atomic size, bond length, and bond strength by holding one atom in the bond constant and varying the other atom within a group or period. For example, the trend in carbon-halogen single bond lengths, C—I > C—Br > C—Cl, parallels the trend in atomic size, I > Br > Cl, and is opposite to the trend in bond energy, C—Cl > C—Br > C—I. Thus, for *single bonds,* longer bonds are usually weaker, and you can see many other examples of this relationship in Table 9.2.

SAMPLE PROBLEM 9.2 Comparing Bond Length and Bond Strength

PROBLEM Without referring to Table 9.2, rank the bonds in each set in order of *decreasing* bond length and bond strength:
(a) S—F, S—Br, S—Cl   **(b)** C=O, C—O, C≡O
PLAN In part (a), S is singly bonded to three different halogen atoms, so all members of the set have a bond order of 1. Bond length increases and bond strength decreases as the halogen's atomic radius increases, and that size trend is clear from the periodic table. In all the bonds in part (b), the same two atoms are involved, but the bond orders differ. In this case, bond strength increases and bond length decreases as bond order increases.
SOLUTION (a) Atomic size increases down a group, so F < Cl < Br.

Bond length: S—Br > S—Cl > S—F

Bond strength: S—F > S—Cl > S—Br

(b) By ranking the bond orders, $C\equiv O > C=O > C-O$, we obtain

Bond length: $C-O > C=O > C\equiv O$

Bond strength: $C\equiv O > C=O > C-O$

CHECK From Table 9.2, we see that the rankings are correct.

COMMENT Remember that for bonds involving pairs of different atoms, as in part (a), *the relationship between length and strength holds* only *for single bonds* and not in every case, so apply it carefully.

FOLLOW-UP PROBLEM 9.2 Rank the bonds in each set in order of *increasing* bond length and bond strength: **(a)** Si—F, Si—C, Si—O; **(b)** N=N, N—N, N≡N.

How the Model Explains the Properties of Covalent Substances

The covalent bonding model proposes that electron sharing between pairs of atoms leads to strong, localized bonds, usually within individual molecules. At first glance, however, it seems that the model is inconsistent with some of the familiar physical properties of covalent substances. After all, most are gases (such as methane and ammonia), liquids (such as benzene and water), or low-melting solids (such as sulfur and paraffin wax). If covalent bonds are so strong (~200 to 500 kJ/mol), why do covalent substances melt and boil at such low temperatures?

To answer this question, we must distinguish between two different sets of forces: (1) the *strong covalent bonding forces* holding the atoms together within the molecule (those we have been discussing), and (2) the *weak intermolecular forces* holding the separate molecules near each other in the macroscopic sample. It is these weak forces *between* the molecules, not the strong covalent bonds *within* each molecule, that are responsible for the physical properties of covalent substances. Consider, for example, what happens when pentane (C_5H_{12}) boils. As Figure 9.14 shows, the weak interactions *between* the pentane molecules are affected, not the strong C—C and C—H covalent bonds *within* each molecule.

Some covalent substances, called *network covalent solids,* do not consist of separate molecules. Rather, they are held together by covalent bonds that extend

Strong covalent bonding forces *within* molecules

Gaseous phase

Liquid phase

Weak intermolecular forces *between* molecules

Figure 9.14 Strong forces within molecules and weak forces between them. When pentane boils, weak forces *between* molecules (intermolecular forces) are overcome, but the strong covalent bonds holding the atoms together within each molecule remain unaffected. Thus, the pentane molecules leave the liquid phase as intact units.

ARIS Presentation Center, Art Library

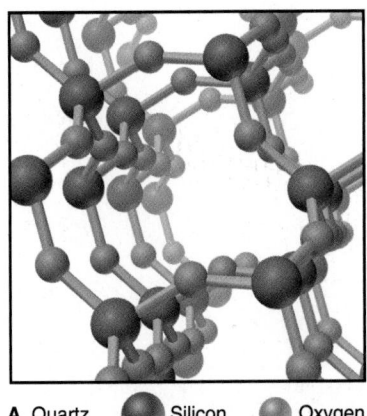

A Quartz   ● Silicon   ● Oxygen

B Diamond   ● Carbon

Figure 9.15 Covalent bonds of network covalent solids. A, In quartz (SiO_2), each Si atom is bonded covalently to four O atoms and each O atom is bonded to two Si atoms in a pattern that extends throughout the sample. Because no separate SiO_2 molecules are present, the melting point of quartz is very high, and it is very hard. **B,** In diamond, each C atom is covalently bonded to four other C atoms throughout the crystal. Diamond is the hardest natural substance known and has an extremely high melting point.

 ARIS Presentation Center, Art Library

in three dimensions *throughout* the sample. The properties of these substances *do* reflect the strength of their covalent bonds. Two examples, quartz and diamond, are shown in Figure 9.15. Quartz (SiO_2) has silicon-oxygen covalent bonds that extend throughout the sample; no separate SiO_2 molecules exist. Quartz is very hard and melts at 1550°C. Diamond has covalent bonds connecting each of its carbon atoms to four others throughout the sample. It is the hardest natural substance known and melts at around 3550°C. Clearly, covalent bonds *are* strong, but because most covalent substances consist of separate molecules with weak forces between them, their physical properties do not reflect this bond strength. (We discuss intermolecular forces in detail in Chapter 12.)

Unlike ionic compounds, most covalent substances are poor electrical conductors, even when melted or when dissolved in water. An electric current is carried by either mobile electrons or mobile ions. In covalent substances, the electrons are localized as either shared or unshared pairs, so they are not free to move, and no ions are present. The Tools of the Laboratory essay describes a tool used widely for studying the types of bonds in covalent substances.

Section Summary

A shared pair of valence electrons attracts the nuclei of two atoms and holds them together in a covalent bond, while filling each atom's outer level. The number of shared pairs between the two atoms is the bond order. • For a given type of bond, the bond energy is the average energy required to completely separate the bonded atoms; the bond length is the average distance between their nuclei. For a given pair of bonded atoms, bond order is directly related to bond energy and inversely related to bond length. • Substances that consist of separate molecules are generally soft and low melting because of the weak forces between molecules. Solids held together by covalent bonds extending throughout the sample are extremely hard and high melting. Most covalent substances have low electrical conductivity because electrons are localized and ions are absent. • The atoms in a covalent bond vibrate, and the energy of these vibrations can be studied with IR spectroscopy.

9.4 BOND ENERGY AND CHEMICAL CHANGE

The relative strengths of the bonds in reactants and products of a chemical change determine whether heat is released or absorbed. In fact, as you'll see in Chapter 20, bond strength is one of two essential factors determining whether the change occurs at all. In this section, we'll discuss the importance of bond energy in chemical change, especially in the combustion of fuels and foods.

Changes in Bond Strengths: Where Does ΔH°_{rxn} Come From?

In Chapter 6, we discussed the heat involved in a chemical change (ΔH°_{rxn}), but we never stopped to ask a central question. When, for example, 1 mol of H_2 and 1 mol of F_2 react at 298 K, 2 mol of HF forms and 546 kJ of heat is released:

$$H_2(g) + F_2(g) \longrightarrow 2HF(g) + 546 \text{ kJ}$$

Where does this heat come from? We find the answer through a very close-up view of the molecules and their energy components.

A system's internal energy has kinetic energy (E_k) and potential energy (E_p) components. Let's examine the contributions to these components to see which one changes during the reaction of H_2 and F_2 to form HF.

Of the various contributions to the kinetic energy, the most important come from the molecules moving through space, rotating, and vibrating and, of course, from the electrons moving within the atoms. Of the various contributions to the

Tools of the Laboratory
Infrared Spectroscopy

nfrared (IR) spectroscopy is an instrumental technique used primarily to study the molecular structure of covalently bonded molecules. Found in most research laboratories, the IR spectrometer is an essential part of the chemist's instrumental toolbox—along with the UV-visible spectrophotometer, mass spectrometer, and NMR spectrometer (described in Chapter 15)—for investigating and identifying organic and biological compounds.

The key components of an IR spectrometer are the same as those of similar instruments (see Figure B7.3, p. 282). The source emits radiation of many wavelengths, and those in the IR region are selected and directed at the sample. For organic compounds, the sample is typically either a pure liquid or a solid mixed with an inorganic salt such as KBr. The sample absorbs certain wavelengths of the IR radiation more than others, and an IR spectrum is generated.

What property of a molecule is displayed in its IR spectrum? All molecules, whether occurring in a gas, a liquid, or a solid, undergo continual rotations and vibrations. Consider, for instance, a sample of ethane gas. The $H_3C—CH_3$ molecules zoom throughout the container, colliding with the walls and each other. If we could look closely at one molecule, however, and disregard its motion through space, we would see the whole molecule rotating and its two CH_3 groups rotating relative to each other about the C—C bond. More important to IR spectroscopy, we would also see each of the bonded atoms vibrating as though the bonds were flexible springs. Figure B9.1 displays the stretching and bending vibrations that diatomic and triatomic molecules undergo. Larger molecules also undergo twisting, wagging, and rocking vibrations. (Thus, the length of a given bond within a molecule is actually the *average* distance between nuclei, analogous to the average length of a spring stretching and compressing.)

The energies of IR photons fall in the same range as the energies of these molecular vibrations. Each vibrational motion has its own natural frequency, which is based on the type of motion, the masses of the atoms, and the strength of the bond between them. These frequencies correspond to wavelengths between 2.5 and 25 μm, a part of the IR region of the electromagnetic spectrum (see Figure 7.3, p. 271). *The energy of each of these vibrations is quantized.* Just as an atom can absorb a photon whose energy corresponds to the difference between two quantized *electron* energy levels, a molecule can absorb an IR photon whose energy corresponds to the difference between two of its quantized *vibrational* energy levels.

The IR spectrum is particularly useful for compound identification because of two related factors. First, *each kind of bond has a characteristic range of IR wavelengths it can absorb.* For example, a C—C bond absorbs IR photons in a different wavelength range from those absorbed by a C=C bond, a C—H bond, a C=O bond, and so forth. Furthermore, groups of atoms that characterize particular types of organic compounds—alcohol, carboxylic acid, ether, and so forth—absorb in slightly different wavelength regions.

(continued)

🐾 Organic chemistry application

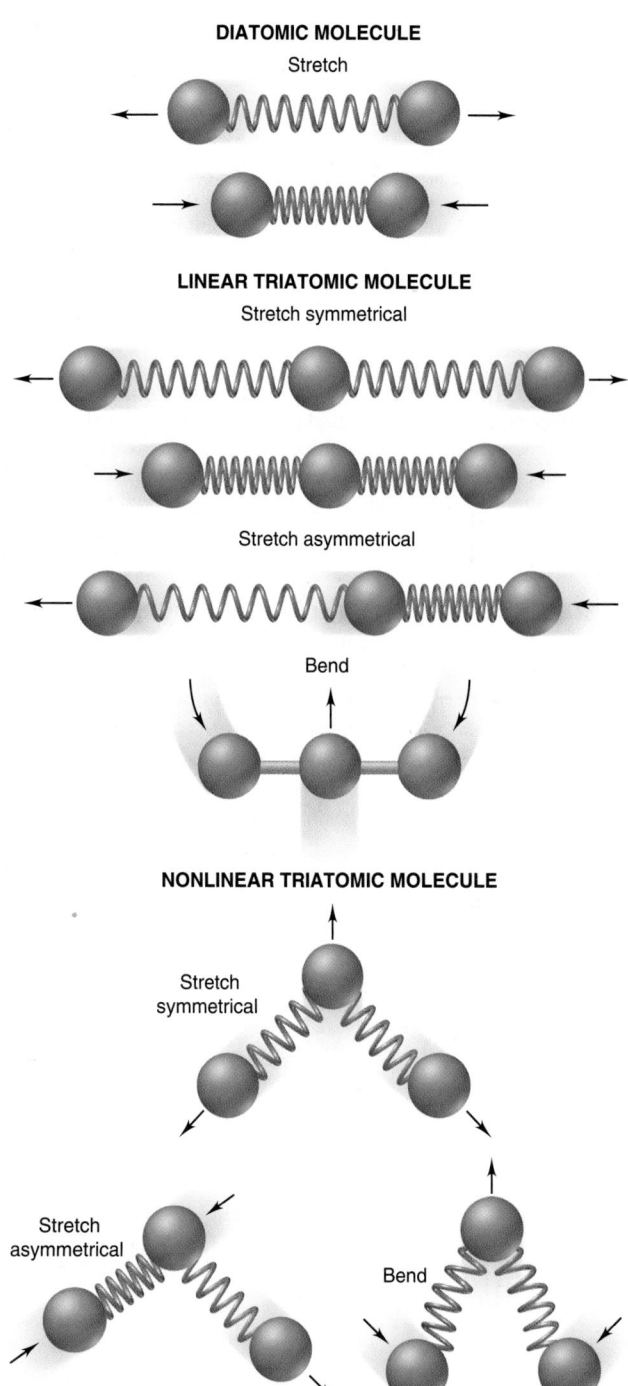

Figure B9.1 Vibrational motions in general diatomic and triatomic molecules. 🔵 ARIS Presentation Center, Art Library

Figure B9.2 The infrared (IR) spectrum of acrylonitrile. The IR spectrum of acrylonitrile is typical of a molecule with several types of covalent bonds. There are many absorption bands (peaks) of differing depths and sharpness. Most peaks correspond to a particular type of vibration (stretch, bend, rock, wag, or twist) involving a particular group of bonded atoms. Some broad peaks (e.g., "combination band") represent several overlapping types of vibrations. The spectrum is reproducible and unique for acrylonitrile. (The bottom axis shows wavenumbers, the inverse of wavelength, so its units are those of length^{-1}. The scale expands to the right of 2000 cm^{-1}.)

ARIS Presentation Center, Art Library

Second, the exact wavelengths and quantity of IR radiation that a molecule absorbs depend on the *overall structure* of the molecule. Combinations of absorptions overlap to create a very characteristic pattern for a given type of compound. This means that *each compound has a characteristic IR spectrum* that can be used to identify it, much as a fingerprint is used to identify a person.

The spectrum appears as a series of downward pointing peaks of varying depth and sharpness. Figure B9.2 shows the IR spectrum of acrylonitrile, a compound used to manufacture synthetic rubber and plastics. No other compound has exactly the same IR spectrum.

Constitutional (structural) isomers are easily distinguished by their IR spectra. These compounds have the same molecular formula but different structural formulas. We might expect very different isomers such as diethyl ether and 2-butanol to have very different IR spectra because their molecular structures are so dissimilar (Figure B9.3). However, even relatively similar compounds, such as 1,3-dimethylbenzene and 1,4-dimethylbenzene, have clearly different spectra (Figure B9.4).

Figure B9.3 The infrared spectra of 2-butanol (*green*) and diethyl ether (*red*). ARIS Presentation Center, Art Library

Figure B9.4 The infrared spectra of 1,3-dimethylbenzene (*green*) and 1,4-dimethylbenzene (*red*). ARIS Presentation Center, Art Library

potential energy, the most important are electrostatic forces between the vibrating atoms, between nucleus and electrons (and between electrons) in each atom, between protons and neutrons in each nucleus, and, of course, between nuclei and the shared electron pair in each bond.

The kinetic energy doesn't change during the reaction because the molecule's motion through space, rotation, and vibration are proportional to the temperature, which is constant at 298 K; and electron motion is not affected by a reaction. Of the potential energy contributions, those within the atoms and nuclei don't change, and vibrational forces vary only slightly as the bonded atoms change. The only significant change in potential energy is in the strength of attraction of the nuclei for the shared electron pair, that is, in the bond energy.

In other words, the answer to "Where does the heat come from?" is that it doesn't really "come from" anywhere: *the energy released or absorbed during a chemical change is due to differences between the reactant bond energies and the product bond energies.*

Using Bond Energies to Calculate ΔH°_{rxn}

Although it does not necessarily occur this way, we can think of any reaction as a two-step process in which *a quantity of heat is absorbed (ΔH° is positive) to break the reactant bonds and form separate atoms and a different quantity is released (ΔH° is negative) when the atoms rearrange to form product bonds.* The sum (symbolized by Σ) of these enthalpy changes is the heat of reaction, ΔH°_{rxn}:

$$\Delta H^\circ_{rxn} = \Sigma \Delta H^\circ_{\text{reactant bonds broken}} + \Sigma \Delta H^\circ_{\text{product bonds formed}} \qquad (9.2)$$

- In an exothermic reaction, the total ΔH° for product bonds formed is *greater* than that for reactant bonds broken, so the sum, ΔH°_{rxn}, is *negative*.
- In an endothermic reaction, the total ΔH° for product bonds formed is *smaller* than that for reactant bonds broken, so the sum, ΔH°_{rxn}, is *positive*.

An equivalent form of Equation 9.2 uses bond energies:

$$\Delta H^\circ_{rxn} = \Sigma BE_{\text{reactant bonds broken}} - \Sigma BE_{\text{product bonds formed}}$$

The minus sign is needed because all bond energies are positive values.

When 1 mol of H—H bonds and 1 mol of F—F bonds absorb energy and break, the 2 mol each of H and F atoms form 2 mol of H—F bonds, which releases energy (Figure 9.16). Recall that *weaker bonds (less stable, more reactive) are easier to break than stronger bonds (more stable, less reactive) because they are higher in energy.* Heat is released when HF forms because the bonds in H_2 and F_2 are weaker (less stable) than the bonds in HF (more stable). Put another way, the sum of the bond energies in 1 mol of H_2 and 1 mol of F_2 is *smaller* than the sum of the bond energies in 2 mol of HF.

ARIS Presentation Center, Art Library

Figure 9.16 **Using bond energies to calculate ΔH°_{rxn}.** Any chemical reaction can be divided conceptually into two hypothetical steps: (1) reactant bonds break to yield separate atoms in a step that absorbs heat (+ sum of BE), and (2) the atoms combine to form product bonds in a step that releases heat (− sum of BE). When the total bond energy of the products is greater than that of the reactants, more energy is released than is absorbed, and the reaction is exothermic (as shown); ΔH°_{rxn} is negative. When the total bond energy of the products is less than that of the reactants, the reaction is endothermic; ΔH°_{rxn} is positive.

We use bond energies to calculate ΔH°_{rxn} by assuming that ***all*** *the reactant bonds break to give individual atoms, from which* ***all*** *the product bonds form.* Even though, typically, only certain bonds break and form, Hess's law allows us to imagine complete bond breakage and then sum the bond energies (with their appropriate signs) to arrive at the overall heat of reaction. (This method assumes all reactants and products in the same physical state; when phase changes occur, additional heat must be taken into account. We address this topic in Chapter 12.)

Let's use bond energies to calculate ΔH°_{rxn} for the combustion of methane. Figure 9.17 shows that all the bonds in CH_4 and O_2 break, and the atoms form the bonds in CO_2 and H_2O. We find the bond energy values in Table 9.2 (p. 353), and use a positive sign for bonds broken and a negative sign for bonds formed:

Bonds broken

$$4 \times C\text{—}H = (4 \text{ mol})(413 \text{ kJ/mol}) = 1652 \text{ kJ}$$
$$\underline{2 \times O_2 = (2 \text{ mol})(498 \text{ kJ/mol}) = 996 \text{ kJ}}$$
$$\Sigma\Delta H^\circ_{\text{reactant bonds broken}} = 2648 \text{ kJ}$$

Bonds formed

$$2 \times C{=}O = (2 \text{ mol})(-799 \text{ kJ/mol}) = -1598 \text{ kJ}$$
$$\underline{4 \times O\text{—}H = (4 \text{ mol})(-467 \text{ kJ/mol}) = -1868 \text{ kJ}}$$
$$\Sigma\Delta H^\circ_{\text{product bonds formed}} = -3466 \text{ kJ}$$

Applying Equation 9.2 gives

$$\Delta H^\circ_{rxn} = \Sigma\Delta H^\circ_{\text{reactant bonds broken}} + \Sigma\Delta H^\circ_{\text{product bonds formed}}$$
$$= 2648 \text{ kJ} + (-3466 \text{ kJ}) = -818 \text{ kJ}$$

It is interesting to compare this value with the value obtained by calorimetry (Section 6.3), which is

$$CH_4(g) + 2O_2(g) \longrightarrow CO_2(g) + 2H_2O(g) \quad \Delta H^\circ_{rxn} = -802 \text{ kJ}$$

Why is there a discrepancy between the bond energy value (-818 kJ) and the calorimetric value (-802 kJ)? Variations in experimental method always intro-

Figure 9.17 Using bond energies to calculate ΔH°_{rxn} for combustion of methane. Treating the combustion of methane as a hypothetical two-step process (see Figure 9.16) means breaking all the bonds in the reactants and forming all the bonds in the products.

duce small discrepancies, but there is a more basic reason in this case. Because bond energies are *average* values obtained from many different compounds, the energy of the bond *in a particular substance* is usually close, but not equal, to this average. For example, the tabulated C—H bond energy of 413 kJ/mol is the average value of C—H bonds in many different molecules. In fact, 415 kJ is actually required to break 1 mol of C—H bonds in methane, or 1660 kJ for 4 mol of these bonds, which gives a ΔH°_{rxn} even closer to the calorimetric value. Thus, it isn't surprising to find a discrepancy between the two ΔH°_{rxn} values. What is surprising—and satisfying in its confirmation of bond theory—is that the values are so close.

SAMPLE PROBLEM 9.3 Using Bond Energies to Calculate ΔH°_{rxn}

PROBLEM Calculate ΔH°_{rxn} for the chlorination of methane to form chloroform:

$$\underset{\underset{\displaystyle H}{|}}{\overset{\overset{\displaystyle H}{|}}{H-C-H}} + 3\,Cl-Cl \longrightarrow \underset{\underset{\displaystyle Cl}{|}}{\overset{\overset{\displaystyle H}{|}}{Cl-C-Cl}} + 3\,H-Cl$$

PLAN We assume that, in the reaction, *all* the reactant bonds break and *all* the product bonds form. We find the bond energies in Table 9.2 (p. 353) and substitute the two sums, with correct signs, into Equation 9.2.

SOLUTION Finding the standard enthalpy changes for bonds broken and for bonds formed: For bonds broken, the bond energy values are

$$4 \times C-H = (4\ mol)(413\ kJ/mol) = 1652\ kJ$$
$$\underline{3 \times Cl-Cl = (3\ mol)(243\ kJ/mol) = 729\ kJ}$$
$$\Sigma\Delta H^{\circ}_{bonds\ broken} = 2381\ kJ$$

For bonds formed, the values are

$$3 \times C-Cl = (3\ mol)(-339\ kJ/mol) = -1017\ kJ$$
$$1 \times C-H = (1\ mol)(-413\ kJ/mol) = -413\ kJ$$
$$\underline{3 \times H-Cl = (3\ mol)(-427\ kJ/mol) = -1281\ kJ}$$
$$\Sigma\Delta H^{\circ}_{bonds\ formed} = -2711\ kJ$$

Calculating ΔH°_{rxn}:

$$\Delta H^{\circ}_{rxn} = \Sigma\Delta H^{\circ}_{bonds\ broken} + \Sigma\Delta H^{\circ}_{bonds\ formed}$$
$$= 2381\ kJ + (-2711\ kJ)$$
$$= -330\ kJ$$

CHECK The signs of the enthalpy changes are correct: $\Sigma\Delta H^{\circ}_{bonds\ broken}$ should be >0, and $\Sigma\Delta H^{\circ}_{bonds\ formed} <0$. More energy is released than absorbed, so ΔH°_{rxn} is negative:

$$\sim 2400\ kJ + [\sim(-2700\ kJ)] = -300\ kJ$$

FOLLOW-UP PROBLEM 9.3 One of the most important industrial reactions is the formation of ammonia from its elements:

$$N\equiv N + 3\,H-H \longrightarrow 2\,\underset{\underset{\displaystyle H}{|}}{H-N-H}$$

Use bond energies to calculate ΔH°_{rxn}.

Relative Bond Strengths in Fuels and Foods

A *fuel* is a material that reacts with atmospheric oxygen to release energy *and* is available at a reasonable cost. The most common fuels for machines are hydrocarbons and coal, and the most common ones for organisms are the fats and

Table 9.4 Heats of Reaction (ΔH_{rxn}) for the Combustion of Some Carbon Compounds

	Two-Carbon Compounds		One-Carbon Compounds	
	Ethane (C_2H_6)	Ethanol (C_2H_5OH)	Methane (CH_4)	Methanol (CH_3OH)
Structural formula	H H H—C—C—H H H	H H H—C—C—O—H H H	H H—C—H H	H H—C—O—H H
Sum of C—C and C—H bonds	7	6	4	3
Sum of C—O and O—H bonds	0	2	0	2
ΔH_{rxn} (kJ/mol)	−1560	−1367	−890	−727
ΔH_{rxn} (kJ/g)	−51.88	−29.67	−55.5	−22.7

Transparency/ARIS Presentation Center, Tables Library

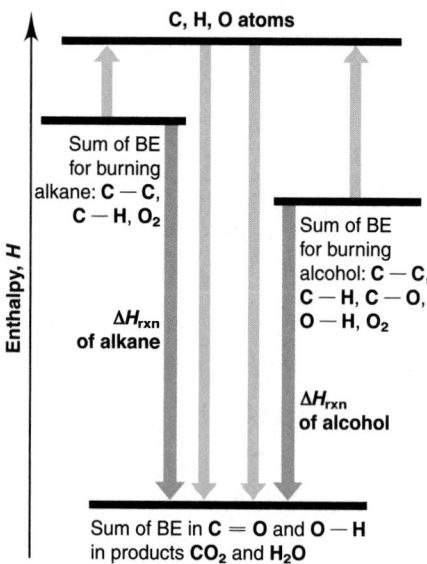

Figure 9.18 Bond strength and the energy from fuels. Fuels with weaker bonds release more energy during combustion than fuels with stronger bonds.

ARIS Presentation Center, Art Library

Environmental application

Biological application

ARIS Presentation Center, Art Library

carbohydrates in foods. Both types of fuels are composed of large organic molecules with mostly C—C and C—H bonds; the foods also contain some C—O and O—H bonds. As with any reaction, the energy released from the combustion of a fuel arises from differences in bond energies between the reactants (fuel plus O_2) and the products (CO_2 and H_2O). When the fuel reacts with O_2, the bonds break, and the C, H, and O atoms form C=O and O—H bonds in the products. Because the reaction is exothermic, we know that the total strength of the bonds in the products is greater than that of the bonds in the reactants. In other words, the bonds in CO_2 and H_2O are stronger (lower in energy, more stable) than those in gasoline (or salad oil) and O_2 (weaker, higher in energy, less stable).

Fuels with more weak bonds yield more energy than fuels with fewer weak bonds. When an alkane burns, C—C and C—H bonds break; when an alcohol burns, C—O and O—H bonds break as well. A look at the bond energy values in Table 9.2 (p. 353) shows that the sum for C—C and C—H bonds (760 kJ/mol) is less than the sum for C—O and O—H bonds (825 kJ/mol). Fuels with relatively more of the weaker bonds will release more energy because it takes less energy to break them apart before the atoms rearrange to form the products (Figure 9.18). Table 9.4 demonstrates this point for some small organic compounds. As the number of C—C and C—H bonds decreases and/or the number of C—O and O—H bonds (shown in red) increases, less energy is released from combustion; that is, ΔH_{rxn} is less negative. As a generalization, we can say that the fewer bonds to O in a fuel, the more energy it releases when burned.

Both fats and carbohydrates serve as high-energy foods—carbohydrates for shorter term release and fats for longer term storage. Fats consist of chains of carbon atoms (C—C bonds) attached to hydrogen atoms (C—H bonds), with a few C—O and O—H bonds (shown in red in the structure below):

Sucrose, a carbohydrate

Triolein, a fat

In contrast, carbohydrates have far fewer C—C and C—H bonds and many more C—O and O—H bonds. Both kinds of foods are metabolized in the body to CO_2 and H_2O. Fats "contain more Calories" per gram than carbohydrates because fats have more weaker bonds and fewer stronger bonds—that is, fewer bonds to O. Therefore, fats should release more energy than carbohydrates, which Table 9.5 confirms.

Section Summary

The only component of internal energy that changes significantly during a reaction is the energy of the bonds in reactants and products; this change in bond energy appears as the heat of reaction, ΔH°_{rxn}. • A reaction involves breaking reactant bonds and forming product bonds. Applying Hess's law, we use tabulated bond energies to calculate ΔH°_{rxn}. • Bonds in fuels and foods are weaker (less stable, higher energy) than bonds in the products of their combustion. Fuels with more weak bonds release more energy than fuels with fewer.

Table 9.5	Heats of Reaction for the Combustion of Some Foods	
Substance	ΔH_{rxn} (kJ/g)
Fats	
Vegetable oil	-37.0
Margarine	-30.1
Butter	-30.0
Carbohydrates	
Table sugar (sucrose)	-16.2
Brown rice	-14.9
Maple syrup	-10.4

ARIS Presentation Center, Tables Library

9.5 | BETWEEN THE EXTREMES: ELECTRONEGATIVITY AND BOND POLARITY

Scientific models are idealized descriptions of reality. As we've discussed them so far, the ionic and covalent bonding models portray compounds as being formed by *either* complete electron transfer *or* complete electron sharing. In most real substances, however, the type of bonding lies somewhere between these extremes. Thus, the great majority of bonds are more accurately thought of as "polar covalent," that is, partially ionic and partially covalent (Figure 9.19).

Electronegativity

One of the most important concepts in chemical bonding is **electronegativity (EN),** the relative ability of a bonded atom to attract the shared electrons.* More than 50 years ago, the American chemist Linus Pauling developed the most common scale of relative EN values for the elements. Here is an example to show the basis of Pauling's approach. We might expect the bond energy of the HF bond to be the average of the energies of an H—H bond (432 kJ/mol) and an F—F bond (159 kJ/mol), or 296 kJ/mol. However, the actual bond energy of H—F is 565 kJ/mol, or 269 kJ/mol *higher* than the average. Pauling reasoned that this difference is due to an *electrostatic (charge) contribution* to the H—F bond energy. If F attracts the shared electron pair more strongly than H, that is, if F is more electronegative than H, the electrons will spend more time closer to F. This unequal sharing of electrons makes the F end of the bond partially negative and the H end partially positive, and the attraction between these partial charges *increases* the energy required to break the bond.

From similar studies with the remaining hydrogen halides and many other compounds, Pauling arrived at the scale of *relative EN values* shown in Figure 9.20 (on the next page). These values are not measured quantities but are based on Pauling's assignment of the highest EN value, 4.0, to fluorine.

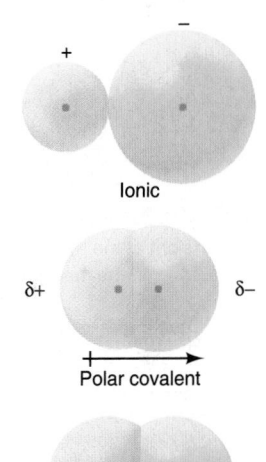

Figure 9.19 The prevalence of electron sharing. The extremes of pure ionic *(top)* and pure covalent *(bottom)* bonding occur rarely, if ever. Much more common is the unequal electron sharing seen in polar covalent bonding.

 ARIS Presentation Center, Art Library

---

*Electronegativity is not the same as electron affinity (EA), although many elements with a high EN also have a highly negative EA. Electronegativity refers to a bonded atom attracting the shared electron pair; electron affinity refers to a separate atom in the gas phase gaining an electron to form a gaseous anion.

Figure 9.20 **The Pauling electronegativity (EN) scale.** The EN is shown by the height of the post with the value on top. The key indicates arbitrary EN cutoffs. In the main groups, EN generally *increases* from left to right and *decreases* from top to bottom. The noble gases are not shown. The transition and inner transition elements show relatively little change in EN. Hydrogen lies near elements of similar EN.

Transparency/ARIS Presentation Center,
Art Library

Literature Sun, Xiaoping. "Redefinition of Electronegativity as the Average Valence Electron Energy: The Third Dimension of the Periodic Table," *Chem. Educator 5* **2000**, *2*, 54–57.

Trends in Electronegativity   Because the nucleus of a smaller atom is closer to the shared pair than that of a larger atom, it attracts the bonding electrons more strongly. So, in general, *electronegativity is inversely related to atomic size*. As Figure 9.21 makes clear for the main-group elements, *electronegativity generally increases up a group and across a period*. The figure also shows that atomic size (on top of each post) decreases with the electronegativity increase. In Pauling's electronegativity scale, and in any of several others that have been devised, *nonmetals are more electronegative than metals*. The most electronegative element is fluorine, with oxygen a close second. Thus, except when it bonds with fluorine, oxygen always pulls bonding electrons toward itself. The least electronegative element (also referred to as the most *electropositive*) is francium, in the lower left corner of the periodic table, but it is radioactive and extremely rare, so for all practical purposes, cesium is the most electropositive.*

Electronegativity and Oxidation Number   One important use of electronegativity is in determining an atom's oxidation number (O.N.; see Section 4.5):

---

*In 1934, the American physicist Robert S. Mulliken developed an approach to electronegativity based solely on atomic properties: EN = (IE − EA)/2. By this approach as well, fluorine, with a high ionization energy (IE) and a large negative electron affinity (EA), has a high EN; and cesium, with a low IE and a small EA, has a low EN.

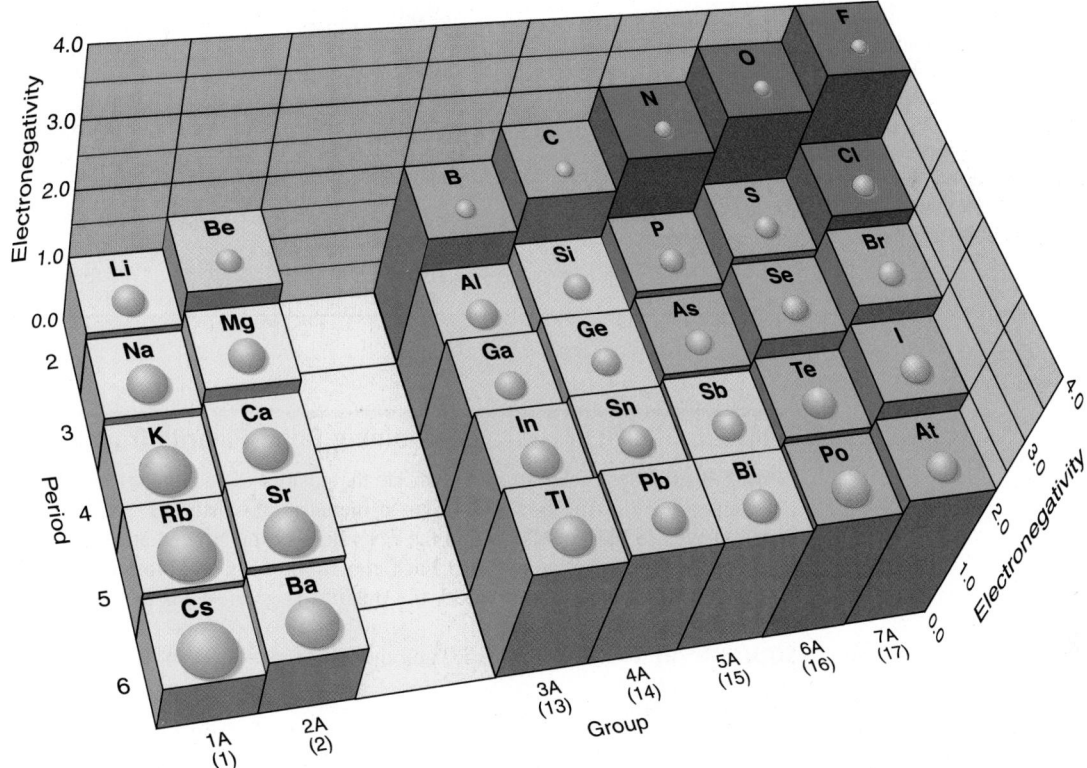

Figure 9.21 Electronegativity and atomic size. The electronegativities of the main-group elements from Periods 2 to 6 (excluding the noble gases) are shown as posts of different heights. On top of each post is a hemisphere showing the element's relative atomic size.

Transparency/ARIS Presentation Center, Art Library

1. The more electronegative atom in a bond is assigned *all* the *shared* electrons; the less electronegative atom is assigned *none*.
2. Each atom in a bond is assigned *all* of its *unshared* electrons.
3. The oxidation number is given by

$$\text{O.N.} = \text{no. of valence e}^- - (\text{no. of shared e}^- + \text{no. of unshared e}^-)$$

In HCl, for example, Cl is more electronegative than H. It has 7 valence electrons but is assigned 8 (2 shared + 6 unshared), so its oxidation number is $7 - 8 = -1$. The H atom has 1 valence electron and is assigned none, so its oxidation number is $1 - 0 = +1$.

Polar Covalent Bonds and Bond Polarity

Whenever atoms of different electronegativities form a bond, as in HF, the bonding pair is shared *un*equally. This unequal distribution of electron density gives the bond partially negative and positive poles. Such a **polar covalent bond** is depicted by a polar arrow ($\longmapsto$) pointing toward the negative pole or by $\delta+$ and $\delta-$ symbols, where the lowercase Greek letter delta (δ) represents a partial charge (see also the discussion of O—H bonds in water, Section 4.1):

H—F̈:   or   H—F̈:

$\delta+$ $\delta-$

In the H—H and F—F bonds, the atoms are identical, so the bonding pair is shared equally, and a **nonpolar covalent bond** results. By knowing the EN values of the atoms in a bond, we can find the direction of the bond polarity. Figure 9.22 (on the next page) compares the distribution of electron density in H_2, F_2, and HF.

Literature Sproul, Gordon. "Electronegativity and Bond Type: Predicting Bond Type," *J. Chem. Educ.* 2001, *78*, 387.

Figure 9.22 Electron density distributions in H₂, F₂, and HF. As these relief maps show, electron density is distributed equally around the two nuclei in the nonpolar covalent molecules H₂ and F₂. (The electron density around the F nuclei is so great that the peaks must be cut off to fit within the figure.) But, in polar covalent HF, the electron density is shifted away from H and toward F.

ARIS Presentation Center, Art Library

ARIS Presentation Center,
Worked Examples Library

ARIS Presentation Center, Art Library

ΔEN	IONIC CHARACTER
>1.7	Mostly ionic
0.4–1.7	Polar covalent
<0.4	Mostly covalent
0	Nonpolar covalent

A

B

Figure 9.23 Boundary ranges for classifying ionic character of chemical bonds. A, The electronegativity difference (ΔEN) between bonded atoms shows cutoff values that act as a general guide to a bond's relative ionic character. **B,** The gradation in ionic character across the entire bonding range is shown as shading from ionic (green) to covalent (yellow).

SAMPLE PROBLEM 9.4 Determining Bond Polarity from EN Values

PROBLEM (a) Use a polar arrow to indicate the polarity of each bond: N—H, F—N, I—Cl. **(b)** Rank the following bonds in order of increasing polarity: H—N, H—O, H—C.
PLAN (a) We use Figure 9.20 to find the EN values of the bonded atoms and point the polar arrow toward the negative end. **(b)** Each choice has H bonded to an atom from Period 2. Since EN increases across a period, the polarity is greatest for the bond whose Period 2 atom is farthest to the right.
SOLUTION (a) The EN of N = 3.0 and the EN of H = 2.1, so N is more electronegative than H: $\overset{\longleftarrow}{N\!-\!H}$

The EN of F = 4.0 and the EN of N = 3.0, so F is more electronegative: $\overset{\longleftarrow}{F\!-\!N}$

The EN of I = 2.5 and the EN of Cl = 3.0, so I is less electronegative: $\overset{\longleftarrow}{I\!-\!Cl}$
(b) The order of increasing EN is C < N < O, and each has a higher EN than H. Therefore, O pulls most on the electron pair shared with H, and C pulls least; so the order of bond polarity is H—C < H—N < H—O.

COMMENT In Chapter 10, you'll see that the polarity of the bonds in a molecule contributes to the overall polarity of the molecule, which is a major factor determining the magnitudes of several physical properties.

FOLLOW-UP PROBLEM 9.4 Arrange each set of bonds in order of increasing polarity, and indicate bond polarity with δ+ and δ− symbols:
(a) Cl—F, Br—Cl, Cl—Cl   **(b)** Si—Cl, P—Cl, S—Cl, Si—Si

The Partial Ionic Character of Polar Covalent Bonds

As we've just seen, if you ask "Is an X—Y bond ionic or covalent?" the answer in almost every case is "*Both, partially!*" A better question is "*To what extent* is the bond ionic or covalent?" The **partial ionic character** of a bond is related directly to the **electronegativity difference (ΔEN),** the difference between the EN values of the bonded atoms: *a greater ΔEN results in larger partial charges and a higher partial ionic character.* Consider these three chlorine-containing molecules: ΔEN for LiCl(g) is 3.0 − 1.0 = 2.0; for HCl(g), it is 3.0 − 2.1 = 0.9; and for Cl₂(g), it is 3.0 − 3.0 = 0. Thus, the bond in LiCl has more ionic character than the H—Cl bond, which has more than the Cl—Cl bond.

Various attempts have been made to classify the ionic character of bonds. All use arbitrary cutoff values, which is inconsistent with the observed gradation, but they are useful for gaining a *relative* sense of a compound's ionic character. One approach uses ΔEN values to divide bonds into ionic, polar covalent, and nonpolar covalent. Based on a range of ΔEN values from 0 (completely nonpolar) to 3.3 (highly ionic), some *approximate* guidelines are given in Figure 9.23A and appear within a gradient in Figure 9.23B.

Figure 9.24 Percent ionic character as a function of electronegativity difference (ΔEN).
A, The percent ionic character is plotted against ΔEN for some gaseous diatomic molecules. Note that, in general, ΔEN correlates with ionic character. (The arbitrary cutoff for an ionic compound is >50% ionic character.) **B,** Even in highly ionic LiF, significant overlap occurs between the ions, indicating some covalent character. The contour map *(left)* and relief map *(right)* depict this overlap.

Another approach calculates the *percent ionic character* of a bond by comparing the actual behavior of a polar molecule in an electric field with the behavior it would show if the electron were transferred completely (a pure ionic bond). A value of 50% ionic character divides substances we call "ionic" from those we call "covalent." Such methods show 43% ionic character for the H—F bond and expected decreases for the other hydrogen halides: H—Cl is 19% ionic, H—Br 11%, and H—I 4%. A plot of percent ionic character vs. ΔEN for a variety of gaseous diatomic molecules is shown in Figure 9.24A. The specific values are not important, but note that *percent ionic character generally increases with ΔEN.* Another point to note is that whereas some molecules, such as $Cl_2(g)$, have 0% ionic character, none has 100% ionic character. That is, *electron sharing occurs to some extent in every bond,* even one between an alkali metal and a halogen, as electron density contour and relief maps show in Figure 9.24B.

The Continuum of Bonding Across a Period

A metal and a nonmetal—elements from opposite sides of the periodic table—have a relatively large ΔEN and typically interact by electron transfer to form an ionic compound. Two nonmetals—elements from the same side of the table—have a small ΔEN and interact by electron sharing to form a covalent compound. When we combine the nonmetal chlorine with each of the other elements in Period 3, starting with sodium, we should observe a steady decrease in ΔEN and a gradation in bond type from ionic through polar covalent to nonpolar covalent.

Figure 9.25 Properties of the Period 3 chlorides. Samples of the compounds formed from each of the Period 3 elements with chlorine are shown in periodic table sequence in the photo. Note the trend in properties displayed in the bar graphs: as ΔEN decreases, both melting point and electrical conductivity (at the melting point) decrease. These trends are consistent with a change in bond type from ionic through polar covalent to nonpolar covalent.

Transparency/ARIS Presentation Center, Art Library

Figure 9.25 shows samples of the common Period 3 chlorides—NaCl, $MgCl_2$, $AlCl_3$, $SiCl_4$, PCl_3, and SCl_2, as well as Cl_2—and some key macroscopic properties, while Figure 9.26 shows an electron density relief map of a bond in each of these substances. Note the steady increase in covalent character, as shown by the height of the electron density *between* the peaks—the bonding region—as we move from ionic NaCl to nonpolar covalent Cl_2.

The first compound in the series is sodium chloride, a white (colorless) crystalline solid with typical ionic properties—high melting point and high electrical conductivity when molten. Having, in addition, a ΔEN of 2.1, NaCl is ionic by any criterion. However, note that, as with LiF (Figure 9.24B), a small but significant covalent region appears in the relief map for NaCl. Magnesium chloride is still considered ionic, with a ΔEN of 1.8, but it has a lower melting point and lower conductivity, as well as a slightly higher bonding region in the relief map.

Aluminum chloride is still less ionic. Rather than having a three-dimensional lattice of Al^{3+} and Cl^- ions, it consists of extended layers of highly polar Al—Cl covalent bonds, as shown by its ΔEN value of 1.5. Weak forces between layers of bonded atoms result in a much lower melting point, and the low electrical conductivity of molten $AlCl_3$ is consistent with a scarcity of ions. And, once again, we see a higher electron density between the nuclei.

ARIS Presentation Center, Art Library

Figure 9.26 Electron density distributions in bonds of the Period 3 chlorides. The relief maps show the electron density for one bond of each chloride. Note the small, but steady, increase in the height of the bonding region between the nuclei, which indicates greater covalent character.

This trend toward less ionic (more covalent) character in the bonds continues through $SiCl_4$, PCl_3, and SCl_2. Each of these compounds occurs as separate molecules, as indicated by the absence of any measurable conductivity. And the forces *between* the molecules are so weak that the melting points are below 0°C. In Cl_2, which is a gas at room temperature, the bond is nonpolar. The relief maps show the increasing height of the electron density in the bonding region.

Thus, *as ΔEN becomes smaller, the bond becomes more covalent,* and the macroscopic properties of the Period 3 chlorides and Cl_2 change from those of a solid consisting of ions to those of a gas consisting of individual molecules.

Section Summary

An atom's electronegativity refers to its ability to pull bonded electrons toward it, which generates partial charges at the ends of the bond. Electronegativity increases across a period and decreases down a group, the reverse of the trends in atomic size. • The greater the ΔEN for the two atoms in a bond, the more polar the bond is and the greater its ionic character. • For chlorides of Period 3 elements, there is a gradation in bond type from ionic to polar covalent to nonpolar covalent.

9.6 AN INTRODUCTION TO METALLIC BONDING

Metallic bonding occurs when large numbers of metal atoms interact. In this section, you'll see a simple, qualitative model for metallic bonding; a more detailed one is presented in Chapter 12.

The Electron-Sea Model

In reactions with nonmetals, reactive metals (such as Na) transfer their outer electrons and form ionic solids (such as NaCl). Two metal atoms can also share their valence electrons in a covalent bond and form gaseous, diatomic molecules (such as Na_2). But what holds the atoms together in a piece of sodium metal? The **electron-sea model** of metallic bonding proposes that all the metal atoms in the sample contribute their valence electrons to form an electron "sea" that is delocalized throughout the piece. The metal ions (the nuclei with their core electrons) are submerged within this electron sea in an orderly array (see Figure 9.2C).

In contrast to ionic bonding, metallic bonding does not hold the metal ions in place as rigidly. In contrast to covalent bonding, no particular pair of metal atoms is bonded through a localized pair of electrons. Rather, *the valence electrons are shared among all the atoms in the sample.* The piece of metal is held together by the mutual attraction of the metal cations for the mobile, highly delocalized valence electrons.

Although there are metallic compounds, two or more metals typically form **alloys,** solid mixtures with variable composition. Many familiar metallic materials are alloys, such as those used for car parts, airplane bodies, building and bridge supports, coins, jewelry, and dental work.

Table 9.6	Melting and Boiling Points of Some Metals	
Element	mp (°C)	bp (°C)
Lithium (Li)	180	1347
Tin (Sn)	232	2623
Aluminum (Al)	660	2467
Barium (Ba)	727	1850
Silver (Ag)	961	2155
Copper (Cu)	1083	2570
Uranium (U)	1130	3930

 ARIS Presentation Center, Tables Library

How the Model Explains the Properties of Metals

Although the physical properties of metals vary over a wide range, most are solids with moderate to high melting points and much higher boiling points (Table 9.6). Metals typically bend or dent rather than crack or shatter. Many can be flattened into sheets (malleable) and pulled into wires (ductile). (See the margin note on page 371.) ● Unlike typical ionic and covalent substances, metals conduct heat and electricity well in *both* the solid and liquid states.

Two features of the electron-sea model that account for these properties are the *regularity,* but not rigidity, of the metal-ion array and the *mobility* of the valence electrons. The melting and boiling points of metals are related to the energy of the metallic bonding. Melting points are only moderately high because the attractions between moveable cations and electrons need not be broken during melting. Boiling a metal requires each cation and its electron(s) to break away from the others, so the boiling points are quite high. Gallium provides a striking example: it melts in your hand (mp 29.8°C) but doesn't boil until over 2400°C.

Periodic trends are also consistent with the electron-sea model. As Figure 9.27 shows, the alkaline earth metals [Group 2A(2)] have higher melting points than the alkali metals [Group 1A(1)]. The 2A metal atoms have two valence electrons and form 2+ cations. Greater attraction between these cations and twice as many valence electrons means stronger metallic bonding than for the 1A metal atoms, so higher temperatures are needed to melt the 2A solids.

Figure 9.27 Melting points of Group 1A(1) and Group 2A(2) metals. Alkaline earth metals *(blue)* have higher melting points than alkali metals *(brown)* because the electron sea has twice as many valence electrons, resulting in stronger attractions among the metal ions.

 ARIS Presentation Center, Art Library

Mechanical and conducting properties are also explained by the model. When a piece of metal is deformed by a hammer, the metal ions slide past each other through the electron sea and end up in new positions. Thus, the metal-ion cores do not repel each other (Figure 9.28). Compare this behavior with the repulsions that occur when an ionic solid is struck (see Figure 9.8, p. 349).

Figure 9.28 The reason metals deform. A, An external force applied to a piece of metal deforms the metal but doesn't break it. **B,** The external force merely moves metal ions past each other through the electron sea.

 ARIS Presentation Center, Photo Library

 Engineering application

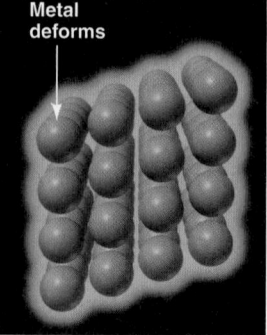

Metals are good conductors of electricity because they have mobile electrons. When a piece of metal is attached to a battery, electrons flow from one terminal into the metal and replace electrons flowing from the metal into the other terminal. Irregularities in the array of metal atoms reduce this conductivity. Ordinary copper wire used to carry an electric current, for example, is more than 99.99% pure because traces of other atoms can drastically restrict the flow of electrons.

Mobile electrons also make metals good conductors of heat. Place your hand on a piece of metal and a piece of wood that are both at room temperature. The metal feels colder because it conducts body heat away from your hand much faster than the wood. The delocalized electrons in the metal disperse the heat from your hand more quickly than the localized electron pairs in the covalent bonds of wood.

Section Summary

According to the electron-sea model, the valence electrons of the metal atoms in a sample are highly delocalized and attract all the metal cations, holding them together. • Metals have only moderately high melting points because the metal ions remain attracted to the electron sea even if their relative positions change. • Boiling requires completely overcoming these bonding attractions, so metals have very high boiling points. • Metals can be deformed because the electron sea prevents repulsions among the cations. • Metals conduct electricity and heat because their electrons are mobile.

Chapter Perspective

Our theme in this chapter has been that the type of chemical bonding—ionic, covalent, metallic, or some blend of these—is governed by the properties of the bonding atoms. This fundamental idea reappears as we investigate the forces that give rise to the properties of liquids, solids, and solutions (Chapters 12 and 13), the behavior of the main-group elements (Chapter 14), and the organic chemistry of carbon (Chapter 15). But first, you'll see in Chapter 10 how the relative placement of atoms and the arrangement of bonding and lone pairs gives a molecule its characteristic shape and how that shape influences the compound's properties. Then, in Chapter 11, you'll see how covalent bonding theory explains the nature of the bond itself and the properties of compounds.

● **The Amazing Malleability of Gold**
All the Group 1B(11) metals—copper, silver, and gold—are soft enough to be machined easily, but gold is in a class by itself. One gram of gold forms a cube 0.37 cm on a side or a sphere the size of a small ball-bearing. It is so ductile that it can be drawn into a wire 20 μm thick and 165 m long, and so malleable that it can be hammered into a 1.0-m^2 sheet that is only 230 atoms (about 70 nm) thick!

 ARIS Presentation Center, Art/Photo Library

CHAPTER REVIEW GUIDE
The following sections provide many aids to help you study this chapter. (Numbers in parentheses refer to pages, unless noted otherwise.)

Learning Objectives These are concepts and skills you should know after studying this chapter.

Relevant section and/or sample problem (SP) numbers appear in parentheses.

Understand These Concepts

1. How differences in atomic properties lead to differences in bond type; the basic distinctions among the three types of bonding (9.1)
2. The essential features of ionic bonding: electron transfer to form ions, and their electrostatic attraction to form a solid (9.2)
3. How lattice energy is ultimately responsible for formation of ionic compounds (9.2)
4. How ionic compound formation is conceptualized as occurring in hypothetical steps (Born-Haber cycle) to calculate lattice energy (9.2)
5. How Coulomb's law explains the periodic trends in lattice energy (9.2)
6. Why ionic compounds are brittle and high melting and conduct electricity only when molten or dissolved in water (9.2)
7. How nonmetal atoms form a covalent bond (9.3)
8. How bonding and lone electron pairs fill the outer (valence) level of each atom in a molecule (9.3)

9. The interrelationships among bond order, bond length, and bond energy (9.3)
10. How the distinction between bonding and nonbonding forces explains the properties of covalent molecules and network covalent solids (9.3)
11. How changes in bond strength account for the heat of reaction (9.4)
12. How a reaction can be divided conceptually into bond-breaking and bond-forming steps (9.4)
13. The periodic trends in electronegativity and the inverse relation of EN values to atomic sizes (9.5)
14. How bond polarity arises from differences in electronegativity of bonded atoms; the direction of bond polarity (9.5)
15. The change in partial ionic character with ΔEN and the change in bonding from ionic to polar covalent to nonpolar covalent across a period (9.5)
16. The role of delocalized electrons in metallic bonding (9.6)
17. How the electron-sea model explains why metals bend, have very high boiling points, and conduct electricity in solid or molten form (9.6)

Learning Objectives (continued)

Master These Skills

1. Using Lewis electron-dot symbols to depict main-group atoms (9.1)
2. Depicting the formation of ions with electron configurations, orbital diagrams, and Lewis symbols, and writing the formula of the ionic compound (SP 9.1)

3. Calculating lattice energy from the enthalpies of the steps to ionic compound formation (9.2)
4. Ranking similar covalent bonds according to their length and strength (SP 9.2)
5. Using bond energies to calculate ΔH°_{rxn} (SP 9.3)
6. Determining bond polarity from EN values (SP 9.4)

Key Terms These important terms appear in boldface in the chapter and are defined again in the Glossary.

Section 9.1
ionic bonding (342)
covalent bonding (342)
metallic bonding (343)
Lewis electron-dot symbol (343)
octet rule (344)

Section 9.2
lattice energy ($\Delta H^\circ_{lattice}$) (346)
Born-Haber cycle (346)

Coulomb's law (348)
ion pair (350)

Section 9.3
covalent bond (351)
bonding (shared) pair (352)
lone (unshared) pair (352)
bond order (352)
single bond (352)
double bond (352)

triple bond (352)
bond energy (BE) (353)
bond length (354)
infrared (IR) spectroscopy (357)

Section 9.5
electronegativity (EN) (363)
polar covalent bond (365)
nonpolar covalent bond (365)

partial ionic character (366)
electronegativity difference (ΔEN) (366)

Section 9.6
electron-sea model (369)
alloy (369)

Key Equations and Relationships Numbered and screened concepts are listed for you to refer to or memorize.

9.1 Relating the energy of attraction to the lattice energy (348):

$$\text{Electrostatic energy} \propto \frac{\text{cation charge} \times \text{anion charge}}{\text{cation radius} + \text{anion radius}} \propto \Delta H^\circ_{lattice}$$

9.2 Calculating heat of reaction from enthalpy changes or bond energies (359):

$$\Delta H^\circ_{rxn} = \Sigma \Delta H^\circ_{reactant\ bonds\ broken} + \Sigma \Delta H^\circ_{product\ bonds\ formed}$$

or $\quad \Delta H^\circ_{rxn} = \Sigma BE_{reactant\ bonds\ broken} - \Sigma BE_{product\ bonds\ formed}$

Highlighted Figures and Tables These figures (F) and tables (T) provide a visual review of key ideas.

These figures (F) and tables (T) provide a quick review of key ideas. Entries in **bold** contain frequently used data.

F9.1 A general comparison of metals and nonmetals (341)
F9.2 The three models of chemical bonding (342)
F9.3 Lewis electron-dot symbols for elements in Periods 2 and 3 (343)
F9.6 The Born-Haber cycle for LiF (347)

F9.7 Trends in lattice energy (348)
F9.11 Covalent bond formation in H_2 (351)
T9.2 Average bond energies and bond lengths (353)
F9.16 Using bond energies to calculate ΔH°_{rxn} (359)
F9.20 The Pauling electronegativity scale (364)
F9.21 Electronegativity and atomic size (365)

Brief Solutions to FOLLOW-UP PROBLEMS Compare your solutions to these calculation steps and answers.

9.1 Mg ([Ne] $3s^2$) + 2Cl ([Ne] $3s^2 3p^5$) $\longrightarrow$

$$\text{Mg}^{2+} \text{([Ne])} + 2\text{Cl}^- \text{([Ne] } 3s^2 3p^6)$$

:Mg· + 2 :Cl: $\longrightarrow$ Mg^{2+} + 2 :Cl:$^-$ Formula: MgCl$_2$

9.2 (a) Bond length: Si—F < Si—O < Si—C
Bond strength: Si—C < Si—O < Si—F
(b) Bond length: N≡N < N=N < N—N
Bond strength: N—N < N=N < N≡N

9.3 N≡N + 3 H—H $\longrightarrow$ 2 H—N—H
$\qquad\qquad\qquad\qquad\qquad\qquad\qquad$ |
$\qquad\qquad\qquad\qquad\qquad\qquad\qquad$ H

$\Sigma \Delta H^\circ_{bonds\ broken}$ = 1 N≡N + 3 H—H
$\qquad\qquad\qquad$ = 945 kJ + 1296 kJ = 2241 kJ
$\Sigma \Delta H^\circ_{bonds\ formed}$ = 6 N—H = −2346 kJ
ΔH°_{rxn} = −105 kJ

9.4 (a) Cl—Cl < $\overset{\delta+}{Br}$—$\overset{\delta-}{Cl}$ < $\overset{\delta+}{Cl}$—$\overset{\delta-}{F}$

(b) Si—Si < $\overset{\delta+}{S}$—$\overset{\delta-}{Cl}$ < $\overset{\delta+}{P}$—$\overset{\delta-}{Cl}$ < $\overset{\delta+}{Si}$—$\overset{\delta-}{Cl}$

PROBLEMS

Problems with colored numbers are answered in Appendix E and worked in detail in the Student Solutions Manual. Problem sections match those in the text and provide the numbers of relevant sample problems. Most offer Concept Review Questions, Skill-Building Exercises (grouped in pairs covering the same concept), and Problems in Context. The Comprehensive Problems are based on material from any section or previous chapter.

Atomic Properties and Chemical Bonds

● Concept Review Questions

9.1 In general terms, how does each of the following atomic properties influence the metallic character of the main-group elements in a period?
(a) Ionization energy
(b) Atomic radius
(c) Number of outer electrons
(d) Effective nuclear charge

9.2 Three solids are represented below. What is the predominant type of intramolecular bonding in each?

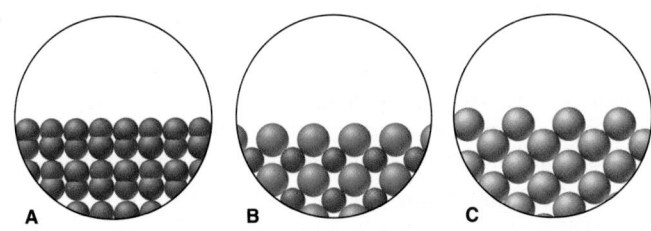

9.3 What is the relationship between the tendency of a main-group element to form a monatomic ion and its position in the periodic table? In what part of the table are the main-group elements that typically form cations? Anions?

● Skill-Building Exercises (grouped in similar pairs)

9.4 Which member of each pair is *more* metallic?
(a) Na or Cs   (b) Mg or Rb   (c) As or N

9.5 Which member of each pair is *less* metallic?
(a) I or O   (b) Be or Ba   (c) Se or Ge

9.6 State the type of bonding—ionic, covalent, or metallic—you would expect in each: (a) $CsF(s)$; (b) $N_2(g)$; (c) $Na(s)$.

9.7 State the type of bonding—ionic, covalent, or metallic—you would expect in each: (a) $ICl_3(g)$; (b) $N_2O(g)$; (c) $LiCl(s)$.

9.8 State the type of bonding—ionic, covalent, or metallic—you would expect in each: (a) $O_3(g)$; (b) $MgCl_2(s)$; (c) $BrO_2(g)$.

9.9 State the type of bonding—ionic, covalent, or metallic—you would expect in each: (a) $Cr(s)$; (b) $H_2S(g)$; (c) $CaO(s)$.

9.10 Draw a Lewis electron-dot symbol for (a) Rb; (b) Si; (c) I.

9.11 Draw a Lewis electron-dot symbol for (a) Ba; (b) Kr; (c) Br.

9.12 Draw a Lewis electron-dot symbol for (a) Sr; (b) P; (c) S.

9.13 Draw a Lewis electron-dot symbol for (a) As; (b) Se; (c) Ga.

9.14 Give the group number and general electron configuration of an element with each electron-dot symbol: (a) $\cdot \ddot{X}$:   (b) $\dot{X}$·

9.15 Give the group number and general electron configuration of an element with each electron-dot symbol: (a) $\cdot \ddot{X}$:   (b) $\dot{X}$·

The Ionic Bonding Model
(Sample Problem 9.1)

● Concept Review Questions

9.16 If energy is required to form monatomic ions from metals and nonmetals, why do ionic compounds exist?

9.17 (a) In general, how does the lattice energy of an ionic compound depend on the charges and sizes of the ions? (b) Ion arrangements of three general salts are represented below. Rank them in order of increasing lattice energy.

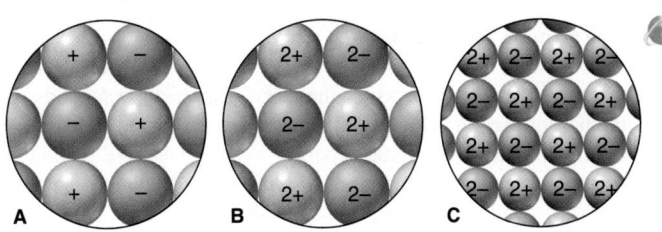

9.18 When gaseous Na^+ and Cl^- ions form gaseous NaCl ion pairs, 548 kJ/mol of energy is released. Why, then, does NaCl occur as a solid under ordinary conditions?

9.19 To form S^{2-} ions from gaseous sulfur atoms requires 214 kJ/mol, but these ions exist in solids such as K_2S. Explain.

● Skill-Building Exercises (grouped in similar pairs)

9.20 Use condensed electron configurations and Lewis electron-dot symbols to depict the ions formed from each of the following atoms, and predict the formula of their compound:
(a) Ba and Cl   (b) Sr and O   (c) Al and F   (d) Rb and O

9.21 Use condensed electron configurations and Lewis electron-dot symbols to depict the ions formed from each of the following atoms, and predict the formula of their compound:
(a) Cs and S   (b) O and Ga   (c) N and Mg   (d) Br and Li

9.22 Identify the main group to which X belongs in each ionic compound formula: (a) XF_2; (b) MgX; (c) X_2SO_4.

9.23 Identify the main group to which X belongs in each ionic compound formula: (a) X_3PO_4; (b) $X_2(SO_4)_3$; (c) $X(NO_3)_2$.

9.24 Identify the main group to which X belongs in each ionic compound formula: (a) X_2O_3; (b) XCO_3; (c) Na_2X.

9.25 Identify the main group to which X belongs in each ionic compound formula: (a) CaX_2; (b) Al_2X_3; (c) XPO_4.

9.26 For each pair, choose the compound with the higher lattice energy, and explain your choice: (a) BaS or CsCl; (b) LiCl or CsCl.

9.27 For each pair, choose the compound with the higher lattice energy, and explain your choice: (a) CaO or CaS; (b) BaO or SrO.

9.28 For each pair, choose the compound with the lower lattice energy, and explain your choice: (a) CaS or BaS; (b) NaF or MgO.

9.29 For each pair, choose the compound with the lower lattice energy, and explain your choice: (a) NaF or NaCl; (b) K_2O or K_2S.

9.30 Use the following to calculate the $\Delta H°_{lattice}$ of NaCl:

$$Na(s) \longrightarrow Na(g) \qquad \Delta H° = 109 \text{ kJ}$$
$$Cl_2(g) \longrightarrow 2Cl(g) \qquad \Delta H° = 243 \text{ kJ}$$
$$Na(g) \longrightarrow Na^+(g) + e^- \qquad \Delta H° = 496 \text{ kJ}$$
$$Cl(g) + e^- \longrightarrow Cl^-(g) \qquad \Delta H° = -349 \text{ kJ}$$
$$Na(s) + \tfrac{1}{2}Cl_2(g) \longrightarrow NaCl(s) \qquad \Delta H°_f = -411 \text{ kJ}$$

Levels of Difficulty: ● Easy ●● Medium ●●● Challenging   This figure is available on the Presentation Center, within the art library.

Compared with the lattice energy of LiF (1050 kJ/mol), is the magnitude of the value for NaCl what you expected? Explain.

9.31 Use the following to calculate the $\Delta H^\circ_{lattice}$ of MgF_2:

$Mg(s) \longrightarrow Mg(g)$	$\Delta H^\circ = \quad 148 \text{ kJ}$
$F_2(g) \longrightarrow 2F(g)$	$\Delta H^\circ = \quad 159 \text{ kJ}$
$Mg(g) \longrightarrow Mg^+(g) + e^-$	$\Delta H^\circ = \quad 738 \text{ kJ}$
$Mg^+(g) \longrightarrow Mg^{2+}(g) + e^-$	$\Delta H^\circ = 1450 \text{ kJ}$
$F(g) + e^- \longrightarrow F^-(g)$	$\Delta H^\circ = -328 \text{ kJ}$
$Mg(s) + F_2(g) \longrightarrow MgF_2(s)$	$\Delta H^\circ_f = -1123 \text{ kJ}$

Compared with the lattice energy of LiF (1050 kJ/mol) or the lattice energy you calculated for NaCl in Problem 9.30, does the relative magnitude of the value for MgF_2 surprise you? Explain.

● **Problems in Context**

9.32 Aluminum oxide (Al_2O_3) is a widely used industrial abrasive (emery, corundum), for which the specific application depends on the hardness of the crystal. What does this hardness imply about the magnitude of the lattice energy? Would you have predicted from the chemical formula that Al_2O_3 is hard? Explain.

9.33 Born-Haber cycles were used to obtain the first reliable values for electron affinity by considering the EA value as the unknown and using a theoretically calculated value for the lattice energy. Use a Born-Haber cycle for KF and the following values to calculate a value for the electron affinity of fluorine:

$K(s) \longrightarrow K(g)$	$\Delta H^\circ = \quad 90 \text{ kJ}$
$K(g) \longrightarrow K^+(g) + e^-$	$\Delta H^\circ = \quad 419 \text{ kJ}$
$F_2(g) \longrightarrow 2F(g)$	$\Delta H^\circ = \quad 159 \text{ kJ}$
$K(s) + \frac{1}{2}F_2(g) \longrightarrow KF(s)$	$\Delta H^\circ_f = -569 \text{ kJ}$
$K^+(g) + F^-(g) \longrightarrow KF(s)$	$\Delta H^\circ = -821 \text{ kJ}$

The Covalent Bonding Model
(Sample Problem 9.2)

● **Concept Review Questions**

9.34 Describe the interactions that occur between individual chlorine atoms as they approach each other and form Cl_2. What combination of forces gives rise to the energy holding the atoms together and to the final internuclear distance?

9.35 Define bond energy using the H—Cl bond as an example. When this bond breaks, is energy absorbed or released? Is the accompanying ΔH value positive or negative? How do the magnitude and sign of this ΔH value relate to the value that accompanies H—Cl bond formation?

9.36 For single bonds between similar types of atoms, how does the strength of the bond relate to the sizes of the atoms? Explain.

9.37 How does the energy of the bond between a given pair of atoms relate to the bond order? Why?

9.38 When liquid benzene (C_6H_6) boils, does the gas consist of molecules, ions, or separate atoms? Explain.

● **Skill-Building Exercises** (grouped in similar pairs)

9.39 Using the periodic table only, arrange the members of each of the following sets in order of increasing bond *strength:*
(a) Br—Br, Cl—Cl, I—I   (b) S—H, S—Br, S—Cl
(c) C≡N, C—N, C≡N

9.40 Using the periodic table only, arrange the members of each of the following sets in order of increasing bond *length:*
(a) H—F, H—I, H—Cl   (b) C—S, C≡O, C—O
(c) N—H, N—S, N—O

● **Problems in Context**

9.41 Formic acid (HCOOH; structural formula shown below) is secreted by certain species of ants when they bite.

Rank the relative strengths of (a) the C—O and C≡O bonds, and (b) the H—C and H—O bonds. Explain these rankings.

9.42 In Figure B9.2, p. 358, the peak labeled "C≡C stretch" occurs at a shorter wavelength than that labeled "C—C stretch," as it does in the IR spectrum of any substance with those bonds. Explain the relative positions of these peaks. In what relative position along the wavelength scale of Figure B9.2 would you expect to find a peak for a C≡C stretch? Explain.

Bond Energy and Chemical Change
(Sample Problem 9.3)

● **Concept Review Questions**

9.43 Write a Plan (without actual numbers, but including the bond energies you would use and how you would combine them algebraically) for calculating the total enthalpy change of the following reaction:
$$H_2(g) + O_2(g) \longrightarrow H_2O_2(g) \quad (H—O—O—H)$$

9.44 The text points out that, for similar types of substances, one with weaker bonds is usually more reactive than one with stronger bonds. Why is this generally true?

9.45 Why is there a discrepancy between a heat of reaction obtained from calorimetry and one obtained from bond energies?

● **Skill-Building Exercises** (grouped in similar pairs)

9.46 Which of the following gases would you expect to have the greater heat of reaction per mole for combustion? Why?

methane   or   formaldehyde

9.47 Which of the following gases would you expect to have the greater heat of reaction per mole for combustion? Why?

ethanol   or   methanol

9.48 Use bond energies to calculate the heat of reaction:

9.49 Use bond energies to calculate the heat of reaction:

● **Problems in Context**

9.50 An important industrial route to extremely pure acetic acid is the reaction of methanol with carbon monoxide:

Use bond energies to calculate the heat of reaction.

• **9.51** Sports trainers treat sprains and soreness with ethyl bromide. It is manufactured by reacting ethylene with hydrogen bromide:

Use bond energies to find the enthalpy change for this reaction.

Between the Extremes: Electronegativity and Bond Polarity
(Sample Problem 9.4)

● Concept Review Questions

• **9.52** Describe the vertical and horizontal trends in electronegativity (EN) among the main-group elements. According to Pauling's scale, what are the two most electronegative elements? The two least electronegative elements?

• **9.53** What is the general relationship between IE_1 and EN for the elements? Why?

• **9.54** Is the H—O bond in water nonpolar covalent, polar covalent, or ionic? Define each term, and explain your choice.

• **9.55** How does electronegativity differ from electron affinity?

• **9.56** How is the partial ionic character of a bond in a diatomic molecule related to ΔEN for the bonded atoms? Why?

● Skill-Building Exercises (grouped in similar pairs)

• **9.57** Using the periodic table only, arrange the elements in each set in order of *increasing* EN: (a) S, O, Si; (b) Mg, P, As.

• **9.58** Using the periodic table only, arrange the elements in each set in order of *increasing* EN: (a) I, Br, N; (b) Ca, H, F.

• **9.59** Using the periodic table only, arrange the elements in each set in order of *decreasing* EN: (a) N, P, Si; (b) Ca, Ga, As.

• **9.60** Using the periodic table only, arrange the elements in each set in order of *decreasing* EN: (a) Br, Cl, P; (b) I, F, O.

• **9.61** Use Figure 9.20, p. 364, to indicate the polarity of each bond with a *polar arrow*: (a) N—B; (b) N—O; (c) C—S; (d) S—O; (e) N—H; (f) Cl—O.

• **9.62** Use Figure 9.20, p. 364, to indicate the polarity of each bond with *partial charges*: (a) Br—Cl; (b) F—Cl; (c) H—O; (d) Se—H; (e) As—H; (f) S—N.

• **9.63** Which is the more polar bond in each of the following pairs from Problem 9.61: (a) or (b); (c) or (d); (e) or (f)?

• **9.64** Which is the more polar bond in each of the following pairs from Problem 9.62: (a) or (b); (c) or (d); (e) or (f)?

• **9.65** Are the bonds in each of the following substances ionic, nonpolar covalent, or polar covalent? Arrange the substances with polar covalent bonds in order of increasing bond polarity:
(a) S_8 (b) RbCl (c) PF_3 (d) SCl_2 (e) F_2 (f) SF_2

• **9.66** Are the bonds in each of the following substances ionic, nonpolar covalent, or polar covalent? Arrange the substances with polar covalent bonds in order of increasing bond polarity:
(a) KCl (b) P_4 (c) BF_3 (d) SO_2 (e) Br_2 (f) NO_2

• **9.67** Rank the members of each set of compounds in order of *increasing* ionic character of their bonds. Use *polar arrows* to indicate the bond polarity of each:

(a) HBr, HCl, HI   (b) H_2O, CH_4, HF   (c) SCl_2, PCl_3, $SiCl_4$

• **9.68** Rank the members of each set of compounds in order of *decreasing* ionic character of their bonds. Use *partial charges* to indicate the bond polarity of each:
(a) PCl_3, PBr_3, PF_3 (b) BF_3, NF_3, CF_4 (c) SeF_4, TeF_4, BrF_3

● Problems in Context

• **9.69** The energy of the C—C bond is 347 kJ/mol, and that of the Cl—Cl bond is 243 kJ/mol. Which of the following values might you expect for the C—Cl bond energy? Explain.
(a) 590 kJ/mol (sum of the values given)
(b) 104 kJ/mol (difference of the values given)
(c) 295 kJ/mol (average of the values given)
(d) 339 kJ/mol (greater than the average of the values given)

An Introduction to Metallic Bonding

● Concept Review Questions

• **9.70** (a) List four physical characteristics of a solid metal.
(b) List two chemical characteristics of a metallic element.

• **9.71** Briefly account for the following relative values:
(a) The melting points of Na and K are 89°C and 63°C, respectively.
(b) The melting points of Li and Be are 180°C and 1287°C, respectively.
(c) Li boils more than 1100°C higher than it melts.

• **9.72** Magnesium metal is easily deformed by an applied force, whereas magnesium fluoride is shattered. Why do these two solids behave so differently?

Comprehensive Problems

•• **9.73** Geologists have a rule of thumb: when molten rock cools and solidifies, crystals of compounds with the smallest lattice energies appear at the bottom of the mass. Suggest a reason for this.

•• **9.74** Acetylene gas (ethyne; HC≡CH) burns in an oxyacetylene torch to produce carbon dioxide and water vapor. The heat of reaction for the combustion of acetylene is 1259 kJ/mol. (a) Calculate the C≡C bond energy, and compare your value with that in Table 9.2, p. 353. (b) When 500.0 g of acetylene burns, how many kilojoules of heat are given off? (c) How many grams of CO_2 form? (d) How many liters of O_2 at 298 K and 18.0 atm are consumed?

••• **9.75** Use Lewis electron-dot symbols to represent the formation of (a) BrF_3 from bromine and fluorine atoms; (b) AlF_3 from aluminum and fluorine atoms.

••• **9.76** Even though so much energy is required to form a metal cation with a 2+ charge, the alkaline earth metals form halides with general formula MX_2, rather than MX.
(a) Use the following data to calculate the ΔH_f° of MgCl:

$Mg(s) \longrightarrow Mg(g)$	$\Delta H^\circ = 148$ kJ
$Cl_2(g) \longrightarrow 2Cl(g)$	$\Delta H^\circ = 243$ kJ
$Mg(g) \longrightarrow Mg^+(g) + e^-$	$\Delta H^\circ = 738$ kJ
$Cl(g) + e^- \longrightarrow Cl^-(g)$	$\Delta H^\circ = -349$ kJ

$\Delta H_{lattice}^\circ$ of MgCl = 783.5 kJ/mol

(b) Is MgCl favored energetically relative to Mg and Cl_2? Explain.
(c) Use Hess's law to calculate ΔH° for the conversion of MgCl to $MgCl_2$ and Mg (ΔH_f° of $MgCl_2 = -641.6$ kJ/mol).
(d) Is MgCl favored energetically relative to $MgCl_2$? Explain.

••• **9.77** Gases react explosively if the heat released when the reaction begins is sufficient to cause more reaction, which leads to a rapid expansion of the gases. Use bond energies to calculate $\Delta H°$ of the following reactions, and predict which occurs explosively:
(a) $H_2(g) + Cl_2(g) \longrightarrow 2HCl(g)$
(b) $H_2(g) + I_2(g) \longrightarrow 2HI(g)$
(c) $2H_2(g) + O_2(g) \longrightarrow 2H_2O(g)$

••• **9.78** By using photons of specific wavelengths, chemists can dissociate gaseous HI to produce H atoms with certain speeds. When HI dissociates, the H atoms move away rapidly, whereas the heavier I atoms move more slowly. (a) What is the longest wavelength (in nm) that can dissociate a molecule of HI? (b) If a photon of 254 nm is used, what is the excess energy (in J) over that needed for dissociation? (c) If this excess energy is carried away by the H atom as kinetic energy, what is its speed (in m/s)?

•• **9.79** Linear, triatomic CO_2 vibrates by symmetric stretch, bend, and asymmetric stretch (Figure B9.1, p. 357), with frequencies of 4.02×10^{13} s^{-1}, 2.00×10^{13} s^{-1}, and 7.05×10^{13} s^{-1}, respectively. (a) In what region of the electromagnetic spectrum are these frequencies? (b) Calculate the energy (in J) of each vibration. Which takes the least energy?

•• **9.80** In developing the concept of electronegativity, Pauling used the term *excess bond energy* for the difference between the actual bond energy of X—Y and the average bond energies of X—X and Y—Y (see text discussion for the case of HF). Based on the values in Figure 9.20, p. 364, which of the following substances contains bonds with *no* excess bond energy?
(a) PH_3 (b) CS_2 (c) $BrCl$ (d) BH_3 (e) Se_8

•• **9.81** Use condensed electron configurations to predict the relative hardness and melting points of rubidium ($Z = 37$), vanadium ($Z = 23$), and cadmium ($Z = 48$).

•• **9.82** Without stratospheric ozone (O_3), harmful solar radiation would cause gene alterations. Ozone forms when the bond in O_2 breaks and each O atom reacts with another O_2 molecule. It is destroyed by reaction with Cl atoms formed when the C—Cl bond in synthetic chemicals breaks. Find the wavelengths of light that can break the C—Cl bond and the bond in O_2.

••• **9.83** "Inert" xenon actually forms many compounds, especially with highly electronegative fluorine. The $\Delta H°_f$ values for xenon difluoride, tetrafluoride, and hexafluoride are -105, -284, and -402 kJ/mol, respectively. Find the average bond energy of the Xe—F bonds in each fluoride.

••• **9.84** The HF bond length is 92 pm, 16% shorter than the sum of the covalent radii of H (37 pm) and F (72 pm). Suggest a reason for this difference. Similar calculations show that the difference becomes smaller down the group from HF to HI. Explain.

••• **9.85** There are two main types of covalent bond breakage. In homolytic breakage (as in Table 9.2, p. 353), each atom in the bond gets one of the shared electrons. In some cases, the electronegativity of adjacent atoms affects the bond energy. In heterolytic breakage, one atom gets both electrons and the other gets none; thus, a cation and an anion form.
(a) Why is the C—C bond in H_3C—CF_3 (423 kJ/mol) stronger than that in H_3C—CH_3 (376 kJ/mol)?
(b) Use bond energy and any other data to calculate the heat of reaction for the heterolytic cleavage of O_2.

•• **9.86** Find the longest wavelengths of light that can cleave the bonds in elemental nitrogen, oxygen, and fluorine.

•• **9.87** The work function (ϕ) of a metal is the minimum energy needed to remove an electron from its surface. (a) Is it easier to remove an electron from a gaseous silver atom or from the surface of solid silver ($\phi = 7.59\times10^{-19}$ J; IE = 731 kJ/mol)? (b) Explain the results in terms of the electron-sea model of metallic bonding.

•• **9.88** Lattice energies can also be calculated for covalent solids using a Born-Haber cycle, and the network solid silicon dioxide has one of the highest $\Delta H°_{lattice}$ values. Silicon dioxide is found in pure crystalline form as transparent rock quartz. Much harder than glass, this material was once prized for making lenses for optical devices and expensive spectacles. Use Appendix B and the following data to calculate $\Delta H°_{lattice}$ of SiO_2:

$Si(s) \longrightarrow Si(g)$	$\Delta H° = 454$ kJ
$Si(g) \longrightarrow Si^{4+}(g) + 4e^-$	$\Delta H° = 9949$ kJ
$O_2(g) \longrightarrow 2O(g)$	$\Delta H° = 498$ kJ
$O(g) + 2e^- \longrightarrow O^{2-}(g)$	$\Delta H° = 737$ kJ

•• **9.89** The average C—H bond energy in CH_4 is 415 kJ/mol. Use Table 9.2 (p. 353) and the following to calculate the average C—H bond energy in ethane (C_2H_6; C—C bond), in ethene (C_2H_4; C=C bond), and in ethyne (C_2H_2; C≡C bond):
$C_2H_6(g) + H_2(g) \longrightarrow 2CH_4(g)$ $\Delta H°_{rxn} = -65.07$ kJ/mol
$C_2H_4(g) + 2H_2(g) \longrightarrow 2CH_4(g)$ $\Delta H°_{rxn} = -202.21$ kJ/mol
$C_2H_2(g) + 3H_2(g) \longrightarrow 2CH_4(g)$ $\Delta H°_{rxn} = -376.74$ kJ/mol

•• **9.90** Carbon-carbon bonds form the "backbone" of nearly every organic and biological molecule. The average bond energy of the C—C bond is 347 kJ/mol. Calculate the frequency and wavelength of the least energetic photon that can break an average C—C bond. In what region of the electromagnetic spectrum is this radiation?

•• **9.91** In a future hydrogen-fuel economy, the cheapest source of H_2 will certainly be water. It takes 467 kJ to produce 1 mol of H atoms from water. What is the frequency, wavelength, and minimum energy of a photon that can free an H atom from water?

•• **9.92** Dimethyl ether (CH_3OCH_3) and ethanol (CH_3CH_2OH) are constitutional isomers (see Table 3.4, p. 103). (a) Use Table 9.2, p. 353, to calculate $\Delta H°_{rxn}$ for the formation of each compound as a gas from methane and oxygen; water vapor also forms. (b) State which reaction is more exothermic. (c) Calculate $\Delta H°_{rxn}$ for the conversion of ethanol to dimethyl ether.

•• **9.93** Heats of reaction calculated from bond energies and from heats of formation are often, but not always, close to each other. (a) Industrial ethanol (CH_3CH_2OH) is produced by a catalytic reaction of ethylene (CH_2=CH_2) with water at high pressures and temperatures. Calculate $\Delta H°_{rxn}$ for this gas-phase hydration of ethylene to ethanol, using bond energies and then using heats of formation. (b) Ethylene glycol is produced by the catalytic oxidation of ethylene to ethylene oxide, which then reacts with water to form ethylene glycol:

$$CH_2-CH_2(l) + H_2O(l) \longrightarrow HOCH_2CH_2OH(l)$$

The $\Delta H°_{rxn}$ for this hydrolysis step, based on heats of formation, is -97 kJ/mol. Calculate $\Delta H°_{rxn}$ for the hydrolysis using bond energies. (c) Why are the two values relatively close for the hydration in part (a) but not close for the hydrolysis in part (b)?

Interconnecting Shapes As in this wooden puzzle, the shapes of many everyday objects fit together to perform a function—key and lock, mortise and tenon, hand and glove. Similarly, an organism's molecules fit together to trigger the processes of life. In this chapter, you learn to depict molecules as two-dimensional structural formulas and then to visualize them as three-dimensional objects.

The Shapes of Molecules

10.1 Depicting Molecules and Ions with Lewis Structures
Using the Octet Rule
Resonance
Formal Charge
Exceptions to the Octet Rule

10.2 Valence-Shell Electron-Pair Repulsion (VSEPR) Theory and Molecular Shape
Electron-Group Arrangements and Molecular Shapes
Molecular Shape with Two Electron Groups
Shapes with Three Electron Groups
Shapes with Four Electron Groups
Shapes with Five Electron Groups
Shapes with Six Electron Groups
Using VSEPR Theory to Determine Molecular Shape
Shapes with More Than One Central Atom

10.3 Molecular Shape and Molecular Polarity
Bond Polarity, Bond Angle, and Dipole Moment
Molecular Polarity and Behavior

Now that you understand how atoms bond, you can explore an idea essential to chemistry—*and* biology! The printed page, covered with atomic symbols, lines, and pairs of dots, makes it easy to forget the amazing, three-dimensional reality of molecular shapes. In any molecule, each atom, bonding pair, and lone pair has its own position in space relative to the others, determined by the attractive and repulsive forces that govern all matter. With definite angles and distances between the nuclei, a molecule has a characteristic minute architecture, extending throughout its tiny volume of space. Whether we consider the details of simple reactions, the properties of synthetic materials, or the intricate life-sustaining processes of living cells, molecular shape is a crucial factor.

IN THIS CHAPTER . . . We first see how to convert the *molecular* formula of a compound into a flat *structural* formula that shows atom attachments and electron-pair location within the molecule. Using the theory that converts these two-dimensional structural formulas into three-dimensional shapes, we describe the five basic classes of shapes that most simple molecules adopt and then see the ways these combine to form the shapes of more complex molecules. You'll learn how shape and bond polarity create a polarity for the entire molecule and glimpse the influence of molecular shape on biological function.

10.1 DEPICTING MOLECULES AND IONS WITH LEWIS STRUCTURES

The first step toward visualizing what a molecule looks like is to convert its molecular formula to its **Lewis structure** (or **Lewis formula**).* This two-dimensional structural formula consists of electron-dot symbols that depict each atom and its neighbors, the bonding pairs that hold them together, and the lone pairs that fill each atom's outer level (valence shell). In many cases, the octet rule (Section 9.1) guides us in allotting electrons to the atoms in a Lewis structure; in many other cases, however, we set the rule aside.

Using the Octet Rule to Write Lewis Structures

To write a Lewis structure, we decide on the relative placement of the atoms in the molecule (or polyatomic ion)—that is, which atoms are adjacent and become bonded to each other—and distribute the total number of valence electrons as bonding and lone pairs. Let's begin by examining Lewis structures for species that "obey" the octet rule—those in which each atom fills its outer level with eight electrons (or two for hydrogen).

Lewis Structures for Molecules with Single Bonds First, we discuss the steps for writing Lewis structures for molecules that have only single bonds, using nitrogen trifluoride, NF_3, as an example. Figure 10.1 lays out the steps.

---

*A Lewis *structure* may be more correctly called a Lewis *formula* because it provides information about the relative placement of atoms in a molecule or ion and shows which atoms are bonded to each other, but it does **not** indicate the three-dimensional shape. Nevertheless, use of the term Lewis *structure* is a convention that we follow.

ARIS Presentation Center, Art Library

Figure 10.1 The steps in converting a molecular formula into a Lewis structure.

Step 1. Place the atoms relative to each other. For compounds of molecular formula AB_n, place the atom with *lower group number* in the center because it needs more electrons to attain an octet; usually, this is also the atom with the *lower electronegativity*. In NF_3, the N (Group 5A; EN = 3.0) has five electrons and so needs three, whereas each F (Group 7A; EN = 4.0) has seven and so needs only one; thus, N goes in the center with the three F atoms around it:

Literature Purser, Gordon H. "Lewis Structures in General Chemistry: Agreement between Electron Density Calculations and Lewis Structures," *J. Chem. Educ.* **2001**, *78*, 981.

$$\begin{array}{ccc} & F & \\ & N & \\ F & & F \end{array}$$

If the atoms have the same group number, as in SO_3 or ClF_3, place the atom with the *higher period number* in the center. H can form only one bond, so it is *never* a central atom.

Step 2. Determine the total number of valence electrons available. For molecules, add up the valence electrons of all the atoms. (Recall that the number of valence electrons equals the A-group number.) In NF_3, N has five valence electrons, and each F has seven:

$$[1 \times N(5e^-)] + [3 \times F(7e^-)] = 5e^- + 21e^- = 26 \text{ valence } e^-$$

For polyatomic ions, *add* one e^- for each negative charge of the ion, or *subtract* one e^- for each positive charge.

Step 3. Draw a single bond from each surrounding atom to the central atom, and subtract two valence electrons for each bond. There must be at least a single bond between bonded atoms:

Literature Ahmad, Wan-Yaacob; Zakaria, Mat B. "Drawing Lewis Structures from Lewis Symbols: A Direct Electron-Pairing Approach," *J. Chem. Educ.* **2000**, *77*, 329.

$$\begin{array}{ccc} & F & \\ & | & \\ F & N & F \end{array}$$

Subtract $2e^-$ for each single bond from the total number of valence electrons available (from step 2) to find the number remaining:

$$3\,N{-}F \text{ bonds} \times 2e^- = 6e^- \quad \text{so} \quad 26e^- - 6e^- = 20e^- \text{ remaining}$$

Step 4. Distribute the remaining electrons in pairs so that each atom ends up with eight electrons (or two for H). First, place lone pairs on the *surrounding (more electronegative) atoms* to give each an octet. If any electrons remain, place them around the central atom. Then check that each atom has $8e^-$:

Literature Purser, Gordon H. "Lewis Structures Are Models for Predicting Molecular Structure, *Not* Electronic Structure," *J. Chem. Educ.* **1999**, *76*, 1013.

$$\begin{array}{ccc} & :\ddot{F}: & \\ & | & \\ :\ddot{F} & N & \ddot{F}: \end{array}$$

This is the Lewis structure for NF_3. Always check that the total number of electrons (bonds plus lone pairs) equals the sum of the valence electrons: $6e^-$ in three bonds plus $20e^-$ in ten lone pairs equals 26 valence electrons.

This particular arrangement of F atoms around an N atom resembles the molecular shape of NF_3 (Section 10.2). But Lewis structures do not indicate shape, so an equally correct depiction of NF_3 is

$$\begin{array}{ccc} & :\ddot{F}: & \\ & | & \\ :\ddot{F}{-}N{-}\ddot{F}: \end{array}$$

or any other that retains the *same connections among the atoms*—a central N atom connected by single bonds to three surrounding F atoms.

Using these four steps, you can write a Lewis structure for any singly bonded molecule whose central atom is C, N, or O, as well as for some molecules with central atoms from higher periods. Remember that, in nearly all their compounds,

- Hydrogen atoms form one bond.
- Carbon atoms form four bonds.
- Nitrogen atoms form three bonds.
- Oxygen atoms form two bonds.
- Halogens form one bond when they are surrounding atoms; fluorine is always a surrounding atom.

Step 1:

          Cl

     F   C   F

          Cl

Step 3:   Cl
            |
        F —— C —— F
            |
            Cl

Step 4:
            :Cl:
             |
        :F —— C —— F:
             |
            :Cl:

SAMPLE PROBLEM 10.1 Writing Lewis Structures for Molecules with One Central Atom

PROBLEM Write a Lewis structure for CCl_2F_2, one of the compounds responsible for the depletion of stratospheric ozone.

SOLUTION *Step 1.* Place the atoms relative to each other. In CCl_2F_2, carbon has the lowest group number and EN, so it is the central atom. The halogen atoms surround it, but their specific positions are not important (see margin).

Step 2. Determine the total number of valence electrons (from A-group numbers): C is in Group 4A, F is in Group 7A, and Cl is in Group 7A, too. Therefore, we have

$$[1 \times C(4e^-)] + [2 \times F(7e^-)] + [2 \times Cl(7e^-)] = 32 \text{ valence } e^-$$

Step 3. Draw single bonds to the central atom and subtract $2e^-$ for each bond (see margin). Four single bonds use $8e^-$, so $32e^- - 8e^-$ leaves $24e^-$ remaining.

Step 4. Distribute the remaining electrons in pairs, beginning with the surrounding atoms, so that each atom has an octet (see margin).

CHECK Counting the electrons shows that each atom has an octet. Remember that bonding electrons are counted as belonging to each atom in the bond. The total number of electrons in bonds (8) and lone pairs (24) equals 32 valence electrons. Note that, as expected, C has four bonds and the surrounding halogens have one each.

FOLLOW-UP PROBLEM 10.1 Write a Lewis structure for each of the following:
(a) H_2S **(b)** OF_2 **(c)** $SOCl_2$

A slightly more complex situation occurs when molecules have two or more central atoms bonded to each other, with the other atoms around them.

Step 1:   H

            H   C   O   H

                 H

Step 3:   H
                 |
            H —— C —— O —— H
                 |
                 H

Step 4:   H
                 |
            H —— C —— Ö —— H
                 |
                 H

SAMPLE PROBLEM 10.2 Writing Lewis Structures for Molecules with More than One Central Atom

PROBLEM Write the Lewis structure for methanol (molecular formula CH_4O), an important industrial alcohol that is being used as a gasoline alternative in car engines.

SOLUTION *Step 1.* Place the atoms relative to each other. The H atoms can have only one bond, so C and O must be adjacent to each other. Recall that C has four bonds and O has two, so we arrange the H atoms to show this (see margin).

Step 2. Find the sum of valence electrons:

$$[1 \times C(4e^-)] + [1 \times O(6e^-)] + [4 \times H(1e^-)] = 14e^-$$

Step 3. Add single bonds and subtract $2e^-$ for each bond (see margin). Five bonds use $10e^-$, so $14e^- - 10e^-$ leaves $4e^-$ remaining.

Step 4. Add the remaining electrons in pairs. Carbon already has an octet, and each H shares two electrons with the C; so the four remaining valence electrons form two lone pairs on O. We now have the Lewis structure for methanol (see margin).

CHECK Each H atom has $2e^-$, and the C and O each have $8e^-$. The total number of valence electrons is $14e^-$, which equals $10e^-$ in bonds plus $4e^-$ in lone pairs. Also note that each H has one bond, C has four, and O has two.

FOLLOW-UP PROBLEM 10.2 Write a Lewis structure for each of the following:
(a) hydroxylamine (NH_3O)   **(b)** dimethyl ether (C_2H_6O; no O—H bonds)

Lewis Structures for Molecules with Multiple Bonds Sometimes, you'll find that, after steps 1 to 4, there are not enough electrons for the central atom (or one of the central atoms) to attain an octet. This usually means that a multiple bond is present, and the following additional step is needed:

Step 5. Cases involving multiple bonds. If, after step 4, a central atom still does not have an octet, make a multiple bond by changing a lone pair from one of the surrounding atoms into a bonding pair to the central atom.

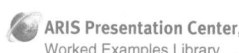
ARIS Presentation Center,
Worked Examples Library

SAMPLE PROBLEM 10.3 Writing Lewis Structures for Molecules with Multiple Bonds

PROBLEM Write Lewis structures for the following:

(a) Ethylene (C_2H_4), the most important reactant in the manufacture of polymers

(b) Nitrogen (N_2), the most abundant atmospheric gas

PLAN We show the structure resulting from steps 1 to 4: placing the atoms, counting the total valence electrons, making single bonds, and distributing the remaining valence electrons in pairs to attain octets. Then we continue with step 5, if needed.

SOLUTION (a) For C_2H_4. After steps 1 to 4, we have

$$\begin{array}{c} H \\ \diagdown \\ C \end{array}\text{—}\begin{array}{c} H \\ \diagup \\ \ddot{C} \\ \diagdown \\ H \end{array}$$

Step 5. Change a lone pair to a bonding pair. The C on the right has an octet, but the C on the left has only $6e^-$, so we convert the lone pair to another bonding pair between the two C atoms,

$$\begin{array}{c} H \\ \diagdown \\ C \end{array}\text{=}\begin{array}{c} H \\ \diagup \\ C \\ \diagdown \\ H \end{array}$$

(b) For N_2. After steps 1 to 4, we have $:\ddot{N}\text{—}\ddot{N}:$

Step 5. Neither N has an octet, so we change a lone pair to a bonding pair, $:\ddot{N}\text{=}N:$ In this case, moving one lone pair to make a double bond still does not give the N on the right an octet, so we move a lone pair from the left N to make a triple bond, $:N{\equiv}N:$

CHECK (a) Each C has four bonds and counts the $4e^-$ in the double bond as part of its own octet. The valence electron total is $12e^-$, all in six bonds. **(b)** Each N has three bonds and counts the $6e^-$ in the triple bond as part of its own octet. The valence electron total is $10e^-$, which equals the electrons in three bonds and two lone pairs.

FOLLOW-UP PROBLEM 10.3 Write Lewis structures for each of the following:
(a) CO (the only common molecule in which C has only three bonds) **(b)** HCN **(c)** CO_2

Resonance: Delocalized Electron-Pair Bonding

Literature Silverstein, Todd P. "The 'Big Dog-Puppy Dog' Analogy for Resonance," *J. Chem. Educ.* **1999**, *76*, 206.

We can often write more than one Lewis structure, each with the same relative placement of atoms, for a molecule or ion with *double bonds next to single bonds.* Consider ozone (O_3), a serious air pollutant at ground level but a life-sustaining absorber of harmful ultraviolet (UV) radiation up in the stratosphere. Two valid Lewis structures (with lettered O atoms for clarity) are

In structure I, oxygen B has a double bond to oxygen A and a single bond to oxygen C. In structure II, the single and double bonds are reversed. These are *not* two different O_3 molecules, just different Lewis structures for the same molecule.

In fact, *neither* Lewis structure depicts O_3 accurately. Bond length and bond energy measurements indicate that the two oxygen-oxygen bonds in O_3 are identical, with properties that lie between those of an O—O bond and an O=O bond, something like a "one-and-a-half" bond. The molecule is shown more correctly with two Lewis structures, called **resonance structures** (or **resonance forms**), and a two-headed resonance arrow ($\longleftrightarrow$) between them. Resonance structures *have the same relative placement of atoms but different locations of bonding and lone electron pairs.* You can convert one resonance form to another by moving lone pairs to bonding positions, and vice versa:

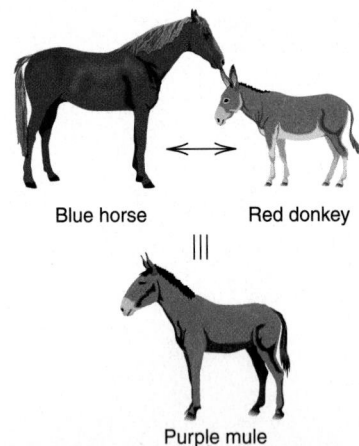

Blue horse   Red donkey

|||

Purple mule

● **A Purple Mule, Not a Blue Horse and a Red Donkey** A mule is a genetic mix, a hybrid, of a horse and a donkey; it is not a horse one instant and a donkey the next. Similarly, the color purple is a mix of two other colors, red and blue, not red one instant and blue the next. In the same sense, a resonance hybrid is one molecular species, not one resonance form this instant and another resonance form the next. The problem is that we cannot depict the hybrid accurately with a single Lewis structure.

ARIS Presentation Center,
Art Library

Organic chemistry application

Resonance structures are not real bonding depictions: O_3 does *not* change back and forth from structure I at one instant to structure II the next. The actual molecule is a **resonance hybrid,** an average of the resonance forms. ●

Our need for more than one Lewis structure to depict the ozone molecule is due to **electron-pair delocalization.** In a single, double, or triple bond, each electron pair is attracted by the nuclei of the two bonded atoms, and the electron density is greatest in the region between the nuclei: each electron pair is *localized.* In the resonance hybrid for O_3, however, two of the electron pairs (one bonding and one lone pair) are *delocalized:* their density is "spread" over the entire molecule. (Recall from Section 9.6 that in a metal, the electrons are delocalized over the entire sample, which is much more extensive than in a resonance hybrid.) In O_3, this results in two identical bonds, each consisting of a single bond (the localized electron pair) and a *partial bond* (the contribution from one of the delocalized electron pairs). We draw the resonance hybrid with a curved dashed line to show the delocalized pairs:

$$\ddot{\text{:O}} \diagdown \overset{\cdot\cdot}{\underset{\cdot\cdot}{\text{O}}} \diagup \ddot{\text{O:}}$$

Resonance is very common, and many molecules (and ions) are best depicted as resonance hybrids. Benzene (C_6H_6), for example, has two important resonance forms in which alternating single and double bonds have different positions. The actual molecule has six identical carbon-carbon bonds because there are six C—C bonds and three electron pairs delocalized over all six C atoms, often shown as a dashed circle (or simply a circle):

resonance forms

or

resonance hybrid

Partial bonding, such as that occurring in resonance hybrids, often leads to fractional bond orders. For O_3, we have

$$\text{Bond order} = \frac{3 \text{ electron pairs}}{2 \text{ bonded-atom pairs}} = 1\tfrac{1}{2}$$

The carbon-to-carbon bond order in benzene is 9 electron pairs/6 bonded-atom pairs, or $1\tfrac{1}{2}$ also. For the carbonate ion, CO_3^{2-}, three resonance structures can be drawn. Each has 4 electron pairs shared among 3 bonded-atom pairs, so the bond order is $\tfrac{4}{3}$, or $1\tfrac{1}{3}$. One of the three resonance structures for CO_3^{2-} is

Note that *the Lewis structure of a polyatomic ion is shown in square brackets, with its charge as a right superscript outside the brackets.*

SAMPLE PROBLEM 10.4 Writing Resonance Structures

PROBLEM Write resonance structures for the nitrate ion, NO_3^-.

PLAN We write a Lewis structure, remembering to add $1e^-$ to the total number of valence electrons because of the $1-$ ionic charge. Then we move lone and bonding pairs to write other resonance forms and connect them with the resonance arrow.

SOLUTION After steps 1 to 4, we have

Step 5. Since N has only $6e^-$, we change one lone pair on an O atom to a bonding pair and form a double bond, which gives each atom an octet. All the O atoms are equivalent, however, so we can move a lone pair from any of the three O atoms and obtain three resonance structures:

CHECK Each structure has the same relative placement of atoms, an octet around each atom, and $24e^-$ (the sum of the valence electron total and $1e^-$ from the ionic charge distributed in four bonds and eight lone pairs).

COMMENT 1. Remember that no double bond actually exists in the NO_3^- ion. The ion is a resonance hybrid of these three structures with a bond order of $1\frac{1}{3}$. (You'll see in the upcoming discussion why N can have four bonds here.)

2. The three resonance forms for NO_3^- contribute equally to the resonance hybrid because all the surrounding atoms are identical. This is not always the case, as you'll see next.

FOLLOW-UP PROBLEM 10.4 One of the three resonance structures for CO_3^{2-} was shown just before Sample Problem 10.4. Draw the other two.

Formal Charge: Selecting the More Important Resonance Structure

As we noted in the Comment in Sample Problem 10.4, the resonance forms for nitrate ion contribute equally to the resonance hybrid because the central atom has surrounding atoms that are all the same. When this is not the case, one resonance form may look more like the hybrid than the others. In other words, because the resonance hybrid is an average of the resonance forms, one form may contribute more and "weight" the average in its favor. We can often select the more important resonance form by determining each atom's **formal charge,** the charge it would have *if the bonding electrons were shared equally.*

An atom's formal charge is its total number of valence electrons minus *all* of its unshared valence electrons and *half* of its shared valence electrons. Thus,

$$\text{Formal charge of atom} = \text{no. of valence } e^- - (\text{no. of unshared valence } e^- + \tfrac{1}{2}\text{ no. of shared valence } e^-) \quad (10.1)$$

For example, in O_3, the formal charge of oxygen A in resonance form I is

$$6 \text{ valence } e^- - (4 \text{ unshared } e^- + \tfrac{1}{2}\text{ of } 4 \text{ shared } e^-) = 6 - 4 - 2 = 0$$

The formal charges of all the atoms in the two O_3 resonance forms are

$$O_A[6 - 4 - \tfrac{1}{2}(4)] = 0 \qquad\qquad O_A[6 - 6 - \tfrac{1}{2}(2)] = -1$$
$$O_B[6 - 2 - \tfrac{1}{2}(6)] = +1 \qquad\qquad O_B[6 - 2 - \tfrac{1}{2}(6)] = +1$$
$$O_C[6 - 6 - \tfrac{1}{2}(2)] = -1 \qquad\qquad O_C[6 - 4 - \tfrac{1}{2}(4)] = 0$$

Forms I and II have the same formal charges but on different O atoms, so they contribute equally to the resonance hybrid. *Formal charges must sum to the actual charge on the species:* zero for a molecule and the ionic charge for an ion.

Literature DeWit, David G. "Using Formal Charges in Teaching Descriptive Inorganic Chemistry," *J. Chem. Educ.* **1994,** *71,* 750.

Note that, in form I, instead of the usual two bonds for oxygen, O_B has three bonds and O_C has one. Only when an atom has a zero formal charge does it have its usual number of bonds; the same holds for C in CO, N in NO_3^-, and so forth.

Three criteria help us choose the more important resonance structures:

- Smaller formal charges (positive or negative) are preferable to larger ones.
- The same nonzero formal charges on adjacent atoms are not preferred.
- A more negative formal charge should reside on a more electronegative atom.

Let's apply these criteria to the cyanate ion, NCO^-, which has two *different* atoms around the central one. Three resonance forms with formal charges are

Formal charges: (−2) (0) (+1)   (−1) (0) (0)   (0) (0) (−1)

Resonance forms: $[:\ddot{N}-C\equiv O:]^- \longleftrightarrow [\ddot{N}=C=\ddot{O}:]^- \longleftrightarrow [:N\equiv C-\ddot{O}:]^-$

I   **II**   **III**

We eliminate form I because it has a larger formal charge on N than the others and a positive formal charge on O, which is more electronegative than N. Forms II and III have the same magnitude of formal charges, but form III has a −1 charge on the more electronegative atom, O. Therefore, II and III are significant contributors to the resonance hybrid of the cyanate ion, but III is the more important.

Note that formal charge (used to examine resonance structures) is *not* the same as oxidation number (used to monitor redox reactions):

Literature Packer, John E.; Woodgate, Sheila D. "Lewis Structures, Formal Charge, and Oxidation Numbers: A More User-Friendly Approach," *J. Chem. Educ.* **1991,** *68,* 456.

- For a *formal charge,* bonding electrons are assigned *equally* to the atoms (as if the bonding were *nonpolar covalent*), so each atom has half of them:

$$\text{Formal charge} = \text{valence e}^- - (\text{lone pair e}^- + \tfrac{1}{2}\text{ bonding e}^-)$$

- For an *oxidation number,* bonding electrons are assigned *completely* to the more electronegative atom (as if the bonding were *ionic*):

$$\text{Oxidation number} = \text{valence e}^- - (\text{lone pair e}^- + \text{bonding e}^-)$$

For the three cyanate ion resonance structures,

Formal charges: (−2) (0) (+1)   (−1) (0) (0)   (0) (0) (−1)

$[:\ddot{N}-C\equiv O:]^- \longleftrightarrow [\ddot{N}=C=\ddot{O}:]^- \longleftrightarrow [:N\equiv C-\ddot{O}:]^-$

Oxidation numbers: −3 +4 −2   −3 +4 −2   −3 +4 −2

Notice that the oxidation numbers *do not* change from one resonance form to another (because the electronegativities *do not* change), but the formal charges *do* change (because the numbers of bonding and lone pairs *do* change).

Lewis Structures for Exceptions to the Octet Rule

The octet rule is a useful guide for most molecules with Period 2 central atoms, but not for every one. Also, many molecules have central atoms from higher periods. As you'll see, some central atoms have fewer than eight electrons around them, and others have more. The most significant octet rule exceptions are for molecules containing electron-deficient atoms, odd-electron atoms, and especially atoms with expanded valence shells. In this discussion, you'll see that formal charge also has limitations when we are selecting the best resonance form.

Electron-Deficient Molecules Gaseous molecules containing either beryllium or boron as the central atom are often **electron deficient;** that is, they have *fewer* than eight electrons around the Be or B atom. The Lewis structures, with formal charges, of gaseous beryllium chloride* and boron trifluoride are

(0) (0) (0)
$:\ddot{Cl}-Be-\ddot{Cl}:$

$:\ddot{F}:$ (0)
$|$
$\underset{(0)}{:\ddot{F}}\overset{B}{\underset{(0)}{}}\underset{(0)}{\ddot{F}:}$

*Even though beryllium is an alkaline earth metal [Group 2A(2)], most of its compounds have properties consistent with covalent, rather than ionic, bonding (Chapter 14). For example, molten $BeCl_2$ does not conduct electricity, indicating the absence of ions.

There are only four electrons around beryllium and six around boron. Why don't lone pairs from the surrounding halogen atoms form multiple bonds to the central atoms, thereby satisfying the octet rule? Because halogens are much more electronegative than beryllium or boron, formal charge rules make the following structures unlikely:

ARIS Presentation Center,
Art Library

(Some data for BF_3 show a shorter than expected B—F bond. Shorter bonds indicate double-bond character, so the structure with the B=F bond may be a minor contributor to a resonance hybrid.) The main way electron-deficient atoms attain an octet is by forming additional bonds in reactions. When BF_3 reacts with ammonia, for instance, a compound forms in which boron attains its octet:*

Odd-Electron Molecules

A few molecules contain a central atom with an odd number of valence electrons, so they cannot possibly have all their electrons in pairs. Such species, called **free radicals,** contain a lone (unpaired) electron, which makes them paramagnetic (Section 8.5) and extremely reactive. ● Most odd-electron molecules have a central atom from an odd-numbered group, such as N [Group 5A(15)] or Cl [Group 7A(17)].

Consider nitrogen dioxide (NO_2) as an example. A major contributor to urban smog, it is formed when the NO in auto exhaust is oxidized. NO_2 has several resonance forms. Two involve an O atom that is doubly bonded, as in the case of ozone. Other resonance forms involve the location of the lone electron, and two of those are shown below. The resonance concept means the lone electron is delocalized over the N and O atoms, but let's see if formal charges can help us decide where it occurs most of the time. The form with the lone electron on the singly bonded O has zero formal charges (*right*):

But, given the way NO_2 reacts, the form with the lone electron on N (*left*) may be more important. Free radicals often react with each other to pair their lone electrons. When two NO_2 molecules collide, the lone electrons pair to form the N—N bond in dinitrogen tetraoxide (N_2O_4) and each N attains an octet:

Apparently, in this case, the lone electron spends most of its time on N, so formal charge is not very useful for picking the most important resonance form; we'll see below other limitations of using formal charge as the sole criterion.

Expanded Valence Shells

Many molecules and ions have more than eight valence electrons around the central atom. An atom must expand its valence shell to hold the additional valence electrons, but this allows the atom to form more bonds, a process that releases energy. Only larger central atoms can accommodate additional pairs, and empty *outer d* orbitals as well as occupied *s* and *p* orbitals are used. Therefore, **expanded valence shells** occur only with a *large central nonmetal atom in which d orbitals are available, that is, one from Period 3 or higher.*

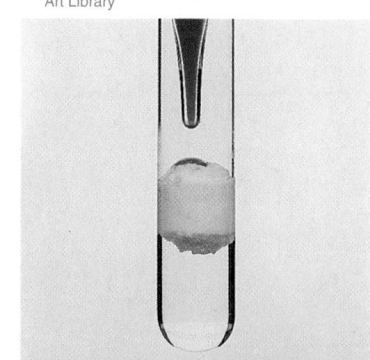

● **Deadly Free-Radical Activity** Free radicals can be extremely dangerous to biological systems because they rupture bonds in cells' biomolecules. If a free radical reacts with a biomolecule, it typically forms a covalent bond to one of the H atoms and removes it, and the biomolecule is left with an unpaired electron, thereby becoming a new free radical. That species repeats the process and creates other species with lone electrons that proliferate to disrupt chromosomes and cell membranes. We benefit from this activity when we apply the disinfectant hydrogen peroxide to a cut because it forms free radicals that destroy bacterial membranes. (The photo shows H_2O_2 reacting vigorously with a drop of blood.) On the other hand, recent studies have suggested that several disease states, including certain forms of cancer, may be attributable to free radicals. Also, vitamin E is believed to interrupt free-radical proliferation.

Biological application

---

*Reactions of the sort shown here, in which one species "donates" an electron pair to another to form a covalent bond, are examples of Lewis acid-base reactions (Chapter 18).

One example is sulfur hexafluoride, SF_6, a remarkably dense and inert gas used as an insulator in electrical equipment. The central sulfur is surrounded by six single bonds, one to each fluorine, for a total of 12 electrons:

Another example is phosphorus pentachloride, PCl_5, a fuming yellow-white solid used to manufacture lacquers and films. PCl_5 forms when phosphorus trichloride, PCl_3, reacts with chlorine gas. The P in PCl_3 has an octet, but two more bonds to chlorine form and P expands its valence shell to 10 electrons in PCl_5. Note that when PCl_5 forms, *one* Cl—Cl bond breaks (left side of the equation), and *two* P—Cl bonds form (right side), for a net increase of one bond:

In SF_6 and PCl_5, the central atom forms bonds to *more than four* atoms. But, by applying the concept of formal charge, many Lewis structures can be drawn with expanded valence shells of central atoms bonded to *four or fewer* atoms. Consider H_2SO_4: two of the resonance forms, with formal charges, are

In form I, sulfur obeys the octet rule, but the molecule has several nonzero formal charges. In form II, sulfur has an expanded valence shell of 12 electrons, but the formal charges are minimized. Thus, based on the formal charge rules, form II contributes more than form I to the resonance hybrid. But, more important than whether or not the rules are followed, form II is also consistent with observation. In gaseous H_2SO_4, the two sulfur-oxygen bonds *with* an H atom attached to the O are 157 pm long, whereas the two sulfur-oxygen bonds *without* an H atom attached to the O are 142 pm long. This shorter bond length indicates double-bond character. Moreover, electron density measurements indicate greater density in the sulfur-oxygen bonds where O has no H attached. Thus, these experimental results favor form II.

This situation is complex, however. As we saw with the lone electron of NO_2, formal charge rules may not always lead to the best resonance form. Indeed, in the case of H_2SO_4 and many other species with central atoms from Period 3 or higher, recent quantum-mechanical calculations indicate that forms with expanded valence shells and zero formal charges may be *less* important than forms that follow the octet rule and have higher formal charges. These findings suggest that the bonds are shorter because the higher formal charges increase bond polarities and, thus, electrostatic attractions. Such considerations favor form I for H_2SO_4.

When sulfuric acid loses two H^+ ions, it forms the sulfate ion, SO_4^{2-}. Measurements indicate all sulfur-oxygen bonds in SO_4^{2-} are 149 pm long, which is between the length of an S=O bond (~142 pm) and an S—O bond (~157 pm). Six of the seven resonance forms consistent with these data have sulfur with an expanded valence shell and zero formal charge. Two of the six are shown at left, and the seventh form, the one that obeys the octet rule, is shown at right:

Measurements show that the sulfur-oxygen bonds in SO_2 and SO_3 are all approximately 142 pm long, indicating S=O bonds. Lewis structures consistent with these data have zero formal charges (two at left), but other structures that obey the octet rule are consistent with the recent calculations (two at right):

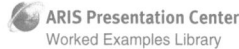

A final word on the limitations of Lewis structures and formal charge: Chemistry has been a central science for more than two centuries, yet controversies often arise over how data should be interpreted, even within such established areas as bonding and structure. Lewis structures have limitations in depicting a molecule, and we require several resonance forms to overcome them. Formal charge is a useful, but far from perfect, tool in choosing the most important resonance form. In the case of expanded valence shells, structures based on calculations differ from those based on formal charge rules. Nevertheless, while keeping these points in mind, we will continue to draw structures based on the rules of formal charge because they provide a simple approach consistent with experimental data.

SAMPLE PROBLEM 10.5 Writing Lewis Structures for Octet-Rule Exceptions

ARIS Presentation Center,
Worked Examples Library

PROBLEM Write a Lewis structure and identify the octet-rule exception for (a) $SClF_5$; (b) H_3PO_4 (draw two resonance forms and select the more important); (c) $BFCl_2$.

PLAN We write each Lewis structure and examine it for exceptions to the octet rule. In (a) and (b), the central atoms are in Period 3, so they can have more than an octet. In (c), the central atom is B, which can have fewer than an octet of electrons.

SOLUTION (a) $SClF_5$ has an *expanded valence shell*. The Lewis structure is

(b) H_3PO_4 has two resonance forms. The structures, with formal charges, are

Structure I obeys the octet rule but has nonzero formal charges. Structure II has an *expanded valence shell* with all formal charges equal to zero. According to formal charge rules, structure II is the more important resonance form.

(c) $BFCl_2$ is an *electron-deficient molecule;* B has only six electrons surrounding it:

COMMENT In (b), structure II is consistent with bond-length measurements, which show one shorter (152 pm) phosphorus-oxygen bond and three longer (157 pm) bonds. But, as we said for H_2SO_4, calculations also indicate the importance of structure I.

FOLLOW-UP PROBLEM 10.5 Write a Lewis structure with minimal formal charges for (a) $POCl_3$; (b) ClO_2; (c) XeF_4.

Section Summary

A stepwise process converts a molecular formula into a Lewis structure, a *two-dimensional* representation of a molecule (or ion) that shows the placement of atoms and distribution of valence electrons among bonding and lone pairs. • When two or more Lewis structures can be drawn for the same relative placement of atoms, the actual structure is a hybrid of those resonance forms. • Formal charges are often useful for determining the most important contributor to the hybrid. • Electron-deficient molecules (central Be or B) and odd-electron species (free radicals) have less than an octet around the central atom but often attain an octet in reactions. • In a molecule (or ion) with a central atom from Period 3 or higher, the atom can hold more than eight electrons because it is larger and uses *d* orbitals to expand its valence shell.

📖 **Literature** Gillespie, Ronald J. "Teaching Molecular Geometry with the VSEPR Model," *J. Chem. Educ.* **2004**, *81*, 298.

📖 **Literature** Hurst, Michael O. "How We Teach Molecular Structure to Freshmen," *J. Chem. Educ.* **2002**, *79*, 763.

📖 **Literature** Pfennig, Brian W.; Frock, Richard L. "The Use of Molecular Modeling and VSEPR Theory in the Undergraduate Curriculum to Predict the Three-Dimensional Structure of Molecules," *J. Chem. Educ.* **1999**, *76*, 1018.

10.2 VALENCE-SHELL ELECTRON-PAIR REPULSION (VSEPR) THEORY AND MOLECULAR SHAPE

Virtually every biochemical process hinges to a great extent on the shapes of interacting molecules. Every medicine you take, odor you smell, or flavor you taste depends on part or all of one molecule fitting physically together with another. This universal importance of molecular shape in the functioning of each organism carries over to the ecosystem. Biologists have found complex interactions regulating behaviors (such as mating, defense, navigation, and feeding) that depend on one molecule's shape matching up with that of another. In this section, we discuss a model for understanding and predicting molecular shape.

The Lewis structure of a molecule is something like the blueprint of a building: a flat drawing showing the relative placement of parts (atom cores), structural connections (groups of bonding valence electrons), and various attachments (nonbonding lone pairs of valence electrons). To construct the molecular shape from the Lewis structure, chemists employ **valence-shell electron-pair repulsion (VSEPR) theory.** Its basic principle is that *each group of valence electrons around a central atom is located as far away as possible from the others in order to minimize repulsions.* We define a "group" of electrons as any number of electrons that occupy a localized region around an atom. Thus, an electron group may consist of a single bond, a double bond, a triple bond, a lone pair, or even a lone electron.* Each group of valence electrons around an atom repels the other groups to maximize the angles between them. It is the three-dimensional arrangement of nuclei joined by these groups that gives rise to the molecular shape.

Electron-Group Arrangements and Molecular Shapes

When two, three, four, five, or six objects attached to a central point maximize the space that each can occupy around that point, five geometric patterns result. Figure 10.2A depicts these patterns with balloons. If the objects are the valence-electron groups of a central atom, their repulsions maximize the space each occupies and give rise to the five *electron-group arrangements* of minimum energy seen in the great majority of molecules and polyatomic ions.

The electron-group arrangement is defined by the valence-electron groups, both bonding and nonbonding, around the central atom. On the other hand, the **molecular shape** is defined by the relative positions of the atomic nuclei. Figure 10.2B shows the molecular shapes that occur when *all* the surrounding electron groups are *bonding* groups. When some are *nonbonding* groups, different molecular shapes occur. Thus, *the same electron-group arrangement can give rise to different molecular shapes:* some with all bonding groups (as in Figure 10.2B) and others with bonding and

📖 **Literature** Gomba, Frank J. "A Comment on Molecular Geometry," *J. Chem. Educ.* **1999**, *76*, 1732.

📖 **Literature** Birk, James P.; Abbassian, Soraya. "Teaching VSEPR: The Plastic Egg Model," *J. Chem. Educ.* **1996**, *73*, 636.

Animation: VSEPR
ARIS Presentation Center
Chapter 10

---

*The two electron pairs in a double bond (or the three pairs in a triple bond) occupy separate orbitals, so they remain near each other and act as one electron group (see Chapter 11).

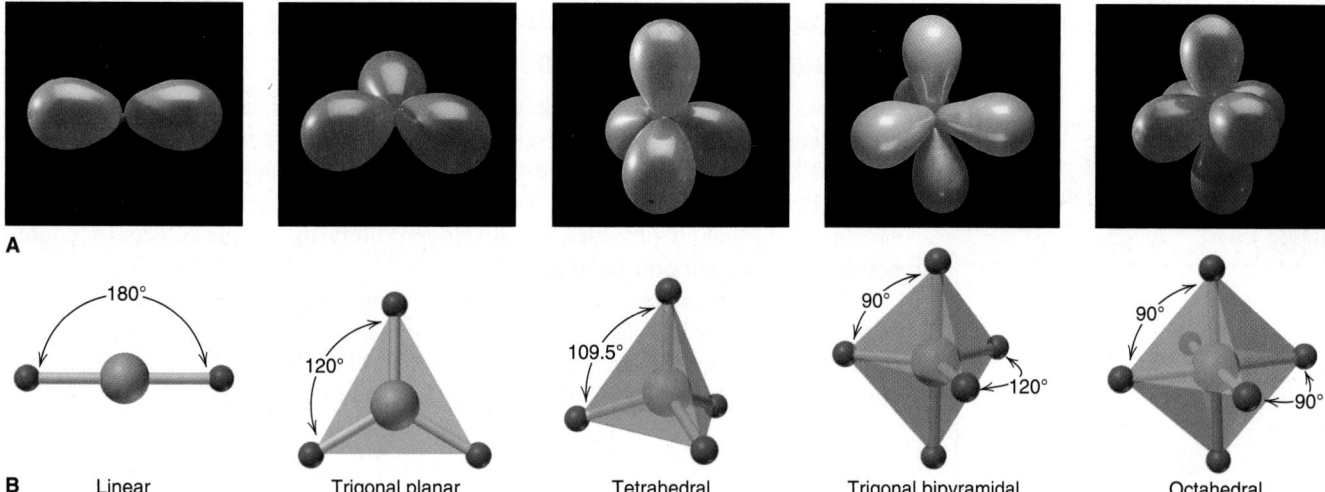

B Linear   Trigonal planar   Tetrahedral   Trigonal bipyramidal   Octahedral

Figure 10.2 **Electron-group repulsions and molecular shapes.** **A,** As an analogy for electron-group arrangements, two to six attached balloons form five geometric orientations in which each balloon occupies as much space as possible. **B,** Mutually repelling electron groups attached to a central atom (*red*) occupy as much space as possible. If each is a bonding group to a surrounding atom (*dark gray*), these molecular shapes and bond angles are observed. The shape has the same name as the electron-group arrangement.

Transparency/ARIS Presentation Center,
Art Library

Animation: VSEPR Theory and the Shape of Molecules
ARIS Presentation Center
Chapter 10

ARIS Presentation Center,
Art Library

nonbonding groups. To classify molecular shapes, we assign each a specific AX_mE_n designation, where m and n are integers, A is the central atom, X is a surrounding atom, and E is a nonbonding valence-electron group (usually a lone pair).

The **bond angle** is the angle formed by the nuclei of two surrounding atoms with the nucleus of the central atom at the vertex. The angles shown for the shapes in Figure 10.2B are *ideal* bond angles, those predicted by simple geometry alone. These are observed when all the bonding electron groups around a central atom are identical and are connected to atoms of the same element. When this is not the case, the bond angles deviate from the ideal angles, as you'll see shortly.

It's important to realize that we use the VSEPR model to account for the molecular shapes observed by means of various laboratory instruments. In almost every case, VSEPR predictions are in accord with actual observations. (We discuss some of these observational methods in Chapter 12.)

The Molecular Shape with Two Electron Groups (Linear Arrangement)

When two electron groups attached to a central atom are oriented as far apart as possible, they point in opposite directions. The **linear arrangement** of electron groups results in a molecule with a **linear shape** and a bond angle of 180°. Figure 10.3 shows the general form (top) and shape (middle) with VSEPR shape class (AX_2), and the formulas of some linear molecules.

Gaseous beryllium chloride ($BeCl_2$) is a linear molecule (AX_2). Gaseous beryllium compounds are electron deficient, with only two electron pairs around the central Be atom:

$$:\ddot{C}l\!\!-\!\!Be\!\!-\!\!\ddot{C}l:$$

In carbon dioxide, the central C atom forms two double bonds with the O atoms:

$$\ddot{O}\!\!=\!\!C\!\!=\!\!\ddot{O}$$

Each double bond acts as one electron group and is oriented 180° away from the other, so CO_2 is linear. Notice that the lone pairs on the O atoms of CO_2 or on the Cl atoms of $BeCl_2$ are not involved in the molecular shape: only electron groups around the *central* atom influence shape.

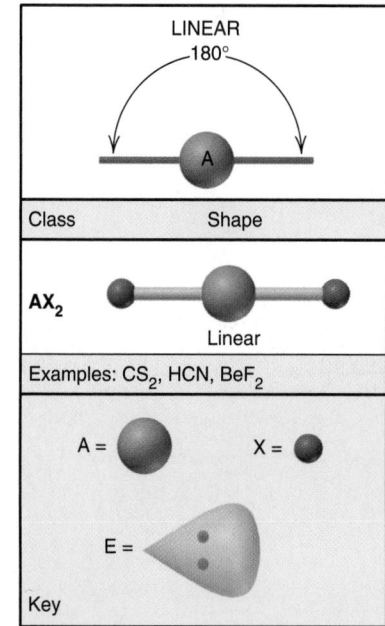

Figure 10.3 **The single molecular shape of the linear electron-group arrangement.** The key (*bottom*) for A, X, and E also refers to Figures 10.4, 10.5, 10.7, and 10.8.

ARIS Presentation Center, Active Art Library

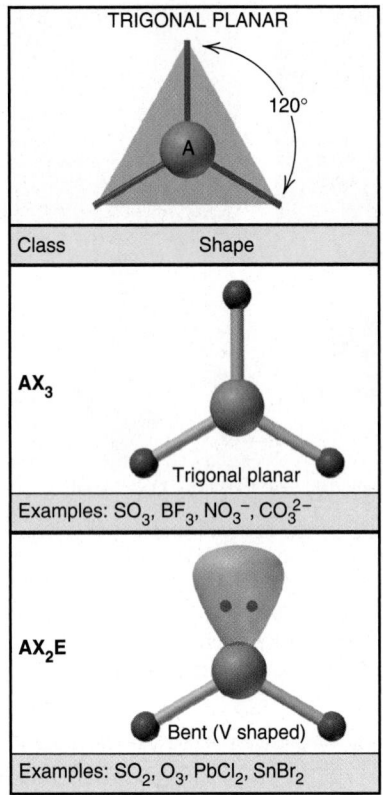

Figure 10.4 The two molecular shapes of the trigonal planar electron-group arrangement.

ARIS Presentation Center,
Active Art Library

Literature Gillespie, Ronald J. "Multiple Bonds and the VSEPR Model," *J. Chem. Educ.* **1992**, *69*, 116.

Molecular Shapes with Three Electron Groups (Trigonal Planar Arrangement)

Three electron groups around the central atom repel each other to the corners of an equilateral triangle, which gives the **trigonal planar arrangement,** shown in Figure 10.4, and an ideal bond angle of 120°. This arrangement has two possible molecular shapes, one with three surrounding atoms and the other with two atoms and one lone pair. It provides our first opportunity to see the effects of double bonds and lone pairs on bond angles.

When the three electron groups are bonding groups, the molecular shape is *trigonal planar* (AX_3). Boron trifluoride (BF_3), another electron-deficient molecule, is an example. It has six electrons around the central B atom in three single bonds to F atoms. The nuclei lie in a plane, and each F—B—F angle is 120°:

The nitrate ion (NO_3^-) is one of several polyatomic ions with the trigonal planar shape. One of three resonance forms of the nitrate ion (Sample Problem 10.4) is

The resonance hybrid has three identical bonds of bond order $1\frac{1}{3}$, so the ideal bond angle is observed.

Effect of Double Bonds
How do bond angles deviate from the ideal angles when the surrounding atoms and electron groups are not identical? Consider formaldehyde (CH_2O), a substance with many uses, including the manufacture of Formica countertops, the production of methanol, and the preservation of cadavers. Its trigonal planar shape is due to two types of surrounding atoms (O and H) and two types of electron groups (single and double bonds):

The actual bond angles deviate from the ideal because *the double bond, with its greater electron density, repels the two single bonds more strongly than they repel each other.* Note that the H—C—H bond angle is less than 120°.

Effect of Lone Pairs
The molecular shape is defined *only* by the positions of the nuclei, so when one of the three electron groups is a lone pair (AX_2E), the shape is **bent,** or **V shaped,** not trigonal planar. Gaseous tin(II) chloride is an example, with the three electron groups in a trigonal plane and the lone pair at one of the triangle's corners. A lone pair can have a major effect on bond angle. Because a lone pair is held by only one nucleus, it is less confined and exerts stronger repulsions than a bonding pair. Thus, *a lone pair repels bonding pairs more strongly than bonding pairs repel each other.* This stronger repulsion *decreases* the angle between bonding pairs. Note the decrease from the ideal 120° angle in $SnCl_2$:

Molecular Shapes with Four Electron Groups (Tetrahedral Arrangement)

The shapes described so far have all been easy to depict in two dimensions, but four electron groups must use three dimensions to achieve maximal separation. Recall that *Lewis structures do **not** depict shape*. Consider methane. The Lewis structure shown below (left) indicates four bonds pointing to the corners of a square, which suggests a 90° bond angle. However, in three dimensions, the four electron groups move farther apart than 90° and point to the vertices of a tetrahedron, a polyhedron with four faces made of identical equilateral triangles. Methane has a bond angle of 109.5°. *Perspective drawings*, such as the one shown below (middle) for methane, indicate depth by using solid and dashed wedges for some of the bonds:

The normal bond lines (blue) represent shared electron groups in the plane of the page; the solid wedge (green) is the bond between the atom in the plane of the page and a group lying toward you above the page; and the dashed wedge (red) is the bond to a group lying away from you below the page. The ball-and-stick model (right) shows the tetrahedral shape clearly.

*All molecules or ions with four electron groups around a central atom adopt the **tetrahedral arrangement*** (Figure 10.5). When all four electron groups are bonding groups, as in the case of methane, the molecular shape is also *tetrahedral* (AX₄), a very common geometry in organic molecules. In Sample Problem 10.1, we drew the Lewis structure for the tetrahedral molecule dichlorodifluoromethane (CCl_2F_2), without regard to how the halogen atoms surround the carbon atom. Because Lewis structures are flat, it may seem as if we can write two different structures for CCl_2F_2, but these actually represent the same molecule, as Figure 10.6 makes clear.

When one of the four electron groups in the tetrahedral arrangement is a lone pair, the molecular shape is that of a **trigonal pyramid** (AX₃E), a tetrahedron with one vertex "missing." Stronger repulsions due to the lone pair make the measured bond angle slightly less than the ideal 109.5°. In ammonia (NH_3), for example, the lone pair forces the N—H bonding pairs closer, and the H—N—H bond angle is 107.3°. NF_3 (see page 379) has a trigonal pyramidal shape also.

Picturing molecular shapes is a great way to visualize what happens during a reaction. For instance, when ammonia accepts the proton from an acid, the lone pair on the N atom of trigonal pyramidal NH_3 forms a covalent bond to the H^+ and yields the ammonium ion (NH_4^+), one of many tetrahedral polyatomic ions.

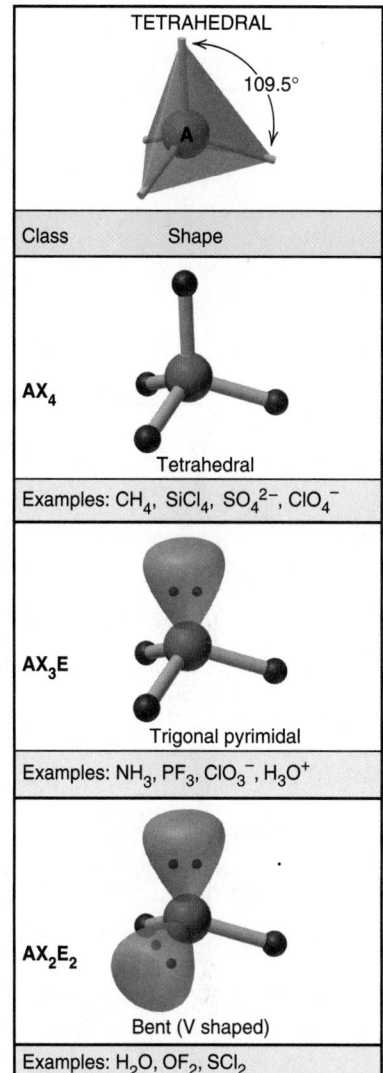

Class	Shape
AX₄	Tetrahedral

Examples: CH_4, $SiCl_4$, SO_4^{2-}, ClO_4^-

| AX₃E | Trigonal pyrimidal |

Examples: NH_3, PF_3, ClO_3^-, H_3O^+

| AX₂E₂ | Bent (V shaped) |

Examples: H_2O, OF_2, SCl_2

Figure 10.5 **The three molecular shapes of the tetrahedral electron-group arrangement.**

ARIS Presentation Center, Active Art Library

same as

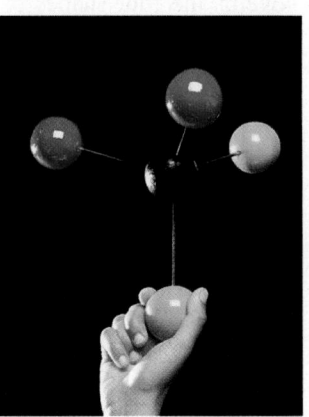

Figure 10.6 **Lewis structures and molecular shapes.** Lewis structures do not indicate geometry. For example, it may seem as if two different Lewis structures can be written for CCl_2F_2, but a twist of the model (Cl, *green;* F, *yellow*) shows that they represent the same molecule.

ARIS Presentation Center, Photo Library

Figure 10.7 **The four molecular shapes of the trigonal bipyramidal electron-group arrangement.**

Note how the H—N—H bond angle expands from 107.3° in NH_3 to 109.5° in NH_4^+, as the lone pair becomes another bonding pair:

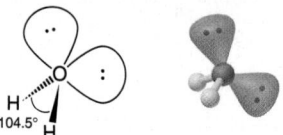

When the four electron groups around the central atom include two bonding and two nonbonding groups, the molecular shape is *bent, or V shaped* (AX_2E_2). [In the trigonal planar arrangement, the shape with two bonding groups and one lone pair (AX_2E) is also called bent, but its ideal bond angle is 120°, not 109.5°.] Water is the most important V-shaped molecule with the tetrahedral arrangement. We might expect the repulsions from its two lone pairs to have a *greater* effect on the bond angle than the repulsions from the single lone pair in NH_3. Indeed, the H—O—H bond angle is 104.5°, even less than the H—N—H angle in NH_3:

Thus, for similar molecules within a given electron-group arrangement, electron-pair repulsions cause deviations from ideal bond angles in the following order:

> Lone pair–lone pair > lone pair–bonding pair > bonding pair–bonding pair   (10.2)

Molecular Shapes with Five Electron Groups (Trigonal Bipyramidal Arrangement)

All molecules with five or six electron groups have a central atom from Period 3 or higher because only these atoms have the *d* orbitals available to expand the valence shell beyond eight electrons.

When five electron groups maximize their separation, they form the **trigonal bipyramidal arrangement.** In a trigonal bipyramid, two trigonal pyramids share a common base, as shown in Figure 10.7. Note that, in a molecule with this arrangement, *there are two types of positions for surrounding electron groups and two ideal bond angles.* Three **equatorial groups** lie in a trigonal plane that includes the central atom, and two **axial groups** lie above and below this plane. Therefore, a 120° bond angle separates equatorial groups, and a 90° angle separates axial from equatorial groups. In general, the greater the bond angle, the weaker the repulsions, so *equatorial-equatorial (120°) repulsions are weaker than axial-equatorial (90°) repulsions.* The tendency of the electron groups to occupy *equatorial* positions, and thus minimize the stronger axial-equatorial repulsions, governs the four shapes of the trigonal bipyramidal arrangement.

With all five positions occupied by bonded atoms, the molecule has the *trigonal bipyramidal* shape (AX_5), as in phosphorus pentachloride (PCl_5):

Three other shapes arise for molecules with lone pairs. Since lone pairs exert stronger repulsions than bonding pairs, *lone pairs occupy equatorial positions.* With one lone pair present at an equatorial position, the molecule has a **seesaw shape** (AX_4E). Sulfur tetrafluoride (SF_4), a powerful fluorinating agent, has this shape, shown here and in Figure 10.7 with the "seesaw" tipped up on an end. Note how the equatorial lone pair repels all four bonding pairs to reduce the bond angles:

The tendency of lone pairs to occupy equatorial positions causes molecules with three bonding groups and two lone pairs to have a **T shape** (AX_3E_2). Bromine trifluoride (BrF_3), one of many compounds with fluorine bonded to a larger halogen, has this shape. Note the predicted decrease from the ideal 90° F—Br—F bond angle:

Molecules with three lone pairs in equatorial positions must have the two bonding groups in axial positions, which gives the molecule a *linear* shape (AX_2E_3) and a 180° axial-to-central-to-axial (X—A—X) bond angle. For example, the triiodide ion (I_3^-), which forms when I_2 dissolves in aqueous I^- solution, is linear:

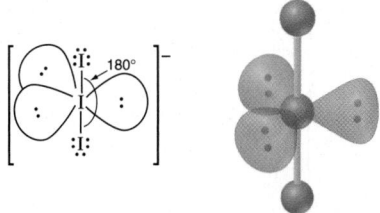

Molecular Shapes with Six Electron Groups (Octahedral Arrangement)

The last of the five major electron-group arrangements is the **octahedral arrangement.** An octahedron is a polyhedron with eight faces made of identical equilateral triangles and six identical vertices, as shown in Figure 10.8. In a molecule (or ion) with this arrangement, six electron groups surround the central atom and each points to one of the six vertices, which gives all the groups a 90° ideal bond angle. Three important molecular shapes occur with this arrangement.

With six bonding groups, the molecular shape is *octahedral* (AX_6). When the seesaw-shaped SF_4 reacts with additional F_2, the central S atom expands its valence shell further to form octahedral sulfur hexafluoride (SF_6):

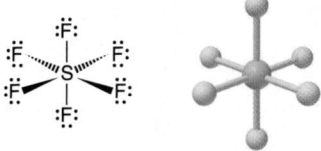

Because all six electron groups have the same ideal bond angle, it makes no difference which position one lone pair occupies. Five bonded atoms and one lone pair define the **square pyramidal shape** (AX_5E), as in iodine pentafluoride (IF_5):

OCTAHEDRAL
90°
A
90°

Class	Shape
AX_6	Octahedral

Examples: SF_6, IOF_5

AX_5E	Square pyramidal

Examples: BrF_5, TeF_5^-, $XeOF_4$

AX_4E_2	Square planar

Examples: XeF_4, ICl_4^-

Figure 10.8 The three molecular shapes of the octahedral electron-group arrangement.

When a molecule has four bonded atoms and two lone pairs, however, the lone pairs always lie at *opposite vertices* to avoid the stronger 90° lone pair–lone pair repulsions. This positioning gives the **square planar shape** (AX_4E_2), as in xenon tetrafluoride (XeF_4):

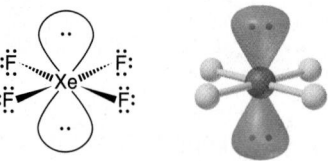

Figure 10.9 is a summary of the molecular shapes we've discussed.

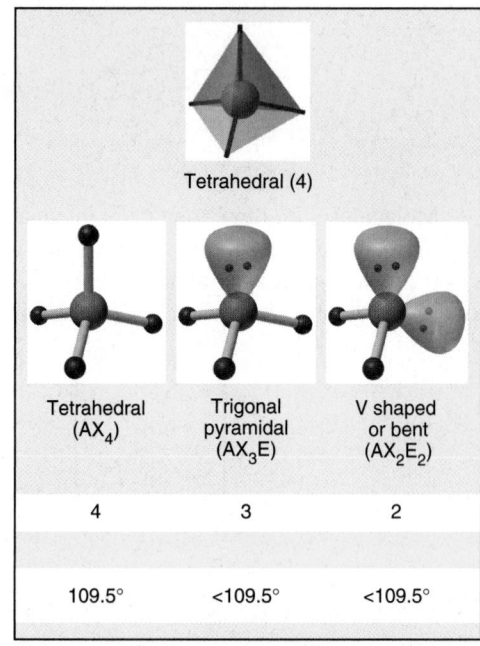

	Linear (2)	Trigonal planar (3)		Tetrahedral (4)		
e⁻ Group arrangement (no. of groups)						
Molecular shape (class)	Linear (AX_2)	Trigonal planar (AX_3)	V shaped or bent (AX_2E)	Tetrahedral (AX_4)	Trigonal pyramidal (AX_3E)	V shaped or bent (AX_2E_2)
No. of bonding groups	2	3	2	4	3	2
Bond angle	180°	120°	<120°	109.5°	<109.5°	<109.5°

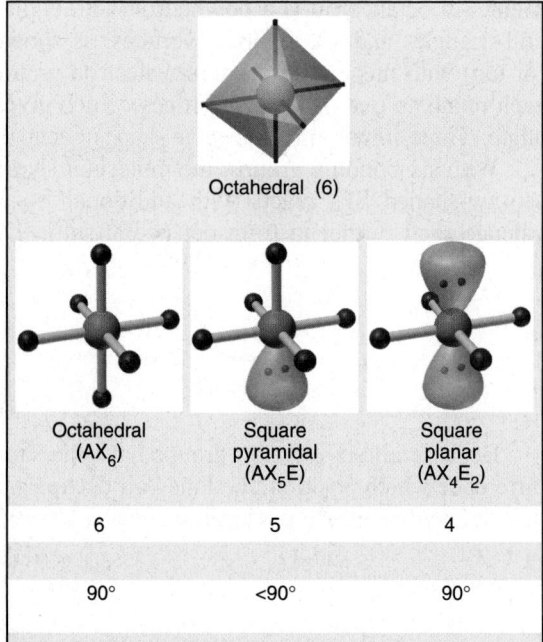

	Trigonal bipyramidal (5)				Octahedral (6)		
e⁻ Group arrangement (no. of groups)							
Molecular shape (class)	Trigonal bipyramidal (AX_5)	Seesaw (AX_4E)	T shaped (AX_3E_2)	Linear (AX_2E_3)	Octahedral (AX_6)	Square pyramidal (AX_5E)	Square planar (AX_4E_2)
No. of bonding groups	5	4	3	2	6	5	4
Bond angle	90° (ax) 120° (eq)	<90° (ax) <120° (eq)	<90° (ax)	180°	90°	<90°	90°

Figure 10.9 A summary of common molecular shapes with two to six electron groups.

Using VSEPR Theory to Determine Molecular Shape

Let's apply a stepwise method for using the VSEPR theory to determine a molecular shape from a molecular formula:

Step 1. Write the Lewis structure from the molecular formula (Figure 10.1, p. 378) to see the relative placement of atoms and the number of electron groups.

Step 2. Assign an electron-group arrangement by counting *all* electron groups around the central atom, bonding *plus* nonbonding.

Step 3. Predict the ideal bond angle from the electron-group arrangement and *the direction of any deviation* caused by lone pairs or double bonds.

Step 4. Draw and name the molecular shape by counting bonding groups and nonbonding groups separately.

Figure 10.10 summarizes these steps, and the next two sample problems apply them.

 Literature Gillespie, Ronald J.; Spencer, James N.; Moog, Richard S. "Bonding and Molecular Geometry without Orbitals—The Electron Domain Model," *J. Chem. Educ.* **1996**, *73*, 622.

Transparency/ARIS Presentation Center, Art Library

Figure 10.10 **The steps in determining a molecular shape.** Four steps are needed to convert a molecular formula to a molecular shape.

SAMPLE PROBLEM 10.6 Predicting Molecular Shapes with Two, Three, or Four Electron Groups

PROBLEM Draw the molecular shapes and predict the bond angles (relative to the ideal angles) of (**a**) PF_3 and (**b**) $COCl_2$.

SOLUTION (**a**) For PF_3.

Step 1. Write the Lewis structure from the formula (see below left).

Step 2. Assign the electron-group arrangement: Three bonding groups plus one lone pair give four electron groups around P and the *tetrahedral arrangement*.

Step 3. Predict the bond angle: For the tetrahedral electron-group arrangement, the ideal bond angle is 109.5°. There is one lone pair, so the actual bond angle should be less than 109.5°.

Step 4. Draw and name the molecular shape: With four electron groups, one of them a lone pair, PF_3 has a trigonal pyramidal shape (AX_3E):

$$:\!\ddot{F}\!-\!\ddot{P}\!-\!\ddot{F}\!: \xrightarrow[\text{groups}]{4\ e^-} \text{Tetrahedral arrangement} \xrightarrow[\text{pair}]{1\ \text{lone}} <109.5° \xrightarrow[\text{groups}]{3\ \text{bonding}}$$

96.3° AX_3E

(**b**) For $COCl_2$.

Step 1. Write the Lewis structure from the formula (see below left).

Step 2. Assign the electron-group arrangement: Two single bonds plus one double bond give three electron groups around C and the *trigonal planar arrangement*.

Step 3. Predict the bond angles: The ideal bond angle is 120°, but the double bond between C and O should compress the Cl—C—Cl angle to less than 120°.

Step 4. Draw and name the molecular shape: With three electron groups and no lone pairs, $COCl_2$ has a trigonal planar shape (AX_3):

$$\xrightarrow[\text{groups}]{3\ e^-} \text{Trigonal planar arrangement} \xrightarrow[\text{bond}]{1\ \text{double}} \begin{array}{l} \text{Cl—C—O} >120° \\ \text{Cl—C—Cl} <120° \end{array} \xrightarrow[\text{groups}]{3\ \text{bonding}}$$

124.5° 111° AX_3

CHECK We compare the answers with the general information in Figure 10.9.

COMMENT Be sure the Lewis structure is correct because it determines the other steps.

FOLLOW-UP PROBLEM 10.6 Draw the molecular shapes and predict the bond angles (relative to the ideal angles) of (**a**) CS_2; (**b**) $PbCl_2$; (**c**) CBr_4; (**d**) SF_2.

ARIS Presentation Center, Worked Examples Library

Literature Myers, Stephanie A. "The Molecular Model Game," *J. Chem. Educ.* **2003**, *80*, 423.

SAMPLE PROBLEM 10.7 Predicting Molecular Shapes with Five or Six Electron Groups

PROBLEM Determine the molecular shapes and predict the bond angles (relative to the ideal angles) of **(a)** SbF_5 and **(b)** BrF_5.

PLAN We proceed as in Sample Problem 10.6, keeping in mind the need to minimize the number of 90° repulsions.

SOLUTION (a) For SbF_5.

Step 1. Lewis structure (see below left).

Step 2. Electron-group arrangement: With five electron groups, this is the *trigonal bipyramidal* arrangement.

Step 3. Bond angles: All the groups and surrounding atoms are identical, so the bond angles are ideal: 120° between equatorial groups and 90° between axial and equatorial groups.

Step 4. Molecular shape: Five electron groups and no lone pairs give the trigonal bipyramidal shape (AX_5):

[Diagram: Lewis structure of SbF_5 → 5 e⁻ groups → Trigonal bipyramidal arrangement → no lone pairs or double bonds → Ideal bond angles → 5 bonding groups → structure showing 120° and 90° angles, labeled AX_5, with ball-and-stick model]

(b) For BrF_5.

Step 1. Lewis structure (see below left).

Step 2. Electron-group arrangement: Six electron groups give the *octahedral* arrangement.

Step 3. Bond angles: The lone pair should make all bond angles less than the ideal 90°.

Step 4. Molecular shape: With six electron groups and one of them a lone pair, BrF_5 has the square pyramidal shape (AX_5E):

[Diagram: Lewis structure of BrF_5 → 6 e⁻ groups → Octahedral arrangement → 1 lone pair → <90° → 5 bonding groups → structure showing 84.8° angle, labeled AX_5E, with ball-and-stick model]

CHECK We compare our answers with Figure 10.9.

COMMENT We will encounter the linear, tetrahedral, square planar, and octahedral shapes in an important group of substances called *coordination compounds,* which we discuss in Chapter 23.

FOLLOW-UP PROBLEM 10.7 Draw the molecular shapes and predict the bond angles (relative to the ideal angles) of **(a)** ICl_2^-; **(b)** ClF_3; **(c)** SOF_4.

Molecular Shapes with More Than One Central Atom

Many molecules, especially those in living systems, have more than one central atom. The shapes of these molecules are combinations of the molecular shapes for each central atom. For these molecules, we find the molecular shape around one central atom at a time. Consider ethane (CH_3CH_3; molecular formula C_2H_6), a component of natural gas (Figure 10.11A). With four bonding groups and no lone pairs around each of the two central carbons, ethane is shaped like two overlapping tetrahedra.

Ethanol (CH_3CH_2OH; molecular formula C_2H_6O), the intoxicating substance in beer and wine, has three central atoms (Figure 10.11B). The CH_3— group is tetrahedrally shaped, and the —CH_2— group has four bonding groups around its central C atom, so it is tetrahedrally shaped also. The O atom has four electron groups and two lone pairs around it, which gives the V shape (AX_2E_2).

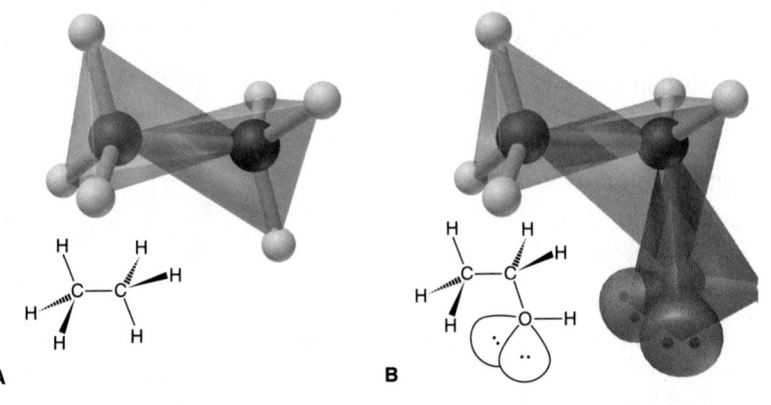

Figure 10.11 The tetrahedral centers of ethane and of ethanol. When a molecule has more than one central atom, the overall shape is a composite of the shape around each center. **A,** Ethane's shape can be viewed as two overlapping tetrahedra. **B,** Ethanol's shape can be viewed as three overlapping tetrahedral arrangements, with the shape around the O atom bent (V shaped) because of its two lone pairs.

 ARIS Presentation Center,
Art Library

 ARIS Presentation Center,
Worked Examples Library

SAMPLE PROBLEM 10.8 **Predicting Molecular Shapes with More Than One Central Atom**

PROBLEM Determine the shape around each of the central atoms in acetone, $(CH_3)_2C{=}O$.
PLAN There are three central atoms, all C, two of which are in $CH_3{-}$ groups. We determine the shape around one central atom at a time.
SOLUTION
Step 1. Lewis structure (see below left).
Step 2. Electron-group arrangement: Each $CH_3{-}$ group has four electron groups around its central C, so its electron-group arrangement is *tetrahedral*. The third C atom has three electron groups around it, so it has the *trigonal planar* arrangement.
Step 3. Bond angles: The H—C—H angle in the $CH_3{-}$ groups should be near the ideal tetrahedral angle of $109.5°$. The C$=$O double bond should compress the C—C—C angle to less than the ideal $120°$.
Step 4. Shapes around central atoms: With four electron groups and no lone pairs, the shapes around the two C atoms in the $CH_3{-}$groups are tetrahedral (AX_4). With three electron groups and no lone pairs, the shape around the middle C atom is trigonal planar (AX_3):

(Lewis structure)	3 e⁻ groups (middle C) → Trigonal planar (middle C)	1 double bond (middle C) → C—C—O >120° C—C—C <120° H—C—H ~109.5° H—C—C ~109.5° → all bonding groups → (ball-and-stick models)
	4 e⁻ groups (end C's) → Tetrahedral (end C's)	

FOLLOW-UP PROBLEM 10.8 Determine the shape around each central atom and predict any deviations from ideal bond angles in the following: **(a)** H_2SO_4; **(b)** propyne (C_3H_4; there is one C$\equiv$C bond); **(c)** S_2F_2.

Organic chemistry application

In recent years, chemists have synthesized some organic molecules with remarkable—some would say beautiful—shapes. Some of their uses may still be speculative, but there is no question about their geometric elegance. A few of these molecules appear in the Gallery on the next page.

Section Summary

The VSEPR theory proposes that each group of electrons (single bond, multiple bond, lone pair, or lone electron) around a central atom remains as far away from the others as possible. • One of five electron-group arrangements results when two, three, four, five, or six electron groups surround a central atom. Each arrangement is associated with one or more molecular shapes. • Ideal bond angles are prescribed by the regular geometric shapes; deviations from these angles occur when the surrounding atoms or electron groups are not identical. • Lone pairs and double bonds exert greater repulsions than single bonds. • Larger molecules have shapes that are composites of the shapes around each of their central atoms.

Molecular Beauty:
Odd Shapes with Useful Functions

Chemists see a strange beauty in the geometric intricacy of nature's most minute objects. The simplicity of spherical atoms disappears when they combine to form pentagons, helices, and countless other shapes. Moreover, many "beautiful" molecules have marvelous practical uses.

Fullerenes Buckminsterfullerene ("bucky ball," named for R. Buckminster Fuller, the architect and philosopher who designed domes based on this shape) is a truncated icosahedron of C atoms, a molecular "soccer ball" with 60 vertices and 32 faces (12 pentagons and 20 hexagons). This C_{60} structure, discovered in soot in 1985 and prepared in bulk in 1990, represents a third crystalline form of carbon (graphite and diamond are the other two). It is the parent of a new family of structures called *fullerenes* and has spawned a new field of synthesis. By inserting atoms, such as potassium, inside the ball (see models) and bonding countless types of chemical groups to the outside, researchers have discovered a whole new range of possible applications, including lubricants, superconductors, rocket fuels, lasers, batteries, magnetic films, cancer and AIDS drugs—the list grows daily!

Nanotubes
These younger cousins of fullerenes consist of extremely long, thin, graphite-like cylinders with fullerene ends. They are often nested within one another (as in the photo behind the model). Despite thicknesses of a few nanometers, these structures are highly conductive along their length and about 40 times stronger than steel! Applications of nanoscale electronic components and medical devices incorporating nanotubes are already appearing.

Dendrimers Named for their structural similarity to the branching of trees, dendrimers form when one molecule with several bonding groups reacts with itself. They have excellent surface properties and are being studied for applications as films, fibers, and paint binders. (The model shows one-third of the circular schematic.)

Cubanes
Compressing carbon's ideal tetrahedral bond angle of 109.5° to 90° requires energy, which becomes stored in the bond.
Thus, cyclobutane, a square of C atoms with two H atoms at each corner, is unstable above 500 K. Much more energy is stored in the bonds of cubane, a cube of C atoms with one H at each corner, synthesized in 1964. The bond-strain energy of cubane can yield enormous explosive power. Tetranitrocubane (shown here) is a very powerful explosive, and octanitrocubane, which was synthesized in 2000, is considered the most powerful non-nuclear explosive known. Other properties of cubanes, which are apparently unrelated to the bond energy, include attacking cancer cells and inactivating an enzyme involved in Parkinson's disease.

Biological application

Literature Martin, James D. "From the Wood-Shop to Crystal Engineering: Teaching Three-Dimensional Chemistry," *J. Chem. Educ.* **2001,** *78,* 1195.

10.3 MOLECULAR SHAPE AND MOLECULAR POLARITY

Knowing the shape of a substance's molecules is a key to understanding its physical and chemical behavior. One of the most important and far-reaching effects of molecular shape is molecular polarity, which can influence melting and boiling points, solubility, chemical reactivity, and even biological function.

In Chapter 9, you learned that a covalent bond is *polar* when it joins atoms of different electronegativities because the atoms share the electrons unequally. In diatomic molecules, such as HF, where there is only one bond, the bond polarity causes the molecule itself to be polar. Molecules with a net imbalance of charge have a **molecular polarity.** In molecules with more than two atoms, *both shape and bond polarity determine molecular polarity.* In an electric field, polar molecules become oriented, on average, with their partial charges pointing toward the oppositely charged electric plates, as shown for HF in Figure 10.12. The **dipole moment (μ)** is the product of these partial charges and the distance between them. It is typically measured in *debye* (D) units; using the SI units of charge (coulomb, C) and length (meter, m), 1 D = 3.34×10^{-30} C·m. [The unit is named for Peter Debye (1884–1966), the Dutch American chemist and physicist who won the Nobel Prize in 1936 for his major contributions to our understanding of molecular structure and solution behavior.]

Bond Polarity, Bond Angle, and Dipole Moment

To determine molecular polarity, we also consider shape because the presence of polar bonds does not *always* lead to a polar molecule. In carbon dioxide, for example, the large electronegativity difference between C (EN = 2.5) and O (EN = 3.5) makes each C=O bond quite polar. However, CO_2 is linear, so its bonds point 180° from each other. As a result, the two identical bond polarities are counterbalanced and give the molecule *no net dipole moment* (μ = 0 D). Note that the electron density model shows regions of high negative charge (*red*) distributed equally on either side of the central region of high positive charge (*blue*):

Water also has identical atoms bonded to the central atom, but it *does* have a significant dipole moment (μ = 1.85 D). In each O—H bond, electron density is pulled toward the more electronegative O atom. Here, the bond polarities are *not* counterbalanced, because the water molecule is V shaped (also see Figure 4.1, p. 141). Instead, the bond polarities are partially reinforced, and the O end of the molecule is more negative than the other end (the region between the H atoms), which the electron density model shows clearly:

(The molecular polarity of water has some amazing effects, from determining the composition of the oceans to supporting life itself, as you'll see in Chapter 12.)

In the two previous examples, molecular shape influences polarity. When different molecules have the same shape, the nature of the atoms surrounding the central atom can have a major effect on polarity. Consider carbon tetrachloride (CCl_4) and chloroform ($CHCl_3$), two tetrahedral molecules with very different polarities. In CCl_4, the surrounding atoms are all Cl atoms. Although each C—Cl bond is polar (ΔEN = 0.5), the molecule is nonpolar (μ = 0 D) because the

Figure 10.12 The orientation of polar molecules in an electric field.
A, A space-filling model of HF (*left*) shows the partial charges of this polar molecule. The electron density model (*right*) shows high electron density (*red*) associated with the F end and low electron density (*blue*) with the H end. **B,** In the absence of an external electric field, HF molecules are oriented randomly. **C,** In the presence of an electric field, the molecules, on average, become oriented with their partial charges pointing toward the oppositely charged plates.

ARIS Presentation Center,
Art Library

Animation: Influence of Shape on Polarity
ARIS Presentation Center
Chapter 10

Literature Mendiara, Sara N.; Perissinotti, L.J. "Tetrahedral Geometry and the Dipole Moment of Molecules," *J. Chem. Educ.* **2002**, *79*, 64.

individual bond polarities counterbalance each other. In $CHCl_3$, H substitutes for one Cl atom, disrupting the balance and giving chloroform a significant dipole moment ($\mu = 1.01$ D):

SAMPLE PROBLEM 10.9 Predicting the Polarity of Molecules

PROBLEM From electronegativity (EN) values and their periodic trends (see Figure 9.20, p. 364), predict whether each of the following molecules is polar and show the direction of bond dipoles and the overall molecular dipole when applicable:
(a) Ammonia, NH_3 **(b)** Boron trifluoride, BF_3
(c) Carbonyl sulfide, COS (atom sequence SCO)
PLAN First, we draw and name the molecular shape. Then, using relative EN values, we decide on the direction of each bond dipole. Finally, we see if the bond dipoles balance or reinforce each other in the molecule as a whole.
SOLUTION **(a)** For NH_3. The molecular shape is trigonal pyramidal. From Figure 9.20, we see that N (EN = 3.0) is more electronegative than H (EN = 2.1), so the bond dipoles point toward N. The bond dipoles partially reinforce each other, and thus the molecular dipole points toward N:

Therefore, ammonia is polar.
(b) For BF_3. The molecular shape is trigonal planar. Because F (EN = 4.0) is farther to the right in Period 2 than B (EN = 2.0), it is more electronegative; thus, each bond dipole points toward F. However, the bond angle is 120°, so the three bond dipoles counterbalance each other, and BF_3 has no molecular dipole:

Therefore, boron trifluoride is nonpolar.
(c) For COS. The molecular shape is linear. With C and S having the same EN, the C=S bond is nonpolar, but the C=O bond is quite polar ($\Delta EN = 1.0$), so there is a net molecular dipole toward the O:

Therefore, carbonyl sulfide is polar.

FOLLOW-UP PROBLEM 10.9 Show the bond dipoles and molecular dipole, if any, for **(a)** dichloromethane (CH_2Cl_2); **(b)** iodine oxide pentafluoride (IOF_5); **(c)** nitrogen tribromide (NBr_3).

The Effect of Molecular Polarity on Behavior

To get a sense of the influence of molecular polarity on physical behavior, consider what effect a molecular dipole might have when many polar molecules lie near each other, as they do in a liquid. How does a molecular property such as dipole moment affect a macroscopic property such as boiling point? A liquid boils when its molecules have enough energy to form bubbles of gas. To enter the bubble, the molecules in the liquid must overcome the weak attractive forces *between* them. A molecular dipole influences the strength of these attractions.

Consider the two dichloroethylenes shown below. These compounds have the *same* molecular formula ($C_2H_2Cl_2$), but they have *different* physical and chemical properties; they are constitutional isomers. In particular, *cis*-1,2-dichloroethylene boils 13°C higher than *trans*-1,2-dichloroethylene. VSEPR theory predicts, and laboratory observation confirms, that all the nuclei lie in the same plane, with a trigonal planar shape around each C atom. The *trans* isomer has no dipole moment ($\mu = 0$ D) because the C—Cl bond polarities balance each other. In contrast, the *cis* isomer *is* polar ($\mu = 1.90$ D) because the bond dipoles partially reinforce each other, with the molecular dipole pointing between the Cl atoms. In the liquid state, the polar *cis* molecules attract each other more strongly than the nonpolar *trans* molecules, so more energy is needed to overcome these stronger forces. Therefore, the *cis* isomer has a higher boiling point. We extend these ideas in Chapter 12. The upcoming Chemical Connections essay discusses some fundamental effects of molecular shape and polarity on biological behavior.

ARIS Presentation Center, Art Library

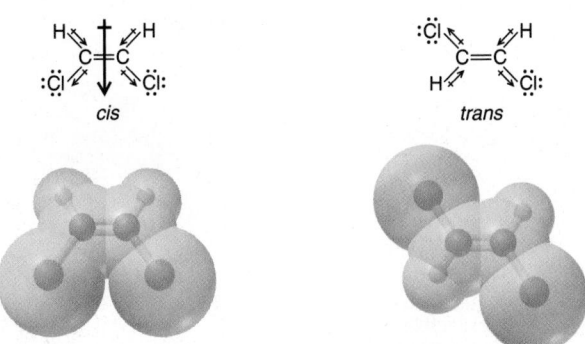

cis　　　trans

Section Summary

Bond polarity and molecular shape determine molecular polarity, which is measured as a dipole moment. • When bond polarities counterbalance each other, the molecule is nonpolar; when they reinforce each other, even partially, the molecule is polar. • Molecular shape and polarity can affect physical properties, such as boiling point, and they play a central role in many aspects of biological function.

Chapter Perspective

In this chapter, you saw how chemists use the simple idea of electrostatic repulsion between electron groups to account for the shapes of molecules and, ultimately, their properties. In Chapter 11, you'll learn how chemists explain molecular shape and behavior through the nature of the orbitals that electrons occupy. The landscape of a molecule—its geometric contours and regions of charge—gives rise to all its other properties. In upcoming chapters, we'll explore the physical behavior of liquids and solids (Chapter 12), the properties of solutions (Chapter 13), and the behavior of the main-group elements (Chapter 14) and organic molecules (Chapter 15), all of which emerge from atomic properties and the resulting molecular properties.

Chemical Connections *to Sensory Physiology*

Molecular Shape, Biological Receptors, and the Sense of Smell

In a very simple sense, a biological cell is a membrane-bound sack filled with molecular shapes interacting in an aqueous fluid. As a result of the magnificent internal organization of cells, many complex processes in an organism begin when a molecular "key" fits into a correspondingly shaped molecular "lock." The key can be a relatively small molecule circulating in a body fluid, whereas the lock is usually a large molecule, known as a *biological receptor,* that is often found embedded in a cell membrane. The receptor contains a precisely shaped cavity, or *receptor site,* that is exposed to the passing fluid. Thousands of molecules collide with this site, but when one with the correct shape (that is, the molecular key) lands on it, the receptor "grabs" it through intermolecular attractions, and the biological response begins.

Let's see how this fitting together of molecular shapes operates in the sense of smell (olfaction). A substance must have certain properties to have an odor. An odorous molecule travels through the air, so it must come from a gas or a volatile liquid or solid. To reach the receptor, it must be soluble, at least to a small extent, in the thin film of aqueous solution that lines the nasal pas-

sages. Most important, the odorous molecule, or a portion of it, must have a shape that fits into one of the olfactory receptor sites that cover the nerve endings deep within the nasal passage (Figure B10.1). When this happens, the resulting nerve impulses travel from these endings to the brain, which interprets the impulses as a specific odor.

In the 1950s, a stereochemical theory of odor (*stereo* means "three-dimensional") was introduced to explain the relationship between odor and molecular shape. Its basic premise is that a molecule's shape (and sometimes its polarity), but *not* its composition, determines its odor. The original theory proposed seven primary odors—camphor-like, musky, floral, minty, ethereal, pungent, and putrid—which correspond to seven different shapes of olfactory receptor sites. Figure B10.2 shows three of the seven receptor sites occupied by a molecule having that odor.

Several predictions of this original theory have been verified. If two substances fit the same receptors, they should have the same odor. The four very different molecules in Figure B10.3 all fit the bowl-shaped camphor-like receptor and all smell like moth

🐾 Organic chemistry application

🧬 Biological application

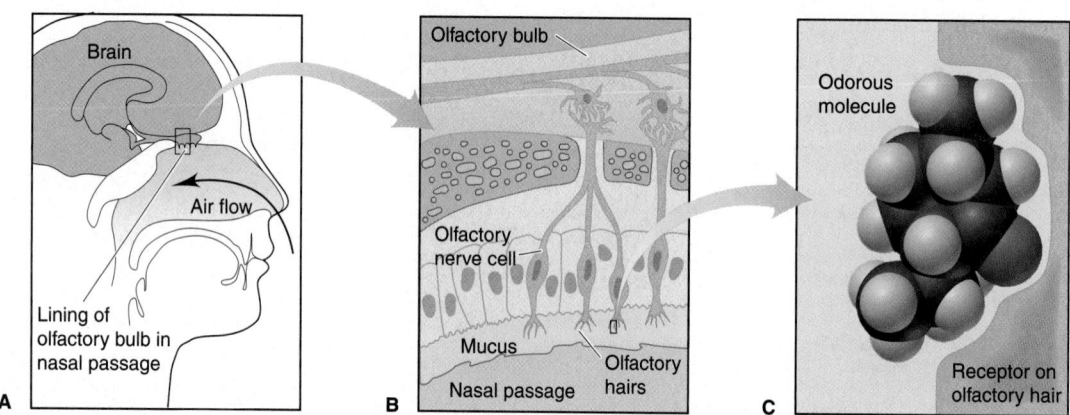

A Brain / Air flow / Lining of olfactory bulb in nasal passage

B Olfactory bulb / Olfactory nerve cell / Mucus / Nasal passage / Olfactory hairs

C Odorous molecule / Receptor on olfactory hair

Figure B10.1 The location of olfactory receptors within the nose. A, The olfactory area lies at the top of the nasal passage very close to the brain. Air containing odorous molecules is sniffed in, warmed, moistened, and channeled toward this region. **B,** A blow-up of the region shows olfactory nerve cells with their hairlike endings protruding into the liquid-coated nasal passage. **C,** A further blow-up shows a receptor site on one of the endings containing an odorous molecule that matches its shape. This particular molecule has a minty odor.

ARIS Presentation Center, Art Library

Floral   Camphor-like   Ethereal

Figure B10.2 Shapes of some olfactory receptor sites. Three of the seven proposed olfactory receptors are shown with a molecule having that odor occupying the site.

ARIS Presentation Center, Art Library

Camphor   **Hexachloroethane**   **Thiophosphoric acid dichloride ethylamide**   **Cyclooctane**

Figure B10.3 Different molecules with the same odor. The theory stating that odor is based on shape, not composition, is sup-ported by these four different substances: all have a camphor-like (moth repellent) odor.

repellent. If different portions of one molecule fit different receptors, the molecule should have a mixed odor. Portions of the benzaldehyde molecule fit the camphor-like, floral, and minty receptors, which gives it an almond odor; other molecules that smell like almonds fit the same three receptors.

Recent evidence suggests, however, that the original model is far too simple. In many cases, the odor predicted from the shape turns out to be wrong when the substance is smelled. One reason for this discrepancy is that a molecule may have a very different shape in the gas phase than it does when it is in solution at the receptor. By the early 1990s, evidence had indicated that there are about 1000 different receptors and that various combinations of stimulation produce the more than 10,000 odors humans can distinguish. In fact, the 2004 Nobel Prize in medicine or physiology was awarded to Richard Axel and Linda B. Buck for this work. Thus, although a key premise of the theory is still accepted—odor depends on molecular shape—the nature of the dependence is

complex and is the focus of active research in the food, cosmetics, and insecticide industries. The sense of smell is so vital for survival that these topics are important areas of study for biologists as well.

Many other biochemical processes are controlled by one molecule fitting into a receptor site on another. Enzymes are proteins that bind cellular reactants and speed their reaction (Figure B10.4). Nerve impulses are transmitted when small molecules released from one nerve fit into receptors on the next. Mind-altering drugs act by chemically disrupting the molecular fit at such nerve receptors in the brain. One type of immune response is triggered when a molecule on a bacterial surface binds to the receptors on "killer" cells in the bloodstream. Hormones regulate energy production and growth by fitting into and activating key receptors. Genes function when certain nucleic acid molecules fit into specific regions of other nucleic acids. Indeed, *no molecular property is more crucial to living systems than shape*.

A

B

Figure B10.4 Molecular shape and enzyme action. A, A small sugar molecule (*bottom*) is about to move up and fit into the receptor site of an enzyme molecule. **B,** The bound sugar molecule about to undergo reaction.

CHAPTER REVIEW GUIDE

The following sections provide many aids to help you study this chapter. (Numbers in parentheses refer to pages, unless noted otherwise.)

Learning Objectives These are concepts and skills you should know after studying this chapter.

Relevant section and/or sample problem (SP) numbers appear in parentheses.

Understand These Concepts

1. How Lewis structures depict the atoms and their bonding and lone electron pairs in a molecule or ion (10.1)
2. How resonance and electron delocalization explain bond properties in many compounds with double bonds adjacent to single bonds (10.1)
3. The meaning of formal charge and how it is used to select the most important resonance structure; the difference between formal charge and oxidation number (10.1)
4. The octet rule and its three major exceptions—molecules with a central atom that has an electron deficiency, an odd number of electrons, or an expanded valence shell (10.1)
5. How electron-group repulsions lead to molecular shapes (10.2)

6. The five electron-group arrangements and their associated molecular shapes (10.2)
7. Why double bonds and lone pairs cause deviations from ideal bond angles (10.2)
8. How bond polarities and molecular shape combine to give a molecule polarity (10.3)

Master These Skills

1. Using a stepwise method for writing a Lewis structure from a molecular formula (SPs 10.1–10.3, 10.5)
2. Writing resonance structures for molecules and ions (SP 10.4)
3. Calculating the formal charge of any atom in a molecule or ion (SP 10.5)
4. Predicting molecular shapes from Lewis structures (SPs 10.6–10.8)
5. Using molecular shape and electronegativity values to predict the direction of a molecular dipole (SP 10.9)

Key Terms These important terms appear in boldface in the chapter and are defined again in the Glossary.

Section 10.1
Lewis structure (Lewis formula) (378)
resonance structure (resonance form) (381)
resonance hybrid (382)
electron-pair delocalization (382)
formal charge (383)
electron deficient (384)

free radical (385)
expanded valence shell (385)

Section 10.2
valence-shell electron-pair repulsion (VSEPR) theory (388)
molecular shape (388)
bond angle (389)
linear arrangement (389)

linear shape (389)
trigonal planar arrangement (390)
bent shape (V shape) (390)
tetrahedral arrangement (391)
trigonal pyramidal shape (391)
trigonal bipyramidal arrangement (392)
equatorial group (392)
axial group (392)

seesaw shape (392)
T shape (393)
octahedral arrangement (393)
square pyramidal shape (393)
square planar shape (394)

Section 10.3
molecular polarity (399)
dipole moment (μ) (399)

Key Equations and Relationships Numbered and screened concepts are listed for you to refer to or memorize.

10.1 Calculating the formal charge on an atom (383):
Formal charge of atom
$$= \text{no. of valence e}^-$$
$$-(\text{no. of unshared valence e}^- + \tfrac{1}{2} \text{ no. of shared valence e}^-)$$

10.2 Ranking the effect of electron-pair repulsions on bond angle (392):
Lone pair–lone pair > lone pair–bonding pair
$$> \text{bonding pair–bonding pair}$$

Highlighted Figures and Tables These figures (F) and tables (T) provide a visual review of key ideas.

F10.1 The steps in writing a Lewis structure (378)
F10.2 Electron-group repulsions and the five basic molecular shapes (389)
F10.3 The single molecular shape of the linear electron-group arrangement (389)
F10.4 The two molecular shapes of the trigonal planar electron-group arrangement (390)
F10.5 The three molecular shapes of the tetrahedral electron-group arrangement (391)

F10.7 The four molecular shapes of the trigonal bipyramidal electron-group arrangement (392)
F10.8 The three molecular shapes of the octahedral electron-group arrangement (393)
F10.9 A summary of molecular shapes (394)
F10.10 The steps in determining a molecular shape (395)

Brief Solutions to FOLLOW-UP PROBLEMS Compare your solutions to these calculation steps and answers.

PROBLEMS

Problems with colored numbers are answered in Appendix E and worked in detail in the Student Solutions Manual. Problem sections match those in the text and provide the numbers of relevant sample problems. Most offer Concept Review Questions, Skill-Building Exercises (grouped in pairs covering the same concept), and Problems in Context. The Comprehensive Problems are based on material from any section or previous chapter.

Depicting Molecules and Ions with Lewis Structures
(Sample Problems 10.1 to 10.5)

● Concept Review Questions

• **10.1** Which of these atoms *cannot* serve as a central atom in a Lewis structure: (a) O; (b) He; (c) F; (d) H; (e) P? Explain.

• **10.2** When is a resonance hybrid needed to adequately depict the bonding in a molecule? Using NO_2 as an example, explain how a resonance hybrid is consistent with the actual bond length, bond strength, and bond order.

• **10.3** In which of these bonding patterns does X obey the octet rule?

(a) (b) (c) (d) (e) (f) (g) (h)

$-\overset{|}{\underset{|}{X}}-$:$\overset{|}{X}-$ $\overset{/}{\underset{\backslash}{X}}$ $\equiv X$: $-\overset{|}{X}=$ $-\overset{..}{X}=$ $\overset{/}{\underset{|}{X}}$:$\overset{..}{\underset{..}{X}}$:$^{2-}$

• **10.4** What is required for an atom to expand its valence shell? Which of the following atoms can expand its valence shell: F, S, H, Al, Se, Cl?

● Skill-Building Exercises *(grouped in similar pairs)*

• **10.5** Draw a Lewis structure for (a) SiF_4; (b) $SeCl_2$; (c) COF_2 (C is central).

• **10.6** Draw a Lewis structure for (a) PH_4^+; (b) C_2F_4; (c) SbH_3.

• **10.7** Draw a Lewis structure for (a) PF_3; (b) H_2CO_3 (both H atoms are attached to O atoms); (c) CS_2.

• **10.8** Draw a Lewis structure for (a) CH_4S; (b) S_2Cl_2; (c) $CHCl_3$.

• **10.9** Draw Lewis structures of all the important resonance forms of (a) NO_2^+; (b) NO_2F (N is central).

• **10.10** Draw Lewis structures of all the important resonance forms of (a) HNO_3 ($HONO_2$); (b) $HAsO_4^{2-}$ ($HOAsO_3^{2-}$).

• **10.11** Draw Lewis structures of all the important resonance forms of (a) N_3^-; (b) NO_2^-.

• **10.12** Draw Lewis structures of all the important resonance forms of (a) HCO_2^- (H is attached to C); (b) $HBrO_4$ ($HOBrO_3$).

• **10.13** Draw the Lewis structure with lowest formal charges, and determine the charge of each atom in (a) IF_5; (b) AlH_4^-.

• **10.14** Draw the Lewis structure with lowest formal charges, and determine the charge of each atom in (a) OCS; (b) NO.

• **10.15** Draw the Lewis structure with lowest formal charges, and determine the charge of each atom in (a) CN^-; (b) ClO^-.

• **10.16** Draw the Lewis structure with lowest formal charges, and determine the charge of each atom in (a) BF_4^-; (b) ClNO.

• **10.17** Draw a Lewis structure for a resonance form of each ion with the lowest possible formal charges, show the charges, and give oxidation numbers of the atoms: (a) BrO_3^-; (b) SO_3^{2-}.

• **10.18** Draw a Lewis structure for a resonance form of each ion with the lowest possible formal charges, show the charges, and give oxidation numbers of the atoms: (a) AsO_4^{3-}; (b) ClO_2^-.

• **10.19** These species do not obey the octet rule. Draw a Lewis structure for each, and state the type of octet-rule exception:
(a) BH_3 (b) AsF_4^- (c) $SeCl_4$

• **10.20** These species do not obey the octet rule. Draw a Lewis structure for each, and state the type of octet-rule exception:
(a) PF_6^- (b) ClO_3 (c) H_3PO_3 (one P—H bond)

• **10.21** These species do not obey the octet rule. Draw a Lewis structure for each, and state the type of octet-rule exception:
(a) BrF_3 (b) ICl_2^- (c) BeF_2

• **10.22** These species do not obey the octet rule. Draw a Lewis structure for each, and state the type of octet-rule exception:
(a) O_3^- (b) XeF_2 (c) SbF_4^-

● Problems in Context

• **10.23** Molten beryllium chloride reacts with chloride ion from molten NaCl to form the $BeCl_4^{2-}$ ion, in which the Be atom attains an octet. Show the net ionic reaction with Lewis structures.

• **10.24** Despite many attempts, the perbromate ion (BrO_4^-) was not prepared in the laboratory until about 1970. (In fact, articles were published explaining theoretically why it could never be prepared!) Draw a Lewis structure for BrO_4^- in which all atoms have lowest formal charges.

• **10.25** Cryolite (Na_3AlF_6) is an indispensable component in the electrochemical manufacture of aluminum. Draw a Lewis structure for the AlF_6^{3-} ion.

• **10.26** Phosgene is a colorless, highly toxic gas employed against troops in World War I and used today as a key reactant in organic syntheses. From the following resonance structures, select the one with the lowest formal charges:

Valence-Shell Electron-Pair Repulsion (VSEPR) Theory and Molecular Shape
(Sample Problems 10.6 to 10.8)

● Concept Review Questions

• **10.27** If you know the formula of a molecule or ion, what is the first step in predicting its shape?

• **10.28** In what situation is the name of the molecular shape the same as the name of the electron-group arrangement?

• **10.29** Which of the following numbers of electron groups can give rise to a bent (V-shaped) molecule: two, three, four, five, six? Draw an example for each case, showing the shape classification (AX_mE_n) and the ideal bond angle.

• **10.30** Name all the molecular shapes that have a tetrahedral electron-group arrangement.

•• **10.31** Consider the following molecular shapes:

(a) Which has the most electron pairs (both shared and unshared) around the central atom? (b) Which has the most unshared pairs around the central atom? (c) Do any have only shared pairs around the central atom?

• **10.32** Use wedge-bond perspective drawings (if necessary) to sketch the atom positions in a general molecule of formula (not shape class) AX_n that has each of the following shapes:
(a) V shaped   (b) trigonal planar   (c) trigonal bipyramidal
(d) T shaped   (e) trigonal pyramidal   (f) square pyramidal

• **10.33** What would you expect to be the electron-group arrangement around atom A in each of the following cases? For each arrangement, give the ideal bond angle and the direction of any expected deviation:

(a)
```
      X
      |
  X—A:
      |
      X
```
(b) X—A≡X

(c)
```
      X
      |
  X=A—X
```
(d) X—Ä—X

(e) X=A=X

(f)
```
      X
      / X
  :A
      \ X
      X
```

● Skill-Building Exercises (grouped in similar pairs)

• **10.34** Determine the electron-group arrangement, molecular shape, and ideal bond angle(s) for each of the following:
(a) O_3 (b) H_3O^+ (c) NF_3

• **10.35** Determine the electron-group arrangement, molecular shape, and ideal bond angle(s) for each of the following:
(a) SO_4^{2-} (b) NO_2^- (c) PH_3

• **10.36** Determine the electron-group arrangement, molecular shape, and ideal bond angle(s) for each of the following:
(a) CO_3^{2-} (b) SO_2 (c) CF_4

• **10.37** Determine the electron-group arrangement, molecular shape, and ideal bond angle(s) for each of the following:
(a) SO_3 (b) N_2O (N is central)   (c) CH_2Cl_2

• **10.38** Name the shape and give the AX_mE_n classification and ideal bond angle(s) for each of the following general molecules:

• **10.39** Name the shape and give the AX_mE_n classification and ideal bond angle(s) for each of the following general molecules:

- **10.40** Determine the shape, ideal bond angle(s), and the direction of any deviation from these angles for each of the following:
(a) ClO_2^- (b) PF_5 (c) SeF_4 (d) KrF_2

- **10.41** Determine the shape, ideal bond angle(s), and the direction of any deviation from these angles for each of the following:
(a) ClO_3^- (b) IF_4^- (c) $SeOF_2$ (d) TeF_5^-

- **10.42** Determine the shape around each central atom in each molecule, and explain any deviation from ideal bond angles:
(a) CH_3OH (b) N_2O_4 (O_2NNO_2)

- **10.43** Determine the shape around each central atom in each molecule, and explain any deviation from ideal bond angles:
(a) H_3PO_4 (no H—P bond)   (b) CH_3—O—CH_2CH_3

- **10.44** Determine the shape around each central atom in each molecule, and explain any deviation from ideal bond angles:
(a) CH_3COOH (b) H_2O_2

- **10.45** Determine the shape around each central atom in each molecule, and explain any deviation from ideal bond angles:
(a) H_2SO_3 (no H—S bond)   (b) N_2O_3 ($ONNO_2$)

- **10.46** Arrange the following AF_n species in order of *increasing* F—A—F bond angles: BF_3, BeF_2, CF_4, NF_3, OF_2.

- **10.47** Arrange the following ACl_n species in order of *decreasing* Cl—A—Cl bond angles: SCl_2, OCl_2, PCl_3, $SiCl_4$, $SiCl_6^{2-}$.

- **10.48** State an ideal value for each of the bond angles in each molecule, and note where you expect deviations:
(a)   (b)   (c)

- **10.49** State an ideal value for each of the bond angles in each molecule, and note where you expect deviations:
(a)   (b)   (c)

● Problems in Context

- **10.50** Because both tin and carbon are members of Group 4A(14), they form structurally similar compounds. However, tin exhibits a greater variety of structures because it forms several ionic species. Predict the shapes and ideal bond angles, including any deviations, for the following:
(a) $Sn(CH_3)_2$ (b) $SnCl_3^-$ (c) $Sn(CH_3)_4$
(d) SnF_5^- (e) SnF_6^{2-}

- **10.51** In the gas phase, phosphorus pentachloride exists as separate molecules. In the solid phase, however, the compound is composed of alternating PCl_4^+ and PCl_6^- ions. What change(s) in molecular shape occur(s) as PCl_5 solidifies? How does the Cl—P—Cl angle change?

Molecular Shape and Molecular Polarity
(Sample Problem 10.9)

● Concept Review Questions

- **10.52** For molecules of general formula AX_n (where $n > 2$), how do you determine if a molecule is polar?

- **10.53** How can a molecule with polar covalent bonds not be polar? Give an example.

- **10.54** Explain in general why the shape of a biomolecule is important to its function.

● Skill-Building Exercises (grouped in similar pairs)

- **10.55** Consider the molecules SCl_2, F_2, CS_2, CF_4, and $BrCl$.
(a) Which has bonds that are the most polar?
(b) Which have a molecular dipole moment?

- **10.56** Consider the molecules BF_3, PF_3, BrF_3, SF_4, and SF_6.
(a) Which has bonds that are the most polar?
(b) Which have a molecular dipole moment?

- **10.57** Which molecule in each pair has the greater dipole moment? Give the reason for your choice.
(a) SO_2 or SO_3 (b) ICl or IF
(c) SiF_4 or SF_4 (d) H_2O or H_2S

- **10.58** Which molecule in each pair has the greater dipole moment? Give the reason for your choice.
(a) ClO_2 or SO_2 (b) HBr or HCl
(c) $BeCl_2$ or SCl_2 (d) AsF_3 or AsF_5

● Problems in Context

- **10.59** There are three different dichloroethylenes (molecular formula $C_2H_2Cl_2$), which we can designate X, Y, and Z. Compound X has no dipole moment, but compound Z does. Compounds X and Z each combine with hydrogen to give the same product:

$$C_2H_2Cl_2 \text{ (X or Z)} + H_2 \longrightarrow ClCH_2-CH_2Cl$$

What are the structures of X, Y, and Z? Would you expect compound Y to have a dipole moment?

- **10.60** Dinitrogen difluoride, N_2F_2, is the only stable, simple inorganic molecule with an N=N bond. The compound occurs in *cis* and *trans* forms.
(a) Draw the molecular shapes of the two forms of N_2F_2.
(b) Predict the polarity, if any, of each form.

Comprehensive Problems

- **10.61** In addition to ammonia, nitrogen forms three other hydrides: hydrazine (N_2H_4), diazene (N_2H_2), and tetrazene (N_4H_4).
(a) Use Lewis structures to compare the strength, length, and order of nitrogen-nitrogen bonds in hydrazine, diazene, and N_2.
(b) Tetrazene (atom sequence H_2NNNNH_2) decomposes above 0°C to hydrazine and nitrogen gas. Draw a Lewis structure for tetrazene, and calculate $\Delta H°_{rxn}$ for this decomposition.

- **10.62** Draw a Lewis structure for each species: (a) PF_5; (b) CCl_4; (c) H_3O^+; (d) ICl_3; (e) BeH_2; (f) PH_2^-; (g) $GeBr_4$; (h) CH_3^-; (i) BCl_3; (j) BrF_4^+; (k) XeO_3; (l) TeF_4.

- **10.63** Give the molecular shape of each species in Problem 10.62.

- **10.64** Consider the following reaction of silicon tetrafluoride:

$$SiF_4 + F^- \longrightarrow SiF_5^-$$

(a) Which depiction below best illustrates the change in molecular shape around Si? (b) Give the name and AX_mE_n designation of each shape in the depiction chosen in part (a).

•• **10.65** Both aluminum and iodine form chlorides, Al_2Cl_6 and I_2Cl_6, with "bridging" Cl atoms. The Lewis structures are

(a) What is the formal charge on each atom? (b) Which of these molecules has a planar shape? Explain.

•• **10.66** The VSEPR model was developed before any xenon compounds had been prepared. Thus, these compounds provided an excellent test of the model's predictive power. What would you have predicted for the shapes of XeF_2, XeF_4, and XeF_6?

•• **10.67** When SO_3 gains two electrons, SO_3^{2-} forms. (a) Which depiction shown below best illustrates the change in molecular shape around S? (b) Does molecular polarity change during this reaction?

•• **10.68** The actual bond angle in NO_2 is 134.3°, and in NO_2^- it is 115.4°, although the ideal bond angle is 120° in both. Explain.

•• **10.69** "Inert" xenon actually forms several compounds, especially with the highly electronegative elements oxygen and fluorine. The simple fluorides XeF_2, XeF_4, and XeF_6 are all formed by direct reaction of the elements. As you might expect from the size of the xenon atom, the Xe—F bond is not a strong one. Calculate the Xe—F bond energy in XeF_6, given that the heat of formation is −402 kJ/mol.

•• **10.70** Propylene oxide is used to make many products, including plastics such as polyurethane. One method for synthesizing it involves oxidizing propene with hydrogen peroxide:

$$CH_3-CH=CH_2 + H_2O_2 \longrightarrow CH_3-CH-CH_2 + H_2O$$
$$\underset{O}{\diagup\diagdown}$$

(a) What is the molecular shape and ideal bond angle around each carbon atom in propylene oxide?
(b) Predict any deviation from the ideal for the actual bond angle (assume the three atoms in the ring form an equilateral triangle).

•• **10.71** Chloral, $Cl_3C-CH=O$, reacts with water to form the sedative and hypnotic agent chloral hydrate, $Cl_3C-CH(OH)_2$. Draw Lewis structures for these substances, and describe the change in molecular shape, if any, that occurs around each of the carbon atoms during the reaction.

•• **10.72** Dichlorine heptaoxide, Cl_2O_7, can be viewed as two ClO_4 groups sharing an O atom. Draw a Lewis structure for Cl_2O_7 with the lowest formal charges, and predict any deviation from the ideal for the Cl—O—Cl bond angle.

•• **10.73** Like several other bonds, carbon-oxygen bonds have lengths and strengths that depend on the bond order. Draw Lewis structures for the following species, and arrange them in order of increasing carbon-oxygen bond length and then by increasing carbon-oxygen bond strength: (a) CO; (b) CO_3^{2-}; (c) H_2CO; (d) CH_4O; (e) HCO_3^- (H attached to O).

•• **10.74** In the 1980s, there was an international agreement to destroy all stockpiles of mustard gas, $ClCH_2CH_2SCH_2CH_2Cl$. When this substance contacts the moisture in eyes, nasal passages, and skin, the —OH groups of water replace the Cl atoms and create high local concentrations of hydrochloric acid, which cause severe blistering and tissue destruction. Write a balanced equation for this reaction, and calculate ΔH_{rxn}°.

•• **10.75** The four bonds of carbon tetrachloride (CCl_4) are polar, but the molecule is nonpolar because the bond polarity is canceled by the symmetric tetrahedral shape. When other atoms substitute for some of the Cl atoms, the symmetry is broken and the molecule becomes polar. Use Figure 9.20 (p. 364) to rank the following molecules from the least polar to the most polar: CH_2Br_2, CF_2Cl_2, CH_2F_2, CH_2Cl_2, CBr_4, CF_2Br_2.

••• **10.76** Ethanol (CH_3CH_2OH) is being used as a gasoline additive or alternative in many parts of the world.
(a) Use bond energies to find ΔH_{rxn}° for the combustion of gaseous ethanol. (Assume H_2O forms as a gas.)
(b) In its standard state at 25°C, ethanol is a liquid. Its vaporization requires 40.5 kJ/mol. Correct the value from part (a) to find the heat of reaction for the combustion of liquid ethanol.
(c) How does the value from part (b) compare with the value you calculate from standard heats of formation (Appendix B)?
(d) "Greener" methods produce ethanol from corn and other plant material, but the main industrial method involves hydrating ethylene from petroleum. Use Lewis structures and bond energies to calculate ΔH_{rxn}° for the formation of gaseous ethanol from ethylene gas with water vapor.

••• **10.77** In the following compounds, the C atoms form a single ring. Draw a Lewis structure for each compound, identify cases for which resonance exists, and determine the carbon-carbon bond order(s): (a) C_3H_4; (b) C_3H_6; (c) C_4H_6; (d) C_4H_4; (e) C_6H_6.

•• **10.78** An oxide of nitrogen is 25.9% N by mass, has a molar mass of 108 g/mol, and contains no nitrogen-nitrogen or oxygen-oxygen bonds. Draw its Lewis structure, and name it.

••• **10.79** An experiment requires 50.0 mL of 0.040 M NaOH for the titration of 1.00 mmol of acid. Mass analysis of the acid shows 2.24% hydrogen, 26.7% carbon, and 71.1% oxygen. Draw the Lewis structure of the acid.

••• **10.80** A gaseous compound has a composition by mass of 24.8% carbon, 2.08% hydrogen, and 73.1% chlorine. At STP, the gas has a density of 4.3 g/L. Draw a Lewis structure that satisfies these facts. Would another structure also satisfy them? Explain.

•• **10.81** Perchlorates are powerful oxidizing agents used in fireworks, flares, and the booster rockets of space shuttles. Lewis structures for the perchlorate ion (ClO_4^-) can be drawn with all single bonds or with one, two, or three double bonds. Draw each of these possible resonance forms, use formal charges to determine the most important, and calculate its average bond order.

•• **10.82** Methane burns in oxygen to form carbon dioxide and water vapor. Hydrogen sulfide burns in oxygen to form sulfur dioxide and water vapor. Use bond energies (Table 9.2, p. 353) to determine the heat of each reaction per mole of O_2 (assume Lewis structures with zero formal charges; BE of S=O is 552 kJ/mol).

•• **10.83** Use Lewis structures to determine which *two* of the following are unstable: (a) SF_2; (b) SF_3; (c) SF_4; (d) SF_5; (e) SF_6.

10.84 A major short-lived, neutral species in flames is OH.
(a) What is unusual about the electronic structure of OH?
(b) Use the standard heat of formation of OH(g) (39.0 kJ/mol) and bond energies to calculate the O—H bond energy in OH(g).
(c) From the average value for the O—H bond energy in Table 9.2 (p. 353) and your value for the O—H bond energy in OH(g), find the energy needed to break the first O—H bond in water.

10.85 Pure HN$_3$ (atom sequence HNNN) is explosive. In aqueous solution, it is a weak acid that yields the azide ion, N$_3^-$. Draw resonance structures to explain why the nitrogen-nitrogen bond lengths are equal in N$_3^-$ but unequal in HN$_3$.

10.86 Except for nitrogen, the elements of Group 5A(15) all form pentafluorides, and most form pentachlorides. The chlorine atoms of PCl$_5$ can be replaced with fluorine atoms one at a time to give, successively, PCl$_4$F, PCl$_3$F$_2$, ... , PF$_5$. (a) Given the sizes of F and Cl, would you expect the first two F substitutions to be at axial or equatorial positions? Explain. (b) Which of the five fluorine-containing molecules have no dipole moment?

10.87 Dinitrogen monoxide (N$_2$O) supports combustion in a manner similar to oxygen, with the nitrogen atoms forming N$_2$. Draw three resonance structures for N$_2$O (one N is central), and use formal charges to decide the relative importance of each. What correlation can you suggest between the most important structure and the observation that N$_2$O supports combustion?

10.88 Oxalic acid (H$_2$C$_2$O$_4$) is found in toxic concentrations in rhubarb leaves. The acid forms two ions, HC$_2$O$_4^-$ and C$_2$O$_4^{2-}$, by the sequential loss of H$^+$ ions. Draw Lewis structures for the three species, and comment on the relative lengths and strengths of their carbon-oxygen bonds. The connections among the atoms are shown below with single bonds only.

$$H-O \quad\quad O-H$$
$$C-C$$
$$O \quad\quad O$$

10.89 The Murchison meteorite that landed in Australia in 1969 contained 92 different amino acids, including 21 found in Earth organisms. A skeleton structure (single bonds only) of one of these extraterrestrial amino acids is shown below. Draw a Lewis structure, and identify any atoms with a nonzero formal charge.

$$H_3N-CH-C-O$$
$$\quad\quad\quad CH_2 \quad O$$
$$\quad\quad\quad CH_3$$

10.90 Hydrazine (N$_2$H$_4$) is used as a rocket fuel because it reacts very exothermically with oxygen to form nitrogen gas and water vapor. The heat released and the increase in number of moles of gas provide thrust. Calculate the heat of reaction.

10.91 A student isolates a product with the molecular shape shown at right (F is orange).
(a) If the species is a neutral compound, can the black sphere represent selenium (Se)?
(b) If the species is an anion, can the black sphere represent N? (c) If the black sphere represents Br, what is the charge of the species?

10.92 When gaseous sulfur trioxide is dissolved in concentrated sulfuric acid, disulfuric acid forms:

$$SO_3(g) + H_2SO_4(l) \longrightarrow H_2S_2O_7(l)$$

Use bond energies (Table 9.2, p. 353) to determine ΔH°_{rxn}. (The S atoms in H$_2$S$_2$O$_7$ are bonded through an O atom. Assume Lewis structures with zero formal charges; BE of S=O is 552 kJ/mol.)

10.93 A molecule of formula AY$_3$ is found experimentally to be polar. Which molecular shapes are possible and which impossible for AY$_3$?

10.94 Consider the following molecular shapes:

(a) Match each shape with one of the following species: XeF$_3^+$, SbBr$_3$, GaCl$_3$.
(b) Which, if any, is polar?
(c) Which has the most valence electrons around the cental atom?

10.95 Hydrogen cyanide can be catalytically reduced with hydrogen to form methylamine. Use Lewis structures and bond energies to determine ΔH°_{rxn} for

$$HCN(g) + 2H_2(g) \longrightarrow CH_3NH_2(g)$$

10.96 Ethylene, C$_2$H$_4$, and tetrafluoroethylene, C$_2$F$_4$, are used to make the polymers polyethylene and polytetrafluoroethylene (Teflon), respectively.
(a) Draw the Lewis structures for C$_2$H$_4$ and C$_2$F$_4$, and give the ideal H—C—H and F—C—F bond angles.
(b) The actual H—C—H and F—C—F bond angles are 117.4° and 112.4°, respectively. Explain these deviations.

10.97 Lewis structures of mescaline, a hallucinogenic compound in peyote cactus, and dopamine, a neurotransmitter in the mammalian brain, appear below. Suggest a reason for mescaline's ability to disrupt nerve impulses.

mescaline   dopamine

10.98 Using bond lengths in Table 9.3 (p. 353) and assuming ideal geometry, calculate each of the following distances:
(a) Between H atoms in C$_2$H$_2$
(b) Between F atoms in SF$_6$ (two answers)
(c) Between equatorial F atoms in PF$_5$

10.99 Phosphorus pentachloride, a key industrial compound with annual world production of about 2×10^7 kg, is used to make other compounds. It reacts with sulfur dioxide to produce phosphorus oxychloride (POCl$_3$) and thionyl chloride (SOCl$_2$). Draw a Lewis structure, and name the molecular shape of each product.

11

Bonding Theory Visualized. How do molecular shapes emerge from interacting atoms? One theory proposes that new types of orbitals—such as these of the central phosphorus atom in PCl_5—arise during bonding. In this chapter, we explore two models that rationalize the rich properties of the covalent bond.

Theories of Covalent Bonding

11.1 Valence Bond (VB) Theory and Orbital Hybridization
Central Themes of VB Theory
Types of Hybrid Orbitals

11.2 The Mode of Orbital Overlap and the Types of Covalent Bonds
Single and Multiple Bonds
Molecular Rotation

11.3 Molecular Orbital (MO) Theory and Electron Delocalization
Central Themes of MO Theory
Homonuclear Diatomic Molecules of the Period 2 Elements
Some Heteronuclear Diatomic Molecules
MO Descriptions of Benzene and Ozone

All scientific models have limitations because they are simplifications of reality. The VSEPR model accounts for molecular shapes by assuming that electron groups minimize their repulsions, and thus occupy as much space as possible around a central atom. But it does *not* explain how the shapes come from the interactions of atomic orbitals. After all, the orbitals we examined in Chapter 7 aren't oriented toward the corners of, for example, a tetrahedron or a trigonal bipyramid. Moreover, knowing the shape doesn't help us explain the magnetic and spectral properties of molecules; only an understanding of their orbitals and energy levels can do that. We'll discuss two theories of covalent bonding to understand shape and other molecular properties. Each theory complements the other, so don't be concerned if we need more than one model to explain a complex phenomenon like bonding. In every science, one model often accounts for some aspect of a topic better than another, and several models are called into service to explain a wider range of phenomena.

IN THIS CHAPTER . . . We introduce two theories of bonding in molecules. Valence bond (VB) theory rationalizes observed molecular shapes through interactions of atomic orbitals that give rise to new "hybrid" orbitals. We apply VB theory to sigma and pi bonding, the two types of covalent bonds. Molecular orbital (MO) theory proposes orbitals associated with the whole molecule to explain molecular energy levels and properties. We apply MO theory to key properties of some diatomic and simple polyatomic molecules.

Concepts & Skills to Review
before you study this chapter

- atomic orbital shapes (Section 7.4)
- the exclusion principle (Section 8.2)
- Hund's rule (Section 8.3)
- writing Lewis structures (Section 10.1)
- resonance in covalent bonding (Section 10.1)
- molecular shapes (Section 10.2)
- molecular polarity (Section 10.3)

Literature Scerri, Eric R. "Have Orbitals Really Been Observed?" *J. Chem. Educ.* **2002**, *79*, 310.

11.1 VALENCE BOND (VB) THEORY AND ORBITAL HYBRIDIZATION

What *is* a covalent bond, and what characteristic gives it strength? And how can we explain *molecular* shapes based on the interactions of *atomic* orbitals? The most useful approach for answering these questions is based on quantum mechanics (Chapter 7) and called **valence bond (VB) theory.**

The Central Themes of VB Theory

The basic principle of VB theory is that *a covalent bond forms when orbitals of two atoms overlap and the overlap region, which is between the nuclei, is occupied by a pair of electrons.* ("Orbital overlap" is another way of saying that the two wave functions are *in phase,* so the amplitude increases between the nuclei.) The central themes of VB theory derive from this principle:

1. *Opposing spins of the electron pair.* As the exclusion principle (Section 8.2) prescribes, the space formed by the overlapping orbitals *has a maximum capacity of two electrons that must have opposite spins.* When a molecule of H_2 forms, for instance, the two $1s$ electrons of two H atoms occupy the overlapping $1s$ orbitals and have opposite spins (Figure 11.1A).

2. *Maximum overlap of bonding orbitals.* The bond strength depends on the attraction of the nuclei for the shared electrons, so *the greater the orbital overlap, the stronger (more stable) the bond.* The extent of overlap depends on the shapes and directions of the orbitals. An *s* orbital is spherical, but *p* and *d* orbitals have more electron density in one direction than in another. Thus, whenever possible, a bond involving *p* or *d* orbitals will be oriented in the direction that maximizes overlap. In the HF bond, for example, the $1s$ orbital of H overlaps the half-filled $2p$ orbital of F *along the long axis* of that orbital (Figure 11.1B). Any other direction would result in less overlap and, thus, a weaker bond. Similarly, in the F—F bond of F_2, the two half-filled $2p$ orbitals interact end to end, that is, *along the long axes* of the orbitals, to maximize overlap (Figure 11.1C).

A Hydrogen, H_2

B Hydrogen fluoride, HF

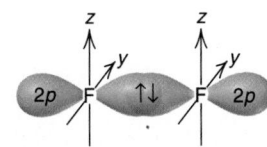

C Fluorine, F_2

Figure 11.1 Orbital overlap and spin pairing in three diatomic molecules.
A, In the H_2 molecule, the two overlapping 1s orbitals are occupied by the two 1s electrons with opposite spins. (The electrons, shown as arrows, spend the most time between the nuclei but move throughout the overlapping orbitals.)
B, To maximize overlap in HF, half-filled H 1s and F 2p orbitals overlap along the long axis of the 2p orbital involved in bonding. (The $2p_x$ orbital is shown bonding; the other two 2p orbitals of F are not shown.)
C, In F_2, the half-filled $2p_x$ orbital on one F points end to end toward the similar orbital on the other F to maximize overlap.

 ARIS Presentation Center, Art Library

Literature Emerson, David W. "A Colorful Demonstration to Simulate Orbital Hybridization," *J. Chem. Educ.* **1988**, *65*, 454.

3. *Hybridization of atomic orbitals*. To account for the bonding in simple diatomic molecules like HF, we picture the direct overlap of *s* and *p* orbitals of isolated atoms. But how can we account for the shapes of so many molecules and polyatomic ions through the overlap of spherical *s* orbitals, dumbbell-shaped *p* orbitals, and cloverleaf-shaped *d* orbitals?

Consider a methane molecule, CH_4. It has four H atoms bonded to a central C atom. An isolated ground-state C atom ([He] $2s^2 2p^2$) has four valence electrons: two in the 2*s* orbital and one each in two of the three 2*p* orbitals. We might easily see how the two half-filled *p* orbitals of C could overlap with the 1*s* orbitals of two H atoms to form *two* C—H bonds with a 90° H—C—H bond angle. But methane is not CH_2 and doesn't have a bond angle of 90°. It's not as easy to see how the orbitals overlap to form the *four* C—H bonds with the 109.5° bond angle that occurs in methane.

To explain such facts, Linus Pauling proposed that *the valence atomic orbitals in the molecule are **different** from those in the isolated atoms*. Indeed, quantum-mechanical calculations show that if we "mix" specific combinations of orbitals mathematically, we obtain new atomic orbitals. The spatial orientations of these new orbitals lead to more stable bonds and are consistent with observed molecular shapes. The process of orbital mixing is called **hybridization,** and the new atomic orbitals are called **hybrid orbitals.** Two key points about the number and type of hybrid orbitals are that

- The *number* of hybrid orbitals obtained *equals* the number of atomic orbitals mixed.
- The *type* of hybrid orbitals obtained *varies* with the types of atomic orbitals mixed.

You can imagine hybridization as a process in which atomic orbitals mix, hybrid orbitals form, and electrons enter them with spins parallel (Hund's rule) to create stable bonds. In truth, though, hybridization is a mathematically derived result from quantum mechanics that accounts for the molecular shapes we observe.

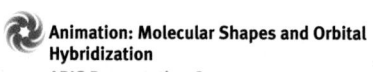

Animation: Molecular Shapes and Orbital Hybridization

ARIS Presentation Center
Chapter 11

Types of Hybrid Orbitals

We postulate the presence of a certain type of hybrid orbital *after* we observe the molecular shape. As we discuss the five common types of hybridization, notice that the spatial orientation of each type of hybrid orbital corresponds with one of the five common electron-group arrangements predicted by VSEPR theory.

***sp* Hybridization** When two electron groups surround the central atom, we observe a linear shape, which means that the bonding orbitals must have a linear orientation. VB theory explains this by proposing that mixing two *nonequivalent* orbitals of a central atom, one *s* and one *p*, gives rise to two *equivalent **sp* hybrid orbitals** that lie 180° apart (Figure 11.2A). Note the shape of the hybrid orbital: with one large and one small lobe, it differs markedly from the shapes of the atomic orbitals that were mixed. The orientations of hybrid orbitals extend electron density in the bonding direction and minimize repulsions between the electrons that occupy them. Thus, *both shape and orientation maximize overlap with the orbital of the other atom in the bond.*

In gaseous $BeCl_2$, for example, the Be atom is said to be *sp* hybridized. The rest of Figure 11.2 depicts the hybridization of Be in $BeCl_2$ in terms of a traditional orbital diagram of the valence level (part B), a similar diagram with orbital contours (C), and the bond formation with Cl (D). The 2*s* and one of the three 2*p* orbitals of Be mix and form two *sp* orbitals. These overlap 3*p* orbitals of two Cl atoms, and the four valence electrons—two from Be and one from each Cl—occupy the overlapped orbitals in pairs with opposite spins. The two unhybridized 2*p* orbitals of Be lie perpendicular to each other and to the bond axes. Thus, through hybridization, the paired 2*s* electrons in the isolated Be atom are distributed into two *sp* orbitals, which form two Be—Cl bonds.

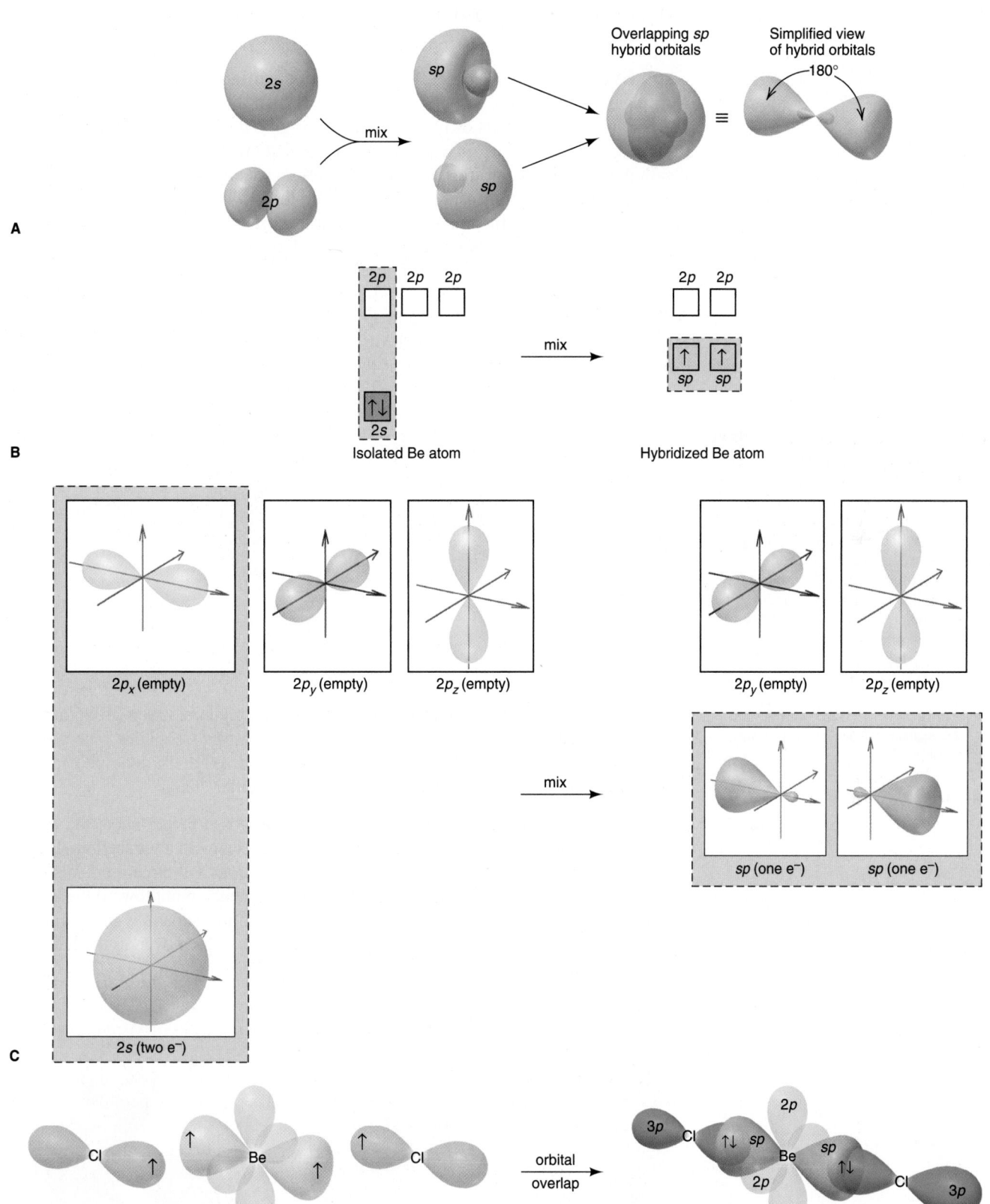

Figure 11.2 The *sp* hybrid orbitals in gaseous BeCl₂. A, One 2*s* and one 2*p* atomic orbital mix to form two *sp* hybrid orbitals (shown above and below one another and turned slightly sideways for ease of viewing). Note the large and small lobes of the hybrid orbitals. In the molecule, the two *sp* orbitals of Be are oriented in opposite directions. For clarity, the simplified hybrid orbitals will be used throughout the text, usually without the small lobe. **B,** The orbital diagram for hybridization in Be is drawn vertically and shows that the 2*s* and one of the three 2*p* orbitals form two *sp* hybrid orbitals, and the two other 2*p* orbitals remain unhybridized. Electrons half-fill the *sp* hybrid orbitals. During bonding, each *sp* orbital fills by sharing an electron from Cl (not shown). **C,** The orbital diagram is shown with orbital contours instead of electron arrows. **D,** BeCl₂ forms by overlap of the two *sp* hybrids with the 3*p* orbitals of two Cl atoms; the two unhybridized Be 2*p* orbitals lie perpendicular to the *sp* hybrids. (For clarity, only the 3*p* orbital involved in bonding is shown for each Cl.)

413

sp^2 Hybridization In order to rationalize the trigonal planar electron-group arrangement and the shapes of molecules based on it, we introduce the mixing of one s and two p orbitals of the central atom to give three hybrid orbitals that point toward the vertices of an equilateral triangle, their axes 120° apart. These are called **sp^2 hybrid orbitals.** (Note that, unlike electron configuration notation, hybrid orbital notation uses superscripts for the *number* of atomic orbitals of a given type that are mixed, *not* for the number of electrons in the orbital: here, one s and two p orbitals were mixed, so we have s^1p^2, or sp^2.)

For example, VB theory proposes that the central B atom in the BF_3 molecule is sp^2 hybridized. Figure 11.3 shows the three sp^2 orbitals in the trigonal plane, with the third $2p$ orbital unhybridized and perpendicular to this plane. Each sp^2 orbital overlaps the $2p$ orbital of an F atom, and the six valence electrons—three from B and one from each of the three F atoms—form three bonding pairs.

Figure 11.3 The sp^2 hybrid orbitals in BF$_3$. A, The orbital diagram shows that the $2s$ and two of the three $2p$ orbitals of the B atom mix to make three sp^2 hybrid orbitals. Three electrons (*up arrows*) half-fill the sp^2 hybrids. The third $2p$ orbital remains empty and unhybridized. **B,** BF$_3$ forms through overlap of $2p$ orbitals on three F atoms with the sp^2 hybrids. During bonding, each sp^2 orbital fills with an electron from one F (*down arrow*). The three sp^2 hybrids of B lie 120° apart, and the unhybridized $2p$ orbital is perpendicular to the trigonal bonding plane.

Transparency/ARIS Presentation Center, Art Library

To account for other molecular shapes within a given electron-group arrangement, we postulate that one or more of the hybrid orbitals contains lone pairs. In ozone (O_3), for example, the central O is sp^2 hybridized and a lone pair fills one of its three sp^2 orbitals, so ozone has a bent molecular shape.

sp^3 Hybridization Now let's return to the question posed earlier about the orbitals in methane, the same question that arises for any species with a tetrahedral electron-group arrangement. VB theory proposes that the one s and all three p orbitals of the central atom mix and form four **sp^3 hybrid orbitals,** which point toward the vertices of a tetrahedron. As shown in Figure 11.4, the C atom in methane is sp^3 hybridized. Its four valence electrons half-fill the four sp^3 hybrids, which overlap the half-filled $1s$ orbitals of four H atoms and form four C—H bonds.

Figure 11.4 The sp^3 hybrid orbitals in CH$_4$. A, The $2s$ and all three $2p$ orbitals of C are mixed to form four sp^3 hybrids. Carbon's four valence electrons half-fill the sp^3 hybrids. **B,** In methane, the four sp^3 orbitals of C point toward the corners of a tetrahedron and overlap the $1s$ orbitals of four H atoms. Each sp^3 orbital fills by addition of an electron from one H (electrons are shown as dots).

 Transparency/ARIS Presentation Center, Art Library

Figure 11.5 The sp^3 hybrid orbitals in NH_3 and H_2O. A, The orbital diagrams show sp^3 hybridization, as in CH_4. In NH_3 (*top*), one sp^3 orbital is filled with a lone pair. In H_2O (*bottom*), two sp^3 orbitals are filled with lone pairs. **B,** Contour diagrams show the tetrahedral orientation of the sp^3 orbitals and the overlap of the bonded H atoms. Each half-filled sp^3 orbital fills by addition of an electron from one H. (Shared pairs and lone pairs are shown as dots.)

Transparency/ARIS Presentation Center, Art Library

Figure 11.5 shows the bonding in two of the other molecular shapes with the tetrahedral electron-group arrangement. The trigonal pyramidal shape of NH_3 arises when a lone pair fills one of the four sp^3 orbitals of N, and the bent shape of H_2O arises when lone pairs fill two of the sp^3 orbitals of O.

sp^3d Hybridization The shapes of molecules with trigonal bipyramidal or octahedral electron-group arrangements are rationalized with VB theory through similar arguments. The only new point is that such molecules have central atoms from Period 3 or higher, so atomic d orbitals, as well as s and p orbitals, are mixed to form the hybrid orbitals.

For the trigonal bipyramidal shape of the PCl_5 molecule, for example, the VB model proposes that the one $3s$, the three $3p$, and one of the five $3d$ orbitals of the central P atom mix and form five **sp^3d hybrid orbitals,** which point to the vertices of a trigonal bipyramid. Each hybrid orbital overlaps a $3p$ orbital of a Cl atom, and the five valence electrons of P, together with one from each of the five Cl atoms, pair up to form five P—Cl bonds, as shown in Figure 11.6. Seesaw, T-shaped, and linear molecules have this electron-group arrangement with lone pairs in, respectively, one, two, or three of the central atom's sp^3d orbitals.

Literature Galbraith, John Morrison. "On the Role of d Orbital Hybridization in the Chemistry Curriculum," *J. Chem. Educ.* **2007,** *84,* 783.

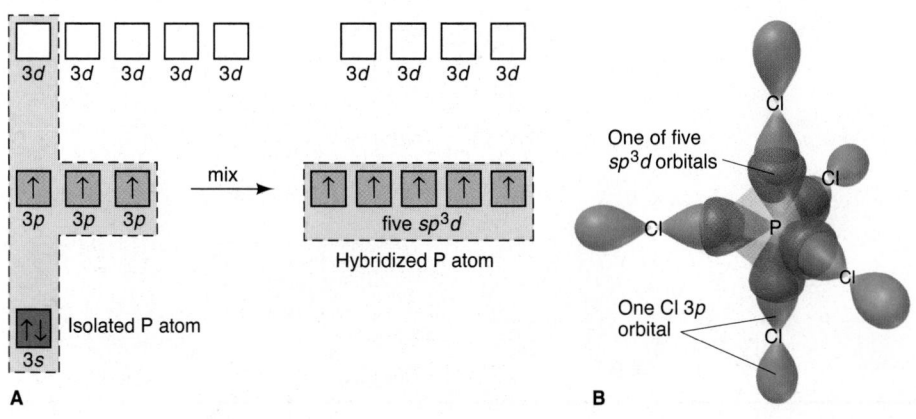

Figure 11.6 The sp^3d hybrid orbitals in PCl_5. A, The orbital diagram shows that one $3s$, three $3p$, and one of the five $3d$ orbitals of P mix to form five sp^3d orbitals that are half-filled. Four $3d$ orbitals are unhybridized and empty. **B,** The trigonal bipyramidal PCl_5 molecule forms by the overlap of a $3p$ orbital from each of the five Cl atoms with the sp^3d hybrid orbitals of P (unhybridized, empty $3d$ orbitals not shown). Each sp^3d orbital fills by addition of an electron from one Cl. (These five bonding pairs are not shown.)

Transparency/ARIS Presentation Center, Art Library

Figure 11.7 The sp^3d^2 hybrid orbitals in SF₆. A, The orbital diagram shows that one 3*s*, three 3*p*, and two 3*d* orbitals of S mix to form six sp^3d^2 orbitals that are half-filled. Three 3*d* orbitals are unhybridized and empty. **B,** The octahedral SF₆ molecule forms from overlap of a 2*p* orbital from each of six F atoms with the sp^3d^2 orbitals of S (unhybridized, empty 3*d* orbitals not shown). Each sp^3d^2 orbital fills by addition of an electron from one F. (These six bonding pairs are not shown.)

sp^3d^2 Hybridization To rationalize the shape of SF₆, the VB model proposes that the one 3*s*, the three 3*p*, and two of the five 3*d* orbitals of the central S atom mix and form six **sp^3d^2 hybrid orbitals,** which point to the vertices of an octahedron. Each hybrid orbital overlaps a 2*p* orbital of an F atom, and the six valence electrons of S, together with one from each of the six F atoms, pair up to form six S—F bonds (Figure 11.7). Square pyramidal and square planar molecules have lone pairs in one and two of the central atom's sp^3d^2 orbitals, respectively.

Table 11.1 summarizes the numbers and types of atomic orbitals that are mixed to obtain the five types of hybrid orbitals. Once again, note the similarities between the orientations of the hybrid orbitals proposed by VB theory and the shapes predicted by VSEPR theory (see Figure 10.2, p. 389). Figure 11.8 shows the three conceptual steps from molecular formula to postulating the hybrid orbitals in the molecule, and Sample Problem 11.1 details the end of that process.

Table 11.1   Composition and Orientation of Hybrid Orbitals

	Linear	Trigonal Planar	Tetrahedral	Trigonal Bipyramidal	Octahedral
Atomic orbitals mixed	one *s* one *p*	one *s* two *p*	one *s* three *p*	one *s* three *p* one *d*	one *s* three *p* two *d*
Hybrid orbitals formed	two *sp*	three *sp²*	four *sp³*	five *sp³d*	six *sp³d²*
Unhybridized orbitals remaining	two *p*	one *p*	none	four *d*	three *d*
Orientation					

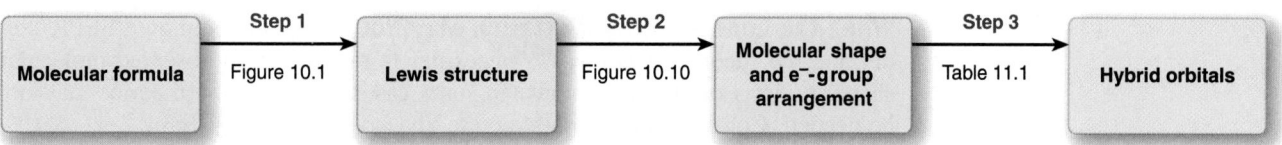

Figure 11.8 **The conceptual steps from molecular formula to the hybrid orbitals used in bonding.** (See Figures 10.1, p. 378, and 10.10, p. 395.) ARIS Presentation Center, Art Library

ARIS Presentation Center, Worked Examples Library

SAMPLE PROBLEM 11.1 Postulating Hybrid Orbitals in a Molecule

PROBLEM Use partial orbital diagrams to describe how mixing of the atomic orbitals of the central atom(s) leads to the hybrid orbitals in each of the following:
(a) Methanol, CH_3OH **(b)** Sulfur tetrafluoride, SF_4

PLAN From the Lewis structure, we determine the number and arrangement of electron groups around each central atom, along with the molecular shape. From that, we postulate the type of hybrid orbitals involved. Then, we write the partial orbital diagram for each central atom before and after the orbitals are hybridized.

SOLUTION **(a)** For CH_3OH. The electron-group arrangement is tetrahedral around both C and O atoms. Therefore, each central atom is sp^3 hybridized. The C atom has four half-filled sp^3 orbitals:

The O atom has two half-filled sp^3 orbitals and two filled with lone pairs:

During bonding, each half-filled C or O orbital becomes filled. The second electron comes from an H atom or, in the C—O bond, from the overlapping C and O orbitals.

(b) For SF_4. The molecular shape is seesaw, which is based on the trigonal bipyramidal electron-group arrangement. Thus, the central S atom is surrounded by five electron groups, which implies sp^3d hybridization. One $3s$ orbital, three $3p$ orbitals, and one $3d$ orbital are mixed. One hybrid orbital is filled with a lone pair, and four are half-filled. Four unhybridized $3d$ orbitals remain empty:

During bonding, each half-filled S orbital becomes filled, with the second electrons coming from F atoms.

FOLLOW-UP PROBLEM 11.1 Use partial orbital diagrams to show how the atomic orbitals of the central atom mix to form hybrid orbitals in **(a)** beryllium fluoride, BeF_2; **(b)** silicon tetrachloride, $SiCl_4$; **(c)** xenon tetrafluoride, XeF_4.

When the Concept of Hybridization May Not Apply We employ VSEPR theory and VB theory *whenever it is necessary* to rationalize an observed molecular shape in terms of atomic orbitals. In some cases, however, the theories may not be needed. Consider the Lewis structure and bond angle of H_2S:

$$H \underset{92°}{\overset{\overset{\displaystyle \ddot{S}}{\diagdown \diagup}}{}} H$$

Based on VSEPR theory, we would predict that, as in H_2O, the four electron groups around H_2S point to the vertices of a tetrahedron, and the two lone pairs compress the H—S—H bond angle somewhat below the ideal 109.5°. Based on VB theory, we would propose that the $3s$ and $3p$ orbitals of the central S atom mix and form four sp^3 hybrids, two of which are filled with lone pairs, while the other two overlap $1s$ orbitals of two H atoms and are filled with bonding pairs.

The problem is that observation does *not* support these arguments. In fact, the H_2S molecule has a bond angle of 92°, close to the 90° angle expected between *unhybridized*, perpendicular, atomic p orbitals. Similar angles occur in the hydrides of other Group 6A(16) elements and also in the hydrides of larger Group 5A(15) elements. It makes no sense to apply a theory when the facts don't warrant it. In the case of H_2S and these other hydrides, neither VSEPR theory nor the concept of hybridization applies. It's important to remember that real factors, such as bond length, atomic size, and electron-electron repulsions, influence molecular shape. Apparently, with these larger central atoms and their longer bonds to H, crowding and electron-electron repulsions decrease, and the simple overlap of unhybridized atomic orbitals rationalizes the observed shapes perfectly well.

Section Summary

VB theory explains that a covalent bond forms when two atomic orbitals overlap and two electrons with paired (opposite) spins occupy the overlapped regions. • Orbital hybridization allows us to explain how atomic orbitals mix and change their characteristics during bonding. • Based on the observed molecular shape (and the related electron-group arrangement), we postulate the type of hybrid orbital needed. In certain cases, we need not invoke hybridization at all.

Organic chemistry application

11.2 THE MODE OF ORBITAL OVERLAP AND THE TYPES OF COVALENT BONDS

In this section, we focus on the *mode* by which orbitals overlap—end to end or side to side—to see the detailed makeup of covalent bonds. These two modes give rise to the two types of covalent bonds—*sigma* bonds and *pi* bonds. We'll use valence bond theory to describe the two types here, but they are essential features of molecular orbital theory as well.

Orbital Overlap in Single and Multiple Bonds

The VSEPR model predicts, and measurements verify, different shapes for ethane (C_2H_6), ethylene (C_2H_4), and acetylene (C_2H_2). Ethane is tetrahedrally shaped at both carbons, with bond angles of about 109.5°. Ethylene is trigonal planar at both carbons, with the double bond acting as one electron group and bond angles near the ideal 120°. Acetylene has a linear shape, with the triple bond acting as one electron group and bond angles of 180°:

ethane   ethylene   acetylene

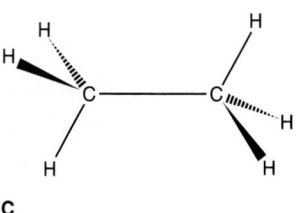

Figure 11.9 **The σ bonds in ethane (C_2H_6). A,** Two sp^3 hybridized C atoms and six H atoms in ethane form one C—C σ bond and six C—H σ bonds. **B,** Electron density is distributed relatively evenly among the σ bonds. **C,** Wedge-bond perspective drawing.

Transparency/ARIS Presentation Center, Art Library

Literature Shusterman, Gwendolyn P.; Shusterman, Alan J. "Teaching Chemistry with Electron Density Models," *J. Chem. Educ.* **1997,** *74,* 771.

A close look at the bonds shows two modes of orbital overlap, which correspond to the two types of covalent bonds:

1. *End-to-end overlap and sigma (σ) bonding.* Both C atoms of ethane are sp^3 hybridized (Figure 11.9). The C—C bond involves overlap of one sp^3 orbital from each C, and each of the six C—H bonds involves overlap of a C sp^3 orbital with an H $1s$ orbital. The bonds in ethane are like all the others described so far in this chapter. Look closely at the C—C bond, for example. It involves the overlap of the end of one orbital with the end of the other. The bond resulting from such *end-to-end* overlap is called a **sigma (σ) bond.** It has its *highest electron density along the bond axis* (an imaginary line joining the nuclei) and is shaped like an ellipse rotated about its long axis (the shape resembles a football). All single bonds, formed by any combination of overlapping hybrid, *s,* or *p* orbitals, have their electron density concentrated along the bond axis, and thus are σ bonds.

2. *Side-to-side overlap and pi (π) bonding.* A close look at Figure 11.10 reveals the double nature of the carbon-carbon bond in ethylene. Here, each C atom is sp^2 hybridized. Each C atom's four valence electrons half-fill its three sp^2 orbitals and its unhybridized $2p$ orbital, which lies perpendicular to the sp^2 plane. Two sp^2 orbitals of each C form C—H σ bonds by overlapping the $1s$ orbitals of two H atoms. The third sp^2 orbital forms a C—C σ bond with an sp^2 orbital of the other C because their orientation allows end-to-end overlap. With the σ-bonded C atoms near each other, their half-filled unhybridized $2p$ orbitals are close enough to overlap *side to side.* Such overlap forms another type of covalent bond called a **pi (π) bond.** It has *two regions of electron density,* one above and one below the σ-bond axis. *One π bond holds two electrons that occupy both regions (lobes) of the bond. A double bond always consists of one σ bond and one π bond.* As shown in Figure 11.10D, the double bond increases electron density between the C atoms.

Now we can answer a question not addressed in our discussion of VSEPR theory in Chapter 10: why do the two electron pairs in a double bond act as one electron group; that is, why don't they push each other apart? The answer is that each electron pair occupies a distinct orbital, a specific region of electron density, so repulsions are reduced.

Transparency/ARIS Presentation Center, Art Library

Figure 11.10 **The σ and π bonds in ethylene (C_2H_4). A,** Two sp^2 hybridized C atoms form one C—C σ bond and four C—H σ bonds. Half-filled unhybridized $2p$ orbitals (shown in stylized form) lie perpendicular to the σ-bond axis. **B,** The $2p$ orbitals (shown in accurate form) actually overlap side to side; the σ bonds are shown by a ball-and-stick model. **C,** The two overlapping regions comprise *one* π bond, which is occupied by two electrons. **D,** With four electrons (one σ bond and one π bond) between the C atoms, electron density (*red shading*) is high. **E,** Wedge-bond perspective drawing.

Two lobes of one π bond

Figure 11.11 The σ bonds and π bonds in acetylene (C₂H₂).

A, The C—C σ bond in acetylene forms when an *sp* orbital of each C overlaps. Two C—H σ bonds form from overlap of the other *sp* orbital of each C and an *s* orbital of an H. Two unhybridized 2*p* orbitals on each C are shown in stylized form (one pair is in *red* for clarity). **B,** The 2*p* orbitals (shown in accurate form) actually overlap side to side; the σ bonds are shown by a ball-and-stick model. **C,** Overlap of pairs of 2*p* orbitals results in two π bonds, one (*black*) with its lobes above and below the C—C σ bond, and the other (*red*) with its lobes in front of and behind the σ bond. **D,** The two π bonds give the molecule cylindrical symmetry. Six electrons (one σ bond and two π bonds) create greater electron density between the C atoms than in ethylene. **E,** Bond-line drawing.

A triple bond, such as the C≡C bond in acetylene, *consists of one σ and two π bonds* (Figure 11.11). To maximize overlap in a linear shape, one *s* and one *p* orbital in *each* C atom form two *sp* hybrids, and two 2*p* orbitals remain unhybridized. The four valence electrons half-fill all four orbitals. Each C uses one of its *sp* orbitals to form a σ bond with an H atom and uses the other to form the C—C σ bond. Side-to-side overlap of one pair of 2*p* orbitals gives one π bond, with electron density above and below the σ bond. Side-to-side overlap of the other pair of 2*p* orbitals gives the other π bond, 90° away from the first, with electron density in front and back of the σ bond. The result is a *cylindrically symmetrical* H—C≡C—H molecule. Note the greater electron density between the C atoms created by the six bonding electrons. Figure 11.12 shows electron density relief maps of the carbon-carbon bonds in these compounds; note the increasing electron density between the nuclei from single to double to triple bond.

Literature Matta, C. F.; Gillespie, R. J. "Understanding and Interpreting Molecular Electron Density Distributions," *J. Chem. Educ.* **2002,** *79,* 1141.

Figure 11.12 Electron density and bond order.
Relief maps of the carbon-carbon bonding region in ethane, ethylene, and acetylene show a large increase in electron density between the nuclei of the C atoms as the bond order increases.

The extent of overlap influences bond strength. Because side-to-side overlap is not as extensive as end-to-end overlap, we might expect a π bond to be weaker than a σ bond, and thus a double bond should be less than twice as strong as a single bond. This expectation *is* borne out for carbon-carbon bonds. However, many factors, such as lone-pair repulsions, bond polarities, and other electrostatic contributions, affect overlap and the relative strength of σ and π bonds between other pairs of atoms. Thus, as a rough approximation, a double bond is about twice as strong as a single bond, and a triple bond is about three times as strong.

ARIS Presentation Center, Worked Examples Library

SAMPLE PROBLEM 11.2 Describing the Types of Bonds in Molecules

PROBLEM Describe the types of bonds and orbitals in acetone, $(CH_3)_2CO$.

PLAN We note, as in Sample Problem 11.1, the shape around each central atom to postulate the hybrid orbitals used, and pay attention to the multiple bonding of the C=O bond.

SOLUTION In Sample Problem 10.8, p. 397, we determined the shapes of the three central atoms of acetone: tetrahedral around each C of the two CH_3 (methyl) groups and trigonal planar around the middle C atom. Thus, the middle C has three sp^2 orbitals and one unhybridized p orbital. Each of the two methyl C atoms has four sp^3 orbitals. Three of these sp^3 orbitals overlap the $1s$ orbitals of the H atoms to form σ bonds; the fourth overlaps an sp^2 orbital of the middle C atom. Thus, two of the three sp^2 orbitals of the middle C form σ bonds to the other two C atoms.

The O atom is also sp^2 hybridized and has an unhybridized p orbital that can form a π bond. Two of the O atom's sp^2 orbitals hold lone pairs, and the third forms a σ bond with the third sp^2 orbital of the middle C atom. The unhybridized, half-filled p orbitals of C and O form a π bond. The σ and π bonds constitute the C=O bond:

σ bonds

π bond
(shown with molecule
rotated 90°)

COMMENT How can we tell if a terminal atom, such as the O atom in acetone, is hybridized? After all, it could use two perpendicular p orbitals for the σ and π bonds with C and leave the other p and the s orbital to hold the two lone pairs. But, having each lone pair in an sp^2 orbital oriented away from the C=O bond, rather than in an s orbital, lowers repulsions.

FOLLOW-UP PROBLEM 11.2 Describe the types of bonds and orbitals in **(a)** hydrogen cyanide, HCN, and **(b)** carbon dioxide, CO_2.

Orbital Overlap and Molecular Rotation

The type of bond influences the ability of one part of a molecule to rotate relative to another part. A σ bond *allows free rotation* of the parts of the molecule with respect to each other because the extent of overlap is not affected. If you could hold one CH_3 group of the ethane molecule, the other CH_3 group could spin like a pinwheel without affecting the C—C σ-bond overlap (see Figure 11.9).

However, p orbitals must be parallel to engage in side-to-side overlap, so a π *bond restricts rotation* around it. Rotating one CH_2 group in ethylene with respect to the other must decrease the side-to-side overlap and break the π bond. You can see why distinct *cis* and *trans* structures can exist for molecules with double bonds, such as 1,2-dichloroethylene (Section 10.3). As Figure 11.13 shows, the π bond allows two *different* arrangements of atoms around the two C atoms, which can have a major effect on molecular polarity (see p. 401). (Rotation around a triple bond is not meaningful: because each triple-bonded C atom is attached to only one other group in a linear arrangement, there can be no difference in the relative positions of the groups.)

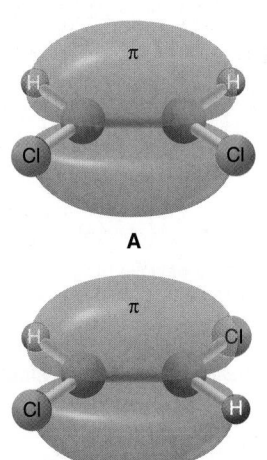

Figure 11.13 Restricted rotation of π-bonded molecules. *Cis-* **(A)** and *trans-*1,2-dichloroethylene **(B)** occur as distinct molecules because the π bond between the C atoms restricts rotation and maintains different relative positions of the H and Cl atoms.

ARIS Presentation Center, Art Library

Section Summary

End-to-end overlap of atomic orbitals forms a σ bond and allows free rotation of the parts of the molecule. • Side-to-side overlap forms a π bond, which restricts rotation. • A multiple bond consists of a σ bond and either one π bond (double bond) or two (triple bond). Multiple bonds have greater electron density between the nuclei than single bonds do.

11.3 MOLECULAR ORBITAL (MO) THEORY AND ELECTRON DELOCALIZATION

Scientists choose the model that best helps them answer a particular question. If the question concerns molecular shape, chemists choose the VSEPR model, followed by hybrid-orbital analysis with VB theory. But VB theory does not adequately explain magnetic and spectral properties, and it understates the importance of electron delocalization. In order to deal with these phenomena, which involve molecular energy levels, chemists choose another model based on quantum mechanics called **molecular orbital (MO) theory.**

In VB theory, a molecule is pictured as a group of atoms bound together through *localized* overlap of valence-shell atomic orbitals. In MO theory, a molecule is pictured as a collection of nuclei with the electron orbitals *delocalized* over the entire molecule. The MO model is a quantum-mechanical treatment for molecules similar to the one for atoms in Chapter 8. Just as an atom has atomic orbitals (AOs) with a given energy and shape that are occupied by the atom's electrons, a molecule has **molecular orbitals (MOs)** with a given energy and shape that are occupied by the molecule's electrons. Despite the great usefulness of MO theory, it too has a drawback: MOs are more difficult to visualize than the easily depicted shapes of VSEPR theory or the hybrid orbitals of VB theory.

The Central Themes of MO Theory

Several key ideas of MO theory appear in its description of the hydrogen molecule and other simple species. These ideas include the formation of MOs, their energy and shape, and how they fill with electrons.

Formation of Molecular Orbitals Because electron motion is so complex, we use approximations to solve the Schrödinger equation (p. 304) for an atom with more than one electron. Similar complications arise even with H_2, the simplest molecule, so we use approximations to solve for the properties of MOs. The most common approximation mathematically *combines* (adds or subtracts) the atomic orbitals (atomic wave functions) of nearby atoms to form MOs (molecular wave functions).

When two H nuclei lie near each other, as in H_2, their AOs overlap. The two ways of combining the AOs are as follows:

- *Adding the wave functions together.* This combination forms a **bonding MO,** which has *a region of high electron density between the nuclei*. Additive overlap is analogous to light waves reinforcing each other, making the resulting amplitude higher and the light brighter. For electron waves, the overlap *increases* the probability that the electrons are between the nuclei (Figure 11.14A).

- *Subtracting the wave functions from each other.* This combination forms an **antibonding MO,** which has *a region of zero electron density (a node) between the nuclei* (Figure 11.14B). Subtractive overlap is analogous to light waves canceling each other, so that the light disappears. With electron waves, the probability that the electrons lie between the nuclei *decreases* to zero.

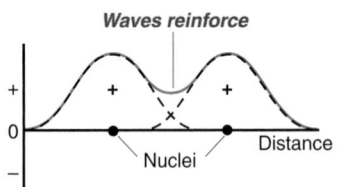

A Amplitudes of wave functions added

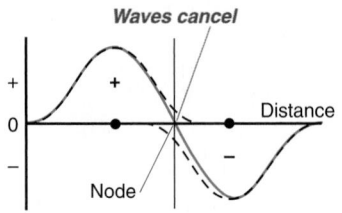

B Amplitudes of wave functions subtracted

Figure 11.14 An analogy between light waves and atomic wave functions. When light waves undergo interference, their amplitudes either add together or subtract. **A,** When the amplitudes of atomic wave functions *(dashed lines)* are added, a bonding molecular orbital (MO) results, and electron density *(red line)* increases between the nuclei. **B,** Conversely, when the amplitudes of the wave functions are subtracted, an antibonding MO results, which has a node (region of zero electron density) between the nuclei.

The two possible combinations for hydrogen atoms H_A and H_B are

AO of H_A + AO of H_B = bonding MO of H_2 (more e^- density between nuclei)

AO of H_A − AO of H_B = antibonding MO of H_2 (less e^- density between nuclei)

Notice that *the number of AOs combined always equals the number of MOs formed:* two H atomic orbitals combine to form two H_2 molecular orbitals.

Energy and Shape of H_2 Molecular Orbitals
The bonding MO is lower in energy and the antibonding MO higher in energy than the AOs that combined to form them. Let's examine Figure 11.15 to see why this is so.

The bonding MO in H_2 is spread mostly *between* the nuclei, with the nuclei attracted to the intervening electrons. An electron in this MO can delocalize its charge over a much larger volume than is possible in an individual AO in H. Because the electron-electron repulsions are reduced, the bonding MO is *lower* in energy than the isolated AOs. Therefore, when electrons occupy this orbital, the molecule is *more stable* than the separate atoms. In contrast, the antibonding MO has a node between the nuclei and most of its electron density *outside* the internuclear region. The electrons do not shield one nucleus from the other, which increases the nucleus-nucleus repulsion and makes the antibonding MO *higher* in energy than the isolated AOs. Therefore, when the antibonding orbital is occupied, the molecule is *less stable* than when this orbital is empty.

Both the bonding and antibonding MOs of H_2 are **sigma (σ) MOs** because they are cylindrically symmetrical about an imaginary line that runs through the two nuclei. The bonding MO is denoted by σ_{1s}, that is, a σ MO formed by combination of $1s$ AOs. Antibonding orbitals are denoted with a superscript star, so the antibonding MO derived from the $1s$ AOs is σ_{1s}^* (spoken "sigma, one ess, star").

To interact effectively and form MOs, *atomic orbitals must have similar energy and orientation.* The $1s$ orbitals on two H atoms have identical energy and orientation, so they interact strongly. This point will be important when we consider molecules composed of atoms with many sublevels.

Filling Molecular Orbitals with Electrons
Electrons fill MOs just as they fill AOs:

- MOs are filled in order of increasing energy (aufbau principle).
- An MO has a maximum capacity of two electrons with opposite spins (exclusion principle).
- Orbitals of equal energy are half-filled, with spins parallel, before any of them is completely filled (Hund's rule).

A **molecular orbital (MO) diagram** shows the relative energy and number of electrons in each MO, as well as the AOs from which they formed. Figure 11.16 is the MO diagram for H_2.

MO theory redefines bond order. In a Lewis structure, bond order is the number of electron pairs per linkage. The **MO bond order** is the number of electrons in bonding MOs minus the number in antibonding MOs, divided by two:

$$\text{Bond order} = \tfrac{1}{2}[(\text{no. of } e^- \text{ in bonding MO}) - (\text{no. of } e^- \text{ in antibonding MO})] \quad \textbf{(11.1)}$$

Thus, for H_2, the bond order is $\tfrac{1}{2}(2-0)=1$. A bond order greater than zero indicates that the molecular species is stable relative to the separate atoms, whereas a bond order of zero implies no net stability and, thus, no likelihood that the species will form. In general, *the higher the bond order, the stronger the bond.*

Another similarity of MO theory to the quantum-mechanical model for atoms is that we can write an electron configuration for a molecule. The symbol of each occupied MO is shown in parentheses, and the number of electrons in it is written outside the parentheses as a superscript: the electron configuration of H_2 is $(\sigma_{1s})^2$.

Literature Smith, Derek W. "The Antibonding Effect," *J. Chem. Educ.* **2000,** *77,* 780.

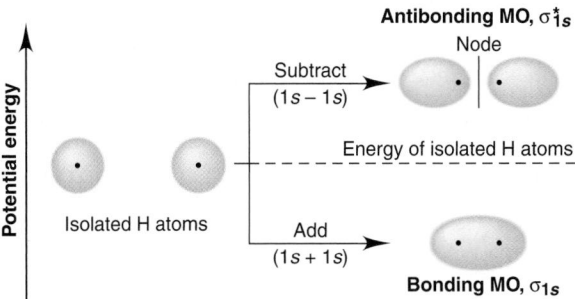

Figure 11.15 Contours and energies of the bonding and antibonding molecular orbitals (MOs) in H_2. When two H $1s$ atomic orbitals (AOs) combine, they form two H_2 MOs. The bonding MO (σ_{1s}) forms from addition of the AOs and is lower in energy than those AOs because most of its electron density lies *between* the nuclei (shown as dots). The antibonding MO (σ_{1s}^*) forms from subtraction of the AOs and is higher in energy because there is a *node* between the nuclei and most of the electron density lies outside the internuclear region.

ARIS Presentation Center, Art Library

ARIS Presentation Center, Art Library

H_2 bond order $= \tfrac{1}{2}(2-0) = 1$

Figure 11.16 The MO diagram for H_2. The positions of the boxes indicate the relative energies and the arrows show the electron occupancy of the MOs and the AOs from which they formed. Two electrons, one from each H atom, fill the lower energy σ_{1s} MO, while the higher energy σ_{1s}^* MO remains empty. Orbital occupancy is also shown by color (darker = full, paler = half-filled, no color = empty).

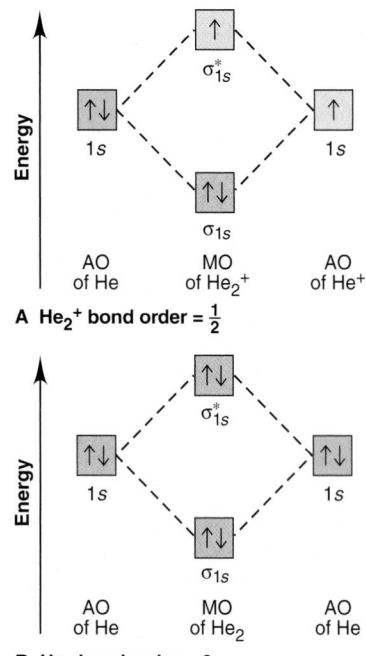

A He₂⁺ bond order = ½

B He₂ bond order = 0

Figure 11.17 MO diagrams for He₂⁺ and He₂. A, In He₂⁺, three electrons enter MOs in order of increasing energy to give a filled σ_{1s} MO and a half-filled σ^*_{1s} MO. The bond order of $\frac{1}{2}$ implies that He₂⁺ exists. **B,** In He₂, the four electrons fill both the σ_{1s} and the σ^*_{1s} MOs, so there is no net stabilization (bond order = 0).

ARIS Presentation Center, Active Art Library

ARIS Presentation Center, Worked Examples Library

One of the early triumphs of MO theory was its ability to *predict* the existence of He₂⁺, the dihelium molecule-ion, which is composed of two He nuclei and three electrons. Let's use MO theory to see why He₂⁺ exists and, at the same time, why He₂ does not. In He₂⁺, the $1s$ atomic orbitals form the molecular orbitals, so the MO diagram, shown in Figure 11.17A, is similar to that for H₂. The three electrons enter the MOs to give a pair in the σ_{1s} MO and a lone electron in the σ^*_{1s} MO. The bond order is $\frac{1}{2}(2 - 1) = \frac{1}{2}$. Thus, He₂⁺ has a relatively weak bond, but it should exist. Indeed, this molecular ionic species has been observed frequently when He atoms collide with He⁺ ions. Its electron configuration is $(\sigma_{1s})^2(\sigma^*_{1s})^1$.

On the other hand, He₂ has four electrons to place in its σ_{1s} and σ^*_{1s} MOs. As Figure 11.17B shows, both the bonding and antibonding orbitals are filled. The stabilization arising from the electron pair in the bonding MO is canceled by the destabilization due to the electron pair in the antibonding MO. From its zero bond order $[\frac{1}{2}(2 - 2) = 0]$, we predict, and experiment has so far confirmed, that a covalent He₂ molecule does not exist.

SAMPLE PROBLEM 11.3 Predicting Stability of Species Using MO Diagrams

PROBLEM Use MO diagrams to find bond orders and predict whether H₂⁺ and H₂⁻ exist. For any that exists, write the electron configuration.

PLAN In these species, the $1s$ orbitals form MOs, so the MO diagrams are similar to that for H₂. We find the number of electrons in each species and distribute the electrons in pairs to the bonding and antibonding MOs in order of increasing energy. We obtain the bond order with Equation 11.1 and write the electron configuration as described in the text.

SOLUTION For H₂⁺. H₂ has two e⁻, so H₂⁺ has only one (see diagram). The bond order is $\frac{1}{2}(1 - 0) = \frac{1}{2}$, so we predict that H₂⁺ does exist. The electron configuration is $(\sigma_{1s})^1$. For H₂⁻. H₂ has two e⁻, so H₂⁻ has three (see diagram). The bond order is $\frac{1}{2}(2 - 1) = \frac{1}{2}$, so we predict that H₂⁻ does exist. The electron configuration is $(\sigma_{1s})^2(\sigma^*_{1s})^1$.

CHECK The number of electrons in the MOs equals the number of electrons in the AOs.

COMMENT Both these species have been detected spectroscopically: H₂⁺ occurs in the hydrogen-containing material around stars; H₂⁻ has been formed in the laboratory.

FOLLOW-UP PROBLEM 11.3 Use an MO diagram to find the bond order and predict whether two hydride ions (H⁻) will form H₂²⁻. If it does exist, write its electron configuration.

Homonuclear Diatomic Molecules of the Period 2 Elements

Homonuclear diatomic molecules are composed of two identical atoms. In addition to H₂ from Period 1, you're familiar with several from Period 2—N₂, O₂, and F₂—as the elemental forms under standard conditions. Others in Period 2—Li₂, Be₂, B₂, C₂, and Ne₂—are observed, if at all, only in high-temperature gas-phase experiments. Molecular orbital descriptions of these species provide some interesting tests of the model. Let's look first at the molecules from the s block, Groups 1A(1) and 2A(2), and then at those from the p block, Groups 3A(13) through 8A(18).

Bonding in the s-Block Homonuclear Diatomic Molecules Both Li and Be occur as metals under normal conditions, but let's see what MO theory predicts for their stability as the diatomic gases dilithium (Li_2) and diberyllium (Be_2).

These atoms have both inner ($1s$) and outer ($2s$) electrons, but the $1s$ orbitals interact negligibly. We ignore the inner electrons here because, in general, *only outer (valence) orbitals interact enough to form molecular orbitals.* Like the MOs formed from $1s$ AOs, those formed from $2s$ AOs are σ orbitals, cylindrically symmetrical around the internuclear axis. Bonding (σ_{2s}) and antibonding (σ_{2s}^*) MOs form, and the two valence electrons fill the bonding MO, with opposing spins (Figure 11.18A). Dilithium has two electrons in bonding MOs and none in antibonding MOs; therefore, its bond order is $\frac{1}{2}(2 - 0) = 1$. In fact, Li_2 *has* been observed, and the MO electron configuration is $(\sigma_{2s})^2$. With its two additional electrons, the MO diagram for Be_2 has filled σ_{2s} and σ_{2s}^* MOs (Figure 11.18B). This is similar to the case of He_2. The bond order is $\frac{1}{2}(2 - 2) = 0$. In keeping with a zero bond order, the ground state of Be_2 has never been observed.

Molecular Orbitals from Atomic p-Orbital Combinations As we move to boron in the p block, atomic $2p$ orbitals become involved, so let's first consider the shapes and energies of the MOs that result from their combinations. Recall that p orbitals can overlap with each other in two different ways, as shown in Figure 11.19. End-to-end combination gives a pair of σ MOs, the σ_{2p} and σ_{2p}^*. Side-to-side combination gives a pair of **pi (π) MOs**, π_{2p} and π_{2p}^*. Similar to MOs formed from s orbitals, bonding MOs from p-orbital combinations have their greatest electron density *between* the nuclei, whereas antibonding MOs from p-orbital combinations have a node between the nuclei and most of their electron density *outside* the internuclear region.

The order of MO energy levels, whether bonding or antibonding, is based on the order of AO energy levels *and* on the mode of the p-orbital combination:

- MOs formed from $2s$ orbitals are lower in energy than MOs formed from $2p$ orbitals because $2s$ AOs are lower in energy than $2p$ AOs.
- Bonding MOs are lower in energy than antibonding MOs, so σ_{2p} is lower in energy than σ_{2p}^* and π_{2p} is lower than π_{2p}^*.
- Atomic p orbitals can interact more extensively end to end than they can side to side. Thus, the σ_{2p} MO is lower in energy than the π_{2p} MO. Similarly, the destabilizing effect of the σ_{2p}^* MO is greater than that of the π_{2p}^* MO.

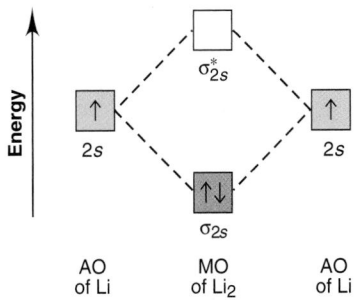

A Li_2 bond order = 1

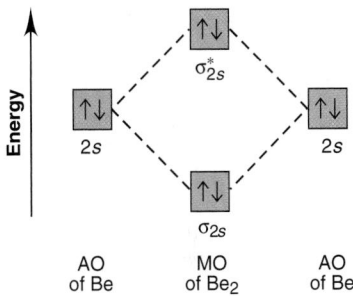

B Be_2 bond order = 0

Figure 11.18 Bonding in s-block homonuclear diatomic molecules.
Only outer (valence) AOs interact enough to form MOs. **A,** Li_2. The two valence electrons from two Li atoms fill the bonding (σ_{2s}) MO, and the antibonding (σ_{2s}^*) remains empty. With a bond order of 1, Li_2 does form. **B,** Be_2. The four valence electrons from two Be atoms fill both MOs to give no net stabilization. Ground-state Be_2 has a zero bond order and has never been observed.

ARIS Presentation Center, Art Library

Figure 11.19 Contours and energies of σ and π MOs from combinations of $2p$ atomic orbitals. A, The p orbitals lying along the line between the atoms (usually designated p_x) undergo end-to-end overlap and form σ_{2p} and σ_{2p}^* MOs. Note the greater electron density between the nuclei for the bonding MO and the node between the nuclei for the antibonding MO. **B,** The p orbitals that lie perpendicular to the internuclear axis (p_y and p_z) undergo side-to-side overlap to form two π MOs. The p_z interactions are the same as those shown here for the p_y orbitals, so a total of four π MOs form. A π_{2p} is a bonding MO with its greatest density above and below the internuclear axis; a π_{2p}^* is an antibonding MO with a node between the nuclei and its electron density outside the internuclear region.

Transparency/ARIS Presentation Center, Active Art Library

Figure 11.20 Relative MO energy levels for Period 2 homonuclear diatomic molecules. A, MO energy levels for O_2, F_2, and Ne_2. The six $2p$ orbitals of the two atoms form six MOs that are higher in energy than the two MOs formed from the two $2s$ orbitals. The AOs forming π orbitals give rise to two bonding MOs (π_{2p}) of equal energy and two antibonding MOs (π^*_{2p}) of equal energy. This sequence of energy levels arises from minimal $2s$-$2p$ orbital mixing. **B,** MO energy levels for B_2, C_2, and N_2. Because of significant $2s$-$2p$ orbital mixing, the energies of σ MOs formed from $2p$ orbitals increase and the energies of those formed from $2s$ orbitals decrease. The major effect of this orbital mixing on the MO sequence is that the σ_{2p} is higher in energy than the π_{2p}. (For clarity, those MOs affected by mixing are shown in purple.)

Transparency/ARIS Presentation Center, Art Library

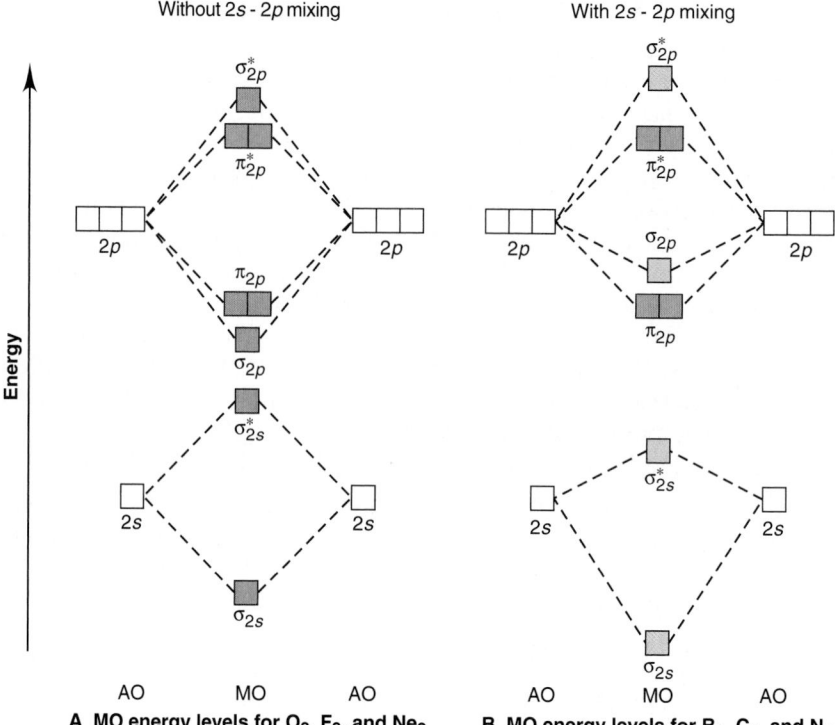

Literature Haim, Albert. "The Relative Energies of Molecular Orbitals for Second-Row Homonuclear Diatomic Molecules: The Effect of *s-p* Mixing," *J. Chem. Educ.* **1991,** *68,* 737.

Thus, the energy order for MOs derived from $2p$ orbitals is

$$\sigma_{2p} < \pi_{2p} < \pi^*_{2p} < \sigma^*_{2p}$$

There are three mutually perpendicular $2p$ orbitals in each atom. When the six p orbitals in two atoms combine, the two orbitals that interact end to end form a σ and a σ^* MO, and the two pairs of orbitals that interact side to side form two π MOs of the same energy and two π^* MOs of the same energy. Combining these orientations with the energy order gives the *expected* MO diagram for the *p*-block Period 2 homonuclear diatomic molecules (Figure 11.20A).

One other factor influences the MO energy order. Recall that only AOs of similar energy interact to form MOs. The order in Figure 11.20A assumes that the s and p AOs are so different in energy that they do not interact with each other: the orbitals do not *mix*. This is true for O, F, and Ne. These atoms are small, so relatively strong repulsions occur as the $2p$ electrons pair up; these repulsions raise the energy of the $2p$ orbitals high enough above the energy of the $2s$ orbitals to minimize orbital mixing. In contrast, B, C, and N atoms are larger, and when the $2p$ AOs are half-filled, repulsions are relatively small; so the $2p$ energies are much closer to the $2s$ energy. As a result, some mixing occurs between the $2s$ orbital of one atom and the end-on $2p$ orbital of the other. This orbital mixing *lowers* the energy of the σ_{2s} and σ^*_{2s} MOs and raises the energy of the σ_{2p} and σ^*_{2p} MOs; the π MOs are not affected. The MO diagram for B_2 through N_2 (Figure 11.20B) reflects this AO mixing. The only change that affects this discussion is the *reverse in energy order* of the σ_{2p} and π_{2p} MOs.

Bonding in the *p*-Block Homonuclear Diatomic Molecules Figure 11.21 shows the MOs, their electron occupancy, and some properties of B_2 through Ne_2. Note how *a higher bond order correlates with a greater bond energy and shorter bond length.* Also note how orbital occupancy correlates with magnetic properties. Recall from Chapter 8 that the spins of unpaired electrons in an atom (or ion) cause the substance to be *paramagnetic*, attracted to an external magnetic field. If all the electron spins are paired, the substance is *diamagnetic*, unaffected

Figure 11.21 MO occupancy and molecular properties for B_2 through Ne_2. The sequence of MOs and their electron populations are shown for the homonuclear diatomic molecules in the p block of Period 2 [Groups 3A(13) to 8A(18)]. The bond energy, bond length, bond order, magnetic properties, and outer (valence) electron configuration appear below the orbital diagrams. Note the correlation between bond order and bond energy, both of which are inversely related to bond length.

Transparency/ARIS Presentation Center, Active Art Library

(or weakly repelled) by the magnetic field. The same observations apply to molecules. These properties are not addressed directly in VSEPR or VB theory.

The B_2 molecule has six outer electrons to place in its MOs. Four of these fill the σ_{2s} and σ^*_{2s} MOs. The remaining two electrons occupy the two π_{2p} MOs, one in each orbital, in keeping with Hund's rule. With four electrons in bonding MOs and two electrons in antibonding MOs, the bond order of B_2 is $\frac{1}{2}(4-2) = 1$. As expected from its MO diagram, B_2 is paramagnetic.

The two additional electrons present in C_2 fill the two π_{2p} MOs. Since C_2 has two more bonding electrons than B_2, it has a bond order of 2 and the expected stronger, shorter bond. All the electrons are paired, and, as the model predicts, C_2 is diamagnetic.

In N_2, the two additional electrons fill the σ_{2p} MO. The resulting bond order is 3, which is consistent with the triple bond in the Lewis structure. As the model predicts, the bond energy is higher and the bond length shorter than for C_2, and N_2 is diamagnetic.

With O_2, we really see the power of MO theory compared to theories based on localized orbitals. For years, it seemed impossible to reconcile bonding theories with the bond strength and magnetic behavior of O_2. On the one hand, the data show a double-bonded molecule that is paramagnetic. On the other hand, we can write two possible Lewis structures for O_2, but neither gives such a molecule. One has a double bond and all electrons paired, the other a single bond and two electrons unpaired:

$$\ddot{O}=\ddot{O} \quad \text{or} \quad :\dot{O}-\dot{O}:$$

MO theory resolves this paradox beautifully. As Figure 11.21 shows, the bond order of O_2 is 2: eight electrons occupy bonding MOs and four occupy antibonding MOs [$\frac{1}{2}(8-4) = 2$]. Note the lower bond energy and greater bond

Figure 11.22 The paramagnetic properties of O_2. Liquid O_2 is attracted to the poles of a magnet because it is paramagnetic, as MO theory predicts. A diamagnetic substance would fall between the poles.

 ARIS Presentation Center, Photo Library

 ARIS Presentation Center, Worked Examples Library

length relative to N_2. The *two* electrons with highest energy occupy the *two* π^*_{2p} MOs with unpaired (parallel) spins, making the molecule paramagnetic. Figure 11.22 shows liquid O_2 suspended between the poles of a powerful magnet.

The two additional electrons in F_2 fill the π^*_{2p} orbitals, which decreases the bond order to 1, and the absence of unpaired electrons makes F_2 diamagnetic. As expected, the bond energy is lower and the bond distance longer than in O_2. Note that the bond energy for F_2 is only about half that for B_2, even though they have the same bond order. F is smaller than B, so we might expect a stronger bond. But the 18 electrons in the smaller volume of F_2 compared with the 10 electrons in B_2 cause greater repulsions and make the F_2 single bond easier to break.

The final member of the series, Ne_2, does not exist for the same reason that He_2 does not: all the MOs are filled, so the stabilization from bonding electrons cancels the destabilization from antibonding electrons, and the bond order is zero.

SAMPLE PROBLEM 11.4 Using MO Theory to Explain Bond Properties

PROBLEM As the following data show, removing an electron from N_2 forms an ion with a weaker, longer bond than in the parent molecule, whereas the ion formed from O_2 has a stronger, shorter bond:

	N_2	N_2^+	O_2	O_2^+
Bond energy (kJ/mol)	945	841	498	623
Bond length (pm)	110	112	121	112

Explain these facts with diagrams that show the sequence and occupancy of MOs.

PLAN We first determine the number of valence electrons in each species. Then, we draw the sequence of MO energy levels for the four species, recalling that they differ for N_2 and O_2 (see Figures 11.20 and 11.21), and fill them with electrons. Finally, we calculate bond orders and compare them with the data. Recall that bond order is related directly to bond energy and inversely to bond length.

SOLUTION Determining the valence electrons:

N has 5 valence e^-, so N_2 has 10 and N_2^+ has 9

O has 6 valence e^-, so O_2 has 12 and O_2^+ has 11

Drawing and filling the MO diagrams:

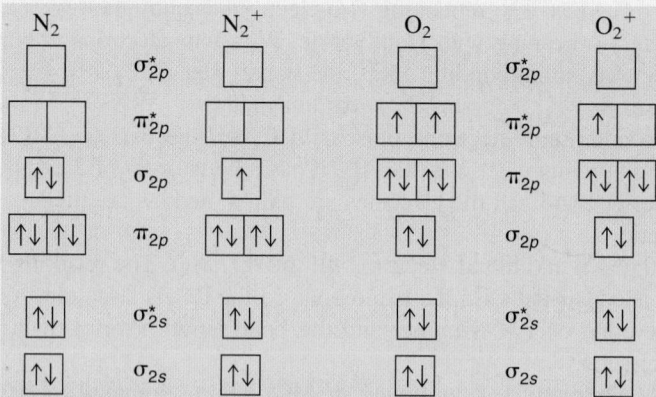

Calculating bond orders:

$$\tfrac{1}{2}(8 - 2) = 3 \qquad \tfrac{1}{2}(7 - 2) = 2.5 \qquad \tfrac{1}{2}(8 - 4) = 2 \qquad \tfrac{1}{2}(8 - 3) = 2.5$$

When N_2 forms N_2^+, a *bonding* electron is removed, so the bond order decreases. Thus, N_2^+ has a weaker, longer bond than N_2. When O_2 forms O_2^+, an *antibonding* electron is removed, so the bond order increases. Thus, O_2^+ has a stronger, shorter bond than O_2.

CHECK The answer makes sense in terms of the relationships among bond order, bond energy, and bond length. Check that the total number of bonding and antibonding electrons equals the number of valence electrons calculated.

FOLLOW-UP PROBLEM 11.4 Determine the bond orders for the following species: F_2^{2-}, F_2^-, F_2, F_2^+, F_2^{2+}. List the species in order of increasing bond energy and in order of increasing bond length.

MO Descriptions of Some Heteronuclear Diatomic Molecules

Heteronuclear diatomic molecules, those composed of two *different* atoms, have asymmetric MO diagrams because the atomic orbitals of the two atoms have unequal energies. Atoms with greater effective nuclear charge (Z_{eff}) draw their electrons closer to the nucleus and thus have atomic orbitals of lower energy (Section 8.2). Greater Z_{eff} also gives these atoms higher electronegativity values. Let's apply MO theory to the bonding in HF and NO.

Bonding in HF To form the MOs in HF, we combine appropriate AOs from isolated H and F atoms. The high effective nuclear charge of F holds all its electrons more tightly than the H nucleus holds its electron. As a result, *all* occupied atomic orbitals of F have lower energy than the 1s orbital of H. Therefore, the H 1s orbital interacts only with the F 2p orbitals, and only one of the three 2p orbitals, say the $2p_z$, leads to end-on overlap. This results in a σ MO and a σ* MO. The two other p orbitals of F ($2p_x$ and $2p_y$) are not involved in bonding and are called **nonbonding MOs;** they have the same energy as the isolated AOs (Figure 11.23).

Because the occupied bonding MO is closer in energy to the AOs of F, we say that the F 2p orbital contributes more to the bonding in HF than the H 1s orbital does. Generally, in polar covalent molecules, *bonding MOs are closer in energy to the AOs of the more electronegative atom.* In effect, fluorine's greater electronegativity lowers the energy of the bonding MO and draws the bonding electrons closer to its nucleus.

Bonding in NO Nitrogen monoxide (nitric oxide) is a highly reactive molecule because it has a lone electron. Two possible Lewis structures for NO, with formal charges (Section 10.1), are

$$\overset{0}{:\!\dot{N}}\!\!=\!\!\overset{0}{\ddot{O}:} \quad \text{or} \quad \overset{-1}{:\!\ddot{N}}\!\!=\!\!\overset{+1}{\ddot{O}:}$$
$$\textbf{I} \qquad\qquad \textbf{II}$$

Both structures show a double bond, but the *measured* bond energy suggests a bond order *higher* than 2. Moreover, it is not clear where the lone electron resides, although the lower formal charges for structure I suggest that it is on the N atom.

MO theory predicts the bond order and indicates lone-electron placement with no difficulty. The MO diagram in Figure 11.24 is asymmetric, with the AOs of the more electronegative O lower in energy. The 11 valence electrons of NO fill MOs in order of increasing energy, leaving the lone electron in one of the π_{2p}^* orbitals. The eight bonding electrons and three antibonding electrons give a bond order of $\frac{1}{2}(8-3) = 2.5$, more in keeping with experiment than either Lewis structure. The bonding electrons lie in MOs closer in energy to the AOs of the O atom. The lone electron occupies an antibonding orbital. Because this orbital receives a greater contribution from the 2p orbitals of N, the lone electron resides closer to the N atom.

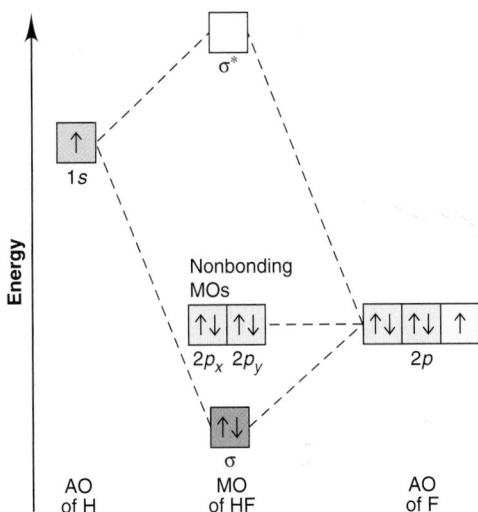

Figure 11.23 The MO diagram for HF. For a polar covalent molecule, the MO diagram is asymmetric because the more electronegative atom has AOs of lower energy. In HF, the bonding MO is closer in energy to the 2p orbital of F. Electrons that are not involved in bonding occupy nonbonding MOs. (The 2s AO of F is not shown.)

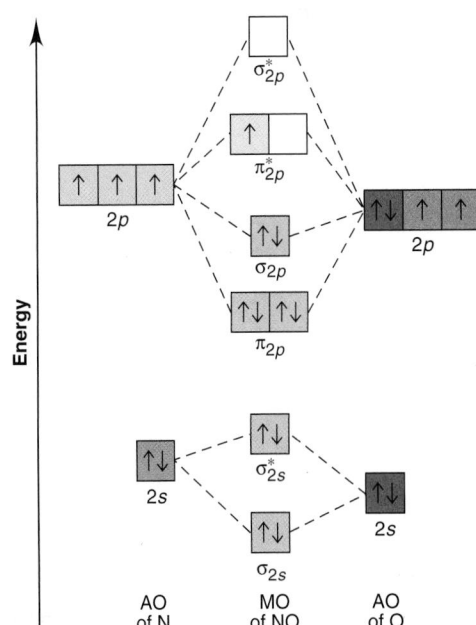

Figure 11.24 The MO diagram for NO. Eleven electrons occupy the MOs in nitrogen monoxide. Note that the lone electron occupies an antibonding MO whose energy is closer to that of the AO of the (less electronegative) N atom.

Figure 11.25 The lowest energy π-bonding MOs in benzene and ozone.
A, The most stable π-bonding MO in C_6H_6 has hexagonal lobes of electron density above and below the σ plane of the six C atoms. **B,** The most stable π-bonding MO in O_3 extends above and below the σ plane (shown as bond lines) of the three O atoms.

ARIS Presentation Center, Art Library

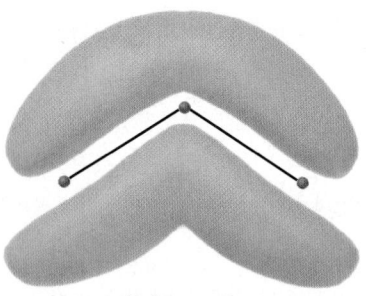

A Benzene, C_6H_6   **B Ozone, O_3**

MO Descriptions of Benzene and Ozone

The orbital shapes and MO diagrams for polyatomic molecules are too complex for a detailed treatment here. However, let's briefly discuss how the model eliminates the need for resonance forms and helps explain the effects of the absorption of energy. Recall that we cannot draw a single Lewis structure that adequately depicts bonding in molecules such as ozone and benzene because the adjacent single and double bonds actually have identical properties. Instead, we draw more than one structure and mentally combine them into a resonance hybrid. The VB model also relies on resonance because it depicts *localized* electron-pair bonds.

In contrast, MO theory pictures a structure of delocalized σ and π bonding and antibonding MOs. Figure 11.25 shows the lowest energy π-bonding MOs in benzene and ozone. Each holds one pair of electrons. The extended electron densities allow delocalization of this π-bonding electron pair over the entire molecule, thus eliminating the need for separate resonance forms. In benzene, the upper and lower hexagonal lobes of this π-bonding MO lie above and below the σ plane of all six carbon nuclei. In ozone, the two lobes of the lowest energy π-bonding MO extend over and under all three oxygen nuclei. Another advantage of MO theory is that it can explain excited states and spectra of molecules: for instance, why O_3 decomposes when it absorbs ultraviolet radiation in the stratosphere (bonding electrons are excited and enter empty antibonding orbitals), and why the ultraviolet spectrum of benzene has its characteristic absorption bands.

Section Summary

Molecular orbital theory treats a molecule as a collection of nuclei with MOs delocalized over the entire structure. • Atomic orbitals of comparable energy can be added and subtracted to obtain bonding and antibonding MOs, respectively. • Bonding MOs have most of their electron density between the nuclei and are lower in energy than the atomic orbitals; most of the electron density of antibonding MOs does not lie between the nuclei, so these MOs are higher in energy. • MOs are filled in order of energy with paired electrons having opposing spins. • MO diagrams show energy levels and orbital occupancy and explain trends in bond energy, bond length, and magnetic behavior. • In heteronuclear diatomic molecules, the more electronegative atom contributes more to the bonding MOs. • MO theory eliminates the need for resonance forms to depict larger molecules.

Chapter Perspective

In this chapter, you've seen that the two most important orbital-based models of covalent bonding, each with its own benefits and limitations, complement each other. Next, you'll see how polarity and electron distribution give rise to the physical behavior of liquids and solids. MO theory will help us understand metallic solids. Then, following Chapter 12 are three chapters that apply what you've learned from Chapter 7 onward to the behavior of solutions and inorganic, organic, and biochemical substances.

CHAPTER REVIEW GUIDE The following sections provide many aids to help you study this chapter. (Numbers in parentheses refer to pages, unless noted otherwise.)

Learning Objectives These are concepts and skills you should know after studying this chapter.

Relevant section and/or sample problem (SP) numbers appear in parentheses.

Understand These Concepts

1. The main ideas of valence bond theory—orbital overlap, opposing electron spins, and hybridization—as a means of rationalizing molecular shapes (11.1)
2. How orbitals mix to form hybrid orbitals with different spatial orientations (11.1)
3. The distinction between end-to-end and side-to-side overlap and the origin of sigma (σ) and pi (π) bonds in simple molecules (11.2)
4. How two modes of orbital overlap lead to single, double, and triple bonds (11.2)
5. Why π bonding restricts rotation around double bonds (11.2)
6. The distinction between the localized bonding of VB theory and the delocalized bonding of MO theory (11.3)

7. How addition or subtraction of AOs forms bonding or antibonding MOs (11.3)
8. The shapes of MOs formed from combinations of two s orbitals and combinations of two p orbitals (11.3)
9. How MO bond order predicts the stability of molecular species (11.3)
10. How MO theory explains the bonding and properties of homonuclear and heteronuclear diatomic molecules of Period 2 (11.3)

Master These Skills

1. Using molecular shape to postulate the hybrid orbitals used by a central atom (SP 11.1)
2. Describing the types of orbitals and bonds in a molecule (SP 11.2)
3. Drawing MO diagrams, writing electron configurations, and calculating bond orders of molecular species (SP 11.3)
4. Explaining bond properties with MO theory (SP 11.4)

Key Terms These important terms appear in boldface in the chapter and are defined again in the Glossary.

Section 11.1
valence bond (VB) theory (411)
hybridization (412)
hybrid orbital (412)
sp hybrid orbital (412)
sp^2 hybrid orbital (414)

sp^3 hybrid orbital (414)
sp^3d hybrid orbital (415)
sp^3d^2 hybrid orbital (416)

Section 11.2
sigma (σ) bond (419)
pi (π) bond (419)

Section 11.3
molecular orbital (MO) theory (422)
molecular orbital (MO) (422)
bonding MO (422)
antibonding MO (422)
sigma (σ) MO (423)

molecular orbital (MO) diagram (423)
MO bond order (423)
homonuclear diatomic molecule (424)
pi (π) MO (425)
nonbonding MO (429)

Key Equations and Relationships Numbered and screened concepts are listed for you to refer to or memorize.

11.1 Calculating the MO bond order (423):   Bond order $= \frac{1}{2}[$(no. of e^- in bonding MO) $-$ (no. of e^- in antibonding MO)$]$

Highlighted Figures and Tables These figures (F) and tables (T) provide a visual review of key ideas.

F11.2 The sp hybrid orbitals in gaseous $BeCl_2$ (413)
T11.1 Composition and orientation of hybrid orbitals (416)
F11.8 The conceptual steps from molecular formula to hybrid orbitals used in bonding (417)
F11.10 The σ and π bonds in ethylene (C_2H_4) (419)
F11.15 Contours and energies of the bonding and antibonding MOs in H_2 (423)

F11.19 Contours and energies of σ and π MOs from combinations of $2p$ atomic orbitals (425)
F11.21 MO occupancy and molecular properties for B_2 through Ne_2 (427)

Brief Solutions to FOLLOW-UP PROBLEMS Compare your solutions to these calculation steps and answers.

11.1 (a) Shape is linear, so Be is *sp* hybridized.

(b) Shape is tetrahedral, so Si is sp^3 hybridized.

(c) Shape is square planar, so Xe is sp^3d^2 hybridized.

11.2 H—C≡N:

HCN is linear, so C is *sp* hybridized. N is also *sp* hybridized. One *sp* of C overlaps the 1*s* of H to form a σ bond. The other *sp* of C overlaps one *sp* of N to form a σ bond. The other *sp* of N holds a lone pair. Two unhybridized *p* orbitals of N and two of C overlap to form two π bonds.

(b) Ö=C=Ö

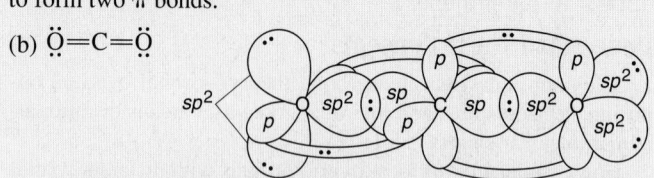

CO_2 is linear, so C is *sp* hybridized. Both O atoms are sp^2 hybridized. Each *sp* of C overlaps one sp^2 of an O to form two σ bonds. Each of the two unhybridized *p* orbitals of C forms a π bond with the unhybridized *p* of one of the two O atoms. Two sp^2 of each O hold lone pairs.

11.3

Does not exist: bond order = $\frac{1}{2}(2-2) = 0$.

11.4 Bond orders: $F_2^{2-} = 0$; $F_2^- = \frac{1}{2}$; $F_2 = 1$; $F_2^+ = 1\frac{1}{2}$
$F_2^{2+} = 2$
Bond energy: $F_2^{2-} < F_2^- < F_2 < F_2^+ < F_2^{2+}$
Bond length: $F_2^{2+} < F_2^+ < F_2 < F_2^-$; F_2^{2-} does not exist

---

PROBLEMS

Problems with **colored** numbers are answered in Appendix E and worked in detail in the Student Solutions Manual. Problem sections match those in the text and provide the numbers of relevant sample problems. Most offer Concept Review Questions, Skill-Building Exercises (grouped in pairs covering the same concept), and Problems in Context. The Comprehensive Problems are based on material from any section or previous chapter.

Valence Bond (VB) Theory and Orbital Hybridization
(Sample Problem 11.1)

● Concept Review Questions

• **11.1** What type of central-atom orbital hybridization corresponds to each electron-group arrangement: (a) trigonal planar; (b) octahedral; (c) linear; (d) tetrahedral; (e) trigonal bipyramidal?

• **11.2** What is the orbital hybridization of a central atom that has one lone pair and bonds to: (a) two other atoms; (b) three other atoms; (c) four other atoms; (d) five other atoms?

• **11.3** How do carbon and silicon differ with regard to the *types* of orbitals available for hybridization? Explain.

• **11.4** How many hybrid orbitals form when four atomic orbitals of a central atom mix? Explain.

● Skill-Building Exercises *(grouped in similar pairs)*

• **11.5** Give the number and type of hybrid orbital that forms when each set of atomic orbitals mixes:
(a) two *d*, one *s*, and three *p*   (b) three *p* and one *s*

• **11.6** Give the number and type of hybrid orbital that forms when each set of atomic orbitals mixes:
(a) one *p* and one *s*   (b) three *p*, one *d*, and one *s*

Levels of Difficulty: • Easy •• Medium ••• Challenging

• **11.7** What is the hybridization of nitrogen in each of the following: (a) NO; (b) NO_2; (c) NO_2^-?

• **11.8** What is the hybridization of carbon in each of the following: (a) CO_3^{2-}; (b) $C_2O_4^{2-}$; (c) NCO^-?

• **11.9** What is the hybridization of chlorine in each of the following: (a) ClO_2; (b) ClO_3^-; (c) ClO_4^-?

• **11.10** What is the hybridization of bromine in each of the following: (a) BrF_3; (b) BrO_2^-; (c) BrF_5?

• **11.11** Which types of atomic orbitals of the central atom mix to form hybrid orbitals in (a) $SiClH_3$; (b) CS_2; (c) SCl_3F; (d) NF_3?

• **11.12** Which types of atomic orbitals of the central atom mix to form hybrid orbitals in (a) Cl_2O; (b) $BrCl_3$; (c) PF_5; (d) SO_3^{2-}?

•• **11.13** Phosphine (PH_3) reacts with borane (BH_3) as follows:
$$PH_3 + BH_3 \longrightarrow H_3P{-}BH_3$$
(a) Which of the illustrations below depicts the change, if any, in the orbital hybridization of P during this reaction? (b) Which depicts the change, if any, in the orbital hybridization of B?

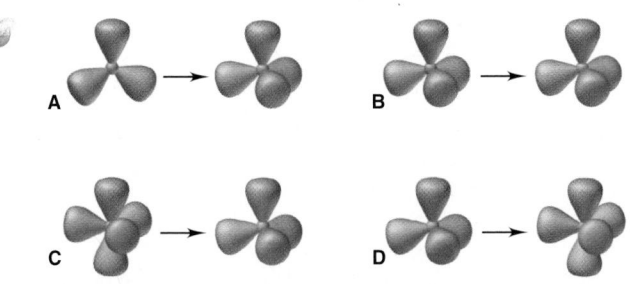

•• **11.14** The illustrations below depict differences in orbital hybridization of some tellurium (Te) fluorides. (a) Which depicts the difference, if any, between TeF_6 (*left*) and TeF_5^- (*right*)? (b) Which depicts the difference, if any, between TeF_4 (*left*) and TeF_6 (*right*)?

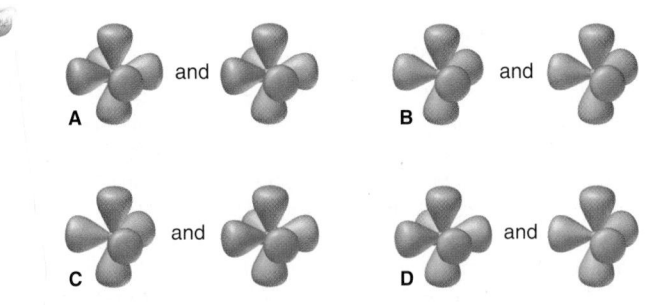

•• **11.15** Use partial orbital diagrams to show how the atomic orbitals of the central atom lead to hybrid orbitals in (a) $GeCl_4$; (b) BCl_3; (c) CH_3^+.

•• **11.16** Use partial orbital diagrams to show how the atomic orbitals of the central atom lead to hybrid orbitals in (a) BF_4^-; (b) PO_4^{3-}; (c) SO_3.

•• **11.17** Use partial orbital diagrams to show how the atomic orbitals of the central atom lead to hybrid orbitals in (a) $SeCl_2$; (b) H_3O^+; (c) IF_4^-.

•• **11.18** Use partial orbital diagrams to show how the atomic orbitals of the central atom lead to hybrid orbitals in (a) $AsCl_3$; (b) $SnCl_2$; (c) PF_6^-.

● Problems in Context

•• **11.19** Methyl isocyanate, $CH_3{-}\ddot{N}{=}C{=}\ddot{O}:$, is an intermediate in the manufacture of many pesticides. It received notoriety in 1984 when a leak from a manufacturing plant resulted in the death of more than 2000 people in Bhopal, India. What are the hybridizations of the N atom and the two C atoms in methyl isocyanate? Sketch the molecular shape.

The Mode of Orbital Overlap and the Types of Covalent Bonds
(Sample Problem 11.2)

● Concept Review Questions

• **11.20** Are these statements true or false? Correct any that are false.
(a) Two σ bonds comprise a double bond.
(b) A triple bond consists of one π bond and two σ bonds.
(c) Bonds formed from atomic *s* orbitals are always σ bonds.
(d) A π bond restricts rotation about the σ-bond axis.
(e) A π bond consists of two pairs of electrons.
(f) End-to-end overlap results in a bond with electron density above and below the bond axis.

● Skill-Building Exercises (grouped in similar pairs)

• **11.21** Describe the hybrid orbitals used by the central atom and the type(s) of bonds formed in (a) NO_3^-; (b) CS_2; (c) CH_2O.

11.22 Describe the hybrid orbitals used by the central atom and the type(s) of bonds formed in (a) O_3; (b) I_3^-; (c) $COCl_2$ (C is central).

• **11.23** Describe the hybrid orbitals used by the central atom(s) and the type(s) of bonds formed in (a) FNO; (b) C_2F_4; (c) $(CN)_2$.

• **11.24** Describe the hybrid orbitals used by the central atom(s) and the type(s) of bonds formed in (a) BrF_3; (b) $CH_3C{\equiv}CH$; (c) SO_2.

● Problems in Context

•• **11.25** 2-Butene ($CH_3CH{=}CHCH_3$) is a starting material in the manufacture of lubricating oils and many other compounds. Draw two different structures for 2-butene, indicating the σ and π bonds in each.

Molecular Orbital (MO) Theory and Electron Delocalization
(Sample Problems 11.3 and 11.4)

● Concept Review Questions

• **11.26** Two *p* orbitals from one atom and two *p* orbitals from another atom are combined to form molecular orbitals for the joined atoms. How many MOs will result from this combination? Explain.

• **11.27** Certain atomic orbitals on two atoms were combined to form the following MOs. Name the atomic orbitals used and the MOs formed, and explain which MO has higher energy:

• **11.28** How do the bonding and antibonding MOs formed from a given pair of AOs compare to each other with respect to (a) energy; (b) presence of nodes; (c) internuclear electron density?

•• **11.29** Antibonding MOs always have at least one node. Can a bonding MO have a node? If so, draw an example.

● **Skill-Building Exercises** *(grouped in similar pairs)*

● **11.30** How many electrons does it take to fill (a) a σ bonding MO; (b) a π antibonding MO; (c) the MOs formed from combination of the 1*s* orbitals of two atoms?

● **11.31** How many electrons does it take to fill (a) the MOs formed from combination of the 2*p* orbitals of two atoms; (b) a σ^*_{2p} MO; (c) the MOs formed from combination of the 2*s* orbitals of two atoms?

●● **11.32** The molecular orbitals depicted below are derived from 2*p* atomic orbitals in F_2^+. (a) Give the orbital designations. (b) Which is occupied by at least one electron in F_2^+? (c) Which is occupied by only one electron in F_2^+?

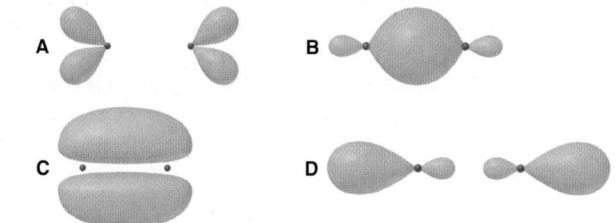

●● **11.33** The molecular orbitals depicted below are derived from *n* = 2 atomic orbitals. (a) Give the orbital designations. (b) Which is highest in energy? (c) Lowest in energy? (d) Rank the MOs in order of increasing energy for B_2.

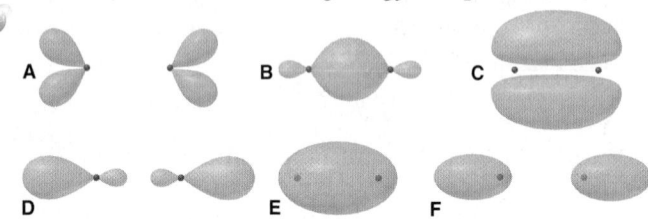

●● **11.34** Use MO diagrams and the bond orders you obtain from them to answer: (a) Is Be_2^+ stable? (b) Is Be_2^+ diamagnetic? (c) What is the outer (valence) electron configuration of Be_2^+?

●● **11.35** Use MO diagrams and the bond orders you obtain from them to answer: (a) Is O_2^- stable? (b) Is O_2^- paramagnetic? (c) What is the outer (valence) electron configuration of O_2^-?

●● **11.36** Use MO diagrams to place C_2^-, C_2, and C_2^+ in order of (a) increasing bond energy; (b) increasing bond length.

●● **11.37** Use MO diagrams to place B_2^+, B_2, and B_2^- in order of (a) decreasing bond energy; (b) decreasing bond length.

Comprehensive Problems

●● **11.38** Predict the shape, state the hybridization of the central atom, and give the ideal bond angle(s) and any expected deviations for:
(a) BrO_3^- (b) $AsCl_4^-$ (c) SeO_4^{2-} (d)BiF_5^{2-}
(e) SbF_4^+ (f) AlF_6^{3-} (g) IF_4^+

●● **11.39** Butadiene *(right)* is a colorless gas used to make synthetic rubber and many other compounds. (a) How many σ bonds and π bonds does the molecule have? (b) Are *cis-trans* arrangements about the double bonds possible? Explain.

●● **11.40** Epinephrine (or adrenaline; *see structure*) is a naturally occurring hormone that is also manufactured commercially for use as a heart stimulant, a nasal decongestant, and to treat glaucoma.

(a) What is the hybridization of each C, O, and N atom? (b) How many σ bonds does the molecule have? (c) How many π electrons are delocalized in the ring?

●● **11.41** Use partial orbital diagrams to show how the atomic orbitals of the central atom lead to the hybrid orbitals in:
(a) IF_2^- (b) ICl_3 (c) $XeOF_4$ (d) BHF_2

●● **11.42** Isoniazid *(below)* is an antibacterial agent that is very useful against many common strains of tuberculosis. (a) How many σ bonds are in the molecule? (b) What is the hybridization of each C and N atom?

●● **11.43** Hydrazine, N_2H_4, and carbon disulfide, CS_2, form a cyclic molecule *(right)*. (a) Draw Lewis structures for N_2H_4 and CS_2. (b) How do electron-group arrangement, molecular shape, and hybridization of N change when N_2H_4 reacts to form the product? (c) How do electron-group arrangement, molecular shape, and hybridization of C change when CS_2 reacts to form the product?

●● **11.44** In each of the following equations, what hybridization change, if any, occurs for the underlined atom?
(a) $\underline{B}F_3 + NaF \longrightarrow Na^+BF_4^-$
(b) $\underline{P}Cl_3 + Cl_2 \longrightarrow PCl_5$
(c) $H\underline{C}{\equiv}CH + H_2 \longrightarrow H_2C{=}CH_2$
(d) $\underline{Si}F_4 + 2F^- \longrightarrow SiF_6^{2-}$
(e) $\underline{S}O_2 + \frac{1}{2}O_2 \longrightarrow SO_3$

●● **11.45** The ionosphere lies about 100 km above Earth's surface. This layer consists mostly of NO, O_2, and N_2, and photoionization creates NO^+, O_2^+, and N_2^+. (a) Use MO theory to compare the bond orders of the molecules and ions. (b) Does the magnetic behavior of each species change when its ion forms?

●● **11.46** Glyphosate *(below)* is a common herbicide that is relatively harmless to animals but deadly to most plants. Describe the shape around and the hybridization of the P, N, and three numbered C atoms.

●● **11.47** The sulfate ion can be represented with four S—O bonds or with two S—O and two S=O bonds. (a) Which representation is better from the standpoint of formal charges? (b) What is the shape of the sulfate ion, and what hybrid orbitals of S are postulated for the σ bonding? (c) In view of the answer to part (b), what orbitals of S must be used for the π bonds? What orbitals of O? (d) Draw a diagram to show how one atomic orbital from S and one from O overlap to form a π bond.

•• **11.48** Tryptophan is one of the amino acids found in proteins:

(a) What is the hybridization of each of the numbered C, N, and O atoms? (b) How many σ bonds are present in tryptophan? (c) Predict the bond angles at points a, b, and c.

•• **11.49** Sulfur forms oxides, oxoanions, and halides. What is the hybridization of the central S in SO_2, SO_3, SO_3^{2-}, SCl_4, SCl_6, and S_2Cl_2 (atom sequence Cl—S—S—Cl)?

••• **11.50** The hydrocarbon allene, $H_2C=C=CH_2$, is obtained indirectly from petroleum and used as a precursor for several types of plastics. What is the hybridization of each C atom in allene? Draw a bonding picture for allene with lines for σ bonds, and show the arrangement of the π bonds. Be sure to represent the geometry of the molecule in three dimensions.

•• **11.51** Some species with two oxygen atoms only are the oxygen molecule, O_2, the peroxide ion, O_2^{2-}, the superoxide ion, O_2^{-}, and the dioxygenyl ion, O_2^{+}. Draw an MO diagram for each, rank them in order of increasing bond length, and find the number of unpaired electrons in each.

••• **11.52** Molecular nitrogen, carbon monoxide, and cyanide ion are isoelectronic. (a) Draw an MO diagram for each. (b) CO and CN^- are toxic. What property may explain why N_2 isn't?

•• **11.53** There is concern in health-related government agencies that the American diet contains too much meat, and numerous recommendations have been made urging people to consume more fruit and vegetables. One of the richest sources of vegetable protein is soy, available in many forms. Among these is soybean curd, or tofu, which is a staple of many Asian diets. Chemists have isolated an anticancer agent called *genistein* from tofu, which may explain the much lower incidence of cancer among people in the Far East. A valid Lewis structure for genistein is

(a) Is the hybridization of each C in the right-hand ring the same? Explain. (b) Is the hybridization of the O atom in the center ring the same as that of the O atoms in OH groups? Explain. (c) How many carbon-oxygen σ bonds are there? How many carbon-oxygen π bonds? (d) Do all the lone pairs on oxygens occupy the same type of hybrid orbital? Explain.

••• **11.54** An organic chemist synthesizes the molecule below:

(a) Which of the orientations of hybrid orbitals shown below are present in the molecule? (b) Are there any present that are not shown below? If so, what are they? (c) How many of each type of hybrid orbital are present?

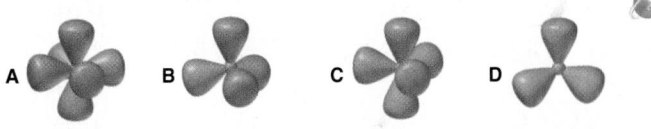

A   **B**   **C**   **D**

•• **11.55** Silicon tetrafluoride reacts with F^- to produce the hexafluorosilicate ion, SiF_6^{2-}; GeF_4 behaves similarly, but CF_4 does not. (a) Draw Lewis structures for SiF_4, GeF_6^{2-}, and CF_4. (b) What is the hybridization of the central atom in each species? (c) Why doesn't CF_4 react with F^- to form CF_6^{2-}?

•• **11.56** Simple proteins consist of amino acids linked together in a long chain; a small portion of such a chain is

Experiment shows that rotation about the C—N bond (indicated by the arrow) is somewhat restricted. Explain with resonance structures, and show the types of bonding involved.

•• **11.57** The compound 2,6-dimethylpyrazine *(below)* gives chocolate its odor and is used in flavorings. (a) Which atomic orbitals mix to form the hybrid orbitals of N? (b) In what type of hybrid orbital do the lone pairs of N reside? (c) Is C in CH_3 hybridized the same as any C in the ring? Explain.

•• **11.58** Acetylsalicylic acid (aspirin), the most widely used medicine in the world, has the Lewis structure at right. (a) What is the hybridization of each C and each O atom? (b) How many localized π bonds are present? (c) How many C atoms have a trigonal planar shape around them? A tetrahedral shape?

••• **11.59** Linoleic acid is an essential fatty acid found in many vegetable oils, such as soy, peanut, and cottonseed. A key structural feature of the molecule is the *cis* orientation around its two double bonds, where R_1 and R_2 represent two different groups that form the rest of the molecule.

(a) How many different compounds are possible, changing only the *cis-trans* arrangements around these two double bonds? (b) How many are possible for a similar compound with three double bonds?

Biological: 11.48, 11.53, 11.56, 11.58, 11.59; Organic: 11.50, 11.54, 11.57

12

All Together The three phases of water appear together in this scene of a misty, snowbound, hot spring in Yellowstone National Park, Wyoming. In this chapter, we discuss the forces among molecules that give rise to the three phases and to the physical changes they undergo.

Intermolecular Forces: Liquids, Solids, and Phase Changes

12.1 An Overview of Physical States and Phase Changes

12.2 Quantitative Aspects of Phase Changes
Heat Involved in Phase Changes
Equilibrium Nature of Phase Changes
Phase Diagrams

12.3 Types of Intermolecular Forces
Ion-Dipole Forces
Dipole-Dipole Forces
The Hydrogen Bond
Charge-Induced Dipole Forces
Dispersion (London) Forces

12.4 Properties of the Liquid State
Surface Tension
Capillarity
Viscosity

12.5 The Uniqueness of Water
Solvent Properties
Thermal Properties
Surface Properties
Density of Solid and Liquid Water

12.6 The Solid State: Structure, Properties, and Bonding
Structural Features of Solids
Crystalline Solids
Amorphous Solids
Bonding in Solids

12.7 Advanced Materials
Electronic Materials
Liquid Crystals
Ceramic Materials
Polymeric Materials
Nanotechnology

All the matter in and around you occurs in one or more of the three physical states—gas, liquid, or solid. Under certain conditions, many pure substances can occur in any of the states. We're all familiar with the three states of water: we inhale and exhale gaseous water; drink, excrete, and wash with liquid water; and shovel, slide on, or cool our drinks with solid water. Look around and you'll find other examples of solids (jewelry, table salt), liquids (gasoline, antifreeze), and gases (air, CO_2 bubbles) that can exist in one or both of the other states, even though the pressures and/or temperatures needed may be unusual.

The three states were introduced in Chapter 1 and their properties compared when we examined gases in Chapter 5; now we turn our attention to liquids and solids. A physical state is one type of **phase,** any physically distinct, homogeneous part of a system. The water in a glass constitutes a single phase. Add some ice and you have two phases; or, if there are air bubbles in the ice, you have three.

In all three phases, electrostatic forces among the particles, called *interparticle forces* or, more commonly, **intermolecular forces,** combine with the particles' kinetic energy to create the properties of each phase as well as **phase changes,** the changes from one phase to another. Liquids and solids are called *condensed* phases (or condensed states) because their particles are extremely close together. Although intermolecular forces have relatively little effect in gases because the particles are so far apart, they have a large effect in liquids and solids.

IN THIS CHAPTER . . . We begin with a kinetic-molecular overview of the states and then take a similarly close-up view of phase changes. We calculate the energy involved in phase changes and highlight the effects of temperature and pressure with *phase diagrams.* We then consider the types of intermolecular forces and apply them first to the properties of liquids. As a special case, we see how the unique physical properties of water arise inevitably from the electron configurations of its atoms. Then we discuss the properties of solids, emphasizing the relationship between type of bonding and intermolecular force. You will learn how the atomic-scale structures of solids are determined and, in the final section, how some exciting advanced materials—semiconductors, liquid crystals, ceramics, polymers, and nanostructures—are produced and used.

Concepts & Skills to Review
before you study this chapter

- properties of gases, liquids, and solids (Section 5.1)
- kinetic-molecular theory of gases (Section 5.6)
- kinetic and potential energy (Section 6.1)
- enthalpy change, heat capacity, and Hess's law (Sections 6.2, 6.3, and 6.5)
- diffraction of light (Section 7.1)
- Coulomb's law (Section 9.2)
- chemical bonding models (Chapter 9)
- molecular polarity (Section 10.3)
- molecular orbital treatment of diatomic molecules (Section 11.3)

12.1 AN OVERVIEW OF PHYSICAL STATES AND PHASE CHANGES

Imagine yourself among the particles of a molecular substance, such as solid iodine, I_2, and you'll discover two types of electrostatic forces at work:

- *Intra*molecular forces (bonding forces) exist *within* each molecule and influence the *chemical* properties of the substance.
- *Inter*molecular forces (nonbonding forces) exist *between* the molecules and influence the *physical* properties of the substance.

Put another way, if you magnified the three states of water a billionfold and focused on just one molecule from each state, they would be identical bent, polar H—O—H molecules. The *chemical* behavior of the three states *is* identical because their molecules are held together by the same *intra*molecular (bonding) forces. But, the *physical* behavior of the states differs greatly because the strengths of the *inter*molecular (nonbonding) forces differ greatly.

A Kinetic-Molecular View of the Three States
Whether a substance is a gas, liquid, or solid depends on the interplay of the potential energy of the intermolecular attractions, which tends to draw the molecules together, and the kinetic energy of the molecules, which tends to disperse them. According to Coulomb's law, the potential energy depends on the charges of the particles and the distances between them (see Section 9.2). The average kinetic energy, which is related to the particles' average speed, is proportional to the absolute temperature.

Table 12.1	A Macroscopic Comparison of Gases, Liquids, and Solids		
State	**Shape and Volume**	**Compressibility**	**Ability to Flow**
Gas	Conforms to shape and volume of container	High	High
Liquid	Conforms to shape of container; volume limited by surface	Very low	Moderate
Solid	Maintains its own shape and volume	Almost none	Almost none

● **Environmental Flow** The environment demonstrates beautifully the differences in the abilities of the three states to flow and diffuse. Atmospheric gases mix so well that the 80 km of air closest to Earth's surface has a uniform composition. Much less mixing occurs in the oceans, and differences in composition at various depths support different species. Rocky solids (see photo) intermingle so little that adjacent strata remain separated for millions of years.

Environmental application

Literature Ramasami, Ponnadurai. "Students as Solids, Liquids, and Gases," *J. Chem. Educ.* 2000, *77*, 485.

In Table 12.1, we distinguish among the three states by focusing on three properties—shape, compressibility, and ability to flow. The following kinetic-molecular view of the three states is an extension of the model we used to understand gases (see Section 5.6):

- *In a gas,* the energy of attraction is small relative to the energy of motion; so, on average, the particles are far apart. This large interparticle distance has several macroscopic consequences. A gas moves randomly throughout its container and fills it. Gases are highly compressible, and they flow and diffuse easily through one another.
- *In a liquid,* the attractions are stronger because the particles are in virtual contact. But their kinetic energy still allows them to tumble randomly over and around each other. Therefore, a liquid conforms to the shape of its container but has a surface. With very little free space between the particles, liquids resist an applied external force and thus compress only very slightly. They flow and diffuse but *much* more slowly than gases.
- *In a solid,* the attractions dominate the motion so much that the particles remain in position relative to one another, jiggling in place. With the positions of the particles fixed, a solid has a specific shape and does not flow significantly. The particles are usually slightly closer together than in a liquid, so solids compress even less than liquids. ●

Types of Phase Changes Phase changes are also determined by the interplay between kinetic energy and intermolecular forces. As the temperature increases, the average kinetic energy increases as well, so the faster moving particles can overcome attractions more easily; conversely, lower temperatures allow the forces to draw the slower moving particles together.

What happens when gaseous water is cooled? A mist appears as the particles form tiny microdroplets that then collect into a bulk sample of liquid with a single surface. The process by which a gas changes into a liquid is called **condensation;** the opposite process, changing from a liquid into a gas, is called **vaporization.** With further cooling, the particles move even more slowly and become fixed in position as the liquid solidifies in the process of **freezing;** the opposite change is called **melting,** or **fusion.** In common speech, the term *freezing* implies low temperature because we typically think of water. But, most metals, for example, freeze (solidify) at much higher temperatures, and this change has many medical, industrial, and artistic applications: gold dental crowns, steel auto bodies, bronze statues, and so forth.

Enthalpy changes accompany phase changes. As the molecules of a gas attract each other and come closer together in the liquid, and then become fixed in the solid, the system of particles loses energy, which is released as heat. Thus, *condensing and freezing are exothermic changes.* On the other hand, energy must be absorbed by the system to overcome the attractive forces that keep the particles in a liquid together and those that keep them fixed in place in a solid. Thus, *melting and vaporizing are endothermic changes.* ●

For a pure substance, each phase change has a specific enthalpy change *per mole* (measured at 1 atm and the temperature of the change). For vaporization, it is called the **heat of vaporization ($\Delta H°_{vap}$),** and for fusion, it is the **heat of fusion ($\Delta H°_{fus}$).** In the case of water, we have

$$H_2O(l) \longrightarrow H_2O(g) \qquad \Delta H = \Delta H°_{vap} = 40.7 \text{ kJ/mol (at } 100°C)$$
$$H_2O(s) \longrightarrow H_2O(l) \qquad \Delta H = \Delta H°_{fus} = 6.02 \text{ kJ/mol (at } 0°C)$$

The reverse processes, condensing and freezing, have enthalpy changes of the *same magnitude but opposite sign:*

$$H_2O(g) \longrightarrow H_2O(l) \qquad \Delta H = -\Delta H°_{vap} = -40.7 \text{ kJ/mol}$$
$$H_2O(l) \longrightarrow H_2O(s) \qquad \Delta H = -\Delta H°_{fus} = -6.02 \text{ kJ/mol}$$

In this regard, water behaves typically in that it takes much less energy to melt 1 mol of the solid than to vaporize 1 mol of the liquid: $\Delta H°_{fus} < \Delta H°_{vap}$. Figure 12.1 shows several examples. The reason is that a phase change is essentially a change in intermolecular distance and freedom of motion. Less energy is needed to overcome the forces holding the molecules in fixed positions (melt a solid) than to separate them completely from each other (vaporize a liquid).

The three states of water are so common because they are stable under ordinary conditions. Carbon dioxide, on the other hand, is familiar as a gas and a solid (dry ice), but liquid CO_2 occurs only at external pressures greater than 5 atm. At ordinary conditions, solid CO_2 becomes a gas without first becoming a liquid. This process is called **sublimation.** On a clear wintry day, you can dry clothes outside on a line, even though it may be too cold for ice to melt, because the ice sublimes. Freeze-dried foods are prepared by sublimation. The opposite process, changing from a gas directly into a solid, is called **deposition**—you may have seen ice crystals form on a cold window from the deposition of water vapor. The **heat of sublimation ($\Delta H°_{subl}$)** is the enthalpy change when 1 mol of a substance sublimes. From Hess's law (Section 6.5), it equals the sum of the heats of fusion and vaporization:

Solid $\longrightarrow$ liquid	$\Delta H°_{fus}$
Liquid $\longrightarrow$ gas	$\Delta H°_{vap}$
Solid $\longrightarrow$ gas	$\Delta H°_{subl}$

● **A Cooling Phase Change** The evaporation of sweat has a cooling effect because heat from your body is used to vaporize the water. Many fur-covered animals achieve this cooling effect, too. Cats (and many rodents) lick themselves, transferring water from inside their bodies to the surface where it can evaporate, and dogs just open their mouths, hang out their tongues, and pant!

ARIS Presentation Center, Photo Library

Biological application

Demo Shakhashiri, Bassam Z. "Evaporation as an Endothermic Process," *Chemical Demonstrations, A Handbook for Teachers of Chemistry,* Vol. 3 (The University of Wisconsin Press, Madison, 1985), 249–251.

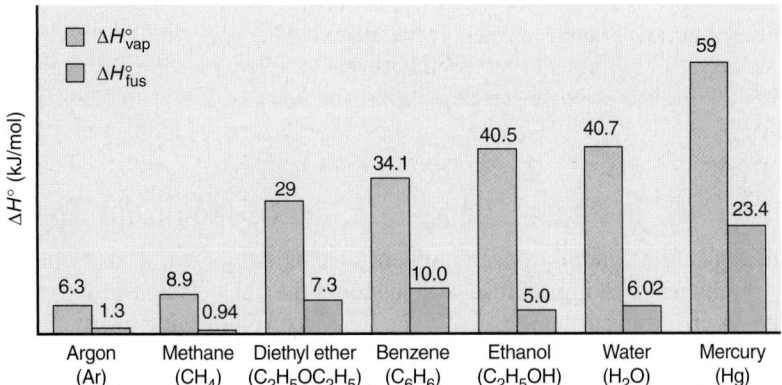

Figure 12.1 Heats of vaporization and fusion for several common substances. $\Delta H°_{vap}$ is always larger than $\Delta H°_{fus}$ because it takes more energy to separate particles completely than just to free them from their fixed positions in the solid.

ARIS Presentation Center, Art Library

Figure 12.2 **Phase changes and their enthalpy changes.** Each type of phase change is shown with its associated enthalpy change. Fusion (or melting), vaporization, and sublimation are endothermic changes (positive $\Delta H°$), whereas freezing, condensation, and deposition are exothermic changes (negative $\Delta H°$).

Transparency/ARIS Presentation Center, Art Library

Animation: Vapor Pressure
ARIS Presentation Center
Chapter 12

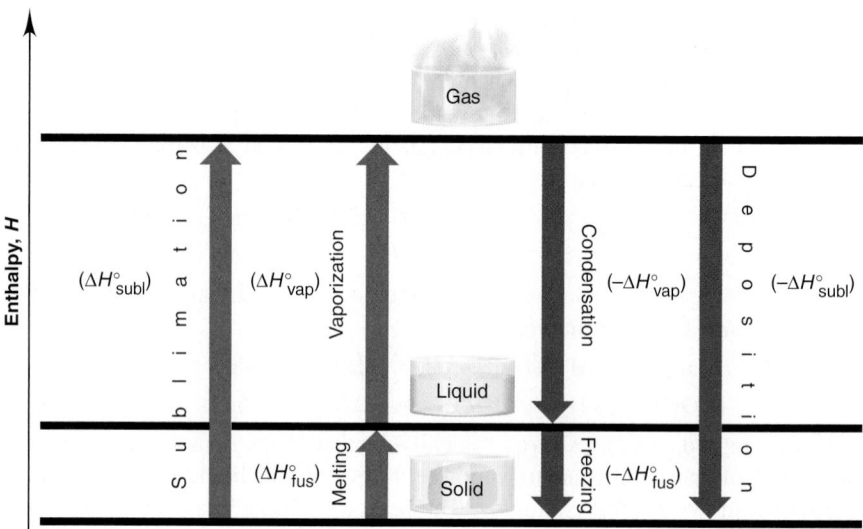

Figure 12.2 summarizes the terminology of the various phase changes and shows the enthalpy changes associated with them.

Section Summary

Because of the relative magnitudes of intermolecular forces and kinetic energy, the particles in a gas are far apart and moving randomly, those in a liquid are in contact but still moving relative to each other, and those in a solid are in contact and fixed relative to one another in a rigid structure. These molecular-level differences in the states of matter account for macroscopic differences in shape, compressibility, and ability to flow. • When a solid becomes a liquid (melting, or fusion) or a liquid becomes a gas (vaporization), energy is absorbed to overcome intermolecular forces and increase the average distance between particles. When particles come closer together in the reverse changes (freezing and condensation), energy is released. Sublimation is the changing of a solid directly into a gas. Each phase change is associated with a given enthalpy change under specified conditions.

12.2 QUANTITATIVE ASPECTS OF PHASE CHANGES

Some of the most spectacular, and familiar, phase changes occur in the weather each day. When it rains, water vapor condenses to a liquid, which changes back to a gas as puddles dry up. In the spring, snow melts and streams fill; in winter, water freezes and falls to earth. These remarkable transformations also take place whenever you make a pot of tea or a tray of ice cubes. In this section, we examine the heat absorbed or released in a phase change and the equilibrium nature of the process.

Heat Involved in Phase Changes: A Kinetic-Molecular Approach

We can apply the kinetic-molecular theory quantitatively to phase changes by means of a **heating-cooling curve,** which shows the changes that occur when heat is added to or removed from a particular sample of matter at a constant rate. As an example, the cooling process is depicted in Figure 12.3 for a 2.50-mol sample of gaseous water in a closed container, with the pressure kept at 1 atm and the temperature changing from 130°C to −40°C. To an observer, the process is continuous, but we can divide it into five heat-releasing (exothermic) stages that correspond to the five portions of the curve—the gas cools, it condenses to a liquid, the liquid cools, it freezes to a solid, and the solid cools a bit further:

Figure 12.3 **A cooling curve for the conversion of gaseous water to ice.** A plot is shown of temperature vs. heat removed as gaseous water at 130°C changes to ice at −40°C. This process occurs in five stages, with a molecular-level depiction shown for each stage. Stage 1: Gaseous water cools. Stage 2: Gaseous water condenses. Stage 3: Liquid water cools. Stage 4: Liquid water freezes. Stage 5: Solid water

cools. The slopes of the lines in stages 1, 3, and 5 reflect the magnitudes of the molar heat capacities of the phases. Although not drawn to scale, the line in stage 2 is longer than the line in stage 4 because ΔH°_{vap} of water is greater than ΔH°_{fus}. A plot of temperature vs. heat added would have the same steps but in reverse order.

Transparency/ARIS Presentation Center, Active Art Library

Stage 1. Gaseous water cools. Picture a collection of water molecules behaving like a typical gas: zooming around chaotically at a range of speeds, smashing into each other and the container walls. At a high enough temperature, the most probable speed, and thus the average kinetic energy (E_k), of the molecules is high enough to overcome the potential energy (E_p) of attractions among them. As the temperature falls, the average E_k decreases, so the attractions become increasingly important.

The change is $H_2O(g)$ [130°C] $\longrightarrow$ $H_2O(g)$ [100°C]. The heat (q) is the product of the amount (number of moles, n) of water, the molar heat capacity of *gaseous water*, $C_{water(g)}$, and the temperature change, ΔT ($T_{final} - T_{initial}$):

$$q = n \times C_{water(g)} \times \Delta T = (2.50 \text{ mol}) (33.1 \text{ J/mol·°C}) (100°C - 130°C)$$
$$= -2482 \text{ J} = -2.48 \text{ kJ}$$

The minus sign indicates that heat is released. (For purposes of canceling, the units for molar heat capacity, C, include °C, rather than K. The magnitude of C is not affected because the kelvin and the Celsius degree represent the same temperature increment.)

Stage 2. Gaseous water condenses. At the condensation point, the slowest of the molecules are near each other long enough for intermolecular attractions to form groups of molecules, which aggregate into microdroplets and then a bulk liquid. Note that while the state is changing from gas to liquid, *the temperature remains constant,* so the average E_k is constant. This means that, even though molecules in a gas certainly move *farther* between collisions than those in a liquid, their average *speed* is the same at a given temperature. Thus, removing heat from the system results in a decrease in the average E_p, as the molecules approach and attract each other more strongly, but not a decrease in the average E_k. In other words, at 100°C, gaseous water and liquid water have the same average E_k, but the liquid has lower E_p.

The change is $H_2O(g)\ [100°C] \longrightarrow H_2O(l)\ [100°C]$. The heat released is the amount (n) times the negative of the heat of vaporization ($-\Delta H°_{vap}$):

$$q = n(-\Delta H°_{vap}) = (2.50\ mol)\ (-40.7\ kJ/mol) = -102\ kJ$$

This step contributes the *greatest portion of the total heat released* because of the decrease in potential energy that occurs with the enormous decrease in distance between molecules in a gas and those in a liquid.

Stage 3. Liquid water cools. The molecules have now condensed to the liquid state. The continued loss of heat appears as a decrease in temperature, that is, as a decrease in the most probable molecular speed and, thus, the average E_k. The temperature decreases as long as the sample remains liquid.

The change is $H_2O(l)\ [100°C] \longrightarrow H_2O(l)\ [0°C]$. The heat depends on amount (n), the molar heat capacity of *liquid water,* and ΔT:

$$q = n \times C_{water(l)} \times \Delta T = (2.50\ mol)\ (75.4\ J/mol·°C)\ (0°C - 100°C)$$
$$= -18,850\ J = -18.8\ kJ$$

Stage 4. Liquid water freezes. At the freezing temperature of water, 0°C, intermolecular attractions overcome the motion of the molecules around one another. Beginning with the slowest, the molecules lose E_p and align themselves into the crystalline structure of ice. Molecular motion continues, but only as vibration of atoms about their fixed positions. As during condensation, the temperature and average E_k remain constant during freezing.

The change is $H_2O(l)\ [0°C] \longrightarrow H_2O(s)\ [0°C]$. The heat released is n times the negative of the heat of fusion ($-\Delta H°_{fus}$):

$$q = n(-\Delta H°_{fus}) = (2.50\ mol)\ (-6.02\ kJ/mol) = -15.0\ kJ$$

Stage 5. Solid water cools. With motion restricted to jiggling in place, further cooling merely reduces the average speed of this jiggling.

The change is $H_2O(s)\ [0°C] \longrightarrow H_2O(s)\ [-40°C]$. The heat released depends on n, the molar heat capacity of *solid water,* and ΔT:

$$q = n \times C_{water(s)} \times \Delta T = (2.50\ mol)\ (37.6\ J/mol·°C)\ (-40°C - 0°C)$$
$$= -3760\ J = -3.76\ kJ$$

According to Hess's law, the total heat released is the sum of the heats released for the individual stages. The sum of q for stages 1 to 5 is $-142\ kJ$.

Two key points stand out in this or any similar process (at constant pressure), whether exothermic or endothermic:

- *Within a phase,* a change in heat is accompanied by *a change in temperature,* which is associated with a change in average E_k as *the most probable speed of the molecules changes.* The heat lost or gained depends on the amount of substance, the molar heat capacity for that phase, and the change in temperature.
- *During a phase change,* a change in heat occurs at a *constant temperature,* which is associated with a change in E_p, as *the average distance between molecules changes.* Both physical states are present during a phase change. The heat lost or gained depends on the amount of substance and the enthalpy of the phase change ($\Delta H°_{vap}$ or $\Delta H°_{fus}$).

Let's work a molecular-scene sample problem to make sure these ideas are clear.

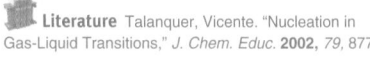
Literature Gesser, H. D. "The Bobbing (Drinking) Bird," *J. Chem. Educ.* **1996,** *73,* 355.

Literature Talanquer, Vicente. "Nucleation in Gas-Liquid Transitions," *J. Chem. Educ.* **2002,** *79,* 877.

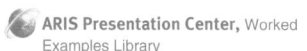

SAMPLE PROBLEM 12.1 Finding the Heat of a Phase Change Depicted by Molecular Scenes

PROBLEM The molecular scenes below represent a phase change of water. Select data from the previous text discussion to find the heat (in kJ) lost or gained when 24.3 g of H_2O undergoes this change.

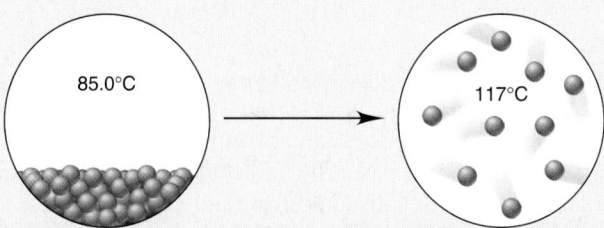

PLAN From the molecular scenes, data in the text, and the given mass (24.3 g) of water, we have to find the heat of the phase change. The scenes show a disorderly, condensed phase at 85°C changing to separate molecules at 117°C. Thus, they represent the vaporization of liquid water. The relevant text data are given per mole, so we first convert the mass (g) of water to amount (mol). There are three endothermic stages (see margin): (1) heating the liquid from 85.0°C to 100.°C, (2) converting the liquid at 100.°C to a gas at 100.°C, and (3) heating the gas from 100.°C to 117°C. We add the values of q for these stages to obtain the total heat of the phase change.

SOLUTION Converting from mass (g) of H_2O to amount (mol):

$$\text{Amount (mol) of } H_2O = 24.3 \text{ g } H_2O \times \frac{1 \text{ mol}}{18.02 \text{ g } H_2O} = 1.35 \text{ mol}$$

Calculating the heat of stage 1, $H_2O(l)$ [85.0°C] ⟶ $H_2O(l)$ [100.°C]:

$$q = n \times C_{water(l)} \times \Delta T = (1.35 \text{ mol}) (75.4 \text{ J/mol·°C}) (100.°C - 85.0°C)$$
$$= 1527 \text{ J} = 1.53 \text{ kJ}$$

Calculating the heat of stage 2, $H_2O(l)$ [100.°C] ⟶ $H_2O(g)$ [100.°C]:

$$q = n(\Delta H^{\circ}_{vap}) = (1.35 \text{ mol}) (40.7 \text{ kJ/mol}) = 54.9 \text{ kJ}$$

Calculating the heat of stage 3, $H_2O(g)$ [100.°C] ⟶ $H_2O(g)$ [117°C]:

$$q = n \times C_{water(g)} \times \Delta T = (1.35 \text{ mol}) (33.1 \text{ J/mol·°C}) (117°C - 100.°C)$$
$$= 759.6 \text{ J} = 0.760 \text{ kJ}$$

Adding the stages together to find the total heat of the phase change:

$$\text{Total heat (kJ)} = 1.53 \text{ kJ} + 54.9 \text{ kJ} + 0.760 \text{ kJ} = \boxed{57.2 \text{ kJ}}$$

CHECK It makes sense that the answer is a positive value because heat is absorbed by the water. Be sure to round off to quickly check each value of q; for example, in stage 1, 1.33 mol × 75 J/mol·°C × 15°C = 1500 J. Note that the phase change itself (stage 2) requires the most energy and, thus, dominates the final answer. The ΔH°_{vap} units include kJ, whereas the molar heat capacity units include J, which is a thousandth as large.

FOLLOW-UP PROBLEM 12.1 The molecular scenes below represent a phase change of water. Select data from the previous text discussion to find the heat (in kJ) lost or gained when 2.25 mol of H_2O undergoes this change.

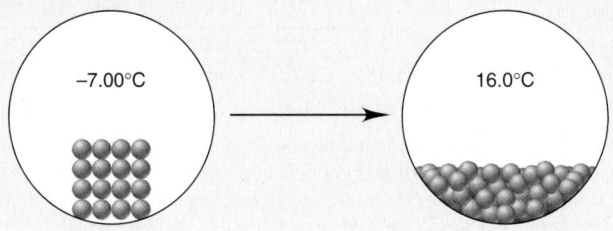

The Equilibrium Nature of Phase Changes

In everyday experience, phase changes take place in open containers—the outdoors, a pot on a stove, the freezer compartment of a refrigerator—so they are not reversible. In a closed container under controlled conditions, however, *phase changes of many substances are reversible and reach equilibrium,* just as chemical changes do.

Liquid-Gas Equilibria Picture an *open* flask containing a pure liquid at constant temperature and focus on the molecules at the surface. Within their range of molecular speeds, some are moving fast enough and in the right direction to overcome attractions, so they vaporize. Nearby molecules immediately fill the gap, and with energy supplied by the constant-temperature surroundings, the process continues until the entire liquid phase is gone.

Now picture starting with a *closed* flask at constant temperature, as in Figure 12.4A, and assume that a vacuum exists above the liquid. As before, some of the molecules at the surface have a high enough E_k to vaporize. As the number of molecules in the vapor phase increases, the pressure of the vapor increases. At the same time, some of the molecules in the vapor that collide with the surface have a low enough E_k to become attracted too strongly to leave the liquid and they condense. For a given surface area, the number of molecules that make up the surface is constant; therefore, the rate of vaporization—the number of molecules leaving the surface per unit time—is also constant. On the other hand, as the vapor becomes more populated, molecules collide with the surface more often, so the rate of condensation slowly increases. As condensation continues to offset vaporization, the increase in the pressure of the vapor slows. Eventually, the rate of condensation equals the rate of vaporization, as depicted in Figure 12.4B. From this time onward, *the pressure of the vapor is constant at that temperature.* Macroscopically, the situation seems static, but at the molecular level, molecules are entering and leaving the liquid surface at equal rates. The system has reached a state of *dynamic equilibrium:*

$$\text{Liquid} \rightleftharpoons \text{gas}$$

Figure 12.4C depicts the entire process graphically.

A Molecules in liquid vaporize.

B Molecules enter and leave liquid at same rate.

C

Figure 12.4 **Liquid-gas equilibrium. A,** In a closed flask at constant temperature with the air removed, the initial pressure is zero. As molecules leave the surface and enter the space above the liquid, the pressure rises. **B,** At equilibrium, the same number of molecules leave as enter the liquid within a given time, so the pressure of the vapor reaches a constant value. **C,** A plot of pressure vs. time shows that the pressure of the vapor increases as long as the rate of vaporization is greater than the rate of condensation. At equilibrium, the rates are equal, so the pressure is constant. The pressure at this point is the *vapor pressure* of the liquid at that temperature.

The pressure exerted by the vapor at equilibrium is called the *equilibrium vapor pressure*, or just the **vapor pressure,** of the liquid at that temperature. If we use a larger flask, more molecules are present in the gas phase at equilibrium; as long as some liquid remains, however, the vapor pressure does not change. Suppose we disturb this equilibrium system by pumping out some of the vapor at constant temperature, thereby lowering the pressure. (If this were a cylinder with piston, we could lower the pressure by moving the piston outward to increase the volume.) The rate of condensation temporarily falls below the rate of vaporization (the forward process is faster). Fewer molecules re-enter the liquid than leave it, so the pressure of the vapor rises until, after a short period, the condensation rate increases enough for equilibrium to be reached again. Similarly, if we disturb the system by pumping in additional vapor (or moving the piston inward to decrease the volume) at constant temperature, thereby raising the pressure, the rate of condensation temporarily exceeds the rate of vaporization. More molecules enter the liquid than leave it (the backward process is faster), but soon the condensation rate decreases, and the pressure again reaches the equilibrium value. This behavior of a pure liquid in contact with its vapor is a general one for any system: *when a system at equilibrium is disturbed, it counteracts the disturbance and eventually re-establishes a state of equilibrium.* We'll return to this key idea often in later chapters.

The Effects of Temperature and Intermolecular Forces on Vapor Pressure The

vapor pressure of a substance depends on the temperature. Raising the temperature of a liquid *increases* the fraction of molecules moving fast enough to escape the liquid and *decreases* the fraction moving slowly enough to be recaptured. This important idea is shown in Figure 12.5. In general, *the higher the temperature is, the higher the vapor pressure.*

Demo Silverstein, Todd P. "Why Do Alcoholic Beverages Have 'Legs' ?" *J. Chem. Educ.* **1998,** *75,* 723.

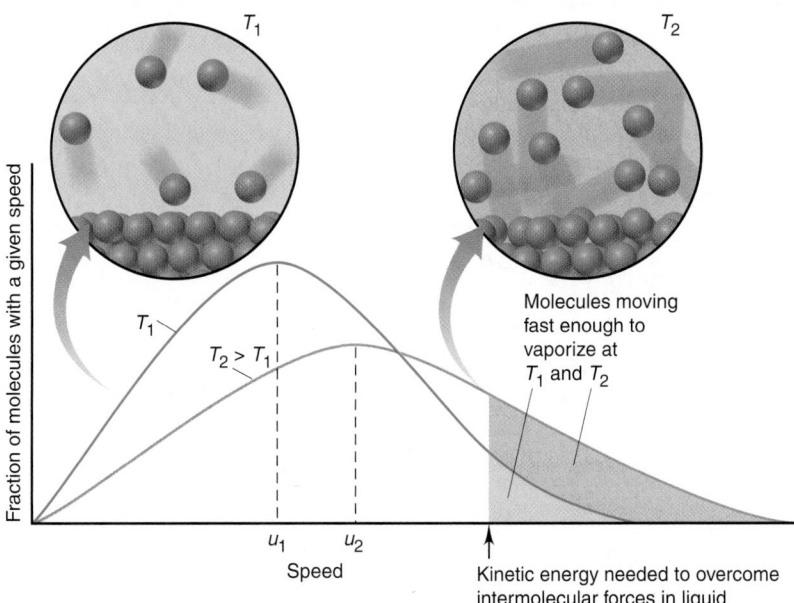

Figure 12.5 **The effect of temperature on the distribution of molecular speeds in a liquid.** With T_1 lower than T_2, the most probable molecular speed u_1 is less than u_2. (Note the similarity to Figure 5.14, p. 211.) The fraction of molecules with enough energy to escape the liquid *(shaded area)* is *greater at the higher temperature.* The molecular views show that at the higher T, equilibrium is reached with more gas molecules in the same volume and thus at a higher vapor pressure.

Transparency/ARIS Presentation Center, Art Library

Figure 12.6 Vapor pressure as a function of temperature and intermolecular forces. At any temperature (vertical dashed line at 20°C), diethyl ether has the highest vapor pressure and water the lowest, because diethyl ether has the weakest intermolecular forces and water the strongest. The horizontal dashed line at 760 torr shows the normal boiling point of each liquid, the temperature at which the vapor pressure equals atmospheric pressure at sea level.

Transparency/ARIS Presentation Center, Art Library

ARIS Presentation Center, Worked Examples Library

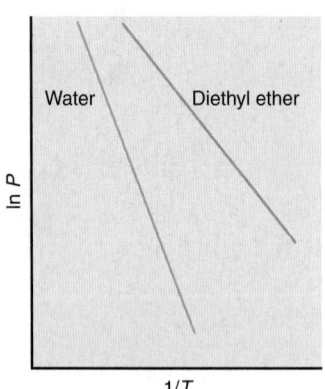

Figure 12.7 A linear plot of the relationship between vapor pressure and temperature. The Clausius-Clapeyron equation gives a straight line when the natural logarithm of the vapor pressure (ln P) is plotted against 1/T. The slope ($-\Delta H_{vap}/R$) allows determination of the ΔH_{vap} of each liquid. The slope is steeper for water because its ΔH_{vap} is greater.

ARIS Presentation Center, Art Library

The vapor pressure also depends on the intermolecular forces present. The average E_k is the same for different substances at a given temperature. Therefore, molecules with weaker intermolecular forces vaporize more easily. In general, *the weaker the intermolecular forces are, the higher the vapor pressure.*

Figure 12.6 shows the vapor pressure of three liquids as a function of temperature. Notice that each curve rises more steeply as the temperature increases. At a given temperature, the substance with the weakest intermolecular forces has the highest vapor pressure; thus, the intermolecular forces in diethyl ether are weaker than those in ethanol, which are weaker than those in water.

The nonlinear relationship between vapor pressure and temperature shown in Figure 12.6 can be expressed as a linear relationship between ln P and 1/T:

$$\ln P = \frac{-\Delta H_{vap}}{R}\left(\frac{1}{T}\right) + C$$
$$y \quad = \quad m \quad\ x\ \ + \ b$$

where ln P is the natural logarithm of the vapor pressure, ΔH_{vap} is the heat of vaporization, R is the universal gas constant (8.314 J/mol·K), T is the absolute temperature, and C is a constant (not related to heat capacity). This is the **Clausius-Clapeyron equation,** which gives us a way of finding the heat of vaporization, the energy needed to vaporize 1 mol of molecules in the liquid state. The blue equation beneath the Clausius-Clapeyron equation is the equation for a straight line, where $y = \ln P$, $x = 1/T$, m (the slope) $= -\Delta H_{vap}/R$, and b (the y-axis intercept) $= C$. Figure 12.7 shows plots for diethyl ether and water. A two-point version of the equation allows a nongraphical determination of ΔH_{vap}:

$$\ln \frac{P_2}{P_1} = \frac{-\Delta H_{vap}}{R}\left(\frac{1}{T_2} - \frac{1}{T_1}\right) \tag{12.1}$$

If ΔH_{vap} and P_1 at T_1 are known, we can calculate the vapor pressure (P_2) at any other temperature (T_2) or the temperature at any other pressure.

SAMPLE PROBLEM 12.2 Using the Clausius-Clapeyron Equation

PROBLEM The vapor pressure of ethanol is 115 torr at 34.9°C. If ΔH_{vap} of ethanol is 40.5 kJ/mol, calculate the temperature (in °C) when the vapor pressure is 760 torr.

PLAN We are given ΔH_{vap}, P_1, P_2, and T_1 and substitute them into Equation 12.1 to solve for T_2. The value of R here is 8.314 J/mol·K, so we must convert T_1 to K to obtain T_2, and then convert T_2 to °C.

SOLUTION Substituting the values into Equation 12.1 and solving for T_2:

$$\ln \frac{P_2}{P_1} = \frac{-\Delta H_{vap}}{R}\left(\frac{1}{T_2} - \frac{1}{T_1}\right)$$
$$T_1 = 34.9°C + 273.15 = 308.0 \text{ K}$$
$$\ln \frac{760 \text{ torr}}{115 \text{ torr}} = \left(-\frac{40.5 \times 10^3 \text{ J/mol}}{8.314 \text{ J/mol·K}}\right)\left(\frac{1}{T_2} - \frac{1}{308.0 \text{ K}}\right)$$
$$1.888 = (-4.87 \times 10^3)\left[\frac{1}{T_2} - (3.247 \times 10^{-3})\right]$$
$$T_2 = 350. \text{ K}$$

Converting T_2 from K to °C:

$$T_2 = 350. \text{ K} - 273.15 = \boxed{77°C}$$

CHECK Round off to check the math. The change is in the right direction: higher P should occur at higher T. As we discuss next, a substance has a vapor pressure of 760 torr at its normal boiling point. Checking the *CRC Handbook of Chemistry and Physics* shows that the boiling point of ethanol is 78.5°C, very close to our answer.

FOLLOW-UP PROBLEM 12.2 At 34.1°C, the vapor pressure of water is 40.1 torr. What is the vapor pressure at 85.5°C? The ΔH_{vap} of water is 40.7 kJ/mol.

Vapor Pressure and Boiling Point In an *open* container, the weight of the atmosphere bears down on the liquid surface. As the temperature rises, molecules leave the surface more often, and they also move more quickly throughout the liquid. At some temperature, the average E_k of the molecules in the liquid is great enough for bubbles of vapor to form *in the interior,* and the liquid boils. At any lower temperature, the bubbles collapse as soon as they start to form because the external pressure is greater than the vapor pressure inside the bubbles. Thus, the **boiling point** is *the temperature at which the vapor pressure equals the external pressure,* which is usually that of the atmosphere. Once boiling begins, the temperature of the liquid remains constant until all of the liquid is gone. Thus, the applied heat is used to overcome intermolecular attractions and allow the molecules to enter the gas phase.

The boiling point varies with elevation because the atmospheric pressure does. At higher elevations, a lower atmospheric pressure is exerted on the liquid surface, and thus molecules in the interior need less kinetic energy to form bubbles. At lower elevations, the opposite is true: the applied pressure is higher and the molecules need more kinetic energy to form bubbles. Thus, *the boiling point depends on the applied pressure.* The *normal boiling point* is observed at standard atmospheric pressure (760 torr, or 101.3 kPa; see the horizontal dashed line in Figure 12.6).

Since the boiling point is the temperature at which the vapor pressure equals the external pressure, we can also interpret the curves in Figure 12.6 as a plot of external pressure vs. boiling point. For instance, the H_2O curve shows that water boils at 100°C at 760 torr (sea level), at 94°C at 610 torr (Boulder, Colorado), and at about 72°C at 270 torr (top of Mt. Everest). ●

Solid-Liquid Equilibria At the molecular level, the particles in a crystal are continually vibrating about their fixed positions. As the temperature rises, the particles vibrate more violently, until some have enough kinetic energy to break free of their positions. At this point, melting begins. As more molecules enter the liquid (molten) phase, some collide with the solid and become fixed in position again. Because the phases remain in contact, a dynamic equilibrium is established when the melting rate equals the freezing rate. The temperature at which this occurs is called the **melting point;** it is the same temperature as the freezing point, differing only in the direction of the energy flow. As we saw with the boiling point, the temperature remains fixed at the melting point as long as both phases are present.

Because liquids and solids are nearly incompressible, a change in pressure has little effect on the rate of movement to or from the solid. Therefore, in contrast to the boiling point, the melting point is affected by pressure only very slightly, and a plot of pressure (*y* axis) vs. temperature (*x* axis) for a solid-liquid phase change is typically a straight, *nearly* vertical line.

Solid-Gas Equilibria Solids have much lower vapor pressures than liquids. Sublimation, the process of a solid changing directly into a gas, is much less familiar than vaporization because the necessary conditions of pressure and temperature are uncommon for most substances. Some solids *do* have high enough vapor pressures to sublime at ordinary conditions, including dry ice (carbon dioxide), iodine (Figure 12.8), and solid room deodorizers.

A substance sublimes rather than melts because the combination of intermolecular attractions and atmospheric pressure is not great enough to keep the particles near one another when they leave the solid state. The pressure vs. temperature plot for the solid-gas transition shows a large effect of temperature on the pressure of the vapor; thus, this curve resembles the liquid-gas line in curving upward at higher temperatures.

📖 **Literature** Vollmer, John J. "Out of 'Thin Air': Exploring Phase Changes," *J. Chem. Educ.* **2000**, *77*, 488A.

🧪 **Demo** Shakhashiri, Bassam Z. "Boiling Liquids at Reduced Pressure," *Chemical Demonstrations, A Handbook for Teachers of Chemistry,* Vol. 2 (The University of Wisconsin Press, Madison, 1985), 81–84.

🧪 **Demo** Summerlin, L. R.; Borgford, C. L.; Ealy, J. B. "Boiling at Reduced Pressure," *Chemical Demonstrations, A Sourcebook for Teachers,* Vol. 2 (American Chemical Society, Washington D.C., 1988), 24–25.

● **Cooking Under Low or High Pressure**
People who live or hike in mountainous regions cook their meals under lower atmospheric pressure. The resulting lower boiling point of the liquid means the food takes *more* time to cook. On the other hand, in a pressure cooker, the pressure exceeds that of the atmosphere, so the temperature rises above the normal boiling point. The food becomes hotter and, thus, takes *less* time to cook.

Figure 12.8 Iodine subliming. At ordinary atmospheric pressure, solid iodine sublimes (changes directly from a solid into a gas). When the I_2 vapor comes in contact with a cold surface, such as the water-filled inner test tube, it deposits I_2 crystals. Sublimation is a means of purification that, along with distillation, was probably discovered by alchemists.

 ARIS Presentation Center, Photo Library

Literature Gramsch, Stephen A. "A Closer Look at Phase Diagrams for the General Chemistry Course," *J. Chem. Educ.* **2000,** *77,* 718.

Literature Kildahl, Nicholas K. "Journey Around a Phase Diagram," *J. Chem. Educ.* **1994,** *71,* 1052.

Literature Harding, Kenn E. "Melting Point Range and Phase Diagrams—Confusing Laboratory Textbook Descriptions," *J. Chem. Educ.* **1999,** *76,* 224.

Phase Diagrams: Effect of Pressure and Temperature on Physical State

To describe the phase changes of a substance at various conditions of temperature and pressure, we construct a **phase diagram,** which combines the liquid-gas, solid-liquid, and solid-gas curves. The shape of the phase diagram for CO_2 (Figure 12.9A) is typical of the diagram for most substances. A phase diagram has these four features:

1. *Regions of the diagram.* Each region corresponds to one phase of the substance. A particular phase is stable for any combination of pressure and temperature within its region. If any of the other phases is placed under those conditions, it will change to the stable phase. In general, the solid is stable at low temperature and high pressure, the gas at high temperature and low pressure, and the liquid at intermediate conditions.

2. *Lines between regions.* The lines separating the regions represent the phase-transition curves discussed earlier. Any point along a line shows the pressure and temperature at which the two phases exist in equilibrium. Note that the solid-liquid line has a *positive* slope (slants to the *right* with increasing pressure) because, for most substances, the solid is more dense than the liquid. Because the liquid occupies slightly more space than the solid, an increase in pressure favors the solid phase. (Water is the major exception, as you'll soon see.)

3. *The critical point.* The liquid-gas line ends at the **critical point.** Picture a liquid in a closed container. As it is heated, it expands, so its density decreases. At the same time, more liquid vaporizes, so the density of the vapor increases. The liquid and vapor densities become closer and closer to each other until, at the *critical temperature* (T_c), the two densities are equal and the phase boundary disappears. The pressure at this temperature is the *critical pressure* (P_c). At this

Figure 12.9 Phase diagrams for CO₂ and H₂O. Each region depicts the temperatures and pressures under which the phase is stable. Lines between two regions show conditions at which the two phases exist in equilibrium. The critical point shows conditions beyond which separate liquid and gas phases no longer exist. At the triple point, the three phases exist in equilibrium. (The axes are not linear.) **A,** The phase di-

agram for CO_2 is typical of most substances in that the solid-liquid line slopes to the right with increasing pressure: the solid is *more* dense than the liquid. **B,** Water is one of the few substances whose solid-liquid line slopes to the left with increasing pressure: the solid is *less* dense than the liquid. (The slopes of the solid-liquid lines in both diagrams are exaggerated.)

 Transparency/ARIS Presentation Center, Art Library

point, the average E_k of the molecules is so high that the vapor cannot be condensed no matter what pressure is applied. The two most common gases in air have critical temperatures far below room temperature: no matter what the pressure, O_2 will not condense above $-119°C$, and N_2 will not condense above $-147°C$. Beyond the critical temperature, a *supercritical fluid* exists rather than separate liquid and gaseous phases. ●

4. *The triple point.* The three phase-transition curves meet at the **triple point:** the pressure and temperature at which three phases are in equilibrium. As strange as it sounds, at the triple point in Figure 12.9A, CO_2 is subliming and depositing, melting and freezing, and vaporizing and condensing simultaneously! Phase diagrams for substances with several solid forms, such as sulfur, have more than one triple point.

The CO_2 phase diagram explains why dry ice (solid CO_2) doesn't melt under ordinary conditions. The triple-point pressure for CO_2 is 5.1 atm; therefore, at around 1 atm, liquid CO_2 does not occur. By following the horizontal dashed line in Figure 12.9A, you can see that when solid CO_2 is heated at 1.0 atm, it sublimes at $-78°C$ to gaseous CO_2 rather than melting. If normal atmospheric pressure were 5.2 atm, liquid CO_2 would be common.

The phase diagram for water differs in one key respect from the general case and reveals an extremely important property (Figure 12.9B). Unlike almost every other substance, solid water is *less dense* than liquid water. Because the solid occupies more space than the liquid, *water expands on freezing*. This behavior results from the unique open crystal structure of ice, which we discuss in Section 12.5. As always, an increase in pressure favors the phase that occupies less space, but in the case of water, this is the *liquid* phase. Therefore, the solid-liquid line for water has a *negative* slope (slants to the *left* with increasing pressure): the higher the pressure, the lower the temperature at which water freezes. In Figure 12.9B, the vertical dashed line at $-1°C$ crosses the solid-liquid line, which means that ice melts at that temperature with only an increase in pressure.

The triple point of water occurs at low pressure (0.006 atm). Therefore, when solid water is heated at 1.0 atm, the horizontal dashed line crosses the solid-liquid line (at 0°C, the normal melting point) and enters the liquid region. Thus, at ordinary pressures, ice melts rather than sublimes. As the temperature rises, the horizontal line crosses the liquid-gas curve (at 100°C, the normal boiling point) and enters the gas region.

Section Summary

A heating-cooling curve depicts the change in temperature with heat gain or loss. Within a phase, temperature (and average E_k) changes as heat is added or removed. During a phase change, temperature (and average E_k) is constant, but E_p changes. The total heat change for the curve is calculated using Hess's law. • In a closed container, equilibrium is established between the liquid and gas phases. Vapor pressure, the pressure of the gas at equilibrium, is related directly to temperature and inversely to the strength of the intermolecular forces. • The Clausius-Clapeyron equation relates the vapor pressure to the temperature through the ΔH_{vap}. • A liquid in an open container boils when its vapor pressure equals the external pressure. • Solid-liquid equilibrium occurs at the melting point. Many solids sublime at low pressures and high temperatures. • A phase diagram shows the phase that exists at a given pressure and temperature and the conditions at the critical point and the triple point of a substance. Water differs from most substances in that its solid phase is less dense than its liquid phase, so its solid-liquid line has a negative slope.

Literature Hawkes, Stephen J. "There Is No Perceptible Inflection at the Triple Point," *J. Chem. Educ.* **1999**, *76*, 226.

● **The Remarkable Behavior of a Supercritical Fluid (SCF)** What material lies beyond the familiar regions of liquid and gas? An SCF expands and contracts like a gas but has the solvent properties of a liquid, which chemists can alter by controlling the density. Supercritical CO_2 has received the most attention so far. It extracts nonpolar ingredients from complex mixtures, such as caffeine from coffee beans, nicotine from tobacco, and fats from potato chips, while leaving taste and aroma ingredients behind, for healthier consumer products. Lower the pressure and the SCF disperses immediately as a harmless gas. Supercritical CO_2 dissolves the fat from meat, along with pesticide and drug residues, which can then be quantified and monitored. It is being studied as an environmentally friendly dry cleaning agent. In an unexpected finding of great potential use, supercritical H_2O was shown to dissolve nonpolar substances even though liquid water does not! Studies are under way to carry out the large-scale removal of nonpolar organic toxins, such as PCBs, from industrial waste by extracting them into supercritical H_2O; after this step, O_2 gas is added to oxidize the toxins to small, harmless molecules.

 Engineering application

Animation: Phase Diagrams of the States of Matter
ARIS Presentation Center
Chapter 12

12.3 TYPES OF INTERMOLECULAR FORCES

As we've seen, the nature of the phases and their changes are due primarily to forces among the molecules. Both bonding (intramolecular) forces and intermolecular forces arise from electrostatic attractions between opposite charges. Bonding forces are due to the attraction between cations and anions (ionic bonding), nuclei and electron pairs (covalent bonding), or metal cations and delocalized valence electrons (metallic bonding). Intermolecular forces, on the other hand, are due to the attraction between molecules as a result of partial charges, or the attraction between ions and molecules. The two types of forces differ in magnitude, and Coulomb's law explains why:

- *Bonding forces are relatively strong* because they involve larger charges that are closer together.
- *Intermolecular forces are relatively weak* because they typically involve smaller charges that are farther apart.

Figure 12.10 **Covalent and van der Waals radii.** As shown here for solid chlorine, the van der Waals (VDW) radius is one-half the distance between adjacent *nonbonded* atoms ($\frac{1}{2}$ × VDW distance), and the covalent radius is one-half the distance between *bonded* atoms ($\frac{1}{2}$ × bond length).

 ARIS Presentation Center, Art Library

Figure 12.11 **Periodic trends in covalent and van der Waals radii (in pm).** Like covalent radii (*blue quarter-circles and top numbers*), van der Waals radii (*red quarter-circles and bottom numbers*) increase down a group and decrease across a period. The covalent radius of an element is *always* less than its van der Waals radius.

 ARIS Presentation Center, Art Library

How far apart are the charges on different molecules that give rise to intermolecular forces between them? Consider Cl_2 as an example. When we measure the distances between two Cl nuclei in a sample of solid Cl_2, we obtain two different values, as shown in Figure 12.10. The shorter distance is between *two **bonded** Cl atoms in the same molecule.* It is, as you know, called the *bond length,* and one-half this distance is the *covalent radius.* The longer distance is between *two **nonbonded** Cl atoms in adjacent molecules.* It is called the *van der Waals distance* (named after the Dutch physicist Johannes van der Waals, who studied the effects of intermolecular forces on the behavior of real gases). This distance is the closest one Cl_2 molecule can approach another, the point at which intermolecular attractions balance electron-cloud repulsions. One-half this distance is the **van der Waals radius,** one-half the closest distance between the nuclei of identical *nonbonded* Cl atoms. *The van der Waals radius of an atom is always larger than its covalent radius,* but van der Waals radii decrease across a period and increase down a group, just as covalent radii do. Figure 12.11 shows these relationships for many of the nonmetals.

There are several types of intermolecular forces: ion-dipole, dipole-dipole, hydrogen bonding, dipole–induced dipole, and dispersion forces. As we discuss these *intermolecular* forces (also called *van der Waals forces*), look at Table 12.2, which compares them with the stronger *intramolecular* (bonding) forces.

Ion-Dipole Forces

When an ion and a nearby polar molecule (dipole) attract each other, an **ion-dipole force** results. The most important example takes place when an ionic compound dissolves in water. The ions become separated because the attractions

Table 12.2   Comparison of Bonding and Nonbonding (Intermolecular) Forces

Force	Model	Basis of Attraction	Energy (kJ/mol)	Example
Bonding				
Ionic		Cation–anion	400–4000	NaCl
Covalent		Nuclei–shared e^- pair	150–1100	H—H
Metallic		Cations–delocalized electrons	75–1000	Fe
Nonbonding (Intermolecular)				
Ion-dipole		Ion charge– dipole charge	40–600	$Na^+\cdots O\big\langle{{H}\atop{H}}$
H bond	$\underset{-A-H}{\overset{\delta-\ \ \delta+}{}}\cdots\underset{:B-}{\overset{\delta-}{}}$	Polar bond to H– dipole charge (high EN of N, O, F)	10–40	$:\!\ddot{O}—H\cdots:\!\ddot{O}—H$
Dipole-dipole		Dipole charges	5–25	I—Cl$\cdots$I—Cl
Ion–induced dipole		Ion charge– polarizable e^- cloud	3–15	$Fe^{2+}\cdots O_2$
Dipole–induced dipole		Dipole charge– polarizable e^- cloud	2–10	H—Cl$\cdots$Cl—Cl
Dispersion (London)		Polarizable e^- clouds	0.05–40	F—F$\cdots$F—F

between the ions and the oppositely charged poles of the H_2O molecules overcome the attractions between the ions themselves. Ion-dipole forces in solutions and their associated energy are discussed fully in Chapter 13.

Dipole-Dipole Forces

In Figure 10.12 (p. 399), you saw how an external electric field can orient gaseous polar molecules. When polar molecules lie near one another, as in liquids and solids, their partial charges act as tiny electric fields that orient them and give rise to **dipole-dipole forces:** the positive pole of one molecule attracts the negative pole of another (Figure 12.12).

These are the forces that give polar *cis*-1,2-dichloroethylene a higher boiling point than the nonpolar *trans* compound (see p. 401). In fact, for molecular compounds of approximately the same size and molar mass, the greater the dipole moment, the greater the dipole-dipole forces between the molecules are, and so

Solid

Liquid

Figure 12.12 Polar molecules and dipole-dipole forces. In a solid or a liquid, the polar molecules are close enough for the partially positive pole of one molecule to attract the partially negative pole of a nearby molecule. The orientation is more orderly in the solid (*top*) than in the liquid (*bottom*) because, at the lower temperatures required for freezing, the average kinetic energy of the particles is lower. (Interparticle spaces are increased for clarity.)

Figure 12.13 Dipole moment and boiling point. For compounds of similar molar mass, the boiling point increases with increasing dipole moment. (Note also the increasing color intensities in the electron-density models.) The greater dipole moment creates stronger dipole-dipole forces, which require higher temperatures to overcome.

Transparency/ARIS Presentation Center, Art Library

Literature Furio, Carlos; Calatayud, Ma Luisa. "Difficulties with the Geometry and Polarity of Molecules: Beyond Misconceptions," *J. Chem. Educ.* **1996**, *73*, 36.

☐ Dipole moment (D x 100)
☐ Boiling point (K)

	Propane $CH_3CH_2CH_3$ 44.09 g/mol	Dimethyl ether CH_3OCH_3 46.07 g/mol	Methyl chloride CH_3Cl 50.48 g/mol	Acetaldehyde CH_3CHO 44.05 g/mol	Acetonitrile CH_3CN 41.05 g/mol
Dipole moment	8	130	187	269	392
Boiling point	231	248	249	294	355

Organic chemistry application

Literature Schultz, Emeric. "The Nature of Hydrogen Bonding," *J. Chem. Educ.* **2005**, *82*, 400A.

the more energy it takes to separate them. Consider the boiling points of the compounds in Figure 12.13, which have similar molar masses. Methyl chloride, for instance, has a smaller dipole moment than acetaldehyde, so less energy is needed to overcome the dipole-dipole forces between its molecules and it boils at a lower temperature.

The Hydrogen Bond

A special type of dipole-dipole force arises between molecules that have *an H atom bonded to a small, highly electronegative atom with lone electron pairs*. The most important atoms that fit this description are N, O, and F. The H—N, H—O, and H—F bonds are very polar, so electron density is withdrawn from H. As a result, the partially positive H of one molecule is attracted to the partially negative lone pair on the N, O, or F of another molecule, and a **hydrogen bond (H bond)** forms. Thus, the atom sequence that allows an H bond (dotted line) to form is —B:····H—A—, where *both* A and B are N, O, or F. Three examples are

$$ —\ddot{\underset{..}{F}}:····H—\ddot{\underset{..}{O}}— \qquad —\ddot{\underset{..}{O}}:····H—\ddot{N}— \qquad —\ddot{N}:····H—\ddot{\underset{..}{F}}: $$

The small sizes of N, O, and F* are essential to H bonding for two reasons:

1. It makes these atoms so electronegative that their covalently bonded H is highly positive.

2. It allows the lone pair on the other N, O, or F to come close to the H.

The Significance of Hydrogen Bonding Hydrogen bonding has a profound impact in many systems. Here we'll examine one major effect on physical properties and preview its enormous importance in biological systems, which we address in Chapters 13 and 15.

---

*H-bond–type interactions occur with the larger atoms P, S, and Cl, but those are so much weaker than the interactions with N, O, and F that we will not consider them.

Figure 12.14 shows the effect of H bonding on the boiling points of the binary hydrides of Groups 4A(14) through 7A(17). For reasons we'll discuss shortly, boiling points typically rise as molar mass increases, as you can see in the Group 4A hydrides, CH₄ through SnH₄. In the other groups, however, the first member in each series—NH₃, H₂O, and HF—deviates enormously from this expected increase. The H bonds between the molecules in these substances require additional energy to break before the molecules can separate and enter the gas phase. For example, on the basis of molar mass alone, we would expect water to boil about 200°C lower than it actually does (dashed line). (In Section 12.5 we'll discuss the effects that H bonds in water have in nature.)

The significance of hydrogen bonding in biological systems cannot be emphasized too strongly. It is a principal feature in the structures of the three major biopolymers—proteins, nucleic acids, and carbohydrates. In addition, it is responsible for the action of many *enzymes*, the specialized proteins that speed all metabolic reactions. In Chapter 13, after we discuss intermolecular forces in solution, you'll be amazed at the essential roles H bonds and the other intermolecular forces play in the structure and function of the molecules of life.

Biological application

Figure 12.14 **Hydrogen bonding and boiling point.** Boiling points of the binary hydrides of Groups 4A(14) to 7A(17) are plotted against period number. H bonds in NH₃, H₂O, and HF give them much higher boiling points than if the trend were based on molar mass, as it is for Group 4A. The red dashed line extrapolates to the boiling point H₂O would have if it formed no H bonds.

Transparency/ARIS Presentation Center, Art Library

ARIS Presentation Center, Worked Examples Library

Literature Schultz, Emeric. "Simple Dynamic Models for Hydrogen Bonding Using Velcro-Polarized Molecular Models," *J. Chem. Educ.* **2005,** *82,* 401.

SAMPLE PROBLEM 12.3   Drawing Hydrogen Bonds Between Molecules of a Substance

PROBLEM Which of the following substances exhibits H bonding? For those that do, draw two molecules of the substance with the H bond(s) between them.

                                                  O
                                                  ‖
(a) C₂H₆   **(b)** CH₃OH   **(c)** CH₃C—NH₂

PLAN We draw each structure to see if it contains N, O, or F covalently bonded to H. If it does, we draw two molecules of the substance in the —B:····H—A— pattern.

SOLUTION **(a)** For C₂H₆. No H bonds are formed.

(b) For CH₃OH. The H covalently bonded to the O in one molecule forms an H bond to the lone pair on the O of an adjacent molecule:

$$
\begin{array}{ccc}
 & \text{H} & \\
 & | & \\
\text{H}-\text{C}-\text{H} & & \text{H} \\
 & | & | \\
\text{H}-\ddot{\text{O}}\text{:}\cdots\text{H}-\ddot{\text{O}}-\text{C}-\text{H} \\
 & & | \\
 & & \text{H}
\end{array}
$$

(c) For CH₃C—NH₂ (with O double-bonded to C). Two of these molecules can form one H bond between an H bonded to N and the O, or they can form two such H bonds:

A third possibility (not shown) could be between an H attached to N in one molecule and the lone pair of N in another molecule.

CHECK The —B:····H—A— sequence (with A and B either N, O, or F) is present.

COMMENT Note that *H covalently bonded to C does not form H bonds* because carbon is not electronegative enough to make the C—H bond very polar.

FOLLOW-UP PROBLEM 12.3 Which of these substances exhibits H bonding? Draw the H bond(s) between two molecules of the substance where appropriate.

(a) CH₃C—OH   (b) CH₃CH₂OH   (c) CH₃CCH₃

Polarizability and Charge-Induced Dipole Forces

Even though electrons are localized in bonding or lone pairs, they are in constant motion, so that we often picture them as "clouds" of negative charge. A nearby electric field can distort a cloud, pulling electron density toward a positive charge or pushing it away from a negative charge. In effect, the field *induces* a distortion in the electron cloud. For a nonpolar molecule, this distortion creates a temporary, induced dipole moment; for a polar molecule, it enhances the dipole moment already present. The source of the electric field can be the electrodes of a battery, the charge of an ion, or the partial charges of a polar molecule.

The ease with which the electron cloud of a particle can be distorted is called its **polarizability.** Smaller atoms (or ions) are less polarizable than larger ones because their electrons are closer to the nucleus and therefore are held more tightly. Thus, we observe several trends:

- *Polarizability increases down a group* because atomic size increases, so the larger electron clouds are farther from the nucleus and, thus, easier to distort.
- *Polarizability decreases from left to right across a period* because the increasing Z_{eff} shrinks atomic size and holds the electrons more tightly.
- Cations are *less* polarizable than their parent atoms because they are smaller; anions are *more* polarizable because they are larger.

Ion–induced dipole and dipole–induced dipole forces are the two types of charge-induced dipole forces; they are most important in solution, so we'll focus on them in Chapter 13. Nevertheless, *polarizability affects all intermolecular forces.*

Dispersion (London) Forces

Polarizability plays the central role in the most universal intermolecular force. Up to this point, we've discussed forces that depend on an existing charge, of either an ion or a polar molecule. But what forces cause nonpolar substances like octane, chlorine, and argon to condense and solidify? Some force must be acting between the particles, or these substances would be gases under any conditions. *The intermolecular force primarily responsible for the condensed states of nonpolar substances is the* **dispersion force** (or **London force,** named for Fritz London, the physicist who explained the quantum-mechanical basis of the attraction).

Dispersion forces are caused by *momentary oscillations of electron charge in atoms and, therefore, are present between all particles (atoms, ions, and molecules).* Picture one atom in a sample of argon gas. Averaged over time, the 18 electrons are distributed uniformly around the nucleus, so the atom is nonpolar. But at any instant, there may be more electrons on one side of the nucleus than on the other, so the atom has an *instantaneous dipole.* When far apart, a pair of argon atoms do not influence each other. But when close together, *the instantaneous dipole in one atom induces a dipole in its neighbor.* The result is a synchronized motion of the electrons in the two atoms, which causes an attraction between them. This process occurs with other nearby atoms and, thus, throughout the sample. At low enough temperatures, the attractions among the dipoles keep all the atoms together. Thus, dispersion forces are *instantaneous dipole–induced dipole forces.* Figure 12.15 depicts the dispersion forces among nonpolar particles.

As we noted, the dispersion force is the *only* force existing between nonpolar particles, but because they exist between *all* particles, *dispersion forces*

📖 **Literature** Laing, Michael. "Boiling Points of the Family of Small Molecules CH₍w₎F₍x₎Cl₍y₎Br₍z₎: How Are They Related to Molecular Mass?" *J. Chem. Educ.* **2001,** *78,* 1544.

📖 **Literature** Kimbrough, Doris R.; DeLorenzo, Ronald. "Solving the Mystery of Fading Fingerprints with London Dispersion Forces," *J. Chem. Educ.* **1998,** *75,* 1300.

📖 **Literature** Wilcox, C. Jayne. "London Dispersion Forces and 'The Wave'," *J. Chem. Educ.* **1998,** *75,* 1301.

💿 **ARIS Presentation Center,** Art Library

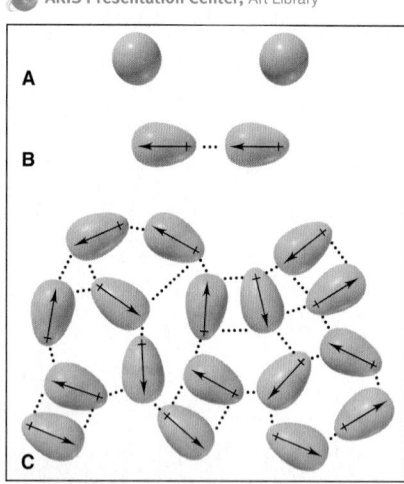

Figure 12.15 Dispersion forces among nonpolar particles. The dispersion force is responsible for the condensed states of noble gases and nonpolar molecules. **A,** Separated Ar atoms are nonpolar. **B,** An instantaneous dipole in one atom induces a dipole in its neighbor. These partial charges attract the atoms together. **C,** This process takes place among atoms throughout the sample.

contribute to the overall energy of attraction of all substances. In fact, except in cases involving small, polar molecules with large dipole moments or those forming strong H bonds, the *dispersion force is the dominant intermolecular force.* Calculations show, for example, that 85% of the total energy of attraction between HCl molecules is due to dispersion forces and only 15% to dipole-dipole forces. Even for water, estimates indicate that 75% of the total energy of attraction comes from the strong H bonds and nearly 25% from dispersion forces!

Dispersion forces are very weak for small particles, like H_2 and He, but much stronger for larger particles, like I_2 and Xe. The relative strength of the dispersion force depends on the polarizability of the particle. *Polarizability depends on the number of electrons, which correlates closely with molar mass* because heavier particles are either larger atoms or composed of more atoms and thus have more electrons. For example, molar mass increases down the halogens and the noble gases, so dispersion forces increase and so do boiling points (Figure 12.16).

For nonpolar substances with the same molar mass, the strength of the dispersion forces is often influenced by molecular shape. Shapes that allow more points of contact have more area over which electron clouds can be distorted, so stronger attractions result. Consider the two five-carbon alkanes, pentane (also called *n-pentane*) and 2,2-dimethylpropane (also called *neopentane*). These isomers have the same molecular formula (C_5H_{12}) but different shapes. *n*-Pentane is shaped like a cylinder, whereas neopentane has a more compact, spherical shape, as shown in Figure 12.17. Thus, two *n*-pentane molecules make more contact than do two neopentane molecules. Greater contact allows the dispersion forces to act at more points, so *n*-pentane has a higher boiling point. Figure 12.18 summarizes how to analyze the intermolecular forces in a sample.

Figure 12.16 **Molar mass and boiling point.** The strength of dispersion forces increases with number of electrons, which usually correlates with molar mass. As a result, boiling points increase down the halogens and the noble gases.

ARIS Presentation Center, Art Library

Organic chemistry application

ARIS Presentation Center, Art Library

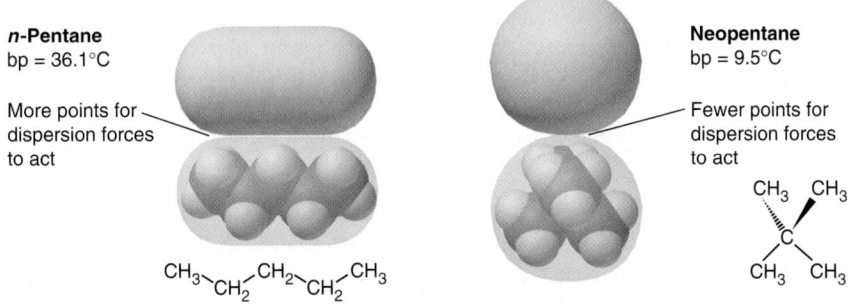

Figure 12.17 **Molecular shape and boiling point.** Spherical neopentane molecules make less contact with each other than do cylindrical *n*-pentane molecules, so neopentane has a lower boiling point.

Figure 12.18 **Summary diagram for analyzing the intermolecular forces in a sample.** *ARIS Presentation Center, Art Library*

ARIS Presentation Center, Worked Examples Library

Literature Murthy, Parvathi S. "Molecular Handshake: Recognition through Weak Noncovalent Interactions," *J. Chem. Educ.* **2006,** *83,* 1010.

(b)

CH₃—N—H⋯⋯:N—CH₃

(c)

CH₃—Ö—H⋯⋯:Ö—CH₃

CH₃CH₂—Ö—H⋯⋯:Ö—CH₂CH₃

(d)

2,2-Dimethylbutane

Hexane

SAMPLE PROBLEM 12.4 Predicting the Types of Intermolecular Force

PROBLEM For each pair of substances, identify the key intermolecular force(s) in each substance, and select the substance with the higher boiling point:
(a) $MgCl_2$ or PCl_3
(b) CH_3NH_2 or CH_3F
(c) CH_3OH or CH_3CH_2OH
(d) Hexane ($CH_3CH_2CH_2CH_2CH_2CH_3$) or 2,2-dimethylbutane (structure: $CH_3CCH_2CH_3$ with two CH_3 groups)

PLAN We examine the formulas and picture (or draw) the structures to identify key differences between members of each pair: Are ions present? Are molecules polar or nonpolar? Is N, O, or F bonded to H? Do the molecules have different masses or shapes? To rank the boiling points, we consult Table 12.2 and Figure 12.18 and remember that

- Bonding forces are stronger than intermolecular forces.
- Hydrogen bonding is a strong type of dipole-dipole force.
- Dispersion forces are always present, but they are decisive when the difference is molar mass or molecular shape.

SOLUTION **(a)** $MgCl_2$ consists of Mg^{2+} and Cl^- ions held together by ionic bonding forces; PCl_3 consists of polar molecules, so intermolecular dipole-dipole forces are present. The forces in $MgCl_2$ are stronger, so it has a higher boiling point.
(b) CH_3NH_2 and CH_3F both consist of polar molecules of about the same molar mass. CH_3NH_2 has N—H bonds, so it can form H bonds (see margin). CH_3F contains a C—F bond but no H—F bond, so dipole-dipole forces occur but not H bonds. Therefore, CH_3NH_2 has the higher boiling point.
(c) CH_3OH and CH_3CH_2OH molecules both contain an O—H bond, so they can form H bonds (see margin). CH_3CH_2OH has an additional —CH_2— group and thus a larger molar mass, which correlates with stronger dispersion forces; therefore, it has a higher boiling point.
(d) Hexane and 2,2-dimethylbutane are nonpolar molecules of the same molar mass but different molecular shapes (see margin). Cylindrical hexane molecules can make more extensive intermolecular contact than the more spherical 2,2-dimethylbutane molecules can, so hexane should have greater dispersion forces and a higher boiling point.

CHECK The actual boiling points show that our predictions are correct:
(a) $MgCl_2$ (1412°C) and PCl_3 (76°C)
(b) CH_3NH_2 (−6.3°C) and CH_3F (−78.4°C)
(c) CH_3OH (64.7°C) and CH_3CH_2OH (78.5°C)
(d) Hexane (69°C) and 2,2-dimethylbutane (49.7°C)

COMMENT Dispersion forces are *always* present, but in parts (a) and (b), they are much less significant than the other forces involved.

FOLLOW-UP PROBLEM 12.4 In each pair, identify all the intermolecular forces present for each substance, and select the substance with the higher boiling point:
(a) CH_3Br or CH_3F **(b)** $CH_3CH_2CH_2OH$ or $CH_3CH_2OCH_3$ **(c)** C_2H_6 or C_3H_8

Section Summary

The van der Waals radius determines the shortest distance over which intermolecular forces operate; it is always larger than the covalent radius. • Intermolecular forces are much weaker than bonding (intramolecular) forces. • Ion-dipole forces occur between ions and polar molecules. • Dipole-dipole forces occur between oppositely charged poles on polar molecules. • Hydrogen bonding, a special type of dipole-dipole force, occurs when H bonded to N, O, or F is attracted to the lone pair of N, O, or F in another molecule. • Electron clouds can be distorted (polarized) in an electric field. • Ion– and dipole–induced dipole forces arise between a charge and the dipole it induces in another molecule. • Dispersion (London) forces are instantaneous dipole–induced dipole forces that occur among all particles and increase with number of electrons (molar mass). Molecular shape determines the extent of contact between molecules and can be a factor in the strength of dispersion forces.

12.4 PROPERTIES OF THE LIQUID STATE

Of the three states of matter, the liquid is the least understood at the molecular level. Because of the *randomness* of the particles in a gas, any region of the sample is virtually identical to any other. As you'll see in Section 12.6, different regions of a crystalline solid are identical because of the *orderliness* of the particles. Liquids, however, have a combination of these attributes that changes continually: a region that is orderly one moment becomes random the next, and vice versa. Despite this complexity at the molecular level, the macroscopic properties of liquids are well understood. In this section, we discuss three liquid properties—surface tension, capillarity, and viscosity.

Surface Tension

In a liquid sample, intermolecular forces exert different effects on a molecule at the surface than on one in the interior (Figure 12.19). Interior molecules are attracted by others on all sides, whereas molecules at the surface have others only below and to the sides. As a result, molecules at the surface experience a *net attraction downward* and move toward the interior to increase attractions and become more stable. Therefore, *a liquid surface tends to have the smallest possible area,* that of a sphere, and behaves like a "taut skin" covering the interior.

To increase the surface area, molecules must move to the surface, thus breaking some attractions in the interior—which requires energy. The **surface tension** is the energy required to increase the surface area by a unit amount; units of J/m^2 are shown in Table 12.3. Comparing these values with those in Table 12.2 (p. 451) shows that, in general, *the stronger the forces are between the particles in a liquid, the greater the surface tension.* Water has a high surface tension because its molecules form multiple H bonds. *Surfactants* (*surf*ace-*act*ive age*nts*), such as soaps, petroleum recovery agents, and biological fat emulsifiers, decrease the surface tension of water by congregating at the surface and disrupting the H bonds.

Figure 12.19 The molecular basis of surface tension. Molecules in the interior of a liquid experience intermolecular attractions in all directions. Molecules at the surface experience a net attraction downward (*red arrow*) and move toward the interior. Thus, a liquid tends to minimize the number of molecules at the surface, which results in surface tension.

ARIS Presentation Center, Art Library

Demo Shakhashiri, Bassam Z. "At the Water's Edge: Surface Spreading and Surface Tension," *Chemical Demonstrations, A Handbook for Teachers of Chemistry,* Vol. 3 (The University of Wisconsin Press, Madison, 1985), 301–304.

Demo Condon, F. E.; Condon, F. E., Jr. "The Floating Needle," *J. Chem. Educ.* **2001**, *78*, 334.

Demo Rosenthal, Andrew J. "Demonstration of Surface Tension," *J. Chem. Educ.* **2001**, *78*, 332.

ARIS Presentation Center, Tables Library

Table 12.3	Surface Tension and Forces Between Particles		
Substance	**Formula**	**Surface Tension (J/m^2) at 20°C**	**Major Force(s)**
Diethyl ether	$CH_3CH_2OCH_2CH_3$	1.7×10^{-2}	Dipole-dipole; dispersion
Ethanol	CH_3CH_2OH	2.3×10^{-2}	H bonding
Butanol	$CH_3CH_2CH_2CH_2OH$	2.5×10^{-2}	H bonding; dispersion
Water	H_2O	7.3×10^{-2}	H bonding
Mercury	Hg	48×10^{-2}	Metallic bonding

Capillarity

The rising of a liquid through a narrow space against the pull of gravity is called *capillary action,* or **capillarity.** In blood-screening tests, a narrow *capillary tube* is held against the skin opening made by pricking a finger (see photo). Capillarity results from a competition between the intermolecular forces within the liquid (cohesive forces) and those between the liquid and the tube walls (adhesive forces).

Picture what occurs at the molecular level when you place a glass capillary tube in water. Glass is mostly silicon dioxide (SiO_2), so the water molecules form H bonds to the oxygen atoms of the tube's inner wall. Because the adhesive forces (H bonding) between the water and the wall are stronger than the cohesive forces (H bonding) within the water, a thin film of water creeps up the wall. At the same time, the cohesive forces that give rise to surface tension pull the liquid surface

Blood being drawn into a capillary tube for a medical test.

ARIS Presentation Center, Photo Library

Figure 12.20 Shape of water or mercury meniscus in glass. A, Water displays a concave meniscus in a glass tube because the adhesive (H-bond) forces between the H_2O molecules and the O—Si—O groups of the glass are *stronger* than the cohesive (H-bond) forces within the water. **B,** Mercury displays a convex meniscus in a glass tube because the cohesive (metallic bonding) forces within the mercury are *stronger* than the adhesive (dispersion) forces between the mercury and the glass.

 ARIS Presentation Center, Art Library

taut. These adhesive and cohesive forces combine to raise the water level and produce the familiar concave meniscus (Figure 12.20A). The liquid rises until gravity pulling down is balanced by the adhesive forces pulling up.

On the other hand, if you place a glass capillary tube in a dish of mercury, the mercury level in the tube drops below that in the dish. Mercury has a higher surface tension than water (see Table 12.3), which means it has stronger cohesive forces (metallic bonding). The cohesive forces among the mercury atoms are much stronger than the adhesive forces (mostly dispersion) between mercury and glass, so the liquid tends to pull away from the walls. At the same time, the surface atoms are being pulled toward the interior of the mercury by its high surface tension, so the level drops. These combined forces produce a convex meniscus (Figure 12.20B, seen in a laboratory barometer or manometer).

Viscosity

When a liquid flows, the molecules slide around and past each other. A liquid's **viscosity,** its resistance to flow, results from intermolecular attractions that impede this movement. Both gases and liquids flow, but liquid viscosities are *much* higher because the intermolecular forces operate over much shorter distances.

Viscosity decreases with heating, as Table 12.4 shows for water. When molecules move faster at higher temperatures, they can overcome intermolecular forces more easily, so the resistance to flow decreases. Next time you heat cooking oil in a pan, watch the oil flow more easily and spread out in a thin layer as it warms.

Molecular shape plays a key role in a liquid's viscosity. Small, spherical molecules make little contact and, like buckshot in a glass, pour easily. Long molecules make more contact and, like spaghetti in a glass, become entangled and pour slowly. Thus, given the same types of forces, liquids containing longer molecules have higher viscosities. A striking example of a change in viscosity occurs during the making of syrup. Even at room temperature, a concentrated aqueous sugar solution has a higher viscosity than water because of H bonding among the many hydroxyl (—OH) groups on the ring-shaped sugar molecules. When the solution is slowly heated to boiling, the sugar molecules react with each other and link covalently, gradually forming long chains. Hydrogen bonds and dispersion forces occur at many points along the chains, and the resulting syrup is a viscous liquid that pours slowly and clings to a spoon. When a viscous syrup is cooled, it may become stiff enough to be picked up and stretched—into taffy candy. The Gallery on page 459 shows other familiar examples of these three liquid properties.

Table 12.4 Viscosity of Water at Several Temperatures	
Temperature (°C)	Viscosity (N·s/m^2)*
20	1.00×10^{-3}
40	0.65×10^{-3}
60	0.47×10^{-3}
80	0.35×10^{-3}

*The units of viscosity are newton-seconds per square meter.

ARIS Presentation Center, Tables Library

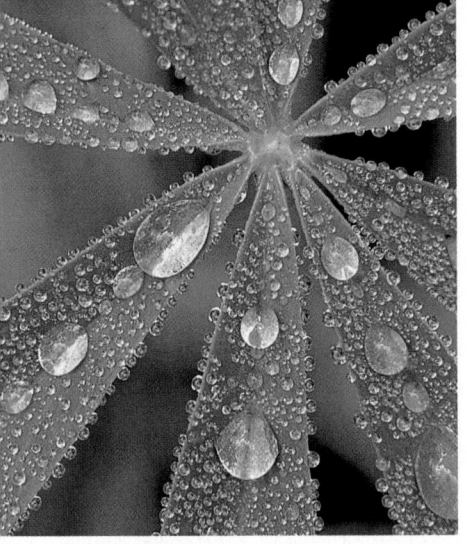

Properties of a Liquid

Of the three states of matter, only liquids combine the ability to flow with the strength that comes from intermolecular contact, and this combination appears in numerous applications.

Beaded droplets on waxy surfaces — Biological application

The adhesive (dipole–induced dipole) forces between water and a nonpolar surface are much weaker than the cohesive (H-bond) forces within water. As a result, water pulls away from a nonpolar surface and forms beaded droplets. You have seen this effect when water beads on a leaf or a freshly waxed car after a rainfall.

Minimizing a surface

In the low-gravity environment of an orbiting space shuttle, the tendency of a liquid to minimize its surface creates perfectly spherical droplets, unlike the flattened drops we see on Earth. For the same reason, bubbles in a soft drink are spherical because the liquid uses the minimum number of molecules needed to surround the gas. A water strider flits across a pond on widespread legs that do not exert enough pressure to exceed the surface tension.

Maintaining motor oil viscosity

To protect engine parts during long drives or in hot weather, when an oil would ordinarily become too thin, motor oils contain additives, called *polymeric viscosity index improvers,* that act as thickeners. As the oil heats up, the additive molecules change shape from compact spheres to spaghetti-like strands and become tangled with the hydrocarbon oil molecules. As a result of the greater dispersion forces, there is an increase in viscosity that compensates for the decrease due to heating. ⊙⊙ **Engineering application**

Capillary action after a shower

Paper and cotton consist of fibers of cellulose, long carbon-containing molecules with many attached hydroxyl (—OH) groups. A towel dries you in two ways:

First, capillary action draws the water molecules away from your body between the closely spaced cellulose molecules.

Second, the water molecules themselves form adhesive H bonds to the —OH groups of cellulose. — Biological application

How a ballpoint pen works ⊙⊙ **Engineering application**

The essential parts of a ballpoint pen are the moving ball and the viscous ink. The material of the ball is chosen for the strong adhesive forces between it and the ink. Cohesive forces within the ink are replaced by those adhesive forces when the ink "wets" the ball. As the ball rolls along the paper, the adhesive forces between ball and ink are overcome by those between ink and paper. The rest of the ink stays in the pen because of its high viscosity.

Ink

Moving ball

Paper surface

459

Section Summary

Surface tension is a measure of the energy required to increase a liquid's surface area. Greater intermolecular forces within a liquid create higher surface tension. • Capillary action, the rising of a liquid through a narrow space, occurs when the forces between a liquid and a solid surface (adhesive) are greater than those within the liquid itself (cohesive). • Viscosity, the resistance to flow, depends on molecular shape and decreases with temperature. Stronger intermolecular forces create higher viscosity.

Figure 12.21 The H-bonding ability of the water molecule. Because it has two O—H bonds and two lone pairs, one H_2O molecule can engage in as many as four H bonds to surrounding H_2O molecules, which are arranged tetrahedrally.

ARIS Presentation Center, Art Library

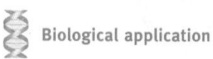
Demo Johnson, J. L. H.; Yalkowsky, S. H. "A Three-Dimensional Model for Water," *J. Chem. Educ.* **2002,** *79,* 1088.

Biological application

Literature Oughton, James; Xu, Silas; Battino, Rubin. "The Purification of Water by Freeze-Thaw or Zone Melting," *J. Chem. Educ.* **2001,** *78,* 1373.

Environmental application

12.5 THE UNIQUENESS OF WATER

Water is absolutely amazing stuff, but it is so familiar that we take it for granted. In fact, water has some of the most unusual properties of any substance. These properties, which are vital to our very existence, arise inevitably from the nature of the H and O atoms that make up the molecule. Each atom attains a filled outer level by sharing electrons in single covalent bonds. With two bonding pairs and two lone pairs around the O atom and a large electronegativity difference in each O—H bond, the H_2O molecule is bent and highly polar. This arrangement is crucial because it allows each water molecule to engage in four H bonds with its neighbors (Figure 12.21). From these fundamental atomic and molecular properties emerges unique and remarkable macroscopic behavior.

Solvent Properties of Water

The *great solvent power* of water is the result of its polarity and exceptional H-bonding ability. It dissolves ionic compounds through ion-dipole forces that separate the ions from the solid and keep them in solution (see Figure 4.2, p. 142). Water dissolves many polar nonionic substances, such as ethanol (CH_3CH_2OH) and glucose ($C_6H_{12}O_6$), by forming H bonds to them. It even dissolves nonpolar gases, such as those in the atmosphere, to a limited extent, through dipole–induced dipole and dispersion forces. Chapter 13 highlights these solvent properties.

Because it can dissolve so many substances, water is the environmental and biological solvent, forming the complex solutions we know as oceans, rivers, lakes, and cellular fluids. Aquatic animals could not survive without dissolved O_2, nor could aquatic plants survive without dissolved CO_2. The coral reefs that girdle the sea bottom in tropical latitudes are composed of carbonates made from dissolved CO_2 and HCO_3^- by tiny marine animals. Life is thought to have begun in a "primordial soup," an aqueous mixture of simple biomolecules from which emerged the larger molecules whose self-sustaining reactions are characteristic of living things. From a chemical point of view, all organisms, from bacteria to humans, can be thought of as highly organized systems of membranes enclosing and compartmentalizing complex aqueous solutions.

Thermal Properties of Water

Water has an exceptionally *high specific heat capacity,* higher than almost any other liquid. Recall from Section 6.3 that heat capacity is a measure of the heat absorbed by a substance for a given temperature rise. When a substance is heated, some of the energy increases the average molecular speed, some increases molecular vibration and rotation, and some is used to overcome intermolecular forces. Because water has so many strong H bonds, it has a high specific heat capacity. With oceans and seas covering 70% of Earth's surface, daytime heat from the Sun causes relatively small changes in temperature, allowing our planet to support life. On the waterless, airless Moon, temperatures can range from 100°C (212°F) to −150°C (−238°F) in the same lunar day. Even in Earth's deserts, where a dense atmosphere tempers these extremes, day-night temperature differences of 40°C (72°F) are common.

Numerous strong H bonds also give water an exceptionally *high heat of vaporization*. A quick calculation shows how essential this property is to our existence. When 1000 g of water absorbs about 4 kJ of heat, its temperature rises 1°C. With a ΔH°_{vap} of 2.3 kJ/g, however, less than 2 g of water must evaporate to keep the temperature constant for the remaining 998 g. The average human adult has 40 kg of body water and generates about 10,000 kJ of heat each day from metabolism. If this heat were used only to increase the average E_k of body water, the temperature rise would mean immediate death. Instead, the heat is converted to E_p as it breaks H bonds and evaporates sweat, resulting in a stable body temperature and minimal loss of body fluid. On a planetary scale, the Sun's energy supplies the heat of vaporization for ocean water. Water vapor, formed in warm latitudes, moves through the atmosphere, and its potential energy is released as heat to warm cooler regions when the vapor condenses to rain. The enormous energy involved in this cycling of water powers storms all over the planet.

Surface Properties of Water

The H bonds that give water its remarkable thermal properties are also responsible for its *high surface tension* and *high capillarity*. Except for some metals and molten salts, water has the highest surface tension of any liquid. This property is vital for surface aquatic life because it keeps plant debris resting on a pond surface, which provides shelter and nutrients for many fish, microorganisms, and insects. Water's high capillarity, a result of its high surface tension, is crucial to land plants. During dry periods, plant roots absorb deep groundwater, which rises by capillary action through the tiny spaces between soil particles.

The Density of Solid and Liquid Water

As you saw in Figure 12.21, through the H bonds of water, each O atom becomes connected to as many as four other O atoms via four H atoms. Continuing this tetrahedral pattern through many molecules in a fixed array leads to the hexagonal, *open structure* of ice shown in Figure 12.22A. The symmetrical beauty of snowflakes, as shown in Figure 12.22B, reflects this hexagonal organization.

This organization explains the negative slope of the solid-liquid line in the phase diagram for water (see Figure 12.9B, p. 448): As pressure is applied, some H bonds break; as a result, some water molecules enter the spaces. The crystal structure breaks down, and the sample liquefies.

The large spaces within ice give *the solid state a lower density than the liquid state*. When the surface of a lake freezes in winter, the ice floats on the liquid water below. If the solid were denser than the liquid, as is true for nearly every other substance, the surface of a lake would freeze and sink repeatedly until the entire lake was solid. Aquatic life would not survive from year to year.

The density of water changes in a complex way. When ice melts at 0°C, the tetrahedral arrangement around each O atom breaks down, and the loosened molecules pack much more closely, filling spaces in the collapsing solid structure. As a result, water is most dense (1.000 g/mL) at around 4°C (3.98°C). With more heating, the density decreases through normal thermal expansion.

This change in density is vital for freshwater life. As lake water becomes colder in the fall and early winter, it becomes more dense *before* it freezes. Similarly, in spring, less dense ice thaws to form more dense water *before* the water expands. During both of these seasonal density changes, the top layer of water reaches the high-density point first and sinks. The next layer of water rises because it is slightly less dense, reaches 4°C, and likewise sinks. This sinking and rising distribute nutrients and dissolved oxygen.

The expansion and contraction of water dramatically affect the land as well. When rain fills the crevices in rocks and then freezes, great outward stress is applied, to be relieved only when the ice melts. In time, this repeated freeze-thaw

A

B

Figure 12.22 The hexagonal structure of ice. A, The geometric arrangement of the H bonds in H_2O leads to the open, hexagonally shaped crystal structure of ice. Thus, when liquid water freezes, the volume *increases*. **B,** The delicate six-pointed beauty of snowflakes reflects the hexagonal crystal structure of ice.

Figure 12.23 The expansion and contraction of water. As water freezes and thaws repeatedly, the stress due to expanding and contracting can break rocks. Over eons of geologic time, this process creates sand and soil.

ARIS Presentation Center, Photo Library

stress cracks the rocks apart (Figure 12.23). Over eons of geologic change, this effect has helped to produce the sand and soil of the planet.

The properties of water and their far-reaching consequences illustrate the theme that runs throughout this book: the world we know is the stepwise, macroscopic outgrowth of the atomic world we seek to know. Figure 12.24 offers a fanciful summary of water's unique properties in the form of a tree whose "roots" (atomic properties) give foundation to the base of the "trunk" (molecular properties), which strengthens the next portion (polarity), from which grows another portion (H bonding). These two portions of the trunk "branch" out to the other properties of water and their global effects.

Section Summary

The atomic properties of hydrogen and oxygen atoms result in the water molecule's bent shape, polarity, and H-bonding ability. • These properties of water enable it to dissolve many ionic and polar compounds. • Water's H bonding results in a high specific heat capacity and a high heat of vaporization, which combine to give Earth and its organisms a narrow temperature range. • H bonds also confer high surface tension and capillarity, which are essential to plant and animal life. • Water expands on freezing because of its H bonds, which lead to an open crystal structure in ice. Seasonal density changes foster nutrient mixing in lakes. Seasonal freeze-thaw stress on rocks results eventually in soil formation.

Figure 12.24 The macroscopic properties of water and their atomic and molecular "roots."

12.6 THE SOLID STATE: STRUCTURE, PROPERTIES, AND BONDING

Stroll through the mineral collection of any school or museum, and you'll be struck by the extraordinary variety and beauty of these solids. In this section, we first discuss the general structural features of crystalline solids and then examine a laboratory method for studying them. We survey the properties of the major types of solids and find the whole range of intermolecular forces at work. We then present a model for bonding in solids that explains many of their properties.

Structural Features of Solids

We can divide solids into two broad categories based on the orderliness of their shapes, which in turn is based on the orderliness of their particles. **Crystalline solids** generally have a well-defined shape, as shown in Figure 12.25, because their particles—atoms, molecules, or ions—occur in an orderly arrangement. **Amorphous solids** have poorly defined shapes because their particles lack long-range ordering throughout the sample. In this discussion, we focus, for the most part, on crystalline solids.

Figure 12.25 **The striking beauty of crystalline solids.** **A,** Wulfanite. **B,** Quartz (amethyst). **C,** Beryl (emerald). **D,** Pyrite. **E,** Barite (*left*) on calcite (*right*).   ARIS Presentation Center, Photo Library

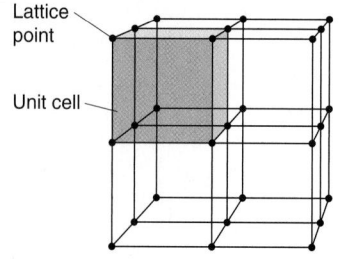

A Portion of 3-D lattice

B 2-D analogy for unit cell and lattice

Figure 12.26 **The crystal lattice and the unit cell. A,** The lattice is an array of points that defines the positions of the particles in a crystal structure. It is shown here as points connected by lines. A unit cell *(colored)* is the simplest array of points that, when repeated in all directions, produces the lattice. A simple cubic unit cell, one of 14 types in nature, is shown. **B,** A checkerboard is a two-dimensional analogy for a lattice.

ARIS Presentation Center, Art Library

Animation: Cubic Unit Cells and Their Origins
ARIS Presentation Center
Chapter 12

Literature Keiter, Richard L.; Puzey, Whitney L.; Blitz, Erin A. "Density Visualization," *J. Chem. Educ.* **2006,** *83,* 1629.

The Crystal Lattice and the Unit Cell If you could see the particles within a crystal, you would find them packed tightly together in an orderly, three-dimensional array. Consider the simplest case, in which all the particles are *identical* spheres, and imagine a point at the same location within each particle in this array, say, at the center. The points form a regular pattern throughout the crystal that is called the crystal **lattice.** Thus, *the lattice consists of all points with identical surroundings.* Put another way, if the rest of each particle were removed, leaving only the lattice point, and you were transported from one point to another, you would not be able to tell that you had moved. Keep in mind that there is no pre-existing array of lattice points; rather, *the arrangement of the points within the particles defines the lattice.*

Figure 12.26A shows a portion of a lattice and the **unit cell,** the *smallest* portion of the crystal that, if repeated in all three directions, gives the crystal. A two-dimensional analogy for a unit cell and a crystal lattice can be seen in a checkerboard (as shown in Figure 12.26B), a section of tiled floor, a strip of wall-paper, or any other pattern that is constructed from a repeating unit. The **coordination number** of a particle in a crystal is the number of nearest neighbors surrounding it.

There are 7 crystal systems and 14 types of unit cells that occur in nature, but we will be concerned primarily with the *cubic system,* which gives rise to the cubic lattice. The solid states of a majority of metallic elements, some covalent compounds, and many ionic compounds occur as cubic lattices. (We also describe the hexagonal unit cell a bit later.) There are three types of cubic unit cells within the cubic system:

1. In the **simple cubic unit cell,** shown in Figure 12.27A, the centers of eight identical particles define the corners of a cube. Attractions pull the particles together, so they touch along the cube's edges; but they do not touch diagonally along the cube's faces or through its center. The coordination number of each particle is 6: four in its own layer, one in the layer above, and one in the layer below.
2. In the **body-centered cubic unit cell,** shown in Figure 12.27B, identical particles lie at each corner *and* in the center of the cube. Those at the corners do not touch each other, but they all touch the one in the center. Each particle is surrounded by eight nearest neighbors, four above and four below, so the coordination number is 8.
3. In the **face-centered cubic unit cell,** shown in Figure 12.27C, identical particles lie at each corner *and* in the center of each face but not in the center of the cube. Those at the corners touch those in the faces but not each other. The coordination number is 12.

One unit cell lies adjacent to another throughout the crystal, with no gaps, so a particle at a corner or face is *shared* by adjacent unit cells. As you can see from Figure 12.27 (third row from the top), in the three cubic unit cells, the particle at each corner is part of eight adjacent cells, so one-eighth of each of these particles belongs to each unit cell (bottom row). There are eight corners in a cube, so each simple cubic unit cell contains $8 \times \frac{1}{8}$ particle = 1 particle. The body-centered cubic unit cell contains one particle from the eight corners and one in the center, for a total of two particles; and the face-centered cubic unit cell contains four particles, one from the eight corners and three from the half-particles in each of the six faces.

A Simple cubic

B Body-centered cubic

C Face-centered cubic

Coordination number = 6

Coordination number = 8

Coordination number = 12

$\frac{1}{8}$ atom
at 8 corners

Atoms/unit cell = $\frac{1}{8}$ × 8 = 1

$\frac{1}{8}$ atom
at 8 corners

1 atom
at center

Atoms/unit cell = ($\frac{1}{8}$ × 8) + 1 = 2

$\frac{1}{8}$ atom
at 8 corners

$\frac{1}{2}$ atom
at 6 faces

Atoms/unit cell = ($\frac{1}{8}$ × 8) + ($\frac{1}{2}$ × 6) = 4

Figure 12.27 **The three cubic unit cells. A,** Simple cubic unit cell. **B,** Body-centered cubic unit cell. **C,** Face-centered cubic unit cell. *Top row:* Cubic arrangements of atoms in expanded view. *Second row:* Space-filling view of these cubic arrangements. All atoms are identical but, for clarity, corner atoms are blue, body-centered atoms pink, and face-centered atoms yellow. *Third row:* A unit cell (*shaded blue*) in a portion of the crystal. The number of nearest neighbors around one particle (*dark blue in center*) is the coordination number. *Bottom row:* The total numbers of atoms in the actual unit cell. The simple cubic has one atom; the body-centered has two; and the face-centered has four.

 Transparency/ARIS Presentation Center, Active Art Library

The efficient packing of fruit.

ARIS Presentation Center, Photo Library

Literature Dungey, Keenan E. "Kixium
Monolayers: A Simple Alternative to the Bubble Raft
Model for Close-Packed Spheres," *J. Chem. Educ.*
2000, *77,* 618.

Packing Efficiency and the Creation of Unit Cells Unit cells result from the various ways atoms pack together, which are very similar to the ways macroscopic spheres—marbles, golf balls, fruit—are packed (see photo).

For particles of the same size, *the higher the coordination number of the crystal is, the greater the number of particles in a given volume.* Therefore, as the coordination numbers in Figure 12.27 indicate, a crystal structure based on the face-centered cubic unit cell has more particles packed into a given volume than one based on the body-centered cubic unit cell, which has more than one based on the simple cubic unit cell. Let's see how to pack *identical* spheres to create these unit cells and the hexagonal unit cell as well:

1. *The simple cubic unit cell.* Suppose we arrange the first layer of spheres as shown in Figure 12.28A. Note the large diamond-shaped spaces (cutaway portion). If we place the next layer of spheres *directly above* the first, as shown in Figure 12.28B, we obtain an arrangement based on the *simple* cubic unit cell. By calculating the **packing efficiency** of this arrangement—the percentage of the total volume occupied by the spheres themselves—we find that only 52% of the available unit-cell volume is occupied by spheres and 48% consists of the empty space between them. This is a very inefficient way to pack spheres, so neither fruit nor atoms are usually packed this way.

2. *The body-centered cubic unit cell.* Rather than placing the second layer directly above the first, we can use space more efficiently by placing the spheres (colored differently for clarity) on the diamond-shaped spaces in the first layer, as shown in Figure 12.28C. Then we pack the third layer onto the spaces in the second so that the first and third layers line up vertically. This arrangement is based on the *body-centered* cubic unit cell, and its packing efficiency is 68%—much higher than for the simple cubic unit cell. Several metallic elements, including chromium, iron, and all the Group 1A(1) elements, have a crystal structure based on the body-centered cubic unit cell.

3. *The hexagonal and face-centered cubic unit cells.* Spheres can be packed even more efficiently. First, we shift rows in the bottom layer so that the large diamond-shaped spaces become smaller triangular spaces. Then we place the second layer over these spaces. Figure 12.28D shows this arrangement, with the first layer labeled *a* (*orange*) and the second layer *b* (*green*).

We can place the third layer of spheres in two different ways, and how we do so gives rise to two different unit cells. If you look carefully at the spaces formed in layer *b* of Figure 12.28D, you'll see that some are orange because they lie above *spheres* in layer *a*, whereas others are white because they lie above *spaces* in layer *a*. If we place the third layer of spheres (*orange*) over the orange spaces (down and left to Figure 12.28E), they lie directly over spheres in layer *a*, and we obtain an *abab. . .* layering pattern because every other layer is placed identically. This gives **hexagonal closest packing,** which is based on the *hexagonal unit cell.* On the other hand, if we place the third layer of spheres (*blue*) over the white spaces (down and right to Figure 12.28F), the spheres lie over spaces in layer *a*. This placement is different from both layers *a* and *b*, so we obtain an *abcabc. . .* pattern. This gives **cubic closest packing,** which is based on the *face-centered* cubic unit cell.

The packing efficiency of both hexagonal and cubic closest packing is 74%, and the coordination number of both is 12. There is no way to pack identical spheres more efficiently. Most metallic elements crystallize in either of these arrangements. Magnesium, titanium, and zinc are some elements that adopt the hexagonal structure; nickel, copper, and lead adopt the cubic structure, as do many ionic compounds and other substances, such as frozen carbon dioxide, methane, and most noble gases.

Our understanding of solids is based on the ability to "see" their crystal structures. Two techniques for doing this are described in the Tools of the Laboratory essay on page 468.

A

B Simple cubic (52%)

C Body-centered cubic (68%)

Layer *a*

Layer *b*

Layer *c*

Next layer *a*

To hexagonal closest packing

To cubic closest packing

D Closest packing of first and second layers

Cutaway side view showing hexagonal unit cell

Expanded side view

Expanded side view

Cutaway side view showing face-centered cubic unit cell

Tilted side view of unit cell

E Hexagonal closest packing (*abab*...) (74%)

F Cubic closest packing (*abcabc*...) (74%)

Figure 12.28 **Packing identical spheres.** **A,** In the first layer, each sphere lies next to another horizontally and vertically; note the large diamond-shaped spaces (*see cutaway*). **B,** If the spheres in the next layer lie *directly* over those in the first, the packing is based on the *simple cubic* unit cell (*pale orange cube, lower right corner*). **C,** If the spheres in the next layer lie in the diamond-shaped spaces of the first layer, the packing is based on the *body-centered cubic* unit cell (*lower right corner*). **D,** The closest possible packing of the first layer (*layer a, orange*) is obtained by shifting every other row in part A, thus reducing the diamond-shaped spaces to smaller triangular spaces. The spheres of the second layer (*layer b, green*) are placed above these spaces; note the orange and white spaces that result. **E,** Follow the left arrow

from part D to obtain hexagonal closest packing. When the third layer (*next layer a, orange*) is placed directly over the first, that is, over the orange spaces, we obtain an *abab.* . . pattern. Rotating the layers 90° produces the side view, with the hexagonal unit cell shown as a cutaway segment, and the expanded side view. **F,** Follow the right arrow from part D to obtain cubic closest packing. When the third layer (*layer c, blue*) covers the white spaces, it lies in a different position from the first *and* second layers to give an *abcabc.* . . pattern. Rotating the layers 90° shows the side view, with the face-centered cubic unit cell as a cutaway, and a further tilt shows the unit cell clearly; finally, we see the expanded view. The packing efficiency for each type of unit cell is given in parentheses.

 Transparency/ARIS Presentation Center, Art Library

Tools of the Laboratory

X-Ray Diffraction Analysis and Scanning Tunneling Microscopy

In this chapter and in Chapter 10, we have discussed crystal structures and molecular shapes as if they have actually been seen. Various tools exist for measuring the dimensions of the molecular world, and two of the most powerful are **x-ray diffraction analysis** and **scanning tunneling microscopy**.

X-Ray Diffraction Analysis

In Chapter 7, we discussed wave diffraction and saw how interference patterns of bright and dark regions appear when light passes through adjacent slits spaced at the distance of the light's wavelength (see Figure 7.5, p. 273). Because x-ray wavelengths are about the same size as the spaces between layers of particles in many solids, the spaces act as "slits" and diffract x-rays.

Let's see how this technique is used to measure a key feature of a crystal structure: the distance (d) between layers of atoms. Figure B12.1 depicts a side view of two layers in a simplified lattice. Two waves impinge on the crystal at an angle θ and are diffracted at the same angle by adjacent layers. When the first wave strikes the top layer and the second strikes the next layer, the waves are *in phase* (peaks aligned with peaks and troughs with troughs). If they are still in phase after being diffracted, a spot appears on a nearby photographic plate. Note that this will occur only if the additional distance traveled by the second wave (DE + EF in the figure) is a whole number of wavelengths, nλ, where n is an integer (1, 2, 3, and so on). From trigonometry, we find that

$$n\lambda = 2d \sin \theta$$

where θ is the known angle of incoming light, λ is its known wavelength, and d is the unknown distance between the layers in the crystal. This relationship is the *Bragg equation,* named for W. H. Bragg and his son W. L. Bragg, who shared the Nobel Prize in physics in 1915 for their work on crystal structure analysis.

Rotating the crystal changes the angle of incoming radiation and produces a different set of spots, eventually yielding a complete diffraction pattern, which is analyzed mathematically to

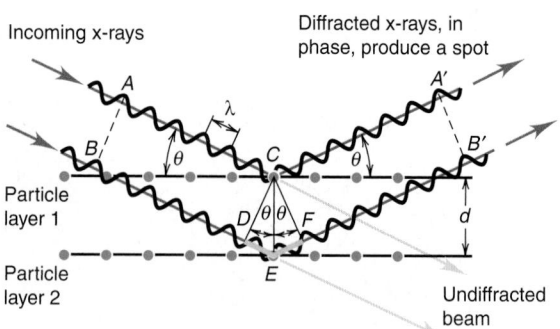

Figure B12.1 Diffraction of x-rays by crystal planes. In-phase x-ray beams *A* and *B* pass into a crystal at angle θ and are diffracted. If the additional distance beam *B* travels (*DE* + *EF*) is a whole number of wavelengths, the beams remain in phase and create a spot. From the pattern of spots, the distance *d* between layers can be calculated.

obtain the dimensions of the crystal and the distances and angles within the lattice (Figure B12.2). Modern x-ray diffraction equipment automatically rotates the crystal and measures thousands of diffractions, and a computer calculates the parameters of interest.

X-ray diffraction analysis is used to answer questions in many branches of chemistry, but its greatest impact has been in biochemistry. It has shown that DNA exists as a double helix, and it is currently helping biochemists learn how a protein's three-dimensional structure is related to its function.

Scanning Tunneling Microscopy

This technique, a much newer method than x-ray diffraction analysis, is used to observe surfaces on the atomic scale. It was invented in the early 1980s by Gerd Binnig and Heinrich Rohrer, two Swiss physicists who won the Nobel Prize in physics in 1986 for their work. The technique is based on the idea that an electron

---

One piece of data that can be obtained from x-ray crystallography is the edge length of the unit cell. Figure 12.29 shows how we use this parameter and apply the Pythagorean theorem and some simple arithmetic to find atomic (or ionic) radii in the three cubic unit cells. Sample Problem 12.5 uses this approach.

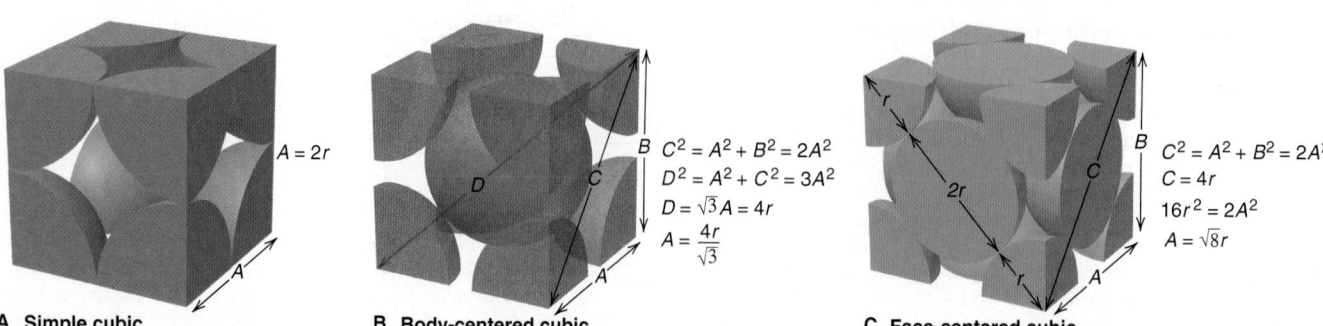

A Simple cubic

$A = 2r$

B Body-centered cubic

$$C^2 = A^2 + B^2 = 2A^2$$
$$D^2 = A^2 + C^2 = 3A^2$$
$$D = \sqrt{3}A = 4r$$
$$A = \frac{4r}{\sqrt{3}}$$

C Face-centered cubic

$$C^2 = A^2 + B^2 = 2A^2$$
$$C = 4r$$
$$16r^2 = 2A^2$$
$$A = \sqrt{8}r$$

Figure 12.29 Edge length and atomic (ionic) radius in the three cubic unit cells. Simple **(A)**, body-centered **(B)**, and face-centered **(C)** cubic unit cells are shown with the geometric derivation needed to

find the edge length (*A*) of each. Since the corners of the unit cells form right angles, the Pythagorean theorem applies. In the body-centered cubic unit cell, the body diagonal (*D*) goes through the central sphere.

Figure B12.2 Formation of an x-ray diffraction pattern of the protein hemoglobin. A, A sample of crystalline hemoglobin is rotated to obtain different angles of incoming and diffracted x-rays. **B,** The diffraction pattern is a complex series of spots. (Large white spot is an artifact.) **C,** Computerized analysis relates the pattern to the crystal structure, and the data are used to generate a picture of the molecule.

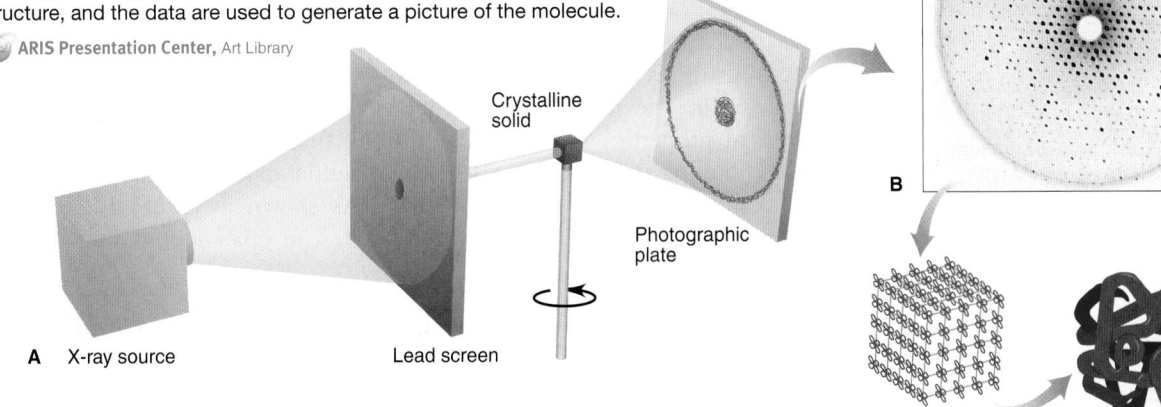

Spots from diffracted x-rays

Crystalline solid

Photographic plate

B

A X-ray source

Lead screen

C Protein crystal   Protein molecule

in an atom has a small probability of existing far from the nucleus, so given the right conditions, it can move ("tunnel") to end up closer to another atom.

In practice, the tunneling electrons create a current that can be used to image the atoms of an adjacent surface. An extremely sharp tungsten-tipped probe, the source of the tunneling electrons, is placed very close (about 0.5 nm) to the surface under study. A small electric potential is applied across this minute gap to increase the probability that the electrons will tunnel across it. The size of the gap is kept constant by maintaining a constant tunneling current generated by the moving electrons. For this to occur, the probe must move tiny distances up and down, thus following the atomic contour of the surface. This movement is electronically monitored, and after many scans, a three-dimensional map of the surface is obtained. The method has revealed magnificent images of atoms and molecules coated on surfaces and is being used to study many aspects of surfaces, such as the nature of defects and the adhesion of films (Figure B12.3).

A   **B**

Figure B12.3 Scanning tunneling micrographs. A, A gold surface. **B,** A chain of cesium atoms (red) on a gallium arsenide surface.

SAMPLE PROBLEM 12.5 Determining Atomic Radius from Crystal Structure

PROBLEM Copper has the most essential industrial applications of the three coinage metals [Group 1B(11)]. Its crystal structure adopts cubic closest packing, and the edge length of the unit cell is 361.5 pm. What is the atomic radius of copper?

PLAN Copper crystallizes in cubic closest packing. This structure has a face-centered cubic unit cell, and we are given the cell edge length. We refer to Figure 12.29C, with edge length $A = 361.5$ pm, and solve for r (see margin).

SOLUTION Using the Pythagorean theorem to find C, the diagonal of the cell's face:

$$C = \sqrt{A^2 + B^2}$$

The unit cell is a cube, so $A = B$. Therefore,

$$C = \sqrt{2A^2} = \sqrt{2(361.5 \text{ pm})^2} = 511.2 \text{ pm}$$

Finding r: $C = 4r$. Therefore, $r = 511.2$ pm/4 = 127.8 pm.

CHECK Rounding and quickly checking the math gives

$$C = \sqrt{2(4\times10^2 \text{ pm})^2} = \sqrt{2(16\times10^4 \text{ pm}^2)}, \text{ or } \sim500\text{–}600 \text{ pm; thus, } r \approx 125\text{–}150 \text{ pm}.$$

The actual value for copper is 128 pm (see Figure 8.15, p. 319).

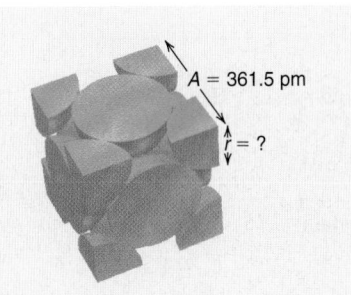

$A = 361.5$ pm

$r = ?$

FOLLOW-UP PROBLEM 12.5 Iron crystallizes in a body-centered cubic structure. If the atomic radius of Fe is 126 pm, find the edge length (in nm) of the unit cell.

469

Types and Properties of Crystalline Solids

Now we can turn to the five most important types of solids, which are summarized in Table 12.5. Each is defined by the type(s) of particle(s) in the crystal, which determines the interparticle forces. You may want to review the bonding models (Chapter 9) to clarify how they relate to the properties of different solids.

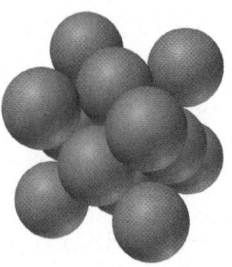

Figure 12.30 Cubic closest packing of frozen argon (face-centered cubic unit cell).

ARIS Presentation Center, Art Library

Atomic Solids Individual atoms that are held together by dispersion forces form an **atomic solid.** The noble gases [Group 8A(18)] are the only examples, and their physical properties reflect the very weak forces among the atoms. Their melting and boiling points and heats of vaporization and fusion are all very low, rising smoothly with increasing molar mass. As shown in Figure 12.30, argon crystallizes in a cubic closest packed structure. The other atomic solids do so as well.

Molecular Solids In the many thousands of **molecular solids,** individual molecules occupy the lattice points. For example, methane crystallizes in a face-centered cubic structure, shown in Figure 12.31 with the center of each carbon as the lattice point.

Various combinations of dipole-dipole, dispersion, and H-bonding forces are at work in molecular solids, which accounts for their wide range of physical properties. Dispersion forces are the principal force acting in nonpolar substances, so melting points generally increase with molar mass (Table 12.5). Among polar molecules, dipole-dipole forces and, where possible, H bonding are important as well. Except for those substances consisting of the simplest molecules, molecular solids have higher melting points than the atomic solids (noble gases). Nevertheless, intermolecular forces are still relatively weak, so the melting points are much lower than those of ionic, metallic, and network covalent solids.

Figure 12.31 Cubic closest packing of frozen methane. Only one CH$_4$ molecule is shown as a space-filling model; the rest are represented as simple spheres.

ARIS Presentation Center, Art Library

Transparency/ARIS Presentation Center, Tables Library

Table 12.5 Characteristics of the Major Types of Crystalline Solids

Type	Particle(s)	Interparticle Forces	Physical Properties	Examples [mp, °C]
Atomic	Atoms	Dispersion	Soft, very low mp, poor thermal and electrical conductors	Group 8A(18) [Ne (-249) to Rn (-71)]
Molecular	Molecules	Dispersion, dipole-dipole, H bonds	Fairly soft, low to moderate mp, poor thermal and electrical conductors	*Nonpolar** O$_2$ [-219], C$_4$H$_{10}$ [-138] Cl$_2$ [-101], C$_6$H$_{14}$ [-95], P$_4$ [44.1] *Polar* SO$_2$ [-73], CHCl$_3$ [-64], HNO$_3$ [-42], H$_2$O [0.0], CH$_3$COOH [17]
Ionic	Positive and negative ions	Ion-ion attraction	Hard and brittle, high mp, good thermal and electrical conductors when molten	NaCl [801] CaF$_2$ [1423] MgO [2852]
Metallic	Atoms	Metallic bond	Soft to hard, low to very high mp, excellent thermal and electrical conductors, malleable and ductile	Na [97.8] Zn [420] Fe [1535]
Network covalent	Atoms	Covalent bond	Very hard, very high mp, usually poor thermal and electrical conductors	SiO$_2$ (quartz) [1610] C (diamond) [~4000]

*Nonpolar molecular solids are arranged in order of increasing molar mass. Note the correlation with increasing melting point (mp).

Na⁺

Cl⁻

Cl⁻ Na⁺

A NaCl unit cell   B

Ionic Solids In crystalline **ionic solids,** the unit cell contains particles with whole, rather than partial, charges. As a result, the interparticle forces (ionic bonds) are *much* stronger than the van der Waals forces in atomic or molecular solids. To maximize attractions, cations are surrounded by as many anions as possible, and vice versa, with *the smaller of the two ions lying in the spaces (holes) formed by the packing of the larger.* Since the unit cell is the smallest portion of the crystal that maintains the overall spatial arrangement, it is also the smallest portion that maintains the overall chemical composition. In other words, *the unit cell has the same cation/anion ratio as the empirical formula.*

Ionic compounds adopt several different crystal structures, but many use cubic closest packing. Let's first consider two structures that have a 1/1 ratio of ions. The *sodium chloride structure* is found in many compounds, including most of the alkali metal [Group 1A(1)] halides and hydrides, the alkaline earth metal [Group 2A(2)] oxides and sulfides, several transition-metal oxides and sulfides, and most of the silver halides. To visualize this structure, first imagine Cl⁻ anions and Na⁺ cations organized separately in face-centered cubic (cubic closest packing) arrays. The crystal structure arises when these two arrays penetrate each other such that the smaller Na⁺ ions end up in the holes between the larger Cl⁻ ions, as shown in Figure 12.32A. Thus, each Na⁺ is surrounded by six Cl⁻, and vice versa (coordination number = 6). Figure 12.32B is a space-filling depiction of the unit cell showing a face-centered cube of Cl⁻ ions with Na⁺ ions between them. Note the four Cl⁻ [$(8 \times \frac{1}{8}) + (6 \times \frac{1}{2}) = 4$ Cl⁻] and four Na⁺ [$(12 \times \frac{1}{4}) + 1$ in the center = 4 Na⁺], giving a 1/1 ion ratio.

Another structure with a 1/1 ion ratio is the *zinc blende (ZnS) structure.* It can be pictured as two face-centered cubic arrays, one of Zn²⁺ ions and the other of S²⁻ ions, interpenetrating such that each ion is tetrahedrally surrounded by four ions of opposite charge (coordination number = 4). Note the 1/1 ratio of ions in the unit cell shown in Figure 12.33. Many other compounds, including AgI, CdS, and the Cu(I) halides adopt the zinc blende structure.

Figure 12.33 The zinc blende structure. Zinc sulfide adopts the zinc blende structure. **A,** The translucent cube shows a face-centered cubic array of four [$(8 \times \frac{1}{8}) + (6 \times \frac{1}{2}) = 4$] S²⁻ ions (*yellow*) tetrahedrally surrounding each of four Zn²⁺ ions (*gray*) to give the 1/1 empirical formula. (Bonds are shown only for clarity.) **B,** The actual unit cell slightly expanded to show the interior ions.

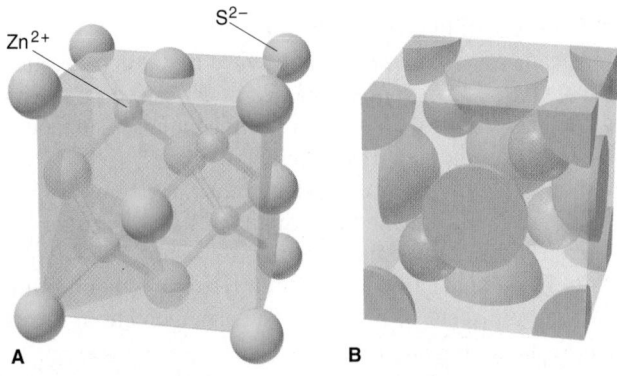

Zn²⁺   S²⁻

A   B

Figure 12.34 The fluorite structure.
Calcium fluoride adopts the fluorite struc-ture. **A,** The translucent cube shows a face-centered cubic array of four Ca^{2+} ions (*blue*) tetrahedrally surrounding each of eight F^- ions (*yellow*) to give the 4/8, or 1/2, ratio. **B,** The actual unit cell (slightly expanded).

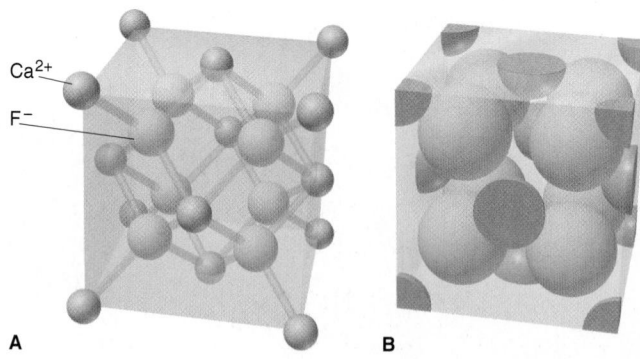

The *fluorite (CaF₂) structure* is common among salts with a 1/2 cation/anion ratio, especially those having relatively large cations and relatively small anions. In the case of CaF_2, the unit cell is a face-centered cubic array of Ca^{2+} ions with F^- ions occupying *all* eight available holes (Figure 12.34). This results in a Ca^{2+}/F^- ratio of 4/8, or 1/2. SrF_2 and $BaCl_2$ have the fluorite structure also. The *antifluorite structure* is often seen in compounds having a cation/anion ratio of 2/1 and a relatively large anion (for example, K_2S). In this structure, the cations occupy all eight holes formed by the cubic closest packing of the anions, just the opposite of the fluorite structure.

The properties of ionic solids are a direct consequence of the *fixed ion positions* and *very strong interionic forces,* which create a high lattice energy. Thus, ionic solids typically have high melting points and low electrical conductivities. When a large amount of heat is supplied and the ions gain enough kinetic energy to break free of their positions, the solid melts and the mobile ions conduct a current. Ionic compounds are hard because only a strong external force can change the relative positions of many trillions of interacting ions. If enough force *is* applied to move them, ions of like charge are brought near each other, and their repulsions crack the crystal (see Figure 9.8, p. 349).

Metallic Solids In contrast to the weak dispersion forces between the atoms in atomic solids, powerful metallic bonding forces hold individual atoms together in **metallic solids.** Most metallic elements crystallize in one of the two closest packed structures (Figure 12.35).

The properties of metals—high electrical and thermal conductivity, luster, and malleability—result from the presence of delocalized electrons, the essential fea-ture of metallic bonding (introduced in Section 9.6). Metals have a wide range of melting points and hardnesses, which are related to the packing efficiency of the crystal structure and the number of valence electrons available for bonding. For example, Group 2A metals are harder and higher melting than Group 1A metals (see Figure 9.27, p. 370), because the 2A metals have closest packed structures (except Ba) and twice as many delocalized valence electrons.

Network Covalent Solids In the final type of crystalline solid, separate particles are not present. Instead, strong covalent bonds link the atoms together through-out a **network covalent solid.** As a consequence of this strong bonding, all these substances have extremely high melting and boiling points, but their conductiv-ity and hardness depend on the details of their bonding.

The two common crystalline forms of elemental carbon are examples of net-work covalent solids. Although graphite and diamond have the same composition, their properties are strikingly different, as Table 12.6 shows. Graphite occurs as stacked flat sheets of hexagonal carbon rings with a strong σ-bond framework and delocalized π bonds, reminiscent of benzene. The arrangement of hexagons looks

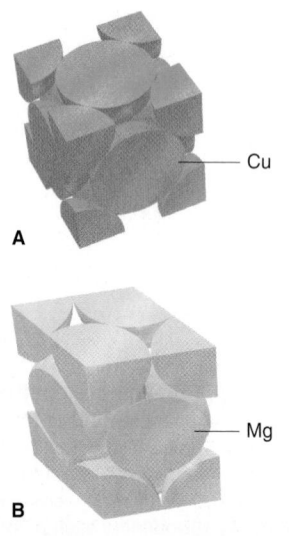

Figure 12.35 Crystal structures of metals. Most metals crystallize in one of the two closest packed arrangements. **A,** Copper adopts cubic closest packing. **B,** Magnesium adopts hexagonal closest packing.

Table 12.6	Comparison of the Properties of Diamond and Graphite		
Property	Graphite		Diamond
Density (g/cm^3)	2.27		3.51
Hardness	<1 (very soft)		10 (hardest)
Melting point (K)	4100		4100
Color	Shiny black		Colorless transparent
Electrical conductivity	High (along sheet)		None
ΔH°_{rxn} for combustion (kJ/mol)	−393.5		−395.4
ΔH°_{f}(kJ/mol)	0 (standard state)		1.90

like chicken wire or honeycomb. Whereas the π-bonding electrons of benzene are delocalized over one ring, those of graphite are delocalized over the entire sheet. These mobile electrons allow graphite to conduct electricity, but only in the plane of the sheets. Graphite is a common electrode material and was once used for lightbulb filaments. The sheets interact via dispersion forces. Common impurities, such as O_2, that lodge between the sheets allow them to slide past each other easily, which explains why graphite is so soft. Diamond crystallizes in a face-centered cubic unit cell, with each carbon atom tetrahedrally surrounded by four others in one virtually endless array. Strong, single bonds throughout the crystal make diamond the hardest natural substance known. Because of its localized bonding electrons, diamond (like most network covalent solids) is unable to conduct electricity.

By far the most important network covalent solids are the *silicates*. They utilize a variety of bonding patterns, but nearly all consist of extended arrays of covalently bonded silicon and oxygen atoms. Quartz (SiO_2) is a common example. We'll discuss silicates, which form the structure of clays, rocks, and many minerals, when we consider the chemistry of silicon in Chapter 14.

Amorphous Solids

Amorphous solids are noncrystalline. Many have small, somewhat ordered regions connected by large disordered regions. Charcoal, rubber, and glass are some familiar examples of amorphous solids.

The process that forms quartz glass is typical of that for many amorphous solids. Crystalline quartz (SiO_2) has a cubic closest packed structure. The crystalline form is melted, and the viscous liquid is cooled rapidly to prevent it from recrystallizing. The chains of silicon and oxygen atoms cannot orient themselves quickly enough into an orderly structure, so they solidify in a distorted jumble containing many gaps and misaligned rows (Figure 12.36). The absence of regularity in the structure confers some properties of a liquid; in fact, glasses are sometimes referred to as *supercooled liquids.*

Literature Hawkes, Stephen J. "Glass Doesn't Flow and Doesn't Crystallize and It Isn't a Liquid," *J. Chem. Educ.* **2000**, *77*, 846.

Figure 12.36 Crystalline and amorphous silicon dioxide. A, The atomic arrangement of cristobalite, one of the many crystalline forms of silica (SiO_2), shows the regularity of cubic closest packing. **B,** The atomic arrangement of a quartz glass is amorphous with a generally disordered structure.

Bonding in Solids: Molecular Orbital Band Theory

Chapter 9 introduced a qualitative model of metallic bonding that pictures metal ions submerged in a "sea" of mobile, delocalized valence electrons. Quantum mechanics offers another model, an extension of molecular orbital (MO) theory, called **band theory.** It is more quantitative than the electron-sea model, and therefore more useful. We'll pay special attention to bonding in metals and differences in electrical conductivity of metals, metalloids, and nonmetals.

Recall that when two atoms form a diatomic molecule, their atomic orbitals (AOs) combine to form an equal number of molecular orbitals (MOs). Let's consider lithium as an example. Figure 12.37 shows the formation of MOs in lithium. In dilithium, Li_2, each atom has four valence orbitals (one $2s$ and three $2p$). (Recall that in Section 11.3, we focused primarily on the $2s$ orbitals.) They combine to form eight MOs, four bonding and four antibonding, spread over both atoms. If two more Li atoms combine, they form Li_4, a slightly larger aggregate, with 16 delocalized MOs. As more Li atoms join the cluster, more MOs are created, their energy levels lying closer and closer together. Extending this process to a 7-g sample of lithium metal (the molar mass) results in 1 mol of Li atoms (Li_{N_A}) combining to form an extremely large number ($4 \times$ Avogadro's number) of delocalized MOs, with *energies so closely spaced that they form a continuum, or band, of MOs.* It is almost as though the entire piece of metal were one enormous Li molecule.

The band model proposes that the lower energy MOs are occupied by the valence electrons and make up the **valence band.** The empty MOs that are higher in energy make up the **conduction band.** In Li metal, the valence band is derived from the $2s$ AOs, and the conduction band is derived mostly from an intermingling of the $2s$ and $2p$ AOs. In Li_2, two valence electrons fill the lowest energy MO and leave the antibonding MO empty. Similarly, in Li metal, 1 mol of valence electrons fills the valence band and leaves the conduction band empty.

The key to understanding metallic properties is that *in metals, the valence and conduction bands are contiguous,* which means that electrons can jump from

Figure 12.37 The band of molecular orbitals in lithium metal. Lithium atoms contain four valence orbitals, one $2s$ and three $2p$ (*left*). When two lithium atoms combine (Li_2), their AOs form eight MOs within a certain range of energy. Four Li atoms (Li_4) form 16 MOs. A mole of Li atoms forms $4N_A$ MOs (N_A = Avogadro's number). The orbital energies are so close together that they form a continuous band. The valence electrons enter the lower energy portion (valence band), while the higher energy portion (conduction band) remains empty. In lithium (and other metals), the valence and conduction bands have no gap between them.

Transparency/ARIS Presentation Center, Art Library

the filled valence band to the unfilled conduction band if they receive even an infinitesimally small quantity of energy. In other words, the electrons are completely delocalized: *they are free to move throughout the piece of metal.* Thus, metals conduct electricity so well because an applied electric field easily excites the highest energy electrons into empty orbitals, and they move through the sample.

Metallic luster (shininess) is another effect of the continuous band of MO energy levels. With so many closely spaced levels available, electrons can absorb and release photons of many frequencies as they move between the valence and conduction bands. Malleability and thermal conductivity also result from the completely delocalized electrons. Under an externally applied force, layers of positive metal ions simply move past each other, always protected from mutual repulsions by the presence of the delocalized electrons (see Figure 9.28B, p. 370). When a metal wire is heated, the highest energy electrons are excited and their extra energy is transferred as kinetic energy along the wire's length.

Large numbers of nonmetal or metalloid atoms can also combine to form bands of MOs. Metals conduct a current well (conductors), whereas most nonmetals do not (insulators), and the conductivity of metalloids lies somewhere in between (semiconductors). Band theory explains these differences in terms of the size of the energy gaps between the valence and conduction bands, as shown in Figure 12.38:

1. *Conductors (metals).* The valence and conduction bands of a **conductor** have no gap between them, so electrons flow when even a tiny electrical potential difference is applied. When the temperature is raised, greater random motion of the atoms hinders electron movement, which *decreases* the conductivity of a metal.
2. *Semiconductors (metalloids).* In a **semiconductor,** a relatively small energy gap exists between the valence and conduction bands. Thermally excited electrons can cross the gap, allowing a small current to flow. Thus, in contrast to a conductor, the conductivity of a semiconductor *increases* when it is heated.
3. *Insulators (nonmetals).* In an **insulator,** the gap between the bands is too large for electrons to jump even when the substance is heated, so no current is observed.

Another type of electrical conductivity, called **superconductivity,** has been generating intense interest for more than two decades. When metals conduct at ordinary temperatures, electron flow is restricted by collisions with atoms vibrating in

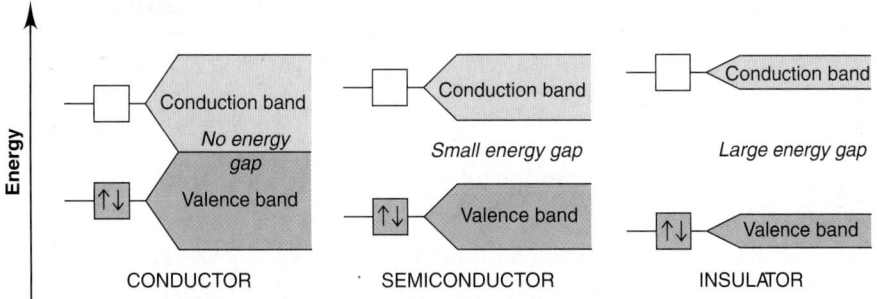

Figure 12.38 Electrical conductivity in a conductor, a semiconductor, and an insulator. Band theory explains differences in electrical conductivity in terms of the size of the energy gap between the material's valence and conduction bands. In conductors (metals), there is no gap. In semiconductors (many metalloids), electrons can jump the small gap if they are given energy, as when the sample is heated. In insulators (most nonmetals), electrons cannot jump the large energy gap.

Figure 12.39 The levitating power of a superconducting oxide. A magnet is suspended above a cooled high-temperature superconductor. Someday, this phenomenon may be used to levitate trains above their tracks for quiet, fast travel.

ARIS Presentation Center, Photo Library

their lattice sites. Such restricted flow appears as resistive heating and represents a loss of energy. To conduct with no energy loss—to superconduct—requires extreme cooling to minimize atom movement. This remarkable phenomenon had been observed in metals only by cooling them to near absolute zero, which can be done only with liquid helium (bp = 4 K; price = $11/L).

In 1986, all this changed with the synthesis of certain ionic oxides that superconduct near the boiling point of liquid nitrogen (bp = 77 K; price = $0.25/L). Like metal conductors, oxide superconductors have no band gap. In the case of $YBa_2Cu_3O_7$, x-ray analysis shows that the Cu ions in the oxide lattice are aligned, which may be related to the superconducting property. In 1989, oxides with Bi and Tl instead of Y and Ba were synthesized and found to superconduct at 125 K; and in 1993, an oxide with Hg, Ba, and Ca, in addition to Cu and O, was shown to superconduct at 133 K. Engineering dreams for these materials include storage and transmission of electricity with no loss of energy (allowing power plants to be located far from cities), ultrasmall microchips for ultrafast computers, electromagnets to levitate superfast railway trains (Figure 12.39), and inexpensive medical diagnostic equipment with remarkable image clarity.

However, manufacturing problems must be overcome. The oxides are very brittle, but some recent developments may remedy this drawback. A more fundamental concern is that when the oxide is warmed or placed in a strong magnetic field, the superconductivity may disappear and not return again on cooling. Clearly, research into superconductivity will involve chemists, physicists, and engineers for many years to come. We consider other remarkable materials in Section 12.7.

Section Summary

The particles in crystalline solids lie at points that form a structure of repeating unit cells. • The three types of unit cells in the cubic system are simple, body-centered, and face-centered. The most efficient packing arrangements are cubic closest packing and hexagonal closest packing. • Bond angles and distances in a crystal structure can be determined with x-ray diffraction analysis and scanning tunneling microscopy. Some of these data can be used to determine atomic radii. • Atomic solids have a closest packed structure, with atoms held together by very weak dispersion forces. • Molecular solids have molecules at the lattice points, often in a cubic closest packed structure. Their intermolecular forces (dispersion, dipole-dipole, H bonding) and resulting physical properties vary greatly. • Ionic solids often crystallize with one type of ion filling holes in a cubic closest packed structure of the other. The high melting points, hardness, and low conductivity of these solids arise from strong ionic attractions. • Most metals have a closest packed structure. • The atoms of network covalent solids are covalently bonded throughout the sample. • Amorphous solids have very little regularity among their particles. • Band theory proposes that orbitals in the atoms of solids combine to form a continuum, or band, of molecular orbitals. Metals are electrical conductors because electrons move freely from the filled (valence band) to the empty (conduction band) portions of this energy continuum. Insulators have a large energy gap between the two portions, and semiconductors have a small gap.

12.7 ADVANCED MATERIALS

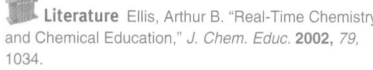
Literature Ellis, Arthur B. "Real-Time Chemistry and Chemical Education," *J. Chem. Educ.* **2002**, *79*, 1034.

In the last few decades, the exciting field of materials science has grown from solid-state chemistry, physics, and engineering, and it is changing our lives in astonishing ways. Objects that were once considered futuristic fantasies of science-fiction writers are realities or soon will be: powerful, ultrafast computers no bigger than this book, connected electronically to millions of others throughout the world; cars powered by sunlight and made of nonmetallic parts stronger than steel and lighter than aluminum; sporting goods made of the materials of space vehicles; ultrasmall machines constructed by manipulating individual atoms and molecules. In this section, we briefly discuss some of these remarkable materials.

Electronic Materials

The ideal of a perfectly ordered crystal is attainable only if the crystal is grown very slowly under carefully controlled conditions. When crystals form more rapidly, **crystal defects** inevitably form. Planes of particles are misaligned, particles are out of place or missing entirely, and foreign particles are lodged in the lattice.

Although crystal defects usually weaken a substance, they are sometimes introduced intentionally to create materials with improved properties, such as increased strength or hardness, or, as you'll see in a moment, to increase a material's conductivity for use in electronic devices. In the process of welding two metals together, for example, *vacancies* form near the surface when atoms vaporize, and then these vacancies move deeper as atoms from lower rows rise to fill the gaps. Welding causes the two types of metal atoms to intermingle and fill each other's vacancies. Metal alloying introduces several kinds of crystal defects, as when some atoms of a second metal occupy lattice sites of the first. Often, the alloy is harder than the pure metal; an example is brass, an alloy of copper with zinc. One reason the welded metals are stronger and the alloy is harder is that the second metal contributes additional valence electrons for metallic bonding.

Doped Semiconductors The manufacture and miniaturization of doped semiconductors have revolutionized the communications, home-entertainment, and information industries. By controlling the number of valence electrons through the creation of specific types of defects, chemists and engineers can greatly increase the conductivity of a semiconductor.

Pure silicon (Si), which lies below carbon in Group 4A(14), conducts poorly at room temperature because an energy gap separates its filled valence band from its conduction band (Figure 12.40A). Its conductivity can be greatly enhanced by **doping,** adding small amounts of other elements to increase or decrease the number of valence electrons in the bands. When Si is doped with phosphorus [or another Group 5A(15) element], P atoms occupy some of the lattice sites. Since P has one more valence electron than Si, this additional electron must enter an empty orbital in the conduction band, thus bridging the energy gap and increasing conductivity. Such doping creates an *n-type semiconductor,* so called because extra *n*egative charges (electrons) are present (Figure 12.40B).

When Si is doped with gallium [or another Group 3A(13) element], Ga atoms occupy some sites (Figure 12.40C). Since Ga has one fewer valence electron than Si, some of the orbitals in the valence band are empty, which creates a positive site. Si electrons can migrate to these empty orbitals, thereby increasing conductivity. Such doping creates a *p-type semiconductor,* so called because the empty orbitals act as *p*ositive holes.

Literature Moore, John W. "Challenges at the Molecular Frontier," *J. Chem. Educ.* **2003,** *80,* 591.

 Engineering application

Figure 12.40 Crystal structures and band representations of doped semiconductors. A, Pure silicon has the same crystal structure as diamond but acts as a semiconductor; the energy gap between its valence and conduction bands keeps conductivity low at room temperature. **B,** Doping silicon with phosphorus (*purple*) adds additional valence electrons, which are free to move through the crystal. They enter the lower portion of the conduction band, which is adjacent to higher energy empty orbitals, thereby increasing conductivity. **C,** Doping silicon with gallium (*orange*) removes electrons from the valence band and introduces positive holes. Nearby Si electrons can enter these empty orbitals, thereby increasing conductivity. The orbitals from which the Si electrons move become vacant; in effect, the holes move.

A Pure silicon crystal

B n-Type doping with phosphorus

C p-Type doping with gallium

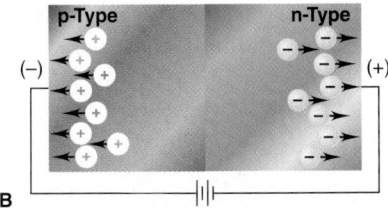

Figure 12.41 The p-n junction. Placing a p-type semiconductor adjacent to an n-type creates a p-n junction. **A,** If the negative battery terminal is connected to the n-type portion, electrons (*yellow sphere with minus sign*) flow toward the p-type portion, which, in effect, moves holes (*white circle with plus sign*) toward the n-type portion. The small arrows indicate the net direction of movement. Note that any hole in the n-type portion lies opposite an electron that moved to the p-type portion. **B,** If the positive battery terminal is connected to the n-type portion, electrons are attracted to it, so no flow takes place across the junction.

ARIS Presentation Center, Art Library

ARIS Presentation Center, Photo Library

● **Solar Cells** One of the more common types of solar cell is, in essence, a p-n junction with an n-type surface exposed to a light source. The light provides the energy to free electrons from the n-type region and accelerate them through an external circuit into the p-type region, thus producing a current to power a calculator, light a bulb, and so on. Arrays of solar cells supply electric power to many residences (*see photo*) and businesses, as well as to space shuttles and most communications satellites.

Green chemistry application

In contact with each other, an n-type and a p-type semiconductor form a *p-n junction.* When the negative terminal of a battery is connected to the n-type portion and the positive terminal to the p-type portion, electrons flow freely in the n-to-p direction, which has the simultaneous effect of moving holes in the p-to-n direction (Figure 12.41). ● No current flows if the terminals are reversed. Such unidirectional current flow makes a p-n junction act as a *rectifier,* a device that converts alternating current into direct current. A p-n junction in a modern integrated circuit can be made smaller than a square 10 μm on a side. Before the p-n junction was created, rectifiers were bulky, expensive vacuum tubes.

A modern computer chip the size of a nickel may incorporate millions of p-n junctions in the form of *transistors*. One of the most common types, an n-p-n transistor, is made by sandwiching a p-type portion between two n-type portions to form adjacent p-n junctions. The current flowing through one junction controls the current flowing through the other and results in an amplified signal. Once again, to accomplish signal amplification before the advent of doped semiconductors required large, and often unreliable, vacuum tubes. Today, minute transistors are found in every radio, TV, and computer and have made possible the multibillion-dollar electronics industry.   Engineering application

Liquid Crystals

Incorporated in the membrane of every cell in your body and in the display of just about every digital watch, pocket calculator, and laptop computer are unique substances known as **liquid crystals.** These materials flow like liquids but, like crystalline solids, pack at the molecular level with a high degree of order.

Properties, Preparation, and Types of Liquid Crystals
To understand the properties of liquid crystals, let's first examine how particles are ordered in the three common physical states and how this affects their properties. The extent of order among particles distinguishes crystalline solids clearly from gases and liquids. Gases have no order, and liquids have little more. Both are considered *isotropic,* which means that their physical properties are the same in every direction within the phase. For example, the viscosity of a gas or a liquid is the same regardless of direction. Glasses and other amorphous solids are also isotropic because they have no regular lattice structure.

In contrast, crystalline solids have a high degree of order among their particles. The properties of a crystal *do* depend on direction, so a crystal is *anisotropic.* The facets in a cut diamond, for instance, arise because the crystal cracks in one direction more easily than in another. Liquid crystals are also anisotropic in that several physical properties, including the electrical and optical properties that lead to their most important applications, differ with direction through the phase.

Organic chemistry application

Like crystalline molecular solids, liquid crystal phases consist of individual molecules. In most cases, the molecules that form liquid crystal phases have two characteristics: a long, cylindrical shape and a structure that allows intermolecular attractions through dispersion and dipole-dipole or H-bonding forces, but that inhibits perfect crystalline packing. Figure 12.42 shows the structures of two molecules that form liquid crystal phases. Note the rodlike shapes and the presence of certain groups—in these cases, flat, benzene-like ring systems—that keep the molecules extended. Many of these types of molecules also have a molecular dipole associated with the long molecular axis. A sufficiently strong electric field can orient large numbers of these polar molecules in approximately the same direction, like compass needles in a magnetic field.

The viscosity of a liquid crystal phase is lowest in the direction parallel to the long axis. Like moistened microscope slides, it is easier for the molecules to slide along each other (because the total attractive force remains the same), than it is for them to pull apart from each other sideways. As a result, the molecules tend to align while the phase flows.

Liquid crystal phases can arise in two general ways, and, sometimes, either way can occur in the same substance. A *thermotropic* phase develops as a result of a change in temperature. As a crystalline solid is heated, the molecules leave their lattice sites, but the intermolecular interactions are still strong enough to keep the molecules aligned with each other along their long axes. Like any other phase, the liquid crystal phase has sharp transition temperatures; however, it exists over a relatively small temperature range. Further heating provides enough kinetic energy for the molecules to become disordered, as in a normal liquid. The typical range for liquid crystal phases of pure substances is from <1°C to around 10°C, but mixing phases of two or more substances can greatly extend this range. For this reason, the liquid crystal phases used within display devices, as well as those within cell membranes, consist of mixtures of molecules.

A *lyotropic* phase occurs in solution as the result of changes in concentration, but the conditions for forming such a phase vary for different substances. For

Figure 12.42 Structures of two typical molecules that form liquid crystal phases. Note the long, extended shapes and the regions of high (*red*) and low (*blue*) electron density.

A Nematic

B Cholesteric

C Smectic

Figure 12.43 The three common types of liquid crystal phases. A, Nematic phase. A rectangular volume of the phase, with expanded view, shows a close-up of the arrangement of molecules. **B,** Cholesteric phase. Note the corkscrew-like arrangement of the layers. **C,** Smectic phase. A rectangular volume of the phase shows the more orderly stacking of layers.

 ARIS Presentation Center, Art Library

example, when purified, some biomolecules that exist naturally in mammalian cell membranes form lyotropic phases in water at the moderate temperature that occurs within the organism. At the other extreme, Kevlar, a fiber used in bulletproof vests and high-performance sports equipment, forms a lyotropic phase at high temperatures in concentrated H_2SO_4 solution.

Molecules that form liquid crystal phases can exhibit various types of order. Three common types are nematic, cholesteric, and smectic:

- In a *nematic* phase, the molecules lie in the same direction but their ends are not aligned, much like a school of fish swimming in synchrony (Figure 12.43A). The nematic phase is the least ordered type of liquid crystal phases.
- In a *cholesteric* phase, which is somewhat more ordered, the molecules lie in layers that each exhibit nematic-type ordering. Rather than lying in parallel fashion, however, each layer is rotated by a fixed angle with respect to the next layer. The result is a helical (corkscrew) arrangement, so a cholesteric phase is often called a *twisted nematic phase* (Figure 12.43B).
- In a *smectic* phase, which is the most ordered, the molecules lie parallel to each other, *with* their ends aligned, in layers that are stacked directly over each other, much like a supermarket display of shelves filled with identical bottles (Figure 12.43C). The long molecular axis has a well-defined angle (shown in the figure as 90°) with respect to the plane of the layer. The molecules in Figure 12.42 are typical of those that form nematic or smectic phases. Liquid crystal–type phases appear in many biological systems (Figure 12.44).

In some cases, a substance that forms a given liquid crystal phase under one set of conditions forms other phases under different conditions. Thus, a given thermotropic liquid crystal substance can pass from disordered liquid through a series of distinct liquid crystal phases to an ordered crystal through a decrease in temperature. A lyotropic substance can undergo similar changes through an increase in concentration.

Figure 12.44 Liquid crystal–type phases in biological systems. A, Nematic arrays of tobacco mosaic virus particles within the fluid of a tobacco leaf. **B,** The orderly arrangement of actin and myosin protein filaments in voluntary muscle cells.

 ARIS Presentation Center, Photo Library

 Biological application

A

B

Applications of Liquid Crystals The ability to control the orientation of the molecules in a liquid crystal allows us to produce materials with high strength or unique optical properties.

High-strength applications involve the use of extremely long molecules called *polymers*. While in a thermotropic liquid crystal phase and during their flow through the processing equipment, these molecules become highly aligned, like the fibers in wood. Cooling solidifies them into fibers, rods, and sheets that can be shaped into materials with superior mechanical properties in the direction of the long molecular axis. Sporting equipment, supersonic aircraft parts, and the sails used in the America's Cup races are fabricated from these polymeric materials. (We discuss the structure and physical behavior of polymers later in this section and their synthesis in Chapter 15.)

Far more important in today's consumer market are the liquid crystal displays (LCDs) used in watches, calculators, cell phones, and computers. All depend on *changes in molecular orientation in an electric field*. The most common type, called a *twisted nematic* display, is shown schematically in Figure 12.45 as a small portion of a wristwatch LCD. The liquid crystal phase consists of layers of nematic phases sandwiched between thin glass plates that incorporate transparent electrodes. As a result of a special coating process, the long molecular axis lies parallel to the plane of the glass plates. The distance between the plates (6–8 μm) is chosen so that the molecular axis within each succeeding layer twists just enough for the molecular orientation at the bottom plate to be 90° from that at the top plate. Above and below this "sandwich" are thin polarizing filters (like those used in Polaroid lenses or camera sunlight filters), which allow light waves oriented in only one direction to pass through. The filters are placed in a "crossed" arrangement, so that light passing through the top filter must twist 90° to pass through the bottom filter. The orientation and optical properties of the molecules twist the orientation of the light by exactly this amount. This whole grouping of filters, plates, and liquid crystal phase lies on a mirror.

A current generated by the watch battery controls the orientation of the molecules within the phase. With the current on in one region of the display, the molecules become oriented *toward* the field, and thus block the light from passing through to the bottom filter, so that region appears dark. With the current off in another region, light passes through the molecules and bottom filter to the mirror and back again, so that region appears bright.

Cholesteric liquid crystals are used in applications that involve color changes with temperature. The twisted molecular orientation in these phases "unwinds" with heating, and the extent of the unwinding determines the color. Liquid crystal thermometers that include a mixture of substances to widen their range of temperatures use this effect. Newer and more important uses include "mapping" the area of a tumor, detecting faulty connections in electronic circuit boards, and nondestructive testing of materials under stress.

Ceramic Materials

First developed by Stone Age people, **ceramics** are defined as nonmetallic, nonpolymeric solids that are hardened by heating to high temperatures. Clay ceramics consist of silicate microcrystals suspended in a glassy cementing medium. In "firing" a ceramic pot, for example, a kiln heats the object made of an aluminosilicate clay, such as kaolinite, to 1500°C and the clay loses water:

$$Si_2Al_2O_5(OH)_4(s) \longrightarrow Si_2Al_2O_7(s) + 2H_2O(g)$$

During the heating process, the structure rearranges to an extended network of Si-centered and Al-centered tetrahedra of O atoms (Section 14.6).

Bricks, porcelain, glazes, and other clay ceramics are useful because of their hardness and resistance to heat and chemicals. Today's high-tech ceramics

Figure 12.45 Schematic of a liquid crystal display (LCD). A close-up of the "2" on a wristwatch LCD reveals two polarizers sandwiching two glass plates, which sandwich a liquid crystal (LC) layer, all lying on a mirror. When light waves oriented in all directions enter the first polarizer, only waves oriented in one direction emerge to enter the LC layer. Enlarging a dark region of the numeral (*top blow-up*) shows that with the current on, the LC molecules are aligned and the light cannot pass through the other polarizer to the mirror; thus, the viewer sees no light. Enlarging a bright region (*bottom blow-up*) shows that with the current off, the LC molecules lie in a twisted nematic arrangement, which rotates the plane of the light waves and allows them to pass through the other polarizer to the mirror. Reflecting and retracing this path (not shown), the light reaches the viewer.

Labels in figure: Polarizer · Top plate · LC layer · Bottom plate · Polarizer · Mirror

Current on: molecules align, light blocked, dark region

Current off: molecules not aligned, light passes, bright region

Table 12.7	Some Uses of Modern Ceramics and Ceramic Mixtures
Ceramic	**Applications**
SiC, Si_3N_4, TiB_2, Al_2O_3	Whiskers (fibers) to strengthen Al and other ceramics
Si_3N_4	Car engine parts; turbine rotors for "turbo" cars; electronic sensor units
Si_3N_4, BN, Al_2O_3	Supports or layering materials (as insulators) in electronic microchips
SiC, Si_3N_4, TiB_2, ZrO_2, Al_2O_3, BN	Cutting tools, edge sharpeners (as coatings and whole devices), scissors, surgical tools, industrial "diamond"
BN, SiC	Armor-plating reinforcement fibers (as in Kevlar composites)
ZrO_2, Al_2O_3	Surgical implants (hip and knee joints)

Engineering application

incorporate these traditional characteristics in addition to superior electrical and magnetic properties (Table 12.7). As just one example, consider the unusual electrical behavior of certain zinc oxide (ZnO) composites. Ordinarily a semiconductor, ZnO can be doped so that it becomes a conductor. Imbedding particles of the doped oxide into an insulating ceramic produces a variable resistor: at low voltage, the material conducts poorly, but at high voltage, it conducts well. Best of all, the changeover voltage can be "preset" by controlling the size of the ZnO particles and the thickness of the insulating medium.

Preparing Modern Ceramics Among the important modern ceramics are silicon carbide (SiC) and nitride (Si_3N_4), boron nitride (BN), and the superconducting oxides. They are prepared by standard chemical methods that involve driving off a volatile component during the reaction.

The SiC ceramics are made from compounds used in silicone polymer manufacture (we'll discuss their structures and uses in Section 14.6):

$$n(CH_3)_2SiCl_2(l) + 2nNa(s) \longrightarrow 2nNaCl(s) + [(CH_3)_2Si]_n(s)$$

This product is heated to 800°C to form the ceramic:

$$[(CH_3)_2Si]_n \longrightarrow nCH_4(g) + nH_2(g) + nSiC(s)$$

Silicon carbide can also be prepared by the direct reaction of silicon and graphite under vacuum:

$$Si(s) + C(graphite) \xrightarrow{\sim 1500°C} SiC(s)$$

The nitride is also prepared by reaction of the elements:

$$3Si(s) + 2N_2(g) \xrightarrow{>1300°C} Si_3N_4(s)$$

The formation of a BN ceramic begins with the reaction of boron trichloride or boric acid with ammonia:

$$B(OH)_3(s) + 3NH_3(g) \longrightarrow B(NH_2)_3(s) + 3H_2O(g)$$

Heat drives off some of the bound nitrogen as NH_3 to yield the ceramic:

$$B(NH_2)_3(s) \xrightarrow{\Delta} 2NH_3(g) + BN(s)$$

One of the common high-temperature superconducting oxides is a ceramic made by heating a mixture of barium carbonate and copper and yttrium oxides, followed by further heating in the presence of O_2:

$$4BaCO_3(s) + 6CuO(s) + Y_2O_3(s) \xrightarrow{\Delta} 2YBa_2Cu_3O_{6.5}(s) + 4CO_2(g)$$

$$YBa_2Cu_3O_{6.5}(s) + \tfrac{1}{4}O_2(g) \xrightarrow{\Delta} YBa_2Cu_3O_7(s)$$

Ceramic Structures and Uses Structures of several ceramic materials are shown in Figure 12.46. Note the diamond-like structure of silicon carbide. Network covalent bonding gives this material great strength. Silicon carbide is made into thin fibers, called *whiskers,* to reinforce other ceramics in a composite structure and prevent cracking, much like steel rods reinforcing concrete. Silicon nitride is virtually inert chemically, retains its strength and wear resistance for extended periods above 1000°C, is dense and hard, and acts as an electrical insulator. Japanese and American automakers are testing it in high-efficiency car and truck engines because it allows an ideal combination of low weight, high operating temperature, and little need for lubrication.

The BN ceramics exist in two structures (Section 14.5), analogous to the common crystalline forms of carbon. In the graphite-like form, BN has extraordinary properties as an electrical insulator. At high temperature and very high pressure (1800°C and 8.5×10^4 atm), it converts to a diamond-like structure, which is extremely hard and durable. Both forms are virtually invisible to radar.

Earlier, we mentioned some potential uses of the superconducting oxides. In nearly every one of these ceramic materials, copper occurs in an unusual oxidation state. In $YBa_2Cu_3O_7$, for instance, assuming oxidation states of $+3$ for Y, $+2$ for Ba, and -2 for O, the three Cu atoms have a total oxidation state of $+7$. This is allocated as $Cu(II)_2Cu(III)$, with one Cu in the unusual $+3$ state. X-ray diffraction analysis indicates that a distortion in the structure makes four of the oxide ions unusually close to the Y^{3+} ion, which aligns the Cu ions into chains within the crystal. It is suspected that a specific half-filled $3d$ orbital in Cu oriented toward a neighboring O^{2-} ion may be associated with superconductivity, although the process is still poorly understood. Because of their brittleness, it has been difficult to fashion these ceramics into wires, but methods for making superconducting films and ribbons have recently been developed.

Research in ceramic processing is beginning to overcome the inherent brittleness of this entire class of materials. The brittleness arises from the strength of the ionic-covalent bonding in these solids and their resulting inability to deform. Under stress, a microfine crystal defect widens and lengthens until the material cracks. One new method forms defect-free ceramics using controlled packing and heat-treating of extremely small, uniform oxide particles coated with organic polymers. Another method is aimed at arresting a widening crack. The ceramics are embedded with zirconia (ZrO_2), whose crystal structure expands up to 5% under the mechanical stress of a crack tip: the moment the advancing crack reaches them, the zirconia particles effectively pinch it shut. Recently, a third method was reported. Japanese researchers prepared a ceramic material made of precisely grown single crystals of Al_2O_3 and $GdAlO_3$, which become entangled during the solidification process. It bends without cracking at temperatures above 1800 K.

Despite the technical difficulties, chemists will continue to develop new ceramic materials and apply their amazing and useful properties well into the 21st century.

Figure 12.46 Expanded view of the atom arrangements in some modern ceramic materials. SiC **(A)** and the high-pressure form of BN **(B)** both have a crystal structure similar to that of diamond and are extremely hard. $YBa_2Cu_3O_7$ **(C)** is one of the high-temperature superconducting oxides.

C **$YBa_2Cu_3O_7$**

A Silicon carbide

B Cubic BN (borazon)

Engineering application

Polymeric Materials

In its simplest form, a **polymer** (Greek, "many parts") is an extremely large molecule, or **macromolecule,** consisting of a covalently linked chain of smaller molecules, called **monomers** (Greek, "one part"). The monomer is the *repeat unit* of the polymer, and a typical polymer may have from hundreds to hundreds of thousands of repeat units. *Synthetic* polymers are created by chemical reactions in the laboratory; *natural* polymers (or *biopolymers*) are created by chemical reactions within organisms. There are many types of monomers, and their chemical structures allow for the complete repertoire of intermolecular forces.

Synthetic polymers such as plastics, rubbers, and specialized glasses have revolutionized everyday life. Virtually every home, car, electronic component, and processed food contains synthetic polymers in its structure or packaging. You interact with dozens of these materials each day—from paints to floor coverings to clothing to the additives and adhesives in this textbook. Some of these materials, like those used in food containers, are very long-lived in the environment and have created a serious waste-disposal problem. Others are being actively recycled into the same or other useful products, such as garbage bags, outdoor furniture, roofing tiles, and even marine pilings and roadside curbs. Still others, such as artificial skin, heart valve components, and hip joints, are designed to have as long a life as possible.

In this section, we'll examine the physical nature of synthetic polymers and explore the role intermolecular forces play in their properties and uses. In Chapter 15, we'll examine the types of monomers, look at the preparation of synthetic polymers, and then focus on the structures and vital functions of the biopolymers.

Dimensions of a Polymer Chain: Mass, Size, and Shape Because of their great lengths, polymers are unlike smaller molecules in several important ways. Let's see how chemists describe the mass, size, and shape of a polymer chain and how the chains exist in a sample. We'll focus throughout on polyethylene, by far the most common synthetic polymer.

1. *Polymer mass.* The molar mass of a polymer chain ($\mathcal{M}_{polymer}$, in g/mol, often referred to as the *molecular weight*) depends on two parameters—the molar mass of the repeat unit ($\mathcal{M}_{repeat}$) and the **degree of polymerization (n),** or the number of repeat units in the chain:

$$\mathcal{M}_{polymer} = \mathcal{M}_{repeat} \times n$$

For example, the molar mass of the ethylene repeat unit is 28 g/mol. If an individual polyethylene chain in a plastic grocery bag has a degree of polymerization of 7100, the molar mass of that particular chain is

$$\mathcal{M}_{polymer} = \mathcal{M}_{repeat} \times n = (28 \text{ g/mol})(7.1 \times 10^3) = 2.0 \times 10^5 \text{ g/mol}$$

Table 12.8 shows some other examples.

ARIS Presentation Center, Tables Library

Table 12.8	Molar Masses of Some Common Polymers		
Name	$\mathcal{M}_{polymer}$ **(g/mol)**	**n**	**Uses**
Acrylates	2×10^5	2×10^3	Rugs, carpets
Polyamide (nylons)	1.5×10^4	1.2×10^2	Tires, fishing line
Polycarbonate	1×10^5	4×10^2	Compact discs
Polyethylene	3×10^5	1×10^4	Grocery bags
Polyethylene (ultrahigh molecular weight)	5×10^6	2×10^5	Hip joints
Poly(ethylene terephthalate)	2×10^4	1×10^2	Soda bottles
Polystyrene	3×10^5	3×10^3	Packing, coffee cups
Poly(vinyl chloride)	1×10^5	1.5×10^3	Plumbing

However, even though any *given* chain within a sample of a polymer has a fixed molar mass, the degree of polymerization often varies considerably from chain to chain. As a result, *all samples of synthetic polymers have a distribution of chain lengths*. For this reason, polymer chemists use various definitions of *average* molar mass, and a common one is the *number-average molar mass*, $\mathcal{M}_n$:

$$\mathcal{M}_n = \frac{\text{total mass of all chains}}{\text{number of moles of chains}}$$

Thus, even though the number-average molar mass of the polyethylene in grocery bags is, say, 1.6×10^5 g/mol, the chains may vary in molar mass from about 7.0×10^4 to 3.0×10^5 g/mol.

2. *Polymer size and shape.* The long axis of a polymer chain is called its *backbone*. The length of an *extended* backbone is simply the number of repeat units (degree of polymerization, n) times the length of each repeat unit (l_0). For instance, the length of an ethylene repeat unit is about 250 pm, so the extended length of our particular grocery-bag polyethylene chain is

Length of extended chain = $n \times l_0 = (7.1 \times 10^3)(2.5 \times 10^2 \text{ pm}) = 1.8 \times 10^6 \text{ pm}$

Comparing this length with the thickness of the chain, which is only about 40 pm, gives a good picture of the threadlike dimensions of the extended chain.

A polymer molecule, however, whether pure or in solution, doesn't exist as an extended chain, but is far more compact. Polymer chemists assume, as a first approximation, that the shape of the chain arises as a result of free rotation around all of its single bonds. Thus, as each repeat unit rotates randomly, the chain continuously changes direction, turning back on itself many times and eventually arriving at the **random coil** shape that most polymers adopt (Figure 12.47). In reality, of course, rotation is not completely free because, as one portion of a chain bends and twists near other portions of the same chain or of other nearby chains, they attract each other. Thus, the nature of intermolecular forces between chain portions, between different chains, and/or between chain and solvent becomes a key factor in establishing the actual shape of a polymer chain.

Polyethylene chain (space-filling)

One of several entangled sections of nearby polyethylene chains

R_g
Radius of gyration

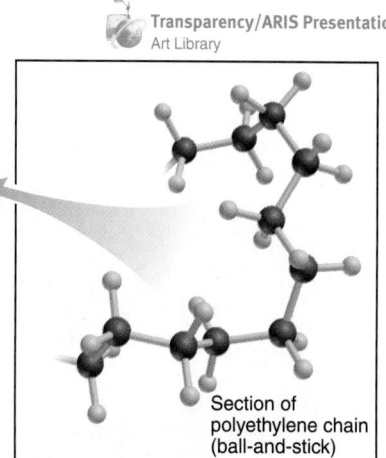

Transparency/ARIS Presentation Center, Art Library

Section of polyethylene chain (ball-and-stick)

Figure 12.47 The random-coil shape of a polymer chain. Note the random coiling of the chain's carbon atoms (*black*). Sections of several nearby chains (*red, green,* and *yellow*) are entangled with this chain, kept near one another by dispersion forces. In reality, entangling chains fill any gaps shown here. The radius of gyration (R_g) represents the average distance from the center of mass of the coiled molecule to its outer edge.

The size of the coiled polymer chain is expressed by its **radius of gyration, R_g,** the average distance from the center of mass of the polymer to the outer edge of the coil (Figure 12.47). Even though R_g is reported as a single value for a given polymer, it represents an average value for many chains. The mathematical expression for the radius of gyration considers the length of each repeat unit and its randomized direction in space, as well as the bond angles between atoms in a unit and between adjacent units:*

$$R_g = \sqrt{\frac{nl_0^2}{6}}$$

As we would expect, the radius of gyration increases with the degree of polymerization, and thus with the molar mass as well. Most importantly, light-scattering experiments and other laboratory measurements correlate with the calculated results, so for many polymers the radius of gyration can be determined experimentally.

For our grocery-bag polyethylene chain, we have

$$R_g = \sqrt{\frac{nl_0^2}{6}} = \sqrt{\frac{(7.1\times10^3)\,(2.5\times10^2\text{ pm})^2}{6}} = 8.6\times10^3\text{ pm}$$

Doubling the radius gives a diameter of 1.7×10^4 pm, less than one-hundredth the length of the extended chain!

3. *Polymer crystallinity.* You may get the impression from the discussion so far that a sample of a given polymer is just a disorderly jumble of chains, but this is often not the case. If the molecular structure allows neighboring chains to pack together and if the chemical groups lead to favorable dipole-dipole, H-bonding, or dispersion forces, portions of the chains can align regularly and exhibit crystallinity.

However, the crystallinity of a polymer is very different from the crystal structures of the simple compounds we discussed earlier. There, the orderly array extends over many molecules, and the unit cell includes at least one molecule. In contrast, the orderly regions of a polymer rarely, if ever, involve even one whole molecule (Figure 12.48). At best, a polymer is *semicrystalline,* because only parts

ARIS Presentation Center, Art Library

Figure 12.48 The semicrystallinity of a polymer chain. This depiction of polyethylene highlights several ordered regions (*darker color*) with random coils between them. Ordered regions from nearby chains (*red* and *yellow*) are shown overlapping those in the main chain.

---

*The mathematical derivation of R_g is beyond the scope of this text, but it is analogous to the two-dimensional "walk of the drunken sailor." With each step, the sailor stumbles in random directions and, given enough time, ends up very close to the starting position. The radius of gyration quantifies how far the end of the polymer chain (the sailor) has gone from the origin.

Overlapping crystalline region from a nearby chain

Two of several crystalline regions in a polyethylene chain

of the molecule align with parts of neighboring molecules (or with other parts of the same chain), while most of the chain remains as a random coil. Thus, the unit cell of a polymer includes only a small part of the molecule.

Flow Behavior of Polymers In Section 12.4, we defined the viscosity of a fluid as its resistance to flow. Some of the most important uses of polymers arise from their ability to change the viscosity of a solvent in which they have been dissolved and to undergo temperature-dependent changes in their own viscosity.

When an appreciable amount of polymer (about 5–15 mass %) dissolves, the viscosity of the solution is much higher than that of the pure solvent. This behavior is put to use by adding polymers to increase the viscosity of many common materials, such as motor oil, paint, and salad dressing. A dissolved polymer increases the viscosity of the solution by interacting with the solvent. As the random coil of a polymer moves through a solution, solvent molecules are attracted to its exterior and interior through intermolecular forces (Figure 12.49). Thus, the polymer coil drags along many solvent molecules that are attracted to other solvent molecules and other coils, and flow is lessened. Increasing the polymer concentration increases the viscosity because the coils are more likely to become entangled in one another. In order for each coil to flow, it must disentangle from its neighbors or drag them along.

Viscosity is a basic property that characterizes the behavior of a particular polymer-solvent pair at a given temperature. Just as it does for a pure polymer, the size (radius of gyration) of a random coil of a polymer in solution increases with molar mass, and so does the viscosity of the solution. To study basic interactions

Transparency/ARIS Presentation Center, Art Library

Figure 12.49 The viscosity of a polymer in solution. A ball-and-stick model of a section (*left panel*) of a poly(ethylene oxide) chain in water shows the H bonds that form between the lone pairs of the chain's O atoms and the H atoms of solvent molecules. The polymer chain, depicted as a blue and red coiled rod (*center panel*), forms many H bonds with solvent molecules. Note the H bonds to water molecules that allow one chain to interact with others nearby. As the concentration of polymer increases (*three right panels*), the viscosity of the solution increases because the movement of each chain is restricted through its interactions with solvent and with other chains.

and to improve polymer manufacture, chemists and other industrial scientists have developed essential quantitative equations to predict the viscosity of polymers of different molar masses in a variety of solvents (see Problem 12.143).

Intermolecular forces also play a major role in the flow of a pure polymer sample. At temperatures high enough to melt them, many polymers exist as viscous liquids, flowing more like honey than water. The forces between the chains, as well as the entangling of chains, hinder the molecules from flowing past each other. As the temperature decreases, the intermolecular attractions exert a greater effect, and the eventual result is a rigid solid. If the chains don't crystallize, the resulting material is called a *polymer glass*. The transition from a liquid to a glass occurs over a narrow (10–20°C) temperature range for a given polymer, but chemists define a single temperature at the midpoint of the range as the *glass transition temperature, T_g.* Like window glass, many polymer glasses are transparent, such as polystyrene in drinking cups and polycarbonate in compact discs.

The flow-related properties of polymers give rise to their familiar **plastic** mechanical behavior. The word "plastic" refers to a material that, when deformed, retains its new shape; in contrast, when an "elastic" object is deformed, it returns to its original shape. Many polymers can be deformed (stretched, bent, twisted) when warm and retain their deformed shape when cooled. In this way, they are made into countless everyday objects—milk bottles, car parts, and so forth.

Molecular Architecture of Polymers

A polymer's architecture—its overall spatial layout and molecular structure—is crucial to its properties. In addition to the linear chains we've discussed so far, chemists create polymers with more complex architectures through the processes of branching and crosslinking.

Branches are smaller chains appended to a polymer backbone. As the number of branches increases, the chains cannot pack together as well, so the degree of crystallinity decreases; as a result, the polymer is less rigid. A small amount of branching occurs as a side reaction in the preparation of high-density polyethylene (HDPE). Because it is still largely linear, though, it is rigid enough for use in milk containers. In contrast, much more branching is intentionally induced to prepare low-density polyethylene (LDPE). The chains cannot pack well, so crystallinity is low. The flexible, transparent material used in food storage bags results.

Dendrimers are the ultimate branched polymers. They are prepared from monomers with three or more attachment points, so each monomer forms branches. In essence, then, dendrimers have no backbone and consist of branches only. As you can see in the Gallery on page 398, a dendrimer has a constantly increasing number of branches and an incredibly large number of end groups at its outer edge. Chemists have used dendrimers to bind one polymer to another in the production of films and fibers and to deliver drug molecules to the desired location in medical applications.

Crosslinks can be thought of as branches that link one chain to another. The extent of crosslinking can result in remarkable differences in properties. In many cases, a small degree of crosslinking yields a *thermoplastic* polymer, one that still flows at high temperatures. But, as the extent of crosslinking increases, a thermoplastic polymer is transformed into a *thermoset* polymer, one that can no longer flow because it has become a single network. Below their glass transition temperatures, some thermosets are extremely rigid and strong, making them ideal as matrix materials in high-strength composites (see photo).

Above their glass transition temperatures, many thermosets become **elastomers,** polymers that can be stretched and immediately spring back to their initial shapes when released, like the net under a trapeze artist or a common rubber band. When you stretch a rubber band, individual polymer chains flow for only a short distance before the connectivity of the network returns them to their original positions. Table 12.9 lists some elastomers.

Bicycle helmet containing a thermoset polymer.
ARIS Presentation Center,
Photo Library

Table 12.9 Some Common Elastomers		
Name	T_g(°C)	Uses
Poly(dimethyl siloxane)	−123	Breast implants
Polybutadiene	−106	Rubber bands
Polyisoprene	−65	Surgical gloves
Polychloroprene (neoprene)	−43	Footwear; medical tubing

Differences in monomer sequences often influence polymer properties as well. A *homopolymer* consists of one type of monomer (A—A—A—A—A— ...), whereas a **copolymer** consists of two or more types. The simplest copolymer is called an *AB block copolymer* because a chain (block) of monomer A and a chain of monomer B are linked at one point:

$$... \text{—A—A—A—A—A—B—B—B—B—B—} ...$$

If the intermolecular forces between the A and B portions of the chain are weaker than those between different regions within each portion, the A and B portions form their own random coils. This ability makes AB block copolymers ideal adhesives for joining two polymer surfaces covalently. An ABA block copolymer has A chains linked at each end of a B chain:

$$... \text{—A—A—A—B—(B)}_n\text{—B—A—A—A—} ...$$

Some of these block copolymers act as *thermoplastic elastomers*, materials shaped at high temperature that become elastomers at room temperature; not surprisingly, some of these materials have revolutionized the footwear industry.

Polymer chemists tailor polymers to control properties such as viscosity, strength, toughness, and flexibility. The effect of intermolecular forces on their physical properties is only part of the story of these remarkably useful materials. Silicone polymers are described in the Gallery in Chapter 14 (p. 593), and organic reactions that form polymer chains from monomers are examined in Chapter 15.

Nanotechnology: Designing Materials Atom by Atom

At the frontier of interdisciplinary science and expanding at an incredible pace, nanotechnology joins scientists from physics, materials science, chemistry, biology, environmental science, medicine, and many branches of engineering. **Nanotechnology** is the science and engineering of nanoscale systems, whose sizes range from 1 to 100 nm. (Recall that 1 nm is 10^{-9} m—about one-billionth the width of your dorm-room desk or one-millionth the width of your pen tip. A nanometer is to a meter as a marble is to Earth!) The field had its conceptual origin in 1959, in the mind of physics Nobel laureate Richard Feynman, who said "The principles of physics . . . do not speak against the possibility of maneuvering things atom by atom. It is not an attempt to violate any laws; it is something, in principle, that can be done; but in practice, it has not been done because we are too big." Until recently, physical scientists had focused on atoms, which are smaller (around 1×10^{-1} nm), or on crystals, which are larger (around 1×10^5 nm and up). As so often happens in science, a technological advance allows a new field to blossom. Now, working between these size extremes, nanotechnologists use the scanning tunneling microscope (see Tools of the Laboratory, p. 468) and the similar scanning probe and atomic force microscopes to examine the chemical and physical properties of nanostructures and build them up, in many cases, atom by atom.

One of the first discoveries in the field was that nanoscale materials behave neither like atoms nor like bulk materials. The reason is that their total surface

Literature Ong, Eddie W.; Razdan, Anshuman; Garcia, Antonio A.; Pizziconi, Vincent; Ramakrishna, B. L.; Glaunsinger, William S. "Interactive Nano-Visualization of Materials over the Internet," *J. Chem. Educ.* 2000, *77*, 1114.

Literature Heinhorst, Sabine; Cannon, Gordon. "Nanotechnology," *J. Chem. Educ.* 1999, *76*, 1472.

Literature Heinhorst, Sabine; Cannon, Gordon C. "Slick Nanoparticles and Other Gleanings from Nature," *J. Chem. Educ.* 1997, *74*, 1026.

Literature Sharma, C. V. Krishnamohan. "Designing Advanced Materials As Simple As Assembling Lego® Blocks!" *J. Chem. Educ.* 2001, *78*, 617.

Literature Sohlberg, Karl. "Introducing the Core Concepts of Nanoscience and Nanotechnology: Two Vignettes," *J. Chem. Educ.* 2006, *83*, 1516.

 Engineering application

area is similar to their total interior volume—for example, a 5-nm particle has about half of its atoms on its surface—and this fact leads to remarkable properties. Cut a piece of ordinary aluminum foil into pieces just large enough to see with a light microscope and they still behave like the original piece. But, if you could cut the original piece into 20–30-nm pieces, they would explode. Indeed, aluminum nanoparticles are used in fireworks and propellants.

Two features of nanoscale engineering occur routinely in nature. The first is *self-assembly*, the ability of smaller, simpler parts to organize themselves into a larger, more complex whole. On the molecular scale, self-assembly refers to atoms or small molecules aggregating through intermolecular forces. Oppositely charged regions on two such particles make contact to form a larger particle, which in turn forms a still larger one. The other feature is *controlled orientation*, the positioning of two molecules near each other long enough for intermolecular forces to result in a change. All biological catalysts (enzymes) function this way. (Catalysts, discussed in Chapter 16, are substances that speed reactions.) Because this new field changes many times faster than a textbook, we can only provide a glimpse of a few exciting research directions in nanotechnology.

Nanoscale Optical Materials *Quantum dots* are nanoparticles of semiconducting materials, such as gallium arsenide (GaAs) or gallium selenide (GaSe), that are smaller than 10 nm. Rather than having a band of energy levels, the dot has discrete energy levels, much like a single atom. As in atoms, the energy levels can be studied with spectroscopy; unlike atoms, quantum dots can be modified chemically. When high-energy (short-wavelength) UV light irradiates a suspension of quantum dots, they become excited and emit radiation of lower energy (longer wavelength). Most importantly, the emitted wavelengths depend on the size of the dots: the larger the particles, the longer the wavelength emitted. Thus, the color of the emitted light can be "tuned" by varying the dimensions of the nanoparticles (Figure 12.50). One use for quantum dots is the imaging of specific cells. The dot is encapsulated in a coating and then bonded to a particular protein essential to the target organ. The cells in the target incorporate the protein, along with its attached quantum dot, and the location of the protein can be viewed spectroscopically.

Nanostructured Materials *Nanostructuring*, the construction of bulk materials from nanoscale building blocks, increases strength, ductility, plasticity, and many other properties. The resulting *nanocomposites* behave as liquid magnets, ductile cements, conducting elastomers, and many other unique materials.

One example mimics a familiar material used by nature. Bones are natural nanocomposites of hydroxyapatite (a type of calcium phosphate) and other minerals. A recently created product, NanOss® Bone Void Filler, is a nanocrystalline hydroxyapatite, with weight-bearing properties identical to bone, that provides a framework for natural bone tissue to grow into and heal a fracture. Another new material, called a *ferrofluid,* consists of magnetic nanoparticles (usually magnetite, Fe_3O_4) dispersed in a viscous fluid. Such materials act like ordinary fluids until they are placed in a magnetic field, upon which the particles align with the field and adopt its shape (Figure 12.51). Ferrofluids find uses in automobile shock absorbers, magnetic plastics, audio speakers, computer hard drives, high-vacuum valves, and many other devices.

High-Surface-Area Materials Carbon nanotubes have remarkable mechanical and electrical properties, and academic and industrial laboratories throughout the world are studying them (see Gallery, p. 398). A major future application involves storing H_2 gas for use in hydrogen-powered vehicles. Currently, this gas is stored in metal cylinders at high pressures (>70 atm). Porous solids, such as carbon nanotubes, can store the same amount of gas at much lower pressures because they have enormous surface area. Carbon nanotubes have been synthesized with surface areas

Engineering application

Literature Chanteau, Stephanie H.; Ruths, Troy; Tour, James M. "Arts and Sciences Reunite in Nanoput: Communicating Synthesis and the Nanoscale to the Layperson," *J. Chem. Educ.* **2003,** *80,* 395.

2.3 ———————→ 5.5
Size (nm)

Figure 12.50 The colors of quantum dots. After being irradiated with UV light, quantum dots emit visible light: the smaller the dots, the shorter the emitted wavelengths.

ARIS Presentation Center, Art Library

Literature Legge, Coulton H. "Chemistry under the Microscope—Lab-on-a-Chip Technologies," *J. Chem. Educ.* **2002,** *79,* 173.

Figure 12.51 The magnetic behavior of a ferrofluid. Nanoparticles of magnetite (Fe_3O_4) dispersed in a viscous fluid are suspended between the poles of a magnet.

ARIS Presentation Center, Photo Library

of 4500 m²/g—about four football fields per gram of material! Thus, a medium-sized container of nanotubes could store a large amount of H_2 fuel under conditions safe enough for a family car.

Other high-surface-area materials being developed are porous membranes for water purification or batteries and multilayer films that incorporate photosynthetic molecules for high-efficiency solar cells. A very exciting development is the molecule-specific biosensor, which can detect as little as 10^{-14} mol of DNA using color changes in gold nanoparticles. Such biosensors could circulate freely in the bloodstream to measure levels of specific disease-related molecules and deliver drugs to individual cells, even individual genes.

Nanomachines Nanoscale machines have been dreamt of by science fiction writers for decades. In nature, the virus is a marvel of nanoscale engineering, designed for delivering genetic material to infect a host cell. For example, the T4 virus latches onto a bacterial cell with its tail fibers, bores through the bacterial membrane with its end plate, and then injects the payload of DNA, packaged in its head, into the bacterium. These complex functions are performed by a biological "machine" measuring 60×200 nm. Researchers are exploring ways for viruses to deliver medicinal agents to specific cells. In other feats of nanoscale engineering, researchers have created nanovalves, nanopropellers, and even a nanocar, complete with a chassis, axles, and buckyball wheels, that is only 4 nm wide and is "driven" on a gold surface under the direction of an atomic force microscope (Figure 12.52).

 Literature Ward, Michael D. "Chemistry and Molecular Electronics: New Molecules as Wires, Switches, and Logic Gates," *J. Chem. Educ.* **2001**, *78*, 321.

Engineering application

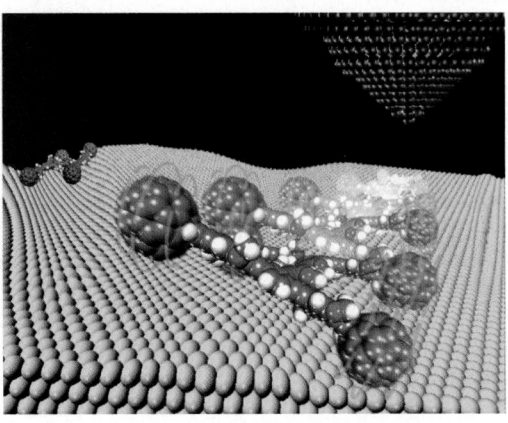

Figure 12.52 Driving a nanocar.
ARIS Presentation Center, Art Library

Demo Campbell, Dean J.; Olson, Joel A.; Calderon, Camilo E.; Doolan, Patrick W.; Mengelt, Elizabeth A.; Ellis, Arthur B.; Lisensky, George C. "Chemistry with Refrigerator Magnets: From Modeling of Nanoscale Characterization to Composite Fabrication," *J. Chem. Educ.* **1999**, *76*, 1205.

Section Summary

Doping increases the conductivity of semiconductors and is essential to modern electronic materials. Doping silicon with Group 5A(15) atoms introduces negative sites (n-type) by adding valence electrons to the conduction band, whereas doping with Group 3A(13) atoms adds positive holes (p-type) by emptying some orbitals in the valence band. Placing these two types of doped Si next to one another forms a p-n junction. Sandwiching a p-type portion between two n-type portions forms a transistor. • Liquid crystal phases flow like liquids but have molecules ordered like crystalline solids. Typically, the molecules have rodlike shapes, and their intermolecular forces keep them aligned. Thermotropic phases are prepared by heating the solid; lyotropic phases form when the solvent concentration is varied. The nematic, cholesteric, and smectic phases of liquid crystals differ in their molecular order. Liquid crystal applications depend on controlling the orientation of the molecules. • Ceramics are very resistant to heat and chemicals. Most are network covalent solids formed at high temperature from simple reactants. They add lightweight strength to other materials. • Polymers are extremely large molecules that adopt the shape of a random coil as a result of intermolecular forces. A polymer sample has an average molar mass because it consists of a range of chain lengths. The high viscosity of a polymer arises from attractions between chains or, in the case of a dissolved polymer, between the chain and the solvent molecules. By varying the degrees of branching, crosslinking, and ordering (crystallinity), chemists tailor polymers with remarkable properties. • Nanoscale materials can be made one atom or molecule at a time through construction processes involving self-assembly and controlled orientation of molecules.

Chapter Perspective

Our central focus—macroscopic behavior resulting from molecular behavior—became still clearer in this chapter, as we built on earlier ideas of bonding, molecular shape, and polarity to understand the intermolecular forces that create the properties of liquids and solids. In Chapter 13, we'll find many parallels to the key ideas in this chapter: the same intermolecular forces create solutions, temperature influences solubility just as it does vapor pressure, and the equilibrium state also underlies the properties of solutions.

CHAPTER REVIEW GUIDE

The following sections provide many aids to help you study this chapter. (Numbers in parentheses refer to pages, unless noted otherwise.)

Learning Objectives
These are concepts and skills you should know after studying this chapter.

Relevant section and/or sample problem (SP) numbers appear in parentheses.

Understand These Concepts

1. How the interplay between kinetic and potential energy underlies the properties of the three states of matter and their phase changes (12.1)
2. The processes involved, both within a phase and during a phase change, when heat is added or removed from a pure substance (12.2)
3. The meaning of vapor pressure and how phase changes are dynamic equilibrium processes (12.2)
4. How temperature and intermolecular forces influence vapor pressure (12.2)
5. The relation between vapor pressure and boiling point (12.2)
6. How a phase diagram shows the phases of a substance at differing conditions of pressure and temperature (12.2)
7. The distinction between bonding and intermolecular forces on the basis of Coulomb's law and the meaning of the van der Waals radius of an atom (12.3)
8. The types and relative strengths of intermolecular forces acting in a substance (dipole-dipole, H-bonding, dispersion), the impact of H bonding on physical properties, and the meaning of polarizability (12.3)
9. The meanings of surface tension, capillarity, and viscosity and how intermolecular forces influence their magnitudes (12.4)
10. How the important macroscopic properties of water arise from atomic and molecular properties (12.5)

11. The meaning of *crystal lattice* and the characteristics of the three types of cubic unit cells (12.6)
12. How packing of spheres gives rise to the hexagonal and cubic unit cells (12.6)
13. Types of crystalline solids and how their intermolecular forces give rise to their properties (12.6)
14. How band theory accounts for the properties of metals and the relative conductivities of metals, nonmetals, and metalloids (12.6)
15. The structures, properties, and functions of modern materials (doped semiconductors, liquid crystals, ceramics, polymers, and nanostructures) on the atomic scale (12.7)

Master These Skills

1. Calculating the overall enthalpy change when heat is gained or lost by a pure substance (12.2 and SP 12.1)
2. Using the Clausius-Clapeyron equation to examine the relationship between vapor pressure and temperature (SP 12.2)
3. Using a phase diagram to predict the physical state and/or phase change of a substance (12.2)
4. Determining whether a substance can form H bonds and drawing the H-bonded structures (SP 12.3)
5. Predicting the types and relative strength of the intermolecular forces acting within a substance from its structure (SP 12.4)
6. Finding the number of particles in a unit cell (12.6)
7. Calculating atomic radius from the crystal structure of an element (SP 12.5)

Key Terms
These important terms appear in boldface in the chapter and are defined again in the Glossary.

phase (437)
intermolecular forces (437)
phase change (437)

Section 12.1
condensation (438)
vaporization (438)
freezing (438)
melting (fusion) (438)
heat of vaporization (ΔH°_{vap}) (439)
heat of fusion (ΔH°_{fus}) (439)
sublimation (439)
deposition (439)
heat of sublimation (ΔH°_{subl}) (439)

Section 12.2
heating-cooling curve (440)
vapor pressure (445)
Clausius-Clapeyron equation (446)
boiling point (447)

melting point (447)
phase diagram (448)
critical point (448)
triple point (449)

Section 12.3
van der Waals radius (450)
ion-dipole force (450)
dipole-dipole force (451)
hydrogen bond (H bond) (452)
polarizability (454)
dispersion (London) force (454)

Section 12.4
surface tension (457)
capillarity (457)
viscosity (458)

Section 12.6
crystalline solid (463)
amorphous solid (463)
lattice (464)
unit cell (464)

coordination number (464)
simple cubic unit cell (464)
body-centered cubic unit cell (464)
face-centered cubic unit cell (464)
packing efficiency (466)
hexagonal closest packing (466)
cubic closest packing (466)
x-ray diffraction analysis (468)
scanning tunneling microscopy (468)
atomic solid (470)
molecular solid (470)
ionic solid (471)
metallic solid (472)
network covalent solid (472)
band theory (474)
valence band (474)
conduction band (474)
conductor (475)

semiconductor (475)
insulator (475)
superconductivity (475)

Section 12.7
crystal defect (477)
doping (477)
liquid crystal (478)
ceramic (481)
polymer (484)
macromolecule (484)
monomer (484)
degree of polymerization (*n*) (484)
random coil (485)
radius of gyration (R_g) (486)
plastic (488)
branch (488)
crosslink (488)
elastomer (488)
copolymer (489)
nanotechnology (489)

Key Equations and Relationships Numbered and screened concepts are listed for you to refer to or memorize.

12.1 Using the vapor pressure at one temperature to find the vapor pressure at another temperature (two-point form of the Clausius-Clapeyron equation) (446):

$$\ln\frac{P_2}{P_1} = \frac{-\Delta H_{vap}}{R}\left(\frac{1}{T_2} - \frac{1}{T_1}\right)$$

Highlighted Figures and Tables These figures (F) and tables (T) provide a visual review of key ideas.

T12.1 Macroscopic comparison of gases, liquids, and solids (438)
F12.2 Phase changes and their enthalpy changes (440)
F12.3 Cooling curve for the conversion of gaseous water to ice (441)
F12.4 Liquid-gas equilibrium (444)
F12.5 Effect of temperature on the distribution of molecular speeds in a liquid (445)
F12.6 Vapor pressure as a function of temperature and intermolecular forces (446)
F12.9 Phase diagrams for CO_2 and H_2O (448)
F12.10 Covalent and van der Waals radii (450)
F12.11 Periodic trends in covalent and van der Waals radii (450)
T12.2 Bonding and nonbonding (intermolecular) forces (451)
F12.14 Hydrogen bonding and boiling point (453)
F12.16 Molar mass and boiling point (455)

F12.18 Summary diagram for analyzing intermolecular forces (455)
F12.21 The H-bonding ability of the water molecule (460)
F12.24 Macroscopic properties of water and their atomic and molecular "roots" (462)
F12.26 The crystal lattice and the unit cell (464)
F12.27 The three cubic unit cells (465)
F12.28 Packing identical spheres (467)
F12.29 Edge length and atomic (ionic) radius in the three cubic unit cells (468)
T12.5 The major types of crystalline solids (470)
F12.32 The sodium chloride structure (471)
F12.37 The band of molecular orbitals in lithium metal (474)
F12.38 Electrical conductivity in a conductor, a semiconductor, and an insulator (475)

Brief Solutions to FOLLOW-UP PROBLEMS Compare your solutions to these calculation steps and answers.

12.1 The scene represents solid water at $-7.00°C$ melting to liquid water at $16.0°C$, so there are three stages.
Stage 1, $H_2O(s)\,[-7.00°C] \longrightarrow H_2O(s)\,[0.00°C]$:
$q = n \times C_{water(s)} \times \Delta T = (2.25\text{ mol})(37.6\text{ J/mol}\cdot°C)(7.00°C)$
$\quad = 592\text{ J} = 0.592\text{ kJ}$
Stage 2, $H_2O(s)\,[0.00°C] \longrightarrow H_2O(l)\,[0.00°C]$:
$q = n(\Delta H°_{fus}) = (2.25\text{ mol})(6.02\text{ kJ/mol}) = 13.5\text{ kJ}$
Stage 3, $H_2O(l)\,[0.00°C] \longrightarrow H_2O(l)\,[16.0°C]$:
$q = n \times C_{water(l)} \times \Delta T = (2.25\text{ mol})(75.4\text{ J/mol}\cdot°C)(16.0°C)$
$\quad = 2714\text{ J} = 2.71\text{ kJ}$

Total heat (kJ) $= 0.592\text{ kJ} + 13.5\text{ kJ} + 2.71\text{ kJ} = 16.8\text{ kJ}$

12.2 $\ln\dfrac{P_2}{P_1} = \left(\dfrac{-40.7\times10^3\text{ J/mol}}{8.314\text{ J/mol}\cdot\text{K}}\right)$

$\qquad \times\left(\dfrac{1}{273.15 + 85.5\text{ K}} - \dfrac{1}{273.15 + 34.1\text{ K}}\right)$
$\qquad = (-4.90\times10^3\text{ K})(-4.66\times10^{-4}\text{ K}^{-1}) = 2.28$
$\dfrac{P_2}{P_1} = 9.8$; thus, $P_2 = 40.1\text{ torr} \times 9.8 = 3.9\times10^2\text{ torr}$

12.3 (a)

(c) No H bonding

(b)

12.4 (a) Dipole-dipole, dispersion; CH_3Br
(b) H bonds, dipole-dipole, dispersion; $CH_3CH_2CH_2OH$
(c) Dispersion; C_3H_8
12.5 From Figure 12.29B,
$A = \dfrac{4r}{\sqrt{3}} = \dfrac{4(126\text{ pm})}{\sqrt{3}} = 291\text{ pm}$

$r = 126$ pm
$\leftarrow A = ? \rightarrow$

PROBLEMS

Problems with colored numbers are answered in Appendix E and worked in detail in the Student Solutions Manual. Problem sections match those in the text and provide the numbers of relevant sample problems. Most offer Concept Review Questions, Skill-Building Exercises (grouped in pairs covering the same concept), and Problems in Context. The Comprehensive Problems are based on material from any section or previous chapter.

An Overview of Physical States and Phase Changes
● Concept Review Questions
• **12.1** How does the energy of attraction between particles compare with their energy of motion in a gas and in a solid? As part of your answer, identify two macroscopic properties that differ between a gas and a solid.
• **12.2** (a) Why are gases more easily compressed than liquids?
(b) Why do liquids have a greater ability to flow than solids?

Levels of Difficulty: • Easy •• Medium ••• Challenging

• **12.3** What type of forces, intramolecular or intermolecular:
   (a) Prevent ice cubes from adopting the shape of their container?
   (b) Are overcome when ice melts?
   (c) Are overcome when liquid water is vaporized?
   (d) Are overcome when gaseous water is converted to hydrogen gas and oxygen gas?
• **12.4** (a) Why is the heat of fusion (ΔH_{fus}) of a substance smaller than its heat of vaporization (ΔH_{vap})?
   (b) Why is the heat of sublimation (ΔH_{subl}) of a substance greater than its ΔH_{vap}?
   (c) At a given temperature and pressure, how does the magnitude of the heat of vaporization of a substance compare with that of its heat of condensation?

● **Skill-Building Exercises** *(grouped in similar pairs)*
• **12.5** Which forces are intramolecular and which intermolecular?
   (a) Those preventing oil from evaporating at room temperature
   (b) Those preventing butter from melting in a refrigerator
   (c) Those allowing silver to tarnish
   (d) Those preventing O_2 in air from forming O atoms
• **12.6** Which forces are intramolecular and which intermolecular?
   (a) Those allowing fog to form on a cool, humid evening
   (b) Those allowing water to form when H_2 is sparked
   (c) Those allowing liquid benzene to crystallize when cooled
   (d) Those responsible for the low boiling point of hexane

• **12.7** Name the phase change in each of these events:
   (a) Dew appears on a lawn in the morning.
   (b) Icicles change into liquid water.
   (c) Wet clothes dry on a summer day.
• **12.8** Name the phase change in each of these events:
   (a) A diamond film forms on a surface from gaseous carbon atoms in a vacuum.
   (b) Mothballs in a bureau drawer disappear over time.
   (c) Molten iron from a blast furnace is cast into ingots ("pigs").

● **Problems in Context**
•• **12.9** Liquid propane, a widely used fuel, is produced by compressing gaseous propane. During the process, approximately 15 kJ of energy is released for each mole of gas liquefied. Where does this energy come from?
• **12.10** Many heat-sensitive and oxygen-sensitive solids, such as camphor, are purified by warming under vacuum. The solid vaporizes directly, and the vapor crystallizes on a cool surface. What phase changes are involved in this method?

Quantitative Aspects of Phase Changes
(Sample Problems 12.1 and 12.2)

● **Concept Review Questions**
• **12.11** Describe the changes (if any) in potential energy and in kinetic energy among the molecules when gaseous PCl_3 condenses to a liquid at a fixed temperature.
•• **12.12** When benzene is at its melting point, two processes occur simultaneously and balance each other. Describe these processes on the macroscopic and molecular levels.
• **12.13** Liquid hexane (bp = 69°C) is placed in a closed container at room temperature. At first, the pressure of the vapor phase increases, but after a short time, it stops changing. Why?
• **12.14** Explain the effect of strong intermolecular forces on each of these parameters: (a) critical temperature; (b) boiling point; (c) vapor pressure; (d) heat of vaporization.

•• **12.15** Match each numbered point in the phase diagram for compound Q with the correct molecular depiction below:

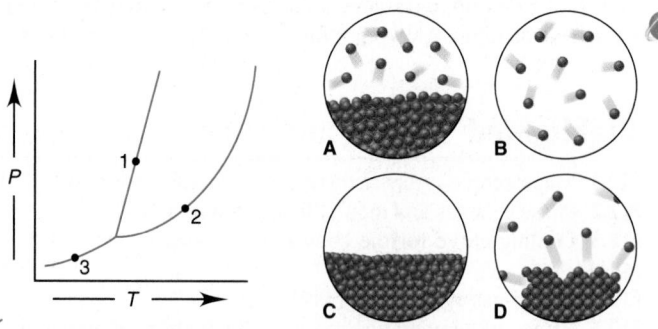

•• **12.16** A liquid is in equilibrium with its vapor in a closed vessel at a fixed temperature. The vessel is connected by a stopcock to an evacuated vessel. When the stopcock is opened, will the final pressure of the vapor be different from the initial value if (a) some liquid remains; (b) all the liquid is first removed? Explain.
• **12.17** The phase diagram for substance A has a solid-liquid line with a positive slope, and that for substance B has a solid-liquid line with a negative slope. What macroscopic property can distinguish A from B?
•• **12.18** Why does water vapor at 100°C cause a more severe burn than liquid water at 100°C?

● **Skill-Building Exercises** *(grouped in similar pairs)*
• **12.19** From the data below, calculate the total heat (in J) needed to convert 22.00 g of ice at −6.00°C to liquid water at 0.500°C:

mp at 1 atm:	0.0°C	ΔH_{fus}°:	6.02 kJ/mol
c_{liquid}:	4.21 J/g·°C	c_{solid}:	2.09 J/g·°C

• **12.20** From the data below, calculate the total heat (in J) needed to convert 0.333 mol of gaseous ethanol at 300°C and 1 atm to liquid ethanol at 25.0°C and 1 atm:

bp at 1 atm:	78.5°C	ΔH_{vap}°:	40.5 kJ/mol
c_{gas}:	1.43 J/g·°C	c_{liquid}:	2.45 J/g·°C

• **12.21** A liquid has a ΔH_{vap}° of 35.5 kJ/mol and a boiling point of 122°C at 1.00 atm. What is its vapor pressure at 113°C?
• **12.22** Diethyl ether has a ΔH_{vap}° of 29.1 kJ/mol and a vapor pressure of 0.703 atm at 25.0°C. What is its vapor pressure at 95.0°C?

• **12.23** What is the ΔH_{vap}° of a liquid that has a vapor pressure of 621 torr at 85.2°C and a boiling point of 95.6°C at 1 atm?
• **12.24** Methane (CH_4) has a boiling point of −164°C at 1 atm and a vapor pressure of 42.8 atm at −100°C. What is the heat of vaporization of CH_4?

• **12.25** Use these data to draw a qualitative phase diagram for ethylene (C_2H_4). Is $C_2H_4(s)$ more or less dense than $C_2H_4(l)$?

bp at 1 atm:	−103.7°C
mp at 1 atm:	−169.16°C
Critical point:	9.9°C and 50.5 atm
Triple point:	−169.17°C and 1.20×10^{-3} atm

• **12.26** Use these data to draw a qualitative phase diagram for H_2. Does H_2 sublime at 0.05 atm? Explain.

mp at 1 atm:	13.96 K
bp at 1 atm:	20.39 K
Triple point:	13.95 K and 0.07 atm
Critical point:	33.2 K and 13.0 atm
Vapor pressure of solid at 10 K:	0.001 atm

This figure is available on the Presentation Center, within the art library.

● **Problems in Context**

• **12.27** Sulfur dioxide is produced in enormous amounts for sulfuric acid production. It melts at $-73°C$ and boils at $-10.°C$. Its $\Delta H°_{fus}$ is 8.619 kJ/mol and its $\Delta H°_{vap}$ is 25.73 kJ/mol. The specific heat capacities of the liquid and gas are 0.995 J/g·K and 0.622 J/g·K, respectively. How much heat is required to convert 2.500 kg of solid SO_2 at the melting point to a gas at $60.°C$?

• **12.28** Butane is a common fuel used in cigarette lighters and camping stoves. Normally supplied in metal containers under pressure, the fuel exists as a mixture of liquid and gas, so high temperatures may cause the container to explode. At $25.0°C$, the vapor pressure of butane is 2.3 atm. What is the pressure in the container at $135°C$ ($\Delta H°_{vap} = 24.3$ kJ/mol)?

•• **12.29** Use Figure 12.9A, p. 448, to answer the following:
(a) Carbon dioxide is sold in steel cylinders under pressures of approximately 20 atm. Is there liquid CO_2 in the cylinder at room temperature (~20°C)? At 40°C? At −40°C? At −120°C?
(b) Carbon dioxide is also sold as solid chunks, called *dry ice,* in insulated containers. If the chunks are warmed by leaving them in an open container at room temperature, will they melt?
(c) If a container is nearly filled with dry ice and then sealed and warmed to room temperature, will the dry ice melt?
(d) If dry ice is compressed at a temperature below its triple point, will it melt?

Types of Intermolecular Forces
(Sample Problems 12.3 and 12.4)

● **Concept Review Questions**

•• **12.30** Why are covalent bonds typically much stronger than intermolecular forces?

•• **12.31** Intermolecular forces are depicted in the scenes below:

A   B
C   D

(a) Name the type of force being depicted in each scene.
(b) Rank the forces in order of increasing strength.

•• **12.32** Oxygen and selenium are members of Group 6A(16). Water forms H bonds, but H_2Se does not. Explain.

•• **12.33** In solid I_2, is the distance between the two I nuclei of one I_2 molecule longer or shorter than the distance between two I nuclei of adjacent I_2 molecules? Explain.

•• **12.34** Polar molecules exhibit dipole-dipole forces. Do they also exhibit dispersion forces? Explain.

•• **12.35** Distinguish between *polarizability* and *polarity*. How does each influence intermolecular forces?

•• **12.36** How can one nonpolar molecule induce a dipole in a nearby nonpolar molecule?

● **Skill-Building Exercises** *(grouped in similar pairs)*

•• **12.37** What is the strongest interparticle force in each substance?
(a) CH_3OH (b) CCl_4 (c) Cl_2

•• **12.38** What is the strongest interparticle force in each substance?
(a) H_3PO_4 (b) SO_2 (c) $MgCl_2$

•• **12.39** What is the strongest interparticle force in each substance?
(a) CH_3Cl (b) CH_3CH_3 (c) NH_3

•• **12.40** What is the strongest interparticle force in each substance?
(a) Kr   (b) BrF   (c) H_2SO_4

•• **12.41** Which member of each pair of compounds forms intermolecular H bonds? Draw the H-bonded structures in each case:
(a) $\underset{\underset{OH}{|}}{CH_3CHCH_3}$ or CH_3SCH_3 (b) HF or HBr

•• **12.42** Which member of each pair of compounds forms intermolecular H bonds? Draw the H-bonded structures in each case:
(a) $(CH_3)_2NH$ or $(CH_3)_3N$ (b) $HOCH_2CH_2OH$ or FCH_2CH_2F

•• **12.43** Which forces oppose vaporization of each substance?
(a) Hexane   (b) Water   (c) $SiCl_4$

•• **12.44** Which forces oppose vaporization of each substance?
(a) Br_2 (b) SbH_3 (c) CH_3NH_2

•• **12.45** Which has the greater polarizability? Explain.
(a) Br^- or I^- (b) $CH_2 = CH_2$ or $CH_3—CH_3$ (c) H_2O or H_2Se

•• **12.46** Which has the greater polarizability? Explain.
(a) Ca^{2+} or Ca   (b) CH_3CH_3 or $CH_3CH_2CH_3$ (c) CCl_4 or CF_4

• **12.47** Which member in each pair of liquids has the *higher* vapor pressure at a given temperature? Explain.
(a) C_2H_6 or C_4H_{10} (b) CH_3CH_2OH or CH_3CH_2F
(c) NH_3 or PH_3

• **12.48** Which member in each pair of liquids has the *lower* vapor pressure at a given temperature? Explain.
(a) $HOCH_2CH_2OH$ or $CH_3CH_2CH_2OH$
(b) CH_3COOH or $(CH_3)_2C=O$ (c) HF or HCl

• **12.49** Which substance has the *lower* boiling point? Explain.
(a) LiCl or HCl   (b) NH_3 or PH_3 (c) Xe or I_2

• **12.50** Which substance has the *higher* boiling point? Explain.
(a) CH_3CH_2OH or $CH_3CH_2CH_3$ (b) NO or N_2 (c) H_2S or H_2Te

• **12.51** Which substance has the *lower* boiling point? Explain.
(a) $CH_3CH_2CH_2CH_3$ or $\underset{\underset{CH_2—CH_2}{|\qquad|}}{CH_2—CH_2}$ (b) NaBr or PBr_3

(c) H_2O or HBr

• **12.52** Which substance has the *higher* boiling point? Explain.
(a) CH_3OH or CH_3CH_3 (b) FNO or ClNO
(c) $\underset{H}{\overset{F}{\diagdown}}C=C\underset{H}{\overset{F}{\diagup}}$ or $\underset{F}{\overset{H}{\diagdown}}C=C\underset{H}{\overset{F}{\diagup}}$

● **Problems in Context**

• **12.53** For pairs of molecules in the gas phase, average H-bond dissociation energies are 17 kJ/mol for NH_3, 22 kJ/mol for H_2O, and 29 kJ/mol for HF. Explain this increase in H-bond strength.

•• **12.54** Dispersion forces are the only intermolecular forces present in motor oil, yet it has a high boiling point. Explain.

• **12.55** Why does the antifreeze ingredient ethylene glycol ($HOCH_2CH_2OH$; $M = 62.07$ g/mol) have a boiling point of $197.6°C$, whereas propanol ($CH_3CH_2CH_2OH$; $M = 60.09$ g/mol), a compound with a similar molar mass, has a boiling point of only $97.4°C$?

Properties of the Liquid State

● **Concept Review Questions**

• **12.56** Before the phenomenon of surface tension was understood, physicists described the surface of water as being covered with a "skin." What causes this skinlike phenomenon?

•• **12.57** Small, equal-sized drops of oil, water, and mercury lie on a waxed floor. How does each liquid behave? Explain.

•• **12.58** Why does an aqueous solution of ethanol (CH_3CH_2OH) have a lower surface tension than water?

• **12.59** Why are units of energy per area (J/m^2) used for surface tension values?

•• **12.60** Does the *strength* of the intermolecular forces in a liquid change as the liquid is heated? Explain. Why does liquid viscosity decrease with rising temperature?

● **Skill-Building Exercises** (grouped in similar pairs)

• **12.61** Rank the following in order of *increasing* surface tension at a given temperature, and explain your ranking:
(a) $CH_3CH_2CH_2OH$ (b) $HOCH_2CH(OH)CH_2OH$
(c) $HOCH_2CH_2OH$

• **12.62** Rank the following in order of *decreasing* surface tension at a given temperature, and explain your ranking:
(a) CH_3OH (b) CH_3CH_3 (c) $H_2C{=}O$

• **12.63** Rank the compounds in Problem 12.61 in order of *decreasing* viscosity at a given temperature; explain your ranking.

• **12.64** Rank the compounds in Problem 12.62 in order of *increasing* viscosity at a given temperature; explain your ranking.

● **Problems in Context**

•• **12.65** Soil vapor extraction (SVE) is used to remove volatile organic pollutants, such as chlorinated solvents, from soil at hazardous waste sites. Vent wells are drilled, and a vacuum pump is applied to the subsurface. (a) How does this remove pollutants? (b) Why does heating combined with SVE speed the process?

• **12.66** Use Figure 12.1, p. 439 to answer the following: (a) Does it take more heat to melt 12.0 g of CH_4 or 12.0 g of Hg? (b) Does it take more heat to vaporize 12.0 g of CH_4 or 12.0 g of Hg? (c) What is the principal intermolecular force in each sample?

• **12.67** Pentanol ($C_5H_{11}OH$; $\mathcal{M} = 88.15$ g/mol) has nearly the same molar mass as hexane (C_6H_{14}; $\mathcal{M} = 86.17$ g/mol) but is more than 12 times as viscous at 20°C. Explain.

The Uniqueness of Water

● **Concept Review Questions**

• **12.68** For what types of substances is water a good solvent? For what types is it a poor solvent? Explain.

• **12.69** A water molecule can engage in as many as four H bonds. Explain.

• **12.70** Warm-blooded animals have a narrow range of body temperature because their bodies have a high water content. Explain.

• **12.71** What property of water keeps plant debris on the surface of lakes and ponds? What is the ecological significance of this?

• **12.72** A drooping plant can be made upright by watering the ground around it. Explain.

•• **12.73** Describe the molecular basis of the property of water responsible for the presence of ice on the surface of a frozen lake.

• **12.74** Describe in molecular terms what occurs when ice melts.

The Solid State: Structure, Properties, and Bonding
(Sample Problem 12.5)

● **Concept Review Questions**

• **12.75** What is the difference between an amorphous solid and a crystalline solid on the macroscopic and molecular levels? Give an example of each.

• **12.76** How are a solid's unit cell and crystal structure related?

• **12.77** For structures consisting of identical atoms, how many atoms are contained in the simple, body-centered, and face-centered cubic unit cells? Explain how you obtained the values.

• **12.78** An element has a crystal structure in which the width of the cubic unit cell equals the diameter of an atom. What type of unit cell does it have?

• **12.79** What specific difference in the positioning of spheres gives a crystal structure based on the face-centered cubic unit cell less empty space than one based on the body-centered cubic unit cell?

• **12.80** Both solid Kr and solid Cu consist of individual atoms. Why do their physical properties differ so much?

• **12.81** What is the energy gap in band theory? Compare its size in superconductors, conductors, semiconductors, and insulators.

• **12.82** Predict the effect (if any) of an increase in temperature on the electrical conductivity of (a) a conductor; (b) a semiconductor; (c) an insulator.

• **12.83** Besides the type of unit cell, what information is needed to find the density of a solid consisting of identical atoms?

● **Skill-Building Exercises** (grouped in similar pairs)

• **12.84** What type of crystal lattice does each metal form? (The number of atoms per unit cell is given in parentheses.)
(a) Ni (4)   (b) Cr (2)   (c) Ca (4)

• **12.85** What is the number of atoms per unit cell for each metal?
(a) Polonium, Po   (b) Manganese, Mn   (c) Silver, Ag

• **12.86** When cadmium oxide reacts to form cadmium selenide, a change in unit cell occurs, as depicted below:

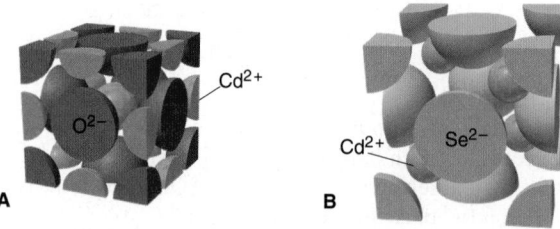

(a) What is the change in unit cell?
(b) Does the coordination number of cadmium change? Explain.

• **12.87** As molten iron cools to 1674 K, it adopts one type of cubic unit cell; then, as the temperature drops below 1181 K, it changes to another, as depicted below:

(a) What is the change in unit cell?
(b) Which crystal structure has the greater packing efficiency?

•• **12.88** Of the five major types of crystalline solid, which does each of the following form, and why: (a) Ni; (b) F_2; (c) CH_3OH; (d) Sn; (e) Si; (f) Xe?

•• **12.89** Of the five major types of crystalline solid, which does each of the following form, and why: (a) SiC; (b) Na_2SO_4; (c) SF_6; (d) cholesterol ($C_{27}H_{45}OH$); (e) KCl; (f) BN?

• **12.90** Zinc oxide adopts the zinc blende crystal structure (Figure P12.90). How many Zn^{2+} ions are in the ZnO unit cell?

• **12.91** Calcium sulfide adopts the sodium chloride crystal structure (Figure P12.91). How many S^{2-} ions are in the CaS unit cell?

Figure P12.90

Figure P12.91

•• **12.92** Zinc selenide (ZnSe) crystallizes in the zinc blende structure (see Figure P12.90) and has a density of 5.42 g/cm^3.
(a) How many Zn and Se ions are in each unit cell?
(b) What is the mass of a unit cell?
(c) What is the volume of a unit cell?
(d) What is the edge length of a unit cell?

•• **12.93** An element crystallizes in a face-centered cubic lattice and has a density of 1.45 g/cm^3. The edge of its unit cell is 4.52×10^{-8} cm.
(a) How many atoms are in each unit cell?
(b) What is the volume of a unit cell?
(c) What is the mass of a unit cell?
(d) Calculate an approximate atomic mass for the element.

• **12.94** Classify each of the following as a conductor, insulator, or semiconductor: (a) phosphorus; (b) mercury; (c) germanium.

• **12.95** Classify each of the following as a conductor, insulator, or semiconductor: (a) carbon (graphite); (b) sulfur; (c) platinum.

•• **12.96** Predict the effect (if any) of an increase in temperature on the electrical conductivity of (a) antimony; (b) tellurium; (c) bismuth.

•• **12.97** Predict the effect (if any) of a decrease in temperature on the electrical conductivity of (a) silicon; (b) lead; (c) germanium.

● **Problems in Context**

•• **12.98** Polonium, the Period 6 member of Group 6A(16), is a rare radioactive metal that is the only element with a crystal structure based on the simple cubic unit cell. If its density is 9.142 g/cm^3, calculate an approximate atomic radius for polonium.

•• **12.99** The coinage metals—copper, silver, and gold—crystallize in a cubic closest packed structure. Use the density of copper (8.95 g/cm^3) and its molar mass (63.55 g/mol) to calculate an approximate atomic radius for copper.

•• **12.100** One of the most important enzymes in the world—nitrogenase, the plant protein that catalyzes nitrogen fixation—contains active clusters of iron, sulfur, and molybdenum atoms. Crystalline molybdenum (Mo) has a body-centered cubic unit cell (d of Mo = 10.28 g/cm^3). (a) Determine the edge length of the unit cell. (b) Calculate the atomic radius of Mo.

•• **12.101** Tantalum (Ta; d = 16.634 g/cm^3 and $\mathcal{M}$ = 180.9479 g/mol) has a body-centered cubic structure with a unit-cell edge length of 3.3058 Å. Use these data to calculate Avogadro's number.

Advanced Materials

● **Concept Review Questions**

•• **12.102** When tin is added to copper, the resulting alloy (bronze) is much harder than copper. Explain.

• **12.103** In the process of doping a semiconductor, certain impurities are added to increase its electrical conductivity. Explain this process for an n-type and a p-type semiconductor.

• **12.104** State two molecular characteristics of substances that typically form liquid crystals. How is each related to function?

• **12.105** Distinguish between isotropic and anisotropic substances. To which category do liquid crystals belong?

• **12.106** How are the properties of high-tech ceramics the same as those of traditional clay ceramics, and how are they different? Refer to specific substances in your answer.

• **12.107** Why is the average molar mass of a polymer sample different from the molar mass of an individual chain?

• **12.108** How does the random coil shape relate to the radius of gyration of a polymer chain?

• **12.109** What factor(s) influence the viscosity of a polymer solution? What factor(s) influence the viscosity of a molten polymer? What is a polymer glass?

• **12.110** Use an example to show how branching and crosslinking can affect the physical behavior of a polymer.

● **Skill-Building Exercises** (grouped in similar pairs)

• **12.111** Silicon and germanium are both semiconducting elements from Group 4A(14) that can be doped to improve their conductivity. Would each of the following form an n-type or a p-type semiconductor: (a) Ge doped with P; (b) Si doped with In?

• **12.112** Would each of the following form an n-type or a p-type semiconductor: (a) Ge doped with As; (b) Si doped with B?

• **12.113** The repeat unit in a polystyrene coffee cup has the formula $C_6H_5CHCH_2$. If the molar mass of the polymer is 3.5×10^5 g/mol, what is the degree of polymerization?

• **12.114** The monomer of poly(vinyl chloride) has the formula C_2H_3Cl. If there are 1565 repeat units in a single chain of the polymer, what is the molecular mass (in amu) of that chain?

• **12.115** The polypropylene (repeat unit CH_3CHCH_2) in a plastic toy has a molar mass of 2.8×10^5 g/mol and a repeat unit length of 0.252 pm. Calculate the radius of gyration.

• **12.116** The polymer that is used to make 2-L soda bottles [poly(ethylene terephthalate)] has a repeat unit with molecular formula $C_{10}H_8O_4$ and length 1.075 nm. Calculate the radius of gyration of a chain with molar mass of 2.30×10^4 g/mol.

Comprehensive Problems

•• **12.117** A 0.75-L bottle is cleaned, dried, and closed in a room where the air is 22°C and 44% relative humidity (that is, the water vapor in the air is 0.44 of the equilibrium vapor pressure at 22°C). The bottle is brought outside and stored at 0.0°C.
(a) What mass of liquid water condenses inside the bottle?
(b) Would liquid water condense at 10°C? (See Table 5.3, p. 207.)

•• **12.118** In an experiment, 5.00 L of N_2 is saturated with water vapor at 22°C and then compressed to half its volume at constant T. (a) What is the partial pressure of H_2O in the compressed gas mixture? (b) What mass of water vapor condenses to liquid?

•• **12.119** Barium is the largest nonradioactive alkaline earth metal. It has a body-centered cubic unit cell and a density of 3.62 g/cm^3.

What is the atomic radius of barium? (Volume of a sphere: $V = \frac{4}{3}\pi r^3$.)

••• 12.120 Two important characteristics used to evaluate risks of fire or explosion are a compound's *lower flammable limit* (LFL) and *flash point*. The LFL is the minimum percentage by volume in air that is ignitable. Below that, the mixture is too "lean" to burn. The flash point is the temperature at which the air over a confined liquid becomes ignitable. *n*-Hexane boils at 68.7°C at 1 atm. At 20.0°C, its vapor pressure is 121 mmHg. The LFL of *n*-hexane is 1.1%. Calculate the flash point of *n*-hexane.

•• 12.121 Bismuth is used to calibrate instruments employed in high-pressure studies because it has several well-characterized crystalline phases. Its phase diagram (*right*) shows the liquid phase and five solid phases that are stable above 1 katm (1000 atm) and up to 300°C. (a) Which solid

phases are stable at 25°C? (b) Which phase is stable at 50 katm and 175°C? (c) As the pressure is reduced from 100 to 1 katm at 200°C, what phase transitions does bismuth undergo? (d) What phases are present at each of the triple points?

•• 12.122 In the photoelectric effect, the work function (φ) is the minimum energy a photon must have to remove an electron from a metal surface (see Problem 7.70). For a given metal, φ depends on how and where the photon strikes the crystal. Copper adopts the face-centered cubic structure. If the photon strikes perpendicular to the unit cell's face, φ is 4.59 eV; if perpendicular to a plane tangent to an edge, φ is 4.48 eV; and if perpendicular to a plane tangent to a corner, φ is 4.94 eV (1 eV = 1.602×10^{-19} J). Find the wavelength (in nm) of the lowest energy photon that can remove an electron from copper, and state how it strikes the unit cell.

•• 12.123 In making computer chips, a 4.00-kg cylindrical ingot of ultrapure n-type doped silicon that is 5.20 inches in diameter is sliced into wafers 1.12×10^{-4} m thick.
(a) Assuming no waste, how many wafers can be made?
(b) What is the mass of a wafer (*d* of Si = 2.34 g/cm³; *V* of a cylinder = $\pi r^2 h$)?
(c) A key step in making p-n junctions for the chip is chemical removal of the oxide layer on the wafer through treatment with gaseous HF. Write a balanced equation for this reaction.
(d) If 0.750% of the Si atoms are removed during the treatment in part (c), how many moles of HF are required per wafer, assuming 100% reaction yield?

•• 12.124 Methyl salicylate, $C_8H_8O_3$, the odorous constituent of oil of wintergreen, has a vapor pressure of 1.00 torr at 54.3°C and 10.0 torr at 95.3°C. (a) What is its vapor pressure at 25°C? (b) What is the minimum number of liters of air that must pass over a sample of the compound at 25°C to vaporize 1.0 mg of it?

•• 12.125 Mercury (Hg) vapor is toxic and readily absorbed from the lungs. At 20.°C, mercury (ΔH_{vap} = 59.1 kJ/mol) has a vapor pressure of 1.20×10^{-3} torr, which is high enough to be hazardous. To reduce the danger to workers in processing plants, Hg is cooled to lower its vapor pressure. At what temperature would the vapor pressure of Hg be at the safer level of 5.0×10^{-5} torr?

••• 12.126 Polytetrafluoroethylene (Teflon) has a repeat unit with the formula F_2C—CF_2. A sample of the polymer consists of fractions with the following distribution of chains:

Fraction	Average no. of repeat units	Amount (mol) of polymer
1	273	0.10
2	330	0.40
3	368	1.00
4	483	0.70
5	525	0.30
6	575	0.10

(a) Determine the molar mass of each fraction.
(b) Determine the number-average molar mass of the sample.
(c) Another type of average molar mass of a polymer sample is called the *weight-average molar mass*, $\mathcal{M}_w$:

$$\mathcal{M}_w = \frac{\Sigma\ (\mathcal{M} \text{ of fraction} \times \text{mass of fraction})}{\text{total mass of all fractions}}$$

Calculate the weight-average molar mass of the above sample.

•• 12.127 A greenhouse contains 256 m³ of air at a temperature of 26°C, and a humidifier in it vaporizes 4.20 L of water. (a) What is the pressure of water vapor in the greenhouse, assuming that none escapes and that the air was originally completely dry (*d* of H_2O = 1.00 g/mL)? (b) What total volume of liquid water would have to be vaporized to saturate the air (i.e., achieve 100% relative humidity)? (See Table 5.4, p. 221.)

•• 12.128 Like most transition metals, tantalum (Ta) exhibits several oxidation states. Give the formula of each tantalum compound whose unit cell is depicted below:

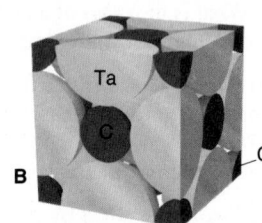

••• 12.129 Iron crystallizes in a body-centered cubic structure. The volume of one Fe atom is 8.38×10^{-24} cm³, and the density of Fe is 7.874 g/cm³. Find an approximate value for Avogadro's number.

••• 12.130 Corn is a valuable source of industrial chemicals. For example, furfural is prepared from corncobs. It is an important solvent in synthetic rubber manufacturing, and it is reduced to furfuryl alcohol, which is used to make polymer resins. Furfural can also be oxidized to 2-furoic acid.

furfuryl alcohol   furfural   2-furoic acid

(a) Which of these compounds can form H bonds? Draw structures in each case.
(b) The molecules of some substances can form an "internal" H bond, that is, an H bond *within* a molecule. This takes the form

of a polygon with atoms as corners and bonds as sides and an H bond as one of the sides. Which of these molecules is (are) likely to form a stable internal H bond? Draw the structure. (*Hint:* Structures with 5 or 6 atoms as corners are most stable.)

•••• **12.131** On a humid day in New Orleans, the temperature is 22.0°C, and the partial pressure of water vapor in the air is 31.0 torr. The 9000-ton air-conditioning system in the Louisiana Superdome maintains an inside air temperature of 22.0°C also, but a partial pressure of water vapor of 10.0 torr. The volume of air in the dome is 2.4×10^6 m^3, and the total pressure inside and outside the dome are both 1.0 atm. (a) What mass of water (in metric tons) must be removed every time the inside air is completely replaced with outside air? (*Hint:* How many moles of gas are in the dome? How many moles of water vapor? How many moles of dry air? How many moles of outside air must be added to the air in the dome to simulate the composition of outside air?) (b) Find the heat released when this mass of water condenses.

•• **12.132** The boiling point of amphetamine, $C_9H_{13}N$, is 201°C at 760 torr and 83°C at 13 torr. What is the concentration (in g/m^3) of amphetamine when it is in contact with 20.°C air?

••• **12.133** Diamond has a face-centered cubic unit cell, with four more C atoms in tetrahedral holes within the cell. Densities of diamonds vary from 3.01 g/cm^3 to 3.52 g/cm^3 because C atoms are missing from some holes. (a) Calculate the unit-cell edge length of the densest diamond. (b) Assuming the cell dimensions are fixed, how many C atoms are in the unit cell of the diamond with the lowest density?

•• **12.134** Is it possible for a salt of formula AB_3 to have a face-centered cubic unit cell of anions with cations in all the eight available holes? Explain.

•• **12.135** The density of solid gallium at its melting point is 5.9 g/cm^3, whereas that of liquid gallium is 6.1 g/cm^3. Is the temperature at the triple point higher or lower than the normal melting point? Is the slope of the solid-liquid line for gallium positive or negative?

••• **12.136** A 4.7-L sealed bottle containing 0.33 g of liquid ethanol, C_2H_6O, is placed in a refrigerator and reaches equilibrium with its vapor at −11°C. (a) What mass of ethanol is present in the vapor? (b) When the container is removed and warmed to room temperature, 20.°C, will all the ethanol vaporize? (c) How much liquid ethanol would be present at 0.0°C? The vapor pressure of ethanol is 10. torr at −2.3°C and 40. torr at 19°C.

••• **12.137** Substance A has the following properties.

mp at 1 atm:	−20.°C	bp at 1 atm:	85°C
ΔH_{fus}:	180. J/g	ΔH_{vap}:	500. J/g
c_{solid}:	1.0 J/g·°C	c_{liquid}:	2.5 J/g·°C
c_{gas}:	0.5 J/g·°C		

At 1 atm, a 25-g sample of A is heated from −40.°C to 100.°C at a constant rate of 450. J/min. (a) How many minutes does it take to heat the sample to its melting point? (b) How many minutes does it take to melt the sample? (c) Perform any other necessary calculations, and draw a curve of temperature vs. time for the entire heating process.

•••• **12.138** An aerospace manufacturer is building a prototype experimental aircraft that cannot be detected by radar. Boron nitride is chosen for incorporation into the body parts, and the boric acid/ammonia method is used to prepare the ceramic material. Given 85.5% and 86.8% yields for the two reaction steps, how much boron nitride can be prepared from 1.00 metric ton of boric acid and 12.5 m^3 of ammonia at 275 K and 3.07×10^3 kPa?

Assume that ammonia does *not* behave ideally under these conditions and is recycled completely in the reaction process.

•• **12.139** The ball-and-stick models below represent three compounds with the same molecular formula, $C_4H_8O_2$:

A   **B**   **C**

(a) Which compound(s) can form intermolecular H bonds?
(b) Which has the highest viscosity?

•• **12.140** The ΔH_f° of gaseous dimethyl ether (CH_3OCH_3) is −185.4 kJ/mol; the vapor pressure is 1.00 atm at −23.7°C and 0.526 atm at −37.8°C. (a) Calculate ΔH_{vap}° of dimethyl ether. (b) Calculate ΔH_f° of liquid dimethyl ether.

•••• **12.141** The crystal structure of sodium is based on the body-centered cubic unit cell. What is the mass of one unit cell of Na?

•••• **12.142** KF has the same type of crystal structure as NaCl. The unit cell of KF has an edge length of 5.39 Å. Find the density of KF.

•••• **12.143** The *intrinsic viscosity* of a solute in a solvent, $[\eta]_{solvent}$, is the portion of the total viscosity due to the solute and is related to solute shape. In polymer science, it relates as well to intermolecular interactions between solvent and polymer: higher $[\eta]_{solvent}$ means stronger interaction. The $[\eta]_{solvent}$ values of polymers in solution are given by the Mark-Houwink equation, $[\eta]_{solvent} = K\mathcal{M}^a$, where $\mathcal{M}$ is the molar mass of the polymer, and K and a are constants specific to the polymer and solvent. Use the data below for 25°C to answer the following questions:

Polymer	Solvent	K (mL/g)	a
Polystyrene	Benzene	9.5×10^{-3}	0.74
	Cyclohexane	8.1×10^{-2}	0.50
Polyisobutylene	Benzene	8.3×10^{-2}	0.50
	Cyclohexane	2.6×10^{-1}	0.70

(a) A polystyrene sample has a molar mass of 104,160 g/mol. Calculate the intrinsic viscosity in benzene and in cyclohexane. Which solvent has stronger interactions with the polymer?
(b) A different polystyrene sample has a molar mass of 52,000 g/mol. Calculate its $[\eta]_{benzene}$. Given a polymer standard of known $\mathcal{M}$, how could you use its measured $[\eta]$ in a given solvent to determine the molar mass of any sample of that polymer?
(c) Compare $[\eta]$ of a polyisobutylene sample [repeat unit $(CH_3)_2CCH_2$] with a molar mass of 104,160 g/mol with that of the polystyrene in part (a). What does this suggest about the solvent-polymer interactions of the two samples?

•• **12.144** At a local convenience store, you purchase a cup of coffee, but, at 98.4°C, it is too hot to drink. You add 23.0 g of ice that is −2.2°C to 248 mL of the coffee. What is the final temperature of the coffee? (Assume the heat capacity and density of the coffee are the same as water and the coffee cup is well insulated.)

•••• **12.145** One way of purifying gaseous H_2 is to pass it under high pressure through the holes of a metal's crystal structure. Palladium, which adopts a cubic closest packed structure, absorbs more H_2 than any other element and is one of the metals currently used for this purpose. Although the metal-hydrogen interaction is unclear, it is estimated that the density of absorbed H_2 approaches that of liquid hydrogen (70.8 g/L). What volume (in L) of gaseous H_2, measured at STP, can be packed into the spaces of 1 dm^3 of palladium metal?

13

Starting the Day with a Mixture Cream mixing into a cup of coffee demonstrates just how familiar mixtures are in everyday life. In this chapter, you'll see why mixtures form and how we quantify their composition and make use of their remarkable physical behavior.

The Properties of Mixtures: Solutions and Colloids

13.1 Types of Solutions: Intermolecular Forces and Solubility
Intermolecular Forces in Solution
Liquid Solutions
Gas Solutions and Solid Solutions

13.2 Intermolecular Forces and Biological Macromolecules
Structures of Proteins
Dual Polarity and Function
Structure of DNA
Structure of Cellulose

13.3 Why Substances Dissolve: Understanding the Solution Process
Heats of Solution and Solution Cycles
Heats of Hydration
The Solution Process and the Change in Entropy

13.4 Solubility as an Equilibrium Process
Effect of Temperature
Effect of Pressure

13.5 Quantitative Ways of Expressing Concentration
Molarity and Molality
Parts of Solute by Parts of Solution
Interconverting Concentration Terms

13.6 Colligative Properties of Solutions
Nonvolatile Nonelectrolyte Solutions
Solute Molar Mass
Volatile Nonelectrolyte Solutions
Strong Electrolyte Solutions

13.7 The Structure and Properties of Colloids

Nearly all the gases, liquids, and solids that make up our world are *mixtures*—two or more substances physically mixed together but not chemically combined. Synthetic mixtures, such as glass and soap, usually contain relatively few components, whereas natural mixtures, such as seawater and soil, are more complex, often containing more than 50 different substances. Living mixtures, such as trees and students, are the most complex—even a simple bacterial cell contains well over 5000 different compounds (Table 13.1).

Concepts & Skills to Review
before you study this chapter

- classification and separation of mixtures (Section 2.9)
- calculations involving mass percent (Section 3.1) and molarity (Section 3.5)
- electrolytes; water as a solvent (Sections 4.1 and 12.5)
- mole fraction and Dalton's law of partial pressures (Section 5.4)
- types of intermolecular forces and the concept of polarizability (Section 12.3)
- distinction between monomer and polymer (Section 12.7)
- vapor pressure of liquids (Section 12.2)

Table 13.1	Approximate Composition of a Bacterium		
Substance	Mass % of Cell	Number of Types	Number of Molecules
Water	~70	1	5×10^{10}
Ions	1	20	?
Sugars*	3	200	3×10^{8}
Amino acids*	0.4	100	5×10^{7}
Lipids*	2	50	3×10^{7}
Nucleotides*	0.4	200	1×10^{7}
Other small molecules	0.2	~200	?
Macromolecules (proteins, nucleic acids, polysaccharides)	23	~5000	6×10^{6}

*Includes precursors and metabolites.

 ARIS Presentation Center, Tables Library

Biological application

Recall from Chapter 2 that a mixture has two defining characteristics: *its composition is variable,* and *it retains some properties of its components.* In this chapter, we focus on two common types of mixtures: solutions and colloids. *A solution is a homogeneous mixture,* one with no boundaries separating its components; thus, a solution exists as one phase. A *heterogeneous mixture* has two or more phases. The pebbles in concrete or the bubbles in champagne are visible indications that these are heterogeneous mixtures. A *colloid* is a heterogeneous mixture in which one component is dispersed as very fine particles in another component, so distinct phases are not easy to see. Smoke and milk are colloids. The essential difference between a solution and a colloid is one of particle *size:*

- In a solution, the particles are individual atoms, ions, or small molecules.
- In a colloid, the particles are typically either macromolecules or aggregations of small molecules that are not large enough to settle out.

As you'll see later in the chapter, these molecular-scale differences result in many observable differences.

IN THIS CHAPTER . . . We examine intermolecular forces between solute and solvent and their role in forming solutions. To predict solubility in different solvents, we focus on the dual polarity of simple organic compounds and then apply this idea to biological macromolecules and the cell membrane. We investigate why a substance dissolves in terms of the enthalpy of solution and the dispersal of matter that occurs during the solution process. To understand the second factor, the concept of entropy is introduced. We examine the equilibrium nature of solubility and see how temperature and pressure affect it. Various concentration units are discussed and interconverted. You'll see why the physical properties of solutions are different from those of pure substances and how we employ those differences. Then we investigate colloids and apply solution and colloid chemistry to the purification of water.

Ion-dipole
(40–600)

H bond
(10–40)

H_2O

Methanol
(CH_3OH)

Dipole-dipole
(5–25)

Ethanal
(CH_3CHO)

Chloroform
($CHCl_3$)

Ion–induced dipole
(3–15)

Cl^-

Hexane
(C_6H_{14})

Dipole–induced dipole
(2–10)

H_2O

Xenon

Dispersion
(0.05–40)

Octane
(C_8H_{18})

C_6H_{14}

Figure 13.1 The major types of intermolecular forces in solutions. Forces are listed in decreasing order of strength (with values in kJ/mol), and an example of each is shown with space-filling models.

13.1 TYPES OF SOLUTIONS: INTERMOLECULAR FORCES AND SOLUBILITY

We often describe solutions in terms of one substance dissolving in another: the **solute** dissolves in the **solvent.** Usually, *the solvent is the most abundant component* of a given solution. In some cases, however, the substances are **miscible,** that is, soluble in each other in any proportion; in such cases, it may not be meaningful to call one the solute and the other the solvent.

The **solubility** (*S*) of a solute is the maximum amount that dissolves in a fixed quantity of a particular solvent at a specified temperature, given that excess solute is present. Different solutes have different solubilities. For example, for sodium chloride (NaCl), $S = 39.12$ g/100. mL water at 100.°C, whereas for silver chloride (AgCl), $S = 0.0021$ g/100. mL water at 100.°C. Obviously, NaCl is much more soluble in water than AgCl is. (Solubility is also expressed in other units, as you'll see later.) Although solubility has a quantitative meaning, *dilute* and *concentrated* are qualitative terms that refer to the *relative amounts of solute:* a dilute solution contains much less dissolved solute than a concentrated one.

From everyday experience, you know that some solvents can dissolve a given solute, whereas others can't. For example, butter doesn't dissolve in water, but it does in cooking oil. A major factor determining whether a solution forms is the relative strength of the intermolecular forces within and between solute and solvent. Experience shows that *substances with similar types of intermolecular forces dissolve in each other.* This fact is summarized in the rule-of-thumb **like dissolves like.** From a knowledge of these forces, we can often predict which solutes will dissolve in which solvents.

Intermolecular Forces in Solution

All the intermolecular forces we discussed in Section 12.3 for pure substances also occur between solute and solvent in solutions (Figure 13.1):

1. *Ion-dipole forces* are the principal force involved in the solubility of ionic compounds in water. When a salt dissolves, each ion on the crystal's surface attracts the oppositely charged end of the water dipole. These attractive forces overcome those between the ions and break down the crystal structure. As each ion becomes separated, more water molecules cluster around it in **hydration shells** (Figure 13.2). Recent evidence shows that water's normal H bonding is altered only among molecules in the closest hydration shell. These molecules are H bonded to others slightly farther away, which are in turn H bonded to still other solvent water molecules. It's important to remember that this process does not lead to a jumble of ions and water molecules; rather, there is less freedom of motion for molecules in the closest hydration shell. For monatomic ions, the number of water molecules in the closest shell depends on the ion's size. Four water molecules can fit tetrahedrally around small ions, such as Li^+, while larger ions, such as Na^+ and F^-, have six water molecules surrounding them octahedrally.

2. *Hydrogen bonding* is especially important in aqueous solution. In fact, it is a primary factor in the solubility in water—and, thus, cell fluid—of many oxygen- and nitrogen-containing organic and biological compounds, such as alcohols, sugars, amines, and amino acids. (Recall that O and N are small and electronegative, so their bound H atoms are partially positive and can get very close to the O of H_2O.)

3. *Dipole-dipole forces,* in the absence of H bonding, account for the solubility of polar organic molecules, such as acetaldehyde (CH_3CHO), in polar, nonaqueous solvents like chloroform ($CHCl_3$).

4. **Ion–induced dipole forces** are one of two types of *charge-induced dipole forces,* which rely on the polarizability of the components. They result when an ion's charge distorts the electron cloud of a nearby nonpolar molecule. This type of force plays an essential biological role that initiates the binding of the Fe^{2+}

Ion-dipole forces

Hydration shells

Hydrogen bonds

Figure 13.2 Hydration shells around an aqueous ion. When an ionic compound dissolves in water, ion-dipole forces orient water molecules around the separated ions to form hydration shells. The cation shown here is octahedrally surrounded by six water molecules, which form H bonds with water molecules in the next hydration shell, and those form H bonds with others farther away.

ARIS Presentation Center, Art Library

ion in hemoglobin and an O_2 molecule in the bloodstream. Because an ion increases the magnitude of any nearby dipole, ion–induced dipole forces also contribute to the solubility of salts in less polar solvents, such as LiCl in ethanol.

5. **Dipole–induced dipole forces,** also based on polarizability, are weaker than the ion–induced dipole forces because the magnitude of the charge is smaller (Coulomb's law). They arise when a polar molecule distorts the electron cloud of a nearby nonpolar molecule. The solubility in water, limited though it is, of atmospheric O_2, N_2, and the noble gases is due in large part to these forces. Paint thinners and grease solvents also function through dipole–induced dipole forces.

6. *Dispersion forces* contribute to the solubility of all solutes in all solvents, but they are the *principal* type of intermolecular force in solutions of nonpolar substances, such as petroleum and gasoline.

As you'll see in Section 13.2, these intermolecular forces are also responsible for keeping cellular macromolecules in their biologically active shapes.

Liquid Solutions and the Role of Molecular Polarity

Solutions can be gaseous, liquid, or solid. In general, *the physical state of the solvent determines the physical state of the solution.* We focus on liquid solutions because they are by far the most common and important.

From cytoplasm to tree sap, gasoline to cleaning fluid, iced tea to urine, solutions in which the solvent is a liquid are familiar in everyday life. Water is the most prominent solvent because it is so common and dissolves so many ionic and polar substances. But there are many other liquid solvents, and their polarity ranges from very polar to nonpolar.

Liquid-Liquid and Solid-Liquid Solutions

Many salts dissolve in water because the strong ion-dipole attractions that water molecules form with the ions are very similar to the strong attractions between the ions themselves and, therefore, can *substitute* for them. The same salts are insoluble in hexane (C_6H_{14}) because the weak ion–induced dipole forces their ions could form with the nonpolar molecules of this solvent *cannot* substitute for attractions between the ions. Similarly, oil does not dissolve in water because the weak dipole–induced dipole forces between oil and water molecules cannot substitute for the strong H bonds between water molecules. Oil *does* dissolve in hexane, however, because the dispersion forces in one substitute readily for the dispersion forces in the other. Thus, for a solution to form, "like dissolves like" means that the forces created between solute and solvent must be *comparable in strength* to the forces destroyed within both the solute and the solvent.

Literature Pinarbasi, Tacettin; Canpolat, Nurtaç. "Students' Understanding of Solution Chemistry Concepts," *J. Chem. Educ.* **2003,** *80,* 1328.

Literature Sanger, Michael J.; Badger, Steven M., II. "Using Computer-Based Visualization Strategies to Improve Students' Understanding of Molecular Polarity and Miscibility," *J. Chem. Educ.* **2001,** *78,* 1412.

Demo Pravia, Katia; Maynard, David F. "Why Don't Water and Oil Mix?" *J. Chem. Educ.* **1996,** *73,* 497.

Organic chemistry application

To examine this idea further, let's compare the solubilities of a series of alcohols in two solvents that act through very different intermolecular forces—water and hexane. Alcohols are organic molecules with a hydroxyl (—OH) group bound to a hydrocarbon group. The simplest type of alcohol has the general formula $CH_3(CH_2)_nOH$; we'll consider alcohols with $n = 0$ to 5. We can view an alcohol molecule as consisting of two portions: the polar —OH group and the nonpolar hydrocarbon chain. The —OH portion forms strong H bonds with water and weak dipole–induced dipole forces with hexane. The hydrocarbon portion interacts through dispersion forces with hexane and through dipole–induced dipole forces with water.

In Table 13.2, the models show the relative change in size of the polar and nonpolar portions of the alcohol molecules. In the smaller alcohols (one to three carbons), the hydroxyl group is a relatively large portion, so the molecules interact with each other through H bonding, just as water molecules do. When the smaller alcohols mix with water, H bonding within solute and within solvent is replaced by H bonding *between* solute and solvent (Figure 13.3). As a result, these smaller alcohols are miscible with water.

Water solubility decreases dramatically for alcohols larger than three carbons, and those with chains longer than six carbons are insoluble in water. For these larger alcohols to dissolve, the nonpolar chains have to move between the water molecules, substituting their weak attractions with those water molecules for strong H bonds among the water molecules themselves. The —OH portion of the alcohol does form H bonds to water, but these don't outweigh the H bonds among water molecules that break to make room for the hydrocarbon portion.

Table 13.2 shows that the opposite trend occurs with hexane. For alcohols in hexane, the major solute-solvent *and* solvent-solvent interactions are dispersion forces. The weak forces between the —OH group of methanol (CH_3OH) and

Table 13.2	Solubility* of a Series of Alcohols in Water and Hexane		
Alcohol	**Model**	**Solubility in Water**	**Solubility in Hexane**
CH_3OH (methanol)		∞	1.2
CH_3CH_2OH (ethanol)		∞	∞
$CH_3(CH_2)_2OH$ (1-propanol)		∞	∞
$CH_3(CH_2)_3OH$ (1-butanol)		1.1	∞
$CH_3(CH_2)_4OH$ (1-pentanol)		0.30	∞
$CH_3(CH_2)_5OH$ (1-hexanol)		0.058	∞

*Expressed in mol alcohol/1000 g solvent at 20°C.

Water · Methanol · A solution of water and methanol

Figure 13.3 Like dissolves like: solubility of methanol in water. The H bonds in water and in methanol are similar in type and strength, so they can substitute for one another. Thus, methanol is soluble in water; in fact, the two substances are miscible.

ARIS Presentation Center, Art Library

hexane cannot substitute for the strong H bonding among CH_3OH molecules, so the solubility of methanol in hexane is relatively low. In any larger alcohol, however, dispersion forces become increasingly more important, and these *can* substitute for dispersion forces in pure hexane; thus, solubility increases. With no strong solvent-solvent forces to be replaced by weak solute-solvent forces, even the two-carbon chain of ethanol has strong enough solute-solvent attractions to be miscible in hexane.

Many other organic molecules have polar and nonpolar portions, and the predominance of one portion or the other determines their solubility in different solvents. For example, carboxylic acids and amines behave very much like alcohols. Thus, methanoic acid (HCOOH, formic acid) and methanamine (CH_3NH_2) are miscible with water and much less soluble in hexane, whereas hexanoic acid [$CH_3(CH_2)_4COOH$] and 1-hexanamine [$CH_3(CH_2)_5NH_2$] are slightly soluble in water and very soluble in hexane. ●

● **"Greener" Organic Solvents** Volatile organic solvents have many uses in paints and other coatings, in cleaning agents, and in propellants for hairspray, cooking oils, lubricants, and so forth. But they also play a role in smog formation and may have adverse health effects. These concerns have led to the development of "greener" processes, such as the use of nonvolatile organic solvents, nonaqueous ionic liquid solvents, supercritical fluids, and solvent-free and water-based methods. With the implementation of these approaches, atmospheric emissions and worker exposure are greatly reduced.

Green chemistry application

ARIS Presentation Center, Worked Examples Library

SAMPLE PROBLEM 13.1 Predicting Relative Solubilities of Substances

PROBLEM Predict which solvent will dissolve more of the given solute:
(a) Sodium chloride in methanol (CH_3OH) or in 1-propanol ($CH_3CH_2CH_2OH$)
(b) Ethylene glycol ($HOCH_2CH_2OH$) in hexane ($CH_3CH_2CH_2CH_2CH_2CH_3$) or in water
(c) Diethyl ether ($CH_3CH_2OCH_2CH_3$) in water or in ethanol (CH_3CH_2OH)

PLAN We examine the formulas of the solute and each solvent to determine the types of forces that could occur. A solute tends to be more soluble in a solvent whose intermolecular forces are similar to, and therefore can substitute for, its own.

SOLUTION (a) Methanol. NaCl is an ionic solid that dissolves through ion-dipole forces. Both methanol and 1-propanol contain a polar —OH group, but 1-propanol's longer hydrocarbon chain can form only weak forces with the ions, so it is less effective at substituting for the ionic attractions in the solute.

(b) Water. Ethylene glycol molecules have two —OH groups, so the molecules interact with each other through H bonding. They are more soluble in H_2O, whose H bonds can substitute for their own H bonds better than the dispersion forces in hexane can.

(c) Ethanol. Diethyl ether molecules interact with each other through dipole-dipole and dispersion forces and can form H bonds to both H_2O and ethanol. The ether is more soluble in ethanol because that solvent can form H bonds *and* substitute for the ether's dispersion forces. Water, on the other hand, can form H bonds with the ether, but it lacks any hydrocarbon portion, so it forms much weaker dispersion forces with that solute.

FOLLOW-UP PROBLEM 13.1 Which solute is more soluble in the given solvent?

(a) 1-Butanol ($CH_3CH_2CH_2CH_2OH$) or 1,4-butanediol ($HOCH_2CH_2CH_2CH_2OH$) in water
(b) Chloroform ($CHCl_3$) or carbon tetrachloride (CCl_4) in water

Table 13.3	Correlation Between Boiling Point and Solubility in Water	
Gas	Solubility (*M*)*	bp (K)
He	4.2×10^{-4}	4.2
Ne	6.6×10^{-4}	27.1
N_2	10.4×10^{-4}	77.4
CO	15.6×10^{-4}	81.6
O_2	21.8×10^{-4}	90.2
NO	32.7×10^{-4}	121.4

*At 273 K and 1 atm.

 ARIS Presentation Center, Tables Library

 Biological application

 Engineering application

Gas-Liquid Solutions Gases that are nonpolar, such as N_2, or are nearly so, such as NO, have low boiling points because their intermolecular attractions are weak. Likewise, they are not very soluble in water because solute-solvent forces are weak. In fact, as Table 13.3 shows, for nonpolar gases, boiling point generally correlates with solubility in water.

In some cases, the small amount of a nonpolar gas that does dissolve is essential to a process. The most important environmental example is the solubility of O_2 in water. At 25°C and 1 atm, the solubility of O_2 is only 3.2 mL/100. mL of water, but aquatic animal life would die without this small amount. In other cases, the solubility of a gas may *seem* high because the gas is not only dissolving but also reacting with the solvent or another component. Oxygen seems much more soluble in blood than in water because O_2 molecules are continually bonding with hemoglobin molecules in red blood cells. Similarly, carbon dioxide, which is essential for aquatic plants and coral-reef growth, seems very soluble in water (~81 mL of CO_2/100. mL of H_2O at 25°C and 1 atm) because it is reacting, in addition to simply dissolving:

$$CO_2(g) + H_2O(l) \longrightarrow H^+(aq) + HCO_3^-(aq)$$

Gas Solutions and Solid Solutions

Despite the central place of liquid solutions in chemistry, gaseous solutions and solid solutions also have vital importance and numerous applications.

Gas-Gas Solutions *All gases are infinitely soluble in one another.* Air is the classic example of a gaseous solution, consisting of about 18 gases in widely differing proportions. Anesthetic gas proportions are finely adjusted to the needs of the patient and the length of the surgical procedure. The ratios of components in many industrial gas mixtures, such as CO/H_2 in syngas production or N_2/H_2 in ammonia production, are controlled to optimize product yield under varying conditions of temperature and pressure.

Gas-Solid Solutions *When a gas dissolves in a solid, it occupies the spaces between the closely packed particles.* Hydrogen gas can be purified by passing an impure sample through a solid metal such as palladium. Only the H_2 molecules are small enough to enter the spaces between the Pd atoms, where they form Pd—H covalent bonds. Under high H_2 pressure, the H atoms are passed along the Pd crystal structure and emerge from the solid as H_2 molecules.

The ability of gases to penetrate a solid also has disadvantages. The electrical conductivity of copper, for example, is drastically reduced by the presence of O_2, which dissolves into the crystal structure and reacts to form copper(I) oxide. High-conductivity copper is prepared by melting and recasting the copper in an O_2-free atmosphere.

Solid-Solid Solutions Because solids diffuse so little, their mixtures are usually heterogeneous; an example is gravel mixed with sand. Some solid-solid solutions can be formed by melting the solids and then mixing them and allowing them to freeze. Many **alloys,** mixtures of elements that have an overall metallic character, are examples of solid-solid solutions (although several common alloys have microscopic heterogeneous regions). In *substitutional* alloys, such as brass (Figure 13.4A) and sterling silver, atoms of another element substitute for some atoms of the main element in the structure. In *interstitial* alloys, atoms of another element (often a nonmetal) fill some of the spaces *(interstices)* between atoms of the main element. For example, some forms of carbon steel (Figure 13.4B) are interstitial alloys of iron with a small amount of carbon.

Waxes are another familiar type of solid-solid solution. Most waxes are amorphous solids that may contain small regions of crystalline regularity. A natural *wax* is a solid of biological origin that is insoluble in water but dissolves in

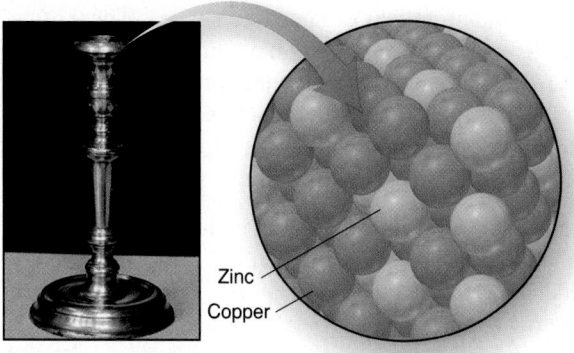

Zinc
Copper

A Brass, a substitutional alloy

Carbon
Iron

B Carbon steel, an interstitial alloy

Figure 13.4 The arrangement of atoms in two types of alloys.
A, Brass is a substitutional alloy in which zinc atoms substitute for copper atoms at many of copper's face-centered cubic packing sites.

B, Some forms of carbon steel are interstitial alloys in which carbon atoms lie in the holes (interstices) of the body-centered cubic crystal structure of iron. **ARIS Presentation Center,** Art Library

nonpolar solvents. Many plants have waxy coatings on their leaves and fruit to prevent evaporation of interior water. ●

Section Summary

Solutions are homogeneous mixtures consisting of a solute dissolved in a solvent through the action of intermolecular forces. • The *solubility* of a solute in a given amount of solvent is the maximum amount that can dissolve at a specified temperature. (For gaseous solutes, the pressure must be specified also.) • In addition to the intermolecular forces that exist in pure substances, ion-dipole, ion–induced dipole, and dipole–induced dipole forces occur in solution. If similar intermolecular forces occur in solute and solvent, a solution will likely form ("like dissolves like"). • When ionic compounds dissolve in water, the ions become surrounded by hydration shells of H-bonded water molecules. • Solubility of organic molecules in various solvents depends on their polarity and the extent of their polar and nonpolar portions. • The solubility of nonpolar gases in water is low because of weak intermolecular forces. Gases are miscible with one another and dissolve in solids by fitting into spaces in the crystal structure. • Solid-solid solutions, such as alloys and waxes, form when the components are mixed while molten.

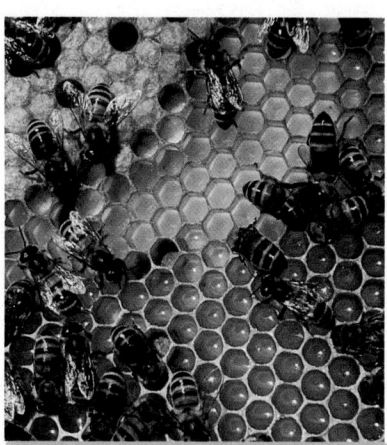

● **Waxes for Home and Auto** Beeswax, the remarkable structural material that bees secrete to build their combs, is a complex mixture of fatty acids, long-chain carboxylic acids, and hydrocarbons in which some of the molecules contain chains more than 40 carbon atoms long. Carnauba wax, from a South American palm, is a mixture of compounds, each consisting of a fatty acid bound to a long-chain alcohol. It is hard but forms a thick gel in nonpolar solvents, perfect for car waxes.

 ARIS Presentation Center, Photo Library

 Biological application

13.2 INTERMOLECULAR FORCES AND BIOLOGICAL MACROMOLECULES

The forces responsible for solutes dissolving in solvents also play a fundamental role in maintaining the complex three-dimensional shapes of cellular biopolymers: proteins, nucleic acids, and polysaccharides. The structures of these molecules, and that of the cell membrane itself, can be understood through two simple ideas involving intermolecular forces: (1) polar and ionic groups interact with the surrounding cellular water, while nonpolar groups do not, and (2) distant groups on a large molecule are often close enough to interact through intermolecular forces.

In this section, we'll see how these forces stabilize the three types of biological macromolecules and the cell membrane; in Chapter 15, we'll see the reactions that form these molecules and how their structure is crucial to their function.

The Structures of Proteins

Proteins are unbranched polymers formed from monomers called **amino acids,** organic compounds with amine ($—NH_2$) and carboxyl ($—COOH$) groups in the same molecule. Proteins range from about 50 amino acids ($\mathcal{M} \approx 5 \times 10^3$ g/mol) to several thousand ($\mathcal{M} \approx 5 \times 10^5$ g/mol). *Protein shapes are determined completely by the sequence of amino acids in the chain.* Some proteins have few amino acids in repeating patterns and, thus, extended helical or sheetlike shapes. These function to give structure to hair, skin, and so forth. Others have many more types of amino acids and more complex, globular shapes. These function as antibodies that defend against infection, enzymes that speed metabolic reactions, membrane gates and pumps that control ion concentrations in the cell, and so forth. Let's see how the amino acid composition influences the shapes of globular proteins.

The Polarity of Amino-Acid Side Chains

In the cell, an individual amino acid has four groups bonded to its α-carbon (Figure 13.5): charged carboxyl ($-COO^-$) and amine ($-NH_3^+$) groups, an H atom, and a *side chain*, which ranges from an H atom to a two-ringed C_9H_8N group. The *peptide* bond is the covalent linkage between amino acids in a protein and is formed between the carboxyl group of one amino acid and the amine group of the next. (We discuss amino acid structures and details of peptide bond formation in Chapter 15.)

In essence, then, the backbone of a protein consists of an α-*carbon bonded to a peptide bond, which is bonded to the next* α-*carbon bonded to the next peptide bond,* and so forth, with the various side chains dangling off the α-carbons on alternate sides of the chain (Figure 13.6). Globular proteins are made up of about 20 different types of amino acids, each characterized by its particular side chain (*gray screens*). The simple classification of amino acids we use here focuses on their polarity and/or charge: nonpolar, polar, and ionic. A few examples are

Figure 13.5 The physiological form of an amino acid. Under physiological conditions, amino acids have charged groups. There are about 20 different side chains. ARIS Presentation Center, Art Library

Figure 13.6 A portion of a polypeptide chain. The peptide bond holds the monomers together in a protein. Three peptide bonds (*orange screens*) joining four amino acids occur in this portion of a polypeptide chain. Note the repeating pattern of the chain: peptide bond—α-carbon—peptide bond—α-carbon—and so on. Also note that the side chains (*gray screens*) dangle off the main chain.

Transparency/ARIS Presentation Center, Art Library

Intermolecular Forces and Protein Shape The forces responsible for protein shapes are *the same bonding and intermolecular forces that operate for all molecules.* A long protein chain has many bends, and distant groups of the same molecule end up near enough to interact.

Figure 13.7 depicts the forces operating in proteins dissolved in the aqueous cell medium. (Membrane-bound proteins have an important structural difference, as we'll see shortly.) The covalent peptide bonds create the chain. The —SH ends of two cysteine side chains often form an —S—S— bond, a covalent *disulfide bridge* that brings together distant parts of the chain. Polar and ionic side chains usually protrude into the surrounding cell fluid, interacting with water through ion-dipole forces and H bonds. But sometimes oppositely charged ends of ionic side chains lie near each other, and the —COO⁻ and —NH₃⁺ groups form an electrostatic *salt link* (or *ion pair*), which secures the chain's bends. Helical and sheetlike segments arise from *H bonds between the C=O of one peptide bond and the N—H of another.* Other H bonds involving atoms in peptide bonds or in side chains act to keep distant portions of the chain near each other. Nonpolar side chains interact through dispersion forces within the nonaqueous protein interior. Thus, through the influence of intermolecular forces among the protein's side chains, *soluble proteins have polar-ionic exteriors and nonpolar interiors.*

A protein's shape is so important because, as we emphasize in Chapter 15, *the amino acid sequence determines the protein's shape, and the shape determines its function.*

H bonds within polypeptide chain

Ion-dipole forces

H bond

H bonds within polypeptide chain

Salt link

Disulfide bridge

Region where dispersion forces predominate

H bond

Figure 13.7 **The forces that maintain protein structure.** A portion of a general protein shows that a combination of covalent, ionic, and intermolecular forces (both within the protein and between the protein and water) is responsible for the overall protein shape. (Water molecules and some amino-acid side chains are shown as ball-and-stick models within a space-filling contour.)

Dual Polarity and Function

The dual polarity that we discussed earlier with reference to solubility can help us understand the function of systems as diverse as soaps, antibiotics, and cell membranes.

Soaps and Detergents A **soap** is the salt of a strong base (metal hydroxide) and a *fatty acid,* a carboxylic acid with a long hydrocarbon chain. A typical soap molecule has an even number of carbon atoms and is usually made up of a nonpolar "tail" 15–19 carbons long and a polar-ionic "head" consisting of the —COO⁻ group and a cation. ● For example, sodium stearate, $CH_3(CH_2)_{16}COONa$, is a major component of many bar soaps:

Sodium stearate $C_{17}H_{35}COONa$

When grease on your hands or clothes is immersed in soapy water, the nonpolar tails of the soap molecules interact with the nonpolar grease molecules through dispersion forces, while the polar-ionic heads interact with the water through ion-dipole forces and H bonds. Agitation creates tiny aggregates of grease molecules with embedded soap molecules, whose polar-ionic heads stick into the water. These aggregates are flushed away by added water (Figure 13.8). Detergents have a similar structure of nonpolar tail and polar-ionic head, but they typically consist of an —SO₃⁻ group (and an inorganic cation) attached to a long hydrocarbon group.

● **The Hardness and Softness of Soaps**
The cation of a soap influences its properties greatly. Lithium soaps are hard and high melting and are used in car lubricants. Softer, more water-soluble sodium soaps are used as common bar soap. Potassium soaps are low melting and used as liquid soaps. The insolubility of calcium soaps explains why it is difficult to clean clothes in hard water, which typically contains Ca^{2+} ions (see Chemical Connections, p. 542).

 Biological application

Figure 13.8 The structure and function of a soap. The tails of soap molecules interact with nonpolar molecules in grease through dispersion forces, while the heads interact with water through H bonds and ion-dipole forces. Agitation breaks up the grease stain into small soap-grease aggregates that rinsing flushes away.

Soap molecule immersed in grease stain

Nonpolar molecule in grease stain

 ARIS Presentation Center, Art Library

The Lipid Bilayer The most abundant molecules in the cell membranes of most species are *phospholipids*. They consist of two long fatty acid chains and a charged organophosphate group all linked to glycerol, a three-carbon trialcohol. Thus, as Figure 13.9 shows, phospholipids, like soaps, have a polar-ionic head and a nonpolar tail.

In water, phospholipids undergo a remarkable self-assembly into an extended sheetlike double layer of molecules, called a **lipid bilayer.** On each surface of the bilayer, the charged heads face the water, and within the interior of the bilayer, the nonpolar tails face each other. So that the tails at the ends of the bilayer are not exposed to the aqueous surroundings, the bilayer tends to close on itself. In the laboratory, bilayers can be made to form spherical vesicles that trap water inside. The result is an energetically favorable structure that optimizes the various intermolecular forces: ion-dipole forces between polar heads and water inside and out, dispersion forces between nonpolar tails within the bilayer, and minimum contact between nonpolar tails and water. Experiments show that the lipid bilayer is impermeable to ions and most polar molecules, including water.

Structure of the Cell Membrane The cell membrane is a perfect example of the effect of intermolecular forces in a complex biological assembly. A typical animal cell membrane is about 8 nm thick and consists of the phospholipid bilayer embedded with various proteins that may comprise as much as 50% by mass of the membrane. Depending on the type of cell, membrane proteins function as pumps that actively move ions and specific molecules in or out, channels or gates that passively allow ions or molecules in or out, or receptors for small biomolecules (see Chemical Connections, pp. 402–403), or they may be involved in energy production or metabolic reactions.

The structures of membrane proteins differ in a fundamental way from those of the soluble proteins we discussed previously. In soluble proteins, in most cases, *polar groups are found on the exterior* and *nonpolar groups on the interior,* away from water. Studies of many systems, from red blood cells to photosynthetic bacteria, show that membrane proteins have this arrangement partly inverted: *the exterior of the protein region that lies within the lipid bilayer consists of nonpolar amino-acid side chains.* Dispersion forces between these groups and the hydrocarbon tails of the membrane lipids stabilize this orientation (Figure 13.10).

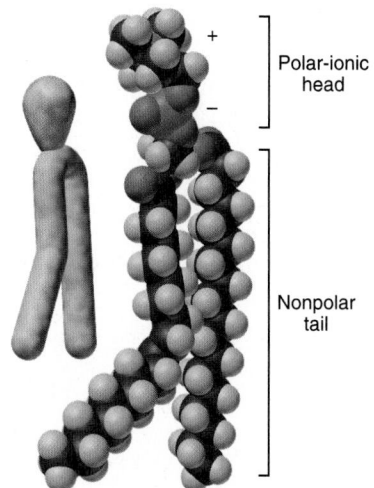

Figure 13.9 Membrane phospholipids. The phospholipids found in cell membranes all have similar portions, with a polar-ionic head and a nonpolar tail; lecithin (phosphatidylcholine) is shown here. The purple-and-gray shape is a simplified representation of the structure.

 Biological application

 ARIS Presentation Center, Art Library

 Transparency/ARIS Presentation Center, Art Library

Figure 13.10 Intermolecular forces and membrane structure.

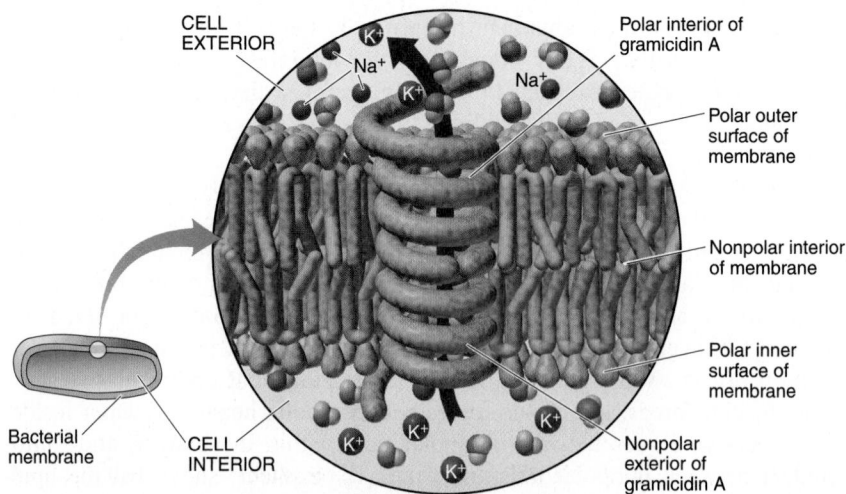

Figure 13.11 The mode of action of an antibiotic. Gramicidin A is a tubelike molecule, nonpolar on the outside and polar on the inside. After dissolving into the nonpolar interior of a bacterial cell membrane via dispersion forces, pairs of gramicidin molecules aligned end to end span the membrane's thickness. Ions diffuse through the polar interior of these molecules, which destroys the cell's ion balance and its segregation of Na^+ outside and K^+ inside.

ARIS Presentation Center, Art Library

Action of Antibiotics A clear example of how this orientation of molecules in the cell membrane functions is seen in channel-forming antibiotics. Antibiotics are produced by one microorganism to destroy another, and many have been adapted by medical science to fight bacterial infection.

One function of the membrane is to balance internal and external ion concentrations: Na^+ is excluded from the cell, and K^+ is kept inside. If this ionic segregation is disrupted, the organism dies. Gramicidin A and similar antibiotics disrupt the ion balance by forming a channel in the bacterial membrane through which ions can flow (Figure 13.11). Two helical gramicidin A molecules—with nonpolar groups outside and polar groups inside—lie end to end and span the bacterial membrane. The nonpolar outside helps the molecule dissolve into the membrane of the target bacterium through dispersion forces, and the polar groups on the inside pass the ions along through ion-dipole forces, like a "bucket brigade." Indeed, over 10^7 ions per second pass through a single channel! Through this polar passageway ions diffuse in and out, and the organism dies.

The Structure of DNA

The chemical information that guides the design, construction, and function of all proteins is contained in **nucleic acids,** unbranched polymers that consist of monomers called **mononucleotides.** Each mononucleotide consists of an N-containing base, a sugar, and a phosphate group (Figure 13.12). In DNA (*deoxyribonucleic acid*), the sugar is *2-deoxyribose,* in which —H substitutes for —OH on the second C atom of the five-carbon sugar ribose.

The repeating pattern of the backbone of a DNA chain is *sugar linked to phosphate linked to sugar linked to phosphate,* and so on. Attached to each sugar is one of four nitrogen-containing bases. The bases are flat ring structures that dangle off the sugar-phosphate chain, similar to the way amino-acid side chains dangle off the protein chain.

Figure 13.12 A short portion of a DNA chain. ARIS Presentation Center, Art Library

Biological application

Figure 13.13 **The double helix of DNA.** In this space-filling segment of DNA (*left*), the polar, negatively charged, sugar-phosphate portion (*pink*) lies on the exterior of the structure and interacts with the aqueous cell fluid through ion-dipole forces and H bonds. Dispersion forces stabilize the stacked, nonpolar bases (*gray*) in the interior. The ball-and-stick model (*right*) shows that specific pairs are attached by H bonds: a guanine-cytosine pair is shown.

Transparency/ARIS Presentation Center, Art Library

Literature Heinhorst, Sabine; Cannon, Gordon. "*Nature:* 'Green' Chemistry, Natural Antioxidants, and a DNA-Fueled Machine," *J. Chem. Educ.* **2001**, *78*, 150.

Demo Bruist, Michael F. "A Simple Demonstration of How Intermolecular Forces Make DNA Helical," *J. Chem. Educ.* **1998**, *75*, 53.

Intermolecular Forces and the Double Helix In the cell nucleus, DNA exists as two chains wrapped around each other in a **double helix** (Figure 13.13). Intermolecular forces play a central role in stabilizing this structure. On the exterior, negatively charged sugar-phosphate chains form ion-dipole and H bonds with the aqueous surroundings. In the interior, the flat, nitrogen-containing bases stack above each other, which allows extensive interaction through dispersion forces. Most important, specific pairs of bases form interchain H bonds; that is, each base in one chain is always H bonded with the same partner base in the other chain. Thus, *the base sequence of one chain is the H-bonded complement of the base sequence of the other.*

A DNA molecule contains millions of H bonds linking bases in these prescribed pairs. The total energy of the H bonds keeps the chains together, but each H bond is weak enough (around 5% of a typical covalent single bond) that a few at a time can break as the chains separate during crucial cellular processes. Linear segments of DNA act as *genes*, chemical blueprints for synthesizing the organism's proteins. In Chapter 15, we'll see how base pairing through H bonding is essential to protein synthesis (the process by which the cell makes its proteins) and DNA replication (the process by which the genes copy themselves during cell division).

The Structure of Cellulose

Strength and energy storage are the two main functions of **polysaccharides,** polymers consisting of monomers called **monosaccharides,** or simple sugars. The three major natural polysaccharides—cellulose, starch, and glycogen—consist entirely of glucose units, but they differ in the way the units are linked and in the extent of crosslinking. In aqueous systems, glucose occurs as a six-membered ring with five —OH groups attached (Figure 13.14).

The same intermolecular forces operate in all three of these polysaccharides, but we'll focus here on cellulose, the most abundant organic chemical on Earth. Cellulose is the only unbranched polysaccharide of the three and consists entirely of long chains of glucose monomers. Picture parallel chains of linked glucose

Figure 13.14 The cyclic structure of glucose in aqueous solution.

ARIS Presentation Center, Art Library

Figure 13.15 The structure of cellulose. The polysaccharide cellulose exists as chains of glucose molecules. Hydrogen bonds within each chain, between chains to form layers, and between layers give cellulose great strength.

 ARIS Presentation Center, Art Library

units lying next to each other in three dimensions, their —OH groups everywhere (Figure 13.15). The opportunities for dispersion forces and extensive interchain H bonding are enormous. In fact, the great strength of wood is due largely to the countless H bonds among cellulose chains. (We compare the structures of the three polysaccharides in Chapter 15.)

Section Summary

Intermolecular forces stabilize the structures of biological macromolecules—proteins, nucleic acids, and polysaccharides—and the cell membrane. • In soluble proteins, polar and ionic amino-acid side chains face the water, while nonpolar side chains interact with each other in the molecular interior. • Soaps have a polar-ionic head and a nonpolar tail that allow them to dissolve grease while interacting with water. • Membranes consist of phospholipids that also have two polarities. They assemble in a self-sealing, water-impermeable lipid bilayer. In a cell membrane, the lipid bilayer has embedded proteins whose nonpolar side chains face the surrounding bilayer. Channel-forming antibiotics also have nonpolar exteriors and polar interiors. • DNA forms a double helix with a sugar-phosphate, polar-ionic exterior. In the interior, bases H bond in specific pairs and stack through dispersion forces. • Cellulose consists of long chains of linked glucose molecules, which interact through extensive H bonding.

13.3 WHY SUBSTANCES DISSOLVE: UNDERSTANDING THE SOLUTION PROCESS

As a qualitative predictive tool, "like dissolves like" is helpful in many cases. As you might expect, this handy *macroscopic* rule is based on the *molecular* interactions that occur between solute and solvent particles. To see *why* like dissolves like, let's break down the solution process conceptually into steps and examine them in terms of changes in *enthalpy* and *entropy* of the system. We discussed enthalpy in Chapter 6 and focus on it first here. The concept of entropy is introduced at the end of this section and treated quantitatively in Chapter 20.

Heats of Solution and Solution Cycles

Picture a general solute and solvent about to form a solution. Both consist of particles attracting each other. For one substance to dissolve in another, three events must occur: (1) solute particles must separate from each other, (2) some solvent

A **Exothermic solution process**   B **Endothermic solution process**

particles must separate to make room for the solute particles, and (3) solute and solvent particles must mix together. No matter what the nature of the attractions within the solute and within the solvent, some energy must be *absorbed* for particles to separate, and some energy is *released* when they mix and attract each other. As a result of these changes, the solution process is typically accompanied by a change in enthalpy. We can divide the process into these three steps, each with its own enthalpy term:

Step 1. Solute particles separate from each other. This step involves overcoming intermolecular attractions, so it is *endothermic:*

$$\text{Solute (aggregated)} + heat \longrightarrow \text{solute (separated)} \qquad \Delta H_{\text{solute}} > 0$$

Step 2. Solvent particles separate from each other. This step also involves overcoming attractions, so it is *endothermic*, too:

$$\text{Solvent (aggregated)} + heat \longrightarrow \text{solvent (separated)} \qquad \Delta H_{\text{solvent}} > 0$$

Step 3. Solute and solvent particles mix. The particles attract each other, so this step is *exothermic:*

$$\text{Solute (separated)} + \text{solvent (separated)} \longrightarrow \text{solution} + heat \qquad \Delta H_{\text{mix}} < 0$$

The total enthalpy change that occurs when a solution forms from solute and solvent is the **heat of solution (ΔH_{soln}),** and we combine the three individual enthalpy changes to find it. The overall process is called a *thermochemical solution cycle* and is yet another application of Hess's law:

$$\Delta H_{\text{soln}} = \Delta H_{\text{solute}} + \Delta H_{\text{solvent}} + \Delta H_{\text{mix}} \qquad \text{(13.1)}$$

(In Section 9.2 we conceptually broke down the heat of formation of an ionic compound into component enthalpies in a similar way using a Born-Haber cycle.) If the sum of the endothermic terms ($\Delta H_{\text{solute}} + \Delta H_{\text{solvent}}$) is smaller than the exothermic term (ΔH_{mix}), ΔH_{soln} is negative; that is, the process is exothermic. Figure 13.16A is an enthalpy diagram for the formation of such a solution. If the sum of the endothermic terms is larger than the exothermic term, ΔH_{soln} is positive; that is, the process is endothermic (Figure 13.16B). However, if ΔH_{soln} is highly positive, the solute may not dissolve to any significant extent in that solvent.

Heats of Hydration: Ionic Solids in Water

We can simplify the solution cycle for aqueous systems. The $\Delta H_{\text{solvent}}$ and ΔH_{mix} components of the heat of solution are difficult to measure individually. Combined, these terms represent the enthalpy change during **solvation,** the process of surrounding a solute particle with solvent particles. Solvation in water is called **hydration.** Thus, enthalpy changes for separating the water molecules ($\Delta H_{\text{solvent}}$)

Figure 13.16 Solution cycles and the enthalpy components of the heat of solution. ΔH_{soln} can be thought of as the sum of three enthalpy changes: $\Delta H_{\text{solvent}}$ (separating the solvent; always >0), ΔH_{solute} (separating the solute; always >0), and ΔH_{mix} (mixing solute and solvent; always <0). **A,** ΔH_{mix} is larger than the sum of ΔH_{solute} and $\Delta H_{\text{solvent}}$, so ΔH_{soln} is negative (exothermic process). **B,** ΔH_{mix} is smaller than the sum of the others, so ΔH_{soln} is positive (endothermic process).

Transparency/ARIS Presentation Center, Art Library

Table 13.4	Trends in Ionic Heats of Hydration	
Ion	**Ionic Radius (pm)**	**ΔH_{hydr} (kJ/mol)**
Group 1A(1)		
Na^+	102	−410
K^+	138	−336
Rb^+	152	−315
Cs^+	167	−282
Group 2A(2)		
Mg^{2+}	72	−1903
Ca^{2+}	100	−1591
Sr^{2+}	118	−1424
Ba^{2+}	135	−1317
Group 7A(17)		
F^-	133	−431
Cl^-	181	−313
Br^-	196	−284
I^-	220	−247

and mixing the solute with them (ΔH_{mix}) are combined into the **heat of hydration** (ΔH_{hydr}). In water, Equation 13.1 becomes

$$\Delta H_{soln} = \Delta H_{solute} + \Delta H_{hydr}$$

The heat of hydration is a key factor in dissolving an ionic solid. Breaking H bonds in water is more than compensated for by forming strong ion-dipole forces, so hydration of an ion is *always* exothermic. The ΔH_{hydr} of an ion is defined as the enthalpy change for the hydration of 1 mol of separated (gaseous) ions:

$$M^+(g) \text{ [or } X^-(g)] \xrightarrow{\;H_2O\;} M^+(aq) \text{ [or } X^-(aq)] \qquad \Delta H_{hydr \text{ of the ion}} \text{ (always} <0)$$

Heats of hydration exhibit trends based on the **charge density** of the ion, the ratio of the ion's charge to its volume. In general, the higher the charge density is, the more negative ΔH_{hydr} is. According to Coulomb's law, the greater the ion's charge is and the closer the ion can approach the oppositely charged end of the water molecule's dipole, the stronger the attraction. Therefore,

- A 2+ ion attracts H_2O molecules more strongly than a 1+ ion of similar size.
- A small 1+ ion attracts H_2O molecules more strongly than a large 1+ ion.

Down a group of ions, for example, from Na^+ to Cs^+ in Group 1A(1), the charge stays the same and the size increases; thus, the charge densities decrease, as do the heats of hydration. Going across the periodic table from Group 1A(1) to Group 2A(2), the 2A ion has a smaller radius *and* a greater charge, so its charge density and ΔH_{hydr} are greater. Table 13.4 shows these relationships for some ions.

The energy required to separate an ionic solute (ΔH_{solute}) into gaseous ions is its lattice energy ($\Delta H_{lattice}$), and ΔH_{solute} is highly positive:

$$M^+X^-(s) \longrightarrow M^+(g) + X^-(g) \qquad \Delta H_{solute} \text{ (always} >0) = \Delta H_{lattice}$$

Thus, the heat of solution for ionic compounds in water combines the lattice energy (always positive) and the combined heats of hydration of cation and anion (always negative),

$$\Delta H_{soln} = \Delta H_{lattice} + \Delta H_{hydr \text{ of the ions}} \qquad (13.2)$$

The sizes of the individual terms determine the sign of the heat of solution.

Figure 13.17 shows enthalpy diagrams for dissolving three ionic solutes in water. The first, NaCl, has a small positive heat of solution ($\Delta H_{soln} = 3.9$ kJ/mol).

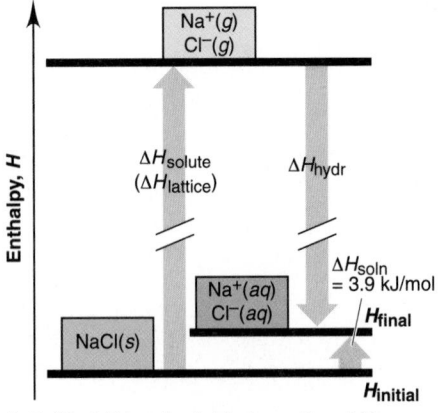

A NaCl. $\Delta H_{lattice}$ is slightly larger than ΔH_{hydr}: ΔH_{soln} is small and positive.

Figure 13.17 Dissolving ionic compounds in water. The enthalpy diagram for an ionic compound in water includes $\Delta H_{lattice}$ (ΔH_{solute}; always positive) and the combined ionic heats of hydration (ΔH_{hydr}; always negative).

B NaOH. ΔH_{hydr} dominates: ΔH_{soln} is large and negative.

C NH₄NO₃. $\Delta H_{lattice}$ dominates: ΔH_{soln} is large and positive.

Its lattice energy is only slightly greater than the combined ionic heats of hydration, so if you dissolve NaCl in water in a flask, you do not notice any temperature change. However, if you dissolve NaOH in water, the flask feels hot. The lattice energy for NaOH is much smaller than the combined ionic heats of hydration, so dissolving NaOH is highly exothermic ($\Delta H_{soln} = -44.5$ kJ/mol). Finally, if you dissolve NH_4NO_3 in water, the flask feels cold. In this case, the lattice energy is much larger than the combined ionic heats of hydration, so the process is quite endothermic ($\Delta H_{soln} = 25.7$ kJ/mol). ●

The Solution Process and the Change in Entropy

The heat of solution (ΔH_{soln}) is one of two factors that determine whether a solute dissolves in a solvent. The other factor concerns the natural tendency of a system to spread out, distribute, or disperse, its energy in as many ways as possible. A thermodynamic variable called **entropy** *(S)* is directly related to the number of ways that a system can distribute its energy, which in turn is closely related to the freedom of motion of the particles and the number of ways they can be arranged.

Let's see what it means for a system to "disperse its energy." We'll first compare the three physical states and then compare solute and solvent with solution. In a solid, the particles are relatively fixed in their positions, but in the liquid, they are free to move around each other. This greater freedom of motion means the particles can distribute their kinetic energy in more ways; thus, the liquid has higher entropy than the solid ($S_{liquid} > S_{solid}$). The gas, in turn, has higher entropy than the liquid because the particles have much more freedom of motion ($S_{gas} > S_{liquid}$). Another way to state this is that the change in entropy when the liquid vaporizes to a gas (ΔS_{vap}) is positive; that is, $\Delta S_{vap} > 0$.

Similarly, *a solution usually has higher entropy than the pure solute and pure solvent.* Here, the number of ways to distribute the energy and the freedom of motion of the particles is related to the different molecules becoming interspersed and, thus, to the number of different interactions between the molecules. There are far more interactions possible when solute and solvent are mixed than when they are pure; thus, $S_{soln} > (S_{solute} + S_{solvent})$, or $\Delta S_{soln} > 0$.

From everyday experience, we know that solutions form naturally, whereas pure solutes and solvents do not. We'll see in Chapter 20 that energy must be expended to reverse this tendency of systems to distribute their energy in more ways—to get "mixed up." Water treatment facilities, oil refineries, metal foundries, and many other industrial processes expend enormous quantities of energy in order to reverse this natural tendency and separate mixtures into pure components.

The solution process involves the interplay of two factors—the change in enthalpy and the change in entropy. *Systems tend toward a state of lower enthalpy and higher entropy.* In many cases, the relative magnitudes of ΔH_{soln} and ΔS_{soln} determine whether a solution forms.

To see this interplay of enthalpy and entropy, let's consider three solute-solvent pairs in which different influences dominate. In our first example, sodium chloride does not dissolve in hexane (C_6H_{14}), as you would predict from the dissimilar intermolecular forces. As Figure 13.18A on page 518 shows, separating the solvent is easy because the dispersion forces are relatively weak, but separating the ionic solute requires supplying $\Delta H_{lattice}$. Mixing releases very little heat because the ion–induced dipole attractions between Na^+ (or Cl^-) ions and hexane molecules are weak. The sum of the endothermic terms is much larger than the exothermic term, so ΔH_{soln} is highly positive. In this case, *a solution does not form because the entropy increase that would accompany the mixing of solute and solvent is much smaller than the enthalpy increase required to separate the ions.*

● **Hot Packs, Cold Packs, and Self-Heating Soup** Strain a muscle or sprain a joint and you may obtain temporary relief by applying the appropriate heat of solution. Hot packs and cold packs consist of a thick outer pouch containing water and a thin inner pouch containing a salt. A squeeze on the outer pouch breaks the inner pouch, and the salt dissolves. Most hot packs use anhydrous $CaCl_2$ ($\Delta H_{soln} = -82.8$ kJ/mol), whereas cold packs use NH_4NO_3 ($\Delta H_{soln} = 25.7$ kJ/mol). The change in temperature can be large—a cold pack, for instance, can bring the solution from room temperature down to 0°C—but a pack's usable time is limited to around half an hour. In Japan, some cans of soup have double walls, with an ionic compound in a packet and water between the walls. When you open the can, the packet breaks, and the exothermic solution process, which can quickly reach about 90°C, warms the soup.

Literature Spencer, J. N.; Lowe, John P. "Entropy: The Effects of Distinguishability," *J. Chem. Educ.* **2003**, *80*, 1417.

Figure 13.18 Enthalpy diagrams for dissolving NaCl and octane in hexane.
A, Because attractions between Na$^+$ (or Cl$^-$) ions and hexane molecules are weak, ΔH_{mix} is *much* smaller than ΔH_{solute}. Thus, ΔH_{soln} is so positive that NaCl does *not* dissolve in hexane.
B, Intermolecular forces in octane and in hexane are so similar that ΔH_{soln} is very small. Octane dissolves in hexane because the solution has greater entropy than the pure components.

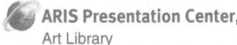
ARIS Presentation Center,
Art Library

The second solute-solvent pair is octane (C_8H_{18}) and hexane. Both consist of nonpolar molecules held together by dispersion forces of comparable strength. We therefore predict that octane is soluble in hexane; in fact, they are infinitely soluble (miscible). But this does not necessarily mean that a great deal of heat is released; in fact, ΔH_{soln} is around zero (Figure 13.18B). *With no enthalpy change driving the process, octane dissolves in hexane because the entropy increases greatly when the pure substances mix.* Thus, the large increase in entropy drives formation of this solution.

In some cases, a large enough increase in entropy can cause a solution to form even when the enthalpy increases significantly ($\Delta H_{soln} > 0$). As Figure 13.17C showed, when ammonium nitrate (NH_4NO_3) dissolves in water, the process is very endothermic. Nevertheless, in that case, *the increase in entropy that occurs when the crystal breaks down and the ions mix with water molecules more than compensates for the increase in enthalpy.* The enthalpy/entropy interplay in physical and chemical systems is covered in depth in Chapter 20.

Section Summary

An overall heat of solution can be obtained from a thermochemical solution cycle as the sum of two endothermic steps (solute separation and solvent separation) and one exothermic step (solute-solvent mixing). • In aqueous solutions, the combination of solvent separation and mixing is called *hydration.* Ionic heats of hydration are always negative because of strong ion-dipole forces. • Most systems have a natural tendency to increase their entropy (disperse their energy in more ways), and a solution has greater entropy than the pure solute and solvent. • The combination of enthalpy and entropy changes determines whether a solution forms. A substance with a positive ΔH_{soln} dissolves *only* if the entropy increase is large enough to outweigh it.

13.4 SOLUBILITY AS AN EQUILIBRIUM PROCESS

When an ionic solid dissolves, ions leave the solid and become dispersed in the solvent. Some dissolved ions collide occasionally with the undissolved solute and recrystallize. As long as the rate of dissolving is greater than the rate of recrystallizing, the concentration of ions rises. Eventually, given enough solid, ions are dissolving at the same rate as ions in the solution are recrystallizing (Figure 13.19).

Figure 13.19 Equilibrium in a saturated solution. In a saturated solution, equilibrium exists between excess solid solute and dissolved solute. At a particular temperature, the number of solute particles dissolving (*white arrows*) per unit time equals the number recrystallizing (*black arrows*).

At this point, even though dissolving and recrystallizing continue, there is no further change in the concentration with time. The system has reached equilibrium; that is, *excess undissolved solute is in equilibrium with the dissolved solute:*

$$\text{Solute (undissolved)} \rightleftharpoons \text{solute (dissolved)}$$

This solution is called **saturated:** it contains the maximum amount of dissolved solute at a given temperature in the presence of undissolved solute. ● Filter off the saturated solution and add more solute to it, and the added solute will not dissolve. A solution that contains less than this concentration of dissolved solute is called **unsaturated:** add more solute, and more will dissolve until the solution becomes saturated.

In some cases, we can prepare a solution that contains more than the equilibrium concentration of dissolved solute. Such a solution is called **supersaturated.** It is unstable relative to the saturated solution: if you add a "seed" crystal of solute, or just tap the container, the excess solute crystallizes immediately, leaving a saturated solution (Figure 13.20). You can often prepare a supersaturated solution of a solute if it has greater solubility at a higher temperature. While warming the contents of the flask, you dissolve more than the amount of solute required to prepare a saturated solution at some lower temperature and then slowly cool the solution. If the excess solute remains dissolved, the solution is supersaturated.

 ARIS Presentation Center, Art Library

● **A Saturated Solution Is Like a Pure Liquid and Its Vapor** Equilibrium processes in a saturated solution are analogous to those for a pure liquid and its vapor in a closed flask (Section 12.2). For the liquid, rates of vaporizing and condensing are equal; in the solution, rates of dissolving and recrystallizing are equal. In the liquid-vapor system, particles leave the liquid to enter the vapor, and their concentration (pressure) increases until, at equilibrium, the available space above the liquid is "saturated" with vapor at a given temperature. In the solution, solute particles leave the crystal to enter the solvent, and their concentration increases until, at equilibrium, the available solvent is saturated with solute at a given temperature.

Demo Ahmad, Jamil. "Crystallization from a Supersaturated Solution of Sodium Acetate," *J. Chem. Educ.* **2000,** *77,* 1446.

Figure 13.20 Sodium acetate crystallizing from a supersaturated solution. When a seed crystal of sodium acetate is added to a supersaturated solution of the compound (**A**), solute begins to crystallize out of solution (**B**) and continues to do so until the remaining solution is saturated (**C**).

 ARIS Presentation Center, Photo Library

Effect of Temperature on Solubility

Temperature affects the solubility of most substances. You may have noticed, for example, that not only does sugar dissolve more quickly in hot tea than in iced tea, but *more* sugar dissolves; in other words, the solubility of sugar in tea is greater at higher temperatures. Let's examine the effects of temperature on the solubility of solids and of gases.

Temperature and the Solubility of Solids Like sugar, *most solids are more soluble at higher temperatures*. Figure 13.21 shows the solubility of several ionic compounds in water as a function of temperature. Note that most of the graphed lines rise from left to right. Cerium sulfate is the only exception shown in the figure, but several other salts, mostly sulfates, behave similarly. Some salts exhibit increasing solubility up to a certain temperature and then decreasing solubility at still higher temperatures.

We might think that the sign of ΔH_{soln} would indicate the effect of temperature. Most ionic solids have a positive ΔH_{soln} because their lattice energies are greater than their heats of hydration. Thus, heat is *absorbed* to form the solution from solute and solvent, and if we think of heat as a reactant, a rise in temperature should increase the rate of the forward process:

$$\text{Solute + solvent} + \textit{heat} \rightleftharpoons \text{saturated solution}$$

Tabulated ΔH_{soln} values refer to the enthalpy change for a solution to reach the standard state of 1 *M*, but in order to understand the effect of temperature, we need to know the sign of the enthalpy change very close to the point of saturation, which may not be the same. To use earlier examples, tabulated values give a negative ΔH_{soln} for $NaOH$ and a positive one for NH_4NO_3, yet both compounds are more soluble at higher temperatures. The point is that solubility is a complex behavior, and, although the effect of temperature ultimately reflects the equilibrium nature of solubility, no single measure can help us predict the effect for a given solute.

Figure 13.21 The relation between solubility and temperature for several ionic compounds. Most ionic compounds have higher solubilities at higher temperatures. Cerium sulfate is one of several exceptions.

Temperature and Gas Solubility in Water The effect of temperature on gas solubility is much more predictable. When a solid dissolves in a liquid, the solute particles must separate, so energy must be added; thus, for a solid, $\Delta H_{solute} > 0$. In contrast, gas particles are already separated, so $\Delta H_{solute} \approx 0$. Because the hydration step is exothermic ($\Delta H_{hydr} < 0$), the sum of these two terms must be negative. Thus, for all gases in water, $\Delta H_{soln} < 0$:

$$\text{Solute}(g) + \text{water}(l) \rightleftharpoons \text{saturated solution}(aq) + \text{heat}$$

This equation means that *gas solubility in water **decreases** with rising temperature.* Gases have weak intermolecular forces, so there are relatively weak intermolecular forces between a gas and water. When the temperature rises, the average kinetic energy of the particles in solution increases, allowing the gas particles to easily overcome these weak forces and re-enter the gas phase.

This behavior can lead to an environmental problem known as *thermal pollution.* During many industrial processes, large amounts of water are taken from a nearby river or lake, pumped through the system to cool liquids, gases, and equipment, and then returned to the body of water at a higher temperature. The metabolic rates of fish and other aquatic animals increase in the warmer water released near the plant outlet; thus, their need for O_2 increases, but the concentration of dissolved O_2 is lower at the higher temperature. This oxygen depletion can harm these aquatic populations. Moreover, the warmer water is less dense, so it floats on the cooler water below and prevents O_2 from reaching it. Thus, creatures living at lower levels become oxygen deprived as well. Farther from the plant, the water temperature returns to ambient levels and the O_2 solubility increases. One way to lessen the problem is with cooling towers, which lower the temperature of the water before it exits the plant (Figure 13.22).

Effect of Pressure on Solubility

Because liquids and solids are almost incompressible, pressure has little effect on their solubility, but it has a major effect on gas solubility. Consider the piston-cylinder assembly in Figure 13.23, with a gas above a saturated aqueous solution of the gas. At a given pressure, the same number of gas molecules enter and leave the solution per unit time; that is, the system is at equilibrium:

$$\text{Gas} + \text{solvent} \rightleftharpoons \text{saturated solution}$$

If you push down on the piston, the gas volume decreases, its pressure increases, and gas particles collide with the liquid surface more often. Thus, more gas particles enter than leave the solution per unit time. Higher gas pressure disturbs the

Environmental application

Figure 13.22 Fighting thermal pollution. Steam billowing from cooling towers is common at large electric power plants. Hot water resulting from the cooling of gases and equipment enters near the top of the tower and cools in contact with air. The cooler water is then released back into the water source.

Demo Shakhashiri, Bassam Z. "Effect of Temperature and Pressure on the Solubility of Gases in Liquids," *Chemical Demonstrations, A Handbook for Teachers of Chemistry,* Vol. 3 (The University of Wisconsin Press, Madison, 1985), 280–282.

Figure 13.23 **The effect of pressure on gas solubility. A,** A saturated solution of a gas is in equilibrium at pressure P_1. **B,** If the pressure is increased to P_2, the volume of the gas decreases. Therefore, the frequency of collisions with the surface increases. **C,** As a result, more gas is in solution when equilibrium is re-established.

● **Scuba Diving and Soda Pop** Although N_2 is barely soluble in water or blood at sea level, high external pressures significantly increase its solubility. Scuba divers who breathe compressed air and dive below about 50 ft have much more N_2 dissolved in their blood. If they ascend quickly to lower pressures, they may be afflicted with decompression sickness (the "bends"), as the dissolved N_2 bubbles out of their blood and causes painful, sometimes fatal, blockage of capillaries. This condition can be avoided by using less soluble gases, such as He, in the breathing mixtures. In principle, the same thing happens when you open a can of a carbonated drink. In the closed can, dissolved CO_2 is at equilibrium with 4 atm of CO_2 gas in the space above it. In the open can, the dissolved CO_2 is under 1 atm of air ($P_{CO_2} = 3 \times 10^{-4}$ atm), so CO_2 bubbles out of solution. With time, the system reaches equilibrium with the P_{CO_2} in air: the drink goes "flat."

 ARIS Presentation Center, Photo Library

 Biological application

balance at equilibrium, so more gas dissolves to reduce this disturbance (a shift to the right in the preceding equation) until the system re-establishes equilibrium. ●

Henry's law expresses the quantitative relationship between gas pressure and solubility: *the solubility of a gas (S_{gas}) is directly proportional to the partial pressure of the gas (P_{gas}) above the solution:*

$$S_{gas} = k_H \times P_{gas} \tag{13.3}$$

where k_H is the Henry's law constant and is specific for a given gas-solvent combination at a given temperature. With S_{gas} in mol/L and P_{gas} in atm, the units of k_H are mol/L·atm.

ARIS Presentation Center, Worked Examples Library

SAMPLE PROBLEM 13.2 Using Henry's Law to Calculate Gas Solubility

PROBLEM The partial pressure of carbon dioxide gas inside a bottle of cola is 4 atm at 25°C. What is the solubility of CO_2? The Henry's law constant for CO_2 dissolved in water is 3.3×10^{-2} mol/L·atm at 25°C.

PLAN We know P_{CO_2} (4 atm) and the value of k_H (3.3×10^{-2} mol/L·atm), so we substitute them into Equation 13.3 to find S_{CO_2}.

SOLUTION $S_{CO_2} = k_H \times P_{CO_2} = (3.3 \times 10^{-2}$ mol/L·atm$)(4$ atm$) = 0.1$ mol/L

CHECK The units are correct for solubility. We rounded the answer to one significant figure because there is only one in the pressure value.

FOLLOW-UP PROBLEM 13.2 If air contains 78% N_2 by volume, what is the solubility of N_2 in water at 25°C and 1 atm (k_H for N_2 in H_2O at 25°C = 7×10^{-4} mol/L·atm)?

Section Summary

A solution that contains the maximum amount of dissolved solute in the presence of excess undissolved solute is saturated. A state of equilibrium exists when a saturated solution is in contact with excess solute, because solute particles are entering and leaving the solution at the same rate. • Most solids are more soluble at higher temperatures. • All gases have a negative ΔH_{soln} in water, so heating lowers gas solubility in water. • Henry's law says that the solubility of a gas is directly proportional to its partial pressure above the solution.

13.5 QUANTITATIVE WAYS OF EXPRESSING CONCENTRATION

Concentration is the *proportion* of a substance in a mixture, so it is an intensive property, one that does not depend on the quantity of mixture present: 1.0 L of 0.1 *M* NaCl has the same concentration as 1.0 mL of 0.1 *M* NaCl. Concentration is most often expressed as the ratio of the quantity of solute to the quantity of *solution*, but sometimes it is the ratio of solute to *solvent*. Because both parts of the ratio can be given in units of mass, volume, or amount (mol), chemists employ several concentration terms, including molarity, molality, and various expressions of "parts of solute per part of solution" (Table 13.5).

Molarity and Molality

Molarity (*M*) is the *number of moles of solute dissolved in 1 L of solution:*

$$\text{Molarity } (M) = \frac{\text{amount (mol) of solute}}{\text{volume (L) of solution}} \tag{13.4}$$

In Chapter 3, you used molarity to convert liters of solution into moles of dissolved solute. Expressing concentration in terms of molarity may have drawbacks, however. Because volume is affected by temperature, so is molarity. A solution expands when heated, so a unit volume of hot solution contains slightly less solute than a unit volume of cold solution. This can be a source of error in very precise

Table 13.5 Concentration Definitions	
Concentration Term	**Ratio**
Molarity (*M*)	$\dfrac{\text{amount (mol) of solute}}{\text{volume (L) of solution}}$
Molality (*m*)	$\dfrac{\text{amount (mol) of solute}}{\text{mass (kg) of solvent}}$
Parts by mass	$\dfrac{\text{mass of solute}}{\text{mass of solution}}$
Parts by volume	$\dfrac{\text{volume of solute}}{\text{volume of solution}}$
Mole fraction (*X*)	$\dfrac{\text{amount (mol) of solute}}{\text{amount (mol) of solute} + \text{amount (mol) of solvent}}$

work. More importantly, because of solute-solvent interactions that are difficult to predict, *solution volumes may not be additive;* that is, adding 500. mL of one solution to 500. mL of another may not give 1000. mL. Therefore, in precise work, a solution with a desired molarity may not be easy to prepare.

A concentration term that does not contain volume in its ratio is **molality (*m*),** *the number of moles of solute dissolved in 1000 g (1 kg) of solvent:*

$$\text{Molality } (m) = \frac{\text{amount (mol) of solute}}{\text{mass (kg) of solvent}} \qquad \textbf{(13.5)}$$

Note that molality includes the quantity of *solvent,* not solution. And, most important, molal solutions are prepared by measuring *masses* of solute and solvent, not solvent or solution *volume.* Mass does not change with temperature, so neither does molality. Moreover, unlike volumes, masses *are* additive: adding 500. g of one solution to 500. g of another *does* give 1000. g of final solution. For these reasons, molality is a preferred unit when temperature, and hence density, may change, as in the examination of solutions' physical properties. For the special case of water, 1 L has a mass of 1 kg, so *molality and molarity are nearly the same for dilute aqueous solutions.*

SAMPLE PROBLEM 13.3 Calculating Molality

PROBLEM What is the molality of a solution prepared by dissolving 32.0 g of $CaCl_2$ in 271 g of water?

PLAN To use Equation 13.5, we convert mass of $CaCl_2$ (32.0 g) to amount (mol) with the molar mass (g/mol) and then divide by the mass of water (271 g), being sure to convert from grams to kilograms.

SOLUTION Converting from grams of solute to moles:

$$\text{Moles of } CaCl_2 = 32.0 \text{ g } CaCl_2 \times \frac{1 \text{ mol } CaCl_2}{110.98 \text{ g } CaCl_2} = 0.288 \text{ mol } CaCl_2$$

Finding molality:

$$\text{Molality} = \frac{\text{mol solute}}{\text{kg solvent}} = \frac{0.288 \text{ mol } CaCl_2}{271 \text{ g} \times \dfrac{1 \text{ kg}}{10^3 \text{ g}}} = 1.06 \text{ } m \text{ } CaCl_2$$

CHECK The answer seems reasonable: the given numbers of moles of $CaCl_2$ and kilograms of H_2O are about the same, so their ratio is about 1.

FOLLOW-UP PROBLEM 13.3 How many grams of glucose ($C_6H_{12}O_6$) must be dissolved in 563 g of ethanol (C_2H_5OH) to prepare a 2.40×10^{-2} *m* solution?

Parts of Solute by Parts of Solution

Several concentration terms are based on the number of solute (or solvent) parts present in a specific number of *solution* parts. The solution parts can be expressed in terms of mass, volume, or amount (mol).

Parts by Mass The most common of the parts-by-mass terms is **mass percent,** which you encountered in Chapter 3. The word *percent* means "per hundred," so mass percent of solute means the mass of solute dissolved in every 100. parts by mass of solution, or the mass fraction times 100:

$$\text{Mass percent} = \frac{\text{mass of solute}}{\text{mass of solute } + \text{ mass of solvent}} \times 100$$

$$= \frac{\text{mass of solute}}{\text{mass of solution}} \times 100 \qquad \textbf{(13.6)}$$

Sometimes mass percent is symbolized as **% (w/w),** indicating that the percentage is a ratio of weights (more accurately, masses). You may have seen mass percent values on jars of solid chemicals to indicate the amounts of impurities present. Two very similar terms are parts per million (ppm) by mass and parts per billion (ppb) by mass: grams of solute per million or per billion grams of solution. For these quantities, in Equation 13.6, you multiply by 10^6 or by 10^9, respectively, instead of by 100. ●

Parts by Volume The most common of the parts-by-volume terms is **volume percent,** the volume of solute in 100. volumes of solution:

$$\text{Volume percent} = \frac{\text{volume of solute}}{\text{volume of solution}} \times 100 \qquad \textbf{(13.7)}$$

A common symbol for volume percent is **% (v/v).** Commercial rubbing alcohol, for example, is an aqueous solution of isopropyl alcohol (a three-carbon alcohol) that contains 70 volumes of isopropyl alcohol per 100. volumes of solution, and the label indicates this as "70% (v/v)." Parts-by-volume concentrations are most often used for liquids and gases. Minor atmospheric components occur in parts per million by volume (ppmv). For example, there are about 0.05 ppmv of the toxic gas carbon monoxide (CO) in clean air, 1000 times as much (about 50 ppmv of CO) in air over urban traffic, and 10,000 times as much (about 500 ppmv of CO) in cigarette smoke. Some extremely tiny concentrations of active substances are found in nature. *Pheromones* are organic compounds secreted by one member of a species to signal other members about food, danger, sexual readiness, and so forth. Many organisms, including dogs and monkeys, release pheromones, and researchers suspect that humans do as well. Some insect pheromones are active even at a few hundred molecules per milliliter of air, about 100 parts per quadrillion by volume (Figure 13.24).

A measure of concentration frequently used for aqueous solutions is % (w/v), a ratio of solute *weight* (actually mass) to solution *volume.* Thus, a 1.5% (w/v)

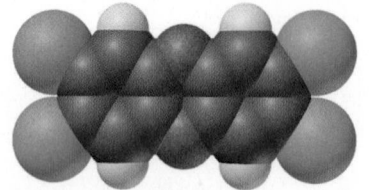

● **Unhealthy Ultralow Concentrations** Environmental toxicologists often measure extremely low concentrations of pollutants. The Centers for Disease Control and Prevention considers TCDD (*te*trachloro*di*benzo*di*oxin, *above*) one of the most toxic substances known. It is a member of the dioxin family of chlorinated hydrocarbons, a byproduct of bleaching paper. TCDD is considered unsafe at soil levels above 1 ppb. From contact with air, water, and soil, most North Americans have an average of 0.01 ppb TCDD in their fatty tissues. The Environmental Protection Agency has assessed TCDD as unsafe at *all* levels measured.

 Environmental application

 Biological application

$$\underset{\displaystyle CH_3-CH-(CH_2)_4-CH-CH-(CH_2)_9-CH_3}{\overset{\displaystyle \overset{CH_3}{|} \qquad\qquad \overset{O}{\diagup\!\!\diagdown}}{}}$$

Figure 13.24 The sex attractant of the gypsy moth. The pheromone secreted by the female gypsy moth (called *disparlure*) can attract a male over 0.5 mi away. Chemists synthesize pheromones for insect pest control.

NaCl solution contains 1.5 g of NaCl per 100. mL of *solution*. This way of expressing concentrations is particularly common in medical labs and other health-related facilities.

Mole Fraction The **mole fraction** (X) of a solute is the ratio of number of solute moles to the total number of moles (solute plus solvent), that is, parts by mole. The *mole percent* is the mole fraction expressed as a percentage:

$$\text{Mole fraction } (X) = \frac{\text{amount (mol) of solute}}{\text{amount (mol) of solute} + \text{amount (mol) of solvent}} \quad \textbf{(13.8)}$$

$$\text{Mole percent (mol \%)} = \text{mole fraction} \times 100$$

We discussed these terms in Chapter 5 in relation to Dalton's law of partial pressures for mixtures of gases, but they apply to liquids and solids as well. Concentrations given as mole fractions provide the clearest picture of the actual proportion of solute (or solvent) particles among all the particles in the solution.

ARIS Presentation Center, Worked Examples Library

SAMPLE PROBLEM 13.4 Expressing Concentrations in Parts by Mass, Parts by Volume, and Mole Fraction

PROBLEM **(a)** Find the concentration of calcium (in ppm) in a 3.50-g pill that contains 40.5 mg of Ca.
(b) The label on a 0.750-L bottle of Italian chianti indicates "11.5% alcohol by volume." How many liters of alcohol does the wine contain?
(c) A sample of rubbing alcohol contains 142 g of isopropyl alcohol (C_3H_7OH) and 58.0 g of water. What are the mole fractions of alcohol and water?
PLAN **(a)** We are given the masses of Ca (40.5 mg) and the pill (3.50 g). We convert the mass of Ca from mg to g, find the ratio of mass of Ca to mass of pill, and multiply by 10^6 to obtain ppm. **(b)** We know the volume % (11.5%, or 11.5 parts by volume of alcohol to 100 parts of chianti) and the total volume (0.750 mL), so we use Equation 13.7 to find the volume of alcohol. **(c)** We know the mass and formula of each component, so we convert masses to amounts (mol) and apply Equation 13.8 to find the mole fractions.
SOLUTION **(a)** Finding parts per million by mass of Ca. Combining the steps, we have

$$\text{ppm Ca} = \frac{\text{mass of Ca}}{\text{mass of pill}} \times 10^6 = \frac{40.5 \text{ mg Ca} \times \dfrac{1 \text{ g}}{10^3 \text{ mg}}}{3.50 \text{ g}} \times 10^6 = 1.16 \times 10^4 \text{ ppm Ca}$$

(b) Finding volume (L) of alcohol:

$$\text{Volume (L) of alcohol} = 0.750 \text{ L chianti} \times \frac{11.5 \text{ L alcohol}}{100. \text{ L chianti}} = 0.0862 \text{ L}$$

(c) Finding mole fractions. Converting from grams to moles:

$$\text{Moles of } C_3H_7OH = 142 \text{ g } C_3H_7OH \times \frac{1 \text{ mol } C_3H_7OH}{60.09 \text{ g } C_3H_7OH} = 2.36 \text{ mol } C_3H_7OH$$

$$\text{Moles of } H_2O = 58.0 \text{ g } H_2O \times \frac{1 \text{ mol } H_2O}{18.02 \text{ g } H_2O} = 3.22 \text{ mol } H_2O$$

Calculating mole fractions:

$$X_{C_3H_7OH} = \frac{\text{moles of } C_3H_7OH}{\text{total moles}} = \frac{2.36 \text{ mol}}{2.36 \text{ mol} + 3.22 \text{ mol}} = 0.423$$

$$X_{H_2O} = \frac{\text{moles of } H_2O}{\text{total moles}} = \frac{3.22 \text{ mol}}{2.36 \text{ mol} + 3.22 \text{ mol}} = 0.577$$

CHECK **(a)** The mass ratio is about 0.04 g/4 g = 10^{-2}, and $10^{-2} \times 10^6 = 10^4$ ppm, so it seems correct. **(b)** The volume % is a bit more than 10%, so the volume of alcohol should be a bit more than 75 mL (0.075 L). **(c)** Always check that the *mole fractions add up to 1*: 0.423 + 0.577 = 1.000.

FOLLOW-UP PROBLEM 13.4 An alcohol solution contains 35.0 g of 1-propanol (C_3H_7OH) and 150. g of ethanol (C_2H_5OH). Calculate the mass percent and the mole fraction of each alcohol.

Interconverting Concentration Terms

All the terms we just discussed represent different ways of expressing concentration, so they are interconvertible. Keep these points in mind:

- To convert a term based on amount (mol) to one based on mass, you need the molar mass. These conversions are similar to the mass-mole conversions you've done earlier.
- To convert a term based on mass to one based on volume, you need the solution *density*. Given the mass of a solution, the density (mass/volume) gives you the volume, or vice versa.
- Molality involves quantity of *solvent*, whereas the other concentration terms involve quantity of *solution*.

 ARIS Presentation Center, Worked Examples Library

SAMPLE PROBLEM 13.5 Converting Concentration Terms

PROBLEM Hydrogen peroxide is a powerful oxidizing agent used in concentrated solution in rocket fuels and in dilute solution as a hair bleach. An aqueous solution of H_2O_2 is 30.0% by mass and has a density of 1.11 g/mL. Calculate its:
(a) Molality   **(b)** Mole fraction of H_2O_2 **(c)** Molarity
PLAN We know the mass % (30.0) and the density (1.11 g/mL). **(a)** For molality, we need the amount (mol) of solute and the mass (kg) of *solvent*. Assuming 100.0 g of H_2O_2 solution allows us to express the mass % directly as grams of substance. We subtract the grams of H_2O_2 to obtain the grams of solvent. To find molality, we convert grams of H_2O_2 to moles and divide by mass of solvent (converting g to kg). **(b)** To find the mole fraction, we use the number of moles of H_2O_2 [from part (a)] and convert the grams of H_2O to moles. Then we divide the moles of H_2O_2 by the total moles. **(c)** To find molarity, we assume 100.0 g of solution and use the given solution density to find the volume. Then we divide the amount (mol) of H_2O_2 [from part (a)] by *solution* volume (in L).
SOLUTION **(a)** From mass % to molality. Finding mass of solvent (assuming 100.0 g of solution):

$$\text{Mass (g) of } H_2O = 100.0 \text{ g solution} - 30.0 \text{ g } H_2O_2 = 70.0 \text{ g } H_2O$$

Converting from grams of H_2O_2 to moles:

$$\text{Moles of } H_2O_2 = 30.0 \text{ g } H_2O_2 \times \frac{1 \text{ mol } H_2O_2}{34.02 \text{ g } H_2O_2} = 0.882 \text{ mol } H_2O_2$$

Calculating molality:

$$\text{Molality of } H_2O_2 = \frac{0.882 \text{ mol } H_2O_2}{70.0 \text{ g} \times \frac{1 \text{ kg}}{10^3 \text{ g}}} = 12.6 \text{ } m \text{ } H_2O_2$$

(b) From mass % to mole fraction:

$$\text{Moles of } H_2O_2 = 0.882 \text{ mol } H_2O_2 \text{ [from part (a)]}$$

$$\text{Moles of } H_2O = 70.0 \text{ g } H_2O \times \frac{1 \text{ mol } H_2O}{18.02 \text{ g } H_2O} = 3.88 \text{ mol } H_2O$$

$$X_{H_2O_2} = \frac{0.882 \text{ mol}}{0.882 \text{ mol} + 3.88 \text{ mol}} = 0.185$$

(c) From mass % and density to molarity. Converting from solution mass to volume:

$$\text{Volume (mL) of solution} = 100.0 \text{ g} \times \frac{1 \text{ mL}}{1.11 \text{ g}} = 90.1 \text{ mL}$$

Calculating molarity:

$$\text{Molarity} = \frac{\text{mol } H_2O_2}{\text{L soln}} = \frac{0.882 \text{ mol } H_2O_2}{90.1 \text{ mL} \times \frac{1 \text{ L soln}}{10^3 \text{ mL}}} = 9.79 \text{ } M \text{ } H_2O_2$$

CHECK Rounding shows the answers seem reasonable: **(a)** The ratio of ~0.9 mol/0.07 kg is greater than 10. **(b)** ~0.9 mol H_2O_2/(1 mol + 4 mol) ≈ 0.2. **(c)** The ratio of moles to liters (0.9/0.09) is around 10.

FOLLOW-UP PROBLEM 13.5 A sample of commercial concentrated hydrochloric acid is 11.8 *M* HCl and has a density of 1.190 g/mL. Calculate the mass %, molality, and mole fraction of HCl.

Section Summary

The concentration of a solution is independent of the amount of solution and can be expressed as molarity (mol solute/L solution), molality (mol solute/kg solvent), parts by mass (mass solute/mass solution), parts by volume (volume solute/volume solution), or mole fraction [mol solute/(mol solute + mol solvent)]. The choice of units depends on convenience or the nature of the solution. • If, in addition to the quantities of solute and solution, the solution density is also known, all ways of expressing concentration are interconvertible.

13.6 COLLIGATIVE PROPERTIES OF SOLUTIONS

We might expect the presence of solute particles to make the physical properties of a solution different from those of the pure solvent. However, what we might *not* expect is that, in the case of four important solution properties, the *number* of solute particles makes the difference, not their chemical identity. These properties, known as **colligative properties** (*colligative* means "collective"), are vapor pressure lowering, boiling point elevation, freezing point depression, and osmotic pressure. Even though most of these effects are small, they have many practical applications, including some that are vital to biological systems.

Historically, colligative properties were measured to explore the nature of a solute in aqueous solution and its extent of dissociation into ions. In Chapter 4, we classified solutes by their ability to conduct an electric current, which requires moving ions to be present. Recall that an aqueous solution of an **electrolyte** conducts an electric current because the solute separates into ions as it dissolves. An electrolyte is classified as strong or weak depending on whether its solution is a good or poor conductor at moderate solute concentrations. Soluble salts, strong acids, and strong bases are *strong electrolytes* because they dissociate completely and, thus, their solutions conduct well. Weak acids and weak bases are *weak electrolytes* because they dissociate very little and, thus, their solutions conduct poorly. Many compounds, such as sugar and alcohol, do not dissociate into ions at all. They are **nonelectrolytes** because their solutions do not conduct a current. Figure 13.25 shows these behaviors.

Literature McCarthy, Sean M.; Gordon-Wylie, Scott W. "A Greener Approach for Measuring Colligative Properties," *J. Chem. Educ.* **2005**, *82*, 116.

Literature Beall, Herbert; Riccardino, Patrick A. "The Colligative Property of Walther Nernst," *J. Chem. Educ.* **2001**, *78*, 511.

A Strong electrolyte   **B Weak electrolyte**   **C Nonelectrolyte**

Figure 13.25 The three types of electrolytes. A, Strong electrolytes conduct a large current because they dissociate completely into ions. **B,** Weak electrolytes conduct a small current because they dissociate very little. **C,** Nonelectrolytes do not conduct a current because they do not dissociate.

ARIS Presentation Center, Photo Library

To predict the magnitude of a colligative property, we refer to the solute formula to find the number of particles in solution. Each mole of nonelectrolyte yields 1 mol of particles in the solution. For example, 0.35 *M* glucose contains 0.35 mol of solute particles per liter. In principle, each mole of strong electrolyte dissociates into the number of moles of ions in the formula unit: 0.4 *M* Na_2SO_4 contains 0.8 mol of Na^+ ions and 0.4 mol of SO_4^{2-} ions, or 1.2 mol of particles, per liter (see Sample Problem 4.1, p. 143). (We examine the equilibrium nature of the dissociation of weak electrolytes in Chapters 18 and 19.)

Colligative Properties of Nonvolatile Nonelectrolyte Solutions

In this section, we focus most of our attention on the simplest case, the colligative properties of solutes that do not dissociate into ions and have negligible vapor pressure even at the boiling point of the solvent. Such solutes are called *nonvolatile nonelectrolytes;* sucrose (table sugar) is an example. Later, we briefly explore the properties of volatile nonelectrolytes and of strong electrolytes.

Vapor Pressure Lowering *The vapor pressure of a solution of a nonvolatile nonelectrolyte is always lower than the vapor pressure of the pure solvent.* We can understand this **vapor pressure lowering (ΔP)** in terms of opposing rates and in terms of changes in entropy. Consider a pure solvent and the opposing rates of vaporization (molecules leaving the liquid) and of condensation (molecules re-entering the liquid). At equilibrium, the two rates are equal. When we add some nonvolatile solute, the number of solvent molecules on the surface is lower, so fewer vaporize per unit time. To maintain equilibrium, fewer gas molecules need to enter the liquid, and this occurs only if the concentration of gas, that is, the vapor pressure, is lowered. In terms of the entropy change, recall that natural processes occur in a direction of increasing entropy. A pure solvent vaporizes because the vapor has a greater entropy than the liquid. However, the solvent in a solution already has a greater entropy than it does as pure solvent, so it has less tendency to vaporize in order to gain entropy. Thus, by either argument, equilibrium is reached at a lower vapor pressure for the solution. Figure 13.26 illustrates this point.

In quantitative terms, we find that the vapor pressure of solvent above the solution ($P_{solvent}$) equals the mole fraction of solvent in the solution ($X_{solvent}$) times the vapor pressure of the pure solvent ($P^{\circ}_{solvent}$). This relationship is expressed by **Raoult's law:**

$$P_{solvent} = X_{solvent} \times P^{\circ}_{solvent} \qquad (13.9)$$

In a solution, $X_{solvent}$ is always less than 1, so $P_{solvent}$ is always less than $P^{\circ}_{solvent}$. An **ideal solution** is one that follows Raoult's law at any concentration. However, just as most gases deviate from ideality, so do most solutions. In practice, Raoult's law gives a good approximation of the behavior of *dilute* solutions only, and it becomes exact at infinite dilution.

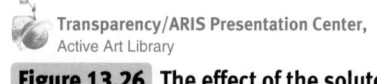

Figure 13.26 **The effect of the solute on the vapor pressure of a solution.** Vaporization occurs because of the tendency of a system to increase its entropy. **A,** Equilibrium is established between a pure liquid and its vapor when the numbers of molecules vaporizing and condensing in a given time are equal. **B,** The presence of a dissolved solute decreases the number of solvent molecules at the surface and increases the entropy of the system, so fewer solvent molecules vaporize in a given time. Therefore, fewer molecules need to condense to balance them, and equilibrium is established at a lower vapor pressure.

Solvent molecules

A

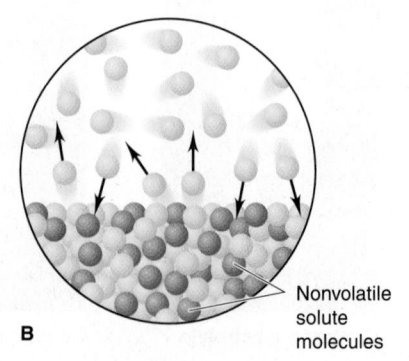

Nonvolatile solute molecules

B

How does the *amount* of solute affect the *magnitude* of the vapor pressure lowering, ΔP? The solution consists of solvent and solute, so the sum of their mole fractions equals 1:

$$X_{\text{solvent}} + X_{\text{solute}} = 1; \quad \text{thus,} \quad X_{\text{solvent}} = 1 - X_{\text{solute}}$$

From Raoult's law, we have

$$P_{\text{solvent}} = X_{\text{solvent}} \times P^{\circ}_{\text{solvent}} = (1 - X_{\text{solute}}) \times P^{\circ}_{\text{solvent}}$$

Multiplying through on the right side gives

$$P_{\text{solvent}} = P^{\circ}_{\text{solvent}} - (X_{\text{solute}} \times P^{\circ}_{\text{solvent}})$$

Rearranging and introducing ΔP gives

$$P^{\circ}_{\text{solvent}} - P_{\text{solvent}} = \Delta P = X_{\text{solute}} \times P^{\circ}_{\text{solvent}} \tag{13.10}$$

Thus, the *magnitude* of ΔP equals the mole fraction of solute times the vapor pressure of the pure solvent—a relationship applied in the next sample problem.

SAMPLE PROBLEM 13.6 Using Raoult's Law to Find Vapor Pressure Lowering

ARIS Presentation Center,
Worked Examples Library

PROBLEM Calculate the vapor pressure lowering, ΔP, when 10.0 mL of glycerol ($C_3H_8O_3$) is added to 500. mL of water at 50.°C. At this temperature, the vapor pressure of pure water is 92.5 torr and its density is 0.988 g/mL. The density of glycerol is 1.26 g/mL.
PLAN To calculate ΔP, we use Equation 13.10. We are given the vapor pressure of pure water ($P^{\circ}_{H_2O} = 92.5$ torr), so we just need the mole fraction of glycerol, X_{glycerol}. We convert the given volume of glycerol (10.0 mL) to mass using the given density (1.26 g/L), find the molar mass from the formula, and convert mass (g) to amount (mol). The same procedure gives amount of H_2O. From these, we find X_{glycerol} and ΔP.
SOLUTION Calculating the amount (mol) of glycerol and of water:

$$\text{Moles of glycerol} = 10.0 \text{ mL glycerol} \times \frac{1.26 \text{ g glycerol}}{1 \text{ mL glycerol}} \times \frac{1 \text{ mol glycerol}}{92.09 \text{ g glycerol}}$$

$$= 0.137 \text{ mol glycerol}$$

$$\text{Moles of } H_2O = 500. \text{ mL } H_2O \times \frac{0.988 \text{ g } H_2O}{1 \text{ mL } H_2O} \times \frac{1 \text{ mol } H_2O}{18.02 \text{ g } H_2O} = 27.4 \text{ mol } H_2O$$

Calculating the mole fraction of glycerol:

$$X_{\text{glycerol}} = \frac{0.137 \text{ mol}}{0.137 \text{ mol} + 27.4 \text{ mol}} = 0.00498$$

Finding the vapor pressure lowering:

$$\Delta P = X_{\text{glycerol}} \times P^{\circ}_{H_2O} = 0.00498 \times 92.5 \text{ torr} = 0.461 \text{ torr}$$

CHECK The amount of each component seems correct: for glycerol, ~10 mL × 1.25 g/mL ÷ 100 g/mol = 0.125 mol; for H_2O, ~500 mL × 1 g/mL ÷ 20 g/mol = 25 mol. The small ΔP is reasonable because the mole fraction of solute is small.
COMMENT The calculation assumes that glycerol is nonvolatile. At 1 atm, glycerol boils at 290.0°C, so the vapor pressure of glycerol at 50°C is so low it can be neglected.

FOLLOW-UP PROBLEM 13.6 Calculate the vapor pressure lowering of a solution of 2.00 g of aspirin ($M = 180.15$ g/mol) in 50.0 g of methanol (CH_3OH) at 21.2°C. Pure methanol has a vapor pressure of 101 torr at this temperature.

Volume (mL) of glycerol (or H_2O)

multiply by density (g/mL)

Mass (g) of glycerol (or H_2O)

divide by M (g/mol)

Amount (mol) of glycerol (or H_2O)

divide by total number of moles

Mole fraction (X) of glycerol

multiply by $P^{\circ}_{H_2O}$

Vapor pressure lowering (ΔP)

Boiling Point Elevation *A solution boils at a higher temperature than the pure solvent.* Let's see why. The boiling point (boiling temperature, T_b) of a liquid is the temperature at which its vapor pressure equals the external pressure. The vapor pressure of a solution is lower than the external pressure at the solvent's boiling point because the vapor pressure of a solution is lower than that of the pure solvent at any temperature. Therefore, the solution does not yet boil. A higher temperature is needed to raise the solution's vapor pressure to equal the external pressure. We can see this **boiling point elevation (ΔT_b)** by superimposing a phase diagram for the solution on a phase diagram for the pure solvent, as shown in

Figure 13.27 Phase diagrams of solvent and solution. Phase diagrams of an aqueous solution (*dashed lines*) and of pure water (*solid lines*) show that, by lowering the vapor pressure (ΔP), a dissolved solute elevates the boiling point (ΔT_b) and depresses the freezing point (ΔT_f).

Transparency/ARIS Presentation Center, Art Library

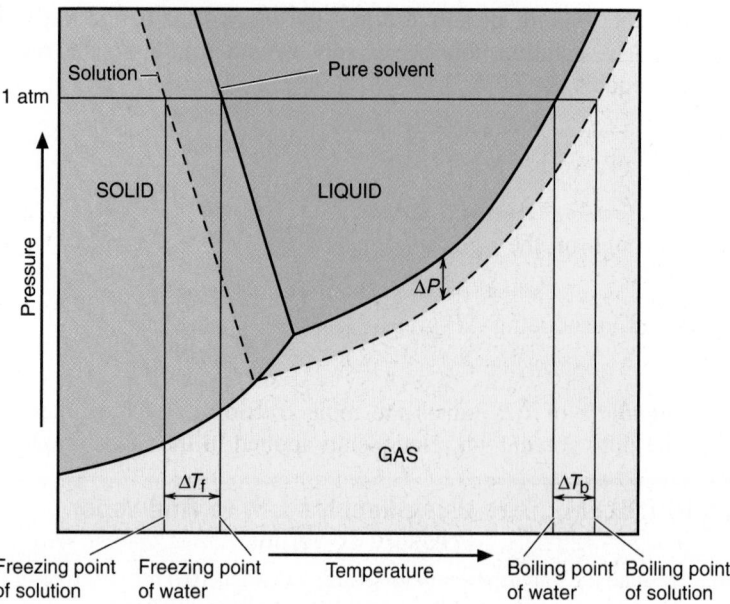

Figure 13.27. Note that the gas-liquid line for the solution lies *below* that for the pure solvent at any temperature and to the right of it at any pressure.

Like the vapor pressure lowering, the magnitude of the boiling point elevation is proportional to the concentration of solute particles:

$$\Delta T_b \propto m \quad \text{or} \quad \Delta T_b = K_b m \tag{13.11}$$

where m is the solution molality, K_b is the *molal boiling point elevation constant,* and ΔT_b is the boiling point elevation. We typically speak of ΔT_b as a positive value, so we subtract the lower temperature from the higher; that is, we subtract the solvent T_b from the solution T_b:

$$\Delta T_b = T_{b(solution)} - T_{b(solvent)}$$

Molality is the concentration term used because it is related to mole fraction and thus to particles of solute. It also involves mass rather than volume of solvent, so it is not affected by temperature changes. The constant K_b has units of degrees Celsius per molal unit (°C/m) and is specific for a given solvent (Table 13.6).

Notice that the K_b for water is only 0.512°C/m, so the changes in boiling point are quite small: if you dissolved 1.00 mol of glucose (180. g; 1.00 mol of particles) in 1.00 kg of water, or 0.500 mol of NaCl (29.2 g; a strong electrolyte, so also 1.00 mol of particles) in 1.00 kg of water, the boiling points of the resulting solutions at 1 atm would be only 100.512°C instead of 100.000°C.

Table 13.6	Molal Boiling Point Elevation and Freezing Point Depression Constants of Several Solvents			
Solvent	**Boiling Point (°C)***	**K_b (°C/m)**	**Melting Point (°C)**	**K_f (°C/m)**
Acetic acid	117.9	3.07	16.6	3.90
Benzene	80.1	2.53	5.5	4.90
Carbon disulfide	46.2	2.34	−111.5	3.83
Carbon tetrachloride	76.5	5.03	−23	30.
Chloroform	61.7	3.63	−63.5	4.70
Diethyl ether	34.5	2.02	−116.2	1.79
Ethanol	78.5	1.22	−117.3	1.99
Water	100.0	0.512	0.0	1.86

*At 1 atm.

Freezing Point Depression As you just saw, only solvent molecules can vaporize from the solution, so molecules of the nonvolatile solute are left behind. Similarly, in many cases, *only solvent molecules can solidify,* again leaving solute molecules behind to form a slightly more concentrated solution. The freezing point of a solution is that temperature at which its vapor pressure equals that of the pure solvent. At this temperature, the two phases—solid solvent and liquid solution—are in equilibrium. Because the vapor pressure of the solution is lower than that of the solvent at any temperature, the solution freezes at a lower temperature than the solvent. In other words, the numbers of solvent particles leaving and entering the solid per unit time become equal at a lower temperature. The **freezing point depression (ΔT_f)** is shown in Figure 13.27; note that the solid-liquid line for the solution lies to the left of that for the pure solvent at any pressure.

Like ΔT_b, the freezing point depression has a magnitude proportional to the molal concentration of solute:

$$\Delta T_f \propto m \quad \text{or} \quad \Delta T_f = K_f m \qquad \text{(13.12)}$$

where K_f is the *molal freezing point depression constant,* which also has units of °C/m (see Table 13.6). Also like ΔT_b, ΔT_f is considered a positive value, so we subtract the lower temperature from the higher; in this case, however, it is the solution T_f from the solvent T_f:

$$\Delta T_f = T_{f(solvent)} - T_{f(solution)}$$

Here, too, the overall effect in aqueous solution is quite small because the K_f value for water is small—only 1.86°C/m. Thus, 1 m glucose, 0.5 m NaCl, and 0.33 m K$_2$SO$_4$, all solutions with 1 mol of particles per kilogram of water, freeze at −1.86°C at 1 atm instead of at 0.00°C.

SAMPLE PROBLEM 13.7   Determining the Boiling Point Elevation and Freezing Point Depression of a Solution

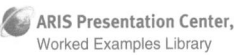
ARIS Presentation Center,
Worked Examples Library

PROBLEM You add 1.00 kg of ethylene glycol (C$_2$H$_6$O$_2$) antifreeze to your car radiator, which contains 4450 g of water. What are the boiling and freezing points of the solution?
PLAN To find the boiling and freezing points of the solution, we first find the molality by converting the given mass of solute (1.00 kg) to amount (mol) and dividing by mass of solvent (4450 g). Then we calculate ΔT_b and ΔT_f from Equations 13.11 and 13.12 (using constants from Table 13.6). We add ΔT_b to the solvent boiling point and subtract ΔT_f from the solvent freezing point. The roadmap shows the steps.
SOLUTION Calculating the molality:

$$\text{Moles of C}_2\text{H}_6\text{O}_2 = 1.00 \text{ kg C}_2\text{H}_6\text{O}_2 \times \frac{10^3 \text{ g}}{1 \text{ kg}} \times \frac{1 \text{ mol C}_2\text{H}_6\text{O}_2}{62.07 \text{ g C}_2\text{H}_6\text{O}_2} = 16.1 \text{ mol C}_2\text{H}_6\text{O}_2$$

$$\text{Molality} = \frac{\text{mol solute}}{\text{kg solvent}} = \frac{16.1 \text{ mol C}_2\text{H}_6\text{O}_2}{4450 \text{ g H}_2\text{O} \times \frac{1 \text{ kg}}{10^3 \text{ g}}} = 3.62 \text{ } m \text{ C}_2\text{H}_6\text{O}_2$$

Finding the boiling point elevation and $T_{b(solution)}$, with $K_b = 0.512$°C/m:

$$\Delta T_b = \frac{0.512\text{°C}}{m} \times 3.62 \text{ } m = 1.85\text{°C}$$

$$T_{b(solution)} = T_{b(solvent)} + \Delta T_b = 100.00\text{°C} + 1.85\text{°C} = \boxed{101.85\text{°C}}$$

Finding the freezing point depression and $T_{f(solution)}$, with $K_f = 1.86$°C/m:

$$\Delta T_f = \frac{1.86\text{°C}}{m} \times 3.62 \text{ } m = 6.73\text{°C}$$

$$T_{f(solution)} = T_{f(solvent)} - \Delta T_f = 0.00\text{°C} - 6.73\text{°C} = \boxed{-6.73\text{°C}}$$

CHECK The changes in boiling and freezing points should be in the same proportion as the constants used. That is, $\Delta T_b/\Delta T_f$ should equal K_b/K_f: 1.85/6.73 = 0.275 = 0.512/1.86.
COMMENT These answers are only approximate because the concentration far exceeds that of a *dilute* solution, for which Raoult's law is most useful.

Mass (g) of solute

divide by $\mathcal{M}$ (g/mol)

Amount (mol) of solute

divide by kg of solvent

Molality (m)

$\Delta T_b = K_b m$ $\Delta T_f = K_f m$

ΔT_b ΔT_f

add *to* $T_{b(solvent)}$ subtract *from* $T_{f(solvent)}$

$T_{b(solution)}$ $T_{f(solution)}$

Osmotic Pressure The fourth colligative property appears when two solutions of different concentrations are separated by a **semipermeable membrane,** one that allows solvent, but *not* solute, molecules to pass through. This phenomenon is called **osmosis.** Many organisms have semipermeable membranes that regulate internal concentrations by osmosis, and synthetic polymers are studied by osmosis.

Consider a simple apparatus in which a semipermeable membrane lies at the curve of a U tube and separates an aqueous sugar solution from pure water. The membrane allows water molecules to pass in *either* direction, but not the larger sugar molecules. Because the solute molecules are present, fewer water molecules touch the membrane on the solution side, so fewer of them leave the solution in a given time than enter it (Figure 13.28A). This *net flow of water into the solution* increases the volume of the solution and thus decreases its concentration.

As the height of the solution rises and that of the solvent falls, the resulting pressure difference pushes some water molecules *from* the solution back through the membrane. At equilibrium, water is pushed out of the solution at the same rate it enters (Figure 13.28B). The pressure difference at equilibrium is the **osmotic pressure (Π),** which is defined as the applied pressure required to *prevent* the net movement of water from solvent to solution (Figure 13.28C).

The osmotic pressure is proportional to the number of solute particles in a given *volume* of solution, that is, to the molarity (*M*):

$$\Pi \propto \frac{n_{solute}}{V_{soln}} \qquad \text{or} \qquad \Pi \propto M$$

The proportionality constant is *R* times the absolute temperature *T*. Thus,

$$\Pi = \frac{n_{solute}}{V_{soln}} RT = MRT \qquad (13.13)$$

The similarity of Equation 13.13 to the ideal gas law ($P = nRT/V$) is not surprising, because both relate the pressure of a system to its concentration and temperature. The Gallery shows some important applications of colligative properties.

Figure 13.28 The development of osmotic pressure. A, In the process of osmosis, a solution and a solvent (or solutions of different concentrations) are separated by a semipermeable membrane, which allows only solvent molecules to pass through. The molecular-scale view (*below*) shows that more solvent molecules enter the solution than leave it in a given time. **B,** As a result, the solution volume increases, so its concentration decreases. At equilibrium, the difference in heights in the two compartments reflects the *osmotic pressure* (Π). The greater height in the solution compartment exerts a backward pressure that eventually equalizes the flow of solvent in both directions. **C,** Osmotic pressure is defined as the applied pressure required to *prevent* this volume change.

Colligative Properties in Industry and Biology

Colligative properties—especially freezing point depression and osmotic pressure—have much practical relevance to everyday life. Some common applications of freezing point depression make life safer during cold winter months. Others are essential to the electronics industry. Applications of osmotic pressure are found throughout nature and in the health and biological sciences—without question, the most important semipermeable membranes surround living cells.

Applications of Freezing Point Depression

Plane de-icing and car antifreezing
Ethylene glycol ($C_2H_6O_2$) is miscible with water through extensive hydrogen bonding and has a high enough boiling point for it to be essentially nonvolatile at 100°C. It is the major ingredient in airplane "de-icers." In "year-round" automobile antifreeze, it lowers the freezing point of water in the radiator in the winter and raises its boiling point in the summer.

ARIS Presentation Center, Photo Library

🧬 Biological application

Biological antifreeze To survive in the Arctic and in northern winters, many fish and insects, including the common housefly, produce large amounts of glycerol ($C_3H_8O_3$)—a substance with a structure very similar to that of ethylene glycol and also miscible with water—which lowers the freezing point of their blood.

Salts for slippery streets Highway crews use salts, such as mixtures of NaCl and $CaCl_2$, to melt ice on streets. A small amount of salt dissolves in the ice by lowering its freezing point and melting it. More salt dissolves, more ice melts, and so forth. An advantage of $CaCl_2$ is that it has a highly negative ΔH_{soln}, so heat is released when it dissolves, which melts more ice.

Engineering application

Silicon rod with impurities | Zone of molten silicon | Inert gas | Heating coil | Purer silicon

Zone refining
Manufacturing computer chips requires extremely pure starting materials. In the process of *zone refining,* a rod of impure silicon (or other metal or metalloid) is passed slowly through a heating coil in an inert atmosphere, and the first narrow zone of impure solid melts. As the next zone melts, the dissolved impurities from the first zone lower the freezing point of the solution, while the purer solvent refreezes. The process continues, zone by zone, throughout the entire rod. After several passes through the coil, in which each zone's impurities move into the adjacent zone, the refrozen silicon at the end of the rod becomes extremely pure—more than 99.999999%. The ultrapure silicon is then sliced into wafers for the production of computer chips (Section 12.7).

(continued)

Applications of Osmotic Pressure

Controlling cell shape The word *tonicity* refers to the tone, or firmness of a biological cell. An *isotonic* solution has the same concentration of particles as in the cell fluid, so water enters the cell at the same rate that it leaves, thereby maintaining the cell's normal shape.

To study cell contents, biochemists rupture membranes by placing the cell in a *hypotonic* solution, one that has a lower concentration of solute particles than the cell fluid. As a result, water enters the cell faster than it leaves, causing the cell to expand and burst.

In a *hypertonic* solution, one that contains a higher concentration of solute particles than the cell fluid, the cell shrinks from the net outward flow of water.

ARIS Presentation Center, Photo Library

Isotonic daily care Contact-lens rinses consist of isotonic saline (0.15 *M* NaCl) to prevent any changes in the volume of corneal cells. Solutions for the intravenous delivery of nutrients or drugs are always isotonic.

Hypotonic watering of trees The dissolved substances in tree sap create a more concentrated solution than the surrounding groundwater. Water enters membranes in the roots and rises into the tree, creating an osmotic pressure that can exceed 20 atm in the tallest trees!

Sodium ion: the extracellular osmoregulator Of the four major biological cations—Na^+, K^+, Mg^{2+}, and Ca^{2+}—Na^+ is essential for all animals (which includes you!) to regulate their fluid volume. The Na^+ ion accounts for more than 90 mol % of all cations *outside* a cell. A high Na^+ concentration draws water out of the cell by osmosis; a low Na^+ concentration leaves more inside. The primary role of Na^+ is to regulate the water volume of the body, and the primary role of the kidneys is to regulate the concentration of Na^+. Changes in blood pressure (volume) activate nerves and hormones to adjust blood flow and alter kidney function.

Hypertonic food preservation Before refrigeration was common, salt was used as a preservative. The salt causes microbes on the food's surface to shrivel and die from loss of water. (Salt was so highly prized for this purpose that Roman soldiers were paid in salt, from which practice comes the word *salary*.) In 1772, Captain James Cook wisely brought on board his ship large amounts of cabbage as food stores and preserved it with salt, thus converting the perishable vegetable into long-lived sauerkraut. The high vitamin-C content of the pickled cabbage prevented the debilitating effects of scurvy and allowed Cook's crew to continue explorations and partake in experiments that would revolutionize longitude measurements and, thus, oceangoing navigation.

The Underlying Theme of Colligative Properties A common theme runs through our explanations of the four colligative properties of nonvolatile solutes. Each property rests on the inability of solute particles to cross between two phases. Solute particles cannot enter the gas phase, which leads to vapor pressure lowering and boiling point elevation. They cannot enter the solid phase, which leads to freezing point depression. They cannot cross a semipermeable membrane, which leads to the development of osmotic pressure. The presence of solute decreases the mole fraction of solvent, which lowers the number of solvent particles leaving the solution per unit time; this lowering requires an adjustment to reach equilibrium again. This adjustment to reach the new balance in numbers of particles crossing between two phases per unit time results in the measured colligative property.

Using Colligative Properties to Find Solute Molar Mass

Each colligative property relates concentration to some measurable quantity—the number of degrees the freezing point is lowered, the magnitude of osmotic pressure created, and so forth. From these measurements, we can determine the amount (mol) of solute particles and, for a known mass of solute, the molar mass of the solute as well.

In principle, any of the colligative properties can be used to find the solute's molar mass, but in practice, some systems provide more accurate data than others. For example, to determine the molar mass of an unknown solute by freezing point depression, you would select a solvent with as large a molal freezing point depression constant as possible (see Table 13.6). If the solute is soluble in acetic acid, for instance, a 1 m concentration of it depresses the freezing point of acetic acid by 3.90°C, more than twice the change in water (1.86°C).

Of the four colligative properties, osmotic pressure creates the largest changes and therefore the most precise measurements. Biological and polymer chemists estimate molar masses as great as 10^5 g/mol by measuring osmotic pressure. Because only a tiny fraction of a mole of a macromolecular solute dissolves, it would create too small a change in the other colligative properties.

SAMPLE PROBLEM 13.8 Determining Molar Mass from Osmotic Pressure

PROBLEM Biochemists have discovered more than 400 mutant varieties of hemoglobin, the blood protein that carries oxygen throughout the body. A physician studying a variety associated with a fatal disease first finds its molar mass ($\mathcal{M}$). She dissolves 21.5 mg of the protein in water at 5.0°C to make 1.50 mL of solution and measures an osmotic pressure of 3.61 torr. What is the molar mass of this variety of hemoglobin?

PLAN We know the osmotic pressure ($\Pi = 3.61$ torr), R, and T (5.0°C). We convert Π from torr to atm, and T from °C to K, and then use Equation 13.13 to solve for molarity (M). Then we calculate the amount (mol) of hemoglobin from the known volume (1.50 mL) and use the known mass (21.5 mg) to find $\mathcal{M}$.

SOLUTION Combining unit conversion steps and solving for molarity from Equation 13.13:

$$M = \frac{\Pi}{RT} = \frac{\dfrac{3.61 \text{ torr}}{760 \text{ torr/1 atm}}}{\left(0.0821 \dfrac{\text{atm·L}}{\text{mol·K}}\right)(273.15 \text{ K} + 5.0)} = 2.08 \times 10^{-4}\ M$$

Finding amount (mol) of solute (after changing mL to L):

$$\text{Moles of solute} = M \times V = \frac{2.08 \times 10^{-4} \text{ mol}}{1 \text{ L soln}} \times 0.00150 \text{ L soln} = 3.12 \times 10^{-7} \text{ mol}$$

Ⅱ (atm)

$M = \Pi/RT$

M (mol/L)

multiply by volume (L) of solution

Amount (mol) of solute

divide *into* mass (g) of solute

𝓜 (g/mol)

 Biological application

Calculating molar mass of hemoglobin (after changing mg to g):

$$\mathcal{M} = \frac{0.0215 \text{ g}}{3.12 \times 10^{-7} \text{ mol}} = 6.89 \times 10^4 \text{ g/mol}$$

CHECK The answers seem reasonable: The small osmotic pressure implies a very low molarity. Hemoglobin is a protein, a biological macromolecule, so we expect a small number of moles [$(\sim 2 \times 10^{-4}$ mol/L) $(1.5 \times 10^{-3}$ L) $= 3 \times 10^{-7}$ mol] and a high molar mass ($\sim 21 \times 10^{-3}$ g/3×10^{-7} mol $= 7 \times 10^4$ g/mol).

FOLLOW-UP PROBLEM 13.8 A 0.30 M solution of sucrose that is at 37°C has approximately the same osmotic pressure as blood does. What is the osmotic pressure of blood?

Figure 13.29 The process of fractional distillation. In the laboratory, a solution of two or more volatile components is attached to a *fractionating column* that is packed with glass beads and connected to a condenser. As the solution is heated, the vapor mixture rises and condenses repeatedly on the beads, each time forming a liquid enriched in the more volatile component. The vapor reaching the condenser consists of the more volatile component only. In industry, this process is used to separate petroleum into many products. A 30-m high fractionating tower can separate components that differ by a few tenths of a degree in their boiling points. (The illustration is a simplified version of a multipart process.)

Engineering application

Colligative Properties of Volatile Nonelectrolyte Solutions

What is the effect on vapor pressure when the solute *is* volatile, that is, when the vapor consists of solute *and* solvent molecules? From Raoult's law (Equation 13.9), we know that

$$P_{\text{solvent}} = X_{\text{solvent}} \times P^{\circ}_{\text{solvent}} \quad \text{and} \quad P_{\text{solute}} = X_{\text{solute}} \times P^{\circ}_{\text{solute}}$$

where X_{solvent} and X_{solute} refer to the mole fractions in the liquid phase. According to Dalton's law of partial pressures, the total vapor pressure is the sum of the partial vapor pressures:

$$P_{\text{total}} = P_{\text{solvent}} + P_{\text{solute}} = (X_{\text{solvent}} \times P^{\circ}_{\text{solvent}}) + (X_{\text{solute}} \times P^{\circ}_{\text{solute}})$$

Just as a nonvolatile solute lowers the vapor pressure of the solvent by making the mole fraction of the solvent less than 1, *the presence of each volatile component lowers the vapor pressure of the other* by making each mole fraction less than 1.

Let's examine this effect in a solution that contains equal amounts (mol) of benzene (C_6H_6) and toluene (C_7H_8): $X_{\text{ben}} = X_{\text{tol}} = 0.500$. At 25°C, the vapor pressure of pure benzene (P°_{ben}) is 95.1 torr and that of pure toluene (P°_{tol}) is 28.4 torr; note that benzene is more volatile than toluene. We find the partial pressures from Raoult's law:

$$P_{\text{ben}} = X_{\text{ben}} \times P^{\circ}_{\text{ben}} = 0.500 \times 95.1 \text{ torr} = 47.6 \text{ torr}$$

$$P_{\text{tol}} = X_{\text{tol}} \times P^{\circ}_{\text{tol}} = 0.500 \times 28.4 \text{ torr} = 14.2 \text{ torr}$$

As you can see, the presence of benzene lowers the vapor pressure of toluene, and vice versa.

Does the composition of the vapor differ from that of the solution? To see, let's calculate the mole fraction of each substance *in the vapor* by applying Dalton's law. Recall from Section 5.4 that $X_A = P_A / P_{\text{total}}$. Therefore, for benzene and toluene in the vapor,

$$X_{\text{ben}} = \frac{P_{\text{ben}}}{P_{\text{total}}} = \frac{47.6 \text{ torr}}{47.6 \text{ torr} + 14.2 \text{ torr}} = 0.770$$

$$X_{\text{tol}} = \frac{P_{\text{tol}}}{P_{\text{total}}} = \frac{14.2 \text{ torr}}{47.6 \text{ torr} + 14.2 \text{ torr}} = 0.230$$

The vapor composition is very different from the solution composition. The essential point to notice is that *the vapor has a higher mole fraction of the **more** volatile solution component.* The 50/50 ratio of benzene to toluene in the liquid created a 77/23 ratio of benzene to toluene in the vapor. Condense this vapor into a separate container, and that new *solution* would have this 77/23 composition, and the new *vapor* above it would be enriched still further in benzene.

In the process of **fractional distillation,** this phenomenon is used to separate a mixture of volatile components. Numerous vaporization-condensation steps continually enrich the vapor, until the vapor reaching the top of the fractionating column consists solely of the most volatile component. Figure 13.29 shows how fractional distillation is used in the industrial process of petroleum refining to separate the hundreds of individual compounds in crude oil into a small number of "fractions" based on boiling point range.

Colligative Properties of Strong Electrolyte Solutions

When we consider colligative properties of strong electrolyte solutions, the solute formula tells us the number of particles. For instance, the boiling point elevation (ΔT_b) of 0.050 m NaCl should be twice that of 0.050 m glucose ($C_6H_{12}O_6$), because NaCl dissociates into two particles per formula unit. Thus, we include a multiplying factor in the equations for the colligative properties of electrolyte solutions. The *van't Hoff factor (i)*, named after the Dutch chemist Jacobus van't Hoff (1852–1911), is the ratio of the *measured* value of the colligative property in the electrolyte solution to the *expected* value for a nonelectrolyte solution:

$$i = \frac{\text{measured value for electrolyte solution}}{\text{expected value for nonelectrolyte solution}}$$

To calculate the colligative properties of strong electrolyte solutions, we incorporate the van't Hoff factor into the equation:

For vapor pressure lowering: $\Delta P = i(X_{\text{solute}} \times P^{\circ}_{\text{solvent}})$

For boiling point elevation: $\Delta T_b = i(K_b m)$

For freezing point depression: $\Delta T_f = i(K_f m)$

For osmotic pressure: $\Pi = i(MRT)$

If strong electrolyte solutions behaved ideally, the factor i would be the amount (mol) of particles in solution divided by the amount (mol) of dissolved solute; that is, i would be 2 for NaCl, 3 for $Mg(NO_3)_2$, and so forth. Careful experiment shows, however, that *most strong electrolyte solutions are **not** ideal*. For example, comparing the boiling point elevation for 0.050 m NaCl solution with that for 0.050 m glucose solution gives a factor i of 1.9, not 2.0:

$$i = \frac{\Delta T_b \text{ of 0.050 } m \text{ NaCl}}{\Delta T_b \text{ of 0.050 } m \text{ glucose}} = \frac{0.049^{\circ}C}{0.026^{\circ}C} = 1.9$$

The measured value of the van't Hoff factor is typically *lower* than that expected from the formula. This deviation implies that the ions are not behaving as independent particles. However, we know from other evidence that soluble salts dissociate completely into ions. The fact that the deviation is greater with multiply charged ions is a strong indication that the ionic charge is somehow involved (Figure 13.30).

To explain this nonideal behavior, we picture ions as separate but near each other. Clustered near a positive ion are, on average, more negative ions than positive ones, and vice versa. Figure 13.31 shows each ion surrounded by an **ionic atmosphere** of net opposite charge. Through these electrostatic associations, each type of ion behaves as if it were "tied up," so its concentration seems *lower* than it actually is. Thus, we often speak of an *effective* concentration, obtained by multiplying i by the *stoichiometric* concentration based on the formula. The greater the charge, the stronger the electrostatic associations; so the deviation from ideal behavior is greater for compounds that dissociate into multiply charged ions.

At ordinary conditions and concentrations, nonideal behavior of solutions is much more common (and the deviations much larger) than nonideal behavior of gases, because the particles in solutions are so much closer together. Nevertheless, the two systems exhibit some interesting similarities. Gases display nearly ideal behavior at low pressures because the distances between particles are large. Similarly, van't Hoff factors (i) approach their ideal values as the solution becomes more dilute, that is, as the distance between ions increases. In gases, attractions between particles cause deviations from the expected pressure. In solutions,

Figure 13.30 Nonideal behavior of strong electrolyte solutions. The van't Hoff factors *(i)* for various ionic solutes in dilute (0.05 m) aqueous solution show that the observed value *(dark blue)* is always *lower* than the expected value *(light blue)*. This deviation is due to ionic interactions that, in effect, reduce the number of free ions in solution. The deviation is greatest for multiply charged ions.

Transparency/ARIS Presentation Center, Art Library

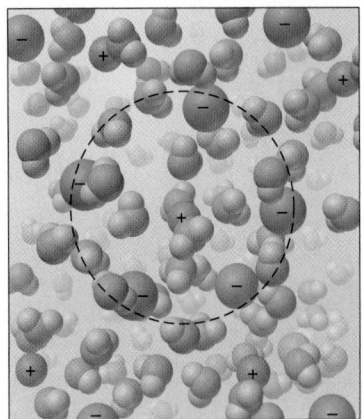

Figure 13.31 An ionic atmosphere model for nonideal behavior of electrolyte solutions. Hydrated anions cluster near cations, and vice versa, to form ionic atmospheres of net opposite charge. Because the ions do not act independently, their concentrations are effectively *less* than expected. Such interactions cause deviations from ideal behavior.

Transparency/ARIS Presentation Center, Art Library

attractions between particles cause deviations from the expected size of a collative property. Finally, for both real gases and real solutions, we use empirically determined numbers (van der Waals constants or van't Hoff factors) to transform theories (the ideal gas law or Raoult's law) into more useful relations.

Transparency/ARIS Presentation Center, Worked Examples Library

SAMPLE PROBLEM 13.9 Finding Colligative Properties from Molecular Scenes

PROBLEM A 0.952-g sample of magnesium chloride is dissolved in 100. g of water in a flask.
(a) Which scene depicts the solution best?

(b) What is the amount (mol) represented by each green sphere?
(c) Assuming the solution is ideal, what is its freezing point (at 1 atm)?
PLAN (a) From the name, we recognize an ionic compound, so we determine the formula to find the numbers of cations and anions per formula unit and compare this result with the three scenes: there is 1 magnesium ion for every 2 chloride ions. **(b)** From the given mass of solute, we find the amount (mol); from part (a), there are twice as many moles of chloride ions (green spheres). Dividing by the total number of green spheres gives the moles/sphere. **(c)** From the moles of solute and the given mass (kg) of water, we find the molality (m). We use K_f for water from Table 13.6 and multiply by m to get ΔT_f, and then subtract that from 0.000°C to get the solution freezing point.
SOLUTION (a) The formula is $MgCl_2$; only scene A has 1 Mg^{2+} for every 2 Cl^-.

(b)
$$\text{Moles of } MgCl_2 = \frac{0.952 \text{ g } MgCl_2}{95.21 \text{ g/mol } MgCl_2} = 0.0100 \text{ mol } MgCl_2$$

Therefore,
$$\text{Moles of } Cl^- = 0.0100 \text{ mol } MgCl_2 \times \frac{2 \text{ Cl}^-}{1 \text{ } MgCl_2} = 0.0200 \text{ mol } Cl^-$$

$$\text{Moles/sphere} = \frac{0.0200 \text{ mol } Cl^-}{8 \text{ spheres}} = 2.50 \times 10^{-3} \text{ mol/sphere}$$

(c)
$$\text{Molality } (m) = \frac{\text{mol of solute}}{\text{kg of solvent}} = \frac{0.0100 \text{ mol } MgCl_2}{100. \text{ g} \times \dfrac{1 \text{ kg}}{1000 \text{ g}}} = 0.100 \text{ } m \text{ } MgCl_2$$

Assuming an ideal solution, the van't Hoff factor, i, is 3 for $MgCl_2$ because there are 3 ions per formula unit, so we have

$$\Delta T_f = i(K_f m) = 3(1.86°C/m \times 0.100 \text{ } m) = 0.558°C$$
and
$$T_f = 0.000°C - 0.558°C = -0.558°C$$

CHECK Let's quickly check part (c): We have 0.01 mol dissolved in 0.1 kg, which gives 0.1 m. Then, rounding K_f, we have about 3(2°C/m × 0.1 m) = 0.6°C.

FOLLOW-UP PROBLEM 13.9 The $MgCl_2$ solution in the sample problem has a density of 1.006 g/mL at 20.0°C. **(a)** What is the osmotic pressure of the solution? **(b)** A U-tube fitted with a semipermeable membrane is filled with this $MgCl_2$ solution in the left arm and a glucose solution of equal molarity in the right arm. After time, which scene depicts the U-tube best?

Section Summary

Colligative properties are related to the number of dissolved solute particles, not their chemical nature. • Compared with the pure solvent, a solution of a nonvolatile non-electrolyte has a lower vapor pressure (Raoult's law), an elevated boiling point, a depressed freezing point, and an osmotic pressure. • Colligative properties can be used to determine the solute molar mass. • When solute *and* solvent are volatile, the vapor pressure of each is lowered by the presence of the other. The vapor pressure of the more volatile component is always higher. • Electrolyte solutions exhibit non-ideal behavior because ionic interactions reduce the effective concentration of the ions.

13.7 THE STRUCTURE AND PROPERTIES OF COLLOIDS

Stir a handful of fine sand into a glass of water, and you'll see the sand particles are suspended at first but gradually settle to the bottom. Sand in water is a **suspension,** a *heterogeneous* mixture containing particles large enough to be seen by the naked eye and clearly distinct from the surrounding fluid. In contrast, stirring sugar into water forms a solution, a *homogeneous* mixture in which the particles are individual molecules distributed evenly throughout the surrounding fluid.

Between the extremes of suspensions and solutions is a large group of mixtures called *colloidal dispersions,* or **colloids,** in which a dispersed (solute-like) substance is distributed throughout a dispersing (solvent-like) substance. Colloidal particles are larger than simple molecules but small enough to remain distributed and not settle out. They range in diameter from 1 to 1000 nm (10^{-9} to 10^{-6} m). A colloidal particle may consist of a single macromolecule (such as a protein or synthetic polymer) or an aggregate of many atoms, ions, or molecules. Whatever their composition, colloidal particles have an enormous total surface area as a result of their small size. A cube with 1-cm sides has a total surface area of 6 cm^2. If it's divided equally into 10^{12} cubes, the cubes are the size of large colloidal particles and have a total surface area of 60,000 cm^2, or 6 m^2. This enormous surface area allows many more interactions to exert a great total adhesive force, which attracts other particles and leads to the practical uses of colloids.

Colloids are classified in Table 13.7 according to whether the dispersed and dispersing substances are gases, liquids, or solids. Many familiar commercial products and natural objects are colloids. Whipped cream is a *foam,* a gas dispersed in a liquid. Firefighting foams are liquid mixtures of proteins in water that are made frothy with fine jets of air. Most biological fluids are aqueous *sols,* solids dispersed in water. Within a typical cell, proteins and nucleic acids are colloidal-size particles dispersed in an aqueous solution of ions and small molecules. The action of soaps and detergents occurs by the formation of an *emulsion,* a liquid (soap dissolved in grease) dispersed in another liquid (water). ●

● **"Soaps" in Your Small Intestine** In the small intestine, fats are digested by *bile salts,* which are soaplike molecules with smaller polar-ionic and larger nonpolar portions. Secreted by the liver, stored in the gallbladder, and released in the intestine, bile salts emulsify fats just as soap emulsifies grease: Fatty aggregates are broken down into particles of colloidal size and dispersed in the watery fluid. The fats are then broken down further and transported into the blood for cellular metabolism.

Biological application

Transparency/ARIS Presentation Center, Tables Library

Table 13.7   Types of Colloids

Colloid Type	Dispersed Substance	Dispersing Medium	Example(s)
Aerosol	Liquid	Gas	Fog
Aerosol	Solid	Gas	Smoke
Foam	Gas	Liquid	Whipped cream
Solid foam	Gas	Solid	Marshmallow
Emulsion	Liquid	Liquid	Milk
Solid emulsion	Liquid	Solid	Butter
Sol	Solid	Liquid	Paint, cell fluid
Solid sol	Solid	Solid	Opal

B

Figure 13.32 Light scattering and the Tyndall effect. A, When a beam of light passes through a solution (*left jar*), its path remains narrow and barely visible. When it passes through a colloid (*right jar*), it is scattered and broadened by the particles and thus easily visible. **B,** Sunlight is scattered as it shines through misty air in a forest.

 ARIS Presentation Center, Photo Library

Most colloids are cloudy or opaque, but some are transparent to the naked eye. When light passes through a colloid, it is scattered randomly by the dispersed particles because their sizes are similar to the wavelengths of visible light (400 to 750 nm). Viewed from the side, the scattered light beam is visibly broader than one passing through a solution. This light-scattering phenomenon is called the **Tyndall effect** (Figure 13.32). Dust in air displays this effect when sunlight shines through it, as does mist pierced by headlights at night.

Under low magnification, you can see colloidal particles exhibit *Brownian motion,* a characteristic movement in which the particles change speed and direction erratically. This motion results because the colloidal particles are being pushed this way and that by molecules of the dispersing medium. These collisions are primarily responsible for keeping colloidal particles from settling out. (Einstein's explanation of Brownian motion in 1905 was a principal factor in the acceptance of the molecular nature of matter.)

When colloidal particles collide, why don't they aggregate into larger particles and settle out? Interparticle forces provide the explanation. Water-dispersed colloidal particles remain dispersed because they have charged surfaces that interact strongly with the water molecules through ion-dipole forces. Soap molecules form spherical *micelles,* with the charged heads forming the micelle exterior and the nonpolar tails interacting via dispersion forces in the interior. Aqueous proteins are typically spherical and mimic this micellar arrangement, with charged amino acid groups facing the water and uncharged groups buried within the molecule. Nonpolar oily particles can be dispersed in water by introducing ions, which are adsorbed onto their surfaces by dispersion forces. Charge repulsions between the adsorbed ions prevent the particles from aggregating.

Despite these forces, various methods can coagulate the particles and "destroy" the colloid. Heating a colloid makes the particles move faster and collide more often and with enough force to coalesce into heavier particles that settle out. Adding an electrolyte solution introduces oppositely charged ions that neutralize the particles' surface charges, which allows the particles to coagulate and settle. Uncharged colloidal particles in smokestack gases are removed by creating ions that become adsorbed on the particles, which are then attracted to the charged plates of a Cottrell precipitator. Such precipitators are installed in the smokestacks of coal-burning power plants (Figure 13.33) and collect about 90% of colloidal smoke particulates, thus preventing their release into the air.

The upcoming Chemical Connections essay applies solution and colloid chemistry to the purification of water for residential and industrial use.

Figure 13.33 A Cottrell precipitator for removing particulates from industrial smokestack gases.

 ARIS Presentation Center, Art Library

① *Hot gases pass through discharge area and some molecules are ionized*

② *Gaseous ions are adsorbed by suspended smoke particles*

Outlet for clean gases

③ *Charged smoke particles are attracted to oppositely charged electrode*

Plate electrode

④ *Charged particles are neutralized, fall, and are discarded*

High DC voltage

Wire

Smoke inlet

Solutions and Colloids in Water Purification

After decades of pollution and waste, we've finally begun to treasure clean water as a priceless, limited resource. Most water destined for human use comes from lakes, rivers, reservoirs, or groundwater sources that may also serve as the final sink after the water is used. Dissolved organic compounds, some of which are toxic, may be present, as well as high concentrations of mineral ions, such as NO_3^- and Fe^{3+}. Clay and microbes are dispersed as colloids, and larger particles and debris are present in suspension. Water must be treated to remove these dissolved, dispersed, and suspended particles to make it clean enough to drink.

Water Treatment Plants

Treating water for human use involves these steps (Figure B13.1):

Step 1. Screening and settling. As water moves into a treatment facility, screens remove large debris and then gravitational settling removes small, denser particles, such as sand.

Step 2. Coagulating. A series of processes must then remove microorganisms and other colloids. These particles typically have negatively charged surfaces, so they repel each other. Coagulating of colloids occurs by adding aluminum sulfate [cake alum; $Al_2(SO_4)_3$] or iron(III) chloride ($FeCl_3$), which supply triply charged cations Al^{3+} or Fe^{3+} to neutralize the surface charges and allow the particles to aggregate through van der Waals forces.

Step 3. Flocculating and sedimenting. In these processes, the water and flocculating agents are carefully mixed in large basins using paddles, which causes a fluffy *floc* to form. Specific cationic polymers added at this point serve as long-chain bridges between floc particles, which then grow bigger and heavier. These particles flow into other basins where they settle out and are removed. Instead of sedimentation, some plants use *dissolved air flotation* (DAF), in which fine air bubbles forced up through the water attach to the floc, and the floating mass is then skimmed.

Step 4. Filtering. Several types of filters are used to remove remaining colloidal particles. In *slow sand filters*, the water passes through a bed of sand and/or gravel of increasing particle size. In *rapid sand filters*, the sand is backwashed by passing water up though it, and the colloidal mass is removed. Membrane filters (not shown in Figure B13.1) with pore sizes of 0.1–10 μm occur as thin tubes bundled together inside a vessel. The water is forced into the tubes and the colloid-free filtrate is collected from a large central core tube. Filtration is very effective at removing microorganisms that are resistant to disinfection.

Step 5. Disinfecting. Nearly all water sources contain harmful microorganisms, and these are killed by one of three disinfecting agents. Chlorine, either as a bleach solution (ClO^-) or as Cl_2 gas, is the most common. But chlorine reacts with dissolved organic compounds to form chlorinated compounds, some of which are carcinogenic. Another agent is UV light emitted by specialized high-intensity fluorescent tubes. It disinfects by disrupting bonds in microorganisms' DNA. A third common disinfectant is ozone (O_3) gas, a powerful oxidizing agent. Often a small amount of Cl_2 is added after UV or O_3 treatment to prevent contamination in the storage tank. Following disinfection, other compounds may be added, such as sodium fluoride (NaF) to prevent tooth decay and phosphate salts to prevent leaching of lead from pipes.

Step 6. Adsorbing onto granular activated carbon (GAC). A common final step in water treatment is the removal of petroleum and other dissolved organic contaminants by adsorption onto GAC. This highly porous material is formed by "activating" wood, coal, or coconut shells through treatment with steam: 1 kg of GAC has a surface area of about 275 acres! When all the active sites are filled, the GAC is discarded or regenerated with steam.

(continued)

Engineering application

Literature Parkinson, Gerald. "Seawater Gets Fresh," *Chem. Eng.*, Vol. 106(3), **1999**, 32–35.

Figure B13.1 The steps in a typical municipal wastewater treatment plant. Before water is sent to users, (1) it is screened to remove large debris, (2) coagulants are added to destabilize colloidal particles, (3) flocculation (slow mixing) is used to cause the particles to aggregate and form fluffy floc, which is removed by sedimentation, (4) filtration is used to remove remaining floc and microorganisms, and (5) chlorine, UV light, or ozone is used to destroy any remaining microorganisms.

Chemical Connections continued

Demo Shakhashiri, Bassam Z. *"Osmosis Through the Membrane of an Egg," Chemical Demonstrations, A Handbook for Teachers of Chem* Vol. 3 (The University of Wisconsin Press, Madison, 1985), 283–285

Water Softening via Ion Exchange

Water that contains large amounts of doubly charged cations, such as Ca^{2+} and Mg^{2+}, is called **hard water.** During cleaning, the cations combine with the fatty-acid anions in soap to produce insoluble deposits on clothes, washing machine parts, and sinks:

$$Ca^{2+}(aq) + 2C_{17}H_{35}COONa(aq) \longrightarrow$$
$$\text{soap} \quad (C_{17}H_{35}COO)_2Ca(s) + 2Na^+(aq)$$
$$\text{insoluble deposit}$$

When a large amount of HCO_3^- is present, the hard-water cations form *scale*, insoluble carbonate deposits in boilers and hot-water pipes that interfere with the transfer of heat and damage plumbing:

$$Ca^{2+}(aq) + 2HCO_3^-(aq) \longrightarrow CaCO_3(s) + CO_2(g) + H_2O(l)$$

The removal of hard-water cations, called **water softening,** is accomplished by exchanging Na^+ ions for Ca^{2+} and Mg^{2+} ions. A typical domestic system for **ion exchange** contains an *ion-exchange resin,* an insoluble polymer that has covalently bonded anion groups, such as $—SO_3^-$ or $—COO^-$, to which Na^+ ions are attached to balance the charges (Figure B13.2). The divalent cations in hard water are attracted to the resin's anionic groups and displace the Na^+ ions into the water. When all the resin sites are occupied, the resin is "regenerated" with a concentrated Na^+ solution, which exchanges Na^+ ions for the bound Ca^{2+} and Mg^{2+}.

ARIS Presentation Center, Art Library

Figure B13.2 Ion exchange for removal of hard-water cations. In a typical ion-exchange resin, negatively charged groups are covalently bonded to resin beads, with Na^+ ions present to neutralize the charges. Hard-water ions, such as Ca^{2+}, exchange with the Na^+ ions, which are displaced into the flowing water.

Membrane Processes and Reverse Osmosis

As we said on p. 541, membranes with pore sizes of 0.1–10 μm are used to filter colloids and microorganisms out of drinking water. In **reverse osmosis,** membranes with pores of 0.0001–0.01 μm are used to remove dissolved ions. Earlier in this chapter (p. 532), you saw how solutions of different concentrations separated by a semipermeable membrane give rise to an osmotic pressure. In reverse osmosis, a pressure *greater* than the osmotic pressure is *applied* to the concentrated solution, forcing the water back through the membrane and filtering out the ions. In domestic water systems, reverse osmosis is used to remove toxic *heavy-metal ions,* such as Pb^{2+}, Cd^{2+}, and Hg^{2+}, present at concentrations too low for removal by ion exchange. On a much larger scale, reverse osmosis is used in the process of **desalination,** which removes large amounts of ions from seawater (Figure B13.3). Seawater containing about 40,000 ppm of dissolved salts can be converted to a solution of about 400 ppm, which is suitable for drinking.

Wastewater Treatment

Used water, or **wastewater,** must be treated before being returned to the groundwater, river, or lake. Industrial wastewater may also contain toxic components. In *primary* wastewater treatment, the water enters a settling basin to remove denser particles. Most municipalities also include *biological* wastewater treatment in which bacteria metabolize organic compounds in the wastewater and are then removed by settling. If needed, biological treatment is supplemented by *advanced* wastewater treatment tailored to a specific pollutant. For example, ammonia in wastewater causes excessive growth of plants and algae in surface water. It can be removed through a two-step process. In a *nitrification* step, specific aerobic bacteria use ammonia as a reducing agent (electron donor) and oxygen as an oxidizing agent (electron acceptor):

$$NH_4^+ + 2O_2 \longrightarrow NO_3^- + 2H^+ + H_2O$$

Then, in a *denitrification* step, an organic compound such as methanol (CH_3OH) is added, and other bacteria oxidize it using the NO_3^- formed in the nitrification step:

$$5CH_3OH + 6NO_3^- \longrightarrow 3N_2 + 5CO_2 + 7H_2O + 6OH^-$$

Thus, the overall process converts NH_3 in wastewater to harmless N_2 gas, which is released to the atmosphere.

ARIS Presentation Center, Art Library

Figure B13.3 Reverse osmosis for the removal of ions. **A,** A section of a coiled reverse-osmosis permeator. **B,** Each permeator contains a bundle of thin, hollow fibers of semipermeable membrane. **C,** Seawater is pumped through the permeator at high pressure. Most of the ions are removed, so much purer water enters the fibers and is collected.

Section Summary

Colloidal particles are smaller than those in a suspension and larger than those in a solution. • Colloids are classified by the physical states of the dispersed and dispersing substances and are formed from many combinations of gas, liquid, and solid. • Colloids have extremely large surface areas, scatter incoming light (Tyndall effect), and exhibit random (Brownian) motion. • Colloidal particles in water have charged surfaces that keep them dispersed, but they can be coagulated by heating or by the addition of ions. ●

Chapter Perspective

In Chapter 12, you saw how pure liquids and solids behave, and here you've seen how their behaviors change when they are mixed together. Two features of these systems reappear in later chapters: their equilibrium nature in Chapters 17 to 19 and the importance of entropy in Chapter 20.

This chapter completes our general discussion of atomic and molecular properties and their influence on the properties of matter. Immediately following is an Interchapter feature that reviews these properties in a pictorial format and helps preview their application in the next two chapters. In Chapter 14, we travel through the main groups of the periodic table, applying the ideas you've learned so far to the chemical and physical behavior of the elements. Then, in Chapter 15, we apply them to carbon and its nearest neighbors to see how their properties are responsible for the fascinating world of organic substances, including synthetic polymers and biomolecules.

● **From Colloid to Civilization** At times, civilizations have been born where colloids were being coagulated by electrolyte solutions. At the mouths of rivers, where salt concentrations increase near an ocean or sea, the clay particles dispersed in the river water come together to form muddy deltas, such as those of the Nile (*reddish area in photo*) and the Mississippi. The ancient Egyptian empire and the city of New Orleans are the results of this global colloid chemistry.

ARIS Presentation Center, Photo Library

CHAPTER REVIEW GUIDE

The following sections provide many aids to help you study this chapter. (Numbers in parentheses refer to pages, unless noted otherwise.)

Learning Objectives These are concepts and skills you should know after studying this chapter.

Relevant section and/or sample problem (SP) numbers appear in parentheses.

Understand These Concepts

1. The quantitative meaning of solubility (13.1)
2. The major types of intermolecular forces in solution and their relative strengths (13.1)
3. How the like-dissolves-like rule depends on intermolecular forces (13.1)
4. Why gases have relatively low solubilities in water (13.1)
5. General characteristics of solutions formed by various combinations of gases, liquids, and solids (13.1)
6. How intermolecular forces stabilize the structures of proteins, the cell membrane, DNA, and cellulose (13.2)
7. The enthalpy components of a solution cycle and their effect on ΔH_{soln} (13.3)
8. The dependence of ΔH_{hydr} on ionic charge density and the factors that determine whether ionic solution processes are exothermic or endothermic (13.3)
9. The meaning of entropy and how the balance between the change in enthalpy and the change in entropy governs the solution process (13.3)
10. The distinctions among saturated, unsaturated, and supersaturated solutions, and the equilibrium nature of a saturated solution (13.4)

11. The relation between temperature and the solubility of solids (13.4)
12. Why the solubility of gases in water decreases with a rise in temperature (13.4)
13. The effect of gas pressure on solubility and its quantitative expression as Henry's law (13.4)
14. The meaning of molarity, molality, mole fraction, and parts by mass or by volume of a solution, and how to convert among them (13.5)
15. The distinction between electrolytes and nonelectrolytes in solution (13.6)
16. The four colligative properties and their dependence on number of dissolved particles (13.6)
17. Ideal solutions and the importance of Raoult's law (13.6)
18. How the phase diagram of a solution differs from that of the pure solvent (13.6)
19. Why the vapor over a solution of volatile nonelectrolyte is richer in the more volatile component (13.6)
20. Why electrolyte solutions are not ideal and the meanings of the van't Hoff factor and ionic atmosphere (13.6)
21. How particle size distinguishes suspensions, colloids, and solutions (13.7)
22. How colloidal behavior is demonstrated by the Tyndall effect and Brownian motion (13.7)

Learning Objectives *(continued)*

Master These Skills

1. Predicting relative solubilities from intermolecular forces (SP 13.1)
2. Using Henry's law to calculate the solubility of a gas (SP 13.2)
3. Expressing concentration in terms of molality, parts by mass, parts by volume, and mole fraction (SPs 13.3, 13.4)
4. Interconverting among the various terms for expressing concentration (SP 13.5)
5. Using Raoult's law to calculate the vapor pressure lowering of a solution (SP 13.6)

6. Determining boiling point elevation and freezing point depression of a solution (SP 13.7)
7. Using a colligative property to calculate the molar mass of a solute (SP 13.8)
8. Calculating the composition of vapor over a solution of volatile nonelectrolyte (13.6)
9. Calculating the van't Hoff factor i from the magnitude of a colligative property (13.6)
10. Using a solution depiction to determine colligative properties (SP 13.9)

Key Terms These important terms appear in boldface in the chapter and are defined again in the Glossary.

Section 13.1
solute (502)
solvent (502)
miscible (502)
solubility (*S*) (502)
like-dissolves-like rule (502)
hydration shell (502)
ion–induced dipole force (502)
dipole–induced dipole force (503)
alloy (506)

Section 13.2
protein (507)
amino acid (507)
soap (510)
lipid bilayer (511)
nucleic acid (512)

mononucleotide (512)
double helix (513)
polysaccharide (513)
monosaccharide (513)

Section 13.3
heat of solution (ΔH_{soln}) (515)
solvation (515)
hydration (515)
heat of hydration (ΔH_{hydr}) (516)
charge density (516)
entropy (*S*) (517)

Section 13.4
saturated solution (519)
unsaturated solution (519)
supersaturated solution (519)
Henry's law (522)

Section 13.5
molality (*m*) (523)
mass percent [% (w/w)] (524)
volume percent [% (v/v)] (524)
mole fraction (*X*) (525)

Section 13.6
colligative property (527)
electrolyte (527)
nonelectrolyte (527)
vapor pressure lowering (ΔP) (528)
Raoult's law (528)
ideal solution (528)
boiling point elevation (ΔT_b) (529)
freezing point depression (ΔT_f) (531)

semipermeable membrane (532)
osmosis (532)
osmotic pressure (Π) (532)
fractional distillation (536)
ionic atmosphere (537)

Section 13.7
suspension (539)
colloid (539)
Tyndall effect (540)
hard water (542)
water softening (542)
ion exchange (542)
reverse osmosis (542)
desalination (542)
wastewater (542)

Key Equations and Relationships Numbered and screened concepts are listed for you to refer to or memorize.

13.1 Dividing the general heat of solution into component enthalpies (515):
$$\Delta H_{soln} = \Delta H_{solute} + \Delta H_{solvent} + \Delta H_{mix}$$
13.2 Dividing the heat of solution of an ionic compound in water into component enthalpies (516):
$$\Delta H_{soln} = \Delta H_{lattice} + \Delta H_{hydr\ of\ the\ ions}$$
13.3 Relating gas solubility to its partial pressure (Henry's law) (522):
$$S_{gas} = k_H \times P_{gas}$$
13.4 Defining concentration in terms of molarity (522):
$$\text{Molarity } (M) = \frac{\text{amount (mol) of solute}}{\text{volume (L) of solution}}$$
13.5 Defining concentration in terms of molality (523):
$$\text{Molality } (m) = \frac{\text{amount (mol) of solute}}{\text{mass (kg) of solvent}}$$
13.6 Defining concentration in terms of mass percent (524):
$$\text{Mass percent } [\% \text{ (w/w)}] = \frac{\text{mass of solute}}{\text{mass of solution}} \times 100$$
13.7 Defining concentration in terms of volume percent (524):
$$\text{Volume percent } [\% \text{ (v/v)}] = \frac{\text{volume of solute}}{\text{volume of solution}} \times 100$$

13.8 Defining concentration in terms of mole fraction (525):
Mole fraction (*X*)
$$= \frac{\text{amount (mol) of solute}}{\text{amount (mol) of solute} + \text{amount(mol) of solvent}}$$
13.9 Expressing the relationship between the vapor pressure of solvent above a solution and its mole fraction in the solution (Raoult's law) (528):
$$P_{solvent} = X_{solvent} \times P°_{solvent}$$
13.10 Calculating the vapor pressure lowering due to solute (529):
$$\Delta P = X_{solute} \times P°_{solvent}$$
13.11 Calculating the boiling point elevation of a solution (530):
$$\Delta T_b = K_b m$$
13.12 Calculating the freezing point depression of a solution (531):
$$\Delta T_f = K_f m$$
13.13 Calculating the osmotic pressure of a solution (532):
$$\Pi = \frac{n_{solute}}{V_{soln}} RT = MRT$$

Highlighted Figures and Tables These figures (F) and tables (T) provide a visual review of key ideas.

F13.1 Major types of intermolecular forces in solutions (502)
F13.2 Hydration shells around an aqueous ion (503)
F13.7 The forces that maintain protein structure (509)
F13.10 Intermolecular forces and membrane structure (511)
F13.13 The double helix of DNA (513)
F13.16 Solution cycles and the enthalpy components of ΔH_{soln} (515)
T13.4 Trends in ionic heats of hydration (516)
F13.17 Dissolving ionic compounds in water (516)

F13.19 Equilibrium in a saturated solution (519)
F13.23 Effect of pressure on gas solubility (521)
T13.5 Concentration definitions (523)
F13.26 Effect of the solute on vapor pressure of a solution (528)
F13.27 Phase diagrams of solvent and solution (530)
F13.28 Development of osmotic pressure (532)
F13.30 Nonideal behavior of strong electrolyte solutions (537)
T13.7 Types of colloids (539)

Brief Solutions to FOLLOW-UP PROBLEMS Compare your solutions to these calculation steps and answers.

13.1 (a) 1,4-Butanediol is more soluble in water because it can form more H bonds.
(b) Chloroform is more soluble in water because of dipole-dipole forces.

13.2 $S_{N_2} = (7\times10^{-4}\ \text{mol/L·atm})\ (0.78\ \text{atm})$
$= 5\times10^{-4}\ \text{mol/L}$

13.3 Mass (g) of glucose = 563 g ethanol

$$\times \frac{1\ \text{kg}}{10^3\ \text{g}} \times \frac{2.40\times10^{-2}\ \text{mol glucose}}{1\ \text{kg ethanol}}$$

$$\times \frac{180.16\ \text{g glucose}}{1\ \text{mol glucose}}$$

$$= 2.43\ \text{g glucose}$$

13.4 Mass% $C_3H_7OH = \dfrac{35.0\ \text{g}}{35.0\ \text{g} + 150.\ \text{g}} \times 100 = 18.9\ \text{mass \%}$

Mass % $C_2H_5OH = 100.0 - 18.9 = 81.1$ mass %

$$X_{C_3H_7OH} = \frac{35.0\ \text{g C}_3\text{H}_7\text{OH} \times \dfrac{1\ \text{mol C}_3\text{H}_7\text{OH}}{60.09\ \text{g C}_3\text{H}_7\text{OH}}}{\left(35.0\ \text{g C}_3\text{H}_7\text{OH} \times \dfrac{1\ \text{mol C}_3\text{H}_7\text{OH}}{60.09\ \text{g C}_3\text{H}_7\text{OH}}\right) + \left(150.\ \text{g C}_2\text{H}_5\text{OH} \times \dfrac{1\ \text{mol C}_2\text{H}_5\text{OH}}{46.07\ \text{g C}_2\text{H}_5\text{OH}}\right)} = 0.152$$

$X_{C_2H_5OH} = 1.000 - 0.152 = 0.848$

13.5 Mass % HCl = $\dfrac{\text{mass of HCl}}{\text{mass of soln}} \times 100$

$$= \frac{\dfrac{11.8\ \text{mol HCl}}{1\ \text{L soln}} \times \dfrac{36.46\ \text{g HCl}}{1\ \text{mol HCl}}}{\dfrac{1.190\ \text{g}}{1\ \text{mL soln}} \times \dfrac{10^3\ \text{mL}}{1\ \text{L}}} \times 100$$

$$= 36.2\ \text{mass \% HCl}$$

Mass (kg) of soln = $1\ \text{L soln} \times \dfrac{1.190\times10^{-3}\ \text{kg soln}}{1\times10^{-3}\ \text{L soln}}$

$$= 1.190\ \text{kg soln}$$

Mass (kg) of HCl = $11.8\ \text{mol HCl} \times \dfrac{36.46\ \text{g HCl}}{1\ \text{mol HCl}} \times \dfrac{1\ \text{kg}}{10^3\ \text{g}}$

$$= 0.430\ \text{kg HCl}$$

Molality of HCl = $\dfrac{\text{mol HCl}}{\text{kg water}} = \dfrac{\text{mol HCl}}{\text{kg soln} - \text{kg HCl}}$

$$= \frac{11.8\ \text{mol HCl}}{0.760\ \text{kg H}_2\text{O}} = 15.5\ m\ \text{HCl}$$

$$X_{HCl} = \frac{\text{mol HCl}}{\text{mol HCl} + \text{mol H}_2\text{O}}$$

$$= \frac{11.8\ \text{mol}}{11.8\ \text{mol} + \left(760\ \text{g H}_2\text{O} \times \dfrac{1\ \text{mol}}{18.02\ \text{g H}_2\text{O}}\right)} = 0.219$$

13.6 $\Delta P = X_{aspirin} \times P^\circ_{methanol}$

$$= \frac{\dfrac{2.00\ \text{g}}{180.15\ \text{g/mol}}}{\dfrac{2.00\ \text{g}}{180.15\ \text{g/mol}} + \dfrac{50.0\ \text{g}}{32.04\ \text{g/mol}}} \times 101\ \text{torr}$$

$$= 0.713\ \text{torr}$$

13.7 Molality of $C_2H_6O_2 = \dfrac{(32°F - 0.00°F)\left(\dfrac{5°C}{9°F}\right)}{1.86°C/m} = 9.56\ m$

13.8 $\Pi = MRT = (0.30\ \text{mol/L})\left(0.0821\ \dfrac{\text{atm·L}}{\text{mol·K}}\right)(37°C + 273.15)$
$= 7.6\ \text{atm}$

13.9 (a) Mass of 0.100 m solution = 1 kg water + 0.100 mol MgCl$_2$
$= 1000\ \text{g} + 9.52\ \text{g} = 1009.52\ \text{g}$

Volume of solution = $1009.52\ \text{g} \times \dfrac{1\ \text{mL}}{1.006\ \text{g}} = 1003\ \text{mL}$

Molarity = $\dfrac{9.52\ \text{g MgCl}_2}{1003\ \text{mL soln}} \times \dfrac{1\ \text{mol}}{95.21\ \text{g MgCl}_2} \times \dfrac{10^3\ \text{mL}}{1\ \text{L}}$
$= 9.97\times10^{-2}\ M$

Osmotic pressure (Π)
$= i(MRT)$
$= 3(9.97\times10^{-2}\ \text{mol/L})\left(0.0821\dfrac{\text{atm·L}}{\text{mol·K}}\right)(293\ \text{K})$
$= 7.19\ \text{atm}$

(b) Scene C

PROBLEMS

Problems with colored numbers are answered in Appendix E and worked in detail in the Student Solutions Manual. Problem sections match those in the text and provide the numbers of relevant sample problems. Most offer Concept Review Questions, Skill-Building Exercises (grouped in pairs covering the same concept), and Problems in Context. The Comprehensive Problems are based on material from any section or previous chapter.

Types of Solutions: Intermolecular Forces and Solubility
(Sample Problem 13.1)

● Concept Review Questions

● **13.1** Describe how properties of seawater illustrate the two characteristics that define mixtures.

● **13.2** What types of intermolecular forces give rise to hydration shells in an aqueous solution of sodium chloride?

● **13.3** Acetic acid is miscible with water. Would you expect carboxylic acids with general formula $CH_3(CH_2)_nCOOH$ to become more or less water soluble as n increases? Explain.

● **13.4** Which would you expect to be more effective as a soap, sodium acetate or sodium stearate? Explain.

● **13.5** Hexane and methanol are miscible as gases but only slightly soluble in each other as liquids. Explain.

● **13.6** Hydrogen chloride (HCl) gas is much more soluble than propane gas (C_3H_8) in water, even though HCl has a lower boiling point. Explain.

● Skill-Building Exercises (grouped in similar pairs)

● **13.7** Which gives the more concentrated solution, (a) KNO_3 in H_2O or (b) KNO_3 in carbon tetrachloride (CCl_4)? Explain.

● **13.8** Which gives the more concentrated solution, stearic acid [$CH_3(CH_2)_{16}COOH$] in (a) H_2O or (b) CCl_4? Explain.

● **13.9** What is the strongest type of intermolecular force between solute and solvent in each solution?
(a) CsCl(s) in H_2O(l)   (b) $CH_3\overset{\text{O}}{\overset{\|}{C}}CH_3$($l$) in H_2O(l)
(c) CH_3OH(l) in CCl_4(l)

● **13.10** What is the strongest type of intermolecular force between solute and solvent in each solution?
(a) Cu(s) in Ag(s)   (b) CH_3Cl(g) in CH_3OCH_3(g)
(c) CH_3CH_3(g) in $CH_3CH_2CH_2NH_2$(l)

● **13.11** What is the strongest type of intermolecular force between solute and solvent in each solution?
(a) CH_3OCH_3(g) in H_2O(l)   (b) Ne(g) in H_2O(l)
(c) N_2(g) in C_4H_{10}(g)

● **13.12** What is the strongest type of intermolecular force between solute and solvent in each solution?
(a) C_6H_{14}(l) in C_8H_{18}(l)   (b) $H_2C{=}O$(g) in CH_3OH(l)
(c) Br_2(l) in CCl_4(l)

● **13.13** Which member of each pair is more soluble in diethyl ether? Why?
(a) NaCl(s) or HCl(g)   (b) H_2O(l) or $CH_3\overset{\text{O}}{\overset{\|}{C}}H$($l$)
(c) $MgBr_2$(s) or CH_3CH_2MgBr(s)

● **13.14** Which member of each pair is more soluble in water? Why?
(a) $CH_3CH_2OCH_2CH_3$(l) or $CH_3CH_2OCH_3$(g)
(b) CH_2Cl_2(l) or CCl_4(l)
(c)

cyclohexane   tetrahydropyran

● Problems in Context

● **13.15** The dictionary defines *homogeneous* as "uniform in composition throughout." River water is a mixture of dissolved compounds, such as calcium bicarbonate, and suspended soil particles. Is river water homogeneous? Explain.

● **13.16** Gluconic acid is a derivative of glucose used in cleaners and in the dairy and brewing industries. Caproic acid is a carboxylic acid used in the flavoring industry. Although both are six-carbon acids (see structures below), gluconic acid is soluble in water and nearly insoluble in hexane, whereas caproic acid has the opposite solubility behavior. Explain.

$$\overset{\text{OH}}{\underset{\ \ \ }{|}}\ \overset{\text{OH}}{\underset{\ \ \ }{|}}\ \overset{\text{OH}}{\underset{\ \ \ }{|}}\ \overset{\text{OH}}{\underset{\ \ \ }{|}}\ \overset{\text{OH}}{\underset{\ \ \ }{|}}$$
$$CH_2{-}CH{-}CH{-}CH{-}CH{-}COOH$$

gluconic acid

$$CH_3{-}CH_2{-}CH_2{-}CH_2{-}CH_2{-}COOH$$

caproic acid

Intermolecular Forces and Biological Macromolecules

● Concept Review Questions

● **13.17** Name three intermolecular forces that stabilize the shape of a soluble, globular protein, and explain how they act.

● **13.18** Name three intermolecular forces that stabilize the shape of DNA, and explain how they act.

● **13.19** Is the sodium salt of propanoic acid as effective a soap as sodium stearate? Explain.

● **13.20** What intermolecular forces stabilize a lipid bilayer?

● **13.21** In what way do proteins embedded in a membrane differ structurally from soluble proteins?

● **13.22** How can wood be so strong if it consists of cellulose chains held together by relatively weak H bonds?

● **13.23** Histones are proteins that control gene function by attaching through salt links to exterior regions of DNA. Name an amino acid whose side chain is often found on the exterior of histones.

Why Substances Dissolve: Understanding the Solution Process

● Concept Review Questions

● **13.24** What is the relationship between solvation and hydration?

● **13.25** For a general solvent, which enthalpy terms in the thermochemical solution cycle are combined to obtain $\Delta H_{\text{solvation}}$?

● **13.26** (a) What is the charge density of an ion, and what two properties of an ion affect it?
(b) Arrange the following in order of increasing charge density:

Levels of Difficulty ● Easy ●● Medium ●●● Challenging   Biological: 13.16–13.23; Organic: 13.14

(c) How do the two properties in part (a) affect the ionic heat of hydration, ΔH_{hydr}?

• **13.27** For ΔH_{soln} to be very small, what quantities must be nearly equal in magnitude? Will their signs be the same or opposite?

• **13.28** Water is added to a flask containing solid NH_4Cl. As the salt dissolves, the solution becomes colder.
(a) Is the dissolving of NH_4Cl exothermic or endothermic?
(b) Is the magnitude of $\Delta H_{lattice}$ of NH_4Cl larger or smaller than the combined ΔH_{hydr} of the ions? Explain.
(c) Given the answer to (a), why does NH_4Cl dissolve in water?

• **13.29** An ionic compound has a highly negative ΔH_{soln} in water. Would you expect it to be very soluble or nearly insoluble in water? Explain in terms of enthalpy and entropy changes.

● Skill-Building Exercises (grouped in similar pairs)

• **13.30** Sketch a qualitative enthalpy diagram for the process of dissolving $KCl(s)$ in H_2O (endothermic).

• **13.31** Sketch a qualitative enthalpy diagram for the process of dissolving $NaI(s)$ in H_2O (exothermic).

• **13.32** Which ion in each pair has greater charge density? Explain.
(a) Na^+ or Cs^+ (b) Sr^{2+} or Rb^+ (c) Na^+ or Cl^-
(d) O^{2-} or F^- (e) OH^- or SH^- (f) Mg^{2+} or Ba^{2+}
(g) Mg^{2+} or Na^+ (h) NO_3^- or CO_3^{2-}

• **13.33** Which ion has the lower ratio of charge to volume? Explain.
(a) Br^- or I^- (b) Sc^{3+} or Ca^{2+} (c) Br^- or K^+
(d) S^{2-} or Cl^- (e) Sc^{3+} or Al^{3+} (f) SO_4^{2-} or ClO_4^-
(g) Fe^{3+} or Fe^{2+} (h) Ca^{2+} or K^+

• **13.34** Which has the *larger* ΔH_{hydr} in each pair of Problem 13.32?

• **13.35** Which has the *smaller* ΔH_{hydr} in each pair of Problem 13.33?

• **13.36** (a) Use the following data to calculate the combined heat of hydration for the ions in potassium bromate ($KBrO_3$):

$$\Delta H_{lattice} = 745 \text{ kJ/mol} \qquad \Delta H_{soln} = 41.1 \text{ kJ/mol}$$

(b) Which ion contributes more to the answer to part (a)? Why?

• **13.37** (a) Use the following data to calculate the combined heat of hydration for the ions in sodium acetate ($NaC_2H_3O_2$):

$$\Delta H_{lattice} = 763 \text{ kJ/mol} \qquad \Delta H_{soln} = 17.3 \text{ kJ/mol}$$

(b) Which ion contributes more to the answer to part (a)? Why?

• **13.38** State whether the entropy of the system increases or decreases in each of the following processes:
(a) Gasoline burns in a car engine.
(b) Gold is extracted and purified from its ore.
(c) Ethanol (CH_3CH_2OH) dissolves in 1-propanol ($CH_3CH_2CH_2OH$).

• **13.39** State whether the entropy of the system increases or decreases in each of the following processes:
(a) Pure gases are mixed to prepare an anesthetic.
(b) Electronic-grade silicon is prepared from sand.
(c) Dry ice (solid CO_2) sublimes.

● Problems in Context

• **13.40** Besides its use in making black-and-white film, silver nitrate ($AgNO_3$) is used similarly in forensic science. The NaCl left behind in the sweat of a fingerprint is treated with $AgNO_3$ solution to form AgCl. This precipitate is developed to show the black-and-white fingerprint pattern. Given $\Delta H_{lattice}$ of $AgNO_3 = 822$ kJ/mol and $\Delta H_{hydr} = -799$ kJ/mol, calculate its ΔH_{soln}.

Solubility as an Equilibrium Process
(Sample Problem 13.2)

● Concept Review Questions

• **13.41** You are given a bottle of solid X and three aqueous solutions of X—one saturated, one unsaturated, and one supersaturated. How would you determine which solution is which?

• **13.42** Potassium permanganate ($KMnO_4$) has a solubility of 6.4 g/100 g of H_2O at 20°C and a curve of solubility vs. temperature that slopes upward to the right. How would you prepare a supersaturated solution of $KMnO_4$?

• **13.43** Why does the solubility of any gas in water decrease with rising temperature?

● Skill-Building Exercises (grouped in similar pairs)

• **13.44** For a saturated aqueous solution of each of the following at 20°C and 1 atm, will the solubility increase, decrease, or stay the same when the indicated change occurs?
(a) $O_2(g)$, increase P (b) $N_2(g)$, increase V

• **13.45** For a saturated aqueous solution of each of the following at 20°C and 1 atm, will the solubility increase, decrease, or stay the same when the indicated change occurs?
(a) $He(g)$, decrease T (b) $RbI(s)$, increase P

• **13.46** The Henry's law constant (k_H) for O_2 in water at 20°C is 1.28×10^{-3} mol/L·atm. (a) How many grams of O_2 will dissolve in 2.50 L of H_2O that is in contact with pure O_2 at 1.00 atm? (b) How many grams of O_2 will dissolve in 2.50 L of H_2O that is in contact with air, where the partial pressure of O_2 is 0.209 atm?

• **13.47** Argon makes up 0.93% by volume of air. Calculate its solubility (mol/L) in water at 20°C and 1.0 atm. The Henry's law constant for Ar under these conditions is 1.5×10^{-3} mol/L·atm.

● Problems in Context

• **13.48** Caffeine is about 10 times as soluble in hot water as in cold water. A chemist puts a hot-water extract of caffeine into an ice bath, and some caffeine crystallizes. Is the remaining solution saturated, unsaturated, or supersaturated?

• **13.49** The partial pressure of CO_2 gas above the liquid in a bottle of champagne at 20°C is 5.5 atm. What is the solubility of CO_2 in champagne? Assume Henry's law constant is the same for champagne as for water: at 20°C, $k_H = 3.7 \times 10^{-2}$ mol/L·atm.

•• **13.50** Respiratory problems are treated with devices that deliver air with a higher partial pressure of O_2 than normal air. Why?

Quantitative Ways of Expressing Concentration
(Sample Problems 13.3 to 13.5)

● Concept Review Questions

• **13.51** Explain the difference between molarity and molality. Under what circumstances would molality be a more accurate measure of the concentration of a prepared solution than molarity? Why?

• **13.52** Which way of expressing concentration includes (a) volume of solution; (b) mass of solution; (c) mass of solvent?

• **13.53** A solute has a solubility in water of 21 g/kg solvent. Is this value the same as 21 g/kg solution? Explain.

• **13.54** You want to convert among molarity, molality, and mole fraction of a solution. You know the masses of solute and solvent and the volume of solution. Is this enough information to carry out all the conversions? Explain.

• **13.55** When a solution is heated, which ways of expressing concentration change in value? Which remain unchanged? Explain.

● **Skill-Building Exercises** (grouped in similar pairs)

• **13.56** Calculate the molarity of each aqueous solution:
   (a) 32.3 g of table sugar ($C_{12}H_{22}O_{11}$) in 100. mL of solution
   (b) 5.80 g of $LiNO_3$ in 505 mL of solution

• **13.57** Calculate the molarity of each aqueous solution:
   (a) 0.82 g of ethanol (C_2H_5OH) in 10.5 mL of solution
   (b) 1.27 g of gaseous NH_3 in 33.5 mL of solution

• **13.58** Calculate the molarity of each aqueous solution:
   (a) 78.0 mL of 0.240 *M* NaOH diluted to 0.250 L with water
   (b) 38.5 mL of 1.2 *M* HNO_3 diluted to 0.130 L with water

• **13.59** Calculate the molarity of each aqueous solution:
   (a) 25.5 mL of 6.25 *M* HCl diluted to 0.500 L with water
   (b) 8.25 mL of 2.00×10^{-2} *M* KI diluted to 12.0 mL with water

• **13.60** How would you prepare the following aqueous solutions?
   (a) 365 mL of 8.55×10^{-2} *M* KH_2PO_4 from solid KH_2PO_4
   (b) 465 mL of 0.335 *M* NaOH from 1.25 *M* NaOH

• **13.61** How would you prepare the following aqueous solutions?
   (a) 2.5 L of 0.65 *M* NaCl from solid NaCl
   (b) 15.5 L of 0.3 *M* urea [$(NH_2)_2C$=O] from 2.1 *M* urea

• **13.62** How would you prepare the following aqueous solutions?
   (a) 1.40 L of 0.288 *M* KBr from solid KBr
   (b) 255 mL of 0.0856 *M* $LiNO_3$ from 0.264 *M* $LiNO_3$

• **13.63** How would you prepare the following aqueous solutions?
   (a) 57.5 mL of 1.53×10^{-3} *M* $Cr(NO_3)_3$ from solid $Cr(NO_3)_3$
   (b) 5.8×10^3 m^3 of 1.45 *M* NH_4NO_3 from 2.50 *M* NH_4NO_3

• **13.64** Calculate the molality of the following:
   (a) A solution containing 85.4 g of glycine (NH_2CH_2COOH) dissolved in 1.270 kg of H_2O
   (b) A solution containing 8.59 g of glycerol ($C_3H_8O_3$) in 77.0 g of ethanol (C_2H_5OH)

• **13.65** Calculate the molality of the following:
   (a) A solution containing 174 g of HCl in 757 g of H_2O
   (b) A solution containing 16.5 g of naphthalene ($C_{10}H_8$) in 53.3 g of benzene (C_6H_6)

• **13.66** What is the molality of a solution consisting of 44.0 mL of benzene (C_6H_6; *d* = 0.877 g/mL) in 167 mL of hexane (C_6H_{14}; *d* = 0.660 g/mL)?

• **13.67** What is the molality of a solution consisting of 2.66 mL of carbon tetrachloride (CCl_4; *d* = 1.59 g/mL) in 76.5 mL of methylene chloride (CH_2Cl_2; *d* = 1.33 g/mL)?

•• **13.68** How would you prepare the following aqueous solutions?
   (a) 3.10×10^2 g of 0.125 *m* ethylene glycol ($C_2H_6O_2$) from ethylene glycol and water
   (b) 1.20 kg of 2.20 mass % HNO_3 from 52.0 mass % HNO_3

•• **13.69** How would you prepare the following aqueous solutions?
   (a) 1.50 kg of 0.0355 *m* ethanol (C_2H_5OH) from ethanol and water
   (b) 445 g of 13.0 mass % HCl from 34.1 mass % HCl

• **13.70** A solution contains 0.35 mol of isopropanol (C_3H_7OH) dissolved in 0.85 mol of water. (a) What is the mole fraction of isopropanol? (b) The mass percent? (c) The molality?

• **13.71** A solution contains 0.100 mol of NaCl dissolved in 8.60 mol of water. (a) What is the mole fraction of NaCl? (b) The mass percent? (c) The molality?

• **13.72** What mass of cesium bromide must be added to 0.500 L of water (*d* = 1.00 g/mL) to produce a 0.400 *m* solution? What are the mole fraction and the mass percent of CsBr?

• **13.73** What are the mole fraction and the mass percent of a solution made by dissolving 0.30 g of KI in 0.400 L of water (*d* = 1.00 g/mL)?

• **13.74** Calculate the molality, molarity, and mole fraction of NH_3 in an 8.00 mass % aqueous solution (*d* = 0.9651 g/mL).

• **13.75** Calculate the molality, molarity, and mole fraction of $FeCl_3$ in a 28.8 mass % aqueous solution (*d* = 1.280 g/mL).

● **Problems in Context**

• **13.76** Wastewater from a cement factory contains 0.25 g of Ca^{2+} ion and 0.056 g of Mg^{2+} ion per 100.0 L of solution. The solution density is 1.001 g/mL. Calculate the Ca^{2+} and Mg^{2+} concentrations in ppm (by mass).

• **13.77** An automobile antifreeze mixture is made by mixing equal volumes of ethylene glycol (*d* = 1.114 g/mL; *$\mathcal{M}$* = 62.07 g/mol) and water (*d* = 1.00 g/mL) at 20°C. The density of the mixture is 1.070 g/mL. Express the concentration of ethylene glycol as:
   (a) Volume percent   (b) Mass percent   (c) Molarity
   (d) Molality   (e) Mole fraction

Colligative Properties of Solutions
(Sample Problems 13.6 to 13.9)

● **Concept Review Questions**

• **13.78** The chemical formula of a solute does *not* affect the extent of the solution's colligative properties. What characteristic of a solute *does* affect these properties? Name a physical property of a solution that *is* affected by the chemical formula of the solute.

• **13.79** What is a nonvolatile nonelectrolyte? Why is this type of solute the simplest case for examining colligative properties?

• **13.80** In what sense is a strong electrolyte "strong"? What property of the substance makes it a strong electrolyte?

• **13.81** Express Raoult's law in words. Is Raoult's law valid for a solution of a volatile solute? Explain.

• **13.82** What are the most important differences between the phase diagram of a pure solvent and the phase diagram of a solution of that solvent?

• **13.83** Is the composition of the vapor at the top of a fractionating column different from the composition at the bottom? Explain.

• **13.84** Is the boiling point of 0.01 *m* KF(*aq*) higher or lower than that of 0.01 *m* glucose(*aq*)? Explain.

• **13.85** Which aqueous solution has a boiling point closer to its predicted value, 0.050 *m* NaF or 0.50 *m* KCl? Explain.

• **13.86** Which aqueous solution has a freezing point closer to its predicted value, 0.01 *m* NaBr or 0.01 *m* $MgCl_2$? Explain.

• **13.87** The freezing point depression constants of the solvents cyclohexane and naphthalene are 20.1°C/*m* and 6.94°C/*m*, respectively. Which solvent would give a more accurate result if you are using freezing point depression to determine the molar mass of a substance that is soluble in either one? Why?

● **Skill-Building Exercises** (grouped in similar pairs)

• **13.88** Classify the following substances as strong electrolytes, weak electrolytes, or nonelectrolytes:
   (a) Hydrogen chloride (HCl)   (b) Potassium nitrate (KNO_3)
   (c) Glucose ($C_6H_{12}O_6$)   (d) Ammonia (NH_3)

• **13.89** Classify the following substances as strong electrolytes, weak electrolytes, or nonelectrolytes:
   (a) Sodium permanganate ($NaMnO_4$)
   (b) Acetic acid (CH_3COOH)
   (c) Methanol (CH_3OH)   (d) Calcium acetate [$Ca(C_2H_3O_2)_2$]

• **13.90** How many moles of solute particles are present in 1 L of each of the following aqueous solutions?
(a) 0.3 M KBr   (b) 0.065 M HNO$_3$
(c) 10^{-4} M KHSO$_4$ (d) 0.06 M ethanol (C$_2$H$_5$OH)

• **13.91** How many moles of solute particles are present in 1 mL of each of the following aqueous solutions?
(a) 0.02 M CuSO$_4$ (b) 0.004 M Ba(OH)$_2$
(c) 0.08 M pyridine (C$_5$H$_5$N)   (d) 0.05 M (NH$_4$)$_2$CO$_3$

• **13.92** Which solution has the lower freezing point?
(a) 11.0 g of CH$_3$OH in 100. g of H$_2$O *or*
22.0 g of CH$_3$CH$_2$OH in 200. g of H$_2$O
(b) 20.0 g of H$_2$O in 1.00 kg of CH$_3$OH *or*
20.0 g of CH$_3$CH$_2$OH in 1.00 kg of CH$_3$OH

• **13.93** Which solution has the higher boiling point?
(a) 38.0 g of C$_3$H$_8$O$_3$ in 250. g of ethanol *or*
38.0 g of C$_2$H$_6$O$_2$ in 250. g of ethanol
(b) 15 g of C$_2$H$_6$O$_2$ in 0.50 kg of H$_2$O *or*
15 g of NaCl in 0.50 kg of H$_2$O

• **13.94** Rank the following aqueous solutions in order of increasing (a) osmotic pressure; (b) boiling point; (c) freezing point; (d) vapor pressure at 50°C:
(I) 0.100 m NaNO$_3$ (II) 0.100 m glucose   (III) 0.100 m CaCl$_2$

• **13.95** Rank the following aqueous solutions in order of decreasing (a) osmotic pressure; (b) boiling point; (c) freezing point; (d) vapor pressure at 298 K:
(I) 0.04 m urea [(NH$_2$)$_2$C=O]
(II) 0.01 m AgNO$_3$
(III) 0.03 m CuSO$_4$

• **13.96** Calculate the vapor pressure of a solution of 34.0 g of glycerol (C$_3$H$_8$O$_3$) in 500.0 g of water at 25°C. The vapor pressure of water at 25°C is 23.76 torr. (Assume ideal behavior.)

• **13.97** Calculate the vapor pressure of a solution of 0.39 mol of cholesterol in 5.4 mol of toluene at 32°C. Pure toluene has a vapor pressure of 41 torr at 32°C. (Assume ideal behavior.)

• **13.98** What is the freezing point of 0.251 m urea in water?

• **13.99** What is the boiling point of 0.200 m lactose in water?

• **13.100** The boiling point of ethanol (C$_2$H$_5$OH) is 78.5°C. What is the boiling point of a solution of 6.4 g of vanillin $\mathcal{M}$ = 152.14 g/mol) in 50.0 g of ethanol (K_b of ethanol = 1.22°C/m)?

• **13.101** The freezing point of benzene is 5.5°C. What is the freezing point of a solution of 5.00 g of naphthalene (C$_{10}$H$_8$) in 444 g of benzene (K_f of benzene = 4.90°C/m)?

•• **13.102** What is the minimum mass of ethylene glycol (C$_2$H$_6$O$_2$) that must be dissolved in 14.5 kg of water to prevent the solution from freezing at −12.0°F? (Assume ideal behavior.)

•• **13.103** What is the minimum mass of glycerol (C$_3$H$_8$O$_3$) that must be dissolved in 11.0 mg of water to prevent the solution from freezing at −15°C? (Assume ideal behavior.)

•• **13.104** Calculate the molality and van't Hoff factor (i) for the following aqueous solutions:
(a) 1.00 mass % NaCl, freezing point = −0.593°C
(b) 0.500 mass % CH$_3$COOH, freezing point = −0.159°C

•• **13.105** Calculate the molality and van't Hoff factor (i) for the following aqueous solutions:
(a) 0.500 mass % KCl, freezing point = −0.234°C
(b) 1.00 mass % H$_2$SO$_4$, freezing point = −0.423°C

● **Problems in Context**

• **13.106** Wastewater discharged into a stream by a sugar refinery contains 3.55 g of sucrose (C$_{12}$H$_{22}$O$_{11}$) per liter. A government-industry project is designed to test the feasibility of removing the sugar by reverse osmosis. What pressure must be applied to the apparatus at 20.°C to produce pure water?

•• **13.107** In a study designed to prepare new gasoline-resistant coatings, a polymer chemist dissolves 6.053 g of poly(vinyl alcohol) in enough water to make 100.0 mL of solution. At 25°C, the osmotic pressure of this solution is 0.272 atm. What is the molar mass of the polymer sample?

• **13.108** The U.S. Food and Drug Administration lists dichloromethane (CH$_2$Cl$_2$) and carbon tetrachloride (CCl$_4$) among the many cancer-causing chlorinated organic compounds. What are the partial pressures of these substances in the vapor above a solution of 1.60 mol of CH$_2$Cl$_2$ and 1.10 mol of CCl$_4$ at 23.5°C? The vapor pressures of pure CH$_2$Cl$_2$ and CCl$_4$ at 23.5°C are 352 torr and 118 torr, respectively. (Assume ideal behavior.)

The Structure and Properties of Colloids

● **Concept Review Questions**

• **13.109** Is the fluid inside a bacterial cell considered a solution, a colloid, or both? Explain.

• **13.110** What type of colloid is each of the following?
(a) Milk   (b) Fog   (c) Shaving cream

• **13.111** What is Brownian motion, and what causes it?

• **13.112** In a movie theater, you can see the beam of projected light. What phenomenon does this exemplify? Why does it occur?

• **13.113** Why don't soap micelles coagulate and form large globules? Is soap more effective in freshwater or in seawater? Why?

Comprehensive Problems

•• **13.114** The three aqueous ionic solutions below have total volumes of 25 mL in A, 50. mL in B, and 100. mL in C. If each sphere represents 0.010 mol of ions, calculate: (a) the total molarity of ions for each solution; (b) the highest molarity of solute; (c) the lowest molality of solute (assuming the solution densities are equal); (d) the highest osmotic pressure (assuming ideal behavior).

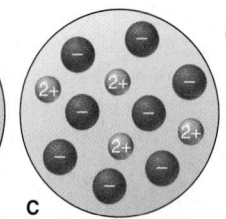

A   **B**   **C**

•• **13.115** An aqueous solution is 10.% glucose by mass (d = 1.039 g/mL at 20°C). Calculate its freezing point, boiling point at 1 atm, and osmotic pressure.

•• **13.116** Because zinc has nearly the same atomic radius as copper (d = 8.95 g/cm^3), zinc atoms substitute for some copper atoms in the many types of brass. Calculate the density of the brass with (a) 10.0 atom % Zn and (b) 38.0 atom % Zn.

•• **13.117** Gold occurs in seawater at an average concentration of 1.1×10^{-2} ppb. How many liters of seawater must be processed to recover 1 troy ounce of gold, assuming 81.5% efficiency (d of seawater = 1.025 g/mL; 1 troy ounce = 31.1 g)?

•• **13.118** Use atomic properties to explain why xenon is 11 times as soluble as helium in water at 0°C on a mole basis.

•• **13.119** Which of the following best represents a molecular-scale view of an ionic compound in aqueous solution? Explain.

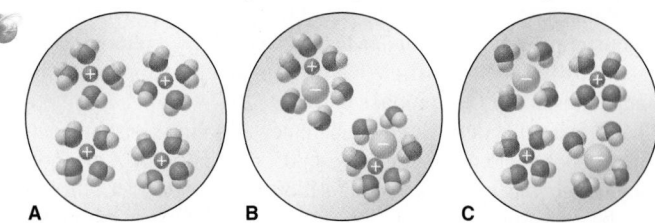

A   B   C

•• **13.120** Four 0.50 *m* aqueous solutions are depicted below. Assume the solutions behave ideally: (a) Which has the highest boiling point? (b) Which has the lowest freezing point? (c) Can you determine which one has the highest osmotic pressure? Explain.

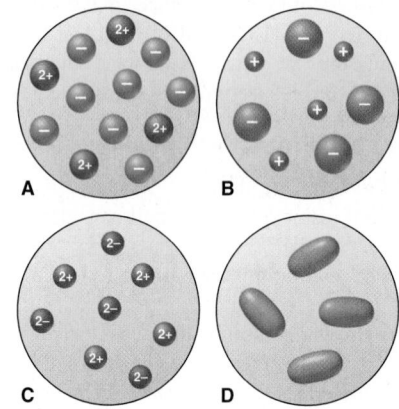

A   B

C   D

•• **13.121** Thermal pollution from industrial wastewater causes the temperature of river or lake water to increase, which can affect fish survival as the concentration of dissolved O_2 decreases. Use the following data to find the molarity of O_2 at each temperature (assume the solution density is the same as water):

Temperature (°C)	Solubility of O_2 (mg/kg H_2O)	Density of H_2O (g/mL)
0.0	14.5	0.99987
20.0	9.07	0.99823
40.0	6.44	0.99224

•• **13.122** Pyridine *(right)* is an essential portion of many biologically active compounds, such as nicotine and vitamin B_6. Like ammonia, it has a nitrogen with a lone pair, which makes it act as a weak base. Because it is miscible in a wide range of solvents, from water to benzene, pyridine is one of the most important bases and solvents in organic syntheses. Account for its solubility behavior in terms of intermolecular forces.

••• **13.123** A chemist is studying small organic compounds for their potential use as an antifreeze. When 0.243 g of a compound is dissolved in 25.0 mL of water, the freezing point of the solution is −0.201°C. (a) Calculate the molar mass of the compound (*d* of water = 1.00 g/mL). (b) Analysis shows that the compound is 53.31 mass % C and 11.18 mass % H, the remainder being O. Calculate the empirical and molecular formulas of the compound. (c) Draw a Lewis structure for a compound with this formula that forms H bonds and another for one that does not.

•• **13.124** Air in a smoky bar contains 4.0×10^{-6} mol/L of CO. What mass of CO is inhaled by a bartender who respires at a rate of 11 L/min during an 8.0-h shift?

•• **13.125** Is 50% by mass of methanol dissolved in ethanol different from 50% by mass of ethanol dissolved in methanol? Explain.

•• **13.126** Three gaseous mixtures of N_2 (blue), Cl_2 (green), and Ne (purple) are depicted below. (a) Which has the smallest mole fraction of N_2? (b) Which have the same mole fraction of Ne? (c) Rank all three in order of increasing mole fraction of Cl_2.

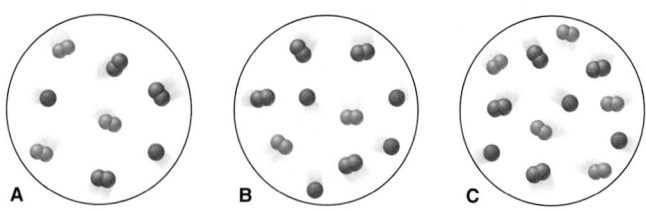

A   B   C

•• **13.127** A water treatment plant needs to attain a fluoride concentration of 4.50×10^{-5} *M*. (a) What mass of NaF must be added to a 5000.-L blending tank of water? (b) What mass per day of fluoride is ingested by a person who drinks 2.0 L of this water?

•• **13.128** Answer each of the following briefly: (a) Why is cake alum $[Al_2(SO_4)_3]$ added during water purification? (b) Why is water that contains large amounts of Ca^{2+} and Mg^{2+} difficult to use for cleaning? (c) What is the meaning of "reverse" in reverse osmosis? (d) Why might a water treatment plant use ozone as a disinfectant instead of chlorine? (e) How does passing a saturated NaCl solution through a "spent" ion-exchange resin regenerate the resin?

•• **13.129** Four U tubes each have distilled water in the right arm, a solution in the left arm, and a semipermeable membrane between arms. (a) If the solute is KCl, which solution is most concentrated? (b) If each solute is different but all the solutions have the same molarity, which contains the smallest number of dissolved ions?

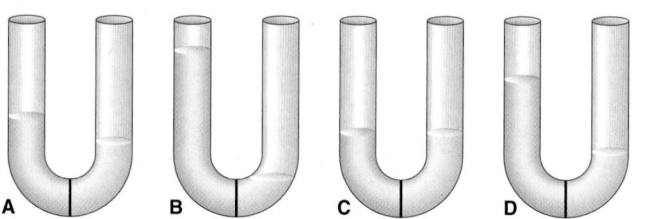

A   B   C   D

•• **13.130** β-Pinene $(C_{10}H_{16})$ and α-terpineol $(C_{10}H_{18}O)$ are used in cosmetics to provide a "fresh pine" scent. At 367 K, the pure substances have vapor pressures of 100.3 torr and 9.8 torr, respectively. What is the composition of the vapor (in terms of mole fractions) above a solution containing equal masses of these compounds at 367 K? (Assume ideal behavior.)

••• **13.131** A solution of 1.50 g of solute dissolved in 25.0 mL of H_2O at 25°C has a boiling point of 100.45°C. (a) What is the molar mass of the solute if it is a nonvolatile nonelectrolyte and the solution behaves ideally (*d* of H_2O at 25°C = 0.997 g/mL)? (b) Conductivity measurements show the solute to be ionic of general formula AB_2 or A_2B. What is the molar mass if the solution behaves ideally? (c) Analysis indicates an empirical formula of CaN_2O_6. Explain the difference between the actual formula mass and that calculated from the boiling point elevation. (d) Find the van't Hoff factor (*i*) for this solution.

••• **13.132** A pharmaceutical preparation made with ethanol (C_2H_5OH) is contaminated with methanol (CH_3OH). A sample of vapor above the liquid mixture contains a 97/1 mass ratio of C_2H_5OH to CH_3OH. What is the mass ratio of these alcohols in the liquid? At the temperature of the liquid, the vapor pressures of C_2H_5OH and CH_3OH are 60.5 torr and 126.0 torr, respectively.

Biological: 13.122, 13.124, 13.127; Environmental: 13.121, 13.127, 13.128; Organic: 13.123, 13.130, 13.132

•• **13.133** Water treatment plants commonly use chlorination to destroy bacteria. A byproduct is chloroform ($CHCl_3$), a suspected carcinogen, produced when HOCl, formed by reaction of Cl_2 and water, reacts with dissolved organic matter. The United States, Canada, and the World Health Organization have set a limit of 100. ppb of $CHCl_3$ in drinking water. Convert this concentration into molarity, molality, mole fraction, and mass percent.

•• **13.134** A saturated Na_2CO_3 solution is prepared, and a small excess of solid is present (white pile in beaker). A seed crystal of $Na_2^{14}CO_3$ (^{14}C is a radioactive isotope of ^{12}C) is added (small red piece), and the radioactivity is measured over time. (a) Would you expect radioactivity in the solution? Explain. (b) Would you expect radioactivity in all the solid or just in the seed crystal? Explain.

Saturated
Na_2CO_3 soln

•• **13.135** A biochemical engineer isolates a bacterial gene fragment and dissolves a 10.0-mg sample in enough water to make 30.0 mL of solution. The osmotic pressure of the solution is 0.340 torr at 25°C. (a) What is the molar mass of the gene fragment? (b) If the solution density is 0.997 g/mL, how large is the freezing point depression for this solution (K_f of water = 1.86°C/m)?

•• **13.136** A river is contaminated with 0.65 mg/L of dichloroethylene ($C_2H_2Cl_2$). What is the concentration (in ng/L) of dichloroethylene at 21°C in the air breathed by a person swimming in the river (k_H for $C_2H_2Cl_2$ in water is 0.033 mol/L·atm)?

•• **13.137** At an air-water interface, fatty acids such as oleic acid lie in a one-molecule-thick layer (*monolayer*), with the heads in the water and the tails perpendicular in the air. When 2.50 mg of oleic acid is placed on a water surface, it forms a circular monolayer 38.6 cm in diameter. Find the surface area (in cm²) occupied by one molecule (M of oleic acid = 283 g/mol).

•• **13.138** A simple device used for estimating the concentration of total dissolved solids in an aqueous solution works by measuring the electrical conductivity of the solution. The method assumes that equal concentrations of different solids give approximately the same conductivity, and that the conductivity is proportional to concentration. The table below gives some actual electrical conductivities (in arbitrary units) for solutions of selected solids at the indicated concentrations (in ppm by mass):

Sample	Conductivity		
	0 ppm	5.00×10^3 ppm	10.00×10^3 ppm
$CaCl_2$	0.0	8.0	16.0
K_2CO_3	0.0	7.0	14.0
Na_2SO_4	0.0	6.0	11.0
Seawater (dil)	0.0	8.0	15.0
Sucrose ($C_{12}H_{22}O_{11}$)	0.0	0.0	0.0
Urea [$(NH_2)_2C{=}O$]	0.0	0.0	0.0

(a) How reliable are these measurements for estimating concentrations of dissolved solids? (b) For what types of substances might this method have a large error? Why? (c) Based on this method, an aqueous $CaCl_2$ solution has a conductivity of 14.0 units. Calculate its mole fraction and molality.

••• **13.139** Two beakers are placed in a closed container (*left*). One beaker contains water, the other a concentrated aqueous sugar

solution. With time, the solution volume increases and the water volume decreases (*right*). Explain on the molecular level.

time

Water | Sugar solution

Sugar solution | Water

•• **13.140** The release of volatile organic compounds into the atmosphere is regulated to limit ozone formation. In a laboratory simulation, 5% of the ethanol in a liquid detergent is released. Thus, a "down-the-drain" factor of 0.05 is used to estimate ethanol emissions from the detergent. The k_H values for ethanol and 2-butoxyethanol ($C_4H_9OCH_2CH_2OH$) are 5×10^{-6} atm·m³/mol and 1.6×10^{-6} atm·m³/mol, respectively. (a) Estimate a "down-the-drain" factor for 2-butoxyethanol in the detergent. (b) What is the k_H for ethanol in units of L·atm/mol? (c) Is this value consistent with a value given as 0.64 Pa·m³/mol?

••• **13.141** Although other solvents are available, dichloromethane (CH_2Cl_2) is still often used to "decaffeinate" drinks because the solubility of caffeine in CH_2Cl_2 is 8.35 times that in water.
(a) A 100.0-mL sample of cola containing 10.0 mg of caffeine is extracted with 60.0 mL of CH_2Cl_2. What mass of caffeine remains in the aqueous phase? (b) A second identical cola sample is extracted with two successive 30.0-mL portions of CH_2Cl_2. What mass of caffeine remains in the aqueous phase after each extraction? (c) Which approach extracts more caffeine?

• **13.142** How do you prepare 250. g of 0.150 m aqueous $NaHCO_3$?

•• **13.143** Tartaric acid occurs in crystalline residues found in wine vats. It is used in baking powders and as an additive in foods. It contains 32.3% by mass carbon and 3.97% by mass hydrogen; the balance is oxygen. When 0.981 g of tartaric acid is dissolved in 11.23 g of water, the solution freezes at −1.26°C. Find the empirical and molecular formulas of tartaric acid.

•• **13.144** Methanol (CH_3OH) and ethanol (C_2H_5OH) are miscible because the major intermolecular force for each is hydrogen bonding. In some methanol-ethanol solutions, the mole fraction of methanol is higher, but the mass percent of ethanol is higher. What is the range of mole fraction of methanol for these solutions?

•• **13.145** A solution of 5.0 g of benzoic acid (C_6H_5COOH) in 100.0 g of carbon tetrachloride has a boiling point of 77.5°C.
(a) Calculate the molar mass of benzoic acid in the solution.
(b) Suggest a reason for the difference between the molar mass based on the formula and that found in part (a).

••• **13.146** Derive a general equation that expresses the relationship between the molarity and the molality of a solution. Why are the numerical values of these two terms approximately equal for very dilute aqueous solutions?

••• **13.147** A florist prepares a solution of nitrogen-phosphorus fertilizer by dissolving 5.66 g of NH_4NO_3 and 4.42 g of $(NH_4)_3PO_4$ in enough water to make 20.0 L of solution. What are the molarities of NH_4^+ and of PO_4^{3-} in the solution?

•• **13.148** Suppose coal-fired power plants used water in scrubbers to remove SO_2 from smokestack gases (see Chemical Connections, p. 256). (a) If the partial pressure of SO_2 in the stack gases is 2.0×10^{-3} atm, what is the solubility of SO_2 in the scrubber liquid (k_H for SO_2 in water is 1.23 mol/L·atm at 200.°C)? (b) From your answer to part (a), why are basic solutions, such as limewater slurries [$Ca(OH)_2$], used in scrubbers?

Biological: 13.135, 13.137; Environmental: 13.133, 13.136, 13.140, 13.148; Industrial: 13.141; Organic: 13.144, 13.145

•• 13.149 Urea is a white crystalline solid used as a fertilizer, in the pharmaceutical industry, and in the manufacture of certain polymer resins. Analysis of urea reveals that, by mass, it is 20.1% carbon, 6.7% hydrogen, 46.5% nitrogen and the balance oxygen. (a) Find the empirical formula of urea. (b) A 5.0 g/L solution of urea in water has an osmotic pressure of 2.04 atm, measured at 25°C. What is the molar mass and molecular formula of urea?

•• 13.150 The total concentration of dissolved particles in blood is 0.30 M. An intravenous (IV) solution must be isotonic with blood, which means it must have the same concentration. (a) To relieve dehydration, a patient is given 100. mL/h of IV glucose ($C_6H_{12}O_6$) for 2.5 h. What mass (g) of glucose did she receive? (b) If isotonic saline (NaCl) is used, what is the molarity of the solution? (c) If the patient is given 150. mL/h of IV saline for 1.5 h, how many grams of NaCl did she receive?

•• 13.151 Deviations from Raoult's law lead to the formation of *azeotropes*, constant boiling mixtures that cannot be separated by distillation, making industrial separations difficult. For components A and B, there is a positive deviation if the A-B attraction is less than A-A and B-B attractions (A and B reject each other), and a negative deviation if the A-B attraction is greater than A-A and B-B attractions. If the A-B attraction is nearly equal to the A-A and B-B attractions, the solution obeys Raoult's law. Explain whether the behavior of each pair will be nearly ideal, have a positive deviation, or a negative deviation: (a) benzene (C_6H_6) and methanol; (b) water and ethyl acetate; (c) hexane and heptane; (d) methanol and water; (e) water and hydrochloric acid.

•• 13.152 Acrylic acid (CH_2=CHCOOH) is a monomer used to make superabsorbent polymers and various compounds for paint and adhesive production. At 1 atm, it boils at 141.5°C but is prone to polymerization. Its vapor pressure at 25°C is 4.1 mbar. What pressure (in mmHg) is needed to distill the pure acid at 65°C?

•• 13.153 To effectively stop polymerization, certain inhibitors require the presence of a small amount of O_2. At equilibrium with 1 atm of air, the concentration of O_2 dissolved in the monomer acrylic acid (CH_2=CHCOOH) is 1.64×10^{-3} M. (a) What is the k_H (mol/L·atm) for O_2 in acrylic acid? (b) If 0.005 atm of O_2 is sufficient to stop polymerization, what is the molarity of O_2? (c) What is the mole fraction? (d) What is the concentration in ppm? (Pure acrylic acid is 14.6 M; P_{O_2} in air is 0.2095 atm.)

•• 13.154 Volatile organic solvents have been implicated in adverse health effects of industrial workers. Greener methods are phasing them out (see the margin note, p. 505). Rank the solvents in Table 13.6 (p. 530) in terms of increasing volatility.

•• 13.155 At ordinary temperatures, water is a poor solvent for organic substances. But at high pressure and above 200°C, water develops many properties of organic solvents. Find the minimum pressure needed to maintain water as a liquid at 200.°C (ΔH_{vap} = 40.7 kJ/mol at 100°C and 1.00 atm; assume it remains constant with temperature).

•• 13.156 In ice-cream making, the ingredients are kept below 0.0°C in an ice-salt bath. (a) Assuming that NaCl dissolves completely and forms an ideal solution, what mass of it is needed to lower the melting point of 5.5 kg of ice to −5.0°C? (b) Given the same assumptions as in part (a), what mass of $CaCl_2$ is needed?

•• 13.157 Perfluorocarbons (PFCs), hydrocarbons with all the H replaced by F, have very weak cohesive forces. One unusual application shows a mouse surviving submerged in O_2-saturated PFCs. (a) At 298 K, perfluorohexane (C_6F_{14}, $\mathcal{M}$ = 338 g/mol, d = 1.674 g/mL) in equilibrium with 101,325 Pa of O_2 has a mole fraction of O_2 of 4.28×10^{-3}. What is the k_H in mol/L·atm? (b) According to one source, the k_H for O_2 in water at 25°C is 756.7 L·atm/mol. What is the solubility of O_2 in water at 25°C in ppm? (c) Rank the k_H values in descending order for O_2 in water, ethanol, C_6F_{14}, and C_6H_{14}. Explain your ranking.

••• 13.158 The solubility of N_2 in blood is a serious problem for divers breathing compressed air (78% N_2 by volume) at depths greater than 50 ft (see the margin note, p. 522). (a) What is the molarity of N_2 in blood at 1.00 atm? (b) What is the molarity of N_2 in blood at a depth of 50. ft? (c) Find the volume (in mL) of N_2, measured at 25°C and 1.00 atm, released per liter of blood when a diver at a depth of 50. ft rises to the surface (k_H for N_2 in water at 25°C is 7.0×10^{-4} mol/L·atm and at 37°C is 6.2×10^{-4} mol/L·atm; assume d of water is 1.00 g/mL).

••• 13.159 Figure 12.9B (p. 448) shows the phase changes of pure water. Consider how the diagram would change if air were present at 1 atm and dissolved in the water. (a) Would the three phases of water still attain equilibrium at some temperature? Explain. (b) In principle, would that temperature be higher, lower, or the same as the triple point for pure water? Explain. (c) Would ice sublime at a few degrees below the freezing point under this pressure? Explain. (d) Would the liquid have the same vapor pressure as that shown in Figure 12.9B at 100°C? At 120°C?

•• 13.160 KNO₃, KClO₃, KCl, and NaCl are recrystallized as follows:
Step 1. A saturated aqueous solution of the compound is prepared at 50°C.
Step 2. The mixture is filtered to remove undissolved compound.
Step 3. The filtrate is cooled to 0°C.
Step 4. The crystals that form are filtered, dried, and weighed.
(a) Which compound has the highest percent recovery and which the lowest (see Figure 13.21, p. 520)? (b) Starting with 100. g of each compound, how many grams of each can be recovered?

•• 13.161 Eighty proof whiskey is 40% ethanol (C_2H_5OH) by volume. A man has 7.0 L of blood and drinks 28 mL of the whiskey, of which 22% of the ethanol goes into his blood. (a) What concentration (in g/mL) of ethanol is in his blood (d of ethanol = 0.789 g/mL)? (b) What volume (in mL) of whiskey would raise his blood alcohol level to 8.0×10^{-4} g/mL, the level at which a person is considered intoxicated?

•• 13.162 Soft drinks are canned under 4 atm of CO_2 and release CO_2 when opened (see the margin note, p. 522). (a) How many moles of CO_2 are dissolved in a 355-mL can of soda before it is opened? (b) After it has gone flat? (c) What volume (in L) would the released CO_2 occupy at 1.00 atm and 25°C (k_H for CO_2 at 25°C is 3.3×10^{-2} mol/L·atm; P_{CO_2} in air is 3×10^{-4} atm)?

•• 13.163 Gaseous O_2 in equilibrium with O_2 dissolved in water at 283 K is depicted at right. (a) Which scene below represents the system at 298 K? (b) Which scene represents the system when the pressure of O_2 is increased by half?

A   B   C

The Basis of Element Behavior The dramatic chemical change that occurs when sodium reacts with bromine is just one example of the countless reactions of the main-group elements. As you have seen, this behavior is based on atomic and molecular properties, which we revisit in the upcoming pages.

INTERCHAPTER
A Perspective on the Properties of the Elements

Topic 1   The Key Atomic Properties

Topic 2   Characteristics of Chemical Bonding

Topic 3   Metallic Behavior

Topic 4   Acid-Base Behavior of the Element Oxides

Topic 5   Redox Behavior of the Elements

Topic 6   Physical States and Phase Changes

Chemistry has a central, underlying principle: *The behavior of a sample of matter emerges from the properties of its component atoms.* This illustrated Interchapter reviews many ideas about these properties from earlier chapters so that you can see in upcoming chapters how they lead to the behavior of the main-group elements. Keep in mind that, despite our categories, clear dividing lines rarely appear in nature; instead, the periodic properties of matter display gradual changes from one substance to another.

The Key Atomic Properties

Four atomic properties are critical to the behavior of an element: *electron configuration, atomic size, ionization energy,* and *electronegativity.*

ARIS Presentation Center, Art Library

Electron configuration ($nl^{\#}$) is the distribution of electrons in the energy levels and sublevels of an atom *(Sections 7.4 and 8.3):*
- The n value (positive integer) indicates the *energy* and *distance* from the nucleus of orbitals in a level: higher value of n means higher energy and greater distance.
- The l value (or its more commonly used letter designation s, p, d, f) indicates the *shape* of orbitals in the sublevel *(right).*
- The superscript ($^{\#}$) tells the *number* of electrons in the sublevel.

The periodic table above shows the sublevel blocks of elements and the ground-state (lowest energy), outer (valence) electron configuration of the main-group (*s-* and *p*-block) elements. Note that
- The ***s* block** lies to the left of the transition elements and the ***p* block** lies to the right. [Although H is not an alkali metal, like the elements in Group 1A(1), and He belongs in Group 8A(18), both are part of the *s* block.]
- Outer electron configurations are *similar within a group.*
- Outer electron configurations are *different within a period.*
- Outer electrons occupy the ns and np sublevels (n = period number).
- Four valence-level orbitals (one ns + three np) occur among the main-group elements.
- Outer electrons *are* the valence electrons of the main-group elements.
- The A-group number (**1A** to **8A**) equals the number of valence electrons.

Outer electrons are *shielded* from the full nuclear charge by electrons in the same level and, especially, by those in lower energy levels. Shielding reduces the attraction they experience to a much lower *effective nuclear charge* (Z_{eff}) *(Section 8.2).* Within a level, electrons that *penetrate* more (spend more time near the nucleus) shield others more effectively. The extent of penetration is in the order $s > p > d > f$. The radial probability distribution curves *(right)* show that
- Inner ($n = 1$) electrons shield outer ($n = 2$) electrons very effectively.
- $2s$ electrons shield $2p$ electrons somewhat because $2s$ electrons spend more time near the nucleus.

Z_{eff} greatly influences atomic properties. In general,
- Z_{eff} increases significantly left to right *across* a period.
- Z_{eff} increases slightly *down* a group.

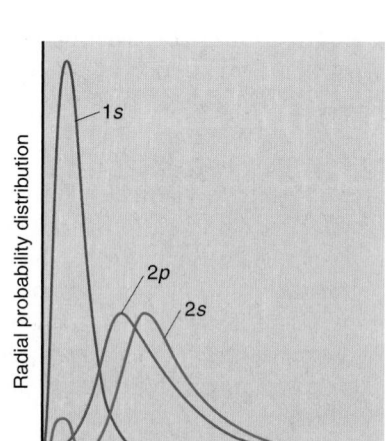

Atomic size is based on atomic radius, one-half the distance between nuclei of identical, bonded atoms *(Section 8.4)*. The small red periodic table shows the trends in atomic size among the main-group elements. Note that

- Atomic size generally *decreases left to right across a period:* increasing Z_{eff} pulls outer electrons closer.
- Atomic size generally *increases down a group:* outer electrons in higher periods lie farther from the nucleus.

Thus, the trends in IE are *opposite* those in atomic size: it is easier to remove an electron (lower IE) that is farther from the nucleus (larger atomic size).

Ionization energy (IE) is the energy required to remove the highest energy electron from 1 mol of gaseous atoms *(Section 8.4)*. The relative magnitude of the IE influences the types of bonds an atom forms: an element with a *low IE* is more likely to *lose* electrons, and one with a *high IE* is more likely to *share (or gain)* electrons (excluding the noble gases). From the small yellow periodic table, note that

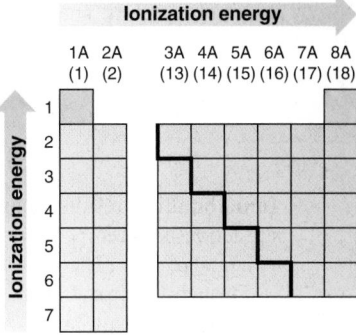

- IE generally *increases left to right across a period:* higher Z_{eff} holds electrons tighter.
- IE generally *decreases down a group:* greater distance from the nucleus lowers the attraction for electrons.

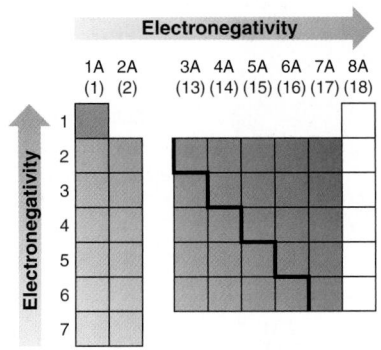

Electronegativity (EN) is a number that refers to the relative ability of an atom in a covalent bond to attract shared electrons *(Section 9.5)*. From the small green periodic table, note that

- EN generally *increases left to right across a period:* higher Z_{eff} and shorter distance from the nucleus strengthen the attraction for the shared pair.
- EN generally *decreases down a group:* greater distance from the nucleus weakens the attraction for the shared pair. (Group 8A is not shaded because the noble gases form few compounds.)

Thus, the trends in EN are *opposite* those in atomic size and the *same* as those in IE.

The graph shows simplified plots of atomic size *(red),* ionization energy *(yellow),* and electronegativity *(green)* versus atomic number. Note that

- Atomic size decreases gradually left to right within a period *(vertical gray band)* and then increases suddenly at the beginning of the next period.
- IE and EN display the opposite pattern, increasing left to right within a period and decreasing at the beginning of the next period.

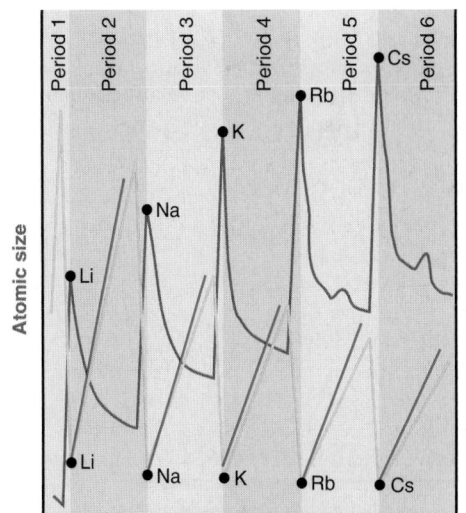

The difference in electronegativity (ΔEN) between the atoms in a bond greatly influences physical and chemical behavior of the compound, as the following diagram shows.

ΔEN → Bond polarity → Molecular polarity → Intermolecular forces → Physical properties / Solubility / Reactivity

Characteristics of Chemical Bonding

Chemical bonds are the forces that hold atoms (or ions) together in an element or compound. The type of bonding, bond properties, nature of orbital overlap, and number of bonds determine physical and chemical behavior.

Types of Bonding

There are three idealized bonding models: *ionic, covalent,* and *metallic*.

Covalent bonding results from the attraction between two nuclei and a localized electron pair. The bond arises through electron sharing between atoms with a small ΔEN (usually two nonmetals) and leads to discrete molecules with specific shapes or to extended networks *(Section 9.3)*.

Ionic bonding results from the attraction between positive and negative ions. The ions arise through electron transfer between atoms with a large ΔEN (from metal to nonmetal). This bonding leads to crystalline solids with ions packed tightly in regular arrays *(Section 9.2)*.

Metallic bonding results from the attraction between the cores of metal atoms (metal cations) and their delocalized valence electrons. This bonding arises through the shared pooling of valence electrons from many atoms and leads to crystalline solids *(Sections 9.6 and 12.6)*.

The actual bonding in real substances usually lies between these distinct models *(Section 9.5)*. The electron density relief maps show a small overlap region even in ionic bonding (NaCl). This region increases in polar covalent bonding (an SiCl bond from $SiCl_4$) and even more in nonpolar covalent bonding (Cl_2).

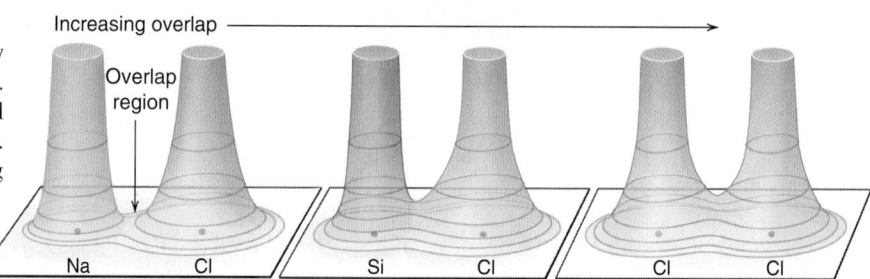

Increasing overlap

Overlap region

Na   Cl   Si   Cl   Cl   Cl

The triangular diagram shows the continuum of bond types among all the Period 3 main-group elements:

- Along the left side of the triangle, compounds of each element with chlorine display a gradual change from ionic to covalent bonding and a *decrease* in bond polarity from bottom to top.

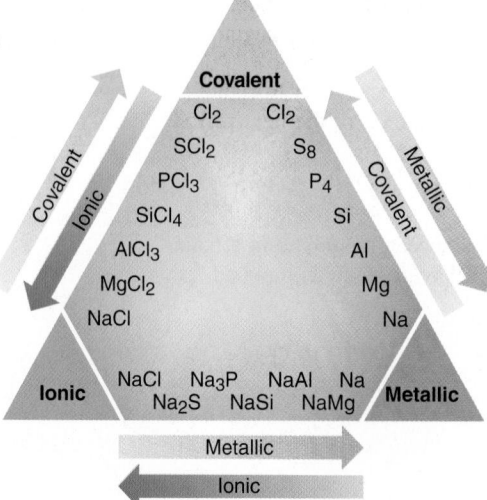

Covalent

Covalent   Metallic
         Ionic   Covalent

Cl_2 Cl_2
SCl_2 S_8
PCl_3 P_4
$SiCl_4$ Si
$AlCl_3$ Al
$MgCl_2$ Mg
NaCl   Na

Ionic   Metallic

NaCl Na_3P NaAl   Na
Na_2S NaSi   NaMg

Metallic

Ionic

- Along the right side, the elements themselves display a gradual change from covalent to metallic bonding.

- Along the base, compounds of each element with sodium display a gradual change from ionic to metallic bonding and, once again, a *decrease* in bond polarity from left to right.

Bond Properties

There are two important properties of a covalent bond *(Section 9.3):*

Bond length is the distance between the nuclei of bonded atoms.
Bond energy (bond strength) is the enthalpy change required to break a given bond in 1 mol of gaseous molecules.

Among similar compounds, these bond properties are related to each other and to reactivity, as shown in the graph for the carbon tetrahalides (CX_4). Note that

- *As bond length increases, bond energy decreases:* shorter bonds are stronger bonds.
- *As bond energy decreases, reactivity increases.*

Nature of Orbital Overlap

In a covalent bond, the shared electrons reside in the entire region composed of the overlapping orbitals of the two atoms. The diagram depicts the bonding in ethylene (C_2H_4).

Bond order is one-half the number of electrons shared. Bond orders of 1 (single bond) and 2 (double bond) are common; a bond order of 3 (triple bond) is much less common. Fractional bond orders occur when there are resonance structures for species with adjacent single and double bonds *(Sections 9.3 and 10.1).*

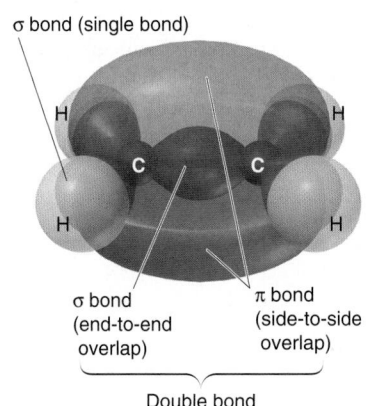

σ bond (single bond)

σ bond (end-to-end overlap)

π bond (side-to-side overlap)

Double bond

Orbitals overlap in two ways, which leads to two types of bonds *(Section 11.2):*

- **End-to-end overlap** (of *s*, *p*, and hybrid atomic orbitals) leads to a sigma (σ) bond, one with electron density distributed symmetrically along the bond axis. A single bond is a σ bond.
- **Side-to-side overlap** (of *p* with *p*, or sometimes *d*, orbitals) leads to a pi (π) bond, one with electron density distributed above and below the bond axis. A double bond consists of one σ bond and one π bond. A π bond restricts rotation around the bond axis, allowing for different spatial arrangements of the atoms and, therefore, different compounds. Pi bonds are often sites of reactivity; for example,

$$CH_2 = CH_2(g) + H - Cl(g) \longrightarrow CH_3 - CH_2 - Cl(g)$$
$$CH_3 - CH_3(g) + H - Cl(g) \longrightarrow \text{no reaction}$$

Number of Bonds and Molecular Shape

The shape of a molecule is defined by the positions of the nuclei of the bonded atoms. According to VSEPR theory *(Section 10.2),* the number of electron groups in the valence level of a central atom, which is based on the number of bonding and lone pairs, is the key factor that determines molecular shape. The small periodic table shows that

- *The elements in Period 2 cannot form more than four bonds* because they have a maximum of four (one *s* and three *p*) valence orbitals. (Only carbon forms four bonds routinely.) Molecular shapes *(small circle)* are based on linear, trigonal planar, and tetrahedral electron-group arrangements.
- *Many elements in Period 3 or higher can form more than four bonds* by using empty *d* orbitals and, thus, expanding their valence levels. Shapes include those above and others based on trigonal bipyramidal and octahedral electron-group arrangements *(large circle).*

Period 2

Periods 3–6

Cannot form more than four bonds

Can form more than four bonds

Metallic Behavior

The elements are often classified as metals, metalloids, or nonmetals. Although exceptions exist, this table contrasts the general atomic, physical, and chemical properties of metals with those of nonmetals *(Section 8.5)*.

	Metals	**Nonmetals**
Atomic Properties	Have fewer valence electrons (in period) Have larger atomic size Have lower ionization energies Have lower electronegativities	Have more valence electrons (in period) Have smaller atomic size Have higher ionization energies Have higher electronegativities
Physical Properties	Occur as solids at room temperature Conduct electricity and heat well Are malleable and ductile	Occur in all three physical states Conduct electricity and heat poorly Are not malleable or ductile
Chemical Properties	Lose electron(s) to become cations React with nonmetals to form ionic compounds Mix with other metals to form solid solutions (alloys)	Gain electron(s) to become anions React with metals to form ionic compounds React with other nonmetals to form covalent compounds

 Transparency/ARIS Presentation Center, Art Library

This small periodic table shows the location of metals, metalloids, and nonmetals among the main-group elements. Note that

- Metals lie in the lower-left portion of the table.
- Nonmetals lie in the upper-right portion of the table.
- Metalloids lie between the metals and the nonmetals. These elements have intermediate values of atomic size, IE, and EN and display an intermediate metallic character: shiny solids with low conductivity; cation-like with nonmetals (e.g., AsF_3); and anion-like with metals (e.g., Na_3As).

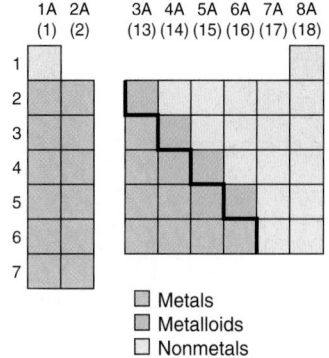

☐ Metals
☐ Metalloids
☐ Nonmetals

From the blue, shaded periodic table, note that
- *Metallic behavior* changes gradually among the elements.
- *Metallic behavior parallels atomic size:* Larger members of a group *(bottom)* or period *(left)* are more metallic; smaller members are less metallic.

Metals and nonmetals typically form crystalline ionic compounds when they react with each other, and ionic size and charge determine the packing in these solids. Monatomic ions exhibit clear size trends:
- Cations are smaller than their parent atoms, and anions are larger.
- Ionic size *increases down* a group.
- Ionic size *decreases left to right across* a period, but anions are much larger than cations.

Within an isoelectronic series of ions, such as the anions of Period 2 and the cations of Period 3, size decreases in the following order:
$$N^{3-} > O^{2-} > F^- > Na^+ > Mg^{2+} > Al^{3+}$$

Acid-Base Behavior of the Element Oxides

Oxides are known for almost every element. The metallic behavior of an element correlates with the acid-base behavior of its oxide in water *(Section 8.5)*. Recall that

- An acid produces H^+ ions when dissolved in water and reacts with a base to form a salt and water.
- A base produces OH^- ions when dissolved in water and reacts with an acid to form a salt and water.

Transparency/ARIS Presentation Center, Art Library

This chart shows that the electronegativity and metallic behavior of an element (E) determine the type of bonding (E-to-O) in its oxide and, therefore, the acid-base behavior of the oxide dissolved in water.

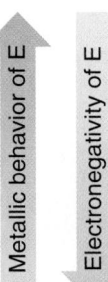

Basic Oxide (Ionic E-to-O)	$+ H_2O \longrightarrow$	more OH^-
	$+ \text{acid} \longrightarrow$	salt (E cation) $+ H_2O$
Amphoteric Oxide	$+ \text{acid} \longrightarrow$	salt (E cation) $+ H_2O$
	$+ \text{base} \longrightarrow$	salt (E oxoanion) $+ H_2O$
Acidic Oxide (Covalent E-to-O)	$+ \text{base} \longrightarrow$	salt (E oxoanion) $+ H_2O$
	$+ H_2O \longrightarrow$	more H^+

Note that, among the main groups,

- *Elements with low EN (metals) form basic oxides.* The E-to-O bonding is ionic.
  The O^{2-} ion is the basic species:
  $$O^{2-}(s) + H_2O(l) \longrightarrow 2OH^-(aq)$$
For example, barium forms the basic oxide BaO:
  Reacts with water: $BaO(s) + H_2O(l) \longrightarrow Ba^{2+}(aq) + 2OH^-(aq)$
  Reacts with acid: $BaO(s) + 2H^+(aq) \longrightarrow Ba^{2+}(aq) + H_2O(l)$
- *Elements with high EN (nonmetals) form acidic oxides.* The E-to-O bonding is covalent.
  Water bonds to E of the oxide to form an acid, which releases H^+. For example, sulfur forms the acidic oxide SO_2:
  Reacts with water: $SO_2(g) + H_2O(l) \rightleftharpoons H_2SO_3(aq) \rightleftharpoons H^+(aq) + HSO_3^-(aq)$
  Reacts with base: $SO_2(g) + 2OH^-(aq) \longrightarrow SO_3^{2-}(aq) + H_2O(l)$
- *Elements with intermediate EN (some metalloids and metals) form amphoteric oxides, which react with acid and base.* For example, Al_2O_3 is an amphoteric oxide:
  Reacts with acid: $Al_2O_3(s) + 6H^+(aq) \longrightarrow 2Al^{3+}(aq) + 3H_2O(l)$
  Reacts with base: $Al_2O_3(s) + 2OH^-(aq) + 3H_2O(l) \longrightarrow 2Al(OH)_4^-(aq)$

	1A (1)	2A (2)	3A (13)	4A (14)	5A (15)	6A (16)	7A (17)	8A (18)
1								
2	Li_2O	BeO	B_2O_3	CO_2	N_2O_5 N_2O_3			
3	Na_2O	MgO	Al_2O_3	SiO_2	P_4O_{10} P_4O_6	SO_3 SO_2	Cl_2O_7 Cl_2O	
4	K_2O	CaO	Ga_2O_3	GeO_2	As_2O_5 As_4O_6	SeO_3 SeO_2	Br_2O	
5	Rb_2O	SrO	In_2O_3 In_2O	SnO_2 SnO	Sb_2O_5 Sb_4O_6	TeO_3 TeO_2	I_2O_5	
6	Cs_2O	BaO	Tl_2O	PbO_2 PbO	Bi_2O_3	PoO_2 PoO		
7	Fr_2O	RaO						

- ☐ Strongly basic
- ☐ Weakly basic
- ☐ Amphoteric
- ☐ Strongly acidic
- ☐ Moderately acidic
- ☐ Weakly acidic

The periodic table shows the acid-base behavior of many main-group oxides in water. Note that

- *Oxide acidity increases left to right across a period and decreases down a group,* which is *opposite* the trend in metallic behavior (and atomic size).
- When an element forms two oxides, *the element has a higher oxidation number in the more acidic oxide* (see also Topic 5). Thus, for example, SO_2 forms the weak acid H_2SO_3, whereas SO_3 forms the strong acid H_2SO_4.

Metallic behavior of E

Electronegativity of E

Topic 4

Redox Behavior of the Elements

The relative ability of an element to lose or gain electrons (or electron charge) when reacting with other elements defines its redox (oxidation-reduction) behavior *(Section 4.5):*

- *The oxidation number, O.N. (or oxidation state)* of an atom in an element is zero. In a compound, the O.N. is the number of electrons that have shifted away from the atom (positive O.N.) or toward it (negative O.N.). O.N. values are determined by a set of rules *(Section 4.5).*
- *An oxidation-reduction (redox) reaction* occurs when the O.N. values of any atoms in the reactants are different from those in the products.
- Among the countless redox reactions are all those that involve *an elemental substance as reactant or product.* These include all combustion reactions and all formation reactions, such as this one between potassium and chlorine: $2K + Cl_2 \longrightarrow 2KCl$

Reducing and Oxidizing Agents

All redox reactions include a reducing *and* an oxidizing agent among the reactants. This chart shows that

- The reducing agent gives electrons *to* (reduces) the oxidizing agent and becomes oxidized (more positive O.N.).
- The oxidizing agent removes electrons *from* (oxidizes) the reducing agent and becomes reduced (more negative O.N.).

Reducing agent	Oxidizing agent
X loses electrons	Y gains electrons
X is oxidized by Y	Y is reduced by X
X oxidation number increases	Y oxidation number decreases

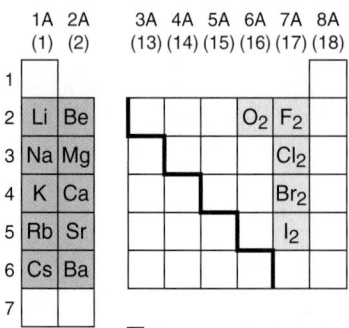

☐ Strong reducing agent
☐ Strong oxidizing agent

The small periodic table shows that oxidizing and reducing ability are related to atomic properties:

- Elements with low IE and low EN [Groups 1A(1) and 2A(2)] are *strong reducing agents.*
- Elements with high IE and high EN [Group 7A(17) and oxygen in Group 6A(16)] are *strong oxidizing agents.*

Oxidation States of the Main-Group Elements

The periodic table below shows some oxidation states (most common states in **bold** type) of the elements in their compounds. Note that

- *The highest (most positive) state in a group equals the A-group number.* It occurs when all of an atom's outer (valence) electrons shift toward a *more* electronegative atom.
- *Among nonmetals, the lowest (most negative) state equals the A-group number minus eight.* It occurs when a nonmetal atom fills its outer level with electrons that shift away from a *less* electronegative atom.
- *Nonmetals have more oxidation states than metals* in the same group or period. (Oxygen and fluorine are exceptions.)
- Odd-numbered oxidation states are the most common ones in odd-numbered groups, and even-numbered states are the most common ones in even-numbered groups. *Oxidation states differ by units of two* because electrons are lost (or gained) in pairs *(Section 14.9).*
- For many metals and metalloids with more than one oxidation state [Groups 3A(13) to 5A(15)], *the lower state becomes more common down the group.* This state results when np electrons only are lost *(Section 8.5).*
- An element with more than one oxidation state exhibits *greater metallic behavior in its lower state.* For example, arsenic(III) oxide is more basic, more like a metal oxide, than is arsenic(V) oxide (see Topic 4).

 Transparency/ARIS Presentation Center, Art Library

□ Metals
□ Metalloids
□ Nonmetals

	1A (1)	2A (2)	3A (13)	4A (14)	5A (15)	6A (16)	7A (17)	8A (18)
1	**H** −1, **+1**							**He**
2	**Li** +1	**Be** +2	**B** +3	**C** −4, **+4,** +2	**N** −3, **+5,** +4, +3, +2, +1	**O** **−1,** −2	**F** **−1**	**Ne**
3	**Na** +1	**Mg** +2	**Al** +3	**Si** −4, **+4,** +2	**P** −3, **+5,** +3	**S** **−2,** +6, +4, +2	**Cl** **−1,** +7, +5, +3, +1	**Ar**
4	**K** +1	**Ca** +2	**Ga** +3, +1	**Ge** +4, +2	**As** −3, **+5,** +3	**Se** **−2,** +6, +4, +2	**Br** **−1,** +7, +5, +3, +1	**Kr** +2
5	**Rb** +1	**Sr** +2	**In** +3, +1	**Sn** +4, +2	**Sb** −3, **+5,** +3	**Te** **−2,** +6, +4, +2	**I** **−1,** +7, +5, +3, +1	**Xe** **+8, +6,** +4, +2
6	**Cs** +1	**Ba** +2	**Tl** +1	**Pb** +4, +2	**Bi** +3	**Po** +4, +2	**At** −1	**Rn** +2
7	**Fr** +1	**Ra** +2						

Physical States and Phase Changes

Physical state and the heat of a phase change reflect the relative strengths of the bonding and/or intermolecular forces between the atoms, ions, or molecules that make up an element or compound *(Sections 12.2, 12.3, 12.4, and 12.6).*

Physical States of the Elements

The large periodic table shows the type of particle or solid structure *(shape in box),* physical state *(color of shape),* and dominant interparticle force *(background color of box)* in the most common form of each of the main-group elements at room temperature. Note that

* *Metals (left and bottom) are solids:* strong metallic bonding holds the atoms in their crystal structures.
* *Metalloids (along staircase line) and carbon are solids:* strong covalent bonding holds the atoms together in extensive networks.
* *Lighter nonmetals and Group 8A(18) (right and top) are gases:* dispersion forces are weak between molecules (H_2, N_2, O_2, F_2, Cl_2) or atoms with smaller, less polarizable electron clouds.
* *Heavier nonmetals are liquid (Br_2) or soft solids (P_4, S_8, and I_2):* dispersion forces are stronger between molecules with larger, more polarizable electron clouds.

Transparency/ARIS Presentation Center, Art Library

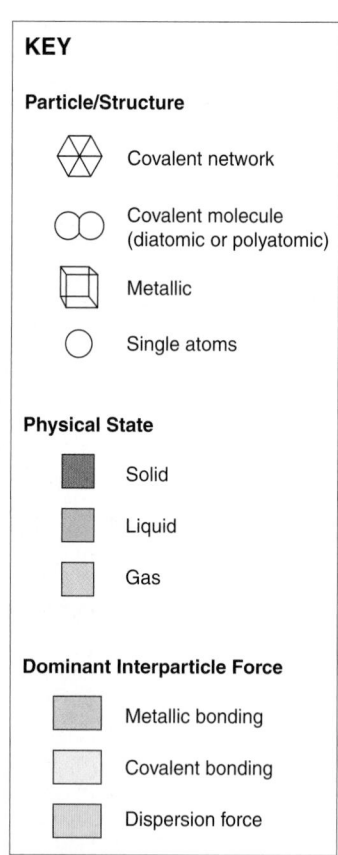

Phase Changes of the Elements

The small periodic table shows trends in melting point, boiling point, ΔH_{fus}, and ΔH_{vap} for Groups 1A(1), 7A(17), and 8A(18):

- In Group 1A(1), these properties generally *increase up* the group: smaller atom cores attract delocalized electrons more strongly, which leads to stronger metallic bonding.
- In Groups 7A(17) and 8A(18), these properties generally *increase down* the group: dispersion forces become stronger with larger, more polarizable atoms.
- In Groups 3A(13) to 6A(16), these properties reflect changes in interparticle forces down the group: lower values for molecular nonmetals, higher values for covalent networks of metalloids (and carbon), and intermediate values for metals.

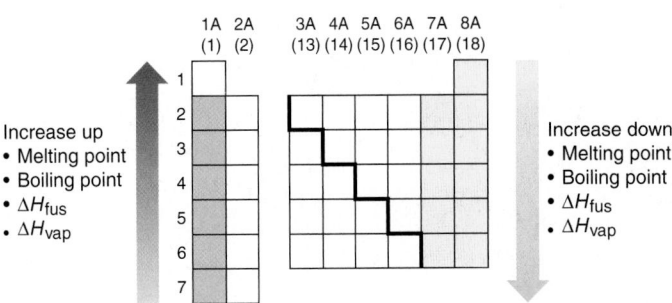

Physical Properties of Compounds

Compounds have physical properties based on the types of bonding and intermolecular forces.

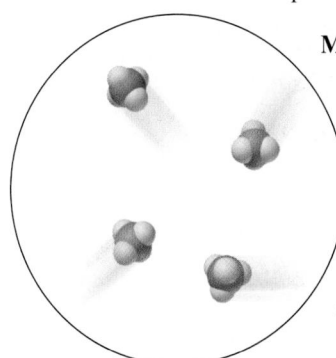

Molecular compounds, such as methane, have a physical state that depends on intermolecular forces. For polar compounds, dipole-dipole forces dominate; for non-polar compounds, dispersion forces are most important. Most of these substances are gases, liquids, or low-melting solids at room temperature.

Hydrogen bonding arises when H is bonded to N, O, or F. It has a major effect on physical properties. For example, despite similar molar masses, H bonding in H_2O (18.02 g/mol) gives it a much higher melting point, boiling point, ΔH_{fus}, and ΔH_{vap} than CH_4 (16.04 g/mol). Water's unusually high specific heat capacity, surface tension, and viscosity also result from H bonding *(Section 12.5)*.

Network covalent compounds, such as silica, have extremely high melting points, boiling points, ΔH_{fus}, and ΔH_{vap}.

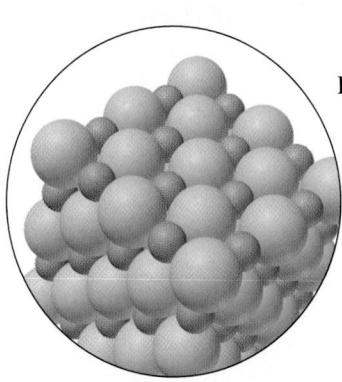

Ionic compounds, such as sodium chloride, have very high melting points, boiling points, ΔH_{fus}, and ΔH_{vap}.

14

Imposing Importance The world's largest periodic table was displayed at the 2006 Chicago Science Expedition in the Richard J. Daley Center to emphasize the importance of science in everyday life. In this chapter, we correlate patterns in the periodic table with patterns in the physical and chemical behavior of the main-group elements.

Periodic Patterns in the Main-Group Elements

14.1 Hydrogen, the Simplest Atom
Highlights of Hydrogen Chemistry

14.2 Trends Across the Periodic Table: The Period 2 Elements

14.3 Group 1A(1): The Alkali Metals

14.4 Group 2A(2): The Alkaline Earth Metals

14.5 Group 3A(13): The Boron Family
Highlights of Boron Chemistry

14.6 Group 4A(14): The Carbon Family
Highlights of Carbon Chemistry
Highlights of Silicon Chemistry

14.7 Group 5A(15): The Nitrogen Family
Highlights of Nitrogen Chemistry
Highlights of Phosphorus Chemistry

14.8 Group 6A(16): The Oxygen Family
Highlights of Oxygen Chemistry
Highlights of Sulfur Chemistry

14.9 Group 7A(17): The Halogens
Highlights of Halogen Chemistry

14.10 Group 8A(18): The Noble Gases

n your study of chemistry so far, you've learned how to name compounds, balance equations, and calculate reaction yields. You've seen how heat is related to chemical and physical change, how electron configuration influences atomic properties, how elements bond to form compounds, and how the arrangement of bonding and lone pairs accounts for molecular shapes. You've learned modern theories of bonding and, most recently, seen how atomic and molecular properties give rise to the macroscopic properties of gases, liquids, solids, and solutions.

The purpose of this knowledge, of course, is to make sense of the magnificent diversity of chemical and physical behavior around you. The periodic table, which organizes much of this diversity, was derived from chemical facts observed in countless hours of 18th- and 19th-century research. One of the greatest achievements in science is 20th-century quantum theory, which provides a theoretical foundation for the periodic table's arrangement. But bear in mind that one theory can rarely account for *all* the facts. Therefore, be amazed by the predictive power of the patterns in the table, but do not be concerned by the occasional exceptions to these patterns. After all, our models are simple, but nature is complex.

IN THIS CHAPTER . . . We apply general ideas of bonding, structure, and reactivity to the main-group elements and see how their behavior correlates with their position in the periodic table. We begin with hydrogen, the simplest and, in some ways, most important of the elements, and then consider Period 2 to survey the general changes across the periodic table. In each of the remaining sections, we deal with one of the eight families of main-group elements. We explore vertical trends in each group and also keep an eye on horizontal trends involving neighboring groups. Especially important elements—boron, carbon, silicon, nitrogen, phosphorus, oxygen, sulfur, and the halogens—receive greater attention. (If you haven't already read the Interchapter preceding this chapter, now is the perfect time.)

Concepts & Skills to Review
before you study this chapter

- see the Interchapter

14.1 HYDROGEN, THE SIMPLEST ATOM

A hydrogen atom consists of a nucleus with a single positive charge, surrounded by a single electron. Despite this simple structure, or perhaps because of it, hydrogen may be the most important element of all. In the Sun, hydrogen (H) nuclei combine to form helium (He) nuclei in a process that provides nearly all Earth's energy. About 90% of all the atoms in the universe are H atoms, making it the most abundant element by far. On Earth, only tiny amounts of the free element occur naturally (as H_2), but hydrogen is abundant in combination with oxygen in water. Because of its simple structure and low molar mass, nonpolar gaseous H_2 is colorless and odorless, and its extremely weak dispersion forces result in very low melting ($-259°C$) and boiling points ($-253°C$).

Where Does Hydrogen Fit in the Periodic Table?

Hydrogen has no perfectly suitable position in the periodic table (Figure 14.1). Because of its single valence electron and common +1 oxidation state, some think it almost fits in Group 1A(1). However, unlike the alkali metals, hydrogen *shares* its electron with nonmetals rather than losing it to them. Moreover, like a nonmetal, it has a relatively high ionization energy (IE = 1311 kJ/mol)—much higher than that of lithium (IE = 520 kJ/mol), whose IE is the highest in Group 1A— and a much higher electronegativity as well (EN of H = 2.1; EN of Li = 1.0).

Others think hydrogen almost fits in Group 7A(17). Like the halogens, it occurs as diatomic molecules and fills its outer (valence) level either by electron sharing or by gaining one electron from a metal to form a monatomic anion, the hydride ion (H⁻; O.N. = −1). However, hydrogen has a lower electronegativity

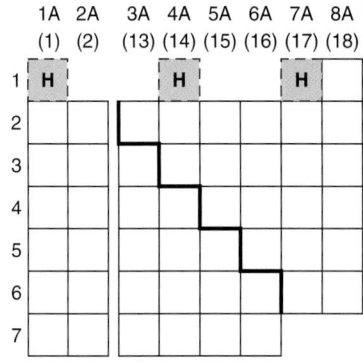

Figure 14.1 Where does hydrogen belong? Hydrogen's small size and single electron confer properties that are not fully consistent with the members of any one periodic group. Depending on the property, hydrogen may fit better in Group 1A(1), 4A(14), or 7A(17).

ARIS Presentation Center, Art Library

Literature Cronyn, Marshall W. "The Proper Place for Hydrogen in the Periodic Table," *J. Chem. Educ.* **2003**, *80*, 947.

(EN = 2.1) than any of the halogens (ENs range from 2.2 to 4.0) and lacks their three valence electron pairs. Moreover, the H^- ion is rare and reactive, whereas halide ions are common and stable.

And, still others think that, because of a half-filled valence level and similarities in ionization energy, electron affinity, electronegativity, and bond energies, hydrogen fits best in Group 4A(14).

Hydrogen's unique behavior is attributable to its tiny size. Hydrogen has a high IE because its electron is very close to the nucleus, with no inner electrons to shield it from the positive charge. It has a low EN (for a nonmetal) because it has only one proton to attract bonding electrons. In this chapter, hydrogen will appear in either Group 1A(1) or 7A(17) depending on the property being considered.

Highlights of Hydrogen Chemistry

In Chapters 12 and 13, we discussed hydrogen bonding and its impact on physical properties (melting and boiling points, heats of fusion and vaporization, and specific heat capacities) and solubilities. You learned that H bonding plays a critical part in stabilizing Earth's climate as well as your body temperature, and you glimpsed its role in the functional structures of biomolecules. In Chapter 6 (see Chemical Connections, p. 258) we considered the possible future use of hydrogen as a fuel, and in Chapter 22 we'll see how hydrogen is used to recover some metals from their ores. Hydrogen is very reactive, combining with nearly every element, so we focus here on the three types of hydrides.

Ionic (Saltlike) Hydrides With very reactive metals, such as those in Group 1A(1) and the larger members of Group 2A(2) (Ca, Sr, and Ba), hydrogen forms *saltlike hydrides*—white, crystalline solids composed of the metal cation and the hydride ion:

$$2Li(s) + H_2(g) \longrightarrow 2LiH(s)$$
$$Ca(s) + H_2(g) \longrightarrow CaH_2(s)$$

In water, H^- is a strong base that pulls H^+ from surrounding H_2O molecules to form H_2 and OH^-:

$$NaH(s) + H_2O(l) \longrightarrow Na^+(aq) + OH^-(aq) + H_2(g)$$

The hydride ion is also a powerful reducing agent; for example, it reduces Ti(IV) to the free metal:

$$TiCl_4(l) + 4LiH(s) \longrightarrow Ti(s) + 4LiCl(s) + 2H_2(g)$$

Covalent (Molecular) Hydrides Hydrogen reacts with nonmetals to form many *covalent hydrides,* such as CH_4, NH_3, H_2O, and HF. Most are gases consisting of small molecules, but many hydrides of boron and carbon are liquids or solids that consist of much larger molecules. In most covalent hydrides, hydrogen has an oxidation number of +1 because the other nonmetal has a higher electronegativity than hydrogen.

Conditions for preparing the covalent hydrides depend on the reactivity of the other nonmetal. For example, with stable, triple-bonded N_2, hydrogen reacts at high temperatures (~400°C) and pressures (~250 atm), and the reaction needs a catalyst to proceed at any practical speed:

$$N_2(g) + 3H_2(g) \xrightarrow{\text{catalyst}} 2NH_3(g) \quad \Delta H^\circ_{rxn} = -91.8\,kJ$$

Industrial facilities throughout the world use this reaction to produce millions of tons of ammonia each year for fertilizers, explosives, and synthetic fibers. On the other hand, hydrogen combines rapidly with reactive, single-bonded F_2, even at extremely low temperatures ($-196°C$):

$$F_2(g) + H_2(g) \longrightarrow 2HF(g) \quad \Delta H°_{rxn} = -546 \text{ kJ}$$

Metallic (Interstitial) Hydrides Many transition elements form *metallic (interstitial) hydrides,* in which H_2 molecules (and H atoms) occupy the holes in the metal's crystal structure. Thus, such hydrides are *not* compounds but gas-solid solutions. Also, unlike ionic and covalent hydrides, interstitial hydrides, such as $TiH_{1.7}$, typically lack a single, stoichiometric formula because the metal can incorporate variable amounts of hydrogen, depending on the pressure and temperature of the gas. ●

14.2 TRENDS ACROSS THE PERIODIC TABLE: THE PERIOD 2 ELEMENTS

Table 14.1 on pages 568 and 569 presents the trends in atomic properties of the Period 2 elements, lithium through neon, and the physical and chemical properties that emerge from them. In general, these trends apply to the other periods as well. Note the following points:

- Electrons fill the one *ns* and the three *np* orbitals according to Pauli's exclusion principle and Hund's rule.
- As a result of increasing nuclear charge and the addition of electrons to orbitals of the same energy level (same *n* value), atomic size generally decreases, whereas first ionization energy and electronegativity generally increase (see bar graphs on p. 569).
- Metallic character decreases with increasing nuclear charge as period members change from metals to metalloids to nonmetals.
- Reactivity is highest at the left and right ends of the period, except for the inert noble gas.
- Bonding between atoms of an element changes from metallic, to covalent in networks, to covalent in individual molecules, to none (noble gases exist as separate atoms). As expected, physical properties, such as melting point, change abruptly at the network/molecule boundary, which occurs in Period 2 between carbon (solid) and nitrogen (gas).
- Bonding between each element and an active nonmetal changes from ionic, to polar covalent, to covalent. Bonding between each element and an active metal changes from metallic to polar covalent to ionic.
- The acid-base behavior of the common element oxide in water changes from basic to amphoteric to acidic as the bond between the element and oxygen becomes more covalent.
- Reducing strength decreases through the metals, and oxidizing strength increases through the nonmetals. In Period 2, common oxidation numbers (O.N.s) equal the A-group number for Li and Be and the A-group number minus eight for O and F. Boron has several O.N.s, Ne has none, and C and N show all possible O.N.s for their groups.

One point that will be coming up frequently but is not shown in Table 14.1 is the *anomalous behavior of the Period 2 elements within their groups.* Later, we'll highlight some particularly clear examples of such *un*representative behavior, which arises from the relatively *small size* and *small number of orbitals* in the outer energy level of these elements.

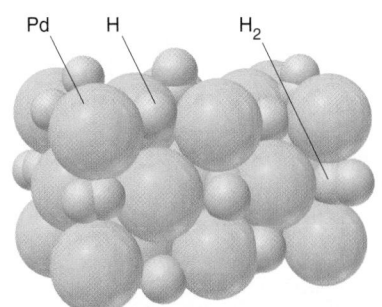

● **Fill 'Er Up with Hydrogen?** Dreams of a future hydrogen economy had included metallic hydrides as "storage containers" for hydrogen fuel in cars. Just drive up to your local refueling station, insert a cylinder of H-packed metal in your car, and zip away again! Some metals, such as palladium (see figure) and niobium, and some alloys, such as $LaNi_5$, store large amounts of hydrogen. However, the metals that store the most hydrogen are expensive and heavy and they hold the gas tightest, releasing it only at very high temperatures. As a result, in 1996, U.S. government funding for research on metallic hydrides stopped. Some researchers are keeping the dream alive, however, through their studies of carbon nanotubes, which store large amounts of H_2, and of an Na/Al alloy that is lightweight, stores 5% H_2 by mass, and releases it below 200°C.

  ARIS Presentation Center, Art Library

Green chemistry application

 Literature Rayner-Canham, Geoffrey. "Periodic Patterns," *J. Chem. Educ.* **2000,** *77,* 1053.

Table 14.1 Trends in Atomic, Physical, and Chemical Properties of the Period 2 Elements

Group: Element/At. No.:	1A(1) Lithium (Li) $Z = 3$	2A(2) Beryllium (Be) $Z = 4$	3A(13) Boron (B) $Z = 5$	4A(14) Carbon (C) $Z = 6$
Atomic Properties				
Condensed electron configuration; partial orbital diagram	[He] $2s^1$ $\uparrow$ 2s 2p	[He] $2s^2$ $\uparrow\downarrow$ 2s 2p	[He] $2s^2 2p^1$ $\uparrow\downarrow$ $\uparrow$ 2s 2p	[He] $2s^2 2p^2$ $\uparrow\downarrow$ $\uparrow$ $\uparrow$ 2s 2p
Physical Properties				
Appearance				
Metallic character	Metal	Metal	Metalloid	Nonmetal
Hardness	Soft	Hard	Very hard	Graphite: soft Diamond: extremely hard
Melting point/ boiling point	Low mp for a metal	High mp	Extremely high mp	Extremely high mp
Chemical Properties				
General reactivity	Reactive	Low reactivity at room temperature	Low reactivity at room temperature	Low reactivity at room temperature; graphite more reactive
Bonding among atoms of element	Metallic	Metallic	Network covalent	Network covalent
Bonding with nonmetals	Ionic	Polar covalent	Polar covalent	Covalent (π bonds common)
Bonding with metals	Metallic	Metallic	Polar covalent	Polar covalent
Acid-base behavior of common oxide	Strongly basic	Amphoteric	Very weakly acidic	Very weakly acidic
Redox behavior (O.N.)	Strong reducing agent (+1)	Moderately strong reducing agent (+2)	Complex hydrides good reducing agents (+3, −3)	Every oxidation state from +4 to −4
Relevance/Uses of Element and Compounds				
	Li soaps for auto grease; thermonuclear bombs; high-voltage, low-weight batteries; treatment of bipolar disorders (Li_2CO_3)	Rocket nose cones; alloys for springs and gears; nuclear reactor parts; x-ray tubes	Cleaning agent (borax); eyewash, antiseptic (boric acid); armor (B_4C); borosilicate glass; plant nutrient	Graphite: lubricant, structural fiber Diamond: jewelry, cutting tools, protective films Limestone ($CaCO_3$) Organic compounds: drugs, fuels, textiles, etc.

5A(15) Nitrogen (N) $Z = 7$	6A(16) Oxygen (O) $Z = 8$	7A(17) Fluorine (F) $Z = 9$	8A(18) Neon (Ne) $Z = 10$
[He] $2s^2 2p^3$	[He] $2s^2 2p^4$	[He] $2s^2 2p^5$	[He] $2s^2 2p^6$
Nonmetal	Nonmetal	Nonmetal	Nonmetal
—	—	—	—
Very low mp and bp	Very low mp and bp	Very low mp and bp	Extremely low mp and bp
Inactive at room temperature	Very reactive	Extremely reactive	Chemically inert
Covalent N_2 molecules	Covalent O_2 (or O_3) molecules	Covalent F_2 molecules	None; separate atoms
Covalent (π bonds common)	Covalent (π bonds common)	Covalent	None
Ionic/polar covalent; anions with active metals	Ionic	Ionic	None
Strongly acidic (NO_2)	—	Acidic	None
Every oxidation state from +5 to −3	O_2 (and O_3) very strong oxidizing agents (−2)	Strongest oxidizing agent (−1)	None
Component of proteins; ammonia for fertilizers, explosives; oxides involved in manufacturing and air pollution (smog, acid rain)	Final oxidizer in residential, industrial, and biological energy production	Manufacture of coatings (Teflon); glass etching (HF); refrigerants involved in ozone depletion (CFCs); dental protection (NaF, SnF_2)	Electrified gas in advertising signs

Atomic radius (pm)

Li 152, Be 112, B 85, C 77, N 75, O 73, F 72, Ne 71

1A(1) 2A(2) 3A(13) 4A(14) 5A(15) 6A(16) 7A(17) 8A(18)

First ionization energy (kJ/mol)

Li 520, Be 899, B 800, C 1086, N 1402, O 1314, F 1681, Ne 2080

1A(1) 2A(2) 3A(13) 4A(14) 5A(15) 6A(16) 7A(17) 8A(18)

Electronegativity

Li 1.0, Be 1.5, B 2.0, C 2.5, N 3.0, O 3.5, F 4.0

1A(1) 2A(2) 3A(13) 4A(14) 5A(15) 6A(16) 7A(17) 8A(18)

14.3 GROUP 1A(1): THE ALKALI METALS

The first group of elements in the periodic table is named for the alkaline (basic) nature of their oxides and for the basic solutions the elements form in water. Group 1A(1) provides the best example of regular trends with no significant exceptions. All the elements in the group—lithium (Li), sodium (Na), potassium (K), rubidium (Rb), cesium (Cs), and rare, radioactive francium (Fr)*—are very reactive metals. The Family Portrait of Group 1A(1), on pages 572 and 573, is the first in a series that provides an overview of each of the main groups. Atomic and physical properties are summarized on the left-hand page; representative reactions and some important compounds appear on the right-hand page. Refer to these family portraits during the discussions for supporting information.

Lithium floating in oil floating on water.

ARIS Presentation Center, Photo Library

Why Are the Alkali Metals Soft, Low Melting, and Lightweight?

Unlike most metals, the alkali metals are soft: Na has the consistency of cold butter, and K can be squeezed like clay. The alkali metals also have lower melting and boiling points than any other group of metals. Except for Li, they all melt below 100°C, and Cs melts a few degrees above room temperature. They have lower densities than most metals: Li floats on lightweight household oil (see photo).

The unusual physical behavior of the alkali metals can be traced to their atomic size, the largest in their respective periods, and to the ns^1 valence electron configuration. Because the single valence electron is relatively far from the nucleus, there is only a weak attraction between the delocalized electrons and the metal-ion cores in the crystal structure. This weak metallic bonding means the alkali metal crystal structure can be easily deformed or broken down, which results in a soft consistency and low melting point. These elements also have the lowest molar masses in their periods; with their large atomic radii, they therefore have relatively low densities.

Why Are the Alkali Metals So Reactive?

The alkali metals are extremely reactive elements. They are *powerful reducing agents* and, thus, always occur in nature as 1+ cations rather than as free metals. (As we discuss in Section 22.4, highly endothermic reduction processes are required to prepare the free metals industrially from their molten salts.) The alkali metals reduce the halogens and form ionic solids in reactions that release large quantities of heat. They reduce the hydrogen in water, reacting vigorously (Rb and Cs explosively) to form H_2 and a metal hydroxide solution. They reduce O_2 in the air, and thus tarnish rapidly. Because of this reactivity, Na and K are usually kept under mineral oil (an unreactive liquid) in the laboratory, and Rb and Cs are handled with gloves under an inert argon atmosphere.

The ns^1 configuration, which is the basis for their physical properties, is also the reason these metals form salts so readily. In a Born-Haber cycle of the reaction between an alkali metal and a nonmetal, for example, the solid metal separates into gaseous atoms, each of the atoms transfers its outer electron to the nonmetal, and the resulting cations and anions attract each other to form an ionic solid (Section 9.2). Properties based on the ns^1 configuration relate to each step.

1. *Low heat of atomization* (ΔH°_{atom}). Consistent with the alkali metals' low melting and boiling points, their weak metallic bonding leads to low values for ΔH°_{atom} (the energy needed to convert the solid into individual gaseous atoms), which decrease down the group:[†]

$$E(s) \longrightarrow E(g) \qquad \Delta H_{atom} \ (\text{Li} > \text{Na} > \text{K} > \text{Rb} > \text{Cs})$$

---

*Francium is so rare (estimates indicate only 15 g of the element in the top kilometer of Earth's crust) that its properties are largely unknown. Therefore, we will mention it only occasionally in the discussion.

[†]Throughout the chapter, we use E to represent any element in a group.

2. *Low IE and small ionic radius.* Each alkali metal has the largest size and the lowest IE in its period. A great *decrease* in size occurs when the outer electron is lost: the volume of the Li^+ ion is less than 13% that of the Li atom! Thus, Group 1A(1) ions are small spheres with considerable charge density.

3. *High lattice energy.* When Group 1A(1) salts crystallize, large amounts of energy are released because the small cations lie close to the anions. Thus, the endothermic atomization and ionization steps are easily outweighed by the highly exothermic formation of the solid. For a given anion, the trend in lattice energy is the inverse of the trend in cation size: *as the cation becomes larger, the lattice energy becomes smaller.* This steady decrease in lattice energy within the Group 1A(1) and 2A(2) chlorides is shown in Figure 14.2.

Despite the strong ionic attractions in the solid, *nearly all Group 1A(1) salts are water soluble.* The attraction between the ions and water molecules creates a highly exothermic heat of hydration (ΔH_{hydr}), and a large increase in entropy occurs when ions in the organized crystal become dispersed and hydrated in solution; together, these factors outweigh the high lattice energy.

The magnitude of the hydration energy *decreases* as ionic size increases:

$$E^+(g) \longrightarrow E^+(aq) \qquad \Delta H = -\Delta H_{hydr}\ (Li^+ > Na^+ > K^+ > Rb^+ > Cs^+)$$

Interestingly, the *smaller* ions attract water molecules strongly enough to form *larger hydrated ions.* This size trend has a major effect on the function of nerves, kidneys, and cell membranes because the *sizes* of $Na^+(aq)$ and $K^+(aq)$, the most common cations in cell fluids, influence their movement in and out of cells.

The Anomalous Behavior of Lithium

As we noted in discussing Table 14.1, all the Period 2 elements display some anomalous (unrepresentative) behavior within their groups. Even within the regular, predictable trends of Group 1A(1), Li has some atypical properties. It is the only member that forms a simple oxide and nitride, Li_2O and Li_3N, on reaction with O_2 and N_2 in air. Only Li forms molecular compounds with hydrocarbon groups from organic halides:

$$2Li(s) + CH_3CH_2Cl(g) \longrightarrow CH_3CH_2Li(s) + LiCl(s)$$

Organolithium compounds, such as CH_3CH_2Li, are liquids or low-melting solids that dissolve in nonpolar solvents and contain polar covalent $^{\delta-}C—Li^{\delta+}$ bonds. They are important reactants in the synthesis of organic compounds.

Because of its small size, Li^+ has a relatively high charge density. Therefore, it can deform nearby electron clouds to a much greater extent than the other 1A ions can, which increases orbital overlap and gives many lithium salts significant covalent character (Figure 14.3). Thus, LiCl, LiBr, and LiI are much more soluble in polar organic solvents, such as ethanol and acetone, than are the halides of Na and K, because the lithium halide dipole interacts with these solvents through dipole-dipole forces. The small, highly positive Li^+ makes dissociation of Li salts into ions more difficult in water; thus, the fluoride, carbonate, hydroxide, and phosphate of Li are much less soluble in water than those of Na and K.

Figure 14.2 Lattice energies of the Group 1A(1) and 2A(2) chlorides. The lattice energy decreases regularly in both groups of metal chlorides as the cations become larger. Lattice energies for the Group 2A chlorides are greater because the 2A cations have higher charge and smaller size.

ARIS Presentation Center, Art Library

Biological application

Figure 14.3 The effect of Li^+ charge density on a nearby electron cloud. The ability of the Li^+ ion to deform nearby electron clouds gives rise to many anomalous properties. In this case, Li^+ polarizes an I^- ion, which gives some covalent character to lithium iodide, LiI, as depicted by a space-filling model **(A)**, an electron density contour **(B)**, and an electron density relief map **(C)**.

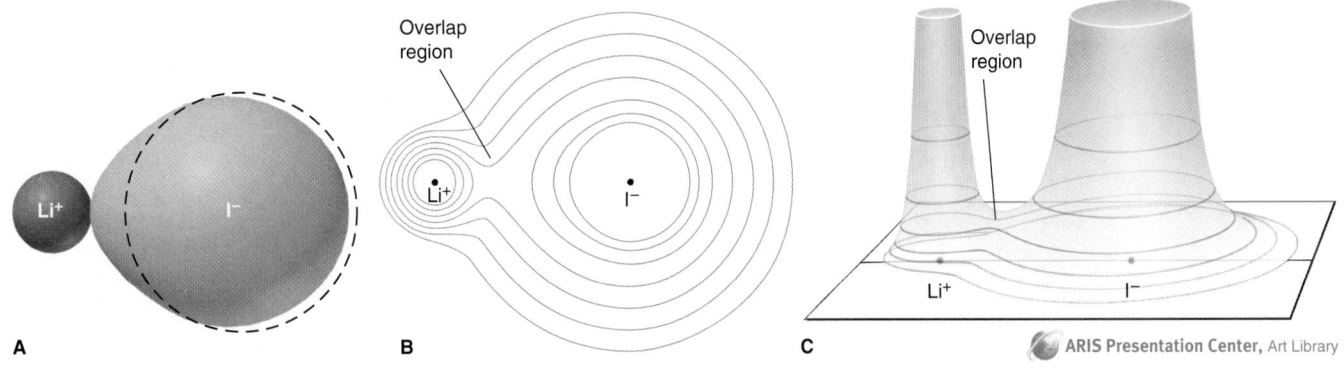

ARIS Presentation Center, Art Library

Group 1A(1): The Alkali Metals

Key Atomic and Physical Properties

KEY

Atomic No.
Symbol
Atomic mass
Valence e⁻ configuration
Common oxidation states

ns^1

GROUP 1A(1)

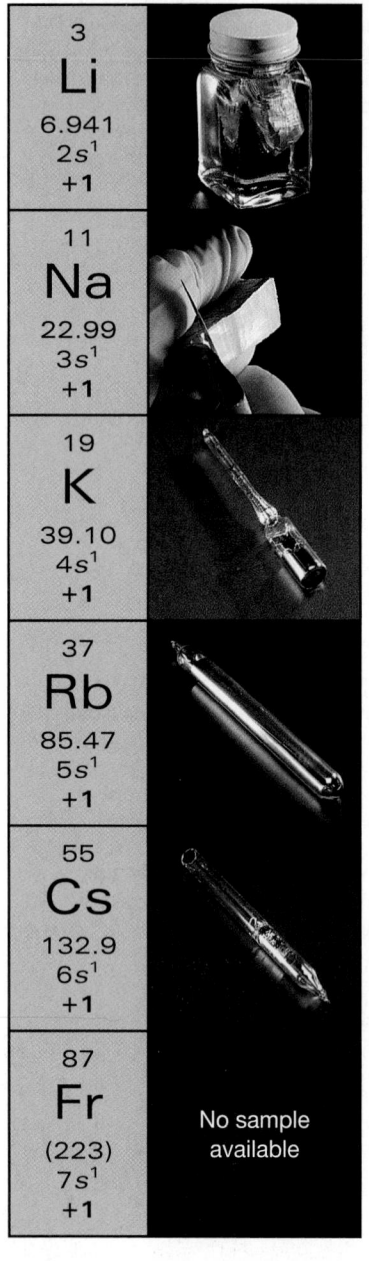

3	
Li	
6.941	
$2s^1$	
+1	
11	
Na	
22.99	
$3s^1$	
+1	
19	
K	
39.10	
$4s^1$	
+1	
37	
Rb	
85.47	
$5s^1$	
+1	
55	
Cs	
132.9	
$6s^1$	
+1	
87	
Fr	No sample available
(223)	
$7s^1$	
+1	

Atomic Properties

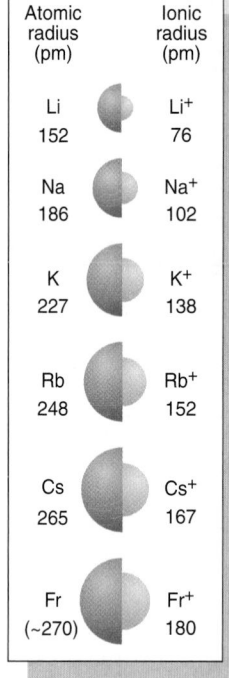

Atomic radius (pm)		Ionic radius (pm)	
Li 152		Li⁺ 76	
Na 186		Na⁺ 102	
K 227		K⁺ 138	
Rb 248		Rb⁺ 152	
Cs 265		Cs⁺ 167	
Fr (~270)		Fr⁺ 180	

Group electron configuration is ns^1. All members have the +1 oxidation state and form an E⁺ ion.

Atoms have the largest size and lowest IE and EN in their periods.

Down the group, atomic and ionic size increase, while IE and EN decrease.

First ionization energy (kJ/mol)

Li 520
Na 496
K 419
Rb 403
Cs 376
Fr ~375

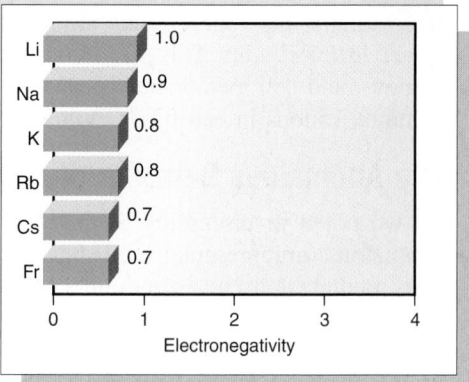

Electronegativity

Li 1.0
Na 0.9
K 0.8
Rb 0.8
Cs 0.7
Fr 0.7

Physical Properties

Metallic bonding is relatively weak because there is only one valence electron. Therefore, these metals are soft with relatively low melting and boiling points. These values decrease down the group because larger atom cores attract delocalized electrons less strongly.

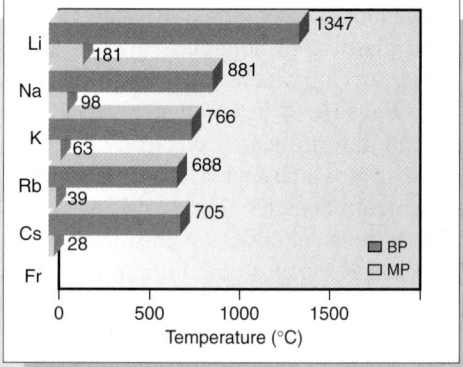

Temperature (°C)

Li 181 / 1347
Na 98 / 881
K 63 / 766
Rb 39 / 688
Cs 28 / 705

■ BP □ MP

Density (g/mL)

Li 0.534
Na 0.968
K 0.856
Rb 1.532
Cs 1.90
Fr

Large atomic size and low atomic mass result in low density; thus density generally increases down the group because mass increases more than size.

Some Reactions and Compounds

Important Reactions

The reducing power of the alkali metals (E) is shown in reactions 1 to 4. Some industrial applications of Group 1A(1) compounds are shown in reactions 5 to 7.

1. The alkali metals reduce H in H_2O from the +1 to the 0 oxidation state:

$$2E(s) + 2H_2O(l) \longrightarrow 2E^+(aq) + 2OH^-(aq) + H_2(g)$$

The reaction becomes more vigorous down the group (*see photo*).

Potassium reacting with water

2. The alkali metals reduce oxygen, but the product depends on the metal. Li forms the oxide, Li_2O; Na forms the peroxide, Na_2O_2; K, Rb, and Cs form the superoxide, EO_2:

$$4Li(s) + O_2(g) \longrightarrow 2Li_2O(s)$$
$$K(s) + O_2(g) \longrightarrow KO_2(s)$$

In emergency breathing units, KO_2 reacts with H_2O and CO_2 in exhaled air to release O_2.

3. The alkali metals reduce hydrogen to form ionic (saltlike) hydrides:

$$2E(s) + H_2(g) \longrightarrow 2EH(s)$$

NaH is an industrial base and reducing agent that is used to prepare other reducing agents, such as $NaBH_4$.

4. The alkali metals reduce halogens to form ionic halides:

$$2E(s) + X_2 \longrightarrow 2EX(s) \qquad (X = F, Cl, Br, I)$$

5. Sodium chloride is the most important alkali metal halide. (a) In the Downs process for the production of sodium metal (*Section 22.4*), reaction 4 is reversed by supplying electricity to molten NaCl:

$$2NaCl(l) \xrightarrow{\text{electricity}} 2Na(l) + Cl_2(g)$$

(b) In the chlor-alkali process (*Section 22.5*), NaCl(*aq*) is electrolyzed to form several key industrial chemicals:

$$2NaCl(aq) + 2H_2O(l) \longrightarrow 2NaOH(aq) + H_2(g) + Cl_2(g)$$

(c) In its reaction with sulfuric acid, NaCl forms two major products:

$$2NaCl(s) + H_2SO_4(aq) \longrightarrow Na_2SO_4(aq) + 2HCl(g)$$

Sodium sulfate is important in the paper industry; HCl is essential in steel, plastics, textile, and food production.

6. Sodium hydroxide is used in the formation of bleaching solutions:

$$2NaOH(aq) + Cl_2(g) \longrightarrow NaClO(aq) + NaCl(aq) + H_2O(l)$$

7. In an ion-exchange process (*Chemical Connections, p. 542*), water is "softened" by removal of dissolved hard-water cations, which displace Na^+ from a resin:

$$M^{2+}(aq) + Na_2Z(s) \longrightarrow MZ(s) + 2Na^+(aq)$$
$$(M = Mg, Ca; Z = resin)$$

Important Compounds

1. Lithium chloride and lithium bromide, LiCl and LiBr. Because the Li^+ ion is so small, Li salts have a high affinity for H_2O and yet a positive heat of solution. Thus, they are used in dehumidifiers and air-cooling units.

2. Lithium carbonate, Li_2CO_3. Used to make porcelain enamels and toughened glasses and as a drug in the treatment of bipolar disorders.

3. Sodium chloride, NaCl. Millions of tons used in the industrial production of Na, NaOH, $Na_2CO_3/NaHCO_3$, Na_2SO_4, and HCl or purified for use as table salt.

4. Sodium carbonate and sodium hydrogen carbonate, Na_2CO_3 and $NaHCO_3$. Carbonate used as an industrial base and to make glass. Hydrogen carbonate, which releases CO_2 at low temperatures (50°C to 100°C), used in baking powder and in fire extinguishers.

5. Sodium hydroxide, NaOH. Most important industrial base; used to make bleach, sodium phosphates, and alcohols.

6. Potassium nitrate, KNO_3. Powerful oxidizing agent used in gunpowder and fireworks (*see photo*).

14.4 GROUP 2A(2): THE ALKALINE EARTH METALS

The Group 2A(2) elements are called *alkaline earth metals* because their oxides give basic (alkaline) solutions and melt at such high temperatures that they remained as solids ("earths") in the alchemists' fires. The group includes a fascinating collection of elements: rare beryllium (Be), common magnesium (Mg) and calcium (Ca), less familiar strontium (Sr) and barium (Ba), and radioactive radium (Ra). The Group 2A(2) Family Portrait (pp. 576 and 577) presents an overview of these elements.

How Do the Physical Properties of the Alkaline Earth and Alkali Metals Compare?

In general terms, the elements in Groups 1A(1) and 2A(2) behave as close cousins. Whatever differences occur between the groups are those of degree, not kind, and are due to the change in outer electron configuration: ns^2 vs. ns^1. Two electrons are available from each 2A atom for metallic bonding, and the nucleus contains one additional positive charge. These factors make the attraction between delocalized electrons and atom cores greater. Consequently, melting and boiling points are much higher for 2A metals than for the corresponding 1A metals; in fact, the 2A elements melt at around the same temperatures as the 1A elements boil! Compared with transition metals, such as iron and chromium, the alkaline earths are soft and lightweight, but they are much harder and more dense than the alkali metals. ●

How Do the Chemical Properties of the Alkaline Earth and Alkali Metals Compare?

The alkaline earth metals display a wider range of chemical behavior than the alkali metals, largely because of the behavior of beryllium, as you'll see shortly. Because the second valence electron lies in the same sublevel as the first, it is not shielded from the additional nuclear charge very well, and so Z_{eff} is greater. Therefore, Group 2A(2) elements have smaller atomic radii and higher ionization energies than Group 1A(1) elements. Despite the higher IEs, *all the alkaline earths (except Be) form ionic compounds* as 2+ cations. Beryllium behaves differently because so much energy is needed to remove two electrons from this tiny atom that it never forms discrete Be^{2+} ions, and its bonds are polar covalent.

Like the alkali metals, the alkaline earth metals are *strong reducing agents*. Each element reduces O_2 in air to form the oxide (Ba also forms the peroxide, BaO_2). Except for Be and Mg, which form adherent oxide coatings, the alkaline earths reduce H_2O at room temperature to form H_2. And, except for Be, they reduce the halogens, N_2, and H_2 to form ionic compounds. The Group 2A oxides are strongly basic (except for amphoteric BeO) and react with acidic oxides to form salts, such as sulfites and carbonates; for example,

$$SrO(s) + CO_2(g) \longrightarrow SrCO_3(s)$$

Natural carbonates, such as limestone and marble, are major structural materials and the commercial sources for most 2A compounds. ●

The alkaline earth metals are reactive because the high lattice energies of their compounds more than compensate for the large total IE needed to form the 2+ cation (Section 9.2). Group 2A salts have much higher lattice energies than Group 1A salts (see Figure 14.2) because the 2A cations are smaller and doubly charged.

One of the main differences between the two groups is the lower solubility of 2A salts in water. Their ions are smaller and more highly charged than 1A ions, resulting in much higher charge densities. Even though this factor increases heats of hydration, it increases lattice energies even more. In fact, most 2A fluorides, carbonates, phosphates, and sulfates are considered insoluble, unlike the corresponding 1A compounds. Nevertheless, the ion-dipole attraction of 2+ ions

● **Versatile Magnesium** Magnesium, second of the alkaline earths, forms a tough, adherent oxide coating that prevents further reaction of the metal in air and confers stability for many uses. The metal can be rolled, forged, welded, and riveted into virtually any shape, and it forms strong, low-density alloys: some weigh 25% as much as steel but are just as strong. Alloyed with aluminum and zinc, magnesium is used for everyday objects, such as camera bodies, luggage, and the "mag" wheels of bicycles and sports cars. Alloyed with the lanthanides (first series of inner transition elements), it is used in objects that require great strength at high temperatures, such as auto engine blocks and missile parts.

Engineering application

Engineering application

● **Lime: The Most Useful Metal Oxide** Calcium oxide (lime) is a slightly soluble, basic oxide that is among the five most heavily produced industrial compounds in the world; it is made by roasting limestone to high temperatures [Group 2A(2) Family Portrait, reaction 7, p. 577]. It has essential roles in steelmaking, water treatment, and smokestack scrubbing. Lime reacts with carbon to form calcium carbide, which is used to make acetylene, and with arsenic acid (H_3AsO_4) to make insecticides for treating cotton, tobacco, and potato plants. Most glass contains about 12% lime by mass. The paper industry uses lime to prepare the bleach [$Ca(ClO)_2$] that whitens paper. Farmers "dust" fields with lime to "sweeten" acidic soil. The food industry uses it to neutralize compounds in milk to make cream and butter and to neutralize impurities in crude sugar.

for water is so strong that many slightly soluble 2A salts crystallize as hydrates; two examples are Epsom salt, $MgSO_4 \cdot 7H_2O$, used as a soak for inflammations, and gypsum, $CaSO_4 \cdot 2H_2O$, used as the bonding material between the paper sheets in wallboard and as the cement in surgical casts.

The Anomalous Behavior of Beryllium

The Period 2 element Be displays far more anomalous behavior than Li in Group 1A(1). If the Be^{2+} ion did exist, its high charge density would polarize nearby electron clouds very strongly and cause extensive orbital overlap. In fact, *all Be compounds exhibit covalent bonding.* Even BeF_2, the most ionic Be compound, has a relatively low melting point and, when melted, a low electrical conductivity.

With only two valence electrons, Be does not attain an octet in its simple gaseous compounds (Section 10.1). When it bonds to an electron-rich atom, however, this electron deficiency is overcome as the gas condenses. Consider beryllium chloride ($BeCl_2$) (Figure 14.4). At temperatures greater than 900°C, it consists of linear molecules in which two *sp* hybrid orbitals hold four electrons around the central Be. As it cools, the molecules bond together, solidifying in long chains, with each Be sp^3 hybridized to finally attain an octet.

Diagonal Relationships: Lithium and Magnesium

One of the clearest ways in which atomic properties influence chemical behavior appears in the **diagonal relationships,** similarities between a Period 2 element and one diagonally down and to the right in Period 3 (Figure 14.5).

A

B

Figure 14.4 Overcoming electron deficiency in beryllium chloride. A, At high temperatures, $BeCl_2$ is a gas with only four electrons around each Be. **B,** Solid $BeCl_2$ occurs in long chains with each Cl bridging two Be atoms, which gives each Be an octet.

ARIS Presentation Center, Art Library

Figure 14.5 Three diagonal relationships in the periodic table. Certain Period 2 elements exhibit behaviors that are very similar to those of the Period 3 elements immediately below and to the right. Three such diagonal relationships exist: Li and Mg, Be and Al, and B and Si.

ARIS Presentation Center, Art Library

The first of three such relationships occurs between Li and Mg and reflects similarities in atomic and ionic size. Note that *one period down increases atomic (or ionic) size and one group to the right decreases it.* Thus, Li has a radius of 152 pm and Mg has a radius of 160 pm; the Li^+ radius is 76 pm and that of Mg^{2+} is 72 pm. From similar atomic properties emerge similar chemical properties. Both elements form nitrides with N_2, hydroxides and carbonates that decompose easily with heat, organic compounds with a polar covalent metal-carbon bond, and salts with similar solubilities. We'll discuss the diagonal relationships between Be and Al and between B and Si in later sections.

Looking Backward and Forward: Groups 1A(1), 2A(2), and 3A(13)

Throughout this chapter, comparing the previous, current, and upcoming groups (Figure 14.6) will help you to keep horizontal trends in mind while examining vertical groups. Not much changes from 1A to 2A, and the elements behave as metals both physically and chemically. With smaller atomic sizes and stronger metallic bonding, 2A elements are harder, higher melting, and denser than those in 1A. Nearly all 1A and most 2A compounds are ionic. The higher ionic charge in Group 2A (2+ vs. 1+) leads to higher lattice energies and less soluble salts. The range of behavior in 2A is wider than that in 1A because of Be, and the range widens much further in Group 3A, from metalloid boron to metallic thallium.

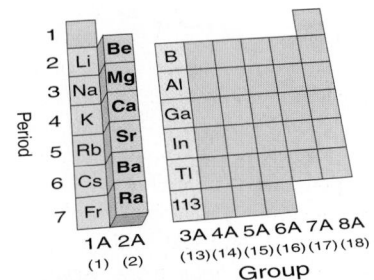

Figure 14.6 Standing in Group 2A(2), looking backward to 1A(1) and forward to 3A(13).

ARIS Presentation Center, Art Library

Group 2A(2): The Alkaline Earth Metals

Key Atomic and Physical Properties

ns^2

GROUP 2A(2)

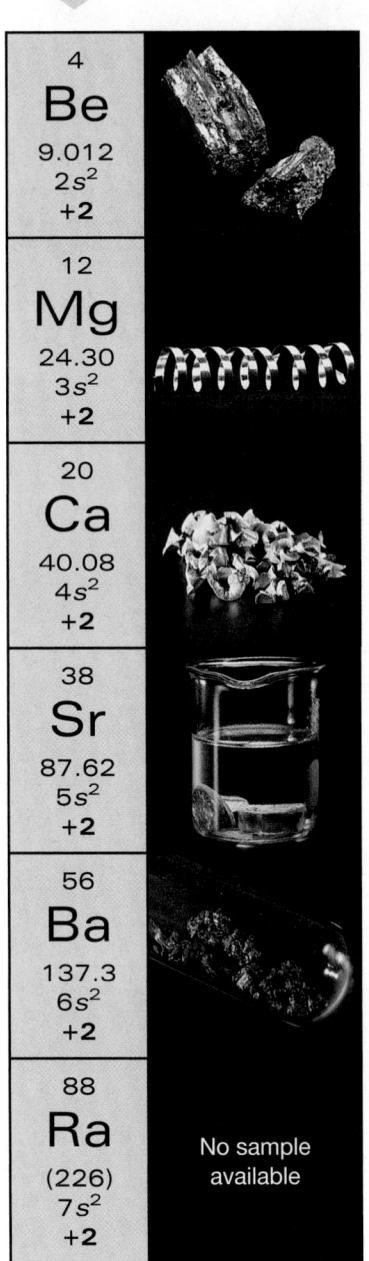

4	
Be	
9.012	
$2s^2$	
+2	

12	
Mg	
24.30	
$3s^2$	
+2	

20	
Ca	
40.08	
$4s^2$	
+2	

38	
Sr	
87.62	
$5s^2$	
+2	

56	
Ba	
137.3	
$6s^2$	
+2	

88	
Ra	
(226)	
$7s^2$	
+2	

No sample
available

Atomic Properties

Atomic radius (pm)	Ionic radius (pm)
Be 112	
Mg 160	Mg^{2+} 72
Ca 197	Ca^{2+} 100
Sr 215	Sr^{2+} 118
Ba 222	Ba^{2+} 135
Ra (~220)	Ra^{2+} 148

Group electron configuration is ns^2 (filled ns sublevel). All members have the +2 oxidation state and, except for Be, form compounds with an E^{2+} ion.

Atomic and ionic sizes increase down the group but are smaller than for the corresponding 1A(1) element.

IE and EN decrease down the group but are higher than for the corresponding 1A(1) element.

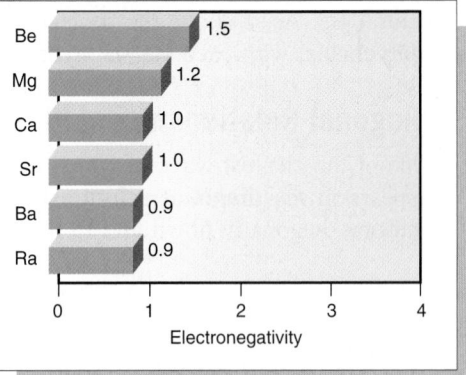

Physical Properties

Metallic bonding involves two valence electrons. These metals are still relatively soft but are much harder than the 1A(1) metals.

Melting and boiling points generally decrease, and densities generally increase down the group. These values are much higher than for 1A(1) elements, and the trend is not as regular.

Some Reactions and Compounds

Important Reactions

The elements (E) act as reducing agents in reactions 1 to 5; note the similarity to reactions of Group 1A(1). Reaction 6 shows the general basicity of the 2A(2) oxides; reaction 7 shows the general instability of their carbonates at high temperature.

1. The metals reduce O_2 to form the oxides:

$$2E(s) + O_2(g) \longrightarrow 2EO(s)$$

Ba also forms the peroxide, BaO_2.

Magnesium ribbon burning

2. The larger metals reduce water to form hydrogen gas:

$$E(s) + 2H_2O(l) \longrightarrow E^{2+}(aq) + 2OH^-(aq) + H_2(g)$$
$$(E = Ca, Sr, Ba)$$

Be and Mg form an adherent oxide coating that allows only slight reaction.

3. The metals reduce halogens to form ionic halides:

$$E(s) + X_2 \longrightarrow EX_2(s) \quad (X = F, Cl, Br, I)$$

4. Most of the elements reduce hydrogen to form ionic hydrides:

$$E(s) + H_2(g) \longrightarrow EH_2(s) \quad (E = \text{all except Be})$$

5. Most of the elements reduce nitrogen to form ionic nitrides:

$$3E(s) + N_2(g) \longrightarrow E_3N_2(s) \quad (E = \text{all except Be})$$

6. Except for amphoteric BeO, the element oxides are basic:

$$EO(s) + H_2O(l) \longrightarrow E^{2+}(aq) + 2OH^-(aq)$$

$Ca(OH)_2$ is a component of cement and mortar.

7. All carbonates undergo thermal decomposition to the oxide:

$$ECO_3(s) \xrightarrow{\Delta} EO(s) + CO_2(g)$$

This reaction is used to produce CaO (lime) in huge amounts from naturally occurring limestone (*margin note, p. 574*).

Important Compounds

1. Beryl, $Be_3Al_2Si_6O_{18}$. Beryl, the industrial source of Be metal, also occurs as a gemstone with a variety of colors (*see photo*). It is chemically identical to emerald, except for the trace of Cr^{3+} that gives emerald its green color.

2. Magnesium oxide, MgO. Because of its high melting point (2852°C), MgO is used as a refractory material for furnace brick (*see photo*) and wire insulation.

Industrial kiln

3. Alkylmagnesium halides, RMgX (R = hydrocarbon group; X = halogen). These compounds are used to synthesize many organic compounds. Organotin agricultural fungicides are made by treating RMgX with $SnCl_4$:

$$3RMgCl + SnCl_4 \longrightarrow 3MgCl_2 + R_3SnCl$$

4. Calcium carbonate, $CaCO_3$. Occurs as enormous natural deposits of limestone, marble, chalk (*see photo*), and coral. Used as a building material, to make lime, and, in high purity, as a toothpaste abrasive and an antacid.

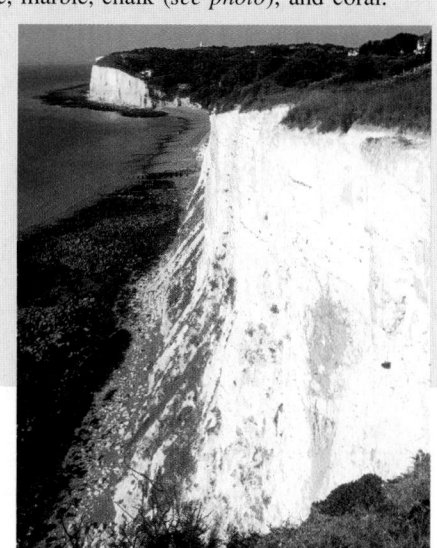

White cliffs, Dover, England

14.5 GROUP 3A(13): THE BORON FAMILY

The third family of main-group elements contains some unusual members and some familiar ones, some exotic bonding, and some strange physical properties. Boron (B) heads the family, but, as you'll see, its properties certainly do not represent the other members. Metallic aluminum (Al) has properties that are more typical of the group, but its great abundance and importance contrast with the rareness of gallium (Ga), indium (In), thallium (Tl), and especially the recently synthesized element 113. The atomic, physical, and chemical properties of these elements are summarized in the Group 3A(13) Family Portrait (pp. 580 and 581).

How Do Transition Elements Influence Group 3A(13) Properties?

Group 3A(13) is the first of the *p* block. If you look at the main groups only, the elements of this group seem to be just one group away from those of Group 2A(2). In Period 4 and higher, however, a large gap separates the two groups (see Figure 8.11, p. 314). The gap holds 10 transition elements (*d* block) each in Periods 4, 5, and 6 and an additional 14 inner transition elements (*f* block) in Period 6. Recall from Section 8.2 that *d* and *f* electrons penetrate very little, and so spend very little time near the nucleus. Thus, the heavier 3A members—Ga, In, and Tl—have nuclei with many more protons, but their outer (*s* and *p*) electrons are only partially shielded from the much higher positive charge; as a result, these elements have greater Z_{eff} values than the two lighter members of the group.

Many properties of these heavier 3A elements are influenced by this stronger nuclear attraction. Figure 14.7 compares several properties of Group 3A(13) with those of Group 3B(3) (the first group of transition elements), in which the additional protons of the *d* block and *f* block have *not* been added. Note the regular changes exhibited by the 3B elements in contrast to the irregular patterns for those in 3A. The deviations for Ga reflect the *d*-block *contraction in size* and can be explained by limited shielding by the *d* electrons of the 10 additional protons of the first transition series. Similarly, the deviations for Tl reflect the *f*-block (lanthanide) contraction and can be explained by limited shielding by the *f* electrons of the 14 additional protons of the first inner transition series.

Physical properties are influenced by the type of bonding that occurs in the element. Boron is a network covalent metalloid—black, hard, and very high melting. The other group members are metals—shiny and relatively soft and low melting. Aluminum's low density and three valence electrons make it an exceptional conductor: for a given mass, aluminum conducts a current twice as effectively as copper. Gallium has the largest liquid temperature range of any element: it melts in your hand (see photo, p. 580) but does not boil until 2403°C. Its metallic bonding is too weak to keep the Ga atoms fixed when the solid is warmed, but strong enough to keep them from escaping the molten metal until it is very hot. ●

What New Features Appear in the Chemical Properties of Group 3A(13)?

Looking down Group 3A(13), we see a wide range of chemical behavior. Boron, the anomalous member from Period 2, is the first metalloid we've encountered so far and the only one in the group. It is much less reactive at room temperature than the other members and forms covalent bonds exclusively. Although aluminum acts like a metal physically, its halides exist in the gas phase as covalent *dimers*— molecules formed by joining two identical smaller molecules (Figure 14.8)—and its oxide is amphoteric rather than basic. Most of the other 3A compounds are ionic, but they have more covalent character than similar 2A compounds. Because the 3A cations are smaller and more highly charged than the 2A cations, they polarize an anion's electron cloud more effectively.

● **Gallium Arsenide: The New Wave of Semiconductors** An important modern use of gallium is in the production of gallium arsenide (GaAs) semiconductors. Electrons move 10 times faster through GaAs than through Si-based chips. GaAs chips also have novel optical properties: a current is created when they absorb light, and, conversely, they emit light when a current is supplied. GaAs devices are used in light-powered calculators, wristwatches, and solar panels. In addition, GaAs-based lasers are much smaller and more powerful than other types.

Engineering application

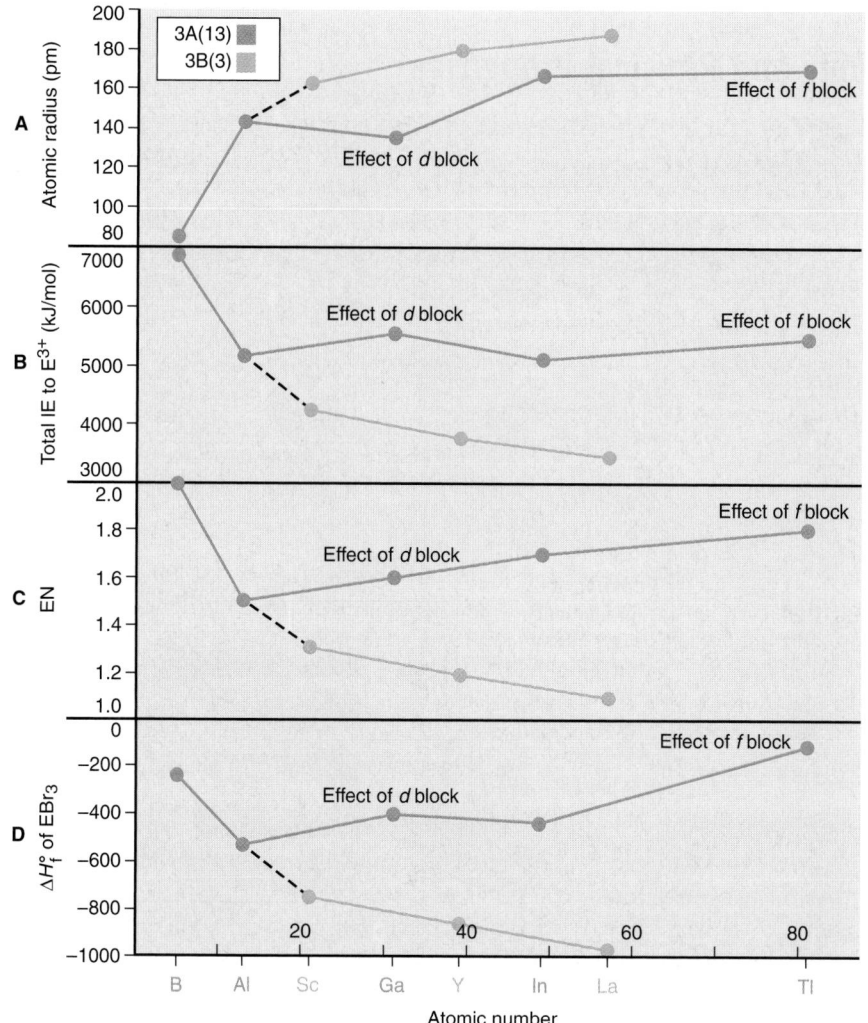

Transparency/ ARIS Presentation Center, Art Library

Figure 14.7 The effect of transition elements on properties: Group 3B(3) vs. Group 3A(13). The additional protons in the nuclei of transition elements exert an exceptionally strong attraction because d and f electrons shield the nuclear charge poorly. This greater effective nuclear charge affects the p-block elements in Periods 4 to 6, as can be seen by comparing properties of Group 3B(3), the first group after the s block, with those of Group 3A(13), the first group after the d block. **A,** Atomic size. In Group 3B, size increases smoothly, whereas in 3A, Ga and Tl are smaller than expected. **B,** Total ionization energy for E^{3+}. The deviations in size lead to deviations in total IE ($IE_1 + IE_2 + IE_3$). Note the regular decrease for Group 3B and the unexpectedly higher values for Ga and Tl in 3A. **C,** Electronegativity. The deviations in size also make Ga and Tl more electronegative than expected. **D,** Heat of formation of EBr_3. The higher IE values for Ga and Tl mean less heat is released upon formation of the ionic compound; thus, the magnitudes of the ΔH_f° values for $GaBr_3$ and $TlBr_3$ are smaller than expected.

The redox behavior of the elements in this group provides a chance to note three general principles that appear in Groups 3A(13) to 6A(16):

1. *Presence of multiple oxidation states.* Many of the larger elements in these groups also have an important oxidation state *two lower than the A-group number.* The lower state occurs when the atoms lose their np electrons only, not their two ns electrons. This fact is often called the *inert-pair effect* (Section 8.5), but it has nothing to do with any inertness of ns electrons; for example, the $6s$ electrons in Tl are lost *more* easily than the $4s$ electrons in Ga. The reason for the common appearance of the lower state involves bond energies of the compounds. Bond energy decreases as atomic size, and therefore bond length, increases. Consider the Tl—Cl bond. It is relatively long and weak, and it takes more energy to remove the two $6s$ electrons from Tl^+ to make Tl^{3+} than is released when two more weak Tl—Cl bonds form in $TlCl_3$. In fact, $TlCl_3$ is so unstable that it decomposes readily to Cl_2 gas and TlCl.

2. *Increasing prominence of the lower oxidation state.* When a group exhibits more than one oxidation state, *the lower state becomes more prominent going down the group.* In Group 3A(13), for instance, all members exhibit the +3 state, but the +1 state first appears with some compounds of gallium and becomes the only important state of thallium.

Figure 14.8 The dimeric structure of gaseous aluminum chloride. Despite its name, aluminum trichloride exists in the gas phase as the dimer, Al_2Cl_6.

ARIS Presentation Center, Art Library

Group 3A(13): The Boron Family

Key Atomic and Physical Properties

ns^2np^1

GROUP 3A(13)

| 5 |
| B |
| 10.81 |
| $2s^22p^1$ |
| +3 |

| 13 |
| Al |
| 26.98 |
| $3s^23p^1$ |
| +3 |

| 31 |
| Ga |
| 69.72 |
| $4s^24p^1$ |
| +3, +1 |

| 49 |
| In |
| 114.8 |
| $5s^25p^1$ |
| +3, +1 |

| 81 |
| Tl |
| 204.4 |
| $6s^26p^1$ |
| +1 |

| 113 |
| |
| (284) |
| $7s^27p^1$ |

Observed in
experiments at
Dubna, Russia,
in 2003

Atomic radius (pm)	Ionic radius (pm)
B 85	
Al 143	Al³⁺ 54
Ga 135	Ga³⁺ 62
In 167	In³⁺ 80
Tl 170	Tl⁺ 150

Atomic Properties

Group electron configuration is ns^2np^1. All except Tl (and presumably element 113) commonly display the +3 oxidation state. The +1 state becomes more common down the group.

Atomic size is smaller and EN is higher than for 2A(2) elements; IE is lower, however, because it is easier to remove an electron from the higher energy p sublevel.

Atomic size, IE, and EN do not change as expected down the group because there are intervening transition and inner transition elements.

First ionization energy (kJ/mol)
- B 800
- Al 577
- Ga 579
- In 558
- Tl 589

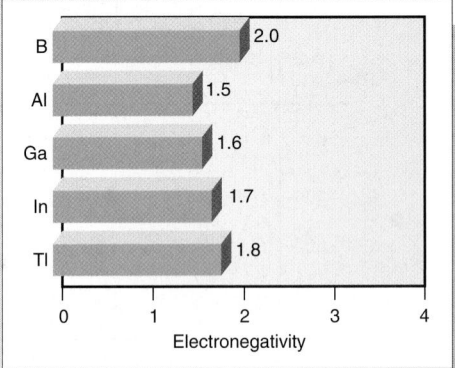

Electronegativity
- B 2.0
- Al 1.5
- Ga 1.6
- In 1.7
- Tl 1.8

Physical Properties

Bonding changes from network covalent in B to metallic in the rest of the group. Thus, B has a much higher melting point than the others, but there is no overall trend. Boiling points decrease down the group.

Temperature (°C) — BP ■ MP □
- B 2180, ~3650
- Al 660, 2467
- Ga 29.8, 2403
- In 157, 2080
- Tl 304, 1457

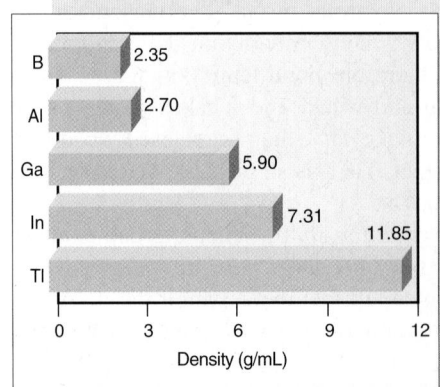

Density (g/mL)
- B 2.35
- Al 2.70
- Ga 5.90
- In 7.31
- Tl 11.85

Densities increase down the group.

Some Reactions and Compounds

Important Reactions

In reactions 1 to 3, the elements (E) usually require higher temperatures to react than those of Groups 1A(1) and 2A(2); note the lower oxidation state of Tl. Reactions 4 to 6 show some compounds in key industrial processes.

1. The elements react sluggishly, if at all, with water:

$$2Ga(s) + 6H_2O(hot) \longrightarrow 2Ga^{3+}(aq) + 6OH^-(aq) + 3H_2(g)$$
$$2Tl(s) + 2H_2O(steam) \longrightarrow 2Tl^+(aq) + 2OH^-(aq) + H_2(g)$$

Al becomes covered with a layer of Al_2O_3 that prevents further reaction (*see photo*).

Electron micrograph of oxide coating on Al surface

2. When strongly heated in pure O_2, all members form oxides:

$$4E(s) + 3O_2(g) \xrightarrow{\Delta} 2E_2O_3(s) \qquad (E = B, Al, Ga, In)$$
$$4Tl(s) + O_2(g) \xrightarrow{\Delta} 2Tl_2O(s)$$

Oxide acidity decreases down the group: B_2O_3 (weakly acidic) > Al_2O_3 > Ga_2O_3 > In_2O_3 > Tl_2O (strongly basic), and the +1 oxide is more basic than the +3 oxide.

3. All members reduce halogens (X_2):

$$2E(s) + 3X_2 \longrightarrow 2EX_3 \qquad (E = B, Al, Ga, In)$$
$$2Tl(s) + X_2 \longrightarrow 2TlX(s)$$

The BX_3 compounds are volatile and covalent.
Trihalides of Al, Ga, and In are (mostly) ionic solids but occur as covalent dimers in the gas phase; in this way, the 3A atom attains a filled outer level.

4. Acid treatment of Al_2O_3 is important in water purification:

$$Al_2O_3(s) + 3H_2SO_4(l) \longrightarrow Al_2(SO_4)_3(s) + 3H_2O(l)$$

In water, $Al_2(SO_4)_3$ and other agents form a colloid that aids in removing suspended particles (*see Chemical Connections, p. 541*).

5. The overall reaction in the production of aluminum metal is a redox process:

$$2Al_2O_3(s) + 3C(s) \longrightarrow 4Al(s) + 3CO_2(g)$$

This process is carried out electrochemically in the presence of cryolite (Na_3AlF_6), which lowers the melting point of the reactant mixture and takes part in the change (*Section 22.4*).

6. A displacement reaction produces gallium arsenide, GaAs (*margin note, p. 578*):

$$(CH_3)_3Ga(g) + AsH_3(g) \longrightarrow 3CH_4(g) + GaAs(s)$$

Important Compounds

1. Boron oxide, B_2O_3. Used in the production of borosilicate glass (*see margin note, p. 582*).

2. Borax, $Na_2[B_4O_5(OH)_4]\cdot8H_2O$. Major mineral source of boron compounds and B_2O_3. Used as a fireproof insulation material and as a washing powder (20-Mule Team Borax).

3. Boric acid, H_3BO_3 [or $B(OH)_3$]. Used as external disinfectant, eyewash, and insecticide.

4. Diborane, B_2H_6. A powerful reducing agent for possible use as a rocket fuel. Used to synthesize higher boranes, compounds that led to new theories of chemical bonding.

5. Aluminum sulfate (cake alum), $Al_2(SO_4)_3$. Used in purifying water, tanning leather, and sizing paper; as a fixative for dyeing cloth (*see photo*); and as an antiperspirant.

6. Aluminum oxide, Al_2O_3. Major compound in natural source (bauxite) of Al metal. Used as abrasive in sandpaper, sanding and cutting tools, and toothpaste. Large crystals with metal ion impurities often of gemstone quality (*see photo*). Inert support for chromatography. In fibrous forms, woven into heat-resistant fabrics; also used to strengthen ceramics and metals.

Ruby

7. $Tl_2Ba_2Ca_2Cu_3O_{10}$. Becomes a high-temperature superconductor at 125 K, which is readily attained with liquid N_2 (*see photo*).

Magnet levitated by superconductor

3. *Relative basicity of oxides.* In general, *oxides with the element in a lower oxidation state are more basic than oxides with the element in a higher oxidation state.* A good example in Group 3A(13) is that In_2O is more basic than In_2O_3. In general, when an element has more than one oxidation state, *it acts more like a metal in its lower state,* and this, too, is related to ionic charge density. In this example, the lower charge of In^+ does not polarize the O^{2-} ion as much as the higher charge of In^{3+} does. Thus, in compounds of general formula E_2O, the E-to-O bonding is more ionic than it is in E_2O_3 compounds, so the O^{2-} ion is more available to act as a base (Interchapter, Topic 4).

Highlights of Boron Chemistry

Like the other Period 2 elements, the chemical behavior of boron is strikingly different from that of the other members of its group. As was pointed out earlier, *all boron compounds are covalent,* and unlike the other Group 3A(13) members, boron forms network covalent compounds or large molecules with metals, H, O, N, and C. The unifying feature of many boron compounds is the element's *electron deficiency,* but boron adopts two strategies to fill its outer level: accepting a bonding pair from an electron-rich atom and forming bridge bonds with electron-poor atoms.

Accepting a Bonding Pair from an Electron-Rich Atom
In gaseous boron trihalides (BX_3), the B atom is electron deficient, with only six electrons around it (Section 10.1). To attain an octet, the B atom accepts a lone pair of electrons from an electron-rich atom (*red*) and forms a covalent bond:

$$BF_3(g) + :NH_3(g) \longrightarrow F_3B\!-\!NH_3(g)$$

(Such reactions, in which one reactant accepts an electron pair from another to form a covalent bond, are very widespread in inorganic, organic, and biochemical processes. They are known as *Lewis acid-base reactions,* and we'll discuss them in Chapters 18 and 23 and see examples of them throughout the second half of the text.)

Similarly, B has only six electrons in boric acid, $B(OH)_3$ (sometimes written as H_3BO_3). In water, the acid itself does not release a proton. Rather, it accepts an electron pair from the O in H_2O, forming a fourth bond and releasing an H^+ ion:

$$B(OH)_3(s) + H_2O(l) \rightleftharpoons B(OH)_4^-(aq) + H^+(aq)$$

Boron's outer shell is filled in the wide variety of borate salts, such as the mineral borax (sodium borate), $Na_2[B_4O_5(OH)_4]{\cdot}8H_2O$, used for decades as a household cleaning agent. ●

Boron also attains an octet in several boron-nitrogen compounds whose crystal or molecular structures are amazingly similar to those of elemental carbon and some of its organic compounds (Figure 14.9). These similarities are due to the fact that the size, IE, and EN of C are between those of B and N (see Table 14.1). Moreover, C has four valence electrons, whereas B has three and N has five, so the $\cdot C\!-\!C\cdot$ and $\cdot B\!-\!\dot N\cdot$ groupings have the same number of valence electrons. Therefore, $H_3C\!-\!CH_3$ (ethane) and $H_3B\!-\!NH_3$ (amine-borane) are isoelectronic, as are benzene and borazine (Figure 14.9A and B).

Boron nitride (Figure 14.9C) has a structure consisting of hexagons fused into sheets, very similar to that of graphite. Moreover, like graphite, its π electrons are highly delocalized through resonance. However, molecular-orbital calculations show that there is a large energy gap between the filled valence band and the empty conductance band in boron nitride, but no such gap in graphite (see Figure 12.38, p. 475). As a result, boron nitride is a white electrical insulator, whereas graphite is a black electrical conductor. At high pressure and temperature, boron nitride forms borazon (Figure 14.9D), which has a crystal structure like that of diamond and is extremely hard and abrasive.

ARIS Presentation Center, Photo Library

● **Borates in Your Labware** Strong heating of boric acid (or borate salts) drives off water molecules and gives molten boron oxide:

$$2B(OH)_3(s) \xrightarrow{\Delta} B_2O_3(l) + 3H_2O(g)$$

The molten oxide dissolves metal oxides to form borate glasses. When mixed with silica (SiO_2), it forms borosilicate glass. Its high transparency and small change in size when heated or cooled make borosilicate glass useful in cookware and in the glassware you use in the lab.

Engineering application

Figure 14.9 Similarities between substances with C—C bonds and those with B—N bonds. Note that in each case, B attains an octet of electrons by bonding with electron-rich N. **A,** Ethane and its BN analog. **B,** Benzene and borazine, which is often referred to as "inorganic" benzene. **C,** Graphite and the similar extended hexagonal sheet structure of boron nitride. **D,** Diamond and borazon have the same crystal structure and are among the hardest substances known.

ARIS Presentation Center, Art Library

Forming Bridge Bonds with Electron-Poor Atoms In elemental boron and its many hydrides (boranes), there is no electron-rich atom to supply boron with electrons. In these substances, boron attains an octet through some unusual bonding. In diborane (B_2H_6) and many larger boranes, for example, two types of B—H bonds exist. The first type is a typical electron-pair bond. The valence bond picture in Figure 14.10 shows an sp^3 orbital of B overlapping a $1s$ orbital of H in each of the four terminal B—H bonds, using two of the three electrons in the valence level of each B atom.

The other type of bond is a hydride **bridge bond** (or three-center, two-electron bond), in which *each B—H—B grouping is held together by only two electrons*. Two sp^3 orbitals, one from *each* B, overlap an H $1s$ orbital between them. Two electrons move through this extended bonding orbital—one from one of the B atoms and the other from the H atom—and join the two B atoms via the H atom bridge. Notice that *each B atom is surrounded by eight electrons:* four from the two normal B—H bonds and four from the two B—H—B bridge bonds.

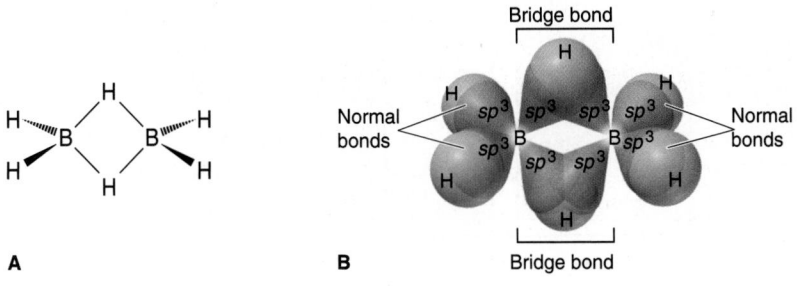

Figure 14.10 The two types of covalent bonding in diborane. A, A wedge-bond perspective drawing of B_2H_6 shows the unusual B—H—B bridge bond and the tetrahedral arrangement around each B atom. **B,** A valence bond depiction shows each sp^3-hybridized B forming normal covalent bonds with two hydrogens and two bridge bonds, in which two electrons bind three atoms, at the two central B—H—B groupings.

ARIS Presentation Center, Art Library

A B₁₂ unit

B B₅H₉

Figure 14.11 The boron icosahedron and one of the boranes. A, The icosahedral structural unit of elemental boron. **B,** The structure of B_5H_9, one of many boranes.

ARIS Presentation Center, Art Library

In many boranes and in elemental boron (Figure 14.11), one B atom bridges two others in a three-center, two-electron B—B—B bond.

Diagonal Relationships: Beryllium and Aluminum

Beryllium in Group 2A(2) and aluminum in Group 3A(13) are another pair of diagonally related elements. Both form oxoanions in strong base: beryllate, $Be(OH)_4^{2-}$, and aluminate, $Al(OH)_4^{-}$. Both have bridge bonds in their hydrides and chlorides. Both form oxide coatings impervious to reaction with water, and both oxides are amphoteric, extremely hard, and high melting. Although the atomic and ionic sizes of these elements differ, the small, highly charged Be^{2+} and Al^{3+} ions polarize nearby electron clouds strongly. Therefore, some Al compounds and all Be compounds have significant covalent character.

14.6 GROUP 4A(14): THE CARBON FAMILY

The whole range of elemental behavior occurs within Group 4A(14): nonmetallic carbon (C) leads off, followed by the metalloids silicon (Si) and germanium (Ge), with metallic tin (Sn) and lead (Pb) next, and newly synthesized element 114 at the bottom of the group. Information about the compounds of C and of Si fills libraries: organic chemistry, most polymer chemistry, and biochemistry are based on carbon, whereas geochemistry and some extremely important polymer and electronic technologies are based on silicon. The Group 4A(14) Family Portrait (pp. 586 and 587) summarizes atomic, physical, and chemical properties.

How Does the Bonding in an Element Affect Physical Properties?

The elements of Group 4A(14) and their neighbors in Groups 3A(13) and 5A(15) illustrate how physical properties, such as melting point and heat of fusion (ΔH_{fus}), depend on the type of bonding in an element (Table 14.2). Within Group 4A, the large decrease in melting point between the network covalent solids C and Si is due to longer, weaker bonds in the Si structure; the large decrease between Ge and Sn is due to the change from covalent network to metallic bonding. Similarly, considering horizontal trends, the large increases in melting point and ΔH_{fus} across a period between Al and Si and between Ga and Ge reflect the change from metallic to covalent network bonding. Note the abrupt rises in the values for these properties from metallic Al, Ga, and Sn to the network covalent metalloids Si, Ge, and Sb, and note the abrupt drops from the covalent networks of C and Si to the individual molecules of N and P in Group 5A.

Table 14.2   Bond Type and the Melting Process in Groups 3A(13) to 5A(15)

Period	Group 3A(13)				Group 4A(14)				Group 5A(15)				Key:	
	Element	Bond Type	Melting Point (°C)	ΔH_{fus} (kJ/mol)	Element	Bond Type	Melting Point (°C)	ΔH_{fus} (kJ/mol)	Element	Bond Type	Melting Point (°C)	ΔH_{fus} (kJ/mol)		Metallic
2	B		2180	23.6	C		4100	Very high	N		−210	0.7		Covalent network
3	Al		660	10.5	Si		1420	50.6	P		44.1	2.5		Covalent molecule
4	Ga		30	5.6	Ge		945	36.8	As		816	27.7		Metal
5	In		157	3.3	Sn		232	7.1	Sb		631	20.0		Metalloid
6	Tl		304	4.3	Pb		327	4.8	Bi		271	10.5		Nonmetal

Allotropism: Different Forms of an Element Striking variations in physical properties often appear among **allotropes,** different crystalline or molecular forms of a substance. One allotrope is usually more stable than another at a particular pressure and temperature. Group 4A(14) provides the first dramatic example of allotropism, in the forms of carbon. It is difficult to imagine two substances made entirely of the same atom that are more different than graphite and diamond. Graphite is a black electrical conductor that is soft and "greasy," whereas diamond is a colorless electrical insulator that is extremely hard. Graphite is the standard state of carbon, the more stable form at ordinary temperature and pressure, as the phase diagram in Figure 14.12 shows. Fortunately for jewelry owners, diamond changes to graphite at a negligible rate under normal conditions.

In the mid-1980s, a newly discovered allotrope of carbon began generating great interest. Mass spectrometric analysis of soot had shown evidence for a soccer ball–shaped molecule of formula C_{60} (Figure 14.13A; see also the Gallery, p. 398). More recently, the molecule has also been found in geological samples formed by meteorite impacts, even the one that occurred around the time the dinosaurs became extinct. The molecule has been dubbed *buckminsterfullerene* (informally called a "buckyball") after the architect-engineer R. Buckminster Fuller, who designed structures with similar shapes. Excitement rose in 1990, when scientists learned how to prepare multigram quantities of C_{60} and related fullerenes, enough to study macroscopic behavior and possible applications. Since then, metal atoms have been incorporated into the structure and many different atoms and groups (fluorine, hydroxyl groups, sugars, etc.) have been attached, resulting in compounds with a range of useful properties.

Then, in 1991, scientists passed an electric discharge through graphite rods sealed with helium gas in a container and obtained extremely thin (~1 nm in diameter) graphite-like tubes with fullerene ends (Figure 14.13B). These *nanotubes* are rigid and, on a mass basis, much stronger than steel along their long axis. They also conduct electricity along this axis because of the delocalized electrons. Reports about fullerenes and nanotubes appear almost daily in scientific journals. With potential applications in nanoscale electronics, catalysis, polymers, and medicine, fullerene and nanotube chemistry will receive increasing attention well into the 21st century.

Tin has two allotropes. White β-tin is stable at room temperature and above, whereas gray α-tin is the more stable form below 13°C (56°F). When white tin is kept for long periods at a low temperature, some converts to microcrystals of gray tin. The random formation and growth of these regions of gray tin, which has a different crystal structure, weaken the metal and make it crumble. In the unheated cathedrals of medieval northern Europe, tin pipes of magnificent organs sometimes crumbled as a result of "tin disease" caused by this allotropic transition from white to gray tin.

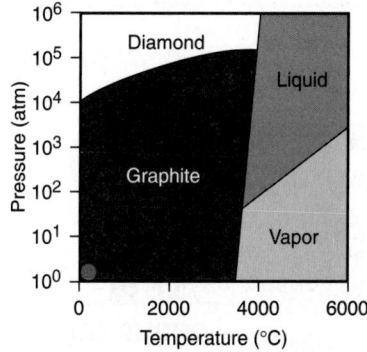

Figure 14.12 Phase diagram of carbon. Graphite is the more stable form of carbon at ordinary conditions (*small red circle at extreme lower left*). Diamond is more stable at very high pressure.

ARIS Presentation Center, Art Library

Engineering application

ARIS Presentation Center, Art Library

Figure 14.13 Buckyballs and nanotubes. A, Crystals of buckminster-fullerene (C_{60}) next to a ball-and-stick model. The parent of the fullerenes, the "buckyball," is a soccer ball–shaped molecule of 60 carbon atoms. **B,** Nanotubes are single or, as shown in this colored model, concentric graphite-like tubes (shown here without the usual fullerene ends; see the Gallery, p. 398).

A

B

Key Atomic and Physical Properties

ns^2np^2

GROUP 4A(14)

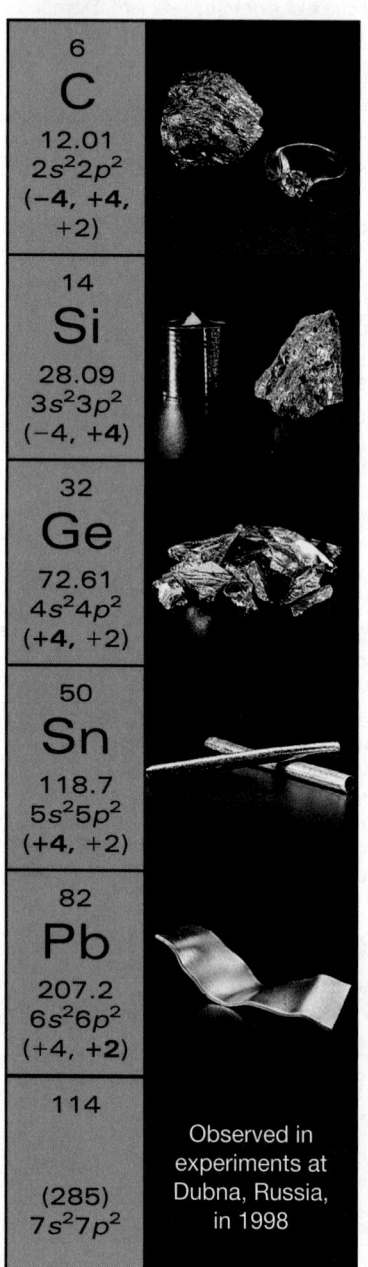

6
C
12.01
$2s^2 2p^2$
(−4, +4,
+2)

14
Si
28.09
$3s^2 3p^2$
(−4, +4)

32
Ge
72.61
$4s^2 4p^2$
(+4, +2)

50
Sn
118.7
$5s^2 5p^2$
(+4, +2)

82
Pb
207.2
$6s^2 6p^2$
(+4, +2)

114
(285)
$7s^2 7p^2$

Observed in
experiments at
Dubna, Russia,
in 1998

Atomic Properties

Atomic radius (pm)		Ionic radius (pm)
C 77		
Si 118		
Ge 122		
Sn 140		Sn^{2+} 118
Pb 146		Pb^{2+} 119

Group electron configuration is ns^2np^2. Down the group, the number of oxidation states decreases, and the lower (+2) state becomes more common.

Down the group, size increases. Because transition and inner transition elements intervene, IE and EN do not decrease smoothly.

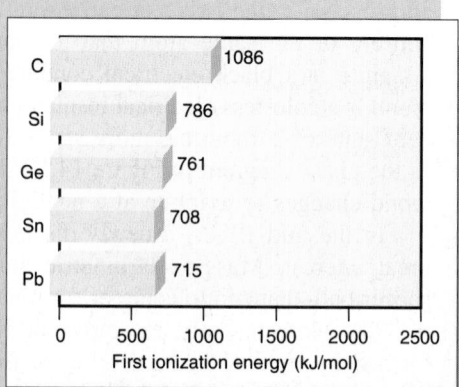

First ionization energy (kJ/mol)

C	1086
Si	786
Ge	761
Sn	708
Pb	715

Electronegativity

C	2.5
Si	1.8
Ge	1.8
Sn	1.8
Pb	1.9

Physical Properties

Trends in properties, such as decreasing hardness and melting point, are due to changes in types of bonding within the solid: network covalent in C, Si, and Ge; metallic in Sn and Pb (*see text*).

Temperature (°C)

	BP	MP
C	Unstable	4100
Si	~3280	1420
Ge	2850	945
Sn	2623	232
Pb	1751	327

■ BP
□ MP

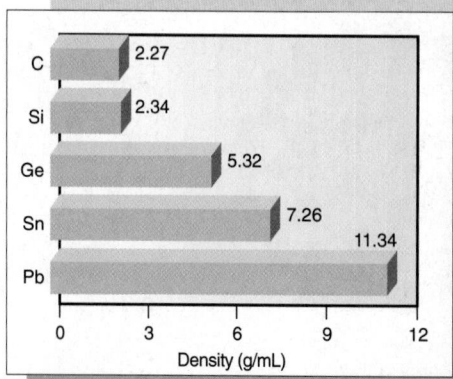

Density (g/mL)

C	2.27
Si	2.34
Ge	5.32
Sn	7.26
Pb	11.34

Down the group, density increases because of several factors, including differences in crystal packing.

Some Reactions and Compounds

Important Reactions

Reactions 1 and 2 pertain to all the elements (E); reactions 3 to 7 concern industrial uses for compounds of C and Si.

1. The elements are oxidized by halogens:

$$E(s) + 2X_2 \longrightarrow EX_4 \quad (E = C, Si, Ge)$$

The +2 halides are more stable for tin and lead, SnX_2 and PbX_2.

2. The elements are oxidized by O_2:

$$E(s) + O_2(g) \longrightarrow EO_2 \quad (E = C, Si, Ge, Sn)$$

Pb forms the +2 oxide, PbO. Oxides become more basic down the group. The reaction of CO_2 and H_2O provides the weak acidity of natural unpolluted waters:

$$CO_2(g) + H_2O(l) \rightleftharpoons [H_2CO_3(aq)]$$
$$\rightleftharpoons H^+(aq) + HCO_3{}^-(aq)$$

3. Air and steam passed over hot coke produce gaseous fuel mixtures (producer gas and water gas):

$$C(s) + air(g) + H_2O(g) \longrightarrow CO(g) + CO_2(g) + N_2(g) + H_2(g)$$
[not balanced]

4. Hydrocarbons react with O_2 to form CO_2 and H_2O. The reaction for methane is adapted to yield heat or electricity:

$$CH_4(g) + 2O_2(g) \longrightarrow CO_2(g) + 2H_2O(g)$$

Oxyacetylene torch

5. Certain metal carbides react with water to produce acetylene:

$$CaC_2(s) + 2H_2O(l) \longrightarrow$$
$$Ca(OH)_2(aq) + C_2H_2(g)$$

The gas is used to make other organic compounds and as a fuel in welding (*see photo*).

6. Freons (chlorofluorocarbons) are formed by fluorinating carbon tetrachloride:

$$CCl_4(l) + HF(g) \longrightarrow CFCl_3(g) + HCl(g)$$

Production of trichlorofluoromethane (Freon-11), the most widely used refrigerant in the world, is being eliminated because of adverse effects on the environment (*margin note, p. 589*).

7. Silica is reduced to form elemental silicon:

$$SiO_2(s) + 2C(s) \longrightarrow Si(s) + 2CO(g)$$

This crude silicon is made ultrapure through zone refining (*Gallery, p. 533*) for the manufacture of computer chips (*see photo*).

Computer chip

Important Compounds

1. Carbon monoxide, CO. Used as a gaseous fuel, as a precursor for organic compounds, and as a reactant in the purification of nickel. Formed in internal combustion engines and released as a toxic air pollutant.

2. Carbon dioxide, CO_2. Atmospheric component used by photosynthetic plants to make carbohydrates and O_2. The final oxidation product of all carbon-based fuels; its increase in the atmosphere is changing Earth's climate. Used industrially as a refrigerant gas, a blanketing gas in fire extinguishers, and an effervescent gas in beverages (*see photo*). Combined with NH_3 to form urea for fertilizers and plastics manufacture.

3. Methane, CH_4. Used as a fuel and in the production of many organic compounds. Major component of natural gas. Formed by anaerobic decomposition of plants (swamp gas) and by microbes in termites and certain mammals. Greenhouse gas that contributes to global warming.

4. Silicon dioxide, SiO_2. Occurs in many amorphous (glassy) and crystalline forms, quartz being the most common. Used to make glass and as an inert chromatography support material.

5. Silicon carbide, SiC. Known as *carborundum*, a major industrial abrasive and a highly refractory ceramic for tough, high-temperature uses. Can be doped to form a high-temperature semiconductor.

6. Organotin compounds, R_4Sn. Used to stabilize PVC [poly(vinyl chloride)] plastics and to cure silicone rubbers. Agricultural biocide for insects, fungi, and weeds.

7. Tetraethyl lead, $(C_2H_5)_4Pb$. Once used as a gasoline additive to improve fuel efficiency, but now banned because of its inactivation of auto catalytic converters. Major source of lead as a toxic air pollutant.

How Does the Type of Bonding Change in Group 4A(14) Compounds?

The Group 4A(14) elements display a wide range of chemical behavior, from the covalent compounds of carbon to the ionic compounds of lead. Carbon's intermediate EN of 2.5 ensures that it virtually always forms covalent bonds, but the larger members of the group form bonds with increasing ionic character. With nonmetals, Si and Ge form strong polar covalent bonds, such as the Si—O bond, one of the strongest of any Period 3 element (BE = 368 kJ/mol). This bond is responsible for the physical and chemical stability of Earth's solid surface, as we discuss later in this section. Although individual Sn or Pb ions rarely exist, the bonding of either element with a nonmetal has considerable ionic character.

The pattern of elements with more than one oxidation state that we observed in Group 3A(13) also appears here. After Si, the shift to greater importance of the lower oxidation state occurs: compounds with Si in the +4 state are much more stable than those with Si in the +2 state, whereas compounds with Pb in the +2 state are more stable than those with Pb in the +4 state. The elements also behave more like metals in the lower oxidation state. Consider the chlorides and oxides of Sn and Pb. The +2 chlorides $SnCl_2$ and $PbCl_2$ are white, relatively high-melting, water-soluble crystals—typical properties of a salt (Figure 14.14). In contrast, $SnCl_4$ is a volatile, benzene-soluble liquid, and $PbCl_4$ is a thermally unstable oil. Both likely consist of individual, tetrahedral molecules. The +2 oxides SnO and PbO are more basic than the +4 oxides SnO_2 and PbO_2: because the +2 metals are less able to polarize the O^{2-} ion, the E-to-O bonding is more ionic.

Figure 14.14 The greater metallic character of tin and lead in the lower oxidation state. Metals with more than one oxidation state exhibit more metallic behavior in the lower state. Tin(II) and lead(II) chlorides are white, crystalline, saltlike solids. In contrast, tin(IV) and lead(IV) chlorides are volatile liquids, indicating the presence of individual molecules.

 ARIS Presentation Center, Photo Library

Highlights of Carbon Chemistry

Like the other Period 2 elements, carbon is an anomaly in its group; indeed, it may be an anomaly in the entire periodic table. Carbon forms bonds with the smaller Group 1A(1) and 2A(2) metals, many transition metals, the halogens, and its neighbors B, Si, N, O, P, and S. It also exhibits every oxidation state possible for its group, from +4 in CO_2 and halides like CCl_4 through −4 in CH_4.

Two features stand out in the chemistry of carbon: its ability to bond to itself to form chains, a process known as *catenation*, and its ability to form multiple bonds. As a result of its small size and its capacity for four bonds, carbon can form chains, branches, and rings that lead to myriad structures. Add a lot of H, some O and N, a bit of S, P, halogens, and a few metals, and you have the whole organic world! Figure 14.15 shows three of the several million organic compounds known. Multiple bonds are common in carbon structures because the C—C bond is short enough for side-to-side overlap of two half-filled 2*p* orbitals to form π bonds. (In Chapter 15, we discuss in detail how the atomic properties of carbon give rise to the diverse structures and reactivities of organic compounds.)

Because the other 4A members are larger, E—E bonds become longer and weaker down the group; thus, in terms of bond strength, C—C > Si—Si > Ge—Ge. The empty *d* orbitals of these larger atoms make the chains much more

Figure 14.15 Three of the several million known organic compounds of carbon. Acrylonitrile, a precursor of acrylic fibers. A typical PCB (one of the polychlorinated biphenyls). Lysine, one of about 20 amino acids that occur in proteins.

ARIS Presentation Center, Art Library

susceptible to chemical attack. Thus, some compounds with long Si chains are known but they are very reactive, and the longest Ge chain has only eight atoms. Moreover, longer bonds do not typically allow sufficient overlap of *p* orbitals for π bonding. Only very recently have compounds with π bonds between Si atoms or between Ge atoms been prepared, and they are also extremely reactive.

In contrast to its organic compounds, carbon's inorganic compounds are simple. Metal carbonates are the main mineral form. Marble, limestone, chalk, coral, and several other types are found in enormous deposits throughout the world. Many of these compounds are remnants of fossilized marine organisms. Carbonates are used in several common antacids because they react with the HCl in stomach acid [see the Group 2A(2) Family Portrait, compound 4, p. 577]:

$$CaCO_3(s) + 2HCl(aq) \longrightarrow CaCl_2(aq) + CO_2(g) + H_2O(l)$$

Identical net ionic reactions with sulfuric and nitric acids protect lakes bounded by limestone deposits from the harmful effects of acid rain.

Unlike the other 4A members, which form only solid network-covalent or ionic oxides, carbon forms two common gaseous oxides, CO_2 and CO. Carbon dioxide is essential to all life: it is the primary source of carbon in plants and animals through photosynthesis. Its aqueous solution is the cause of acidity in natural waters. However, its atmospheric buildup from deforestation and excessive use of fossil fuels is severely affecting the global climate. Carbon monoxide forms when carbon or its compounds burn in an inadequate supply of O_2:

$$2C(s) + O_2(g) \longrightarrow 2CO(g)$$

Carbon monoxide is a key component of syngas fuels (see Chemical Connections, p. 256) and is widely used in the production of methanol, formaldehyde, and other major industrial compounds.

CO binds strongly to many transition metals. When inhaled in cigarette smoke or polluted air, it enters the blood and binds strongly to the Fe(II) in hemoglobin, preventing the normal binding of O_2, and to other iron-containing proteins. The cyanide ion (CN^-) is *isoelectronic* with CO:

$$[:C≡N:]^- \text{ same electronic structure as } :C≡O:$$

Biological application

Cyanide binds to many of the same iron-containing proteins and is also toxic.

Monocarbon halides (or halomethanes) are tetrahedral molecules whose stability to heat and light decreases as the size of the halogen atom increases and the C—X bond becomes longer (weaker) (see the Interchapter, Topic 2). The chlorofluorocarbons (CFCs, or Freons) have important industrial uses; their short, strong carbon-halogen bonds are exceptionally stable, however, and the environmental persistence of these compounds has created major problems. ●

● CFCs: The Good, the Bad, and the Strong The strengths of the C—F and C—Cl bonds make CFCs, such as Freon-12 (CCl_2F_2), shown here, both useful and harmful. Chemically and thermally stable, nontoxic, and nonflammable, they are excellent cleaners for electronic parts, coolants in refrigerators and air conditioners, and propellants in aerosol cans. However, their bond strengths also mean that CFCs decompose *very* slowly near Earth's surface. In the lower atmosphere, they contribute to climate warming, absorbing infrared radiation 16,000 times as effectively as CO_2. Once in the stratosphere, however, they are bombarded by ultraviolet (UV) radiation, which breaks the otherwise stable C—Cl bonds, releasing free Cl atoms that initiate ozone-destroying reactions (Chapter 16). Legal production of CFCs has ended in the United States, but international production and smuggling are widespread.

Environmental application

Figure 14.16 The impact of *p,d–π* bonding on the structure of trisilylamine. A, The molecular shape of trimethylamine is trigonal *pyramidal*, like that of ammonia. **B,** Trisilylamine, the silicon analog, has a trigonal *planar* shape because of the formation of a double bond between N and Si. The ball-and-stick model shows one of three resonance forms. **C,** The lone pair in an unhybridized *p* orbital of N overlaps with an empty *d* orbital of Si to give a *p,d–π* bond. In the resonance hybrid, a *d* orbital on each Si is involved in the π bond.

ARIS Presentation Center, Art Library

Highlights of Silicon Chemistry

Silicon halides are more reactive than carbon halides because Si has empty *d* orbitals that are available for bond formation. Surprisingly, the silicon halides have longer yet *stronger* bonds than the corresponding carbon halides (see Table 9.2, p. 353). One explanation for this unusual strength is that the Si—X bond has some double-bond character because of the presence of a σ bond *and* a different type of π bond, called a *p,d–π* bond. This bond arises from side-to-side overlap of an Si *d* orbital and a halogen *p* orbital. A similar type of *p,d–π* bonding leads to the trigonal planar molecular shape of trisilylamine, in contrast to the trigonal pyramidal shape of trimethylamine (Figure 14.16).

To a great extent, the chemistry of silicon is the chemistry of the *silicon-oxygen bond*. Just as carbon forms unending C—C chains, the —Si—O— grouping repeats itself endlessly in a wide variety of **silicates,** the most important minerals on the planet, and in **silicones,** synthetic polymers that have many applications:

1. *Silicate minerals.* From common sand and clay to semiprecious amethyst and carnelian, silicate minerals are the dominant form of matter in the nonliving world. Oxygen, the most abundant element on Earth, and silicon, the next most abundant, compose these minerals and thus account for four of every five atoms on the surface of the planet!

The silicate building unit is the *orthosilicate* grouping, —SiO$_4$—, a tetrahedral arrangement of four oxygens around a central silicon. Several well-known minerals contain SiO$_4^{4-}$ ions or small groups of them linked together. The gemstone zircon (ZrSiO$_4$) contains one unit; hemimorphite [Zn$_4$(OH)$_2$Si$_2$O$_7$·H$_2$O] contains two units linked through an oxygen corner; and beryl (Be$_3$Al$_2$Si$_6$O$_{18}$), the major source of beryllium, contains six units joined into a cyclic ion (Figure 14.17). As you'll see shortly, in addition to these separate ions, SiO$_4$ units are linked more extensively to create much of the planet's mineral structure.

2. *Silicone polymers.* Unlike the naturally occurring silicates, silicone polymers are manufactured substances, consisting of alternating Si and O atoms with two organic groups also bonded to each Si atom. A key starting material is formed by the reaction of silicon with methyl chloride:

$$Si(s) + 2CH_3Cl(g) \xrightarrow[\sim 300°C]{Cu \text{ catalyst}} (CH_3)_2SiCl_2(g)$$

Although the corresponding carbon compound [(CH$_3$)$_2$CCl$_2$] is inert in water, the empty *d* orbitals of Si make (CH$_3$)$_2$SiCl$_2$ reactive:

$$(CH_3)_2SiCl_2(l) + 2H_2O(l) \longrightarrow (CH_3)_2Si(OH)_2(l) + 2HCl(g)$$

Figure 14.17 **Structures of the silicate anions in some minerals.**

The product reacts with itself to form water and a silicone chain molecule called a *poly(dimethyl siloxane)*:

$$n(CH_3)_2Si(OH)_2 \longrightarrow \left[\begin{array}{c} CH_3 \\ | \\ O-Si \\ | \\ CH_3 \end{array}\right]_n + nH_2O$$

Silicones have properties of both plastics and minerals. The organic groups give them the flexibility and weak intermolecular forces between chains that are characteristic of a plastic, while the O—Si—O backbone confers the thermal stability and nonflammability of a mineral. In the Gallery on the following pages, note the structural parallels between silicates and silicones.

Diagonal Relationships: Boron and Silicon

The final diagonal relationship that we consider occurs between the metalloids boron and silicon. Both of these elements exhibit the electrical properties of a semiconductor (Section 12.6). Both B and Si and their mineral oxoanions—borates and silicates—occur in extended covalent networks. Both boric acid [B(OH)$_3$] and silicic acid [Si(OH)$_4$] are weakly acidic solids that occur as layers held together by widespread H bonding. Their hydrides—the compact boranes and the extended silanes—are both flammable, low-melting compounds that act as reducing agents.

Looking Backward and Forward: Groups 3A(13), 4A(14), and 5A(15)

Standing in Group 4A(14), we look back at Group 3A(13) as the transition from the *s* block of metals to the *p* block of mostly metalloids and nonmetals (Figure 14.18). Major changes in physical behavior occur moving across a period from metals to covalent networks. Changes in chemical behavior occur as well, as cations give way to covalent tetrahedra. Looking ahead to Group 5A(15), we note the more frequent occurrence of covalent bonding and (as electronegativity increases) the expanded valence shells of nonmetals, as well as the first appearance of monatomic anions.

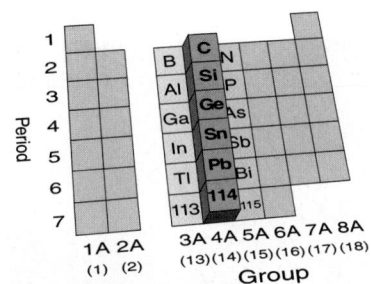

Figure 14.18 Standing in Group 4A(14), looking backward to 3A(13) and forward to 5A(15).

Silicate Minerals and Silicone Polymers

The silicates and silicones show beautifully how organization at the molecular level manifests itself in the properties of macroscopic substances. Interestingly, both of these types of materials exhibit the same three structural classes: chains, sheets, and three-dimensional frameworks.

Silicate Minerals

Chain Silicates

In the simplest structural class of silicates, each SiO_4 unit shares two of its O corners with other SiO_4 units, forming a chain. Two chains can link laterally into a ribbon; the most common ribbon has repeating $Si_4O_{11}^{6-}$ units. Metal ions bind the polyanionic ribbons together into neutral sheaths. With only weak intermolecular forces between sheaths, the material occurs in fibrous strands, as in the family of asbestos minerals.

Portion of ribbon

$= Si$

$Si_4O_{11}^{6-}$ unit

Asbestos

Sheet (Layer) Silicates

In the next structural class of silicates, each SiO_4 unit shares three of its four O corners with other SiO_4 units to form a sheet; double sheets arise when the fourth O is shared with another sheet. In talc, the softest mineral, the sheets interact through weak forces, so talcum powder feels slippery. If Al substitutes for some Si, or if $Al(OH)_3$ layers interleave with silicate sheets, an aluminosilicate results, such as the clay kaolinite. Different substitutions for Si and/or interlayers of other ions give the micas. In muscovite mica, ions lie between aluminosilicate double layers. Mica flakes when the ionic attractions are overcome.

SiO_4
K^+
SiO_4
OH^-
Al^{3+}
OH^-
SiO_4
K^+
SiO_4

Muscovite mica

GALLERY

Framework Silicates

In the final structural class of silicates, SiO_4 units share all four O corners to give a three-dimensional framework silicate, such as silica (SiO_2), which occurs most often as α-quartz. Some of silica's 12 crystalline forms exist as semiprecious gems. Feldspars, which comprise 60% of Earth's crust, result when some Si in a framework silicate is replaced by Al; granite consists of microcrystals of feldspar, mica, and quartz. Eons of weathering convert feldspars into clays. Zeolites have open frameworks of polyhedra that give rise to minute tunnels. Synthetic zeolites are made with cavities of specific sizes to trap particular molecules; they are used to dry gas mixtures, separate hydrocarbons, and prepare catalysts.

Natural zeolite

Quartz

Silicone Polymers

Polymer Chemistry

The design, synthesis, and production of polymers form one of the largest branches of modern chemical science. Nearly half of all industrial chemists and chemical engineers are involved in this pursuit. A major branch of polymer chemistry is devoted to exploring the properties and countless uses of silicones.

Repeat unit

Chain-terminating unit

Chain Silicones: Oils and Greases

The simplest structural class of silicones consists of the poly(dimethyl siloxane) chain, in which each $(CH_3)_2 Si(OH)_2$ unit uses both OH groups to link to two others. A chain-terminating compound with a third organic group, such as $(CH_3)_3SiOH$, is added to control chain length. These polymers are unreactive, oily liquids with high viscosity and low surface tension. They are used as hydraulic oils and lubricants, as antifoaming agents for frying potato chips, and as components of suntan oil, car polish, digestive aids, and makeup.

Organic chemistry application

(continued)

Sheet Silicones: Elastomers

In the next class of silicones, the third OH group of an added bridging compound, such as $CH_3Si(OH)_3$, reacts to condense chains laterally into gummy sheets that are made into elastomers (rubbers), which are flexible, elastic, and stable from $-100°C$ to $250°C$. These are used in gaskets, rollers, cable insulation, space suits, contact lenses, and dentures.

⊙⊙ **Engineering application**

Makeup consisting of chain and sheet silicones being applied to create the character Freddie Krueger

Framework Silicones: Resins

In the final structural class of silicones, reactions that free OH groups, while substituting larger organic groups for some CH_3 groups, interlink sheets to produce strong, thermally stable resins. These are used as insulating laminates on printed circuit boards and as nonstick coatings on cookware. Sheet and framework silicones have revolutionized modern surgical practice by allowing the fabrication of numerous parts that can be implanted permanently in a patient to replace damaged ones. Some of these (indicated by dots on Leonardo's man) are artificial skin, bone, joints, blood vessels, and organ parts.

≋ **Biological application**

14.7 GROUP 5A(15): THE NITROGEN FAMILY

For the first two elements of Group 5A(15), gaseous nonmetallic nitrogen (N) and solid nonmetallic phosphorus (P), the industrial and environmental significance of their compounds is matched only by their importance in the structures and functions of biomolecules. Below these nonmetals are two metalloids, arsenic (As) and antimony (Sb), followed by the metal bismuth (Bi), the last nonradioactive element in the periodic table, and the newly synthesized element 115. The Group 5A(15) Family Portrait (pp. 596 and 597) provides an overview.

What Accounts for the Wide Range of Physical Behavior in Group 5A(15)?

Group 5A(15) displays the widest range of physical behavior we've seen so far because of large changes in bonding and intermolecular forces. Nitrogen occurs as a gas consisting of N_2 molecules, which interact through such weak dispersion forces that the element *boils* more than 200°C below room temperature. Elemental phosphorus exists most commonly as tetrahedral P_4 molecules. However, because P is heavier and more polarizable than N, stronger dispersion forces are present, and the element melts about 25°C above room temperature. Arsenic consists of extended, puckered sheets in which each As atom is covalently bonded to three others and forms nonbonding interactions with three nearest neighbors in adjacent sheets. This arrangement gives As the highest melting point in the group. A similar covalent network for Sb gives it a much higher melting point than the next member of the group, Bi, which has metallic bonding.

Phosphorus has several allotropes. The white and red forms have very different properties because of the way the atoms are linked. The highly reactive white form is prepared as a gas and condensed to a whitish, waxy solid under cold water to prevent it from igniting in air. It consists of tetrahedral molecules of four P atoms bonded *to each other,* with no atom in the center (Figure 14.19A). Each P atom uses its half-filled $3p$ orbitals to bond to the other three. Whereas the $3p$ orbitals of an isolated P atom lie 90° apart, the bond angles in P_4 are 60°. With the smaller angle comes poor orbital overlap and a P—P bond strength only about 80% that of a normal P—P bond (Figure 14.19B). Weaker bonds break more easily, contributing to the white allotrope's high reactivity. Heating the white form in the absence of air breaks one of the P—P bonds in each tetrahedron, and those $3p$ orbitals overlap with others to form the chains of P_4 units that make up the red form (Figure 14.19C). The individual molecules of white P make it highly reactive, low melting (44.1°C), and soluble in nonpolar solvents; the chains of red P make it much less reactive, high melting (~600°C), and insoluble.

ARIS Presentation Center, Art Library

Figure 14.19 Two allotropes of phosphorus. A, White phosphorus exists as individual P_4 molecules, with the P—P bonds forming the edges of a tetrahedron. **B,** The reactivity of P_4 is due in part to the bond strain that arises from the 60° bond angle. Note how overlap of the $3p$ orbitals is decreased because they do not meet directly end to end (overlap is shown here for only three of the P—P bonds), which makes the bonds easier to break. **C,** In red phosphorus, one of the P—P bonds of the white form has broken and links the P_4 units together into long chains. Lone pairs (not shown) reside in s orbitals in both allotropes.

A White phosphorus (P_4)   **B** Strained bonds in P_4 **C** Red phosphorus

Group 5A(15): The Nitrogen Family

Key Atomic and Physical Properties

ns^2np^3

GROUP 5A(15)

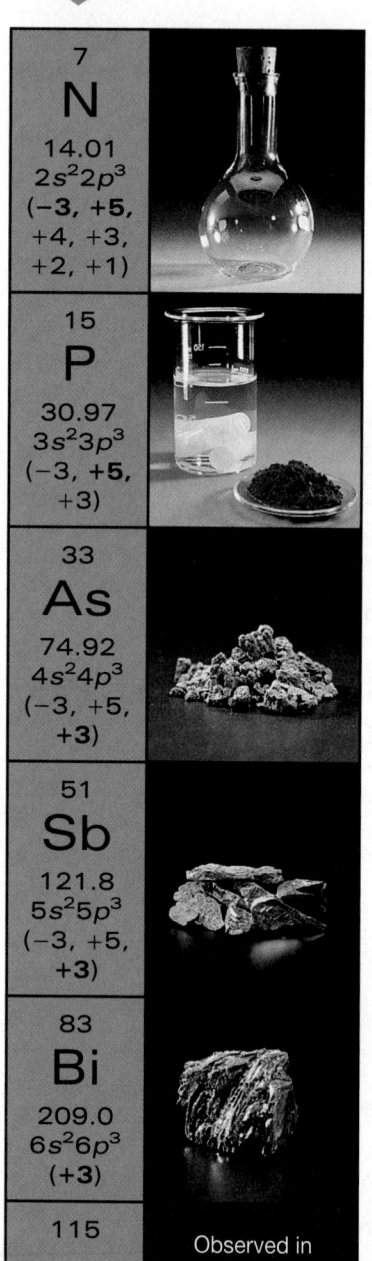

7	
N	
14.01	
$2s^22p^3$	
(−3, **+5**, +4, +3, +2, +1)	
15	
P	
30.97	
$3s^23p^3$	
(−3, **+5**, +3)	
33	
As	
74.92	
$4s^24p^3$	
(−3, +5, **+3**)	
51	
Sb	
121.8	
$5s^25p^3$	
(−3, +5, **+3**)	
83	
Bi	
209.0	
$6s^26p^3$	
(**+3**)	
115	
(288)	
$7s^27p^3$	

Observed in experiments at Dubna, Russia, in 2003

Atomic Properties

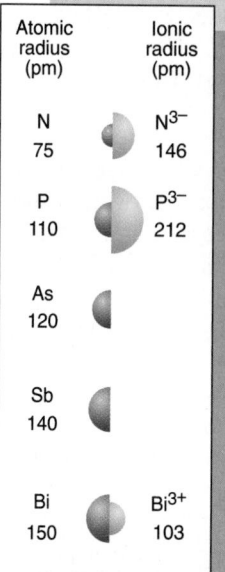

Atomic radius (pm)		Ionic radius (pm)
N 75		N^{3-} 146
P 110		P^{3-} 212
As 120		
Sb 140		
Bi 150		Bi^{3+} 103

Group electron configuration is ns^2np^3. The np sublevel is half-filled, with each p orbital containing one electron. The number of oxidation states decreases down the group, and the lower (+3) state becomes more common.

Atomic properties follow generally expected trends. The large (~50%) increases in size from N to P correlates with the much lower IE and EN of P.

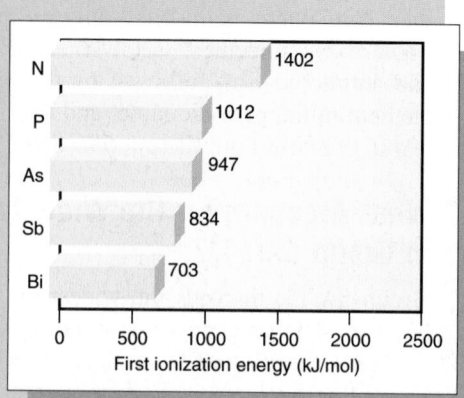

First ionization energy (kJ/mol): N 1402, P 1012, As 947, Sb 834, Bi 703

Electronegativity: N 3.0, P 2.1, As 2.0, Sb 1.9, Bi 1.9

Physical Properties

Physical properties reflect the change from individual molecules (N, P) to network covalent solid (As, Sb) to metal (Bi). Thus, melting points increase and then decrease.

Temperature (°C): N −196 / −210; P 280 / 44.1; As 615 (subl) / 816 (at 39 atm); Sb 1587 / 631; Bi 1564 / 271. (BP, MP)

Large atomic size and low atomic mass result in low density. Because mass increases more than size down the group, the density of the elements as solids increases. The dramatic increase from P to As is due to the intervening transition elements.

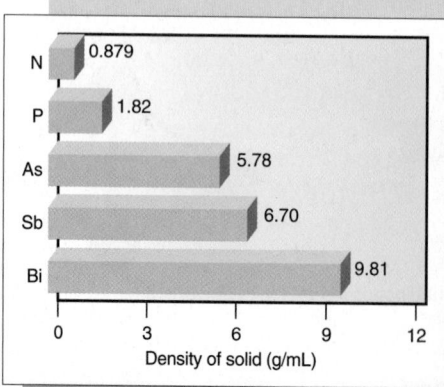

Density of solid (g/mL): N 0.879, P 1.82, As 5.78, Sb 6.70, Bi 9.81

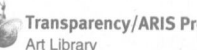

Transparency/ARIS Presentation Center, Art Library

Some Reactions and Compounds

Important Reactions

General group behavior is shown in reactions 1 to 3, whereas phosphorus chemistry is the theme in reactions 4 and 5.

1. Nitrogen is "fixed" industrially in the Haber process:

$$N_2(g) + 3H_2(g) \rightleftharpoons 2NH_3(g)$$

Further reactions convert NH_3 to NO, NO_2, and HNO_3 (*Highlights of Nitrogen Chemistry*). Hydrides of some other group members are formed from reaction in water (or with H_3O^+) of a metal phosphide, arsenide, and so forth:

$$Ca_3P_2(s) + 6H_2O(l) \longrightarrow 2PH_3(g) + 3Ca(OH)_2(aq)$$

2. Halides are formed by direct combination of the elements:

$$2E(s) + 3X_2 \longrightarrow 2EX_3 \qquad \text{(E = all except N)}$$

$$EX_3 + X_2 \longrightarrow EX_5 \qquad \text{(E = all except N and Bi}$$
$$\text{with X = F and Cl, but no}$$
$$\text{BiCl}_5; \text{E = P for X = Br)}$$

3. Oxoacids are formed from the halides in a reaction with water that is common to many nonmetal halides:

$$EX_3 + 3H_2O(l) \longrightarrow H_3EO_3(aq) + 3HX(aq)$$
$$\text{(E = all except N)}$$

$$EX_5 + 4H_2O(l) \longrightarrow H_3EO_4(aq) + 5HX(aq)$$
$$\text{(E = all except N and Bi)}$$

Note that the oxidation number of E does *not* change.

4. Phosphate ions are dehydrated to form polyphosphates:

$$3NaH_2PO_4(s) \longrightarrow Na_3P_3O_9(s) + 3H_2O(g)$$

5. When P_4 reacts in basic solution, its oxidation number both decreases *and* increases (disproportionation):

$$P_4(s) + 3OH^-(aq) + 3H_2O(l) \longrightarrow PH_3(g) + 3H_2PO_2^-(aq)$$

Analogous reactions are typical of many nonmetals, such as S_8 and X_2 (halogens).

Important Compounds

1. Ammonia, NH_3. First substance formed when atmospheric N_2 is used to make N-containing compounds. Annual multimillion-ton production for use in fertilizers, explosives, rayon, and polymers such as nylon, urea-formaldehyde resins, and acrylics.

2. Hydrazine, N_2H_4 (*margin note, p. 598*).

3. Nitric oxide (NO), nitrogen dioxide (NO_2), and nitric acid (HNO_3). Oxides are intermediates in HNO_3 production. This acid is used in fertilizer manu-

Detonation of explosives

facture, nylon production, metal etching, and the explosives industry (*Highlights of Nitrogen Chemistry*).

4. Amino acids, H_3N^+—CH(R)—COO⁻ (R = one of 20 different organic groups). Occur in every organism, both free and linked together into proteins. Essential to the growth and function of all cells. Synthetic amino acids are used as dietary supplements.

5. Phosphorus trichloride, PCl_3. Used to form many organic phosphorus compounds, including oil and fuel additives, plasticizers, flame retardants, and insecticides. Also used to make PCl_5, $POCl_3$, and other important P-containing compounds.

6. Tetraphosphorus decaoxide (P_4O_{10}) and phosphoric acid (H_3PO_4) (*Highlights of Phosphorus Chemistry*).

7. Sodium tripolyphosphate, $Na_5P_3O_{10}$. As a water-softening agent (Calgon), it combines with hard-water Mg^{2+} and Ca^{2+} ions, preventing them from reacting with soap anions, and thus

Algal growth in polluted lake

improving cleaning action. Its use has been curtailed in the United States because it pollutes lakes and streams by causing excessive algal growth (*see photo*).

8. Adenosine triphosphate (ATP) and other biophosphates. ATP acts to transfer chemical energy in the cell; it is necessary for all biological processes requiring energy. Phosphate groups also occur in sugars, fats, proteins, and nucleic acids.

9. Bismuth subsalicylate, $BiO(C_7H_5O_3)$. The active ingredient in Pepto-Bismol (*see photo*), a widely used remedy for diarrhea and nausea. (The pink color is not due to this white compound.)

What Patterns Appear in the Chemical Behavior of Group 5A(15)?

The same general pattern of chemical behavior that we discussed for Group 4A(14) appears again in this group, reflected in the change from nonmetallic N to metallic Bi. The overwhelming majority of Group 5A(15) compounds have *covalent bonds*. Whereas N can form no more than four bonds, the next three members can expand their valence shells by using empty d orbitals.

For a 5A element to form an ion with a noble gas electron configuration, it must *gain* three electrons, the last two in endothermic steps. Nevertheless, the enormous lattice energy released when such highly charged anions attract cations drives their formation. However, the $3-$ anion of N occurs only in compounds with active metals, such as Li_3N and Mg_3N_2 (and that of P may occur in Na_3P). Metallic Bi forms mostly covalent compounds but exists as a cation in a few compounds, such as BiF_3 and $Bi(NO_3)_3 \cdot 5H_2O$, through *loss* of its three valence p electrons.

Again, as we move down this group, we see the patterns first seen in Groups 3A and 4A. Fewer oxidation states occur, with the lower state becoming more prominent: N exhibits every state possible for a 5A element, from $+5$ to -3; only the $+5$ and $+3$ states are common for P, As, and Sb; and $+3$ is the only common state of Bi. The oxides change from acidic to amphoteric to basic, reflecting the increase in the metallic character of the elements. In addition, the lower oxide of an element is more basic than the higher oxide, reflecting the greater ionic character of the E-to-O bonding in the lower oxide.

All the Group 5A(15) elements form gaseous hydrides of formula EH_3. Except for NH_3, these are extremely reactive and poisonous and are synthesized by reaction of a metal phosphide, arsenide, and so forth, which acts as a strong base in water or aqueous acid. For example,

$$Ca_3As_2(s) + 6H_2O(l) \longrightarrow 2AsH_3(g) + 3Ca(OH)_2(aq)$$

Ammonia is made industrially by direct combination of the elements at high pressure and moderately high temperature:

$$N_2(g) + 3H_2(g) \rightleftharpoons 2NH_3(g)$$

Molecular properties of the Group 5A(15) hydrides reveal some interesting bonding and structural patterns:

- Despite its much lower molar mass, NH_3 melts and boils at higher temperatures than the other 5A hydrides as a result of *H bonding*.
- Bond angles decrease from $107.3°$ for NH_3 to around $90°$ for the other hydrides, which suggests that the larger atoms use unhybridized p orbitals.
- E—H bond lengths increase down the group, so bond strength and thermal stability decrease: AsH_3 decomposes at $250°C$, SbH_3 at $20°C$, and BiH_3 at $-45°C$.

We'll see these features—H bonding for the smallest member, change in bond angles, change in bond energies—in the hydrides of Group 6A(16) as well. ●

The Group 5A(15) elements all form trihalides (EX_3). All except nitrogen form pentafluorides (EF_5), but only a few other pentahalides (PCl_5, PBr_5, $AsCl_5$, and $SbCl_5$) are known. Nitrogen cannot form pentahalides because it cannot expand its valence shell. Most trihalides are prepared by direct combination:

$$P_4(s) + 6Cl_2(g) \longrightarrow 4PCl_3(l)$$

The pentahalides form with excess halogen:

$$PCl_3(l) + Cl_2(g) \longrightarrow PCl_5(s)$$

As with the hydrides, the thermal stability of the halides decreases as the E—X bond becomes longer. Among the nitrogen halides, for example, NF_3 is a stable, rather unreactive gas. NCl_3 is explosive and reacts rapidly with water. (The chemist who first prepared it lost three fingers and an eye!) NBr_3 can only be made below $-87°C$. NI_3 has never been prepared, but an ammoniated product ($NI_3 \cdot NH_3$) explodes at the slightest touch.

ARIS Presentation Center, Art Library

95°

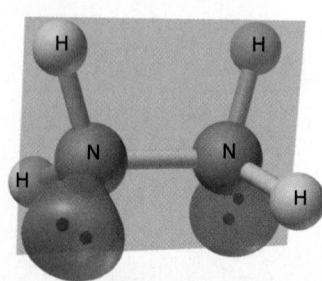

● **Hydrazine, Nitrogen's Other Hydride**
Aside from NH_3, the most important Group 5A(15) hydride is hydrazine, N_2H_4. Like NH_3, it is a weak base, forming $N_2H_5^+$ and $N_2H_6^{2+}$ ions with acids. Hydrazine is also used to make antituberculin drugs, plant growth regulators, and fungicides, and organic hydrazine derivatives are used in rocket and missile fuels. We might expect the lone pairs in the molecule to lie opposite each other and give a symmetrical structure with no dipole moment. However, repulsions between each lone pair and the bonding pairs on the other atom force the lone pairs to lie below the N—N bond (see model), resulting in a large dipole moment ($\mu = 1.85$ D in the gas phase). A similar arrangement occurs in hydrogen peroxide (H_2O_2), the analogous hydride of oxygen.

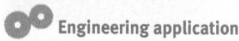 Engineering application

🐾 Organic chemistry application

In an aqueous reaction pattern *typical of many nonmetal halides,* each 5A halide reacts with water to yield the hydrogen halide and the oxoacid, in which E has the *same* oxidation number as it had in the original halide. For example, PX_5 (O.N. of P = +5) produces phosphoric acid (O.N. of P = +5) and HX:

$$PCl_5(s) + 4H_2O(l) \longrightarrow H_3PO_4(l) + 5HCl(g)$$

Highlights of Nitrogen Chemistry

Surely the most striking highlight of nitrogen chemistry is the inertness of N_2 itself. Nearly four-fifths of the atmosphere consists of N_2, and the other fifth is nearly all O_2, a very strong oxidizing agent. Nevertheless, the searing temperature of a lightning bolt is required for significant amounts of atmospheric nitrogen oxides to form. Thus, even though N_2 is inert at moderate temperatures, it reacts at high temperatures with H_2, Li, Group 2A(2) members, B, Al, C, Si, Ge, O_2, and many transition elements. In fact, nearly every element in the periodic table forms bonds to N. Here we focus on the oxides and the oxoacids and their salts.

Nitrogen Oxides Nitrogen is remarkable for having six stable oxides, each with a *positive* heat of formation because of the great strength of the N≡N bond (BE = 945 kJ/mol). Their structures and some properties are shown in Table 14.3. Unlike the hydrides and halides of nitrogen, the oxides are planar. Nitrogen displays all its positive oxidation states in these compounds, and in N_2O and N_2O_3, the two N atoms have different states.

Transparency/ARIS Presentation Center, Tables Library

Table 14.3 Structures and Properties of the Nitrogen Oxides

Formula	Name	Space-filling Model	Lewis Structure	Oxidation State of N	ΔH_f° (kJ/mol) at 298 K	Comment
N_2O	Dinitrogen monoxide (dinitrogen oxide; nitrous oxide)		:N≡N—Ö:	+1 (0, +2)	82.0	Colorless gas; used as dental anesthetic ("laughing gas") and aerosol propellant
NO	Nitrogen monoxide (nitrogen oxide; nitric oxide)		:Ṅ=Ö:	+2	90.3	Colorless, paramagnetic gas; biochemical messenger; air pollutant
N_2O_3	Dinitrogen trioxide		structure	+3 (+2, +4)	83.7	Reddish brown gas (reversibly dissociates to NO and NO_2)
NO_2	Nitrogen dioxide		:Ö—Ṅ=Ö:	+4	33.2	Orange-brown, paramagnetic gas formed during HNO_3 manufacture; poisonous air pollutant
N_2O_4	Dinitrogen tetraoxide		structure	+4	9.16	Colorless to yellow liquid (reversibly dissociates to NO_2)
N_2O_5	Dinitrogen pentaoxide		structure	+5	11.3	Colorless, volatile solid consisting of NO_2^+ and NO_3^-; gas consists of N_2O_5 molecules

Dinitrogen monoxide (N_2O; also called dinitrogen oxide or nitrous oxide) is the dental anesthetic "laughing gas" and the propellant in canned whipped cream. It is a linear molecule with an electronic structure best described by three resonance forms (note formal charges; see Problem 10.87):

$$\underset{\text{most important}}{\overset{0\ \ +1\ \ -1}{:N\equiv N-\ddot{O}:}} \longleftrightarrow \overset{-1\ \ +1\ \ \ 0}{:\ddot{N}=N=\ddot{O}:} \longleftrightarrow \underset{\text{least important}}{\overset{-2\ \ +1\ \ +1}{:\ddot{N}-N\equiv O:}}$$

● **Nitrogen Monoxide: A Biochemical Surprise** The tiny inorganic free radical NO is a remarkable player in an enormous range of biochemical systems. NO is biosynthesized in animal species from barnacles to humans. In the late 1980s and early 1990s, it was shown to play various roles in neurotransmission, blood clotting, blood pressure, and the destruction of cancer cells. The list of effects is now much longer and includes the harmful and beneficial ones of the NO-derived species NO_2, N_2O_3, $ONOO^-$, and HNO. Living systems are full of surprises!

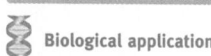

Biological application

Nitrogen monoxide (NO; also called nitrogen oxide or nitric oxide) is an odd-electron molecule with a vital biochemical function. ● In Section 11.3, we used MO theory to explain its bonding. The commercial preparation of NO through the oxidation of ammonia occurs as a first step in the production of nitric acid:

$$4NH_3(g) + 5O_2(g) \longrightarrow 4NO(g) + 6H_2O(g)$$

Nitrogen monoxide is also produced whenever air is heated to high temperatures, as in a car engine or a lightning storm:

$$N_2(g) + O_2(g) \xrightarrow{\text{high } T} 2NO(g)$$

Heating converts NO to two other oxides:

$$3NO(g) \xrightarrow{\Delta} N_2O(g) + NO_2(g)$$

This type of redox reaction is called a **disproportionation.** It occurs when a substance *acts as both an oxidizing and a reducing agent in a reaction.* In the process, an atom with an intermediate oxidation state in the reactant occurs in the products in both lower and higher states: the oxidation state of N in NO ($+2$) is intermediate between that in N_2O ($+1$) and that in NO_2 ($+4$).

Nitrogen dioxide (NO_2), a brown poisonous gas, forms to a small extent when NO reacts with additional oxygen:

$$2NO(g) + O_2(g) \rightleftharpoons 2NO_2(g)$$

Like NO, NO_2 is also an odd-electron molecule, but the electron is more localized on the N atom. Thus, NO_2 dimerizes reversibly to dinitrogen tetraoxide:

$$O_2N\cdot(g) + \cdot NO_2(g) \rightleftharpoons O_2N-NO_2(g) \quad \text{(or } N_2O_4\text{)}$$

Environmental application

$$\text{(graph: Concentration (ppm) vs. Time of day)}$$

Thunderstorms form NO and NO_2 and carry them down to the soil, where they act as natural fertilizers. In urban traffic, however, their formation leads to *photochemical smog* (Figure 14.20). During the morning commute, NO forms in car and truck (especially diesel) engines. It is then oxidized to NO_2 by free radicals ($RO_2\cdot$), which form by oxidation of unburned hydrocarbons in gasoline fumes. As the Sun climbs later in the day, radiant energy breaks down some NO_2:

$$NO_2(g) \xrightarrow{\text{sunlight}} NO(g) + O(g)$$

The O atoms collide with O_2 molecules and form ozone, O_3, a powerful oxidizing agent that damages synthetic rubber, plastics, and plant and animal tissue:

$$O_2(g) + O(g) \longrightarrow O_3(g)$$

A complex series of reactions involving NO, NO_2, O_3, unburned gasoline, and various free-radical species forms *peroxyacylnitrates* (PANs), a group of potent nose and eye irritants. The outcome of this fascinating atmospheric chemistry is choking, brown smog.

Figure 14.20 The formation of photochemical smog. Localized atmospheric concentrations of the precursor components of smog change during the day. In early morning traffic, exhaust from motor vehicles raises NO and hydrocarbon levels, followed by increases in NO_2 as NO is oxidized by free radicals. Ozone peaks later, as NO_2 breaks down in stronger sunlight and releases O atoms that react with O_2. By midafternoon, all the components for the production of peroxyacylnitrates (PANs) are present, and photochemical smog results.

Nitrogen Oxoacids and Oxoanions

The two common nitrogen oxoacids are nitric acid and nitrous acid (Figure 14.21). The first two steps in the *Ostwald process* for the production of nitric acid—the oxidations of NH_3 to NO and of NO to NO_2—have been shown. The final step is a disproportionation, as the oxidation numbers show:

$$\overset{+4}{3NO_2}(g) + H_2O(l) \longrightarrow \overset{+5}{2HNO_3}(aq) + \overset{+2}{NO}(g)$$

The NO is recycled to make more NO_2.

Figure 14.21 The structures of nitric and nitrous acids and their oxoanions. A, Nitric acid loses a proton (H^+) to form the trigonal planar nitrate ion (one of three resonance forms is shown). **B,** Nitrous acid, a much weaker acid, forms the planar nitrite ion (one of two resonance forms is shown). Note the effect of nitrogen's lone pair (the lone pairs of the oxygens are not shown) in reducing the ideal 120° bond angle to 115°.

 ARIS Presentation Center, Art Library

In nitric acid, as in all oxoacids, *the acidic H is attached to one of the O atoms.* In the laboratory, nitric acid is used as a strong oxidizing acid. The products of its reactions with metals vary with the metal's reactivity and the acid's concentration. In the following examples, notice from the net ionic equations that *the NO_3^- ion is the oxidizing agent.* Nitrate ion that is not reduced is a spectator ion and does not appear in the net ionic equations.

- With an active metal, such as Al, and dilute acid, N is reduced from the $+5$ state all the way to the -3 state in the ammonium ion, NH_4^+:

$$8Al(s) + 30HNO_3(aq; 1\ M) \longrightarrow 8Al(NO_3)_3(aq) + 3NH_4NO_3(aq) + 9H_2O(l)$$
$$8Al(s) + 30H^+(aq) + 3NO_3^-(aq) \longrightarrow 8Al^{3+}(aq) + 3NH_4^+(aq) + 9H_2O(l)$$

- With a less reactive metal, such as Cu, and more concentrated acid, N is reduced to the $+2$ state in NO:

$$3Cu(s) + 8HNO_3(aq; 3\ to\ 6\ M) \longrightarrow 3Cu(NO_3)_2(aq) + 4H_2O(l) + 2NO(g)$$
$$3Cu(s) + 8H^+(aq) + 2NO_3^-(aq) \longrightarrow 3Cu^{2+}(aq) + 4H_2O(l) + 2NO(g)$$

- With still more concentrated acid, N is reduced only to the $+4$ state in NO_2:

$$Cu(s) + 4HNO_3(aq; 12\ M) \longrightarrow Cu(NO_3)_2(aq) + 2H_2O(l) + 2NO_2(g)$$
$$Cu(s) + 4H^+(aq) + 2NO_3^-(aq) \longrightarrow Cu^{2+}(aq) + 2H_2O(l) + 2NO_2(g)$$

Nitrates form when HNO_3 reacts with metals or with their hydroxides, oxides, or carbonates. *All nitrates are soluble in water.*

Nitrous acid, HNO_2, a much weaker acid than HNO_3, forms when metal nitrites are treated with a strong acid:

$$NaNO_2(aq) + HCl(aq) \longrightarrow HNO_2(aq) + NaCl(aq)$$

These two acids reveal a *general pattern in relative acid strength among oxoacids:* the more O atoms bonded to the central nonmetal, the stronger the acid. Thus, HNO_3 is stronger than HNO_2. The O atoms pull electron density from the N atom, which in turn pulls electron density from the O of the O—H bond, facilitating the release of the H^+ ion. The O atoms also act to stabilize the resulting oxoanion by delocalizing its negative charge. The same pattern occurs in the oxoacids of sulfur and the halogens; we'll discuss the pattern quantitatively in Chapter 18.

A P_4O_6

B P_4O_{10}

Figure 14.22 Important oxides of phosphorus. A, P_4O_6**. B,** P_4O_{10}.

ARIS Presentation Center, Art Library

Demo Taylor, Larry C. "Burning Phosphorus under Water Safely," *J. Chem. Educ.* **1997**, *74*, 1074.

Demo Keiter, Richard L.; Gamage, Chaminda P. "Combustion of White Phosphorus," *J. Chem. Educ.* **2001**, *78*, 908.

● **The Countless Uses of Phosphates**
Phosphates have an amazing array of applications in home and industry. Na_3PO_4 is a paint stripper and grease remover and is still used in cleaning powders in other countries. Na_2HPO_4 is a laxative ingredient and is used to adjust the acidity of boiler water. The potassium salt K_3PO_4 is used to stabilize latex for synthetic rubber, and K_2HPO_4 is a radiator corrosion inhibitor. Ammonium phosphates are used as fertilizers and as flame retardants on curtains and paper costumes. Calcium phosphates are used in baking powders and toothpastes, as mineral supplements in livestock feed, and (on the hundred-million-ton scale) as fertilizers throughout the world.

Engineering application

Highlights of Phosphorus Chemistry: Oxides and Oxoacids

Phosphorus forms two important oxides, P_4O_6 and P_4O_{10}. Tetraphosphorus hexaoxide, P_4O_6, has P in its $+3$ oxidation state and forms when white P_4 reacts with limited oxygen:

$$P_4(s) + 3O_2(g) \longrightarrow P_4O_6(s)$$

P_4O_6 has the tetrahedral orientation of the P atoms found in P_4, with an O atom between each pair of P atoms (Figure 14.22A). It reacts with water to form phos*phorous* acid (note the spelling):

$$P_4O_6(s) + 6H_2O(l) \longrightarrow 4H_3PO_3(l)$$

The formula H_3PO_3 is misleading because the acid has only two acidic H atoms; the third is bonded to the central P and does not dissociate. Phosphorous acid is a weak acid in water but reacts completely in two steps with excess strong base:

Salts of phosphorous acid contain the phosphite ion, HPO_3^{2-}.

In tetraphosphorus decaoxide, P_4O_{10}, P is in the $+5$ oxidation state. Commonly known as "phosphorus pentoxide" from the empirical formula (P_2O_5), it forms when P_4 burns in excess O_2:

$$P_4(s) + 5O_2(g) \longrightarrow P_4O_{10}(s)$$

Its structure can be viewed as that of P_4O_6 with another O atom bonded to each of the four corner P atoms (Figure 14.22B). P_4O_{10} is a powerful drying agent and, in a vigorous exothermic reaction with water, forms phosphoric acid (H_3PO_4), one of the "top-10" most important compounds in chemical manufacturing:

$$P_4O_{10}(s) + 6H_2O(l) \longrightarrow 4H_3PO_4(l)$$

The presence of many H bonds makes pure H_3PO_4 syrupy, more than 75 times as viscous as water. The laboratory-grade concentrated acid is an 85% by mass aqueous solution. H_3PO_4 is a weak triprotic acid; in water, it loses one proton in the following equilibrium reaction:

$$H_3PO_4(l) + H_2O(l) \rightleftharpoons H_2PO_4^-(aq) + H_3O^+(aq)$$

In excess strong base, however, the three protons dissociate completely in three steps to give the three phosphate oxoanions:

dihydrogen phosphate ion   hydrogen phosphate ion   phosphate ion

Phosphoric acid has a central role in fertilizer production, and it is also used as a polishing agent for aluminum car trim and as an additive in soft drinks to give a touch of tartness. The various phosphate salts have numerous essential applications. ●

Polyphosphates are formed by heating hydrogen phosphates, which lose water as they form P—O—P linkages. This type of reaction, in which an H_2O molecule is lost for every pair of OH groups that join, is called a **dehydration-condensation;** it occurs frequently in the formation of polyoxoanion chains and other polymeric structures, both synthetic and natural. For example, sodium diphosphate, $Na_4P_2O_7$, is prepared by heating sodium hydrogen phosphate:

$$2Na_2HPO_4(s) \xrightarrow{\Delta} Na_4P_2O_7(s) + H_2O(g)$$

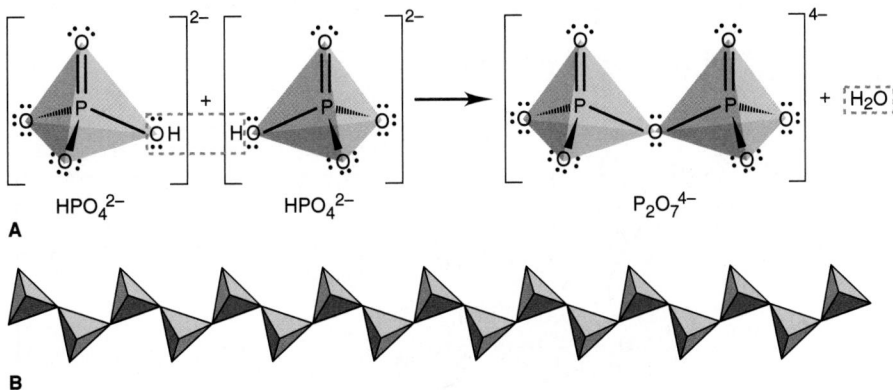

A

B

Figure 14.23 The diphosphate ion and polyphosphates. A, When two hydrogen phosphate ions undergo a dehydration-condensation reaction, they lose a water molecule and join through a shared O atom to form a diphosphate ion. **B,** Polyphosphates are chains of many such tetrahedral PO_4 units. Chain shapes depend on the orientation of the individual units. Note the similarity to silicate chains (see the Gallery, p. 592).

The diphosphate ion, $P_2O_7^{4-}$, the smallest of the polyphosphates, consists of two PO_4 units linked through a common oxygen corner (Figure 14.23A). Its reaction with water, the reverse of the previous reaction, generates heat:

$$P_2O_7^{4-}(aq) + H_2O(l) \longrightarrow 2HPO_4^-(aq) + heat$$

A similar process is put to vital use by organisms, when a third PO_4 unit linked to diphosphate creates the triphosphate grouping, part of the all-important high-energy biomolecule adenosine triphosphate (ATP). In Chapters 20 and 21, we discuss the central role of ATP in biological energy production. Extended polyphosphate chains consist of many tetrahedral PO_4 units (Figure 14.23B) and are structurally similar to silicate chains. As a final look at the chemical versatility of phosphorus, the margin note describes some of the numerous important compounds with sulfur and with nitrogen. ●

● **Match Heads, Bug Sprays, and O-Rings** Phosphorus forms many sulfides and nitrides. P_4S_3 is used in "strike-anywhere" match heads, and P_4S_{10} is used in the manufacture of organophosphorus pesticides, such as malathion. Polyphosphazenes have properties similar to those of silicones. Indeed, the $—(R_2)P{=}N—$ unit is isoelectronic with the silicone unit, $—(R_2)Si—O—$. Sheets, films, fibers, and foams of polyphosphazene are water repellent, flame resistant, solvent resistant, and flexible at low temperatures—perfect for the gaskets and O-rings in spacecraft and polar vehicles.

Engineering application

Literature Quinsey, Carolyn S. "Oxygen—Abundant and Essential," *J. Chem. Educ.* **2003**, *80*, 1124.

Literature Wright, Stephen W. "A Method for Generating Oxygen from Consumer Chemicals," *J. Chem. Educ.* **2003**, *80*, 1158.

Demo Shakhashiri, Bassam Z. "Preparation and Properties of Oxygen," *Chemical Demonstrations, A Handbook for Teachers of Chemistry*, Vol. 2 (The University of Wisconsin Press, Madison, 1985), 137–141.

ARIS Presentation Center, Art Library

14.8 GROUP 6A(16): THE OXYGEN FAMILY

The first two members of this family—gaseous nonmetallic oxygen (O) and solid nonmetallic sulfur (S)—are among the most important elements in industry, the environment, and living things. Two metalloids, selenium (Se) and tellurium (Te), appear below them, then comes the lone metal, radioactive polonium (Po), and newly synthesized element 116 ends the group. The Group 6A(16) Family Portrait (pp. 604 and 605) displays the features of these elements.

How Do the Oxygen and Nitrogen Families Compare Physically?

Group 6A(16) resembles Group 5A(15) in many respects, so let's look at some common themes. The pattern of physical properties we saw in Group 5A appears again in this group. Like nitrogen, oxygen occurs as a low-boiling diatomic gas. Like phosphorus, sulfur occurs as a polyatomic molecular solid. Like arsenic, selenium commonly occurs as a gray metalloid. Like antimony, tellurium is slightly more metallic than the preceding group member but still displays network covalent bonding. Finally, like bismuth, polonium has a metallic crystal structure. As we expect by comparison with the Group 5A elements, electrical conductivities increase steadily down Group 6A as bonding changes from individual molecules (insulators) to metalloid networks (semiconductors) to a metallic solid (conductor).

Group 6A(16): The Oxygen Family

Key Atomic and Physical Properties

ns^2np^4

GROUP 6A(16)

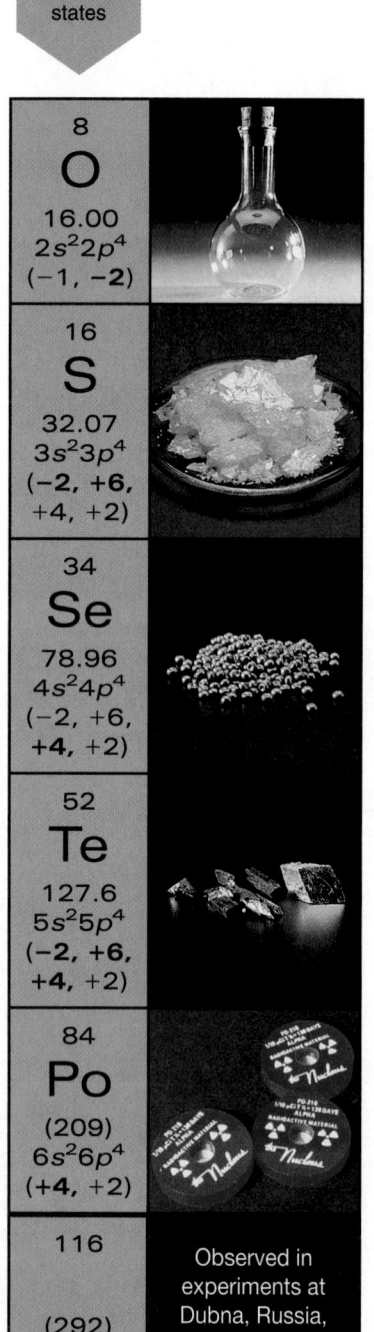

8
O
16.00
$2s^22p^4$
(−1, **−2**)

16
S
32.07
$3s^23p^4$
(**−2, +6,** +4, +2)

34
Se
78.96
$4s^24p^4$
(**−2, +6, +4,** +2)

52
Te
127.6
$5s^25p^4$
(**−2, +6, +4,** +2)

84
Po
(209)
$6s^26p^4$
(**+4,** +2)

116
(292)
$7s^27p^4$

Observed in experiments at Dubna, Russia, in 2004

Atomic Properties

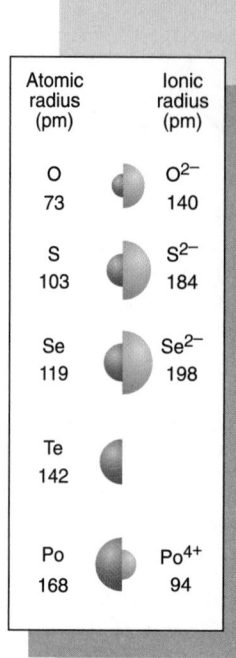

Atomic radius (pm)	Ionic radius (pm)
O 73	O^{2-} 140
S 103	S^{2-} 184
Se 119	Se^{2-} 198
Te 142	
Po 168	Po^{4+} 94

Group electron configuration is ns^2np^4. As in Groups 3A(13) and 5A(15), a lower (+4) oxidation state becomes more common down the group.

Down the group, atomic and ionic size increase, and IE and EN decrease.

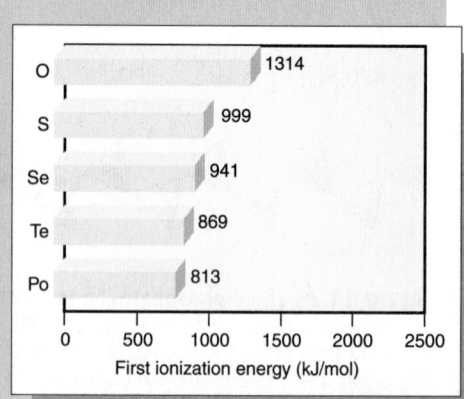

First ionization energy (kJ/mol)
O 1314
S 999
Se 941
Te 869
Po 813

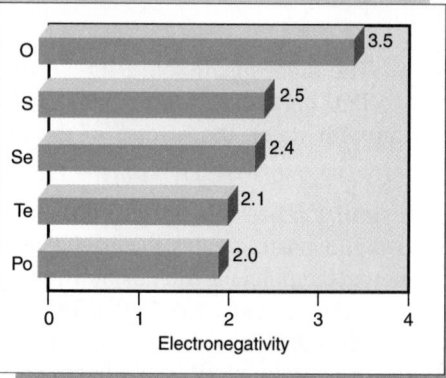

Electronegativity
O 3.5
S 2.5
Se 2.4
Te 2.1
Po 2.0

Physical Properties

Melting points increase through Te, which has covalent bonding, and then decrease for Po, which has metallic bonding.

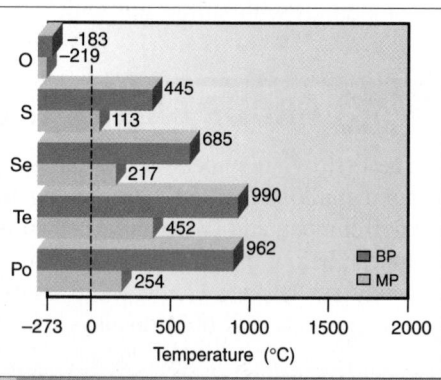

Temperature (°C)
O −183 / −219
S 445 / 113
Se 685 / 217
Te 990 / 452
Po 962 / 254
BP / MP

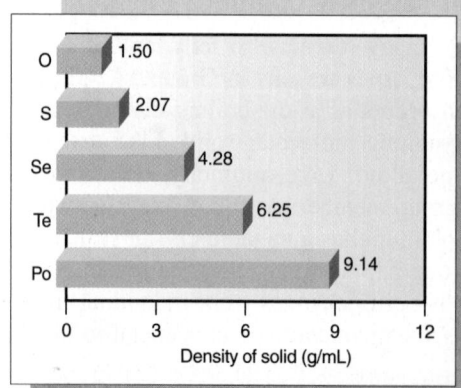

Density of solid (g/mL)
O 1.50
S 2.07
Se 4.28
Te 6.25
Po 9.14

Densities of the elements as solids increase steadily.

Some Reactions and Compounds

Important Reactions

Halogenation and oxidation of the elements (E) appear in reactions 1 and 2, and sulfur chemistry in reactions 3 and 4.

1. Halides are formed by direct combination:

$$E(s) + X_2(g) \longrightarrow \text{various halides}$$
$$(E = S, Se, Te; X = F, Cl)$$

2. The other elements in the group are oxidized by O_2:

$$E(s) + O_2(g) \longrightarrow EO_2 \quad (E = S, Se, Te, Po)$$

SO_2 is oxidized further, and the product is used in the final step of H_2SO_4 manufacture (*Highlights of Sulfur Chemistry*):

$$2SO_2(g) + O_2(g) \longrightarrow 2SO_3(g)$$

3. Sulfur is recovered when hydrogen sulfide is oxidized:

$$8H_2S(g) + 4O_2(g) \longrightarrow S_8(s) + 8H_2O(g)$$

This reaction is used to obtain sulfur when natural deposits are not available.

4. The thiosulfate ion is formed when an alkali metal sulfite reacts with sulfur, as in the preparation of "hypo," photographer's developing solution:

$$S_8(s) + 8Na_2SO_3(aq) \longrightarrow 8Na_2S_2O_3(aq)$$

Important Compounds

1. Water, H_2O. The single most important compound on Earth (*Section 12.5*).

2. Hydrogen peroxide, H_2O_2. Used as an oxidizing agent, disinfectant, and bleach, and in the production of peroxy compounds for polymerization (*margin note, p. 607*).

3. Hydrogen sulfide, H_2S. Vile-smelling toxic gas formed during anaerobic decomposition of plant and animal matter, in volcanoes, and in deep-sea thermal vents. Used as a source of sulfur and in the manufacture of paper. Atmospheric traces cause silver to tarnish through formation of black Ag_2S (*see photo*).

5. Sulfur trioxide (SO_3) and sulfuric acid (H_2SO_4). SO_3, formed from SO_2 over a K_2O/V_2O_5 catalyst, is then converted to H_2SO_4. The acid is the cheapest strong acid and is so widely used in industry that its production level is an indicator of a nation's economic strength. It is a strong dehydrating agent that removes water from any organic source (*Highlights of Sulfur Chemistry*).

6. Sulfur hexafluoride, SF_6. Extremely inert gas used as an electrical insulator.

Untarnished and tarnished silver spoons

4. Sulfur dioxide, SO_2. Colorless, choking gas formed in volcanoes (*see photo*) or whenever an S-containing material (coal, oil, metal sulfide ores, and so on) is burned. More than 90% of SO_2 produced is used to make sulfuric acid. Also used as a fumigant and a preservative of fruit, syrups, and wine. As a reducing agent, removes excess Cl_2 from industrial wastewater, removes O_2 from petroleum handling tanks, and prepares ClO_2 for bleaching paper. Atmospheric pollutant in acid rain.

Allotropism is more common in Group 6A(16) than in Group 5A(15). Oxygen has two allotropes: life-giving dioxygen (O_2), and poisonous triatomic ozone (O_3). Oxygen gas is colorless, odorless, paramagnetic, and thermally stable. In contrast, ozone gas is bluish, has a pungent odor, is diamagnetic, and decomposes in heat and especially in ultraviolet (UV) light:

$$2O_3(g) \xrightarrow{\text{UV}} 3O_2(g)$$

This ability to absorb high-energy photons makes stratospheric ozone vital to life. A thinning of the ozone layer, observed above the North Pole and especially the South Pole, means that more UV light will reach Earth's surface, with potentially hazardous effects. (We'll discuss the chemical causes of ozone depletion in Chapter 16.)

Sulfur is the allotrope "champion" of the periodic table, with more than 10 forms. The S atom's ability to bond to other S atoms creates numerous rings and chains, with S—S bond lengths that range from 180 pm to 260 pm and bond angles from 90° to 180°. At room temperature, the sulfur molecule is a crown-shaped ring of eight atoms, called *cyclo-S$_8$* (Figure 14.24). The most stable allotrope is orthorhombic α-S_8, which consists entirely of these molecules; all other S allotropes eventually revert to this one.

Demo Shakhashiri, Bassam Z. "Combining Volume of Oxygen with Sulfur," *Chemical Demonstrations, A Handbook for Teachers of Chemistry,* Vol. 2 (The University of Wisconsin Press, Madison, 1985), 190–192.

Figure 14.24 The cyclo-S$_8$ molecule. A, Top view of a space-filling model of the cyclo-S$_8$ molecule. **B,** Side view of a ball-and-stick model of the molecule; note the crownlike shape.

ARIS Presentation Center, Art Library

● **Selenium and Xerography** Photocopying was invented in the 1940s as a rapid, inexpensive, dry means of copying documents (*xerography*, Greek "dry writing"). It is based on the ability of Se to conduct a current when illuminated. A film of amorphous Se is deposited on an aluminum drum and electrostatically charged. Exposure to a document produces an "image" of low and high positive charges corresponding to the document's bright and dark areas. Negatively charged black, dry ink (toner) particles are attracted to the regions of high charge more than to those of low charge. This pattern of black particles is transferred electrostatically to paper, and the particles are fused to the paper's surface by heat or solvent. Excess toner is removed from the Se film, the charges are "erased" by exposure to light, and the film is ready for the next page.

Selenium also has several allotropes, some consisting of crown-shaped Se$_8$ molecules. Gray Se is composed of layers of helical chains. Its electrical conductivity in visible light revolutionized the photocopying industry. ● When molten glass, cadmium sulfide, and gray Se are mixed and heated in the absence of air, a ruby-red glass forms, which you see whenever you stop at a traffic light.

How Do the Oxygen and Nitrogen Families Compare Chemically?

Changes in chemical behavior in Group 6A(16) are also similar to those in the previous group. Even though O and S occur as anions much more often than do N and P, like N and P, they bond covalently with almost every other nonmetal. Covalent bonds appear in the compounds of Se and Te (as in those of As and Sb), whereas Po behaves like a metal (as does Bi) in some of its saltlike compounds. In contrast to nitrogen, oxygen has few common oxidation states, but the earlier pattern returns with the other members: the +6, +4, and −2 states occur most often, with the lower positive (+4) state becoming more common in Te and Po [as the lower positive (+3) state does in Sb and Bi].

The range in atomic properties is wider in this group than in Group 5A(15) because of oxygen's high EN (3.5) and great oxidizing strength, second only to that of fluorine. As in 5A and earlier groups, the behavior of the Period 2 element stands out. In fact, aside from a similar outer electron configuration, the larger

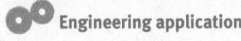
Engineering application

members of Group 6A behave very little like oxygen: they are much less electronegative, form anions much less often (S^{2-} occurs with active metals), and their hydrides exhibit no H bonding.

Except for O, all the 6A elements form foul-smelling, poisonous, gaseous hydrides (H_2E) on treatment with acid of the metal sulfide, selenide, and so forth. For example,

$$FeSe(s) + 2HCl(aq) \longrightarrow H_2Se(g) + FeCl_2(aq)$$

Hydrogen sulfide also forms naturally in swamps from the breakdown of organic matter. It is as toxic as HCN, and even worse, it anesthetizes your olfactory nerves, so that as its concentration increases, you smell it less! The other hydrides are about 100 times *more* toxic.

In their bonding and thermal stability, these Group 6A hydrides have several features in common with those of Group 5A:

- Only water can form H bonds, so it melts and boils much higher than the other H_2E compounds (see Figure 12.14, p. 453). (Oxygen's other hydride, H_2O_2, is also extensively H bonded.) ●
- Bond angles drop from the nearly tetrahedral value for H_2O (104.5°) to around 90° for the larger 6A hydrides, suggesting that the central atom uses unhybridized *p* orbitals.
- E—H bond length increases (bond energy decreases) down the group. Thus, H_2Te decomposes above 0°C, and H_2Po can be made only in extreme cold because thermal energy from the radioactive Po decomposes it. Another result of longer (weaker) bonds is that the 6A hydrides are acids in water, and their acidity increases from H_2S to H_2Po.

Except for O, the Group 6A elements form a wide range of halides, whose structure and reactivity patterns depend on the *sizes of the central atom and the surrounding halogens:*

- Sulfur forms many fluorides, a few chlorides, one bromide, but no stable iodides.
- As the central atom becomes larger, the halides become more stable. Thus, tetrachlorides and tetrabromides of Se, Te, and Po are known, as are tetraiodides of Te and Po. Hexafluorides are known only for S, Se, and Te.

The inverse relationship between bond length and bond strength that we've seen previously does not account for this pattern. Rather, it is based on the effect of electron repulsions due to crowding of lone pairs and halogen (X) atoms around the central Group 6A atom. With S, the larger X atoms would be too crowded, which explains why sulfur iodides do not occur. With increasing size of E, and therefore increasing length of E—X bonds, however, lone pairs and X atoms do not crowd each other as much, and a greater number of stable halides form.

Two sulfur fluorides illustrate clearly how crowding and orbital availability affect reactivity. Sulfur tetrafluoride (SF_4) is extremely reactive. It forms SO_2 and HF when exposed to moisture and is commonly used to fluorinate many compounds:

$$3SF_4(g) + 4BCl_3(g) \longrightarrow 4BF_3(g) + 3SCl_2(l) + 3Cl_2(g)$$

In contrast, sulfur hexafluoride (SF_6) is almost as inert as a noble gas! It is odorless, tasteless, nonflammable, nontoxic, and insoluble. Hot metals, boiling HCl, molten KOH, and high-pressure steam have no effect on it. It is used as an insulating gas in high-voltage generators, withstanding over 10^6 volts across electrodes only 50 mm apart. A look at the structures of these two fluorides provides

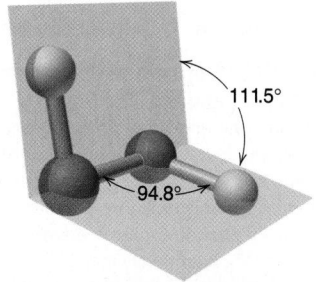

111.5°

94.8°

● **Hydrogen Peroxide: Hydrazine's Cousin** In addition to water, oxygen forms another hydride, called hydrogen peroxide, H_2O_2 (HO—OH). Like hydrazine (H_2N—NH_2), the related hydride of nitrogen, hydrogen peroxide has a skewed molecular shape. It is a colorless liquid with a high density, viscosity, and boiling point because of extensive H bonding. In peroxides, O has the −1 oxidation state, midway between that in O_2 (zero) and that in oxides (−2); thus, H_2O_2 readily disproportionates:

$$H_2O_2(l) \longrightarrow H_2O(l) + \tfrac{1}{2}O_2(g)$$

Aside from the familiar use of H_2O_2 as a hair bleach and disinfectant, more than 70% of the 0.5 million tons produced each year is used to bleach paper pulp, textiles, straw, and leather and to make other chemicals. H_2O_2 is also used in sewage treatment to oxidize bacteria.

ARIS Presentation Center, Art Library

SF$_4$
Sulfur tetrafluoride

SF$_6$
Sulfur hexafluoride

Figure 14.25 Structural differences between SF$_4$ and SF$_6$. The S atom in SF$_4$ has a lone pair and empty d orbitals that can become involved in bonding. In SF$_6$, the central S atom already has the maximum number of bonds that it can form, and closely packed F atoms envelop it, rendering SF$_6$ chemically inert.

ARIS Presentation Center, Art Library

ARIS Presentation Center, Photo Library

● **Acid from the Sky** Awareness of SO$_2$ as a major air pollutant has been widespread for years. Even though enormous amounts of SO$_2$ form during volcanic and other geothermal activity, they are dwarfed by the amounts emitted from human sources: coal-burning power plants, petroleum refineries, and metal-ore smelters. In the atmosphere, SO$_2$ undergoes several reaction steps to form sulfuric acid, which falls in rain, snow, and dust on animals, plants, buildings, and lakes (Chapter 19).

Environmental application

the key to these astounding differences (Figure 14.25). Sulfur in SF$_4$ can form bonds to another atom either by donating its lone electron pair or by accepting a lone pair into one of its empty d orbitals. On the other hand, the central S atom in SF$_6$ has no lone pair, and the six F atoms around it form an octahedral sheath that blocks chemical attack.

Highlights of Oxygen Chemistry: Range of Oxide Properties

Oxygen is the most abundant element on Earth's surface, occurring both as the free element and in innumerable oxides, silicates, carbonates, and phosphates, as well as in water. Virtually all free O$_2$ has a biological origin, having been formed for billions of years by photosynthetic algae and multicellular plants in an over-all equation that looks deceptively simple:

$$n\text{H}_2\text{O}(l) + n\text{CO}_2(g) \xrightarrow{\text{light}} n\text{O}_2(g) + (\text{CH}_2\text{O})_n \text{ (carbohydrates)}$$

The reverse process occurs during combustion and respiration. Through these O$_2$-forming and O$_2$-utilizing processes, the 1.5×10^9 km^3 of water on Earth is, on average, used and remade every 2 million years!

Every element (except He, Ne, and Ar) forms at least one oxide, many by direct combination. A broad spectrum of properties characterizes these compounds. Some oxides are gases that condense at very low temperatures, such as CO (bp = $-192°$C); others are solids that melt at extremely high temperatures, such as BeO (mp = $2530°$C). Oxides cover the full range of conductivity: insulators (MgO), semiconductors (NiO), conductors (ReO$_3$), and superconductors (YBa$_2$Cu$_3$O$_7$). Some oxides have endothermic heats of formation (ΔH_f° of NO = $+90.3$ kJ/mol), whereas other oxides have exothermic ones (ΔH_f° of CO$_2$ = -393.5 kJ/mol). They may be thermally stable (CaO) or unstable (HgO), as well as chemically reactive (Li$_2$O) or inert (Fe$_2$O$_3$).

Given this vast range of behavior, another useful way to classify element oxides is by their acid-base properties (see Interchapter, Topic 4). The oxides of Group 6A(16) exhibit expected trends in acidity, with SO$_3$ the most acidic and PoO$_2$ the most basic.

Highlights of Sulfur Chemistry: Oxides, Oxoacids, and Sulfides

Like phosphorus, sulfur forms two important oxides, sulfur dioxide (SO$_2$) and sulfur trioxide (SO$_3$). Sulfur is in its $+4$ oxidation state in SO$_2$, a colorless, choking gas that forms whenever S, H$_2$S, or a metal sulfide burns in air: ●

$$2\text{H}_2\text{S}(g) + 3\text{O}_2(g) \longrightarrow 2\text{H}_2\text{O}(g) + 2\text{SO}_2(g)$$
$$4\text{FeS}_2(s) + 11\text{O}_2(g) \longrightarrow 2\text{Fe}_2\text{O}_3(s) + 8\text{SO}_2(g)$$

Most SO$_2$ is used to produce sulfuric acid.

In water, sulfur dioxide forms sulfurous acid, which exists in equilibrium with hydrated SO$_2$ rather than as isolable H$_2$SO$_3$ molecules:

$$\text{SO}_2(aq) + \text{H}_2\text{O}(l) \rightleftharpoons [\text{H}_2\text{SO}_3(aq)] \rightleftharpoons \text{H}^+(aq) + \text{HSO}_3^-(aq)$$

(Similarly, carbonic acid occurs in equilibrium with hydrated CO$_2$ and cannot be isolated as H$_2$CO$_3$ molecules.) Sulfurous acid is weak and has two acidic protons, forming the hydrogen sulfite (bisulfite, HSO$_3^-$) and sulfite (SO$_3^{2-}$) ions with strong base. Because the S in SO$_3^{2-}$ is in the $+4$ state and is easily oxidized to the $+6$ state, sulfites are good reducing agents and are used to preserve foods and wine by eliminating undesirable products of air oxidation.

En route to sulfuric acid, SO$_2$ is first oxidized to SO$_3$ (S in the $+6$ state) by heating in O$_2$ over a catalyst:

$$\text{SO}_2(g) + \tfrac{1}{2}\text{O}_2(g) \underset{}{\overset{\text{V}_2\text{O}_5/\text{K}_2\text{O catalyst}}{\rightleftharpoons}} \text{SO}_3(g)$$

Figure 14.26 The dehydration of carbohydrates by sulfuric acid. The protons of concentrated H_2SO_4 combine exothermically with water, dehydrating many organic materials. When table sugar is treated with sulfuric acid, the components of water are removed from the carbohydrate molecules $(CH_2O)_n$. Steam forms, and the remaining carbon expands to a porous mass.

(We discuss how catalysts work in Chapter 16 and how H_2SO_4 is produced in Chapter 22.) The SO_3 is absorbed into concentrated H_2SO_4 and treated with additional H_2O:

$$SO_3(\text{in concentrated } H_2SO_4) + H_2O(l) \longrightarrow H_2SO_4(l)$$

With more than 40 million tons produced each year in the United States alone, H_2SO_4 ranks first among all industrial chemicals. Fertilizer production, metal, pigment, and textile processing, and soap and detergent manufacturing are just a few of the major industries that depend on sulfuric acid.

Concentrated laboratory-grade sulfuric acid is a viscous, colorless liquid that is 98% H_2SO_4 by mass. Like other strong acids, H_2SO_4 dissociates completely in water, forming the hydrogen sulfate (or bisulfate) ion, a much weaker acid:

hydrogen sulfate ion   sulfate ion

Most common hydrogen sulfates and sulfates are water soluble, but those of Group 2A(2) members (except $MgSO_4$), Pb^{2+}, and Hg_2^{2+} are not.

Concentrated sulfuric acid is an excellent dehydrating agent. Its loosely held proton transfers to water in a highly exothermic formation of hydronium (H_3O^+) ions. This process can occur even when the reacting substance contains no free water. For example, H_2SO_4 dehydrates wood, natural fibers, and many other organic substances by removing the components of water from the molecular structure, leaving behind a carbonaceous mass (Figure 14.26).

Thiosulfuric acid ($H_2S_2O_3$) is a structural analog of sulfuric acid in which a second S substitutes for one of the O atoms (*thio-* means "containing sulfur in place of oxygen"). The thiosulfate ion ($S_2O_3^{2-}$) is an important reducing agent in chemical analysis. However, its largest commercial use is in photography, where sodium thiosulfate pentahydrate ($Na_2S_2O_3 \cdot 5H_2O$), known as "hypo," is involved in fixing the image (Section 23.3).

Many metals combine directly with S to form *metal sulfides*. Indeed, naturally occurring sulfides are ores, which are mined for the extraction of many metals, including copper, zinc, lead, and silver. Aside from the sulfides of Groups 1A(1) and 2A(2), most metal sulfides do not have discrete S^{2-} ions. Several transition metals, such as chromium, iron, and nickel, form covalent, alloy-like, nonstoichiometric compounds with S, such as $Cr_{0.88}S$ or $Fe_{0.86}S$. Some important minerals contain S_2^{2-} ions; an example is iron pyrite, or "fool's gold" (FeS_2) (Figure 14.27). We discuss the metallurgy of ores in Chapter 22.

Figure 14.27 A common sulfide mineral. Pyrite (FeS_2), or fool's gold, is a beautiful but relatively cheap mineral that contains the S_2^{2-} ion, but no gold.

Figure 14.28 Standing in Group 6A(16), looking backward to Group 5A(15) and forward to Group 7A(17).

ARIS Presentation Center, Art Library

Looking Backward and Forward: Groups 5A(15), 6A(16), and 7A(17)

Groups 5A(15) and 6A(16) are very similar in their physical and chemical trends and in the versatility of phosphorus and sulfur (Figure 14.28). Their greatest difference is the sluggish behavior of N_2 compared with the striking reactivity of O_2. In both groups, metallic character appears only in the largest members. From here on, metals and even metalloids are left behind: all Group 7A(17) elements are reactive nonmetals. Anion formation, which was rare in 5A and more common in 6A, is one of the dominant features of 7A. Another feature of the 7A elements is the number of covalent compounds they form with oxygen *and* with each other.

14.9 GROUP 7A(17): THE HALOGENS

Our last chance to view elements of great reactivity occurs in Group 7A(17). The halogens begin with fluorine (F), the strongest electron "grabber" of all. Chlorine (Cl), bromine (Br), and iodine (I) also form compounds with most elements, and even extremely rare astatine (At) is thought to be reactive. The key features of Group 7A(17) are presented in the Family Portrait on pages 612 and 613.

What Accounts for the Regular Changes in the Halogens' Physical Properties?

Like the alkali metals at the other end of the periodic table, the halogens display regular trends in their physical properties. However, whereas melting and boiling points and heats of fusion and vaporization *decrease* down Group 1A(1), these properties *increase* down Group 7A(17) (see the Interchapter, Topic 6). The reason for these opposite trends is the different type of bonding in the elements. The alkali metals consist of atoms held together by metallic bonding, which *decreases* in strength as the atoms become larger. The halogens, on the other hand, exist as diatomic molecules that interact through dispersion forces, which *increase* in strength as the atoms become larger and more easily polarized. Thus, F_2 is a very pale yellow gas, Cl_2 a yellow-green gas, Br_2 a brown-orange liquid, and I_2 a purple-black solid.

Why Are the Halogens So Reactive?

The Group 7A(17) elements react with most metals and nonmetals to form many ionic and covalent compounds: metal and nonmetal halides, halogen oxides, and oxoacids. The reason for halogen reactivity is the same as that for alkali metal reactivity—an electron configuration one electron away from that of a noble gas. Whereas a 1A metal atom must *lose* one electron to attain a filled outer level, *a 7A nonmetal atom must gain one electron to fill its outer level*. It accomplishes this filling in either of two ways:

1. Gaining an electron from a metal atom, thus forming a negative ion as the metal forms a positive one
2. Sharing an electron pair with a nonmetal atom, thus forming a covalent bond

Down the group, reactivity reflects the decrease in electronegativity: F_2 is the most reactive and I_2 the least. The exceptional reactivity of elemental F_2 is also related to the weakness of the F—F bond. Because F is small, the bond is short; however, the lone pairs on each F atom repel those on the other, which weakens the bond (Figure 14.29). As a result of these factors, F_2 reacts with every element (except He, Ne, and Ar), in many cases, explosively.

The halogens display the largest range in electronegativity of any group, but all are electronegative enough to behave as nonmetals. They act as *oxidizing*

Figure 14.29 Bond energies and bond lengths of the halogens. A, In keeping with the increase in atomic size down the group, bond lengths increase steadily. **B,** The halogens show a general decrease in bond energy as bond length increases. However, F_2 deviates from this trend because its small, close, electron-rich atoms repel each other, thereby lowering its bond energy.

agents in the majority of their reactions, and halogens higher in the group can oxidize halide ions lower down:

$$F_2(g) + 2X^-(aq) \longrightarrow 2F^-(aq) + X_2(aq) \quad (X = Cl, Br, I)$$

Thus, the oxidizing ability of X_2 *decreases* down the group: the lower the EN, the less strongly each X atom pulls electrons. Similarly, the reducing ability of X^- *increases* down the group: the larger the ion, the more easily it gives up its electron (Figure 14.30).

The halogens undergo some important aqueous redox chemistry. Fluorine is such a powerful oxidizing agent that it tears water apart, oxidizing the O to produce O_2, some O_3, and HFO (hypofluorous acid). The other halogens undergo disproportionations (note the oxidation numbers):

$$\overset{0}{X_2} + H_2O(l) \rightleftharpoons \overset{-1}{H}X(aq) + \overset{+1}{H}XO(aq) \qquad (X = Cl, Br, I)$$

At equilibrium, very little product is present unless excess OH^- ion is added, which reacts with the HX and HXO and drives the reaction to completion:

$$X_2 + 2OH^-(aq) \longrightarrow X^-(aq) + XO^-(aq) + H_2O(l)$$

When X is Cl, the product mixture acts as a bleach: household bleach is a dilute solution of sodium hypochlorite (NaClO). Heating causes XO^- to disproportionate further, creating oxoanions with X in a higher oxidation state:

$$3\overset{+1}{X}O^-(aq) \overset{\Delta}{\longrightarrow} 2\overset{-1}{X}^-(aq) + \overset{+5}{X}O_3^-(aq)$$

$$Cl_2(aq) + 2I^-(aq) \longrightarrow 2Cl^-(aq) + I_2 \text{ (in } CCl_4)$$

Figure 14.30 The relative oxidizing ability of the halogens. A, Halogen redox behavior is based on atomic properties such as electron affinity, ionic charge density, and electronegativity. A halogen (X_2) higher in the group can oxidize a halide ion (X^-) lower down. **B,** As an example, when aqueous Cl_2 is added to a solution of I^- (*top layer*), it oxidizes the I^- to I_2, which dissolves in the CCl_4 solvent (*bottom layer*) to give a purple solution.

ARIS Presentation Center, Photo Library

Key Atomic and Physical Properties

KEY

Atomic No.
Symbol
Atomic mass
Valence e⁻ configuration
Common oxidation states

ns^2np^5

GROUP 7A(17)

9	
F	Photograph not available
19.00	
$2s^22p^5$	
(−1)	

17	
Cl	
35.45	
$3s^23p^5$	
(−1, +7, +5, +3, +1)	

35	
Br	
79.90	
$4s^24p^5$	
(−1, +7, +5, +3, +1)	

53	
I	
126.9	
$5s^25p^5$	
(−1, +7, +5, +3, +1)	

85	
At	Extremely rare, no sample available
(210)	
$6s^26p^5$	
(−1)	

Atomic Properties

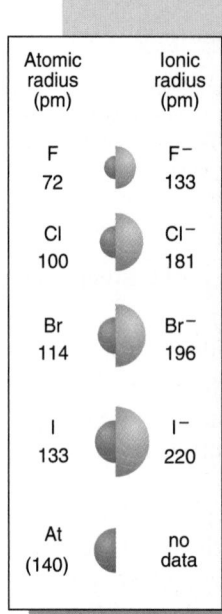

Atomic radius (pm)		Ionic radius (pm)
F 72		F⁻ 133
Cl 100		Cl⁻ 181
Br 114		Br⁻ 196
I 133		I⁻ 220
At (140)		no data

Group electron configuration is ns^2np^5; elements lack one electron to complete their outer level. The −1 oxidation state is the most common for all members. Except for F, the halogens exhibit all odd-numbered states (+7 through −1).

Down the group, atomic and ionic size increase steadily, as IE and EN decrease.

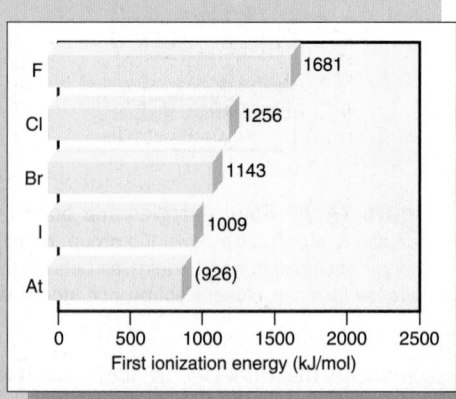

First ionization energy (kJ/mol)

F	1681
Cl	1256
Br	1143
I	1009
At	(926)

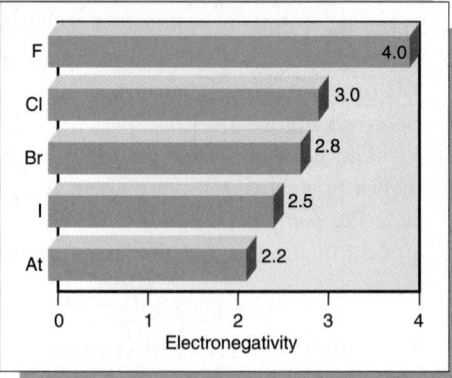

Electronegativity

F	4.0
Cl	3.0
Br	2.8
I	2.5
At	2.2

Physical Properties

Down the group, melting and boiling points increase smoothly as a result of stronger dispersion forces between heavier molecules.

Temperature (°C)

F	−188 / −219
Cl	−34.0 / −101
Br	59.5 / −7.2
I	185 / 114
At	

■ BP
□ MP

Density of liquid (g/mL)

F	1.51 (−188°C)
Cl	1.66 (−70°C)
Br	3.19 (0°C)
I	3.96 (120°C)
At	

The densities of the elements as liquids (at given T) increase steadily with molar mass.

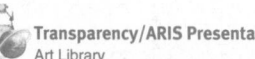

Some Reactions and Compounds

Important Reactions

Oxidizing strength and aqueous redox chemistry of the halogens are shown in reactions 1 and 2, and industrial processes involving fluorine in reactions 3 and 4.

1. The halogens (X_2) oxidize many metals and nonmetals. The reaction with hydrogen, although not used commercially for HX production (except for high-purity HCl), is characteristic of these strong oxidizing agents:

$$X_2 + H_2(g) \longrightarrow 2HX(g)$$

2. The halogens disproportionate in water:

$$X_2 + H_2O(l) \rightleftharpoons HX(aq) + HXO(aq)$$
$$(X = Cl, Br, I)$$

In aqueous base, the reaction goes to completion to form hypohalites (*see text*) and, at higher temperatures, halates; for example:

$$3Cl_2(g) + 6OH^-(aq) \xrightarrow{\Delta} ClO_3^-(aq) + 5Cl^-(aq) + 3H_2O(l)$$

3. F_2 is produced electrolytically at moderate temperature:

$$2HF \text{ (as a soln of KF in HF)} \xrightarrow[90°C]{electrolysis} H_2(g) + F_2(g)$$

A major use of F_2 is in the preparation of UF_6 for nuclear fuel.

4. Glass (amorphous silica) is etched with HF:

$$SiO_2(s) + 6HF(g) \longrightarrow H_2SiF_6(aq) + 2H_2O(l)$$

Space shuttle

Important Compounds

1. Fluorspar (fluorite), CaF_2. Widely distributed mineral used as a flux in steelmaking and in the production of HF (*see photo*).

Fluorite

2. Hydrogen fluoride, HF. Colorless, extremely toxic gas used to make F_2, organic fluorine compounds, and polymers. Also used in aluminum manufacture and in glass etching (*margin note, p. 614*).

3. Hydrogen chloride, HCl. Extremely water-soluble gas that forms hydrochloric acid, which occurs naturally in the stomach fluids of mammals (humans produce 1.5 L of 0.1 *M* HCl daily) and in volcanic gases (from reaction of H_2O and sea salt). Made by reaction of NaCl and H_2SO_4 and as a byproduct of plastics (PVC) production. Used in the "pickling" of steel (removal of adhering oxides) and in the production of syrups, rayon, and plastic.

4. Sodium hypochlorite, NaClO, and calcium hypochlorite, $Ca(ClO)_2$. Oxidizing agents used to bleach wood pulp and textiles and to disinfect swimming pools, foods, and sewage (also used to disinfect the Apollo 11 on return from the Moon). Household bleach is 5.25% NaClO by mass in water.

5. Ammonium perchlorate, NH_4ClO_4. Strong oxidizing agent used in the space shuttle program (*see photo*).

6. Potassium iodide, KI. Most common soluble iodide. Added to table salt to prevent thyroid disease (goiter). Used in chemical analysis because it is easily oxidized to I_2, which provides a colored end point.

7. Polychlorinated biphenyls, PCBs. Mixture of chlorinated organic compounds used as nonflammable insulating liquids in electrical transformers. Production discontinued because of persistence in the environment, where they become concentrated in the tissues of fish, birds, and mammals, and cause reproductive disturbances and possibly cancer.

ARIS Presentation Center, Art Library

Highlights of Halogen Chemistry

In this section, we examine the compounds the halogens form with hydrogen and with each other, as well as their oxides, oxoanions, and oxoacids.

The Hydrogen Halides The halogens form gaseous hydrogen halides (HX) through direct combination with H_2 or through the action of a concentrated acid on the metal halide (a nonoxidizing acid is used for HBr and HI): ●

$$CaF_2(s) + H_2SO_4(l) \longrightarrow CaSO_4(s) + 2HF(g)$$

$$3NaBr(s) + H_3PO_4(l) \longrightarrow Na_3PO_4(s) + 3HBr(g)$$

Commercially, most HCl is formed as a byproduct in the chlorination of hydrocarbons for plastics production:

$$CH_2{=}CH_2(g) + Cl_2(g) \longrightarrow ClCH_2CH_2Cl(l) \xrightarrow{500°C} CH_2{=}CHCl(g) + HCl(g)$$
$$\text{vinyl chloride}$$

In this case, the vinyl chloride reacts in a separate process to form poly(vinyl chloride), or PVC, a polymer used extensively in modern plumbing pipes.

In water, gaseous HX molecules form a *hydrohalic acid*. Only HF, with its relatively short, strong bond, forms a weak acid:

$$HF(g) + H_2O(l) \rightleftharpoons H_3O^+(aq) + F^-(aq)$$

The others dissociate completely to form the stoichiometric amount of H_3O^+ ions:

$$HBr(g) + H_2O(l) \longrightarrow H_3O^+(aq) + Br^-(aq)$$

(We saw reactions similar to these in Chapter 4. They involve transfer of a proton from acid to H_2O and are classified as *Brønsted-Lowry acid-base reactions*. In Chapter 18, we discuss them thoroughly and examine the relation between bond length and acidity of the larger HX molecules.)

Interhalogen Compounds: The "Halogen Halides" Halogens react exothermically with one another to form many **interhalogen compounds**. The simplest are diatomic molecules, such as ClF or BrCl. Every binary combination of the four common halogens is known. The more electronegative halogen is in the -1 oxidation state, and the less electronegative is in the $+1$ state. It has been proposed that interhalogens of general formula XY_n ($n = 3, 5, 7$) form when the larger members of the group (X) use d orbitals to expand their valence shells. In every case, the central atom has the *lower electronegativity* and a positive oxidation state.

The commercially useful interhalogens are powerful *fluorinating agents,* some of which react with metals, nonmetals, and oxides—even wood and asbestos:

$$Sn(s) + ClF_3(l) \longrightarrow SnF_2(s) + ClF(g)$$

$$P_4(s) + 5ClF_3(l) \longrightarrow 4PF_3(g) + 3ClF(g) + Cl_2(g)$$

$$2B_2O_3(s) + 4BrF_3(l) \longrightarrow 4BF_3(g) + 2Br_2(l) + 3O_2(g)$$

Their reactions with water are nearly explosive and yield HF and *the oxoacid in which the central halogen has the same oxidation state.* For example,

$$3H_2O(l) + BrF_5(l) \longrightarrow 5HF(g) + \overset{+5}{H}BrO_3(aq)$$

The Oddness and Evenness of Oxidation States In Topic 5 of the Interchapter, we saw that odd-numbered groups exhibit odd-numbered oxidation states and even-numbered groups exhibit even-numbered states. The reason for this general behavior is that *almost all stable molecules have paired electrons,* either as bonding or lone pairs. Therefore, *when bonds form or break, two electrons are involved, so the oxidation state changes by 2.*

Consider the interhalogens. Four general formulas are XY, XY_3, XY_5, and XY_7; examples are shown in Figure 14.31. With Y in the -1 state, X must be in the $+1$, $+3$, $+5$, and $+7$ state, respectively. The -1 state arises when Y fills its valence level; the $+7$ state arises when the central halogen (X) is completely

CIF

CIF₃

BrF₅

IF₇

| Linear, XY | T-shaped, XY₃ | Square pyramidal, XY₅ | Pentagonal bipyramidal, XY₇ |

Figure 14.31 Molecular shapes of the main types of interhalogen compounds.

oxidized, that is, when all seven valence electrons have shifted away from it to the more electronegative Y atoms around it.

Let's examine the iodine fluorides to see why the oxidation states jump by two units. When I_2 reacts with F_2, IF forms (note the oxidation number of I):

$$I_2 + F_2 \longrightarrow 2\overset{+1}{I}F$$

In IF_3, I uses *two* more valence electrons to form *two* more bonds:

$$\overset{+1}{I}F + F_2 \longrightarrow \overset{+3}{I}F_3$$

Otherwise, an unstable lone-electron species containing two fluorines would form. With more fluorine, another jump of two units occurs and the pentafluoride forms:

$$\overset{+3}{I}F_3 + F_2 \longrightarrow \overset{+5}{I}F_5$$

With still more fluorine, the heptafluoride forms:

$$\overset{+5}{I}F_5 + F_2 \longrightarrow \overset{+7}{I}F_7$$

An element in an even-numbered group, such as sulfur in Group 6A(16), shows the same tendency for its compounds to have paired electrons. Elemental sulfur (oxidation number, O.N. = 0) gains or shares two electrons to complete its shell (O.N. = −2). It uses two electrons to react with fluorine, for example, and form SF_2 (O.N. = +2), two more electrons for SF_4 (O.N. = +4), and two more for SF_6 (O.N. = +6). Thus, an element with one even state typically has all even states, and an element with one odd state typically has all odd states. To reiterate the main point, *successive oxidation states differ by two units because stable molecules have electrons in pairs around their atoms.*

Halogen Oxides, Oxoacids, and Oxoanions

The Group 7A(17) elements form many oxides that are *powerful oxidizing agents and acids in water.* Dichlorine monoxide (Cl_2O) and especially chlorine dioxide (ClO_2) are used to bleach paper (Figure 14.32). ClO_2 is unstable to heat and shock, so it is prepared on site, and more than 100,000 tons are used annually:

$$2NaClO_3(s) + SO_2(g) + H_2SO_4(aq) \longrightarrow 2ClO_2(g) + 2NaHSO_4(aq)$$

The dioxide has an unpaired electron and Cl in the unusual +4 oxidation state.

Lone e⁻

Dichlorine monoxide (Cl_2O)

Chlorine dioxide (ClO_2)

Dichlorine heptaoxide (Cl_2O_7)

Figure 14.32 Chlorine oxides.
Dichlorine monoxide and chlorine dioxide are bent molecules; note the lone electron in ClO_2. Dichlorine heptaoxide can be viewed as two ClO_4 tetrahedra joined through an O corner. (Each ball-and-stick model shows the structure in which each atom has its lowest formal charge. Lone pairs are shown on the central atoms only.)

● **Pyrotechnic Perchlorates** Thousands of tons of perchlorates are made each year for use in explosives and fireworks. The white flash of a fireworks display is caused by $KClO_4$ reacting with powdered sulfur and aluminum. At rock concerts and other theatrical productions, mixtures of $KClO_4$ and Mg are often used for special effects.

Chlorine is in its highest $(+7)$ oxidation state in dichlorine heptaoxide, Cl_2O_7, which is a symmetrical molecule formed when two $HClO_4$ $(HO—ClO_3)$ molecules undergo a dehydration-condensation reaction:

$$O_3Cl—O\boxed{H + HO}—ClO_3 \longrightarrow O_3Cl—O—ClO_3(l) + H_2O(l)$$

The halogen oxoacids and oxoanions are produced by reaction of the halogens and their oxides with water. Most of the oxoacids are stable only in solution. Table 14.4 shows ball-and-stick models of the acids in which each atom has its lowest formal charge; note the formulas, which emphasize that H is bonded to O. The hypohalites (XO^-), halites (XO_2^-), and halates (XO_3^-) are oxidizing agents formed by aqueous disproportionation reactions [see the Group 7A(17) Family Portrait, reaction 2]. You may have heated solid alkali chlorates in the laboratory to form small amounts of O_2:

$$2MClO_3(s) \xrightarrow{\Delta} 2MCl(s) + 3O_2(g)$$

Potassium chlorate is the oxidizer in "safety" matches.

Several perhalates are also strong oxidizing agents. ● Ammonium perchlorate, prepared from sodium perchlorate, is the oxidizing agent for the aluminum powder in the solid-fuel booster rockets of space shuttles; each launch uses more than 700 tons of NH_4ClO_4:

$$10Al(s) + 6NH_4ClO_4(s) \longrightarrow 4Al_2O_3(s) + 12H_2O(g) + 3N_2(g) + 2AlCl_3(g)$$

The relative strengths of the halogen oxoacids depend on two factors:

1. *Electronegativity of the halogen.* Among oxoacids with the halogen in the same oxidation state, such as the halic acids, HXO_3 (or $HOXO_2$), acid strength decreases as the halogen's EN decreases:

$$HOClO_2 > HOBrO_2 > HOIO_2$$

The more electronegative the halogen, the more electron density it removes from the O—H bond, and the more easily the proton is lost.

2. *Oxidation state of the halogen.* Among oxoacids of a given halogen, such as chlorine, acid strength decreases as the oxidation state of the halogen decreases:

$$HOClO_3 > HOClO_2 > HOClO > HOCl$$

The higher the oxidation state (number of attached O atoms) of the halogen, the more electron density it pulls from the O—H bond. We consider these trends quantitatively in Chapter 18.

Table 14.4 **The Known Halogen Oxoacids***

Central Atom	Hypohalous Acid (HOX)	Halous Acid (HOXO)	Halic Acid (HOXO$_2$)	Perhalic Acid (HOXO$_3$)
Fluorine	HOF	—	—	—
Chlorine	HOCl	HOClO	HOClO$_2$	HOClO$_3$
Bromine	HOBr	(HOBrO)?	HOBrO$_2$	HOBrO$_3$
Iodine	HOI	—	HOIO$_2$	HOIO$_3$, (HO)$_5$IO
Oxoanion	Hypohalite	Halite	Halate	Perhalate

*Lone pairs are shown only on the halogen atom, and each atom has its lowest formal charge.

14.10 GROUP 8A(18): THE NOBLE GASES

The last main group consists of individual atoms too "noble" to interact with others. The Group 8A(18) elements display regular trends in physical properties and very low, if any, reactivity. The group consists of helium (He), the second most abundant element in the universe, neon (Ne) and argon (Ar), krypton (Kr) and xenon (Xe), the only members for which compounds have been well studied, and finally, radioactive radon (Rn). The noble gases make up about 1% by volume of the atmosphere, primarily due to the abundance of Ar. Their properties appear in the Group 8A(18) Family Portrait on page 618.

How Can Noble Gases Form Compounds?

Lying at the far right side of the periodic table, the Group 8A(18) elements consist of individual atoms with filled outer levels and the smallest radii in their periods: even Li, the smallest alkali metal (152 pm), is bigger than Rn, the largest noble gas (140 pm). These elements come as close to behaving as ideal gases as any other substances. Only at very low temperatures do they condense and solidify. In fact, He is the only substance that does *not* solidify by a reduction in temperature alone; it requires an increase in pressure as well. Helium has the lowest melting point known ($-272.2°C$ at 25 atm), only one degree above 0 K ($-273.15°C$), and it boils only about three degrees higher. Weak dispersion forces hold these elements in condensed states, with melting and boiling points that increase, as expected, with molar mass.

Ever since their discovery in the late 19th century, these elements had been considered, and even formerly named, the "inert" gases. Atomic theory and, more important, all experiments had supported this idea. Then, in 1962, all this changed when the first noble gas compound was prepared. How, with filled outer levels and extremely high ionization energies, *can* noble gases react?

The discovery of noble gas reactivity is a classic example of clear thinking in the face of an unexpected event. At the time, a young inorganic chemist named Neil Bartlett was studying platinum fluorides, known to be strong oxidizing agents. When he accidentally exposed PtF_6 to air, its deep-red color lightened slightly, and analysis showed that the PtF_6 had oxidized O_2 to form the ionic compound $[O_2]^+[PtF_6]^-$. Knowing that the ionization energy of the oxygen molecule ($O_2 \longrightarrow O_2^+ + e^-$; IE = 1175 kJ/mol) is very close to IE_1 of xenon (1170 kJ/mol), Bartlett reasoned that PtF_6 might be able to oxidize xenon. Shortly thereafter, he prepared $XePtF_6$, an orange-yellow solid. Within a few months, the white crystalline XeF_2 and XeF_4 (Figure 14.33) were also prepared. In addition to its +2 and +4 oxidation states, Xe has the +6 state in several compounds, such as XeF_6, and the +8 state in the unstable oxide, XeO_4. A few compounds of Kr have also been made. The xenon fluorides react rapidly in water to form HF and various other products, including different xenon compounds.

Looking Backward and Forward: Groups 7A(17), 8A(18), and 1A(1)

With the disappearance of metallic behavior, the Group 7A(17) elements form a host of anions, covalent oxides, and oxoanions, which change oxidation states readily in a rich aqueous chemistry. The vigorous reactivity of the halogens is in stark contrast to the inertness of their 8A neighbors. Filled outer levels render the noble gas atoms largely inert, despite a limited ability to react with the most electronegative elements, fluorine and oxygen. The least reactive family in the periodic table stands between the two most reactive: the halogens, which need one more electron to fill their outer level, and the alkali metals, which need one fewer (Figure 14.34). As you know, atomic, physical, and chemical properties change dramatically from Group 8A(18) to Group 1A(1).

Figure 14.33 Crystals of xenon tetra-fluoride (XeF₄).

ARIS Presentation Center, Photo Library

Figure 14.34 Standing in Group 8A(18), looking backward at the halogens, Group 7A(17), and ahead to the alkali metals, Group 1A(1).

ARIS Presentation Center, Art Library

Key Atomic and Physical Properties

KEY

Atomic No.
Symbol
Atomic mass
Valence e⁻ configuration
Common oxidation states

ns^2np^6

GROUP 8A(18)

2	**He**
	4.003
	$1s^2$
	(none)

10	**Ne**
	20.18
	$2s^22p^6$
	(none)

18	**Ar**
	39.95
	$3s^23p^6$
	(none)

36	**Kr**
	83.80
	$4s^24p^6$
	(+2)

54	**Xe**
	131.3
	$5s^25p^6$
	(+8, +6, +4, +2)

86	**Rn**
	(222)
	$6s^26p^6$
	(+2)

Mass spectral peak

Atomic Properties

Group electron configuration is $1s^2$ for He and ns^2np^6 for the others. The valence shell is filled. Only Kr and Xe (and perhaps Rn) are known to form compounds. The more reactive Xe exhibits all even oxidation states (+2 to +8).

This group contains the smallest atoms with the highest IEs in their periods. Down the group, atomic size increases and IE decreases steadily. (EN values are given only for Kr and Xe.)

Atomic radius (pm)

He 31
Ne 71
Ar 98
Kr 112
Xe 131
Rn (140)

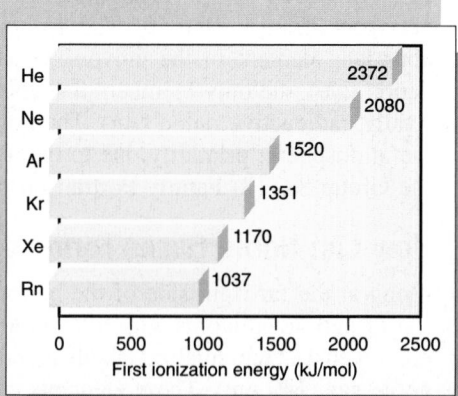

First ionization energy (kJ/mol)

He	2372
Ne	2080
Ar	1520
Kr	1351
Xe	1170
Rn	1037

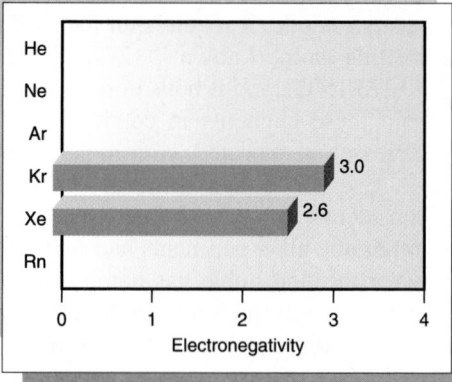

Electronegativity

Kr	3.0
Xe	2.6

Physical Properties

Melting and boiling points of these gaseous elements are extremely low but increase down the group because of stronger dispersion forces. Note the extremely small liquid ranges.

Temperature (°C)

	BP	MP
He	−269	−270
Ne	−246	−249
Ar	−186	−189
Kr	−153	−157
Xe	−108	−112
Rn	−62	−71

■ BP □ MP

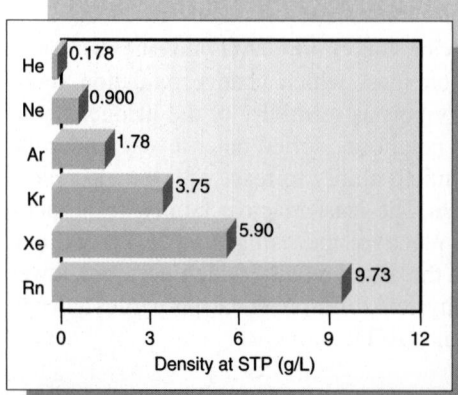

Density at STP (g/L)

He	0.178
Ne	0.900
Ar	1.78
Kr	3.75
Xe	5.90
Rn	9.73

Densities (at STP) increase steadily, as expected.

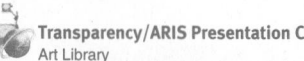

Transparency/ARIS Presentation Center, Art Library

Chapter Perspective

Our excursion through the bonding and reactivity patterns of the elements has come full circle, but we've only been able to touch on some of the most important features. Clearly, the properties of the atoms and the resultant physical and chemical behaviors of the elements are magnificent testimony to nature's diversity. In Chapter 15, we continue with our theme of macroscopic behavior emerging from atomic properties in an investigation of the marvelously complex organic compounds of carbon. In Chapter 22, we revisit the most important main-group elements to see how they occur in nature and how we isolate and use them.

CHAPTER REVIEW GUIDE
The following sections provide many aids to help you study this chapter. (Numbers in parentheses refer to pages, unless noted otherwise.)

Learning Objectives These are concepts and skills you should know after studying this chapter.

Relevant section numbers appear in parentheses.

Understand These Concepts

Note: Many characteristic reactions appear in the "Important Reactions" section of each group's Family Portrait.

1. How hydrogen is similar to, yet different from, alkali metals and halogens; the differences between ionic, covalent, and metallic hydrides (Section 14.1)

2. Key horizontal trends in atomic properties, types of bonding, oxide acid-base properties, and redox behavior as the elements change from metals to nonmetals (Section 14.2)

3. How the ns^1 configuration accounts for the physical and chemical properties of the alkali metals (Section 14.3)

4. How small atomic size and limited number of valence orbitals account for the anomalous behavior of the Period 2 member of each group (see each section for details)

5. How the ns^2 configuration accounts for the key differences between Groups 1A(1) and 2A(2) (Section 14.4)

6. The basis of the three important diagonal relationships (Li/Mg, Be/Al, B/Si) (Sections 14.4–14.6)

7. How the presence of inner $(n-1)d$ electrons affects properties in Group 3A(13) (Section 14.5)

8. Patterns among larger members of Groups 3A(13) to 6A(16): two common oxidation states (inert-pair effect), lower state more important down the group, and more basic lower oxide (Sections 14.5–14.8)

9. How boron attains an octet of electrons (Section 14.5)

10. The effect of bonding on the physical behavior of Groups 4A(14) to 6A(16) (Sections 14.6–14.8)

11. Allotropism in carbon, phosphorus, and sulfur (Sections 14.6–14.8)

12. How atomic properties lead to catenation and multiple bonds in organic compounds (Section 14.6)

13. Structures and properties of the silicates and silicones (Section 14.6)

14. Patterns of behavior among hydrides and halides of Groups 5A(15) and 6A(16) (Sections 14.7, 14.8)

15. The meaning of disproportionation (Section 14.7) and its importance in halogen aqueous chemistry (Section 14.9)

16. Structure and chemistry of the nitrogen oxides and oxoacids (Section 14.7)

17. Structure and chemistry of the phosphorus oxides and oxoacids (Section 14.7)

18. Dehydration-condensation reactions and polyphosphate structures (Section 14.7)

19. Structure and chemistry of the sulfur oxides and oxoacids (Section 14.8)

20. How the ns^2np^5 configuration accounts for halogen reactivity with metals (Section 14.9)

21. Why the oxidation states of an element change by two units (Section 14.9)

22. Structure and chemistry of the halogen oxides and oxoacids (Section 14.9)

23. How the ns^2np^6 configuration accounts for the relative inertness of the noble gases (Section 14.10)

Key Terms These important terms appear in boldface in the chapter and are defined again in the Glossary.

Section 14.4	Section 14.6	Section 14.7	Section 14.9
diagonal relationship (575)	allotrope (585)	disproportionation reaction (600)	interhalogen compound (614)
Section 14.5	silicate (590)		
bridge bond (583)	silicone (590)	dehydration-condensation reaction (602)	

Highlighted Figures and Tables These figures (F) and tables (T) provide a visual review of key ideas.

T14.1 Trends in properties of the Period 2 elements (568–569)
Family Portrait Group 1A(1) (572–573)
F14.5 Three diagonal relationships (575)
Family Portrait Group 2A(2) (576–577)
Family Portrait Group 3A(13) (580–581)
T14.2 Bond type and melting process in Groups 3A(13) to 5A(15) (584)

Family Portrait Group 4A(14) (586–587)
Family Portrait Group 5A(15) (596–597)
T14.3 Structures and properties of the nitrogen oxides (599)
Family Portrait Group 6A(16) (604–605)
Family Portrait Group 7A(17) (612–613)
Family Portrait Group 8A(18) (618)

PROBLEMS

Problems with colored numbers are answered in Appendix E and worked in detail in the Student Solutions Manual. Problem sections match those in the text and provide the numbers of relevant sample problems. Most offer Concept Review Questions, Skill-Building Exercises (grouped in pairs covering the same concept), and Problems in Context. The Comprehensive Problems are based on material from any section or previous chapter.

Note: This set begins with problems based on material from the Interchapter.

Interchapter Review

● Concept Review Questions

•**14.1** (a) Define the four atomic properties that are reviewed in the Interchapter.
(b) To what do the three parts of the electron configuration ($nl^{\#}$) correlate?
(c) Which part of the electron configuration is primarily associated with the size of the atom?
(d) What is the major distinction between the outer electron configurations within a group and those within a period?
(e) What correlation, if any, exists between the group number and the number of valence electrons?

•**14.2** (a) What trends, if any, exist for Z_{eff} across a period and down a group? (b) How does Z_{eff} influence atomic size, IE_1, and EN across a period?

•**14.3** Iodine monochloride and elemental bromine have nearly the same molar mass and liquid density but very different boiling points. (a) What molecular property is primarily responsible for this difference in boiling point? What atomic property gives rise to it? Explain. (b) Which substance has a higher boiling point? Why?

•**14.4** How does the trend in atomic size differ from the trend in ionization energy? Explain.

•**14.5** How are bond energy, bond length, and reactivity related for similar compounds?

•**14.6** How are covalent and metallic bonding similar? How are they different?

•**14.7** If the leftmost element in a period combined with each of the others in the period, how would the type of bonding change from left to right? Explain in terms of atomic properties.

•**14.8** Why is rotation about the bond axis possible for single-bonded atoms but not double-bonded atoms?

•**14.9** Explain the horizontal irregularity in size of the most common ions of Period 3 elements.

•**14.10** Is S^{2-} larger or smaller than Cl^-? Is Mg^{2+} larger or smaller than Na^+? Explain.

•**14.11** How do each of the following correlate with the electronegativity of the elements: (a) the type of bonding in element oxides; (b) the acid-base behavior of element oxides?

•**14.12** (a) How does the metallic character of an element correlate with the *acidity* of its oxide? (b) What trends, if any, exist in oxide *basicity* across a period and down a group?

•**14.13** How are atomic size, IE_1, and EN related to redox behavior of the elements in Groups 1A(1), 2A(2), 6A(16), and 7A(17)?

•**14.14** How do the physical properties of a network covalent solid and a molecular covalent solid differ? Why?

● Skill-Building Exercises (grouped in similar pairs)

•**14.15** Rank the following elements in order of *increasing*
(a) atomic size: Ba, Mg, Sr   (b) IE_1: P, Na, Al
(c) EN: Br, Cl, Se   (d) no. of valence electrons: Bi, Ga, Sn

•**14.16** Rank the following elements in order of *decreasing*
(a) atomic size: N, Si, P   (b) IE_1: Kr, K, Ar
(c) EN: In, Rb, I   (d) no. of valence electrons: Sb, S, Cs

•**14.17** Which of the following pairs react to form *ionic* compounds: (a) Cl and Br; (b) Na and Br; (c) P and Se; (d) H and Ba?

•**14.18** Which of the following pairs react to form *covalent* compounds: (a) Be and C; (b) Sr and O; (c) Ca and Cl; (d) P and F?

•**14.19** Draw a Lewis structure for a compound with a bond order of 2 throughout the molecule.

•**14.20** Draw a Lewis structure for a polyatomic ion with a fractional bond order throughout the molecule.

•**14.21** Rank the following in order of *increasing* bond length: $SiCl_4$, CF_4, $GeBr_4$.

•**14.22** Rank the following in order of *decreasing* bond energy: NF_3, NI_3, NCl_3.

•**14.23** Rank the following in order of *increasing* radius:
(a) O^{2-}, F^-, Na^+
(b) S^{2-}, P^{3-}, Cl^-

•**14.24** Rank the following in order of *decreasing* radius:
(a) Ca^{2+}, K^+, Ga^{3+}
(b) Br^-, Sr^{2+}, Rb^+

•**14.25** Rank O^-, O^{2-}, and O in order of *increasing* size.

•**14.26** Rank Tl^{3+}, Tl, and Tl^+ in order of *decreasing* size.

••**14.27** Which member of each pair gives the more *basic* solution in water: (a) CaO or SO_3; (b) BeO or BaO; (c) CO_2 or SO_2; (d) P_4O_{10} or K_2O? For the pair in part (a), write an equation for the dissolving of each oxide to support your answer.

••**14.28** Which member of each pair gives the more *acidic* solution in water: (a) CO_2 or SrO; (b) SnO or SnO_2; (c) Cl_2O or Na_2O; (d) SO_2 or MgO? For the pair in part (a), write an equation for the dissolving of each oxide to support your answer.

Levels of Difficulty: • Easy •• Medium ••• Challenging

• **14.29** Which member of each pair has more *covalent* character in its bonds: (a) LiCl or KCl; (b) $AlCl_3$ or PCl_3; (c) NCl_3 or $AsCl_3$?

• **14.30** Which member of each pair has more *ionic* character in its bonds: (a) BeF_2 or CaF_2; (b) PbF_2 or PbF_4; (c) GeF_4 or PF_3?

•• **14.31** Rank the following in order of *increasing*
(a) Melting point: Na, Si, Ar   (b) ΔH_{fus}: Rb, Cs, Li

•• **14.32** Rank the following in order of *decreasing*
(a) Boiling point: O_2, Br_2, As(s)   (b) ΔH_{vap}: Cl_2, Ar, I_2

Hydrogen, the Simplest Atom

● Concept Review Questions

• **14.33** Hydrogen has only one proton, but its IE_1 is much greater than that of lithium, which has three protons. Explain.

• **14.34** Sketch a periodic table, and label the areas containing elements that give rise to the three types of hydrides.

● Skill-Building Exercises *(grouped in similar pairs)*

• **14.35** Draw Lewis structures for the following compounds, and predict which member of each pair will form hydrogen bonds:
(a) NF_3 or NH_3 (b) CH_3OCH_3 or CH_3CH_2OH

• **14.36** Draw Lewis structures for the following compounds, and predict which member of each pair will form hydrogen bonds:
(a) NH_3 or AsH_3 (b) CH_4 or H_2O

• **14.37** Complete and balance the following equations:
(a) An active metal reacting with acid,
$$Al(s) + HCl(aq) \longrightarrow$$
(b) A saltlike (alkali metal) hydride reacting with water,
$$LiH(s) + H_2O(l) \longrightarrow$$

• **14.38** Complete and balance the following equations:
(a) A saltlike (alkaline earth metal) hydride reacting with water,
$$CaH_2(s) + H_2O(l) \longrightarrow$$
(b) Reduction of a metal halide by hydrogen to form a metal,
$$PdCl_2(aq) + H_2(g) \longrightarrow$$

● Problems in Context

• **14.39** Compounds such as $NaBH_4$, $Al(BH_4)_3$, and $LiAlH_4$ are complex hydrides used as reducing agents in many syntheses.
(a) Give the oxidation state of each element in these compounds.
(b) Write a Lewis structure for the polyatomic anion in $NaBH_4$, and predict its shape.

•• **14.40** Unlike the F^- ion, which has an ionic radius close to 133 pm in all alkali metal fluorides, the ionic radius of H^- varies from 137 pm in LiH to 152 pm in CsH. Suggest an explanation for the large variability in the size of H^- but not F^-.

Trends Across the Periodic Table

● Concept Review Questions

• **14.41** How does the maximum oxidation number vary across a period in the main groups? Is the pattern in Period 2 different?

• **14.42** What correlation, if any, exists for the Period 2 elements between group number and the number of covalent bonds the element typically forms? How is the correlation different for elements in Periods 3 to 6?

• **14.43** Each of the chemically active Period 2 elements forms stable compounds that have bonds to fluorine.
(a) What are the names and formulas of these compounds?
(b) Does ΔEN increase or decrease left to right across the period?
(c) Does percent ionic character increase or decrease left to right?
(d) Draw Lewis structures for these compounds.

• **14.44** Period 6 is unusual in several ways.
(a) It is the longest period in the table. How many elements belong to Period 6? How many metals?
(b) It contains no metalloids. Where is the metal/nonmetal boundary in Period 6?

•• **14.45** An element forms an oxide, E_2O_3, and a fluoride, EF_3.
(a) Of which two groups might E be a member?
(b) How does the group to which E belongs affect the properties of the oxide and the fluoride?

• **14.46** Fluorine lies between oxygen and neon in Period 2. Whereas atomic sizes and ionization energies of these three elements change smoothly, their electronegativities display a dramatic change. What is this change, and how do their electron configurations explain it?

Group 1A(1): The Alkali Metals

● Concept Review Questions

• **14.47** Lithium salts are often much less soluble in water than the corresponding salts of other alkali metals. For example, at 18°C, the concentration of a saturated LiF solution is 1.0×10^{-2} M, whereas that of a saturated KF solution is 1.6 M. How would you explain this behavior?

• **14.48** The alkali metals play virtually the same general chemical role in all their reactions. (a) What is this role? (b) How is it based on atomic properties? (c) Using sodium, write two balanced equations that illustrate this role.

• **14.49** How do atomic properties account for the low densities of the Group 1A(1) elements?

● Skill-Building Exercises *(grouped in similar pairs)*

• **14.50** Each of the following properties shows a regular trend in Group 1A(1). Predict whether each increases or decreases *down* the group: (a) density; (b) ionic size; (c) E—E bond energy; (d) IE_1; (e) magnitude of ΔH_{hydr} of E^+ ion.

• **14.51** Each of the following properties shows a regular trend in Group 1A(1). Predict whether each increases or decreases *up* the group: (a) melting point; (b) E—E bond length; (c) hardness; (d) molar volume; (e) lattice energy of EBr.

• **14.52** Write a balanced equation for the formation from its elements of sodium peroxide, an industrial bleach.

• **14.53** Write a balanced equation for the formation of rubidium bromide through a reaction of a strong acid and a strong base.

● Problems in Context

• **14.54** Although the alkali metal halides can be prepared directly from the elements, the far less expensive industrial route is treatment of the carbonate or hydroxide with aqueous hydrohalic acid (HX) followed by recrystallization. Balance the reaction between potassium carbonate and aqueous hydriodic acid.

• **14.55** Lithium forms several useful organolithium compounds. Calculate the mass percent of Li in the following:
(a) Lithium stearate ($C_{17}H_{35}COOLi$), a water-resistant grease used in cars because it does not harden at cold temperatures
(b) Butyllithium (LiC_4H_9), a reagent in organic syntheses

Group 2A(2): The Alkaline Earth Metals

● Concept Review Questions

• **14.56** How do Groups 1A(1) and 2A(2) compare with respect to reaction of the metals with water?

14.57 Alkaline earth metals are involved in two key diagonal relationships in the periodic table. (a) Give the two pairs of elements in these diagonal relationships. (b) For each pair, cite two similarities that demonstrate the relationship. (c) Why are the members of each pair so similar in behavior?

14.58 The melting points of alkaline earth metals are many times higher than those of the alkali metals. Explain this difference on the basis of atomic properties. Name three other physical properties for which Group 2A(2) metals have higher values than the corresponding 1A(1) metals.

● **Skill-Building Exercises** *(grouped in similar pairs)*

14.59 Write a balanced equation for each reaction:
(a) "Slaking" of lime (treatment with water)
(b) Combustion of calcium in air

14.60 Write a balanced equation for each reaction:
(a) Thermal decomposition of witherite (barium carbonate)
(b) Neutralization of stomach acid (HCl) by milk of magnesia (magnesium hydroxide)

● **Problems in Context**

14.61 Lime (CaO) is one of the most abundantly produced chemicals in the world. Write balanced equations for
(a) The preparation of lime from natural sources
(b) The use of slaked lime to remove SO_2 from flue gases
(c) The reaction of lime with arsenic acid (H_3AsO_4) to manufacture the insecticide calcium arsenate
(d) The regeneration of NaOH in the paper industry by reaction of lime with aqueous sodium carbonate

14.62 In some reactions, Be behaves like a typical alkaline earth metal; in others, it does not. Complete and balance the following equations:
(a) $BeO(s) + H_2O(l) \longrightarrow$
(b) $BeCl_2(l) + Cl^-(l;$ from molten NaCl$) \longrightarrow$
In which reaction does Be behave like the other Group 2A(2) members?

Group 3A(13): The Boron Family

● **Concept Review Questions**

14.63 How do the transition metals in Period 4 affect the pattern of ionization energies in Group 3A(13)? How does this pattern compare with that in Group 3B(3)?

14.64 How do the acidities of aqueous solutions of Tl_2O and Tl_2O_3 compare with each other? Explain.

14.65 Despite the expected decrease in atomic size, there is an unexpected drop in the first ionization energy between Groups 2A(2) and 3A(13) in Periods 2 through 4. Explain this pattern in terms of electron configurations and orbital energies.

14.66 Many compounds of Group 3A(13) elements have chemical behavior that reflects an electron deficiency.
(a) What is the meaning of *electron deficiency?*
(b) Give two reactions that illustrate this behavior.

14.67 Boron's chemistry is not typical of its group.
(a) Cite three ways in which boron and its compounds differ significantly from the other 3A(13) members and their compounds.
(b) What is the reason for these differences?

● **Skill-Building Exercises** *(grouped in similar pairs)*

14.68 Rank the following oxides in order of increasing aqueous *acidity:* Ga_2O_3, Al_2O_3, In_2O_3.

14.69 Rank the following hydroxides in order of increasing aqueous *basicity:* $Al(OH)_3$, $B(OH)_3$, $In(OH)_3$.

14.70 Thallium forms the compound TlI_3. What is the apparent oxidation state of Tl in this compound? Given that the anion is I_3^-, what is the actual oxidation state of Tl? Draw the shape of the anion, giving its VSEPR class and bond angles. Propose a reason why the compound does not exist as $(Tl^{3+})(I^-)_3$.

14.71 Very stable dihalides of the Group 3A(13) metals are known. What is the apparent oxidation state of Ga in $GaCl_2$? Given that $GaCl_2$ consists of a Ga^+ cation and a $GaCl_4^-$ anion, what are the actual oxidation states of Ga? Draw the shape of the anion, giving its VSEPR class and bond angles.

● **Problems in Context**

14.72 Give the name and symbol or formula of a Group 3A(13) element or compound that fits each description or use:
(a) Component of heat-resistant (Pyrex-type) glass
(b) Manufacture of high-speed computer chips
(c) Largest temperature range for liquid state of an element
(d) Elemental substance with three-center, two-electron bonds
(e) Metal protected from oxidation by adherent oxide coat
(f) Mild antibacterial agent (e.g., for eye infections)
(g) Toxic metal that lies between two other toxic metals

14.73 Indium (In) reacts with HCl to form a diamagnetic solid with the formula $InCl_2$. (a) Write condensed electron configurations for In, In^+, In^{2+}, and In^{3+}. (b) Which of these species is (are) diamagnetic and which paramagnetic? (c) What is the apparent oxidation state of In in $InCl_2$? (d) Given your answers to parts (b) and (c), explain how $InCl_2$ can be diamagnetic.

14.74 Use VSEPR theory to draw structures, with ideal bond angles, for boric acid and the anion it forms in reaction with water.

14.75 Boron nitride (BN) has a structure similar to graphite, but is a white insulator rather than a black conductor. It is synthesized by heating diboron trioxide with ammonia at about 1000°C.
(a) Write a balanced equation for the formation of BN; water forms also.
(b) Calculate ΔH°_{rxn} for the production of BN (ΔH°_f of BN is -254 kJ/mol).
(c) Boron is obtained from the mineral borax, $Na_2B_4O_7 \cdot 10H_2O$. How much borax is needed to produce 1.0 kg of BN, assuming 72% yield?

Group 4A(14): The Carbon Family

● **Concept Review Questions**

14.76 How does the basicity of SnO_2 in water compare with that of CO_2? Explain.

14.77 Nearly every compound of silicon has the element in the $+4$ oxidation state. In contrast, most compounds of lead have the element in the $+2$ state.
(a) What general observation do these facts illustrate?
(b) Explain in terms of atomic and molecular properties.
(c) Give an analogous example from Group 3A(13).

14.78 The sum of IE_1 through IE_4 for Group 4A(14) elements shows a decrease from C to Si, a slight increase from Si to Ge, a decrease from Ge to Sn, and an increase from Sn to Pb.
(a) What is the expected trend for IEs down a group?
(b) Suggest a reason for the deviations in Group 4A(14).
(c) Which group might show even greater deviations?

14.79 Give explanations for the large drops in melting point from C to Si and from Ge to Sn.

14.80 What is an allotrope? Name two Group 4A(14) elements that exhibit allotropism, and identify two of their allotropes.

● **14.81** Even though EN values vary relatively little down Group 4A(14), the elements change from nonmetal to metal. Explain.

● **14.82** How do atomic properties account for the enormous number of carbon compounds? Why don't other Group 4A(14) elements behave similarly?

● **Skill-Building Exercises** (grouped in similar pairs)

●● **14.83** Draw a Lewis structure for
(a) The cyclic silicate ion $Si_4O_{12}^{8-}$
(b) A cyclic hydrocarbon with formula C_4H_8

●● **14.84** Draw a Lewis structure for
(a) The cyclic silicate ion $Si_6O_{18}^{12-}$
(b) A cyclic hydrocarbon with formula C_6H_{12}

●● **14.85** Show three units of a linear silicone polymer made from $(CH_3)_2Si(OH)_2$ with some $(CH_3)_3SiOH$ added to end the chain.

●● **14.86** Show two chains of three units each of a sheet silicone polymer made from $(CH_3)_2Si(OH)_2$ with $CH_3Si(OH)_3$ added to crosslink the chains.

● **Problems in Context**

●● **14.87** Zeolite A, $Na_{12}[(AlO_2)_{12}(SiO_2)_{12}] \cdot 27H_2O$, is used to soften water by replacing Ca^{2+} and Mg^{2+} with Na^+. Hard water from a certain source is 4.5×10^{-3} M Ca^{2+} and 9.2×10^{-4} M Mg^{2+}, and a pipe delivers 25,000 L of this hard water per day. What mass (in kg) of zeolite A is needed to soften a week's supply of the water? (Assume zeolite A loses its capacity to exchange ions when 85 mol % of its Na^+ has been lost.)

● **14.88** Give the name and symbol or formula of a Group 4A(14) element or compound that fits each description or use:
(a) Hardest known natural substance
(b) Medicinal antacid
(c) Atmospheric gas implicated in the greenhouse effect
(d) Waterproofing polymer based on silicon
(e) Synthetic abrasive composed entirely of elements in the group
(f) Product formed when coke burns in a limited supply of air
(g) Toxic metal found in plumbing and paints

●● **14.89** One similarity between B and Si is the explosive combustion of their hydrides in air. Write balanced equations for the combustion of B_2H_6 and of Si_4H_{10}.

Group 5A(15): The Nitrogen Family

● **Concept Review Questions**

● **14.90** Which Group 5A(15) elements form trihalides? Pentahalides? Explain.

● **14.91** As you move down Group 5A(15), the melting points of the elements increase and then decrease. Explain.

● **14.92** (a) What is the range of oxidation states shown by the elements of Group 5A(15) as you move down the group? (b) How does this range illustrate the general rule for the range of oxidation states in groups on the right side of the periodic table?

● **14.93** Bismuth(V) compounds are such powerful oxidizing agents that they have not been prepared in pure form. How is this fact consistent with the location of Bi in the periodic table?

● **14.94** Rank the following oxides in order of increasing acidity in water: Sb_2O_3, Bi_2O_3, P_4O_{10}, Sb_2O_5.

● **Skill-Building Exercises** (grouped in similar pairs)

● **14.95** Assuming acid strength relates directly to electronegativity of the central atom, rank H_3PO_4, HNO_3, and H_3AsO_4 in order of *increasing* acid strength.

● **14.96** Assuming acid strength relates directly to number of O atoms bonded to the central atom, rank $H_2N_2O_2$ [or $(HON)_2$]; HNO_3 (or $HONO_2$); and HNO_2 (or $HONO$) in order of *decreasing* acid strength.

● **14.97** Complete and balance the following:
(a) $As(s) + \text{excess } O_2(g) \longrightarrow$
(b) $Bi(s) + \text{excess } F_2(g) \longrightarrow$
(c) $Ca_3As_2(s) + H_2O(l) \longrightarrow$

● **14.98** Complete and balance the following:
(a) $\text{Excess } Sb(s) + Br_2(l) \longrightarrow$
(b) $HNO_3(aq) + MgCO_3(s) \longrightarrow$
(c) $K_2HPO_4(s) \xrightarrow{\Delta}$

● **14.99** Complete and balance the following:
(a) $N_2(g) + Al(s) \xrightarrow{\Delta}$ (b) $PF_5(g) + H_2O(l) \longrightarrow$

●● **14.100** Complete and balance the following:
(a) $AsCl_3(l) + H_2O(l) \longrightarrow$ (b) $Sb_2O_3(s) + NaOH(aq) \longrightarrow$

●● **14.101** Based on the relative sizes of F and Cl, predict the structure of PF_2Cl_3.

●● **14.102** Use the VSEPR model to predict the structure of the cyclic ion $P_3O_9^{3-}$.

● **Problems in Context**

●● **14.103** The pentafluorides of the larger members of Group 5A(15) have been prepared, but N can have only eight electrons. A claim has been made that, at low temperatures, a compound with the empirical formula NF_5 forms. Draw a possible Lewis structure for this compound. (*Hint:* NF_5 is ionic.)

● **14.104** Give the name and symbol or formula of a Group 5A(15) element or compound that fits each description or use:
(a) Hydride produced at multimillion-ton level
(b) Element(s) essential in plant nutrition
(c) Hydride used to manufacture rocket propellants and drugs
(d) Odd-electron molecule (two examples)
(e) Amphoteric hydroxide with a 5A element in its +3 state
(f) Phosphorus-containing water softener
(g) Element that is an electrical conductor

●● **14.105** In addition to those in Table 14.3, other less stable nitrogen oxides exist. Draw a Lewis structure for each of the following:
(a) N_2O_2, a dimer of nitrogen monoxide with an N—N bond
(b) N_2O_2, a dimer of nitrogen monoxide with no N—N bond
(c) N_2O_3 with no N—N bond
(d) NO^+ and NO_3^-, products of the ionization of liquid N_2O_4

●● **14.106** Nitrous oxide (N_2O), the "laughing gas" used as an anesthetic by dentists, is made by thermal decomposition of solid NH_4NO_3. Write a balanced equation for this reaction. What are the oxidation states of N in NH_4NO_3 and in N_2O?

●● **14.107** Write balanced equations for the thermal decomposition of potassium nitrate (O_2 is also formed in both cases): (a) at low temperature to the nitrite; (b) at high temperature to the metal oxide and nitrogen.

Group 6A(16): The Oxygen Family

● **Concept Review Questions**

● **14.108** Rank the following in order of increasing electrical conductivity, and explain your ranking: Po, S, Se.

● **14.109** The oxygen and nitrogen families have some obvious similarities and differences.
(a) State two general physical similarities between Group 5A(15) and 6A(16) elements.

(b) State two general chemical similarities between Group 5A(15) and 6A(16) elements.

(c) State two chemical similarities between P and S.

(d) State two physical similarities between N and O.

(e) State two chemical differences between N and O.

•14.110 A molecular property of the Group 6A(16) hydrides changes abruptly down the group. This change has been explained in terms of a change in orbital hybridization.

(a) Between what periods does the change occur?

(b) What is the change in the molecular property?

(c) What is the change in hybridization?

(d) What other group displays a similar change?

● **Skill-Building Exercises** (grouped in similar pairs)

••14.111 Complete and balance the following:

(a) $NaHSO_4(aq) + NaOH(aq) \longrightarrow$

(b) $S_8(s) + \text{excess } F_2(g) \longrightarrow$

(c) $FeS(s) + HCl(aq) \longrightarrow$

(d) $Te(s) + I_2(s) \longrightarrow$

••14.112 Complete and balance the following:

(a) $H_2S(g) + O_2(g) \longrightarrow$

(b) $SO_3(g) + H_2O(l) \longrightarrow$

(c) $SF_4(g) + H_2O(l) \longrightarrow$

(d) $Al_2Se_3(s) + H_2O(l) \longrightarrow$

•14.113 Is each oxide basic, acidic, or amphoteric in water: (a) SeO_2; (b) N_2O_3; (c) K_2O; (d) BeO; (e) BaO?

•14.114 Is each oxide basic, acidic, or amphoteric in water: (a) MgO; (b) N_2O_5; (c) CaO; (d) CO_2; (e) TeO_2?

•14.115 Rank the following hydrides in order of *increasing* acid strength: H_2S, H_2O, H_2Te.

•14.116 Rank the following species in order of *decreasing* acid strength: H_2SO_4, H_2SO_3, HSO_3^-.

● **Problems in Context**

••14.117 Describe the physical changes observed when solid sulfur is heated from room temperature to 440°C and then poured quickly into cold water. Explain the molecular changes responsible for the macroscopic changes.

•14.118 Give the name and symbol or formula of a Group 6A(16) element or compound that fits each description or use:

(a) Unreactive gas used as an electrical insulator

(b) Unstable allotrope of oxygen

(c) Oxide having sulfur with the same O.N. as in sulfuric acid

(d) Air pollutant produced by burning sulfur-containing coal

(e) Powerful dehydrating agent

(f) Compound used in solution in the photographic process

(g) Gas in trace amounts in air that tarnishes silver

•14.119 Give the oxidation state of sulfur in (a) S_8; (b) SF_4; (c) SF_6; (d) H_2S; (e) FeS_2; (f) H_2SO_4; (g) $Na_2S_2O_3 \cdot 5H_2O$.

•14.120 Disulfur decafluoride is intermediate in reactivity between SF_4 and SF_6. It disproportionates at 150°C to these monosulfur fluorides. Write a balanced equation for this reaction, and give the oxidation state of S in each compound.

Group 7A(17): The Halogens

● **Concept Review Questions**

•14.121 (a) Give the physical state and color of each halogen at STP. (b) Explain the change in physical state down Group 7A(17) in terms of molecular properties.

•14.122 (a) What are the common oxidation states of the halogens? (b) Give an explanation based on electron configuration for the range and values of the oxidation states of chlorine. (c) Why is fluorine an exception to the pattern of oxidation states found for the other group members?

•14.123 How many electrons does a halogen atom need to complete its octet? Give examples of the ways a Cl atom can do so.

•14.124 Select the stronger bond in each pair:

(a) Cl—Cl or Br—Br   (b) Br—Br or I—I

(c) F—F or Cl—Cl. Why doesn't the F—F bond strength follow the group trend?

••14.125 In addition to interhalogen compounds, many interhalogen ions exist. Would you expect interhalogen ions with a 1+ or a 1− charge to have an even or odd number of atoms? Explain.

••14.126 (a) A halogen (X_2) disproportionates in base in several steps to X^- and XO_3^-. Write the overall equation for the disproportionation of Br_2 to Br^- and BrO_3^-. (b) Write a balanced equation for the reaction of ClF_5 with aqueous base (see the reaction of BrF_5 shown on p. 614).

● **Skill-Building Exercises** (grouped in similar pairs)

••14.127 Complete and balance the following equations. If no reaction occurs, write NR:

(a) $Rb(s) + Br_2(l) \longrightarrow$ (b) $I_2(s) + H_2O(l) \longrightarrow$

(c) $Br_2(l) + I^-(aq) \longrightarrow$ (d) $CaF_2(s) + H_2SO_4(l) \longrightarrow$

••14.128 Complete and balance the following equations. If no reaction occurs, write NR:

(a) $H_3PO_4(l) + NaI(s) \longrightarrow$ (b) $Cl_2(g) + I^-(aq) \longrightarrow$

(c) $Br_2(l) + Cl^-(aq) \longrightarrow$ (d) $ClF(g) + F_2(g) \longrightarrow$

•14.129 Rank the following acids in order of *increasing* acid strength: $HClO$, $HClO_2$, $HBrO$, HIO.

•14.130 Rank the following acids in order of *decreasing* acid strength: $HBrO_3$, $HBrO_4$, HIO_3, $HClO_4$.

● **Problems in Context**

•14.131 Give the name and symbol or formula of a Group 7A(17) element or compound that fits each description or use:

(a) Used in etching glass

(b) Naturally occurring source (ore) of fluorine

(c) Oxide used in bleaching paper pulp and textiles

(d) Weakest hydrohalic acid

(e) Compound used to prevent goiter (thyroid disorder)

(f) Element that is produced in the largest quantity

(g) Organic chloride used to make plastics

••14.132 An industrial chemist treats solid NaCl with concentrated H_2SO_4 and obtains gaseous HCl and $NaHSO_4$. When she substitutes solid NaI for NaCl, gaseous H_2S, solid I_2, and S_8 are obtained but no HI. (a) What type of reaction did the H_2SO_4 undergo with NaI? (b) Why does NaI, but not NaCl, cause this type of reaction? (c) To produce HI(g) by the reaction of NaI with an acid, how does the acid have to differ from sulfuric acid?

•14.133 Rank the halogens Cl_2, Br_2, and I_2 in order of increasing oxidizing strength based on their products with metallic Re: $ReCl_6$, $ReBr_5$, ReI_4. Explain your ranking.

Group 8A(18): The Noble Gases

● **Concept Review Questions**

•14.134 Which noble gas is the most abundant in the universe? In Earth's atmosphere?

•14.135 What oxidation states does Xe show in its compounds?

•14.136 Why do the noble gases have such low boiling points?

14.137 Explain why Xe, and to a limited extent Kr, form compounds, whereas He, Ne, and Ar do not.

14.138 (a) Why do stable xenon fluorides have an even number of F atoms? (b) Why do the ionic species XeF_3^+ and XeF_7^- have odd numbers of F atoms? (c) Predict the shape of XeF_3^+.

Comprehensive Problems

14.139 Xenon tetrafluoride reacts with antimony pentafluoride to form the following ionic complex: $[XeF_3]^+[SbF_6]^-$. (a) Which of the following illustrates the molecular shapes of the reactants and product? (b) How, if at all, does the hybridization of xenon change in the reaction?

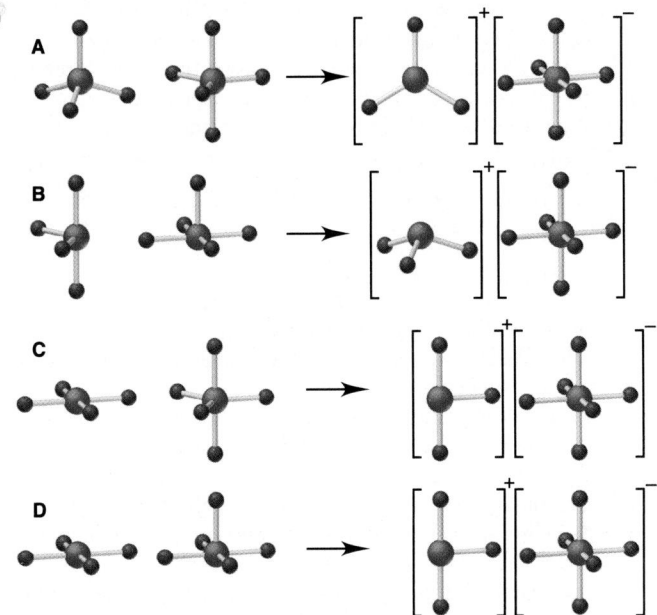

A

B

C

D

14.140 Given the following information,

$$H^+(g) + H_2O(g) \longrightarrow H_3O^+(g) \qquad \Delta H = -720 \text{ kJ}$$
$$H^+(g) + H_2O(l) \longrightarrow H_3O^+(aq) \qquad \Delta H = -1090 \text{ kJ}$$
$$H_2O(l) \longrightarrow H_2O(g) \qquad \Delta H = 40.7 \text{ kJ}$$

calculate the heat of solution of the hydronium ion:

$$H_3O^+(g) \xrightarrow{H_2O} H_3O^+(aq)$$

14.141 The electronic transition in Na from $3p^1$ to $3s^1$ gives rise to a bright yellow-orange emission at 589.2 nm. What is the energy of this transition?

14.142 Unlike other Group 2A(2) metals, beryllium reacts like aluminum and zinc with concentrated aqueous base to release hydrogen gas and form oxoanions of formula $M(OH)_4^{n-}$. Write equations for the reactions of these three metals with NaOH.

14.143 The interhalogen IF undergoes the reaction depicted below (I is purple and F is green):

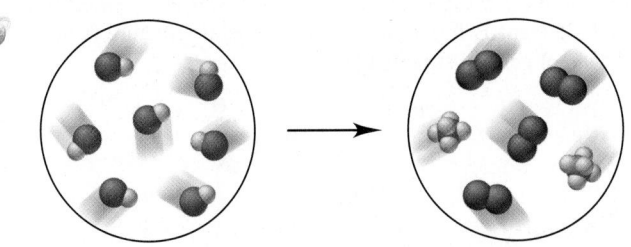

(a) Write the balanced equation. (b) Name the interhalogen product. (c) What type of reaction is shown? (d) If each molecule of IF represents 2.50×10^{-3} mol, what mass of each product forms?

14.144 Cyclopropane (C_3H_6), the smallest cyclic hydrocarbon, is reactive for the same reason white phosphorus is. Use bond properties and valence bond theory to explain its high reactivity.

14.145 Thionyl chloride ($SOCl_2$) is a sulfur oxohalide used industrially to dehydrate metal halide hydrates. (a) Write a balanced equation for its reaction with magnesium chloride hexahydrate, in which SO_2 and HCl form along with the metal halide. (b) Draw a Lewis structure of $SOCl_2$ with minimal formal charges.

14.146 The main reason alkali metal dihalides (MX_2) do *not* form is the high IE_2 of the metal. (a) Why is IE_2 so high for alkali metals? (b) The IE_2 for Cs is 2255 kJ/mol, low enough for CsF_2 to form exothermically ($\Delta H_f^\circ = -125$ kJ/mol). This compound cannot be synthesized, however, because CsF forms with a much greater release of heat ($\Delta H_f^\circ = -530$ kJ/mol). Thus, the breakdown of CsF_2 to CsF happens readily. Write the equation for this breakdown, and calculate the heat of reaction per mole of CsF.

14.147 Semiconductors made from elements in Groups 3A(13) and 5A(15) are typically prepared by direct reaction of the elements at high temperature. An engineer treats 32.5 g of molten gallium with 20.4 L of white phosphorus vapor at 515 K and 195 kPa. If purification losses are 7.2% by mass, how many grams of gallium phosphide will be prepared?

14.148 Two substances with empirical formula HNO are hyponitrous acid ($\mathcal{M} = 62.04$ g/mol) and nitroxyl ($\mathcal{M} = 31.02$ g/mol).
(a) What is the molecular formula of each species?
(b) For each species, draw the Lewis structure having the lowest formal charges. (*Hint:* Hyponitrous acid has an N=N bond.)
(c) Predict the shape around the N atoms of each species.
(d) When hyponitrous acid loses two protons, it forms the hyponitrite ion. Draw *cis* and *trans* forms of this ion.

14.149 The species CO, CN^-, and C_2^{2-} are isoelectronic.
(a) Draw their Lewis structures.
(b) Draw their MO diagrams (assume $2s$-$2p$ mixing, as in N_2), and give the bond order and electron configuration for each.

14.150 The Ostwald process is a series of three reactions used for the industrial production of nitric acid from ammonia.
(a) Write a series of balanced equations for the Ostwald process.
(b) If NO is *not* recycled, how many moles of NH_3 are consumed per mole of HNO_3 produced?
(c) In a typical industrial unit, the process is very efficient, with a 96% yield for the first step. Assuming 100% yields for the subsequent steps, what volume of concentrated aqueous nitric acid (60.% by mass; $d = 1.37$ g/mL) can be prepared for each cubic meter of a gas mixture that is 90.% air and 10.% NH_3 by volume at the industrial conditions of 5.0 atm and 850.°C?

14.151 Buckminsterfullerene (C_{60}) is a soccer-ball–shaped molecule with a C atom at each vertex. It crystallizes in a face-centered cubic unit cell that contains 240 C atoms and has a density of only 1.693 g/cm³, less than half that of diamond. Calculate the volume and edge length of the unit cell, and suggest a reason for the low density despite the closest packing.

14.152 Perhaps surprisingly, some pure liquid interhalogens that contain fluorine have high electrical conductivity. The explanation is that one molecule transfers a fluoride ion to another molecule. Write the equation that would explain the high electrical conductivity of bromine trifluoride.

•• **14.153** All common plant fertilizers contain nitrogen compounds. Determine the mass % of N in (a) ammonia; (b) ammonium nitrate; (c) ammonium hydrogen phosphate.

•• **14.154** Producer gas is a fuel formed by passing air over red-hot coke (amorphous carbon). What mass of a producer gas that consists of 25% CO, 5.0% CO_2, and 70.% N_2 by mass can be formed from 1.75 metric tons of coke, assuming an 87% yield?

•• **14.155** Gaseous F_2 reacts with water to form HF and O_2. In NaOH solution, F_2 forms F^-, water, and oxygen difluoride (OF_2), a highly toxic gas and powerful oxidizing agent. The OF_2 reacts with excess OH^-, forming O_2, water, and F^-. (a) For each reaction, write a balanced equation, give the oxidation state of O in all compounds, and identify the oxidizing and reducing agents. (b) Draw a Lewis structure for OF_2, and predict its shape.

•• **14.156** What is a disproportionation reaction, and which of the following fit the description?
(a) $I_2(s) + KI(aq) \longrightarrow KI_3(aq)$
(b) $2ClO_2(g) + H_2O(l) \longrightarrow HClO_3(aq) + HClO_2(aq)$
(c) $Cl_2(g) + 2NaOH(aq) \longrightarrow$
$$NaCl(aq) + NaClO(aq) + H_2O(l)$$
(d) $NH_4NO_2(s) \longrightarrow N_2(g) + 2H_2O(g)$
(e) $3MnO_4^{2-}(aq) + 2H_2O(l) \longrightarrow$
$$2MnO_4^-(aq) + MnO_2(s) + 4OH^-(aq)$$
(f) $3AuCl(s) \longrightarrow AuCl_3(s) + 2Au(s)$

•• **14.157** Explain the following observations:
(a) In reactions with Cl_2, phosphorus forms PCl_5 in addition to the expected PCl_3, but nitrogen forms only NCl_3.
(b) Carbon tetrachloride is unreactive toward water, but silicon tetrachloride reacts rapidly and completely. (To give what?)
(c) The sulfur-oxygen bond in SO_4^{2-} is shorter than expected for an S—O single bond.
(d) Chlorine forms ClF_3 and ClF_5, but ClF_4 is unknown.

•• **14.158** Which group(s) of the periodic table is (are) described by each of the following general statements?
(a) The elements form compounds of VSEPR class AX_3E.
(b) The free elements are strong oxidizing agents and form monatomic ions and oxoanions.
(c) The atoms form compounds by combining with two other atoms that donate one electron each.
(d) The free elements are strong reducing agents, show only one nonzero oxidation state, and form mainly ionic compounds.
(e) The elements can form stable compounds with only three bonds, but as a central atom, they can accept a pair of electrons from a fourth atom without expanding their valence shell.
(f) Only larger members of the group are chemically active.

•• **14.159** Diiodine pentaoxide (I_2O_5) was discovered by Joseph Gay-Lussac in 1813, but its structure was unknown until 1970! Like Cl_2O_7, it can be prepared by the dehydration-condensation of the corresponding oxoacid.
(a) Name the precursor oxoacid, write a reaction for formation of the oxide, and draw a likely Lewis structure.
(b) Data show that the bonds to the terminal O are shorter than the bonds to the bridging O. Why?
(c) I_2O_5 is one of the few chemicals that can oxidize CO rapidly and completely; elemental iodine forms in the process. Write a balanced equation for this reaction.

•• **14.160** An important starting material for the manufacture of polyphosphazenes is the cyclic molecule $(NPCl_2)_3$. The molecule has a symmetrical six-membered ring of alternating N and

P atoms, with the Cl atoms bonded to the P atoms. The nitrogen-phosphorus bond length is significantly less than that expected for an N—P single bond.
(a) Draw a likely Lewis structure for the molecule.
(b) How many lone pairs of electrons do the ring atoms have?
(c) What is the order of the nitrogen-phosphorus bond?

•• **14.161** Potassium fluotitanate (K_2TiF_6), a starting material for producing other titanium salts, is made by dissolving titanium(IV) oxide in concentrated HF, adding aqueous KF, and then heating to dryness to drive off excess HF. The impure solid is recrystallized by dissolving in hot water and then cooling to 0°C, which leaves excess KF in solution. For the reaction of 5.00 g of TiO_2, 250. mL of hot water is used to recrystallize the product. How many grams of purified K_2TiF_6 are obtained? (Solubility of K_2TiF_6 in water at 0°C is 0.60 g/100 mL.)

•• **14.162** Bromine monofluoride (BrF) disproportionates to bromine gas and bromine tri- and pentafluorides. Use the following to find ΔH°_{rxn} for the decomposition of BrF to its elements:
$$3BrF(g) \longrightarrow Br_2(g) + BrF_3(l) \quad \Delta H_{rxn} = -125.3 \text{ kJ}$$
$$5BrF(g) \longrightarrow 2Br_2(g) + BrF_5(l) \quad \Delta H_{rxn} = -166.1 \text{ kJ}$$
$$BrF_3(l) + F_2(g) \longrightarrow BrF_5(l) \quad \Delta H_{rxn} = -158.0 \text{ kJ}$$

•• **14.163** Account for the following facts:
(a) Ca^{2+} and Na^+ have very nearly the same radii.
(b) CaF_2 is insoluble in water, but NaF is quite soluble.
(c) Molten $BeCl_2$ is a poor electrical conductor, whereas molten $CaCl_2$ is an excellent one.

•• **14.164** Cake alum (aluminum sulfate) is used as a *flocculating agent* in water purification. The "floc" is a gelatinous precipitate of aluminum hydroxide that carries small suspended particles and bacteria out of solution for removal by filtration.
(a) The hydroxide is formed by the reaction of the aluminum ion with water. Write the equation for this reaction.
(b) The acidity from this reaction is partially neutralized by the sulfate ion. Write an equation for this reaction.

•• **14.165** Carbon tetrachloride is made by passing Cl_2 in the presence of a catalyst through liquid CS_2 near its boiling point. Disulfur dichloride (S_2Cl_2) also forms, in addition to the byproducts SCl_2 and CCl_3SCl. How many grams of CS_2 reacted if excess Cl_2 yielded 50.0 g of CCl_4 and 0.500 g of CCl_3SCl? Is it possible to determine how much S_2Cl_2 and SCl_2 also formed? Explain.

•• **14.166** H_2 may act as a reducing agent or an oxidizing agent, depending on the substance reacting with it. Using, in turn, sodium or chlorine as the other reactant, write balanced equations for the two reactions, and characterize the redox role of hydrogen.

•• **14.167** P_4 is prepared by heating phosphate rock [principally $Ca_3(PO_4)_2$] with sand and coke:
$$Ca_3(PO_4)_2(s) + SiO_2(s) + C(s) \longrightarrow$$
$$CaSiO_3(s) + CO(g) + P_4(g) \text{ [unbalanced]}$$
How many kilograms of phosphate rock are needed to produce 315 mol of P_4, assuming that the conversion is 90.% efficient?

•• **14.168** Element E forms an oxide of general structure A and a chloride of general structure B:

For the anion EF_5^-, what is (a) the molecular shape; (b) the hybridization of E; (c) the O.N. of E?

•• **14.169** From its formula, one might expect CO to be quite polar, but its dipole moment is actually low (0.11 D). (a) Draw the Lewis structure for CO. (b) Calculate the formal charges. (c) Based on your answers to parts (a) and (b), explain why the dipole moment is so low.

••• **14.170** In addition to Al_2Cl_6, aluminum forms other species with bridging halide ions to two aluminum atoms. One such species is the ion $Al_2Cl_7^-$. The ion is symmetrical, with a 180° Al—Cl—Al bond angle.
(a) What orbitals does Al use to bond with the Cl atoms?
(b) What is the shape around each Al?
(c) What is the hybridization of the central Cl?
(d) What do the shape and hybridization suggest about the presence of lone pairs of electrons on the central Cl?

•• **14.171** When an alkaline earth carbonate is heated, it releases CO_2, leaving the metal oxide. The temperature at which each Group 2A(2) carbonate yields a CO_2 partial pressure of 1 atm is

Carbonate	Temperature (°C)
$MgCO_3$	542
$CaCO_3$	882
$SrCO_3$	1155
$BaCO_3$	1360

(a) Suggest a reason for this trend. (b) Mixtures of $CaCO_3$ and MgO are used to absorb dissolved silicates from boiler water. How would you prepare a mixture of $CaCO_3$ and MgO from dolomite, which contains $CaCO_3$ and $MgCO_3$?

•• **14.172** The bond angles in the nitrite ion, nitrogen dioxide, and the nitronium ion (NO_2^+) are 115°, 134°, and 180°, respectively. Explain these values using Lewis structures and VSEPR theory.

•• **14.173** A common method for producing a gaseous hydride is to treat a salt containing the anion of the volatile hydride with a strong acid.
(a) Write an equation for each of the following examples: (1) the production of HF from CaF_2; (2) the production of HCl from NaCl; (3) the production of H_2S from FeS.
(b) In some cases even a weak acid such as water will suffice if the anion of the salt has a sufficiently strong attraction for protons. An example is the production of PH_3 from Ca_3P_2 and water. Write the equation for this reaction.
(c) By analogy, predict the products and write the equation for the reaction of Al_4C_3 with water.

•• **14.174** Chlorine trifluoride was formerly used in the production of uranium hexafluoride for the U.S. nuclear industry:

$$U(s) + 3ClF_3(l) \longrightarrow UF_6(l) + 3ClF(g)$$

How many grams of UF_6 can form from 1.00 metric ton of uranium ore that is 1.55% by mass uranium and 12.75 L of chlorine trifluoride ($d = 1.88$ g/mL)?

••• **14.175** Chlorine is used to make bleach solutions containing 5.25% NaClO (by mass). Assuming 100% yield in the reaction producing NaClO from Cl_2, how many liters of $Cl_2(g)$ at STP will be needed to make 1000. L of bleach solution ($d = 1.07$ g/mL)?

•• **14.176** The triatomic molecular ion H_3^+ was first detected and characterized by J. J. Thomson using mass spectrometry. Use the bond energy of H_2 (432 kJ/mol) and the proton affinity of H_2 ($H_2 + H^+ \longrightarrow H_3^+$; $\Delta H = -337$ kJ/mol) to calculate the heat of reaction for

$$H + H + H^+ \longrightarrow H_3^+$$

•• **14.177** An atomic hydrogen torch is used for cutting and welding thick sheets of metal. When H_2 passes through an electric arc, the molecules decompose into atoms, which react with O_2 at the end of the torch. Temperatures in excess of 5000°C are reached, which can melt all metals. Write equations for the breakdown of H_2 to H atoms and for the subsequent overall reaction of the H atoms with oxygen. Use Appendix B to find the standard heat of each reaction per mole of product.

•• **14.178** In aqueous hydrogen fluoride solution, an important species is the ion HF_2^-, which has the bonding arrangement FHF^-. Draw the Lewis structure for this ion, and explain how it arises.

•• **14.179** Which of the following oxygen ions are paramagnetic: O^+, O^-, O^{2-}, O^{2+}?

••• **14.180** Copper(II) hydrogen arsenite ($CuHAsO_3$) is a green pigment once used in wallpaper. In damp conditions, mold metabolizes this compound to trimethylarsine [$(CH_3)_3As$], a highly toxic gas. (a) Calculate the mass percent of As in each compound. (b) How much $CuHAsO_3$ must react to reach a toxic level in a room that measures 12.35 m × 7.52 m × 2.98 m (arsenic is toxic at 0.50 mg/m³)?

••• **14.181** Hydrogen peroxide can act as either an oxidizing agent or a reducing agent. (a) When H_2O_2 is treated with aqueous KI, I_2 forms. In which role is H_2O_2 acting? What is the oxygen-containing product formed? (b) When H_2O_2 is treated with aqueous $KMnO_4$, the purple color of MnO_4^- disappears and a gas forms. In which role is H_2O_2 acting? What is the oxygen-containing product formed?

15

Herbal Essence This colored scanning electron micrograph of a freeze-fractured sage leaf reveals the round, reddish oil glands that secrete the numerous organic compounds responsible for the herb's taste and odor. Ultimately, the behavior of a living system is derived from that of its organic chemicals, which, as you'll see in this chapter, emerges from the behavior of the carbon atom.

Organic Compounds and the Atomic Properties of Carbon

15.1 The Special Nature of Carbon and the Characteristics of Organic Molecules
Structural Complexity
Chemical Diversity

15.2 The Structures and Classes of Hydrocarbons
Carbon Skeletons and Hydrogen Skins
Alkanes
Constitutional Isomers
Optical Isomers
Alkenes and Geometric Isomers
Alkynes

Aromatic Hydrocarbons
Catenated Inorganic Hydrides

15.3 Some Important Classes of Organic Reactions
Types of Organic Reactions
Organic Redox Reactions

15.4 Properties and Reactivities of Common Functional Groups
Groups with Only Single Bonds
Groups with Double Bonds
Groups with Single and Double Bonds
Groups with Triple Bonds

15.5 The Monomer-Polymer Theme I: Synthetic Macromolecules
Addition Polymers
Condensation Polymers

15.6 The Monomer-Polymer Theme II: Biological Macromolecules
Sugars and Polysaccharides
Amino Acids and Proteins
Nucleotides and Nucleic Acids

Organic chemistry application (entire chapter covers essential topics in organic chemistry, with many biological applications)

The most remarkable chemical system known is a living cell. Through delicately controlled mechanisms, it oxidizes food for energy, maintains the concentrations of thousands of aqueous components, interacts continuously with its environment, synthesizes both simple and complex molecules, and even reproduces itself! For all our technological prowess, no human-made system even approaches the cell in its complexity and sheer elegance of function.

This amazing chemical machine consumes, creates, and consists largely of *organic compounds*. Except for a few inorganic salts and ever-present water, nearly everything you put into or on your body—food, medicine, cosmetics, and clothing—consists of organic compounds. Organic fuels warm our homes, cook our meals, and power our society. Major industries are devoted to producing organic compounds, such as polymers, pharmaceuticals, and insecticides.

What *is* an organic compound? Dictionaries define it as "a compound of carbon," but that definition is too general because it includes carbonates, cyanides, carbides, cyanates, and other carbon-containing ionic compounds that most chemists classify as inorganic. Here is a more specific definition: all **organic compounds** contain carbon, nearly always bonded to other carbons and hydrogen, and often to other elements.

The word *organic* has a biological connotation arising from a major misconception that stifled research into the chemistry of living systems for many decades. In the early 19th century, many prominent thinkers believed that an unobservable spiritual energy, a "vital force," existed *within* the compounds of living things, making them impossible to synthesize and fundamentally different from compounds of the mineral world. This idea of *vitalism* was challenged in 1828, when the young German chemist Friedrich Wöhler heated ammonium cyanate, a "mineral-world" compound, and produced urea, a "living-world" compound:

$$NH_4OCN \xrightarrow{\Delta} H_2N-\overset{\overset{\textstyle O}{\|}}{C}-NH_2$$

Although Wöhler did not appreciate the significance of this reaction—he was more interested in the fact that two compounds can have the same molecular formula—his experiment is considered a key event in the origin of organic chemistry. Chemists soon synthesized methane, acetic acid, acetylene, and many other organic compounds from inorganic sources. ● Today, we know that *the same chemical principles govern organic and inorganic systems* because the behavior of a compound arises from the properties of its elements, no matter how marvelous that behavior may be.

IN THIS CHAPTER . . . We apply the text's central theme to the enormous field of organic chemistry to see how the structure and reactivity of organic molecules emerge naturally from the properties of their component atoms. First, we review the special atomic properties of carbon and see how they lead to the complex structure and reactivity of organic molecules. Next, we focus on writing and naming hydrocarbons as a prelude to naming other types of organic compounds. We classify the main types of organic reactions and apply them to various families of organic compounds. Finally, we extend these ideas to the giant molecules of commerce and life—synthetic and natural polymers.

15.1 THE SPECIAL NATURE OF CARBON AND THE CHARACTERISTICS OF ORGANIC MOLECULES

Although there is nothing mystical about organic molecules, their indispensable role in biology and industry leads us to ask if carbon has some extraordinary attributes that give it a special chemical "personality." Of course, each element

Concepts & Skills to Review
before you study this chapter

- naming simple organic compounds; the functional-group concept (Section 2.8)
- constitutional isomerism (Section 3.2)
- ΔEN and bond polarity (Section 9.5)
- resonance structures (Section 10.1)
- VSEPR theory (Section 10.2)
- orbital hybridization (Section 11.1)
- σ and π bonding (Section 11.2)
- types of intermolecular forces and the shapes of biological macromolecules (Sections 12.3, 12.7, and 13.2)
- properties of the Period 2 elements (Section 14.2)
- properties of the Group 4A(14) elements (Section 14.6)

urea

● **Organic Chemistry Is Enough to Drive One Mad** The vitalists did not change their beliefs overnight. Indeed, organic compounds *do* seem different from inorganic compounds because of their complex structures and compositions. Imagine the consternation of the 19th-century chemist who was accustomed to studying compounds with formulas such as $CaCO_3$, $Ba(NO_3)_2$, and $CuSO_4 \cdot 5H_2O$ and then isolated white crystals from a gallstone and found their empirical formula to be $C_{27}H_{46}O$ (cholesterol). Even Wöhler later said, "Organic chemistry . . . is enough to drive one mad. It gives me the impression of a primeval tropical forest, full of the most remarkable things, a monstrous and boundless thicket, with no way of escape, into which one may well dread to enter."

ARIS Presentation Center, Art Library

has its own specific properties, and carbon is no more unique than sodium, hafnium, or any other element. But the atomic properties of carbon do give it bonding capabilities beyond those of any other element, which in turn lead to the two obvious characteristics of organic molecules—structural complexity and chemical diversity.

The Structural Complexity of Organic Molecules

Most organic molecules have much more complex structures than most inorganic molecules, and a quick review of carbon's atomic properties and bonding behavior shows why:

1. *Electron configuration, electronegativity, and covalent bonding.* Carbon's ground-state electron configuration of [He] $2s^2 2p^2$—four electrons more than He and four fewer than Ne—means that the formation of carbon ions is energetically impossible under ordinary conditions: to form a C^{4+} cation requires energy equal to the sum of IE_1 through IE_4, and to form a C^{4-} anion requires the sum of EA_1 through EA_4, the last three steps of which are endothermic. Carbon's position in the periodic table (Figure 15.1) and its electronegativity are midway between the most metallic and the most nonmetallic elements of Period 2: Li = 1.0, C = 2.5, F = 4.0. Therefore, *carbon shares electrons to attain a filled outer (valence) level,* bonding covalently in all its elemental forms and compounds.

2. *Bond properties, catenation, and molecular shape.* The *number* and *strength* of carbon's bonds lead to its outstanding ability to *catenate* (form chains of atoms), which allows it to form a multitude of chemically and thermally stable chain, ring, and branched compounds. Through the process of orbital hybridization (Section 11.1), *carbon forms four bonds in virtually all its compounds,* and they point in as many as four different directions. The small size of carbon allows close approach to another atom and thus greater orbital overlap, meaning that *carbon forms relatively short, strong bonds.* The C—C bond is short enough to allow side-to-side overlap of half-filled, unhybridized *p* orbitals and the formation of *multiple bonds,* which restrict rotation of attached groups (see Figure 11.13, p. 421). These features add more possibilities for the molecular shapes of carbon compounds.

3. *Molecular stability.* Although silicon and several other elements also catenate, none can compete with carbon. Atomic and bonding properties confer three crucial differences between C and Si chains that explain why C chains are so stable and, therefore, so common:

- *Atomic size and bond strength.* As atomic size increases down Group 4A(14), bonds between identical atoms become longer and weaker. Thus, a C—C bond (347 kJ/mol) is much stronger than an Si—Si bond (226 kJ/mol).
- *Relative heats of reaction.* A C—C bond (347 kJ/mol) and a C—O bond (358 kJ/mol) have nearly the same energy, so relatively little heat is released when a C chain reacts and one bond replaces the other. In contrast, an Si—O bond (368 kJ/mol) is much stronger than an Si—Si bond (226 kJ/mol), so a large quantity of heat is released when an Si chain reacts.
- *Orbitals available for reaction.* Unlike C, Si has low-energy *d* orbitals that can be attacked (occupied) by the lone pairs of incoming reactants. Thus, for example, ethane (CH_3—CH_3) is stable in water and does not react in air unless sparked, whereas disilane (SiH_3—SiH_3) breaks down in water and ignites spontaneously in air.

Figure 15.1 The position of carbon in the periodic table. Lying at the center of Period 2, carbon has an intermediate electronegativity (EN), and its position at the top of Group 4A(14) means it is relatively small. Other elements common in organic compounds are H, N, O, P, S, and the halogens.

ARIS Presentation Center, Art Library

The Chemical Diversity of Organic Molecules

In addition to their elaborate geometries, organic compounds are noted for their sheer number and diverse chemical behavior. Look, for example, at the *CRC Handbook of Chemistry and Physics,* and you'll find that the number of organic compounds of carbon dwarfs the number of inorganic compounds of all the elements combined! Several million organic compounds are known, and thousands more are discovered or synthesized each year.

This incredible diversity is also founded on atomic and bonding behavior and is due to three interrelated factors:

1. *Bonding to heteroatoms.* Many organic compounds contain **heteroatoms,** atoms other than C or H. The most common heteroatoms are N and O, but S, P, and the halogens often occur, and organic compounds with other elements are known as well. Figure 15.2 shows that 23 different molecular structures are possible from various arrangements of four C atoms singly bonded to each other, the necessary number of H atoms, and just one O atom (either singly or doubly bonded).

2. *Electron density and reactivity.* Most reactions start—that is, a new bond begins to form—*when a region of high electron density on one molecule meets a region of low electron density on another.* These regions may be due to the presence of a multiple bond or to the partial charges that occur in carbon-heteroatom bonds. For example, consider four bonds commonly found in organic molecules:

- *The C—C bond.* When C is singly bonded to another C, as occurs in portions of nearly every organic molecule, the EN values are equal and the bond is nonpolar. Therefore, in general, *C—C bonds are unreactive.*
- *The C—H bond.* This bond, which also occurs in nearly every organic molecule, is very nearly nonpolar because it is short (109 pm) and the EN values of H (2.1) and C (2.5) are close. Thus, *C—H bonds are largely unreactive* as well.
- *The C—O bond.* This bond, which occurs in many types of organic molecules, is highly polar ($\Delta EN = 1.0$), with the O end of the bond electron rich and the C end electron poor. As a result of this imbalance in electron density, *the C—O bond is reactive* and, given appropriate conditions, a reaction will occur there.
- *Bonds to other heteroatoms.* Even when a carbon-heteroatom bond has a small ΔEN, such as that for C—Br ($\Delta EN = 0.3$), or none at all, as for C—S ($\Delta EN = 0$), heteroatoms like these are large, and so their bonds to carbon are long, weak, and thus reactive.

3. *Nature of functional groups.* One of the most important ideas in organic chemistry is that of the **functional group,** a specific combination of bonded atoms that reacts in a *characteristic* way, no matter what molecule it occurs in (we mentioned this idea briefly in Section 2.8). In nearly every case, *the reaction of an organic compound takes place at the functional group.* In fact, as you'll see, we often substitute a general symbol for the remainder of the molecule because it usually stays the same while the functional group reacts. Functional groups vary from carbon-carbon multiple bonds to several combinations of carbon-heteroatom bonds, and each has its own pattern of reactivity. A particular bond may be a functional group itself or *part* of one or more functional groups. For example, the C—O bond occurs in four functional groups. We will discuss the reactivity of three of these groups in this chapter:

   alcohol group   carboxylic acid group   ester group

Figure 15.2 The chemical diversity of organic compounds. Different arrangements of chains, branches, rings, and heteroatoms give rise to many structures. There are 23 different compounds possible from just four C atoms joined by single bonds, one O atom, and the necessary H atoms.

Transparency/ARIS Presentation Center, Art Library

Section Summary

The structural complexity of organic compounds arises from carbon's small size, intermediate EN, four valence electrons, ability to form multiple bonds, and absence of *d* orbitals in the valence level. These factors lead to chains, branches, and rings of C atoms joined by strong, chemically resistant bonds that point in as many as four directions from each C. • The chemical diversity of organic compounds arises from carbon's ability to bond to many other elements, including O and N, which creates polar bonds and greater reactivity. These factors lead to compounds that contain functional groups, specific portions of molecules that react in characteristic ways.

15.2 THE STRUCTURES AND CLASSES OF HYDROCARBONS

A fanciful, anatomical analogy can be made between an organic molecule and an animal. The carbon-carbon bonds form the skeleton: the longest continual chain is the backbone, and any branches are the limbs. Covering the skeleton is a skin of hydrogen atoms, with functional groups protruding at specific locations, like chemical fingers ready to grab an incoming reactant.

In this section, we "dissect" one group of compounds down to their skeletons and see how to name and draw them. **Hydrocarbons,** the simplest type of organic compound, are a large group of substances containing only H and C atoms. Some common fuels, such as natural gas and gasoline, are hydrocarbon mixtures. Hydrocarbons are also important *feedstocks,* precursor reactants used to make other compounds. Ethylene, acetylene, and benzene, for example, are feedstocks for hundreds of other substances.

Carbon Skeletons and Hydrogen Skins

Let's begin by examining the possible bonding arrangements of C atoms only (we'll leave off the H atoms at first) in simple skeletons without multiple bonds or rings. To distinguish different skeletons, focus on the *arrangement* of C atoms (that is, the successive linkages of one to another) and keep in mind that *groups joined by single (sigma) bonds are relatively free to rotate* (Section 11.2).

Structures with one, two, or three carbons can be arranged in only one way. Whether you draw three C atoms in a line or with a bend, the arrangement is the same. Four C atoms, however, have two possible arrangements—a four-C chain or a three-C chain with a one-C branch at the central C:

Even when we show the chain more realistically, with the bends due to the tetrahedral shape around the C atoms (as in the ball-and-stick models), the situation is the same. Notice that if the branch is added to either end of the three-C chain, it is simply a bend in a four-C chain, *not* a different arrangement. Similarly, if the branch points down instead of up, it represents the same arrangement because groups joined by single bonds rotate.

As the total number of C atoms increases, the number of different arrangements increases as well. Five C atoms have 3 possible arrangements; 6 C atoms can be arranged in 5 ways, 7 C atoms in 9 ways, 10 C atoms in 75 ways, and

Figure 15.3 Some five-carbon skeletons. A, Three five-C skeletons are possible with only single bonds. **B,** Five more skeletons are possible with one C=C bond present. **C,** Five more skeletons are possible with one ring present. Even more would be possible with a ring *and* a double bond.

20 C atoms in more than 300,000 ways! If we include multiple bonds and rings, the number of arrangements increases further. For example, including one C=C bond in the five-C skeletons creates 5 more arrangements, and including one ring creates 5 more (Figure 15.3).

When determining the number of different skeletons for a given number of C atoms, remember that

- Each C atom can form a *maximum* of four single bonds, or two single and one double bond, or one single and one triple bond.
- The *arrangement* of C atoms determines the skeleton, so a straight chain and a bent chain represent the same skeleton.
- Groups joined by single bonds can *rotate,* so a branch pointing down is the same as one pointing up. (Recall that a double bond restricts rotation.)

If we put a hydrogen "skin" on a carbon skeleton, we obtain a hydrocarbon. Figure 15.4 shows that the skeleton has the correct number of H atoms when each C has four bonds. Sample Problem 15.1 provides practice drawing hydrocarbons.

A C atom single-bonded to one other atom gets three H atoms.

A C atom single-bonded to two other atoms gets two H atoms.

A C atom single-bonded to three other atoms gets one H atom.

A C atom single-bonded to four other atoms is already fully bonded (no H atoms).

A double-bonded C atom is treated as if it were bonded to two other atoms.

A double- and single-bonded C atom or a triple-bonded C atom is treated as if it were bonded to three other atoms.

Figure 15.4 Adding the H-atom skin to the C-atom skeleton. In a hydrocarbon molecule, each carbon atom bonds to as many hydrogen atoms as needed to give the carbon a total of four bonds.

SAMPLE PROBLEM 15.1 Drawing Hydrocarbons

PROBLEM Draw structures that have different atom arrangements for hydrocarbons with:
(a) Six C atoms, no multiple bonds, and no rings
(b) Four C atoms, one double bond, and no rings
(c) Four C atoms, no multiple bonds, and one ring
PLAN In each case, we draw the longest carbon chain first and then work down to smaller chains with branches at different points along them. The process typically involves trial and error. Then, we add H atoms to give each C a total of four bonds.
SOLUTION **(a)** Compounds with six C atoms:

6-C chain:

5-C chains:

4-C chains:

(b) Compounds with four C atoms and one double bond:

4-C chains: 3-C chain:

(c) Compounds with four C atoms and one ring:

4-C ring: 3-C ring:

CHECK Be sure each skeleton has the correct number of C atoms, multiple bonds, and/or rings, and no arrangements are repeated or omitted; remember a double bond counts as two bonds.
COMMENT Avoid some *common mistakes:*
In (a): C—C—C—C—C is the same skeleton as C—C—C—C—C

C—C—C—C is the same skeleton as C—C—C—C

In (b): C—C—C=C is the same skeleton as C=C—C—C
The double bond restricts rotation, so, in addition to the *cis* form shown in part (b), another possibility is the *trans* form:

(We discuss *cis-trans* isomers fully later in this section.)

Also, there are too many bonds to one C in

$$\text{5 bonds}$$
$$H-\overset{\overset{\displaystyle H}{|}}{\underset{\underset{\displaystyle H}{|}}{C}}-\overset{\displaystyle H}{\underset{\displaystyle \overset{|}{C}\!-\!H}{C}}=\overset{\displaystyle H}{C}-H$$
$$\overset{\displaystyle H-\overset{|}{C}-H}{\underset{\displaystyle H}{|}}$$

In (c): Too many bonds to one C in

$$\text{5 bonds}$$
$$H-\overset{\overset{\displaystyle H}{|}}{\underset{\underset{\displaystyle H}{|}}{C}}-\overset{\overset{\displaystyle H\ H}{\diagdown\diagup}}{\underset{\underset{\displaystyle H}{|}}{C}}-\overset{\overset{\displaystyle H}{|}}{\underset{\underset{\displaystyle H}{|}}{C}}-H$$

FOLLOW-UP PROBLEM 15.1 Draw all hydrocarbons that have different atom arrangements with:
(a) Seven C atoms, no multiple bonds, and no rings (nine arrangements)
(b) Five C atoms, one triple bond, and no rings (three arrangements)

Hydrocarbons can be classified into four main groups. In the remainder of this section, we examine the names and some structural features and physical properties of each group. Later, we discuss the chemical behavior of the hydrocarbons.

Alkanes: Hydrocarbons with Only Single Bonds

A hydrocarbon that contains only single bonds is an **alkane** (general formula C_nH_{2n+2}, where n is a positive integer). For example, if $n = 5$, the formula is $C_5H_{[(2\times5)+2]}$, or C_5H_{12}. The alkanes comprise a **homologous series,** one in which each member differs from the next by a $-CH_2-$ (methylene) group. In an alkane, each C is sp^3 hybridized. Because each C is bonded to the *maximum number of other atoms* (C or H), alkanes are referred to as **saturated hydrocarbons.**

Naming Alkanes You learned how to name simple alkanes in Section 2.8. Here we discuss general rules for naming any alkane and, by extension, other organic compounds as well. The key point is that *each chain, branch, or ring has a name based on the **number** of C atoms*. The name of a compound has three portions:

PREFIX + ROOT + SUFFIX

- *Root:* The root tells the number of C atoms in the longest *continuous* chain in the molecule. The roots for the ten smallest alkanes are shown in Table 15.1. As you can see, there are special roots for compounds of one to four C atoms; roots of longer chains are based on Greek numbers.
- *Suffix:* The suffix tells the *type of organic compound* the molecule represents; that is, it identifies the key functional group the molecule possesses. The suffix is placed *after* the root.
- *Prefix:* Each prefix identifies a *group attached to the main chain* and the number of the carbon to which it is attached. Prefixes identifying hydrocarbon branches are the same as root names (Table 15.1) but have *-yl* as their ending. Each prefix is placed *before* the root.

For example, in the name 2-methylbutane, *2-methyl-* is the prefix (a one-carbon branch is attached to C-2 of the main chain), *-but-* is the root (the main chain has four C atoms), and *-ane* is the suffix (the compound is an alkane).

To obtain the systematic name of a compound,

1. Name the longest chain (root).
2. Add the compound type (suffix).
3. Name any branches (prefix).

Table 15.1 Numerical Roots for Carbon Chains and Branches

Roots	Number of C Atoms
meth-	1
eth-	2
prop-	3
but-	4
pent-	5
hex-	6
hept-	7
oct-	8
non-	9
dec-	10

 ARIS Presentation Center, Tables Library

Table 15.2 Rules for Naming an Organic Compound

1. Naming the longest chain (root)
   (a) Find the longest *continuous* chain of C atoms.
   (b) Select the root that corresponds to the number of C atoms in this chain.

2. Naming the compound type (suffix)
   (a) For alkanes, add the suffix *-ane* to the chain root. (Other suffixes appear in Table 15.5 with their functional group and compound type.)
   (b) If the chain forms a ring, the name is preceded by *cyclo-*.

3. Naming the branches (prefixes) (If the compound has no branches, the name consists of the root and suffix.)
   (a) Each branch name consists of a subroot (number of C atoms) and the ending *-yl* to signify that it is not part of the main chain.
   (b) Branch names precede the chain name. When two or more branches are present, their names appear in *alphabetical* order.
   (c) To specify where the branch occurs along the chain, number the main-chain C atoms consecutively, starting at the end *closer* to a branch, to achieve the *lowest* numbers for the branches. Precede each branch name with the number of the main-chain C to which that branch is attached.

6 carbons $\Longrightarrow$ hex-

hex- + -ane $\Longrightarrow$ hexane

CH_3 methyl

CH_3—CH—CH—CH_2—CH_2—CH_3

CH_2—CH_3 ethyl

ethylmethylhexane

3-ethyl-2-methylhexane

Table 15.2 presents the rules for naming any organic compound and applies them to an alkane component of gasoline. Other organic compounds are named with a variety of other prefixes and suffixes (see Table 15.5, p. 651). In addition to these *systematic* names, we'll also note important *common* names still in use.

Depicting Alkanes with Formulas and Models Chemists have several ways to depict organic compounds. Expanded, condensed, and carbon-skeleton formulas are easy to draw; ball-and-stick and space-filling models show the actual shapes.

The *expanded formula* shows each atom and bond. One type of *condensed formula* groups each C atom with its H atoms. *Carbon-skeleton formulas* show only carbon-carbon bonds and appear as zig-zag lines, often with branches. *Each end or bend of a zig-zag line or branch represents a C atom attached to the number of H atoms that gives it a total of four bonds:*

propane CH_3—CH_2—CH_3

2,3-dimethylbutane CH_3—CH—CH—CH_3 (with CH_3 CH_3 branches)

Figure 15.5 Ways of depicting an alkane.

Figure 15.5 shows these types of formulas, together with ball-and-stick and space-filling models, of the compound named in Table 15.2.

| Expanded formula | Condensed formula | Carbon-skeleton formula | Ball-and-stick model | Space-filling model |

ARIS Presentation Center, Art Library

A Cyclopropane   B Cyclobutane   C Cyclopentane   D Cyclohexane

Figure 15.6 Depicting cycloalkanes. Cycloalkanes are usually drawn as regular polygons. Each side is a C—C bond, and each corner represents a C atom with its required number of H atoms. The expanded formulas show each bond in the molecule. The ball-and-stick and space-filling models show that, except for cyclopropane, the rings are not planar. These conformations minimize electron repulsions between adjacent H atoms. Cyclohexane **(D)** is shown in its more stable chair conformation.

Cyclic Hydrocarbons A **cyclic hydrocarbon** contains one or more rings in its structure. When a straight-chain alkane (C_nH_{2n+2}) forms a ring, two H atoms are lost as the C—C bond forms to join the two ends of the chain. Thus, *cycloalkanes* have the general formula C_nH_{2n}. Cyclic hydrocarbons are often drawn with carbon-skeleton formulas, as shown at the top of Figure 15.6. Except for three-carbon rings, *cycloalkanes are nonplanar*. This structural feature arises from the tetrahedral shape around each C atom and the need to minimize electron repulsions between adjacent H atoms. As a result, orbital overlap of adjacent C atoms is maximized. The most stable form of cyclohexane, called the *chair conformation,* is shown in Figure 15.6D.

Constitutional Isomerism and the Physical Properties of Alkanes

Recall from Section 3.2 that two or more compounds with the same molecular formula but different properties are called **isomers.** Those with *different arrangements of bonded atoms* are **constitutional (or structural) isomers;** alkanes with the same number of C atoms but different skeletons are examples. The smallest alkane to exhibit constitutional isomerism has four C atoms: two different compounds have the formula C_4H_{10}, as shown in Table 15.3, on page 638. The unbranched one is butane (common name, *n*-butane; *n*- stands for "normal," or having a straight chain), and the other is 2-methylpropane (common name, *iso*butane). Similarly, three compounds have the formula C_5H_{12} (shown in Table 15.3). The unbranched isomer is pentane (common name, *n*-pentane); the one with a

Table 15.3	The Constitutional Isomers of C_4H_{10} and C_5H_{12}				
Systematic Name (Common Name)	**Expanded Formula**	**Condensed and Skeleton Formulas**	**Space-Filling Model**	**Density (g/mL)**	**Boiling Point (°C)**
Butane (*n*-butane)		$CH_3-CH_2-CH_2-CH_3$		0.579	−0.5
2-Methylpropane (isobutane)		$CH_3-CH-CH_3$ $\quad\quad CH_3$		0.549	−11.
Pentane (*n*-pentane)		$CH_3-CH_2-CH_2-CH_2-CH_3$		0.626	36.1
2-Methylbutane (isopentane)		$CH_3-CH-CH_2-CH_3$ $\quad\quad CH_3$		0.620	27.8
2,2-Dimethylpropane (neopentane)		CH_3-C-CH_3 with CH_3 above and below		0.614	9.5

methyl group at C-2 of a four-C chain is 2-methylbutane (common name, *iso*pentane). The third isomer has two methyl branches on C-2 of a three-C chain, so its name is 2,2-dimethylpropane (common name, *neo*pentane).

Because alkanes are nearly nonpolar, we expect their physical properties to be determined by dispersion forces, and the boiling points in Table 15.3 certainly bear this out. The four-C alkanes boil lower than the five-C compounds. Moreover, within each group of isomers, the more spherical member (isobutane or neopentane) boils lower than the more elongated one (*n*-butane or *n*-pentane). As you saw in Chapter 12, this trend occurs because a spherical shape leads to less intermolecular contact, and thus lower total dispersion forces, than does an elongated shape.

A particularly clear example of the effect of dispersion forces on physical properties occurs among the unbranched alkanes (*n*-alkanes). Among these compounds, boiling points increase steadily with chain length: the longer the chain, the greater the intermolecular contact, the stronger the dispersion forces, and the higher the boiling point (Figure 15.7). Pentane (five C atoms) is the smallest *n*-alkane that exists as a liquid at room temperature. The solubility of alkanes, and of all hydrocarbons, is easy to predict from the like-dissolves-like rule (Section 13.1). Alkanes are miscible in each other and in other nonpolar solvents, such as benzene, but are nearly insoluble in water. The solubility of pentane in water, for example, is only 0.36 g/L at room temperature.

methane (CH_4; $\mathcal{M}$ = 16.04)	CH_4	−164
ethane (C_2H_6; $\mathcal{M}$ = 30.07)	CH_3CH_3	−89
propane (C_3H_8; $\mathcal{M}$ = 44.09)	$CH_3CH_2CH_3$	−42
butane (C_4H_{10}; $\mathcal{M}$ = 58.12)	$CH_3CH_2CH_2CH_3$	−0.5
pentane (C_5H_{12}; $\mathcal{M}$ = 72.15)	$CH_3CH_2CH_2CH_2CH_3$	36
hexane (C_6H_{14}; $\mathcal{M}$ = 86.17)	$CH_3CH_2CH_2CH_2CH_2CH_3$	69
heptane (C_7H_{16}; $\mathcal{M}$ = 100.20)	$CH_3CH_2CH_2CH_2CH_2CH_2CH_3$	98
octane (C_8H_{18}; $\mathcal{M}$ = 114.22)	$CH_3CH_2CH_2CH_2CH_2CH_2CH_2CH_3$	126
nonane (C_9H_{20}; $\mathcal{M}$ = 128.25)	$CH_3CH_2CH_2CH_2CH_2CH_2CH_2CH_2CH_3$	151
decane ($C_{10}H_{22}$; $\mathcal{M}$ = 142.28)	$CH_3CH_2CH_2CH_2CH_2CH_2CH_2CH_2CH_2CH_3$	174

Temperature (°C): −273   −200   −100   0   100   200

Figure 15.7 Boiling points of the first 10 unbranched alkanes. Boiling point increases smoothly with chain length because dispersion forces increase. Each entry includes the name, molecular formula, molar mass ($\mathcal{M}$, in g/mol), condensed formula, space-filling model, and boiling point at 1 atm pressure.

ARIS Presentation Center, Art Library

Chiral Molecules and Optical Isomerism

Another type of isomerism exhibited by some alkanes and many other organic (as well as some inorganic) compounds is called *stereoisomerism.* **Stereoisomers** are molecules with the same arrangement of atoms *but different orientations of groups in space.* *Optical isomerism* is one type of stereoisomerism: *when two objects are mirror images of each other and cannot be superimposed, they are* **optical isomers,** also called *enantiomers.* To use a familiar example, your right hand is an optical isomer of your left. Look at your right hand in a mirror, and you will see that the *image* is identical to your left hand (Figure 15.8). No matter how you twist your arms around, however, your hands cannot lie on top of each other with all parts superimposed. They are not superimposable because each is *asymmetric:* there is no plane of symmetry that divides your hand into two identical parts.

An asymmetric molecule is called **chiral** (Greek *cheir,* "hand"). Typically, an organic molecule *is chiral if it contains a carbon atom that is bonded to four **different** groups.* This C atom is called a *chiral center* or an asymmetric carbon. In 3-methylhexane, for example, C-3 is a chiral center because it is bonded to four different groups: H—, CH_3—, CH_3—CH_2—, and CH_3—CH_2—CH_2— (Figure 15.9A). Like your two hands, the two forms are mirror images and cannot be superimposed on each other: when two of the groups are superimposed, the other two are opposite each other. Thus, the two forms are optical isomers. The central C atom in the amino acid alanine is also a chiral center (Figure 15.9B).

Figure 15.8 An analogy for optical isomers. The reflection of your right hand looks like your left hand. Each hand is asymmetric, so you cannot superimpose them with your palms facing in the same direction.

Literature Yee, Gordon T. "Through the Looking Glass and What Alice Ate There," *J. Chem. Educ.* **2002,** *79,* 569.

A Optical isomers of 3-methylhexane

B Optical isomers of alanine

Figure 15.9 Two chiral molecules. A, 3-Methylhexane is chiral because C-3 is bonded to four different groups. These two models are optical isomers (enantiomers). **B,** The central C in the amino acid alanine is also bonded to four different groups.

ARIS Presentation Center, Art Library

Figure 15.10 The rotation of plane-polarized light by an optically active substance. The source emits light oscillating in all planes. When the light passes through the first polarizing filter, light oscillating in only one plane emerges. The plane-polarized light enters the sample compartment, which contains a known concentration of an optical isomer, and the plane rotates as the light passes through the solution. For the experimenter to see the light, it must pass through a second polarizing filter (called the *analyzer*), which is attached to a movable ring that is calibrated in degrees and measures the angle of rotation.

ARIS Presentation Center, Art Library

Literature Luján-Upton, Hannia. "Introducing Stereochemistry to Non-science Majors," *J. Chem. Educ.* **2001**, *78*, 475.

ARIS Presentation Center, Art Library

Chiral center

● **Chiral Medicines** Many drugs are chiral molecules of which one optical isomer is biologically active and the other has either a different type of activity or none at all. Naproxen, the pain reliever and anti-inflammatory agent, is an example; one isomer (see the model) is active as an anti-arthritic agent, and the other is a potent liver toxin that must be removed from the mixture during synthesis. The notorious drug thalidomide is another example. One optical isomer is active against depression, whereas the other causes fetal mutations and deaths. Tragically, the drug was sold in the 1950s as the racemic mixture and caused limb malformations in many children whose mothers had it prescribed to relieve "morning sickness" during pregnancy.

Biological application

Unlike constitutional isomers, optical isomers are identical in all but two respects:

1. In their physical properties, *optical isomers differ only in the direction that each isomer rotates the plane of polarized light.* A **polarimeter** is used to measure the angle that the plane is rotated (Figure 15.10). A beam of light consists of waves that oscillate in all planes. A polarizing filter blocks all waves except those in one plane, so the light emerging through the filter is *plane-polarized.* An optical isomer is **optically active** because it rotates the plane of this polarized light. (Liquid crystal displays incorporate polarizing filters and optically active compounds; see Figure 12.45, p. 481.) The *dextrorotatory* isomer (designated *d* or +) rotates the plane of light clockwise; the *levorotatory* isomer (designated *l* or −) is the mirror image of the *d* isomer and rotates the plane counterclockwise. An equimolar mixture of the two isomers (called a *racemic mixture*) does not rotate the plane at all because the dextrorotation cancels the levorotation. The *specific rotation* is a characteristic, measurable property of the isomer at a certain temperature, concentration, and wavelength of light.

2. In their chemical properties, *optical isomers differ only in a chiral (asymmetric) chemical environment,* one that distinguishes "right-handed" from "left-handed" molecules. As an analogy, your right hand fits well in your right glove but not in your left glove. Typically, one isomer of an optically active reactant is added to a mixture of optical isomers of another compound. The products of the reaction have different properties and can be separated. For example, when the anion *d*-lactate is added to a mixture of the cationic forms *d*-alanine and *l*-alanine, the *d* isomer of alanine crystallizes as the *d*-lactate/*d*-alanine salt, but the *d*-lactate/*l*-alanine salt is more soluble and remains in solution.

Optical isomerism plays a vital role in living cells. Nearly all carbohydrates and amino acids are optically active, but only one of the isomers is biologically usable. For example, *d*-glucose is metabolized for energy, but *l*-glucose is excreted unused. Similarly, *l*-alanine is incorporated naturally into proteins, but *d*-alanine is not. ● The organism distinguishes one optical isomer from the other through its enzymes, large molecules that speed virtually every reaction in the cell by binding to the reactants. A specific portion of the enzyme provides the asymmetric environment, so only one of the isomers can bind there and undergo reaction (Figure 15.11).

Alkenes: Hydrocarbons with Double Bonds

A hydrocarbon that contains at least one C=C bond is called an **alkene.** With two H atoms removed to make the double bond, alkenes have the general formula C_nH_{2n}. The double-bonded C atoms are sp^2 hybridized. Because their carbon atoms are bonded to fewer than the maximum of four atoms each, alkenes are considered **unsaturated hydrocarbons.**

Alkene names differ from those of alkanes in two respects:

1. The main chain (root) *must* contain both C atoms of the double bond, even if it is not the longest chain. The chain is numbered from the end *closer* to the C=C bond, and the position of the bond is indicated by the number of the *first* C atom in it.

2. The suffix for alkenes is *-ene*.

For example, there are three four-C alkenes (C_4H_8), two unbranched and one branched (see Sample Problem 15.1b). The branched isomer is 2-methylpropene; the unbranched isomer with the C=C bond between C-1 and C-2 is 1-butene; the unbranched isomer with the C=C bond between C-2 and C-3 is 2-butene. As you'll see next, there are two isomers of 2-butene, but they are of a different sort (discussed briefly in Sections 10.3 and 11.2).

The C=C Bond and Geometric (*cis-trans*) Isomerism There are two major structural differences between alkenes and alkanes. First, alkanes have a *tetrahedral* geometry (bond angles of ~109.5°) around each C atom, whereas the double-bonded C atoms in alkenes are *trigonal planar* (~120°). Second, the C—C bond *allows* rotation of bonded groups, so the atoms in an alkane continually change their relative positions. In contrast, the π bond of the C=C bond *restricts* rotation, which fixes the relative positions of the atoms bonded to it.

This rotational restriction leads to another type of stereoisomerism. **Geometric isomers** (also called ***cis-trans*** **isomers**) have different orientations of groups around a double bond (or similar structural feature). Table 15.4 shows the two geometric isomers of 2-butene (also see Comment, Sample Problem 15.1). One isomer, *cis*-2-butene, has the CH_3 groups on the *same* side of the C=C bond, whereas the other isomer, *trans*-2-butene, has them on *opposite* sides of the C=C bond. In general, the *cis* isomer has the *larger portions of the main chain* (in this case, two CH_3 groups) *on the same side* of the double bond, and the *trans* isomer has them on opposite sides. (*Cis-trans* isomerism also occurs in many transition metal compounds, which have a different structural feature that restricts rotation, as we discuss in Chapter 23.) For a molecule to have geometric isomers, *each C atom in the C=C bond must be bonded to two different groups.* Like other isomers, geometric isomers have different properties. Note in Table 15.4 that the two 2-butenes differ in molecular shape *and* physical properties. The *cis* isomer has a bend in the chain that the *trans* isomer lacks. In Chapters 10 and 12, you saw how such a difference affects molecular polarity and physical properties, which arise from differing strengths of intermolecular attractions. The upcoming Chemical Connections essay shows how this simple difference in geometry has profound effects in biological systems as well.

Figure 15.11 The binding site of an enzyme. Organisms can utilize only one of a pair of optical isomers because their enzymes have binding sites with shapes that are chiral (asymmetric). The shape of the mirror image of this molecule does not allow it to bind at this site.

ARIS Presentation Center, Art Library

Biological application

Demo Hart, Harold. "'Dishing Out' Stereochemical Principles," *J. Chem. Educ.* **2001,** *78,* 1632.

Literature Haudrechy, Arnaud. "A Simple Method of Drawing Stereoisomers from Complicated Symmetrical Structures," *J. Chem. Educ.* **2000,** *77,* 864.

Table 15.4	**The Geometric Isomers of 2-Butene**			
Systematic Name	**Condensed and Skeleton Formulas**	**Space-Filling Model**	**Density (g/mL)**	**Boiling Point (°C)**
cis-2-Butene			0.621	3.7
trans-2-Butene			0.604	0.9

Chemical Connections to Sensory Physiology

Geometric Isomers and the Chemistry of Vision

Of all our senses, sight provides the most information about the external world. Light bouncing off objects enters the lens of the eye and is focused on the retina. From there, molecular signals are converted to mechanical signals and then to electrical signals that are transmitted to the brain. This remarkable sequence is one of the few physiological processes that we understand thoroughly at the molecular level. The first step relies on the different shapes of geometric isomers.

The molecule responsible for receiving the light energy is *retinal.* It is derived from retinol (vitamin A), which we obtain mostly from β-carotene in yellow and green vegetables. Retinal is a 20-C compound consisting of a 15-C chain and five 1-C branches. As you can see from the ball-and-stick model in Figure B15.1, the chain includes five C=C bonds, a six-C ring at one end, and a C=O bond at the other. There are two biologically occurring isomers of retinal. The all-*trans* isomer, shown on the right, has a *trans* orientation around all five double bonds. The 11-*cis* isomer, shown on the left, has a *cis* orientation around the C=C bond between C-11 and C-12. Note the significant difference in shape between the two isomers.

Certain cells of the retina are densely packed with *rhodopsin,* which consists of the relatively small 11-*cis*-retinal covalently bonded to a large protein. The initial chemical event in vision occurs when rhodopsin absorbs a photon of visible light. The energy of visible photons (between 165 and 293 kJ/mol) lies in the range needed to break a C=C π bond (~250 kJ/mol). Retinal is bonded to the protein in such a way that the 11-*cis* π bond is the most susceptible to breakage. The photon is absorbed in a few trillionths of a second, the *cis* π bond breaks, the groups attached to the intact σ bond rotate, and a π bond re-forms to produce all-*trans*-retinal in another few millionths of a second. In effect, light energy is converted into the mechanical energy of the moving molecular chain.

This rapid and relatively large change in the shape of retinal causes the protein portion of rhodopsin to change shape as well—a process that breaks the bond between the protein portion and retinal. The change in protein shape triggers a flow of ions into the retina cells, initiating electrical impulses to the optic nerve, which conducts them to the brain. Meanwhile, the free all-*trans*-retinal diffuses away and is changed back to the *cis* form, which then binds to the protein portion again.

Retinal must be ideally suited to its function because it has been selected through evolution as a photon absorber in organisms as different as purple bacteria, mollusks, insects, and vertebrates. The features that make it the perfect choice are its strong absorption in the visible region, the efficiency with which light converts the *cis* to the *trans* form, and the large structural change it undergoes when this takes place.

Figure B15.1 The initial chemical event in vision. In rhodopsin, the geometric isomer 11-*cis*-retinal is bonded to (*shown here as lying on*) a large protein. A photon absorbed by the retinal breaks the 11-*cis* π bond. The two parts of the retinal chain rotate around the σ bond (*shown in circle*), and the π bond forms again but with a *trans* orientation. This change in the shape of retinal changes the shape of the protein, causing it to release the all-*trans*-retinal.

 ARIS Presentation Center, Art Library

Alkynes: Hydrocarbons with Triple Bonds

Hydrocarbons that contain at least one C≡C bond are called **alkynes.** Their general formula is C_nH_{2n-2} because they have two H atoms fewer than alkenes with the same number of carbons. Because a carbon in a C≡C bond can bond to only one other atom, the geometry around each C atom is linear (180°): each C is *sp* hybridized. Alkynes are named in the same way as alkenes, except that the suffix is *-yne*. Because of their localized π electrons, C=C and C≡C bonds are electron rich and act as functional groups. Thus, alkenes and alkynes are much more reactive than alkanes, as we'll discuss in Section 15.4.

 ARIS Presentation Center, Worked Examples Library

SAMPLE PROBLEM 15.2 Naming Alkanes, Alkenes, and Alkynes

PROBLEM Give the systematic name for each of the following, indicate the chiral center in part (d), and draw two geometric isomers for part (e):

(a)
$$CH_3-\underset{\underset{CH_3}{|}}{\overset{\overset{CH_3}{|}}{C}}-CH_2-CH_3$$

(b)
$$CH_3-CH_2-\underset{\underset{\underset{CH_3}{|}}{CH_2}}{\overset{\overset{CH_3}{|}}{CH}}-CH-CH_3$$

(c)

(d)
$$CH_3-CH_2-\underset{\underset{CH_3}{|}}{CH}-CH=CH_2$$

(e)
$$CH_3-CH_2-CH=\underset{\underset{CH_3}{|}}{C}-CH-CH_3$$

PLAN For (a) to (c), we refer to Table 15.2, p. 636. We first name the longest chain (*root-* + *-ane*). Then we find the *lowest* branch numbers by counting C atoms from the end *closer* to a branch. Finally, we name each branch (*root-* + *-yl*) and put the names alphabetically before the chain name. For (d) and (e), the longest chain that *includes* the multiple bond is numbered from the end closer to it. For (d), the chiral center is the C atom bonded to four different groups. In (e), the *cis* isomer has larger groups on the same side of the double bond, and the *trans* isomer has them on opposite sides.

SOLUTION

(a)

2,2-dimethylbutane

When a type of branch appears more than once, we group the chain numbers and indicate the number of branches with a numerical prefix, such as 2,2-*dimethyl*.

(b)

3,4-dimethylhexane

In this case, we can number the chain from either end because the branches are the same and are attached to the two central C atoms.

(c)

1-ethyl-2-methylcyclopentane

We number the ring C atoms so that a branch is attached to C-1.

(d)

3-methyl-1-pentene

(e)

cis-2,3-dimethyl-3-hexene

trans-2,3-dimethyl-3-hexene

CHECK A good check (and excellent practice) is to reverse the process by drawing structures for the names to see if you come up with the structures given in the problem.

COMMENT In part (b), C-3 and C-4 are chiral centers, as are C-1 and C-2 in part (c). However, in (b) the molecule is not chiral: it has a plane of symmetry between C-3 and C-4, so half of the molecule rotates light opposite to the other half. Avoid these common mistakes: In (b), 2-ethyl-3-methylpentane is wrong: the longest chain is *hexane*. In (c), 1-methyl-2-ethylcyclopentane is wrong: the branch names appear *alphabetically*.

FOLLOW-UP PROBLEM 15.2 Draw condensed formulas for the following compounds: **(a)** 3-ethyl-3-methyloctane; **(b)** 1-ethyl-3-propylcyclohexane (also draw a carbon-skeleton formula for this compound); **(c)** 3,3-diethyl-1-hexyne; **(d)** *trans*-3-methyl-3-heptene.

Aromatic Hydrocarbons: Cyclic Molecules with Delocalized π Electrons

Unlike the cycloalkanes, **aromatic hydrocarbons** are planar molecules, usually with one or more rings of six C atoms, and are often drawn with alternating single and double bonds. As you learned for benzene (Section 10.1), however, all the ring bonds are identical, with values of length and strength *between* those of a C—C and a C=C bond. To indicate this, benzene is also shown as a resonance hybrid, with a circle (or dashed circle) representing the delocalized character of the π electrons (Figure 15.12A). An orbital picture shows the two lobes of the delocalized π cloud above and below the hexagonal plane of the σ-bonded C atoms (Figure 15.12B).

Figure 15.12 Representations of benzene. A, Benzene is often drawn with two resonance forms, showing alternating single and double bonds in different positions. The molecule is more correctly depicted as the resonance hybrid, with the delocalized electrons shown as an unbroken or dashed circle. **B,** The delocalized π cloud and the σ-bond plane of the benzene ring.

ARIS Presentation Center, Art Library

The systematic naming of simple aromatic compounds is quite straightforward. Usually, benzene is the parent compound, and attached groups, or *substituents,* are named as prefixes. However, many common names are still in use. For example, benzene with one methyl group attached is systematically named *methylbenzene* but is better known by its common name, *toluene.* With only one substituent present in toluene, we do not number the ring C atoms; when two or more groups are attached, however, we number in such a way that one of the groups is attached to ring C-1. Thus, toluene and the three structural isomers with two methyl groups attached are

methylbenzene
(toluene)
bp = 110.6°C

1,2-dimethylbenzene
(*o*-xylene)
bp = 144.4°C

1,3-dimethylbenzene
(*m*-xylene)
bp = 139.1°C

1,4-dimethylbenzene
(*p*-xylene)
bp = 138.3°C

ARIS Presentation Center, Art Library

In common names, the positions of two groups are indicated by *o-* (ortho) for groups on adjacent ring C atoms, *m-* (meta) for groups separated by one ring C atom, and *p-* (para) for groups on opposite ring C atoms. The dimethylbenzenes (commonly known as *xylenes*) are important solvents and feedstocks for polyester fibers and dyes. (See the margin note on page 645.) ●

The number of isomers increases with more than two attached groups. For example, there are six isomers for a compound with one methyl and three nitro

(—NO$_2$) groups attached to a benzene ring; the explosive TNT, shown below, is only one:

2,4,6-trinitromethylbenzene
(**trin**itro**t**oluene, TNT)

One of the most important methods for determining the structures of organic molecules is discussed in the upcoming Tools of the Laboratory essay.

Variations on a Theme: Catenated Inorganic Hydrides

In short discussions called *Variations on a Theme* found throughout this chapter, we examine similarities between organic and inorganic compounds. From this perspective, you'll see that the behavior of carbon is remarkable, but not unique, in the chemistry of the elements.

Although no element approaches carbon in the variety and complexity of its hydrides, catenation occurs frequently in the periodic table, and many ring, chain, and cage structures are known. Some of the most fascinating belong to the boron hydrides, or boranes (Section 14.5). Although their shapes rival even those of the hydrocarbons, the weakness of their unusual bridge bonds renders most of them thermally and chemically unstable.

An obvious structural similarity exists between alkanes and the silicon hydrides, or silanes (Section 14.6). Silanes even have an analogous general formula (Si$_n$H$_{2n+2}$). Branched silanes are also known, but no cyclic or unsaturated (containing an Si=Si bond) compounds had been prepared until very recently. Unlike alkanes, silanes are unstable thermally and ignite spontaneously in air.

Sulfur's ability to catenate is second only to carbon's, and many chains and rings occur among its allotropes (Section 14.8). A large series of sulfur hydrides, or polysulfanes, is known. However, these molecules are unbranched chains with H atoms at the ends only (H—S$_n$—H). Like the silanes, the polysulfanes are oxidized easily and decompose readily to sulfur's only stable hydride, H$_2$S, and its most stable allotrope, cyclo-S$_8$.

Section Summary

Hydrocarbons contain only C and H atoms, so their physical properties depend on the strength of their dispersion forces. • Names of organic compounds have a root for the longest chain, a prefix for any attached group, and a suffix for the type of compound. • Alkanes (C$_n$H$_{2n+2}$) have only single bonds. Cycloalkanes (C$_n$H$_{2n}$) have ring structures that are typically nonplanar. Alkenes (C$_n$H$_{2n}$) have at least one C=C bond. Alkynes (C$_n$H$_{2n-2}$) have at least one C≡C bond. Aromatic hydrocarbons have at least one planar ring with delocalized π electrons.

Isomers are compounds with the same molecular formula but different properties. • Structural isomers have different atom arrangements. • Stereoisomers (optical and geometric) have the same arrangement of atoms, but their atoms are oriented differently in space. Optical isomers cannot be superimposed on each other because they are asymmetric, with four different groups bonded to the C that is the chiral center. They have identical physical and chemical properties except in their rotation of plane-polarized light and their reaction with chiral reactants. Geometric (*cis-trans*) isomers have groups oriented differently around a C=C bond, which restricts rotation. • Light converts a *cis* isomer of retinal to the all-*trans* form, which initiates the visual response. • ^{1}H-NMR spectroscopy indicates the relative numbers of H atoms in the various environments within an organic molecule.

Naphthalene

Benzo[*a*]pyrene

● **Aromatic Carcinogens** *Polycyclic aromatic compounds* have "fused" ring systems, in which two or more rings share one or more sides. The simplest is naphthalene, a feedstock for dyes. Many of these compounds (and benzene itself) have been shown to have carcinogenic (cancer-causing) activity. One of the most potent, benzo[*a*]pyrene, is found in soot, cigarette smoke, car exhaust, and even the smoke from barbecue grills!

 Environmental application

Tools of the Laboratory

Nuclear Magnetic Resonance (NMR) Spectroscopy

Chemists rely on a battery of instrumental methods to determine the complex structures of organic molecules. We discussed the methods of mass spectrometry in Chapter 2 and infrared (IR) spectroscopy in Chapter 9. One of the most useful tools for analyzing organic and biochemical structures is **nuclear magnetic resonance (NMR) spectroscopy,** which measures environments of certain nuclei in a molecule to elucidate its structure.

In Chapter 8, we saw that an electron can spin in either of two directions, each of which creates a tiny magnetic field. Several types of nuclei behave in a similar way. The most important of these are ^1H, ^{13}C, ^{19}F, and ^{31}P. In this discussion we focus primarily on the proton, ^1H, the nucleus of the most common isotope of hydrogen, and therefore refer to ^1H-NMR spectroscopy. Generally, the magnetic fields of all the protons in a sample of compound are oriented randomly. When placed in a strong external magnetic field (B_0), however, the proton fields become aligned with it (parallel) or against it (antiparallel). The parallel orientation is slightly lower in energy, with the energy difference (ΔE) between the two energy states (spin states) lying in the radio-frequency (rf) region of the spectrum.

The protons in the sample oscillate rapidly between the two spin states, and in a process known as *resonance,* the rf value is varied until it matches the frequency of this oscillation. At this point, the protons are "in resonance" with the rf radiation and they absorb and re-emit the energy, which is detected by the rf receiver of the ^1H-NMR spectrometer (Figure B15.2). If all the protons in a sample required the same ΔE for resonance, a ^1H-NMR spectrum would have only one peak and be useless. However, the ΔE between the two states depends on the *actual* magnetic field felt by each proton, which is affected by the tiny magnetic fields of the *electrons* of atoms adjacent to that proton. Thus, the ΔE of each proton depends on the electrons in the adjacent atoms—C atoms, electronegative atoms, multiple bonds, and aromatic rings—in

Figure B15.2 The basis of proton spin resonance. The randomly oriented magnetic fields of protons in a sample become aligned in a strong external field (B_0), with slightly more protons in the lower spin state. In a process called *resonance,* the spin flips when the proton absorbs a photon with energy equal to ΔE (radio-frequency region).

other words, on the specific molecular environment. Therefore, the ^1H-NMR spectrum is *unique* to that compound.

The ^1H-NMR spectrum of a compound is a series of peaks that represents the resonance of each proton as a function of the changing magnetic field. The *chemical shift* of the protons in a given environment is where a peak appears; it represents the ratio of the frequency at which resonance occurs for that proton to the magnitude of the external magnetic field. The chemical shifts are shown relative to that of an added standard, tetramethylsilane [$(CH_3)_4Si$, TMS], which has 12 protons bonded to four C atoms that are bonded to one Si atom. Because the shape of TMS is tetrahedral around Si, all 12 protons are in identical environments and so produce one peak. Figure B15.3 shows the ^1H-NMR spectrum of acetone. The six protons of acetone also have identical environments—bonded to two C atoms that are each bonded to the C atom in a $C=O$ bond—so they, too, produce one peak, but at a different position from that for TMS. (The axes need not concern us.) The spectrum of dimethoxymethane in Figure B15.4 shows *two* peaks in addition to the TMS peak. The taller one is due

---

15.3 SOME IMPORTANT CLASSES OF ORGANIC REACTIONS

In Chapter 4, we classified chemical reactions based on the chemical process involved (precipitation, acid-base, or redox) and then briefly included a classification based on the number of reactants and products (combination, decomposition, or displacement). We take a similar approach here with organic reactions.

From here on, we use the notation of an uppercase R with a single bond, R—, to signify a general organic group attached to one of the atoms shown; you can usually picture R— as an **alkyl group,** a saturated hydrocarbon chain with one bond available to link to another atom. Thus, R—CH_2—Br has an alkyl group attached to a CH_2 group bearing a Br atom; R—$CH=CH_2$ is an alkene with an alkyl group attached to one of the carbons in the double bond; and so forth. (Often, when more than one R group is present, we write R, R′, R″, and so forth, to indicate that these groups may be different.)

Types of Organic Reactions

Most organic reactions are examples of three broad types that can be identified by comparing the *number of bonds to C* in reactants and products:

Figure B15.3 The ¹H-NMR spectrum of acetone. The six CH_3 protons are in identical environments, so they produce one peak; it has a different chemical shift from that for the 12 protons of the TMS standard.

ARIS Presentation Center, Art Library

Figure B15.4 The ¹H-NMR spectrum of dimethoxymethane. The two peaks show chemical shifts of protons in two different environments. The area under each peak is proportional to the number of protons.

to the six CH_3 protons, and the shorter is due to the two CH_2 protons. The area under each peak (given here in units of chart-paper spaces) is proportional to *the number of protons in a given environment*. Note that the area ratio is 20.3/6.8 ≈ 3/1, the same as the ratio of six CH_3 protons to two CH_2 protons. Thus, by analyzing the chemical shifts and peak areas, the chemist learns the type and number of hydrogens in the compound.

Newer NMR methods measure the resonance of protons and other nuclei and have many applications in biochemistry and medicine. In fact, Kurt Wüthrich shared the 2002 Nobel Prize in chemistry for developing NMR methods to study biopolymer structures in solution. ¹³C-NMR is used to monitor changes in protein and nucleic acid shape and function, and ³¹P-NMR can determine the health of various organs. For example, the extent of damage from a heart attack can be learned by using ³¹P-NMR to measure the concentrations of specific phosphate-containing molecules involved in energy utilization by cardiac muscle tissue.

Computer-aided magnetic resonance imaging (MRI) has allowed visualization of damage to organs and greatly assists physicians in medical diagnosis. For example, an MRI scan of the head (Figure B15.5) can show various levels of metabolic activity in different regions of the brain. The 2003 Nobel Prize in physiology or medicine was awarded to Paul C. Lauterbur and Peter Mansfield for their discoveries applying magnetic resonance to imaging of organs and other body parts.

Figure B15.5 Magnetic resonance imaging (MRI) of a brain tumor

ARIS Presentation Center, Photo Library

Biological application

1. An **addition reaction** occurs when an unsaturated reactant becomes a saturated product:

$$R—CH{=}CH—R \ + \ X—Y \longrightarrow R—\overset{\overset{X}{|}}{C}H—\overset{\overset{Y}{|}}{C}H—R$$

Note the C atoms are bonded to *more* atoms in the product than in the reactant.

The C=C and C≡C bonds and the C=O bond commonly undergo addition reactions. In each case, the π bond breaks, leaving the σ bond intact. In the product, the two C atoms (or C and O) form two additional σ bonds. Let's examine the standard heat of reaction (ΔH°_{rxn}) for a typical addition reaction to see why these reactions occur. Consider the reaction between ethene (common name, ethylene) and HCl:

$$CH_2{=}CH_2 \ + \ H—Cl \longrightarrow H—CH_2—CH_2—Cl$$

Reactants (bonds broken)	Product (bonds formed)
1 C=C = 614 kJ	1 C—C = −347 kJ
4 C—H = 1652 kJ	5 C—H = −2065 kJ
1 H—Cl = 427 kJ	1 C—Cl = −339 kJ
Total = 2693 kJ	Total = −2751 kJ

$\Delta H^\circ_{rxn} = \Sigma \Delta H^\circ_{bonds\ broken} + \Sigma \Delta H^\circ_{bonds\ formed} = 2693\ kJ + (-2751\ kJ) = -58\ kJ$

A   **B**

Figure 15.13 A color test for C=C bonds. A, Br$_2$ (in pipet) reacts with a compound that has a C=C bond (in beaker), and its orange-brown color disappears:

B, The compound in this beaker has no C=C bond, so the Br$_2$ does not react, and the orange-brown color remains.

ARIS Presentation Center, Photo Library

The reaction is exothermic. By looking at the *net* change in bonds, we see that the driving force for many additions is the formation of two σ bonds (in this case, C—H and C—Cl) from one σ bond (in this case, H—Cl) and one relatively weak π bond. An addition reaction is the basis of a color test for the presence of C=C bonds (Figure 15.13).

2. **Elimination reactions** are the opposite of addition reactions. They occur when a saturated reactant becomes an unsaturated product:

$$\overset{\overset{\displaystyle Y}{|}}{R{-}CH}{-}\overset{\overset{\displaystyle X}{|}}{CH_2} \longrightarrow R{-}CH{=}CH_2 \ + \ X{-}Y$$

Note that the C atoms are bonded to *fewer* atoms in the product than in the reactant. A pair of halogen atoms, an H atom and a halogen atom, or an H atom and an —OH group are typically eliminated, but C atoms are not. Thus, the driving force for many elimination reactions is the formation of a small, stable molecule, such as HCl(g) or H$_2$O, which increases the entropy of the system (Section 13.3):

$$CH_3{-}\overset{\overset{\displaystyle OH}{|}}{CH}{-}\overset{\overset{\displaystyle H}{|}}{CH_2} \xrightarrow{H_2SO_4} CH_3{-}CH{=}CH_2 \ + \ H{-}OH$$

3. A **substitution reaction** occurs when an atom (or group) from an added reagent substitutes for one in the organic reactant:

$$\overset{|}{\underset{|}{R{-}C}}{-}X \ + \ :Y \longrightarrow \overset{|}{\underset{|}{R{-}C}}{-}Y \ + \ :X$$

Note that the C atom is bonded to the *same number* of atoms in the product as in the reactant. The C atom may be saturated or unsaturated, and X and Y can be many different atoms, but generally *not* C. The main flavor ingredient in banana oil, for instance, forms through a substitution reaction; note that the O substitutes for the Cl:

$$CH_3{-}\overset{\overset{\displaystyle O}{\|}}{C}{-}Cl \ + \ HO{-}CH_2{-}CH_2{-}\overset{\overset{\displaystyle CH_3}{|}}{CH}{-}CH_3 \longrightarrow$$

$$CH_3{-}\overset{\overset{\displaystyle O}{\|}}{C}{-}O{-}CH_2{-}CH_2{-}\overset{\overset{\displaystyle CH_3}{|}}{CH}{-}CH_3 \ + \ H{-}Cl$$

ARIS Presentation Center, Worked Examples Library

SAMPLE PROBLEM 15.3 Recognizing the Type of Organic Reaction

PROBLEM State whether each reaction is an addition, elimination, or substitution:

(a) CH$_3$—CH$_2$—CH$_2$—Br $\longrightarrow$ CH$_3$—CH=CH$_2$ + HBr

(b) + H$_2$ $\longrightarrow$

(c) CH$_3$$\overset{\overset{\displaystyle O}{\|}}{C}$—Br + CH$_3CH_2$OH $\longrightarrow$ CH$_3$$\overset{\overset{\displaystyle O}{\|}}{C}$—OCH$_2CH_3$ + HBr

PLAN We determine the type of reaction by looking for any change in the number of atoms bonded to C:

• More atoms bonded to C is an *addition*.
• Fewer atoms bonded to C is an *elimination*.
• Same number of atoms bonded to C is a *substitution*.

SOLUTION

(a) Elimination: two bonds in the reactant, C—H and C—Br, are absent in the product, so fewer atoms are bonded to C.

(b) Addition: two more C—H bonds have formed in the product, so more atoms are bonded to C.

(c) Substitution: the reactant C—Br bond becomes a C—O bond in the product, so the same number of atoms are bonded to C.

FOLLOW-UP PROBLEM 15.3 Write a balanced equation for each of the following:
(a) An addition reaction between 2-butene and Cl_2
(b) A substitution reaction between CH_3—CH_2—CH_2—Br and OH^-
(c) The elimination of H_2O from $(CH_3)_3C$—OH

The Redox Process in Organic Reactions

An important process in many organic reactions is *oxidation-reduction.* But chemists do not usually monitor the change in oxidation numbers of the various C atoms in a reaction. Rather, they note the movement of electron density around a C atom by counting the number of bonds to more electronegative atoms (usually O) or to less electronegative atoms (usually H). A more electronegative atom takes some electron density from the C, whereas a less electronegative atom gives some electron density to the C. Moreover, even though a redox reaction always involves both an oxidation and a reduction, organic chemists typically *focus on the organic reactant only.* Therefore,

• When a C atom in the organic reactant forms more bonds to O or fewer bonds to H, the reactant is oxidized and the reaction is called an *oxidation.*
• When a C atom in the organic reactant forms fewer bonds to O or more bonds to H, the reactant is reduced and the reaction is called a *reduction.*

The most dramatic redox reactions are combustion reactions. All organic compounds contain C and H atoms and burn in excess O_2 to form CO_2 and H_2O. For ethane, the reaction is

$$2CH_3-CH_3 + 7O_2 \longrightarrow 4CO_2 + 6H_2O$$

Obviously, when ethane is converted to CO_2 and H_2O, each of its C atoms has more bonds to O and fewer bonds to H. Thus, ethane is oxidized, and this reaction is referred to as an *oxidation,* even though O_2 is reduced as well.

Most oxidations do not involve such a total breaking apart of the molecule, however. When 2-propanol reacts with potassium dichromate in acidic solution (a common oxidizing agent in organic reactions), the organic compound that forms is 2-propanone:

Note that C-2 has one fewer bond to H and one more bond to O in 2-propanone than it does in 2-propanol. Thus, 2-propanol is oxidized, so this is an *oxidation.* Don't forget, however, that the dichromate ion is reduced at the same time:

$$Cr_2O_7^{2-} + 14H^+ + 6e^- \longrightarrow 2Cr^{3+} + 7H_2O$$

The addition of H_2 to an alkene is referred to as a *reduction:*

$$CH_2{=}CH_2 + H_2 \xrightarrow{Pd} CH_3-CH_3$$

Note that each C has more bonds to H in ethane than it has in ethene, so the ethene is reduced. (The H_2 is oxidized in the process, and the palladium shown over the arrow acts as a catalyst to speed up the reaction.)

Section Summary

In an addition reaction, a π bond breaks and the two C atoms bond to more atoms. • In an elimination, a π bond forms and the two C atoms bond to fewer atoms. • In a substitution, one atom replaces another, but the total number of atoms bonded to C does not change. • In an organic redox process, the organic reactant is oxidized if a C atom in the compound forms more bonds to O atoms (or fewer bonds to H atoms), and it is reduced if a C atom forms more bonds to H (or fewer bonds to O).

15.4 PROPERTIES AND REACTIVITIES OF COMMON FUNCTIONAL GROUPS

The central organizing principle of organic reaction chemistry is the *functional group*. To predict how an organic compound might react, we narrow our focus there because *the distribution of electron density in a functional group affects the reactivity*. The electron density can be high, as in the C=C and C≡C bonds, or it can be low at one end of a bond and high at the other, as in the C—Cl and C—O bonds. Such bond sites enhance a dipole in the other reactant. As a result, the reactants attract each other and begin a sequence of bond-forming and bond-breaking steps that lead to product. Thus, *the intermolecular forces that affect physical properties and solubility also affect reactivity*. Table 15.5 lists some of the important functional groups in organic compounds.

When we classify functional groups by bond order (single, double, and so forth), they tend to follow certain patterns of reactivity:

- Functional groups with only single bonds undergo substitution or elimination.
- Functional groups with double or triple bonds undergo addition.
- Functional groups with both single and double bonds undergo substitution.

Functional Groups with Only Single Bonds

The most common functional groups with only single bonds are alcohols, haloalkanes, and amines.

Alcohols The **alcohol** functional group consists of carbon bonded to an —OH group, $-\overset{|}{\underset{|}{C}}-\ddot{O}-H$, and the general formula of an alcohol is R—OH. Alcohols are named by dropping the final -*e* from the parent hydrocarbon name and adding the suffix -*ol*. Thus, the two-carbon alcohol is ethanol (ethan- + -ol). The common name is the hydrocarbon *root*- + -*yl*, followed by "alcohol"; thus, the common name of ethanol is ethyl alcohol. (This substance, obtained from fermented grain, has been consumed by people as an intoxicant in beverages since ancient times; today, it is recognized as the most abused drug in the world.) Alcohols are important laboratory reagents, and the functional group occurs in many biomolecules, including carbohydrates, sterols, and some amino acids. Figure 15.14 shows the names, structures, and uses of some important compounds that contain the alcohol group.

You can think of an alcohol as a water molecule with an R group in place of one of the H atoms. In fact, alcohols react with very active metals, such as the alkali metals [Group 1A(1)], in a manner similar to water:

$$2Na + 2HOH \longrightarrow 2NaOH + H_2$$
$$2Na + 2CH_3{-}CH_2{-}OH \longrightarrow 2CH_3{-}CH_2{-}ONa + H_2$$

Water forms a solution of the strongly basic hydroxide ion (HO^-), and an alcohol forms a solution of the strongly basic *alkoxide ion* (RO^-): ethanol forms the ethoxide ion ($CH_3{-}CH_2{-}O^-$). The physical properties of the smaller alcohols are also similar to those of water. They have high melting and boiling points as a result of hydrogen bonding, and they dissolve polar molecules and some salts. (Substitution of R groups for both H atoms of water gives *ethers*, R—O—R.)

Alcohols undergo elimination and substitution reactions. Dehydration, the elimination of H and OH, requires acid and forms alkenes:

Methanol (methyl alcohol)
Byproduct in coal gasification; de-icing agent; gasoline substitute; precursor of organic compounds

1,2-Ethanediol (ethylene glycol)
Main component of auto antifreeze

Serine
Amino acid found in most proteins

Cholesterol
Major sterol in animals; essential for cell membranes; precursor of steroid hormones

Figure 15.14 Some molecules with the alcohol functional group.

cyclohexanol   cyclohexene

Table 15.5 | **Important Functional Groups in Organic Compounds**

Functional Group	Compound Type	Prefix or Suffix of Name	Example		
			Lewis Structure	Ball-and-Stick Model	Systematic Name (Common Name)
C=C	alkene	-ene			ethene (ethylene)
$-\text{C}\equiv\text{C}-$	alkyne	-yne	$\text{H}-\text{C}\equiv\text{C}-\text{H}$		ethyne (acetylene)
$-\overset{\|}{\underset{\|}{\text{C}}}-\overset{..}{\underset{..}{\text{O}}}-\text{H}$	alcohol	-ol			methanol (methyl alcohol)
$-\overset{\|}{\underset{\|}{\text{C}}}-\overset{..}{\underset{..}{\text{X}}}:$ (X = halogen)	haloalkane	halo-			chloromethane (methyl chloride)
$-\overset{\|}{\underset{\|}{\text{C}}}-\overset{..}{\underset{\|}{\text{N}}}-$	amine	-amine			ethanamine (ethylamine)
$-\overset{:\text{O}:}{\underset{}{\text{C}}}-\text{H}$	aldehyde	-al			ethanal (acetaldehyde)
$-\overset{\|}{\underset{\|}{\text{C}}}-\overset{:\text{O}:}{\underset{}{\text{C}}}-\overset{\|}{\underset{\|}{\text{C}}}-$	ketone	-one			2-propanone (acetone)
$-\overset{:\text{O}:}{\underset{}{\text{C}}}-\overset{..}{\underset{..}{\text{O}}}-\text{H}$	carboxylic acid	-oic acid			ethanoic acid (acetic acid)
$-\overset{:\text{O}:}{\underset{}{\text{C}}}-\overset{..}{\underset{..}{\text{O}}}-\overset{\|}{\underset{\|}{\text{C}}}-$	ester	-oate			methyl ethanoate (methyl acetate)
$-\overset{:\text{O}:}{\underset{}{\text{C}}}-\overset{..}{\underset{\|}{\text{N}}}-$	amide	-amide			ethanamide (acetamide)
$-\text{C}\equiv\text{N}:$	nitrile	-nitrile	$\text{H}-\overset{\text{H}}{\underset{\text{H}}{\text{C}}}-\text{C}\equiv\text{N}:$		ethanenitrile (acetonitrile, methyl cyanide)

Elimination of two H atoms requires inorganic oxidizing agents, such as $K_2Cr_2O_7$ in aqueous H_2SO_4. As we saw in Section 15.3, the reaction is an oxidation and produces the $C=O$ group (shown with condensed and carbon-skeleton formulas):

$$CH_3-CH_2-\underset{\underset{OH}{|}}{CH}-CH_3 \xrightarrow[H_2SO_4]{K_2Cr_2O_7} CH_3-CH_2-\underset{\overset{O}{\|}}{C}-CH_3$$

2-butanol   2-butanone

For alcohols with an OH group at the end of the chain ($R-CH_2-OH$), another oxidation occurs. Wine turns sour, for example, when the ethanol in contact with air is oxidized to acetic acid:

$$CH_3-\underset{\underset{CH_2}{|}}{\overset{OH}{|}} \xrightarrow[-H_2O]{\frac{1}{2} O_2} CH_3-\overset{O}{\overset{\|}{CH}} \xrightarrow{\frac{1}{2} O_2} CH_3-\overset{O}{\overset{\|}{C}}-OH$$

Substitution yields products with other single-bonded functional groups. With hydrohalic acids, many alcohols give haloalkanes:

$$R_2CH-OH + HBr \longrightarrow R_2CH-Br + HOH$$

As you'll see below, *the C atom undergoing the change in a substitution is bonded to a more electronegative element*, which makes it partially positive and, thus, a target for a negatively charged or electron-rich group of an incoming reactant.

Haloalkanes A *halogen* atom (X) bonded to C gives the **haloalkane** functional group, $-\overset{|}{\underset{|}{C}}-\ddot{\underset{..}{X}}:$, and compounds with the general formula $R-X$. Haloalkanes (common name, **alkyl halides**) are named by adding the halogen as a prefix to the hydrocarbon name and numbering the C atom to which the halogen is attached, as in bromomethane, 2-chloropropane, or 1,3-diiodohexane.

Just as many alcohols undergo substitution to alkyl halides when treated with halide ions in acid, many halides undergo substitution to alcohols in base. For example, OH^- attacks the positive C end of the $C-X$ bond and displaces X^-:

$$CH_3-CH_2-CH_2-CH_2-Br + OH^- \longrightarrow CH_3-CH_2-CH_2-CH_2-OH + Br^-$$

1-bromobutane   1-butanol

Substitution by groups such as $-CN$, $-SH$, $-OR$, and $-NH_2$ allows chemists to convert alkyl halides to a host of other compounds.

Just as addition of HX *to* an alkene produces haloalkanes, elimination of HX *from* a haloalkane by reaction with a strong base, such as potassium ethoxide, produces an alkene:

$$CH_3-\underset{\underset{Cl}{|}}{\overset{\overset{CH_3}{|}}{C}}-CH_3 + CH_3-CH_2-OK \longrightarrow CH_3-\overset{\overset{CH_3}{|}}{C}=CH_2 + KCl + CH_3-CH_2-OH$$

2-chloro-2-methylpropane   potassium ethoxide   2-methylpropene

Haloalkanes have many important uses, but many are carcinogenic in mammals, have severe neurological effects in humans, and, to make matters worse, are very stable and accumulate in the environment. ●

Amines The **amine** functional group is $-\overset{|}{\underset{|}{C}}-\overset{..}{N}-$. Chemists classify amines as derivatives of ammonia, with R groups in place of one or more H atoms. *Primary* (1°) amines are RNH_2, *secondary* (2°) amines are R_2NH, and *tertiary* (3°) amines are R_3N. Like ammonia, amines have trigonal pyramidal shapes and a lone pair

ARIS Presentation Center, Art Library

● **Pollutants in the Food Chain** Until recently, halogenated aromatic hydrocarbons, such as the *polychlorinated biphenyls* (*PCBs;* one of 209 different compounds is shown above), were used as insulating fluids in electrical transformers and then discharged in wastewater. Because of their low solubility and high stability, they accumulate for decades in river and lake sediment and are eaten by microbes and invertebrates. Fish eat the invertebrates, and birds and mammals, including humans, eat the fish. PCBs become increasingly concentrated in body fat at each stage. As a result of their health risks, PCBs in natural waters present an enormous cleanup problem.

Environmental application

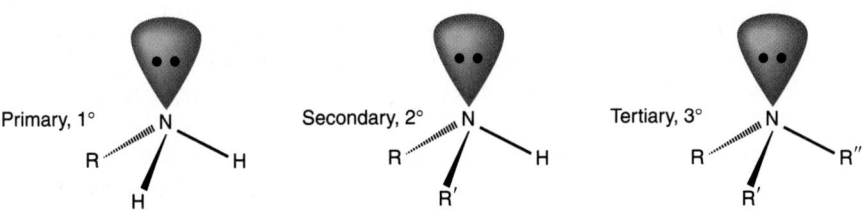

Figure 15.15 General structures of amines. Amines have a trigonal pyramidal shape and are classified by the number of R groups bonded to N. The lone pair on the nitrogen atom is the key to amine reactivity.

of electrons on a partially negative N atom (Figure 15.15). Systematic names drop the final -*e* of the alkane and add the suffix -*amine*, as in ethanamine. However, there is still wide usage of common names, in which the suffix -*amine* follows the name of the alkyl group; thus, methylamine has one methyl group attached to N, diethylamine has two ethyl groups attached, and so forth. Figure 15.16 shows that the amine functional group occurs in many biomolecules.

Primary and secondary amines can form H bonds, so they have higher melting and boiling points than hydrocarbons and alkyl halides of similar molar mass. For example, dimethylamine ($\mathcal{M}$ = 45.09 g/mol) boils 45°C higher than ethyl fluoride ($\mathcal{M}$ = 48.06 g/mol). Trimethylamine has a greater molar mass than dimethylamine, but it melts more than 20°C *lower* because trimethylamine molecules are not H bonded.

Amines of low molar mass are fishy smelling, water soluble, and weakly basic. The reaction with water proceeds only slightly to the right to reach equilibrium:

$$CH_3-\ddot{N}H_2 + H_2O \rightleftharpoons CH_3-\overset{+}{N}H_3 + OH^-$$

Amines undergo substitution reactions in which the lone pair on N attacks the partially positive C in alkyl halides to displace X⁻ and form a larger amine:

$$2CH_3-CH_2-\ddot{N}H_2 + CH_3-CH_2-Cl \longrightarrow \underset{\overset{|}{CH_3-CH_2}}{CH_3-CH_2-\ddot{N}H} + CH_3-CH_2-\overset{+}{N}H_3Cl^-$$

$\quad\quad$ ethylamine $\quad\quad\quad\quad$ chloroethane $\quad\quad\quad\quad$ diethylamine $\quad\quad\quad$ ethylammonium chloride

(Two molecules of ethylamine are needed: while one attacks the chloroethane, the other binds the released H⁺ to prevent it from remaining on the diethylamine.)

Lysine (primary amine)
Amino acid found in
most proteins

Adenine (primary amine)
Component of nucleic
acids

Epinephrine (adrenaline;
secondary amine)
Neurotransmitter in brain;
hormone released
during stress

Cocaine (tertiary amine)
Brain stimulant; widely
abused drug

Figure 15.16 Some biomolecules with the amine functional group.

Biological application

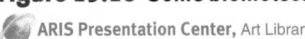

Variations on a Theme: Inorganic Compounds with Single Bonds to O, X, and N

The —OH group occurs frequently in inorganic compounds. All oxoacids contain at least one —OH, usually bonded to a relatively electronegative nonmetal atom, which in most cases is bonded to other O atoms. Oxoacids are acidic in water because these additional O atoms pull electron density from the central nonmetal, which pulls electron density from the O—H bond, releasing an H^+ ion and stabilizing the oxoanion through resonance. Alcohols are *not* acidic in water because they lack the additional O atoms and the electronegative nonmetal.

Halides of nearly every nonmetal are known, and many undergo substitution reactions in base. As in the case of an alkyl halide, the process involves an attack on the partially positive central atom by OH^-:

Thus, alkyl halides undergo the same general reaction as other nonmetal halides, such as BCl_3, SiF_4, and PCl_5.

The bonds between nitrogen and larger nonmetals, such as Si, P, and S, have significant double-bond character, which affects structure and reactivity. For example, trisilylamine, the Si analog of trimethylamine (see Figure 14.16, p. 590), is planar, rather than trigonal pyramidal, partially as a result of p,d–π bonding. The lone pair on N is delocalized in this π bond, so trisilylamine is not basic.

ARIS Presentation Center, Worked Examples Library

SAMPLE PROBLEM 15.4 Predicting the Reactions of Alcohols, Alkyl Halides, and Amines

PROBLEM Determine the reaction type and predict the product(s) for each reaction:

(a) CH_3—CH_2—CH_2—I + NaOH $\longrightarrow$

(b) CH_3—CH_2—Br + 2CH_3—CH_2—CH_2—NH_2 $\longrightarrow$ (c) CH_3—$\overset{\displaystyle |}{\underset{\displaystyle OH}{CH}}$—$CH_3$ $\xrightarrow[\text{H}_2\text{SO}_4]{\text{Cr}_2\text{O}_7{}^{2-}}$

PLAN We first determine the functional group(s) of the reactant(s) and then examine any inorganic reagent(s) to decide on the reaction type, keeping in mind that, in general, these functional groups undergo substitution or elimination. In (a), the reactant is an alkyl halide, so the OH^- of the inorganic reagent substitutes for the I. In (b), the reactants are an amine and an alkyl halide, so the N of the amine substitutes for the Br. In (c), the reactant is an alcohol, the inorganic reagents form a strong oxidizing agent, and the alcohol undergoes elimination to a carbonyl compound.

SOLUTION (a) Substitution: The products are CH_3—CH_2—CH_2—OH + NaI

(b) Substitution: The products are CH_3—CH_2—CH_2—$\overset{\displaystyle |}{\underset{\displaystyle CH_2—CH_3}{NH}}$ + CH_3—CH_2—CH_2—$\overset{+}{NH_3}Br^-$

(c) Elimination: (oxidation): The product is CH_3—$\overset{\displaystyle ||}{\underset{\displaystyle O}{C}}$—$CH_3$

CHECK The only changes should be at the functional group.

FOLLOW-UP PROBLEM 15.4 Fill in the blank in each reaction. (*Hint:* Examine any inorganic compounds and the organic product to determine the organic reactant.)

(a) _____ + CH_3—ONa $\longrightarrow$ CH_3—$CH{=}\overset{\displaystyle \overset{\displaystyle CH_3}{|}}{C}$—$CH_3$ + NaCl + CH_3—OH

(b) _____ $\xrightarrow[\text{H}_2\text{SO}_4]{\text{Cr}_2\text{O}_7{}^{2-}}$ CH_3—CH_2—$\overset{\displaystyle \overset{\displaystyle O}{||}}{C}$—$OH$

Functional Groups with Double Bonds

The most important functional groups with double bonds are the C=C of alkenes and the C=O of aldehydes and ketones. Both appear in many organic and biological molecules. *Their most common reaction type is addition.*

Comparing the Reactivity of Alkenes and Aromatic Compounds

The C=C bond is the essential portion of the alkene functional group, $\ce{C=C}$. Although they can be further unsaturated to alkynes, *alkenes typically undergo addition.* The electron-rich double bond is readily attracted to the partially positive H atoms of hydronium ions and hydrohalic acids, yielding alcohols and alkyl halides, respectively:

$$CH_3-\underset{\underset{\displaystyle CH_2}{||}}{C}-CH_3 \ + \ H_3O^+ \ \longrightarrow \ CH_3-\underset{\underset{\displaystyle OH}{|}}{\overset{\overset{\displaystyle CH_3}{|}}{C}}-CH_3 \ + \ H^+$$

2-methylpropene

2-methyl-2-propanol

$$CH_3-CH=CH_2 \ + \ HCl \ \longrightarrow \ CH_3-\underset{\underset{\displaystyle Cl}{|}}{CH}-CH_3$$

propene

2-chloropropane

The *localized* unsaturation of alkenes is very different from the *delocalized* unsaturation of benzene. Despite the way we depict its resonance forms, benzene does *not* have double bonds. Thus, for example, benzene does not decolorize Br_2.

In general, aromatic rings are much *less* reactive than alkenes because they have delocalized π electrons. For example, let's compare the heats of reaction for the addition of H_2. Hydrogenation of cyclohexene, with one C=C bond, has a ΔH°_{rxn} of -120 kJ/mol. If we extrapolate to the imaginary molecule "cyclohexatriene," that is, the structure with three C=C bonds, its hypothetical ΔH°_{rxn} for hydrogenation is three times as much, or -360 kJ/mol. Hydrogenation of benzene has a ΔH°_{rxn} of -208 kJ/mol. Thus, the ΔH°_{rxn} for hydrogenation of benzene is 152 kJ/mol *less* than that for "cyclohexatriene," an energy difference attributed to the aromatic stabilization of benzene (Figure 15.17).

This extra energy needed to break up the delocalized π system means *addition* reactions with benzene occur rarely. But, benzene does undergo many *substitution* reactions in which the delocalization is retained when an H atom attached to a ring C is replaced by another group:

benzene   bromobenzene

Aldehydes and Ketones

The C=O bond, or **carbonyl group,** is one of the most chemically versatile. In the **aldehyde** functional group, the carbonyl C is bonded to H (and often to another C), so it occurs *at the end of a chain,* $R-\overset{\overset{\displaystyle H}{|}}{C}=O$. Aldehyde names drop the final *-e* from the alkane name and add *-al;* thus, the three-C aldehyde is propanal. In the **ketone** functional group, the carbonyl C is bonded to two other C atoms, $-\overset{\overset{\displaystyle :O:}{||}}{\underset{\underset{\displaystyle |}{|}}{C}}-\overset{\overset{\displaystyle }{|}}{\underset{\underset{\displaystyle |}{|}}{C}}-$, so it occurs *within the chain.*

ARIS Presentation Center, Art Library

Figure 15.17 The stability of benzene. Because of the resonance stabilization caused by the aromatic π-electron system in benzene, the molecule releases 152 kJ/mol *less* energy when it is hydrogenated than would "cyclohexatriene," the analogous molecule with three separate double bonds.

Methanal (formaldehyde)
Used to make resins in plywood, dishware, countertops; biological preservative

Ethanal (acetaldehyde)
Narcotic product of ethanol metabolism; used to make perfumes, flavors, plastics, other chemicals

Benzaldehyde
Artificial almond flavoring

2-Propanone (acetone)
Solvent for fat, rubber, plastic, varnish, lacquer; chemical feedstock

2-Butanone (methyl ethyl ketone)
Important solvent

Figure 15.18 Some common aldehydes and ketones.

 ARIS Presentation Center, Active Art Library

A

B

Figure 15.19 The carbonyl group.
A, The σ and π bonds that make up the C=O bond of the carbonyl group. **B,** The charged resonance form shows that the C=O bond is polar (ΔEN = 1.0).

ARIS Presentation Center, Art Library

Ketones, R—C—R′, are named by numbering the carbonyl C, dropping the final -e from the alkane name, and adding -one. For example, the unbranched, five-C ketone with the carbonyl C as C-2 in the chain is named 2-pentanone. Figure 15.18 shows some common carbonyl compounds.

Like the C=C bond, the C=O bond is *electron rich;* unlike the C=C bond, it is *highly polar* (ΔEN = 1.0). Figure 15.19 emphasizes this polarity with an electron density model and a charged resonance form. Aldehydes and ketones are formed by the oxidation of alcohols:

$$CH_3—CH_2—OH \xrightarrow{\text{oxidation}} CH_3—\overset{O}{\overset{\|}{C}}—H$$
ethanol   ethanal (common name, acetaldehyde)

3-pentanol $\xrightarrow{\text{oxidation}}$ 3-pentanone (common name, diethylketone)

Conversely, as a result of their unsaturation, carbonyl compounds can undergo *addition* and be reduced to alcohols:

cyclobutanone $\xrightarrow{\text{reduction}}$ cyclobutanol

As a result of the bond polarity, addition often occurs with an electron-rich group bonding to the carbonyl C and an electron-poor group bonding to the carbonyl O. **Organometallic compounds,** which have a metal atom (usually Li or Mg) attached to an R group through a polar covalent bond (Sections 14.3 and 14.4), take part in this type of reaction. In a two-step sequence, they convert carbonyl compounds to alcohols with *different carbon skeletons:*

$$\overset{\delta+}{R}—\overset{\delta-}{CH}=\overset{\delta-}{O} + \overset{\delta-}{R'}—\overset{\delta+}{Li} \longrightarrow \xrightarrow{H_2O} R—\overset{OH}{\underset{|}{CH}}—R' + LiOH$$

In the following reaction steps, for example, the electron-rich C bonded to Li in ethyllithium, CH_3CH_2—Li, attacks the electron-poor carbonyl C of 2-propanone, adding its ethyl group; at the same time, the Li adds to the carbonyl O. Treating the mixture with water forms the C—OH group:

$$CH_3—\overset{\delta+}{\underset{\delta-}{CH_2}} + \overset{\delta+}{\underset{|}{C}}—CH_3 \longrightarrow CH_3—CH_2—\overset{O^-Li^+}{\underset{\underset{CH_3}{|}}{C}}—CH_3 \xrightarrow{H_2O} CH_3—CH_2—\overset{OH}{\underset{\underset{CH_3}{|}}{C}}—CH_3 + LiOH$$

Note that the product skeleton combines the two reactant skeletons. The field of *organic synthesis* often employs organometallic compounds to create molecules with different skeletons and, thus, synthesize new compounds.

SAMPLE PROBLEM 15.5 Predicting the Steps in a Reaction Sequence

ARIS Presentation Center, Worked Examples Library

PROBLEM Fill in the blanks in the following reaction sequence:

$$CH_3-CH_2-\overset{\displaystyle Br}{\underset{\displaystyle |}{CH}}-CH_3 \xrightarrow{\text{OH}^-} \underline{\quad} \xrightarrow[\text{H}_2\text{SO}_4]{\text{Cr}_2\text{O}_7{}^{2-}} \underline{\quad} \xrightarrow{\text{CH}_3-\text{Li}} \xrightarrow{\text{H}_2\text{O}} \underline{\quad}$$

PLAN For each step, we examine the functional group of the reactant and the reagent above the yield arrow to decide on the most likely product.

SOLUTION The sequence starts with an alkyl halide reacting with OH^-. Substitution gives an alcohol. Oxidation of this alcohol with acidic dichromate gives a ketone. Finally, a two-step reaction of a ketone with CH_3-Li and then water forms an alcohol with a carbon skeleton that has the CH_3- group attached to the carbonyl C:

2-bromobutane   2-butanol   2-butanone   2-methyl-2-butanol

CHECK In this case, make sure that the first two reactions alter the functional group only and that the final steps change the C skeleton.

FOLLOW-UP PROBLEM 15.5 Choose reactants to obtain the following products:

(a) $\underline{\quad} \xrightarrow[\text{H}_2\text{SO}_4]{\text{Cr}_2\text{O}_7{}^{2-}}$

(b) $\underline{\quad} \xrightarrow{\text{CH}_3-\text{CH}_2-\text{Li}} \xrightarrow{\text{H}_2\text{O}}$

Variations on a Theme: Inorganic Compounds with Double Bonds Homonuclear (same kind of atom) double bonds are rare for atoms other than C. However, we've seen many double bonds between O and other nonmetals, for example, in the oxides of S, N, and the halogens. Like carbonyl compounds, these substances undergo addition reactions. For example, the partially negative O of water attacks the partially positive S of SO_3 to form sulfuric acid:

S bonded to 3 atoms   S bonded to 4 atoms

Functional Groups with Both Single and Double Bonds

A family of three functional groups contains C double bonded to O (a carbonyl group) *and* single bonded to O or N. The parent of the family is the **carboxylic acid** group, $-\overset{\displaystyle :\text{O}:}{\underset{\displaystyle \|}{\text{C}}}-\overset{..}{\text{O}}\text{H}$, also called the *carboxyl group* and written —COOH. *The most important reaction type of this family is substitution from one member to another.* Substitution for the —OH by the —OR of an alcohol gives the **ester** group, $-\overset{\displaystyle :\text{O}:}{\underset{\displaystyle \|}{\text{C}}}-\overset{..}{\text{O}}-\text{R}$; substitution by the $-\overset{\displaystyle |}{\underset{\displaystyle |}{\overset{..}{\text{N}}}}-$ of an amine gives the **amide** group, $-\overset{\displaystyle :\text{O}:}{\underset{\displaystyle \|}{\text{C}}}-\overset{\displaystyle |}{\underset{\displaystyle |}{\overset{..}{\text{N}}}}-$.

Methanoic acid (formic acid)
An irritating component
of ant and bee stings

Butanoic acid (butyric acid)
Odor of rancid butter;
suspected component of
monkey sex attractant

Benzoic acid
Calorimetric standard;
used in preserving food,
dyeing fabric, curing tobacco

Octadecanoic acid (stearic acid)
Found in animal fats; used in
making candles and soaps

Figure 15.20 Some molecules with the carboxylic acid functional group.

 ARIS Presentation Center, Active Art Library

 Biological application

Literature Doyle, Ellin. "Trans Fatty Acids," *J. Chem. Educ.* **1997,** *74,* 1030.

Carboxylic Acids Carboxylic acids, R—C(=O)—OH, are named by dropping the *-e* from the alkane name and adding *-oic acid;* however, many common names are used. For example, the four-C acid is butanoic acid (the carboxyl C is counted when choosing the root); its common name is butyric acid. Figure 15.20 shows some important carboxylic acids. The carboxyl C already has three bonds, so it forms only one other. In formic acid (methanoic acid), the carboxyl C bonds to an H, but in all other carboxylic acids it bonds to a chain or ring.

Carboxylic acids are weak acids in water:

$$CH_3-C(=O)-OH(l) + H_2O(l) \rightleftharpoons CH_3-C(=O)-O^-(aq) + H_3O^+(aq)$$
ethanoic acid
(acetic acid)

At equilibrium in acid solutions of typical concentration, more than 99% of the acid molecules are undissociated at any given moment. In strong base, however, they react completely to form a salt and water:

$$CH_3-C(=O)-OH(l) + NaOH(aq) \longrightarrow CH_3-C(=O)-O^-(aq) + Na^+(aq) + H_2O(l)$$

The anion is the *carboxylate ion,* named by dropping *-oic acid* and adding *-oate;* the sodium salt of butanoic acid, for instance, is sodium butanoate.

Carboxylic acids with long hydrocarbon chains are **fatty acids,** an essential group of compounds found in all cells. Animal fatty acids have saturated chains, whereas many from vegetable sources are unsaturated, usually with the C=C bonds in the *cis* configuration. The double bond makes them much easier to metabolize. Nearly all fatty acid skeletons have an even number of C atoms—16 and 18 carbons are very common—because cells use two-carbon units in synthesizing them. Fatty acid salts are soaps, with the cation usually from Group 1A(1) or 2A(2) (Section 13.2).

Substitution of carboxylic acids and other members of this family occurs through a two-step sequence: *addition plus elimination equals substitution.* Addition to the trigonal planar shape of the carbonyl group gives an unstable tetrahedral intermediate, which immediately undergoes elimination to revert to a trigonal planar product (in this case, X is OH):

$$R-C(=O)-X + Z-Y \underset{addition}{\rightleftharpoons} \left[R\overset{O-Z}{\underset{Y}{\overset{|}{\underset{|}{C}}}}-X \right] \underset{elimination}{\rightleftharpoons} R-C(=O)-Y + Z-X$$

Strong heating of carboxylic acids forms an **acid anhydride** through a type of substitution called a *dehydration-condensation reaction* (Section 14.7), in which two molecules condense into one with loss of water:

$$R-C(=O)-\boxed{OH + H}O-C(=O)-R \xrightarrow{\Delta} R-C(=O)-O-C(=O)-R + HOH$$

Esters *An alcohol and a carboxylic acid form an ester.* The first part of an ester name designates the alcohol portion and the second the acid portion (named in the same way as the carboxylate ion). For example, the ester formed between ethanol and ethanoic acid is ethyl ethanoate (common name, ethyl acetate), a solvent for nail polish and model glue.

The ester group occurs commonly in **lipids,** a large group of fatty biological substances. Most dietary fats are *triglycerides,* esters composed of three fatty acids linked to the alcohol 1,2,3-trihydroxypropane (common name, glycerol) that

Cetyl palmitate The most common lipid in whale blubber

Lecithin Phospholipid found in all cell membranes

Figure 15.21 Some lipid molecules with the ester functional group. ARIS Presentation Center, Art Library

Tristearin Typical dietary fat used as an energy store in animals

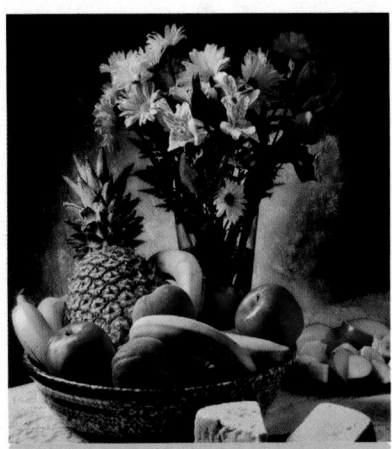

Biological application

function as energy stores. Some important lipids are shown in Figure 15.21; lecithin is one of several phospholipids that make up the lipid bilayer in all cell membranes (Section 13.2).

Esters, like acid anhydrides, form through a dehydration-condensation reaction; in this case, it is called an *esterification*:

$$R-C\overset{O}{\|}-\boxed{OH + H}-O-R' \underset{\overset{H^+}{\rightleftharpoons}}{} R-C\overset{O}{\|}-O-R' + HOH$$

In Chapter 16, we will discuss the role of H^+ in increasing the rate of this multi-step reaction. ●

Note that the esterification reaction is reversible. The opposite of dehydration-condensation is called **hydrolysis,** in which the O atom of water is attracted to the partially positive C atom of the ester, cleaving (lysing) the molecule into two parts. One part receives water's —OH, and the other part receives water's other H. In the process of soap manufacture, or *saponification* (Latin *sapon,* "soap"), begun in ancient times, ester bonds in animal or vegetable fats are hydrolyzed with strong base:

a triglyceride + 3NaOH $\overset{\Delta}{\longrightarrow}$ 3 soaps + glycerol

Amides The product of a substitution between an amine (or NH_3) and an ester is an amide. The partially negative N of the amine is attracted to the partially positive C of the ester, an alcohol (ROH) is lost, and an amide forms:

$$CH_3-C\overset{O}{\|}-O-CH_3 + H\ddot{N}-CH_2-CH_3 \longrightarrow CH_3-C\overset{O}{\|}-N-CH_2-CH_3 + CH_3-OH$$

methyl ethanoate (methyl acetate)   ethanamine (ethylamine)   *N*-ethylethanamide (*N*-ethylacetamide)   methanol

Amides are named by denoting the amine portion with *N-* and replacing *-oic acid* from the parent carboxylic acid with *-amide*. In the amide from the previous

● **A Pungent, Pleasant Banquet** Many organic compounds have strong odors, but none can rival the diverse odors of carboxylic acids and esters. From the pungent, vinegary odors of the one-C and two-C acids to the cheesy stench of slightly larger ones, carboxylic acids possess some awful odors. When butanoic acid reacts with ethanol, however, its rancid-butter smell becomes the peachy pineapple scent of ethyl butanoate, and when pentanoic acid reacts with pentanol, its Limburger cheese odor becomes the fresh apple aroma of pentyl pentanoate. Naturally occurring and synthetic esters are used to add fruity, floral, and herbal odors to foods, cosmetics, household deodorizers, and medicines.

 ARIS Presentation Center, Photo Library

Acetaminophen
Active ingredient in nonaspirin
pain relievers; used to make dyes
and photographic chemicals

***N,N*-Dimethylmethanamide
(dimethylformamide)**
Major organic solvent; used in
production of synthetic fibers

Lysergic acid diethylamide (LSD-25) A potent hallucinogen

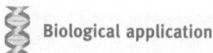

**Figure 15.22 Some molecules with
the amide functional group.**

ARIS Presentation Center, Art Library

Biological application

ARIS Presentation Center, Worked Examples
Library

reaction, the ethyl group comes from the amine, and the acid portion comes from
ethanoic acid (acetic acid). Some amides are shown in Figure 15.22.

Amides are hydrolyzed in hot water (or base) to a carboxylic acid and an
amine. Thus, even though amides are not normally formed in the following way,
they can be viewed as the result of a reversible dehydration-condensation:

The most important example of the amide group is the *peptide bond* (discussed
in Sections 13.2 and 15.6), which links amino acids in a protein.

The carboxylic acid family also undergoes reduction to form other functional
groups. For example, certain inorganic reducing agents convert acids or esters to
alcohols and convert amides to amines:

$$R-\overset{O}{\underset{||}{C}}-OH \text{ (or } R-\overset{O}{\underset{||}{C}}-O-R') \xrightarrow{reduction} R-CH_2-OH + HOH \text{ (or } R'-OH)$$

$$R-\overset{O}{\underset{||}{C}}-NH-R' \xrightarrow{reduction} R-CH_2-NH-R' + H_2O$$

SAMPLE PROBLEM 15.6 **Predicting Reactions of the Carboxylic Acid
Family**

PROBLEM Predict the product(s) of the following reactions:

(a) $CH_3-CH_2-CH_2-\overset{O}{\underset{||}{C}}-OH + CH_3-\overset{OH}{\underset{|}{CH}}-CH_3 \overset{H^+}{\rightleftharpoons}$

(b) $CH_3-\overset{CH_3}{\underset{|}{CH}}-CH_2-CH_2-\overset{O}{\underset{||}{C}}-NH-CH_2-CH_3 \xrightarrow[H_2O]{NaOH}$

PLAN We discussed substitution reactions (including addition-elimination and dehydration-condensation) and hydrolysis. (a) A carboxylic acid and an alcohol react, so the reaction must be a substitution to form an ester and water. (b) An amide reacts with OH⁻, so it is hydrolyzed to an amine and a sodium carboxylate.

SOLUTION (a) Formation of an ester:

$$CH_3-CH_2-CH_2-\overset{O}{\underset{||}{C}}-O-\overset{CH_3}{\underset{|}{CH}}-CH_3 + H_2O$$

(b) Basic hydrolysis of an amide:

$$CH_3-\overset{CH_3}{\underset{|}{CH}}-CH_2-CH_2-\overset{O}{\underset{||}{C}}-O^- + Na^+ + H_2N-CH_2-CH_3$$

CHECK Note that in part (b), the carboxylate ion forms, rather than the acid, because the aqueous NaOH that is present reacts with the carboxylic acid.

FOLLOW-UP PROBLEM 15.6 Fill in the blanks in the following reactions:

(a) _____ + CH₃—OH $\xrightleftharpoons{H^+}$ ⬡—CH₂—$\overset{\displaystyle O}{\overset{\|}{C}}$—O—CH₃ + H₂O

(b) _____ + _____ $\longrightarrow$ CH₃—CH₂—CH₂—$\overset{\displaystyle O}{\overset{\|}{C}}$—NH—CH₂—CH₃ + CH₃—OH

Variations on a Theme: Oxoacids, Esters, and Amides of Other Nonmetals

A nonmetal that is both double and single bonded to O occurs in most inorganic oxoacids, such as phosphoric, sulfuric, and chlorous acids. Those with additional O atoms are stronger acids than carboxylic acids.

Diphosphoric and disulfuric acids are acid anhydrides formed by dehydration-condensation reactions, just as a carboxylic acid anhydride is formed (Figure 15.23). Inorganic oxoacids form esters and amides that are part of many biological molecules. We already saw that certain lipids are phosphate esters (see Figure 15.21). The first compound formed when glucose is digested is a phosphate ester (Figure 15.24A); a similar phosphate ester is a major structural feature of nucleic acids, as we'll see shortly. Amides of organic sulfur-containing oxoacids, called *sulfonamides*, are potent antibiotics; the simplest of these is depicted in Figure 15.24B. More than 10,000 different sulfonamides have been synthesized.

Figure 15.23 The formation of carboxylic, phosphoric, and sulfuric acid anhydrides.

🌐 ARIS Presentation Center, Art Library

A Glucose-6-phosphate

B Sulfanilamide

Figure 15.24 An ester and an amide with another nonmetal. A, Glucose-6-phosphate contains a phosphate ester group. **B,** Sulfanilamide contains a sulfonamide group.

🌐 ARIS Presentation Center, Art Library

🧬 Biological application

Functional Groups with Triple Bonds

There are only two important functional groups with triple bonds. *Alkynes,* with an electron-rich —C≡C— group, undergo addition (by H₂O, H₂, HX, X₂, and so forth) to form double-bonded or saturated compounds:

CH₃—C≡CH $\xrightarrow{H_2}$ CH₃—CH=CH₂ $\xrightarrow{H_2}$ CH₃—CH₂—CH₃
propyne   propene   propane

Nitriles (R—C≡N) contain the **nitrile** group (—C≡N:) and are made by substituting a CN⁻ (cyanide) ion for X⁻ in a reaction with an alkyl halide:

CH₃—CH₂—Cl + NaCN $\longrightarrow$ CH₃—CH₂—C≡N + NaCl

This reaction is useful because it *increases the hydrocarbon chain by one C atom.* Nitriles are versatile because once they are formed, they can be reduced to amines or hydrolyzed to carboxylic acids:

CH₃—CH₂—CH₂—NH₂ $\xleftarrow{\text{reduction}}$ CH₃—CH₂—C≡N $\xrightarrow[\text{hydrolysis}]{H_3O^+, H_2O}$ CH₃—CH₂—$\overset{\displaystyle O}{\overset{\|}{C}}$—OH + NH₄⁺

Variations on a Theme: Inorganic Compounds with Triple Bonds Triple bonds are as scarce in the inorganic world as in the organic world. Carbon monoxide (:C≡O:), elemental nitrogen (:N≡N:), and the cyanide ion ([:C≡N:]⁻) are the only common examples.

You've seen quite a few functional groups by this time, and it is especially important that you can recognize them in a complex organic molecule. Sample Problem 15.7 provides some practice.

SAMPLE PROBLEM 15.7   Recognizing Functional Groups

PROBLEM Circle and name the functional groups in the following molecules:

PLAN We use Table 15.5 (p. 651) to identify the various functional groups.
SOLUTION

FOLLOW-UP PROBLEM 15.7 Circle and name the functional groups:

Section Summary

Organic reactions are initiated when regions of high and low electron density of different reactant molecules attract each other. • Groups containing only single bonds—alcohols, amines, and alkyl halides—take part in substitution and elimination reactions. • Groups with double or triple bonds—alkenes, aldehydes, ketones, alkynes, and nitriles—generally take part in addition reactions. • Aromatic compounds typically undergo substitution, rather than addition, because delocalization of the π electrons stabilizes the ring. • Groups with both double and single bonds—carboxylic acids, esters, and amides—generally take part in substitution reactions. • Many reactions change one functional group to another, but some, especially reactions with organometallic compounds and with the cyanide ion, change the C skeleton.

Engineering application

15.5 THE MONOMER-POLYMER THEME I: SYNTHETIC MACROMOLECULES

In our survey of advanced materials in Chapter 12, you saw that polymers are extremely large molecules that consist of many monomeric repeat units. In that chapter, we focused on the mass, shape, and physical properties of polymers. Now we'll see how polymers are named and discuss the two types of reactions that link monomers covalently into a chain. To name a polymer, just add the prefix *poly-* to the monomer name, as in *polyethylene* or *polystyrene*. When the monomer has a two-word name, parentheses are used, as in *poly(vinyl chloride)*.

Because synthetic polymers consist of petroleum-based monomers, a major shortage of raw materials for their manufacture is expected in the coming decades.

To avoid this and create renewable materials, green chemists are devising polymers based on corn, sugarcane, and even cashew-nut shells. Before such a method is embraced as a "green" technology, however, the total impact of growing, transporting, and manufacturing must be assessed.

The two major types of reaction processes that form synthetic polymers lend their names to the resulting classes of polymer—addition and condensation.

🌎 Green chemistry application

Addition Polymers

Addition polymers form when monomers undergo an addition reaction with one another. These are also called *chain-reaction* (or *chain-growth*) *polymers* because as each monomer adds to the chain, it forms a new reactive site to continue the process. The monomers of most addition polymers have the ⟩C═C⟨ grouping.

As you can see from Table 15.6, the essential differences between an acrylic sweater, a plastic grocery bag, and a bowling ball are due to the different chemical groups that are attached to the double-bonded C atoms of the monomer.

📖 **Literature** Jensen, William B. "The Origin of Vinyl," *J. Chem. Educ.* **2004,** *81,* 464.

Table 15.6 Some Major Addition Polymers

Monomer	Polymer	Applications
	polyethylene	Plastic bags; bottles; toys
	polytetrafluoroethylene	Cooking utensils (e.g., Teflon)
	polypropylene	Carpeting (indoor-outdoor); bottles
	poly(vinyl chloride)	Plastic wrap; garden hose; indoor plumbing
	polystyrene	Insulation; furniture; packing materials
	polyacrylonitrile	Yarns, fabrics, and wigs (e.g., Orlon, Acrilan)
	poly(vinyl acetate)	Adhesives; paints; textile coatings; computer disks
	poly(vinylidene chloride)	Food wrap (e.g., Saran)
	poly(methyl methacrylate)	Glass substitute (e.g., Lucite, Plexiglas); bowling balls; paint

(Peroxide initiator)

Y—O—O—Y

Step 1
Formation of
free radical

2Y—O•

$H_2C=CH_2$

Step 2
Addition of
monomer

Y—O—CH_2—CH_2•

$(n+1)$

$(n+1)$ $H_2C=CH_2$

Step 3
Addition of
more monomer

Y—O—CH_2—CH_2—[CH_2—CH_2]$_n$—CH_2—CH_2•

•CH_2—CH_2—[CH_2—CH_2]$_m$—O—Y

Step 4
Chain termination by
joining of two free radicals

Y—O—CH_2—CH_2—[CH_2—CH_2]$_{n+1}$—[CH_2—CH_2]$_{m+1}$—O—Y

The *free-radical polymerization* of ethene (ethylene, $CH_2=CH_2$) to polyethylene is a simple example of the addition process. In Figure 15.25, the monomer reacts to form a *free radical,* a species with an unpaired electron, that seeks an electron from another monomer to form a covalent bond. The process begins when an *initiator,* usually a peroxide, generates a free radical that attacks the π bond of an ethylene unit, forming a σ bond with one of the *p* electrons and leaving the other unpaired. This new free radical then attacks the π bond of another ethylene, joining it to the chain end, and the backbone of the polymer grows. As each ethylene adds, it leaves an unpaired electron on the growing end to find an electron "mate" and make the chain one repeat unit longer. This process stops when two free radicals form a covalent bond or when a very stable free radical is formed by addition of an *inhibitor* molecule. Recent progress in controlling the high reactivity of free-radical species promises an even wider range of polymers. In a similar method, polymerization is initiated by the formation of a cation (or anion) instead of a free radical. The cationic (or anionic) reactive end of the chain attacks the π bond of another monomer to form a new cationic (or anionic) end, and the process continues.

The most important polymerization reactions take place under relatively mild conditions through the use of catalysts that incorporate transition metals. In 1963, Karl Ziegler and Giulio Natta received the Nobel Prize in chemistry for developing *Ziegler-Natta catalysts,* which employ an organoaluminum compound, such as $Al(C_2H_5)_3$, and the tetrachloride of titanium or vanadium. Today, chemists use organometallic catalysts that are *stereoselective* to create polymers whose repeat units have groups spatially oriented in particular ways. Through the use of these catalysts, polyethylene chains with molar masses of 10^4 to 10^5 g/mol are made by varying conditions and reagents.

Similar methods are used to make polypropylenes, —[CH_2—CH]—$_n$, that have

$$CH_3$$

all the CH_3 groups of the repeat units oriented either on one side of the chain or on alternating sides. The different orientations lead to different packing efficiencies of the chains and, thus, different degrees of crystallinity, which lead to differences in such physical properties as density, rigidity, and elasticity (Section 12.7).

Condensation Polymers

The monomers of **condensation polymers** must have *two functional groups;* we can designate such a monomer as A—**R**—B (where **R** is the rest of the molecule). Most commonly, the monomers link when an A group on one undergoes a *dehydration-condensation reaction* with a B group on another:

$$\tfrac{1}{2}nH—A—R—B—OH + \tfrac{1}{2}nH—A—R—B—OH \xrightarrow{-(n-1)HOH}$$

$$H—[A—R—B]_n—OH$$

Many condensation polymers are *copolymers,* those consisting of two or more different repeat units (Section 12.7). For example, condensation of carboxylic acid and amine monomers forms *polyamides (nylons),* whereas carboxylic acid and alcohol monomers form *polyesters.*

One of the most common polyamides is *nylon-66* (see the Gallery, p. 74), manufactured by mixing equimolar amounts of a six-C diamine (1,6-diaminohexane)

Figure 15.25 Steps in the free-radical polymerization of ethylene. In this polymerization method, free radicals initiate, propagate, and terminate the formation of an addition polymer. An initiator (Y—O—O—Y) is split to form two molecules of a free radical (Y—O•). The free radical attacks the π bond of a monomer and creates another free radical (Y—O—CH_2—CH_2•). The process continues, and the chain grows (propagates) until the system runs out of monomer and two free radicals combine.

ARIS Presentation Center, Active Art Library

and a six-C diacid (1,6-hexanedioic acid). The basic amine reacts with the acid to form a "nylon salt." Heating drives off water and forms the amide bonds:

$$n\text{HO}-\overset{\overset{\text{O}}{\|}}{\text{C}}-(\text{CH}_2)_4-\overset{\overset{\text{O}}{\|}}{\text{C}}-\text{OH} \; + \; n\text{H}_2\text{N}-(\text{CH}_2)_6-\text{NH}_2 \xrightarrow[-(2n-1)\text{H}_2\text{O}]{\Delta} \text{HO}\left[\overset{\overset{\text{O}}{\|}}{\text{C}}-(\text{CH}_2)_4-\overset{\overset{\text{O}}{\|}}{\text{C}}-\text{NH}-(\text{CH}_2)_6-\text{NH}\right]_n\text{H}$$

In the laboratory, this nylon is made without heating by using a more reactive acid component (Figure 15.26). Covalent bonds within the chains give nylons great strength, and H bonds between chains give them great flexibility. About half of all nylons are made to reinforce automobile tires; the others are used for rugs, clothing, fishing line, and so forth.

Dacron, a popular polyester fiber, is woven from polymer strands formed when equimolar amounts of 1,4-benzenedicarboxylic acid and 1,2-ethanediol react. Blending these polyester fibers with various amounts of cotton gives fabrics that are durable, easily dyed, and crease resistant. Extremely thin Mylar films, used for recording tape and food packaging, are also made from this polymer.

Variations on a Theme: Inorganic Polymers You already know that some synthetic polymers have inorganic backbones. In Chapter 14, we discussed the silicones, polymers with the repeat unit $-(\text{R}_2)\text{Si}-\text{O}-$. Depending on the chain crosslinks and the R groups, silicones range from oily liquids to elastic sheets to rigid solids and have applications that include artificial limbs and spacesuits. We also mentioned the polyphosphazenes in Chapter 14, which exist as flexible chains even at low temperatures and have the repeat unit $-(\text{R}_2)\text{P}{=}\text{N}-$.

Section Summary

Polymers are extremely large molecules that are made of repeat units called monomers. • Addition polymers are formed from unsaturated monomers that commonly link through free-radical reactions. • Most condensation polymers are formed by linking two types of monomer through a dehydration-condensation reaction. • Reaction conditions, catalysts, and monomers can be varied to produce polymers with different properties.

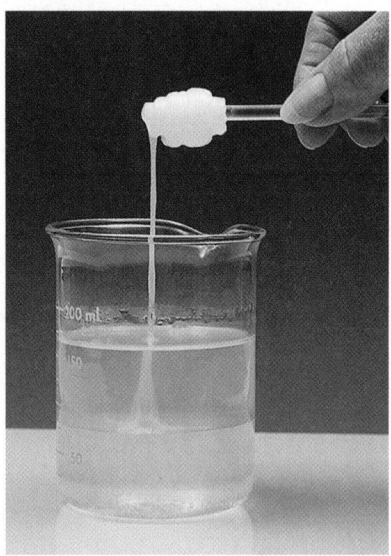

Figure 15.26 The formation of nylon-66. Nylons are formed industrially by the reaction of diamine and diacid monomers. In this laboratory demonstration, the six-C diacid chloride, which is more reactive than the diacid, is used as one monomer; the polyamide forms between the two liquid phases.

ARIS Presentation Center, Photo Library

 Biological application

15.6 THE MONOMER-POLYMER THEME II: BIOLOGICAL MACROMOLECULES

The monomer-polymer theme was being played out in nature eons before humans employed it to such great advantage. Biological macromolecules are nothing more than condensation polymers created by nature's reaction chemistry and improved through evolution. These remarkable molecules are the greatest proof of the versatility of carbon and its handful of atomic partners.

Natural polymers are the "stuff of life"—polysaccharides, proteins, and nucleic acids. Some have structures that make wood strong, hair curly, fingernails hard, and wool flexible. Others speed up the myriad reactions that occur in every cell or defend the body against infection. Still others possess the genetic information organisms need to forge other biomolecules. Remarkable as these giant molecules are, the functional groups of their monomers and the reactions that link them are identical to those of other, smaller organic molecules. Moreover, as you saw in Section 13.2, the same intermolecular forces that dissolve smaller molecules stabilize these giant molecules in the aqueous medium of the cell.

Sugars and Polysaccharides

In essence, the same chemical change occurs when you burn a piece of wood or eat a piece of bread. Wood and bread are mixtures of *carbohydrates*, substances that provide energy through oxidation.

A

Cyclic form of glucose

Figure 15.27 **The structure of glucose in aqueous solution and the formation of a disaccharide. A,** A molecule of glucose undergoes an internal addition reaction between the aldehyde group of C-1 and the alcohol group of C-5 to form a cyclic monosaccharide. **B,** Two monosaccharides, glucose and fructose, undergo a dehydration-condensation reaction to form the disaccharide sucrose (table sugar) and a water molecule.

ARIS Presentation Center, Art Library

B Glucose   Fructose $-H_2O$ Sucrose

Monomer Structure and Linkage

Glucose and other simple sugars, from the three-C *trioses* to the seven-C *heptoses,* are called **monosaccharides** and consist of carbon chains with attached hydroxyl and carbonyl groups. In addition to their roles as individual molecules engaged in energy metabolism, they serve as the monomer units of **polysaccharides.** Most natural polysaccharides are formed from five- and six-C units. In aqueous solution, *an alcohol group and the aldehyde (or ketone) group of a given monosaccharide react with each other to form a cyclic molecule with either a five- or six-membered ring* (Figure 15.27A).

When two monosaccharides undergo a dehydration-condensation reaction, a **disaccharide** forms. For example, sucrose (table sugar) is a disaccharide of glucose (linked at C-1) and fructose (linked at C-2) (Figure 15.27B); lactose (milk sugar) is a disaccharide of glucose (C-1) and galactose (C-4); and maltose, used in brewing and as a sweetener, is a disaccharide of two glucose units (C-1 to C-4).

Types of Polysaccharides

A polysaccharide consists of *many* monosaccharide units linked together. The three major natural polysaccharides—cellulose, starch, and glycogen—consist entirely of glucose units, but they differ in the ring positions of the links, in the orientation of certain bonds, and in the extent of crosslinking. Some other polysaccharides contain nitrogen in their attached groups. ●

Cellulose is the most abundant organic chemical on Earth. More than 50% of the carbon in plants occurs in the cellulose of stems and leaves; wood is largely cellulose, and cotton is more than 90% cellulose (see Section 13.2). Cellulose consists of long chains of glucose. The great strength of wood is due largely to the H bonds between cellulose chains. The monomers are linked in a particular way from C-1 in one unit to C-4 in the next. Humans lack the enzymes to break this link, so we cannot digest cellulose (unfortunately!); however, microorganisms in the digestive tracts of some animals, such as cows, sheep, and termites, can.

Starch is a mixture of polysaccharides of glucose and serves as an *energy store* in plants. When a plant needs energy, some starch is broken down by hydrolysis of the bonds between units, and the released glucose is oxidized through a multistep metabolic pathway. Starch occurs in plant cells as insoluble granules of amylose, a helical molecule of several thousand glucose units, and amylopectin, a highly branched, bushlike molecule of up to a million glucose units. Most of the glucose units are linked by C-1 to C-4 bonds, as in cellulose, but a different

● **Polysaccharide Skeletons of Lobsters and Roaches** The variety of polysaccharide properties that arises from simple changes in the monomers is amazing. For example, substituting an —NH_2 group for the —OH group at C-2 in glucose produces *glucosamine.* The amide that is formed from glucosamine and acetic acid (*N*-acetylglucosamine) is the monomer of *chitin* (pronounced "KY-tin"), a polysaccharide that is the main component of the tough, brittle, external skeletons of insects and crustaceans.

ARIS Presentation Center, Photo Library

orientation around the chiral C-1 allows our digestive enzymes to break starch down into monomers. A C-6 to C-1 crosslink joins chains every 24 to 30 units.

Glycogen functions as the energy storage molecule in animals. It occurs in liver and muscle cells as large, insoluble granules consisting of glycogen molecules made from 1000 to more than 500,000 glucose units. In glycogen, the units are also linked by C-1 to C-4 bonds, but the molecule is more highly crosslinked than starch, with C-6 to C-1 crosslinks every 8 to 12 units.

Amino Acids and Proteins

As you saw in Section 15.5, synthetic polyamides (such as nylon-66) are formed from two monomers, one with a carboxyl group at each end and the other with an amine group at each end. **Proteins,** the polyamides of nature, are unbranched polymers formed from monomers called **amino acids.** As we discussed in Section 13.2, *each amino acid has a carboxyl group and an amine group.*

Monomer Structure and Linkage An amino acid has both its carboxyl group and its amine group attached to the α-*carbon,* the second C atom in the chain. Proteins are made up of about 20 different types of amino acids, each with its own particular R group, ranging from an H atom to a polycyclic N-containing aromatic structure (Figure 15.28).

Transparency/ARIS Presentation Center, Art Library

Figure 15.28 The common amino acids. About 20 different amino acids occur in proteins. The R groups are screened gray, and the α-carbons (*boldface*), with carboxyl and amino groups, are screened yellow. Here the amino acids are shown with the charges they have under physiological conditions. They are grouped by polarity, acid-base character, and presence of an aromatic ring. The R groups play a major role in the shape and function of the protein.

Nonpolar amino acids

Glycine   Alanine   Valine   Leucine   Isoleucine   Methionine   Proline

Polar amino acids

Serine   Cysteine   Threonine

Aromatic amino acids

Phenylalanine   Tyrosine   Tryptophan

Basic amino acids

Histidine   Lysine   Arginine

Acidic amino acids and amide derivatives

Aspartic acid   Glutamic acid   Asparagine   Glutamine

In the aqueous cell fluid, the NH_2 and COOH groups of amino acids are charged because the carboxyl group transfers an H^+ ion to H_2O to form H_3O^+, which transfers the H^+ to the amine group. The overall process is, in effect, an intramolecular acid-base reaction:

An H atom is the third group bonded to the α-carbon, and the fourth is the R group (also called the *side chain*).

Each amino acid is linked to the next through a *peptide (amide) bond* formed by a dehydration-condensation reaction in which the carboxyl group of one monomer reacts with the amine group of the next. Therefore, as noted in Section 13.2, the polypeptide chain—the backbone of the protein—has a repeating structure that consists of an *α-carbon bonded to an amide group bonded to the next α-carbon bonded to the next amide group,* and so forth (see Figure 13.6, p. 508). The various R groups dangle from the α-carbons on alternate sides of the chain.

The Hierarchy of Protein Structure *Each type of protein has its own amino acid composition,* a specific number and proportion of various amino acids. However, it is not the composition that defines the protein's role in the cell; rather, *the sequence of amino acids determines the protein's shape and function.* Proteins range from about 50 to several thousand amino acids, yet from a purely mathematical point of view, even a small protein of 100 amino acids has a virtually limitless number of possible sequences of the 20 types of amino acids ($20^{100} \approx 10^{130}$). In fact, though, only a tiny fraction of these possibilities occur in actual proteins. For example, even in an organism as complex as a human being, there are only about 10^5 different types of protein.

A protein folds into its *native shape* as it is being synthesized in the cell. Some shapes are simple—long rods or undulating sheets. Others are far more complex—baskets, Y shapes, spheroid blobs, and countless other globular forms. Biochemists define a hierarchy for the overall structure of a protein (Figure 15.29):

Figure 15.29 The structural hierarchy of proteins. A typical protein's structure can be viewed at different levels. Primary structure (shown as a long string of balls leaving and returning to the picture frame) is the sequence of amino acids. Secondary structure consists of highly ordered regions that occur as an α-helix or a β-sheet. Tertiary structure combines these regions with random coil sections. In many proteins, several tertiary units interact to give the quaternary structure.

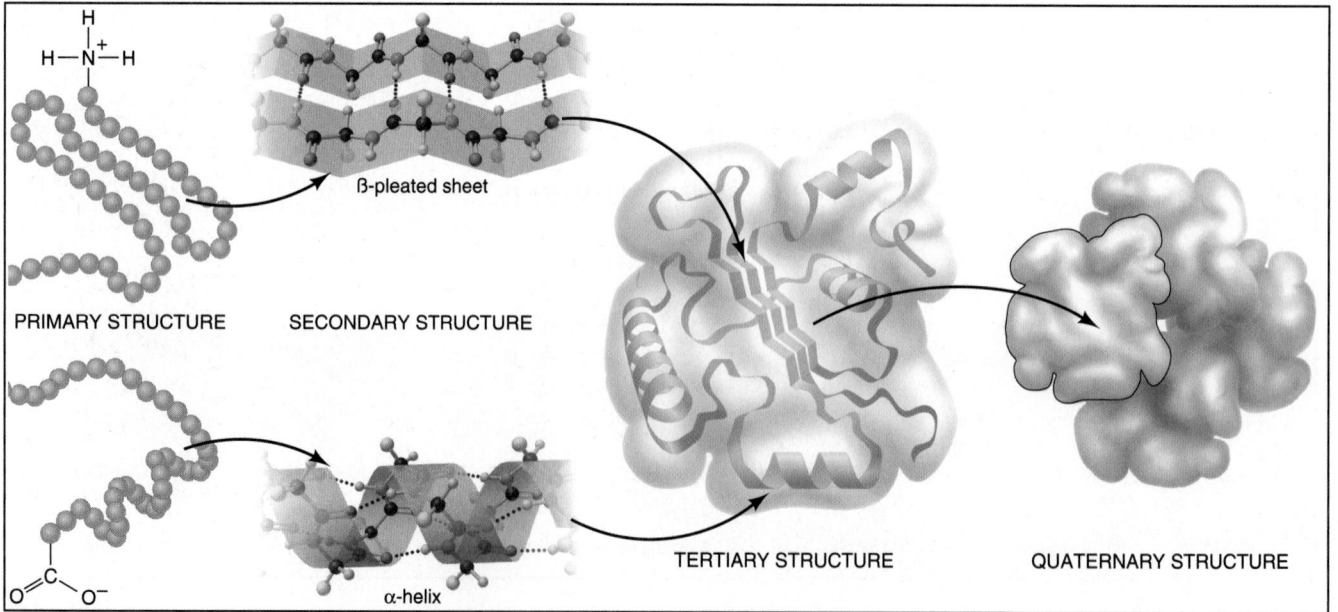

PRIMARY STRUCTURE   SECONDARY STRUCTURE   TERTIARY STRUCTURE   QUATERNARY STRUCTURE

β-pleated sheet

α-helix

1. *Primary (1°) structure*, the most basic level, refers to the sequence of co-valently bonded amino acids in the polypeptide chain.
2. *Secondary (2°) structure* refers to sections of the chain that, as a result of H bonding between nearby peptide groupings, adopt shapes called α-helices and β-pleated sheets.
3. *Tertiary (3°) structure* refers to the three-dimensional folding of the whole polypeptide chain. In some proteins, certain folding patterns form characteristic regions that play a role in the protein's function—binding a hormone, attaching to a membrane, forming a polar channel, and so forth.
4. *Quaternary (4°) structure*, the most complex level, occurs in proteins made up of several polypeptide chains (subunits) and refers to the way the chains assemble into the overall protein.

Note that *only the 1° structure involves covalent bonds; the 2°, 3°, and 4° structures rely primarily on intermolecular forces*, as discussed in Section 13.2.

The Relation Between Structure and Function

Two broad classes of proteins differ in the complexity of their amino acid compositions and sequences and, therefore, in their structure and function:

1. *Fibrous proteins* have relatively simple amino acid compositions and correspondingly simple structures. They are key components of hair, wool, skin, and connective tissue—materials that require strength and flexibility. Like synthetic polymers, these proteins have a small number of different amino acids in a repeating sequence. Consider collagen, the most common animal protein, which makes up as much as 40% of human body weight. More than 30% of its amino acids are glycine (G) and another 20% are proline (P). It exists as three chains, each an extended helix, that wind tightly around each other as the peptide C=O groups in one chain form H bonds to the peptide N—H groups in another. The result is a long, triple-helical cable with the sequence —G—X—P—G—X—P— and so on (where X is another amino acid), as shown in Figure 15.30A. As the main component of tendons, skin, and blood vessels, collagen has a high tensile strength; in fact, a 1-mm thick strand can support a 10-kg weight!

In silk fibroin, secreted by the silk moth caterpillar, more than 85% of the amino acids are glycine, alanine, and serine (Figure 15.30B). Fibroin chain segments bend back and forth, running alongside each other, and form interchain H bonds to create a *pleated sheet*. Stacks of sheets interact through dispersion forces, which make fibroin strong and flexible but not very extendable—perfect for a silkworm's cocoon.

2. *Globular proteins* have much more complex compositions, often containing all 20 amino acids in varying proportions. As the name implies, globular proteins are typically more rounded and compact, with a wide variety of shapes and a correspondingly wide range of functions: defenders against bacterial invasion, messengers that trigger cell actions, catalysts of chemical change, membrane gatekeepers that maintain aqueous concentrations, and many others. The locations of particular amino-acid R groups are crucial to the protein's function. For example, in catalytic proteins, a few R groups form a crevice that closely matches the shapes of reactant molecules. These groups typically hold the reactants through intermolecular forces and speed their reaction to products by bringing them together and twisting and stretching their bonds. Experiment has shown repeatedly that a slight change in one of these critical R groups decreases function dramatically. This fact supports the essential idea that *the protein's amino acid sequence determines its structure, which in turn determines its function*:

$$\text{SEQUENCE} \Longrightarrow \text{STRUCTURE} \Longrightarrow \text{FUNCTION}$$

Next, we'll see how the amino acid sequence of every protein in every organism is prescribed by the genetic information that is held within the organism's nucleic acids.

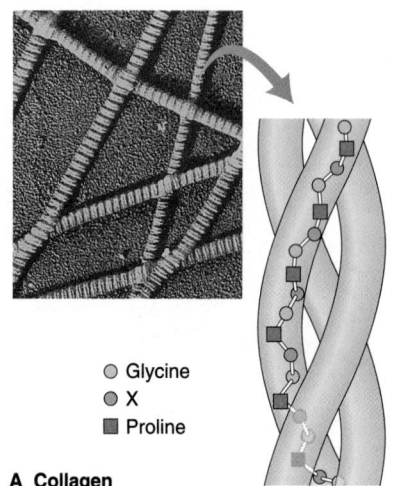

○ Glycine
● X
■ Proline

A Collagen

Glycine

B Silk fibroin — Alanine (or serine)

Figure 15.30 The shapes of fibrous proteins. Fibrous proteins have largely structural roles in an organism, and their amino acid sequence is relatively simple. **A,** The triple helix of a collagen molecule has a glycine as every third amino acid along the chain. **B,** The pleated sheet structure of silk fibroin has a repeating sequence of glycine alternating with either alanine or serine. The sheets interact through dispersion forces.

ARIS Presentation Center, Art Library

Nucleotides and Nucleic Acids

An organism's nucleic acids construct its proteins. And, given that the proteins determine how the organism looks and behaves, no job could be more essential.

Monomer Structure and Linkage **Nucleic acids** are unbranched polymers that consist of linked monomer units called **mononucleotides,** which consist of an N-containing base, a sugar, and a phosphate group. The two types of nucleic acid, *ribonucleic acid* (RNA) and *deoxyribonucleic acid* (DNA), differ in the sugar portions of their mononucleotides. RNA contains *ribose,* a five-C sugar, and DNA contains *deoxyribose,* in which —H substitutes for —OH on the 2′ position of ribose. (Carbon atoms in the sugar portion are given numbers with a prime to distinguish them from carbon atoms in the base.)

The cellular precursors that form a nucleic acid are *nucleoside triphosphates* (Figure 15.31A). Dehydration-condensation reactions between them release inorganic diphosphate ($H_2P_2O_7^{2-}$) and create *phosphodiester* bonds to form a polynucleotide chain. Therefore, the repeating pattern of the nucleic acid backbone is —*sugar—phosphate—sugar—phosphate—*, and so on (Figure 15.31B). Attached to each sugar is one of four N-containing bases, either a pyrimidine (six-membered ring) or a purine (six- and five-membered rings sharing a side). The pyrimidines are thymine (T) and cytosine (C); the purines are guanine (G) and adenine (A); in RNA, uracil (U) substitutes for thymine. The bases dangle off the sugar-phosphate chain, much as R groups dangle off the polypeptide chain of a protein.

Figure 15.31 Nucleic acid precursors and their linkage. A, In the cell, nucleic acids are constructed from nucleoside triphosphates, precursors of the mononucleotide units. Each one consists of an N-containing base (structure not shown), a sugar, and a triphosphate group. In RNA (*top*), the sugar is ribose; in DNA, it is 2′-deoxyribose (note the absence of an —OH group on C-2 of the ring). **B,** A tiny segment of the polynucleotide chain of DNA (with one, mononucleotide unit boxed) shows the phosphodiester bonds that link the 5′-OH group of one sugar to the 3′-OH group of the next and are formed through dehydration-condensation reactions (which also release diphosphate ion). The bases dangle off the chain.

Transparency/ARIS Presentation Center, Art Library

Figure 15.32 The double helix of DNA. A segment of DNA is shown as a space-filling model (*left*). The boxed area is expanded (*center*) to show how the polar sugar-phosphate (—S—P—) backbone faces the watery outside, and the nonpolar bases form H bonds to each other in the DNA core. The boxed area is expanded (*right*) to show how a pyrimidine and a purine always form H-bonded base pairs to maintain the double helix width. The members of the pairs are always the same: A pairs with T, and G pairs with C.

Transparency/ARIS Presentation Center, Art Library

Literature Heinhorst, Sabine; Cannon, Gordon. "*Nature:* What Matters in Nature—DNA, Light-Driven Ion Pumps, and Antimatter," *J. Chem. Educ.* **2003,** *80,* 466.

The Central Importance of Base Pairing In the nucleus of the cell, DNA exists as two chains wrapped around each other in a **double helix** (Figure 15.32). Each base in one chain "pairs" with a base in the other through H bonding. A double-helical DNA molecule may contain many millions of H-bonded bases in specified pairs. Two features of these **base pairs** are crucial to the structure and function of DNA:

- A pyrimidine and a purine are always paired, which gives the double helix a constant diameter.
- Each base is always paired with the same partner: A with T and G with C. Thus, *the base sequence on one chain is the complement of the sequence on the other.* For example, the sequence A—C—T on one chain is *always* paired with T—G—A on the other: A with T, C with G, and T with A.

Each DNA molecule is folded into a tangled mass that forms one of the cell's *chromosomes.* The DNA molecule is amazingly long and thin: if the largest human chromosome were stretched out, it would be 4 cm long; in the cell nucleus, however, it is wound into a structure only 5 nm long—8 million times shorter! Segments of the DNA chains are the *genes* that contain the chemical information for synthesizing the organism's proteins.

An Outline of Protein Synthesis *The information content of a gene resides in its base sequence.* In the **genetic code,** each base acts as a "letter," each three-base sequence as a "word," and *each word codes for a specific amino acid.* For example, the sequence C—A—C codes for the amino acid histidine, A—A—G codes for lysine, and so on. Through a complex series of interactions, greatly simplified

Figure 15.33 Key stages in protein synthesis.

Transparency/ARIS Presentation Center, Art Library

in the overview in Figure 15.33, one amino acid at a time is positioned and linked to the next in the process of protein synthesis. To fully appreciate this aspect of the chemical basis of biology, keep in mind that *this amazingly complex process occurs largely through H bonding between base pairs.*

Here is an outline of *protein synthesis.* DNA occurs in the cell nucleus, but the genetic message is decoded outside it, so the information must be sent to the synthesis site. RNA serves in this messenger role, as well as in several others. A portion of the DNA is temporarily unwound and one chain segment acts as a *template* for the formation of a complementary chain of *messenger RNA* (mRNA) made by individual mononucleoside triphosphates linking together; thus, *the DNA code words are transcribed into RNA code words* through base pairing. The mRNA leaves the nucleus and binds, again through base pairing, to an RNA-rich particle in the cell called a *ribosome.* The words (three-base sequences) in the mRNA are then decoded by molecules of *transfer RNA* (tRNA). These smaller nucleic acid "shuttles" have two key portions on opposite ends of their structures:

- A three-base sequence that is the complement of a word on the mRNA
- A binding site for the amino acid coded by that word

The ribosome moves along the bound mRNA, one word at a time, while tRNAs bind to the mRNA and position their amino acids near one another in preparation for peptide bond formation and synthesis of the protein.

In essence, then, protein synthesis involves the DNA message of three-base words being transcribed into the RNA message of three-base words, which is then translated into a sequence of amino acids that are linked to make a protein:

DNA BASE SEQUENCE ⇒ RNA BASE SEQUENCE ⇒ PROTEIN AMINO ACID SEQUENCE

An Outline of DNA Replication Another complex series of interactions allows *DNA replication,* the ability of DNA to copy itself. When a cell divides, its chromosomes are replicated, or reproduced, ensuring that the new cells have the same number and types of chromosomes. In this process, a small portion of the double helix is

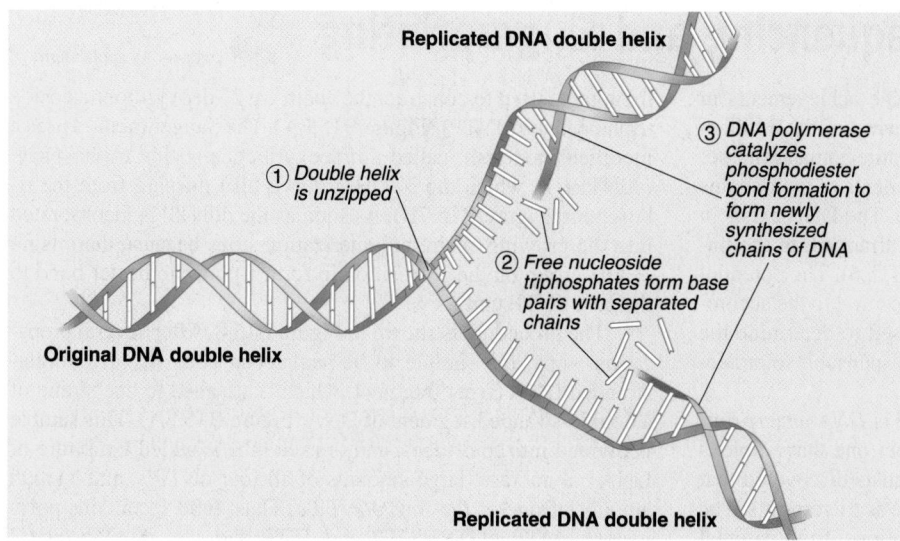

Figure 15.34 Key stages in DNA replication.

ARIS Presentation Center, Art Library

① Double helix is unzipped

② Free nucleoside triphosphates form base pairs with separated chains

③ DNA polymerase catalyzes phosphodiester bond formation to form newly synthesized chains of DNA

Replicated DNA double helix

Original DNA double helix

Replicated DNA double helix

"unzipped," and each DNA chain acts as a template for the base pairing of its mononucleotide monomers with free mononucleoside triphosphates (Figure 15.34). These new units, H-bonded to their complements, are linked through phosphodiester bonds into a chain by an enzyme called *DNA polymerase*. Gradually, each of the unzipped chains forms the complementary half of a double helix, leading to *two* double helices. Because each base always pairs with its complement, the original double helix is copied and the genetic makeup of the cell preserved. (Base pairing is central to methods for learning the sequence of nucleotides in our genes, as the upcoming Chemical Connections essay shows.)

The biopolymers provide striking evidence for the folly of vitalism—that some sort of vital force exists only in substances from living systems. No unknowable force is required to explain the marvels of the living world. The same atomic properties that give rise to covalent bonds, molecular shape, and intermolecular forces provide the means for all life forms to flourish.

Section Summary

The three types of natural polymers—polysaccharides, proteins, and nucleic acids—are formed by dehydration-condensation reactions. • Polysaccharides are formed from cyclic monosaccharides, such as glucose. Cellulose, starch, and glycogen have structural or energy-storage roles. • Proteins are polyamides formed from as many as 20 different types of amino acids. Fibrous proteins have extended shapes and play structural roles. Globular proteins have compact shapes and play metabolic, immunologic, and hormonal roles. The amino acid sequence of a protein determines its shape and function. • Nucleic acids (DNA and RNA) are polynucleotides consisting of four different mononucleotides. The base sequence of the DNA chain determines the sequence of amino acids in an organism's proteins. Hydrogen bonding between specific base pairs is the key to protein synthesis and DNA replication. • DNA sequencing is used to determine the identity and order of nucleotides; DNA fingerprinting is used to identify a person from the characteristic pattern of gene fragments in a sample of cells.

Chapter Perspective

The amazing diversity of organic compounds highlights the importance of atomic properties. While gaining an appreciation for the special properties of carbon, you've seen similar types of bonding, molecular shapes, intermolecular forces, and reactivities in organic and inorganic compounds. Upcoming chapters focus on the dynamic aspects of all reactions—their speed, extent, and direction—as we explore the kinetics, equilibrium, and thermodynamics of chemical and physical change. We'll have many opportunities to exemplify these topics with organic compounds and biochemical systems.

Chemical Connections to Genetics and Forensics

DNA Sequencing and Fingerprinting

Biological application

Engineering application

As a result of one of the most remarkable achievements in modern science, we now know the sequence of the 3 billion nucleotide base pairs in the DNA of the entire human genome! The *genome* consists of DNA molecules; segments of their chains are the genes, the functional units of heredity. The DNA exists in each cell's nucleus as tightly coiled threads arranged, in the human, into 23 pairs of chromosomes (Figure B15.6). The potential benefits of this knowledge are profound, and central to the accomplishment is *DNA sequencing,* the process used to determine the identity and order of bases. Sequencing is indispensable to molecular biology and biochemical genetics.

Just as indispensable to forensic science is *DNA fingerprinting* (or *DNA profiling*), a different process but one that employs some similar methods. In 1985, British scientists discovered that portions of an individual's DNA are as unique as fingerprints. The technique has been applied in countless situations, from parental custody claims to crime scene investigations to identifying the remains of victims of the September 11, 2001, terrorist attacks.

An Outline of DNA Sequencing

A given chromosome may have 100 million nucleotide bases, but the sequencing process can handle, at one time, DNA fragments only about 2000 bases long. Therefore, the chromosome is first broken into pieces by enzymes that cleave at specific sites. Then, to obtain enough sample for analysis, the DNA is replicated through a variety of "amplification" methods, which make many copies of the individual DNA "target" fragments.

The most popular sequencing method is the *Sanger chain-termination method*, which uses chemically altered bases to stop the growth of a complementary DNA chain at specific locations. As you've seen in the previous text discussion, the chain consists of linked 2'-deoxyribonucleoside monophosphate units (dNMP, where N represents A, T, G, or C). The link is a phosphodiester bond from the 3'-OH of one unit to the 5'-OH of the next. The free

monomers used to construct the chain are 2'-deoxyribonucleoside triphosphates (dNTP) (Figure B15.7A). The Sanger method uses a modified monomer, called a *di*deoxyribonucleoside triphosphate (ddNTP), in which the 3'-OH group is also missing from the ribose unit (Figure B15.7B). As soon as the ddNTP is incorporated into the growing chain, polymerization stops because there is no —OH group on the 3' position to form a phosphodiester bond to the next dNTP unit.

The procedure is shown in Figure B15.8. After several preparation steps, the sample to be sequenced consists of a single-stranded DNA target fragment, which is attached to one strand of a double-stranded segment of DNA (Figure B15.8A). This sample is divided into four tubes, and to each tube is added a mixture of DNA polymerase, large amounts of all four dNTP's, and a small amount of one of the four ddNTP's. Thus, tube 1 contains polymerase, dATP, dGTP, dCTP, and dTTP, and, say, ddATP; tube 2 contains the same, except ddGTP instead of ddATP; and so forth. After the polymerization reaction is complete, each tube contains the original target fragment paired to complementary chains of varying length (Figure B15.8B). The chain lengths vary because in tube 1, each complementary chain ends in ddA (designated A in the figure); in tube 2, each ends in ddG (G); in tube 3, each ends in ddC (C); and in tube 4, each ends in ddT (T).

Each double-stranded product is divided into single strands, and then the complementary chains are separated by means of *high-resolution polyacrylamide-gel electrophoresis*. This technique applies an electric field to separate charged species through differences in their rate of migration through pores in a gel: the smaller the species, the faster it moves. Polynucleotide fragments are commonly separated by electrophoresis because they have charged phosphate groups. High-resolution gels can be made with pores that vary so slightly in size that they can separate fragments differing by only a single nucleotide.

In this step, each sample is applied to its own "lane" on a gel, and, after electrophoresis, the gel is scanned to locate the chains,

Figure B15.6 DNA, the genetic material. In the cell nucleus, each chromosome consists of a DNA molecule wrapped around globular proteins called histones. Segments of the DNA chains are genes.

$N = A, G, C,$ or T

Base N

Deoxynucleoside triphosphate (dNTP)

Base N

Dideoxynucleoside triphosphate (ddNTP)

no 3'-OH

Figure B15.7 Nucleoside triphosphate monomers. The normal *deoxy* monomer (dNTP; *top*) has no 2'-OH group but *does* have a 3'-OH group to continue growth of the polynucleotide chain. The modified *dideoxy* monomer in the Sanger method (ddNTP; *bottom*) also lacks the 3'-OH group.

Chemical Connections *(continued)*

A

B

C

D

Figure B15.8 Steps in the Sanger method of DNA sequencing. (See text for full discussion.)

which appear in bands. Because all four ddNTP's were used, all possible chain fragments are formed, so the sequence of the original DNA fragment can be determined (Figure B15.8C).

An automated approach begins with each ddNTP tagged with a different fluorescent dye that emits light of a distinct color. The entire mixture of complementary chains is introduced onto the gel and separated. The gel is scanned by a laser, which activates the dyes. The fluorescent intensity versus distance along the chain is detected and plotted by a computer (Figure B15.8D).

An Outline of DNA Fingerprinting

Biological samples collected at a crime scene are often very small, so, as with sequencing, the first step to obtain enough DNA for forensic analysis involves amplification. Modern techniques require less than 1 ng of DNA. Thus, in addition to blood and semen, licked stamps and envelopes, gum, and clothing containing dead skin cells can also be used as sources for the DNA.

Currently, the most successful and widely used DNA profiling procedure is called *short tandem repeat (STR) analysis*. STRs are specific areas on a chromosome that contain short (3–7 bases) sequences that repeat themselves within the DNA molecule. There are hundreds of different types of STRs, but each person has a unique number and assortment of types. Therefore, the more STRs that are characterized, the more discriminating the analysis. Once the STRs have been amplified, they are separated by electrophoresis, and an analyst determines the number of base repeats within each STR (Figure B15.9).

ARIS Presentation Center, Art Library

Figure B15.9 STR analysis of DNA in a blood stain. Stained gels containing the STRs of DNA in blood samples from several suspects are compared with those from the blood found at the crime scene. Suspect 3 has an identical pattern of STRs to those in the blood stain.

Literature Helser, Terry L. "Genetic Code Wordsearch," *J. Chem. Educ.* **2003,** *80,* 417.

CHAPTER REVIEW GUIDE

The following sections provide many aids to help you study this chapter. (Numbers in parentheses refer to pages, unless noted otherwise.)

Learning Objectives
These are concepts and skills you should know after studying this chapter.

Relevant section and/or sample problem (SP) numbers appear in parentheses.

Understand These Concepts

1. How carbon's atomic properties give rise to its ability to form four strong covalent bonds, multiple bonds, and chains, which results in the great structural diversity of organic compounds (15.1)
2. How carbon's atomic properties give rise to its ability to bond to various heteroatoms, which creates regions of charge imbalance that result in functional groups (15.1)
3. Structures and names of alkanes, alkenes, and alkynes (15.2)
4. The distinctions among constitutional, optical, and geometric isomers (15.2)
5. The importance of optical isomerism in organisms (15.2)
6. The effect of restricted rotation around a π bond on the structures and properties of alkenes (15.2)
7. The nature of organic addition, elimination, and substitution reactions (15.3)
8. The properties and reaction types of the various functional groups (15.4):
   - Substitution and elimination for alcohols, alkyl halides, and amines
   - Addition for alkenes, alkynes, and aldehydes and ketones
   - Substitution for the carboxylic acid family (acids, esters, and amides)
9. Why delocalization of electrons causes aromatic rings to have lower reactivity than alkenes (15.4)
10. The polarity of the carbonyl bond and the importance of organometallic compounds in addition reactions of carbonyl compounds (15.4)
11. How addition plus elimination lead to substitution in the reactions of the carboxylic acid family (15.4)
12. How addition and condensation polymers form (15.5)
13. The three types of biopolymers and their monomers (15.6)

14. How amino acid sequence determines protein shape, which determines function (15.6)
15. How complementary base pairing controls the processes of protein synthesis and DNA replication (15.6)
16. How DNA base sequence determines RNA base sequence, which determines amino acid sequence (15.6)

Master These Skills

1. Drawing hydrocarbon structures given the number(s) of C atoms, multiple bonds, and rings (SP 15.1)
2. Naming hydrocarbons and drawing expanded, condensed, and carbon skeleton formulas (15.2 and SP 15.2)
3. Drawing geometric isomers and identifying chiral centers of molecules (SP 15.2)
4. Recognizing the type of reaction from the structures of reactants and products (SP 15.3)
5. Recognizing a reaction as an oxidation or reduction from the structures of reactants and products (15.3)
6. Determining the reactants and products of the reactions of alcohols, alkyl halides, and amines (SP 15.4 and Follow-up)
7. Determining the products in a stepwise reaction sequence (SP 15.5)
8. Determining the reactants of the reactions of aldehydes and ketones (Follow-up Problem 15.5)
9. Determining the reactants and products of the reactions of the carboxylic acid family (SP 15.6 and Follow-up)
10. Recognizing and naming the functional groups in an organic molecule (SP 15.7)
11. Drawing an abbreviated synthetic polymer structure based on monomer structures (15.5)
12. Drawing small peptides from amino acid structures (15.6)
13. Using the base-paired sequence of one DNA strand to predict the sequence of the other (15.6)

Key Terms
These important terms appear in boldface in the chapter and are defined again in the Glossary.

organic compound (629)

Section 15.1
heteroatom (631)
functional group (631)

Section 15.2
hydrocarbon (632)
alkane (C_nH_{2n+2}) (635)
homologous series (635)
saturated hydrocarbon (635)
cyclic hydrocarbon (637)
isomers (637)
constitutional (structural) isomers (637)
stereoisomers (639)
optical isomers (639)

chiral molecule (639)
polarimeter (640)
optically active (640)
alkene (C_nH_{2n}) (640)
unsaturated hydrocarbon (640)
geometric (*cis-trans*) isomers (641)
alkyne (C_nH_{2n-2}) (643)
aromatic hydrocarbon (644)
nuclear magnetic resonance (NMR) spectroscopy (646)

Section 15.3
alkyl group (646)
addition reaction (647)
elimination reaction (648)
substitution reaction (648)

Section 15.4
alcohol (650)
haloalkane (alkyl halide) (652)
amine (652)
carbonyl group (655)
aldehyde (655)
ketone (655)
organometallic compound (656)
carboxylic acid (657)
ester (657)
amide (657)
fatty acid (658)
acid anhydride (658)
lipid (658)
hydrolysis (659)

nitrile (661)

Section 15.5
addition polymer (663)
condensation polymer (664)

Section 15.6
monosaccharide (666)
polysaccharide (666)
disaccharide (666)
protein (667)
amino acid (667)
nucleic acid (670)
mononucleotide (670)
double helix (671)
base pair (671)
genetic code (671)

Highlighted Figures and Tables These figures (F) and tables (T) provide a visual review of key ideas.

F15.4 Adding H-atom skin to C-atom skeleton (633)
T15.1 Numerical roots for carbon chains and branches (635)
T15.2 Rules for naming an organic compound (636)

T15.5 Important organic functional groups (651)
F15.31 Nucleic acid precursors and their linkage (670)

Brief Solutions to FOLLOW-UP PROBLEMS Compare your solutions to these calculation steps and answers.

PROBLEMS

Problems with **colored** numbers are answered in Appendix E and worked in detail in the Student Solutions Manual. Problem sections match those in the text and provide the numbers of relevant sample problems. Most offer Concept Review Questions, Skill-Building Exercises (grouped in pairs covering the same concept), and Problems in Context. The Comprehensive Problems are based on material from any section or previous chapter.

The Special Nature of Carbon and the Characteristics of Organic Molecules

● Concept Review Questions

15.1 Give the names and formulas of two carbon compounds that are organic and two that are inorganic.

15.2 Through the first quarter of the 19th century, the concept of vitalism—that a key difference existed between substances isolated from animate sources and those from inanimate sources—was widely accepted. What was the central notion of vitalism? Give an example of a finding that led to its eventual downfall.

15.3 Explain each statement in terms of atomic properties:
(a) Carbon engages in covalent rather than ionic bonding.
(b) Carbon has four bonds in all its organic compounds.
(c) Carbon forms neither stable cations, like many metals, nor stable anions, like many nonmetals.
(d) Carbon bonds to itself more extensively than does any other element.
(e) Carbon forms stable multiple bonds.

15.4 Carbon bonds to many elements other than itself.
(a) Name six elements that commonly bond to carbon in organic compounds.
(b) Which of these elements are heteroatoms?
(c) Which of these elements are more electronegative than carbon? Less electronegative?
(d) How does bonding of carbon to heteroatoms increase the number of organic compounds?

15.5 Silicon lies just below carbon in Group 4A(14) and also forms four covalent bonds. Why aren't there as many silicon compounds as carbon compounds?

15.6 What is the range of oxidation states for carbon? Name a compound in which carbon has its highest oxidation state and one in which it has its lowest.

15.7 Which of these bonds to carbon would you expect to be relatively reactive: C—H, C—C, C—I, C=O, C—Li? Explain.

The Structures and Classes of Hydrocarbons
(Sample Problems 15.1 and 15.2)

● Concept Review Questions

15.8 (a) What structural feature is associated with each type of hydrocarbon: an alkane; a cycloalkane; an alkene; an alkyne?
(b) Give the general formula for each type.
(c) Which hydrocarbons are considered saturated?

15.9 Define each type of isomer: (a) constitutional; (b) geometric; (c) optical. Which types of isomers are stereoisomers?

15.10 Among alkenes, alkynes, and aromatic hydrocarbons, only alkenes exhibit *cis-trans* isomerism. Why don't the others?

15.11 Which objects are asymmetric (have no plane of symmetry): (a) a circular clock face; (b) a football; (c) a dime; (d) a brick; (e) a hammer; (f) a spring?

15.12 Explain how a polarimeter works and what it measures.

15.13 How does an aromatic hydrocarbon differ from a cycloalkane in terms of its bonding? How does this difference affect structure?

● Skill-Building Exercises (grouped in similar pairs)

15.14 Draw all possible skeletons for a 7-C compound with
(a) A 6-C chain and 1 double bond
(b) A 5-C chain and 1 double bond
(c) A 5-C ring and no double bonds

15.15 Draw all possible skeletons for a 6-C compound with
(a) A 5-C chain and 2 double bonds
(b) A 5-C chain and 1 triple bond
(c) A 4-C ring and no double bonds

15.16 Add the correct number of hydrogens to each of the skeletons in Problem 15.14.

15.17 Add the correct number of hydrogens to each of the skeletons in Problem 15.15.

15.18 Draw correct structures, by making a single change, for any that are incorrect:

15.19 Draw correct structures, by making a single change, for any that are incorrect:

15.20 Draw the structure or give the name of each compound:
(a) 2,3-dimethyloctane   (b) 1-ethyl-3-methylcyclohexane

15.21 Draw the structure or give the name of each compound:

(c) 1,2-diethylcyclopentane   (d) 2,4,5-trimethylnonane

15.22 Each of the following names is wrong. Draw structures based on them, and correct the names:
(a) 4-methylhexane   (b) 2-ethylpentane
(c) 2-methylcyclohexane   (d) 3,3-methyl-4-ethyloctane

Levels of Difficulty: ● Easy ●● Medium ●●● Challenging

15.23 Each of the following names is wrong. Draw structures based on them, and correct the names:
(a) 3,3-dimethylbutane   (b) 1,1,1-trimethylheptane
(c) 1,4-diethylcyclopentane   (d) 1-propylcyclohexane

15.24 Each of the following compounds can exhibit optical activity. Circle the chiral center(s) in each:
(a)   (b)

15.25 Each of the following compounds can exhibit optical activity. Circle the chiral center(s) in each:
(a)   (b)

15.26 Draw structures from the following names, and determine which compounds are optically active:
(a) 3-bromohexane   (b) 3-chloro-3-methylpentane
(c) 1,2-dibromo-2-methylbutane

15.27 Draw structures from the following names, and determine which compounds are optically active:
(a) 1,3-dichloropentane   (b) 3-chloro-2,2,5-trimethylhexane
(c) 1-bromo-1-chlorobutane

15.28 Which of the following structures exhibit geometric isomerism? Draw and name the two isomers in each case:
(a) CH₃—CH₂—CH=CH—CH₃   (b)

(c) CH₃—C=CH—CH—CH₂—CH₃ (with CH₃ groups)

15.29 Which of the following structures exhibit geometric isomerism? Draw and name the two isomers in each case:
(a) CH₃—C—CH=CH—CH₃ (with two CH₃ groups)   (b)

(c) Cl—CH₂—CH=C—CH₂—CH₂—CH₂—CH₃ (with CH₃ group)

15.30 Which compounds exhibit geometric isomerism? Draw and name the two isomers in each case:
(a) propene   (b) 3-hexene
(c) 1,1-dichloroethene   (d) 1,2-dichloroethene

15.31 Which compounds exhibit geometric isomerism? Draw and name the two isomers in each case:
(a) 1-pentene   (b) 2-pentene
(c) 1-chloropropene   (d) 2-chloropropene

15.32 Draw and name all the constitutional isomers of dichlorobenzene.

15.33 Draw and name all the constitutional isomers of trimethylbenzene.

Problems in Context

15.34 Butylated hydroxytoluene (BHT) is a common preservative added to cereals and other dry foods. Its systematic name is 1-hydroxy-2,6-di-*tert*-butyl-4-methylbenzene (where "*tert*-butyl" is 1,1-dimethylethyl). Draw the structure of BHT.

15.35 There are two compounds with the name 2-methyl-3-hexene, but only one with the name 2-methyl-2-hexene. Explain with structures.

15.36 Any tetrahedral atom with four different groups attached can be a chiral center. Which of these species is optically active?
(a) CHClBrF
(b) NBrCl₂H⁺
(c) PFClBrI⁺
(d) SeFClBrH

15.37 The number of peaks in the ¹H-NMR spectrum of a compound depends on the number of different kinds (environments) of H atoms. How many peaks appear in the spectrum of each isomer of C_4H_{10} and of C_5H_{12} (see Table 15.3, p. 638)?

Some Important Classes of Organic Reactions
(Sample Problem 15.3)

Concept Review Questions

15.38 In terms of numbers of reactant and product substances, which organic reaction type corresponds to (a) a combination reaction, (b) a decomposition reaction, (c) a displacement reaction?

15.39 The same type of bond is broken in an addition reaction and formed in an elimination reaction. Name the type.

15.40 Can a redox reaction also be an addition, elimination, or substitution reaction? Explain with examples.

Skill-Building Exercises *(grouped in similar pairs)*

15.41 Determine the type of each of the following reactions:

(a) CH₃—CH₂—CH—CH₃ (with Br) $\xrightarrow[\Delta]{NaOH}$
       CH₃—CH=CH—CH₃ + NaBr + H₂O

(b) CH₃—CH=CH—CH₂—CH₃ + H₂ $\xrightarrow{Pt}$
       CH₃—CH₂—CH₂—CH₂—CH₃

15.42 Determine the type of each of the following reactions:

(a) CH₃—CH (with O) + HCN $\longrightarrow$ CH₃—CH—CN (with OH)

(b) CH₃—C—O—CH₃ (with O) + CH₃—NH₂ $\xrightarrow{H^+}$
       CH₃—C—NH—CH₃ (with O) + CH₃—OH

15.43 Write equations for the following: (a) an addition reaction between H₂O and 3-hexene (H⁺ is a catalyst); (b) an elimination reaction between 2-bromopropane and hot potassium ethoxide, CH₃—CH₂—OK (KBr and ethanol are also products); (c) a light-induced substitution reaction between Cl₂ and ethane to form 1,1-dichloroethane.

15.44 Write equations for the following: (a) a substitution reaction between 2-bromopropane and KI; (b) an addition reaction between cyclohexene and Cl₂; (c) an addition reaction between 2-propanone and H₂ (Ni metal is a catalyst).

15.45 Based on the number of bonds and the nature of the bonded atoms, state whether each of the following changes is an oxidation or a reduction:

(a) $=CH_2$ becomes $-CH_2-OH$

(b) $=CH-$ becomes $-CH_2-$

(c) $\equiv C-$ becomes $-CH_2-$

15.46 Based on the number of bonds and the nature of the bonded atoms, state whether each of the following changes is an oxidation or a reduction:

(a) $-\overset{|}{\underset{|}{C}}-OH$ becomes $-C=O$

(b) $-CH_2-OH$ becomes $=CH_2$

(c) $-\overset{O}{\overset{||}{C}}-\overset{O}{\overset{||}{C}}-$ becomes $-\overset{O}{\overset{||}{C}}-O-$

15.47 Is the organic reactant oxidized, reduced, or neither in each of the following reactions?

(a) 2-hexene $\xrightarrow[\text{cold OH}^-]{\text{KMnO}_4}$ 2,3-dihydroxyhexane

(b) cyclohexane $\xrightarrow[\text{catalyst}]{\Delta}$ benzene + $3H_2$

15.48 Is the organic reactant oxidized, reduced, or neither in each of the following reactions?

(a) 1-butyne + H_2 $\xrightarrow{\text{Pt}}$ 1-butene

(b) toluene $\xrightarrow[\text{H}_3\text{O}^+, \Delta]{\text{KMnO}_4}$ benzoic acid

● **Problems in Context**

15.49 Phenylethylamine is a natural substance that is structurally similar to amphetamine. It is found in sources as diverse as almond oil and human urine, where it occurs at elevated concentrations as a result of stress and certain forms of schizophrenia. One method of synthesizing the compound for pharmacological and psychiatric studies involves two steps:

Classify each step as an addition, elimination, or substitution.

Properties and Reactivities of Common Functional Groups

(Sample Problems 15.4 to 15.7)

● **Concept Review Questions**

15.50 Compounds with nearly identical molar masses often have very different physical properties. Choose the compound with the higher value for each of the following properties, and explain your choice.
(a) Solubility in water: chloroethane or methylethylamine
(b) Melting point: diethyl ether or 1-butanol
(c) Boiling point: trimethylamine or propylamine

15.51 Fill in each blank with a general formula for the type of compound formed:

15.52 Of the three major types of organic reactions, which do *not* occur readily with benzene? Why?

15.53 Why does the C=O group react differently from the C=C group? Show an example of the difference.

15.54 Many substitution reactions involve an initial electrostatic attraction between reactants. Show where this attraction arises in the formation of an amide from an amine and an ester.

15.55 Although both carboxylic acids and alcohols contain an —OH group, one is acidic in water and the other is not. Explain.

15.56 What reaction type is common to the formation of esters and acid anhydrides? What is the other product?

15.57 Both alcohols and carboxylic acids undergo substitution, but the processes are very different. Explain.

● **Skill-Building Exercises** (grouped in similar pairs)

15.58 Name the type of organic compound from the following description of its functional group: (a) polar group that has only single bonds and does not include O or N; (b) group that is polar and has a triple bond; (c) group that has single and double bonds and is acidic in water; (d) group that has a double bond and must be at the end of a C chain.

15.59 Name the type of organic compound from the following description of its functional group: (a) N-containing group with single and double bonds; (b) group that is not polar and has a double bond; (c) polar group that has a double bond and cannot be at the end of a C chain; (d) group that has only single bonds and is basic in water.

15.60 Circle and name the functional group(s) in each compound:

15.61 Circle and name the functional group(s) in each compound:

15.62 Draw all alcohols with the formula $C_5H_{12}O$.

15.63 Draw all aldehydes and ketones with the formula $C_5H_{10}O$.

15.64 Draw all amines with the formula $C_4H_{11}N$.

15.65 Draw all carboxylic acids with the formula $C_5H_{10}O_2$.

15.66 Draw the product resulting from mild oxidation of (a) 2-butanol; (b) 2-methylpropanal; (c) cyclopentanol.

15.67 Draw the alcohol you'd oxidize to produce (a) 2-methylpropanal; (b) 2-pentanone; (c) 3-methylbutanoic acid.

15.68 Draw the organic product formed when the following compounds undergo a substitution reaction: (a) acetic acid and methylamine; (b) butanoic acid and 2-propanol; (c) formic acid and 2-methyl-1-propanol.

• **15.69** Draw the organic product formed when the following compounds undergo a substitution reaction: (a) acetic acid and 1-hexanol; (b) propanoic acid and dimethylamine; (c) ethanoic acid and diethylamine.

• **15.70** Draw condensed formulas for the carboxylic acid and alcohol that form the following esters:

(a)

(b)

(c) CH_3—CH_2—O—$\overset{\overset{\textstyle O}{\|}}{C}$—$CH_2$—$CH_2$—

• **15.71** Draw condensed formulas for the carboxylic acid and amine that form the following amides:

(a) H_3C— —CH_2—$\overset{\overset{\textstyle O}{\|}}{C}$—$NH_2$

(b)   (c) $HC\overset{\overset{\textstyle O}{\|}}{}$—$NH$—

• **15.72** Fill in the expected organic substances:

(a) CH_3—CH_2—$Br \xrightarrow{OH^-}$ ____ $\xrightarrow{CH_3-CH_2-\overset{\overset{\textstyle O}{\|}}{C}-OH}{H^+}$ ____

(b) CH_3—CH_2—$\overset{\overset{\textstyle Br}{|}}{CH}$—$CH_3 \xrightarrow{CN^-}$ ____ $\xrightarrow{H_3O^+, H_2O}$ ____

•• **15.73** Fill in the expected organic substances:

(a) CH_3—CH_2—CH=$CH_2 \xrightarrow{H^+, H_2O}$ ____ $\xrightarrow{Cr_2O_7^{2-}, H^+}$ ____

(b) CH_3—CH_2—$\overset{\overset{\textstyle O}{\|}}{C}$—$CH_3 \xrightarrow{CH_3-CH_2-Li}$ $\xrightarrow{H_2O}$ ____

• **15.74** Supply the missing organic and/or inorganic substances:

(a) CH_3—CH_2—OH + ____ $\xrightarrow{H_3O^+, H_2O}$

CH_3—CH_2—O—$\overset{\overset{\textstyle O}{\|}}{C}$—$CH_2$—$CH_3$

(b) CH_3—$\overset{\overset{\textstyle O}{\|}}{C}$—O—$CH_3 \xrightarrow{?}$ CH_3—CH_2—NH—$\overset{\overset{\textstyle O}{\|}}{C}$—$CH_3$

• **15.75** Supply the missing organic and/or inorganic substances:

(a) CH_3—$\overset{\overset{\textstyle Cl}{|}}{CH}$—$CH_3 \xrightarrow{?}$ CH_3—CH=$CH_2 \xrightarrow{?}$ CH_3—$\overset{\overset{\textstyle Br}{|}}{CH}$—$\overset{\overset{\textstyle Br}{|}}{CH_2}$

(b) CH_3—CH_2—CH_2—$OH \xrightarrow{?}$ CH_3—CH_2—$\overset{\overset{\textstyle O}{\|}}{C}$—$OH$ + ____ $\xrightarrow{?}$ $\xrightarrow{?}$

CH_3—CH_2—$\overset{\overset{\textstyle O}{\|}}{C}$—O—$CH_2$—

● **Problems in Context**

•• **15.76** (a) Draw the four isomers of $C_5H_{12}O$ that can be oxidized to an aldehyde. (b) Draw the three isomers of $C_5H_{12}O$ that can be oxidized to a ketone. (c) Draw the isomers of $C_5H_{12}O$ that cannot be easily oxidized to an aldehyde or ketone. (d) Name any isomer that is an alcohol.

•• **15.77** Ethyl formate ($H\overset{\overset{\textstyle O}{\|}}{C}$—O—$CH_2$—$CH_3$) is added to foods to give them the flavor of rum. How would you synthesize ethyl formate from ethanol, methanol, and any inorganic reagents?

The Monomer-Polymer Theme I: Synthetic Macromolecules

● **Concept Review Questions**

• **15.78** Name the reaction processes that lead to the two types of synthetic polymers.

• **15.79** Which functional group occurs in the monomers of addition polymers? How are these polymers different from one another?

• **15.80** What is a free radical? How is it involved in polymer formation?

• **15.81** Which intermolecular force is primarily responsible for the different types of polyethylene? Explain.

• **15.82** Which of the two types of synthetic polymer is more similar chemically to biopolymers? Explain.

• **15.83** Which functional groups react to form nylons? Polyesters?

● **Skill-Building Exercises** (grouped in similar pairs)

• **15.84** Draw an abbreviated structure for the following polymers, with brackets around the repeat unit:

(a) Poly(vinyl chloride) (PVC) from

(b) Polypropylene from

• **15.85** Draw an abbreviated structure for the following polymers, with brackets around the repeat unit:

(a) Teflon from   (b) Polystyrene from

● **Problems in Context**

•• **15.86** Polyethylene terephthalate (PET) is used to make synthetic fibers, such as Dacron, thin films, such as Mylar, and bottles for carbonated beverages.

PET

PET is produced from ethylene glycol and either of two monomers, depending on whether the reaction proceeds by dehydration-condensation or by displacement. Write equations for the two syntheses. (*Hint:* The displacement is reversed by adding methanol to PET at high *T* and *P*.)

•• **15.87** Write a balanced equation for the reaction of dihydroxydimethylsilane (*right*) to form the condensation polymer known as Silly Putty.

$\overset{\overset{\textstyle CH_3}{|}}{HO-Si-OH}\underset{\underset{\textstyle CH_3}{|}}{}$

The Monomer-Polymer Theme II: Biological Macromolecules

● Concept Review Questions

● **15.88** Which type of polymer is formed from each of the following monomers: (a) amino acids; (b) alkenes; (c) simple sugars; (d) mononucleotides?

● **15.89** What is the key structural difference between fibrous and globular proteins? How is it related, in general, to the proteins' amino acid composition?

● **15.90** Protein shape, function, and amino acid sequence are interrelated. Which determines which?

● **15.91** What linkage joins the monomers in each strand of DNA?

● **15.92** What is base pairing? How does it pertain to DNA structure?

● **15.93** RNA base sequence, protein amino acid sequence, and DNA base sequence are interrelated. Which determines which in the process of protein synthesis?

● Skill-Building Exercises *(grouped in similar pairs)*

● **15.94** Draw the R group of (a) alanine; (b) histidine; (c) methionine.

● **15.95** Draw the R group of (a) glycine; (b) isoleucine; (c) tyrosine.

●● **15.96** Draw the structure of each of the following tripeptides:
(a) Aspartic acid-histidine-tryptophan
(b) Glycine-cysteine-tyrosine with the charges existing in cell fluid

●● **15.97** Draw the structure of each of the following tripeptides:
(a) Lysine-phenylalanine-threonine
(b) Alanine-leucine-valine with the charges that exist in cell fluid

● **15.98** Write the sequence of the complementary DNA strand that pairs with each of the following DNA base sequences:
(a) TTAGCC   (b) AGACAT

● **15.99** Write the sequence of the complementary DNA strand that pairs with each of the following DNA base sequences:
(a) GGTTAC   (b) CCCGAA

● **15.100** Write the base sequence of the DNA template from which this RNA sequence was derived: UGUUACGGA. How many amino acids are coded for in this sequence?

● **15.101** Write the base sequence of the DNA template from which this RNA sequence was derived: GUAUCAAUGAACUUG. How many amino acids are coded for in this sequence?

● Problems in Context

●● **15.102** Protein shapes are maintained by a variety of forces that arise from interactions between the amino-acid R groups. Name the amino acid that possesses each R group and the force that could arise in each of the following interactions:
(a) —CH₂—SH with HS—CH₂—

(b) —(CH₂)₄—NH₃⁺ with ⁻O—C(=O)—CH₂—

(c) —CH₂—C(=O)—NH₂ with HO—CH₂—

(d) —CH(CH₃)—CH₃ with ⟨benzene ring⟩—CH₂—

●● **15.103** Amino acids have an average molar mass of 100 g/mol. How many bases on a single strand of DNA are needed to code for a protein with a molar mass of 5×10^5 g/mol?

Comprehensive Problems

●● **15.104** A synthesis of 2-butanol was performed by treating 2-bromobutane with hot sodium hydroxide solution. The yield

was 60%, indicating that a significant portion of the reactant was converted into a second product. Predict what this other product might be. What simple test would support your prediction?

●● **15.105** Ethers (general formula R—O—R') have many important uses. Until recently, methyl *tert*-butyl ether (MTBE, *right*) was used as an octane booster and fuel additive for gasoline. It increases the oxygen content of the fuel, which reduces CO emissions. MTBE is synthesized by the catalyzed reaction of 2-methylpropene with methanol.
(a) Write a balanced equation for the synthesis of MTBE. (*Hint: Alcohols add to alkenes similarly to the way water does.*)
(b) If the government required that auto fuel mixtures contain 2.7% oxygen by mass to reduce CO emissions, how many grams of MTBE would have to be added to each 100. g of gasoline?
(c) How many liters of MTBE would be in each liter of fuel mixture? (The density of both gasoline and MTBE is 0.740 g/mL.)
(d) How many liters of air (21% O₂ by volume) are needed at 24°C and 1.00 atm to fully combust 1.00 L of MTBE?

●●● **15.106** An alcohol is oxidized to a carboxylic acid, and 0.2003 g of the acid is titrated with 45.25 mL of 0.03811 *M* NaOH.
(a) What is the molar mass of the acid? (b) What are the molar mass and molecular formula of the alcohol?

●● **15.107** Some of the most useful compounds for organic synthesis are Grignard reagents (general formula R—MgX, where X is a halogen), which are made by combining an alkyl halide, R—X, with Mg. They are used to change the carbon skeleton of a starting carbonyl compound in a reaction similar to that with R—Li:

$$R'-\overset{O}{\underset{}{C}}-R'' + R-MgBr \longrightarrow R'-\overset{OMgBr}{\underset{R}{C}}-R'' \overset{H_2O}{\longrightarrow}$$

$$R'-\overset{OH}{\underset{R}{C}}-R'' + Mg(OH)Br$$

(a) What is the product, after a final step with water, of the reaction between ethanal and the Grignard reagent of bromobenzene? (b) What is the product, after a final step with water, of the reaction between 2-butanone and the Grignard reagent of 2-bromopropane? (c) There are often two (or more) combinations of Grignard reagent and carbonyl compound that will give the same product. Choose another pair of reactants to give the product in (a). (d) What carbonyl compound must react with a Grignard reagent to yield a product with the —OH group at the *end* of the carbon chain? (e) What Grignard reagent and carbonyl compound would you use to prepare 2-methyl-2-butanol?

●● **15.108** Starting with the given organic reactant and any necessary inorganic reagents, explain how you would perform each of the following syntheses:
(a) From CH₃—CH₂—CH₂—OH, make CH₃—CH(Br)—CH₂—Br

(b) From CH₃—CH₂—OH, make CH₃—C(=O)—O—CH₂—CH₃

●●● **15.109** Compound A, composed of C, H, and O, is heated in a 1.00-L flask to 160.°C until all of the A has vaporized and displaced the air. The flask is then cooled, and 2.48 g of A remains. When 0.500 g of A burns in O₂, 0.409 g of H₂O and 1.00 g of CO₂ are produced. Compound A is not acidic, but it can be oxidized to compound B, which is weakly acidic: 1.000 g of B is neutralized with 33.9 mL of 0.5 M sodium hydroxide. When B is heated to 260°C, it gives off water and forms C, whose solution in $CDCl_3$ (where D is deuterium, ^2H) has one peak in its ^1H-NMR spectrum. (a) What are the structures of A, B, and C? (b) Compound A is a controlled substance because it is metabolized to the weakly acidic "date rape" drug GHB, $C_4H_8O_3$. What are the structure and name of GHB?

••• **15.110** Cadaverine (1,5-diaminopentane) and putrescine (1,4-diaminobutane) are two compounds that are formed by bacterial action and are responsible for the odor of rotting flesh. Draw their structures. Suggest a series of reactions to synthesize putrescine from 1,2-dibromoethane and any inorganic reagents.

•• **15.111** Pyrethrins, such as jasmolin II *(below)*, are a group of natural compounds synthesized by flowers of the genus *Chrysanthemum* (known as pyrethrum flowers) to act as insecticides.
(a) Circle and name the functional groups in jasmolin II.
(b) What is the hybridization of the numbered carbons?
(c) Which, if any, of the numbered carbons are chiral centers?

••• **15.112** Compound A is branched and optically active and contains C, H, and O. (a) A 0.500-g sample burns in excess O_2 to yield 1.25 g of CO_2 and 0.613 g of H_2O. Determine the empirical formula. (b) When 0.225 g of compound A vaporizes at 755 torr and 97°C, the vapor occupies 78.0 mL. Determine the molecular formula. (c) Careful oxidation of the compound yields a ketone. Name and draw compound A, and circle the chiral center.

••• **15.113** Ibuprofen is one of the most common anti-inflammatory drugs. (a) Identify the functional group(s) and chiral center(s) in ibuprofen. (b) Write a four-step synthesis of a racemic mixture of ibuprofen from 4-isobutylbenzaldehyde, using inorganic reactants and one organometallic reactant (see Problem 15.107).

ibuprofen   4-isobutylbenzaldehyde

•• **15.114** Combustion of gasoline releases almost all of the carbon it contains as CO_2, the major greenhouse gas. As a simplification, assume that gasoline consists of eight-C alkanes (average $d = 0.74$ g/mL). How many metric tons of CO_2 are released in a year by an SUV that gets 15 mpg and is driven 18,000 miles?

•• **15.115** Which features of retinal make it so useful as a photon absorber in the visual systems of organisms?

•• **15.116** The polypeptide chain in proteins does not exhibit free rotation because of the partial double-bond character of the peptide bond. Explain this fact with resonance structures.

••• **15.117** Other nonmetals form compounds that are structurally analogous to those of carbon, but these inorganic compounds are usually more reactive. Predict any missing products and write balanced equations for each reaction: (a) the decomposition and chlorination of diborane to boron trichloride; (b) the combustion of pentaborane (B_5H_9) in O_2; (c) the addition of water to bor-

azine's double bonds ($B_3N_3H_6$, Figure 14.9B, p. 583) *(Hint: The —OH group bonds to B)*; (d) the hydrolysis of trisilane (Si_3H_8) to silica (SiO_2) and H_2; (e) the complete halogenation of disilane with Cl_2; (f) the thermal decomposition of H_2S_5 to hydrogen sulfide and sulfur molecules; (g) the hydrolysis of PCl_5.

••• **15.118** The genetic code consists of a series of three-base words that each code for a given amino acid.
(a) Using the selections from the genetic code shown below, determine the amino acid sequence coded by the following segment of RNA: UCCACAGCCUAUAUGGCAAACUUGAAG

AUG = methionine	CCU = proline	CAU = histidine
UGG = tryptophan	AAG = lysine	UAU = tyrosine
GCC = alanine	UUG = leucine	CGG = arginine
UGU = cysteine	AAC = asparagine	ACA = threonine
UCC = serine	GCA = alanine	UCA = serine

(b) What is the complementary DNA sequence from which this RNA sequence was made? (c) If you were sequencing the DNA fragment in part (b), how many complementary chain pieces would you obtain in the tube containing ddATP?

••• **15.119** In addition to their use in water treatment, ion-exchange resins are used to extract Au, Ag, and Pt ions from solution. One of the most common resins consists of a polymer with a benzene-containing backbone to which sulfonic acid groups (—SO_3H) are then added. (a) What monomer can be used to prepare the polymer backbone? (b) This polymer typically contains 4–16% crosslinking. Draw the structure of the benzene-containing monomer used to crosslink the polymer.

••• **15.120** Complete hydrolysis of a 100.00-g sample of a peptide gave the following amounts of individual amino acids (molar masses, in g/mol, appear in parentheses):

3.00 g of glycine (75.07)	0.90 g of alanine (89.10)
3.70 g of valine (117.15)	6.90 g of proline (115.13)
7.30 g of serine (105.10)	86.00 g of arginine (174.21)

(a) Why does the total mass of amino acids exceed the mass of peptide? (b) What are the relative numbers of amino acids in the peptide? (c) What is the minimum molar mass of the peptide?

•• **15.121** Aircraft de-icing agents, such as propylene glycol [1,2-propanediol; $CH_3CH(OH)CH_2OH$; $d = 1.036$ g/mL], are used to remove ice and snow prior to winter flights, but their presence in runoff depletes O_2 in natural waters. (a) Write a balanced equation for the complete oxidation of propylene glycol by O_2. (b) What is the "theoretical oxygen demand," in grams of O_2 required per gram of propylene glycol oxidized? (c) What is the theoretical oxygen demand, in grams of O_2 required per liter of solution, of 90.0% by volume aqueous propylene glycol?

••• **15.122** 2-Butanone is reduced by hydride ion donors, such as sodium borohydride ($NaBH_4$), to the alcohol 2-butanol. Even though the alcohol has a chiral center, the product isolated from the redox reaction is not optically active. Explain.

•• **15.123** Biphenyl consists of two benzene rings connected by a single bond. When Cl atoms substitute for some of the H atoms, polychlorinated biphenyls (PCBs) are formed. There are over 200 different PCBs, but only six with one Cl on each ring. Draw these constitutional isomers.

••• **15.124** Wastewater from a cheese factory has the following composition: 8.0 g/L protein ($C_{16}H_{24}O_5N_4$); 12 g/L carbohydrate (CH_2O); and 2.0 g/L fat ($C_8H_{16}O$). What is the total organic carbon (TOC) of the wastewater in g of C/L?

16

Rate and Survival Evidence mounts that, unlike modern reptiles, a dinosaur such as the mighty *Tyrannosaurus rex* could generate enough metabolic heat to sustain its biochemical reactions at high rates. In this chapter, you'll see how reaction rate is influenced by several factors, including temperature, and how chemists can control those factors.

Kinetics: Rates and Mechanisms of Chemical Reactions

16.1 Factors That Influence Reaction Rate

16.2 Expressing the Reaction Rate
Average, Instantaneous, and Initial Reaction Rates
Rate and Concentration

16.3 The Rate Law and Its Components
Determining the Initial Rate
Reaction Order Terminology
Determining Reaction Orders
Determining the Rate Constant

16.4 Integrated Rate Laws: Concentration Changes over Time
First-, Second-, and Zero-Order Reactions
Reaction Order
Reaction Half-Life

16.5 The Effect of Temperature on Reaction Rate

16.6 Explaining the Effects of Concentration and Temperature
Collision Theory
Transition State Theory

16.7 Reaction Mechanisms: Steps in the Overall Reaction
Elementary Reactions
The Rate-Determining Step
The Mechanism and the Rate Law

16.8 Catalysis: Speeding Up a Chemical Reaction
Homogeneous Catalysis
Heterogeneous Catalysis

Until now we've taken a rather simple approach to chemical change: reactants mix and products form. A balanced equation is an essential quantitative tool for calculating product yields from reactant amounts, but it tells us nothing about three dynamic aspects of the reaction, which are essential to understanding chemical change and which we examine in the next several chapters:

- How fast is the reaction proceeding at a given moment?
- What will reactant and product concentrations be when the reaction is complete?
- Will the reaction proceed by itself and release energy, or will it require energy to proceed?

This chapter addresses the first of these questions and focuses on the field of *kinetics*, which deals with the speed of a reaction and its mechanism, the stepwise changes that reactants undergo in their conversion to products. Chapters 17 through 19 are concerned with *equilibrium*, the dynamic balance between forward and reverse reactions and how external influences alter reactant and product concentrations. Chapters 20 and 21 present *thermodynamics* and its application to electrochemistry; there, we investigate why reactions occur and how we can put them to use. These central subdisciplines of chemistry apply to all physical and chemical change and, thus, are crucial to our understanding of modern technology, the environment, and the reactions in living things.

Chemical kinetics is the study of *reaction rates,* the changes in concentrations of reactants (or products) as a function of time (Figure 16.1). Reactant concentrations decrease quickly in a reaction with a high rate and slowly in a reaction with a low rate.

Reactions occur at a wide range of rates (Figure 16.2). Some, like a neutralization, a precipitation, or an explosive redox process, seem to be over as soon as the reactants make contact—in a fraction of a second. Others, such as the reactions involved in cooking or rusting, take a moderate length of time, from minutes to months. Still others take much longer: the reactions that make up the human aging process continue for decades, and those involved in the formation of coal from dead plants take hundreds of millions of years.

Figure 16.1 Reaction rate: the central focus of chemical kinetics. The rate at which reactant becomes product is the underlying theme of chemical kinetics. As time elapses, reactant *(purple)* decreases and product *(green)* increases.

Transparency/ARIS Presentation Center, Art Library

ARIS Presentation Center, Photo Library

Figure 16.2 The wide range of reaction rates. Reactions proceed at a wide range of rates. An explosion **(A)** is much faster than the process of ripening **(B),** which is much faster than the process of rusting **(C),** which is much faster than the process of human aging **(D).**

Knowing how fast a chemical change occurs can be essential. How quickly a medicine acts or blood clots can make the difference between life and death. How long it takes for cement to harden or polyethylene to form can make the difference between profit and loss. In general, the rates of these diverse processes depend on the same variables, most of which chemists can manipulate to maximize yields within a given time or to slow down an unwanted reaction.

IN THIS CHAPTER . . . We first overview three key factors that affect reaction rate. Then we see how to express rate in the form of a *rate law* and how the components of a rate law are determined experimentally. We focus on the effects of concentration and temperature on rate and examine the models that explain those effects. Only then can we develop a reaction mechanism by noting the steps the reaction goes through and picturing the structure that exists as reactant bonds are breaking and product bonds are forming. Finally, we discuss how catalysts increase reaction rates, highlighting two vital areas—the reactions of a living cell and the depletion of atmospheric ozone.

16.1 FACTORS THAT INFLUENCE REACTION RATE

Let's begin our study of kinetics with a qualitative look at the key factors that affect how fast a reaction proceeds. Under any given set of conditions, *each reaction has its own characteristic rate,* which is determined by the chemical nature of the reactants. At room temperature, for example, hydrogen reacts explosively with fluorine but extremely slowly with nitrogen:

$$H_2(g) + F_2(g) \longrightarrow 2HF(g) \quad \text{[very fast]}$$
$$3H_2(g) + N_2(g) \longrightarrow 2NH_3(g) \quad \text{[very slow]}$$

We can control four factors that affect the rate of a given reaction: the concentrations of the reactants, the physical state of the reactants, the temperature at which the reaction occurs, and the use of a catalyst. We'll consider the first three factors here and discuss the fourth in Section 16.8.

1. *Concentration: molecules must collide to react.* A major factor influencing the rate of a given reaction is reactant concentration. Consider the reaction between ozone and nitrogen monoxide (nitric oxide) that occurs in the stratosphere, where the oxide is released in the exhaust gases of supersonic aircraft:

$$NO(g) + O_3(g) \longrightarrow NO_2(g) + O_2(g)$$

Imagine what this might look like at the molecular level with the reactants confined in a reaction vessel. Nitrogen monoxide and ozone molecules zoom every which way, crashing into each other and the vessel walls. A reaction between NO and O_3 can occur only when the molecules collide. The more molecules present in the container, the more frequently they collide, and the more often a reaction occurs. Thus, *reaction rate is proportional to the concentration of reactants:*

$$\text{Rate} \propto \text{collision frequency} \propto \text{concentration}$$

In this case, we're looking at a very simple reaction, one in which reactant molecules collide and form product molecules in one step, but even the rates of complex reactions depend on reactant concentration.

2. *Physical state: molecules must mix to collide.* The frequency of collisions between molecules also depends on the physical states of the reactants. When the reactants are in the same phase, as in an aqueous solution, random thermal motion brings them into contact. When they are in different phases, contact occurs only at the interface, so vigorous stirring and grinding may be needed. In these cases, *the more finely divided a solid or liquid reactant, the greater its surface area per unit volume, the more contact it makes with the other reactant, and the faster the reaction occurs.* Figure 16.3A shows a steel nail heated in oxygen glowing feebly; in Figure 16.3B, the same mass of steel wool bursts into flame. For the same reason, you start a campfire with wood chips and thin branches, not logs.

A

B

Figure 16.3 The effect of surface area on reaction rate. A, A hot nail glows in O_2. **B,** The same mass of hot steel wool bursts into flame in O_2. The greater surface area per unit volume of the steel wool means that more metal makes contact with O_2, so the reaction is faster.

 ARIS Presentation Center, Photo Library

3. *Temperature: molecules must collide with enough energy to react.* Temperature usually has a major effect on the speed of a reaction. Two familiar kitchen appliances employ this effect: a refrigerator slows down chemical processes that spoil food, whereas an oven speeds up other chemical processes that cook it.

Recall that molecules in a sample of gas have a range of speeds, with the most probable speed dependent on the temperature (see Figure 5.14, p. 211). Thus, *at a higher temperature, more collisions occur in a given time.* Even more important, however, is the fact that temperature affects the kinetic energy of the molecules, and thus the *energy* of the collisions. In the jumble of molecules in the reaction of NO and O_3, mentioned previously, most collisions result in the molecules simply recoiling, like billiard balls, with no reaction taking place. However, some collisions occur with sufficient energy for the molecules to react (Figure 16.4). And, at a higher temperature, more of these sufficiently energetic collisions occur. Thus, *raising the temperature increases the reaction rate by increasing the number and, especially, the energy of the collisions:*

$$\text{Rate} \propto \text{collision energy} \propto \text{temperature}$$

The qualitative idea that reaction rate is influenced by the frequency and energy of reactant collisions leads to several quantitative questions: How can we describe the dependence of rate on reactant concentration mathematically? Do all changes in concentration affect the rate to the same extent? Do all rates increase to the same extent with a given rise in temperature? How do reactant molecules use the energy of collision to form product molecules, and is there a way to determine this energy? What do the reactants look like as they are turning into products? We address these questions in the following sections.

Section Summary

Chemical kinetics deals with reaction rates and the stepwise molecular events by which a reaction occurs. • Under a given set of conditions, each reaction has its own rate. • Concentration affects rate by influencing the frequency of collisions between reactant molecules. Physical state affects rate by determining the surface area per unit volume of reactant(s). Temperature affects rate by influencing the frequency and, even more importantly, the energy of the reactant collisions.

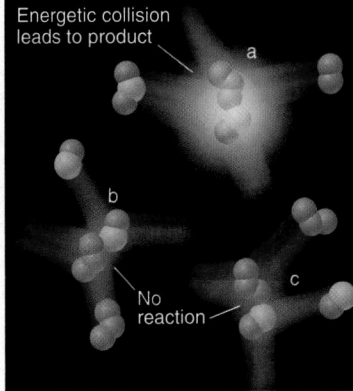

Figure 16.4 Collision energy and reaction rate. The reaction equation is shown in the panel. Although many collisions between NO and O_3 molecules occur, relatively few have enough energy to cause reaction. At this temperature, only collision *a* is energetic enough to lead to product; the reactant molecules in collisions *b* and *c* just bounce off each other.

ARIS Presentation Center, Art Library

Literature "Chemistry Time: Factors Affecting the Rate of a Chemical Reaction," *J. Chem. Educ.* **1998,** *75,* 1120A.

16.2 EXPRESSING THE REACTION RATE

Before we can deal quantitatively with the effects of concentration and temperature on reaction rate, we must express the rate mathematically. A *rate* is a change in some variable per unit of time. The most common examples relate to the rate of motion (speed) of an object, which is the change in its position divided by the change in time. Suppose, for instance, we measure a runner's starting position, x_1, at time t_1 and final position, x_2, at time t_2. The runner's average speed is

$$\text{Rate of motion (speed)} = \frac{\text{change in position}}{\text{change in time}} = \frac{x_2 - x_1}{t_2 - t_1} = \frac{\Delta x}{\Delta t}$$

In the case of a chemical change, we are concerned with the **reaction rate,** the changes in concentrations of reactants or products per unit time: *reactant concentrations decrease while product concentrations increase.* Consider a general reaction, A $\longrightarrow$ B. We quickly measure the starting reactant concentration (conc A_1) at t_1, allow the reaction to proceed, and then quickly measure the reactant concentration again (conc A_2) at t_2. The change in concentration divided by the change in time gives the *average* rate:

$$\text{Rate of reaction} = -\frac{\text{change in concentration of A}}{\text{change in time}} = -\frac{\text{conc } A_2 - \text{conc } A_1}{t_2 - t_1} = -\frac{\Delta(\text{conc A})}{\Delta t}$$

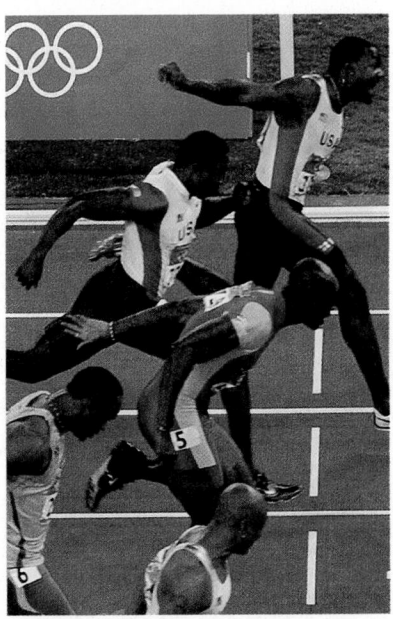

Runners changing position with time.

ARIS Presentation Center, Photo Library

Note the minus sign. By convention, reaction rate is a *positive* number, but conc A_2 will always be *lower* than conc A_1, so the *change in concentration (final − initial) of reactant A is always negative.* We use the minus sign simply to convert the negative change in reactant concentration to a positive value for the rate. Suppose the concentration of A changes from 1.2 mol/L (conc A_1) to 0.75 mol/L (conc A_2) over a 125-s period. The average rate is

$$\text{Rate} = -\frac{0.75\ \text{mol/L} - 1.2\ \text{mol/L}}{125\ \text{s} - 0\ \text{s}} = 3.6 \times 10^{-3}\ \text{mol/L·s}$$

We use *square brackets,* [ ], *to express concentration in moles per liter.* That is, [A] is the concentration of A in mol/L, so the rate expressed in terms of A is

$$\text{Rate} = -\frac{\Delta[\text{A}]}{\Delta t} \tag{16.1}$$

The rate has units of moles per liter per second (mol L^{-1} s^{-1}, or mol/L·s), or any time unit convenient for the particular reaction (minutes, years, and so on).

If instead we measure the *product* to determine the reaction rate, we find its concentration *increasing* over time. That is, conc B_2 is always *higher* than conc B_1. Thus, the *change* in product concentration, $\Delta[\text{B}]$, is *positive,* and the reaction rate for A $\longrightarrow$ B expressed in terms of B is

$$\text{Rate} = \frac{\Delta[\text{B}]}{\Delta t}$$

Average, Instantaneous, and Initial Reaction Rates

In most reactions, not only the concentration changes, but *the rate itself varies with time as the reaction proceeds.* Consider the reversible gas-phase reaction between ethylene and ozone, one of many reactions that can be involved in the formation of photochemical smog:

$$\text{C}_2\text{H}_4(g) + \text{O}_3(g) \rightleftharpoons \text{C}_2\text{H}_4\text{O}(g) + \text{O}_2(g)$$

For now, we consider only reactant concentrations. You can see from the equation coefficients that for every molecule of C_2H_4 that reacts, a molecule of O_3 reacts with it. In other words, the concentrations of both reactants decrease at the same rate in this particular reaction:

$$\text{Rate} = -\frac{\Delta[\text{C}_2\text{H}_4]}{\Delta t} = -\frac{\Delta[\text{O}_3]}{\Delta t}$$

By measuring the concentration of either reactant, we can follow the reaction rate.

Suppose we have a known concentration of O_3 in a closed reaction vessel kept at 30°C (303 K). Table 16.1 shows the concentration of O_3 at various times during the first minute after we introduce C_2H_4 gas. The rate over the entire 60.0 s is the total change in concentration divided by the change in time:

$$\text{Rate} = -\frac{\Delta[\text{O}_3]}{\Delta t} = -\frac{(1.10 \times 10^{-5}\ \text{mol/L}) - (3.20 \times 10^{-5}\ \text{mol/L})}{60.0\ \text{s} - 0.0\ \text{s}} = 3.50 \times 10^{-7}\ \text{mol/L·s}$$

This calculation gives us the **average rate** over that period; that is, during the first 60.0 s of the reaction, ozone concentration decreases an *average* of 3.50×10^{-7} mol/L each second. However, the average rate does not show that the rate is changing, and it tells us nothing about how fast the ozone concentration is decreasing *at any given instant.*

We can see the rate change during the reaction by calculating the average rate over two shorter periods—one earlier and one later. Between the starting time 0.0 s and 10.0 s, the average rate is

$$\text{Rate} = -\frac{\Delta[\text{O}_3]}{\Delta t} = -\frac{(2.42 \times 10^{-5}\ \text{mol/L}) - (3.20 \times 10^{-5}\ \text{mol/L})}{10.0\ \text{s} - 0.0\ \text{s}} = 7.80 \times 10^{-7}\ \text{mol/L·s}$$

Table 16.1 Concentration of O_3 at Various Times in Its Reaction with C_2H_4 at 303 K

Time (s)	Concentration of O_3 (mol/L)
0.0	3.20×10^{-5}
10.0	2.42×10^{-5}
20.0	1.95×10^{-5}
30.0	1.63×10^{-5}
40.0	1.40×10^{-5}
50.0	1.23×10^{-5}
60.0	1.10×10^{-5}

ARIS Presentation Center, Tables Library

During the last 10.0 s, between 50.0 s and 60.0 s, the average rate is

$$\text{Rate} = -\frac{\Delta[O_3]}{\Delta t} = -\frac{(1.10 \times 10^{-5} \text{ mol/L}) - (1.23 \times 10^{-5} \text{ mol/L})}{60.0 \text{ s} - 50.0 \text{ s}} = 1.30 \times 10^{-7} \text{ mol/L·s}$$

The earlier rate is six times as fast as the later rate. Thus, *the rate decreases during the course of the reaction*. This makes perfect sense from a molecular point of view: as O_3 molecules are used up, fewer of them are present to collide with C_2H_4 molecules, so the rate, the change in their concentration over time, decreases.

The change in rate can also be seen by plotting the concentrations vs. the times at which they were measured (Figure 16.5). A curve is obtained, which means that the rate changes. *The slope of the straight line ($\Delta y/\Delta x$, that is, $\Delta[O_3]/\Delta t$) joining any two points gives the average rate over that period.*

The shorter the time period we choose, the closer we come to the **instantaneous rate,** the rate at a particular instant during the reaction. *The slope of a line tangent to the curve at a particular point gives the instantaneous rate at that time.* For example, the rate of the reaction of C_2H_4 and O_3 35.0 s after it began is 2.50×10^{-7} mol/L·s, the slope of the line drawn tangent to the curve through the point at which $t = 35.0$ s (line *d* in Figure 16.5). In general, we use the term *reaction rate* to mean the *instantaneous* reaction rate.

As a reaction continues, the product concentrations increase, and so the reverse reaction (reactants ⟵ products) proceeds more quickly. To find the overall (net) rate, we would have to take both forward and reverse reactions into account and calculate the difference between their rates. A common way to avoid this complication for many reactions is to measure the **initial rate,** the instantaneous rate at the moment the reactants are mixed. Under these conditions, the product concentrations are negligible, so the reverse rate is negligible. Moreover, we know the reactant concentrations from the concentrations and volumes of the solutions we mix together. The initial rate is measured by determining the slope of the line tangent to the curve at $t = 0$ s. In Figure 16.5, the initial rate is 10.0×10^{-7} mol/L·s (line *a*). Unless stated otherwise, we will use initial rate data to determine other kinetic parameters.

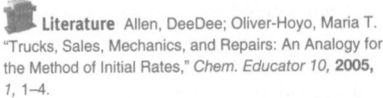
Literature Allen, DeeDee; Oliver-Hoyo, Maria T. "Trucks, Sales, Mechanics, and Repairs: An Analogy for the Method of Initial Rates," *Chem. Educator 10*, **2005**, *1*, 1–4.

Line	Rate (mol/L·s)
a	10.0×10^{-7}
b	3.50×10^{-7}
c	7.80×10^{-7}
d	2.50×10^{-7}
e	1.30×10^{-7}

Figure 16.5 The concentration of O_3 vs. time during its reaction with C_2H_4. Plotting the data in Table 16.1 gives a curve because the rate changes during the reaction. The *average* rate over a given period is the slope of a line joining two points along the curve. The slope of line *b* is the average rate over the first 60.0 s of the reaction. The slopes of lines *c* and *e* give the average rate over the first and last 10.0-s intervals, respectively. Line *c* is steeper than line *e* because the average rate over the earlier period is higher. The *instantaneous* rate at 35.0 s is the slope of line *d*, the tangent to the curve at $t = 35.0$ s. The *initial* rate is the slope of line *a*, the tangent to the curve at $t = 0$ s.

Transparency/ARIS Presentation Center, Art Library

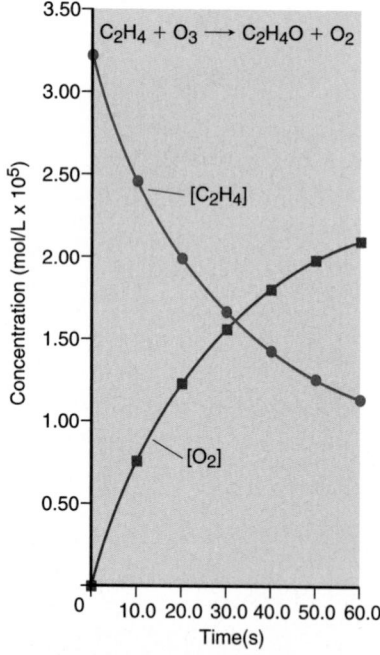

Figure 16.6 Plots of [C₂H₄] and [O₂] vs. time. Measuring reactant concentration, [C₂H₄], and product concentration, [O₂], gives curves of identical shapes but changing in opposite directions. The steep upward (positive) slope of [O₂] early in the reaction mirrors the steep downward (negative) slope of [C₂H₄] because the faster C₂H₄ is used up, the faster O₂ is formed.

ARIS Presentation Center, Art Library

Literature Sanger, Michael J. "Flipping Pennies and Burning Candles: Adventures in Kinetics," *J. Chem. Educ.* 2003, 80, 304A.

ARIS Presentation Center, Worked Examples Library

Expressing Rate in Terms of Reactant and Product Concentrations

So far, in our discussion of the reaction of C_2H_4 and O_3, we've expressed the rate in terms of the decreasing concentration of O_3. The rate is the same in terms of C_2H_4, but it is exactly the opposite in terms of the products because their concentrations are *increasing*. From the balanced equation, we see that one molecule of C_2H_4O and one of O_2 appear for every molecule of C_2H_4 and of O_3 that disappear. We can express the rate in terms of any of the four substances involved:

$$\text{Rate} = -\frac{\Delta[C_2H_4]}{\Delta t} = -\frac{\Delta[O_3]}{\Delta t} = +\frac{\Delta[C_2H_4O]}{\Delta t} = +\frac{\Delta[O_2]}{\Delta t}$$

Again, note the negative values for the reactants and the positive values for the products (usually written without the plus sign). Figure 16.6 shows a plot of the simultaneous monitoring of one reactant and one product. Because, in this case, product concentration increases at the same rate that reactant concentration decreases, the curves have the same shapes but are inverted.

In many other cases, though, the reactants disappear and the products appear at different rates. Consider the reaction between hydrogen and iodine to form hydrogen iodide:

$$H_2(g) + I_2(g) \longrightarrow 2HI(g)$$

For every molecule of H_2 that disappears, one molecule of I_2 disappears and *two* molecules of HI appear. In other words, the rate of [H_2] decrease is the same as the rate of [I_2] decrease, but both are only half the rate of [HI] increase. By referring the change in [I_2] and [HI] to the change in [H_2], we have

$$\text{Rate} = -\frac{\Delta[H_2]}{\Delta t} = -\frac{\Delta[I_2]}{\Delta t} = \frac{1}{2}\frac{\Delta[HI]}{\Delta t}$$

If we refer the change in [H_2] and [I_2] to the change in [HI] instead, we obtain

$$\text{Rate} = \frac{\Delta[HI]}{\Delta t} = -2\frac{\Delta[H_2]}{\Delta t} = -2\frac{\Delta[I_2]}{\Delta t}$$

Notice that this expression is just a rearrangement of the previous one; also note that it gives a numerical value for the rate that is double the previous value. Thus, the mathematical expression for the rate of a particular reaction and *the numerical value of the rate depend on which substance serves as the reference.*

We can summarize these results for any reaction,

$$aA + bB \longrightarrow cC + dD$$

where *a*, *b*, *c,* and *d* are coefficients of the balanced equation. In general, the rate is related to reactant or product concentrations as follows:

$$\text{Rate} = -\frac{1}{a}\frac{\Delta[A]}{\Delta t} = -\frac{1}{b}\frac{\Delta[B]}{\Delta t} = \frac{1}{c}\frac{\Delta[C]}{\Delta t} = \frac{1}{d}\frac{\Delta[D]}{\Delta t} \qquad (16.2)$$

SAMPLE PROBLEM 16.1 Expressing Rate in Terms of Changes in Concentration with Time

PROBLEM Because it has a nonpolluting combustion product (water vapor), hydrogen gas is used for fuel aboard the space shuttles and in earthbound cars with prototype engines:

$$2H_2(g) + O_2(g) \longrightarrow 2H_2O(g)$$

(a) Express the rate in terms of changes in [H_2], [O_2], and [H_2O] with time.
(b) When [O_2] is decreasing at 0.23 mol/L·s, at what rate is [H_2O] increasing?
PLAN (a) Of the three substances in the equation, let's choose O_2 as the reference because its coefficient is 1. For every molecule of O_2 that disappears, two molecules of H_2 disappear, so the rate of [O_2] decrease is one-half the rate of [H_2] decrease. By similar reasoning, we see that the rate of [O_2] decrease is one-half the rate of [H_2O] increase.
(b) Because [O_2] is decreasing, the change in its concentration must be negative. We substitute the negative value into the expression and solve for $\Delta[H_2O]/\Delta t$.

SOLUTION (a) Expressing the rate in terms of each component:

$$\text{Rate} = -\frac{1}{2}\frac{\Delta[H_2]}{\Delta t} = -\frac{\Delta[O_2]}{\Delta t} = \frac{1}{2}\frac{\Delta[H_2O]}{\Delta t}$$

(b) Calculating the rate of change of $[H_2O]$:

$$\frac{1}{2}\frac{\Delta[H_2O]}{\Delta t} = -\frac{\Delta[O_2]}{\Delta t} = -(-0.23 \text{ mol/L·s})$$

$$\frac{\Delta[H_2O]}{\Delta t} = 2(0.23 \text{ mol/L·s}) = 0.46 \text{ mol/L·s}$$

CHECK (a) A good check is to use the rate expression to obtain the balanced equation: $[H_2]$ changes twice as fast as $[O_2]$, so two H_2 molecules react for each O_2. $[H_2O]$ changes twice as fast as $[O_2]$, so two H_2O molecules form from each O_2. From this reasoning, we get $2H_2 + O_2 \longrightarrow 2H_2O$. The $[H_2]$ and $[O_2]$ decrease, so they take minus signs; $[H_2O]$ increases, so it takes a plus sign. Another check is to use Equation 16.2, with $A = H_2$, $a = 2$; $B = O_2$, $b = 1$; $C = H_2O$, $c = 2$. Thus,

$$\text{Rate} = -\frac{1}{a}\frac{\Delta[A]}{\Delta t} = -\frac{1}{b}\frac{\Delta[B]}{\Delta t} = \frac{1}{c}\frac{\Delta[C]}{\Delta t}$$

or

$$\text{Rate} = -\frac{1}{2}\frac{\Delta[H_2]}{\Delta t} = -\frac{\Delta[O_2]}{\Delta t} = \frac{1}{2}\frac{\Delta[H_2O]}{\Delta t}$$

(b) Given the rate expression, it makes sense that the numerical value of the rate of $[H_2O]$ increase is twice that of $[O_2]$ decrease.

COMMENT Thinking through this type of problem at the molecular level is the best approach, but use Equation 16.2 to confirm your answer.

FOLLOW-UP PROBLEM 16.1 (a) Balance the following equation and express the rate in terms of the change in concentration with time for each substance:

$$NO(g) + O_2(g) \longrightarrow N_2O_3(g)$$

(b) How fast is $[O_2]$ decreasing when $[NO]$ is decreasing at a rate of 1.60×10^{-4} mol/L·s?

Demo Wellman, Whitney E.; Noble, Mark E. "Greening the Blue Bottle," *J. Chem. Educ.* **2003**, *80*, 537.

Demo Noble, Mark E. "Out of the Blue," *J. Chem. Educ.* **2003**, *80*, 536A.

Section Summary

The average reaction rate is the change in reactant (or product) concentration over a change in time, Δt. The rate slows as reactants are used up. The instantaneous rate at time t is obtained from the slope of the tangent to a concentration vs. time curve at time t. The initial rate, the instantaneous rate at $t = 0$, occurs when reactants are just mixed and before any product accumulates. • The expression for a reaction rate and its numerical value depend on which reaction component is being monitored.

16.3 THE RATE LAW AND ITS COMPONENTS

The centerpiece of any kinetic study is the **rate law** (or **rate equation**) for the reaction in question. The rate law expresses the rate as a function of reactant concentrations, product concentrations, and temperature. Any hypothesis we make about how the reaction occurs on the molecular level must conform to the rate law because it is based on experimental fact.

In this discussion, we generally consider reactions for which the products do not appear in the rate law. In these cases, *the reaction rate depends only on reactant concentrations and temperature.* First, we look at the effect of concentration on rate for reactions occurring at a fixed temperature. For a general reaction,

$$aA + bB + \cdots \longrightarrow cC + dD + \cdots$$

the rate law has the form

$$\text{Rate} = k[A]^m[B]^n \cdots \qquad (16.3)$$

Aside from the concentration terms, [A] and [B], the other parameters in Equation 16.3 require some definition. The proportionality constant k, called the **rate constant,** is specific for a given reaction at a given temperature; it does *not* change as the reaction proceeds. (As you'll see in Section 16.5, k *does* change with temperature and therefore determines how temperature affects the rate.) The exponents m and n, called the **reaction orders,** define how the rate is affected by reactant concentration. Thus, if the rate doubles when [A] doubles, the rate depends on [A] raised to the first power, $[A]^1$, so $m = 1$. Similarly, if the rate quadruples when [B] doubles, the rate depends on [B] raised to the second power, $[B]^2$, so $n = 2$. In another reaction, the rate may not change at all when [A] doubles; in that case, the rate does *not* depend on [A] or, to put it another way, the rate depends on [A] raised to the zero power, $[A]^0$, so $m = 0$. Keep in mind that the coefficients a and b in the general balanced equation are *not* necessarily related in any way to the reaction orders m and n.

A key point to remember is that *the components of the rate law—rate, reaction orders, and rate constant—must be found by experiment;* they cannot be

---

Tools of the Laboratory

Measuring Reaction Rates

Speculation about how a reaction occurs at the molecular level must be based on measurements of reaction rates. There are many experimental approaches, but all must obtain the results quickly and reproducibly. We consider four common methods, with specific examples.

Spectrometric Methods

These methods are used to measure the concentration of a reactant or product that absorbs (or emits) light of a narrow range of wavelengths. The reaction is typically performed *within* the sample compartment of a spectrometer that has been set to measure a wavelength characteristic of one of the species (Figure B16.1; see also Tools of the Laboratory, pp. 281–282). For example, in the reaction of NO and O_3, only NO_2 has a color:

$$NO(g, \text{colorless}) + O_3(g, \text{colorless}) \longrightarrow$$
$$O_2(g, \text{colorless}) + NO_2(g, \text{brown})$$

Known amounts of the reactants are injected into a gas sample tube of known volume, and the rate of NO_2 formation is measured by monitoring the color over time. Reactions in aqueous solution are studied similarly.

Conductometric Methods

When nonionic reactants form ionic products, or vice versa, the change in conductivity of the solution over time can be used to measure the rate. Electrodes are immersed in the reaction mixture, and the increase (or decrease) in conductivity correlates with the formation of product (Figure B16.2). Consider the reaction between an organic halide, such as 2-bromo-2-methylpropane, and water:

$$(CH_3)_3C-Br(l) + H_2O(l) \longrightarrow$$
$$(CH_3)_3C-OH(l) + H^+(aq) + Br^-(aq)$$

The HBr that forms is a strong acid in water, so it dissociates completely into ions. As time passes, more ions form, so the conductivity of the reaction mixture increases.

Figure B16.1 Spectrometric monitoring of a reaction. The investigator adds the reactant(s) to the sample tube and immediately places it in the spectrometer. For a reactant (or product) that is colored, rate data are determined from a plot of absorbance (of light of particular wavelength) vs. time. The rate of disappearance of a blue dye is being studied here.

deduced from the reaction stoichiometry. So let's take an experimental approach to finding the components by

1. Using concentration measurements to find the *initial rate*
2. Using initial rates from several experiments to find the *reaction orders*
3. Using these values to calculate the *rate constant*

Once we know the rate law, we can use it to predict the rate for any initial reactant concentrations.

Determining the Initial Rate

In the last section, we showed how initial rates are determined from a plot of concentration vs. time. Because we use initial rate data to determine the reaction orders and rate constant, an accurate experimental method for measuring concentration at various times during the reaction is essential to finding the rate law. A few of the many techniques used for measuring changes in concentration over time are presented in the Tools of the Laboratory essay, which begins on p. 692.

Figure B16.2 Conductometric monitoring of a reaction. When a reactant mixture differs in conductivity from the product mixture, the change in conductivity is proportional to the reaction rate. This method is usually used when nonionic reactants form ionic products.

Figure B16.3 Manometric monitoring of a reaction. When a reaction results in a change in the number of moles of gas, the change in pressure with time corresponds to a change in reaction rate. The rate of formation of NO_2 from N_2O_4 is being studied here.

Manometric Methods

If a reaction involves a change in the number of moles of a gaseous reactant or product, the rate can be determined from the change in pressure (at constant volume and temperature) over time. In practice, a manometer is attached to a reaction vessel of known volume that is immersed in a constant-temperature bath. For example, the reaction between zinc and acetic acid can be monitored by this method:

$$Zn(s) + 2CH_3COOH(aq) \longrightarrow$$
$$Zn^{2+}(aq) + 2CH_3COO^-(aq) + H_2(g)$$

As H_2 forms, the gas pressure increases. Thus, the reaction rate is directly proportional to the rate of increase of the H_2 gas pressure. In Figure B16.3, this method is being used to study a reaction involving nitrogen dioxide.

Direct Chemical Methods

Rates of slow reactions, or of those that can be easily slowed, are often studied by direct chemical methods. A small, measured portion (called an *aliquot*) of the reaction mixture is removed, and the reaction in this portion is stopped, usually by rapid cooling. The concentration of reactant or product in the aliquot is measured, while the bulk of the reaction mixture continues to react and is sampled later. For example, the reaction between an organic halide and water can be studied by titration, as well as by conductometric monitoring. The reaction rate in an aliquot is slowed by quickly transferring it to a chilled flask in an ice bath. HBr concentration in the aliquot is determined by titrating with standardized NaOH solution. To determine the change in HBr concentration with time, the procedure is repeated at regular intervals during the reaction.

Reaction Order Terminology

Before we see how reaction orders are determined from initial rate data, let's discuss the meaning of reaction order and some important terminology. We speak of a reaction as having an *individual* order "with respect to" or "in" each reactant as well as an *overall* order, which is simply the sum of the individual orders.

In the simplest case, a reaction with a single reactant A, the reaction is *first order* overall if the rate is directly proportional to [A]:

$$\text{Rate} = k[\text{A}]$$

It is *second order* overall if the rate is directly proportional to the square of [A]:

$$\text{Rate} = k[\text{A}]^2$$

And it is *zero order* overall if the rate is *not* dependent on [A] at all, a situation that is quite common in metal-catalyzed and biochemical processes:

$$\text{Rate} = k[\text{A}]^0 = k(1) = k$$

Here are some real examples. For the reaction between nitrogen monoxide and ozone,

$$\text{NO}(g) + \text{O}_3(g) \longrightarrow \text{NO}_2(g) + \text{O}_2(g)$$

the rate law has been experimentally determined to be

$$\text{Rate} = k[\text{NO}][\text{O}_3]$$

This reaction is first order with respect to NO (or first order in NO), which means that the rate depends on NO concentration raised to the first power, that is, $[\text{NO}]^1$ (an exponent of 1 is generally omitted). It is also first order with respect to O_3, or $[\text{O}_3]^1$. This reaction is second order overall $(1 + 1 = 2)$.

Now consider a different gas-phase reaction:

$$2\text{NO}(g) + 2\text{H}_2(g) \longrightarrow \text{N}_2(g) + 2\text{H}_2\text{O}(g)$$

The rate law for this reaction has been determined to be

$$\text{Rate} = k[\text{NO}]^2[\text{H}_2]$$

The reaction is second order in NO and first order in H_2, so it is third order overall.

Finally, for the hydrolysis of 2-bromo-2-methylpropane, which we just considered in the Tools of the Laboratory essay,

$$(\text{CH}_3)_3\text{C}-\text{Br}(l) + \text{H}_2\text{O}(l) \longrightarrow (\text{CH}_3)_3\text{C}-\text{OH}(l) + \text{H}^+(aq) + \text{Br}^-(aq)$$

the rate law has been found to be

$$\text{Rate} = k[(\text{CH}_3)_3\text{CBr}]$$

This reaction is first order in 2-bromo-2-methylpropane. Note that the concentration of H_2O does not even appear in the rate law. Thus, the reaction is zero order with respect to H_2O $([\text{H}_2\text{O}]^0)$. This means that the rate does not depend on the concentration of H_2O. We can also write the rate law as

$$\text{Rate} = k[(\text{CH}_3)_3\text{CBr}][\text{H}_2\text{O}]^0$$

Overall, this is a first-order reaction.

These examples demonstrate a major point: *reaction orders* **cannot be** *deduced from the balanced equation*. For the reaction between NO and H_2 and for the hydrolysis of 2-bromo-2-methylpropane, the reaction orders in the rate laws do *not* correspond to the coefficients of the balanced equations. Reaction orders *must* be determined from rate data.

Reaction orders are usually positive integers or zero, but they can also be fractional or negative. For the reaction

$$\text{CHCl}_3(g) + \text{Cl}_2(g) \longrightarrow \text{CCl}_4(g) + \text{HCl}(g)$$

Literature Arnáiz, Francisco J. "Mice in the Box for Zero-Order Kinetics," *J. Chem. Educ.* **1999**, *76*, 1458.

a fractional order appears in the rate law:

$$\text{Rate} = k[CHCl_3][Cl_2]^{1/2}$$

This reaction order means that the rate depends on the square root of the Cl_2 concentration. For example, if the initial Cl_2 concentration is increased by a factor of 4, while the initial $CHCl_3$ concentration is kept the same, the rate increases by a factor of 2, the square root of the change in $[Cl_2]$. A negative exponent means that the rate *decreases* when the concentration of that component increases. Negative orders are often seen for reactions whose rate laws include products. For example, for the atmospheric reaction

$$2O_3(g) \rightleftharpoons 3O_2(g)$$

the rate law has been shown to be

$$\text{Rate} = k[O_3]^2[O_2]^{-1} = k\frac{[O_3]^2}{[O_2]}$$

If the O_2 concentration doubles, the reaction proceeds half as fast.

SAMPLE PROBLEM 16.2 Determining Reaction Order from Rate Laws

PROBLEM For each of the following reactions, use the given rate law to determine the reaction order with respect to each reactant and the overall order:
(a) $2NO(g) + O_2(g) \longrightarrow 2NO_2(g)$; rate $= k[NO]^2[O_2]$
(b) $CH_3CHO(g) \longrightarrow CH_4(g) + CO(g)$; rate $= k[CH_3CHO]^{3/2}$
(c) $H_2O_2(aq) + 3I^-(aq) + 2H^+(aq) \longrightarrow I_3^-(aq) + 2H_2O(l)$; rate $= k[H_2O_2][I^-]$
PLAN We inspect the exponents in the rate law, *not* the coefficients of the balanced equation, to find the individual orders, and then take their sum to find the overall reaction order.
SOLUTION (a) The exponent of [NO] is 2, so the reaction is second order with respect to NO, first order with respect to O_2, and third order overall.

(b) The reaction is $\frac{3}{2}$ order in CH_3CHO and $\frac{3}{2}$ order overall.

(c) The reaction is first order in H_2O_2, first order in I^-, and second order overall.

The reactant H^+ does not appear in the rate law, so the reaction is zero order in H^+.
CHECK Be sure that each reactant has an order and that the sum of the individual orders gives the overall order.

FOLLOW-UP PROBLEM 16.2 Experiment shows that the reaction

$$5Br^-(aq) + BrO_3^-(aq) + 6H^+(aq) \longrightarrow 3Br_2(l) + 3H_2O(l)$$

obeys this rate law: rate $= k[Br^-][BrO_3^-][H^+]^2$. What are the reaction orders in each reactant and the overall reaction order?

ARIS Presentation Center, Worked Examples Library

Determining Reaction Orders Experimentally

Sample Problem 16.2 shows how to find the reaction orders from a known rate law. Now let's see how they are found from data *before* the rate law is known. Consider the reaction between oxygen and nitrogen monoxide, a key step in the formation of acid rain and in the industrial production of nitric acid:

$$O_2(g) + 2NO(g) \longrightarrow 2NO_2(g)$$

The rate law, expressed in general form, is

$$\text{Rate} = k[O_2]^m[NO]^n$$

To find the reaction orders, *we run a series of experiments, starting each one with a different set of reactant concentrations and obtaining an initial rate in each case.*

Literature Marzzacco, Charles J. "An Analogy to Help Students Understand Reaction Orders," *J. Chem. Educ.* **1998,** *75,* 482.

Table 16.2	Initial Rates for a Series of Experiments with the Reaction Between O_2 and NO		
Experiment	Initial Reactant Concentrations (mol/L)		Initial Rate (mol/L·s)
	O_2	NO	
1	1.10×10^{-2}	1.30×10^{-2}	3.21×10^{-3}
2	2.20×10^{-2}	1.30×10^{-2}	6.40×10^{-3}
3	1.10×10^{-2}	2.60×10^{-2}	12.8×10^{-3}
4	3.30×10^{-2}	1.30×10^{-2}	9.60×10^{-3}
5	1.10×10^{-2}	3.90×10^{-2}	28.8×10^{-3}

Table 16.2 shows experiments that change one reactant concentration while keeping the other constant. If we compare experiments 1 and 2, we see the effect of doubling $[O_2]$ on the rate. First, we take the ratio of their rate laws:

$$\frac{\text{Rate 2}}{\text{Rate 1}} = \frac{k[O_2]_2^m[NO]_2^n}{k[O_2]_1^m[NO]_1^n}$$

where $[O_2]_2$ is the O_2 concentration for experiment 2, $[NO]_1$ is the NO concentration for experiment 1, and so forth. Because k is a constant and $[NO]$ does not change between these two experiments, these quantities cancel:

$$\frac{\text{Rate 2}}{\text{Rate 1}} = \frac{[O_2]_2^m}{[O_2]_1^m} = \left(\frac{[O_2]_2}{[O_2]_1}\right)^m$$

Substituting the values from Table 16.2, we obtain

$$\frac{6.40 \times 10^{-3}\ \text{mol/L·s}}{3.21 \times 10^{-3}\ \text{mol/L·s}} = \left(\frac{2.20 \times 10^{-2}\ \text{mol/L}}{1.10 \times 10^{-2}\ \text{mol/L}}\right)^m$$

Dividing, we obtain

$$1.99 = (2.00)^m$$

Rounding to one significant figure gives

$$2 = 2^m$$

therefore, $m = 1$

The reaction is first order in O_2: when $[O_2]$ doubles, the rate doubles.

When the exponent is not easy to determine by inspection, you can solve for m as follows: $a = b^x$, so $x = \log a/\log b$. Thus, we have $m = \log 1.99/\log 2.00 = 0.993$, which rounds to 1.

To find the order with respect to NO, we compare experiments 3 and 1, in which $[O_2]$ is held constant and $[NO]$ is doubled:

$$\frac{\text{Rate 3}}{\text{Rate 1}} = \frac{k[O_2]_3^m[NO]_3^n}{k[O_2]_1^m[NO]_1^n}$$

As before, k is constant, and in this pair of experiments $[O_2]$ does not change, so these quantities cancel:

$$\frac{\text{Rate 3}}{\text{Rate 1}} = \left(\frac{[NO]_3}{[NO]_1}\right)^n$$

The actual values give

$$\frac{12.8 \times 10^{-3}\ \text{mol/L·s}}{3.21 \times 10^{-3}\ \text{mol/L·s}} = \left(\frac{2.60 \times 10^{-2}\ \text{mol/L}}{1.30 \times 10^{-2}\ \text{mol/L}}\right)^n$$

Dividing, we obtain

$$3.99 = (2.00)^n$$

Solving for n:

$$n = \log 3.99/\log 2.00 = 2.00$$

therefore, $n = 2$

The reaction is second order in NO: when [NO] doubles, the rate quadruples. Thus, the rate law is

$$\text{Rate} = k[O_2][NO]^2$$

You may want to use experiment 1 in combination with experiments 4 and 5 to check this result.

SAMPLE PROBLEM 16.3 Determining Reaction Orders from Initial Rate Data

ARIS Presentation Center, Worked Examples Library

Literature Sanger, Michael J.; Wiley, Russell A., Jr.; Richter, Erwin W.; Phelps, Amy J. "Rate Law Determination of Everyday Processes," *J. Chem. Educ.* **2002,** *79,* 989.

PROBLEM Many gaseous reactions occur in car engines and exhaust systems. One of these is

$$NO_2(g) + CO(g) \longrightarrow NO(g) + CO_2(g) \qquad \text{rate} = k[NO_2]^m[CO]^n$$

Use the following data to determine the individual and overall reaction orders:

Experiment	Initial Rate (mol/L·s)	Initial [NO$_2$] (mol/L)	Initial [CO] (mol/L)
1	0.0050	0.10	0.10
2	0.080	0.40	0.10
3	0.0050	0.10	0.20

depends on [NO$_2$]

PLAN We need to solve the general rate law for the reaction orders *m* and *n*. To solve for each exponent, we proceed as in the text, taking the ratio of the rate laws for two experiments in which only the reactant in question changes.

SOLUTION Calculating *m* in [NO$_2$]^m: We take the ratio of the rate laws for experiments 1 and 2, in which [NO$_2$] varies but [CO] is constant:

$$\frac{\text{Rate 2}}{\text{Rate 1}} = \frac{k[NO_2]_2^m[CO]_2^n}{k[NO_2]_1^m[CO]_1^n} = \left(\frac{[NO_2]_2}{[NO_2]_1}\right)^m \quad \text{or} \quad \frac{0.080 \text{ mol/L·s}}{0.0050 \text{ mol/L·s}} = \left(\frac{0.40 \text{ mol/L}}{0.10 \text{ mol/L}}\right)^m$$

This gives $16 = (4.0)^m$, so we have $m = \log 16/\log 4.0 = 2.0$. The reaction is second order in NO$_2$.

Calculating *n* in [CO]^n: We take the ratio of the rate laws for experiments 1 and 3, in which [CO] varies but [NO$_2$] is constant:

$$\frac{\text{Rate 3}}{\text{Rate 1}} = \frac{k[NO_2]_3^2[CO]_3^n}{k[NO_2]_1^2[CO]_1^n} = \left(\frac{[CO]_3}{[CO]_1}\right)^n \quad \text{or} \quad \frac{0.0050 \text{ mol/L·s}}{0.0050 \text{ mol/L·s}} = \left(\frac{0.20 \text{ mol/L}}{0.10 \text{ mol/L}}\right)^n$$

We have $1.0 = (2.0)^n$, so $n = 0$. The rate does not change when [CO] varies, so the reaction is zero order in CO.

Therefore, the rate law is

$$\text{Rate} = k[NO_2]^2[CO]^0 = k[NO_2]^2(1) = k[NO_2]^2$$

The reaction is second order overall.

CHECK A good check is to reason through the orders. If $m = 1$, quadrupling [NO$_2$] would quadruple the rate; but the rate *more* than quadruples, so $m > 1$. If $m = 2$, quadrupling [NO$_2$] would increase the rate by a factor of 16 (4^2). The ratio of rates is 0.080/0.005 = 16, so $m = 2$. In contrast, increasing [CO] has no effect on the rate, which can happen only if [CO]$^n = 1$, so $n = 0$.

FOLLOW-UP PROBLEM 16.3 Find the rate law and the overall reaction order for the reaction $H_2 + I_2 \longrightarrow 2HI$ from the following data at 450°C:

Experiment	Initial Rate (mol/L·s)	Initial [H$_2$] (mol/L)	Initial [I$_2$] (mol/L)
1	1.9×10^{-23}	0.0113	0.0011
2	1.1×10^{-22}	0.0220	0.0033
3	9.3×10^{-23}	0.0550	0.0011
4	1.9×10^{-22}	0.0220	0.0056

To provide a final overview of this idea, we'll go through one more sample problem, this time using molecular scenes to determine reaction orders.

SAMPLE PROBLEM 16.4 Determining Reaction Orders from a Series of Molecular Scenes

PROBLEM At a particular temperature, two gases, A (red) and B (blue), react to form products. The following molecular scenes represent four experiments run at the same volume, labeled 1 through 4 with their initial rates (in mol/L·s):

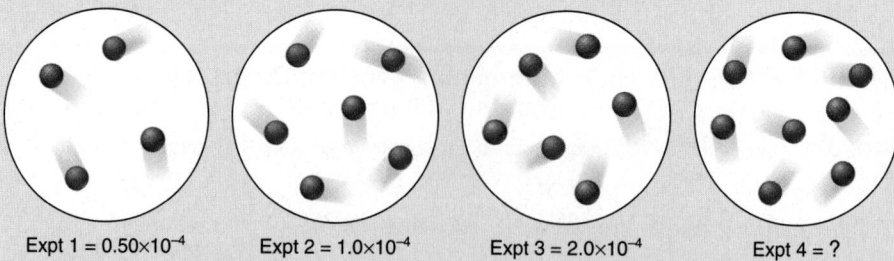

| Expt 1 = 0.50×10⁻⁴ | Expt 2 = 1.0×10⁻⁴ | Expt 3 = 2.0×10⁻⁴ | Expt 4 = ? |

Expt 1 = 0.50×10^{-4} Expt 2 = 1.0×10^{-4} Expt 3 = 2.0×10^{-4} Expt 4 = ?

(a) What is the reaction order with respect to A? With respect to B? The overall order?
(b) Write the rate law for the reaction.
(c) Predict the initial rate of experiment 4.

PLAN (a) As in previous situations, we find the individual reaction orders by seeing how a change in each reactant changes the rate. The difference here is that, instead of using concentration data, we count numbers of spheres. As before, if an increase in one reactant has no effect on the rate, the order with respect to that reactant is 0. Similarly, if the increase causes an increase in the rate by the same factor, the order is 1. And, if the increase in a reactant causes a square of that increase in the rate, the order is 2. The sum of the individual orders is the overall order. **(b)** To write the rate law, we use the orders from part (a) as exponents in the general rate law. **(c)** Using the results from Expts 1 through 3 and the rate law from part (b), we find the initial rate of Expt 4.

SOLUTION (a) Finding the individual and overall orders: For reactant A (red), from Expts 1 and 2, the number of A doubles (from 2 to 4), while B is constant (at 2), and the rate doubles (from 0.5×10^{-4} mol/L·s to 1.0×10^{-4} mol/L·s), so the order with respect to A is 1. For reactant B (blue), from Expts 1 and 3, the number of B doubles (from 2 to 4), while the number of A is constant (at 2), and the rate quadruples (from 0.5×10^{-4} mol/L·s to 2.0×10^{-4} mol/L·s), so the order with respect to B is 2. The overall order is $1 + 2 = 3$.
(b) Writing the rate law: The general rate law is rate $= k[\text{A}]^m[\text{B}]^n$, so we have

$$\text{Rate} = k[\text{A}][\text{B}]^2$$

(c) Finding the initial rate of Expt 4: Comparing, for example, Expts 3 and 4, we see that the number of A doubles (from 2 to 4). Furthermore, the rate law shows that the reaction is first order in A. Therefore, the initial rate in Expt 4 should be 4.0×10^{-4} mol/L·s.
CHECK A good check is to compare other pairs of experiments. **(a)** Comparing Expts 2 and 3 shows that the number of B doubles, which causes the rate to quadruple, and A decreases by half, which causes the rate to halve; so the overall rate change should double (from 1.0×10^{-4} mol/L·s to 2.0×10^{-4} mol/L·s), which it does. **(c)** Comparing Expts 2 and 4, with A constant, shows that the number of B doubles, so the rate should quadruple, which means the initial rate of Expt 4 would be 4.0×10^{-4} mol/L·s, as we found.

FOLLOW-UP PROBLEM 16.4 The scenes below show three experiments at a given temperature involving reactants X (black) and Y (green), with their initial rates (in mol/L·s):

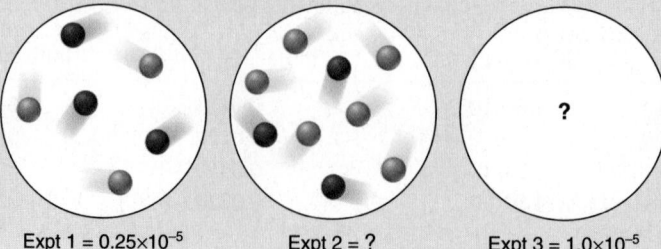

Expt 1 = 0.25×10^{-5} Expt 2 = ?   Expt 3 = 1.0×10^{-5}

If the rate law for the reaction is rate $= k[\text{X}]^2$: **(a)** What is the initial rate of Expt 2? **(b)** Draw a scene for Expt 3 that involves a single change of the scene for Expt 1.

Determining the Rate Constant

With the rate, reactant concentrations, and reaction orders known, the sole remaining unknown in the rate law is the rate constant, k. The rate constant is specific for a particular reaction *at a particular temperature*. The experiments with the reaction of O_2 and NO were run at the same temperature, so we can use data from any to solve for k. From experiment 1 in Table 16.2, for instance, we obtain

$$k = \frac{\text{rate 1}}{[O_2]_1[NO]_1^2} = \frac{3.21\times10^{-3}\ \text{mol/L·s}}{(1.10\times10^{-2}\ \text{mol/L})(1.30\times10^{-2}\ \text{mol/L})^2}$$

$$= \frac{3.21\times10^{-3}\ \text{mol/L·s}}{1.86\times10^{-6}\ \text{mol}^3/\text{L}^3} = 1.73\times10^3\ \text{L}^2/\text{mol}^2\text{·s}$$

Always check that the values of k for a series are constant within experimental error. To three significant figures, the average value of k for the five experiments in Table 16.2 is $1.72\times10^3\ \text{L}^2/\text{mol}^2\text{·s}$.

Note the units for the rate constant. With concentrations in mol/L and the reaction rate in units of mol/L·time, the units for k depend on the order of the reaction and, of course, the time unit. The units for k in our example, $\text{L}^2/\text{mol}^2\text{·s}$, are required to give a rate with units of mol/L·s:

$$\frac{\text{mol}}{\text{L·s}} = \frac{\text{L}^2}{\text{mol}^2\text{·s}} \times \frac{\text{mol}}{\text{L}} \times \left(\frac{\text{mol}}{\text{L}}\right)^2$$

The rate constant will *always* have these units for an overall third-order reaction with the time unit in seconds. Table 16.3 shows the units of k for some common overall reaction orders, but you can always determine the units mathematically.

Section Summary

An experimentally determined rate law shows how the rate of a reaction depends on concentration. If we consider only initial rates, the rate law often takes this form: rate = $k[A]^m[B]^n \cdots$. • With an accurate method for obtaining initial rates, reaction orders are determined experimentally by comparing rates for different initial concentrations, that is, by performing several experiments and varying the concentration of one reactant at a time to see its effect on the rate. • With rate, concentrations, and reaction orders known, only the rate constant remains to be calculated.

16.4 INTEGRATED RATE LAWS: CONCENTRATION CHANGES OVER TIME

Notice that the rate laws we've developed so far do not include time as a variable. They tell us the rate or concentration at a given instant, allowing us to answer a critical question, "How fast is the reaction proceeding at the moment when y moles per liter of A are reacting with z moles per liter of B?" However, by employing different forms of the rate laws, called **integrated rate laws,** we can consider the time factor and answer other questions, such as "How long will it take for x moles per liter of A to be used up?" and "What is the concentration of A after y minutes of reaction?"

Integrated Rate Laws for First-, Second-, and Zero-Order Reactions

Consider a simple first-order reaction, $A \longrightarrow B$. (Because first- and second-order reactions are more common, we'll discuss them before zero-order reactions.) As we discussed previously, the rate can be expressed as the change in the concentration of A divided by the change in time:

$$\text{Rate} = -\frac{\Delta[A]}{\Delta t}$$

Table 16.3 Units of the Rate Constant k for Several Overall Reaction Orders

Overall Reaction Order	Units of k (t in seconds)
0	mol/L·s (or mol L^{-1} s^{-1})
1	1/s (or s^{-1})
2	L/mol·s (or L mol^{-1} s^{-1})
3	$\text{L}^2/\text{mol}^2\text{·s}$ (or L^2 mol^{-2} s^{-1})

General formula:

$$\text{Units of } k = \frac{\left(\dfrac{\text{L}}{\text{mol}}\right)^{order-1}}{\text{unit of } t}$$

ARIS Presentation Center, Tables Library

Demo Creary, Xavier; Morris, Karen M. "A New Twist on the Iodine Clock Reaction: Determining the Order of a Reaction," *J. Chem. Educ.* **1999,** *76*, 530.

Demo Shakhashiri, Bassam Z. "Hydrogen Peroxide Iodine Clock: Oxidation of Potassium Iodide by Hydrogen Peroxide," *Chemical Demonstrations, A Handbook for Teachers of Chemistry*, Vol. 4 (The University of Wisconsin Press, Madison, 1985), 37–43.

Demo Shakhashiri, Bassam Z. "Color Variations of the Landolt Reaction," *Chemical Demonstrations, A Handbook for Teachers of Chemistry*, Vol. 4 (The University of Wisconsin Press, Madison, 1985), 22–28.

Literature Wright, Stephen W. "The Vitamin C Clock Reaction," *J. Chem. Educ.* **2002,** *79*, 41.

It can also be expressed in terms of the rate law:

$$\text{Rate} = k[A]$$

Setting these different expressions equal to each other gives

$$-\frac{\Delta[A]}{\Delta t} = k[A]$$

Through the methods of calculus, this expression is integrated over time to obtain the integrated rate law for a first-order reaction:

$$\ln \frac{[A]_0}{[A]_t} = kt \qquad (\text{first-order reaction; rate} = k[A]) \qquad \textbf{(16.4)}$$

where ln is the natural logarithm, $[A]_0$ is the concentration of A at $t = 0$, and $[A]_t$ is the concentration of A at any time t during an experiment. In mathematical terms, $\ln \frac{a}{b} = \ln a - \ln b$, so we have

$$\ln [A]_0 - \ln [A]_t = kt$$

For a general second-order reaction, the expression including time is quite complex, so let's consider the simple case in which the rate law contains only one reactant. Setting the rate expressions equal to each other gives

$$\text{Rate} = -\frac{\Delta[A]}{\Delta t} = k[A]^2$$

Integrating over time gives the integrated rate law for a second-order reaction involving one reactant:

$$\frac{1}{[A]_t} - \frac{1}{[A]_0} = kt \qquad (\text{second-order reaction; rate} = k[A]^2) \qquad \textbf{(16.5)}$$

For a zero-order reaction, we have

$$\text{Rate} = -\frac{\Delta[A]}{\Delta t} = k[A]^0$$

Integrating over time gives the integrated rate law for a zero-order reaction:

$$[A]_t - [A]_0 = -kt \qquad (\text{zero-order reaction; rate} = k[A]^0 = k) \qquad \textbf{(16.6)}$$

Sample Problem 16.5 shows one way integrated rate laws are applied.

SAMPLE PROBLEM 16.5 Determining the Reactant Concentration at a Given Time

PROBLEM At 1000°C, cyclobutane (C_4H_8) decomposes in a first-order reaction, with the very high rate constant of 87 s^{-1}, to two molecules of ethylene (C_2H_4).
(a) If the initial C_4H_8 concentration is 2.00 M, what is the concentration after 0.010 s?
(b) What fraction of C_4H_8 has decomposed in this time?
PLAN (a) We must find the concentration of cyclobutane at time t, $[C_4H_8]_t$. The problem tells us this is a first-order reaction, so we use the integrated first-order rate law:

$$\ln \frac{[C_4H_8]_0}{[C_4H_8]_t} = kt$$

We know k (87 s^{-1}), t (0.010 s), and $[C_4H_8]_0$ (2.00 M), so we can solve for $[C_4H_8]_t$.
(b) The fraction decomposed is the concentration that has decomposed divided by the initial concentration:

$$\text{Fraction decomposed} = \frac{[C_4H_8]_0 - [C_4H_8]_t}{[C_4H_8]_0}$$

SOLUTION (a) Substituting the data into the integrated rate law:

$$\ln \frac{2.00 \text{ mol/L}}{[C_4H_8]_t} = (87 \text{ s}^{-1})(0.010 \text{ s}) = 0.87$$

Taking the antilog of both sides:

$$\frac{2.00 \text{ mol/L}}{[C_4H_8]_t} = e^{0.87} = 2.4$$

Solving for $[C_4H_8]_t$:

$$[C_4H_8]_t = \frac{2.00 \text{ mol/L}}{2.4} = 0.83 \text{ mol/L}$$

(b) Finding the fraction that has decomposed after 0.010 s:

$$\frac{[C_4H_8]_0 - [C_4H_8]_t}{[C_4H_8]_0} = \frac{2.00 \text{ mol/L} - 0.83 \text{ mol/L}}{2.00 \text{ mol/L}} = 0.58$$

CHECK The concentration remaining after 0.010 s (0.83 mol/L) is less than the starting concentration (2.00 mol/L), which makes sense. Raising e to an exponent slightly less than 1 should give a number (2.4) slightly less than the value of e (2.718). Moreover, the final result makes sense: a high rate constant indicates a fast reaction, so it's not surprising that so much decomposes in such a short time.

COMMENT Integrated rate laws are also used to solve for the time it takes to reach a certain reactant concentration, as in the follow-up problem.

FOLLOW-UP PROBLEM 16.5 At 25°C, hydrogen iodide breaks down very slowly to hydrogen and iodine: rate $= k[HI]^2$. The rate constant at 25°C is 2.4×10^{-21} L/mol·s. If 0.0100 mol of HI(g) is placed in a 1.0-L container, how long will it take for the concentration of HI to reach 0.00900 mol/L (10.0% reacted)?

Determining the Reaction Order from the Integrated Rate Law

Suppose you don't know the rate law for a reaction and don't have the initial rate data needed to determine the reaction orders (which we did have in Sample Problem 16.3). Another method for finding reaction orders is a graphical technique that uses concentration and time data directly.

An integrated rate law can be rearranged into the form of an equation for a straight line, $y = mx + b$, where m is the slope and b is the y-axis intercept. For a first-order reaction, we have

$$\ln [A]_0 - \ln [A]_t = kt$$

Rearranging and changing signs gives

$$\ln [A]_t = -kt + \ln [A]_0$$
$$y \quad = \quad mx + \quad b$$

Therefore, a plot of $\ln [A]_t$ vs. time gives a straight line with slope $= -k$ and y intercept $= \ln [A]_0$ (Figure 16.7A).

For a simple second-order reaction, we have

$$\frac{1}{[A]_t} - \frac{1}{[A]_0} = kt$$

Rearranging gives

$$\frac{1}{[A]_t} = kt + \frac{1}{[A]_0}$$
$$y \quad = \quad mx + \quad b$$

In this case, a plot of $1/[A]_t$ vs. time gives a straight line with slope $= k$ and y intercept $= 1/[A]_0$ (Figure 16.7B).

For a zero-order reaction, we have

$$[A]_t - [A]_0 = -kt$$

Rearranging gives

$$[A]_t = -kt + [A]_0$$
$$y \quad = \quad mx + \quad b$$

Thus, a plot of $[A]_t$ vs. time gives a straight line with slope $= -k$ and y intercept $= [A]_0$ (Figure 16.7C).

Transparency/ARIS Presentation Center, Art Library

A First order

B Second order

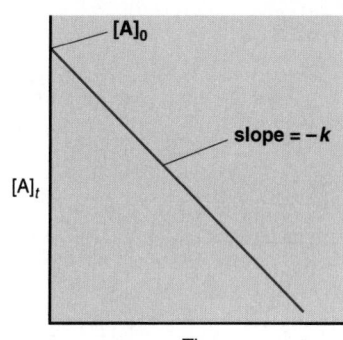

C Zero order

Figure 16.7 Integrated rate laws and reaction orders. A, Plot of ln $[A]_t$ vs. time gives a straight line for a reaction that is first order in A. **B,** Plot of $1/[A]_t$ vs. time gives a straight line for a reaction that is second order in A. **C,** Plot of $[A]_t$ vs. time gives a straight line for a reaction that is zero order in A.

Time (min)	$[N_2O_5]$	$\ln [N_2O_5]$	$1/[N_2O_5]$
0	0.0165	−4.104	60.6
10	0.0124	−4.390	80.6
20	0.0093	−4.68	1.1×10^2
30	0.0071	−4.95	1.4×10^2
40	0.0053	−5.24	1.9×10^2
50	0.0039	−5.55	2.6×10^2
60	0.0029	−5.84	3.4×10^2

Figure 16.8 **Graphical determination of the reaction order for the decomposition of N_2O_5.** A table of time and concentration data for determining reaction order appears below the graphs. **A,** A plot of $[N_2O_5]$ vs. time is curved, indicating that the reaction is *not* zero order in N_2O_5. **B,** A plot of $\ln [N_2O_5]$ vs. time gives a straight line, indicating that the reaction *is* first order in N_2O_5. **C,** A plot of $1/[N_2O_5]$ vs. time is curved, indicating that the reaction is *not* second order in N_2O_5. Plots A and C support the conclusion from plot B.

Transparency/ARIS Presentation Center, Art Library

Literature Urbansky, Edward. "Don't Be Tricked by Your Integrated Rate Plot," *J. Chem. Educ.* **2004,** *81,* 32.

Therefore, some trial-and-error graphical plotting is required to find the reaction order from the concentration and time data:

- If you obtain a straight line when you plot ln [reactant] vs. time, the reaction is *first order* with respect to that reactant.
- If you obtain a straight line when you plot 1/[reactant] vs. time, the reaction is *second order* with respect to that reactant.
- If you obtain a straight line when you plot [reactant] vs. time, the reaction is *zero order* with respect to that reactant.

Figure 16.8 shows how this approach is used to determine the order for the decomposition of N_2O_5. Since the plot of $\ln [N_2O_5]$ *is* linear and the plot of $1/[N_2O_5]$ *is not,* the decomposition of N_2O_5 must be first order in N_2O_5.

Reaction Half-Life

The **half-life** $(t_{1/2})$ of a reaction is the time required for the reactant concentration to reach half its initial value. A half-life is expressed in time units appropriate for a given reaction and is characteristic of that reaction at a given temperature.

At fixed conditions, *the half-life of a first-order reaction is a constant, independent of reactant concentration.* For example, the half-life for the first-order decomposition of N_2O_5 at 45°C is 24.0 min. The meaning of this value is that if we start with, say, 0.0600 mol/L of N_2O_5 at 45°C, after 24 min (one half-life), 0.0300 mol/L has been consumed and 0.0300 mol/L remains; after 48 min (two half-lives), 0.0150 mol/L remains; after 72 min (three half-lives), 0.0075 mol/L remains, and so forth (Figure 16.9).

We can see from the integrated rate law why the half-life of a first-order reaction is independent of concentration:

$$\ln \frac{[A]_0}{[A]_t} = kt$$

After one half-life, $t = t_{1/2}$, and $[A]_t = \frac{1}{2}[A]_0$. Substituting, we obtain

$$\ln \frac{[A]_0}{\frac{1}{2}[A]_0} = kt_{1/2} \qquad \text{or} \qquad \ln 2 = kt_{1/2}$$

Then, solving for $t_{1/2}$, we have

$$t_{1/2} = \frac{\ln 2}{k} = \frac{0.693}{k} \qquad \text{(first-order process; rate} = k[A]) \qquad \text{(16.7)}$$

Figure 16.9 **A plot of [N$_2$O$_5$] vs. time for three half-lives.** During each half-life, the concentration is halved (T = 45°C and [N$_2$O$_5$]$_0$ = 0.0600 mol/L). The blow-up volumes, with N$_2$O$_5$ molecules as colored spheres, show that after three half-lives, $\frac{1}{2} \times \frac{1}{2} \times \frac{1}{2} = \frac{1}{8}$ of the original concentration remains.

 Transparency/ARIS Presentation Center, Active Art Library

As you can see, *the time to reach one-half the starting concentration in a first-order reaction does not depend on what that starting concentration is.*

Radioactive decay of an unstable nucleus is another example of a first-order process. For example, the half-life for the decay of uranium-235 is 7.1×10^8 yr. After 710 million years, a sample containing 1 kg of uranium-235 will contain 0.5 kg of uranium-235, and a sample containing 1 mg of uranium-235 will contain 0.5 mg. (We discuss the kinetics of radioactive decay thoroughly in Chapter 24.) Whether we consider a molecule or a radioactive nucleus, the *decomposition of each particle in a first-order process is independent of the number of other particles present.*

We'll go through two sample problems to solidify this idea: the first generalizes the concept with molecular scenes, and the second uses real substances.

ARIS Presentation Center,
Worked Examples Library

SAMPLE PROBLEM 16.6 Using Molecular Scenes to Determine Half-Life

PROBLEM Compound A (red) converts to compound B (black) in a first-order process. The following scenes represent the reaction mixture at the start of the reaction (at 0.0 s) and after 30.0 s:

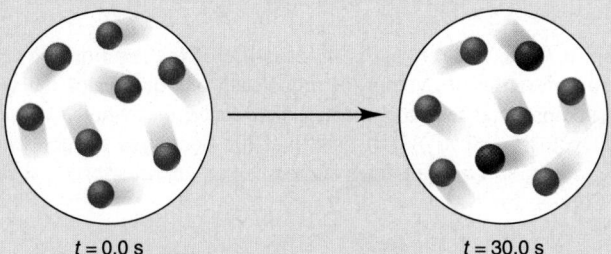

(a) Find the half-life, $t_{1/2}$, of the reaction. **(b)** Calculate the rate constant, k. **(c)** Draw a scene that represents the reaction mixture after 2.00 min.

PLAN (a) Given that the reaction is first order, we know the half-life is a constant. Counting the numbers of red and black spheres at the start and at the later time tells us how fast the reaction is proceeding. When half of the red spheres have become black, the elapsed time is the half-life. **(b)** We substitute the value of $t_{1/2}$ from part (a) into Equation 16.7 and solve for k. **(c)** Once we know how long it takes half of A to become B, we can calculate how many red and black spheres would be in the mixture at 2.00 min (120. s).

SOLUTION (a) Examining the scenes to find $t_{1/2}$: At $t = 0.0$ s, there are 8 molecules of A (red) and no B (black). At $t = 30.0$ s, 6 of A and 2 of B are present, so one-quarter of A has reacted. Therefore, it takes twice as long (2×30.0 s) for half of the original 8 molecules of A to react: $t_{1/2} = 60.0$ s

(b) Finding the rate constant:

$$t_{1/2} = 0.693/k$$

therefore, $$k = 0.693/t_{1/2} = 0.693/60.0 \text{ s} = \boxed{1.16 \times 10^{-2} \text{ s}^{-1}}$$

(c) If half of the original number of molecules react in 60.0 s, half of the remaining number react in two half-lives, or 120. s. The scene representing the mixture after 120. s is

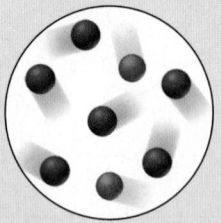

CHECK For **(b)**, rounding gives $0.7/60$ s^{-1}, which is slightly greater than 1×10^{-2} s^{-1}, so the answer seems correct.

FOLLOW-UP PROBLEM 16.6 Compound X (blue) undergoes a slow first-order transformation to Y (orange), as represented by the scenes below:

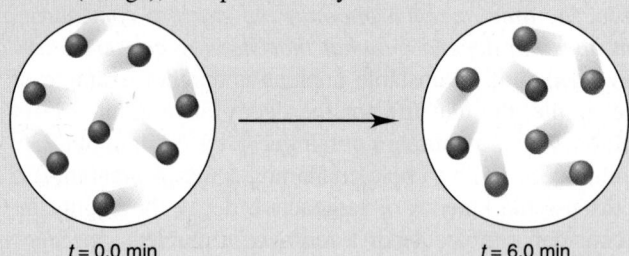

$t = 0.0$ min $t = 6.0$ min

Determine $t_{1/2}$ and k of the reaction.

SAMPLE PROBLEM 16.7 Determining the Half-Life of a First-Order Reaction

PROBLEM Cyclopropane is the smallest cyclic hydrocarbon. Because its 60° bond angles reduce orbital overlap, its bonds are weak. As a result, it is thermally unstable and rearranges to propene at 1000°C via the following first-order reaction:

$$\underset{\displaystyle H_2C\!-\!CH_2(g)}{\overset{\displaystyle CH_2}{\diagup\diagdown}} \quad \overset{\Delta}{\longrightarrow} \quad CH_3\!-\!CH\!=\!CH_2(g)$$

The rate constant is 9.2 s^{-1}. **(a)** What is the half-life of the reaction? **(b)** How long does it take for the concentration of cyclopropane to reach one-quarter of the initial value?

PLAN (a) The cyclopropane rearrangement is first order, so to find $t_{1/2}$ we use Equation 16.7 and substitute for k (9.2 s^{-1}). **(b)** Each half-life decreases the concentration to one-half of its initial value, so two half-lives decrease it to one-quarter.

SOLUTION (a) Solving for $t_{1/2}$:

$$t_{1/2} = \frac{\ln 2}{k} = \frac{0.693}{9.2 \text{ s}^{-1}} = \boxed{0.075 \text{ s}}$$

It takes 0.075 s for half the cyclopropane to form propene at this temperature.

(b) Finding the time to reach one-quarter of the initial concentration:

$$\text{Time} = 2(t_{1/2}) = 2(0.075 \text{ s}) = \boxed{0.15 \text{ s}}$$

CHECK For **(a)**, rounding gives $0.7/9$ s^{-1} = 0.08 s, so the answer seems correct.

FOLLOW-UP PROBLEM 16.7 Iodine-123 is used to study thyroid gland function. This radioactive isotope breaks down in a first-order process with a half-life of 13.1 h. What is the rate constant for the process?

Transparency/ARIS Presentation Center, Worked Examples Library

Demo Hughes, Enrique A.; Zalts, Anita. "Radioactivity in the Classroom," *J. Chem. Educ.* **2000**, *77*, 613.

Literature Schultz, Emeric. "Dice Shaking as an Analogy for Radioactive Decay and First Order Kinetics," *J. Chem. Educ.* **1997**, *74*, 505.

Table 16.4 **An Overview of Zero-Order, First-Order, and Simple Second-Order Reactions**

	Zero Order	First Order	Second Order
Rate law	rate $= k$	rate $= k[A]$	rate $= k[A]^2$
Units for k	mol/L·s	1/s	L/mol·s
Integrated rate law in straight-line form	$[A]_t =$ $-kt + [A]_0$	$\ln [A]_t =$ $-kt + \ln [A]_0$	$1/[A]_t =$ $kt + 1/[A]_0$
Plot for straight line	$[A]_t$ vs. time	$\ln [A]_t$ vs. time	$1/[A]_t$ vs. time
Slope, y intercept	$-k$, $[A]_0$	$-k$, $\ln [A]_0$	k, $1/[A]_0$
Half-life	$[A]_0/2k$	$(\ln 2)/k$	$1/k[A]_0$

Transparency/ARIS Presentation Center, Tables Library

In contrast to the half-life of a first-order reaction, the half-life of a second-order reaction *does* depend on reactant concentration:

$$t_{1/2} = \frac{1}{k[A]_0} \qquad \text{(second-order process; rate } = k[A]^2)$$

Note that here *the half-life is **inversely** proportional to the initial reactant concentration*. This relationship means that a second-order reaction with a high initial reactant concentration has a shorter half-life, and one with a low initial reactant concentration has a longer half-life. Therefore, *as a second-order reaction proceeds, the half-life increases*.

In contrast to the half-life of a second-order reaction, *the half-life of a zero-order reaction is **directly** proportional to the initial reactant concentration*:

$$t_{1/2} = \frac{[A]_0}{2k} \qquad \text{(zero-order process; rate } = k)$$

Thus, if a zero-order reaction begins with a high reactant concentration, it has a longer half-life than if it begins with a low reactant concentration. Table 16.4 summarizes the essential features of zero-, first-, and second-order reactions.

Section Summary

Integrated rate laws are used to find either the time needed to reach a certain concentration of reactant or the concentration present after a given time. Rearrangements of the integrated rate laws allow us to determine reaction orders and rate constants graphically. • The half-life is the time needed for the reaction to consume half the reactant; for first-order reactions, it is independent of concentration.

16.5 THE EFFECT OF TEMPERATURE ON REACTION RATE

Temperature often has a major effect on reaction rate. As Figure 16.10A shows for a common organic reaction (hydrolysis, or reaction with water, of an ester), when reactant concentrations are held constant, the rate nearly doubles with each rise in temperature of 10 K (or 10°C). In fact, for many reactions near room temperature, an increase of 10°C causes a doubling or tripling of the rate.

ARIS Presentation Center, Art Library

Figure 16.10 **Dependence of the rate constant on temperature. A,** In the hydrolysis of the ester ethyl acetate,

$CH_3COOCH_2CH_3 + H_2O \rightleftharpoons$
$\qquad CH_3COOH + CH_3CH_2OH$

when reactant concentrations are held constant and temperature increases, the rate and rate constant increase. Note the near doubling of k with each rise of 10 K (10°C). **B,** A plot of rate constant vs. temperature for this reaction shows an exponentially increasing curve.

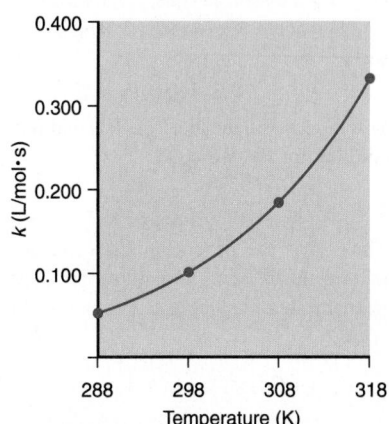

Expt	[Ester]	[H₂O]	T (K)	Rate (mol/L·s)	k (L/mol·s)
1	0.100	0.200	288	1.04×10^{-3}	0.0521
2	0.100	0.200	298	2.02×10^{-3}	0.101
3	0.100	0.200	308	3.68×10^{-3}	0.184
4	0.100	0.200	318	6.64×10^{-3}	0.332

A

B

Literature Leenson, Ilya A. "The Arrhenius Law and Storage of Food in a Freezer," *J. Chem. Educ.* **1999,** 76, 504.

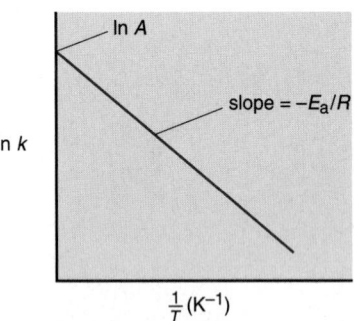

Figure 16.11 Graphical determination of the activation energy. A plot of ln k vs. $1/T$ gives a straight line with slope = $-E_a/R$.

ARIS Presentation Center,
Art Library

ARIS Presentation Center,
Worked Examples Library

● **The Significance of R** The importance of R extends beyond the study of gases (or osmotic pressure). By expressing pressure and volume in more fundamental quantities, we obtain dimensions for energy, E:

$$P = \frac{\text{force}}{\text{area}} = \frac{\text{force}}{(\text{length})^2} \quad V = (\text{length})^3$$

Thus,

$$PV = \frac{\text{force}}{(\text{length})^2} \times (\text{length})^3$$
$$= \text{force} \times \text{length}$$

Energy is used when a force moves an object over a distance. Thus,

$$E = \text{force} \times \text{distance (or length)}$$

so

$$PV = \text{force} \times \text{length} = E$$

Solving for R in the ideal gas law and substituting for PV, we obtain

$$R = \frac{PV}{nT} = \frac{E}{\text{amount} \times T}$$

Thus, R is the proportionality constant that relates the energy, amount (mol), and temperature of any chemical system.

Aside from mammals and birds, most animals—all reptiles and many species of amphibians, insects, and fish—absorb heat from the external environment to maintain metabolic and physical activity. For example, lizards and crocodiles bask in the sun or lie in the shade to regulate their internal reaction rates.

How does the rate law express this effect of temperature? If we collect concentration and time data for the same reaction run at *different* temperatures (T), and then solve each rate expression for k, we find that k increases as T increases. In other words, *temperature affects the rate by affecting the rate constant.* A plot of k vs. T gives a curve that increases exponentially (Figure 16.10B).

These results are consistent with studies made in 1889 by the Swedish chemist Svante Arrhenius, who discovered a key relationship between T and k. In its modern form, the **Arrhenius equation** is

$$k = Ae^{-E_a/RT} \qquad \textbf{(16.8)}$$

where k is the rate constant, e is the base of natural logarithms, T is the absolute temperature, and R is the universal gas constant. We'll discuss the meaning of the constant A, which is related to the orientation of the colliding molecules, in the next section. The E_a term is the **activation energy** of the reaction, which Arrhenius considered the *minimum energy* the molecules must have to react; we'll explore its meaning in the next section as well. This negative exponential relationship between T and k means that *as T increases, the negative exponent becomes smaller, so the value of k becomes larger, which means that the rate increases:*

$$\text{Higher } T \Longrightarrow \text{larger } k \Longrightarrow \text{increased rate}$$

We can calculate E_a from the Arrhenius equation by taking the natural logarithm of both sides and recasting the equation into one for a straight line:

$$\ln k = \ln A - \frac{E_a}{R}\left(\frac{1}{T}\right)$$
$$y \;=\; b \;+\; mx$$

A plot of ln k vs. $1/T$ gives a straight line whose slope is $-E_a/R$ and whose y intercept is ln A (Figure 16.11). Therefore, with the constant R known, we can determine E_a graphically from a series of k values at different temperatures.

Since the relationship between ln k and $1/T$ is linear, we can use a simpler method to find E_a if we know the rate constants at two temperatures, T_2 and T_1:

$$\ln k_2 = \ln A - \frac{E_a}{R}\left(\frac{1}{T_2}\right) \qquad \ln k_1 = \ln A - \frac{E_a}{R}\left(\frac{1}{T_1}\right)$$

When we subtract ln k_1 from ln k_2, the term ln A drops out and the other terms can be rearranged to give

$$\ln \frac{k_2}{k_1} = -\frac{E_a}{R}\left(\frac{1}{T_2} - \frac{1}{T_1}\right) \qquad \textbf{(16.9)}$$

From this, we can solve for E_a. ●

SAMPLE PROBLEM 16.8 Determining the Energy of Activation

PROBLEM The decomposition of hydrogen iodide,

$$2HI(g) \longrightarrow H_2(g) + I_2(g)$$

has rate constants of 9.51×10^{-9} L/mol·s at 500. K and 1.10×10^{-5} L/mol·s at 600. K. Find E_a.

PLAN We are given the rate constants, k_1 and k_2, at two temperatures, T_1 and T_2, so we substitute into Equation 16.9 and solve for E_a.

SOLUTION Rearranging Equation 16.9 to solve for E_a:

$$\ln \frac{k_2}{k_1} = -\frac{E_a}{R}\left(\frac{1}{T_2} - \frac{1}{T_1}\right)$$

$$E_a = -R\left(\ln \frac{k_2}{k_1}\right)\left(\frac{1}{T_2} - \frac{1}{T_1}\right)^{-1}$$

$$= -(8.314 \text{ J/mol·K})\left(\ln \frac{1.10 \times 10^{-5}\text{ L/mol·s}}{9.51 \times 10^{-9}\text{ L/mol·s}}\right)\left(\frac{1}{600. \text{ K}} - \frac{1}{500. \text{ K}}\right)^{-1}$$

$$= 1.76 \times 10^5 \text{ J/mol}$$

$$= 1.76 \times 10^2 \text{ kJ/mol}$$

COMMENT Be sure to retain the same number of significant figures in $1/T$ as you have in T, or a significant error could be introduced. Round to the correct number of significant figures only at the final answer. On most pocket calculators, the expression $(1/T_2 - 1/T_1)$ is entered as follows: $(T_2)(1/x) - (T_1)(1/x) =$

FOLLOW-UP PROBLEM 16.8 The reaction

$$2NOCl(g) \longrightarrow 2NO(g) + Cl_2(g)$$

has an E_a of 1.00×10^2 kJ/mol and a rate constant of 0.286 L/mol·s at 500. K. What is the rate constant at 490. K?

In this section and Sections 16.3 and 16.4, we discussed a series of experimental and mathematical methods for the study of reaction kinetics. Figure 16.12 presents a useful summary of this information. Notice that the integrated rate law provides an alternative method for obtaining reaction orders and the rate constant.

Section Summary

As the Arrhenius equation shows, rate increases with temperature because a temperature rise increases the rate constant. • The activation energy, E_a, the minimum energy needed for a reaction to occur, can be determined graphically from k values at different T values.

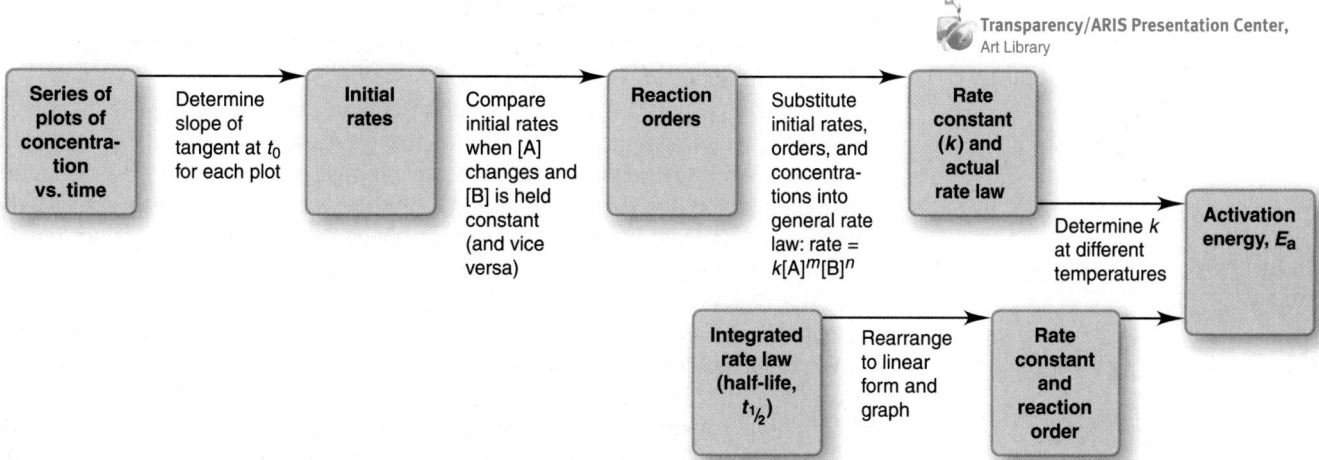

Figure 16.12 **Information sequence to determine the kinetic parameters of a reaction.** Note that the integrated rate law does not depend on the experimental comparison of initial rates and that it is also used to determine reaction orders and the rate constant.

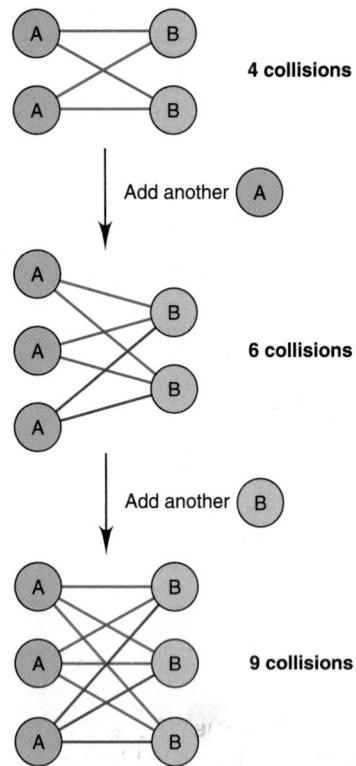

Figure 16.13 The dependence of number of possible collisions on the product of reactant concentrations. Concentrations are multiplied, not added, in the rate law because the number of possible collisions is the *product,* not the sum, of the numbers of particles present.

ARIS Presentation Center,
Art Library

16.6 EXPLAINING THE EFFECTS OF CONCENTRATION AND TEMPERATURE

The Arrhenius equation was developed empirically from the observations of many reactions. The two major models that explain the observed effects of concentration and temperature on reaction rate highlight different aspects of the reaction process but are completely compatible. *Collision theory* views the reaction rate as the result of particles colliding with a certain frequency and minimum energy. *Transition state theory* offers a close-up view of how the energy of a collision converts reactant to product.

Collision Theory: Basis of the Rate Law

The basic tenet of **collision theory** is that reactant particles—atoms, molecules, and ions—must collide with each other to react. Therefore, the number of collisions per unit time provides an upper limit on how fast a reaction can take place. The model restricts itself to simple one-step reactions in which two particles collide and form products: $A + B \longrightarrow$ products. With its emphasis on collisions between three-dimensional particles, this model explains why reactant concentrations are multiplied together in the rate law, how temperature affects the rate, and what influence molecular structure has on rate.

Why Concentrations Are Multiplied in the Rate Law
If particles must collide to react, the laws of probability tell us why the rate depends on the *product* of the reactant concentrations, not their sum. Imagine that you have only two particles of A and two of B confined in a reaction vessel. Figure 16.13 shows that four A-B collisions are possible. If you add another particle of A, there can be six A-B collisions (3×2), not just five ($3 + 2$); add another particle of B, and there can be nine A-B collisions (3×3), not just six ($3 + 3$). Thus, collision theory is consistent with the observation that concentrations are *multiplied* in the rate law.

How Temperature Affects Rate: The Importance of Activation Energy
Increasing the temperature of a reaction increases the average speed of particles and therefore their collision frequency. But collision frequency cannot be the only factor affecting rate, or every gaseous reaction would be almost instantaneous: after all, at 1 atm and 20°C, the molecules in a milliliter of gas experience about 10^{27} collisions per second, or about 4×10^7 collisions/molecule, so their reaction would be over in less than 1 ten-millionth of a second! In fact, *in the vast majority of collisions, the molecules rebound without reacting.*

Arrhenius proposed that every reaction has an *energy threshold* that the colliding molecules must exceed in order to react. (An analogy might be an athlete who must exceed the height of the bar to accomplish a high jump.) This minimum collision energy is the *activation energy* (E_a), the energy required to activate the molecules into a state from which reactant bonds can change into product bonds. Recall that at any given temperature, molecules have a range of kinetic energies; thus, their collisions have a range of energies as well. According to collision theory, *only those collisions with enough energy to exceed E_a can lead to reaction.*

We noted earlier that many reactions near room temperature approximately double or triple their rates with a 10°C rise in temperature. Is the rate increase due to a higher number of collisions? Actually, this has only a minor effect. Calculations show that a 10°C rise increases the average molecular speed by only 2%. If an increase in speed is the only effect of temperature and if the speed of each colliding molecule increases by 2%, we should observe at most a 4% increase in rate. Far more important is that *the temperature rise enlarges the fraction of collisions with enough energy to exceed the activation energy.* This key point is shown in Figure 16.14.

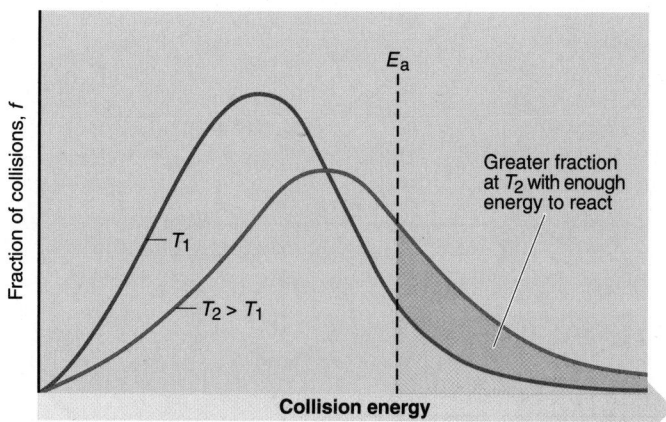

Figure 16.14 **The effect of temperature on the distribution of collision energies.** At the higher temperature, T_2, a larger fraction of collisions occur with enough energy to exceed E_a.

At a given temperature, the fraction f of molecular collisions with energy greater than or equal to the activation energy, E_a, is given by

$$f = e^{-E_a/RT}$$

where e is the base of natural logarithms, T is the absolute temperature, and R is the universal gas constant. [Notice that the right side of this equation is the central component in the Arrhenius equation (Equation 16.8).] *The magnitudes of both E_a and T affect the fraction of sufficiently energetic collisions.* In the top portion of Table 16.5, you can see the effect on this fraction of increasing E_a at a fixed temperature. Note how much the fraction shrinks with a 25-kJ/mol increase in activation energy. (As the height of the bar is raised, fewer athletes can accomplish the jump.) In the bottom portion of the table, you can see the effect on the fraction of increasing T at a fixed E_a of 50 kJ/mol, a typical value for many reactions. Note that the fraction nearly doubles for a 10°C increase. Doubling the fraction doubles the rate constant, which doubles the reaction rate.

A reversible reaction has two activation energies (Figure 16.15). The activation energy for the forward reaction, $E_{a(fwd)}$, is the energy difference between the activated state and the reactants; the activation energy for the reverse reaction, $E_{a(rev)}$, is the energy difference between the activated state and the products. The figure shows an energy-level diagram for an exothermic reaction, so the products are at a lower energy than the reactants, and $E_{a(fwd)}$ is less than $E_{a(rev)}$.

Table 16.5 **The Effect of E_a and T on the Fraction (f) of Collisions with Sufficient Energy to Allow Reaction**

E_a (kJ/mol)	f (at $T = 298$ K)
50	1.70×10^{-9}
75	7.03×10^{-14}
100	2.90×10^{-18}
T	**f (at $E_a = 50$ kJ/mol)**
25°C (298 K)	1.70×10^{-9}
35°C (308 K)	3.29×10^{-9}
45°C (318 K)	6.12×10^{-9}

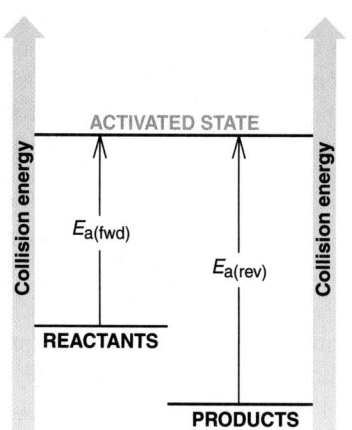

Figure 16.15 **Energy-level diagram for a reaction.** For molecules to react, they must collide with enough energy to reach an activated state. This minimum collision energy is the energy of activation, E_a. A reaction can occur in either direction, so the diagram shows two activation energies. Here, the forward reaction is exothermic, and $E_{a(fwd)} < E_{a(rev)}$.

Figure 16.16 An energy-level diagram of the fraction of collisions whose energy exceeds E_a. When Figure 16.14 is aligned vertically on the left and right axes of Figure 16.15, we see that

• In either direction, the fraction of collisions with energy exceeding E_a is larger at the higher T.
• In an *exothermic* reaction at any temperature, the fraction of collisions with energy exceeding $E_{a(fwd)}$ is larger than the fraction exceeding $E_{a(rev)}$.

Literature Piepgrass, Kent W. "Audience-Appropriate Analogies: Collision Theory," *J. Chem. Educ.* **1998,** *75,* 724.

By turning the collision-energy distribution curve of Figure 16.14 vertically and aligning it on both sides of Figure 16.15, we obtain Figure 16.16, which shows how temperature affects the fraction of collisions that have energy exceeding the activation energy for both the forward and reverse reactions. Several ideas are illustrated in this composite figure:

• In both reaction directions, a larger fraction of collisions exceeds the activation energy at the higher temperature, T_2: higher T increases reaction rate.
• For an *exothermic* process (forward reaction here) at any temperature, the fraction of reactant collisions with energy exceeding $E_{a(fwd)}$ is *larger* than the fraction of product collisions with energy exceeding $E_{a(rev)}$; thus, the forward reaction is faster. On the other hand, in an *endothermic* process (reverse reaction here), $E_{a(fwd)}$ is greater than $E_{a(rev)}$, so the fraction of product collisions with energy exceeding $E_{a(rev)}$ is larger, and the reverse reaction is faster.

These conclusions are consistent with the Arrhenius equation; that is, *the smaller the E_a (or the higher the temperature), the larger the value of k, and the faster the reaction:*

Smaller E_a (or higher T) $\Longrightarrow$ larger k $\Longrightarrow$ increased rate

Conversely, *the larger the E_a (or the lower the temperature), the smaller the value of k, and the slower the reaction:*

Larger E_a (or lower T) $\Longrightarrow$ smaller k $\Longrightarrow$ decreased rate

How Molecular Structure Affects Rate You've seen that the enormous number of collisions per second is greatly reduced when we count only those with enough energy to react. However, even this tiny fraction of the total collisions does not reveal the true number of **effective collisions,** those that actually lead to product. In addition to colliding with enough energy, *the molecules must collide so that the reacting atoms make contact.* In other words, to be effective, a collision must have enough energy *and* a particular *molecular orientation.*

In the Arrhenius equation, the effect of molecular orientation is contained in the term A:

$$k = Ae^{-E_a/RT}$$

This term is called the **frequency factor,** the product of the collision frequency Z and an *orientation probability factor, p,* which is specific for each reaction: $A = pZ$. The factor p is related to the structural complexity of the colliding particles. You can think of it as the ratio of effectively oriented collisions to all possible collisions. For example, Figure 16.17 shows a few of the possible collision orientations for the following simple gaseous reaction:

$$NO(g) + NO_3(g) \longrightarrow 2NO_2(g)$$

Of the five collisions shown, only one has an orientation in which the N of NO makes contact with an O of NO_3. Actually, the orientation probability factor (*p* value) for this reaction is 0.006: only 6 collisions in every 1000 (1 in 167) have an orientation that can lead to reaction.

Collisions between individual atoms have *p* values near 1: almost no matter how they hit, as long as the collision has enough energy, the particles react. In such cases, the rate constant depends only on the frequency and energy of the collisions. At the other extreme are biochemical reactions, in which the reactants are often two small molecules that can react only when they collide with a specific tiny region of a giant molecule—a protein or nucleic acid. The orientation probability factor for these reactions is often less than 10^{-6}: fewer than one in a million sufficiently energetic collisions lead to product. The fact that countless such biochemical reactions are occurring right now, as you read this sentence, helps make the point that the number of collisions per second is truly astounding.

Transition State Theory: Molecular Nature of the Activated Complex

Collision theory is a simple, easy to visualize model, but it provides no insight about why the activation energy is crucial and how the activated molecules look. To understand these aspects of the process, we turn to **transition state theory,** which focuses on the high-energy species that forms through an effective collision.

Visualizing the Transition State Recall from our discussion of energy changes (Chapter 6) that the internal energy of a system is the sum of its kinetic and potential energies. When two molecules approach each other, some kinetic energy is converted to potential energy as the electron clouds repel each other. At the moment of a head-on collision, the molecules stop, and their kinetic energy is converted to the potential energy of the collision. *If this potential energy is less than the activation energy, the molecules recoil,* bouncing off each other like billiard balls; the molecules zoom apart without reacting.

The tiny fraction of molecules that are oriented effectively *and* moving at the highest speed behave differently. *Their kinetic energy pushes them together with enough force to overcome repulsions and react.* Nuclei in one atom attract electrons in another; atomic orbitals overlap and electron densities shift; some bonds lengthen and weaken while others start to form. At some point during this smooth transformation, what exists is *neither reactant nor product but a transitional species with partial bonds.* This extremely unstable species, which is called the **transition state,** or **activated complex,** exists only at the instant when the reacting system is highest in potential energy. Thus, *the activation energy is the quantity needed to stretch and deform bonds in order to reach the transition state.*

Because transition states *cannot* be isolated, our knowledge of them, until recently, came from reasoning and studies of analogous, more stable species. This situation is changing rapidly due to the pioneering work of Ahmed H. Zewail, who received the 1999 Nobel Prize in chemistry. He used lasers pulsing on the same femtosecond (10^{-15}) timescale as bond vibrations to observe the detailed changes occurring when transition states form and decompose.

Consider the reaction between methyl bromide and hydroxide ion:

$$CH_3Br + OH^- \longrightarrow CH_3OH + Br^-$$

The electronegative bromine makes the carbon of methyl bromide partially positive. If the reactants are moving toward each other fast enough and are oriented effectively when they collide, the negatively charged oxygen in OH^- approaches the carbon with enough energy to begin forming a C—O bond, which causes the

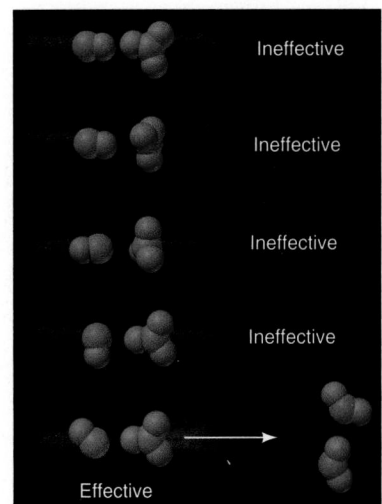

Ineffective

Ineffective

Ineffective

Ineffective

Effective

Figure 16.17 The importance of molecular orientation to an effective collision. Only one of the five orientations shown for the collision between NO and NO_3 has the correct orientation to lead to product. In the effective orientation, contact occurs between the atoms that will become bonded in the product.

Transparency/ARIS Presentation Center, Art Library

Figure 16.18 Nature of the transition state in the reaction between CH₃Br and OH⁻. Note the partial (elongated) C—O and C—Br bonds and the trigonal bipyramidal shape of the transition state of this reaction.

ARIS Presentation Center, Art Library

Organic chemistry application

C—Br bond to weaken. In the transition state (Figure 16.18), C is surrounded by five atoms (trigonal bipyramidal), which never occurs in its stable compounds. This high-energy species has three normal C—H bonds and two partial bonds, one from C to O and the other from C to Br. Reaching this transition state is no guarantee that the reaction will proceed to products. A transition state can change in either direction: if the C—O bond continues to shorten and strengthen, products form; however, if the C—Br bond becomes shorter and stronger again, the transition state reverts to reactants.

Depicting the Change with Reaction Energy Diagrams

A useful way to depict the events we just described is with a **reaction energy diagram,** which shows the potential energy of the system during the reaction as a smooth curve. Figure 16.19 shows the reaction energy diagram for the reaction of CH_3Br and OH^-, and also includes electron density relief maps, structural formulas, and molecular-scale views at various points during the change.

The horizontal axis, labeled "Reaction progress," indicates that reactants change to products from left to right. This reaction is exothermic, so reactants are higher in

Figure 16.19 Reaction energy diagram for the reaction between CH₃Br and OH⁻. A plot of potential energy vs. reaction progress shows the relative energy levels of reactants, products, and transition state joined by a curved line, as well as the activation energies of the forward and reverse reactions and the heat of reaction. The electron density relief maps, structural formulas, and molecular-scale views depict the change at five points. Note the gradual bond forming and bond breaking as the system goes through the transition state.

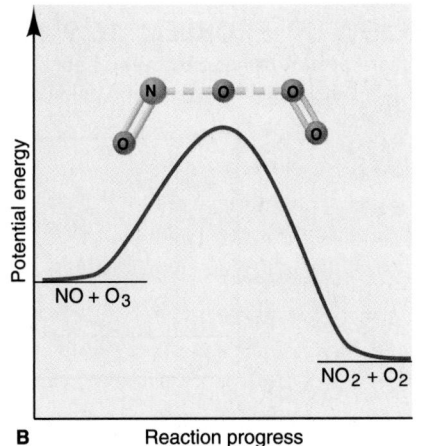

A Reaction progress

B Reaction progress

energy than products. The diagram also shows activation energies for the forward and reverse reactions; in this case, $E_{a(fwd)}$ is less than $E_{a(rev)}$. This difference, which reflects the change in bond energies, equals the heat of reaction, ΔH_{rxn}:

$$\Delta H_{rxn} = E_{a(fwd)} - E_{a(rev)} \qquad (16.10)$$

Transition state theory proposes that *every reaction (and every step in an overall reaction) goes through its own transition state,* from which it can continue in either direction. We imagine how a transition state might look by examining the reactant and product bonds that change. Figure 16.20 depicts reaction energy diagrams for two simple reactions. Note that the shape of the postulated transition state in each case is based on a specific collision orientation between the atoms that become bonded to form the product.

SAMPLE PROBLEM 16.9 Drawing Reaction Energy Diagrams and Transition States

PROBLEM A key reaction in the upper atmosphere is

$$O_3(g) + O(g) \longrightarrow 2O_2(g)$$

The $E_{a(fwd)}$ is 19 kJ, and the ΔH_{rxn} for the reaction as written is −392 kJ. Draw a reaction energy diagram for this reaction, postulate a transition state, and calculate $E_{a(rev)}$.

PLAN The reaction is highly exothermic ($\Delta H_{rxn} = -392$ kJ), so the products are much lower in energy than the reactants. The small $E_{a(fwd)}$ (19 kJ) means the energy of the reactants lies slightly below that of the transition state. We use Equation 16.10 to calculate $E_{a(rev)}$. To postulate the transition state, we sketch the species and note that one of the bonds in O$_3$ weakens, and this partially bonded O begins forming a bond to the separate O atom.

SOLUTION Solving for $E_{a(rev)}$:
$$\Delta H_{rxn} = E_{a(fwd)} - E_{a(rev)}$$

So, $\qquad E_{a(rev)} = E_{a(fwd)} - \Delta H_{rxn} = 19 \text{ kJ} - (-392 \text{ kJ}) = 411 \text{ kJ}$

The reaction energy diagram (not drawn to scale), with transition state, is

CHECK Rounding to find $E_{a(rev)}$ gives ∼20 + 390 = 410.

FOLLOW-UP PROBLEM 16.9 The following reaction energy diagram depicts another key atmospheric reaction. Label the axes, identify $E_{a(fwd)}$, $E_{a(rev)}$, and ΔH_{rxn}, draw and label the transition state, and calculate $E_{a(rev)}$ for the reaction.

Section Summary

According to collision theory, reactant particles must collide to react, and the number of collisions depends on the product of the reactant concentrations. • At higher temperatures, more collisions have enough energy to exceed the activation energy (E_a). • The relative E_a values for the forward and the reverse reactions depend on whether the overall reaction is exothermic or endothermic. • Molecules must collide with an effective orientation for them to react, so structural complexity decreases rate. • Transition state theory pictures the kinetic energy of the particles changing to potential energy during a collision. • Given a sufficiently energetic collision and an effective molecular orientation, the reactant species become an unstable transition state, which either forms product(s) or reverts to reactant(s). • A reaction energy diagram depicts the changing energy of the chemical system as it progresses from reactants through transition state(s) to products.

Literature Jennings, Laura D.; Keller, Steven W. "An Interactive Classroom Activity Demonstrating Reaction Mechanisms and Rate-Determining Steps," *J. Chem. Educ.* **2005,** *82,* 549.

16.7 REACTION MECHANISMS: STEPS IN THE OVERALL REACTION

Imagine trying to figure out how a car works just by examining the body, wheels, and dashboard. It can't be done—you need to look under the hood and inside the engine to see how the parts fit together and function. Similarly, because our main purpose is to know how a reaction works *at the molecular level,* examining the overall balanced equation is not much help—we must "look under the yield arrow and inside the reaction" to see how reactants change into products.

When we do so, we find that most reactions occur through a **reaction mechanism,** a sequence of single reaction steps that sum to the overall reaction. For example, a possible mechanism for the overall reaction

$$2A + B \longrightarrow E + F$$

might involve these three simpler steps:

(1) A + B $\longrightarrow$ C
(2) C + A $\longrightarrow$ D
(3)   D $\longrightarrow$ E + F

Adding them together and canceling common substances, we obtain the overall equation:

$$A + B + \cancel{C} + A + \cancel{D} \longrightarrow \cancel{C} + \cancel{D} + E + F \quad \text{or} \quad 2A + B \longrightarrow E + F$$

Note what happens to C and to D in this mechanism. C is a product in step 1 and a reactant in step 2, and D is a product in 2 and a reactant in 3. Each functions as a **reaction intermediate,** a substance that is formed and used up during the overall reaction. Reaction intermediates do not appear in the overall balanced equation but are absolutely essential for the reaction to occur. They are usually unstable relative to the reactants and products but are far more stable than transition states (activated complexes). Reaction intermediates are molecules with normal bonds and are sometimes stable enough to be isolated.

Chemists *propose* a reaction mechanism to explain how a particular reaction might occur, and then they *test* the mechanism. This section focuses on the nature of the individual steps and how they fit together to give a rate law consistent with experimental results.

Elementary Reactions and Molecularity

The individual steps, which together make up a proposed reaction mechanism, are called **elementary reactions** (or **elementary steps**). Each describes a *single molecular event,* such as one particle decomposing or two particles colliding and combining. *An elementary step is **not** made up of simpler steps.*

An elementary step is characterized by its **molecularity,** the number of *reactant* particles involved in the step. Consider the mechanism for the breakdown of ozone in the stratosphere. The overall reaction is

$$2O_3(g) \longrightarrow 3O_2(g)$$

A two-step mechanism has been proposed for this reaction. Notice that the two steps sum to the overall reaction. The first elementary step is a **unimolecular reaction,** one that involves the decomposition or rearrangement of a single particle:

(1) $O_3(g) \longrightarrow O_2(g) + O(g)$

The second step is a **bimolecular reaction,** one in which two particles react:

(2) $O_3(g) + O(g) \longrightarrow 2O_2(g)$

Some *termolecular* elementary steps occur, but they are extremely rare because the probability of three particles colliding simultaneously with enough energy and with an effective orientation is very small. Higher molecularities are not known. Unless evidence exists to the contrary, it makes good chemical sense to propose only unimolecular or bimolecular reactions as the elementary steps in a reaction mechanism.

The rate law for an elementary reaction, unlike that for an overall reaction, *can* be deduced from the reaction stoichiometry. An elementary reaction occurs in one step, so its rate must be proportional to the product of the reactant concentrations. Therefore, *we use the equation coefficients as the reaction orders in the rate law for an elementary step; that is, reaction order equals molecularity* (Table 16.6). Remember that this statement holds *only* when we know that the reaction is elementary; you've already seen that for an overall reaction, the reaction orders must be determined experimentally.

Table 16.6   Rate Laws for General Elementary Steps

Elementary Step	Molecularity	Rate Law
A $\longrightarrow$ product	Unimolecular	Rate = $k[A]$
2A $\longrightarrow$ product	Bimolecular	Rate = $k[A]^2$
A + B $\longrightarrow$ product	Bimolecular	Rate = $k[A][B]$
2A + B $\longrightarrow$ product	Termolecular	Rate = $k[A]^2[B]$

 ARIS Presentation Center, Tables Library

SAMPLE PROBLEM 16.10 Determining Molecularity and Rate Laws for Elementary Steps

PROBLEM The following two reactions are proposed as elementary steps in the mechanism for an overall reaction:

(1) $\qquad NO_2Cl(g) \longrightarrow NO_2(g) + Cl(g)$

(2) $NO_2Cl(g) + Cl(g) \longrightarrow NO_2(g) + Cl_2(g)$

(a) Write the overall balanced equation.

(b) Determine the molecularity of each step.

(c) Write the rate law for each step.

PLAN We find the overall equation from the sum of the elementary steps. The molecularity of each step equals the total number of reactant particles. We write the rate law for each step using the molecularities as reaction orders.

SOLUTION **(a)** Writing the overall balanced equation:

$$NO_2Cl(g) \longrightarrow NO_2(g) + Cl(g)$$
$$\underline{NO_2Cl(g) + Cl(g) \longrightarrow NO_2(g) + Cl_2(g)}$$
$$NO_2Cl(g) + NO_2Cl(g) + \cancel{Cl(g)} \longrightarrow NO_2(g) + \cancel{Cl(g)} + NO_2(g) + Cl_2(g)$$
$$2NO_2Cl(g) \longrightarrow 2NO_2(g) + Cl_2(g)$$

(b) Determining the molecularity of each step: The first elementary step has only one reactant, NO_2Cl, so it is unimolecular. The second elementary step has two reactants, NO_2Cl and Cl, so it is bimolecular.

(c) Writing rate laws for the elementary reactions:

(1) $Rate_1 = k_1[NO_2Cl]$

(2) $Rate_2 = k_2[NO_2Cl][Cl]$

CHECK In part (a), be sure the equation is balanced; in part (c), be sure the substances in brackets are the reactants of each elementary step.

FOLLOW-UP PROBLEM 16.10 The following elementary steps constitute a proposed mechanism for a reaction:

(1) $\qquad 2NO(g) \longrightarrow N_2O_2(g)$

(2) $\qquad 2H_2(g) \longrightarrow 4H(g)$

(3) $N_2O_2(g) + H(g) \longrightarrow N_2O(g) + HO(g)$

(4) $2HO(g) + 2H(g) \longrightarrow 2H_2O(g)$

(5) $H(g) + N_2O(g) \longrightarrow HO(g) + N_2(g)$

(a) Write the balanced equation for the overall reaction.

(b) Determine the molecularity of each step.

(c) Write the rate law for each step.

The Rate-Determining Step of a Reaction Mechanism

All the elementary steps in a mechanism do not have the same rate. Usually, one of the elementary steps is much slower than the others, so it limits how fast the overall reaction proceeds. This step is called the **rate-determining step** (or **rate-limiting step**).

Because the rate-determining step limits the rate of the overall reaction, its rate law represents the rate law for the overall reaction. Consider the reaction between nitrogen dioxide and carbon monoxide:

$$NO_2(g) + CO(g) \longrightarrow NO(g) + CO_2(g)$$

If the overall reaction were an elementary reaction—that is, if the mechanism consisted of only one step—we could immediately write the overall rate law as

$$Rate = k[NO_2][CO]$$

However, as you saw in Sample Problem 16.3, experiment shows that the actual rate law is

$$Rate = k[NO_2]^2$$

Based on this observation, we know immediately that the reaction shown cannot be elementary.

A proposed two-step mechanism is

(1) $NO_2(g) + NO_2(g) \longrightarrow NO_3(g) + NO(g)$ [slow; rate determining]

(2) $NO_3(g) + CO(g) \longrightarrow NO_2(g) + CO_2(g)$ [fast]

Note that NO_3 functions as a reaction intermediate in the mechanism. Rate laws for these elementary steps are

(1) $Rate_1 = k_1[NO_2][NO_2] = k_1[NO_2]^2$

(2) $Rate_2 = k_2[NO_3][CO]$

Note that if $k_1 = k$, *the rate law for the rate-determining step (step 1) is identical to the experimental rate law.* Because the first step is slow, $[NO_3]$ is very low; thus, the fast second step cannot increase the overall rate, and the reaction takes essentially as long as the first step. Here you can see one reason a reactant (in this case, CO) has a reaction order of zero is that it takes part in the reaction only *after* the rate-determining step.

Correlating the Mechanism with the Rate Law

Conjuring up a reasonable reaction mechanism is one of the most exciting aspects of chemical kinetics and can be a classic example of the use of the scientific method. We use observations and data from rate experiments to hypothesize what the individual steps might be and then test our hypothesis by gathering further evidence. If the evidence supports it, we continue to apply that mechanism; if not, we propose a new one. However, *we can never prove, just from data, that a particular mechanism represents the actual chemical change, only that it is consistent with it.*

Regardless of the elementary steps proposed for a mechanism, they must meet three criteria:

1. *The elementary steps must add up to the overall balanced equation.* We cannot wind up with more (or fewer) reactants or products than are present in the balanced equation.
2. *The elementary steps must be physically reasonable.* As we noted, most steps should involve one reactant particle (unimolecular) or two (bimolecular). Steps with three reactant particles (termolecular) are very unlikely.
3. *The mechanism must correlate with the rate law.* Most importantly, a mechanism must support the experimental facts shown by the rate law, not the other way around.

Let's see how the mechanisms of several reactions conform to these criteria and how the elementary steps fit together.

Mechanisms with a Slow Initial Step We've already seen one mechanism with a rate-determining first step—that for the reaction of NO_2 and CO. Another example is the reaction between nitrogen dioxide and fluorine gas:

$$2NO_2(g) + F_2(g) \longrightarrow 2NO_2F(g)$$

The experimental rate law is first order in NO_2 and in F_2:

$$Rate = k[NO_2][F_2]$$

The accepted mechanism for the reaction is

(1) $NO_2(g) + F_2(g) \longrightarrow NO_2F(g) + F(g)$ [slow; rate determining]

(2) $NO_2(g) + F(g) \longrightarrow NO_2F(g)$ [fast]

Molecules of reactant and product appear in both elementary steps. The free fluorine atom is a reaction intermediate.

Does this mechanism meet the three crucial criteria?

1. The elementary reactions sum to the balanced equation:

$$NO_2(g) + NO_2(g) + F_2(g) + \cancel{F(g)} \longrightarrow NO_2F(g) + NO_2F(g) + \cancel{F(g)}$$

or $2NO_2(g) + F_2(g) \longrightarrow 2NO_2F(g)$

2. Both steps are bimolecular, so they are chemically reasonable.

3. The mechanism gives the rate law for the overall equation. To show this, we write the rate laws for the elementary steps:

(1) $Rate_1 = k_1[NO_2][F_2]$

(2) $Rate_2 = k_2[NO_2][F]$

Step 1 is the rate-determining step and therefore gives the overall rate law, with $k_1 = k$. Because the second molecule of NO_2 appears in the step that follows the rate-determining step, it does not appear in the overall rate law. Thus, we see that *the overall rate law includes only species active in the reaction up to and including those in the rate-determining step.* This point was also illustrated by the mechanism for NO_2 and CO shown earlier. Carbon monoxide was absent from the overall rate law because it appeared *after* the rate-determining step.

Figure 16.21 is a reaction energy diagram for the reaction of NO_2 and F_2. Note that

- *Each step in the mechanism has its own transition state.* (Note that only one molecule of NO_2 reacts in step 1, and only the first transition state is depicted.)
- The F atom intermediate is a reactive, unstable species (as you know from halogen chemistry), so it is higher in energy than the reactants or product.
- The first step is slower (rate limiting), so its activation energy is *larger* than that of the second step.
- The overall reaction is exothermic, so the product is lower in energy than the reactants.

Figure 16.21 Reaction energy diagram for the two-step reaction of NO_2 and F_2. Each step in the mechanism has its own transition state. The proposed transition state is shown for step 1. Reactants for the second step are the F atom intermediate and the second molecule of NO_2. Note that the first step is slower (higher E_a). The overall reaction is exothermic ($\Delta H_{rxn} < 0$).

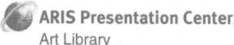

ARIS Presentation Center, Art Library

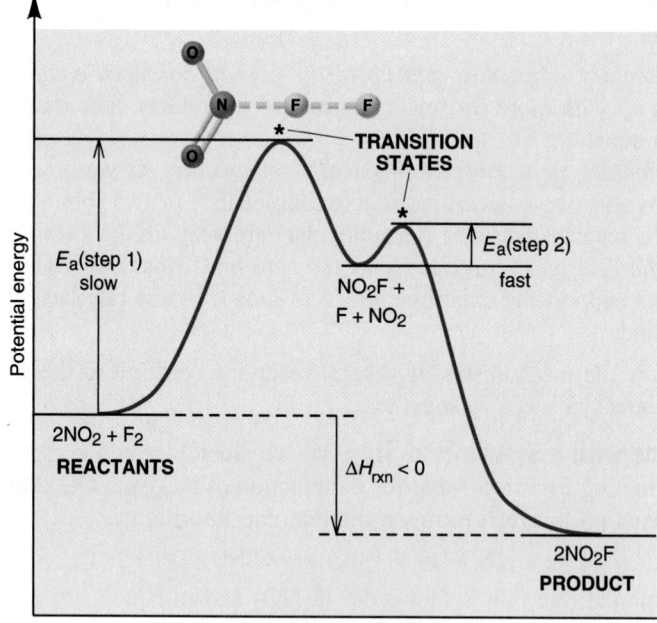

Reaction progress

Mechanisms with a Fast Initial Step

If the rate-limiting step in a mechanism is *not* the initial step, it acts as a bottleneck later in the reaction sequence. As a result, the product of a fast initial step builds up and starts reverting to reactant, while waiting for the slow step to remove it. With time, the product of the initial step is changing back to reactant as fast as it is forming. In other words, the *fast initial step reaches equilibrium.* As you'll see, this situation allows us to fit the mechanism to the overall rate law.

Consider once again the oxidation of nitrogen monoxide:

$$2NO(g) + O_2(g) \longrightarrow 2NO_2(g)$$

The experimentally determined rate law is

$$\text{Rate} = k[NO]^2[O_2]$$

and a proposed mechanism is

(1) $NO(g) + O_2(g) \rightleftharpoons NO_3(g)$ [fast, reversible]

(2) $NO_3(g) + NO(g) \longrightarrow 2NO_2(g)$ [slow; rate determining]

Note that, with cancellation of the reaction intermediate NO_3, the first criterion is met because the sum of the steps gives the overall equation. Also note that the second criterion is met because both steps are bimolecular.

To meet the third criterion (that the mechanism conforms to the overall rate law), we first write rate laws for the elementary steps:

(1) $\text{Rate}_{1(\text{fwd})} = k_1[NO][O_2]$

$\text{Rate}_{1(\text{rev})} = k_{-1}[NO_3]$

where k_{-1} is the rate constant for the reverse reaction.

(2) $\text{Rate}_2 = k_2[NO_3][NO]$

Now we must show that the rate law for the rate-determining step (step 2) gives the overall rate law. As written, it does not, because it contains the intermediate NO_3, and *an overall rate law can include only reactants (and products)*. Therefore, we must eliminate $[NO_3]$ from the step 2 rate law. To do so, we express $[NO_3]$ in terms of reactants. Step 1 reaches equilibrium when the forward and reverse rates are equal:

$$\text{Rate}_{1(\text{fwd})} = \text{Rate}_{1(\text{rev})} \quad \text{or} \quad k_1[NO][O_2] = k_{-1}[NO_3]$$

To express $[NO_3]$ in terms of reactants, we isolate it algebraically:

$$[NO_3] = \frac{k_1}{k_{-1}}[NO][O_2]$$

Then, substituting for $[NO_3]$ in the rate law for step 2, we obtain

$$\text{Rate}_2 = k_2[NO_3][NO] = k_2\left(\frac{k_1}{k_{-1}}[NO][O_2]\right)[NO] = \frac{k_2 k_1}{k_{-1}}[NO]^2[O_2]$$

This rate law is identical to the overall rate law, with $k = \dfrac{k_2 k_1}{k_{-1}}$.

Thus, to test the validity of a mechanism with a fast initial, reversible step:

1. Write rate laws for both directions of the fast step and for the slow step.
2. Show that the slow step's rate law is equivalent to the overall rate law, by *expressing [intermediate] in terms of [reactant]:* set the forward rate law of the fast, reversible step equal to the reverse rate law, and solve for [intermediate].
3. Substitute the expression for [intermediate] into the rate law for the slow step to obtain the overall rate law.

Several end-of-chapter problems, including Problems 16.72 and 16.73, provide additional examples of this approach.

Section Summary

The mechanisms of most common reactions consist of two or more elementary steps, reactions that occur in one step and depict a single chemical change. • The molecularity of an elementary step equals the number of reactant particles and is the same as the reaction order of its rate law. Unimolecular and bimolecular steps are common. • The rate-determining, or rate-limiting (slowest), step determines how fast the overall reaction occurs, and its rate law represents the overall rate law. • Reaction intermediates are species that form in one step and react in a later one. • The steps in a proposed mechanism must add up to the overall reaction, be physically reasonable, and conform to the overall rate law. If a fast step precedes a slow step, the fast step reaches equilibrium, and the concentrations of intermediates in the rate law of the slow step must be expressed in terms of reactants.

16.8 CATALYSIS: SPEEDING UP A CHEMICAL REACTION

There are many situations in which the rate of a reaction must be increased for it to be useful. In an industrial process, for example, a higher rate often determines whether a new product can be made economically. Sometimes, we can speed up a reaction sufficiently with a higher temperature, but energy is costly and many substances are heat sensitive and easily decomposed. Alternatively, we can often employ a **catalyst,** a substance that increases the rate *without* being consumed in the reaction. Because catalysts are not consumed, only very small, nonstoichiometric quantities are generally required. Nevertheless, these substances are employed in so many important processes that several million tons of industrial catalysts are produced annually in the United States alone! Nature is the master designer and user of catalysts. Even the simplest bacterium employs thousands of biological catalysts, known as *enzymes,* to speed up its cellular reactions. Every organism relies on enzymes to sustain life.

Each catalyst has its own specific way of functioning, but in general, *a catalyst causes a lower activation energy, which in turn makes the rate constant larger and the rate higher.* Two important points stand out in Figure 16.22:

- A catalyst speeds up the forward *and* reverse reactions. A reaction with a catalyst *does not yield more product* than one without a catalyst, but it yields the product *more quickly.*
- A catalyst causes a *lower activation energy* by providing a *different mechanism* for the reaction, and thus a new, lower energy pathway.

Consider a general *uncatalyzed* reaction that proceeds by a one-step mechanism involving a bimolecular collision:

$$A + B \longrightarrow \text{product} \qquad \text{[slower]}$$

In the *catalyzed* reaction, a reactant molecule interacts with the catalyst, so the mechanism might involve a two-step pathway:

$$A + \text{catalyst} \longrightarrow C \qquad \qquad \text{[faster]}$$
$$C + B \longrightarrow \text{product} + \text{catalyst} \qquad \text{[faster]}$$

Literature White, Colin. "A Reaction That Takes Place in Beakers but Not in Conical Flasks: A Catalysis-Related Demonstration," *J. Chem. Educ.* **2004,** *81,* 364.

Figure 16.22 **Reaction energy diagram for a catalyzed and an uncatalyzed process.** A catalyst speeds a reaction by providing a new, lower energy pathway, in this case by replacing the one-step mechanism with a two-step mechanism. Both forward and reverse rates are increased to the same extent, so *a catalyst does not affect the overall reaction yield.* (The only activation energy shown for the catalyzed reaction is the larger one for the forward direction.)

Transparency/ARIS Presentation Center, Art Library

Reaction progress

Note that *the catalyst is not consumed,* as its definition requires. Rather, it is used and then regenerated, and the activation energies of both steps are lower than the activation energy of the uncatalyzed pathway.

Chemists recognize two general categories of catalyst—homogeneous and heterogeneous—based on whether the catalyst occurs in the same phase as the reactant and product.

Homogeneous Catalysis

A **homogeneous catalyst** exists in solution with the reaction mixture. All homogeneous catalysts are gases, liquids, or soluble solids. Some industrial processes that employ these catalysts are shown in Table 16.7 *(top).*

A thoroughly studied example of homogeneous catalysis is the hydrolysis of an organic ester (RCOOR′), a reaction we examined in Section 15.4:

$$\text{R}-\overset{\displaystyle\text{O}}{\overset{\|}{\text{C}}}-\text{O}-\text{R}' + \text{H}_2\text{O} \rightleftharpoons \text{R}-\overset{\displaystyle\text{O}}{\overset{\|}{\text{C}}}-\text{OH} + \text{R}'-\text{OH}$$

Here R and R′ are hydrocarbon groups, $\text{R}-\overset{\displaystyle\text{O}}{\overset{\|}{\text{C}}}-\text{OH}$ is a carboxylic acid, and R′—OH is an alcohol. The reaction rate is low at room temperature but can be increased greatly by adding a small amount of strong acid, which provides H^+ ion, the catalyst in the reaction; strong bases, which supply OH^- ions, also speed ester hydrolysis, but by a slightly different mechanism.

In the first step of the acid-catalyzed reaction (Figure 16.23), the H^+ of a hydronium ion forms a bond to the double-bonded O atom. From the resonance forms, we see that the bonding of H^+ then makes the C atom more positive, which *increases its attraction* for the partially negative O atom of water. In effect, H^+ increases the likelihood that the bonding of water, which is the rate-determining step, will take place. Several steps later, a water molecule, acting as a base, removes the H^+ and returns it to solution. Thus, H^+ acts as a catalyst because it speeds up the reaction but is not itself consumed: it is used up in one step and re-formed in another.

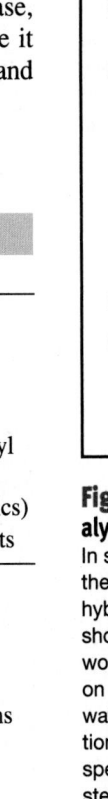

Figure 16.23 Mechanism for the catalyzed hydrolysis of an organic ester. In step 1, the catalytic H^+ ion binds to the electron-rich oxygen. The resonance hybrid of this product *(see gray panel)* shows the C atom is more positive than it would ordinarily be. The enhanced charge on C attracts the partially negative O of water more strongly, increasing the fraction of effective collisions and thus speeding up step 2, the rate-determining step. Loss of R′OH and removal of H^+ by water occur in a final series of fast steps.

Table 16.7	Some Modern Processes That Employ Catalysts		
Reactants	**Catalyst**	**Product**	**Use**
Homogeneous			
Propylene, oxidizer	Mo(VI) complexes	Propylene oxide	Polyurethane foams; polyesters
Methanol, CO	$[\text{Rh(CO)}_2\text{I}_2]^-$	Acetic acid	Poly(vinyl acetate) coatings; poly(vinyl alcohol)
Butadiene, HCN	Ni/P compounds	Adiponitrile	Nylons (fibers, plastics)
α-Olefins, CO, H_2	Rh/P compounds	Aldehydes	Plasticizers, lubricants
Heterogeneous			
Ethylene, O_2	Silver, cesium chloride on alumina	Ethylene oxide	Polyesters, ethylene glycol, lubricants
Propylene, NH_3, O_2	Bismuth molybdates	Acrylonitrile	Plastics, fibers, resins
Ethylene	Organochromium and titanium halides on silica	High-density polyethylene	Molded products

Biological application

Many digestive enzymes, which catalyze the hydrolysis of proteins, fats, and carbohydrates during the digestion of foods, employ very similar mechanisms. The difference is that the acids or bases that speed these reactions are not the strong inorganic reagents used in the lab, but rather specific amino-acid side chains of the enzymes that release or abstract H^+ ions.

Heterogeneous Catalysis

A **heterogeneous catalyst** speeds up a reaction that occurs in a separate phase. The catalyst is most often a solid interacting with gaseous or liquid reactants. Because reaction occurs on the solid's surface, heterogeneous catalysts usually have enormous surface areas for contact, between 1 and 500 m^2/g. Interestingly, many reactions that occur on a metal surface, such as the decomposition of HI on gold and the decomposition of N_2O on platinum, are zero order because the rate-determining step occurs on the surface itself. Thus, despite an enormous surface area, once the reactant gas covers the surface, increasing the reactant concentration cannot increase the rate. Table 16.7 *(bottom)* lists some polymer manufacturing processes that employ heterogeneous catalysts.

One of the most important examples of heterogeneous catalysis is the addition of H_2 to the C=C bonds of organic compounds to form C—C bonds. The petroleum, plastics, and food industries frequently use catalytic **hydrogenation.** The conversion of vegetable oil into margarine is one example.

The simplest hydrogenation converts ethylene (ethene) to ethane:

$$H_2C=CH_2(g) + H_2(g) \longrightarrow H_3C-CH_3(g)$$

In the absence of a catalyst, the reaction occurs very slowly. At high H_2 pressure in the presence of finely divided Ni, Pd, or Pt, the reaction becomes rapid even at ordinary temperatures. These Group 8B(10) metals catalyze by *chemically adsorbing the reactants onto their surface* (Figure 16.24). The H_2 lands and splits into separate H atoms chemically bound to the solid catalyst's metal atoms (catM):

$$H-H(g) + 2catM(s) \longrightarrow 2catM-H \text{ (H atoms bound to metal surface)}$$

Then, C_2H_4 adsorbs and reacts with two H atoms, one at a time, to form C_2H_6. The H—H bond breakage is the rate-determining step in the overall process, and interaction with the catalyst's surface provides the low-E_a step as part of an alternative reaction mechanism. ●

Two Chemical Connections essays follow. The first introduces the remarkable catalytic abilities of the enzymes inside you, and the second discusses how catalysis operates in the depletion of stratospheric ozone.

Green chemistry application

● **Catalytically Cleaning Your Car's Exhaust** A heterogeneous catalyst you use every day, but rarely see, is in the catalytic converter in your car's exhaust system. This device is designed to convert polluting exhaust gases into nontoxic ones. In a single pass through the catalyst bed, CO and unburned gasoline are oxidized to CO_2 and H_2O, while NO is reduced to N_2. As in the mechanism for catalytic hydrogenation of an alkene, the catalyst lowers the activation energy of the rate-determining step by adsorbing the molecules, thereby weakening their bonds. Mixtures of transition metals and their oxides embedded in inert supports convert as much exhaust gas as possible in the shortest time. It is estimated, for example, that an NO molecule is adsorbed and split into catalyst-bound N and O atoms in less than 2×10^{-12} s!

Figure 16.24 The metal-catalyzed hydrogenation of ethene.

ARIS Presentation Center,
Active Art Library

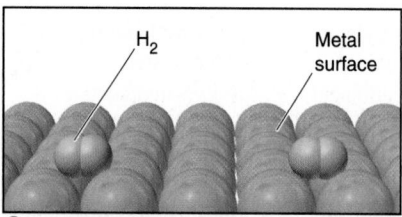
① *H_2 adsorbs to metal surface*

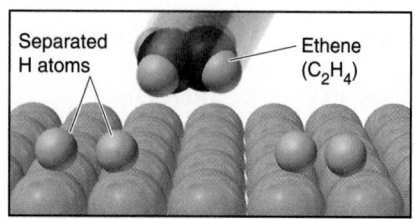
② *Rate-limiting step is H—H bond breakage.*

③ *After C_2H_4 adsorbs, one C—H forms.*

④ *Another C—H bond forms; C_2H_6 leaves surface.*

Kinetics and Function of Biological Catalysts

Within every living cell, thousands of individual reactions occur. Many involve complex chemical changes, yet they take place in dilute solution at ordinary temperatures and pressures. The rate of each reaction responds smoothly to momentary or permanent changes in other reaction rates, various signals from other cells, and environmental stresses. Virtually every reaction in this marvelous chemical harmony is catalyzed by its own specific **enzyme,** a protein catalyst whose function has been perfected through evolution.

Enzymes have complex three-dimensional shapes (see Section 15.6) and molar masses ranging from about 15,000 to 1,000,000 g/mol. At a specific region on an enzyme's surface is its

active site, a crevice whose shape results from the shapes of the amino acid side chains (R groups) directly involved in catalyzing the reaction (also see Figure B10.4, p. 403). When the reactant molecules, called the **substrates** of the reaction, collide with the active site, the chemical change is initiated. The active site makes up only a small part of the enzyme's surface—like a tiny hollow carved into a mountainside—and often includes amino acid R groups from distant regions of the molecular chain that lie near each other because of the folding of the protein's backbone, as depicted in Figure B16.4. In most cases, *substrates bind to the active site through intermolecular forces:* H bonds, dipole forces, and other weak attractions.

(continued)

Literature Woodyer, Ryan; Chen, Wilfred; Zhao, Huimin. "Outrunning Nature: Directed Evolution of Superior Biocatalysts," *J. Chem. Educ.* **2004,** *81,* 126.

Demo Johnson, A. Kristin. "A Simple Method for Demonstrating Enzyme Kinetics Using Catalase from Beef Liver Extract," *J. Chem. Educ.* **2000,** *77,* 1451.

Biological application

Figure B16.4 The widely separated amino acid groups that form an active site. The amino acids in chymotrypsin, a digestive enzyme, are shown as linked spheres and numbered consecutively from the beginning of the chain. The R groups of the active-site amino acids 57 (histidine, His), 102 (aspartic acid, Asp), and 195 (serine, Ser) are shown because they play a crucial role in the enzyme-catalyzed reaction. Part of the substrate molecule lies within the active site's crevice (binding pocket) to anchor it during reaction. (Illustration by Irving Geis. Rights owned by Howard Hughes Medical Institute. Not to be used without permission.)

 ARIS Presentation Center,
Art Library

An enzyme has features of both a homogeneous and a heterogeneous catalyst. Most enzymes are enormous compared with their substrates, and they are often found embedded within membranes of larger cell parts. Thus, like a heterogeneous catalyst, an enzyme provides an active surface on which a reactant is immobilized temporarily, waiting for its reaction partner to land nearby. Like a homogeneous catalyst, the enzyme's amino acid R groups interact actively with the substrates in multistep sequences involving intermediates.

Enzymes are incredibly *efficient* catalysts. Consider the hydrolysis of urea as an example:

$$(NH_2)_2C{=}O(aq) + 2H_2O(l) + H^+(aq) \longrightarrow$$
$$2NH_4^+(aq) + HCO_3^-(aq)$$

In water at room temperature, the rate constant for the uncatalyzed reaction is approximately 3×10^{-10} s^{-1}. Under the same conditions in the presence of the enzyme *urease* (pronounced "*yur*-ee-ase"), the rate constant is 3×10^4 s^{-1}, a 10^{14}-fold increase! Enzymes increase rates by 10^8 to 10^{20} times, values that industrial chemists who design catalysts can only dream of.

Enzymes are also extremely *specific:* each reaction is generally catalyzed by a particular enzyme. Urease catalyzes *only* the hydrolysis of urea, and none of the several thousand other enzymes present in the cell catalyzes that reaction. This remarkable specificity results from the particular groups that comprise the active site. Two models of enzyme action are illustrated in Figure B16.5. According to the **lock-and-key model,** when the "key" (substrate) fits the "lock" (active site), the chemical change begins. However, x-ray crystallographic and spectroscopic methods show that, in many cases, *the enzyme changes shape when the substrate lands at the active site.* This **induced-fit model** of enzyme action pictures the substrate inducing the active site to adopt a perfect fit. Rather than a rigidly shaped lock and key, therefore, we should picture a hand in a glove, in which the "glove" (active site) does not attain its functional shape until the "hand" (substrate) moves into place.

The kinetics of enzyme catalysis has many features in common with that of ordinary catalysis. In an uncatalyzed reaction, the rate is affected by the concentrations of the reactants; in a catalyzed reaction, the rate is affected by the *concentration of reactant bound to catalyst.* In the enzyme-catalyzed case, substrate (S) and enzyme (E) form an intermediate **enzyme-substrate complex (ES),** whose concentration determines the rate of product (P) formation. The steps common to virtually all enzyme-catalyzed reactions are

(1) $E + S \rightleftharpoons ES$ [fast, reversible]
(2) $ES \longrightarrow E + P$ [slow; rate determining]

Thus, rate = $k[ES]$. When all the available enzyme molecules are bound to substrate and exist as ES, increasing [S] has no effect on rate and the process is zero order. To cite one very common example, the breakdown of ethanol in the body is zero order.

Enzymes employ a variety of catalytic mechanisms. In some cases, the active-site R groups bring the reacting atoms of the bound substrates closer together. In other cases, the groups move apart slightly, stretching the substrate bond that is to be broken in the process. Some R groups are acidic and thus are able to provide H$^+$ ions that increase the speed of a rate-determining step; others act as bases and remove H$^+$ ions at a critical step. *Hydrolases* are a class of enzymes that cleave bonds by such *acid-base catalysis.* For example, lysozyme, an enzyme found in tears, hydrolyzes bacterial cell walls, thus protecting the eyes from microbes, and chymotrypsin, an enzyme found in the small intestine, hydrolyzes proteins into smaller molecules during digestion.

No matter what their specific mode of action, *all enzymes function by stabilizing the reaction's transition state.* For instance, in the lysozyme-catalyzed reaction, the transition state is a sugar molecule whose bonds are twisted and stretched into unusual lengths and angles so that it fits the lysozyme active site perfectly. By stabilizing the transition state through binding it effectively, lysozyme lowers the activation energy of the reaction and thus increases the rate.

A Lock-and-key model

B Induced-fit model

Figure B16.5 Two models of enzyme action. A, In the lock-and-key model, the active site is thought to be an exact fit for the substrate shapes. **B,** In the induced-fit model, the active site is thought to change shape to fit the substrates. Most enzyme-catalyzed reactions proceed through the fast, reversible formation of an enzyme-substrate complex, followed by a slow conversion to product and free enzyme.

ARIS Presentation Center,
Art Library

Chemical Connections *to Atmospheric Science*

Depletion of Earth's Ozone Layer

ARIS Presentation Center,
Art Library

Both homogeneous and heterogeneous catalysts play key roles in one of today's most serious environmental concerns—the depletion of ozone from the stratosphere. The stratospheric ozone layer absorbs UV radiation with wavelengths between 280 and 320 nm emitted by the Sun that would otherwise reach Earth's surface. This radiation (called UV-B) has enough energy to break bonds in deoxyribonucleic acid (DNA) and thereby damage genes. Depletion of stratospheric ozone could significantly increase risks to human health from UV-B, particularly a higher incidence of skin cancer and cataracts (loss of transparency of the lens of the eye). Plant life, especially the simpler forms at the base of the food chain, may also be damaged.

Before 1976, stratospheric ozone concentration varied seasonally but remained nearly constant from year to year through a series of atmospheric reactions. Oxygen atoms formed from dissociation of O_2 by UV radiation with wavelengths less than 242 nm react with O_2 to form O_3, and with O_3 to regenerate O_2:

$$O_2 \xrightarrow{UV} 2O$$
$$O + O_2 \longrightarrow O_3 \quad \text{[ozone formation]}$$
$$O + O_3 \longrightarrow 2O_2 \quad \text{[ozone breakdown]}$$

However, research by Paul J. Crutzen, Mario J. Molina, and F. Sherwood Rowland, for which they received the Nobel Prize in chemistry in 1995, revealed that industrially produced chlorofluorocarbons (CFCs) shifted this balance by catalyzing the breakdown reaction.

CFCs were widely used as aerosol propellants, foam-blowing agents, and air-conditioning coolants, so large quantities were released into the atmosphere. Unreactive in the troposphere, CFC molecules gradually reach the stratosphere, where they encounter UV radiation that can split them and release chlorine atoms:

$$CF_2Cl_2 \xrightarrow{UV} CF_2Cl\cdot + Cl\cdot$$

(The dots are unpaired electrons resulting from bond cleavage.)

Like many species with unpaired electrons (free radicals), atomic Cl is very reactive. The Cl atoms react with stratospheric ozone to produce an intermediate, chlorine monoxide ($ClO\cdot$), which then reacts with free O atoms to regenerate Cl atoms:

$$O_3 + Cl\cdot \longrightarrow ClO\cdot + O_2$$
$$ClO\cdot + O \longrightarrow Cl\cdot + O_2$$

The sum of these steps is the ozone breakdown reaction:

$$O_3 + \cancel{Cl\cdot} + \cancel{ClO\cdot} + O \longrightarrow \cancel{ClO\cdot} + O_2 + \cancel{Cl\cdot} + O_2$$
$$\text{or} \qquad O_3 + O \longrightarrow 2O_2$$

Note that the Cl atom acts as a homogeneous catalyst: it exists in the same phase as the reactants, speeds up the process via a different mechanism, and is regenerated. The problem is that each Cl atom has a stratospheric half-life of about 2 years, during which time it speeds the breakdown of about 100,000 ozone molecules. Bromine is a more effective catalyst, but much less ends up in the stratosphere. Halons, used as fire suppressants, and methyl bromide, an agricultural insecticide, are the main anthropogenic (human-made) sources of stratospheric bromine.

In 1985, British scientists reported their decade-long observation of the development of an Antarctic *ozone hole*, the reduction of stratospheric ozone over the Antarctic region when the Sun rises after the long winter darkness. Later measurements of high [$ClO\cdot$] over Antarctica are consistent with the breakdown mecha-

Sept 1979   Sept 2006

Total ozone

100   DU   500

Figure B16.6 The increasing size of the Antarctic ozone hole. Satellite images show the large increase in the size of the "hole" in the ozone layer (indicated by the leftmost colors in the key) as a result of the large decrease in ozone concentration over the South Pole from September 1979 to September 2006. Since 1995, data show a similar thinning, though not yet as severe, over the North Pole.

nism. Expansion of the ozone hole as CFCs accumulated is shown in Figure B16.6. The ozone hole enlarges by heterogeneous catalysis involving polar stratospheric clouds. The clouds provide a surface for reactions that convert inactive chlorine compounds, such as HCl and chlorine nitrate ($ClONO_2$), to substances, such as Cl_2, that are cleaved by UV radiation to Cl atoms. Heterogeneous catalysis also occurs on fine particles in the stratosphere. Dust from the eruption of Mt. Pinatubo in 1991 reduced stratospheric ozone for 2 years. There is ozone thinning over the North Pole, and NASA has also documented the loss of stratospheric ozone at middle latitudes, such as over the United States. So far, both of these losses occur to a lower extent and at a lower rate.

The Montreal Protocol in 1987 and later amendments curtailed growth in production of CFCs and set dates for phasing out these and other chlorinated and brominated compounds, such as CCl_4, CCl_3CH_3, and CH_3Br. Another change was replacement of CFC aerosol propellants by hydrocarbons, such as isobutane, $(CH_3)_3CH$. Replacement of CFCs in refrigeration units is more difficult, because it will require changes in equipment. The first replacements for CFCs were hydrochlorofluorocarbons (HCFCs), such as CHF_2Cl. These deplete ozone less than CFCs do because they have relatively short tropospheric lifetimes due to abstraction of H atoms by hydroxyl radicals ($\cdot OH$):

$$CHF_2Cl + \cdot OH \longrightarrow H_2O + \cdot CF_2Cl$$

After this initial attack, further degradation occurs rapidly.

HCFCs will be phased out by 2040 and replaced by hydrofluorocarbons (HFCs, which contain only C, H, and F) in devices such as car air conditioners. HFCs are preferable to CFCs or HCFCs because fluorine is a poor catalyst of ozone breakdown. Nevertheless, because of the long lifetimes of CFCs and the ongoing emissions of HCFCs until their phase-out, full recovery of the ozone layer may take another century! The good news is that tropospheric halogen levels have already begun to fall.

Literature Manning, Thomas J.; Little, Brian; Purcell, Jerry; Feldman, Amy; Parker, William; Register, Katrice; Sumner, Brandy; Schibner, Clint. "Ozone Decomposition Data for Kinetics Exercises," *Chem. Educator* 7, **2002**, *5*, 278–283.

Section Summary

A catalyst is a substance that increases the rate of a reaction without being consumed. It accomplishes this by providing an alternative mechanism with a lower activation energy. • Homogeneous catalysts function in the same phase as the reactants. Heterogeneous catalysts act in a different phase from the reactants. • The hydrogenation of carbon-carbon double bonds takes place on a solid catalyst, which speeds the breakage of the H—H bond in H_2. • Enzymes are biological catalysts with spectacular efficiency and specificity. • Chlorine atoms derived from CFC molecules catalyze the breakdown of stratospheric ozone.

Chapter Perspective

With this introduction to chemical kinetics, we have begun to explore the dynamic inner workings of chemical change. Variations in reaction rate are observed with concentration and temperature changes, which operate on the molecular level through the frequency and energetics of particle collisions and the details of reactant structure. Kinetics allows us to speculate about the molecular pathway of a reaction. Modern industry and biochemistry depend on its principles. However, speed and yield are very different aspects of a reaction. In Chapter 17, we'll see how opposing reaction rates give rise to the equilibrium state and examine how much product has formed once the net reaction stops.

CHAPTER REVIEW GUIDE

The following sections provide many aids to help you study this chapter. (Numbers in parentheses refer to pages, unless noted otherwise.)

Learning Objectives These are concepts and skills you should know after studying this chapter.

Relevant section and/or sample problem (SP) numbers appear in parentheses.

Understand These Concepts

1. How reaction rate depends on concentration, physical state, and temperature (Section 16.1)
2. The meaning of reaction rate in terms of changing concentrations over time (Section 16.2)
3. How the rate can be expressed in terms of reactant or product concentrations (Section 16.2)
4. The distinction between average and instantaneous rate and why the instantaneous rate changes during the reaction (Section 16.2)
5. The interpretation of reaction rate in terms of reactant and product concentrations (Section 16.2)
6. The experimental basis of the rate law and the information needed to determine it—initial rate data, reaction orders, and rate constant (Section 16.3)
7. The importance of reaction order in determining the rate (Section 16.3)
8. How reaction order is determined from initial rates at different concentrations (Section 16.3)
9. How integrated rate laws show the dependence of concentration on time (Section 16.4)
10. What reaction half-life means and why it is constant for a first-order reaction (Section 16.4)
11. Activation energy and the effect of temperature on the rate constant (Arrhenius equation) (Section 16.5)
12. Why concentrations are multiplied in the rate law (Section 16.6)
13. How temperature affects rate by influencing collision energy and, thus, the fraction of collisions with energy exceeding the activation energy (Section 16.6)

14. Why molecular orientation and complexity influence the number of effective collisions and the rate (Section 16.6)
15. The transition state as the momentary species between reactants and products whose formation requires the activation energy (Section 16.6)
16. How an elementary step represents a single molecular event and its molecularity equals the number of colliding particles (Section 16.7)
17. How a reaction mechanism consists of several elementary steps, with the slowest step determining the overall rate (Section 16.7)
18. The criteria for a valid reaction mechanism (Section 16.7)
19. How a catalyst speeds a reaction by lowering the activation energy (Section 16.8)
20. The distinction between homogeneous and heterogeneous catalysis (Section 16.8)

Master These Skills

1. Calculating instantaneous rate from the slope of a tangent to a concentration vs. time plot (Section 16.2)
2. Expressing reaction rate in terms of changes in concentration over time (SP 16.1)
3. Determining reaction order from a known rate law (SP 16.2)
4. Determining reaction order from changes in initial rate with concentration (SPs 16.3, 16.4)
5. Calculating the rate constant and its units (Section 16.3)
6. Using an integrated rate law to find concentration at a given time or the time to reach a given concentration (SP 16.5)
7. Determining reaction order graphically with a rearranged integrated rate law (Section 16.4)
8. Determining the half-life of a reaction (SPs 16.6, 16.7)
9. Using a form of the Arrhenius equation to calculate the activation energy (SP 16.8)

Learning Objectives *(continued)*

10. Using reaction energy diagrams to depict the energy changes during a reaction (SP 16.9)
11. Postulating a transition state for a simple reaction (SP 16.9)

12. Determining the molecularity and rate law for an elementary step (SP 16.10)
13. Constructing a mechanism with either a slow or a fast initial step (Section 16.7)

Key Terms These important terms appear in boldface in the chapter and are defined again in the Glossary.

chemical kinetics (685)

Section 16.2
reaction rate (687)
average rate (688)
instantaneous rate (689)
initial rate (689)

Section 16.3
rate law (rate equation) (691)
rate constant (692)
reaction orders (692)

Section 16.4
integrated rate law (699)

half-life ($t_{1/2}$) (702)

Section 16.5
Arrhenius equation (706)
activation energy (E_a) (706)

Section 16.6
collision theory (708)
effective collision (710)
frequency factor (710)
transition state theory (711)
transition state (activated complex) (711)

reaction energy diagram (712)

Section 16.7
reaction mechanism (714)
reaction intermediate (715)
elementary reaction (elementary step) (715)
molecularity (715)
unimolecular reaction (715)
bimolecular reaction (715)
rate-determining (rate-limiting) step (716)

Section 16.8
catalyst (720)
homogeneous catalyst (721)
heterogeneous catalyst (722)
hydrogenation (722)
enzyme (723)
active site (723)
substrate (723)
lock-and-key model (724)
induced-fit model (724)
enzyme-substrate complex (ES) (724)

Key Equations and Relationships Numbered and screened concepts are listed for you to refer to or memorize.

16.1 Expressing reaction rate in terms of reactant A (688):

$$\text{Rate} = -\frac{\Delta[A]}{\Delta t}$$

16.2 Expressing the rate of a general reaction (690):

$$aA + bB \longrightarrow cC + dD$$

$$\text{Rate} = -\frac{1}{a}\frac{\Delta[A]}{\Delta t} = -\frac{1}{b}\frac{\Delta[B]}{\Delta t} = \frac{1}{c}\frac{\Delta[C]}{\Delta t} = \frac{1}{d}\frac{\Delta[D]}{\Delta t}$$

16.3 Writing a general rate law (in which products do not appear) (691):

$$\text{Rate} = k[A]^m[B]^n \cdots$$

16.4 Calculating the time to reach a given [A] in a first-order reaction (rate = k[A]) (700):

$$\ln\frac{[A]_0}{[A]_t} = kt$$

16.5 Calculating the time to reach a given [A] in a simple second-order reaction (rate = $k[A]^2$) (700):

$$\frac{1}{[A]_t} - \frac{1}{[A]_0} = kt$$

16.6 Calculating the time to reach a given [A] in a zero-order reaction (rate = k) (700):

$$[A]_t - [A]_0 = -kt$$

16.7 Finding the half-life of a first-order process (702):

$$t_{1/2} = \frac{\ln 2}{k} = \frac{0.693}{k}$$

16.8 Relating the rate constant to the temperature (Arrhenius equation) (706):

$$k = Ae^{-E_a/RT}$$

16.9 Calculating the activation energy (rearranged form of Arrhenius equation) (706):

$$\ln\frac{k_2}{k_1} = -\frac{E_a}{R}\left(\frac{1}{T_2} - \frac{1}{T_1}\right)$$

16.10 Relating the heat of reaction to the forward and reverse activation energies (713):

$$\Delta H_{rxn} = E_{a(fwd)} - E_{a(rev)}$$

Highlighted Figures and Tables These figures (F) and tables (T) provide a visual review of key ideas.

F16.1 Reaction rate (685)
F16.5 Concentration of O_3 vs. time during reaction with C_2H_4 (689)
T16.3 Units of k for overall reaction orders (699)
F16.7 Integrated rate laws and reaction orders (701)
F16.8 Graphical determination of reaction order (702)
F16.9 [N_2O_5] vs. time for three half-lives (703)
T16.4 Zero-order, first-order, and simple second-order reactions (705)
F16.10 Dependence of the rate constant on temperature (705)
F16.11 Graphical determination of E_a (706)
F16.12 Determining the kinetic parameters of a reaction (707)

F16.14 Effect of temperature on distribution of collision energies (709)
T16.5 Effect of E_a and T on the fraction (f) of sufficiently energetic collisions (709)
F16.15 Energy-level diagram for a reaction (709)
F16.16 Fraction of collisions with energy exceeding E_a (710)
F16.19 Reaction energy diagram for the reaction of CH_3Br and OH^- (712)
T16.6 Rate laws for general elementary steps (715)
F16.22 Reaction energy diagram for catalyzed and uncatalyzed processes (720)

Brief Solutions to FOLLOW-UP PROBLEMS Compare your solutions to these calculation steps and answers.

16.1 (a) $4NO(g) + O_2(g) \longrightarrow 2N_2O_3(g)$;

$$\text{rate} = -\frac{\Delta[O_2]}{\Delta t} = -\frac{1}{4}\frac{\Delta[NO]}{\Delta t} = \frac{1}{2}\frac{\Delta[N_2O_3]}{\Delta t}$$

(b) $-\dfrac{\Delta[O_2]}{\Delta t} = -\dfrac{1}{4}\dfrac{\Delta[NO]}{\Delta t}$

$$= -\frac{1}{4}(-1.60\times10^{-4}\,\text{mol/L·s}) = 4.00\times10^{-5}\,\text{mol/L·s}$$

16.2 First order in Br^-, first order in BrO_3^-, second order in H^+, fourth order overall

16.3 Rate $= k[H_2]^m[I_2]^n$. From experiments 1 and 3, $m = 1$. From experiments 2 and 4, $n = 1$.
Therefore, rate $= k[H_2][I_2]$; second order overall.

16.4 (a) The rate law shows the reaction is zero order in Y, so the rate is not affected by doubling Y:
rate of Expt 2 $= 0.25\times10^{-5}$ mol/L·s
(b) The rate of Expt 3 is four times that of Expt 1, so [X] doubles.

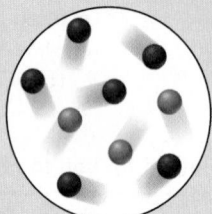

16.5 $1/[HI]_1 - 1/[HI]_0 = kt$;
111 L/mol $- 100.$ L/mol $= (2.4\times10^{-21}$ L/mol·s)(t)
$t = 4.6\times10^{21}$ s (or 1.5×10^{14} yr)

16.6 By inspection, we see that one-quarter of the original number remain after 6.0 min, so one-half remain after 3.0 min: $t_{1/2} = 3.0$ min; $k = 0.693/3.0$ min $= 0.23$ min^{-1}

16.7 $t_{1/2} = (\ln 2)/k$; $k = 0.693/13.1$ h $= 5.29\times10^{-2}$ h^{-1}

16.8 $\ln\dfrac{0.286\ \text{L/mol·s}}{k_1} = -\dfrac{1.00\times10^5\ \text{J/mol}}{8.314\ \text{J/mol·K}}$

$$\times\left(\frac{1}{500.\ \text{K}} - \frac{1}{490.\ \text{K}}\right)$$

$$= 0.491$$

$$k_1 = 0.175\ \text{L/mol·s}$$

16.9

16.10 (a) Balanced equation:
$2NO(g) + 2H_2(g) \longrightarrow N_2(g) + 2H_2O(g)$
(b) Step 2 is unimolecular. All others are bimolecular.
(c) Rate$_1 = k_1[NO]^2$; rate$_2 = k_2[H_2]$; rate$_3 = k_3[N_2O_2][H]$; rate$_4 = k_4[HO][H]$; rate$_5 = k_5[H][N_2O]$.

PROBLEMS

Problems with colored numbers are answered in Appendix E and worked in detail in the Student Solutions Manual. Problem sections match those in the text and provide the numbers of relevant sample problems. Most offer Concept Review Questions, Skill-Building Exercises (grouped in pairs covering the same concept), and Problems in Context. The Comprehensive Problems are based on material from any section or previous chapter.

Factors That Influence Reaction Rate

● Concept Review Questions

• **16.1** What variable of a chemical reaction is measured over time to obtain the reaction rate?

• **16.2** How does an increase in pressure affect the rate of a gas-phase reaction? Explain.

• **16.3** A reaction is carried out with water as the solvent. How does the addition of more water to the reaction vessel affect the rate of the reaction? Explain.

• **16.4** A gas reacts with a solid that is present in large chunks. Then the reaction is run again with the solid pulverized. How does the increase in the surface area of the solid affect the rate of its reaction with the gas? Explain.

• **16.5** How does an increase in temperature affect the rate of a reaction? Explain the two factors involved.

• **16.6** In a kinetics experiment, a chemist places crystals of iodine in a closed reaction vessel, introduces a given quantity of hydrogen gas, and obtains data to calculate the rate of hydrogen iodide formation. In a second experiment, she uses the same amounts of iodine and hydrogen, but first warms the flask to 130°C, a temperature above the sublimation point of iodine. In which of these experiments does the reaction proceed at a higher rate? Explain.

Expressing the Reaction Rate
(Sample Problem 16.1)

● Concept Review Questions

• **16.7** Define *reaction rate*. Assuming constant temperature and a closed reaction vessel, why does the rate change with time?

• **16.8** (a) What is the difference between an average rate and an instantaneous rate? (b) What is the difference between an initial rate and an instantaneous rate?

• **16.9** Give two reasons to measure initial rates in a kinetics study.

•• **16.10** For the reaction $A(g) \longrightarrow B(g)$, sketch two curves on the same set of axes that show
(a) The formation of product as a function of time
(b) The consumption of reactant as a function of time

Levels of Difficulty: • Easy •• Medium ••• Challenging

16.11 For the reaction $C(g) \longrightarrow D(g)$, [C] vs. time is plotted:

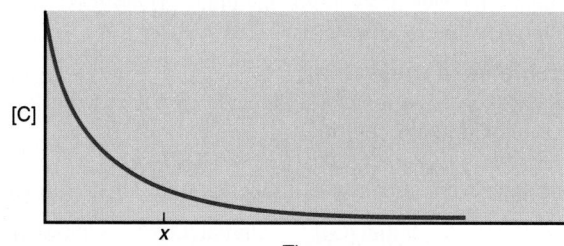

Time

How do you determine each of the following?
(a) The average rate over the entire experiment
(b) The reaction rate at time x
(c) The initial reaction rate
(d) Would the values in parts (a), (b), and (c) be different if you plotted [D] vs. time? Explain.

● **Skill-Building Exercises** *(grouped in similar pairs)*

16.12 The compound AX_2 decomposes according to the equation $2AX_2(g) \longrightarrow 2AX(g) + X_2(g)$. In one experiment, $[AX_2]$ was measured at various times and these data were obtained:

Time (s)	$[AX_2]$ (mol/L)
0.0	0.0500
2.0	0.0448
6.0	0.0300
8.0	0.0249
10.0	0.0209
20.0	0.0088

(a) Find the average rate over the entire experiment.
(b) Is the initial rate higher or lower than the rate in part (a)? Use graphical methods to estimate the initial rate.

16.13 (a) Use the data from Problem 16.12 to calculate the average rate from 8.0 to 20.0 s.
(b) Is the rate at exactly 5.0 s higher or lower than the rate in part (a)? Use graphical methods to estimate the rate at 5.0 s.

16.14 Express the rate of reaction in terms of the change in concentration of each of the reactants and products:
$$2A(g) \longrightarrow B(g) + C(g)$$
When [C] is increasing at 2 mol/L·s, how fast is [A] decreasing?

16.15 Express the rate of reaction in terms of the change in concentration of each of the reactants and products:
$$D(g) \longrightarrow \tfrac{3}{2}E(g) + \tfrac{5}{2}F(g)$$
When [E] is increasing at 0.25 mol/L·s, how fast is [F] increasing?

16.16 Express the rate of reaction in terms of the change in concentration of each of the reactants and products:
$$A(g) + 2B(g) \longrightarrow C(g)$$
When [B] is decreasing at 0.5 mol/L·s, how fast is [A] decreasing?

16.17 Express the rate of reaction in terms of the change in concentration of each of the reactants and products:
$$2D(g) + 3E(g) + F(g) \longrightarrow 2G(g) + H(g)$$
When [D] is decreasing at 0.1 mol/L·s, how fast is [H] increasing?

16.18 Reaction rate is expressed in terms of changes in concentration of reactants and products. Write a balanced equation for
$$\text{Rate} = -\frac{1}{2}\frac{\Delta[N_2O_5]}{\Delta t} = \frac{1}{4}\frac{\Delta[NO_2]}{\Delta t} = \frac{\Delta[O_2]}{\Delta t}$$

16.19 Reaction rate is expressed in terms of changes in concentration of reactants and products. Write a balanced equation for
$$\text{Rate} = -\frac{\Delta[CH_4]}{\Delta t} = -\frac{1}{2}\frac{\Delta[O_2]}{\Delta t} = \frac{1}{2}\frac{\Delta[H_2O]}{\Delta t} = \frac{\Delta[CO_2]}{\Delta t}$$

● **Problems in Context**

16.20 The decomposition of NOBr is studied manometrically because the number of moles of gas changes; it cannot be studied colorimetrically because both NOBr and Br_2 are reddish brown:
$$2NOBr(g) \longrightarrow 2NO(g) + Br_2(g)$$
Use the data below to answer the following:
(a) Determine the average rate over the entire experiment.
(b) Determine the average rate between 2.00 and 4.00 s.
(c) Use graphical methods to estimate the initial reaction rate.
(d) Use graphical methods to estimate the rate at 7.00 s.
(e) At what time does the instantaneous rate equal the average rate over the entire experiment?

Time (s)	[NOBr] (mol/L)
0.00	0.0100
2.00	0.0071
4.00	0.0055
6.00	0.0045
8.00	0.0038
10.00	0.0033

16.21 The formation of ammonia is one of the most important processes in the chemical industry:
$$N_2(g) + 3H_2(g) \longrightarrow 2NH_3(g)$$
Express the rate in terms of changes in $[N_2]$, $[H_2]$, and $[NH_3]$.

16.22 Although the depletion of stratospheric ozone threatens life on Earth today, its accumulation was one of the crucial processes that allowed life to develop in prehistoric times:
$$3O_2(g) \longrightarrow 2O_3(g)$$
(a) Express the reaction rate in terms of $[O_2]$ and $[O_3]$.
(b) At a given instant, the reaction rate in terms of $[O_2]$ is 2.17×10^{-5} mol/L·s. What is it in terms of $[O_3]$?

The Rate Law and Its Components
(Sample Problems 16.2 to 16.4)

● **Concept Review Questions**

16.23 The rate law for the general reaction
$$a A + b B + \cdots \longrightarrow c C + d D + \cdots$$
is rate $= k[A]^m[B]^n \cdots$.
(a) Explain the meaning of k. (b) Explain the meanings of m and n. Does $m = a$ and $n = b$? Explain. (c) If the reaction is first order in A and second order in B, and time is measured in minutes (min), what are the units for k?

16.24 You are studying the reaction
$$A_2(g) + B_2(g) \longrightarrow 2AB(g)$$
to determine its rate law. Assuming that you have a valid experimental procedure for obtaining $[A_2]$ and $[B_2]$ at various times, explain how you determine (a) the initial rate, (b) the reaction orders, and (c) the rate constant.

16.25 By what factor does the rate change in each of the following cases (assuming constant temperature)?
(a) A reaction is first order in reactant A, and [A] is doubled.
(b) A reaction is second order in reactant B, and [B] is halved.
(c) A reaction is second order in reactant C, and [C] is tripled.

● **Skill-Building Exercises** (grouped in similar pairs)

● **16.26** Give the individual reaction orders for all substances and the overall reaction order from the following rate law:

$$Rate = k[BrO_3^-][Br^-][H^+]^2$$

● **16.27** Give the individual reaction orders for all substances and the overall reaction order from the following rate law:

$$Rate = k\frac{[O_3]^2}{[O_2]}$$

● **16.28** By what factor does the rate in Problem 16.26 change if each of the following changes occurs: (a) $[BrO_3^-]$ is doubled; (b) $[Br^-]$ is halved; (c) $[H^+]$ is quadrupled?

● **16.29** By what factor does the rate in Problem 16.27 change if each of the following changes occurs: (a) $[O_3]$ is doubled; (b) $[O_2]$ is doubled; (c) $[O_2]$ is halved?

● **16.30** Give the individual reaction orders for all substances and the overall reaction order from this rate law:

$$Rate = k[NO_2]^2[Cl_2]$$

● **16.31** Give the individual reaction orders for all substances and the overall reaction order from this rate law:

$$Rate = k\frac{[HNO_2]^4}{[NO]^2}$$

● **16.32** By what factor does the rate in Problem 16.30 change if each of the following changes occurs: (a) $[NO_2]$ is tripled; (b) $[NO_2]$ and $[Cl_2]$ are doubled; (c) $[Cl_2]$ is halved?

● **16.33** By what factor does the rate in Problem 16.31 change if each of the following changes occurs: (a) $[HNO_2]$ is doubled; (b) $[NO]$ is doubled; (c) $[HNO_2]$ is halved?

● **16.34** For the reaction

$$4A(g) + 3B(g) \longrightarrow 2C(g)$$

the following data were obtained at constant temperature:

Experiment	Initial [A] (mol/L)	Initial [B] (mol/L)	Initial Rate (mol/L·min)
1	0.100	0.100	5.00
2	0.300	0.100	45.0
3	0.100	0.200	10.0
4	0.300	0.200	90.0

(a) What is the order with respect to each reactant? (b) Write the rate law. (c) Calculate k (using the data from experiment 1).

● **16.35** For the reaction

$$A(g) + B(g) + C(g) \longrightarrow D(g)$$

the following data were obtained at constant temperature:

Expt	Initial [A] (mol/L)	Initial [B] (mol/L)	Initial [C] (mol/L)	Initial Rate (mol/L·s)
1	0.0500	0.0500	0.0100	6.25×10^{-3}
2	0.1000	0.0500	0.0100	1.25×10^{-2}
3	0.1000	0.1000	0.0100	5.00×10^{-2}
4	0.0500	0.0500	0.0200	6.25×10^{-3}

(a) What is the order with respect to each reactant? (b) Write the rate law. (c) Calculate k (using the data from experiment 1).

● **16.36** Without consulting Table 16.3, give the units of the rate constants for reactions with the following overall orders: (a) first order; (b) second order; (c) third order; (d) $\frac{5}{2}$ order.

● **16.37** Give the overall reaction order that corresponds to a rate constant with each of the following units: (a) mol/L·s; (b) yr^{-1}; (c) $(mol/L)^{1/2} \cdot s^{-1}$; (d) $(mol/L)^{-5/2} \cdot min^{-1}$.

● **Problems in Context**

● **16.38** Phosgene is a toxic gas prepared by the reaction of carbon monoxide with chlorine:

$$CO(g) + Cl_2(g) \longrightarrow COCl_2(g)$$

These data were obtained in a kinetics study of its formation:

Experiment	Initial [CO] (mol/L)	Initial [Cl₂] (mol/L)	Initial Rate (mol/L·s)
1	1.00	0.100	1.29×10^{-29}
2	0.100	0.100	1.33×10^{-30}
3	0.100	1.00	1.30×10^{-29}
4	0.100	0.0100	1.32×10^{-31}

(a) Write the rate law for the formation of phosgene.
(b) Calculate the average value of the rate constant.

Integrated Rate Laws: Concentration Changes over Time
(Sample Problems 16.5 to 16.7)

● **Concept Review Questions**

● **16.39** How are integrated rate laws used to determine reaction order? What is the order in reactant if a plot of
(a) The natural logarithm of [reactant] vs. time is linear?
(b) The inverse of [reactant] vs. time is linear?
(c) [Reactant] vs. time is linear?

● **16.40** Define the *half-life* of a reaction. Explain on the molecular level why the half-life of a first-order reaction is constant.

● **Skill-Building Exercises** (grouped in similar pairs)

● **16.41** For the simple decomposition reaction

$$AB(g) \longrightarrow A(g) + B(g)$$

rate $= k[AB]^2$ and $k = 0.2$ L/mol·s. How long will it take for $[AB]$ to reach one-third of its initial concentration of 1.50 M?

16.42 For the reaction in Problem 16.41, what is $[AB]$ after 10.0 s?

● **16.43** In a first-order decomposition reaction, 50.0% of a compound decomposes in 10.5 min. (a) What is the rate constant of the reaction? (b) How long does it take for 75.0% of the compound to decompose?

● **16.44** A decomposition reaction has a rate constant of 0.0012 yr^{-1}. (a) What is the half-life of the reaction? (b) How long does it take for [reactant] to reach 12.5% of its original value?

● **Problems in Context**

● **16.45** In a study of ammonia production, an industrial chemist discovers that the compound decomposes to its elements N_2 and H_2 in a first-order process. She collects the following data:

Time (s)	0	1.000	2.000
[NH₃] (mol/L)	4.000	3.986	3.974

(a) Use graphical methods to determine the rate constant.
(b) What is the half-life for ammonia decomposition?

The Effect of Temperature on Reaction Rate
(Sample Problem 16.8)

● **Concept Review Questions**

● **16.46** Use the exponential term in the Arrhenius equation to explain how temperature affects reaction rate.

● **16.47** How is the activation energy determined from the Arrhenius equation?

16.48 (a) Graph the relationship between k (y axis) and T (x axis). (b) Graph the relationship between ln k (y axis) and $1/T$ (x axis). How is the activation energy determined from this graph?

● Skill-Building Exercises (grouped in similar pairs)

16.49 The rate constant of a reaction is 4.7×10^{-3} s^{-1} at 25°C, and the activation energy is 33.6 kJ/mol. What is k at 75°C?

16.50 The rate constant of a reaction is 4.50×10^{-5} L/mol·s at 195°C and 3.20×10^{-3} L/mol·s at 258°C. What is the activation energy of the reaction?

● Problems in Context

16.51 Understanding the high-temperature formation and break-down of the nitrogen oxides is essential for controlling the pollutants generated from power plants and cars. The first-order breakdown of dinitrogen monoxide to its elements has rate constants of 0.76/s at 727°C and 0.87/s at 757°C. What is the activation energy of this reaction?

Explaining the Effects of Concentration and Temperature
(Sample Problem 16.9)

● Concept Review Questions

16.52 What is the central idea of collision theory? How does this idea explain the effect of concentration on reaction rate?

16.53 Is collision frequency the only factor affecting rate? Explain.

16.54 Arrhenius proposed that each reaction has an energy threshold that must be reached for the particles to react. The kinetic theory of gases proposes that the average kinetic energy of the particles is proportional to the absolute temperature. How do these concepts relate to the effect of temperature on rate?

16.55 (a) For a reaction with a given E_a, how does an increase in T affect the rate? (b) For a reaction at a given T, how does a decrease in E_a affect the rate?

16.56 In the reaction AB + CD $\rightleftharpoons$ EF, 4×10^{-5} mol of AB molecules collide with 4×10^{-5} mol of CD molecules. Will 4×10^{-5} mol of EF form? Explain.

16.57 Assuming the activation energies are equal, which of the following reactions will occur at a higher rate at 50°C? Explain.

$$NH_3(g) + HCl(g) \longrightarrow NH_4Cl(s)$$
$$N(CH_3)_3(g) + HCl(g) \longrightarrow (CH_3)_3NHCl(s)$$

● Skill-Building Exercises (grouped in similar pairs)

16.58 For the reaction A(g) + B(g) $\longrightarrow$ AB(g), how many unique collisions between A and B are possible if there are four particles of A and three particles of B present in the vessel?

16.59 For the reaction A(g) + B(g) $\longrightarrow$ AB(g), how many unique collisions between A and B are possible if 1.01 mol of A(g) and 2.12 mol of B(g) are present in the vessel?

16.60 At 25°C, what is the fraction of collisions with energy equal to or greater than an activation energy of 100. kJ/mol?

16.61 If the temperature in Problem 16.60 is increased to 50.°C, by what factor does the fraction of collisions with energy equal to or greater than the activation energy change?

16.62 For the reaction ABC + D $\rightleftharpoons$ AB + CD, $\Delta H^{\circ}_{rxn} = -55$ kJ/mol and $E_{a(fwd)} = 215$ kJ/mol. Assuming a one-step reaction, (a) draw a reaction energy diagram; (b) calculate $E_{a(rev)}$; and (c) sketch a possible transition state if ABC is V-shaped.

16.63 For the reaction $A_2 + B_2 \longrightarrow$ 2AB, $E_{a(fwd)} = 125$ kJ/mol and $E_{a(rev)} = 85$ kJ/mol. Assuming the reaction occurs in one

step, (a) draw a reaction energy diagram; (b) calculate ΔH°_{rxn}; and (c) sketch a possible transition state.

● Problems in Context

16.64 Aqua regia, a mixture of HCl and HNO$_3$, has been used since alchemical times to dissolve many metals, including gold. Its orange color is due to the presence of nitrosyl chloride. Consider this one-step reaction for the formation of this compound:

$$NO(g) + Cl_2(g) \longrightarrow NOCl(g) + Cl(g) \qquad \Delta H^{\circ} = 83 \text{ kJ}$$

(a) Draw a reaction energy diagram, given $E_{a(fwd)}$ is 86 kJ/mol. (b) Calculate $E_{a(rev)}$. (c) Sketch a possible transition state for the reaction. (*Note:* The atom sequence of nitrosyl chloride is Cl—N—O.)

Reaction Mechanisms: Steps in the Overall Reaction
(Sample Problem 16.10)

● Concept Review Questions

16.65 Is the rate of an overall reaction lower, higher, or equal to the average rate of the individual steps? Explain.

16.66 Explain why the coefficients of an elementary step equal the reaction orders of its rate law but those of an overall reaction do not.

16.67 Is it possible for more than one mechanism to be consistent with the rate law of a given reaction? Explain.

16.68 What is the difference between a reaction intermediate and a transition state?

16.69 Why is a bimolecular step more reasonable physically than a termolecular step?

16.70 If a slow step precedes a fast step in a two-step mechanism, do the substances in the fast step appear in the rate law? Explain.

16.71 If a fast step precedes a slow step in a two-step mechanism, how is the fast step affected? How is this effect used to determine the validity of the mechanism?

● Skill-Building Exercises (grouped in similar pairs)

16.72 The proposed mechanism for a reaction is

(1) A(g) + B(g) $\rightleftharpoons$ X(g)	[fast]
(2) X(g) + C(g) $\longrightarrow$ Y(g)	[slow]
(3) Y(g) $\longrightarrow$ D(g)	[fast]

(a) What is the overall equation?
(b) Identify the intermediate(s), if any.
(c) What are the molecularity and the rate law for each step?
(d) Is the mechanism consistent with the actual rate law: rate = k[A][B][C]?
(e) Is the following one-step mechanism equally valid: A(g) + B(g) + C(g) $\longrightarrow$ D(g)?

16.73 Consider the following mechanism:

(1) ClO$^-$(aq) + H$_2$O(l) $\rightleftharpoons$ HClO(aq) + OH$^-$(aq)	[fast]
(2) I$^-$(aq) + HClO(aq) $\longrightarrow$ HIO(aq) + Cl$^-$(aq)	[slow]
(3) OH$^-$(aq) + HIO(aq) $\longrightarrow$ H$_2$O(l) + IO$^-$(aq)	[fast]

(a) What is the overall equation?
(b) Identify the intermediate(s), if any.
(c) What are the molecularity and the rate law for each step?
(d) Is the mechanism consistent with the actual rate law: rate = k[ClO$^-$][I$^-$]?

● Problems in Context

16.74 In a study of nitrosyl halides, a chemist proposes the following mechanism for the synthesis of nitrosyl bromide:

$$NO(g) + Br_2(g) \rightleftharpoons NOBr_2(g) \quad \text{[fast]}$$
$$NOBr_2(g) + NO(g) \longrightarrow 2NOBr(g) \quad \text{[slow]}$$

If the rate law is rate = $k[NO]^2[Br_2]$, is the proposed mechanism valid? If so, show that it satisfies the three criteria for validity.

•• **16.75** The rate law for $2NO(g) + O_2(g) \longrightarrow 2NO_2(g)$ is rate = $k[NO]^2[O_2]$. In addition to the mechanism in the text (p. 719), the following ones have been proposed:

I $2NO(g) + O_2(g) \longrightarrow 2NO_2(g)$

II $2NO(g) \rightleftharpoons N_2O_2(g)$ [fast]
    $N_2O_2(g) + O_2(g) \longrightarrow 2NO_2(g)$ [slow]

III $2NO(g) \rightleftharpoons N_2(g) + O_2(g)$ [fast]
    $N_2(g) + 2O_2(g) \longrightarrow 2NO_2(g)$ [slow]

(a) Which of these mechanisms is consistent with the rate law?
(b) Which of these mechanisms is most reasonable? Why?

Catalysis: Speeding Up a Chemical Reaction

● Concept Review Questions

•• **16.76** Consider the reaction $N_2O(g) \xrightarrow{Au} N_2(g) + \frac{1}{2}O_2(g)$.
(a) Is the gold a homogeneous or a heterogeneous catalyst?
(b) On the same set of axes, sketch the reaction energy diagrams for the catalyzed and the uncatalyzed reaction.

•• **16.77** Does a catalyst increase reaction rate by the same means as a rise in temperature does? Explain.

•• **16.78** In a classroom demonstration, hydrogen gas and oxygen gas are mixed in a balloon. The mixture is stable under normal conditions, but if a spark is applied to it or some powdered metal is added, the mixture explodes. (a) Is the spark acting as a catalyst? Explain. (b) Is the metal acting as a catalyst? Explain.

•• **16.79** A principle of green chemistry is that the energy needs of industrial processes should have minimal environmental impact. How can the use of catalysts lead to "greener" technologies?

Comprehensive Problems

•• **16.80** Experiments show that each of the following redox reactions is second order overall:

Reaction 1: $NO_2(g) + CO(g) \longrightarrow NO(g) + CO_2(g)$
Reaction 2: $NO(g) + O_3(g) \longrightarrow NO_2(g) + O_2(g)$

(a) When $[NO_2]$ in reaction 1 is doubled, the rate quadruples. Write the rate law for this reaction.
(b) When $[NO]$ in reaction 2 is doubled, the rate doubles. Write the rate law for this reaction.
(c) In each reaction, the initial concentrations of the reactants are equal. For each reaction, what is the ratio of the initial rate to the rate when the reaction is 50% complete?
(d) In reaction 1, the initial $[NO_2]$ is twice the initial $[CO]$. What is the ratio of the initial rate to the rate at 50% completion?
(e) In reaction 2, the initial $[NO]$ is twice the initial $[O_3]$. What is the ratio of the initial rate to the rate at 50% completion?

•• **16.81** Consider the following reaction energy diagram:

Reaction progress

(a) How many elementary steps are in the reaction mechanism?
(b) Which step is rate limiting?
(c) Is the overall reaction exothermic or endothermic?

•• **16.82** Reactions between certain organic (alkyl) halides and water produce alcohols. Consider the overall reaction for *t*-butyl bromide (2-bromo-2-methylpropane):

$(CH_3)_3CBr(aq) + H_2O(l) \longrightarrow$
$\qquad\qquad (CH_3)_3COH(aq) + H^+(aq) + Br^-(aq)$

The experimental rate law is rate = $k[(CH_3)_3CBr]$. The accepted mechanism for the reaction is

(1) $(CH_3)_3C—Br(aq) \longrightarrow (CH_3)_3C^+(aq) + Br^-(aq)$ [slow]
(2) $(CH_3)_3C^+(aq) + H_2O(l) \longrightarrow (CH_3)_3C—OH_2^+(aq)$ [fast]
(3) $(CH_3)_3C—OH_2^+(aq) \longrightarrow H^+(aq) + (CH_3)_3C—OH(aq)$
$\qquad\qquad\qquad\qquad\qquad\qquad\qquad\qquad$ [fast]

(a) Why doesn't H_2O appear in the rate law?
(b) Write rate laws for the elementary steps.
(c) What reaction intermediates appear in the mechanism?
(d) Show that the mechanism is consistent with the experimental rate law.

•• **16.83** The catalytic destruction of ozone occurs via a two-step mechanism, where X can be any of several species:

(1) $X + O_3 \longrightarrow XO + O_2$ [slow]
(2) $XO + O \longrightarrow X + O_2$ [fast]

(a) Write the overall reaction.
(b) Write the rate law for each step.
(c) X acts as _____, and XO acts as _____.
(d) High-flying aircraft release NO into the stratosphere, which catalyzes this process. When O_3 and NO concentrations are 5×10^{12} molecule/cm³ and 1.0×10^9 molecule/cm³, respectively, what is the rate of O_3 depletion (k for the rate-determining step is 6×10^{-15} cm³/molecule·s)?

•• **16.84** Archeologists can determine the age of artifacts made of wood or bone by measuring the amount of the radioactive isotope ^{14}C present in the object. The amount of isotope decreases in a first-order process. If 15.5% of the original amount of ^{14}C is present in a wooden tool at the time of analysis, what is the age of the tool? The half-life of ^{14}C is 5730 yr.

•• **16.85** A slightly bruised apple will rot extensively in about 4 days at room temperature (20°C). If it is kept in the refrigerator at 0°C, the same extent of rotting takes about 16 days. What is the activation energy for the rotting reaction?

•• **16.86** Benzoyl peroxide, the substance most widely used against acne, has a half-life of 9.8×10^3 days when refrigerated. How long will it take to lose 5% of its potency (95% remaining)?

•• **16.87** The rate law for the reaction

$NO_2(g) + CO(g) \longrightarrow NO(g) + CO_2(g)$

is rate = $k[NO_2]^2$; one possible mechanism is shown on p. 717.
(a) Draw a reaction energy diagram for that mechanism, given that $\Delta H^\circ_{overall} = -226$ kJ/mol.
(b) Consider the following alternative mechanism:

(1) $2NO_2(g) \longrightarrow N_2(g) + 2O_2(g)$ [slow]
(2) $2CO(g) + O_2(g) \longrightarrow 2CO_2(g)$ [fast]
(3) $N_2(g) + O_2(g) \longrightarrow 2NO(g)$ [fast]

Is the alternative mechanism consistent with the rate law? Is one mechanism more reasonable physically? Explain.

•• **16.88** Consider the following general reaction and data:

$2A + 2B + C \longrightarrow D + 3E$

Expt	Initial [A] (mol/L)	Initial [B] (mol/L)	Initial [C] (mol/L)	Initial Rate (mol/L·s)
1	0.024	0.085	0.032	6.0×10^{-6}
2	0.096	0.085	0.032	9.6×10^{-5}
3	0.024	0.034	0.080	1.5×10^{-5}
4	0.012	0.170	0.032	1.5×10^{-6}

(a) What is the reaction order with respect to each reactant?
(b) Calculate the rate constant.
(c) Write the rate law for this reaction.
(d) Express the rate in terms of changes in concentration with time for each of the components.

•• **16.89** In acidic solution, the breakdown of sucrose into glucose and fructose has this rate law: rate = $k[H^+][\text{sucrose}]$. The initial rate of sucrose breakdown is measured in a solution that is 0.01 M H^+, 1.0 M sucrose, 0.1 M fructose, and 0.1 M glucose. How does the rate change if
(a) [Sucrose] is changed to 2.5 M?
(b) [Sucrose], [fructose], and [glucose] are all changed to 0.5 M?
(c) $[H^+]$ is changed to 0.0001 M?
(d) [Sucrose] and $[H^+]$ are both changed to 0.1 M?

•• **16.90** Enzymes are remarkably efficient catalysts that can increase reaction rates by as many as 20 orders of magnitude.
(a) How does an enzyme affect the transition state of a reaction, and how does this effect increase the reaction rate?
(b) What characteristics of enzymes give them this tremendous effectiveness as catalysts?

•• **16.91** The following molecular scenes represent starting mixtures I and II for the reaction of A (black) with B (orange):

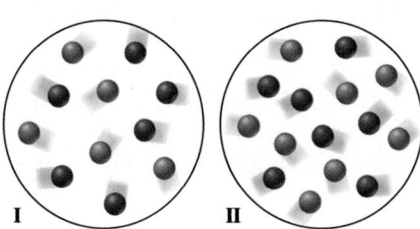

I   **II**

Each sphere counts as 0.010 mol, and the volume is 0.50 L. If the initial rate in I is 8.3×10^{-4} mol/L·min, what is the initial rate in II?

•• **16.92** Biochemists consider the citric acid cycle to be the central reaction sequence in metabolism. One of the key steps is an oxidation catalyzed by the enzyme isocitrate dehydrogenase and the oxidizing agent NAD^+. Under certain conditions, the reaction in yeast obeys 11th-order kinetics:
Rate = $k[\text{enzyme}][\text{isocitrate}]^4[\text{AMP}]^2[\text{NAD}^+]^m[\text{Mg}^{2+}]^2$
What is the order with respect to NAD^+?

•• **16.93** At body temperature (37°C), k of an enzyme-catalyzed reaction is 2.3×10^{14} times greater than k of the uncatalyzed reaction. Assuming that the frequency factor A is the same for both reactions, by how much does the enzyme lower the E_a?

•• **16.94** Enzymes in the human liver catalyze a large number of reactions that degrade ingested toxic chemicals. By what factor is the rate of a detoxification reaction changed if a liver enzyme lowers the activation energy by 5 kJ/mol at 37°C?

•• **16.95** Experiment shows that the rate of formation of carbon tetrachloride from chloroform,
$$CHCl_3(g) + Cl_2(g) \longrightarrow CCl_4(g) + HCl(g)$$

is first order in $CHCl_3$, $\frac{1}{2}$ order in Cl_2, and $\frac{3}{2}$ order overall. Show that the following mechanism is consistent with the rate law:
(1) $Cl_2(g) \rightleftharpoons 2Cl(g)$ [fast]
(2) $Cl(g) + CHCl_3(g) \longrightarrow HCl(g) + CCl_3(g)$ [slow]
(3) $CCl_3(g) + Cl(g) \longrightarrow CCl_4(g)$ [fast]

•• **16.96** A biochemist studying breakdown of the insecticide DDT finds that it decomposes by a first-order reaction with a half-life of 12 yr. How long does it take DDT in a soil sample to decompose from 275 ppbm to 10. ppbm (parts per billion by mass)?

•• **16.97** Insulin is a polypeptide hormone released from the pancreas that stimulates fat and muscle to take up glucose. In a certain patient, it has a first-order half-life in the blood of 3.5 min. To maintain an adequate blood concentration of insulin, it must be replenished in a time interval equal to $1/k$. How long is the time interval in this patient?

•• **16.98** For the reaction $A(g) + B(g) \longrightarrow AB(g)$, the rate is 0.20 mol/L·s, when $[A]_0 = [B]_0 = 1.0$ mol/L. If the reaction is first order in B and second order in A, what is the rate when $[A]_0 = 2.0$ mol/L and $[B]_0 = 3.0$ mol/L?

•• **16.99** The acid-catalyzed hydrolysis of sucrose occurs by the following overall reaction whose kinetic data are given below:

$$\underset{\text{sucrose}}{C_{12}H_{22}O_{11}(s)} + H_2O(l) \longrightarrow \underset{\text{glucose}}{C_6H_{12}O_6(aq)} + \underset{\text{fructose}}{C_6H_{12}O_6(aq)}$$

[Sucrose] (mol/L)	Time (h)
0.501	0.00
0.451	0.50
0.404	1.00
0.363	1.50
0.267	3.00

(a) Determine the rate constant and the half-life of the reaction.
(b) How long does it take to hydrolyze 75% of the sucrose?
(c) Other studies have shown that this reaction is actually second order overall but appears to follow first-order kinetics. (Such a reaction is called a *pseudo–first-order reaction.*) Suggest a reason for this apparent first-order behavior.

•• **16.100** Is each of these statements true? If not, explain why.
(a) At a given T, all molecules have the same kinetic energy.
(b) Halving the P of a gaseous reaction doubles the rate.
(c) A higher activation energy gives a lower reaction rate.
(d) A temperature rise of 10°C doubles the rate of any reaction.
(e) If reactant molecules collide with greater energy than the activation energy, they change into product molecules.
(f) The activation energy of a reaction depends on temperature.
(g) The rate of a reaction increases as the reaction proceeds.
(h) Activation energy depends on collision frequency.
(i) A catalyst increases the rate by increasing collision frequency.
(j) Exothermic reactions are faster than endothermic reactions.
(k) Temperature has no effect on the frequency factor (A).
(l) The activation energy of a reaction is lowered by a catalyst.
(m) For most reactions, ΔH_{rxn} is lowered by a catalyst.
(n) The orientation probability factor (p) is near 1 for reactions between single atoms.
(o) The initial rate of a reaction is its maximum rate.
(p) A bimolecular reaction is generally twice as fast as a unimolecular reaction.
(q) The molecularity of an elementary reaction is proportional to the molecular complexity of the reactant(s).

•• **16.101** For the decomposition of gaseous dinitrogen pentaoxide, $2N_2O_5(g) \longrightarrow 4NO_2(g) + O_2(g)$, the rate constant is $k = 2.8 \times 10^{-3}$ s^{-1} at 60°C. The initial concentration of N_2O_5 is 1.58 mol/L. (a) What is $[N_2O_5]$ after 5.00 min? (b) What fraction of the N_2O_5 has decomposed after 5.00 min?

•• **16.102** Even when a mechanism is consistent with the rate law, later work may show it to be incorrect. For example, the reaction between hydrogen and iodine has this rate law: rate = $k[H_2][I_2]$. The long-accepted mechanism had a single bimolecular step; that is, the overall reaction was thought to be elementary:

$$H_2(g) + I_2(g) \longrightarrow 2HI(g)$$

In the 1960s, however, spectroscopic evidence showed the presence of free I atoms during the reaction. Kineticists have since proposed a three-step mechanism:

(1) $I_2(g) \rightleftharpoons 2I(g)$ [fast]
(2) $H_2(g) + I(g) \rightleftharpoons H_2I(g)$ [fast]
(3) $H_2I(g) + I(g) \longrightarrow 2HI(g)$ [slow]

Show that this mechanism is consistent with the rate law.

•• **16.103** Suggest an experimental method for measuring the change in concentration with time for each of the following reactions:
(a) $CH_3CH_2Br(l) + H_2O(l) \longrightarrow CH_3CH_2OH(l) + HBr(aq)$
(b) $2NO(g) + Cl_2(g) \longrightarrow 2NOCl(g)$

•• **16.104** An atmospheric chemist fills a container with gaseous N_2O_5 to a pressure of 125 kPa, and the gas decomposes to NO_2 and O_2. What is the partial pressure of NO_2, P_{NO_2} (in kPa), when the total pressure is 178 kPa?

••• **16.105** Many drugs decompose in blood by a first-order process.
(a) Two tablets of aspirin supply 0.60 g of the active compound. After 30 min, this compound reaches a maximum concentration of 2 mg/100 mL of blood. If the half-life for its breakdown is 90 min, what is its concentration (in mg/100 mL) 2.5 h after it reaches its maximum concentration?
(b) For the decomposition of an antibiotic in a person with a normal temperature (98.6°F), $k = 3.1 \times 10^{-5}$ s^{-1}; for a person with a fever at 101.9°F, $k = 3.9 \times 10^{-5}$ s^{-1}. If the person with the fever must take another pill when $\frac{2}{3}$ of the first pill has decomposed, how many hours should she wait to take a second pill? A third pill? (Assume the pill is effective immediately.)
(c) Calculate E_a for decomposition of the antibiotic in part (b).

•• **16.106** While developing a catalytic process to make ethylene glycol from synthesis gas (CO + H_2), a chemical engineer finds the rate is fourth order in gas pressure. The uncertainty in the pressure reading is 5%. When the catalyst is modified, the rate increases by 10%. If you were the company patent attorney, would you file for a patent on this catalyst modification? Explain.

•• **16.107** Iodide ion reacts with chloromethane to displace chloride ion in a common organic substitution reaction:

$$I^- + CH_3Cl \longrightarrow CH_3I + Cl^-$$

(a) Draw a wedge-bond structure of chloroform and indicate the most effective direction of I^- attack.
(b) The analogous reaction with 2-chlorobutane [Figure P16.107(b)] results in a major change in specific rotation as measured by polarimetry. Explain, showing a wedge-bond structure of the product.
(c) Under different conditions, 2-chlorobutane loses Cl^- in a rate-determining step to form a planar intermediate [Figure P16.107(c)]. This cationic species reacts with HI and then loses H^+ to form a product that exhibits no optical activity. Explain, showing a wedge-bond structure.

Figure P16.107(b)   Figure P16.107(c)

•• **16.108** Assume water boils at 100.0°C in Houston (near sea level), and at 90.0°C in Cripple Creek, Colorado (near 9500 ft). If it takes 4.8 min to cook an egg in Cripple Creek and 4.5 min in Houston, what is E_a for this process?

•• **16.109** Sulfonation of benzene has the following mechanism:
(1) $2H_2SO_4 \longrightarrow H_3O^+ + HSO_4^- + SO_3$ [fast]
(2) $SO_3 + C_6H_6 \longrightarrow H(C_6H_5^+)SO_3^-$ [slow]
(3) $H(C_6H_5^+)SO_3^- + HSO_4^- \longrightarrow C_6H_5SO_3^- + H_2SO_4$ [fast]
(4) $C_6H_5SO_3^- + H_3O^+ \longrightarrow C_6H_5SO_3H + H_2O$ [fast]
(a) Write an overall equation for the reaction.
(b) Write the overall rate law for the initial rate of the reaction.

•• **16.110** In the lower troposphere, ozone is one of the components of photochemical smog. It is generated in air when nitrogen dioxide, formed by the oxidation of nitrogen monoxide from car exhaust, reacts by the following mechanism:
(1) $NO_2(g) \xrightarrow[hv]{k_1} NO(g) + O(g)$
(2) $O(g) + O_2(g) \xrightarrow{k_2} O_3(g)$
Assuming the rate of formation of atomic oxygen in step 1 equals the rate of its consumption in step 2, use the data below to calculate (a) the concentration of atomic oxygen [O]; (b) the rate of ozone formation.

$k_1 = 6.0 \times 10^{-3}$ s^{-1} $[NO_2] = 4.0 \times 10^{-9}$ M
$k_2 = 1.0 \times 10^6$ L/mol·s $[O_2] = 1.0 \times 10^{-2}$ M

•• **16.111** Chlorine is commonly used to disinfect drinking water, and inactivation of pathogens by chlorine follows first-order kinetics. The following data show *E. coli* inactivation:

Contact time (min)	Percent (%) inactivation
0.00	0.0
0.50	68.3
1.00	90.0
1.50	96.8
2.00	99.0
2.50	99.7
3.00	99.9

(a) Determine the first-order inactivation constant, k. [*Hint*: % inactivation = $100 \times (1 - [A]_t/[A]_0)$.]
(b) How much contact time is required for 95% inactivation?

•• **16.112** The reaction and rate law for the gas-phase decomposition of dinitrogen pentaoxide are

$$2N_2O_5(g) \longrightarrow 4NO_2(g) + O_2(g) \qquad \text{rate} = k[N_2O_5]$$

Which of the following can be considered valid mechanisms for the reaction?

I One-step collision

II $2N_2O_5(g) \longrightarrow 2NO_3(g) + 2NO_2(g)$ [slow]
   $2NO_3(g) \longrightarrow 2NO_2(g) + 2O(g)$ [fast]
   $2O(g) \longrightarrow O_2(g)$ [fast]

III $N_2O_5(g) \rightleftharpoons NO_3(g) + NO_2(g)$ [fast]
$NO_2(g) + N_2O_5(g) \longrightarrow 3NO_2(g) + O(g)$ [slow]
$NO_3(g) + O(g) \longrightarrow NO_2(g) + O_2(g)$ [fast]

IV $2N_2O_5(g) \rightleftharpoons 2NO_2(g) + N_2O_3(g) + 3O(g)$ [fast]
$N_2O_3(g) + O(g) \longrightarrow 2NO_2(g)$ [slow]
$2O(g) \longrightarrow O_2(g)$ [fast]

V $2N_2O_5(g) \longrightarrow N_4O_{10}(g)$ [slow]
$N_4O_{10}(g) \longrightarrow 4NO_2(g) + O_2(g)$ [fast]

••• 16.113 Nitrification is a biological process for removing NH_3 from wastewater as NH_4^+:

$$NH_4^+ + 2O_2 \longrightarrow NO_3^- + 2H^+ + H_2O$$

The first-order rate constant is given as

$$k_1 = 0.47e^{0.095\,(T-15°C)}$$

where k_1 is in day^{-1} and T is in °C.
(a) If the initial concentration of NH_3 is 3.0 mol/m³, how long will it take to reduce the concentration to 0.35 mol/m³ in the spring ($T = 20°C$)? (b) In the winter ($T = 10°C$)? (c) Using your answer to part (a), what is the rate of O_2 consumption?

•• 16.114 Carbon disulfide, a poisonous, flammable liquid, is an excellent solvent for phosphorus, sulfur, and some other nonmetals. A kinetic study of its gaseous decomposition reveals these data:

Experiment	Initial [CS₂] (mol/L)	Initial Rate (mol/L·s)
1	0.100	2.7×10^{-7}
2	0.080	2.2×10^{-7}
3	0.055	1.5×10^{-7}
4	0.044	1.2×10^{-7}

(a) Write the rate law for the decomposition of CS_2.
(b) Calculate the average value of the rate constant.

•• 16.115 Like any catalyst, palladium, platinum, and nickel catalyze both directions of a reaction: addition of hydrogen to (hydrogenation) and its elimination from (dehydrogenation) carbon double bonds.
(a) Which variable determines whether an alkene will be hydrogenated or dehydrogenated?
(b) Which reaction requires a higher temperature?
(c) How can all-*trans* fats arise during hydrogenation of fats that contain some *cis*- double bonds?

•• 16.116 In a *clock reaction,* a dramatic color change occurs at a time determined by concentration and temperature. Consider the iodine clock reaction, whose overall equation is

$$2I^-(aq) + S_2O_8^{2-}(aq) \longrightarrow I_2(aq) + 2SO_4^{2-}(aq)$$

As I_2 forms, it is immediately consumed by its reaction with a fixed amount of added $S_2O_3^{2-}$:

$$I_2(aq) + 2S_2O_3^{2-}(aq) \longrightarrow 2I^-(aq) + S_4O_6^{2-}(aq)$$

Once the $S_2O_3^{2-}$ is consumed, the excess I_2 forms a blue-black product with starch present in solution:

$$I_2 + \text{starch} \longrightarrow \text{starch} \cdot I_2 \text{ (blue-black)}$$

The rate of the reaction is also influenced by the total concentration of ions, so KCl and $(NH_4)_2SO_4$ are added to maintain a constant value. Use the data below, obtained at 23°C, to determine
(a) The average rate for each trial
(b) The order with respect to each reactant
(c) The rate constant
(d) The rate law for the overall reaction

	Expt 1	Expt 2	Expt 3
0.200 M KI (mL)	10.0	20.0	20.0
0.100 M Na₂S₂O₈ (mL)	20.0	20.0	10.0
0.0050 M Na₂S₂O₃ (mL)	10.0	10.0	10.0
0.200 M KCl (mL)	10.0	0.0	0.0
0.100 M (NH₄)₂SO₄ (mL)	0.0	0.0	10.0
Time to color (s)	29.0	14.5	14.5

•• 16.117 Heat transfer to and from a reaction flask is often a critical factor in controlling reaction rate. The heat transferred (q) depends on a heat transfer coefficient (h) for the flask material, the temperature difference (ΔT) across the flask wall, and the commonly "wetted" area (A) of the flask and bath, $q = h \cdot A \cdot \Delta T$. When an exothermic reaction is run at a given T, there is a bath temperature at which the reaction can no longer be controlled, and the reaction "runs away" suddenly. A similar problem is often seen when a reaction is "scaled up" from, say, a half-filled small flask to a half-filled large flask. Explain these behaviors.

•• 16.118 The molecular scenes below represent the first-order reaction as cyclopropane (red) is converted to propene (green):

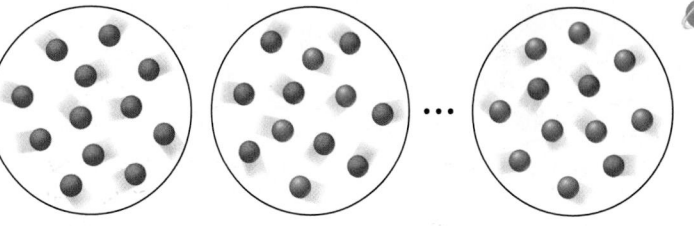

Determine (a) the half-life and (b) the first-order rate constant.

•• 16.119 The growth of *Pseudomonas* bacteria is modeled as a first-order process with $k = 0.035$ min^{-1} at 37°C. The initial *Pseudomonas* population density is 1.0×10^3 cells/L. (a) What is the population density after 2 h? (b) What is the time required for the population to go from 1.0×10^3 to 2.0×10^3 cells/ L?

••• 16.120 Consider the following organic reaction, in which one halogen replaces another in an alkyl halide:

$$CH_3CH_2Br + KI \longrightarrow CH_3CH_2I + KBr$$

In acetone, this particular reaction goes to completion because KI is soluble in acetone but KBr is not. In the mechanism, I^- approaches the carbon *opposite* to the Br (see Figure 16.19, with I^- instead of OH^-). After Br^- has been replaced by I^- and precipitates as KBr, other I^- ions react with the ethyl iodide by the same mechanism.
(a) If we designate the carbon bonded to the halogen as C-1, what is the shape around C-1 and the hybridization of C-1 in ethyl iodide?
(b) In the transition state, one of the two lobes of the unhybridized $2p$ orbital of C-1 overlaps a p orbital of I, while the other lobe overlaps a p orbital of Br. What is the shape around C-1 and the hybridization of C-1 in the transition state?
(c) The deuterated reactant, CH_3CHDBr (where D is deuterium, ^2H), has two optical isomers because C-1 is chiral. If the reaction is run with one of the isomers, the ethyl iodide is *not* optically active. Explain.

•• 16.121 Another radioisotope of iodine, ^{131}I, is also used to study thyroid function (see Follow-up Problem 16.7). A patient is given a sample that is 1.7×10^{-4} M ^{131}I. If the half-life is 8.04 days, what fraction of the radioactivity remains after 30. days?

Biological: 16.119, 16.121; Environmental: 16.113; Organic: 16.115, 16.117, 16.118, 16.120

••• **16.122** The effect of substrate concentration on the first-order microbial growth rate follows the Monod equation:

$$\mu = \frac{\mu_{max} S}{K_s + S}$$

where μ is the first-order growth rate (s^{-1}), μ_{max} is the maximum growth rate (s^{-1}), S is the substrate concentration (kg/m^3), and K_s is the half-saturation coefficient (kg/m^3). For $\mu_{max} = 1.5\times10^{-4}$ s^{-1} and $K_s = 0.03$ kg/m^3:
(a) Prepare a plot of μ vs. S for S between 0.0 and 1.0 kg/m^3.
(b) The initial population density is 5.0×10^3 cells/m^3. What is the density after 1.0 h, if the initial S is 0.30 kg/m^3?
(c) What is it if the initial S is 0.70 kg/m^3?

•• **16.123** The scenes depict four initial reaction mixtures for the reaction of A (blue) and B (yellow), with and without a solid present (gray cubes). The initial rate, $-\Delta[A]/\Delta t$ (in mol/L·s), is shown for each sphere representing 0.010 mol and the container volume at 0.50 L.

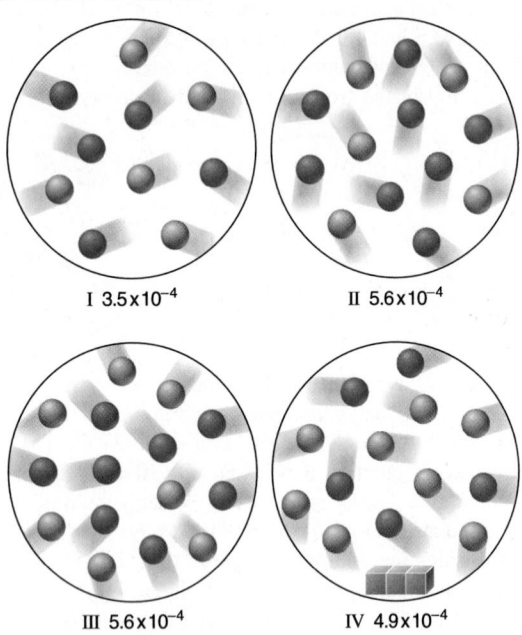

I 3.5x10^{-4} II 5.6x10^{-4}

III 5.6x10^{-4} IV 4.9x10^{-4}

(a) What is the rate law in the absence of a catalyst?
(b) What is the overall reaction order?
(c) Find the rate constant.
(d) Do the gray cubes have a catalytic effect? Explain.

•• **16.124** The mathematics of the first-order rate law can be applied to any situation in which a quantity decreases by a constant fraction per unit of time (or any other variable).

(a) As light moves through a solution, its intensity decreases per unit distance traveled in the solution. Show that

$$\ln\left(\frac{\text{intensity of light leaving the solution}}{\text{intensity of light entering the solution}}\right)$$
$$= -\text{fraction of light removed per unit of length} \times \text{distance traveled in solution}$$

(b) The value of your savings declines under conditions of constant inflation. Show that

$$\ln\left(\frac{\text{value remaining}}{\text{initial value}}\right)$$
$$= -\text{fraction lost per unit of time} \times \text{savings time interval}$$

16.125 Figure 16.24 (p. 722) shows key steps in the metal-catalyzed (M) hydrogenation of ethylene:

$$C_2H_4(g) + H_2(g) \xrightarrow{M} C_2H_6(g)$$

Use the following symbols to write a mechanism that gives the overall equation:
H$_2$(ads)   adsorbed hydrogen molecules
M-H   hydrogen atoms bonded to metal atoms
C$_2$H$_4$(ads)   adsorbed ethylene molecules
C$_2$H$_5$(ads)   adsorbed ethyl radicals

•• **16.126** Acetone is one of the most important solvents in organic chemistry, used to dissolve everything from fats and waxes to airplane glue and nail polish. At high temperatures, it decomposes in a first-order process to methane and ketene (CH$_2$=C=O). At 600°C, the rate constant is 8.7×10^{-3} s^{-1}.
(a) What is the half-life of the reaction?
(b) How long does it take for 40.% of a sample of acetone to decompose?
(c) How long does it take for 90.% of a sample of acetone to decompose?

•• **16.127** A (green), B (blue), and C (red) are structural isomers. The molecular filmstrip depicts them undergoing a chemical change as time proceeds.

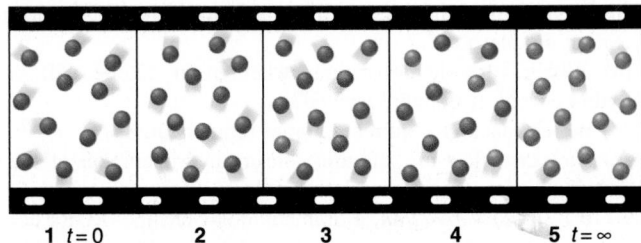

1 $t=0$ 2   3   4   5 $t=\infty$

(a) Write a mechanism for the reaction.
(b) What role does C play?

Juggling Equilibrium Like the opposing steps of a reversible reaction, balls passing back and forth between the hands of a juggler move at equal rates in both directions. This chapter—the first of three to examine dynamic equilibrium, which is the universal condition of chemical change—focuses primarily on equilibrium in gaseous systems.

17

Equilibrium: The Extent of Chemical Reactions

17.1 The Equilibrium State and the Equilibrium Constant

17.2 The Reaction Quotient and the Equilibrium Constant
Writing the Reaction Quotient
Variations in the Form of Q

17.3 Expressing Equilibria with Pressure Terms: Relation Between K_c and K_p

17.4 Reaction Direction: Comparing Q and K

17.5 How to Solve Equilibrium Problems
Using Quantities to Determine K
Using K to Determine Quantities
Determining Reaction Direction

17.6 Reaction Conditions and the Equilibrium State: Le Châtelier's Principle
Change in Concentration
Change in Pressure (Volume)
Change in Temperature
Lack of Effect of a Catalyst

Concepts & Skills to Review
before you study this chapter

- reversibility of reactions (Section 4.7)
- equilibrium vapor pressure (Section 12.2)
- equilibrium nature of a saturated solution (Section 13.4)
- dependence of rate on concentration (Sections 16.2 and 16.6)
- rate laws for elementary reactions (Section 16.7)
- function of a catalyst (Section 16.8)

O ur study of kinetics in Chapter 16 addressed one of three central questions in reaction chemistry, that of reaction rate, which also concerns mechanism. Now we turn to the second question, which we previewed in Chapter 4: how much product forms under a given set of starting concentrations and conditions? The principles of kinetics and equilibrium apply to different aspects of a reaction:

- Kinetics applies to the *speed* (or rate) of a reaction, the concentration of reactant that disappears (or of product that appears) per unit time.
- Equilibrium applies to the *extent* of a reaction, the concentrations of reactant and product present after an unlimited time, or once no further change occurs.

Just as reactions vary greatly in their speed, they also vary in their extent. A fast reaction may go almost completely or barely at all toward products. Consider the dissociation of an acid in water. In 1 *M* HCl, virtually all the hydrogen chloride molecules are dissociated into ions. In contrast, in 1 *M* CH_3COOH, fewer than 1% of the acetic acid molecules are dissociated at any given time. Yet both reactions take less than a second to reach completion. Similarly, some slow reactions eventually yield a large amount of product, whereas others yield very little. After a few years at ordinary temperatures, a steel water-storage tank will rust, and it will do so completely given enough time; but no matter how long you wait, the water inside will not decompose to hydrogen and oxygen.

Knowing the extent of a given reaction is crucial. How much product— medicine, polymer, or fuel—can you obtain from a particular reaction mixture? How can you adjust conditions to obtain more? If a reaction is slow but has a good yield, will a catalyst speed it up enough to make it useful?

IN THIS CHAPTER . . . We consider equilibrium principles in systems of gases and pure liquids and solids; we'll discuss various solution equilibria in Chapters 18 and 19. Throughout these chapters, you'll learn how to solve the major kinds of equilibrium problems. We first examine the equilibrium state at the macroscopic and molecular levels. Then we focus on the relation between the reaction quotient, which changes as the reaction proceeds, and the equilibrium constant, which applies to the system at equilibrium. We express the equilibrium condition in terms of concentrations or pressures. Then we see how to determine whether a system is proceeding toward products or reactants as it approaches equilibrium. We examine how reaction conditions affect the equilibrium state and end with applications of equilibrium in metabolism and in industrial production.

Literature Tyson, Louise; Treagust, David F.; Bucat, Robert B. "The Complexity of Teaching and Learning Chemical Equilibrium," *J. Chem. Educ.* **1999**, *76*, 554.

Literature Dewprashad, Brahmadeo; Kosky, Charles; Vaz, Geraldine S.; Martin, Charlotte L. "Using Clinical Cases to Teach General Chemistry," *J. Chem. Educ.* **2004**, *81*, 1471.

Demo Hanson, Robert M. "Playing-Card Equilibrium," *J. Chem. Educ.* **2003**, *80*, 1271.

Literature Howald, Reed. "The Fizz Keeper, a Case Study in Chemical Education, Equilibrium, and Kinetics," *J. Chem. Educ.* **1999**, *76*, 208.

Demo Pacer, Richard A. "The Coupling of Related Demonstrations to Illustrate Principles in Chemical Kinetics and Equilibrium," *J. Chem. Educ.* **1997**, *74*, 543.

17.1 THE EQUILIBRIUM STATE AND THE EQUILIBRIUM CONSTANT

Countless experiments with chemical systems have shown that, given sufficient time, *the concentrations of reactants and products no longer change*. This apparent cessation of chemical activity occurs because *all reactions are reversible and reach a state of equilibrium*. Let's examine a chemical system at the macroscopic and molecular levels to see how the equilibrium state arises. The system consists of two gases, colorless dinitrogen tetraoxide and brown nitrogen dioxide:

$$N_2O_4(g; \text{colorless}) \rightleftharpoons 2NO_2(g; \text{brown})$$

When we introduce some $N_2O_4(l)$ into a sealed flask kept at 200°C, a change occurs immediately. The liquid vaporizes (bp = 21°C) and the gas begins to turn pale brown. The color darkens, and after a few moments, the color stops changing, as shown in Figure 17.1.

Figure 17.1 Reaching the equilibrium state for the reaction N₂O₄ ⇌ 2NO₂ on the macroscopic and molecular levels.
A, When the experiment begins, the reaction mixture consists mostly of colorless N_2O_4. **B,** As N_2O_4 decomposes to reddish brown NO_2, the color of the mixture becomes pale brown. **C,** When equilibrium is reached, the concentrations of NO_2 and N_2O_4 are constant, and the color reaches its final intensity. **D,** Because the reaction continues in the forward and reverse directions at equal rates, the concentrations (and color) remain constant.

Transparency/ARIS Presentation Center, Active Art Library

On the molecular level, a much more dynamic scene unfolds. The N_2O_4 molecules fly wildly throughout the flask, a few splitting into two NO_2 molecules. As time passes, more N_2O_4 molecules decompose and the concentration of NO_2 rises. As observers in the macroscopic world, we see the flask contents darken, because NO_2 is reddish brown. As the number of N_2O_4 molecules decreases, N_2O_4 decomposition slows. At the same time, increasing numbers of NO_2 molecules collide and combine, so re-formation of N_2O_4 speeds up. Eventually, N_2O_4 molecules decompose into NO_2 molecules as fast as the NO_2 molecules combine into N_2O_4. The system has reached equilibrium: *reactant and product concentrations stop changing because the forward and reverse rates have become equal:*

Literature Leenson, I. A. "Approaching Equilibrium in the N₂O₄-NO₂ System: A Common Mistake in Textbooks," *J. Chem. Educ.* **2000,** *77,* 1652.

$$\text{At equilibrium: rate}_{fwd} = \text{rate}_{rev} \qquad (17.1)$$

Thus, a system at equilibrium continues to be dynamic at the molecular level, but we observe *no further **net** change because changes in one direction are balanced by changes in the other.*

At a particular temperature, when the system reaches equilibrium, product and reactant concentrations are constant. Therefore, their ratio must be a constant. We'll use the N_2O_4-NO_2 system to derive this constant. At equilibrium, we have

$$\text{rate}_{fwd} = \text{rate}_{rev}$$

In this case, both forward and reverse reactions are elementary steps (Section 16.7), so we can write their rate laws directly from the balanced equation:

$$k_{fwd}[N_2O_4]_{eq} = k_{rev}[NO_2]^2_{eq}$$

where k_{fwd} and k_{rev} are the forward and reverse rate constants, respectively, and the subscript "eq" refers to concentrations at equilibrium. By rearranging, we set the ratio of the rate constants equal to the ratio of the concentration terms:

$$\frac{k_{fwd}}{k_{rev}} = \frac{[NO_2]^2_{eq}}{[N_2O_4]_{eq}}$$

The ratio of constants gives rise to a new overall constant called the **equilibrium constant (K):**

$$K = \frac{k_{\text{fwd}}}{k_{\text{rev}}} = \frac{[NO_2]^2_{\text{eq}}}{[N_2O_4]_{\text{eq}}} \qquad (17.2)$$

Demo Shakhashiri, Bassam Z. "Equilibrium between Nitrogen Dioxide and Dinitrogen Tetroxide," *Chemical Demonstrations, A Handbook for Teachers of Chemistry*, Vol. 2 (The University of Wisconsin Press, Madison, 1985), 180–183.

The equilibrium constant K is a number equal to a particular ratio of equilibrium concentrations of product and reactant at a particular temperature. We examine this idea closely in the next section and show that it holds as well for overall reactions made up of several elementary steps.

The magnitude of K is an indication of how far a reaction proceeds toward product at a given temperature. Remember, it is the opposing *rates* that are equal at equilibrium, not necessarily the concentrations. Indeed, different reactions, even at the same temperature, have a wide range of concentrations at equilibrium—from almost all reactant to almost all product—and, therefore, they have a wide range of equilibrium constants (Figure 17.2). Here are three examples of different magnitudes of K:

1. *Small K.* If a reaction yields very little product before reaching equilibrium, it has a small K, and we may even say it looks like there is "no reaction." For example, the oxidation of nitrogen barely proceeds at 1000 K:*

$$N_2(g) + O_2(g) \rightleftharpoons 2NO(g) \qquad K = 1 \times 10^{-30}$$

2. *Large K.* Conversely, if a reaction reaches equilibrium with very little reactant remaining, it has a large K, and we say it "goes to completion." The oxidation of carbon monoxide goes to completion at 1000 K:

$$2CO(g) + O_2(g) \rightleftharpoons 2CO_2(g) \qquad K = 2.2 \times 10^{22}$$

3. *Intermediate K.* When significant amounts of both reactant and product are present at equilibrium, K has an intermediate value, as when bromine monochloride breaks down to its elements at 1000 K:

$$2BrCl(g) \rightleftharpoons Br_2(g) + Cl_2(g) \qquad K = 5$$

---

*To distinguish the equilibrium constant from the Kelvin temperature unit, the equilibrium constant is written with an uppercase italic *K*, whereas the kelvin is an uppercase roman K. Also, since the kelvin is a unit, it always follows a number.

A

B

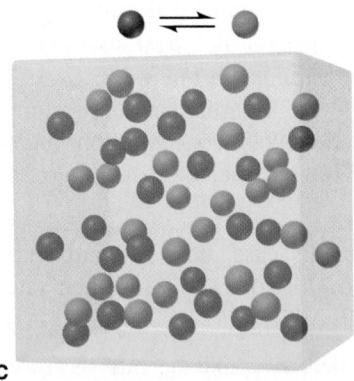

C

Figure 17.2 **The range of equilibrium constants. A,** A system that reaches equilibrium with very little product has a small K. For this reaction, $K = 1/49 = 0.020$. **B,** A system that reaches equilibrium with nearly all product has a large K. For this reaction, $K = 49/1 = 49$.

C, A system that reaches equilibrium with significant concentrations of reactant and product has an intermediate K. For this reaction, $K = 25/25 = 1.0$.

Section Summary

Kinetics and equilibrium are distinct aspects of a reaction system, and the rate and extent of a reaction are not necessarily related. • When the forward and reverse reactions occur at the same rate, the system has reached dynamic equilibrium and concentrations no longer change. • The equilibrium constant (K) is a number based on a particular ratio of product and reactant concentrations: K is small for reactions that reach equilibrium with a high concentration of reactant(s) and large for reactions that reach equilibrium with a low concentration of reactant(s).

17.2 THE REACTION QUOTIENT AND THE EQUILIBRIUM CONSTANT

Our derivation of the equilibrium constant in Section 17.1 was based on kinetics. But the fundamental observation of equilibrium studies was stated many years before the principles of kinetics were developed. In 1864, two Norwegian chemists, Cato Guldberg and Peter Waage, observed that *at a given temperature, a chemical system reaches a state in which a particular ratio of reactant and product concentrations has a constant value.* This is one way of stating the **law of chemical equilibrium,** or the **law of mass action.** Note that there is no mention of rates.

In their formulating of the law of mass action, Guldberg and Waage found that, *for a particular system and temperature, the same equilibrium state is attained regardless of how the reaction is run.* For example, in the N_2O_4-NO_2 system at 200°C, we can start with pure reactant (N_2O_4), pure product (NO_2), or any mixture of NO_2 and N_2O_4, and the ratio of concentrations will attain the same equilibrium value (within experimental error).

The particular ratio of concentration terms that we write for a given reaction is called the **reaction quotient** (Q, also known as the *mass-action expression*). For the breakdown of N_2O_4 to form NO_2, the reaction quotient, which is based directly on the balanced equation, is

$$N_2O_4(g) \longrightarrow 2NO_2(g)$$

$$Q = \frac{[NO_2]^2}{[N_2O_4]}$$

As the reaction proceeds toward the equilibrium state, there is a continual, smooth change in the concentrations of reactants and products. Thus, the ratio of concentrations also changes: at the beginning of the reaction, the concentrations have initial values, and Q has an initial value; a moment later, after the reaction has proceeded a bit, the concentrations have slightly different values, and so does Q; another moment into the reaction, and there is more change in the concentrations and more change in Q; and on and on, *until the reacting system reaches equilibrium.* At that point, at a given temperature, the reactant and product concentrations have reached their equilibrium levels and no longer change. And the value of Q has reached its equilibrium value and no longer changes. It equals K at that temperature:

$$\text{At equilibrium: } Q = K \qquad \text{(17.3)}$$

Therefore, monitoring Q tells whether the system has reached equilibrium, how far away it is if it has not, and, as we discuss later, in which direction it is changing to reach equilibrium. The data from four experiments are presented in

Literature Garritz, Adoni. "The Painting-Sponging Analogy for Chemical Equilibrium," *J. Chem. Educ.* **1997**, *74*, 544.

Literature Wilson, Audrey H. "Equilibrium: A Teaching/Learning Activity," *J. Chem. Educ.* **1998**, *75*, 1176.

Table 17.1 Initial and Equilibrium Concentration Ratios for the N_2O_4-NO_2 System at 200°C (473 K)

	Initial		Ratio (Q)	Equilibrium		Ratio (K)
Expt	$[N_2O_4]$	$[NO_2]$	$[NO_2]^2/[N_2O_4]$	$[N_2O_4]_{eq}$	$[NO_2]_{eq}$	$[NO_2]^2_{eq}/[N_2O_4]_{eq}$
1	0.1000	0.0000	0.0000	0.00357	0.193	10.4
2	0.0000	0.1000	∞	9.24×10^{-4}	9.82×10^{-2}	10.4
3	0.0500	0.0500	0.0500	0.00204	0.146	10.4
4	0.0750	0.0250	0.00833	0.00275	0.170	10.5

ARIS Presentation Center, Tables Library

Figure 17.3 The change in Q during the N_2O_4-NO_2 reaction. The curved plots and the darkening brown screen above them show that $[N_2O_4]$ and $[NO_2]$, and therefore the value of Q, change with time. Before equilibrium is reached, the concentrations are changing continuously, so $Q \neq K$. Once equilibrium is reached (*vertical line*) and at any time thereafter, $Q = K$.

Transparency/ARIS Presentation Center, Art Library

Table 17.1. Each is a different run of the N_2O_4-NO_2 reaction at 200°C. There are two essential points to note:

- The ratio of *initial* concentrations varies widely but always gives the same ratio of *equilibrium* concentrations.
- The *individual* equilibrium concentrations are different in each case, but the *ratio* of these equilibrium concentrations is constant.

The curves in Figure 17.3 show experiment 1 in Table 17.1. Note that $[N_2O_4]$ and $[NO_2]$ change smoothly during the course of the reaction and, thus, so does the value of Q. Once the system reaches equilibrium, as indicated by the constant brown color at the top, the concentrations no longer change and Q equals K. In other words, for any given chemical system, *K is a special value of Q that occurs when the reactant and product terms have their equilibrium values.*

Writing the Reaction Quotient

In Chapter 16, you saw that the rate law for an overall reaction cannot be written from the balanced equation, but must be determined from rate data. In contrast, the reaction quotient *can* be written directly from the balanced equation: *Q is a ratio made up of product concentration terms multiplied together and divided by reactant concentration terms multiplied together, with each term raised to the power of its stoichiometric coefficient in the balanced equation.*

The most common form of the reaction quotient shows reactant and product terms as molar concentrations, which are designated by square brackets, [ ]. In the cases you've seen so far, K is the *equilibrium constant based on concentrations*, designated from now on as K_c. Similarly, we designate the reaction quotient based on concentrations as Q_c. For the general balanced equation

$$aA + bB \rightleftharpoons cC + dD$$

where a, b, c, and d are the stoichiometric coefficients, the reaction quotient is

$$Q_c = \frac{[C]^c[D]^d}{[A]^a[B]^b} \tag{17.4}$$

(Another form of the reaction quotient that we discuss later shows gaseous reactant and product terms as pressures.)

To construct the reaction quotient for any reaction, write the balanced equation first. For the formation of ammonia from its elements, for example, the balanced equation (with colored coefficients for easy reference) is

$$N_2(g) + 3H_2(g) \rightleftharpoons 2NH_3(g)$$

To construct the reaction quotient, we place the product term in the numerator and the reactant terms in the denominator, multiplied by each other, and raise each term to the power of its balancing coefficient (colored as in the equation above):

$$Q_c = \frac{[NH_3]^2}{[N_2][H_2]^3}$$

Let's practice this essential skill.

SAMPLE PROBLEM 17.1 Writing the Reaction Quotient from the Balanced Equation

PROBLEM Write the reaction quotient, Q_c, for each of the following reactions:
(a) The decomposition of dinitrogen pentaoxide, $N_2O_5(g) \rightleftharpoons NO_2(g) + O_2(g)$
(b) The combustion of propane gas, $C_3H_8(g) + O_2(g) \rightleftharpoons CO_2(g) + H_2O(g)$

PLAN We balance the equations and then construct the reaction quotient (Equation 17.4).

SOLUTION (a) $2N_2O_5(g) \rightleftharpoons 4NO_2(g) + O_2(g)$

$$Q_c = \frac{[NO_2]^4[O_2]}{[N_2O_5]^2}$$

(b) $C_3H_8(g) + 5O_2(g) \rightleftharpoons 3CO_2(g) + 4H_2O(g)$

$$Q_c = \frac{[CO_2]^3[H_2O]^4}{[C_3H_8][O_2]^5}$$

CHECK Always be sure that the exponents in Q are the same as the balancing coefficients. A good check is to reverse the process: turn the numerator into products and the denominator into reactants, and change the exponents to coefficients.

FOLLOW-UP PROBLEM 17.1 Write the reaction quotient, Q_c, for each of the following reactions (unbalanced):
(a) The first step in nitric acid production, $NH_3(g) + O_2(g) \rightleftharpoons NO(g) + H_2O(g)$
(b) The disproportionation of nitric oxide, $NO(g) \rightleftharpoons N_2O(g) + NO_2(g)$

Variations in the Form of the Reaction Quotient

As you'll see in the upcoming discussion, the reaction quotient Q is a collection of terms based on the balanced equation *exactly as written* for a given reaction. Therefore, the value of Q, which varies during the reaction, and the value of K, which is the constant value that Q attains when the system has reached equilibrium, also depend on how the balanced equation is written.

A Word about Units for *Q* and *K* In this text (and most others), *the values of Q and K are shown as unitless numbers.* This is because each term in the reaction quotient represents the *ratio* of the measured quantity of the substance (molar concentration or pressure) to the thermodynamic standard-state quantity of the substance. Recall from Section 6.6 that these standard states are 1 M for a substance in solution, 1 atm for gases, and the pure substance for a liquid or solid. Thus, a concentration of 1.20 M becomes $\frac{1.20\ M}{1\ M} = 1.20$; similarly, a pressure of 0.53 atm becomes $\frac{0.53\ \text{atm}}{1\ \text{atm}} = 0.53$. With these quantity terms unitless, the ratio of terms we use to find the value of Q (or K) is also unitless.

Form of *Q* for an Overall Reaction Notice that we've been writing reaction quotients without knowing whether an equation represents an individual reaction step or an overall multistep reaction. We can do this because we obtain the same expression for the overall reaction as we do when we combine the expressions for the individual steps. That is, *if an overall reaction is the **sum** of two or more reactions, the overall reaction quotient (or equilibrium constant) is the **product** of the reaction quotients (or equilibrium constants) for the steps:*

$$Q_{\text{overall}} = Q_1 \times Q_2 \times Q_3 \times \cdots$$

and

$$K_{\text{overall}} = K_1 \times K_2 \times K_3 \times \cdots \qquad (17.5)$$

Sample Problem 17.2 demonstrates this point.

ARIS Presentation Center, Worked Examples Library

Green chemistry application

SAMPLE PROBLEM 17.2 Writing the Reaction Quotient and Finding K for an Overall Reaction

PROBLEM Understanding reactions involving normal components of air is essential for solving problems dealing with atmospheric pollution. Here is a reaction sequence involving N_2 and O_2, the most abundant gases in air. Nitrogen dioxide is a toxic pollutant that contributes to photochemical smog.

(1) $N_2(g) + O_2(g) \rightleftharpoons 2NO(g)$ $K_{c1} = 4.3 \times 10^{-25}$
(2) $2NO(g) + O_2(g) \rightleftharpoons 2NO_2(g)$ $K_{c2} = 6.4 \times 10^9$

(a) Show that the overall Q_c for this reaction sequence is the same as the product of the Q_c's for the individual reactions.
(b) Given that the K_c's occur at the same temperature, find K_c for the overall reaction.
PLAN (a) We first write the overall reaction by adding the individual reactions and then write the overall Q_c. Next, we write the Q_c for each reaction. Because we *add* the individual steps, we *multiply* their Q_c's and cancel common terms to obtain the overall Q_c.
(b) We are given the individual K_c values (4.3×10^{-25} and 6.4×10^9), so we multiply them to find $K_{c(overall)}$.
SOLUTION (a) Writing the overall reaction and its reaction quotient:

$$
\begin{array}{ll}
(1) & N_2(g) + O_2(g) \rightleftharpoons 2NO(g) \\
(2) & 2NO(g) + O_2(g) \rightleftharpoons 2NO_2(g) \\
\hline
\text{Overall:} & N_2(g) + 2O_2(g) \rightleftharpoons 2NO_2(g)
\end{array}
$$

$$Q_{c(overall)} = \frac{[NO_2]^2}{[N_2][O_2]^2}$$

Writing the reaction quotients for the individual steps:

For step 1, $Q_{c1} = \dfrac{[NO]^2}{[N_2][O_2]}$

For step 2, $Q_{c2} = \dfrac{[NO_2]^2}{[NO]^2[O_2]}$

Multiplying the individual reaction quotients and canceling:

$$Q_{c1} \times Q_{c2} = \frac{[NO]^2}{[N_2][O_2]} \times \frac{[NO_2]^2}{[NO]^2[O_2]} = \frac{[NO_2]^2}{[N_2][O_2]^2} = Q_{c(overall)}$$

(b) Calculating the overall K_c:

$$K_{c(overall)} = K_{c1} \times K_{c2} = (4.3 \times 10^{-25})(6.4 \times 10^9) = 2.8 \times 10^{-15}$$

CHECK Round off and check the calculation in part (b):

$$K_c \approx (4 \times 10^{-25})(6 \times 10^9) = 24 \times 10^{-16} = 2.4 \times 10^{-15}$$

FOLLOW-UP PROBLEM 17.2 The following sequence of individual steps has been proposed for the overall reaction between H_2 and Br_2 to form HBr:

(1) $Br_2(g) \rightleftharpoons 2Br(g)$
(2) $Br(g) + H_2(g) \rightleftharpoons HBr(g) + H(g)$
(3) $H(g) + Br(g) \rightleftharpoons HBr(g)$

Write the overall balanced equation and show that the overall Q_c is the product of the Q_c's for the individual steps.

Form of Q for a Forward and a Reverse Reaction
The form of the reaction quotient depends on the *direction* in which the balanced equation is written. Consider, for example, the oxidation of sulfur dioxide to sulfur trioxide. This reaction is a key step in acid rain formation and sulfuric acid production. The balanced equation is

$$2SO_2(g) + O_2(g) \rightleftharpoons 2SO_3(g)$$

The reaction quotient for this equation *as written* is

$$Q_{c(fwd)} = \frac{[SO_3]^2}{[SO_2]^2[O_2]}$$

If we had written the reverse reaction, the decomposition of sulfur trioxide,

$$2SO_3(g) \rightleftharpoons 2SO_2(g) + O_2(g)$$

the reaction quotient would be the *reciprocal* of $Q_{c(fwd)}$:

$$Q_{c(rev)} = \frac{[SO_2]^2[O_2]}{[SO_3]^2} = \frac{1}{Q_{c(fwd)}}$$

Thus, *a reaction quotient (or equilibrium constant) for a forward reaction is the* ***reciprocal*** *of the reaction quotient (or equilibrium constant) for the reverse reaction:*

$$Q_{c(fwd)} = \frac{1}{Q_{c(rev)}} \quad \text{and} \quad K_{c(fwd)} = \frac{1}{K_{c(rev)}} \tag{17.6}$$

The K_c values for the forward and reverse reactions at 1000 K are

$$K_{c(fwd)} = 261 \quad \text{and} \quad K_{c(rev)} = \frac{1}{K_{c(fwd)}} = \frac{1}{261} = 3.83 \times 10^{-3}$$

These values make sense: if the forward reaction goes far to the right (high K_c), the reverse reaction does not (low K_c).

Form of Q for a Reaction with Coefficients Multiplied by a Common Factor

Multiplying all the coefficients of the equation by some factor also changes the form of Q. For example, multiplying all the coefficients in the equation for the formation of SO_3 (bottom of p. 744) by $\frac{1}{2}$ gives

$$SO_2(g) + \tfrac{1}{2}O_2(g) \rightleftharpoons SO_3(g)$$

For this equation, the reaction quotient is

$$Q'_{c(fwd)} = \frac{[SO_3]}{[SO_2][O_2]^{1/2}}$$

Notice that Q_c for the halved equation equals Q_c for the original equation raised to the $\frac{1}{2}$ power:

$$Q'_{c(fwd)} = Q_{c(fwd)}^{1/2} = \left(\frac{[SO_3]^2}{[SO_2]^2[O_2]}\right)^{1/2} = \frac{[SO_3]}{[SO_2][O_2]^{1/2}}$$

Once again, the same property holds for the equilibrium constants. Relating the halved reaction to the original, we have

$$K'_{c(fwd)} = K_{c(fwd)}^{1/2} = (261)^{1/2} = 16.2$$

Similarly, if you double coefficients, the reaction quotient is the original expression squared; if you triple coefficients, it is the original expression cubed; and so on. It may seem that we have changed the extent of the reaction, as indicated by a change in K, merely by changing the balancing coefficients of the equation, but this clearly cannot be true. *A particular K has meaning only in relation to a particular balanced equation.* In this case, $K_{c(fwd)}$ and $K'_{c(fwd)}$ relate to different equations and thus cannot be compared directly.

In general, *if all the coefficients of the balanced equation are multiplied by some factor, that factor becomes the exponent for relating the reaction quotients and the equilibrium constants.* For a multiplying factor n, which we can write as

$$n(aA + bB \rightleftharpoons cC + dD)$$

the reaction quotient and equilibrium constant are

$$Q' = Q^n = \left(\frac{[C]^c[D]^d}{[A]^a[B]^b}\right)^n \quad \text{and} \quad K' = K^n \tag{17.7}$$

SAMPLE PROBLEM 17.3 Finding the Equilibrium Constant for an Equation Multiplied by a Common Factor

PROBLEM For the ammonia-formation reaction,

$$N_2(g) + 3H_2(g) \rightleftharpoons 2NH_3(g)$$

the equilibrium constant, K_c, is 2.4×10^{-3} at 1000 K. If we change the coefficients of this equation, which we'll call the reference (ref) equation, what are the values of K_c for the following balanced equations?

(a) $\frac{1}{3}N_2(g) + H_2(g) \rightleftharpoons \frac{2}{3}NH_3(g)$ **(b)** $NH_3(g) \rightleftharpoons \frac{1}{2}N_2(g) + \frac{3}{2}H_2(g)$

PLAN We compare each equation with the reference equation to see how the direction and coefficients have changed. **(a)** The equation is the reference equation multiplied by $\frac{1}{3}$, so K_c equals $K_{c(ref)}$ (2.4×10^{-3}) raised to the $\frac{1}{3}$ power. **(b)** The equation is one-half the *reverse* of the reference equation, so K_c is the reciprocal of $K_{c(ref)}$ raised to the $\frac{1}{2}$ power.

SOLUTION The reaction quotient for the reference equation is $Q_{c(ref)} = \dfrac{[NH_3]^2}{[N_2][H_2]^3}$.

(a)
$$Q_c = Q_{c(ref)}^{1/3} = \left(\frac{[NH_3]^2}{[N_2][H_2]^3}\right)^{1/3} = \frac{[NH_3]^{2/3}}{[N_2]^{1/3}[H_2]}$$

Thus,
$$K_c = K_{c(ref)}^{1/3} = (2.4 \times 10^{-3})^{1/3} = 0.13$$

(b)
$$Q_c = \left(\frac{1}{Q_{c(ref)}}\right)^{1/2} = \left(\frac{1}{\dfrac{[NH_3]^2}{[N_2][H_2]^3}}\right)^{1/2} = \frac{[N_2]^{1/2}[H_2]^{3/2}}{[NH_3]}$$

Thus,
$$K_c = \left(\frac{1}{K_{c(ref)}}\right)^{1/2} = \left(\frac{1}{2.4 \times 10^{-3}}\right)^{1/2} = 20.$$

CHECK A good check is to work the math backward. **(a)** $(0.13)^3 = 2.2 \times 10^{-3}$, within rounding of 2.4×10^{-3}. The reaction goes in the same direction, so at equilibrium, there should be mostly reactants, as $K_c < 1$ indicates. **(b)** $1/(20.)^2 = 2.5 \times 10^{-3}$, again within rounding. At equilibrium, the reverse reaction should yield mostly products, as $K_c > 1$ indicates.

FOLLOW-UP PROBLEM 17.3 At 1200 K, the reaction of hydrogen and chlorine to form hydrogen chloride is

$$H_2(g) + Cl_2(g) \rightleftharpoons 2HCl(g) \qquad K_c = 7.6 \times 10^8$$

Calculate K_c for the following reactions:

(a) $\frac{1}{2}H_2(g) + \frac{1}{2}Cl_2(g) \rightleftharpoons HCl(g)$ **(b)** $\frac{4}{3}HCl(g) \rightleftharpoons \frac{2}{3}H_2(g) + \frac{2}{3}Cl_2(g)$

Literature Paiva, João C. M.; Gil, Victor M. S. "The State of Division of Solids and Chemical Equilibria," *J. Chem. Educ.* **2001**, *78*, 222.

Form of Q for a Reaction Involving Pure Liquids and Solids Until now, we've looked at *homogeneous* equilibria, systems in which all the components of the reaction are in the same phase, such as a system of reacting gases. When the components are in different phases, the system reaches *heterogeneous* equilibrium.

Consider the decomposition of limestone to lime and carbon dioxide, in which a gas and two solids make up the reaction components:

$$CaCO_3(s) \rightleftharpoons CaO(s) + CO_2(g)$$

Based on the rules for writing the reaction quotient, we have

$$Q_c = \frac{[CaO][CO_2]}{[CaCO_3]}$$

A pure solid, however, such as $CaCO_3$ or CaO, always has the same "concentration" at a given temperature, that is, the same number of moles per liter of solid, just as it has the same density (g/cm^3). Therefore, the concentration of a pure solid is constant, as is the concentration of a pure liquid (but not that of a dissolved solute).

Figure 17.4 The reaction quotient for a heterogeneous system. Even though the two containers have different amounts of the two solids CaO and CaCO$_3$, as long as both solids are present, at a given temperature, the containers have the same [CO$_2$] at equilibrium.

Because we are concerned only with concentrations that *change* as they approach equilibrium, *we eliminate the terms for pure liquids and solids from the reaction quotient.* We do this by multiplying both sides of the above equation by [CaCO$_3$] and dividing both sides by [CaO] and get a new reaction quotient, Q'_c. Thus, the only substance whose concentration can change is the gas CO$_2$:

$$Q'_c = Q_c \frac{[\text{CaCO}_3]}{[\text{CaO}]} = [\text{CO}_2]$$

No matter how much CaO and CaCO$_3$ are in the reaction vessel, *as long as some of each is present,* the reaction quotient for the reaction equals the CO$_2$ concentration (Figure 17.4).

Table 17.2 summarizes the ways of writing Q and calculating K.

Table 17.2 Ways of Expressing Q and Calculating K

Form of Chemical Equation	Form of Q	Value of K
Reference reaction: A $\rightleftharpoons$ B	$Q_{(\text{ref})} = \dfrac{[\text{B}]}{[\text{A}]}$	$K_{(\text{ref})} = \dfrac{[\text{B}]_{\text{eq}}}{[\text{A}]_{\text{eq}}}$
Reverse reaction: B $\rightleftharpoons$ A	$Q = \dfrac{1}{Q_{(\text{ref})}} = \dfrac{[\text{A}]}{[\text{B}]}$	$K = \dfrac{1}{K_{(\text{ref})}}$
Reaction as sum of two steps: (1) A $\rightleftharpoons$ C (2) C $\rightleftharpoons$ B	$Q_1 = \dfrac{[\text{C}]}{[\text{A}]}; Q_2 = \dfrac{[\text{B}]}{[\text{C}]}$ $Q_{\text{overall}} = Q_1 \times Q_2 = Q_{(\text{ref})}$ $= \dfrac{[\cancel{\text{C}}]}{[\text{A}]} \times \dfrac{[\text{B}]}{[\cancel{\text{C}}]} = \dfrac{[\text{B}]}{[\text{A}]}$	$K_{\text{overall}} = K_1 \times K_2$ $= K_{(\text{ref})}$
Coefficients multiplied by n	$Q = Q^n_{(\text{ref})}$	$K = K^n_{(\text{ref})}$
Reaction with pure solid or liquid component, such as A(s)	$Q = Q_{(\text{ref})}[\text{A}] = [\text{B}]$	$K = K_{(\text{ref})}[\text{A}] = [\text{B}]$

Section Summary

The reaction quotient, Q, is a particular ratio of product to reactant terms. Substituting experimental values into this expression gives the value of Q, which changes as the reaction proceeds. When the system reaches equilibrium at a particular temperature, $Q = K$. • If a reaction is the sum of two or more steps, the overall Q (or K) is the product of the individual Q's (or K's). • The *form* of Q is based directly on the balanced equation for the reaction exactly as written, so it changes if the equation is reversed or multiplied by some factor, and K changes accordingly. • Pure liquids or solids do not appear in the expression for Q because their concentrations are constant.

17.3 EXPRESSING EQUILIBRIA WITH PRESSURE TERMS: RELATION BETWEEN K_c and K_p

It is easier to measure the pressure of a gas than its concentration and, as long as the gas behaves nearly ideally under the conditions of the experiment, the ideal gas law (Section 5.3) allows us to relate these variables to each other:

$$PV = nRT, \quad \text{so} \quad P = \frac{n}{V}RT \quad \text{or} \quad \frac{P}{RT} = \frac{n}{V}$$

where P is the pressure of a gas and n/V is its molar concentration (M). Thus, with R a constant and if T is kept constant, *pressure is directly proportional to molar concentration.* When the substances in the reaction are gases, we can express the reaction quotient and calculate its value in terms of partial pressures instead of concentrations. For example, in the reaction between gaseous NO and O_2,

$$2NO(g) + O_2(g) \rightleftharpoons 2NO_2(g)$$

the reaction quotient based on partial pressures, Q_p, is

$$Q_p = \frac{P_{NO_2}^2}{P_{NO}^2 \times P_{O_2}}$$

(In later chapters, you'll see cases where some reaction components are expressed as concentrations and others as partial pressures.) The equilibrium constant obtained when all components are present at their equilibrium partial pressures is designated K_p, the *equilibrium constant based on pressures.* In many cases, K_p has a value different from K_c, but if you know one, you can calculate the other by noting the *change in amount (mol) of gas*, Δn_{gas}, from the balanced equation. Let's see this relationship by converting the terms in Q_c for the reaction of NO and O_2 to those in Q_p:

$$2NO(g) + O_2(g) \rightleftharpoons 2NO_2(g)$$

As the balanced equation shows,

$$3 \text{ mol } (2 \text{ mol} + 1 \text{ mol}) \text{ gaseous reactants} \rightleftharpoons 2 \text{ mol gaseous products}$$

With Δ meaning final *minus* initial (products *minus* reactants), we have

$$\Delta n_{gas} = \text{moles of gaseous product} - \text{moles of gaseous reactant} = 2 - 3 = -1$$

Keep this value of Δn_{gas} in mind because it appears in the algebraic conversion that follows. The reaction quotient based on concentrations is

$$Q_c = \frac{[NO_2]^2}{[NO]^2[O_2]}$$

Rearranging the ideal gas law to $n/V = P/RT$, we express concentrations as n/V and convert them to partial pressures, P; then we collect the RT terms and cancel:

$$Q_c = \frac{\dfrac{n_{NO_2}^2}{V^2}}{\dfrac{n_{NO}^2}{V^2} \times \dfrac{n_{O_2}}{V}} = \frac{\dfrac{P_{NO_2}^2}{(RT)^2}}{\dfrac{P_{NO}^2}{(RT)^2} \times \dfrac{P_{O_2}}{RT}} = \frac{P_{NO_2}^2}{P_{NO}^2 \times P_{O_2}} \times \frac{\dfrac{1}{(RT)^2}}{\dfrac{1}{(RT)^2} \times \dfrac{1}{RT}} = \frac{P_{NO_2}^2}{P_{NO}^2 \times P_{O_2}} \times RT$$

The far right side of the previous expression is Q_p multiplied by RT: $Q_c = Q_p(RT)$. Also, at equilibrium, $K_c = K_p(RT)$; thus, $K_p = \dfrac{K_c}{RT}$, or $K_c(RT)^{-1}$.

Notice that *the exponent of the RT term equals the change in the amount (mol) of gas (Δn_{gas}) from the balanced equation,* -1. Thus, in general,

$$K_p = K_c(RT)^{\Delta n_{gas}} \tag{17.8}$$

The units generally used for the partial pressure terms in K_p are atmospheres, pascals, or torr, raised to some power, and the units of R must be consistent with those units. As Equation 17.8 shows, for those reactions in which the amount (mol) of gas does not change, we have $\Delta n_{gas} = 0$, so the RT term drops out and $K_p = K_c$.

ARIS Presentation Center, Worked Examples Library

Engineering application

SAMPLE PROBLEM 17.4 Converting Between K_c and K_p

PROBLEM A chemical engineer injects limestone ($CaCO_3$) into the hot flue gas of a coal-burning power plant to form lime (CaO), which scrubs SO_2 from the gas and forms gypsum ($CaSO_4 \cdot 2H_2O$). Find K_c for the following reaction, if CO_2 pressure is in atmospheres:

$$CaCO_3(s) \rightleftharpoons CaO(s) + CO_2(g) \qquad K_p = 2.1 \times 10^{-4} \text{ (at 1000. K)}$$

PLAN We know K_p (2.1×10^{-4}), so to convert between K_p and K_c, we must first determine Δn_{gas} from the balanced equation. Then we rearrange Equation 17.8. With gas pressure in atmospheres, R is 0.0821 atm·L/mol·K.

SOLUTION Determining Δn_{gas}: There is 1 mol of gaseous product and no gaseous reactant, so $\Delta n_{gas} = 1 - 0 = 1$.

Rearranging Equation 17.8 and calculating K_c:

$$K_p = K_c(RT)^1$$

so
$$K_c = K_p(RT)^{-1}$$
$$= (2.1 \times 10^{-4})(0.0821 \times 1000.)^{-1}$$
$$= 2.6 \times 10^{-6}$$

CHECK Work backward to see whether you obtain the given K_p:

$$K_p = (2.6 \times 10^{-6})(0.0821 \times 1000.)$$
$$= 2.1 \times 10^{-4}$$

FOLLOW-UP PROBLEM 17.4 Calculate K_p for the following reaction:

$$PCl_3(g) + Cl_2(g) \rightleftharpoons PCl_5(g) \qquad K_c = 1.67 \text{ (at 500. K)}$$

Section Summary

The reaction quotient and the equilibrium constant are most often expressed in terms of concentrations (Q_c and K_c). For gases, they can also be expressed in terms of partial pressures (Q_p and K_p). • The values of K_p and K_c are related by a derivation that relies on the ideal gas law: $K_p = K_c(RT)^{\Delta n_{gas}}$.

17.4 REACTION DIRECTION: COMPARING Q and K

Suppose you start a reaction with a mixture of reactants and products and you know the equilibrium constant at the temperature of the reaction. How do you know if the reaction has reached equilibrium? By comparing the value of Q at a particular time with the known K, you can tell whether the reaction has attained equilibrium or, if not, in which direction it is progressing. With product terms in the numerator of Q and reactant terms in the denominator, *more product makes Q larger, and more reactant makes Q smaller.*

Reactants ──→ Products   Equilibrium: no net change   Reactants ←── Products

Figure 17.5 Reaction direction and the relative sizes of *Q* and *K*. When Q_c is smaller than K_c, the equilibrium of the reaction system shifts to the right, that is, toward products. When Q_c is larger than K_c, the equilibrium of the reaction system shifts to the left. Both shifts continue until $Q_c = K_c$. Note that the size of K_c remains the same throughout.

The three possible relative sizes of Q and K are shown in Figure 17.5.

- $Q < K$. If the value of Q is smaller than K, the denominator (reactants) is large relative to the numerator (products). For Q to become equal to K, the reactants must decrease and the products increase. In other words, the reaction will progress to the right, toward products, until equilibrium is reached:

$$\text{If } Q < K, \text{ reactants} \longrightarrow \text{products}$$

- $Q > K$. If Q is larger than K, the numerator (products) will decrease and the denominator (reactants) increase until equilibrium is reached. Therefore, the reaction will progress to the left, toward reactants:

$$\text{If } Q > K, \text{ reactants} \longleftarrow \text{products}$$

- $Q = K$. This situation exists only when the reactant and product concentrations (or pressures) have attained their equilibrium values. Thus, despite the processes continuing at the molecular level, no further net change appears:

$$\text{If } Q = K, \text{ reactants} \rightleftharpoons \text{products}$$

The next two sample problems focus on this idea: the first relies on molecular scenes and the second on concentration data to determine reaction direction.

SAMPLE PROBLEM 17.5 Using Molecular Scenes to Determine Reaction Direction

PROBLEM During one run of the reaction $A(g) \rightleftharpoons B(g)$, the equilibrium mixture at 175°C had the following composition: $[A] = 2.8\times10^{-4}\ M$ and $[B] = 1.2\times10^{-4}\ M$. The molecular scenes below represent mixtures at various times during runs 1–4 of the reaction at 175°C (A is red; B is blue). In which direction does the reaction shift for each mixture to reach equilibrium?

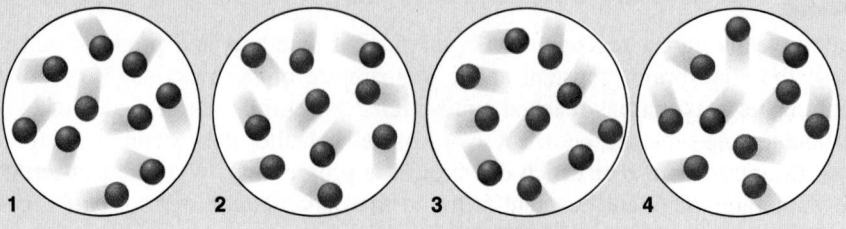

PLAN We must compare Q_c with K_c to determine the reaction direction, so we first use the given equilibrium concentrations to find K_c. Then we count spheres and calculate Q_c for each mixture. If $Q_c < K_c$, the reaction shifts to the right (reactants to products); if $Q_c > K_c$, the reaction shifts to the left (products to reactants); and if $Q_c = K_c$, there is no change.

SOLUTION Writing the reaction quotient and using the data to find K_c:

$$Q_c = \frac{[B]}{[A]} = \frac{1.2 \times 10^{-4}}{2.8 \times 10^{-4}} = 0.43 = K_c$$

Counting red (A) and blue (B) spheres to calculate Q_c for each mixture:

1. $Q_c = 8/2 = 4.0$ 2. $Q_c = 3/7 = 0.43$ 3. $Q_c = 4/6 = 0.67$ 4. $Q_c = 2/8 = 0.25$

Comparing Q_c with K_c to determine reaction direction:

1. $Q_c > K_c$: left   2. $Q_c = K_c$: no change   3. $Q_c > K_c$: left   4. $Q_c < K_c$: right

CHECK Making an error in the calculation for K_c would lead to incorrect conclusions throughout, so check that step: the exponents are the same, and 1.2/2.8 is a bit less than 0.5, as is the calculated K_c. You can check the final answers by inspection; for example, for the number of B (8) in mixture 1 to equal the number at equilibrium (3), more B must change to A, so the reaction must shift to the left.

FOLLOW-UP PROBLEM 17.5 At 338 K, the reaction $X(g) \rightleftharpoons Y(g)$ has a K_c of 1.4. The scenes below represent different mixtures at 338 K, with X orange and Y green. In which direction does the reaction shift (if at all) for each mixture to reach equilibrium?

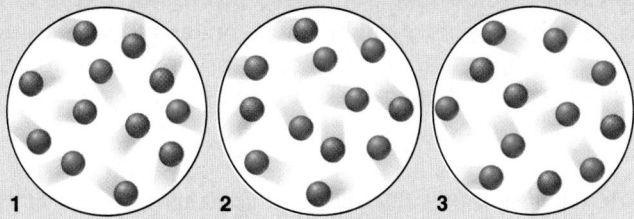

SAMPLE PROBLEM 17.6 Comparing Q and K to Determine Reaction Direction

ARIS Presentation Center, Worked Examples Library

PROBLEM For the reaction $N_2O_4(g) \rightleftharpoons 2NO_2(g)$, $K_c = 0.21$ at 100°C. At a point during the reaction, $[N_2O_4] = 0.12$ M and $[NO_2] = 0.55$ M. Is the reaction at equilibrium? If not, in which direction is it progressing?

PLAN We write the expression for Q_c, find its value by substituting the given concentrations, and then compare its value with the given K_c.

SOLUTION Writing the reaction quotient and solving for Q_c:

$$Q_c = \frac{[NO_2]^2}{[N_2O_4]} = \frac{0.55^2}{0.12} = 2.5$$

With $Q_c > K_c$, the reaction is not at equilibrium and will proceed to the left until $Q_c = K_c$.

CHECK With $[NO_2] > [N_2O_4]$, we expect to obtain a value for Q_c that is greater than 0.21. If $Q_c > K_c$, the numerator will decrease and the denominator will increase until $Q_c = K_c$; that is, this reaction will proceed toward reactants.

FOLLOW-UP PROBLEM 17.6 Chloromethane forms by the reaction

$$CH_4(g) + Cl_2(g) \rightleftharpoons CH_3Cl(g) + HCl(g)$$

At 1500 K, $K_p = 1.6 \times 10^4$. In the reaction mixture, $P_{CH_4} = 0.13$ atm, $P_{Cl_2} = 0.035$ atm, $P_{CH_3Cl} = 0.24$ atm, and $P_{HCl} = 0.47$ atm. Is CH_3Cl or CH_4 forming?

Section Summary

We compare the values of Q and K to determine the direction in which a reaction will proceed toward equilibrium.

• If $Q_c < K_c$, more product forms.
• If $Q_c > K_c$, more reactant forms.
• If $Q_c = K_c$, there is no net change.

Literature Silverstein, Todd P. "Graphing Calculator Strategies for Solving Chemical Equilibrium Problems," *J. Chem. Educ.* **2000,** *77,* 1120.

17.5 HOW TO SOLVE EQUILIBRIUM PROBLEMS

As you've seen in previous sections, three criteria define a system at equilibrium:

- Reactant and product concentrations are constant over time.
- The forward reaction rate equals the reverse reaction rate.
- The reaction quotient equals the equilibrium constant: $Q = K$.

Now, let's apply equilibrium principles quantitatively. Many kinds of equilibrium problems arise in the real world—as well as on chemistry exams—but we can group most of them into two types:

1. We are given equilibrium quantities (concentrations or partial pressures) and solve for K.
2. We are given K and initial quantities and solve for the equilibrium quantities.

Using Quantities to Determine the Equilibrium Constant

There are two common variations on the type of equilibrium problem in which we solve for K: one involves a straightforward substitution of quantities, and the other requires first finding some of the quantities.

Substituting Given Equilibrium Quantities into Q to Find K In the straightforward case, we are given the equilibrium quantities and we must calculate K.

Suppose, for example, that equal amounts of gaseous hydrogen and iodine are injected into a 1.50-L reaction flask at a fixed temperature. In time, the following equilibrium is attained:

$$H_2(g) + I_2(g) \rightleftharpoons 2HI(g)$$

At equilibrium, analysis shows that the flask contains 1.80 mol of H_2, 1.80 mol of I_2, and 0.520 mol of HI. We calculate K_c by finding the concentrations and substituting them into the reaction quotient. Given the balanced equation, we then write the reaction quotient:

$$Q_c = \frac{[HI]^2}{[H_2][I_2]}$$

We first have to convert the amounts (mol) to concentrations (mol/L), using the flask volume of 1.50 L:

$$[H_2] = \frac{1.80 \text{ mol}}{1.50 \text{ L}} = 1.20 \ M$$

Similarly, $[I_2] = 1.20 \ M$, and $[HI] = 0.347 \ M$. Substituting these values into the expression for Q_c gives K_c:

$$K_c = \frac{(0.347)^2}{(1.20)(1.20)} = 8.36 \times 10^{-2}$$

Using a Reaction Table to Determine Equilibrium Quantities and Find K When some quantities are not given, we determine them first from the reaction stoichiometry and then find K. In the following example, pay close attention to a valuable tool being applied: *the reaction table.*

In a study of carbon oxidation, an evacuated vessel containing a small amount of powdered graphite is heated to 1080 K, and then CO_2 is added to a pressure of 0.458 atm. Once the CO_2 is added, the system starts to produce CO. After equilibrium has been reached, the total pressure inside the vessel is 0.757 atm. Calculate K_p.

As always, we start to solve the problem by writing the balanced equation and the reaction quotient:

$$CO_2(g) + C(\text{graphite}) \rightleftharpoons 2CO(g)$$

The data are given in atmospheres and we must find K_p, so we write Q in terms of partial pressures (note the absence of a term for the solid, C):

$$Q_p = \frac{P_{CO}^2}{P_{CO_2}}$$

We are given the initial P_{CO_2} and P_{total} at equilibrium. To find K_p, we must find the equilibrium pressures of CO_2 and CO, which requires solving a stoichiometry problem, and then substitute them into the expression for Q_p.

Let's think through what happened in the vessel. An unknown portion of the CO_2 reacted with graphite to form an unknown amount of CO. We already know the *relative* amounts of CO_2 and CO from the balanced equation: for each mole of CO_2 that reacts, 2 mol of CO forms, which means that when x atm of CO_2 reacts, $2x$ atm of CO forms:

$$x \text{ atm } CO_2 \longrightarrow 2x \text{ atm } CO$$

From the data given, we can predict something about the size of K_p. If it were very large, almost all the CO_2 would be converted to CO, so the final (total) pressure would be twice the initial pressure $[2(0.458 \text{ atm}) \approx 0.9 \text{ atm}]$. On the other hand, if K_p were very small, almost no CO would form, so the final pressure would be close to the initial pressure, 0.458 atm. However, the final pressure (0.757 atm) is between these extremes, so K_p must have an intermediate value. The pressure of CO_2 at equilibrium, $P_{CO_2(eq)}$, is the initial pressure, $P_{CO_2(init)}$, *minus* the change in CO_2 pressure, x, that is, the CO_2 that reacts:

$$P_{CO_2(init)} - x = P_{CO_2(eq)}$$

Similarly, the pressure of CO at equilibrium, $P_{CO(eq)}$, is the initial pressure, $P_{CO(init)}$, *plus* the change in CO pressure, which is the CO that forms, $2x$. Because $P_{CO(init)}$ is zero at the beginning of the reaction, we have

$$P_{CO(init)} + 2x = 0 + 2x = 2x = P_{CO(eq)}$$

To summarize this information, we'll use a reaction table similar to those in Chapter 3. It shows the balanced equation and what we know about

- the *initial* quantities (concentrations or pressures) of reactants and products
- the *changes* in these quantities during the reaction
- the *equilibrium* quantities

Pressure (atm)	$CO_2(g)$	$+$	C(graphite)	$\rightleftharpoons$	$2CO(g)$
Initial	0.458		—		0
Change	$-x$		—		$+2x$
Equilibrium	$0.458 - x$		—		$2x$

As in an arithmetic calculation, in each column, we *add the initial value to the change to obtain the "sum," the equilibrium value*. Note that we include data *only for those substances whose concentrations change;* thus, in this case, the column for C(graphite) is blank. Reaction tables appear in many of the equilibrium problems in this and later chapters.

To solve for K_p, we must substitute equilibrium values into the reaction quotient, so we first have to find x. To do this, we use the other piece of data given, P_{total}. According to Dalton's law of partial pressures and using the quantities from the bottom (equilibrium) row of the reaction table,

$$P_{total} = 0.757 \text{ atm} = P_{CO_2(eq)} + P_{CO} = (0.458 \text{ atm} - x) + 2x$$

Thus, $0.757 \text{ atm} = 0.458 \text{ atm} + x$ and $x = 0.299 \text{ atm}$

With x known, we determine the equilibrium partial pressures:

$$P_{CO_2(eq)} = 0.458 \text{ atm} - x = 0.458 \text{ atm} - 0.299 \text{ atm} = 0.159 \text{ atm}$$

$$P_{CO(eq)} = 2x = 2(0.299 \text{ atm}) = 0.598 \text{ atm}$$

Now, we substitute these values into the expression for Q_p to find K_p:

$$Q_p = \frac{P^2_{CO(eq)}}{P_{CO_2(eq)}} = \frac{0.598^2}{0.159} = 2.25 = K_p$$

As we predicted, K_p is neither very large nor very small.

(From now on, the subscripts "init" and "eq" appear only when it is not clear whether a concentration or pressure is an initial or equilibrium value.)

SAMPLE PROBLEM 17.7 Calculating K_c from Concentration Data

PROBLEM In order to study hydrogen halide decomposition, a researcher fills an evacuated 2.00-L flask with 0.200 mol of HI gas and allows the reaction to proceed at 453°C:

$$2HI(g) \rightleftharpoons H_2(g) + I_2(g)$$

At equilibrium, [HI] = 0.078 M. Calculate K_c.

PLAN To calculate K_c, we need the equilibrium concentrations. We can find the initial [HI] from the amount (0.200 mol) and the flask volume (2.00 L), and we are given [HI] at equilibrium (0.078 M). From the balanced equation, when $2x$ mol of HI reacts, x mol of H_2 and x mol of I_2 form. We set up a reaction table, use the known [HI] at equilibrium to solve for x (the change in [H_2] or [I_2]), and substitute the concentrations into Q_c.

SOLUTION Calculating initial [HI]:

$$[HI] = \frac{0.200 \text{ mol}}{2.00 \text{ L}} = 0.100 \, M$$

Setting up the reaction table, with $x =$ [H_2] or [I_2] that forms and $2x =$ [HI] that reacts:

Concentration (M)	2HI(g) $\rightleftharpoons$	H$_2$(g) +	I$_2$(g)
Initial	0.100	0	0
Change	$-2x$	$+x$	$+x$
Equilibrium	$0.100 - 2x$	x	x

Solving for x, using the known [HI] at equilibrium:

$$[HI] = 0.100 \, M - 2x = 0.078 \, M$$
$$x = 0.011 \, M$$

Therefore, the equilibrium concentrations are

$$[H_2] = [I_2] = 0.011 \, M \quad \text{and} \quad [HI] = 0.078 \, M$$

Substituting into the reaction quotient:

$$Q_c = \frac{[H_2][I_2]}{[HI]^2}$$

Thus,
$$K_c = \frac{(0.011)(0.011)}{0.078^2} = 0.020$$

CHECK Rounding gives $\sim 0.01^2/0.08^2 = 0.02$. Because the initial [HI] of 0.100 M fell slightly at equilibrium to 0.078 M, relatively little product formed; so we expect $K_c < 1$.

FOLLOW-UP PROBLEM 17.7 The atmospheric oxidation of nitrogen monoxide, $2NO(g) + O_2(g) \rightleftharpoons 2NO_2(g)$, was studied at 184°C with initial pressures of 1.000 atm of NO and 1.000 atm of O_2. At equilibrium, $P_{O_2} = 0.506$ atm. Calculate K_p.

Using the Equilibrium Constant to Determine Quantities

Like the type of problem that involves finding K, the type that involves finding equilibrium concentrations (or pressures) has several variations. Sample Problem 17.8 is one variation, in which we know K and some of the equilibrium concentrations and must find another equilibrium concentration.

SAMPLE PROBLEM 17.8 Determining Equilibrium Concentrations from K_c

PROBLEM In a study of the conversion of methane to other fuels, a chemical engineer mixes gaseous CH_4 and H_2O in a 0.32-L flask at 1200 K. At equilibrium, the flask contains 0.26 mol of CO, 0.091 mol of H_2, and 0.041 mol of CH_4. What is [H_2O] at equilibrium? $K_c = 0.26$ for the equation

$$CH_4(g) + H_2O(g) \rightleftharpoons CO(g) + 3H_2(g)$$

PLAN First, we use the balanced equation to write the reaction quotient. We can calculate the equilibrium concentrations from the given numbers of moles and the flask volume (0.32 L). Substituting these into Q_c and setting it equal to the given K_c (0.26), we solve for the unknown equilibrium concentration, [H_2O].

SOLUTION Writing the reaction quotient:

$$CH_4(g) + H_2O(g) \rightleftharpoons CO(g) + 3H_2(g) \qquad Q_c = \frac{[CO][H_2]^3}{[CH_4][H_2O]}$$

Determining the equilibrium concentrations:

$$[CH_4] = \frac{0.041 \text{ mol}}{0.32 \text{ L}} = 0.13 \, M$$

Similarly, $[CO] = 0.81 \, M$ and $[H_2] = 0.28 \, M$.
Calculating $[H_2O]$ at equilibrium: Since $Q_c = K_c$, rearranging gives

$$[H_2O] = \frac{[CO][H_2]^3}{[CH_4]K_c} = \frac{(0.81)(0.28)^3}{(0.13)(0.26)} = 0.53 \, M$$

CHECK Always check by substituting the concentrations into Q_c to confirm K_c:

$$Q_c = \frac{[CO][H_2]^3}{[CH_4][H_2O]} = \frac{(0.81)(0.28)^3}{(0.13)(0.53)} = 0.26 = K_c$$

FOLLOW-UP PROBLEM 17.8 Nitrogen monoxide decomposes by the following equation: $2NO(g) \rightleftharpoons N_2(g) + O_2(g)$; $K_c = 2.3 \times 10^{30}$ at 298 K. In the atmosphere, $P_{O_2} = 0.209$ atm and $P_{N_2} = 0.781$ atm. What is the equilibrium partial pressure of NO in the air we breathe? [*Hint*: You need K_p to find the partial pressure.]

In a somewhat more involved variation, we know K and *initial* quantities and must find *equilibrium* quantities, for which we use a reaction table. In Sample Problem 17.9, the amounts were chosen to simplify the math, allowing us to focus more easily on the overall approach.

ARIS Presentation Center,
Worked Examples Library

Engineering application

SAMPLE PROBLEM 17.9 Determining Equilibrium Concentrations from Initial Concentrations and K_c

PROBLEM Fuel engineers use the extent of the change from CO and H_2O to CO_2 and H_2 to regulate the proportions of synthetic fuel mixtures. If 0.250 mol of CO and 0.250 mol of H_2O are placed in a 125-mL flask at 900 K, what is the composition of the equilibrium mixture? At this temperature, K_c is 1.56 for the equation

$$CO(g) + H_2O(g) \rightleftharpoons CO_2(g) + H_2(g)$$

PLAN We have to find the "composition" of the equilibrium mixture, in other words, the equilibrium concentrations. As always, we use the balanced equation to write the reaction quotient. We find the initial [CO] and $[H_2O]$ from the given amounts (0.250 mol of each) and volume (0.125 L), use the balanced equation to define x and set up a reaction table, substitute into Q_c, and solve for x, from which we calculate the concentrations.

SOLUTION Writing the reaction quotient:

$$CO(g) + H_2O(g) \rightleftharpoons CO_2(g) + H_2(g) \qquad Q_c = \frac{[CO_2][H_2]}{[CO][H_2O]}$$

Calculating initial reactant concentrations:

$$[CO] = [H_2O] = \frac{0.250 \text{ mol}}{0.125 \text{ L}} = 2.00 \, M$$

Setting up the reaction table, with $x = [CO]$ and $[H_2O]$ that react:

Concentration (*M*)	CO(*g*)	+	H₂O(*g*)	⇌	CO₂(*g*)	+	H₂(*g*)
Initial	2.00		2.00		0		0
Change	$-x$		$-x$		$+x$		$+x$
Equilibrium	$2.00 - x$		$2.00 - x$		x		x

Substituting into the reaction quotient and solving for x:

$$Q_c = \frac{[CO_2][H_2]}{[CO][H_2O]} = \frac{(x)(x)}{(2.00 - x)(2.00 - x)} = \frac{x^2}{(2.00 - x)^2}$$

At equilibrium, we have

$$Q_c = K_c = 1.56 = \frac{x^2}{(2.00 - x)^2}$$

We can apply the following math shortcut in this case *but not in general:* Because the right side of the equation is a perfect square, we take the square root of both sides:

$$\sqrt{1.56} = \frac{x}{2.00 - x} = \pm 1.25$$

A positive number (1.56) has a positive *and* a negative square root, but *only the positive root has any chemical meaning,* so we ignore the negative root:*

$$1.25 = \frac{x}{2.00 - x} \quad \text{or} \quad 2.50 - 1.25x = x$$

So $2.50 = 2.25x;$ therefore, $x = 1.11\,M$

Calculating equilibrium concentrations:

$$[CO] = [H_2O] = 2.00\,M - x = 2.00\,M - 1.11\,M = 0.89\,M$$
$$[CO_2] = [H_2] = x = 1.11\,M$$

CHECK From the intermediate size of K_c, it makes sense that the changes in concentration are moderate. It's a good idea to check that the sign of x in the reaction table makes sense—only reactants were initially present, so the change had to proceed to the right: x is the change in concentration, so it has a negative sign for reactants and a positive sign for products. Also check that the equilibrium concentrations give the known K_c:
$$\frac{(1.11)(1.11)}{(0.89)(0.89)} = 1.56.$$

FOLLOW-UP PROBLEM 17.9 The decomposition of HI at low temperature was studied by injecting 2.50 mol of HI into a 10.32-L vessel at 25°C. What is $[H_2]$ at equilibrium for the reaction $2HI(g) \rightleftharpoons H_2(g) + I_2(g)$; $K_c = 1.26 \times 10^{-3}$?

Using the Quadratic Formula to Solve for the Unknown The shortcut that we used to simplify the math in Sample Problem 17.9 is a special case that occurs when both numerator and denominator of the reaction quotient are perfect squares. It worked out that way because we started with equal concentrations of the two reactants, but that is not ordinarily the case.

Suppose, for example, we instead start the reaction in the sample problem with 2.00 M CO and 1.00 M H$_2$O. Now, the reaction table becomes

Concentration (M)	CO(g)	+	H₂O(g)	⇌	CO₂(g)	+	H₂(g)
Initial	2.00		1.00		0		0
Change	−x		−x		+x		+x
Equilibrium	2.00 − x		1.00 − x		x		x

Substituting these values into Q_c, we obtain

$$Q_c = \frac{[CO_2][H_2]}{[CO][H_2O]} = \frac{(x)(x)}{(2.00 - x)(1.00 - x)} = \frac{x^2}{x^2 - 3.00x + 2.00}$$

At equilibrium, we have

$$1.56 = \frac{x^2}{x^2 - 3.00x + 2.00}$$

----

*The negative root gives $-1.25 = \dfrac{x}{2.00 - x}$, or $-2.50 + 1.25x = x$.

So $-2.50 = -0.25x$, and $x = 10.\,M$

This value has no chemical meaning because we started with 2.00 M of each reactant, so it is impossible for 10. M to react. Moreover, the square root of an equilibrium constant is another equilibrium constant, which cannot have a negative value.

To solve for x in this case, we rearrange the previous expression into the form of a *quadratic equation:*

$$a\ x^2 + \ b\ x + \ c\ = 0$$
$$0.56x^2 - 4.68x + 3.12 = 0$$

where $a = 0.56$, $b = -4.68$, and $c = 3.12$. Then we find x with the quadratic formula (Appendix A):

$$x = \frac{-b \pm \sqrt{b^2 - 4ac}}{2a}$$

The $\pm$ sign means that we obtain two possible values for x:

$$x = \frac{4.68 \pm \sqrt{(-4.68)^2 - 4(0.56)(3.12)}}{2(0.56)}$$

$$x = 7.6\ M \quad \text{and} \quad x = 0.73\ M$$

Note that only one of the values for x makes sense chemically. The larger value gives negative concentrations at equilibrium (for example, $2.00\ M - 7.6\ M = -5.6\ M$), which have no meaning. Therefore, $x = 0.73\ M$, and we have

$$[CO] = 2.00\ M - x = 2.00\ M - 0.73\ M = 1.27\ M$$
$$[H_2O] = 1.00\ M - x = 0.27\ M$$
$$[CO_2] = [H_2] = x = 0.73\ M$$

Checking to see if these values give the known K_c, we have

$$K_c = \frac{(0.73)(0.73)}{(1.27)(0.27)} = 1.6 \text{ (within rounding of 1.56)}$$

Simplifying Assumption for Finding an Unknown Quantity In many cases, we can use chemical "common sense" to make an assumption that avoids the use of the quadratic formula to find x. In general, *if a reaction has a relatively small K and a relatively large initial reactant concentration, the concentration change (x) can often be neglected* without introducing significant error. This assumption does not mean that $x = 0$, because then there would be no reaction at all. It means that if a reaction proceeds very little (small K) and and starts out with a high initial reactant concentration, only a very small proportion will be used up; therefore, at equilibrium, the reactant concentration will have hardly changed:

$$[\text{reactant}]_{\text{init}} - x = [\text{reactant}]_{\text{eq}} \approx [\text{reactant}]_{\text{init}}$$

Here's a similar situation from everyday life. On a bathroom scale, you weigh 158 lb. Take off your wristwatch, and you still weigh 158 lb. Within the precision of the measurement, the weight of the wristwatch is so small compared with your body weight that it can be neglected:

Initial body weight − weight of watch = final body weight $\approx$ initial body weight

Similarly, if the initial concentration of A is, for example, 0.500 M and, because of a small K_c, the concentration of A that reacts is 0.002 M, we can assume that

$$0.500\ M - 0.002\ M = 0.498\ M \approx 0.500\ M$$

that is,

$$[A]_{\text{init}} - [A]_{\text{reacting}} = [A]_{\text{eq}} \approx [A]_{\text{init}} \tag{17.9}$$

To justify the assumption that x, the concentration reacting, is negligible, you must check that the error introduced is not significant. One common criterion for "significant" is the 5% rule: *if the assumption results in a change that is less than 5% of the initial concentration, the error is not significant, and the assumption is justified.* Let's make this assumption in a sample problem to see how it simplifies the math and if it is justified in the cases of two different initial concentrations. (We make this assumption often in Chapters 18 and 19.)

SAMPLE PROBLEM 17.10 Calculating Equilibrium Concentrations with a Simplifying Assumption

PROBLEM Phosgene is a potent chemical warfare agent that is now outlawed by international agreement. It decomposes by the reaction

$$COCl_2(g) \rightleftharpoons CO(g) + Cl_2(g) \qquad K_c = 8.3\times10^{-4} \text{ (at 360°C)}$$

Calculate [CO], [Cl$_2$], and [COCl$_2$] when each of the following amounts of phosgene decomposes and reaches equilibrium in a 10.0-L flask:
(a) 5.00 mol of COCl$_2$ **(b)** 0.100 mol of COCl$_2$

PLAN We know from the balanced equation that when x mol of COCl$_2$ decomposes, x mol of CO and x mol of Cl$_2$ form. We use the volume (10.0 L) to convert amount (5.00 mol or 0.100 mol) to concentration, define x and set up the reaction table, and substitute the values into Q_c. Before using the quadratic formula, we assume that x is negligibly small. After solving for x, we check the assumption and find the concentrations. If the assumption is not justified, we must use the quadratic formula to find x.

SOLUTION (a) For 5.00 mol of COCl$_2$. Writing the reaction quotient:

$$Q_c = \frac{[CO][Cl_2]}{[COCl_2]}$$

Calculating initial [COCl$_2$]:

$$[COCl_2]_{init} = \frac{5.00 \text{ mol}}{10.0 \text{ L}} = 0.500 \, M$$

Setting up the reaction table, with x in the "Change" row equal to [COCl$_2$]$_{reacting}$:

Concentration (*M*)	COCl$_2$(*g*) $\rightleftharpoons$	CO(*g*) +	Cl$_2$(*g*)
Initial	0.500	0	0
Change	$-x$	$+x$	$+x$
Equilibrium	$0.500 - x$	x	x

If we use the equilibrium values in Q_c, we obtain

$$Q_c = \frac{[CO][Cl_2]}{[COCl_2]} = \frac{x^2}{0.500 - x} = K_c = 8.3\times10^{-4}$$

Because K_c is small, the reaction does not proceed very far to the right, so let's simplify the math by assuming that x ([COCl$_2$]$_{reacting}$) is so small that the equilibrium concentration is nearly the same as the initial concentration, 0.500 M. Therefore,

$$0.500 \, M - x \approx 0.500 \, M$$

Using this assumption, we substitute and solve for x:

$$K_c = 8.3\times10^{-4} \approx \frac{x^2}{0.500}$$

$$x^2 \approx (8.3\times10^{-4})(0.500) \qquad \text{so} \qquad x \approx 2.0\times10^{-2}$$

Checking the assumption by finding the percent error:

$$\frac{[\text{Change}]}{[\text{Initial}]} = \frac{2.0\times10^{-2}}{0.500} \times 100 = 4\% \text{ (less than 5\%, so the assumption is justified)}$$

Solving for the equilibrium concentrations:

$$[CO] = [Cl_2] = x = 2.0\times10^{-2} \, M$$

$$[COCl_2] = 0.500 \, M - x = 0.480 \, M$$

(b) For 0.100 mol of COCl$_2$. The calculation in this case is the same as the calculation in part (a), except that [COCl$_2$]$_{init}$ = 0.100 mol/10.0 L = 0.0100 M. Thus, at equilibrium,

$$Q_c = \frac{[CO][Cl_2]}{[COCl_2]} = \frac{x^2}{0.0100 - x}$$

$$= K_c = 8.3\times10^{-4}$$

Making the assumption that 0.0100 $M - x \approx 0.0100 \, M$ and solving for x:

$$K_c = 8.3\times10^{-4} \approx \frac{x^2}{0.0100} \qquad x \approx 2.9\times10^{-3}$$

Checking the assumption:

$$\frac{2.9\times10^{-3}}{0.0100} \times 100 = 29\% \qquad \text{(more than 5\%, so the assumption is \textit{not} justified)}$$

We must solve the quadratic equation, $x^2 + (8.3\times10^{-4})x - (8.3\times10^{-6}) = 0$, for which the only meaningful value of x is 2.5×10^{-3} (see Appendix A).

Solving for the equilibrium concentrations:

$$[CO] = [Cl_2] = 2.5\times10^{-3}\,M$$

$$[COCl_2] = 1.00\times10^{-2}\,M - x = 7.5\times10^{-3}\,M$$

CHECK Once again, the best check is to use the calculated values to be sure you obtain the given K_c.

COMMENT Note that the simplifying assumption was justified at the high $[COCl_2]_{init}$ but *not* at the low $[COCl_2]_{init}$. To summarize, we assume that x ($[A]_{reacting}$) can be neglected if K_c is small relative to $[A]_{init}$. The same holds for K_p and $P_{A(init)}$. Here's a benchmark for determining when the assumption is justified:

- If $\dfrac{[A]_{init}}{K_c} > 400$, the assumption is justified: neglecting x introduces an error $< 5\%$.

- If $\dfrac{[A]_{init}}{K_c} < 400$, the assumption is *not* justified; neglecting x introduces an error $> 5\%$,

so we solve a quadratic equation to find x.

FOLLOW-UP PROBLEM 17.10 In a study of the effect of temperature on halogen decomposition, 0.50 mol of I_2 was heated in a 2.5-L vessel, and the following reaction occurred: $I_2(g) \rightleftharpoons 2I(g)$.
(a) Calculate $[I_2]$ and $[I]$ at equilibrium at 600 K; $K_c = 2.94\times10^{-10}$.
(b) Calculate $[I_2]$ and $[I]$ at equilibrium at 2000 K; $K_c = 0.209$.

Mixtures of Reactants and Products: Determining Reaction Direction

In the problems we've worked so far, the direction of the reaction was obvious: with only reactants present at the start, the reaction *had* to go toward products. Thus, in the reaction tables, we knew that the unknown change in reactant concentration had a negative sign $(-x)$ and the change in product concentration had a positive sign $(+x)$. Suppose, however, we start with a *mixture* of reactants and products. Whenever the reaction direction is not obvious, we first *compare the value of Q with K to find the direction* in which the reaction proceeds to reach equilibrium. This tells us the sign of x, the unknown change in concentration. (In order to focus on this idea, the next sample problem eliminates the need for the quadratic formula.)

SAMPLE PROBLEM 17.11 Predicting Reaction Direction and Calculating Equilibrium Concentrations

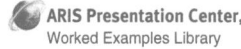

PROBLEM The research and development unit of a chemical company is studying the reaction of CH_4 and H_2S, two components of natural gas:

$$CH_4(g) + 2H_2S(g) \rightleftharpoons CS_2(g) + 4H_2(g)$$

In one experiment, 1.00 mol of CH_4, 1.00 mol of CS_2, 2.00 mol of H_2S, and 2.00 mol of H_2 are mixed in a 250-mL vessel at 960°C. At this temperature, $K_c = 0.036$.
(a) In which direction will the reaction proceed to reach equilibrium?
(b) If $[CH_4] = 5.56\,M$ at equilibrium, what are the equilibrium concentrations of the other substances?

PLAN (a) To find the direction, we convert the given initial amounts and volume (0.250 L) to concentrations, calculate Q_c, and compare it with K_c. (b) Based on the results from (a), we determine the sign of each concentration change for the reaction table and then use the known $[CH_4]$ at equilibrium (5.56 M) to determine x and the other equilibrium concentrations.

SOLUTION (a) Calculating the initial concentrations:

$$[CH_4] = \frac{1.00 \text{ mol}}{0.250 \text{ L}} = 4.00 \text{ } M$$

Similarly, $[H_2S] = 8.00 \text{ } M$, $[CS_2] = 4.00 \text{ } M$, and $[H_2] = 8.00 \text{ } M$.
Calculating the value of Q_c:

$$Q_c = \frac{[CS_2][H_2]^4}{[CH_4][H_2S]^2} = \frac{(4.00)(8.00)^4}{(4.00)(8.00)^2} = 64.0$$

Comparing Q_c and K_c: $Q_c > K_c$ (64.0 > 0.036), so the reaction goes to the left. Therefore, concentrations of reactants increase and those of products decrease.
(b) Setting up a reaction table, with $x = [CS_2]$ that reacts, which equals $[CH_4]$ that forms:

Concentration (*M*)	CH₄(*g*)	+	2H₂S(*g*)	⇌	CS₂(*g*)	+	4H₂(*g*)
Initial	4.00		8.00		4.00		8.00
Change	+x		+2x		−x		−4x
Equilibrium	4.00 + x		8.00 + 2x		4.00 − x		8.00 − 4x

Solving for x: At equilibrium,

$$[CH_4] = 5.56 \text{ } M = 4.00 \text{ } M + x$$

So,

$$x = 1.56 \text{ } M$$

Thus,

$$[H_2S] = 8.00 \text{ } M + 2x = 8.00 \text{ } M + 2(1.56 \text{ } M) = \boxed{11.12 \text{ } M}$$

$$[CS_2] = 4.00 \text{ } M - x = \boxed{2.44 \text{ } M}$$

$$[H_2] = 8.00 \text{ } M - 4x = \boxed{1.76 \text{ } M}$$

CHECK The comparison of Q_c and K_c showed the reaction proceeding to the left. The given data from part (b) confirm this because $[CH_4]$ increases from $4.00 \text{ } M$ to $5.56 \text{ } M$ during the reaction. Check that the concentrations give the known K_c:

$$\frac{(2.44)(1.76)^4}{(5.56)(11.12)^2} = 0.0341, \text{ which is close to } 0.036$$

FOLLOW-UP PROBLEM 17.11 An inorganic chemist studying the reactions of phosphorus halides mixes 0.1050 mol of PCl_5 with 0.0450 mol of Cl_2 and 0.0450 mol of PCl_3 in a 0.5000-L flask at 250°C: $PCl_5(g) \rightleftharpoons PCl_3(g) + Cl_2(g)$; $K_c = 4.2 \times 10^{-2}$.
(a) In which direction will the reaction proceed?
(b) If $[PCl_5] = 0.2065 \text{ } M$ at equilibrium, what are the equilibrium concentrations of the other components?

By this time, you've seen quite a few variations on the type of equilibrium problem in which you know K and some initial quantities and must find the equilibrium quantities. Figure 17.6 presents a useful summary of the steps involved in solving these types of equilibrium problems. A good way to organize the steps is to group them into three overall parts, as the figure shows.

Section Summary

In equilibrium problems, we typically use quantities (concentrations or pressures) of reactants and products to find K, or we use K to find quantities. • Reaction tables summarize the initial quantities, how they change, and the equilibrium quantities. To simplify calculations, we assume that if K is small and the initial quantity of reactant is large, the unknown change in reactant (x) can be neglected. If this assumption is not justified (that is, if the error is greater than 5%), we use the quadratic formula to find x. • To determine reaction direction, we compare the values of Q and K.

SOLVING EQUILIBRIUM PROBLEMS

PRELIMINARY SETTING UP

1. Write the balanced equation
2. Write the reaction quotient, Q
3. Convert all amounts into the correct units (M or atm)

WORKING ON THE REACTION TABLE

4. When reaction direction is not known, compare Q with K
5. Construct a reaction table

✓ Check the sign of x, the change in the concentration (or pressure)

SOLVING FOR x AND EQUILIBRIUM QUANTITIES

6. Substitute the quantities into Q
7. To simplify the math, assume that x is negligible ($[A]_{init} - x = [A]_{eq} \approx [A]_{init}$)
8. Solve for x

✓ Check that assumption is justified (<5% error). If not, solve quadratic equation for x.

9. Find the equilibrium quantities

✓ Check to see that calculated values give the known K

Figure 17.6 **Steps in solving equilibrium problems.** These nine steps, grouped into three tasks, provide a useful approach to calculating equilibrium quantities, given initial quantities and K.

Transparency/ARIS Presentation Center, Art Library

17.6 REACTION CONDITIONS AND THE EQUILIBRIUM STATE: LE CHÂTELIER'S PRINCIPLE

The most remarkable feature of a system at equilibrium is its ability to return to equilibrium after a change in conditions moves it away from that state. This drive to reattain equilibrium is stated in **Le Châtelier's principle:** when a chemical system at equilibrium is disturbed, it reattains equilibrium by undergoing a net reaction that reduces the effect of the disturbance.

Two phrases in this statement need further explanation. First, what does it mean to "disturb" a system? At equilibrium, Q equals K. When a change in conditions forces the system temporarily out of equilibrium ($Q \neq K$), we say that the system has been stressed, or disturbed. Three common disturbances are a change in concentration of a component (that appears in Q), a change in pressure (caused by a change in volume), or a change in temperature. We'll discuss each of these changes below.

The other phrase, "net reaction," is often referred to as a shift in the *equilibrium position* of the system to the right or left. The equilibrium position is just the specific equilibrium concentrations (or pressures). A shift in the equilibrium position to the right means that there is a net reaction to the right (reactant to product) until equilibrium is reattained; a shift to the left means that there is a net reaction to the left (product to reactant). Thus, when a disturbance occurs, we say that the equilibrium position shifts, which means that *concentrations (or pressures) change in a way that reduces the disturbance, and the system attains a new equilibrium position ($Q = K$ again).*

Le Châtelier's principle allows us to predict the direction of the shift in equilibrium position. Most importantly, it helps research and industrial chemists create conditions that maximize yields. For the remainder of this section, we examine each of the three kinds of disturbances—concentration, pressure (volume), and temperature—to see how a system at equilibrium responds; then, we'll note whether a catalyst has any effect.

In the following discussions, we focus on the reversible gaseous reaction between phosphorus trichloride and chlorine to produce phosphorus pentachloride:

$$PCl_3(g) + Cl_2(g) \rightleftharpoons PCl_5(g)$$

However, the basis of Le Châtelier's principle holds for any system at equilibrium, whether in the natural or social sciences. ●

The Effect of a Change in Concentration

When a system at equilibrium is disturbed by a change in concentration of one of the components, the system reacts in the direction that reduces the change:

- If the concentration increases, the system reacts to consume some of it.
- If the concentration decreases, the system reacts to produce some of it.

Of course, the component must be one that appears in Q; thus, pure liquids and solids, which do not appear in Q because their concentrations are constant, are not involved.

At 523 K, the PCl_3-Cl_2-PCl_5 system reaches equilibrium when

$$Q_c = \frac{[PCl_5]}{[PCl_3][Cl_2]} = 24.0 = K_c$$

What happens if we now inject some Cl_2 gas, one of the reactants? The system will always act to reduce the disturbance, so it will reduce the increase in reactant

● **The Universality of Le Châtelier's Principle** Although Le Châtelier limited the scope of his principle to physical and chemical systems, its application is much wider. Ecologists and economists, for example, often see it at work. On the African savannah, the numbers of herbivores (antelope, wildebeest, zebra) and carnivores (lion, cheetah) are in delicate balance. Any disturbance (drought, disease) causes shifts in the relative numbers until they attain a new balance. The economic principle of supply-and-demand is another example of this equilibrium principle. A given demand-supply balance for a product establishes a given price. If either demand or supply is disturbed, the disturbance causes a shift in the other and, therefore, in the prevailing price until a new balance is attained.

ARIS Presentation Center, Photo Library

Literature Novak, Igor. "Microscopic Description of Le Châtelier's Principle," *J. Chem. Educ.* **2005,** *82,* 1190.

Literature Huddle, Benjamin P. "Conceptual Questions on Le Châtelier's Principle," *J. Chem. Educ.* **1998,** *75,* 1175.

by proceeding toward the product side, thereby consuming some of the additional Cl_2. In terms of the reaction quotient, when we add Cl_2, the $[Cl_2]$ term increases, so the value of Q_c immediately falls as the denominator becomes larger; thus, the system is no longer at equilibrium. As some of the added Cl_2 reacts with some of the PCl_3 present and produces more PCl_5, the denominator becomes smaller once again and the numerator larger, until eventually Q_c again equals K_c. The concentrations of the components have changed, however: the concentrations of Cl_2 and PCl_5 are higher than in the original equilibrium position, and the concentration of PCl_3 is lower. Nevertheless, the ratio of values gives the same K_c. We describe this change by saying that *the equilibrium position shifts to the right when a component on the left is added:*

$$PCl_3 + Cl_2(\text{added}) \longrightarrow PCl_5$$

What happens if, instead of adding Cl_2, we remove some PCl_3, the other reactant? In this case, the system reduces the disturbance (the decrease in reactant), by proceeding toward the reactant side, thereby consuming some PCl_5. Once again, thinking in terms of Q_c, when we remove PCl_3, the $[PCl_3]$ term decreases, the denominator becomes smaller, and the value of Q_c rises above K_c. As some of the PCl_5 decomposes to PCl_3 and Cl_2, the numerator decreases and the denominator increases until Q_c equals K_c once again. Here, too, the concentrations are different from those of the original equilibrium position, but the value of K_c is not. We say that *the equilibrium position shifts to the left when a component on the left is removed:*

$$PCl_3(\text{removed}) + Cl_2 \longleftarrow PCl_5$$

The same points we just made for adding or removing a reactant also hold for adding or removing a product. If we add PCl_5, its concentration rises and the equilibrium position shifts to the left, just as it did when we removed some PCl_3; if we remove some PCl_5, the equilibrium position shifts to the right, just as it did when we added some Cl_2. In other words, no matter how the disturbance in concentration comes about, the system responds to make Q_c and K_c equal again. To summarize the effects of concentration changes (Figure 17.7),

- The equilibrium position shifts to the *right* if a reactant is added or a product is removed: [reactant] increases or [product] decreases.
- The equilibrium position shifts to the *left* if a reactant is removed or a product is added: [reactant] decreases or [product] increases.

In general, whenever the concentration of a component changes, *the equilibrium system reacts to consume some of the added substance or produce some of the removed substance.* In this way, the system "reduces the effect of the disturbance." The effect is not completely eliminated, however, as you'll see next from a quantitative comparison of original and new equilibrium positions.

Consider the case in which we added Cl_2 to the system at equilibrium. Suppose the original equilibrium position was established with the following concentrations: $[PCl_3] = 0.200\ M$, $[Cl_2] = 0.125\ M$, and $[PCl_5] = 0.600\ M$. Thus,

$$Q_c = \frac{[PCl_5]}{[PCl_3][Cl_2]} = \frac{0.600}{(0.200)(0.125)} = 24.0 = K_c$$

Now we add enough Cl_2 to increase its concentration by $0.075\ M$. Before any reaction occurs, this addition creates a new set of initial concentrations. Then the system reacts and comes to a new equilibrium position. From Le Châtelier's principle, we predict that adding more reactant will produce more product, that is, shift the equilibrium position to the right. Experiment shows that the new $[PCl_5]$ at equilibrium is $0.637\ M$.

ANY OF THESE CHANGES CAUSES A SHIFT TO THE *RIGHT*

increase   increase   decrease

$$PCl_3 + Cl_2 \rightleftharpoons PCl_5$$

decrease decrease   increase

ANY OF THESE CHANGES CAUSES A SHIFT TO THE *LEFT*

Figure 17.7 **The effect of a change in concentration.**

Table 17.3 The Effect of Added Cl_2 on the PCl_3-Cl_2-PCl_5 System

Concentration (M)	$PCl_3(g)$	+	$Cl_2(g)$	$\rightleftharpoons$	$PCl_5(g)$
Original equilibrium	0.200		0.125	·	0.600
Disturbance			+0.075		
New initial	0.200		0.200		0.600
Change	−x		−x		+x
New equilibrium	0.200 − x		0.200 − x		0.600 + x
					(0.637)*

*Experimentally determined value.

Transparency/ARIS Presentation Center, Tables Library

Table 17.3 shows a reaction table of the entire process: the original equilibrium position, the disturbance, the (new) initial concentrations, the size and direction of the change needed to reattain equilibrium, and the new equilibrium position. Figure 17.8 depicts the process.

From Table 17.3,

$$[PCl_5] = 0.600\ M + x = 0.637\ M, \quad \text{so} \quad x = 0.037\ M$$

Also, $\quad [PCl_3] = [Cl_2] = 0.200\ M - x = 0.163\ M$

Therefore, at equilibrium,

$$K_{c(\text{original})} = \frac{0.600}{(0.200)(0.125)} = 24.0$$

$$K_{c(\text{new})} = \frac{0.637}{(0.163)(0.163)} = 24.0$$

There are several key points to notice about the new equilibrium concentrations that exist after Cl_2 is added:

- As we predicted, $[PCl_5]$ (0.637 M) is higher than its original concentration (0.600 M).
- $[Cl_2]$ (0.163 M) is higher than its original equilibrium concentration (0.125 M), but lower than its initial concentration just after the addition (0.200 M); thus, the disturbance (addition of Cl_2) is *reduced but not eliminated.*
- $[PCl_3]$ (0.163 M), the other left-side component, is lower than its original concentration (0.200 M) because some reacted with the added Cl_2.
- Most importantly, although the position of equilibrium shifted to the right, K_c *remains the same.*

Be sure to note that the system adjusts by changing concentrations, but the value of K_c at a given temperature does **not** change with a change in concentration.

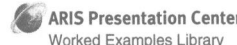

Figure 17.8 The effect of added Cl_2 on the PCl_3-Cl_2-PCl_5 system. In the original equilibrium (*gray region*), all concentrations are constant. When Cl_2 (*yellow curve*) is added, its concentration jumps and then starts to fall as Cl_2 reacts with some PCl_3 to form more PCl_5. After a period of time, equilibrium is re-established at new concentrations (*blue region*) but with the same K.

Transparency/ARIS Presentation Center, Art Library

ARIS Presentation Center, Worked Examples Library

Environmental application

SAMPLE PROBLEM 17.12 Predicting the Effect of a Change in Concentration on the Equilibrium Position

PROBLEM To improve air quality and obtain a useful product, chemists often remove sulfur from coal and natural gas by treating the fuel contaminant hydrogen sulfide with O_2:

$$2H_2S(g) + O_2(g) \rightleftharpoons 2S(s) + 2H_2O(g)$$

What happens to
(a) $[H_2O]$ if O_2 is added?   **(b)** $[H_2S]$ if O_2 is added?
(c) $[O_2]$ if H_2S is removed?   **(d)** $[H_2S]$ if sulfur is added?
PLAN We write the reaction quotient to see how Q_c is affected by each disturbance, relative to K_c. This effect tells us the direction in which the reaction proceeds for the system to reattain equilibrium and how each concentration changes.

SOLUTION Writing the reaction quotient: $Q_c = \dfrac{[H_2O]^2}{[H_2S]^2[O_2]}$

(a) When O_2 is added, the denominator of Q_c increases, so $Q_c < K_c$. The reaction proceeds to the right until $Q_c = K_c$ again, so $[H_2O]$ increases.

(b) As in part (a), when O_2 is added, $Q_c < K_c$. Some H_2S reacts with the added O_2 as the reaction proceeds to the right, so $[H_2S]$ decreases.

(c) When H_2S is removed, the denominator of Q_c decreases, so $Q_c > K_c$. As the reaction proceeds to the left to re-form H_2S, more O_2 is produced as well, so $[O_2]$ increases.

(d) The concentration of solid S is unchanged as long as some is present, so it does not appear in the reaction quotient. Adding more S has no effect, so $[H_2S]$ is unchanged (but see Comment 2 below).

CHECK Apply Le Châtelier's principle to see that the reaction proceeds in the direction that lowers the increased concentration or raises the decreased concentration.

COMMENT 1. As you know, sulfur exists most commonly as S_8. How would this change in formula affect the answers? The balanced equation and Q_c would be

$$8H_2S(g) + 4O_2(g) \rightleftharpoons S_8(s) + 8H_2O(g) \qquad Q_c = \dfrac{[H_2O]^8}{[H_2S]^8[O_2]^4}$$

The value of K_c is different for this equation, but the changes described in the problem have the same effects. For example, in (a), if O_2 were added, the denominator of Q_c would increase, so $Q_c < K_c$. As above, the reaction would proceed to the right until $Q_c = K_c$ again. In other words, changes predicted by Le Châtelier's principle for a given reaction are not affected by a change in the balancing coefficients.

2. In (d), you saw that adding a solid has no effect on the concentrations of other components: because *the **concentration** of the solid cannot change*, it does not appear in Q. But *the **amount** of solid can change*. Adding H_2S shifts the reaction to the right, and more S forms.

FOLLOW-UP PROBLEM 17.12 In a study of the chemistry of glass etching, an inorganic chemist examines the reaction between sand (SiO_2) and hydrogen fluoride at a temperature above the boiling point of water:

$$SiO_2(s) + 4HF(g) \rightleftharpoons SiF_4(g) + 2H_2O(g)$$

Predict the effect on $[SiF_4]$ when **(a)** $H_2O(g)$ is removed; **(b)** some liquid water is added; **(c)** HF is removed; **(d)** some sand is removed.

The Effect of a Change in Pressure (Volume)

Changes in pressure have significant effects only on equilibrium systems with gaseous components. Aside from phase changes, a change in pressure has a negligible effect on liquids and solids because they are nearly incompressible. Pressure changes can occur in three ways:

- Changing the concentration of a gaseous component
- Adding an inert gas (one that does not take part in the reaction)
- Changing the volume of the reaction vessel

We just considered the effect of changing the concentration of a component, and that reasoning holds here. Next, let's see why *adding an inert gas has no effect on the equilibrium position.* Adding an inert gas does not change the volume, so all *reactant and product concentrations remain the same.* In other words, the volume and the number of moles of the reactant and product gases do not change, so their *partial pressures do not change.* Because we use these (unchanged) partial pressures in the reaction quotient, the equilibrium position cannot change. Moreover, the inert gas does not appear in Q, so it cannot have an effect.

On the other hand, changing the pressure by changing the volume often causes a large shift in the equilibrium position. Suppose we let the PCl_3-Cl_2-PCl_5 system come to equilibrium in a cylinder-piston assembly. Then, we press down on the piston to halve the volume: the gas pressure immediately doubles. To reduce this increase in gas pressure, the system responds by reducing the number of gas molecules. And it does so in the only possible way—by shifting the reaction toward the side with *fewer moles of gas,* in this case, toward the product side:

$$PCl_3(g) + Cl_2(g) \longrightarrow PCl_5(g)$$
$$\text{2 mol gas} \longrightarrow \text{1 mol gas}$$

Notice that *a change in volume results in a change in concentration:* a decrease in container volume raises the concentration, and an increase in volume lowers the concentration. Recall that $Q_c = \dfrac{[PCl_5]}{[PCl_3][Cl_2]}$. When the volume is halved, the concentrations double, but the denominator of Q_c is the product of two concentrations, so it quadruples while the numerator only doubles. Thus, Q_c becomes less than K_c. As a result, the system forms more PCl_5 and a new equilibrium position is reached. Because it is just another way to change the concentration, *a change in pressure due to a change in volume does **not** alter K_c.*

Thus, for a system that contains gases at equilibrium, in which the amount (mol) of gas, n_{gas}, changes during the reaction (Figure 17.9):

- If the volume becomes smaller (pressure is higher), the reaction shifts so that the total number of gas molecules decreases.
- If the volume becomes larger (pressure is lower), the reaction shifts so that the total number of gas molecules increases.

In many cases, however, n_{gas} does not change ($\Delta n_{gas} = 0$). For example,

$$H_2(g) + I_2(g) \rightleftharpoons 2HI(g)$$
$$\text{2 mol gas} \longrightarrow \text{2 mol gas}$$

Q_c has the same number of terms in the numerator and denominator:

$$Q_c = \frac{[HI]^2}{[H_2][I_2]} = \frac{[HI][HI]}{[H_2][I_2]}$$

Therefore, a change in volume has the same effect on the numerator and denominator. Thus, *if $\Delta n_{gas} = 0$, there is no effect on the equilibrium position.*

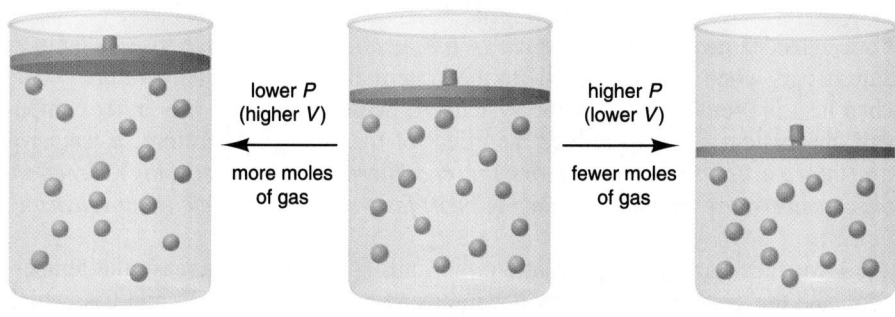

Transparency/ARIS Presentation Center, Art Library

Figure 17.9 **The effect of pressure (volume) on a system at equilibrium.** The system of gases *(center)* is at equilibrium. For the reaction

● + ○ ⇌ ●

an increase in pressure *(right)* decreases the volume, so the reaction shifts to the right to make fewer molecules. A decrease in pressure *(left)* increases the volume, so the reaction shifts to the left to make more molecules.

SAMPLE PROBLEM 17.13 Predicting the Effect of a Change in Volume (Pressure) on the Equilibrium Position

PROBLEM How would you change the volume of each of the following reactions to *increase* the yield of the products?

(a) $CaCO_3(s) \rightleftharpoons CaO(s) + CO_2(g)$ (b) $S(s) + 3F_2(g) \rightleftharpoons SF_6(g)$

(c) $Cl_2(g) + I_2(g) \rightleftharpoons 2ICl(g)$

PLAN Whenever gases are present, a change in volume causes a change in concentration. For reactions in which the number of moles of gas changes, if the volume decreases (pressure increases), the equilibrium position shifts to relieve the pressure by reducing the number of moles of gas. A volume increase (pressure decrease) has the opposite effect.

SOLUTION (a) The only gas is the product CO_2. To make the system produce more CO_2, we increase the volume (decrease the pressure).

(b) With 3 mol of gas on the left and only 1 mol on the right, we decrease the volume (increase the pressure) to form more SF_6.

(c) The number of moles of gas is the same on both sides of the equation, so a change in volume (pressure) will have no effect on the yield of ICl.

CHECK Let's predict the relative values of Q_c and K_c. In (a), $Q_c = [CO_2]$, so increasing the volume will make $Q_c < K_c$, and the system will make more CO_2. In (b), $Q_c = [SF_6]/[F_2]^3$. Lowering the volume increases $[F_2]$ and $[SF_6]$ proportionately, but Q_c decreases because of the exponent 3 in the denominator. To make $Q_c = K_c$ again, $[SF_6]$ must increase. In (c), $Q_c = [ICl]^2/[Cl_2][I_2]$. A change in volume (pressure) affects the numerator (2 mol) and denominator (2 mol) equally, so it will have no effect.

FOLLOW-UP PROBLEM 17.13 Would you increase or decrease the pressure (via a volume change) of each of the following reaction mixtures to *decrease* the yield of products?

(a) $2SO_2(g) + O_2(g) \rightleftharpoons 2SO_3(g)$ (b) $4NH_3(g) + 5O_2(g) \rightleftharpoons 4NO(g) + 6H_2O(g)$

(c) $CaC_2O_4(s) \rightleftharpoons CaCO_3(s) + CO(g)$

The Effect of a Change in Temperature

Of the three types of disturbances—a change in concentration, in pressure, or in temperature—*only temperature changes alter K*. To see why, we must take the heat of reaction into account:

$$PCl_3(g) + Cl_2(g) \rightleftharpoons PCl_5(g) \qquad \Delta H^\circ_{rxn} = -111 \text{ kJ}$$

The forward reaction is exothermic (releases heat; $\Delta H^\circ < 0$), so the reverse reaction is endothermic (absorbs heat; $\Delta H^\circ > 0$):

$$PCl_3(g) + Cl_2(g) \longrightarrow PCl_5(g) + \textbf{\textit{heat}} \text{ (exothermic)}$$
$$PCl_3(g) + Cl_2(g) \longleftarrow PCl_5(g) + \textbf{\textit{heat}} \text{ (endothermic)}$$

If we consider *heat as a component of the equilibrium system,* a rise in temperature occurs when heat is "added" to the system and a drop in temperature occurs when heat is "removed" from the system. As with a change in any other component, the system shifts to reduce the effect of the change. Therefore, *a temperature increase (adding heat) favors the endothermic (heat-absorbing) direction, and a temperature decrease (removing heat) favors the exothermic (heat-releasing) direction.*

If we start with the system at equilibrium, Q equals K. Increase the temperature, and the system responds by decomposing some PCl_5 to PCl_3 and Cl_2, which absorbs the added heat. The denominator of Q becomes larger and the numerator smaller, so the system reaches a new equilibrium position at a smaller ratio of concentration terms, that is, a lower K. Similarly, the system responds to a drop in temperature by forming more PCl_5 from some PCl_3 and Cl_2, which releases more heat. The numerator of Q becomes larger, the denominator smaller, and the new equilibrium position has a higher K. Thus,

- *A temperature rise will increase K for a system with a positive ΔH°_{rxn}.*
- *A temperature rise will decrease K for a system with a negative ΔH°_{rxn}.*

Let's review these ideas with a sample problem.

SAMPLE PROBLEM 17.14 Predicting the Effect of a Change in Temperature on the Equilibrium Position

PROBLEM How does an *increase* in temperature affect the equilibrium concentration of the underlined substance and K for each of the following reactions?

(a) $CaO(s) + H_2O(l) \rightleftharpoons \underline{Ca(OH)_2}(aq)$ $\Delta H° = -82 \text{ kJ}$

(b) $CaCO_3(s) \rightleftharpoons CaO(s) + \underline{CO_2}(g)$ $\Delta H° = 178 \text{ kJ}$

(c) $\underline{SO_2}(g) \rightleftharpoons S(s) + O_2(g)$ $\Delta H° = 297 \text{ kJ}$

PLAN We write each equation to show heat as a reactant or product. The temperature increases when we add heat, so the system shifts to absorb the heat; that is, the endothermic reaction occurs. Thus, K will increase if the forward reaction is endothermic and decrease if it is exothermic.

SOLUTION (a) $CaO(s) + H_2O(l) \rightleftharpoons Ca(OH)_2(aq) + \textbf{\textit{heat}}$

Adding heat shifts the system to the left: $[Ca(OH)_2]$ and K will decrease.

(b) $CaCO_3(s) + \textbf{\textit{heat}} \rightleftharpoons CaO(s) + CO_2(g)$

Adding heat shifts the system to the right: $[CO_2]$ and K will increase.

(c) $SO_2(g) + \textbf{\textit{heat}} \rightleftharpoons S(s) + O_2(g)$

Adding heat shifts the system to the right: $[SO_2]$ will decrease and K will increase.

CHECK You can check your answers by going through the reasoning for a *decrease* in temperature: heat is removed and the exothermic direction is favored. All the answers should be opposite.

FOLLOW-UP PROBLEM 17.14 How does a *decrease* in temperature affect the partial pressure of the underlined substance and the value of K for each of the following reactions?

(a) $C(graphite) + 2\underline{H_2}(g) \rightleftharpoons CH_4(g)$ $\Delta H° = -75 \text{ kJ}$

(b) $\underline{N_2}(g) + O_2(g) \rightleftharpoons 2NO(g)$ $\Delta H° = 181 \text{ kJ}$

(c) $P_4(s) + 10Cl_2(g) \rightleftharpoons 4\underline{PCl_5}(g)$ $\Delta H° = -1528 \text{ kJ}$

The van't Hoff Equation: The Effect of *T* on *K*

The *van't Hoff equation* shows how the equilibrium constant is affected by changes in temperature:

$$\ln \frac{K_2}{K_1} = -\frac{\Delta H°_{rxn}}{R}\left(\frac{1}{T_2} - \frac{1}{T_1}\right) \qquad (17.10)$$

where K_1 is the equilibrium constant at T_1, K_2 is the equilibrium constant at T_2, and R is the universal gas constant (8.314 J/mol·K). If we know $\Delta H°_{rxn}$ and K at one temperature, the van't Hoff equation allows us to find K at any other temperature (or to find $\Delta H°_{rxn}$, given the two K's at two T's).

Here's a typical problem that requires the van't Hoff equation. Coal gasification processes usually begin with the formation of syngas from carbon and steam:

$$C(s) + H_2O(g) \rightleftharpoons CO(g) + H_2(g) \qquad \Delta H°_{rxn} = 131 \text{ kJ/mol}$$

An engineer knows that K_p is only 9.36×10^{-17} at 25°C and therefore wants to find a temperature that allows a much higher yield. Calculate K_p at 700°C.

$$\ln \frac{K_2}{K_1} = -\frac{\Delta H°_{rxn}}{R}\left(\frac{1}{T_2} - \frac{1}{T_1}\right)$$

The temperatures must be converted, and we make the units of $\Delta H°$ and R consistent:

$$\ln\left(\frac{K_{p2}}{9.36 \times 10^{-17}}\right) = -\frac{131 \times 10^3 \text{ J/mol}}{8.314 \text{ J/mol·K}}\left(\frac{1}{973 \text{ K}} - \frac{1}{298 \text{ K}}\right)$$

$$\frac{K_{p2}}{9.36 \times 10^{-17}} = 8.51 \times 10^{15}$$

$$K_{p2} = 0.797$$

● **Temperature-Dependent System** There is a striking similarity among expressions for the temperature dependence of K, k (rate constant), and P (equilibrium vapor pressure):

$$\ln \frac{K_2}{K_1} = -\frac{\Delta H^\circ_{rxn}}{R}\left(\frac{1}{T_2} - \frac{1}{T_1}\right)$$

$$\ln \frac{k_2}{k_1} = -\frac{E_a}{R}\left(\frac{1}{T_2} - \frac{1}{T_1}\right)$$

$$\ln \frac{P_2}{P_1} = -\frac{\Delta H_{vap}}{R}\left(\frac{1}{T_2} - \frac{1}{T_1}\right)$$

Each concentration-related term (K, k, or P) is dependent on T through an energy term (ΔH°_{rxn}, E_a, or ΔH_{vap}, respectively) divided by R. The similarity arises because the equations express the same relationship. K equals a ratio of rate constants, and $E_{a(fwd)} - E_{a(rev)} = \Delta H^\circ_{rxn}$. In the vapor pressure case, for the phase change $A(l) \rightleftharpoons A(g)$, the heat of reaction is equal to the heat of vaporization: $\Delta H_{vap} = \Delta H^\circ_{rxn}$. Also, the equilibrium constant equals the equilibrium vapor pressure: $K_p = P_A$.

Transparency/ARIS Presentation Center, Tables Library

ARIS Presentation Center, Art Library

● **Catalyzed Perpetual Motion?** This engine consists of a piston attached to a flywheel, whose rocker arm holds a catalyst and moves in and out of the reaction in the cylinder. What if the catalyst *could* increase the rate of PCl_5 breakdown but not of its formation? In the cylinder, the catalyst would speed up the breakdown of PCl_5 to PCl_3 and Cl_2—1 mol of gas to 2 mol—increasing gas pressure and pushing the piston out. With the catalyst out of the cylinder, PCl_3 and Cl_2 would re-form PCl_5, lowering gas pressure and moving the piston in. The process would supply power with no input of external energy. If a catalyst could change the rate in only one direction, it would change K, and we could design some amazing machines!

Engineering application

Equation 17.10 confirms the qualitative prediction from Le Châtelier's principle: for a temperature rise, we have

$$T_2 > T_1 \quad \text{and} \quad 1/T_2 < 1/T_1, \quad \text{so } 1/T_2 - 1/T_1 < 0$$

Therefore,

- For an endothermic reaction ($\Delta H^\circ_{rxn} > 0$), the $-(\Delta H^\circ_{rxn}/R)$ term is < 0. With $1/T_2 - 1/T_1 < 0$, the right side of the equation is > 0. Thus, $\ln(K_2/K_1) > 0$, so $K_2 > K_1$.
- For an exothermic reaction ($\Delta H^\circ_{rxn} < 0$), the $-(\Delta H^\circ_{rxn}/R)$ term is > 0. With $1/T_2 - 1/T_1 < 0$, the right side of the equation is < 0. Thus, $\ln(K_2/K_1) < 0$, so $K_2 < K_1$. ●

(For further practice with the van't Hoff equation, see Problems 17.73 and 17.74.)

The Lack of Effect of a Catalyst

Let's briefly consider a final external change to the reacting system: adding a catalyst. Recall from Chapter 16 that a catalyst speeds up a reaction by providing an alternative mechanism with a lower activation energy, thereby increasing the forward *and* reverse rates to the same extent. In other words, it shortens the time needed to attain the *final* concentrations. Thus, *a catalyst shortens the time it takes to reach equilibrium but has **no** effect on the equilibrium position.* ●

If, for instance, we add a catalyst to a mixture of PCl_3 and Cl_2 at 523 K, the system will attain the *same* equilibrium concentrations of PCl_3, Cl_2, and PCl_5 *more quickly* than it did without the catalyst. As you'll see in a moment, however, catalysts often play key roles in optimizing reaction systems.

Table 17.4 summarizes the effects of changing conditions on the position of equilibrium. Note that many changes alter the equilibrium *position*, but only temperature changes alter the value of the equilibrium *constant*. Sample Problem 17.15 shows how to visualize equilibrium at the molecular level. Then, two Chemical Connections essays apply equilibrium principles: the first examines metabolic processes in organisms, and the second describes a major industrial process.

Table 17.4 Effects of Various Disturbances on a System at Equilibrium

Disturbance	Net Direction of Reaction	Effect on Value of K
Concentration		
Increase [reactant]	Toward formation of product	None
Decrease [reactant]	Toward formation of reactant	None
Increase [product]	Toward formation of reactant	None
Decrease [product]	Toward formation of product	None
Pressure		
Increase P (decrease V)	Toward formation of fewer moles of gas	None
Decrease P (increase V)	Toward formation of more moles of gas	None
Increase P (add inert gas, no change in V)	None; concentrations unchanged	None
Temperature		
Increase T	Toward absorption of heat	Increases if $\Delta H^\circ_{rxn} > 0$ Decreases if $\Delta H^\circ_{rxn} < 0$
Decrease T	Toward release of heat	Increases if $\Delta H^\circ_{rxn} < 0$ Decreases if $\Delta H^\circ_{rxn} > 0$
Catalyst added	None; forward and reverse equilibrium attained sooner; rates increase equally	None

SAMPLE PROBLEM 17.15 Determining Equilibrium Parameters from Molecular Scenes

PROBLEM For the reaction,

$$X(g) + Y_2(g) \rightleftharpoons XY(g) + Y(g) \qquad \Delta H > 0$$

the following molecular scenes depict different reaction mixtures (X is green, Y is purple):

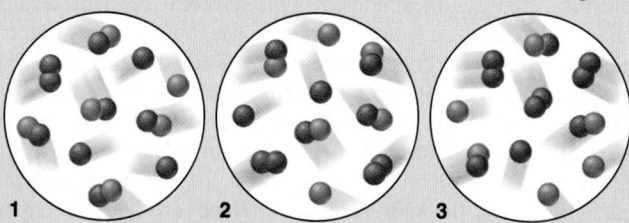

1   2   3

(a) If $K = 2$ at the temperature of the reaction, which scene represents the mixture at equilibrium? **(b)** Will the reaction mixtures in the other two scenes proceed toward reactants or toward products to reach equilibrium? **(c)** For the mixture at equilibrium, how will a rise in temperature affect $[Y_2]$?

PLAN **(a)** We are given the balanced equation and the value of K and must choose the scene representing the mixture at equilibrium. We write the expression for Q, and for each scene, count particles and plug in the numbers to solve for the value of Q. Whichever scene gives a Q equal to K represents the mixture at equilibrium. **(b)** To determine the direction each reaction proceeds in the other two scenes, we compare the value of Q with the given K. If $Q > K$, the numerator (product side) is too high, so the reaction proceeds toward reactants; if $Q < K$, the reaction proceeds toward products. **(c)** We are given the sign of ΔH and must see whether a rise in T (corresponding to supplying heat) will increase or decrease the amount of the reactant Y_2. We treat heat as a reactant or product and see whether adding heat shifts the reaction right or left.

SOLUTION **(a)** For the reaction, we have $Q = \dfrac{[XY][Y]}{[X][Y_2]}$. Thus,

$$\text{scene 1: } Q = \frac{5 \times 3}{1 \times 1} = 15 \qquad \text{scene 2: } Q = \frac{4 \times 2}{2 \times 2} = 2 \qquad \text{scene 3: } Q = \frac{3 \times 1}{3 \times 3} = \frac{1}{3}$$

For scene 2, $Q = K$, so scene 2 represents the mixture at equilibrium.

(b) For scene 1, Q (15) $> K$ (2), so the reaction proceeds toward reactants.

For scene 3, Q ($\frac{1}{3}$) $< K$ (2), so the reaction proceeds toward products.

(c) The reaction is endothermic, so heat acts as a reactant:

$$X(g) + Y_2(g) + heat \rightleftharpoons XY(g) + Y(g)$$

Therefore, adding heat to the left shifts the reaction to the right, so $[Y_2]$ decreases.

CHECK **(a)** Remember that quantities in the numerator (or denominator) of Q are multiplied, not added. For example, the denominator for scene 1 is $1 \times 1 = 1$, not $1 + 1 = 2$.

(c) A good check is to imagine that $\Delta H < 0$ and see if you get the opposite result:

$$X(g) + Y_2(g) \rightleftharpoons XY(g) + Y(g) + heat$$

If $\Delta H < 0$, adding heat would shift the reaction to the left and increase $[Y_2]$.

FOLLOW-UP PROBLEM 17.15 For the reaction

$$C_2(g) + D_2(g) \rightleftharpoons 2CD(g) \qquad \Delta H < 0$$

these molecular scenes depict different reaction mixtures (C is red, D is blue):

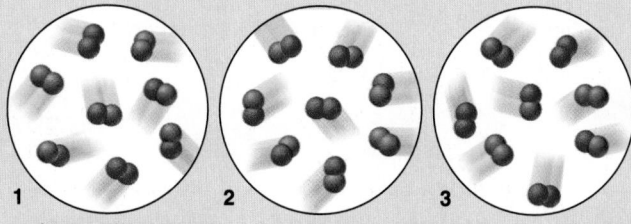

1   2   3

At equilibrium

(a) Calculate the value of K_p. **(b)** In which direction will the reaction proceed for the mixtures *not* at equilibrium? **(c)** For the mixture at equilibrium, what effect will a rise in T have on the total moles of gas (increase, decrease, no effect)? Explain.

Design and Control of a Metabolic Pathway

Biological cells are microscopic wizards of chemical change. From the simplest bacterium to the most specialized neuron, every cell performs thousands of individual reactions that allow it to grow and reproduce, feed and excrete, move and communicate. Taken together, these chemical reactions constitute the cell's *metabolism*. These myriad feats of biochemical breakdown, synthesis, and energy flow are organized into reaction sequences called **metabolic pathways.** Some pathways disassemble the biopolymers in food into monomer building blocks: sugars, amino acids, and nucleotides. Others extract energy from these small molecules and then use much of the energy to assemble molecules with other functions, such as hormones, neurotransmitters, defensive toxins, and the cell's own biopolymers.

In principle, *each step in a metabolic pathway is a reversible reaction catalyzed by a specific enzyme* (Section 16.8). However, *equilibrium is never reached* in a pathway. As in other reaction sequences, the product of the first reaction becomes the reactant of the second, the product of the second becomes the reactant of the third, and so on. Consider the five-step pathway shown in Figure B17.1, in which one amino acid, threonine, is converted into another, isoleucine, in the cells of a bread mold. Threonine, supplied from a different region of the cell, forms ketobutyrate through the catalytic action of the first enzyme (reaction 1). The equilibrium position of reaction 1 continuously shifts to the right because the ketobutyrate is continuously removed as the reactant in reaction 2. Similarly, reaction 2 shifts to the right as its product is used in reaction 3. Each subsequent reaction shifts the equilibrium position of the previous reaction in the direction of product. The final product, isoleucine, is typically removed to make proteins elsewhere in the cell. Thus, *the entire pathway operates in one direction.*

This continuous shift in equilibrium position results in two major features of metabolic pathways. First, *each step proceeds with nearly 100% yield:* Virtually every molecule of threonine that enters this region of the cell eventually changes to ketobutyrate, every molecule of ketobutyrate to the next product, and so on.

Second, *reactant and product concentrations remain constant* or vary within extremely narrow limits, even though the system never attains equilibrium. This situation, called a *steady state,* is different in a basic way from those we have studied so far. In equilibrium systems, equal reaction rates in *opposing* directions give rise to constant concentrations of reactants and products. In steady-state systems, on the other hand, the rates of reactions in *one* direction—into, through, and out of the system—give rise to nearly constant concentrations of intermediates. Ketobutyrate, for example, is formed via reaction 1 just as fast as it is used via reaction 2, so its concentration is kept constant. (You can establish a steady-state amount of water by filling a sink and then adjusting the faucet and drain so that water enters just as fast as it leaves.)

It is critical that a cell regulate the concentration of the final product of a pathway. Recall from Chapter 16 that an enzyme catalyzes a reaction when the substrate (reactant) occupies the *active site.* However, substrate concentrations are so much higher than enzyme concentrations that, if no other factors were involved, the active sites on all the enzyme molecules would always be occupied and all cellular reactions would occur at their maximum rates. This might be ideal for an industrial process, but it could be catastrophic for an organism. To regulate overall product formation, the rates of certain key steps are controlled by *regulatory enzymes,* which contain an *inhibitor site* on their surface in addition to an active site. The enzyme's three-dimensional structure is such that when the inhibitor site is occupied, the shape of the active site is deformed and the reaction cannot be catalyzed (Figure B17.2).

In the simplest case of metabolic control (Figure B17.1), *the final product of the pathway is also the inhibitor molecule, and the regulatory enzyme catalyzes the first step.* Suppose, for instance, that a bread mold cell is temporarily making less protein, so the isoleucine synthesized by the pathway in Figure B17.1 is not being removed as quickly. As the concentration of isoleucine rises, the chance increases that an isoleucine molecule will land on the inhibitor site of the first enzyme in the pathway, thereby inhibiting its own production. This process is called *end-product feedback inhibition.* More complex pathways have more elaborate regulatory schemes, but many operate through similar types of inhibitory feedback (see Problem 17.113). The exquisitely detailed regulation of a cell's numerous metabolic pathways arises directly from the principles of chemical kinetics and equilibrium.

Substrate in active site when inhibitor is absent: catalysis occurs

Active site is deformed when inhibitor is present: substrate cannot bind so no catalysis occurs

Portion of enzyme

Portion of enzyme

A   Inhibitor binding site   **B**   Inhibitor bound to site

Figure B17.2 Effect of inhibitor binding on shape of active site. **A,** If inhibitor site is not occupied, the enzyme catalyzes the reaction. **B,** If inhibitor site is occupied, the enzyme molecule does not function.

Enzyme 1 Threonine Dehydratase   Enzyme 2   Enzyme 3   Enzyme 4   Enzyme 5

THREONINE ⟶ Ketobutyrate ⇌ Acetohydroxybutyrate ⇌ Dihydroxymethylvalerate ⇌ Ketomethylvalerate ⇌ ISOLEUCINE

INHIBITION

Figure B17.1 The biosynthesis of isoleucine from threonine. Isoleucine is synthesized from threonine in a sequence of five enzyme-catalyzed reactions. The unequal equilibrium arrows show that each step is reversible but is shifted toward product as each product becomes the reactant of the subsequent step. When the cell's need for isoleucine is temporarily satisfied, the concentration of isoleucine, the end product, builds up and inhibits the catalytic activity of threonine dehydratase, the first enzyme in the pathway.

Chemical Connections to Industrial Production

The Haber Process for the Synthesis of Ammonia ∞ Engineering application

Nitrogen occurs in many essential natural and synthetic compounds. By far the richest source of nitrogen is the atmosphere, where four of every five molecules are N_2. Despite this abundance, the supply of *usable* nitrogen for biological and manufacturing processes is limited because of the low chemical reactivity of N_2. Because of the strong triple bond holding the two N atoms together, the nitrogen atom is very difficult to "fix," that is, to combine with other atoms.

Natural nitrogen fixation occurs either through the elegant specificity of enzymes found in bacteria that live on plant roots or through the brute force of lightning. Nearly 13% of nitrogen fixation on Earth is accomplished industrially through the **Haber process** for the formation of ammonia from its elements:

$$N_2(g) + 3H_2(g) \rightleftharpoons 2NH_3(g) \qquad \Delta H^\circ_{rxn} = -91.8 \text{ kJ}$$

The process was developed by the German chemist Fritz Haber and first used in 1913. From its humble beginnings in a plant with a capacity of 12,000 tons a year, world production of ammonia has exploded to a current level of more than 110 million tons a year. On a mole basis, more ammonia is produced industrially than any other compound. Over 80% of this ammonia is used in fertilizer applications. In fact, the most common form of fertilizer is compressed anhydrous, liquid NH_3 injected directly into the soil (Figure B17.3). Other uses of NH_3 include the production of explosives, via the formation of HNO_3, and the making of nylons and other polymers. Smaller amounts are used as refrigerants, rubber stabilizers, and household cleaners and in the synthesis of pharmaceuticals and other organic chemicals.

The Haber process provides an excellent opportunity to apply equilibrium principles and see the compromises needed to make an industrial process economically worthwhile. From inspection of the balanced equation, we can see three ways to maximize the yield of ammonia:

1. *Decrease [NH_3].* Ammonia is the product, so removing it as it forms will make the system produce more in a continual drive to reattain equilibrium.

Table B17.1	Effect of Temperature on K_c for Ammonia Synthesis
T (K)	K_c
200.	7.17×10^{15}
300.	2.69×10^{8}
400.	3.94×10^{4}
500.	1.72×10^{2}
600.	4.53×10^{0}
700.	2.96×10^{-1}
800.	3.96×10^{-2}

ARIS Presentation Center, Tables Library

2. *Decrease volume (increase pressure).* Because 4 mol of gas reacts to form 2 mol of gas, decreasing the volume will shift the equilibrium position toward fewer moles of gas, that is, toward ammonia formation.

3. *Decrease temperature.* Because the formation of ammonia is exothermic, decreasing the temperature (removing heat) will shift the equilibrium position toward the product, thereby increasing K_c (Table B17.1).

Therefore, the ideal conditions for maximizing the yield of ammonia are continual removal of NH_3 as it forms, high pressure, and low temperature. Figure B17.4 shows the percent yield of ammonia at various conditions of pressure and temperature. Note the almost complete conversion (98.3%) to ammonia at 1000 atm and the relatively low temperature of 473 K (200.°C).

Unfortunately, a problem arises that highlights the distinction between the principles of equilibrium and kinetics. Although the *yield* is favored by low temperature, the *rate* of formation is not. In fact, ammonia forms so slowly at low temperatures that the process becomes uneconomical. In practice, a compromise is achieved that optimizes yield *and* rate. High pressure and continuous removal are used to increase yield, but the temperature is

(continued)

Figure B17.3 Liquid ammonia used as fertilizer.

ARIS Presentation Center, Photo Library

Figure B17.4 Percent yield of ammonia vs. temperature (°C) at five different operating pressures. At very high pressure and low temperature (*top left*), the yield is high, but the rate of formation is low. Industrial conditions (*circle*) are between 200 and 300 atm at about 400°C.

ARIS Presentation Center, Art Library

raised to a moderate level and a catalyst is used to increase the rate. Achieving the same rate without employing a catalyst would require much higher temperatures and would result in a much lower yield.

The stages in the industrial production of ammonia are shown schematically in Figure B17.5. To extend equipment life and minimize cost, modern ammonia plants operate at pressures of about 200 to 300 atm and temperatures of around 673 K (400.°C). The catalyst consists of 5-mm to 10-mm chunks of iron crystals embedded in a fused mixture of MgO, Al_2O_3, and SiO_2. The stoi-

chiometric ratio of compressed reactant gases ($N_2/H_2 = 1/3$ by volume) is injected into the heated, pressurized reaction chamber, where it flows over the catalyst beds. Some of the heat needed is supplied by the enthalpy change of the reaction. The emerging equilibrium mixture, which contains about 35% NH_3 by volume, is cooled by refrigeration coils until the NH_3 (boiling point, $-33.4°C$) condenses and is removed. Because N_2 and H_2 have much lower boiling points than NH_3, they remain gaseous and are recycled by pumps back into the reaction chamber to continue the process.

ARIS Presentation Center, Art Library

Figure B17.5 Key stages in the Haber process for synthesizing ammonia.

Section Summary

Le Châtelier's principle states that if a system at equilibrium is disturbed, the system undergoes a net reaction that reduces the disturbance and allows equilibrium to be reattained. • Changes in concentration cause a net reaction away from the added component or toward the removed component. For a reaction that involves a change in number of moles of gas, an increase in pressure (decrease in volume) causes a net reaction toward fewer moles of gas, and a decrease in pressure causes the opposite change. Although the equilibrium concentrations of components change as a result of concentration and volume changes, K does not change. • A temperature change *does* change K: higher T increases K for an endothermic reaction (positive $\Delta H°_{rxn}$) and decreases K for an exothermic reaction (negative $\Delta H°_{rxn}$). • A catalyst causes the system to reach equilibrium more quickly by speeding forward and reverse reactions equally, but it does not affect the equilibrium position. • A metabolic pathway is a cellular reaction sequence in which each step is shifted completely toward product. Its overall yield is controlled by feedback inhibition of certain key enzymes. • Ammonia is produced in a process favored by high pressure, low temperature, and continual removal of product. To make the process economical, an intermediate temperature and a catalyst are used.

Chapter Perspective

The equilibrium phenomenon is central to all natural systems. In this introduction, which focused primarily on gaseous systems, we discussed the nature of equilibrium on the observable and molecular levels, ways to solve relevant problems, and how conditions affect systems at equilibrium. In Chapters 18 and 19, we apply these ideas to acids and bases and to other aqueous ionic systems. Then we examine the fundamental relationship between the equilibrium constant and the energy change that accompanies a reaction to understand why chemical reactions occur.

CHAPTER REVIEW GUIDE
The following sections provide many aids to help you study this chapter. (Numbers in parentheses refer to pages, unless noted otherwise.)

Learning Objectives
These are concepts and skills you should know after studying this chapter.

Relevant section and/or sample problem (SP) numbers appear in parentheses.

Understand These Concepts

1. The distinction between the speed (rate) and the extent of a reaction (Introduction)
2. Why a system attains dynamic equilibrium when forward and reverse reaction rates are equal (Section 17.1)
3. The equilibrium constant as a number that is equal to a particular ratio of rate constants and of concentration terms (Section 17.1)
4. How the magnitude of K is related to the extent of the reaction (Section 17.1)
5. Why the same equilibrium state is reached no matter what the starting concentrations of the reacting system (Section 17.2)
6. How the reaction quotient (Q) changes continuously until the system reaches equilibrium, at which point $Q = K$ (Section 17.2)
7. Why the form of Q is based exactly on the balanced equation *as written* (Section 17.2)
8. How the *sum* of reaction steps gives the overall reaction, and the *product* of Q's (or K's) gives the overall Q (or K) (Section 17.2)
9. Why pure solids and liquids do not appear in Q (Section 17.2)
10. How the interconversion of K_c and K_p is based on the ideal gas law and Δn_{gas} (Section 17.3)
11. How the reaction direction depends on the relative values of Q and K (Section 17.4)
12. How a reaction table is used to find an unknown quantity (concentration or pressure) (Section 17.5)
13. How assuming that the change in [reactant] is relatively small simplifies finding equilibrium quantities (Section 17.5)
14. How Le Châtelier's principle explains the effects of a change in concentration, pressure (volume), or temperature on a system at equilibrium and on K (Section 17.6)
15. Why a change in temperature *does* affect K (Section 17.6)
16. Why the addition of a catalyst does *not* affect K (Section 17.6)

Master These Skills

1. Writing the reaction quotient (Q) from a balanced equation (SP 17.1)

2. Writing Q and calculating K for a reaction consisting of more than one step (SP 17.2)
3. Writing Q and finding K for a reaction multiplied by a common factor (SP 17.3)
4. Writing Q for heterogeneous equilibria (Section 17.2)
5. Converting between K_c and K_p (SP 17.4)
6. Comparing Q and K to determine reaction direction (SPs 17.5, 17.6)
7. Substituting quantities (concentrations or pressures) into Q to find K (Section 17.5)
8. Using a reaction table to determine quantities and find K (SP 17.7)
9. Finding one equilibrium quantity from other equilibrium quantities and K (SP 17.8)
10. Finding an equilibrium quantity from initial quantities and K (SP 17.9)
11. Solving a quadratic equation for an unknown equilibrium quantity (Section 17.5)
12. Assuming that the change in [reactant] is relatively small to find equilibrium quantities and checking the assumption (SP 17.10)
13. Comparing the values of Q and K to find reaction direction and x, the unknown change in a quantity (SP 17.11)
14. Using the relative values of Q and K to predict the effect of a change in concentration on the equilibrium position and on K (SP 17.12)
15. Using Le Châtelier's principle and Δn_{gas} to predict the effect of a change in pressure (volume) on the equilibrium position (SP 17.13)
16. Using Le Châtelier's principle and $\Delta H°$ to predict the effect of a change in temperature on the equilibrium position and on K (SP 17.14)
17. Using the van't Hoff equation to calculate K at one temperature given K at another temperature (Section 17.6)
18. Using molecular scenes to find equilibrium parameters (SP 17.15)

Key Terms
These important terms appear in boldface in the chapter and are defined again in the Glossary.

Section 17.1
equilibrium constant (K) (740)

Section 17.2
law of chemical equilibrium (law of mass action) (741)
reaction quotient (Q) (741)

Section 17.6
Le Châtelier's principle (761)
metabolic pathway (770)
Haber process (771)

Key Equations and Relationships
Numbered and screened concepts are listed for you to refer to or memorize.

17.1 Defining equilibrium in terms of reaction rates (739):

$$\text{At equilibrium: rate}_{fwd} = \text{rate}_{rev}$$

17.2 Defining the equilibrium constant for the reaction A $\rightleftharpoons$ 2B (740):

$$K = \frac{k_{fwd}}{k_{rev}} = \frac{[B]^2_{eq}}{[A]_{eq}}$$

17.3 Defining the equilibrium constant in terms of the reaction quotient (741):

$$\text{At equilibrium: } Q = K$$

17.4 Expressing Q_c for the reaction aA + bB $\rightleftharpoons$ cC + dD (742):

$$Q_c = \frac{[C]^c[D]^d}{[A]^a[B]^b}$$

Key Equations and Relationships (continued)

17.5 Finding the overall K for a reaction sequence (743):
$$K_{overall} = K_1 \times K_2 \times K_3 \times \cdots$$

17.6 Finding K of a reaction from K of the reverse reaction (745):
$$K_{fwd} = \frac{1}{K_{rev}}$$

17.7 Finding K of a reaction multiplied by a factor n (745):
$$K' = K^n$$

17.8 Relating K based on pressures to K based on concentrations (749):
$$K_p = K_c(RT)^{\Delta n_{gas}}$$

17.9 Assuming that ignoring the concentration that reacts introduces no significant error (757):
$$[A]_{init} - [A]_{reacting} = [A]_{eq} \approx [A]_{init}$$

17.10 Finding K at one temperature given K at another (van't Hoff equation) (767):
$$\ln \frac{K_2}{K_1} = -\frac{\Delta H^{\circ}_{rxn}}{R}\left(\frac{1}{T_2} - \frac{1}{T_1}\right)$$

Highlighted Figures and Tables These figures (F) and tables (T) provide a visual review of key ideas.

F17.2 The range of equilibrium constants (740)
F17.3 The change in Q during a reaction (742)
T17.2 Ways of expressing Q and calculating K (747)
F17.5 Reaction direction and the relative sizes of Q and K (750)
F17.6 Steps in solving equilibrium problems (760)

F17.7 Effect of a change in concentration (762)
T17.3 Effect of added Cl_2 on the PCl_3-Cl_2-PCl_5 system (763)
F17.9 Effect of pressure (volume) on equilibrium (765)
T17.4 Effects of disturbances on equilibrium (768)

Brief Solutions to FOLLOW-UP PROBLEMS Compare your solutions to these calculation steps and answers.

17.1 (a) $Q_c = \dfrac{[NO]^4[H_2O]^6}{[NH_3]^4[O_2]^5}$ (b) $Q_c = \dfrac{[N_2O][NO_2]}{[NO]^3}$

17.2 $H_2(g) + Br_2(g) \rightleftharpoons 2HBr(g)$;

$$Q_{c(overall)} = \frac{[HBr]^2}{[H_2][Br_2]}$$

$$Q_{c(overall)} = Q_{c1} \times Q_{c2} \times Q_{c3}$$

$$= \frac{[\cancel{Br}]^2}{[Br_2]} \times \frac{[HBr]\cancel{[H]}}{\cancel{[Br]}[H_2]} \times \frac{[HBr]}{\cancel{[H]}\cancel{[Br]}} = \frac{[HBr]^2}{[H_2][Br_2]}$$

17.3 (a) $K_c = K_{c(ref)}^{1/2} = 2.8 \times 10^4$

(b) $K_c = \left(\dfrac{1}{K_{c(ref)}}\right)^{2/3} = 1.2 \times 10^{-6}$

17.4 $K_p = K_c(RT)^{-1} = 1.67\left(0.0821\dfrac{atm \cdot L}{mol \cdot K} \times 500.\ K\right)^{-1}$

$$= 4.07 \times 10^{-2}$$

17.5 $K_c = \dfrac{[Y]}{[X]} = 1.4$

1. $Q = 0.33$, right
2. $Q = 1.4$, no change
3. $Q = 2.0$, left

17.6 $Q_p = \dfrac{(P_{CH_3Cl})(P_{HCl})}{(P_{CH_4})(P_{Cl_2})} = \dfrac{(0.24)(0.47)}{(0.13)(0.035)} = 25$;

$Q_p < K_p$, so CH_3Cl is forming.

17.7 From the reaction table for $2NO + O_2 \rightleftharpoons 2NO_2$,
$$P_{O_2} = 1.000\ atm - x = 0.506\ atm \qquad x = 0.494\ atm$$
Also, $P_{NO} = 0.012\ atm$ and $P_{NO_2} = 0.988\ atm$, so

$$K_p = \frac{0.988^2}{0.012^2(0.506)} = 1.3 \times 10^4$$

17.8 Since $\Delta n_{gas} = 0$, $K_p = K_c = 2.3 \times 10^{30} = \dfrac{(0.781)(0.209)}{P_{NO}^2}$

Thus, $P_{NO} = 2.7 \times 10^{-16}\ atm$

17.9 From the reaction table, $[H_2] = [I_2] = x$; $[HI] = 0.242 - 2x$.

Thus, $K_c = 1.26 \times 10^{-3} = \dfrac{x^2}{(0.242 - 2x)^2}$

Taking the square root of both sides, ignoring the negative root, and solving gives $x = [H_2] = 8.02 \times 10^{-3}\ M$.

17.10 (a) Based on the reaction table, and assuming that $0.20\ M - x \approx 0.20\ M$,

$$K_c = 2.94 \times 10^{-10} \approx \frac{4x^2}{0.20} \qquad x \approx 3.8 \times 10^{-6}$$

Error $= 1.9 \times 10^{-3}\%$, so assumption is justified; therefore, at equilibrium, $[I_2] = 0.20\ M$ and $[I] = 7.6 \times 10^{-6}\ M$.

(b) Based on the same reaction table and assumption, $x \approx 0.10$; error is 50%, so assumption is *not* justified. Solve equation:
$$4x^2 + 0.209x - 0.042 = 0 \qquad x = 0.080\ M$$

Therefore, at equilibrium, $[I_2] = 0.12\ M$ and $[I] = 0.16\ M$.

17.11 (a) $Q_c = \dfrac{(0.0900)(0.0900)}{0.2100} = 3.86 \times 10^{-2}$

$Q_c < K_c$, so reaction proceeds to the right.
(b) From the reaction table,

$[PCl_5] = 0.2100\ M - x = 0.2065\ M \qquad so \qquad x = 0.0035\ M$

So, $[Cl_2] = [PCl_3] = 0.0900\ M + x = 0.0935\ M$.

17.12 (a) $[SiF_4]$ increases; (b) decreases; (c) decreases; (d) no effect.

17.13 (a) Decrease P; (b) increase P; (c) increase P.

17.14 (a) P_{H_2} will decrease; K will increase; (b) P_{N_2} will increase; K will decrease; (c) P_{PCl_5} will increase; K will increase.

17.15 (a) Since $P = \dfrac{n}{V}RT$ and, in this case, V, R, and T cancel,

$$K_p = \frac{n_{CD}^2}{n_{C_2} \times n_{D_2}} = \frac{16}{(2)(2)} = 4$$

(b) Scene 2, to the left; scene 3, to the right. (c) There are 2 mol of gas on each side of the balanced equation, so there is no effect on total moles of gas.

PROBLEMS

Problems with colored numbers are answered in Appendix E and worked in detail in the Student Solutions Manual. Problem sections match those in the text and provide the numbers of relevant sample problems. Most offer Concept Review Questions, Skill-Building Exercises (grouped in pairs covering the same concept), and Problems in Context. The Comprehensive Problems are based on material from any section or previous chapter.

The Equilibrium State and the Equilibrium Constant

● Concept Review Questions

● **17.1** A change in reaction conditions increases the rate of a certain forward reaction more than that of the reverse reaction. What is the effect on the equilibrium constant and the concentrations of reactants and products at equilibrium?

● **17.2** When a chemical company employs a new reaction to manufacture a product, the chemists consider its rate (kinetics) and yield (equilibrium). How do each of these affect the usefulness of a manufacturing process?

● **17.3** If there is no change in concentrations, why is the equilibrium state considered dynamic?

● **17.4** Is K very large or very small for a reaction that goes essentially to completion? Explain.

● **17.5** White phosphorus, P_4, is produced by the reduction of phosphate rock, $Ca_3(PO_4)_2$. If exposed to oxygen, the waxy, white solid smokes, bursts into flames, and releases a large quantity of heat. Does the reaction

$$P_4(g) + 5O_2(g) \rightleftharpoons P_4O_{10}(s)$$

have a large or small equilibrium constant? Explain.

The Reaction Quotient and the Equilibrium Constant

(Sample Problems 17.1 to 17.3).

● Concept Review Questions

● **17.6** For a given reaction at a given temperature, the value of K is constant. Is the value of Q also constant? Explain.

● **17.7** In a study of the thermal decomposition of lithium peroxide,

$$2Li_2O_2(s) \rightleftharpoons 2Li_2O(s) + O_2(g)$$

a chemist finds that, as long as some Li_2O_2 is present at the end of the experiment, the amount of O_2 obtained in a given container at a given T is the same. Explain.

● **17.8** In a study of the formation of HI from its elements,

$$H_2(g) + I_2(g) \rightleftharpoons 2HI(g)$$

equal amounts of H_2 and I_2 were placed in a container, which was then sealed and heated.
(a) On one set of axes, sketch concentration vs. time curves for H_2 and HI, and explain how Q changes as a function of time.
(b) Is the value of Q different if $[I_2]$ is plotted instead of $[H_2]$?

● **17.9** Explain the difference between a heterogeneous and a homogeneous equilibrium. Give an example of each.

● **17.10** Does Q for the formation of 1 mol of NO from its elements differ from Q for the decomposition of 1 mol of NO to its elements? Explain and give the relationship between the two Q's.

● **17.11** Does Q for the formation of 1 mol of NH_3 from H_2 and N_2 differ from Q for the formation of NH_3 from H_2 and 1 mol of N_2? Explain and give the relationship between the two Q's.

● Skill-Building Exercises (grouped in similar pairs)

● **17.12** Balance each reaction and write its reaction quotient, Q_c:
(a) $NO(g) + O_2(g) \rightleftharpoons N_2O_3(g)$
(b) $SF_6(g) + SO_3(g) \rightleftharpoons SO_2F_2(g)$
(c) $SClF_5(g) + H_2(g) \rightleftharpoons S_2F_{10}(g) + HCl(g)$

● **17.13** Balance each reaction and write its reaction quotient, Q_c:
(a) $C_2H_6(g) + O_2(g) \rightleftharpoons CO_2(g) + H_2O(g)$
(b) $CH_4(g) + F_2(g) \rightleftharpoons CF_4(g) + HF(g)$
(c) $SO_3(g) \rightleftharpoons SO_2(g) + O_2(g)$

● **17.14** Balance each reaction and write its reaction quotient, Q_c:
(a) $NO_2Cl(g) \rightleftharpoons NO_2(g) + Cl_2(g)$
(b) $POCl_3(g) \rightleftharpoons PCl_3(g) + O_2(g)$
(c) $NH_3(g) + O_2(g) \rightleftharpoons N_2(g) + H_2O(g)$

● **17.15** Balance each reaction and write its reaction quotient, Q_c:
(a) $O_2(g) \rightleftharpoons O_3(g)$
(b) $NO(g) + O_3(g) \rightleftharpoons NO_2(g) + O_2(g)$
(c) $N_2O(g) + H_2(g) \rightleftharpoons NH_3(g) + H_2O(g)$

● **17.16** At a particular temperature, $K_c = 1.6 \times 10^{-2}$ for

$$2H_2S(g) \rightleftharpoons 2H_2(g) + S_2(g)$$

Calculate K_c for each of the following reactions:
(a) $\frac{1}{2}S_2(g) + H_2(g) \rightleftharpoons H_2S(g)$
(b) $5H_2S(g) \rightleftharpoons 5H_2(g) + \frac{5}{2}S_2(g)$

● **17.17** At a particular temperature, $K_c = 6.5 \times 10^2$ for

$$2NO(g) + 2H_2(g) \rightleftharpoons N_2(g) + 2H_2O(g)$$

Calculate K_c for each of the following reactions:
(a) $NO(g) + H_2(g) \rightleftharpoons \frac{1}{2}N_2(g) + H_2O(g)$
(b) $2N_2(g) + 4H_2O(g) \rightleftharpoons 4NO(g) + 4H_2(g)$

● **17.18** Balance each of the following examples of heterogeneous equilibria and write its reaction quotient, Q_c:
(a) $Na_2O_2(s) + CO_2(g) \rightleftharpoons Na_2CO_3(s) + O_2(g)$
(b) $H_2O(l) \rightleftharpoons H_2O(g)$
(c) $NH_4Cl(s) \rightleftharpoons NH_3(g) + HCl(g)$

● **17.19** Balance each of the following examples of heterogeneous equilibria and write its reaction quotient, Q_c:
(a) $H_2O(l) + SO_3(g) \rightleftharpoons H_2SO_4(aq)$
(b) $KNO_3(s) \rightleftharpoons KNO_2(s) + O_2(g)$
(c) $S_8(s) + F_2(g) \rightleftharpoons SF_6(g)$

● **17.20** Balance each of the following examples of heterogeneous equilibria and write its reaction quotient, Q_c:
(a) $NaHCO_3(s) \rightleftharpoons Na_2CO_3(s) + CO_2(g) + H_2O(g)$
(b) $SnO_2(s) + H_2(g) \rightleftharpoons Sn(s) + H_2O(g)$
(c) $H_2SO4(l) + SO_3(g) \rightleftharpoons H_2S_2O_7(l)$

● **17.21** Balance each of the following examples of heterogeneous equilibria and write its reaction quotient, Q_c:
(a) $Al(s) + NaOH(aq) + H_2O(l) \rightleftharpoons$
$$Na[Al(OH)_4](aq) + H_2(g)$$
(b) $CO_2(s) \rightleftharpoons CO_2(g)$
(c) $N_2O_5(s) \rightleftharpoons NO_2(g) + O_2(g)$

● Problems in Context

● **17.22** Write Q_c for each of the following:
(a) Hydrogen chloride gas reacts with oxygen gas to produce chlorine gas and water vapor.
(b) Solid diarsenic trioxide reacts with fluorine gas to produce liquid arsenic pentafluoride and oxygen gas.

Levels of Difficulty ● Easy ●● Medium ●●● Challenging

(c) Gaseous sulfur tetrafluoride reacts with liquid water to produce gaseous sulfur dioxide and hydrogen fluoride gas.

(d) Solid molybdenum(VI) oxide reacts with gaseous xenon difluoride to form liquid molybdenum(VI) fluoride, xenon gas, and oxygen gas.

• **17.23** The interhalogen ClF_3 is prepared in a two-step fluorination of chlorine gas:

$$Cl_2(g) + F_2(g) \rightleftharpoons ClF(g)$$
$$ClF(g) + F_2(g) \rightleftharpoons ClF_3(g)$$

(a) Balance each step and write the overall equation.

(b) Show that the overall Q_c equals the product of the Q_c's for the individual steps.

Expressing Equilibria with Pressure Terms: Relation Between K_c and K_p
(Sample Problem 17.4)

● **Concept Review Questions**

• **17.24** Guldberg and Waage proposed the definition of the equilibrium constant as a certain ratio of *concentrations*. What relationship allows us to use a particular ratio of *partial pressures* (for a gaseous reaction) to express an equilibrium constant? Explain.

• **17.25** When are K_c and K_p equal, and when are they not?

●● **17.26** A certain reaction at equilibrium has more moles of gaseous products than of gaseous reactants.

(a) Is K_c larger or smaller than K_p?

(b) Write a statement about the relative sizes of K_c and K_p for any gaseous equilibrium.

● **Skill-Building Exercises** *(grouped in similar pairs)*

• **17.27** Determine Δn_{gas} for each of the following reactions:

(a) $2KClO_3(s) \rightleftharpoons 2KCl(s) + 3O_2(g)$

(b) $2PbO(s) + O_2(g) \rightleftharpoons 2PbO_2(s)$

(c) $I_2(s) + 3XeF_2(s) \rightleftharpoons 2IF_3(s) + 3Xe(g)$

• **17.28** Determine Δn_{gas} for each of the following reactions:

(a) $MgCO_3(s) \rightleftharpoons MgO(s) + CO_2(g)$

(b) $2H_2(g) + O_2(g) \rightleftharpoons 2H_2O(l)$

(c) $HNO_3(l) + ClF(g) \rightleftharpoons ClONO_2(g) + HF(g)$

• **17.29** Calculate K_c for each of the following equilibria:

(a) $CO(g) + Cl_2(g) \rightleftharpoons COCl_2(g)$; $K_p = 3.9 \times 10^{-2}$ at 1000. K

(b) $S_2(g) + C(s) \rightleftharpoons CS_2(g)$; $K_p = 28.5$ at 500. K

• **17.30** Calculate K_c for each of the following equilibria:

(a) $H_2(g) + I_2(g) \rightleftharpoons 2HI(g)$; $K_p = 49$ at 730. K

(b) $2SO_2(g) + O_2(g) \rightleftharpoons 2SO_3(g)$; $K_p = 2.5 \times 10^{10}$ at 500. K

• **17.31** Calculate K_p for each of the following equilibria:

(a) $N_2O_4(g) \rightleftharpoons 2NO_2(g)$; $K_c = 6.1 \times 10^{-3}$ at 298 K

(b) $N_2(g) + 3H_2(g) \rightleftharpoons 2NH_3(g)$; $K_c = 2.4 \times 10^{-3}$ at 1000. K

• **17.32** Calculate K_p for each of the following equilibria:

(a) $H_2(g) + CO_2(g) \rightleftharpoons H_2O(g) + CO(g)$; $K_c = 0.77$ at 1020. K

(b) $3O_2(g) \rightleftharpoons 2O_3(g)$; $K_c = 1.8 \times 10^{-56}$ at 570. K

Reaction Direction: Comparing Q and K
(Sample Problems 17.5 and 17.6)

● **Concept Review Questions**

• **17.33** When the numerical value of Q is less than K, in which direction does the reaction proceed to reach equilibrium? Explain.

●● **17.34** The following molecular scenes depict the aqueous reaction $2D \rightleftharpoons E$, with D red and E blue. Each sphere represents 0.0100 mol, but the volume in scene A is 1.00 L, whereas in scenes B and C, it is 0.500 L.

A   **B**   **C**

(a) If the reaction in scene A is at equilibrium, calculate K_c.

(b) Are the reactions in scenes B and C at equilibrium? Which, if either, is not, and in which direction will it proceed?

● **Skill-Building Exercises** *(grouped in similar pairs)*

• **17.35** At 425°C, $K_p = 4.18 \times 10^{-9}$ for the reaction

$$2HBr(g) \rightleftharpoons H_2(g) + Br_2(g)$$

In one experiment, 0.20 atm of $HBr(g)$, 0.010 atm of $H_2(g)$, and 0.010 atm of $Br_2(g)$ are introduced into a container. Is the reaction at equilibrium? If not, in which direction will it proceed?

• **17.36** At 100°C, $K_p = 60.6$ for the reaction

$$2NOBr(g) \rightleftharpoons 2NO(g) + Br_2(g)$$

In a given experiment, 0.10 atm of each component is placed in a container. Is the system at equilibrium? If not, in which direction will the reaction proceed?

● **Problems in Context**

●● **17.37** The water-gas shift reaction plays a central role in the chemical methods for obtaining cleaner fuels from coal:

$$CO(g) + H_2O(g) \rightleftharpoons CO_2(g) + H_2(g)$$

At a given temperature, $K_p = 2.7$. If 0.13 mol of CO, 0.56 mol of H_2O, 0.62 mol of CO_2, and 0.43 mol of H_2 are put in a 2.0-L flask, in which direction does the reaction proceed ?

How to Solve Equilibrium Problems
(Sample Problems 17.7 to 17.11)

● **Concept Review Questions**

●● **17.38** In the 1980s, CFC-11 was one of the most heavily produced chlorofluorocarbons. The last step in its formation is

$$CCl_4(g) + HF(g) \rightleftharpoons CFCl_3(g) + HCl(g)$$

If you start the reaction with equal concentrations of CCl_4 and HF, you obtain equal concentrations of $CFCl_3$ and HCl at equilibrium. Are the final concentrations of $CFCl_3$ and HCl equal if you start with unequal concentrations of CCl_4 and HF? Explain.

• **17.39** For a problem involving the catalyzed reaction of methane and steam, the following reaction table was prepared:

Pressure (atm)	$CH_4(g)$ +	$2H_2O(g)$ $\rightleftharpoons$	$CO_2(g)$ +	$4H_2(g)$
Initial	0.30	0.40	0	0
Change	$-x$	$-2x$	$+x$	$+4x$
Equilibrium	$0.30 - x$	$0.40 - 2x$	x	$4x$

Explain the entries in the "Change" and "Equilibrium" rows.

• **17.40** (a) What is the basis of the approximation that avoids using the quadratic formula to find an equilibrium concentration?

(b) When should this approximation *not* be made?

● **Skill-Building Exercises** *(grouped in similar pairs)*

• **17.41** In an experiment to study the formation of $HI(g)$,

$$H_2(g) + I_2(g) \rightleftharpoons 2HI(g)$$

$H_2(g)$ and $I_2(g)$ were placed in a sealed container at a certain temperature. At equilibrium, $[H_2] = 6.50\times10^{-5}$ M, $[I_2] = 1.06\times10^{-3}$ M, and $[HI] = 1.87\times10^{-3}$ M. Calculate K_c for the reaction at this temperature.

• **17.42** Gaseous ammonia was introduced into a sealed container and heated to a certain temperature:

$$2NH_3(g) \rightleftharpoons N_2(g) + 3H_2(g)$$

At equilibrium, $[NH_3] = 0.0225$ M, $[N_2] = 0.114$ M, and $[H_2] = 0.342$ M. Calculate K_c for the reaction at this temperature.

• **17.43** Gaseous PCl_5 decomposes according to the reaction

$$PCl_5(g) \rightleftharpoons PCl_3(g) + Cl_2(g)$$

In one experiment, 0.15 mol of $PCl_5(g)$ was introduced into a 2.0-L container. Construct the reaction table for this process.

• **17.44** Hydrogen fluoride, HF, can be made from the reaction

$$H_2(g) + F_2(g) \rightleftharpoons 2HF(g)$$

In one experiment, 0.10 mol of $H_2(g)$ and 0.050 mol of $F_2(g)$ are added to a 0.50-L flask. Write a reaction table for this process.

• **17.45** For the following reaction, $K_p = 6.5\times10^4$ at 308 K:

$$2NO(g) + Cl_2(g) \rightleftharpoons 2NOCl(g)$$

At equilibrium, $P_{NO} = 0.35$ atm and $P_{Cl_2} = 0.10$ atm. What is the equilibrium partial pressure of $NOCl(g)$?

• **17.46** For the following reaction, $K_p = 0.262$ at 1000°C:

$$C(s) + 2H_2(g) \rightleftharpoons CH_4(g)$$

At equilibrium, P_{H_2} is 1.22 atm. What is the equilibrium partial pressure of $CH_4(g)$?

• **17.47** Ammonium hydrogen sulfide decomposes according to the following reaction, for which $K_p = 0.11$ at 250°C:

$$NH_4HS(s) \rightleftharpoons H_2S(g) + NH_3(g)$$

If 55.0 g of $NH_4HS(s)$ is placed in a sealed 5.0-L container, what is the partial pressure of $NH_3(g)$ at equilibrium?

• **17.48** Hydrogen sulfide decomposes according to the following reaction, for which $K_c = 9.30\times10^{-8}$ at 700°C:

$$2H_2S(g) \rightleftharpoons 2H_2(g) + S_2(g)$$

If 0.45 mol of H_2S is placed in a 3.0-L container, what is the equilibrium concentration of $H_2(g)$ at 700°C?

• **17.49** Even at high T, the formation of NO is not favored:

$$N_2(g) + O_2(g) \rightleftharpoons 2NO(g) \qquad K_c = 4.10\times10^{-4} \text{ at } 2000°C$$

What is [NO] when a mixture of 0.20 mol of $N_2(g)$ and 0.15 mol of $O_2(g)$ reach equilibrium in a 1.0-L container at 2000°C?

• **17.50** Nitrogen dioxide decomposes according to the reaction

$$2NO_2(g) \rightleftharpoons 2NO(g) + O_2(g)$$

where $K_p = 4.48\times10^{-13}$ at a certain temperature. If 0.75 atm of NO_2 is added to a container and allowed to come to equilibrium, what are the equilibrium partial pressures of $NO(g)$ and $O_2(g)$?

• **17.51** Hydrogen iodide decomposes according to the reaction

$$2HI(g) \rightleftharpoons H_2(g) + I_2(g)$$

A sealed 1.50-L container initially holds 0.00623 mol of H_2, 0.00414 mol of I_2, and 0.0244 mol of HI at 703 K. When equilibrium is reached, the concentration of $H_2(g)$ is 0.00467 M. What are the concentrations of $HI(g)$ and $I_2(g)$?

• **17.52** Compound A decomposes according to the equation

$$A(g) \rightleftharpoons 2B(g) + C(g)$$

A sealed 1.00-L container initially contains 1.75×10^{-3} mol of $A(g)$, 1.25×10^{-3} mol of $B(g)$, and 6.50×10^{-4} mol of $C(g)$ at 100°C. At equilibrium, [A] is 2.15×10^{-3} M. Find [B] and [C].

● Problems in Context

• **17.53** In an analysis of interhalogen reactivity, 0.500 mol of ICl was placed in a 5.00-L flask, where it decomposed at a high T:

$$2ICl(g) \rightleftharpoons I_2(g) + Cl_2(g)$$

Calculate the equilibrium concentrations of I_2, Cl_2, and ICl ($K_c = 0.110$ at this temperature).

• **17.54** A toxicologist studying mustard gas, $S(CH_2CH_2Cl)_2$, a blistering agent, prepares a mixture of 0.675 M SCl_2 and 0.973 M C_2H_4 and allows it to react at room temperature (20.0°C):

$$SCl_2(g) + 2C_2H_4(g) \rightleftharpoons S(CH_2CH_2Cl)_2(g)$$

At equilibrium, $[S(CH_2CH_2Cl)_2] = 0.350$ M. Calculate K_p.

• **17.55** The first step in HNO_3 production is the catalyzed oxidation of NH_3. Without a catalyst, a different reaction predominates:

$$4NH_3(g) + 3O_2(g) \rightleftharpoons 2N_2(g) + 6H_2O(g)$$

When 0.0150 mol of $NH_3(g)$ and 0.0150 mol of $O_2(g)$ are placed in a 1.00-L container at a certain temperature, the N_2 concentration at equilibrium is 1.96×10^{-3} M. Calculate K_c.

• **17.56** A key step in the extraction of iron from its ore is

$$FeO(s) + CO(g) \rightleftharpoons Fe(s) + CO_2(g) \qquad K_p = 0.403 \text{ at } 1000°C$$

This step occurs in the 700°C to 1200°C zone within a blast furnace. What are the equilibrium partial pressures of $CO(g)$ and $CO_2(g)$ when 1.00 atm of $CO(g)$ and excess $FeO(s)$ react in a sealed container at 1000°C?

Reaction Conditions and the Equilibrium State: Le Châtelier's Principle
(Sample Problems 17.12 to 17.15)

● Concept Review Questions

• **17.57** What does "disturbance" mean in Le Châtelier's principle?

• **17.58** What is the difference between the equilibrium position and the equilibrium constant of a reaction? Which changes as a result of a change in reactant concentration?

• **17.59** Scenes A, B, and C below depict this reaction at three temperatures:

$$NH_4Cl(s) \rightleftharpoons NH_3(g) + HCl(g) \qquad \Delta H°_{rxn} = 176 \text{ kJ}$$

A   B   C

(a) Which best represents the reaction mixture at the highest temperature? Explain. (b) Which best represents the reaction mixture at the lowest temperature? Explain.

• **17.60** What is implied by the word "constant" in the term *equilibrium constant?* Give two reaction parameters that can be changed without changing the value of an equilibrium constant.

• **17.61** Le Châtelier's principle is related ultimately to the rates of the forward and reverse steps in a reaction. Explain (a) why an increase in reactant concentration shifts the equilibrium position to the right but does not change K; (b) why a decrease in V shifts the equilibrium position toward fewer moles of gas but does not change K; (c) why a rise in T shifts the equilibrium position of an exothermic reaction toward reactants and also changes K; and (d) why a rise in temperature of an endothermic reaction from T_1 to T_2 results in K_2 being larger than K_1.

17.62 An equilibrium mixture of two solids and a gas, in the reaction $XY(s) \rightleftharpoons X(g) + Y(s)$, is depicted at right (X is green and Y is black). Does scene A, B, or C best represent the system at equilibrium after two formula units of Y(s) is added? Explain.

A   **B**   **C**

● **Skill-Building Exercises** (grouped in similar pairs)

17.63 Consider this equilibrium system:
$$CO(g) + Fe_3O_4(s) \rightleftharpoons CO_2(g) + 3FeO(s)$$
How does the equilibrium position shift as a result of each of the following disturbances? (a) CO is added. (b) CO_2 is removed by adding solid NaOH. (c) Additional $Fe_3O_4(s)$ is added to the system. (d) Dry ice is added at constant temperature.

17.64 Sodium bicarbonate undergoes thermal decomposition according to the reaction
$$2NaHCO_3(s) \rightleftharpoons Na_2CO_3(s) + CO_2(g) + H_2O(g)$$
How does the equilibrium position shift as a result of each of the following disturbances? (a) 0.20 atm of argon gas is added. (b) $NaHCO_3(s)$ is added. (c) $Mg(ClO_4)_2(s)$ is added as a drying agent to remove H_2O. (d) Dry ice is added at constant T.

17.65 Predict the effect of *increasing* the container volume on the amounts of each reactant and product in the following reactions:
(a) $F_2(g) \rightleftharpoons 2F(g)$
(b) $2CH_4(g) \rightleftharpoons C_2H_2(g) + 3H_2(g)$

17.66 Predict the effect of *increasing* the container volume on the amounts of each reactant and product in the following reactions:
(a) $CH_3OH(l) \rightleftharpoons CH_3OH(g)$
(b) $CH_4(g) + NH_3(g) \rightleftharpoons HCN(g) + 3H_2(g)$

17.67 Predict the effect of *decreasing* the container volume on the amounts of each reactant and product in the following reactions:
(a) $H_2(g) + Cl_2(g) \rightleftharpoons 2HCl(g)$
(b) $2H_2(g) + O_2(g) \rightleftharpoons 2H_2O(l)$

17.68 Predict the effect of *decreasing* the container volume on the amounts of each reactant and product in the following reactions:
(a) $C_3H_8(g) + 5O_2(g) \rightleftharpoons 3CO_2(g) + 4H_2O(l)$
(b) $4NH_3(g) + 3O_2(g) \rightleftharpoons 2N_2(g) + 6H_2O(g)$

17.69 How would you adjust the *volume* of the container in order to maximize product yield in each of the following reactions?
(a) $Fe_3O_4(s) + 4H_2(g) \rightleftharpoons 3Fe(s) + 4H_2O(g)$
(b) $2C(s) + O_2(g) \rightleftharpoons 2CO(g)$

17.70 How would you adjust the *volume* of the container in order to maximize product yield in each of the following reactions?
(a) $Na_2O_2(s) \rightleftharpoons 2Na(l) + O_2(g)$
(b) $C_2H_2(g) + 2H_2(g) \rightleftharpoons C_2H_6(g)$

17.71 Predict the effect of *increasing* the temperature on the amounts of products in the following reactions:
(a) $CO(g) + 2H_2(g) \rightleftharpoons CH_3OH(g)$ $\Delta H°_{rxn} = -90.7$ kJ
(b) $C(s) + H_2O(g) \rightleftharpoons CO(g) + H_2(g)$ $\Delta H°_{rxn} = 131$ kJ

(c) $2NO_2(g) \rightleftharpoons 2NO(g) + O_2(g)$ (endothermic)
(d) $2C(s) + O_2(g) \rightleftharpoons 2CO(g)$ (exothermic)

17.72 Predict the effect of *decreasing* the temperature on the amounts of reactants in the following reactions:
(a) $C_2H_2(g) + H_2O(g) \rightleftharpoons CH_3CHO(g)$ $\Delta H°_{rxn} = -151$ kJ
(b) $CH_3CH_2OH(l) + O_2(g) \rightleftharpoons CH_3CO_2H(l) + H_2O(g)$
$$\Delta H°_{rxn} = -451 \text{ kJ}]$$
(c) $2C_2H_4(g) + O_2(g) \rightleftharpoons 2CH_3CHO(g)$ (exothermic)
(d) $N_2O_4(g) \rightleftharpoons 2NO_2(g)$ (endothermic)

17.73 The molecule D_2 (where D, deuterium, is ^2H) undergoes a reaction with ordinary H_2 that leads to isotopic equilibrium:
$$D_2(g) + H_2(g) \rightleftharpoons 2DH(g) \qquad K_p = 1.80 \text{ at } 298 \text{ K}$$
If $\Delta H°_{rxn}$ is 0.32 kJ/mol DH, calculate K_p at 500. K.

17.74 The formation of methanol is important to the processing of new fuels. At 298 K, $K_p = 2.25 \times 10^4$ for the reaction
$$CO(g) + 2H_2(g) \rightleftharpoons CH_3OH(l)$$
If $\Delta H°_{rxn} = -128$ kJ/mol CH_3OH, calculate K_p at 0°C.

● **Problems in Context**

17.75 The minerals hematite (Fe_2O_3) and magnetite (Fe_3O_4) exist in equilibrium with atmospheric oxygen:
$$4Fe_3O_4(s) + O_2(g) \rightleftharpoons 6Fe_2O_3(s) \quad K_p = 2.5 \times 10^{87} \text{ at } 298 \text{ K}$$
(a) Determine P_{O_2} at equilibrium. (b) Given that P_{O_2} in air is 0.21 atm, in which direction will the reaction proceed to reach equilibrium? (c) Calculate K_c at 298 K.

17.76 The oxidation of SO_2 is the key step in H_2SO_4 production:
$$SO_2(g) + \tfrac{1}{2}O_2(g) \rightleftharpoons SO_3(g) \qquad \Delta H°_{rxn} = -99.2 \text{ kJ}$$
(a) What qualitative combination of T and P maximizes SO_3 yield?
(b) How does addition of O_2 affect Q? K?
(c) Why is catalysis used for this reaction?

Comprehensive Problems

17.77 The "filmstrip" represents five molecular scenes of a gaseous mixture as it reaches equilibrium over time:

A   **B**   **C**   **D**   **E**

X is purple and Y is orange: $X_2(g) + Y_2(g) \rightleftharpoons 2XY(g)$.
(a) Write the reaction quotient, Q, for this reaction.
(b) If each particle represents 0.1 mol, find Q for each scene.
(c) If $K > 1$, is time progressing to the right or to the left? Explain.
(d) Calculate K at this temperature.
(e) If $\Delta H°_{rxn} < 0$, which scene, if any, best represents the mixture at a higher temperature? Explain.
(f) Which scene, if any, best represents the mixture at a higher pressure (lower volume)? Explain.

17.78 The powerful chlorinating agent sulfuryl dichloride (SO_2Cl_2) can be prepared by the following two-step sequence:
$$H_2S(g) + O_2(g) \rightleftharpoons SO_2(g) + H_2O(g)$$
$$SO_2(g) + Cl_2(g) \rightleftharpoons SO_2Cl_2(g)$$
(a) Balance each step, and write the overall equation.

(b) Show that the overall Q_c equals the product of the Q_c's for the individual steps.

•• **17.79** A mixture of 5.00 volumes of N_2 and 1.00 volume of O_2 passes slowly through a heated furnace. Assuming it reaches equilibrium at 1850 K and 5.00 atm, the reaction is

$$N_2(g) + O_2(g) \rightleftharpoons 2NO(g) \qquad K_p = 1.47 \times 10^{-4}$$

(a) What is the partial pressure of NO? (b) What is the concentration in micrograms per liter (μg/L) of NO in the mixture?

•• **17.80** For the following equilibrium system, which of the changes will form more $CaCO_3$?

$$CO_2(g) + Ca(OH)_2(s) \rightleftharpoons CaCO_3(s) + H_2O(l)$$
$$\Delta H° = -113 \text{ kJ}$$

(a) Decrease temperature at constant pressure (no phase change)
(b) Increase volume at constant temperature
(c) Increase partial pressure of CO_2
(d) Remove one-half of the initial $CaCO_3$

•• **17.81** Ammonium carbamate (NH_2COONH_4) is a salt of carbamic acid that is found in the blood and urine of mammals. At 250.°C, $K_c = 1.58 \times 10^{-8}$ for the following equilibrium:

$$NH_2COONH_4(s) \rightleftharpoons 2NH_3(g) + CO_2(g)$$

If 7.80 g of NH_2COONH_4 is put into a 0.500-L evacuated container, what is the total pressure at equilibrium?

•• **17.82** Isolation of Group 8B(10) elements, used as industrial catalysts, involves a series of steps. For nickel, the sulfide ore is roasted in air: $Ni_3S_2(s) + O_2(g) \rightleftharpoons NiO(s) + SO_2(g)$. The metal oxide is reduced by the H_2 in water gas (CO + H_2) to impure Ni: $NiO(s) + H_2(g) \rightleftharpoons Ni(s) + H_2O(g)$. The CO in water gas then reacts with the metal in the Mond process to form gaseous nickel carbonyl, $Ni(s) + CO(g) \rightleftharpoons Ni(CO)_4(g)$, which is subsequently decomposed to the metal. (a) Balance each of the three steps, and obtain an overall balanced equation for the conversion of Ni_3S_2 to $Ni(CO)_4$. (b) Show that the overall Q_c is the product of the Q_c's for the individual reactions.

•• **17.83** Consider the formation of ammonia in two experiments.
(a) To a 1.00-L container at 727°C, 1.30 mol of N_2 and 1.65 mol of H_2 are added. At equilibrium, 0.100 mol of NH_3 is present. Calculate the equilibrium concentrations of N_2 and H_2, and find K_c for the reaction: $2NH_3(g) \rightleftharpoons N_2(g) + 3H_2(g)$.
(b) In a different 1.00-L container at the same temperature, equilibrium is established with 8.34×10^{-2} mol of NH_3, 1.50 mol of N_2, and 1.25 mol of H_2 present. Calculate K_c for the reaction:
$$NH_3(g) \rightleftharpoons \tfrac{1}{2}N_2(g) + \tfrac{3}{2}H_2(g)$$
(c) What is the relationship between the K_c values in parts (a) and (b)? Why aren't these values the same?

•• **17.84** An important industrial source of ethanol is the reaction, catalyzed by H_3PO_4, of steam with ethylene derived from oil:

$$C_2H_4(g) + H_2O(g) \rightleftharpoons C_2H_5OH(g)$$
$$\Delta H°_{rxn} = -47.8 \text{ kJ} \qquad K_c = 9 \times 10^3 \text{ at } 600. \text{ K}$$

(a) At equilibrium, $P_{C_2H_5OH} = 200.$ atm and $P_{H_2O} = 400.$ atm. Calculate $P_{C_2H_4}$. (b) Is the highest yield of ethanol obtained at high or low P? High or low T? (c) Calculate K_c at 450. K. (d) In NH_3 manufacture, the yield is increased by condensing the NH_3 to a liquid and removing it. Would condensing the C_2H_5OH have the same effect in ethanol production? Explain.

•• **17.85** Which of the following situations represent equilibrium?
(a) Migratory birds fly north in summer and south in winter.
(b) In a grocery store, some carts are kept inside and some outside. Customers bring carts out to their cars, while store clerks bring carts in to replace those taken out.

(c) In a tug o' war, a ribbon tied to the center of the rope moves back and forth until one side loses, after which the ribbon goes all the way to the winner's side.
(d) As a stew is cooking, water in the stew vaporizes, and the vapor condenses on the lid to droplets that drip into the stew.

•• **17.86** An industrial chemist introduces 2.0 atm of H_2 and 2.0 atm of CO_2 into a 1.00-L container at 25.0°C and then raises the temperature to 700.°C, at which $K_c = 0.534$:

$$H_2(g) + CO_2(g) \rightleftharpoons H_2O(g) + CO(g)$$

How many grams of H_2 are present at equilibrium?

•• **17.87** As an EPA scientist studying catalytic converters and urban smog, you want to find K_c for the following reaction:

$$2NO_2(g) \rightleftharpoons N_2(g) + 2O_2(g) \qquad K_c = ?$$

Use the following data to find the unknown K_c:

$$\tfrac{1}{2}N_2(g) + \tfrac{1}{2}O_2(g) \rightleftharpoons NO(g) \qquad K_c = 4.8 \times 10^{-10}$$
$$2NO_2(g) \rightleftharpoons 2NO(g) + O_2(g) \qquad K_c = 1.1 \times 10^{-5}$$

•• **17.88** An inorganic chemist places 1 mol of BrCl in container A and 0.5 mol of Br_2 and 0.5 mol of Cl_2 in container B. She seals the containers and heats them to 300°C. With time, both containers hold identical mixtures of BrCl, Br_2, and Cl_2.
(a) Write a balanced equation for the reaction in container A.
(b) Write the reaction quotient, Q, for this reaction.
(c) How do the values of Q in A and in B compare over time?
(d) Explain on the molecular level how it is possible for both containers to end up with identical mixtures.

•• **17.89** An engineer examining the oxidation of SO_2 in the manufacture of sulfuric acid determines that $K_c = 1.7 \times 10^8$ at 600. K:

$$2SO_2(g) + O_2(g) \rightleftharpoons 2SO_3(g)$$

(a) At equilibrium, $P_{SO_3} = 300.$ atm and $P_{O_2} = 100.$ atm. Calculate P_{SO_2}. (b) The engineer places a mixture of 0.0040 mol of $SO_2(g)$ and 0.0028 mol of $O_2(g)$ in a 1.0-L container and raises the temperature to 1000 K. At equilibrium, 0.0020 mol of $SO_3(g)$ is present. Calculate K_c and P_{SO_2} for this reaction at 1000. K.

•• **17.90** Phosgene ($COCl_2$) is a toxic substance that forms readily from carbon monoxide and chlorine at elevated temperatures:

$$CO(g) + Cl_2(g) \rightleftharpoons COCl_2(g)$$

If 0.350 mol of each reactant is placed in a 0.500-L flask at 600 K, what are the concentrations of all three substances at equilibrium ($K_c = 4.95$ at this temperature)?

•• **17.91** When 0.100 mol of $CaCO_3(s)$ and 0.100 mol of $CaO(s)$ are placed in an evacuated sealed 10.0-L container and heated to 385 K, $P_{CO_2} = 0.220$ atm after equilibrium is established:

$$CaCO_3(s) \rightleftharpoons CaO(s) + CO_2(g)$$

An additional 0.300 atm of $CO_2(g)$ is pumped in. What is the total mass (in g) of $CaCO_3$ after equilibrium is re-established?

•• **17.92** Use each of the following reaction quotients to write the balanced equation:

(a) $Q = \dfrac{[CO_2]^2[H_2O]^2}{[C_2H_4][O_2]^3}$
(b) $Q = \dfrac{[NH_3]^4[O_2]^7}{[NO_2]^4[H_2O]^6}$

•• **17.93** Hydrogenation of carbon-carbon π bonds is important in the petroleum and food industries. The conversion of acetylene to ethylene is a simple example of the process:

$$C_2H_2(g) + H_2(g) \rightleftharpoons C_2H_4(g)$$

The calculated K_c at 2000. K is 2.9×10^8. But the process is run at lower temperatures with the aid of a catalyst to prevent decomposition. Use $\Delta H°_f$ values to calculate the K_c at 300. K.

Biological: 17.81; Environmental: 17.87; Industrial: 17.82–17.84, 1789; Organic: 17.93

•• **17.94** For the reaction $M_2 + N_2 \rightleftharpoons 2MN$, scene A represents the mixture at equilibrium, with M black and N orange. If each molecule represents 0.10 mol and the volume is 1.0 L, how many moles of each substance will be present in scene B when that mixture reaches equilibrium?

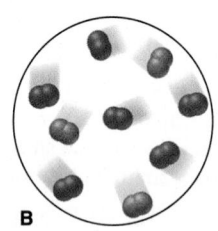

A   B

•• **17.95** Highly toxic disulfur decafluoride decomposes by a free-radical process: $S_2F_{10}(g) \rightleftharpoons SF_4(g) + SF_6(g)$. In a study of the decomposition, S_2F_{10} was placed in a 2.0-L flask and heated to 100°C; $[S_2F_{10}]$ was 0.50 M at equilibrium. More S_2F_{10} was added, and when equilibrium was reattained, $[S_2F_{10}]$ was 2.5 M. How did $[SF_4]$ and $[SF_6]$ change from the original to the new equilibrium position after the addition of more S_2F_{10}?

•• **17.96** A study of the water-gas shift reaction (see Problem 17.37) was made in which equilibrium was reached with $[CO] = [H_2O] = [H_2] = 0.10\ M$ and $[CO_2] = 0.40\ M$. After 0.60 mol of H_2 is added to the 2.0-L container and equilibrium is re-established, what are the new concentrations of all the components?

•• **17.97** A gaseous mixture of 10.0 volumes of CO_2, 1.00 volume of unreacted O_2, and 50.0 volumes of unreacted N_2 leaves an engine at 4.0 atm and 800. K. Assuming that the mixture reaches equilibrium, what are (a) the partial pressure and (b) the concentration (in picograms per liter, pg/L) of CO in this exhaust gas?
$$2CO_2(g) \rightleftharpoons 2CO(g) + O_2(g) \qquad K_p = 1.4\times10^{-28} \text{ at 800. K}$$
(The actual concentration of CO in exhaust gas is much higher because the gases do *not* reach equilibrium in the short transit time through the engine and exhaust system.)

•• **17.98** Consider the following reaction:
$$3Fe(s) + 4H_2O(g) \rightleftharpoons Fe_3O_4(s) + 4H_2(g)$$
(a) What is the apparent oxidation state of Fe in Fe_3O_4?
(b) Actually, Fe has two oxidation states in Fe_3O_4. What are they?
(c) At 900°C, K_c for the reaction is 5.1. If 0.050 mol of $H_2O(g)$ and 0.100 mol of $Fe(s)$ are placed in a 1.0-L container at 900°C, how many grams of Fe_3O_4 are present at equilibrium?

Note: The synthesis of ammonia is a major process throughout the industrialized world. Problems 17.99 to 17.105 refer to various aspects of this all-important reaction:
$$N_2(g) + 3H_2(g) \rightleftharpoons 2NH_3(g) \qquad \Delta H°_{rxn} = -91.8 \text{ kJ}$$

••• **17.99** When ammonia is made industrially, the mixture of N_2, H_2, and NH_3 that emerges from the reaction chamber is far from equilibrium. Why does the plant supervisor use reaction conditions that produce less than the maximum yield of ammonia?

•• **17.100** The following reaction can be used to make H_2 for the synthesis of ammonia from the greenhouse gases carbon dioxide and methane:
$$CH_4(g) + CO_2(g) \rightleftharpoons 2CO(g) + 2H_2(g)$$
(a) What is the percent yield of H_2 when an equimolar mixture of CH_4 and CO_2 with a total pressure of 20.0 atm reaches equilibrium at 1200. K, at which $K_p = 3.548\times10^6$?

(b) What is the percent yield of H_2 for this system at 1300. K, at which $K_p = 2.626\times10^7$?
(c) Use the van't Hoff equation to find $\Delta H°_{rxn}$.

•• **17.101** The methane used to obtain H_2 for NH_3 manufacture is impure and usually contains other hydrocarbons, such as propane, C_3H_8. Imagine the reaction of propane occurring in two steps:
$$C_3H_8(g) + 3H_2O(g) \rightleftharpoons 3CO(g) + 7H_2(g)$$
$$K_p = 8.175\times10^{15} \text{ at 1200. K}$$
$$CO(g) + H_2O(g) \rightleftharpoons CO_2(g) + H_2(g)$$
$$K_p = 0.6944 \text{ at 1200. K}$$
(a) Write the overall equation for the reaction of propane and steam to produce carbon dioxide and hydrogen.
(b) Calculate K_p for the overall process at 1200. K.
(c) When 1.00 volume of C_3H_8 and 4.00 volumes of H_2O, each at 1200. K and 5.0 atm, are mixed in a container, what is the final pressure? Assume the total volume remains constant, that the reaction is essentially complete, and that the gases behave ideally.
(d) What percentage of the C_3H_8 remains unreacted?

•• **17.102** Using CH_4 and steam as a source of H_2 for NH_3 synthesis requires high temperatures. Rather than burning CH_4 separately to heat the mixture, it is more efficient to inject some O_2 into the reaction mixture. All of the H_2 is thus released for the synthesis, and the heat of reaction for the combustion of CH_4 helps maintain the required temperature. Imagine the reaction occurring in two steps:
$$2CH_4(g) + O_2(g) \rightleftharpoons 2CO(g) + 4H_2(g)$$
$$K_p = 9.34\times10^{28} \text{ at 1000. K}$$
$$CO(g) + H_2O(g) \rightleftharpoons CO_2(g) + H_2(g)$$
$$K_p = 1.374 \text{ at 1000. K}$$
(a) Write the overall equation for the reaction of methane, steam, and oxygen to form carbon dioxide and hydrogen.
(b) What is K_p for the overall reaction?
(c) What is K_c for the overall reaction?
(d) A mixture of 2.0 mol of CH_4, 1.0 mol of O_2, and 2.0 mol of steam with a total pressure of 30. atm reacts at 1000. K at constant volume. Assuming that the reaction is complete and the ideal gas law is a valid approximation, what is the final pressure?

••• **17.103** A mixture of 3.00 volumes of H_2 and 1.00 volume of N_2 reacts at 344°C to form ammonia. The equilibrium mixture at 110. atm contains 41.49% NH_3 by volume. Calculate K_p for the reaction, assuming that the gases behave ideally.

•• **17.104** One mechanism for the synthesis of ammonia proposes that N_2 and H_2 molecules catalytically dissociate into atoms:
$$N_2(g) \rightleftharpoons 2N(g) \qquad \log K_p = -43.10$$
$$H_2(g) \rightleftharpoons 2H(g) \qquad \log K_p = -17.30$$
(a) Find the partial pressure of N in N_2 at 1000. K and 200. atm.
(b) Find the partial pressure of H in H_2 at 1000. K and 600. atm.
(c) How many N atoms and H atoms are present per liter?
(d) Based on these answers, which of the following is a more reasonable step to continue the mechanism after the catalytic dissociation? Explain.
$$N(g) + H(g) \longrightarrow NH(g)$$
$$N_2(g) + H(g) \longrightarrow NH(g) + N(g)$$

••• **17.105** You are a member of a research team of chemists discussing the plans to operate an ammonia processing plant:
$$N_2(g) + 3H_2(g) \rightleftharpoons 2NH_3(g)$$
(a) The plant operates at close to 700 K, at which K_p is 1.00×10^{-4}, and employs the stoichiometric 1/3 ratio of N_2/H_2.

At equilibrium, the partial pressure of NH_3 is 50. atm. Calculate the partial pressures of each reactant and P_{total}.

(b) One member of the team suggests the following: since the partial pressure of H_2 is cubed in the reaction quotient, the plant could produce the same amount of NH_3 if the reactants were in a 1/6 ratio of N_2/H_2 and could do so at a lower pressure, which would cut operating costs. Calculate the partial pressure of each reactant and P_{total} under these conditions, assuming an unchanged partial pressure of 50. atm for NH_3. Is the suggestion valid?

•• **17.106** The molecular scenes below depict the reaction $Y \rightleftharpoons 2Z$ at four different times, out of sequence, as it reaches equilibrium. Each sphere (Y is red and Z is green) represents 0.025 mol and the volume is 0.40 L. (a) Which scene(s) represent(s) equilibrium? (b) List the scenes in the correct sequence. (c) Calculate K_c.

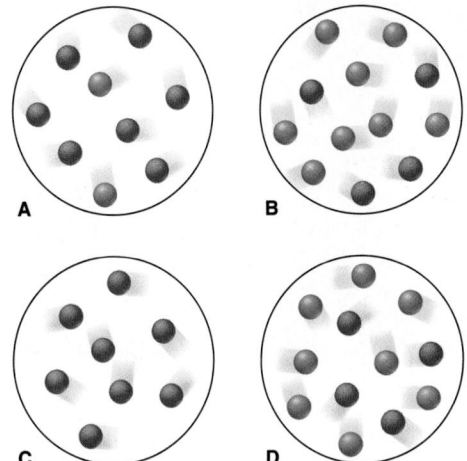

••• **17.107** For the equilibrium

$$H_2S(g) \rightleftharpoons 2H_2(g) + S_2(g) \qquad K_c = 9.0 \times 10^{-8} \text{ at } 700°C$$

the initial concentrations of the three gases are 0.300 M H_2S, 0.300 M H_2, and 0.150 M S_2. Determine the equilibrium concentrations of the gases.

•• **17.108** The two most abundant atmospheric gases react to a tiny extent at 298 K in the presence of a catalyst:

$$N_2(g) + O_2(g) \rightleftharpoons 2NO(g) \qquad K_p = 4.35 \times 10^{-31}$$

(a) What are the equilibrium pressures of the three gases when the atmospheric partial pressures of O_2 (0.210 atm) and of N_2 (0.780 atm) are put into an evacuated 1.00-L flask at 298 K with catalyst? (b) What is P_{total} in the container? (c) Find K_c at 298 K.

•• **17.109** The oxidation of nitrogen monoxide is favored at 457 K:

$$2NO(g) + O_2(g) \rightleftharpoons 2NO_2(g) \qquad K_p = 1.3 \times 10^4$$

(a) Calculate K_c at 457 K. (b) Find $\Delta H°_{rxn}$ from standard heats of formation. (c) At what temperature does $K_c = 6.4 \times 10^9$?

••• **17.110** The kinetics and equilibrium of the decomposition of hydrogen iodide have been studied extensively:

$$2HI(g) \rightleftharpoons H_2(g) + I_2(g)$$

(a) At 298 K, $K_c = 1.26 \times 10^{-3}$ for this reaction. Calculate K_p.
(b) Calculate K_c for the *formation* of HI at 298 K.
(c) Calculate $\Delta H°_{rxn}$ for HI decomposition from $\Delta H°_f$ values.
(d) At 729 K, $K_c = 2.0 \times 10^{-2}$ for HI decomposition. Calculate ΔH_{rxn} for this reaction from the van't Hoff equation.

•• **17.111** Isopentyl alcohol reacts with pure acetic acid to form isopentyl acetate, the essence of banana oil:

$$C_5H_{11}OH + CH_3COOH \rightleftharpoons CH_3COOC_5H_{11} + H_2O$$

A student adds a drying agent to remove H_2O and thus increase the yield of banana oil. Is this approach reasonable? Explain.

•• **17.112** Isomers Q (blue) and R (yellow) interconvert. They are depicted in an equilibrium mixture in scene A. Scene B represents the mixture after addition of more Q. How many molecules of each isomer are present when the mixture in scene B attains equilibrium again?

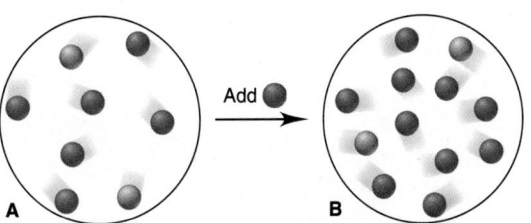

•• **17.113** Many essential metabolites are products in branched pathways, such as the one shown below:

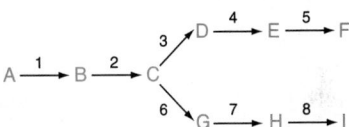

One method of control of these pathways occurs through inhibition of the first enzyme specific for a branch. (a) Which enzyme is inhibited by F? (b) Which enzyme is inhibited by I? (c) What is the disadvantage if F inhibited enzyme 1? (d) What is the disadvantage if F inhibited enzyme 6?

••• **17.114** Glauber's salt, $Na_2SO_4 \cdot 10H_2O$, was used by J. R. Glauber in the 17th century as a medicinal agent. At 25°C, $K_p = 4.08 \times 10^{-25}$ for the loss of water of hydration from Glauber's salt:

$$Na_2SO_4 \cdot 10H_2O(s) \rightleftharpoons Na_2SO_4(s) + 10H_2O(g)$$

(a) What is the vapor pressure of water at 25°C in a closed container holding a sample of $Na_2SO_4 \cdot 10H_2O(s)$?
(b) How do the following changes affect the ratio (higher, lower, same) of hydrated form to anhydrous form for the system above?
(1) Add more $Na_2SO_4(s)$ (2) Reduce the container volume
(3) Add more water vapor   (4) Add N_2 gas

••• **17.115** In a study of synthetic fuels, 0.100 mol of CO and 0.100 mol of water vapor are added to a 20.00-L container at 900.°C, and they react to form CO_2 and H_2. At equilibrium, [CO] is 2.24×10^{-3} M. (a) Calculate K_c at this temperature. (b) Calculate P_{total} in the flask at equilibrium. (c) How many moles of CO must be added to double this pressure? (d) After P_{total} is doubled and the system reattains equilibrium, what is $[CO]_{eq}$?

•• **17.116** Synthetic diamonds are made under conditions of high temperature (2000 K) and high pressure (10^{10} Pa; 10^5 atm) in the presence of catalysts. Carbon's phase diagram is useful for finding the conditions for formation of natural and synthetic diamonds. Along the diamond-graphite line, the two allotropes are in equilibrium.
(a) At point A, what is the sign of ΔH for the formation of diamond from graphite? Explain.
(b) Which allotrope is denser? Explain.

18

Gardening with Acid-Base Equilibria Hydrangea flowers contain a substance in their extracellular fluids that changes color depending on the acidity of the soil: Acidic soil produces reddish flowers, and basic soil yields bluish ones. In this chapter, we explore the essential nature of acids and bases and their extremely important equilibria.

Acid-Base Equilibria

18.1 Acids and Bases in Water
The Arrhenius Acid-Base Definition
The Acid-Dissociation Constant (K_a)
Relative Strengths of Acids and Bases

18.2 Autoionization of Water and the pH Scale
Autoionization and K_w
The pH Scale

18.3 Proton Transfer and the Brønsted-Lowry Acid-Base Definition
The Conjugate Acid-Base Pair
Net Direction of Acid-Base Reactions

18.4 Solving Problems Involving Weak-Acid Equilibria
Finding K_a Given Concentrations

Finding Concentrations Given K_a
Extent of Acid Dissociation
Polyprotic Acids

18.5 Weak Bases and Their Relation to Weak Acids
Ammonia and the Amines
Anions of Weak Acids
The Relation Between K_a and K_b

18.6 Molecular Properties and Acid Strength
Nonmetal Hydrides
Oxoacids
Acidity of Hydrated Metal Ions

18.7 Acid-Base Properties of Salt Solutions
Salts That Yield Neutral Solutions
Salts That Yield Acidic Solutions
Salts That Yield Basic Solutions
Salts of Weakly Acidic Cations and Weakly Basic Anions
Salts of Amphiprotic Anions

18.8 Generalizing the Brønsted-Lowry Concept: The Leveling Effect

18.9 Electron-Pair Donation and the Lewis Acid-Base Definition
Molecules as Lewis Acids
Metal Cations as Lewis Acids
Overview of Acid-Base Definitions

A cids and bases have been used as laboratory chemicals since the time of the alchemists, and it is likely that their properties were being applied for at least a millenium before then. They remain indispensable, not only in academic and industrial labs, but in the home as well (Table 18.1).

Notice that some of the acids (e.g., acetic and citric) have a sour taste. In fact, sourness had been a defining property since the 17th century: an acid was any substance that had a sour taste; reacted with active metals, such as aluminum and zinc, to produce hydrogen gas; and turned certain organic compounds characteristic colors. (We discuss *indicators* later and in Chapter 19.) A base was any substance that had a bitter taste and slippery feel and turned the same organic compounds different characteristic colors. (Please remember *NEVER* to taste or touch laboratory chemicals; instead, try some acetic acid in the form of vinegar on your next salad.) Moreover, it was known that *when acids and bases react, each cancels the properties of the other in a process called neutralization.* But limited definitions in science are replaced by broader ones as new phenomena are discovered. Although the early definitions of acids and bases described distinctive properties, they inevitably gave way to those based on molecular behavior.

IN THIS CHAPTER . . . We develop three definitions of acids and bases that allow us to understand ever-increasing numbers of reactions. In the process, we apply the principles of chemical equilibrium to this essential group of substances. After presenting the Arrhenius acid-base definition, we examine acid dissociation to see why acids vary in strength. The pH scale is introduced as a means of comparing the acidity or basicity of aqueous solutions. Then, we'll see that the Brønsted-Lowry acid-base definition greatly expands the meaning of "base," along with the scope of acid-base reactions. We use molecular structures of acids and bases to rationalize variations in their strengths and see that the very designations "acid" and "base" depend on relative strengths and on the solvent. Finally, we'll see that the Lewis acid-base definition expands the meaning of "acid" and acid-base behavior even further.

Concepts & Skills to Review
before you study this chapter

- role of water as solvent (Section 4.1)
- writing ionic equations (Section 4.2)
- acids, bases, and acid-base reactions (Section 4.4)
- proton transfer in acid-base reactions (Section 4.4)
- properties of an equilibrium constant (Section 17.2)
- solving equilibrium problems (Section 17.5)

ARIS Presentation Center, Tables Library

Table 18.1	Some Common Acids and Bases and Their Household Uses	
Substance	**Use**	
Acids		
Acetic acid, CH_3COOH	Flavoring, preservative	
Citric acid, $H_3C_6H_5O_7$	Flavoring	
Phosphoric acid, H_3PO_4	Rust remover	
Boric acid, H_3BO_3	Mild antiseptic, insecticide	
Aluminum salts, $NaAl(SO_4)_2 \cdot 12H_2O$	In baking powder, with sodium hydrogen carbonate	
Hydrochloric acid (muriatic acid), HCl	Brick and ceramic tile cleaner	
Bases		
Sodium hydroxide (lye), NaOH	Oven and drain cleaners	
Ammonia, NH_3	Household cleaner	
Sodium carbonate, Na_2CO_3	Water softener, grease remover	
Sodium hydrogen carbonate, $NaHCO_3$	Fire extinguisher, rising agent in cake mixes (baking soda), mild antacid	
Sodium phosphate, Na_3PO_4	Cleaner for surfaces before painting or wallpapering	

18.1 ACIDS AND BASES IN WATER

Although water is not an essential participant in all modern acid-base definitions, most laboratory work with acids and bases involves water, as do most environmental, biological, and industrial applications. Recall from our discussion in Chapter 4 that *water is a product in all reactions between strong acids and strong bases:*

$$HCl(aq) + NaOH(aq) \longrightarrow NaCl(aq) + H_2O(l)$$

Indeed, as the net ionic equation of this reaction shows, water is *the* product:

$$H^+(aq) + OH^-(aq) \longrightarrow H_2O(l)$$

Furthermore, whenever an acid dissociates in water, solvent molecules participate in the reaction:

$$HA(g \text{ or } l) + H_2O(l) \longrightarrow A^-(aq) + H_3O^+(aq)$$

As you saw in that earlier discussion, water molecules surround the proton to form H-bonded species with the general formula $H(H_2O)_n^+$. Because the proton is so small, its charge density is very high, so its attraction to water is especially strong. The proton bonds covalently to one of the lone electron pairs of a water molecule's O atom to form a **hydronium ion, H_3O^+,** which forms H bonds to several other water molecules (see Figure 4.7, p. 151). To emphasize the active role of water and the nature of the proton-water interaction, the hydrated proton is usually shown in the text as $H_3O^+(aq)$, although in some cases this hydrated species is shown more simply as $H^+(aq)$.

Release of H^+ or OH^- and the Arrhenius Acid-Base Definition

The earliest and simplest definition of acids and bases that reflects their molecular nature was suggested by Svante Arrhenius, whose work on the rate constant we encountered in Chapter 16. In the **Arrhenius acid-base definition,** acids and bases are classified in terms of their formulas and their behavior *in water:*

- An *acid* is a substance that has H in its formula and dissociates in water to yield H_3O^+.
- A *base* is a substance that has OH in its formula and dissociates in water to yield OH^-.

Some typical Arrhenius acids are HCl, HNO_3, and HCN, and some typical bases are NaOH, KOH, and $Ba(OH)_2$. Although Arrhenius bases contain discrete OH^- ions in their structures, Arrhenius acids *never* contain discrete H^+ ions. On the contrary, these acids contain *covalently bonded H atoms that ionize when the molecule dissolves in water.*

When an acid and a base react, they undergo **neutralization.** The meaning of acid-base reactions has changed along with the definitions of acid and base, but in the Arrhenius sense, neutralization occurs when *the H^+ ion from the acid and the OH^- ion from the base combine to form H_2O.* This description explains an observation that puzzled many of Arrhenius's colleagues. They observed that all neutralization reactions between what we now call strong acids and strong bases (those that dissociate completely in water) had the same heat of reaction. No matter which strong acid and base reacted, and no matter which salt formed, ΔH_{rxn}° was -55.9 kJ per mole of water formed. Arrhenius suggested that the heat of reaction was always the same because the actual reaction was always the same—a hydrogen ion and a hydroxide ion formed water:

$$H^+(aq) + OH^-(aq) \longrightarrow H_2O(l) \qquad \Delta H_{rxn}^\circ = -55.9 \text{ kJ}$$

Literature Jensen, William B. "The Origin of the Term 'Base'," *J. Chem. Educ.* **2006,** *83,* 1130.

Literature de Vos, Wobbe; Pilot, Albert. "Acids and Bases in Layers: The Stratal Structure of an Ancient Topic," *J. Chem. Educ.* **2001,** *78,* 494.

Literature Summerfield, John H. "An Acid-Base Chemistry Example: Conversion of Nicotine," *J. Chem. Educ.* **1999,** *76,* 1397.

Literature Ault, Addison. "The Acid-Base Chemistry of Nicotine: Extensions, Analogies, and a Generalization," *J. Chem. Educ.* **2001,** *78,* 500.

The dissolved salt that formed along with the water, for example, NaCl in the reaction of sodium hydroxide with hydrochloric acid,

$$Na^+(aq) + OH^-(aq) + H^+(aq) + Cl^-(aq) \longrightarrow Na^+(aq) + Cl^-(aq) + H_2O(l)$$

did not affect the ΔH°_{rxn} but existed as hydrated spectator ions.

Despite its importance at the time, limitations in the Arrhenius definition soon became apparent. Arrhenius and many others realized that even though some substances do *not* have discrete OH$^-$ ions, they still behave as bases. For example, NH$_3$ and K$_2$CO$_3$ also yield OH$^-$ in water. As you'll see shortly, broader acid-base definitions are required to include these species.

Variation in Acid Strength: The Acid-Dissociation Constant (K_a)

Acids and bases differ greatly in their *strength* in water, that is, in the amount of H$_3$O$^+$ or OH$^-$ produced per mole of substance dissolved. We generally classify acids and bases as either strong or weak, according to the extent of their dissociation into ions in water (see Table 4.2, p. 151). Remember, however, that a *gradation* in strength exists, as we'll examine quantitatively in a moment. Acids and bases are electrolytes in water, so this classification of acid and base strength correlates with our earlier classification of electrolyte strength: *strong electrolytes dissociate completely, and weak electrolytes dissociate partially.*

• *Strong acids dissociate completely into ions in water* (Figure 18.1):

$$HA(g \text{ or } l) + H_2O(l) \longrightarrow H_3O^+(aq) + A^-(aq)$$

In a dilute solution of a strong acid, *virtually no HA molecules are present;* that is, $[H_3O^+] \approx [HA]_{init}$. In other words, $[HA]_{eq} \approx 0$, so the value of K_c is extremely large:

$$Q_c = \frac{[H_3O^+][A^-]}{[HA][H_2O]} \quad \text{(at equilibrium, } Q_c = K_c \gg 1\text{)}$$

Because the reaction is essentially complete, it is not very useful to express it as an equilibrium process. In a dilute aqueous nitric acid solution, for example, there are virtually no undissociated nitric acid molecules:

$$HNO_3(l) + H_2O(l) \longrightarrow H_3O^+(aq) + NO_3^-(aq)$$

Literature Silverstein, Todd P. "Weak vs. Strong Acids and Bases: The Football Analogy," *J. Chem. Educ.* **2000,** *77,* 849.

Literature Moran, Michael J. "Factors That Influence Relative Acid Strength in Water: A Simple Model," *J. Chem. Educ.* **2006,** *83,* 800.

Animation: Dissociation of Strong and Weak Acids
ARIS Presentation Center
Chapter 18

Strong acid: HA(*g* or *l*) + H$_2$O(*l*) $\longrightarrow$ H$_3$O$^+$(*aq*) + A$^-$(*aq*)

Figure 18.1 The extent of dissociation for strong acids. The bar graphs show the relative numbers of moles of species before *(left)* and after *(right)* acid dissociation occurs. When a strong acid dissolves in water, it dissociates completely, yielding H$_3$O$^+$(*aq*) and A$^-$(*aq*) ions; virtually no HA molecules are present.

Transparency/ARIS Presentation Center, Art Library

• *Weak acids dissociate very slightly into ions in water* (Figure 18.2):

$$HA(aq) + H_2O(l) \rightleftharpoons H_3O^+(aq) + A^-(aq)$$

In a dilute solution of a weak acid, *the great majority of HA molecules are undissociated.* Thus, $[H_3O^+] \ll [HA]_{init}$. In other words, $[HA]_{eq} \approx [HA]_{init}$, so the value of K_c is very small. Hydrocyanic acid is an example of a weak acid:

$$HCN(aq) + H_2O(l) \rightleftharpoons H_3O^+(aq) + CN^-(aq)$$

$$Q_c = \frac{[H_3O^+][CN^-]}{[HCN][H_2O]} \qquad \text{(at equilibrium, } Q_c = K_c \ll 1\text{)}$$

Figure 18.2 The extent of dissociation for weak acids. In contrast to a strong acid in water (see Figure 18.1), a weak acid dissociates very little, remaining mostly as intact acid molecules and, thus, yielding relatively few $H_3O^+(aq)$ and $A^-(aq)$ ions.

Transparency/ARIS Presentation Center, Art Library

Weak acid: $HA(aq) + H_2O(l) \rightleftharpoons H_3O^+(aq) + A^-(aq)$

(As in Chapter 17, brackets with no subscript mean molar concentration *at equilibrium;* that is, $[X]$ means $[X]_{eq}$. In this chapter, we are dealing with systems *at equilibrium,* so instead of writing Q and stating that Q equals K at equilibrium, we'll express K directly as a collection of equilibrium concentration terms.)

Because of the effect of concentration on rate (Section 16.6), this dramatic difference in $[H_3O^+]$ causes a dramatic difference in the reaction rate when an equal concentration of a strong or a weak acid reacts with an active metal, such as zinc (Figure 18.3):

$$Zn(s) + 2H_3O^+(aq) \longrightarrow Zn^{2+}(aq) + 2H_2O(l) + H_2(g)$$

In a strong acid, such as 1 *M* HCl, zinc reacts rapidly, forming bubbles of H_2 vigorously. In a weak acid, such as 1 *M* CH_3COOH, zinc reacts slowly, forming bubbles of H_2 sluggishly on the metal piece. The strong acid has a much higher $[H_3O^+]$; much more H_3O^+ ion is available for reaction, so the rate is much higher.

The Meaning of K_a   There is a *specific* equilibrium constant for acid dissociation that highlights only those species whose concentrations change to any significant extent. The equilibrium expression for the dissociation of a general *weak acid, HA,* in water is

$$K_c = \frac{[H_3O^+][A^-]}{[HA][H_2O]}$$

In general, the concentration of water, $[H_2O]$, is so much larger than $[HA]$ that it changes negligibly when HA dissociates; thus, it is treated as a constant. Therefore, as you saw for solids in Section 17.2, we simplify the equilibrium expression by multiplying $[H_2O]$ by K_c to define a new equilibrium constant, the **acid-dissociation constant** (or **acid-ionization constant**), K_a:

$$K_c[H_2O] = K_a = \frac{[H_3O^+][A^-]}{[HA]} \qquad \textbf{(18.1)}$$

ARIS Presentation Center, Photo Library

Strong acid (1 *M* HCl)   Weak acid (1 *M* CH_3COOH)

Figure 18.3   Reaction of zinc with a strong and a weak acid. In the reaction of zinc with a strong acid (*left*), the higher concentration of H_3O^+ results in rapid formation of H_2 (bubbles). In a weak acid (*right*), the H_3O^+ concentration is much lower, so formation of H_2 is much slower.

Like any equilibrium constant, K_a is a number whose magnitude is temperature dependent and tells how far to the right the reaction has proceeded to reach equilibrium. Thus, *the stronger the acid, the higher the [H_3O^+] at equilibrium, and the larger the K_a:*

$$\text{Stronger acid} \Longrightarrow \text{higher [H}_3\text{O}^+] \Longrightarrow \text{larger } K_a$$

The range of values for the acid-dissociation constants of weak acids extends over many orders of magnitude. Listed below are some benchmark K_a values for typical weak acids to give you a general idea of the fraction of HA molecules that dissociate into ions:

- For a weak acid with a relatively high K_a ($\sim 10^{-2}$), a 1 M solution has $\sim 10\%$ of the HA molecules dissociated. The K_a of chlorous acid ($HClO_2$) is 1.1×10^{-2}, and 1 M $HClO_2$ is 10.% dissociated.
- For a weak acid with a moderate K_a ($\sim 10^{-5}$), a 1 M solution has $\sim 0.3\%$ of the HA molecules dissociated. The K_a of acetic acid (CH_3COOH) is 1.8×10^{-5}, and 1 M CH_3COOH is 0.42% dissociated.
- For a weak acid with a relatively low K_a ($\sim 10^{-10}$), a 1 M solution has $\sim 0.001\%$ of the HA molecules dissociated. The K_a of HCN is 6.2×10^{-10}, and 1 M HCN is 0.0025% dissociated.

Thus, for solutions of the same initial HA concentration, *the smaller the K_a, the lower the percent dissociation of HA:*

$$\text{Smaller } K_a \Longrightarrow \text{lower \% dissociation of HA} \Longrightarrow \text{weaker acid}$$

Table 18.2 lists K_a values in water of some weak *monoprotic* acids, those with one ionizable proton. (A more extensive list appears in Appendix C.) Strong acids are absent because they do not have meaningful K_a values. Note that the ionizable proton in organic acids is attached to oxygen in the —COOH group; the H atoms bonded to C atoms do *not* ionize. (In Section 18.4, we discuss the behavior of polyprotic acids, those with more than one ionizable proton.)

Table 18.2 K_a Values for Some Monoprotic Acids at 25°C		
Name (Formula)	Lewis Structure*	K_a
Chlorous acid ($HClO_2$)	H—Ö—Cl̈=Ö	1.1×10^{-2}
Nitrous acid (HNO_2)	H—Ö—N̈=Ö	7.1×10^{-4}
Hydrofluoric acid (HF)	H—F̈:	6.8×10^{-4}
Formic acid (HCOOH)	$\overset{:O:}{\underset{\parallel}{H-C-\ddot{O}-H}}$	1.8×10^{-4}
Acetic acid (CH_3COOH)	$H-\overset{H}{\underset{H}{C}}-\overset{:O:}{C}-\ddot{O}-H$	1.8×10^{-5}
Propanoic acid (CH_3CH_2COOH)	$H-\overset{H}{\underset{H}{C}}-\overset{H}{\underset{H}{C}}-\overset{:O:}{C}-\ddot{O}-H$	1.3×10^{-5}
Hypochlorous acid (HClO)	H—Ö—Cl̈:	2.9×10^{-8}
Hydrocyanic acid (HCN)	H—C≡N:	6.2×10^{-10}

ACID STRENGTH ↑

*Red type indicates the ionizable proton; all atoms have zero formal charge.

Classifying the Relative Strengths of Acids and Bases

Using a table of acid-dissociation constants is the surest way to quantify relative strengths of weak acids, but you can often classify acids and bases qualitatively as strong or weak just from their formulas:

- *Strong acids.* Two types of strong acids, with examples that *you should memorize,* are
  1. The hydrohalic acids HCl, HBr, and HI
  2. Oxoacids in which the number of O atoms exceeds the number of ionizable protons by two or more, such as HNO_3, H_2SO_4, and $HClO_4$; for example, in H_2SO_4, 4 O's − 2 H's = 2

- *Weak acids.* There are many *more* weak acids than strong ones. Four types, with examples, are
  1. The hydrohalic acid HF
  2. Acids in which H is not bonded to O or to a halogen, such as HCN and H_2S
  3. Oxoacids in which the number of O atoms equals or exceeds by one the number of ionizable protons, such as HClO, HNO_2, and H_3PO_4
  4. Carboxylic acids (general formula RCOOH, with the ionizable proton shown in red), such as CH_3COOH and C_6H_5COOH

- *Strong bases.* Water-soluble compounds containing O^{2-} or OH^- ions are strong bases. The cations are usually those of the most active metals:
  1. M_2O or MOH, where M = Group 1A(1) metal (Li, Na, K, Rb, Cs)
  2. MO or $M(OH)_2$, where M = Group 2A(2) metal (Ca, Sr, Ba)
     [MgO and $Mg(OH)_2$ are only slightly soluble in water, but the soluble portion dissociates completely.]

- *Weak bases.* Many compounds with an electron-rich nitrogen atom are weak bases (none are Arrhenius bases). The common structural feature is an N atom with a lone electron pair (shown here in blue in the formulas):
  1. Ammonia ($\ddot{N}H_3$)
  2. Amines (general formula $R\ddot{N}H_2$, $R_2\ddot{N}H$, or $R_3\ddot{N}$), such as $CH_3CH_2\ddot{N}H_2$, $(CH_3)_2\ddot{N}H$, and $(C_3H_7)_3\ddot{N}$

ARIS Presentation Center, Worked Examples Library

SAMPLE PROBLEM 18.1 Classifying Acid and Base Strength from the Chemical Formula

PROBLEM Classify each of the following compounds as a strong acid, weak acid, strong base, or weak base:
(a) H_2SeO_4 (b) $(CH_3)_2CHCOOH$
(c) KOH   (d) $(CH_3)_2CHNH_2$

PLAN We examine the formula and classify each acid or base, using the text descriptions. Particular points to note for acids are the numbers of O atoms relative to H atoms and the presence of the —COOH group. For bases, note the nature of the cation or the presence of an N atom that has a lone pair.

SOLUTION (a) Strong acid: H_2SeO_4 is an oxoacid in which the number of O atoms exceeds the number of ionizable protons by two.

(b) Weak acid: $(CH_3)_2CHCOOH$ is a carboxylic acid, as indicated by the —COOH group.

(c) Strong base: KOH is one of the Group 1A(1) hydroxides.

(d) Weak base: $(CH_3)_2CHNH_2$ has a lone pair on the N and is an amine.

FOLLOW-UP PROBLEM 18.1 Which member of each pair is the stronger acid or base?
(a) HClO or $HClO_3$ (b) HCl or CH_3COOH (c) NaOH or CH_3NH_2

Section Summary

In aqueous solution, water binds the proton released from an acid to form the hydrated species represented by $H_3O^+(aq)$. • In the Arrhenius definition, acids contain H and yield H_3O^+ in water, bases contain OH and yield OH^- in water, and an acid-base reaction (neutralization) is the reaction of H^+ and OH^- to form H_2O. • Acid strength depends on $[H_3O^+]$ relative to $[HA]$ in aqueous solution. Strong acids dissociate completely and weak acids slightly. • The extent of dissociation is expressed by the acid-dissociation constant, K_a. Weak acids have K_a values ranging from about 10^{-1} to 10^{-12}. • Many acids and bases can be classified qualitatively as strong or weak based on their formulas.

18.2 AUTOIONIZATION OF WATER AND THE pH SCALE

Before we discuss the next major definition of acid-base behavior, let's examine a crucial property of water that enables us to quantify $[H_3O^+]$ in any aqueous system: *water is an extremely weak electrolyte.* The electrical conductivity of tap water is due almost entirely to dissolved ions, but even water that has been repeatedly distilled and deionized exhibits a tiny conductance. The reason is that water itself dissociates into ions very slightly in an equilibrium process known as **autoionization** (or self-ionization):

$$H_2O(l) \quad + \quad H_2O(l) \quad \rightleftharpoons \quad H_3O^+(aq) \quad + \quad OH^-(aq)$$

ARIS Presentation Center, Art Library

The Equilibrium Nature of Autoionization: The Ion-Product Constant for Water (K_w)

Like any equilibrium process, the autoionization of water is described quantitatively by an equilibrium constant:

$$K_c = \frac{[H_3O^+][OH^-]}{[H_2O]^2}$$

Because the concentration of H_2O is essentially constant here, we simplify this equilibrium expression by including the constant $[H_2O]^2$ term with the value of K_c to obtain a new equilibrium constant, the **ion-product constant for water, K_w:**

$$K_c[H_2O]^2 = K_w = [H_3O^+][OH^-] = 1.0 \times 10^{-14} \text{ (at 25°C)} \qquad (18.2)$$

Notice that *one H_3O^+ ion and one OH^- ion appear for each H_2O molecule that dissociates.* Therefore, in pure water, we find that

$$[H_3O^+] = [OH^-] = \sqrt{1.0 \times 10^{-14}} = 1.0 \times 10^{-7} \, M \text{ (at 25°C)}$$

Pure water has a concentration of about 55.5 M $\left(\text{that is, } \dfrac{1000 \text{ g/L}}{18.02 \text{ g/mol}} \right)$, so these equilibrium concentrations are attained when only 1 in 555 million water molecules dissociates reversibly into ions!

Autoionization of water has two major consequences for aqueous acid-base chemistry:

1. *A change in $[H_3O^+]$ causes an inverse change in $[OH^-]$,* and vice versa:

Higher $[H_3O^+] \Longrightarrow$ lower $[OH^-]$ **and**   Higher $[OH^-] \Longrightarrow$ lower $[H_3O^+]$

790 **Chapter 18** Acid-Base Equilibria

Recall from our discussion of Le Châtelier's principle (Section 17.6) that a change in concentration of either ion shifts the equilibrium position, but it does *not* change the equilibrium constant. Therefore, if some acid is added, $[H_3O^+]$ increases, and so $[OH^-]$ must decrease; if some base is added, $[OH^-]$ increases, and so $[H_3O^+]$ must decrease. Note that the addition of H_3O^+ or OH^- leads to the formation of H_2O, so the value of K_w is maintained.

2. *Both ions are present in all aqueous systems.* Thus, all acidic solutions contain a low concentration of OH^- ions, and all basic solutions contain a low concentration of H_3O^+ ions. The equilibrium nature of autoionization allows us to define "acidic" and "basic" solutions in terms of relative magnitudes of $[H_3O^+]$ and $[OH^-]$:

In an *acidic* solution, $[H_3O^+] > [OH^-]$

In a *neutral* solution, $[H_3O^+] = [OH^-]$

In a *basic* solution, $[H_3O^+] < [OH^-]$

Figure 18.4 summarizes these relationships and the relative solution acidity. Moreover, if you know the value of K_w at a particular temperature and the concentration of one of these ions, you can easily calculate the concentration of the other ion by solving for it from the K_w expression:

$$[H_3O^+] = \frac{K_w}{[OH^-]} \quad \text{or} \quad [OH^-] = \frac{K_w}{[H_3O^+]}$$

Figure 18.4 The relationship between $[H_3O^+]$ and $[OH^-]$ and the relative acidity of solutions.

Transparency/ARIS Presentation Center, Active Art Library

ARIS Presentation Center, Worked Examples Library

SAMPLE PROBLEM 18.2 Calculating $[H_3O^+]$ or $[OH^-]$ in Aqueous Solution

PROBLEM A research chemist adds a measured amount of HCl gas to pure water at 25°C and obtains a solution with $[H_3O^+] = 3.0\times10^{-4}\ M$. Calculate $[OH^-]$. Is the solution neutral, acidic, or basic?

PLAN We use the known value of K_w at 25°C (1.0×10^{-14}) and the given $[H_3O^+]$ ($3.0\times10^{-4}\ M$) to solve for $[OH^-]$. Then we compare $[H_3O^+]$ with $[OH^-]$ to determine whether the solution is acidic, basic, or neutral (see Figure 18.4).

SOLUTION Calculating $[OH^-]$:

$$[OH^-] = \frac{K_w}{[H_3O^+]} = \frac{1.0\times10^{-14}}{3.0\times10^{-4}}$$
$$= 3.3\times10^{-11}\ M$$

Because $[H_3O^+] > [OH^-]$, the solution is acidic.

CHECK It makes sense that adding an acid to water results in an acidic solution. Moreover, because $[H_3O^+]$ is greater than $10^{-7}\ M$, $[OH^-]$ must be less than $10^{-7}\ M$ to give a constant K_w.

FOLLOW-UP PROBLEM 18.2 Calculate $[H_3O^+]$ in a solution that is at 25°C and has $[OH^-] = 6.7\times10^{-2}\ M$. Is the solution neutral, acidic, or basic?

Expressing the Hydronium Ion Concentration: The pH Scale

Literature Ault, Addison. "Do pH in Your Head," *J. Chem. Educ.* **1999**, *76*, 936.

Literature Plumsky, Roger. "A pHorseshoe," *J. Chem. Educ.* **1999**, *76*, 935.

In aqueous solutions, $[H_3O^+]$ can vary from about 10 *M* to 10^{-15} *M*. To handle numbers with negative exponents more conveniently in calculations, we convert them to positive numbers using a numerical system called a *p-scale*, the negative of the common (base-10) logarithm of the number. Applying this numerical system to $[H_3O^+]$ gives **pH**, the negative logarithm of $[H^+]$ (or $[H_3O^+]$):

$$pH = -\log [H_3O^+] \qquad (18.3)$$

What is the pH of 10^{-12} *M* H_3O^+ solution?

$$pH = -\log [H_3O^+] = -\log 10^{-12} = (-1)(-12) = 12$$

Similarly, a 10^{-3} *M* H_3O^+ solution has a pH of 3, and a 5.4×10^{-4} *M* H_3O^+ solution has a pH of 3.27:

$$pH = -\log [H_3O^+] = (-1)(\log 5.4 + \log 10^{-4}) = 3.27$$

As with any measurement, the number of significant figures in a pH value reflects the precision with which the concentration is known. However, a pH value is a logarithm, so the number of significant figures in the concentration equals the number of digits *to the right of the decimal point in the pH value* (see Appendix A). In the preceding example, 5.4×10^{-4} *M* has two significant figures, so its negative logarithm, 3.27, has two digits to the right of the decimal point.

Note in particular that *the higher the pH, the lower the $[H_3O^+]$*. Therefore, *an acidic solution has a lower pH (higher $[H_3O^+]$) than a basic solution*. At 25°C in pure water, $[H_3O^+]$ is 1.0×10^{-7} *M*, so

pH of a neutral solution	$= 7.00$
pH of an acidic solution	< 7.00
pH of a basic solution	> 7.00

Figure 18.5 shows that the pH values of some familiar aqueous solutions fall within a range of 0 to 14.

Another important point arises when we compare $[H_3O^+]$ in different solutions. Because the pH scale is logarithmic, a solution of pH 1.0 has an $[H_3O^+]$ that is 10 times higher than that of a pH 2.0 solution, 100 times higher than that of a pH 3.0 solution, and so forth. To find the $[H_3O^+]$ from the pH, you perform the opposite arithmetic process; that is, you find the negative antilog of pH:

$$[H_3O^+] = 10^{-pH}$$

A p-scale is used to express other quantities as well:

- Hydroxide ion concentration can be expressed as pOH:

$$pOH = -\log [OH^-]$$

Acidic solutions have a higher pOH (lower $[OH^-]$) than basic solutions.

- Equilibrium constants can be expressed as p*K*:

$$pK = -\log K$$

A low pK corresponds to a high K. A reaction that reaches equilibrium with mostly products present (proceeds far to the right) has a low p*K* (high *K*), whereas one that has mostly reactants present at equilibrium has a high p*K* (low *K*). Table 18.3 shows this relationship for aqueous equilibria of some weak acids.

ARIS Presentation Center, Tables Library

Table 18.3	The Relationship Between K_a and pK_a	
Acid Name (Formula)	K_a **at 25°C**	**pK_a**
Hydrogen sulfate ion (HSO_4^-)	1.0×10^{-2}	1.99
Nitrous acid (HNO_2)	7.1×10^{-4}	3.15
Acetic acid (CH_3COOH)	1.8×10^{-5}	4.74
Hypobromous acid ($HBrO$)	2.3×10^{-9}	8.64
Phenol (C_6H_5OH)	1.0×10^{-10}	10.00

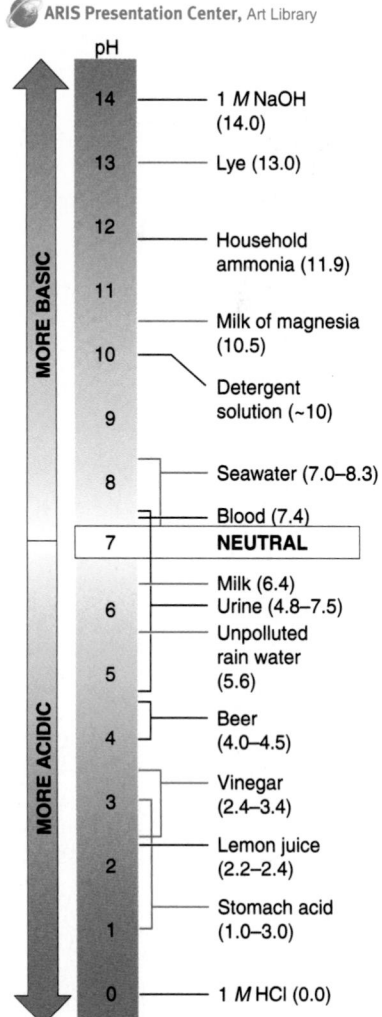
ARIS Presentation Center, Art Library

Figure 18.5 The pH values of some familiar aqueous solutions.

Figure 18.6 The relations among $[H_3O^+]$, pH, $[OH^-]$, and pOH. Because K_w is constant, $[H_3O^+]$ and $[OH^-]$ are interdependent, and they change in opposite directions as the acidity or basicity of the aqueous solution increases. The pH and pOH are interdependent in the same way. Note that at 25°C, the product of $[H_3O^+]$ and $[OH^-]$ is 1.0×10^{-14}, and the sum of pH and pOH is 14.00.

Transparency/ARIS Presentation Center, Art Library

	$[H_3O^+]$	pH	$[OH^-]$	pOH
MORE BASIC	1.0×10^{-15}	15.00	1.0×10^{1}	−1.00
	1.0×10^{-14}	14.00	1.0×10^{0}	0.00
	1.0×10^{-13}	13.00	1.0×10^{-1}	1.00
BASIC	1.0×10^{-12}	12.00	1.0×10^{-2}	2.00
	1.0×10^{-11}	11.00	1.0×10^{-3}	3.00
	1.0×10^{-10}	10.00	1.0×10^{-4}	4.00
	1.0×10^{-9}	9.00	1.0×10^{-5}	5.00
	1.0×10^{-8}	8.00	1.0×10^{-6}	6.00
NEUTRAL	1.0×10^{-7}	7.00	1.0×10^{-7}	7.00
	1.0×10^{-6}	6.00	1.0×10^{-8}	8.00
	1.0×10^{-5}	5.00	1.0×10^{-9}	9.00
MORE ACIDIC	1.0×10^{-4}	4.00	1.0×10^{-10}	10.00
	1.0×10^{-3}	3.00	1.0×10^{-11}	11.00
ACIDIC	1.0×10^{-2}	2.00	1.0×10^{-12}	12.00
	1.0×10^{-1}	1.00	1.0×10^{-13}	13.00
	1.0×10^{0}	0.00	1.0×10^{-14}	14.00
	1.0×10^{1}	−1.00	1.0×10^{-15}	15.00

The Relations Among pH, pOH, and pK_w Taking the negative log of both sides of the K_w expression gives a very useful relationship among pK_w, pH, and pOH:

$$K_w = [H_3O^+][OH^-] = 1.0 \times 10^{-14} \text{ (at 25°C)}$$

$$-\log K_w = (-\log [H_3O^+]) + (-\log [OH^-]) = -\log (1.0 \times 10^{-14})$$

$$p K_w = pH + pOH = 14.00 \qquad \text{(at 25°C)} \qquad \textbf{(18.4)}$$

The sum of pH and pOH is pK_w for any aqueous solution at any temperature, with the value of pK_w equal to 14.00 at 25°C. With pH, pOH, $[H_3O^+]$, and $[OH^-]$ interrelated through K_w, knowing any one of the values allows us to determine the others (Figure 18.6).

ARIS Presentation Center, Worked Examples Library

SAMPLE PROBLEM 18.3 Calculating $[H_3O^+]$, pH, $[OH^-]$, and pOH

PROBLEM In an art restoration project, a conservator prepares copper-plate etching solutions by diluting concentrated HNO_3 to 2.0 M, 0.30 M, and 0.0063 M HNO_3. Calculate $[H_3O^+]$, pH, $[OH^-]$, and pOH of the three solutions at 25°C.

PLAN We know from its formula that HNO_3 is a strong acid, so it dissociates completely; thus, $[H_3O^+] = [HNO_3]_{init}$. We use the given concentrations and the value of K_w at 25°C (1.0×10^{-14}) to find $[H_3O^+]$ and $[OH^-]$ and then use them to calculate pH and pOH.

SOLUTION Calculating the values for 2.0 M HNO_3:

$$[H_3O^+] = 2.0 \, M$$

$$pH = -\log [H_3O^+] = -\log 2.0 = \boxed{-0.30}$$

$$[OH^-] = \frac{K_w}{[H_3O^+]} = \frac{1.0 \times 10^{-14}}{2.0} = \boxed{5.0 \times 10^{-15} \, M}$$

$$pOH = -\log (5.0 \times 10^{-15}) = \boxed{14.30}$$

Calculating the values for 0.30 M HNO_3:

$$[H_3O^+] = 0.30 \, M$$

$$pH = -\log [H_3O^+] = -\log 0.30 = \boxed{0.52}$$

$$[OH^-] = \frac{K_w}{[H_3O^+]} = \frac{1.0 \times 10^{-14}}{0.30} = \boxed{3.3 \times 10^{-14} \, M}$$

$$pOH = -\log (3.3 \times 10^{-14}) = \boxed{13.48}$$

Calculating the values for 0.0063 M HNO_3:

$$[H_3O^+] = 6.3 \times 10^{-3} M$$

$$pH = -\log [H_3O^+] = -\log (6.3 \times 10^{-3}) = 2.20$$

$$[OH^-] = \frac{K_w}{[H_3O^+]} = \frac{1.0 \times 10^{-14}}{6.3 \times 10^{-3}} = 1.6 \times 10^{-12} M$$

$$pOH = -\log (1.6 \times 10^{-12}) = 11.80$$

CHECK As the solution becomes more dilute, $[H_3O^+]$ decreases, so pH increases, as we expect. An $[H_3O^+]$ greater than 1.0 M, as in 2.0 M HNO_3, gives a positive log, so it results in a negative pH. The arithmetic seems correct because pH + pOH = 14.00 in each case.
COMMENT On most calculators, finding the pH requires several keystrokes. For example, to find the pH of 6.3×10^{-3} M HNO_3 solution, you enter: 6.3, EXP, 3, +/−, log, +/−.
FOLLOW-UP PROBLEM 18.3 A solution of NaOH has a pH of 9.52. What is its pOH, $[H_3O^+]$, and $[OH^-]$ at 25°C?

Measuring pH In the laboratory, pH values are usually obtained with an acid-base indicator or, more precisely, with an instrument called a pH meter. **Acid-base indicators** are organic molecules whose colors depend on the acidity or basicity of the solution in which they are dissolved. The pH of a solution is estimated quickly with *pH paper*, a paper strip impregnated with one or a mixture of indicators. A drop of test solution is placed on the paper strip, and the color of the strip is compared with a color chart, as shown in Figure 18.7A.

The *pH meter* measures $[H_3O^+]$ by means of two electrodes immersed in the test solution. One electrode provides a stable reference voltage; the other has an extremely thin, conducting, glass membrane that separates a known internal $[H_3O^+]$ from the unknown external $[H_3O^+]$. The difference in $[H_3O^+]$ creates a voltage difference across the membrane, which is measured and displayed in pH units (Figure 18.7B). We examine this device further in Chapter 21.

A

B

Section Summary

Pure water has a low conductivity because it autoionizes to a small extent. This process is described by an equilibrium reaction whose equilibrium constant is the ion-product constant for water, K_w (1.0×10^{-14} at 25°C). Thus, $[H_3O^+]$ and $[OH^-]$ are inversely related: in acidic solution, $[H_3O^+]$ is greater than $[OH^-]$; the reverse is true in basic solution; and the two are equal in neutral solution. • To express small values of $[H_3O^+]$ more simply, we use the pH scale (pH = $-\log [H_3O^+]$). A high pH represents a low $[H_3O^+]$. Similarly, pOH = $-\log [OH^-]$, and pK = $-\log K$. In acidic solutions, pH < 7.00; in basic solutions, pH > 7.00; and in neutral solutions, pH = 7.00. The sum of pH and pOH equals pK_w (14.00 at 25°C).

18.3 PROTON TRANSFER AND THE BRØNSTED-LOWRY ACID-BASE DEFINITION

Earlier we noted a major shortcoming of the Arrhenius acid-base definition: many substances that yield OH^- ions when they dissolve in water do not contain OH in their formulas. Examples include ammonia, the amines, and many salts of weak acids, such as NaF. Another limitation of the Arrhenius definition was that water had to be the solvent for acid-base reactions. In the early 20th century, J. N. Brønsted and T. M. Lowry suggested definitions that remove these limitations. (Recall that we discussed their ideas briefly in Section 4.4.) According to the **Brønsted-Lowry acid-base definition,**

• *An acid is a **proton donor**,* any species that donates an H^+ ion. An acid must contain H in its formula; HNO_3 and $H_2PO_4^-$ are two of many examples. All Arrhenius acids are Brønsted-Lowry acids.

Figure 18.7 Methods for measuring the pH of an aqueous solution. A, A few drops of the solution are placed on a strip of pH paper, and the color is compared with the color chart. **B,** The electrodes of a pH meter immersed in the test solution measure $[H_3O^+]$. (In this instrument, the two electrodes are housed in one probe.)

ARIS Presentation Center, Photo Library

Literature Adcock, Jamie L. "Teaching Brønsted-Lowry Acid-Base Theory in a Direct Comprehensive Way," *J. Chem. Educ.* **2001,** *78,* 1495.

- *A base is a **proton acceptor,** any species that accepts an H^+ ion.* A base must contain a lone pair of electrons to bind the H^+ ion; a few examples are NH_3, CO_3^{2-}, and F^-, as well as OH^- itself. Brønsted-Lowry bases are not Arrhenius bases, but all Arrhenius bases contain the Brønsted-Lowry base OH^-.

From the Brønsted-Lowry perspective, the only requirement for an acid-base reaction is that *one species donates a proton and another species accepts it: an acid-base reaction is a proton-transfer process.* Acid-base reactions can occur between gases, in nonaqueous solutions, and in heterogeneous mixtures, as well as in aqueous solutions.

An acid and a base always work together in the transfer of a proton. In other words, one species behaves as an acid and the other species behaves as a base *simultaneously.* Even when an acid or a base merely dissolves in water, an acid-base reaction occurs because water acts as the other partner. Consider two typical acidic and basic solutions:

1. *Acid donates a proton to water* (Figure 18.8A). When HCl dissolves in water, an H^+ ion (a proton) is transferred from HCl to H_2O, where it becomes attached to a lone pair of electrons on the O atom, forming H_3O^+. In effect, HCl (the acid) has *donated* the H^+, and H_2O (the base) has *accepted* it:

$$HCl(g) + H_2\ddot{\underset{..}{O}}(l) \longrightarrow Cl^-(aq) + H_3\ddot{O}^+(aq)$$

2. *Base accepts a proton from water* (Figure 18.8B). Proton transfer also occurs in an aqueous solution of ammonia. An H^+ from H_2O attaches to the lone pair of N, forming NH_4^+. Having transferred an H^+, the H_2O becomes an OH^- ion:

$$\ddot{N}H_3(aq) + H_2O(l) \rightleftharpoons NH_4^+(aq) + OH^-(aq)$$

In this case, H_2O (the acid) has *donated* the H^+, and NH_3 (the base) has *accepted* it. Thus, H_2O is *amphiprotic:* it acts as a base (accepts an H^+) in one case and as an acid (donates an H^+) in the other. As you'll see, many other species are amphiprotic as well.

Figure 18.8 Proton transfer as the essential feature of a Brønsted-Lowry acid-base reaction. A, When HCl dissolves in water, it acts as an acid by donating a proton to water, which acts as a base by accepting it. **B,** In aqueous solution, NH_3 acts as a base by accepting a proton from water, which acts as an acid by donating it. Thus, in the Brønsted-Lowry sense, an acid-base reaction occurs in both cases.

Transparency/ARIS Presentation Center, Art Library

The Conjugate Acid-Base Pair

The Brønsted-Lowry definition provides a new way to look at acid-base reactions because it focuses on the reactants *and* the products. For example, let's examine the reaction between hydrogen sulfide and ammonia:

$$H_2S + NH_3 \rightleftharpoons HS^- + NH_4^+$$

In the forward reaction, H_2S acts as an acid by donating an H^+ to NH_3, which acts as a base by accepting. The reverse reaction involves another acid-base pair. The ammonium ion, NH_4^+, acts as an acid by donating an H^+ to the hydrogen sulfide ion, HS^-, which acts as a base. Notice that the acid, H_2S, becomes a base, HS^-, and the base, NH_3, becomes an acid, NH_4^+.

In Brønsted-Lowry terminology, H_2S and HS^- are a **conjugate acid-base pair**: HS^- is the conjugate base of the acid H_2S. Similarly, NH_3 and NH_4^+ form a conjugate acid-base pair: NH_4^+ is the conjugate acid of the base NH_3. *Every acid has a conjugate base, and every base has a conjugate acid.* Note that, for any conjugate acid-base pair,

- The conjugate base has one *fewer* H and one *more* minus charge than the acid.
- The conjugate acid has one *more* H and one *fewer* minus charge than the base.

A Brønsted-Lowry acid-base reaction occurs when *an acid and a base react to form their conjugate base and conjugate acid, respectively:*

$$\text{acid}_1 + \text{base}_2 \rightleftharpoons \text{base}_1 + \text{acid}_2$$

Table 18.4 shows some Brønsted-Lowry acid-base reactions. Notice that

- Each reaction has an acid and a base as reactants *and* as products, and these comprise two conjugate acid-base pairs.
- Acids and bases can be neutral, cationic, or anionic.
- The same species can be an acid or a base (amphiprotic), depending on the other species reacting. Water behaves this way in reactions 1 and 4, and HPO_4^{2-} does so in reactions 4 and 6.

Transparency/ARIS Presentation Center, Tables Library

Table 18.4   The Conjugate Pairs in Some Acid-Base Reactions

	Acid	+	Base	→ Base	+	Acid
Reaction 1	HF	+	H_2O	F^-	+	H_3O^+
Reaction 2	HCOOH	+	CN^-	$HCOO^-$	+	HCN
Reaction 3	NH_4^+	+	CO_3^{2-}	NH_3	+	HCO_3^-
Reaction 4	$H_2PO_4^-$	+	OH^-	HPO_4^{2-}	+	H_2O
Reaction 5	H_2SO_4	+	$N_2H_5^+$	HSO_4^-	+	$N_2H_6^{2+}$
Reaction 6	HPO_4^{2-}	+	SO_3^{2-}	PO_4^{3-}	+	HSO_3^-

(Conjugate Pair brackets span Acid + Base; Conjugate Pair brackets span Base + Acid)

SAMPLE PROBLEM 18.4 Identifying Conjugate Acid-Base Pairs

ARIS Presentation Center, Worked Examples Library

PROBLEM The following reactions are important environmental processes. Identify the conjugate acid-base pairs.
(a) $H_2PO_4^-(aq) + CO_3^{2-}(aq) \rightleftharpoons HCO_3^-(aq) + HPO_4^{2-}(aq)$
(b) $H_2O(l) + SO_3^{2-}(aq) \rightleftharpoons OH^-(aq) + HSO_3^-(aq)$

PLAN To find the conjugate pairs, we find the species that donated an H^+ (acid) and the species that accepted it (base). The acid (or base) on the left becomes its conjugate base (or conjugate acid) on the right. Remember, the conjugate acid has one more H and one fewer minus charge than its conjugate base.

SOLUTION (a) $H_2PO_4^-$ has one more H^+ than HPO_4^{2-}; CO_3^{2-} has one fewer H^+ than HCO_3^-. Therefore, $H_2PO_4^-$ and HCO_3^- are the acids, and HPO_4^{2-} and CO_3^{2-} are the bases. The conjugate acid-base pairs are $H_2PO_4^-/HPO_4^{2-}$ and HCO_3^-/CO_3^{2-}.

(b) H_2O has one more H^+ than OH^-; SO_3^{2-} has one fewer H^+ than HSO_3^-. The acids are H_2O and HSO_3^-; the bases are OH^- and SO_3^{2-}. The conjugate acid-base pairs are H_2O/OH^- and HSO_3^-/SO_3^{2-}.

FOLLOW-UP PROBLEM 18.4 Identify the conjugate acid-base pairs:
(a) $CH_3COOH(aq) + H_2O(l) \rightleftharpoons CH_3COO^-(aq) + H_3O^+(aq)$
(b) $H_2O(l) + F^-(aq) \rightleftharpoons OH^-(aq) + HF(aq)$

Literature Moran, Michael J. "Factors That Influence Relative Acid Strength in Water: A Simple Model," *J. Chem. Educ.* **2006**, *83*, 800.

Relative Acid-Base Strength and the Net Direction of Reaction

The net *direction* of an acid-base reaction depends on the relative strengths of the acids and bases involved. *A reaction proceeds to the greater extent in the direction in which a stronger acid and stronger base form a weaker acid and weaker base.* If the stronger acid and base are written on the left, the net direction is to the right, so $K_c > 1$.

The net direction of the reaction of H_2S and NH_3 is to the right ($K_c > 1$) because H_2S is a stronger acid than NH_4^+, the other acid present, and NH_3 is a stronger base than HS^-, the other base:

$$H_2S \; + \; NH_3 \; \rightleftharpoons \; HS^- \; + \; NH_4^+$$

stronger acid + stronger base ⟶ weaker base + weaker acid

You might think of the process as *a competition for the proton between the two bases,* NH_3 and HS^-, in which NH_3 wins.

In the same sense, the extent of acid (HA) dissociation in water depends on a competition for the proton between the two bases, A^- and H_2O, and strong acids and weak acids give very different results. When the strong acid HNO_3 dissolves, it completely transfers an H^+ to the base, H_2O, forming the conjugate base of HNO_3, which is NO_3^-, and the conjugate acid of H_2O, which is H_3O^+:

$$HNO_3 \; + \; H_2O \; \longrightarrow \; NO_3^- \; + \; H_3O^+$$

stronger acid + stronger base ⟶ weaker base + weaker acid

(In this case, the net direction is so far to the right that it would be inappropriate to show an equilibrium arrow.) HNO_3 is a stronger acid than H_3O^+, and H_2O is a stronger base than NO_3^-. Thus, with strong acids, the H_2O wins the competition for the proton because A^- (NO_3^-) is a much weaker base. On the other hand, with weak acids such as HF, the A^- (F^-) wins the competition because it is a stronger base than H_2O:

$$HF \; + \; H_2O \; \rightleftharpoons \; F^- \; + \; H_3O^+$$

weaker acid + weaker base ⟵ stronger base + stronger acid

Based on evidence gathered from the results of many such reactions, we can rank conjugate pairs in terms of the ability of the acid to transfer its proton (Figure 18.9). Note, especially, that *a weaker acid has a stronger conjugate base.* This makes perfect sense: the acid gives up its proton less readily because its conjugate base holds it more strongly.

We can use this list to predict the direction of a reaction between any two pairs, in other words, whether the equilibrium position lies predominantly to the right ($K_c > 1$) or to the left ($K_c < 1$). *An acid-base reaction proceeds to the right if the acid reacts with a base that is lower on the list* because this combination produces a weaker conjugate base and a weaker conjugate acid. Here are two sample problems, one using formulas and the other molecular scenes, that demonstrate this key idea.

ARIS Presentation Center, Worked Examples Library

SAMPLE PROBLEM 18.5 Predicting the Net Direction of an Acid-Base Reaction

PROBLEM Predict the net direction and whether K_c is greater or less than 1 for each of the following reactions (assume equal initial concentrations of all species):

(a) $H_2PO_4^-(aq) + NH_3(aq) \rightleftharpoons NH_4^+(aq) + HPO_4^{2-}(aq)$

(b) $H_2O(l) + HS^-(aq) \rightleftharpoons OH^-(aq) + H_2S(aq)$

PLAN We first identify the conjugate acid-base pairs. To predict the direction, we consult Figure 18.9 to see which acid and base are stronger. The stronger acid and base form the weaker acid and base, so the reaction proceeds in that net direction. If the reaction *as written* proceeds to the right, then [products] is higher than [reactants], and $K_c > 1$.

SOLUTION (a) The conjugate pairs are $H_2PO_4^-/HPO_4^{2-}$ and NH_4^+/NH_3. $H_2PO_4^-$ is higher on the list of acids, so it is stronger than NH_4^+; and NH_3 is lower on the list of bases, so it is stronger than HPO_4^{2-}. Therefore,

$$H_2PO_4^-(aq) + NH_3(aq) \rightleftharpoons NH_4^+(aq) + HPO_4^{2-}(aq)$$
stronger acid + stronger base ⟶ weaker acid + weaker base

The net direction is right, so $K_c > 1$.

(b) The conjugate pairs are H_2O/OH^- and H_2S/HS^-. H_2S is higher on the list of acids, and OH^- is lower on the list of bases. Thus, we have

$$H_2O(l) + HS^-(aq) \rightleftharpoons OH^-(aq) + H_2S(aq)$$
weaker acid + weaker base ⟵ stronger base + stronger acid

The net direction is to the left, so $K_c < 1$.

FOLLOW-UP PROBLEM 18.5 Explain with balanced equations, in which you show the net direction of the reaction, each of the following observations:
(a) You smell ammonia when NH_3 dissolves in water.
(b) The odor goes away when you add an excess of HCl to the solution in part (a).
(c) The odor returns when you add an excess of NaOH to the solution in part (b).

Figure 18.9 **Strengths of conjugate acid-base pairs.** The stronger the acid is, the weaker its conjugate base. The strongest acid appears at top left and the strongest base at bottom right. When an acid reacts with a base farther down the list, the reaction proceeds to the right ($K_c > 1$).

Transparency/ARIS Presentation Center, Art Library

SAMPLE PROBLEM 18.6 Using Molecular Scenes to Predict the Net Direction of an Acid-Base Reaction

PROBLEM Given that a 0.10 M solution of HX (blue and green) has a pH of 2.88 and a 0.10 M solution of HY (blue and orange) has a pH of 3.52, which scene best represents the equilibrium mixture after equimolar solutions of HX and Y^- (orange) are mixed?

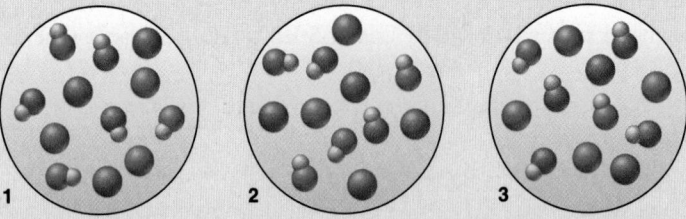

PLAN A stronger acid and base yield a weaker acid and base, so we have to determine the relative acid strengths of HX and HY in order to choose the correct molecular scene. The concentrations of the acid solutions are equal, so we can pick the stronger acid directly from the pH values of the two solutions. The stronger acid will react to a greater extent, so fewer molecules of it will be in the scene than molecules of the weaker acid.

SOLUTION The HX solution has a lower pH (2.88) than the HY solution (3.52), so we know right away that HX is the stronger acid and Y^- is the stronger base. Therefore, the reaction of HX and Y^- has a $K_c > 1$, which means the equilibrium mixture will have more HY than HX. Scene 1 has equal numbers of HX and HY, which would occur if the acids were of equal strength, and scene 2 shows fewer HY than HX, which would occur if HY were stronger. Therefore, only scene 3 is consistent with the relative acid strengths.

FOLLOW-UP PROBLEM 18.6 The following scene represents the reaction mixture after 0.10 M solutions of HA (blue and red) and B^- (black) react according to

$$HA(aq) + B^-(aq) \rightleftharpoons A^-(aq) + HB(aq)$$

and the reaction reaches equilibrium. Does the reaction have a K_c that is >1 or <1?

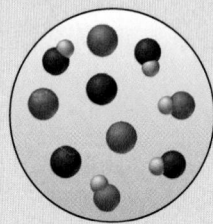

Section Summary

The Brønsted-Lowry acid-base definition does not require that bases contain OH or that acid-base reactions occur in aqueous solution. • An acid is a species that donates a proton and a base is one that accepts it. • An acid and a base act together in proton transfer. When an acid donates a proton, it becomes the conjugate base; when a base accepts a proton, it becomes the conjugate acid. In an acid-base reaction, acids and bases form their conjugates. A stronger acid has a weaker conjugate base, and vice versa. • An acid-base reaction proceeds in the net direction in which a stronger acid and base form a weaker base and acid.

18.4 SOLVING PROBLEMS INVOLVING WEAK-ACID EQUILIBRIA

Just as you saw in Chapter 17 for equilibrium problems in general, there are two types of equilibrium problems involving weak acids and their conjugate bases:

1. Given equilibrium concentrations, find K_a.
2. Given K_a and some concentration information, find the other equilibrium concentrations.

For all of these problems, we'll apply the same problem-solving approach, notation system, and assumptions:

- *The problem-solving approach.* As always, start with what is given in the problem statement and move logically toward what you want to find. Make a habit of applying the following steps:
  1. Write the balanced equation and K_a expression; these tell you what to find.
  2. Define x as the unknown change in concentration that occurs during the reaction. Frequently, $x = [HA]_{dissoc}$, the concentration of HA that dissociates, which, through the use of certain assumptions, also equals $[H_3O^+]$ and $[A^-]$ at equilibrium.
  3. Construct a reaction table that incorporates the unknown.
  4. Make assumptions (see below) that simplify the calculations, usually that x is very small relative to the initial concentration.
  5. Substitute the values into the K_a expression, and solve for x.
  6. Check that the assumptions are justified. (Apply the 5% test that was first used in Sample Problem 17.10, p. 758.) If they are not justified, use the quadratic formula to find x.

Literature Deng, Yuanjian. "Weak Electrolytes: When Should We Solve the Quadratic Equation?" *Chem. Educator* 8 **2003**, *5*, 295–297.

- *The notation system.* As always, the molar concentration of each species is shown with brackets. A subscript refers to where the species comes from or when it occurs in the reaction process. For example, $[H_3O^+]_{from\ HA}$ is the molar concentration of H_3O^+ that comes from the dissociation of HA; $[HA]_{init}$ is the initial molar concentration of HA, that is, before the dissociation occurs; $[HA]_{dissoc}$ is the molar concentration of HA that dissociates; and so forth. Recall that *brackets with no subscript refer to the molar concentration of the species at equilibrium.*

- *The assumptions.* We make two assumptions to simplify the arithmetic:
  1. The $[H_3O^+]$ from the autoionization of water is negligible. It is so much smaller than the $[H_3O^+]$ from the dissociation of HA that we can neglect it in these problems:

$$[H_3O^+] = [H_3O^+]_{from\ HA} + [H_3O^+]_{from\ H_2O} \approx [H_3O^+]_{from\ HA}$$

  Indeed, Le Châtelier's principle tells us that $[H_3O^+]_{from\ HA}$ *decreases* the extent of autoionization of water, so $[H_3O^+]_{from\ H_2O}$ in the HA solution is even less than $[H_3O^+]$ in pure water. Note that each molecule of HA that dissociates forms one H_3O^+ and one A^-, so $[A^-] = [H_3O^+]$.
  2. A weak acid has a small K_a. Therefore, it dissociates to such a small extent that we can neglect the change in its concentration to find its equilibrium concentration:

$$[HA] = [HA]_{init} - [HA]_{dissoc} \approx [HA]_{init}$$

Finding K_a Given Concentrations

This type of problem involves finding K_a of a weak acid from the concentration of one of the species in solution, usually $[H_3O^+]$ from a given pH:

$$HA(aq) + H_2O(l) \rightleftharpoons H_3O^+(aq) + A^-(aq)$$

$$K_a = \frac{[H_3O^+][A^-]}{[HA]}$$

A common approach is to prepare an aqueous solution of HA and measure its pH. Because you prepared the solution, you know $[HA]_{init}$. You can calculate $[H_3O^+]$ from the measured pH and then determine $[A^-]$ and $[HA]$ at equilibrium. Then, you substitute these values into the K_a expression and solve for K_a. You'll see how to make the assumptions discussed above and go through the approach in Sample Problem 18.7.

Phenylalanine, one of the amino acids that make up aspartame, is metabolized to phenylacetic acid (model).

SAMPLE PROBLEM 18.7 Finding K_a of a Weak Acid from the Solution pH

PROBLEM Phenylacetic acid ($C_6H_5CH_2COOH$, simplified here to HPAc; see model) builds up in the blood of persons with phenylketonuria, an inherited disorder that, if untreated, causes mental retardation and death. A study of the acid shows that the pH of 0.12 M HPAc is 2.62. What is the K_a of phenylacetic acid?

PLAN We are given $[HPAc]_{init}$ (0.12 M) and the pH (2.62) and must find K_a. We first write the equation for HPAc dissociation and the expression for K_a to see which values we need to find:

$$HPAc(aq) + H_2O(l) \rightleftharpoons H_3O^+(aq) + PAc^-(aq) \qquad K_a = \frac{[H_3O^+][PAc^-]}{[HPAc]}$$

- To find $[H_3O^+]$: We know the pH, so we can find $[H_3O^+]$. Because a pH of 2.62 is more than four pH units (10^4-fold) *lower* than the pH of pure water itself (pH = 7.0), we can assume that $[H_3O^+]_{from\ HPAc} \gg [H_3O^+]_{from\ H_2O}$. Therefore, $[H_3O^+]_{from\ HPAc} + [H_3O^+]_{from\ H_2O} \approx [H_3O^+]_{from\ HPAc} \approx [H_3O^+]$. Note that the initial $[H_3O^+]$ in the reaction table is zero.
- To find $[PAc^-]$: Because each HPAc that dissociates forms one H_3O^+ and one PAc^-, $[H_3O^+] \approx [PAc^-]$.
- To find $[HPAc]$: We know $[HPAc]_{init}$. Because HPAc is a weak acid, we assume that very little dissociates, so $[HPAc]_{init} - [HPAc]_{dissoc} = [HPAc] \approx [HPAc]_{init}$.

We set up a reaction table, make the assumptions, substitute the equilibrium values, solve for K_a, and then check the assumptions.

SOLUTION Calculating $[H_3O^+]$:

$$[H_3O^+] = 10^{-pH} = 10^{-2.62} = 2.4 \times 10^{-3}\ M$$

Setting up the reaction table, with $x = [HPAc]_{dissoc} = [H_3O^+]_{from\ HPAc} = [PAc^-] \approx [H_3O^+]$:

Concentration (M)	HPAc(aq)	+	H₂O(l)	⇌	H₃O⁺(aq)	+	PAc⁻(aq)
Initial	0.12		—		0		0
Change	$-x$		—		$+x$		$+x$
Equilibrium	$0.12 - x$		—		x		x

Making the assumptions:
1. The calculated $[H_3O^+]$ ($2.4 \times 10^{-3}\ M$) $\gg [H_3O^+]_{from\ H_2O}$, so we assume that $[H_3O^+] \approx [H_3O^+]_{from\ HPAc} = x$ (the change in $[HPAc]$).
2. HPAc is a weak acid, so we assume that $[HPAc] = 0.12\ M - x \approx 0.12\ M$.

Solving for the equilibrium concentrations:

$$x \approx [H_3O^+] = [PAc^-] = 2.4 \times 10^{-3}\ M$$
$$[HPAc] = 0.12\ M - x = 0.12\ M - (2.4 \times 10^{-3}\ M) \approx 0.12\ M\ (\text{to 2 sf})$$

Substituting these values into K_a:

$$K_a = \frac{[H_3O^+][PAc^-]}{[HPAc]} \approx \frac{(2.4 \times 10^{-3})(2.4 \times 10^{-3})}{0.12} = 4.8 \times 10^{-5}$$

Checking the assumptions by finding the percent error in concentration:

1. For $[H_3O^+]_{from\ H_2O}$: $\dfrac{1 \times 10^{-7}\ M}{2.4 \times 10^{-3}\ M} \times 100 = 4 \times 10^{-3}\%$ (<5%; assumption is justified).

2. For $[HPAc]_{dissoc}$: $\dfrac{2.4 \times 10^{-3}\ M}{0.12\ M} \times 100 = 2.0\%$ (<5%; assumption is justified).

CHECK The $[H_3O^+]$ makes sense: pH 2.62 should give $[H_3O^+]$ between 10^{-2} and $10^{-3}\ M$. The K_a calculation also seems in the correct range: $(10^{-3})^2/10^{-1} = 10^{-5}$, and this value seems reasonable for a weak acid.

COMMENT $[H_3O^+]_{from\ H_2O}$ is so small relative to $[H_3O^+]_{from\ HA}$ that, as shown in this problem, we will enter it as zero in all subsequent reaction tables.

FOLLOW-UP PROBLEM 18.7 The conjugate acid of ammonia is NH_4^+, a weak acid. If a 0.2 M NH_4Cl solution has a pH of 5.0, what is the K_a of NH_4^+?

Finding Concentrations Given K_a

The second type of equilibrium problem involving weak acids gives some concentration data and the K_a value and asks for the equilibrium concentration of some component. Such problems are very similar to those we solved in Chapter 17 in which a substance with a given initial concentration reacted to an unknown extent (see Sample Problems 17.9 to 17.11).

 ARIS Presentation Center, Worked Examples Library

 Biological application

SAMPLE PROBLEM 18.8 Determining Concentrations from K_a and Initial [HA]

PROBLEM Propanoic acid (CH_3CH_2COOH, which we simplify as HPr) is a carboxylic acid whose salts are used to retard mold growth in foods. What is the $[H_3O^+]$ of 0.10 M HPr ($K_a = 1.3\times10^{-5}$)?

PLAN We know the initial concentration (0.10 M) and K_a (1.3×10^{-5}) of HPr, and we need to find $[H_3O^+]$. First, we write the balanced equation and the expression for K_a:

$$HPr(aq) + H_2O(l) \rightleftharpoons H_3O^+(aq) + Pr^-(aq) \qquad K_a = \frac{[H_3O^+][Pr^-]}{[HPr]} = 1.3\times10^{-5}$$

We know $[HPr]_{init}$ but not [HPr]. If we let $x = [HPr]_{dissoc}$, x is also $[H_3O^+]_{from\ HPr}$ and $[Pr^-]$ because each HPr that dissociates yields one H_3O^+ and one Pr^-. With this information, we can set up a reaction table. In solving for x, we assume that, because HPr has a small K_a, it dissociates very little; therefore, $[HPr]_{init} - x = [HPr] \approx [HPr]_{init}$. After we find x, we check the assumption.

SOLUTION Setting up a reaction table, with $x = [HPr]_{dissoc} = [H_3O^+]_{from\ HPr} = [Pr^-] = [H_3O^+]$:

Concentration (M)	HPr(aq)	+	H₂O(l)	$\rightleftharpoons$	H₃O⁺(aq)	+	Pr⁻(aq)
Initial	0.10		—		0		0
Change	$-x$		—		$+x$		$+x$
Equilibrium	$0.10 - x$		—		x		x

Making the assumption: K_a is small, so x is small compared with $[HPr]_{init}$; therefore, $0.10\ M - x \approx 0.10\ M$. Substituting into the K_a expression and solving for x:

$$K_a = \frac{[H_3O^+][Pr^-]}{[HPr]} = 1.3\times10^{-5} \approx \frac{(x)(x)}{0.10}$$

$$x \approx \sqrt{(0.10)(1.3\times10^{-5})}$$
$$= 1.1\times10^{-3}\ M = [H_3O^+]$$

Checking the assumption:

For $[HPr]_{dissoc}$: $\dfrac{1.1\times10^{-3}\ M}{0.10\ M} \times 100 = 1.1\%$ (<5%; assumption is justified).

CHECK The $[H_3O^+]$ seems reasonable for a dilute solution of a weak acid with a moderate K_a. By reversing the calculation, we can check the math: $(1.1\times10^{-3})^2/0.10 = 1.2\times10^{-5}$, which is within rounding of the given K_a.

COMMENT We assume the concentration of HA that dissociates ($[HA]_{dissoc} = x$) can be neglected if K_a is small. But that also depends on $[HA]_{init}$. We'll use the same benchmark introduced in Sample Problem 17.10 (see Comment) to see if the assumption is justified:

- If $\dfrac{[HA]_{init}}{K_a} > 400$, the assumption is justified: neglecting x introduces an error <5%.

- If $\dfrac{[HA]_{init}}{K_a} < 400$, the assumption is *not* justified; neglecting x introduces an error >5%, so we solve a quadratic equation to find x.

The latter situation occurs in the follow-up problem.

FOLLOW-UP PROBLEM 18.8 Cyanic acid (HOCN) is an extremely acrid, unstable substance. What is the $[H_3O^+]$ and pH of 0.10 M HOCN ($K_a = 3.5\times10^{-4}$)?

The Effect of Concentration on the Extent of Acid Dissociation

If we repeat the calculation in Sample Problem 18.8, but start with a lower [HPr], we observe a very interesting fact about the extent of dissociation of a weak acid. Suppose the initial concentration of HPr is one-tenth as much, 0.010 M rather than 0.10 M. After filling in the reaction table and making the same assumptions, we find that

$$x = [\text{HPr}]_{\text{dissoc}} = 3.6 \times 10^{-4}\, M$$

Now, let's compare the percentages of HPr molecules dissociated at the two different initial acid concentrations, using the relationship

$$\text{Percent HA dissociated} = \frac{[\text{HA}]_{\text{dissoc}}}{[\text{HA}]_{\text{init}}} \times 100 \qquad (18.5)$$

Case 1: $[\text{HPr}]_{\text{init}} = 0.10\, M$

$$\text{Percent dissociated} = \frac{1.1 \times 10^{-3}\, M}{1.0 \times 10^{-1}\, M} \times 100 = 1.1\%$$

Case 2: $[\text{HPr}]_{\text{init}} = 0.010\, M$

$$\text{Percent dissociated} = \frac{3.6 \times 10^{-4}\, M}{1.0 \times 10^{-2}\, M} \times 100 = 3.6\%$$

As the initial acid concentration decreases, the percent dissociation of the acid increases. Don't confuse the *concentration* of HA dissociated with the *percent* HA dissociated. The concentration, $[\text{HA}]_{\text{dissoc}}$, is lower in the diluted HA solution because the actual *number* of dissociated HA molecules is smaller. It is the *fraction* (or the *percent*) of dissociated HA molecules that increases with dilution.

This phenomenon is analogous to a change in container volume (pressure) for gases at equilibrium (see the discussion in Section 17.6). In that situation, an increase in volume shifts the equilibrium position to favor more moles of gas. In the case of HA dissociation, the available volume increases as solvent is added and the solution is diluted; this increase in volume shifts the equilibrium position to favor more moles of ions. Sample Problem 18.9 uses molecular scenes to highlight this idea. (Note that, in order to depict the scenes practically, the acid has a much higher percent dissociation than actually occurs for any weak acid.)

SAMPLE PROBLEM 18.9 Using Molecular Scenes to Determine the Extent of HA Dissociation

PROBLEM A 0.15 M solution of acid HA (blue and green) is 33% dissociated. Which scene represents a sample of that solution after it is diluted with water?

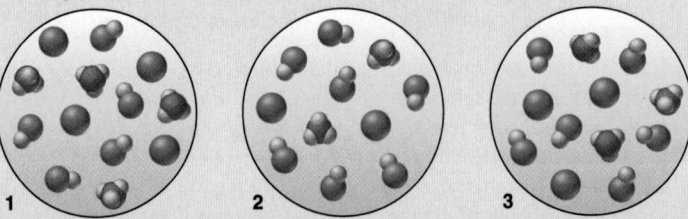

PLAN We are given the percent dissociation of the original HA solution (33%), and we know that the percent dissociation increases as the acid is diluted. Thus, we calculate the percent dissociation of each diluted sample and see which is greater than 33%. To determine percent dissociation, we apply Equation 18.5, with $\text{HA}_{\text{dissoc}}$ equal to the number of H_3O^+ (or A^-) and HA_{total} equal to the number of HA *plus* the number of H_3O^+ (or A^-).
SOLUTION Calculating the percent dissociation of each diluted solution with Equation 18.5:

Solution 1. Percent dissociated = $4/(5 + 4) \times 100 = 44\%$
Solution 2. Percent dissociated = $2/(7 + 2) \times 100 = 22\%$
Solution 3. Percent dissociated = $3/(6 + 3) \times 100 = 33\%$

Therefore, scene 1 represents the diluted solution.

CHECK Let's confirm our choice by examining the other scenes: in scene 2, HA is *less* dissociated than originally, so that scene must represent a more concentrated HA solution; scene 3 represents another solution with the same percent dissociation as the original.

FOLLOW-UP PROBLEM 18.9 The scene in the margin represents a sample of a weak acid HB (purple) dissolved in water. Draw a scene that represents the same volume after the solution has been diluted with water.

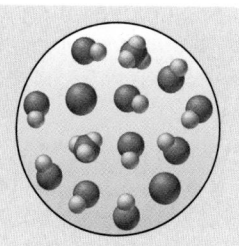

The Behavior of Polyprotic Acids

Acids with more than one ionizable proton are **polyprotic acids.** In a solution of a polyprotic acid, one proton at a time dissociates from the acid molecule, and each dissociation step has a different K_a. For example, phosphoric acid is a triprotic acid (three ionizable protons), so it has three K_a values:

$$H_3PO_4(aq) + H_2O(l) \rightleftharpoons H_2PO_4^-(aq) + H_3O^+(aq)$$

$$K_{a1} = \frac{[H_2PO_4^-][H_3O^+]}{[H_3PO_4]} = 7.2 \times 10^{-3}$$

$$H_2PO_4^-(aq) + H_2O(l) \rightleftharpoons HPO_4^{2-}(aq) + H_3O^+(aq)$$

$$K_{a2} = \frac{[HPO_4^{2-}][H_3O^+]}{[H_2PO_4^-]} = 6.3 \times 10^{-8}$$

$$HPO_4^{2-}(aq) + H_2O(l) \rightleftharpoons PO_4^{3-}(aq) + H_3O^+(aq)$$

$$K_{a3} = \frac{[PO_4^{3-}][H_3O^+]}{[HPO_4^{2-}]} = 4.2 \times 10^{-13}$$

As you can see from the relative K_a values, H_3PO_4 is a much stronger acid than $H_2PO_4^-$, which is much stronger than HPO_4^{2-}. Table 18.5 lists some common polyprotic acids and their K_a values. (More are listed in Appendix C.) Notice that, in every case, the first proton comes off to a much greater extent than the second and, where applicable, the second does to a much greater extent than the third:

$$K_{a1} > K_{a2} > K_{a3}$$

 ARIS Presentation Center, Tables Library

Table 18.5	Successive K_a Values for Some Polyprotic Acids at 25°C			
Name (Formula)	**Lewis Structure***	K_{a1}	K_{a2}	K_{a3}
Oxalic acid ($H_2C_2O_4$)	H—Ö—C—C—Ö—H (with :O: :O: double bonds)	5.6×10^{-2}	5.4×10^{-5}	
Sulfurous acid (H_2SO_3)	H—Ö—S—Ö—H (with :O:)	1.4×10^{-2}	6.5×10^{-8}	
Phosphoric acid (H_3PO_4)	H—Ö—P—Ö—H (with :O: and :Ö—H)	7.2×10^{-3}	6.3×10^{-8}	4.2×10^{-13}
Arsenic acid (H_3AsO_4)	H—Ö—As—Ö—H (with :O: and :Ö—H)	6×10^{-3}	1.1×10^{-7}	3×10^{-12}
Carbonic acid (H_2CO_3)	H—Ö—C—Ö—H (with :O:)	4.5×10^{-7}	4.7×10^{-11}	
Hydrosulfuric acid (H_2S)	H—S̈—H	9×10^{-8}	1×10^{-17}	

*Red type indicates the ionizable protons.

ACID STRENGTH (upward arrow)

This trend makes sense: it is more difficult for an H^+ ion to leave a singly charged anion (such as $H_2PO_4^-$) than to leave a neutral molecule (such as H_3PO_4), and more difficult still for it to leave a doubly charged anion (HPO_4^{2-}). Successive acid-dissociation constants typically differ by several orders of magnitude. This fact greatly simplifies pH calculations involving polyprotic acids because *we can usually neglect the H_3O^+ coming from the subsequent dissociations.*

ARIS Presentation Center, Worked Examples Library

Biological application

SAMPLE PROBLEM 18.10 Calculating Equilibrium Concentrations for a Polyprotic Acid

PROBLEM Ascorbic acid ($H_2C_6H_6O_6$; H_2Asc for this problem), known as vitamin C, is a diprotic acid ($K_{a1} = 1.0 \times 10^{-5}$ and $K_{a2} = 5 \times 10^{-12}$) found in citrus fruit. Calculate $[H_2Asc]$, $[HAsc^-]$, $[Asc^{2-}]$, and the pH of 0.050 M H_2Asc.

PLAN We know the initial concentration (0.050 M) and both K_a's for H_2Asc, and we have to calculate the equilibrium concentrations of all species and convert $[H_3O^+]$ to pH. We first write the equations and K_a expressions:

$$H_2Asc(aq) + H_2O(l) \rightleftharpoons HAsc^-(aq) + H_3O^+(aq)$$

$$K_{a1} = \frac{[HAsc^-][H_3O^+]}{[H_2Asc]} = 1.0 \times 10^{-5}$$

$$HAsc^-(aq) + H_2O(l) \rightleftharpoons Asc^{2-}(aq) + H_3O^+(aq)$$

$$K_{a2} = \frac{[Asc^{2-}][H_3O^+]}{[HAsc^-]} = 5 \times 10^{-12}$$

Then we assume the following:

1. Because $K_{a1} \gg K_{a2}$, dissociation of the first proton produces almost all the H_3O^+: $[H_3O^+]_{from\ H_2Asc} \gg [H_3O^+]_{from\ HAsc^-}$.

2. Because K_{a1} is small, the amount of H_2Asc that dissociates can be neglected relative to the initial concentration: $[H_2Asc]_{init} \approx [H_2Asc]$.

We set up a reaction table for the first dissociation, with x equal to the concentration of H_2Asc that dissociates, and then solve for $[H_3O^+]$ and $[HAsc^-]$. Because the second dissociation occurs to a much lesser extent than the first, we can substitute values from the first dissociation directly to find $[Asc^{2-}]$ of the second.

SOLUTION Setting up a reaction table with $x = [H_2Asc]_{dissoc} = [HAsc^-] \approx [H_3O^+]$:

Concentration (*M*)	$H_2Asc(aq)$	$+$	$H_2O(l)$	$\rightleftharpoons$	$H_3O^+(aq)$	$+$	$HAsc^-(aq)$
Initial	0.050		—		0		0
Change	$-x$		—		$+x$		$+x$
Equilibrium	$0.050 - x$		—		x		x

Making the assumptions:

1. Because $K_{a2} \ll K_{a1}$, $[H_3O^+]_{from\ HAsc^-} \ll [H_3O^+]_{from\ H_2Asc}$. Therefore,

$$[H_3O^+]_{from\ H_2Asc} \approx [H_3O^+]$$

2. Because K_{a1} is small, $[H_2Asc]_{init} - x = [H_2Asc] \approx [H_2Asc]_{init}$. Thus,

$$[H_2Asc] = 0.050\ M - x \approx 0.050\ M$$

Substituting into the expression for K_{a1} and solving for x:

$$K_{a1} = \frac{[H_3O^+][HAsc^-]}{[H_2Asc]} = 1.0 \times 10^{-5} \approx \frac{x^2}{0.050}$$

$$x = [HAsc^-] \approx [H_3O^+] \approx 7.1 \times 10^{-4}\ M$$

$$pH = -\log [H_3O^+] = -\log (7.1 \times 10^{-4}) = 3.15$$

Checking the assumptions:

1. $[H_3O^+]_{from\ HAsc^-} \ll [H_3O^+]_{from\ H_2Asc}$: For any second dissociation that does occur,

$$[H_3O^+]_{from\ HAsc^-} \approx \sqrt{[HAsc^-](K_{a2})} = \sqrt{(7.1 \times 10^{-4})(5 \times 10^{-12})} = 6 \times 10^{-8}\ M$$

This is even less than $[H_3O^+]_{from\ H_2O}$, so the assumption is justified.

2. $[H_2Asc]_{dissoc} \ll [H_2Asc]_{init}$: $\dfrac{7.1 \times 10^{-4}\ M}{0.050\ M} \times 100 = 1.4\%$ ($<5\%$; assumption is justified).

Also, from the Comment in Sample Problem 18.8, note that

$$\frac{[H_2Asc]_{init}}{K_{a1}} = \frac{0.050}{1.0\times10^{-5}} = 5000 > 400$$

Calculating $[Asc^{2-}]$:

$$K_{a2} = \frac{[H_3O^+][Asc^{2-}]}{[HAsc^-]} \quad \text{and} \quad [Asc^{2-}] = \frac{(K_{a2})[HAsc^-]}{[H_3O^+]}$$

$$[Asc^{2-}] = \frac{(5\times10^{-12})(7.1\times10^{-4})}{7.1\times10^{-4}} = 5\times10^{-12}\,M$$

CHECK $K_{a1} \gg K_{a2}$, so it makes sense that $[HAsc^-] \gg [Asc^{2-}]$ because Asc^{2-} is produced only in the second (much weaker) dissociation. Both K_a's are small, so all concentrations except $[H_2Asc]$ should be much lower than the original 0.050 M.

FOLLOW-UP PROBLEM 18.10 Oxalic acid (HOOC—COOH, or $H_2C_2O_4$) is the simplest diprotic carboxylic acid. Its commercial uses include bleaching straw and leather and removing rust and ink stains. Calculate $[H_2C_2O_4]$, $[HC_2O_4^-]$, $[C_2O_4^{2-}]$, and the pH of a 0.150 M $H_2C_2O_4$ solution. Use K_a values from Appendix C.

Section Summary

Two common types of weak-acid equilibrium problems involve finding K_a from a concentration and finding a concentration from K_a. • We summarize the information in a reaction table, and we simplify the arithmetic by assuming (1) $[H_3O^+]_{from\ H_2O}$ is so small relative to $[H_3O^+]_{from\ HA}$ that it can be neglected, and (2) weak acids dissociate so little that $[HA]_{init} \approx [HA]$ at equilibrium. • The *fraction* of weak acid molecules that dissociates is greater in a more dilute solution, even though the total $[H_3O^+]$ is less. • Polyprotic acids have more than one ionizable proton, but we assume that the first dissociation provides virtually all the H_3O^+.

18.5 WEAK BASES AND THEIR RELATION TO WEAK ACIDS

By focusing on where the proton comes from and goes to, the Brønsted-Lowry concept expands the definition of a base to encompass a host of species that the Arrhenius definition excludes: a base is any species that accepts a proton; to do so, *the base must have a lone electron pair.* (The lone electron pair also plays the central role in the Lewis acid-base definition, as you'll see later in this chapter.)

Now let's examine the equilibrium system of a weak base and focus, as we did for weak acids, on the base (B) dissolving in water. When B dissolves, it accepts a proton from H_2O, which acts as an acid, leaving behind an OH^- ion:

$$B(aq) + H_2O(l) \rightleftharpoons BH^+(aq) + OH^-(aq)$$

This general reaction for a base in water is described by the equilibrium expression

$$K_c = \frac{[BH^+][OH^-]}{[B][H_2O]}$$

Based on our earlier reasoning that $[H_2O]$ is treated as a constant in aqueous reactions, we incorporate $[H_2O]$ in the value of K_c and obtain the **base-dissociation constant** (or **base-ionization constant**), K_b:

$$K_b = \frac{[BH^+][OH^-]}{[B]} \tag{18.6}$$

Despite the name "base-dissociation constant," *no base dissociates in the process,* as you can see from the reaction.

As in the relation between pK_a and K_a, we know that pK_b, the negative logarithm of the base-dissociation constant, decreases with increasing K_b (that is, increasing base strength). In aqueous solution, the two large classes of weak bases are ammonia and the amines, and the anions of weak acids.

Literature Dostál, Jirí. "Two Faces of Alkaloids," *J. Chem. Educ.* **2000,** *77,* 993.

Molecules as Weak Bases: Ammonia and the Amines

Ammonia is the simplest nitrogen-containing compound that acts as a weak base in water:

$$NH_3(aq) + H_2O(l) \rightleftharpoons NH_4^+(aq) + OH^-(aq) \qquad K_b = 1.76 \times 10^{-5} \text{ (at 25°C)}$$

Despite labels on reagent bottles that read "ammonium hydroxide," an aqueous solution of ammonia consists largely of *unprotonated* NH_3 molecules, as its small K_b indicates. In a 1.0 M NH_3 solution, for example, $[OH^-] = [NH_4^+] = 4.2 \times 10^{-3}$ M, so about 99.58% of the NH_3 is not ionized. Table 18.6 shows the K_b values for a few common molecular bases. (A more extensive list appears in Appendix C.)

Table 18.6 K_b **Values for Some Molecular (Amine) Bases at 25°C**

Name (Formula)	Lewis Structure*	K_b
Diethylamine [$(CH_3CH_2)_2NH$]		8.6×10^{-4}
Dimethylamine [$(CH_3)_2NH$]		5.9×10^{-4}
Methylamine (CH_3NH_2)		4.4×10^{-4}
Ammonia (NH_3)		1.76×10^{-5}
Pyridine (C_5H_5N)		1.7×10^{-9}
Aniline ($C_6H_5NH_2$)		4.0×10^{-10}

BASE STRENGTH →

*Blue type indicates the basic nitrogen and its lone pair.

If one or more of the hydrogen atoms in ammonia is replaced by an organic group (designated as R), an *amine* results: RNH_2, R_2NH, or R_3N (Section 15.4; see Figure 15.15). The key structural feature of these organic compounds, as in all Brønsted-Lowry bases, is *a lone pair of electrons that can bind the proton donated by the acid.* Figure 18.10 depicts this process for methylamine, the simplest amine.

Figure 18.10 Abstraction of a proton from water by methylamine. The amines are organic derivatives of ammonia. Methylamine, the simplest amine, acts as a base in water by accepting a proton, thereby increasing [OH^-].

To find the pH of a solution of a molecular weak base, we use an approach very similar to that for a weak acid: write the equilibrium expression, set up a reaction table to find [base]$_{reacting}$, make the usual assumptions, and then solve for [OH$^-$]. The main difference is that we must convert [OH$^-$] to [H$_3$O$^+$] in order to calculate pH.

ARIS Presentation Center, Worked Examples Library

SAMPLE PROBLEM 18.11 Determining pH from K_b and Initial [B]

PROBLEM Dimethylamine, (CH$_3$)$_2$NH (see margin), a key intermediate in detergent manufacture, has a K_b of 5.9×10^{-4}. What is the pH of 1.5 M (CH$_3$)$_2$NH?

PLAN We know the initial concentration (1.5 M) and K_b (5.9×10^{-4}) of (CH$_3$)$_2$NH and have to find the pH. The amine reacts with water to form OH$^-$, so we have to find [OH$^-$] and then calculate [H$_3$O$^+$] and pH. The balanced equation and K_b expression are

$$(CH_3)_2NH(aq) + H_2O(l) \rightleftharpoons (CH_3)_2NH_2^+(aq) + OH^-(aq)$$

$$K_b = \frac{[(CH_3)_2NH_2^+][OH^-]}{[(CH_3)_2NH]}$$

Because $K_b \gg K_w$, the [OH$^-$] from the autoionization of water is negligible, and we disregard it. Therefore,

$$[OH^-]_{from\ base} = [(CH_3)_2NH_2^+] = [OH^-]$$

Because K_b is small, we assume that the amount of amine reacting is small, so

$$[(CH_3)_2NH]_{init} - [(CH_3)_2NH]_{reacting} = [(CH_3)_2NH] \approx [(CH_3)_2NH]_{init}$$

As usual, we set up a reaction table, make the assumption, and solve for x. Then we check the assumption and convert [OH$^-$] to [H$_3$O$^+$] using K_w; finally, we calculate pH.

SOLUTION Setting up the reaction table, with

$$x = [(CH_3)_2NH]_{reacting} = [(CH_3)_2NH_2^+] = [OH^-]$$

Concentration (M)	(CH$_3$)$_2$NH(aq)	+ H$_2$O(l)	$\rightleftharpoons$	(CH$_3$)$_2$NH$_2^+$(aq)	+ OH$^-$(aq)
Initial	1.5	—		0	0
Change	$-x$	—		$+x$	$+x$
Equilibrium	$1.5 - x$	—		x	x

Making the assumption:
K_b is small, so [(CH$_3$)$_2$NH]$_{init} \approx$ [(CH$_3$)$_2$NH]; thus, 1.5 M $- x \approx$ 1.5 M.
Substituting into the K_b expression and solving for x:

$$K_b = \frac{[(CH_3)_2NH_2^+][OH^-]}{[(CH_3)_2NH]} = 5.9 \times 10^{-4} \approx \frac{x^2}{1.5}$$

$$x = [OH^-] \approx 3.0 \times 10^{-2}\ M$$

Checking the assumption:

$$\frac{3.0 \times 10^{-2}\ M}{1.5\ M} \times 100 = 2.0\% \ (<5\%;\ \text{assumption is justified})$$

Note that the Comment in Sample Problem 18.8 applies to weak bases as well:

$$\frac{[B]_{init}}{K_b} = \frac{1.5}{5.9 \times 10^{-4}} = 2.5 \times 10^3 > 400$$

Calculating pH:

$$[H_3O^+] = \frac{K_w}{[OH^-]} = \frac{1.0 \times 10^{-14}}{3.0 \times 10^{-2}} = 3.3 \times 10^{-13}\ M$$

$$pH = -\log(3.3 \times 10^{-13}) = \underline{12.48}$$

CHECK The value of x seems reasonable: $\sqrt{(\sim 6 \times 10^{-4})(1.5)} = \sqrt{9 \times 10^{-4}} = 3 \times 10^{-2}$.
Because (CH$_3$)$_2$NH is a weak base, the pH should be several pH units greater than 7.

FOLLOW-UP PROBLEM 18.11 Pyridine (C$_5$H$_5$N, see margin) serves as a solvent and base in organic syntheses. It has a pK_b of 8.77. What is the pH of 0.10 M pyridine?

Dimethylamine

ARIS Presentation Center, Art Library

ARIS Presentation Center, Art Library

Pyridine

Anions of Weak Acids as Weak Bases

The other large group of Brønsted-Lowry bases consists of anions of weak acids:*

$$A^-(aq) + H_2O(l) \rightleftharpoons HA(aq) + OH^-(aq) \qquad K_b = \frac{[HA][OH^-]}{[A^-]}$$

For example, F^-, the anion of the weak acid HF, acts as a weak base:

$$F^-(aq) + H_2O(l) \rightleftharpoons HF(aq) + OH^-(aq) \qquad K_b = \frac{[HF][OH^-]}{[F^-]}$$

Why is a solution of HA acidic and a solution of A^- basic? Let's approach the question by examining the relative concentrations of species present in 1 M HF and in 1 M NaF:

1. *The acidity of HA (aq).* Because HF is a weak acid, most of it is undissociated. The small fraction of HF that does dissociate yields small concentrations of H_3O^+ and F^-. The equilibrium position of the system lies far to the left:

$$HF(aq) + H_2O(l) \overset{\longleftarrow}{\rightleftharpoons} H_3O^+(aq) + F^-(aq)$$

Water molecules also contribute H_3O^+ and OH^-, but these concentrations are extremely small:

$$2H_2O(l) \overset{\longleftarrow}{\rightleftharpoons} H_3O^+(aq) + OH^-(aq)$$

Of all the species present—HF, H_2O, H_3O^+, F^-, and OH^-—the two that can influence the acidity of the solution are H_3O^+, predominantly from HF, and OH^- from water. The solution is acidic because $[H_3O^+]_{from\ HF} >> [OH^-]_{from\ H_2O}$.

2. *The basicity of A^-(aq).* Now, consider the species present in 1 M NaF. The salt dissociates completely to yield a stoichiometric concentration of F^-. The Na^+ ion behaves as a spectator, and some F^- reacts as a weak base to produce very small amounts of HF and OH^-:

$$F^-(aq) + H_2O(l) \overset{\longleftarrow}{\rightleftharpoons} HF(aq) + OH^-(aq)$$

As before, water dissociation contributes minute amounts of H_3O^+ and OH^-. Thus, in addition to the Na^+ ion, the species present are the same as in the HF solution: HF, H_2O, H_3O^+, F^-, and OH^-. The two species that affect the acidity are OH^-, predominantly from the F^- reaction with water, and H_3O^+ from water. In this case, $[OH^-]_{from\ F^-} >> [H_3O^+]_{from\ H_2O}$, so the solution is basic.

To summarize, *the relative concentrations of HA and A^- determine the acidity or basicity of the solution:*

- In an HA solution, $[HA] >> [A^-]$ and $[H_3O^+]_{from\ HA} >> [OH^-]_{from\ H_2O}$, so the solution is acidic.
- In an A^- solution, $[A^-] >> [HA]$ and $[OH^-]_{from\ A^-} >> [H_3O^+]_{from\ H_2O}$, so the solution is basic.

The Relation Between K_a and K_b of a Conjugate Acid-Base Pair

An important relationship exists between the K_a of HA and the K_b of A^-, which we can see by treating the two dissociation reactions as a reaction sequence and adding them together:

$$HA + H_2O \rightleftharpoons H_3O^+ + A^-$$
$$\underline{A^- + H_2O \rightleftharpoons HA + OH^-}$$
$$2H_2O \rightleftharpoons H_3O^+ + OH^-$$

---

*This equation and equilibrium expression are sometimes referred to as a *hydrolysis reaction* and a *hydrolysis constant*, K_h, because water is dissociated (hydrolyzed), and its parts end up in the products. Actually, except for the charge on the base, this process is the same as the proton-abstraction process with molecular bases such as ammonia, so a new term and equilibrium constant are unnecessary. That is, K_h is just another symbol for K_b, so we'll use K_b throughout.

The sum of the two dissociation reactions is the autoionization of water. Recall from Chapter 17 that, for a reaction that is the *sum* of two or more reactions, the overall equilibrium constant is the *product* of the individual equilibrium constants. Therefore, writing the expressions for each reaction gives

$$\frac{[H_3O^+][A^-]}{[HA]} \times \frac{[HA][OH^-]}{[A^-]} = [H_3O^+][OH^-]$$

or

$$K_a \quad \times \quad K_b \quad = \quad K_w \tag{18.7}$$

This relationship allows us to find K_a of the acid in a conjugate pair given K_b of the base, and vice versa. Reference tables typically have K_a and K_b values for *molecular species only*. The K_b for F^- or the K_a for $CH_3NH_3^+$, for example, does not appear in standard tables, but you can calculate either value simply by looking up the value for the molecular conjugate species and relating it to K_w. To find the K_b value for F^-, for instance, we look up the K_a value for HF and apply Equation 18.7:

$$K_a \text{ of HF} = 6.8 \times 10^{-4} \text{ (from Appendix C)}$$

So, we have

$$K_a \text{ of HF} \times K_b \text{ of } F^- = K_w$$

or

$$K_b \text{ of } F^- = \frac{K_w}{K_a \text{ of HF}} = \frac{1.0 \times 10^{-14}}{6.8 \times 10^{-4}} = 1.5 \times 10^{-11}$$

We can use this calculated K_b value to finish solving the problem.

SAMPLE PROBLEM 18.12 Determining the pH of a Solution of A^-

ARIS Presentation Center, Worked Examples Library

PROBLEM Sodium acetate (CH_3COONa, or NaAc for this problem) is used in photographic development and textile dyeing. What is the pH of 0.25 *M* NaAc? K_a of acetic acid (HAc) is 1.8×10^{-5}.

PLAN From the formula (NaAc) and the fact that all sodium salts are water soluble, we know that the initial concentration of acetate ion, Ac^-, is 0.25 *M*. We also know the K_a of the parent acid, HAc (1.8×10^{-5}). We have to find the pH of the solution of Ac^-, which acts as a base in water:

$$Ac^-(aq) + H_2O(l) \rightleftharpoons HAc(aq) + OH^-(aq) \qquad K_b = \frac{[HAc][OH^-]}{[Ac^-]}$$

If we calculate $[OH^-]$, we can find $[H_3O^+]$ and convert it to pH. To solve for $[OH^-]$, we need the K_b of Ac^-, which we obtain from the K_a of HAc and K_w. All sodium salts are soluble, so we know that $[Ac^-] = 0.25$ *M*. Our usual assumption is that $[Ac^-]_{init} \approx [Ac^-]$.

SOLUTION Setting up the reaction table, with $x = [Ac^-]_{reacting} = [HAc] = [OH^-]$:

Concentration (*M*)	$Ac^-(aq)$	+	$H_2O(l)$	$\rightleftharpoons$	$HAc(aq)$	+	$OH^-(aq)$
Initial	0.25		—		0		0
Change	$-x$		—		$+x$		$+x$
Equilibrium	$0.25 - x$		—		x		x

Solving for K_b:

$$K_b = \frac{K_w}{K_a} = \frac{1.0 \times 10^{-14}}{1.8 \times 10^{-5}} = 5.6 \times 10^{-10}$$

Making the assumption: Because K_b is small, $0.25 M - x \approx 0.25 M$.
Substituting into the expression for K_b and solving for x:

$$K_b = \frac{[HAc][OH^-]}{[Ac^-]} = 5.6 \times 10^{-10} \approx \frac{x^2}{0.25} \qquad x = [OH^-] \approx 1.2 \times 10^{-5} M$$

Checking the assumption:

$$\frac{1.2 \times 10^{-5} M}{0.25 M} \times 100 = 4.8 \times 10^{-3}\% \ (<5\%; \text{ assumption is justified})$$

Note that

$$\frac{0.25}{5.6 \times 10^{-10}} = 4.5 \times 10^8 > 400$$

Solving for pH:

$$[H_3O^+] = \frac{K_w}{[OH^-]} = \frac{1.0 \times 10^{-14}}{1.2 \times 10^{-5}} = 8.3 \times 10^{-10} \, M$$

$$pH = -\log(8.3 \times 10^{-10}) = 9.08$$

CHECK The K_b calculation seems reasonable: $\sim 10 \times 10^{-15}/2 \times 10^{-5} = 5 \times 10^{-10}$. Because Ac^- is a weak base, $[OH^-] > [H_3O^+]$; thus, pH > 7, which makes sense.

FOLLOW-UP PROBLEM 18.12 Sodium hypochlorite (NaClO) is the active ingredient in household laundry bleach. What is the pH of 0.20 M NaClO?

Section Summary

The extent to which a weak base accepts a proton from water to form OH^- is expressed by a base-dissociation constant, K_b. • Brønsted-Lowry bases include NH_3 and amines and the anions of weak acids. All produce basic solutions by accepting H^+ from water, which yields OH^- and thus makes $[H_3O^+] < [OH^-]$. • A solution of HA is acidic because $[HA] \gg [A^-]$, so $[H_3O^+] > [OH^-]$. A solution of A^- is basic because $[A^-] \gg [HA]$, so $[OH^-] > [H_3O^+]$. • By multiplying the expressions for K_a of HA and K_b of A^-, we obtain K_w. This relationship allows us to calculate either K_a of BH^+, the cationic conjugate acid of a molecular weak base B, or K_b of A^-, the anionic conjugate base of a molecular weak acid HA.

Literature Schmid, Roland; Miah, Arzu M. "The Strength of the Hydrohalic Acids," *J. Chem. Educ.* **2001**, *78*, 116.

18.6 MOLECULAR PROPERTIES AND ACID STRENGTH

The strength of an acid depends on its ability to donate a proton, which depends in turn on the strength of the bond to the acidic proton. In this section, we apply trends in atomic and bond properties to determine the trends in acid strength of nonmetal hydrides and oxoacids and discuss the acidity of hydrated metal ions.

Trends in Acid Strength of Nonmetal Hydrides

Two factors determine how easily a proton is released from a nonmetal hydride: the electronegativity of the central nonmetal (E) and the strength of the E—H bond. Figure 18.11 displays two periodic trends:

1. *Across a period, nonmetal hydride acid strength increases.* Across a period, the electronegativity of the nonmetal E determines the trend. As E becomes more electronegative, electron density around H is withdrawn, and the E—H bond becomes more polar. As a result, H^+ is released more easily to an O atom of a surrounding water molecule. In aqueous solution, the hydrides of Groups 3A(13) to 5A(15) do not behave as acids, but an increase in acid strength is seen in Groups 6A(16) and 7A(17). Thus, HCl is a stronger acid than H_2S because Cl is more electronegative (EN = 3.0) than S (EN = 2.5). The same relationship holds across each period.

2. *Down a group, nonmetal hydride acid strength increases.* Down a group, E—H bond strength determines the trend. As E becomes larger, the E—H bond becomes longer and weaker, so H^+ comes off more easily.* Thus, the hydrohalic acids increase in strength down the group:

$$HF \ll HCl < HBr < HI$$

A similar trend in increasing acid strength is seen down Group 6A(16). (The trend in hydrohalic acid strength is not seen in aqueous solution, where HCl, HBr, and HI are all equally strong; we discuss how this trend is observed in Section 18.8.)

Electronegativity increases, acidity increases →

6A(16)	7A(17)
H_2O	HF
H_2S	HCl
H_2Se	HBr
H_2Te	HI

← Bond strength decreases, acidity increases

Figure 18.11 The effect of atomic and molecular properties on nonmetal hydride acidity. As the electronegativity of the nonmetal (E) bonded to the ionizable proton increases *(left to right),* the acidity increases. As the length of the E—H bond increases *(top to bottom),* the bond strength decreases, so the acidity increases. (In water, HCl, HBr, and HI are equally strong, for reasons discussed in Section 18.8.)

---

*Actually, bond energy refers to bond breakage that forms an H atom, whereas acidity refers to bond breakage that forms an H^+ ion, so the two processes are not the same. Nevertheless, the magnitudes of the two types of bond breakage parallel each other.

Transparency/ARIS Presentation Center, Art Library

Trends in Acid Strength of Oxoacids

All oxoacids have the acidic H atom bonded to an O atom, so bond strength (length) is not a factor in their acidity, though it is with the nonmetal hydrides. Rather, as you saw in Section 14.8, two factors determine the acid strength of oxoacids: the electronegativity of the central nonmetal (E) and the number of O atoms.

1. *For oxoacids with the **same** number of oxygens around E, acid strength increases with the electronegativity of E.* Consider the hypohalous acids (written here as HOE, where E is a halogen atom). The more electronegative E is, the more electron density it pulls from the O—H bond; the more polar the O—H bond becomes, the more easily H^+ is lost (Figure 18.12A). Since electronegativity decreases down the group, we predict that acid strength decreases: HOCl > HOBr > HOI. Our prediction is confirmed by the K_a values:

$$K_a \text{ of HOCl} = 2.9 \times 10^{-8} \qquad K_a \text{ of HOBr} = 2.3 \times 10^{-9} \qquad K_a \text{ of HOI} = 2.3 \times 10^{-11}$$

We also predict (correctly) that in Group 6A(16), H_2SO_4 is stronger than H_2SeO_4; in Group 5A(15), H_3PO_4 is stronger than H_3AsO_4, and so forth.

2. *For oxoacids with **different** numbers of oxygens around a given E, acid strength increases with number of O atoms.* The electronegative O atoms pull electron density away from E, which makes the O—H bond more polar. The more O atoms present, the greater the shift in electron density, and the more easily the H^+ ion comes off (Figure 18.12B). Therefore, we predict, for instance, that chlorine oxoacids (written here as $HOCl_n$, with n from 0 to 3) increase in strength in the order HOCl < $HOClO$ < $HOClO_2$ < $HOClO_3$. Once again, the K_a values support the prediction:

$$K_a \text{ of HOCl (hypochlorous acid)} = 2.9 \times 10^{-8}$$
$$K_a \text{ of HOClO (chlorous acid)} = 1.12 \times 10^{-2}$$
$$K_a \text{ of HOClO}_2 \text{ (chloric acid)} \approx 1$$
$$K_a \text{ of HOClO}_3 \text{ (perchloric acid)} = >10^7$$

It follows from this that HNO_3 is stronger than HNO_2, that H_2SO_4 is stronger than H_2SO_3, and so forth.

Figure 18.12 The relative strengths of oxoacids. A, Among these hypohalous acids, HOCl is the strongest and HOI the weakest. Because Cl is the most electronegative of the halogens shown here, it withdraws electron density (indicated by thickness of green arrow) from the O—H bond most effectively, making that bond most polar in HOCl (indicated by the relative sizes of the δ symbols). **B,** Among the chlorine oxoacids, the additional O atoms in $HOClO_3$ pull electron density from the O—H bond, making the bond much more polar than that in HOCl.

Acidity of Hydrated Metal Ions

The aqueous solutions of certain metal ions are acidic because the *hydrated* metal ion transfers an H^+ ion to water. Consider a general metal nitrate, $M(NO_3)_n$, as it dissolves in water. The ions separate and become bonded to a specific number of surrounding H_2O molecules. This equation shows the hydration of the cation (M^{n+}) with H_2O molecules; hydration of the anion (NO_3^-) is indicated by (*aq*):

$$M(NO_3)_n(s) + xH_2O(l) \longrightarrow M(H_2O)_x^{n+}(aq) + nNO_3^-(aq)$$

Table 18.7 K_a Values of Some Hydrated Metal Ions at 25°C		
Free Ion	**Hydrated Ion**	K_a
Fe^{3+}	$Fe(H_2O)_6^{3+}(aq)$	6×10^{-3}
Sn^{2+}	$Sn(H_2O)_6^{2+}(aq)$	4×10^{-4}
Cr^{3+}	$Cr(H_2O)_6^{3+}(aq)$	1×10^{-4}
Al^{3+}	$Al(H_2O)_6^{3+}(aq)$	1×10^{-5}
Cu^{2+}	$Cu(H_2O)_6^{2+}(aq)$	3×10^{-8}
Pb^{2+}	$Pb(H_2O)_6^{2+}(aq)$	3×10^{-8}
Zn^{2+}	$Zn(H_2O)_6^{2+}(aq)$	1×10^{-9}
Co^{2+}	$Co(H_2O)_6^{2+}(aq)$	2×10^{-10}
Ni^{2+}	$Ni(H_2O)_6^{2+}(aq)$	1×10^{-10}

ACID STRENGTH

ARIS Presentation Center, Tables Library

Demo Koubek, Edward. "Acid-Base Chemistry of the Aluminum Ion in Aqueous Solution," *J. Chem. Educ.* 1998, *75*, 60.

Figure 18.13 The acidic behavior of the hydrated Al^{3+} ion. When a metal ion enters water, it is hydrated as water molecules bond to it. If the ion is small and multiply charged, as is the Al^{3+} ion, it pulls sufficient electron density from the O—H bonds of the attached water molecules to make the bonds more polar, and an H^+ ion is transferred to a nearby water molecule.

ARIS Presentation Center, Art Library

If the metal ion, M^{n+}, is *small and highly charged,* it has a high charge density and withdraws sufficient electron density from the O—H bonds of these bonded water molecules for a proton to be released. That is, the hydrated cation, $M(H_2O)_x^{n+}$, acts as a typical Brønsted-Lowry acid. In the process, the bound H_2O molecule that releases the proton becomes a bound OH^- ion:

$$M(H_2O)_x^{n+}(aq) + H_2O(l) \rightleftharpoons M(H_2O)_{x-1}OH^{(n-1)+}(aq) + H_3O^+(aq)$$

Each type of hydrated metal ion that releases a proton has a characteristic K_a value. Table 18.7 shows some common examples (also see Appendix C).

Aluminum ion, for example, has the small size and high positive charge needed to produce an acidic solution. When an aluminum salt, such as $Al(NO_3)_3$, dissolves in water, the following steps occur:

$$Al(NO_3)_3(s) + 6H_2O(l) \longrightarrow Al(H_2O)_6^{3+}(aq) + 3NO_3^-(aq)$$
[dissolution and hydration]

$$Al(H_2O)_6^{3+}(aq) + H_2O(l) \rightleftharpoons Al(H_2O)_5OH^{2+}(aq) + H_3O^+(aq)$$
[dissociation of weak acid]

Note the formulas of the hydrated metal ions in the last step. When H^+ is released, the number of bound H_2O molecules decreases by 1 (from 6 to 5) and the number of bound OH^- ions increases by 1 (from 0 to 1), which reduces the ion's positive charge by 1 (from 3 to 2) (Figure 18.13).

Through its ability to withdraw electron density from the O—H bonds of the bonded water molecules, a small, highly charged central metal ion behaves like a central electronegative atom in an oxoacid. Salts of most M^{2+} and M^{3+} ions yield acidic aqueous solutions.

Electron density drawn toward Al^{3+} Nearby H_2O acts as base

$Al(H_2O)_6^{3+}$ H_2O $Al(H_2O)_5OH^{2+}$ H_3O^+

Section Summary

The strength of an acid depends on the ease with which the ionizable proton is released. • For nonmetal hydrides, acid strength increases across a period, with the electronegativity of the nonmetal (E), and down a group, with the length of the E—H bond. • For oxoacids with the same number of O atoms, acid strength increases with electronegativity of E; for oxoacids with the same E, acid strength increases with number of O atoms. • Small, highly charged metal ions are acidic in water because they withdraw electron density from the O—H bonds of bound H_2O molecules, releasing an H^+ ion to the solution.

18.7 ACID-BASE PROPERTIES OF SALT SOLUTIONS

When a salt dissolves, one or both of its ions may react with the water and affect the pH of the solution. Up to now you've seen that cations of weak bases (such as NH_4^+) are acidic, anions of weak acids (such as CN^-) are basic, and small, highly charged metal cations (such as Al^{3+}) are acidic. In addition, certain ions (such as $H_2PO_4^-$ and HCO_3^-) can act as an acid or a base, and we must calculate whether their solutions are acidic or basic. In this section, we'll classify the acid-base behavior of the various types of salt solutions.

Demo Shakhashiri, Bassam Z. "Hydrolysis: Acidic and Basic Properties of Salts," *Chemical Demonstrations, A Handbook for Teachers of Chemistry*, Vol. 3 (The University of Wisconsin Press, Madison, 1989), 103–108.

Salts That Yield Neutral Solutions

A salt consisting of the anion of a strong acid and the cation of a strong base yields a neutral solution because the ions do not react with water. To see why the ions don't react, let's consider the dissociation of the parent acid and base. When a strong acid such as HNO_3 dissolves, complete dissociation takes place:

$$HNO_3(l) + H_2O(l) \longrightarrow NO_3^-(aq) + H_3O^+(aq)$$

H_2O is a much stronger base than NO_3^-, so the reaction proceeds essentially to completion. The same argument can be made for any strong acid: *the anion of a strong acid is a much weaker base than water.* Therefore, a strong acid anion is hydrated, but nothing further happens.

Now consider the dissociation of a strong base, such as NaOH:

$$NaOH(s) \xrightarrow{H_2O} Na^+(aq) + OH^-(aq)$$

When Na^+ enters water, it becomes hydrated but nothing further happens. *The cations of all strong bases behave this way.*

The anions of strong acids are the halide ions, except F^-, and those of strong oxoacids, such as NO_3^- and ClO_4^-. The cations of strong bases are those from Group 1A(1) and Ca^{2+}, Sr^{2+}, and Ba^{2+} from Group 2A(2). Salts containing only these ions, such as NaCl and $Ba(NO_3)_2$, *yield neutral solutions because no reaction takes place between the ions and water.*

Salts That Yield Acidic Solutions

A salt consisting of the anion of a strong acid and the cation of a weak base yields an acidic solution because the cation acts as a weak acid, and the anion does not react. For example, NH_4Cl produces an acidic solution because the NH_4^+ ion, the cation that forms from the weak base NH_3, is a weak acid, and the Cl^- ion, the anion of a strong acid, does not react:

$$NH_4Cl(s) \xrightarrow{H_2O} NH_4^+(aq) + Cl^-(aq) \quad \text{[dissolution and hydration]}$$
$$NH_4^+(aq) + H_2O(l) \rightleftharpoons NH_3(aq) + H_3O^+(aq) \quad \text{[dissociation of weak acid]}$$

As you saw earlier, *small, highly charged metal ions* make up another group of cations that yield H_3O^+ in solution. For example, $Fe(NO_3)_3$ produces an acidic solution because the hydrated Fe^{3+} ion acts as a weak acid, whereas the NO_3^- ion, the anion of a strong acid, does not react:

$$Fe(NO_3)_3(s) + 6H_2O(l) \xrightarrow{H_2O} Fe(H_2O)_6^{3+}(aq) + 3NO_3^-(aq)$$
$$\text{[dissolution and hydration]}$$
$$Fe(H_2O)_6^{3+}(aq) + H_2O(l) \rightleftharpoons Fe(H_2O)_5OH^{2+}(aq) + H_3O^+(aq)$$
$$\text{[dissociation of weak acid]}$$

Salts That Yield Basic Solutions

A salt consisting of the anion of a weak acid and the cation of a strong base yields a basic solution in water because the anion acts as a weak base, and the cation does not react. The anion of a weak acid accepts a proton from water to yield OH^- ion. Sodium acetate, for example, yields a basic solution because the Na^+ ion, the cation of a strong base, does not react with water, and the CH_3COO^- ion, the anion of the weak acid CH_3COOH, acts as a weak base:

$$CH_3COONa(s) \xrightarrow{H_2O} Na^+(aq) + CH_3COO^-(aq) \quad \text{[dissolution and hydration]}$$
$$CH_3COO^-(aq) + H_2O(l) \rightleftharpoons CH_3COOH(aq) + OH^-(aq) \quad \text{[reaction of weak base]}$$

SAMPLE PROBLEM 18.13 Predicting Relative Acidity of Salt Solutions

PROBLEM Predict whether aqueous solutions of the following are acidic, basic, or neutral, and write an equation for the reaction of any ion with water:
(a) Potassium perchlorate, $KClO_4$
(b) Sodium benzoate, C_6H_5COONa
(c) Chromium(III) nitrate, $Cr(NO_3)_3$

PLAN We examine the formulas to determine the cations and anions. Depending on the nature of these ions, the solution will be neutral (strong-acid anion and strong-base cation), acidic (weak-base cation and strong-acid anion or highly charged metal cation), or basic (weak-acid anion and strong-base cation).

SOLUTION (a) Neutral. The ions are K^+ and ClO_4^-. The K^+ ion is from the strong base KOH, and the ClO_4^- anion is from the strong acid $HClO_4$. Neither ion reacts with water.
(b) Basic. The ions are Na^+ and $C_6H_5COO^-$. Na^+ is the cation of the strong base $NaOH$ and does not react with water. The benzoate ion, $C_6H_5COO^-$, is from the weak acid benzoic acid, so it reacts with water to produce OH^- ion:

$$C_6H_5COO^-(aq) + H_2O(l) \rightleftharpoons C_6H_5COOH(aq) + OH^-(aq)$$

(c) Acidic. The ions are Cr^{3+} and NO_3^-. The NO_3^- ion is the anion of the strong acid HNO_3, so it does not react with water. Cr^{3+} is a small metal ion with a high positive charge, so the hydrated ion, $Cr(H_2O)_6^{3+}$, reacts with water to produce H_3O^+:

$$Cr(H_2O)_6^{3+}(aq) + H_2O(l) \rightleftharpoons Cr(H_2O)_5OH^{2+}(aq) + H_3O^+(aq)$$

FOLLOW-UP PROBLEM 18.13 Write equations to predict whether solutions of the following salts are acidic, basic, or neutral: **(a)** $KClO_2$; **(b)** $CH_3NH_3NO_3$; **(c)** CsI.

Salts of Weakly Acidic Cations and Weakly Basic Anions

If a salt consists of a cation that can act as a weak acid *and* an anion that can act as a weak base, both ions react with water. In these cases, the overall acidity of the solution depends on the relative acid strength (K_a) or base strength (K_b) of the separated ions. For example, will an aqueous solution of ammonium cyanide, NH_4CN, be acidic or basic? First, we write equations for any reactions that occur between the separated ions and water. Ammonium ion is the conjugate acid of a weak base, so it acts as a weak acid:

$$NH_4^+(aq) + H_2O(l) \rightleftharpoons NH_3(aq) + H_3O^+(aq)$$

Cyanide ion is the anion of the weak acid HCN and it acts as a weak base:

$$CN^-(aq) + H_2O(l) \rightleftharpoons HCN(aq) + OH^-(aq)$$

The reaction that goes farther to the right will have the greater influence on the pH of the solution, so we must compare the K_a of NH_4^+ with the K_b of CN^-. Recall that only molecular compounds are listed in K_a and K_b tables, so we first have to calculate these values for the ions:

$$K_a \text{ of } NH_4^+ = \frac{K_w}{K_b \text{ of } NH_3} = \frac{1.0\times10^{-14}}{1.76\times10^{-5}} = 5.7\times10^{-10}$$

$$K_b \text{ of } CN^- = \frac{K_w}{K_a \text{ of } HCN} = \frac{1.0\times10^{-14}}{6.2\times10^{-10}} = 1.6\times10^{-5}$$

The difference in magnitude of the equilibrium constants ($K_b \approx 3\times10^4 K_a$) indicates that the acceptance of a proton from H_2O by CN^- proceeds much further than the donation of a proton to H_2O by NH_4^+. Thus, because K_b of $CN^- > K_a$ of NH_4^+, the NH_4CN solution is basic.

Salts of Amphiprotic Anions

The only salts left to consider are those in which the cation comes from a strong base and the anion comes from a polyprotic acid and still has one or more ionizable protons attached. These anions are amphiprotic—they can release a proton *to* water, thereby acting as an acid, or they can abstract a proton *from* water, thereby acting as a base. As with the previous type of salt, we have to compare the magnitudes of K_a and K_b, but here, we compare K_a and K_b of the anion.

For example, Na_2HPO_4 consists of Na^+, the cation of a strong base, which does not react with water, and HPO_4^{2-}, the second anion of the weak polyprotic acid H_3PO_4. Dissolved in water, the salt can undergo these three reactions. Does reaction 2 or 3 go farther to the right?

1. $$Na_2HPO_4(s) \xrightarrow{\ H_2O\ } 2Na^+(aq) + HPO_4^{2-}(aq) \qquad \text{[dissolution and hydration]}$$

2. $HPO_4^{2-}(aq) + H_2O(l) \rightleftharpoons PO_4^{3-}(aq) + H_3O^+(aq) \qquad \text{[acting as a weak acid]}$

From Appendix C, we find that the K_a of HPO_4^{2-} is 4.2×10^{-13}

3. $HPO_4^{2-}(aq) + H_2O(l) \rightleftharpoons H_2PO_4^-(aq) + OH^-(aq) \qquad \text{[acting as a weak base]}$

$$K_b \text{ of } HPO_4^{2-} = \frac{K_w}{K_a \text{ of } H_2PO_4^-} \qquad \text{or} \qquad \frac{1.0 \times 10^{-14}}{6.3 \times 10^{-8}} = 1.6 \times 10^{-7}$$

Because $K_b > K_a$, HPO_4^{2-} acts as a base, and a solution of Na_2HPO_4 is basic.

Table 18.8 displays the acid-base behavior of the various types of salts in water.

Transparency/ARIS Presentation Center, Tables Library

Table 18.8   The Behavior of Salts in Water

Salt Solution: Examples	pH	Nature of Ions	Ion That Reacts with Water: Example
Neutral: NaCl, KBr, Ba(NO$_3$)$_2$	7.0	Cation of strong base Anion of strong acid	None
Acidic: NH$_4$Cl, NH$_4$NO$_3$, CH$_3$NH$_3$Br	<7.0	Cation of weak base Anion of strong acid	Cation: $NH_4^+ + H_2O \rightleftharpoons NH_3 + H_3O^+$
Acidic: Al(NO$_3$)$_3$, CrBr$_3$, FeCl$_3$	<7.0	Small, highly charged cation Anion of strong acid	Cation: $Al(H_2O)_6^{3+} + H_2O \rightleftharpoons Al(H_2O)_5OH^{2+} + H_3O^+$
Acidic/Basic: NH$_4$ClO$_2$, NH$_4$CN, Pb(CH$_3$COO)$_2$	<7.0 if $K_{a(cation)} > K_{b(anion)}$ >7.0 if $K_{b(anion)} > K_{a(cation)}$	Cation of weak base (or small, highly charged cation) Anion of weak acid	Cation and anion: $NH_4^+ + H_2O \rightleftharpoons NH_3 + H_3O^+$ $CN^- + H_2O \rightleftharpoons HCN + OH^-$
Acidic/Basic: NaH$_2$PO$_4$, KHCO$_3$, NaHSO$_3$	<7.0 if $K_{a(anion)} > K_{b(anion)}$ >7.0 if $K_{b(anion)} > K_{a(anion)}$	Cation of strong base Anion of polyprotic acid	Anion: $HSO_3^- + H_2O \rightleftharpoons SO_3^{2-} + H_3O^+$ $HSO_3^- + H_2O \rightleftharpoons H_2SO_3 + OH^-$

ARIS Presentation Center, Worked Examples Library

SAMPLE PROBLEM 18.14 Predicting the Relative Acidity of a Salt Solution from K_a and K_b of the Ions

PROBLEM Determine whether an aqueous solution of zinc formate, $Zn(HCOO)_2$, is acidic, basic, or neutral.

PLAN The formula consists of the small, highly charged, and therefore weakly acidic, Zn^{2+} cation and the weakly basic $HCOO^-$ anion of the weak acid HCOOH. To determine the relative acidity of the solution, we write equations that show the reactions of the ions with water, and then find K_a of Zn^{2+} (from Appendix C) and calculate K_b of $HCOO^-$ (from K_a of HCOOH in Appendix C) to see which ion reacts to a greater extent.

SOLUTION Writing the reactions with water:

$$Zn(H_2O)_6^{2+}(aq) + H_2O(l) \rightleftharpoons Zn(H_2O)_5OH^+(aq) + H_3O^+(aq)$$

$$HCOO^-(aq) + H_2O(l) \rightleftharpoons HCOOH(aq) + OH^-(aq)$$

Obtaining K_a and K_b of the ions: The K_a of $Zn(H_2O)_6^{2+}(aq)$ is 1×10^{-9}. We obtain K_a of HCOOH and solve for K_b of $HCOO^-$:

$$K_b \text{ of } HCOO^- = \frac{K_w}{K_a \text{ of } HCOOH} = \frac{1.0 \times 10^{-14}}{1.8 \times 10^{-4}} = 5.6 \times 10^{-11}$$

K_a of $Zn(H_2O)_6^{2+} > K_b$ of $HCOO^-$, so the solution is acidic.

FOLLOW-UP PROBLEM 18.14 Determine whether solutions of the following salts are acidic, basic, or neutral: **(a)** $Cu(CH_3COO)_2$; **(b)** NH_4F; **(c)** $KHSO_3$.

Section Summary

Salts that always yield a neutral solution consist of ions that do not react with water. • Salts that always yield an acidic solution contain an unreactive anion and a cation that releases a proton to water. • Salts that always yield a basic solution contain an unreactive cation and an anion that accepts a proton from water. • If both cation and anion react with water, the ion that reacts to the greater extent (higher K) determines the acidity or basicity of the salt solution. • If the anion is amphiprotic (first anion of a polyprotic acid), the strength of the anion as an acid (K_a) or as a base (K_b) determines the acidity of the salt.

18.8 GENERALIZING THE BRØNSTED-LOWRY CONCEPT: THE LEVELING EFFECT

We conclude our focus on the Brønsted-Lowry concept with an important principle that holds for acid-base behavior in any solvent. Notice that, in H_2O, all Brønsted-Lowry acids yield H_3O^+ and all Brønsted-Lowry bases yield OH^- —the ions that form when the solvent autoionizes. In general, *an acid yields the cation and a base yields the anion of solvent autoionization.*

This idea lets us examine a question you may have been wondering about: why are all strong acids and strong bases *equally* strong in water? The answer is that *in water, the strongest acid possible is H_3O^+ and the strongest base possible is OH^-.* The moment we put some gaseous HCl in water, it reacts with the base H_2O and forms H_3O^+. The same holds for HNO_3, H_2SO_4, and any strong acid. All strong acids are equally strong in water because they dissociate *completely* to form H_3O^+. Given that the strong acid is no longer present, we are actually observing the acid strength of H_3O^+.

Similarly, strong bases, such as $Ba(OH)_2$, dissociate completely in water to yield OH^-. Even those that do not contain hydroxide ions in the solid, such as K_2O, do so. The oxide ion, which is a stronger base than OH^-, immediately takes a proton from water to form OH^-:

$$2K^+(aq) + O^{2-}(aq) + H_2O(l) \longrightarrow 2K^+(aq) + 2OH^-(aq)$$

No matter what species we try, any acid stronger than H_3O^+ simply donates its proton to H_2O, and any base stronger than OH^- accepts a proton from H_2O. Thus, water exerts a **leveling effect** on any strong acid or base by reacting with it to form the products of water's autoionization. Acting as a base, water levels the strength of all strong acids by making them appear equally strong, and acting as an acid, it levels the strength of all strong bases as well.

To rank strong acids in terms of relative strength, we must dissolve them in a solvent that is a *weaker* base than water, one that accepts their protons less readily. For example, you saw in Figure 18.11 that the hydrohalic acids increase in strength as the halogen becomes larger, as a result of the longer, weaker H—X bond. In water, HF is weaker than the other hydrogen halides, but HCl, HBr, and HI appear equally strong because water causes them to dissociate completely. When we dissolve them in pure acetic acid, however, *the acetic acid acts as the base* and accepts a proton from the acids:

$$\begin{array}{cccc} \text{acid} & \text{base} & \text{base} & \text{acid} \\ HCl(g) + CH_3COOH(l) & \rightleftharpoons & Cl^-(acet) + CH_3COOH_2^+(acet) \\ HBr(g) + CH_3COOH(l) & \rightleftharpoons & Br^-(acet) + CH_3COOH_2^+(acet) \\ HI(g) + CH_3COOH(l) & \rightleftharpoons & I^-(acet) + CH_3COOH_2^+(acet) \end{array}$$

[The use of (*acet*) instead of (*aq*) indicates solvation by CH_3COOH.] However, because acetic acid is a *weaker base* than water, the three acids protonate it to *different* extents. Measurements show that HI protonates the solvent to a greater extent than HBr, and HBr does so more than HCl; that is, in pure acetic acid, $K_{HI} > K_{HBr} > K_{HCl}$. Therefore, HCl is a weaker acid than HBr, which is weaker than HI. Similarly, the relative strength of strong bases is determined in a solvent that is a weaker acid than H_2O, such as liquid NH_3.

Section Summary

Strong acids (or strong bases) dissociate completely to yield H_3O^+ (or OH^-) in water; in effect, water equalizes (levels) their strengths. Acids that are equally strong in water show differences in strength when dissolved in a solvent that is a weaker base than water, such as acetic acid.

18.9 ELECTRON-PAIR DONATION AND THE LEWIS ACID-BASE DEFINITION

Literature Shaffer, Alan A. "Let Us Give Lewis Acid-Base Theory the Priority It Deserves," *J. Chem. Educ.* **2006**, *83*, 1746.

The final acid-base concept we consider was developed by Gilbert N. Lewis, whose contribution to understanding the importance of valence electron pairs in molecular bonding we discussed in Chapter 9. Whereas the Brønsted-Lowry concept focuses on the proton in defining a species as an acid or a base, the Lewis concept highlights the role of the *electron pair*. The **Lewis acid-base definition** holds that

- A *base* is any species that *donates* an electron pair.
- An *acid* is any species that *accepts* an electron pair.

The Lewis definition, like the Brønsted-Lowry definition, requires that a base have an electron pair to donate, so it does not expand the classes of bases. However, *it greatly expands the classes of acids.* Many species, such as CO_2 and Cu^{2+}, that do not contain H in their formula (and thus cannot be Brønsted-Lowry acids) function as Lewis acids by accepting an electron pair in their reactions. Lewis stated his objection to the proton as the defining feature of an acid this way: "To restrict the group of acids to those substances which contain hydrogen interferes as seriously with the systematic understanding of chemistry as would the restriction of the term oxidizing agent to those substances containing oxygen." Moreover, in the Lewis sense, the proton itself functions as an acid because it accepts the electron pair donated by a base:

$$B\!:\!\overset{\frown}{+}\,H^+ \rightleftharpoons B\!-\!H^+$$

Thus, *all Brønsted-Lowry acids donate H^+, a Lewis acid.*

The product of any Lewis acid-base reaction is called an **adduct,** *a single species that contains a **new** covalent bond:*

$$A\overset{\frown}{+}\!:\!B \rightleftharpoons A\!-\!B \text{ (adduct)}$$

Thus, the Lewis concept radically broadens the idea of acid-base reactions. What to Arrhenius was the formation of H_2O from H^+ and OH^- became, to Brønsted and Lowry, the transfer of a proton from a stronger acid to a stronger base to form a weaker base and weaker acid. To Lewis, the same process became *the donation and acceptance of an electron pair to form a covalent bond in an adduct*.

As we've seen, the key feature of a *Lewis base is a lone pair of electrons to donate*. The key feature of a *Lewis acid is a vacant orbital* (or the ability to rearrange its bonds to form one) to accept that lone pair and form a new bond. In the upcoming discussion, you'll encounter a variety of neutral molecules and positively charged ions that satisfy this requirement.

Molecules as Lewis Acids

Many neutral molecules function as Lewis acids. In every case, the atom that accepts the electron pair is low in electron density because of either an electron deficiency or a polar multiple bond.

Lewis Acids with Electron-Deficient Atoms Some molecular Lewis acids contain a central atom that is *electron deficient*, one surrounded by fewer than eight valence electrons. The most important of these acids are covalent compounds of the Group 3A(13) elements boron and aluminum. As noted in Chapters 10 and 14, these compounds react to complete their octet. For example, boron trifluoride accepts an electron pair from ammonia to form a covalent bond in a gaseous Lewis acid-base reaction:

Unexpected solubility behavior is sometimes due to adduct formation. Aluminum chloride, for instance, dissolves freely in relatively nonpolar diethyl ether because of a Lewis acid-base reaction, in which the ether's O atom donates an electron pair to Al to form a covalent bond:

Demo Shakhashiri, Bassam Z. "Lewis Acid-Base Properties of Aluminum Chloride and Hydrogen Chloride," *Chemical Demonstrations, A Handbook for Teachers of Chemistry*, Vol. 3 (The University of Wisconsin Press, Madison, 1989), 192–196.

This acidic behavior of boron and aluminum halides is put to use in many organic syntheses. For example, toluene, an important solvent and organic reagent, can be made by the action of CH_3Cl on benzene in the presence of $AlCl_3$. The Lewis acid $AlCl_3$ abstracts the Lewis base Cl^- from CH_3Cl to form an adduct that has a reactive $CH_3{}^+$ group, which attacks the benzene ring:

$$CH_3Cl \; + \; AlCl_3 \; \rightleftharpoons \; [CH_3]^+[Cl{-}AlCl_3]^-$$
$$\text{base} \qquad \text{acid} \qquad\qquad \text{adduct}$$

$$C_6H_6 \; + \; [CH_3]^+[Cl{-}AlCl_3]^- \; \rightleftharpoons \; C_6H_5CH_3 \; + \; AlCl_3 \; + \; HCl$$
$$\text{benzene} \qquad\qquad\qquad\qquad \text{toluene}$$

Lewis Acids with Polar Multiple Bonds Molecules that contain a polar double bond also function as Lewis acids. As the electron pair on the Lewis base approaches the partially positive end of the double bond, one of the bonds breaks to form the new bond in the adduct. For example, consider the reaction that occurs when SO_2 dissolves in water. The electronegative O atoms in SO_2 withdraw electron density from the central S, so it is partially positive. The O atom of water donates a lone pair to the S, breaking one of the π bonds and forming an S—O bond, and a proton is transferred from water to that O. The resulting adduct is sulfurous acid, and the overall process is

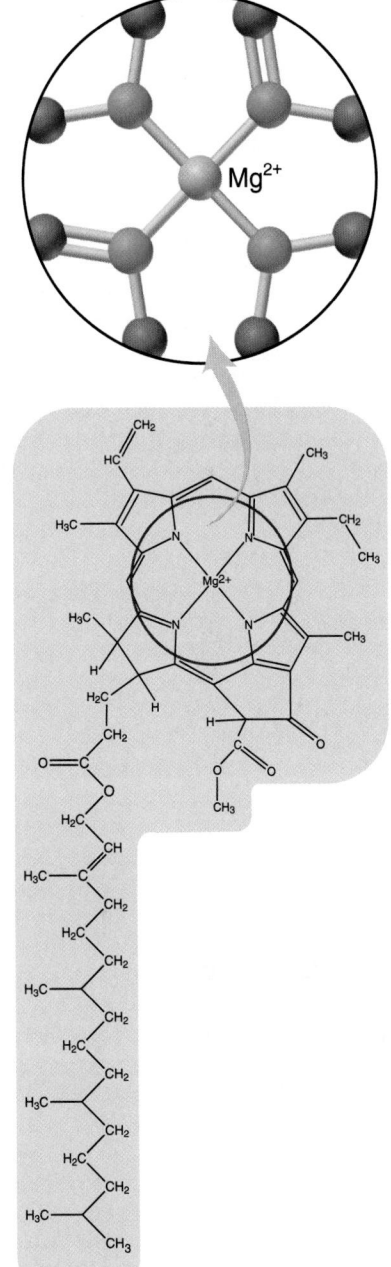

The formation of carbonates from a metal oxide and carbon dioxide is an analogous reaction that occurs in a nonaqueous heterogeneous system. The O^{2-} ion (shown below from CaO) donates an electron pair to the partially positive C in CO_2, a π bond breaks, and the CO_3^{2-} ion forms as the adduct:

Metal Cations as Lewis Acids

Earlier we saw that certain hydrated metal ions act as Brønsted-Lowry acids. In the Lewis sense, the hydration process itself is an acid-base reaction. The hydrated cation is the adduct, as lone electron pairs on the O atoms of water form covalent bonds to the positively charged ion; thus, *any metal ion acts as a Lewis acid when it dissolves in water:*

Ammonia is a stronger Lewis base than water because it displaces H_2O from a hydrated ion when aqueous NH_3 is added:

$$Ni(H_2O)_6^{2+}(aq) + 6NH_3(aq) \rightleftharpoons Ni(NH_3)_6^{2+}(aq) + 6H_2O(l)$$

We discuss the equilibrium nature of these acid-base reactions in greater detail in Chapter 19, and we investigate the structures of these ions in Chapter 23.

Many biomolecules are Lewis adducts with central metal ions. Most often, O and N atoms of organic groups, with their lone pairs, serve as the Lewis bases. Chlorophyll is a Lewis adduct of a central Mg^{2+} ion and the four N atoms of an organic tetrapyrrole ring system (Figure 18.14). Vitamin B_{12} has a similar structure with a central Co^{3+}, and so does heme, but with a central Fe^{2+}. Several other metal ions, such as Zn^{2+}, Mo^{2+}, and Cu^{2+}, are bound at the active sites of enzymes and function as Lewis acids in the catalytic action. **Biological application**

Figure 18.14   The Mg^{2+} ion as a Lewis acid in the chlorophyll molecule. Many biomolecules contain metal ions that act as Lewis acids. In chlorophyll, Mg^{2+} accepts electron pairs from surrounding N atoms that are part of the large organic portion of the molecule.

SAMPLE PROBLEM 18.15 Identifying Lewis Acids and Bases

PROBLEM   Identify the Lewis acids and Lewis bases in the following reactions:
(a) $H^+ + OH^- \rightleftharpoons H_2O$
(b) $Cl^- + BCl_3 \rightleftharpoons BCl_4^-$
(c) $K^+ + 6H_2O \rightleftharpoons K(H_2O)_6^+$
PLAN   We examine the formulas to see which species accepts the electron pair (Lewis acid) and which donates it (Lewis base) in forming the adduct.

SOLUTION (a) The H^+ ion accepts an electron pair from the OH^- ion in forming a bond. H^+ is the acid and OH^- is the base.
(b) The Cl^- ion has four lone pairs and uses one to form a new bond to the central B. Therefore, BCl_3 is the acid and Cl^- is the base.
(c) The K^+ ion does not have any valence electrons to provide, so the bond is formed when electron pairs from O atoms of water enter empty orbitals on K^+. Thus, K^+ is the acid and H_2O is the base.
CHECK The Lewis acids (H^+, BCl_3, and K^+) each have an unfilled valence shell that can accept an electron pair from the Lewis bases (OH^-, Cl^-, and H_2O).

FOLLOW-UP PROBLEM 18.15 Identify the Lewis acids and Lewis bases in each of the following reactions:
(a) $OH^- + Al(OH)_3 \rightleftharpoons Al(OH)_4^-$
(b) $SO_3 + H_2O \rightleftharpoons H_2SO_4$
(c) $Co^{3+} + 6NH_3 \rightleftharpoons Co(NH_3)_6^{3+}$

An Overview of Acid-Base Definitions

By looking closely at the essential chemical change involved, chemists can see a common theme in reactions as diverse as a standardized base being used to analyze an unknown fatty acid, baking soda being used in breadmaking, and even oxygen binding to hemoglobin in a blood cell. From this broader chemical perspective, the diversity of acid-base reactions takes on more unity. Let's stand back and survey the scope of the three acid-base definitions and see how they fit together.

The *Arrhenius definition,* which was the first attempt at describing acids and bases on the molecular level, is the most limited and narrow of the three definitions. It applies only to species whose structures include an H atom or an OH group that is released as an ion when the species dissolves in water. Because relatively few species have these structural prerequisites, Arrhenius acid-base reactions are relatively few in number, and all such reactions occur in H_2O and result in the formation of H_2O.

The *Brønsted-Lowry definition* is more general, seeing acid-base reactions as proton-transfer processes and eliminating the requirement that they occur in water. Whereas a Brønsted-Lowry acid, like an Arrhenius acid, still must have an H, a Brønsted-Lowry base is defined as any species with an electron pair available to accept a transferred proton. This definition includes a great many more species as bases. Furthermore, it defines the acid-base reaction in terms of conjugate acid-base pairs, with an acid and a base on both sides of the reaction. The system reaches an equilibrium state based on the relative strengths of the acid, the base, and their conjugates.

The *Lewis definition* has the widest scope of the three. The defining event of a Lewis acid-base reaction is the donation and acceptance of an electron pair to form a new covalent bond. Lewis bases still must have an electron pair to donate, but Lewis acids—as electron-pair acceptors—include many species not encompassed by the other two definitions, including molecules with electron-deficient atoms or with polar double bonds, metal ions, and the proton itself.

Section Summary

The Lewis acid-base definition focuses on the donation or acceptance of an electron pair to form a new covalent bond in an adduct, the product of an acid-base reaction. Lewis bases donate the electron pair, and Lewis acids accept it. • Thus, many species that do not contain H are Lewis acids. Molecules with polar double bonds act as Lewis acids, as do those with electron-deficient atoms. • Metal ions act as Lewis acids when they dissolve in water, which acts as a Lewis base, to form the adduct, a hydrated cation. • Many metal ions function as Lewis acids in biomolecules.

Chapter Perspective

In this chapter, we extended the principles of equilibrium to acids and bases. We also investigated one direction in which the science of chemistry matured, as narrow definitions of acids and bases progressively widened to encompass different species, physical states, solvent systems, and reaction types. Acids and bases, by whatever definition, are an extremely important group of substances. In Chapter 19, we continue our discussion of acid-base behavior and apply many of the ideas developed here to other aqueous equilibria.

CHAPTER REVIEW GUIDE

The following sections provide many aids to help you study this chapter. (Numbers in parentheses refer to pages, unless noted otherwise.)

Learning Objectives
These are concepts and skills you should know after studying this chapter.

Relevant section and/or sample problem (SP) numbers appear in parentheses.

Understand These Concepts

1. Why the proton is bonded to a water molecule, as H_3O^+, in all aqueous acid-base systems (Section 18.1)
2. The Arrhenius definitions of an acid and a base (Section 18.1)
3. Why all reactions of a strong acid and a strong base have the same ΔH°_{rxn} (Section 18.1)
4. How the strength of an acid (or base) relates to the extent of its dissociation into ions in water (Section 18.1)
5. How relative acid strength is expressed by the acid-dissociation constant K_a (Section 18.1)
6. Why water is a very weak electrolyte and how its autoionization is expressed by K_w (Section 18.2)
7. Why $[H_3O^+]$ is inversely related to $[OH^-]$ in any aqueous solution (Section 18.2)
8. How the relative magnitudes of $[H_3O^+]$ and $[OH^-]$ define whether a solution is acidic, basic, or neutral (Section 18.2)
9. The Brønsted-Lowry definitions of an acid and a base and how an acid-base reaction can be viewed as a proton-transfer process (Section 18.3)
10. How water acts as a base (or as an acid) when an acid (or base) dissolves in it (Section 18.3)
11. How a conjugate acid-base pair differs by one proton (Section 18.3)
12. How a Brønsted-Lowry acid-base reaction involves two conjugate acid-base pairs (Section 18.3)
13. Why a stronger acid and base react ($K_c > 1$) to form a weaker base and acid (Section 18.3)
14. How the percent dissociation of a weak acid increases as its concentration decreases (Section 18.4)
15. How a polyprotic acid dissociates in two or more steps and why only the first step supplies significant $[H_3O^+]$ (Section 18.4)
16. How weak bases in water accept a proton rather than dissociate; the meaning of K_b and pK_b (Section 18.5)
17. How ammonia, amines, and weak-acid anions act as weak bases in water (Section 18.5)
18. Why relative concentrations of HA and A^- determine the acidity or basicity of their solution (Section 18.5)
19. The relationship of the K_a and K_b of a conjugate acid-base pair to K_w (Section 18.5)
20. The effects of electronegativity, bond polarity, and bond strength on acid strength (Section 18.6)
21. Why aqueous solutions of small, highly charged metal ions are acidic (Section 18.6)
22. The various combinations of cations and anions that lead to acidic, basic, or neutral salt solutions (Section 18.7)
23. Why the strengths of strong acids are leveled in water but differentiated in a less basic solvent (Section 18.8)
24. The Lewis definitions of an acid and a base and how a Lewis acid-base reaction involves the donation and acceptance of an electron pair to form a covalent bond (Section 18.9)
25. How molecules with electron-deficient atoms, molecules with polar multiple bonds, and metal cations act as Lewis acids (Section 18.9)

Master These Skills

1. Classifying strong and weak acids and bases from their formulas (SP 18.1)
2. Using K_w to calculate $[H_3O^+]$ or $[OH^-]$ in an aqueous solution (SP 18.2)
3. Using p-scales to express $[H_3O^+]$, $[OH^-]$, and K (Section 18.2)
4. Calculating $[H_3O^+]$, pH, $[OH^-]$, and pOH (SP 18.3)
5. Identifying conjugate acid-base pairs (SP 18.4)
6. Using relative acid strengths to predict the net direction of an acid-base reaction (SPs 18.5, 18.6)
7. Calculating K_a of a weak acid from pH (SP 18.7)
8. Calculating $[H_3O^+]$ (and, thus, pH) from K_a and $[HA]_{init}$ (SP 18.8)
9. Applying the quadratic equation to find a concentration (Follow-up Problem 18.8)
10. Calculating the percent dissociation of a weak acid (Section 18.4 and SP 18.9)
11. Calculating $[H_3O^+]$ and other concentrations for a polyprotic acid (SP 18.10)
12. Calculating pH from K_b and $[B]_{init}$ (SP 18.11)
13. Finding K_b of A^- from K_a of HA and K_w (Section 18.5 and SP 18.12)
14. Calculating pH from K_b of A^- and $[A^-]_{init}$ (SP 18.12)
15. Predicting relative acid strengths of nonmetal hydrides and oxoacids (Section 18.6)
16. Predicting the relative acidity of a salt solution from the nature of the cation and anion (SPs 18.13, 18.14)
17. Identifying Lewis acids and bases (SP 18.15)

Key Terms These important terms appear in boldface in the chapter and are defined again in the Glossary.

Section 18.1
hydronium ion, H_3O^+ (784)
Arrhenius acid-base definition (784)
neutralization (784)
acid-dissociation (acid-ionization) constant (K_a) (786)

Section 18.2
autoionization (789)
ion-product constant for water (K_w) (789)
pH (791)
acid-base indicator (793)

Section 18.3
Brønsted-Lowry acid-base definition (793)

proton donor (793)
proton acceptor (794)
conjugate acid-base pair (795)

Section 18.4
polyprotic acid (803)

Section 18.5
base-dissociation (base-ionization) constant (K_b) (805)

Section 18.8
leveling effect (816)

Section 18.9
Lewis acid-base definition (817)
adduct (817)

Key Equations and Relationships Numbered and screened concepts are listed for you to refer to or memorize.

18.1 Defining the acid-dissociation constant (786):
$$K_a = \frac{[H_3O^+][A^-]}{[HA]}$$

18.2 Defining the ion-product constant for water (789):
$$K_w = [H_3O^+][OH^-] = 1.0 \times 10^{-14} \text{ (at 25°C)}$$

18.3 Defining pH (791):
$$pH = -\log [H_3O^+]$$

18.4 Relating pK_w to pH and pOH (792):
$$pK_w = pH + pOH = 14.00 \text{ (at 25°C)}$$

18.5 Finding the percent dissociation of HA (802):
$$\text{Percent HA dissociated} = \frac{[HA]_{\text{dissoc}}}{[HA]_{\text{init}}} \times 100$$

18.6 Defining the base-dissociation constant (805):
$$K_b = \frac{[BH^+][OH^-]}{[B]}$$

18.7 Expressing the relationship among K_a, K_b, and K_w (809):
$$K_a \times K_b = K_w$$

Highlighted Figures and Tables These figures (F) and tables (T) provide a visual review of key ideas.

Entries in **bold** contain frequently used data.

F18.1 Extent of dissociation for strong acids (785)
F18.2 Extent of dissociation for weak acids (786)
F18.4 Relationship between $[H_3O^+]$ and $[OH^-]$ and relative acidity of solutions (790)
F18.6 Relations among $[H_3O^+]$, pH, $[OH^-]$, and pOH (792)
F18.8 Proton transfer in Brønsted-Lowry acid-base reactions (794)

T18.4 Some conjugate acid-base pairs (795)
F18.9 **Strengths of conjugate acid-base pairs (797)**
F18.11 Effects of atomic and molecular properties on nonmetal hydride acidity (810)
F18.12 Relative strengths of oxoacids (811)
T18.8 Behavior of salts in water (815)

Brief Solutions to FOLLOW-UP PROBLEMS Compare your solutions to these calculation steps and answers.

18.1 (a) $HClO_3$; (b) HCl; (c) NaOH

18.2 $[H_3O^+] = \dfrac{1.0 \times 10^{-14}}{6.7 \times 10^{-2}} = 1.5 \times 10^{-13}$ M; basic

18.3 pOH $= 14.00 - 9.52 = 4.48$
$[H_3O^+] = 10^{-9.52} = 3.0 \times 10^{-10}$ M
$[OH^-] = \dfrac{1.0 \times 10^{-14}}{3.0 \times 10^{-10}} = 3.3 \times 10^{-5}$ M

18.4 (a) CH_3COOH/CH_3COO^- and H_3O^+/H_2O
(b) H_2O/OH^- and HF/F^-

18.5 (a) $NH_3(g) + H_2O(l) \rightleftharpoons NH_4^+(aq) + OH^-(aq)$
(b) $NH_3(g) + H_3O^+(aq;\text{ from HCl}) \longrightarrow NH_4^+(aq) + H_2O(l)$
(c) $NH_4^+(aq) + OH^-(aq;\text{ from NaOH}) \longrightarrow NH_3(g) + H_2O(l)$

18.6 There are more HB molecules than HA, so $K_c > 1$.

18.7 $NH_4^+(aq) + H_2O(l) \rightleftharpoons NH_3(aq) + H_3O^+(aq)$
$[H_3O^+] = 10^{-pH} = 10^{-5.0} = 1 \times 10^{-5}$ $M = [NH_3]$
From reaction table, $K_a = \dfrac{[NH_3][H_3O^+]}{[NH_4^+]} = \dfrac{(x)(x)}{0.2 - x}$
$\approx \dfrac{(1 \times 10^{-5})^2}{0.2} = 5 \times 10^{-10}$

18.8 $K_a = \dfrac{[H_3O^+][OCN^-]}{[HOCN]} = \dfrac{(x)(x)}{0.10 - x} = 3.5 \times 10^{-4}$
Since $\dfrac{[HOCN]_{\text{init}}}{K_a} = \dfrac{0.10}{3.5 \times 10^{-4}} = 286 < 400$, you must solve a quadratic equation: $x^2 + (3.5 \times 10^{-4})x - (3.5 \times 10^{-5}) = 0$.
$x = [H_3O^+] = 5.7 \times 10^{-3}$ M; pH $= 2.24$

18.9 There is no single correct scene: any scene in which the total number of $HB + H_3O^+$ (or $HB + B^-$) is less than in the original solution, yet the number of HB dissociated is greater, would be correct. For example,

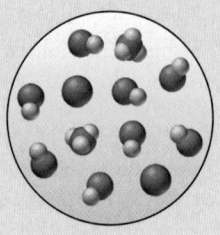

18.10 $K_{a1} = \dfrac{[HC_2O_4^-][H_3O^+]}{[H_2C_2O_4]} = \dfrac{x^2}{0.150 - x} = 5.6 \times 10^{-2}$

Since $\dfrac{[H_2C_2O_4]_{init}}{K_{a1}} < 400$, you must solve a quadratic equation:

$x^2 + (5.6 \times 10^{-2})x - (8.4 \times 10^{-3}) = 0$

$x = [H_3O^+] = 0.068\ M;\ \text{pH} = 1.17$

$x = [HC_2O_4^-] = 0.068\ M;\ [H_2C_2O_4] = 0.150\ M - x = 0.082\ M$

$[C_2O_4^{2-}] = \dfrac{(K_{a2})[HC_2O_4^-]}{[H_3O^+]} = \dfrac{(5.4 \times 10^{-5})(0.068)}{0.068}$

$\qquad\qquad = 5.4 \times 10^{-5}\ M$

18.11 $K_b = \dfrac{[C_5H_5NH^+][OH^-]}{[C_5H_5N]} = 10^{-8.77} = 1.7 \times 10^{-9}$

Assuming $0.10\ M - x \approx 0.10\ M,\ K_b = 1.7 \times 10^{-9} \approx \dfrac{(x)(x)}{0.10}$;

$x = [OH^-] \approx 1.3 \times 10^{-5}\ M;\ [H_3O^+] = 7.7 \times 10^{-10}\ M;$
pH = 9.11

18.12 K_b of $ClO^- = \dfrac{K_w}{K_a \text{ of } HClO} = \dfrac{1.0 \times 10^{-14}}{2.9 \times 10^{-8}} = 3.4 \times 10^{-7}$

Assuming $0.20\ M - x \approx 0.20\ M$,

$K_b = 3.4 \times 10^{-7} = \dfrac{[HClO][OH^-]}{[ClO^-]} \approx \dfrac{x^2}{0.20}$;

$x = [OH^-] \approx 2.6 \times 10^{-4}\ M;\ [H_3O^+] = 3.8 \times 10^{-11}\ M;$
pH = 10.42

18.13 (a) Basic:
$\quad ClO_2^-(aq) + H_2O(l) \rightleftharpoons HClO(aq) + OH^-(aq)$
K^+ is from strong base KOH.
(b) Acidic:
$\quad CH_3NH_3^+(aq) + H_2O(l) \rightleftharpoons CH_3NH_2(aq) + H_3O^+(aq)$
NO_3^- is from strong acid HNO_3.
(c) Neutral: Cs^+ is from strong base CsOH; I^- is from strong acid HI.

18.14 (a) K_a of $Cu(H_2O)_6^{2+} = 3 \times 10^{-8}$

K_b of $CH_3COO^- = \dfrac{K_w}{K_a \text{ of } CH_3COOH} = 5.6 \times 10^{-10}$

Since $K_a > K_b$, $Cu(CH_3COO)_2(aq)$ is acidic.

(b) K_a of $NH_4^+ = \dfrac{K_w}{K_b \text{ of } NH_3} = 5.7 \times 10^{-10}$

K_b of $F^- = \dfrac{K_w}{K_a \text{ of } HF} = 1.5 \times 10^{-11}$

Because $K_a > K_b$, $NH_4F(aq)$ is acidic.

(c) From Appendix C, K_a of $HSO_3^- = 6.5 \times 10^{-8}$

K_b of $HSO_3^- = \dfrac{K_w}{K_a \text{ of } H_2SO_3} = 7.1 \times 10^{-13}$

Because $K_a > K_b$, $KHSO_3(aq)$ is acidic

18.15 (a) OH^- is the Lewis base; $Al(OH)_3$ is the Lewis acid.
(b) H_2O is the Lewis base; SO_3 is the Lewis acid.
(c) NH_3 is the Lewis base; Co^{3+} is the Lewis acid.

PROBLEMS

Problems with colored numbers are answered in Appendix E and worked in detail in the Student Solutions Manual. Problem sections match those in the text and provide the numbers of relevant sample problems. Most offer Concept Review Questions, Skill-Building Exercises (grouped in pairs covering the same concept), and Problems in Context. The Comprehensive Problems are based on material from any section or previous chapter.

Note: Unless stated otherwise, all problems refer to aqueous solutions at 298 K (25°C).

Acids and Bases in Water
(Sample Problem 18.1)

● Concept Review Questions

- **18.1** What is the role of water in the Arrhenius acid-base definition?
- **18.2** What do Arrhenius acids have in common? What do Arrhenius bases have in common? Explain neutralization in terms of the Arrhenius acid-base definition. What data led Arrhenius to propose this idea of neutralization?
- **18.3** Why is the Arrhenius acid-base definition too limited? Give an example for which the Arrhenius definition does not apply.
- **18.4** What do "strong" and "weak" mean for acids and bases? K_a values of weak acids vary over more than 10 orders of magnitude. What do the acids have in common that makes them "weak"?

● Skill-Building Exercises (grouped in similar pairs)

- **18.5** Which of the following are Arrhenius acids?
  (a) H_2O (b) $Ca(OH)_2$ (c) H_3PO_3 (d) HI

- **18.6** Which of the following are Arrhenius acids?
  (a) $NaHSO_4$ (b) CH_4 (c) NaH   (d) H_3N

- **18.7** Which of the following are Arrhenius bases?
  (a) H_3AsO_4 (b) $Ba(OH)_2$ (c) HClO   (d) KOH

- **18.8** Which of the following are Arrhenius bases?
  (a) CH_3COOH (b) HOH   (c) CH_3OH (d) H_2NNH_2

- **18.9** Write the K_a expression for each of the following in water:
  (a) HCN   (b) HCO_3^- (c) HCOOH

- **18.10** Write the K_a expression for each of the following in water:
  (a) $CH_3NH_3^+$ (b) HClO   (c) H_2S

- **18.11** Write the K_a expression for each of the following in water:
  (a) HNO_2 (b) CH_3COOH (c) $HBrO_2$

- **18.12** Write the K_a expression for each of the following in water:
  (a) $H_2PO_4^-$ (b) H_3PO_2 (c) HSO_4^-

- **18.13** Use Appendix C to rank the following in order of *increasing* acid strength: HIO_3, HI, CH_3COOH, HF.

- **18.14** Use Appendix C to rank the following in order of *decreasing* acid strength: HClO, HCl, HCN, HNO_2.

- **18.15** Classify each as a strong or weak acid or base:
  (a) H_3AsO_4 (b) $Sr(OH)_2$ (c) HIO   (d) $HClO_4$

- **18.16** Classify each as a strong or weak acid or base:
  (a) CH_3NH_2 (b) K_2O (c) HI   (d) HCOOH

- **18.17** Classify each as a strong or weak acid or base:
  (a) RbOH   (b) HBr   (c) H_2Te (d) HClO

- **18.18** Classify each as a strong or weak acid or base:
  (a) $HOCH_2CH_2NH_2$ (b) H_2SeO_4 (c) HS^- (d) $B(OH)_3$

Autoionization of Water and the pH Scale
(Sample Problems 18.2 and 18.3)

● Concept Review Questions

18.19 What is an autoionization reaction? Write equations for the autoionization reactions of H_2O and of H_2SO_4.

18.20 What is the difference between K_c and K_w for the autoionization of water?

18.21 (a) What is the change in pH when $[OH^-]$ increases by a factor of 10? (b) What is the change in $[H_3O^+]$ when the pH decreases by 3 units?

18.22 Which solution has the higher pH? Explain.
(a) A 0.1 M solution of an acid with $K_a = 1 \times 10^{-4}$ or one with $K_a = 4 \times 10^{-5}$
(b) A 0.1 M solution of an acid with $pK_a = 3.0$ or one with $pK_a = 3.5$
(c) A 0.1 M solution or a 0.01 M solution of a weak acid
(d) A 0.1 M solution of a weak acid or a 0.1 M solution of a strong acid
(e) A 0.1 M solution of an acid or a 0.01 M solution of a base
(f) A solution of pOH 6.0 or one of pOH 8.0

● Skill-Building Exercises (grouped in similar pairs)

18.23 (a) What is the pH of 0.0111 M NaOH? Is the solution neutral, acidic, or basic? (b) What is the pOH of 1.35×10^{-3} M HCl? Is the solution neutral, acidic, or basic?

18.24 (a) What is the pH of 0.0333 M HNO_3? Is the solution neutral, acidic, or basic? (b) What is the pOH of 0.0347 M KOH? Is the solution neutral, acidic, or basic?

18.25 (a) What is the pH of 6.14×10^{-3} M HI? Is the solution neutral, acidic, or basic? (b) What is the pOH of 2.55 M $Ba(OH)_2$? Is the solution neutral, acidic, or basic?

18.26 (a) What is the pH of 7.52×10^{-4} M CsOH? Is the solution neutral, acidic, or basic? (b) What is the pOH of 1.59×10^{-3} M $HClO_4$? Is the solution neutral, acidic, or basic?

18.27 (a) What are $[H_3O^+]$, $[OH^-]$, and pOH in a solution with a pH of 9.85? (b) What are $[H_3O^+]$, $[OH^-]$, and pH in a solution with a pOH of 9.43?

18.28 (a) What are $[H_3O^+]$, $[OH^-]$, and pOH in a solution with a pH of 3.47? (b) What are $[H_3O^+]$, $[OH^-]$, and pH in a solution with a pOH of 4.33?

18.29 (a) What are $[H_3O^+]$, $[OH^-]$, and pOH in a solution with a pH of 4.77? (b) What are $[H_3O^+]$, $[OH^-]$, and pH in a solution with a pOH of 5.65?

18.30 (a) What are $[H_3O^+]$, $[OH^-]$, and pOH in a solution with a pH of 8.97? (b) What are $[H_3O^+]$, $[OH^-]$, and pH in a solution with a pOH of 11.27?

18.31 How many moles of H_3O^+ or OH^- must you add per liter of HA solution to adjust its pH from 3.15 to 3.65? Assume a negligible volume change.

18.32 How many moles of H_3O^+ or OH^- must you add per liter of HA solution to adjust its pH from 9.33 to 9.07? Assume a negligible volume change.

18.33 How many moles of H_3O^+ or OH^- must you add to 5.6 L of HA solution to adjust its pH from 4.52 to 5.25? Assume a negligible volume change.

18.34 How many moles of H_3O^+ or OH^- must you add to 87.5 mL of HA solution to adjust its pH from 8.92 to 6.33? Assume a negligible volume change.

● Problems in Context

18.35 The two molecular scenes below depict the relative concentrations of H_3O^+ (purple) in solutions of the same volume (with counter ions and solvent molecules omitted for clarity). If the pH in scene A is 4.8, what is the pH in scene B?

A   **B**

18.36 Like any equilibrium constant, K_w changes with temperature. (a) Given that autoionization is endothermic, how does K_w change with rising T? Explain with a reaction that includes heat as reactant or product. (b) In many medical applications, the value of K_w at 37°C (body temperature) may be more appropriate than the value at 25°C, 1.0×10^{-14}. The pH of pure water at 37°C is 6.80. Calculate K_w, pOH, and $[OH^-]$ at this temperature.

Proton Transfer and the Brønsted-Lowry Acid-Base Definition
(Sample Problems 18.4 to 18.6)

● Concept Review Questions

18.37 How are the Arrhenius and Brønsted-Lowry acid-base definitions different? How are they similar? Name two Brønsted-Lowry bases that are not Arrhenius bases. Can you do the same for acids? Explain.

18.38 What is a conjugate acid-base pair? What is the relationship between the two members of the pair?

18.39 (a) A Brønsted-Lowry acid-base reaction proceeds in the net direction in which a stronger acid and stronger base form a weaker acid and weaker base. Explain.

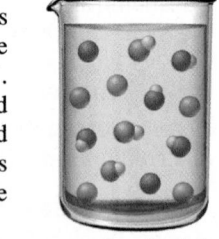

(b) The molecular scene at right depicts an aqueous solution of two conjugate acid-base pairs: HA/A^- and HB/B^-. The base in the first pair is represented by red spheres and the base in the second pair by green spheres; solvent molecules are omitted for clarity. Which is the stronger acid? Stronger base? Explain.

18.40 What is an amphiprotic species? Name one and write balanced equations that show why it is amphiprotic.

● Skill-Building Exercises (grouped in similar pairs)

18.41 Write balanced equations and K_a expressions for these Brønsted-Lowry acids in water:
(a) H_3PO_4 (b) C_6H_5COOH (c) HSO_4^-

18.42 Write balanced equations and K_a expressions for these Brønsted-Lowry acids in water:
(a) HCOOH   (b) $HClO_3$ (c) $H_2AsO_4^-$

18.43 Give the formula of the conjugate base:
(a) HCl   (b) H_2CO_3 (c) H_2O

18.44 Give the formula of the conjugate base:
(a) HPO_4^{2-} (b) NH_4^+ (c) HS^-

18.45 Give the formula of the conjugate acid:
(a) NH_3 (b) NH_2^- (c) nicotine, $C_{10}H_{14}N_2$

18.46 Give the formula of the conjugate acid:
(a) O^{2-} (b) SO_4^{2-} (c) H_2O

18.47 In each equation, label the acids, bases, and conjugate pairs:
(a) $HCl + H_2O \rightleftharpoons Cl^- + H_3O^+$
(b) $HClO_4 + H_2SO_4 \rightleftharpoons ClO_4^- + H_3SO_4^+$
(c) $HPO_4^{2-} + H_2SO_4 \rightleftharpoons H_2PO_4^- + HSO_4^-$

18.48 In each equation, label the acids, bases, and conjugate pairs:
(a) $NH_3 + HNO_3 \rightleftharpoons NH_4^+ + NO_3^-$
(b) $O^{2-} + H_2O \rightleftharpoons OH^- + OH^-$
(c) $NH_4^+ + BrO_3^- \rightleftharpoons NH_3 + HBrO_3$

18.49 In each equation, label the acids, bases, and conjugate pairs:
(a) $NH_3 + H_3PO_4 \rightleftharpoons NH_4^+ + H_2PO_4^-$
(b) $CH_3O^- + NH_3 \rightleftharpoons CH_3OH + NH_2^-$
(c) $HPO_4^{2-} + HSO_4^- \rightleftharpoons H_2PO_4^- + SO_4^{2-}$

18.50 In each equation, label the acids, bases, and conjugate pairs:
(a) $NH_4^+ + CN^- \rightleftharpoons NH_3 + HCN$
(b) $H_2O + HS^- \rightleftharpoons OH^- + H_2S$
(c) $HSO_3^- + CH_3NH_2 \rightleftharpoons SO_3^{2-} + CH_3NH_3^+$

18.51 Write balanced net ionic equations for the following reactions, and label the conjugate acid-base pairs:
(a) $NaOH(aq) + NaH_2PO_4(aq) \rightleftharpoons H_2O(l) + Na_2HPO_4(aq)$
(b) $KHSO_4(aq) + K_2CO_3(aq) \rightleftharpoons K_2SO_4(aq) + KHCO_3(aq)$

18.52 Write balanced net ionic equations for the following reactions, and label the conjugate acid-base pairs:
(a) $HNO_3(aq) + Li_2CO_3(aq) \rightleftharpoons LiNO_3(aq) + LiHCO_3(aq)$
(b) $2NH_4Cl(aq) + Ba(OH)_2(aq) \rightleftharpoons$
$\qquad\qquad 2H_2O(l) + BaCl_2(aq) + 2NH_3(aq)$

18.53 The following aqueous species constitute two conjugate acid-base pairs. Use them to write one acid-base reaction with $K_c > 1$ and another with $K_c < 1$: HS^-, Cl^-, HCl, H_2S.

18.54 The following aqueous species constitute two conjugate acid-base pairs. Use them to write one acid-base reaction with $K_c > 1$ and another with $K_c < 1$: NO_3^-, F^-, HF, HNO_3.

18.55 Use Figure 18.9 (p. 797) to determine whether $K_c > 1$ for
(a) $HCl + NH_3 \rightleftharpoons NH_4^+ + Cl^-$
(b) $H_2SO_3 + NH_3 \rightleftharpoons HSO_3^- + NH_4^+$

18.56 Use Figure 18.9 (p. 797) to determine whether $K_c > 1$ for
(a) $OH^- + HS^- \rightleftharpoons H_2O + S^{2-}$
(b) $HCN + HCO_3^- \rightleftharpoons H_2CO_3 + CN^-$

18.57 Use Figure 18.9 (p. 797) to determine whether $K_c < 1$ for
(a) $NH_4^+ + HPO_4^{2-} \rightleftharpoons NH_3 + H_2PO_4^-$
(b) $HSO_3^- + HS^- \rightleftharpoons H_2SO_3 + S^{2-}$

18.58 Use Figure 18.9 (p. 797) to determine whether $K_c < 1$ for
(a) $H_2PO_4^- + F^- \rightleftharpoons HPO_4^{2-} + HF$
(b) $CH_3COO^- + HSO_4^- \rightleftharpoons CH_3COOH + SO_4^{2-}$

Solving Problems Involving Weak-Acid Equilibria
(Sample Problems 18.7 to 18.10)

● Concept Review Questions

18.59 In each of the following cases, is the concentration of acid before and after dissociation nearly the same or very different? Explain your reasoning: (a) a concentrated solution of a strong acid; (b) a concentrated solution of a weak acid; (c) a dilute solution of a weak acid; (d) a dilute solution of a strong acid.

18.60 A sample of $0.0001\ M$ HCl has $[H_3O^+]$ close to that of a sample of $0.1\ M$ CH_3COOH. Are acetic acid and hydrochloric acid equally strong in these samples? Explain.

18.61 In which of the following solutions will $[H_3O^+]$ be approximately equal to $[CH_3COO^-]$: (a) $0.1\ M$ CH_3COOH; (b) $1\times10^{-7}\ M$ CH_3COOH; (c) a solution containing both $0.1\ M$ CH_3COOH and $0.1\ M$ CH_3COONa? Explain.

18.62 Why do successive K_a's decrease for all polyprotic acids?

● Skill-Building Exercises (grouped in similar pairs)

18.63 A $0.15\ M$ solution of butanoic acid, $CH_3CH_2CH_2COOH$, contains $1.51\times10^{-3}\ M$ H_3O^+. What is the K_a of butanoic acid?

18.64 A $0.035\ M$ solution of a weak acid (HA) has a pH of 4.88. What is the K_a of the acid?

18.65 Nitrous acid, HNO_2, has a K_a of 7.1×10^{-4}. What are $[H_3O^+]$, $[NO_2^-]$, and $[OH^-]$ in $0.60\ M$ HNO_2?

18.66 Hydrofluoric acid, HF, has a K_a of 6.8×10^{-4}. What are $[H_3O^+]$, $[F^-]$, and $[OH^-]$ in $0.75\ M$ HF?

18.67 Chloroacetic acid, $ClCH_2COOH$, has a pK_a of 2.87. What are $[H_3O^+]$, pH, $[ClCH_2COO^-]$, and $[ClCH_2COOH]$ in $1.25\ M$ $ClCH_2COOH$?

18.68 Hypochlorous acid, HClO, has a pK_a of 7.54. What are $[H_3O^+]$, pH, $[ClO^-]$, and $[HClO]$ in $0.115\ M$ HClO?

18.69 In a $0.20\ M$ solution, a weak acid is 3.0% dissociated.
(a) Calculate the $[H_3O^+]$, pH, $[OH^-]$, and pOH of the solution.
(b) Calculate K_a of the acid.

18.70 In a $0.735\ M$ solution, a weak acid is 12.5% dissociated.
(a) Calculate the $[H_3O^+]$, pH, $[OH^-]$, and pOH of the solution.
(b) Calculate K_a of the acid.

18.71 A 0.250-mol sample of HX is dissolved in enough H_2O to form 655 mL of solution. If the pH of the solution is 3.54, what is the K_a of HX?

18.72 A 4.85×10^{-3}-mol sample of HY is dissolved in enough H_2O to form 0.095 L of solution. If the pH of the solution is 2.68, what is the K_a of HY?

18.73 The weak acid HZ has a K_a of 2.55×10^{-4}.
(a) Calculate the pH of $0.075\ M$ HZ.
(b) Calculate the pOH of $0.045\ M$ HZ.

18.74 The weak acid HQ has a pK_a of 4.89.
(a) Calculate the $[H_3O^+]$ of $3.5\times10^{-2}\ M$ HQ.
(b) Calculate the $[OH^-]$ of $0.65\ M$ HQ.

18.75 (a) Calculate the pH of $0.175\ M$ HY, if $K_a = 1.50\times10^{-4}$.
(b) Calculate the pOH of $0.175\ M$ HX, if $K_a = 2.00\times10^{-2}$.

18.76 (a) Calculate the pH of $0.55\ M$ HCN, if $K_a = 6.2\times10^{-10}$.
(b) Calculate the pOH of $0.044\ M$ HIO_3, if $K_a = 0.16$.

18.77 Use Appendix C to calculate the percent dissociation of $0.55\ M$ benzoic acid, C_6H_5COOH.

18.78 Use Appendix C to calculate the percent dissociation of $0.050\ M$ CH_3COOH.

18.79 Use Appendix C to calculate $[H_2S]$, $[HS^-]$, $[S^{2-}]$, $[H_3O^+]$, pH, $[OH^-]$, and pOH in a $0.10\ M$ solution of the diprotic acid hydrosulfuric acid.

18.80 Use Appendix C to calculate $[H_2C_2O_4]$, $[HC_2O_4^-]$, $[C_2O_4^{2-}]$, $[H_3O^+]$, pH, $[OH^-]$, and pOH in a $0.200\ M$ solution of the diprotic acid oxalic acid.

● Problems in Context

18.81 Acetylsalicylic acid (aspirin), $HC_9H_7O_4$, is the most widely used pain reliever and fever reducer. Find the pH of $0.018\ M$ aqueous aspirin at body temperature (K_a at 37°C $= 3.6\times10^{-4}$).

• **18.82** Formic acid, HCOOH, the simplest carboxylic acid, is used in the textile and rubber industries and is secreted as a defense by many species of ants (family *Formicidae*). Calculate the percent dissociation of 0.75 *M* HCOOH.

Weak Bases and Their Relation to Weak Acids
(Sample Problems 18.11 and 18.12)

● Concept Review Questions

• **18.83** What is the key structural feature of all Brønsted-Lowry bases? How does this feature function in an acid-base reaction?

• **18.84** Why are most anions basic in H_2O? Give formulas of four anions that are not basic.

• **18.85** Except for the Na^+ spectator ion, aqueous solutions of CH_3COOH and CH_3COONa contain the same species. (a) What are the species (other than H_2O)? (b) Why is 0.1 *M* CH_3COOH acidic and 0.1 *M* CH_3COONa basic?

● Skill-Building Exercises (grouped in similar pairs)

• **18.86** Write balanced equations and K_b expressions for these Brønsted-Lowry bases in water:
(a) Pyridine, C_5H_5N (b) CO_3^{2-}

• **18.87** Write balanced equations and K_b expressions for these Brønsted-Lowry bases in water:
(a) Benzoate ion, $C_6H_5COO^-$ (b) $(CH_3)_3N$

• **18.88** Write balanced equations and K_b expressions for these Brønsted-Lowry bases in water:
(a) Hydroxylamine, HO—NH_2 (b) HPO_4^{2-}

• **18.89** Write balanced equations and K_b expressions for these Brønsted-Lowry bases in water:
(a) Guanidine, $(H_2N)_2C$=NH (double-bonded N is most basic)
(b) Acetylide ion, HC≡C$^-$

• **18.90** What is the pH of 0.070 *M* dimethylamine?
• **18.91** What is the pH of 0.12 *M* diethylamine?

• **18.92** What is the pH of 0.25 *M* ethanolamine?
• **18.93** What is the pH of 0.26 *M* aniline?

• **18.94** (a) What is the K_b of the acetate ion?
(b) What is the K_a of the anilinium ion, $C_6H_5NH_3^+$?
• **18.95** (a) What is the K_b of the benzoate ion, $C_6H_5COO^-$?
(b) What is the K_a of the 2-hydroxyethylammonium ion, $HOCH_2CH_2NH_3^+$ (pK_b of $HOCH_2CH_2NH_2$ = 4.49)?

• **18.96** (a) What is the pK_b of ClO_2^-?
(b) What is the pK_a of the dimethylammonium ion, $(CH_3)_2NH_2^+$?
• **18.97** (a) What is the pK_b of NO_2^-?
(b) What is the pK_a of the hydrazinium ion, H_2N—NH_3^+ (K_b of hydrazine = 8.5×10^{-7})?

• **18.98** (a) What is the pH of 0.150 *M* KCN?
(b) What is the pH of 0.40 *M* triethylammonium chloride, $(CH_3CH_2)_3NHCl$?
• **18.99** (a) What is the pH of 0.100 *M* sodium phenolate, C_6H_5ONa, the sodium salt of phenol?
(b) What is the pH of 0.15 *M* methylammonium bromide, CH_3NH_3Br (K_b of CH_3NH_2 = 4.4×10^{-4})?

• **18.100** (a) What is the pH of 0.65 *M* potassium formate, HCOOK?
(b) What is the pH of 0.85 *M* NH_4Br?
• **18.101** (a) What is the pH of 0.75 *M* NaF?
(b) What is the pH of 0.88 *M* pyridinium chloride, C_5H_5NHCl?

● Problems in Context

•• **18.102** Sodium hypochlorite solution, sold as "chlorine bleach," is potentially dangerous because of the basicity of ClO^-, the active bleaching ingredient. What is $[OH^-]$ in an aqueous solution that is 6.5% NaClO by mass? What is the pH of the solution? (Assume *d* of solution = 1.0 g/mL.)

• **18.103** Codeine ($C_{18}H_{21}NO_3$) is a narcotic pain reliever that forms a salt with HCl. What is the pH of 0.050 *M* codeine hydrochloride (pK_b of codeine = 5.80)?

Molecular Properties and Acid Strength

● Concept Review Questions

• **18.104** Across a period, how does the electronegativity of a nonmetal affect the acidity of its binary hydride?

• **18.105** How does the atomic size of a nonmetal affect the acidity of its binary hydride?

• **18.106** A strong acid has a weak bond to its acidic proton, whereas a weak acid has a strong bond to its acidic proton. Explain.

• **18.107** Perchloric acid, $HClO_4$, is the strongest of the halogen oxoacids, and hypoiodous acid, HIO, is the weakest. What two factors govern this difference in acid strength?

● Skill-Building Exercises (grouped in similar pairs)

• **18.108** Choose the *stronger* acid in each of the following pairs:
(a) H_2SeO_3 or H_2SeO_4 (b) H_3PO_4 or H_3AsO_4 (c) H_2S or H_2Te
• **18.109** Choose the *weaker* acid in each of the following pairs:
(a) HBr or H_2Se (b) $HClO_4$ or H_2SO_4 (c) H_2SO_3 or H_2SO_4

• **18.110** Choose the *stronger* acid in each of the following pairs:
(a) H_2Se or H_3As (b) $B(OH)_3$ or $Al(OH)_3$ (c) $HBrO_2$ or HBrO
• **18.111** Choose the *weaker* acid in each of the following pairs:
(a) HI or HBr   (b) H_3AsO_4 or H_2SeO_4 (c) HNO_3 or HNO_2

•• **18.112** Use Appendix C to choose the solution with the *lower* pH:
(a) 0.5 *M* $CuSO_4$ or 0.25 *M* $Al_2(SO_4)_3$
(b) 0.3 *M* $ZnCl_2$ or 0.3 *M* $PbCl_2$
•• **18.113** Use Appendix C to choose the solution with the *lower* pH:
(a) 0.1 *M* $FeCl_3$ or 0.1 *M* $AlCl_3$
(b) 0.1 *M* $BeCl_2$ or 0.1 *M* $CaCl_2$

•• **18.114** Use Appendix C to choose the solution with the *higher* pH:
(a) 0.2 *M* $Ni(NO_3)_2$ or 0.2 *M* $Co(NO_3)_2$
(b) 0.35 *M* $Al(NO_3)_3$ or 0.35 *M* $Cr(NO_3)_3$
•• **18.115** Use Appendix C to choose the solution with the *higher* pH:
(a) 0.1 *M* $NiCl_2$ or 0.1 *M* NaCl
(b) 0.1 *M* $Sn(NO_3)_2$ or 0.1 *M* $Co(NO_3)_2$

Acid-Base Properties of Salt Solutions
(Sample Problems 18.13 and 18.14)

● Concept Review Questions

• **18.116** What determines whether an aqueous solution of a salt will be acidic, basic, or neutral? Give an example of each type of salt.

• **18.117** Why is aqueous NaF basic but aqueous NaCl neutral?

• **18.118** The NH_4^+ ion forms acidic solutions, and the CH_3COO^- ion forms basic solutions. However, a solution of ammonium acetate is almost neutral. Do all of the ammonium salts of weak acids form neutral solutions? Explain your answer.

● Skill-Building Exercises (grouped in similar pairs)

•• **18.119** Explain with equations and calculations, when necessary, whether an aqueous solution of each of these salts is acidic, basic, or neutral: (a) KBr; (b) NH_4I; (c) KCN.

•• **18.120** Explain with equations and calculations, when necessary, whether an aqueous solution of each of these salts is acidic, basic, or neutral: (a) $Cr(NO_3)_3$; (b) NaHS; (c) $Zn(CH_3COO)_2$.

•• **18.121** Explain with equations and calculations, when necessary, whether an aqueous solution of each of these salts is acidic, basic, or neutral: (a) Na_2CO_3; (b) $CaCl_2$; (c) $Cu(NO_3)_2$.

•• **18.122** Explain with equations and calculations, when necessary, whether an aqueous solution of each of these salts is acidic, basic, or neutral: (a) CH_3NH_3Cl; (b) $KClO_4$; (c) CoF_2.

•• **18.123** Explain with equations and calculations, when necessary, whether an aqueous solution of each of these salts is acidic, basic, or neutral: (a) $SrBr_2$; (b) $Ba(CH_3COO)_2$; (c) $(CH_3)_2NH_2Br$.

•• **18.124** Explain with equations and calculations, when necessary, whether an aqueous solution of each of these salts is acidic, basic, or neutral: (a) $Fe(HCOO)_3$; (b) $KHCO_3$; (c) K_2S.

•• **18.125** Explain with equations and calculations, when necessary, whether an aqueous solution of each of these salts is acidic, basic, or neutral: (a) $(NH_4)_3PO_4$; (b) Na_2SO_4; (c) LiClO.

•• **18.126** Explain with equations and calculations, when necessary, whether an aqueous solution of each of these salts is acidic, basic, or neutral: (a) $Pb(CH_3COO)_2$; (b) $Cr(NO_2)_3$; (c) CsI.

•• **18.127** Rank the following salts in order of *increasing* pH of their 0.1 M aqueous solutions:
(a) KNO_3, K_2SO_3, K_2S, $Fe(NO_3)_2$
(b) NH_4NO_3, $NaHSO_4$, $NaHCO_3$, Na_2CO_3

•• **18.128** Rank the following salts in order of *decreasing* pH of their 0.1 M aqueous solutions:
(a) $FeCl_2$, $FeCl_3$, $MgCl_2$, $KClO_2$
(b) NH_4Br, $NaBrO_2$, NaBr, $NaClO_2$

Generalizing the Brønsted-Lowry Concept: The Leveling Effect

● Concept Review Questions

• **18.129** The methoxide ion, CH_3O^-, and amide ion, NH_2^-, are very strong bases that are "leveled" by water. What does this mean? Write the reactions that occur in the leveling process. What species do the two leveled solutions have in common?

• **18.130** Explain the differing extents of dissociation of H_2SO_4 in CH_3COOH, H_2O, and NH_3.

•• **18.131** In H_2O, HF is weak and the other hydrohalic acids are equally strong. In NH_3, however, all the hydrohalic acids are equally strong. Explain.

Electron-Pair Donation and the Lewis Acid-Base Definition
(Sample Problem 18.15)

● Concept Review Questions

• **18.132** What feature must a molecule or ion have for it to act as a Lewis base? A Lewis acid? Explain the roles of these features.

• **18.133** How do Lewis acids differ from Brønsted-Lowry acids? How are they similar? Do Lewis bases differ from Brønsted-Lowry bases? Explain.

•• **18.134** (a) Is a weak Brønsted-Lowry base necessarily a weak Lewis base? Explain with an example.
(b) Identify the Lewis bases in the following reaction:
$Cu(H_2O)_4^{2+}(aq) + 4CN^-(aq) \rightleftharpoons Cu(CN)_4^{2-}(aq) + 4H_2O(l)$
(c) Given that $K_c > 1$ for the reaction in part (b), which Lewis base is stronger?

• **18.135** In which of the three acid-base concepts can water be a product of an acid-base reaction? In which is it the only product?

• **18.136** (a) Give an example of a *substance* that is a base in two of the three acid-base definitions, but not in the third.
(b) Give an example of a *substance* that is an acid in one of the three acid-base definitions, but not in the other two.

● Skill-Building Exercises (grouped in similar pairs)

• **18.137** Which are Lewis acids and which are Lewis bases?
(a) Cu^{2+} (b) Cl^- (c) $SnCl_2$ (d) OF_2

• **18.138** Which are Lewis acids and which are Lewis bases?
(a) Na^+ (b) NH_3 (c) CN^- (d) BF_3

• **18.139** Which are Lewis acids and which are Lewis bases?
(a) BF_3 (b) S^{2-} (c) SO_3^{2-} (d) SO_3

• **18.140** Which are Lewis acids and which are Lewis bases?
(a) Mg^{2+} (b) OH^- (c) SiF_4 (d) $BeCl_2$

• **18.141** Identify the Lewis acid and Lewis base in each equation:
(a) $Na^+ + 6H_2O \rightleftharpoons Na(H_2O)_6^+$
(b) $CO_2 + H_2O \rightleftharpoons H_2CO_3$
(c) $F^- + BF_3 \rightleftharpoons BF_4^-$

• **18.142** Identify the Lewis acid and Lewis base in each equation:
(a) $Fe^{3+} + 2H_2O \rightleftharpoons FeOH^{2+} + H_3O^+$
(b) $H_2O + H^- \rightleftharpoons OH^- + H_2$
(c) $4CO + Ni \rightleftharpoons Ni(CO)_4$

• **18.143** Classify the following as Arrhenius, Brønsted-Lowry, or Lewis acid-base reactions. A reaction may fit all, two, one, or none of the categories:
(a) $Ag^+ + 2NH_3 \rightleftharpoons Ag(NH_3)_2^+$
(b) $H_2SO_4 + NH_3 \rightleftharpoons HSO_4^- + NH_4^+$
(c) $2HCl \rightleftharpoons H_2 + Cl_2$
(d) $AlCl_3 + Cl^- \rightleftharpoons AlCl_4^-$

• **18.144** Classify the following as Arrhenius, Brønsted-Lowry, or Lewis acid-base reactions. A reaction may fit all, two, one, or none of the categories:
(a) $Cu^{2+} + 4Cl^- \rightleftharpoons CuCl_4^{2-}$
(b) $Al(OH)_3 + 3HNO_3 \rightleftharpoons Al^{3+} + 3H_2O + 3NO_3^-$
(c) $N_2 + 3H_2 \rightleftharpoons 2NH_3$
(d) $CN^- + H_2O \rightleftharpoons HCN + OH^-$

Comprehensive Problems

•• **18.145** Pantothenic acid ($C_9H_{17}NO_5$; vitamin B_3) is shown below. This biologically active molecule is an optical isomer that behaves like a monoprotic Brønsted-Lowry acid in water.

(a) Use the molecular formula to write the equation for the reaction of pantothenic acid with water and the K_a expression.
(b) Which C atom is the chiral center?

•• **18.146** Chloral ($Cl_3C—CH{=}O$) forms a monohydrate, chloral hydrate, the sleep-inducing depressant called "knockout drops" in old movies. (a) Write two possible structures for chloral hydrate, one involving hydrogen bonding and one that is a Lewis adduct. (b) What spectroscopic method could be used to identify the real structure? Explain.

•• **18.147** In humans, blood pH is maintained within a narrow range: *acidosis* occurs if the blood pH is below 7.35, and *alkalosis* occurs if the pH is above 7.45. Given that the pK_w of blood is 13.63 at 37°C (body temperature), what is the normal range of $[H_3O^+]$ and of $[OH^-]$ in blood?

•• **18.148** The disinfectant phenol, C_6H_5OH, has a pK_a of 10.0 in water, but 14.4 in methanol. (a) Why are the values different? (b) Is methanol is a stronger or weaker base than water? (c) Write the dissociation reaction of phenol in methanol. (d) Write an expression for the autoionization constant of methanol.

•• **18.149** When carbon dioxide dissolves in water, it undergoes a multistep equilibrium process, with $K_{overall} = 4.5 \times 10^{-7}$, which is simplified to the following:

$$CO_2(g) + H_2O(l) \rightleftharpoons H_2CO_3(aq)$$
$$H_2CO_3(aq) + H_2O(l) \rightleftharpoons HCO_3^-(aq) + H_3O^+(aq)$$

(a) Classify each step as a Lewis or a Brønsted-Lowry reaction. (b) What is the pH of nonpolluted rainwater in equilibrium with clean air (P_{CO_2} in clean air $= 3.2 \times 10^{-4}$ atm; Henry's law constant for CO_2 at 25°C is 0.033 mol/L·atm)? (c) What is $[CO_3^{2-}]$ in rainwater (K_a of $HCO_3^- = 4.7 \times 10^{-11}$)? (d) If the partial pressure of CO_2 in clean air doubles in the next few decades, what will the pH of rainwater become?

•• **18.150** Seashells are mostly calcium carbonate, which reacts with H_3O^+ according to the equation

$$CaCO_3(s) + H_3O^+(aq) \rightleftharpoons Ca^{2+}(aq) + HCO_3^-(aq) + H_2O(l)$$

If K_w increases at higher pressure, will seashells dissolve more rapidly near the surface of the ocean or at great depths? Explain.

•• **18.151** Many molecules with central atoms from Period 3 or higher take part in Lewis acid-base reactions in which the central atom expands its valence shell. $SnCl_4$ reacts with $(CH_3)_3N$ as follows:

(a) Identify the Lewis acid and the Lewis base in the reaction. (b) Give the *nl* designation of the sublevel of the central atom in the acid before it accepts the lone pair.

•• **18.152** A chemist makes four successive 1:10 dilutions of 1.0×10^{-5} *M* HCl. Calculate the pH of the original solution and of each diluted solution (through 1.0×10^{-9} *M* HCl).

•• **18.153** Chlorobenzene, C_6H_5Cl, is a key intermediate in the manufacture of dyes and pesticides. It is made by the chlorination of benzene, catalyzed by $FeCl_3$, in this series of steps:

(1) $Cl_2 + FeCl_3 \rightleftharpoons FeCl_5$ (or $Cl^+FeCl_4^-$)
(2) $C_6H_6 + Cl^+FeCl_4^- \rightleftharpoons C_6H_6Cl^+ + FeCl_4^-$
(3) $C_6H_6Cl^+ \rightleftharpoons C_6H_5Cl + H^+$
(4) $H^+ + FeCl_4^- \rightleftharpoons HCl + FeCl_3$

(a) Which of the step(s) is (are) Lewis acid-base reactions? (b) Identify the Lewis acids and bases in each of those steps.

•• **18.154** The beakers shown contain 0.300 L of aqueous solutions of a moderately weak acid HY. Each particle represents 0.010 mol; solvent molecules are omitted for clarity. (a) The reaction in

beaker A is at equilibrium. Calculate *Q* for B, C, and D to determine which, if any, is also at equilibrium. (b) For any not at equilibrium, in which direction does the reaction proceed? (c) Does dilution affect the extent of dissociation of a weak acid? Explain.

A   B   C   D

•• **18.155** Hydrogen peroxide, H_2O_2 ($pK_a = 11.75$), is a bleaching agent and antiseptic. It is sold as 3% H_2O_2 by mass and contains 0.001% H_3PO_4 by mass to stabilize the solution. Which contributes more H_3O^+ to the solution, the H_2O_2 or the H_3PO_4?

•• **18.156** The strength of an acid or base is related to its strength as an electrolyte. (a) Is the electrical conductivity of 0.1 *M* HCl higher, lower, or the same as that of 0.1 *M* CH_3COOH? Explain. (b) Is the electrical conductivity of 1×10^{-7} *M* HCl higher, lower, or the same as that of 1×10^{-7} *M* CH_3COOH? Explain.

•• **18.157** Esters, RCOOR′, are formed by the reaction of carboxylic acids, RCOOH, and alcohols, R′OH, where R and R′ are hydrocarbon groups. Many esters are responsible for the odors of fruit and, thus, have important uses in the food and cosmetics industries. The first two steps in the mechanism of ester formation are

$$(1)\ R-\overset{\overset{\displaystyle :O:}{\|}}{C}-\ddot{O}H + H^+ \rightleftharpoons R-\overset{\overset{\displaystyle :\ddot{O}H}{|}}{\underset{+}{C}}-\ddot{O}H$$

$$(2)\ R-\overset{\overset{\displaystyle :\ddot{O}H}{|}}{\underset{+}{C}}-\ddot{O}H + R'-\ddot{O}H \rightleftharpoons R-\overset{\overset{\displaystyle :\ddot{O}H}{|}}{\underset{\underset{\displaystyle H}{\overset{\displaystyle |}{\underset{+}{:O}-R'}}}{C}}-\ddot{O}H$$

Identify the Lewis acids and Lewis bases in these two steps.

•• **18.158** Three beakers contain 100. mL of 0.10 *M* acid, either HCl, $HClO_2$, or HClO. (a) Find the pH of each. (b) Describe quantitatively how to make the pH equal in the beakers through the addition of water only.

•• **18.159** Human urine has a normal pH of 6.2. If a person eliminates an average of 1250. mL of urine per day, how many H^+ ions are eliminated per week?

•• **18.160** Liquid ammonia autoionizes like water:

$$2NH_3(l) \longrightarrow NH_4^+(am) + NH_2^-(am)$$

where (*am*) represents solvation by NH_3.
(a) Write the ion-product constant expression, K_{am}.
(b) What are the strongest acid and base that can exist in $NH_3(l)$?
(c) HNO_3 and HCOOH are leveled in $NH_3(l)$. Explain with equations.
(d) At the boiling point ($-33°C$), $K_{am} = 5.1 \times 10^{-27}$. Calculate $[NH_4^+]$ at this temperature.
(e) Pure sulfuric acid also autoionizes. Write the ion-product constant expression, K_{sulf}, and find the concentration of the conjugate base at 20°C ($K_{sulf} = 2.7 \times 10^{-4}$ at 20°C).

••• **18.161** Autoionization (see Problem 18.160) occurs in methanol (CH_3OH) and in ethylenediamine ($NH_2CH_2CH_2NH_2$).
(a) The autoionization constant of methanol (K_{met}) is 2×10^{-17}. What is $[CH_3O^-]$ in pure CH_3OH?

(b) The concentration of $NH_2CH_2CH_2NH_3^+$ in pure $NH_2CH_2CH_2NH_2$ is 2×10^{-8} M. What is the autoionization constant of ethylenediamine (K_{en})?

•• **18.162** Thiamine hydrochloride ($C_{12}H_{18}ON_4SCl_2$) is a water-soluble form of thiamine (vitamin B_1; $K_a = 3.37\times10^{-7}$). How many grams of the hydrochloride must be dissolved in 10.00 mL of water to give a pH of 3.50?

•• **18.163** Tris(hydroxymethyl)aminomethane, known as TRIS or THAM, is a water-soluble base used to synthesize surfactants and pharmaceuticals, as an emulsifying agent in cosmetics, and in cleaning mixtures for textiles and leather. In biomedical research, solutions of TRIS are used to maintain nearly constant pH for the study of enzymes and other cellular components. Given that the pK_b is 5.91, calculate the pH of 0.075 M TRIS.

•• **18.164** When Fe^{3+} salts are dissolved in water, the solution becomes acidic due to formation of $Fe(H_2O)_5OH^{2+}$ and H_3O^+. The overall process involves both Lewis and Brønsted-Lowry acid-base reactions. Write the equations for the process.

•• **18.165** What is the pH of a vinegar with 5.0% (w/v) acetic acid in water?

•• **18.166** How would you differentiate between a strong and a weak monoprotic acid from the results of the following procedures? (a) Electrical conductivity of an equimolar solution of each acid is measured. (b) Equal molarities of each are tested with pH paper. (c) Zinc metal is added to solutions of equal concentration.

•• **18.167** At 50°C and 1 atm, $K_w = 5.19\times10^{-14}$. At these conditions, find (a) $[H_3O^+]$ in pure water, (b) $[H_3O^+]$ in 0.010 M NaOH, and (c) $[OH^-]$ in 0.0010 M HClO$_4$. (d) Find $[H_3O^+]$ in 0.0100 M KOH at 100°C and 1000 atm ($K_w = 1.10\times10^{-12}$). (e) Find the pH of pure water at 100°C and 1000 atm.

•• **18.168** The catalytic efficiency of an enzyme is called its *activity* and refers to the rate at which it catalyzes the reaction. Most enzymes have optimum activity over a relatively narrow pH range, which is related to the pH of the local cellular fluid. The pH profiles of three digestive enzymes are shown.

Salivary amylase begins digestion of starches in the mouth and has optimum activity at a pH of 6.8; pepsin begins protein digestion in the stomach and has optimum activity at a pH of 2.0; and trypsin, released in pancreatic juices, continues protein digestion in the small intestine and has optimum activity at a pH of 9.5. Calculate $[H_3O^+]$ in the local cellular fluid for each enzyme.

•• **18.169** Acetic acid has a K_a of 1.8×10^{-5}, and ammonia has a K_b of 1.8×10^{-5}. Find $[H_3O^+]$, $[OH^-]$, pH, and pOH for (a) 0.240 M acetic acid and (b) 0.240 M ammonia.

•• **18.170** The uses of sodium phosphate include clarifying crude sugar, manufacturing paper, removing boiler scale, and washing concrete. What is the pH of a solution containing 33 g of Na$_3$PO$_4$ per liter? What is $[OH^-]$ of this solution?

•• **18.171** The Group 5A(15) hydrides react with boron trihalides in a reversible Lewis acid-base reaction. When 0.15 mol of PH$_3$BCl$_3$(s) is introduced into a 3.0-L container at a certain temperature, 8.4×10^{-3} mol of PH$_3$ is present at equilibrium: PH$_3$BCl$_3$(s) $\rightleftharpoons$ PH$_3$(g) + BCl$_3$(g). (a) Find K_c for the reaction at this temperature. (b) Draw a Lewis structure for the reactant.

••• **18.172** A 1.000 m solution of chloroacetic acid (ClCH$_2$COOH) freezes at −1.93°C. Find the K_a of chloroacetic acid. (Assume the molarities equal the molalities.)

•• **18.173** Sodium stearate ($C_{17}H_{35}$COONa) is a major component of bar soap. The K_a of the stearic acid is 1.3×10^{-5}. What is the pH of 10.0 mL of a solution containing 0.42 g of sodium stearate?

•• **18.174** Calcium propionate [Ca(CH$_3$CH$_2$COO)$_2$; calcium propanoate] is a mold inhibitor used in food, tobacco, and pharmaceuticals. (a) Use balanced equations to show whether aqueous calcium propionate is acidic, basic, or neutral. (b) Use Appendix C to find the resulting pH when 8.75 g of Ca(CH$_3$CH$_2$COO)$_2$ dissolves in enough water to give 0.500 L of solution.

••• **18.175** A site in Pennsylvania receives a total annual deposition of 2.688 g/m^2 of sulfate from fertilizer and acid rain. The ratio by mass of ammonium sulfate/ammonium bisulfate/sulfuric acid is 3.0/5.5/1.0. (a) How much acid, expressed as kg of sulfuric acid, is deposited over an area of 10. km^2? (b) How many pounds of CaCO$_3$ are needed to neutralize this acid? (c) If 10. km^2 is the area of an unpolluted lake 3 m deep and there is no loss of acid, what pH would be attained in the year? (Assume constant volume.)

•• **18.176** (a) If $K_w = 1.139\times10^{-15}$ at 0°C and 5.474×10^{-14} at 50°C, find $[H_3O^+]$ and pH of water at 0°C and 50°C.
(b) The autoionization constant for heavy water (deuterium oxide, D$_2$O) is 3.64×10^{-16} at 0°C and 7.89×10^{-15} at 50°C. Find $[D_3O^+]$ and pD of heavy water at 0°C and 50°C.
(c) Suggest a reason for these differences.

••• **18.177** HX ($\mathcal{M} = 150.$ g/mol) and HY ($\mathcal{M} = 50.0$ g/mol) are weak acids. A solution of 12.0 g/L of HX has the same pH as one containing 6.00 g/L of HY. Which is the stronger acid? Why?

•• **18.178** The beakers below depict the aqueous dissociations of weak acids HA (blue and green) and HB (blue and yellow); solvent molecules are omitted for clarity. If the HA solution is 0.50 L, and the HB solution is 0.25 L, and each particle represents 0.010 mol, find the K_a of each acid. Which acid, if either, is stronger?

A   **B**

•• **18.179** In his acid-base studies, Arrhenius discovered an important fact involving reactions like the following:
$$KOH(aq) + HNO_3(aq) \longrightarrow ?$$
$$NaOH(aq) + HCl(aq) \longrightarrow ?$$
(a) Complete the reactions and use the data for the individual ions in Appendix B to calculate each ΔH°_{rxn}.
(b) Explain your results and use them to predict ΔH°_{rxn} for
$$KOH(aq) + HCl(aq) \longrightarrow ?$$

•• **18.180** Putrescine [NH$_2$(CH$_2$)$_4$NH$_2$], found in rotting animal tissue, is now known to be in all cells and essential for normal and abnormal (cancerous) growth. It also plays a key role in the formation of GABA, a neurotransmitter. A 0.10 M aqueous solution of putrescine has $[OH^-] = 2.1\times10^{-3}$. What is the K_b?

Biological: 18.162, 18.163, 18.168, 18.180; Environmental: 18.175; Industrial: 18.170

•• **18.181** The molecular scene depicts the relative concentrations of H_3O^+ (purple) and OH^- (green) in an aqueous solution at 25°C. (Counter ions and solvent molecules are omitted for clarity.) (a) Calculate the pH. (b) How many H_3O^+ ions would you have to draw for every OH^- ion to depict a solution of pH 4?

•• **18.182** Polymers are not very soluble in water, but their solubility increases if they have charged groups. (a) Casein, a milk protein, contains many —COO⁻ groups on its side chains. How does the solubility of casein vary with pH? (b) Histones are proteins essential to the function of DNA. They are weakly basic due to the presence of side chains with —NH₂ and =NH groups. How does the solubility of a histone vary with pH?

•• **18.183** Hemoglobin (Hb) transports oxygen in the blood:

$$HbH^+(aq) + O_2(aq) + H_2O(l) \longrightarrow HbO_2(aq) + H_3O^+(aq)$$

In blood, $[H_3O^+]$ is held nearly constant at $4\times10^{-8}\ M$.
(a) How does the equilibrium position change in the lungs?
(b) How does it change in O_2-deficient cells?
(c) Excessive vomiting may lead to metabolic *alkalosis*, in which $[H_3O^+]$ in blood *decreases*. How does this condition affect the ability of Hb to transport O_2?
(d) Diabetes mellitus may lead to metabolic *acidosis*, in which $[H_3O^+]$ in blood *increases*. How does this condition affect the ability of Hb to transport O_2?

••• **18.184** Nitrogen is discharged from wastewater treatment facilities into rivers and streams, usually as NH_3 and NH_4^+:

$$NH_3(aq) + H_2O(l) \rightleftharpoons NH_4^+(aq) + OH^-(aq)\quad K_b = 1.76\times10^{-5}$$

One strategy for removing it is to raise the pH and "strip" the NH_3 from solution by bubbling air through the water. (a) At pH 7.00, what fraction of the total nitrogen in solution is NH_3, defined as $[NH_3]/([NH_3] + [NH_4^+])$? (b) What is the fraction at pH 10.00? (c) Explain the basis of ammonia stripping.

•• **18.185** Kevlar®, a polymeric aromatic amide, is synthesized from the monomers *p*-phenylenediamine *(left)* and terephthaloyl chloride *(right)*. The polymer strands are initially aligned randomly, but they are dissolved and spun to give a lightweight, flexible product that is five times stronger than steel of the same weight.

The first step in the reaction of an amine with an acid chloride is

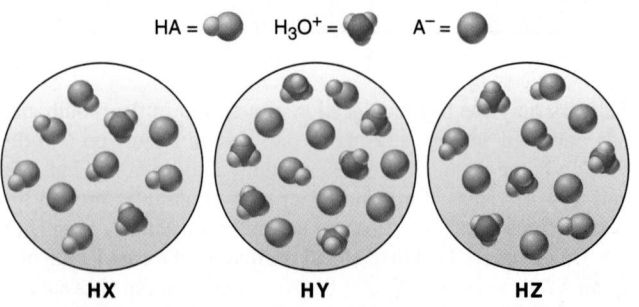

(a) Describe this step in terms of a Lewis acid-base reaction.
(b) Would you expect water or ammonia to react faster with an acid chloride? Explain.

•••**18.186** A solution of propanoic acid (CH_3CH_2COOH), made by dissolving 7.500 g in sufficient water to make 100.0 mL, has a freezing point of $-1.890°C$.
(a) Calculate the molarity of the solution.
(b) Calculate the molarity of the propanoate ion. (Assume the molarity of the solution equals the molality.)
(c) Calculate the percent dissociation of propanoic acid.

•••**18.187** The antimalarial properties of quinine ($C_{20}H_{24}N_2O_2$) saved thousands of lives during construction of the Panama Canal. This substance is a classic example of the medicinal wealth of tropical forests. Both N atoms are basic, but the N (colored) of the 3° amine group is far more basic ($pK_b = 5.1$) than the N within the aromatic ring system ($pK_b = 9.7$).

(a) A saturated solution of quinine in water is only $1.6\times10^{-3}\ M$. What is the pH of this solution?
(b) Show that the aromatic N contributes negligibly to the pH of the solution.
(c) Because of its low solubility, quinine is given as the salt quinine hydrochloride ($C_{20}H_{24}N_2O_2\cdot HCl$), which is 120 times more soluble than quinine. What is the pH of 0.33 M quinine hydrochloride?
(d) An antimalarial concentration in water is 1.5% quinine hydrochloride by mass ($d = 1.0$ g/mL). What is the pH?

•••**18.188** Drinking water is often disinfected with Cl_2, which hydrolyzes to form $HClO$, a weak acid but powerful disinfectant:

$$Cl_2(aq) + 2H_2O(l) \longrightarrow HClO(aq) + H_3O^+(aq) + Cl^-(aq)$$

The fraction of HClO in solution is defined as

$$\frac{[HClO]}{[HClO] + [ClO^-]}$$

(a) What is the fraction of HClO at pH 7.00 (K_a of HClO = 2.9×10^{-8})? (b) What is the fraction at pH 10.00?

•••**18.189** The following scenes represent three weak acids HA (where A = X, Y, or Z) dissolved in water (H_2O is not shown):

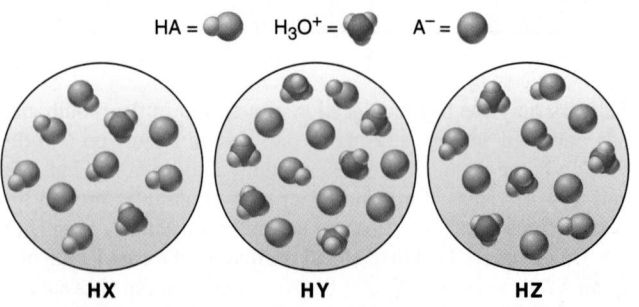

(a) Rank the acids in order of increasing K_a. (b) Rank the acids in order of increasing pK_a. (c) Rank the conjugate bases in order of increasing pK_b. (d) What is the percent dissociation of HX? (e) If equimolar amounts of the sodium salts of the acids (NaX, NaY, and NaZ) were dissolved in water, which solution would have the highest pOH? The lowest pH?

19

Magnificent Shifts in Balance Awesome, subterranean limestone formations, such as these in the caves of Nerja on Spain's Costa del Sol, result from gradual shifts in the equilibrium positions of carbonate systems over eons. In this chapter, we explore the quantitative basis of three major types of aqueous ionic equilibria—those involving acid-base buffers, slightly soluble salts, and complex ions.

Ionic Equilibria in Aqueous Systems

19.1 Equilibria of Acid-Base Buffer Systems
The Common-Ion Effect
The Henderson-Hasselbalch Equation
Buffer Capacity and Range
Preparing a Buffer

19.2 Acid-Base Titration Curves
Acid-Base Indicators
Strong Acid–Strong Base Titrations
Weak Acid–Strong Base Titrations
Weak Base–Strong Acid Titrations
Polyprotic Acid Titrations
Amino Acids as Polyprotic Acids

19.3 Equilibria of Slightly Soluble Ionic Compounds
The Solubility-Product Constant (K_{sp})
Calculations Involving K_{sp}
The Effect of a Common Ion
The Effect of pH
Q_{sp} vs. K_{sp}
Selective Precipitation of Ions

19.4 Equilibria Involving Complex Ions
Formation of Complex Ions
Complex Ions and Solubility
Amphoteric Hydroxides

Concepts & Skills to Review
before you study this chapter

- solubility rules for ionic compounds (Section 4.3)
- effect of concentration on equilibrium position (Section 17.6)
- conjugate acid-base pairs (Section 18.3)
- calculations for weak-acid and weak-base equilibria (Sections 18.4 and 18.5)
- acid-base properties of salt solutions (Section 18.7)
- Lewis acids and bases (Section 18.9)

Europa, one of Jupiter's moons, has an icy surface with hints of vast oceans of liquid water beneath. Is there life on Europa? If so, perhaps some Europan astronomer viewing Earth would be asking the same question, because liquid water is essential for the aqueous systems that maintain life. Every astronaut has felt awe at seeing our "beautiful blue orb" from space. A biologist peering at the fabulous watery world of a living cell probably feels the same way. A chemist is awed by the principles of equilibrium and their universal application to aqueous solutions wherever they occur.

Consider just a few cases of aqueous equilibria. The magnificent formations in limestone caves and the vast expanses of oceanic coral reefs result from subtle shifts in carbonate solubility equilibria. Carbonates also influence soil pH and prevent acidification of lakes by acid rain. Equilibria involving carbon dioxide and phosphates help organisms maintain cellular pH within narrow limits. Equilibria involving clays in soils control the availability of ionic nutrients for plants. The principles of ionic equilibrium also govern how water is softened, how substances are purified by precipitation of unwanted ions, and even how the weak acids in wine and vinegar influence the delicate taste of a fine French sauce.

IN THIS CHAPTER . . . We explore three aqueous ionic equilibrium systems: acid-base buffers, slightly soluble salts, and complex ions. Our discussion of buffers introduces the common-ion effect, an important phenomenon in many ionic equilibria. We discuss why buffers are important, how they work, and how to prepare them, and we then examine the various types of acid-base titrations and how buffered solutions are involved in them. Slightly soluble salts have major roles in the laboratory and in nature, and we'll see how conditions influence their solubility equilibria. Finally, we investigate how complex ions form and change from one type to another.

19.1 EQUILIBRIA OF ACID-BASE BUFFER SYSTEMS

Why do some lakes become acidic when showered by acid rain, while others remain unaffected? How does blood maintain a constant pH in contact with countless cellular acid-base reactions? How can a chemist sustain a nearly constant $[H_3O^+]$ in reactions that consume or produce H_3O^+ or OH^-? The answer in each case depends on the action of a buffer. In everyday language, a buffer is something that lessens the impact of an external force. An **acid-base buffer** is a solution that *lessens the impact on pH from the addition of acid or base.* Figure 19.1 shows that a small amount of H_3O^+ or OH^- added to an unbuffered solution (or just water) changes the pH by several units. Note that, because of the logarithmic nature of pH, *this change is several orders of magnitude larger* than the change that results from the same addition to a buffered solution, shown in Figure 19.2. To withstand the addition of strong acid or strong base without signifi-

Figure 19.1 The effect of adding acid or base to an unbuffered solution.
A, A 100-mL sample of dilute HCl is adjusted to pH 5.00. **B,** After the addition of 1 mL of 1 *M* HCl (*left*) or of 1 *M* NaOH (*right*), the pH changes by several units.

Figure 19.2 The effect of adding acid or base to a buffered solution.
A, A 100-mL sample of a buffered solution, made by mixing 1 M CH_3COOH with 1 M CH_3COONa, is adjusted to pH 5.00. **B,** After the addition of 1 mL of 1 M HCl (*left*) or of 1 M NaOH (*right*), the pH change is negligibly small. Compare these changes with those in Figure 19.1.

ARIS Presentation Center,
Photo Library

cantly changing its pH, a buffer must contain an acidic component that can react with the added OH^- ion *and* a basic component that can react with added H_3O^+ ion. However, these buffer components cannot be just any acid and base because they would neutralize each other. Most often, *the components of a buffer are the conjugate acid-base pair of a weak acid (or base)*. The buffer in Figure 19.2, for example, is a mixture of acetic acid (CH_3COOH) and acetate ion (CH_3COO^-).

How a Buffer Works: The Common-Ion Effect

Buffers work through a phenomenon known as the **common-ion effect.** An example of this effect occurs when acetic acid dissociates in water and some sodium acetate is added. As you know, acetic acid dissociates only slightly in water:

$$CH_3COOH(aq) + H_2O(l) \rightleftharpoons H_3O^+(aq) + CH_3COO^-(aq)$$

From Le Châtelier's principle (Section 17.6), we know that if some CH_3COO^- ion is added (from the soluble sodium acetate), the equilibrium position shifts to the left; thus, $[H_3O^+]$ decreases, in effect lowering the extent of acid dissociation:

$$CH_3COOH(aq) + H_2O(l) \overset{\longleftarrow}{\rightleftharpoons} H_3O^+(aq) + CH_3COO^-(aq; \text{ added})$$

Similarly, if we dissolve acetic acid *in* a sodium acetate solution, acetate ion and H_3O^+ ion from the acid enter the solution. The acetate ion already present in the solution acts to suppress the acid from dissociating as much as it would in pure water, thus keeping the $[H_3O^+]$ lower. The effect again is to lower the acid dissociation. Acetate ion is called *the common ion* in this case because it is "common" to both the acetic acid and sodium acetate solutions; that is, acetate ion from the acid enters a solution in which it is already present. *The common-ion effect occurs when a given ion is added to an equilibrium mixture that already contains that ion, and the position of equilibrium shifts away from forming more of it.*

Table 19.1 shows the percent dissociation and the pH of an acetic acid solution containing various concentrations of acetate ion (supplied from solid sodium acetate). Note that the *common ion, CH₃COO⁻, suppresses the dissociation of CH₃COOH*, which makes the solution less acidic (higher pH).

Demo Summerlin, L.R., Ealy, J.B. "The Common Ion Effect: Ammonium Hydroxide and Ammonium Acetate," *Chemical Demonstrations, A Sourcebook for Teachers*, Vol. 1 (American Chemical Society, Washington, DC, 1988), 95.

Table 19.1 **The Effect of Added Acetate Ion on the Dissociation of Acetic Acid**

$[CH_3COOH]_{init}$	$[CH_3COO^-]_{added}$	% Dissociation*	pH
0.10	0.00	1.3	2.89
0.10	0.050	0.036	4.44
0.10	0.10	0.018	4.74
0.10	0.15	0.012	4.92

*% Dissociation $= \dfrac{[CH_3COOH]_{dissoc}}{[CH_3COOH]_{init}} \times 100$

ARIS Presentation Center,
Tables Library

Demo Shakhashiri, Bassam Z. "Buffering Action of Alka-Seltzer," *Chemical Demonstrations, A Handbook for Teachers of Chemistry,* Vol. 3 (The University of Wisconsin Press, Madison, 1989), 186–187.

The Essential Feature of a Buffer In the previous case, we prepared a buffer by mixing a weak acid (CH_3COOH) and its conjugate base (CH_3COO^-). *How* does this solution resist pH changes when H_3O^+ or OH^- is added? The essential feature of a buffer is that *it consists of high concentrations of the acidic (HA) and basic (A^-) components.* When small amounts of H_3O^+ or OH^- ions are added to the buffer, they cause *a small amount of one buffer component to convert into the other,* which changes the relative concentrations of the two components. As long as the amount of H_3O^+ or OH^- added is much smaller than the amounts of HA and A^- present originally, *the added ions have little effect on the pH because they are consumed by one or the other buffer component:* a large excess of A^- reacts with any added H_3O^+, and a large excess of HA reacts with any added OH^-.

Consider what happens quantitatively to a solution containing high $[CH_3COOH]$ and high $[CH_3COO^-]$ when we add small amounts of strong acid or base. The expression for HA dissociation at equilibrium is

$$K_a = \frac{[CH_3COO^-][H_3O^+]}{[CH_3COOH]}$$

Solving for $[H_3O^+]$ gives

$$[H_3O^+] = K_a \times \frac{[CH_3COOH]}{[CH_3COO^-]}$$

Note that because K_a is constant, *the $[H_3O^+]$ of the solution depends directly on the buffer-component concentration ratio,* $\dfrac{[CH_3COOH]}{[CH_3COO^-]}$:

- If the ratio $[HA]/[A^-]$ goes up, $[H_3O^+]$ goes up.
- If the ratio $[HA]/[A^-]$ goes down, $[H_3O^+]$ goes down.

When we add a small amount of strong acid, the increased amount of H_3O^+ ion reacts with a *stoichiometric amount* of acetate ion from the buffer to form more acetic acid:

$$H_3O^+(aq;\ added) + CH_3COO^-(aq;\ from\ buffer) \longrightarrow CH_3COOH(aq) + H_2O(l)$$

As a result, $[CH_3COO^-]$ goes down by that amount and $[CH_3COOH]$ goes up by that amount, which increases the buffer-component concentration ratio (Figure 19.3, *left*). The $[H_3O^+]$ also increases but only very slightly.

Transparency/ARIS Presentation Center, Art Library

Figure 19.3 How a buffer works.
A buffer consists of high concentrations of a conjugate acid-base pair, in this case, acetic acid (CH_3COOH) and acetate ion (CH_3COO^-). When a small amount of H_3O^+ is added (*left*), that same amount of CH_3COO^- combines with it, which increases the amount of CH_3COOH slightly. Similarly, when a small amount of OH^- is added (*right*), that amount of CH_3COOH combines with it, which increases the amount of CH_3COO^- slightly. In both cases, the relative changes in amounts of the buffer components are small, so the changes in their concentration ratio, and therefore the pH, are small as well.

Adding a small amount of strong base produces the opposite result. It supplies OH^- ions, which react with a *stoichiometric amount* of CH_3COOH from the buffer, forming that much more CH_3COO^- (Figure 19.3, *right*):

$$CH_3COOH(aq;\ from\ buffer) + OH^-(aq;\ added) \longrightarrow CH_3COO^-(aq) + H_2O(l)$$

The buffer-component concentration ratio decreases, which decreases $[H_3O^+]$, but once again, the change is very slight.

Thus, the buffer components consume virtually all the added H_3O^+ or OH^-. To reiterate, as long as the amount of added H_3O^+ or OH^- is small compared with the amounts of the buffer components, *the conversion of one component into the other produces a small change in the buffer-component concentration ratio and, consequently, a small change in $[H_3O^+]$ and in pH.* Sample Problem 19.1 demonstrates how small these pH changes typically are. Note that the latter two parts of the problem combine a stoichiometry portion, like the problems in Chapter 3, and a weak-acid dissociation portion, like those in Chapter 18.

 ARIS Presentation Center, Worked Examples Library

Literature Jensen, William B. "The Symbol for pH," *J. Chem. Educ.* **2004,** *81,* 21.

SAMPLE PROBLEM 19.1 Calculating the Effect of Added H_3O^+ or OH^- on Buffer pH

PROBLEM Calculate the pH:
(a) Of a buffer solution consisting of 0.50 *M* CH_3COOH and 0.50 *M* CH_3COONa
(b) After adding 0.020 mol of solid NaOH to 1.0 L of the buffer solution in part (a)
(c) After adding 0.020 mol of HCl to 1.0 L of the buffer solution in part (a)
K_a of $CH_3COOH = 1.8 \times 10^{-5}$. (Assume the additions cause negligible volume changes.)
PLAN In each case, we know, or can find, $[CH_3COOH]_{init}$ and $[CH_3COO^-]_{init}$ and the K_a of CH_3COOH (1.8×10^{-5}), and we need to find $[H_3O^+]$ at equilibrium and convert it to pH. **(a)** We use the given concentrations of buffer components (each 0.50 *M*) as the initial values. As in earlier problems, we assume that x, the $[CH_3COOH]$ that dissociates, which equals $[H_3O^+]$, is so small relative to $[CH_3COOH]_{init}$ that it can be neglected. We set up a reaction table, solve for x, and check the assumption. **(b)** and **(c)** We assume that the added OH^- or H_3O^+ reacts completely with the buffer components to yield new $[CH_3COOH]_{init}$ and $[CH_3COO^-]_{init}$, which then dissociate to an unknown extent. We set up two reaction tables. The first summarizes the stoichiometry of adding strong base (0.020 mol) or acid (0.020 mol). The second summarizes the dissociation of the new HA concentrations, so we proceed as in part (a) to find the new $[H_3O^+]$.
SOLUTION **(a)** The original pH: $[H_3O^+]$ in the original buffer.
Setting up a reaction table with $x = [CH_3COOH]_{dissoc} = [H_3O^+]$ (as in Chapter 18, we assume that $[H_3O^+]$ from H_2O is negligible and disregard it):

Concentration (*M*)	$CH_3COOH(aq)$	$+$	$H_2O(l)$	$\rightleftharpoons$	$CH_3COO^-(aq)$	$+$	$H_3O^+(aq)$
Initial	0.50		—		0.50		0
Change	$-x$		—		$+x$		$+x$
Equilibrium	$0.50 - x$		—		$0.50 + x$		x

Making the assumption and finding the equilibrium $[CH_3COOH]$ and $[CH_3COO^-]$: With K_a small, x is small, so we assume

$$[CH_3COOH] = 0.50\ M - x \approx 0.50\ M \quad \text{and} \quad [CH_3COO^-] = 0.50\ M + x \approx 0.50\ M$$

Solving for x ($[H_3O^+]$ at equilibrium):

$$x = [H_3O^+] = K_a \times \frac{[CH_3COOH]}{[CH_3COO^-]} \approx (1.8 \times 10^{-5}) \times \frac{0.50}{0.50} = 1.8 \times 10^{-5}\ M$$

Checking the assumption:

$$\frac{1.8 \times 10^{-5} M}{0.50\ M} \times 100 = 3.6 \times 10^{-3}\% < 5\%$$

The assumption is justified, and we will use the same assumption in parts (b) and (c). Calculating pH:

$$pH = -\log [H_3O^+] = -\log (1.8 \times 10^{-5}) = 4.74$$

(b) The pH after adding base (0.020 mol of NaOH to 1.0 L of buffer). Finding $[OH^-]_{added}$:

$$[OH^-]_{added} = \frac{0.020 \text{ mol } OH^-}{1.0 \text{ L soln}} = 0.020 \, M \, OH^-$$

Setting up a reaction table for the *stoichiometry* of adding OH^- to CH_3COOH:

Concentration (M)	$CH_3COOH(aq)$ +	$OH^-(aq)$	$\longrightarrow$	$CH_3COO^-(aq)$ +	$H_2O(l)$
Initial	0.50	0.020		0.50	—
Change	−0.020	−0.020		+0.020	—
Final	0.48	0		0.52	—

Setting up a reaction table for the *acid dissociation,* using these new initial concentrations. As in part (a), $x = [CH_3COOH]_{dissoc} = [H_3O^+]$:

Concentration (M)	$CH_3COOH(aq)$ +	$H_2O(l)$	$\rightleftharpoons$	$CH_3COO^-(aq)$ +	$H_3O^+(aq)$
Initial	0.48	—		0.52	0
Change	−x	—		+x	+x
Equilibrium	0.48 − x	—		0.52 + x	x

Making the assumption that x is small, and solving for x:

$$[CH_3COOH] = 0.48 \, M - x \approx 0.48 \, M \quad \text{and} \quad [CH_3COO^-] = 0.52 \, M + x \approx 0.52 \, M$$

$$x = [H_3O^+] = K_a \times \frac{[CH_3COOH]}{[CH_3COO^-]} \approx (1.8 \times 10^{-5}) \times \frac{0.48}{0.52} = 1.7 \times 10^{-5} \, M$$

Calculating the pH:

$$pH = -\log [H_3O^+] = -\log (1.7 \times 10^{-5})$$
$$= 4.77$$

The addition of strong base increased the concentration of the basic buffer component at the expense of the acidic buffer component. Note especially that the pH *increased only slightly,* from 4.74 to 4.77.

(c) The pH after adding acid (0.020 mol of HCl to 1.0 L of buffer). Finding $[H_3O^+]_{added}$:

$$[H_3O^+]_{added} = \frac{0.020 \text{ mol } H_3O^+}{1.0 \text{ L soln}} = 0.020 \, M \, H_3O^+$$

Now we proceed as in part (b), by first setting up a reaction table for the *stoichiometry* of adding H_3O^+ to CH_3COO^-:

Concentration (M)	$CH_3COO^-(aq)$ +	$H_3O^+(aq)$	$\longrightarrow$	$CH_3COOH(aq)$ +	$H_2O(l)$
Initial	0.50	0.020		0.50	—
Change	−0.020	−0.020		+0.020	—
Final	0.48	0		0.52	—

The reaction table for the *acid dissociation,* with $x = [CH_3COOH]_{dissoc} = [H_3O^+]$ is

Concentration (M)	$CH_3COOH(aq)$ +	$H_2O(l)$	$\rightleftharpoons$	$CH_3COO^-(aq)$ +	$H_3O^+(aq)$
Initial	0.52	—		0.48	0
Change	−x	—		+x	+x
Equilibrium	0.52 − x	—		0.48 + x	x

Making the assumption that x is small, and solving for x:

$$[CH_3COOH] = 0.52 \, M - x \approx 0.52 \, M \quad \text{and} \quad [CH_3COO^-] = 0.48 \, M + x \approx 0.48 \, M$$

$$x = [H_3O^+] = K_a \times \frac{[CH_3COOH]}{[CH_3COO^-]} \approx (1.8 \times 10^{-5}) \times \frac{0.52}{0.48} = 2.0 \times 10^{-5} \, M$$

Calculating the pH:

$$pH = -\log [H_3O^+] = -\log (2.0 \times 10^{-5})$$
$$= 4.70$$

The addition of strong acid increased the concentration of the acidic buffer component at the expense of the basic buffer component and *lowered* the pH only slightly, from 4.74 to 4.70.

CHECK The changes in $[CH_3COOH]$ and $[CH_3COO^-]$ occur in opposite directions in parts (b) and (c), which makes sense. The additions were of equal amounts, so the pH increase in (b) should equal the pH decrease in (c), within rounding.

COMMENT In part (a), we justified our assumption that x can be neglected. Therefore, in parts (b) and (c), we could have used the "Final" values from the last line of the stoichiometry tables directly for the ratio of buffer components; that would have allowed us to dispense with a reaction table for the dissociation. In subsequent problems in this chapter, we will follow this simplified approach.

FOLLOW-UP PROBLEM 19.1 Calculate the pH of a buffer consisting of 0.50 M HF and 0.45 M F$^-$ **(a)** before and **(b)** after addition of 0.40 g of NaOH to 1.0 L of the buffer (K_a of HF = 6.8×10^{-4}).

The Henderson-Hasselbalch Equation

For any weak acid, HA, the dissociation equation and K_a expression are

$$HA + H_2O \rightleftharpoons H_3O^+ + A^-$$

$$K_a = \frac{[H_3O^+][A^-]}{[HA]}$$

The key variable that determines $[H_3O^+]$ is the concentration *ratio* of acid species to base species, so rearranging to isolate $[H_3O^+]$ gives

$$[H_3O^+] = K_a \times \frac{[HA]}{[A^-]}$$

Taking the negative common logarithm (base 10) of both sides gives

$$-\log [H_3O^+] = -\log K_a - \log \left(\frac{[HA]}{[A^-]} \right)$$

from which we obtain

$$pH = pK_a + \log \left(\frac{[A^-]}{[HA]} \right)$$

(Note the inversion of the buffer-component concentration ratio when the sign of the logarithm is changed.) A key point we'll emphasize again later is that when $[A^-] = [HA]$, their ratio becomes 1; the log term then becomes 0, and thus $pH = pK_a$.

Generalizing the previous equation for any conjugate acid-base pair gives the **Henderson-Hasselbalch equation:**

$$pH = pK_a + \log \left(\frac{[base]}{[acid]} \right) \qquad (19.1)$$

This relationship allows us to solve directly for pH instead of having to calculate $[H_3O^+]$ first. For instance, by applying the Henderson-Hasselbalch equation in part (b) of Sample Problem 19.1, we could have found the pH of the buffer after the addition of NaOH as follows:

$$pH = pK_a + \log \left(\frac{[CH_3COO^-]}{[CH_3COOH]} \right)$$

$$= 4.74 + \log \left(\frac{0.52}{0.48} \right) = 4.77$$

(We just derived the Henderson-Hasselbalch equation from fundamental definitions and simple algebra. Whenever possible, it's a good idea to derive simple relationships rather than having to memorize them.)

Literature Po, Henry N.; Senozan, N. M. "Henderson-Hasselbalch Equation: Its History and Limitations," *J. Chem. Educ.* **2001**, *78*, 1499.

Figure 19.4   **The relation between buffer capacity and pH change.**
The four bars in the graph represent CH_3COOH/CH_3COO^- buffers with the same initial pH (4.74) but different component concentrations (labeled on or near each bar). When a given amount of strong base is added to each buffer, the pH increases. The length of the bar corresponds to the pH increase. Note that the more concentrated the buffer, the greater its capacity, and the smaller the pH change.

ARIS Presentation Center, Art Library

Demo Shakhashiri, Bassam Z. "Buffering Action and Capacity," *Chemical Demonstrations, A Handbook for Teachers of Chemistry,* Vol. 3 (The University of Wisconsin Press, Madison, 1989), 173–185.

Literature Urbansky, Edward T.; Schock, Michael R. "Understanding, Deriving, and Computing Buffer Capacity," *J. Chem. Educ.* **2000,** *77,* 1640.

Buffer Capacity and Buffer Range

As you've seen, a buffer resists a pH change as long as the concentrations of buffer components are *large* compared with the amount of strong acid or base added. **Buffer capacity** is a measure of this ability to resist pH change and depends on both the absolute and relative component concentrations.

In absolute terms, *the more concentrated the components of a buffer, the greater the buffer capacity.* In other words, you must add more H_3O^+ or OH^- to a high-capacity (concentrated) buffer than to a low-capacity (dilute) buffer to obtain the same pH change. Conversely, adding the same amount of H_3O^+ or OH^- to buffers of different capacities produces a smaller pH change in the higher capacity buffer (Figure 19.4). It's important to realize that *the pH of a buffer is distinct from its buffer capacity.* A buffer made of equal volumes of 1.0 *M* CH_3COOH and 1.0 *M* CH_3COO^- has the same pH (4.74) as a buffer made of equal volumes of 0.10 *M* CH_3COOH and 0.10 *M* CH_3COO^-, but the more concentrated buffer has a much larger capacity for resisting a pH change.

Buffer capacity is also affected by the *relative* concentrations of the buffer components. As a buffer functions, the concentration of one component increases relative to the other. Because the ratio of these concentrations determines the pH, the less the ratio changes, the less the pH changes. *For a given addition of acid or base, the buffer-component concentration ratio changes less when the concentrations are similar than when they are different.* Suppose that a buffer has [HA] = [A$^-$] = 1.000 *M*. When we add 0.010 mol of OH^- to 1.00 L of buffer, [A$^-$] becomes 1.010 *M* and [HA] becomes 0.990 *M*:

$$\frac{[A^-]_{init}}{[HA]_{init}} = \frac{1.000\ M}{1.000\ M} = 1.000 \qquad \frac{[A^-]_{final}}{[HA]_{final}} = \frac{1.010\ M}{0.990\ M} = 1.02$$

$$\text{Percent change} = \frac{1.02 - 1.000}{1.000} \times 100 = 2\%$$

Now let us suppose that the component concentrations are [HA] = 0.250 *M* and [A$^-$] = 1.750 *M*. The same addition of 0.010 mol of OH^- to 1.00 L of buffer gives [HA] = 0.240 *M* and [A$^-$] = 1.760 *M*, so the ratios are

$$\frac{[A^-]_{init}}{[HA]_{init}} = \frac{1.750\ M}{0.250\ M} = 7.00 \qquad \frac{[A^-]_{final}}{[HA]_{final}} = \frac{1.760\ M}{0.240\ M} = 7.33$$

$$\text{Percent change} = \frac{7.33 - 7.00}{7.00} \times 100 = 4.7\%$$

As you can see, the change in the buffer-component concentration ratio is much larger when the initial concentrations of the components are very different.

It follows that *a buffer has the highest capacity when the component concentrations are equal,* that is, when [A$^-$]/[HA] = 1:

$$pH = pK_a + \log\left(\frac{[A^-]}{[HA]}\right) = pK_a + \log 1 = pK_a + 0 = pK_a$$

Note this important result: for a given concentration, *a buffer whose pH is equal to or near the pK$_a$ of its acid component has the highest buffer capacity.*

The **buffer range** is the pH range over which the buffer acts effectively, and it is related to the relative component concentrations. The further the buffer-component concentration ratio is from 1, the less effective the buffering action (that is, the lower the buffer capacity). In practice, if the [A$^-$]/[HA] ratio is greater than 10 or less than 0.1—that is, if one component concentration is more than 10 times the other—buffering action is poor. Recalling that log 10 = +1 and log 0.1 = −1, we find that *buffers have a usable range within ±1 pH unit of the* pK$_a$ *of the acid component:*

$$pH = pK_a + \log\left(\frac{10}{1}\right) = pK_a + 1 \qquad \text{and} \qquad pH = pK_a + \log\left(\frac{1}{10}\right) = pK_a - 1$$

SAMPLE PROBLEM 19.2 Using Molecular Scenes to Examine Buffers

PROBLEM The molecular scenes below represent samples of four HA/A^- buffer solutions. (HA is blue and green, A^- is green, and other ions and water are not shown.)

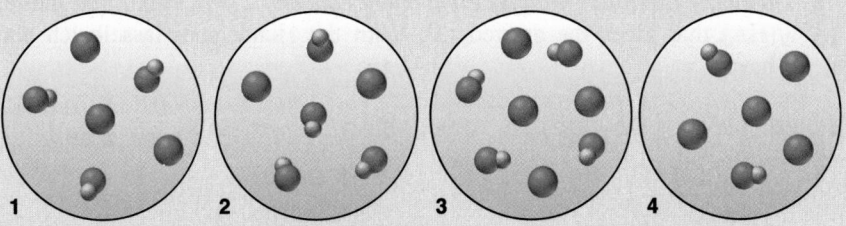

1   2   3   4

(a) Which buffer has the highest pH?

(b) Which buffer has the greatest capacity?

(c) Should we add a small amount of concentrated strong acid or strong base to convert sample 1 to sample 2 (assuming no volume change)?

PLAN The molecular scenes show varying numbers of weak acid molecules (HA) and the conjugate base anion (A^-). Numbers correspond to moles and, because the volumes are equal, the scenes represent molarities as well. **(a)** As the pH rises, more HA donates its H^+ and becomes A^-, so the $[A^-]/[HA]$ ratio will increase. We examine the scenes to see which has the highest ratio. **(b)** Buffer capacity depends on buffer-component concentration *and* ratio. We examine the scenes to see which has a high concentration and a ratio close to 1. **(c)** Adding strong acid converts some A^- to HA, and adding strong base does the opposite. Comparing the $[A^-]/[HA]$ ratios in samples 1 and 2 tells us which to add.

SOLUTION (a) The $[A^-]/[HA]$ ratios are as follows: For sample 1, $[A^-]/[HA] = 3/3 = 1$. Similarly, sample 2 = 0.5; sample 3 = 1; sample 4 = 2. Sample 4 has the highest pH because it has the highest $[A^-]/[HA]$ ratio.

(b) Samples 1 and 3 have a $[A^-]/[HA]$ ratio of 1, but sample 3 has the greater capacity because it has a higher concentration.

(c) Sample 2 has a lower $[A^-]/[HA]$ ratio than sample 1, so we would add strong acid to sample 1 to convert some A^- to HA.

FOLLOW-UP PROBLEM 19.2 The molecular scene (see margin) shows a sample of an HB/B^- buffer solution. (HB is blue and yellow, B^- is yellow, and other ions and water are not shown.) **(a)** Would you add a small amount of concentrated strong acid or strong base to increase the buffer capacity? **(b)** Assuming no volume change, draw a scene that represents the buffer with the highest possible capacity after the addition in part (a).

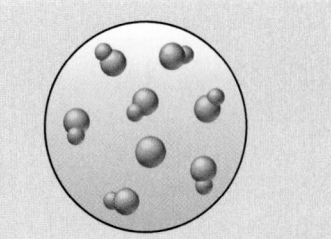

Preparing a Buffer

Any large chemical supply-house catalog lists many common buffers in a variety of pH values and concentrations. So, you might ask, why learn how to prepare a buffer? In many cases, a common buffer is simply not available with the desired pH or concentration, and you have to make it yourself. In many environmental or biomedical research applications, a buffer of unusual composition may be required to simulate an ecosystem or stabilize a fragile biological macromolecule. Even the most sophisticated, automated laboratory frequently relies on personnel with a good knowledge of basic chemistry to prepare a buffer. Several steps are required to prepare a buffer of a desired pH:

1. *Choose the conjugate acid-base pair.* First, decide on the chemical composition of the buffer, that is, the conjugate acid-base pair. This choice is determined to a large extent by the desired pH. Remember that a buffer is most effective when the ratio of its component concentrations is close to 1, in which case the pH $\approx$ pK_a of the acid.

Suppose that, for a biochemical experiment, you need a buffer whose pH is 3.90: the pK_a of the acid component should be as close to 3.90 as possible; or $K_a = 10^{-3.90} = 1.3 \times 10^{-4}$. A table of acid-dissociation constants (see Appendix C) shows that lactic acid (pK_a = 3.86), glycolic acid (pK_a = 3.83), and formic acid (pK_a = 3.74) are good choices. To avoid common biological species, you use formic acid. Therefore, the buffer components will be formic acid, HCOOH, and formate ion, HCOO$^-$, supplied by a soluble salt, such as sodium formate, HCOONa.

2. *Calculate the ratio of buffer component concentrations.* Next, find the ratio of [A$^-$]/[HA] that gives the desired pH. With the Henderson-Hasselbalch equation, we have

$$pH = pK_a + \log\left(\frac{[A^-]}{[HA]}\right) \quad \text{or} \quad 3.90 = 3.74 + \log\left(\frac{[HCOO^-]}{[HCOOH]}\right)$$

$$\log\left(\frac{[HCOO^-]}{[HCOOH]}\right) = 0.16 \quad \text{so} \quad \left(\frac{[HCOO^-]}{[HCOOH]}\right) = 10^{0.16} = 1.4$$

Thus, for every 1.0 mol of HCOOH in a given volume of solution, you need 1.4 mol of HCOONa.

3. *Determine the buffer concentration.* Next, decide how concentrated the buffer should be. Remember that the higher the concentrations of components, the greater the buffer capacity. For most laboratory-scale applications, concentrations of about 0.5 *M* are suitable, but the decision is often based on availability of stock solutions.

Suppose you have a large stock of 0.40 *M* HCOOH and you need approximately 1.0 L of final buffer. First, find the moles of sodium formate that will give the needed 1.4/1.0 ratio, and then convert to grams:

$$\text{Moles of HCOOH} = 1.0\,\text{L soln} \times \frac{0.40\,\text{mol HCOOH}}{1.0\,\text{L soln}} = 0.40\,\text{mol HCOOH}$$

$$\text{Moles of HCOONa} = 0.40\,\text{mol HCOOH} \times \frac{1.4\,\text{mol HCOONa}}{1.0\,\text{mol HCOOH}} = 0.56\,\text{mol HCOONa}$$

$$\text{Mass (g) of HCOONa} = 0.56\,\text{mol HCOONa} \times \frac{68.01\,\text{g HCOONa}}{1\,\text{mol HCOONa}} = 38\,\text{g HCOONa}$$

4. *Mix the solution and adjust the pH.* Thoroughly dissolve 38 g of solid sodium formate in 0.40 *M* HCOOH to a total volume of 1.0 L. Finally, note that because of the behavior of nonideal solutions (Section 13.6), a buffer prepared in this way may vary from the desired pH by a few tenths of a pH unit. Therefore, after making up the solution, *adjust the buffer pH* to the desired value by adding strong acid or strong base, while monitoring the solution with a pH meter. (The following sample problem does not refer to the Henderson-Hasselbalch equation.)

ARIS Presentation Center, Worked Examples Library

Environmental application

SAMPLE PROBLEM 19.3 Preparing a Buffer

PROBLEM An environmental chemist needs a carbonate buffer of pH 10.00 to study the effects of the acid rain on limestone-rich soils. How many grams of Na$_2$CO$_3$ must she add to 1.5 L of freshly prepared 0.20 *M* NaHCO$_3$ to make the buffer? K_a of HCO$_3^-$ is 4.7×10^{-11}.

PLAN The conjugate pair is already chosen, HCO$_3^-$ (acid) and CO$_3^{2-}$ (base), as are the volume (1.5 L) and concentration (0.20 *M*) of HCO$_3^-$, so we must find the buffer-component concentration ratio that gives pH 10.00 and the mass of Na$_2$CO$_3$ to dissolve. We first convert pH to [H$_3$O$^+$] and use the K_a expression from the equation to solve for the required [CO$_3^{2-}$]. Multiplying by the volume of solution gives the amount (mol) of CO$_3^{2-}$ required, and then we use the molar mass to find the mass (g) of Na$_2$CO$_3$.

SOLUTION Calculating [H$_3$O$^+$]: [H$_3$O$^+$] = 10^{-pH} = $10^{-10.00}$ = 1.0×10^{-10} *M*

Solving for [CO$_3^{2-}$] in the concentration ratio:

$$HCO_3^-(aq) + H_2O(l) \rightleftharpoons H_3O^+(aq) + CO_3^{2-}(aq) \qquad K_a = \frac{[H_3O^+][CO_3^{2-}]}{[HCO_3^-]}$$

So

$$[CO_3^{2-}] = K_a\frac{[HCO_3^-]}{[H_3O^+]} = \frac{(4.7\times10^{-11})(0.20)}{1.0\times10^{-10}} = 0.094\,M$$

Calculating the amount (mol) of CO_3^{2-} needed for the given volume:

$$\text{Moles of } CO_3^{2-} = 1.5 \text{ L soln} \times \frac{0.094 \text{ mol } CO_3^{2-}}{1 \text{ L soln}} = 0.14 \text{ mol } CO_3^{2-}$$

Calculating the mass (g) of Na_2CO_3 needed:

$$\text{Mass (g) of } Na_2CO_3 = 0.14 \text{ mol } Na_2CO_3 \times \frac{105.99 \text{ g } Na_2CO_3}{1 \text{ mol } Na_2CO_3} = 15 \text{ g } Na_2CO_3$$

We dissolve 15 g of Na_2CO_3 into 1.3 L of 0.20 M $NaHCO_3$ and add 0.20 M $NaHCO_3$ to make 1.5 L. Using a pH meter, we adjust the pH to 10.00 with strong acid or base.
CHECK For a useful buffer range, the concentration of the acidic component, $[HCO_3^-]$, must be within a factor of 10 of the concentration of the basic component, $[CO_3^{2-}]$. We have (1.5 L)(0.20 M HCO_3^-), or 0.30 mol of HCO_3^-, and 0.14 mol of CO_3^{2-}; $0.30/0.14 = 2.1$, which seems fine. Make sure the relative amounts of components seem reasonable: since we want a pH lower than the pK_a of HCO_3^- (10.33), it makes sense that we have more of the acidic than the basic species.

FOLLOW-UP PROBLEM 19.3 How would you prepare a benzoic acid/benzoate buffer with pH = 4.25, starting with 5.0 L of 0.050 M sodium benzoate (C_6H_5COONa) solution and adding the acidic component? K_a of benzoic acid (C_6H_5COOH) is 6.3×10^{-5}.

Another way to prepare a buffer is to form one of the components during the final mixing step by *partial neutralization* of the other component. For example, you can prepare an $HCOOH/HCOO^-$ buffer by mixing appropriate amounts of $HCOOH$ solution and $NaOH$ solution. As the OH^- ions react with the $HCOOH$ molecules, neutralization of part of the total $HCOOH$ present produces the $HCOO^-$ needed:

$HCOOH$ (HA total) + OH^- (amt added) $\longrightarrow$

$HCOOH$ (HA total $-$ OH^- amt added) + $HCOO^-$ (OH^- amt added) + H_2O

This method is based on the same chemical process that occurs when a weak acid is titrated with a strong base, as you'll see in Section 19.2.

Section Summary

A buffered solution exhibits a much smaller change in pH when H_3O^+ or OH^- is added than does an unbuffered solution. • A buffer consists of relatively high concentrations of the components of a conjugate weak acid–base pair. The buffer-component concentration ratio determines the pH, and the ratio and pH are related by the Henderson-Hasselbalch equation. As H_3O^+ or OH^- is added, one buffer component reacts with it and is converted into the other component; therefore, the buffer-component concentration ratio, and consequently the free $[H_3O^+]$ (and pH), changes only slightly. • A concentrated buffer undergoes smaller changes in pH than a dilute buffer. When the buffer pH equals the pK_a of the acid component, the buffer has its highest capacity. • A buffer has an effective range of $pK_a \pm 1$ pH unit. • To prepare a buffer, choose the conjugate acid-base pair, calculate the ratio of buffer components, determine buffer concentration, and adjust the final solution to the desired pH.

19.2 ACID-BASE TITRATION CURVES

In Chapter 4, we discussed the acid-base titration as an analytical method. Let's re-examine it, this time tracking the change in pH with an **acid-base titration curve,** a plot of pH vs. volume of titrant added. The behavior of an acid-base indicator and its role in the titration are described first. To better understand the titration process, we apply the principles of the acid-base behavior of salt solutions (Section 18.7) and, later in the section, the principles of buffer action.

Figure 19.5 Colors and approximate pH range of some common acid-base indicators. Most have a range of about 2 pH units, in keeping with the useful buffer range of 2 pH units ($pK_a \pm 1$). (pH range depends to some extent on the solvent used to prepare the indicator.)

Literature Kooser, Ara S.; Jenkins, Judith L.; Welch, Lawrence E. "Acid-Base Indicators: A New Look at an Old Topic," *J. Chem. Educ.* **2001,** *78,* 1504.

Demo Epp, Dianne N. "Teas as Natural Indicators," *J. Chem. Educ.* **1993,** *70,* 326.

Figure 19.6 The color change of the indicator bromthymol blue. The acidic form of bromthymol blue is yellow (*left*) and the basic form is blue (*right*). Over the pH range in which the indicator is changing, both forms are present, so the mixture appears greenish (*center*).

Monitoring pH with Acid-Base Indicators

The two common devices for measuring pH in the laboratory are pH meters and acid-base indicators. (We discuss the operation of pH meters in Chapter 21.) An *acid-base indicator* is a weak organic acid (denoted here as HIn) that has a different color than its conjugate base (In$^-$), with the color change occurring over a specific and relatively narrow pH range. Typically, one or both of the forms are intensely colored, so only a tiny amount of indicator is needed, far too little to affect the pH of the solution being studied.

Indicators are used for estimating the pH of a solution and for monitoring the pH in acid-base titrations and in reactions. Figure 19.5 shows the color changes and their pH ranges for some common acid-base indicators. Selecting an indicator requires that you know the approximate pH of the titration end point, which in turn requires that you know which ionic species are present. You'll see how to identify those in our discussion of acid-base titration curves. Because the indicator molecule is a weak acid, the ratio of the two forms is governed by the [H$_3$O$^+$] of the test solution:

$$HIn(aq) + H_2O(l) \rightleftharpoons H_3O^+(aq) + In^-(aq) \qquad K_a \text{ of HIn} = \frac{[H_3O^+][In^-]}{[HIn]}$$

Therefore,

$$\frac{[HIn]}{[In^-]} = \frac{[H_3O^+]}{K_a}$$

How we perceive colors has a major influence on the use of indicators. Typically, the experimenter will see the HIn color if the [HIn]/[In$^-$] ratio is 10/1 or greater and the In$^-$ color if the [HIn]/[In$^-$] ratio is 1/10 or less. Between these extremes, the colors of the two forms merge into an intermediate hue. Therefore, an indicator has a *color range* that reflects a 10^2-fold range in the [HIn]/[In] ratio, which means that an *indicator changes color over a range of about 2 pH units.* For example, as you can see in Figure 19.5, bromthymol blue has a pH range of about 6.0 to 7.6 and, as Figure 19.6 shows, it is yellow below that range, blue above it, and greenish in between.

Strong Acid–Strong Base Titration Curves

A typical curve for the titration of a strong acid with a strong base appears in Figure 19.7, along with the data used to construct it.

Features of the Curve There are three distinct regions of the curve, which correspond to three major changes in slope:

1. The pH starts out low, reflecting the high $[H_3O^+]$ of the strong acid, and increases slowly as acid is gradually neutralized by the added base.
2. Suddenly, the pH rises steeply. This rise begins when the moles of OH^- that have been added nearly equal the moles of H_3O^+ originally present in the acid. An additional drop or two of base neutralizes the final tiny excess of acid and introduces a tiny excess of base, so the pH jumps 6 to 8 units.
3. Beyond this steep portion, the pH increases slowly as more base is added.

The **equivalence point,** which occurs within the nearly vertical portion of the curve, is the point at which the *number of moles of added OH⁻ equals the number of moles of H₃O⁺ originally present.* At the equivalence point of a strong acid–strong base titration, *the solution consists of the anion of the strong acid and the cation of the strong base.* Recall from Chapter 18 that *these ions do not react with water, so the solution is neutral: pH = 7.00.* The volume and concentration of base needed to reach the equivalence point allow us to calculate the amount of acid originally present (see Sample Problem 4.6, p. 154).

Literature Barnum, Dennis W. "Predicting Acid-Base Titration Curves Without Calculations," *J. Chem. Educ.* **1999,** *76,* 938.

Transparency/ARIS Presentation Center, Art Library

Volume of NaOH added (mL)	pH
00.00	1.00
10.00	1.22
20.00	1.48
30.00	1.85
35.00	2.18
39.00	2.89
39.50	3.20
39.75	3.50
39.90	3.90
39.95	4.20
39.99	4.90
40.00	7.00
40.01	9.10
40.05	9.80
40.10	10.10
40.25	10.50
40.50	10.79
41.00	11.09
45.00	11.76
50.00	12.05
60.00	12.30
70.00	12.43
80.00	12.52

Titration of 40.00 mL of 0.1000 *M* HCl with 0.1000 *M* NaOH

Phenolphthalein

pH = 7.00 at equivalence point

Methyl red

Total volume (mL)   40 + 0 = 40   40 + 20 = 60   40 + 40 = 80   40 + 60 = 100

Figure 19.7 Curve for a strong acid–strong base titration. Data obtained from the titration of 40.00 mL of 0.1000 *M* HCl with 0.1000 *M* NaOH are shown at left. The titration curve shows pH increasing gradually at first and then rapidly near and at the equivalence point, which occurs when amount (mol) of OH^- added = amount (mol) of H_3O^+ originally present. In a strong acid–strong base titration, pH = 7.00 at the equivalence point, and an indicator that changes color on the steep portion of the curve is used (methyl red and phenolphthalein are shown in the photos). Beyond this point, added OH^- causes a gradual pH increase again. Each blow-up circle depicts the relative numbers of species present, other than solvent, in a tiny portion of the solution at key points during the titration. The circles increase in size in proportion to the total volume of the solution.

Before the titration begins, we add a few drops of an appropriate indicator to the acid solution to signal when we reach the equivalence point. The **end point** of the titration occurs when the indicator changes color. *We choose an indicator with an end point close to the equivalence point,* one that changes color in the pH range on the steep vertical portion of the curve. Figure 19.7 shows the color changes for two indicators that are suitable for a strong acid–strong base titration. Methyl red changes from red at pH 4.2 to yellow at pH 6.3, whereas phenolphthalein changes from colorless at pH 8.3 to pink at pH 10.0. Even though neither color change occurs *at* the equivalence point (pH 7.00), both occur on the vertical portion of the curve, where a single drop of base causes a large pH change: when methyl red turns yellow, or when phenolphthalein turns pink, we know we are within a drop or two of the equivalence point. For example, in going from 39.90 to 39.99 mL, one to two drops, the pH changes one whole unit. For all practical purposes, then, the *visible* change in color of the indicator (end point) signals the *invisible* point at which moles of added base equal the original moles of acid (equivalence point).

Calculating the pH By knowing the chemical species present during the titration, we can calculate the pH at various points along the way:

1. *Original solution of strong HA.* In Figure 19.7, 40.00 mL of 0.1000 M HCl is titrated with 0.1000 M NaOH. Because a strong acid is completely dissociated, $[HCl] = [H_3O^+] = 0.1000\ M$. Therefore, the initial pH is[*]

$$pH = -\log[H_3O^+] = -\log(0.1000) = 1.00$$

2. *Before the equivalence point.* As soon as we start adding base, two changes occur that affect the pH calculations: (1) some acid is neutralized, and (2) the volume of solution increases. To find the pH at various points up to the equivalence point, we find the initial *amount (mol)* of H_3O^+ present, subtract the amount reacted, *which equals the amount (mol) of OH^- added,* and then use the change in volume to calculate the *concentration,* $[H_3O^+]$, and convert to pH. For example, after adding 20.00 mL of 0.1000 M NaOH:

- *Find the moles of H_3O^+ remaining.* Subtracting the number of moles of H_3O^+ reacted from the number initially present gives the number remaining. Moles of H_3O^+ reacted equals moles of OH^- added, so

$$\begin{array}{ll}
\text{Initial moles of } H_3O^+ = 0.04000\ L \times 0.1000\ M = & 0.004000\ \text{mol}\ H_3O^+ \\
-\text{Moles of } OH^-\ \text{added} = 0.02000\ L \times 0.1000\ M = & 0.002000\ \text{mol}\ OH^- \\
\hline
\text{Moles of } H_3O^+\ \text{remaining} = & 0.002000\ \text{mol}\ H_3O^+
\end{array}$$

- *Calculate $[H_3O^+]$, taking the total volume into account.* To find the ion concentrations, we use the *total volume* because the water of one solution dilutes the ions of the other:

$$[H_3O^+] = \frac{\text{amount (mol) of } H_3O^+ \text{ remaining}}{\text{original volume of acid} + \text{volume of added base}}$$

$$= \frac{0.002000\ \text{mol}\ H_3O^+}{0.04000\ L + 0.02000\ L} = 0.03333\ M \qquad pH = 1.48$$

Given the moles of OH^- added, we are halfway to the equivalence point; but we are still on the initial slow rise of the curve, so the pH is still very low. Similar calculations give values up to the equivalence point.

3. *At the equivalence point.* After 40.00 mL of 0.1000 M NaOH has been added, the equivalence point is reached. All the H_3O^+ from the acid has been

---

[*]In acid-base titrations, volumes and concentrations are usually known to four significant figures, but pH is generally reported to no more than two digits to the right of the decimal point.

neutralized, and the solution contains Na^+ and Cl^-, neither of which reacts with water. Because of the autoionization of water, however,

$$[H_3O^+] = 1.0 \times 10^{-7}\ M \qquad pH = 7.00$$

In this example, 0.004000 mol of OH^- reacted with 0.004000 mol of H_3O^+ to reach the equivalence point.

4. *After the equivalence point.* From the equivalence point on, the pH calculation is based on the moles of *excess* OH^- present. For example, after adding 50.00 mL of NaOH, we have

$$
\begin{array}{ll}
\text{Total moles of } OH^- \text{ added} = 0.05000\ L \times 0.1000\ M = 0.005000\ \text{mol } OH^- \\
-\text{Moles of } H_3O^+ \text{ consumed} = 0.04000\ L \times 0.1000\ M = 0.004000\ \text{mol } H_3O^+ \\
\hline
\text{Moles of excess } OH^- = \hspace{4.5cm} 0.001000\ \text{mol } OH^-
\end{array}
$$

$$[OH^-] = \frac{0.001000\ \text{mol } OH^-}{0.04000\ L + 0.05000\ L} = 0.01111\ M \qquad pOH = 1.95$$

$$pH = pK_w - pOH = 14.00 - 1.95 = 12.05$$

Weak Acid–Strong Base Titration Curves

Now let's turn to the titration of a weak acid with a strong base. Figure 19.8 shows the curve obtained when we use 0.1000 M NaOH to titrate 40.00 mL of 0.1000 M propanoic acid, a weak organic acid (CH_3CH_2COOH; $K_a = 1.3 \times 10^{-5}$). (We abbreviate the acid as HPr and the conjugate base, $CH_3CH_2COO^-$, as Pr^-.)

Transparency/ARIS Presentation Center, Active Art Library

Figure 19.8 **Curve for a weak acid–strong base titration.** The curve for the titration of 40.00 M CH_3CH_2COOH (HPr) with 0.1000 M NaOH is compared with that for the strong acid HCl (*dotted portion*). Note the buffer region and the equivalence point, which is above pH 7, as indicated by the change in color of phenolphthalein (*photo*). The blow-up circles depict the relative numbers of species present at key points during the titration.

Features of the Curve When we compare this weak acid–strong base titration curve with the strong acid–strong base titration curve (dotted curve portion in Figure 19.8 corresponds to bottom half of curve in Figure 19.7), four key regions appear, and the first three differ from the strong acid case:

1. *The initial pH is higher.* Because the weak acid (HPr) dissociates slightly, much less H_3O^+ is present than with the strong acid.
2. *A gradually rising portion of the curve, called the buffer region, appears before the steep rise to the equivalence point.* As HPr reacts with the strong base, more and more conjugate base (Pr^-) forms, which creates an HPr/Pr^- buffer. At the midpoint of the buffer region, *half the original HPr has reacted,* so $[HPr] = [Pr^-]$, or $[Pr^-]/[HPr] = 1$. Therefore, *the pH equals the pK_a:*

$$pH = pK_a + \log\left(\frac{[Pr^-]}{[HPr]}\right) = pK_a + \log 1 = pK_a + 0 = pK_a$$

Observing the pH at the midpoint of the buffer region is a common method for estimating the pK_a of an unknown acid.
3. *The pH at the equivalence point is greater than 7.00.* The solution contains the strong-base cation Na^+, which does not react with water, and the weak-acid anion Pr^-, which acts as a weak base to accept a proton from H_2O and yield OH^-.
4. Beyond the equivalence point, the pH increases slowly as *excess OH^-* is added.

Our choice of indicator is more limited here than for a strong acid–strong base titration because the steep rise occurs over a smaller pH range. Phenolphthalein is suitable because its color change lies within this range (Figure 19.8). However, the figure shows that methyl red, our other choice for the strong acid–strong base titration, changes color earlier and slowly over a large volume (~10 mL) of titrant, thereby giving a vague and false indication of the equivalence point.

Calculating the pH The calculation procedure for the weak acid–strong base titration is different from that for the strong acid–strong base titration because we have to consider the partial dissociation of the weak acid and the reaction of the conjugate base with water. There are four key regions of the titration curve, each of which requires a different type of calculation to find $[H_3O^+]$:

1. *Solution of HA.* Before base is added, the $[H_3O^+]$ is that of a weak-acid solution, so we find $[H_3O^+]$ as in Section 18.4: we set up a reaction table with $x = [HPr]_{dissoc}$, assume $[H_3O^+] = [HPr]_{dissoc} \ll [HPr]_{init}$, and solve for x:

$$K_a = \frac{[H_3O^+][Pr^-]}{[HPr]} \approx \frac{x^2}{[HPr]_{init}} \qquad \text{therefore,} \qquad x = [H_3O^+] \approx \sqrt{K_a \times [HPr]_{init}}$$

2. *Solution of HA and added base.* As soon as we add NaOH, it reacts with HPr to form Pr^-. This means that up to the equivalence point, we have a mixture of acid and conjugate base, and an HPr/Pr^- buffer solution exists over much of that interval. Therefore, we find $[H_3O^+]$ from the relationship

$$[H_3O^+] = K_a \times \frac{[HPr]}{[Pr^-]}$$

(Of course, we can find pH directly with the Henderson-Hasselbalch equation, which is just an alternative form of this relationship.) Note that in this calculation we do *not* have to consider the new total volume because *the volumes cancel in the ratio of concentrations.* That is, $[HPr]/[Pr^-]$ = moles of HPr/moles of Pr^-, so we need not calculate concentrations.

3. *Equivalent amounts of HA and added base.* At the equivalence point, the original amount of HPr has reacted, so the flask contains a solution of Pr^-, a weak base that reacts with water to form OH^-:

$$Pr^-(aq) + H_2O(l) \rightleftharpoons HPr(aq) + OH^-(aq)$$

Therefore, as mentioned previously, in a weak acid–strong base titration, the solution at the equivalence point is slightly basic, pH > 7.00. We calculate $[H_3O^+]$ as in Section 18.5: we first find K_b of Pr^- from K_a of HPr, set up a reaction table (assume $[Pr^-] \gg [Pr^-]_{reacting}$), and solve for $[OH^-]$. We need a single concentration, $[Pr^-]$, to solve for $[OH^-]$, so we *do* need the total volume. Then, we convert to $[H_3O^+]$. These two steps are

(1) $[OH^-] \approx \sqrt{K_b \times [Pr^-]}$, where $K_b = \dfrac{K_w}{K_a}$ and $[Pr^-] = \dfrac{\text{moles of HPr}_{init}}{\text{total volume}}$

(2) $[H_3O^+] = \dfrac{K_w}{[OH^-]}$

Combining them into one step gives

$$[H_3O^+] \approx \frac{K_w}{\sqrt{K_b \times [Pr^-]}}$$

4. *Solution of excess added base.* Beyond the equivalence point, we are just adding excess OH^- ion, so the calculation is the same as for the strong acid–strong base titration:

$$[H_3O^+] = \frac{K_w}{[OH^-]}, \qquad \text{where } [OH^-] = \frac{\text{moles of excess } OH^-}{\text{total volume}}$$

Sample Problem 19.4 shows the overall approach.

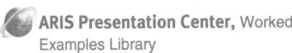

SAMPLE PROBLEM 19.4 Finding the pH During a Weak Acid–Strong Base Titration

PROBLEM Calculate the pH during the titration of 40.00 mL of 0.1000 *M* propanoic acid (HPr; $K_a = 1.3 \times 10^{-5}$) after each of the following volumes of 0.1000 *M* NaOH has been added:
(a) 0.00 mL   **(b)** 30.00 mL
(c) 40.00 mL   **(d)** 50.00 mL

PLAN **(a)** 0.00 mL: No base has been added yet, so this is a weak-acid solution. Thus, we calculate the pH as we did in Section 18.4. **(b)** 30.00 mL: A mixture of Pr^- and HPr is present. We find the amount (mol) of each, substitute into the K_a expression to solve for $[H_3O^+]$, and convert to pH. **(c)** 40.00 mL: The amount (mol) of NaOH added equals the initial amount (mol) of HPr, so a solution of Na^+ and the weak base Pr^- exists. We calculate the pH as we did in Section 18.5, except that we need *total* volume to find $[Pr^-]$. **(d)** 50.00 mL: Excess NaOH is added, so we calculate the amount (mol) of excess OH^- in the total volume and convert to $[H_3O^+]$ and then pH.

SOLUTION **(a)** 0.00 mL of 0.1000 *M* NaOH added. Following the approach used in Sample Problem 18.8 and just described in the text, we obtain

$$[H_3O^+] \approx \sqrt{K_a \times [HPr]_{init}} = \sqrt{(1.3 \times 10^{-5})(0.1000)} = 1.1 \times 10^{-3} \, M$$
$$pH = 2.96$$

(b) 30.00 mL of 0.1000 *M* NaOH added. Calculating the ratio of moles of HPr to Pr^-:

Original moles of HPr = 0.04000 L × 0.1000 *M* = 0.004000 mol HPr
Moles of NaOH added = 0.03000 L × 0.1000 *M* = 0.003000 mol OH^-

For 1 mol of NaOH that reacts, 1 mol of Pr^- forms, so we construct the following reaction table for the stoichiometry:

Amount (mol)	HPr(*aq*)	+	OH⁻(*aq*)	⟶	Pr⁻(*aq*)	+	H₂O(*l*)
Initial	0.004000		0.003000		0		—
Change	−0.003000		−0.003000		+0.003000		—
Final	0.001000		0		0.003000		—

The last line of this table shows the new initial amounts of HPr and Pr^- that will react to attain a new equilibrium. However, with *x* very small, we assume that the [HPr]/[Pr⁻]

ratio at equilibrium is essentially equal to the ratio of these new initial amounts (see Comment in Sample Problem 19.1). Thus,

$$\frac{[\text{HPr}]}{[\text{Pr}^-]} = \frac{0.001000 \text{ mol}}{0.003000 \text{ mol}} = 0.3333$$

Solving for $[\text{H}_3\text{O}^+]$:
$$[\text{H}_3\text{O}^+] = K_\text{a} \times \frac{[\text{HPr}]}{[\text{Pr}^-]} = (1.3 \times 10^{-5})(0.3333) = 4.3 \times 10^{-6} M$$

$$\text{pH} = 5.37$$

(c) 40.00 mL of 0.1000 M NaOH added. Calculating $[\text{Pr}^-]$ after all HPr has reacted:

$$[\text{Pr}^-] = \frac{0.004000 \text{ mol}}{0.04000 \text{ L} + 0.04000 \text{ L}} = 0.05000 M$$

Calculating K_b:
$$K_\text{b} = \frac{K_\text{w}}{K_\text{a}} = \frac{1.0 \times 10^{-14}}{1.3 \times 10^{-5}} = 7.7 \times 10^{-10}$$

Solving for $[\text{H}_3\text{O}^+]$ as described in the text:

$$[\text{H}_3\text{O}^+] \approx \frac{K_\text{w}}{\sqrt{K_\text{b} \times [\text{Pr}^-]}} = \frac{1.0 \times 10^{-14}}{\sqrt{(7.7 \times 10^{-10})(0.05000)}} = 1.6 \times 10^{-9} M$$

$$\text{pH} = 8.80$$

(d) 50.00 mL of 0.1000 M NaOH added.

$$\text{Moles of excess OH}^- = (0.1000 M)(0.05000 \text{ L} - 0.04000 \text{ L}) = 0.001000 \text{ mol}$$

$$[\text{OH}^-] = \frac{\text{moles of excess OH}^-}{\text{total volume}} = \frac{0.001000 \text{ mol}}{0.09000 \text{ L}} = 0.01111 M$$

$$[\text{H}_3\text{O}^+] = \frac{K_\text{w}}{[\text{OH}^-]} = \frac{1.0 \times 10^{-14}}{0.01111} = 9.0 \times 10^{-13} M$$

$$\text{pH} = 12.05$$

CHECK As expected from the continuous addition of base, the pH increases through the four stages. Be sure to round off and check the arithmetic along the way.

FOLLOW-UP PROBLEM 19.4 A chemist titrates 20.00 mL of 0.2000 M HBrO ($K_\text{a} = 2.3 \times 10^{-9}$) with 0.1000 M NaOH. What is the pH **(a)** before any base is added; **(b)** when $[\text{HBrO}] = [\text{BrO}^-]$; **(c)** at the equivalence point; **(d)** when the moles of OH^- added are twice the moles of HBrO originally present? **(e)** Sketch the titration curve.

Weak Base–Strong Acid Titration Curves

In the previous case, we titrated a weak acid with a strong base. The opposite process is the titration of a weak base (NH_3) with a strong acid (HCl), shown in Figure 19.9. Note that *the curve has the same shape as the weak acid–strong base curve* (Figure 19.8, p. 845), *but it is inverted*. Thus, the regions of the curve have the same features, but *the pH decreases* throughout the process:

1. The initial solution is that of a weak base, so *the pH starts out above 7.00*.
2. The pH decreases gradually in the buffer region, where significant amounts of base (NH_3) and conjugate acid (NH_4^+) are present. At the midpoint of the buffer region, *the pH equals the pK_a of the ammonium ion*.
3. After the buffer region, the curve drops vertically to the equivalence point, at which all the NH_3 has reacted and the solution contains only NH_4^+ and Cl^-. Note that *the pH at the equivalence point is below 7.00* because Cl^- does not react with water and NH_4^+ is acidic:

$$\text{NH}_4^+(aq) + \text{H}_2\text{O}(l) \rightleftharpoons \text{NH}_3(aq) + \text{H}_3\text{O}^+(aq)$$

4. Beyond the equivalence point, the pH decreases slowly as *excess H_3O^+* is added.

Titration of 40.00 mL of 0.1000 M NH₃ with 0.1000 M HCl

Figure 19.9 Curve for a weak base–strong acid titration. Titrating 40.00 mL of 0.1000 *M* NH₃ with a solution of 0.1000 *M* HCl leads to a curve whose shape is the same as that of the weak acid–strong base curve in Figure 19.8 but inverted. The midpoint of the buffer region occurs when [NH₃] = [NH₄⁺]; the pH at this point equals the pK_a of the acid, NH₄⁺. In this case, methyl red (*photo*) is a suitable indicator because it changes color on the steep portion. The blow-up circles depict the relative numbers of species present at key points during the titration.

 Transparency/ARIS Presentation Center, Active Art Library

For this titration also, we must be more careful in choosing the indicator than we would be for a strong acid–strong base titration. Phenolphthalein changes color too soon and too slowly to indicate the equivalence point; but methyl red lies on the steep portion of the curve and straddles the equivalence point, so it is a perfect choice.

Titration Curves for Polyprotic Acids

As we discussed in Section 18.4, polyprotic acids have more than one ionizable proton. Except for sulfuric acid, the common polyprotic acids are all weak. The successive K_a values for a polyprotic acid differ by several orders of magnitude, which means that the first H⁺ is lost much more easily than subsequent ones, as these values for sulfurous acid show:

Literature Ault, Addison. "The pH at the First Equivalence Point in the Titration of a Diprotic Acid," *J. Chem. Educ.* **2003,** *80,* 1382.

$$H_2SO_3(aq) + H_2O(l) \rightleftharpoons HSO_3^-(aq) + H_3O^+(aq)$$
$$K_{a1} = 1.4 \times 10^{-2} \quad \text{and} \quad pK_{a1} = 1.85$$

$$HSO_3^-(aq) + H_2O(l) \rightleftharpoons SO_3^{2-}(aq) + H_3O^+(aq)$$
$$K_{a2} = 6.5 \times 10^{-8} \quad \text{and} \quad pK_{a2} = 7.19$$

In a titration of a diprotic acid such as H_2SO_3, two OH^- ions are required to completely remove both H^+ ions from each molecule of acid. Figure 19.10 shows the titration curve for sulfurous acid with strong base. Because of the large difference in K_a values, we assume that each mole of H^+ is titrated separately; that is, all H_2SO_3 molecules lose one H^+ before any HSO_3^- ions lose one:

$$H_2SO_3 \xrightarrow{\text{1 mol OH}^-} HSO_3^- \xrightarrow{\text{1 mol OH}^-} SO_3^{2-}$$

As you can see from the curve in Figure 19.10, the loss of each mole of H^+ shows up as a separate equivalence point and buffer region. As on the curve for a weak monoprotic acid (Figure 19.8, p. 845), the pH at the midpoint of the buffer region is equal to the pK_a of that acidic species. Note also that the same volume of added base (in this case, 40.00 mL of 0.1000 M OH^-) is required to remove each mole of H^+.

Figure 19.10 Curve for the titration of a weak polyprotic acid. Titrating 40.00 mL of 0.1000 M H_2SO_3 with 0.1000 M NaOH leads to a curve with two buffer regions and two equivalence points. Because the K_a values are separated by several orders of magnitude, the titration curve looks like two weak acid–strong base curves joined end to end. The pH of the first equivalence point is below 7.00 because the solution contains HSO_3^-, which is a stronger acid than it is a base (K_a of $HSO_3^- = 6.5\times10^{-8}$; K_b of $HSO_3^- = 7.1\times10^{-13}$).

ARIS Presentation Center, Art Library

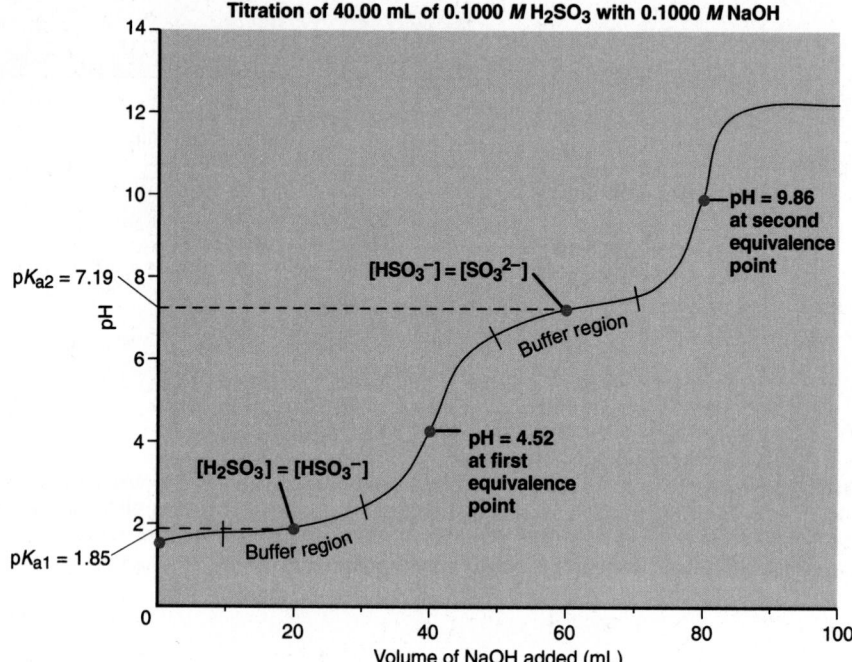

Biological application

Amino Acids as Biological Polyprotic Acids

Amino acids, the molecules that link together to form proteins (Sections 13.2 and 15.6), have the general formula NH_2—$CH(R)$—$COOH$, where R can be one of about 20 different groups. At low pH, both the amino group (—NH_2) and the acid group (—$COOH$) are protonated: $^+NH_3$—$CH(R)$—$COOH$. Thus, in this form the amino acid behaves like a polyprotic acid. In the case of glycine (R = H), the simplest amino acid, the dissociation reactions and pK_a values are

$$^+NH_3CH_2COOH(aq) + H_2O(l) \rightleftharpoons {}^+NH_3CH_2COO^-(aq) + H_3O^+(aq) \quad pK_{a1} = 2.35$$

$$^+NH_3CH_2COO^-(aq) + H_2O(l) \rightleftharpoons NH_2CH_2COO^-(aq) + H_3O^+(aq) \quad pK_{a2} = 9.78$$

The pK_a values show that the —$COOH$ group is much more acidic than the —NH_3^+ group. As we saw with H_2SO_3, the protons are titrated separately, so virtually all the —$COOH$ protons are removed before any —NH_3^+ protons are:

protonated   zwitterion   unprotonated
(low pH)   (neutral pH)   (high pH)

Thus, at physiological pH (~7), *glycine exists predominantly as a zwitterion* (German *zwitter*, "double"), a species with opposite charges on the same molecule: $^+NH_3CH_2COO^-$. Among the 20 different R groups of amino acids in proteins, several have *additional* —COO^- or —NH_3^+ groups at pH 7 (see Figure 15.28, p. 667).

When amino acids link to form a protein, their charged R groups give the protein its overall charge and often play a role in its function. A widely studied example of the effect of charge on function occurs in the hereditary disease sickle cell anemia. Normal red blood cells are packed tightly with molecules of hemoglobin. Two of the amino acids in this protein—both glutamic acid—are critical to the mobility of hemoglobin molecules, and this mobility is critical to the shape of the red blood cells (Figure 19.11). Each glutamic acid has a negatively charged R group (—$CH_2CH_2COO^-$). The abnormal hemoglobin molecules in sickle cell anemia have uncharged (—CH_3) groups instead of these charged ones. This change in just two of hemoglobin's 574 amino acids lowers the charge repulsions between hemoglobin molecules. As a result, they clump together in fiber-like structures, which leads to the sickle shape of the red blood cells. The misshapen cells block capillaries, and the painful course of sickle cell anemia usually ends in early death.

Figure 19.11 Sickle shape of red blood cells in sickle cell anemia.

 ARIS Presentation Center, Photo Library

Section Summary

An acid-base (pH) indicator is a weak acid that has differently colored acidic and basic forms and changes color over about 2 pH units. • In a strong acid–strong base titration, the pH starts out low, rises slowly, then shoots up near the equivalence point (pH = 7). • In a weak acid–strong base titration, the pH starts out higher than in the strong acid titration, rises slowly in the buffer region (pH = pK_a at the midpoint), then rises more quickly near the equivalence point (pH > 7). • A weak base–strong acid titration has a shape that is the inverse of the weak acid–strong base case, with the pH decreasing to the equivalence point (pH < 7). • Polyprotic acids have two or more acidic protons, each with its own K_a value. Because the K_a's differ by several orders of magnitude, each proton is titrated separately. • Amino acids exist in different charged forms that depend on the pH of the solution.

19.3 EQUILIBRIA OF SLIGHTLY SOLUBLE IONIC COMPOUNDS

In this section, we explore the aqueous equilibria of slightly soluble ionic compounds, which up to now we've called "insoluble." In Chapter 13, we found that most solutes, even those called "soluble," have a limited solubility in a particular solvent. Add more than this amount, and some solute remains undissolved. In a saturated solution at a particular temperature, equilibrium exists between the undissolved and dissolved solute. Slightly soluble ionic compounds have a relatively low solubility, so they reach equilibrium with relatively little solute dissolved. At this point, it would be a good idea for you to review the solubility rules listed in Table 4.1 (p. 148).

When a soluble ionic compound dissolves in water, it dissociates completely into ions. In this discussion, we will assume that the small amount of a slightly soluble ionic compound that does dissolve in water also dissociates completely into ions. In reality, however, this is not the case. Many slightly soluble salts, particularly those of transition metals and heavy main-group metals, have metal-nonmetal bonds with significant covalent character, and their solutions often contain other species that are partially dissociated or even undissociated. For example, when lead(II) chloride is thoroughly stirred in water (see photo), a small amount dissolves, and the solution contains not only the $Pb^{2+}(aq)$ and $Cl^-(aq)$ ions expected from complete dissociation, but also undissociated $PbCl_2(aq)$

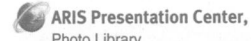 ARIS Presentation Center, Photo Library

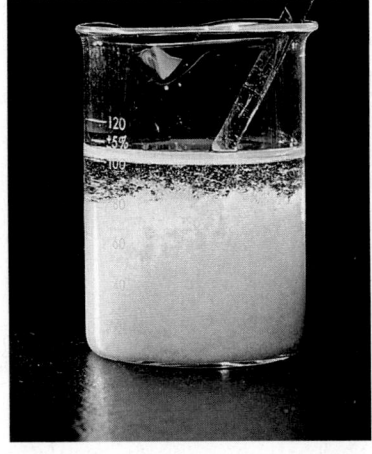

$PbCl_2$, a slightly soluble ionic compound.

molecules and $PbCl^+(aq)$ ions. In solutions of some other salts, such as $CaSO_4$, there are no molecules, but pairs of ions exist, such as $Ca^{2+}SO_4{}^{2-}(aq)$. These species increase the solubility above what we calculate assuming complete dissociation. More advanced courses address these complexities, but we will only hint at some of them in the Comments of several sample problems. For these reasons, it is best to treat our calculations here as first approximations.

The Ion-Product Expression (Q_{sp}) and the Solubility-Product Constant (K_{sp})

Literature Raviolo, Andrés. "Assessing Students' Conceptual Understanding of Solubility Equilibrium," *J. Chem. Educ.* **2001**, *78*, 629.

If we make the assumption that there is complete dissociation of a slightly soluble ionic compound into its component ions, then *equilibrium exists between solid solute and aqueous ions*. Thus, for example, for a saturated solution of lead(II) sulfate in water, we have

$$PbSO_4(s) \rightleftharpoons Pb^{2+}(aq) + SO_4{}^{2-}(aq)$$

As with all the other equilibrium systems we've looked at, this one can be expressed by a reaction quotient:

Literature Hawkes, Stephen J. "What Should We Teach Beginners about Solubility and Solubility Products?" *J. Chem. Educ.* **1998**, *75*, 1179.

$$Q_c = \frac{[Pb^{2+}][SO_4{}^{2-}]}{[PbSO_4]}$$

As in previous cases, we incorporate the constant concentration of the solid, $[PbSO_4]$, into the value of Q_c, which gives the *ion-product expression*, Q_{sp}:

$$Q_{sp} = Q_c[PbSO_4]$$
$$= [Pb^{2+}][SO_4{}^{2-}]$$

Literature Clark, Roy W.; Bonicamp, Judith M. "The K_{sp}-Solubility Conundrum," *J. Chem. Educ.* **1998**, *75*, 1182.

And, when solid $PbSO_4$ attains equilibrium with Pb^{2+} and $SO_4{}^{2-}$ ions, that is, when the solution reaches saturation, the numerical value of Q_{sp} attains a constant value called the **solubility-product constant**, K_{sp}. The K_{sp} for $PbSO_4$ at 25°C, for example, is 1.6×10^{-8}.

As we've seen with other equilibrium constants, a given K_{sp} value depends *only* on the temperature, not on the individual ion concentrations. Suppose, for example, you add some lead(II) nitrate, a soluble lead salt, to increase the solution's $[Pb^{2+}]$. The equilibrium position shifts to the left, and $[SO_4{}^{2-}]$ goes down as more $PbSO_4$ precipitates; so the K_{sp} value is maintained.

The form of Q_{sp} is identical to that of the other reaction quotients we have written: each ion concentration is raised to an exponent equal to the coefficient in the balanced equation, which in this case also *equals the subscript of each ion in the compound's formula*. Thus, in general, for a saturated solution of a slightly soluble ionic compound, M_pX_q, composed of the ions M^{n+} and X^{z-}, the equilibrium condition is

$$Q_{sp} = [M^{n+}]^p[X^{z-}]^q = K_{sp} \qquad (19.2)$$

At saturation, the concentration terms have their equilibrium values, so we can write the ion-product expression directly with the symbol K_{sp}. For example, the equation and ion-product expression that describe a saturated solution of $Cu(OH)_2$ are

$$Cu(OH)_2(s) \rightleftharpoons Cu^{2+}(aq) + 2OH^-(aq)$$
$$K_{sp} = [Cu^{2+}][OH^-]^2$$

Insoluble metal sulfides present a slightly different case. The sulfide ion, S^{2-}, is so basic that it is not stable in water and reacts completely to form the hydrogen sulfide ion (HS^-) and the hydroxide ion (OH^-):

$$S^{2-}(aq) + H_2O(l) \longrightarrow HS^-(aq) + OH^-(aq)$$

For instance, when manganese(II) sulfide is shaken with water, the solution contains Mn^{2+}, HS^-, and OH^- ions. Although the sulfide ion does not exist as such in water, you can imagine the dissolution process as the sum of two steps, with S^{2-} occurring as an intermediate that is consumed immediately:

$$MnS(s) \rightleftharpoons Mn^{2+}(aq) + \cancel{S^{2-}(aq)}$$
$$\underline{\cancel{S^{2-}(aq)} + H_2O(l) \longrightarrow HS^-(aq) + OH^-(aq)}$$
$$MnS(s) + H_2O(l) \rightleftharpoons Mn^{2+}(aq) + HS^-(aq) + OH^-(aq)$$

Therefore, the ion-product expression is

$$K_{sp} = [Mn^{2+}][HS^-][OH^-]$$

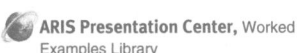
ARIS Presentation Center, Worked Examples Library

SAMPLE PROBLEM 19.5 Writing Ion-Product Expressions for Slightly Soluble Ionic Compounds

PROBLEM Write the ion-product expression for each compound: **(a)** magnesium carbonate; **(b)** iron(II) hydroxide; **(c)** calcium phosphate; **(d)** silver sulfide.

PLAN We write an equation that describes a saturated solution and then write the ion-product expression, K_{sp}, according to Equation 19.2, noting the sulfide in part (d).

SOLUTION **(a)** Magnesium carbonate:

$$MgCO_3(s) \rightleftharpoons Mg^{2+}(aq) + CO_3^{2-}(aq) \qquad K_{sp} = [Mg^{2+}][CO_3^{2-}]$$

(b) Iron(II) hydroxide:

$$Fe(OH)_2(s) \rightleftharpoons Fe^{2+}(aq) + 2OH^-(aq) \qquad K_{sp} = [Fe^{2+}][OH^-]^2$$

(c) Calcium phosphate:

$$Ca_3(PO_4)_2(s) \rightleftharpoons 3Ca^{2+}(aq) + 2PO_4^{3-}(aq) \qquad K_{sp} = [Ca^{2+}]^3[PO_4^{3-}]^2$$

(d) Silver sulfide:

$$Ag_2S(s) \rightleftharpoons 2Ag^+(aq) + \cancel{S^{2-}(aq)}$$
$$\underline{\cancel{S^{2-}(aq)} + H_2O(l) \longrightarrow HS^-(aq) + OH^-(aq)}$$
$$Ag_2S(s) + H_2O(l) \rightleftharpoons 2Ag^+(aq) + HS^-(aq) + OH^-(aq) \qquad K_{sp} = [Ag^+]^2[HS^-][OH^-]$$

CHECK Except for part (d), you can check by reversing the process to see if you obtain the formula of the compound from K_{sp}.

COMMENT In part (d), we include H_2O as reactant to obtain a balanced equation.

FOLLOW-UP PROBLEM 19.5 Write the ion-product expression for each compound: **(a)** calcium sulfate; **(b)** chromium(III) carbonate; **(c)** magnesium hydroxide; **(d)** arsenic(III) sulfide.

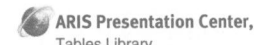
ARIS Presentation Center, Tables Library

The value of K_{sp} indicates how far to the right the dissolution proceeds at equilibrium (saturation). Table 19.2 presents some representative K_{sp} values. (Appendix C includes a much more extensive list.) Even though the values are all quite low, they range over many orders of magnitude.

Calculations Involving the Solubility-Product Constant

In Chapters 17 and 18, we described two types of equilibrium problems. In one type, we use concentrations to find K, and in the other, we use K to find concentrations. Here we encounter the same two types.

Determining K_{sp} from Solubility The solubilities of ionic compounds are determined experimentally, and several chemical handbooks tabulate them. Most solubility values are given in units of grams of solute dissolved in 100. grams of H_2O. Because the mass of compound in solution is small, a negligible error is introduced if we assume that "100. g of water" is equal to "100. mL of solution." We then convert the solubility from grams of solute per 100. mL of solution to **molar solubility**, the amount (mol) of solute dissolved per liter of solution (that is, the molarity of the solute). Next, we use the equation for the dissolution of the solute to find the molarity of each ion and substitute into the ion-product expression to find the value of K_{sp}.

Table 19.2 Solubility-Product Constants (K_{sp}) of Selected Ionic Compounds at 25°C

Name, Formula	K_{sp}
Aluminum hydroxide, $Al(OH)_3$	3×10^{-34}
Cobalt(II) carbonate, $CoCO_3$	1.0×10^{-10}
Iron(II) hydroxide, $Fe(OH)_2$	4.1×10^{-15}
Lead(II) fluoride, PbF_2	3.6×10^{-8}
Lead(II) sulfate, $PbSO_4$	1.6×10^{-8}
Mercury(I) iodide, Hg_2I_2	4.7×10^{-29}
Silver sulfide, Ag_2S	8×10^{-48}
Zinc iodate, $Zn(IO_3)_2$	3.9×10^{-6}

SAMPLE PROBLEM 19.6 Determining K_{sp} from Solubility

PROBLEM (a) Lead(II) sulfate ($PbSO_4$) is a key component in lead-acid car batteries. Its solubility in water at 25°C is 4.25×10^{-3} g/100 mL solution. What is the K_{sp} of $PbSO_4$?
(b) When lead(II) fluoride (PbF_2) is shaken with pure water at 25°C, the solubility is found to be 0.64 g/L. Calculate the K_{sp} of PbF_2.

PLAN We are given the solubilities in various units and must find K_{sp}. For each compound, we write an equation for its dissolution to see the number of moles of each ion, and then write the ion-product expression. We convert the solubility to molar solubility, find the molarity of each ion, and substitute into the ion-product expression to calculate K_{sp}.

SOLUTION (a) For $PbSO_4$. Writing the equation and ion-product (K_{sp}) expression:

$$PbSO_4(s) \rightleftharpoons Pb^{2+}(aq) + SO_4^{2-}(aq) \qquad K_{sp} = [Pb^{2+}][SO_4^{2-}]$$

Converting solubility to molar solubility:

$$\text{Molar solubility of } PbSO_4 = \frac{0.00425 \text{ g } PbSO_4}{100 \text{ mL soln}} \times \frac{1000 \text{ mL}}{1 \text{ L}} \times \frac{1 \text{ mol } PbSO_4}{303.3 \text{ g } PbSO_4}$$

$$= 1.40 \times 10^{-4} M \, PbSO_4$$

Determining molarities of the ions: Because 1 mol of Pb^{2+} and 1 mol of SO_4^{2-} form when 1 mol of $PbSO_4$ dissolves, $[Pb^{2+}] = [SO_4^{2-}] = 1.40 \times 10^{-4} M$.

Calculating K_{sp}:

$$K_{sp} = [Pb^{2+}][SO_4^{2-}] = (1.40 \times 10^{-4})^2 = 1.96 \times 10^{-8}$$

(b) For PbF_2. Writing the equation and K_{sp} expression:

$$PbF_2(s) \rightleftharpoons Pb^{2+}(aq) + 2F^-(aq) \qquad K_{sp} = [Pb^{2+}][F^-]^2$$

Converting solubility to molar solubility:

$$\text{Molar solubility of } PbF_2 = \frac{0.64 \text{ g } PbF_2}{1 \text{ L soln}} \times \frac{1 \text{ mol } PbF_2}{245.2 \text{ g } PbF_2} = 2.6 \times 10^{-3} M \, PbF_2$$

Determining molarities of the ions: Since 1 mol of Pb^{2+} and 2 mol of F^- form when 1 mol of PbF_2 dissolves,

$$[Pb^{2+}] = 2.6 \times 10^{-3} M \qquad \text{and} \qquad [F^-] = 2(2.6 \times 10^{-3} M) = 5.2 \times 10^{-3} M$$

Calculating K_{sp}:

$$K_{sp} = [Pb^{2+}][F^-]^2 = (2.6 \times 10^{-3})(5.2 \times 10^{-3})^2 = 7.0 \times 10^{-8}$$

CHECK The low solubilities are consistent with K_{sp} values being small. (a) The molar solubility seems about right: $\sim \dfrac{4 \times 10^{-2} \text{ g/L}}{3 \times 10^2 \text{ g/mol}} \approx 1.3 \times 10^{-4} M$. Squaring this number gives 1.7×10^{-8}, close to the calculated K_{sp}. (b) We check the final step as follows: $\sim(3 \times 10^{-3})(5 \times 10^{-3})^2 = 7.5 \times 10^{-8}$, close to the calculated K_{sp}.

COMMENT 1. In part (b), the formula PbF_2 means that $[F^-]$ is twice $[Pb^{2+}]$. Then we square this value of $[F^-]$. Always follow the ion-product expression explicitly.
2. The tabulated K_{sp} values for these compounds (Table 19.2) are lower than our calculated values. For PbF_2, for instance, the tabulated value is 3.6×10^{-8}, but we calculated 7.0×10^{-8} from solubility data. The discrepancy arises because we assumed that the PbF_2 in solution dissociates completely to Pb^{2+} and F^-. Here is an example of the complexity pointed out at the beginning of this section. Actually, about a third of the PbF_2 dissolves as $PbF^+(aq)$ and a small amount as undissociated $PbF_2(aq)$. The solubility (0.64 g/L) is determined experimentally and includes these other species, which we did not include in our simple calculation. This is why we treat such calculated K_{sp} values as approximations.

FOLLOW-UP PROBLEM 19.6 When powdered fluorite (CaF_2; see photo) is shaken with pure water at 18°C, 1.5×10^{-4} g dissolves for every 10.0 mL of solution. Calculate the K_{sp} of CaF_2 at 18°C.

Fluorite

Determining Solubility from K_{sp} The reverse of Sample Problem 19.6 involves finding the solubility of a compound based on its formula and K_{sp} value. An approach similar to the one we used for weak acids in Sample Problem 18.8 is to define the unknown amount dissolved—molar solubility—as S. Then we define the ion concentrations in terms of this unknown in a reaction table, and solve for S.

ARIS Presentation Center, Worked Examples Library

SAMPLE PROBLEM 19.7 Determining Solubility from K_{sp}

PROBLEM Calcium hydroxide (slaked lime) is a major component of mortar, plaster, and cement, and solutions of $Ca(OH)_2$ are used in industry as a cheap, strong base. Calculate the solubility of $Ca(OH)_2$ in water if the K_{sp} is 6.5×10^{-6}.

PLAN We write the dissolution equation and the ion-product expression. We know K_{sp} (6.5×10^{-6}); to find molar solubility (S), we set up a reaction table that expresses $[Ca^{2+}]$ and $[OH^-]$ in terms of S, substitute into the ion-product expression, and solve for S.

SOLUTION Writing the equation and ion-product expression:

$$Ca(OH)_2(s) \rightleftharpoons Ca^{2+}(aq) + 2OH^-(aq) \qquad K_{sp} = [Ca^{2+}][OH^-]^2 = 6.5 \times 10^{-6}$$

Setting up a reaction table, with S = molar solubility:

Concentration (M)	$Ca(OH)_2(s)$	$\rightleftharpoons$	$Ca^{2+}(aq)$	+	$2OH^-(aq)$
Initial	—		0		0
Change	—		$+S$		$+2S$
Equilibrium	—		S		$2S$

Substituting into the ion-product expression and solving for S:

$$K_{sp} = [Ca^{2+}][OH^-]^2 = (S)(2S)^2 = (S)(4S^2) = 4S^3 = 6.5 \times 10^{-6}$$

$$S = \sqrt[3]{\frac{6.5 \times 10^{-6}}{4}} = \boxed{1.2 \times 10^{-2} \ M}$$

CHECK We expect a low solubility from a slightly soluble salt. If we reverse the calculation, we should obtain the given K_{sp}: $4(1.2 \times 10^{-2})^3 = 6.9 \times 10^{-6}$, close to 6.5×10^{-6}.

COMMENT 1. Note that we did not double and *then* square $[OH^-]$. $2S$ *is* the $[OH^-]$, so we just squared it, as the ion-product expression required.

2. Once again, we assumed that the solid dissociates completely. Actually, the solubility is increased to about $2.0 \times 10^{-2} \ M$ by the presence of $CaOH^+(aq)$ formed in the reaction $Ca(OH)_2(s) \rightleftharpoons CaOH^+(aq) + OH^-(aq)$. Our calculated answer is only approximate because we did not take this other species into account.

FOLLOW-UP PROBLEM 19.7 A suspension of $Mg(OH)_2$ in water is marketed as "milk of magnesia," which alleviates minor symptoms of indigestion by neutralizing stomach acid. The $[OH^-]$ is too low to harm the mouth and throat, but the suspension dissolves in the acidic stomach juices. What is the molar solubility of $Mg(OH)_2$ ($K_{sp} = 6.3 \times 10^{-10}$) in pure water?

Using K_{sp} Values to Compare Solubilities The K_{sp} values provide a guide to *relative* solubility, as long as we compare compounds whose formulas contain the *same total number of ions*. In such cases, *the higher the K_{sp}, the greater the solubility*. Table 19.3 shows this point for several compounds. Note that for compounds that form three ions, the relationship holds whether the cation/anion ratio is 1/2 or 2/1, because the mathematical expression containing S is the same ($4S^3$) in the calculation (see Sample Problem 19.7).

Table 19.3 Relationship Between K_{sp} and Solubility at 25°C				
No. of Ions	**Formula**	**Cation/Anion**	**K_{sp}**	**Solubility (M)**
2	$MgCO_3$	1/1	3.5×10^{-8}	1.9×10^{-4}
2	$PbSO_4$	1/1	1.6×10^{-8}	1.3×10^{-4}
2	$BaCrO_4$	1/1	2.1×10^{-10}	1.4×10^{-5}
3	$Ca(OH)_2$	1/2	6.5×10^{-6}	1.2×10^{-2}
3	BaF_2	1/2	1.5×10^{-6}	7.2×10^{-3}
3	CaF_2	1/2	3.2×10^{-11}	2.0×10^{-4}
3	Ag_2CrO_4	2/1	2.6×10^{-12}	8.7×10^{-5}

Transparency/ARIS Presentation Center, Tables Library

The Effect of a Common Ion on Solubility

The presence of a common ion decreases the solubility of a slightly soluble ionic compound. As in acid-base systems, Le Châtelier's principle helps explain this effect. Let's examine the equilibrium condition for a saturated solution of lead(II) chromate:

$$PbCrO_4(s) \rightleftharpoons Pb^{2+}(aq) + CrO_4^{2-}(aq) \qquad K_{sp} = [Pb^{2+}][CrO_4^{2-}] = 2.3 \times 10^{-13}$$

At a given temperature, K_{sp} depends on the product of the ion concentrations. If the concentration of either ion goes up, the other must go down to maintain K_{sp}. Suppose we add Na_2CrO_4, a soluble salt, to the saturated $PbCrO_4$ solution. The concentration of the common ion, CrO_4^{2-}, increases, and some of it combines with Pb^{2+} ion to form more solid $PbCrO_4$ (Figure 19.12). The overall effect is a shift in the position of equilibrium to the left:

$$PbCrO_4(s) \overset{\longleftarrow}{\rightleftharpoons} Pb^{2+}(aq) + CrO_4^{2-}(aq; \text{added})$$

After the addition, $[CrO_4^{2-}]$ is higher, but $[Pb^{2+}]$ is lower. In this case, $[Pb^{2+}]$ represents the amount of $PbCrO_4$ dissolved; thus, in effect, the solubility of $PbCrO_4$ has decreased. The same result is obtained if we dissolve $PbCrO_4$ in a Na_2CrO_4 solution. We also obtain this result by adding a soluble lead(II) salt, such as $Pb(NO_3)_2$. The added Pb^{2+} ion combines with some $CrO_4^{2-}(aq)$, thereby lowering the amount of dissolved $PbCrO_4$.

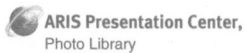
ARIS Presentation Center,
Photo Library

Figure 19.12 The effect of a common ion on solubility. When a common ion is added to a saturated solution of an ionic compound, the solubility is lowered and more of the compound precipitates. **A,** Lead(II) chromate, a slightly soluble salt, forms a saturated aqueous solution. **B,** When Na_2CrO_4 solution is added, the amount of $PbCrO_4(s)$ increases. Thus, $PbCrO_4$ is less soluble in the presence of the common ion CrO_4^{2-}.

A $PbCrO_4(s) \rightleftharpoons Pb^{2+}(aq) + CrO_4^{2-}(aq)$

B $PbCrO_4(s) \overset{\longleftarrow}{\rightleftharpoons} Pb^{2+}(aq) + CrO_4^{2-}(aq; \text{added})$

ARIS Presentation Center, Worked Examples Library

SAMPLE PROBLEM 19.8   Calculating the Effect of a Common Ion on Solubility

PROBLEM In Sample Problem 19.7, we calculated the solubility of $Ca(OH)_2$ in water. What is its solubility in 0.10 M $Ca(NO_3)_2$? K_{sp} of $Ca(OH)_2$ is 6.5×10^{-6}.

PLAN Addition of Ca^{2+}, the common ion, should lower the solubility. We write the equation and ion-product expression and set up a reaction table, with $[Ca^{2+}]_{init}$ coming from $Ca(NO_3)_2$ and S equal to $[Ca^{2+}]_{from\ Ca(OH)_2}$. To simplify the math, we assume that, because K_{sp} is low, S is so small relative to $[Ca^{2+}]_{init}$ that it can be neglected. Then we solve for S and check the assumption.

SOLUTION Writing the equation and ion-product expression:

$$Ca(OH)_2(s) \rightleftharpoons Ca^{2+}(aq) + 2OH^-(aq) \qquad K_{sp} = [Ca^{2+}][OH^-]^2 = 6.5 \times 10^{-6}$$

Setting up the reaction table, with $S = [Ca^{2+}]_{from\ Ca(OH)_2}$:

Concentration (M)	$Ca(OH)_2(s)$	$\rightleftharpoons$	$Ca^{2+}(aq)$	+	$2OH^-(aq)$
Initial	—		0.10		0
Change	—		$+S$		$+2S$
Equilibrium	—		$0.10 + S$		$2S$

Making the assumption: K_{sp} is small, so $S \ll 0.10$ M; thus, 0.10 M + $S \approx 0.10$ M.

Substituting into the ion-product expression and solving for S:

$$K_{sp} = [Ca^{2+}][OH^-]^2 = 6.5 \times 10^{-6} \approx (0.10)(2S)^2$$

Therefore, $4S^2 \approx \dfrac{6.5 \times 10^{-6}}{0.10}$ so $S \approx \sqrt{\dfrac{6.5 \times 10^{-5}}{4}} = \boxed{4.0 \times 10^{-3} \, M}$

Checking the assumption: $\dfrac{4.0 \times 10^{-3} \, M}{0.10 \, M} \times 100 = 4.0\% < 5\%$

CHECK In Sample Problem 19.7, the solubility of $Ca(OH)_2$ was 0.012 M; here it is 0.0040 M, so the solubility *decreased* in the presence of added Ca^{2+}, the common ion, as expected.

FOLLOW-UP PROBLEM 19.8 To improve the quality of x-ray images used in the diagnosis of intestinal disorders, the patient drinks an aqueous suspension of $BaSO_4$ before the x-ray procedure (see photo). The Ba^{2+} in the suspension is opaque to x-rays, but it is also toxic; thus, the Ba^{2+} concentration is lowered by the addition of dilute Na_2SO_4. What is the solubility of $BaSO_4$ ($K_{sp} = 1.1 \times 10^{-10}$) in (a) pure water and in (b) 0.10 M Na_2SO_4?

$BaSO_4$ imaging of a human large intestine.

 ARIS Presentation Center,
Photo Library

Biological application

The Effect of pH on Solubility

The hydronium ion concentration can have a profound effect on the solubility of an ionic compound. *If the compound contains the anion of a weak acid, addition of H_3O^+ (from a strong acid) increases its solubility.* Once again, Le Châtelier's principle explains why. In a saturated solution of calcium carbonate, for example,

$$CaCO_3(s) \rightleftharpoons Ca^{2+}(aq) + CO_3^{2-}(aq)$$

Adding some strong acid introduces a large amount of H_3O^+, which immediately reacts with CO_3^{2-} to form the weak acid HCO_3^-:

$$CO_3^{2-}(aq) + H_3O^+(aq) \longrightarrow HCO_3^-(aq) + H_2O(l)$$

If enough H_3O^+ is added, further reaction occurs to form carbonic acid, which decomposes immediately to H_2O and CO_2, and the gas escapes the container:

$$HCO_3^-(aq) + H_3O^+(aq) \longrightarrow H_2CO_3(aq) + H_2O(l) \longrightarrow CO_2(g) + 2H_2O(l)$$

As this sequence of changes shows, the net effect of added H_3O^+ is a shift in the equilibrium position to the right, and more $CaCO_3$ dissolves:

$$CaCO_3(s) \xrightleftharpoons{\hspace{1cm}} Ca^{2+} + CO_3^{2-} \xrightarrow{H_3O^+} HCO_3^- \xrightarrow{H_3O^+} H_2CO_3 \longrightarrow$$
$$CO_2(g) + H_2O + Ca^{2+}$$

In fact, this example illustrates a qualitative field test for carbonate minerals because the CO_2 bubbles vigorously (Figure 19.13).

In contrast, adding H_3O^+ to a saturated solution of a compound with a strong-acid anion, such as silver chloride, has no effect on the equilibrium position:

$$AgCl(s) \rightleftharpoons Ag^+(aq) + Cl^-(aq)$$

Because Cl^- ion is the conjugate base of a strong acid (HCl), it can coexist in solution with high $[H_3O^+]$. The Cl^- does not leave the system, so the equilibrium position is not affected.

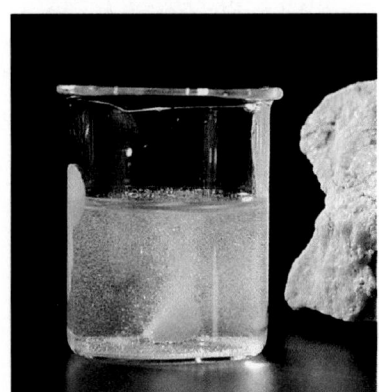

Figure 19.13 Test for the presence of a carbonate. When a mineral that contains carbonate ion is treated with strong acid, the added H_3O^+ shifts the equilibrium position of the carbonate solubility. More carbonate dissolves, and the carbonic acid that is formed breaks down to water and gaseous CO_2.

 ARIS Presentation Center,
Photo Library

ARIS Presentation Center,
Worked Examples Library

SAMPLE PROBLEM 19.9 Predicting the Effect on Solubility of Adding Strong Acid

PROBLEM Write balanced equations to explain whether addition of H_3O^+ from a strong acid affects the solubility of these ionic compounds:
(a) Lead(II) bromide   (b) Copper(II) hydroxide   (c) Iron(II) sulfide
PLAN We write the balanced dissolution equation and note the anion: Weak-acid anions react with H_3O^+ and shift the equilibrium position toward more dissolution. Strong-acid anions do not react, so added H_3O^+ has no effect.
SOLUTION (a) $PbBr_2(s) \rightleftharpoons Pb^{2+}(aq) + 2Br^-(aq)$
 No effect. Br^- is the anion of HBr, a strong acid, so it does not react with H_3O^+.
(b) $Cu(OH)_2(s) \rightleftharpoons Cu^{2}(aq) + 2OH^-(aq)$ Increases solubility. OH^- is the anion of H_2O, a very weak acid, so it reacts with the added H_3O:

$$\boxed{OH^-(aq) + H_3O^+(aq) \longrightarrow 2H_2O(l)}$$

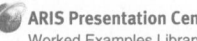 **Demo** Shakhashiri, Bassam Z. "Fizzing and Foaming: Reactions of Acids with Carbonates," *Chemical Demonstrations, A Handbook for Teachers of Chemistry,* Vol. 3 (The University of Wisconsin Press, Madison, 1989), 96–99.

(c) $FeS(s) + H_2O(l) \rightleftharpoons Fe^{2+}(aq) + HS^-(aq) + OH^-(aq)$ Increases solubility. We noted earlier that the S^{2-} ion reacts immediately with water to form HS^- and OH^-. The added H_3O^+ reacts with both of these weak-acid anions:

$$HS^-(aq) + H_3O^+(aq) \longrightarrow H_2S(aq) + H_2O(l)$$
$$OH^-(aq) + H_3O^+(aq) \longrightarrow 2H_2O(l)$$

FOLLOW-UP PROBLEM 19.9 Write balanced equations to show how addition of $HNO_3(aq)$ affects the solubility of these ionic compounds:
(a) Calcium fluoride   **(b)** Zinc sulfide   **(c)** Silver iodide

The principles of ionic equilibria are illustrated in the next Chemical Connections essay, on the carbonate equilibria that give rise to limestone caves.

Predicting the Formation of a Precipitate: Q_{sp} vs. K_{sp}

In Chapter 17, we compared the values of Q and K to see if a reaction had reached equilibrium and, if not, in which net direction it would move until it did. We use the same approach here to see if a precipitate will form and, if not, what changes in ion concentrations will cause it to do so. As you know, $Q_{sp} = K_{sp}$ when the solution is saturated. If Q_{sp} is greater than K_{sp}, the solution is momentarily supersaturated, and some solid precipitates until the remaining solution becomes saturated ($Q_{sp} = K_{sp}$). If Q_{sp} is less than K_{sp}, the solution is unsaturated, and no precipitate forms at that temperature (more solid can dissolve). To summarize,

- $Q_{sp} = K_{sp}$: solution is saturated and no change occurs.
- $Q_{sp} > K_{sp}$: precipitate forms until solution is saturated.
- $Q_{sp} < K_{sp}$: solution is unsaturated and no precipitate forms.

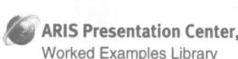
SAMPLE PROBLEM 19.10 Predicting Whether a Precipitate Will Form

PROBLEM A common laboratory method for preparing a precipitate is to mix solutions containing the component ions. Does a precipitate form when 0.100 L of 0.30 *M* $Ca(NO_3)_2$ is mixed with 0.200 L of 0.060 *M* NaF?

PLAN First, we decide which slightly soluble salt could form and look up its K_{sp} value in Appendix C. To see whether mixing these solutions forms the precipitate, we find the initial ion concentrations by calculating the amount (mol) of each ion from its concentration and volume, and then dividing by the *total* volume because each solution dilutes the other. Finally, we write the ion-product expression, calculate Q_{sp}, and compare Q_{sp} with K_{sp}.

SOLUTION The ions present are Ca^{2+}, Na^+, F^-, and NO_3^-. All sodium and all nitrate salts are soluble (Table 4.1, p. 148), so the only possibility is CaF_2 ($K_{sp} = 3.2 \times 10^{-11}$). Calculating the ion concentrations:

$$\text{Moles of } Ca^{2+} = 0.30\ M\ Ca^{2+} \times 0.100\ L = 0.030\ \text{mol } Ca^{2+}$$

$$[Ca^{2+}]_{init} = \frac{0.030\ \text{mol } Ca^{2+}}{0.100\ L + 0.200\ L} = 0.10\ M\ Ca^{2+}$$

$$\text{Moles of } F^- = 0.060\ M\ F^- \times 0.200\ L = 0.012\ \text{mol } F^-$$

$$[F^-]_{init} = \frac{0.012\ \text{mol } F^-}{0.100\ L + 0.200\ L} = 0.040\ M\ F^-$$

Substituting into the ion-product expression and comparing Q_{sp} with K_{sp}:

$$Q_{sp} = [Ca^{2+}]_{init}[F^-]^2_{init} = (0.10)(0.040)^2 = 1.6 \times 10^{-4}$$

Because $Q_{sp} > K_{sp}$, CaF_2 will precipitate until $Q_{sp} = 3.2 \times 10^{-11}$.

CHECK Make sure you round off and quickly check the math. For example, $Q_{sp} = (1 \times 10^{-1})(4 \times 10^{-2})^2 = 1.6 \times 10^{-4}$. With K_{sp} so low, CaF_2 must have a low solubility, and given the sizable concentrations being mixed, we would expect CaF_2 to precipitate.

FOLLOW-UP PROBLEM 19.10 As a result of mineral erosion and biological activity, phosphate ion is common in natural waters, where it often precipitates as insoluble salts such as $Ca_3(PO_4)_2$. If $[Ca^{2+}]_{init} = [PO_4^{3-}]_{init} = 1.0 \times 10^{-9}\ M$ in a given river, will $Ca_3(PO_4)_2$ precipitate? K_{sp} of $Ca_3(PO_4)_2$ is 1.2×10^{-29}.

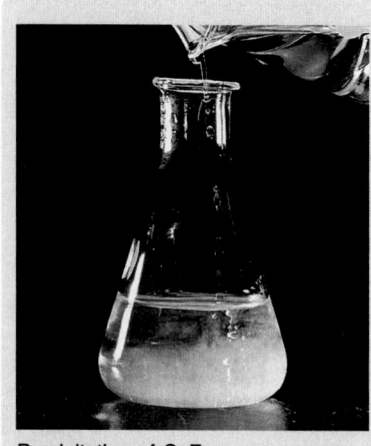
Precipitation of CaF_2.

Chemical Connections to Geology

Creation of a Limestone Cave

Limestone caves and the intricate structures within them provide striking evidence of the workings of aqueous ionic equilibria (Figure B19.1). The spires and vaults of these natural cathedrals are the products of reactions between carbonate rocks and the water that has run through them for millennia. Limestone is predominantly calcium carbonate ($CaCO_3$), a slightly soluble ionic compound with a K_{sp} of 3.3×10^{-9}. This rocky material began accumulating in the Earth over 400 million years ago, and a relatively young cave, such as Howe Caverns in eastern New York State, began forming about 800,000 years ago.

Two key facts help us understand how limestone caves form:

1. Gaseous CO_2 is in equilibrium with aqueous CO_2 in natural waters:

$$CO_2(g) \underset{}{\overset{H_2O(l)}{\rightleftharpoons}} CO_2(aq) \qquad \textbf{(equation 1)}$$

The concentration of CO_2 in the water is proportional to the partial pressure of $CO_2(g)$ in contact with the water (Henry's law; Section 13.4):

$$[CO_2(aq)] \propto P_{CO_2}$$

Because of the continual release of CO_2 from within the Earth (outgassing), the P_{CO_2} in soil-trapped air is *higher* than the P_{CO_2} in the atmosphere.

2. As discussed in the text, the presence of $H_3O^+(aq)$ increases the solubility of ionic compounds that contain the anion of a weak acid. The reaction of CO_2 with water produces H_3O^+:

$$CO_2(aq) + 2H_2O(l) \rightleftharpoons H_3O^+(aq) + HCO_3^-(aq)$$

Thus, the presence of $CO_2(aq)$ leads to the formation of H_3O^+, which increases the solubility of $CaCO_3$:

$$CaCO_3(s) + CO_2(aq) + H_2O(l) \rightleftharpoons$$
$$Ca^{2+}(aq) + 2HCO_3^-(aq) \qquad \textbf{(equation 2)}$$

Here is an overview of the principal cave-forming process in most limestone caves: As surface water trickles through cracks in the ground, it meets soil-trapped air with its high P_{CO_2}. As a result, $[CO_2(aq)]$ increases (equation 1 shifts to the right), and the solution becomes more acidic. When this CO_2-rich water contacts limestone, more $CaCO_3$ dissolves (equation 2 shifts to the right). As a result, more rock is carved out, more water flows in, and so on. Centuries pass and a cave slowly begins to form.

Eating its way through underground tunnels, some of the aqueous solution, largely dilute $Ca(HCO_3)_2$, passes through the ceiling of the growing cave. As it drips, it meets air, which has a lower P_{CO_2} than the soil, so some $CO_2(aq)$ comes out of solution (equation 1 shifts to the left). This causes some $CaCO_3$ to precipitate on the ceiling and on the floor below, where the drops land (equation 2 shifts to the left). Decades pass, and the ceiling bears an "icicle" of $CaCO_3$, called a *stalactite,* while a spike of $CaCO_3$, called a *stalagmite,* grows upward from the cave floor. With time, they meet to form a column of precipitated limestone.

The same chemical process can lead to different shapes. Standing pools of $Ca(HCO_3)_2$ solution form limestone "lily pads" or "corals." Cascades of solution form delicate limestone "draperies" on a cave wall, with fabulous colors arising from trace metal ions, such as iron (reddish brown) or copper (bluish green).

Recently, vast, spectacularly shaped deposits of gypsum ($CaSO_4 \cdot 2H_2O$; Figure B19.2) found in some limestone caves in Mexico point strongly to another process of cave formation. Certain bacteria found in abundance in these caves can survive at very low pH and consume hydrogen sulfide (H_2S), which leaks into the cave from oil deposits deep below it. The evidence shows that these *acidophilic* microbes use O_2 to oxidize the H_2S to sulfuric acid (H_2SO_4), a strong acid:

$$H_2S(g) + 2O_2(g) + H_2O(l) \xrightarrow{\text{bacteria}} H_3O^+(aq) + HSO_4^-(aq)$$

The sulfuric acid contributes greatly to the formation of the cave.

Literature Weston, Ralph E., Jr. "Climate Change and Its Effect on Coral Reefs," *J. Chem. Educ.* **2000,** *77,* 1574.

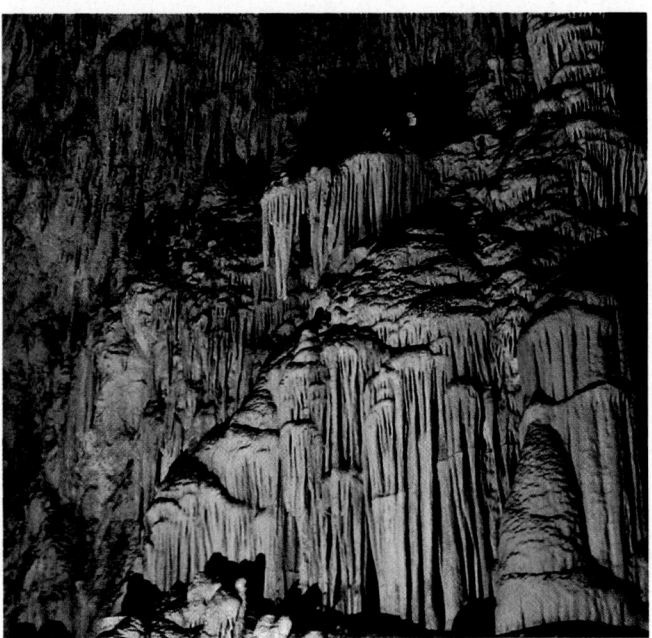

Figure B19.1 Limestone cave. The formations within this cave at Nerja, Málaga, Spain, result from subtle shifts in carbonate equilibria acting over millions of years.

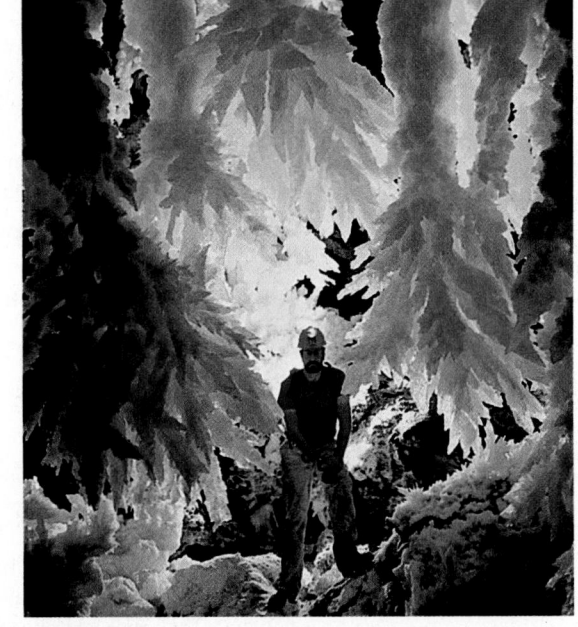

Figure B19.2 Gypsum deposits. Formations of gypsum in the Chandelier Ballroom of Lechuguilla Cave in Mexico.

SAMPLE PROBLEM 19.11 Using Molecular Scenes to Predict Whether a Precipitate Will Form

PROBLEM The following molecular scenes represent four solutions of silver (gray) and carbonate (black and red) ions that are above solid silver carbonate. (For clarity, the solid, other ions, and water are not shown.)

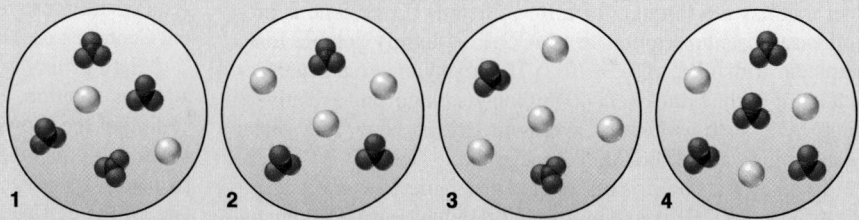

(a) Which scene best represents the solution at equilibrium with the solid?
(b) In which, if any, other scene(s) will additional solid be formed?
(c) Explain how, if at all, addition of a small volume of concentrated strong acid affects the $[Ag^+]$ in scene 4 and the mass of solid present.

PLAN (a) The solution of silver and carbonate ions in equilibrium with the solid (Ag_2CO_3) should have the same relative numbers of cation and anion as appear in the formula. We examine the scenes to see which maintains a ratio of 2 Ag^+ to 1 CO_3^{2-}. **(b)** A solid forms whenever the ion product exceeds the K_{sp}. We first calculate the value of Q_{sp} in each scene according to the ion-product expression and then see which, if any, exceeds the value for the solution from part (a). **(c)** The CO_3^{2-} ion reacts with added H_3O^+, so adding strong acid will shift the equilibrium to the right. We write the equations and determine how a shift to the right affects $[Ag^+]$ and the mass of solid Ag_2CO_3.

SOLUTION (a) Scene 3 is the only one with 2 Ag^+ to 1 CO_3^{2-} as in the solid's formula.
(b) Calculating the ion products:

$$Ag_2CO_3(s) \rightleftharpoons 2Ag^+(aq) + CO_3^{2-}(aq) \qquad Q_{sp} = [Ag^+]^2[CO_3^{2-}]$$

Scene 1: $Q_{sp} = (2)^2(4) = 16$ Scene 2: $Q_{sp} = (3)^2(3) = 27$
Scene 3: $Q_{sp} = (4)^2(2) = 32$ Scene 4: $Q_{sp} = (3)^2(4) = 36$

Therefore, from scene 3, $K_{sp} = 32$; the Q_{sp} value for scene 4 is the only other one that equals or exceeds 32, so a precipitate of Ag_2CO_3 will form.
(c) Writing the equations:
1. $Ag_2CO_3(s) \rightleftharpoons 2Ag^+(aq) + CO_3^{2-}(aq)$
2. $CO_3^{2-}(aq) + 2H_3O^+(aq) \longrightarrow H_2CO_3(aq) + 2H_2O(l) \longrightarrow 3H_2O(l) + CO_2(g)$

The CO_2 leaves as a gas, so adding H_3O^+ shifts the equilibrium position of reaction 2 to the right. This change lowers the $[CO_3^{2-}]$ in reaction 1, thereby causing more CO_3^{2-} to form. As a result, more solid dissolves, which means that the $[Ag^+]$ increases and the mass of Ag_2CO_3 decreases.

CHECK (a) In scene 1, the formula has two CO_3^{2-} per formula unit, not two Ag^+. **(b)** Even though scene 4 has fewer Ag^+ ions than scene 3, its Q_{sp} value is higher and exceeds the K_{sp}.

FOLLOW-UP PROBLEM 19.11 The following scenes represent four solutions of nickel(II) (black) and hydroxide (red and blue) ions that are above solid nickel(II) hydroxide. (For clarity, the solid, other ions, and water are not shown.)

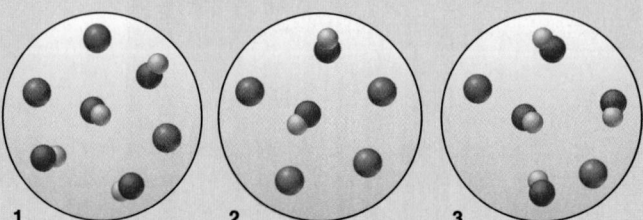

(a) Which scene best depicts the solution at equilibrium with the solid?
(b) In which, if any, other scene(s) will additional solid be formed?
(c) Will addition of a small amount of concentrated strong acid or strong base affect the mass of solid present in any scene? Explain.

Selective Precipitation of Ions

We can often separate one ion in a solution from another by exploiting differences in the solubility of their compounds with a given precipitating ion. In the process of **selective precipitation**, we add a solution of precipitating ion until the Q_{sp} value of the *more soluble* compound is almost equal to its K_{sp} value. This method ensures that the K_{sp} value of the *less soluble* compound is exceeded as much as possible. As a result, the maximum amount of the less soluble compound precipitates, but none of the more soluble compound does.

ARIS Presentation Center, Worked
Examples Library

SAMPLE PROBLEM 19.12 Separating Ions by Selective Precipitation

PROBLEM A solution consists of 0.20 M $MgCl_2$ and 0.10 M $CuCl_2$. Calculate the $[OH^-]$ that would separate the metal ions as their hydroxides. K_{sp} of $Mg(OH)_2$ is 6.3×10^{-10}; K_{sp} of $Cu(OH)_2$ is 2.2×10^{-20}.

PLAN The two hydroxides have the same formula type (1/2) (see p. 855), so we can compare their K_{sp} values to see which is more soluble: $Mg(OH)_2$ is about 10^{10} times more soluble than $Cu(OH)_2$. Thus, $Cu(OH)_2$ precipitates first. We want to precipitate as much $Cu(OH)_2$ as possible without precipitating any $Mg(OH)_2$. We solve for the $[OH^-]$ that will just give a saturated solution of $Mg(OH)_2$ because this $[OH^-]$ will precipitate the greatest amount of Cu^{2+} ion. As confirmation, we calculate the $[Cu^{2+}]$ remaining to see if the separation was accomplished.

SOLUTION Writing the equations and ion-product expressions:

$$Mg(OH)_2(s) \rightleftharpoons Mg^{2+}(aq) + 2OH^-(aq) \qquad K_{sp} = [Mg^{2+}][OH^-]^2$$
$$Cu(OH)_2(s) \rightleftharpoons Cu^{2+}(aq) + 2OH^-(aq) \qquad K_{sp} = [Cu^{2+}][OH^-]^2$$

Calculating the $[OH^-]$ that gives a saturated $Mg(OH)_2$ solution:

$$[OH^-] = \sqrt{\frac{K_{sp}}{[Mg^{2+}]}} = \sqrt{\frac{6.3 \times 10^{-10}}{0.20}} = 5.6 \times 10^{-5}\ M$$

This is the maximum $[OH^-]$ that will *not* precipitate Mg^{2+} ion.
Calculating the $[Cu^{2+}]$ remaining in the solution with this $[OH^-]$:

$$[Cu^{2+}] = \frac{K_{sp}}{[OH^-]^2} = \frac{2.2 \times 10^{-20}}{(5.6 \times 10^{-5})^2} = 7.0 \times 10^{-12}\ M$$

Since the initial $[Cu^{2+}]$ was 0.10 M, virtually all the Cu^{2+} ion is precipitated.

CHECK Rounding, we find that the $[OH^-]$ seems right: $\sim \sqrt{(6 \times 10^{-10})/0.2} = 5 \times 10^{-5}$. The $[Cu^{2+}]$ remaining also seems correct: $(200 \times 10^{-22})/(5 \times 10^{-5})^2 = 8 \times 10^{-12}$.

COMMENT Often, more than one approach will accomplish the same analytical step. Another possibility in this case is to add excess ammonia to make the solution basic enough to precipitate both hydroxides:

$$Mg^{2+}(aq) + 2NH_3(aq) + 2H_2O(l) \rightleftharpoons Mg(OH)_2(s) + 2NH_4^+(aq)$$
$$Cu^{2+}(aq) + 2NH_3(aq) + 2H_2O(l) \rightleftharpoons Cu(OH)_2(s) + 2NH_4^+(aq)$$

Then, the $Cu(OH)_2$ dissolves in excess NH_3 by forming a soluble complex ion:

$$Cu(OH)_2(s) + 4NH_3(aq) \rightleftharpoons Cu(NH_3)_4^{2+}(aq) + 2OH^-(aq)$$

FOLLOW-UP PROBLEM 19.12 A solution containing two alkaline earth metal ions is made from 0.050 M $BaCl_2$ and 0.025 M $CaCl_2$. What concentration of SO_4^{2-} must be present to leave 99.99% of only one of the cations in solution? K_{sp} of $BaSO_4$ is 1.1×10^{-10}, and K_{sp} of $CaSO_4$ is 2.4×10^{-5}.

Sometimes two or more types of ionic equilibria are controlled simultaneously to precipitate ions selectively. This approach is used commonly to separate ions as their sulfides, so the HS^- ion is the precipitating ion. In the manner used in Sample Problem 19.12, we control the $[HS^-]$ to exceed the K_{sp} value of one metal sulfide but not another. We exert this control on the $[HS^-]$ by controlling H_2S dissociation because H_2S is the source of HS^-; we control H_2S dissociation by adjusting the $[H_3O^+]$:

$$H_2S(aq) + H_2O(l) \rightleftharpoons H_3O^+(aq) + HS^-(aq)$$

These interactions are controlled in the following way: If we add strong acid to the solution, $[H_3O^+]$ is high, so H_2S dissociation shifts to the left, which decreases $[HS^-]$. With a low $[HS^-]$, the *less* soluble sulfide precipitates. Conversely, if we add strong base, $[H_3O^+]$ is low, so H_2S dissociation shifts to the right, which increases $[HS^-]$, and then the *more* soluble sulfide precipitates. In essence, what we do is shift one equilibrium system (H_2S dissociation) by adjusting a second (H_2O ionization) to control a third (metal sulfide solubility).

As the upcoming Chemical Connections essay demonstrates, the principles of ionic equilibria often help us understand the chemical basis of complex environmental problems and may provide ways to solve them.

Section Summary

As a first approximation, the dissolved portion of a slightly soluble salt dissociates completely into ions. • In a saturated solution, the ions are in equilibrium with the solid, and the product of the ion concentrations, each raised to the power of its subscript in the compound's formula, has a constant value ($Q_{sp} = K_{sp}$). • The value of K_{sp} can be obtained from the solubility, and vice versa. • Adding a common ion lowers an ionic compound's solubility. • Adding H_3O^+ (lowering the pH) increases a compound's solubility if the anion of the compound is that of a weak acid. • If $Q_{sp} > K_{sp}$ for an ionic compound, a precipitate forms when two solutions, each containing one of the compound's ions, are mixed. • Limestone caves result from shifts in the $CaCO_3/CO_2$ equilibrium system. • Lakes bounded by limestone-rich soils form buffer systems that prevent harmful acidification. • Ions are precipitated selectively by adding a precipitating ion until the K_{sp} of one compound is exceeded as much as possible without exceeding the K_{sp} of the other. An extension of this approach simultaneously controls an equilibrium system that contains the precipitating ion itself.

19.4 EQUILIBRIA INVOLVING COMPLEX IONS

The final type of aqueous ionic equilibrium we consider involves a different kind of ion than we've examined up to now. A simple ion, such as Na^+ or SO_4^{2-}, consists of one or a few bound atoms, with an excess or deficit of electrons. A **complex ion** consists of a central metal ion covalently bonded to two or more anions or molecules, called **ligands.** Hydroxide, chloride, and cyanide ions are some ionic ligands; water, carbon monoxide, and ammonia are some molecular ligands. In the complex ion $Cr(NH_3)_6^{3+}$, for example, Cr^{3+} is the central metal ion and six NH_3 molecules are the ligands, giving an overall $3+$ charge (Figure 19.14).

As we noted in Section 18.9, *all complex ions are Lewis adducts.* The metal ion acts as a Lewis acid (accepts an electron pair) and the ligand acts as a Lewis base (donates an electron pair). The acidic hydrated metal ions that we discussed in Section 18.6 are complex ions with water molecules as ligands. In Chapter 23, we discuss the transition metals and the structures and properties of the numerous complex ions they form. Our focus here is on equilibria of hydrated ions with ligands other than water.

Formation of Complex Ions

Whenever a metal ion enters water, a complex ion forms, with water as the ligand. In many cases, when we treat this hydrated cation with a solution of another ligand, the bound water molecules exchange for the other ligand. For example, a hydrated M^{2+} ion, $M(H_2O)_4^{2+}$, forms $M(NH_3)_4^{2+}$ in aqueous NH_3:

$$M(H_2O)_4^{2+}(aq) + 4NH_3(aq) \rightleftharpoons M(NH_3)_4^{2+}(aq) + 4H_2O(l)$$

At equilibrium, this system is expressed by a ratio of concentration terms whose form follows that of any other equilibrium expression:

$$K_c = \frac{[M(NH_3)_4^{2+}][H_2O]^4}{[M(H_2O)_4^{2+}][NH_3]^4}$$

NH_3

Cr^{3+}

Figure 19.14 $Cr(NH_3)_6^{3+}$, a typical **complex ion.** A complex ion consists of a central metal ion, such as Cr^{3+}, covalently bonded to a specific number of ligands, such as NH_3.

ARIS Presentation Center, Art Library

Chemical Connections to Environmental Science

The Acid-Rain Problem

The adverse effect of an industrial society on the environment is clear in the problem of *acid rain,* the deposition of acids, formed as a result of human activity, that occurs in wet form as rain, snow, and fog, or in dry form on solid particles. Acidic precipitation has been recorded in all parts of the United States, in Canada, Mexico, the Amazon basin, throughout Europe and Russia and many parts of Asia, and even at the North and South Poles. We've addressed several aspects of this problem earlier; now, let's examine some effects of acidic precipitation on biological and mineral systems and see how to prevent them. The strong acids H_2SO_4 and HNO_3 cause the greatest concern (Figure B19.3):

- *Formation of sulfuric acid.* Sulfur dioxide (SO_2) formed primarily by the burning of high-sulfur coal, forms sulfurous acid (H_2SO_3) in contact with water. Oxidants, such as hydrogen peroxide (H_2O_2) and ozone (O_3), that are present as pollutants in the atmosphere also dissolve in the water in clouds and oxidize the sulfurous acid to sulfuric acid:

$$H_2O_2(aq) + H_2SO_3(aq) \longrightarrow H_2SO_4(aq) + H_2O(l)$$

SO_2 is also oxidized directly by atmospheric hydroxyl radicals to sulfur trioxide (SO_3), which forms H_2SO_4 in contact with water.

- *Formation of nitric acid.* Nitrogen oxides (collectively known as NO_x) form through the reaction of N_2 and O_2. NO is formed during high-temperature combustion, primarily in car engines and electric power plants, and it forms NO_2 and HNO_3 in air in the process that creates smog (p. 600). At night, NO_x is converted to N_2O_5, which hydrolyzes to HNO_3 in the presence of water. Ammonium salts, deposited as NH_4HSO_4 or NH_4NO_3, produce HNO_3 in soil through biochemical oxidation.

How does the pH of acidic precipitation compare with that of natural waters? Normal rainwater is weakly acidic because it contains dissolved CO_2 from the air:

$$CO_2(g) + 2H_2O(l) \rightleftharpoons H_3O^+(aq) + HCO_3^-(aq)$$

Based on the volume percent of CO_2 in air, the solubility of CO_2 in water, and the K_{a1} of H_2CO_3, the pH of normal rainwater is about 5.6 (see Problem 19.148 at the end of the chapter). In stark contrast, the average pH of rainfall in many parts of the United States was 4.2 in 1984, representing 25 times as much H_3O^+. Worldwide, rain in Sweden had a pH of 2.7, about the same as vinegar. Rain in Wheeling, West Virginia, had a pH of 1.8, less than that of lemon juice. Acidic fog in California sometimes has a pH of 1.6 as evaporation from the particles concentrates the acid.

Such high $[H_3O^+]$ is very destructive to living things. Some fish and shellfish die at pH values between 4.5 and 5.0. The young of most species are generally most sensitive. At pH 5, most fish eggs cannot hatch. With tens of thousands of rivers and lakes around the world becoming acidified, the loss of fish became a major concern long ago. In addition, acres of forest have been harmed by the acid, which removes nutrients and releases toxic substances from the soil (Figure B19.4 on the next page).

Many principles of aqueous equilibria bear directly on the effects of acid rain. The aluminosilicates that make up most soils are extremely insoluble in water. In these materials, the Al^{3+} ion is bonded to OH^- and O^{2-} ions (see the Gallery in Chapter 14, p. 592). Continual contact with water at pH < 5 causes these and other Al-bearing substances to react through a complex series of equilibria to release some of the bound Al^{3+}, which is toxic to fish and plants. Along with dissolved Al^{3+} ions, acid rain dissolves and carries away Ca^{2+} and Mg^{2+} ions from the soil, which serve as nutrients for plants and animals.

(continued)

Demo Goss, Lisa M. "A Demonstration of Acid Rain and Lake Acidification: Wet Deposition of Sulfur Dioxide," *J. Chem. Educ.* **2003,** *80,* 39.

Figure B19.3 Formation of acidic precipitation. A complex interplay of human activities, atmospheric chemistry, and environmental distribution leads to acidic precipitation and its harmful effects. Car exhaust and power plant waste gases contain lower oxides of nitrogen and sulfur. These are oxidized in the atmosphere to higher oxides (NO_2, SO_3), which react with moisture to form acidic rain, snow, and fog. In contact with acidic precipitation, many lakes become acidified, whereas limestone-bounded lakes form a carbonate buffer that prevents acidification.

Transparency/ARIS Presentation Center, art Library   Environmental application

Demo Shakhashiri, Bassam Z. "Acid-Neutralization Capacity of Lake Beds," *Chemical Demonstrations, A Handbook for Teachers of Chemistry,* Vol. 3 (The University of Wisconsin Press, Madison, 1989), 103–108.

Figure B19.4 **A forest damaged by acid rain.**

Marble and limestone (both primarily $CaCO_3$) in buildings and monuments react with sulfuric acid to form gypsum ($CaSO_4 \cdot 2H_2O$), which flakes off (Figure B19.5). Ironically, the same chemical process that destroys these structures is responsible for saving those lakes that lie on and are bounded by limestone-rich soil. Limestone dissolves sufficiently in lake water to form an HCO_3^-/CO_3^{2-} buffer solution capable of absorbing the incoming H_3O^+ and maintaining a mildly basic pH:

$$CO_3^{2-}(aq) + H_3O^+(aq) \rightleftharpoons HCO_3^-(aq) + H_2O(l)$$

Acidified lakes and rivers that are in contact with granite and other weathering-resistant bedrock can be remediated by *liming* (treating with limestone). This approach is expensive and, at best, only a stopgap because the lakes are acidic again within several years, so it is more cost effective to treat the problem at its source.

As we pointed out earlier (see Chemical Connections, Chapter 6, p. 256), the principal means of controlling sulfur dioxide is by "scrubbing" the emissions from power plants with limestone.

Figure B19.5 **The effect of acid rain on marble statuary.** Calcium carbonate, which is the major component of marble, decomposes when exposed to acid rain. These photos, taken many years apart, of a gargoyle at Notre Dame Cathedral in Paris demonstrate this decomposition.

Both dry and wet scrubbers are used. Another method reduces some of the SO_2 with methane, coal, or H_2S, and the mixture is converted catalytically to sulfur, which is sold as a byproduct:

$$16H_2S(g) + 8SO_2(g) \longrightarrow 3S_8(s) + 16H_2O(l)$$

Burning low-sulfur coal to reduce SO_2 formation is sometimes an option, but such coal deposits are rare and expensive to mine. Alternatively, coal can be converted into gaseous and liquid low-sulfur fuels (see Chemical Connections, p. 256). The sulfur is removed (as H_2S) in an acid-gas scrubber after gasification.

Through the use of a catalytic converter in an auto exhaust system, NO_x species are reduced to N_2. In power plants, the amount of NO_x is decreased by adjusting combustion conditions, but it also can be removed from the hot stack gases by treatment with ammonia in the presence of a heterogeneous catalyst:

$$4NO(g) + 4NH_3(g) + O_2(g) \longrightarrow 4N_2(g) + 6H_2O(g)$$

Progress in North America and Europe, primarily by reducing power-plant emissions, has resulted in increased pH of rainfall, and some surface waters have recovered. Further progress is expected under current legislation, but, in much of eastern North America and the sensitive regions of northern Europe, additional measures are needed for full recovery.

$$M(H_2O)_4^{2+}(aq) + NH_3(aq) \rightleftharpoons M(H_2O)_3(NH_3)^{2+}(aq) \xrightarrow[\text{3 more steps}]{3NH_3} M(NH_3)_4^{2+}(aq) + 4H_2O(l)$$

 Transparency/ARIS Presentation Center, Art Library

Once again, because the concentration of water is essentially constant in aqueous reactions, we incorporate it into K_c and obtain the expression for a new equilibrium constant, the **formation constant, K_f**:

$$K_f = \frac{K_c}{[H_2O]^4} = \frac{[M(NH_3)_4^{2+}]}{[M(H_2O)_4^{2+}][NH_3]^4}$$

At the molecular level (Figure 19.15), the actual process is stepwise, with ammonia molecules replacing water molecules one at a time to give a series of intermediate species, each with its own formation constant:

$$M(H_2O)_4^{2+}(aq) + NH_3(aq) \rightleftharpoons M(H_2O)_3(NH_3)^{2+}(aq) + H_2O(l)$$

$$K_{f1} = \frac{[M(H_2O)_3(NH_3)^{2+}]}{[M(H_2O)_4^{2+}][NH_3]}$$

$$M(H_2O)_3(NH_3)^{2+}(aq) + NH_3(aq) \rightleftharpoons M(H_2O)_2(NH_3)_2^{2+}(aq) + H_2O(l)$$

$$K_{f2} = \frac{[M(H_2O)_2(NH_3)_2^{2+}]}{[M(H_2O)_3(NH_3)^{2+}][NH_3]}$$

$$M(H_2O)_2(NH_3)_2^{2+}(aq) + NH_3(aq) \rightleftharpoons M(H_2O)(NH_3)_3^{2+}(aq) + H_2O(l)$$

$$K_{f3} = \frac{[M(H_2O)(NH_3)_3^{2+}]}{[M(H_2O)_2(NH_3)_2^{2+}][NH_3]}$$

$$M(H_2O)(NH_3)_3^{2+}(aq) + NH_3(aq) \rightleftharpoons M(NH_3)_4^{2+}(aq) + H_2O(l)$$

$$K_{f4} = \frac{[M(NH_3)_4^{2+}]}{[M(H_2O)(NH_3)_3^{2+}][NH_3]}$$

The *sum* of the equations gives the overall equation, so the *product* of the individual formation constants gives the overall formation constant:

$$K_f = K_{f1} \times K_{f2} \times K_{f3} \times K_{f4}$$

In this case, the K_f for each step is much larger than 1 because ammonia is a stronger Lewis base than water. Therefore, if we add excess ammonia to the $M(H_2O)_4^{2+}$ solution, the H_2O ligands are replaced and essentially all the M^{2+} ion exists as $M(NH_3)_4^{2+}$.

Table 19.4 (and Appendix C) shows the formation constants of some complex ions. Notice that the K_f values are all 10^6 or greater, which means that these ions form readily. Because of this behavior, complex-ion formation is used to retrieve a metal from its ore, eliminate a toxic or unwanted metal ion from a solution, or convert a metal ion to a different form, as Sample Problem 19.13 shows for the zinc ion.

Figure 19.15 The stepwise exchange of NH_3 for H_2O in $M(H_2O)_4^{2+}$. The ligands of a complex ion can exchange for other ligands. When ammonia is added to a solution of the hydrated M^{2+} ion, $M(H_2O)_4^{2+}$, NH_3 molecules replace the bound H_2O molecules one at a time to form the $M(NH_3)_4^{2+}$ ion. The molecular-scale views show the first exchange and the fully ammoniated ion.

Demo Summerlin, L. R., Ealy, J. B. "Colorful Complex Ions in Ammonia," *Chemical Demonstrations, A Sourcebook for Teachers*, Vol. 2 (American Chemical Society, Washington, DC, 1988), 75–76.

ARIS Presentation Center, Tables Library

Table 19.4	Formation Constants (K_f) of Some Complex Ions at 25°C
Complex Ion	**K_f**
$Ag(CN)_2^-$	3.0×10^{20}
$Ag(NH_3)_2^+$	1.7×10^7
$Ag(S_2O_3)_2^{3-}$	4.7×10^{13}
AlF_6^{3-}	4×10^{19}
$Al(OH)_4^-$	3×10^{33}
$Be(OH)_4^{2-}$	4×10^{18}
CdI_4^{2-}	1×10^6
$Co(OH)_4^{2-}$	5×10^9
$Cr(OH)_4^-$	8.0×10^{29}
$Cu(NH_3)_4^{2+}$	5.6×10^{11}
$Fe(CN)_6^{4-}$	3×10^{35}
$Fe(CN)_6^{3-}$	4.0×10^{43}
$Hg(CN)_4^{2-}$	9.3×10^{38}
$Ni(NH_3)_6^{2+}$	2.0×10^8
$Pb(OH)_3^-$	8×10^{13}
$Sn(OH)_3^-$	3×10^{25}
$Zn(CN)_4^{2-}$	4.2×10^{19}
$Zn(NH_3)_4^{2+}$	7.8×10^8
$Zn(OH)_4^{2-}$	3×10^{15}

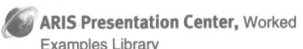

ARIS Presentation Center, Worked
Examples Library

SAMPLE PROBLEM 19.13 Calculating the Concentration of a Complex Ion

PROBLEM An industrial chemist converts $Zn(H_2O)_4^{2+}$ to the more stable $Zn(NH_3)_4^{2+}$ by mixing 50.0 L of 0.0020 M $Zn(H_2O)_4^{2+}$ and 25.0 L of 0.15 M NH_3. What is the final $[Zn(H_2O)_4^{2+}]$? K_f of $Zn(NH_3)_4^{2+}$ is 7.8×10^8.

PLAN We write the equation and the K_f expression and use a reaction table to calculate the equilibrium concentrations. To set up the table, we must first find $[Zn(H_2O)_4^{2+}]_{init}$ and $[NH_3]_{init}$. We are given the individual volumes and molar concentrations, so we find the moles and divide by the *total* volume, because the solutions are mixed. With the large excess of NH_3 and high K_f, we assume that almost all the $Zn(H_2O)_4^{2+}$ is converted to $Zn(NH_3)_4^{2+}$. Because $[Zn(H_2O)_4^{2+}]$ at equilibrium is very small, we use x to represent it.

SOLUTION Writing the equation and the K_f expression:

$$Zn(H_2O)_4^{2+}(aq) + 4NH_3(aq) \rightleftharpoons Zn(NH_3)_4^{2+}(aq) + 4H_2O(l)$$

$$K_f = \frac{[Zn(NH_3)_4^{2+}]}{[Zn(H_2O)_4^{2+}][NH_3]^4}$$

Finding the initial reactant concentrations:

$$[Zn(H_2O)_4^{2+}]_{init} = \frac{50.0 \text{ L} \times 0.0020 \, M}{50.0 \text{ L} + 25.0 \text{ L}} = 1.3 \times 10^{-3} \, M$$

$$[NH_3]_{init} = \frac{25.0 \text{ L} \times 0.15 \, M}{50.0 \text{ L} + 25.0 \text{ L}} = 5.0 \times 10^{-2} \, M$$

Setting up a reaction table: We assume that nearly all the $Zn(H_2O)_4^{2+}$ is converted to $Zn(NH_3)_4^{2+}$, so we set up the table with $x = [Zn(H_2O)_4^{2+}]$ at equilibrium. Because 4 mol of NH_3 is needed per mole of $Zn(H_2O)_4^{2+}$, the change in $[NH_3]$ is

$$[NH_3]_{reacted} \approx 4(1.3 \times 10^{-3} \, M) = 5.2 \times 10^{-3} \, M \quad \text{and} \quad [Zn(NH_3)_4^{2+}] \approx 1.3 \times 10^{-3} \, M$$

Concentration (M)	$Zn(H_2O)_4^{2+}(aq)$	+	$4NH_3(aq)$	$\rightleftharpoons$	$Zn(NH_3)_4^{2+}(aq)$	+	$4H_2O(l)$
Initial	1.3×10^{-3}		5.0×10^{-2}		0		—
Change	$\sim(-1.3 \times 10^{-3})$		$\sim(-5.2 \times 10^{-3})$		$\sim(+1.3 \times 10^{-3})$		—
Equilibrium	x		4.5×10^{-2}		1.3×10^{-3}		—

Solving for x, the $[Zn(H_2O)_4^{2+}]$ remaining at equilibrium:

$$K_f = \frac{[Zn(NH_3)_4^{2+}]}{[Zn(H_2O)_4^{2+}][NH_3]^4} = 7.8 \times 10^8 \approx \frac{1.3 \times 10^{-3}}{x(4.5 \times 10^{-2})^4}$$

$$x = [Zn(H_2O)_4^{2+}] \approx 4.1 \times 10^{-7} \, M$$

CHECK The K_f is large, so we expect the $[Zn(H_2O)_4^{2+}]$ remaining to be very low.

FOLLOW-UP PROBLEM 19.13 Cyanide ion is toxic because it forms stable complex ions with the Fe^{3+} ion in certain iron-containing proteins engaged in energy production. To study this effect, a biochemist mixes 25.5 mL of $3.1 \times 10^{-2} \, M$ $Fe(H_2O)_6^{3+}$ with 35.0 mL of 1.5 M NaCN. What is the final $[Fe(H_2O)_6^{3+}]$? K_f of $Fe(CN)_6^{3-}$ is 4.0×10^{43}.

 Biological application

Complex Ions and the Solubility of Precipitates

In Section 19.3, you saw that H_3O^+ increases the solubility of a slightly soluble ionic compound if its anion is that of a weak acid. Similarly, *a ligand increases the solubility of a slightly soluble ionic compound if it forms a complex ion with the cation.* For example, zinc sulfide is very slightly soluble:

$$ZnS(s) + H_2O(l) \rightleftharpoons Zn^{2+}(aq) + HS^-(aq) + OH^-(aq) \qquad K_{sp} = 2.0 \times 10^{-22}$$

When we add some 1.0 M NaCN, the CN^- ions act as ligands and react with the small amount of $Zn^{2+}(aq)$ to form the complex ion $Zn(CN)_4^{2-}$:

$$Zn^{2+}(aq) + 4CN^-(aq) \rightleftharpoons Zn(CN)_4^{2-}(aq) \qquad K_f = 4.2 \times 10^{19}$$

To see the effect of complex-ion formation on the solubility of ZnS, we add the equations and, therefore, multiply their equilibrium constants:

$$ZnS(s) + 4CN^-(aq) + H_2O(l) \rightleftharpoons Zn(CN)_4^{2-}(aq) + HS^-(aq) + OH^-(aq)$$

$$K_{overall} = K_{sp} \times K_f = (2.0 \times 10^{-22})(4.2 \times 10^{19}) = 8.4 \times 10^{-3}$$

The overall equilibrium constant increased by more than a factor of 10^{19} in the presence of the ligand; this reflects the increased amount of ZnS in solution.

SAMPLE PROBLEM 19.14 Calculating the Effect of Complex-Ion Formation on Solubility

ARIS Presentation Center, Worked Examples Library

PROBLEM In black-and-white film developing (see photo), excess AgBr is removed from the film negative by "hypo," an aqueous solution of sodium thiosulfate ($Na_2S_2O_3$), which forms the complex ion $Ag(S_2O_3)_2^{3-}$. Calculate the solubility of AgBr in (a) H_2O; (b) 1.0 M hypo. K_f of $Ag(S_2O_3)_2^{3-}$ is 4.7×10^{13} and K_{sp} of AgBr is 5.0×10^{-13}.

PLAN (a) After writing the equation and the ion-product expression, we use the given K_{sp} to solve for S, the molar solubility of AgBr. (b) In hypo, Ag^+ forms a complex ion with $S_2O_3^{2-}$, which shifts the equilibrium and dissolves more AgBr. We write the complex-ion equation and add it to the equation for dissolving AgBr to obtain the overall equation for dissolving AgBr in hypo. We multiply K_{sp} by K_f to find $K_{overall}$. To find the solubility of AgBr in hypo, we set up a reaction table, with $S = [Ag(S_2O_3)_2^{3-}]$, substitute into the expression for $K_{overall}$, and solve for S.

SOLUTION (a) Solubility in water. Writing the equation for the saturated solution and the ion-product expression:

$$AgBr(s) \rightleftharpoons Ag^+(aq) + Br^-(aq) \qquad K_{sp} = [Ag^+][Br^-]$$

Solving for solubility (S) directly from the equation: We know that

$$S = [AgBr]_{dissolved} = [Ag^+] = [Br^-]$$

Thus,

$$K_{sp} = [Ag^+][Br^-] = S^2 = 5.0 \times 10^{-13}$$

so

$$S = 7.1 \times 10^{-7} M$$

Developing the image in "hypo."

ARIS Presentation Center, Photo Library

Demo Summerlin, L. R., Ealy, J. B. "Solubility of Some Silver Compounds," *Chemical Demonstrations, A Sourcebook for Teachers*, Vol. 2 (American Chemical Society, Washington, DC, 1988), 83.

(b) Solubility in 1.0 M hypo. Writing the overall equation:

$$AgBr(s) \rightleftharpoons Ag^+(aq) + Br^-(aq)$$
$$Ag^+(aq) + 2S_2O_3^{2-}(aq) \rightleftharpoons Ag(S_2O_3)_2^{3-}(aq)$$
$$\overline{AgBr(s) + 2S_2O_3^{2-}(aq) \rightleftharpoons Ag(S_2O_3)_2^{3-}(aq) + Br^-(aq)}$$

Calculating $K_{overall}$:

$$K_{overall} = \frac{[Ag(S_2O_3)_2^{3-}][Br^-]}{[S_2O_3^{2-}]^2} = K_{sp} \times K_f = (5.0 \times 10^{-13})(4.7 \times 10^{13}) = 24$$

Setting up a reaction table, with $S = [AgBr]_{dissolved} = [Ag(S_2O_3)_2^{3-}]$:

Concentration (*M*)	AgBr(s) +	2S₂O₃²⁻(aq) ⇌	Ag(S₂O₃)₂³⁻(aq) +	Br⁻(aq)
Initial	—	1.0	0	0
Change	—	$-2S$	$+S$	$+S$
Equilibrium	—	$1.0 - 2S$	S	S

Substituting the values into the expression for $K_{overall}$ and solving for S:

$$K_{overall} = \frac{[Ag(S_2O_3)_2^{3-}][Br^-]}{[S_2O_3^{2-}]^2} = \frac{S^2}{(1.0\,M - 2S)^2} = 24$$

Taking the square root of both sides gives

$$\frac{S}{1.0\,M - 2S} = \sqrt{24} = 4.9 \qquad [Ag(S_2O_3)_2^{3-}] = S = 0.45\,M$$

CHECK (a) From the number of ions in the formula of AgBr, we know that $S = \sqrt{K_{sp}}$, so the order of magnitude seems right: $\sim \sqrt{10^{-14}} \approx 10^{-7}$. (b) The $K_{overall}$ seems correct: the exponents cancel, and $5 \times 5 = 25$. Most importantly, the answer makes sense because the photographic process requires the remaining AgBr to be washed off the film and the large $K_{overall}$ confirms that. We can check S by rounding and working backward to find $K_{overall}$: from the reaction table, we find that

$$[(S_2O_3)^{2-}] = 1.0\,M - 2S = 1.0\,M - 2(0.45\,M) = 1.0\,M - 0.90\,M = 0.1\,M$$

so $K_{overall} \approx (0.45)^2/(0.1)^2 = 20$, within rounding of the calculated value.

FOLLOW-UP PROBLEM 19.14 How does the solubility of AgBr in 1.0 M NH₃ compare with its solubility in 1.0 M hypo? K_f of $Ag(NH_3)_2^+$ is 1.7×10^7.

Demo Shakhashiri, Bassam Z. "Amphoteric Properties of Hydroxides of Aluminum, Zinc, Chromium, and Lead," *Chemical Demonstrations, A Handbook for Teachers of Chemistry*, Vol. 3 (The University of Wisconsin Press, Madison, 1989), 128–135.

Complex Ions of Amphoteric Hydroxides

Many of the same metals that form amphoteric oxides (Chapter 8, p. 327, and Interchapter Topic 4, p. 559) also form slightly soluble *amphoteric hydroxides*. These compounds dissolve very little in water, but they dissolve to a much greater extent in both acidic and basic solutions. Aluminum hydroxide is one of several examples:

$$Al(OH)_3(s) \rightleftharpoons Al^{3+}(aq) + 3OH^-(aq)$$

It is insoluble in water ($K_{sp} = 3\times10^{-34}$), but

- It dissolves in acid because H_3O^+ reacts with the OH^- anion (Sample Problem 19.9),

$$3H_3O^+(aq) + 3OH^-(aq) \longrightarrow 6H_2O(l)$$

giving the overall equation

$$Al(OH)_3(s) + 3H_3O^+(aq) \longrightarrow Al^{3+}(aq) + 6H_2O(l)$$

- It dissolves in base through the formation of a complex ion:

$$Al(OH)_3(s) + OH^-(aq) \longrightarrow Al(OH)_4^-(aq)$$

Figure 19.16 illustrates this amphoteric behavior but includes some unexpected formulas for the species. Let's see how these species arise. When we dissolve a soluble aluminum salt, such as $Al(NO_3)_3$, in water and then slowly add a strong base, a white precipitate first forms and then dissolves as more base is added.

What reactions are occurring? The formula for the hydrated Al^{3+} ion is $Al(H_2O)_6^{3+}(aq)$. It acts as a weak polyprotic acid and reacts with added OH^- ions in a stepwise manner. In each step, one of the bound H_2O molecules loses a proton and becomes a bound OH^- ion, so the number of bound H_2O molecules is reduced by 1:

$$Al(H_2O)_6^{3+}(aq) + OH^-(aq) \rightleftharpoons Al(H_2O)_5OH^{2+}(aq) + H_2O(l)$$

$$Al(H_2O)_5OH^{2+}(aq) + OH^-(aq) \rightleftharpoons Al(H_2O)_4(OH)_2^{+}(aq) + H_2O(l)$$

$$Al(H_2O)_4(OH)_2^{+}(aq) + OH^-(aq) \rightleftharpoons Al(H_2O)_3(OH)_3(s) + H_2O(l)$$

After three protons have been removed from each $Al(H_2O)_6^{3+}$, the white precipitate has formed, which is the insoluble hydroxide $Al(H_2O)_3(OH)_3(s)$, often written more simply as $Al(OH)_3(s)$. Now you can see that the precipitate actually consists of the hydrated Al^{3+} ion with an H^+ removed from each of three bound H_2O molecules. Addition of H_3O^+ protonates the OH^- ions and re-forms the hydrated Al^{3+} ion.

Further addition of OH^- removes a fourth H^+ and the precipitate dissolves as the soluble ion $Al(H_2O)_2(OH)_4^-(aq)$ forms, which we usually write with the formula $Al(OH)_4^-(aq)$:

$$Al(H_2O)_3(OH)_3(s) + OH^-(aq) \rightleftharpoons Al(H_2O)_2(OH)_4^-(aq) + H_2O(l)$$

In other words, this complex ion is not created by ligands substituting for bound water molecules but through an acid-base reaction in which added OH^- ions titrate bound water molecules.

Several other slightly soluble hydroxides, including those of cadmium, chromium(III), cobalt(III), lead(II), tin(II), and zinc, are amphoteric and exhibit similar reactions:

$$Zn(H_2O)_2(OH)_2(s) + OH^-(aq) \rightleftharpoons Zn(H_2O)(OH)_3^-(aq) + H_2O(l)$$

In contrast to the preceding hydroxides, the slightly soluble hydroxides of iron(II), iron(III), and calcium dissolve in acid, but do *not* dissolve in base, because the

$$3H_2O(l) + Al(H_2O)_6{}^{3+}(aq) \xleftarrow{\quad 3H_3O^+ \quad} Al(H_2O)_3(OH)_3(s) \xrightarrow{\quad OH^- \quad} Al(H_2O)_2(OH)_4{}^-(aq) + H_2O(l)$$

Figure 19.16 The amphoteric behavior of aluminum hydroxide.
When solid Al(OH)$_3$ is treated with H$_3$O$^+$ (*left*) or with OH$^-$ (*right*), it dissolves as a result of the formation of soluble complex ions. The molecular views show that Al(OH)$_3$ is actually the hydrated species Al(H$_2$O)$_3$(OH)$_3$. (The extensive OH bridging that occurs between Al^{3+} ions throughout the solid is not shown.) Addition of OH$^-$ (*right*) forms the soluble Al(H$_2$O)$_2$(OH)$_4{}^-$ ion; addition of H$_3$O$^+$ (*left*) forms the soluble Al(H$_2$O)$_6{}^{3+}$ ion.

ARIS **Presentation Center,** Active Art Library

three remaining bound water molecules are not acidic enough to lose any of their protons:

$$Fe(H_2O)_3(OH)_3(s) + 3H_3O^+(aq) \longrightarrow Fe(H_2O)_6{}^{3+}(aq) + 3H_2O(l)$$

$$Fe(H_2O)_3(OH)_3(s) + OH^-(aq) \longrightarrow \text{no reaction}$$

This difference in solubility in base between Al(OH)$_3$ and Fe(OH)$_3$ is the key to an important separation step in the production of aluminum metal, so we'll consider it again in Section 22.4.

Section Summary

A complex ion consists of a central metal ion covalently bonded to two or more negatively charged or neutral ligands. Its formation is described by a formation constant, K_f. • A hydrated metal ion is a complex ion with water molecules as ligands. Other ligands can displace the water in a stepwise process. In most cases, the K_f value of each step is large, so the fully substituted complex ion forms almost completely in the presence of excess ligand. • Adding a solution containing a ligand increases the solubility of an ionic precipitate if the cation forms a complex ion with the ligand. • Amphoteric metal hydroxides dissolve in acid and base due to acid-base reactions that form soluble complex ions.

Chapter Perspective

This chapter is the last of three that explore the nature and variety of equilibrium systems. In Chapter 17, we discussed the central ideas of equilibrium in the context of gaseous systems. In Chapter 18, we extended our understanding to acid-base equilibria. In this chapter, we highlighted three types of aqueous ionic systems and examined their role in the laboratory and the environment. The equilibrium constant, in all its forms, is a number that provides a limit to changes in a system, whether a chemical reaction, a physical change, or the dissolution of a substance. You now have the skills to predict whether a change will take place and to calculate its result, but you still do not know why the change occurs in the first place, or why it stops when it does. In Chapter 20, you'll find out.

CHAPTER REVIEW GUIDE
The following sections provide many aids to help you study this chapter. (Numbers in parentheses refer to pages, unless noted otherwise.)

Learning Objectives These are concepts and skills you should know after studying this chapter.

Relevant section and/or sample problem (SP) numbers appear in parentheses.

Understand These Concepts

1. How the presence of a common ion suppresses a reaction that forms it (Section 19.1)
2. Why the concentrations of buffer components must be high to minimize the change in pH from addition of small amounts of H_3O^+ or OH^- (Section 19.1)
3. How buffer capacity depends on buffer concentration and on the pK_a of the acid component; why buffer range is within ± 1 pH unit of the pK_a (Section 19.1)
4. The nature of an acid-base indicator as a conjugate acid-base pair with differently colored acidic and basic forms (Section 19.2)
5. The distinction between equivalence point and end point in an acid-base titration (Section 19.2)
6. Why the shapes of strong acid–strong base, weak acid–strong base, and strong acid–weak base titration curves differ (Section 19.2)
7. How the pH at the equivalence point is determined by the species present; why the pH at the midpoint of the buffer region equals the pK_a of the acid (Section 19.2)
8. How the titration curve of a polyprotic acid has a buffer region and equivalence point for each ionizable proton (Section 19.2)
9. How a slightly soluble ionic compound reaches equilibrium in water, expressed by an equilibrium (solubility-product) constant, K_{sp} (Section 19.3)
10. Why incomplete dissociation of an ionic compound leads to approximate calculated values for K_{sp} and solubility (Section 19.3)
11. Why a common ion in a solution decreases the solubility of its compounds (Section 19.3)
12. How pH affects the solubility of a compound that contains a weak-acid anion (Section 19.3)
13. How precipitate formation depends on the relative values of Q_{sp} and K_{sp} (Section 19.3)
14. How selective precipitation and simultaneous equilibria are used to separate ions (Section 19.3)

15. How complex-ion formation is stepwise and expressed by an overall equilibrium (formation) constant, K_f (Section 19.4)
16. Why addition of a ligand increases the solubility of a compound whose metal ion forms a complex ion (Section 19.4)
17. How the aqueous chemistry of amphoteric hydroxides involves precipitation, complex-ion formation, and acid-base equilibria (Section 19.4)

Master These Skills

1. Using stoichiometry and equilibrium problem-solving techniques to calculate the effect of added H_3O^+ or OH^- on buffer pH (SP 19.1)
2. Using the Henderson-Hasselbalch equation to calculate buffer pH (Section 19.1)
3. Using molecular scenes to depict buffer capacity (SP 19.2)
4. Choosing the components of a buffer with a given pH and calculating their quantities (SP 19.3)
5. Calculating the pH at any point in an acid-base titration (Section 19.2 and SP 19.4)
6. Choosing an appropriate indicator based on the pH at various points in a titration (Section 19.2)
7. Writing K_{sp} expressions for slightly soluble ionic compounds (SP 19.5)
8. Using K_{sp} values to compare solubilities for compounds with the same total number of ions (Section 19.3)
9. Calculating a K_{sp} value from solubility data (SP 19.6)
10. Calculating solubility from a K_{sp} value (SP 19.7)
11. Calculating the decrease in solubility caused by the presence of a common ion (SP 19.8)
12. Predicting the effect of added H_3O^+ on solubility (SP 19.9)
13. Using ion concentrations to calculate Q_{sp} and compare it with K_{sp} to predict whether a precipitate forms (SPs 19.10, 19.11)
14. Comparing K_{sp} values in order to separate ions by selective precipitation (SP 19.12)
15. Calculating the concentration of metal ion remaining after addition of excess ligand forms a complex ion (SP 19.13)
16. Using an overall equilibrium constant ($K_{sp} \times K_f$) to calculate the effect of complex-ion formation on solubility (SP 19.14)

Key Terms These important terms appear in boldface in the chapter and are defined again in the Glossary.

Section 19.1
acid-base buffer (832)
common-ion effect (833)
Henderson-Hasselbalch
equation (837)

buffer capacity (838)
buffer range (838)
Section 19.2
acid-base titration curve (841)
equivalence point (843)
end point (844)

Section 19.3
solubility-product constant
(K_{sp}) (852)
molar solubility (853)
selective precipitation (861)

Section 19.4
complex ion (862)
ligand (862)
formation constant (K_f) (865)

Key Equations and Relationships Numbered and screened concepts are listed for you to refer to or memorize.

19.1 Finding the pH from known concentrations of a conjugate acid-base pair (Henderson-Hasselbalch equation) (837):

$$pH = pK_a + \log\left(\frac{[base]}{[acid]}\right)$$

19.2 Defining the equilibrium condition for a saturated solution of a slightly soluble compound, M_pX_q, composed of M^{n+} and X^{z-} ions (852):

$$Q_{sp} = [M^{n+}]^p[X^{z-}]^q = K_{sp}$$

Highlighted Figures and Tables These figures (F) and tables (T) provide a visual review of key ideas.

F19.3 How a buffer works (834)
F19.4 Buffer capacity and pH change (838)
F19.5 Colors and pH ranges of acid-base indicators (842)

F19.7 A strong acid–strong base titration curve (843)
F19.8 A weak acid–strong base titration curve (845)
F19.9 A weak base–strong acid titration curve (849)

Brief Solutions to FOLLOW-UP PROBLEMS Compare your solutions to these calculation steps and answers.

19.1 (a) Before addition:
Assuming x is small enough to be neglected,
[HF] = 0.50 M and   [F⁻] = 0.45 M

$$[H_3O^+] = K_a \times \frac{[HF]}{[F^-]} \approx (6.8 \times 10^{-4})\left(\frac{0.50}{0.45}\right) = 7.6 \times 10^{-4} M$$

pH = 3.12
(b) After addition of 0.40 g of NaOH (0.010 mol of NaOH) to 1.0 L of buffer,
[HF] = 0.49 M and   [F⁻] = 0.46 M

$$[H_3O^+] \approx (6.8 \times 10^{-4})\left(\frac{0.49}{0.46}\right) = 7.2 \times 10^{-4} M; \text{ pH} = 3.14$$

19.2 (a) Strong base would convert HB to B⁻, thereby making the ratio closer to 1.
(b)

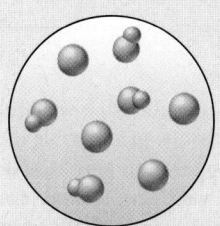

19.3 $[H_3O^+] = 10^{-pH} = 10^{-4.25} = 5.6 \times 10^{-5}$

$$[C_6H_5COOH] = \frac{[H_3O^+][C_6H_5COO^-]}{K_a}$$

$$= \frac{(5.6 \times 10^{-5})(0.050)}{6.3 \times 10^{-5}} = 0.044 M$$

Mass (g) of C_6H_5COOH

$$= 5.0 \text{ L soln} \times \frac{0.044 \text{ mol } C_6H_5COOH}{1 \text{ L soln}}$$

$$\times \frac{122.12 \text{ g } C_6H_5COOH}{1 \text{ mol } C_6H_5COOH}$$

$$= 27 \text{ g } C_6H_5COOH$$

Dissolve 27 g of C_6H_5COOH in 4.9 L of 0.050 M C_6H_5COONa and add solution to make 5.0 L. Adjust pH to 4.25 with strong acid or base.
19.4 (a) $[H_3O^+] \approx \sqrt{(2.3 \times 10^{-9})(0.2000)} = 2.1 \times 10^{-5} M$
pH = 4.68

(b) $[H_3O^+] = K_a \times \frac{[HBrO]}{[BrO^-]} = (2.3 \times 10^{-9})(1) = 2.3 \times 10^{-9} M$

pH = 8.64

(c) $[BrO^-] = \frac{\text{moles of } BrO^-}{\text{total volume}} = \frac{0.004000 \text{ mol}}{0.06000 \text{ L}} = 0.06667 M$

K_b of $BrO^- = \frac{K_w}{K_a \text{ of HBrO}} = 4.3 \times 10^{-6}$

$$[H_3O^+] = \frac{K_w}{\sqrt{K_b \times [BrO^-]}}$$

$$\approx \frac{1.0 \times 10^{-14}}{\sqrt{(4.3 \times 10^{-6})(0.06667)}} = 1.9 \times 10^{-11} M$$

pH = 10.72

(d) Moles of OH⁻ added = 0.008000 mol
Volume (L) of OH⁻ soln = 0.08000 L

$$[OH^-] = \frac{\text{moles of OH}^- \text{ unreacted}}{\text{total volume}}$$

$$= \frac{0.008000 \text{ mol} - 0.004000 \text{ mol}}{(0.02000 + 0.08000) \text{ L}} = 0.04000 M$$

$[H_3O^+] = \frac{K_w}{[OH^-]} = 2.5 \times 10^{-13}$

pH = 12.60

Brief Solutions to FOLLOW-UP PROBLEMS *(continued)*

(e)

pH vs. Volume of added base (mL): points at 4.68, 8.64, 10.72, 12.60

19.5 (a) $K_{sp} = [Ca^{2+}][SO_4^{2-}]$
(b) $K_{sp} = [Cr^{3+}]^2[CO_3^{2-}]^3$
(c) $K_{sp} = [Mg^{2+}][OH^-]^2$
(d) $K_{sp} = [As^{3+}]^2[HS^-]^3[OH^-]^3$

19.6 $[CaF_2] = \dfrac{1.5\times10^{-4}\ g\ CaF_2}{10.0\ mL\ soln} \times \dfrac{1000\ mL}{1\ L} \times \dfrac{1\ mol\ CaF_2}{78.08\ g\ CaF_2}$
$= 1.9\times10^{-4}\ M$
$CaF_2(s) \rightleftharpoons Ca^{2+}(aq) + 2F^-(aq)$
$[Ca^{2+}] = 1.9\times10^{-4}\ M$ and $[F^-] = 3.8\times10^{-4}\ M$
$K_{sp} = [Ca^{2+}][F^-]^2 = (1.9\times10^{-4})(3.8\times10^{-4})^2 = 2.7\times10^{-11}$
19.7 From the reaction table, $[Mg^{2+}] = S$ and $[OH^-] = 2S$
$K_{sp} = [Mg^{2+}][OH^-]^2 = 4S^3 = 6.3\times10^{-10};\ S = 5.4\times10^{-4}\ M$
19.8 (a) In water: $K_{sp} = [Ba^{2+}][SO_4^{2-}] = S^2 = 1.1\times10^{-10}$;
$S = 1.0\times10^{-5}$
(b) In 0.10 M Na_2SO_4: $[SO_4^{2-}] = 0.10\ M$
$K_{sp} = 1.1\times10^{-10} \approx S \times 0.10;\ S = 1.1\times10^{-9}\ M$
S decreases in presence of the common ion SO_4^{2-}.
19.9 (a) Increases solubility.
$$CaF_2(s) \rightleftharpoons Ca^{2+}(aq) + 2F^-(aq)$$
$$F^-(aq) + H_3O^+(aq) \longrightarrow HF(aq) + H_2O(l)$$

(b) Increases solubility.
$$ZnS(s) + H_2O(l) \rightleftharpoons Zn^{2+}(aq) + HS^-(aq) + OH^-(aq)$$
$$HS^-(aq) + H_3O^+(aq) \longrightarrow H_2S(aq) + H_2O(l)$$
$$OH^-(aq) + H_3O^+(aq) \longrightarrow 2H_2O(l)$$
(c) No effect. $I^-(aq)$ is the conjugate base of strong acid HI.
19.10 $Ca_3(PO_4)_2(s) \rightleftharpoons 3Ca^{2+}(aq) + 2PO_4^{3-}(aq)$
$Q_{sp} = [Ca^{2+}]^3[PO_4^{3-}]^2 = (1.0\times10^{-9})^5 = 1.0\times10^{-45}$
$Q_{sp} < K_{sp}$, so $Ca_3(PO_4)_2$ will not precipitate.
19.11 (a) Scene 3 has the same relative numbers of ions as in the formula. (b) Based on $Ni(OH)_2(s) \rightleftharpoons Ni^{2+}(aq) + 2OH^-(aq)$ and $Q_{sp} = [Ni^{2+}][OH^-]^2$, the ion products are $(3)(4)^2 = 48$ in scene 1; $(4)(2)^2 = 16$ in scene 2; and $(2)(4)^2 = 32 = K_{sp}$ in scene 3. Q_{sp} of scene 1 exceeds K_{sp} of scene 3. (c) Addition of acid will decrease mass of $Ni(OH)_2(s)$ by reacting with OH^-, thereby causing more solid to dissolve; addition of base will increase mass of $Ni(OH)_2(s)$ due to common-ion effect.
19.12 Both are 1/1 salts, and K_{sp} values show $CaSO_4$ is more soluble:
$$[SO_4^{2-}] = \frac{K_{sp}}{[Ca^{2+}]} = \frac{2.4\times10^{-5}}{(0.025)(0.9999)} = 9.6\times10^{-4}\ M$$
19.13 $[Fe(H_2O)_6^{3+}]_{init} = \dfrac{(0.0255\ L)(3.1\times10^{-2}\ M)}{0.0255\ L + 0.0350\ L}$
$= 1.3\times10^{-2}\ M$
Similarly, $[CN^-]_{init} = 0.87\ M$. From the reaction table,
$$K_f = \frac{[Fe(CN)_6^{3-}]}{[Fe(H_2O)_6^{3+}][CN^-]^6} = 4.0\times10^{43} \approx \frac{1.3\times10^{-2}}{x(0.79)^6}$$
$x = [Fe(H_2O)_6^{3+}] \approx 1.3\times10^{-45}$
19.14 $AgBr(s) + 2NH_3(aq) \rightleftharpoons Ag(NH_3)_2^+(aq) + Br^-(aq)$
$K_{overall} = K_{sp}$ of $AgBr \times K_f$ of $Ag(NH_3)_2^+$
$= 8.5\times10^{-6}$

From the reaction table, $\dfrac{S}{1.0 - 2S} = \sqrt{8.5\times10^{-6}} = 2.9\times10^{-3}$

$$S = [Ag(NH_3)_2^+] = 2.9\times10^{-3}\ M$$

Solubility is greater in 1.0 M hypo than in 1.0 M NH_3.

PROBLEMS

Problems with colored numbers are answered in Appendix E and worked in detail in the Student Solutions Manual. Problem sections match those in the text and provide the numbers of relevant sample problems. Most offer Concept Review Questions, Skill-Building Exercises (grouped in pairs covering the same concept), and Problems in Context. The Comprehensive Problems are based on material from any section or previous chapter.

Note: Unless stated otherwise, all of the problems for this chapter refer to aqueous solutions at 298 K (25°C).

Equilibria of Acid-Base Buffer Systems
(Sample Problems 19.1 to 19.3)

● **Concept Review Questions**

• **19.1** What is the purpose of an acid-base buffer?

• **19.2** How do the acid and base components of a buffer function? Why are they often a conjugate acid-base pair of a weak acid?

• **19.3** What is the common-ion effect? How is it related to Le Châtelier's principle? Explain with equations that include HF and NaF.

• **19.4** The scenes below depict solutions of the same HA/A⁻ buffer (with counterions and water molecules omitted for clarity). (a) Which solution has the greatest buffer capacity? (b) Explain how the pH ranges of the buffers compare. (c) Which solution can react with the largest amount of added strong acid?

• **19.5** What is the difference between buffers with high and low capacities? Will adding 0.01 mol of HCl produce a greater pH change in a buffer with a high or a low capacity? Explain.

• **19.6** Which of these factors influence buffer capacity? How?
(a) Conjugate acid-base pair   (b) pH of the buffer
(c) Concentration of buffer components   (d) Buffer range
(e) pK_a of the acid component

• **19.7** What is the relationship between the buffer range and the buffer-component concentration ratio?

• **19.8** A chemist needs a pH 3.5 buffer. Should she use NaOH with formic acid ($K_a = 1.8 \times 10^{-4}$) or with acetic acid ($K_a = 1.8 \times 10^{-5}$)? Why? What is the disadvantage of choosing the other acid? What is the role of the NaOH?

• **19.9** State and explain the relative change in the pH and in the buffer-component concentration ratio, [NaA]/[HA], for each of the following additions:
(a) Add 0.1 M NaOH to the buffer
(b) Add 0.1 M HCl to the buffer
(c) Dissolve pure NaA in the buffer
(d) Dissolve pure HA in the buffer

• **19.10** Does the pH increase or decrease, and does it do so to a large or small extent, with each of the following additions?
(a) 5 drops of 0.1 M NaOH to 100 mL of 0.5 M acetate buffer
(b) 5 drops of 0.1 M HCl to 100 mL of 0.5 M acetate buffer
(c) 5 drops of 0.1 M NaOH to 100 mL of 0.5 M HCl
(d) 5 drops of 0.1 M NaOH to distilled water

● **Skill-Building Exercises** (grouped in similar pairs)

• **19.11** What are the [H_3O^+] and the pH of a propanoic acid–propanoate buffer that consists of 0.35 M CH_3CH_2COONa and 0.15 M CH_3CH_2COOH (K_a of propanoic acid = 1.3×10^{-5})?

• **19.12** What are the [H_3O^+] and the pH of a benzoic acid–benzoate buffer that consists of 0.33 M C_6H_5COOH and 0.28 M C_6H_5COONa (K_a of benzoic acid = 6.3×10^{-5})?

• **19.13** What are the [H_3O^+] and the pH of a buffer that consists of 0.55 M HNO_2 and 0.75 M KNO_2 (K_a of HNO_2 = 7.1×10^{-4})?

• **19.14** What are the [H_3O^+] and the pH of a buffer that consists of 0.20 M HF and 0.25 M KF (K_a of HF = 6.8×10^{-4})?

• **19.15** Find the pH of a buffer that consists of 0.45 M HCOOH and 0.63 M HCOONa (pK_a of HCOOH = 3.74).

• **19.16** Find the pH of a buffer that consists of 0.95 M HBrO and 0.68 M KBrO (pK_a of HBrO = 8.64).

• **19.17** Find the pH of a buffer that consists of 1.3 M sodium phenolate (C_6H_5ONa) and 1.2 M phenol (C_6H_5OH) (pK_a of phenol = 10.00).

• **19.18** Find the pH of a buffer that consists of 0.12 M boric acid (H_3BO_3) and 0.82 M sodium borate (NaH_2BO_3) (pK_a of boric acid = 9.24).

• **19.19** Find the pH of a buffer that consists of 0.25 M NH_3 and 0.15 M NH_4Cl (pK_b of NH_3 = 4.75).

• **19.20** Find the pH of a buffer that consists of 0.50 M methylamine (CH_3NH_2) and 0.60 M CH_3NH_3Cl (pK_b of CH_3NH_2 = 3.35).

• **19.21** A buffer consists of 0.22 M $KHCO_3$ and 0.37 M K_2CO_3. Carbonic acid is a diprotic acid with $K_{a1} = 4.5 \times 10^{-7}$ and $K_{a2} = 4.7 \times 10^{-11}$. (a) Which K_a value is more important to this buffer? (b) What is the buffer pH?

• **19.22** A buffer consists of 0.50 M NaH_2PO_4 and 0.40 M Na_2HPO_4. Phosphoric acid is a triprotic acid ($K_{a1} = 7.2 \times 10^{-3}$, $K_{a2} = 6.3 \times 10^{-8}$, and $K_{a3} = 4.2 \times 10^{-13}$). (a) Which K_a value is most important to this buffer? (b) What is the buffer pH?

• **19.23** What is the component concentration ratio, [Pr⁻]/[HPr], of a buffer that has a pH of 5.44 (K_a of HPr = 1.3×10^{-5})?

• **19.24** What is the component concentration ratio, [NO_2^-]/[HNO_2], of a buffer that has a pH of 2.95 (K_a of HNO_2 = 7.1×10^{-4})?

• **19.25** What is the component concentration ratio, [BrO⁻]/[HBrO], of a buffer that has a pH of 7.95 (K_a of HBrO = 2.3×10^{-9})?

• **19.26** What is the component concentration ratio, [CH_3COO^-]/[CH_3COOH], of a buffer that has a pH of 4.39 (K_a of CH_3COOH = 1.8×10^{-5})?

• **19.27** A buffer containing 0.2000 M of acid, HA, and 0.1500 M of its conjugate base, A⁻, has a pH of 3.35. What is the pH after 0.0015 mol of NaOH is added to 0.5000 L of this solution?

• **19.28** A buffer that contains 0.40 M base, B, and 0.25 M of its conjugate acid, BH^+, has a pH of 8.88. What is the pH after 0.0020 mol of HCl is added to 0.25 L of this solution?

• **19.29** A buffer that contains 0.110 M HY and 0.220 M Y⁻ has a pH of 8.77. What is the pH after 0.0015 mol of Ba(OH)$_2$ is added to 0.350 L of this solution?

• **19.30** A buffer that contains 1.05 M B and 0.750 M BH^+ has a pH of 9.50. What is the pH after 0.0050 mol of HCl is added to 0.500 L of this solution?

•• **19.31** A buffer is prepared by mixing 204 mL of 0.452 M HCl and 0.500 L of 0.400 M sodium acetate. (See Appendix C.) (a) What is the pH? (b) How many grams of KOH must be added to 0.500 L of the buffer to change the pH by 0.15 units?

•• **19.32** A buffer is prepared by mixing 50.0 mL of 0.050 M sodium bicarbonate and 10.7 mL of 0.10 M NaOH. (See Appendix C.) (a) What is the pH? (b) How many grams of HCl must be added to 25.0 mL of the buffer to change the pH by 0.07 units?

• **19.33** Choose specific acid-base conjugate pairs to make the following buffers: (a) pH ≈ 4.5; (b) pH ≈ 7.0. (See Appendix C.)

• **19.34** Choose specific acid-base conjugate pairs to make the following buffers: (a) [H_3O^+] ≈ 1×10^{-9} M; (b) [OH⁻] ≈ 3×10^{-5} M. (See Appendix C.)

• **19.35** Choose specific acid-base conjugate pairs to make the following buffers: (a) pH ≈ 3.5; (b) pH ≈ 5.5. (See Appendix C.)

• **19.36** Choose specific acid-base conjugate pairs to make the following buffers: (a) [OH⁻] ≈ 1×10^{-6} M; (b) [H_3O^+] ≈ 4×10^{-4} M. (See Appendix C.)

● **Problems in Context**

• **19.37** An industrial chemist studying bleaching and sterilizing prepares several hypochlorite buffers. Find the pH of (a) 0.100 M HClO and 0.100 M NaClO; (b) 0.100 M HClO and 0.150 M NaClO; (c) 0.150 M HClO and 0.100 M NaClO; (d) 1.0 L of the solution in part (a) after 0.0050 mol of NaOH has been added.

• **19.38** Oxoanions of phosphorus are buffer components in blood. For a KH_2PO_4/Na_2HPO_4 solution with pH = 7.40 (pH of normal arterial blood), what is the buffer-component concentration ratio?

Acid-Base Titration Curves
(Sample Problem 19.4)

● **Concept Review Questions**

• **19.39** How can you estimate the pH range of an indicator's color change? Why do some indicators have two separate pH ranges?

• **19.40** Why does the color change of an indicator take place over a range of about 2 pH units?

• **19.41** Why doesn't the addition of an acid-base indicator affect the pH of the test solution?

• **19.42** What is the difference between the end point of a titration and the equivalence point? Is the equivalence point always reached first? Explain.

•• **19.43** The scenes below depict the relative concentrations of H_3PO_4, $H_2PO_4^-$, and HPO_4^{2-} during a titration with aqueous NaOH, but they are out of order. (Phosphate groups are purple, hydrogens are blue, and Na^+ ions and water molecules are not shown.) (a) List the scenes in the correct order. (b) What is the pH in the correctly ordered second scene (see Appendix C)? (c) If it requires 10.00 mL of the NaOH solution to reach this scene, how much more is needed to reach the last scene?

⬤ = H_3PO_4 ⬤ = $H_2PO_4^-$ ⬤ = HPO_4^{2-}

A   B   C   D

• **19.44** Explain how *strong acid*–strong base, *weak acid*–strong base, and *weak base*–strong acid titrations using the same concentrations differ in terms of (a) the initial pH and (b) the pH at the equivalence point. (The component in italics is in the flask.)

• **19.45** What species are in the buffer region of a weak acid–strong base titration? How are they different from the species at the equivalence point? How are they different from the species in the buffer region of a weak base–strong acid titration?

• **19.46** Why is the center of the buffer region of a weak acid–strong base titration significant?

• **19.47** How does the titration curve of a monoprotic acid differ from that of a diprotic acid?

● **Skill-Building Exercises** *(grouped in similar pairs)*

• **19.48** The indicator cresol red has $K_a = 3.5 \times 10^{-9}$. Over what approximate pH range does it change color?

• **19.49** The indicator ethyl red has $K_a = 3.8 \times 10^{-6}$. Over what approximate pH range does it change color?

• **19.50** Use Figure 19.5 to find an indicator for these titrations:
(a) 0.10 M HCl with 0.10 M NaOH
(b) 0.10 M HCOOH (Appendix C) with 0.10 M NaOH

• **19.51** Use Figure 19.5 to find an indicator for these titrations:
(a) 0.10 M CH_3NH_2 (Appendix C) with 0.10 M HCl
(b) 0.50 M HI with 0.10 M KOH

• **19.52** Use Figure 19.5 to find an indicator for these titrations:
(a) 0.5 M $(CH_3)_2NH$ (Appendix C) with 0.5 M HBr
(b) 0.2 M KOH with 0.2 M HNO_3

• **19.53** Use Figure 19.5 to find an indicator for these titrations:
(a) 0.25 M C_6H_5COOH (Appendix C) with 0.25 M KOH
(b) 0.50 M NH_4Cl (Appendix C) with 0.50 M NaOH

•• **19.54** Calculate the pH during the titration of 40.00 mL of 0.1000 M HCl with 0.1000 M NaOH solution after the following additions of base:
(a) 0 mL   (b) 25.00 mL   (c) 39.00 mL   (d) 39.90 mL
(e) 40.00 mL   (f) 40.10 mL   (g) 50.00 mL

•• **19.55** Calculate the pH during the titration of 30.00 mL of 0.1000 M KOH with 0.1000 M HBr solution after the following additions of acid:
(a) 0 mL   (b) 15.00 mL   (c) 29.00 mL   (d) 29.90 mL
(e) 30.00 mL   (f) 30.10 mL   (g) 40.00 mL

•• **19.56** Find the pH during the titration of 20.00 mL of 0.1000 M butanoic acid, $CH_3CH_2CH_2COOH$ ($K_a = 1.54 \times 10^{-5}$), with 0.1000 M NaOH solution after the following additions of titrant:
(a) 0 mL   (b) 10.00 mL   (c) 15.00 mL   (d) 19.00 mL
(e) 19.95 mL   (f) 20.00 mL   (g) 20.05 mL   (h) 25.00 mL

•• **19.57** Find the pH during the titration of 20.00 mL of 0.1000 M triethylamine, $(CH_3CH_2)_3N$ ($K_b = 5.2 \times 10^{-4}$), with 0.1000 M HCl solution after the following additions of titrant:
(a) 0 mL   (b) 10.00 mL   (c) 15.00 mL   (d) 19.00 mL
(e) 19.95 mL   (f) 20.00 mL   (g) 20.05 mL   (h) 25.00 mL

•• **19.58** Find the pH of the equivalence point(s) and the volume (mL) of 0.0372 M NaOH needed to reach it in titrations of
(a) 42.2 mL of 0.0520 M CH_3COOH
(b) 28.9 mL of 0.0850 M H_2SO_3 (two equivalence points)

•• **19.59** Find the pH of the equivalence point(s) and the volume (mL) of 0.0588 M KOH needed to reach it in titrations of
(a) 23.4 mL of 0.0390 M HNO_2
(b) 17.3 mL of 0.130 M H_2CO_3 (two equivalence points)

•• **19.60** Find the pH of the equivalence point(s) and the volume (mL) of 0.125 M HCl needed to reach it in titrations of
(a) 65.5 mL of 0.234 M NH_3
(b) 21.8 mL of 1.11 M CH_3NH_2

•• **19.61** Find the pH and volume (mL) of 0.447 M HNO_3 needed to reach the equivalence point(s) in titrations of
(a) 2.65 L of 0.0750 M pyridine (C_5H_5N)
(b) 0.188 L of 0.250 M ethylenediamine ($H_2NCH_2CH_2NH_2$)

Equilibria of Slightly Soluble Ionic Compounds
(Sample Problems 19.5 to 19.12)

● **Concept Review Questions**

• **19.62** The molar solubility of M_2X is 5×10^{-5} M. What is the molarity of each ion? How do you set up the calculation to find K_{sp}? What assumption must you make about the dissociation of M_2X into ions? Why is the calculated K_{sp} higher than the actual value?

• **19.63** Why does pH affect the solubility of BaF_2 but not of $BaCl_2$?

• **19.64** A list of K_{sp} values like that in Appendix C can be used to compare the solubility of silver chloride directly with that of silver bromide but not with that of silver chromate. Explain.

• **19.65** In a gaseous equilibrium, the reverse reaction occurs when $Q_c > K_c$. What occurs in aqueous solution when $Q_{sp} > K_{sp}$?

● **Skill-Building Exercises** *(grouped in similar pairs)*

• **19.66** Write the ion-product expressions for (a) silver carbonate; (b) barium fluoride; (c) copper(II) sulfide.

• **19.67** Write the ion-product expressions for (a) iron(III) hydroxide; (b) barium phosphate; (c) tin(II) sulfide.

• **19.68** Write the ion-product expressions for (a) calcium chromate; (b) silver cyanide; (c) nickel(II) sulfide.

• **19.69** Write the ion-product expressions for (a) lead(II) iodide; (b) strontium sulfate; (c) cadmium sulfide.

• **19.70** The solubility of silver carbonate is 0.032 M at 20°C. Calculate its K_{sp}.

• **19.71** The solubility of zinc oxalate is 7.9×10^{-3} M at 18°C. Calculate its K_{sp}.

• **19.72** The solubility of silver dichromate at 15°C is 8.3×10^{-3} g/100 mL solution. Calculate its K_{sp}.

• **19.73** The solubility of calcium sulfate at 30°C is 0.209 g/100 mL solution. Calculate its K_{sp}.

•• **19.74** Find the molar solubility of $SrCO_3$ ($K_{sp} = 5.4 \times 10^{-10}$) in (a) pure water and (b) 0.13 M $Sr(NO_3)_2$.

•• **19.75** Find the molar solubility of $BaCrO_4$ ($K_{sp} = 2.1 \times 10^{-10}$) in (a) pure water and (b) 1.5×10^{-3} M Na_2CrO_4.

•• **19.76** Calculate the molar solubility of $Ca(IO_3)_2$ in (a) 0.060 M $Ca(NO_3)_2$ and (b) 0.060 M $NaIO_3$. (See Appendix C.)

•• **19.77** Calculate the molar solubility of Ag_2SO_4 in (a) 0.22 M $AgNO_3$ and (b) 0.22 M Na_2SO_4. (See Appendix C.)

• **19.78** Which compound in each pair is more soluble in water?
(a) Magnesium hydroxide or nickel(II) hydroxide
(b) Lead(II) sulfide or copper(II) sulfide
(c) Silver sulfate or magnesium fluoride

• **19.79** Which compound in each pair is more soluble in water?
(a) Strontium sulfate or barium chromate
(b) Calcium carbonate or copper(II) carbonate
(c) Barium iodate or silver chromate

• **19.80** Which compound in each pair is more soluble in water?
(a) Barium sulfate or calcium sulfate
(b) Calcium phosphate or magnesium phosphate
(c) Silver chloride or lead(II) sulfate

• **19.81** Which compound in each pair is more soluble in water?
(a) Manganese(II) hydroxide or calcium iodate
(b) Strontium carbonate or cadmium sulfide
(c) Silver cyanide or copper(I) iodide

• **19.82** Write equations to show whether the solubility of either of the following is affected by pH: (a) AgCl; (b) $SrCO_3$.

• **19.83** Write equations to show whether the solubility of either of the following is affected by pH: (a) CuBr; (b) $Ca_3(PO_4)_2$.

• **19.84** Write equations to show whether the solubility of either of the following is affected by pH: (a) $Fe(OH)_2$; (b) CuS.

• **19.85** Write equations to show whether the solubility of either of the following is affected by pH: (a) PbI_2; (b) $Hg_2(CN)_2$.

• **19.86** Does any solid $Cu(OH)_2$ form when 0.075 g of KOH is dissolved in 1.0 L of 1.0×10^{-3} M $Cu(NO_3)_2$?

• **19.87** Does any solid $PbCl_2$ form when 3.5 mg of NaCl is dissolved in 0.250 L of 0.12 M $Pb(NO_3)_2$?

• **19.88** Does any solid $Ba(IO_3)_2$ form when 7.5 mg of $BaCl_2$ is dissolved in 500. mL of 0.023 M $NaIO_3$?

• **19.89** Does any solid Ag_2CrO_4 form when 2.7×10^{-5} g of $AgNO_3$ is dissolved in 15.0 mL of 4.0×10^{-4} M K_2CrO_4?

● **Problems in Context**

•• **19.90** When blood is donated, sodium oxalate solution is used to precipitate Ca^{2+}, which triggers clotting. A 104-mL sample of

blood contains 9.7×10^{-5} g Ca^{2+}/mL. A technologist treats the sample with 100.0 mL of 0.1550 M $Na_2C_2O_4$. Calculate $[Ca^{2+}]$ after the treatment. (See Appendix C for K_{sp} of $CaC_2O_4 \cdot H_2O$.)

•• **19.91** A 50.0-mL volume of 0.50 M $Fe(NO_3)_3$ is mixed with 125 mL of 0.25 M $Cd(NO_3)_2$.
(a) If aqueous NaOH is added, which ion precipitates first? (See Appendix C.)
(b) Describe how the metal ions can be separated using NaOH.
(c) Calculate the $[OH^-]$ that will accomplish the separation.

Equilibria Involving Complex Ions
(Sample Problems 19.13 and 19.14)

● **Concept Review Questions**

• **19.92** How can a metal cation be at the center of a complex anion?

• **19.93** Write equations to show the stepwise reaction of $Cd(H_2O)_4^{2+}$ in an aqueous solution of KI to form CdI_4^{2-}. Show that $K_{f(overall)} = K_{f1} \times K_{f2} \times K_{f3} \times K_{f4}$.

•• **19.94** Consider the dissolution of PbS in water:
$$PbS(s) + H_2O(l) \rightleftharpoons Pb^{2+}(aq) + HS^-(aq) + OH^-(aq)$$
Adding aqueous NaOH causes more PbS to dissolve. Does this violate Le Châtelier's principle? Explain.

● **Skill-Building Exercises** (grouped in similar pairs)

• **19.95** Write a balanced equation for the reaction of $Hg(H_2O)_4^{2+}$ in aqueous KCN.

• **19.96** Write a balanced equation for the reaction of $Zn(H_2O)_4^{2+}$ in aqueous NaCN.

• **19.97** Write a balanced equation for the reaction of $Ag(H_2O)_2^+$ in aqueous $Na_2S_2O_3$.

• **19.98** Write a balanced equation for the reaction of $Al(H_2O)_6^{3+}$ in aqueous KF.

•• **19.99** What is $[Ag^+]$ when 25.0 mL each of 0.044 M $AgNO_3$ and 0.57 M $Na_2S_2O_3$ are mixed [K_f of $Ag(S_2O_3)_2^{3-} = 4.7 \times 10^{13}$]?

•• **19.100** Potassium thiocyanate, KSCN, is often used to detect the presence of Fe^{3+} ions in solution through the formation of the red $Fe(H_2O)_5SCN^{2+}$ (or, more simply, $FeSCN^{2+}$). What is $[Fe^{3+}]$ when 0.50 L each of 0.0015 M $Fe(NO_3)_3$ and 0.20 M KSCN are mixed? K_f of $FeSCN^{2+} = 8.9 \times 10^2$.

•• **19.101** Find the solubility of $Cr(OH)_3$ in a buffer of pH 13.0 [K_{sp} of $Cr(OH)_3 = 6.3 \times 10^{-31}$; K_f of $Cr(OH)_4^- = 8.0 \times 10^{29}$].

•• **19.102** Find the solubility of AgI in 2.5 M NH_3 [K_{sp} of AgI = 8.3×10^{-17}; K_f of $Ag(NH_3)_2^+ = 1.7 \times 10^7$].

•• **19.103** When 0.84 g of $ZnCl_2$ is dissolved in 245 mL of 0.150 M NaCN, what are $[Zn^{2+}]$, $[Zn(CN)_4^{2-}]$, and $[CN^-]$ [K_f of $Zn(CN)_4^{2-} = 4.2 \times 10^{19}$]?

•• **19.104** When 2.4 g of $Co(NO_3)_2$ is dissolved in 0.350 L of 0.22 M KOH, what are $[Co^{2+}]$, $[Co(OH)_4^{2-}]$, and $[OH^-]$ [K_f of $Co(OH)_4^{2-} = 5 \times 10^9$]?

Comprehensive Problems

•• **19.105** What volumes of 0.200 M HCOOH and 2.00 M NaOH would make 500. mL of a buffer with the pH of one made from 475 mL of 0.200 M benzoic acid and 25 mL of 2.00 M NaOH?

•• **19.106** A microbiologist is preparing a medium on which to culture *E. coli* bacteria. She buffers the medium at pH 7.00 to minimize

Biological: 19.90, 19.106

the effect of acid-producing fermentation. What volumes of equimolar aqueous solutions of K_2HPO_4 and KH_2PO_4 must she combine to make 100. mL of the pH 7.00 buffer?

•• **19.107** As an FDA physiologist, you need 0.700 L of formic acid–formate buffer with a pH of 3.74.
(a) What is the required buffer-component concentration ratio?
(b) How do you prepare this solution from stock solutions of 1.0 M HCOOH and 1.0 M NaOH?
(c) What is the final concentration of HCOOH in this solution?

•• **19.108** Tris(hydroxymethyl)aminomethane [$(HOCH_2)_3CNH_2$, known as TRIS] is a weak base used in biochemical experiments to make buffer solutions in the pH range of 7 to 9. A certain TRIS buffer has a pH of 8.10 at 25°C and a pH of 7.80 at 37°C. Why does the pH change with temperature?

•• **19.109** Water flowing through pipes of carbon steel must be kept at pH 5 or greater to limit corrosion. If an 8.0×10^3 lb/hr water stream contains 10 ppm sulfuric acid and 0.015% acetic acid, how many pounds per hour of sodium acetate trihydrate must be added to maintain that pH?

•• **19.110** Gout is caused by an error in metabolism that leads to a buildup of uric acid in body fluids, which is deposited as slightly soluble sodium urate ($C_5H_3N_4O_3Na$) in the joints. If the extracellular [Na^+] is 0.15 M and the solubility of sodium urate is 0.085 g/100. mL, what is the minimum urate ion concentration (abbreviated [Ur^-]) that will cause a deposit of sodium urate?

•• **19.111** Cadmium ion in solution is analyzed by precipitation as the sulfide, a yellow compound used as a pigment in everything from artists' oil paints to glass and rubber. Calculate the molar solubility of cadmium sulfide at 25°C.

•• **19.112** In the discussion of cave formation in the Chemical Connections essay (p. 859), the dissolution of CO_2 (equation 1) has a K_{eq} of 3.1×10^{-2}, and the formation of aqueous $Ca(HCO_3)_2$ (equation 2) has a K_{eq} of 1×10^{-12}. The fraction by volume of atmospheric CO_2 is 3×10^{-4}. (a) Find [$CO_2(aq)$] in equilibrium with atmospheric CO_2. (b) Determine [Ca^{2+}] arising from (equation 2) given current levels of atmospheric CO_2. (c) Calculate [Ca^{2+}] if atmospheric CO_2 doubles.

•• **19.113** Phosphate systems form essential buffers in organisms. Calculate the pH of a buffer made by dissolving 0.80 mol of NaOH in 0.50 L of 1.0 M H_3PO_4.

•• **19.114** The solubility of KCl is 3.7 M at 20°C. Two beakers contain 100. mL of saturated KCl solution: 100. mL of 6.0 M HCl is added to the first beaker and 100. mL of 12 M HCl to the second.
(a) Find the ion-product constant of KCl at 20°C.
(b) What mass, if any, of KCl will precipitate from each beaker?

•• **19.115** It is possible to detect NH_3 gas over 10^{-2} M NH_3. To what pH must 0.15 M NH_4Cl be raised to form detectable NH_3?

•• **19.116** Manganese(II) sulfide is one of the compounds found in the nodules on the ocean floor that may eventually be a primary source of many transition metals. The solubility of MnS is 4.7×10^{-4} g/100 mL solution. Estimate the K_{sp} of MnS.

•• **19.117** The normal pH of blood is 7.40 ± 0.05 and is controlled in part by the H_2CO_3/HCO_3^- buffer system.
(a) Assuming that the K_a value for carbonic acid at 25°C applies to blood, what is the [H_2CO_3]/[HCO_3^-] ratio in normal blood?
(b) In a condition called *acidosis*, the blood is too acidic. What is the [H_2CO_3]/[HCO_3^-] ratio in a patient whose blood pH is 7.20?

•• **19.118** A bioengineer preparing cells for cloning bathes a small piece of rat epithelial tissue in a TRIS buffer (see Problem 19.108). The buffer is made by dissolving 43.0 g of TRIS ($pK_b = 5.91$) in enough 0.095 M HCl to make 1.00 L of solution. What is the molarity of TRIS and the pH of the buffer?

•• **19.119** Sketch a qualitative curve for the titration of ethylenediamine, $H_2NCH_2CH_2NH_2$, with 0.1 M HCl.

•• **19.120** A solution contains 0.10 M $ZnCl_2$ and 0.020 M $MnCl_2$. Given the following information, how would you adjust the pH to separate the ions as their sulfides ([H_2S] of a saturated aqueous solution at 25°C = 0.10 M; $K_w = 1.0\times10^{-14}$ at 25°C)?

$$MnS + H_2O \rightleftharpoons Mn^{2+} + HS^- + OH^- \quad K_{sp} = 3\times10^{-11}$$
$$ZnS + H_2O \rightleftharpoons Zn^{2+} + HS^- + OH^- \quad K_{sp} = 2\times10^{-22}$$
$$H_2S + H_2O \rightleftharpoons H_3O^+ + HS^- \quad K_{a1} = 9\times10^{-8}$$

••• **19.121** Amino acids [general formula $NH_2CH(R)COOH$] can be considered polyprotic acids. In many cases, the R group contains additional amine and carboxyl groups.
(a) Can an amino acid dissolved in pure water have a protonated COOH group and an unprotonated NH_2 group (K_a of COOH group = 4.47×10^{-3}; K_b of NH_2 group = 6.03×10^{-5})? Use glycine, NH_2CH_2COOH, to explain why.
(b) Calculate [$^+NH_3CH_2COO^-$]/[$^+NH_3CH_2COOH$] at pH 5.5.
(c) The R group of lysine is $-CH_2CH_2CH_2CH_2NH_2$ (pK_b = 3.47). Draw the structure of lysine at pH 1, physiological pH (~7), and pH 13.
(d) The R group of glutamic acid is $-CH_2CH_2COOH$ (pK_a = 4.07). Of the forms of glutamic acid that are shown below, which predominates at (1) pH 1, (2) physiological pH (~7), and (3) pH 13?

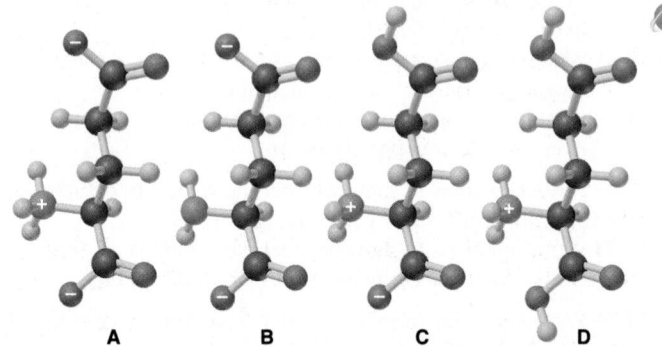

A   **B**   **C**   **D**

•• **19.122** The scene below depicts a saturated solution of $MCl_2(s)$ in the presence of dilute aqueous NaCl; each sphere represents 1.0×10^{-6} mol of ion, and the volume is 250.0 mL (solid MCl_2 is shown as green chunks; Na^+ ions and water molecules are not shown).

(a) Calculate the K_{sp} of MCl_2.

(b) If $M(NO_3)_2(s)$ is added, is there an increase, decrease, or no change in the number of Cl^- particles? In the K_{sp}? In the mass of $MCl_2(s)$?

•• **19.123** Tooth enamel consists of hydroxyapatite, $Ca_5(PO_4)_3OH$ ($K_{sp} = 6.8\times10^{-37}$). Fluoride ion added to drinking water reacts with $Ca_5(PO_4)_3OH$ to form the more tooth decay–resistant fluorapatite, $Ca_5(PO_4)_3F$ ($K_{sp} = 1.0\times10^{-60}$). Fluoridated water has dramatically decreased cavities among children. Calculate the solubility of $Ca_5(PO_4)_3OH$ and of $Ca_5(PO_4)_3F$ in water.

•• **19.124** The acid-base indicator ethyl orange turns from red to yellow over the pH range 3.4 to 4.8. Estimate K_a for ethyl orange.

•• **19.125** Use the values obtained in Problem 19.54 to sketch a curve of $[H_3O^+]$ vs. mL of added titrant. Are there advantages or disadvantages to viewing the results in this form? Explain.

••• **19.126** Instrumental acid-base titrations use a pH meter to monitor the changes in pH and volume. The equivalence point is found from the volume at which the curve has the steepest slope.

(a) Use Figure 19.7 to calculate the slope $\Delta pH/\Delta V$ for all pairs of adjacent points and to calculate the average volume (V_{avg}) for each interval.

(b) Plot $\Delta pH/\Delta V$ vs. V_{avg} to find the steepest slope, and thus the volume at the equivalence point. (For example, the first pair of points gives $\Delta pH = 0.22$, $\Delta V = 10.00$ mL; hence, $\Delta pH/\Delta V = 0.022$ mL^{-1}, and $V_{avg} = 5.00$ mL.)

•• **19.127** What is the pH of a solution of 6.5×10^{-9} mol of $Ca(OH)_2$ in 10.0 L of water [K_{sp} of $Ca(OH)_2 = 6.5\times10^{-6}$]?

•• **19.128** The Henderson-Hasselbalch equation gives a relationship for obtaining the pH of a buffer solution consisting of HA and A^-. Derive an analogous relationship for obtaining the pOH of a buffer solution consisting of B and BH^+.

•• **19.129** Muscle physiologists study the accumulation of lactic acid [$CH_3CH(OH)COOH$] during exercise. Food chemists study its occurrence in sour milk, beer, wine, and fruit. Industrial microbiologists study its formation by various bacterial species from carbohydrates. A biochemist prepares a lactic acid–lactate buffer by mixing 225 mL of 0.85 M lactic acid ($K_a = 1.38\times10^{-4}$) with 435 mL of 0.68 M sodium lactate. What is the buffer pH?

•• **19.130** A student wants to dissolve the maximum amount of CaF_2 ($K_{sp} = 3.2\times10^{-11}$) to make 1 L of aqueous solution.

(a) Into which of the following solvents should she dissolve the salt?

(I) Pure water   (II) 0.01 M HF

(III) 0.01 M NaOH   (IV) 0.01 M HCl

(V) 0.01 M $Ca(OH)_2$

(b) Which would dissolve the least amount of salt?

•• **19.131** A 500.-mL solution consists of 0.050 mol of solid NaOH and 0.13 mol of hypochlorous acid (HClO; $K_a = 3.0\times10^{-8}$) dissolved in water.

(a) Aside from water, what is the concentration of each species that is present?

(b) What is the pH of the solution?

(c) What is the pH after adding 0.0050 mol of HCl to the flask?

•• **19.132** Calcium ion present in water supplies is easily precipitated as calcite ($CaCO_3$):

$$Ca^{2+}(aq) + CO_3^{2-}(aq) \rightleftharpoons CaCO_3(s)$$

Because the K_{sp} decreases with temperature, heating hard water forms a calcite "scale," which clogs pipes and water heaters. Find the solubility of calcite in water (a) at 10°C ($K_{sp} = 4.4\times10^{-9}$) and (b) at 30°C ($K_{sp} = 3.1\times10^{-9}$).

•• **19.133** Calculate the molar solubility of $Hg_2C_2O_4$ ($K_{sp} = 1.75\times10^{-13}$) in 0.13 M $Hg_2(NO_3)_2$.

•• **19.134** The well water in an area is "hard" because it is in equilibrium with $CaCO_3$ in the surrounding rocks. What is the concentration of Ca^{2+} in the well water (assuming the water's pH is such that the CO_3^{2-} ion is not significantly protonated)? (See Appendix C for K_{sp} of $CaCO_3$.)

•• **19.135** Environmental engineers use *alkalinity* as a measure of the capacity of carbonate buffering systems in water samples:

Alkalinity (mol/L) = $[HCO_3^-] + 2[CO_3^{2-}] + [OH^-] - [H^+]$

Find the alkalinity of a water sample that has a pH of 9.5, 26.0 mg/L CO_3^{2-}, and 65.0 mg/L HCO_3^-.

•• **19.136** Human blood contains one buffer system based on phosphate species and one on carbonate species. Assuming that blood has a normal pH of 7.4, what are the principal phosphate and carbonate species present? What is the ratio of the two phosphate species? (In the presence of the dissolved ions and other species in blood, K_{a1} of $H_3PO_4 = 1.3\times10^{-2}$, $K_{a2} = 2.3\times10^{-7}$, and $K_{a3} = 6\times10^{-12}$; K_{a1} of $H_2CO_3 = 8\times10^{-7}$ and $K_{a2} = 1.6\times10^{-10}$.)

•• **19.137** Litmus is an organic dye extracted from lichens. It is red below pH 4.5 and blue above pH 8.3. One drop of either 0.1 M HCl (pH 1) or a pH 3 buffer changes blue litmus paper to red, but a drop of 0.001 M HCl (also pH 3) does not. Explain.

•• **19.138** An environmental technician collects a sample of rainwater. Back in the lab, her pH meter isn't working, so she uses indicator solutions to estimate the pH. A piece of litmus paper turns red, indicating acidity, so she divides the sample into thirds and obtains the following results: thymol blue turns yellow; bromphenol blue turns green; and methyl red turns red. Estimate the pH of the rainwater.

••• **19.139** A 0.050 M H_2S solution contains 0.15 M $NiCl_2$ and 0.35 M $Hg(NO_3)_2$. What pH is required to precipitate the maximum amount of HgS but none of the NiS? (See Appendix C.)

••• **19.140** Quantitative analysis of Cl^- ion is often performed by a titration with silver nitrate, using sodium chromate as an indicator. As standardized $AgNO_3$ is added, both white AgCl and red Ag_2CrO_4 precipitate, but so long as some Cl^- remains, the Ag_2CrO_4 redissolves as the mixture is stirred. When the red color is permanent, the equivalence point has been reached.

(a) Calculate the equilibrium constant for the reaction

$$2AgCl(s) + CrO_4^{2-}(aq) \rightleftharpoons Ag_2CrO_4(s) + 2Cl^-(aq)$$

(b) Explain why the silver chromate redissolves.

(c) If 25.00 cm^3 of 0.1000 M NaCl is mixed with 25.00 cm^3 of 0.1000 M $AgNO_3$, what is the concentration of Ag^+ remaining in solution? Is this sufficient to precipitate any silver chromate?

•• **19.141** An ecobotanist separates the components of a tropical bark extract by chromatography. She discovers a large proportion of quinidine, a dextrorotatory isomer of quinine used for control of arrhythmic heartbeat. Quinidine has two basic nitrogens ($K_{b1} = 4.0\times10^{-6}$ and $K_{b2} = 1.0\times10^{-10}$). To measure the concentration, she carries out a titration. Because of the low

---

Biological: 19.123, 19.129, 19.136, 19.141; Environmental: 19.134, 19.135, 19.138; Industrial: 19.132

solubility of quinidine, she first protonates both nitrogens with excess HCl and titrates the acidified solution with standardized base. A 33.85-mg sample of quinidine ($\mathcal{M}$ = 324.41 g/mol) is acidified with 6.55 mL of 0.150 M HCl.
(a) How many milliliters of 0.0133 M NaOH are needed to titrate the excess HCl?
(b) How many additional milliliters of titrant are needed to reach the first equivalence point of quinidine dihydrochloride?
(c) What is the pH at the first equivalence point?

•• **19.142** Some kidney stones form by the precipitation of calcium oxalate monohydrate ($CaC_2O_4 \cdot H_2O$, $K_{sp} = 2.3 \times 10^{-9}$). The pH of urine varies from 5.5 to 7.0, and the average $[Ca^{2+}]$ in urine is 2.6×10^{-3} M.
(a) If the [oxalic acid] in urine is 3.0×10^{-13} M, will kidney stones form at pH = 5.5?
(b) At pH = 7.0?
(c) Vegetarians have a urine pH above 7. Are they more or less likely to form kidney stones?

•• **19.143** A biochemist needs a medium for acid-producing bacteria. The pH of the medium must not change by more than 0.05 pH units for every 0.0010 mol of H_3O^+ generated by the organisms per liter of medium. A buffer consisting of 0.10 M HA and 0.10 M A^- is included in the medium to control its pH. What volume of this buffer must be included in 1.0 L of medium?

•• **19.144** A 35.00-mL solution of 0.2500 M HF is titrated with a standardized 0.1532 M solution of NaOH at 25°C.
(a) What is the pH of the HF solution before titrant is added?
(b) How many milliliters of titrant are required to reach the equivalence point?
(c) What is the pH at 0.50 mL before the equivalence point?
(d) What is the pH at the equivalence point?
(e) What is the pH at 0.50 mL after the equivalence point?

••• **19.145** Because of the toxicity of mercury compounds, mercury(I) chloride is used in antibacterial salves. The mercury(I) ion (Hg_2^{2+}) consists of two bound Hg^+ ions.
(a) What is the empirical formula of mercury(I) chloride?
(b) Calculate $[Hg_2^{2+}]$ in a saturated solution of mercury(I) chloride ($K_{sp} = 1.5 \times 10^{-18}$).
(c) A seawater sample contains 0.20 lb of NaCl per gallon. Find $[Hg_2^{2+}]$ if the seawater is saturated with mercury(I) chloride.
(d) How many grams of mercury(I) chloride are needed to saturate 4900 km^3 of pure water (the volume of Lake Michigan)?
(e) How many grams of mercury(I) chloride are needed to saturate 4900 km^3 of seawater?

••• **19.146** A lake that has a surface area of 10.0 acres (1 acre = 4.840×10^3 yd^2) receives 1.00 in. of rain of pH 4.20. (Assume that the acidity of the rain is due to a strong, monoprotic acid.)
(a) How many moles of H_3O^+ are in the rain falling on the lake?
(b) If the lake is unbuffered (pH = 7.00) and its average depth is 10.0 ft before the rain, find the pH after the rain has been mixed with lake water. (Ignore runoff from the surrounding land.)
(c) If the lake contains hydrogen carbonate ions (HCO_3^-), what mass of HCO_3^- would neutralize the acid in the rain?

•• **19.147** A 35.0-mL solution of 0.075 M $CaCl_2$ is mixed with 25.0 mL of 0.090 M $BaCl_2$.
(a) If aqueous KF is added, which fluoride precipitates first?
(b) Describe how the metal ions can be separated using KF to form the fluorides.

(c) Calculate the fluoride ion concentration that will accomplish the separation.

••• **19.148** Even before the industrial age, rainwater was slightly acidic due to dissolved CO_2. Use the following data to calculate pH of unpolluted rainwater at 25°C: vol % in air of CO_2 = 0.033 vol %; solubility of CO_2 in pure water at 25°C and 1 atm = 88 mL CO_2/100 mL H_2O; K_{a1} of H_2CO_3 = 4.5×10^{-7}.

••• **19.149** Seawater at the surface has a pH of about 8.5.
(a) Which of the following species has the highest concentration at this pH: H_2CO_3; HCO_3^-; CO_3^{2-}? Explain.
(b) What are the concentration ratios $[CO_3^{2-}]/[HCO_3^-]$ and $[HCO_3^-]/[H_2CO_3]$ at this pH?
(c) In the deep sea, light levels are low, and the pH is around 7.5. Suggest a reason for the lower pH at the greater ocean depth. (*Hint:* Consider the presence or absence of plant and animal life, and the effects on carbon dioxide concentrations.)

••• **19.150** Ethylenediaminetetraacetic acid (abbreviated H_4EDTA) is a tetraprotic acid. Its salts are used to treat toxic metal poisoning by forming soluble complex ions that are then excreted. Because $EDTA^{4-}$ also binds essential calcium ions, it is often administered as the calcium disodium salt. For example, when $Na_2Ca(EDTA)$ is given to a patient, the $[Ca(EDTA)]^{2-}$ ions react with circulating Pb^{2+} ions and the metal ions are exchanged:

$$[Ca(EDTA)]^{2-}(aq) + Pb^{2+}(aq) \rightleftharpoons$$
$$[Pb(EDTA)]^{2-}(aq) + Ca^{2+}(aq) \qquad K_c = 2.5 \times 10^7$$

A child has a dangerous blood lead level of 120 μg/100 mL. If the child is administered 100. mL of 0.10 M $Na_2Ca(EDTA)$, assuming the exchange reaction and excretion process are 100% efficient, what is the final concentration of Pb^{2+} in μg/100 mL blood? (Total blood volume is 1.5 L.)

•• **19.151** Buffers that are based on 3-morpholinopropanesulfonic acid (MOPS) are often used in RNA analysis. The useful pH range of a MOPS buffer is 6.5 to 7.9. Estimate the K_a of MOPS.

•• **19.152** NaCl is purified by adding HCl to a saturated solution of NaCl (317 g/L). When 28.5 mL of 8.65 M HCl is added to 0.100 L of saturated solution, what mass (g) of pure NaCl precipitates?

•• **19.153** Scenes A to D represent tiny portions of 0.10 M aqueous solutions of a weak acid HA (red and blue; $K_a = 4.5 \times 10^{-5}$), its conjugate base A^- (red), or a mixture of the two (only these species are shown):

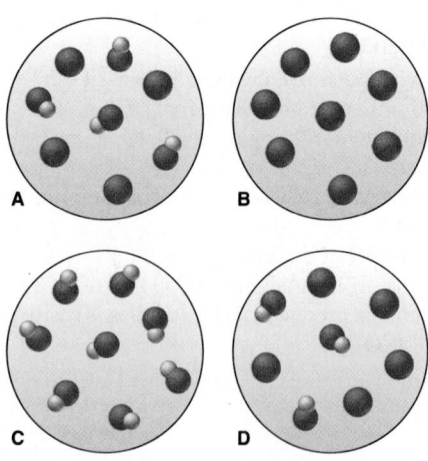

(a) Which scene(s) show(s) a buffer?

(b) What is the pH of each solution?

(c) Arrange the scenes in sequence, assuming that they represent stages in a weak acid–strong base titration.

(d) Which scene represents the titration at its equivalence point?

•• **19.154** Scenes A to C represent aqueous solutions of the slightly soluble salt MZ (only the ions of this salt are shown):

$$MZ(s) \rightleftharpoons M^{2+}(aq) + Z^{2-}(aq)$$

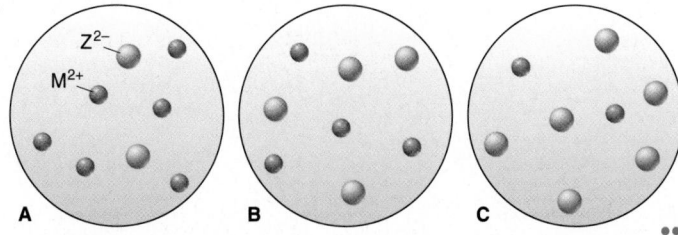

(a) Which scene represents the solution just after solid MZ is stirred thoroughly in distilled water?

(b) If each sphere represents $2.5 \times 10^{-6}\ M$ of ions, what is the K_{sp} of MZ?

(c) Which scene represents the solution after $Na_2Z(aq)$ is added?

(d) If Z^{2-} is CO_3^{2-}, which scene represents the solution after the pH has been lowered?

•• **19.155** The solubility of Ag(I) in aqueous solutions containing different concentrations of Cl^- is based on the following equilibria:

$$Ag^+(aq) + Cl^-(aq) \rightleftharpoons AgCl(s) \qquad K_{sp} = 1.8 \times 10^{-10}$$
$$Ag^+(aq) + 2Cl^-(aq) \rightleftharpoons AgCl_2^-(aq) \qquad K_f = 1.8 \times 10^5$$

When solid AgCl is shaken with a solution containing Cl^-, Ag(I) is present as both Ag^+ and $AgCl_2^-$. The solubility of AgCl is the sum of the concentrations of Ag^+ and $AgCl_2^-$.

(a) Show that $[Ag^+]$ in solution is given by

$$[Ag^+] = 1.8 \times 10^{-10}/[Cl^-]$$

and that $[AgCl_2^-]$ in solution is given by

$$[AgCl_2^-] = (3.2 \times 10^{-5})([Cl^-])$$

(b) Find the $[Cl^-]$ at which $[Ag^+] = [AgCl_2^-]$.

(c) Explain the shape of a plot of AgCl solubility vs. $[Cl^-]$.

(d) Find the solubility of AgCl at the $[Cl^-]$ of part (b), which is the minimum solubility of AgCl in the presence of Cl^-.

••• **19.156** EDTA binds metal ions to form complex ions (see Problem 19.150), so it is used to determine the concentrations of metal ions in solution:

$$M^{n+}(aq) + EDTA^{4-}(aq) \longrightarrow MEDTA^{n-4}(aq)$$

A 50.0-mL sample of $0.048\ M$ Co^{2+} is titrated with $0.050\ M$ $EDTA^{4-}$. Find $[Co^{2+}]$ and $[EDTA^{4-}]$ after (a) 25.0 mL and (b) 75.0 mL of $EDTA^{4-}$ are added (log K_f of $CoEDTA^{2-} = 16.31$).

20

Thermodynamics and the Steam Engine The relationship between heat and work was first applied most practically in the 19th century to understand and improve the newly invented steam engine. In this chapter, we examine this relationship to learn the real-world limit on the energy that can be obtained from any chemical or physical change. In the process, we see *why* reactions occur.

Thermodynamics: Entropy, Free Energy, and the Direction of Chemical Reactions

20.1 The Second Law of Thermodynamics: Predicting Spontaneous Change
Limitations of the First Law
The Sign of ΔH and Spontaneous Change
Freedom of Motion and Dispersal of Energy
Entropy and the Number of Microstates
Entropy and the Second Law
Standard Molar Entropies and the Third Law

20.2 Calculating the Change in Entropy of a Reaction
The Standard Entropy of Reaction
Entropy Changes in the Surroundings
Entropy Change and the Equilibrium State
Spontaneous Exothermic and Endothermic Reactions

20.3 Entropy, Free Energy, and Work
Free Energy Change and Reaction Spontaneity
Standard Free Energy Changes
ΔG and Work
Temperature and Reaction Spontaneity
Coupling of Reactions

20.4 Free Energy, Equilibrium, and Reaction Direction

In the last few chapters, we've posed and answered some essential questions about chemical and physical change: How fast does the change occur, and how is this rate affected by concentration and temperature? How much product will be present when the net change ceases, and how is this yield affected by concentration and temperature? We've explored these questions for systems ranging from the stratosphere to a limestone cave to the cells of your body.

But why does the change occur in the first place? From everyday experience, it seems that some changes happen by themselves—that is, spontaneously—almost as if a force were driving them in one direction and not the other. Turn on a gas stove, for example, and the methane mixes with oxygen and burns immediately to yield carbon dioxide and water vapor. But those products will not remake methane and oxygen no matter how long they mix. A steel shovel left outside slowly rusts, but put a rusty one outside and it won't become shiny. A cube of sugar dissolves in a cup of coffee after a few seconds of stirring, but stir for another century and the dissolved sugar won't ever reappear as a cube.

Chemists speak of a process that occurs by itself as being *spontaneous*. Some absorb energy, whereas others release it. The principles of thermodynamics, which were developed almost 200 years ago to help utilize the power derived from steam, allow us to understand the nature of spontaneous change. Despite their narrow historical purpose, these principles apply, as far as we know, to every system in the universe!

IN THIS CHAPTER . . . We begin by looking for a criterion to predict the direction of a spontaneous change. A review of the first law of thermodynamics shows that it accounts for the quantity of energy in a change but not the direction of the change. The sign of the enthalpy change is *not* a criterion for predicting direction either. Rather, we find this criterion in the second law of thermodynamics and its quantitative application of entropy (S), the state function that relates to the natural tendency of a system's energy to become more dispersed. We examine these ideas for both exothermic and endothermic changes. The concept of free energy we develop gives a simplified criterion for spontaneous change, and we see how it relates to the work a system can do. Finally, we consider the key relationship between the free energy change and the equilibrium constant of a reaction.

Concepts & Skills to Review
before you study this chapter

- internal energy, heat, and work (Section 6.1)
- state functions (Section 6.1) and standard states (Section 6.6)
- enthalpy, ΔH, and Hess's law (Sections 6.2 and 6.5)
- entropy and solution formation (Section 13.3)
- comparing Q and K to find reaction direction (Section 17.4)

20.1 THE SECOND LAW OF THERMODYNAMICS: PREDICTING SPONTANEOUS CHANGE

In a formal sense, a **spontaneous change** of a system, whether a chemical or physical change, or just a change in location, is one that occurs by itself under specified conditions, without a continuous input of energy from outside the system. The freezing of water, for example, is spontaneous for the system at 1 atm and −5°C. A spontaneous process such as burning or falling may need a little "push" to get started—a spark to ignite gasoline, a shove to knock a book off your desk—but once the process begins, it keeps going without the need for any continuous external input of energy.

In contrast, for a nonspontaneous change to occur, the surroundings must supply the system with a continuous input of energy. A book falls spontaneously, but it rises only if something else, such as a human hand (or a strong wind), supplies energy in the form of work. Under a given set of conditions, *if a change is spontaneous in one direction, it is **not** spontaneous in the other*.

The term *spontaneous* does not mean *instantaneous* or have anything to say about how long a process takes to occur; it means that, given enough time, the process will happen by itself. Many processes are spontaneous but slow—ripening, rusting, and (happily!) aging.

Literature Galley, William C. "Mass–Elastic Band Thermodynamics: A Visual Teaching Aid at the Introductory Level," *J. Chem. Educ.* **2007,** *84,* 1147.

Literature Gislason, Eric A.; Craig, Norman C. "Why Do Two Objects at Different Temperatures Come to a Common Intermediate Temperature When Put in Contact? Entropy Is Maximized," *J. Chem. Educ.* **2006,** *83,* 885.

Literature Holman, John; Pilling, Gwen. "Thermodynamics in Context: A Case Study of Contextualized Teaching for Undergraduates," *J. Chem. Educ.* **2004,** *81,* 373.

A chemical reaction proceeding toward equilibrium is an example of a spontaneous change. As you learned in Chapter 17, we can predict the net direction of the reaction—its spontaneous direction—by comparing the reaction quotient (Q) with the equilibrium constant (K). But *why* is there a drive to attain equilibrium? And what determines the value of the equilibrium constant? Can we tell the direction of a spontaneous change in cases that are not as obvious as burning gasoline or falling books? Because energy changes seem to be involved, let's begin by reviewing the idea of conservation of energy to see whether it can help uncover the criterion for spontaneity.

Limitations of the First Law of Thermodynamics

In Chapter 6, we discussed the first law of thermodynamics (the law of conservation of energy) and its application to chemical and physical systems. The first law states that the internal energy (E) of a system, the sum of the kinetic and potential energy of all its particles, changes when heat (q) and/or work (w) are gained or lost:

$$\Delta E = q + w$$

Whatever is not part of the system (sys) is part of the surroundings (surr), so the system and surroundings together constitute the universe (univ):

$$E_{univ} = E_{sys} + E_{surr}$$

Heat and/or work gained by the system is lost by the surroundings, and vice versa:

$$(q + w)_{sys} = -(q + w)_{surr}$$

It follows from these ideas that *the total energy of the universe is constant:*[*]

$$\Delta E_{sys} = -\Delta E_{surr} \qquad \text{therefore} \qquad \Delta E_{sys} + \Delta E_{surr} = 0 = \Delta E_{univ}$$

Literature Liberko, Charles A. "Using Science Fiction To Teach Thermodynamics: Vonnegut, Icenine, and Global Warming," *J. Chem. Educ.* **2004**, *81*, 509.

Is the first law sufficient to explain why a natural process takes place as it does? It certainly accounts for the energy involved. When gasoline burns in your car's engine, the first law states that the potential energy difference between the chemical bonds in the fuel mixture and those in the exhaust gases is converted to the kinetic energy of the moving car and its parts plus the heat released to the environment. When an ice cube melts in your hand, the first law tells that energy from your hand was transferred to the ice to change it to a liquid. If you could measure the work and heat involved in each case, you would find that energy is conserved as it is converted from one form to another.

However, the first law does not help us make sense of the *direction* of the change. Why doesn't the heat released in the car engine convert exhaust fumes back into gasoline and oxygen? Why doesn't the pool of water in your cupped hand transfer the heat back to your hand and refreeze? Neither of these events would violate the first law—energy would still be conserved—but they never happen. The first law by itself tells nothing about the direction of a spontaneous change, so we must look elsewhere for a way to predict that direction.

The Sign of ΔH Cannot Predict Spontaneous Change

In the mid-19th century, some scientists thought that the sign of the enthalpy change (ΔH), the heat gained or lost at constant pressure (q_P), was the criterion for spontaneity. They thought that exothermic processes ($\Delta H < 0$) were spontaneous and endothermic ones ($\Delta H > 0$) were nonspontaneous. This hypothesis has a lot of support from observation; after all, many spontaneous processes *are*

---

[*]Any modern statement of conservation of energy must take into account mass-energy equivalence and the processes in stars, which convert enormous amounts of matter into energy. These can be included by stating that the total *mass-energy* of the universe is constant.

exothermic. All combustion reactions, such as methane burning, are spontaneous and exothermic:

$$CH_4(g) + 2O_2(g) \longrightarrow CO_2(g) + 2H_2O(g) \qquad \Delta H^\circ_{rxn} = -802 \text{ kJ}$$

Iron metal oxidizes spontaneously and exothermically:

$$2Fe(s) + \tfrac{3}{2}O_2(g) \longrightarrow Fe_2O_3(s) \qquad \Delta H^\circ_{rxn} = -826 \text{ kJ}$$

Ionic compounds, such as NaCl, form spontaneously and exothermically from their elements:

$$Na(s) + \tfrac{1}{2}Cl_2(g) \longrightarrow NaCl(s) \qquad \Delta H^\circ_{rxn} = -411 \text{ kJ}$$

However, in many other cases, the sign of ΔH is no help. An exothermic process occurs spontaneously under certain conditions, whereas the opposite, endothermic, process occurs spontaneously under other conditions. Consider the following examples of phase changes, dissolving salts, and chemical changes.

At ordinary pressure, water freezes below 0°C but melts above 0°C. Both changes are spontaneous, but the first is exothermic and the second endothermic:

$$H_2O(l) \longrightarrow H_2O(s) \qquad \Delta H^\circ_{rxn} = -6.02 \text{ kJ } (exo\text{thermic; spontaneous at } T < 0°C)$$
$$H_2O(s) \longrightarrow H_2O(l) \qquad \Delta H^\circ_{rxn} = +6.02 \text{ kJ } (endo\text{thermic; spontaneous at } T > 0°C)$$

At ordinary pressure and room temperature, liquid water vaporizes spontaneously in dry air, another endothermic change:

$$H_2O(l) \longrightarrow H_2O(g) \qquad \Delta H^\circ_{rxn} = +44.0 \text{ kJ}$$

In fact, all melting and vaporizing are endothermic changes that are spontaneous under proper conditions.

Recall from Chapter 13 that most water-soluble salts have a positive ΔH°_{soln} yet dissolve spontaneously:

$$NaCl(s) \xrightarrow{H_2O} Na^+(aq) + Cl^-(aq) \qquad \Delta H^\circ_{soln} = +3.9 \text{ kJ}$$
$$RbClO_3(s) \xrightarrow{H_2O} Rb^+(aq) + ClO_3^-(aq) \qquad \Delta H^\circ_{soln} = +47.7 \text{ kJ}$$
$$NH_4NO_3(s) \xrightarrow{H_2O} NH_4^+(aq) + NO_3^-(aq) \qquad \Delta H^\circ_{soln} = +25.7 \text{ kJ}$$

Some endothermic chemical changes are also spontaneous:

$$N_2O_5(s) \longrightarrow 2NO_2(g) + \tfrac{1}{2}O_2(g) \qquad \Delta H^\circ_{rxn} = +109.5 \text{ kJ}$$
$$Ba(OH)_2 \cdot 8H_2O(s) + 2NH_4NO_3(s) \longrightarrow$$
$$Ba^{2+}(aq) + 2NO_3^-(aq) + 2NH_3(aq) + 10H_2O(l) \qquad \Delta H^\circ_{rxn} = +62.3 \text{ kJ}$$

In the second reaction, the released waters of hydration solvate the ions, but the reaction mixture cannot absorb heat from the surroundings quickly enough, and the container becomes so cold that a wet block of wood freezes to it (Figure 20.1).

Freedom of Particle Motion and Dispersal of Particle Energy

What features common to the previous endothermic processes can help us see why they occur spontaneously? In each case, the particles that make up the matter have more freedom of motion after the change occurs. And this means that their energy of motion becomes more dispersed. As you'll see shortly, "dispersed" means spread over more quantized energy levels.

Phase changes lead from a solid, in which particle motion is restricted, to a liquid, in which the particles have more freedom to move around each other, to a gas, with its much greater freedom of particle motion. Along with this greater freedom of motion, the energy of the particles becomes dispersed over more levels. Dissolving a salt leads from a crystalline solid and pure liquid to ions and solvent molecules moving and interacting throughout the solution; their energy of motion, therefore, is much more dispersed. In the chemical reactions shown above, *fewer* moles of crystalline solids produce *more* moles of gases and/or solvated ions. In these cases, there is not only more freedom of motion, but more particles to disperse their energy over more levels.

A

B

Figure 20.1 A spontaneous endothermic chemical reaction. A, When the crystalline solid reactants barium hydroxide octahydrate and ammonium nitrate are mixed, a slurry soon forms as waters of hydration are released. **B,** The reaction mixture absorbs heat so quickly from the surroundings that the beaker becomes covered with frost, and a moistened block of wood freezes to it.

ARIS Presentation Center, Photo Library

Literature Hanson, Robert M.; Michalek, Bridget. "Give Them Money: The Boltzmann Game, a Classroom or Laboratory Activity Modeling Entropy Changes and the Distribution of Energy in Chemical Systems," *J. Chem. Educ.* **2006,** *83,* 581.

Thus, in each process, the particles have more freedom of motion and, therefore, their energy of motion has more levels over which to be dispersed:

less freedom of particle motion $\longrightarrow$ more freedom of particle motion
localized energy of motion $\longrightarrow$ dispersed energy of motion

Phase change: solid $\longrightarrow$ liquid $\longrightarrow$ gas
Dissolving of salt: crystalline solid + liquid $\longrightarrow$ ions in solution
Chemical change: crystalline solids $\longrightarrow$ gases + ions in solution

In thermodynamic terms, a change in the freedom of motion of particles in a system, that is, in the dispersal of their energy of motion, is a key factor determining the direction of a spontaneous process.

Entropy and the Number of Microstates

Literature Kozliak, Evguenii I.; Lambert, Frank L. "'Order-to-Disorder' for Entropy Change? Consider the Numbers!" *Chem. Educator 10* **2005**, *1*, 24–25.

Literature Lambert, Frank L. "Disorder—A Cracked Crutch for Supporting Entropy Discussions," *J. Chem. Educ.* **2002**, *79*, 187.

Literature Lambert, Frank L. "Entropy Is Simple, Qualitatively," *J. Chem. Educ.* **2002**, *79*, 1241.

Literature Lambert, Frank L. "'Disorder' in Unstretched Rubber Bands?" *J. Chem. Educ.* **2003**, *80*, 145.

Demo Lechner, Joseph H. "Visualizing Entropy," *J. Chem. Educ.* **1999**, *76*, 1382.

Let's see how freedom of motion and dispersal of energy relate to spontaneous change. In earlier chapters, we discussed the quantization of energy, not just quantized *electronic* energy levels of an atom or molecule, but also quantized *kinetic* energy levels—vibrational, rotational, and translational—of a molecule and its atoms (see Tools of the Laboratory, p. 357).

Picture a system of, say, 1 mol of N_2 gas and focus on one molecule. At any instant, it is moving through space (translating) at a specific speed, it is rotating at a specific frequency, and its atoms are vibrating at a specific frequency. In the next instant, the molecule collides with another or with the container, and these motional (kinetic) energy states change to different values. The complete quantum state of the molecule at any instant is given by these specific electronic, translational, rotational, and vibrational states.

Each quantized state of the whole system of molecules is called a *microstate*, and at any one instant, the total energy of the system is dispersed throughout one microstate. In the next instant, the energy of the system is dispersed throughout a different microstate. Many quantum states are possible for a single molecule, and the number of microstates possible for the system of 1 mol of molecules is staggering—on the order of $10^{10^{23}}$. At a given set of conditions, each microstate has the same total energy as any other. Therefore, each microstate is equally possible for the system, and the laws of probability say that, over time, all microstates are equally likely. If we focus only on microstates associated with thermal energy (that is, not electron energy states), the number of microstates for a system is the number of ways it can disperse its thermal (kinetic) energy among the various modes of motion of all its molecules.

In 1877, the Austrian mathematician and physicist Ludwig Boltzmann related the number of microstates (W) to the **entropy (S)** of the system:

$$S = k \ln W \qquad (20.1)$$

where k, the *Boltzmann constant,* is the universal gas constant (R) divided by Avogadro's number (N_A), or R/N_A, and equals 1.38×10^{-23} J/K. The term W is just a number of microstates, so it has no units; therefore, S has units of joules/kelvin (J/K). The system has its total energy in only one microstate at any one instant; therefore, the total number of microstates possible defines the extent of energy dispersal and, thus, entropy. Thus,

- A system with fewer microstates (smaller W) has *lower entropy (lower S).*
- A system with more microstates (larger W) has *higher entropy (higher S).*

For our earlier examples,

lower entropy (fewer microstates) $\longrightarrow$ higher entropy (more microstates)

Phase change: solid $\longrightarrow$ liquid $\longrightarrow$ gas
Dissolving of salt: crystalline solid + liquid $\longrightarrow$ ions in solution
Chemical change: crystalline solids $\longrightarrow$ gases + ions in solution

(In Chapter 13, we used some of these ideas to explain solution behavior.)

Changes in Entropy If a physical or chemical change results in a greater number of microstates, there are more ways to disperse the energy of the system. Thus, the entropy increases:

$$S_{\text{more microstates}} > S_{\text{fewer microstates}}$$

If the change results in a lower number of microstates, the entropy decreases.

Like internal energy (E) and enthalpy (H), entropy is a state function and, thus, depends only on the present state of the system, not on the path it took to arrive at that state (see Chapter 6, p. 241). Therefore, the change in entropy of the system (ΔS_{sys}) depends only on the difference between its final and initial values:

$$\Delta S_{\text{sys}} = S_{\text{final}} - S_{\text{initial}}$$

Like any state function, $\Delta S_{\text{sys}} > 0$ when its value increases during a change. For example, when dry ice sublimes to gaseous carbon dioxide, we have

$$CO_2(s) \longrightarrow CO_2(g) \qquad \Delta S_{\text{sys}} = S_{\text{final}} - S_{\text{initial}} = S_{\text{gaseous } CO_2} - S_{\text{solid } CO_2} > 0$$

Similarly, $\Delta S_{\text{sys}} < 0$ means the entropy decreases during a change, as when water vapor condenses:

$$H_2O(g) \longrightarrow H_2O(l) \qquad \Delta S_{\text{sys}} = S_{\text{liquid } H_2O} - S_{\text{gaseous } H_2O} < 0$$

Or, consider the decomposition of dinitrogen tetraoxide (written as $O_2N{-}NO_2$):

$$O_2N{-}NO_2(g) \longrightarrow 2NO_2(g)$$

When the N—N bond in 1 mol of dinitrogen tetraoxide molecules breaks, the 2 mol of NO_2 molecules have many more available rotational, vibrational, and translational motions; thus, in any given instant, their energy disperses throughout any one of a greater number of microstates:

$$\Delta S_{\text{sys}} = \Delta S_{\text{rxn}} = S_{\text{final}} - S_{\text{initial}} = S_{\text{products}} - S_{\text{reactants}} = 2S_{NO_2} - S_{N_2O_4} > 0$$

Quantitative Meaning of an Entropy Change Two approaches for quantifying an entropy change are different but give the same result. The first is a statistical approach based on the number of microstates possible for a system at any given instant. The second is based on the heat absorbed (or released) by the system. We'll explore both for a system of 1 mol of a gas expanding from 1 L to 2 L and behaving ideally, much as neon does at 298 K:

$$1 \text{ mol neon (initial: 1 L and 298 K)} \longrightarrow 1 \text{ mol neon (final: 2 L and 298 K)}$$

1. *The approach based on number of microstates.* We use this approach to find ΔS_{sys} by applying Equation 20.1. Figure 20.2 shows a container consisting of two identical flasks connected by a stopcock, with 1 mol of neon in the left flask and an evacuated right flask. We know that when we open the stopcock, the gas will expand to fill both flasks to 0.5 mol each—but *why?* Opening the stopcock increases the volume, which increases the number of translational energy levels available as it increases the number of particle locations. As a result, the particles will occupy more energy levels, so the number of possible microstates—and the entropy—increases.

Figure 20.3 presents this idea with particles on energy levels in a box of changeable volume. Note that when the stopcock opens, there are more energy levels, and they are closer together, so more distributions of particles are possible.

Figure 20.3 The entropy increase due to expansion of a gas. The energy levels available to a group of 21 gas particles are depicted as horizontal lines in a vertical box of narrow width. Each distribution of energies for the 21 particles represents one microstate. When the stopcock is opened, the volume increases, as indicated by the wider box on the right, and the particles have more freedom of motion and more energy levels available. At constant temperature, the total energy of the particles remains constant, but their energy has more ways to be distributed, so there are more microstates possible, and the entropy increases.

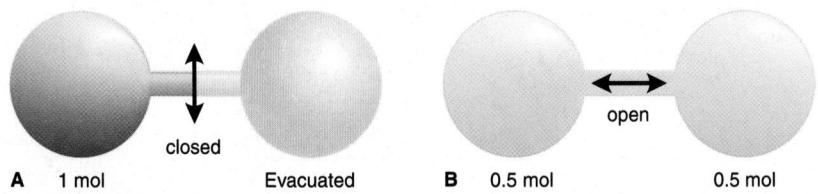

Figure 20.2 Spontaneous expansion of a gas. The container consists of two identical flasks connected by a stopcock. **A,** Initially, 1 mol of neon gas occupies one flask, and the other is evacuated. **B,** Open the stopcock, and the gas expands spontaneously until each flask contains 0.5 mol.

In Figure 20.4, we symbolically represent the number of microstates by the locations of particles in two attached flasks. Let's start with one neon atom and think through what happens as we add more atoms to the system, and open the stopcock. At a given instant, an atom in the left flask has its energy in one of some number (W) of possible microstates, and it has the same number possible in the right flask. Opening the stopcock increases the volume, which increases the number of possible locations and translational energy levels. As a result, the system has 2^1, or 2 *times* as many microstates available when the atom moves through both flasks (final state, W_{final}) as when it is confined to one (initial state, $W_{initial}$).

With more atoms, different sets of them can occupy various energy levels, and each distribution of occupied levels represents a microstate. Thus, as Figure 20.4 shows for two atoms, A and B, moving through both flasks, there are 2^2, or 4, times as many alternative microstates as when they are confined initially to one flask—some number of microstates with A and B in the left, the same number with A in the left and B in the right, that number with B in the left and A in the right, and that number with A and B in the right. Add another atom, and there are 2^3, or 8, times as many alternative microstates when the stopcock is open. With 10 Ne atoms, there are 2^{10}, or 1024, times as many microstates possible for the gas in both flasks. Finally, with 1 mol (N_A) of Ne, there are 2^{N_A} times as many microstates possible for the atoms in the larger volume (W_{final}) as in the smaller ($W_{initial}$). Thus, for 1 mol, we have

$$\frac{W_{final}}{W_{initial}} = 2^{N_A}$$

Now let's find ΔS_{sys} through the Boltzmann equation, $S = k \ln W$. From the properties of logarithms (Appendix A), we know that $\ln A - \ln B = \ln A/B$. Thus,

$$\Delta S_{sys} = S_{final} - S_{initial} = k \ln W_{final} - k \ln W_{initial} = k \ln \frac{W_{final}}{W_{initial}}$$

Figure 20.4 **Expansion of a gas and the increase in number of microstates.** When a gas confined to one flask is allowed to spread through two flasks, the energy of the system of particles is dispersed over more microstates, and so the entropy is higher. Each set of locations of particles in the available volume symbolically represents a different microstate. When the volume increases, the number of possible microstates increases by a factor of 2^n, where n is the number of particles.

Transparency/ARIS Presentation Center, Art Library

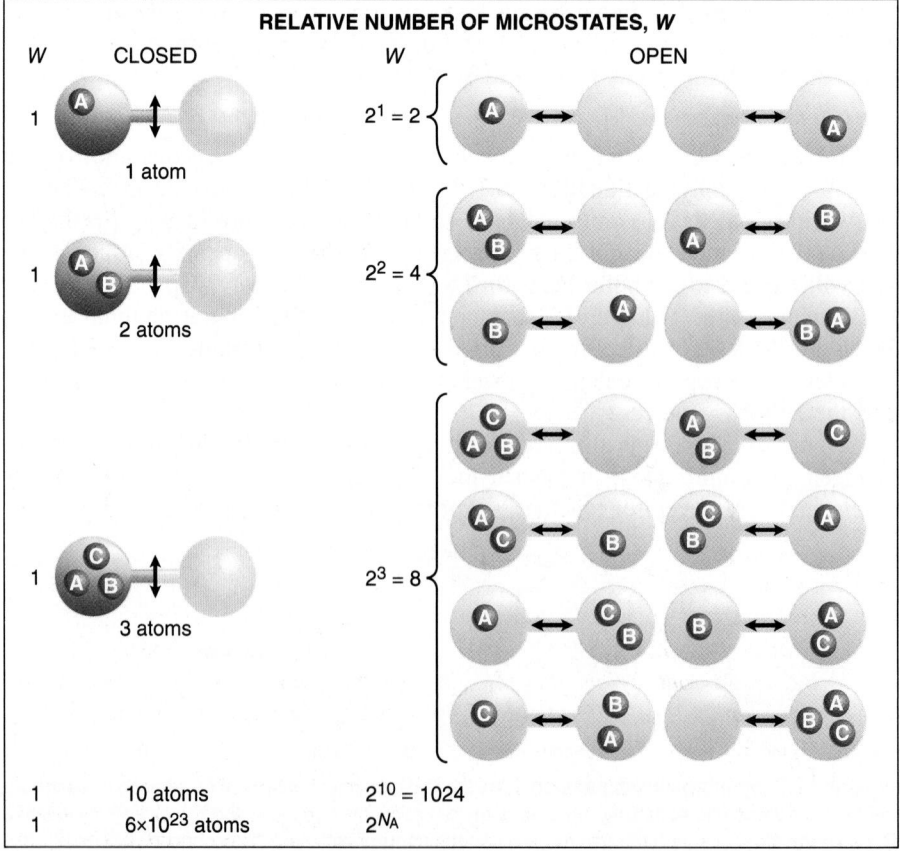

Also, from Appendix A, $\ln A^y = y \ln A$; thus, with $k = R/N_A$, we have

$$\Delta S_{sys} = \frac{R}{N_A} \ln 2^{N_A} = \left(\frac{R}{N_A}\right) N_A \ln 2 = R \ln 2$$

$$= (8.314 \text{ J/mol·K})(0.693) = 5.76 \text{ J/mol·K}$$

2. *The approach based on heat changes.* As you know, when a sample of gas is heated, it expands. Let's compare the entropy change we found from the statistical approach—a gas expanding into a vacuum—to the entropy change that occurs when a gas is heated and does work on the surroundings. This approach for finding the entropy change uses the relationship

$$\Delta S_{sys} = \frac{q_{rev}}{T} \qquad (20.2)$$

where T is the temperature at which the heat change occurs and q is the heat absorbed. The subscript "rev" refers to a *reversible process*, one that occurs slowly enough for equilibrium to be maintained, so that the direction of the change can be reversed by an *infinitesimal* reversal of conditions.

We can approximate the reversible expansion of the same sample of neon gas by placing it in a piston-cylinder assembly surrounded by a heat reservoir to maintain constant temperature, and confining the gas by the "pressure" of a beaker of sand resting on the piston (Figure 20.5). We remove one grain of sand (an "infinitesimal" change in pressure) with a pair of tweezers, and the gas expands a tiny amount, raising the piston and doing work, $-w$. If the neon behaves ideally, it absorbs from the heat reservoir a tiny increment of heat q, equivalent to $-w$. We remove another grain of sand, and the gas expands a tiny bit more and absorbs another tiny increment of heat. This expansion is very close to being reversible because we can reverse it at any point by putting a grain of sand back into the beaker, which causes a tiny quantity of work done by the surroundings compressing the gas and a tiny release of heat into the reservoir. If we continue this expansion process to 2 L and apply calculus through the process of integration, which adds together all the tiny increments of heat, we find q_{rev} is 1718 J. Thus, applying Equation 20.2, the entropy change is

$$\Delta S_{sys} = \frac{q_{rev}}{T} = \frac{1718 \text{ J}}{298 \text{ K}} = 5.76 \text{ J/K}$$

Notice that this is the same value we obtained previously. That first approach, which is often limited to simple systems like ideal gases, shows how a system's entropy changes when the volume occupied by the particles increases, and it allows us to visualize the changes in terms of energy dispersal. This second approach, which involves incremental heat changes, is not as easy to visualize, but is applicable to more complex systems.

Figure 20.5 A "reversible" change.
At constant temperature, changing the pressure by tiny increments approximates a reversible change in volume. Remove a grain of sand, and the gas does a tiny quantity of work, $-w$, to the surroundings and absorbs a tiny quantity of heat, q. The sum of all the quantities of heat divided by T is the change in entropy.

ARIS Presentation Center, Art Library

Demo Laird, Brian B. "Entropy, Disorder, and Freezing," *J. Chem. Educ.* **1999,** *76,* 1388.

Entropy and the Second Law of Thermodynamics

Now back to our earlier question: what criterion determines the direction of a spontaneous process? The change in entropy is the key, but to evaluate that change correctly, we must consider more than just the system. After all, some processes, such as ice melting or a crystal dissolving, occur spontaneously, and the system ends up with higher entropy; other processes, such as water freezing or a crystal forming, occur spontaneously, and the system ends up with lower entropy. But if we consider changes in *both* the system *and* its surroundings, we find that *all real processes occur spontaneously in the direction that increases the entropy of the universe (system plus surroundings).* This is one way to state the **second law of thermodynamics.**

The second law places no limitations on the entropy change of the system *or* the surroundings: either may be negative; that is, either system *or* surroundings may have lower entropy after the process. The law does state, however, that for a spontaneous process, the *sum* of the entropy changes must be positive. If the entropy of the system decreases, the entropy of the surroundings must increase even more to offset the system's decrease, so that the entropy of the universe (system *plus* surroundings) increases. A quantitative statement of the second law is, for any real spontaneous process,

$$\Delta S_{univ} = \Delta S_{sys} + \Delta S_{surr} > 0 \qquad (20.3)$$

Standard Molar Entropies and the Third Law

Both entropy and enthalpy are state functions, but the nature of their values differs in a fundamental way. Recall that we cannot determine absolute *enthalpies* because we have no easily measurable starting point, no baseline value for the enthalpy of a substance. Therefore, we measure only enthalpy *changes*.

In contrast, we *can* determine the absolute *entropy* of a substance. To do so, we apply the **third law of thermodynamics,** which states that *a perfect crystal has zero entropy at a temperature of absolute zero:* $S_{sys} = 0$ at 0 K. "Perfect" means that all the particles are aligned flawlessly in the crystal structure, with no defects of any kind. At absolute zero, all particles in the crystal have the minimum energy, so there is only one microstate, only one way the energy can be dispersed: thus, in Equation 20.1, $W = 1$, so $S = k \ln 1 = 0$. When we warm the crystal, its total energy increases and thus can be dispersed into more than one microstate. Thus, $W > 1$, $\ln W > 0$, and $S > 0$.

To obtain a value for S at a given temperature, we first cool a crystalline sample of the substance as close to 0 K as possible. Then we heat it in small increments, dividing q by T to get the increase in S for each increment, and add up all the entropy increases to the temperature of interest, usually 298 K. The entropy of a substance at a given temperature is therefore an *absolute* value that is equal to the entropy increase obtained when the substance is heated from 0 K to that temperature.

As with other thermodynamic variables, we usually compare entropy values for substances in their *standard states* at the temperature of interest: *1 atm for gases, 1 M for solutions, and the pure substance in its most stable form for solids or liquids.* Because entropy is an *extensive* property, that is, one that depends on the amount of substance, we are interested in the **standard molar entropy ($S°$)** in units of J/mol·K (or Jmol^{-1}K^{-1}). The $S°$ values at 298 K for many elements, compounds, and ions appear, with other thermodynamic variables, in Appendix B.

Predicting Relative $S°$ Values of a System Based on an understanding of systems at the molecular level and the effects of heat absorbed, we can see how the entropy of a substance is affected by conditions. Let's consider the effects of temperature, the most fundamental condition, and also physical state, dissolution, and atomic or molecular complexity. (All $S°$ values in the following discussion have units of J/mol·K and, unless stated otherwise, refer to the system at 298 K.)

1. *Temperature changes.* For a given substance, $S°$ *increases as the temperature rises.* Consider these typical values for copper metal:

T (K):	273	295	298
$S°$:	31.0	32.9	33.2

The temperature increases as heat is absorbed ($q > 0$), which represents an increase in the average kinetic energy of the particles. Recall from Figure 5.14 (p. 211) that the kinetic energies of gas particles are distributed over a range, which becomes wider as the temperature rises, and a similar behavior holds for liquids and solids. With more microstates available in which the energy can be dispersed at any instant, the entropy of the substance goes up. In other words, raising the temperature allows more possible microstates. Figure 20.6 provides three ways to view the effect of temperature on entropy.

Figure 20.6 **Visualizing the effect of temperature on entropy. A,** These computer simulations show the paths of particle centers in a crystal. At any $T > 0$ K, each particle moves about its lattice position: the higher the temperature, the greater the movement. Adding heat increases the total energy. Therefore, the particles have greater freedom of motion, and their energy is more dispersed; thus, the entropy increases. **B,** At any T, there is a range of occupied energy levels and, thus, a certain number of available microstates. As heat is added, the total energy increases, and the range of occupied energy levels becomes greater: lower T means fewer microstates (lower S); higher T means more microstates (higher S). **C,** The energy levels available to a group of 21 gas particles are shown as horizontal lines in a vertical box of fixed width. The height of the box represents the total energy and number of energy levels available: at lower T, the box is short, and at higher T, it is tall. Thus, when heat is added, the total energy increases *and* becomes more dispersed, so entropy increases.

Figure 20.7 **The increase in entropy from solid to liquid to gas.** A plot of entropy vs. temperature for O_2 includes selected entropy values (in J/mol·K). Note the gradual increase in entropy within a phase and the abrupt increase at a phase change. The molecular-scale views depict the increase in freedom of motion as the solid melts and, even more so, as the liquid vaporizes.

Transparency/ARIS Presentation Center, Active Art Library

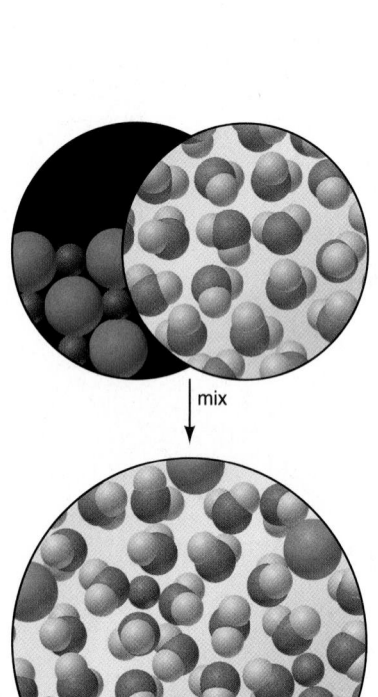

Figure 20.8 **The entropy change accompanying the dissolution of a salt.** When a crystalline salt and liquid water form a solution, the entropy change has two contributions: a positive contribution as the crystal separates into ions and the pure liquid disperses them, and a negative contribution as water molecules become organized around each ion. The relative magnitudes of these contributions determine the overall entropy change. The entropy of a salt solution is usually *greater* than that of the solid and water.

ARIS Presentation Center, Art Library

2. *Physical states and phase changes.* For a phase change such as melting or vaporizing, heat is absorbed ($q > 0$). The particles have more freedom of motion and their energy is more dispersed, so the entropy change is positive. Thus, $S°$ *increases for a substance as it changes from a solid to a liquid to a gas:*

	Na	**H_2O**	**C(graphite)**
$S°(s$ or $l)$:	51.4(s)	69.9(l)	5.7(s)
$S°(g)$:	153.6	188.7	158.0

Figure 20.7 shows how the entropy of molecular oxygen changes as it is heated from solid to liquid to gas, and the behavior of O_2 is typical of many other substances. Note the gradual increase within a phase as the temperature rises and the large, sudden increase at the phase change. Particles in the solid vibrate about their positions but, on average, remain fixed. The energy of the solid has the fewest microstates available and, thus, the least dispersed energy and the lowest entropy. As the temperature rises, the entropy gradually increases with the increase in the particles' kinetic energy. When the solid melts, the particles move much more freely between and around each other, so there is an abrupt increase in entropy. Further heating increases the speed of the particles in the liquid, and the entropy increases gradually. Finally, the particles undergo another, and much larger, abrupt entropy increase as a gas. That is, *the increase in entropy from liquid to gas is much larger than that from solid to liquid:* $\Delta S°_{vap} >> \Delta S°_{fus}$.

3. *Dissolving a solid or liquid.* The entropy of a dissolved molecular solid or liquid is *greater* than the entropy of the pure solute. For ionic solutes, however, the nature of solute *and* solvent affects the overall change:

	NaCl	**$AlCl_3$**	**CH_3OH**
$S°(s$ or $l)$:	72.1(s)	167(s)	127(l)
$S°(aq)$:	115.1	−148	132

When an ionic solid dissolves in water, the crystal breaks down, and the ions experience a great increase in freedom of motion as they separate, with their energy dispersed into more possible microstates. We expect the entropy of the ions themselves to be greater in the solution than in the crystal. However, some of the water molecules become arranged around the ions (Figure 20.8), which limits

A Ethanol   **B Water**   **C Solution of water and ethanol**

Figure 20.9 The small increase in entropy when ethanol dissolves in water. Pure ethanol **(A)** and pure water **(B)** have many intermolecular H bonds. **C,** In a solution of these two substances, the molecules form H bonds to one another, so their freedom of motion does not change significantly. Thus, the entropy increase is relatively small and is due mostly to identical molecules being separated and intermixed.

ARIS Presentation Center, Art Library

their freedom of motion (see Figure 13.2, p. 503). In fact, for small, multiply charged ions, some solvent molecules become so organized around the ions that their energy of motion becomes less dispersed. This negative portion of the total entropy change can lead to *negative S°* values for the ions in solution. For example, in the case of $AlCl_3$ above, the $Al^{3+}(aq)$ ion has such a negative S° value (-313 J/mol·K) that when $AlCl_3$ dissolves in water, even though S° of $Cl^-(aq)$ is positive, the overall entropy of aqueous $AlCl_3$ is lower than that of solid $AlCl_3$.*

For molecular solutes, the increase in entropy upon dissolving is typically much smaller than for ionic solutes. After all, for a solid such as glucose, there is no separation into ions, and for a liquid such as methanol or ethanol (Figure 20.9), the breakdown of a crystal structure is absent as well. Furthermore, in these small alcohols and in pure water, the molecules form many H bonds, so there is relatively little change in their freedom of motion when they are mixed. The small increase in the entropy of dissolved methanol or ethanol arises from the separation of identical molecules from each other in the pure solute and in water when the molecules are mixed.

4. *Dissolving a gas.* The particles in a gas already have so much freedom of motion—and, thus, such highly dispersed energy—that they lose some when they dissolve in a liquid or solid. Therefore, the entropy of a solution of a gas in a liquid or in a solid is always *less* than the entropy of the gas itself. For instance, when gaseous O_2 [$S°(g) = 205.0$ J/mol·K] dissolves in water, its entropy decreases dramatically [$S°(aq) = 110.9$ J/mol·K] (Figure 20.10). When a gas dissolves in another gas, however, the entropy increases as a result of the separation and mixing of the molecules.

5. *Atomic size or molecular complexity.* In general, differences in entropy values for substances in the same phase are based on atomic size and molecular complexity. Within a periodic group, energy levels become closer together for heavier atoms, which means the number of possible microstates, and thus molar entropy, increases down the group:

Figure 20.10 The large *decrease* in entropy of a gas when it dissolves in a liquid. The wide range of motion and high entropy of molecules of O_2 are reduced greatly when the gas dissolves in water.

ARIS Presentation Center, Art Library

	Li	Na	K	Rb	Cs
Atomic radius (pm):	152	186	227	248	265
Molar mass (g/mol):	6.941	22.99	39.10	85.47	132.9
$S°(s)$:	29.1	51.4	64.7	69.5	85.2

The same trend of increasing entropy down a group holds for similar compounds:

	HF	HCl	HBr	HI
Molar mass (g/mol):	20.01	36.46	80.91	127.9
$S°(g)$:	173.7	186.8	198.6	206.3

*An S° value for a hydrated ion can be negative because it is relative to the S° value for the hydrated proton, $H^+(aq)$, which is assigned a value of 0. In other words, $Al^{3+}(aq)$ has a lower entropy than $H^+(aq)$.

For an element that occurs in different forms (allotropes), the entropy is *higher* in the form that allows the atoms more freedom of motion, which disperses their energy because there are more microstates. For example, the $S°$ of graphite is 5.69 J/mol·K, whereas the $S°$ of diamond is 2.44 J/mol·K. In diamond, covalent bonds extend in three dimensions, allowing the atoms little movement; in graphite, covalent bonds extend only within a two-dimensional sheet, and motion of the sheets relative to each other is relatively easy.

For compounds, entropy increases with chemical complexity, that is, with the number of atoms in a formula unit or molecule of the compound. This trend holds for both ionic and covalent substances, as long as they are in the same phase:

	NaCl	AlCl$_3$	P$_4$O$_{10}$	NO	NO$_2$	N$_2$O$_4$
$S°(s)$:	72.1	167	229			
$S°(g)$:				211	240	304

The trend is based on the types of movement, and thus number of microstates, possible for the atoms (or ions) in each compound. For example, among the nitrogen oxides listed above, the two atoms of NO can vibrate only toward and away from each other. The three atoms of NO$_2$ have more vibrational motions, and the six atoms of N$_2$O$_4$ have even more (Figure 20.11; see also Figure B9.1, p. 357).

Figure 20.11 Entropy and vibrational motion. A diatomic molecule, such as NO, can vibrate in only one way. NO$_2$ can vibrate in more ways, and N$_2$O$_4$ in even more. Thus, as the number of atoms increases, there is more freedom of motion and there are more possible microstates into which a molecule can disperse its vibrational energy, so it has higher entropy.

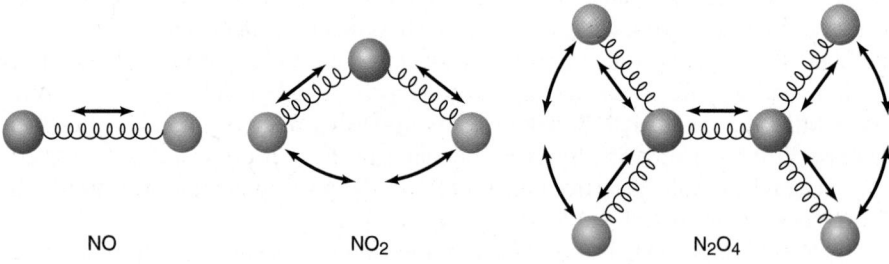

NO   NO$_2$ N$_2$O$_4$

For larger molecules, we also consider how one part of a molecule moves relative to other parts. A long hydrocarbon chain can rotate and vibrate in more ways than a short one, so entropy increases with chain length. A ring compound, such as cyclopentane (C$_5$H$_{10}$), has lower entropy than the corresponding chain compound, pentene (C$_5$H$_{10}$), because the ring structure restricts freedom of motion:

	CH$_4$(g)	C$_2$H$_6$(g)	C$_3$H$_8$(g)	C$_4$H$_{10}$(g)	C$_5$H$_{10}$(g)	C$_5$H$_{10}$(cyclo, g)	C$_2$H$_5$OH(l)
$S°$:	186	230	270	310	348	293	161

Remember, these trends hold only for *substances in the same physical state*. Gaseous methane (CH$_4$) has higher entropy than liquid ethanol (C$_2$H$_5$OH), even though ethanol molecules are more complex. When gases are compared with liquids, *the effect of physical state usually dominates that of molecular complexity*.

SAMPLE PROBLEM 20.1   Predicting Relative Entropy Values

PROBLEM   Choose the member with the higher entropy in each of the following pairs, and justify your choice [assume constant temperature, except in part (e)]:
(a) 1 mol of SO$_2$(g) or 1 mol of SO$_3$(g)
(b) 1 mol of CO$_2$(s) or 1 mol of CO$_2$(g)
(c) 3 mol of O$_2$(g) or 2 mol of O$_3$(g)
(d) 1 mol of KBr(s) or 1 mol of KBr(aq)

(e) Seawater at 2°C or at 23°C

(f) 1 mol of $CF_4(g)$ or 1 mol of $CCl_4(g)$

PLAN In general, we know that particles with more freedom of motion have more microstates into which to disperse their kinetic energy, so they have higher entropy, and that raising the temperature increases entropy. We apply the general categories described in the text to choose the member with the higher entropy.

SOLUTION (a) 1 mol of $SO_3(g)$. For equal numbers of moles of substances with the same types of atoms in the same physical state, the more atoms in the molecule, the more types of motion available, and thus the higher the entropy.

(b) 1 mol of $CO_2(g)$. For a given substance, entropy increases in the sequence $s < l < g$.

(c) 3 mol of $O_2(g)$. The two samples contain the same number of oxygen atoms but different numbers of molecules. Despite the greater complexity of O_3, the greater number of molecules dominates in this case because there are many more microstates possible for three moles of particles than for two moles.

(d) 1 mol of $KBr(aq)$. The two samples have the same number of ions, but their motion is more limited and their energy less dispersed in the solid than in the solution.

(e) Seawater at 23°C. Entropy increases with rising temperature.

(f) 1 mol of $CCl_4(g)$. For similar compounds, entropy increases with molar mass.

FOLLOW-UP PROBLEM 20.1 For 1 mol of substance at a given temperature, select the member with the higher entropy in each of the following pairs, and give the reason for your choice:

(a) $PCl_3(g)$ or $PCl_5(g)$ **(b)** $CaF_2(s)$ or $BaCl_2(s)$ **(c)** $Br_2(g)$ or $Br_2(l)$

Section Summary

A change is spontaneous under specified conditions if it occurs without a continuous input of energy. • Neither the first law of thermodynamics nor the sign of ΔH predicts the direction of a spontaneous change. • All spontaneous processes involve an increase in the dispersion of energy. • Entropy is a state function that measures the extent of energy dispersal into the number of microstates possible for a system, which is related to the freedom of motion of its particles. • The second law of thermodynamics states that, in a spontaneous process, the entropy of the universe (system plus surroundings) increases. • Absolute entropy values can be found because perfect crystals have zero entropy at 0 K (third law). • Standard molar entropy, $S°$ (J/mol·K), is affected by temperature, phase changes, dissolution, and atomic size or molecular complexity.

20.2 CALCULATING THE CHANGE IN ENTROPY OF A REACTION

In addition to understanding trends in $S°$ values for different substances or for the same substance in different phases, chemists are especially interested in learning how to predict the sign *and* calculate the value of the entropy change that occurs during a reaction.

Entropy Changes in the System: Standard Entropy of Reaction ($\Delta S°_{rxn}$)

Based on the ideas we just discussed, we can often predict the sign of the **standard entropy of reaction, $\Delta S°_{rxn}$,** the entropy change that occurs when all reactants and products are in their standard states. A deciding event is usually a change in the number of moles of gas. Because gases have such great freedom of motion and thus high molar entropies, *if the number of moles of gas increases, $\Delta S°_{rxn}$ is usually positive; if the number decreases, $\Delta S°_{rxn}$ is usually negative.*

For example, when $H_2(g)$ and $I_2(s)$ form $HI(g)$, the total number of moles of *substance* stays the same, but we predict that the entropy increases because the number of moles of *gas* increases:

$$H_2(g) + I_2(s) \longrightarrow 2HI(g) \qquad \Delta S°_{rxn} = S°_{products} - S°_{reactants} > 0$$

When ammonia forms from its elements, 4 mol of gas produces 2 mol of gas, so we predict that the entropy decreases:

$$N_2(g) + 3H_2(g) \rightleftharpoons 2NH_3(g) \qquad \Delta S°_{rxn} = S°_{products} - S°_{reactants} < 0$$

Sometimes, even when the amount (mol) of gases stays the same, another factor related to freedom of motion may help predict the sign of the entropy change. For example, when cyclopropane is heated to 500°C, the ring opens and propene forms. The chain has more freedom of motion than the ring, so the standard molar entropy of the product is greater than that of the reactant:

$$\underset{\underset{H_2C-CH_2(g)}{\diagup \diagdown}}{CH_2} \longrightarrow CH_3-CH=CH_2(g) \qquad \Delta S° = S°_{products} - S°_{reactants} > 0$$

Keep in mind, however, that in general *we cannot predict the sign of the entropy change unless the reaction involves a change in number of moles of gas.*

Recall that by applying Hess's law (Chapter 6), we can combine $\Delta H°_f$ values to find the standard heat of reaction, $\Delta H°_{rxn}$. Similarly, we combine $S°$ values to find the standard entropy of reaction, $\Delta S°_{rxn}$:

$$\Delta S°_{rxn} = \Sigma m S°_{products} - \Sigma n S°_{reactants} \qquad \text{(20.4)}$$

where m and n are the amounts (mol) of the individual species, given by their coefficients in the balanced equation. For the formation of ammonia, we have the following:

$$\Delta S°_{rxn} = [(2\text{ mol }NH_3)(S° \text{ of }NH_3)] - [(1\text{ mol }N_2)(S° \text{ of }N_2) + (3\text{ mol }H_2)(S° \text{ of }H_2)]$$

From Appendix B, we find the appropriate $S°$ values:

$$\Delta S°_{rxn} = [(2\text{ mol})(193\text{ J/mol·K})] - [(1\text{ mol})(191.5\text{ J/mol·K}) + (3\text{ mol})(130.6\text{ J/mol·K})]$$
$$= -197\text{ J/K}$$

As we predicted from the decrease in number of moles of gas, $\Delta S°_{rxn} < 0$.

ARIS Presentation Center, Worked Examples Library

SAMPLE PROBLEM 20.2   Calculating the Standard Entropy of Reaction, $\Delta S°_{rxn}$

PROBLEM Calculate $\Delta S°_{rxn}$ for the combustion of 1 mol of propane at 25°C:

$$C_3H_8(g) + 5O_2(g) \longrightarrow 3CO_2(g) + 4H_2O(l)$$

PLAN To determine $\Delta S°_{rxn}$, we apply Equation 20.4. We predict the sign of $\Delta S°_{rxn}$ from the change in the number of moles of gas: 6 mol of gas yields 3 mol of gas, so the entropy will probably decrease ($\Delta S°_{rxn} < 0$).

SOLUTION Calculating $\Delta S°_{rxn}$. Using Appendix B values,

$$\Delta S°_{rxn} = [(3\text{ mol }CO_2)(S° \text{ of }CO_2) + (4\text{ mol }H_2O)(S° \text{ of }H_2O)]$$
$$- [(1\text{ mol }C_3H_8)(S° \text{ of }C_3H_8) + (5\text{ mol }O_2)(S° \text{ of }O_2)]$$
$$= [(3\text{ mol})(213.7\text{ J/mol·K}) + (4\text{ mol})(69.9\text{ J/mol·K})]$$
$$- [(1\text{ mol})(269.9\text{ J/mol·K}) + (5\text{ mol})(205.0\text{ J/mol·K})]$$
$$= -374\text{ J/K}$$

CHECK $\Delta S° < 0$, so our prediction is correct. Rounding gives $[3(200) + 4(70)] - [270 + 5(200)] = 880 - 1270 = -390$, close to the calculated value.

FOLLOW-UP PROBLEM 20.2
Balance the following equations, predict the sign of $\Delta S°_{rxn}$ if possible, and calculate its value at 25°C:
(a) $NaOH(s) + CO_2(g) \longrightarrow Na_2CO_3(s) + H_2O(l)$
(b) $Fe(s) + H_2O(g) \longrightarrow Fe_2O_3(s) + H_2(g)$

Entropy Changes in the Surroundings: The Other Part of the Total

In many spontaneous reactions, such as the synthesis of ammonia and the combustion of propane, we see that the entropy of the reacting system decreases ($\Delta S^{\circ}_{rxn} < 0$). The second law dictates that, for a spontaneous process, *decreases in the entropy of the system can occur only if **increases** in the entropy of the surroundings outweigh them.* Let's examine the influence of the surroundings—in particular, the addition (or removal) of heat and the temperature at which this heat change occurs—on the *total* entropy change.

The essential role of the surroundings is to *either add heat to the system or remove heat from it.* In essence, the surroundings function as an enormous heat source or heat sink, one so large that its temperature remains constant, even though its entropy changes through the loss or gain of heat. The surroundings participate in the two possible types of enthalpy changes as follows:

1. *Exothermic change.* Heat lost by the system is gained by the surroundings. This heat gain increases the freedom of motion of the particles, which allows their energy to be more dispersed, so the entropy of the surroundings increases:

    For an exothermic change: $q_{sys} < 0$, $q_{surr} > 0$,   and $\Delta S_{surr} > 0$

2. *Endothermic change.* Heat gained by the system is lost by the surroundings. This heat loss reduces the freedom of motion of the particles, which makes their energy less dispersed, so the entropy of the surroundings decreases:

    For an endothermic change: $q_{sys} > 0$, $q_{surr} < 0$,   and $\Delta S_{surr} < 0$

The *temperature* of the surroundings when the heat is transferred also affects ΔS_{surr}. Consider the effect of an exothermic reaction at low or high temperature:

- At a lower temperature, such as 20 K, there is very little random motion in the surroundings and, thus, relatively little energy. This means there are fewer energy levels in each microstate, and fewer available microstates, in which to disperse the energy. Therefore, transferring heat to the surroundings has a large effect on how much the energy is dispersed.
- At a higher temperature, such as 298 K, the surroundings already have a relatively large quantity of energy dispersed. This means there are more energy levels in each microstate, and a greater number of available microstates. Thus, transferring the same amount of heat has a smaller effect on the total dispersal of energy. ●

In other words, the change in entropy of the surroundings is greater when heat is added at a lower temperature. Putting these ideas together, *the change in entropy of the surroundings is directly related to an opposite change in the heat of the system and inversely related to the temperature at which the heat is transferred.* Combining these relationships gives an equation that is closely related to Equation 20.2:

$$\Delta S_{surr} = -\frac{q_{sys}}{T}$$

Recall that for a process at *constant pressure,* the heat (q_P) is ΔH, so

$$\Delta S_{surr} = -\frac{\Delta H_{sys}}{T} \qquad (20.5)$$

Thus, we can calculate ΔS_{surr} by measuring ΔH_{sys} and the temperature T at which the change takes place.

To restate the central point, if a spontaneous reaction has a negative ΔS_{sys} (fewer microstates into which energy is dispersed), ΔS_{surr} must be positive enough (*many* more microstates into which energy is dispersed) for ΔS_{univ} to be positive (net more microstates into which energy is dispersed). Sample Problem 20.3 illustrates this situation for one of the reactions we considered earlier.

● **A Checkbook Analogy for Heating the Surroundings** A monetary analogy may clarify the relative sizes of the changes that arise from heating the surroundings at different initial temperatures. If you have $10 in your checking account, a $10 deposit represents a 100% increase in your net worth; that is, a given change to a low initial state has a large impact. If, however, you have a $1000 balance, a $10 deposit represents only a 1% increase. That is, the same change to a high initial state has a smaller impact.

SAMPLE PROBLEM 20.3   Determining Reaction Spontaneity

PROBLEM At 298 K, the formation of ammonia has a negative ΔS°_{sys}:

$$N_2(g) + 3H_2(g) \longrightarrow 2NH_3(g) \qquad \Delta S^{\circ}_{sys} = -197 \text{ J/K}$$

Calculate ΔS_{univ}, and state whether the reaction occurs spontaneously at this temperature.

PLAN For the reaction to occur spontaneously, $\Delta S_{univ} > 0$, and so ΔS_{surr} must be greater than $+197$ J/K. To find ΔS_{surr}, we need ΔH°_{sys}, which is the same as ΔH°_{rxn}. We use ΔH°_{f} values from Appendix B to find ΔH°_{rxn}. Then, we use ΔH°_{rxn} and the given T (298 K) to find ΔS_{surr}. To find ΔS_{univ}, we add the calculated ΔS_{surr} to the given ΔS°_{sys} (-197 J/K).

SOLUTION Calculating ΔH°_{sys}:

$$\begin{aligned} \Delta H^{\circ}_{sys} &= \Delta H^{\circ}_{rxn} \\ &= [(2 \text{ mol NH}_3)(-45.9 \text{ kJ/mol})] - [(3 \text{ mol H}_2)(0 \text{ kJ/mol}) + (1 \text{ mol N}_2)(0 \text{ kJ/mol})] \\ &= -91.8 \text{ kJ} \end{aligned}$$

Calculating ΔS_{surr}:

$$\Delta S_{surr} = -\frac{\Delta H^{\circ}_{sys}}{T} = -\frac{-91.8 \text{ kJ} \times \dfrac{1000 \text{ J}}{1 \text{ kJ}}}{298 \text{ K}} = 308 \text{ J/K}$$

Determining ΔS_{univ}:

$$\Delta S_{univ} = \Delta S^{\circ}_{sys} + \Delta S_{surr} = -197 \text{ J/K} + 308 \text{ J/K} = \boxed{111 \text{ J/K}}$$

$\Delta S_{univ} > 0$, so $\boxed{\text{the reaction occurs spontaneously at 298 K}}$ (see figure in margin).

CHECK Rounding to check the math, we have

$$\Delta H^{\circ}_{rxn} \approx 2(-45 \text{ kJ}) = -90 \text{ kJ}$$
$$\Delta S_{surr} \approx -(-90{,}000 \text{ J})/300 \text{ K} = 300 \text{ J/K}$$
$$\Delta S_{univ} \approx -200 \text{ J/K} + 300 \text{ J/K} = 100 \text{ J/K}$$

Given the negative ΔH°_{rxn}, Le Châtelier's principle predicts that low temperature should favor NH_3 formation, and so the answer is reasonable (see the Chemical Connections essay on the Haber process, pp. 771–772).

COMMENT 1. Note that ΔH° has units of kJ, whereas ΔS has units of J/K. Don't forget to convert kJ to J, or you'll introduce a large error.

2. This example highlights the distinction between thermodynamic and kinetic considerations. Even though NH_3 forms spontaneously, it does so slowly, and the chemical industry must expend great effort to catalyze the formation at a practical rate.

FOLLOW-UP PROBLEM 20.3 Does the oxidation of FeO(*s*) to Fe_2O_3(*s*) occur spontaneously at 298 K?

As Sample Problem 20.3 shows, taking the surroundings into account is crucial to determining reaction spontaneity. Moreover, it clarifies the relevance of thermodynamics to biology, as the Chemical Connections essay shows.

The Entropy Change and the Equilibrium State

For a process spontaneously approaching equilibrium, $\Delta S_{univ} > 0$. When the process reaches equilibrium, there is no longer any force driving it to proceed further and, thus, no net change in either direction; that is, $\Delta S_{univ} = 0$. At that point, any entropy change in the system is exactly balanced by an opposite entropy change in the surroundings:

At equilibrium: $\Delta S_{univ} = \Delta S_{sys} + \Delta S_{surr} = 0$ or $\Delta S_{sys} = -\Delta S_{surr}$

For example, let's calculate ΔS_{univ} for a phase change. For the vaporization-condensation of 1 mol of water at 100°C (373 K),

$$H_2O(l; 373 \text{ K}) \rightleftharpoons H_2O(g; 373 \text{ K})$$

First, we find ΔS_{univ} for the forward change (vaporization) by calculating ΔS°_{sys}:

$$\begin{aligned} \Delta S^{\circ}_{sys} &= \Sigma m S^{\circ}_{products} - \Sigma n S^{\circ}_{reactants} = S^{\circ} \text{ of } H_2O(g; 373 \text{ K}) - S^{\circ} \text{ of } H_2O(l; 373 \text{ K}) \\ &= 195.9 \text{ J/K} - 86.8 \text{ J/K} = 109.1 \text{ J/K} \end{aligned}$$

Do Living Things Obey the Laws of Thermodynamics?

Organisms can be thought of as chemical machines that evolved by extracting energy as efficiently as possible from the environment. Such a mechanical view holds that all processes, whether they involve living or nonliving systems, are consistent with thermodynamic principles. Let's examine the first and second laws of thermodynamics to see if they apply to living systems.

Organisms certainly comply with the first law. The chemical bond energy in food is converted into the mechanical energy of sprouting, crawling, swimming, and countless other movements; the electrical energy of nerve conduction; the thermal energy of warming the body; and so forth. Many experiments have demonstrated that in all these energy conversions, the total energy is conserved. Some of the earliest studies of this question were performed by Lavoisier, who included animal respiration in his new theory of combustion (Figure B20.1). He was the first to show that "animal heat" was produced by a slow combustion process occurring continually in the body. In experiments with guinea pigs, he measured the intake of food and O_2 and the output of CO_2 and heat, for which he invented a calorimeter based on the melting of ice, and he established the principles and methods for measuring the basal rate of metabolism. Modern experiments using a calorimeter large enough to contain an exercising human continue to confirm the conservation of energy (Figure B20.2).

It may not seem as clear, however, that an organism, and even the whole parade of life, complies with the second law. Mature humans are far more complex than the simple egg and sperm cells from which they develop, and modern organisms are far more complex than the one-celled ancestral specks from which they evolved. Energy must be highly localized and macromolecules constrained to carry out myriad reactions that synthesize biopolymers from their monomers. Are the growth of an organism and the evolution of life exceptions to the spontaneous tendency of natural processes to disperse their energy and increase molecular free-

dom? Does biology violate the second law? Not at all, if we examine the system *and* its surroundings. For an organism to grow or a species to evolve, countless moles of oxygen and food molecules—carbohydrates, proteins, and fats—undergo exothermic combustion reactions to form many more moles of gaseous CO_2 and H_2O. The formation and discharge of these waste gases, as well as the heat released, represent a tremendous net increase in the entropy of the surroundings. Thus, the localization of energy and restrictions on molecular freedom apparent in the growth and evolution of organisms occur at the expense of a far greater dispersal of energy and freedom of motion in the Earth-Sun surroundings. When system and surroundings are considered together, the entropy of the universe, as always, increases.

Figure B20.2 A whole-body calorimeter. In this room-sized apparatus, a subject exercises while respiratory gases, energy input and output, and other physiological variables are monitored.

Figure B20.1 Lavoisier studying human respiration as a form of combustion. Lavoisier (standing at right) measured substances consumed and excreted in addition to changes in heat to understand the chemical nature of respiration. The artist (Mme. Lavoisier) shows herself taking notes (seated at right).

As we expect, the entropy of the system increases ($\Delta S^\circ_{sys} > 0$) as the liquid absorbs heat and changes to a gas.

For ΔS_{surr}, we have

$$\Delta S_{surr} = -\frac{\Delta H^\circ_{sys}}{T}$$

where $\Delta H^\circ_{sys} = \Delta H^\circ_{vap}$ at 373 K $= 40.7$ kJ/mol $= 40.7 \times 10^3$ J/mol. For 1 mol of water, we have

$$\Delta S_{surr} = -\frac{\Delta H^\circ_{vap}}{T} = -\frac{40.7 \times 10^3 \text{ J}}{373 \text{ K}} = -109 \text{ J/K}$$

The surroundings lose heat, and the negative sign means that the entropy of the surroundings decreases. The two entropy changes have the same magnitude but opposite signs, so they cancel:

$$\Delta S_{univ} = 109 \text{ J/K} + (-109 \text{ J/K}) = 0$$

For the reverse change (condensation), ΔS_{univ} also equals zero, but ΔS°_{sys} and ΔS_{surr} have signs opposite those for vaporization. A similar treatment of a chemical change shows the same result: the entropy change of the forward reaction is *equal in magnitude but opposite in sign* to the entropy change of the reverse reaction. Thus, *when a system reaches equilibrium, neither the forward nor the reverse reaction is spontaneous,* and so there is no net reaction in either direction.

Spontaneous Exothermic and Endothermic Reactions: A Summary

We can now see why exothermic and endothermic spontaneous reactions occur. No matter what its *enthalpy* change, a reaction occurs because the total *entropy* of the reacting system *and* its surroundings increases. The two possibilities are

1. *For an exothermic reaction* ($\Delta H_{sys} < 0$), heat is released by the system, which increases the freedom of motion and energy dispersed and, thus, the entropy of the surroundings ($\Delta S_{surr} > 0$).

• If the reacting system yields products whose entropy is greater than that of the reactants ($\Delta S_{sys} > 0$), the total entropy change ($\Delta S_{sys} + \Delta S_{surr}$) will be positive (Figure 20.12A). For example, in the oxidation of glucose, an essential reaction for all higher organisms,

$$C_6H_{12}O_6(s) + 6O_2(g) \longrightarrow 6CO_2(g) + 6H_2O(g) + \textbf{heat}$$

6 mol of gas yields 12 mol of gas and heat; thus, $\Delta S_{sys} > 0$, $\Delta S_{surr} > 0$, and $\Delta S_{univ} > 0$.

• If, on the other hand, the entropy of the system decreases as the reaction occurs ($\Delta S_{sys} < 0$), the entropy of the surroundings must increase even more ($\Delta S_{surr} \gg 0$) to make the total ΔS positive (Figure 20.12B). For example, when calcium oxide and carbon dioxide form calcium carbonate,

$$CaO(s) + CO_2(g) \longrightarrow CaCO_3(s) + \textbf{heat}$$

the entropy of the system decreases because the amount (mol) of gas decreases. However, the heat released increases the entropy of the surroundings even more; thus, $\Delta S_{sys} < 0$, but $\Delta S_{surr} \gg 0$, so $\Delta S_{univ} > 0$.

2. *For an endothermic reaction* ($\Delta H_{sys} > 0$), the heat lost by the surroundings decreases molecular freedom of motion and dispersal of energy and, thus, decreases the entropy of the surroundings ($\Delta S_{surr} < 0$). Therefore, the only way an endothermic reaction can occur spontaneously is if ΔS_{sys} is positive and large enough to outweigh the negative ΔS_{surr} (Figure 20.12C).

• In the solution process for many ionic compounds, heat is absorbed to form the solution, so the entropy of the surroundings decreases ($\Delta S_{surr} < 0$). However, when the crystalline solid becomes freely moving ions, the entropy increase is so large ($\Delta S_{sys} \gg 0$) that it outweighs the negative ΔS_{surr}. Thus, ΔS_{univ} is positive.

A   B   C

Figure 20.12 Components of ΔS_{univ} for spontaneous reactions. For a reaction to occur spontaneously, ΔS_{univ} must be positive. **A,** An exothermic reaction in which ΔS_{sys} increases; the size of ΔS_{surr} is not important. **B,** An exothermic reaction in which ΔS_{sys} decreases; ΔS_{surr} must be larger than ΔS_{sys}. **C,** An endothermic reaction in which ΔS_{sys} increases; ΔS_{surr} must be smaller than ΔS_{sys}.

Transparency/ARIS Presentation Center, Art Library

- Spontaneous endothermic reactions have similar features. For example, consider the reaction between barium hydroxide octahydrate and ammonium nitrate (see Figure 20.1, p. 883),

$$heat + Ba(OH)_2 \cdot 8H_2O(s) + 2NH_4NO_3(s) \longrightarrow$$
$$Ba^{2+}(aq) + 2NO_3^-(aq) + 2NH_3(aq) + 10H_2O(l)$$

In this endothermic reaction, 3 mol of crystalline solids absorbs heat from the surroundings ($\Delta S_{surr} < 0$) and yields 15 mol of dissolved ions and molecules, which have much more freedom of motion and, therefore, much greater entropy ($\Delta S_{sys} >> 0$).

Section Summary

The standard entropy of reaction, ΔS_{rxn}°, is calculated from S° values. • When the amount (mol) of gas (Δn_{gas}) increases in a reaction, usually $\Delta S_{rxn}^\circ > 0$. • The value of ΔS_{surr} is related directly to ΔH_{sys}° and inversely to the T at which the change occurs. • In a spontaneous change, the entropy of the system can decrease only if the entropy of the surroundings increases even more. • For a system at equilibrium, $\Delta S_{univ} = 0$.

20.3 ENTROPY, FREE ENERGY, AND WORK

By making *two* separate measurements, ΔS_{sys} and ΔS_{surr}, we can predict whether a reaction will be spontaneous at a particular temperature. It would be useful, however, to have *one* criterion for spontaneity that we can find by examining the system only. The Gibbs free energy, or simply **free energy (G),** is a function that combines the system's enthalpy and entropy:

$$G = H - TS$$

Named for Josiah Willard Gibbs, the physicist who proposed it and laid much of the foundation for chemical thermodynamics, this function provides the criterion for spontaneity we've been seeking. ●

Free Energy Change and Reaction Spontaneity

The free energy change (ΔG) is a measure of the spontaneity of a process and of the useful energy available from it. Let's see how the free energy change is derived from the second law of thermodynamics. Recall that by definition, the entropy change of the universe is the sum of the entropy changes of the system and the surroundings:

$$\Delta S_{univ} = \Delta S_{sys} + \Delta S_{surr}$$

● **The Greatness and Obscurity of J. Willard Gibbs** Even today, one of the greatest minds in science is barely known outside chemistry and physics. In 1878, Josiah Willard Gibbs (1839–1903), a professor of mathematical physics at Yale, completed a 323-page paper that virtually established the science of chemical thermodynamics and major principles governing chemical equilibria, phase-change equilibria, and the energy changes in electrochemical cells. The European scientists James Clerk Maxwell, Wilhelm Ostwald, and Henri Le Châtelier realized the significance of Gibbs's achievements long before most of his American colleagues did. In fact, he was elected to the Hall of Fame of Distinguished Americans as late as 1950 because, embarrassingly, he had not received enough votes until then!

 ARIS Presentation Center, Photo Library

At constant pressure,

$$\Delta S_{surr} = -\frac{\Delta H_{sys}}{T}$$

Substituting for ΔS_{surr} gives a relationship that lets us focus solely on the system:

$$\Delta S_{univ} = \Delta S_{sys} - \frac{\Delta H_{sys}}{T}$$

Multiplying both sides by $-T$ gives

$$-T\Delta S_{univ} = \Delta H_{sys} - T\Delta S_{sys}$$

Now we can introduce the new free energy quantity to replace the enthalpy and entropy terms. From $G = H - TS$, the *Gibbs equation* shows us the change in the free energy of the system (ΔG_{sys}) at constant temperature and pressure:

$$\Delta G_{sys} = \Delta H_{sys} - T\Delta S_{sys} \tag{20.6}$$

Combining this equation with the previous one shows that

$$-T\Delta S_{univ} = \Delta H_{sys} - T\Delta S_{sys} = \Delta G_{sys}$$

*The **sign** of ΔG tells if a reaction is spontaneous.* The second law dictates

- $\Delta S_{univ} > 0$ for a spontaneous process
- $\Delta S_{univ} < 0$ for a nonspontaneous process
- $\Delta S_{univ} = 0$ for a process at equilibrium

Of course, absolute temperature is always positive, so

$$T\Delta S_{univ} > 0 \quad \text{or} \quad -T\Delta S_{univ} < 0 \text{ for a spontaneous process}$$

Because $\Delta G = -T\Delta S_{univ}$, we know that

- $\Delta G < 0$ for a spontaneous process
- $\Delta G > 0$ for a nonspontaneous process
- $\Delta G = 0$ for a process at equilibrium

An important point to keep in mind is that if a process is *nonspontaneous* in one direction ($\Delta G > 0$), it is *spontaneous* in the opposite direction ($\Delta G < 0$). By using ΔG, we have not incorporated any new ideas, but we can predict reaction spontaneity from one variable (ΔG_{sys}) rather than two (ΔS_{sys} and ΔS_{surr}).

As we noted at the beginning of the chapter, the degree of spontaneity of a reaction—that is, the sign *and* magnitude of ΔG—tells us nothing about the rate of the reaction. Remember that some spontaneous reactions are extremely slow. For example, the reaction between $H_2(g)$ and $O_2(g)$ at room temperature is highly spontaneous—ΔG has a large negative value—but in the absence of a catalyst or a flame, the reaction doesn't occur to a measurable extent because its rate is so low.

Calculating Standard Free Energy Changes

Because free energy (G) combines three state functions, H, S, and T, it is also a state function. As with enthalpy, we focus on the free energy *change* (ΔG).

The Standard Free Energy Change As we did previously with the other thermodynamic variables, to compare the free energy changes of different reactions we calculate the **standard free energy change ($\Delta G°$),** which occurs when all components of the system are in their standard states. Adapting the Gibbs equation (20.6), we have

$$\Delta G°_{sys} = \Delta H°_{sys} - T\Delta S°_{sys} \tag{20.7}$$

This important relationship is used frequently to find any one of these three central thermodynamic variables, given the other two, as in Sample Problem 20.4.

Literature Ferguson, Arthur. "The Gibbs Free Energy of a Chemical Reaction System As a Function of the Extent of Reaction and the Prediction of Spontaneity," *J. Chem. Educ.* **2004,** *81,* 606.

Literature Weiss, Hilton M. "Developing an Intuitive Understanding of Free Energy," *J. Chem. Educ.* **2001,** *78,* 1362.

Literature Kozliak, Evguenii I. "Energy as Money, Chemical Bonding as Business, and Negative ΔH and ΔG as Investment," *J. Chem. Educ.* **2002,** *79,* 1435.

SAMPLE PROBLEM 20.4 Calculating ΔG°_{rxn} from Enthalpy and Entropy Values

PROBLEM Potassium chlorate, a common oxidizing agent in fireworks (see photo) and matchheads, undergoes a solid-state disproportionation reaction when heated:

$$4KClO_3(s) \xrightarrow{\Delta} 3\overset{+7}{KClO_4}(s) + \overset{-1}{KCl}(s)$$

Use ΔH°_f and S° values to calculate ΔG°_{sys} (ΔG°_{rxn}) at 25°C for this reaction.

PLAN To solve for ΔG°, we need values from Appendix B. We use ΔH°_f values to calculate ΔH°_{rxn} (ΔH°_{sys}), use S° values to calculate ΔS°_{rxn} (ΔS°_{sys}), and then apply Equation 20.7.

SOLUTION Calculating ΔH°_{sys} from ΔH°_f values (with Equation 6.8):

$$\Delta H^{\circ}_{sys} = \Delta H^{\circ}_{rxn} = \Sigma m \Delta H^{\circ}_{f(products)} - \Sigma n \Delta H^{\circ}_{f(reactants)}$$

$$= [(3 \text{ mol } KClO_4)(\Delta H^{\circ}_f \text{ of } KClO_4) + (1 \text{ mol } KCl)(\Delta H^{\circ}_f \text{ of } KCl)]$$
$$- [(4 \text{ mol } KClO_3)(\Delta H^{\circ}_f \text{ of } KClO_3)]$$

$$= [(3 \text{ mol})(-432.8 \text{ kJ/mol}) + (1 \text{ mol})(-436.7 \text{ kJ/mol})]$$
$$- [(4 \text{ mol})(-397.7 \text{ kJ/mol})]$$

$$= -144 \text{ kJ}$$

Calculating ΔS°_{sys} from S° values (with Equation 20.4):

$$\Delta S^{\circ}_{sys} = \Delta S^{\circ}_{rxn} = \Sigma m S^{\circ}_{products} - \Sigma n S^{\circ}_{reactants}$$

$$= [(3 \text{ mol } KClO_4)(S^{\circ} \text{ of } KClO_4)$$
$$+ (1 \text{ mol } KCl)(S^{\circ} \text{ of } KCl)] - [(4 \text{ mol } KClO_3)(S^{\circ} \text{ of } KClO_3)]$$

$$= [(3 \text{ mol})(151.0 \text{ J/mol·K}) + (1 \text{ mol})(82.6 \text{ J/mol·K})]$$
$$- [(4 \text{ mol})(143.1 \text{ J/mol·K})]$$

$$= -36.8 \text{ J/K}$$

Calculating ΔG°_{sys} at 298 K:

$$\Delta G^{\circ}_{sys} = \Delta H^{\circ}_{sys} - T\Delta S^{\circ}_{sys} = -144 \text{ kJ} - \left[(298 \text{ K})(-36.8 \text{ J/K})\left(\frac{1 \text{ kJ}}{1000 \text{ J}}\right)\right] = -133 \text{ kJ}$$

CHECK Rounding to check the math:

$$\Delta H^{\circ} \approx [3(-433 \text{ kJ}) + (-440 \text{ kJ})] - [4(-400 \text{ kJ})] = -1740 \text{ kJ} + 1600 \text{ kJ} = -140 \text{ kJ}$$

$$\Delta S^{\circ} \approx [3(150 \text{ J/K}) + 85 \text{ J/K}] - [4(145 \text{ J/K})] = 535 \text{ J/K} - 580 \text{ J/K} = -45 \text{ J/K}$$

$$\Delta G^{\circ} \approx -140 \text{ kJ} - 300 \text{ K}(-0.04 \text{ kJ/K}) = -140 \text{ kJ} + 12 \text{ kJ} = -128 \text{ kJ}$$

All values are close to the calculated ones. Another way to calculate ΔG° for this reaction is presented in Sample Problem 20.5.

COMMENT 1. For a spontaneous reaction under *any* conditions, the free energy change, ΔG, is negative. Under standard-state conditions, a spontaneous reaction has a negative *standard* free energy change; that is, $\Delta G^{\circ} < 0$.

2. This reaction is spontaneous, but the rate is very low in the solid. When $KClO_3$ is heated slightly above its melting point, the ions are free to move and the reaction occurs readily.

FOLLOW-UP PROBLEM 20.4 Determine the standard free energy change at 298 K for the reaction $2NO(g) + O_2(g) \longrightarrow 2NO_2(g)$.

Potassium chlorate is the oxidizing agent in fireworks.

The Standard Free Energy of Formation

Another way to calculate ΔG°_{rxn} is with values for the **standard free energy of formation (ΔG°_f)** of the components; ΔG°_f is the free energy change that occurs when 1 mol of compound is made *from its elements,* with all components in their standard states. Because free energy is a state function, we can combine ΔG°_f values of reactants and products to calculate ΔG°_{rxn} no matter how the reaction takes place:

$$\Delta G^{\circ}_{rxn} = \Sigma m \Delta G^{\circ}_{f(products)} - \Sigma n \Delta G^{\circ}_{f(reactants)} \qquad (20.8)$$

ΔG°_f values have properties similar to ΔH°_f values:

- ΔG°_f of an element in its standard state is zero.
- An equation coefficient (*m* or *n* above) multiplies ΔG°_f by that number.
- Reversing a reaction changes the sign of ΔG°_f.

Many ΔG°_f values appear along with those for ΔH°_f and S° in Appendix B.

SAMPLE PROBLEM 20.5 Calculating ΔG°_{rxn} from ΔG°_f Values

PROBLEM Use ΔG°_f values to calculate ΔG°_{rxn} for the reaction in Sample Problem 20.4:

$$4KClO_3(s) \longrightarrow 3KClO_4(s) + KCl(s)$$

PLAN We apply Equation 20.8 to calculate ΔG°_{rxn}.

SOLUTION

$$\begin{aligned}
\Delta G^{\circ}_{rxn} &= \Sigma m \Delta G^{\circ}_{f(products)} - \Sigma n \Delta G^{\circ}_{f(reactants)} \\
&= [(3 \text{ mol } KClO_4)(\Delta G^{\circ}_f \text{ of } KClO_4) + (1 \text{ mol } KCl)(\Delta G^{\circ}_f \text{ of } KCl)] \\
&\quad - [(4 \text{ mol } KClO_3)(\Delta G^{\circ}_f \text{ of } KClO_3)] \\
&= [(3 \text{ mol})(-303.2 \text{ kJ/mol}) + (1 \text{ mol})(-409.2 \text{ kJ/mol})] \\
&\quad - [(4 \text{ mol})(-296.3 \text{ kJ/mol})] \\
&= -134 \text{ kJ}
\end{aligned}$$

CHECK Rounding to check the math:

$$\Delta G^{\circ}_{rxn} \approx [3(-300 \text{ kJ}) + 1(-400 \text{ kJ})] - 4(-300 \text{ kJ})$$
$$= -1300 \text{ kJ} + 1200 \text{ kJ} = -100 \text{ kJ}$$

COMMENT The slight discrepancy between this answer and that obtained in Sample Problem 20.4 is within experimental error. As you can see, when ΔG°_f values are available for a reaction taking place at 25°C, this method is simpler than that in Sample Problem 20.4.

FOLLOW-UP PROBLEM 20.5 Use ΔG°_f values to calculate the free energy change at 25°C for each of the following reactions:
(a) $2NO(g) + O_2(g) \longrightarrow 2NO_2(g)$ (from Follow-up Problem 20.4)
(b) $2C(graphite) + O_2(g) \longrightarrow 2CO(g)$

ΔG and the Work a System Can Do

Recall that the science of thermodynamics was born soon after the invention of the steam engine, and one of the most practical relationships in the field is that between the free energy change and the work a system can do:

- For a spontaneous process ($\Delta G < 0$) at constant T and P, ΔG is the *maximum useful work obtainable **from** the system as the process takes place:*

$$\Delta G = w_{max} \qquad (20.9)$$

- For a nonspontaneous process ($\Delta G > 0$) at constant T and P, ΔG is the *minimum work that must be done **to** the system to make the process take place.*

Gas at $V_{initial}$

A

Gas at V_{final}

B

Figure 20.13 An expanding gas lifting a weight.

The phrase "maximum useful work obtainable" requires some explanation. First, what do we mean by the "maximum work obtainable," and what determines this limit? The free energy change is the maximum work the system can *possibly* do. But the work the system *actually* does depends on how that change in free energy is released. Let's first consider a nonchemical process. Suppose a gas is confined within a cylinder, at $V_{initial}$, by a piston attached to a 1-kg brass weight (Figure 20.13A). As the gas expands and lifts the weight, its pressure becomes balanced by the weight at some final volume, V_{final} (Figure 20.13B). The gas lifted the weight in one step, thus doing a certain quantity of work. *As the number of steps increases, the quantity of work done by the gas increases.* Thus, the gas can do more work in two steps, by lifting a 2-kg weight to one-half V_{final} and then lifting the 1-kg weight the rest of the way to V_{final}. Lifting the weight in three steps would do even more work. The gas would get closer to performing the maximum work if the brass weight were replaced by a container of sand so that the weight could be lifted grain by grain, that is, almost infinitesimally. (We described a similar method for finding the entropy change on p. 887.) In this way, the gas would lift the container in a very high number of steps.

However, only in a hypothetical process in which the work is done in an infinite number of steps, and an infinitesimal increase in the weight reverses the expansion, can the maximum work be done: *the maximum work is done by a spontaneous process only if the work is carried out reversibly*. Of course, in any *real* process, work is performed *irreversibly*, that is, in a finite number of steps, so *we can never obtain the maximum work*. The free energy not used for work is lost to the surroundings as heat. This "unharnessed" energy is a consequence of any real process.

Next, what do we mean by "useful work"? As the gas expands and raises the weights, some of the work just pushes back the atmosphere. That work increases the atmosphere's energy, but it doesn't do anything useful, such as turn a turbine, run a car, etc. By the same token, if we wanted to compress the gas, the atmosphere would do part of the work by pushing on the piston, but that part of the work doesn't come from our muscles or from gasoline or from the power company. The ΔG function ignores the work done to or by the atmosphere and tells us about the work we can use or have to do ourselves.

Let's consider two examples of systems that use a chemical reaction to do useful work—a car engine and a battery. When gasoline (represented here by octane) is burned in a car engine,

$$C_8H_{18}(l) + \tfrac{25}{2}O_2(g) \longrightarrow 8CO_2(g) + 9H_2O(g)$$

a large amount of energy is given off as heat ($\Delta H_{sys} < 0$), and because the number of moles of gas increases, the entropy of the system increases ($\Delta S_{sys} > 0$). Therefore, the reaction is spontaneous ($\Delta G_{sys} < 0$). The free energy available does work turning wheels, moving belts, playing a CD, and so on. However, only if it is released reversibly, that is, in a series of infinitesimal steps, do we obtain the maximum work available from this reaction. In reality, the process is carried out irreversibly and much of the total free energy just warms the engine, the passenger compartment, and the outside air, which increases the freedom of motion of the particles in the universe, in accord with the second law.

As you'll see in Chapter 21, a battery is essentially a packaged spontaneous redox reaction that releases free energy to the surroundings (flashlight, radio, motor, etc.). If we connect the battery terminals to each other through a short piece of wire, the free energy change is released all at once but does no work—it just heats the wire and battery. If we connect the terminals to a motor, the free energy is released more slowly, and a sizeable portion of it runs the motor, but some is still converted to heat in the battery and the motor. If we discharge the battery still more slowly, more of the free energy change does work and less is converted to heat, but only when the battery discharges infinitely slowly can we obtain the maximum quantity of work.

The compromise that all engineers and machine designers must face is that, *in the real world, some free energy is always changed to heat and is, thus, unharnessed: no **real** process uses all the available free energy to do work.* ●

Let's summarize the relationship between the free energy change of a reaction and the work it can actually do:

- A spontaneous reaction ($\Delta G_{sys} < 0$) will occur *and* can do work on the surroundings ($-w$). In any real machine, however, the actual work done by the reaction is *always less than the maximum possible* because some of the ΔG released is lost as heat.
- A nonspontaneous reaction ($\Delta G_{sys} > 0$) will not occur unless the surroundings do work on it ($+w$). In any real machine, however, the actual work done on the reaction is *always more than the minimum* because some of the ΔG added is lost as heat.
- A reaction at equilibrium ($\Delta G_{sys} = 0$) can no longer do any work.

 Engineering application

● **The Wide Range of Energy Efficiency** One definition for the efficiency of a device is the percentage of the energy input that results in work output. The range of efficiencies is enormous for the common "energy-conversion systems" of society. For instance, a common incandescent lightbulb converts only 5% of incoming electrical energy to light, the rest is given off as heat. At the other extreme, a large electrical generator converts 99% of the incoming mechanical energy to electricity. Although improvements are continually being made, here are some efficiency values for other devices: home oil furnace, 65%; hand tool motor, 63%; liquid fuel rocket, 50%; car engine, <30%; fluorescent lamp, 20%; solar cell, ~15%.

The Effect of Temperature on Reaction Spontaneity

In most cases, the enthalpy contribution (ΔH) to the free energy change (ΔG) is much *larger* than the entropy contribution ($T\Delta S$). For this reason, most exothermic reactions are spontaneous: the negative ΔH helps make ΔG negative. However, the *temperature of a reaction influences the magnitude of the $T\Delta S$ term,* so, for many reactions, the overall spontaneity depends on the temperature.

By scrutinizing the signs of ΔH and ΔS, we can predict the effect of temperature on the sign of ΔG. The values for the thermodynamic variables in this discussion are based on standard state values from Appendix B, but we show them without the degree sign to emphasize that the relationships among ΔG, ΔH, and ΔS are valid at any conditions. Also, we assume that ΔH and ΔS change little with temperature, which is true as long as no phase changes occur.

Let's examine the four combinations of positive and negative ΔH and ΔS—two combinations that are independent of temperature and two that are dependent on temperature:

- *Temperature-independent cases.* When ΔH and ΔS have *opposite* signs, the reaction occurs spontaneously either at all temperatures or at none.

  1. *Reaction is spontaneous at all temperatures:* $\Delta H < 0$, $\Delta S > 0$. Both contributions favor the spontaneity of the reaction. ΔH is negative and ΔS is positive, so $-T\Delta S$ is negative; thus, ΔG is always negative. Most combustion reactions are in this category. The decomposition of hydrogen peroxide, a common disinfectant, is also spontaneous at all temperatures:

  $$2H_2O_2(l) \longrightarrow 2H_2O(l) + O_2(g)$$
  $$\Delta H = -196 \text{ kJ} \quad \text{and} \quad \Delta S = 125 \text{ J/K}$$

  2. *Reaction is nonspontaneous at all temperatures:* $\Delta H > 0$, $\Delta S < 0$. Both contributions oppose the spontaneity of the reaction. ΔH is positive and ΔS is negative, so $-T\Delta S$ is positive; thus, ΔG is always positive. The formation of ozone from oxygen is not spontaneous at any temperature:

  $$3O_2(g) \longrightarrow 2O_3(g)$$
  $$\Delta H = 286 \text{ kJ} \quad \text{and} \quad \Delta S = -137 \text{ J/K}$$

- *Temperature-dependent cases.* When ΔH and ΔS have the *same* sign, the relative magnitudes of the $-T\Delta S$ and ΔH terms determine the sign of ΔG. In these cases, the magnitude of T is crucial to reaction spontaneity.

  3. *Reaction is spontaneous at higher temperatures:* $\Delta H > 0$ *and* $\Delta S > 0$. Here, ΔS favors spontaneity ($-T\Delta S < 0$), but ΔH does not. For example,

  $$2N_2O(g) + O_2(g) \longrightarrow 4NO(g)$$
  $$\Delta H = 197.1 \text{ kJ} \quad \text{and} \quad \Delta S = 198.2 \text{ J/K}$$

  With a positive ΔH, the reaction will occur spontaneously only when $-T\Delta S$ is large enough to make ΔG negative, which will happen at higher temperatures. The oxidation of N_2O occurs spontaneously at $T > 994$ K.

  4. *Reaction is spontaneous at lower temperatures:* $\Delta H < 0$ *and* $\Delta S < 0$. Now, ΔH favors spontaneity, but ΔS does not ($-T\Delta S > 0$). For example,

  $$4Fe(s) + 3O_2(g) \longrightarrow 2Fe_2O_3(s)$$
  $$\Delta H = -1651 \text{ kJ} \quad \text{and} \quad \Delta S = -549.4 \text{ J/K}$$

  With a negative ΔH, the reaction will occur spontaneously only if the $-T\Delta S$ term is smaller than the ΔH term, and this happens at lower temperatures. The production of iron(III) oxide occurs spontaneously at any $T < 3005$ K.

Table 20.1 summarizes these four possible combinations of ΔH and ΔS, and Sample Problem 20.6 applies them to physical and chemical changes.

Table 20.1	Reaction Spontaneity and the Signs of ΔH, ΔS, and ΔG			
ΔH	ΔS	$-T\Delta S$	ΔG	Description
−	+	−	−	Spontaneous at all T
+	−	+	+	Nonspontaneous at all T
+	+	−	+ or −	Spontaneous at higher T; nonspontaneous at lower T
−	−	+	+ or −	Spontaneous at lower T; nonspontaneous at higher T

Transparency/ARIS Presentation Center, Tables Library

SAMPLE PROBLEM 20.6 Using Molecular Scenes to Determine the Signs of ΔH, ΔS, and ΔG

ARIS Presentation Center, Worked Examples Library

PROBLEM The following scenes represent a familiar phase change for water (blue spheres):

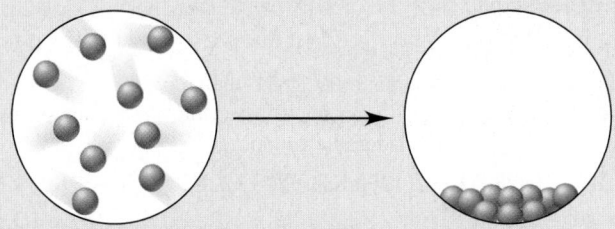

(a) What are the signs of ΔH and ΔS for this process? Explain.
(b) Is the process spontaneous at all T, no T, low T, or high T? Explain.
PLAN The scenes depict water vapor on the left and liquid water on the right. **(a)** From the scenes, we determine any change in amount of gas, which indicates the sign of ΔS, and any change in the freedom of motion of the particles, which indicates whether heat is absorbed or released, and thus the sign of ΔH. **(b)** The question refers to the sign of ΔG (+ or −) at the different temperature possibilities, so we apply Equation 20.6 and refer to the previous text discussion and Table 20.1.
SOLUTION (a) The scene represents the condensation of water vapor, so the amount of gas decreases dramatically, and the separated molecules give up energy as they come closer together. Therefore, $\Delta S < 0$ and $\Delta H < 0$. **(b)** According to Equation 20.6, with ΔS negative, the $-T\Delta S$ term is positive. In order for $\Delta G < 0$, the magnitude of T must be small. Therefore, the process is spontaneous at low T.
CHECK The answer in part (b) seems reasonable based on our analysis in part (a). The answer makes sense because we know from everyday experience that water condenses spontaneously, and it does so at low temperatures.

FOLLOW-UP PROBLEM 20.6 The following molecular scenes represent the gas-phase decomposition of X_2Y_2 to X_2 (red) and Y_2 (blue):

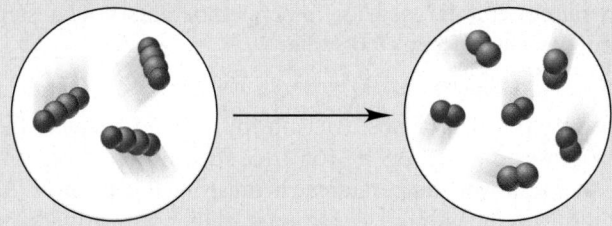

(a) What is the sign of ΔS for the reaction?
(b) If the reaction is spontaneous only above 325°C, what is the sign of ΔH? Explain.

As you saw in Sample Problem 20.4, one way to calculate ΔG is from enthalpy and entropy changes. Because ΔH and ΔS usually change little with temperature if no phase changes occur, we use their values at 298 K in Sample Problem 20.7 to examine the effect of T on ΔG and, thus, on reaction spontaneity.

SAMPLE PROBLEM 20.7 Determining the Effect of Temperature on ΔG

PROBLEM A key step in the production of sulfuric acid is the oxidation of $SO_2(g)$ to $SO_3(g)$:

$$2SO_2(g) + O_2(g) \longrightarrow 2SO_3(g)$$

At 298 K, $\Delta G = -141.6$ kJ; $\Delta H = -198.4$ kJ; and $\Delta S = -187.9$ J/K.
(a) Use the data to decide if this reaction is spontaneous at 25°C, and predict how ΔG will change with increasing T.
(b) Assuming that ΔH and ΔS are constant with T, is the reaction spontaneous at 900.°C?
PLAN (a) We note the sign of ΔG to see if the reaction is spontaneous and the signs of ΔH and ΔS to see the effect of T. **(b)** We use Equation 20.7 to calculate ΔG from the given ΔH and ΔS at the higher T (in K).
SOLUTION (a) $\Delta G < 0$, so the reaction is spontaneous at 298 K: SO_2 and O_2 will form SO_3 spontaneously. With $\Delta S < 0$, the term $-T\Delta S > 0$, and this term will become more positive at higher T. Therefore,

ΔG will become less negative, and the reaction less spontaneous, with increasing T.

(b) Calculating ΔG at 900.°C ($T = 273 + 900. = 1173$ K):

$$\Delta G = \Delta H - T\Delta S = -198.4 \text{ kJ} - [(1173 \text{ K})(-187.9 \text{ J/K})(1 \text{ kJ}/1000 \text{ J})] = 22.0 \text{ kJ}$$

$\Delta G > 0$, so the reaction is nonspontaneous at the higher T.
CHECK The answer in part (b) seems reasonable based on our prediction in part (a). The arithmetic seems correct, given considerable rounding:

$$\Delta G \approx -200 \text{ kJ} - [(1200 \text{ K})(-200 \text{ J/K})/1000 \text{ J}] = +40 \text{ kJ}$$

FOLLOW-UP PROBLEM 20.7 A reaction is nonspontaneous at room temperature but *is* spontaneous at −40°C. What can you say about the signs and relative magnitudes of ΔH, ΔS, and $-T\Delta S$?

The Temperature at Which a Reaction Becomes Spontaneous As you have just seen, when the signs of ΔH and ΔS are the same, some reactions that are nonspontaneous at one temperature become spontaneous at another, and vice versa. It would certainly be useful to know the temperature at which a reaction becomes spontaneous. This is the temperature at which a positive ΔG switches to a negative ΔG because of the changing magnitude of the $-T\Delta S$ term. We find this "crossover" temperature by setting ΔG equal to zero and solving for T:

$$\Delta G = \Delta H - T\Delta S = 0$$

Therefore,

$$\Delta H = T\Delta S \qquad \text{and} \qquad T = \frac{\Delta H}{\Delta S} \tag{20.10}$$

Consider the reaction of copper(I) oxide with carbon, which does *not* occur at lower temperatures but is used at higher temperatures in a step during the extraction of copper metal from chalcocite:

$$Cu_2O(s) + C(s) \longrightarrow 2Cu(s) + CO(g)$$

We predict this reaction has a positive entropy change because the number of moles of gas increases; in fact, $\Delta S = 165$ J/K. Furthermore, because the reaction is *non*spontaneous at lower temperatures, it must have a positive ΔH (58.1 kJ). As the $-T\Delta S$ term becomes more negative at higher temperatures, it will eventually outweigh the positive ΔH term, and the reaction will occur spontaneously.

Let's calculate ΔG for this reaction at 25°C and then find the temperature above which the reaction is spontaneous. At 25°C (298 K),

$$\Delta G = \Delta H - T\Delta S = 58.1 \text{ kJ} - \left(298 \text{ K} \times 165 \text{ J/K} \times \frac{1 \text{ kJ}}{1000 \text{ J}}\right) = 8.9 \text{ kJ}$$

Because ΔG is positive, the reaction will not proceed on its own at 25°C. At the crossover temperature, $\Delta G = 0$, so

$$T = \frac{\Delta H}{\Delta S} = \frac{58.1 \text{ kJ} \times \dfrac{1000 \text{ J}}{1 \text{ kJ}}}{165 \text{ J/K}} = 352 \text{ K}$$

At any temperature above 352 K (79°C), a moderate one for recovering a metal from its ore, the reaction occurs spontaneously. Figure 20.14 depicts this result. The line for $T\Delta S$ rises steadily (and thus the $-T\Delta S$ term becomes more negative) with rising temperature. This line crosses the relatively constant ΔH line at 352 K. At any higher temperature, the $-T\Delta S$ term is greater than the ΔH term, so ΔG is negative.

Coupling of Reactions to Drive a Nonspontaneous Change

When considering the spontaneity of a complex, multistep reaction, we often find that a nonspontaneous step is driven by a spontaneous step in a **coupling of reactions.** *One step supplies enough free energy for the other to occur,* just as the combustion of gasoline supplies enough free energy to move a car.

Look again at the reduction of copper(I) oxide by carbon. Previously, we found that the *overall* reaction becomes spontaneous at any temperature above 352 K. Dividing the reaction into two steps, however, we find that even at a higher temperature, such as 375 K, copper(I) oxide does not spontaneously decompose to its elements:

$$\text{Cu}_2\text{O}(s) \longrightarrow 2\text{Cu}(s) + \tfrac{1}{2}\text{O}_2(g) \qquad \Delta G_{375} = 140.0 \text{ kJ}$$

However, the oxidation of carbon to CO at 375 K is quite spontaneous:

$$\text{C}(s) + \tfrac{1}{2}\text{O}_2(g) \longrightarrow \text{CO}(g) \qquad \Delta G_{375} = -143.8 \text{ kJ}$$

Coupling these reactions means having the carbon in contact with the Cu_2O, which allows the reaction with the larger negative ΔG to "drive" the one with the smaller positive ΔG. Adding the reactions together and canceling the common substance ($\tfrac{1}{2}\text{O}_2$) gives an overall reaction with a negative ΔG:

$$\text{Cu}_2\text{O}(s) + \text{C}(s) \longrightarrow 2\text{Cu}(s) + \text{CO}(g) \qquad \Delta G_{375} = -3.8 \text{ kJ}$$

Many biochemical reactions are also nonspontaneous. Key steps in the synthesis of proteins and nucleic acids, the formation of fatty acids, the maintenance of ion balance, and the breakdown of nutrients are among the many essential processes with positive ΔG values. Driving these energetically unfavorable steps by coupling them to a spontaneous one is a life-sustaining strategy common to all organisms—animals, plants, and microbes—as you'll see in the Chemical Connections essay.

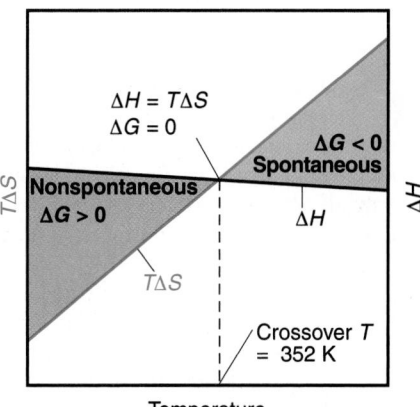

Figure 20.14 The effect of temperature on reaction spontaneity. The two terms that make up ΔG are plotted against T. The figure shows a relatively constant ΔH and a steadily increasing $T\Delta S$ (and thus more negative $-T\Delta S$) for the reaction between Cu_2O and C. At low T, the reaction is nonspontaneous ($\Delta G > 0$) because the positive ΔH term has a greater magnitude than the negative $T\Delta S$ term. At 352 K, $\Delta H = T\Delta S$, so $\Delta G = 0$. At any higher T, the reaction becomes spontaneous ($\Delta G < 0$) because the $-T\Delta S$ term dominates.

 Transparency/ARIS Presentation Center, Art Library

 Biological application

Section Summary

The sign of the free energy change, $\Delta G = \Delta H - T\Delta S$, is directly related to reaction spontaneity: a negative ΔG corresponds to a positive ΔS_{univ}. • We use the standard free energy of formation (ΔG_f°) to calculate $\Delta G_{\text{rxn}}^\circ$ at 25°C. • The maximum work a system can do is never obtained from a real (irreversible) process because some free energy is always converted to heat. • The magnitude of T influences the spontaneity of temperature-dependent reactions (same signs of ΔH and ΔS) by affecting the size of $T\Delta S$. For such reactions, the T at which the reaction becomes spontaneous can be estimated by setting $\Delta G = 0$. • A nonspontaneous reaction ($\Delta G > 0$) can be coupled to a more spontaneous one ($\Delta G \ll 0$) to make it occur. For example, in organisms, the hydrolysis of ATP drives many reactions with a positive ΔG.

Chemical Connections *to Biological Energetics*

The Universal Role of ATP

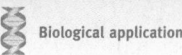

One of the most remarkable features of organisms, as well as a strong indication of a common ancestry, is the utilization of the same few biomolecules for all the reactions of life. Despite their incredible diversity of appearance and behavior, virtually all organisms use the same amino acids to make their proteins, the same nucleotides to make their nucleic acids, and the same carbohydrate (glucose) to provide energy.

In addition, *all organisms use the same spontaneous reaction to provide the free energy needed to drive a wide variety of nonspontaneous ones*. This reaction is the hydrolysis of a high-energy molecule called **adenosine triphosphate (ATP)** to adenosine diphosphate (ADP):

$$ATP^{4-} + H_2O \rightleftharpoons ADP^{3-} + HPO_4^{2-} + H^+$$
$$\Delta G^{\circ\prime} = -30.5 \text{ kJ}$$

In biochemical systems, the standard-state concentration of H^+ is $10^{-7}\ M$, not the usual $1\ M$, and the standard free energy change has the symbol $\Delta G^{\circ\prime}$. In the metabolic breakdown of glucose, for example, the initial step is the addition of a phosphate group to a glucose molecule in a dehydration-condensation reaction:

$$\text{Glucose} + HPO_4^{2-} + H^+ \rightleftharpoons [\text{glucose phosphate}]^- + H_2O$$
$$\Delta G^{\circ\prime} = 13.8 \text{ kJ}$$

Coupling this nonspontaneous reaction to ATP hydrolysis makes the overall process spontaneous. If we add the two reactions, HPO_4^{2-}, H^+, and H_2O cancel, and we obtain

$$\text{Glucose} + ATP^{4-} \rightleftharpoons [\text{glucose phosphate}]^- + ADP^{3-}$$
$$\Delta G^{\circ\prime} = -16.7 \text{ kJ}$$

Like the reactions to extract copper that we just discussed, these two reactions cannot affect each other if they are physically separated. Coupling of the reactions is accomplished through an enzyme, a biological catalyst (Section 16.8) that simultaneously binds glucose and ATP such that the phosphate group of ATP to be transferred lies next to the particular —OH group of glucose that will accept it (Figure B20.3). Enzymes play similar catalytic roles in all the reactions driven by ATP hydrolysis.

The ADP formed in these energy-releasing reactions is combined with phosphate to regenerate ATP in energy-absorbing reactions catalyzed by other enzymes. In fact, one major biochemical reason an organism eats and breathes is to make ATP, so that it has the energy to move, grow, reproduce—and study chemistry, of course. Thus, there is a continuous cycling of ATP to ADP and back to ATP again to supply energy to the cells (Figure B20.4).

Figure B20.4 The cycling of metabolic free energy through ATP. Processes that release free energy are coupled to the formation of ATP from ADP, whereas those that require free energy are coupled to the hydrolysis of ATP to ADP.

Figure B20.3 The coupling of a nonspontaneous reaction to the hydrolysis of ATP. The glucose molecule must lie next to the ATP molecule (shown as ADP—O—PO₃H) in the enzyme's active site for the correct atoms to form bonds. ADP (shown as ADP—OH) and glucose phosphate are released.

What makes ATP hydrolysis such a good supplier of free energy? By examining the phosphate portions of ATP, ADP, and HPO_4^{2-}, we can see two reasons (Figure B20.5). The first is that, at physiological pH (~7), the triphosphate portion of the ATP molecule has an average of four negative charges grouped closely together. As a result, the ATP molecule has a *high charge repulsion* built into its structure. In ADP, however, this charge repulsion is reduced.

The second reason relates to the greater delocalization of π electrons in the hydrolysis products, which we can see from resonance structures. For example, once ATP is hydrolyzed, the π electrons in HPO_4^{2-} can be more readily delocalized, which stabilizes the ion. Thus, greater charge repulsion and less electron delocalization make ATP higher in energy than the sum of ADP and HPO_4^{2-}. When ATP is hydrolyzed, some of this additional energy is released, to be harnessed by the organism to drive the metabolic reactions that could not otherwise take place.

Figure B20.5 Why is ATP a high-energy molecule? The ATP molecule releases a large amount of free energy when it is hydrolyzed because **(A)** the high charge repulsion in the triphosphate portion of ATP is reduced, and **(B)** the free HPO_4^{2-} ion is stabilized by delocalization of its π electrons, as shown by these resonance forms.

ARIS Presentation Center, Art Library

20.4 FREE ENERGY, EQUILIBRIUM, AND REACTION DIRECTION

The sign of ΔG allows us to predict reaction direction, but it is not the only way to do so. In Chapter 17, we predicted reaction direction by comparing the values of the reaction quotient (Q) and the equilibrium constant (K). Recall that

Literature Hirsch, Warren. "Rubber Bands, Free Energy, and Le Châtelier's Principle," *J. Chem. Educ.* 2002, *79*, 200A.

- If $Q < K$ ($Q/K < 1$), the reaction as written proceeds to the right.
- If $Q > K$ ($Q/K > 1$), the reaction as written proceeds to the left.
- If $Q = K$ ($Q/K = 1$), the reaction has reached equilibrium, and there is no net reaction in either direction.

As you might expect, these two ways of predicting reaction spontaneity—the sign of ΔG and the magnitude of Q/K—are related. Their relationship emerges when we compare the signs of $\ln Q/K$ with ΔG:

- If $Q/K < 1$, then $\ln Q/K < 0$: reaction proceeds to the right ($\Delta G < 0$).
- If $Q/K > 1$, then $\ln Q/K > 0$: reaction proceeds to the left ($\Delta G > 0$).
- If $Q/K = 1$, then $\ln Q/K = 0$: reaction is at equilibrium ($\Delta G = 0$).

Note that the signs of ΔG and $\ln Q/K$ are identical for a given reaction direction. In fact, ΔG and $\ln Q/K$ are proportional to each other and made equal through the constant RT:

$$\Delta G = RT \ln \frac{Q}{K} = RT \ln Q - RT \ln K \qquad (20.11)$$

What does this central relationship mean? As you know, Q represents the concentrations (or pressures) of a system's components at any time during the reaction, whereas K represents them when the reaction has reached equilibrium. Therefore, Equation 20.11 says that ΔG depends on how different the ratio of concentrations, Q, is from the equilibrium ratio, K.

The last term in Equation 20.11 is very important. By choosing standard-state values for Q, we obtain the standard free energy change ($\Delta G°$). When all concentrations are 1 M (or all pressures 1 atm), ΔG is $\Delta G°$ and Q equals 1:

$$\Delta G° = RT \ln 1 - RT \ln K$$

We know that $\ln 1 = 0$, so the $RT \ln Q$ term drops out, and we have

$$\Delta G° = -RT \ln K \qquad (20.12)$$

Table 20.2   The Relationship Between $\Delta G°$ and K at 298 K

$\Delta G°$ (kJ)	K	Significance
200	9×10^{-36}	Essentially no forward reaction; reverse reaction goes to completion
100	3×10^{-18}	
50	2×10^{-9}	
10	2×10^{-2}	
1	7×10^{-1}	Forward and reverse reactions proceed to same extent
0	1	
−1	1.5	
−10	5×10^1	
−50	6×10^8	
−100	3×10^{17}	Forward reaction goes to completion; essentially no reverse reaction
−200	1×10^{35}	

(Arrow indicating FORWARD REACTION downward and REVERSE REACTION upward)

This relationship allows us to calculate the standard free energy change of a reaction ($\Delta G°$) from its equilibrium constant, or vice versa. Because $\Delta G°$ is related logarithmically to K, even a small change in the value of $\Delta G°$ has a large effect on the value of K. Table 20.2 shows the K values that correspond to a range of $\Delta G°$ values. Note that as $\Delta G°$ becomes more positive, K becomes smaller, which means the reaction reaches equilibrium with less product and more reactant. Similarly, as $\Delta G°$ becomes more negative, K becomes larger. For example, if $\Delta G° = +10$ kJ, $K \approx 0.02$, which means that the product terms are about $\frac{1}{50}$ those of the reactant terms; whereas, if $\Delta G° = -10$ kJ, they are 50 times larger. Of course, reactions do not usually begin with all components in their standard states. By substituting the relationship between $\Delta G°$ and K (Equation 20.12) into the expression for ΔG (Equation 20.11), we obtain a relationship that applies to any starting concentrations:

$$\Delta G = \Delta G° + RT \ln Q \qquad (20.13)$$

Sample Problems 20.8 and 20.9 illustrate some ways to apply Equations 20.12 and 20.13; the first employs molecular scenes to examine these ideas, while the second highlights an industrial reaction.

SAMPLE PROBLEM 20.8   Using Molecular Scenes to Find ΔG for a Reaction at Nonstandard Conditions

PROBLEM The molecular scenes below represent three mixtures of A_2 (black) and B_2 (green) forming AB. Each molecule represents 0.10 mol, and the volume is 1 L. The equation is

$$A_2(g) + B_2(g) \rightleftharpoons 2AB(g) \qquad \Delta G° = -3.4 \text{ kJ/mol}$$

(Three molecular scene illustrations labeled 1, 2, 3)

(a) Mixture 1 is at equilibrium; calculate K.
(b) Which reaction mixture has the lowest (most negative) ΔG, and which has the highest (most positive)?
(c) Is the reaction spontaneous when $[A_2] = [B_2] = [AB] = 1.0 \; M$?

PLAN (a) Mixture 1 is at equilibrium, so we first write the expression for Q and then find the molarity of each substance from the numbers of molecules to calculate K. **(b)** To find ΔG, we apply Equation 20.13. We are given $\Delta G°$ (−3.4 kJ/mol), and we know R (8.314 J/mol·K). We calculate T from Equation 20.12 with the value of K that we found in part (a), and substitute the molarity of each substance to get Q. **(c)** The concentration 1.0 M represents the standard state, and we substitute that concentration into Q in Equation 20.13 and solve for ΔG, which in this case equals $\Delta G°$.

SOLUTION (a) Writing the expression for Q and calculating K:

$$A_2(g) + B_2(g) \rightleftharpoons 2AB(g) \qquad Q = \frac{[AB]^2}{[A_2][B_2]} \qquad K = \frac{(0.40)^2}{(0.20)(0.20)} = 4.0$$

(b) Calculating T from Equation 20.12 for use in Equation 20.13:

$$\Delta G° = -RT \ln K = -\frac{3.4 \text{ kJ}}{\text{mol}} = -\left(\frac{8.314 \text{ J}}{\text{mol·K}}\right) T \ln 4.0$$

$$T = \frac{\dfrac{-3.4 \text{ kJ}}{\text{mol}}\left(\dfrac{1000 \text{ J}}{1 \text{ kJ}}\right)}{-\left(\dfrac{8.314 \text{ J}}{\text{mol·K}}\right) \ln 4.0} = 295 \text{ K}$$

Calculating the value of ΔG for each mixture from Equation 20.13:

For mixture 1:

$$\Delta G = \Delta G° + RT \ln Q = -3.4 \text{ kJ} + RT \ln 4.0$$

$$= -3.4 \text{ kJ}\left(\frac{1000 \text{ J}}{1 \text{ kJ}}\right) + \left(\frac{8.314 \text{ J}}{\text{mol·K}}\right)(295 \text{ K}) \ln 4.0$$

$$= -3400 \text{ J} + 3400 \text{ J} = 0.0 \text{ J}$$

For mixture 2:

$$\Delta G = -3.4\text{kJ} + RT \ln \frac{(0.20)^2}{(0.30)(0.30)}$$

$$= -3.4 \text{ kJ}\left(\frac{1000 \text{ J}}{1 \text{ kJ}}\right) + \left(\frac{8.314 \text{ J}}{\text{mol·K}}\right)(295 \text{ K}) \ln 0.44 = -5.4 \times 10^3 \text{ J}$$

For mixture 3:

$$\Delta G = -3.4 \text{ kJ/mol} + RT \ln \frac{(0.60)^2}{(0.10)(0.10)}$$

$$= -3.4 \text{ kJ/mol}\left(\frac{1000 \text{ J}}{1 \text{ kJ}}\right) + \left(\frac{8.314 \text{ J}}{\text{mol·K}}\right)(295 \text{ K}) \ln 36 = 5.4 \times 10^3 \text{ J}$$

Mixture 2 has the most negative ΔG, and mixture 3 has the most positive ΔG.

(c) Finding ΔG when $[A_2] = [B_2] = [AB] = 1.0 \ M$:

$$\Delta G = \Delta G° + RT \ln Q = -3.4 \text{ kJ/mol} + RT \ln \frac{(1.0)^2}{(1.0)(1.0)}$$

$$= -3.4 \text{ kJ/mol} + RT \ln 1.0 = -3.4 \text{ kJ/mol}$$

Yes, the reaction is spontaneous when the components are in their standard states.

CHECK In (b), round to check the arithmetic; for example, for mixture 3,

$$\Delta G \approx -3000 \text{ J} + (8 \text{ J/mol·K})(300 \text{ K})4 \approx 7000 \text{ J}$$

which is in the correct ballpark.

COMMENT 1. By using the properties of logarithms, we didn't have to calculate T and ΔG in (b). In mixture 2, $Q < 1$, so $\ln Q$ is negative, which makes ΔG more negative. Also, note that $Q(0.44) < K(4.0)$, so $\Delta G < 0$. In mixture 3, $Q > 1$ (and is greater than it is in mixture 1), so $\ln Q$ is positive, which makes ΔG positive. Also, $Q(36) > K(4.0)$, so $\Delta G > 0$.
2. In (b), the value of zero for ΔG of the equilibrium mixture (mixture 1) makes sense, because a system at equilibrium has released all of its free energy.

FOLLOW-UP PROBLEM 20.8 The scenes below depict mixtures of X_2 (tan) and Y_2 (blue) forming XY_2. Each molecule represents 0.10 mol, and the volume is 1 L. The equation is $X_2(g) + 2Y_2(g) \rightleftharpoons 2XY_2(g)$; $\Delta G° = -1.3$ kJ/mol.

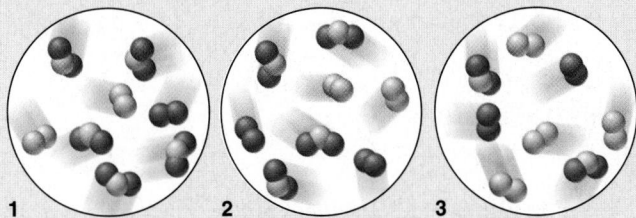

(a) If $K = 2$, which mixture is at equilibrium? (b) Rank the three mixtures from the lowest (most negative) ΔG to highest (most positive) ΔG. (c) What is the sign of ΔG for the change that occurs as each nonequilibrium mixture approaches equilibrium?

ARIS Presentation Center, Worked Examples Library

Engineering application

SAMPLE PROBLEM 20.9 Calculating ΔG at Nonstandard Conditions

PROBLEM The oxidation of $SO_2(g)$ (which we considered in Sample Problem 20.7),

$$2SO_2(g) + O_2(g) \longrightarrow 2SO_3(g)$$

is too slow at 298 K to be useful in the manufacture of sulfuric acid. To overcome this low rate, the process is conducted at an elevated temperature.

(a) Calculate K at 298 K and at 973 K. ($\Delta G^{\circ}_{298} = -141.6$ kJ/mol for reaction as written; using ΔH° and ΔS° values at 973 K, $\Delta G^{\circ}_{973} = -12.12$ kJ/mol for reaction as written.)

(b) In experiments to determine the effect of temperature on reaction spontaneity, two sealed containers are filled with 0.500 atm of SO_2, 0.0100 atm of O_2, and 0.100 atm of SO_3 and kept at 25°C and at 700.°C. In which direction, if any, will the reaction proceed to reach equilibrium at each temperature?

(c) Calculate ΔG for the system in part (b) at each temperature.

PLAN **(a)** We know ΔG°, T, and R, so we can calculate the K's from Equation 20.12. **(b)** To determine if a net reaction will occur at the given pressures, we calculate Q with the given partial pressures and compare it with each K from part (a). **(c)** Because these are not standard-state pressures, we calculate ΔG at each T from Equation 20.13 with the values of ΔG° (given) and Q [found in part (b)].

SOLUTION **(a)** Calculating K at the two temperatures:

$$\Delta G^{\circ} = -RT \ln K$$

so

$$K = e^{-(\Delta G^{\circ}/RT)}$$

At 298 K, the exponent is

$$-(\Delta G^{\circ}/RT) = -\left(\frac{-141.6 \text{ kJ/mol} \times \dfrac{1000 \text{ J}}{1 \text{ kJ}}}{8.314 \text{ J/mol·K} \times 298 \text{ K}} \right) = 57.2$$

so

$$K = e^{-(\Delta G^{\circ}/RT)} = e^{57.2} = 7 \times 10^{24}$$

At 973 K, the exponent is

$$-(\Delta G^{\circ}/RT) = -\left(\frac{-12.12 \text{ kJ/mol} \times \dfrac{1000 \text{ J}}{1 \text{ kJ}}}{8.314 \text{ J/mol·K} \times 973 \text{ K}} \right) = 1.50$$

so

$$K = e^{-(\Delta G^{\circ}/RT)} = e^{1.50} = 4.5$$

(b) Calculating the value of Q:

$$Q = \frac{P^2_{SO_3}}{P^2_{SO_2} \times P_{O_2}} = \frac{0.100^2}{0.500^2 \times 0.0100} = 4.00$$

Because $Q < K$ at both temperatures, the denominator will decrease and the numerator increase—more SO_3 will form—until Q equals K. However, the reaction will go far to the right at 298 K while approaching equilibrium, whereas it will move only slightly to the right at 973 K.

(c) Calculating ΔG, the nonstandard free energy change, at 298 K:

$$\Delta G_{298} = \Delta G^{\circ} + RT \ln Q$$

$$= -141.6 \text{ kJ/mol} + \left(8.314 \text{ J/mol·K} \times \frac{1 \text{ kJ}}{1000 \text{ J}} \times 298 \text{ K} \times \ln 4.00 \right)$$

$$= -138.2 \text{ kJ/mol}$$

Calculating ΔG at 973 K:

$$\Delta G_{973} = \Delta G^{\circ} + RT \ln Q$$

$$= -12.12 \text{ kJ/mol} + \left(8.314 \text{ J/mol·K} \times \frac{1 \text{ kJ}}{1000 \text{ J}} \times 973 \text{ K} \times \ln 4.00 \right)$$

$$= -0.9 \text{ kJ/mol}$$

CHECK Note that in parts (a) and (c) we made the energy units in free energy changes (kJ) consistent with those in R (J). Based on the rules for significant figures in addition and subtraction, we retain one digit to the right of the decimal place in part (c).

COMMENT For these starting gas pressures at 973 K, the process is barely spontaneous ($\Delta G = -0.9$ kJ/mol), so why use a higher temperature? Like the synthesis of NH_3 (Chemical Connections, p. 771), this process is carried out at a higher temperature *with a catalyst* to attain a higher *rate*, even though the *yield* is greater at a lower temperature. We discuss the details of the industrial production of sulfuric acid in Chapter 22.

FOLLOW-UP PROBLEM 20.9 At 298 K, hypobromous acid (HBrO) dissociates in water with a K_a of 2.3×10^{-9}.
(a) Calculate $\Delta G°$ for the dissociation of HBrO.
(b) Calculate ΔG if $[H_3O^+] = 6.0\times10^{-4}$ M, $[BrO^-] = 0.10$ M, and $[HBrO] = 0.20$ M.

Another Look at the Meaning of Spontaneity At this point, let's discuss some terminology related to, but distinct from, the terms *spontaneous* and *nonspontaneous*. Consider the general reaction A $\rightleftharpoons$ B, for which $K = [B]/[A] > 1$, and therefore, the reaction proceeds largely from left to right (Figure 20.15A). From pure A to the equilibrium point, $Q < K$ and the curved *green* arrow indicates that the reaction is spontaneous ($\Delta G < 0$). From there on, the curved *red* arrow shows that the reaction is nonspontaneous ($\Delta G > 0$). From pure B to the equilibrium point, $Q > K$ and the reaction is also spontaneous ($\Delta G < 0$), but not thereafter. In either case, *the free energy decreases as the reaction proceeds, until it reaches a minimum at the equilibrium mixture: $Q = K$ and $\Delta G = 0$.* For the overall reaction A $\rightleftharpoons$ B (starting with all components in their standard states), $G_B°$ is smaller than $G_A°$, so $\Delta G°$ is negative, which corresponds to $K > 1$. We call this a *product-favored* reaction because the final state of the system contains mostly product.

Now consider the opposite situation, a general reaction C $\rightleftharpoons$ D, for which $K = [D]/[C] < 1$: the reaction proceeds only slightly from left to right (Figure 20.15B). Here, too, whether we start with pure C or pure D, the reaction is spontaneous ($\Delta G < 0$) until the equilibrium point. But here, the equilibrium mixture contains mostly C (the reactant), so we say the reaction is *reactant favored*. In this case, $G_D°$ is *larger* than $G_C°$, so $\Delta G°$ is *positive*, which corresponds to $K < 1$. The point is that *spontaneous* refers to that portion of a reaction in which the free energy is decreasing, that is, from some starting mixture to the equilibrium mixture, whereas *product-favored* refers to a reaction that goes predominantly, but not necessarily completely, to product (see Table 20.2).

Transparency/ARIS Presentation Center, Art Library

Figure 20.15 The relation between free energy and the extent of reaction. The free energy of the system is plotted against the extent of reaction. Each reaction proceeds spontaneously ($Q \neq K$ and $\Delta G < 0$; *curved green arrows*) from either pure reactants (A or C) or pure products (B or D) *to* the equilibrium mixture, at which point $\Delta G = 0$. The reaction *from* the equilibrium mixture to either pure reactants or products is nonspontaneous ($\Delta G > 0$; *curved red arrows*). **A,** For the product-favored reaction A $\rightleftharpoons$ B, $G_A° > G_B°$, so $\Delta G° < 0$ and $K > 1$. **B,** For the reactant-favored reaction C $\rightleftharpoons$ D, $G_D° > G_C°$, so $\Delta G° > 0$ and $K < 1$.

Section Summary

Two ways of predicting reaction direction are from the value of ΔG and from the relation of Q to K. These variables represent different aspects of the same phenomenon and are related to each other by $\Delta G = RT \ln Q/K$. When $Q = K$, the system can release no more free energy. • Beginning with Q at the standard state, the free energy change is $\Delta G°$, and it is related to the equilibrium constant by $\Delta G° = -RT \ln K$. • Any nonequilibrium mixture of reactants and products moves spontaneously ($\Delta G < 0$) toward the equilibrium mixture. • A product-favored reaction goes predominantly toward product and, thus, has $K > 1$ and $\Delta G° < 0$.

Chapter Perspective

As processes move toward equilibrium, some of the energy released becomes degraded to unusable heat, and the entropy of the universe increases. This unalterable necessity built into all natural processes has led to speculation about the "End of Everything," when all possible processes have stopped occurring, no free energy remains in any system, and nothing but waste heat is dispersed evenly throughout the universe—the final equilibrium! Even if this grim future is in store, however, it is many billions of years away, so you have plenty of time to appreciate more hopeful applications of thermodynamics. In Chapter 21, you'll see how spontaneous reactions can generate electricity and how electricity supplies free energy to drive nonspontaneous ones.

▌CHAPTER REVIEW GUIDE   The following sections provide many aids to help you study this chapter. (Numbers in parentheses refer to pages, unless noted otherwise.)

Learning Objectives   These are concepts and skills you should know after studying this chapter.

Relevant section and/or sample problem (SP) numbers appear in parentheses.

Understand These Concepts

1. How the tendency of a process to occur by itself is distinct from how long it takes to occur (Introduction)
2. The distinction between a spontaneous and a nonspontaneous change (Section 20.1)
3. Why the first law of thermodynamics and the sign of $\Delta H°$ cannot predict the direction of a spontaneous process (Section 20.1)
4. How the entropy (S) of a system is defined by the number of microstates over which its energy is dispersed (Section 20.1)
5. How entropy is alternatively defined by the heat absorbed (or released) at constant T in a reversible process (Section 20.1)
6. The criterion for spontaneity according to the second law of thermodynamics: that a change increases S_{univ} (Section 20.1)
7. How absolute values of standard molar entropies ($S°$) can be obtained because the third law of thermodynamics provides a "zero point" (Section 20.1)
8. How temperature, physical state, dissolution, atomic size, and molecular complexity influence $S°$ values (Section 20.1)
9. How $\Delta S°_{rxn}$ is based on the difference between the summed $S°$ values for the reactants and those for products (Section 20.2)
10. How the surroundings add heat to or remove heat from the system and how ΔS_{surr} influences overall $\Delta S°_{rxn}$ (Section 20.2)
11. The relationship between ΔS_{surr} and ΔH_{sys} (Section 20.2)
12. How reactions proceed spontaneously toward equilibrium ($\Delta S_{univ} > 0$) but proceed no further at equilibrium ($\Delta S_{univ} = 0$) (Section 20.2)
13. How the free energy change (ΔG) combines the system's entropy and enthalpy changes (Section 20.3)

14. How the expression for the free energy change is derived from the second law (Section 20.3)
15. The relationship between ΔG and the maximum work a system can perform and why this quantity of work is never performed in a real process (Section 20.3)
16. How temperature determines spontaneity for reactions in which ΔS and ΔH have the same sign (Section 20.3)
17. Why the temperature at which a reaction becomes spontaneous occurs when $\Delta G = 0$ (Section 20.3)
18. How a spontaneous change can be coupled to a nonspontaneous change to make it occur (Section 20.3)
19. How ΔG is related to the ratio of Q to K (Section 20.4)
20. The meaning of $\Delta G°$ and its relation to K (Section 20.4)
21. The relation of ΔG to $\Delta G°$ and Q (Section 20.4)
22. Why G decreases, no matter what the starting concentrations, as the reacting system moves toward equilibrium (Section 20.4)

Master These Skills

1. Predicting relative $S°$ values of systems (Section 20.1 and SP 20.1)
2. Calculating $\Delta S°_{rxn}$ for a chemical change (SP 20.2)
3. Finding reaction spontaneity from ΔS_{surr} and $\Delta H°_{sys}$ (SP 20.3)
4. Calculating $\Delta G°_{rxn}$ from $\Delta H°_f$ and $S°$ values (SP 20.4)
5. Calculating $\Delta G°_{rxn}$ from $\Delta G°_f$ values (SP 20.5)
6. Predicting the signs of ΔH, ΔS, and ΔG (SP 20.6)
7. Calculating the effect of temperature on ΔG (SP 20.7)
8. Calculating the temperature at which a reaction becomes spontaneous (Section 20.3)
9. Calculating K from $\Delta G°$ (Section 20.4 and SP 20.9)
10. Using $\Delta G°$ and Q to calculate ΔG at any conditions (SPs 20.8, 20.9)

Key Terms These important terms appear in boldface in the chapter and are defined again in the Glossary.

Section 20.1
spontaneous change (881)
entropy (S) (884)
second law of thermodynamics (887)
third law of thermodynamics (888)

standard molar entropy ($S°$) (888)

Section 20.2
standard entropy of reaction ($\Delta S°_{rxn}$) (893)

Section 20.3
free energy (G) (899)
standard free energy change ($\Delta G°$) (900)
standard free energy of formation ($\Delta G°_f$) (901)

coupling of reactions (907)
adenosine triphosphate (ATP) (908)

Key Equations and Relationships Numbered and screened concepts are listed for you to refer to or memorize.

20.1 Quantifying entropy in terms of the number of microstates (W) over which the energy of a system can be distributed (884):
$$S = k \ln W$$

20.2 Quantifying the entropy change in terms of heat absorbed (or released) in a reversible process (887)
$$\Delta S_{sys} = \frac{q_{rev}}{T}$$

20.3 Stating the second law of thermodynamics, for a spontaneous process (888):
$$\Delta S_{univ} = \Delta S_{sys} + \Delta S_{surr} > 0$$

20.4 Calculating the standard entropy of reaction from the standard molar entropies of reactants and products (894):
$$\Delta S°_{rxn} = \Sigma m S°_{products} - \Sigma n S°_{reactants}$$

20.5 Relating the entropy change in the surroundings to the enthalpy change of the system and the temperature (895):
$$\Delta S_{surr} = -\frac{\Delta H_{sys}}{T}$$

20.6 Expressing the free energy change of the system in terms of its component enthalpy and entropy changes (Gibbs equation) (900):
$$\Delta G_{sys} = \Delta H_{sys} - T\Delta S_{sys}$$

20.7 Calculating the standard free energy change from standard enthalpy and entropy changes (900):
$$\Delta G°_{sys} = \Delta H°_{sys} - T\Delta S°_{sys}$$

20.8 Calculating the standard free energy change from the standard free energies of formation (901):
$$\Delta G°_{rxn} = \Sigma m \Delta G°_{f(products)} - \Sigma n \Delta G°_{f(reactants)}$$

20.9 Relating the free energy change to the maximum work a process can perform (902):
$$\Delta G = w_{max}$$

20.10 Finding the temperature at which a reaction becomes spontaneous (906):
$$T = \frac{\Delta H}{\Delta S}$$

20.11 Expressing the free energy change in terms of Q and K (909):
$$\Delta G = RT \ln \frac{Q}{K} = RT \ln Q - RT \ln K$$

20.12 Expressing the free energy change when Q is at standard state conditions (909):
$$\Delta G° = -RT \ln K$$

20.13 Expressing the free energy change for nonstandard initial conditions (910):
$$\Delta G = \Delta G° + RT \ln Q$$

Highlighted Figures and Tables These figures (F) and tables (T) provide a visual review of key ideas.

F20.3 Entropy and expansion of a gas (885)
F20.4 Expansion of a gas and number of microstates (886)
F20.6 Entropy and temperature (889)
F20.7 Entropy and phase changes (890)

F20.12 Components of ΔS_{univ} for spontaneous reactions (899)
T20.1 Reaction spontaneity and the signs of ΔH, ΔS, and ΔG (905)
T20.2 The relationship of $\Delta G°$ and K (910)
F20.15 Free energy and extent of reaction (913)

Brief Solutions to FOLLOW-UP PROBLEMS Compare your solutions to these calculation steps and answers.

20.1 (a) $PCl_5(g)$: higher molar mass and more complex molecule; (b) $BaCl_2(s)$: higher molar mass; (c) $Br_2(g)$: gases have more freedom of motion and dispersal of energy than liquids.

20.2 (a) $2NaOH(s) + CO_2(g) \longrightarrow Na_2CO_3(s) + H_2O(l)$
$\Delta n_{gas} = -1$, so $\Delta S°_{rxn} < 0$
$\Delta S°_{rxn} = [(1 \text{ mol } H_2O)(69.9 \text{ J/mol·K})$
$\quad + (1 \text{ mol } Na_2CO_3)(139 \text{ J/mol·K})]$
$\quad - [(1 \text{ mol } CO_2)(213.7 \text{ J/mol·K})$
$\quad + (2 \text{ mol } NaOH)(64.5 \text{ J/mol·K})]$
$\quad = -134 \text{ J/K}$

(b) $2Fe(s) + 3H_2O(g) \longrightarrow Fe_2O_3(s) + 3H_2(g)$
$\Delta n_{gas} = 0$, so cannot predict sign of $\Delta S°_{rxn}$
$\Delta S°_{rxn} = [(1 \text{ mol } Fe_2O_3)(87.4 \text{ J/mol·K})$
$\quad + (3 \text{ mol } H_2)(130.6 \text{ J/mol·K})]$
$\quad - [(2 \text{ mol } Fe)(27.3 \text{ J/mol·K})$
$\quad + (3 \text{ mol } H_2O)(188.7 \text{ J/mol·K})]$
$\quad = -141.5 \text{ J/K}$

20.3 $2FeO(s) + \frac{1}{2}O_2(g) \longrightarrow Fe_2O_3(s)$
$\Delta S°_{sys} = (1 \text{ mol } Fe_2O_3)(87.4 \text{ J/mol·K})$
$\quad - [(2 \text{ mol } FeO)(60.75 \text{ J/mol·K})$
$\quad + (\frac{1}{2} \text{ mol } O_2)(205.0 \text{ J/mol·K})]$
$\quad = -136.6 \text{ J/K}$
$\Delta H°_{sys} = (1 \text{ mol } Fe_2O_3)(-825.5 \text{ kJ/mol})$
$\quad - [(2 \text{ mol } FeO)(-272.0 \text{ kJ/mol}) + (\frac{1}{2} \text{ mol } O_2)(0 \text{ kJ/mol})]$
$\quad = -281.5 \text{ kJ}$
$\Delta S_{surr} = -\frac{\Delta H°_{sys}}{T} = -\frac{(-281.5 \text{ kJ} \times 1000 \text{ J/kJ})}{298 \text{ K}} = +945 \text{ J/K}$
$\Delta S_{univ} = \Delta S°_{sys} + \Delta S_{surr} = -136.6 \text{ J/K} + 945 \text{ J/K}$
$\quad = 808 \text{ J/K}$; reaction is spontaneous at 298 K.

20.4 Using $\Delta H°_f$ and $S°$ values from Appendix B,
$\Delta H°_{rxn} = -114.2 \text{ kJ}$ and $\Delta S°_{rxn} = -146.5 \text{ J/K}$
$\Delta G°_{rxn} = \Delta H°_{rxn} - T\Delta S°_{rxn}$
$\quad = -114.2 \text{ kJ} - [(298 \text{ K})(-146.5 \text{ J/K})(1 \text{ kJ}/1000 \text{ J})]$
$\quad = -70.5 \text{ kJ}$

Brief Solutions to FOLLOW-UP PROBLEMS (continued)

20.5 (a) ΔG°_{rxn} = (2 mol NO_2)(51 kJ/mol)
    $-$ [(2 mol NO)(86.60 kJ/mol)
    + (1 mol O_2)(0 kJ/mol)]
    = -71 kJ
(b) ΔG°_{rxn} = (2 mol CO)(-137.2 kJ/mol) $-$ [(2 mol C)(0 kJ/mol)
    + (1 mol O_2)(0 kJ/mol)]
    = -274.4 kJ

20.6 (a) More moles of gas are present after the reaction, so $\Delta S > 0$.
(b) The problem says the reaction is spontaneous ($\Delta G < 0$) only above 325°C, which implies high T. If $\Delta S > 0$, $-T\Delta S < 0$, so ΔG will become negative at higher T only if $\Delta H > 0$.

20.7 ΔG becomes negative at lower T, so $\Delta H < 0$, $\Delta S < 0$, and $-T\Delta S > 0$. At lower T, the negative ΔH value becomes larger than the positive $-T\Delta S$ value.

20.8 (a) Mixture 2 is at equilibrium. (b) 3 (most negative) < 2 < 1 (most positive). (c) Any reaction mixture moves spontaneously toward equilibrium, so both changes have a negative ΔG.

20.9 (a) $\Delta G^\circ = -RT \ln K = -8.314$ J/mol·K $\times \dfrac{1 \text{ kJ}}{1000 \text{ J}} \times 298$ K
    $\times \ln (2.3 \times 10^{-9})$
    = 49 kJ/mol
(b) $Q = \dfrac{[H_3O^+][BrO^-]}{[HBrO]} = \dfrac{(6.0 \times 10^{-4})(0.10)}{0.20} = 3.0 \times 10^{-4}$
$\Delta G = \Delta G^\circ + RT \ln Q$
    = 49 kJ/mol
    $+ \left[8.314 \text{ J/mol·K} \times \dfrac{1 \text{ kJ}}{1000 \text{ J}} \times 298 \text{ K} \times \ln (3.0 \times 10^{-4}) \right]$
    = 29 kJ/mol

PROBLEMS

Problems with **colored** numbers are answered in Appendix E and worked in detail in the Student Solutions Manual. Problem sections match those in the text and provide the numbers of relevant sample problems. Most offer Concept Review Questions, Skill-Building Exercises (grouped in pairs covering the same concept), and Problems in Context. The Comprehensive Problems are based on material from any section or previous chapter.

Note: Unless stated otherwise, problems refer to systems at 298 K (25°C). Solving these problems may require values from Appendix B.

The Second Law of Thermodynamics: Predicting Spontaneous Change
(Sample Problem 20.1)

● **Concept Review Questions**

• **20.1** Distinguish between the terms *spontaneous* and *instantaneous*. Give an example of a process that is spontaneous but very slow, and one that is very fast but not spontaneous.

• **20.2** Distinguish between the terms *spontaneous* and *nonspontaneous*. Can a nonspontaneous process occur? Explain.

• **20.3** State the first law of thermodynamics in terms of (a) the energy of the universe; (b) the creation or destruction of energy; (c) the energy change of system and surroundings. Does the first law reveal the direction of spontaneous change? Explain.

• **20.4** State qualitatively the relationship between entropy and freedom of particle motion. Use this idea to explain why you will probably never (a) be suffocated because all the air near you has moved to the other side of the room; (b) see half the water in your cup of tea freeze while the other half boils.

• **20.5** Why is ΔS_{vap} of a substance always larger than ΔS_{fus}?

• **20.6** How does the entropy of the surroundings change during an exothermic reaction? An endothermic reaction? Other than the examples in text, describe a spontaneous endothermic process.

• **20.7** (a) What is the entropy of a perfect crystal at 0 K?

(b) Does entropy increase or decrease as the temperature rises?
(c) Why is $\Delta H^\circ_f = 0$ but $S^\circ > 0$ for an element?
(d) Why does Appendix B list ΔH°_f values but not ΔS°_f values?

● **Skill-Building Exercises** (grouped in similar pairs)

• **20.8** Which of the following processes are spontaneous?
(a) Water evaporates from a puddle. (b) A lion chases an antelope.
(c) An isotope undergoes radioactive disintegration.

• **20.9** Which of the following processes are spontaneous?
(a) Earth moves around the Sun. (b) A boulder rolls up a hill.
(c) Sodium metal and chlorine gas form solid sodium chloride.

• **20.10** Which of the following processes are spontaneous?
(a) Methane burns in air. (b) A teaspoonful of sugar dissolves in a cup of hot coffee. (c) A soft-boiled egg becomes raw.

• **20.11** Which of the following processes are spontaneous?
(a) A satellite falls to Earth. (b) Water decomposes to H_2 and O_2 at 298 K and 1 atm. (c) Average car prices increase.

• **20.12** Predict the sign of ΔS_{sys} for each process: (a) A piece of wax melts. (b) Silver chloride precipitates from solution. (c) Dew forms on a lawn in morning.

• **20.13** Predict the sign of ΔS_{sys} for each process: (a) Gasoline vapors mix with air in a car engine. (b) Hot air expands. (c) Humidity condenses in cold air.

• **20.14** Predict the sign of ΔS_{sys} for each process: (a) Alcohol evaporates. (b) A solid explosive converts to a gas. (c) Perfume vapors diffuse through a room.

• **20.15** Predict the sign of ΔS_{sys} for each process: (a) A pond freezes in winter. (b) Atmospheric CO_2 dissolves in the ocean. (c) An apple tree bears fruit.

• **20.16** Without using Appendix B, predict the sign of ΔS° for
(a) $2K(s) + F_2(g) \longrightarrow 2KF(s)$
(b) $NH_3(g) + HBr(g) \longrightarrow NH_4Br(s)$
(c) $NaClO_3(s) \longrightarrow Na^+(aq) + ClO_3^-(aq)$

• **20.17** Without using Appendix B, predict the sign of ΔS° for
(a) $H_2S(g) + \frac{1}{2}O_2(g) \longrightarrow \frac{1}{8}S_8(s) + H_2O(g)$
(b) $HCl(aq) + NaOH(aq) \longrightarrow NaCl(aq) + H_2O(l)$
(c) $2NO_2(g) \longrightarrow N_2O_4(g)$

Levels of Difficulty • Easy •• Medium ••• Challenging

- **20.18** Without using Appendix B, predict the sign of $\Delta S°$ for
  (a) $CaCO_3(s) + 2HCl(aq) \longrightarrow CaCl_2(aq) + H_2O(l) + CO_2(g)$
  (b) $2NO(g) + O_2(g) \longrightarrow 2NO_2(g)$
  (c) $2KClO_3(s) \longrightarrow 2KCl(s) + 3O_2(g)$
- **20.19** Without using Appendix B, predict the sign of $\Delta S°$ for
  (a) $Ag^+(aq) + Cl^-(aq) \longrightarrow AgCl(s)$
  (b) $KBr(s) \longrightarrow KBr(aq)$

  (c) $CH_3CH{=}CH_2(g) \longrightarrow \underset{\text{H}_2\text{C}{-}\text{CH}_2(g)}{\overset{\text{CH}_2}{\diagup \diagdown}}$

- **20.20** Predict the sign of ΔS for each process:
  (a) $C_2H_5OH(g)$ (350 K and 500 torr) $\longrightarrow$
  $\qquad\qquad\qquad\qquad C_2H_5OH(g)$ (350 K and 250 torr)
  (b) $N_2(g)$ (298 K and 1 atm) $\longrightarrow N_2(aq)$ (298 K and 1 atm)
  (c) $O_2(aq)$ (303 K and 1 atm) $\longrightarrow O_2(g)$ (303 K and 1 atm)
- **20.21** Predict the sign of ΔS for each process:
  (a) $O_2(g)$ (1.0 L at 1 atm) $\longrightarrow O_2(g)$ (0.10 L at 10 atm)
  (b) $Cu(s)$ (350°C and 2.5 atm) $\longrightarrow Cu(s)$ (450°C and 2.5 atm)
  (c) $Cl_2(g)$ (100°C and 1 atm) $\longrightarrow Cl_2(g)$ (10°C and 1 atm)
- **20.22** Predict which substance has greater molar entropy. Explain.
  (a) Butane $CH_3CH_2CH_2CH_3(g)$ or 2-butene $CH_3CH{=}CHCH_3(g)$
  (b) $Ne(g)$ or $Xe(g)$ (c) $CH_4(g)$ or $CCl_4(l)$
- **20.23** Predict which substance has greater molar entropy. Explain.
  (a) $NO_2(g)$ or $N_2O_4(g)$ (b) $CH_3OCH_3(l)$ or $CH_3CH_2OH(l)$
  (c) $HCl(g)$ or $HBr(g)$
- **20.24** Predict which substance has greater molar entropy. Explain.
  (a) $CH_3OH(l)$ or $C_2H_5OH(l)$ (b) $KClO_3(s)$ or $KClO_3(aq)$
  (c) $Na(s)$ or $K(s)$
- **20.25** Predict which substance has greater molar entropy. Explain.
  (a) $P_4(g)$ or $P_2(g)$ (b) $HNO_3(aq)$ or $HNO_3(l)$
  (c) $CuSO_4(s)$ or $CuSO_4{\cdot}5H_2O(s)$
- **20.26** Without consulting Appendix B, arrange each group in order of *increasing* standard molar entropy ($S°$). Explain.
  (a) Graphite, diamond, charcoal
  (b) Ice, water vapor, liquid water
  (c) O_2, O_3, O atoms
- **20.27** Without consulting Appendix B, arrange each group in order of *increasing* standard molar entropy ($S°$). Explain.
  (a) Glucose ($C_6H_{12}O_6$), sucrose ($C_{12}H_{22}O_{11}$), ribose ($C_5H_{10}O_5$)
  (b) $CaCO_3$, $Ca + C + \frac{3}{2}O_2$, $CaO + CO_2$
  (c) $SF_6(g)$, $SF_4(g)$, $S_2F_{10}(g)$
- **20.28** Without consulting Appendix B, arrange each group in order of *decreasing* standard molar entropy ($S°$). Explain.
  (a) $ClO_4^-(aq)$, $ClO_2^-(aq)$, $ClO_3^-(aq)$
  (b) $NO_2(g)$, $NO(g)$, $N_2(g)$
  (c) $Fe_2O_3(s)$, $Al_2O_3(s)$, $Fe_3O_4(s)$
- **20.29** Without consulting Appendix B, arrange each group in order of *decreasing* standard molar entropy ($S°$). Explain.
  (a) Mg metal, Ca metal, Ba metal
  (b) Hexane (C_6H_{14}), benzene (C_6H_6), cyclohexane (C_6H_{12})
  (c) $PF_2Cl_3(g)$, $PF_5(g)$, $PF_3(g)$

Calculating the Change in Entropy of a Reaction
(Sample Problems 20.2 and 20.3)

● **Concept Review Questions**

- **20.30**. For the reaction depicted in the molecular scenes, X is red and Y is green.

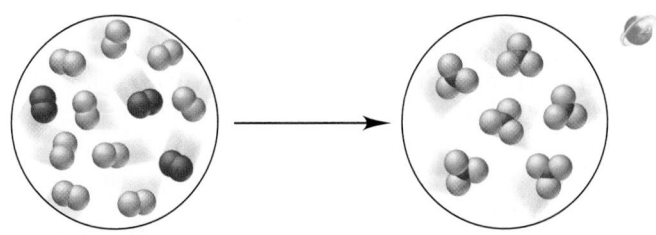

  (a) Write a balanced equation.
  (b) Determine the sign of ΔS.
  (c) Which species has the highest molar entropy?
- **20.31** Describe the equilibrium condition in terms of the entropy changes of a system and its surroundings. What does this description mean about the entropy change of the universe?
- **20.32** For the reaction $H_2O(g) + Cl_2O(g) \longrightarrow 2HClO(g)$, you know $\Delta S°_{rxn}$ and $S°$ of $HClO(g)$ and of $H_2O(g)$. Write an expression that can be used to determine $S°$ of $Cl_2O(g)$.

● **Skill-Building Exercises** (grouped in similar pairs)

- **20.33** For each reaction, predict the sign and find the value of $\Delta S°$:
  (a) $3NO(g) \longrightarrow N_2O(g) + NO_2(g)$
  (b) $3H_2(g) + Fe_2O_3(s) \longrightarrow 2Fe(s) + 3H_2O(g)$
  (c) $P_4(s) + 5O_2(g) \longrightarrow P_4O_{10}(s)$
- **20.34** For each reaction, predict the sign and find the value of $\Delta S°$:
  (a) $3NO_2(g) + H_2O(l) \longrightarrow 2HNO_3(l) + NO(g)$
  (b) $N_2(g) + 3F_2(g) \longrightarrow 2NF_3(g)$
  (c) $C_6H_{12}O_6(s) + 6O_2(g) \longrightarrow 6CO_2(g) + 6H_2O(g)$

- **20.35** Find $\Delta S°$ for the combustion of ethane (C_2H_6) to carbon dioxide and gaseous water. Is the sign of $\Delta S°$ as expected?
- **20.36** Find $\Delta S°$ for the combustion of methane to carbon dioxide and liquid water. Is the sign of $\Delta S°$ as expected?

- **20.37** Find $\Delta S°$ for the reaction of nitrogen monoxide with hydrogen to form ammonia and water vapor. Is the sign of $\Delta S°$ as expected?
- **20.38** Find $\Delta S°$ for the combustion of ammonia to nitrogen dioxide and water vapor. Is the sign of $\Delta S°$ as expected?

- **20.39** Find $\Delta S°$ for the formation of $Cu_2O(s)$ from its elements.
- **20.40** Find $\Delta S°$ for the formation of $HI(g)$ from its elements.

- **20.41** Find $\Delta S°$ for the formation of $CH_3OH(l)$ from its elements.
- **20.42** Find $\Delta S°$ for the formation of $PCl_5(g)$ from its elements.

● **Problems in Context**
- **20.43** Sulfur dioxide is released in the combustion of coal. Scrubbers use aqueous slurries of calcium hydroxide to remove the SO_2 from flue gases. Write a balanced equation for this reaction and calculate $\Delta S°$ at 298 K [$S°$ of $CaSO_3(s) = 101.4$ J/mol·K].
- **20.44** Oxyacetylene welding is used to repair metal structures, including bridges, buildings, and even the Statue of Liberty. Calculate $\Delta S°$ for the combustion of 1 mol of acetylene (C_2H_2).

Entropy, Free Energy, and Work
(Sample Problems 20.4 to 20.7)

● **Concept Review Questions**
- **20.45** What is the advantage of calculating free energy changes rather than entropy changes to determine reaction spontaneity?
- **20.46** Given that $\Delta G_{sys} = -T\Delta S_{univ}$, explain how the sign of ΔG_{sys} correlates with reaction spontaneity.
- **20.47** (a) Is an endothermic reaction more likely to be spontaneous at higher temperatures or lower temperatures? Explain.

(b) The change depicted below occurs at constant pressure. Explain your answers to each of the following: (1) What is the sign of ΔH? (2) What is the sign of ΔS? (3) What is the sign of ΔS_{surr}? (4) How does the sign of ΔG vary with temperature?

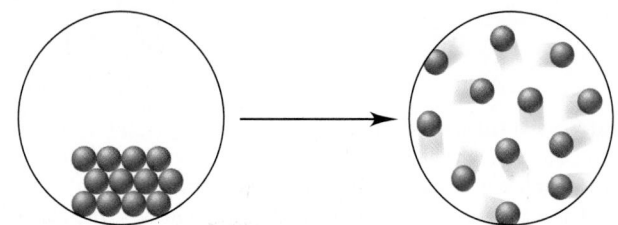

• 20.48 With its components in their standard states, a certain reaction is spontaneous only at high T. What do you know about the signs of $\Delta H°$ and $\Delta S°$? Describe a process for which this is true.

• 20.49 How can $\Delta S°$ be relatively independent of T if $S°$ of each reactant and product increases with T?

● **Skill-Building Exercises** (grouped in similar pairs)

• 20.50 Calculate $\Delta G°$ for each reaction using $\Delta G_f°$ values:
(a) $2Mg(s) + O_2(g) \longrightarrow 2MgO(s)$
(b) $2CH_3OH(g) + 3O_2(g) \longrightarrow 2CO_2(g) + 4H_2O(g)$
(c) $BaO(s) + CO_2(g) \longrightarrow BaCO_3(s)$

• 20.51 Calculate $\Delta G°$ for each reaction using $\Delta G_f°$ values:
(a) $H_2(g) + I_2(s) \longrightarrow 2HI(g)$
(b) $MnO_2(s) + 2CO(g) \longrightarrow Mn(s) + 2CO_2(g)$
(c) $NH_4Cl(s) \longrightarrow NH_3(g) + HCl(g)$

• 20.52 Find $\Delta G°$ for the reactions in Problem 20.50 using $\Delta H_f°$ and $S°$ values.

• 20.53 Find $\Delta G°$ for the reactions in Problem 20.51 using $\Delta H_f°$ and $S°$ values.

• 20.54 Consider the oxidation of carbon monoxide:
$$CO(g) + \tfrac{1}{2}O_2(g) \longrightarrow CO_2(g)$$
(a) Predict the signs of $\Delta S°$ and $\Delta H°$. Explain.
(b) Calculate $\Delta G°$ by two different methods.

• 20.55 Consider the combustion of butane gas:
$$C_4H_{10}(g) + \tfrac{13}{2}O_2(g) \longrightarrow 4CO_2(g) + 5H_2O(g)$$
(a) Predict the signs of $\Delta S°$ and $\Delta H°$. Explain.
(b) Calculate $\Delta G°$ by two different methods.

• 20.56 For the gaseous reaction of xenon and fluorine to form xenon hexafluoride:
(a) Calculate $\Delta S°$ at 298 K ($\Delta H° = -402$ kJ/mol and $\Delta G° = -280.$ kJ/mol).
(b) Assuming that $\Delta S°$ and $\Delta H°$ change little with temperature, calculate $\Delta G°$ at 500. K.

• 20.57 For the gaseous reaction of carbon monoxide and chlorine to form phosgene ($COCl_2$):
(a) Calculate $\Delta S°$ at 298 K ($\Delta H° = -220.$ kJ/mol and $\Delta G° = -206$ kJ/mol).
(b) Assuming that $\Delta S°$ and $\Delta H°$ change little with temperature, calculate $\Delta G°$ at 450. K.

• 20.58 One reaction used to produce small quantities of pure H_2 is
$$CH_3OH(g) \rightleftharpoons CO(g) + 2H_2(g)$$
(a) Determine $\Delta H°$ and $\Delta S°$ for the reaction at 298 K.
(b) Assuming that these values are relatively independent of temperature, calculate $\Delta G°$ at 28°C, 128°C, and 228°C.
(c) What is the significance of the different values of $\Delta G°$?

• 20.59 A reaction that occurs in the internal combustion engine is
$$N_2(g) + O_2(g) \rightleftharpoons 2NO(g)$$
(a) Determine $\Delta H°$ and $\Delta S°$ for the reaction at 298 K.
(b) Assuming that these values are relatively independent of temperature, calculate $\Delta G°$ at 100.°C, 2560.°C, and 3540.°C.
(c) What is the significance of the different values of $\Delta G°$?

• 20.60 Use $\Delta H°$ and $\Delta S°$ values for the following process at 1 atm to find the normal boiling point of Br_2:
$$Br_2(l) \rightleftharpoons Br_2(g)$$

• 20.61 Use $\Delta H°$ and $\Delta S°$ values to find the temperature at which these sulfur allotropes reach equilibrium at 1 atm:
$$S(rhombic) \rightleftharpoons S(monoclinic)$$

● **Problems in Context**

• 20.62 As a fuel, $H_2(g)$ produces only nonpolluting $H_2O(g)$ when it burns. Moreover, it combines with $O_2(g)$ in a fuel cell (Chapter 21) to provide electrical energy.
(a) Calculate $\Delta H°$, $\Delta S°$, and $\Delta G°$ per mole of H_2 at 298 K.
(b) Is the spontaneity of this reaction dependent on T? Explain.
(c) At what temperature does the reaction become spontaneous?

• 20.63 The U.S. government requires automobile fuels to contain a renewable component. Fermentation of glucose from corn yields ethanol, which is added to gasoline to fulfill this requirement:
$$C_6H_{12}O_6(s) \longrightarrow 2C_2H_5OH(l) + 2CO_2(g)$$
Calculate $\Delta H°$, $\Delta S°$, and $\Delta G°$ for the reaction at 25°C. Is the spontaneity of this reaction dependent on T? Explain.

Free Energy, Equilibrium, and Reaction Direction
(Sample Problems 20.8 and 20.9)

● **Concept Review Questions**

• 20.64 (a) If $K << 1$ for a reaction, what do you know about the sign and magnitude of $\Delta G°$? (b) If $\Delta G° << 0$ for a reaction, what do you know about the magnitude of K? Of Q?

• 20.65 How is the free energy change of a process related to the work that can be obtained from the process? Is this quantity of work obtainable in practice? Explain.

•• 20.66 The scenes and the graph relate to the reaction of $X_2(g)$ (black) with $Y_2(g)$ (orange) to form $XY(g)$.
(a) If reactants and products are in their standard states, what quantity is represented on the graph by x?
(b) Which scene represents point 1? Explain. (c) Which scene represents point 2? Explain.

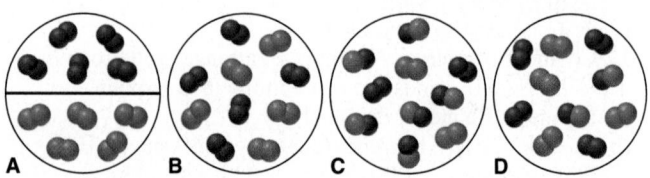

• 20.67 What is the difference between $\Delta G°$ and ΔG? Under what circumstances does $\Delta G = \Delta G°$?

● **Skill-Building Exercises** (grouped in similar pairs)

• 20.68 Calculate K at 298 K for each reaction:
(a) $NO(g) + \tfrac{1}{2}O_2(g) \rightleftharpoons NO_2(g)$

(b) $2HCl(g) \rightleftharpoons H_2(g) + Cl_2(g)$

(c) $2C(graphite) + O_2(g) \rightleftharpoons 2CO(g)$

• **20.69** Calculate K at 298 K for each reaction:

(a) $MgCO_3(s) \rightleftharpoons Mg^{2+}(aq) + CO_3^{2-}(aq)$

(b) $2HCl(g) + Br_2(l) \rightleftharpoons 2HBr(g) + Cl_2(g)$

(c) $H_2(g) + O_2(g) \rightleftharpoons H_2O_2(l)$

• **20.70** Calculate K at 298 K for each reaction:

(a) $2H_2S(g) + 3O_2(g) \rightleftharpoons 2H_2O(g) + 2SO_2(g)$

(b) $H_2SO_4(l) \rightleftharpoons H_2O(l) + SO_3(g)$

(c) $HCN(aq) + NaOH(aq) \rightleftharpoons NaCN(aq) + H_2O(l)$

• **20.71** Calculate K at 298 K for each reaction:

(a) $SrSO_4(s) \rightleftharpoons Sr^{2+}(aq) + SO_4^{2-}(aq)$

(b) $2NO(g) + Cl_2(g) \rightleftharpoons 2NOCl(g)$

(c) $Cu_2S(s) + O_2(g) \rightleftharpoons 2Cu(s) + SO_2(g)$

• **20.72** Use Appendix B to determine the K_{sp} of Ag_2S.

• **20.73** Use Appendix B to determine the K_{sp} of CaF_2.

• **20.74** For the reaction $I_2(g) + Cl_2(g) \rightleftharpoons 2ICl(g)$, calculate K_p at 25°C [ΔG_f° of $ICl(g)$ = −6.075 kJ/mol].

• **20.75** For the reaction $CaCO_3(s) \rightleftharpoons CaO(s) + CO_2(g)$, calculate the equilibrium P_{CO_2} at 25°C.

• **20.76** The K_{sp} of $PbCl_2$ is 1.7×10^{-5} at 25°C. What is ΔG°? Is it possible to prepare a solution that contains $Pb^{2+}(aq)$ and $Cl^-(aq)$, at their standard-state concentrations?

• **20.77** The K_{sp} of ZnF_2 is 3.0×10^{-2} at 25°C. What is ΔG°? Is it possible to prepare a solution that contains $Zn^{2+}(aq)$ and $F^-(aq)$ at their standard-state concentrations?

• **20.78** The equilibrium constant for the reaction

$$2Fe^{3+}(aq) + Hg_2^{2+}(aq) \rightleftharpoons 2Fe^{2+}(aq) + 2Hg^{2+}(aq)$$

is $K_c = 9.1 \times 10^{-6}$ at 298 K.

(a) What is ΔG° at this temperature?

(b) If standard-state concentrations of the reactants and products are mixed, in which direction does the reaction proceed?

(c) Calculate ΔG when $[Fe^{3+}]$ = 0.20 M, $[Hg_2^{2+}]$ = 0.010 M, $[Fe^{2+}]$ = 0.010 M, and $[Hg^{2+}]$ = 0.025 M. In which direction will the reaction proceed to achieve equilibrium?

• **20.79** The formation constant for the reaction

$$Ni^{2+}(aq) + 6NH_3(aq) \rightleftharpoons Ni(NH_3)_6^{2+}(aq)$$

is $K_f = 5.6 \times 10^8$ at 25°C.

(a) What is ΔG° at this temperature?

(b) If standard-state concentrations of the reactants and products are mixed, in which direction does the reaction proceed?

(c) Determine ΔG when $[Ni(NH_3)_6^{2+}]$ = 0.010 M, $[Ni^{2+}]$ = 0.0010 M, and $[NH_3]$ = 0.0050 M. In which direction will the reaction proceed to achieve equilibrium?

● **Problems in Context**

• **20.80** High levels of ozone (O_3) cause rubber to deteriorate, green plants to turn brown, and many people to have difficulty breathing.

(a) Is the formation of O_3 from O_2 favored at all T, no T, high T, or low T?

(b) Calculate ΔG° for this reaction at 298 K.

(c) Calculate ΔG at 298 K for this reaction in urban smog where $[O_2]$ = 0.21 atm and $[O_3]$ = 5×10^{-7} atm.

• **20.81** A $BaSO_4$ slurry is ingested before the gastrointestinal tract is x-rayed because it is opaque to x-rays and defines the contours of the tract. Ba^{2+} ion is toxic, but the compound is nearly insoluble. If ΔG° at 37°C (body temperature) is 59.1 kJ/mol for the process

$$BaSO_4(s) \rightleftharpoons Ba^{2+}(aq) + SO_4^{2-}(aq)$$

what is $[Ba^{2+}]$ in the intestinal tract? (Assume that the only source of SO_4^{2-} is the ingested slurry.)

Comprehensive Problems

•• **20.82** According to advertisements, "a diamond is forever."

(a) Calculate ΔH°, ΔS°, and ΔG° at 298 K for the phase change

$$\text{Diamond} \longrightarrow \text{graphite}$$

(b) Given the conditions under which diamond jewelry is normally kept, argue for and against the statement in the ad.

(c) Given the answers in part (a), what would need to be done to make synthetic diamonds from graphite?

(d) Assuming ΔH° and ΔS° do not change with temperature, can graphite be converted to diamond spontaneously at 1 atm?

•• **20.83** Replace each question mark with the correct information:

	ΔS_{rxn}	ΔH_{rxn}	ΔG_{rxn}	Comment
(a)	+	−	−	?
(b)	?	0	−	Spontaneous
(c)	−	+	?	Not spontaneous
(d)	0	?	−	Spontaneous
(e)	?	0	+	?
(f)	+	+	?	$T\Delta S > \Delta H$

•• **20.84** Among the many complex ions of cobalt are the following: $Co(NH_3)_6^{3+}(aq) + 3en(aq) \rightleftharpoons Co(en)_3^{3+}(aq) + 6NH_3(aq)$ where "en" stands for ethylenediamine, $H_2NCH_2CH_2NH_2$. Six Co—N bonds are broken and six Co—N bonds are formed in this reaction, so $\Delta H_{rxn}^\circ \approx 0$; yet $K > 1$. What are the signs of ΔS° and ΔG°? What drives the reaction?

•• **20.85** What is the change in entropy when 0.200 mol of potassium freezes at 63.7°C (ΔH_{fus} = 2.39 kJ/mol)?

•• **20.86** Is each statement true or false? If false, correct it.

(a) All spontaneous reactions occur quickly.

(b) The reverse of a spontaneous reaction is nonspontaneous.

(c) All spontaneous processes release heat.

(d) The boiling of water at 100°C and 1 atm is spontaneous.

(e) If a process increases the freedom of motion of the particles of a system, the entropy of the system decreases.

(f) The energy of the universe is constant; the entropy of the universe decreases toward a minimum.

(g) All systems disperse their energy spontaneously.

(h) Both ΔS_{sys} and ΔS_{surr} equal zero at equilibrium.

••• **20.87** Hemoglobin carries O_2 from the lungs to tissue cells, where the O_2 is released. The protein is represented as Hb in its unoxygenated form and as Hb·O_2 in its oxygenated form. One reason CO is toxic is that it competes with O_2 in binding to Hb:

$$Hb·O_2(aq) + CO(g) \rightleftharpoons Hb·CO(aq) + O_2(g)$$

(a) If $\Delta G^\circ \approx -14$ kJ at 37°C (body temperature), what is the ratio of [Hb·CO] to [Hb·O_2] at 37°C with $[O_2]$ = [CO]?

(b) How is Le Châtelier's principle used to treat CO poisoning?

•• **20.88** An important ore of lithium is spodumene, $LiAlSi_2O_6$. To extract the metal, the α form of spodumene is first converted into the less dense β form in preparation for subsequent leaching and washing steps. Use the following data to calculate the lowest temperature at which the $\alpha \longrightarrow \beta$ conversion is feasible:

	ΔH_f° (kJ/mol)	S° (J/mol·K)
α-spodumene	−3055	129.3
β-spodumene	−3027	154.4

•• **20.89** Magnesia (MgO) is used for fire brick, crucibles, and furnace linings because of its high melting point. It is produced by decomposing magnesite ($MgCO_3$) at around 1200°C.
(a) Write a balanced equation for magnesite decomposition.
(b) Use $\Delta H°$ and $S°$ values to find $\Delta G°$ at 298 K.
(c) Assuming $\Delta H°$ and $S°$ do not change with temperature, find the minimum temperature at which the reaction is spontaneous.
(d) Calculate the equilibrium P_{CO_2} above $MgCO_3$ at 298 K.
(e) Calculate the equilibrium P_{CO_2} above $MgCO_3$ at 1200 K.

•• **20.90** To prepare nuclear fuel, U_3O_8 ("yellow cake") is converted to $UO_2(NO_3)_2$, which is then converted to UO_3 and finally UO_2. The fuel is enriched (the proportion of the ^{235}U is increased) by a two-step conversion of UO_2 into UF_6, a volatile solid, followed by a gaseous-diffusion separation of the ^{235}U and ^{238}U isotopes:

$$UO_2(s) + 4HF(g) \longrightarrow UF_4(s) + 2H_2O(g)$$
$$UF_4(s) + F_2(g) \longrightarrow UF_6(s)$$

Calculate $\Delta G°$ for the overall process at 85°C:

	$\Delta H_f°$ (kJ/mol)	$S°$ (J/mol·K)	$\Delta G_f°$ (kJ/mol)
$UO_2(s)$	−1085	77.0	−1032
$UF_4(s)$	−1921	152	−1830.
$UF_6(s)$	−2197	225	−2068

•• **20.91** Methanol, a major industrial feedstock, is made by several catalyzed reactions, such as $CO(g) + 2H_2(g) \longrightarrow CH_3OH(l)$.
(a) Show that this reaction is thermodynamically feasible.
(b) Is it favored at low or at high temperatures?
(c) One concern about using CH_3OH as an auto fuel is its oxidation in air to yield formaldehyde, $CH_2O(g)$, which poses a health hazard. Calculate $\Delta G°$ at 100.°C for this oxidation.

•• **20.92** (a) Write a balanced equation for the gaseous reaction between N_2O_5 and F_2 to form NF_3 and O_2. (b) Determine $\Delta G_{rxn}°$. (c) Find ΔG_{rxn} at 298 K if $P_{N_2O_5} = P_{F_2} = 0.20$ atm, $P_{NF_3} = 0.25$ atm, and $P_{O_2} = 0.50$ atm.

•• **20.93** Consider the following reaction:

$$2NOBr(g) \rightleftharpoons 2NO(g) + Br_2(g) \qquad K = 0.42 \text{ at } 373 \text{ K}$$

Given that $S°$ of $NOBr(g) = 272.6$ J/mol·K and that $\Delta S_{rxn}°$ and $\Delta H_{rxn}°$ are constant with temperature, find
(a) $\Delta S_{rxn}°$ at 298 K (b) $\Delta G_{rxn}°$ at 373 K
(c) $\Delta H_{rxn}°$ at 373 K (d) $\Delta H_f°$ of NOBr at 298 K
(e) $\Delta G_{rxn}°$ at 298 K (f) $\Delta G_f°$ of NOBr at 298 K

•• **20.94** Calculate the equilibrium constants for decomposition of the hydrogen halides at 298 K:

$$2HX(g) \rightleftharpoons H_2(g) + X_2(g)$$

What do these values indicate about the extent of decomposition of HX at 298 K? Suggest a reason for this trend.

•• **20.95** Hydrogenation is the addition of H_2 to double (or triple) carbon-carbon bonds. Peanut butter and most commercial baked goods include hydrogenated oils. Find $\Delta H°$, $\Delta S°$, and $\Delta G°$ for the hydrogenation of ethene (C_2H_4) to ethane (C_2H_6) at 25°C.

••• **20.96** Styrene is produced by catalytic dehydrogenation of ethylbenzene at high temperature in the presence of superheated steam.
(a) Find $\Delta H_{rxn}°$, $\Delta G_{rxn}°$, and $\Delta S_{rxn}°$, given these data at 298 K:

Compound	$\Delta H_f°$ (kJ/mol)	$\Delta G_f°$ (kJ/mol)	$S°$ (J/mol·K)
Ethylbenzene, $C_6H_5—CH_2CH_3$	−12.5	119.7	255
Styrene, $C_6H_5—CH=CH_2$	103.8	202.5	238

(b) At what temperature is the reaction spontaneous?
(c) What are $\Delta G_{rxn}°$ and K at 600.°C?
(d) With 5.0 parts steam to 1.0 part ethylbenzene in the reactant mixture and the total pressure kept constant at 1.3 atm, what is ΔG at 50.% conversion, that is, when 50.% of the ethylbenzene has reacted?

•• **20.97** Bromine monochloride is formed from the elements:

$$Cl_2(g) + Br_2(g) \longrightarrow 2BrCl(g)$$
$$\Delta H_{rxn}° = -1.35 \text{ kJ/mol} \qquad \Delta G_f° = -0.88 \text{ kJ/mol}$$

Calculate (a) $\Delta H_f°$ and (b) $S°$ of $BrCl(g)$.

••• **20.98** Propylene (propene; $CH_3CH=CH_2$) is used to produce polypropylene and many other chemicals. Although most is obtained from the cracking of petroleum, about 2% is produced by catalytic dehydrogenation of propane ($CH_3CH_2CH_3$):

$$CH_3CH_2CH_3 \xrightarrow{Pt/Al_2O_3} CH_3CH=CH_2 + H_2$$

Because this reaction is endothermic, heaters are placed between the reactor vessels to maintain the required temperature.
(a) If the molar entropy, $S°$, of propylene is 267.1 J/mol·K, find its entropy of formation, $S_f°$.
(b) Find $\Delta G_f°$ of propylene ($\Delta H_f°$ for propylene = 20.4 kJ/mol).
(c) Calculate $\Delta H_{rxn}°$ and $\Delta G_{rxn}°$ for the dehydrogenation.
(d) What is the theoretical yield of propylene at 580°C if the initial pressure of propane is 1.00 atm?
(e) Would the yield change if the reactor walls were permeable to H_2? Explain.
(f) At what temperature is the dehydrogenation spontaneous, with all substances in the standard state?

Note: Problems 20.99 to 20.102 relate to the thermodynamics of adenosine triphosphate (ATP). Refer to the Chemical Connections essay on pp. 908–909.

•• **20.99** Find K for (a) the hydrolysis of ATP, (b) the dehydration-condensation to form glucose phosphate, and (c) the coupled reaction between ATP and glucose. (d) How does each K change when T changes from 25°C to 37°C?

•• **20.100** The oxidation of 1 mol of glucose supplies enough metabolic energy to form 36 mol of ATP. Oxidation of 1 mol of a typical dietary fat like tristearin ($C_{57}H_{116}O_6$) yields enough energy to form 458 mol of ATP. (a) How many molecules of ATP can form per gram of glucose? (b) Per gram of tristearin?

•• **20.101** Nonspontaneous processes like muscle contraction, protein building, and nerve conduction are coupled to the spontaneous hydrolysis of ATP to ADP. ATP is then regenerated by coupling its synthesis to other energy-yielding reactions such as

Creatine phosphate $\longrightarrow$ creatine + phosphate
$$\Delta G°' = -43.1 \text{ kJ/mol}$$
ADP + phosphate $\longrightarrow$ ATP $\qquad \Delta G°' = +30.5 \text{ kJ/mol}$

Calculate $\Delta G°'$ for the overall reaction that regenerates ATP.

••• **20.102** Energy from ATP hydrolysis drives many nonspontaneous cell reactions:

$ATP^{4-}(aq) + H_2O(l) \rightleftharpoons$
$\quad ADP^{3-}(aq) + HPO_4^{2-}(aq) + H^+(aq) \qquad \Delta G°' = -30.5 \text{ kJ}$

Energy for the reverse process comes ultimately from glucose metabolism:

$$C_6H_{12}O_6(s) + 6O_2(g) \longrightarrow 6CO_2(g) + 6H_2O(l)$$

(a) Find K for the hydrolysis of ATP at 37°C.
(b) Find $\Delta G_{rxn}°'$ for metabolism of 1 mol of glucose.

(c) How many moles of ATP can be produced by metabolism of 1 mol of glucose?

(d) If 36 mol of ATP is formed, what is the actual yield?

•• **20.103** From the following reaction and data, find (a) $S°$ of $SOCl_2$ and (b) T at which the reaction becomes nonspontaneous:

$$SO_3(g) + SCl_2(l) \longrightarrow SOCl_2(l) + SO_2(g) \qquad \Delta G°_{rxn} = -75.2 \text{ kJ}$$

	$SO_3(g)$	$SCl_2(l)$	$SOCl_2(l)$	$SO_2(g)$
$\Delta H°_f$ (kJ/mol)	−396	−50.0	−245.6	−296.8
$S°$ (J/mol·K)	256.7	184	—	248.1

•• **20.104** Write equations for the oxidation of Fe and of Al. Use $\Delta G°_f$ to determine whether either process is spontaneous at 25°C.

•• **20.105**. The molecular scene depicts a gaseous equilibrium mixture at 460°C for the reaction of H_2 (blue) and I_2 (purple) to form HI. Each molecule represents 0.010 mol and the container volume is 1.0 L. (a) Is $K_c >$, =, or < 1? (b) Is $K_p >$, =, or < K_c? (c) Calculate $\Delta G°_{rxn}$. (d) How would the value of $\Delta G°_{rxn}$ change if the purple molecules represented H_2 and the blue I_2? Explain.

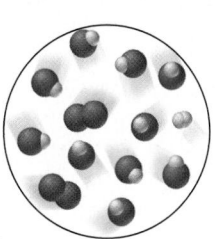

•• **20.106** A key step in the metabolism of glucose for energy is the isomerization of glucose-6-phosphate (G6P) to fructose-6-phosphate (F6P): G6P $\rightleftharpoons$ F6P; $K = 0.510$ at 298 K.

(a) Calculate $\Delta G°$ at 298 K.

(b) Calculate ΔG when Q, the [F6P]/[G6P] ratio, equals 10.0.

(c) Calculate ΔG when $Q = 0.100$.

(d) Calculate Q if $\Delta G = -2.50$ kJ/mol.

•• **20.107** A chemical reaction, such as HI forming from its elements, can reach equilibrium at many temperatures. In contrast, a phase change, such as ice melting, is in equilibrium at a given pressure only at the melting point. Each of the graphs below depicts G_{sys} vs. extent of change. (a) Which graph depicts how G_{sys} changes for the formation of HI? Explain. (b) Which graph depicts how G_{sys} changes as ice melts at 1°C and 1 atm? Explain.

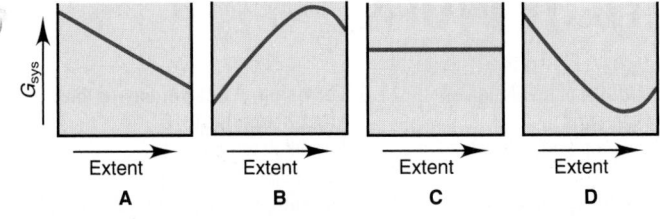

•• **20.108** When heated, the DNA double helix separates into two random-coil single strands. When cooled, the random coils reform the double helix: double helix $\rightleftharpoons$ 2 random coils.

(a) What is the sign of ΔS for the forward process? Why?

(b) Energy must be added to overcome H bonds and dispersion forces between the strands. What is the sign of ΔG for the forward process when $T\Delta S$ is smaller than ΔH?

(c) Write an expression that shows T in terms of ΔH and ΔS when the reaction is at equilibrium. (This temperature is called the *melting temperature* of the nucleic acid.)

•••**20.109** In the process of respiration, glucose is oxidized completely. In fermentation, O_2 is absent and glucose is broken down to ethanol and CO_2. Ethanol is oxidized to CO_2 and H_2O.

(a) Balance the following equations for these processes:

Respiration: $C_6H_{12}O_6(s) + O_2(g) \longrightarrow CO_2(g) + H_2O(l)$

Fermentation: $C_6H_{12}O_6(s) \longrightarrow C_2H_5OH(l) + CO_2(g)$

Ethanol oxidation: $C_2H_5OH(l) + O_2(g) \longrightarrow CO_2(g) + H_2O(l)$

(b) Calculate $\Delta G°_{rxn}$ for respiration of 1.00 g of glucose.

(c) Calculate $\Delta G°_{rxn}$ for fermentation of 1.00 g of glucose.

(d) Calculate $\Delta G°_{rxn}$ for oxidation of the ethanol from part (c).

•••**20.110** Consider the formation of ammonia:

$$N_2(g) + 3H_2(g) \rightleftharpoons 2NH_3(g)$$

(a) Assuming that $\Delta H°$ and $\Delta S°$ are constant with temperature, find the temperature at which $K_p = 1.00$.

(b) Find K_p at 400.°C, a typical temperature for NH_3 production.

(c) Given the lower K_p at the higher temperature, why are these conditions used industrially?

•••**20.111** Kyanite, sillimanite, and andalusite all have the formula Al_2SiO_5. Each is stable under different conditions (see the graph at right). At the point where the three phases intersect:

(a) Which mineral, if any, has the lowest free energy?

(b) Which mineral, if any, has the lowest enthalpy?

(c) Which mineral, if any, has the highest entropy?

(d) Which mineral, if any, has the lowest density?

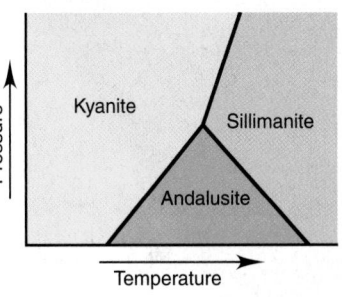

•••**20.112** Acetylene is produced commercially by the partial oxidation of methane. At 1500°C and pressures of 1–10 bar, the yield of acetylene is about 20%. The major side product is carbon monoxide, and some soot and carbon dioxide also form.

(a) At what temperature is the desired reaction spontaneous:

$$2CH_4 + \tfrac{1}{2}O_2 \longrightarrow C_2H_2 + 2H_2 + H_2O$$

(b) Acetylene can also be made by the reaction of its elements, carbon (graphite) and hydrogen. At what temperature is this formation reaction spontaneous? (c) Why must this reaction mixture be immediately cooled?

•••**20.113** Synthesis gas, a mixture that includes the fuels CO and H_2, is used to produce liquid hydrocarbons and methanol. It is made at pressures up to 100 atm by oxidation of methane followed by the steam re-forming and water-gas shift reactions. Because the process is exothermic, temperatures reach 950–1100°C, and the conditions are such that the amounts of H_2, CO, CO_2, CH_4, and H_2O leaving the reactor are close to the equilibrium amounts for the steam re-forming and water-gas shift reactions:

$CH_4(g) + H_2O(g) \rightleftharpoons CO(g) + 3H_2(g)$ (steam re-forming)

$CO(g) + H_2O(g) \rightleftharpoons CO_2(g) + H_2(g)$ (water-gas shift)

(a) At 1000.°C, what are $\Delta G°$ and $\Delta H°$ for the steam re-forming reaction and for the water-gas shift reaction?

(b) By doubling the steam re-forming step and adding it to the water-gas shift step, we obtain the following combined reaction:

$$2CH_4(g) + 3H_2O(g) \rightleftharpoons CO_2(g) + CO(g) + 7H_2(g)$$

Is this reaction spontaneous at 1000.°C in the standard state?

(c) Is it spontaneous at 98 atm and 50.% conversion (when 50.% of the starting materials have reacted)?

(d) Is it spontaneous at 98 atm and 90.% conversion?

21

Electrochemical Cells for Every Occasion Perhaps near you right now are a laptop, an MP3 player, and a cell phone—just a few of the devices whose energy comes from an electrochemical cell. In this chapter, you'll look at the two faces of electrochemistry—reactions in cells, such as those in batteries, that do electrical work and reactions in cells that require electrical work to occur. Both are indispensable to our modern way of life.

Electrochemistry: Chemical Change and Electrical Work

21.1 Redox Reactions and Electrochemical Cells
Review of Oxidation-Reduction Concepts
Half-Reaction Method for Balancing Redox Reactions
Electrochemical Cells

21.2 Voltaic Cells: Using Spontaneous Reactions to Generate Electrical Energy
Construction and Operation
Cell Notation
Why Does the Cell Work?

21.3 Cell Potential: Output of a Voltaic Cell
Standard Cell Potentials
Strengths of Oxidizing and Reducing Agents

21.4 Free Energy and Electrical Work
Standard Cell Potential and K
Effect of Concentration on E_{cell}
Changes in E_{cell} During Cell Operation
Concentration Cells

21.5 Electrochemical Processes in Batteries
Primary (Nonrechargeable) Batteries
Secondary (Rechargeable) Batteries
Fuel Cells

21.6 Corrosion: A Case of Environmental Electrochemistry
Corrosion of Iron
Protecting Against Corrosion

21.7 Electrolytic Cells: Using Electrical Energy to Drive Nonspontaneous Reactions
Construction and Operation
Predicting Electrolysis Products
Stoichiometry of Electrolysis

Thermodynamics has many applications other than expanding gases inside steam engines. In fact, some applications are probably within your reach right now, in the form of battery-operated devices—laptop computer, palm organizer, DVD remote, and, of course, wristwatch and calculator—or in the form of metal-plated jewelry or silverware. The operation and creation of these objects, and the many similar ones you use daily, involve the principles of electrochemistry, one of the most important areas of applied thermodynamics.

Electrochemistry is the study of the relationship between chemical change and electrical work. It is typically investigated through the use of **electrochemical cells,** systems that incorporate a redox reaction to produce or utilize electrical energy. The common objects just mentioned highlight the essential difference between the two types of electrochemical cells:

- One type of cell *does work by releasing free energy from a spontaneous reaction to produce electricity.* A battery houses this type of cell. ●
- The other type of cell *does work by absorbing free energy from a source of electricity to drive a nonspontaneous reaction.* Such cells are used to plate a thin layer of metal on objects and, in major industrial processes, produce some compounds and nonmetals and recover many metals from their ores.

IN THIS CHAPTER . . . We review redox concepts and highlight a method for balancing redox equations (mentioned briefly in Chapter 4) that is particularly useful for electrochemical cells. An overview of the two cell types follows. The first type releases free energy to do electrical work, and we see how and why it operates. The cell's electrical output and its relation to the relative strengths of the redox species concern us next. We then examine the free energy change and equilibrium nature of the cell reaction and how they relate to the cell output. The concentration cells and batteries we consider are useful examples of this type of cell, and corrosion is a destructive electrochemical process that operates by the same principle. Next, we focus on cells that absorb free energy to do electrical work and see how they are used to isolate elements from their compounds and how to determine the identity and amount of product formed. Finally, we examine the redox system that generates energy in living cells.

21.1 REDOX REACTIONS AND ELECTROCHEMICAL CELLS

Whether an electrochemical process releases or absorbs free energy, it always involves the *movement of electrons from one chemical species to another* in an oxidation-reduction (redox) reaction. In this section, we review the redox process and describe the half-reaction method of balancing redox reactions that was mentioned in Section 4.5. Then we see how such reactions are used in the two types of electrochemical cells.

A Quick Review of Oxidation-Reduction Concepts

In electrochemical reactions, as in any redox process, *oxidation* is the loss of electrons, and *reduction* is the gain of electrons. An *oxidizing agent* is the species that does the oxidizing, taking electrons from the substance being oxidized. A *reducing agent* is the species that does the reducing, giving electrons to the substance being reduced. After the reaction, the oxidized substance has a higher (more positive or less negative) oxidation number (O.N.), and the reduced substance has a lower (less positive or more negative) one. Keep in mind three key points:

- Oxidation (electron loss) always accompanies reduction (electron gain).
- The oxidizing agent is reduced, and the reducing agent is oxidized.
- The total number of electrons gained by the atoms/ions of the oxidizing agent always equals the total number lost by the atoms/ions of the reducing agent.

Concepts & Skills to Review
before you study this chapter

- redox terminology (Section 4.5 and Interchapter Topic 5)
- balancing redox reactions (Section 4.5)
- activity series of the metals (Section 4.6)
- free energy, work, and equilibrium (Sections 20.3 and 20.4)
- Q vs. K (Section 17.4) and ΔG vs. $\Delta G°$ (Section 20.4)

● **The Electrochemical Future Is Here** As the combustion products of coal and gasoline continue to threaten our atmosphere, a new generation of electrochemical devices is being developed. Battery-gasoline hybrid cars are already common, achieving much higher mileage than traditional gasoline-powered cars. Soon, electric cars, powered by banks of advanced fuel cells, may reduce the need for gasoline to a minimum. Every major car company has a fuel-cell prototype in operation. And these devices, once used principally on space missions, are now being produced for everyday residential and industrial applications as well.

 ARIS Presentation Center, Photo Library

 Engineering application

Figure 21.1 A summary of redox terminology. In the reaction between zinc and hydrogen ion, Zn is oxidized and H^+ is reduced.

 ARIS Presentation Center, Art Library

PROCESS	$Zn(s) + 2H^+(aq) \longrightarrow Zn^{2+}(aq) + H_2(g)$	
OXIDATION • One reactant loses electrons. • Reducing agent is oxidized. • Oxidation number increases.	Zinc **loses** electrons. Zinc is the reducing agent and becomes **oxidized**. The oxidation number of Zn **increases** from 0 to +2.	
REDUCTION • Other reactant gains electrons. • Oxidizing agent is reduced. • Oxidation number decreases.	Hydrogen ion **gains** electrons. Hydrogen ion is the oxidizing agent and becomes **reduced**. The oxidation number of H **decreases** from +1 to 0.	

Figure 21.1 presents these ideas for the aqueous reaction between zinc metal and a strong acid. Be sure you can identify the oxidation and reduction parts of a redox process. If you're having trouble, you may want to review the full discussion in Chapter 4 and the summary in Topic 5 of the Interchapter.

Half-Reaction Method for Balancing Redox Reactions

In Chapter 4, two methods for balancing redox reactions were mentioned—the oxidation number method and the half-reaction method—but only the first was discussed in detail (see Sample Problem 4.10, p. 163). The essential difference between the two methods is that the **half-reaction method** *divides the overall redox reaction into oxidation and reduction half-reactions.* Each half-reaction is balanced for atoms and charge. Then, one or both are multiplied by some integer to make electrons gained equal electrons lost, and the half-reactions are recombined to give the balanced redox equation. The half-reaction method offers several advantages for studying electrochemistry:

- It separates the oxidation and reduction steps, which reflects their actual physical separation in electrochemical cells.
- It is readily applied to redox reactions that take place in acidic or basic solution, which is common in these cells.
- It (usually) does *not* require assigning O.N.s. (In cases where the half-reactions are not obvious, we assign O.N.s to determine which atoms undergo a change and write half-reactions with the species that contain those atoms.)

In general, we begin with a "skeleton" ionic reaction, which shows only the species that are oxidized and reduced. *If the oxidized form of a species is on the left side of the skeleton reaction, the reduced form of that species must be on the right, and vice versa.* Unless H_2O, H^+, and OH^- are being oxidized or reduced, they do not appear in the skeleton reaction. The following steps are used in balancing a redox reaction by the half-reaction method:

Step 1. Divide the skeleton reaction into two half-reactions, each of which contains the oxidized and reduced forms of one of the species. (Which half-reaction is the oxidation and which the reduction becomes clear in the next step.)

Step 2. Balance the atoms and charges in each half-reaction.
- Atoms are balanced *in order:* atoms other than O and H, then O, and then H.
- Charge is balanced by *adding electrons* (e^-). They are added *to the left in the reduction half-reaction* because the reactant gains them; they are added *to the right in the oxidation half-reaction* because the reactant loses them.

Step 3. If necessary, multiply one or both half-reactions by an integer to make the number of e^- gained in the reduction equal the number lost in the oxidation.

Step 4. Add the balanced half-reactions, and include states of matter.

Step 5. Check that the atoms and charges are balanced.

We'll balance a redox reaction that occurs in acidic solution first and then go through Sample Problem 21.1 to balance one in basic solution.

Balancing Redox Reactions in Acidic Solution When a redox reaction occurs in acidic solution, H_2O molecules and H^+ ions are available for balancing. Even though we've usually used H_3O^+ to indicate the proton in water, we use H^+ in this chapter because it makes the balanced equations less complex. However, after using H^+ in this example, we'll balance the same equation with H_3O^+ just to show you that the only difference is in the number of water molecules needed.

Let's balance the redox reaction between dichromate ion and iodide ion to form chromium(III) ion and solid iodine, which occurs in acidic solution (Figure 21.2). The skeleton ionic reaction shows only the oxidized and reduced species:

$$Cr_2O_7^{2-}(aq) + I^-(aq) \longrightarrow Cr^{3+}(aq) + I_2(s) \quad \text{[acidic solution]}$$

Step 1. Divide the reaction into half-reactions, each of which contains the oxidized and reduced forms of one species. The two chromium species make up one half-reaction, and the two iodine species make up the other:

$$Cr_2O_7^{2-} \longrightarrow Cr^{3+}$$
$$I^- \longrightarrow I_2$$

Step 2. Balance atoms and charges in each half-reaction. We use H_2O to balance O atoms, H^+ to balance H atoms, and e^- to balance positive charges.

- For the $Cr_2O_7^{2-}/Cr^{3+}$ half-reaction:
  a. *Balance atoms other than O and H.* We balance the two Cr on the left with a coefficient 2 on the right:
  $$Cr_2O_7^{2-} \longrightarrow 2Cr^{3+}$$
  b. *Balance O atoms by adding H_2O molecules.* Each H_2O has one O atom, so we add seven H_2O on the right to balance the seven O in $Cr_2O_7^{2-}$:
  $$Cr_2O_7^{2-} \longrightarrow 2Cr^{3+} + 7H_2O$$
  c. *Balance H atoms by adding H^+ ions.* Each H_2O contains two H, and we added seven H_2O, so we add 14 H^+ ions on the left:
  $$14H^+ + Cr_2O_7^{2-} \longrightarrow 2Cr^{3+} + 7H_2O$$
  d. *Balance charge by adding electrons.* Each H^+ ion has a 1+ charge, and 14 H^+ plus $Cr_2O_7^{2-}$ gives 12+ on the left. Two Cr^{3+} give 6+ on the right. There is an excess of 6+ on the left, so we add six e^- on the left:
  $$6e^- + 14H^+ + Cr_2O_7^{2-} \longrightarrow 2Cr^{3+} + 7H_2O$$

This half-reaction is balanced, and we see it is the *reduction* because electrons appear on the *left, as reactants:* the reactant $Cr_2O_7^{2-}$ gains electrons (is reduced), so $Cr_2O_7^{2-}$ is the *oxidizing agent.* (Note that the O.N. of Cr decreases from +6 on the left to +3 on the right.)

- For the I^-/I_2 half-reaction:
  a. *Balance atoms other than O and H.* Two I atoms on the right require a coefficient 2 on the left:
  $$2I^- \longrightarrow I_2$$
  b. *Balance O atoms with H_2O.* Not needed; there are no O atoms.
  c. *Balance H atoms with H^+.* Not needed; there are no H atoms.
  d. *Balance charge with e^-.* To balance the 2− on the left, we add two e^- on the right:
  $$2I^- \longrightarrow I_2 + 2e^-$$

This half-reaction is balanced, and it is the *oxidation* because electrons appear on the *right, as products:* the reactant I^- loses electrons (is oxidized), so I^- is the *reducing agent.* (Note that the O.N. of I increases from −1 to 0.)

Figure 21.2 The redox reaction between dichromate ion and iodide ion. When $Cr_2O_7^{2-}$ (*left*) and I^- (*center*) are mixed in acid solution, they react to form Cr^{3+} and I_2 (*right*).

ARIS Presentation Center, Photo Library

Step 3. Multiply each half-reaction, if necessary, by an integer so that the number of e^- lost in the oxidation equals the number of e^- gained in the reduction. Two e^- are lost in the oxidation and six e^- are gained in the reduction, so we multiply the oxidation by 3:

$$3(2I^- \longrightarrow I_2 + 2e^-)$$
$$6I^- \longrightarrow 3I_2 + 6e^-$$

Step 4. Add the half-reactions together, canceling substances that appear on both sides, and include states of matter. In this example, only the electrons cancel:

$$6e^- + 14H^+ + Cr_2O_7^{2-} \longrightarrow 2Cr^{3+} + 7H_2O$$
$$6I^- \longrightarrow 3I_2 + 6e^-$$
$$\overline{6I^-(aq) + 14H^+(aq) + Cr_2O_7^{2-}(aq) \longrightarrow 3I_2(s) + 7H_2O(l) + 2Cr^{3+}(aq)}$$

Step 5. Check that atoms and charges balance:

$$\text{Reactants (6I, 14H, 2Cr, 7O; 6+)} \longrightarrow \text{products (6I, 14H, 2Cr, 7O; 6+)}$$

Balancing a Redox Reaction Using H_3O^+ Now let's balance the reduction half-reaction with H_3O^+ as the supplier of H atoms. In step 2, balancing Cr atoms and balancing O atoms with H_2O are the same as before, so we start with

$$Cr_2O_7^{2-} \longrightarrow 2Cr^{3+} + 7H_2O$$

Now we use H_3O^+ instead of H^+ to balance H atoms. Because each H_3O^+ is an H^+ bonded to an H_2O, we balance the 14 H on the right with 14 H_3O^+ on the left and immediately balance the H_2O that are part of those 14 H_3O^+ by adding 14 more H_2O on the right:

$$14H_3O^+ + Cr_2O_7^{2-} \longrightarrow 2Cr^{3+} + 7H_2O + 14H_2O$$

Simplifying the right side by taking the sum of the H_2O molecules gives

$$14H_3O^+ + Cr_2O_7^{2-} \longrightarrow 2Cr^{3+} + 21H_2O$$

None of this affects the redox change, so balancing the charge still requires six e^- on the left, and we obtain the balanced reduction half-reaction:

$$6e^- + 14H_3O^+ + Cr_2O_7^{2-} \longrightarrow 2Cr^{3+} + 21H_2O$$

Adding the balanced oxidation half-reaction gives the balanced redox equation:

$$6I^-(aq) + 14H_3O^+(aq) + Cr_2O_7^{2-}(aq) \longrightarrow 3I_2(s) + 21H_2O(l) + 2Cr^{3+}(aq)$$

Note, as mentioned earlier, that the only difference is the number of H_2O molecules: there are 14 more H_2O (21 instead of 7) from the 14 H_3O^+.

To depict the reaction even more accurately, we could show the metal ion in its hydrated form, which would also affect only the total number of H_2O:

$$6I^-(aq) + 14H_3O^+(aq) + Cr_2O_7^{2-}(aq) \longrightarrow 3I_2(s) + 9H_2O(l) + 2Cr(H_2O)_6^{3+}(aq)$$

Although these added steps present the species in solution more accurately, they change only the number of water molecules and make balancing more difficult; thus, they detract somewhat from the key chemical event—the redox change. Therefore, we'll employ the method that uses H^+ to balance H atoms.

Balancing Redox Reactions in Basic Solution As you just saw, in acidic solution, H_2O molecules and H^+ (or H_3O^+) ions are available for balancing. As Sample Problem 21.1 shows, in basic solution, H_2O molecules and OH^- ions are available. Only one additional step is needed to balance a redox equation that takes place in basic solution. It appears after both half-reactions have first been balanced *as if they took place in acidic solution* (steps 1 and 2), the e^- lost have been made equal to the e^- gained (step 3), and the half-reactions have been combined (step 4). At this point, *we add one OH^- ion **to both sides of the equation** for every H^+ ion present.* (We label this step "4 Basic.") The H^+ ions on one side are combined with the added OH^- ions to form H_2O, and OH^- ions appear on the other side of the equation. Excess H_2O molecules are canceled, and states of matter are identified. Finally, we check that atoms and charges balance (step 5).

SAMPLE PROBLEM 21.1 Balancing Redox Reactions by the Half-Reaction Method

PROBLEM Permanganate ion is a strong oxidizing agent, and its deep purple color makes it useful as an indicator in redox titrations (see Figure 4.15, p. 165). It reacts in basic solution with the oxalate ion to form carbonate ion and solid manganese dioxide. Balance the skeleton ionic equation for the reaction between $NaMnO_4$ and $Na_2C_2O_4$ in basic solution:

$$MnO_4^-(aq) + C_2O_4^{2-}(aq) \longrightarrow MnO_2(s) + CO_3^{2-}(aq) \quad \text{[basic solution]}$$

PLAN We proceed through step 4 as if this took place in acidic solution. Then, we add the appropriate number of OH^- ions and cancel excess H_2O molecules (step 4 Basic).

SOLUTION

1. Divide into half-reactions.

$$MnO_4^- \longrightarrow MnO_2 \qquad\qquad\qquad C_2O_4^{2-} \longrightarrow CO_3^{2-}$$

2. Balance.

   a. Atoms other than O and H,   a. Atoms other than O and H,

      Not needed $C_2O_4^{2-} \longrightarrow 2CO_3^{2-}$

   b. O atoms with H_2O,   b. O atoms with H_2O,

      $MnO_4^- \longrightarrow MnO_2 + 2H_2O$ $2H_2O + C_2O_4^{2-} \longrightarrow 2CO_3^{2-}$

   c. H atoms with H^+,   c. H atoms with H^+,

      $4H^+ + MnO_4^- \longrightarrow MnO_2 + 2H_2O$ $2H_2O + C_2O_4^{2-} \longrightarrow 2CO_3^{2-} + 4H^+$

   d. Charge with e^-,   d. Charge with e^-,

      $3e^- + 4H^+ + MnO_4^- \longrightarrow MnO_2 + 2H_2O$ $2H_2O + C_2O_4^{2-} \longrightarrow 2CO_3^{2-} + 4H^+ + 2e^-$

                       [reduction]   [oxidation]

3. Multiply each half-reaction, if necessary, by some integer to make e^- lost equal e^- gained.

$2(3e^- + 4H^+ + MnO_4^- \longrightarrow MnO_2 + 2H_2O$) $3(2H_2O + C_2O_4^{2-} \longrightarrow 2CO_3^{2-} + 4H^+ + 2e^-)$

$6e^- + 8H^+ + 2MnO_4^- \longrightarrow 2MnO_2 + 4H_2O$ $6H_2O + 3C_2O_4^{2-} \longrightarrow 6CO_3^{2-} + 12H^+ + 6e^-$

4. Add half-reactions, and cancel substances appearing on both sides.

The six e^- cancel, eight H^+ cancel to leave four H^+ on the right, and four H_2O cancel to leave two H_2O on the left:

$$\cancel{6e^-} + 8H^{\cancel{+}} + 2MnO_4^- \longrightarrow 2MnO_2 + 4\cancel{H_2O}$$
$$\underline{2\,6H_2O + 3C_2O_4^{2-} \longrightarrow 6CO_3^{2-} + 4\,\cancel{12}H^+ + \cancel{6e^-}}$$
$$2MnO_4^- + 2H_2O + 3C_2O_4^{2-} \longrightarrow 2MnO_2 + 6CO_3^{2-} + 4H^+$$

4 Basic. Add OH^- to both sides to neutralize H^+, and cancel H_2O.

Adding four OH^- to both sides forms four H_2O on the right, two of which cancel the two H_2O on the left, leaving two H_2O on the right:

$$2MnO_4^- + 2H_2O + 3C_2O_4^{2-} + 4OH^- \longrightarrow 2MnO_2 + 6CO_3^{2-} + [4H^+ + 4OH^-]$$
$$2MnO_4^- + 2H_2O + 3C_2O_4^{2-} + 4OH^- \longrightarrow 2MnO_2 + 6CO_3^{2-} + 2\,4H_2O$$

Including states of matter gives the final balanced equation:

$$2MnO_4^-(aq) + 3C_2O_4^{2-}(aq) + 4OH^-(aq) \longrightarrow 2MnO_2(s) + 6CO_3^{2-}(aq) + 2H_2O(l)$$

5. Check that atoms and charges balance.

$$(2Mn, 24O, 6C, 4H; 12-) \longrightarrow (2Mn, 24O, 6C, 4H; 12-)$$

COMMENT As a final step, we can obtain the balanced *molecular* equation for this reaction by noting the number of moles of each anion in the balanced ionic equation and adding the correct number of moles of spectator ions (in this case, Na^+) to obtain neutral compounds. Thus, for instance, balancing the charge of 2 mol of MnO_4^- requires 2 mol of Na^+, so we have $2NaMnO_4$. The balanced molecular equation is

$$2NaMnO_4(aq) + 3Na_2C_2O_4(aq) + 4NaOH(aq) \longrightarrow$$
$$2MnO_2(s) + 6Na_2CO_3(aq) + 2H_2O(l)$$

FOLLOW-UP PROBLEM 21.1 Write a balanced molecular equation for the reaction between $KMnO_4$ and KI in basic solution. The skeleton ionic reaction is

$$MnO_4^-(aq) + I^-(aq) \longrightarrow MnO_4^{2-}(aq) + IO_3^-(aq) \quad \text{[basic solution]}$$

The half-reaction method reveals a great deal about redox processes and is essential to understanding electrochemical cells. The major points are

- Any redox reaction can be treated as the sum of a reduction and an oxidation half-reaction.
- Atoms and charge are conserved in each half-reaction.
- Electrons lost in one half-reaction are gained in the other.
- Although the half-reactions are treated separately, electron loss and electron gain occur simultaneously.

An Overview of Electrochemical Cells

We distinguish two types of electrochemical cells based on the general thermodynamic nature of the reaction:

1. A **voltaic cell** (or **galvanic cell**) uses a spontaneous reaction ($\Delta G < 0$) to generate electrical energy. In the cell reaction, the difference in chemical potential energy between higher energy reactants and lower energy products is converted into electrical energy. This energy is used to operate the load—flashlight bulb, CD player, car starter motor, or other electrical device. In other words, *the system does work on the surroundings*. All batteries contain voltaic cells.
2. An **electrolytic cell** uses electrical energy to drive a nonspontaneous reaction ($\Delta G > 0$). In the cell reaction, electrical energy from an external power supply converts lower energy reactants into higher energy products. Thus, *the surroundings do work on the system*. Electroplating and recovering metals from ores involve electrolytic cells.

The two types of cell have certain design features in common (Figure 21.3). Two **electrodes,** which conduct the electricity between cell and surroundings, are dipped into an **electrolyte,** a mixture of ions (usually in aqueous solution) that

Demo Summerlin, L.R.; Borgford, C.L.; Ealy, J.B. "Visible Oxidation-Reduction in Electrochemical Cells," *Chemical Demonstrations, A Sourcebook for Teachers,* Vol. 2 (American Chemical Society, Washington, DC, 1988), 205–206.

Figure 21.3 General characteristics of voltaic and electrolytic cells. A voltaic cell **(A)** generates energy from a spontaneous reaction ($\Delta G < 0$), whereas an electrolytic cell **(B)** requires energy to drive a nonspontaneous reaction ($\Delta G > 0$). In both types of cell, two electrodes dip into electrolyte solutions, and an external circuit provides the means for electrons to flow between them. Most important, notice that oxidation takes place at the anode and reduction takes place at the cathode, but the relative electrode charges are opposite in the two cells.

Transparency/ARIS Presentation Center, Art Library

A   B

are involved in the reaction or that carry the charge. An electrode is identified as either **anode** or **cathode** depending on the half-reaction that takes place there:

- *The oxidation half-reaction occurs at the anode.* Electrons are lost by the substance being oxidized (reducing agent) and *leave the cell* at the anode.
- *The reduction half-reaction occurs at the cathode.* Electrons are gained by the substance being reduced (oxidizing agent) and *enter the cell* at the cathode. ●

As shown in Figure 21.3, the relative charges of the electrodes are *opposite* in the two types of cell. As you'll see in the following sections, these opposite charges result from the different phenomena that cause the electrons to flow.

Section Summary

An oxidation-reduction (redox) reaction involves the transfer of electrons from a reducing agent to an oxidizing agent. • The half-reaction method of balancing divides the overall reaction into half-reactions that are balanced separately and then recombined. • The two types of electrochemical cells are based on redox reactions. In a voltaic cell, a spontaneous reaction generates electricity and does work on the surroundings; in an electrolytic cell, the surroundings supply electricity that does work to drive a nonspontaneous reaction. In both types, two electrodes dip into electrolyte solutions; oxidation occurs at the anode, and reduction occurs at the cathode.

● **Which Half-Reaction Occurs at Which Electrode?** Here are some memory aids to help you remember which half-reaction occurs at which electrode:

1. The words *anode* and *oxidation* start with vowels; the words *cathode* and *reduction* start with consonants.
2. Alphabetically, the *A* in anode comes before the *C* in cathode, and the *O* in oxidation comes before the *R* in reduction.
3. Look at the first syllables and use your imagination:

ANode, OXidation; REDuction, CAThode
⇒ AN OX and a RED CAT

21.2 VOLTAIC CELLS: USING SPONTANEOUS REACTIONS TO GENERATE ELECTRICAL ENERGY

If you put a strip of zinc metal in a solution of Cu^{2+} ion, the blue color of the solution fades as a brown-black crust of Cu metal forms on the Zn strip (Figure 21.4). Judging from what we see, the reaction involves the reduction of Cu^{2+} ion to Cu metal, which must be accompanied by the oxidation of Zn metal to Zn^{2+} ion. The overall reaction consists of two half-reactions:

$$Cu^{2+}(aq) + 2e^- \longrightarrow Cu(s) \qquad \text{[reduction]}$$
$$Zn(s) \longrightarrow Zn^{2+}(aq) + 2e^- \qquad \text{[oxidation]}$$
$$\overline{Zn(s) + Cu^{2+}(aq) \longrightarrow Zn^{2+}(aq) + Cu(s)} \qquad \text{[overall reaction]}$$

Figure 21.4 The spontaneous reaction between zinc and copper(II) ion. When a strip of zinc metal is placed in a solution of Cu^{2+} ion, a redox reaction begins (*left*), in which the zinc is oxidized to Zn^{2+} and the Cu^{2+} is reduced to copper metal. As the reaction proceeds (*right*), the deep blue color of the solution of hydrated Cu^{2+} ion lightens, and the Cu "plates out" on the Zn and falls off in chunks. (The Cu appears black because it is very finely divided.) At the atomic scale, each Zn atom loses two electrons, which are gained by a Cu^{2+} ion. The process is summarized with symbols in the balanced equation.

 ARIS Presentation Center, Art Library

In the remainder of this section, we examine this spontaneous reaction as the basis of a voltaic (galvanic) cell.

Construction and Operation of a Voltaic Cell

Animation: Operation of a Voltaic Cell
ARIS Presentation Center
Chapter 21

Electrons are being transferred in the Zn/Cu^{2+} reaction (Figure 21.4), but the system does not generate electrical energy because the oxidizing agent (Cu^{2+}) and the reducing agent (Zn) are in the same beaker. If, however, the half-reactions are physically separated and connected by an external circuit, the electrons are transferred by traveling through the circuit and an electric current is produced.

This separation of half-reactions is the essential idea behind a voltaic cell (Figure 21.5A). The components of each half-reaction are placed in a separate container, or **half-cell,** which consists of one electrode dipping into an electrolyte solution. The two half-cells are joined by the circuit, which consists of a wire and a salt bridge (the inverted U tube in the figure; we will discuss its function shortly). In order to measure the voltage generated by the cell, a voltmeter is inserted in the path of the wire connecting the electrodes. A switch (not shown) closes (completes) or opens (breaks) the circuit. By convention, *the oxidation half-cell (anode compartment) is shown on the left and the reduction half-cell (cathode compartment) on the right.* Here are the key points about the Zn/Cu^{2+} voltaic cell:

Literature "Silver to Black—and Back," *J. Chem. Educ.* **2000,** *77,* 328A.

1. *The oxidation half-cell.* In this case, the anode compartment consists of a zinc bar (the anode) immersed in a Zn^{2+} electrolyte (such as a solution of zinc sulfate, $ZnSO_4$). The zinc bar is the reactant in the oxidation half-reaction, and it conducts the released electrons *out* of its half-cell.

2. *The reduction half-cell.* In this case, the cathode compartment consists of a copper bar (the cathode) immersed in a Cu^{2+} electrolyte [such as a solution of copper(II) sulfate, $CuSO_4$]. Copper metal is the product in the reduction half-reaction, and the bar conducts electrons *into* its half-cell.

Literature Sanger, Michael J.; Greenbowe, Thomas J. "Students' Misconceptions in Electrochemistry Regarding Current Flow in Electrolyte Solutions and the Salt Bridge," *J. Chem. Educ.* **1997,** *74,* 819.

3. *Relative charges on the electrodes.* The electrode charges are determined by the *source of electrons* and the *direction of electron flow* through the circuit. In this cell, zinc atoms are oxidized at the anode to Zn^{2+} ions and electrons. The Zn^{2+} ions enter the solution, while the electrons enter the bar and then the wire. *The electrons flow left to right* through the wire to the cathode, where Cu^{2+} ions in the solution accept them and are reduced to Cu atoms. As the cell operates, electrons are continuously generated at the anode and consumed at the cathode. Therefore, the anode has an excess of electrons and a negative charge *relative* to the cathode. *In any **voltaic** cell, the anode is negative and the cathode is positive.*

4. *The purpose of the salt bridge.* The cell cannot operate unless the circuit is complete. The oxidation half-cell originally contains a neutral solution of Zn^{2+} and SO_4^{2-} ions, but as Zn atoms in the bar lose electrons, the solution would develop a net positive charge from the Zn^{2+} ions entering. Similarly, in the reduction half-cell, the neutral solution of Cu^{2+} and SO_4^{2-} ions would develop a net negative charge as Cu^{2+} ions leave the solution to form Cu atoms. If the half-cells do not remain neutral, the resulting charge imbalance would stop cell operation. To avoid this situation and enable the cell to operate, the two half-cells are joined by a **salt bridge,** which acts as a "liquid wire," allowing ions to flow through both compartments and complete the circuit. The salt bridge shown in Figure 21.5A is an inverted U tube containing a solution of the nonreacting ions Na^+ and SO_4^{2-} in a gel. The solution cannot pour out, but ions can diffuse through it into or out of the half-cells.

To maintain neutrality in the reduction half-cell (right; cathode compartment) as Cu^{2+} ions change to Cu atoms, Na^+ ions move from the salt bridge into the solution (and some SO_4^{2-} ions move from the solution into the salt bridge). Similarly, to maintain neutrality in the oxidation half-cell (left; anode compartment)

Oxidation half-reaction
$$Zn(s) \longrightarrow Zn^{2+}(aq) + 2e^-$$

Reduction half-reaction
$$Cu^{2+}(aq) + 2e^- \longrightarrow Cu(s)$$

Overall (cell) reaction
$$Zn(s) + Cu^{2+}(aq) \longrightarrow Zn^{2+}(aq) + Cu(s)$$

A

B

Figure 21.5 A voltaic cell based on the zinc-copper reaction. A, The anode half-cell (oxidation) consists of a Zn electrode dipping into a Zn^{2+} solution. The two electrons generated in the oxidation of each Zn atom move through the Zn bar and the wire, and into the Cu electrode, which dips into a Cu^{2+} solution in the cathode half-cell (reduction). There, the electrons reduce Cu^{2+} ions. Thus, electrons flow left to right through electrodes and wire. A salt bridge contains unreactive Na^+ and SO_4^{2-} ions that maintain neutral charge in the electrolyte solutions: anions in the salt bridge flow to the left, and cations flow to the right. The voltmeter registers the electrical output of the cell. **B,** After the cell runs for several hours, the Zn anode weighs less because Zn atoms have been oxidized to aqueous Zn^{2+} ions, and the Cu cathode weighs more because aqueous Cu^{2+} ions have been reduced to Cu metal.

Transparency/ARIS Presentation Center, Active Art Library

as Zn atoms change to Zn^{2+} ions, SO_4^{2-} ions move from the salt bridge into that solution (and some Zn^{2+} ions move from the solution into the salt bridge). Thus, as Figure 21.5A shows, the circuit is completed *as electrons move left to right through the wire, while anions move right to left and cations move left to right through the salt bridge.*

5. *Active vs. inactive electrodes.* The electrodes in the Zn/Cu^{2+} cell are *active* because the metal bars themselves are components of the half-reactions. As the cell operates, the mass of the zinc electrode gradually decreases, and the $[Zn^{2+}]$ in the anode half-cell increases. At the same time, the mass of the copper electrode increases, and the $[Cu^{2+}]$ in the cathode half-cell decreases; we say that the Cu^{2+} "plates out" on the electrode. Look at Figure 21.5B to see how the electrodes look, removed from their half-cells, after several hours of operation.

For many redox reactions, there are no reactants or products capable of serving as electrodes, so *inactive* electrodes are used. Most commonly, inactive electrodes are rods of *graphite* or *platinum*: they conduct electrons into or out of the half-cells but cannot take part in the half-reactions. In a voltaic cell based on the following half-reactions, for instance, the reacting species cannot act as electrodes:

$$2I^-(aq) \longrightarrow I_2(s) + 2e^- \qquad \text{[anode; oxidation]}$$
$$MnO_4^-(aq) + 8H^+(aq) + 5e^- \longrightarrow Mn^{2+}(aq) + 4H_2O(l) \qquad \text{[cathode; reduction]}$$

Oxidation half-reaction

$2I^-(aq) \longrightarrow I_2(s) + 2e^-$

Reduction half-reaction

$MnO_4^-(aq) + 8H^+(aq) + 5e^- \longrightarrow$
$Mn^{2+}(aq) + 4H_2O(l)$

Overall (cell) reaction

$2MnO_4^-(aq) + 16H^+(aq) + 10I^-(aq) \longrightarrow$
$2Mn^{2+}(aq) + 5I_2(s) + 8H_2O(l)$

Figure 21.6 A voltaic cell using inactive electrodes. The reaction between I^- and MnO_4^- in acidic solution does not have species that can be used as electrodes, so inactive graphite (C) electrodes are used.

Therefore, each half-cell consists of inactive electrodes immersed in an electrolyte solution that contains *all the reactant species involved in that half-reaction* (Figure 21.6). In the anode half-cell, I^- ions are oxidized to solid I_2. The electrons that are released flow into the graphite anode, through the wire, and into the graphite cathode. From there, the electrons are consumed by MnO_4^- ions, which are reduced to Mn^{2+} ions. (A KNO_3 salt bridge is used.)

As Figures 21.5A and 21.6 show, there are certain consistent features in the *diagram* of any voltaic cell. The physical arrangement includes the half-cell containers, electrodes, wire, and salt bridge, and the following details appear:

- Components of the half-cells: electrode materials, electrolyte ions, and other substances involved in the reaction
- Electrode name (anode or cathode) and charge. By convention, the anode compartment always appears *on the left*.
- Each half-reaction with its half-cell and the overall cell reaction
- Direction of electron flow in the external circuit
- Nature of ions and direction of ion flow in the salt bridge

You'll see how to specify these details and diagram a cell shortly.

Notation for a Voltaic Cell

A useful shorthand notation describes the components of a voltaic cell. For example, the notation for the Zn/Cu^{2+} cell is

$$Zn(s)\,|\,Zn^{2+}(aq)\,\|\,Cu^{2+}(aq)\,|\,Cu(s)$$

Key parts of the notation are

- The components of the anode compartment (oxidation half-cell) are written *to the left* of the components of the cathode compartment (reduction half-cell).
- A single vertical line represents a phase boundary. For example, $Zn(s)\,|\,Zn^{2+}(aq)$ indicates that the *solid* Zn is a *different* phase from the *aqueous* Zn^{2+}. A comma separates the half-cell components that are in the *same* phase. For example, the notation for the voltaic cell housing the reaction between I^- and MnO_4^- shown in Figure 21.6 is

$$graphite\,|\,I^-(aq)\,|\,I_2(s)\,\|\,H^+(aq),\,MnO_4^-(aq),\,Mn^{2+}(aq)\,|\,graphite$$

That is, in the cathode compartment, H^+, MnO_4^-, and Mn^{2+} ions are all in aqueous solution with solid graphite immersed in it. Often, we specify the concentrations of dissolved components; for example, if the concentrations of Zn^{2+} and Cu^{2+} are 1 *M*, we write

$$Zn(s)\,|\,Zn^{2+}(1\,M)\,\|\,Cu^{2+}(1\,M)\,|\,Cu(s)$$

- Half-cell components usually appear in the same order as in the half-reaction, and electrodes appear at the far left and far right of the notation.
- A double vertical line indicates the separated half-cells and represents the phase boundary on either side of the salt bridge (the ions in the salt bridge are omitted because they are not part of the reaction).

SAMPLE PROBLEM 21.2 Describing a Voltaic Cell with a Diagram and Notation

PROBLEM Draw a diagram, show balanced equations, and write the notation for a voltaic cell that consists of one half-cell with a Cr bar in a $Cr(NO_3)_3$ solution, another half-cell with an Ag bar in an $AgNO_3$ solution, and a KNO_3 salt bridge. Measurement indicates that the Cr electrode is negative relative to the Ag electrode.

PLAN From the given contents of the half-cells, we can write the half-reactions. We must determine which is the anode compartment (oxidation) and which is the cathode (reduction). To do so, we must find the direction of the spontaneous redox reaction, which is given by the relative electrode charges. Electrons are released into the anode during oxidation, so it has a negative charge. We are told that Cr is negative, so it must be the anode; and, therefore, Ag is the cathode.

SOLUTION Writing the balanced half-reactions. The Ag electrode is positive, so the half-reaction consumes e^-:

$$Ag^+(aq) + e^- \longrightarrow Ag(s) \qquad [\text{reduction; cathode}]$$

The Cr electrode is negative, so the half-reaction releases e^-:

$$Cr(s) \longrightarrow Cr^{3+}(aq) + 3e^- \qquad [\text{oxidation; anode}]$$

Writing the balanced overall cell reaction. We triple the reduction half-reaction in order to balance e^-, and then we combine the half-reactions to obtain the overall reaction:

$$Cr(s) + 3Ag^+(aq) \longrightarrow Cr^{3+}(aq) + 3Ag(s)$$

Determining direction of electron and ion flow. The released e^- in the Cr electrode (negative) flow through the external circuit to the Ag electrode (positive). As Cr^{3+} ions enter the anode electrolyte, NO_3^- ions enter from the salt bridge to maintain neutrality. As Ag^+ ions leave the cathode electrolyte and plate out on the Ag electrode, K^+ ions enter from the salt bridge to maintain neutrality. The diagram of this cell is shown in the margin. Writing the cell notation:

$$Cr(s) \mid Cr^{3+}(aq) \parallel Ag^+(aq) \mid Ag(s)$$

CHECK Always be sure that the half-reactions and cell reaction are balanced, the half-cells contain *all* components of the half-reactions, and the electron and ion flow are shown. You should be able to write the half-reactions from the cell notation as a check.

COMMENT The key to diagramming a voltaic cell is to use the direction of the spontaneous reaction to identify the oxidation (anode; negative) and reduction (cathode; positive) half-reactions.

FOLLOW-UP PROBLEM 21.2 In one compartment of a voltaic cell, a graphite rod dips into an acidic solution of $K_2Cr_2O_7$ and $Cr(NO_3)_3$; in the other compartment, a tin bar dips into a $Sn(NO_3)_2$ solution. A KNO_3 salt bridge joins them. The tin electrode is negative relative to the graphite. Draw a diagram of the cell, show the balanced equations, and write the cell notation.

Oxidation half-reaction
$Cr(s) \longrightarrow Cr^{3+}(aq) + 3e^-$

Reduction half-reaction
$Ag^+(aq) + e^- \longrightarrow Ag(s)$

Overall (cell) reaction
$Cr(s) + 3Ag^+(aq) \longrightarrow Cr^{3+}(aq) + 3Ag(s)$

ARIS Presentation Center, Art Library

Why Does a Voltaic Cell Work?

By placing a lightbulb in the circuit or looking at the voltmeter, we can see that the Zn/Cu^{2+} cell generates electrical energy. But what principle explains *how* the reaction takes place, and *why* do electrons flow in the direction shown?

Let's examine what happens when the switch is open and no reaction is occurring. In each half-cell, we can consider the metal electrode to be in equilibrium with the metal ions in the electrolyte and the electrons residing in the metal:

$$Zn(s) \rightleftharpoons Zn^{2+}(aq) + 2e^- \text{(in Zn metal)}$$
$$Cu(s) \rightleftharpoons Cu^{2+}(aq) + 2e^- \text{(in Cu metal)}$$

From the direction of the overall spontaneous reaction, we know that Zn gives up its electrons more easily than Cu does; thus, Zn is a stronger reducing agent. Therefore, the equilibrium position of the Zn half-reaction lies farther to the right: Zn produces more electrons than Cu does. You might think of the electrons in the Zn electrode as being subject to a greater electron "pressure" than those in the Cu electrode, a greater potential energy (referred to as *electrical potential*) ready to "push" them through the circuit. Close the switch, and electrons flow from the Zn to the Cu electrode to equalize this difference in electrical potential. The flow disturbs the equilibrium at each electrode. The Zn half-reaction shifts to the right to restore the electrons flowing out, and the Cu half-reaction shifts to the left to remove the electrons flowing in. Thus, *the spontaneous reaction occurs as a result of the different abilities of these metals to give up their electrons and the ability of the electrons to flow through the circuit.* ●

● **Electron Flow and Water Flow** Consider this analogy between electron "pressure" and water pressure. A U tube (cell) is separated into two arms (two half-cells) by a stopcock (switch), and the two arms contain water at different heights (the half-cells contain half-reactions with different electrical potentials). Open the stopcock (close the switch), and the different heights (potential difference) become equal as water flows (electrons flow and current is generated).

Section Summary

A voltaic cell consists of oxidation (anode) and reduction (cathode) half-cells, connected by a wire to conduct electrons and a salt bridge to maintain charge neutrality as the cell operates. • Electrons move from anode (left) to cathode (right), while cations move from the salt bridge into the cathode half-cell and anions from the salt bridge into the anode half-cell. • The cell notation shows the species and their phases in each half-cell, as well as the direction of current flow. • A voltaic cell operates because species in the two half-cells differ in their tendency to lose electrons.

21.3 CELL POTENTIAL: OUTPUT OF A VOLTAIC CELL

The purpose of a voltaic cell is to convert the free energy change of a spontaneous reaction into the kinetic energy of electrons moving through an external circuit (electrical energy). This electrical energy is proportional to the *difference in electrical potential between the two electrodes*, which is called the **cell potential (E_{cell})**; it is also called the **voltage** of the cell or the **electromotive force (emf)**.

Demo Bolton, Ed; Richter, Mark M. "Light Emission at Electrodes: An Electrochemiluminescence Demonstration," *J. Chem. Educ.* **2001**, *78*, 641.

Electrons flow spontaneously from the negative to the positive electrode, that is, toward the electrode with the more positive electrical potential. Thus, when the cell operates *spontaneously,* there is a *positive* cell potential:

$$E_{cell} > 0 \text{ for a spontaneous process} \qquad (21.1)$$

The more positive E_{cell} is, the more work the cell can do, and the farther the reaction proceeds to the right as written. A *negative* cell potential, on the other hand, is associated with a *nonspontaneous* cell reaction. If $E_{cell} = 0$, the reaction has reached equilibrium and the cell can do no more work. (There is a clear relationship between E_{cell}, K, and ΔG that we'll discuss in Section 21.4.)

How are the units of cell potential related to those of energy available to do work? As you've seen, work is done when charge moves between electrode compartments that differ in electrical potential. The SI unit of electrical potential is the **volt (V),** and the SI unit of electrical charge is the **coulomb (C).** By definition, for two electrodes that differ by 1 volt of electrical potential, 1 joule of energy is released (that is, 1 joule of work can be done) for each coulomb of charge that moves between the electrodes. Thus,

$$1 \text{ V} = 1 \text{ J/C} \qquad (21.2)$$

Table 21.1 lists the voltages of some commercial and natural voltaic cells. Next, we'll see how to measure cell potential.

Table 21.1	Voltages of Some Voltaic Cells
Voltaic Cell	**Voltage (V)**
Common alkaline flashlight battery	1.5
Lead-acid car battery (6 cells = 12 V)	2.0
Calculator battery (mercury)	1.3
Lithium-ion laptop battery	3.7
Electric eel (~5000 cells in 6-ft eel = 750 V)	0.15
Nerve of giant squid (across cell membrane)	0.070

Standard Cell Potentials

The measured potential of a voltaic cell is affected by changes in concentration as the reaction proceeds and by energy losses due to heating of the cell and the external circuit. Therefore, in order to compare the output of different cells, we obtain a **standard cell potential** ($E°_{cell}$), the potential measured at a specified temperature (usually 298 K) with no current flowing* and *all components in their standard states:* 1 atm for gases, 1 *M* for solutions, the pure solid for electrodes. When the zinc-copper cell that we diagrammed in Figure 21.5 begins operating under standard state conditions, that is, when $[Zn^{2+}] = [Cu^{2+}] = 1$ *M*, the cell produces 1.10 V at 298 K (Figure 21.7):

$$Zn(s) + Cu^{2+}(aq; 1\,M) \longrightarrow Zn^{2+}(aq; 1\,M) + Cu(s) \qquad E°_{cell} = 1.10\text{ V}$$

Standard Electrode (Half-Cell) Potentials Just as each half-reaction makes up part of the overall reaction, the potential of each half-cell makes up a part of the overall cell potential. The **standard electrode potential** ($E°_{half\text{-}cell}$) is the potential associated with a given half-reaction (electrode compartment) when all the components are in their standard states.

By convention, *a standard electrode potential always refers to the half-reaction written as a* ***reduction.*** For the zinc-copper reaction, for example, the standard electrode potentials for the zinc half-reaction ($E°_{zinc}$, anode compartment) and for the copper half-reaction ($E°_{copper}$, cathode compartment) refer to the processes written as reductions:

$$Zn^{2+}(aq) + 2e^- \longrightarrow Zn(s) \qquad E°_{zinc}\ (E°_{anode}) \qquad [\text{reduction}]$$
$$Cu^{2+}(aq) + 2e^- \longrightarrow Cu(s) \qquad E°_{copper}\ (E°_{cathode}) \qquad [\text{reduction}]$$

The overall cell reaction involves the *oxidation* of zinc at the anode, not the *reduction* of Zn^{2+}, so we reverse the zinc half-reaction:

$$Zn(s) \longrightarrow Zn^{2+}(aq) + 2e^- \qquad [\text{oxidation}]$$
$$Cu^{2+}(aq) + 2e^- \longrightarrow Cu(s) \qquad [\text{reduction}]$$

The overall redox reaction is the sum of these half-reactions:

$$Zn(s) + Cu^{2+}(aq) \longrightarrow Zn^{2+}(aq) + Cu(s)$$

Because electrons flow spontaneously toward the copper electrode (cathode), it must have a more positive $E°_{half\text{-}cell}$ than the zinc electrode (anode). Therefore, to obtain a positive $E°_{cell}$, we subtract $E°_{zinc}$ from $E°_{copper}$:

$$E°_{cell} = E°_{copper} - E°_{zinc}$$

We can generalize this result for any voltaic cell: *the standard cell potential is the difference between the standard electrode potential of the cathode (reduction) half-cell and the standard electrode potential of the anode (oxidation) half-cell:*

$$\boxed{E°_{cell} = E°_{cathode\ (reduction)} - E°_{anode\ (oxidation)}} \qquad (21.3)$$

For a spontaneous reaction at standard conditions, $E°_{cell} > 0$.

Determining $E°_{half\text{-}cell}$: The Standard Hydrogen Electrode What portion of $E°_{cell}$ for the zinc-copper reaction is contributed by the anode half-cell (oxidation of Zn) and what portion by the cathode half-cell (reduction of Cu^{2+})? That is, how can we know half-cell potentials if we can only measure the potential of the complete cell? Half-cell potentials, such as $E°_{zinc}$ and $E°_{copper}$, are not absolute quantities, but rather are values *relative* to that of a standard. *This standard reference*

Figure 21.7 Measurement of a standard cell potential. The zinc-copper cell, operating at 298 K under standard-state conditions, produces a voltage of 1.10 V.

ARIS Presentation Center, Photo Library

Animation: Galvanic Cell
ARIS Presentation Center
Chapter 21

Literature Özkaya, Ali Riza; Üce, Musa; Sariçair, Hakan; Şahin, Musa. "Effectiveness of Conceptual Change-Oriented Teaching Strategy to Improve Students' Understanding of Galvanic Cells," *J. Chem. Educ.* **2006,** *83,* 1719.

---

*The current required to operate modern digital voltmeters makes a negligible difference in the value of $E°_{cell}$.

half-cell has its standard electrode potential defined as zero ($E°_{reference} \equiv 0.00$ V). The **standard reference half-cell** is a **standard hydrogen electrode,** which consists of a specially prepared platinum electrode immersed in a 1 *M* aqueous solution of a strong acid, $H^+(aq)$ [or $H_3O^+(aq)$], through which H_2 gas at 1 atm is bubbled. Thus, the reference half-reaction is

$$2H^+(aq; 1 M) + 2e^- \rightleftharpoons H_2(g; 1 \text{ atm}) \qquad E°_{reference} = 0.00 \text{ V}$$

Now we can construct a voltaic cell consisting of this reference half-cell and another half-cell whose potential we want to determine. With $E°_{reference}$ defined as zero, the overall $E°_{cell}$ allows us to find the unknown standard electrode potential, $E°_{unknown}$. When H_2 is oxidized, the reference half-cell is the anode, and so reduction occurs at the unknown half-cell:

$$E°_{cell} = E°_{cathode} - E°_{anode} = E°_{unknown} - E°_{reference} = E°_{unknown} - 0.00 \text{ V} = E°_{unknown}$$

When H^+ is reduced, the reference half-cell is the cathode, and so oxidation occurs at the unknown half-cell:

$$E°_{cell} = E°_{cathode} - E°_{anode} = E°_{reference} - E°_{unknown} = 0.00 \text{ V} - E°_{unknown} = -E°_{unknown}$$

Figure 21.8 shows a voltaic cell that has the Zn/Zn^{2+} half-reaction in one compartment and the H^+/H_2 (or H_3O^+/H_2) half-reaction in the other. The zinc electrode is negative relative to the hydrogen electrode, so we know that the zinc is being oxidized and is the anode. The measured $E°_{cell}$ is +0.76 V, and we use this value to find the unknown standard electrode potential, $E°_{zinc}$:

$$
\begin{array}{lll}
2H^+(aq) + 2e^- \longrightarrow H_2(g) & E°_{reference} = 0.00 \text{ V} & \text{[cathode; reduction]} \\
Zn(s) \longrightarrow Zn^{2+}(aq) + 2e^- & E°_{zinc} = ? \text{ V} & \text{[anode; oxidation]} \\
\hline
Zn(s) + 2H^+(aq) \longrightarrow Zn^{2+}(aq) + H_2(g) & E°_{cell} = 0.76 \text{ V}
\end{array}
$$

$$E°_{cell} = E°_{cathode} - E°_{anode} = E°_{reference} - E°_{zinc}$$
$$E°_{zinc} = E°_{reference} - E°_{cell} = 0.00 \text{ V} - 0.76 \text{ V} = -0.76 \text{ V}$$

Oxidation half-reaction
$$Zn(s) \longrightarrow Zn^{2+}(aq) + 2e^-$$

Reduction half-reaction
$$2H_3O^+(aq) + 2e^- \longrightarrow H_2(g) + 2H_2O(l)$$

Overall (cell) reaction
$$Zn(s) + 2H_3O^+(aq) \longrightarrow Zn^{2+}(aq) + H_2(g) + 2H_2O(l)$$

Figure 21.8 Determining an unknown $E°_{half-cell}$ with the standard reference (hydrogen) electrode. A voltaic cell has the Zn/Zn^{2+} half-reaction in one half-cell and the hydrogen reference half-reaction in the other. The magnified view of the hydrogen half-reaction shows two H_3O^+ ions being reduced to two H_2O molecules and an H_2 molecule, which enters the H_2 bubble. The Zn/Zn^{2+} half-cell potential is negative (anode), and the cell potential is 0.76 V. The potential of the standard reference electrode is defined as 0.00 V, so the cell potential equals the negative of the anode potential; that is,

$$0.76 \text{ V} = 0.00 \text{ V} - E°_{zinc} \qquad \text{so} \qquad E°_{zinc} = -0.76 \text{ V}$$

Now let's return to the zinc-copper cell and use the measured value of $E°_{cell}$ (1.10 V) and the value we just found for $E°_{zinc}$ to calculate $E°_{copper}$:

$$E°_{cell} = E°_{cathode} - E°_{anode} = E°_{copper} - E°_{zinc}$$
$$E°_{copper} = E°_{cell} + E°_{zinc} = 1.10 \text{ V} + (-0.76 \text{ V}) = 0.34 \text{ V}$$

By continuing this process of constructing cells with one known and one unknown electrode potential, we can find many other standard electrode potentials. Let's go over these ideas once more with a sample problem.

SAMPLE PROBLEM 21.3 Calculating an Unknown $E°_{half-cell}$ from $E°_{cell}$

PROBLEM A voltaic cell houses the reaction between aqueous bromine and zinc metal:

$$Br_2(aq) + Zn(s) \longrightarrow Zn^{2+}(aq) + 2Br^-(aq) \qquad E°_{cell} = 1.83 \text{ V}$$

Calculate $E°_{bromine}$, given $E°_{zinc} = -0.76$ V.

PLAN $E°_{cell}$ is positive, so the reaction is spontaneous as written. By dividing the reaction into half-reactions, we see that Br_2 is reduced and Zn is oxidized; thus, the zinc half-cell contains the anode. We use Equation 21.3 to find $E°_{unknown}$ ($E°_{bromine}$).

SOLUTION Dividing the reaction into half-reactions:

$$Br_2(aq) + 2e^- \longrightarrow 2Br^-(aq) \qquad E°_{unknown} = E°_{bromine} = ? \text{ V}$$
$$Zn(s) \longrightarrow Zn^{2+}(aq) + 2e^- \qquad E°_{zinc} = -0.76 \text{ V}$$

Calculating $E°_{bromine}$:

$$E°_{cell} = E°_{cathode} - E°_{anode} = E°_{bromine} - E°_{zinc}$$
$$E°_{bromine} = E°_{cell} + E°_{zinc} = 1.83 \text{ V} + (-0.76 \text{ V}) = 1.07 \text{ V}$$

CHECK A good check is to make sure that calculating $E°_{bromine} - E°_{zinc}$ gives $E°_{cell}$: 1.07 V − (−0.76 V) = 1.83 V.

COMMENT Keep in mind that, whichever is the unknown half-cell, reduction is the cathode half-reaction and oxidation is the anode half-reaction. Always subtract $E°_{anode}$ from $E°_{cathode}$ to get $E°_{cell}$.

FOLLOW-UP PROBLEM 21.3 A voltaic cell based on the reaction between aqueous Br_2 and vanadium(III) ions has $E°_{cell} = 1.39$ V:

$$Br_2(aq) + 2V^{3+}(aq) + 2H_2O(l) \longrightarrow 2VO^{2+}(aq) + 4H^+(aq) + 2Br^-(aq)$$

What is $E°_{vanadium}$, the standard electrode potential for the reduction of VO^{2+} to V^{3+}?

Relative Strengths of Oxidizing and Reducing Agents

One of the things we can learn from measuring potentials of voltaic cells is the relative strengths of the oxidizing and reducing agents involved. Three oxidizing agents present in the voltaic cell just discussed are Cu^{2+}, H^+, and Zn^{2+}. We can rank their relative oxidizing strengths by writing each half-reaction as a gain of electrons (reduction), with its corresponding standard electrode potential:

$$Cu^{2+}(aq) + 2e^- \longrightarrow Cu(s) \qquad E° = 0.34 \text{ V}$$
$$2H^+(aq) + 2e^- \longrightarrow H_2(g) \qquad E° = 0.00 \text{ V}$$
$$Zn^{2+}(aq) + 2e^- \longrightarrow Zn(s) \qquad E° = -0.76 \text{ V}$$

The more positive the $E°$ value, the more readily the reaction (as written) occurs; thus, Cu^{2+} gains two electrons more readily than H^+, which gains them more readily than Zn^{2+}. In terms of strength as an oxidizing agent, therefore, $Cu^{2+} > H^+ > Zn^{2+}$. Moreover, this listing also ranks the strengths of the reducing agents: $Zn > H_2 > Cu$. Notice that this list of half-reactions in order of *decreasing* half-cell potential shows, *from top to bottom*, the oxidizing agents (reactants) *decreasing* in strength and the reducing agents (products) *increasing* in strength; that is, Cu^{2+} (top left) is the strongest oxidizing agent, and Zn (bottom right) is the strongest reducing agent.

By combining many pairs of half-cells into voltaic cells, we can create a list of reduction half-reactions and arrange them in *decreasing* order of standard electrode potential (from most positive to most negative). Such a list, called an *emf series* or a *table of standard electrode potentials,* appears in Appendix D, with a few examples in Table 21.2. There are several key points to keep in mind:

- All values are relative to the standard hydrogen (reference) electrode:

$$2H^+(aq; 1\ M) + 2e^- \rightleftharpoons H_2(g; 1\ atm) \qquad E^\circ_{reference} = 0.00\ V$$

- By convention, the half-reactions are written as *reductions,* which means that *only reactants are oxidizing agents and only products are reducing agents.*
- The more positive the $E^\circ_{half\text{-}cell}$, the more readily the half-reaction occurs.
- Half-reactions are shown with an equilibrium arrow because each can occur as a reduction or an oxidation (that is, take place at the cathode or anode, respectively), depending on the $E^\circ_{half\text{-}cell}$ of the other half-reaction.
- As Appendix D (and Table 21.2) is arranged, the strength of the oxidizing agent (reactant) *increases going up (bottom to top),* and the strength of the reducing agent (product) *increases going down (top to bottom).*

ARIS Presentation Center, Tables Library

Table 21.2   Selected Standard Electrode Potentials (298 K)

Half-Reaction	$E^\circ_{half\text{-}cell}$ (V)
$F_2(g) + 2e^- \rightleftharpoons 2F^-(aq)$	+2.87
$Cl_2(g) + 2e^- \rightleftharpoons 2Cl^-(aq)$	+1.36
$MnO_2(s) + 4H^+(aq) + 2e^- \rightleftharpoons Mn^{2+}(aq) + 2H_2O(l)$	+1.23
$NO_3^-(aq) + 4H^+(aq) + 3e^- \rightleftharpoons NO(g) + 2H_2O(l)$	+0.96
$Ag^+(aq) + e^- \rightleftharpoons Ag(s)$	+0.80
$Fe^{3+}(aq) + e^- \rightleftharpoons Fe^{2+}(aq)$	+0.77
$O_2(g) + 2H_2O(l) + 4e^- \rightleftharpoons 4OH^-(aq)$	+0.40
$Cu^{2+}(aq) + 2e^- \rightleftharpoons Cu(s)$	+0.34
$2H^+(aq) + 2e^- \rightleftharpoons H_2(g)$	0.00
$N_2(g) + 5H^+(aq) + 4e^- \rightleftharpoons N_2H_5^+(aq)$	−0.23
$Fe^{2+}(aq) + 2e^- \rightleftharpoons Fe(s)$	−0.44
$Zn^{2+}(aq) + 2e^- \rightleftharpoons Zn(s)$	−0.76
$2H_2O(l) + 2e^- \rightleftharpoons H_2(g) + 2OH^-(aq)$	−0.83
$Na^+(aq) + e^- \rightleftharpoons Na(s)$	−2.71
$Li^+(aq) + e^- \rightleftharpoons Li(s)$	−3.05

Strength of oxidizing agent (increasing upward). Strength of reducing agent (increasing downward).

Thus, $F_2(g)$ is the strongest oxidizing agent (has the largest positive E°), which means $F^-(aq)$ is the weakest reducing agent. Similarly, $Li^+(aq)$ is the weakest oxidizing agent (has the most negative E°), which means $Li(s)$ is the strongest reducing agent. (You may have noticed an analogy to conjugate acid-base pairs: a strong acid forms a weak conjugate base, and vice versa, just as a *strong oxidizing agent forms a weak reducing agent,* and vice versa.)

If you forget the ranking in the table, just rely on your chemical knowledge of the elements. You know that F_2 is very electronegative and typically occurs as F^-. It is easily reduced (gains electrons), so it must be a strong oxidizing agent (high, positive E°). Similarly, Li metal has a low ionization energy and typically occurs as Li^+. Therefore, it is easily oxidized (loses electrons), so it must be a strong reducing agent (low, negative E°).

Writing Spontaneous Redox Reactions

Appendix D can be used to write spontaneous redox reactions, which is useful for constructing voltaic cells.

Every redox reaction is the sum of two half-reactions, so there is a reducing agent and an oxidizing agent on each side. In the zinc-copper reaction, for instance, Zn and Cu are the reducing agents, and Cu^{2+} and Zn^{2+} are the oxidizing agents. The stronger oxidizing and reducing agents react spontaneously to form the weaker oxidizing and reducing agents:

$$Zn(s) \ + \ Cu^{2+}(aq) \longrightarrow Zn^{2+}(aq) \ + \ Cu(s)$$

stronger	stronger	weaker	weaker
reducing agent	oxidizing agent	oxidizing agent	reducing agent

Here, too, note the similarity to acid-base chemistry. The stronger acid and base spontaneously form the weaker base and acid, respectively. The members of a conjugate acid-base pair differ by a proton: the acid has the proton and the base does not. The members of a redox pair, or *redox couple,* such as Zn and Zn^{2+}, differ by one or more electrons: the reduced form (Zn) has the electrons and the oxidized form (Zn^{2+}) does not. In acid-base reactions, we compare acid and base strength using K_a and K_b values. In redox reactions, we compare oxidizing and reducing strength using $E°$ values.

Based on the order of the $E°$ values in Appendix D, *the stronger oxidizing agent (species on the left) has a half-reaction with a larger (more positive or less negative) $E°$ value, and the stronger reducing agent (species on the right) has a half-reaction with a smaller (less positive or more negative) $E°$ value.* Therefore, a spontaneous reaction ($E°_{cell} > 0$) will occur between an oxidizing agent and any reducing agent that lies *below* it in the list. For instance, Zn (right) lies below Cu^{2+} (left), and Cu^{2+} and Zn react spontaneously. In other words, *for a spontaneous reaction to occur, the half-reaction higher in the list proceeds at the cathode as written, and the half-reaction lower in the list proceeds at the anode in reverse.* This pairing ensures that the stronger oxidizing agent (higher on the left) and stronger reducing agent (lower on the right) will be the reactants.

However, if we know the electrode potentials, we can write a spontaneous redox reaction even if Appendix D is not available. Let's choose a pair of half-reactions from the appendix and, without referring to their relative positions in the list, use them to write a spontaneous redox reaction:

$$Ag^+(aq) + e^- \longrightarrow Ag(s) \qquad E°_{silver} = 0.80 \text{ V}$$
$$Sn^{2+}(aq) + 2e^- \longrightarrow Sn(s) \qquad E°_{tin} = -0.14 \text{ V}$$

There are two steps involved:

1. Reverse one of the half-reactions into an oxidation step such that the difference of the electrode potentials (cathode *minus* anode) gives a *positive* $E°_{cell}$. Note that when we reverse the half-reaction, we need *not* reverse the sign of $E°_{half-cell}$ because the minus sign in Equation 21.3 ($E°_{cell} = E°_{cathode} - E°_{anode}$) will do that.

2. Add the rearranged half-reactions to obtain a balanced overall equation. Be sure to multiply by coefficients so that e^- lost equals e^- gained and to cancel species common to both sides.

(You may be tempted in this particular case to add the two half-reactions as written, because you obtain a positive $E°_{cell}$, but you would then have two oxidizing agents forming two reducing agents, which cannot occur.)

We want to pair the stronger oxidizing and reducing agents as reactants. The larger (more positive) $E°$ value for the silver half-reaction means that Ag^+ is a stronger oxidizing agent (gains electrons more readily) than Sn^{2+}, and the smaller (more negative) $E°$ value for the tin half-reaction means that Sn is a stronger reducing agent (loses electrons more readily) than Ag. Therefore, we reverse the tin half-reaction (but *not* the sign of $E°_{tin}$):

$$Sn(s) \longrightarrow Sn^{2+}(aq) + 2e^- \qquad E°_{tin} = -0.14 \text{ V}$$

Subtracting $E°_{half-cell}$ of the tin half-reaction (anode, oxidation) from $E°_{half-cell}$ of the silver half-reaction (cathode, reduction) gives a positive $E°_{cell}$; that is, 0.80 V − (−0.14 V) = 0.94 V.

With the half-reactions written in the correct direction, we must next make sure that the *number of electrons lost in the oxidation equals the number gained in the reduction*. In this case, we double the silver (reduction) half-reaction. Adding the half-reactions and applying Equation 21.3 gives the balanced equation and E_{cell}:

$$2Ag^+(aq) + 2e^- \longrightarrow 2Ag(s) \qquad\qquad E^\circ_{silver} = 0.80 \text{ V} \qquad \text{[reduction]}$$
$$Sn(s) \longrightarrow Sn^{2+}(aq) + 2e^- \qquad\qquad E^\circ_{tin} = -0.14 \text{ V} \qquad \text{[oxidation]}$$
$$\overline{Sn(s) + 2Ag^+(aq) \longrightarrow Sn^{2+}(aq) + 2Ag(s) \qquad E^\circ_{cell} = E^\circ_{silver} - E^\circ_{tin} = 0.94 \text{ V}}$$

With the reaction spontaneous as written, the stronger oxidizing and reducing agents are reactants, which confirms that Sn is a stronger reducing agent than Ag, and Ag^+ is a stronger oxidizing agent than Sn^{2+}.

A very important point to note is that, when we doubled the coefficients of the silver half-reaction to balance the number of electrons, we did *not* double its E° value—it remained 0.80 V. That is, *changing the balancing coefficients of a half-reaction does **not** change its E° value*. The reason is that a standard electrode potential is an *intensive* property, one that does *not* depend on the amount of substance present. The potential is the *ratio* of energy to charge. When we change the coefficients, thus changing the amount of substance, the energy *and* the charge change proportionately, so their ratio stays the same. (Recall that density, which is also an intensive property, does not change with the amount of substance because the mass *and* the volume change proportionately.)

 ARIS Presentation Center, Worked Examples Library

SAMPLE PROBLEM 21.4 Writing Spontaneous Redox Reactions and Ranking Oxidizing and Reducing Agents by Strength

PROBLEM (a) Combine the following three half-reactions into three balanced equations (A, B, and C) for spontaneous reactions, and calculate E°_{cell} for each. **(b)** Rank the relative strengths of the oxidizing and reducing agents.

$$\text{(1) } NO_3^-(aq) + 4H^+(aq) + 3e^- \longrightarrow NO(g) + 2H_2O(l) \qquad E^\circ = 0.96 \text{ V}$$
$$\text{(2) } N_2(g) + 5H^+(aq) + 4e^- \longrightarrow N_2H_5^+(aq) \qquad E^\circ = -0.23 \text{ V}$$
$$\text{(3) } MnO_2(s) + 4H^+(aq) + 2e^- \longrightarrow Mn^{2+}(aq) + 2H_2O(l) \qquad E^\circ = 1.23 \text{ V}$$

PLAN (a) To write the redox equations, we combine the possible pairs of half-reactions: (1) and (2), (1) and (3), and (2) and (3). They are all written as reductions, so the oxidizing agents appear as reactants and the reducing agents appear as products. In each pair, we reverse the reduction half-reaction that has the smaller (less positive or more negative) E° value to an oxidation to obtain a positive E°_{cell}. We make e^- lost equal e^- gained, without changing the magnitude of the E° value, add the half-reactions together, and then apply Equation 21.3 to find E°_{cell}. **(b)** Because each reaction is spontaneous as written, the stronger oxidizing and reducing agents are the reactants. To obtain the overall ranking, we first rank the relative strengths within each equation and then compare them.

SOLUTION (a) Combining half-reactions (1) and (2) gives equation (A). The E° value for half-reaction (1) is larger (more positive) than that for (2), so we reverse (2) to obtain a positive E°_{cell}:

$$\text{(1) } NO_3^-(aq) + 4H^+(aq) + 3e^- \longrightarrow NO(g) + 2H_2O(l) \qquad E^\circ = 0.96 \text{ V}$$
$$\text{(rev 2) } \qquad N_2H_5^+(aq) \longrightarrow N_2(g) + 5H^+(aq) + 4e^- \qquad E^\circ = -0.23 \text{ V}$$

To make e^- lost equal e^- gained, we multiply (1) by four and the reversed (2) by three; then add half-reactions and cancel appropriate numbers of common species (H^+ and e^-):

$$4NO_3^-(aq) + 16H^+(aq) + 12e^- \longrightarrow 4NO(g) + 8H_2O(l) \qquad E^\circ = 0.96 \text{ V}$$
$$\overline{\qquad\qquad 3N_2H_5^+(aq) \longrightarrow 3N_2(g) + 15H^+(aq) + 12e^- \qquad E^\circ = -0.23 \text{ V}}$$

$$\text{(A) } 3N_2H_5^+(aq) + 4NO_3^-(aq) + H^+(aq) \longrightarrow 3N_2(g) + 4NO(g) + 8H_2O(l)$$

$$E^\circ_{cell} = 0.96 \text{ V} - (-0.23 \text{ V}) = 1.19 \text{ V}$$

Combining half-reaction (1) and half-reaction (3) gives equation (B). Half-reaction (1) must be reversed:

(rev 1) $NO(g) + 2H_2O(l) \longrightarrow NO_3^-(aq) + 4H^+(aq) + 3e^-$ $E° = 0.96$ V
(3) $MnO_2(s) + 4H^+(aq) + 2e^- \longrightarrow Mn^{2+}(aq) + 2H_2O(l)$ $E° = 1.23$ V

We multiply reversed (1) by two and (3) by three, then add and cancel:

$2NO(g) + 4H_2O(l) \longrightarrow 2NO_3^-(aq) + 8H^+(aq) + 6e^-$ $E° = 0.96$ V
$3MnO_2(s) + 12H^+(aq) + 6e^- \longrightarrow 3Mn^{2+}(aq) + 6H_2O(l)$ $E° = 1.23$ V

(B) $3MnO_2(s) + 4H^+(aq) + 2NO(g) \longrightarrow 3Mn^{2+}(aq) + 2H_2O(l) + 2NO_3^-(aq)$

$$E°_{cell} = 1.23 \text{ V} - 0.96 \text{ V} = 0.27 \text{ V}$$

Combining half-reaction (2) and half-reaction (3) gives equation (C). Half-reaction (2) must be reversed:

(rev 2) $N_2H_5^+(aq) \longrightarrow N_2(g) + 5H^+(aq) + 4e^-$ $E° = -0.23$ V
(3) $MnO_2(s) + 4H^+(aq) + 2e^- \longrightarrow Mn^{2+}(aq) + 2H_2O(l)$ $E° = 1.23$ V

We multiply reaction (3) by two, add the half-reactions, and cancel:

$N_2H_5^+(aq) \longrightarrow N_2(g) + 5H^+(aq) + 4e^-$ $E° = -0.23$ V
$2MnO_2(s) + 8H^+(aq) + 4e^- \longrightarrow 2Mn^{2+}(aq) + 4H_2O(l)$ $E° = 1.23$ V

(C) $N_2H_5^+(aq) + 2MnO_2(s) + 3H^+(aq) \longrightarrow N_2(g) + 2Mn^{2+}(aq) + 4H_2O(l)$

$$E°_{cell} = 1.23 \text{ V} - (-0.23 \text{ V}) = 1.46 \text{ V}$$

(b) Ranking the oxidizing and reducing agents within each equation:

Equation (A): Oxidizing agents: $NO_3^- > N_2$ Reducing agents: $N_2H_5^+ > NO$
Equation (B): Oxidizing agents: $MnO_2 > NO_3^-$ Reducing agents: $NO > Mn^{2+}$
Equation (C): Oxidizing agents: $MnO_2 > N_2$ Reducing agents: $N_2H_5^+ > Mn^{2+}$

Determining the overall ranking of oxidizing and reducing agents. Comparing the relative strengths from the three balanced equations gives

$$\text{Oxidizing agents: } MnO_2 > NO_3^- > N_2$$

$$\text{Reducing agents: } N_2H_5^+ > NO > Mn^{2+}$$

CHECK As always, check that atoms and charge balance on each side of the equation. A good way to check the ranking and equations is to list the given half-reactions in order of decreasing $E°$ value:

$MnO_2(s) + 4H^+(aq) + 2e^- \longrightarrow Mn^{2+}(aq) + 2H_2O(l)$ $E° = 1.23$ V
$NO_3^-(aq) + 4H^+(aq) + 3e^- \longrightarrow NO(g) + 2H_2O(l)$ $E° = 0.96$ V
$N_2(g) + 5H^+(aq) + 4e^- \longrightarrow N_2H_5^+(aq)$ $E° = -0.23$ V

Then the oxidizing agents (reactants) decrease in strength going down the list, so the reducing agents (products) decrease in strength going up. Moreover, each of the three spontaneous reactions (A, B, and C) should combine a reactant with a product that is lower down on this list.

FOLLOW-UP PROBLEM 21.4   Is the following reaction spontaneous as written?

$$3Fe^{2+}(aq) \longrightarrow Fe(s) + 2Fe^{3+}(aq)$$

If not, write the equation for the spontaneous reaction, calculate $E°_{cell}$, and rank the three species of iron in order of decreasing reducing strength.

Relative Reactivities of Metals   In Chapter 4, we discussed the activity series of the metals (see Figure 4.21, p. 171), which ranks metals by their ability to "displace" one another from aqueous solution. Now you'll see *why* this displacement occurs, as well as why many, but not all, metals react with acid to form H_2, and why a few metals form H_2 even in water.

1. *Metals that can displace H_2 from acid.* The standard hydrogen half-reaction represents the reduction of H^+ ions from an acid to H_2:

$$2H^+(aq) + 2e^- \longrightarrow H_2(g) \qquad E^\circ = 0.00 \text{ V}$$

To see which metals reduce H^+ (referred to as "displacing H_2") from acids, choose a metal, write its half-reaction as an oxidation, combine this half-reaction with the hydrogen half-reaction, and see if E°_{cell} is positive. What you find is that the metals Li through Pb, those that lie *below* the standard hydrogen (reference) half-reaction in Appendix D, give a positive E°_{cell} when reducing H^+. Iron, for example, reduces H^+ from an acid to H_2:

$Fe(s) \longrightarrow Fe^{2+}(aq) + 2e^-$	$E^\circ = -0.44$ V	[anode; oxidation]
$2H^+(aq) + 2e^- \longrightarrow H_2(g)$	$E^\circ = 0.00$ V	[cathode; reduction]
$Fe(s) + 2H^+(aq) \longrightarrow H_2(g) + Fe^{2+}(aq)$	$E^\circ_{cell} = 0.00$ V $- (-0.44$ V$) = 0.44$ V	

The lower the metal in the list, the stronger it is as a reducing agent; therefore, the more positive its half-cell potential when the half-reaction is reversed, and the higher the E°_{cell} for its reduction of H^+ to H_2. *If E°_{cell} for the reduction of H^+ is more positive for metal A than it is for metal B, metal A is a stronger reducing agent than metal B and a more* **active** *metal.*

2. *Metals that cannot displace H_2 from acid.* Metals that are *above* the standard hydrogen (reference) half-reaction *cannot* reduce H^+ from acids. When we reverse the metal half-reaction, the E°_{cell} is negative, so the reaction does not occur. For example, the coinage metals—copper, silver, and gold, which are in Group 1B(11)—are not strong enough reducing agents to reduce H^+ from acids:

$Ag(s) \longrightarrow Ag^+(aq) + e^-$	$E^\circ = 0.80$ V	[anode; oxidation]
$2H^+(aq) + 2e^- \longrightarrow H_2(g)$	$E^\circ = 0.00$ V	[cathode; reduction]
$2Ag(s) + 2H^+(aq) \longrightarrow 2Ag^+(aq) + H_2(g)$	$E^\circ_{cell} = 0.00$ V $- 0.80$ V $= -0.80$ V	

The higher the metal in the list, the more negative is its E°_{cell} for the reduction of H^+ to H_2, the lower is its reducing strength, and the less active it is. Thus, gold is less active than silver, which is less active than copper.

3. *Metals that can displace H_2 from water.* Metals active enough to reduce H_2O lie *below that half-reaction:*

$$2H_2O(l) + 2e^- \longrightarrow H_2(g) + 2OH^-(aq) \qquad E = -0.42 \text{ V}$$

(The value shown here is the *nonstandard* electrode potential because, in pure water, $[OH^-]$ is 1.0×10^{-7} M, not the standard-state value of 1 M.) For example, consider the reaction of sodium in water (with the Na^+/Na half-reaction reversed and doubled):

$2Na(s) \longrightarrow 2Na^+(aq) + 2e^-$	$E^\circ = -2.71$ V	[anode; oxidation]
$2H_2O(l) + 2e^- \longrightarrow H_2(g) + 2OH^-(aq)$	$E = -0.42$ V	[cathode; reduction]
$2Na(s) + 2H_2O(l) \longrightarrow 2Na^+(aq) + H_2(g) + 2OH^-(aq)$		
	$E_{cell} = -0.42$ V $- (-2.71$ V$) = 2.29$ V	

The alkali metals [Group 1A(1)] and the larger alkaline earth metals [Group 2A(2)] can reduce water, or displace H_2 from H_2O (Figure 21.9).

4. *Metals that can displace other metals from solution.* We can also predict whether one metal can reduce the aqueous ion of another metal. Any metal that is lower in the list in Appendix D can reduce the ion of a metal that is higher up, and thus displace that metal from solution. For example, zinc can displace iron from solution:

$Zn(s) \longrightarrow Zn^{2+}(aq) + 2e^-$	$E^\circ = -0.76$ V	[anode; oxidation]
$Fe^{2+}(aq) + 2e^- \longrightarrow Fe(s)$	$E^\circ = -0.44$ V	[cathode; reduction]
$Zn(s) + Fe^{2+}(aq) \longrightarrow Zn^{2+}(aq) + Fe(s)$	$E^\circ_{cell} = -0.44$ V $- (-0.76$ V$) = 0.32$ V	

Oxidation half-reaction
$Ca(s) \longrightarrow Ca^{2+}(aq) + 2e^-$

Reduction half-reaction
$2H_2O(l) + 2e^- \longrightarrow H_2(g) + 2OH^-(aq)$

Overall (cell) reaction
$Ca(s) + 2H_2O(l) \longrightarrow Ca(OH)_2(aq) + H_2(g)$

Figure 21.9 The reaction of calcium in water. Calcium is one of the metals active enough to displace H_2 from H_2O.

This particular reaction has tremendous economic importance in protecting iron from rusting, as you'll see shortly. The reducing power of metals has other, more personal, consequences, as the margin note points out. ●

Section Summary

The output of a cell is called the cell potential (E_{cell}) and is measured in volts (1 V = 1 J/C). ● When all substances are in their standard states, the output is the standard cell potential (E°_{cell}). $E^\circ_{cell} > 0$ for a spontaneous reaction at standard-state conditions. ● By convention, a standard electrode potential ($E^\circ_{half\text{-}cell}$) refers to the *reduction* half-reaction. E°_{cell} equals $E^\circ_{half\text{-}cell}$ of the cathode *minus* $E^\circ_{half\text{-}cell}$ of the anode. ● Using a standard hydrogen (reference) electrode ($E^\circ_{reference} = 0$ V), $E^\circ_{half\text{-}cell}$ values can be measured and used to rank oxidizing (or reducing) agents (see Appendix D). ● Spontaneous redox reactions combine stronger oxidizing and reducing agents to form weaker ones. ● A metal can reduce another species (H^+, H_2O, or an ion of another metal) if E°_{cell} for the reaction is positive.

21.4 FREE ENERGY AND ELECTRICAL WORK

In Chapter 20, we discussed the relationship of useful work, free energy, and the equilibrium constant. In this section, we examine this relationship in the context of electrochemical cells and see the effect of concentration on cell potential.

Standard Cell Potential and the Equilibrium Constant

As you know from Section 20.3, a spontaneous reaction has a *negative* free energy change ($\Delta G < 0$), and you've just seen that a spontaneous electrochemical reaction has a *positive* cell potential ($E_{cell} > 0$). Note that *the signs of ΔG and E_{cell} are opposite for a spontaneous reaction.* These two indications of spontaneity are proportional to each other:

$$\Delta G \propto -E_{cell}$$

Let's determine this proportionality constant by focusing on the electrical work done (w, in joules), which is the product of the potential (E_{cell}, in volts) and the amount of charge that flows (in coulombs). The value used for E_{cell} is measured with no current flowing and, therefore, no energy lost to heating the cell. Thus, E_{cell} is the maximum voltage possible for the cell, and the work is the maximum work possible (w_{max}). For work done *by* the cell *on* the surroundings, this quantity is negative:*

$$w_{max} = -E_{cell} \times \text{charge}$$

Equation 20.9, p. 902, shows that the maximum work done *on* the surroundings is ΔG:

$$w_{max} = -E_{cell} \times \text{charge} = \Delta G \quad \text{and so} \quad \Delta G = -E_{cell} \times \text{charge}$$

The charge that flows through the cell equals the number of moles of electrons (n) transferred times the charge of 1 mol of electrons (symbol F):

$$\text{Charge} = \text{moles of } e^- \times \frac{\text{charge}}{\text{mol } e^-} \quad \text{or} \quad \text{charge} = nF$$

The charge of 1 mol of electrons is the **Faraday constant (F)**, named in honor of Michael Faraday, the 19th-century British scientist who pioneered the study of electrochemistry:

$$F = \frac{96,485 \text{ C}}{\text{mol } e^-}$$

---

*Recall from Chapter 20 that only a reversible process can do maximum work. For no current to flow and the process to be reversible, E_{cell} must be opposed by an equal potential in the measuring circuit: if the opposing potential is infinitesimally smaller, the cell reaction goes forward; if it is infinitesimally larger, the reaction goes backward.

 ARIS Presentation Center, Art Library

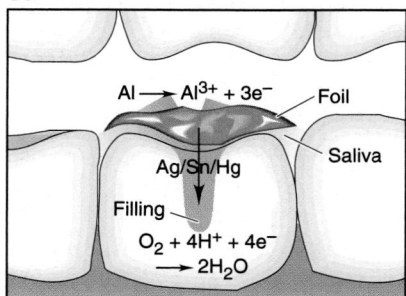

● **The Pain of a Dental Voltaic Cell**
Have you ever felt a jolt of pain when biting down with a filled tooth on a scrap of foil left on a piece of food? Here's the reason. The aluminum foil acts as an active anode (E° of Al = -1.66 V), saliva as the electrolyte, and the filling (usually a silver/tin/mercury alloy) as an inactive cathode. O_2 is reduced to water, and the short circuit between the foil in contact with the filling creates a current that is sensed by the nerve of the tooth.

 Biological application

Literature Bindel, Thomas H. "Understanding Electrochemical Thermodynamics through Entropy Analysis," *J. Chem. Educ.* 2000, *77*, 1031.

Because 1 V = 1 J/C, we have 1 C = 1 J/V, and

$$F = 9.65 \times 10^4 \, \frac{J}{V \cdot \text{mol e}^-} \quad \text{(3 sf)} \quad \text{(21.4)}$$

Substituting for charge, the proportionality constant is nF:

$$\Delta G = -nFE_{\text{cell}} \quad \text{(21.5)}$$

When all of the components are in their standard states, we have

$$\Delta G^\circ = -nFE^\circ_{\text{cell}} \quad \text{(21.6)}$$

Using this relationship, we can relate the standard cell potential to the equilibrium constant of the redox reaction. Recall from Equation 20.12 that

$$\Delta G^\circ = -RT \ln K$$

Substituting for ΔG° from Equation 21.6 gives

$$-nFE^\circ_{\text{cell}} = -RT \ln K$$

Solving for E°_{cell} gives

$$E^\circ_{\text{cell}} = \frac{RT}{nF} \ln K \quad \text{(21.7)}$$

Figure 21.10 summarizes the interconnections among the standard free energy change, the equilibrium constant, and the standard cell potential. Notice that, even though these relationships hold under *standard-state conditions*, they are still extremely useful. Recall from Chapter 20 that determining K required that we know ΔG°, either from ΔH° and ΔS° values or from ΔG°_f values. Now, for redox reactions, we have a direct experimental method for determining K *and* ΔG°: measuring E°_{cell}.

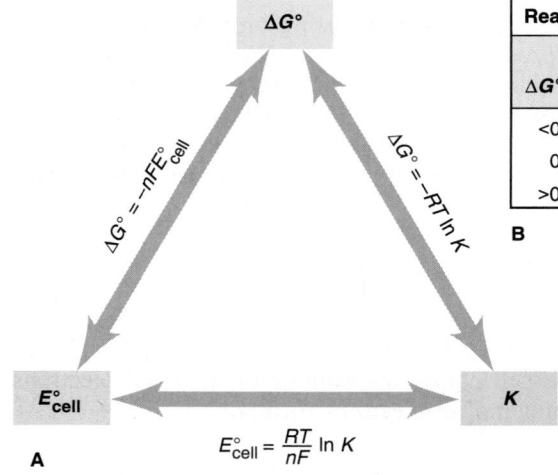

Reaction Parameters at the Standard State			
ΔG°	K	E°_{cell}	Reaction at standard-state conditions
<0	>1	>0	Spontaneous
0	1	0	At equilibrium
>0	<1	<0	Nonspontaneous

B

Figure 21.10 **The interrelationship of ΔG°, E°_{cell}, and K. A,** Any one of these three central thermodynamic parameters can be used to find the other two. **B,** The signs of ΔG° and E°_{cell} determine the reaction direction at standard-state conditions.

Transparency/ARIS Presentation Center, Art Library

It is common practice to simplify Equation 21.7 in calculations by

- Substituting the known value of 8.314 J/(mol rxn·K) for the constant R
- Substituting the known value of 9.65×10^4 J/(V·mol e$^-$) for the constant F
- Substituting the standard temperature of 298.15 K for T, but keeping in mind that the cell can run at other temperatures.
- Multiplying by 2.303 to convert from natural to common (base-10) logarithms. This conversion shows that a *10-fold change in K makes E°_{cell} change by 1.*

Thus, when n moles of e^- are transferred per mole of reaction in the balanced equation, this simplified relation between E°_{cell} and K gives

$$E^{\circ}_{cell} = \frac{RT}{nF} \ln K = 2.303 \, \frac{RT}{nF} \log K = 2.303 \times \frac{8.314 \, \frac{J}{\text{mol rxn·K}} \times 298.15 \text{ K}}{\frac{n \, \text{mol e}^-}{\text{mol rxn}} \left(9.65 \times 10^4 \, \frac{J}{\text{V·mol e}^-}\right)} \log K$$

And, we have

$$E^{\circ}_{cell} = \frac{0.0592 \text{ V}}{n} \log K \quad \text{or} \quad \log K = \frac{nE^{\circ}_{cell}}{0.0592 \text{ V}} \quad \text{(at 298.15 K)} \quad \textbf{(21.8)}$$

SAMPLE PROBLEM 21.5 Calculating K and ΔG° from E°_{cell}

PROBLEM Lead can displace silver from solution:

$$Pb(s) + 2Ag^+(aq) \longrightarrow Pb^{2+}(aq) + 2Ag(s)$$

As a consequence, silver is a valuable byproduct in the industrial extraction of lead from its ore. Calculate K and ΔG° at 298.15 K for this reaction.

PLAN We divide the spontaneous redox equation into the half-reactions and use values from Appendix D to calculate E°_{cell}. Then, we substitute this result into Equation 21.8 to find K and into Equation 21.6 to find ΔG°.

SOLUTION Writing the half-reactions with their E° values:

(1) $Ag^+(aq) + e^- \longrightarrow Ag(s)$ $E^{\circ} = 0.80$ V
(2) $Pb^{2+}(aq) + 2e^- \longrightarrow Pb(s)$ $E^{\circ} = -0.13$ V

Calculating E°_{cell}: We double (1), reverse (2), add the half-reactions, and subtract E°_{lead} from E°_{silver}:

$$2Ag^+(aq) + 2e^- \longrightarrow 2Ag(s) \qquad\qquad E^{\circ} = 0.80 \text{ V}$$
$$Pb(s) \longrightarrow Pb^{2+}(aq) + 2e^- \qquad\qquad E^{\circ} = -0.13 \text{ V}$$

$$Pb(s) + 2Ag^+(aq) \longrightarrow Pb^{2+}(aq) + 2Ag(s) \quad E^{\circ}_{cell} = 0.80 \text{ V} - (-0.13 \text{ V}) = 0.93 \text{ V}$$

Calculating K with Equations 21.7 and 21.8:

$$E^{\circ}_{cell} = \frac{RT}{nF} \ln K = 2.303 \, \frac{RT}{nF} \log K$$

The adjusted half-reactions show that 2 mol of e^- are transferred per mole of reaction as written, so $n = 2$. Then, performing the substitutions for R and F that we just discussed with the cell running at 25°C (298.15 K), we have

$$E^{\circ}_{cell} = \frac{0.0592 \text{ V}}{2} \log K = 0.93 \text{ V}$$

So, $\log K = \dfrac{0.93 \text{ V} \times 2}{0.0592 \text{ V}} = 31.42$ and $K = 2.6 \times 10^{31}$

Calculating ΔG° (Equation 21.6):

$$\Delta G^{\circ} = -nFE^{\circ}_{cell} = -\frac{2 \text{ mol e}^-}{\text{mol rxn}} \times \frac{96.5 \text{ kJ}}{\text{V·mol e}^-} \times 0.93 \text{ V} = \boxed{-1.8 \times 10^2 \text{ kJ/mol rxn}}$$

CHECK The three variables are consistent with the reaction being spontaneous at standard-state conditions: $E^{\circ}_{cell} > 0$, $\Delta G^{\circ} < 0$, and $K > 1$. Be sure to round and check the order of magnitude: in the ΔG° calculation, for instance, $\Delta G^{\circ} \approx -2 \times 100 \times 1 = -200$, so the overall math seems right. Another check would be to obtain ΔG° directly from its relation with K:

$$\Delta G^{\circ} = -RT \ln K = -8.314 \text{ J/mol rxn·K} \times 298.15 \text{ K} \times \ln (2.6 \times 10^{31})$$
$$= -1.8 \times 10^5 \text{ J/mol rxn} = -1.8 \times 10^2 \text{ kJ/mol rxn}$$

FOLLOW-UP PROBLEM 21.5 When cadmium metal reduces Cu^{2+} in solution, Cd^{2+} forms in addition to copper metal. Given that $\Delta G^{\circ} = -143$ kJ/mol rxn, calculate K at 25°C. What is E°_{cell} of a voltaic cell that uses this reaction?

ARIS Presentation Center, Worked Examples Library

Engineering application

The Effect of Concentration on Cell Potential

So far, we've considered cells with all components in their standard states and found standard cell potential (E°_{cell}) from standard half-cell potentials ($E^\circ_{half-cell}$). However, most cells do not start with all components in their standard states, and even if they did, the concentrations change after a few moments of operation. Moreover, in all practical voltaic cells, such as batteries, reactant concentrations are far from standard-state values. Clearly, we must be able to determine E_{cell}, the cell potential under nonstandard conditions.

To do so, let's derive an expression for the relation between cell potential and concentration based on the relation between free energy and concentration. Recall from Chapter 20 (Equation 20.13) that ΔG equals ΔG° (the free energy change when the system moves from standard-state concentrations to equilibrium) *plus* $RT \ln Q$ (the free energy change when the system moves from nonstandard-state to standard-state concentrations):

$$\Delta G = \Delta G^\circ + RT \ln Q$$

ΔG is related to E_{cell} and ΔG° to E°_{cell} (Equations 21.5 and 21.6), so we substitute for them and get

$$-nFE_{cell} = -nFE^\circ_{cell} + RT \ln Q$$

Dividing both sides by $-nF$, we obtain the **Nernst equation,** developed by the German chemist Walther Hermann Nernst in 1889: ●

$$E_{cell} = E^\circ_{cell} - \frac{RT}{nF} \ln Q \qquad (21.9)$$

The Nernst equation says that a cell potential under any conditions depends on the potential at standard-state concentrations *and* a term for the potential at nonstandard-state concentrations. How do changes in Q affect cell potential? From Equation 21.9, we see that

- When $Q < 1$ and thus [reactant] > [product], $\ln Q < 0$, so $E_{cell} > E^\circ_{cell}$.
- When $Q = 1$ and thus [reactant] = [product], $\ln Q = 0$, so $E_{cell} = E^\circ_{cell}$.
- When $Q > 1$ and thus [reactant] < [product], $\ln Q > 0$, so $E_{cell} < E^\circ_{cell}$.

As before, to obtain a simplified form of the Nernst equation for use in calculations, let's substitute known values of R and F, operate the cell at 298.15 K, and convert to common (base-10) logarithms:

$$E_{cell} = E^\circ_{cell} - \frac{RT}{nF} \ln Q = E^\circ_{cell} - 2.303 \frac{RT}{nF} \log Q$$

$$= E^\circ_{cell} - 2.303 \times \frac{8.314 \frac{J}{mol\ rxn \cdot K} \times 298.15\ K}{\frac{n\ mol\ e^-}{mol\ rxn}\left(9.65 \times 10^4 \frac{J}{V \cdot mol\ e^-}\right)} \log Q$$

And we obtain:

$$E_{cell} = E^\circ_{cell} - \frac{0.0592\ V}{n} \log Q \quad (\text{at } 298.15\ K) \qquad (21.10)$$

Remember that the expression for Q *contains only those species with concentrations (and/or pressures) that can vary;* thus, solids do not appear, even when they are the electrodes. For example, in the reaction between cadmium and silver ion, the Cd and Ag electrodes do not appear in the expression for Q:

$$Cd(s) + 2Ag^+(aq) \longrightarrow Cd^{2+}(aq) + 2Ag(s) \qquad Q = \frac{[Cd^{2+}]}{[Ag^+]^2}$$

● **Walther Hermann Nernst (1864–1941)** This great German physical chemist was only 25 years old when he developed the equation for the relationship between cell voltage and concentration. His career, which culminated in the Nobel Prize in chemistry in 1920, included formulating the third law of thermodynamics, establishing the principle of the solubility product, and contributing key ideas to photochemistry and the Haber process. He is shown here in later years making a point to a few of his celebrated colleagues: from left to right, Nernst, Albert Einstein, Max Planck, Robert Millikan, and Max von Laue.

SAMPLE PROBLEM 21.6 Using the Nernst Equation to Calculate E_{cell}

ARIS Presentation Center, Worked Examples Library

PROBLEM In a test of a new reference electrode, a chemist constructs a voltaic cell consisting of a Zn/Zn^{2+} half-cell and an H_2/H^+ half-cell under the following conditions:

$$[Zn^{2+}] = 0.010\ M \qquad [H^+] = 2.5\ M \qquad P_{H_2} = 0.30\ atm$$

Calculate E_{cell} at 298.15 K.

PLAN To apply the Nernst equation and determine E_{cell}, we must know E°_{cell} and Q. We write the spontaneous reaction, calculate E°_{cell} from standard electrode potentials (Appendix D), and use the given pressure and concentrations to find Q. (Recall that the ideal gas law allows us to use P at constant T as another way of writing concentration, n/V.) Then we substitute into Equation 21.10.

SOLUTION Determining the cell reaction and E°_{cell}:

$$\begin{array}{ll} 2H^+(aq) + 2e^- \longrightarrow H_2(g) & E^\circ = 0.00\ V \\ \underline{Zn(s) \longrightarrow Zn^{2+}(aq) + 2e^-} & \underline{E^\circ = -0.76\ V} \\ 2H^+(aq) + Zn(s) \longrightarrow H_2(g) + Zn^{2+}(aq) & E^\circ_{cell} = 0.00\ V - (-0.76\ V) = 0.76\ V \end{array}$$

Calculating Q:

$$Q = \frac{P_{H_2} \times [Zn^{2+}]}{[H^+]^2} = \frac{0.30 \times 0.010}{2.5^2} = 4.8 \times 10^{-4}$$

Solving for E_{cell} at 25°C (298.15 K), with $n = 2$:

$$E_{cell} = E^\circ_{cell} - 2.303 \frac{RT}{nF} \log Q = E^\circ_{cell} - \frac{0.0592\ V}{n} \log Q$$

$$= 0.76\ V - \left[\frac{0.0592\ V}{2} \log (4.8 \times 10^{-4}) \right] = 0.76\ V - (-0.0982\ V) = \boxed{0.86\ V}$$

CHECK After you check the arithmetic, reason through the answer: $E_{cell} > E^\circ_{cell}$ (0.86 > 0.76) because the log Q term was negative, which is consistent with $Q < 1$.

FOLLOW-UP PROBLEM 21.6 Consider a voltaic cell based on the following reaction: $Fe(s) + Cu^{2+}(aq) \longrightarrow Fe^{2+}(aq) + Cu(s)$. If $[Cu^{2+}] = 0.30\ M$, what must $[Fe^{2+}]$ be to increase E_{cell} by 0.25 V above E°_{cell} at 25°C?

Changes in Potential During Cell Operation

Let's see how the potential of the zinc-copper cell changes as concentrations change during cell operation. In this case, the only concentrations that change are [reactant] = $[Cu^{2+}]$ and [product] = $[Zn^{2+}]$:

$$Zn(s) + Cu^{2+}(aq) \longrightarrow Zn^{2+}(aq) + Cu(s) \qquad Q = \frac{[Zn^{2+}]}{[Cu^{2+}]}$$

Literature Fishtik, Ilie; Berka, Ladislav H. "Variability of the Cell Potential of a Given Chemical Reaction," *J. Chem. Educ.* **2004**, *81*, 584.

The positive E°_{cell} (1.10 V) means that this reaction proceeds *spontaneously* from standard-state conditions, at which $[Zn^{2+}] = [Cu^{2+}] = 1\ M$ ($Q = 1$), to some point at which $[Zn^{2+}] > [Cu^{2+}]$ ($Q > 1$). Now, suppose that we start the cell when $[Zn^{2+}] < [Cu^{2+}]$ ($Q < 1$), for example, when $[Zn^{2+}] = 1.0 \times 10^{-4}\ M$ and $[Cu^{2+}] = 2.0\ M$. Thus, the cell potential is *higher* than the standard cell potential:

$$E_{cell} = E^\circ_{cell} - \frac{0.0592\ V}{2} \log \frac{[Zn^{2+}]}{[Cu^{2+}]} = 1.10\ V - \left(\frac{0.0592\ V}{2} \log \frac{1.0 \times 10^{-4}}{2.0} \right)$$

$$= 1.10\ V - \left[\frac{0.0592\ V}{2} (-4.30) \right] = 1.10\ V + 0.127\ V = 1.23\ V$$

As the cell operates, $[Zn^{2+}]$ increases (as the Zn electrode deteriorates) and $[Cu^{2+}]$ decreases (as Cu plates out on the Cu electrode). Although the changes during this process occur smoothly, if we keep Equation 21.10 in mind, we can

$$Q = \frac{[Zn^{2+}]}{[Cu^{2+}]}$$

Figure 21.11 **The relation between E_{cell} and log Q for the zinc-copper cell.** **A,** A plot of E_{cell} vs. Q (on a logarithmic scale) for the zinc-copper cell shows a linear decrease. When $Q < 1$ (*left*), [reactant] is relatively high, and the cell can do relatively more work. When $Q = 1$, $E_{cell} = E_{cell}°$. When $Q > 1$ (*right*), [product] is relatively high, and the cell can do relatively less work. **B,** A summary of the changes in E_{cell} as the cell operates, including the changes in $[Zn^{2+}]$, denoted [P] for [product], and $[Cu^{2+}]$, denoted [R] for [reactant].

Transparency/ARIS Presentation Center, Art Library

identify four general stages of operation. Figure 21.11A shows the first three. The main point to note is that *as the cell operates, its potential decreases:*

Stage 1. $E_{cell} > E_{cell}°$ when $Q < 1$: When the cell begins operation, $[Cu^{2+}] > [Zn^{2+}]$, so the $[(0.0592\ V/n)\ \log Q]$ term < 0 and $E_{cell} > E_{cell}°$.

As cell operation continues, $[Zn^{2+}]$ increases and $[Cu^{2+}]$ decreases; thus, Q becomes larger, the $[(0.0592\ V/n)\ \log Q]$ term becomes less negative (more positive), and E_{cell} decreases.

Stage 2. $E_{cell} = E_{cell}°$ when $Q = 1$: At the point when $[Cu^{2+}] = [Zn^{2+}]$, $Q = 1$, so the $[(0.0592\ V/n)\ \log Q]$ term $= 0$ and $E_{cell} = E_{cell}°$.

Stage 3. $E_{cell} < E_{cell}°$ when $Q > 1$: As the $[Zn^{2+}]/[Cu^{2+}]$ ratio continues to increase, the $[(0.0592\ V/n)\ \log Q]$ term > 0, so $E_{cell} < E_{cell}°$.

Stage 4. $E_{cell} = 0$ when $Q = K$: Eventually, the $[(0.0592\ V/n)\ \log Q]$ term becomes so large that it equals $E_{cell}°$, which means that E_{cell} is zero. This occurs *when the system reaches* **equilibrium:** *no more free energy is released, so the cell can do no more work.* At this point, we say that a battery is "dead."

Figure 21.11B summarizes these four key stages in the operation of a voltaic cell. Let's find K for the zinc-copper cell. At equilibrium, Equation 21.10 becomes

$$0 = E_{cell}° - \left(\frac{0.0592\ V}{n}\right) \log K, \text{ which rearranges to } E_{cell}° = \frac{0.0592\ V}{n} \log K$$

Note that this result is identical to Equation 21.8, which we obtained from $\Delta G°$. Solving for K of the zinc-copper cell ($E_{cell}° = 1.10\ V$),

$$\log K = \frac{2 \times E_{cell}°}{0.0592\ V}, \quad \text{so} \quad K = 10^{(2 \times 1.10\ V)/0.0592\ V} = 10^{37.16} = 1.4 \times 10^{37}$$

Thus, the zinc-copper cell does work until the $[Zn^{2+}]/[Cu^{2+}]$ ratio is *very* high. To conclude, let's examine cell potential in terms of the *starting Q/K* ratio:

- If $Q/K < 1$, E_{cell} is positive for the reaction *as written.* The smaller the Q/K ratio, the greater the value of E_{cell}, and the more electrical work the cell can do.
- If $Q/K = 1$, $E_{cell} = 0$. The cell is at equilibrium and can no longer do work.
- If $Q/K > 1$, E_{cell} is negative for the reaction as written. The reverse reaction will take place, and the cell will do work until Q/K equals 1 at equilibrium.

Concentration Cells

If you mix a concentrated solution of a salt with a dilute solution of the salt, the final concentration equals some intermediate value. A **concentration cell** employs this phenomenon to generate electrical energy. The two solutions are in separate half-cells, so they do not mix; rather, their concentrations become equal as the cell operates.

How a Concentration Cell Works Suppose both compartments of a voltaic cell house the Cu/Cu^{2+} half-reaction. The cell reaction is the sum of identical half-reactions, written in opposite directions, so the *standard* half-cell potentials cancel ($E°_{copper} - E°_{copper}$) and $E°_{cell}$ is zero. This occurs because standard electrode potentials are based on concentrations of 1 *M*. *In a concentration cell, however, the half-reactions are the same but the* **concentrations** *are different.* As a result, even though $E°_{cell}$ equals zero, the *nonstandard* cell potential, E_{cell}, does *not* equal zero because it depends on the *ratio of concentrations*.

In Figure 21.12A, a concentration cell has 0.10 *M* Cu^{2+} in the anode half-cell and 1.0 *M* Cu^{2+}, a 10-fold higher concentration, in the cathode half-cell:

$$Cu(s) \longrightarrow Cu^{2+}(aq; 0.10\,M) + 2e^- \qquad \text{[anode; oxidation]}$$
$$Cu^{2+}(aq; 1.0\,M) + 2e^- \longrightarrow Cu(s) \qquad \text{[cathode; reduction]}$$

The overall cell reaction is the sum of the half-reactions:

$$Cu^{2+}(aq; 1.0\,M) \longrightarrow Cu^{2+}(aq; 0.10\,M) \qquad E_{cell} = ?$$

The cell potential at the initial concentrations of 0.10 *M* (dilute) and 1.0 *M* (concentrated), with $n = 2$, is obtained from the Nernst equation:

$$E_{cell} = E°_{cell} - \frac{0.0592\,V}{2} \log \frac{[Cu^{2+}]_{dil}}{[Cu^{2+}]_{conc}} = 0\,V - \left(\frac{0.0592\,V}{2} \log \frac{0.10\,M}{1.0\,M} \right)$$

$$= 0\,V - \left[\frac{0.0592\,V}{2} (-1.00) \right] = 0.0296\,V$$

As you can see, because $E°_{cell}$ for a concentration cell equals zero, E_{cell} for nonstandard conditions depends entirely on the $[(0.0592\,V/n) \log Q]$ term.

What is actually going on as this cell operates? In the half-cell with dilute electrolyte (anode), the Cu atoms in the electrode give up electrons and become Cu^{2+} ions, which enter the solution and make it *more* concentrated. The electrons released at the anode flow to the cathode compartment. There, Cu^{2+} ions in the concentrated solution pick up the electrons and become Cu atoms, which plate out on the electrode, so that solution becomes *less* concentrated. As in any voltaic cell, E_{cell} decreases until equilibrium is attained, which happens when $[Cu^{2+}]$ is the same in both half-cells (Figure 21.12B). The same final concentration would result if we mixed the two solutions, but no electrical work would be done.

Oxidation half-reaction
$$Cu(s) \longrightarrow Cu^{2+}(aq; 0.10\,M) + 2e^-$$

Reduction half-reaction
$$Cu^{2+}(aq; 1.0\,M) + 2e^- \longrightarrow Cu(s)$$

Overall (cell) reaction
$$Cu^{2+}(aq; 1.0\,M) \longrightarrow Cu^{2+}(aq; 0.10\,M)$$

A

Figure 21.12 A concentration cell based on the Cu/Cu^{2+} half-reaction. A, The half-reactions are the same, so $E°_{cell} = 0$. The cell operates because the half-cell concentrations are different, which makes $E_{cell} > 0$ in this case. **B,** The cell operates until the half-cell concentrations are equal. Note the change in electrodes (exaggerated here for clarity) and the identical color of the solutions.

SAMPLE PROBLEM 21.7 Calculating the Potential of a Concentration Cell

PROBLEM A concentration cell consists of two Ag/Ag^+ half-cells. In half-cell A, electrode A dips into 0.010 M $AgNO_3$; in half-cell B, electrode B dips into 4.0×10^{-4} M $AgNO_3$. What is the cell potential at 298.15 K? Which electrode has a positive charge?

PLAN The standard half-cell reactions are identical, so E°_{cell} is zero, and we calculate E_{cell} from the Nernst equation. Because half-cell A has a higher $[Ag^+]$, Ag^+ ions will be reduced and plate out on electrode A. In half-cell B, Ag will be oxidized and Ag^+ ions will enter the solution. As in all voltaic cells, reduction occurs at the cathode, which is positive.

SOLUTION Writing the spontaneous reaction: The $[Ag^+]$ decreases in half-cell A and increases in half-cell B, so the spontaneous reaction is

$$Ag^+(aq;\, 0.010\, M) \text{ [half-cell A]} \longrightarrow Ag^+(aq;\, 4.0 \times 10^{-4}\, M) \text{ [half-cell B]}$$

Calculating E_{cell}, with $n = 1$:

$$E_{cell} = E^{\circ}_{cell} - \frac{0.0592\, V}{1} \log \frac{[Ag^+]_{dil}}{[Ag^+]_{conc}} = 0\, V - \left(0.0592\, V \log \frac{4.0 \times 10^{-4}}{0.010} \right)$$

$$= 0.0828\, V$$

Reduction occurs at the cathode, electrode A: $Ag^+(aq;\, 0.010\, M) + e^- \longrightarrow Ag(s)$. Thus, electrode A has a positive charge due to a relative electron deficiency.

FOLLOW-UP PROBLEM 21.7 A concentration cell is built using two Au/Au^{3+} half-cells. In half-cell A, $[Au^{3+}] = 7.0 \times 10^{-4}\, M$, and in half-cell B, $[Au^{3+}] = 2.5 \times 10^{-2}\, M$. What is E_{cell}, and which electrode is negative?

● **Concentration Cells in Your Nerve Cells** The nerve cells that mediate every thought, movement, and other bodily process function by the principle of a concentration cell. The nerve membrane is imbedded with a host of specialized enzyme "gates" that use energy from ATP hydrolysis to separate an *interior* solution of low $[Na^+]$ and high $[K^+]$ from an *exterior* solution of high $[Na^+]$ and low $[K^+]$. As a result of the differences in $[Na^+]$ and $[K^+]$, the membrane is more positive outside than inside. One-third of the body's ATP is used to create and maintain these concentration differences. (The 1997 Nobel Prize in chemistry was shared by Jens C. Skou for elucidation of this enzyme mechanism.) When the nerve membrane is stimulated, Na^+ ions spontaneously rush in, and in 0.001 s, the inside of the membrane becomes more positive than the outside. This event is followed by K^+ ions spontaneously rushing out; after another 0.001 s, the membrane is again more positive outside. These large changes in charge in one region of the membrane stimulate the neighboring region and the electrical impulse moves down the length of the cell.

Biological application

Applications of Concentration Cells Chemists, biologists, and environmental scientists apply the principle of a concentration cell in a host of applications. ● The most important is the measurement of unknown ion concentrations, particularly $[H^+]$. Suppose we construct a concentration cell based on the H_2/H^+ half-reaction, in which the cathode compartment houses the standard hydrogen electrode and the anode compartment has the same apparatus dipping into an unknown $[H^+]$ in solution. The half-reactions and overall reaction are

$$H_2(g;\, 1\, atm) \longrightarrow 2H^+(aq;\, unknown) + 2e^- \quad \text{[anode; oxidation]}$$
$$2H^+(aq;\, 1\, M) + 2e^- \longrightarrow H_2(g;\, 1\, atm) \quad \text{[cathode; reduction]}$$
$$\overline{2H^+(aq;\, 1\, M) \longrightarrow 2H^+(aq;\, unknown) \quad E_{cell} = ?}$$

As for the Cu/Cu^{2+} concentration cell, E°_{cell} is zero; however, the half-cells differ in $[H^+]$, so E_{cell} is *not* zero. From the Nernst equation, with $n = 2$, we obtain

$$E_{cell} = E^{\circ}_{cell} - \frac{0.0592\, V}{2} \log \frac{[H^+]^2_{unknown}}{[H^+]^2_{standard}}$$

Substituting 1 M for $[H^+]_{standard}$ and 0 V for E°_{cell} gives

$$E_{cell} = 0\, V - \frac{0.0592\, V}{2} \log \frac{[H^+]^2_{unknown}}{1^2} = -\frac{0.0592\, V}{2} \log [H^+]^2_{unknown}$$

Because $\log x^2 = 2 \log x$ (see Appendix A), we obtain

$$E_{cell} = -\left[\frac{0.0592\, V}{2} (2 \log [H^+]_{unknown}) \right] = -0.0592\, V \times \log [H^+]_{unknown}$$

Substituting $-\log [H^+] = pH$, we have

$$E_{cell} = 0.0592\, V \times pH$$

Thus, by measuring E_{cell}, we can find the pH.

In the routine measurement of pH, a concentration cell incorporating two hydrogen electrodes is too bulky and difficult to maintain. Instead, as was pointed out in Chapter 18, a pH meter is used. As shown in Figure 21.13A, two separate electrodes dip into the solution being tested. One of them is a *glass electrode*, which consists of an Ag/AgCl half-reaction immersed in an HCl solution of fixed concentration (usually 1.000 *M*) and enclosed by a thin (~0.05 mm) membrane made of a special glass that is highly sensitive to the presence of H$^+$ ions. The other electrode is a reference electrode, typically a *saturated calomel electrode.* It consists of a platinum wire immersed in a paste of Hg$_2$Cl$_2$ (calomel), liquid Hg, and saturated KCl solution. The glass electrode monitors the solution's [H$^+$] relative to its own fixed internal [H$^+$], and the instrument converts the potential difference between the glass and reference electrodes into a measure of pH. In modern instruments, a combination electrode is used, which houses both electrodes in one tube (Figure 21.13B).

The pH electrode is one example of an *ion-selective (or ion-specific) electrode.* Electrodes have been designed with highly specialized membranes in order to selectively measure the concentrations of many different ions in industrial, environmental, and biological samples. Recent advances allow measurement in the picomolar-to-femtomolar ($10^{-12}-10^{-15}$ *M*) range. Table 21.3 shows a few of the many ions for which ion-selective electrodes are available. ●

● **Minimicroanalysis** Electronic miniaturization has greatly broadened the applications of electrochemistry. Environmental chemists use pocket-sized pH meters in the field to study natural waters and soil. Physiologists and biochemists employ electrodes so small that they can be surgically implanted in a single biological cell to study ion channels and receptors. In the photo, a microelectrode is recording the electrical impulses of a single neuron in a monkey's visual cortex.

Table 21.3 **Some Ions Measured with Ion-Specific Electrodes**

Species Detected	Typical Sample
NH$_3$/NH$_4^+$	Industrial wastewater, seawater
CO$_2$/HCO$_3^-$	Blood, groundwater
F$^-$	Drinking water, urine, soil, industrial stack gases
Br$^-$	Grain, plant tissue
I$^-$	Milk, pharmaceuticals
NO$_3^-$	Soil, fertilizer, drinking water
K$^+$	Blood serum, soil, wine
H$^+$	Laboratory solutions, soil, natural waters

Section Summary

A spontaneous process is indicated by a negative ΔG or a positive E_{cell}, which are related: $\Delta G = -nFE_{cell}$. The ΔG of the cell reaction represents the maximum amount of electrical work the cell can do. • Because the standard free energy change, $\Delta G°$, is related to $E°_{cell}$ and to K, we can use $E°_{cell}$ to determine K. • For nonstandard conditions, the Nernst equation shows that E_{cell} depends on $E°_{cell}$ and a correction term based on Q. E_{cell} is high when Q is small (high [reactant]), and it decreases as the cell operates. At equilibrium, ΔG and E_{cell} are zero, which means that $Q = K$. • Concentration cells have identical half-reactions, but solutions of differing concentration; thus, they generate electrical energy as the concentrations become equal. • Ion-specific electrodes, such as the pH electrode, measure the concentration of one species.

Engineering application

Literature Smith, Michael J.; Vincent, Colin A. "Why Do Some Batteries Last Longer Than Others?" *J. Chem. Educ.* **2002**, *79*, 851.

Literature Smith, Michael J.; Vincent, Colin A. "Structure and Content of Some Primary Batteries," *J. Chem. Educ.* **2001**, *78*, 519.

21.5 ELECTROCHEMICAL PROCESSES IN BATTERIES

Because of their compactness and mobility, batteries have a major influence on our way of life. In industrialized countries, each person uses an average of 10 batteries per year! A **battery,** strictly speaking, is a self-contained group of voltaic cells arranged in series (plus-to-minus-to-plus, and so on), so that their individual voltages are added together. In everyday speech, however, the term may also be applied to a single voltaic cell. Batteries are ingeniously engineered devices that house rather unusual half-reactions and half-cells, but they operate through the same electrochemical principles we've been discussing. In this section, we examine the three categories of batteries—primary, secondary, and fuel cells— and note important examples, including some newer designs, of each.

Primary (Nonrechargeable) Batteries

A *primary battery* cannot be recharged, so it is discarded when the components have reached their equilibrium concentrations, that is, when the cell is "dead." We'll discuss the alkaline battery, with a brief historical reference to the dry cell, the mercury and silver "button" batteries, and the primary lithium battery.

Alkaline Battery The precursor of today's ubiquitous alkaline battery, the common dry cell, or *Leclanché cell,* was invented in the 1860s and became a familiar household item during the first three-quarters of the 20th century. The anode of the dry cell was a zinc can that housed a mixture of MnO_2 and a weakly acidic electrolyte paste, consisting of NH_4Cl, $ZnCl_2$, H_2O, and starch. Powdered graphite was used to improve conductivity, and the cathode was an inactive graphite rod. Even today, the cathode half-reaction is not completely understood, but it involved the reduction of $MnO_2(s)$ to $Mn_2O_3(s)$ and an acid-base reaction between NH_4^+ and OH^-. At a high current drain, this reaction generated ammonia gas, which could build up and cause a serious voltage drop. Moreover, because the zinc anode reacted with the acidic NH_4^+ ions, dry cells had a short shelf life.

Although more expensive than the dry cell, the alkaline battery avoids these drawbacks. The same electrode materials, zinc and manganese dioxide, are used, but the electrolyte is a paste of KOH and water. The half-reactions are essentially the same, but use of the KOH paste eliminates the NH_3 gas buildup and maintains the Zn electrode (Figure 21.14):

Anode (oxidation): $\qquad\qquad Zn(s) + 2OH^-(aq) \longrightarrow ZnO(s) + H_2O(l) + 2e^-$

Cathode (reduction): $\quad MnO_2(s) + 2H_2O(l) + 2e^- \longrightarrow Mn(OH)_2(s) + 2OH^-(aq)$

Overall (*cell*) *reaction*:

$\qquad\quad Zn(s) + MnO_2(s) + H_2O(l) \longrightarrow ZnO(s) + Mn(OH)_2(s) \qquad E_{cell} = 1.5\ V$

Like the dry cell, the alkaline battery powers portable radios, toys, flashlights, and so on, is safe, and comes in many sizes. It has no voltage drop, longer shelf life, and better performance in terms of power capability and stored energy.

Figure 21.14 Alkaline battery.

ARIS Presentation Center, Photo Library

Positive button
Steel case
MnO_2 in KOH paste
Zn (anode)
Graphite rod (cathode)
Absorbent/separator
Negative end cap

Mercury and Silver (Button) Batteries Mercury and silver batteries are quite similar. Both use a zinc container as the anode (reducing agent) in a basic medium. The mercury battery employs HgO as the oxidizing agent, the silver uses Ag_2O, and both use a steel can around the cathode. The solid reactants are compacted with KOH and separated with moist paper. The half-reactions are

Anode (oxidation):　　　　　　$Zn(s) + 2OH^-(aq) \longrightarrow ZnO(s) + H_2O(l) + 2e^-$
Cathode (reduction) (mercury): $HgO(s) + H_2O(l) + 2e^- \longrightarrow Hg(l) + 2OH^-(aq)$
Cathode (reduction) (silver):　$Ag_2O(s) + H_2O(l) + 2e^- \longrightarrow 2Ag(s) + 2OH^-(aq)$
Overall (cell) reaction (mercury):
　　　　　$Zn(s) + HgO(s) \longrightarrow ZnO(s) + Hg(l)$　　$E_{cell} = 1.3$ V
Overall (cell) reaction (silver):
　　　　　$Zn(s) + Ag_2O(s) \longrightarrow ZnO(s) + 2Ag(s)$　　$E_{cell} = 1.6$ V

Both cells are manufactured as small button-sized batteries. The mercury cell is used in calculators (Figure 21.15). Because of its very steady output, the silver cell is used in watches, cameras, heart pacemakers, and hearing aids. Their major disadvantages are toxicity of discarded mercury and high cost of silver cells.

Anode cap
Cathode can
Zn in KOH gel (anode) (−)
Gasket
Separator
Pellet of Ag_2O in graphite (cathode) (+)

Figure 21.15 Silver button battery.
ARIS Presentation Center, Art Library

Primary Lithium Batteries The primary lithium battery is widely used in watches and implanted medical devices. It offers an extremely high energy/mass ratio, producing 1 mol of e^- (1 F) from less than 7 g of metal ($\mathcal{M}$ of Li = 6.941 g/mol). The anode is lithium metal foil, which requires a nonaqueous electrolyte. The cathode is one of several metal oxides in which lithium ions lie between oxide layers. Some implantable medical devices have a silver vanadium oxide (SVO; $AgV_2O_{5.5}$) cathode and can provide power for several years, but at a low rate because energy storage is limited (Figure 21.16). The half-reactions are

Anode (oxidation):　　　　　$3.5Li(s) \longrightarrow 3.5Li^+ + 3.5e^-$
Cathode (reduction):　$AgV_2O_{5.5} + 3.5Li^+ + 3.5e^- \longrightarrow Li_{3.5}AgV_2O_{5.5}$
Overall (cell) reaction:　$AgV_2O_{5.5} + 3.5Li(s) \longrightarrow Li_{3.5}AgV_2O_{5.5}$

Secondary (Rechargeable) Batteries

In contrast to primary batteries, a *secondary,* or *rechargeable, battery* is recharged when it runs down *by supplying electrical energy to reverse the cell reaction* and re-form reactant. In other words, in this type of battery, the voltaic cells are periodically converted to electrolytic cells to restore *nonequilibrium* concentrations of the cell components. By far the most widely used secondary battery is the common car battery. Two newer types are the nickel–metal hydride battery and the lithium-ion battery, a secondary version of a lithium battery.

Lead-Acid Battery A typical lead-acid car battery has six cells connected in series, each of which delivers about 2.1 V for a total of about 12 V. Each cell contains two lead grids loaded with the electrode materials: high-surface-area (spongy) Pb in the anode and high-surface-area PbO_2 in the cathode. The grids are immersed in an electrolyte solution of ~4.5 M H_2SO_4. Fiberglass sheets between the grids prevent

ARIS Presentation Center, Photo Library

Heart pacemaker

Stainless steel case (−)
Feedthrough pin (+)
Insulative seal
Multiplate cell stack
Electrolyte fill hole
Cathode lead bridge
Polypropylene separator
Lithium (anode)　SVO (cathode)

Figure 21.16 Lithium battery.

Figure 21.17 Lead-acid battery.

Cathode (positive): lead grids filled with PbO_2

Anode (negative): similar grids filled with spongy lead

H_2SO_4 electrolyte

ARIS Presentation Center, Art Library

Literature Treptow, Richard S. "The Lead-Acid Battery: Its Voltage in Theory and in Practice," *J. Chem. Educ.* **2002**, *79*, 334.

shorting due to physical contact (Figure 21.17). When the cell discharges, it generates electrical energy as a voltaic cell:

Anode (oxidation): $\quad\quad\quad Pb(s) + HSO_4^-(aq) \longrightarrow PbSO_4(s) + H^+ + 2e^-$

Cathode (reduction):

$$PbO_2(s) + 3H^+(aq) + HSO_4^-(aq) + 2e^- \longrightarrow PbSO_4(s) + 2H_2O(l)$$

Notice that both half-reactions produce Pb^{2+} ions, one through the oxidation of Pb, the other through the reduction of PbO_2. The Pb^{2+} forms $PbSO_4(s)$ at both electrodes by reacting with HSO_4^-.

Overall (cell) reaction (discharge):

$$PbO_2(s) + Pb(s) + 2H_2SO_4(aq) \longrightarrow 2PbSO_4(s) + 2H_2O(l) \quad\quad E_{cell} = 2.1 \text{ V}$$

When the cell is recharged, it uses electrical energy as an electrolytic cell, and the half-cell and overall reactions are reversed.

Overall (cell) reaction (recharge):

$$2PbSO_4(s) + 2H_2O(l) \longrightarrow PbO_2(s) + Pb(s) + 2H_2SO_4(aq)$$

For more than a century, car and truck owners have relied on the lead-acid battery to provide the large burst of current to the starter motor needed to start the engine—and to do so for years in both hot and cold weather. Nevertheless, there are problems with the lead-acid battery, mainly loss of capacity and safety concerns. Loss of capacity arises from several factors, including corrosion of the positive (Pb) grid, detachment of active material as a result of normal mechanical bumping, and the formation of large crystals of $PbSO_4$, which make recharging more difficult. Most of the safety concerns have been remedied in modern batteries. Older batteries had a cap on each cell for monitoring electrolyte density and replacing water lost on overcharging. During recharging, some water could be electrolyzed to H_2 and O_2, which could explode if sparked, and splatter H_2SO_4. Modern batteries are sealed, so they don't require addition of water during normal operation, and they use flame attenuators to reduce the explosion hazard.

Nickel–Metal Hydride (Ni-MH) Battery

Concerns about the toxicity of the nickel-cadmium (nicad) battery are leading to its replacement by the nickel–metal hydride battery. The anode half-reaction oxidizes the hydrogen absorbed within a metal alloy (designated M; e.g., $LaNi_5$) in a basic (KOH) electrolyte, while nickel(III) in the form of NiO(OH) is reduced at the cathode (Figure 21.18):

Anode (oxidation): $\quad\quad MH(s) + OH^-(aq) \longrightarrow M(s) + H_2O(l) + e^-$

Cathode (reduction): $NiO(OH)(s) + H_2O(l) + e^- \longrightarrow Ni(OH)_2(s) + OH^-(aq)$

Overall (cell) reaction: $\quad MH(s) + NiO(OH)(s) \longrightarrow M(s) + Ni(OH)_2(s)$

$$E_{cell} = 1.4 \text{ V}$$

The cell reaction is reversed during recharging. The Ni-MH battery is common in cordless razors, camera flash units, and power tools. It is lightweight, has high power, and is nontoxic, but it may discharge excessively during long-term storage.

Figure 21.18 Nickel–metal hydride battery.

ARIS Presentation Center, Art Library

NiO(OH) (anode)

Separator

MH (cathode)

Insulator

Gasket

(−)

(+)

Heat shrink tube

Lithium-Ion Battery The secondary lithium-ion battery has an anode of Li atoms that lie between sheets of graphite (designated Li_xC_6). The cathode is a lithium metal oxide, such as $LiMn_2O_4$ or $LiCoO_2$, and a typical electrolyte is 1 M $LiPF_6$ in an organic solvent, such as dimethyl carbonate (often mixed with methylethyl carbonate). Electrons flow through the circuit, while solvated Li^+ ions flow from anode to cathode within the cell, as shown in Figure 21.19. The cell reactions are

Anode (oxidation): $\qquad\qquad\qquad Li_xC_6 \longrightarrow xLi^+ + xe^- + C_6(s)$

Cathode (reduction): $Li_{1-x}Mn_2O_4(s) + xLi^+ + xe^- \longrightarrow LiMn_2O_4(s)$

Overall (cell) reaction: $\qquad Li_xC_6 + Li_{1-x}Mn_2O_4(s) \longrightarrow LiMn_2O_4(s) + C_6(s)$

$$E_{cell} = 3.7\ V$$

The cell reaction is reversed during recharging. The lithium-ion battery powers countless laptop computers, cell phones, and camcorders. Its key drawbacks are expense and the flammability of the organic solvent.

Fuel Cells

In contrast to primary and secondary batteries, a **fuel cell,** sometimes called a *flow battery,* is not self-contained. The reactants (usually a combustible fuel and oxygen) enter the cell, and the products leave the cell, generating electricity through the controlled oxidation of the fuel. In other words, *fuel cells use combustion to produce electricity.* The fuel does not burn because, as in other batteries, the half-reactions are separated, and the electrons are transferred through an external circuit.

The most common fuel cell being developed for use in cars is the *proton exchange membrane (PEM) cell,* which uses H_2 as the fuel and has an operating temperature of around 80°C (Figure 21.20). The cell reactions are

Anode (oxidation): $\qquad\qquad 2H_2(g) \longrightarrow 4H^+(aq) + 4e^-$

Cathode (reduction): $O_2(g) + 4H^+(aq) + 4e^- \longrightarrow 2H_2O(g)$

Overall (cell) reaction: $\qquad 2H_2(g) + O_2(g) \longrightarrow 2H_2O(g)$ $\qquad E_{cell} = 1.2\ V$

The reactions in fuel cells have much lower rates than those in other batteries, so they require an *electrocatalyst* to decrease the activation energy (Section 16.8). The PEM cell electrodes are composites consisting of nanoparticles of a Pt-based catalyst deposited on graphite. These are embedded in a polymer electrolyte membrane having a perfluoroethylene backbone ($-\!\!-\!\!(F_2C\!-\!\!CF_2)_n\!\!-$) with attached sulfonic acid groups (RSO_3^-) that play a key role in ferrying protons from anode to cathode. At the anode, two H_2 molecules adsorb onto the catalyst and are split and oxidized. From each H_2, two e^- travel through the wire to the cathode, while two H^+ become hydrated and migrate through the electrolyte as H_3O^+. According to a proposed multistep mechanism, at the cathode, an O_2 molecule adsorbs onto the catalyst, which provides an e^- to form O_2^-. One H_3O^+ donates its H^+ to the O_2^-, forming HO_2 (that is, HO—O). The O—O bond stretches and breaks as another H_3O^+ gives its H^+ and the catalyst provides another e^-: the first H_2O has formed. In similar fashion, a third H^+ and e^- attach to the freed O atom to form OH, and a fourth H^+ and e^- are transferred to form the second H_2O. Both water molecules desorb and leave the cell.

Hydrogen fuel cells have been used for years to provide electricity and pure water during space flights. In the very near future, similar ones will supply electric power for transportation, residential, and commercial needs. Already, every major car manufacturer has a fuel-cell prototype. By themselves, these cells produce no pollutants, and they convert about 75% of the fuel's bond energy into

Figure 21.19 Lithium-ion battery.

ARIS Presentation Center, Art Library

Literature Treptow, Richard S. "Lithium Batteries: A Practical Application of Chemical Principles," *J. Chem. Educ.* **2003,** *80,* 1015.

Figure 21.20 Hydrogen fuel cell.

ARIS Presentation Center, Art Library

 Green chemistry application

Literature Wald, Matthew. "Questions About a Hydrogen Economy," *Scientific American* **2004**, *May*, 66.

useable power, in contrast to 40% for a coal-fired power plant and 25% for a gasoline-powered car engine. Of course, their overall environmental impact will depend on how the H_2 is obtained; for example, electrolyzing water with solar power will have a negligible impact, whereas electrolyzing it with electricity from a coal-fired plant will have a sizeable one. Despite steady progress, current fuel-cell research remains focused on lowering costs by improving membrane conductivity and developing more efficient electrocatalysts.

Section Summary

Batteries contain several voltaic cells in series and are classified as primary (e.g., alkaline, mercury, silver, and lithium), secondary (e.g., lead-acid, nickel–metal hydride, and lithium-ion), or fuel cell. • Supplying electricity to a rechargeable (secondary) battery reverses the redox reaction, forming more reactant for further use. • Fuel cells generate a current through the controlled oxidation of a fuel such as H_2.

Green chemistry application

Demo Tejada, Silvia; Guevara, Estela; Olivares, Esperanza. "Slide Projector Corrosion Cell," *J. Chem. Educ.* **1998**, *75*, 747.

21.6 CORROSION: A CASE OF ENVIRONMENTAL ELECTROCHEMISTRY

By now, you may be thinking that spontaneous electrochemical processes are always beneficial, but consider the problem of **corrosion,** the natural redox process that oxidizes metals to their oxides and sulfides. In chemical terms, corrosion is the reverse of isolating a metal from its oxide or sulfide ore; in electrochemical terms, the process shares many similarities with the operation of a voltaic cell. Damage from corrosion to cars, ships, buildings, and bridges runs into tens of billions of dollars annually, so it is a major problem in much of the world. Although we focus here on the corrosion of iron, many other metals, such as copper and silver, also corrode.

The Corrosion of Iron

The most common and economically destructive form of corrosion is the rusting of iron. About 25% of the steel produced in the United States is made just to replace steel already in use that has corroded. Contrary to the simplified equation presented earlier, rust is *not* a direct product of the reaction between iron and oxygen but arises through a complex electrochemical process. Let's look at the facts of iron corrosion and then use the features of a voltaic cell to explain them:

1. Iron does not rust in dry air: moisture must be present.
2. Iron does not rust in air-free water: oxygen must be present.
3. The loss of iron and the depositing of rust often occur at *different* places on the *same* object.
4. Iron rusts more quickly at low pH (high $[H^+]$).
5. Iron rusts more quickly in contact with ionic solutions.
6. Iron rusts more quickly in contact with a less active metal (such as Cu) and more slowly in contact with a more active metal (such as Zn).

Picture the magnified surface of a piece of iron or steel (Figure 21.21). Strains, ridges, and dents in contact with water are typically the sites of iron loss (fact 1). These sites are called *anodic regions* because the following half-reaction occurs there:

$$Fe(s) \longrightarrow Fe^{2+}(aq) + 2e^- \qquad \text{[anodic region; oxidation]}$$

Once the iron atoms lose electrons, the damage to the object has been done, and a pit forms where the iron is lost.

The freed electrons move through the external circuit—the piece of iron itself—until they reach a region of relatively high O_2 concentration (fact 2), near

A   **B**

Figure 21.21 The corrosion of iron. A, A close-up view of an iron surface. Corrosion usually occurs at a surface irregularity. **B,** A sche-matic depiction of a small area of the surface, showing the steps in the corrosion process.

Transparency/ARIS Presentation Center, Art Library

the surface of a surrounding water droplet, for instance. At this *cathodic region,* the electrons released from the iron atoms reduce O_2 molecules:

$$O_2(g) + 4H^+(aq) + 4e^- \longrightarrow 2H_2O(l) \qquad \text{[cathodic region; reduction]}$$

Notice that this overall redox process is complete; thus, the iron loss has occurred without any rust forming:

$$2Fe(s) + O_2(g) + 4H^+(aq) \longrightarrow 2Fe^{2+}(aq) + 2H_2O(l)$$

Rust forms through another redox reaction in which the reactants make direct contact. The Fe^{2+} ions formed originally at the anodic region disperse through the surrounding water and react with O_2, often at some distance from the pit (fact 3). The overall reaction for this step is

$$2Fe^{2+}(aq) + \tfrac{1}{2}O_2(g) + (2+n)H_2O(l) \longrightarrow Fe_2O_3 \cdot nH_2O(s) + 4H^+(aq)$$

[The inexact coefficient n for H_2O in the above equation appears because rust, $Fe_2O_3 \cdot nH_2O$, is a form of iron(III) oxide with a variable number of waters of hydration.] The rust deposit is really incidental to the damage caused by loss of iron—a chemical insult added to the original injury.

Adding the previous two equations together shows the overall equation for the rusting of iron:

$$2Fe(s) + \tfrac{3}{2}O_2(g) + nH_2O(l) + \cancel{4H^+(aq)} \longrightarrow Fe_2O_3 \cdot nH_2O(s) + \cancel{4H^+(aq)}$$

The canceled H^+ ions are shown to emphasize that they act as a catalyst; that is, they speed the process as they are used up in one step of the overall reaction and created in another. As a result of this action, rusting is faster at low pH (high $[H^+]$) (fact 4). Ionic solutions speed rusting by improving the conductivity of the aqueous medium near the anodic and cathodic regions (fact 5). The effect of ions is especially evident on ocean-going vessels (Figure 21.22) and on the under-bodies and around the wheel wells of cars driven in cold climates, where salts are used to melt ice on slippery roads.

In many ways, the components of the corrosion process resemble those of a voltaic cell:

- Anodic and cathodic regions are separated in space.
- The regions are connected via an external circuit through which the electrons travel.
- In the anodic region, iron behaves like an active electrode, whereas in the cathodic region, it is inactive.
- The moisture surrounding the pit functions somewhat like a salt bridge, a means for ions to ferry back and forth and keep the solution neutral.

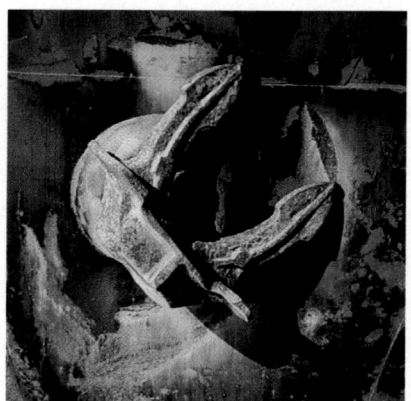

Figure 21.22 Enhanced corrosion at sea. The high ion concentration of seawater leads to its high conductivity, which en-hances the corrosion of iron in the hulls and anchors of ocean-going vessels.

ARIS Presentation Center, Photo Library

Figure 21.23 The effect of metal-metal contact on the corrosion of iron. A, When iron is in contact with a less active metal, such as copper, the iron loses electrons more readily (is more anodic), so it corrodes faster. **B,** When iron is in contact with a more active metal, such as zinc, the zinc acts as the anode and loses electrons. Therefore, the iron is cathodic, so it does not corrode. The process is known as *cathodic protection.*

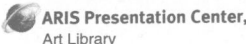 ARIS Presentation Center, Art Library

② *Electrons from Fe reduce O_2 to H_2O*

Water droplet

$2H_2O$ $2Fe^{2+}$

$4H^+$ O_2

$4e^-$ Cu (inactive cathode)

2Fe (anode)

① *Fe gives up electrons to Cu cathode*

A Enhanced corrosion

② *Electrons from Zn reduce O_2 to H_2O*

$2H_2O$
$2Zn^{2+}$

$4H^+$ O_2

2Zn (anode)

$4e^-$

Fe (inactive cathode)

① *Zn gives up electrons to Fe cathode*

B Cathodic protection

Figure 21.24 The use of sacrificial anodes to prevent iron corrosion.
In cathodic protection, an active metal, such as magnesium or aluminum, is connected to underground iron pipes to prevent their corrosion. The active metal is sacrificed instead of the iron.

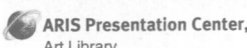 ARIS Presentation Center, Art Library

Protecting Against the Corrosion of Iron

A common approach to preventing or limiting corrosion is to eliminate contact with the corrosive factors. The simple act of washing off road salt removes the ionic solution from auto bodies. Iron objects are frequently painted to keep out O_2 and moisture, but if the paint layer chips, rusting proceeds. More permanent coatings include chromium plated on plumbing fixtures. In the "blueing" of gun barrels, wood stoves, and other steel objects, an adherent coating of Fe_3O_4 (magnetite) is bonded to the surface.

The only fact regarding corrosion that we have not yet addressed concerns the relative activity of other metals in contact with iron (fact 6), which leads to the most effective way to prevent corrosion. The essential idea is that *iron functions as both anode and cathode in the rusting process, but it is lost only at the anode.* Therefore, anything that makes iron behave more like the anode increases corrosion. As you can see in Figure 21.23A, when iron is in contact with a *less* active metal (weaker reducing agent), such as copper, its anodic function is enhanced. As a result, when iron plumbing is connected directly to copper plumbing, the iron pipe corrodes rapidly.

On the other hand, anything that makes iron behave more like the cathode prevents corrosion. In *cathodic protection,* iron makes contact with a *more* active metal (stronger reducing agent), such as zinc. The iron becomes cathodic and remains intact, while the zinc acts as the anode and loses electrons (Figure 21.23B). Coating steel with a "sacrificial" layer of zinc is the basis of the *galvanizing* process. In addition to blocking physical contact with H_2O and O_2, the zinc is "sacrificed" (oxidized) instead of the iron.

Sacrificial anodes are employed to protect iron and steel structures (pipes, tanks, oil rigs, and so on) in marine and moist underground environments. The metals that are most frequently used for this purpose are magnesium and aluminum because they are much more active than iron. As a result, they act as the anode while iron acts as the cathode (Figure 21.24). Another advantage of these metals is that they form adherent oxide coatings, which slows their own corrosion.

Section Summary

Corrosion damages metal structures through a natural electrochemical change. • Iron corrosion occurs in the presence of oxygen and moisture and is increased by high $[H^+]$, high [ion], or contact with a less active metal, such as Cu. Fe is oxidized and O_2 is reduced in one redox reaction, while rust (hydrated form of Fe_2O_3) is formed in another reaction that often takes place at a different location. Because Fe functions as both anode and cathode in the process, an iron or steel object can be protected by physically covering its surface or joining it to a more active metal (such as Zn, Mg, or Al), which acts as the anode in place of the Fe.

21.7 ELECTROLYTIC CELLS: USING ELECTRICAL ENERGY TO DRIVE NONSPONTANEOUS REACTIONS

Up to now, we've been considering voltaic cells, those that generate electrical energy from a spontaneous redox reaction. The principle of an electrolytic cell is exactly the opposite: *electrical energy from an external source drives a nonspontaneous reaction.*

Construction and Operation of an Electrolytic Cell

Let's examine the operation of an electrolytic cell by constructing one from a voltaic cell. Consider the tin-copper voltaic cell in Figure 21.25A. The Sn anode will gradually become oxidized to Sn^{2+} ions, and the Cu^{2+} ions will gradually be reduced and plate out on the Cu cathode because the cell reaction is spontaneous in that direction:

For the voltaic cell

$$Sn(s) \longrightarrow Sn^{2+}(aq) + 2e^- \quad \text{[anode; oxidation]}$$
$$Cu^{2+}(aq) + 2e^- \longrightarrow Cu(s) \quad \text{[cathode; reduction]}$$
$$Sn(s) + Cu^{2+}(aq) \longrightarrow Sn^{2+}(aq) + Cu(s) \quad E°_{cell} = 0.48 \text{ V and } \Delta G° = -93 \text{ kJ}$$

Therefore, the *reverse* cell reaction is *non*spontaneous and never happens of its own accord, as the negative $E°_{cell}$ and positive $\Delta G°$ indicate:

$$Cu(s) + Sn^{2+}(aq) \longrightarrow Cu^{2+}(aq) + Sn(s) \quad E°_{cell} = -0.48 \text{ V and } \Delta G° = 93 \text{ kJ}$$

However, we can make this process happen by supplying from an external source an electric potential *greater than* $E°_{cell}$. In effect, we have converted the voltaic cell into an electrolytic cell and changed the nature of the electrodes—anode is now cathode, and cathode is now anode (Figure 21.25B):

For the electrolytic cell

$$Cu(s) \longrightarrow Cu^{2+}(aq) + 2e^- \quad \text{[anode; oxidation]}$$
$$Sn^{2+}(aq) + 2e^- \longrightarrow Sn(s) \quad \text{[cathode; reduction]}$$
$$Cu(s) + Sn^{2+}(aq) \longrightarrow Cu^{2+}(aq) + Sn(s) \quad \text{[overall (cell) reaction]}$$

Demo Shakhashiri, Bassam Z. "Electrolytic Cells in Series: A Red, White, and Blue Electrolysis," *Chemical Demonstrations, A Handbook for Teachers of Chemistry*, Vol. 4 (The University of Wisconsin Press, Madison, 1992), 170–173.

A Voltaic cell

B Electrolytic cell

Figure 21.25 The tin-copper reaction as the basis of a voltaic and an electrolytic cell. A, At standard conditions, the spontaneous reaction between Sn and Cu^{2+} generates 0.48 V in a voltaic cell. **B,** If more than 0.48 V is supplied, the same apparatus and components become an electrolytic cell, and the nonspontaneous reaction between Cu and Sn^{2+} occurs. Note the changes in electrode charges and direction of electron flow.

Transparency/ARIS Presentation Center, Active Art Library

Note that in an electrolytic cell, as in a voltaic cell, *oxidation takes place at the anode and reduction takes place at the cathode, but the direction of electron flow and the signs of the electrodes are reversed.*

To understand these changes, keep in mind the *cause* of the electron flow:

- In a voltaic cell, electrons are generated at the anode, so it is negative, and electrons are consumed at the cathode, so it is positive.
- In an electrolytic cell, the electrons come from the external power source, which *supplies* them *to* the cathode, so it is negative, and *removes* them *from* the anode, so it is positive.

A rechargeable battery functions as a voltaic cell when it is discharging and as an electrolytic cell when it is recharging, so it provides a good way to compare these two cell types and the changes in the processes and charges of the electrodes. Figure 21.26 shows these changes in the lead-acid battery. In the discharge mode (voltaic cell), oxidation occurs at electrode I, thus making the *negative* electrode the anode. In the recharge mode (electrolytic cell), oxidation occurs at electrode II, thus making the *positive* electrode the anode. Similarly, the cathode is *positive* during discharge (electrode II) and *negative* during recharge (electrode I). To reiterate, regardless of whether the cell is discharging or recharging, oxidation occurs at the anode and reduction occurs at the cathode.

Engineering application

Transparency/ARIS Presentation Center, Art Library

Figure 21.26 The processes occurring during the discharge and recharge of a lead-acid battery. When the lead-acid battery is discharging (*top*), it behaves like a voltaic cell: the anode is negative (electrode I), and the cathode is positive (electrode II). When it is recharging (*bottom*), it behaves like an electrolytic cell: the anode is positive (electrode II), and the cathode is negative (electrode I).

Table 21.4 Comparison of Voltaic and Electrolytic Cells

Cell Type	ΔG	E_{cell}	Electrode		
			Name	Process	Sign
Voltaic	<0	>0	Anode	Oxidation	−
Voltaic	<0	>0	Cathode	Reduction	+
Electrolytic	>0	<0	Anode	Oxidation	+
Electrolytic	>0	<0	Cathode	Reduction	−

Table 21.4 summarizes the processes and signs in the two types of electrochemical cells.

Predicting the Products of Electrolysis

Electrolysis, the splitting (lysing) of a substance by the input of electrical energy, is often used to decompose a compound into its elements. Electrolytic cells are involved in key industrial production steps for some of the most commercially important elements, including chlorine, aluminum, and copper, as you'll see in Chapter 22. The first laboratory electrolysis of H_2O to H_2 and O_2 was performed in 1800, and the process is still used to produce these gases in ultrahigh purity. The electrolyte in an electrolytic cell can be the pure compound (such as H_2O or a molten salt), a mixture of molten salts, or an aqueous solution of a salt. The products obtained depend on atomic properties and several other factors, so let's examine some actual cases.

Electrolysis of Pure Molten Salts Many electrolytic applications involve isolating a metal or nonmetal from a molten salt. Predicting the product at each electrode is simple if the salt is pure because *the cation will be reduced and the anion oxidized.* The electrolyte is the molten salt itself, and the ions move through the cell because they are attracted by the oppositely charged electrodes.

Consider the electrolysis of molten (fused) calcium chloride. The two species present are Ca^{2+} and Cl^-, so Ca^{2+} ion is reduced and Cl^- ion is oxidized:

$$2Cl^-(l) \longrightarrow Cl_2(g) + 2e^- \qquad \text{[anode; oxidation]}$$
$$\underline{Ca^{2+}(l) + 2e^- \longrightarrow Ca(s) \qquad \text{[cathode; reduction]}}$$
$$Ca^{2+}(l) + 2Cl^-(l) \longrightarrow Ca(s) + Cl_2(g) \qquad \text{[overall]}$$

Metallic calcium is prepared industrially this way, as are several other active metals, such as Na and Mg, and the halogens Cl_2 and Br_2. We examine the details of these and several other electrolytic processes in Chapter 22.

Electrolysis of Mixed Molten Salts More typically, the electrolyte is a mixture of molten salts, which is then electrolyzed to obtain a particular metal. When we have a choice of product, how can we tell which species will react at which electrode? The general rule for all electrolytic cells is that *the more easily oxidized species (stronger reducing agent) reacts at the anode, and the more easily reduced species (stronger oxidizing agent) reacts at the cathode.*

It's important to realize that for electrolysis of mixtures of molten salts, we *cannot* use tabulated $E°$ values to tell the relative strength of the oxidizing and reducing agents. Those values refer to the *change from aqueous ion to free element,* $M^{n+}(aq) + ne^- \longrightarrow M(s)$, under standard-state conditions, but there are no aqueous ions in a molten salt. Instead, we rely on our knowledge of periodic atomic trends to predict which of the ions present gains or loses electrons more easily (Sections 8.4 and 9.5).

Demo Summerlin, L.R.; Borgford, C.L.; Ealy, J.B. "Electroplating Copper," *Chemical Demonstrations, A Sourcebook for Teachers,* Vol. 2 (American Chemical Society, Washington, DC, 1988), 199–200.

ARIS Presentation Center, Worked Examples Library

SAMPLE PROBLEM 21.8 Predicting the Electrolysis Products of a Molten Salt Mixture

PROBLEM A chemical engineer melts a naturally occurring mixture of NaBr and $MgCl_2$ and decomposes it in an electrolytic cell. Predict the substance formed at each electrode, and write balanced half-reactions and the overall cell reaction.

PLAN We have to determine which metal and nonmetal will form more easily at the electrodes. We first list the ions as oxidizing or reducing agents. If one metal holds its electrons more tightly than another, it has a higher ionization energy (IE). Therefore, as a cation, it gains electrons more easily; it is the stronger oxidizing agent and is reduced at the cathode. Similarly, if one nonmetal holds its electrons less tightly than another, it has a lower electronegativity (EN). Therefore, as an anion, it loses electrons more easily; it is the stronger reducing agent and is oxidized at the anode.

SOLUTION Listing the ions as oxidizing or reducing agents:

The possible oxidizing agents are Na^+ and Mg^{2+}.

The possible reducing agents are Br^- and Cl^-.

Determining the cathode product (more easily reduced cation): Mg is to the right of Na in Period 3. IE increases from left to right, so Mg has a higher IE. It takes more energy to remove an e^- from Mg than from Na, so it follows that Mg^{2+} has a greater attraction for e^- and thus is more easily reduced (stronger oxidizing agent):

$$Mg^{2+}(l) + 2e^- \longrightarrow Mg(l) \qquad \text{[cathode; reduction]}$$

Determining the anode product (more easily oxidized anion): Br is below Cl in Group 7A(17). EN decreases down the group, so Br has a lower EN than Cl. Therefore, it follows that Br^- holds its e^- less tightly than Cl^-, so Br^- is more easily oxidized (stronger reducing agent):

$$2Br^-(l) \longrightarrow Br_2(g) + 2e^- \qquad \text{[anode; oxidation]}$$

Writing the overall cell reaction:

$$Mg^{2+}(l) + 2Br^-(l) \longrightarrow Mg(l) + Br_2(g) \qquad \text{[overall]}$$

COMMENT The cell temperature must be high enough to keep the salt mixture molten. In this case, the temperature is greater than the melting point of Mg, so it appears as a liquid in the equation, and greater than the boiling point of Br_2, so it appears as a gas.

FOLLOW-UP PROBLEM 21.8 A sample of $AlBr_3$ contaminated with KF is melted and electrolyzed. Determine the electrode products and write the overall cell reaction.

Oxidation half-reaction
$2H_2O(l) \longrightarrow O_2(g) + 4H^+(aq) + 4e^-$

Reduction half-reaction
$2H_2O(l) + 2e^- \longrightarrow H_2(g) + 2OH^-(aq)$

Overall (cell) reaction
$2H_2O(l) \longrightarrow 2H_2(g) + O_2(g)$

Figure 21.27 The electrolysis of water. A certain volume of oxygen forms through the oxidation of H_2O at the anode (*right*), and twice that volume of hydrogen forms through the reduction of H_2O at the cathode (*left*).

ARIS Presentation Center, Photo Library

Electrolysis of Water and Nonstandard Half-Cell Potentials
Before we can analyze the electrolysis products of aqueous salt solutions, we must examine the electrolysis of water itself. Extremely pure water is difficult to electrolyze because very few ions are present to conduct a current. If we add a small amount of a salt that cannot be electrolyzed in water (such as Na_2SO_4), however, electrolysis proceeds rapidly. A glass electrolytic cell with separated gas compartments is used to keep the H_2 and O_2 gases from mixing (Figure 21.27). At the anode, water is oxidized as the O.N. of O changes from -2 to 0:

$$2H_2O(l) \longrightarrow O_2(g) + 4H^+(aq) + 4e^- \qquad E = 0.82 \text{ V} \qquad \text{[anode; oxidation]}$$

At the cathode, water is reduced as the O.N. of H changes from $+1$ to 0:

$$2H_2O(l) + 2e^- \longrightarrow H_2(g) + 2OH^-(aq) \qquad E = -0.42 \text{ V} \qquad \text{[cathode; reduction]}$$

Doubling the cathode half-reaction to equate e^- loss and gain, adding the half-reactions (which involves combining the H^+ and OH^- into H_2O and canceling e^- and excess H_2O), and calculating E_{cell} gives the overall reaction:

$$2H_2O(l) \longrightarrow 2H_2(g) + O_2(g) \qquad E_{cell} = -0.42 \text{ V} - 0.82 \text{ V} = -1.24 \text{ V} \qquad \text{[overall]}$$

Notice that these electrode potentials are not written with a standard-state symbol (degree sign) because they are *not* standard electrode potentials. The $[H^+]$

and $[OH^-]$ are 1.0×10^{-7} M rather than the standard-state value of 1 M. These E values are obtained by applying the Nernst equation. For example, the calculation for the anode potential (with $n = 4$) is

$$E_{cell} = E°_{cell} - \frac{0.0592 \text{ V}}{4} \log(P_{O_2} \times [H^+]^4)$$

The standard potential for the *oxidation* of water is -1.23 V (from Appendix D) and $P_{O_2} \approx 1$ atm in the half-cell, so we have

$$E_{cell} = -1.23 \text{ V} - \left\{ \frac{0.0592 \text{ V}}{4} \times [\log 1 + 4 \log(1.0 \times 10^{-7})] \right\} = -0.82 \text{ V}$$

In aqueous ionic solutions, $[H^+]$ and $[OH^-]$ are approximately 10^{-7} M also, so we use these nonstandard E_{cell} values to predict electrode products.

Electrolysis of Aqueous Salt Solutions and the Phenomenon of Overvoltage

Aqueous salt solutions are mixtures of ions *and* water, so we have to compare the various electrode potentials to predict the electrode products. When two half-reactions are possible at an electrode,

- *The reduction with the less negative (more positive) electrode potential occurs.*
- *The oxidation with the less positive (more negative) electrode potential occurs.*

What happens, for instance, when a solution of potassium iodide is electrolyzed? The possible oxidizing agents are K^+ and H_2O, and their reduction half-reactions are

$$K^+(aq) + e^- \longrightarrow K(s) \qquad\qquad E° = -2.93 \text{ V}$$
$$2H_2O(l) + 2e^- \longrightarrow H_2(g) + 2OH^-(aq) \qquad E = -0.42 \text{ V} \qquad \text{[reduction]}$$

The less *negative* electrode potential for water means that it is much easier to reduce than K^+, so H_2 forms at the cathode. The possible reducing agents are I^- and H_2O, and their oxidation half-reactions are

$$2I^-(aq) \longrightarrow I_2(s) + 2e^- \qquad\qquad E° = 0.53 \text{ V} \qquad \text{[oxidation]}$$
$$2H_2O(l) \longrightarrow O_2(g) + 4H^+(aq) + 4e^- \qquad E = 0.82 \text{ V}$$

The less *positive* electrode potential for I^- means that a lower potential is needed to oxidize it than to oxidize H_2O, so I_2 forms at the anode.

However, the products predicted from this type of comparison of electrode potentials are not always the actual products. For gases such as $H_2(g)$ and $O_2(g)$ to be produced at metal electrodes, an additional voltage is required. This increment above the expected voltage is called the **overvoltage**, and it is 0.4 to 0.6 V for these gases. The overvoltage results from kinetic factors, such as the large activation energy (Section 16.5) required for gases to form at the electrode.

Overvoltage has major practical significance. A multibillion-dollar example is the industrial production of chlorine from concentrated NaCl solution. Water is easier to reduce than Na^+, so H_2 forms at the cathode, even *with* an overvoltage of 0.6 V:

$$Na^+(aq) + e^- \longrightarrow Na(s) \qquad\qquad E° = -2.71 \text{ V}$$
$$2H_2O(l) + 2e^- \longrightarrow H_2(g) + 2OH^-(aq) \qquad E = -0.42 \text{ V } (\approx -1 \text{ V with overvoltage})$$
$$\text{[reduction]}$$

But Cl_2 *does* form at the anode, even though the electrode potentials themselves would lead us to predict that O_2 should form:

$$2H_2O(l) \longrightarrow O_2(g) + 4H^+(aq) + 4e^- \qquad E = 0.82 \text{ V } (\sim 1.4 \text{ V with overvoltage})$$
$$2Cl^-(aq) \longrightarrow Cl_2(g) + 2e^- \qquad\qquad E° = 1.36 \text{ V} \qquad \text{[oxidation]}$$

An overvoltage of ~ 0.6 V makes the potential needed to form O_2 slightly above that for Cl_2. Keeping the $[Cl^-]$ high also favors Cl_2 formation. Thus, chlorine, which is one of the 10 most heavily produced industrial chemicals, can be formed from plentiful natural sources of aqueous sodium chloride (see Chapter 22).

From these and other examples, we can determine which elements can be prepared electrolytically from aqueous solutions of their salts:

1. Cations of less active metals *are* reduced to the metal, including gold, silver, copper, chromium, platinum, and cadmium.
2. Cations of more active metals *are not* reduced, including those in Groups 1A(1) and 2A(2), and Al from 3A(13). Water is reduced to H_2 and OH^- instead.
3. Anions that *are* oxidized, because of overvoltage from O_2 formation, include the halides ($[Cl^-]$ must be high), except for F^-.
4. Anions that *are not* oxidized include F^- and common oxoanions, such as SO_4^{2-}, CO_3^{2-}, NO_3^-, and PO_4^{3-}, because the central nonmetal in these oxoanions is already in its highest oxidation state. Water is oxidized to O_2 and H^+ instead.

SAMPLE PROBLEM 21.9 Predicting the Electrolysis Products of Aqueous Salt Solutions

PROBLEM What products form during electrolysis of aqueous solutions of the following salts: **(a)** KBr; **(b)** $AgNO_3$; **(c)** $MgSO_4$?

PLAN We identify the reacting ions and compare their electrode potentials with those of water, taking the 0.4 to 0.6 V overvoltage into consideration. The reduction half-reaction with the less negative electrode potential occurs at the cathode, and the oxidation half-reaction with the less positive electrode potential occurs at the anode.

SOLUTION

(a)
$$K^+(aq) + e^- \longrightarrow K(s) \qquad\qquad E° = -2.93 \text{ V}$$
$$2H_2O(l) + 2e^- \longrightarrow H_2(g) + 2OH^-(aq) \qquad E = -0.42 \text{ V}$$

Despite the overvoltage, which makes E for the reduction of water between -0.8 and -1.0 V, H_2O is still easier to reduce than K^+, so $H_2(g)$ forms at the cathode.

$$2Br^-(aq) \longrightarrow Br_2(l) + 2e^- \qquad\qquad E° = 1.07 \text{ V}$$
$$2H_2O(l) \longrightarrow O_2(g) + 4H^+(aq) + 4e^- \qquad E = 0.82 \text{ V}$$

Because of the overvoltage, which makes E for the oxidation of water between 1.2 and 1.4 V, Br^- is easier to oxidize than water, so $Br_2(l)$ forms at the anode (see photo).

(b)
$$Ag^+(aq) + e^- \longrightarrow Ag(s) \qquad\qquad E° = 0.80 \text{ V}$$
$$2H_2O(l) + 2e^- \longrightarrow H_2(g) + 2OH^-(aq) \qquad E = -0.42 \text{ V}$$

As the cation of an inactive metal, Ag^+ is a better oxidizing agent than H_2O, so $Ag(s)$ forms at the cathode. NO_3^- cannot be oxidized, because N is already in its highest ($+5$) oxidation state. Thus, $O_2(g)$ forms at the anode:

$$2H_2O(l) \longrightarrow O_2(g) + 4H^+(aq) + 4e^-$$

(c)
$$Mg^{2+}(aq) + 2e^- \longrightarrow Mg(s) \qquad E° = -2.37 \text{ V}$$

Like K^+ in part (a), Mg^{2+} cannot be reduced in the presence of water, so $H_2(g)$ forms at the cathode. The SO_4^{2-} ion cannot be oxidized because S is in its highest ($+6$) oxidation state. Thus, H_2O is oxidized, and $O_2(g)$ forms at the anode:

$$2H_2O(l) \longrightarrow O_2(g) + 4H^+(aq) + 4e^-$$

FOLLOW-UP PROBLEM 21.9 Write half-reactions showing the products you predict will form in the electrolysis of aqueous $AuBr_3$.

Electrolysis of aqueous KBr.

The Stoichiometry of Electrolysis: The Relation Between Amounts of Charge and Product

As you've seen, the charge flowing through an electrolytic cell yields products at the electrodes. In the electrolysis of molten NaCl, for example, the power source supplies electrons to the cathode, where Na^+ ions migrate to pick them up and become Na metal. At the same time, the power source pulls from the anode the electrons that Cl^- ions release as they become Cl_2 gas. It follows that the more electrons picked up by Na^+ ions and released by Cl^- ions, the greater the amounts of Na and Cl_2 that form. This relationship was first determined experimentally by Michael Faraday and is referred to as *Faraday's law of electrolysis: the amount of substance produced at each electrode is directly proportional to the quantity of charge flowing through the cell.* ●

Each balanced half-reaction shows the amounts (mol) of reactant, electrons, and product involved in the change, so it contains the information we need to answer such questions as "How much material will form as a result of a given quantity of charge?" or, conversely, "How much charge is needed to produce a given amount of material?" To apply Faraday's law,

1. Balance the half-reaction to find the number of moles of electrons needed per mole of product.
2. Use the Faraday constant ($F = 9.65 \times 10^4$ C/mol e^-) to find the corresponding charge.
3. Use the molar mass to find the charge needed for a given mass of product.

In practice, to supply the correct amount of electricity, we need some means of finding the charge flowing through the cell. We cannot measure charge directly, but we *can* measure current, the charge flowing per unit time. The SI unit of current is the **ampere (A),** which is defined as a charge of 1 coulomb flowing through a conductor in 1 second:

$$1 \text{ ampere} = 1 \text{ coulomb/second} \quad \text{or} \quad 1 \text{ A} = 1 \text{ C/s} \quad (21.11)$$

Thus, the current multiplied by the time gives the charge:

$$\text{Current} \times \text{time} = \text{charge} \quad \text{or} \quad A \times s = \frac{C}{s} \times s = C$$

Therefore, we find the charge by measuring the current *and* the time during which the current flows. This, in turn, relates to the amount of product formed. Figure 21.28 summarizes these relationships.

● **The Father of Electrochemistry and Much More** The investigations by Michael Faraday (1791–1867) into the mass of an element that is equivalent to a given amount of charge established electrochemistry as a quantitative science, but his breakthroughs in physics are even more celebrated. In 1821, Faraday developed the precursor of the electric motor, and his later studies of currents induced by electric and magnetic fields eventually led to the electric generator and the transformer. After his death, this self-educated blacksmith's son was rated by Albert Einstein as the peer of Newton, Galileo, and Maxwell. He is shown here delivering one of his famous lectures, "Chemical History of a Candle."

ARIS Presentation Center, Photo Library

MASS (g) of substance oxidized or reduced		AMOUNT (mol) of substance oxidized or reduced		AMOUNT (mol) of electrons transferred		CHARGE (C)		CURRENT (A)
	$\mathcal{M}$ (g/mol)		balanced half-reaction		Faraday constant (C/mol e^-)		time (s)	

Figure 21.28 A summary diagram for the stoichiometry of electrolysis.

Transparency/ARIS Presentation Center, Art Library

Problems based on Faraday's law often ask you to calculate current, mass of material, or time. The electrode half-reaction provides the key to solving these problems because it is related to the mass for a certain quantity of charge.

As an example, let's consider a typical problem in practical electrolysis: how long does it take to produce 3.0 g of $Cl_2(g)$ by electrolysis of aqueous NaCl using a power supply with a current of 12 A? The problem asks for the time needed to produce a certain mass, so let's first relate mass to number of moles of electrons to find the charge needed. Then, we'll relate the charge to the current to find the time.

We know the mass of Cl_2 produced, so we can find the amount (mol) of Cl_2. The half-reaction tells us that the loss of 2 mol of electrons produces 1 mol of chlorine gas:

$$2Cl^-(aq) \longrightarrow Cl_2(g) + 2e^-$$

We use this relationship as a conversion factor, and multiplying by the Faraday constant gives us the total charge:

$$\text{Charge (C)} = 3.0 \text{ g Cl}_2 \times \frac{1 \text{ mol Cl}_2}{70.90 \text{ g Cl}_2} \times \frac{2 \text{ mol e}^-}{1 \text{ mol Cl}_2} \times \frac{9.65 \times 10^4 \text{ C}}{1 \text{ mol e}^-} = 8.2 \times 10^3 \text{ C}$$

Now we use the relationship between charge and current to find the time needed:

$$\text{Time (s)} = \frac{\text{charge(C)}}{\text{current (A, or C/s)}} = 8.2 \times 10^3 \text{ C} \times \frac{1 \text{ s}}{12 \text{ C}} = 6.8 \times 10^2 \text{ s } (\sim 11 \text{ min})$$

Note that the entire calculation follows Figure 21.28 until the last step:

$$\text{grams of Cl}_2 \Rightarrow \text{moles of Cl}_2 \Rightarrow \text{moles of e}^- \Rightarrow \text{coulombs} \Rightarrow \text{seconds}$$

Sample Problem 21.10 demonstrates the steps as they appear in Figure 21.28.

SAMPLE PROBLEM 21.10   Applying the Relationship Among Current, Time, and Amount of Substance

PROBLEM A technician is plating a faucet with 0.86 g of chromium from an electrolytic bath containing aqueous $Cr_2(SO_4)_3$. If 12.5 min is allowed for the plating, what current is needed?

PLAN To find the current, we divide the charge by the time; therefore, we need to find the charge. First, we write the half-reaction for Cr^{3+} reduction. From it, we know the number of moles of e^- required per mole of Cr. As the roadmap shows, to find the charge, we convert the mass of Cr needed (0.86 g) to amount (mol) of Cr. The balanced half-reaction gives the amount (mol) of e^- transferred. Then, we use the Faraday constant (9.65×10^4 C/mol e^-) to find the charge and divide by the time (12.5 min, converted to seconds) to obtain the current.

SOLUTION Writing the balanced half-reaction:

$$Cr^{3+}(aq) + 3e^- \longrightarrow Cr(s)$$

Combining steps to find amount (mol) of e^- transferred for mass of Cr needed:

$$\text{Moles of e}^- \text{ transferred} = 0.86 \text{ g Cr} \times \frac{1 \text{ mol Cr}}{52.00 \text{ g Cr}} \times \frac{3 \text{ mol e}^-}{1 \text{ mol Cr}} = 0.050 \text{ mol e}^-$$

Calculating the charge:

$$\text{Charge (C)} = 0.050 \text{ mol e}^- \times \frac{9.65 \times 10^4 \text{ C}}{1 \text{ mol e}^-} = 4.8 \times 10^3 \text{ C}$$

Calculating the current:

$$\text{Current (A)} = \frac{\text{charge (C)}}{\text{time (s)}} = \frac{4.8 \times 10^3 \text{ C}}{12.5 \text{ min}} \times \frac{1 \text{ min}}{60 \text{ s}} = 6.4 \text{ C/s} = 6.4 \text{ A}$$

CHECK Rounding gives

$$(\sim 0.9 \text{ g})(1 \text{ mol Cr}/50 \text{ g})(3 \text{ mol e}^-/1 \text{ mol Cr}) = 5 \times 10^{-2} \text{ mol e}^-$$

then $(5 \times 10^{-2} \text{ mol e}^-)(\sim 1 \times 10^5 \text{ C/mol e}^-) = 5 \times 10^3 \text{ C}$

and $(5 \times 10^3 \text{ C}/12 \text{ min})(1 \text{ min}/60 \text{ s}) = 7 \text{ A}$

COMMENT For the sake of introducing Faraday's law, the details of the electroplating process have been simplified here. Actually, electroplating chromium is only 30% to 40% efficient and must be run at a particular temperature range for the plate to appear bright. Nearly 10,000 metric tons (2×10^8 mol) of chromium is used annually for electroplating.

FOLLOW-UP PROBLEM 21.10 Using a current of 4.75 A, how many minutes does it take to plate onto a sculpture 1.50 g of Cu from a $CuSO_4$ solution?

The following Chemical Connections essay links several themes of this chapter in the setting of a living cell.

Chemical Connections to *Biological Energetics*

Cellular Electrochemistry and the Production of ATP

Biological cells apply the principles of electrochemical cells to generate energy. The complex multistep process can be divided into two parts:

1. Bond energy in food is used to generate an electrochemical potential.
2. The potential is used to create the bond energy of the high-energy molecule adenosine triphosphate (ATP; see Chemical Connections, pp. 908–909).

The redox species that accomplish these steps are part of the *electron-transport chain* (ETC), which lies on the inner membranes of *mitochondria,* the subcellular particles that produce the cell's energy (Figure B21.1).

The ETC is a series of large molecules (mostly proteins), each of which contains a *redox couple* (the oxidized and reduced forms of a species), such as Fe^{3+}/Fe^{2+}, that passes electrons down the chain. At three points along the chain, large potential differences supply enough free energy to convert adenosine diphosphate (ADP) into ATP.

Bond Energy to Electrochemical Potential

Cells utilize the energy in food by releasing it in controlled steps rather than all at once. The reaction that ultimately powers the ETC is the oxidation of hydrogen to form water:

$$H_2 + \tfrac{1}{2}O_2 \longrightarrow H_2O$$

However, instead of H_2 gas, which does not occur in organisms, the hydrogen takes the form of two H^+ ions and two e^-. A bio-logical oxidizing agent called NAD^+ (*n*icotinamide *a*denine *d*inucleotide) acquires these protons and electrons in the process of oxidizing the molecules in food. To show this process, we use the following half-reaction (without canceling an H^+ on both sides):

$$NAD^+(aq) + 2H^+(aq) + 2e^- \longrightarrow NADH(aq) + H^+(aq)$$

At the mitochondrial inner membrane, the NADH and H^+ transfer the two e^- to the first redox couple of the ETC and release the two H^+. The electrons are transported down the chain of redox couples, where they finally reduce O_2 to H_2O. The overall process, with standard electrode potentials,* is

$$NADH(aq) + H^+(aq) \longrightarrow NAD^+(aq) + 2H^+(aq) + 2e^-$$
$$E^{\circ\prime} = -0.315 \text{ V}$$
$$\tfrac{1}{2}O_2(aq) + 2e^- + 2H^+(aq) \longrightarrow H_2O(l) \qquad E^{\circ\prime} = 0.815 \text{ V}$$

$$NADH(aq) + H^+(aq) + \tfrac{1}{2}O_2(aq) \longrightarrow NAD^+(aq) + H_2O(l)$$
$$E^{\circ\prime}_{\text{overall}} = 0.815 \text{ V} - (-0.315 \text{ V}) = 1.130 \text{ V}$$

Thus, for each mole of NADH that enters the ETC, the free-energy equivalent of 1.14 V is available:

$$\Delta G^{\circ\prime} = -nFE^{\circ\prime}$$
$$= -(2 \text{ mol } e^-/\text{mol NADH})(96.5 \text{ kJ/V·mol } e^-)(1.130 \text{ V})$$
$$= -218 \text{ kJ/mol NADH}$$

(continued)

---

*In biological systems, standard potentials are designated $E^{\circ\prime}$ and the standard states include a pH of 7.0 ($[H^+] = 1 \times 10^{-7}$ M).

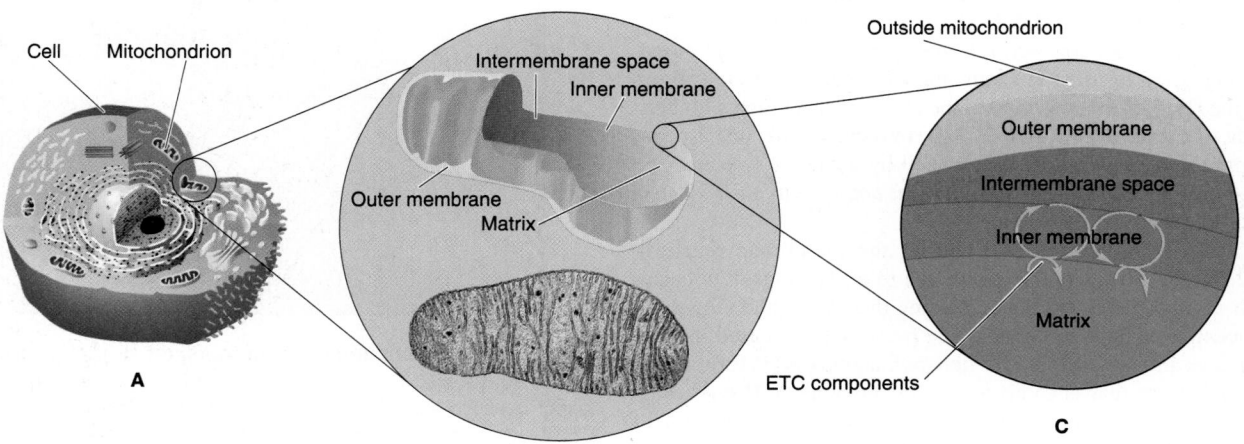

Figure B21.1 The mitochondrion. A, Mitochondria are subcellular particles outside the cell nucleus. **B,** They have a smooth outer membrane and a highly folded inner membrane, shown schematically and in an electron micrograph. **C,** The components of the electron-transport chain are attached to the inner membrane.

 ARIS Presentation Center, Art Library

Note that this aspect of the process functions like a voltaic cell: a spontaneous reaction, the reduction of O_2 to H_2O, is used to generate a potential. In contrast to a laboratory voltaic cell, in which the overall process occurs in one step, this process occurs in many small steps. Figure B21.2 is a greatly simplified diagram of the three key steps in the ETC that generate the high potential to produce ATP.

In the cell, each of these steps is part of a complex consisting of several components, most of which are proteins. Electrons are passed from one redox couple to the next down the chain, such that the reduced form of the first couple reduces the oxidized form of the second, and so forth. Most of the ETC components are iron-containing proteins, and the redox change consists of the oxidation of Fe^{2+} to Fe^{3+} in one component accompanied by the reduction of Fe^{3+} to Fe^{2+} in another:

$$Fe^{2+} \text{ (in A)} + Fe^{3+} \text{ (in B)} \longrightarrow Fe^{3+} \text{ (in A)} + Fe^{2+} \text{ (in B)}$$

In other words, metal ions within the ETC proteins are the actual species undergoing the redox reactions.

Electrochemical Potential to Bond Energy

At the three points shown in Figure B21.2, the large potential difference is used to form ATP:

$$ADP^{3-}(aq) + HPO_4^{2-}(aq) + H^+(aq) \longrightarrow ATP^{4-}(aq) + H_2O(l)$$
$$\Delta G^{\circ\prime} = 30.5 \text{ kJ/mol}$$

Note that the free energy that is *released* at each of the three ATP-producing points exceeds 30.5 kJ, the free energy that must be *absorbed* to form ATP. Thus, just as in an electrolytic cell, an electrochemical potential is supplied to drive a nonspontaneous reaction.

So far, we've followed the flow of *electrons* through the members of the ETC, but where have the released *protons* gone? The answer is the key to *how* electrochemical potential is converted to bond energy in ATP. As the electrons flow and the redox couples change oxidation states, free energy released at the three key steps is used to force H^+ ions into the intermembrane space, so that the $[H^+]$ of the intermembrane space soon becomes higher than that of the matrix (Figure B21.3). In other words, this aspect of the process acts as an electrolytic cell by using the free energy supplied by the three steps to *create what amounts to an H^+ concentration cell across the membrane.*

When the $[H^+]$ difference across the membrane reaches about 2.5-fold, it triggers the membrane to let H^+ ions flow back through spontaneously (in effect, closing the switch and allowing the concentration cell to operate). The free energy released in this spontaneous process drives the nonspontaneous ATP formation via a mechanism that is catalyzed by the enzyme ATP synthase. (Paul D. Boyer and John E. Walker shared half of the 1997 Nobel Prize in chemistry for elucidating this mechanism.)

Thus, the mitochondrion uses the "electron-motive force" of redox couples on the membrane to generate a "proton-motive force" across the membrane, which converts a potential difference to bond energy.

Figure B21.2 The main energy-yielding steps in the electron-transport chain (ETC). The electron carriers in the ETC undergo oxidation and reduction as they pass electrons to one another along the chain. At the three points shown, the difference in potential $E^{\circ\prime}$ (or free energy, $\Delta G^{\circ\prime}$) is large enough to be used for ATP production. (A complex consists of many components, mostly proteins; Q is a large organic molecule; and Cyt is the abbreviation for a cytochrome, a protein that contains a metal-ion redox couple, such as Fe^{3+}/Fe^{2+}, which is the actual electron carrier.)

Figure B21.3 Coupling electron transport to proton transport to ATP synthesis. The purpose of the ETC is to convert the free energy released from food molecules into the stored free energy of ATP. It accomplishes this by transporting electrons along the chain (*curved yellow line*), while protons are pumped out of the mitochondrial matrix. This pumping creates an $[H^+]$ difference and generates a potential across the inner membrane (in effect, a concentration cell). When this potential reaches a "trigger" value, H^+ flows back into the inner space, and the free energy released drives the formation of ATP.

Section Summary

An electrolytic cell uses electrical energy to drive a nonspontaneous reaction. • Oxidation occurs at the anode and reduction at the cathode, but the direction of electron flow and the charges of the electrodes are opposite those in voltaic cells. • When two products can form at each electrode, the more easily oxidized substance reacts at the anode and the more easily reduced at the cathode. • The reduction or oxidation of water takes place at nonstandard conditions. • Overvoltage causes the actual voltage required to be unexpectedly high and can affect the electrode product that forms. • The amount of product that forms depends on the quantity of charge flowing through the cell, which is related to the magnitude of the current and the time it flows. • Biological redox systems combine aspects of voltaic, concentration, and electrolytic cells to convert bond energy in food into electrochemical potential and then into the bond energy of ATP.

Chapter Perspective

The field of electrochemistry is one of the many areas in which the principles of thermodynamics lead to practical benefits. As you've seen, electrochemical cells can use a reaction to generate energy or use energy to drive a reaction. Such processes are central not only to our mobile way of life, but also to our biological existence. In Chapter 22, we examine the electrochemical (and other) methods used by industry to convert raw natural resources into some of the materials modern society finds indispensable.

CHAPTER REVIEW GUIDE   The following sections provide many aids to help you study this chapter. (Numbers in parentheses refer to pages, unless noted otherwise.)

Learning Objectives   These are concepts and skills you should know after studying this chapter.

Relevant section and/or sample problem (SP) numbers appear in parentheses.

Understand These Concepts

1. The meanings of *oxidation* and *reduction;* why an oxidizing agent is reduced and a reducing agent is oxidized (Section 21.1; also Section 4.5)

2. How the half-reaction method is used to balance redox reactions in acidic or basic solution (Section 21.1)

3. The distinction between voltaic and electrolytic cells in terms of the sign of ΔG (Section 21.1)

4. How voltaic cells use a spontaneous reaction to release electrical energy (Section 21.2)

5. The physical makeup of a voltaic cell: arrangement and composition of half-cells, relative charges of electrodes, and purpose of a salt bridge (Section 21.2)

6. How the difference in reducing strength of the electrodes determines the direction of electron flow (Section 21.2)

7. The correspondence between a positive E_{cell} and a spontaneous cell reaction (Section 21.3)

8. The usefulness and significance of standard electrode potentials ($E^{\circ}_{\text{half-cell}}$) (Section 21.3)

9. How $E^{\circ}_{\text{half-cell}}$ values are combined to give E°_{cell} (Section 21.3)

10. How the standard reference electrode is used to find an unknown $E^{\circ}_{\text{half-cell}}$ (Section 21.3)

11. How an emf series (e.g., Table 21.2 or Appendix D) is used to write spontaneous redox reactions (Section 21.3)

12. How the relative reactivity of a metal is determined by its reducing power and is related to the negative of its $E^{\circ}_{\text{half-cell}}$ (Section 21.3)

13. How E_{cell} (the nonstandard cell potential) is related to ΔG (maximum work) and the charge (moles of electrons times the Faraday constant) flowing through the cell (Section 21.4)

14. The interrelationship of ΔG°, E°_{cell}, and K (Section 21.4)

15. How E_{cell} changes as the cell operates (Q changes) (Section 21.4)

16. Why a voltaic cell can do work until $Q = K$ (Section 21.4)

17. How a concentration cell does work until the half-cell concentrations are equal (Section 21.4)

18. The distinction between primary (nonrechargeable) and secondary (rechargeable) batteries and the nature of fuel cells (Section 21.5)

19. How corrosion occurs and is prevented; the similarities between a corroding metal and a voltaic cell (Section 21.6)

20. How electrolytic cells use nonspontaneous redox reactions driven by an external source of electricity (Section 21.7)

21. How atomic properties (ionization energy and electronegativity) determine the products of the electrolysis of molten salt mixtures (Section 21.7)

22. How the electrolysis of water influences the products of aqueous electrolysis; the importance of overvoltage (Section 21.7)

23. The relationship between the quantity of charge flowing through the cell and the amount of product formed (Section 21.7)

Learning Objectives *(continued)*

Master These Skills

1. Balancing redox reactions by the half-reaction method (Section 21.1 and SP 21.1)
2. Diagramming and notating a voltaic cell (Section 21.2 and SP 21.2)
3. Combining $E^\circ_{\text{half-cell}}$ values to obtain E°_{cell} (Section 21.3)
4. Using E°_{cell} and a known $E^\circ_{\text{half-cell}}$ to find an unknown $E^\circ_{\text{half-cell}}$ (SP 21.3)
5. Manipulating half-reactions to write a spontaneous redox reaction and calculate its E°_{cell} (SP 21.4)
6. Ranking the relative strengths of oxidizing and reducing agents in a redox reaction (SP 21.4)

7. Predicting whether a metal can displace hydrogen or another metal from solution (Section 21.3)
8. Using the interrelationship of ΔG°, E°_{cell}, and K to calculate one of the three given the other two (Section 21.4 and SP 21.5)
9. Using the Nernst equation to calculate the nonstandard cell potential (E_{cell}) (SP 21.6)
10. Calculating E_{cell} of a concentration cell (SP 21.7)
11. Predicting the products of the electrolysis of a mixture of molten salts (SP 21.8)
12. Predicting the products of the electrolysis of aqueous salt solutions (SP 21.9)
13. Calculating the current (or time) needed to produce a given amount of product by electrolysis (SP 21.10)

Key Terms These important terms appear in boldface in the chapter and are defined again in the Glossary.

electrochemistry (923)
electrochemical cell (923)

Section 21.1
half-reaction method (924)
voltaic (galvanic) cell (928)
electrolytic cell (928)
electrode (928)
electrolyte (928)
anode (929)
cathode (929)

Section 21.2
half-cell (930)
salt bridge (930)

Section 21.3
cell potential (E_{cell}) (934)
voltage (934)
electromotive force (emf) (934)
volt (V) (934)
coulomb (C) (934)

standard cell potential (E°_{cell}) (935)
standard electrode (half-cell) potential ($E^\circ_{\text{half-cell}}$) (935)
standard reference half-cell (standard hydrogen electrode) (936)

Section 21.4
Faraday constant (F) (943)
Nernst equation (946)
concentration cell (948)

Section 21.5
battery (952)
fuel cell (955)

Section 21.6
corrosion (956)

Section 21.7
electrolysis (961)
overvoltage (963)
ampere (A) (965)

Key Equations and Relationships Numbered and screened concepts are listed for you to refer to or memorize.

21.1 Relating a spontaneous process to the sign of the cell potential (934):
$$E_{\text{cell}} > 0 \text{ for a spontaneous process}$$

21.2 Relating electric potential to energy and charge in SI units (934):
$$\text{Potential} = \text{energy/charge} \quad \text{or} \quad 1\text{ V} = 1\text{ J/C}$$

21.3 Relating standard cell potential to standard electrode potentials in a voltaic cell (935):
$$E^\circ_{\text{cell}} = E^\circ_{\text{cathode (reduction)}} - E^\circ_{\text{anode (oxidation)}}$$

21.4 Defining the Faraday constant (944):
$$F = 9.65 \times 10^4 \frac{\text{J}}{\text{V} \cdot \text{mol e}^-} \quad (3\text{ sf})$$

21.5 Relating the free energy change to the cell potential (944):
$$\Delta G = -nFE_{\text{cell}}$$

21.6 Finding the standard free energy change from the standard cell potential (944):
$$\Delta G^\circ = -nFE^\circ_{\text{cell}}$$

21.7 Finding the equilibrium constant from the standard cell potential (944):
$$E^\circ_{\text{cell}} = \frac{RT}{nF} \ln K$$

21.8 Substituting known values of R, F, and T into Equation 21.7 and converting to common logarithms (945):
$$E^\circ_{\text{cell}} = \frac{0.0592\text{ V}}{n} \log K \quad \text{or} \quad \log K = \frac{nE^\circ_{\text{cell}}}{0.0592\text{ V}} \quad (\text{at } 298.15\text{ K})$$

21.9 Calculating the nonstandard cell potential (Nernst equation) (946):
$$E_{\text{cell}} = E^\circ_{\text{cell}} - \frac{RT}{nF} \ln Q$$

21.10 Substituting known values of R, F, and T into the Nernst equation and converting to common logarithms (946):
$$E_{\text{cell}} = E^\circ_{\text{cell}} - \frac{0.0592\text{ V}}{n} \log Q \quad (\text{at } 298.15\text{ K})$$

21.11 Relating current to charge and time (965):
$$\text{Current} = \text{charge/time} \quad \text{or} \quad 1\text{ A} = 1\text{ C/s}$$

Highlighted Figures and Tables These figures (F) and tables (T) provide a visual review of key ideas.

F21.1 Summary of redox terminology (924)
F21.3 Voltaic and electrolytic cells (928)
F21.5 A voltaic cell based on the zinc-copper reaction (931)
F21.10 The interrelationship of $\Delta G°$, $E°_{cell}$, and K (944)
F21.11 E_{cell} and log Q for the zinc-copper cell (948)

F21.21 The corrosion of iron (957)
F21.25 Tin-copper reaction in voltaic and electrolytic cells (959)
T21.4 Comparison of voltaic and electrolytic cells (961)
F21.28 Summary of the stoichiometry of electrolysis (965)

Brief Solutions to FOLLOW-UP PROBLEMS Compare your solutions to these calculation steps and answers.

21.1 $6KMnO_4(aq) + 6KOH(aq) + KI(aq) \longrightarrow$
$$6K_2MnO_4(aq) + KIO_3(aq) + 3H_2O(l)$$

21.2
$$Sn(s) \longrightarrow Sn^{2+}(aq) + 2e^-$$
[anode; oxidation]
$$6e^- + 14H^+(aq) + Cr_2O_7{}^{2-}(aq) \longrightarrow 2Cr^{3+}(aq) + 7H_2O(l)$$
[cathode; reduction]

$$3Sn(s) + Cr_2O_7{}^{2-}(aq) + 14H^+(aq) \longrightarrow$$
$$3Sn^{2+}(aq) + 2Cr^{3+}(aq) + 7H_2O(l) \quad \text{[overall]}$$

Cell notation:
$Sn(s) \mid Sn^{2+}(aq) \parallel H^+(aq), Cr_2O_7{}^{2-}(aq), Cr^{3+}(aq) \mid$ graphite

21.3 $Br_2(aq) + 2e^- \longrightarrow 2Br^-(aq)$ $E°_{bromine} = 1.07$ V
[cathode]
$2V^{3+}(aq) + 2H_2O(l) \longrightarrow 2VO^{2+}(aq) + 4H^+(aq) + 2e^-$
$E°_{vanadium} = ?$ [anode]
$E°_{vanadium} = E°_{bromine} - E°_{cell} = 1.07$ V $- 1.39$ V $= -0.32$ V

21.4 $Fe^{2+}(aq) + 2e^- \longrightarrow Fe(s)$ $E° = -0.44$ V
$2[Fe^{2+}(aq) \longrightarrow Fe^{3+}(aq) + e^-]$ $E° = 0.77$ V

$3Fe^{2+}(aq) \longrightarrow 2Fe^{3+}(aq) + Fe(s)$

$E°_{cell} = -0.44$ V $- 0.77$ V $= -1.21$ V
The reaction is nonspontaneous. The spontaneous reaction is
$2Fe^{3+}(aq) + Fe(s) \longrightarrow 3Fe^{2+}(aq)$ $E°_{cell} = 1.21$ V
$Fe > Fe^{2+} > Fe^{3+}$

21.5 $Cd(s) + Cu^{2+}(aq) \longrightarrow Cd^{2+}(aq) + Cu(s)$
$\Delta G° = -RT \ln K = -8.314$ J/mol rxn·K $\times 298$ K $\times \ln K$
     $= -143$ kJ/mol rxn; $K = 1.2 \times 10^{25}$
$E°_{cell} = \dfrac{0.0592 \text{ V}}{2} \log (1.2 \times 10^{25}) = 0.742$ V

21.6
$$Fe(s) \longrightarrow Fe^{2+}(aq) + 2e^- \quad E° = -0.44 \text{ V}$$
$$Cu^{2+}(aq) + 2e^- \longrightarrow Cu(s) \quad E° = 0.34 \text{ V}$$

$Fe(s) + Cu^{2+}(aq) \longrightarrow Fe^{2+}(aq) + Cu(s)$ $E°_{cell} = 0.78$ V
So $E_{cell} = 0.78$ V $+ 0.25$ V $= 1.03$ V
$$1.03 \text{ V} = 0.78 \text{ V} - \frac{0.0592 \text{ V}}{2} \log \frac{[Fe^{2+}]}{[Cu^{2+}]}$$
$$\frac{[Fe^{2+}]}{[Cu^{2+}]} = 3.6 \times 10^{-9}$$
$[Fe^{2+}] = 3.6 \times 10^{-9} \times 0.30$ $M = 1.1 \times 10^{-9}$ M

21.7 $Au^{3+}(aq; 2.5 \times 10^{-2}$ $M)$ [B] $\longrightarrow$
$$Au^{3+}(aq; 7.0 \times 10^{-4} \text{ } M) \text{ [A]}$$
$$E_{cell} = 0 \text{ V} - \left(\frac{0.0592 \text{ V}}{3} \times \log \frac{7.0 \times 10^{-4}}{2.5 \times 10^{-2}} \right) = 0.0306 \text{ V}$$

The electrode in A is negative, so it is the anode.

21.8 Oxidizing agents: K^+ and Al^{3+}.
Reducing agents: F^- and Br^-.
Al is above and to the right of K in the periodic table, so it has a higher IE:
$Al^{3+}(l) + 3e^- \longrightarrow Al(s)$ [cathode; reduction]
Br is below F in Group 7A(17), so it has a lower EN:
$2Br^-(l) \longrightarrow Br_2(g) + 2e^-$ [anode; oxidation]
$2Al^{3+}(l) + 6Br^-(l) \longrightarrow 2Al(s) + 3Br_2(g)$ [overall]

21.9 The reduction with the more positive electrode potential is
$Au^{3+}(aq) + 3e^- \longrightarrow Au(s); E° = 1.50$ V
[cathode; reduction]
Because of overvoltage, O_2 will not form at the anode, so Br_2 will form:
$2Br^-(aq) \longrightarrow Br_2(l) + 2e^-; E° = 1.07$ V
[cathode; oxidation]

21.10 $Cu^{2+}(aq) + 2e^- \longrightarrow Cu(s)$; therefore,
2 mol e^-/1 mol Cu = 2 mol e^-/63.55 g Cu
$$\text{Time (min)} = 1.50 \text{ g Cu} \times \frac{2 \text{ mol } e^-}{63.55 \text{ g Cu}}$$
$$\times \frac{9.65 \times 10^4 \text{ C}}{1 \text{ mol } e^-} \times \frac{1 \text{ s}}{4.75 \text{ C}} \times \frac{1 \text{ min}}{60 \text{ s}}$$
$$= 16.0 \text{ min}$$

PROBLEMS

Problems with **colored** numbers are answered in Appendix E and worked in detail in the Student Solutions Manual. Problem sections match those in the text and provide the numbers of relevant sample problems. Most offer Concept Review Questions, Skill-Building Exercises (grouped in pairs covering the same concept), and Problems in Context. The Comprehensive Problems are based on material from any section or previous chapter.

Note: Unless stated otherwise, all problems refer to systems at 298 K (25°C).

Redox Reactions and Electrochemical Cells
(Sample Problem 21.1)

● Concept Review Questions

• **21.1** Define *oxidation* and *reduction* in terms of electron transfer and change in oxidation number.

• **21.2** Why must an electrochemical process involve a redox reaction?

• **21.3** Can one half-reaction in a redox process take place independently of the other? Explain.

• **21.4** Water is used to balance O atoms in the half-reaction method. Why can't O^{2-} ions be used instead?

• **21.5** During the redox balancing process, what step is taken to ensure that e^- loss equals e^- gain?

• **21.6** How are protons removed when balancing a redox reaction in basic solution?

• **21.7** Are spectator ions used to balance the half-reactions of a redox reaction? At what stage might spectator ions enter the balancing process?

• **21.8** Which type of electrochemical cell has $\Delta G_{sys} < 0$? Which type shows an increase in free energy?

• **21.9** Which statements are true? Correct any that are false.
(a) In a voltaic cell, the anode is negative relative to the cathode.
(b) Oxidation occurs at the anode of a voltaic or electrolytic cell.
(c) Electrons flow into the cathode of an electrolytic cell.
(d) In a voltaic cell, the surroundings do work on the system.
(e) A metal that plates out of an electrolytic cell appears on the cathode.
(f) The cell electrolyte provides a solution of mobile electrons.

● Skill-Building Exercises (grouped in similar pairs)

• **21.10** Consider the following balanced redox reaction:
$$16H^+(aq) + 2MnO_4^-(aq) + 10Cl^-(aq) \longrightarrow$$
$$2Mn^{2+}(aq) + 5Cl_2(g) + 8H_2O(l)$$
(a) Which species is being oxidized?
(b) Which species is being reduced?
(c) Which species is the oxidizing agent?
(d) Which species is the reducing agent?
(e) From which species to which does electron transfer occur?
(f) Write the balanced molecular equation, with K^+ and SO_4^{2-} as the spectator ions.

• **21.11** Consider the following balanced redox reaction:
$$2CrO_2^-(aq) + 2H_2O(l) + 6ClO^-(aq) \longrightarrow$$
$$2CrO_4^{2-}(aq) + 3Cl_2(g) + 4OH^-(aq)$$
(a) Which species is being oxidized?
(b) Which species is being reduced?
(c) Which species is the oxidizing agent?

(d) Which species is the reducing agent?
(e) From which species to which does electron transfer occur?
(f) Write the balanced molecular equation, with Na^+ as the spectator ion.

• **21.12** Balance the following skeleton reactions and identify the oxidizing and reducing agents:
(a) $ClO_3^-(aq) + I^-(aq) \longrightarrow I_2(s) + Cl^-(aq)$ [acidic]
(b) $MnO_4^-(aq) + SO_3^{2-}(aq) \longrightarrow$
$$MnO_2(s) + SO_4^{2-}(aq) \text{ [basic]}$$
(c) $MnO_4^-(aq) + H_2O_2(aq) \longrightarrow Mn^{2+}(aq) + O_2(g)$ [acidic]

• **21.13** Balance the following skeleton reactions and identify the oxidizing and reducing agents:
(a) $O_2(g) + NO(g) \longrightarrow NO_3^-(aq)$ [acidic]
(b) $CrO_4^{2-}(aq) + Cu(s) \longrightarrow$
$$Cr(OH)_3(s) + Cu(OH)_2(s) \text{ [basic]}$$
(c) $AsO_4^{3-}(aq) + NO_2^-(aq) \longrightarrow$
$$AsO_2^-(aq) + NO_3^-(aq) \text{ [basic]}$$

• **21.14** Balance the following skeleton reactions and identify the oxidizing and reducing agents:
(a) $Cr_2O_7^{2-}(aq) + Zn(s) \longrightarrow Zn^{2+}(aq) + Cr^{3+}(aq)$ [acidic]
(b) $Fe(OH)_2(s) + MnO_4^-(aq) \longrightarrow$
$$MnO_2(s) + Fe(OH)_3(s) \text{ [basic]}$$
(c) $Zn(s) + NO_3^-(aq) \longrightarrow Zn^{2+}(aq) + N_2(g)$ [acidic]

• **21.15** Balance the following skeleton reactions and identify the oxidizing and reducing agents:
(a) $BH_4^-(aq) + ClO_3^-(aq) \longrightarrow$
$$H_2BO_3^-(aq) + Cl^-(aq) \text{ [basic]}$$
(b) $CrO_4^{2-}(aq) + N_2O(g) \longrightarrow Cr^{3+}(aq) + NO(g)$ [acidic]
(c) $Br_2(l) \longrightarrow BrO_3^-(aq) + Br^-(aq)$ [basic]

• **21.16** Balance the following skeleton reactions and identify the oxidizing and reducing agents:
(a) $Sb(s) + NO_3^-(aq) \longrightarrow Sb_4O_6(s) + NO(g)$ [acidic]
(b) $Mn^{2+}(aq) + BiO_3^-(aq) \longrightarrow$
$$MnO_4^-(aq) + Bi^{3+}(aq) \text{ [acidic]}$$
(c) $Fe(OH)_2(s) + Pb(OH)_3^-(aq) \longrightarrow$
$$Fe(OH)_3(s) + Pb(s) \text{ [basic]}$$

• **21.17** Balance the following skeleton reactions and identify the oxidizing and reducing agents:
(a) $NO_2(g) \longrightarrow NO_3^-(aq) + NO_2^-(aq)$ [basic]
(b) $Zn(s) + NO_3^-(aq) \longrightarrow Zn(OH)_4^{2-}(aq) + NH_3(g)$ [basic]
(c) $H_2S(g) + NO_3^-(aq) \longrightarrow S_8(s) + NO(g)$ [acidic]

• **21.18** Balance the following skeleton reactions and identify the oxidizing and reducing agents:
(a) $As_4O_6(s) + MnO_4^-(aq) \longrightarrow$
$$AsO_4^{3-}(aq) + Mn^{2+}(aq) \text{ [acidic]}$$
(b) $P_4(s) \longrightarrow HPO_3^{2-}(aq) + PH_3(g)$ [acidic]
(c) $MnO_4^-(aq) + CN^-(aq) \longrightarrow$
$$MnO_2(s) + CNO^-(aq) \text{ [basic]}$$

• **21.19** Balance the following skeleton reactions and identify the oxidizing and reducing agents:
(a) $SO_3^{2-}(aq) + Cl_2(g) \longrightarrow SO_4^{2-}(aq) + Cl^-(aq)$ [basic]
(b) $Fe(CN)_6^{3-}(aq) + Re(s) \longrightarrow$
$$Fe(CN)_6^{4-}(aq) + ReO_4^-(aq) \text{ [basic]}$$

Levels of Difficulty: ● Easy ●● Medium ●●● Challenging

(c) $MnO_4^-(aq) + HCOOH(aq) \longrightarrow$
$$Mn^{2+}(aq) + CO_2(g) \text{ [acidic]}$$

● Problems in Context

● **21.20** In many residential water systems, the aqueous Fe^{3+} concentration is high enough to stain sinks and turn drinking water light brown. The iron content is analyzed by first reducing the Fe^{3+} to Fe^{2+} and then titrating with MnO_4^- in acidic solution. Balance the skeleton reaction of the titration step:
$$Fe^{2+}(aq) + MnO_4^-(aq) \longrightarrow Mn^{2+}(aq) + Fe^{3+}(aq)$$

● **21.21** *Aqua regia,* a mixture of concentrated HNO_3 and HCl, was developed by alchemists as a means to "dissolve" gold. The process is a redox reaction with the simplified skeleton reaction:
$$Au(s) + NO_3^-(aq) + Cl^-(aq) \longrightarrow AuCl_4^-(aq) + NO_2(g)$$
(a) Balance the reaction by the half-reaction method.
(b) What are the oxidizing and reducing agents?
(c) What is the function of HCl in aqua regia?

Voltaic Cells: Using Spontaneous Reactions to Generate Electrical Energy
(Sample Problem 21.2)

● Concept Review Questions

● **21.22** Consider the following general voltaic cell:

Identify the (a) anode, (b) cathode, (c) salt bridge, (d) electrode at which e^- leave the cell, (e) electrode with a positive charge, and (f) electrode that gains mass as the cell operates (assuming that a metal plates out).

● **21.23** Why does a voltaic cell not operate unless the two compartments are connected through an external circuit?

● **21.24** What purpose does the salt bridge serve in a voltaic cell, and how does it accomplish this purpose?

● **21.25** What is the difference between an active and an inactive electrode? Why are inactive electrodes used? Name two substances commonly used for inactive electrodes.

● **21.26** When a piece of metal A is placed in a solution containing ions of metal B, metal B plates out on the piece of A.
(a) Which metal is being oxidized?
(b) Which metal is being displaced?
(c) Which metal would you use as the anode in a voltaic cell incorporating these two metals?
(d) If bubbles of H_2 form when B is placed in acid, will they form if A is placed in acid? Explain.

● Skill-Building Exercises *(grouped in similar pairs)*

● **21.27** A voltaic cell is constructed with an Sn/Sn^{2+} half-cell and a Zn/Zn^{2+} half-cell. The zinc electrode is negative.
(a) Write balanced half-reactions and the overall reaction.
(b) Diagram the cell, labeling electrodes with their charges and showing the directions of electron flow in the circuit and of cation and anion flow in the salt bridge.

● **21.28** A voltaic cell is constructed with an Ag/Ag^+ half-cell and a Pb/Pb^{2+} half-cell. The silver electrode is positive.
(a) Write balanced half-reactions and the overall reaction.
(b) Diagram the cell, labeling electrodes with their charges and showing the directions of electron flow in the circuit and of cation and anion flow in the salt bridge.

● **21.29** Consider the following voltaic cell:

(a) In which direction do electrons flow in the external circuit?
(b) In which half-cell does oxidation occur?
(c) In which half-cell do electrons enter the cell?
(d) At which electrode are electrons consumed?
(e) Which electrode is negatively charged?
(f) Which electrode decreases in mass during cell operation?
(g) Suggest a solution for the cathode electrolyte.
(h) Suggest a pair of ions for the salt bridge.
(i) For which electrode could you use an inactive material?
(j) In which direction do anions within the salt bridge move to maintain charge neutrality?
(k) Write balanced half-reactions and an overall cell reaction.

● **21.30** Consider the following voltaic cell:

(a) In which direction do electrons flow in the external circuit?
(b) In which half-cell does reduction occur?
(c) In which half-cell do electrons leave the cell?
(d) At which electrode are electrons generated?
(e) Which electrode is positively charged?
(f) Which electrode increases in mass during cell operation?
(g) Suggest a solution for the anode electrolyte.
(h) Suggest a pair of ions for the salt bridge.
(i) For which electrode could you use an inactive material?
(j) In which direction do cations within the salt bridge move to maintain charge neutrality?
(k) Write balanced half-reactions and an overall cell reaction.

● **21.31** A voltaic cell is constructed with an Fe/Fe^{2+} half-cell and an Mn/Mn^{2+} half-cell. The iron electrode is positive.
(a) Write balanced half-reactions and the overall reaction.
(b) Diagram the cell, labeling electrodes with their charges and showing the directions of electron flow in the circuit and of cation and anion flow in the salt bridge.

21.32 A voltaic cell is constructed with a Cu/Cu^{2+} half-cell and an Ni/Ni^{2+} half-cell. The nickel electrode is negative.
(a) Write balanced half-reactions and the overall reaction.
(b) Diagram the cell, labeling electrodes with their charges and showing the directions of electron flow in the circuit and of cation and anion flow in the salt bridge.

21.33 Write the cell notation for the voltaic cell that incorporates each of the following redox reactions:
(a) $Al(s) + Cr^{3+}(aq) \longrightarrow Al^{3+}(aq) + Cr(s)$
(b) $Cu^{2+}(aq) + SO_2(g) + 2H_2O(l) \longrightarrow$
$$Cu(s) + SO_4^{2-}(aq) + 4H^+(aq)$$

21.34 Write a balanced equation from each cell notation:
(a) $Mn(s) \mid Mn^{2+}(aq) \parallel Cd^{2+}(aq) \mid Cd(s)$
(b) $Fe(s) \mid Fe^{2+}(aq) \parallel NO_3^-(aq) \mid NO(g) \mid Pt(s)$

Cell Potential: Output of a Voltaic Cell
(Sample Problems 21.3 and 21.4)

● **Concept Review Questions**

21.35 How is a standard reference electrode used to determine unknown $E^\circ_{half-cell}$ values?

21.36 What does a negative E°_{cell} indicate about a redox reaction? What does it indicate about the reverse reaction?

21.37 The standard cell potential is a thermodynamic state function. How are E° values treated similarly to ΔH°, ΔG°, and S° values? How are they treated differently?

● **Skill-Building Exercises** (grouped in similar pairs)

21.38 In basic solution, Se^{2-} and SO_3^{2-} ions react spontaneously:
$$2Se^{2-}(aq) + 2SO_3^{2-}(aq) + 3H_2O(l) \longrightarrow$$
$$2Se(s) + 6OH^-(aq) + S_2O_3^{2-}(aq) \qquad E^\circ_{cell} = 0.35 \text{ V}$$
(a) Write balanced half-reactions for the process.
(b) If $E^\circ_{sulfite}$ is -0.57 V, calculate $E^\circ_{selenium}$.

21.39 In acidic solution, O_3 and Mn^{2+} ion react spontaneously:
$$O_3(g) + Mn^{2+}(aq) + H_2O(l) \longrightarrow$$
$$O_2(g) + MnO_2(s) + 2H^+(aq) \qquad E^\circ_{cell} = 0.84 \text{ V}$$
(a) Write the balanced half-reactions.
(b) Using Appendix D to find E°_{ozone}, calculate $E^\circ_{manganese}$.

21.40 Use the emf series (Appendix D) to arrange the species.
(a) In order of *decreasing* strength as *oxidizing* agents: Fe^{3+}, Br_2, Cu^{2+}
(b) In order of *increasing* strength as *oxidizing* agents: Ca^{2+}, $Cr_2O_7^{2-}$, Ag^+

21.41 Use the emf series (Appendix D) to arrange the species.
(a) In order of *decreasing* strength as *reducing* agents: SO_2, $PbSO_4$, MnO_2
(b) In order of *increasing* strength as *reducing* agents: Hg, Fe, Sn

21.42 Balance each skeleton reaction, calculate E°_{cell}, and state whether the reaction is spontaneous:
(a) $Co(s) + H^+(aq) \longrightarrow Co^{2+}(aq) + H_2(g)$
(b) $Mn^{2+}(aq) + Br_2(l) \longrightarrow MnO_4^-(aq) + Br^-(aq)$ [acidic]
(c) $Hg_2^{2+}(aq) \longrightarrow Hg^{2+}(aq) + Hg(l)$

21.43 Balance each skeleton reaction, calculate E°_{cell}, and state whether the reaction is spontaneous:
(a) $Cl_2(g) + Fe^{2+}(aq) \longrightarrow Cl^-(aq) + Fe^{3+}(aq)$
(b) $Mn^{2+}(aq) + Co^{3+}(aq) \longrightarrow MnO_2(s) + Co^{2+}(aq)$ [acidic]
(c) $AgCl(s) + NO(g) \longrightarrow$
$$Ag(s) + Cl^-(aq) + NO_3^-(aq) \text{ [acidic]}$$

21.44 Balance each skeleton reaction, calculate E°_{cell}, and state whether the reaction is spontaneous:
(a) $Ag(s) + Cu^{2+}(aq) \longrightarrow Ag^+(aq) + Cu(s)$
(b) $Cd(s) + Cr_2O_7^{2-}(aq) \longrightarrow Cd^{2+}(aq) + Cr^{3+}(aq)$
(c) $Ni^{2+}(aq) + Pb(s) \longrightarrow Ni(s) + Pb^{2+}(aq)$

21.45 Balance each skeleton reaction, calculate E°_{cell}, and state whether the reaction is spontaneous:
(a) $Cu^+(aq) + PbO_2(s) + SO_4^{2-}(aq) \longrightarrow$
$$PbSO_4(s) + Cu^{2+}(aq) \text{ [acidic]}$$
(b) $H_2O_2(aq) + Ni^{2+}(aq) \longrightarrow O_2(g) + Ni(s)$ [acidic]
(c) $MnO_2(s) + Ag^+(aq) \longrightarrow MnO_4^-(aq) + Ag(s)$ [basic]

21.46 Use the following half-reactions to write three spontaneous reactions, calculate E°_{cell} for each reaction, and rank the strengths of the oxidizing and reducing agents:
(1) $Al^{3+}(aq) + 3e^- \longrightarrow Al(s) \qquad E^\circ = -1.66 \text{ V}$
(2) $N_2O_4(g) + 2e^- \longrightarrow 2NO_2^-(aq) \qquad E^\circ = 0.867 \text{ V}$
(3) $SO_4^{2-}(aq) + H_2O(l) + 2e^- \longrightarrow SO_3^{2-}(aq) + 2OH^-(aq)$
$$E^\circ = 0.93 \text{ V}$$

21.47 Use the following half-reactions to write three spontaneous reactions, calculate E°_{cell} for each reaction, and rank the strengths of the oxidizing and reducing agents:
(1) $Au^+(aq) + e^- \longrightarrow Au(s) \qquad E^\circ = 1.69 \text{ V}$
(2) $N_2O(g) + 2H^+(aq) + 2e^- \longrightarrow N_2(g) + H_2O(l)$
$$E^\circ = 1.77 \text{ V}$$
(3) $Cr^{3+}(aq) + 3e^- \longrightarrow Cr(s) \qquad E^\circ = -0.74 \text{ V}$

21.48 Use the following half-reactions to write three spontaneous reactions, calculate E°_{cell} for each reaction, and rank the strengths of the oxidizing and reducing agents:
(1) $2HClO(aq) + 2H^+(aq) + 2e^- \longrightarrow Cl_2(g) + 2H_2O(l)$
$$E^\circ = 1.63 \text{ V}$$
(2) $Pt^{2+}(aq) + 2e^- \longrightarrow Pt(s) \qquad E^\circ = 1.20 \text{ V}$
(3) $PbSO_4(s) + 2e^- \longrightarrow Pb(s) + SO_4^{2-}(aq) \qquad E^\circ = -0.31 \text{ V}$

21.49 Use the following half-reactions to write three spontaneous reactions, calculate E°_{cell} for each reaction, and rank the strengths of the oxidizing and reducing agents:
(1) $I_2(s) + 2e^- \longrightarrow 2I^-(aq) \qquad E^\circ = 0.53 \text{ V}$
(2) $S_2O_8^{2-}(aq) + 2e^- \longrightarrow 2SO_4^{2-}(aq) \qquad E^\circ = 2.01 \text{ V}$
(3) $Cr_2O_7^{2-}(aq) + 14H^+(aq) + 6e^- \longrightarrow$
$$2Cr^{3+}(aq) + 7H_2O(l) \qquad E^\circ = 1.33 \text{ V}$$

● **Problems in Context**

21.50 When metal A is placed in a solution of a salt of metal B, the surface of metal A changes color. When metal B is placed in acid solution, gas bubbles form on the surface of the metal. When metal A is placed in a solution of a salt of metal C, no change is observed in the solution or on the metal A surface. Will metal C cause formation of H_2 when placed in acid solution? Rank metals A, B, and C in order of *decreasing* reducing strength.

21.51 When a clean iron nail is placed in an aqueous solution of copper(II) sulfate, the nail becomes coated with a brownish black material.
(a) What is the material coating the iron?
(b) What are the oxidizing and reducing agents?
(c) Can this reaction be made into a voltaic cell?
(d) Write the balanced equation for the reaction.
(e) Calculate E°_{cell} for the process.

Free Energy and Electrical Work
(Sample Problems 21.5 to 21.7)

● **Concept Review Questions**

● **21.52** (a) How do the relative magnitudes of Q and K relate to the signs of ΔG and E_{cell}? Explain.
(b) Can a cell do work when $Q/K > 1$ or $Q/K < 1$? Explain.

● **21.53** A voltaic cell consists of a metal A/A^+ electrode and a metal B/B^+ electrode, with the A/A^+ electrode negative. The initial $[A^+]/[B^+]$ is such that $E_{cell} > E°_{cell}$.
(a) How do $[A^+]$ and $[B^+]$ change as the cell operates?
(b) How does E_{cell} change as the cell operates?
(c) What is $[A^+]/[B^+]$ when $E_{cell} = E°_{cell}$? Explain.
(d) Is it possible for E_{cell} to be less than $E°_{cell}$? Explain.

● **21.54** Explain whether E_{cell} of a voltaic cell will increase or decrease with each of the following changes:
(a) Decrease in cell temperature
(b) Increase in [active ion] in the anode compartment
(c) Increase in [active ion] in the cathode compartment
(d) Increase in pressure of a gaseous reactant in the cathode compartment

● **21.55** In a concentration cell, is the more concentrated electrolyte in the cathode or the anode compartment? Explain.

● **Skill-Building Exercises** (grouped in similar pairs)

● **21.56** What is the value of the equilibrium constant for the reaction between each pair at 25°C?
(a) $Ni(s)$ and $Ag^+(aq)$ (b) $Fe(s)$ and $Cr^{3+}(aq)$

● **21.57** What is the value of the equilibrium constant for the reaction between each pair at 25°C?
(a) $Al(s)$ and $Cd^{2+}(aq)$ (b) $I_2(s)$ and $Br^-(aq)$

● **21.58** What is the value of the equilibrium constant for the reaction between each pair at 25°C?
(a) $Ag(s)$ and $Mn^{2+}(aq)$ (b) $Cl_2(g)$ and $Br^-(aq)$

● **21.59** What is the value of the equilibrium constant for the reaction between each pair at 25°C?
(a) $Cr(s)$ and $Cu^{2+}(aq)$ (b) $Sn(s)$ and $Pb^{2+}(aq)$

● **21.60** Calculate $\Delta G°$ for each of the reactions in Problem 21.56.
● **21.61** Calculate $\Delta G°$ for each of the reactions in Problem 21.57.

● **21.62** Calculate $\Delta G°$ for each of the reactions in Problem 21.58.
● **21.63** Calculate $\Delta G°$ for each of the reactions in Problem 21.59.

● **21.64** What are $E°_{cell}$ and $\Delta G°$ of a redox reaction at 25°C for which $n = 1$ and $K = 5.0 \times 10^4$?

● **21.65** What are $E°_{cell}$ and $\Delta G°$ of a redox reaction at 25°C for which $n = 1$ and $K = 5.0 \times 10^{-6}$?

● **21.66** What are $E°_{cell}$ and $\Delta G°$ of a redox reaction at 25°C for which $n = 2$ and $K = 65$?

● **21.67** What are $E°_{cell}$ and $\Delta G°$ of a redox reaction at 25°C for which $n = 2$ and $K = 0.065$?

● **21.68** A voltaic cell consists of a standard hydrogen electrode in one half-cell and a Cu/Cu^{2+} half-cell. Calculate $[Cu^{2+}]$ when E_{cell} is 0.22 V.

● **21.69** A voltaic cell consists of an Mn/Mn^{2+} half-cell and a Pb/Pb^{2+} half-cell. Calculate $[Pb^{2+}]$ when $[Mn^{2+}]$ is 1.4 M and E_{cell} is 0.44 V.

●● **21.70** A voltaic cell with Ni/Ni^{2+} and Co/Co^{2+} half-cells has the following initial concentrations: $[Ni^{2+}] = 0.80$ M; $[Co^{2+}] = 0.20$ M.

(a) What is the initial E_{cell}?
(b) What is $[Ni^{2+}]$ when E_{cell} reaches 0.03 V?
(c) What are the equilibrium concentrations of the ions?

●● **21.71** A voltaic cell with Mn/Mn^{2+} and Cd/Cd^{2+} half-cells has the following initial concentrations: $[Mn^{2+}] = 0.090$ M; $[Cd^{2+}] = 0.060$ M.
(a) What is the initial E_{cell}?
(b) What is E_{cell} when $[Cd^{2+}]$ reaches 0.050 M?
(c) What is $[Mn^{2+}]$ when E_{cell} reaches 0.055 V?
(d) What are the equilibrium concentrations of the ions?

● **21.72** A concentration cell consists of two H_2/H^+ half-cells. Half-cell A has H_2 at 0.95 atm bubbling into 0.10 M HCl. Half-cell B has H_2 at 0.60 atm bubbling into 2.0 M HCl. Which half-cell houses the anode? What is the voltage of the cell?

● **21.73** A concentration cell consists of two Sn/Sn^{2+} half-cells. The electrolyte in compartment A is 0.13 M $Sn(NO_3)_2$. The electrolyte in B is 0.87 M $Sn(NO_3)_2$. Which half-cell houses the cathode? What is the voltage of the cell?

Electrochemical Processes in Batteries

● **Concept Review Questions**

● **21.74** What is the direction of electron flow with respect to the anode and the cathode in a battery? Explain.

● **21.75** In the everyday batteries used for flashlights, toys, etc., no salt bridge is evident. What is used in these cells to separate the anode and cathode compartments?

●● **21.76** Both a D-sized and an AAA-sized alkaline battery have an output of 1.5 V. What property of the cell potential allows this to occur? What is different about these two batteries?

● **Problems in Context**

● **21.77** Many common electrical devices require the use of more than one battery.
(a) How many alkaline batteries must be placed in series to light a flashlight with a 6.0-V bulb?
(b) What is the voltage requirement of a camera that uses six silver batteries?
(c) How many volts can a car battery deliver if two of its anode/cathode cells are shorted?

Corrosion: A Case of Environmental Electrochemistry

● **Concept Review Questions**

● **21.78** During reconstruction of the Statue of Liberty, Teflon spacers were placed between the iron skeleton and the copper plates that cover the statue. What purpose do these spacers serve?

● **21.79** Why do steel bridge-supports rust at the waterline but not above or below it?

● **21.80** After the 1930s, chromium replaced nickel for corrosion resistance and appearance on car bumpers and trim. How does chromium protect steel from corrosion?

● **21.81** Which of the following metals are suitable for use as sacrificial anodes to protect against corrosion of underground iron pipes? If any are not suitable, explain why:
(a) Aluminum
(b) Magnesium
(c) Sodium
(d) Lead
(e) Nickel
(f) Zinc
(g) Chromium

Electrolytic Cells: Using Electrical Energy to Drive Nonspontaneous Reactions

(Sample Problems 21.8 to 21.10)

● **Concept Review Questions**

Note: Unless stated otherwise, assume that the electrolytic cells in the following problems operate at 100% efficiency.

● **21.82** Consider the following general electrolytic cell:

(a) At which electrode does oxidation occur?
(b) At which electrode does elemental M form?
(c) At which electrode are electrons being released by ions?
(d) At which electrode are electrons entering the cell?

●● **21.83** A voltaic cell consists of Cr/Cr^{3+} and Cd/Cd^{2+} half-cells with all components in their standard states. After 10 minutes of operation, a thin coating of cadmium metal has plated out on the cathode. Describe what will happen if you attach the negative terminal of a dry cell (1.5 V) to the cell cathode and the positive terminal to the cell anode.

● **21.84** Why are $E_{half\text{-}cell}$ values for the oxidation and reduction of water different from $E^{\circ}_{half\text{-}cell}$ values for the same processes?

● **21.85** In an aqueous electrolytic cell, nitrate ions never react at the anode, but nitrite ions do. Explain.

● **21.86** How does overvoltage influence the products in the electrolysis of aqueous salts?

● **Skill-Building Exercises** (grouped in similar pairs)

● **21.87** In the electrolysis of molten NaBr,
(a) What product forms at the anode?
(b) What product forms at the cathode?

● **21.88** In the electrolysis of molten BaI_2,
(a) What product forms at the negative electrode?
(b) What product forms at the positive electrode?

● **21.89** In the electrolysis of a molten mixture of KI and MgF_2, identify the product that forms at the anode and at the cathode.

● **21.90** In the electrolysis of a molten mixture of CsBr and $SrCl_2$, identify the product that forms at the negative electrode and at the positive electrode.

● **21.91** In the electrolysis of a molten mixture of NaCl and $CaBr_2$, identify the product that forms at the anode and at the cathode.

● **21.92** In the electrolysis of a molten mixture of RbF and $CaCl_2$, identify the product that forms at the negative electrode and at the positive electrode.

● **21.93** Identify those elements that can be prepared by electrolysis of their aqueous salts: copper, barium, aluminum, bromine.

● **21.94** Identify those elements that can be prepared by electrolysis of their aqueous salts: strontium, gold, tin, chlorine.

● **21.95** Identify those elements that can be prepared by electrolysis of their aqueous salts: lithium, iodine, zinc, silver.

● **21.96** Identify those elements that can be prepared by electrolysis of their aqueous salts: fluorine, manganese, iron, cadmium.

● **21.97** What product forms at each electrode in the aqueous electrolysis of the following salts: (a) LiF; (b) $SnSO_4$?

● **21.98** What product forms at each electrode in the aqueous electrolysis of the following salts: (a) $ZnBr_2$; (b) $Cu(HCO_3)_2$?

● **21.99** What product forms at each electrode in the aqueous electrolysis of the following salts: (a) $Cr(NO_3)_3$; (b) $MnCl_2$?

● **21.100** What product forms at each electrode in the aqueous electrolysis of the following salts: (a) FeI_2; (b) K_3PO_4?

● **21.101** Electrolysis of molten $MgCl_2$ is the final production step in the isolation of magnesium from seawater by the Dow process (Section 22.4). Assuming that 45.6 g of Mg metal forms,
(a) How many moles of electrons are required?
(b) How many coulombs are required?
(c) How many amps will produce this amount in 3.50 h?

● **21.102** Electrolysis of molten NaCl in a Downs cell is the major isolation step in the production of sodium metal (Section 22.4). Assuming that 215 g of Na metal forms,
(a) How many moles of electrons are required?
(b) How many coulombs are required?
(c) How many amps will produce this amount in 9.50 h?

● **21.103** How many grams of radium can form by passing 235 C through an electrolytic cell containing a molten radium salt?

● **21.104** How many grams of aluminum can form by passing 305 C through an electrolytic cell containing a molten aluminum salt?

● **21.105** How many seconds does it take to deposit 65.5 g of Zn on a steel gate when 21.0 A is passed through a $ZnSO_4$ solution?

● **21.106** How many seconds does it take to deposit 1.63 g of Ni on a decorative drawer handle when 13.7 A is passed through a $Ni(NO_3)_2$ solution?

● **Problems in Context**

● **21.107** A professor adds Na_2SO_4 to water to facilitate its electrolysis in a lecture demonstration. (a) What is the purpose of the Na_2SO_4? (b) Why is the water electrolyzed instead of the salt?

● **21.108** Subterranean brines in parts of the United States are rich in iodides and bromides and serve as an industrial source of these elements. In one recovery method, the brines are evaporated to dryness and then melted and electrolyzed. Which halogen is more likely to form from this treatment? Why?

● **21.109** Zinc plating (galvanizing) is an important means of corrosion protection. Although the process is done customarily by dipping the object into molten zinc, the metal can also be electroplated from aqueous solutions. How many grams of zinc can be deposited on a steel tank from a $ZnSO_4$ solution when a 0.855-A current flows for 2.50 days?

Comprehensive Problems

●● **21.110** The MnO_2 used in alkaline batteries can be produced by an electrochemical process of which one half-reaction is

$$Mn^{2+}(aq) + 2H_2O(l) \longrightarrow MnO_2(s) + 4H^+(aq) + 2e^-$$

If a current of 25.0 A is used, how many hours are needed to produce 1.00 kg of MnO_2? At which electrode is the MnO_2 formed?

●● **21.111** Car manufacturers are developing engines that use H_2 as fuel. In Iceland, Sweden, and other parts of Scandinavia, where hydroelectric plants produce inexpensive electric power, the H_2 can be made industrially by the electrolysis of water.

(a) How many coulombs are needed to produce 3.5×10^6 L of H_2 gas at 12.0 atm and 25°C? (Assume the ideal gas law applies.)

(b) If the coulombs are supplied at 1.44 V, how many joules are produced?

(c) If the combustion of oil yields 4.0×10^4 kJ/kg, what mass of oil must be burned to yield the number of joules in part (b)?

•• **21.112** The overall cell reaction occurring in an alkaline battery is

$$Zn(s) + MnO_2(s) + H_2O(l) \longrightarrow ZnO(s) + Mn(OH)_2(s)$$

(a) How many moles of electrons flow per mole of reaction?

(b) If 4.50 g of zinc is oxidized, how many grams of manganese dioxide and of water are consumed?

(c) What is the total mass of reactants consumed in part (b)?

(d) How many coulombs are produced in part (b)?

(e) In practice, voltaic cells of a given capacity (coulombs) are heavier than the calculation in part (c) indicates. Explain.

•• **21.113** An inexpensive and accurate method of measuring the quantity of electricity passing through a circuit is to pass it through a solution of a metal ion and weigh the metal deposited. A silver electrode immersed in an Ag^+ solution weighs 1.7854 g before the current passes and weighs 1.8016 g after the current has passed. How many coulombs have passed?

••• **21.114** Brass, an alloy of copper and zinc, can be produced by simultaneously electroplating the two metals from a solution containing their 2+ ions. If 65.0% of the total current is used to plate copper, while 35.0% goes to plating zinc, what is the mass percent of copper in the brass?

•• **21.115** Compare and contrast a voltaic cell and an electrolytic cell with respect to each of the following:

(a) Sign of the free energy change

(b) Nature of the half-reaction at the anode

(c) Nature of the half-reaction at the cathode

(d) Charge on the electrode labeled "anode"

(e) Electrode from which electrons leave the cell

••• **21.116** A thin circular-disk earring 4.00 cm in diameter is plated with a coating of gold 0.25 mm thick from an Au^{3+} bath.

(a) How many days does it take to deposit the gold on one side of one earring if the current is 0.013 A (d of gold = 19.3 g/cm³)?

(b) How many days does it take to deposit the gold on both sides of the pair of earrings?

(c) If the price of gold is $320 per troy ounce (31.10 g), what is the total cost of the gold plating?

•• **21.117** (a) How many minutes does it take to form 10.0 L of O_2 measured at 99.8 kPa and 28°C from water if a current of 1.3 A passes through the electrolytic cell? (b) What mass of H_2 forms?

••• **21.118** Trains powered by electricity, including subways, use direct current. One conductor is the overhead wire (or "third rail" for subways), and the other is the rails upon which the wheels run. The rails are on supports in contact with the ground. To minimize corrosion, should the overhead wire or the rails be connected to the positive terminal? Explain.

•• **21.119** A silver button battery used in a watch contains 0.75 g of zinc and can run until 80% of the zinc is consumed. (a) How many days can the battery run at a current of 0.85 microamps (10^{-6} amps)? (b) When the battery dies, 95% of the Ag_2O has been consumed. How many grams of Ag was used to make the battery? (c) If Ag costs $13.00 per troy ounce (31.10 g), what is the cost of the Ag consumed each day the watch runs?

•• **21.120** Like any piece of apparatus, an electrolytic cell operates at less than 100% efficiency. A cell depositing Cu from a Cu^{2+} bath

operates for 10 h with an average current of 5.8 A. If 53.4 g of copper is deposited, at what efficiency is the cell operating?

•• **21.121** Commercial electrolysis is performed on both molten NaCl and aqueous NaCl solutions. Identify the anode product, cathode product, species reduced, and species oxidized for the (a) molten electrolysis and (b) aqueous electrolysis.

•• **21.122** To examine the effect of ion removal on cell voltage, a chemist constructs two voltaic cells, each with a standard hydrogen electrode in one compartment. One cell also contains a Pb/Pb^{2+} half-cell; the other contains a Cu/Cu^{2+} half-cell.

(a) What is $E°$ of each cell at 298 K?

(b) Which electrode in each cell is negative?

(c) When Na_2S solution is added to the Pb^{2+} electrolyte, solid PbS forms. What happens to the cell voltage?

(d) When sufficient Na_2S is added to the Cu^{2+} electrolyte, CuS forms and $[Cu^{2+}]$ drops to 1×10^{-16} M. Find the cell voltage.

•• **21.123** Electrodes used in electrocardiography are disposable, and many incorporate silver. The metal is deposited in a thin layer on a small plastic "button," and then some is converted to AgCl:

$$Ag(s) + Cl^-(aq) \rightleftharpoons AgCl(s) + e^-$$

(a) If the surface area of the button is 2.0 cm² and the thickness of the silver layer is 7.5×10^{-6} m, calculate the volume (in cm³) of Ag used in one electrode.

(b) The density of silver metal is 10.5 g/cm³. How many grams of silver are used per electrode?

(c) If Ag is plated on the button from an Ag^+ solution with a current of 12.0 mA, how many minutes does the plating take?

(d) If bulk silver costs $13.00 per troy ounce (31.10 g), what is the cost (in cents) of the silver in one disposable electrode?

•• **21.124** Commercial aluminum production is done by the electrolysis of a bath containing Al_2O_3 dissolved in molten Na_3AlF_6. Why isn't it done by electrolysis of an aqueous $AlCl_3$ solution?

••• **21.125** Comparing the standard electrode potentials ($E°$) of the Group 1A(1) metals Li, Na, and K with the negative of their first ionization energies reveals a discrepancy:

Ionization process reversed: $M^+(g) + e^- \rightleftharpoons M(g)$ (−IE)

Electrode reaction: $M^+(aq) + e^- \rightleftharpoons M(s)$ ($E°$)

Metal	−IE (kJ/mol)	$E°$ (V)
Li	−520	−3.05
Na	−496	−2.71
K	−419	−2.93

Note that the electrode potentials do not decrease smoothly down the group, as the ionization energies do. You might expect that if it is more difficult to remove an electron from an atom to form a gaseous ion (larger IE), then it would be less difficult to add an electron to an aqueous ion to form an atom (smaller $E°$), yet $Li^+(aq)$ is *more* difficult to reduce than $Na^+(aq)$. Applying Hess's law, use an approach similar to that for a Born-Haber cycle to break down the process occurring at the electrode into three steps and label the energy involved in each step. How can you account for the discrepancy?

•• **21.126** To improve conductivity in the electroplating of automobile bumpers, a thin coating of copper separates the steel from a heavy coating of chromium.

(a) What mass of Cu is deposited on an automobile trim piece if plating continues for 1.25 h at a current of 5.0 A?

(b) If the area of the trim piece is 50.0 cm², what is the thickness of the Cu coating (d of Cu = 8.95 g/cm³)?

21.127 In Appendix D, standard electrode potentials range from about +3 to −3 V. Thus, it might seem possible to use a half-cell from each end of this range to construct a cell with a voltage of approximately 6 V. However, most commercial aqueous voltaic cells have $E°$ values of 1.5 to 2 V. Why are there no aqueous cells with significantly higher potentials?

21.128 Tin is used to coat "tin" cans used for food storage. If the tin is scratched and the iron of the can exposed, will the iron corrode more or less rapidly than if the tin were not present? Inside the can, the tin itself is coated with a clear varnish. Explain.

21.129 Commercial electrolytic cells for producing aluminum operate at 5.0 V and 100,000 A.
(a) How long does it take to produce exactly 1 metric ton (1000 kg) of aluminum?
(b) How much electrical power (in kilowatt-hours, kW·h) is used [1 W = 1 J/s; 1 kW·h = $3.6×10^3$ kJ]?
(c) If electricity costs 0.90¢ per kW·h and cell efficiency is 90.%, what is the cost of electricity to produce exactly 1 lb of aluminum?

21.130 Magnesium bars are connected electrically to underground iron pipes to serve as sacrificial anodes.
(a) Do electrons flow from the bar to the pipe or the reverse?
(b) A 12-kg Mg bar is attached to an iron pipe, and it takes 8.5 yr for the Mg to be consumed. What is the average current flowing between the Mg and the Fe during this period?

21.131 Bubbles of H_2 form when metal D is placed in hot H_2O. No reaction occurs when D is placed in a solution of a salt of metal E, but D is discolored and coated immediately when placed in a solution of a salt of metal F. What happens if E is placed in a solution of a salt of metal F? Rank metals D, E, and F in order of *increasing* reducing strength.

21.132 Calcium is obtained industrially by electrolysis of molten $CaCl_2$ and is used in aluminum alloys. How many coulombs are needed to produce 10.0 g of Ca metal? If a cell runs at 15 A, how many minutes will it take to produce 10.0 g of Ca(s)?

21.133 In addition to reacting with gold (see Problem 21.21), aqua regia is used to bring other precious metals into solution. Balance the skeleton equation for the reaction with Pt:

$$Pt(s) + NO_3^-(aq) + Cl^-(aq) \longrightarrow PtCl_6^{2-}(aq) + NO(g)$$

21.134 The following reactions are used in batteries:

I $2H_2(g) + O_2(g) \longrightarrow 2H_2O(l)$ $E_{cell} = 1.23$ V

II $Pb(s) + PbO_2(s) + 2H_2SO_4(aq) \longrightarrow$
$2PbSO_4(s) + 2H_2O(l)$ $E_{cell} = 2.04$ V

III $2Na(l) + FeCl_2(s) \longrightarrow 2NaCl(s) + Fe(s)$ $E_{cell} = 2.35$ V

Reaction I is used in fuel cells, II in the automobile lead-acid battery, and III in an experimental high-temperature battery for powering electric vehicles. The aim is to obtain as much work as possible from a cell, while keeping its weight to a minimum.
(a) In each cell, find the moles of electrons transferred and ΔG.
(b) Calculate the ratio, in kJ/g, of w_{max} to mass of reactants for each of the cells. Which has the highest ratio, which the lowest, and why? (*Note:* For simplicity, ignore the masses of cell components that do not appear in the cell as reactants, including electrode materials, electrolytes, separators, cell casing, wiring, etc.)

21.135 A current is applied to two electrolytic cells in series. In the first, silver is deposited; in the second, a zinc electrode is consumed. How much Ag is plated out if 1.2 g of Zn dissolves?

21.136 You are investigating a particular chemical reaction. State all the types of data available in standard tables that enable you to calculate the equilibrium constant for the reaction at 298 K.

21.137 In an electric power plant, personnel monitor the O_2 content of boiler feed water to prevent corrosion of the boiler tubes. Why does Fe corrode faster in steam and hot water than in cold water?

21.138 A voltaic cell using Cu/Cu²⁺ and Sn/Sn²⁺ half-cells is set up at standard conditions, and each compartment has a volume of 345 mL. The cell delivers 0.17 A for 48.0 h. (a) How many grams of Cu(s) are deposited? (b) What is the [Cu²⁺] remaining?

21.139 The $E_{half-cell}$ for the reduction of water is very different from the $E°_{half-cell}$ value. Calculate the $E_{half-cell}$ value when H_2 is in its standard state.

21.140 From the skeleton equations below, create a list of balanced half-reactions in which the strongest oxidizing agent is on top and the weakest is on the bottom:

$$U^{3+}(aq) + Cr^{3+}(aq) \longrightarrow Cr^{2+}(aq) + U^{4+}(aq)$$
$$Fe(s) + Sn^{2+}(aq) \longrightarrow Sn(s) + Fe^{2+}(aq)$$
$$Fe(s) + U^{4+}(aq) \longrightarrow \text{no reaction}$$
$$Cr^{3+}(aq) + Fe(s) \longrightarrow Cr^{2+}(aq) + Fe^{2+}(aq)$$
$$Cr^{2+}(aq) + Sn^{2+}(aq) \longrightarrow Sn(s) + Cr^{3+}(aq)$$

21.141 You are given the following three half-reactions:
(1) $Fe^{3+}(aq) + e^- \rightleftharpoons Fe^{2+}(aq)$
(2) $Fe^{2+}(aq) + 2e^- \rightleftharpoons Fe(s)$
(3) $Fe^{3+}(aq) + 3e^- \rightleftharpoons Fe(s)$
(a) Use $E°_{half-cell}$ values for (1) and (2) to find $E°_{half-cell}$ for (3).
(b) Calculate $\Delta G°$ for (1) and (2) from their $E°_{half-cell}$ values.
(c) Calculate $\Delta G°$ for (3) from (1) and (2).
(d) Calculate $E°_{half-cell}$ for (3) from its $\Delta G°$.
(e) What is the relationship between the $E°_{half-cell}$ values for (1) and (2) and the $E°_{half-cell}$ value for (3)?

21.142 Use the half-reaction method to balance the equation for the conversion of ethanol to acetic acid in acid solution:

$$CH_3CH_2OH + Cr_2O_7^{2-} \longrightarrow CH_3COOH + Cr^{3+}$$

21.143 When zinc is refined by electrolysis, the desired half-reaction at the cathode is

$$Zn^{2+}(aq) + 2e^- \longrightarrow Zn(s)$$

A competing reaction, which lowers the yield, is the formation of hydrogen gas:

$$2H^+(aq) + 2e^- \longrightarrow H_2(g)$$

If 91.50% of the current flowing results in zinc being deposited, while 8.50% produces hydrogen gas, how many liters of H_2, measured at STP, form per kilogram of zinc?

21.144 A chemist designs an ion-specific probe for measuring [Ag⁺] in an NaCl solution saturated with AgCl. One half-cell has an Ag-wire electrode immersed in the unknown AgCl-saturated NaCl solution. It is connected through a salt bridge to the other half-cell, which has a calomel reference electrode [a platinum wire immersed in a paste of mercury and calomel (Hg_2Cl_2)] in a saturated KCl solution. The measured E_{cell} is 0.060 V.
(a) Given the following standard half-reactions, calculate [Ag⁺].
Calomel: $Hg_2Cl_2(s) + 2e^- \longrightarrow$
$2Hg(l) + 2Cl^-(aq)$ $E° = 0.24$ V
Silver: $Ag^+(aq) + e^- \longrightarrow Ag(s)$ $E° = 0.80$ V
(*Hint:* Assume [Cl⁻] is so high that it is essentially constant.)
(b) A mining engineer wants an ore sample analyzed with the Ag⁺-selective probe. After pretreating the ore sample, the chemist measures the cell voltage as 0.53 V. What is [Ag⁺]?

•• **21.145** Use Appendix D to calculate the K_{sp} of AgCl.

••• **21.146** Black-and-white photographic film is coated with silver halides. Because silver is expensive, the manufacturer monitors the Ag^+ content of the waste stream, $[Ag^+]_{waste}$, from the plant with an Ag^+-selective electrode at 25°C. A stream of known Ag^+ concentration, $[Ag^+]_{standard}$, is passed over the electrode in turn with the waste stream and the data recorded by a computer.
(a) Write the equations relating the nonstandard cell potential to the standard cell potential and $[Ag^+]$ for each solution.
(b) Combine these into a single equation to find $[Ag^+]_{waste}$.
(c) Rewrite the equation from part (b) to find $[Ag^+]_{waste}$ in ng/L.
(d) If E_{waste} is 0.003 V higher than $E_{standard}$, and the standard solution contains 1000. ng/L, what is $[Ag^+]_{waste}$?
(e) Rewrite the equation in part (b) to find $[Ag^+]_{waste}$ for a system in which T changes and T_{waste} and $T_{standard}$ may be different.

•• **21.147** Calculate the K_f of $Ag(NH_3)_2^+$ from

$$Ag^+(aq) + e^- \rightleftharpoons Ag(s) \qquad\qquad E° = 0.80 \text{ V}$$
$$Ag(NH_3)_2^+(aq) + e^- \rightleftharpoons Ag(s) + 2NH_3(aq) \qquad E° = 0.37 \text{ V}$$

•• **21.148** Even though the toxicity of cadmium has become a concern, nickel-cadmium (nicad) batteries are still used commonly in many devices. The overall cell reaction is

$$Cd(s) + 2NiO(OH)(s) + 2H_2O(l) \longrightarrow$$
$$2Ni(OH)(s) + Cd(OH)_2(s)$$

A certain nicad battery weighs 18.3 g and has a capacity of 300. mA·h (that is, the cell can store charge equivalent to a current of 300. mA flowing for 1 h).
(a) What is the capacity of this cell in coulombs?
(b) What mass of reactants is needed to deliver 300. mA·h?
(c) What percentage of the cell mass consists of reactants?

•• **21.149** The zinc-air battery is a less expensive alternative for silver batteries in hearing aids. The cell reaction is

$$2Zn(s) + O_2(g) \longrightarrow 2ZnO(s)$$

A new battery weighs 0.275 g. The zinc accounts for exactly $\frac{1}{10}$ of the mass, and the oxygen does not contribute to the mass because it is supplied by the air.
(a) How much electricity (in C) can the battery deliver?
(b) How much free energy (in J) is released if E_{cell} is 1.3 V?

•• **21.150** Use Appendix D to create an activity series of Mn, Fe, Ag, Sn, Cr, Cu, Ba, Al, Na, Hg, Ni, Li, Au, Zn, and Pb. Rank these metals in order of decreasing reducing strength, and divide them into three groups: those that displace H_2 from water, those that displace H_2 from acid, and those that cannot displace H_2.

•• **21.151** Both Ti and V are reactive enough to displace H_2 from water. The difference in their $E°_{half-cell}$ values is 0.43 V. Given

$$V(s) + Cu^{2+}(aq) \longrightarrow V^{2+}(aq) + Cu(s) \quad \Delta G° = -298 \text{ kJ/mol}$$

use Appendix D to calculate the $E°_{half-cell}$ values for V and Ti.

•• **21.152** For the reaction

$$S_4O_6^{2-}(aq) + 2I^-(aq) \longrightarrow I_2(s) + S_2O_3^{2-}(aq)$$
$$\Delta G° = 87.8 \text{ kJ/mol}$$

(a) Identify the oxidizing and reducing agents. (b) Calculate $E°_{cell}$. (c) For the reduction half-reaction, write a balanced equation, give the oxidation number of each element, and calculate $E°_{half-cell}$.

••• **21.153** Two concentration cells are prepared, both with 90.0 mL of 0.0100 M $Cu(NO_3)_2$ and a Cu bar in each half-cell.
(a) In the first concentration cell, 10.0 mL of 0.500 M NH_3 is added to one half-cell; the complex ion $Cu(NH_3)_4^{2+}$ forms, and E_{cell} is 0.129 V. Calculate K_f for the formation of the complex ion.

(b) Calculate E_{cell} when an additional 10.0 mL of 0.500 M NH_3 is added.
(c) In the second concentration cell, 10.0 mL of 0.500 M NaOH is added to one half-cell; the precipitate $Cu(OH)_2$ forms ($K_{sp} = 2.2 \times 10^{-20}$). Calculate $E°_{cell}$.
(d) What would the molarity of NaOH have to be for the addition of 10.0 mL to result in an $E°_{cell}$ of 0.340 V?

••• **21.154** Two voltaic cells are to be joined so that one will run the other as an electrolytic cell. In the first cell, one half-cell has Au foil in 1.00 M $Au(NO_3)_3$, and the other half-cell has a Cr bar in 1.00 M $Cr(NO_3)_3$. In the second cell, one half-cell has a Co bar in 1.00 M $Co(NO_3)_2$, and the other half-cell has a Zn bar in 1.00 M $Zn(NO_3)_2$. (a) Calculate $E°_{cell}$ for each cell. (b) Calculate the total potential if the two cells are connected as voltaic cells in series. (c) When the electrode wires are switched in one of the cells, which cell will run as the voltaic cell and which as the electrolytic cell? (d) Which metal ion is being reduced in each cell? (e) If 2.00 g of metal plates out in the voltaic cell, how much metal ion plates out in the electrolytic cell?

•• **21.155** A voltaic cell has one half-cell with a Cu bar in a 1.00 M Cu^{2+} salt, and the other half-cell with a Cd bar in the same volume of a 1.00 M Cd^{2+} salt. (a) Find $E°_{cell}$, $\Delta G°$, and K. (b) As the cell operates, $[Cd^{2+}]$ increases; find E_{cell} and ΔG when $[Cd^{2+}]$ is 1.95 M. (c) Find E_{cell}, ΔG, and $[Cu^{2+}]$ at equilibrium.

•• **21.156** Gasoline is a mixture of hydrocarbons, but the heat released when it burns is close to that of octane, $C_8H_{18}(l)$ ($\Delta H°_f = -250.1$ kJ/mol). As an alternative to gasoline, research is underway to use H_2 from the electrolysis of water in fuel cells to power cars.
(a) Calculate $\Delta H°$ when 1.00 gal of gasoline ($d = 0.7028$ g/mL) burns to produce carbon dioxide gas and water vapor.
(b) How many liters of H_2 at 25°C and 1.00 atm must burn to produce this quantity of energy?
(c) How long would it take to produce this amount of H_2 by electrolysis with a current of 1.00×10^3 A at 6.00 V?
(d) How much power in kilowatt·hours (kW·h) is required to generate this amount of H_2? (1 W = 1 J/s, 1 J = 1 C·V, and 1 kW·h = 3.6×10^6 J.)
(e) If the cell is 88.0% efficient and electricity costs 0.950¢ per kW·h, what is the cost of producing the amount of H_2 equivalent to 1.00 gal of gasoline?

•• **21.157** Balance the following redox reactions:
(a) In acidic solution, xenon trioxide reacts with iodide ion to produce xenon gas, triiodide ion (I_3^-), and water.
(b) In basic solution, the hydrogen xenate ion ($HXeO_4^-$) disproportionates to xenon gas, perxenate ion (XeO_6^{4-}), water, and oxygen gas.
(c) In basic solution, bismuthate ion (BiO_3^-) reacts with manganese(II) ion to produce bismuth(III) and permanganate ions.
(d) In basic solution, oxygen difluoride reacts to produce fluoride ion, oxygen gas, and water.
(e) In acidic solution, manganese(IV) oxide reacts with sulfite ion to form manganese(II) ion, water, and dithionate ion ($S_2O_6^{2-}$).
(f) In one of the few reactions known for astatine, the oxidation of astinide ion by chlorine gas in acid solution produces astinate ion (AtO_3^-) and chloride ion.

•• **21.158** If the E_{cell} of the following cell is 0.915 V, what is the pH in the anode compartment?

$$Pt(s) \mid H_2(1.00 \text{ atm}) \mid H^+(aq) \parallel Ag^+(0.100 \, M) \mid Ag(s)$$

22

Essential Products from Industry In industrial plants around the world, concentrated NaCl solutions are used to produce millions of metric tons of two of the most important chemicals, Cl_2 and NaOH. In this chapter, we apply concepts from several previous chapters to the practical side of chemistry—the ways in which we obtain and utilize the elements and their compounds.

The Elements in Nature and Industry

22.1 How the Elements Occur in Nature
Earth's Structure and the Abundance of the Elements
Sources of the Elements

22.2 The Cycling of Elements Through the Environment
The Carbon Cycle
The Nitrogen Cycle
The Phosphorus Cycle

22.3 Metallurgy: Extracting a Metal from Its Ore
Pretreating the Ore
Converting Mineral to Element
Refining and Alloying

22.4 Tapping the Crust: Isolation and Uses of Selected Elements
Sodium and Potassium
Iron, Copper, and Aluminum
Magnesium and Bromine
Hydrogen

22.5 Chemical Manufacturing: Two Case Studies
Sulfuric Acid
The Chlor-Alkali Process

In this chapter, we return to the periodic table, but from a new perspective. In Chapter 14, we considered the atomic, physical, and chemical behavior of the main-group elements. Here, we examine various elements from a practical viewpoint to learn how much of an element is present in nature, in what form it occurs, how organisms affect its distribution, and how we isolate it for our own use.

Remember that chemistry is, above all, a practical science, and its concepts were developed to address real-life problems: How do we isolate and recycle aluminum? How do we make iron stronger? How do we lower energy costs and increase yield in the production of sulfuric acid? How does fossil-fuel combustion affect the environmental cycling of carbon? Here, we apply concepts you learned from the chapters on kinetics, equilibrium, thermodynamics, and electrochemistry to explain chemical processes in nature and industry that influence our lives.

IN THIS CHAPTER . . . We first discuss the abundances and sources of elements on Earth, then consider how three essential elements—carbon, nitrogen, and phosphorus—cycle through the environment. After seeing where and in what forms elements are found, we focus on their isolation and utilization. We discuss general redox and metallurgical procedures for extracting an element from its ore and examine in detail the isolation and uses of certain key elements. The chapter ends with a close look at two of the most important processes in chemical manufacturing: the production of sulfuric acid and the isolation of chlorine.

Concepts & Skills to Review
before you study this chapter

- catalysts and reaction rate (Section 16.8)
- Le Châtelier's principle (Section 17.6)
- acid-base equilibria (Sections 18.3 and 18.9)
- solubility and complex-ion equilibria (Sections 19.3 and 19.4)
- temperature and reaction spontaneity (Section 20.3)
- free energy and equilibrium (Section 20.4)
- standard electrode potentials (Section 21.3)
- electrolysis of molten salts and aqueous salt solutions (Section 21.7)

22.1 HOW THE ELEMENTS OCCUR IN NATURE

To begin our examination of how we use the elements, let's take inventory of our elemental stock—the distribution and relative amounts of the elements on Earth, especially that thin outer portion of the planet that we can reach.

Earth's Structure and the Abundance of the Elements

Any attempt to isolate an element must begin with a knowledge of its **abundance,** the amount of the element in a particular region of the natural world. The abundances of the elements on Earth and in its various regions are the result of the specific details of our planet's evolution.

Literature Dutch, Steven I. "Periodic Tables of Elemental Abundance," *J. Chem. Educ.* **1999**, *76*, 356.

Formation and Layering of Planet Earth About 4.5 billion years ago, vast clouds of cold gases and interstellar debris from exploded older stars gradually coalesced into the Sun and planets. At first, Earth was a cold, solid sphere of uniformly distributed elements and simple compounds. In the next billion years or so, heat from radioactive decay and virtually continuous meteor impacts raised the planet's temperature to around 10^4 K, sufficient to form an enormous molten mass. Any remaining gaseous elements, such as the cosmically abundant hydrogen and helium, were ejected into space.

As Earth cooled, chemical and physical processes resulted in its **differentiation,** the formation of regions of different composition and density. Differentiation gave Earth its layered internal structure, which consists of a dense (10–15 g/cm^3) **core,** composed of a molten outer core and a solid Moon-sized inner core. (Recent evidence has revealed remarkable properties of the inner core: it is nearly as hot as the surface of the Sun and spins within the molten outer core slightly faster than does Earth itself!) Around the core lies a thick homogeneous **mantle,** consisting of lower and upper portions, with an overall density of 4–6 g/cm^3. All the comings and goings of life take place on the thin heterogeneous **crust** (average density = 2.8 g/cm^3). Figure 22.1 on page 982 shows these layers and compares the abundances of some key elements in the universe, the whole Earth (actually core plus mantle only, which account for more than 99%

ELEMENT	UNIVERSE	EARTH	CRUST 3–72 km (2.8 g/cm³)	MANTLE 2880 km (4–6 g/cm³)	CORE 3440 km (10–15 g/cm³)
O	1.07	29.5	49.5	43.7	—
Si	0.06	15.2	25.7	21.6	—
Al	—	1.1	7.5	1.8	—
Fe	0.19	34.6	4.7	13.3	88.6
Ca	0.007	1.1	3.4	2.1	—
Mg	0.06	12.7	2.8	16.6	—
Na	—	0.6	2.6	0.8	—
K	—	0.07	2.4	0.2	—
H	73.9	—	0.87	—	—
Ti	—	0.05	0.58	0.18	0.01
Cl	—	—	0.19	—	—
P	—	—	0.12	—	—
Mn	—	—	0.09	—	—
C	0.46	—	0.08	—	—
S	0.04	1.9	0.06	>2	—
Ni	0.006	2.4	0.008	0.3	8.5
Co	—	0.13	0.003	0.04	0.6

Core

Mantle

Crust

Figure 22.1 Cosmic and terrestrial abundances of selected elements (mass percent). The mass percents of the elements are superimposed on a cutaway of Earth. The internal structure consists of a large core, thick mantle, and thin crust. The densities of these regions are indicated in the column heads. Blank abundance values mean either that reliable data are not available or that the value is less than 0.001% by mass. (Helium is not listed because it is not abundant on Earth, but it accounts for 24.0% by mass of the universe.)

Transparency/ARIS Presentation Center, Art Library

of its mass), and Earth's three regions. (Because the deepest terrestrial sampling can penetrate only a few kilometers into the crust, some of these data represent extrapolations from meteor samples and from seismic studies of earthquakes.) Several points stand out:

1. Cosmic and whole-Earth abundances are very different, particularly for H.
2. The elements O, Si, Fe, and Mg are abundant both cosmically and on Earth. Together, they account for more than 90% of Earth's mass.
3. The core is particularly rich in the dense Group 8B(8) through 8B(10) metals: Co, Ni, and especially Fe, the most abundant element in the whole Earth.
4. Crustal abundances are very different from whole-Earth abundances. The crust makes up only 0.4% of Earth's mass, but has the largest share of nonmetals, metalloids, and light, active metals: Al, Ca, Na, and K. The mantle contains smaller proportions of these, and the core has none. Oxygen is the most abundant element in the crust and mantle but is absent from the core.

These compositional differences in Earth's major layers, or *phases,* arose from the effects of thermal energy. When Earth was molten, gravity and convection caused more dense materials to sink and less dense materials to rise. Most of the Fe sank to form the core, or *iron phase.* In the light outer phase, oxygen combined with Si, Al, Mg, and some Fe to form silicates, the material of rocks. This *silicate phase* later separated into the mantle and crust. The *sulfide phase,* intermediate in density and insoluble in the other two, consisted mostly of iron sulfide and mixed with parts of the silicate phase above and the iron phase below. *Outgassing,* the expulsion of trapped gases, produced a thin, primitive atmosphere, probably a mixture of water vapor (which gave rise to the oceans), carbon monoxide, and nitrogen (or ammonia).

The distribution of the remaining elements (discussed here using only the new group numbers) was controlled by their chemical affinity for one of the three phases. In general terms, as Figure 22.2 shows, elements with low or high electronegativity—active metals (Groups 1 through 5, Cr, and Mn) and nonmetals (O, lighter members of Groups 13 to 15, and all of Group 17)—tended to congregate in the silicate phase as ionic compounds. Metals with intermediate electronegativities (many from Groups 6 to 10) dissolved in the iron core. Lower-melting transition metals and many metals and metalloids in Groups 11 to 16 became concentrated in the sulfide phase.

Figure 22.2 Geochemical differentiation of the elements. During Earth's formation, elements became concentrated primarily in the atmosphere or one of three phases: silicate, sulfide, or iron.

Transparency/ARIS Presentation Center, Art Library

Biological application

The Impact of Life on Crustal Abundances

At present, the crust is the only physically accessible portion of Earth, so only crustal abundances have practical significance. The crust is divided into solid, liquid, and gaseous portions called the **lithosphere, hydrosphere,** and **atmosphere.** Over billions of years, weathering and volcanic upsurges have dramatically altered the composition of the crust. The **biosphere,** which consists of the living systems that have inhabited the planet, has been another *major influence on crustal element composition and distribution.*

When the earliest rocks were forming and the ocean basins filling with water, binary inorganic molecules in the atmosphere were reacting to form first simple, and then more complex, organic molecules. The energy for these (mostly) endothermic changes was supplied by lightning, solar radiation, geologic heating, and meteoric impact. In an amazingly short period of time, probably no more than 500 million years, the first organisms appeared. It took less than another billion years for these to evolve into simple algae that could derive metabolic energy from photosynthesis, converting CO_2 and H_2O into organic molecules and releasing O_2 as a byproduct. The importance of this process to crustal chemistry cannot be overstated. Over the next 300 million years, the atmosphere gradually became richer in O_2, and *oxidation became the major source of free energy in the crust and biosphere.* Geologists mark this period by the appearance of Fe(III)-containing minerals in place of the Fe(II)-containing minerals that had predominated (Figure 22.3A). Paleontologists see in it an explosion of O_2-utilizing lifeforms, a few of which evolved into the organisms of today (Figure 22.3B).

Figure 22.3 Ancient effects of an O_2-rich atmosphere. A, With the appearance of photosynthetic organisms and the resulting oxidizing environment, much of the iron(II) in minerals was oxidized to iron(III). These ancient banded-iron formations (red beds) from Michigan contain hematite, Fe_2O_3. **B,** A fossil of *Hallucigenia sparsa*, one of the multicellular organisms found in the Burgess shale, British Columbia, Canada, whose appearance coincided with the increase in O_2. Many of these organisms have no known modern descendants, having disappeared in a mass extinction at the end of the Cambrian Period about 505 million years ago.

ARIS Presentation Center, Photo Library

In addition to creating an oxidizing environment, organisms have had profound effects on the *distribution* of specific elements. Why, for instance, is the K^+ concentration of the oceans so much lower than the Na^+ concentration? Forming over eons from outgassed water vapor condensing into rain and streaming over the land, the oceans became complex ionic solutions of dissolved minerals with 30 times as much Na^+ as K^+. There are two principal reasons for this difference. Clays, which make up a large portion of many soils, bind K^+ preferentially through an ion-exchange process. Even more importantly, plants require K^+ for growth, absorbing dissolved K^+ that would otherwise wash down to the sea.

As another example, consider the enormous subterranean deposits of organic carbon. Buried deeply and decomposing under high pressure and temperature in the absence of free O_2, ancient plants gradually turned into coal, and animals buried under similar conditions turned into petroleum. Crustal deposits of this organic carbon provide the fuels that move our cars, heat our homes, and electrify our cities. Moreover, carbon *and* calcium, the fifth most abundant element in the crust, occur together in vast sedimentary deposits all over the world as limestone, dolomite, marble, and chalk, the fossilized skeletal remains of early marine organisms (see photo).

Table 22.1 compares selected elemental abundances in the whole crust, the three crustal regions, and the human body, a representative portion of the biosphere. Note the quantities of the four major elements of life—O, C, H, and N. Oxygen is either the first or second most abundant element in all cases. The biosphere contains large amounts of carbon in its biomolecules and large amounts of hydrogen as water. Nitrogen is abundant in the atmosphere as free N_2 and in organisms as part of their proteins. Phosphorus and sulfur occur in exceptionally high amounts in organisms, too.

One striking difference among the lithosphere, hydrosphere, and biosphere (human) is in the abundances of the transition metals vanadium through zinc.

Dolomite Mountains, Italy.

ARIS Presentation Center, Photo Library

Transparency/ARIS Presentation Center, Tables Library

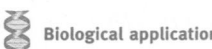

Biological application

Table 22.1   Abundances of Selected Elements in the Crust, Its Regions, and the Human Body as Representative of the Biosphere (Mass %)

| Element | Crust | Crustal Regions | | | Human |
		Lithosphere	Hydrosphere	Atmosphere	
O	49.5	45.5	85.8	23.0	65.0
C	0.08	0.018	—	0.01	18.0
H	0.87	0.15	10.7	0.02	10.0
N	0.03	0.002	—	75.5	3.0
P	0.12	0.11	—	—	1.0
Mg	1.9	2.76	0.13	—	0.50
K	2.4	1.84	0.04	—	0.34
Ca	3.4	4.66	0.05	—	2.4
S	0.06	0.034	—	—	0.26
Na	2.6	2.27	1.1	—	0.14
Cl	0.19	0.013	2.1	—	0.15
Fe	4.7	6.2	—	—	0.005
Zn	0.013	0.008	—	—	0.003
Cr	0.02	0.012	—	—	3×10^{-6}
Co	0.003	0.003	—	—	3×10^{-6}
Cu	0.007	0.007	—	—	4×10^{-4}
Mn	0.09	0.11	—	—	1×10^{-4}
Ni	0.008	0.010	—	—	3×10^{-6}
V	0.015	0.014	—	—	3×10^{-6}

Their relatively insoluble oxides and sulfides make them much scarcer in water than on land. Yet organisms, which evolved in the seas, developed the ability to concentrate the trace amounts of these elements present in their aqueous environment. In every case, the biological concentration shows an increase of at least 100-fold, with Mn increasing about 1000-fold, and Cu, Zn, and Fe even more. Each of these elements performs an essential role in living systems (see Chemical Connections, pp. 1055–1056).

Sources of the Elements

Given an element's abundance in a region of the crust, we must next determine its **occurrence,** or **source,** the form(s) in which the element exists. Practical considerations often determine the commercial source. Oxygen, for example, is abundant in all three crustal regions, but the atmosphere is its primary industrial source because it occurs there as the free element. Nitrogen and the noble gases (except helium) are also obtained from the atmosphere. Several other elements occur uncombined, formed in large deposits by prehistoric biological action; examples are sulfur in caprock salt domes and nearly pure carbon in coal. The relatively unreactive elements gold and platinum also occur in an uncombined (*native*) state.

The overwhelming majority of elements, however, occur in **ores,** natural compounds or mixtures of compounds from which an element can be extracted by economically feasible means (see photo). The phrase "economically feasible" refers to an important point: *the financial costs of mining, isolating, and purifying an element must be considered when choosing a process to obtain it.*

Figure 22.4 shows the most useful sources of the elements. Alkali metal halides are ores for both of their component groups of elements. One example of how cost influences recovery concerns the feasibility of using silicates as ores. Even though many elements occur as silicates, most of these sources are very stable thermodynamically. Thus, the cost of the energy required to process them prohibits their use—aside from silicon, only lithium and beryllium are obtained from their silicates. The Group 2A(2) metals occur as carbonates in the marble and limestone of mountain ranges, although magnesium's great abundance in seawater makes that the preferred source, as we discuss later.

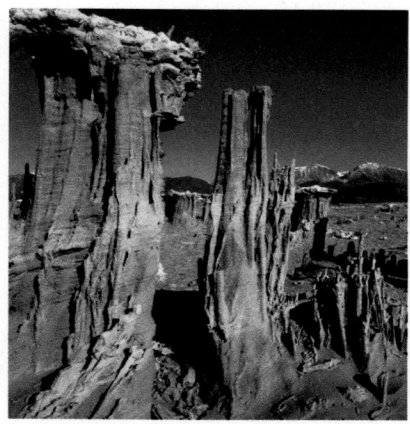

Deposit of borate (tufa), the ore of boron.

ARIS Presentation Center, Photo Library

Figure 22.4 **Sources of the elements.** The major sources of most of the elements are shown. Although a few elements occur uncombined, most are obtained from their oxide or sulfide ores.

Transparency/ARIS Presentation Center, Art Library

The ores of most industrially important metals are either *oxides*, which dominate the left half of the transition series, or *sulfides*, which dominate the right half of the transition series and a few of the main groups beyond. The reasons for the prominence of oxides and sulfides are complex and include processes of weathering, selective precipitation, and relative solubilities. Certain atomic properties are relevant as well. Elements toward the left side of the periodic table have lower ionization energies and electronegativities, so they tend to give up electrons or hold them loosely in bonds. The O^{2-} ion is small enough to approach a metal cation closely, which results in a high lattice energy for the oxide. In contrast, elements toward the right side have higher ionization energies and electronegativities. Thus, they tend to form bonds that are more covalent, which suits the larger, more polarizable S^{2-} ion.

Section Summary

As the young Earth cooled, the elements became differentiated into a dense, metallic core, a silicate-rich mantle, and a low-density crust. • High abundances of light metals, metalloids, and nonmetals are concentrated in the crust, which consists of the lithosphere (solid), hydrosphere (liquid), and atmosphere (gaseous). • The biosphere (living systems) has affected crustal chemistry primarily by producing free O_2, and thus an oxidizing environment. • Some elements occur in their native state, but most are combined in ores. The ores of most important metallic elements are oxides or sulfides.

Biological application

Environmental application

22.2 THE CYCLING OF ELEMENTS THROUGH THE ENVIRONMENT

The distributions of many elements are in flux, changing at widely differing rates. The physical, chemical, and biological paths that the atoms of an element take on their journey through regions of the crust constitute the element's **environmental cycle**. In this section, we consider the cycles for three important elements—carbon, nitrogen, and phosphorus—and highlight the effects of humans on them.

The Carbon Cycle

Carbon is one of a handful of elements that appear in all three portions of Earth's crust. In the lithosphere, it occurs in elemental form as graphite and diamond, in fully oxidized form in carbonate minerals, in fully reduced form in petroleum hydrocarbons, and in complex mixtures, such as coal and living matter. In the hydrosphere, it occurs in living matter, in carbonate minerals formed by the action of coral-reef organisms, and in dissolved CO_2. In the atmosphere, it occurs principally in gaseous CO_2, a minor but essential component that exists in equilibrium with the aqueous fraction.

Figure 22.5 depicts the complex interplay of the carbon sources and the considerable effect of the biosphere on the element's environmental cycle. It provides an estimate of the size of each source and, where reliable numbers exist, the amount of carbon moving annually between sources. Some key points to note are

- The portions of the cycle are linked by the atmosphere—one link between oceans and air, the other between land and air. The 2.6×10^{12} metric tons of CO_2 in air cycles through the oceans and atmosphere about once every 300 years but spends a much longer time in carbonate minerals.
- Atmospheric CO_2 accounts for a tiny fraction (0.003%) of crustal carbon, but the atmosphere is the prime mover of carbon through the other regions. Estimates indicate that a CO_2 molecule spends, on average, 3.5 years in the atmosphere.

Figure 22.5 The carbon cycle. The major sources and changes in carbon distribution are shown. Note that the atmosphere is the major conduit of carbon between land and sea and that CO_2 becomes fixed through the action of organisms. Numbers in boxes refer to the size of the source; numbers along arrows refer to the annual movement of the element from source to source. Values are in 10^9 metric tons of C (1 metric ton = 1000 kg = 1.1 tons).

Transparency/ARIS Presentation Center, Art Library

- The land and oceans are also in contact with the largest sources of carbon, which are effectively immobilized as carbonates, coal, and oil in the rocky sediment beneath the soil.
- The biological processes of *photosynthesis, respiration,* and *decay* are major factors in the carbon cycle. **Fixation** is the process of converting a gaseous substance into a condensed, more usable form. Via photosynthesis, marine plankton and terrestrial plants use sunlight to fix atmospheric CO_2 into carbohydrates. Plants release CO_2 by respiration at night and, much more slowly, when they decay. Animals eat the plants and release CO_2 by respiration and in their decay.
- Natural fires and volcanoes also release CO_2 into the air.

For hundreds of millions of years, the cycle has maintained a relatively constant amount of atmospheric CO_2. The tremendous growth of human industry over the past century and a half, and especially since World War II, has shifted this natural balance by increasing atmospheric CO_2. The principal cause of this increase is the combustion of coal, wood, and oil for fuel and for the thermal decomposition of limestone to make cement, coupled with the simultaneous clearing of vast stretches of forest and jungle for lumber, paper, and agriculture. With the 1990s the hottest decade ever recorded, evidence is mounting that these activities are creating one of the most dire environmental changes in human history—global warming through the greenhouse effect (see Chemical Connections, p. 257). Higher temperatures, alterations in dry and rainy seasons, melting of polar ice, and increasing ocean acidity, with associated changes in carbonate equilibria, are some of the results currently being observed. Nearly all environmental and science policy experts are pressing for the immediate adoption of a program that combines conservation of carbon-based fuels, an end to deforestation, and development of alternative energy sources.

The Nitrogen Cycle

In contrast to the carbon cycle, the nitrogen cycle includes a direct interaction of land and sea (Figure 22.6). All nitrites and nitrates are soluble, so rain and runoff contribute huge amounts of nitrogen to lakes, rivers, and oceans. Human activity, in the form of fertilizer production and use, plays a major role in this cycle.

Like carbon dioxide, atmospheric nitrogen must be fixed to be used by organisms. However, whereas CO_2 can be incorporated by plants in either its gaseous or aqueous form, the great stability of N_2 prevents plants from using it directly. Fixation of N_2 requires a great deal of energy and occurs through atmospheric, industrial, and biological processes:

1. *Atmospheric fixation.* Lightning brings about the high-temperature endothermic reaction of N_2 and O_2 to form NO, which is then oxidized by ozone exothermically to NO_2:

$$N_2(g) + O_2(g) \longrightarrow 2NO(g) \qquad \Delta H° = 180.6 \text{ kJ}$$
$$NO(g) + O_3(g) \longrightarrow NO_2(g) + O_2(g) \qquad \Delta H° = -200.1 \text{ kJ}$$

During the day, NO_2 reacts with hydroxyl radical to form HNO_3:

$$NO_2(g) + HO\cdot(g) \longrightarrow HNO_3(g)$$

At night, a multistep reaction between NO_2 and ozone is involved. In either case, rain dissociates the nitric acid, and $NO_3^-(aq)$ enters both sea and land to be utilized by plants.

2. *Industrial fixation.* Most human-caused fixation occurs industrially during ammonia synthesis via the Haber process (see Chemical Connections, pp. 771–772). The process takes place on an enormous scale: on a mole basis, NH_3 ranks first among compounds in amount produced. Some of this NH_3 is converted to HNO_3 in the Ostwald process (Chapter 14, p. 600), but most is used as fertilizer, either directly or in the form of urea and ammonium salts (sulfate, phosphate, and nitrate), which enter the biosphere as they are taken up by plants (item 3 below). In recent decades, high-temperature combustion in electric power plants and in car, truck, and plane engines has become an important contributor to total fixed nitrogen, mainly because of the enormous rise in vehicle traffic. High engine operating temperatures mimic lightning to form NO from the air taken in to burn the hydrocarbon fuel. The NO in exhaust gases reacts to form nitric acid in the atmosphere, which adds to the nitrate load that reaches the ground.

The overuse of fertilizers and automobiles presents an increasingly serious water pollution problem in many areas. Leaching of the land by rain causes nitrate from fertilizer use and vehicle operation to enter natural waters (lakes, rivers, and

Environmental application

Engineering application

Figure 22.6 **The nitrogen cycle.** The major source of nitrogen, N_2, is fixed atmospherically, industrially, and biologically. Various types of bacteria are indispensable throughout the cycle. Numbers in boxes are in 10^9 metric tons of N and refer to the size of the source; numbers along arrows are in 10^6 metric tons of N and refer to the annual movement of the element between sources.

coastal estuaries) and cause *eutrophication,* the depletion of O_2 and the death of aquatic animal life from excessive algal and plant growth and decay. Excess nitrate also spoils nearby drinkable water.

3. *Biological fixation.* The biological fixation of atmospheric N_2 occurs in blue-green algae and in nitrogen-fixing bacteria that live on the roots of leguminous plants (such as peas, alfalfa, and clover). On a planetary scale, these microbial processes dwarf the previous two, fixing more than seven times as much nitrogen as the atmosphere and six times as much as industry. Root bacteria fix N_2 by reducing it to NH_3 and NH_4^+ through the action of enzymes that contain the transition metal molybdenum at their active site. Enzymes in other soil bacteria catalyze the multistep oxidation of NH_4^+ to NO_2^- and finally NO_3^-, which the plants reduce again to make their proteins. When the plants die, still other soil bacteria oxidize the proteins to NO_3^-.

Animals eat the plants, use the plant proteins to make their own proteins, and excrete nitrogenous wastes, such as urea [$(H_2N)_2C{=}O$]. The nitrogen in the proteins is released when the animals die and decay, and is converted by soil bacteria to NO_2^- and NO_3^- again. This central pool of inorganic nitrate has three main fates: some enters marine and terrestrial plants, some enters the enormous sediment store of mineral nitrates, and some is reduced by denitrifying bacteria to NO_2^- and then reduced further to N_2O and N_2, which re-enter the atmosphere to complete the cycle.

The Phosphorus Cycle

Virtually all the mineral sources of phosphorus contain the phosphate group, PO_4^{3-}. The most commercially important ores are **apatites,** compounds of general formula $Ca_5(PO_4)_3X$, where X is usually F, Cl, or OH. The cycling of phosphorus through the environment involves three interlocking subcycles, illustrated in Figure 22.7. Two rapid biological cycles—a land-based cycle completed in a matter of years and a water-based cycle completed in weeks to years—are superimposed on an inorganic cycle that takes millions of years to complete. Unlike the carbon and nitrogen cycles, the phosphorus cycle has no gaseous component and thus does *not* involve the atmosphere.

The Inorganic Cycle Most phosphate rock formed when Earth's lithosphere solidified. ● The most important components of this material are nearly insoluble phosphate salts:

$$K_{sp} \text{ of } Ca_3(PO_4)_2 \approx 10^{-29}$$
$$K_{sp} \text{ of } Ca_5(PO_4)_3OH \approx 10^{-51}$$
$$K_{sp} \text{ of } Ca_5(PO_4)_3F \approx 10^{-60}$$

Weathering slowly leaches phosphates from the soil and carries the ions through rivers to the sea. Plants in the land-based biological cycle speed this process. In the ocean, some phosphate is absorbed by organisms in the water-based biological cycle, but the majority is precipitated again by Ca^{2+} ion and deposited on the continental shelf. Geologic activity lifts the continental shelves, returning the phosphate to the land.

The Land-Based Biological Cycle The biological cycles involve the incorporation of phosphate into organisms (biomolecules, bones, teeth, and so forth) and the release of phosphate through their excretion and decay. In the land-based cycle, plants continually remove phosphate from the inorganic cycle. Recall that there are three phosphate oxoanions, and they exist in equilibrium in the aqueous environment:

$$H_2PO_4^- \rightleftharpoons HPO_4^{2-} \rightleftharpoons PO_4^{3-}$$
$$\quad\quad + H^+ \quad\quad + H^+$$

In topsoil, phosphates occur as insoluble compounds of Ca^{2+}, Fe^{3+}, and Al^{3+}. Because plants can absorb only the water-soluble dihydrogen phosphate, they have evolved the ability to secrete acids near their roots to convert the insoluble salts gradually into the soluble ion:

$$Ca_3(PO_4)_2(s; \text{ soil}) + 4H^+(aq; \text{ secreted by plants}) \longrightarrow$$
$$3Ca^{2+}(aq) + 2H_2PO_4^-(aq; \text{ absorbed by plants})$$

Animals that eat the plants excrete soluble phosphate, which is used in turn by newly growing plants. As the plants and animals excrete, die, and decay, some phosphate is washed into rivers and from there to the ocean. Most of the 2 million metric tons of phosphate that washes from land to sea each year comes from biological decay. Thus, the biosphere greatly increases the movement of phosphate from lithosphere to hydrosphere.

The Water-Based Biological Cycle Various phosphate oxoanions continually enter the aquatic environment. Studies that incorporate trace amounts of radioactive phosphorus into $H_2PO_4^-$ and monitor its uptake show that, within 1 minute, 50% of the ion is taken up by photosynthetic algae, which use it to synthesize

Figure 22.7 **The phosphorus cycle.** The inorganic subcycle *(purple arrows)*, water-based biological subcycle *(blue arrows)*, and land-based biological subcycle *(yellow arrows)* interact through erosion and through the growth and decay of organisms. The phosphorus cycle has no gaseous component.

 Transparency/ARIS Presentation Center, Art Library

their biomolecules (this synthesis is denoted by the top curved arrow in the upcoming equation). The overall process might be represented by the following equation:

$$106CO_2(g) + 16NO_3^-(aq) + H_2PO_4^-(aq) + 122H_2O(l) + 17H^+(aq) \underset{\text{decay}}{\overset{\text{synthesis}}{\rightleftharpoons}}$$

$$C_{106}H_{263}O_{110}N_{16}P(aq; \text{ in algal cell fluid}) + 138O_2(g)$$

(The complex formula on the right represents the total composition of algal biomolecules, not some particular compound.)

As on land, animals eat the plants and are eaten by other animals, and they all excrete phosphate, die, and decay (denoted by the bottom curved arrow in the preceding equation). Some of the released phosphate is used by other aquatic organisms, and some returns to the land when fish-eating animals (mostly humans and birds) excrete phosphate, die, and decay. In addition, some aquatic phosphate precipitates with Ca^{2+}, sinks to the seabed, and returns to the long-term mineral deposits of the inorganic cycle.

Human Effects on Phosphorus Movement From prehistoric through preindustrial times, these interlocking phosphorus cycles were balanced. Modern human activity, however, alters the movement of phosphorus considerably. Our influence on the inorganic cycle is minimal, despite our annual removal of more than

Figure 22.8 Industrial uses of phosphorus. The major end products from phosphate rock are shown, with box heights proportional to the amounts (mass %) of phosphate rock used annually to make the end product. Over 100 million metric tons of phosphate rock is mined each year (100%).

Engineering application

Demeton, an insecticide

V agent, a lethal nerve gas

● **Phosphorus Nerve Poisons** Because phosphorus is essential for all organisms, its compounds have many biological actions. In addition to their widespread use as fertilizer, phosphorus compounds are also used as pesticides *and* as military nerve gases. In fact, these two types of substances often have similar structures (see above) and modes of action. Both types affect the target organism by inactivating an enzyme involved in muscle contraction. They bind an amino acid critical to the enzyme's function, allowing the muscle to contract but not relax; the effect on the organism—insect or human—is often fatal. Chemists have synthesized a nerve-gas antidote whose shape mimics the binding site of the enzyme. The drug binds the poison, removing it from the enzyme, and is excreted.

Biological application

100 million metric tons of phosphate rock for a variety of uses (Figure 22.8). The end products of this removal, however, have unbalanced the two biological cycles.

The imbalance arises from our overuse of *soluble phosphate fertilizers*, such as $Ca(H_2PO_4)_2$ and $NH_4H_2PO_4$, the end products of about 85% of phosphate rock mined. Some fertilizer finds its way into rivers, lakes, and oceans to enter the water-based cycle. Much larger pollution sources, however, are the crops grown with the fertilizer and the detergents made from phosphate rock. The great majority of this phosphate arrives eventually in cities as crops, as animals that were fed crops, and as consumer products, such as the tripolyphosphates in detergents. Human garbage, excrement, wash water with detergents, and industrial wastewater containing phosphate return through sewers to the aquatic system. This human contribution equals the natural contribution—another 2 million metric tons of phosphate per year. The increased concentration in rivers and lakes causes eutrophication, which robs the water of O_2 so that it cannot support life. Such "dead" rivers and lakes are no longer usable for fishing, drinking, or recreation.

As early as 1912, a chemical solution to this problem was tried in Switzerland, where the enormous Lake Zurich had been choked with algae and become devoid of fish because of the release of human sewage. Treatment with $FeCl_3$ precipitated enough phosphate to return the lake gradually to its natural state. Soluble aluminum compounds have the same effect:

$$FeCl_3(aq) + PO_4^{3-}(aq) \longrightarrow FePO_4(s) + 3Cl^-(aq)$$
$$KAl(SO_4)_2 \cdot 12H_2O(aq) + PO_4^{3-}(aq) \longrightarrow$$
$$AlPO_4(s) + K^+(aq) + 2SO_4^{2-}(aq) + 12H_2O(l)$$

After several lakes in the United States became polluted in this way, phosphates in detergents were drastically reduced, and stringent requirements for sewage treatment, including tertiary treatments to remove phosphate (see Chemical Connections, pp. 541–542) were put into effect. Lake Erie, which was severely polluted by the late 1960s and early 1970s, was the target of a program to reduce phosphates and is today relatively clean and restocked with fish. ●

Section Summary

The environmental distribution of many elements changes cyclically with time and is affected in major ways by organisms. • Carbon occurs in all three regions of Earth's crust, with atmospheric CO_2 linking the other two regions. Photosynthesis and decay of organisms alter the amount of carbon in the land and oceans. Human activity has increased atmospheric CO_2. • Nitrogen is fixed by lightning, by industry, and primarily by microorganisms. When plants decay, bacteria decompose organic nitrogen and eventually return N_2 to the atmosphere. Through extensive use of fertilizers, humans have added excess nitrogen to freshwaters. • The phosphorus cycle has no gaseous component. Inorganic phosphates leach slowly into land and water, where biological cycles interact. When plants and animals decay, they release phosphate to natural waters, where it is then absorbed by other plants and animals. Through overuse of phosphate fertilizers and detergents, humans double the amount of phosphorus entering aqueous systems.

22.3 METALLURGY: EXTRACTING A METAL FROM ITS ORE

Metallurgy is the branch of materials science concerned with the extraction and utilization of metals. In this section, we discuss the general procedures for isolating metals and, occasionally, the nonmetals that are found in their ores. The extraction process applies one or more of these three types of metallurgy: *pyrometallurgy* uses heat to obtain the metal, *electrometallurgy* employs an electrochemical step, and *hydrometallurgy* relies on the metal's aqueous solution chemistry.

The extraction of an element begins with *mining the ore*. Most ores consist of mineral and gangue. The **mineral** contains the element; it is defined as a naturally occurring, homogeneous, crystalline inorganic solid, with a well-defined composition. The **gangue** is the portion of the ore with no commercial value, such as sand, rock, and clay attached to the mineral. Table 22.2 lists the common mineral sources of some metals.

Humans have been removing metals from Earth's crust for thousands of years, so most of the known concentrated sources are long gone. In many cases, ores containing very low mass percents of a metal are all that remain, and these require elaborate—and ingenious—extraction methods. Nevertheless, the general procedure for extracting most metals (and many nonmetals) involves a few basic steps, described in the upcoming subsections and previewed in Figure 22.9.

Table 22.2 Common Mineral Sources of Some Elements

Element	Mineral, Formula
Al	Gibbsite (in bauxite), $Al(OH)_3$
Ba	Barite, $BaSO_4$
Be	Beryl, $Be_3Al_2Si_6O_{18}$
Ca	Limestone, $CaCO_3$
Fe	Hematite, Fe_2O_3
Hg	Cinnabar, HgS
Na	Halite, $NaCl$
Pb	Galena, PbS
Sn	Cassiterite, SnO_2
Zn	Sphalerite, ZnS

Figure 22.9 Steps in metallurgy.

Pretreating the Ore

Following a crushing, grinding, or pulverizing step, which can be very expensive, pretreatment usually takes advantage of some physical or chemical difference to separate mineral from gangue. For magnetic minerals, such as magnetite (Fe_3O_4), a magnet can remove the mineral and leave the gangue behind. Where large density differences exist, a *cyclone separator* is used; it blows high-pressure air through the pulverized mixture to separate the particles. The lighter silicate-rich

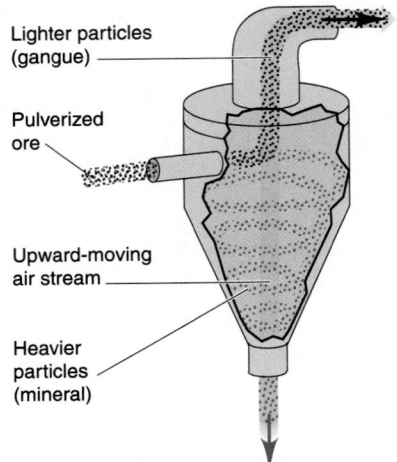

Lighter particles (gangue)

Pulverized ore

Upward-moving air stream

Heavier particles (mineral)

Figure 22.10 The cyclone separator. The pulverized ore enters at an angle, resulting in the more dense particles (mineral) hitting the walls and spiraling downward, while the less dense particles (silicate-rich gangue) are carried upward in a stream of air.

🌐 ARIS Presentation Center, Art Library

Stirrer
Ore and oil-detergent mixture
Compressed air
Froth of detergent-coated mineral

Gangue (rock, sand)

Figure 22.11 The flotation process. The pulverized ore is treated with an oil-detergent mixture, then stirred and aerated to create a froth that separates the detergent-coated mineral from the gangue. The mineral-rich froth is collected for further processing.

🌐 ARIS Presentation Center, Art Library

gangue is blown away, while the denser mineral-rich particles hit the walls of the separator and fall through the open bottom (Figure 22.10).

In **flotation,** an oil-detergent mixture is stirred with the pulverized ore in water to form a slurry (Figure 22.11). Flotation takes advantage of the different abilities of the surfaces of mineral and of gangue to become wet in contact with water and detergent. Rapid mixing moves air bubbles through the mixture and produces an oily, mineral-rich froth that floats, while the silicate particles sink. Skimming, followed by solvent removal of the oil-detergent mixture, isolates the concentrated mineral fraction. Flotation is a key step in copper recovery.

Leaching is a hydrometallurgical process that selectively extracts the metal, usually by forming a complex ion. The modern extraction of gold is a good example of this technique, because nuggets of the precious metal are found only in museums nowadays. ● The crushed ore, which often contains as little as 25 ppm of gold, is contained in a plastic-lined pool, treated with a cyanide ion solution, and aerated. In the presence of CN^-, the O_2 in air oxidizes gold metal to gold(I) ion, Au^+, which forms the soluble complex ion, $Au(CN)_2^-$:

$$4Au(s) + O_2(g) + 8CN^-(aq) + 2H_2O(l) \longrightarrow 4Au(CN)_2^-(aq) + 4OH^-(aq)$$

(The method has become controversial in some areas because cyanide lost from the contained area enters streams and lakes, where it poisons fish and birds.)

● **Panning and Fleecing for Gold** The picture of the prospector, alone but for his faithful mule, hunched over the bank of a mountain stream patiently panning for gold, is mostly a relic of old movies. However, the process is valid. It is based on the very large density difference between gold ($d = 19.3$ g/cm^3) and sand ($d \approx 2.5$ g/cm^3). In ancient times, river sands were washed over a sheep's fleece to trap the gold grains, a practice historians think represents the origin of the Golden Fleece of Greek mythology.

🌐 ARIS Presentation Center, Photo Library

Converting Mineral to Element

After the mineral has been freed of debris and concentrated, it may undergo several chemical steps during its conversion to the element.

Converting the Mineral to Another Compound First, the mineral is often converted to another compound, one that has more appropriate solubility properties, is easier to reduce, or is free of a troublesome impurity. Conversion to an oxide is common because oxides can be reduced easily. Carbonates are heated to convert them to the oxide:

$$CaCO_3(s) \xrightarrow{\Delta} CaO(s) + CO_2(g)$$

Metal sulfides, such as ZnS, can be converted to oxides by **roasting** in air:

$$2ZnS(s) + 3O_2(g) \xrightarrow{\Delta} 2ZnO(s) + 2SO_2(g)$$

In most countries, hydrometallurgical methods are now used to avoid the atmospheric release of SO_2 during roasting. One way of processing copper, for example, is by bubbling air through an acidic slurry of insoluble Cu_2S, which completes both the pretreatment and conversion steps:

$$2Cu_2S(s) + 5O_2(g) + 4H^+(aq) \longrightarrow 4Cu^{2+}(aq) + 2SO_4^{2-}(aq) + 2H_2O(l)$$

🌐 Environmental application

Converting the Compound to the Element Through Chemical Redox The next step converts the new mineral form (usually an oxide) to the free element by either chemical or electrochemical methods. Let's consider the chemical methods first. In these processes, a reducing agent reacts directly with the compound. The most common reducing agents are carbon, hydrogen, and an active metal.

1. *Reduction with carbon.* Carbon, in the form of coke (a porous residue from incomplete combustion of coal) or charcoal, is a very common reducing agent. Heating an oxide with a reducing agent such as coke to obtain the metal is called **smelting.** Many metal oxides, such as zinc oxide and tin(IV) oxide, are smelted with carbon to free the metal, which may need to be condensed and solidified:

$$ZnO(s) + C(s) \longrightarrow Zn(g) + CO(g)$$
$$SnO_2(s) + 2C(s) \longrightarrow Sn(l) + 2CO(g)$$

Several nonmetals that occur with positive oxidation states in minerals can be reduced with carbon as well. Phosphorus, for example, is produced from calcium phosphate:

$$2Ca_3(PO_4)_2(s) + 10C(s) + 6SiO_2(s) \longrightarrow 6CaSiO_3(s) + 10CO(g) + P_4(s)$$

(Metallic calcium is a much stronger reducing agent than carbon, so it is not formed.)

Low cost and ready availability explain why carbon is such a common reducing agent, and thermodynamic principles explain why it is such an effective one in many cases. Consider the standard free energy change ($\Delta G°$) for the reduction of tin(IV) oxide:

$$SnO_2(s) + 2C(s) \longrightarrow Sn(s) + 2CO(g) \qquad \Delta G° = 245 \text{ kJ at } 25°C \text{ (298 K)}$$

The magnitude and sign of $\Delta G°$ indicate a highly *non*spontaneous reaction for standard conditions at 25°C. However, because solid C becomes gaseous CO, the standard molar entropy change is very positive ($\Delta S° \approx 380$ J/K). Therefore, *the $-T\Delta S°$ term of $\Delta G°$ becomes more negative with higher temperature* (Section 20.3). As the temperature increases, $\Delta G°$ decreases, and at a high enough temperature, the reaction becomes spontaneous ($\Delta G° < 0$). At 1000°C (1273 K), for example, $\Delta G° = -62.8$ kJ. The actual process is carried out at a temperature of around 1250°C (1523 K), so $\Delta G°$ is even more negative.

As you know, overall reactions often mask several intermediate steps. For instance, the reduction of tin(IV) oxide may begin with formation of tin(II) oxide:

$$SnO_2(s) + C(s) \longrightarrow \cancel{SnO(s)} + CO(g)$$
$$\cancel{SnO(s)} + C(s) \longrightarrow Sn(l) + CO(g)$$

[Molten tin (mp = 232 K) is obtained at this temperature.] The second step may even be composed of others, in which CO, not C, is the actual reducing agent:

$$SnO(s) + \cancel{CO(g)} \longrightarrow Sn(l) + \cancel{CO_2(g)}$$
$$\cancel{CO_2(g)} + C(s) \rightleftharpoons 2CO(g)$$

Other pyrometallurgical processes are similarly complex. Shortly, you'll see that the overall reaction for the smelting of iron is

$$2Fe_2O_3(s) + 3C(s) \longrightarrow 4Fe(l) + 3CO_2(g)$$

However, the actual process is a multistep one with CO as the reducing agent. The advantage of CO over C is the far greater contact the gaseous reducing agent can make with the other reactant, which speeds the process.

2. *Reduction with hydrogen.* For oxides of some metals, especially some members of Groups 6B(6) and 7B(7), reduction with carbon forms metal carbides. These carbides are difficult to convert further, so other reducing agents are used. Hydrogen gas is used for less active metals, such as tungsten:

$$WO_3(s) + 3H_2(g) \longrightarrow W(s) + 3H_2O(g)$$

One step in the purification of the metalloid germanium uses hydrogen:

$$GeO_2(s) + 2H_2(g) \longrightarrow Ge(s) + 2H_2O(g)$$

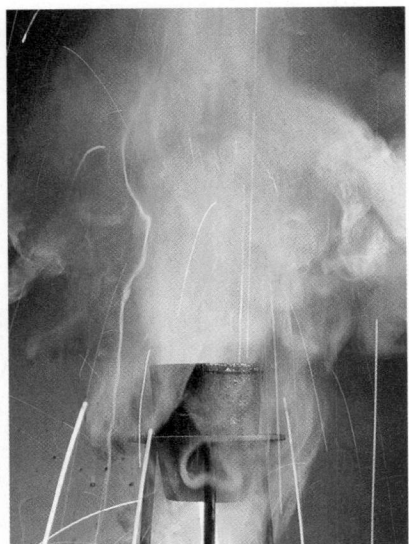

Figure 22.12 The thermite reaction.
This highly exothermic redox reaction
uses powdered aluminum to reduce
metal oxides.

ARIS Presentation Center, Photo Library

Animation: Thermite Reaction
ARIS Presentation Center
Chapter 22

Demo Arnáiz, Francisco J.; Aguado, Rafael;
Arnáiz, Susana. "Microscale Thermite Reactions,"
J. Chem. Educ. **1998**, *75*, 1630.

3. *Reduction with an active metal.* When a metal might form an undesirable hydride, its oxide is reduced by a more active metal. In the *thermite reaction,* aluminum powder reduces the metal oxide in a spectacular exothermic reaction to give the molten metal (Figure 22.12). The reaction for chromium is

$$Cr_2O_3(s) + 2Al(s) \longrightarrow 2Cr(l) + Al_2O_3(s) \qquad \Delta H° \ll 0$$

Reduction by a more active metal is also used in situations that do not involve an oxide. In the extraction of gold, after the pretreatment by leaching, the gold(I) complex ion is reduced with zinc to form the metal and a zinc complex ion:

$$2Au(CN)_2^-(aq) + Zn(s) \longrightarrow 2Au(s) + Zn(CN)_4^{2-}(aq)$$

In some cases, an active metal, such as calcium, is used to recover an even more active metal, such as rubidium, from its molten salt:

$$Ca(l) + 2RbCl(l) \longrightarrow CaCl_2(l) + 2Rb(g)$$

A similar process is used to recover sodium, as detailed in Section 22.4.

4. *Oxidation with an active nonmetal.* Just as chemical *reduction* of a mineral is used to obtain the metal, chemical *oxidation* of a mineral is sometimes used to obtain a nonmetal. A stronger oxidizing agent is used to remove electrons from the nonmetal anion to give the free nonmetal, as in the industrial production of iodine from concentrated brines by oxidation with chlorine gas:

$$2I^-(aq) + Cl_2(g) \longrightarrow 2Cl^-(aq) + I_2(s)$$

Converting the Compound to the Element Through Electrochemical Redox
In these processes, the mineral components are converted to the elements in an electrolytic cell (Section 21.7). Sometimes, the *pure* mineral, in the form of the molten halide or oxide, is used to prevent unwanted side reactions. The cation is reduced to the metal at the cathode, and the anion is oxidized to the nonmetal at the anode:

$$BeCl_2(l) \longrightarrow Be(s) + Cl_2(g)$$

High-purity hydrogen gas is prepared by electrochemical reduction:

$$2H_2O(l) \longrightarrow 2H_2(g) + O_2(g)$$

Specially designed cells separate the products to prevent recombination. Cost is a major factor in the use of electrolysis, and an inexpensive source of electricity is essential for large-scale methods. The current and voltage requirements depend on the electrochemical potential and any overvoltage effects. Figure 22.13 shows the most common step for obtaining each free element from its mineral.

Figure 22.13 The redox step in converting a mineral to the element. Most elements are reduced from the compound with carbon, a more active metal, or H_2.

Transparency/ARIS Presentation Center,
Art Library

☐ Uncombined in nature
☐ Reduction of molten halide (or oxide) electrolytically
☐ Reduction of halide with active metal (e.g., Na, Mg, Ca)
☐ Reduction of oxide/halide with Al or H_2
☐ Reduction of oxide with C (coke or charcoal)
☐ Oxidation of anion (or oxoanion) chemically and/or electrolytically

Refining and Alloying the Element

At this point in the isolation process, the element typically contains impurities from previous steps, so it must be refined. Then, once purified, metallic elements are often alloyed to improve their properties.

Refining (Purifying) the Element Refining is a purification procedure, often carried out by one of three common methods. An electrolytic cell is employed in **electrorefining,** in which the impure metal acts as the anode and a sample of the pure metal acts as the cathode. As the reaction proceeds, the anode slowly disintegrates, and the metal ions are reduced to deposit on the cathode. In some cases, the impurities, which fall beneath the disintegrating anode, are the source of several valuable and less abundant elements. (We examine the electrorefining of copper in Section 22.4.)

Metals with relatively low boiling points, such as zinc and mercury, are refined by *distillation.*

In the process of **zone refining** (see the Gallery, p. 533), impurities are removed from a bar of the element by concentrating them in a thin molten zone, while the purified element recrystallizes. Metalloids used in electronic semiconductors, such as silicon and germanium, must be zone refined to greater than 99.999999% purity.

Alloying the Purified Element An **alloy** is a metal-like mixture consisting of solid phases of two or more pure elements, a solid solution, or, in some cases, distinct intermediate phases (these alloys are sometimes referred to as *intermetallic compounds*). The separate phases are often so finely divided that they can only be distinguished microscopically. Alloying a metal with other metals (and, in some cases, nonmetals) is done to alter the metal's melting point and to enhance properties such as luster, conductivity, malleability, ductility, and strength.

Iron, probably the most important metal, is used only when it is alloyed. In pure form, it is soft and corrodes easily. However, when alloyed with carbon and other metals, such as Mo for hardness and Cr and Ni for corrosion resistance, it forms the various steels. Copper stiffens when zinc is added to make brass. Mercury solidifies when alloyed with sodium, and vanadium becomes extremely tough with some carbon added. Table 22.3 shows the composition and uses of some common alloys.

The simplest alloys, called *binary alloys,* contain only two elements. In some cases, *the added element enters interstices,* spaces between the parent metal atoms

ARIS Presentation Center, Tables Library

Table 22.3	Some Familiar Alloys and Their Composition	
Name	**Composition (Mass %)**	**Uses**
Stainless steel	73–79 Fe, 14–18 Cr, 7–9 Ni	Cutlery, instruments
Nickel steel	96–98 Fe, 2–4 Ni	Cables, gears
High-speed steels	80–94 Fe, 14–20 W (or 6–12 Mo)	Cutting tools
Permalloy	78 Ni, 22 Fe	Ocean cables
Bronzes	70–95 Cu, 1–25 Zn, 1–18 Sn	Statues, castings
Brasses	50–80 Cu, 20–50 Zn	Plating, ornamental objects
Sterling silver	92.5 Ag, 7.5 Cu	Jewelry, tableware
14-Carat gold	58 Au, 4–28 Ag, 14–28 Cu	Jewelry
18-Carat white gold	75 Au, 12.5 Ag, 12.5 Cu	Jewelry
Typical tin solder	67 Pb, 33 Sn	Electrical connections
Dental amalgam	69 Ag, 18 Sn, 12 Cu, 1 Zn (dissolved in Hg)	Dental fillings

A Vanadium carbide

B β - Brass

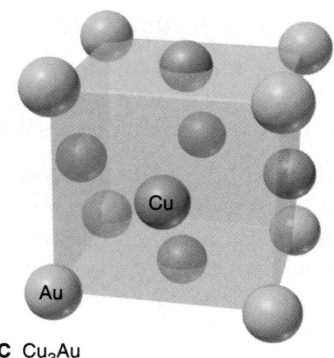

C Cu₃Au

Figure 22.14 Three binary alloys.

ARIS Presentation Center, Art Library

in the crystal structure (Section 12.6). In vanadium carbide, for instance, carbon atoms occupy holes of a face-centered cubic vanadium unit cell (Figure 22.14A). In other cases, the *added element substitutes for parent atoms* in the unit cell. In many types of brass, for example, relatively few zinc atoms substitute randomly for copper atoms in the face-centered cubic copper unit cell (see Figure 13.4A, p. 507). On the other hand, β-brass is an intermediate phase that crystallizes as the Zn/Cu ratio approaches 1/1 in a structure with a Zn atom surrounded by eight Cu atoms (Figure 22.14B). In one alloy of copper and gold, Cu atoms occupy the faces of a face-centered cube, and Au atoms lie at the corners (Figure 22.14C).

Atomic properties, including size, electron configuration, and number of valence electrons, determine which metals form stable alloys. Thus, transition metals having few *d* electrons often form stable alloys with transition metals having many *d* electrons. Thus, metals from the left half of the *d* block [Groups 3B(3) to 5B(5)] often form alloys with metals from the right half [Groups 8B(9) to 1B(11)]. For example, electron-poor Zr, Nb, and Ta form strong alloys with electron-rich Ir, Pt, or Au; an example is the very stable ZrPt₃.

Section Summary

Metallurgy involves mining an ore, separating it from debris, pretreating it to concentrate the mineral source, converting the mineral to another compound that is easier to process further, reducing this compound to the metal, purifying the metal, and in many cases, alloying it to obtain a more useful material.

22.4 TAPPING THE CRUST: ISOLATION AND USES OF SELECTED ELEMENTS

Once an element's abundance and sources are known and the metallurgical methods chosen, the isolation process begins. Obviously, the process depends on the physical and chemical properties of the source: we isolate an element that occurs uncombined in the air differently from one dissolved in the sea or one found in a rocky ore. In this section, we detail methods for recovering some important elements; pay special attention to the application of ideas from earlier chapters to these key industrial processes.

Producing the Alkali Metals: Sodium and Potassium

The alkali metals are among the most reactive elements and thus are always found as ions in nature, either in solid minerals or in aqueous solution. In the laboratory, the free elements must be protected from contact with air (for example, by being stored under mineral oil) to prevent their immediate oxidation and to minimize the possibility of fires. The two most important alkali metals are sodium and potassium. Their abundant, water-soluble compounds are used throughout industry and research, and the Na$^+$ and K$^+$ ions are essential to organisms.

Industrial Production of Sodium and Potassium The sodium ore is *halite* (largely NaCl), which is obtained either by evaporation of concentrated salt solutions (brines) or by mining vast salt deposits formed from the evaporation of prehistoric seas. The Cheshire salt field in Britain, for example, is 60 km by 24 km by 400 m thick and contains more than 90% NaCl. Other large deposits occur in New Mexico, Michigan, New York, and Kansas. ●

The brine is evaporated to dryness, and the solid is crushed and fused (melted) for use in an electrolytic apparatus called the **Downs cell** (Figure 22.15). To reduce heating costs, the NaCl (mp = 801°C) is mixed with $1\frac{1}{2}$ parts CaCl$_2$ to form a mixture that melts at only 580°C. Reduction of the metal ions to Na and Ca takes place at a cylindrical steel cathode, with the molten metals floating on the denser molten salt mixture. As they rise through a short collecting pipe, the liquid Na is siphoned off, while a higher melting Na/Ca alloy solidifies and falls back into the molten electrolyte. Chloride ions are oxidized to Cl$_2$ gas at a large anode within an inverted cone-shaped chamber. The cell design separates the metals from the Cl$_2$ to prevent their explosive recombination. The Cl$_2$ gas is collected, purified, and sold as a valuable byproduct.

Sylvite (mostly KCl) is the major ore of potassium. The metal is too soluble in molten KCl to be prepared by a method similar to that for sodium. Instead, chemical reduction of K$^+$ ions by liquid Na is used. The method is based on atomic properties and the nature of equilibrium systems. Recall that an Na atom is smaller than a K atom, so it holds its outer electron more tightly, as the first ionization energies indicate: IE$_1$ of Na = 496 kJ/mol; IE$_1$ of K = 419 kJ/mol. Thus, if the isolation method for K were based solely on this atomic property, Na would not be very effective at reducing K$^+$, and the reaction would not move very far toward products. An ingenious approach, which applies Le Châtelier's

● **A Plentiful Oceanic Supply of NaCl** Will we ever run out of sodium chloride? Not likely. The rock-salt (halite) equivalent of NaCl in the world's oceans has been estimated at 1.9×10^7 km^3, about 150% of the volume of North America above sea level. Or, looking at this amount another way: a row of 1-km^3 cubes of NaCl would stretch from Earth to the Moon 47 times! The photo shows evaporative salt beds near San Francisco Bay.

ARIS Presentation Center, Photo Library

Engineering application

Figure 22.15 The Downs cell for production of sodium. The mixture of solid NaCl and CaCl$_2$ forms the molten electrolyte. Sodium and calcium are formed at the cathode and float, but an Na/Ca alloy solidifies and falls back into the bath. Chlorine gas forms at the anode.

Inlet for NaCl

Cl$_2$(g)

Na/Ca alloy

Na(l)

Molten electrolyte (NaCl/CaCl$_2$, 2/3)

Molten Na

(−)

(+)

Cathode (reduction)
2Na$^+$(l) + 2e$^-$ ⟶ 2Na(l)

Anode (oxidation)
2Cl$^-$(l) ⟶ Cl$_2$(g) + 2e$^-$

ARIS Presentation Center, Art Library

Engineering application

principle, overcomes this drawback. The reduction is carried out at 850°C, which is above the boiling point of K, so the equilibrium mixture contains *gaseous* K:

$$Na(l) + K^+(l) \rightleftharpoons Na^+(l) + K(g)$$

As the K gas is removed, the system shifts to produce more K. The gas is then condensed and purified by fractional distillation. The same general method (with Ca as the reducing agent) is used to produce rubidium and cesium.

Uses of Sodium and Potassium The compounds of Na and K (and indeed, of all the alkali metals) have many more uses than the elements themselves [see the Group 1A(1) Family Portrait, p. 573]. Nevertheless, there are some interesting uses of the metals that take advantage of their strong reducing power. Large amounts of Na were used as an alloy with lead to make gasoline antiknock additives, such as tetraethyllead:

$$4C_2H_5Cl(g) + 4Na(s) + Pb(s) \longrightarrow (C_2H_5)_4Pb(l) + 4NaCl(s)$$

The toxic effects of environmental lead have limited this use to aviation fuel in the United States. Moreover, although leaded gasoline is still used in several developing countries, the global market for tetraethyllead is declining by 15% annually.

If certain types of nuclear reactors, called *breeder reactors* (Chapter 24), become a practical way to generate energy in the United States, Na production would increase enormously. Its low melting point, viscosity, and absorption of neutrons, combined with its high thermal conductivity and heat capacity, make it perfect for cooling the reactor and exchanging heat to the steam generator.

The major use of potassium at present is in an alloy with sodium for use as a heat exchanger in chemical and nuclear reactors. Another application relies on production of its superoxide, which it forms by direct contact with O_2:

$$K(s) + O_2(g) \longrightarrow KO_2(s)$$

This material is employed as an emergency source of O_2 in breathing masks for miners, divers, submarine crews, and firefighters (see photo):

$$4KO_2(s) + 4CO_2(g) + 2H_2O(g) \longrightarrow 4KHCO_3(s) + 3O_2(g)$$

The Indispensable Three: Iron, Copper, and Aluminum

Their familiar presence, innumerable applications, and enormous production levels make three metals—iron (in the form of steel), copper, and aluminum—stand out as indispensable materials of industrial society.

Metallurgy of Iron and Steel Although people have practiced iron smelting for more than 3000 years, it was only a little over 235 years ago that iron assumed its current dominant role. In 1773, an inexpensive process to convert coal to carbon in the form of coke was discovered, and the material was used in a blast furnace. The coke process made iron smelting cheap and efficient and led to large-scale iron production, which ushered in the Industrial Revolution.

Modern society rests, quite literally, on the various alloys of iron known as **steel.** Although steel production has grown enormously since the 18th century—more than 700 million tons are produced annually—the process of recovering iron from its ores still employs the same general approach: *reduction by carbon in a blast furnace.* The most important minerals of iron are listed in Table 22.4. The first four are used for steelmaking, but the sulfide minerals cannot be used because traces of sulfur make the steel brittle.

The conversion of iron ore to iron metal involves a series of overall redox and acid-base reactions that appear quite simple, although the detailed chemistry is not entirely understood even today. A modern **blast furnace** (Figure 22.16), such as those used in South Korea and Japan, is a tower about 14 m wide by 40 m high, made of a brick that can withstand intense heat. The *charge,* which consists of an iron ore (usually hematite) containing the mineral, coke, and lime-

Firefighter with KO_2 breathing mask.

Table 22.4	Important Minerals of Iron
Mineral Type	**Mineral, Formula**
Oxide	Hematite, Fe_2O_3
	Magnetite, Fe_3O_4
	Ilmenite, $FeTiO_3$
Carbonate	Siderite, $FeCO_3$
Sulfide	Pyrite, FeS_2
	Pyrrhotite, FeS

Charge (ore, limestone, coke)

Waste gases
(CO, CO₂, NO₂)

200°C	Raw materials preheated
	Partial reduction of iron ore
	$3Fe_2O_3(s) + CO(g) \rightarrow 2Fe_3O_4(s) + CO_2(g)$
	$CaCO_3(s) \rightarrow CaO(s) + CO_2(g)$
	$Fe_3O_4(s) + CO(g) \rightarrow 3FeO(s) + CO_2(g)$
700°C	Final reduction
	$C(s) + CO_2(g) \rightarrow 2CO(g)$
	$FeO(s) + CO(g) \rightarrow Fe(l) + CO_2(g)$
1200°C	Melting of impure Fe and slag
1500°C	Phosphates and silicates reduced;
	P, Si, and other impurities enter molten Fe
2000°C	$2C(s) + O_2(g) \rightarrow 2CO(g)$

Airblast
~900°C

Slag

Hearth

Human

Molten Fe
(pig iron)

Figure 22.16 The major reactions in a blast furnace. Iron is produced from iron ore, coke, and limestone in a complex process that depends on the temperature at different heights in the furnace. (A human is shown for scale.)

Transparency/ARIS Presentation Center, Active Art Library

Animation: Iron Smelting
ARIS Presentation Center
Chapter 22

stone, is fed through the top. More coke is burned in air at the bottom. The charge falls and meets a *blast* of rapidly rising hot air created by the burning coke:

$$2C(s) + O_2(g) \longrightarrow 2CO(g) + \textbf{\textit{heat}}$$

At the bottom of the furnace, the temperature exceeds 2000°C, while at the top, it reaches only 200°C. As a result, different stages of the overall reaction process occur at different heights as the charge descends:

1. In the upper part of the furnace (200°C to 700°C), the charge is preheated, and a *partial reduction* step occurs. The hematite is reduced to magnetite and then to iron(II) oxide (FeO) by CO, the actual reducing agent for the iron oxides. Carbon dioxide is formed as well. Because the blast of hot air passes through the entire furnace in only 10 s, the various gas-solid reactions do *not* reach equilibrium, and many intermediate products form. Limestone also decomposes to form CO₂ and the basic oxide CaO, which is important later in the process.

2. Lower down, at 700°C to 1200°C, a *final reduction* step occurs, as some of the coke reduces CO₂ to form more CO, which reduces the FeO to Fe.

3. At still higher temperatures, between 1200°C and 1500°C, the iron melts and drips to the bottom of the furnace. Acidic silica particles from the gangue react with basic calcium oxide in a Lewis acid-base reaction to form a molten waste product called **slag:**

$$CaO(s) + SiO_2(s) \longrightarrow CaSiO_3(l)$$

The siliceous slag drips down and floats on the denser iron (much as Earth's silicate-rich mantle and crust float on its molten iron core). ●

4. Some unwanted reactions occur at this hottest stage, between 1500°C and 2000°C. Any remaining phosphates and silicates are reduced to P and Si, and some Mn and traces of S dissolve into the molten iron along with carbon. The resulting impure product is called *pig iron* and contains about 3% to 4% C. A small amount of pig iron is used to make *cast iron,* but most is purified and alloyed to make various kinds of steel.

● Was It Slag That Made the Great Ship Go Down? Direct observation of the sunken luxury liner *Titanic* shows that six small slits caused by collision with an iceberg, rather than an enormous gash, sank the ship. The slits appear between sections of steel plates in the lower hull and suggest weakness in the rivets that connected the plates. Metallurgical analysis of the rivets shows three times as much slag content as modern rivets. Too much of this silicate-rich material makes the wrought iron used in the rivets brittle and easier to break under stress. Moreover, rather than being evenly distributed, the slag particles within the iron, viewed microscopically, appear to be clumped, which causes weak spots. Could low-grade rivets with too much slag have caused this tragedy? Follow-up analysis of more rivets and a historical search of rivet standards in the early 1900s may finally solve the mystery of what brought down the unsinkable *Titanic* and caused the deaths of 1500 passengers.

Engineering application

A

Figure 22.17 The basic-oxygen process for making steel. A, Jets of pure O_2 in combination with a basic flux (CaO) are used to remove impurities and lower the carbon content of pig iron in the manufacture of steel. **B,** Impure molten iron is added to a basic-oxygen furnace.

B

Pig iron is converted to steel in a separate furnace by means of the **basic-oxygen process,** as shown in Figure 22.17. High-pressure O_2 is blown over and through the molten iron, so that impurities (C, Si, P, Mn, S) are oxidized rapidly. The highly negative standard heats of formation of their oxides (such as ΔH_f° of $CO_2 = -394$ kJ/mol and ΔH_f° of $SiO_2 = -911$ kJ/mol) raise the temperature, which speeds the reaction. A lime (CaO) flux is added, which converts the oxides to a molten slag [primarily $CaSiO_3$ and $Ca_3(PO_4)_2$] that is decanted from the molten steel. The product is **carbon steel,** which contains 1% to 1.5% C and other impurities. It is alloyed with metals that prevent corrosion and increase its strength or flexibility.

Isolation and Electrorefining of Copper

After many centuries of being mined to make bronze and brass articles, copper ores have become less plentiful and less rich in copper, so the metal is more expensive to extract. Despite this, more than 2.5 billion pounds of copper is produced in the United States annually. The most common copper ore is chalcopyrite, $CuFeS_2$, a mixed sulfide of FeS and CuS (see photo). Most remaining deposits contain less than 0.5% Cu by mass. To "win" this small amount of copper from the ore requires several metallurgical steps, including a final refining to achieve the 99.99% purity needed for electrical wiring, copper's most important application.

The low copper content in chalcopyrite must be enriched by removing the iron. The first step in copper extraction is pretreatment by flotation (see Figure 22.11, p. 994), which concentrates the ore to around 15% Cu by mass. The next step in many processing plants is a controlled roasting step, which oxidizes the FeS but not the CuS:

$$2FeCuS_2(s) + 3O_2(g) \longrightarrow 2CuS(s) + 2FeO(s) + 2SO_2(g)$$

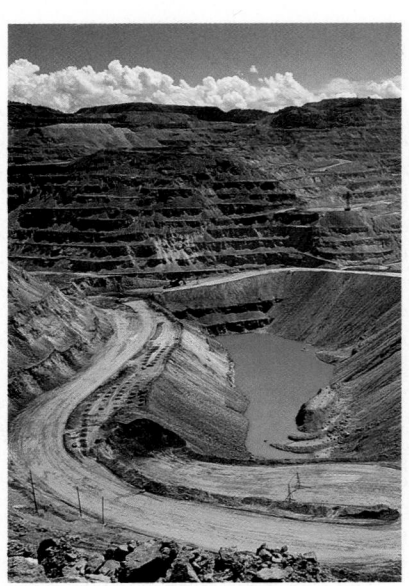

Mining chalcopyrite.

To remove the FeO and convert the CuS to a more convenient form, the mixture is heated to 1100°C with sand and more of the concentrated ore. Several reactions occur in this step. The FeO reacts with sand to form a molten slag:

$$FeO(s) + SiO_2(s) \longrightarrow FeSiO_3(l)$$

The CuS is thermodynamically unstable at the elevated temperature and decomposes to yield Cu_2S, which is drawn off as a liquid.

In the final smelting step, the Cu_2S is roasted in air, which converts some of it to Cu_2O:

$$2Cu_2S(s) + 3O_2(g) \longrightarrow 2Cu_2O(s) + 2SO_2(g)$$

The two copper(I) compounds then react, with sulfide ion acting as the reducing agent:

$$Cu_2S(s) + 2Cu_2O(s) \longrightarrow 6Cu(l) + SO_2(g)$$

The copper obtained at this stage is usable for plumbing, but it must be purified further for electrical applications by removing unwanted impurities (Fe and Ni) as well as valuable ones (Ag, Au, and Pt). Purification is accomplished by *electrorefining,* which involves the oxidation of Cu and the formation of Cu^{2+} ions in solution, followed by their reduction and the plating out of Cu metal (Figure 22.18). The impure copper obtained from smelting is cast into plates to be used as anodes, and cathodes are made from already purified copper. The electrodes are immersed in acidified $CuSO_4$ solution, and a controlled voltage is applied that accomplishes two tasks simultaneously:

1. Copper and the more active impurities (Fe, Ni) are oxidized to their cations, while the less active ones (Ag, Au, Pt) are not. As the anode slabs react, these unoxidized metals fall off as a valuable "anode mud" and are purified separately. Sale of the precious metals in the anode mud nearly offsets the cost of electricity to operate the cell, making Cu wire inexpensive.
2. Because Cu is much less active than the Fe and Ni impurities, Cu^{2+} ions are reduced at the cathode, but Fe^{2+} and Ni^{2+} ions remain in solution:

$$Cu^{2+}(aq) + 2e^- \longrightarrow Cu(s) \qquad E° = 0.34 \text{ V}$$
$$Ni^{2+}(aq) + 2e^- \longrightarrow Ni(s) \qquad E° = -0.25 \text{ V}$$
$$Fe^{2+}(aq) + 2e^- \longrightarrow Fe(s) \qquad E° = -0.44 \text{ V}$$

Figure 22.18 The electrorefining of copper. A, Copper is refined electrolytically, using impure slabs of copper as anodes and sheets of pure copper as cathodes. The Cu^{2+} ions released from the anode are reduced to Cu metal and plate out at the cathode. The "anode mud" contains valuable metal byproducts. **B,** A small section of an industrial facility for electrorefining copper.

Isolation, Uses, and Recycling of Aluminum

Aluminum is the most abundant metal in Earth's crust by mass and the third most abundant element (after O and Si). It is found in numerous aluminosilicate minerals (Gallery, pp. 592–594), such as feldspars, micas, and clays, and in the rare gems garnet, beryl, spinel, and turquoise. Corundum, pure aluminum oxide (Al_2O_3), is extremely hard; mixed with traces of transition metals, it exists as ruby and sapphire. Impure Al_2O_3 is used in sandpaper and other abrasives.

Through eons of weathering, certain clays became *bauxite,* the major ore of aluminum. This mixed oxide-hydroxide occurs in enormous surface deposits in Mediterranean and tropical regions (see photo), but with world aluminum production approaching 100 million tons annually, it may someday be scarce. In addition to hydrated Al_2O_3 (about 75%), industrial-grade bauxite also contains Fe_2O_3, SiO_2, and TiO_2, which are removed during the extraction.

The overall two-step process combines hydro- and electrometallurgical techniques. In the first step, Al_2O_3 is separated from bauxite; in the second, it is converted to the metal.

1. *Isolating Al_2O_3 from bauxite.* After mining, bauxite is pretreated by extended boiling in 30% NaOH in the *Bayer process,* which involves acid-base, solubility, and complex-ion equilibria. The acidic SiO_2 and the amphoteric Al_2O_3 dissolve in the base, but the basic Fe_2O_3 and TiO_2 do not:

$$SiO_2(s) + 2NaOH(aq) + 2H_2O(l) \longrightarrow Na_2Si(OH)_6(aq)$$
$$Al_2O_3(s) + 2NaOH(aq) + 3H_2O(l) \longrightarrow 2NaAl(OH)_4(aq)$$
$$Fe_2O_3(s) + NaOH(aq) \longrightarrow \text{no reaction}$$
$$TiO_2(s) + NaOH(aq) \longrightarrow \text{no reaction}$$

Further heating slowly precipitates the $Na_2Si(OH)_6$ as an aluminosilicate, which is filtered out with the insoluble Fe_2O_3 and TiO_2 ("red mud").

Acidifying the filtrate precipitates Al^{3+} as $Al(OH)_3$. Recall from our discussion of complex-ion equilibria that the aluminate ion, $Al(OH)_4^-(aq)$, is actually the complex ion $Al(H_2O)_2(OH)_4^-$, in which four of the six water molecules surrounding Al^{3+} have each lost a proton (see Figure 19.16, p. 869). Weakly acidic CO_2 is added to produce a small amount of H^+ ion, which reacts with this complex ion. Cooling supersaturates the solution, and the solid that forms is filtered out:

$$CO_2(g) + H_2O(l) \rightleftharpoons H^+(aq) + HCO_3^-(aq)$$
$$Al(H_2O)_2(OH)_4^-(aq) + H^+(aq) \longrightarrow Al(H_2O)_3(OH)_3(s)$$

[Recall that we usually write $Al(H_2O)_3(OH)_3$ more simply as $Al(OH)_3$.]

Drying at high temperature converts the hydroxide to the oxide:

$$2Al(H_2O)_3(OH)_3(s) \xrightarrow{\Delta} Al_2O_3(s) + 9H_2O(g)$$

2. *Converting Al_2O_3 to the free metal.* Aluminum is an active metal, much too strong a reducing agent to be formed at the cathode from aqueous solution (Section 21.7), so the oxide itself must be electrolyzed. However, the melting point of Al_2O_3 is very high (2030°C), so it is dissolved in molten *cryolite* (Na_3AlF_6) to give a mixture that is electrolyzed at ~1000°C. Obviously, the use of cryolite provides a major energy (and cost) savings. The only sizeable cryolite mines are in Greenland, however, and they cannot supply enough natural mineral to meet the demand. Therefore, production of synthetic cryolite has become a major subsidiary industry in aluminum manufacture.

The electrolytic step, called the *Hall-Heroult process,* takes place in a graphite-lined furnace, with the lining itself acting as the cathode. Anodes of graphite dip into the molten Al_2O_3-Na_3AlF_6 mixture (Figure 22.19). The cell typically operates at a moderate voltage of 4.5 V, but with an enormous current flow of 1.0×10^5 to 2.5×10^5 A.

Mining bauxite.

ARIS Presentation Center, Photo Library

Animation: Aluminum Production
ARIS Presentation Center
Chapter 22

The process is complex and its details are still not entirely known. Therefore, the specific reactions shown below are chosen from among several other possibilities. Molten cryolite contains several ions (including AlF_6^{3-}, AlF_4^{-}, and F^{-}), which react with Al_2O_3 to form fluoro-oxy ions (including $AlOF_3^{2-}$, $Al_2OF_6^{2-}$, and $Al_2O_2F_4^{2-}$) that dissolve in the mixture. For example,

$$2Al_2O_3(s) + 2AlF_6^{3-}(l) \longrightarrow 3Al_2O_2F_4^{2-}(l)$$

Al forms at the cathode (reduction), shown here with AlF_6^{3-} as reactant:

$$AlF_6^{3-}(l) + 3e^- \longrightarrow Al(l) + 6F^-(l) \quad \text{[cathode; reduction]}$$

The graphite anodes are oxidized and form carbon dioxide gas. Using one of the fluoro-oxy species as an example, the anode reaction is

$$Al_2O_2F_4^{2-}(l) + 8F^-(l) + C(\text{graphite}) \longrightarrow 2AlF_6^{3-}(l) + CO_2(g) + 4e^-$$

$$\text{[anode; oxidation]}$$

Thus, the anodes themselves are consumed in this half-reaction and must be replaced frequently.

Combining the three previous equations and making sure that e^- gained at the cathode equal e^- lost at the anode gives the overall reaction:

$$2Al_2O_3(\text{in } Na_3AlF_6) + 3C(\text{graphite}) \longrightarrow 4Al(l) + 3CO_2(g) \quad \text{[overall (cell) reaction]}$$

The Hall-Heroult process uses an enormous quantity of energy: aluminum production accounts for more than 5% of total U.S. electrical usage. ●

Aluminum is a superb decorative, functional, and structural metal. It is lightweight, attractive, easy to work, and forms strong alloys. Although Al is very active, it does not corrode readily because of an adherent oxide layer that forms rapidly in air and prevents more O_2 from penetrating. Nevertheless, when it is in

● **Energy Received and Returned**
Aluminum manufacture in the United States uses more electricity in 1 day than a city of 100,000 uses in 1 year! The reason for the high energy needs of Al manufacture is the electron configuration of Al ([Ne] $3s^23p^1$). Each Al^{3+} ion needs $3e^-$ to form an Al atom, and the atomic mass of Al is so low (~27 g/mol) that 1 mol of e^- produces only 9 g of Al. Compare this with the masses of Mg or Ca (two other lightweight structural metals) that 1 mol of e^- produces:

$$\tfrac{1}{3}Al^{3+} + e^- = \tfrac{1}{3}Al(s), \sim 9\text{ g}$$
$$\tfrac{1}{2}Mg^{2+} + e^- = \tfrac{1}{2}Mg(s), \sim 12\text{ g}$$
$$\tfrac{1}{2}Ca^{2+} + e^- = \tfrac{1}{2}Ca(s), \sim 20\text{ g}$$

An Al battery can turn this disadvantage around. Once produced, Al represents a concentrated form of electrical energy that can deliver 1 mol of e^- (96,500 C) for every 9 g of Al consumed in the battery. Thus, its electrical output per gram of metal is high. In fact, aluminum-air batteries are now being produced.

Engineering application

Graphite rods
Anodes (+): $Al_2O_2F_4^{2-} + 8F^- + C \longrightarrow 2AlF_6^{3-} + CO_2 + 4e^-$

Al_2O_3 dissolved in molten Na_3AlF_6

Bubbles of CO_2

Molten Al

Power source (−) (+)

Graphite furnace lining
Cathode (−): $AlF_6^{3-} + 3e^- \longrightarrow Al + 6F^-$

Figure 22.19 The electrolytic cell in the manufacture of aluminum. Purified Al_2O_3 is mixed with cryolite (Na_3AlF_6) and melted. Reduction at the graphite furnace lining (cathode) gives molten Al. Oxidation at the graphite rods (anodes) slowly converts them to CO_2, so they must be replaced periodically.

 ARIS Presentation Center, Art Library

Figure 22.20 The many familiar and essential uses of aluminum.

ARIS Presentation Center, Art Library

Environmental application

contact with less active metals such as Fe, Cu, and Pb, aluminum becomes the anode and deteriorates rapidly (Section 21.6). To prevent this, aluminum objects are often *anodized,* that is, made to act as the anode in an electrolysis that coats them with an oxide layer. The object is immersed in a 20% H_2SO_4 bath and connected to a graphite cathode:

$$6H^+(aq) + 6e^- \longrightarrow 3H_2(g) \qquad \text{[cathode; reduction]}$$
$$\underline{2Al(s) + 3H_2O(l) \longrightarrow Al_2O_3(s) + 6H^+(aq) + 6e^- \qquad \text{[anode; oxidation]}}$$
$$2Al(s) + 3H_2O(l) \longrightarrow Al_2O_3(s) + 3H_2(g) \qquad \text{[overall (cell) reaction]}$$

The Al_2O_3 layer deposited is typically from 10 μ to 100 μ thick, depending on the object's intended use. Figure 22.20 is a pie chart showing the countless uses of aluminum.

More than 3.5 billion pounds (1.5 million metric tons) of aluminum cans and packaging are discarded each year—a waste of one of the most useful materials in the world *and* the energy used to make it. A quick calculation of the energy needed to prepare 1 mol of Al from *purified* Al_2O_3, compared with the energy needed for recycling, conveys a clear message. The overall cell reaction in the Hall-Heroult process has a $\Delta H°$ of 2272 kJ and a $\Delta S°$ of 635.4 J/K. If we consider *only* the standard free energy change of the reaction at 1000.°C, for 1 mol of Al, we obtain

$$\Delta G° = \Delta H° - T\Delta S° = \frac{2272 \text{ kJ}}{4 \text{ mol Al}} - \left(1273 \text{ K} \times \frac{0.6354 \text{ kJ/K}}{4 \text{ mol Al}}\right) = 365.8 \text{ kJ/mol Al}$$

The molar mass of Al is nearly twice the mass of a soft-drink or beer can, so the electrolysis step requires nearly 200 kJ of energy for each can!

When aluminum is recycled, the major energy input (which is for melting the cans and foil) has been calculated as ~26 kJ/mol Al. The ratio of these energy inputs is

$$\frac{\text{Energy to recycle 1 mol Al}}{\text{Energy for electrolysis of 1 mol Al}} = \frac{26 \text{ kJ}}{365.8 \text{ kJ}} = 0.071$$

Based on just these portions of the process, recycling uses about 7% as much energy as electrolysis. Recent energy estimates for the entire manufacturing process (including mining, pretreating, maintaining operating conditions, electrolyzing, and so forth) are about 6000 kJ/mol Al, which means recycling requires less than 1% as much energy as manufacturing! The economic advantages, not to mention the environmental ones, are obvious, and recycling of aluminum has become common in the United States (see photo).

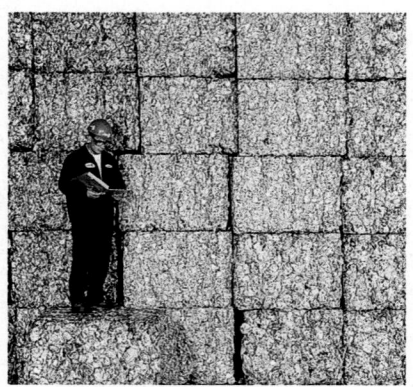

Crushed aluminum cans ready for recycling.

ARIS Presentation Center, Photo Library

Mining the Sea: Magnesium and Bromine

In the not too distant future, as terrestrial sources of certain elements become scarce or too costly to mine, the oceans will become an important source. Despite the abundant distribution of magnesium on land, it is already being obtained from the sea, and its ocean-based production is a good example of the approach we might one day use for other elements. Bromine, too, is recovered from the sea as well as from inland brines.

Isolation and Uses of Magnesium The *Dow process* for the isolation of magnesium from the sea involves steps that are similar to the procedures used for rocky ores (Figure 22.21):

1. *Mining.* Intake of seawater and straining the debris are the "mining" steps. No pretreatment is needed.
2. *Converting to mineral.* The dissolved Mg^{2+} ion is converted to the mineral $Mg(OH)_2$ with $Ca(OH)_2$, which is generated on-site (at the plant). Seashells ($CaCO_3$) are crushed, decomposed with heat to CaO, and mixed with water to make slaked lime [$Ca(OH)_2$]. This is pumped into the seawater intake tank to precipitate the dissolved Mg^{2+} as the hydroxide ($K_{sp} \approx 10^{-9}$):

$$Ca(OH)_2(aq) + Mg^{2+}(aq) \longrightarrow Mg(OH)_2(s) + Ca^{2+}(aq)$$

3. *Converting to compound.* The solid $Mg(OH)_2$ is filtered and mixed with excess HCl, which is also made on-site, to form aqueous $MgCl_2$:

$$Mg(OH)_2(s) + 2HCl(aq) \longrightarrow MgCl_2(aq) + 2H_2O(l)$$

The water is evaporated in stages to give solid, hydrated $MgCl_2 \cdot nH_2O$.

4. *Electrochemical redox.* Heating above 700°C drives off the water of hydration and melts the $MgCl_2$. Electrolysis gives chlorine gas and the molten metal, which floats on the denser molten salt:

$$MgCl_2(l) \longrightarrow Mg(l) + Cl_2(g)$$

The Cl_2 that forms is recycled to make the HCl used in step 3.

Figure 22.21 The production of elemental Mg from seawater. The Dow process involves obtaining Mg^{2+} ion from seawater; the Mg^{2+} is converted to $Mg(OH)_2$ and then to aqueous $MgCl_2$, which is evaporated to dryness and then electrolytically reduced to Mg metal and Cl_2 gas.

 ARIS Presentation Center, Active Art Library

 Engineering application

Jet engine assembly.

Magnesium is the lightest structural metal available (about one-half the density of Al and one-fifth that of steel). Although Mg is quite reactive, it forms an extremely adherent, high-melting oxide layer (MgO); thus, it finds many uses in metal alloys and can be machined into any form. Magnesium alloys occur in everything from aircraft bodies to camera bodies, and from luggage to auto engine blocks. The pure metal is a strong reducing agent, which makes it useful for sacrificial anodes (Section 21.6) and in the metallurgical extraction of other metals, such as Be, Ti, Zr, Hf, and U. For example, titanium, an essential component of jet engines (see photo), is made from its major ore, ilmenite, in two steps:

(1) $\quad 2FeTiO_3(s) + 7Cl_2(g) + 6C(s) \longrightarrow 2TiCl_4(l) + 2FeCl_3(s) + 6CO(g)$

(2) $\quad\quad\quad\quad\quad TiCl_4(l) + 2Mg(l) \longrightarrow Ti(s) + 2MgCl_2(l)$

Isolation and Uses of Bromine The largest source of bromine is the oceans, where it occurs as Br^- at a concentration of 0.065 g/L (65 ppm). However, the cost of concentrating such a dilute solution shifts the choice of source, when available, to much more concentrated salt lakes and natural brines. Arkansas brines, the most important source in the United States, contain about 4.5 g Br^-/L (4500 ppm). At standard-state conditions, the Br^- is readily oxidized to Br_2 with aqueous Cl_2:

$$2Br^-(aq) + Cl_2(aq) \longrightarrow Br_2(l) + 2Cl^-(aq) \qquad \Delta G^\circ = -61.5 \text{ kJ}$$

The Br_2 is removed by passing steam through the mixture and then cooling and drying the liquid bromine. Iodide is present in these brines also, and it is oxidized to I_2 by aqueous Cl_2 in a similar way.

World production of Br_2 is about 1% that of Cl_2. Its major use is in the preparation of organic chemicals, such as ethylene dibromide, which was added to leaded gasoline to prevent lead oxides from forming and depositing on engine parts. Newer products requiring bromine include the flame retardants in rugs and textiles. Bromine is also needed in the synthesis of inorganic chemicals, especially silver bromide for photographic emulsions.

The Many Sources and Uses of Hydrogen

Although hydrogen accounts for 90% of the atoms in the universe, it makes up only 15% of the atoms in Earth's crust. Moreover, whereas it occurs in the universe mostly as H_2 molecules and free H atoms, virtually all hydrogen in the crust is combined with other elements, either oxygen in natural waters or carbon in biomass, petroleum, and coal.

Properties of the Hydrogen Isotopes Hydrogen has three naturally occurring isotopes. Ordinary hydrogen (^1H), or protium, is the most abundant and has no neutrons. Deuterium (^2H or D), the next in abundance, has one neutron, and rare, radioactive tritium (^3H or T) has two. All three occur as diatomic molecules, H_2, D_2 (or $^2H_2$), and T_2 (or $^3H_2$), as well as HD, DT, and HT. Table 22.5 compares

Table 22.5	Some Molecular and Physical Properties of Diatomic Protium, Deuterium, and Tritium		
Property	**H_2**	**D_2**	**T_2**
Molar mass (g/mol)	2.016	4.028	6.032
Bond length (pm)	74.14	74.14	74.14
Melting point (K)	13.96	18.73	20.62
Boiling point (K)	20.39	23.67	25.04
ΔH°_{fus} (kJ/mol)	0.117	0.197	0.250
ΔH°_{vap} (kJ/mol)	0.904	1.226	1.393
Bond energy (kJ/mol at 298 K)	432	443	447

some molecular and physical properties of H_2, D_2, and T_2. As you can see, the heavier the isotope is, the higher the molar mass of the molecule, and the higher its melting point, boiling point, and heats of phase change; it also effuses at a lower rate (Section 5.6).

Because hydrogen is so light, the relative difference in mass of its isotopes is enormous compared to that of isotopes of other common elements. (For example, the mass of ^{13}C is only 8% greater than that of ^{12}C.) Note, especially, that the mass difference leads to different bond energies, which affect reactivity. Because the mass of D is twice the mass of H, H atoms bonded to a given atom vibrate at a higher frequency than do D atoms, so their bonds are higher in energy. As a result, the *bond to H is weaker and thus quicker to break.* Therefore, any reaction that includes breaking a bond to hydrogen in the rate-determining step occurs *faster with H than with D.* This phenomenon is called a *kinetic isotope effect;* we'll see an example shortly. No other element displays a kinetic isotope effect nearly as large.

Industrial Production of Hydrogen Hydrogen gas (H_2) is produced on an industrial scale worldwide: 250,000 metric tons (3×10^{12} L at STP; equivalent to $\sim 1 \times 10^{11}$ mol) are produced annually in the United States alone. All production methods are energy intensive, so the choice is determined by energy costs. In Scandinavia, where hydroelectric power is plentiful, electrolysis is the chosen method; on the other hand, where natural gas from oil refineries is plentiful, as in the United States and Great Britain, thermal methods are used to produce hydrogen.

The most common thermal methods use water and a simple hydrocarbon in two steps. Modern U.S. plants use methane, which has the highest H/C ratio of any hydrocarbon. In the first step, the reactants are heated to around 1000°C over a nickel-based catalyst in the endothermic *steam-reforming process:*

$$CH_4(g) + H_2O(g) \longrightarrow CO(g) + 3H_2(g) \qquad \Delta H° = 206 \text{ kJ}$$

Heat is supplied by burning methane at the refinery. To generate more H_2, the product mixture (called *water gas*) is heated with steam at 400°C over an iron or cobalt oxide catalyst in the exothermic *water-gas shift reaction,* and the CO reacts:

$$H_2O(g) + CO(g) \rightleftharpoons CO_2(g) + H_2(g) \qquad \Delta H° = -41 \text{ kJ}$$

The reaction mixture is recycled several times, which decreases the CO to around 0.2% by volume. Passing the mixture through liquid water removes the more soluble CO_2 (solubility = 0.034 mol/L) from the H_2 (solubility < 0.001 mol/L). Calcium oxide can also be used to remove CO_2 by formation of $CaCO_3$. By removing CO_2, these steps shift the equilibrium position to the right and produce H_2 that is about 98% pure. To attain greater purity ($\sim 99.9\%$), the gas mixture is passed through a *synthetic zeolite* (see the Gallery, p. 593) selected to filter out nearly all molecules larger than H_2.

In regions where inexpensive electricity is available, very pure H_2 is prepared through electrolysis of water with Pt (or Ni) electrodes:

$$2H_2O(l) + 2e^- \longrightarrow H_2(g) + 2OH^-(aq) \qquad E = -0.42 \text{ V} \quad \text{[cathode; reduction]}$$
$$\underline{H_2O(l) \longrightarrow \tfrac{1}{2}O_2(g) + 2H^+(aq) + 2e^- \qquad E = 0.82 \text{ V} \quad \text{[anode; oxidation]}}$$
$$H_2O(l) \longrightarrow H_2(g) + \tfrac{1}{2}O_2(g) \qquad E_{cell} = -0.42 \text{ V} - 0.82 \text{ V} = -1.24 \text{ V}$$

Overvoltage makes the cell potential about -2 V (Section 21.7). Therefore, under typical operating conditions, it takes about 400 kJ of energy to produce 1 mol of H_2:

$$\Delta G = -nFE = (-2 \text{ mol e}^-)\left(\frac{96.5 \text{ kJ}}{\text{V·mol e}^-}\right)(-2 \text{ V}) = 4 \times 10^2 \text{ kJ/mol } H_2$$

High-purity O_2 is a valuable byproduct of this process that offsets some of the costs of isolating H_2.

Calcium reacts with water to form H_2.

Demo Shakhashiri, Bassam Z. "Preparation and Properties of Hydrogen," *Chemical Demonstrations, A Handbook for Teachers of Chemistry*, Vol. 2 (The University of Wisconsin Press, Madison, 1985), 128–130.

ARIS Presentation Center, Photo Library

Laboratory Production of Hydrogen You may already have produced small amounts of H_2 in the lab by one of several methods. A characteristic reaction of the very active metals in Groups 1A(1) and 2A(2) is the reduction of water to H_2 and OH^- (see photo):

$$Ca(s) + 2H_2O(l) \longrightarrow Ca^{2+}(aq) + 2OH^-(aq) + H_2(g)$$

Alternatively, the strongly reducing hydride ion can be used:

$$NaH(s) + H_2O(l) \longrightarrow Na^+(aq) + OH^-(aq) + H_2(g)$$

Less active metals reduce H^+ in acids:

$$Zn(s) + 2H^+(aq) \longrightarrow Zn^{2+}(aq) + H_2(g)$$

In view of hydrogen's great potential as a fuel, chemists are seeking ways to decompose water by lowering the overall activation energy of the process. More than 10,000 thermodynamically *and* kinetically favorable *water-splitting* schemes have been devised.

However, at 298 K, $\Delta G°$ for the decomposition of 1 mol of $H_2O(g)$ is 229 kJ, so all water-splitting schemes require energy. Given the increase in the number of moles of gas, $\Delta S°$ is positive; therefore, this is an endothermic process ($\Delta H° > 0$) and, at standard-state conditions, requires heat to occur spontaneously. One promising source of this heat is focused sunlight.

Industrial Uses of Hydrogen Typically, a plant produces H_2 to make some other end product. In fact, more than 95% of H_2 produced industrially is consumed on-site in ammonia or petrochemical facilities. In a plant that synthesizes NH_3 from N_2 and H_2, the reactant gases are formed through a series of reactions that involve methane, including the steam-reforming and water-gas shift reactions discussed previously. For this reason, the cost of NH_3 is closely correlated with that of CH_4.

Here is a typical example of how the reaction series in an ammonia plant works. The steam-reforming reaction is performed with excess CH_4, which depletes the reaction mixture of H_2O:

$$CH_4(g; excess) + H_2O(g) \longrightarrow CO(g) + 3H_2(g)$$

In the next step, an excess of the product mixture (CH_4, CO, and H_2) is burned in an amount of air ($N_2 + O_2$) that is insufficient to effect complete combustion, but is just enough to consume the O_2, heat the mixture to 1100°C, and form additional H_2O:

$$4CH_4(g; excess) + 7O_2(g) \longrightarrow 2CO_2(g) + 2CO(g) + 8H_2O(g)$$
$$2H_2(g; excess) + O_2(g) \longrightarrow 2H_2O(g)$$
$$2CO(g; excess) + O_2(g) \longrightarrow 2CO_2(g)$$

Any remaining CH_4 reacts by the steam-reforming reaction, the remaining CO reacts by the water-gas shift reaction ($H_2O + CO \rightleftharpoons CO_2 + H_2$) to form more H_2, and then the CO_2 is removed with CaO. The amounts are carefully adjusted to produce a final mixture that contains a 1/3 ratio of N_2 (from the added air) to H_2 (with traces of CH_4, Ar, and CO). This mixture is used directly in the synthesis of ammonia, described in Chapter 17 (see the Chemical Connections essay, pp. 771–772).

A second major use of H_2 is in *hydrogenation* of the C=C bonds in liquid oils to form the C—C bonds in solid fats and margarine. The process uses H_2 in contact with transition metal catalysts, such as powdered nickel (see Figure 16.24, p. 722). Solid fats are also used in commercial baked goods. Look at the list of ingredients on most packages of bread, cake, and cookies, and you'll see the "partially hydrogenated vegetable oils" made by this process.

Hydrogen is also essential in the manufacture of numerous "bulk" chemicals, those that are produced in large amounts because they have many further uses. One application that has been gaining great attention is the production of methanol. In this process, carbon monoxide reacts with hydrogen over a copper–zinc oxide catalyst:

$$CO(g) + 2H_2(g) \xrightarrow{\text{Cu-ZnO catalyst}} CH_3OH(l)$$

Many automotive engineers expect methanol to be used as a gasoline additive.

Of course, as less expensive sources of hydrogen become available, it will be used directly in fuel cells on a major scale (Section 21.5).

Production and Uses of Deuterium Deuterium and its compounds are produced from D_2O (heavy water), which is present as a minor component (0.016 mol % D_2O) in normal water and is isolated on the multiton scale by *electrolytic enrichment*. This process is based on the kinetic isotope effect for hydrogen noted earlier, specifically on the *higher rate of bond breaking for O—H bonds compared to O—D bonds*, and thus on the higher rate of electrolysis of H_2O compared with D_2O.

For example, with Pt electrodes, H_2O is electrolyzed about 14 times faster than D_2O. As some of the liquid decomposes to the elemental gases, the remainder becomes enriched in D_2O. Thus, by the time the volume of water has been reduced to 1/20,000 of its original volume, the remaining water is around 99% D_2O. By combining samples and repeating the electrolysis, more than 99.9% D_2O is obtained.

Deuterium gas is produced by electrolysis of D_2O or by any of the chemical reactions that produce hydrogen gas from water, such as

$$2Na(s) + 2D_2O(l) \longrightarrow 2Na^+(aq) + 2OD^-(aq) + D_2(g)$$

Compounds containing deuterium (or tritium) are produced from reactions that give rise to the corresponding hydrogen-containing compound; for example,

$$SiCl_4(l) + 2D_2O(l) \longrightarrow SiO_2(s) + 4DCl(g)$$

Compounds with acidic protons undergo hydrogen/deuterium exchange:

$$CH_3COOH(l) + D_2O(l; \text{ excess}) \longrightarrow CH_3COOD(l) + DHO(l; \text{ small amount})$$

Notice that only the acidic H atom, the one in the COOH group, is exchanged, not any of those attached to carbon. We discuss the natural and synthetic formation of tritium in Chapter 24.

Section Summary

Production highlights of key elements are as follows:
- Na is isolated by electrolysis of molten NaCl in the Downs process; Cl_2 is a byproduct.
- K is produced by reduction with Na in a thermal process.
- Fe is produced through a multistep high-temperature process in a blast furnace. The crude pig iron is converted to carbon steel in the basic-oxygen process and then alloyed with other metals to make different steels.
- Cu is produced by concentration of the ore through flotation, reduction to the metal by smelting, and purification by electrorefining. The metal has extensive electrical and plumbing uses.
- Al is extracted from bauxite by pretreating the ore with concentrated base, followed by electrolysis of the Al_2O_3 in molten cryolite. Al alloys are widely used in homes and industry. The total energy required to extract Al from its ore is over 100 times that needed for recycling it.
- The Mg^{2+} in seawater is converted to $Mg(OH)_2$ and then to $MgCl_2$, which is electrolyzed to obtain the metal; Mg forms strong, lightweight alloys.
- Br_2 is obtained from brines by oxidizing Br^- with Cl_2.
- H_2 is produced by electrolysis of water or in the formation of gaseous fuels from hydrocarbons. It is used in NH_3 production and in hydrogenation of vegetable oils. The isotopes of hydrogen differ significantly in atomic mass and thus in the rate at which their bonds to other atoms break. This difference is used to obtain D_2O from water.

22.5 CHEMICAL MANUFACTURING: TWO CASE STUDIES

From laboratory-scale endeavors of the first half of the 19th century, today's chemical industries have grown to multinational corporations producing materials that define modern life: polymers, electronics, pharmaceuticals, consumer goods, and fuels. In this final section, we examine the interplay of theory and practice in two of the most important processes in the inorganic chemical industry: (1) the contact process for the production of sulfuric acid, and (2) the chlor-alkali process for the production of chlorine.

Sulfuric Acid, the Most Important Chemical

The manufacture of sulfuric acid began more than 400 years ago, when the acid was known as *oil of vitriol* and was distilled from "green vitriol" ($FeSO_4 \cdot 7H_2O$). Considering its countless uses, it is not surprising that today sulfuric acid is produced throughout the world on a gigantic scale—more than 150 million tons a year. The green vitriol method and the many production methods used since have all been superseded by the modern **contact process,** which is based on the *catalyzed oxidation of SO₂.* Here are the key steps.

    1. *Obtaining sulfur.* In most countries today, the production of sulfuric acid starts with the production of elemental sulfur, often by the *Claus process,* in which the H_2S in "sour" natural gas is chemically separated and then oxidized:

$$2H_2S(g) + 2O_2(g) \xrightarrow{\text{low temperature}} \tfrac{1}{8}S_8(g) + SO_2(g) + 2H_2O(g)$$

$$2H_2S(g) + SO_2(g) \xrightarrow{\text{Fe}_2\text{O}_3 \text{ catalyst}} \tfrac{3}{8}S_8(g) + 2H_2O(g)$$

    Where natural gas is not abundant but natural underground deposits of the element are, sulfur is obtained by the *Frasch process,* a nonchemical method that taps these deposits. A hole is drilled to the deposit, and superheated water (about 160°C) is pumped down two outer concentric pipes to melt the sulfur (Figure 22.22A). Then, a combination of the hydrostatic pressure in the outermost pipe and the pressure of compressed air sent through a narrow inner pipe forces the sulfur to the surface (Figure 22.22B and C). The costs of drilling, pumping, and supplying water (5×10^6 gallons per day) are balanced somewhat by the fact that the product is very pure (~99.7% S).

    2. *From sulfur to sulfur dioxide.* Once obtained, the sulfur is burned in air to form SO_2:

$$\tfrac{1}{8}S_8(s) + O_2(g) \longrightarrow SO_2(g) \qquad \Delta H° = -297 \text{ kJ}$$

Some SO_2 is also obtained from the roasting of metal sulfide ores. About 90% of processed sulfur is used in making sulfur dioxide for production of the all-important sulfuric acid. Indeed, this end product from sulfur is central to so many chemical industries that a nation's level of sulfur production is a reliable indicator of its overall industrial capacity: the United States, Russia, Japan, and Germany are the top four sulfur producers.

    3. *From sulfur dioxide to trioxide.* The contact process oxidizes SO_2 with O_2 to SO_3:

$$SO_2(g) + \tfrac{1}{2}O_2(g) \rightleftharpoons SO_3(g) \qquad \Delta H° = -99 \text{ kJ}$$

The reaction is *exothermic* and very *slow* at room temperature. From Le Châtelier's principle (Section 17.6), we know that the yield of SO_3 can be increased by (1) changing the temperature, (2) increasing the pressure (more moles of gas are on the left than on the right), or (3) adjusting the concentrations (adding excess O_2 and removing SO_3).

    First, let's examine the temperature effect. Adding heat (raising the temperature) increases the frequency of SO_2-O_2 collisions and thus increases the *rate* of SO_3 formation. However, because the formation of SO_3 is exothermic, removing heat (lowering the temperature) shifts the equilibrium position to the right and

Figure 22.22 The Frasch process for mining elemental sulfur. This nonchemical process uses superheated water and compressed air to melt underground sulfur and bring it to the surface. **A,** Superheated water is forced down a drilled hole into the subterranean sulfur deposit, and the sulfur melts. **B,** Compressed air is sent down an inner pipe in the drilling apparatus to force up the molten sulfur. **C,** Sulfur obtained from the Frasch process.

 ARIS Presentation Center, Art Library

thus increases the *yield* of SO_3. This is a classic situation that calls for use of a catalyst. By lowering the activation energy, *a catalyst allows equilibrium to be reached more quickly and at a lower temperature;* thus, rate *and* yield are optimized (Section 16.8). The catalyst in the contact process is V_2O_5 on inert silica, which is active between 400°C and 600°C.

The pressure effect is small and economically not worth exploiting. The concentration effects are controlled by providing an excess of O_2 in the form of a 5/1 mixture of air to SO_2, or about 1/1 O_2 to SO_2, about twice as much O_2 as is called for by the reaction stoichiometry. The mixture is passed over catalyst beds in four stages, and the SO_3 is removed at several points to favor more SO_3 formation. The overall yield of SO_3 is 99.5%.

4. *From sulfur trioxide to acid.* Sulfur trioxide is the anhydride of sulfuric acid, so a hydration step is next. However, SO_3 cannot be added to water because, at the operating temperature, it would first meet water vapor, which catalyzes its polymerization to $(SO_3)_x$, and results in a smoke of solid particles that makes poor contact with water and yields little acid. To prevent this, previously formed H_2SO_4 absorbs the SO_3 and forms pyrosulfuric acid (or disulfuric acid, $H_2S_2O_7$; see margin), which is then hydrolyzed with sufficient water:

disulfuric acid

$$SO_3(g) + H_2O(l) \longrightarrow H_2SO_4(l) \text{ [low yield]}$$
$$SO_3(g) + H_2SO_4(l) \longrightarrow H_2S_2O_7(l)$$
$$H_2S_2O_7(l) + H_2O(l) \longrightarrow 2H_2SO_4(l)$$

Figure 22.23 The many indispensable applications of sulfuric acid.

ARIS Presentation Center, Art Library

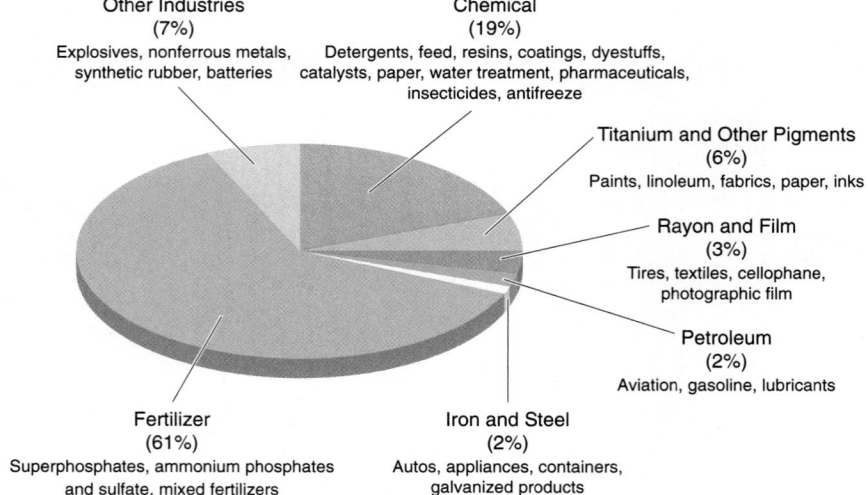

Other Industries (7%)
Explosives, nonferrous metals, synthetic rubber, batteries

Chemical (19%)
Detergents, feed, resins, coatings, dyestuffs, catalysts, paper, water treatment, pharmaceuticals, insecticides, antifreeze

Titanium and Other Pigments (6%)
Paints, linoleum, fabrics, paper, inks

Rayon and Film (3%)
Tires, textiles, cellophane, photographic film

Petroleum (2%)
Aviation, gasoline, lubricants

Iron and Steel (2%)
Autos, appliances, containers, galvanized products

Fertilizer (61%)
Superphosphates, ammonium phosphates and sulfate, mixed fertilizers

The uses of sulfuric acid are legion, as Figure 22.23 indicates.

Sulfuric acid is remarkably inexpensive (about \$150/ton), largely because each step in the process is exothermic—burning S ($\Delta H^\circ = -297$ kJ/mol), oxidizing SO_2 ($\Delta H^\circ = -99$ kJ/mol), hydrating SO_3 ($\Delta H^\circ = -132$ kJ/mol)—and the heat is a valuable byproduct. Three-quarters of the heat is sold as steam, and the rest of it is used to pump gases through the plant. A typical plant making 825 tons of H_2SO_4 per day produces enough steam to generate 7×10^6 watts of electric power.

The Chlor-Alkali Process

Chlorine is produced and used in amounts many times greater than all the other halogens combined, ranking among the top 10 chemicals produced in the United States. Its production methods depend on *the oxidation of Cl⁻ ion from NaCl.*

Earlier, we discussed the Downs process for isolating sodium, which yields Cl_2 gas as the other product (Section 22.4). The **chlor-alkali process,** which forms the basis of one of the largest inorganic chemical industries, electrolyzes concentrated aqueous NaCl to produce Cl_2 and several other important chemicals. As you learned in Section 21.7, the electrolysis of aqueous NaCl does not yield both of the component elements. Because of the effects of overvoltage, Cl⁻ ions rather than H_2O molecules are oxidized at the anode. However, Na^+ ions are not reduced at the cathode because the half-cell potential (-2.71 V) is much more negative than that for reduction of H_2O (-0.42 V), even with the normal overvoltage (around -0.6 V). Therefore, the half-reactions for electrolysis of aqueous NaCl are

$$2Cl^-(aq) \longrightarrow Cl_2(g) + 2e^- \qquad E^\circ = 1.36 \text{ V} \quad \text{[anode; oxidation]}$$
$$2H_2O(l) + 2e^- \longrightarrow 2OH^-(aq) + H_2(g) \qquad E \approx -1.0 \text{ V} \quad \text{[cathode; reduction]}$$
$$\overline{2Cl^-(aq) + 2H_2O(l) \longrightarrow 2OH^-(aq) + H_2(g) + Cl_2(g) \qquad E_{cell} = -1.0 \text{ V} - 1.36 \text{ V} = -2.4 \text{ V}}$$

To obtain commercially meaningful amounts of Cl_2, however, a voltage almost twice this value and a current in excess of 3×10^4 A are used.

When we include the spectator ion Na^+, the total ionic equation shows another important product made by the process:

$$2Na^+(aq) + 2Cl^-(aq) + 2H_2O(l) \longrightarrow 2Na^+(aq) + 2OH^-(aq) + H_2(g) + Cl_2(g)$$

As Figure 22.24 shows, the sodium salts in the cathode compartment exist as an aqueous mixture of NaCl and NaOH; the NaCl is removed by fractional crystallization. Thus, in this version of the chlor-alkali process, which uses an *asbestos*

Figure 22.24 A diaphragm cell for the chlor-alkali process. This process uses concentrated aqueous NaCl to make NaOH, Cl_2, and H_2 in an electrolytic cell. The difference in liquid level between compartments keeps a net movement of solution into the cathode compartment, which prevents reaction between OH^- and Cl_2. The cathode electrolyte is concentrated and fractionally crystallized to give industrial-grade NaOH.

ARIS Presentation Center, Active Art Library

diaphragm to separate the anode and cathode compartments, electrolysis of NaCl brines yields Cl_2, H_2, and industrial-grade NaOH, an important base. Like other reactive products, H_2 and Cl_2 are kept apart to prevent explosive recombination. Note the higher liquid level in the anode compartment. This slight hydrostatic pressure difference minimizes backflow of NaOH, which prevents the disproportionation (self–oxidation-reduction) reactions of Cl_2 that occur in the presence of OH^- (Section 14.9), such as

$$Cl_2(g) + 2OH^-(aq) \longrightarrow Cl^-(aq) + ClO^-(aq) + H_2O(l)$$

If high-purity NaOH is desired, a slightly different version, called the chlor-alkali *mercury-cell* process, is employed. Mercury is used as the cathode, which creates such a large overvoltage for reduction of H_2O to H_2 that the process *does* favor reduction of Na^+. The sodium dissolves in the mercury to form sodium amalgam, Na(Hg). In the mercury-cell version, the half-reactions are

$$2Cl^-(aq) \longrightarrow Cl_2(g) + 2e^- \qquad \text{[anode; oxidation]}$$
$$2Na^+(aq) + 2e^- \xrightarrow{\text{Hg}} 2Na(Hg) \qquad \text{[cathode; reduction]}$$

To obtain sodium hydroxide, the sodium amalgam is pumped out of the system and treated with H_2O, which is reduced by the Na:

$$2Na(Hg) + 2H_2O(l) \xrightarrow{-Hg} 2Na^+(aq) + 2OH^-(aq) + H_2(g)$$

The mercury released in this step is recycled back to the electrolysis bath. Therefore, *the products are the same in both versions,* but the purity of NaOH from the mercury-cell version is much higher.

Despite the formation of purer NaOH, the mercury-cell method is being steadily phased out in the United States and has been eliminated in Japan for more than a decade. Cost is not the major reason for the phase-out, although the method does consume about 15% more electricity than the diaphragm-cell version. Rather, the problem is that as the mercury is recycled, some is lost in the industrial wastewater. On average, 200 g of Hg is lost per ton of Cl_2 produced. In the 1980s, annual U.S. production via the mercury-cell method was 2.75 million tons of Cl_2; thus, during that decade, 550,000 kg of this toxic heavy metal was flowing into U.S. waterways each year!

The newer chlor-alkali *membrane-cell* process replaces the diaphragm with a polymeric membrane to separate the cell compartments. The membrane allows only cations to move through it and only from anode to cathode compartments. Thus, as Cl^- ions are removed at the anode through oxidation to Cl_2, Na^+ ions

Green chemistry application

in the anode compartment move through the membrane to the cathode compartment and form an NaOH solution. In addition to forming purer NaOH than the older diaphragm-cell method, the membrane-cell process uses less electricity and eliminates the problem of Hg pollution. As a result, it has been adopted throughout much of the industrialized world.

Section Summary

Sulfuric acid production starts with the extraction of sulfur, either by the oxidation of H_2S or the mining of sulfur deposits. The sulfur is roasted to SO_2, which is oxidized to SO_3 by the catalyzed contact process, which optimizes the yield at lower temperatures. Absorption of the SO_3 into H_2SO_4, followed by hydration, forms sulfuric acid. • In the diaphragm-cell chlor-alkali process, aqueous NaCl is electrolyzed to form Cl_2, H_2, and low-purity NaOH. The mercury-cell method produces high-purity NaOH but has been almost completely phased out because of mercury pollution. The membrane-cell method requires less electricity and does not use Hg.

Chapter Perspective

In this chapter, we took a realistic approach to the chemistry of the elements that allowed a glimpse at the intertwining natural cycles in which various elements take part, as well as at the ingenious methods that have been developed to extract them. The impact of living systems in general, and of humans in particular, on the origin and distribution of elements reveals a complex, evolving relationship. Understanding of and respect for that relationship are required for a healthy, productive balance between society's needs and the environment.

In Chapter 23, we conclude our exploration of the elements with a close look at the transition metals, a fascinating and extremely useful collection of metals. Then, in Chapter 24, we dive down to the atom's nuclear core and investigate its properties and its great potential for energy production and medical science.

CHAPTER REVIEW GUIDE

The following sections provide many aids to help you study this chapter. (Numbers in parentheses refer to pages, unless noted otherwise.)

Learning Objectives These are concepts and skills you should know after studying this chapter.

Relevant section numbers appear in parentheses.

Understand These Concepts

1. How gravity, thermal convection, and elemental properties led to the silicate, sulfide, and iron phases and the predominance of certain elements in Earth's crust, mantle, and core (Section 22.1)
2. How organisms affect crustal abundances of elements, especially oxygen, carbon, calcium, and some transition metals; the onset of oxidation as an energy source (Section 22.1)
3. How atomic properties influence which elements have oxide ores and which have sulfide ores (Section 22.1)
4. The central role of CO_2 and the importance of photosynthesis, respiration, and decay in the carbon cycle (Section 22.2)
5. The central role of N_2 and the importance of atmospheric, industrial, and biological fixation in the nitrogen cycle (Section 22.2)
6. The absence of a gaseous component, the interactions of the inorganic and biological cycles, and the impact of humans on the phosphorus cycle (Section 22.2)
7. How pyro-, electro-, and hydrometallurgical processes are employed to extract a metal from its ore; the importance of the reduction step from compound to metal; refining and alloying processes (Section 22.3)

8. Functioning of the Downs cell for Na production and the application of Le Châtelier's principle for K production (Section 22.4)
9. How iron ore is reduced in a blast furnace and how pig iron is purified by the basic-oxygen process (Section 22.4)
10. How Fe is removed from copper ore and impure Cu is electrorefined (Section 22.4)
11. The importance of amphoterism in the Bayer process for isolating Al_2O_3 from bauxite; the significance of cryolite in the electrolytic step; the energy advantage of Al recycling (Section 22.4)
12. The steps in the Dow process for the extraction of Mg from seawater (Section 22.4)
13. How H_2 production, whether by chemical or electrolytic means, is tied to NH_3 production (Section 22.4)
14. How the kinetic isotope effect is applied to produce deuterium (Section 22.4)
15. How the Frasch process is used to obtain sulfur from natural deposits (Section 22.5)
16. The importance of equilibrium and kinetic factors in H_2SO_4 production (Section 22.5)
17. How overvoltage allows electrolysis of aqueous NaCl to form Cl_2 gas in the chlor-alkali process; coproduction of NaOH; comparison of diaphragm-cell, mercury-cell, and membrane-cell methods (Section 22.5)

Key Terms These important terms appear in boldface in the chapter and are defined again in the Glossary.

Section 22.1
abundance (981)
differentiation (981)
core (981)
mantle (981)
crust (981)
lithosphere (983)
hydrosphere (983)
atmosphere (983)
biosphere (983)

occurrence (source) (985)
ore (985)

Section 22.2
environmental cycle (986)
fixation (987)
apatites (990)

Section 22.3
metallurgy (993)
mineral (993)

gangue (993)
flotation (994)
leaching (994)
roasting (994)
smelting (995)
electrorefining (997)
zone refining (997)
alloy (997)

Section 22.4
Downs cell (999)

steel (1000)
blast furnace (1000)
slag (1001)
basic-oxygen process (1002)
carbon steel (1002)

Section 22.5
contact process (1012)
chlor-alkali process (1014)

Highlighted Figures and Tables These figures (F) and tables (T) provide a visual review of key ideas.

F22.1 Cosmic and terrestrial abundances of selected elements (982)
F22.2 Geochemical differentiation of the elements (983)
T22.1 Abundances of elements in crust and biosphere (984)

F22.4 Sources of the elements (985)
F22.5 The carbon cycle (987)
F22.6 The nitrogen cycle (989)
F22.7 The phosphorus cycle (991)

PROBLEMS

Problems with colored numbers are answered in Appendix E and worked in detail in the Student Solutions Manual. Problem sections match those in the text and provide the numbers of relevant sample problems. Most offer Concept Review Questions, Skill-Building Exercises (grouped in pairs covering the same concept), and Problems in Context. The Comprehensive Problems are based on material from any section or previous chapter.

How the Elements Occur in Nature

● Concept Review Questions

22.1 Hydrogen is by far the most abundant element cosmically. In interstellar space, it exists mainly as H_2. In contrast, on Earth, it exists very rarely as H_2 and is ninth in abundance in the crust. Why is hydrogen so abundant in the universe? Why is hydrogen so rare as a diatomic gas in Earth's atmosphere?

22.2 Metallic elements can be recovered from ores that are oxides, carbonates, halides, or sulfides. Give an example of each type.

22.3 The location of elements in the regions of Earth has enormous practical importance. (a) Define the term *differentiation*, and explain which physical property of a substance is primarily responsible for this process. (b) What are the four most abundant elements in the crust? (c) Which element is abundant in the crust and mantle but not the core?

22.4 How does the position of a metal in the periodic table relate to whether it occurs primarily as an oxide or as a sulfide?

● Problems in Context

22.5 What material is the source for commercial production of each of the following elements: (a) aluminum; (b) nitrogen; (c) chlorine; (d) calcium; (e) sodium?

22.6 Aluminum is widely distributed throughout the world in the form of aluminosilicates. What property of these minerals prevents them from being a source of aluminum?

22.7 Describe two ways in which the biosphere has influenced the composition of Earth's crust.

The Cycling of Elements Through the Environment

● Concept Review Questions

22.8 Use atomic and molecular properties to explain why life is based on carbon, rather than some other element such as silicon.

22.9 Define the term *fixation*. Name two elements that undergo environmental fixation. What natural forms of them are fixed?

22.10 Carbon dioxide enters the atmosphere by natural processes and from human activity. Why is the latter a cause of concern?

22.11 Diagrams of environmental cycles are simplified to omit minor contributors. For example, the production of lime from limestone is not shown in the cycle for carbon (Figure 22.5, p. 987). Which labeled category in the figure includes this process? Name two other processes that contribute to this category.

22.12 Describe three pathways for the utilization of atmospheric nitrogen. Is human activity a significant factor? Explain.

22.13 Why don't the N-containing species in Figure 22.6 (p. 989) include rings or long chains with N—N bonds?

22.14 (a) Which region of Earth's crust is not involved in the phosphorus cycle? (b) Name two roles organisms play in the cycle.

● Problems in Context

22.15 Nitrogen fixation requires a great deal of energy because the N_2 bond is strong. (a) How do the processes of atmospheric and industrial fixation reflect this energy requirement? (b) How do the thermodynamics of the two processes differ? (*Hint:* Examine the respective heats of formation.) (c) In view of the mild conditions for biological fixation, what must be the source of the "great deal of energy"? (d) What would be the most obvious environmental result of a low activation energy for N_2 fixation?

22.16 The following steps are *unbalanced* half-reactions involved in the nitrogen cycle. Balance each half-reaction to show the number of electrons lost or gained, and state whether it is an oxidation or a reduction (all occur in acidic conditions):
(a) $N_2(g) \longrightarrow NO(g)$
(b) $N_2O(g) \longrightarrow NO_2(g)$
(c) $NH_3(aq) \longrightarrow NO_2^-(aq)$
(d) $NO_3^-(aq) \longrightarrow NO_2^-(aq)$
(e) $N_2(g) \longrightarrow NO_3^-(aq)$

Levels of Difficulty ● Easy ●● Medium ●●● Challenging   Biological: 22.8; Environmental: 22.7, 22.9, 22.10–12, 22.14. Note: Nearly all other problems have Engineering/Industrial relevance.

•• **22.17** The use of silica to form slag in the production of phosphorus from phosphate rock was introduced by Robert Boyle more than 300 years ago. When fluorapatite [$Ca_5(PO_4)_3F$] is used in phosphorus production, most of the fluorine atoms appear in the slag, but some end up in toxic and corrosive $SiF_4(g)$.
(a) If 15% by mass of the fluorine in 100. kg of $Ca_5(PO_4)_3F$ forms SiF_4, what volume of this gas is collected at 1.00 atm and the industrial furnace temperature of 1450.°C?
(b) In some facilities, the SiF_4 is used to produce sodium hexafluorosilicate (Na_2SiF_6) which is sold for water fluoridation:

$$2SiF_4(g) + Na_2CO_3(s) + H_2O(l) \longrightarrow$$
$$Na_2SiF_6(aq) + SiO_2(s) + CO_2(g) + 2HF(aq)$$

How many cubic meters of drinking water can be fluoridated to a level of 1.0 ppm of F^- using the SiF_4 produced in part (a)?

•• **22.18** An impurity sometimes found in $Ca_3(PO_4)_2$ is Fe_2O_3, which is removed during the production of phosphorus as *ferrophosphorus* (Fe_2P). (a) Why is this impurity troubling from an economic standpoint? (b) If 50. metric tons of crude $Ca_3(PO_4)_2$ contains 2.0% Fe_2O_3 by mass and the overall yield of phosphorus is 90.%, how many metric tons of P_4 can be isolated?

Metallurgy: Extracting a Metal from Its Ore

● Concept Review Questions

• **22.19** Define: (a) ore; (b) mineral; (c) gangue; (d) brine.

• **22.20** Define: (a) roasting; (b) smelting; (c) flotation; (d) refining.

• **22.21** What factors determine which reducing agent is selected for producing a specific metal?

• **22.22** Use atomic properties to explain the reduction of a less active metal by a more active one: (a) in aqueous solution; (b) in the molten state. Give a specific example of each process.

• **22.23** What class of element is obtained by oxidation of a mineral? What class of element is obtained by reduction of a mineral?

● Problems in Context

• **22.24** Which group of elements gives each of the following alloys: (a) brass; (b) stainless steel; (c) bronze; (d) sterling silver?

1. Cu, Ag	2. Cu, Sn, Zn	3. Ag, Au
4. Fe, Cr, Ni	5. Fe, V	6. Cu, Zn

Tapping the Crust: Isolation and Uses of the Elements

● Concept Review Questions

• **22.25** How are each of the following involved in iron metallurgy: (a) slag; (b) pig iron; (c) steel; (d) basic-oxygen process?

• **22.26** What are the distinguishing features of each extraction process: pyrometallurgy, electrometallurgy, and hydrometallurgy? Explain briefly how the types of metallurgy are used in the production of (a) Fe; (b) Na; (c) Au; (d) Al.

• **22.27** What property allows copper to be purified in the presence of iron and nickel impurities? Explain.

• **22.28** Why is cryolite used in the electrolysis of aluminum oxide?

• **22.29** (a) What is a kinetic isotope effect? (b) Do compounds of hydrogen exhibit a relatively large or small kinetic isotope effect? Explain. (c) Carbon compounds also exhibit a kinetic isotope effect. How do you expect it to compare in magnitude with that for hydrogen compounds? Why?

• **22.30** How is Le Châtelier's principle involved in the production of elemental potassium?

● Problems in Context

•• **22.31** Elemental Li and Na are prepared by electrolysis of a molten salt, whereas K, Rb, and Cs are prepared by chemical reduction.

(a) In general terms, explain why the alkali metals cannot be prepared by electrolysis of their aqueous salt solutions. (b) Use ionization energies (see the Family Portraits, pp. 572 and 576) to explain why calcium should *not* be able to isolate Rb from molten RbX (X = halide). (c) Use physical properties to explain why calcium *is* used to isolate Rb from molten RbX. (d) Can Ca be used to isolate Cs from molten CsX? Explain.

• **22.32** A Downs cell operating at 77.0 A produces 31.0 kg of Na.
(a) What volume of $Cl_2(g)$ is produced at 1.0 atm and 540.°C?
(b) How many coulombs were passed through the cell?
(c) How long did the cell operate?

• **22.33** (a) In the industrial production of iron, what is the reducing substance loaded into the blast furnace? (b) In addition to furnishing the reducing power, what other function does this substance serve? (c) What is the formula of the active reducing agent in the process? (d) Write equations for the stepwise reduction of Fe_2O_3 to iron in the furnace.

•• **22.34** One of the substances loaded into a blast furnace is limestone, which produces lime in the furnace. (a) Give the chemical equation for the reaction forming lime. (b) Explain the purpose of lime in the furnace. The term *flux* is often used as a label for a substance acting as the lime does. What is the derivation of this word, and how does it relate to the function of the lime? (c) Write a chemical equation describing the action of lime flux.

• **22.35** The last step in the Dow process for the production of magnesium metal involves electrolysis of molten $MgCl_2$.
(a) Why isn't the electrolysis carried out with aqueous $MgCl_2$? What are the products of this aqueous electrolysis?
(b) Do the high temperatures required to melt $MgCl_2$ favor products or reactants? (*Hint:* Consider the ΔH_f° of $MgCl_2$.)

•• **22.36** Iodine is the only halogen that occurs in a positive oxidation state, in $NaIO_3$ impurities within Chile saltpeter, $NaNO_3$.
(a) Is this mode of occurrence consistent with iodine's location in the periodic table? Explain.
(b) In the production of I_2, IO_3^- reacts with HSO_3^-:
$$IO_3^-(aq) + HSO_3^-(aq) \longrightarrow$$
$$HSO_4^-(aq) + SO_4^{2-}(aq) + H_2O(l) + I_2(s) \quad \text{[unbalanced]}$$
Identify the oxidizing and reducing agents.
(c) If 0.78 mol % of an $NaNO_3$ deposit is $NaIO_3$, how much I_2 (in g) can be obtained from 1.000 ton (2000. lb) of the deposit?

•• **22.37** Selenium is prepared by the reaction of H_2SeO_3 with gaseous SO_2. (a) What redox process does the sulfur dioxide undergo? What is the oxidation state of sulfur in the product? (b) Given that the reaction occurs in acidic aqueous solution, what is the formula of the sulfur-containing species? (c) Write the balanced redox equation for the process.

•• **22.38** F_2 and Cl_2 are produced by electrolytic oxidation, whereas Br_2 and I_2 are produced by chemical oxidation of the halide ions in a concentrated aqueous solution (brine) by a more electronegative halogen. Give two reasons why Cl_2 isn't prepared this way.

• **22.39** Silicon is prepared by the reduction of K_2SiF_6 with Al. Write the equation for this reaction. (*Hint:* Can F^- be oxidized in this reaction? Can K^+ be reduced?)

• **22.40** What is the mass percent of iron in each of the following iron ores: Fe_2O_3, Fe_3O_4, FeS_2?

•• **22.41** Phosphorus is one of the impurities present in pig iron that is removed in the basic-oxygen process. Assuming phosphorus is present as P atoms, write equations for its oxidation and subsequent reaction in the basic slag.

22.42 The final step in the smelting of $FeCuS_2$ is

$$Cu_2S(s) + 2Cu_2O(s) \longrightarrow 6Cu(l) + SO_2(g)$$

(a) Give the oxidation states of copper in Cu_2S, Cu_2O, and Cu.
(b) What are the oxidizing and reducing agents in this reaction?

22.43 Use equations to show how acid-base properties are used to separate Fe_2O_3 and TiO_2 from Al_2O_3 in the Bayer process.

22.44 A piece of Al with a surface area of 2.5 m^2 is anodized to produce a film of Al_2O_3 that is 23 μ (23×10^{-6} m) thick.
(a) How many coulombs flow through the cell in this process (assume that the density of the Al_2O_3 layer is 3.97 g/cm^3)?
(b) If it takes 18 min to produce this film, what current must flow through the cell?

22.45 The production of H_2 gas by the electrolysis of water typically requires about 400 kJ of energy per mole.
(a) Use the relationship between work and cell potential (Section 21.4) to calculate the minimum work needed to form 1.0 mol of H_2 gas at a cell potential of 1.24 V.
(b) What is the energy efficiency of the cell operation?
(c) Find the cost of producing 500. mol of H_2 if electricity is $0.06 per kilowatt·hour (1 watt·second = 1 joule).

22.46 (a) What are the components of the reaction mixture following the water-gas shift reaction? (b) Explain how zeolites (Gallery, p. 593) are used to purify the H_2 formed.

22.47 Metal sulfides are often first converted to oxides by roasting in air and then reduced with carbon to produce the metal. Why aren't the metal sulfides reduced directly by carbon to yield CS_2? Give a thermodynamic analysis of both processes for ZnS.

Chemical Manufacturing: Two Case Studies

● Concept Review Questions

22.48 Explain in detail why a catalyst is used to produce SO_3.

22.49 Among the exothermic steps in the manufacture of sulfuric acid is the process of hydrating SO_3. (a) Write two chemical reactions that show this process. (b) Why is the direct reaction of SO_3 with water not feasible?

22.50 Why is commercial H_2SO_4 so inexpensive?

22.51 (a) What are the three commercial products formed in the chlor-alkali process? (b) State an advantage and a disadvantage of the mercury-cell method for this process.

● Problems in Context

22.52 For the reaction of SO_2 to SO_3 at standard conditions, (a) Calculate $\Delta G°$ at 25°C. Is the reaction spontaneous? (b) Why is the reaction not performed at 25°C? (c) Is the reaction spontaneous at 500.°C? (Assume $\Delta H°$ and $\Delta S°$ are constant with changing T.) (d) Compare K at 500.°C and at 25°C. (e) What is the highest T at which the reaction is spontaneous?

22.53 If a chlor-alkali cell used a current of 3×10^4 A, how many pounds of Cl_2 would be produced in a typical 8-h operating day?

22.54 The products of the chlor-alkali process, Cl_2 and NaOH, are kept separated. (a) Why is this necessary when Cl_2 is the desired product? (b) ClO^- or ClO_3^- may form by disproportionation of Cl_2 in basic solution. What determines which product forms? (c) What mole ratio of Cl_2 to OH^- will produce ClO^-? ClO_3^-?

Comprehensive Problems

22.55 The key step in the manufacture of sulfuric acid is the oxidation of sulfur dioxide in the presence of a catalyst, such as V_2O_5. At 727°C, 0.010 mol of SO_2 is injected into an empty

2.00-L container (K_p = 3.18). (a) What is the equilibrium pressure of O_2 that is needed to maintain a 1/1 mole ratio of SO_3 to SO_2? (b) What is the equilibrium pressure of O_2 needed to maintain a 95/5 mole ratio of SO_3 to SO_2?

22.56 Many metal oxides are converted to the free metal by reduction with other elements, such as C or Si. For each of the following reactions, calculate the temperature at which the reduction occurs spontaneously:
(a) $MnO_2(s) \longrightarrow Mn(s) + O_2(g)$
(b) $MnO_2(s) + 2C(graphite) \longrightarrow Mn(s) + 2CO(g)$
(c) $MnO_2(s) + C(graphite) \longrightarrow Mn(s) + CO_2(g)$
(d) $MnO_2(s) + Si(s) \longrightarrow Mn(s) + SiO_2(s)$

22.57 Tetraphosphorus decaoxide (P_4O_{10}) is made from phosphate rock and used as a drying agent in the laboratory.
(a) Write a balanced equation for its reaction with water.
(b) What is the pH of a solution formed from the addition of 8.5 g of P_4O_{10} in sufficient water to form 0.750 L? (See Table 18.5, p. 803, for additional information.)

22.58 Heavy water (D_2O) is used to make deuterated chemicals.
(a) What major species, aside from the starting compounds, do you expect to find in a solution of CH_3OH and D_2O?
(b) Write equations to explain how these various species arise. (*Hint:* Consider the autoionization of both components.)

22.59 A blast furnace uses Fe_2O_3 to produce 8400. t of Fe per day.
(a) What mass of CO_2 is produced each day?
(b) Compare this amount of CO_2 with that produced by 1.0 million automobiles, each burning 5.0 gal of gasoline a day. Assume that gasoline has the formula C_8H_{18} and a density of 0.74 g/mL, and that it burns completely. (Note that U.S. gasoline consumption is over 4×10^8 gal/day.)

22.60 A major use of Cl_2 is in the manufacture of vinyl chloride, the monomer of poly(vinyl chloride). The two-step sequence for formation of vinyl chloride is depicted below.

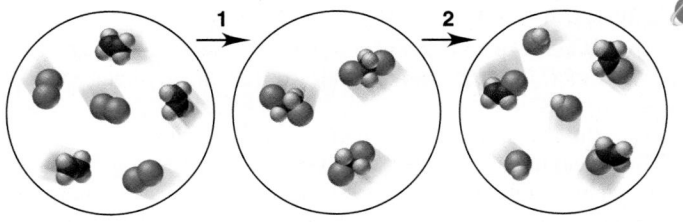

(a) Write a balanced equation for each step. (b) Write the overall equation. (c) What type of organic reaction is shown in step 1? (d) What type of organic reaction is shown in step 2? (e) If each molecule depicted in the initial reaction mixture represents 0.15 mol of substance, what mass (in g) of vinyl chloride forms?

22.61 In the production of magnesium, $Mg(OH)_2$ is precipitated by using $Ca(OH)_2$, which itself is "insoluble." (a) Use K_{sp} values to show that $Mg(OH)_2$ can be precipitated from seawater in which $[Mg^{2+}]$ is initially 0.052 M. (b) If the seawater is saturated with $Ca(OH)_2$, what fraction of the Mg^{2+} is precipitated?

22.62 The Ostwald process for the production of HNO_3 is
(1) $4NH_3(g) + 5O_2(g) \xrightarrow{\text{Pt/Rh catalyst}} 4NO(g) + 6H_2O(g)$
(2) $2NO(g) + O_2(g) \longrightarrow 2NO_2(g)$
(3) $3NO_2(g) + H_2O(l) \longrightarrow 2HNO_3(aq) + NO(g)$
(a) Describe the nature of the change that occurs in step 3.
(b) Write an overall equation that includes NH_3 and HNO_3 as the only nitrogen-containing species.
(c) Calculate $\Delta H°_{rxn}$ (in kJ/mol N atoms) for this reaction at 25°C.

•• **22.63** Step 1 of the Ostwald process for nitric acid production is

$$4NH_3(g) + 5O_2(g) \xrightarrow{\text{Pt/Rh catalyst}} 4NO(g) + 6H_2O(g)$$

An unwanted side reaction for this step is

$$4NH_3(g) + 3O_2(g) \longrightarrow 2N_2(g) + 6H_2O(g)$$

(a) Calculate K_p for these two NH_3 oxidations at 25°C.
(b) Calculate K_p for these two NH_3 oxidations at 900.°C.
(c) The Pt/Rh catalyst is one of the most efficient in the chemical industry, achieving 96% yield in 1 millisecond of contact with the reactants. However, at normal operating conditions (5 atm and 850°C), about 175 mg of Pt is lost per metric ton (t) of HNO_3 produced. If the annual U.S. production of HNO_3 is 1.01×10^7 t and the market price of Pt is \$1260/troy oz, what is the annual cost of the lost Pt (1 kg = 32.15 troy oz)?
(d) Because of the high price of Pt, a filtering unit composed of ceramic fiber is often installed, which recovers as much as 75% of the lost Pt. What is the value of the Pt captured by a recovery unit with 72% efficiency?

•• **22.64** The compounds $(NH_4)_2HPO_4$ and $NH_4(H_2PO_4)$ are water-soluble fertilizers. (a) Which has the higher mass % of P? (b) Why might the one with the lower mass % of P be preferred?

•• **22.65** Several transition metals are prepared by reduction of the metal halide with magnesium. Titanium is prepared by the Kroll method, in which ore (ilmenite) is converted to the gaseous chloride, which is then reduced to Ti metal by molten Mg (see p. 1008). Assuming yields of 84% for step 1 and 93% for step 2, and an excess of the other reactants, what mass of Ti metal can be prepared from 21.5 metric tons of ilmenite?

•• **22.66** The production of S_8 from the $H_2S(g)$ found in natural gas deposits occurs through the Claus process (Section 22.5):
(a) Use these two unbalanced steps to write an overall balanced equation for this process:
(1) $H_2S(g) + O_2(g) \longrightarrow S_8(g) + SO_2(g) + H_2O(g)$
(2) $H_2S(g) + SO_2(g) \longrightarrow S_8(g) + H_2O(g)$
(b) Write the overall reaction with Cl_2 as the oxidizing agent instead of O_2. Use thermodynamic data to show whether $Cl_2(g)$ can be used to oxidize $H_2S(g)$.
(c) Why is oxidation by O_2 preferred to oxidation by Cl_2?

•• **22.67** Acid mine drainage (AMD) occurs when geologic deposits containing pyrite (FeS_2) are exposed to oxygen and moisture. AMD is generated in a multistep process catalyzed by acidophilic (acid-loving) bacteria. Balance each step and identify those that increase acidity:
(1) $FeS_2(s) + O_2(g) \longrightarrow Fe^{2+}(aq) + SO_4^{2-}(aq)$
(2) $Fe^{2+}(aq) + O_2(g) \longrightarrow Fe^{3+}(aq) + H_2O(l)$
(3) $Fe^{3+}(aq) + H_2O(l) \longrightarrow Fe(OH)_3(s) + 12H^+(aq)$
(4) $FeS_2(s) + Fe^{3+}(aq) \longrightarrow Fe^{2+}(aq) + SO_4^{2-}(aq)$

••• **22.68** Based on a concentration of 0.065 g Br^-/L, the ocean is estimated to have \$391,000,000 of Br_2/mi^3 at recent prices. (a) What is the price per gram of Br_2? (b) Find the mass (in g) of $AgNO_3$ needed to precipitate 90.0% of the Br^- in 1.00 L of ocean water. (c) Use Appendix C to find how low $[Cl^-]$ must be to prevent the unwanted precipitation of AgCl. (d) Given a $[Cl^-]$ of about 0.3 M, is precipitation with silver ion a practical way to obtain bromine from the ocean? Explain.

••• **22.69** Below 912°C, pure iron crystallizes in a body-centered cubic structure (ferrite) with a density of 7.86 g/cm³; from 912°C to 1394°C, it adopts a face-centered cubic structure (austenite) with a density of 7.40 g/cm³. Both types of iron form interstitial alloys with carbon. The maximum amount of carbon is 0.0218 mass% in ferrite and 2.08 mass% in austenite. Calculate the density of each alloy.

•• **22.70** Why isn't nitric acid produced by oxidizing N_2 as follows?
(1) $N_2(g) + 2O_2(g) \longrightarrow 2NO_2(g)$
(2) $3NO_2(g) + H_2O(l) \longrightarrow 2HNO_3(aq) + NO(g)$
(3) $\underline{2NO(g) + O_2(g) \longrightarrow 2NO_2(g)}$
   $3N_2(g) + 6O_2(g) + 2H_2O(l) \longrightarrow 4HNO_3(aq) + 2NO(g)$
(*Hint:* Evaluate the thermodynamics of each step.)

••• **22.71** Before the development of the Downs cell, the Castner cell was used for the industrial production of Na metal. The Castner cell was based on the electrolysis of molten NaOH.
(a) Write balanced cathode and anode half-reactions for this cell.
(b) A major problem with this cell was that the water produced at one electrode diffused to the other and reacted with the Na. If all the water produced reacted with Na, what would be the maximum efficiency of the Castner cell expressed as moles of Na produced per mole of electrons flowing through the cell?

••• **22.72** When gold ores are leached with CN^- solutions, gold forms a complex ion, $Au(CN)_2^-$. (a) Find E_{cell} for the oxidation in air ($P_{O_2} = 0.21$) of Au to Au^+ in basic (pH 13.55) solution with $[Au^+] = 0.50$ M. Is the reaction $Au^+(aq) + e^- \longrightarrow Au(s)$, $E° = 1.68$ V, spontaneous? (b) How does formation of the complex ion change $E°$ so that the oxidation can be accomplished?

•• **22.73** Nitric oxide occurs in the tropospheric nitrogen cycle, but it destroys ozone in the stratosphere. (a) Write equations for its reaction with ozone and for the reverse reaction. (b) Given that the forward and reverse steps are first order in each component, write general rate laws for them. (c) Calculate $\Delta G°$ for this reaction at 280. K, the average temperature in the stratosphere. (Assume that the $\Delta H°$ and $S°$ values in Appendix B do not change with temperature.) (d) What ratio of rate constants is consistent with K at this temperature?

•• **22.74** A key part of the carbon cycle is the fixation of CO_2 by photosynthesis to produce carbohydrates and oxygen gas.
(a) Using the formula $(CH_2O)_n$ to represent a carbohydrate, write a balanced equation for the photosynthetic reaction.
(b) If a tree fixes 48 g of CO_2 per day, what volume of O_2 gas measured at 1.0 atm and 78°F does the tree produce per day?
(c) What volume of air (0.035 mol % CO_2) at the same conditions contains this amount of CO_2?

•• **22.75** Farmers use ammonium sulfate as a fertilizer. In the soil, nitrifying bacteria oxidize NH_4^+ to NO_3^-, a groundwater contaminant that causes methemoglobinemia ("blue baby" syndrome). The World Health Organization standard for maximum $[NO_3^-]$ in groundwater is 45 mg/L. A farmer adds 210. kg of $(NH_4)_2SO_4$ to a field and 37% is oxidized to NO_3^-. What is the groundwater $[NO_3^-]$ (in mg/L) if 1000. m^3 of the water is contaminated?

•• **22.76** In the 1980s, U.S. Fish and Wildlife Service researchers found high mortality among newborn coots and ducks in parts of California. U.S. Geological Survey scientists later found the cause to be a high selenium concentration in agricultural drainage water. Balance these half-reactions in the selenium cycle:
(a) $Se^{2-}(aq) \longrightarrow Se(s)$
(b) $SeO_3^{2-}(aq) \longrightarrow SeO_4^{2-}(aq)$
(c) $SeO_4^{2-}(aq) \longrightarrow Se(s)$

•• **22.77** The key reaction (unbalanced) in the manufacture of synthetic cryolite for aluminum electrolysis is

$$HF(g) + Al(OH)_3(s) + NaOH(aq) \longrightarrow Na_3AlF_6(aq) + H_2O(l)$$

Assuming a 95.6% yield of dried, crystallized product, what mass (in kg) of cryolite can be obtained from the reaction of 365 kg of Al(OH)$_3$, 1.20 m^3 of 50.0% by mass aqueous NaOH (d = 1.53 g/mL), and 265 m^3 of gaseous HF at 305 kPa and 91.5°C? (Assume that the ideal gas law holds.)

•• **22.78** Because of their different molar masses, H$_2$ and D$_2$ effuse at different rates (Section 5.6). (a) If it takes 16.5 min for 0.10 mol of H$_2$ to effuse, how long does it take for 0.10 mol of D$_2$ to do so in the same apparatus at the same T and P? (b) How many effusion steps does it take to separate an equimolar mixture of D$_2$ and H$_2$ to 99 mol % purity?

•• **22.79** The disproportionation of CO to graphite and CO$_2$ is thermodynamically favored but slow.
(a) What does this mean in terms of the magnitudes of the equilibrium constant (K), rate constant (k), and activation energy (E_a)?
(b) Write a balanced equation for the disproportionation of CO.
(c) Calculate K_c at 298 K. (d) Calculate K_p at 298 K.

•• **22.80** The overall cell reaction for aluminum production is

$$2Al_2O_3(\text{in } Na_3AlF_6) + 3C(\text{graphite}) \longrightarrow 4Al(l) + 3CO_2(g)$$

(a) Assuming 100% efficiency, how many metric tons (t) of Al$_2$O$_3$ are consumed per metric ton of Al produced?
(b) Assuming 100% efficiency, how many metric tons of the graphite anode are consumed per metric ton of Al produced?
(c) Actual conditions in an aluminum plant require 1.89 t of Al$_2$O$_3$ and 0.45 t of graphite per metric ton of Al. What is the percent yield of Al with respect to Al$_2$O$_3$?
(d) What is the percent yield of Al with respect to graphite?
(e) What volume of CO$_2$ (in m^3) is produced per metric ton of Al at operating conditions of 960.°C and exactly 1 atm?

•• **22.81** World production of chromite (FeCr$_2$O$_4$), the main ore of chromium, was 1.5×10^7 metric tons in 2003. To isolate chromium, a mixture of chromite and sodium carbonate is heated in air to form sodium chromate, iron(III) oxide, and carbon dioxide. The sodium chromate is dissolved in water, and this solution is acidified with sulfuric acid to produce the less soluble sodium dichromate. The sodium dichromate is filtered out and reduced with carbon to produce chromium(III) oxide, sodium carbonate, and carbon monoxide. The chromium(III) oxide is then reduced to chromium with aluminum metal. (a) Write balanced equations for each step. (b) What mass of chromium (in kg) could be prepared from the 2003 world production of chromite?

•• **22.82** Like heavy water (D$_2$O), so-called "semi-heavy water" (HDO) undergoes H/D exchange. The scenes below depict an initial mixture of HDO and H$_2$ reaching equilibrium.

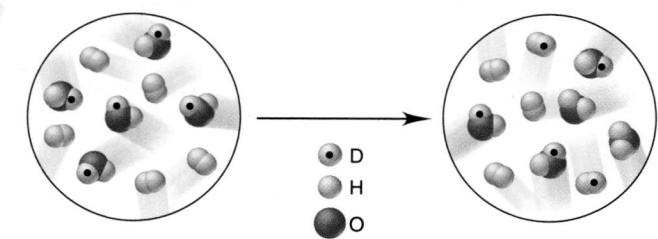

(a) Write the balanced equation for the reaction. (b) Is the value of K greater or less than 1? (c) If each molecule depicted represents 0.10 M, calculate K.

••• **22.83** Even though most metal sulfides are sparingly soluble in water, their solubilities differ by several orders of magnitude. This difference is sometimes used to separate the metals by con-

trolling the pH. Use the following data to find the pH at which you can separate 0.10 M Cu^{2+} and 0.10 M Ni^{2+}:
Saturated H$_2$S = 0.10 M

K_{a1} of H$_2$S = 9×10^{-8} K_{a2} of H$_2$S = 1×10^{-17}
K_{sp} of NiS = 1.1×10^{-18} K_{sp} of CuS = 8×10^{-34}

•• **22.84** Ores with as little as 0.25% by mass of copper are used as sources of the metal. (a) How many kilograms of such an ore would be needed for another Statue of Liberty, which contains 2.0×10^5 lb of copper? (b) If the mineral in the ore is chalcopyrite (FeCuS$_2$), what is the mass % of chalcopyrite in the ore?

••• **22.85** How does acid rain affect the leaching of phosphate into groundwater from terrestrial phosphate rock? Calculate the solubility of Ca$_3$(PO$_4$)$_2$ in each of the following: (a) Pure water, pH 7.0 (Assume that PO$_4^{3-}$ does not react with water.) (b) Moderately acidic rainwater, pH 4.5 (*Hint:* Assume that all the phosphate exists in the form that predominates at this pH.)

•• **22.86** The lead(IV) oxide used in car batteries is prepared by coating the electrode plate with PbO and then oxidizing it to lead dioxide (PbO$_2$). Despite its name, PbO$_2$ has a nonstoichiometric ratio of lead to oxygen of about 1/1.98. In fact, the holes in the PbO$_2$ crystal structure due to missing O atoms are responsible for the oxide's conductivity. (a) What is the mole % of O missing from the PbO$_2$ structure? (b) What is the molar mass of the nonstoichiometric compound?

•• **22.87** Chemosynthetic bacteria reduce CO$_2$ by "splitting" H$_2$S(g) instead of the H$_2$O(g) used by photosynthetic organisms. Compare the free energy change for splitting H$_2$S with that for splitting H$_2$O. Is there an advantage to using H$_2$S instead of H$_2$O?

•• **22.88** Silver has a face-centered cubic structure with a unit cell edge length of 408.6 pm. Sterling silver is a substitutional alloy that contains 7.5% copper atoms. Assuming the unit cell remains the same, find the density of silver and of sterling silver.

•• **22.89** Earth's mass is estimated to be 5.98×10^{24} kg, and titanium represents 0.05% by mass of this total. (a) How many moles of Ti are present? (b) If half of the Ti is found as ilmenite (FeTiO$_3$), what mass of ilmenite is present? (c) If the airline and auto industries use 1.00×10^5 tons of Ti per year, how many years would it take to use up all the Ti (1 ton = 2000 lb)?

•• **22.90** In 1790, Nicolas Leblanc found a way to form Na$_2$CO$_3$ from NaCl. His process, now obsolete, consisted of three steps:

$$2NaCl(s) + H_2SO_4(aq) \longrightarrow Na_2SO_4(aq) + 2HCl(g)$$
$$Na_2SO_4(s) + 2C(s) \longrightarrow Na_2S(s) + 2CO_2(g)$$
$$Na_2S(s) + CaCO_3(s) \longrightarrow Na_2CO_3(s) + CaS(s)$$

(a) Write a balanced overall equation for the process.
(b) Calculate the ΔH_f° of CaS if ΔH_{rxn}° is 351.8 kJ/mol.
(c) Is the overall process spontaneous at standard-state conditions and 298 K?
(d) How many grams of Na$_2$CO$_3$ form from 250. g of NaCl if the process is 73% efficient?

•• **22.91** Limestone (CaCO$_3$) is the second most abundant mineral on Earth after SiO$_2$. For many uses, it is first decomposed thermally to quicklime (CaO). MgO is prepared similarly from MgCO$_3$. (a) At what T is each decomposition spontaneous? (b) Quicklime reacts with SiO$_2$ to form a slag (CaSiO$_3$), a byproduct of steelmaking. In 2003, the total steelmaking capacity of the U.S. steel industry was 2,370,000 tons per week, but only 84% of this capacity was utilized. If 50. kg of slag is produced per ton of steel, what mass (in kg) of limestone was used to make slag in 2003?

23

Traces of Ions and the Beauty of Gems Much of the color in precious and semiprecious stones comes from trace amounts of transition-metal ions in mineral structures. The Cr^{3+} ion within beryllium aluminum silicate makes emeralds green, and the same ion in aluminum oxide makes rubies red; a mixture of Fe^{2+}, Fe^{3+}, and Ti^{4+} ions in aluminum oxide makes sapphires blue. In this chapter, we examine the transition metals and some of their numerous applications as well as the remarkable structures and properties of their compounds.

The Transition Elements and Their Coordination Compounds

23.1 Properties of the Transition Elements
Electron Configurations
Atomic and Physical Properties
Chemical Properties

23.2 The Inner Transition Elements
The Lanthanides
The Actinides

23.3 Highlights of Selected Transition Metals
Chromium
Manganese
Silver
Mercury

23.4 Coordination Compounds
Structures of Complex Ions
Formulas and Names

Alfred Werner and Coordination Theory
Isomerism

23.5 Theoretical Basis for the Bonding and Properties of Complexes
Valence Bond Theory
Crystal Field Theory

O ur exploration of the elements to this point is far from complete; in fact, we have skirted the majority of them and some of the most familiar. Whereas most important uses of the main-group elements involve their compounds, the transition elements are remarkably useful in their uncombined form. Figure 23.1 shows that the **transition elements** *(transition metals)* make up the *d* block (B groups) and *f* block *(inner transition elements)*.

In Chapter 22, you saw how two of the most important—copper and iron— are extracted from their ores. Aside from those two, with their countless essential uses, many other transition elements are indispensable as well: chromium in automobile parts, gold and silver in jewelry, tungsten in lightbulb filaments, platinum in automobile catalytic converters, titanium in bicycle frames and aircraft parts, and zinc in batteries, to mention just a few of the better known elements. You may be less aware of zirconium in nuclear-reactor liners, vanadium in axles and crankshafts, molybdenum in boiler plates, nickel in coins, tantalum in organ-replacement parts, palladium in telephone-relay contacts—the list goes on and on. As ions, they provide fascinating insights into chemical bonding and structure, while many also play vital roles in living organisms.

IN THIS CHAPTER . . . We first discuss some atomic, physical, and chemical properties of the transition elements and then focus on the chemistry of four familiar ones: chromium, manganese, silver, and mercury. Next, we concentrate on the most distinctive feature of transition element chemistry, the formation of coordination compounds, substances that contain complex ions. We consider two models that explain the striking colors of these compounds, as well as their magnetic properties and structures, and then end with some essential biochemical functions of transition metal ions.

Concepts & Skills to Review
before you study this chapter

- properties of light (Section 7.1)
- electron shielding of nuclear charge (Section 8.2)
- electron configuration, ionic size, and magnetic behavior (Sections 8.3 to 8.5)
- valence bond theory (Section 11.1)
- constitutional, geometric, and optical isomerism (Section 15.2)
- Lewis acid-base concepts (Section 18.9)
- complex-ion formation (Section 19.4)
- redox behavior and standard electrode potentials (Section 21.3)

Figure 23.1 The transition elements (*d* block) and inner transition elements (*f* block) in the periodic table.

Scandium, Sc; 3B(3)

Titanium, Ti; 4B(4)

Vanadium, V; 5B(5)

Chromium, Cr; 6B(6)

Manganese, Mn; 7B(7)

Figure 23.2 The Period 4 transition metals. Samples of all ten elements appear as pure metals, in chunk or powder form, in periodic-table order on this and the facing page.

 ARIS Presentation Center, Photo Library

23.1 PROPERTIES OF THE TRANSITION ELEMENTS

The transition elements differ considerably in physical and chemical behavior from the main-group elements. In some ways, they are more uniform: main-group elements in each period change from metal to nonmetal, but *all transition elements are metals.* In other ways, the transition elements are more diverse: most main-group ionic compounds are colorless and diamagnetic, but *many transition metal compounds are highly colored and paramagnetic.* We first discuss electron configurations of the atoms and ions, and then examine certain key properties of transition elements, with an occasional comparison to the main-group elements.

Electron Configurations of the Transition Metals and Their Ions

As with any of the elements, the properties of the transition elements and their compounds arise largely from the electron configurations of their atoms (Section 8.3) and ions (Section 8.5). The *d*-block (B-group) elements occur in four series that lie within Periods 4 through 7 between the last *ns*-block element [Group 2A(2)] and the first *np*-block element [Group 3A(13)]. Each series represents the filling of five *d* orbitals and, thus, contains ten elements. In 1996 and 1997, elements 110 through 112 were synthesized in particle accelerators, so the Period 7 series is complete; thus, all 40 *d*-block transition elements are known. Lying between the first and second members of the *d*-block transition series in Periods 6 and 7 are the inner transition elements, whose *f* orbitals are being filled.

Even though there are several exceptions, in general, the *condensed* ground-state electron configuration for the elements in each *d*-block series is

$$[\text{noble gas}]\ ns^2(n-1)d^x, \text{ with } n = 4 \text{ to } 7 \text{ and } x = 1 \text{ to } 10$$

In Periods 6 and 7, the condensed configuration includes the *f* sublevel:

$$[\text{noble gas}]\ ns^2(n-2)f^{14}(n-1)d^x, \text{ with } n = 6 \text{ or } 7$$

The *partial* (valence-level) electron configuration for the *d*-block elements excludes the noble gas core and the filled inner *f* sublevel:

$$ns^2(n-1)d^x$$

The first transition series occurs in Period 4 and consists of scandium (Sc) through zinc (Zn) (Figure 23.2 and Table 23.1). Scandium has the electron configuration [Ar] $4s^2 3d^1$, and the addition of one electron at a time (along with one proton in the nucleus) first half-fills, then fills, the 3*d* orbitals across the periodic table to zinc. Recall that chromium and copper are two exceptions to this general pattern: the 4*s* and 3*d* orbitals in Cr are both half-filled to give [Ar] $4s^1 3d^5$, and the 4*s* in Cu is half-filled to give [Ar] $4s^1 3d^{10}$. The reasons for these exceptions involve the change in relative energies of the 4*s* and 3*d* orbitals as electrons are added across the series and the unusual stability of half-filled and filled sublevels.

Transition metal ions form through *the loss of the ns electrons before the (n − 1)d electrons.* Therefore, the electron configuration of Ti^{2+} is [Ar] $3d^2$, *not* [Ar] $4s^2$, and Ti^{2+} is referred to as a d^2 ion. Ions of different metals with the same configuration often have similar properties. For example, both Mn^{2+} and Fe^{3+} are d^5 ions; both have pale colors in aqueous solution and form complex ions with similar magnetic properties.

Table 23.1 shows a general pattern in number of unpaired electrons (or half-filled orbitals) across the Period 4 transition series. Note that the number increases

Iron, Fe; 8B(8)

Cobalt, Co; 8B(9)

Nickel, Ni; 8B(10)

Copper, Cu; 1B(11)

Zinc, Zn; 2B(12)

Table 23.1 Orbital Occupancy of the Period 4 Transition Metals

ARIS Presentation Center, Tables Library

Element	4s	3d					4p			Unpaired Electrons
Sc	↑↓	↑								1
Ti	↑↓	↑	↑							2
V	↑↓	↑	↑	↑						3
Cr	↑	↑	↑	↑	↑	↑				6
Mn	↑↓	↑	↑	↑	↑	↑				5
Fe	↑↓	↑↓	↑	↑	↑	↑				4
Co	↑↓	↑↓	↑↓	↑	↑	↑				3
Ni	↑↓	↑↓	↑↓	↑↓	↑	↑				2
Cu	↑	↑↓	↑↓	↑↓	↑↓	↑↓				1
Zn	↑↓	↑↓	↑↓	↑↓	↑↓	↑↓				0

in the first half of the series and, when pairing begins, decreases through the second half. As you'll see, it is the electron configuration of the transition metal *atom* that correlates with physical properties of the *element,* such as density and magnetic behavior, whereas it is the electron configuration of the *ion* that determines the properties of the *compounds.*

SAMPLE PROBLEM 23.1 Writing Electron Configurations of Transition Metal Atoms and Ions

ARIS Presentation Center, Worked Examples Library

PROBLEM Write *condensed* electron configurations for the following: **(a)** Zr; **(b)** V^{3+}; **(c)** Mo^{3+}. (Assume that elements in higher periods behave like those in Period 4.)

PLAN We locate the element in the periodic table and count its position in the respective transition series. These elements are in Periods 4 and 5, so the general configuration is [noble gas] $ns^2(n-1)d^x$. For the ions, we recall that ns electrons are lost first.

SOLUTION (a) Zr is the second element in the $4d$ series: [Kr] $5s^2 4d^2$.

(b) V is the third element in the $3d$ series: [Ar] $4s^2 3d^3$. In forming V^{3+}, three electrons are lost (two $4s$ and one $3d$), so V^{3+} is a d^2 ion: [Ar] $3d^2$.

(c) Mo lies below Cr in Group 6B(6), so we expect the same exception as for Cr. Thus, Mo is [Kr] $5s^1 4d^5$. To form the ion, Mo loses the one $5s$ and two of the $4d$ electrons, so Mo^{3+} is a d^3 ion: [Kr] $4d^3$.

CHECK Figure 8.11 (p. 314) shows we're correct for the atoms. Be sure that charge plus number of d electrons in the ion equals the sum of outer s and d electrons in the atom.

FOLLOW-UP PROBLEM 23.1 Write *partial* electron configurations for the following: **(a)** Ag^+; **(b)** Cd^{2+}; **(c)** Ir^{3+}.

Atomic and Physical Properties of the Transition Elements

The atomic properties of the transition elements contrast in several ways with those of a comparable set of main-group elements.

Literature Laing, Michael. "Periodic Patterns," *J. Chem. Educ.* **2001,** *78,* 877.

Trends Across a Period

Consider the variations in atomic size, electronegativity, and ionization energy across Period 4 (Figure 23.3):

A Atomic radius (pm)

• *Atomic size.* Atomic size decreases overall across the period (Figure 23.3A). However, there is a smooth, steady decrease across the main groups because the electrons are added to *outer* orbitals, which shield the increasing nuclear charge poorly. This steady decrease is suspended throughout the transition series, where *atomic size decreases at first but then remains fairly constant.* Recall that the *d* electrons fill *inner* orbitals, so they shield outer electrons from the increasing nuclear charge very efficiently. As a result, the outer 4*s* electrons are not pulled closer.

B Electronegativity

• *Electronegativity.* Electronegativity generally increases across the period but, once again, the transition elements exhibit a relatively *small change in electronegativity* (Figure 23.3B), consistent with the relatively small change in size. In contrast, the main groups show a steady, much steeper increase between the metal potassium (0.8) and the nonmetal bromine (2.8). The transition elements all have intermediate electronegativity values, much like the large, metallic members of Groups 3A(13) to 5A(15).

C First ionization energy (kJ/mol)

Figure 23.3 Horizontal trends in key atomic properties of the Period 4 elements. The atomic radius (**A**), electronegativity (**B**), and first ionization energy (**C**) of the elements in Period 4 are shown as posts of different heights, with darker shading for the transition series. The transition elements exhibit smaller, less regular changes for these properties than do the main-group elements.

Transparency/ARIS Presentation Center, Art Library

• *Ionization energy.* The ionization energies of the Period 4 main-group elements rise steeply from left to right, more than tripling from potassium (419 kJ/mol) to krypton (1351 kJ/mol), as electrons become more difficult to remove from the poorly shielded, increasing nuclear charge. In the transition metals, however, the *first ionization energies increase relatively little* because the inner 3*d* electrons shield effectively (Figure 23.3C); thus, the outer 4*s* electron experiences only a slightly higher effective nuclear charge. [Recall from Section 8.4 that the drop at Group 3A(13) occurs because it is relatively easy to remove the first electron from the outer *np* orbital.]

Trends Within a Group

Vertical trends for transition elements are also different from those for the main groups.

• *Atomic size.* As expected, atomic size increases from Period 4 to 5, as it does for the main-group elements, but there is virtually *no size increase from Period 5 to 6* (Figure 23.4A). Remember that the lanthanides, with their buried 4*f* sublevel, appear between the 4*d* (Period 5) and 5*d* (Period 6) series. Therefore, an element in Period 6 is separated from the one above it in Period 5 by 32 elements (ten 4*d*, six 5*p*, two 6*s*, and fourteen 4*f* orbitals) instead of just 18. The extra shrinkage that results from the increase in nuclear charge due to the addition of 14 protons is called the **lanthanide contraction.** By coincidence, this *decrease* is about equal to the normal *increase* between periods, so the Periods 5 and 6 transition elements have about the same atomic sizes.

• *Electronegativity.* The vertical trend in electronegativity seen in most transition groups is opposite the trend in main groups. Here, we see an *increase* in electronegativity from Period 4 to Period 5, but then no further increase in Period 6 (Figure 23.4B). The heavier elements, especially gold (EN = 2.4), become quite electronegative, with values exceeding those of most metalloids and even some nonmetals (e.g., EN of Te and of P = 2.1). (In fact, gold forms the saltlike CsAu and the Au⁻ ion, which exists in liquid ammonia.) Although the

Figure 23.4 **Vertical trends in key properties within the transition elements.** The trends are unlike those for the main-group elements in several ways: **A,** The second and third members of a transition metal group are nearly the same size. **B,** Electronegativity increases down a transition group. **C,** First ionization energies are highest at the bottom of a transition group. **D,** Densities increase down a transition group because mass increases faster than volume.

Transparency/ARIS Presentation Center, Art Library

atomic size increases slightly from the top to the bottom of a group, the nuclear charge increases much more. Therefore, the heavier transition metals exhibit more covalent character in their bonds and attract electrons more strongly than do main-group metals.

- *Ionization energy.* The relatively small increase in size combined with the relatively large increase in nuclear charge also explains why *the first ionization energy generally increases* down a transition group (Figure 23.4C). This trend also runs counter to the pattern in the main groups, in which heavier members are so much larger that their outer electron is easier to remove.

- *Density.* Atomic size, and therefore volume, is inversely related to density. Across a period, densities increase, then level off, and finally dip a bit at the end of a series (Figure 23.4D). Down a transition group, densities increase dramatically because atomic volumes change little from Period 5 to 6, but atomic masses increase significantly. As a result, the Period 6 series contains some of the densest elements known: tungsten, rhenium, osmium, iridium, platinum, and gold have densities about 20 times that of water and twice that of lead.

Chemical Properties of the Transition Metals

Like their atomic and physical properties, the chemical properties of the transition elements are very different from those of the main-group elements. Let's examine the key properties in the Period 4 transition series and then see how behavior changes within a group.

A

B

Figure 23.5 Aqueous oxoanions of transition elements. A, Often, a given transition element has multiple oxidation states. Here, Mn is shown in the +2 (Mn^{2+}, *left*), the +6 (MnO_4^{2-}, *middle*), and the +7 states (MnO_4^-, *right*). **B,** The highest possible oxidation state equals the group number in these oxoanions: +5 in VO_4^{3-} (*left*), +6 in $Cr_2O_7^{2-}$ (*middle*), and +7 in MnO_4^- (*right*).

 ARIS Presentation Center, Photo Library

Oxidation States One of the most characteristic chemical properties of the transition metals is the occurrence of *multiple oxidation states*. For example, in their compounds, vanadium exhibits two common positive oxidation states, chromium three, and manganese three (Figure 23.5A), and many other oxidation states are seen less often. Since the ns and $(n-1)d$ electrons are so close in energy, transition elements can use all or most of these electrons in bonding. This behavior is markedly different from that of the main-group metals, which display one or, at most, two oxidation states in their compounds.

The highest oxidation state of elements in Groups 3B(3) through 7B(7) is equal to the group number, as shown in Table 23.2. These oxidation states are seen when the elements combine with highly electronegative oxygen or fluorine. For instance, in the oxoanion solutions shown in Figure 23.5B, vanadium occurs as the vanadate ion (VO_4^{3-}; O.N. of V = +5), chromium occurs as the dichromate ion ($Cr_2O_7^{2-}$; O.N. of Cr = +6), and manganese occurs as the permanganate ion (MnO_4^-; O.N. of Mn = +7). Elements in Groups 8B(8), 8B(9), and 8B(10) exhibit fewer oxidation states, and the highest state is less common and never equal to the group number. For example, we never encounter iron in the +8 state and only rarely in the +6 state. The +2 and +3 states are the most common ones for iron* and cobalt, and the +2 state is most common for nickel, copper, and zinc. *The +2 oxidation state is common because ns^2 electrons are readily lost.*

Copper, silver, and gold (the coinage metals) in Group 1B(11) are unusual. Although copper has a fairly common oxidation state of +1, which results from loss of its single 4s electron, its most common state is +2. Silver behaves more predictably, exhibiting primarily the +1 oxidation state. Gold exhibits the +3 state and, less often, the +1 state. Zinc, cadmium, and mercury in Group 2B(12) exhibit the +2 state, but mercury also exhibits the +1 state in the Hg_2^{2+} ion, with its Hg—Hg bond.†

---

*Iron may seem to have unusual oxidation states in the common ores magnetite (Fe_3O_4) and pyrite (FeS_2), but this is not really the case. In magnetite, one-third of the metal ions are Fe^{2+} and two-thirds are Fe^{3+}, which is equivalent to a 1/1 ratio of FeO to Fe_2O_3 and gives an overall formula of Fe_3O_4. Pyrite contains Fe^{2+} combined with the disulfide ion, S_2^{2-}.

†Some evidence suggests that the Hg_2^{2+} ion may exist as an Hg atom bonded to an Hg^{2+} ion, with the atom donating its two 6s electrons for the covalent bond.

 Transparency/ARIS Presentation Center, Tables Library

Literature Steinborn, Dirk. "The Concept of Oxidation States in Metal Complexes," *J. Chem. Educ.* 2004, *81*, 1148.

Table 23.2   Oxidation States and *d*-Orbital Occupancy of the Period 4 Transition Metals*

Oxidation State	3B (3) Sc	4B (4) Ti	5B (5) V	6B (6) Cr	7B (7) Mn	8B (8) Fe	8B (9) Co	8B (10) Ni	1B (11) Cu	2B (12) Zn
0	d^1	d^2	d^3	d^5	d^5	d^6	d^7	d^8	d^{10}	d^{10}
+1			d^3	d^5	d^5	d^6	d^7	d^8	d^{10}	
+2		d^2	d^3	d^4	d^5	d^6	d^7	d^8	d^9	d^{10}
+3	d^0	d^1	d^2	d^3	d^4	d^5	d^6	d^7	d^8	
+4		d^0	d^1	d^2	d^3	d^4	d^5	d^6		
+5			d^0	d^1	d^2		d^4			
+6				d^0	d^1	d^2				
+7					d^0					

*The most important orbital occupancies are in color.

Metallic Behavior and Reducing Strength Atomic size and oxidation state have a major effect on the nature of bonding in transition metal compounds. Like the metals in Groups 3A(13), 4A(14), and 5A(15), the transition elements in their *lower* oxidation states behave chemically more like metals. That is, *ionic bonding is more prevalent for the lower oxidation states, and covalent bonding is more prevalent for the higher states.* For example, at room temperature, $TiCl_2$ is an ionic solid, whereas $TiCl_4$ is a molecular liquid. In the higher oxidation states, the atoms have higher charge densities, so they polarize the electron clouds of the nonmetal ions more strongly and the bonding becomes more covalent. For the same reason, the oxides become less basic as the oxidation state increases: TiO is weakly basic in water, whereas TiO_2 is amphoteric (reacts with both acid and base).

Table 23.3 shows the standard electrode potentials of the Period 4 transition metals in their +2 oxidation state in acid solution. Note that, in general, reducing strength decreases across the series. All the Period 4 transition metals, except copper, are active enough to reduce H^+ from aqueous acid to form hydrogen gas. In contrast to the rapid reaction at room temperature of the Group 1A(1) and 2A(2) metals with water, however, the transition metals have an oxide coating that allows rapid reaction only with hot water or steam.

Color and Magnetism of Compounds *Most main-group ionic compounds are colorless* because the metal ion has a filled outer level (noble gas electron configuration). With only much higher energy orbitals available to receive an excited electron, the ion does not absorb visible light. In contrast, electrons in a partially filled *d* sublevel can absorb visible wavelengths and move to slightly higher energy *d* orbitals. As a result, *many transition metal compounds have striking colors*. Exceptions are the compounds of scandium, titanium(IV), and zinc, which are colorless because their metal ions have either an empty *d* sublevel (Sc^{3+} or Ti^{4+}: [Ar] $3d^0$) or a filled one (Zn^{2+}: [Ar] $3d^{10}$) (Figure 23.6).

Magnetic properties are also related to sublevel occupancy (Section 8.5). Recall that a *paramagnetic* substance has atoms or ions with unpaired electrons, which cause it to be attracted to an external magnetic field. A *diamagnetic* substance has only paired electrons, so it is unaffected (or slightly repelled) by a magnetic field. *Most main-group metal ions are diamagnetic* for the same reason they are colorless: all their electrons are paired. In contrast, *many transition metal compounds are paramagnetic because of their unpaired d electrons*. For example, $MnSO_4$ is paramagnetic, but $CaSO_4$ is diamagnetic. The Ca^{2+} ion has the electron configuration of argon, whereas Mn^{2+} has a d^5 configuration. Transition metal ions with a d^0 or d^{10} configuration are also colorless and diamagnetic.

Table 23.3 Standard Electrode Potentials of Period 4 M^{2+} Ions	
Half-Reaction	***E°* (V)**
$Ti^{2+}(aq) + 2e^- \rightleftharpoons Ti(s)$	-1.63
$V^{2+}(aq) + 2e^- \rightleftharpoons V(s)$	-1.19
$Cr^{2+}(aq) + 2e^- \rightleftharpoons Cr(s)$	-0.91
$Mn^{2+}(aq) + 2e^- \rightleftharpoons Mn(s)$	-1.18
$Fe^{2+}(aq) + 2e^- \rightleftharpoons Fe(s)$	-0.44
$Co^{2+}(aq) + 2e^- \rightleftharpoons Co(s)$	-0.28
$Ni^{2+}(aq) + 2e^- \rightleftharpoons Ni(s)$	-0.25
$Cu^{2+}(aq) + 2e^- \rightleftharpoons Cu(s)$	0.34
$Zn^{2+}(aq) + 2e^- \rightleftharpoons Zn(s)$	-0.76

ARIS Presentation Center, Tables Library

Figure 23.6 Colors of representative compounds of the Period 4 transition metals. Staggered from left to right, the compounds are scandium oxide *(white)*, titanium(IV) oxide *(white)*, vanadyl sulfate dihydrate *(light blue)*, sodium chromate *(yellow)*, manganese(II) chloride tetrahydrate *(light pink)*, potassium ferricyanide *(red-orange)*, cobalt(II) chloride hexahydrate *(violet)*, nickel(II) nitrate hexahydrate *(green)*, copper(II) sulfate pentahydrate *(blue)*, and zinc sulfate heptahydrate *(white)*.

 ARIS Presentation Center, Photo Library

Chemical Behavior Within a Group The *increase* in reactivity going down a group of main-group metals, as shown by the *decrease* in first ionization energy (IE_1), does *not* occur going down a group of transition metals. Consider the chromium (Cr) group [6B(6)], which shows a typical pattern (Table 23.4). IE_1 *increases* down the group, which makes the two heavier metals *less* reactive than the lightest one. Chromium is also a much stronger reducing agent than molybdenum (Mo) or tungsten (W), as shown by the standard electrode potentials.

Table 23.4	Some Properties of Group 6B(6) Elements		
Element	Atomic Radius (pm)	IE_1 (kJ/mol)	$E°$ (V) for $M^{3+}(aq)\|M(s)$
Cr	128	653	−0.74
Mo	139	685	−0.20
W	139	770	−0.11

The similarity in atomic size of Period 5 and 6 members also leads to similar chemical behavior, a fact that has some important practical consequences. Because Mo and W compounds behave similarly, for example, their ores often occur together in nature, which makes the elements very difficult to separate from each other. The same situation occurs with zirconium and hafnium in Group 4B(4) and with niobium and tantalum in Group 5B(5).

Section Summary

All transition elements are metals. • Atoms of the *d*-block elements have $(n − 1)d$ orbitals being filled, and their ions have an empty *ns* orbital. • Unlike the trends in the main-group elements, atomic size, electronegativity, and first ionization energy change relatively little across a transition series. Because of the lanthanide contraction, atomic size changes little from Period 5 to 6 in a transition metal group; thus, electronegativity, first ionization energy, and density *increase* down a group. • Transition metals typically have several oxidation states, with the +2 state most common. The elements exhibit more metallic behavior in their lower states. • Most Period 4 transition metals are active enough to reduce hydrogen ion from acid solution. • Many transition metal compounds are colored and paramagnetic because the metal ion has unpaired *d* electrons.

23.2 THE INNER TRANSITION ELEMENTS

The 14 **lanthanides**—cerium (Ce; $Z = 58$) through lutetium (Lu; $Z = 71$)—lie between lanthanum ($Z = 57$) and hafnium ($Z = 72$) in the third *d*-block transition series. Below them are the 14 radioactive **actinides,** thorium (Th; $Z = 90$) through lawrencium (Lr; $Z = 103$), which lie between actinium (Ac; $Z = 89$) and rutherfordium (Rf; $Z = 104$). The lanthanides and actinides are called **inner transition elements** because, in most cases, their seven inner 4*f* or 5*f* orbitals are being filled.

The Lanthanides

The lanthanides are sometimes called the *rare earth elements,* a term referring to their presence in unfamiliar oxides, but they are actually not rare at all. Cerium (Ce), for instance, ranks 26th in natural abundance (by mass %) and is five times more abundant than lead. All the lanthanides are silvery, high-melting (800°C to 1600°C) metals. Their chemical properties represent an extreme case of the small variations typical of transition elements in a period or a group, which makes the lanthanides very difficult to separate. ●

● **A Remarkable Laboratory Feat** The discovery of the lanthanides is a testimony to the remarkable laboratory skills of 19th-century chemists. The two mineral sources, ceria and yttria, are mixtures of compounds of all 14 lanthanides; yttria also includes oxides of Sc, Y, and La. Purification of these elements was so difficult that many claims of new elements turned out to be mixtures of similar elements. To confuse matters, the periodic table of the time had space for only one element between Ba and Hf, which turned out to be La. Not until 1913, when the atomic-number basis of the table was established, did chemists realize that 14 new elements could fit, and in 1918, Niels Bohr proposed that the $n = 4$ level be expanded to include the *f* sublevel.

The natural co-occurrence of the lanthanides arises because they exist as M^{3+} ions of very similar radii in their common ores. Most lanthanides have the ground-state electron configuration [Xe] $6s^2 4f^x 5d^0$, where x varies across the series. The three exceptions (Ce, Gd, and Lu) have a single electron in one of their $5d$ orbitals: Ce ([Xe] $6s^2 4f^1 5d^1$) forms a stable 4+ ion with an empty (f^0) sublevel, and the Gd^{3+} and Lu^{3+} ions have a stable half-filled (f^7) or filled (f^{14}) sublevel.

Lanthanide compounds and their mixtures have many uses. Several oxides are used for tinting sunglasses and welder's goggles and for adding color to the fluorescent powder coatings in TV screens. High-quality camera lenses incorporate La_2O_3 because of its extremely high index of refraction. Samarium forms an alloy with cobalt, $SmCo_5$, that is used in the strongest known permanent magnet (see Sample Problem 23.2). Two industrial uses account for more than 60% of rare earth applications. In gasoline refining, the zeolite catalysts (Gallery, p. 593) used to "crack" hydrocarbon components into smaller molecules contain 5% by mass rare earth oxides. In steelmaking, a mixture of lanthanides, called *misch metal,* is used to remove carbon impurities from molten iron and steel.

SAMPLE PROBLEM 23.2 Finding the Number of Unpaired Electrons

PROBLEM The alloy $SmCo_5$ forms a permanent magnet because both samarium and cobalt have unpaired electrons. How many unpaired electrons are in the Sm atom ($Z = 62$)?

PLAN We write the condensed electron configuration of Sm and then, using Hund's rule and the aufbau principle, place the electrons in a partial orbital diagram and count the unpaired electrons.

SOLUTION Samarium is the eighth element after Xe. Two electrons go into the $6s$ sublevel. In general, the $4f$ sublevel fills before the $5d$ (among the lanthanides, only Ce, Gd, and Lu have $5d$ electrons), so the remaining electrons go into the $4f$. Thus, Sm is [Xe] $6s^2 4f^6$. There are seven f orbitals, so each of the six f electrons enters a separate orbital:

Thus, Sm has six unpaired electrons.

CHECK Six $4f$ e$^-$ plus two $6s$ e$^-$ plus the 54 e$^-$ in Xe gives 62, the atomic number of Sm.

FOLLOW-UP PROBLEM 23.2 How many unpaired electrons are in the Er^{3+} ion?

The Actinides

All actinides are radioactive. Like the lanthanides, they share very similar physical and chemical properties. Thorium and uranium occur in nature, but the transuranium elements, those with Z greater than 92, have been synthesized in high-energy particle accelerators, some in only tiny amounts (Section 24.3). In fact, macroscopic samples of mendelevium (Md), nobelium (No), and lawrencium (Lr) have never been seen. The actinides that have been isolated are silvery and chemically reactive and, like the lanthanides, form highly colored compounds. The actinides and lanthanides have similar outer-electron configurations. Although the +3 oxidation state is characteristic of the actinides, as it is for the lanthanides, other states also occur. For example, uranium exhibits +3 through +6 states, with the +6 state the most prevalent; thus, the most common oxide of uranium is UO_3.

Literature Karol, Paul J. "The Mendeleev-Seaborg Periodic Table: Through $Z = 138$ and Beyond," *J. Chem. Educ.* **2002,** *79,* 60.

Literature Hoffman, Darleane C.; Lee, Diana M. "Chemistry of the Heaviest Elements—One Atom at a Time," *J. Chem. Educ.* **1999,** *76,* 331.

Section Summary

There are two series of inner transition elements. The lanthanides ($4f$ series) have a common +3 oxidation state and exhibit very similar properties. • The actinides ($5f$ series) are radioactive. All actinides have a +3 oxidation state; several, including uranium, have higher states as well.

23.3 HIGHLIGHTS OF SELECTED TRANSITION METALS

Let's investigate several transition elements, focusing on the general patterns in the aqueous chemistry of each. We examine chromium and manganese from Period 4, silver from Period 5, and mercury from Period 6.

Chromium

Chromium is a very shiny, silvery metal, whose name (from the Greek *chroma*, "color") refers to its many colorful compounds. A solution of Cr^{3+} is deep violet, for example, and a trace of Cr^{3+} in the Al_2O_3 crystal structure gives ruby its beautiful red hue. Chromium readily forms a thin, adherent, transparent coating of Cr_2O_3 in air, making the metal extremely useful as an attractive protective coating on easily corroded metals, such as iron. "Stainless" steels often contain as much as 18% chromium by mass and are highly resistant to corrosion.

With six valence electrons ($[Ar]\,4s^1 3d^5$), chromium occurs in all possible positive oxidation states, but the three most important are +2, +3, and +6 (see Table 23.2). In its three common oxides, chromium exhibits the pattern seen in many elements: *nonmetallic character and oxide acidity increase with metal oxidation state*. Chromium(II) oxide (CrO) is basic and largely ionic. It forms an insoluble hydroxide in neutral or basic solution but dissolves in acidic solution to yield the Cr^{2+} ion:

$$CrO(s) + 2H^+(aq) \longrightarrow Cr^{2+}(aq) + H_2O(l)$$

Chromium(III) oxide (Cr_2O_3) is amphoteric, dissolving in acid to yield the violet Cr^{3+} ion,

$$Cr_2O_3(s) + 6H^+(aq) \longrightarrow 2Cr^{3+}(aq) + 3H_2O(l)$$

and in base to form the green $Cr(OH)_4^-$ ion:

$$Cr_2O_3(s) + 3H_2O(l) + 2OH^-(aq) \longrightarrow 2Cr(OH)_4^-(aq)$$

Thus, chromium in its +3 state is similar to the main-group metal aluminum in several respects, including its amphoterism (Section 19.4).

Deep-red chromium(VI) oxide (CrO_3) is covalent and acidic, forming chromic acid (H_2CrO_4) in water,

$$CrO_3(s) + H_2O(l) \longrightarrow H_2CrO_4(aq)$$

which yields the yellow chromate ion (CrO_4^{2-}) in base:

$$H_2CrO_4(aq) + 2OH^-(aq) \longrightarrow CrO_4^{2-}(aq) + 2H_2O(l)$$

In acidic solution, the chromate ion immediately forms the orange dichromate ion ($Cr_2O_7^{2-}$):

$$2CrO_4^{2-}(aq) + 2H^+(aq) \rightleftharpoons Cr_2O_7^{2-}(aq) + H_2O(l)$$

Because both ions contain chromium(VI), this is not a redox reaction; rather, it is a dehydration-condensation, as you can see from the structures in Figure 23.7. Hydrogen-ion concentration controls the equilibrium position: yellow CrO_4^{2-} predominates at high pH, and orange $Cr_2O_7^{2-}$ at low pH. The bright colors of chromium(VI) compounds lead to their wide use in pigments for artist's paints and ceramic glazes. Lead chromate (chrome yellow) is also used in the yellow stripes that delineate traffic lanes.

Chromium metal and the Cr^{2+} ion are potent reducing agents. The metal displaces hydrogen from dilute acids to form blue $Cr^{2+}(aq)$, which reduces O_2 in air within minutes to form the violet Cr^{3+} ion:

$$4Cr^{2+}(aq) + O_2(g) + 4H^+(aq) \longrightarrow 4Cr^{3+}(aq) + 2H_2O(l) \qquad E^\circ_{overall} = 1.64\ V$$

Chromium(VI) compounds in acid solution are strong oxidizing agents (concentrated solutions are *extremely* corrosive!), the chromium(VI) being readily reduced to chromium(III):

$$Cr_2O_7^{2-}(aq) + 14H^+(aq) + 6e^- \longrightarrow 2Cr^{3+}(aq) + 7H_2O(l) \qquad E^\circ = 1.33\ V$$

CrO_4^{2-}

$126°$

$Cr_2O_7^{2-}$

A

B

Figure 23.7 The bright colors of chromium(VI) compounds.
A, Structures of the chromate (CrO_4^{2-}) and dichromate ($Cr_2O_7^{2-}$) ions. **B,** Samples of K_2CrO_4 (*yellow*) and $K_2Cr_2O_7$ (*orange*).

This reaction is often used to determine the iron content of a water or soil sample by oxidizing Fe^{2+} to Fe^{3+} ion. In basic solution, the CrO_4^{2-} ion, which is a much weaker oxidizing agent, predominates:

$$CrO_4^{2-}(aq) + 4H_2O(l) + 3e^- \longrightarrow Cr(OH)_3(s) + 5OH^-(aq) \qquad E° = -0.13 \text{ V}$$

The Concept of Valence-State Electronegativity Why does oxide acidity increase with oxidation state? And how can a metal, like chromium, form an oxoanion? To answer such questions, we must apply the concept of electronegativity to the various oxidation states of an element. A metal in a higher oxidation state is more positively charged, which increases its attraction for electrons; in effect, its *electronegativity increases*. This effective electronegativity, called *valence-state electronegativity,* also has numerical values. The electronegativity of chromium metal is 1.6, close to that of aluminum (1.5), another active metal. For chromium(III), the value increases to 1.7, still characteristic of a metal. However, the electronegativity of chromium(VI) is 2.3, close to the values of some nonmetals, such as phosphorus (2.1), selenium (2.4), and carbon (2.5). Thus, like P in PO_4^{3-}, Cr in chromium(VI) compounds often occurs covalently bonded at the center of the oxoanion of a relatively strong acid.

Manganese

Elemental manganese is hard and shiny and, like vanadium and chromium, is used mostly to make steel alloys. A small amount of Mn ($<1\%$) makes steel easier to roll, forge, and weld. Steel made with 12% Mn is tough enough to be used for naval armor, front-end loader buckets (see photo), and other extremely hard steel objects. Small amounts of manganese are added to aluminum beverage cans and bronze alloys to make them stiffer and tougher as well.

The chemistry of manganese resembles that of chromium in some respects. The free metal is quite reactive and readily reduces H^+ from acids, forming the pale-pink Mn^{2+} ion:

$$Mn(s) + 2H^+(aq) \longrightarrow Mn^{2+}(aq) + H_2(g) \qquad E° = 1.18 \text{ V}$$

Like chromium, manganese can use all its valence electrons in its compounds, exhibiting every possible positive oxidation state, with the +2, +4, and +7 states most common (Table 23.5). As the oxidation state of manganese rises, its valence-state electronegativity increases and its oxides change from basic to acidic. Manganese(II) oxide (MnO) is basic, and manganese(III) oxide (Mn_2O_3) is amphoteric. Manganese(IV) oxide (MnO_2) is insoluble and shows no acid-base properties. [It is used in dry cells and alkaline batteries as the oxidizing agent in a redox reaction with zinc (Chapter 21, p. 952).] Manganese(VII) oxide (Mn_2O_7),

Front-end loader buckets are made of steel containing manganese.

ARIS Presentation Center, Photo Library

Demo Kolb, Doris. "Oxidation States of Manganese," *J. Chem. Educ.* **1988,** *65,* 1004.

ARIS Presentation Center, Tables Library

Table 23.5 **Some Oxidation States of Manganese**

Oxidation state*	**Mn(II)**	Mn(III)	**Mn(IV)**	Mn(VI)	**Mn(VII)**
Example	Mn^{2+}	Mn_2O_3	MnO_2	MnO_4^{2-}	MnO_4^-
Orbital occupancy	d^5	d^4	d^3	d^1	d^0
Oxide acidity	BASIC				ACIDIC

*Most common states in **boldface.**

● **Sharing the Ocean's Wealth** At their current rate of usage, known reserves of many key transition metals will be depleted in less than 50 years. Other sources must be found. One promising source is nodules strewn over large portions of the ocean floor. Varying from a few millimeters to a few meters in diameter, these chunks consist mainly of manganese and iron oxides, with oxides of other elements present in smaller amounts. Billions of tons exist, but mining them presents major technical and political challenges. Global agreements have designated the ocean floor as international property, so cooperation will be required to mine the nodules and share the mineral rewards.

ARIS Presentation Center, Photo Library

which forms by reaction of Mn with pure O_2, reacts with water to form permanganic acid ($HMnO_4$), which is as strong as perchloric acid ($HClO_4$).

All manganese species with oxidation states greater than $+2$ act as oxidizing agents, but the purple permanganate (MnO_4^-) ion is particularly powerful. Like ions with chromium in its highest oxidation state, MnO_4^- is a much stronger oxidizing agent in acidic than in basic solution:

$$MnO_4^-(aq) + 4H^+(aq) + 3e^- \longrightarrow MnO_2(s) + 2H_2O(l) \qquad E° = 1.68 \text{ V}$$
$$MnO_4^-(aq) + 2H_2O(l) + 3e^- \longrightarrow MnO_2(s) + 4OH^-(aq) \qquad E° = 0.59 \text{ V}$$

Unlike Cr^{2+} and Fe^{2+}, the Mn^{2+} ion resists oxidation in air. The Cr^{2+} ion is a d^4 species and readily loses a $3d$ electron to form the d^3 ion Cr^{3+}, which is more stable. The Fe^{2+} ion is a d^6 species, and removing a $3d$ electron yields the stable, half-filled d^5 configuration of Fe^{3+}. Removing an electron from Mn^{2+} disrupts its stable d^5 configuration. ●

Silver

Silver, the second member of the coinage metals [Group 1B(11)], has been admired for thousands of years and is still treasured for use in jewelry and fine flatware. Because the pure metal is too soft for these purposes, however, it is alloyed with copper to form the harder sterling silver. In former times, silver was used in coins, but it has been replaced almost universally by copper-nickel alloys. Silver has the *highest electrical conductivity of any element* but is not used in wiring because copper is cheaper and more plentiful. In the past, silver was found in nuggets and veins of rock, often mixed with gold, because both elements are chemically inert enough to exist uncombined. Nearly all of those deposits have been mined, so most silver is now obtained from the anode mud formed during the electrorefining of copper (Section 22.4).

The only important oxidation state of silver is +1. Its most important *soluble* compound is silver nitrate, used for electroplating and in the manufacture of the halides used for photographic film. Although silver forms no oxide in air, it tarnishes to black Ag_2S by reaction with traces of sulfur-containing compounds. Some polishes remove the Ag_2S, along with some silver, by physically abrading the surface. An alternative "home remedy" that removes the tarnish and restores the metal involves heating the object in a solution of table salt or baking soda ($NaHCO_3$) in an aluminum pan. Aluminum, a strong reducing agent, reduces the Ag^+ ions back to the metal:

$$2Al(s) + 3Ag_2S(s) + 6H_2O(l) \longrightarrow 2Al(OH)_3(s) + 6Ag(s) + 3H_2S(g) \quad E° = 0.86 \text{ V}$$

The Chemistry of Black-and-White Photography
The most widespread use of silver compounds—particularly the three halides AgCl, AgBr, and AgI—is in black-and-white photography, an art that applies transition metal chemistry and solution kinetics. The photographic film itself is simply a flexible plastic support for the light-sensitive emulsion, which consists of AgBr microcrystals dispersed in gelatin. The five steps in obtaining a final photograph are exposing the film, developing the image, fixing the image, washing the negative, and printing the image. Figure 23.8 summarizes the first four of these steps. The process depends on several key chemical properties of silver and its compounds:

- Silver halides undergo a redox reaction when exposed to visible light.
- Silver chloride, bromide, and iodide are *not* water soluble.
- Ag^+ is easily reduced: $Ag^+(aq) + e^- \longrightarrow Ag(s); E° = 0.80 \text{ V}$.
- Ag^+ forms several stable, water-soluble complex ions.

1. *Exposing the film.* Light reflected from the objects in a scene—more light from bright objects than from dark ones—enters the camera lens and strikes the film. Exposed AgBr crystals absorb photons ($h\nu$) in a very localized redox reac-

① **Expose.** *Photons hit film: Br⁻ is oxidized, Ag⁺ is reduced.*

② **Develop.** *Additional Ag⁺ is reduced.*

③ **Fix.** *Further reduction of Ag⁺ is prevented by forming Ag(S₂O₃)₂³⁻(aq).*

④ **Wash.** *Soluble species are removed, leaving Ag granules in place on film.*

AgBr crystals before developing

AgBr crystals after developing

Negative

Figure 23.8 Steps in producing a black-and-white negative.

ARIS Presentation Center, Active Art Library

tion. A Br^- ion is excited by a photon and oxidized, and the released electron almost immediately reduces a nearby Ag^+ ion:

$$\begin{aligned} Br^- &\xrightarrow{h\nu} Br + e^- \\ Ag^+ + e^- &\longrightarrow Ag \\ \hline Ag^+ + Br^- &\xrightarrow{h\nu} Ag + Br \end{aligned}$$

Wherever more light strikes a microcrystal, more Ag atoms form. The exposed crystals are called a *latent image* because the few scattered atoms of photoreduced Ag are not yet visible. Nevertheless, their presence as crystal defects within the AgBr crystal makes it highly susceptible to further reduction.

2. *Developing the image.* The latent image is developed into the actual image by reducing more of the silver ions in the crystal in a controlled manner. Developing is a rate-dependent step: crystals with many photoreduced Ag atoms react more quickly than those with only a few. The developer is a weak reducing agent, such as the organic substance hydroquinone ($H_2C_6H_4O_2$; H_2Q) (see margin):

$$2Ag^+(s) + H_2Q(aq; \text{reduced form}) \longrightarrow 2Ag(s) + Q(aq; \text{oxidized form}) + 2H^+(aq)$$

The reaction rate depends on H_2Q concentration, solution temperature, and the length of time the emulsion is bathed in the solution. After developing, approximately 10^6 as many Ag atoms are present on the film as there were in the latent image, and they form very small, black clusters of silver.

3. *Fixing the image.* After the image is developed, it must be "fixed"; that is, the reduction of Ag^+ must be stopped, or the entire film will blacken on exposure to more light. Fixing involves removing the remaining Ag^+ chemically by converting it to a soluble complex ion with sodium thiosulfate solution ("hypo"):

$$AgBr(s) + 2S_2O_3{}^{2-}(aq) \longrightarrow Ag(S_2O_3)_2{}^{3-}(aq) + Br^-(aq)$$

4. *Washing the negative.* The water-soluble ions are washed away in water. Washing is the final step in producing a photographic *negative,* in which dark objects in the scene appear bright in the image, and vice versa.

hydroquinone   quinone

+ 2H⁺ + 2e⁻

5. *Printing the image.* Through the use of an enlarger, the image on the negative is projected onto print paper coated with emulsion (silver halide in gelatin) and exposed to light, and the previous chemical steps are repeated to produce a "positive" of the image. Bright areas on a negative, such as car tires, allow a great deal of light to pass through and reduce many Ag^+ ions on the print paper. Dark areas on a negative, such as clouds, allow much less light to pass through and, thus, much less Ag^+ reduction. Print-paper emulsion usually contains silver chloride, which reacts more slowly than silver bromide, giving finer control of the print image. High-speed film incorporates silver iodide, the most light-sensitive of the three halides.

Mercury

Mercury has been known since ancient times because cinnabar (HgS), its principal ore, is a naturally occurring red pigment (vermilion) that readily undergoes a redox reaction in the heat of a fire. Sulfide ion, the reducing agent for the process, is already present as part of the ore:

$$HgS(s) + O_2(g) \longrightarrow Hg(g) + SO_2(g)$$

The gaseous Hg condenses on cool nearby surfaces.

The Latin name *hydrargyrum* ("liquid silver") is a good description of mercury, the only metal that is liquid at room temperature. Two factors account for this unusual property. First, because of a distorted crystal structure, each mercury atom is surrounded by 6 rather than 12 nearest neighbors. Second, a filled, tightly held *d* sublevel leaves only the two 6*s* electrons available for metallic bonding. Thus, interactions among mercury atoms are relatively few and relatively weak and, as a result, the solid form breaks down at $-38.9°C$.

Many of mercury's uses arise from its unusual physical properties. Its liquid range ($-39°C$ to $357°C$) encompasses most everyday temperatures, so Hg is commonly used in thermometers. As you might expect from its position in Period 6 following the lanthanide contraction, mercury is quite dense (13.5 g/mL), which makes it convenient for use in barometers and manometers. Mercury's fluidity and conductivity make it useful for "silent" switches in thermostats. At high pressures, mercury vapor can be excited electrically to emit the bright white light seen in sports stadium and highway lights.

Mercury is a good solvent for other metals, and many *amalgams* (alloys of mercury) exist. In mercury batteries, a zinc amalgam acts as the anode and mercury(II) oxide acts as the cathode (Chapter 21, p. 953). In the mercury-cell version of the chlor-alkali process (Chapter 22, p. 1015), mercury acts as the cathode and as the solvent for the sodium metal that forms. Recall that, when treated with water, the resulting sodium amalgam, Na(Hg), forms two important byproducts:

$$2Na(Hg) + 2H_2O(l) \xrightarrow{-Hg} 2NaOH(aq) + H_2(g)$$

The mercury is released in this step and reused in the electrolytic cell. In view of mercury's toxicity, which we discuss shortly, a newer method involving a polymeric membrane is replacing the mercury-cell method. Nevertheless, past contamination of wastewater from the chlor-alkali process and the disposal of old mercury batteries remain serious environmental concerns.

The chemical properties of mercury are unique within Group 2B(12). The other members, zinc and cadmium, occur in the +2 oxidation state as d^{10} ions, but mercury occurs in the +1 state as well, with the condensed electron configuration [Xe] $6s^1 4f^{14} 5d^{10}$. We can picture the unpaired 6*s* electron allowing two Hg(I) species to form the *diatomic ion* $[Hg—Hg]^{2+}$ (written Hg_2^{2+}), one of the first species known with a covalent metal-metal bond (but also see the footnote on p. 1028). Mercury's more common oxidation state is +2. Whereas HgF_2 is largely ionic, many other compounds, such as $HgCl_2$, contain bonds that are predominantly covalent. Most mercury(II) compounds are insoluble in water.

The common ions Zn^{2+}, Cd^{2+}, and Hg^{2+} are biopoisons. Zinc oxide is used as an external antiseptic ointment. The Cd^{2+} and Hg^{2+} ions are two of the so-called toxic heavy-metal ions. The cadmium in solder may be more responsible than the lead for solder's high toxicity. Mercury compounds have been used in agriculture as fungicides and pesticides and in medicine as internal drugs, but these uses have been largely phased out.

Because most mercury(II) compounds are insoluble in water, they were once thought to be harmless in the environment; but we now know otherwise. Microorganisms in sludge and river sediment convert mercury atoms and ions to the methyl mercury ion, $CH_3—Hg^+$, and then to organomercury compounds, such as dimethylmercury, $CH_3—Hg—CH_3$. These toxic compounds are nonpolar and, like chlorinated hydrocarbons, become increasingly concentrated in fatty tissues, as they move up a food chain from microorganisms to worms to fish and, finally, to birds and mammals. Indeed, fish living in a mercury-polluted lake or coastal estuary can have a mercury concentration thousands of times higher than that of the water itself.

The mechanism for the toxicity of mercury and other heavy-metal ions is not fully understood. It is thought that the ions migrate from fatty tissue and bind strongly to thiol (—SH) groups of amino acids in proteins, thereby disrupting the proteins' structure and function. Because the brain has a high fat content, small amounts of lead, cadmium, and mercury ions circulating in the blood become deposited in the brain's fatty tissue, interact with its proteins, and often cause devastating neurological and psychological effects. ●

Section Summary

Chromium and manganese add corrosion resistance and hardness to steels. They are typical of transition metals in having several oxidation states. • Valence-state electronegativity refers to the ability of an element in its various oxidation states to attract bonding electrons. It increases with oxidation number, which is the reason elements act more metallic (more ionic compounds, more basic oxides) in lower states and more nonmetallic (more acidic oxides, oxoanions of acids) in higher states. • Cr and Mn produce H_2 in acid. Cr(VI) undergoes a pH-sensitive dehydration-condensation reaction. Both Cr(VI) and Mn(VII) are stronger oxidizing agents in acid than in base. • The only important oxidation state for silver is +1. The silver halides are light sensitive and are used in photography. • Mercury, the only metal that is liquid at room temperature, dissolves many other metals in important applications. The mercury(I) ion is diatomic and has a metal-metal covalent bond. The element and its compounds are toxic and become concentrated as they move up a food chain.

23.4 COORDINATION COMPOUNDS

The most distinctive aspect of transition metal chemistry is the formation of **coordination compounds** (also called *complexes*). These are substances that contain at least one **complex ion,** a species consisting of *a central metal cation (either a transition metal or a main-group metal) that is bonded to molecules and/or anions called **ligands.*** In order to maintain charge neutrality in the coordination compound, the complex ion is typically associated with other ions, called **counter ions.**

Biological application

Environmental application

● **Mad as a Hatter** The felt top hat and zany manner of the Mad Hatter in *Alice in Wonderland* are literary references to the toxicity of mercury compounds. Mercury(II) nitrate and chloride were used with HNO_3 to make felt for hats from animal hair. Inhalation of the dust generated by the process led to "hatter's shakes," abnormal behavior, and other neurological and psychological symptoms.

ARIS Presentation Center, Art Library

Figure 23.9 Components of a coordination compound. Coordination compounds, shown here as models (*top*), wedge-bond perspective drawings (*middle*), and chemical formulas (*bottom*), typically consist of a complex ion and counter ions to neutralize the charge. The complex ion has a central metal ion surrounded by ligands. **A,** When solid [Co(NH₃)₆]Cl₃ dissolves in water, the complex ion and the counter ions separate, but the ligands remain bound to the metal ion. Six ligands around the metal ion give the complex ion an octahedral geometry. **B,** Complex ions with a central d^8 metal ion have four ligands and a square planar geometry.

Transparency/ARIS Presentation Center, Art Library

Literature Venkataraman, D.; Du, Yuhua; Wilson, Scott R.; Hirsch, Keith A.; Zhang, Peng; Moore, Jeffrey S. "A Coordination Geometry Table of the *d*-Block Elements and Their Ions," *J. Chem. Educ.* **1997,** *74,* 915.

A typical coordination compound appears in Figure 23.9A: the coordination compound is [Co(NH₃)₆]Cl₃, the complex ion (always enclosed in square brackets) is [Co(NH₃)₆]³⁺, the six NH₃ molecules bonded to the central Co³⁺ are ligands, and the three Cl⁻ ions are counter ions. *A coordination compound behaves like an electrolyte in water:* the complex ion and counter ions separate from each other. But the complex ion behaves like a polyatomic ion: *the ligands and central metal ion remain attached.* Thus, as Figure 23.9A shows, 1 mol of [Co(NH₃)₆]Cl₃ yields 1 mol of [Co(NH₃)₆]³⁺ ions and 3 mol of Cl⁻ ions.

We discussed the Lewis acid-base properties of hydrated metal ions, which are a type of complex ion, in Section 18.9, and we examined complex-ion equilibria in Section 19.4. In this section, we consider the bonding, structure, and properties of complex ions.

Complex Ions: Coordination Numbers, Geometries, and Ligands

A complex ion is described by the metal ion and the number and types of ligands attached to it. Its structure is related to three characteristics—coordination number, geometry, and number of donor atoms per ligand:

- *Coordination number.* The **coordination number** is the *number of ligand atoms* that are bonded directly to the central metal ion and is *specific* for a given metal ion in a particular oxidation state and compound. The coordination number of the Co³⁺ ion in [Co(NH₃)₆]³⁺ is 6 because six ligand atoms (N of NH₃) are bonded to it. The coordination number of the Pt²⁺ ion in many of its complexes is 4, whereas that of the Pt⁴⁺ ion in its complexes is 6. Copper(II) may have a coordination number of 2, 4, or 6 in different complex ions. In general, *the most common coordination number in complex ions is 6,* but 2 and 4 are often seen, and some higher ones are also known.

- *Geometry.* The geometry (shape) of a complex ion depends on the coordination number and nature of the metal ion. Table 23.6 shows the geometries associated with the coordination numbers 2, 4, and 6, with some examples of each. A complex ion whose metal ion has a coordination number of 2, such as [Ag(NH₃)₂]⁺, is *linear.* The coordination number 4 gives rise to either of two geometries—square planar or tetrahedral. Most d^8 metal ions form *square planar* complex ions, depicted in Figure 23.9B. The d^{10} ions are among those that

Transparency/ARIS Presentation Center, Tables Library

Table 23.6 Coordination Numbers and Shapes of Some Complex Ions

Coordination Number	Shape		Examples
2	Linear		$[CuCl_2]^-$, $[Ag(NH_3)_2]^+$, $[AuCl_2]^-$
4	Square planar		$[Ni(CN)_4]^{2-}$, $[PdCl_4]^{2-}$, $[Pt(NH_3)_4]^{2+}$, $[Cu(NH_3)_4]^{2+}$
4	Tetrahedral		$[Cu(CN)_4]^{3-}$, $[Zn(NH_3)_4]^{2+}$, $[CdCl_4]^{2-}$, $[MnCl_4]^{2-}$
6	Octahedral		$[Ti(H_2O)_6]^{3+}$, $[V(CN)_6]^{4-}$, $[Cr(NH_3)_4Cl_2]^+$, $[Mn(H_2O)_6]^{2+}$, $[FeCl_6]^{3-}$, $[Co(en)_3]^{3+}$

form *tetrahedral* complex ions. A coordination number of 6 results in an *octahedral* geometry, as shown by $[Co(NH_3)_6]^{3+}$ in Figure 23.9A. Note the similarity with some of the molecular shapes in VSEPR theory (Section 10.2).

• *Donor atoms per ligand.* The ligands of complex ions are *molecules or anions* with one or more **donor atoms** that each *donate a lone pair of electrons* to the metal ion to form a covalent bond. Because they have at least one lone pair, donor atoms often come from Group 5A(15), 6A(16), or 7A(17).

Ligands are classified in terms of the number of donor atoms, or "teeth," that each uses to bond to the central metal ion. *Monodentate* (Latin, "one-toothed") ligands, such as Cl^- and NH_3, use a single donor atom. *Bidentate* ligands have two donor atoms, each of which bonds to the metal ion. *Polydentate* ligands have more than two donor atoms. Table 23.7 shows some common ligands in coordination compounds; note that each ligand has one or more donor atoms (colored type), each with a lone pair of electrons to donate. Bidentate and polydentate

Transparency/ARIS Presentation Center, Tables Library

Table 23.7 Some Common Ligands in Coordination Compounds

Ligand Type	Examples
Monodentate	$H_2\ddot{O}$: water $:\ddot{F}:^-$ fluoride ion $[:C≡N:]^-$ cyanide ion $[:\ddot{O}—H]^-$ hydroxide ion :NH_3 ammonia $:\ddot{C}l:^-$ chloride ion $[:\ddot{S}=C=\ddot{N}:]^-$ thiocyanate ion $[:\ddot{O}—N=\ddot{O}:]^-$ nitrite ion
Bidentate	ethylenediamine (en) oxalate ion
Polydentate	diethylenetriamine triphosphate ion ethylenediaminetetraacetate ion (EDTA^{4-})

[Pb(EDTA)]$^{2-}$

● **Grabbing Ions** Because it has 6 donor atoms, the ethylenediaminetetraacetate (EDTA^{4-}) ion forms very stable complexes with many metal ions. This property makes EDTA useful in treating heavy-metal poisoning. Once ingested by the patient, the ion acts as a scavenger to remove lead and other heavy-metal ions from the blood and other body fluids.

ARIS Presentation Center, Art Library

Literature Haworth, Daniel T. "Some Linguistic Detail on Chelation," *J. Chem. Educ.* **1998,** *75,* 47.

ligands give rise to *rings* in the complex ion. For instance, ethylenediamine (abbreviated *en* in formulas) has a chain of four atoms (:N—C—C—N:), so it forms a five-membered ring, with the two electron-donating N atoms bonding to the metal atom. Such ligands seem to grab the metal ion like claws, so a complex ion that contains them is also called a **chelate** (pronounced "KEY-late"; Greek *chela,* "crab's claw"). ●

Formulas and Names of Coordination Compounds

There are three important rules for writing the formulas of coordination compounds, the first two being the same for writing formulas of any ionic compound:

1. *The cation is written before the anion.*
2. *The charge of the cation(s) is balanced by the charge of the anion(s).*
3. *In the complex ion, neutral ligands are written before anionic ligands, and the formula for the whole ion is placed in brackets.*

Let's apply these rules as we examine the combinations of ions in coordination compounds. *The whole complex ion may be a cation or an anion.* A complex cation has anionic counter ions, and a complex anion has cationic counter ions. It's easy to find the charge of the central metal ion. For example, in $K_2[Co(NH_3)_2Cl_4]$, two K^+ counter ions balance the charge of the complex anion $[Co(NH_3)_2Cl_4]^{2-}$, which contains two NH_3 molecules and four Cl^- ions as ligands. The two NH_3 are neutral, the four Cl^- have a total charge of $4-$, and the entire complex ion has a charge of $2-$, so the central metal ion must be Co^{2+}:

$$\text{Charge of complex ion} = \text{Charge of metal ion} + \text{total charge of ligands}$$
$$2- = \text{Charge of metal ion} + [(2 \times 0) + (4 \times 1-)]$$

So, $\text{Charge of metal ion} = (2-) - (4-) = 2+$

In the compound $[Co(NH_3)_4Cl_2]Cl$, the complex ion is $[Co(NH_3)_4Cl_2]^+$ and one Cl^- is the counter ion. The four NH_3 ligands are neutral, the two Cl^- ligands have a total charge of $2-$, and the complex cation has a charge of $1+$, so the central metal ion must be Co^{3+} [that is, $1+ = (3+) + (2-)$]. Some coordination compounds have a complex cation *and* a complex anion, as in $[Co(NH_3)_5Br]_2[Fe(CN)_6]$. In this compound, the complex cation is $[Co(NH_3)_5Br]^{2+}$, with Co^{3+}, and the complex anion is $[Fe(CN)_6]^{4-}$, with Fe^{2+}.

Coordination compounds were originally named after the person who first prepared them or from their color, and some of these common names are still used, but most coordination compounds are named systematically through a set of rules:

1. *The cation is named before the anion.* In naming $[Co(NH_3)_4Cl_2]Cl$, for example, we name the $[Co(NH_3)_4Cl_2]^+$ ion before the Cl^- ion. Thus, the name is

    tetraamminedichlorocobalt(III) chloride

    The only space in the name appears between the cation and the anion.
2. *Within the complex ion, the ligands are named, in alphabetical order,* ***before*** *the metal ion.* Note that in the $[Co(NH_3)_4Cl_2]^+$ ion of the compound named in rule 1, the four NH_3 and two Cl^- are named before the Co^{3+}.
3. *Neutral ligands generally have the molecule name,* but there are a few exceptions (Table 23.8). *Anionic ligands drop the -ide and add -o after the root name;* thus, the name *fluoride* for the F^- ion becomes the ligand name *fluoro.* The two ligands in $[Co(NH_3)_4Cl_2]^+$ are *ammine* (NH_3) and *chloro* (Cl^-) with *ammine* coming before *chloro* alphabetically.
4. *A numerical prefix indicates the number of ligands of a particular type.* For example, *tetra*ammine denotes *four* NH_3, and *di*chloro denotes *two* Cl^-. Other prefixes are *tri-, penta-,* and *hexa-.* These prefixes do *not* affect the alphabet-

Table 23.8   Names of Some Neutral and Anionic Ligands

Neutral		Anionic	
Name	Formula	Name	Formula
Aqua	H_2O	Fluoro	F^-
Ammine	NH_3	Chloro	Cl^-
Carbonyl	CO	Bromo	Br^-
Nitrosyl	NO	Iodo	I^-
		Hydroxo	OH^-
		Cyano	CN^-

ical order; thus, *tetraammine* comes before *dichloro*. Because some ligand names already contain a numerical prefix (such as ethylene*di*amine), we use *bis* (2), *tris* (3), or *tetrakis* (4) to indicate the number of such ligands, followed by the ligand name in parentheses. For example, a complex ion that has two ethylenediamine ligands has *bis(ethylenediamine)* in its name.

5. *The oxidation state of the central metal ion is given by a Roman numeral (in parentheses) only* if the metal ion can have more than one state, as in the compound named in rule 1.

6. *If the complex ion is an anion, we drop the ending of the metal name and add -ate.* Thus, the name for $K[Pt(NH_3)Cl_5]$ is

potassium amminepentachloroplatinate(IV)

(Note that there is one K^+ counter ion, so the complex anion has a charge of $1-$. The five Cl^- ligands have a total charge of $5-$, so Pt must be in the $+4$ oxidation state.) For some metals, we use the Latin root with the *-ate* ending, shown in Table 23.9. For example, the name for $Na_4[FeBr_6]$ is

sodium hexabromoferrate(II)

Table 23.9   Names of Some Metal Ions in Complex Anions

Metal	Name in Anion
Iron	Ferrate
Copper	Cuprate
Lead	Plumbate
Silver	Argentate
Gold	Aurate
Tin	Stannate

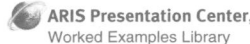

SAMPLE PROBLEM 23.3   Writing Names and Formulas of Coordination Compounds

PROBLEM (a) What is the systematic name of $Na_3[AlF_6]$?
(b) What is the systematic name of $[Co(en)_2Cl_2]NO_3$?
(c) What is the formula of tetraamminebromochloroplatinum(IV) chloride?
(d) What is the formula of hexaamminecobalt(III) tetrachloroferrate(III)?

PLAN We use the rules that were presented above and refer to Tables 23.8 and 23.9.

SOLUTION (a) The complex ion is $[AlF_6]^{3-}$. There are six *(hexa-)* F^- ions *(fluoro)* as ligands, so we have *hexafluoro*. The complex ion is an anion, so the ending of the metal ion (aluminum) must be changed to *-ate:* hexafluoroaluminate. Aluminum has only the $+3$ oxidation state, so we do *not* use a Roman numeral. The positive counter ion is named first and separated from the anion by a space: sodium hexafluoroaluminate.

(b) Listed alphabetically, there are two Cl^- *(dichloro)* and two en [*bis(ethylenediamine)*] as ligands. The complex ion is a cation, so the metal name is unchanged, but we specify its oxidation state because cobalt can have several. One NO_3^- balances the $1+$ cation charge: with $2-$ for two Cl^- and 0 for two en, the metal must be *cobalt(III)*. The word *nitrate* follows a space: dichlorobis(ethylenediamine)cobalt(III) nitrate.

(c) The central metal ion is written first, followed by the neutral ligands and then (in alphabetical order) by the negative ligands. *Tetraammine* is four NH_3, *bromo* is one Br^-, *chloro* is one Cl^-, and *platinate(IV)* is Pt^{4+}, so the complex ion is $[Pt(NH_3)_4BrCl]^{2+}$. Its $2+$ charge is the sum of $4+$ for Pt^{4+}, 0 for four NH_3, $1-$ for one Br^-, and $1-$ for one Cl^-. To balance the $2+$ charge, we need two Cl^- counter ions: $[Pt(NH_3)_4BrCl]Cl_2$.

(d) This compound consists of two different complex ions. In the cation, *hexaammine* is six NH_3 and *cobalt(III)* is Co^{3+}, so the cation is $[Co(NH_3)_6]^{3+}$. The 3+ charge is the sum of 3+ for Co^{3+} and 0 for six NH_3. In the anion, *tetrachloro* is four Cl^-, and *ferrate(III)* is Fe^{3+}, so the anion is $[FeCl_4]^-$. The 1− charge is the sum of 3+ for Fe^{3+} and 4− for four Cl^-. In the neutral compound, one 3+ cation is balanced by three 1− anions: $[Co(NH_3)_6][FeCl_4]_3$.

CHECK Reverse the process to be sure you obtain the name or formula asked for in the problem.

FOLLOW-UP PROBLEM 23.3 **(a)** What is the name of $[Cr(H_2O)_5Br]Cl_2$? **(b)** What is the formula of barium hexacyanocobaltate(III)?

A Historical Perspective: Alfred Werner and Coordination Theory

The substances we now call coordination compounds had been known for almost 200 years when the young Swiss chemist Alfred Werner began studying them in the 1890s. He investigated a series of compounds such as the cobalt series shown in Table 23.10, each of which contains one cobalt(III) ion, three chloride ions, and a given number of ammonia molecules. At the time, which was 30 years before the idea of atomic orbitals was proposed, no structural theory could explain how compounds with similar, even identical, formulas could have widely different properties.

Werner measured the conductivity of each compound in aqueous solution to determine the total number of ions that became dissociated. He treated the solutions with excess $AgNO_3$ to precipitate released Cl^- ions as AgCl and thus determine the number of free Cl^- ions per formula unit. Previous studies had established that the NH_3 molecules were not free in solution. Werner's data, summarized in Table 23.10, could not be explained by the accepted, traditional formulas of the compounds. Other chemists had proposed "chain" structures, like those of organic compounds, to explain such data. For example, a proposed structure for $[Co(NH_3)_6]Cl_3$ was

$$
\begin{array}{l}
NH_3-Cl \\
\quad | \\
Co-NH_3-Cl \\
\quad | \\
NH_3-NH_3-NH_3-NH_3-Cl
\end{array}
$$

However, these models proved inadequate.

Werner's novel idea was the coordination complex, a central metal ion surrounded by a *constant total number* of covalently bonded molecules and/or anions. The coordination complex could be neutral or charged; if charged, it combined with oppositely charged counter ions, in this case Cl^-, to form the neutral compound.

Table 23.10 **Some Coordination Compounds of Cobalt Studied by Werner**

Traditional Formula	Werner's Data*		Modern Formula	Charge of Complex Ion
	Total Ions	Free Cl^-		
$CoCl_3 \cdot 6NH_3$	4	3	$[Co(NH_3)_6]Cl_3$	3+
$CoCl_3 \cdot 5NH_3$	3	2	$[Co(NH_3)_5Cl]Cl_2$	2+
$CoCl_3 \cdot 4NH_3$	2	1	$[Co(NH_3)_4Cl_2]Cl$	1+
$CoCl_3 \cdot 3NH_3$	0	0	$[Co(NH_3)_3Cl_3]$	—

*Moles per mole of compound.

Werner proposed two types of valence, or *combining ability,* for metal ions. *Primary valence,* now called *oxidation state,* is the positive charge on the metal ion that must be satisfied by an equivalent negative charge. In Werner's cobalt series, the primary valence is +3, and it is always balanced by three Cl$^-$ ions. These anions can be bonded covalently to Co as part of the complex ion and/or associated with it as counter ions. *Secondary valence,* now called *coordination number,* is the constant total number of connections (anionic or neutral ligands) *within* the complex ion. The secondary valence in this series of Co compounds is 6.

As you can see, Werner's data are satisfied if the total number of ligands remains the same for each compound, even though the numbers of Cl$^-$ ions and NH_3 molecules in the different complex ions vary. For example, the first compound, [Co(NH_3)$_6$]Cl$_3$, has a total of four ions: one [Co(NH_3)$_6$]$^{3+}$ and three Cl$^-$. All three Cl$^-$ ions are free to form AgCl. The last compound, [Co(NH_3)$_3$Cl$_3$], contains no separate ions.

Surprisingly, Werner was an organic chemist, and his work on coordination compounds, which virtually revolutionized his contemporaries' understanding of chemical bonding, was his attempt to demonstrate the unity of chemistry. For these pioneering studies, especially his prediction of optical isomerism (discussed shortly), Werner received the Nobel Prize in chemistry in 1913.

Isomerism in Coordination Compounds

Isomers are compounds with the same chemical formula but different properties. We discussed many aspects of isomerism in the context of organic compounds in Section 15.2; it may be helpful to review that section now. Figure 23.10 presents an overview of the most common types of isomerism in coordination compounds.

Transparency/ARIS Presentation Center, Art Library

Figure 23.10 **Important types of isomerism in coordination compounds.**

ISOMERS
Same chemical formula, but different properties

Constitutional (structural) isomers
Atoms connected differently

Stereoisomers
Different spatial arrangement

Coordination isomers
Ligand and counter-ion exchange

Linkage isomers
Different donor atom

Geometric (*cis-trans*) isomers (diastereomers)
Different arrangement around metal ion

Optical isomers (enantiomers)
Nonsuperimposable mirror images

Constitutional Isomers: Same Atoms Connected Differently Two compounds with the same formula, but with the atoms connected differently, are called **constitutional (structural) isomers.** Coordination compounds exhibit the following two types of constitutional isomers: one involves a difference in the composition of the complex ion, the other in the donor atom of the ligand.

1. **Coordination isomers** occur when the composition of the complex ion changes but not that of the compound. One way this type of isomerism occurs is when ligand and counter ion exchange positions, as in [Pt(NH_3)$_4$Cl$_2$](NO$_2$)$_2$ and [Pt(NH_3)$_4$(NO$_2$)$_2$]Cl$_2$. In the first compound, the Cl$^-$ ions are the ligands, and the NO$_2^-$ ions are counter ions; in the second, the roles are reversed. Another way that this type of isomerism occurs is in compounds of two complex ions in which the two sets of ligands in one compound are reversed in the other, as in [Cr(NH_3)$_6$][Co(CN)$_6$] and [Co(NH_3)$_6$][Cr(CN)$_6$]; note that NH_3 is a ligand of Cr^{3+} in one compound and of Co^{3+} in the other.

Demo Summerlin, L. R.; Ealy, J. B. "Cobalt Complexes: Changing Coordination Numbers," *Chemical Demonstrations, A Sourcebook for Teachers,* Vol. 1 (American Chemical Society, Washington, DC, 1988), 41–42.

2. Linkage isomers occur when the composition of the complex ion remains the same but the attachment of the ligand donor atom changes. Some ligands can bind to the metal ion through *either of two donor atoms*. For example, the nitrite ion can bind through a lone pair on either the N atom (*nitro*, O₂N:) or one of the O atoms (*nitrito*, ONO:) to give linkage isomers, as in the orange compound penta-ammine*nitro*cobalt(III) chloride $[Co(NH_3)_5(NO_2)]Cl_2$ *(below left)* and its reddish linkage isomer pentaammine*nitrito*cobalt(III) chloride $[Co(NH_3)_5(ONO)]Cl_2$ *(below right)*:

Nitro isomer, $[Co(NH_3)_5(NO_2)]Cl_2$

Nitrito isomer, $[Co(NH_3)_5(ONO)]Cl_2$

Another example is the cyanate ion, which can attach via a lone pair on the O atom (*cyanato*, NCO:) or the N atom (*isocyanato*, OCN:); the thiocyanate ion behaves similarly, attaching via the S atom or the N atom:

nitrite   cyanate   thiocyanate

Stereoisomers: Different Spatial Arrangements of Atoms

Stereoisomers are compounds that have the same atomic connections but different spatial arrangements of the atoms. The two types we discussed for organic compounds, called *geometric* and *optical* isomers, are seen with coordination compounds as well:

Literature Luján-Upton, Hannia. "Introducing Stereochemistry to Non-science Majors," *J. Chem. Educ.* **2001**, *78*, 475.

1. Geometric isomers (also called **cis-trans** isomers and, sometimes, *diastereomers*) occur when atoms or groups of atoms are arranged differently in space relative to the central metal ion. For example, the square planar $[Pt(NH_3)_2Cl_2]$ has two arrangements, which give rise to two different compounds (Figure 23.11A). The isomer with identical ligands *next* to each other is *cis*-diamminedichloroplatinum(II), and the one with identical ligands *across* from each other is *trans*-diamminedichloroplatinum(II); their biological behaviors are remarkably different. ● Octahedral complexes also exhibit *cis-trans* isomerism (Figure 23.11B). The *cis* isomer of the $[Co(NH_3)_4Cl_2]^+$ ion has the two Cl⁻ ligands next to each other and is violet, whereas the *trans* isomer has these two ligands across from each other and is green.

2. Optical isomers (also called *enantiomers*) occur when a molecule and its mirror image cannot be superimposed (see Figures 15.8 to 15.10, pp. 639 and 640). Unlike other types of isomers, which have distinct physical properties, optical isomers are physically identical in all ways but one: *the direction in which they rotate the plane of polarized light*. Octahedral complex ions show many examples of optical isomerism, which we can observe by rotating one isomer and seeing if it is superimposable on the other isomer (its mirror image). For example, as you can see in Figure 23.12A, the two structures (I and II) of $[Co(en)_2Cl_2]^+$, the *cis*-dichlorobis(ethylenediamine)cobalt(III) ion, are mirror images of each other. Rotate structure I 180° around a vertical axis, and you obtain III. The Cl⁻ li-

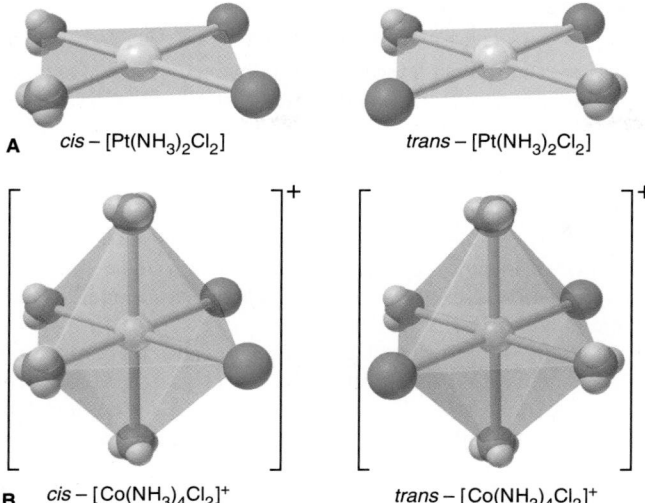

Figure 23.11 Geometric *(cis-trans)* isomerism. A, The *cis* and *trans* isomers of the square planar coordination compound [Pt(NH₃)₂Cl₂]. **B,** The *cis* and *trans* isomers of the octahedral complex ion [Co(NH₃)₄Cl₂]⁺. The colored shapes represent the actual colors of the species.

$cis - [Pt(NH_3)_2Cl_2]$

$trans - [Pt(NH_3)_2Cl_2]$

A

B $cis - [Co(NH_3)_4Cl_2]^+$ $trans - [Co(NH_3)_4Cl_2]^+$

ARIS Presentation Center, Art Library

gands of III match those of II, but the en ligands do not: II and III (rotated I) are not superimposable; therefore, they are optical isomers. One isomer is designated d-[Co(en)₂Cl₂]⁺ and the other is l-[Co(en)₂Cl₂]⁺, depending on whether it rotates polarized light to the right (d- for "dextro-") or to the left (l- for "levo-"). (The d- or l- designation can only be determined experimentally, *not* by examination of the structure.) In contrast, as shown in Figure 23.12B, the two structures of the *trans*-dichlorobis(ethylenediamine)cobalt(III) ion are *not* optical isomers: rotate I 90° around a vertical axis and you obtain III, which *is* superimposable on II.

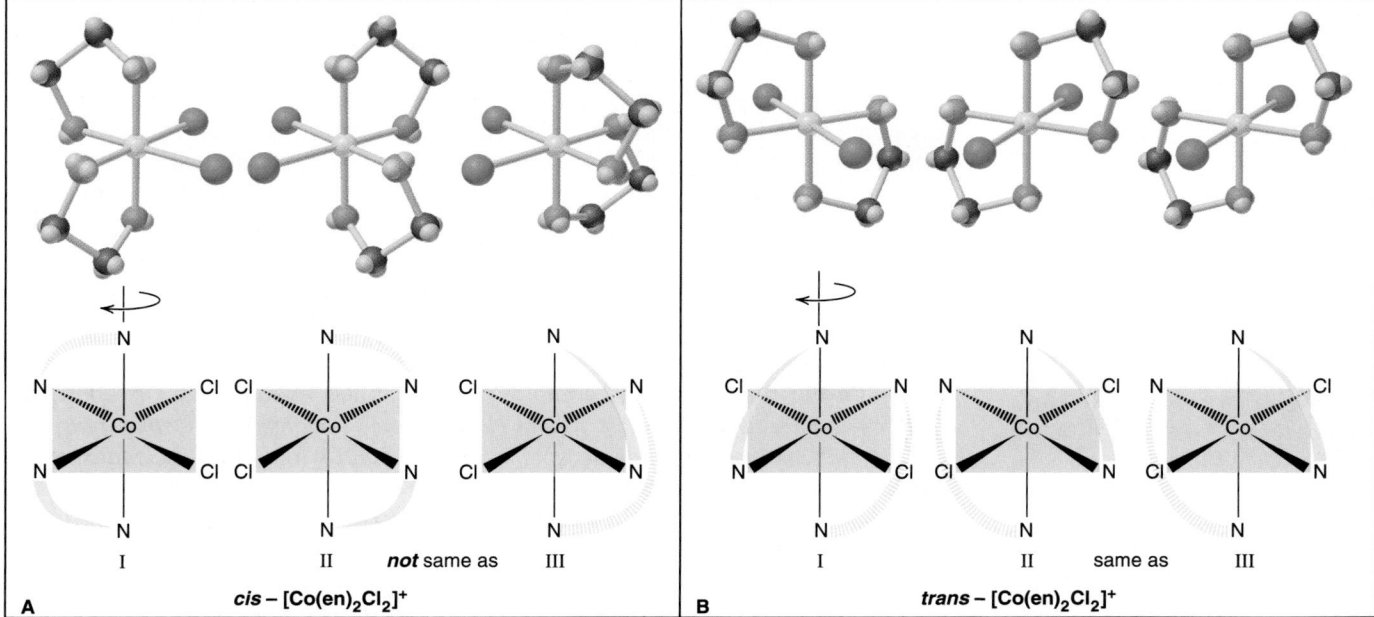

A   I   II *not* same as III

$cis - [Co(en)_2Cl_2]^+$

B   I   II same as III

$trans - [Co(en)_2Cl_2]^+$

Figure 23.12 Optical isomerism in an octahedral complex ion. A, Structure I and its mirror image, structure II, are optical isomers of *cis*-[Co(en)₂Cl₂]⁺. Rotating structure I gives structure III, which is *not* the same as structure II. (The curved wedges represent the bidentate ligand ethylenediamine, H₂N—CH₂—CH₂—NH₂.) **B,** The *trans* isomer does *not* have optical isomers. Rotating structure I gives III, which is *identical* to II, the mirror image of I.

(a)
trans   cis
Mirror

(b)

ARIS Presentation Center,
Worked Examples Library

SAMPLE PROBLEM 23.4 Determining the Type of Stereoisomerism

PROBLEM Draw all stereoisomers for each of the following and state the type of isomerism:
(a) $[Pt(NH_3)_2Br_2]$ (square planar)   **(b)** $[Cr(en)_3]^{3+}$ (en = $H_2NCH_2CH_2NH_2$)
PLAN We first determine the geometry around each metal ion and the nature of the ligands. If there are two different ligands that can be placed in different positions relative to each other, geometric *(cis-trans)* isomerism occurs. Then, we see whether the mirror image of an isomer is superimposable on the original. If it is *not*, optical isomerism occurs.
SOLUTION **(a)** The Pt(II) complex is square planar, and there are two different monodentate ligands. Each pair of ligands can lie next to or across from each other (see structures in margin). Thus, geometric isomerism occurs. Each isomer *is* superimposable on its mirror image, so there is no optical isomerism.
(b) Ethylenediamine (en) is a bidentate ligand. The Cr^{3+} has a coordination number of 6 and an octahedral geometry, like Co^{3+}. The three bidentate ligands are identical, so there is no geometric isomerism. However, the complex ion has a nonsuperimposable mirror image (see structures in margin). Thus, optical isomerism occurs.

FOLLOW-UP PROBLEM 23.4 What stereoisomers, if any, are possible for the $[Co(NH_3)_2(en)Cl_2]^+$ ion?

Section Summary

Coordination compounds consist of a complex ion and charge-balancing counter ions. The complex ion has a central metal ion bonded to neutral and/or anionic ligands, which have one or more donor atoms that each provide a lone pair of electrons. • The most common geometry is octahedral (six ligand atoms bonding). • Formulas and names of coordination compounds follow systematic rules. • Werner established the structural basis of coordination compounds. • These compounds can exhibit constitutional isomerism (coordination and linkage) and stereoisomerism (geometric and optical).

23.5 THEORETICAL BASIS FOR THE BONDING AND PROPERTIES OF COMPLEXES

In this section, we consider models that address, in different ways, several key features of complexes: how metal-ligand bonds form, why certain geometries are preferred, and why these complexes are brightly colored and often paramagnetic. As you saw with covalent bonding in other compounds (Chapter 11), more than one model is often needed to tell the whole story.

Application of Valence Bond Theory to Complex Ions

Valence bond (VB) theory, which helped explain bonding and structure in main-group compounds (Section 11.1), is also used to describe bonding in complex ions. In the formation of a complex ion, the filled ligand orbital overlaps the empty metal-ion orbital. *The ligand (Lewis base) donates the electron pair, and the metal ion (Lewis acid) accepts it to form one of the covalent bonds of the complex ion (Lewis adduct)* (Section 18.9). Such a bond, in which one atom in the bond contributes both electrons, is called a **coordinate covalent bond,** although, once formed, it is identical to any covalent single bond. Recall that the VB concept of hybridization proposes the mixing of particular combinations of s, p, and d orbitals to give sets of hybrid orbitals, which have specific geometries. Similarly, for coordination compounds, the model proposes that *the number and type of metal-ion hybrid orbitals occupied by ligand lone pairs determine the geometry of the complex ion.* Let's discuss the orbital combinations that lead to octahedral, square planar, and tetrahedral geometries.

Figure 23.13 Hybrid orbitals and bonding in the octahedral [Cr(NH₃)₆]³⁺ ion. A, VB depiction of the Cr(NH₃)₆³⁺ ion. **B,** The partial orbital diagrams depict the mixing of two 3d, one 4s, and three 4p orbitals in Cr³⁺ to form six d^2sp^3 hybrid orbitals, which are filled with six NH₃ lone pairs (red).

ARIS Presentation Center, Art Library

Octahedral Complexes The hexaamminechromium(III) ion, $[Cr(NH_3)_6]^{3+}$, illustrates the application of VB theory to an *octahedral complex* (Figure 23.13). The six lowest energy empty orbitals of the Cr^{3+} ion—two 3d, one 4s, and three 4p—mix and become six equivalent d^2sp^3 hybrid orbitals that point toward the corners of an octahedron.* Six NH₃ molecules donate lone pairs from their nitrogens to form six metal-ligand bonds. The three unpaired 3d electrons of the central Cr^{3+} ion ([Ar] $3d^3$), which make the complex ion paramagnetic, remain in unhybridized orbitals.

Square Planar Complexes Metal ions with a d^8 configuration usually form *square planar complexes* (Figure 23.14). In the $[Ni(CN)_4]^{2-}$ ion, for example, the model proposes that one 3d, one 4s, and two 4p orbitals of Ni^{2+} mix and form four dsp^2 hybrid orbitals, which point to the corners of a square and accept one electron pair from each of four CN^- ligands.

A look at the ground-state electron configuration of the Ni^{2+} ion, however, raises a key question: how can the Ni^{2+} ion ([Ar] $3d^8$) offer an empty 3d orbital for accepting a lone pair, if its eight 3d electrons lie in three filled and two half-filled orbitals? Apparently, in the d^8 configuration of Ni^{2+}, electrons in the half-filled orbitals pair up and leave one 3d orbital empty. This explanation is consistent with the fact that the complex is diamagnetic (no unpaired electrons). Moreover, it requires that the energy *gained* by using a 3d orbital for bonding in the hybrid orbital is greater than the energy *required* to overcome repulsions from pairing the 3d electrons.

---

*Note the distinction between the hybrid-orbital designation here and that for octahedral molecules like SF₆. The designation gives the orbitals in energy order within a given *n* value. In the $[Cr(NH_3)_6]^{3+}$ complex ion, the d orbitals have a *lower n* value than the s and p orbitals, so the hybrid orbitals are d^2sp^3. For the orbitals in SF₆, the d orbitals have the *same n* value as the s and p, so the hybrid orbitals are sp^3d^2.

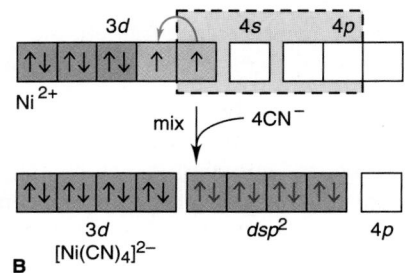

Figure 23.14 Hybrid orbitals and bonding in the square planar [Ni(CN)₄]²⁻ ion. A, VB depiction of [Ni(CN)₄]²⁻. **B,** Two lone 3d electrons pair up and free one 3d orbital for hybridization with the 4s and two of the 4p orbitals to form four dsp^2 orbitals, which become occupied with lone pairs (red) from four CN^- ligands.

ARIS Presentation Center, Art Library

Figure 23.15 Hybrid orbitals and bonding in the tetrahedral $[Zn(OH)_4]^{2-}$ ion. A, VB depiction of $[Zn(OH)_4]^{2-}$. **B,** Mixing one 4s and three 4p orbitals gives four sp^3 hybrid orbitals available for accepting lone pairs (*red*) from OH$^-$ ligands.

ARIS Presentation Center, Art Library

Tetrahedral Complexes Metal ions that have a filled d sublevel, such as Zn^{2+} ([Ar] $3d^{10}$), often form *tetrahedral complexes* (Figure 23.15). For the complex ion $[Zn(OH)_4]^{2-}$, for example, VB theory proposes that the lowest available Zn^{2+} orbitals—one 4s and three 4p—mix to become four sp^3 hybrid orbitals that point to the corners of a tetrahedron and are occupied by four lone pairs, one from each of four OH$^-$ ligands.

Crystal Field Theory

The VB model is easy to picture and rationalizes bonding and shape, but it treats the orbitals as little more than empty "slots" for accepting electron pairs. Consequently, it gives no insight into the colors of coordination compounds and sometimes predicts their magnetic properties incorrectly. In contrast to the VB approach, **crystal field theory** provides little insight about metal-ligand bonding but explains color and magnetism clearly. To do so, it highlights the *effects on the d-orbital energies of the metal ion as the ligands approach.* Before we discuss this theory, let's consider what causes a substance to be colored.

What Is Color? White light is electromagnetic radiation consisting of all wavelengths (λ) in the visible range (Section 7.1). It can be dispersed into a spectrum of colors, each of which has a narrower range of wavelengths. Objects appear colored in white light because they absorb certain wavelengths and reflect or transmit others: an opaque object *reflects* light, whereas a clear one *transmits* it. The reflected or transmitted light enters the eye and the brain perceives a color. If an object absorbs all visible wavelengths, it appears black; if it reflects all, it appears white.

Each color has a *complementary* color. For example, green and red are complementary colors. A mixture of complementary colors absorbs all visible wavelengths and appears black. Figure 23.16 shows these relationships on an artist's color wheel, where complementary colors appear as wedges opposite each other.

An object has a particular color for one of two reasons:

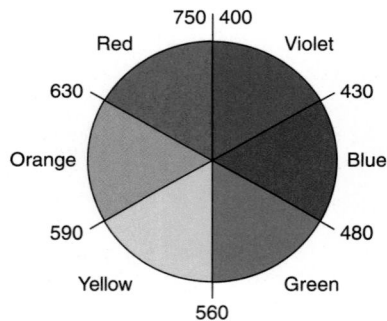

Figure 23.16 An artist's wheel. Colors, with approximate wavelength ranges (in nm), are shown as wedges. Complementary colors, such as red and green, lie opposite each other.

ARIS Presentation Center, Art Library

• It reflects (or transmits) light of *that* color. Thus, if an object absorbs all wavelengths *except* green, the reflected (or transmitted) light enters our eyes and is interpreted as green.
• It absorbs light of the *complementary* color. Thus, if the object absorbs only red, the *complement* of green, the remaining mixture of reflected (or transmitted) wavelengths enters our eyes and is interpreted as green also.

Biological application

Table 23.11 lists the color absorbed and the resulting color perceived. A familiar example of this phenomenon results in the seasonal colors of deciduous trees. In the spring and summer, leaves contain high concentrations of the photosynthetic pigment *chlorophyll* and lower concentrations of other pigments called *xanthophylls*. Chlorophyll absorbs strongly in the blue and red regions, reflecting mostly green wavelengths into your eyes. In the fall, photosynthesis slows, so the leaf no longer makes chlorophyll. Gradually, the green color fades as the chlorophyll decomposes, revealing the xanthophylls that were present all along but masked

Literature Poncini, Laurence; Wimmer, Franz L. "Color Classification of Coordination Compounds," *J. Chem. Educ.* **1987,** *64,* 1001.

Table 23.11 **Relation Between Absorbed and Observed Colors**

Absorbed Color	λ (nm)	Observed Color	λ (nm)
Violet	400	Green-yellow	560
Blue	450	Yellow	600
Blue-green	490	Red	620
Yellow-green	570	Violet	410
Yellow	580	Dark blue	430
Orange	600	Blue	450
Red	650	Green	520

 ARIS Presentation Center, Tables Library

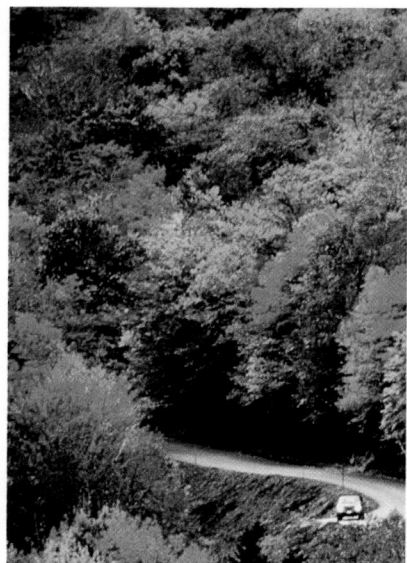

Foliage changing color in autumn.

ARIS Presentation Center, Photo Library

by the chlorophyll. Xanthophylls absorb green and blue strongly, reflecting the bright yellows and reds of autumn (see photo).

Splitting of *d* Orbitals in an Octahedral Field of Ligands

The crystal field model explains that the properties of complexes result from the splitting of *d*-orbital energies, which arises from electrostatic interactions between metal ion and ligands. The model assumes that a complex ion forms as a result of *electrostatic attractions between the metal cation and the negative charge of the ligands*. This negative charge is either partial, as in a polar neutral ligand like NH_3, or full, as in an anionic ligand like Cl^-. The ligands approach the metal ion along the mutually perpendicular x, y, and z axes, which minimizes the overall energy of the system.

Picture what happens as the ligands approach. Figure 23.17A shows six ligands moving toward a metal ion to form an octahedral complex. Let's see how the various *d* orbitals of the metal ion are affected as the complex forms. As the ligands approach, their electron pairs repel electrons in the five *d* orbitals. In the isolated metal ion, the *d* orbitals have equal energies despite their different orientations. In the electrostatic field of ligands, however, the *d* electrons are *repelled unequally because their orbitals have different orientations*. Because the ligands move along the x, y, and z axes, they approach *directly toward* the lobes of the $d_{x^2-y^2}$ and d_{z^2} orbitals (Figure 23.17B and C) but *between* the lobes of the d_{xy}, d_{xz}, and d_{yz} orbitals (Figure 23.17D to F). Thus, electrons in the $d_{x^2-y^2}$ and d_{z^2} orbitals experience *stronger* repulsions than those in the d_{xy}, d_{xz}, and d_{yz} orbitals.

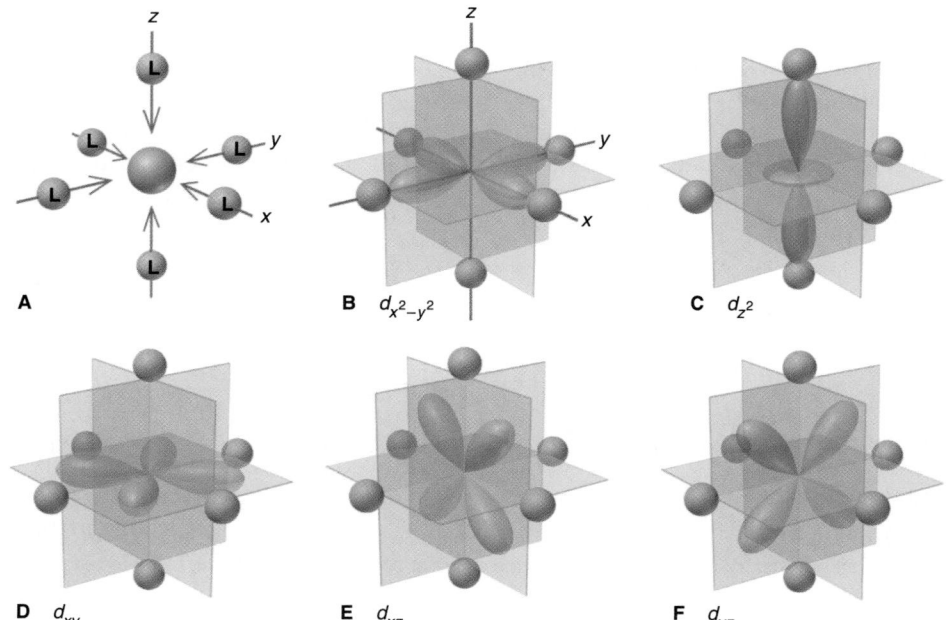

Figure 23.17 **The five *d* orbitals in an octahedral field of ligands.** The direction of ligand approach influences the strength of repulsions of electrons in the five metal ion *d* orbitals. **A,** We assume that ligands approach a metal ion along the three linear axes in an octahedral orientation. **B and C,** Lobes of the $d_{x^2-y^2}$ and d_{z^2} orbitals lie *directly in line* with the approaching ligands, so repulsions are stronger. **D to F,** Lobes of the d_{xy}, d_{xz}, and d_{yz} orbitals lie *between* the approaching ligands, so repulsions are weaker.

Transparency/ARIS Presentation Center, Art Library

Figure 23.18 **Splitting of *d*-orbital energies by an octahedral field of ligands.** Electrons in the *d* orbitals of the free metal ion experience an *average* net repulsion in the negative ligand field that increases all *d*-orbital energies. Electrons in the d_{xy}, d_{yz}, and d_{xz} orbitals, which form the t_{2g} set, are repelled less than those in the $d_{x^2-y^2}$, and d_{z^2} orbitals, which form the e_g set. The energy difference between these two sets is the crystal field splitting energy, Δ.

Transparency/ARIS Presentation Center, Active Art Library

Figure 23.19 **The effect of the ligand on splitting energy.** Ligands interacting strongly with metal-ion *d* orbitals, such as CN^-, produce a larger Δ than those interacting weakly, such as H_2O.

ARIS Presentation Center, Art Library

An energy diagram of the orbitals shows that all five *d* orbitals are higher in energy in the forming complex than in the free metal ion because of repulsions from the approaching ligands, but *the orbital energies split, with two d orbitals higher in energy than the other three* (Figure 23.18). The two higher energy orbitals are called ***e_g* orbitals,** and the three lower energy ones are ***t_{2g}* orbitals.** (These designations refer to features of the orbitals that need not concern us here.)

The splitting of orbital energies is called the *crystal field effect,* and the difference in energy between the e_g and t_{2g} sets of orbitals is the **crystal field splitting energy (Δ).** Different ligands create crystal fields of different strength and, thus, cause the *d*-orbital energies to split to different extents. **Strong-field ligands** lead to a *larger* splitting energy (larger Δ); **weak-field ligands** lead to a *smaller* splitting energy (smaller Δ). For instance, H_2O is a weak-field ligand, and CN^- is a strong-field ligand (Figure 23.19). The magnitude of Δ relates directly to the color and magnetic properties of a complex.

Explaining the Colors of Transition Metals The remarkably diverse colors of coordination compounds are determined by the energy difference (Δ) between the t_{2g} and e_g orbital sets in their complex ions. When the ion absorbs light in the visible range, electrons are excited ("jump") from the lower energy t_{2g} level to the higher e_g level. Recall that the *difference* between two electronic energy levels in the ion is equal to the energy (and inversely related to the wavelength) of the absorbed photon:

$$\Delta E_{\text{electron}} = E_{\text{photon}} = h\nu = hc/\lambda$$

The substance has a color because only certain wavelengths of the incoming white light are absorbed.

Consider the $[Ti(H_2O)_6]^{3+}$ ion, which appears purple in aqueous solution (Figure 23.20). Hydrated Ti^{3+} is a d^1 ion, with the *d* electron in one of the three

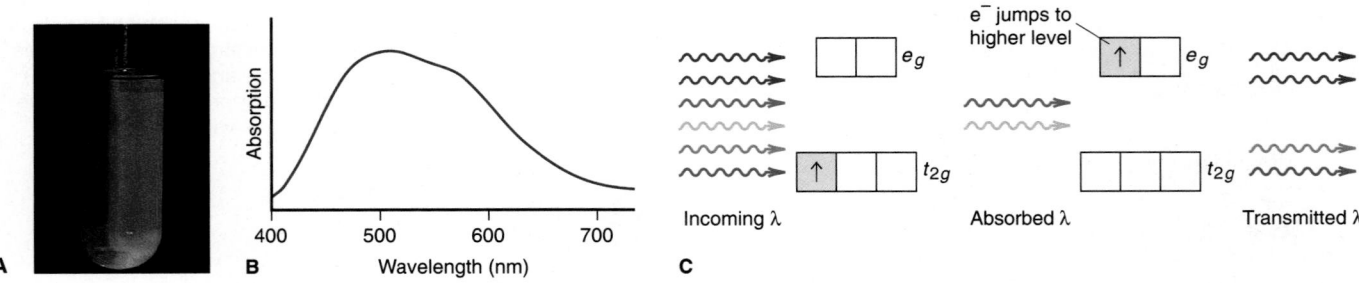

Figure 23.20 **The color of $[Ti(H_2O)_6]^{3+}$. A,** The hydrated Ti^{3+} ion is purple in aqueous solution. **B,** An absorption spectrum shows that incoming wavelengths corresponding to green and yellow light are absorbed, whereas other wavelengths are transmitted. **C,** An orbital diagram depicts the colors absorbed in the excitation of the *d* electron to the higher level. ARIS Presentation Center, Art Library

lower energy t_{2g} orbitals. The energy difference (Δ) between the t_{2g} and e_g orbitals in this ion corresponds to the energy of photons spanning the green and yellow range. When white light shines on the solution, these colors of light are absorbed, and the electron jumps to one of the e_g orbitals. Red, blue, and violet light are transmitted, so the solution appears purple.

Absorption spectra show the wavelengths absorbed by a given metal ion with different ligands and by different metal ions with the same ligand. From such data, we relate the energy of the absorbed light to the Δ values, and two important observations emerge:

1. *For a given ligand, the color depends on the oxidation state of the metal ion.* A solution of $[V(H_2O)_6]^{2+}$ ion is violet, and a solution of $[V(H_2O)_6]^{3+}$ ion is yellow (Figure 23.21A).
2. *For a given metal ion, the color depends on the ligand.* Even a single ligand substitution can have a major effect on the wavelengths absorbed and, thus, the color, as you can see for two Cr^{3+} complex ions in Figure 23.21B.

The second observation allows us to rank ligands into a **spectrochemical series** with regard to their ability to split *d*-orbital energies. An abbreviated series, moving from weak-field ligands (small splitting, small Δ) to strong-field ligands (large splitting, large Δ), is shown in Figure 23.22. Using this series, we can predict the *relative* size of Δ for a series of octahedral complexes of the same metal ion. Although it is difficult to predict the actual color of a given complex, we can determine whether a complex will absorb longer or shorter wavelengths than other complexes in the series.

Figure 23.21 Effects of the metal oxidation state and of ligand identity on color. A, Solutions of $[V(H_2O)_6]^{2+}$ (*left*) and $[V(H_2O)_6]^{3+}$ (*right*) ions have different colors. **B,** A change in even a single ligand can influence the color. The $[Cr(NH_3)_6]^{3+}$ ion is yellow-orange (*left*); the $[Cr(NH_3)_5Cl]^{2+}$ ion is purple (*right*).

ARIS Presentation Center, Photo Library

SAMPLE PROBLEM 23.5   Ranking Crystal Field Splitting Energies for Complex Ions of a Given Metal

PROBLEM Rank the ions $[Ti(H_2O)_6]^{3+}$, $[Ti(NH_3)_6]^{3+}$, and $[Ti(CN)_6]^{3-}$ in terms of the relative value of Δ and of the energy of visible light absorbed.

PLAN The formulas show that titanium's oxidation state is $+3$ in the three ions. From Figure 23.22, we rank the ligands in terms of crystal field strength: the stronger the ligand, the greater the splitting, and the higher the energy of light absorbed.

SOLUTION The ligand field strength is in the order $CN^- > NH_3 > H_2O$, so the relative size of Δ and energy of light absorbed is

$$Ti(CN)_6^{3-} > Ti(NH_3)_6^{3+} > Ti(H_2O)_6^{3+}$$

FOLLOW-UP PROBLEM 23.5 Which complex ion absorbs visible light of higher energy, $[V(H_2O)_6]^{3+}$ or $[V(NH_3)_6]^{3+}$?

ARIS Presentation Center,
Worked Examples Library

Animation: Formation of an Ionic Compound
ARIS Presentation Center
Chapter 23

$$I^- < Cl^- < F^- < OH^- < H_2O < SCN^- < NH_3 < en < NO_2^- < CN^- < CO$$

| WEAKER FIELD | STRONGER FIELD |

| SMALLER Δ | LARGER Δ |

| LONGER λ | SHORTER λ |

Figure 23.22   The spectrochemical series. As the crystal field strength of the ligand increases, the splitting energy (Δ) increases, so shorter wavelengths (λ) of light must be absorbed to excite electrons. Water is usually a weak-field ligand.

Transparency/ARIS Presentation Center,
Art Library

Figure 23.23 High-spin and low-spin complex ions of Mn²⁺. **A,** The free Mn^{2+} ion has five unpaired electrons. **B,** Bonded to weak-field ligands (smaller Δ), Mn^{2+} still has five unpaired electrons (high-spin complex). **C,** Bonded to strong-field ligands (larger Δ), Mn^{2+} has only one unpaired electron (low-spin complex).

Explaining the Magnetic Properties of Transition Metal Complexes The splitting of energy levels influences magnetic properties by affecting the number of *unpaired* electrons in the metal ion's d orbitals. Based on Hund's rule, electrons occupy orbitals one at a time as long as orbitals of equal energy are available. When all lower energy orbitals are half-filled, the next electron can

- enter a half-filled orbital and pair up by overcoming a repulsive *pairing energy* ($E_{pairing}$), or
- enter an empty, higher energy orbital by overcoming the crystal field splitting energy (Δ).

Thus, *the relative sizes of $E_{pairing}$ and Δ determine the occupancy of the d orbitals.* The orbital occupancy, in turn, determines the number of unpaired electrons and, thus, the paramagnetic behavior of the ion.

As an example, the isolated Mn^{2+} ion ([Ar] $3d^5$) has five unpaired electrons in $3d$ orbitals of equal energy (Figure 23.23A). In an octahedral field of ligands, the orbital energies split. The orbital occupancy is affected by the ligand in one of two ways:

- *Weak-field ligands and high-spin complexes.* Weak-field ligands, such as H_2O in $[Mn(H_2O)_6]^{2+}$, cause a *small* splitting energy, so it takes *less* energy for d electrons to jump to the e_g set than to pair up in the t_{2g} set. Therefore, the d electrons remain unpaired (Figure 23.23B). Thus, with weak-field ligands, the pairing energy is *greater* than the splitting energy ($E_{pairing} > \Delta$); therefore, *the number of unpaired electrons in the complex ion is the **same** as in the free ion.* Weak-field ligands create **high-spin complexes,** those with the *maximum* number of unpaired electrons.
- *Strong-field ligands and low-spin complexes.* In contrast, strong-field ligands, such as CN^- in $[Mn(CN)_6]^{4-}$, cause a *large* splitting of the d-orbital energies, so it takes *more* energy for electrons to jump to the e_g set than to pair up in the t_{2g} set (Figure 23.23C). With strong-field ligands, the pairing energy is *smaller* than the splitting energy ($E_{pairing} < \Delta$); therefore, *the number of unpaired electrons in the complex ion is **less** than in the free ion.* Strong-field ligands create **low-spin complexes,** those with *fewer* unpaired electrons.

Orbital diagrams for the d^1 through d^9 ions in octahedral complexes show that both high-spin and low-spin options are possible only for d^4, d^5, d^6, and d^7 ions (Figure 23.24). With three lower energy t_{2g} orbitals available, the d^1, d^2, and d^3 ions always form high-spin complexes because there is no need to pair up. Similarly, d^8 and d^9 ions always form high-spin complexes: because the t_{2g} set is filled with six electrons, the two e_g orbitals *must* have either two (d^8) or one (d^9) unpaired electron.

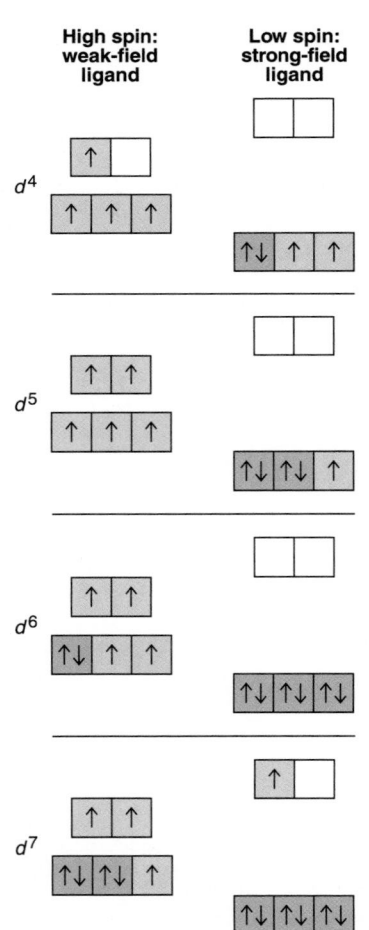

Figure 23.24 Orbital occupancy for high-spin and low-spin complexes of d^4 through d^7 metal ions.

SAMPLE PROBLEM 23.6 Identifying Complex Ions as High Spin or Low Spin

PROBLEM Iron(II) forms an essential complex in hemoglobin. For each of the two octahedral complex ions $[Fe(H_2O)_6]^{2+}$ and $[Fe(CN)_6]^{4-}$, draw an orbital splitting diagram, predict the number of unpaired electrons, and identify the ion as low spin or high spin.

PLAN The Fe^{2+} electron configuration gives us the number of d electrons, and the spectrochemical series in Figure 23.22 shows the relative strengths of the two ligands. We draw the diagrams, separating the t_{2g} and e_g sets by a greater distance for the strong-field ligand. Then we add electrons, noting that a weak-field ligand gives the *maximum* number of unpaired electrons and a high-spin complex, whereas a strong-field ligand leads to electron pairing and a low-spin complex.

SOLUTION Fe^{2+} has the $[Ar]\, 3d^6$ configuration. According to Figure 23.22, H_2O produces smaller splitting than CN^-. The diagrams are shown in the margin. The $[Fe(H_2O)_6]^{2+}$ ion has four unpaired electrons (high spin), and the $[Fe(CN)_6]^{4-}$ ion has no unpaired electrons (low spin).

COMMENT 1. H_2O is a weak-field ligand, so it almost always forms high-spin complexes. **2.** These results are correct, but we cannot confidently predict the spin of a complex without having actual values for Δ and $E_{pairing}$. **3.** Cyanide ions and carbon monoxide are highly toxic because they interact with the iron cations in proteins.

FOLLOW-UP PROBLEM 23.6 How many unpaired electrons do you expect for $[Mn(CN)_6]^{3-}$? Is this a high-spin or low-spin complex ion?

Crystal Field Splitting in Tetrahedral and Square Planar Complexes

Four ligands around a metal ion also cause d-orbital splitting, but the magnitude and pattern of the splitting depend on whether the ligands are in a tetrahedral or a square planar arrangement.

- *Tetrahedral complexes.* With the ligands approaching from the corners of a tetrahedron, none of the five d orbitals is directly in their paths (Figure 23.25). Thus, splitting of d-orbital energies is *less* in a tetrahedral complex than in an octahedral complex having the same ligands:

$$\Delta_{\text{tetrahedral}} < \Delta_{\text{octahedral}}$$

Minimal repulsions arise if the ligands approach the d_{xy}, d_{xz}, and d_{yz} orbitals closer than they approach the $d_{x^2-y^2}$ and d_{z^2} orbitals. This situation is the *opposite of the octahedral case,* and the relative d-orbital energies are reversed: the d_{xy}, d_{xz}, and d_{yz} orbitals become *higher* in energy than the $d_{x^2-y^2}$ and d_{z^2} orbitals. *Only high-spin tetrahedral complexes are known* because the magnitude of Δ is so small.

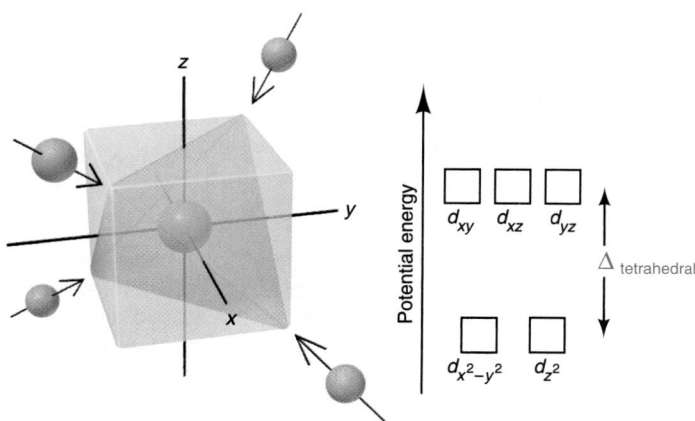

Figure 23.25 Splitting of d-orbital energies by a tetrahedral field of ligands. Electrons in d_{xy}, d_{xz}, and d_{yz} orbitals experience greater repulsions than those in $d_{x^2-y^2}$ and d_{z^2} so the tetrahedral splitting pattern is the opposite of the octahedral pattern.

Figure 23.26 **Splitting of *d*-orbital energies by a square planar field of ligands.** In a square planar field, the energies of d_{xz}, d_{yz}, and especially d_{z^2} orbitals decrease relative to the octahedral pattern.

ARIS Presentation Center, Art Library

• *Square planar complexes.* The effects of the ligand field in the square planar case are easier to picture if we imagine starting with an octahedral geometry and then remove the two ligands along the *z*-axis, as depicted in Figure 23.26. With no *z*-axis interactions present, the d_{z^2} orbital energy decreases greatly, and the energies of the other orbitals with a *z*-axis component, the d_{xz} and d_{yz}, also decrease. As a result, the two *d* orbitals in the *xy* plane interact most strongly with the ligands, and because the $d_{x^2-y^2}$ orbital has its lobes *on* the axes, its energy is highest. As a consequence of this splitting pattern, square planar complexes with d^8 metal ions, such as $[PdCl_4]^{2-}$, are diamagnetic, with four pairs of *d* electrons filling the four lowest energy orbitals. Thus, as a general rule, *square planar complexes are low spin.*

At this point, a final word about bonding theories may be helpful. As you have seen in our discussions of several other topics, no one model is satisfactory in every respect. The VB approach offers a simple picture of bond formation but does not even attempt to explain color. The crystal field model predicts color and magnetic behavior beautifully but treats the metal ion and ligands as points of opposite charge and thus offers no insight about the covalent nature of metal-ligand bonding.

Despite its complexity, chemists now rely on a more refined model, called *ligand field–molecular orbital theory.* This theory combines aspects of the previous two models with MO theory (Section 11.3). We won't explore this model here, but it is a powerful predictive tool, yielding information on bond properties that result from the overlap of metal ion and ligand orbitals as well as information on the spectral and magnetic properties that result from the splitting of a metal ion's *d* orbitals.

In addition to their important chemical applications, complexes of the transition elements play vital roles in living systems, as the following Chemical Connections essay describes.

Section Summary

Valence bond theory pictures bonding in complex ions as arising from coordinate covalent bonding between Lewis bases (ligands) and Lewis acids (metal ions). • Ligand lone pairs occupy hybridized metal-ion orbitals to form complex ions with characteristic shapes. • Crystal field theory explains the color and magnetism of complexes. • As the result of a surrounding field of ligands, the *d*-orbital energies of the metal ion split. The magnitude of this crystal field splitting energy (Δ) depends on the charge of the metal ion and the crystal field strength of the ligand. • In turn, Δ influences the energy of the photon absorbed (color) and the number of unpaired *d* electrons (paramagnetism). Strong-field ligands create a large Δ and produce low-spin complexes that absorb light of higher energy (shorter λ); the reverse is true of weak-field ligands. • Several transition metals, such as iron and zinc, are essential dietary trace elements that function in complexes within proteins.

Chapter Perspective

Our study of the transition elements, a large group of metals with many essential industrial and biological roles, points out once again that macroscopic properties, such as color and magnetism, have their roots at the atomic and molecular levels, which in turn have such a crucial influence on structure. In the next, and final, chapter we explore the core of the atom and learn how we can apply its enormous power.

Chemical Connections to Nutritional Science

Transition Metals as Essential Dietary Trace Elements

Biological application

Living things consist primarily of water and complex organic compounds of four building-block elements: carbon, oxygen, hydrogen, and nitrogen (see Figure 2.17, p. 63). All organisms also contain seven other elements, known as *macro*nutrients because they occur in fairly high concentrations. In order of increasing atomic number, they are sodium, magnesium, phosphorus, sulfur, chlorine, potassium, and calcium. In addition, organisms contain a surprisingly large number of other elements in much lower concentrations, and most of these *micro*nutrients, or *trace elements,* are transition metals.

With the exception of scandium and titanium, all Period 4 transition elements are essential for organisms, and plants require molybdenum (from Period 5) as well. The transition metal ion usually occurs at a bend of a protein chain covalently bonded to surrounding amino acid groups whose N and O atoms act as ligands. Despite the structural complexity of biomolecules, the principles of bonding and *d*-orbital splitting are the same as in simple inorganic systems. Table B23.1 (on page 1056) is a list of the Period 4 transition metals known, or thought, to be essential in human nutrition. In this discussion we focus on iron and zinc.

Iron plays a crucial role in oxygen transport in all vertebrates. The oxygen-transporting protein hemoglobin (Figure B23.1A) consists of four folded protein chains called *globins,* each cradling the iron-containing complex *heme.* Heme is a porphyrin, a complex derived from a metal ion and the tetradentate ring ligand known as *porphin.* Iron(II) is centered in the plane of the porphin ring, forming coordinate covalent bonds with four N lone pairs, resulting in a square planar complex. When heme is bound in hemoglobin (Figure B23.1B), the complex is *octahedral,* with the fifth ligand of iron(II) being an N atom from a nearby amino acid (histidine), and the sixth an O atom from either an O_2 (shown) or an H_2O molecule.

Hemoglobin exists in two forms, depending on the nature of the sixth ligand. In the blood vessels of the lungs, where O_2 concentration is high, heme binds O_2 to form *oxyhemoglobin,* which is transported in the arteries to O_2-depleted tissues. At these tissues, the O_2 is released and replaced by an H_2O molecule to form *deoxyhemoglobin,* which is transported in the veins back to the lungs. The H_2O is a weak-field ligand, so the d^6 ion Fe^{2+} in deoxyhemoglobin is part of a high-spin complex. Because of the relatively small *d*-orbital splitting, deoxyhemoglobin absorbs light at the red (low-energy) end of the spectrum and looks purplish blue, which accounts for the dark color of venous blood. On the other hand, O_2 is a strong-field ligand, so it increases the splitting energy, which gives rise to a low-spin complex. For this reason, oxyhemoglobin absorbs at the blue (high-energy) end of the spectrum, which accounts for the bright red color of arterial blood.

The position of the Fe^{2+} ion relative to the plane of the porphin ring also depends on this sixth ligand. Bound to O_2, Fe^{2+} is *in* the porphin plane; bound to H_2O, it moves *out of* the plane slightly. This tiny (~60 pm = 6×10^{-11} m) change in the position of Fe^{2+} on release or attachment of O_2 influences the shape of its globin chain, which in turn alters the shape of a neighboring globin chain, triggering the release or attachment of *its* O_2, and so on to the other two globin chains. The very survival of vertebrate life is the result of this cooperative "teamwork" by the four globin

(continued)

Figure B23.1 Hemoglobin and the octahedral complex in heme. A, Hemoglobin consists of four protein chains, each with a bound heme. (Illustration by Irving Geis. Rights owned by Howard Hughes Medical Institute. Not to be used without permission.) **B,** In oxyhemoglobin, the octahedral complex in heme has iron(II) at the center surrounded by the four N atoms of the porphin ring, a fifth N from histidine *(below),* and an O_2 molecule *(above).*

ARIS Presentation Center, Art Library

Table B23.1	Some Transition Metal Trace Elements in Humans	
Element	**Biomolecule Containing Element**	**Function of Biomolecule**
Vanadium	Protein (?)	Redox couple in fat metabolism (?)
Chromium	Glucose tolerance factor	Glucose utilization
Manganese	Isocitrate dehydrogenase	Cell respiration
Iron	Hemoglobin and myoglobin Cytochrome c Catalase	Oxygen transport Cell respiration; ATP formation Decomposition of H_2O_2
Cobalt	Cobalamin (vitamin B_{12})	Development of red blood cells
Copper	Ceruloplasmin Cytochrome oxidase	Hemoglobin synthesis Cell respiration; ATP formation
Zinc	Carbonic anhydrase Carboxypeptidase A Alcohol dehydrogenase	Elimination of CO_2 Protein digestion Metabolism of ethanol

ARIS Presentation Center, Tables Library

chains because it allows hemoglobin to pick up O_2 rapidly from the lungs and unload it rapidly in the tissues.

Carbon monoxide is highly toxic because it binds to the Fe^{2+} ion in heme about 200 times more tightly than O_2, thereby eliminating the heme group from functioning in the circulation. Like O_2, CO is a strong-field ligand and produces a bright red, "healthy" look in the individual. Because heme binding is an equilibrium process, CO poisoning can be reversed by breathing extremely high concentrations of O_2, which effectively displaces CO from the heme:

$$\text{heme}-\text{CO} + O_2 \rightleftharpoons \text{heme}-O_2 + \text{CO}$$

Porphin rings are among the most common biological ligands. Chlorophyll, the photosynthetic pigment of green plants, is a porphyrin with Mg^{2+} at the center of the porphin ring, and vitamin B_{12} has Co^{3+} at the center of a very similar ring system. Heme itself is found not only in hemoglobin, but also in proteins called *cytochromes* that are involved in energy metabolism (see Chemical Connections, pp. 967–968).

The zinc ion occurs in many enzymes, the protein catalysts of cells (see Chemical Connections, pp. 723–724). With its d^{10} configuration, Zn^{2+} has a tetrahedral geometry, typically with the N atoms of three amino acid groups at three of the positions, and the fourth position free to interact with the molecule whose reaction is being catalyzed (Figure B23.2). In every case studied, the Zn^{2+} ion acts as a Lewis acid, accepting a lone pair from the reactant as a key step in the catalytic process. Consider the enzyme carbonic anhydrase, which catalyzes the essential reaction between H_2O and CO_2 during respiration:

$$CO_2(g) + H_2O(l) \rightleftharpoons H^+(aq) + HCO_3^-(aq)$$

The Zn^{2+} ion at the enzyme's active site binds three histidine N atoms and the H_2O reactant as the fourth ligand. By withdrawing electron density from the O—H bonds, the Zn^{2+} makes the H_2O acidic enough to lose a proton. In the rate-determining step, the resulting bound OH^- ion attacks the partially positive C atom of CO_2 much more vigorously than could the lone pair of a free water molecule; thus, the reaction rate is higher. One reason the Cd^{2+} ion is toxic is that it competes with Zn^{2+} for fitting into the carbonic anhydrase active site.

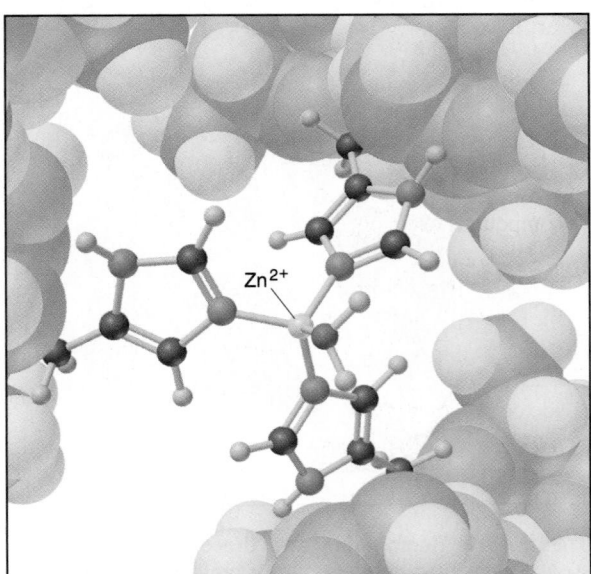

Figure B23.2 The tetrahedral Zn^{2+} complex in carbonic anhydrase. ARIS Presentation Center, Art Library

CHAPTER REVIEW GUIDE
The following sections provide many aids to help you study this chapter. (Numbers in parentheses refer to pages, unless noted otherwise.)

Learning Objectives
These are concepts and skills you should know after studying this chapter.

Relevant section and/or sample problem (SP) numbers appear in parentheses.

Understand These Concepts

1. The positions of the *d*- and *f*-block elements and the general forms of their atomic and ionic electron configurations (Section 23.1)
2. How atomic size, ionization energy, and electronegativity vary across a period and down a group of transition elements and how these trends differ from those of the main-group elements; why the densities of Period 6 transition elements are so high (Section 23.1)
3. Why the transition elements often have multiple oxidation states and why the +2 state is common (Section 23.1)
4. Why metallic behavior (prevalence of ionic bonding and basic oxides) decreases as oxidation state increases (Section 23.1)
5. Why many transition metal compounds are colored and paramagnetic (Section 23.1)
6. The common +3 oxidation state of lanthanides and the similarity in their M^{3+} radii; the radioactivity of actinides (Section 23.2)
7. How valence-state electronegativity explains why oxides become more covalent and acidic as the O.N. of the metal increases (Section 23.3)
8. Why Cr and Mn oxoanions are stronger oxidizing agents in acidic than in basic solutions (Section 23.3)
9. The role silver halides play in black-and-white photography (Section 23.3)
10. How the high density and low melting point of mercury account for its common uses; the toxicity of organomercury compounds (Section 23.3)
11. The coordination numbers, geometries, and ligand structures of complex ions (Section 23.4)
12. How coordination compounds are named and their formulas written (Section 23.4)
13. How Werner correlated the properties and structures of coordination compounds (Section 23.4)
14. The types of constitutional isomerism (coordination and linkage) and stereoisomerism (geometric and optical) of coordination compounds (Section 23.4)
15. How valence bond theory uses hybridization to account for the shapes of octahedral, square planar, and tetrahedral complexes (Section 23.5)
16. How crystal field theory explains that approaching ligands cause *d*-orbital energies to split (Section 23.5)
17. How the relative crystal-field strength of ligands (spectrochemical series) affects the *d*-orbital splitting energy (Δ) (Section 23.5)
18. How the magnitude of Δ accounts for the energy of light absorbed and, thus, the color of a complex (Section 23.5)
19. How the relative sizes of pairing energy and Δ determine the occupancy of *d* orbitals and, thus, the magnetic properties of complexes (Section 23.5)
20. How *d*-orbital splitting in tetrahedral and square planar complexes differs from that in octahedral complexes (Section 23.5)

Master These Skills

1. Writing electron configurations of transition metal atoms and ions (SP 23.1)
2. Using a partial orbital diagram to determine the number of unpaired electrons in a transition-metal atom or ion (SP 23.2)
3. Recognizing the structural components of complex ions (Section 23.4)
4. Naming and writing formulas of coordination compounds (SP 23.3)
5. Determining the type of stereoisomerism in complexes (SP 23.4)
6. Correlating a complex ion's shape with the number and type of hybrid orbitals of the central metal ion (Section 23.5)
7. Using the spectrochemical series to rank complex ions in terms of Δ and the energy of light absorbed (SP 23.5)
8. Using the spectrochemical series to determine if a complex ion is high or low spin (SP 23.6)

Key Terms
These important terms appear in boldface in the chapter and are defined again in the Glossary.

transition elements (1023)

Section 23.1
lanthanide contraction (1026)

Section 23.2
lanthanides (1030)
actinides (1030)
inner transition elements (1030)

Section 23.4
coordination compound (1037)
complex ion (1037)
ligand (1037)
counter ion (1037)
coordination number (1038)
donor atom (1039)
chelate (1040)
isomer (1043)

constitutional (structural) isomers (1043)
coordination isomers (1043)
linkage isomers (1044)
stereoisomers (1044)
geometric (*cis-trans*) isomers (1044)
optical isomers (1044)

Section 23.5
coordinate covalent bond (1046)

crystal field theory (1048)
e_g orbital (1050)
t_{2g} orbital (1050)
crystal field splitting energy (Δ) (1050)
strong-field ligand (1050)
weak-field ligand (1050)
spectrochemical series (1051)
high-spin complex (1052)
low-spin complex (1052)

Highlighted Figures and Tables
These figures (F) and tables (T) provide a visual review of key ideas.

T23.1 Orbital occupancy of Period 4 transition metals (1025)

F23.3 Trends in atomic properties of Period 4 elements (1026)

F23.4 Vertical trends in key properties of the transition elements (1027)

T23.2 Oxidation states and *d*-orbital occupancy of Period 4 transition metals (1028)

T23.6 Coordination numbers and shapes of complex ions (1039)

T23.7 Common ligands in coordination compounds (1039)

T23.8 Names of neutral and anionic ligands (1041)

T23.9 Names of metal ions in complex anions (1041)

F23.10 Isomerism in coordination compounds (1043)

F23.17 *d* Orbitals in an octahedral field of ligands (1049)

F23.18 Splitting of *d*-orbital energies by an octahedral field of ligands (1050)

F23.22 The spectrochemical series (1051)

F23.24 Orbital occupancy for high-spin and low-spin complexes (1052)

F23.25 Splitting of *d*-orbital energies by a tetrahedral field of ligands (1053)

F23.26 Splitting of *d*-orbital energies by a square planar field of ligands (1054)

Brief Solutions to FOLLOW-UP PROBLEMS
Compare your solutions to these calculations steps and answers.

23.1 (a) Ag^+: $4d^{10}$; (b) Cd^{2+}: $4d^{10}$; (c) Ir^{3+}: $5d^6$

23.2 Three; Er^{3+} is $[Xe]\,4f^{11}$:

23.3 (a) Pentaaquabromochromium(III) chloride;
(b) $Ba_3[Co(CN)_6]_2$

23.4 Two sets of *cis-trans* isomers, and the two *cis* isomers are optical isomers.

23.5 Both metal ions are V^{3+}; in terms of ligand field energy, $NH_3 > H_2O$, so $[V(NH_3)_6]^{3+}$ absorbs light of higher energy.

23.6 The metal ion is Mn^{3+}: $[Ar]\,3d^4$.

large Δ

Two unpaired *d* electrons; low-spin complex

PROBLEMS

Problems with **colored** numbers are answered in Appendix E and worked in detail in the Student Solutions Manual. Problem sections match those in the text and provide the numbers of relevant sample problems. Most offer Concept Review Questions, Skill-Building Exercises (grouped in pairs covering the same concept), and Problems in Context. The Comprehensive Problems are based on material from any section or previous chapter.

Note: In these problems, the term *electron configuration* refers to the condensed, ground-state electron configuration.

Properties of the Transition Elements
(Sample Problem 23.1)

● Concept Review Questions

• **23.1** How is the *n* value of the *d* sublevel of a transition element related to the period number of the element?

• **23.2** Write the general electron configuration of a transition element (a) in Period 5; (b) in Period 6.

• **23.3** What is the general rule concerning the order in which electrons are removed from a transition metal atom to form an ion? Give an example from Group 5B(5). Name two types of measurements used to study electron configurations of ions.

• **23.4** What is the maximum number of unpaired *d* electrons that an atom or ion can possess? Give an example of an atom and an ion that have this number.

• **23.5** How does the variation in atomic size across a transition series contrast with the change across the main-group elements of the same period? Why?

• **23.6** (a) What is the lanthanide contraction? (b) How does it affect atomic size down a group of transition elements? (c) How does it influence the densities of the Period 6 transition elements?

• **23.7** (a) What is the range in electronegativity across the first (3*d*) transition series? (b) What is the range across Period 4 of main-group elements? (c) Explain the difference.

• **23.8** (a) Explain the major difference between the number of oxidation states of most transition elements and that of most main-

Levels of Difficulty: • Easy   •• Medium   ••• Challenging

group elements. (b) Why is the +2 oxidation state so common among transition elements?

• **23.9** (a) What behavior distinguishes paramagnetic and diamagnetic substances? (b) Why are paramagnetic ions common among transition elements but not main-group elements? (c) Why are colored solutions of metal ions common among transition elements but not main-group elements?

● **Skill-Building Exercises** (grouped in similar pairs)

• **23.10** Using the periodic table to locate each element, write the electron configuration of (a) V; (b) Y; (c) Hg.

• **23.11** Using the periodic table to locate each element, write the electron configuration of (a) Ru; (b) Cu; (c) Ni.

• **23.12** Using the periodic table to locate each element, write the electron configuration of (a) Os; (b) Co; (c) Ag.

• **23.13** Using the periodic table to locate each element, write the electron configuration of (a) Zn; (b) Mn; (c) Re.

• **23.14** Give the electron configuration and the number of unpaired electrons for (a) Sc^{3+}; (b) Cu^{2+}; (c) Fe^{3+}; (d) Nb^{3+}.

• **23.15** Give the electron configuration and the number of unpaired electrons for: (a) Cr^{3+}; (b) Ti^{4+}; (c) Co^{3+}; (d) Ta^{2+}.

• **23.16** What is the highest oxidation state for (a) Ta; (b) Zr; (c) Mn?

• **23.17** What is the highest oxidation state for (a) Nb; (b) Y; (c) Tc?

• **23.18** Which transition metals have a maximum O.N. of +6?

• **23.19** Which transition metals have a maximum O.N. of +4?

• **23.20** In which compound does Cr exhibit greater metallic behavior, CrF_2 or CrF_6? Explain.

• **23.21** VF_5 is a liquid that boils at 48°C, whereas VF_3 is a solid that melts above 800°C. Explain this difference in properties.

• **23.22** Is it more difficult to oxidize Cr or Mo? Explain.

• **23.23** Is MnO_4^- or ReO_4^- a stronger oxidizing agent? Explain.

• **23.24** Which oxide, CrO_3 or CrO, is more acidic in water? Why?

• **23.25** Which oxide, Mn_2O_3 or Mn_2O_7, is more basic in water? Why?

● **Problems in Context**

• **23.26** The green patina of Cu-alloy roofs results from corrosion in the presence of O_2, H_2O, CO_2, and sulfur compounds. The other members of Group 1B(11), Ag and Au, do not form a patina. Corrosion of Cu and Ag in the presence of sulfur compounds leads to a black tarnish, but Au does not tarnish. This pattern is different from that in Group 1A(1), where ease of oxidation *increases* down the group. Explain these different group patterns.

The Inner Transition Elements
(Sample Problem 23.2)

● **Concept Review Questions**

• **23.27** What atomic property of the lanthanides leads to their remarkably similar chemical properties?

• **23.28** (a) What is the maximum number of unpaired electrons in a lanthanide ion? (b) How does this number relate to occupancy of the 4f subshell?

• **23.29** Which of the actinides are radioactive?

● **Skill-Building Exercises** (grouped in similar pairs)

• **23.30** Give the electron configuration of (a) La; (b) Ce^{3+}; (c) Es; (d) U^{4+}.

• **23.31** Give the electron configuration of (a) Pm; (b) Lu^{3+}; (c) Th; (d) Fm^{3+}.

•• **23.32** Only a few lanthanides show an oxidation state other than +3. Two of these, europium (Eu) and terbium (Tb), are found near the middle of the series, and their unusual oxidation states can be associated with a half-filled f subshell. (a) Write the electron configurations of Eu^{2+}, Eu^{3+}, and Eu^{4+}. Why is Eu^{2+} a common ion, whereas Eu^{4+} is unknown? (b) Write the electron configurations of Tb^{2+}, Tb^{3+}, and Tb^{4+}. Would you expect Tb to show a +2 or a +4 oxidation state? Explain.

•• **23.33** Cerium (Ce) and ytterbium (Yb) exhibit some oxidation states in addition to +3. (a) Write the electron configurations of Ce^{2+}, Ce^{3+}, and Ce^{4+}. (b) Write the electron configurations of Yb^{2+}, Yb^{3+}, and Yb^{4+}. (c) In addition to the 3+ ions, the ions Ce^{4+} and Yb^{2+} are stable. Suggest a reason for this stability.

● **Problems in Context**

• **23.34** Which lanthanide has the maximum number of unpaired electrons in both its atom and 3+ ion? Give the number of unpaired electrons in the atom and ion.

Highlights of Selected Transition Metals

● **Concept Review Questions**

• **23.35** Why is chromium so useful for decorative electroplating?

• **23.36** What is valence-state electronegativity? Explain the change in acidity of the oxides of Mn with changing O.N. of the metal.

• **23.37** What property does manganese confer to steel?

• **23.38** Why is silver used in jewelry and other decorative objects?

• **23.39** How is a photographic latent image different from the image you see on a piece of developed film?

• **23.40** Why is mercury a liquid? Why does it have a 1+ ion?

● **Problems in Context**

• **23.41** When a basic solution of $Cr(OH)_4^-$ ion is slowly acidified, solid $Cr(OH)_3$ precipitates and then redissolves in excess acid. If $Cr(OH)_4^-$ exists as $Cr(H_2O)_2(OH)_4^-$, write equations that represent these two reactions.

• **23.42** Use the following data to determine if $Cr^{2+}(aq)$ can be prepared by the reaction of $Cr(s)$ with $Cr^{3+}(aq)$:

$$Cr^{3+}(aq) + e^- \longrightarrow Cr^{2+}(aq) \quad E° = -0.41 \text{ V}$$
$$Cr^{3+}(aq) + 3e^- \longrightarrow Cr(s) \quad E° = -0.74 \text{ V}$$
$$Cr^{2+}(aq) + 2e^- \longrightarrow Cr(s) \quad E° = -0.91 \text{ V}$$

•• **23.43** When solid CrO_3 is dissolved in water, the solution is orange rather than the yellow of H_2CrO_4. How does this observation indicate that CrO_3 is an acidic oxide?

•• **23.44** Solutions of $KMnO_4$ are used in redox titrations because dark purple MnO_4^- oxidizing agent changes to the faint pink Mn^{2+} as it is reduced. The end point occurs when a slight purple remains as more $KMnO_4$ is added. If a sample that has reached this end point stands for a long time, the purple fades and a suspension of a small amount of brown, muddy MnO_2 appears. Use standard electrode potentials to explain this result.

Coordination Compounds
(Sample Problems 23.3 and 23.4)

● **Concept Review Questions**

• **23.45** Describe the makeup of a complex ion, including the nature of the ligands and their interaction with the central metal ion. Explain how a complex ion can be positive or negative and how it occurs as part of a neutral coordination compound.

• **23.46** What electronic feature must a donor atom of a ligand have?

• **23.47** What is the coordination number of a metal ion in a complex ion? How does it differ from oxidation number?

23.48 What structural feature is characteristic of a chelate?

23.49 What geometries are associated with the coordination numbers 2, 4, and 6?

23.50 What are the coordination numbers of cobalt(III), platinum(II), and platinum(IV) in complexes?

23.51 How is a complex ion a Lewis adduct?

23.52 What does the ending -ate signify in a complex ion name?

23.53 In what order are the metal ion and ligands given in the name of a complex ion?

23.54 Is a linkage isomer a type of constitutional isomer or stereoisomer? Explain.

● **Skill-Building Exercises** (grouped in similar pairs)

23.55 Give systematic names for the following formulas:
(a) $[Ni(H_2O)_6]Cl_2$ (b) $[Cr(en)_3](ClO_4)_3$ (c) $K_4[Mn(CN)_6]$

23.56 Give systematic names for the following formulas:
(a) $[Co(NH_3)_4(NO_2)_2]Cl$ (b) $[Cr(NH_3)_6][Cr(CN)_6]$
(c) $K_2[CuCl_4]$

23.57 What are the charge and coordination number of the central metal ion(s) in each compound of Problem 23.55?

23.58 What are the charge and coordination number of the central metal ion(s) in each compound of Problem 23.56?

23.59 Give systematic names for the following formulas:
(a) $K[Ag(CN)_2]$ (b) $Na_2[CdCl_4]$ (c) $[Co(NH_3)_4(H_2O)Br]Br_2$

23.60 Give systematic names for the following formulas:
(a) $K[Pt(NH_3)Cl_5]$ (b) $[Cu(en)(NH_3)_2][Co(en)Cl_4]$
(c) $[Pt(en)_2Br_2](ClO_4)_2$

23.61 What are the charge and coordination number of the central metal ion(s) in each compound of Problem 23.59?

23.62 What are the charge and coordination number of the central metal ion(s) in each compound of Problem 23.60?

23.63 Give formulas corresponding to the following names:
(a) Tetraamminezinc sulfate
(b) Pentaamminechlorochromium(III) chloride
(c) Sodium bis(thiosulfato)argentate(I)

23.64 Give formulas corresponding to the following names:
(a) Dibromobis(ethylenediamine)cobalt(III) sulfate
(b) Hexaamminechromium(III) tetrachlorocuprate(II)
(c) Potassium hexacyanoferrate(II)

23.65 What is the coordination number of the metal ion and the number of individual ions per formula unit in each of the compounds in Problem 23.63?

23.66 What is the coordination number of the metal ion and the number of individual ions per formula unit in each of the compounds in Problem 23.64?

23.67 Give formulas corresponding to the following names:
(a) Hexaaquachromium(III) sulfate
(b) Barium tetrabromoferrate(III)
(c) Bis(ethylenediamine)platinum(II) carbonate

23.68 Give formulas corresponding to the following names:
(a) Potassium tris(oxalato)chromate(III)
(b) Tris(ethylenediamine)cobalt(III) pentacyanoiodomanganate(II)
(c) Diamminediaquabromochloroaluminum nitrate

23.69 Give the coordination number of the metal ion and the number of ions per formula unit in each compound in Problem 23.67.

23.70 Give the coordination number of the metal ion and the number of ions per formula unit in each compound in Problem 23.68.

23.71 Which of these ligands can participate in linkage isomerism: (a) NO_2^-; (b) SO_2; (c) NO_3^-? Explain with Lewis structures.

23.72 Which of these ligands can participate in linkage isomerism: (a) SCN^-; (b) $S_2O_3^{2-}$ (thiosulfate); (c) HS^-? Explain with Lewis structures.

23.73 For any of the following that can exist as isomers, state the type of isomerism and draw the structures:
(a) $[Pt(CH_3NH_2)_2Br_2]$ (b) $[Pt(NH_3)_2FCl]$ (c) $[Pt(H_2O)(NH_3)FCl]$

23.74 For any of the following that can exist as isomers, state the type of isomerism and draw the structures:
(a) $[Zn(en)F_2]$ (b) $[Zn(H_2O)(NH_3)FCl]$ (c) $[Pd(CN)_2(OH)_2]^{2-}$

23.75 For any of the following that can exist as isomers, state the type of isomerism and draw the structures:
(a) $[PtCl_2Br_2]^{2-}$ (b) $[Cr(NH_3)_5(NO_2)]^{2+}$ (c) $[Pt(NH_3)_4I_2]^{2+}$

23.76 For any of the following that can exist as isomers, state the type of isomerism and draw the structures:
(a) $[Co(NH_3)_5Cl]Br_2$ (b) $[Pt(CH_3NH_2)_3Cl]Br$
(c) $[Fe(H_2O)_4(NH_3)_2]^{2+}$

● **Problems in Context**

23.77 Chromium(III), like cobalt(III), has a coordination number of 6 in many of its complex ions. Compounds are known that have the traditional formula $CrCl_3 \cdot nNH_3$, where $n = 3$ to 6. Which of the compounds has an electrical conductivity in aqueous solution similar to that of an equimolar NaCl solution?

23.78 When $MCl_4(NH_3)_2$ is dissolved in water and treated with $AgNO_3$, 2 mol of AgCl precipitates immediately for each mole of $MCl_4(NH_3)_2$. Give the coordination number of M in the complex.

23.79 Palladium, like its group neighbor platinum, forms four-coordinate Pd(II) and six-coordinate Pd(IV) complexes. Write modern formulas for the complexes with these compositions:
(a) $PdK(NH_3)Cl_3$ (b) $PdCl_2(NH_3)_2$
(c) PdK_2Cl_6 (d) $Pd(NH_3)_4Cl_4$

Theoretical Basis for the Bonding and Properties of Complexes

(Sample Problems 23.5 and 23.6)

● **Concept Review Questions**

23.80 (a) What is a coordinate covalent bond?
(b) Is it involved when $FeCl_3$ dissolves in water? Explain.
(c) Is it involved when HCl gas dissolves in water? Explain.

23.81 According to valence bond theory, what set of orbitals is used by a Period 4 metal ion in forming (a) a square planar complex; (b) a tetrahedral complex?

23.82 A metal ion uses d^2sp^3 orbitals when forming a complex. What is its coordination number and the shape of the complex?

23.83 A complex in solution absorbs green light. What is the color of the solution?

23.84 In terms of the theory of color absorption, explain two ways that a solution can be blue.

23.85 (a) What is the crystal field splitting energy (Δ)?
(b) How does it arise for an octahedral field of ligands?
(c) How is it different for a tetrahedral field of ligands?

23.86 What is the distinction between a weak-field ligand and a strong-field ligand? Give an example of each.

23.87 Is a complex with the same number of unpaired electrons as the free gaseous metal ion termed high spin or low spin?

23.88 How do the relative magnitudes of $E_{pairing}$ and Δ affect the paramagnetism of a complex?

- **23.89** Why are there both high-spin and low-spin octahedral complexes but only high-spin tetrahedral complexes?

● **Skill-Building Exercises** (*grouped in similar pairs*)

- **23.90** Give the number of d electrons (n of d^n) for the central metal ion in (a) $[TiCl_6]^{2-}$; (b) $K[AuCl_4]$; (c) $[RhCl_6]^{3-}$.
- **23.91** Give the number of d electrons (n of d^n) for the central metal ion in (a) $[Cr(H_2O)_6](ClO_3)_2$; (b) $[Mn(CN)_6]^{2-}$; (c) $[Ru(NO)(en)_2Cl]Br$.
- **23.92** How many d electrons (n of d^n) are in the central metal ion in (a) $Ca[IrF_6]$; (b) $[HgI_4]^{2-}$; (c) $[Co(EDTA)]^{2-}$?
- **23.93** How many d electrons (n of d^n) are in the central metal ion in (a) $[Ru(NH_3)_5Cl]SO_4$; (b) $Na_2[Os(CN)_6]$; (c) $[Co(NH_3)_4CO_3I]$?

- **23.94** Sketch the orientation of the orbitals relative to the ligands in an octahedral complex to explain the splitting and the relative energies of the d_{xy} and the $d_{x^2-y^2}$ orbitals.
- **23.95** The two e_g orbitals are identical in energy in an octahedral complex but have different energies in a square planar complex, with the d_{z^2} orbital being much lower in energy than the $d_{x^2-y^2}$. Explain with orbital sketches.

- **23.96** Which of these ions *cannot* form both high- and low-spin octahedral complexes: (a) Ti^{3+}; (b) Co^{2+}; (c) Fe^{2+}; (d) Cu^{2+}?
- **23.97** Which of these ions *cannot* form both high- and low-spin octahedral complexes: (a) Mn^{3+}; (b) Nb^{3+}; (c) Ru^{3+}; (d) Ni^{2+}?

- **23.98** Draw orbital-energy splitting diagrams and use the spectrochemical series to show the orbital occupancy for each of the following (assuming that H_2O is a weak-field ligand):
  (a) $[Cr(H_2O)_6]^{3+}$ (b) $[Cu(H_2O)_4]^{2+}$ (c) $[FeF_6]^{3-}$
- **23.99** Draw orbital-energy splitting diagrams and use the spectrochemical series to show the orbital occupancy for each of the following (assuming that H_2O is a weak-field ligand):
  (a) $[Cr(CN)_6]^{3-}$ (b) $[Rh(CO)_6]^{3+}$ (c) $[Co(OH)_6]^{4-}$

- **23.100** Draw orbital-energy splitting diagrams and use the spectrochemical series to show the orbital occupancy for each of the following (assuming that H_2O is a weak-field ligand):
  (a) $[MoCl_6]^{3-}$ (b) $[Ni(H_2O)_6]^{2+}$ (c) $[Ni(CN)_4]^{2-}$
- **23.101** Draw orbital-energy splitting diagrams and use the spectrochemical series to show the orbital occupancy for each of the following (assuming that H_2O is a weak-field ligand):
  (a) $[Fe(C_2O_4)_3]^{3-}$ ($C_2O_4^{2-}$ creates a weaker field than H_2O does.)
  (b) $[Co(CN)_6]^{4-}$ (c) $[MnCl_6]^{4-}$

- **23.102** Rank the following in order of *increasing* Δ and energy of light absorbed: $[Cr(NH_3)_6]^{3+}$, $[Cr(H_2O)_6]^{3+}$, $[Cr(NO_2)_6]^{3-}$.
- **23.103** Rank the following in order of *decreasing* Δ and energy of light absorbed: $[Cr(en)_3]^{3+}$, $[Cr(CN)_6]^{3-}$, $[CrCl_6]^{3-}$.

- **23.104** A complex, ML_6^{2+}, is violet. The same metal forms a complex with another ligand, Q, that creates a weaker field. What color might MQ_6^{2+} be expected to show? Explain.
- **23.105** $[Cr(H_2O)_6]^{2+}$ is violet. Another CrL_6 complex is green. Can ligand L be CN^-? Can it be Cl^-? Explain.

● **Problems in Context**

- **23.106** Octahedral $[Ni(NH_3)_6]^{2+}$ is paramagnetic, whereas planar $[Pt(NH_3)_4]^{2+}$ is diamagnetic, even though both metal ions are d^8 species. Explain.
- **23.107** The hexaaqua complex $[Ni(H_2O)_6]^{2+}$ is green, whereas the hexaammonia complex $[Ni(NH_3)_6]^{2+}$ is violet. Explain.

- **23.108** Three of the complex ions that are formed by Co^{3+} are $[Co(H_2O)_6]^{3+}$, $[Co(NH_3)_6]^{3+}$, and $[CoF_6]^{3-}$. These ions have the observed colors (listed in arbitrary order) yellow-orange, green, and blue. Match each complex with its color. Explain.

Comprehensive Problems

- **23.109** When neptunium (Np) and plutonium (Pu) were discovered, the periodic table did not include the actinides, so these elements were placed in Group 7B(7) and 8B(8). When americium (Am) and curium (Cm) were synthesized, they were placed in Group 8B(9) and 8B(10). However, during chemical isolation procedures, Glenn Seaborg and his group, who had synthesized these elements, could not find their compounds among other compounds of these groups, which led Seaborg to suggest they were part of a new inner transition series. (a) How do the electron configurations of these elements support Seaborg's suggestion? (b) The highest fluorides of Np and Pu are hexafluorides, as is the highest fluoride of uranium. How does this chemical evidence support the placement of Np and Pu as *inner* transition elements rather than transition elements?

- **23.110** How many different formulas are there for octahedral complexes with a metal M and four ligands A, B, C, and D? Give the number of isomers for each formula and describe the isomers.

- **23.111** At one time, it was common to write the formula for copper(I) chloride as Cu_2Cl_2, instead of $CuCl$, analogously to Hg_2Cl_2 for mercury(I) chloride. Use electron configurations to explain why Hg_2Cl_2 and $CuCl$ are both correct.

- **23.112** Correct each name that has an error:
  (a) $Na[FeBr_4]$, sodium tetrabromoferrate(II)
  (b) $[Ni(NH_3)_6]^{2+}$, nickel hexaammine ion
  (c) $[Co(NH_3)_3I_3]$, triamminetriiodocobalt(III)
  (d) $[V(CN)_6]^{3-}$, hexacyanovanadium(III) ion
  (e) $K[FeCl_4]$, potassium tetrachloroiron(III)

- **23.113** For the compound $[Co(en)_2Cl_2]Cl$, give:
  (a) The coordination number of the metal ion
  (b) The oxidation number of the central metal ion
  (c) The number of individual ions per formula unit
  (d) The moles of AgCl that precipitate immediately when 1 mol of compound is dissolved in water and treated with $AgNO_3$

- **23.114** Hexafluorocobaltate(III) ion is a high-spin complex. Draw the orbital-energy splitting diagram for its d orbitals.

- **23.115** A salt of each of the ions in Table 23.3 (p. 1029) is dissolved in water. A Pt electrode is immersed in each solution and connected to a 0.38-V battery. All of the electrolytic cells are run for the same amount of time with the same current.
  (a) In which cell(s) will a metal plate out? Explain.
  (b) Which cell will plate out the least mass of metal? Explain.

- **23.116** Criticize and correct the following statement: strong-field ligands always give rise to low-spin complexes.

- **23.117** Two major bidentate ligands used in analytical chemistry are bipyridyl (bipy) and *ortho*-phenanthroline (*o*-phen):

bipyridyl

o-phenanthroline

Draw structures and discuss the possibility of isomers for
(a) $[Pt(bipy)Cl_2]$ (b) $[Fe(o\text{-phen})_3]^{3+}$
(c) $[Co(bipy)_2F_2]^+$ (d) $[Co(o\text{-phen})(NH_3)_3Cl]^{2+}$

•• **23.118** The following reaction is a key step in black-and-white photography (see p. 1035):

$$AgBr(s) + 2S_2O_3^{2-}(aq) \longrightarrow Ag(S_2O_3)_2^{3-}(aq) + Br^-(aq)$$

During fixing, 258 mL of hypo (sodium thiosulfate) was used. The hypo concentration was 0.1052 M before the AgBr reacted, and 0.0378 M afterward. How many grams of AgBr reacted?

•• **23.119** The metal ion in platinum(IV) complexes, like that in cobalt(III) complexes, has a coordination number of 6 and often has Cl^- ions and NH_3 molecules as ligands. For the traditional (before the work of Werner) formulas $PtCl_4 \cdot 6NH_3$ and $PtCl_4 \cdot 4NH_3$, (a) give the modern formula and charge of the complex ion, and (b) predict the moles of ions per mole of compound dissolved and moles of AgCl formed with excess $AgNO_3$.

•• **23.120** In 1940, when the elements Np and Pu were prepared, a controversy arose about whether the elements from Ac on were analogs of the transition elements (and related to Y through Mo) or analogs of the lanthanides (and related to La through Nd). The arguments hinged primarily on comparing observed oxidation states to those of the earlier elements. (a) If the actinides were analogs of the transition elements, what would you predict about the maximum oxidation state for U? For Np? (b) If the actinides were analogs of the lanthanides, what would you predict about the maximum oxidation state for U? For Pu?

•• **23.121** In many species, a transition metal has an unusually high or low oxidation state. Write balanced equations for the following and find the oxidation state of the transition metal in the product:
(a) Iron(III) ion reacts with hypochlorite ion in basic solution to form ferrate ion (FeO_4^{2-}), Cl^-, and water.
(b) Potassium hexacyanomanganate(II) reacts with K metal to form $K_6[Mn(CN)_6]$.
(c) Heating sodium superoxide (NaO_2) with Co_3O_4 produces Na_4CoO_4 and O_2 gas.
(d) Vanadium(III) chloride reacts with Na metal under a CO atmosphere to produce $Na[V(CO)_6]$ and NaCl.
(e) Barium peroxide reacts with nickel(II) ions in basic solution to produce $BaNiO_3$.
(f) Bubbling CO through a basic solution of cobalt(II) ion produces $[Co(CO)_4]^-$, CO_3^{2-}, and water.
(g) Heating cesium tetrafluorocuprate(II) with F_2 gas under pressure gives Cs_2CuF_6.
(h) Heating tantalum(V) chloride with Na metal produces NaCl and Ta_6Cl_{15}, in which half of the Ta is in the $+2$ state.
(i) Potassium tetracyanonickelate(II) reacts with hydrazine (N_2H_4) in basic solution to form $K_4[Ni_2(CN)_6]$ and N_2 gas.

•• **23.122** Draw a Lewis structure with lowest formal charges for MnO_4^-.

•• **23.123** The coordination compound $[Pt(NH_3)_2(SCN)_2]$ displays two types of isomerism. Name the types and give names and structures for the six possible isomers.

•• **23.124** In the sepia "toning" of a black-and-white photograph, the image is converted to a rich brownish violet by placing the finished photograph in a solution of gold(III) ions, in which metallic gold replaces the metallic silver. Use Appendix D to explain the chemistry of this process.

•• **23.125** An octahedral complex with three different ligands (A, B, and C) can have formulas with three different ratios of the ligands:
$[MA_4BC]^{n+}$,   such as $[Co(NH_3)_4(H_2O)Cl]^{2+}$
$[MA_3B_2C]^{n+}$,   such as $[Cr(H_2O)_3Br_2Cl]$
$[MA_2B_2C_2]^{n+}$,   such as $[Cr(NH_3)_2(H_2O)_2Br_2]^+$

For each example, give the name, state the type(s) of isomerism present, and draw all isomers.

•• **23.126** In black-and-white photography, what are the major chemical changes involved in exposing, developing, and fixing?

•• **23.127** In $[Cr(NH_3)_6]Cl_3$, the $[Cr(NH_3)_6]^{3+}$ ion absorbs visible light in the blue-violet range, and the compound is yellow-orange. In $[Cr(H_2O)_6]Br_3$, the $[Cr(H_2O)_6]^{3+}$ ion absorbs visible light in the red range, and the compound is blue-gray. Explain these differences in light absorbed and color of the compound.

•• **23.128** Dark green manganate salts contain the MnO_4^{2-} ion. The ion is stable in basic solution but disproportionates in acid to $MnO_2(s)$ and MnO_4^-. (a) What is the oxidation state of Mn in MnO_4^{2-}, MnO_4^-, and MnO_2? (b) Write a balanced equation for the reaction of MnO_4^{2-} in acidic solution.

•• **23.129** Aqueous electrolysis of potassium manganate (K_2MnO_4) in basic solution is used for the production of tens of thousands of tons of potassium permanganate ($KMnO_4$) annually. (a) Write a balanced equation for the electrolysis reaction. (Water is reduced also.) (b) How many moles of MnO_4^- can be formed if 12 A flows through a tank of aqueous MnO_4^{2-} for 96 h?

•• **23.130** The actinides Pa, U, and Np form a series of complex ions, such as the anion in the compound $Na_3[UF_8]$, in which the central metal ion has an unusual geometry and oxidation state. In the crystal structure, the complex ion can be pictured as resulting from interpenetration of simple cubic arrays of uranium and fluoride ions. (a) What is the coordination number of the metal ion in the complex ion? (b) What is the oxidation state of uranium in the compound? (c) Sketch the complex ion.

•• **23.131** Consider the square planar complex shown at right. Which of the structures below are geometric isomers of it?

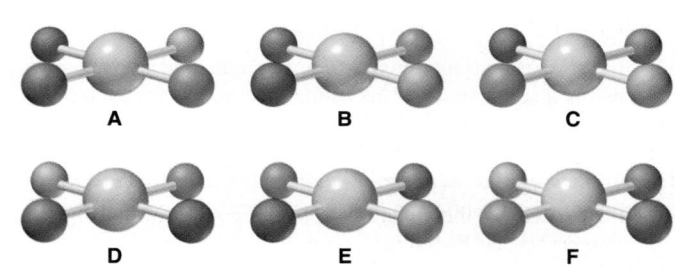

A   **B**   **C**

D   **E**   **F**

•• **23.132** (a) What are the central metal ions in chlorophyll, heme, and vitamin B_{12}? (b) What similarity in structure do these compounds have?

••• **23.133** Several coordination isomers, with both Co and Cr as $3+$ ions, have the molecular formula $CoCrC_6H_{18}N_{12}$. (a) Give the name and formula of the isomer in which the Co complex ion has six NH_3 groups. (b) Give the name and formula of the isomer in which the Co complex ion has one CN and five NH_3 groups.

•• **23.134** Several coordination isomers have the molecular formula $Pt_2H_{12}Cl_6N_4$. (a) Give the name and standard formula of the isomer in which the Pt complex ion has six Cl groups and a charge of $2-$. (b) Give the name and standard formula of the isomer in which the Pt complex ion has four Cl groups and a charge of $2-$.

••• **23.135** A shortcut to finding optical isomers is to see if the complex has a *plane of symmetry*—a plane passing through the metal atom such that every atom on one side of the plane is matched by an identical one at the same distance from the plane on the other side. Any planar complex has a plane of symmetry, since all atoms lie in one plane. Use this approach to determine whether these exist as optical isomers: (a) $[Zn(NH_3)_2Cl_2]$ (tetrahedral); (b) $[Pt(en)_2]^{2+}$; (c) *trans*-$[PtBr_4Cl_2]^{2-}$; (d) *trans*-$[Co(en)_2F_2]^+$; (e) *cis*-$[Co(en)_2F_2]^+$.

•• **23.136** Werner prepared two compounds by heating a solution of $PtCl_2$ with triethyl phosphine, $P(C_2H_5)_3$, which is an excellent ligand for Pt. The two compounds gave the same analysis: Pt, 38.8%; Cl, 14.1%; C, 28.7%; P, 12.4%; and H, 6.02%. Write formulas, structures, and systematic names for the two isomers.

•• **23.137** Some octahedral complexes have distorted shapes. In some, two metal-ligand bonds that are 180° apart are shorter than the other four. In $[Cu(NH_3)_6]^{2+}$, for example, two Cu—N bonds are 207 pm long, and the other four are 262 pm long. (a) Calculate the longest distance between two N atoms in this complex. (b) Calculate the shortest distance between two N atoms.

•• **23.138** Mercury has a small but significant vapor pressure, and Hg poisoning can occur in poorly ventilated rooms where Hg has been spilled. What is the partial pressure (in torr) of Hg if the maximum concentration at 25°C is 20.0 mg/m^3?

•• **23.139** Normal carbonic anhydrase has zinc at its active site (see Figure B23.2, p. 1056). Suggest a structural reason why carbonic anhydrase synthesized with Ni^{2+}, Fe^{2+}, or Mn^{2+} in place of Zn^{2+} gives an enzyme with less catalytic efficiency.

••• **23.140** The effect of entropy on reactions is evident in the stabilities of certain complexes. (a) Using the criterion of number of product particles, predict which of the following will be favored in terms of $\Delta S°_{rxn}$:

$[Cu(NH_3)_4]^{2+}(aq) + 4H_2O(l) \longrightarrow$
$\qquad\qquad [Cu(H_2O)_4]^{2+}(aq) + 4NH_3(aq)$

$[Cu(H_2NCH_2CH_2NH_2)_2]^{2+}(aq) + 4H_2O(l) \longrightarrow$
$\qquad\qquad [Cu(H_2O)_4]^{2+}(aq) + 2en(aq)$

(b) Given that the Cu—N bond strength is approximately the same in both complexes, which complex will be more stable with respect to ligand exchange in water? Explain.

•• **23.141** The extent of crystal field splitting is often determined from spectra. (a) Given the wavelength (λ) of maximum absorption, find the crystal field splitting energy (Δ), in kJ/mol, for each of the following complex ions:

Ion	λ (nm)	Ion	λ (nm)
$[Cr(H_2O)_6]^{3+}$	562	$[Fe(H_2O)_6]^{2+}$	966
$[Cr(CN)_6]^{3-}$	381	$[Fe(H_2O)_6]^{3+}$	730
$[CrCl_6]^{3-}$	735	$[Co(NH_3)_6]^{3+}$	405
$[Cr(NH_3)_6]^{3+}$	462	$[Rh(NH_3)_6]^{3+}$	295
$[Ir(NH_3)_6]^{3+}$	244		

(b) Construct a spectrochemical series for the ligands in the Cr complexes. (c) Use the Fe data to state how oxidation state affects Δ. (d) Use the Co, Rh, and Ir data to state how period number affects Δ.

••• **23.142** You know the following about a coordination compound:
(1) The partial empirical formula is $KM(CrO_4)Cl_2(NH_3)_4$.
(2) It has A (red) and B (blue) crystal forms.
(3) When 1.0 mol of A or B reacts with 1.0 mol of $AgNO_3$, 0.50 mol of a red precipitate forms immediately.
(4) After the reaction in (3), 1.0 mol of A reacts very slowly with 1.0 mol of silver oxalate ($Ag_2C_2O_4$) to form 2.0 mol of a white precipitate. (Oxalate can displace other ligands.)
(5) After the reaction in (3), 1.0 mol of B does not react further with 1.0 mol of $AgNO_3$.
From this information, determine the following:
(a) The coordination number of M
(b) The group(s) bonded to M ionically and covalently
(c) The stereochemistry of the red and blue forms

•• **23.143** Ionic liquids have many new applications in engineering and materials science. The dissolution of the metavanadate ion in chloroaluminate ionic liquids has been studied:

$$VO_3^- + AlCl_4^- \longrightarrow VO_2Cl_2^- + AlOCl_2^-$$

(a) What is the oxidation number of V and Al in each ion?
(b) In reactions of V_2O_5 with HCl, acid concentration affects the product. At low acid concentration, $VO_2Cl_2^-$ and VO_3^- form:

$$V_2O_5 + HCl \longrightarrow VO_2Cl_2^- + VO_3^- + H^+$$

At high acid concentration, $VOCl_3$ forms:

$$V_2O_5 + HCl \longrightarrow VOCl_3 + H_2O$$

Balance each equation, and state which, if either, involves a redox process.
(c) What mass of $VO_2Cl_2^-$ or $VOCl_3$ can form from 12.5 g of V_2O_5 and the appropriate concentration of acid?

•• **23.144** The orbital occupancies for the *d* orbitals of several complex ions are diagrammed below.

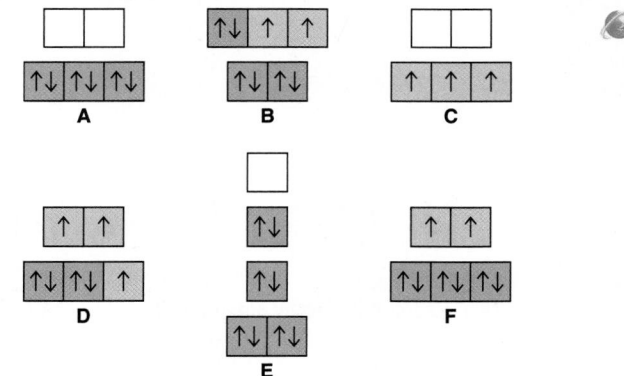

(a) Which diagram corresponds to the orbital occupancy of the cobalt ion in $[Co(CN)_6]^{3-}$?
(b) If diagram D depicts the orbital occupancy of the cobalt ion in $[CoF_6]^n$, what is the value of n?
(c) $[NiCl_4]^{2-}$ is paramagnetic and $[Ni(CN)_4]^{2-}$ is diamagnetic. Which diagrams correspond to the orbital occupancies of the nickel ions in these species?
(d) Diagram C shows the orbital occupancy of V^{2+} in the octahedral complex VL_6. Can you determine whether L is a strong- or weak-field ligand? Explain.

24

You Are Made of Stars In this false-color composite, the halo of gas known as Cassiopeia A and the tiny *neutron star (blue dot at center, white arrow)* are all that remain of a massive star that exploded 325 years ago. Its freshly made atomic nuclei were expelled throughout interstellar space, only to be incorporated into new stars, their planets, and the elements and compounds (and creatures) they consist of. In this chapter, we examine atomic nuclei to understand their composition and how we make use of their breakdown and the enormous quantities of energy released when they undergo change.

Nuclear Reactions and Their Applications

24.1 Radioactive Decay and Nuclear Stability
Components of the Nucleus
Types of Radioactive Emissions
Types of Radioactive Decay; Nuclear Equations
Nuclear Stability and Mode of Decay

24.2 The Kinetics of Radioactive Decay
Rate of Radioactive Decay
Radioisotopic Dating

24.3 Nuclear Transmutation: Induced Changes in Nuclei
Early Transmutation Experiments; Nuclear Shorthand Notation
Particle Accelerators

24.4 The Effects of Nuclear Radiation on Matter
Effects of Ionizing Radiation on Living Matter
Sources of Ionizing Radiation
Assessing the Risk from Ionizing Radiation

24.5 Applications of Radioisotopes
Radioactive Tracers
Additional Applications of Ionizing Radiation

24.6 The Interconversion of Mass and Energy
The Mass Difference Between a Nucleus and Its Nucleons
Nuclear Binding Energy and Binding Energy per Nucleon

24.7 Applications of Fission and Fusion
Nuclear Fission
Nuclear Fusion

F ar below the outer fringes of its cloud of electrons lies the atom's tiny, dense core, held together by the strongest force in the universe. For nearly the entire text, we have focused on an atom's electrons, treating the nucleus as their electrostatic anchor, examining the effect of its positive charge on atomic properties and, ultimately, chemical behavior. But, for the scientists probing the structure and behavior of the nucleus itself, *there* is the scene of real action, one that holds enormous potential benefit and great mystery and wonder.

Society is ambivalent about the applications of nuclear research, however. The promise of abundant energy and treatments for disease comes hand-in-hand with the threat of nuclear waste contamination, reactor accidents, and unimaginable destruction from nuclear war or terrorism. Can the power of the nucleus be harnessed for our benefit, or are the risks too great? In this chapter, we discuss the principles that can help you consider this vital question knowledgably.

The changes that occur in atomic nuclei are strikingly different from chemical changes. In chemical reactions, electrons are shared or transferred to form *compounds,* while nuclei sit by passively, never changing their identities. In nuclear reactions, the roles are reversed, as electrons in their orbitals take part much less often, while the nuclei undergo changes that, in nearly every case, form different *elements*. Nuclear reactions are often accompanied by energy changes a million times greater than those in chemical reactions, energy changes so great that changes in mass *are* detectable. Moreover, nuclear reaction yields and rates are *not* subject to the effects of pressure, temperature, and catalysis. Table 24.1 summarizes these general differences.

Concepts & Skills to Review
before you study this chapter

- discovery of the atomic nucleus (Section 2.4)
- protons, neutrons, mass number, and the $^A_Z X$ notation (Section 2.5)
- half-life and first-order reaction rate (Section 16.4)

 Transparency/ARIS Presentation Center, Tables Library

Table 24.1 Comparison of Chemical and Nuclear Reactions

Chemical Reactions	Nuclear Reactions
1. One substance is converted into another, but atoms never change identity.	1. Atoms of one element typically are converted into atoms of another element.
2. Orbital electrons are involved as bonds break and form; nuclear particles do not take part.	2. Protons, neutrons, and other particles are involved; orbital electrons take part much less often.
3. Reactions are accompanied by relatively small changes in energy and no measurable changes in mass.	3. Reactions are accompanied by relatively large changes in energy and measurable changes in mass.
4. Reaction rates are influenced by temperature, concentration, catalysts, and the compound in which an element occurs.	4. Reaction rates depend on number of nuclei, but are not affected by temperature, catalysts, or, except on rare occasions, the compound in which an element occurs.

IN THIS CHAPTER . . . We survey the field of nuclear chemistry, beginning with an investigation of nuclear stability—why some nuclei are stable, whereas others are unstable and undergo radioactive decay. You'll see how radioactivity is detected and how the kinetics of decay is applied. We explore how nuclei synthesized in particle accelerators have extended the periodic table. Then, we consider the effects of radioactive emissions on matter, especially living matter, and focus on some major applications of radioisotopes in science, technology, and medicine. Next, we discuss the central idea of the mass difference that arises when a nucleus forms from its subatomic particles and the energy equivalent to this difference. We then discuss current and future attempts to harness the energy released when heavy nuclei split and lighter ones fuse. Finally, we look at the nuclear processes that create the chemical elements in the stars.

24.1 RADIOACTIVE DECAY AND NUCLEAR STABILITY

A stable nucleus remains intact indefinitely, but *the great majority of nuclei are unstable.* An unstable nucleus exhibits **radioactivity:** it spontaneously disintegrates, or *decays,* by emitting radiation. In Section 24.2, you'll see that each type of unstable nucleus has its own characteristic *rate* of radioactive decay, which can range from less than a billionth of a second to billions of years. In this section, we consider important terms and notation for nuclei, discuss some of the key events in the discovery of radioactivity, and describe the various types of radioactive decay and how to predict which type occurs for a given nucleus.

The Components of the Nucleus: Terms and Notation

Recall from Chapter 2 that the nucleus contains essentially all the atom's mass but is only about 10^{-5} times its radius (or 10^{-15} times its volume). Obviously, the nucleus is incredibly dense: about 10^{14} g/mL. ● *Protons* and *neutrons,* the elementary particles that make up the nucleus, are called **nucleons.** The term **nuclide** refers to a nucleus with a particular composition, that is, with specific numbers of the two types of nucleons. Most elements occur in nature as a mixture of **isotopes,** atoms with the characteristic number of protons of the element but different numbers of neutrons. Therefore, each isotope of an element is a particular nuclide that we identify by its numbers of protons and neutrons. For example, oxygen has three naturally occurring isotopes—the most abundant contains eight protons and eight neutrons, whereas the least abundant contains eight protons and nine neutrons.

The relative mass and charge of a particle—a nucleon, another elementary particle, or a nucleus—is described by the notation $^A_Z X$, where X is the *symbol* for the particle, A is the *mass number,* or the total number of nucleons, and Z is the *charge* of the particle; for nuclei, A is the *sum of protons and neutrons* and Z is the *number of protons* (atomic number). In this notation, the three subatomic elementary particles are

$$_{-1}^{0}e \text{ (electron)}, \ _1^1p \text{ (proton)}, \text{ and } _0^1n \text{ (neutron)}$$

(A proton is also sometimes represented as $_1^1H^+$.) The number of neutrons (N) in a nucleus is the mass number (A) minus the atomic number (Z): $N = A - Z$. The two naturally occurring stable isotopes of chlorine, for example, have 17 protons ($Z = 17$), but one has 18 neutrons ($_{17}^{35}Cl$, also written ^{35}Cl) and the other has 20 ($_{17}^{37}Cl$, or ^{37}Cl). Nuclides can also be designated with the element name followed by the mass number, for example, chlorine-35 and chlorine-37. In naturally occurring samples of an element or its compounds, *the isotopes of the element are present in specific proportions* that can vary only very slightly. Thus, in a sample of sodium chloride (or any Cl-containing substance), 75.77% of the Cl atoms are chlorine-35 and the remaining 24.23% are chlorine-37.

To understand this chapter, it's very important that you are comfortable with nuclear notations, so please take a moment to review Sample Problem 2.4 (p. 54) and Problems 2.39 to 2.46 (pp. 82–83).

The Discovery of Radioactivity and the Types of Emissions

In 1896, the French physicist Antoine-Henri Becquerel discovered, quite by accident, that uranium minerals, even when wrapped in paper and stored in the dark, emit a penetrating radiation that can expose a photographic plate. Becquerel also found that the radiation creates an electric discharge in air, thus providing a means for measuring its intensity. Two years later, a young doctoral student named Marie Sklodowska Curie began a search for other minerals that behaved like uranium in this way. She found that thorium minerals also emit radiation and, most importantly, that *the intensity of the radiation is directly proportional to the concen-*

● **The Tiny, Massive Nucleus** If you could strip the electrons from the atoms in an object and compress the nuclei together, the object would lose only a fraction of a percent of its mass, but it would shrink to 0.0000000000001% ($10^{-13}\%$) of its volume. An atom the size of the Houston Astrodome would have a nucleus the size of a marble, which would contain virtually all the atom's mass.

tration of the element in the mineral, not to the nature of the mineral or compound in which the element occurs. Curie named the emissions *radioactivity* and showed that they are *unaffected by temperature, pressure, or other physical and chemical conditions.*

To her surprise, Curie found that certain minerals containing uranium were even more radioactive than pure uranium, which implied that they contained traces of one or more as yet unknown, highly radioactive elements. She and her husband, the physicist Pierre Curie, set out to isolate all the radioactive components in pitchblende, the principal ore of uranium. After months of painstaking chemical work, they isolated two extremely small, highly radioactive fractions, one that precipitated with bismuth compounds and another that precipitated with alkaline earth compounds. Marie Curie was able to show that these fractions contained two new elements, which she named polonium (after her native Poland) and radium. Polonium (Po; $Z = 84$), the most metallic member of Group 6A(16), lies to the right of bismuth in Period 6. Radium (Ra; $Z = 88$), which is the heaviest alkaline earth metal, lies under barium in Group 2A(2).

Purifying radium proved to be arduous also. Starting with several tons of pitchblende residues from which the uranium had been extracted, Curie prepared compounds of the larger Group 2A(2) elements, continually separating minuscule amounts of radium compounds from enormously larger amounts of chemically similar barium compounds. It took her four years to isolate 0.1 g of radium chloride, which she melted and electrolyzed to obtain pure metallic radium. ●

During the next few years, Henri Becquerel, the Curies, and P. Villard in France and Ernest Rutherford and his coworkers in England studied the nature of radioactive emissions. A key finding occurred when Rutherford and his colleague Frederick Soddy observed that elements other than radium were formed when radium decayed. In 1902, they proposed that radioactive emission results in the change of one element into another. This explanation sounded like a resurrection of alchemy and was met with disbelief and ridicule. We now know this idea to be true: *when a nuclide of one element decays, it emits radiation and, under most circumstances, changes into a nuclide of a different element.* The three natural types of radioactive emission are

- **Alpha particles** (symbolized α, $^4_2\alpha$, or $^4_2\text{He}^{2+}$) are identical to helium-4 nuclei.
- **Beta particles** (symbolized β, β^-, or sometimes $^0_{-1}\beta$) are high-speed electrons. (The emission of electrons from the nucleus may seem strange, but as you'll see shortly, they result from a nuclear reaction.)
- **Gamma rays** (symbolized γ, or sometimes $^0_0\gamma$) are very high-energy photons.

Figure 24.1 illustrates the behavior of these emissions in an electric field: α particles curve to a small extent toward the negative plate, β particles curve to a greater extent toward the positive plate, and γ rays are not affected by the electric field.

Types of Radioactive Decay; Balancing Nuclear Equations

When a nuclide decays, it forms a nuclide of lower energy, and the excess energy is carried off by the emitted radiation and the recoiling nucleus. The decaying, or reactant, nuclide is called the *parent;* the product nuclide is called the *daughter.* Nuclides can decay in several ways. As each of the major types of decay is introduced (summarized in Table 24.2 on the next page), we'll show examples of that type and apply the key principle used to balance nuclear reactions: *the total Z (charge, number of protons) and the total A (sum of protons and neutrons) of the reactants equal those of the products:*

$$\text{Total } A \atop \text{Total } Z \text{ Reactants} = {\text{Total } A \atop \text{Total } Z} \text{ Products} \qquad (24.1)$$

● **Her Brilliant Career** Marie Curie (1867–1934) is the only person to be awarded Nobel Prizes in two *different* sciences, one in physics in 1903 for her research into radioactivity and the other in chemistry in 1911 for the discovery of polonium and the discovery, isolation, and study of radium and its compounds.

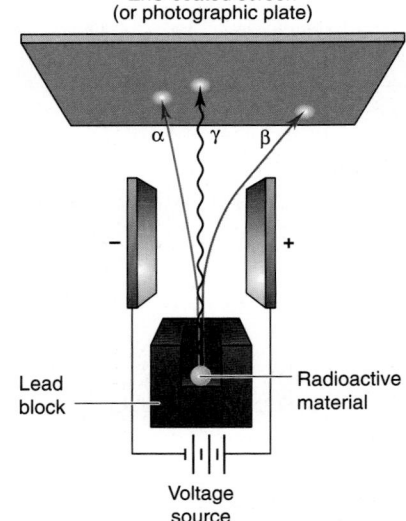

Figure 24.1 Three types of radioactive emissions in an electric field. Positively charged α particles curve toward the negative plate; negatively charged β particles curve toward the positive plate. (Later, we'll give these the symbol β^-.) The curvature is greater for β particles because they have much lower mass. The γ rays, uncharged high-energy photons, are unaffected by the field.

Table 24.2   Modes of Radioactive Decay*

Mode	Emission	Decay Process	Change in		
			A	Z	N
α Decay	α ($^4_2\text{He}^{2+}$)	α expelled	-4	-2	-2
β^- Decay[†]	β^-($^0_{-1}\beta$)	nucleus with xp^+ and yn^0 → nucleus with $(x+1)p^+$ and $(y-1)n^0$ + $^0_{-1}\beta$ β^- expelled Net: ^1_0n in nucleus → ^1_1p in nucleus + $^0_{-1}\beta$ β^- expelled	0	$+1$	-1
Positron (β^+) emission[†]	β^+($^0_1\beta$)	nucleus with xp^+ and yn^0 → nucleus with $(x-1)p^+$ and $(y+1)n^0$ + $^0_1\beta$ β^+ expelled Net: ^1_1p in nucleus → ^1_0n in nucleus + $^0_1\beta$ β^+ expelled	0	-1	$+1$
Electron (e^-) capture (EC)[†]	x-ray	low-energy orbital; nucleus with xp^+ and yn^0 → nucleus with $(x-1)p^+$ and $(y+1)n^0$ Net: $^0_{-1}e$ absorbed from low-energy orbital + ^1_1p in nucleus → ^1_0n in nucleus	0	-1	$+1$
Gamma (γ) emission	γ	excited nucleus → stable nucleus + γ γ photon radiated	0	0	0

*Nuclear chemists consider β^- decay, positron emission, and electron capture to be three decay modes of the more general process known as beta decay (see text).

[†]Neutrinos (ν) or antineutrinos ($\bar{\nu}$) are also formed during the three types of beta decay (see the margin note on page 1069). Although we will not include them in other equations in the chapter, keep in mind that antineutrinos are always expelled during β^- decay, and neutrinos are expelled during β^+ emission and e^- capture.

Transparency/ARIS Presentation Center, Tables Library

1. **Alpha (α) decay** involves the loss of an α particle from a nucleus. For each α particle emitted by the parent, *A decreases by 4 and Z decreases by 2* in the daughter. Every element beyond bismuth (Bi; $Z = 83$) is radioactive and exhibits α decay. Thus, α *decay is the most common means for a heavy, unstable nucleus to become more stable.* In Rutherford's experiment that established the existence of the atomic nucleus (Section 2.4, pp. 50–51), radium was the source of the α particles that were used as projectiles. Radium undergoes α decay to yield radon (Rn; $Z = 86$):

$$^{226}_{88}\text{Ra} \longrightarrow {}^{222}_{86}\text{Rn} + {}^{4}_{2}\alpha$$

Note that the *A* value for Ra equals the sum of the *A* values for Rn and α ($226 = 222 + 4$), and that the *Z* value for Ra equals the sum of the *Z* values for Rn and α ($88 = 86 + 2$).

2. **Beta (β) decay** is a more general class of radioactive decay that includes three types: β^- decay, β^+ emission, and electron capture.
   - **β^- decay** (or *negatron emission*) involves the ejection of a β^- particle from the nucleus. This change does not involve expulsion of a β^- particle that was in the nucleus; rather, *a neutron is converted into a proton, which remains in the nucleus, and a β^- particle, which is expelled immediately:* ●

   $$^{1}_{0}\text{n} \longrightarrow {}^{1}_{1}\text{p} + {}^{0}_{-1}\beta$$

   As always, the totals of the *A* and the *Z* values for reactant and products are equal. Radioactive nickel-63 becomes stable copper-63 through β^- decay:

   $$^{63}_{28}\text{Ni} \longrightarrow {}^{63}_{29}\text{Cu} + {}^{0}_{-1}\beta$$

   Another example is the β^- decay of carbon-14, used in radiocarbon dating:

   $$^{14}_{6}\text{C} \longrightarrow {}^{14}_{7}\text{N} + {}^{0}_{-1}\beta$$

   Note that β^- *decay results in a product nuclide with the same A but with Z one higher (one more proton) than in the reactant nuclide.* In other words, an atom of the element with the next *higher* atomic number is formed.
   - **Positron (β^+) emission** involves the emission of a β^+ particle from the nucleus. A key idea of modern physics is that most fundamental particles have corresponding *antiparticles* with the same mass but opposite charge. (The neutrino and antineutrino are an example.) The **positron** is the antiparticle of the electron. Positron emission occurs through a process in which *a proton in the nucleus is converted into a neutron, and a positron is expelled.* In terms of the effect on *A* and *Z*, *positron emission has the opposite effect of β^- decay: the daughter has the same A but Z is one lower (one fewer proton) than the parent.* Thus, an atom of the element with the next *lower* atomic number forms. Carbon-11, a synthetic radioisotope, decays to a stable boron isotope through β^+ emission:

   $$^{11}_{6}\text{C} \longrightarrow {}^{11}_{5}\text{B} + {}^{0}_{1}\beta$$

   - **Electron (e^-) capture (EC)** occurs when the nucleus interacts with an electron in an orbital from a low atomic energy level. The net effect is that *a proton is transformed into a neutron:*

   $$^{1}_{1}\text{p} + {}^{0}_{-1}\text{e} \longrightarrow {}^{1}_{0}\text{n}$$

   (We use the symbol "e" to distinguish an orbital electron from a beta particle, β.) The orbital vacancy is quickly filled by an electron that moves down from a higher energy level, and that process continues through still higher energy levels, with x-ray photons and neutrinos carrying off the energy difference in each step. Radioactive iron forms stable manganese through electron capture:

   $$^{55}_{26}\text{Fe} + {}^{0}_{-1}\text{e} \longrightarrow {}^{55}_{25}\text{Mn} + h\nu \text{ (x-rays and neutrinos)}$$

   Even though the processes are different, *electron capture has the same net effect as positron emission: Z lower by 1, A unchanged.*

● **The Little Neutral One** A neutral particle called an *antineutrino* ($\overline{\nu}$) is also emitted during β^- decay:

$$^{1}_{0}\text{n} \longrightarrow {}^{1}_{1}\text{p} + {}^{0}_{-1}\beta + \overline{\nu}$$

Its antiparticle, the *neutrino* (ν), is emitted during the other two types of beta decay. Theory suggests that these particles have masses much less than 10^{-4} times that of an electron, and that at least 10^9 of them exist for every proton. Neutrinos and antineutrinos interact with matter so slightly that it would take a piece of lead 1 light-year thick to absorb them. We will not discuss them further, except to mention that experiments in Japan in the 1990s detected neutrinos and obtained evidence that they do indeed have mass. Using a cathedral-sized pool containing 50,000 tons of ultra-pure water buried 1 mile underground in a zinc mine, an international team of scientists obtained results suggesting that these particles may account for a portion of the "missing" matter and dark energy in the universe.

3. **Gamma (γ) emission** involves the radiation of high-energy γ photons from an excited nucleus. Just as an atom in an excited *electronic* state reduces its energy by emitting photons, usually in the UV and visible ranges (see Section 7.2), a nucleus in an excited state lowers its energy by emitting γ photons, which are of much higher energy (much shorter wavelength) than UV photons. Many nuclear processes leave the nucleus in an excited state, so γ *emission accompanies many other (but mostly β) types of decay.* Several γ photons (also called γ rays) of different energies can be emitted from an excited nucleus as it returns to the ground state. Some of Marie Curie's experiments involved the release of γ rays, such as

$$\ce{^{215}_{84}Po \longrightarrow ^{211}_{82}Pb + ^{4}_{2}\alpha} \text{ (several γ emitted)}$$

Gamma emission is common subsequent to β^{-} decay, as in the following:

$$\ce{^{99}_{43}Tc \longrightarrow ^{99}_{44}Ru + ^{0}_{-1}\beta} \text{ (several γ emitted)}$$

Because γ rays have no mass or charge, *γ emission does not change A or Z.* Two gamma rays are emitted when a particle and an antiparticle annihilate each other. In the medical technique known as *positron-emission tomography* (which we discuss in Section 24.5), a positron and an electron annihilate each other (with all *A* and *Z* values shown):

$$\ce{^{0}_{1}\beta + ^{0}_{-1}e \longrightarrow 2^{0}_{0}\gamma}$$

SAMPLE PROBLEM 24.1   Writing Equations for Nuclear Reactions

PROBLEM Write balanced equations for the following nuclear reactions:
(a) Naturally occurring thorium-232 undergoes α decay.
(b) Zirconium-86 undergoes electron capture.

PLAN We first write a skeleton equation that includes the mass numbers, atomic numbers, and symbols of all the particles on the correct sides of the equation, showing the unknown product particle as $^{A}_{Z}X$. Then, because the total of mass numbers and the total of charges on the left side and the right side must be equal, we solve for *A* and *Z*, and use *Z* to determine X from the periodic table.

SOLUTION (a) Writing the skeleton equation, with the α particle as a product:

$$\ce{^{232}_{90}Th \longrightarrow ^{A}_{Z}X + ^{4}_{2}\alpha}$$

Solving for *A* and *Z* and balancing the equation: For *A*, $232 = A + 4$, so $A = 228$. For *Z*, $90 = Z + 2$, so $Z = 88$. From the periodic table, we see that the element with $Z = 88$ is radium (Ra). Thus, the balanced equation is

$$\ce{^{232}_{90}Th \longrightarrow ^{228}_{88}Ra + ^{4}_{2}\alpha}$$

(b) Writing the skeleton equation, with the captured electron as a reactant:

$$\ce{^{86}_{40}Zr + ^{0}_{-1}e \longrightarrow ^{A}_{Z}X}$$

Solving for *A* and *Z* and balancing the equation: For *A*, $86 + 0 = A$, so $A = 86$. For *Z*, $40 + (-1) = Z$, so $Z = 39$. The element with $Z = 39$ is yttrium (Y), so we have

$$\ce{^{86}_{40}Zr + ^{0}_{-1}e \longrightarrow ^{86}_{39}Y}$$

CHECK Always read across superscripts and then across subscripts, with the yield arrow as an equal sign, to check your arithmetic. In part (a), for example, $232 = 228 + 4$, and $90 = 88 + 2$.

FOLLOW-UP PROBLEM 24.1 Write a balanced equation for the reaction in which a nuclide undergoes β^{-} decay and produces cesium-133.

Nuclear Stability and the Mode of Decay

There are several ways that an unstable nuclide might decay, but can we predict how it will decay? Indeed, can we predict whether a given nuclide will decay at all? Our knowledge of the nucleus is much less complete than that of the whole atom, but some patterns emerge by observing the naturally occurring nuclides.

Literature Atwood, Charles H.; Paul, Kimberly M.; Todd, Stefani D. "Simulating and Visualizing Nuclear Reactions," *J. Chem. Educ.* **1995,** *72,* 515.

The Band of Stability Two key factors determine the stability of a nuclide. The first is the number of neutrons (N), the number of protons (Z), and their ratio (N/Z), which we calculate from $(A - Z)/Z$. This factor relates primarily to nuclides that undergo one of the three modes of β decay. The second factor affecting stability is the total mass of the nuclide, which mostly relates to nuclides that undergo α decay.

Figure 24.2A is a plot of number of neutrons vs. number of protons for all stable nuclides. Note the following:

- The points form a narrow **band of stability** that gradually curves above the line for $N = Z$ ($N/Z = 1$).
- Very few stable nuclides exist with $N/Z < 1$; the only two are $^1_1$H and $^3_2$He.
- Many lighter nuclides with $N = Z$ are stable, such as $^4_2$He, $^{12}_6$C, $^{16}_8$O, and $^{20}_{10}$Ne; the heaviest of these is $^{40}_{20}$Ca. Thus, for lighter nuclides, one neutron for each proton ($N = Z$) is enough to provide stability.
- The N/Z ratio of stable nuclides gradually increases as Z increases. A few examples appear on the figure: for $^{56}_{26}$Fe, $N/Z = 1.15$; for $^{107}_{47}$Ag, $N/Z = 1.28$; for $^{184}_{74}$W, $N/Z = 1.49$, and, finally, for $^{209}_{83}$Bi, $N/Z = 1.52$. Thus, for heavier stable nuclides, $N > Z$ ($N/Z > 1$), and N increases faster than Z. As we discuss below, if N/Z of a nuclide is either too high (above the band) or not high enough (below the band), the nuclide is unstable and undergoes one of the three modes of beta decay.

Literature Olbris, Donald J.; Herzfeld, Judith. "*Nucleogenesis!* A Game with Natural Rules for Teaching Nuclear Synthesis and Decay," *J. Chem. Educ.* **1999**, *76*, 349.

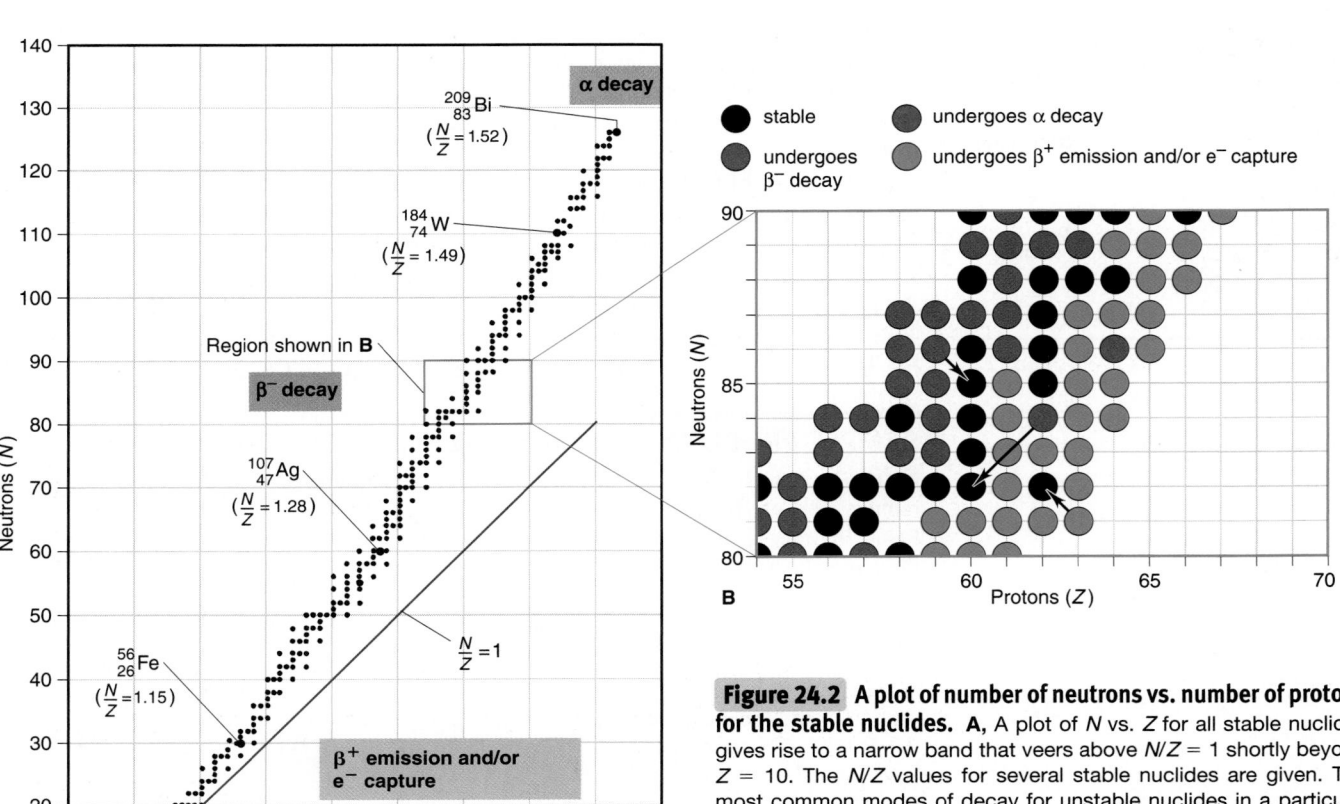

Figure 24.2 A plot of number of neutrons vs. number of protons for the stable nuclides. A, A plot of N vs. Z for all stable nuclides gives rise to a narrow band that veers above $N/Z = 1$ shortly beyond $Z = 10$. The N/Z values for several stable nuclides are given. The most common modes of decay for unstable nuclides in a particular region are shown: nuclides with a high N/Z ratio often undergo β^- decay; those with a low ratio undergo e^- capture or β^+ emission; heavy nuclei beyond the stable band undergo α decay. **B,** The blue box in part A is expanded to show the stable *and* many of the unstable nuclides in that area. Note the modes of decay: α decay decreases both N and Z by 2; β^- decay decreases N and increases Z by 1; β^+ emission and e^- capture increase N and decrease Z by 1.

- All nuclides with $Z > 83$ are unstable. This means that all nuclides with a higher Z—the largest members of main groups 1A(1) through 8A(18), actinium and the actinides ($Z = 89$–103), and the other elements of the fourth ($6d$) transition series ($Z = 104$–112)—are radioactive, and, as we point out below, undergo α decay.

Stability and Nuclear Structure Given that protons are positively charged and neutrons are uncharged, what holds the nucleus together? Nuclear scientists answer this question and explain the importance of N and Z values in terms of two opposing forces. Electrostatic repulsive forces between protons would break the nucleus apart if not for the presence of an attractive force called the **strong force,** which exists between all nucleons (protons and neutrons). This force is 137 times stronger than the repulsive force but *operates only over the short distances within the nucleus.* Competition between the *attractive* strong force and the *repulsive* electrostatic force determines nuclear stability.

The oddness or evenness of N and Z values is related to some important patterns of nuclear stability. Two interesting points become apparent when we classify the known stable nuclides:

- Elements with an even Z (number of protons) usually have a larger number of stable nuclides than elements with an odd Z. Table 24.3 demonstrates this point for cadmium ($Z = 48$) through xenon ($Z = 54$).
- Well over half the stable nuclides have *both* even N and even Z (Table 24.4). Only four nuclides with odd N and odd Z are stable: $^{2}_{1}$H, $^{6}_{3}$Li, $^{10}_{5}$B, and $^{14}_{7}$N.

One model of nuclear structure that attempts to explain the stability of even values of N and Z postulates that protons and neutrons lie in *nucleon energy levels,* and that greater stability results from the *pairing of spins* of like nucleons. (Note the analogy to electron energy levels and the stability from pairing of electron spins.)

Just as noble gases—with 2, 10, 18, 36, 54, and 86 electrons—are exceptionally stable because they have filled *electron* levels, nuclides with N or Z values of 2, 8, 20, 28, 50, 82 (and $N = 126$) are exceptionally stable. These so-called *magic numbers* are thought to correspond to the numbers of protons or neutrons in filled *nucleon* levels. A few examples are $^{50}_{22}$Ti ($N = 28$), $^{88}_{38}$Sr ($N = 50$), and the ten stable nuclides of tin ($Z = 50$). Some extremely stable nuclides have doubly magic numbers: $^{4}_{2}$He, $^{16}_{8}$O, $^{40}_{20}$Ca, and $^{208}_{82}$Pb ($N = 126$).

Table 24.3 **Number of Stable Nuclides for Elements 48 to 54***

Element	Atomic No. (Z)	No. of Nuclides
Cd	**48**	**8**
In	49	2
Sn	**50**	**10**
Sb	51	2
Te	**52**	**8**
I	53	1
Xe	**54**	**9**

**Even Z shown in boldface.*

 ARIS Presentation Center, Tables Library

Table 24.4 **An Even-Odd Breakdown of the Stable Nuclides**

Z	N	No. of Nuclides
Even	Even	157
Even	Odd	53
Odd	Even	50
Odd	Odd	4
	TOTAL	264

 ARIS Presentation Center, Tables Library

 ARIS Presentation Center, Worked Examples Library

SAMPLE PROBLEM 24.2 **Predicting Nuclear Stability**

PROBLEM Which of the following nuclides would you predict to be stable and which radioactive: **(a)** $^{18}_{10}$Ne; **(b)** $^{32}_{16}$S; **(c)** $^{236}_{90}$Th; **(d)** $^{123}_{56}$Ba? Explain.

PLAN In order to evaluate the stability of each nuclide, we determine the N and Z values, the N/Z ratio from $(A - Z)/Z$, the value of Z, stable N/Z ratios (from Figure 24.2), and whether Z and N are even or odd.

SOLUTION **(a)** Radioactive. This nuclide has $N = 8$ ($18 - 10$) and $Z = 10$, so $N/Z = \dfrac{18 - 10}{10} = 0.8$. Except for hydrogen-1 and helium-3, no nuclides with $N < Z$ are stable; despite even N and Z, this nuclide has too few neutrons to be stable.

(b) Stable. This nuclide has $N = Z = 16$, so $N/Z = 1.0$. With $Z < 20$ and even N and Z, this nuclide is most likely stable.

(c) Radioactive. This nuclide has $Z = 90$, and every nuclide with $Z > 83$ is radioactive.

(d) Radioactive. This nuclide has $N = 67$ and $Z = 56$, so $N/Z = 1.20$. For Z values of 55 to 60, Figure 24.2A shows $N/Z \geq 1.3$, so this nuclide has too few neutrons to be stable.

CHECK By consulting a table of isotopes, such as the one in the *CRC Handbook of Chemistry and Physics,* we find that our predictions are correct.

FOLLOW-UP PROBLEM 24.2 Why is $^{31}_{15}$P stable but $^{30}_{15}$P unstable?

Predicting the Mode of Decay An unstable nuclide generally decays in a mode that shifts its N/Z ratio toward the band of stability. This fact is illustrated in Figure 24.2B, which expands a small region of Figure 24.2A to show the stable *and* many of the unstable nuclides in that region, as well as their modes of decay. Note the following points:

1. *Neutron-rich nuclides.* Nuclides with too many neutrons for stability (a high N/Z) lie above the band of stability. They undergo β^- *decay,* which converts a neutron into a proton, thus reducing the value of N/Z.
2. *Proton-rich nuclides.* Nuclides with too many protons for stability (a low N/Z) lie below the band. They undergo β^+ *emission* and/or e^- *capture,* both of which convert a proton into a neutron, thus increasing the value of N/Z. (The rate of e^- capture increases with Z, so β^+ emission is more common among lighter elements and e^- capture more common among heavier elements.)
3. *Heavy nuclides.* Nuclides with $Z > 83$ are too heavy to be stable and undergo α *decay,* which reduces their Z and N values by two units per emission.

If we have the information in Figure 24.2, predicting the mode of decay of an unstable nuclide is just a matter of comparing its N/Z ratio with those in the nearby region of the band of stability. But, even when Figure 24.2 is not available, we can often make an educated guess at the mode of decay. The atomic mass of an element represents the weighted average of its naturally occurring isotopes, so we would expect the mass number A of a *stable* nuclide to be relatively close to the atomic mass. If an *unstable* nuclide of the element (given Z) has an A value higher than the atomic mass, it is neutron rich and will probably decay by β^- emission. If, on the other hand, the unstable nuclide has an A value lower than the atomic mass, it is proton rich and will probably decay by β^+ emission and/or e^- capture. In the next sample problem, we compare the mass number with the atomic mass to help us predict the mode of decay of an unstable nuclide and then check the prediction with the N/Z values in Figure 24.2.

SAMPLE PROBLEM 24.3 Predicting the Mode of Nuclear Decay

PROBLEM Use the atomic mass of the element to predict the mode(s) of decay of the following radioactive nuclides: **(a)** $^{12}_{5}B$; **(b)** $^{234}_{92}U$; **(c)** $^{81}_{33}As$; **(d)** $^{127}_{57}La$.

PLAN If the nuclide is too heavy to be stable ($Z > 83$), it undergoes α decay. For other cases, we use the Z value to locate the element in the periodic table and obtain its atomic mass. If the mass number of the nuclide is higher than the atomic mass, the nuclide has too many neutrons: N too high $\Rightarrow \beta^-$ decay. If the mass number is lower than the atomic mass, the nuclide has too many protons: Z too high $\Rightarrow \beta^+$ emission and/or e^- capture.

SOLUTION **(a)** This nuclide has $Z = 5$, which is boron (B), and the atomic mass is 10.81. The nuclide's A value of 12 is higher than its atomic mass, so this nuclide is neutron rich. It will probably undergo β^- decay lowering N to 6 and raising Z to 6 to form the stable $^{12}_{6}C$.

(b) This nuclide has $Z = 92$, so it will undergo α decay and decrease its total mass.

(c) This nuclide has $Z = 33$, which is arsenic (As), and the atomic mass is 74.92. The A value of 81 is higher, so this nuclide is neutron rich and will probably undergo β^- decay.

(d) This nuclide has $Z = 57$, which is lanthanum (La), and the atomic mass is 138.9. The A value of 127 is lower, so this nuclide is proton rich and will probably undergo β^+ emission and/or e^- capture.

CHECK To confirm our predictions in (a), (c), and (d), let's compare each nuclide's N/Z ratio to those in the band of stability. **(a)** This nuclide has $N = 7$ and $Z = 5$, so $N/Z = 1.40$, which is too high for this region of the band, so it will undergo β^- decay. **(c)** This nuclide has $N = 48$ and $Z = 33$, so $N/Z = 1.45$, which is too high for this region of the band, so it undergoes β^- decay. **(d)** This nuclide has $N = 70$ and $Z = 57$, so $N/Z = 1.23$, which is too low for this region of the band, so it undergoes β^+ emission and/or e^- capture. Our predictions based on N/Z *values* were the same as those based on atomic mass.

COMMENT Both possible modes of decay are observed for the nuclide in part (d).

Figure 24.3 The ^{238}U decay series.
Uranium-238 (*top right*) undergoes a series of 14 steps involving α or β⁻ decay until lead-206 forms (*bottom left*).

Transparency/ARIS Presentation Center, Art Library

Decay Series A parent nuclide may undergo a series of decay steps before a stable daughter nuclide forms. The succession of steps is called a **decay series,** or **disintegration series,** and is typically depicted on a gridlike display. Figure 24.3 shows the decay series from uranium-238 to lead-206. Numbers of neutrons (N) are plotted against numbers of protons (Z) to form the grid, which displays a series of α and β⁻ decays. The typical zigzag pattern arises because $N > Z$, which means that α decay, which reduces both N and Z by two units, decreases Z by a slightly greater percentage than it does N. Therefore, α decays result in neutron-rich daughters, which undergo β⁻ decay to gain more stability. Note that a given nuclide can undergo both types of decay. (Gamma emission accompanies many of these steps but does not affect the type of nuclide.) This series is one of three that occur in nature. All end with isotopes of lead whose nuclides all have one ($Z = 82$) or two ($N = 126, Z = 82$) magic numbers. A second series begins with uranium-235 and ends with lead-207, and a third begins with thorium-232 and ends with lead-208. (Neptunium-237 began a fourth series, but its half-life is so much less than the age of Earth that only traces of it remain today.)

Section Summary

Nuclear reactions are normally not affected by reaction conditions or chemical composition and release much more energy than chemical reactions. • To become more stable, a radioactive nuclide may emit α particles ($^4_2$He nuclei), β particles (β⁻ or $^0_{-1}$β; high-speed electrons), positrons (β⁺ or $^0_1$β), or γ rays (high-energy photons) or may capture an orbital electron ($^0_{-1}$e). • A narrow band of neutron-to-proton ratios (N/Z) includes those of all the stable nuclides. • Certain "magic numbers" of neutrons and protons are associated with very stable nuclides. • By comparing a nuclide's mass number with the atomic mass and its N/Z ratio with those in the band of stability, we can predict that, in general, neutron-rich nuclides undergo β⁻ decay and proton-rich nuclides undergo β⁺ emission and/or e⁻ capture. Heavy nuclides ($Z > 83$) undergo α decay. • Three naturally occurring decay series all end in isotopes of lead.

24.2 THE KINETICS OF RADIOACTIVE DECAY

Chemical and nuclear systems both tend toward maximum stability. Just as the concentrations in a chemical system change in a predictable direction to give a stable equilibrium ratio, the type and number of nucleons in an unstable nucleus change in a predictable direction to give a stable N/Z ratio. As you know, however, the tendency of a chemical system to become more stable tells nothing about how long that process will take, and the same holds true for nuclear systems. In this section, we examine the kinetics of nuclear change; later, we'll examine the energetics of nuclear change. To begin, a Tools of the Laboratory essay on page 1075 describes how radioactivity is detected and measured.

The Rate of Radioactive Decay

Radioactive nuclei decay at a characteristic rate, regardless of the chemical substance in which they occur. The *decay rate,* or **activity** ($\mathscr{A}$), of a radioactive sample is the change in number of nuclei ($\mathscr{N}$) divided by the change in time (t). As we saw with chemical reaction rates, because the number of nuclei is *decreasing,* a minus sign precedes the expression for the decay rate:

$$\text{Decay rate } (\mathscr{A}) = -\frac{\Delta\mathscr{N}}{\Delta t}$$

Animation: Radioactive Decay ARIS Presentation Center Chapter 24

Literature Schultz, Emeric. "Dice Shaking as an Analogy for Radioactive Decay and First Order Kinetics," *J. Chem. Educ.* 1997, 74, 505.

Tools of the Laboratory

Counters for the Detection of Radioactive Emissions

Radioactive emissions interact with atoms in surrounding materials. To determine the rate of nuclear decay, we measure the radioactivity of a sample by observing the effects of these interactions over time. Because these effects can be electrically amplified billions of times, it is even possible to detect the decay of a single nucleus. Ionization counters and scintillation counters are two devices used to measure radioactive emissions.

An *ionization counter* detects radioactive emissions as they ionize a gas. Ionization produces free electrons and gaseous cations, which are attracted to electrodes that conduct a current to a recording device. The most common type of ionization counter is a **Geiger-Müller counter** (Figure B24.1). It consists of a tube filled with an argon-methane mixture; the tube housing acts as the cathode, and a thin wire in the center of the tube acts as the anode. Emissions from the sample enter the tube through a thin window and strike argon atoms, producing free electrons that are accelerated toward the anode. These electrons collide with other argon atoms and free more electrons in an *avalanche effect*. The current created is amplified and appears as a meter reading and/or an audible click. An initial release of 1 electron can release 10^{10} electrons in a microsecond, giving the Geiger-Müller counter great sensitivity.

In a **scintillation counter,** radioactive emissions are detected by their ability to excite atoms and cause them to emit light. The light-emitting substance in the counter, called a *phosphor,* is coated onto part of a *photomultiplier tube,* a device that increases the original electrical signal. Incoming radioactive particles strike the phosphor, which emits photons. Each photon, in turn, strikes a cathode, releasing an electron through the photoelectric effect (Section 7.1). This electron hits other portions of the tube that release increasing numbers of electrons, and the resulting current is recorded. Liquid scintillation counters employ an organic mixture that contains a phosphor and a solvent (Figure B24.2). This "cocktail" dissolves the sample *and* emits light when excited by the emission. These counters are often used to measure β^- emissions from dissolved biological samples, particularly those containing compounds of ^{3}H and ^{14}C.

Figure B24.2 Vials of a scintillation "cocktail" emitting light. A radioactive substance dissolved in an organic mixture (cocktail) emits particles that cause excitation of a phosphor, which emits pulses of light. The number of pulses is proportional to the concentration of the radioactive substance. **ARIS Presentation Center,** Photo Library

Figure B24.1 Detection of radioactivity by an ionization counter. When an Ar atom absorbs the energy of a radioactive particle *(red),* it is ionized to an Ar$^+$ ion *(purple)* and an electron *(yellow).* The free electron collides with and ionizes another Ar atom. As the process continues, the Ar$^+$ ions migrate to the negative electrode, and the electrons migrate to the positive electrode, resulting in a current. **ARIS Presentation Center,** Art Library

Literature Hughes, Enrique A.; Zalts, Anita. "Radioactivity in the Classroom," *J. Chem. Educ.* **2000**, *77*, 613.

The SI unit of radioactivity is the **becquerel (Bq)**; it is defined as one disintegration per second (d/s): 1 Bq = 1 d/s. A much larger and more common unit of radioactivity is the **curie (Ci).** Originally, the curie was defined as the number of disintegrations per second in 1 g of radium-226, but it is now a fixed quantity:

$$1 \text{ Ci} = 3.70 \times 10^{10} \text{ d/s} \qquad \textbf{(24.2)}$$

Because the curie is so large, the millicurie (mCi) and microcurie (μCi) are commonly used. We often express the radioactivity of a sample in terms of *specific activity,* the decay rate per gram.

An activity is meaningful only when we consider the large number of nuclei in a macroscopic sample. Suppose there are 1×10^{15} radioactive nuclei of a particular type in a sample and they decay at a rate of 10% per hour. Although any particular nucleus in the sample might decay in a microsecond or in a million hours, the *average* of all decays results in 10% of the entire collection of nuclei disintegrating each hour. During the first hour, 10% of the *original* number, or 1×10^{14} nuclei, will decay. During the next hour, 10% of the remaining 9×10^{14} nuclei, or 9×10^{13} nuclei, will decay. During the next hour, 10% of those remaining will decay, and so forth. Thus, for a large collection of radioactive nuclei, *the number decaying per unit time is proportional to the number present:*

$$\text{Decay rate } (\mathcal{A}) \propto \mathcal{N} \qquad \text{or} \qquad \mathcal{A} = k\mathcal{N}$$

where k is called the **decay constant** and is characteristic of each type of nuclide. The larger the value of k, the higher is the decay rate: larger $k \Rightarrow$ higher $\mathcal{A}$.

Combining the two rate expressions just given, we obtain

$$\mathcal{A} = -\frac{\Delta \mathcal{N}}{\Delta t} = k\mathcal{N} \qquad \textbf{(24.3)}$$

Note that the activity depends only on $\mathcal{N}$ raised to the first power (and on the constant value of k). Therefore, *radioactive decay is a first-order process* (see Section 16.4). The only difference in the case of nuclear decay is that we consider the *number* of nuclei rather than their concentration.

Half-Life of Radioactive Decay Decay rates are also commonly expressed in terms of the fraction of nuclei that decay over a given time interval. The **half-life ($t_{1/2}$)** of a nuclide is the time it takes for half the nuclei present in a sample to decay. *The number of nuclei remaining is halved after each half-life.* Thus, half-life has the same meaning for a nuclear change as for a chemical change (Section 16.4). Figure 24.4 shows the decay of carbon-14, which has a half-life of 5730 years, in terms of number of ^{14}C nuclei remaining:

$$^{14}_{6}\text{C} \longrightarrow ^{14}_{7}\text{N} + ^{0}_{-1}\beta$$

Animation: Half-Life ARIS Presentation Center Chapter 24

Figure 24.4 Decrease in number of ^{14}C nuclei over time. A plot of number of ^{14}C nuclei vs. time gives a decreasing curve. In each half-life (5730 years), half the ^{14}C nuclei present undergo decay. A plot of mass of ^{14}C (or any other variable proportional to number of nuclei) vs. time has the same shape. (As you'll see in Figure 24.5, p. 1079, a plot of the natural logarithm of the variable on the y-axis vs. time gives a straight line.)

Transparency/ARIS Presentation Center, Art Library

$$\mathcal{N}_t = \mathcal{N}_0 \times \left(\frac{1}{2}\right)^n$$

Number of nuclei at time t | Initial number of nuclei | Number of half-lives

After 1st half-life (5730 yr)
After 2nd half-life (11,460 yr)
After 3rd half-life (17,190 yr)

We can also consider the half-life in terms of mass of substance. As ^{14}C decays, the mass of ^{14}C decreases while the mass of ^{14}N increases. If we start with 1.0 g of ^{14}C, half that mass of ^{14}C (0.50 g) will be left after 5730 years, half of that mass (0.25 g) after another 5730 years, and so on. The activity depends on the number of nuclei, so the activity is halved after each succeeding half-life as well.

We determine the half-life of a nuclear reaction from its rate constant. Rearranging Equation 24.3 and integrating over time gives an expression for finding the number of nuclei remaining after a given time t, $\mathcal{N}_t$:

$$\ln \frac{\mathcal{N}_t}{\mathcal{N}_0} = -kt \quad \text{or} \quad \mathcal{N}_t = \mathcal{N}_0 e^{-kt} \quad \text{and} \quad \ln \frac{\mathcal{N}_0}{\mathcal{N}_t} = kt \quad (24.4)$$

where $\mathcal{N}_0$ is the number of nuclei at $t = 0$. (Note the similarity to Equation 16.4, p. 700.) To calculate the half-life ($t_{1/2}$), we set $\mathcal{N}_t$ equal to $\frac{1}{2}\mathcal{N}_0$ and solve for $t_{1/2}$:

$$\ln \frac{\mathcal{N}_0}{\frac{1}{2}\mathcal{N}_0} = kt_{1/2} \quad \text{so} \quad t_{1/2} = \frac{\ln 2}{k} \quad (24.5)$$

Exactly analogous to the half-life of a first-order chemical change, *this half-life is **not** dependent on the number of nuclei and is inversely related to the decay constant:*

large $k \Rightarrow$ short $t_{1/2}$ and small $k \Rightarrow$ long $t_{1/2}$

The decay constants and half-lives of radioactive nuclides vary over a very wide range, even for the nuclides of a given element (Table 24.5).

Table 24.5 Decay Constants (k) and Half-Lives ($t_{1/2}$) of Beryllium Isotopes

Nuclide	k	$t_{1/2}$
$^{7}_{4}$Be	1.30×10^{-2}/day	53.3 days
$^{8}_{4}$Be	1.0×10^{16}/s	6.7×10^{-17} s
$^{9}_{4}$Be	Stable	
$^{10}_{4}$Be	4.3×10^{-7}/yr	1.6×10^{6} yr
$^{11}_{4}$Be	5.02×10^{-2}/s	13.8 s

 ARIS Presentation Center, Tables Library

 ARIS Presentation Center, Worked Examples Library

 Biological application

SAMPLE PROBLEM 24.4 Finding the Number of Radioactive Nuclei

PROBLEM Strontium-90 is a radioactive byproduct of nuclear reactors that behaves biologically like calcium, the element above it in Group 2A(2). When ^{90}Sr is ingested by mammals, it is found in their milk and eventually in the bones of those drinking the milk. If a sample of ^{90}Sr has an activity of 1.2×10^{12} d/s, what are the activity and the fraction of nuclei that have decayed after 59 yr ($t_{1/2}$ of ^{90}Sr = 29 yr)?

PLAN The fraction of nuclei that have decayed is the change in number of nuclei, expressed as a fraction of the starting number. The activity of the sample ($\mathcal{A}$) is proportional to the number of nuclei ($\mathcal{N}$), so we know that

$$\text{Fraction decayed} = \frac{\mathcal{N}_0 - \mathcal{N}_t}{\mathcal{N}_0} = \frac{\mathcal{A}_0 - \mathcal{A}_t}{\mathcal{A}_0}$$

We are given $\mathcal{A}_0$ (1.2×10^{12} d/s), so we find $\mathcal{A}_t$ from the integrated form of the first-order rate equation (Equation 24.4), in which t is 59 yr. To solve that equation, we first need k, which we can calculate from the given $t_{1/2}$ (29 yr).

SOLUTION Calculating the decay constant k:

$$t_{1/2} = \frac{\ln 2}{k} \quad \text{so} \quad k = \frac{\ln 2}{t_{1/2}} = \frac{0.693}{29 \text{ yr}} = 0.024 \text{ yr}^{-1}$$

Applying Equation 24.4 to calculate $\mathcal{A}_t$, the activity remaining at time t:

$$\ln \frac{\mathcal{N}_0}{\mathcal{N}_t} = \ln \frac{\mathcal{A}_0}{\mathcal{A}_t} = kt \quad \text{or} \quad \ln \mathcal{A}_0 - \ln \mathcal{A}_t = kt$$

So, $\ln \mathcal{A}_t = -kt + \ln \mathcal{A}_0 = -(0.024 \text{ yr}^{-1} \times 59 \text{ yr}) + \ln (1.2\times10^{12} \text{ d/s})$
$\ln \mathcal{A}_t = -1.4 + 27.81 = 26.4$
$\mathcal{A}_t = 2.9\times10^{11}$ d/s

(All the data contain two significant figures, so we retained two in the answer.) Calculating the fraction decayed:

$$\text{Fraction decayed} = \frac{\mathcal{A}_0 - \mathcal{A}_t}{\mathcal{A}_0} = \frac{1.2\times10^{12} \text{ d/s} - 2.9\times10^{11} \text{ d/s}}{1.2\times10^{12} \text{ d/s}} = 0.76$$

CHECK The answer is reasonable: t is about 2 half-lives, so $\mathcal{A}_t$ should be about $\frac{1}{4}\mathcal{A}_0$, or about 0.3×10^{12}; therefore, the activity should have decreased by about $\frac{3}{4}$.

COMMENT 1. A useful substitution of Equation 24.4 for finding $\mathcal{A}_t$, the activity at time t, is $\mathcal{A}_t = \mathcal{A}_0\,e^{-kt}$.

2. Another way to find the fraction of activity (or nuclei) remaining incorporates the number of half-lives $(t/t_{1/2})$. By combining Equations 24.4 and 24.5 and substituting $(\ln 2)/t_{1/2}$ for k, we obtain

$$\ln\frac{\mathcal{N}_0}{\mathcal{N}_t} = \left(\frac{\ln 2}{t_{1/2}}\right)t = \frac{t}{t_{1/2}}\ln 2 = \ln 2^{t/t_{1/2}}$$

Inverting the ratio gives

$$\ln\frac{\mathcal{N}_t}{\mathcal{N}_0} = \ln\left(\frac{1}{2}\right)^{t/t_{1/2}}$$

Taking the antilog gives

$$\text{Fraction remaining} = \frac{\mathcal{N}_t}{\mathcal{N}_0} = \left(\frac{1}{2}\right)^{t/t_{1/2}} = \left(\frac{1}{2}\right)^{59/29} = 0.24$$

So,   Fraction decayed $= 1.00 - 0.24 = 0.76$

FOLLOW-UP PROBLEM 24.4 Sodium-24 ($t_{1/2} = 15$ h) is used to study blood circulation. If a patient is injected with an aqueous solution of ^{24}NaCl whose activity is 2.5×10^9 d/s, how much of the activity is present in the patient's body and excreted fluids after 4.0 days?

Radioisotopic Dating

The historical record fades rapidly with time and virtually disappears for events of more than a few thousand years ago. Much of our understanding of prehistory comes from a technique called **radioisotopic dating**, which uses **radioisotopes** to determine the age of an object. The method supplies data in fields as diverse as art history, archeology, geology, and paleontology.

The technique of *radiocarbon dating*, for which the American chemist Willard F. Libby won the Nobel Prize in chemistry in 1960, is based on measuring the amounts of ^{14}C and ^{12}C in materials of biological origin. The accuracy of the method falls off after about six half-lives of ^{14}C ($t_{1/2} = 5730$ yr), so it is used to date objects up to about 36,000 years old.

Here is how the method works. High-energy cosmic rays, consisting mainly of protons, enter the atmosphere from outer space and initiate a cascade of nuclear reactions, some of which produce neutrons that bombard ordinary ^{14}N atoms to form ^{14}C:

$$^{14}_{7}\text{N} + ^{1}_{0}\text{n} \longrightarrow ^{14}_{6}\text{C} + ^{1}_{1}\text{p}$$

Through the competing processes of formation and radioactive decay, the amount of ^{14}C in the atmosphere has remained nearly constant.*

The ^{14}C atoms combine with O_2, diffuse throughout the lower atmosphere, and enter the total carbon pool as gaseous $^{14}CO_2$ and aqueous $H^{14}CO_3^-$. They mix with ordinary $^{12}CO_2$ and $H^{12}CO_3^-$, reaching a constant ^{12}C/^{14}C ratio of about $10^{12}/1$. Carbon atoms in the CO_2 are taken up by plants during photosynthesis, and then taken up and excreted by animals that eat the plants. Thus, the ^{12}C/^{14}C ratio of a living organism is the same as the ratio in the environment. When an organism dies, however, it no longer absorbs or releases CO_2, so the ^{12}C/^{14}C ratio steadily increases because *the amount of ^{14}C decreases as it decays*:

$$^{14}_{6}\text{C} \longrightarrow ^{14}_{7}\text{N} + ^{0}_{-1}\beta$$

The difference between the ^{12}C/^{14}C ratio in a dead organism and the ratio in living organisms reflects the time elapsed since the organism died. ●

Biological application

ARIS Presentation Center, Photo Library

● **The Case of the Shroud of Turin**
One of the holiest Christian relics is the famed Shroud of Turin. It is a piece of linen that bears a faint image of a man's body and was thought to be the burial cloth used to wrap the body of Jesus Christ. In 1988, the Vatican allowed scientific testing of the cloth by radiocarbon dating. Three labs in Europe and the United States independently measured the ^{12}C/^{14}C ratio of a 50-mg piece of the linen and determined that the flax from which the cloth was made was grown between 1260 AD and 1390 AD. Despite this evidence, the shroud lost none of its fascination: 10 years later, in 1998, when the shroud was again put on display, about 2 million people lined up to view it.

---

*Cosmic ray intensity does vary slightly with time, which affects the amount of atmospheric ^{14}C. From ^{14}C activity in ancient trees, we know the amount fell slightly about 3000 years ago to current levels. Recently, nuclear testing and fossil fuel combustion have also altered the fraction of ^{14}C slightly. Taking these factors into account improves the accuracy of the dating method.

Natural logarithm of specific activity

Charcoal from earliest Polynesian culture in Hawaii
Linen wrappings from Book of Isaiah, Dead Sea Scrolls
Charcoal from earliest settlement in Japan
Burned tree from eruption that created Crater Lake, Oregon
Burned bones of sloth in Chilean cave. Earliest trace of human presence at tip of South America
Burned bison bones associated with Folsom Man, found at Lubbock, Texas
Mesolithic-Neolithic transition site, Belt Cave, Iran
Charcoal from Lascaux Caves, France, site of extensive cave paintings (see background)

Age (years)

Figure 24.5 Radiocarbon dating for determining the age of artifacts. The natural logarithms of the specific activities of ^{14}C (activity/g ^{14}C) for various artifacts are projected onto a line whose slope equals $-k$, the negative of the ^{14}C decay constant. The age (in years) of an artifact is determined from the horizontal axis.

ARIS Presentation Center, Art Library

As you saw in Sample Problem 24.4, the first-order rate equation can be expressed in terms of a ratio of activities:

$$\ln \frac{\mathcal{N}_0}{\mathcal{N}_t} = \ln \frac{\mathcal{A}_0}{\mathcal{A}_t} = kt$$

We use this expression in radiocarbon dating, where $\mathcal{A}_0$ is the activity in a living organism and $\mathcal{A}_t$ is the activity in the object whose age is unknown. Solving for t gives the age of the object:

$$t = \frac{1}{k} \ln \frac{\mathcal{A}_0}{\mathcal{A}_t} \tag{24.6}$$

A useful graphical method in radioisotopic dating shows a plot of the natural logarithm of the specific activity vs. time, which gives a straight line with a slope of $-k$, the negative of the decay constant. Using such a plot and measuring the ^{14}C specific activity of an object, we can determine its age; several examples appear in Figure 24.5. To determine the ages of more ancient objects or of objects that do not contain carbon, different radioisotopes are measured. ●

● **How Old Is the Solar System?** By comparing the ratio of ^{238}U ($t_{1/2} = 4.5 \times 10^9$ yr) to its final decay product, ^{206}Pb, geochemists found that the oldest known surface rocks on Earth—granite in western Greenland—are about 3.7 billion years old. The ratio of ^{238}U/^{206}Pb in meteorites gives 4.65 billion years for the age of the Solar System, and therefore Earth. From this and other isotope ratios, such as ^{40}K/^{40}Ar ($t_{1/2}$ of ^{40}K = 1.3×10^9 yr) as well as ^{87}Rb/^{87}Sr ($t_{1/2}$ of ^{87}Rb = 4.9×10^{10} yr), Moon rocks collected by Apollo astronauts have been shown to be 4.2 billion years old. They also provide evidence for volcanic activity on the Moon's surface about 3.3 billion years ago, just about the time that, according to these methods, the first organisms were evolving on Earth.

ARIS Presentation Center, Worked Examples Library

 Biological application

SAMPLE PROBLEM 24.5 Applying Radiocarbon Dating

PROBLEM The charred bones of a sloth in a cave in Chile represent the earliest evidence of human presence at the southern tip of South America. A sample of the bone has a specific activity of 5.22 disintegrations per minute per gram of carbon (d/min·g). If the ^{12}C/^{14}C ratio for living organisms results in a specific activity of 15.3 d/min·g, how old are the bones ($t_{1/2}$ of ^{14}C = 5730 yr)?

PLAN We first calculate k from the given $t_{1/2}$ (5730 yr). Then we apply Equation 24.6 to find the age (t) of the bones, using the given activities of the bones ($\mathcal{A}_t$ = 5.22 d/min·g) and of a living organism ($\mathcal{A}_0$ = 15.3 d/min·g).

SOLUTION Calculating k for ^{14}C decay:

$$k = \frac{\ln 2}{t_{1/2}} = \frac{0.693}{5730 \text{ yr}} = 1.21 \times 10^{-4} \text{ yr}^{-1}$$

Calculating the age (t) of the bones:

$$t = \frac{1}{k} \ln \frac{\mathscr{A}_0}{\mathscr{A}_t} = \frac{1}{1.21 \times 10^{-4} \text{ yr}^{-1}} \ln \left(\frac{15.3 \text{ d/min·g}}{5.22 \text{ d/min·g}} \right) = 8.89 \times 10^3 \text{ yr}$$

The bones are about 8900 years old.

CHECK The activity of the bones is between $\frac{1}{2}$ and $\frac{1}{4}$ the activity of a living organism, so the age should be between one and two half-lives (5730 to 11,460 yr).

FOLLOW-UP PROBLEM 24.5 A sample of wood from an Egyptian mummy case has a specific activity of 9.41 d/min·g. How old is the case?

Section Summary

Ionization and scintillation counters measure the number of emissions from a radioactive sample. • The decay rate (activity) of a sample is proportional to the number of radioactive nuclei. Nuclear decay is a first-order process, so the half-life is a constant for each nuclide. • Radioisotopic methods, such as ^{14}C dating, determine the ages of objects by measuring the ratio of specific isotopes in the sample.

24.3 NUCLEAR TRANSMUTATION: INDUCED CHANGES IN NUCLEI

The alchemists' dream of changing base metals into gold was never realized, but in the early 20th century, nuclear physicists found that they *could* change one element into another. Research into **nuclear transmutation,** the *induced* conversion of one nucleus into another, was closely linked with research into atomic structure and led to the discovery of the neutron and to the production of artificial radioisotopes. Later, high-energy bombardment of nuclei in particle accelerators began a scientific endeavor, which continues to this day, of creating many new nuclides and a growing number of new elements.

Early Transmutation Experiments; Nuclear Shorthand Notation

The first recognized transmutation occurred in 1919, when Ernest Rutherford showed that α particles emitted from radium bombarded atmospheric nitrogen to form a proton and oxygen-17:

$$^{14}_{7}\text{N} + ^{4}_{2}\alpha \longrightarrow ^{1}_{1}\text{p} + ^{17}_{8}\text{O}$$

By 1926, experimenters had found that α bombardment transmuted most elements with low atomic numbers to the next higher element, with ejection of a proton.

A shorthand notation used specifically for nuclear transmutation reactions shows the reactant (target) nucleus to the left and the product nucleus to the right of a set of parentheses, within which a comma separates the projectile particle from the ejected particle(s):

reactant nucleus (particle in, particle(s) out) product nucleus   (24.7)

Using this notation, the previous reaction is ^{14}N (α,p) ^{17}O.

An unexpected finding in a transmutation experiment led to *the discovery of the neutron.* When lithium, beryllium, and boron were bombarded with α particles, they emitted highly penetrating radiation that could not be deflected by a magnetic or electric field. Unlike γ radiation, these emissions were massive enough to eject protons from the substances they penetrated. In 1932, James Chadwick, a student of Rutherford, proposed that these emissions consisted of neutral particles with a mass similar to that of a proton, and he named them *neutrons.* Chadwick received the Nobel Prize in physics in 1935 for his discovery.

In 1933, Irene and Frederic Joliot-Curie (see photo), daughter and son-in-law of Marie and Pierre Curie, created the first artificial radioisotope, phosphorus-30.

The Joliot-Curies in their laboratory.

When they bombarded aluminum foil with α particles, phosphorus-30 and neutrons were formed:

$$^{27}_{13}\text{Al} + ^{4}_{2}\alpha \longrightarrow ^{1}_{0}\text{n} + ^{30}_{15}\text{P} \qquad \text{or} \qquad ^{27}\text{Al}\,(\alpha,\text{n})\,^{30}\text{P}$$

Since then, other techniques for producing artificial radioisotopes have been developed. In fact, the majority of the nearly 1000 known radionuclides have been produced artificially.

Particle Accelerators and the Transuranium Elements

During the 1930s and 1940s, researchers probing the nucleus bombarded elements with neutrons, α particles, protons, and **deuterons** (nuclei of the stable hydrogen isotope deuterium, ^{2}H). Neutrons are especially useful as projectiles because they have no charge and thus are not repelled as they approach a target nucleus. The other particles are all positive, so early researchers found it difficult to give them enough energy to overcome their repulsion by the target nuclei. Beginning in the 1930s, however, **particle accelerators** were invented to impart high kinetic energies to particles by placing them in an electric field, usually in combination with a magnetic field. In the simplest and earliest design, protons are introduced at one end of a tube and attracted to the other end by a potential difference.

A major advance was the *linear accelerator,* a series of separated tubes of increasing length that, through a source of alternating voltage, change their charge from positive to negative in synchrony with the movement of the particle through them (Figure 24.6).

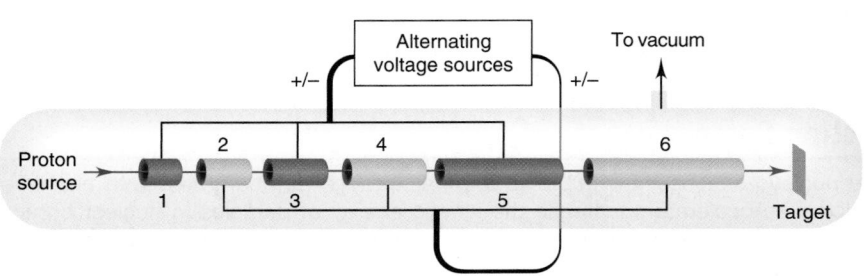

Figure 24.6 Schematic of a linear accelerator. The voltage of each tubular section is alternated, so that the positively charged particle (a proton here) is repelled from the section it is leaving and attracted to the section it is entering. As a result, the particle's speed is continually increased.

A proton, for example, exits the first tube just when that tube becomes positive and the next tube negative. Repelled by the first tube and attracted by the second, the proton accelerates across the gap between them. The process is performed in stages to achieve high particle energies without having to apply a single high voltage. A 40-ft linear accelerator with 46 tubes, built in California after World War II, accelerated protons to speeds several million times faster than earlier accelerators.

The *cyclotron* (Figure 24.7), invented by E. O. Lawrence in 1930, applies the principle of the linear accelerator but uses electromagnets to give the particle a spiral path, thus saving space. The magnets lie within an evacuated chamber above and below two "dees," open, D-shaped electrodes that act like the tubes in the linear design. The particle is accelerated as it passes from one dee, which is momentarily positive, to the other, which is momentarily negative. Its speed and radius increase until it is deflected toward the target nucleus.

Figure 24.7 Schematic of a cyclotron accelerator.

Table 24.6	Formation of Some Transuranium Nuclides*				
Reaction					**Half-life of Product**
$^{239}_{94}\text{Pu}$	$+$ 2^{1}_{0}n	$\longrightarrow$	$^{241}_{95}\text{Am}$	$+$ $^{0}_{-1}\beta$	432 yr
$^{239}_{94}\text{Pu}$	$+$ $^{4}_{2}\alpha$	$\longrightarrow$	$^{242}_{96}\text{Cm}$	$+$ $^{1}_{0}\text{n}$	163 days
$^{241}_{95}\text{Am}$	$+$ $^{4}_{2}\alpha$	$\longrightarrow$	$^{243}_{97}\text{Bk}$	$+$ 2^{1}_{0}n	4.5 h
$^{242}_{96}\text{Cm}$	$+$ $^{4}_{2}\alpha$	$\longrightarrow$	$^{245}_{98}\text{Cf}$	$+$ $^{1}_{0}\text{n}$	45 min
$^{253}_{99}\text{Es}$	$+$ $^{4}_{2}\alpha$	$\longrightarrow$	$^{256}_{101}\text{Md}$	$+$ $^{1}_{0}\text{n}$	76 min
$^{243}_{95}\text{Am}$	$+$ $^{18}_{8}\text{O}$	$\longrightarrow$	$^{256}_{103}\text{Lr}$	$+$ 5^{1}_{0}n	28 s

*Like chemical reactions, nuclear reactions may occur in several steps. For example, the first reaction here is actually an overall process that occurs in three steps:
(1) $^{239}_{94}\text{Pu} + ^{1}_{0}\text{n} \longrightarrow ^{240}_{94}\text{Pu}$ (2) $^{240}_{94}\text{Pu} + ^{1}_{0}\text{n} \longrightarrow ^{241}_{94}\text{Pu}$ (3) $^{241}_{94}\text{Pu} \longrightarrow ^{241}_{95}\text{Am} + ^{0}_{-1}\beta$

● **The Powerful Bevatron** The *bevatron,* used to study the physics of high-energy particle collisions, includes a linear section and a synchrotron section. The instrument at the Lawrence Berkeley Laboratory in California increases the kinetic energy of the particles by a factor of more than 6 billion. A beam of 10^{10} protons makes more than 4 million revolutions, a distance of 300,000 miles, in 1.8 s, attaining a final speed about 90% of the speed of light! Even more powerful accelerators are in use at the Fermi Lab near Chicago and at CERN, near Geneva, Switzerland.

The *synchrotron* uses a synchronously increasing magnetic field to make the particle's path circular rather than spiral. ●

Scientists use accelerators for many applications, from producing radioisotopes used in medical applications to studying the fundamental nature of matter. Perhaps the most specific application for chemists is the synthesis of **transuranium elements,** those with atomic numbers higher than uranium, the heaviest naturally occurring element. Some reactions that were used to form several of these elements appear in Table 24.6. The transuranium elements include the remaining actinides ($Z = 93$ to 103), in which the $5f$ sublevel is being filled, and the elements in the fourth transition series ($Z = 104$ to 112), in which the $6d$ sublevel is being filled.) ●

Section Summary

One nucleus can be transmuted to another through bombardment with high-energy particles. • Accelerators increase the kinetic energy of particles in nuclear bombardment experiments and are used to produce transuranium elements and radioisotopes for medical use.

● **Naming the Transuranium Elements** The last naturally occurring element was named after Uranus, thought at the time to be the outermost planet, so the first two artificial elements were named after the more recently discovered Neptune and Pluto. The next few elements were named after famous scientists, as in curium, and places, as in americium. But conflicting claims of discovery by scientists in the U.S. and Soviet Union led to controversies about names for some of the elements. To provide interim names until the disputes could be settled, the International Union of Pure and Applied Chemistry (IUPAC) adopted a system that uses the atomic number as the basis for a Latin name. Thus, for example, element 104 was named unnilquadium (un = 1, nil = 0, quad = 4, ium = metal suffix), with the symbol Unq. After much compromise, the IUPAC has finalized these names: 104, rutherfordium (Rf); 105, dubnium (Db); 106, seaborgium (Sg); 107, bohrium (Bh); 108, hassium (Hs); 109, meitnerium (Mt); 110, darmstadtium (Ds), and 111, roentgenium (Rg). Elements with atomic numbers 112 and higher have not yet been named.

24.4 THE EFFECTS OF NUCLEAR RADIATION ON MATTER

In 1986, an accident at the Chernobyl nuclear facility in the former Soviet Union released radioactivity that, according to the World Health Organization, will cause thousands of cancer deaths. In the same year, isotopes used in medical treatment emitted radioactivity that prevented thousands of cancer deaths. In this section and Section 24.5, we examine radioactivity's harmful and beneficial effects.

The key to both of these effects is that *nuclear changes cause chemical changes in surrounding matter.* In other words, even though the nucleus of an atom may undergo a reaction with little or no involvement of the atom's electrons, the emissions from that reaction *do* affect the electrons of nearby atoms.

Virtually all radioactivity causes **ionization** in surrounding matter, as the emissions collide with atoms and dislodge electrons:

$$\text{Atom} \xrightarrow{\text{ionizing radiation}} \text{ion}^{+} + \text{e}^{-}$$

From each ionization event, a cation and a free electron result, and the number of such *cation-electron pairs* produced is directly related to the energy of the incoming **ionizing radiation.**

Many applications of ionizing radiation depend not only on the ionizing event itself, but also on secondary processes. For example in a Geiger-Müller counter, the free electron of the cation-electron pair often collides with another atom, which ejects a second electron, and in a scintillation counter, the inital ionization eventually results in emission of light (see Tools of the Laboratory essay, p. 1075). In nuclear power plants (Section 24.7), the inital process as well as several secondary processes cause the heating that makes steam to create electricity.

Effects of Ionizing Radiation on Living Matter

Ionizing radiation has a destructive effect on living tissue, and if the ionized atom is part of a key biological macromolecule or cell membrane, the results can be devastating to the cell and perhaps the organism.

The danger from a radionuclide depends on three factors: the type of radiation, its half-life, and, most importantly, its biochemical behavior. For example, both ^{235}U and ^{239}Pu emit α particles and have long half-lives, but uranium is rapidly excreted by the body, wheras plutonium behaves like calcium and, thus, is incorporated into bones and teeth. One of the most dangerous radionuclides is $^{90}Sr^{2+}$ (formed during a nuclear explosion) because it behaves very similarly to Ca^{2+} and is absorbed strongly by bones. Let's first see how to measure radiation dose and then delve into various aspects of these factors in more detail.

Units of Radiation Dose and Its Effects To measure the effects of ionizing radiation, we need a unit for radiation dose. Units of radioactive decay, such as the becquerel and curie, measure the number of decay events in a given time but not their energy or absorption by matter. The number of cation-electron pairs produced in a given amount of living tissue is a measure of the energy absorbed by the tissue. The SI unit for such energy absorption is the **gray (Gy)**; it is equal to 1 joule of energy absorbed per kilogram of body tissue: 1 Gy = 1 J/kg. A more widely used unit is the **rad (radiation-absorbed dose),** which is equal to 0.01 Gy:

$$1 \text{ rad} = 0.01 \text{ J/kg} = 0.01 \text{ Gy}$$

To measure actual tissue damage, we must account for differences in the strength of the radiation, the exposure time, and the type of tissue. To do this, we multiply the number of rads by a *relative biological effectiveness* (RBE) factor, which depends on the effect of a given type of radiation on a given tissue or body part. The product is the **rem (roentgen equivalent for man),** the unit of radiation dosage equivalent to a given amount of tissue damage in a human:

$$\text{no. of rems} = \text{no. of rads} \times \text{RBE}$$

Doses are often expressed in millirems (10^{-3} rem). The SI unit for dosage equivalent is the **sievert (Sv).** It is defined in the same way as the rem but with absorbed dose in grays; thus, 1 rem = 0.01 Sv.

Penetrating Power of Emissions The effect on living tissue of a radiation dose depends on the penetrating power *and* ionizing ability of the radiation. Figure 24.8 depicts the differences in penetrating power of the three common emissions. Note, in general, that *penetrating power is inversely related to the mass, charge, and energy of the emission.* In other words, if a particle interacts strongly with matter, it penetrates only slightly, and vice versa:

- α *Particles.* Alpha particles are massive and highly charged, which means that they interact with matter most strongly of the three common types of emissions. As a result, they penetrate so little that a piece of paper, light clothing, or the outer layer of skin can stop α radiation from an external source. Internally, however, such as from ingestion, an α emitter can cause grave localized damage through extensive ionization. ●

Figure 24.8 Penetrating power of radioactive emissions. Penetrating power is often measured in terms of the depth of water that stops 50% of the incoming radiation. (Water is the main component of living tissue.) Alpha particles, with the highest mass and charge, have the lowest penetrating power, and γ rays have the highest. (Average values of actual penetrating distances are shown.)

ARIS Presentation Center, Art Library

● **A Tragic Way to Tell Time in the Dark**
In the early 20th century, wristwatch and clock dials were painted by hand with paint containing radium so they would glow in the dark. To write numbers clearly, the young women hired to apply the paint "tipped" fine brushes repeatedly between their lips. Small amounts of ingested $^{226}Ra^{2+}$ were incorporated into the bones of the women, along with normal Ca^{2+}, which led to numerous cases of bone fracture and jaw cancer.

 Biological application

● **A Tale of Two Agents** The relative danger from α- and β⁻-emitting radionuclides is dramatically revealed in the poisoning of two former Russian intelligence agents. In 1957, an agent was poisoned by thallium-204, a β⁻ emitter, but he survived because β⁻ particles do relatively little radiation damage. In contrast, in 2006, another agent was poisoned with polonium-210, an α emitter, and died in three weeks because of the much greater radiation damage. The clandestine nature of this action was enhanced by the fact that ^{210}Po emits few γ rays, so it cannot be detected by airport and hotel screening devices.

Biological application

- *Beta Particles and Positrons.* Beta particles (β^-) and positrons (β^+) have less charge and much less mass than α particles, so they interact less strongly with matter. Even though a given particle has less chance of causing ionization, a β^- (or β^+) emitter is a more destructive external source because the particles penetrate deeper. Specialized heavy clothing or a thick (0.5 cm) piece of metal is required to stop these particles. ●
- *γ Rays.* Neutral, massless γ rays interact least with matter and, thus, penetrate most. A block of lead several inches thick is needed to stop them. Therefore, an external γ ray source is the most dangerous because the energy can ionize many layers of living tissue.

A more variable case occurs with emission of *neutrons,* which are neutral and massive. Outside a nucleus, neutrons decay with a half-life of 10.6 min into a proton and an electron (and an antineutrino). With energies ranging from very low to very high, they may be scattered or captured by interaction with nuclei.

Molecular Interactions How does the damage take place on the molecular level? When ionizing radiation interacts with a molecule, it causes the *loss of an electron from a bond or a lone pair.* The resulting charged species go on to form **free radicals,** molecular or atomic species with one or more unpaired electrons. As you've seen several times in earlier chapters, species with lone electrons are very reactive. They form electron pairs by attacking bonds in other molecules, sometimes forming more free radicals.

When γ radiation strikes biological tissue, for instance, the most likely molecule to absorb it is water, which forms an electron and a water ion-radical:

$$H_2O + \gamma \longrightarrow H_2O\cdot^+ + e^-$$

The $H_2O\cdot^+$ and e^- collide with more water to form more free radicals:

$$H_2O\cdot^+ + H_2O \longrightarrow H_3O^+ + \cdot OH \qquad \text{and} \qquad e^- + H_2O \longrightarrow H\cdot + OH^-$$

These free radicals go on to attack more water molecules and surrounding biomolecules, whose bonding and structure, as you know, are intimately connected with their function (Section 15.6).

The double bonds in membrane lipids are particularly susceptible to free-radical attack:

$$H\cdot + RCH{=}CHR' \longrightarrow RCH_2{-}\dot{C}HR'$$

In this reaction, one electron of the π bond forms a C—H bond between one of the double-bonded carbons and the H·, and the other electron resides on the other carbon to form a free radical. Changes to lipid structure cause changes in membrane fluidity and other damage that, in turn, result in leakage through cell membranes and destruction of the protective fatty tissue around organs. Changes to critical bonds in enzymes lead to their malfunction as catalysts of metabolic reactions. Changes in the nucleic acids and proteins that govern the rate of cell division can cause cancer. Genetic damage and mutations may occur when bonds in the DNA of sperm and egg cells are altered by free radicals.

Sources of Ionizing Radiation

We are continuously exposed to ionizing radiation from natural and artificial sources (Table 24.7). Indeed, life evolved in the presence of natural ionizing radiation, called **background radiation.** The same radiation that alters bonds in DNA and causes harmful mutations also causes beneficial mutations that, over evolutionary time, allow species to change.

Background radiation has several sources. One source is *cosmic radiation,* which increases with altitude because of decreased absorption by the atmosphere. Thus, people in Denver absorb twice as much cosmic radiation as people in Los

Table 24.7	Typical Radiation Doses from Natural and Artificial Sources	
Source of Radiation		**Average Adult Exposure**
Natural		
Cosmic radiation		30–50 mrem/yr
Radiation from the ground		
From clay soil and rocks		~25–170 mrem/yr
In wooden houses		10–20 mrem/yr
In brick houses		60–70 mrem/yr
In concrete (cinder block) houses		60–160 mrem/yr
Radiation from the air (mainly radon)		
Outdoors, average value		20 mrem/yr
In wooden houses		70 mrem/yr
In brick houses		130 mrem/yr
In concrete (cinder block) houses		260 mrem/yr
Internal radiation from minerals in tap water and daily intake of food (^{40}K, ^{14}C, Ra)		~40 mrem/yr
Artificial		
Diagnostic x-ray methods		
Lung (local)		0.04–0.2 rad/film
Kidney (local)		1.5–3 rad/film
Dental (dose to the skin)		≤1 rad/film
Therapeutic radiation treatment		Locally ≤ 10,000 rad
Other sources		
Jet flight (4 h)		~1 mrem
Nuclear testing		<4 mrem/yr
Nuclear power industry		<1 mrem/yr
Total average value		100–200 mrem/yr

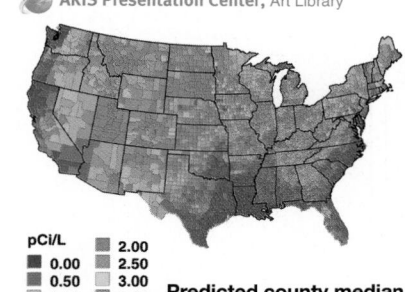

pCi/L 2.00
0.00 2.50
0.50 3.00
1.00 3.50   **Predicted county median**
1.50 4.00   **concentration**

Angeles; even a jet flight involves measurable absorption. The sources of most background radiation are thorium and uranium minerals present in rocks and soil. Radon, the heaviest noble gas [Group 8A(18)], is a radioactive product of uranium and thorium decay. Its concentration in the air we breathe is associated with certain common building materials and types of local soil and rocks. Its decay products cause most of the damage. ● About 150 g of K^+ ions is dissolved in the water in the tissues of an average adult, and 0.0118% of these ions are radioactive ^{40}K. The presence of these substances and of atmospheric $^{14}CO_2$ makes all food, water, clothing, and building materials slightly radioactive.

The largest artificial source of radiation, and the one that is easiest to control, is from medical diagnostic techniques, especially x-rays. The radiation dosage from nuclear testing and radioactive waste disposal is miniscule for most people, but exposures for those living near test sites or disposal areas may be much higher.

Assessing the Risk from Ionizing Radiation

How much radiation is too much? To approach this question, we must ask several others: How strong is the exposure? How long is the exposure? Which tissue is exposed? Are children affected? One reason we lack clear data to answer these questions is that scientific ethical standards forbid the intentional exposure of humans in an experimental setting. However, accidentally exposed radiation workers and Japanese atomic bomb survivors have been studied extensively.

● **Risk of Radon** Radon (Rn; $Z = 86$), the largest noble gas, is a natural decay product of uranium. Therefore, the uranium content of the local soil and rocks is critical, but radon occurs everywhere in varying concentrations. It exists in especially high amounts in cinder blocks made from the waste ash sold by power plants that burn coal high in uranium. Radon itself serves as the volatile medium for entering the body, where it poses a serious potential hazard as it decays to radioactive nuclides of Po, Pb, and Bi, through α, $β^-$, and γ emission. The emissions damage lung tissue, and the heavy-metal atoms formed aggravate the problem. Recent EPA estimates indicate that radon contributes to 15% of annual lung cancer deaths.

 Environmental application

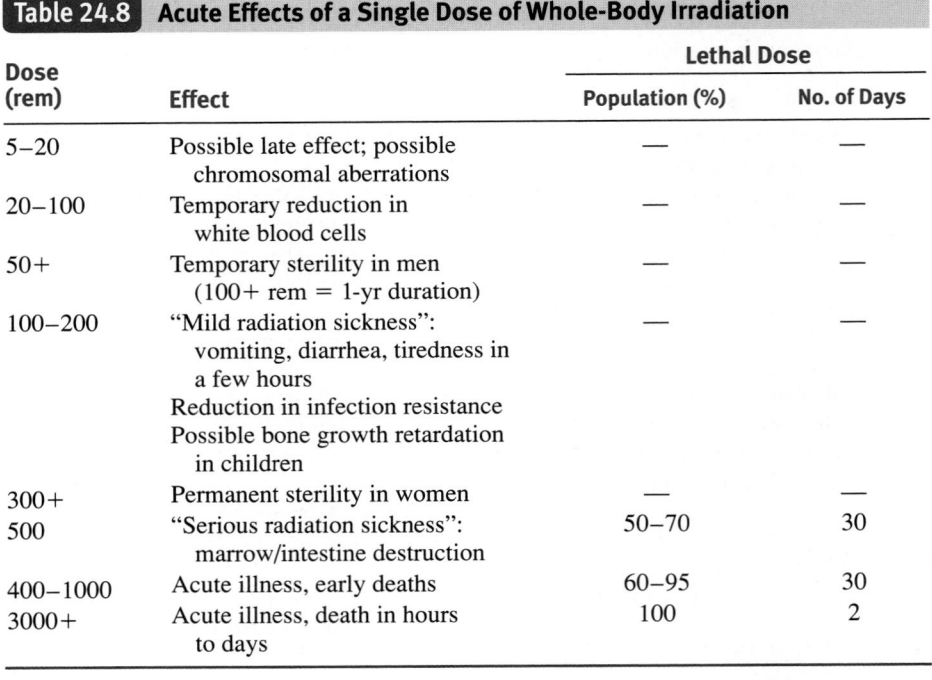

Table 24.8	Acute Effects of a Single Dose of Whole-Body Irradiation		
		Lethal Dose	
Dose (rem)	**Effect**	**Population (%)**	**No. of Days**
5–20	Possible late effect; possible chromosomal aberrations	—	—
20–100	Temporary reduction in white blood cells	—	—
50+	Temporary sterility in men (100+ rem = 1-yr duration)	—	—
100–200	"Mild radiation sickness": vomiting, diarrhea, tiredness in a few hours Reduction in infection resistance Possible bone growth retardation in children	—	—
300+	Permanent sterility in women	—	—
500	"Serious radiation sickness": marrow/intestine destruction	50–70	30
400–1000	Acute illness, early deaths	60–95	30
3000+	Acute illness, death in hours to days	100	2

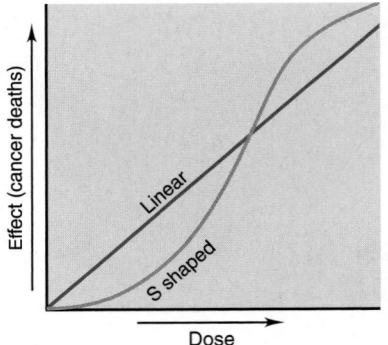

● **Modeling Radiation Risk** There are two current models of effect vs. dose. The *linear response* model proposes that radiation effects, such as cancer risks, accumulate over time regardless of dose and that populations should not be exposed to any radiation above background levels. The *S-shaped response* model implies there is a threshhold above which the effects are more significant: very low risk at low dose and high risk at high dose. If the linear model is more accurate, we should limit all excess exposure, but this would severely restrict medical diagnosis, military testing, and energy production.

Table 24.8 summarizes the immediate tissue effects on humans of an acute single dose of ionizing radiation to the whole body. The severity of the effects increases with dose; a dose of 500 rem will kill about 50% of the exposed population within a month.

Most data come from laboratory animals, whose biological systems may differ greatly from ours. Nevertheless, studies with mice and dogs show that lesions and cancers appear after massive whole-body exposure, with rapidly dividing cells affected first. In an adult animal, these are cells of the bone marrow, organ linings, and reproductive organs, but many other tissues are affected in an immature animal or fetus. Studies in both animals and humans show an increase in the incidence of cancer from either a high, single exposure or a low, chronic exposure.

Reliable data on genetic effects are few. Pioneering studies on fruit flies show a linear increase in genetic defects with both dose and exposure time. However, in the mouse, whose genetic system is obviously much more similar to ours than is the fruit fly's, a total dose given over a long period created one-third as many genetic defects as the same dose given over a short period. Therefore, rate of exposure is a key factor. The children of atomic bomb survivors show higher-than-normal childhood cancer rates, implying that their parents' reproductive systems were affected. ●

Section Summary

All radioactive emissions cause ionization. • The effect of ionizing radiation on living matter depends on the quantity of energy absorbed and the extent of ionization in a given type of tissue. Radiation dose for the human body is measured in rem. • Ionization forms free radicals, some of which proliferate and destroy biomolecular function. • All organisms are exposed to varying quantities of natural ionizing radiation. • Studies show that either a large acute dose or a chronic small dose is harmful.

24.5 APPLICATIONS OF RADIOISOTOPES

Our ability to detect minute amounts of radioisotopes makes them powerful tools for studying processes in biochemistry, medicine, materials science, environmental studies, and many other scientific and industrial fields. Such uses depend on the fact that *isotopes of an element exhibit **very** similar chemical and physical behavior.* In other words, except for having a less stable nucleus, a radioisotope has nearly the same chemical properties as a nonradioactive isotope of that element.* For example, the fact that $^{14}CO_2$ is utilized by a plant in the same way as $^{12}CO_2$ forms the basis of radiocarbon dating.

Radioactive Tracers

Just think how useful it could be to follow a substance through a complex process or from one region of a system to another. A tiny amount of a radioisotope mixed with a large amount of the stable isotope can act as a **tracer,** a chemical "beacon" emitting radiation that signals the presence of the substance.

Reaction Pathways Tracers help us choose from among possible reaction pathways. One well-studied example is the reaction between periodate and iodide ions:

$$IO_4^-(aq) + 2I^-(aq) + H_2O(l) \longrightarrow I_2(s) + IO_3^-(aq) + 2OH^-(aq)$$

Is IO_3^- the result of IO_4^- reduction or I^- oxidation? When we add "cold" (nonradioactive) IO_4^- to a solution of I^- that contains some "hot" (radioactive, indicated in red) $^{131}I^-$, we find that the I_2 is radioactive, not the IO_3^-:

$$IO_4^-(aq) + 2^{131}I^-(aq) + H_2O(l) \longrightarrow {}^{131}I_2(s) + IO_3^-(aq) + 2OH^-(aq)$$

These results show that IO_3^- forms through the reduction of IO_4^-, and that I_2 forms through the oxidation of I^-. To confirm this pathway, we add IO_4^- containing some $^{131}IO_4^-$ to a solution of I^-. As we expected, the IO_3^- is radioactive, not the I_2:

$$^{131}IO_4^-(aq) + 2I^-(aq) + H_2O(l) \longrightarrow I_2(s) + {}^{131}IO_3^-(aq) + 2OH^-(aq)$$

Thus, tracers act like "handles" we can "hold" to follow the changing reactants.

Far more complex pathways can be followed with tracers as well. The photosynthetic pathway, the most essential and widespread metabolic process on Earth, in which energy from sunlight is used to form the chemical bonds of glucose, has an overall reaction that looks quite simple:

$$6CO_2(g) + 6H_2O(l) \xrightarrow[\text{chlorophyll}]{\text{light}} C_6H_{12}O_6(s) + 6O_2(g)$$

However, the actual process is extremely complex: 13 enzyme-catalyzed steps are required to incorporate each C atom from CO_2, so the six CO_2 molecules incorporated to form a molecule of $C_6H_{12}O_6$ require six repetitions of the pathway. Melvin Calvin and his coworkers took seven years to determine the pathway, using ^{14}C in CO_2 as the tracer and paper chromatography as the means of separating the products formed after different times of light exposure. Calvin won the Nobel Prize in chemistry in 1961 for this remarkable achievement.

Tracers are used in many studies of biological function. Recently, future opportunities for living in outer space have required answers to new questions. In an animal study of red blood cell loss during extended space flight, blood plasma

Literature Yalow, Rosalyn S. "Development and Proliferation of Radioimmunoassay Technology," *J. Chem. Educ.* **1999,** *76,* 767.

Biological application

---

*Although this statement is generally correct, differences in isotopic mass *can* influence bond strengths and therefore reaction rates. Such behavior is called a *kinetic isotope effect* and is particularly important for isotopes of hydrogen—^1H, ^2H, and ^3H—because their masses differ by such large proportions. Section 22.4 discussed how the kinetic isotope effect is employed in the industrial production of heavy water, D_2O.

volume was measured with ^{125}I-labeled albumin (a blood protein), and red blood cells labeled with ^{51}Cr as a trace substitute for Fe were used to assess survival of blood cells. In another study, blood flow in skin under long periods of microgravity was monitored using injected ^{133}Xe.

Engineering application

Material Flow Tracers are used in studies of solid surfaces and the flow of materials. Metal atoms hundreds of layers deep within a solid have been shown to exchange with metal ions from the surrounding solution within a matter of minutes. Chemists and engineers use tracers to study material movement in semiconductor chips, paint, and metal plating, in detergent action, and in the process of corrosion, to mention just a few of many applications.

Hydrologic engineers use tracers to study the volume and flow of large bodies of water. By following radionuclides that formed during atmospheric nuclear bomb tests (^{3}H in H_2O, $^{90}Sr^{2+}$, and $^{137}Cs^{+}$), scientists have mapped the flow of water from land to lakes and streams to oceans. They also use tracers to study the surface and deep ocean currents that circulate around the globe, as well as the mechanisms of hurricane formation and the mixing of the troposphere and stratosphere. Industries employ tracers to study material flow during the manufacturing process, such as the flow of ore pellets in smelting kilns, the paths of wood chips and bleach in paper mills, the diffusion of fungicide into lumber, and in a particularly important application, the porosity and leakage of oil and gas wells in geological formations.

Activation Analysis Another use of tracers is in *neutron activation analysis* (NAA). In this method, neutrons bombard a nonradioactive sample, converting a small fraction of its atoms to radioisotopes, which exhibit characteristic decay patterns, such as γ-ray spectra, that reveal the elements present. Unlike chemical analysis, NAA leaves the sample virtually intact, so the method can be used to determine the composition of a valuable object or a very small sample. For example, a painting thought to be a 16th-century Dutch masterpiece was shown through NAA to be a 20th-century forgery, because a microgram-sized sample of its pigment contained much less silver and antimony than the pigments used by the Dutch masters. Forensic chemists use NAA to detect traces of ammunition on a suspect's hand or traces of arsenic in the hair of a victim of poisoning.

Engineering application

Automotive engineers employ NAA and γ-ray detectors to measure friction and wear of moving parts without having to take the engine apart. For example, when a steel surface that has been neutron-activated to form some radioactive ^{59}Fe moves against a second steel surface, the amount of radioactivity on the second surface indicates the amount of material rubbing off. The radioactivity appearing in a lubricant placed between the surfaces can demonstrate the lubricant's ability to reduce wear.

Biological application

Medical Diagnosis The largest use of radioisotopes is in medical science. In fact, over 25% of U.S. hospital admissions are for diagnoses based on data from radioisotopes. Tracers with half-lives of a few minutes to a few days are employed to observe specific organs and body parts. For example, a healthy thyroid gland incorporates dietary I$^-$ into iodine-containing hormones at a known rate. To assess thyroid function, the patient drinks a solution containing a trace amount of Na^{131}I, and a scanning monitor follows the uptake of ^{131}I$^-$ into the thyroid (Figure 24.9A). Technetium-99 ($Z = 43$) is also used for imaging the thyroid (Figure 24.9B), as well as the heart, lungs, and liver. Technetium does not occur naturally, so the radioisotope (actually a metastable form, ^{99m}Tc) is prepared from radioactive molybdenum just before use:

$$^{99}_{42}\text{Mo} \longrightarrow \, ^{99m}_{43}\text{Tc} + \, ^{0}_{-1}\beta$$

Figure 24.9 The use of radioisotopes to image the thyroid gland. Thyroid scanning is used to assess nutritional deficiencies, inflammation, tumor growth, and other thyroid-related ailments. **A,** In ^{131}I scanning, the thyroid gland absorbs ^{131}I$^-$ ions, which undergo β^- decay that exposes a photographic film. The asymmetric image indicates disease. **B,** A ^{99}Tc scan of a healthy thyroid.

Biological application

ARIS Presentation Center, Photo Library

Tracers are also used to measure physiological processes, such as blood flow. The rate at which the heart pumps blood, for example, can be observed by injecting ^{59}Fe, which becomes incorporated into the hemoglobin of blood cells. Several radioisotopes used in medical diagnosis are listed in Table 24.9.

Positron-emission tomography (PET) is a powerful imaging method for observing brain structure and function. A biological substance is synthesized with one of its atoms replaced by an isotope that emits positrons. The substance is injected into a patient's bloodstream, from which it is taken up into the brain. The isotope emits positrons, each of which annihilates a nearby electron. In this process, two γ photons are emitted simultaneously 180° from each other:

$$^{0}_{1}\beta + ^{0}_{-1}e \longrightarrow 2^{0}_{0}\gamma$$

An array of detectors around the patient's head pinpoints the sites of γ emission, and the image is analyzed by computer. Two of the isotopes used are ^{15}O, injected as H$_2$^{15}O to measure blood flow, and ^{18}F bonded to a glucose analog to measure glucose uptake, which is an indicator of energy metabolism. Among many fascinating PET findings are those that show how changes in blood flow and glucose uptake accompany normal or abnormal brain activity (Figure 24.10). In a recent nonmedical development, substances incorporating ^{11}C and ^{15}O are being investigated using PET to learn how molecules interact with and move along the surface of a catalyst.

Table 24.9   Some Radioisotopes Used as Medical Tracers

Isotope	Body Part or Process
^{11}C, ^{18}F, ^{13}N, ^{15}O	PET studies of brain, heart
^{60}Co, ^{192}Ir	Cancer therapy
^{64}Cu	Metabolism of copper
^{59}Fe	Blood flow, spleen
^{67}Ga	Tumor imaging
^{123}I, ^{131}I	Thyroid
^{111}In	Brain, colon
^{42}K	Blood flow
^{81m}Kr	Lung
^{99m}Tc	Heart, thyroid, liver, lung, bone
^{201}Tl	Heart muscle
^{90}Y	Cancer, arthritis

ARIS Presentation Center, Tables Library

Animation: Nuclear Medicine
ARIS Presentation Center
Chapter 24

Figure 24.10 PET and brain activity. These PET scans show brain activity in a normal person (*left*) and in a patient with Alzheimer's disease (*right*). Red and yellow indicate relatively high activity within a region.

ARIS Presentation Center, Photo Library

Additional Applications of Ionizing Radiation

Biological application

To be used as a tracer, a radioisotope need only emit radiation that can be detected. Many other uses of radioisotopes depend on ionizing radiation of higher energy.

Cancer cells divide more rapidly than normal cells, so radioisotopes that interfere with cell division kill more cancer cells than normal ones. Implants of ^{198}Au or of ^{90}Sr, which decays to the γ-emitting ^{90}Y, have been used to destroy pituitary and breast tumor cells, and γ rays from ^{60}Co have been used to destroy tumors of the brain and other body parts.

Irradiation of food increases shelf life by killing microorganisms that cause rotting or spoilage (Figure 24.11), but the practice is quite controversial. Advocates point to the benefits of preserving fresh foods, grains, and seeds for long periods, whereas opponents suggest that irradiation might lower the food's nutritional content or produce harmful byproducts. Irradiation also provides a way to destroy newer, more resistant bacterial strains that survive the increasing use of the more common antibiotics in animal feed. The United Nations has approved irradiation for potatoes, wheat, chicken, and strawberries, and, in 2003, the U.S. Food and Drug Administration approved it as well.

Ionizing radiation has been used to control harmful insects. Captured males are sterilized by radiation and released to mate, thereby reducing the number of offspring. This method has been used to control the Mediterranean fruit fly in California and disease-causing insects, such as the tsetse fly and malarial mosquito, in other parts of the world.

A recent, nonharmful use of ionizing radiation relies on its secondary processes. Most spacecraft use solar energy to provide power for instruments. In deep-space missions, when solar energy is too weak, however, a radioisotope heater unit (RHU) is used. It consists of a ^{238}PuO$_2$ fuel pellet the size of a pencil eraser that is clad within a multilayered graphite and metal shell a little smaller than a flashlight battery. The whole RHU weighs only 40 g.

Nonirradiated   Irradiated

Figure 24.11 The increased shelf life of irradiated food.

ARIS Presentation Center, Photo Library

Section Summary

Radioisotopic tracers have been used to study reaction mechanisms, material flow, elemental composition, and medical conditions. • Ionizing radiation has been used in devices that destroy cancer tissue, kill organisms that spoil food, control insect populations, and power spacecraft instruments.

Literature Treptow, Richard S. "*E = mc²* for the Chemist: When Is Mass Conserved?" *J. Chem. Educ.* 2005, *82*, 1636.

24.6 THE INTERCONVERSION OF MASS AND ENERGY

Most of the nuclear processes we've considered so far have involved radioactive decay, in which a nucleus emits one or a few small particles or photons to become a more stable, slightly lighter nucleus. Two other nuclear processes cause much greater mass changes. In nuclear **fission,** a heavy nucleus splits into two much lighter nuclei, emitting several small particles at the same time. In nuclear **fusion,** the opposite process occurs: two lighter nuclei combine to form a heavier one. Both fission and fusion release enormous quantities of energy. Let's take a look at the origins of this energy by first examining the change in mass that accompanies the breakup of a nucleus into its nucleons and then considering the energy that is equivalent to this mass change.

The Mass Difference Between a Nucleus and Its Nucleons

For most of the 20th century, we have known that mass and energy are interconvertible. Thus, the separate mass and energy conservation laws are combined to state that *the total quantity of mass-energy in the universe is constant.* Therefore, when *any* reacting system releases or absorbs energy, it also loses or gains mass.

This relation between mass and energy is not important for chemical reactions because the energy changes involved in breaking or forming chemical bonds are so small that the mass changes are negligible. For example, when 1 mol of water breaks up into its atoms, heat is absorbed and we have:

$$H_2O(g) \longrightarrow 2H(g) + O(g) \qquad \Delta H^{\circ}_{rxn} = 2 \times \text{BE of O—H} = 934 \text{ kJ}$$

We find the mass that is equivalent to this energy from *Einstein's equation:*

$$E = mc^2 \qquad \text{or} \qquad \Delta E = \Delta mc^2 \qquad \text{so} \qquad \Delta m = \frac{\Delta E}{c^2} \qquad (24.8)$$

where Δm is the mass difference between reactants and products:

$$\Delta m = m_{products} - m_{reactants}$$

Substituting the heat of reaction (in J/mol) for ΔE and the numerical value for c (2.9979×10^8 m/s), we obtain

$$\Delta m = \frac{9.34 \times 10^5 \text{ J/mol}}{(2.9979 \times 10^8 \text{ m/s})^2} = 1.04 \times 10^{-11} \text{ kg/mol} = 1.04 \times 10^{-8} \text{ g/mol}$$

(Units of kg/mol are obtained because the joule includes the kilogram: 1 J = 1 kg·m²/s².) The mass of 1 mol of H_2O molecules (reactant) is about 10 ng *less* than the combined masses of 2 mol of H atoms and 1 mol of O atoms (products), a change difficult to measure with even the most sophisticated balance. Such minute mass changes when bonds break or form allow us to assume that, for all practical purposes, mass is conserved in *chemical* reactions.

The much larger mass change that accompanies a *nuclear* process is related to the enormous energy required to bind the nucleus together from its parts. In an analogy with the calculation above involving the water molecule, consider the change in mass that occurs when one ^{12}C nucleus breaks apart into its nucleons—six protons and six neutrons:

$$^{12}C \longrightarrow 6^1_1p + 6^1_0n$$

We calculate this mass difference in a special way. By combining the mass of six H *atoms* and six neutrons and then subtracting the mass of one ^{12}C *atom,* the masses of the electrons cancel: six e⁻ (in six ^{1}H atoms) cancel six e⁻ (in one ^{12}C atom). The mass of one ^{1}H atom is 1.007825 amu, and the mass of one neutron is 1.008665 amu, so

$$\text{Mass of six } ^1\text{H atoms} = 6 \times 1.007825 \text{ amu} = 6.046950 \text{ amu}$$
$$\underline{\text{Mass of six neutrons} = 6 \times 1.008665 \text{ amu} = 6.051990 \text{ amu}}$$
$$\text{Total mass} = 12.098940 \text{ amu}$$

The mass of the reactant, one ^{12}C atom, is 12 amu (exactly). The mass difference (Δm) we obtain is the total mass of the nucleons minus the mass of the nucleus:

$$\Delta m = 12.098940 \text{ amu} - 12.000000 \text{ amu}$$
$$= 0.098940 \text{ amu}/^{12}\text{C} = 0.098940 \text{ g/mol } ^{12}\text{C}$$

First, and most important, note that *the mass of the nucleus is **less** than the combined masses of its nucleons:* there is *always* a mass decrease when nucleons form a nucleus. Second, note that the mass change of this nuclear process (9.89×10^{-2} g/mol) is nearly 10 million times that of the chemical process (10.4×10^{-9} g/mol) we saw earlier and easily observed on any laboratory balance.

Nuclear Binding Energy and the Binding Energy per Nucleon

Once again, Einstein's equation for the relation between mass and energy allows us to find the energy equivalent of the mass change. For ^{12}C, after converting grams to kilograms, we have

$$\Delta E = \Delta mc^2 = (9.8940 \times 10^{-5} \text{ kg/mol})(2.9979 \times 10^8 \text{ m/s})^2$$
$$= 8.8921 \times 10^{12} \text{ J/mol} = 8.8921 \times 10^9 \text{ kJ/mol}$$

This quantity of energy is called the **nuclear binding energy** for carbon-12, and the positive value means *energy is absorbed*. The nuclear binding energy is the energy required to break 1 mol of atoms into neutrons and hydrogen atoms, which equals the energy to *break 1 mol of nuclei into individual nucleons:*

$$\text{Nucleus} + \text{nuclear binding energy} \longrightarrow \text{nucleons}$$

Thus, the nuclear binding energy is qualitatively analogous to the sum of bond energies of a covalent compound or the lattice energy of an ionic compound. But, quantitatively, nuclear binding energies are typically several million times greater.

We use joules to express the binding energy per mole of nuclei, but the joule is much too large a unit to express the binding energy of a single nucleus. Instead, nuclear scientists use the **electron volt (eV),** the energy an electron acquires when it moves through a potential difference of 1 volt:

$$1 \text{ eV} = 1.602 \times 10^{-19} \text{ J}$$

Binding energies are commonly expressed in millions of electron volts, that is, in *mega–electron volts* (MeV):

$$1 \text{ MeV} = 10^{6} \text{ eV} = 1.602 \times 10^{-13} \text{ J}$$

A particularly useful factor converts the atomic mass unit to its energy equivalent in electron volts:

$$1 \text{ amu} = 931.5 \times 10^{6} \text{ eV} = 931.5 \text{ MeV} \qquad \textbf{(24.9)}$$

Earlier we found the mass change when ^{12}C breaks apart into its nucleons to be 0.098940 amu. The binding energy per ^{12}C nucleus, expressed in MeV, is

$$\frac{\text{Binding energy}}{^{12}\text{C nucleus}} = 0.098940 \text{ amu} \times \frac{931.5 \text{ MeV}}{1 \text{ amu}} = 92.16 \text{ MeV}$$

We can compare the stability of nuclides of different elements by determining the *binding energy per nucleon.* For ^{12}C, we have

$$\text{Binding energy per nucleon} = \frac{\text{binding energy}}{\text{no. of nucleons}} = \frac{92.16 \text{ MeV}}{12 \text{ nucleons}} = 7.680 \text{ MeV/nucleon}$$

SAMPLE PROBLEM 24.6 Calculating the Binding Energy per Nucleon

PROBLEM Iron-56 is an extremely stable nuclide. Compute the binding energy per nucleon for ^{56}Fe and compare it with that for ^{12}C (mass of ^{56}Fe atom = 55.934939 amu; mass of ^{1}H atom = 1.007825 amu; mass of neutron = 1.008665 amu).

PLAN Iron-56 has 26 protons and 30 neutrons. We calculate the mass difference, Δm, when the nucleus forms by subtracting the given mass of one ^{56}Fe atom from the sum of the masses of 26 ^{1}H atoms and 30 neutrons. To find the binding energy per nucleon, we multiply Δm by the equivalent in MeV (931.5 MeV/amu) and divide by the number of nucleons (56).

SOLUTION Calculating the mass difference, Δm:

$$
\begin{aligned}
\text{Mass difference} &= [(26 \times \text{mass } ^{1}\text{H atom}) + (30 \times \text{mass neutron})] - \text{mass } ^{56}\text{Fe atom} \\
&= [(26)(1.007825 \text{ amu}) + (30)(1.008665 \text{ amu})] - 55.934939 \text{ amu} \\
&= 0.52846 \text{ amu}
\end{aligned}
$$

Calculating the binding energy per nucleon:

$$\text{Binding energy per nucleon} = \frac{0.52846 \text{ amu} \times 931.5 \text{ MeV/amu}}{56 \text{ nucleons}} = 8.790 \text{ MeV/nucleon}$$

An ^{56}Fe nucleus would require more energy per nucleon to break up into its nucleons than would ^{12}C (7.680 MeV/nucleon), so ^{56}Fe is more stable than ^{12}C.

CHECK The answer is consistent with the great stability of ^{56}Fe. Given the number of decimal places in the values, rounding to check the math is useful only to find a *major* error. The number of nucleons (56) is an exact number, so we retain four significant figures.

FOLLOW-UP PROBLEM 24.6 Uranium-235 is an essential component of the fuel in nuclear power plants. Calculate the binding energy per nucleon for ^{235}U. Is this nuclide more or less stable than ^{12}C (mass of ^{235}U atom = 235.043924 amu)?

Figure 24.12 The variation in binding energy per nucleon. A plot of the binding energy per nucleon vs. mass number shows that nuclear stability is greatest in the region near ^{56}Fe. Lighter nuclei may undergo fusion to become more stable; heavier ones may undergo fission. Note the exceptional stability of ^{4}He among extremely light nuclei.

Transparency/ARIS Presentation Center, Art Library

Fission or Fusion: Means of Increasing the Binding Energy Per Nucleon Calculations similar to those in Sample Problem 24.6 for other nuclides show that the binding energy per nucleon varies considerably. The essential point is that *the greater the binding energy per nucleon, the more stable the nuclide.*

Figure 24.12 shows a plot of the binding energy per nucleon vs. mass number. It provides information about nuclide stability and the two possible processes nuclides can undergo to form more stable nuclides. Most nuclides with fewer than 10 nucleons have a relatively small binding energy per nucleon. The ^{4}He nucleus is an exception—it is stable enough to be emitted intact as an α particle. Above $A = 12$, the binding energy per nucleon varies from about 7.6 to 8.8 MeV.

The most important observation is that *the binding energy per nucleon peaks at elements with $A \approx 60$.* In other words, nuclides become more stable with increasing mass number up to around 60 nucleons and then become less stable with higher numbers of nucleons. The existence of a peak of stability suggests that there are two ways nuclides can increase their binding energy per nucleon:

- *Fission.* A heavier nucleus can *split into lighter ones (closer to $A \approx 60$)* by undergoing fission. The product nuclei have greater binding energy per nucleon (are more stable) than the reactant nucleus, and the difference in *energy is released.* Nuclear power plants generate energy through fission, as do atomic bombs (Section 24.7).
- *Fusion.* Lighter nuclei, on the other hand, can *combine to form a heavier one (closer to $A \approx 60$)* by undergoing fusion. Once again, the product is more stable than the reactants, and *energy is released.* The Sun and other stars generate energy through fusion, as do thermonuclear (hydrogen) bombs. In these examples and in all current research efforts for developing fusion as a useful energy source, hydrogen nuclei fuse to form the very stable helium-4 nucleus.

In Section 24.7, we examine fission and fusion and the industrial energy facilities designed to utilize them.

Section Summary

The mass of a nucleus is less than the sum of the masses of its nucleons. The energy equivalent to this mass difference is the nuclear binding energy, often expressed in units of MeV. • The binding energy per nucleon is a measure of nuclide stability and varies with the number of nucleons. Nuclides with $A \approx 60$ are most stable. • Lighter nuclides join (fusion) or heavier nuclides split (fission) to create more stable products.

24.7 APPLICATIONS OF FISSION AND FUSION

Of the many beneficial applications of nuclear reactions, the greatest is the potential for almost limitless amounts of energy, which is based on the multimillion-fold increase in energy yield of nuclear reactions over chemical reactions. Our experience with nuclear energy from power plants in the late 20th century, however, has shown that we must improve ways to tap this energy source safely and economically and deal with the waste generated. In this section, we discuss how fission and fusion occur and how we are applying them.

The Process of Nuclear Fission

During the mid-1930s, Enrico Fermi and coworkers bombarded uranium ($Z = 92$) with neutrons in an attempt to synthesize transuranium elements. Many of the unstable nuclides produced were tentatively identified as having $Z > 92$, but other scientists were skeptical. Four years later, the German chemist Otto Hahn and his associate F. Strassmann showed that one of these unstable nuclides was an isotope of barium ($Z = 56$). The Austrian physicist Lise Meitner, a coworker of Hahn, and her nephew Otto Frisch proposed that barium resulted from the *splitting* of the uranium nucleus into *smaller* nuclei, a process that they called *fission* as an analogy to cell division in biology. ●

The ^{235}U nucleus can split in many different ways, giving rise to various daughter nuclei, but all routes have the same general features. Figure 24.13 depicts one of these fission patterns. Neutron bombardment results in a highly excited ^{236}U nucleus, which splits apart in 10^{-14} s. The products are two nuclei of unequal mass, two or three neutrons (average of 2.4), and a large quantity of energy. A single ^{235}U nucleus releases 3.5×10^{-11} J when it splits; 1 mol of ^{235}U (about $\frac{1}{2}$ lb) releases 2.1×10^{13} J—a billion times as much energy as burning $\frac{1}{2}$ lb of coal (about 2×10^4 J)!

We harness the energy of nuclear fission, much of which eventually appears as heat, by means of a **chain reaction,** illustrated in Figure 24.14: the two to three neutrons that are released by the fission of one nucleus collide with other fissionable nuclei and cause them to split, releasing more neutrons, which then col-

● **Lise Meitner (1878–1968)** Until very recently, this extraordinary physicist received little of the acclaim she deserved. Meitner worked in the laboratory of the chemist Otto Hahn, and she was responsible for the discovery of protactinium (Pa; $Z = 91$) and numerous radioisotopes. After leaving Germany in advance of the Nazi domination, Meitner proposed the correct explanation of nuclear fission. In 1944 Hahn received the Nobel Prize in chemistry, but he did not even acknowledge Meitner in his acceptance speech. Today, most physicists believe Meitner should have received the prize. Despite controversy over names for elements 104 to 109, it was widely agreed that element 109 should be named meitnerium.

ARIS Presentation Center, Photo Library

Literature Sime, Ruth Lewin. "Lise Meitner and the Discovery of Fission," *J. Chem. Educ.* **1989,** *66,* 373.

ARIS Presentation Center, Active Art Library

Figure 24.13 Induced fission of ^{235}U. A neutron bombarding a ^{235}U nucleus results in an extremely unstable ^{236}U nucleus, which becomes distorted in the act of splitting. In this case, which shows one of many possible splitting patterns, the products are ^{92}Kr and ^{141}Ba. Three neutrons and a great deal of energy are released also.

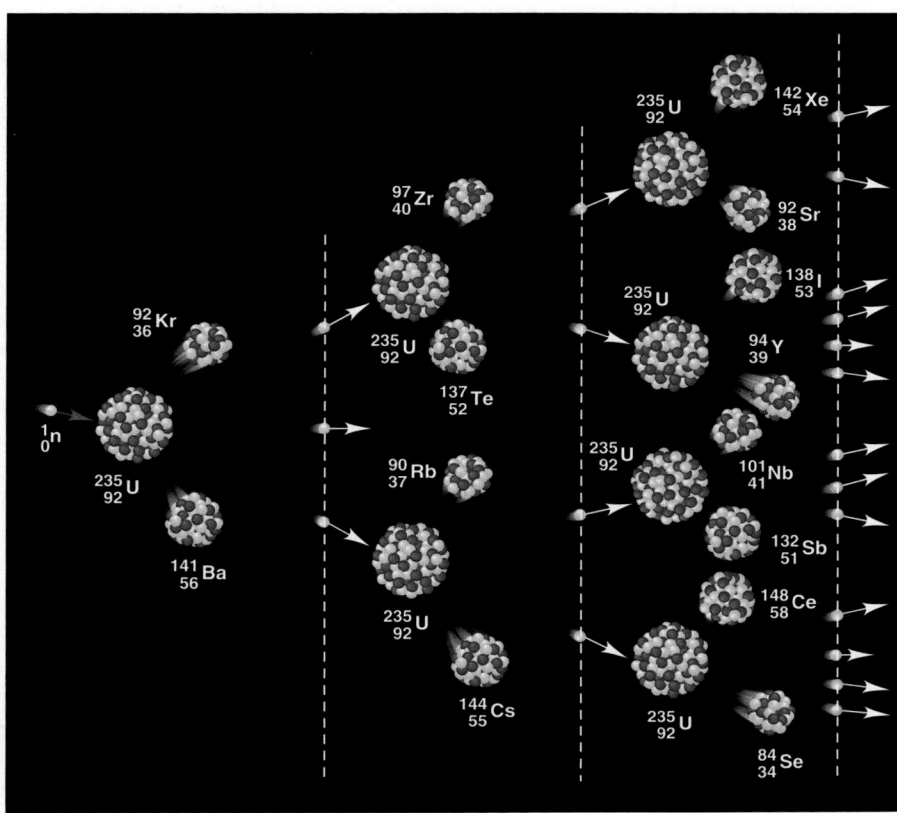

Figure 24.14 A chain reaction of ^{235}U. If a sample exceeds the critical mass, neutrons produced by the first fission event collide with other nuclei, causing their fission and the production of more neutrons to continue the process. Note that various product nuclei form. The vertical dashed lines identify succeeding "generations" of neutrons.

Transparency/ARIS Presentation Center, Art Library

lide with other nuclei, and so on, in a self-sustaining process. In this manner, the energy released increases rapidly because each fission event in a chain reaction releases two to three times as much energy as the preceding one.

Whether a chain reaction occurs depends on the mass (and thus the volume) of the fissionable sample. If the piece of uranium is large enough, the product neutrons strike another fissionable nucleus *before* flying out of the sample, and a chain reaction takes place. The mass required to achieve a chain reaction is called the **critical mass.** If the sample has less than the critical mass (a *subcritical mass*), too many product neutrons leave the sample before they collide with and cause the fission of another ^{235}U nucleus, and thus a chain reaction does not occur.

Uncontrolled Fission: The Atomic Bomb An uncontrolled chain reaction can be adapted to make an extremely powerful explosive, as several of the world's leading atomic physicists suspected just prior to the beginning of World War II. In August 1939, Albert Einstein wrote the president of the United States, Franklin Delano Roosevelt, to this effect, warning of the danger of allowing the Nazi government to develop this power first. It was this concern that led to the Manhattan Project, an enormous scientific effort to develop a bomb based on nuclear fission, which was initiated in 1941.* In August 1945, the United States detonated two atomic bombs over Japan, and the horrible destructive power of these bombs was a major factor in the surrender of the Japanese a few days later.

Literature Klotz, Irving M. "Captives of Their Fantasies: The German Atomic Bomb Scientists," *J. Chem. Educ.* **1997,** *74,* 204.

*For an excellent scientific and historical account of the development of the atomic bomb, see R. Rhodes, *The Making of the Atomic Bomb,* New York, Simon and Schuster, 1986.

Figure 24.15 An atomic bomb based on ^{235}U. Small TNT explosions bring subcritical masses together, and the chain reaction occurs.

Transparency/ARIS Presentation Center, Art Library

ARIS Presentation Center, Photo Library

Figure 24.16 A light-water nuclear reactor. A, Photo of a facility showing the concrete containment shell and nearby water source. **B,** Schematic of a light-water reactor.

Engineering application

In an atomic bomb, small explosions of trinitrotoluene (TNT) bring subcritical masses of fissionable material together to exceed the critical mass, and the ensuing chain reaction brings about the explosion (Figure 24.15). The proliferation of nuclear power plants, which use fissionable materials to generate energy for electricity, has increased concern that more countries (and unscrupulous individuals) may have access to such material for making bombs. Since the devastating terrorist attacks of September 11, 2001 in the United States, this concern has been heightened.

Controlled Fission: Nuclear Energy Reactors Like a coal-fired power plant, *a nuclear power plant generates heat to produce steam, which turns a turbine attached to an electric generator.* But a nuclear plant has the potential to produce electric power much more cleanly than can the combustion of coal.

Heat generation takes place in the **reactor core** of a nuclear plant (Figure 24.16). The core contains the *fuel rods,* which consist of fuel enclosed in tubes of a corrosion-resistant zirconium alloy. The fuel is uranium(IV) oxide (UO$_2$) that has been *enriched* from 0.7% ^{235}U, the natural abundance of this fissionable isotope, to the 3% to 4% ^{235}U required to sustain a chain reaction in a practical vol-

A

Containment shell

③ *Extremely hot water under high pressure passes into steam generator*

④ *Steam produced operates turbine-generator*

→ Electric power

Steam generator

Turbine   Generator

Reactor core

② *Control rods regulate rate of chain reaction*

Steam condenser

Moderator

⑤ *Cool water from nearby source condenses steam and is warmed*

① *Enriched uranium in fuel rods releases energy from fission*

Pump

Coolant water out

Pump

Pump

Coolant water in

B

ume. (Enrichment of nuclear fuel is the most important application of Graham's law; see the margin note, p. 215.) Sandwiched between the fuel rods are movable *control rods* made of cadmium or boron (or, in nuclear submarines, hafnium), substances that absorb neutrons very efficiently. When the control rods are moved between the fuel rods, the chain reaction slows because fewer neutrons are available to bombard uranium atoms; when they are removed, the chain reaction speeds up. Neutrons that leave the fuel-rod assembly collide with a *reflector,* usually made of a beryllium alloy, which absorbs very few neutrons. Reflecting the neutrons back to the fuel rods speeds the chain reaction.

Flowing around the fuel and control rods in the reactor core is the *moderator,* a substance that slows the neutrons, making them much better at causing fission than the fast ones emerging directly from the fission event. In most modern reactors, the moderator also acts as the *coolant,* the fluid that transfers the released heat to the steam-producing region. *Light-water reactors* use H_2O as the moderator because ^1H absorbs neutrons to some extent; in heavy-water reactors, D_2O is used. The advantage of D_2O is that it absorbs very few neutrons, leaving more available for fission, so heavy-water reactors can use uranium that has been *less enriched.* As the coolant flows around the encased fuel, pumps circulate it through coils that transfer its heat to the water reservoir. Steam formed in the reservoir turns the turbine that runs the generator. The steam is then condensed in large cooling towers (see Figure 13.22, p. 521) using water from a lake or river and returned to the water reservoir. ●

Some major accidents at nuclear plants have caused decidedly negative public reactions. In 1979, malfunctions of coolant pumps and valves at the Three-Mile Island facility in Pennsylvania led to melting of some of the fuel and damage to the reactor core, but the release of only a very small amount (about 1 Ci) of radioactive gases into the atmosphere. In 1986, a million times as much radioactivity (1 MCi) was released when a cooling system failure at the Chernobyl plant in Ukraine caused a much greater melting of fuel and an uncontrolled reaction. High-pressure steam and ignited graphite moderator rods caused the reactor building to explode and expel radioactive debris. Carried by prevailing winds, the radioactive particles contaminated vegetables and milk in much of Europe. And, as we said earlier, health officials have evidence that thousands of people living near the accident have already developed or may eventually develop cancer from radiation exposure. The design of the Chernobyl plant was particularly unsafe because, unlike reactors in the United States and western Europe, the reactor was not contained in a massive, concrete building.*

Despite potential safety problems, nuclear power remains an important source of electricity. In the late 1990s, nearly every European country employed nuclear power, and it is the major power source in some countries—Sweden creates 50% of its electricity this way and France almost 80%. Currently, the United States obtains about 20% of its electricity from nuclear power, and Canada slightly less. As our need for energy grows and climate change from fossil-fuel consumption continues, safer reactors are being designed and are likely to be built.

However, even a smoothly operating plant has certain inherent problems. The problem of *thermal pollution* is common to all power plants. Water used to condense the steam is several degrees warmer when returned to its source, which can harm aquatic organisms (Section 13.4). A more serious problem is *nuclear waste disposal.* Many of the fission products formed in nuclear reactors have long half-lives, and no satisfactory plan for their permanent disposal has yet been devised. Proposals to place the waste in containers and bury them in deep bedrock cannot

● **"Breeding" Nuclear Fuel** Uranium-235 is not an abundant isotope. One solution to a potential fuel shortage is a *breeder reactor,* designed to consume one type of nuclear fuel as it produces another. Outside the moderator, fuel rods are surrounded by natural U_3O_8, which contains 99.3% *nonfissionable* ^{238}U atoms. As neutrons formed during ^{235}U fission escape the fuel rod, they collide with ^{238}U, transmuting it into ^{239}Pu, another fissionable nucleus:

$$^{238}_{92}U + {}^1_0n \longrightarrow {}^{239}_{92}U$$
$$(t_{1/2} \text{ of } {}^{239}_{92}U = 23.5 \text{ min})$$

$$^{239}_{92}U \longrightarrow {}^{239}_{93}Np + {}^{0}_{-1}\beta$$
$$(t_{1/2} \text{ of } {}^{239}_{93}Np = 2.35 \text{ days})$$

$$^{239}_{93}Np \longrightarrow {}^{239}_{94}Pu + {}^{0}_{-1}\beta$$
$$(t_{1/2} \text{ of } {}^{239}_{94}Pu = 2.4\times10^4 \text{ yr})$$

Although breeder reactors can make fuel as they operate, they are difficult and expensive to build, and ^{239}Pu is extremely toxic and long lived. Breeder reactors are not used in the United States, although several are operating in Europe and Japan.

🌐 **Environmental application**

🏛 **Literature** Zevos, Nick. "Radioactivity, Radiation, and the Chemistry of Nuclear Waste," *J. Chem. Educ.* **2002,** *79,* 692.

🏛 **Literature** Krow, Grant R.; Krow, Jessica B. "Low-Level Radioactive Waste Disposal: An Exercise in Dealing with Pollution," *J. Chem. Educ.* **1998,** *75,* 1583.

---

*An even greater release of radioactivity (1.1 MCi), involving several radionuclides including the extremely dangerous ^{90}Sr, occurred in 1957 in Kyshtym, a town in the South Ural Mountains of Russia. As a result, radioactivity spread globally, but the accident was kept secret by the former Soviet government until 1980.

possibly be field-tested for the thousands of years the material will remain harmful. Leakage of radioactive material into groundwater is a danger, and earthquakes can occur even in geologically stable regions. Despite studies indicating that the proposed disposal site at Yucca Mountain, Nevada, is too geologically active, the U.S. government recently approved the site. It remains to be seen whether we can operate fission reactors *and* dispose of the waste safely and economically.

The Promise of Nuclear Fusion

Nuclear fusion is the ultimate source of nearly all the energy on Earth because almost all other sources depend, directly or indirectly, on the energy produced by nuclear fusion in the Sun. But the Sun and other stars generate more than energy; in fact, *all the elements larger than hydrogen were formed in fusion and decay processes within stars,* as the next Chemical Connections essay describes.

Much research is being devoted to making nuclear fusion a practical, direct source of energy on Earth. To understand the advantages of fusion, let's consider one of the most discussed fusion reactions, in which deuterium and tritium react:

$$^{2}_{1}H + {}^{3}_{1}H \longrightarrow {}^{4}_{2}He + {}^{1}_{0}n$$

This reaction produces 1.7×10^{9} kJ/mol, an enormous quantity of energy with no radioactive byproducts. Moreover, the reactant nuclei are relatively easy to come by. We obtain deuterium from the electrolysis of water (Section 22.4). In nature, tritium forms through the cosmic (neutron) irradiation of ^{14}N:

$$^{14}_{7}N + {}^{1}_{0}n \longrightarrow {}^{3}_{1}H + {}^{12}_{6}C$$

However, this process results in a natural abundance of only $10^{-7}\%$ ^{3}H. More practically, tritium is produced in nuclear accelerators by bombarding lithium-6 or by surrounding the fusion reactor itself with material containing lithium-6:

$$^{6}_{3}Li + {}^{1}_{0}n \longrightarrow {}^{3}_{1}H + {}^{4}_{2}He$$

Thus, in principle, fusion seems very promising and may represent an ideal source of power. However, some extremely difficult technical problems remain. Fusion requires enormous energy in the form of heat to give the positively charged nuclei enough kinetic energy to force themselves together. The fusion of deuterium and tritium, for example, occurs at practical rates at about 10^{8} K, a temperature hotter than the Sun's core! How can such conditions be achieved? Two current research approaches have promise. In one, atoms are stripped of their electrons at high temperatures, which results in a gaseous *plasma*, a neutral mixture of positive nuclei and electrons. Because of the extreme temperatures needed for fusion, no *material* can contain the plasma. The most successful approach to date has been to enclose the plasma within a magnetic field. The *tokamak* design has a donut-shaped container in which a helical magnetic field confines the plasma and prevents it from contacting the walls (Figure 24.17). Scientists at the Princeton University Plasma Physics facility have achieved some success in generating energy from fusion this way. In another approach, the high temperature is reached by using many focused lasers to compress and heat the fusion reactants. In any event, one or more major breakthroughs are needed before fusion will be realized as a practical, everyday source of energy.

Section Summary

In nuclear fission, neutron bombardment causes a nucleus to split into two smaller nuclei and release neutrons that split other nuclei, giving rise to a chain reaction. • A nuclear power plant controls the rate of the chain reaction to produce heat that creates steam, which is used to generate electricity. • Potential hazards, such as radiation leaks, thermal pollution, and disposal of nuclear waste, remain current concerns. • Nuclear fusion holds great promise as a source of clean abundant energy, but it requires extremely high temperatures and is not yet practical. • The elements were formed through a complex series of nuclear reactions in evolving stars.

Literature Jensen, Torkil H. "Fusion—A Potential Power Source," *J. Chem. Educ.* **1994**, *71*, 820.

Engineering application

Vacuum container
for plasma   Plasma   Magnets

Figure 24.17 The tokamak design for magnetic containment of a fusion plasma. The donut-shaped chamber of the tokamak (photo, *top;* schematic, *bottom*) contains the plasma within a helical magnetic field.

ARIS Presentation Center, Art Library

Chemical Connections to Cosmology

Origin of the Elements in the Stars

How did the universe begin? Where did matter come from? How were the elements formed? Every culture has creation myths that deal with similar questions, but only recently have astronomers, physicists, and chemists begun to offer a scientific explanation. The most accepted current model proposes that a sphere of unimaginable properties—diameter of 10^{-28} cm, density of 10^{96} g/mL (density of a nucleus $\approx 10^{14}$ g/mL), and temperature of 10^{32} K—exploded in a "Big Bang," for reasons not yet known, and distributed its contents through the void of space. Cosmologists consider this moment the beginning of time.

One second later, the universe was an expanding mixture of neutrons, protons, and electrons, denser than rock and hotter than an exploding hydrogen bomb (about 10^{10} K). In the next few minutes, it became a gigantic fusion reactor creating the first atomic nuclei: ^{2}H, ^{3}He, and ^{4}He. After 10 min, more than 25% of the mass of the universe existed as ^{4}He, and only about 0.025% as ^{2}H. About 100 million years later, or about 15 billion years ago, gravitational forces pulled this cosmic mixture into primitive, contracting stars.

This account of the origin of the universe is based on the observation of spectra from the Sun, other stars, nearby galaxies, and cosmic (interstellar) dust. Spectral analysis of planets and chemical analysis of Earth and Moon rocks, meteorites, and cosmic-ray particles furnish data about isotope abundance. From these, a model has been developed for **stellar nucleosynthesis,** the creation of the elements in stars. The overall process occurs in several stages during a star's evolution, and the entire sequence of steps occurs only in very massive stars, having 10 to 100 times the mass of the Sun. Each step involves a contraction of the star that produces higher temperature and heavier nuclei. Such events are forming elements in stars today. The key stages in the process are shown in Figure B24.3 and described below:

1. *Hydrogen burning produces He.* The initial contraction of a star heats its core to about 10^7 K, at which point a fusion process called *hydrogen burning* begins. Through three possible reaction cycles, helium is produced from the abundant protons:

$$4\,^1_1\text{H} \longrightarrow\,^4_2\text{He} + 2\,^0_1\beta + 2\nu + \text{energy}$$

2. *Helium burning produces C, O, Ne, and Mg.* After several billion years of hydrogen burning, about 10% of the ^{1}H is consumed, and the star contracts further. The ^{4}He forms a dense core, hot enough (2×10^8 K) to fuse ^{4}He. The energy released during *helium burning* expands the remaining ^{1}H into a vast envelope: the star becomes a *red giant,* more than 100 times its original diameter. Within its core, pairs of ^{4}He nuclei (α particles) fuse into unstable ^{8}Be nuclei ($t_{1/2} = 7\times10^{-17}$ s), which play the role of an activated complex in chemical reactions. A ^{8}Be nucleus collides with another ^{4}He to form ^{12}C. Then, further fusion with ^{4}He creates nuclei up to ^{24}Mg:

$$^{12}\text{C} \xrightarrow{\alpha}\,^{16}\text{O} \xrightarrow{\alpha}\,^{20}\text{Ne} \xrightarrow{\alpha}\,^{24}\text{Mg}$$

3. *Elements through Fe and Ni form.* For another 10 million years, ^{4}He is consumed, and the heavier nuclei created form a core. This core contracts and heats, expanding the star into a *supergiant.* Within the hot core (7×10^8 K), *carbon and oxygen burning* occur:

$$^{12}\text{C} + \,^{12}\text{C} \longrightarrow\,^{23}\text{Na} + \,^1\text{H}$$
$$^{12}\text{C} + \,^{16}\text{O} \longrightarrow\,^{28}\text{Si} + \gamma$$

Absorption of α particles forms nuclei up to ^{40}Ca:

$$^{12}\text{C} \xrightarrow{\alpha}\,^{16}\text{O} \xrightarrow{\alpha}\,^{20}\text{Ne} \xrightarrow{\alpha}\,^{24}\text{Mg} \xrightarrow{\alpha}$$
$$^{28}\text{Si} \xrightarrow{\alpha}\,^{32}\text{S} \xrightarrow{\alpha}\,^{36}\text{Ar} \xrightarrow{\alpha}\,^{40}\text{Ca}$$

Further contraction and heating to a temperature of 3×10^9 K allow reactions in which nuclei release neutrons, protons, and α particles and then recapture them. As a result, nuclei with lower binding energies supply nucleons to create those with higher binding energies. In stars of moderate mass, less than 10 times the mass of the Sun, this process stops at iron-56 and nickel-58, the nuclei with the highest binding energies.

4. *Heavier elements form.* In very massive stars, the next stage is the most spectacular. With all the fuel consumed, the core collapses within a second. Many Fe and Ni nuclei break down into neutrons and protons. Protons capture electrons to form neutrons, and the entire core forms an incredibly dense *neutron star.* (An Earth-sized star that became a neutron star would fit in the Houston Astrodome!) As the core implodes, the outer layers explode in a *supernova,* which expels material throughout space. A supernova occurs an average of every few hundred years in each galaxy or once every second somewhere in the universe. Figure B24.4 shows a few recently detected supernovae. The heavier elements are formed during supernova events and are found in *second-generation stars,* those that coalesce from interstellar ^{1}H and ^{4}He and the debris of exploded first-generation stars.

Heavier elements form through *neutron-capture* processes. In the *s-process* (*slow* neutron-capture process), a nucleus captures a neutron, sometime over a period of 10 to 1000 years. If the resulting nucleus is unstable, it undergoes β^- decay to form the next element, as in this conversion of ^{68}Zn to ^{70}Ge:

$$^{68}\text{Zn} \xrightarrow{n}\,^{69}\text{Zn} \xrightarrow{\beta^-}\,^{69}\text{Ga} \xrightarrow{n}\,^{70}\text{Ga} \xrightarrow{\beta^-}\,^{70}\text{Ge}$$

The stable isotopes of most heavy elements up to ^{209}Bi form by the s-process.

Less stable isotopes and those with A greater than 230 cannot form by the s-process because their half-lives are too short. These form by the *r-process* (*rapid* neutron-capture process) during the fury of the supernova. Many neutron captures, followed by many β^- decays, occur in a second, as when ^{56}Fe is converted to ^{79}Br:

$$^{56}_{26}\text{Fe} + 23\,^1_0\text{n} \longrightarrow\,^{79}_{26}\text{Fe} \longrightarrow\,^{79}_{35}\text{Br} + 9\,^0_{-1}\beta$$

We know from the heavy elements present in the Sun that it is at least a second-generation star presently undergoing hydrogen burning. Together with its planets, it was formed from the dust of exploded stars about 4.6×10^9 years ago. This means that many of the atoms on Earth, including some within you, came from exploded stars and are older than the Solar System itself!

Any theory of element formation must be consistent with the element abundances we observe (Section 22.1). Although local compositions, such as those of Earth and the Sun, differ, and mineral grains in meteorites may have abnormal isotopic compositions, large regions of the universe have, on average, similar compositions. Scientists believe element forming reaches a dynamic equilibrium, resulting in *relatively constant amounts of the isotopes.*

(continued)

Coalescing cosmic dust

Cosmic dust

⑤ *Debris from supernova eventually ends up in second-generation stars.*

① *Primitive coalescing star burns hydrogen at 10^7K.*

Neutron star

② *Star begins burning He, causing expansion into red giant. Core of red giant burns He to form ^{12}C, ^{16}O, ^{20}Ne, ^{24}Mg at 2×10^8 K.*

④ *Core implodes to form a neutron star, while outer layers explode into a supernova. Neutron-capture processes form heavier nuclei.*

③ *Red supergiant core burns carbon and oxygen to form nuclei through ^{40}Ca at 7×10^8 K. Further heating to 3×10^9 K forms nuclei through ^{56}Fe and ^{58}Ni.*

Figure B24.3 Element synthesis in the life cycle of a star.

Transparency/ARIS Presentation Center, Art Library

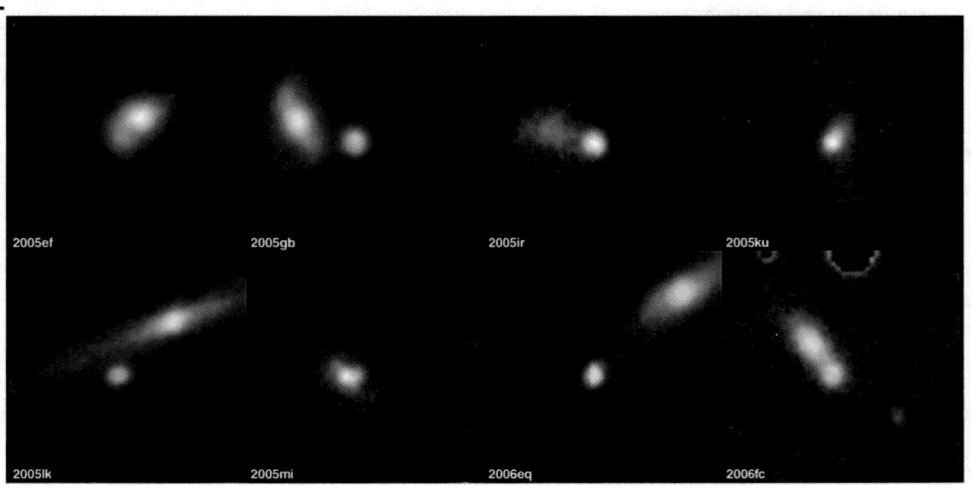

Figure B24.4 Images of 8 of the more than 100 supernovae seen by the Supernovae Survey from 2005 to 2006.

Chapter Perspective and Epilog

With this chapter, our earlier picture of the nucleus as a static point of positive mass at the atom's core has changed radically. Now we picture a dynamic body, capable of a host of changes that involve incredible quantities of energy. Our attempts to apply the behavior of this minute system to benefit society have created some of the most fascinating and challenging fields in science today.

We began our investigation of chemistry 24 chapters ago, by seeing how the chemical elements and the products we make from them influence nearly every aspect of our material existence. Now we have come full circle to learn that these elements, whose patterns of behavior we have become familiar with, yet still marvel at, are continually being born in countless infernos twinkling in the night sky.

For you, the end of this course is a beginning—a chance to apply your new abilities to visualize molecular events and solve problems in whatever field you choose. For the science of chemistry, future challenges are great: What greener energy sources can satisfy our needs while minimizing climate change? What new products can feed, clothe, and house the world's people and maintain precious resources? How can we apply our new genetic insight to defend against cancer, AIDS, and other dread diseases? What new materials and technologies can make life more productive and meaningful? The questions are many, but the science of chemistry will always be one of our most powerful means of answering them.

CHAPTER REVIEW GUIDE

The following sections provide many aids to help you study this chapter. (Numbers in parentheses refer to pages, unless noted otherwise.)

Learning Objectives
These are concepts and skills you should know after studying this chapter.

Relevant section and/or sample problem (SP) numbers appear in parentheses.

Understand These Concepts

1. How nuclear changes differ, in general, from chemical changes (Introduction)
2. The meanings of *radioactivity, nucleon, nuclide,* and *isotope* (Section 24.1)
3. Characteristics of three types of radioactive emissions: α, β, and γ (Section 24.1)
4. The various modes of radioactive decay and how each changes the values of A and Z (Section 24.1)
5. How the N/Z ratio, the even-odd nature of N and Z, and magic numbers correlate with nuclear stability (Section 24.1)
6. How an unstable nuclide's mass number or N/Z ratio correlates with its mode of decay (Section 24.1)
7. How a decay series combines numerous decay steps and ends with a stable nuclide (Section 24.1)
8. Why radioactive decay is a first-order process; the meanings of *decay rate* and *specific activity* (Section 24.2)
9. The meaning of *half-life* in the context of radioactive decay (Section 24.2)
10. How the specific activity of an isotope in an object is used to determine the object's age (Section 24.2)
11. How particle accelerators and reactors are used to synthesize new nuclides (Section 24.3)
12. The units of radiation dose; the effects on living tissue of various dosage levels; the inverse relationship between the mass and charge of an emission and its penetrating power (Section 24.4)
13. How ionizing radiation creates free radicals that damage tissue; sources and risks of ionizing radiation (Section 24.4)
14. How radioisotopes are used in research, analysis, and medical diagnosis (Section 24.5)

15. Why the mass of a nucleus is less than the total mass of its nucleons and how this mass difference is related to the nuclear binding energy (Section 24.6)
16. How nuclear stability is related to binding energy per nucleon (Section 24.6)
17. How heavy nuclides undergo fission and lighter ones undergo fusion to increase the binding energy per nucleon (Section 24.6)
18. The current application of fission and potential application of fusion to produce energy (Section 24.7)

Master These Skills

1. Expressing the mass and charge of a particle with the $_Z^A X$ notation (Section 24.1; see also Section 2.5)
2. Using changes in the values of A and Z to write and balance nuclear equations (SP 24.1)
3. Using the N/Z ratio, the even-odd nature of N and Z, and the presence of magic numbers to predict nuclear stability (SP 24.2)
4. Using the atomic mass of the element or the N/Z ratio to predict the mode of decay of an unstable nuclide (SP 24.3)
5. Converting units of radioactivity (Section 24.2)
6. Calculating specific activity, decay constant, half-life, and number of nuclei (Section 24.2 and SP 24.4)
7. Estimating the age of an object from the specific activity and half-life of carbon-14 (SP 24.5)
8. Writing notations and balancing equations for nuclear transmutation (Section 24.3)
9. Calculating radiation dose and converting units (Section 24.4)
10. Calculating the mass difference between a nucleus and its nucleons and the energy equivalent (Section 24.6)
11. Calculating the binding energy per nucleon and using it to compare stabilities of nuclides (SP 24.6)

Key Terms These important terms appear in boldface in the chapter and are defined again in the Glossary.

Section 24.1
radioactivity (1066)
nucleon (1066)
nuclide (1066)
isotope (1066)
alpha (α) particle (1067)
beta (β) particle (1067)
gamma (γ) ray (1067)
α decay (1069)
β⁻ decay (1069)
positron emission (1069)
positron (β⁺) (1069)
electron (e⁻) capture (EC) (1069)
gamma emission (1070)

band of stability (1071)
strong force (1072)
decay (disintegration) series (1074)

Section 24.2
activity ($\mathscr{A}$) (1074)
Geiger-Müller counter (1075)
scintillation counter (1075)
becquerel (Bq) (1076)
curie (Ci) (1076)
decay constant (1076)
half-life ($t_{1/2}$) (1076)
radioisotopic dating (1078)
radioisotope (1078)

Section 24.3
nuclear transmutation (1080)
deuteron (1081)
particle accelerator (1081)
transuranium element (1082)

Section 24.4
ionization (1082)
ionizing radiation (1082)
gray (Gy) (1083)
rad (radiation-absorbed dose) (1083)
rem (roentgen equivalent for man) (1083)
sievert (Sv) (1083)

free radical (1084)
background radiation (1084)

Section 24.5
tracer (1087)

Section 24.6
fission (1090)
fusion (1090)
nuclear binding energy (1092)
electron volt (eV) (1092)

Section 24.7
chain reaction (1094)
critical mass (1095)
reactor core (1096)
stellar nucleosynthesis (1099)

Key Equations and Relationships Numbered and screened concepts are listed for you to refer to or memorize.

24.1 Balancing a nuclear equation (1067):
$$\text{Total } A_{\text{Total } Z}\text{Reactants} = \text{Total } A_{\text{Total } Z}\text{Products}$$

24.2 Defining the unit of radioactivity (curie, Ci) (1076):
$$1 \text{ Ci} = 3.70 \times 10^{10} \text{ disintegrations per second (d/s)}$$

24.3 Expressing the decay rate (activity) for radioactive nuclei (1076):
$$\text{Decay rate } (\mathscr{A}) = -\frac{\Delta \mathscr{N}}{\Delta t} = k\mathscr{N}$$

24.4 Finding the number of nuclei remaining after a given time, $\mathscr{N}_t$ (1077):
$$\ln\frac{\mathscr{N}_t}{\mathscr{N}_0} = -kt \quad \text{or} \quad \mathscr{N}_t = \mathscr{N}_0 e^{-kt}$$

24.5 Finding the half-life of a radioactive nuclide (1077):
$$t_{1/2} = \frac{\ln 2}{k}$$

24.6 Calculating the time to reach a given specific activity (age of an object in radioisotopic dating) (1079):
$$t = \frac{1}{k} \ln\frac{\mathscr{A}_0}{\mathscr{A}_t}$$

24.7 Writing a transmutation in shorthand notation (1080):
reactant nucleus (particle in, particle(s) out) product nucleus

24.8 Adapting Einstein's equation to calculate mass difference and/or nuclear binding energy (1091):
$$\Delta m = \frac{\Delta E}{c^2} \quad \text{or} \quad \Delta E = \Delta mc^2$$

24.9 Relating the atomic mass unit to its energy equivalent in MeV (1092):
$$1 \text{ amu} = 931.5 \times 10^6 \text{ eV} = 931.5 \text{ MeV}$$

Highlighted Figures and Tables These figures (F) and tables (T) provide a visual review of key ideas.

T24.1 Chemical vs. nuclear reactions (1065)
F24.1 Radioactive emissions in an electric field (1067)
T24.2 Modes of radioactive decay (1068)

F24.2 N vs. Z for the stable nuclides (1071)
F24.4 Decrease in number of ^{14}C nuclei over time (1076)
F24.12 The variation in binding energy per nucleon (1093)

Brief Solutions to FOLLOW-UP PROBLEMS Compare your solutions to these calculation steps and answers.

24.1 $^{133}_{54}\text{Xe} \longrightarrow {}^{133}_{55}\text{Cs} + {}^{0}_{-1}\beta$

24.2 ^{31}P has an even N (16), but ^{30}P has both N and Z odd. ^{31}P also has a slightly higher N/Z ratio that is closer to the band of stability.

24.3 (a) $Z = 26$ is iron (Fe), and the atomic mass is 55.85. The A value of 61 is higher: β⁻ decay; (b) $Z > 83$, which is too high for stability: α decay

24.4 $\ln \mathscr{A}_t = -kt + \ln \mathscr{A}_0$
$$= -\left(\frac{\ln 2}{15 \text{ h}} \times 4.0 \text{ days} \times \frac{24 \text{ h}}{1 \text{ day}}\right) + \ln (2.5 \times 10^9)$$
$$= 17.20$$
$$\mathscr{A}_t = 3.0 \times 10^7 \text{ d/s}$$

24.5 $t = \frac{1}{k} \ln\frac{\mathscr{A}_0}{\mathscr{A}_t} = \frac{5730 \text{ yr}}{\ln 2} \ln\left(\frac{15.3 \text{ d/min·g}}{9.41 \text{ d/min·g}}\right) = 4.02 \times 10^3 \text{ yr}$

The mummy case is about 4000 years old.

24.6 ^{235}U has 92 $^1_1$p and 143 $^1_0$n.
$$\Delta m = [(92 \times 1.007825 \text{ amu}) + (143 \times 1.008665 \text{ amu})]$$
$$- 235.043924 \text{ amu} = 1.9151 \text{ amu}$$

$$\frac{\text{Binding energy}}{\text{nucleon}} = \frac{1.9151 \text{ amu} \times \dfrac{931.5 \text{ MeV}}{1 \text{ amu}}}{235 \text{ nucleons}}$$
$$= 7.591 \text{ MeV/nucleon}$$

Therefore, ^{235}U is less stable than ^{12}C.

PROBLEMS

Problems with colored numbers are answered in Appendix E and worked in detail in the Student Solutions Manual. Problem sections match those in the text and provide the numbers of relevant sample problems. Most offer Concept Review Questions, Skill-Building Exercises (grouped in pairs covering the same concept), and Problems in Context. The Comprehensive Problems are based on material from any section or previous chapter.

Radioactive Decay and Nuclear Stability
(Sample Problems 24.1 to 24.3)

● Concept Review Questions

• **24.1** How do chemical and nuclear reactions differ in
(a) Magnitude of the energy change?
(b) Effect on rate of increasing temperature?
(c) Effect on rate of higher reactant concentration?
(d) Effect on yield of higher reactant concentration?

• **24.2** Sulfur has four naturally occurring stable isotopes. The one with the lowest mass number is sulfur-32, which is also the most abundant (95.02%).
(a) What percentage of the S atoms in a matchhead are ^{32}S?
(b) The isotopic mass of ^{32}S is 31.972070 amu. Is the atomic mass of S larger, smaller, or equal to this mass? Explain.

• **24.3** What led Marie Curie to draw the following conclusions?
(a) Radioactivity is a property of the element and not the compound in which it is found. (b) A highly radioactive element, aside from uranium, occurs in pitchblende.

• **24.4** Which of the following produce an atom of a *different* element: (a) α decay; (b) β$^-$ decay; (c) γ emission; (d) β$^+$ emission; (e) e$^-$ capture? Show how Z and N change, if at all, with each process.

• **24.5** Why is $^3_2$He stable, but $^2_2$He has never been detected?

• **24.6** How do the modes of decay differ for a neutron-rich nuclide and a proton-rich nuclide?

• **24.7** Why might it be difficult to use only a nuclide's N/Z ratio to predict whether it will decay by β$^+$ emission or by e$^-$ capture? What other factor is important?

● Skill-Building Exercises *(grouped in similar pairs)*

• **24.8** Write balanced nuclear equations for the following:
(a) Alpha decay of $^{234}_{92}$U
(b) Electron capture by neptunium-232
(c) Positron emission by $^{12}_7$N

• **24.9** Write balanced nuclear equations for the following:
(a) β$^-$ decay of sodium-26
(b) β$^-$ decay of francium-223
(c) Alpha decay of $^{212}_{83}$Bi

• **24.10** Write balanced nuclear equations for the following:
(a) β$^-$ emission by magnesium-27
(b) β$^+$ emission by $^{23}_{12}$Mg
(c) Electron capture by $^{103}_{46}$Pd

• **24.11** Write balanced nuclear equations for the following:
(a) β$^-$ decay of silicon-32
(b) Alpha decay of polonium-218
(c) Electron capture by $^{110}_{49}$In

• **24.12** Write balanced nuclear equations for the following:
(a) Formation of $^{48}_{22}$Ti through positron emission

(b) Formation of silver-107 through electron capture
(c) Formation of polonium-206 through α decay

• **24.13** Write balanced nuclear equations for the following:
(a) Formation of $^{241}_{95}$Am through β$^-$ decay
(b) Formation of $^{228}_{89}$Ac through β$^-$ decay
(c) Formation of $^{203}_{83}$Bi through α decay

• **24.14** Write balanced nuclear equations for the following:
(a) Formation of ^{186}Ir through electron capture
(b) Formation of francium-221 through α decay
(c) Formation of iodine-129 through β$^-$ decay

• **24.15** Write balanced nuclear equations for the following:
(a) Formation of ^{52}Mn through positron emission
(b) Formation of polonium-215 through α decay
(c) Formation of ^{81}Kr through electron capture

• **24.16** Which nuclide(s) would you predict to be stable? Why?
(a) $^{20}_8$O (b) $^{59}_{27}$Co (c) $^9_3$Li

• **24.17** Which nuclide(s) would you predict to be stable? Why?
(a) $^{146}_{60}$Nd (b) $^{114}_{48}$Cd (c) $^{88}_{42}$Mo

• **24.18** Which nuclide(s) would you predict to be stable? Why?
(a) ^{127}I (b) tin-106 (c) ^{68}As

• **24.19** Which nuclide(s) would you predict to be stable? Why?
(a) ^{48}K (b) ^{79}Br (c) argon-32

• **24.20** What is the most likely mode of decay for each?
(a) $^{238}_{92}$U (b) $^{48}_{24}$Cr (c) $^{50}_{25}$Mn

• **24.21** What is the most likely mode of decay for each?
(a) $^{111}_{47}$Ag (b) $^{41}_{17}$Cl (c) $^{110}_{44}$Ru

• **24.22** What is the most likely mode of decay for each?
(a) ^{15}C (b) ^{120}Xe (c) ^{224}Th

• **24.23** What is the most likely mode of decay for each?
(a) ^{106}In (b) ^{141}Eu (c) ^{241}Am

• **24.24** Why is $^{52}_{24}$Cr the most stable isotope of chromium?

• **24.25** Why is $^{40}_{20}$Ca the most stable isotope of calcium?

● Problems in Context

•• **24.26** ^{237}Np is the parent nuclide of a decay series that starts with α emission, followed by β$^-$ emission, and then two more α emissions. Write a balanced nuclear equation for each step.

• **24.27** Why is helium found in deposits of uranium and thorium ores? What kind of radioactive emission produces it?

•• **24.28** In a natural decay series, how many α and β$^-$ emissions per atom of uranium-235 result in an atom of lead-207?

The Kinetics of Radioactive Decay
(Sample Problems 24.4 and 24.5)

● Concept Review Questions

• **24.29** What electronic process is the basis for detecting radioactivity in (a) a scintillation counter; (b) a Geiger-Müller counter?

• **24.30** What is the reaction order of radioactive decay? Explain.

•• **24.31** After 1 minute, three radioactive nuclei remain from an original sample of six. Is it valid to conclude that $t_{1/2}$ equals 1 minute? Is this conclusion valid if the original sample contained 6×10^{12} nuclei and 3×10^{12} remain after 1 minute? Explain.

•• **24.32** Radioisotopic dating depends on the constant rate of decay and formation of various nuclides in a sample. How is the proportion of ^{14}C kept relatively constant in living organisms?

Levels of Difficulty • Easy •• Medium ••• Challenging

● **Skill-Building Exercises** *(grouped in similar pairs)*

• **24.33** What is the specific activity (in Ci/g) if 1.65 mg of an isotope emits 1.56×10^6 α particles per second?

• **24.34** What is the specific activity (in Ci/g) if 2.6 g of an isotope emits 4.13×10^8 β^- particles per hour?

• **24.35** What is the specific activity (in Bq/g) if 8.58 μg of an isotope emits 7.4×10^4 α particles per minute?

• **24.36** What is the specific activity (in Bq/g) if 1.07 kg of an isotope emits 3.77×10^7 β^- particles per minute?

• **24.37** If one-trillionth of the atoms of a radioactive isotope disintegrate each day, what is the decay constant of the process?

• **24.38** If $2.8 \times 10^{-10}\%$ of the atoms of a radioactive isotope disintegrate in 1.0 yr, what is the decay constant of the process?

• **24.39** If 1.00×10^{-12} mol of ^{135}Cs emits 1.39×10^5 β^- particles in 1.00 yr, what is the decay constant?

• **24.40** If 6.40×10^{-9} mol of ^{176}W emits 1.07×10^{15} β^+ in 1.00 h, what is the decay constant?

• **24.41** The isotope $^{212}_{83}$Bi has a half-life of 1.01 yr. What mass (in mg) of a 2.00-mg sample will remain after 3.75×10^3 h?

• **24.42** The half-life of radium-226 is 1.60×10^3 yr. How many hours will it take for a 2.50-g sample to decay to the point where 0.185 g of the isotope remains?

• **24.43** A rock contains 270 μmol of ^{238}U ($t_{1/2} = 4.5 \times 10^9$ yr) and 110 μmol of ^{206}Pb. Assuming that all the ^{206}Pb comes from decay of the ^{238}U, estimate the rock's age.

• **24.44** A fabric remnant from a burial site has a ^{14}C/^{12}C ratio of 0.735 of the original value. How old is the fabric?

● **Problems in Context**

•• **24.45** Due to decay of ^{40}K, cow's milk has a specific activity of about 6×10^{-11} mCi per milliliter. How many disintegrations of ^{40}K nuclei are there per minute in an 8.0-oz glass of milk?

•• **24.46** Plutonium-239 ($t_{1/2} = 2.41 \times 10^4$ yr) represents a serious nuclear waste hazard. If seven half-lives are required to reach a tolerable level of radioactivity, how long must ^{239}Pu be stored?

•• **24.47** A rock that contains 3.1×10^{-15} mol of ^{232}Th ($t_{1/2} = 1.4 \times 10^{10}$ yr) has 9.5×10^4 fission tracks, each track representing the fission of one atom of ^{232}Th. How old is the rock?

•• **24.48** A volcanic eruption melts a large area of rock, and all gases are expelled. After cooling, $^{40}_{18}$Ar accumulates from the ongoing decay of $^{40}_{19}$K in the rock ($t_{1/2} = 1.25 \times 10^9$ yr). When a piece of rock is analyzed, it is found to contain 1.38 mmol of ^{40}K and 1.14 mmol of ^{40}Ar. How long ago did the rock cool?

Nuclear Transmutation: Induced Changes in Nuclei

● **Concept Review Questions**

• **24.49** Irene and Frederic Joliot-Curie converted $^{27}_{13}$Al to $^{30}_{15}$P in 1933. Why was this transmutation significant?

• **24.50** Early workers mistakenly thought neutron beams were γ radiation. Why? What evidence led to the correct conclusion?

• **24.51** Why must the electrical polarity of the tubes in a linear accelerator be reversed at very short time intervals?

• **24.52** Why does bombardment with protons usually require higher energies than bombardment with neutrons?

● **Skill-Building Exercises** *(grouped in similar pairs)*

• **24.53** Determine the missing species in these transmutations, and write a full nuclear equation from the shorthand notation:
(a) ^{10}B (α,n) __

(b) ^{28}Si (d,__) ^{29}P (where d is a deuteron, ^{2}H$^+$)
(c) __ (α,2n) ^{244}Cf

•• **24.54** Name the unidentified species, and write each transmutation process in shorthand notation: (a) gamma irradiation of a nuclide yields a proton, a neutron, and ^{29}Si; (b) bombardment of ^{252}Cf with ^{10}B yields five neutrons and a nuclide; (c) bombardment of ^{238}U with a particle yields three neutrons and ^{239}Pu.

● **Problems in Context**

•• **24.55** Elements 104, 105, and 106 have been named rutherfordium (Rf), dubnium (Db), and seaborgium (Sg), respectively. These elements are synthesized from californium-249 by bombarding with carbon-12, nitrogen-15, and oxygen-18 nuclei, respectively. Four neutrons are formed in each reaction as well. (a) Write balanced nuclear equations for the formation of these elements. (b) Write the equations in shorthand notation.

The Effects of Nuclear Radiation on Matter

● **Concept Review Questions**

• **24.56** The effects on matter of γ rays and α particles differ. Explain.

• **24.57** What is a cation-electron pair, and how does it form?

• **24.58** Why is ionizing radiation more harmful to children than adults?

• **24.59** Why is ·OH more dangerous than OH$^-$ in an organism?

● **Skill-Building Exercises** *(grouped in similar pairs)*

•• **24.60** A 135-lb person absorbs 3.3×10^{-7} J of energy from radioactive emissions. (a) How many rads does she receive? (b) How many grays (Gy) does she receive?

•• **24.61** A 3.6-kg laboratory animal receives a single dose of 8.92×10^{-4} Gy. (a) How many rads did the animal receive? (b) How many joules did the animal absorb?

•• **24.62** A 70.-kg person exposed to ^{90}Sr absorbs 6.0×10^5 β^- particles, each with an energy of 8.74×10^{-14} J. (a) How many grays does the person receive? (b) If the RBE is 1.0, how many millirems is this? (c) What is the equivalent dose in sieverts (Sv)?

•• **24.63** A laboratory rat weighs 265 g and absorbs 1.77×10^{10} β^- particles, each with an energy of 2.20×10^{-13} J. (a) How many rads does the animal receive? (b) What is this dose in Gy? (c) If the RBE is 0.75, what is the equivalent dose in Sv?

● **Problems in Context**

•• **24.64** If 2.50 pCi [1 pCi (picocurie) $= 1 \times 10^{-12}$ Ci] of radioactivity from ^{239}Pu is emitted in a 95-kg human for 65 h, and each disintegration has an energy of 8.25×10^{-13} J, how many grays does the person receive?

•• **24.65** A small region of a cancer patient's brain is exposed for 24.0 min to 475 Bq of radioactivity from ^{60}Co for treatment of a tumor. If the brain mass exposed is 1.858 g and each β^- particle emitted has an energy of 5.05×10^{-14} J, what is the dose in rads?

Applications of Radioisotopes

● **Concept Review Questions**

• **24.66** What two ways are radioactive tracers used in organisms?

• **24.67** Why is neutron activation analysis (NAA) useful to art historians and criminologists?

• **24.68** Positrons cannot penetrate matter more than a few atomic diameters, but positron emission of radiotracers can be monitored in medical diagnosis. Explain.

•• **24.69** A steel part is treated to form some iron-59. Oil used to lubricate the part emits 298 β^- particles (with the energy characteristic of ^{59}Fe) per minute per milliliter of oil. What other information would you need to calculate the rate of removal of the steel from the part during use?

● **Problem in Context**

•• **24.70** The oxidation of methanol to formaldehyde can be accomplished by reaction with chromic acid:

$$6H^+(aq) + 3CH_3OH(aq) + 2H_2CrO_4(aq) \longrightarrow$$
$$3CH_2O(aq) + 2Cr^{3+}(aq) + 8H_2O(l)$$

The reaction can be studied with the stable isotope tracer ^{18}O and mass spectrometry. When a small amount of $CH_3{}^{18}OH$ is present in the alcohol reactant, $CH_2{}^{18}O$ forms. When a small amount of $H_2Cr^{18}O_4$ is present, $H_2{}^{18}O$ forms. Does chromic acid or methanol supply the O atom to the aldehyde? Explain.

The Interconversion of Mass and Energy
(Sample Problem 24.6)

Note: Data for solving problems in this section: mass of ^{1}H atom = 1.007825 amu; mass of neutron = 1.008665 amu.

● **Concept Review Questions**

• **24.71** Many scientists at first reacted skeptically to Einstein's equation, $E = mc^2$. Why?

• **24.72** How does a change in mass arise when a nuclide forms from nucleons?

• **24.73** When a nucleus forms from nucleons, is energy absorbed or released? Why?

• **24.74** What is the binding energy per nucleon? Why is the binding energy per nucleon, rather than per nuclide, used to compare nuclide stability?

● **Skill-Building Exercises** (grouped in similar pairs)

• **24.75** A ^{3}H nucleus decays with an energy of 0.01861 MeV. Convert this energy into (a) electron volts; (b) joules.

• **24.76** Arsenic-84 decays with an energy of 1.57×10^{-15} kJ per nucleus. Convert this energy into (a) eV; (b) MeV.

• **24.77** How many joules are released when 1.5 mol of ^{239}Pu decays, if each nucleus releases 5.243 MeV?

• **24.78** How many MeV are released per nucleus when 3.2×10^{-3} mol of chromium-49 releases 8.11×10^5 kJ?

•• **24.79** Oxygen-16 is one of the most stable nuclides. The mass of a ^{16}O atom is 15.994915 amu. Calculate the binding energy (a) per nucleon in MeV; (b) per atom in MeV; (c) per mole in kJ.

•• **24.80** Lead-206 is the end product of ^{238}U decay. One ^{206}Pb atom has a mass of 205.974440 amu. Calculate the binding energy (a) per nucleon in MeV; (b) per atom in MeV; (c) per mole in kJ.

•• **24.81** Cobalt-59 is the only stable isotope of this transition metal. One ^{59}Co atom has a mass of 58.933198 amu. Calculate the binding energy (a) per nucleon in MeV; (b) per atom in MeV; (c) per mole in kJ.

•• **24.82** Iodine-131 is one of the most important isotopes used in the diagnosis of thyroid cancer. One atom has a mass of 130.906114 amu. Calculate the binding energy (a) per nucleon in MeV; (b) per atom in MeV; (c) per mole in kJ.

● **Problem in Context**

•• **24.83** The ^{80}Br nuclide decays either by β^- decay or by electron capture. (a) What is the product of each process? (b) Which process releases more energy? (Masses of atoms: ^{80}Br = 79.918528 amu; ^{80}Kr = 79.916380 amu; ^{80}Se =

79.916520 amu; neglect the mass of electrons involved because these are atomic, not nuclear, masses.)

Applications of Fission and Fusion

● **Concept Review Questions**

• **24.84** What is the minimum number of neutrons from each fission event that must be absorbed by other nuclei for a chain reaction to be sustained?

• **24.85** In what main way is fission different from radioactive decay? Are all fission events in a chain reaction identical? Explain.

• **24.86** What is the purpose of enrichment in the preparation of fuel rods? How is it accomplished?

• **24.87** Describe the nature and purpose of these components of a nuclear reactor: (a) control rods; (b) moderator; (c) reflector.

• **24.88** State an advantage and a disadvantage of heavy-water reactors compared to light-water reactors.

• **24.89** What are the expected advantages of fusion reactors over fission reactors?

• **24.90** Why is iron the most abundant element in Earth's core?

• **24.91** Compare the *s*- and *r*-processes in stellar nucleosynthesis.

• **24.92** What is the cosmic importance of unstable ^{8}Be?

● **Problem in Context**

• **24.93** The reaction that will probably power the first commercial fusion reactor is

$$^3_1H + {}^2_1H \longrightarrow {}^4_2He + {}^1_0n$$

How much energy would be produced per mole of reaction? (Masses of atoms: ^3_1H = 3.01605 amu; ^2_1H = 2.0140 amu; ^4_2He = 4.00260 amu; mass of ^1_0n = 1.008665 amu.)

Comprehensive Problems

•• **24.94** Some $^{243}_{95}$Am was present when Earth formed, but it all decayed in the next billion years. The first three steps in this decay series are emission of an α particle, a β^- particle, and another α particle. What other isotopes were present on the young Earth in a rock that contained some $^{243}_{95}$Am?

•• **24.95** Curium-243 undergoes α decay to plutonium-239:

$$^{243}Cm \longrightarrow {}^{239}Pu + \alpha$$

(a) Find the change in mass, Δm (in kg). (Masses: ^{243}Cm = 243.0614 amu; ^{239}Pu = 239.0522 amu; ^{4}He = 4.0026 amu; 1 amu = 1.661×10^{-24} g.) (b) Find the energy released in joules. (c) Find the energy released in kJ/mol of reaction, and comment on the difference between this value and a typical heat of reaction for a chemical change, which is a few hundred kJ/mol.

•• **24.96** Plutonium "triggers" for nuclear weapons were manufactured at the Rocky Flats plant in Colorado. An 85-kg worker inhaled a dust particle containing 1.00 μg of $^{239}_{94}$Pu, which resided in his body for 16 h ($t_{1/2}$ of ^{239}Pu = 2.41×10^4 yr; each disintegration released 5.15 MeV). (a) How many rads did he receive? (b) How many grays?

•• **24.97** Archeologists removed some charcoal from a Native American campfire, burned it in O_2, and bubbled the CO_2 formed into $Ca(OH)_2$ solution (limewater). The $CaCO_3$ that precipitated was filtered and dried. If 4.58 g of the $CaCO_3$ had a radioactivity of 3.2 d/min, how long ago was the campfire?

•• **24.98** A 5.4-μg sample of ^{226}RaCl$_2$ has a radioactivity of 1.5×10^5 Bq. Calculate $t_{1/2}$ of ^{226}Ra.

•• **24.99** How many rads does a 65-kg human receive each year from the approximately 10^{-8} g of $^{14}_6$C naturally present in her body ($t_{1/2}$ = 5730 yr; each disintegration releases 0.156 MeV)?

•• **24.100** The overall reaction taking place during hydrogen burning in a young star is

$$4^1_1H \longrightarrow {}^4_2He + 2^0_1\beta + 2\nu + \text{energy}$$

How much energy (in MeV) is released per He nucleus formed? Per mole of He? (Masses: ^1_1H atom = 1.007825 amu; ^4_2He atom = 4.00260 amu; positron = 5.48580×10^{-4} amu.)

•• **24.101** A sample of AgCl emits 175 nCi/g. A saturated solution prepared from the solid emits 1.25×10^{-2} Bq/mL due to radioactive Ag^+ ions. What is the molar solubility of AgCl?

•• **24.102** The scene below depicts a neutron bombarding ^{235}U:

(a) Is this an example of fission or of fusion? (b) Identify the other nuclide formed. (c) What is the most likely mode of decay of the nuclide with Z = 55?

•• **24.103** What fraction of the ^{235}U ($t_{1/2} = 7.0 \times 10^8$ yr) created when Earth was formed would remain after 2.8×10^9 yr?

••• **24.104** ^{238}U ($t_{1/2} = 4.5 \times 10^9$ yr) begins a decay series that ultimately forms ^{206}Pb. The scene below depicts the relative number of ^{238}U atoms (red) and ^{206}Pb atoms (green) in a mineral. If all the Pb comes from ^{238}U, calculate the age of the sample.

•• **24.105** Cosmologists modeling the origin of the elements postulate nuclides with very short half-lives. One of these nuclides, ^8Be ($t_{1/2} = 7 \times 10^{-17}$ s), plays a key role in stellar nucleosynthesis (p. 1099) because it must fuse with ^4He to form ^{12}C before decaying. Another explanation involves the simultaneous fusion of three ^4He nuclei to form ^{12}C. Comment on the validity of this alternative mechanism.

••• **24.106** Technetium-99m is a metastable nuclide used in numerous cancer diagnostic and treatment programs. It is prepared just before use because it decays rapidly through γ emission:

$$^{99m}Tc \longrightarrow {}^{99}Tc + \gamma$$

Use the data below to determine (a) the half-life of ^{99m}Tc; (b) the percentage of the isotope that is lost if it takes 2.0 h to prepare and administer the dose.

Time (h)	γ Emission (photons/s)
0	5000.
4	3150.
8	2000.
12	1250.
16	788
20	495

•• **24.107** How many curies are produced by 1.0 mol of ^{40}K ($t_{1/2} = 1.25 \times 10^9$ yr)? How many becquerels?

•• **24.108** The fraction of a radioactive isotope remaining at time t is $(\frac{1}{2})^{t/t_{1/2}}$, where $t_{1/2}$ is the half-life. If the half-life of carbon-14 is 5730 yr, what fraction of carbon-14 in a piece of charcoal remains after (a) 10.0 yr; (b) 10.0×10^3 yr; (c) 10.0×10^4 yr?

(d) Why is radiocarbon dating more reliable for the fraction remaining in part (b) than that in part (a) or in part (c)?

••• **24.109** The isotopic mass of $^{210}_{86}Rn$ is 209.989669 amu. When this nuclide decays by electron capture, it emits 2.368 MeV. What is the isotopic mass of the resulting nuclide?

•• **24.110** Exactly 0.1 of the radioactive nuclei in a sample decay per hour. Thus, after n hours, the fraction of nuclei remaining is $(0.900)^n$. Find the value of n equal to one half-life.

•• **24.111** In neutron activation analysis (NAA), stable isotopes are bombarded with neutrons. Depending on the isotope and the energy of the neutron, various emissions are observed. What are the products when the following neutron-activated species decay? Write an overall equation in shorthand notation for the reaction starting with the stable isotope before neutron activation.
(a) $^{52}_{23}V^* \longrightarrow [\beta^-$ emission]   (b) $^{64}_{29}Cu^* \longrightarrow [\beta^+$ emission]
(c) $^{28}_{13}Al^* \longrightarrow [\beta^-$ emission]

•• **24.112** In the 1950s, radioactive material was spread over the land from aboveground nuclear tests. A woman drinks some contaminated milk and ingests 0.0500 g of ^{90}Sr, which is taken up by bones and teeth and not eliminated. (a) How much ^{90}Sr ($t_{1/2} = 29$ yr) is present in her body after 10 yr? (b) How long will it take for 99.9% of the ^{90}Sr to decay?

•• **24.113** The scene below represents a reaction (with neutrons gray and protons purple) that occurs during the lifetime of a star. (a) Write a balanced nuclear equation for the reaction. (b) If the mass difference is 7.7×10^{-2} amu, find the energy (kJ) released.

••• **24.114** What volume of radon will be produced per hour at STP from 1.000 g of ^{226}Ra ($t_{1/2} = 1599$ yr; 1 yr = 8766 h; mass of one ^{226}Ra atom = 226.025402 amu)?

•• **24.115** ^{90}Kr ($t_{1/2} = 32$ s) is used to study respiration. How soon after being made must a sample be administered to the patient if the activity must be at least 90% of the original activity?

•• **24.116** Which isotope in each pair is more stable? Why?
(a) $^{140}_{55}Cs$ or $^{133}_{55}Cs$ (b) $^{79}_{35}Br$ or $^{78}_{35}Br$
(c) $^{28}_{12}Mg$ or $^{24}_{12}Mg$ (d) $^{14}_{7}N$ or $^{18}_{7}N$

•• **24.117** A bone sample containing strontium-90 ($t_{1/2} = 29$ yr) emits 7.0×10^4 β^- particles per month. How long will it take for the emission to decrease to 1.0×10^4 particles per month?

•• **24.118** The 23rd-century starship *Enterprise* uses a substance called "dilithium crystals" as its fuel.
(a) Assuming this material is the result of fusion, what is the product of the fusion of two ^6Li nuclei?
(b) How much energy is released per kilogram of dilithium formed? (Mass of one ^6Li atom is 6.015121 amu.)
(c) When four ^1H atoms fuse to form ^4He, how many positrons are released?
(d) To determine the energy potential of the fusion processes in parts (b) and (c), compare the changes in mass per kilogram of dilithium and of ^4He.
(e) Compare the change in mass per kilogram in part (b) to that for the formation of ^4He by the method used in current fusion reactors (see p. 1098). (For masses, see Problem 24.93.)
(f) Using early 21st-century fusion technology, how much tritium can be produced per kilogram of ^6Li in the following reaction: $^6_3Li + ^1_0n \longrightarrow {}^4_2He + ^3_1H$? When this amount of tritium is

fused with deuterium, what is the change in mass? How does this quantity compare with the use of dilithium in part (b)?

•••24.119 Uranium and radium are found in many rocky soils throughout the world. Both undergo radioactive decay, and one of the products is radon-222, the heaviest noble gas ($t_{1/2}$ = 3.82 days). Inhalation of this gas contributes to many lung cancers. According to Environmental Protection Agency recommendations, the level of radioactivity from radon in homes should not exceed 4.0 pCi/L of air.
(a) What is the safe level of radon in Bq/L of air?
(b) A home has a radon measurement of 41.5 pCi/L. The owner vents the basement in such a way that no more radon enters the living area. What is the activity of the radon remaining in the room air (in Bq/L) after 9.5 days?
(c) How many more days does it take to reach the EPA recommended level?

••24.120 Nuclear disarmament could be accomplished if weapons were not "replenished." The tritium in warheads decays to helium with a half-life of 12.26 yr and must be replaced or the weapon is useless. What fraction of the tritium is lost in 5.50 yr?

••24.121 A decay series starts with the synthetic isotope $^{239}_{92}$U. The first four steps are emissions of a β^- particle, another β^-, an α particle, and another α. Write a balanced nuclear equation for each step. Which natural series could start by this sequence?

••24.122 How long can a 54-lb child be exposed to 1.0 mCi of radiation from ^{222}Rn before accumulating 1.0 mrad if the energy of each disintegration is 5.59 MeV?

••24.123 An earthquake in the area of present-day San Francisco is to be dated by measuring the ^{14}C activity ($t_{1/2}$ = 5730 yr) of parts of a tree uprooted during the event. The tree parts have an activity of 12.9 d/min·g C, and a living tree has an activity of 15.3 d/min·g C. How long ago did the earthquake occur?

••24.124 Were organisms a billion years ago exposed to more or less ionizing radiation than similar organisms today? Explain.

••24.125 Tritium (^{3}H; $t_{1/2}$ = 12.26 yr) is continually formed in the upper troposphere by interaction of solar particles with nitrogen. As a result, natural waters contain a small amount of tritium. Two samples of wine are analyzed, one known to be made in 1941 and another made earlier. The water in the 1941 wine has 2.23 times as much tritium as the water in the other. When was the other wine produced?

••24.126 Even though plutonium-239 ($t_{1/2}$ = 2.41×10⁴ yr) is one of the main fission fuels, it is still a radiation hazard present in spent uranium fuel from nuclear power plants. How many years does it take for 99% of the plutonium-239 in spent fuel to decay?

••24.127 Carbon from the remains of an extinct Australian marsupial, called *Diprotodon,* has a specific activity of 0.61 pCi/g. Modern carbon has a specific activity of 6.89 pCi/g. How long ago did the *Diprotodon* apparently become extinct?

••24.128 The reaction that allows for radiocarbon dating is the continual formation of carbon-14 in the upper atmosphere:
$$^{14}_{7}N + ^1_0n \longrightarrow ^{14}_6C + ^1_1H$$
What is the energy change associated with this process in eV/reaction and in kJ/mol reaction? (Masses of atoms: $^{14}_7$N = 14.003074 amu; $^{14}_6$C = 14.003241 amu; $^1_1$H = 1.007825 amu; mass of $^1_0$n = 1.008665 amu.)

••24.129 What is the nuclear binding energy of a lithium-7 nucleus in units of kJ/mol and eV/nucleus? (Mass of a lithium-7 atom = 7.016003 amu.)

••24.130 Gadolinium-146 undergoes electron capture. Identify the product, and use Figure 24.2 to find the modes of decay and the other two nuclides in the series below:

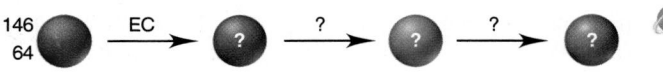

•••24.131 Using 21ˢᵗ-century technology, hydrogen fusion requires temperatures around 10⁸ K. But, lower initial temperatures are used if the hydrogen is compressed. In the late 24ᵗʰ century, the starship *Leinad* uses such methods to fuse hydrogen at 10⁶ K.
(a) What is the kinetic energy of an H atom at 1.00×10⁶ K?
(b) How many H atoms are heated to 1.00×10⁶ K from the energy of one H and one anti-H atom annihilating each other?
(c) If these H atoms fuse into ^{4}He atoms (with the loss of two positrons per ^{4}He formed), how much energy (in J) is generated?
(d) How much more energy is generated by the fusion in (c) than by the hydrogen-antihydrogen collision in (b)?
(e) Should the captain of the *Leinad* change the technology and produce ^{3}He (mass = 3.01603 amu) instead of ^{4}He?

•••24.132 A metastable (excited) form of ^{50}Sc changes to its stable form by emitting γ radiation with a wavelength of 8.73 pm. What is the change in mass of 1 mol of the isotope when it undergoes this change?

••24.133 A sample of cobalt-60 ($t_{1/2}$ = 5.27 yr), a powerful γ emitter used to treat cancer, was purchased by a hospital on March 1, 2007. The sample must be replaced when its activity reaches 70.% of the original value. On what date must it be replaced?

•••24.134 Uranium-233 decays to thorium-229 by α decay, but the emissions have different energies and products: 83% emit an α particle with energy of 4.816 MeV and give ^{229}Th in its ground state; 15% emit an α particle of 4.773 MeV and give ^{229}Th in excited state I; and 2% emit a lower energy α particle and give ^{229}Th in the higher excited state II. Excited state II emits a γ ray of 0.060 MeV to reach excited state I. (a) Find the γ-ray energy and wavelength that would convert excited state I to the ground state. (b) Find the energy of the α particle that would convert ^{233}U to excited state II.

•••24.135 Uranium-238 undergoes a slow decay step ($t_{1/2}$ = 4.5×10⁹ yr) followed by a series of fast steps to form the stable isotope ^{206}Pb. Thus, on a time scale of billions of years, ^{238}U effectively decays "directly" to ^{206}Pb, and the relative amounts of these isotopes are used to find the age of some rocks (see margin note, p. 1079). Two students derive equations relating number of half-lives (n) since the rock formed to the amounts of the isotopes:

Student 1: $\quad (\frac{1}{2})^n = \dfrac{^{238}_{92}U}{^{206}_{82}Pb}$

Student 2: $\quad (\frac{1}{2})^n = \dfrac{^{238}_{92}U}{^{238}_{92}U + ^{206}_{82}Pb}$

(a) Which equation is correct, and why?
(b) If a rock contains exactly twice as much ^{238}U as ^{206}Pb, what is its age in years?

••24.136 In the naturally occurring thorium-232 decay series, the steps emit this sequence of particles: α, β^-, β^-, α, α, α, α, β^-, β^-, and α. Write a balanced equation for each step.

••24.137 At death, a nobleman in ancient Egypt was mummified and his body contained 1.4×10⁻³ g of ^{40}K ($t_{1/2}$ = 1.25×10⁹ yr), 1.2×10⁻⁸ g of ^{14}C ($t_{1/2}$ = 5730 yr), and 4.8×10⁻¹⁴ g of ^{3}H

($t_{1/2}$ = 12.26 yr). Which nuclide would give the most accurate estimate of the mummy's age? Explain.

•• **24.138** Assuming that many radioactive nuclides can be considered safe after 20 half-lives, how long will it take for each of the following nuclides to be safe: (a) ^{242}Cm ($t_{1/2}$ = 163 days); (b) ^{214}Po ($t_{1/2}$ = 1.6×10^{-4} s); (c) ^{232}Th ($t_{1/2}$ = 1.39×10^{10} yr)?

•• **24.139** An ancient sword has a blade from the early Roman Empire, around 100 AD, but the wooden handle, inlaid wooden decorations, leather ribbon, and leather sheath have different styles. Given the following activities, estimate the age of each part. Which part was made near the time of the blade ($t_{1/2}$ of ^{14}C = 5730 yr; $\mathscr{A}_0$ = 15.3 d/min·g)?

Part	$\mathscr{A}_t$ (d/min·g)
Handle	10.1
Inlaid wood	13.8
Ribbon	12.1
Sheath	15.0

••• **24.140** The starship *Voyager,* like many other vessels of the newly designed 24th-century fleet, uses antimatter as fuel.
(a) How much energy is released when 1.00 kg each of antimatter and matter annihilate each other?
(b) When the antimatter is atomic antihydrogen, a small amount of it is mixed with excess atomic hydrogen (gathered from interstellar space during flight). The annihilation releases so much heat that the remaining hydrogen nuclei fuse to form ^{4}He. If each hydrogen-antihydrogen collision releases enough heat to fuse 1.00×10^5 hydrogen atoms, how much energy (in kJ) is released per kilogram of antihydrogen?
(c) Which produces more energy per kilogram of antihydrogen, the procedure in part (a) or that in part (b)?

••• **24.141** Use Einstein's equation, the mass in grams of 1 amu, and the relation between electron volts and joules to find the energy equivalent (in MeV) of a mass difference of 1 amu.

•• **24.142** Determine the age of a rock containing 0.065 g of uranium-238 ($t_{1/2}$ = 4.5×10^9 yr) and 0.023 g of lead-206. (Assume all the lead-206 came from ^{238}U decay.)

••• **24.143** Plutonium-242 decays to uranium-238 by emission of an α particle with an energy of 4.853 MeV. The ^{238}U that forms is unstable and emits a γ ray (λ = 0.02757 nm). (a) Write balanced equations for these reactions. (b) What would be the energy of the α particle if ^{242}Pu decayed directly to the more stable ^{238}U?

•• **24.144** Seaborgium-263 (Sg), the first isotope of element 106 synthesized, was made, together with four neutrons, by bombarding californium-249 with oxygen-18. It then underwent a series of decays starting with three α emissions. Write balanced equations for the synthesis and the three α emissions of ^{263}Sg.

•• **24.145** Some nuclear power plants use plutonium-239, which is produced in breeder reactors (see margin note, p. 1097). The rate-determining step is the second β^- emission. How long does it take to make 1.00 kg of ^{239}Pu if the reaction is complete when the product is 90.% ^{239}Pu?

••• **24.146** A random-number generator can be used to simulate the probability of a given atom decaying over a given time. For example, the formula "=RAND()" in the Excel spreadsheet returns a random number between 0 and 1; thus, for one radioactive atom and a time of one half-life, a number less than 0.5 means the atom decays and a number greater than 0.5 means it doesn't.
(a) Place the "=RAND()" formula in cells A1 to A10 of an Excel spreadsheet. In cell B1, place "=IF(A1<0.5, 0, 1)." This formula returns 0 if A1 is <0.5 (the atom decays) and 1 if A1 is >0.5 (the atom does not decay). Place analogous formulas in cells B2 to B10 (using the "Fill Down" procedure in Excel). To determine the number of atoms remaining after one half-life, sum cells B1 to B10 by placing "=SUM(B1:B10)" in cell B12. To create a new set of random numbers, click on an empty cell (e.g., B13) and hit "Delete." Perform 10 simulations, each time recording the total number of atoms remaining. Do half of the atoms remain after each half-life? If not, why not?
(b) Increase the number of atoms to 100 by placing suitable formulas in cells A1 to A100, B1 to B100, and B102. Perform 10 simulations, and record the number of atoms remaining each time. Are these results more realistic for radioactive decay? Explain.

••• **24.147** In the following Excel-based simulation, the fate of 256 atoms is followed over five half-lives. Set up formulas in columns A and B, as in Problem 24.146, and simulate the fate of the sample of 256 atoms over one half-life. Cells B1 to B256 should contain 1 or 0. In cell C1, enter "=IF(B1=0, 0, RAND())." This returns 0 if the original atom decayed in the previous half-life or a random number between 0 and 1 if it did not. Fill the formula in C1 down to cell C256. Column D should have formulas similar to those in B, but with modified references, as should columns F, H, and J. Columns E, G, and I should have formulas similar to those in C, but with modified references. In cell B258, enter "=SUM(B1:B256)." This records the number of atoms remaining after the first half-life. Put formulas in cells D258, F258, H258, and J258 to record atoms remaining after subsequent half-lives.
(a) Ideally, how many atoms should remain after each half-life?
(b) Make a table of the atoms remaining after each half-life in four separate simulations. Compare these outcomes to the ideal outcome. How would you make the results more realistic?

•• **24.148** Representations of three nuclei (with neutrons gray and protons purple) are shown below. Nucleus 1 is stable, but 2 and 3 are not. (a) Write the symbol for each isotope. (b) What is (are) the most likely mode(s) of decay for 2 and 3?

1   2   3

COMMON MATHEMATICAL OPERATIONS IN CHEMISTRY

In addition to basic arithmetic and algebra, four mathematical operations are used frequently in general chemistry: manipulating logarithms, using exponential notation, solving quadratic equations, and graphing data. Each is discussed briefly below.

MANIPULATING LOGARITHMS

Meaning and Properties of Logarithms

A *logarithm* is an exponent. Specifically, if $x^n = A$, we can say that the logarithm to the base x of the number A is n, and we can denote it as

$$\log_x A = n$$

Because logarithms are exponents, they have the following properties:

$$\log_x 1 = 0$$
$$\log_x (A \times B) = \log_x A + \log_x B$$
$$\log_x \frac{A}{B} = \log_x A - \log_x B$$
$$\log_x A^y = y \log_x A$$

Types of Logarithms

Common and natural logarithms are used frequently in chemistry and the other sciences. For *common* logarithms, the base (x in the examples above) is 10, but they are written without specifying the base; that is, $\log_{10} A$ is written more simply as $\log A$. The common logarithm of 1000 is 3; in other words, you must raise 10 to the 3rd power to obtain 1000:

$$\log 1000 = 3 \qquad \text{or} \qquad 10^3 = 1000$$

Similarly, we have

$$\log 10 = 1 \qquad \text{or} \qquad 10^1 = 10$$
$$\log 1{,}000{,}000 = 6 \qquad \text{or} \qquad 10^6 = 1{,}000{,}000$$
$$\log 0.001 = -3 \qquad \text{or} \qquad 10^{-3} = 0.001$$
$$\log 853 = 2.931 \qquad \text{or} \qquad 10^{2.931} = 853$$

The last example illustrates an important point about significant figures with all logarithms: the number of significant figures in the number equals the number of digits to the right of the decimal point in the logarithm. That is, the number 853 has three significant figures, and the logarithm 2.931 has three digits to the right of the decimal point.

To find a common logarithm with an electronic calculator, you simply enter the number and press the LOG button.

For *natural* logarithms, the base is the number e, which is 2.71828 . . . , and $\log_e A$ is written $\ln A$. The relationship between the common and natural logarithms is easily obtained: because

$$\log 10 = 1 \qquad \text{and} \qquad \ln 10 = 2.303$$

Therefore, we have

$$\ln A = 2.303 \log A$$

To find a natural logarithm with an electronic calculator, you simply enter the number and press the LN button. If your calculator does not have an LN button, enter the number, press the LOG button, and multiply by 2.303.

Antilogarithms

The *antilogarithm* is the number you obtain when you raise the base to the logarithm:

$$\text{antilogarithm (antilog) of } n \text{ is } 10^n$$

Using two of the earlier examples, the antilog of 3 is 1000, and the antilog of 2.931 is 853. To obtain the antilog with a calculator, you enter the number and press the 10^x button. Similarly, to obtain the natural antilogarithm, you enter the number and press the e^x button. [On some calculators, you enter the number and first press INV and then the LOG (or LN) button.]

USING EXPONENTIAL (SCIENTIFIC) NOTATION

Many quantities in chemistry are very large or very small. For example, in the conventional way of writing numbers, the number of gold atoms in 1 gram of gold is

59,060,000,000,000,000,000,000 atoms (to four significant figures)

As another example, the mass in grams of one gold atom is

0.00000000000000000000003272 g (to four significant figures)

Exponential (scientific) notation provides a much more practical way of writing such numbers. In exponential notation, we express numbers in the form

$$A \times 10^n$$

where A (the coefficient) is greater than or equal to 1 and less than 10 (that is, $1 \le A < 10$), and n (the exponent) is an integer.

If the number we want to express in exponential notation is larger than 1, the exponent is positive ($n > 0$); if the number is smaller than 1, the exponent is negative ($n < 0$). The size of n tells the number of places the decimal point (in conventional notation) must be moved to obtain a coefficient A greater than or equal to 1 and less than 10 (in exponential notation). In exponential notation, 1 gram of gold contains 5.906×10^{22} atoms, and each gold atom has a mass of 3.272×10^{-22} g.

Changing Between Conventional and Exponential Notation

In order to use exponential notation, you must be able to convert to it from conventional notation, and vice versa.

1. To change a number from conventional to exponential notation, move the decimal point to the left for numbers equal to or greater than 10 and to the right for numbers between 0 and 1:

    75,000,000 changes to 7.5×10^7 (decimal point 7 places to the left)
    0.006042 changes to 6.042×10^{-3} (decimal point 3 places to the right)

2. To change a number from exponential to conventional notation, move the decimal point the number of places indicated by the exponent to the right for numbers with positive exponents and to the left for numbers with negative exponents:

    1.38×10^5 changes to 138,000 (decimal point 5 places to the right)
    8.41×10^{-6} changes to 0.00000841 (decimal point 6 places to the left)

3. An exponential number with a coefficient greater than 10 or less than 1 can be changed to the standard exponential form by converting the coefficient to the standard form and adding the exponents:

$$582.3 \times 10^6 \text{ changes to } 5.823 \times 10^2 \times 10^6 = 5.823 \times 10^{(2+6)} = 5.823 \times 10^8$$
$$0.0043 \times 10^{-4} \text{ changes to } 4.3 \times 10^{-3} \times 10^{-4} = 4.3 \times 10^{[(-3)+(-4)]} = 4.3 \times 10^{-7}$$

Using Exponential Notation in Calculations

In calculations, you can treat the coefficient and exponents separately and apply the properties of exponents (see earlier section on logarithms).

1. To multiply exponential numbers, multiply the coefficients, add the exponents, and reconstruct the number in standard exponential notation:

$$(5.5 \times 10^3)(3.1 \times 10^5) = (5.5 \times 3.1) \times 10^{(3+5)} = 17 \times 10^8 = 1.7 \times 10^9$$
$$(9.7 \times 10^{14})(4.3 \times 10^{-20}) = (9.7 \times 4.3) \times 10^{[14+(-20)]} = 42 \times 10^{-6} = 4.2 \times 10^{-5}$$

2. To divide exponential numbers, divide the coefficients, subtract the exponents, and reconstruct the number in standard exponential notation:

$$\frac{2.6 \times 10^6}{5.8 \times 10^2} = \frac{2.6}{5.8} \times 10^{(6-2)} = 0.45 \times 10^4 = 4.5 \times 10^3$$
$$\frac{1.7 \times 10^{-5}}{8.2 \times 10^{-8}} = \frac{1.7}{8.2} \times 10^{[(-5)-(-8)]} = 0.21 \times 10^3 = 2.1 \times 10^2$$

3. To add or subtract exponential numbers, change all numbers so that they have the same exponent, then add or subtract the coefficients:

$$(1.45 \times 10^4) + (3.2 \times 10^3) = (1.45 \times 10^4) + (0.32 \times 10^4) = 1.77 \times 10^4$$
$$(3.22 \times 10^5) - (9.02 \times 10^4) = (3.22 \times 10^5) - (0.902 \times 10^5) = 2.32 \times 10^5$$

SOLVING QUADRATIC EQUATIONS

A *quadratic equation* is one in which the highest power of x is 2. The general form of a quadratic equation is

$$ax^2 + bx + c = 0$$

where a, b, and c are numbers. For given values of a, b, and c, the values of x that satisfy the equation are called *solutions* of the equation. We calculate x with the quadratic formula:

$$x = \frac{-b \pm \sqrt{b^2 - 4ac}}{2a}$$

We commonly require the quadratic formula when solving for some concentration in an equilibrium problem. For example, we might have an expression that is rearranged into the quadratic equation

$$4.3x^2 + 0.65x - 8.7 = 0$$

Applying the quadratic formula, with $a = 4.3$, $b = 0.65$, and $c = -8.7$, gives

$$x = \frac{-0.65 \pm \sqrt{(0.65)^2 - 4(4.3)(-8.7)}}{2(4.3)}$$

The "plus or minus" sign ($\pm$) indicates that there are always two possible values for x. In this case, they are

$$x = 1.3 \quad \text{and} \quad x = -1.5$$

In any real physical system, however, only one of the values will have any meaning. For example, if x were $[H_3O^+]$, the negative value would mean a negative concentration, which has no meaning.

GRAPHING DATA IN THE FORM OF A STRAIGHT LINE

Visualizing changes in variables by means of a graph is a very useful technique in science. In many cases, it is most useful if the data can be graphed in the form of a straight line. Any equation will appear as a straight line if it has, or can be rearranged to have, the following general form:

$$y = mx + b$$

where y is the dependent variable (typically plotted along the vertical axis), x is the independent variable (typically plotted along the horizontal axis), m is the slope of the line, and b is the intercept of the line on the y axis. The intercept is the value of y when $x = 0$:

$$y = m(0) + b = b$$

The slope of the line is the change in y for a given change in x:

$$\text{Slope } (m) = \frac{y_2 - y_1}{x_2 - x_1} = \frac{\Delta y}{\Delta x}$$

The *sign* of the slope tells the *direction* of the line. If y increases as x increases, m is positive, and the line slopes upward with higher values of x; if y decreases as x increases, m is negative, and the line slopes downward with higher values of x. The *magnitude* of the slope indicates the *steepness* of the line. A line with $m = 3$ is three times as steep (y changes three times as much for a given change in x) as a line with $m = 1$.

Consider the linear equation $y = 2x + 1$. A graph of this equation is shown in Figure A.1. In practice, you can find the slope by drawing a right triangle to the line, using the line as the hypotenuse. Then, one leg gives Δy, and the other gives Δx. In the figure, $\Delta y = 8$ and $\Delta x = 4$.

At several places in the text, an equation is rearranged into the form of a straight line in order to determine information from the slope and/or the intercept. For example, in Chapter 16, we obtained the following expression:

$$\ln \frac{[A]_0}{[A]_t} = kt$$

Based on the properties of logarithms, we have

$$\ln [A]_0 - \ln [A]_t = kt$$

Rearranging into the form of an equation for a straight line gives

$$\ln [A]_t = -kt + \ln [A]_0$$
$$ y = mx + b$$

Thus, a plot of $\ln [A]_t$ vs. t is a straight line, from which you can see that the slope is $-k$ (the negative of the rate constant) and the intercept is $\ln [A]_0$ (the natural logarithm of the initial concentration of A).

At many other places in the text, linear relationships occur that were not shown in graphical terms. For example, the conversion of temperature scales in Chapter 1 can also be expressed in the form of a straight line:

$$°F = \tfrac{9}{5} °C + 32$$
$$ y = mx + b$$

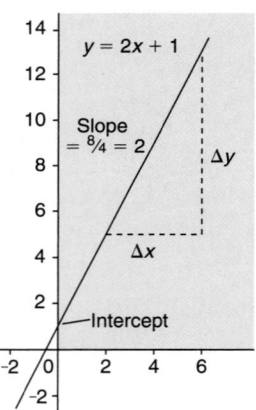

Figure A.1

STANDARD THERMODYNAMIC VALUES FOR SELECTED SUBSTANCES*

Substance or Ion	ΔH°_f (kJ/mol)	ΔG°_f (kJ/mol)	S° (J/mol·K)	Substance or Ion	ΔH°_f (kJ/mol)	ΔG°_f (kJ/mol)	S° (J/mol·K)
$e^-(g)$	0	0	20.87	$CaCO_3(s)$	−1206.9	−1128.8	92.9
Aluminum				$CaO(s)$	−635.1	−603.5	38.2
$Al(s)$	0	0	28.3	$Ca(OH)_2(s)$	−986.09	−898.56	83.39
$Al^{3+}(aq)$	−524.7	−481.2	−313	$Ca_3(PO_4)_2(s)$	−4138	−3899	263
$AlCl_3(s)$	−704.2	−628.9	110.7	$CaSO_4(s)$	−1432.7	−1320.3	107
$Al_2O_3(s)$	−1676	−1582	50.94	**Carbon**			
Barium				$C(graphite)$	0	0	5.686
$Ba(s)$	0	0	62.5	$C(diamond)$	1.896	2.866	2.439
$Ba(g)$	175.6	144.8	170.28	$C(g)$	715.0	669.6	158.0
$Ba^{2+}(g)$	1649.9	—	—	$CO(g)$	−110.5	−137.2	197.5
$Ba^{2+}(aq)$	−538.36	−560.7	13	$CO_2(g)$	−393.5	−394.4	213.7
$BaCl_2(s)$	−806.06	−810.9	126	$CO_2(aq)$	−412.9	−386.2	121
$BaCO_3(s)$	−1219	−1139	112	$CO_3^{2-}(aq)$	−676.26	−528.10	−53.1
$BaO(s)$	−548.1	−520.4	72.07	$HCO_3^-(aq)$	−691.11	587.06	95.0
$BaSO_4(s)$	−1465	−1353	132	$H_2CO_3(aq)$	−698.7	−623.42	191
Boron				$CH_4(g)$	−74.87	−50.81	186.1
$B(\beta$-rhombo-	0	0	5.87	$C_2H_2(g)$	227	209	200.85
hedral)				$C_2H_4(g)$	52.47	68.36	219.22
$BF_3(g)$	−1137.0	−1120.3	254.0	$C_2H_6(g)$	−84.667	−32.89	229.5
$BCl_3(g)$	−403.8	−388.7	290.0	$C_3H_8(g)$	−105	−24.5	269.9
$B_2H_6(g)$	35	86.6	232.0	$C_4H_{10}(g)$	−126	−16.7	310
$B_2O_3(s)$	−1272	−1193	53.8	$C_6H_6(l)$	49.0	124.5	172.8
$H_3BO_3(s)$	−1094.3	−969.01	88.83	$CH_3OH(g)$	−201.2	−161.9	238
Bromine				$CH_3OH(l)$	−238.6	−166.2	127
$Br_2(l)$	0	0	152.23	$HCHO(g)$	−116	−110	219
$Br_2(g)$	30.91	3.13	245.38	$HCOO^-(aq)$	−410	−335	91.6
$Br(g)$	111.9	82.40	174.90	$HCOOH(l)$	−409	−346	129.0
$Br^-(g)$	−218.9	—	—	$HCOOH(aq)$	−410	−356	164
$Br^-(aq)$	−120.9	−102.82	80.71	$C_2H_5OH(g)$	−235.1	−168.6	282.6
$HBr(g)$	−36.3	−53.5	198.59	$C_2H_5OH(l)$	−277.63	−174.8	161
Cadmium				$CH_3CHO(g)$	−166	−133.7	266
$Cd(s)$	0	0	51.5	$CH_3COOH(l)$	−487.0	−392	160
$Cd(g)$	112.8	78.20	167.64	$C_6H_{12}O_6(s)$	−1273.3	−910.56	212.1
$Cd^{2+}(aq)$	−72.38	−77.74	−61.1	$C_{12}H_{22}O_{11}(s)$	−2221.7	−1544.3	360.24
$CdS(s)$	−144	−141	71	$CN^-(aq)$	151	166	118
Calcium				$HCN(g)$	135	125	201.7
$Ca(s)$	0	0	41.6	$HCN(l)$	105	121	112.8
$Ca(g)$	192.6	158.9	154.78	$HCN(aq)$	105	112	129
$Ca^{2+}(g)$	1934.1	—	—	$CS_2(g)$	117	66.9	237.79
$Ca^{2+}(aq)$	−542.96	−553.04	−55.2	$CS_2(l)$	87.9	63.6	151.0
$CaF_2(s)$	−1215	−1162	68.87	$CH_3Cl(g)$	−83.7	−60.2	234
$CaCl_2(s)$	−795.0	−750.2	114	$CH_2Cl_2(l)$	−117	−63.2	179

*All values at 298 K.

(*continued*)

Substance or Ion	ΔH_f° (kJ/mol)	ΔG_f° (kJ/mol)	S° (J/mol·K)	Substance or Ion	ΔH_f° (kJ/mol)	ΔG_f° (kJ/mol)	S° (J/mol·K)
$CHCl_3(l)$	−132	−71.5	203	$Fe^{2+}(aq)$	−87.9	−84.94	113
$CCl_4(g)$	−96.0	−53.7	309.7	$FeCl_2(s)$	−341.8	−302.3	117.9
$CCl_4(l)$	−139	−68.6	214.4	$FeCl_3(s)$	−399.5	−334.1	142
$COCl_2(g)$	−220	−206	283.74	$FeO(s)$	−272.0	−251.4	60.75
Cesium				$Fe_2O_3(s)$	−825.5	−743.6	87.400
$Cs(s)$	0	0	85.15	$Fe_3O_4(s)$	−1121	−1018	145.3
$Cs(g)$	76.7	49.7	175.5	**Lead**			
$Cs^+(g)$	458.5	427.1	169.72	$Pb(s)$	0	0	64.785
$Cs^+(aq)$	−248	−282.0	133	$Pb^{2+}(aq)$	1.6	−24.3	21
$CsF(s)$	−554.7	−525.4	88	$PbCl_2(s)$	−359	−314	136
$CsCl(s)$	−442.8	−414	101.18	$PbO(s)$	−218	−198	68.70
$CsBr(s)$	−395	−383	121	$PbO_2(s)$	−276.6	−219.0	76.6
$CsI(s)$	−337	−333	130	$PbS(s)$	−98.3	−96.7	91.3
Chlorine				$PbSO_4(s)$	−918.39	−811.24	147
$Cl_2(g)$	0	0	223.0	**Lithium**			
$Cl(g)$	121.0	105.0	165.1	$Li(s)$	0	0	29.10
$Cl^-(g)$	−234	−240	153.25	$Li(g)$	161	128	138.67
$Cl^-(aq)$	−167.46	−131.17	55.10	$Li^+(g)$	687.163	649.989	132.91
$HCl(g)$	−92.31	−95.30	186.79	$Li^+(aq)$	−278.46	−293.8	14
$HCl(aq)$	−167.46	−131.17	55.06	$LiF(s)$	−616.9	−588.7	35.66
$ClO_2(g)$	102	120	256.7	$LiCl(s)$	−408	−384	59.30
$Cl_2O(g)$	80.3	97.9	266.1	$LiBr(s)$	−351	−342	74.1
Chromium				$LiI(s)$	−270	−270	85.8
$Cr(s)$	0	0	23.8	**Magnesium**			
$Cr^{3+}(aq)$	−1971	—	—	$Mg(s)$	0	0	32.69
$CrO_4^{2-}(aq)$	−863.2	−706.3	38	$Mg(g)$	150	115	148.55
$Cr_2O_7^{2-}(aq)$	−1461	−1257	214	$Mg^{2+}(g)$	2351	—	—
Copper				$Mg^{2+}(aq)$	−461.96	−456.01	118
$Cu(s)$	0	0	33.1	$MgCl_2(s)$	−641.6	−592.1	89.630
$Cu(g)$	341.1	301.4	166.29	$MgCO_3(s)$	−1112	−1028	65.86
$Cu^+(aq)$	51.9	50.2	−26	$MgO(s)$	−601.2	−569.0	26.9
$Cu^{2+}(aq)$	64.39	64.98	−98.7	$Mg_3N_2(s)$	−461	−401	88
$Cu_2O(s)$	−168.6	−146.0	93.1	**Manganese**			
$CuO(s)$	−157.3	−130	42.63	$Mn(s, \alpha)$	0	0	31.8
$Cu_2S(s)$	−79.5	−86.2	120.9	$Mn^{2+}(aq)$	−219	−223	−84
$CuS(s)$	−53.1	−53.6	66.5	$MnO_2(s)$	−520.9	−466.1	53.1
Fluorine				$MnO_4^-(aq)$	−518.4	−425.1	190
$F_2(g)$	0	0	202.7	**Mercury**			
$F(g)$	78.9	61.8	158.64	$Hg(l)$	0	0	76.027
$F^-(g)$	−255.6	−262.5	145.47	$Hg(g)$	61.30	31.8	174.87
$F^-(aq)$	−329.1	−276.5	−9.6	$Hg^{2+}(aq)$	171	164.4	−32
$HF(g)$	−273	−275	173.67	$Hg_2^{2+}(aq)$	172	153.6	84.5
Hydrogen				$HgCl_2(s)$	−230	−184	144
$H_2(g)$	0	0	130.6	$Hg_2Cl_2(s)$	−264.9	−210.66	196
$H(g)$	218.0	203.30	114.60	$HgO(s)$	−90.79	−58.50	70.27
$H^+(aq)$	0	0	0	**Nitrogen**			
$H^+(g)$	1536.3	1517.1	108.83	$N_2(g)$	0	0	191.5
Iodine				$N(g)$	473	456	153.2
$I_2(s)$	0	0	116.14	$N_2O(g)$	82.05	104.2	219.7
$I_2(g)$	62.442	19.38	260.58	$NO(g)$	90.29	86.60	210.65
$I(g)$	106.8	70.21	180.67	$NO_2(g)$	33.2	51	239.9
$I^-(g)$	−194.7	—	—	$N_2O_4(g)$	9.16	97.7	304.3
$I^-(aq)$	−55.94	−51.67	109.4	$N_2O_5(g)$	11	118	346
$HI(g)$	25.9	1.3	206.33	$N_2O_5(s)$	−43.1	114	178
Iron				$NH_3(g)$	−45.9	−16	193
$Fe(s)$	0	0	27.3	$NH_3(aq)$	−80.83	26.7	110
$Fe^{3+}(aq)$	−47.7	−10.5	−293	$N_2H_4(l)$	50.63	149.2	121.2

(*continued*)

Substance or Ion	ΔH°_f (kJ/mol)	ΔG°_f (kJ/mol)	S° (J/mol·K)	Substance or Ion	ΔH°_f (kJ/mol)	ΔG°_f (kJ/mol)	S° (J/mol·K)
$NO_3^-(aq)$	−206.57	−110.5	146	$Ag(g)$	289.2	250.4	172.892
$HNO_3(l)$	−173.23	−79.914	155.6	$Ag^+(aq)$	105.9	77.111	73.93
$HNO_3(aq)$	−206.57	−110.5	146	$AgF(s)$	−203	−185	84
$NF_3(g)$	−125	−83.3	260.6	$AgCl(s)$	−127.03	−109.72	96.11
$NOCl(g)$	51.71	66.07	261.6	$AgBr(s)$	−99.51	−95.939	107.1
$NH_4Cl(s)$	−314.4	−203.0	94.6	$AgI(s)$	−62.38	−66.32	114
Oxygen				$AgNO_3(s)$	−45.06	19.1	128.2
$O_2(g)$	0	0	205.0	$Ag_2S(s)$	−31.8	−40.3	146
$O(g)$	249.2	231.7	160.95	Sodium			
$O_3(g)$	143	163	238.82	$Na(s)$	0	0	51.446
$OH^-(aq)$	−229.94	−157.30	−10.54	$Na(g)$	107.76	77.299	153.61
$H_2O(g)$	−241.826	−228.60	188.72	$Na^+(g)$	609.839	574.877	147.85
$H_2O(l)$	−285.840	−237.192	69.940	$Na^+(aq)$	−239.66	−261.87	60.2
$H_2O_2(l)$	−187.8	−120.4	110	$NaF(s)$	−575.4	−545.1	51.21
$H_2O_2(aq)$	−191.2	−134.1	144	$NaCl(s)$	−411.1	−384.0	72.12
Phosphorus				$NaBr(s)$	−361	−349	86.82
$P_4(s, white)$	0	0	41.1	$NaOH(s)$	−425.609	−379.53	64.454
$P(g)$	314.6	278.3	163.1	$Na_2CO_3(s)$	−1130.8	−1048.1	139
$P(s, red)$	−17.6	−12.1	22.8	$NaHCO_3(s)$	−947.7	−851.9	102
$P_2(g)$	144	104	218	$NaI(s)$	−288	−285	98.5
$P_4(g)$	58.9	24.5	280	Strontium			
$PCl_3(g)$	−287	−268	312	$Sr(s)$	0	0	54.4
$PCl_3(l)$	−320	−272	217	$Sr(g)$	164	110	164.54
$PCl_5(g)$	−402	−323	353	$Sr^{2+}(g)$	1784	—	—
$PCl_5(s)$	−443.5	—	—	$Sr^{2+}(aq)$	−545.51	−557.3	−39
$P_4O_{10}(s)$	−2984	−2698	229	$SrCl_2(s)$	−828.4	−781.2	117
$PO_4^{3-}(aq)$	−1266	−1013	−218	$SrCO_3(s)$	−1218	−1138	97.1
$HPO_4^{2-}(aq)$	−1281	−1082	−36	$SrO(s)$	−592.0	−562.4	55.5
$H_2PO_4^-(aq)$	−1285	−1135	89.1	$SrSO_4(s)$	−1445	−1334	122
$H_3PO_4(aq)$	−1277	−1019	228	Sulfur			
Potassium				$S(rhombic)$	0	0	31.9
$K(s)$	0	0	64.672	$S(monoclinic)$	0.3	0.096	32.6
$K(g)$	89.2	60.7	160.23	$S(g)$	279	239	168
$K^+(g)$	514.197	481.202	154.47	$S_2(g)$	129	80.1	228.1
$K^+(aq)$	−251.2	−282.28	103	$S_8(g)$	101	49.1	430.211
$KF(s)$	−568.6	−538.9	66.55	$S^{2-}(aq)$	41.8	83.7	22
$KCl(s)$	−436.7	−409.2	82.59	$HS^-(aq)$	−17.7	12.6	61.1
$KBr(s)$	−394	−380	95.94	$H_2S(g)$	−20.2	−33	205.6
$KI(s)$	−328	−323	106.39	$H_2S(aq)$	−39	−27.4	122
$KOH(s)$	−424.8	−379.1	78.87	$SO_2(g)$	−296.8	−300.2	248.1
$KClO_3(s)$	−397.7	−296.3	143.1	$SO_3(g)$	−396	−371	256.66
$KClO_4(s)$	−432.75	−303.2	151.0	$SO_4^{2-}(aq)$	−907.51	−741.99	17
Rubidium				$HSO_4^-(aq)$	−885.75	−752.87	126.9
$Rb(s)$	0	0	69.5	$H_2SO_4(l)$	−813.989	−690.059	156.90
$Rb(g)$	85.81	55.86	169.99	$H_2SO_4(aq)$	−907.51	−741.99	17
$Rb^+(g)$	495.04	—	—	Tin			
$Rb^+(aq)$	−246	−282.2	124	$Sn(white)$	0	0	51.5
$RbF(s)$	−549.28	—	—	$Sn(gray)$	3	4.6	44.8
$RbCl(s)$	−435.35	−407.8	95.90	$SnCl_4(l)$	−545.2	−474.0	259
$RbBr(s)$	−389.2	−378.1	108.3	$SnO_2(s)$	−580.7	−519.7	52.3
$RbI(s)$	−328	−326	118.0	Zinc			
Silicon				$Zn(s)$	0	0	41.6
$Si(s)$	0	0	18.0	$Zn(g)$	130.5	94.93	160.9
$SiF_4(g)$	−1614.9	−1572.7	282.4	$Zn^{2+}(aq)$	−152.4	−147.21	−106.5
$SiO_2(s)$	−910.9	−856.5	41.5	$ZnO(s)$	−348.0	−318.2	43.9
Silver				$ZnS(s, zinc$	−203	−198	57.7
$Ag(s)$	0	0	42.702	blende)			

APPENDIX C

EQUILIBRIUM CONSTANTS FOR SELECTED SUBSTANCES*

Dissociation (Ionization) Constants (K_a) of Selected Acids

Name and Formula	Lewis Structure[†]	K_{a1}	K_{a2}	K_{a3}
Acetic acid CH_3COOH		1.8×10^{-5}		
Acetylsalicylic acid[††] $CH_3COOC_6H_4COOH$		3.6×10^{-4}		
Adipic acid $HOOC(CH_2)_4COOH$		3.8×10^{-5}	3.8×10^{-6}	
Arsenic acid H_3AsO_4		6×10^{-3}	1.1×10^{-7}	3×10^{-12}
Ascorbic acid $H_2C_6H_6O_6$		1.0×10^{-5}	5×10^{-12}	
Benzoic acid C_6H_5COOH		6.3×10^{-5}		
Carbonic acid H_2CO_3		4.5×10^{-7}	4.7×10^{-11}	
Chloroacetic acid $ClCH_2COOH$		1.4×10^{-3}		
Chlorous acid $HClO_2$		1.1×10^{-2}		

*All values at 298 K, except for acetylsalicylic acid.
[†]Acidic (ionizable) proton(s) shown in red. Structures shown have lowest formal charges.
[††]At 37°C in 0.15 M NaCl.

Dissociation (Ionization) Constants (K_a) of Selected Acids

Name and Formula	Lewis Structure†	K_{a1}	K_{a2}	K_{a3}
Citric acid $HOOCCH_2C(OH)(COOH)CH_2COOH$		7.4×10^{-4}	1.7×10^{-5}	4.0×10^{-7}
Formic acid HCOOH		1.8×10^{-4}		
Glyceric acid $HOCH_2CH(OH)COOH$		2.9×10^{-4}		
Glycolic acid $HOCH_2COOH$		1.5×10^{-4}		
Glyoxylic acid HC(O)COOH		3.5×10^{-4}		
Hydrocyanic acid HCN		6.2×10^{-10}		
Hydrofluoric acid HF		6.8×10^{-4}		
Hydrosulfuric acid H_2S		9×10^{-8}	1×10^{-17}	
Hypobromous acid HBrO		2.3×10^{-9}		
Hypochlorous acid HClO		2.9×10^{-8}		
Hypoiodous acid HIO		2.3×10^{-11}		
Iodic acid HIO_3		1.6×10^{-1}		
Lactic acid $CH_3CH(OH)COOH$		1.4×10^{-4}		
Maleic acid $HOOCCH{=}CHCOOH$		1.2×10^{-2}	4.7×10^{-7}	

(*continued*)

Dissociation (Ionization) Constants (K_a) of Selected Acids (*continued*)

Name and Formula	Lewis Structure[†]	K_{a1}	K_{a2}	K_{a3}
Malonic acid $HOOCCH_2COOH$		1.4×10^{-3}	2.0×10^{-6}	
Nitrous acid HNO_2		7.1×10^{-4}		
Oxalic acid $HOOCCOOH$		5.6×10^{-2}	5.4×10^{-5}	
Phenol C_6H_5OH		1.0×10^{-10}		
Phenylacetic acid $C_6H_5CH_2COOH$		4.9×10^{-5}		
Phosphoric acid H_3PO_4		7.2×10^{-3}	6.3×10^{-8}	4.2×10^{-13}
Phosphorous acid $HPO(OH)_2$		3×10^{-2}	1.7×10^{-7}	
Propanoic acid CH_3CH_2COOH		1.3×10^{-5}		
Pyruvic acid $CH_3C(O)COOH$		2.8×10^{-3}		
Succinic acid $HOOCCH_2CH_2COOH$		6.2×10^{-5}	2.3×10^{-6}	
Sulfuric acid H_2SO_4		Very large	1.0×10^{-2}	
Sulfurous acid H_2SO_3		1.4×10^{-2}	6.5×10^{-8}	

(*continued*)

Dissociation (Ionization) Constants (K_b) of Selected Amine Bases

Name and Formula	Lewis Structure[†]	K_{b1}	K_{b2}
Ammonia NH_3		1.76×10^{-5}	
Aniline $C_6H_5NH_2$		4.0×10^{-10}	
Diethylamine $(CH_3CH_2)_2NH$		8.6×10^{-4}	
Dimethylamine $(CH_3)_2NH$		5.9×10^{-4}	
Ethanolamine $HOCH_2CH_2NH_2$		3.2×10^{-5}	
Ethylamine $CH_3CH_2NH_2$		4.3×10^{-4}	
Ethylenediamine $H_2NCH_2CH_2NH_2$		8.5×10^{-5}	7.1×10^{-8}
Methylamine CH_3NH_2		4.4×10^{-4}	
tert-Butylamine $(CH_3)_3CNH_2$		4.8×10^{-4}	
Piperidine $C_5H_{10}NH$		1.3×10^{-3}	
n-Propylamine $CH_3CH_2CH_2NH_2$		3.5×10^{-4}	

[†]Blue type indicates the basic nitrogen and its lone pair.

(*continued*)

Dissociation (Ionization) Constants (K_b) of Selected Amine Bases (*continued*)

Name and Formula	Lewis Structure[†]	K_{b1}	K_{b2}
Isopropylamine $(CH_3)_2CHNH_2$		4.7×10^{-4}	
1,3-Propylenediamine $H_2NCH_2CH_2CH_2NH_2$		3.1×10^{-4}	3.0×10^{-6}
Pyridine C_5H_5N		1.7×10^{-9}	
Triethylamine $(CH_3CH_2)_3N$		5.2×10^{-4}	
Trimethylamine $(CH_3)_3N$		6.3×10^{-5}	

Dissociation (Ionization) Constants (K_a) of Some Hydrated Metal Ions

Free Ion	Hydrated Ion	K_a
Fe^{3+}	$Fe(H_2O)_6^{3+}(aq)$	6×10^{-3}
Sn^{2+}	$Sn(H_2O)_6^{2+}(aq)$	4×10^{-4}
Cr^{3+}	$Cr(H_2O)_6^{3+}(aq)$	1×10^{-4}
Al^{3+}	$Al(H_2O)_6^{3+}(aq)$	1×10^{-5}
Cu^{2+}	$Cu(H_2O)_6^{2+}(aq)$	3×10^{-8}
Pb^{2+}	$Pb(H_2O)_6^{2+}(aq)$	3×10^{-8}
Zn^{2+}	$Zn(H_2O)_6^{2+}(aq)$	1×10^{-9}
Co^{2+}	$Co(H_2O)_6^{2+}(aq)$	2×10^{-10}
Ni^{2+}	$Ni(H_2O)_6^{2+}(aq)$	1×10^{-10}

Formation Constants (K_f) of Some Complex Ions

Complex Ion	K_f
$Ag(CN)_2^-$	3.0×10^{20}
$Ag(NH_3)_2^+$	1.7×10^7
$Ag(S_2O_3)_2^{3-}$	4.7×10^{13}
AlF_6^{3-}	4×10^{19}
$Al(OH)_4^-$	3×10^{33}
$Be(OH)_4^{2-}$	4×10^{18}
CdI_4^{2-}	1×10^6
$Co(OH)_4^{2-}$	5×10^9
$Cr(OH)_4^-$	8.0×10^{29}
$Cu(NH_3)_4^{2+}$	5.6×10^{11}
$Fe(CN)_6^{4-}$	3×10^{35}
$Fe(CN)_6^{3-}$	4.0×10^{43}
$Hg(CN)_4^{2-}$	9.3×10^{38}
$Ni(NH_3)_6^{2+}$	2.0×10^8
$Pb(OH)_3^-$	8×10^{13}
$Sn(OH)_3^-$	3×10^{25}
$Zn(CN)_4^{2-}$	4.2×10^{19}
$Zn(NH_3)_4^{2+}$	7.8×10^8
$Zn(OH)_4^{2-}$	3×10^{15}

Solubility-Product Constants (K_{sp}) of Slightly Soluble Ionic Compounds

Name, Formula	K_{sp}	Name, Formula	K_{sp}
Carbonates		Cobalt(II) hydroxide, $Co(OH)_2$	1.3×10^{-15}
Barium carbonate, $BaCO_3$	2.0×10^{-9}	Copper(II) hydroxide, $Cu(OH)_2$	2.2×10^{-20}
Cadmium carbonate, $CdCO_3$	1.8×10^{-14}	Iron(II) hydroxide, $Fe(OH)_2$	4.1×10^{-15}
Calcium carbonate, $CaCO_3$	3.3×10^{-9}	Iron(III) hydroxide, $Fe(OH)_3$	1.6×10^{-39}
Cobalt(II) carbonate, $CoCO_3$	1.0×10^{-10}	Magnesium hydroxide, $Mg(OH)_2$	6.3×10^{-10}
Copper(II) carbonate, $CuCO_3$	3×10^{-12}	Manganese(II) hydroxide, $Mn(OH)_2$	1.6×10^{-13}
Lead(II) carbonate, $PbCO_3$	7.4×10^{-14}	Nickel(II) hydroxide, $Ni(OH)_2$	6×10^{-16}
Magnesium carbonate, $MgCO_3$	3.5×10^{-8}	Zinc hydroxide, $Zn(OH)_2$	3×10^{-16}
Mercury(I) carbonate, Hg_2CO_3	8.9×10^{-17}	**Iodates**	
Nickel(II) carbonate, $NiCO_3$	1.3×10^{-7}	Barium iodate, $Ba(IO_3)_2$	1.5×10^{-9}
Strontium carbonate, $SrCO_3$	5.4×10^{-10}	Calcium iodate, $Ca(IO_3)_2$	7.1×10^{-7}
Zinc carbonate, $ZnCO_3$	1.0×10^{-10}	Lead(II) iodate, $Pb(IO_3)_2$	2.5×10^{-13}
Chromates		Silver iodate, $AgIO_3$	3.1×10^{-8}
Barium chromate, $BaCrO_4$	2.1×10^{-10}	Strontium iodate, $Sr(IO_3)_2$	3.3×10^{-7}
Calcium chromate, $CaCrO_4$	1×10^{-8}	Zinc iodate, $Zn(IO_3)_2$	3.9×10^{-6}
Lead(II) chromate, $PbCrO_4$	2.3×10^{-13}	**Oxalates**	
Silver chromate, Ag_2CrO_4	2.6×10^{-12}	Barium oxalate dihydrate, $BaC_2O_4\cdot2H_2O$	1.1×10^{-7}
Cyanides		Calcium oxalate monohydrate, $CaC_2O_4\cdot H_2O$	2.3×10^{-9}
Mercury(I) cyanide, $Hg_2(CN)_2$	5×10^{-40}	Strontium oxalate monohydrate,	
Silver cyanide, $AgCN$	2.2×10^{-16}	$SrC_2O_4\cdot H_2O$	5.6×10^{-8}
Halides		**Phosphates**	
Fluorides		Calcium phosphate, $Ca_3(PO_4)_2$	1.2×10^{-29}
Barium fluoride, BaF_2	1.5×10^{-6}	Magnesium phosphate, $Mg_3(PO_4)_2$	5.2×10^{-24}
Calcium fluoride, CaF_2	3.2×10^{-11}	Silver phosphate, Ag_3PO_4	2.6×10^{-18}
Lead(II) fluoride, PbF_2	3.6×10^{-8}	**Sulfates**	
Magnesium fluoride, MgF_2	7.4×10^{-9}	Barium sulfate, $BaSO_4$	1.1×10^{-10}
Strontium fluoride, SrF_2	2.6×10^{-9}	Calcium sulfate, $CaSO_4$	2.4×10^{-5}
Chlorides		Lead(II) sulfate, $PbSO_4$	1.6×10^{-8}
Copper(I) chloride, $CuCl$	1.9×10^{-7}	Radium sulfate, $RaSO_4$	2×10^{-11}
Lead(II) chloride, $PbCl_2$	1.7×10^{-5}	Silver sulfate, Ag_2SO_4	1.5×10^{-5}
Silver chloride, $AgCl$	1.8×10^{-10}	Strontium sulfate, $SrSO_4$	3.2×10^{-7}
Bromides		**Sulfides**	
Copper(I) bromide, $CuBr$	5×10^{-9}	Cadmium sulfide, CdS	1.0×10^{-24}
Silver bromide, $AgBr$	5.0×10^{-13}	Copper(II) sulfide, CuS	8×10^{-34}
Iodides		Iron(II) sulfide, FeS	8×10^{-16}
Copper(I) iodide, CuI	1×10^{-12}	Lead(II) sulfide, PbS	3×10^{-25}
Lead(II) iodide, PbI_2	7.9×10^{-9}	Manganese(II) sulfide, MnS	3×10^{-11}
Mercury(I) iodide, Hg_2I_2	4.7×10^{-29}	Mercury(II) sulfide, HgS	2×10^{-50}
Silver iodide, AgI	8.3×10^{-17}	Nickel(II) sulfide, NiS	3×10^{-16}
Hydroxides		Silver sulfide, Ag_2S	8×10^{-48}
Aluminum hydroxide, $Al(OH)_3$	3×10^{-34}	Tin(II) sulfide, SnS	1.3×10^{-23}
Cadmium hydroxide, $Cd(OH)_2$	7.2×10^{-15}	Zinc sulfide, ZnS	2.0×10^{-22}
Calcium hydroxide, $Ca(OH)_2$	6.5×10^{-6}		

APPENDIX D

STANDARD ELECTRODE (HALF-CELL) POTENTIALS*

Half-Reaction	$E°$ (V)
$F_2(g) + 2e^- \rightleftharpoons 2F^-(aq)$	+2.87
$O_3(g) + 2H^+(aq) + 2e^- \rightleftharpoons O_2(g) + H_2O(l)$	+2.07
$Co^{3+}(aq) + e^- \rightleftharpoons Co^{2+}(aq)$	+1.82
$H_2O_2(aq) + 2H^+(aq) + 2e^- \rightleftharpoons 2H_2O(l)$	+1.77
$PbO_2(s) + 3H^+(aq) + HSO_4^-(aq) + 2e^- \rightleftharpoons PbSO_4(s) + 2H_2O(l)$	+1.70
$Ce^{4+}(aq) + e^- \rightleftharpoons Ce^{3+}(aq)$	+1.61
$MnO_4^-(aq) + 8H^+(aq) + 5e^- \rightleftharpoons Mn^{2+}(aq) + 4H_2O(l)$	+1.51
$Au^{3+}(aq) + 3e^- \rightleftharpoons Au(s)$	+1.50
$Cl_2(g) + 2e^- \rightleftharpoons 2Cl^-(aq)$	+1.36
$Cr_2O_7^{2-}(aq) + 14H^+(aq) + 6e^- \rightleftharpoons 2Cr^{3+}(aq) + 7H_2O(l)$	+1.33
$MnO_2(s) + 4H^+(aq) + 2e^- \rightleftharpoons Mn^{2+}(aq) + 2H_2O(l)$	+1.23
$O_2(g) + 4H^+(aq) + 4e^- \rightleftharpoons 2H_2O(l)$	+1.23
$Br_2(l) + 2e^- \rightleftharpoons 2Br^-(aq)$	+1.07
$NO_3^-(aq) + 4H^+(aq) + 3e^- \rightleftharpoons NO(g) + 2H_2O(l)$	+0.96
$2Hg^{2+}(aq) + 2e^- \rightleftharpoons Hg_2^{2+}(aq)$	+0.92
$Hg_2^{2+}(aq) + 2e^- \rightleftharpoons 2Hg(l)$	+0.85
$Ag^+(aq) + e^- \rightleftharpoons Ag(s)$	+0.80
$Fe^{3+}(aq) + e^- \rightleftharpoons Fe^{2+}(aq)$	+0.77
$O_2(g) + 2H^+(aq) + 2e^- \rightleftharpoons H_2O_2(aq)$	+0.68
$MnO_4^-(aq) + 2H_2O(l) + 3e^- \rightleftharpoons MnO_2(s) + 4OH^-(aq)$	+0.59
$I_2(s) + 2e^- \rightleftharpoons 2I^-(aq)$	+0.53
$O_2(g) + 2H_2O(l) + 4e^- \rightleftharpoons 4OH^-(aq)$	+0.40
$Cu^{2+}(aq) + 2e^- \rightleftharpoons Cu(s)$	+0.34
$AgCl(s) + e^- \rightleftharpoons Ag(s) + Cl^-(aq)$	+0.22
$SO_4^{2-}(aq) + 4H^+(aq) + 2e^- \rightleftharpoons SO_2(g) + 2H_2O(l)$	+0.20
$Cu^{2+}(aq) + e^- \rightleftharpoons Cu^+(aq)$	+0.15
$Sn^{4+}(aq) + 2e^- \rightleftharpoons Sn^{2+}(aq)$	+0.13
$2H^+(aq) + 2e^- \rightleftharpoons H_2(g)$	0.00
$Pb^{2+}(aq) + 2e^- \rightleftharpoons Pb(s)$	−0.13
$Sn^{2+}(aq) + 2e^- \rightleftharpoons Sn(s)$	−0.14
$N_2(g) + 5H^+(aq) + 4e^- \rightleftharpoons N_2H_5^+(aq)$	−0.23
$Ni^{2+}(aq) + 2e^- \rightleftharpoons Ni(s)$	−0.25
$Co^{2+}(aq) + 2e^- \rightleftharpoons Co(s)$	−0.28
$PbSO_4(s) + H^+(aq) + 2e^- \rightleftharpoons Pb(s) + HSO_4^-(aq)$	−0.31
$Cd^{2+}(aq) + 2e^- \rightleftharpoons Cd(s)$	−0.40
$Fe^{2+}(aq) + 2e^- \rightleftharpoons Fe(s)$	−0.44
$Cr^{3+}(aq) + 3e^- \rightleftharpoons Cr(s)$	−0.74
$Zn^{2+}(aq) + 2e^- \rightleftharpoons Zn(s)$	−0.76
$2H_2O(l) + 2e^- \rightleftharpoons H_2(g) + 2OH^-(aq)$	−0.83
$Mn^{2+}(aq) + 2e^- \rightleftharpoons Mn(s)$	−1.18
$Al^{3+}(aq) + 3e^- \rightleftharpoons Al(s)$	−1.66
$Mg^{2+}(aq) + 2e^- \rightleftharpoons Mg(s)$	−2.37
$Na^+(aq) + e^- \rightleftharpoons Na(s)$	−2.71
$Ca^{2+}(aq) + 2e^- \rightleftharpoons Ca(s)$	−2.87
$Sr^{2+}(aq) + 2e^- \rightleftharpoons Sr(s)$	−2.89
$Ba^{2+}(aq) + 2e^- \rightleftharpoons Ba(s)$	−2.90
$K^+(aq) + e^- \rightleftharpoons K(s)$	−2.93
$Li^+(aq) + e^- \rightleftharpoons Li(s)$	−3.05

*All values at 298 K. Written as reductions; $E°$ value refers to all components in their standard states: 1 M for dissolved species; 1 atm pressure for the gas behaving ideally; the pure substance for solids and liquids.

ANSWERS TO SELECTED PROBLEMS

Chapter 1

1.2 Gas molecules fill the entire container; the volume of a gas is the volume of the container. Solids and liquids have a definite volume. The volume of the container does not affect the volume of a solid or liquid.   (a) gas   (b) liquid   (c) liquid   **1.4** Physical property: a characteristic shown by a substance itself, without any interaction with or change into other substances. Chemical property: a characteristic of a substance that appears as it interacts with, or transforms into, other substances.   (a) Color (yellow-green and silvery to white) and physical state (gas and metal to crystals) are physical properties. The interaction between chlorine gas and sodium metal is a chemical property.   (b) Color and magnetism are physical properties. No chemical changes.   **1.6**(a) Physical change; there is only a temperature change.   (b) Chemical change; the change in appearance indicates an irreversible chemical change.   (c) Physical change; there is only a change in size, not composition.   (d) Chemical change; the wood (and air) become different substances with different compositions.   **1.8**(a) fuel   (b) wood   **1.13** Lavoisier measured the total mass of the reactants and products, not just the mass of the solids. The total mass of the reactants and products remained constant. His measurements showed that a gas was involved in the reaction. He called this gas *oxygen* (one of his key discoveries).   **1.16** A well-designed experiment must have the following essential features: (1) There must be at least two variables that are expected to be related; (2) there must be a way to control all the variables, so that only one at a time may be changed; (3) the results must be reproducible.   **1.19**(a) $(2.54 \text{ cm})^2/(1 \text{ in})^2$ and $(1 \text{ m})^2/(100 \text{ cm})^2$ (b) $(1000 \text{ m})^2/(1 \text{ km})^2$ and $(100 \text{ cm})^2/(1 \text{ m})^2$ (c) $(1.609 \times 10^3 \text{ m/mi})$ and $(1 \text{ h}/3600 \text{ s})$ (d) $(1000 \text{ g}/2.205 \text{ lb})$ and $(3.531 \times 10^{-5} \text{ ft}^3/\text{cm}^3)$ **1.21** An extensive property depends on the amount of material present. An intensive property is the same regardless of how much material is present.   (a) extensive property   (b) intensive property   (c) extensive property   (d) intensive property   **1.23**(a) increases   (b) remains the same   (c) decreases   (d) increases   (e) remains the same   **1.26** 1.43 nm   **1.28** 3.94×10^3 in   **1.30**(a) $2.07 \times 10^{-9} \text{ km}^2$ (b) $10.43   **1.32** Answers will vary, depending on the person's mass.   **1.34**(a) $5.52 \times 10^3 \text{ kg/m}^3$ (b) 345 lb/ft^3 **1.36**(a) $2.56 \times 10^{-9} \text{ mm}^3$ (b) 10^{-10} L   **1.38**(a) 9.626 cm^3 (b) 64.92 g   **1.40** 2.70 g/cm^3 **1.42**(a) 20.°C; 293 K   (b) 109 K; -263°F   (c) -273°C; $-460.$°F   **1.45**(a) 2.47×10^{-7} m   (b) 67.6 Å   **1.52**(a) none   (b) none   (c) 0.041$\underline{0}$ (d) 4.$\underline{0}$1$\underline{00}\times10^4$ **1.54**(a) 0.00036   (b) 35.83   (c) 22.5   **1.56** 6×10^2 **1.58**(a) 1.34 m   (b) 72.391 m^3 (c) 443 cm   **1.60**(a) 1.310000×10^5 (b) 4.7×10^{-4} (c) 2.10006×10^5 (d) 2.1605×10^3 **1.62**(a) 5550   (b) 10,070.   (c) 0.000000885   (d) 0.003004   **1.64**(a) 8.025×10^4 (b) 1.0098×10^{-3} (c) 7.7×10^{-11} **1.66**(a) 4.06×10^{-19} J

(b) 1.61×10^{24} molecules   (c) 1.82×10^5 J/mol   **1.68**(a) Height measured, not exact.   (b) Planets counted, exact.   (c) Number of grams in a pound is not a unit definition, not exact.   (d) Definition of "millimeter," exact.   **1.70** 7.50 ± 0.05 cm   **1.72**(a) $I_{avg} = 8.72$ g; $II_{avg} = 8.72$ g; $III_{avg} = 8.50$ g; $IV_{avg} = 8.56$ g; sets I and II are most accurate.   (b) Set III is the most precise, but is the least accurate.   (c) Set I has the best combination of high accuracy and high precision.   (d) Set IV has both low accuracy and low precision.

1.74(a)

compressed spring
less stable—energy
stored in spring

(b)

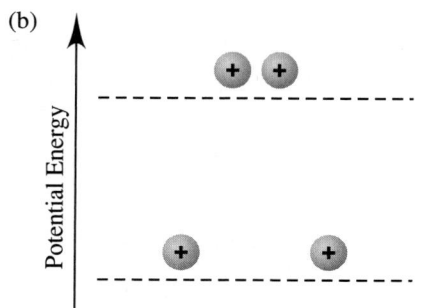

two charges near each other
less stable—repulsion of like
charges

1.76 7.7/1   **1.78**(a) 0.21 g/L, will float   (b) CO_2 is denser than air, will sink   (c) 0.30 g/L, will float   (d) O_2 is denser than air, will sink   (e) 1.38 g/L, will sink   (f) 0.55 g for empty ball; 0.50 g for ball filled with hydrogen   **1.81**(a) 8.0×10^{12} g   (b) 4.1×10^5 m^3 (c) 9.5×10^{13} dollars   **1.83**(a) -195.79°C   (b) -320.42°F   (c) 5.05 L   **1.86**(a) 2.6 m/s   (b) 16 km   (c) 12:45 pm   **1.90** freezing point $= -3.7$°X; boiling point $= 63.3$°X   **1.91** 2.3×10^{25} g oxygen; 1.4×10^{25} g silicon; 5×10^{15} g ruthenium (and rhodium)

Chapter 2

2.1 Compounds contain different types of atoms; there is only one type of atom in an element.   **2.4**(a) The presence of more than one element makes pure calcium chloride a compound. (b) There is only one kind of atom, so sulfur is an element. (c) The presence of more than one compound makes baking powder a mixture.   (d) The presence of more than one type of atom means cytosine cannot be an element. The specific, not variable, arrangement means it is a compound.   **2.12**(a) elements, compounds, and mixtures   (b) compounds   (c) compounds **2.14**(a) Law of definite composition: the composition is the same regardless of its source.   (b) Law of mass conservation: the total quantity of matter does not change.   (c) Law of multiple proportions: two elements can combine to form two different compounds that have different proportions of those elements.   **2.16**(a) No, the percent by mass of each element in a compound is fixed. (b) Yes, the *mass* of each element in a compound depends on the mass of the compound.   **2.18** The two experiments demonstrate the law of definite composition. The unknown compound decomposes the same way both times. The experiments also demonstrate the law of conservation of mass since the total mass before reaction equals the total mass after reaction.   **2.20**(a) 1.34 g F (b) 0.514 Ca; 0.486 F   (c) 51.4 mass % Ca; 48.6 mass % F **2.22**(a) 0.603   (b) 322 g Mg   **2.24** 3.498×10^6 g Cu; 1.766×10^6 g S **2.26** compound 1: 0.905 S/Cl; compound 2: 0.451 S/Cl; ratio: 2.00/1.00   **2.29** Coal A   **2.31** Dalton postulated that atoms of an element are identical and that compounds result from the chemical combination of specific ratios of different elements. **2.32** If you know the ratio of any two quantities and the value of one of them, the other can always be calculated; in this case, the charge and the charge-to-mass ratio were known.   **2.36** The identity of an element is based on the number of protons, not the number of neutrons. The mass number can vary (by a change in number of neutrons) without changing the identity of the element. **2.39** All three isotopes have 18 protons and 18 electrons. Their respective mass numbers are 36, 38, and 40, with the respective numbers of neutrons being 18, 20, and 22.   **2.41**(a) These have the same number of protons and electrons, but different numbers of neutrons; same Z.   (b) These have the same number of neutrons, but different numbers of protons and electrons; same N. (c) These have different numbers of protons, neutrons, and electrons; same A.   **2.43**(a) $^{38}_{18}Ar$ (b) $^{55}_{25}Mn$ (c) $^{109}_{47}Ag$ **2.45**(a) $^{48}_{22}Ti$ (b) $^{79}_{34}Se$ (c) $^{11}_{5}B$

22e⁻ — 22p⁺ 26n⁰   34e⁻ — 34p⁺ 45n⁰   5e⁻ — 5p⁺ 6n⁰

2.47 69.72 amu   **2.49** $^{35}Cl = 75.774\%$, $^{37}Cl = 24.226\%$ **2.52**(a) In the modern periodic table, the elements are arranged in order of increasing atomic number.   (b) Elements in a group (or family) have similar chemical properties.   (c) Elements can be classified as metals, metalloids, or nonmetals.   **2.55** The alkali metals [Group 1A(1)] are metals and readily lose one electron to form cations; the halogens [Group 7A(17)] are nonmetals and

readily gain one electron to form anions.   **2.56**(a) germanium; Ge; 4A(14); metalloid   (b) phosphorus; P; 5A(15); nonmetal (c) helium; He; 8A(18); nonmetal   (d) lithium; Li; 1A(1); metal (e) molybdenum; Mo; 6B(6); metal   **2.58**(a) Ra; 88   (b) Si; 14 (c) Cu; 63.55   (d) Br; 79.90   **2.60** Atoms of these two kinds of substances will form ionic bonds, in which one or more electrons are transferred from the metal atom to the nonmetal atom to form a cation and an anion, respectively.   **2.63** Coulomb's law states the energy of attraction in an ionic bond is directly proportional to the product of charges and inversely proportional to the distance between charges. The product of charges in MgO $[(2+) (2-)]$ is greater than that in LiF $[(1+) (1-)]$. Thus, MgO has stronger ionic bonding.   **2.66** The Group 1A(1) elements form cations, and the Group 7A(17) elements form anions. **2.68** Each potassium atom loses one electron to form an ion with a 1+ charge. Each sulfur atom gains two electrons to form an ion with a 2− charge. Two potassiums, giving one electron each, are required for each sulfur, which takes two electrons. The oppositely charged ions attract each other to form an ionic solid, K_2S. **2.70** K^+; I^- **2.72**(a) oxygen; 17; 6A(16); 2   (b) fluorine; 19; 7A(17); 2   (c) calcium; 40; 2A(2); 4   **2.74** Lithium forms the Li^+ ion; oxygen forms the O^{2-} ion. Number of O^{2-} ions = 4.2×10^{21} O^{2-} ions.   **2.76** NaCl   **2.78** An empirical formula shows the simplest ratio of atoms of each element present in a compound, whereas a molecular formula describes the type and actual number of atoms of each element in a molecule of the compound. The empirical formula and the molecular formula can be the same. The molecular formula is always a whole-number multiple of the empirical formula.   **2.80** The two samples are similar in that both contain 20 billion oxygen atoms and 20 billion hydrogen atoms. They differ in that they contain different types of molecules: H_2O_2 molecules in the hydrogen peroxide sample, and H_2 and O_2 molecules in the mixture. In addition, the mixture contains 20 billion molecules (10 billion H_2 and 10 billion O_2), while the hydrogen peroxide sample contains 10 billion molecules.   **2.84**(a) NH_2 (b) CH_2O **2.86**(a) Na_3N, sodium nitride   (b) SrO, strontium oxide   (c) $AlCl_3$, aluminum chloride **2.88**(a) MgF_2, magnesium fluoride   (b) ZnS, zinc sulfide (c) $SrCl_2$, strontium chloride   **2.90**(a) $SnCl_4$ (b) iron(III) bromide   (c) CuBr   (d) manganese(III) oxide   **2.92**(a) cobalt(II) oxide   (b) Hg_2Cl_2 (c) lead(II) acetate trihydrate   (d) Cr_2O_3 **2.94**(a) BaO   (b) $Fe(NO_3)_2$ (c) MgS   **2.96**(a) H_2SO_4; sulfuric acid   (b) HIO_3; iodic acid   (c) HCN; hydrocyanic acid   (d) H_2S; hydrosulfuric acid   **2.98**(a) ammonium ion, NH_4^+; ammonia, NH_3 (b) magnesium sulfide, MgS; magnesium sulfite, $MgSO_3$; magnesium sulfate, $MgSO_4$ (c) hydrochloric acid, HCl; chloric acid, $HClO_3$; chlorous acid, $HClO_2$ (d) cuprous bromide, CuBr; cupric bromide, $CuBr_2$ **2.100** Disulfur tetrafluoride, S_2F_4 **2.102**(a) calcium chloride   (b) copper(I) oxide   (c) stannic fluoride   (d) hydrochloric acid   **2.104**(a) 12 oxygen atoms; 342.2 amu   (b) 9 hydrogen atoms; 132.06 amu   (c) 8 oxygen atoms; 344.6 amu   **2.106**(a) $(NH_4)_2SO_4$; 132.15 amu   (b) NaH_2PO_4; 119.98 amu   (c) $KHCO_3$; 100.12 amu   **2.108**(a) 108.02 amu (b) 331.2 amu   (c) 72.08 amu   **2.110**(a) SO_3; sulfur trioxide; 80.07 amu   (b) C_3H_8; propane; 44.09 amu   **2.112** disulfur dichloride; SCl; 135.04 amu   **2.116** Separating the components of a mixture requires physical methods only; that is, no chemical changes (no changes in composition) take place, and the components maintain their chemical identities and properties throughout. Separating the components of a compound requires a chemical change (change in composition). **2.119**(a) compound (b) homogeneous mixture (c) heterogeneous mixture (d) homogeneous mixture (e) homogeneous mixture **2.121**(a) Salt dissolves in water and pepper does not. Procedure: Add water to mixture; filter to remove solid pepper. Evaporate water to recover solid salt. (b) Sugar dissolves in water and sand does not. Procedure: Add water to mixture; filter to remove sand. Evaporate water to recover sugar. (c) Oil is less volatile (lower tendency to vaporize to a gas) than water. Procedure: Separate the components by distillation (boiling), and condense the water from vapor thus leaving the oil in the distillation flask. (d) Vegetable oil and vinegar (solution of acetic acid in water) are not soluble in each other. Procedure: allow mixture to separate into two layers and drain off the lower (denser) layer. **2.123**(a) filtration (b) extraction or chromatography (c) extraction and filtration **2.125**(a) fraction of volume = 5.2×10^{-13} (b) mass of nucleus = 6.64466×10^{-24} g; fraction of mass = 0.999726 **2.126** strongest ionic bonding: MgO; weakest ionic bonding: RbI **2.130** 2.7 mass % Fe; 2.4 mass % Mg; 43.2 mass % Si; 51.6 mass % O **2.133**(a) I = NO; II = N_2O_3; III = N_2O_5 (b) I has 1.14 g O per 1.00 g N; II, 1.71 g O; III, 2.86 g O **2.139**(a) Cl^-, 1.898 mass %; Na^+, 1.056 mass %; SO_4^{2-}, 0.265 mass %; Mg^{2+}, 0.127 mass %; Ca^{2+}, 0.04 mass %; K^+, 0.038 mass %; HCO_3^-, 0.014 mass % (b) 30.72% (c) Alkaline earth metal ions, total mass % = 0.17%; alkali metal ions, total mass % = 1.094% (d) Anions (2.177 mass %) make up a larger mass fraction than cations (1.26 mass %). **2.142** Molecular, $C_4H_6O_4$; empirical, $C_2H_3O_2$; molecular mass, 118.09 amu; 40.68% by mass C; 5.122% by mass H; 54.20% by mass O **2.144** B, C, D; 44.05 amu **2.146**(a) Formulas and masses in amu: $^{15}N_2{}^{18}O$, 48; $^{15}N_2{}^{16}O$, 46; $^{14}N_2{}^{18}O$, 46; $^{14}N_2{}^{16}O$, 44; $^{15}N^{14}N^{18}O$, 47; $^{15}N^{14}N^{16}O$, 45 (b) $^{15}N_2{}^{18}O$, least common; $^{14}N_2{}^{16}O$, most common **2.149** 58.091 amu **2.151** nitroglycerin, 39.64 mass % NO; isoamyl nitrate, 22.54 mass % NO. **2.152** 0.370 lb C; 0.0222 lb H; 0.423 lb O; 0.185 lb N **2.156**(a) 20.39 mass % Mg (b) 12.84 mass % Mg (c) $x = 2$ **2.159**(1) chemical change (2) physical change (3) chemical change (4) chemical change (5) physical change

Chapter 3

3.2(a) 12 mol C atoms (b) 1.445×10^{25} C atoms **3.7**(a) left (b) left (c) left (d) neither **3.8**(a) 121.64 g/mol (b) 76.02 g/mol (c) 106.44 g/mol (d) 152.00 g/mol **3.10**(a) 134.7 g/mol (b) 175.3 g/mol (c) 342.17 g/mol (d) 125.84 g/mol **3.12**(a) 1.1×10^2 g $KMnO_4$ (b) 0.188 mol O atoms (c) 1.5×10^{20} O atoms **3.14**(a) 9.73 g $MnSO_4$ (b) 44.6 mol $Fe(ClO_4)_3$ (c) 1.74×10^{21} N atoms **3.16**(a) 1.56×10^3 g Cu_2CO_3 (b) 0.0725 g N_2O_5 (c) 0.644 mol $NaClO_4$; 3.88×10^{23} formula units $NaClO_4$ (d) 3.88×10^{23} Na^+ ions; 3.88×10^{23} ClO_4^- ions; 3.88×10^{23} Cl atoms; 1.55×10^{24} O atoms **3.18**(a) 6.375 mass % H (b) 71.52 mass % O **3.20**(a) 0.1252 mass fraction C (b) 0.3428 mass fraction O **3.23**(a) 0.9507 mol cisplatin (b) 3.5×10^{24} H atoms **3.25**(a) 195 mol rust (b) 195 mol Fe_2O_3 (c) 2.18×10^4 g Fe **3.27** $CO(NH_2)_2 > NH_4NO_3 > (NH_4)_2SO_4 > KNO_3$ **3.28**(a) 883 mol PbS (b) 1.88×10^{25} Pb atoms **3.31**(b) From the mass percent, determine the empirical formula. Add up the total number of atoms in the empirical formula, and divide that

number into the total number of atoms in the molecule. The result is the multiplier that makes the empirical formula into the molecular formula. (c) (mass % expressed directly in grams) (1/molar mass) = moles of each element (e) Count the numbers of the various types of atoms in the structural formula and put these into a molecular formula. **3.33**(a) CH_2; 14.03 g/mol (b) CH_3O; 31.03 g/mol (c) N_2O_5; 108.02 g/mol (d) $Ba_3(PO_4)_2$; 601.8 g/mol (e) TeI_4; 635.2 g/mol **3.35**(a) C_3H_6 (b) N_2H_4 (c) N_2O_4 (d) $C_5H_5N_5$ **3.37**(a) Cl_2O_7 (b) $SiCl_4$ (c) CO_2 **3.39**(a) NO_2 (b) N_2O_4 **3.41**(a) 1.20 mol F (b) 24.0 g M (c) calcium **3.44** $C_{21}H_{30}O_5$ **3.46** $C_{10}H_{20}O$ **3.47** The balanced equation provides information on the identities of reactants and products, the physical states of reactants and products, and the molar ratios by which reactants form products. **3.50** b

3.51(a) $16Cu(s) + S_8(s) \longrightarrow 8Cu_2S(s)$
(b) $P_4O_{10}(s) + 6H_2O(l) \longrightarrow 4H_3PO_4(l)$
(c) $B_2O_3(s) + 6NaOH(aq) \longrightarrow 2Na_3BO_3(aq) + 3H_2O(l)$
(d) $4CH_3NH_2(g) + 9O_2(g) \longrightarrow$
$$4CO_2(g) + 10H_2O(g) + 2N_2(g)$$

3.53(a) $2SO_2(g) + O_2(g) \longrightarrow 2SO_3(g)$
(b) $Sc_2O_3(s) + 3H_2O(l) \longrightarrow 2Sc(OH)_3(s)$
(c) $H_3PO_4(aq) + 2NaOH(aq) \longrightarrow Na_2HPO_4(aq) + 2H_2O(l)$
(d) $C_6H_{10}O_5(s) + 6O_2(g) \longrightarrow 6CO_2(g) + 5H_2O(g)$

3.55(a) $4Ga(s) + 3O_2(g) \longrightarrow 2Ga_2O_3(s)$
(b) $2C_6H_{14}(l) + 19O_2(g) \longrightarrow 12CO_2(g) + 14H_2O(g)$
(c) $3CaCl_2(aq) + 2Na_3PO_4(aq) \longrightarrow$
$$Ca_3(PO_4)_2(s) + 6NaCl(aq)$$

3.61 Balance the equation for the reaction: $aA + bB \longrightarrow cC$. Since A is the limiting reactant, A is used to determine the amount of C:

$$\text{Mass C} = (25 \text{ g A})\left(\frac{1 \text{ mol A}}{\text{molar mass A}}\right)\left(\frac{c \text{ mol C}}{a \text{ mol A}}\right)\left(\frac{\text{molar mass C}}{1 \text{ mol C}}\right)$$
$$= \text{g C}$$

3.63(a) 0.455 mol Cl_2 (b) 32.3 g Cl_2 **3.65**(a) 1.42×10^3 mol KNO_3 (b) 1.43×10^5 g KNO_3 **3.67** 195.8 g H_3BO_3; 19.16 g H_2 **3.69** 2.60×10^3 g Cl_2

3.71(a) $I_2(s) + Cl_2(g) \longrightarrow 2ICl(s)$
$ICl(s) + Cl_2(g) \longrightarrow ICl_3(s)$
(b) $I_2(s) + 3Cl_2(g) \longrightarrow 2ICl_3(s)$
(c) 1.33×10^3 g I_2

3.73(a) 0.105 mol CaO (b) 0.175 mol CaO (c) calcium (d) 5.88 g CaO **3.75** 1.36 mol HIO_3, 239 g HIO_3; 44.9 g H_2O in excess **3.77** 4.40 g CO_2; 4.80 g O_2 in excess **3.79** 12.2 g $Al(NO_2)_3$, no NH_4Cl, 48.7 g $AlCl_3$, 30.7 g N_2, 39.5 g H_2O **3.81** 50.% **3.83** 90.5% **3.85** 25.6 g CH_3Cl **3.87** 39.7 g CF_4 **3.88** A **3.92**(a) C (b) B (c) C (d) B **3.95** No, it should read: "Take 100.0 mL of the 10.0 M solution and add water until the total volume is 1000. mL." **3.96**(a) 7.85 g $Ca(C_2H_3O_2)_2$ (b) 0.254 M KI (c) 124 mol NaCN **3.98**(a) 4.65 g K_2SO_4 (b) 0.0653 M $CaCl_2$ (c) 1.11×10^{20} Mg^{2+} ions **3.100**(a) 0.0617 M KCl (b) 0.00363 M $(NH_4)_2SO_4$ (c) 0.138 M Na^+ **3.102**(a) 987 g HNO_3/L (b) 15.7 M HNO_3 **3.104** 845 mL **3.106** 0.88 g $BaSO_4$ **3.109**(a) Instructions: Be sure to wear goggles to protect your eyes! Pour approximately 2.0 gal of water into the container. Add to the water, slowly and with mixing, 0.90 gal of concentrated HCl. Dilute to 3.0 gal with more water. (b) 22.6 mL **3.114** $x = 3$ **3.115** ethane > propane > cetyl palmitate > ethanol > benzene **3.120**(a) $Fe_2O_3(s) + 3CO(g) \longrightarrow 2Fe(s) + 3CO_2(g)$

(b) 3.39×10^7 g CO   **3.122** 89.8%   **3.124**(a) $2AB_2 + B_2 \longrightarrow$ $2AB_3$ (b) AB_2 (c) 5.0 mol AB_3 (d) 0.5 mol B_2
3.127 44.3%   **3.130**(a) C   (b) B   (c) D   **3.134** 0.071 M KBr
3.138(a) 586 g CO_2 (b) 10.5% CH_4 by mass   **3.140** 10/0.66/1.0
3.146(a) 192.12 g/mol; $C_6H_8O_7$ (b) 0.549 mol
3.147(a) $N_2(g) + O_2(g) \longrightarrow 2NO(g)$
$\quad\quad\quad 2NO(g) + O_2(g) \longrightarrow 2NO_2(g)$
$\quad\quad\quad 3NO_2(g) + H_2O(l) \longrightarrow 2HNO_3(g) + NO(g)$
(b) $2N_2(g) + 5O_2(g) + 2H_2O(l) \longrightarrow 4HNO_3(g)$
(c) 6.07×10^3 t HNO_3
3.149 A   **3.152**(a) 0.039 g heme   (b) 6.3×10^{-5} mol heme
(c) 3.5×10^{-3} g Fe   (d) 4.1×10^{-2} g hemin   **3.155**(a) 46.65
mass % N in urea; 31.98 mass % N in arginine; 21.04 mass %
N in ornithine   (b) 28.45 g N   **3.157** 29.54%   **3.159**(a) 89.3%
(b) 1.47 g ethylene   **3.161**(a) 125 g salt   (b) 65.6 L H_2O

Chapter 4

4.2 Ionic or polar covalent compounds   **4.3** Ions must be present,
and they come from ionic compounds or from electrolytes
such as acids and bases.   **4.6**(2)   **4.10**(a) Benzene is likely to
be insoluble in water because it is nonpolar and water is polar.
(b) Sodium hydroxide, an ionic compound, is likely to be very
soluble in water.   (c) Ethanol (CH_3CH_2OH) is likely to be
soluble in water because the alcohol group (—OH) will hydrogen bond with a water molecule.   (d) Potassium acetate,
an ionic compound, is likely to be very soluble in water.
4.12(a) Yes, CsBr is a soluble salt.   (b) Yes, HI is a strong acid.
4.14(a) 0.64 mol   (b) 0.242 mol   (c) 1.18×10^{-4} mol
4.16(a) 3.0 mol   (b) 7.57×10^{-5} mol   (c) 0.148 mol
4.18(a) 0.058 mol Al^{3+}; 3.5×10^{22} Al^{3+} ions; 0.18 mol Cl^-;
1.1×10^{23} Cl^- ions   (b) 4.62×10^{-4} mol Li^+; 2.78×10^{20} Li^+
ions; 2.31×10^{-4} mol SO_4^{2-}; 1.39×10^{20} SO_4^{2-} ions
(c) 1.50×10^{-2} mol K^+; 9.02×10^{21} K^+ ions; 1.50×10^{-2} mol
Br^-; 9.02×10^{21} Br^- ions   **4.20**(a) 0.35 mol H^+ (b) 6.3×10^{-3}
mol H^+ (c) 0.22 mol H^+ **4.24** Spectator ions do not appear
because they are not involved in the reaction and are present
only to balance charge.   **4.28** Assuming that the left beaker
contains $AgNO_3$ (because it has gray Ag^+ ion), the right must
contain NaCl. Then, NO_3^- is blue, Na^+ is brown, and Cl^- is
green.
Molecular equation: $AgNO_3(aq) + NaCl(aq) \longrightarrow$
$\quad\quad\quad\quad\quad\quad\quad\quad\quad\quad AgCl(s) + NaNO_3(aq)$
Total ionic equation: $Ag^+(aq) + NO_3^-(aq) + Na^+(aq) +$
$\quad\quad\quad Cl^-(aq) \longrightarrow AgCl(s) + Na^+(aq) + NO_3^-(aq)$
Net ionic equation: $Ag^+(aq) + Cl^-(aq) \longrightarrow AgCl(s)$
4.29(a) Molecular: $Hg_2(NO_3)_2(aq) + 2KI(aq) \longrightarrow$
$\quad\quad\quad\quad\quad\quad\quad\quad\quad\quad Hg_2I_2(s) + 2KNO_3(aq)$
$\quad\quad$ Total ionic: $Hg_2^{2+}(aq) + 2NO_3^-(aq) + 2K^+(aq) +$
$\quad\quad\quad 2I^-(aq) \longrightarrow Hg_2I_2(s) + 2K^+(aq) + 2NO_3^-(aq)$
$\quad\quad$ Net ionic: $Hg_2^{2+}(aq) + 2I^-(aq) \longrightarrow Hg_2I_2(s)$
$\quad\quad$ Spectator ions are K^+ and NO_3^-.
(b) Molecular: $FeSO_4(aq) + Sr(OH)_2(aq) \longrightarrow$
$\quad\quad\quad\quad\quad\quad\quad\quad\quad\quad Fe(OH)_2(s) + SrSO_4(s)$
$\quad\quad$ Total ionic: $Fe^{2+}(aq) + SO_4^{2-}(aq) + Sr^{2+}(aq) +$
$\quad\quad\quad 2OH^-(aq) \longrightarrow Fe(OH)_2(s) + SrSO_4(s)$
$\quad\quad$ Net ionic: This is the same as the total ionic equation, be-
cause there are no spectator ions.

4.31(a) No precipitate will form.   (b) A precipitate will form
because silver ions, Ag^+, and bromide ions, Br^-, will combine
to form a solid salt, silver bromide, AgBr. The ammonium and
nitrate ions do not form a precipitate.
Molecular: $NH_4Br(aq) + AgNO_3(aq) \longrightarrow$
$\quad\quad\quad\quad\quad\quad\quad\quad\quad\quad AgBr(s) + NH_4NO_3(aq)$
Total ionic: $NH_4^+(aq) + Br^-(aq) + Ag^+(aq) + NO_3^-(aq) \longrightarrow$
$\quad\quad\quad\quad\quad\quad AgBr(s) + NH_4^+(aq) + NO_3^-(aq)$
Net ionic: $Ag^+(aq) + Br^-(aq) \longrightarrow AgBr(s)$
4.33(a) No precipitate will form.
$\quad$(b) $BaSO_4$ will precipitate.
$\quad\quad$ Molecular: $(NH_4)_2SO_4(aq) + BaCl_2(aq) \longrightarrow$
$\quad\quad\quad\quad\quad\quad\quad\quad\quad\quad BaSO_4(s) + 2NH_4Cl(aq)$
$\quad\quad$ Total ionic: $2NH_4^+(aq) + SO_4^{2-}(aq) + Ba^{2+}(aq) +$
$\quad\quad\quad 2Cl^-(aq) \longrightarrow BaSO_4(s) + 2NH_4^+(aq) + Cl^-(aq)$
$\quad\quad$ Net ionic: $SO_4^{2-}(aq) + Ba^{2+}(aq) \longrightarrow BaSO_4(s)$
4.35 0.0354 M Pb^{2+} **4.39** 2.206 mass % Cl^- **4.45**(a) Formation of a gas, $SO_2(g)$, and of a nonelectrolyte (water) will
cause the reaction to go to completion.   (b) Formation of a
precipitate $Ba_3(PO_4)_2(s)$, and of a nonelectrolyte (water) will
cause the reaction to go to completion.
4.47(a) Molecular equation: $KOH(aq) + HBr(aq) \longrightarrow$
$\quad\quad\quad\quad\quad\quad\quad\quad\quad\quad KBr(aq) + H_2O(l)$
$\quad\quad$ Total ionic equation: $K^+(aq) + OH^-(aq) + H^+(aq) +$
$\quad\quad\quad Br^-(aq) \longrightarrow K^+(aq) + Br^-(aq) + H_2O(l)$
$\quad\quad$ Net ionic equation: $OH^-(aq) + H^+(aq) \longrightarrow H_2O(l)$
$\quad\quad$ The spectator ions are $K^+(aq)$ and $Br^-(aq)$.
$\quad$(b) Molecular equation: $NH_3(aq) + HCl(aq) \longrightarrow$
$\quad\quad\quad\quad\quad\quad\quad\quad\quad\quad NH_4Cl(aq)$
$\quad\quad$ Total ionic equation: $NH_3(aq) + H^+(aq) + Cl^-(aq) \longrightarrow$
$\quad\quad\quad\quad\quad\quad\quad\quad\quad\quad NH_4^+(aq) + Cl^-(aq)$
$\quad\quad NH_3$, a weak base, is written in the molecular (undissociated) form. HCl, a strong acid, is written as dissociated ions. NH_4Cl is a soluble compound, because all
ammonium compounds are soluble.
$\quad\quad$ Net ionic equation: $NH_3(aq) + H^+(aq) \longrightarrow NH_4^+(aq)$
$\quad\quad Cl^-$ is the only spectator ion.
4.49 Total ionic equation: $CaCO_3(s) + 2H^+(aq) + 2Cl^-(aq) \longrightarrow$
$\quad\quad\quad\quad\quad Ca^{2+}(aq) + 2Cl^-(aq) + H_2O(l) + CO_2(g)$
$\quad\quad$ Net ionic equation: $CaCO_3(s) + 2H^+(aq) \longrightarrow$
$\quad\quad\quad\quad\quad\quad\quad Ca^{2+}(aq) + H_2O(l) + CO_2(g)$
4.51 0.05839 M CH_3COOH **4.62**(a) S has O.N. = +6 in
SO_4^{2-} (i.e., H_2SO_4), and O.N. = +4 in SO_2, so S has been
reduced (and I^- oxidized); H_2SO_4 acts as an oxidizing agent.
(b) The oxidation numbers remain constant throughout; H_2SO_4
transfers a proton to F^- to produce HF, so it acts as an acid.
4.64(a) +4   (b) +3   (c) +4   (d) −3   **4.66**(a) −1   (b) +2
(c) −3   (d) +3   **4.68**(a) −3   (b) +5   (c) +3   **4.70**(a) +6
(b) +3   (c) +7   **4.72**(a) MnO_4^- is the oxidizing agent;
$H_2C_2O_4$ is the reducing agent.   (b) Cu is the reducing agent;
NO_3^- is the oxidizing agent.   **4.74**(a) Oxidizing agent is
NO_3^-; reducing agent is Sn.   (b) Oxidizing agent is MnO_4^-;
reducing agent is Cl^-.   **4.76** S is in Group 6A(16), so its
highest possible O.N. is +6 and its lowest possible O.N. is
$6 - 8 = -2$.   (a) S = −2. The S can only increase its O.N.
(oxidize), so S^{2-} can function only as a reducing agent.
(b) S = +6. The S can only decrease its O.N. (reduce), so SO_4^{2-}
can function only as an oxidizing agent.   (c) S = +4.

The S can increase or decrease its O.N. Therefore, SO_2 can function as either an oxidizing or reducing agent.
4.78(a) $8HNO_3(aq) + K_2CrO_4(aq) + 3Fe(NO_3)_2(aq) \longrightarrow$
$2KNO_3(aq) + 3Fe(NO_3)_3(aq) + Cr(NO_3)_3(aq) + 4H_2O(l)$
Oxidizing agent is K_2CrO_4; reducing agent is $Fe(NO_3)_2$.
(b) $8HNO_3(aq) + 3C_2H_6O(l) + K_2Cr_2O_7(aq) \longrightarrow$
$2KNO_3(aq) + 3C_2H_4O(l) + 7H_2O(l) + 2Cr(NO_3)_3(aq)$
Oxidizing agent is $K_2Cr_2O_7$; reducing agent is C_2H_6O.
(c) $6HCl(aq) + 2NH_4Cl(aq) + K_2Cr_2O_7(aq) \longrightarrow$
$2KCl(aq) + 2CrCl_3(aq) + N_2(g) + 7H_2O(l)$
Oxidizing agent is $K_2Cr_2O_7$; reducing agent is NH_4Cl.
(d) $KClO_3(aq) + 6HBr(aq) \longrightarrow$
$3Br_2(l) + 3H_2O(l) + KCl(aq)$
Oxidizing agent is $KClO_3$; reducing agent is HBr.
4.80(a) 4.54×10^{-3} mol MnO_4^- (b) 0.0113 mol H_2O_2
(c) 0.386 g H_2O_2 (d) 2.61 mass % H_2O_2 (e) H_2O_2
4.86 A combination reaction that is also a redox reaction is $2Mg(s) + O_2(g) \longrightarrow 2MgO(s)$. A combination reaction that is not a redox reaction is $CaO(s) + H_2O(l) \longrightarrow Ca(OH)_2(aq)$.
4.87(a) $Ca(s) + 2H_2O(l) \longrightarrow$
$Ca(OH)_2(aq) + H_2(g)$; displacement
(b) $2NaNO_3(s) \longrightarrow 2NaNO_2(s) + O_2(g)$; decomposition
(c) $C_2H_2(g) + 2H_2(g) \longrightarrow C_2H_6(g)$; combination
4.89(a) $2Sb(s) + 3Cl_2(g) \longrightarrow 2SbCl_3(s)$; combination
(b) $2AsH_3(g) \longrightarrow 2As(s) + 3H_2(g)$; decomposition
(c) $Zn(s) + Fe(NO_3)_2(aq) \longrightarrow$
$Zn(NO_3)_2(aq) + Fe(s)$; displacement
4.91(a) $Sr(s) + Br_2(l) \longrightarrow SrBr_2(s)$
(b) $2Ag_2O(s) \xrightarrow{\Delta} 4Ag(s) + O_2(g)$
(c) $Mn(s) + Cu(NO_3)_2(aq) \longrightarrow Mn(NO_3)_2(aq) + Cu(s)$
4.93(a) $N_2(g) + 3H_2(g) \longrightarrow 2NH_3(g)$
(b) $2NaClO_3(s) \xrightarrow{\Delta} 2NaCl(s) + 3O_2(g)$
(c) $Ba(s) + 2H_2O(l) \longrightarrow Ba(OH)_2(aq) + H_2(g)$
4.95(a) $2Cs(s) + I_2(s) \longrightarrow 2CsI(s)$
(b) $2Al(s) + 3MnSO_4(aq) \longrightarrow Al_2(SO_4)_3(aq) + 3Mn(s)$
(c) $2SO_2(g) + O_2(g) \longrightarrow 2SO_3(g)$
(d) $2C_4H_{10}(g) + 13O_2(g) \longrightarrow 8CO_2(g) + 10H_2O(g)$
(e) $2Al(s) + 3Mn^{2+}(aq) \longrightarrow 2Al^{3+}(aq) + 3Mn(s)$
4.97 315 g O_2; 3.95 kg Hg   **4.99**(a) O_2 is in excess.
(b) 0.117 mol Li_2O (c) 0 g Li, 3.49 g Li_2O, and 4.63 g O_2
4.101 67.2 mass % $KClO_3$ **4.103** 2.79 kg Fe   **4.104** 99.9 g compound B, which is $FeCl_2$ **4.108** The reaction is $2NO + Br_2 \rightleftharpoons 2NOBr$, which can proceed in either direction. If NO and Br_2 are placed in a container, they will react to form NOBr, and NOBr will decompose to form NO and Br_2. Eventually, the concentrations of NO, Br_2, and NOBr adjust so that the rates of the forward and reverse reactions become equal, and equilibrium is reached.
4.110(a) $Fe(s) + 2H^+(aq) \longrightarrow Fe^{2+}(aq) + H_2(g)$
O.N.:   0   +1   +2   0
(b) 3.1×10^{21} Fe^{2+} ions
4.112 5.11 g C_2H_5OH; 2.49 L CO_2
4.114(a) $Ca^{2+}(aq) + C_2O_4^{2-}(aq) \longrightarrow CaC_2O_4(s)$
(b) $5H_2C_2O_4(aq) + 2MnO_4^-(aq) + 6H^+(aq) \longrightarrow$
$10CO_2(g) + 2Mn^{2+}(aq) + 8H_2O(l)$
(c) $KMnO_4$ (d) $H_2C_2O_4$ (e) 55.06 mass % $CaCl_2$
4.117 2.809 mass % $CaMg(CO_3)_2$ **4.119**(a) Step 1: oxidizing agent is O_2; reducing agent is NH_3. Step 2: oxidizing agent is O_2; reducing agent is NO. Step 3: oxidizing agent is NO_2; reducing agent is NO_2.   (b) 1.2×10^4 kg NH_3 **4.124** 627 L air

4.127 0.75 kg SiO_2; 0.26 kg Na_2CO_3; 0.18 kg $CaCO_3$
4.130(a) 4 mol IO_3^- (b) 12 mol I_2; IO_3^- is oxidizing agent and I^- is reducing agent.   (c) 12.87 mass % thyroxine
4.134 3.0×10^{-3} mol CO_2 (b) 0.11 L CO_2 **4.136**(a) $C_7H_5O_4Bi$
(b) $C_{21}H_{15}O_{12}Bi_3$ (c) $Bi(OH)_3(s) + 3HC_7H_5O_3(aq) \longrightarrow$
$Bi(C_7H_5O_3)_3(s) + 3H_2O(l)$ (d) 0.490 mg $Bi(OH)_3$
4.138(a) Ethanol: $C_2H_5OH(l) + 3O_2(g) \longrightarrow 2CO_2(g) + 3H_2O(l)$
Gasoline: $2C_8H_{18}(l) + 25O_2(g) \longrightarrow$
$16CO_2(g) + 18H_2O(g)$
(b) 2.50×10^3 g O_2 (c) 1.75×10^3 L O_2 (d) 8.38×10^3 L air
4.140 yes   **4.142**(a) Reaction 2 is a redox process.
(b) 2.00×10^5 g Fe_2O_3; 4.06×10^5 g $FeCl_3$ (c) 2.09×10^5 g Fe; 4.75×10^5 g $FeCl_2$ (d) 0.313

Chapter 5

5.1(a) The volume of the liquid remains constant, but the volume of the gas increases to the volume of the larger container.   (b) The volume of the container holding the gas sample increases when heated, but the volume of the container holding the liquid sample remains essentially constant when heated.   (c) The volume of the liquid remains essentially constant, but the volume of the gas is reduced.   **5.6** 990 cmH_2O **5.8**(a) 566 mmHg   (b) 1.32 bar
(c) 3.60 atm   (d) 107 kPa   **5.10** 0.9408 atm   **5.12** 0.966 atm
5.18 At constant temperature and volume, the pressure of a gas is directly proportional to number of moles of the gas.
5.20(a) Volume decreases to one-third of the original volume.
(b) Volume increases by a factor of 3.0.   (c) Volume increases by a factor of 4.   **5.22**(a) Volume decreases by a factor of 2.
(b) Volume increases by a factor of 1.48.   (c) Volume decreases by a factor of 3.   **5.24** $-144°C$ **5.26** 35.3 L   **5.28** 0.085 mol Cl_2 **5.30** 1.16 g ClF_3 **5.33** yes   **5.35** Beaker is inverted for H_2 and upright for CO_2. The molar mass of CO_2 is greater than the molar mass of air, which, in turn, has a greater molar mass than H_2.   **5.39** 5.86 g/L   **5.41** 1.78×10^{-3} mol AsH_3; 3.48 g/L
5.43 51.1 g/mol   **5.45** 1.33 atm   **5.49** C_5H_{12} **5.51**(a) 0.90 mol
(b) 6.76 torr   **5.52** 39.3 g P_4 **5.54** 41.2 g PH_3 **5.56** 0.0249 g Al
5.58 286 mL SO_2 **5.60** 0.0997 atm SiF_4 **5.65** At STP, the volume occupied by a mole of any gas is the same. At the same temperature, all gases have the same average kinetic energy, resulting in the same pressure.   **5.68**(a) $P_A > P_B > P_C$ (b) $E_A = E_B = E_C$ (c) rate$_A$ > rate$_B$ > rate$_C$ (d) total E_A > total E_B > total E_C (e) $d_A = d_B = d_C$ (f) collision frequency in A > collision frequency in B > collision frequency in C   **5.69** 13.21
5.71(a) curve 1   (b) curve 1   (c) curve 1; fluorine and argon have about the same molar mass   **5.73** 14.9 min   **5.75** 4 atoms per molecule   **5.78** negative deviations; $N_2 < Kr < CO_2$ **5.80** At 1 atm; because the pressure is lower.   **5.83** 6.81×10^4 g/mol
5.86(a) 22.1 atm   (b) 20.9 atm   **5.90**(a) 597 torr N_2; 159 torr O_2; 0.3 torr CO_2; 3.5 torr H_2O (b) 74.9 mol % N_2; 13.7 mol % O_2; 5.3 mol % CO_2; 6.2 mol % H_2O (c) 1.6×10^{21} molecules O_2 **5.92**(a) 4×10^2 mL   (b) 0.013 mol N_2 **5.94** 35.7 L NO_2
5.99 Al_2Cl_6 **5.102** 1.52×10^{-2} mol SO_3 **5.107**(a) 1.95×10^3 g Ni   (b) 3.5×10^4 g Ni   (c) 63 m^3 CO   **5.109**(a) 9 volumes of $O_2(g)$ (b) CH_5N **5.114** The lungs would expand by a factor of 4.86; the diver can safely ascend 52.5 ft to a depth of 73 ft.
5.116 6.07 g H_2O_2 **5.121** 6.53×10^{-3} g N_2 **5.125**(a) xenon
(b) water vapor   (c) mercury   (d) water vapor   **5.129** 17.2 g CO_2; 17.8 g Kr   **5.134** Ne, 676 m/s; Ar, 481 m/s; He, 1.52×10^3 m/s
5.136(a) 0.052 g   (b) 1.1 mL   **5.143**(a) 16.5 L CO_2

(b) P_{H_2O} = 48.8 torr; 3.7×10^2 torr P_{O_2} = P_{CO_2} **5.148** 332 steps   **5.150** 1.4   **5.154** P_{total} = 0.323 atm; P_{I_2} = 0.0332 atm

Chapter 6

6.4 Increase: eating food, lying in the sun, taking a hot bath. Decrease: exercising, taking a cold bath, going outside on a cold day.   **6.6** The amount of the change in internal energy is the same for heater and air conditioner. By the law of energy conservation, the change in energy of the universe is zero.   **6.8** 0 J   **6.10** 1.54×10^3 J   **6.12**(a) 6.6×10^7 kJ   (b) 1.6×10^7 kcal   (c) 6.3×10^7 Btu   **6.15** 8.8 h   **6.17** Measuring the heat transfer at constant pressure is more convenient than measuring at constant volume.   **6.19**(a) exothermic   (b) endothermic   (c) exothermic   (d) exothermic   (e) endothermic   (f) endothermic   (g) exothermic

6.22

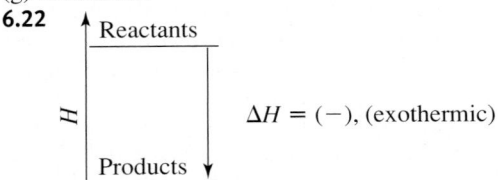

6.24(a) Combustion of ethane: $2C_2H_6(g) + 7O_2(g) \longrightarrow$
$$4CO_2(g) + 6H_2O(g) + heat$$

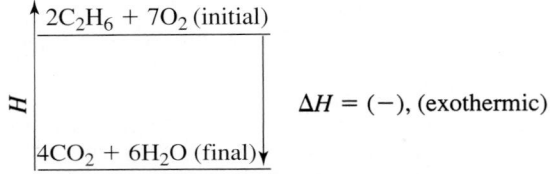

(b) Freezing of water: $H_2O(l) \longrightarrow H_2O(s) + heat$

6.26(a) $2CH_3OH(l) + 3O_2(g) \longrightarrow 2CO_2(g) + 4H_2O(g) + heat$

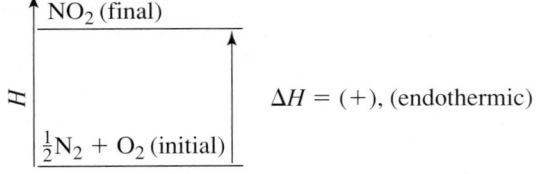

(b) $\frac{1}{2}N_2(g) + O_2(g) + heat \longrightarrow NO_2(g)$

6.28(a) This is a phase change from the solid phase to the gas phase. Heat is absorbed by the system so q_{sys} is positive (+).
(b) The volume of the system is expanding as more moles of gas are present after the phase change than were present before the phase change. So the system has done work of expansion, and w

is negative. Since $\Delta E_{sys} = q + w$, q is positive, and w is negative, the sign of ΔE_{sys} cannot be predicted. It will be positive if $q > w$ and negative if $q < w$.   (c) $\Delta E_{univ} = 0$. If the system loses energy, the surroundings gain an equal amount of energy. The sum of the energy of the system and the energy of the surroundings remains constant.   **6.31** To determine the specific heat capacity of a substance, you need its mass, the heat added (or lost), and the change in temperature.   **6.33** Specific heat capacity is the quantity of heat required to raise one gram of a substance by one kelvin. Heat capacity is also the quantity of heat required for a one-kelvin temperature change, but it applies to an object instead of a specified amount of a substance.   (a) Heat capacity, because the fixture is a combination of substances   (b) Specific heat capacity, because the copper wire is a pure substance   (c) Specific heat capacity, because the water is a pure substance   **6.35** 6.9×10^3 J   **6.37** 295°C   **6.39** 77.5°C   **6.41** 45°C   **6.43** 36.6°C   **6.50** The reaction has a positive ΔH_{rxn}, because this reaction requires the input of energy to break the oxygen-oxygen bond.   **6.51** ΔH is negative; it is opposite in sign and half of the value for the vaporization of 2 mol of H_2O.   **6.52**(a) exothermic   (b) 20.2 kJ   (c) -4.2×10^2 kJ   (d) -15.7 kJ   **6.54**(a) $\frac{1}{2}N_2(g) + \frac{1}{2}O_2(g) \longrightarrow NO(g)$; ΔH = 90.29 kJ
   (b) -10.5 kJ
6.56 -1.88×10^6 kJ
6.60(a) $C_2H_4(g) + 3O_2(g) \longrightarrow 2CO_2(g) + 2H_2O(l)$;
   $\Delta H_{rxn} = -1411$ kJ
   (b) 1.39 g C_2H_4
6.64 -110.5 kJ   **6.65** -813.4 kJ   **6.69** 44.0 kJ   **6.72** The standard heat of reaction, $\Delta H°_{rxn}$, is the enthalpy change for any reaction where all substances are in their standard states. The standard heat of formation, $\Delta H°_f$, is the enthalpy change that accompanies the formation of one mole of a compound in its standard state from elements in their standard states.
6.74(a) $\frac{1}{2}Cl_2(g) + Na(s) \longrightarrow NaCl(s)$
   (b) $H_2(g) + \frac{1}{2}O_2(g) \longrightarrow H_2O(g)$
   (c) no changes
6.75(a) $Ca(s) + Cl_2(g) \longrightarrow CaCl_2(s)$
   (b) $Na(s) + \frac{1}{2}H_2(g) + C(graphite) + \frac{3}{2}O_2(g) \longrightarrow$
$$NaHCO_3(s)$$
   (c) $C(graphite) + 2Cl_2(g) \longrightarrow CCl_4(l)$
   (d) $\frac{1}{2}H_2(g) + \frac{1}{2}N_2(g) + \frac{3}{2}O_2(g) \longrightarrow HNO_3(l)$
6.77(a) -1036.8 kJ   (b) -433 kJ   **6.79** -157.3 kJ/mol
6.81(a) 503.9 kJ   (b) $-\Delta H_1 + 2\Delta H_2 = 504$ kJ
6.82(a) $C_{18}H_{36}O_2(s) + 26O_2(g) \longrightarrow 18CO_2(g) + 18H_2O(g)$
   (b) $-10,488$ kJ   (c) -36.9 kJ; -8.81 kcal
   (d) 8.81 kcal/g $\times$ 11.0 g = 96.9 kcal   **6.84**(a) initial = 23.6 L/mol; final = 24.9 L/mol   (b) 187 J   (c) -1.2×10^2 J   (d) 3.1×10^2 J   (e) 310 J   (f) $\Delta H = \Delta E + P\Delta V = \Delta E - w = (q + w) - w = q_P$
6.93(a) 1.1×10^3 J   (b) 1.1×10^8 J   (c) 1.3×10^2 mol CH_4 (d) \$0.0050/mol   (e) \$0.85   **6.102**(a) III > II > I   (b) I > III > II   **6.104**(a) $\Delta H°_{rxn1} = -657.0$ kJ; $\Delta H°_{rxn2} = 32.9$ kJ   (b) -106.6 kJ   **6.107**(a) 6.81×10^3 J   (b) $+243$°C
6.108 -22.2 kJ   **6.109**(a) 34 kJ/mol   (b) -757 kJ
6.111(a) -1.25×10^3 kJ   (b) 2.24×10^3°C

Chapter 7

7.2(a) x-ray < ultraviolet < visible < infrared < microwave < radio waves   (b) radio < microwave < infrared < visible < ultraviolet < x-ray   (c) radio < microwave < infrared < visible

< ultraviolet < x-ray   **7.7** 316 m; 3.16×10^{11} nm; 3.16×10^{12} Å
7.9 2.5×10^{-23} J   **7.11** b < c < a   **7.13** 1.3483×10^9 nm;
1.3483×10^8 Å   **7.16**(a) 1.24×10^{15} s^{-1}; 8.21×10^{-19} J
(b) 1.4×10^{15} s^{-1}; 9.0×10^{-19} J   **7.18** Bohr's key assumption
was that the electron in an atom does not radiate energy while in
a stationary state, and it can move to a different orbit only by absorbing or emitting a photon whose energy is equal to the
difference in energy between two states. These differences in energy correspond to the wavelengths in the known line spectra for
the hydrogen atom.   **7.20**(a) absorption   (b) emission   (c) emission   (d) absorption   **7.22** Yes, the predicted line spectra are
accurate. The energies could be predicted from

$$E_n = \frac{-(Z^2)(2.18\times10^{-18}\ \text{J})}{n^2},$$ where Z is the atomic number for

the atom or ion. The energy levels for Be^{3+} will be greater by a
factor of 16 ($Z = 4$) than those for the hydrogen atom. This means
that the pattern of lines will be similar, but at different wavelengths.
7.23 434.17 nm   **7.25** 1875.6 nm   **7.27** -2.76×10^5 J/mol
7.29 d < a < c < b   **7.31** $n = 4$ **7.37** Macroscopic objects do
exhibit a wavelike motion, but the wavelength is too small for
humans to perceive.   **7.39**(a) 7.10×10^{-37} m   (b) 1×10^{-35} m
7.41 2.2×10^{-26} m/s   **7.43** 3.75×10^{-36} kg   **7.47** The total
probability of finding the 1s electron in any distance r from the
nucleus is greatest when the value of r is 0.529 Å.   **7.48**(a) principal determinant of the electron's energy or distance from the
nucleus   (b) determines the shape of the orbital   (c) determines
the orientation of the orbital in three-dimensional space
7.49(a) one   (b) five   (c) three   (d) nine   **7.51**(a) m_l: $-2, -1$,
$0, +1, +2$ (b) m_l: 0 (if $n = 1$, then $l = 0$)   (c) m_l: $-3, -2$,
$-1, 0, +1, +2, +3$
7.53(a)   (b)

                    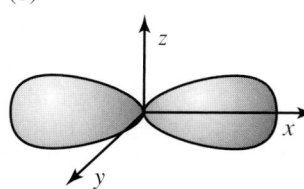

7.55

Sublevel	Allowable m_l	No. of orbitals
(a) d ($l = 2$)	$-2, -1, 0, +1, +2$	5
(b) p ($l = 1$)	$-1, 0, +1$	3
(c) f ($l = 3$)	$-3, -2, -1, 0, +1, +2, +3$	7

7.57(a) $n = 5$ and $l = 0$; one orbital   (b) $n = 3$ and $l = 1$; three
orbitals   (c) $n = 4$ and $l = 3$; seven orbitals   **7.59**(a) no; $n = 2$,
$l = 1$, $m_l = -1$; $n = 2$, $l = 0$, $m_l = 0$ (b) allowed   (c) allowed
(d) no; $n = 5$, $l = 3$, $m_l = +3$; $n = 5$, $l = 2$, $m_l = 0$
7.62(a) $E = -(2.180\times10^{-18}$ J)$(1/n^2)$. This is identical to the
expression from Bohr's theory.   (b) 3.028×10^{-19} J   (c) 656.1 nm
7.63(a) The attraction of the nucleus for the electrons must
be overcome.   (b) The electrons in silver are more tightly held
by the nucleus.   (c) silver   (d) Once the electron is freed from
the atom, its energy increases in proportion to the frequency of
the light.   **7.66** Li^{2+} **7.69**(a) $2 \longrightarrow 1$ (b) $5 \longrightarrow 2$
(c) $4 \longrightarrow 2$ (d) $3 \longrightarrow 2$ (e) $6 \longrightarrow 3$ **7.71**(a) Ba; 462 nm
(b) 278 to 292 nm   **7.74**(a) 2.7×10^2 s   (b) 3.6×10^8 m

7.75(a) $l = 1$ or 2   (b) $l = 1$ or 2   (c) $l = 3, 4, 5$, or 6
(d) $l = 2$ or 3

7.77(a) $\Delta E = (-2.18\times10^{-18}\ \text{J})\left(\dfrac{1}{\infty^2} - \dfrac{1}{n_{\text{initial}}^2}\right)Z^2\left(\dfrac{6.022\times10^{23}}{1\ \text{mol}}\right)$

(b) 3.28×10^7 J/mol   (c) 205 nm   (d) 22.8 nm
7.79(a) 5.293×10^{-11} m   (b) 5.293×10^{-9} m   **7.81** 6.4×10^{27}
photons   **7.83**(a) no overlap   (b) overlap   (c) two   (d) At
longer wavelengths, the hydrogen spectrum begins to become a
continuous band.   **7.86**(a) 7.56×10^{-18} J; 2.63×10^{-8} m
(b) 5.122×10^{-17} J; 3.881×10^{-9} m   (c) 1.2×10^{-18} J;
1.66×10^{-7} m   **7.88**(a) 1.87×10^{-19} J   (b) 3.58×10^{-19} J
7.90(a) red; green   (b) 5.89 kJ (Sr); 5.83 kJ (Ba)   **7.92**(a) This
is the wavelength of maximum absorbance, so it gives the highest sensitivity.   (b) ultraviolet region   (c) 1.93×10^{-2} g
vitamin A/g oil   **7.97** 1.0×10^{18} photons/s   **7.100** $3s \longrightarrow 2p$;
$3d \longrightarrow 2p$; $4s \longrightarrow 2p$; $3p \longrightarrow 2s$

Chapter 8

8.1 Elements are listed in the periodic table in an ordered, systematic way that correlates with a periodicity of their chemical
and physical properties. The theoretical basis for the table in
terms of atomic number and electron configuration does not allow for a "new element" between Sn and Sb.   **8.3**(a) predicted
atomic mass = 54.23 amu   (b) predicted melting point = 6.3°C
8.6 The quantum number m_s relates to just the electron; all the
others describe the orbital.   **8.9** Shielding occurs when inner
electrons protect, or shield, outer electrons from the full nuclear
attraction. The effective nuclear charge is the nuclear charge an
electron actually experiences. As the number of inner electrons
increases, the effective nuclear charge decreases.   **8.11**(a) 6
(b) 10   (c) 2   **8.13**(a) 6   (b) 2   (c) 14   **8.13** Degenerate orbitals (those of identical energy) will be filled in such a way that
the electron spins will have the same value.
N: $1s^2 2s^2 2p^3$

↑↓	↑↓	↑	↑	↑
1s	2s		2p	

8.18 Main-group elements from the same group have similar
outer electron configurations, and the (old) group number equals
the number of outer electrons. Outer electron configurations vary
in a periodic manner within a period, with each succeeding element having an additional electron.   **8.20** The maximum
number of electrons in any energy level n is $2n^2$, so the $n = 4$
energy level holds a maximum of $2(4^2) = 32$ electrons.
8.21(a) $n = 5, l = 0, m_l = 0$, and $m_s = +\frac{1}{2}$ (b) $n = 3, l = 1$,
$m_l = +1$, and $m_s = -\frac{1}{2}$ (c) $n = 5, l = 0, m_l = 0$, and
$m_s = +\frac{1}{2}$ (d) $n = 2, l = 1, m_l = +1$, and $m_s = -\frac{1}{2}$
8.23(a) Rb: $1s^2 2s^2 2p^6 3s^2 3p^6 4s^2 3d^{10} 4p^6 5s^1$
  (b) Ge: $1s^2 2s^2 2p^6 3s^2 3p^6 4s^2 3d^{10} 4p^2$
  (c) Ar: $1s^2 2s^2 2p^6 3s^2 3p^6$
8.25(a) Cl: $1s^2 2s^2 2p^6 3s^2 3p^5$ (b) Si: $1s^2 2s^2 2p^6 3s^2 3p^2$
(c) Sr: $1s^2 2s^2 2p^6 3s^2 3p^6 4s^2 3d^{10} 4p^6 5s^2$
8.27(a) Ti: [Ar] $4s^2 3d^2$

↓	↑	↑								
s		3d					4p			

(b) Cl: [Ne] $3s^2 3p^5$

↓	↑↓	↑↓	↑
s		3p	

(c) V: [Ar] $4s^2 3d^3$

↓	↑	↑	↑						

s $3d$ $4p$

8.29(a) Mn: [Ar] $4s^2 3d^5$

↑↓	↑	↑	↑	↑	↑			

$4s$ $3d$ $4p$

(b) P: [Ne] $3s^2 3p^3$

↑↓	↑	↑	↑

$3s$ $3p$

(c) Fe: [Ar] $4s^2 3d^6$

↑↓	↑↓	↑	↑	↑	↑			

$4s$ $3d$ $4p$

8.31(a) O; Group 6A(16); Period 2

↑↓	↑↓	↑	↑

$2s$ $2p$

(b) P; Group 5A(15); Period 3

↑↓	↑	↑	↑

$3s$ $3p$

8.33(a) Cl; Group 7A(17); Period 3

↑↓	↑↓	↑↓	↑

$3s$ $3p$

(b) As; Group 5A(15); Period 4

↑↓	↑↓	↑↓	↑↓	↑↓	↑↓	↑	↑	↑

$4s$ $3d$ $4p$

8.35(a) [Ar] $4s^2 3d^{10} 4p^1$; Group 3A(13)   (b) [He] $2s^2 2p^6$; Group 8A(18)

8.37

	Inner Electrons	Outer Electrons	Valence Electrons
(a) O	2	6	6
(b) Sn	46	4	4
(c) Ca	18	2	2
(d) Fe	18	2	8
(e) Se	28	6	6

8.39(a) B; Al, Ga, In, and Tl   (b) S; O, Se, Te, and Po   (c) La; Sc, Y, and Ac   **8.41**(a) C; Si, Ge, Sn, and Pb   (b) V; Nb, Ta, and Db   (c) P; N, As, Sb, and Bi

8.43 Na (first excited state): $1s^2 2s^2 2p^6 3p^1$

↑↓	↑↓	↑↓	↑↓	↑↓			↑		

$1s$ $2s$ $2p$ $3s$ $3p$

8.50 A high IE$_1$ and a very negative EA$_1$ suggest that the elements are halogens, in Group 7A(17), which form 1− ions.
8.53(a) K < Rb < Cs   (b) O < C < Be   (c) Cl < S < K
(d) Mg < Ca < K   **8.55**(a) Ba < Sr < Ca   (b) B < N < Ne
(c) Rb < Se < Br   (d) Sn < Sb < As   **8.57** $1s^2 2s^2 2p^1$ (boron, B)
8.59(a) Na   (b) Na   (c) Be   **8.61**(1) Metals conduct electricity; nonmetals do not. (2) When they form stable ions, metal ions tend to have a positive charge; nonmetal ions tend to have a negative charge. (3) Metal oxides are ionic and act as bases; nonmetal oxides are covalent and act as acids.   **8.62** Metallic character increases down a group and decreases toward the right across a period. These trends are the same as those for atomic size and opposite those for ionization energy.   **8.64** Possible

ions are 2+ and 4+. The 2+ ions form by loss of the outermost two p electrons, while the 4+ ions form by loss of these and the outermost two s electrons.   **8.68**(a) Rb   (b) Ra   (c) I
8.70(a) As   (b) P   (c) Be   **8.72** acidic solution; $SO_2(g) + H_2O(l) \longrightarrow H_2SO_3(aq)$ **8.74**(a) $Cl^-: 1s^2 2s^2 2p^6 3s^2 3p^6$
(b) $Na^+: 1s^2 2s^2 2p^6$ (c) $Ca^{2+}: 1s^2 2s^2 2p^6 3s^2 3p^6$
8.76(a) $Al^{3+}: 1s^2 2s^2 2p^6$ (b) $S^{2-}: 1s^2 2s^2 2p^6 3s^2 3p^6$ (c) $Sr^{2+}:$ $1s^2 2s^2 2p^6 3s^2 3p^6 4s^2 3d^{10} 4p^6$ **8.78**(a) 0   (b) 3   (c) 0   (d) 1
8.80 a, b, and d   **8.82**(a) V^{3+}: [Ar] $3d^2$, paramagnetic
(b) Cd^{2+}: [Kr] $4d^{10}$, diamagnetic   (c) Co^{3+}: [Ar] $3d^6$, paramagnetic   (d) Ag^+: [Kr] $4d^{10}$, diamagnetic   **8.84** For palladium to be diamagnetic, all of its electrons must be paired. (a) You might first write the condensed electron configuration for Pd as [Kr] $5s^2 4d^8$. However, the partial orbital diagram is not consistent with diamagnetism.

↑↓	↑↓	↑↓	↑↓	↑	↑			

$5s$ $4d$ $5p$

(b) This is the only configuration that supports diamagnetism, [Kr] $4d^{10}$.

	↑↓	↑↓	↑↓	↑↓	↑↓			

$5s$ $4d$ $5p$

(c) Promoting an s electron into the d sublevel still leaves two electrons unpaired.

↑	↑↓	↑↓	↑↓	↑↓	↑			

$5s$ $4d$ $5p$

8.86(a) $Li^+ < Na^+ < K^+$ (b) $Rb^+ < Br^- < Se^{2-}$ (c) $F^- < O^{2-} < N^{3-}$ **8.90** Ce: [Xe] $6s^2 4f^1 5d^1$; Ce^{4+}: [Xe]; Eu: [Xe] $6s^2 4f^7$; Eu^{2+}: [Xe] $4f^7$. Ce^{4+} has a noble-gas configuration; Eu^{2+} has a half-filled f subshell.   **8.92**(a) Cl_2O, dichlorine monoxide   (b) Cl_2O_3, dichlorine trioxide   (c) Cl_2O_5, dichlorine pentaoxide   (d) Cl_2O_7, dichlorine heptaoxide   (e) SO_3, sulfur trioxide   (f) SO_2, sulfur dioxide   (g) N_2O_5, dinitrogen pentaoxide   (h) N_2O_3, dinitrogen trioxide   (i) CO_2, carbon dioxide (j) P_2O_5, diphosphorus pentaoxide   **8.95**(a) $SrBr_2$, strontium bromide   (b) CaS, calcium sulfide   (c) ZnF_2, zinc fluoride (d) LiF, lithium fluoride   **8.96**(a) 2009 kJ/mol   (b) −549 kJ/mol
8.98 All ions except Fe^{8+} and Fe^{14+} are paramagnetic; Fe^+ and Fe^{3+} would be most attracted.

Chapter 9

9.1(a) Greater ionization energy decreases metallic character.
(b) Larger atomic radius increases metallic character.   (c) Higher number of outer electrons decreases metallic character.   (d) Larger effective nuclear charge decreases metallic character.
9.4(a) Cs   (b) Rb   (c) As   **9.6**(a) ionic   (b) covalent
(c) metallic   **9.8**(a) covalent   (b) ionic   (c) covalent

9.10(a) Rb·   (b) ·Si·   (c) :Ï·

9.12(a) ·Sr·   (b) :P̈·   (c) :S̈·

9.14(a) 6A(16); [noble gas] $ns^2 np^4$ (b) 3(A)13; [noble gas] $ns^2 np^1$ **9.20**(a) Ba^{2+}, [Xe]; Cl^-, [Ne] $3s^2 3p^6$, :C̈l:⁻; $BaCl_2$

(b) Sr^{2+}, [Kr]; O^{2-}, [He] $2s^2 2p^6$, :Ö:²⁻; SrO   (c) Al^{3+}, [Ne]; F^-, [He] $2s^2 2p^6$, :F̈:⁻; AlF_3 (d) Rb^+, [Kr]; O^{2-}, [He] $2s^2 2p^6$, :Ö:²⁻; Rb_2O.   **9.22**(a) 2A(2)   (b) 6A(16)   (c) 1A(1)   **9.24**(a) 3A(13) (b) 2A(2)   (c) 6A(16)   **9.26**(a) BaS; the charge on each ion is twice the charge on the ions in CsCl.   (b) LiCl; Li^+ is smaller than Cs^+. **9.28**(a) BaS; Ba^{2+} is larger than Ca^{2+}. (b) NaF; the charge on each ion is less than the charge on Mg and O. **9.30** 788 kJ; the lattice energy for NaCl is less than that for LiF, because the Na^+ and Cl^- ions are larger than Li^+ and F^- ions. **9.33** -336 kJ **9.34** When two chlorine atoms are far apart, there is no interaction between them. As the atoms move closer together, the nucleus of each atom attracts the electrons of the other atom. The closer the atoms, the greater this attraction; however, the repulsions of the two nuclei and two electrons also increase at the same time. The final internuclear distance is the distance at which maximum attraction is achieved in spite of the repulsion. **9.35** The bond energy is the energy required to break the bond between H atoms and Cl atoms in one mole of HCl molecules in the gaseous state. Energy is needed to break bonds, so bond energy is always endothermic and $\Delta H^\circ_{\text{bond breaking}}$ is positive. The amount of energy needed to break the bond is released upon its formation, so $\Delta H^\circ_{\text{bond forming}}$ has the same magnitude as $\Delta H^\circ_{\text{bond breaking}}$ but is opposite in sign (always exothermic and negative). **9.39**(a) I—I < Br—Br < Cl—Cl (b) S—Br < S—Cl < S—H (c) C—N < C=N < C≡N **9.41**(a) C—O < C=O; the C=O bond (bond order = 2) is stronger than the C—O bond (bond order = 1). (b) C—H < O—H; O is smaller than C so the O—H bond is shorter and stronger than the C—H bond. **9.44** Less energy is required to break weak bonds. **9.46** Both are one-carbon molecules. Since methane contains fewer C=O bonds, it will have the greater heat of reaction per mole for combustion. **9.48** -168 kJ **9.50** -22 kJ **9.51** -59 kJ **9.52** Electronegativity increases from left to right and increases from bottom to top within a group. Fluorine and oxygen are the two most electronegative elements. Cesium and francium are the two least electronegative elements. **9.54** The H—O bond in water is polar covalent. A nonpolar covalent bond occurs between two atoms with identical electronegativities. A polar covalent bond occurs when the atoms have differing electronegativities. Ionic bonds result from electron transfer between atoms. **9.57**(a) Si < S < O (b) Mg < As < P **9.59**(a) N > P > Si (b) As > Ga > Ca

9.61 (a) N—B (b) N—O (c) C—S (none)

(d) S—O (e) N—H (f) Cl—O

9.63 a, d, and e **9.65**(a) nonpolar covalent (b) ionic (c) polar covalent (d) polar covalent (e) nonpolar covalent (f) polar covalent; $SCl_2 < SF_2 < PF_3$

9.67(a) H—I < H—Br < H—Cl

(b) H—C < H—O < H—F

(c) S—Cl < P—Cl < Si—Cl

9.70(a) A shiny solid that conducts heat, is malleable, and melts at high temperatures. (b) Metals lose electrons to form positive ions, and metals form basic oxides. **9.74**(a) 800. kJ/mol, which is lower than the value in Table 9.2 (b) -2.417×10^4 kJ (c) 1690. g CO_2 (d) 65.2 L O_2 **9.76**(a) -125 kJ (b) Yes, since ΔH°_f is negative (c) -392 kJ (d) No, ΔH°_f for $MgCl_2$ is

much more negative than that for MgCl. **9.78**(a) 406 nm (b) 2.93×10^{-19} J (c) 1.87×10^4 m/s **9.82** C—Cl: 3.53×10^{-7} m; bond in O_2: 2.40×10^{-7} m **9.83** XeF_2: 132 kJ/mol; XeF_4: 150. kJ/mol; XeF_6: 146 kJ/mol **9.85**(a) The presence of the very electronegative fluorine atoms bonded to one of the carbons makes the C—C bond polar. This polar bond will tend to undergo heterolytic rather than homolytic cleavage. More energy is required to achieve heterolytic cleavage. (b) 1420 kJ **9.88** $-13,286$ kJ **9.90** 8.70×10^{14} s^{-1}; 3.45×10^{-7} m, which is in the ultraviolet region of the electromagnetic spectrum. **9.92**(a) $CH_3OCH_3(g)$: -326 kJ; $CH_3CH_2OH(g)$: -369 kJ (b) The formation of gaseous ethanol is more exothermic. (c) 43 kJ

Chapter 10

10.1 He and H cannot serve as central atoms in a Lewis structure. Both can have no more than two valence electrons. Fluorine needs only one electron to complete its valence level, and it does not have d orbitals available to expand its valence level. Thus, it can bond to only one other atom. **10.3** All the structures obey the octet rule except c and g.

10.5(a) [Lewis structure of SiF_4: F bonded to Si with three other F]   (b) [Lewis structure: Cl—Se—Cl]   (c) [Lewis structure: F—C—F with double bond to O]

10.7(a) [Lewis structure: F—P—F with F below]   (b) [Lewis structure: H—O—C—O—H with double bond to O]   (c) [Lewis structure: S=C=S]

10.9(a) $\left[O{=}N{=}O \right]^+$

(b) [Two resonance structures of NO_2F with N central, F top, two O bottom]

10.11(a) $[N{=}N{=}N]^- \longleftrightarrow [N{-}N{\equiv}N]^- \longleftrightarrow [N{\equiv}N{-}N]^-$

(b) $[O{=}N{-}O]^- \longleftrightarrow [O{-}N{=}O]^-$

10.13(a) [Lewis structure of IF_4 with four F around I]   formal charges: I = 0, F = 0

(b) $\left[H{-}Al{-}H \text{ with H top and bottom} \right]^-$ formal charges: H = 0, Al = -1

10.15(a) $[C{\equiv}N]^-$ formal charges: C = -1, N = 0

(b) $[Cl{-}O]^-$ formal charges: Cl = 0, O = -1

10.17(a) $\left[O{=}Br{=}O \text{ with O below} \right]^-$ formal charges: Br = 0, doubly bonded O = 0, singly bonded O = -1 O.N.: Br = $+5$; O = -2

(b) $\left[O{-}S{-}O \text{ with O below} \right]^{2-}$ formal charges: S = 0, singly bonded O = -1, doubly bonded O = 0 O.N.: S = $+4$; O = -2

10.19(a) B has 6 valence electrons, so the molecule BH_3 is electron deficient.   (b) As has an expanded valence level with 10 electrons. (c) Se has an expanded valence level with 10 electrons.

(a)   (b)   (c)

10.21(a) Br expands its valence level to 10 electrons.   (b) I has an expanded valence level of 10 electrons.   (c) Be has only 4 valence shell electrons, so the molecule BeF_2 is electron deficient.

(a) :F—Br—F: (b) [:Cl—I—Cl:]⁻   (c) :F—Be—F:
         |
        :F:

10.23

10.26 structure A   **10.28** The molecular shape and the electron-group arrangement are the same when no lone pairs are present on the central atom.   **10.30** tetrahedral, AX_4; trigonal pyramidal, AX_3E; bent or V shaped, AX_2E_2

10.32(a) (b) (c) (d) (e) (f)

10.34(a) trigonal planar, bent, 120°   (b) tetrahedral, trigonal pyramidal, 109.5°   (c) tetrahedral, trigonal pyramidal, 109.5°
10.36(a) trigonal planar, trigonal planar, 120°   (b) trigonal planar, bent, 120°   (c) tetrahedral, tetrahedral, 109.5°
10.38(a) trigonal planar, AX_3, 120°   (b) trigonal pyramidal, AX_3E, 109.5°   (c) trigonal bipyramidal, AX_5, 90° and 120°
10.40(a) bent, 109.5°, less than 109.5°   (b) trigonal bipyramidal, 90° and 120°, angles are ideal   (c) seesaw, 90° and 120°, less than ideal   (d) linear, 180°, angle is ideal   **10.42**(a) C: tetrahedral, 109.5°; O: bent, < 109.5°   (b) N: trigonal planar, 120°
10.44(a) C in CH_3: tetrahedral, 109.5°; C with C=O: trigonal planar, 120°; O with H: bent, < 109.5°   (b) O: bent, < 109.5°
10.46 $OF_2 < NF_3 < CF_4 < BF_3 < BeF_2$ **10.48**(a) The C and N each have three groups, so the ideal angles are 120°; the O has four groups, so the ideal angle is 109.5°. The N and O have lone pairs, so the angles are less than ideal.   (b) All central atoms have four pairs, so the ideal angles are 109.5°. The lone pairs on the O reduce this value.   (c) The B has three groups and an ideal bond angle of 120°. All the O's have four groups (ideal bond angles of 109.5°), two of which are lone pairs that reduce the angle.
10.51

In the gas phase, PCl_5 is AX_5, so the shape is trigonal bipyramidal, and the bond angles are 120° and 90°. The PCl_4^+ ion is AX_4, so the shape is tetrahedral, and the bond angles are 109.5°. The PCl_6^- ion is AX_6, so the shape is octahedral, and the bond angles are 90°.   **10.52** Molecules are polar if they have polar bonds that are not arranged to cancel each other. A polar bond is present any time there is a bond between elements with differing electronegativities.   **10.55**(a) CF_4 (b) BrCl and SCl_2
10.57(a) SO_2, because it is polar and SO_3 is not.   (b) IF has a greater electronegativity difference between its atoms.   (c) SF_4, because it is polar and SiF_4 is not.   (d) H_2O has a greater electronegativity difference between its atoms.
10.59

X   Y   Z

Yes, compound Y has a dipole moment.
10.61(a)

Hydrazine   Diazene   Nitrogen

The single N—N bond (bond order = 1) is weaker and longer than the others. The triple bond (bond order = 3) is stronger and shorter than the others. The double bond (bond order = 2) has an intermediate strength and length.
(b) $\Delta H^\circ_{rxn} = -367$ kJ

10.65(a) formal charges: Al = −1, end Cl = 0, bridging Cl = +1; I = −1, end Cl = 0, bridging Cl = +1   (b) The iodine atoms are each AX_4E_2, and the shape around each is square planar. Placing these square planar portions adjacent gives a planar molecule.
10.70

(a) In propylene oxide, the C's have ideal angles of 109.5°.
(b) The C that is not part of the three-membered ring should have close to the ideal angle. The atoms in the ring form an equilateral triangle, so the angles around the two C's in the ring are reduced from the ideal 109.5° to 60°.
10.72

The Cl—O—Cl bond angle is less than 109.5° because of the lone pairs.   **10.76**(a) −1267 kJ/mol   (b) −1226 kJ/mol
(c) −1234.8 kJ/mol. The two answers differ by less than 10 kJ/mol. This is very good agreement since average bond energies were used to calculate answers a and b.   (d) −37 kJ
10.79

10.82 CH_4: −409 kJ/mol O_2; H_2S: −398 kJ/mol O_2

10.84(a) The O in the OH species only has 7 valence electrons, which is less than an octet, and 1 electron is unpaired. (b) 426 kJ (c) 508 kJ **10.86**(a) The F atoms will substitute at the axial positions first. (b) PF_5 and PCl_3F_2

10.88 $H_2C_2O_4$:

HC_2O_4^−:

C_2O_4^{2−}:

In $H_2C_2O_4$, there are two shorter and stronger C=O bonds and two longer and weaker C—O bonds. In $HC_2O_4^-$, the carbon-oxygen bonds on the side retaining the H remain as one long, weak C—O and one short, strong C=O. The carbon-oxygen bonds on the other side of the molecule have resonance forms with a bond order of 1.5, so they are intermediate in length and strength. In $C_2O_4^{2-}$, all the carbon-oxygen bonds have a bond order of 1.5. **10.92** 22 kJ **10.93** Trigonal planar molecules are nonpolar, so AY_3 cannot be that shape. Trigonal pyramidal molecules and T-shaped molecules are polar, so either could represent AY_3. **10.98**(a) 339 pm (b) 316 pm and 223 pm (c) 271 pm

Chapter 11

11.1(a) sp^2 (b) sp^3d^2 (c) sp (d) sp^3 (e) sp^3d **11.3** C has only $2s$ and $2p$ atomic orbitals, allowing for a maximum of four hybrid orbitals. Si has $3s$, $3p$, and $3d$ atomic orbitals, allowing it to form more than four hybrid orbitals. **11.5**(a) six, sp^3d^2 (b) four, sp^3 **11.7**(a) sp^2 (b) sp^2 (c) sp^2 **11.9**(a) sp^3 (b) sp^3 (c) sp^3 **11.11**(a) Si: one s and three p atomic orbitals form sp^3 hybrid orbitals. (b) C: one s and one p atomic orbitals form sp hybrid orbitals. (c) S: one s, three p, and one d atomic orbital mix to form sp^3d hybrid orbitals. (d) N: one s and three p atomic orbitals mix to form sp^3 hybrid orbitals.
11.13(a) B ($sp^3 \longrightarrow sp^3$) (b) A ($sp^2 \longrightarrow sp^3$)

11.15 (a) ... 4s ... 4p ... sp^3
(b) ... 2s ... 2p ... sp^2 ... 2p
(c) ... 2s ... 2p ... sp^2 ... 2p ... +e^−

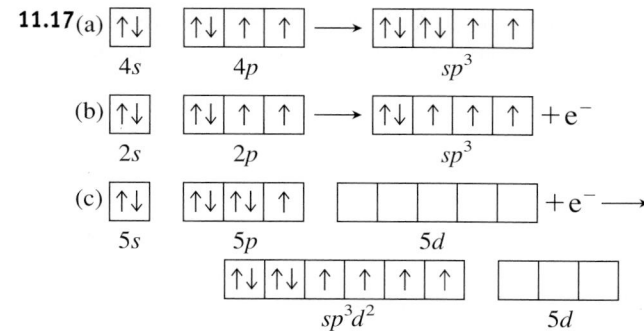

11.17(a) ... 4s ... 4p ... sp^3
(b) ... 2s ... 2p ... sp^3 ... +e^−
(c) ... 5s ... 5p ... 5d ... sp^3d^2 ... 5d

11.20(a) False. A double bond is one σ and one π bond. (b) False. A triple bond consists of one σ and two π bonds. (c) True (d) True (e) False. A π bond consists of a second pair of electrons after a σ bond has been previously formed. (f) False. End-to-end overlap results in a bond with electron density along the bond axis. **11.21**(a) Nitrogen uses sp^2 to form three σ bonds and one π bond. (b) Carbon uses sp to form two σ bonds and two π bonds. (c) Carbon uses sp^2 to form three σ bonds and one π bond.
11.23(a) N: sp^2, forming 2 σ bonds and 1 π bond

:F̈—N̈=Ö:

(b) C: sp^2, forming 3 σ bonds and 1 π bond

(c) C: sp, forming 2 σ bonds and 2 π bonds

:N≡C—C≡N:

11.25

cis   *trans*

The single bonds are all σ bonds. The double bond is one σ bond and one π bond. **11.26** Four molecular orbitals form from the four p atomic orbitals. The total number of molecular orbitals must equal the number of atomic orbitals. **11.28**(a) Bonding MOs have lower energy than antibonding MOs. Lower energy means more stable. (b) Bonding MOs do not have a nodal plane perpendicular to the bond. (c) Bonding MOs have higher electron density between nuclei than antibonding MOs. **11.30**(a) two (b) two (c) four **11.32**(a) A is π^*_{2p}, B is σ_{2p}, C is π_{2p}, and D is σ^*_{2p}. (b) π^*_{2p} (A), σ_{2p} (B), and π_{2p} (C) have at least one electron. (c) π^*_{2p} (A) has only one electron. **11.34**(a) stable (b) paramagnetic (c) $(\sigma_{2s})^2(\sigma^*_{2s})^1$ **11.36**(a) $C_2^+ < C_2 < C_2^-$ (b) $C_2^- < C_2 < C_2^+$ **11.40**(a) C (ring): sp^2; C (all others): sp^3; O (all): sp^3; N: sp^3 (b) 26 (c) 6 **11.42**(a) 17 (b) All carbons are sp^2; the ring N is sp^2, the other N's are sp^3. **11.44**(a) B changes from sp^2 to sp^3. (b) P changes from sp^3 to sp^3d. (c) C changes from sp to sp^2. Two electron groups surround C in C_2H_2, and three electron groups surround C in C_2H_4. (d) Si changes from sp^3 to sp^3d^2. (e) no change for S **11.46** P: tetrahedral, sp^3; N: trigonal pyramid, sp^3; C_1 and C_2: tetrahedral, sp^3; C_3: trigonal planar, sp^2 **11.50** The central C is sp hybridized, and the other two C atoms are sp^2 hybridized.

11.54(a) B and D are present.　(b) Yes, *sp* hybrid orbitals.
(c) Two sets of *sp* orbitals, four sets of *sp*^{2} orbitals, and three sets
of *sp*^{3} orbitals.　**11.56** Through resonance, the C—N bond gains
some double-bond character, which hinders rotation about that
bond.

11.58(a) C in —CH$_3$: *sp*^{3}; all other C atoms: *sp*^{2}; O in two C—O
bonds: *sp*^{3}; O in two C=O bonds: *sp*^{2}　(b) two　(c) eight; one
11.59(a) four　(b) eight

Chapter 12

12.1 In a solid, the energy of attraction of the particles is greater
than their energy of motion; in a gas, it is less. Gases have high
compressibility and the ability to flow, while solids have neither.
12.4(a) Because the intermolecular forces are only partially overcome when fusion occurs but need to be totally overcome in
vaporization.　(b) Because solids have greater intermolecular
forces than liquids do.　(c) $\Delta H_{vap} = -\Delta H_{cond}$　**12.5**(a) intermolecular　(b) intermolecular　(c) intramolecular
(d) intramolecular　**12.7**(a) condensation　(b) fusion
(c) vaporization　**12.9** The gas molecules slow down as the gas
is compressed. Therefore, much of the kinetic energy lost by the
propane molecules is released to the surroundings.　**12.13** At
first, the vaporization of liquid molecules from the surface predominates, which increases the number of gas molecules and
hence the vapor pressure. As more molecules enter the gas phase,
gas molecules hit the surface of the liquid and "stick" more frequently, so the condensation rate increases. When the
vaporization and condensation rates become equal, the vapor
pressure becomes constant.　**12.14** As intermolecular forces increase,　(a) critical temperature increases,　(b) boiling point
increases,　(c) vapor pressure decreases, and　(d) heat of
vaporization increases.　**12.18** because the condensation of the
vapor supplies an additional 40.7 kJ/mol　**12.19** 7.67×10^3 J
12.21 0.777 atm　**12.23** 2×10^4 J/mol
12.25

Solid ethylene is more dense than liquid ethylene.
12.28 32 atm　**12.32** O is smaller and more electronegative than
Se; so the electron density on O is greater, which attracts H more
strongly.　**12.34** All particles (atoms and molecules) exhibit
dispersion forces, but the total force is weak for small molecules.
Dipole-dipole forces between small polar molecules dominate

the dispersion forces.　**12.37**(a) hydrogen bonding　(b) dispersion forces　(c) dispersion forces　**12.39**(a) dipole-dipole forces
(b) dispersion forces　(c) hydrogen bonding
12.41(a)

$$CH_3—CH—\overset{..}{\underset{..}{O}}:\cdots H—\overset{..}{\underset{..}{O}}—CH—CH_3$$

(with CH_3 on the left carbon and H, CH_3 on the right)

(b)

12.43(a) dispersion forces　(b) hydrogen bonding　(c) dispersion forces　**12.45**(a) I$^-$　(b) CH$_2$=CH$_2$　(c) H$_2$Se. In (a) and
(c) the larger particle has the higher polarizability. In (b), the less
tightly held π electron clouds are more easily distorted.
12.47(a) C$_2$H$_6$; it is a smaller molecule exhibiting weaker dispersion forces than C$_4$H$_{10}$.　(b) CH$_3$CH$_2$F; it has no H—F bonds,
so it only exhibits dipole-dipole forces, which are weaker than
the hydrogen bonds of CH$_3$CH$_2$OH.　(c) PH$_3$; it has weaker intermolecular forces (dipole-dipole) than NH$_3$ (hydrogen
bonding).　**12.49**(a) HCl; it has dipole-dipole forces, and there
are stronger ionic bonds in LiCl.　(b) PH$_3$; it has dipole-dipole
forces, and there is stronger hydrogen bonding in NH$_3$.　(c) Xe;
it exhibits weaker dispersion forces since its smaller size results
in lower polarizability than the larger I$_2$ molecules.　**12.51**(a) C$_4$H$_8$
(cyclobutane), because it is more compact than C$_4$H$_{10}$.　(b) PBr$_3$;
the dipole-dipole forces in PBr$_3$ are weaker than the ionic bonds
in NaBr.　(c) HBr; the dipole-dipole forces in HBr are weaker
than the hydrogen bonds in water.　**12.53** As atomic size decreases and electronegativity increases, the electron density of an
atom increases. High electron density strengthens the attraction
to an H atom on another molecule and makes its attached H atom
more positive. Fluorine is the smallest of the three and the most
electronegative, so the H bonds in hydrogen fluoride are the
strongest. Oxygen is smaller and more electronegative than
nitrogen, so H bonds in water are stronger than H bonds in
ammonia.　**12.57** The cohesive forces in water and mercury are
stronger than the adhesive forces to the nonpolar wax on the
floor. Weak adhesive forces result in spherical drops. The
adhesive forces overcome the even weaker cohesive forces in the
oil, and so the oil drop spreads out.　**12.59** Surface tension is
defined as the energy needed to increase the surface area by a
given amount, so units of energy per area are appropriate.
12.61 CH$_3$CH$_2$CH$_2$OH < HOCH$_2$CH$_2$OH <
HOCH$_2$CH(OH)CH$_2$OH. More hydrogen bonding means more
attraction between molecules so more energy is needed to increase surface area.　**12.63** HOCH$_2$CH(OH)CH$_2$OH >
HOCH$_2$CH$_2$OH > CH$_3$CH$_2$CH$_2$OH. More hydrogen bonding
means more attraction between molecules, so they flow less easily.
12.68 Water is a good solvent for polar and ionic substances and
a poor solvent for nonpolar substances. Water is a polar molecule
and dissolves polar substances because their intermolecular
forces are of similar strength.　**12.69** A single water molecule
can form four H bonds. The two hydrogen atoms each form one
H bond to oxygen atoms on neighboring water molecules. The
two lone pairs on the oxygen atom form H bonds with two hydrogen atoms on neighboring molecules.　**12.72** Water exhibits
strong capillary action, which allows it to be easily absorbed by
the narrow spaces in the plant's roots and transported upward to
the leaves.　**12.78** simple cubic　**12.81** The energy gap is the
energy difference between the highest filled energy level

(valence band) and the lowest unfilled energy level (conduction band). In conductors and superconductors, the energy gap is zero because the valence band overlaps the conduction band. In semiconductors, the energy gap is small. In insulators, the gap is large. **12.83** atomic mass and atomic radius **12.84**(a) face-centered cubic (b) body-centered cubic (c) face-centered cubic **12.86**(a) The change in unit cell is from a sodium chloride structure in CdO to a zinc blende structure in CdSe. (b) Yes, the coordination number of Cd changes from 6 in CdO to 4 in CdSe. **12.88**(a) Nickel forms a metallic solid since it is a metal whose atoms are held together by metallic bonds. (b) Fluorine forms a molecular solid since the F_2 molecules are held together by dispersion forces. (c) Methanol forms a molecular solid since the CH_3OH molecules are held together by hydrogen bonds. (d) Tin forms a metallic solid since it is a metal whose atoms are held together by metallic bonds. (e) Silicon is in the same group as carbon, so it exhibits similar bonding properties. Since diamond and graphite are both network covalent solids, it makes sense that Si forms a network covalent solid as well. (f) Xe is an atomic solid since its individual atoms are held together by dispersion forces. **12.90** four **12.92**(a) four Se^{2-} ions, four Zn^{2+} ions (b) 577.48 amu (c) 1.77×10^{-22} cm^3 (d) 5.61×10^{-8} cm **12.94**(a) insulator (b) conductor (c) semiconductor **12.96**(a) Conductivity increases. (b) Conductivity increases. (c) Conductivity decreases. **12.98** 1.68×10^{-8} cm **12.105** A substance whose properties are the same in all directions is isotropic; otherwise, the substance is anisotropic. Liquid crystals have a degree of order only in certain directions, so they are anisotropic. **12.111**(a) n-type semiconductor (b) p-type semiconductor **12.113** $n = 3.4 \times 10^3$ **12.115** 8.4 pm **12.118**(a) 19.8 torr (b) 0.0485 g **12.122** 277 nm; strikes the cubic unit cell on an edge **12.125** 259°K **12.127**(a) 2.23×10^{-2} atm (b) 6.26 L **12.130**(a) furfuryl alcohol

2-furoic acid

(b) furfuryl alcohol   2-furoic acid

12.131(a) 49.3 metric tons H_2O (b) -1.11×10^8 kJ **12.132** 2.9 g/m^3 **12.137**(a) 1.1 min (b) 10. min

(c)

[graph: T (°C) vs Time (min), heating curve]

12.138 2.98×10^5 g BN **12.141** 45.98 amu

Chapter 13

13.2 When a salt such as NaCl dissolves, ion-dipole forces cause the ions to separate, and many water molecules cluster around each ion in hydration shells. **13.4** Sodium stearate would be a more effective soap because the hydrocarbon chain in the stearate ion is longer than that in the acetate ion. The longer chain allows for more dispersion forces with grease molecules. **13.7** KNO_3 is an ionic compound and is therefore more soluble in water. **13.9**(a) ion-dipole forces (b) hydrogen bonding (c) dipole–induced dipole forces **13.11**(a) hydrogen bonding (b) dipole–induced dipole forces (c) dispersion forces **13.13**(a) HCl(g), because the molecular interactions (dipole-dipole forces) in ether are like those in HCl but not like the ionic bonding in NaCl. (b) $CH_3CHO(l)$, because the molecular interactions with ether (dipole-dipole) can replace those between CH_3CHO, but not the H bonds in water. (c) $CH_3CH_2MgBr(s)$, because the molecular interactions (dispersion forces) are greater than between ether and the ions in $MgBr_2$. **13.16** Gluconic acid is soluble in water due to extensive hydrogen bonding by the —OH groups attached to five of its carbons. The dispersion forces in the nonpolar tail of caproic acid are more similar to the dispersion forces in hexane; thus, caproic acid is soluble in hexane. **13.18** The nitrogen-containing bases form H bonds to their complementary bases. The flat, N-containing bases stack above each other, which allows extensive dispersion forces. The exterior negatively charged sugar-phosphate chains form ion-dipole and H bonds with water molecules in the aqueous surroundings, which also stabilizes the structure. **13.20** Dispersion forces are present between the nonpolar tails of the lipid molecules within the bilayer. Polar heads form H bonds and ion-dipole forces with the aqueous surroundings. **13.25** The energy change needed to separate the solvent ($\Delta H_{solvent}$) and that needed to mix the solvent and solute (ΔH_{mix}) combine to obtain $\Delta H_{solution}$. **13.29** Very soluble because a decrease in enthalpy and an increase in entropy both favor the formation of a solution. **13.30**

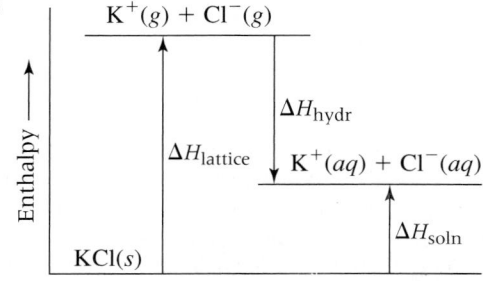

13.32(a) The volume of Na^+ is smaller, so it has the greater charge density.   (b) Sr^{2+} has a larger ionic charge and a smaller volume, so it has the greater charge density.   (c) Na^+ is smaller than Cl^-, so it has the greater charge density.   (d) O^{2-} has a larger ionic charge with a similar ion volume, so it has the greater charge density.   (e) OH^- has a smaller volume than SH^-, so it has the greater charge density.   (f) Mg^{2+} has a smaller volume, so it has the greater charge density.   (g) Mg^{2+} has both a smaller volume and a larger ionic charge, so it has the greater charge density.   (h) CO_3^{2-} has both a smaller volume and a larger ionic charge, so it has the greater charge density.
13.34(a) Na^+ (b) Sr^{2+} (c) Na^+ (d) O^{2-} (e) OH^-
(f) Mg^{2+} (g) Mg^{2+} (h) CO_3^{2-} **13.36**(a) -704 kJ/mol
(b) The K^+ ion contributes more because it is smaller and, therefore, has a greater charge density.   **13.38**(a) increases
(b) decreases   (c) increases   **13.41** Add a pinch of the solid solute to each solution. Addition of a "seed" crystal of solute to a supersaturated solution causes the excess solute to crystallize immediately, leaving behind a saturated solution. The solution in which the added solid solute dissolves is the unsaturated solution. The solution in which the added solid solute remains undissolved is the saturated solution.   **13.44**(a) increase
(b) decrease   **13.46**(a) 0.102 g O_2 (b) 0.0214 g O_2
13.49 0.20 mol/L   **13.52**(a) molarity and % w/v or % v/v
(b) parts-by-mass (% w/w)   (c) molality   **13.54** With just this information, you can convert between molality and molarity, but you need to know the molar mass of the solvent to convert to mole fraction.   **13.56**(a) 0.944 M $C_{12}H_{22}O_{11}$ (b) 0.167 M $LiNO_3$
13.58(a) 0.0749 M NaOH   (b) 0.36 M HNO_3 **13.60**(a) Add 4.25 g KH_2PO_4 to enough water to make 365 mL of aqueous solution.
(b) Add 125 mL of 1.25 M NaOH to enough water to make 465 mL of solution.   **13.62**(a) Weigh out 48.0 g KBr, dissolve it in about 1 L distilled water, and then dilute to 1.40 L with distilled water.   (b) Measure 82.7 mL of the 0.264 M solution and add distilled water to make a total of 255 mL.   **13.64**(a) 0.896 m glycine   (b) 1.21 m glycerol   **13.66** 4.48 m C_6H_6 **13.68**(a) Add 2.39 g $C_2H_6O_2$ to 308 g H_2O.   (b) Add 0.0508 kg of 52.0% HNO_3 by mass to 1.15 kg H_2O to make 1.20 kg of 2.20% HNO_3 by mass.   **13.70**(a) 0.29   (b) 58 mass %   (c) 23 m C_3H_7OH
13.72 42.6 g CsBr; mole fraction $= 7.82\times10^{-3}$; 8.51% by mass
13.74 5.11 m NH_3; 4.53 M NH_3; mole fraction $= 0.0843$
13.76 2.5 ppm Ca^{2+}, 0.56 ppm Mg^{2+} **13.80** It creates a "strong" current. A strong electrolyte dissociates completely in solution.
13.82 The boiling point is higher and the freezing point is lower for the solution compared to the solvent.   **13.85** A dilute solution of an electrolyte behaves more ideally than a concentrated one. With increasing concentration, the effective concentration deviates from the molar concentration because of ionic attractions. Thus, 0.050 m NaF has a boiling point closer to its predicted value.   **13.88**(a) strong electrolyte   (b) strong electrolyte
(c) nonelectrolyte   (d) weak electrolyte   **13.90**(a) 0.6 mol of solute particles   (b) 0.13 mol   (c) 2×10^{-4} mol   (d) 0.06 mol
13.92(a) CH_3OH in H_2O (b) H_2O in CH_3OH solution
13.94(a) $\Pi_{II} < \Pi_I < \Pi_{III}$ (b) $bp_{II} < bp_I < bp_{III}$
(c) $fp_{III} < fp_I < fp_{II}$ (d) $vp_{III} < vp_I < vp_{II}$ **13.96** 23.4 torr
13.98 $-0.467°C$ **13.100** 79.5°C   **13.102** 1.18×10^4 g $C_2H_6O_2$
13.104(a) NaCl: 0.173 m and $i = 1.84$ (b) CH_3COOH: 0.0837 m and $i = 1.02$ **13.106** 0.249 atm   **13.108** 209 torr for CH_2Cl_2;
48.1 torr for CCl_4 **13.109** The fluid inside a bacterial cell is both a solution and a colloid. It is a solution of ions and small

molecules and a colloid of large molecules (proteins and nucleic acids).   **13.113** Soap micelles have nonpolar tails pointed inward and anionic heads pointed outward. The charges on the heads of one micelle repel those on the heads of a neighboring micelle because the charges are the same. This repulsion between micelles keeps them from coagulating. Soap is more effective in freshwater than in seawater because the divalent cations in seawater combine with the anionic heads to form a precipitate.
13.117 3.4×10^9 L   **13.121** 0.0°C: 4.53×10^{-4} M O_2;
20.0°C: 2.83×10^{-4} M O_2; 40.0°C: 2.00×10^{-4} M O_2
13.123(a) 89.9 g/mol   (b) C_2H_5O; $C_4H_{10}O_2$
(c) Forms H bonds:

$$HO-\overset{\overset{\displaystyle H}{|}}{\underset{\underset{\displaystyle H}{|}}{C}}-\overset{\overset{\displaystyle H}{|}}{\underset{\underset{\displaystyle H}{|}}{C}}-\overset{\overset{\displaystyle H}{|}}{\underset{\underset{\displaystyle H}{|}}{C}}-\overset{\overset{\displaystyle H}{|}}{\underset{\underset{\displaystyle H}{|}}{C}}-OH$$

Does not form H bonds:

$$H-\overset{\overset{\displaystyle H}{|}}{\underset{\underset{\displaystyle H}{|}}{C}}-O-\overset{\overset{\displaystyle H}{|}}{\underset{\underset{\displaystyle H}{|}}{C}}-O-\overset{\overset{\displaystyle H}{|}}{\underset{\underset{\displaystyle H}{|}}{C}}-\overset{\overset{\displaystyle H}{|}}{\underset{\underset{\displaystyle H}{|}}{C}}-H$$

13.127(a) 9.45 g NaF   (b) 0.0017 g F^- **13.131**(a) 68 g/mol
(b) 2.1×10^2 g/mol   (c) The molar mass of CaN_2O_6 is 164.10 g/mol. This value is less than the 2.1×10^2 g/mol calculated when the compound is assumed to be a strong electrolyte and is greater than the 68 g/mol calculated when the compound is assumed to be a nonelectrolyte. Thus, the compound forms a nonideal solution because the ions interact but do not dissociate completely in solution.   (d) $i = 2.4$ **13.135**(a) 1.82×10^4 g/mol
(b) $3.41\times10^{-5}°C$ **13.136** 9.4×10^5 ng/L   **13.140**(a) 0.02
(b) 5×10^{-3} atm·L/mol   (c) 6.3×10^{-6} atm·m^3/mol
13.142 Weigh 3.11 g $NaHCO_3$ and dissolve in 247 g water.
13.146 $M = m$ (kg solvent/L solution) $= m \times d_{solution}$. Thus, for very dilute solutions molality $\times$ density $=$ molarity. In an aqueous solution, the liters of solution have approximately the same value as the kg of solvent because the density of water is close to 1 kg/L, so $m = M$.   **13.148**(a) 2.5×10^{-3} M SO_2 (b) The base reacts with the sulfur dioxide to produce calcium sulfite. The reaction of sulfur dioxide makes "room" for more sulfur dioxide to dissolve.   **13.153** Mass percents: iodine in chloroform: 2.6% I_2; iodine in carbon tetrachloride: 2.4% I_2; iodine in carbon disulfide: 14% I_2; Mole fractions: iodine in chloroform: 0.012; iodine in carbon tetrachloride: 0.015; iodine in carbon disulfide: 0.046; Molality: iodine in chloroform: 0.11 m I_2; iodine in carbon tetrachloride: 0.098 m I_2; iodine in carbon disulfide: 0.63 m I_2
13.157(a) 0.0212 mol/L·atm   (b) 8.90 ppm   (c) k_H: $C_6F_{14} >$ $C_6H_{14} >$ ethanol $>$ water. To dissolve oxygen in a solvent, the solvent molecules must be moved apart to make room for the gas. The stronger the intermolecular forces in the solvent, the more difficult it is to separate solvent particles and the lower the solubility of the gas. Both C_6F_{14} and C_6H_{14} have weak dispersion forces, with C_6F_{14} having the weaker forces due to the electronegative fluorine atoms repelling each other. Both ethanol and water are held together by strong hydrogen bonds with those bonds being stronger in water, as the boiling point indicates.
13.159(a) Yes, the phases of water can still coexist at some temperature and can therefore establish equilibrium.   (b) The triple point would occur at a lower pressure and lower temperature because the dissolved air would lower the vapor pressure of the

solvent.   (c) Yes, this is possible because the gas-solid phase boundary exists below the new triple point.   (d) No; at both temperatures, the presence of the solute lowers the vapor pressure of the liquid.   **13.161**(a) 2.8×10^{-4} g/mL   (b) 81 mL

Chapter 14

14.2(a) Z_{eff}, the effective nuclear charge, increases significantly across a period and decreases slightly down a group.   (b) Moving across a period, atomic size decreases, IE_1 increases, and EN increases, all because of the increased Z_{eff}.   **14.3**(a) Polarity is the molecular property that is responsible for the difference in boiling points between iodine monochloride (polar) and bromine (nonpolar). It arises from different EN values of the bonded atoms.   (b) The boiling point of polar ICl is higher than the boiling point of Br_2.   **14.7** The leftmost element in a period is an alkali metal [except for hydrogen in Group 1A(1)]. As we move to the right, the bonding to the other atom changes from metallic to polar covalent to ionic. The IE, EA, and EN of the two atoms become increasingly different, causing changes in the character of the bond.   **14.9** The elements on the left side in Period 3 form cations. The loss of electrons leads to an ion smaller than the atom. The elements on the right side form anions. The gain of electrons leads to an ion larger than the atom.   **14.11**(a) The greater the electronegativity of the element, the more covalent the bonding is in its oxide.   (b) The more electronegative the element, the more acidic the oxide is.   **14.14** In general, network solids have very high melting and boiling points and are very hard, while molecular solids have low melting and boiling points and are soft. The properties of network solids reflect the necessity of breaking chemical bonds throughout the substances, whereas the properties of molecular solids reflect the weaker forces between individual molecules.   **14.15**(a) Mg < Sr < Ba (b) Na < Al < P   (c) Se < Br < Cl   (d) Ga < Sn < Bi   **14.17** b and d   **14.19** $\ddot{O}=C=\ddot{O}$ **14.21** $CF_4 < SiCl_4 < GeBr_4$ **14.23**(a) $Na^+ < F^- < O^{2-}$ (b) $Cl^- < S^{2-} < P^{3-}$ **14.25** $O < O^- < O^{2-}$

14.27(a) $CaO(s) + H_2O(l) \longrightarrow$
$\qquad Ca(OH)_2(s) \rightleftharpoons Ca^{2+}(aq) + 2OH^-(aq)$(more basic)
$SO_3(g) + H_2O(l) \longrightarrow H_2SO_4(aq) \longrightarrow H^+(aq) + HSO_4^-(aq)$
(b) BaO   (c) CO_2 (d) K_2O **14.29**(a) LiCl   (b) PCl_3 (c) NCl_3 **14.31**(a) Ar < Na < Si   (b) Cs < Rb < Li   **14.33** The outermost electron is attracted by a smaller effective nuclear charge in Li because of shielding by the inner electrons, and it is farther from the nucleus in Li. Both of these factors lead to a lower ionization energy.
14.35(a) NH_3 will hydrogen bond.

$$:\ddot{F}-\ddot{N}-\ddot{F}: \qquad H-\ddot{N}-H$$
$$\quad\ \ |\qquad\qquad\qquad\ \ |$$
$$\ \ :\ddot{F}:\qquad\qquad\qquad H$$

(b) CH_3CH_2OH will hydrogen bond.

$$H \qquad H \qquad\qquad H\ \ H$$
$$|\qquad\ |\qquad\qquad\qquad |\ \ \ |$$
$$H-C-\ddot{O}-C-H \qquad H-C-C-\ddot{O}-H$$
$$|\qquad\ |\qquad\qquad\qquad |\ \ \ |$$
$$H \qquad H \qquad\qquad H\ \ H$$

14.37(a) $2Al(s) + 6HCl(aq) \longrightarrow 2AlCl_3(aq) + 3H_2(g)$
(b) $LiH(s) + H_2O(l) \longrightarrow LiOH(aq) + H_2(g)$
14.39(a) $NaBH_4$: +1 for Na, +3 for B, -1 for H
$\qquad Al(BH_4)_3$: +3 for Al, +3 for B, -1 for H
$\qquad LiAlH_4$: +1 for Li, +3 for Al, -1 for H

(b) tetrahedral

$$\left[\begin{array}{c} H \\ | \\ H-B-H \\ | \\ H \end{array} \right]^-$$

14.42 For Groups 1A(1) to 4A(14), the number of covalent bonds equals the (old) group number. For Groups 5A(15) to 7A(17), it equals 8 minus the (old) group number. There are exceptions in Period 3 to Period 6 because it is possible for the 3A(13) to 7A(17) elements to use d orbitals and form more bonds.
14.45(a) Group 3A(13) or 3B(3)   (b) If E is in Group 3B(3), the oxide and fluoride will be more ionic.   **14.48**(a) reducing agent (b) Alkali metals have relatively low ionization energies, which means they easily lose the outermost electron.
(c) $2Na(s) + 2H_2O(l) \longrightarrow 2Na^+(aq) + 2OH^-(aq) + H_2(g)$
$\qquad 2Na(s) + Cl_2(g) \longrightarrow 2NaCl(s)$
14.50 Density and ionic size increase down a group; the other three properties decrease down a group.
14.52 $2Na(s) + O_2(g) \longrightarrow Na_2O_2(s)$
14.54 $K_2CO_3(s) + 2HI(aq) \longrightarrow 2KI(aq) + H_2O(l) + CO_2(g)$
14.58 Group 2A(2) metals have an additional bonding electron to increase the strength of metallic bonding, which leads to higher melting points, higher boiling points, greater hardness, and greater density.
14.59(a) $CaO(s) + H_2O(l) \longrightarrow Ca(OH)_2(s)$
$\qquad$(b) $2Ca(s) + O_2(g) \longrightarrow 2CaO(s)$
14.62(a) $BeO(s) + H_2O(l) \longrightarrow$ no reaction
$\qquad$(b) $BeCl_2(l) + 2Cl^-$(solvated) $\longrightarrow BeCl_4^{2-}$(solvated).
Be behaves like other Group 2A(2) elements in reaction (b).
14.65 The electron removed from Group 2A(2) atoms occupies the outer s orbital, whereas in Group 3A(13) atoms, the electron occupies the outer p orbital. For example, the electron configuration for Be is $1s^2 2s^2$ and for B it is $1s^2 2s^2 2p^1$. It is easier to remove the p electron of B than an s electron of Be, because the energy of a p orbital is higher than that of the s orbital of the same level. Even though atomic size decreases because of increasing Z_{eff}, IE decreases from 2A(2) to 3A(13).
14.66(a) Most atoms form stable compounds when they complete their outer shell (octet). Some compounds of Group 3A(13) elements, like boron, have only six electrons around the central atom. Having fewer than eight electrons is called *electron deficiency.*
(b) $BF_3(g) + NH_3(g) \longrightarrow F_3B-NH_3(g)$
$\qquad B(OH)_3(aq) + OH^-(aq) \longrightarrow B(OH)_4^-(aq)$
14.68 $In_2O_3 < Ga_2O_3 < Al_2O_3$ **14.70** Apparent O.N., +3; actual O.N., +1. $\left[:\ddot{I}-\ddot{I}-\ddot{I}: \right]^-$ The anion I_3^- has the general formula AX_2E_3 and bond angles of 180°. $(Tl^{3+})(I^-)_3$ does not exist because of the low strength of the Tl—I bond.
14.75(a) $B_2O_3(s) + 2NH_3(g) \longrightarrow 2BN(s) + 3H_2O(g)$
(b) 1.30×10^2 kJ   (c) 5.3 kg borax   **14.76** Basicity in water is greater for the oxide of a metal. Tin(IV) oxide is more basic in water than carbon dioxide because tin has more metallic character than carbon.   **14.78**(a) Ionization energy generally decreases down a group.   (b) The deviations (increases) from the expected trend are due to the presence of the first transition series between Si and Ge and of the lanthanides between Sn and Pb.   (c) Group 3A(13)   **14.81** Atomic size increases down a group. As atomic size increases, ionization energy decreases and so it is easier to form a positive ion. An atom that is easier to ionize exhibits greater metallic character.

14.83(a)

(b)

14.85

14.88(a) diamond, C   (b) calcium carbonate, $CaCO_3$ (c) carbon dioxide, CO_2 (d) poly(dimethyl siloxane), $[(CH_3)_2SiO]_n$ (e) silicon carbide, SiC   (f) carbon monoxide, CO   (g) lead, Pb   **14.92**(a) -3 to $+5$ (b) For a group of nonmetals, the oxidation states range from the lowest, group number -8, or $5-8 = -3$ for Group 5A, to the highest, equal to the group number, or $+5$ for Group 5A.   **14.95** $H_3AsO_4 < H_3PO_4 < HNO_3$
14.97(a) $4As(s) + 5O_2(g) \longrightarrow 2As_2O_5(s)$
(b) $2Bi(s) + 3F_2(g) \longrightarrow 2BiF_3(s)$
(c) $Ca_3As_2(s) + 6H_2O(l) \longrightarrow 3Ca(OH)_2(s) + 2AsH_3(g)$
14.99(a) $N_2(g) + 2Al(s) \xrightarrow{\Delta} 2AlN(s)$
(b) $PF_5(g) + 4H_2O(l) \longrightarrow H_3PO_4(aq) + 5HF(g)$
14.101 trigonal bipyramidal, with axial F atoms and equatorial Cl atoms
14.105(a)

(b)

(c)

(d)

14.107(a) $2KNO_3(s) \xrightarrow{\Delta} 2KNO_2(s) + O_2(g)$ (b) $4KNO_3(s)$ $\xrightarrow{\Delta} 2K_2O(s) + 2N_2(g) + 5O_2(g)$ **14.109**(a) Boiling point and conductivity vary in similar ways down both groups. (b) Degree of metallic character and types of bonding vary in similar ways down both groups.   (c) Both P and S have allotropes, and both bond covalently with almost every other nonmetal.   (d) Both N and O are diatomic gases at normal temperatures and pressures.   (e) O_2 is a reactive gas, whereas N_2 is not. Nitrogen can have any of six oxidation states, whereas oxygen has two.
14.111(a) $NaHSO_4(aq) + NaOH(aq) \longrightarrow Na_2SO_4(aq) + H_2O(l)$
(b) $S_8(s) + 24F_2(g) \longrightarrow 8SF_6(g)$
(c) $FeS(s) + 2HCl(aq) \longrightarrow H_2S(g) + FeCl_2(aq)$
(d) $Te(s) + 2I_2(s) \longrightarrow TeI_4(s)$
14.113(a) acidic   (b) acidic   (c) basic   (d) amphoteric   (e) basic   **14.115** $H_2O < H_2S < H_2Te$ **14.118**(a) SF_6, sulfur hexafluoride   (b) O_3, ozone   (c) SO_3, sulfur trioxide   (d) SO_2, sulfur dioxide   (e) H_2SO_4, sulfuric acid   (f) $Na_2S_2O_3 \cdot 5H_2O$, sodium thiosulfate pentahydrate   (g) H_2S, hydrogen sulfide   **14.120** $S_2F_{10}(g) \longrightarrow SF_4(g) + SF_6(g)$; O.N. of S in S_2F_{10} is $+5$; O.N. of S in SF_4 is $+4$; O.N. of S in SF_6 is $+6$.   **14.122**(a) -1, $+1$, $+3$, $+5$, $+7$ (b) The electron configuration for Cl is $[Ne]\,3s^23p^5$. By gaining one electron, Cl achieves an octet. By forming covalent bonds, Cl completes or expands its valence level by maintaining electron pairs in bonds or as lone pairs.   (c) Fluorine has only the -1 oxidation state because its small

size and absence of d orbitals prevent it from forming more than one covalent bond.   **14.124**(a) Cl—Cl bond is stronger than Br—Br bond.   (b) Br—Br bond is stronger than I—I bond.   (c) Cl—Cl bond is stronger than F—F bond. The fluorine atoms are so small that electron-electron repulsion of the lone pairs decreases the strength of the bond.
14.126(a) $3Br_2(l) + 6OH^-(aq) \longrightarrow$
$$5Br^-(aq) + BrO_3^-(aq) + 3H_2O(l)$$
(b) $ClF_5(l) + 6OH^-(aq) \longrightarrow$
$$5F^-(aq) + ClO_3^-(aq) + 3H_2O(l)$$
14.127(a) $2Rb(s) + Br_2(l) \longrightarrow 2RbBr(s)$
(b) $I_2(s) + H_2O(l) \longrightarrow HI(aq) + HIO(aq)$
(c) $Br_2(l) + 2I^-(aq) \longrightarrow I_2(s) + 2Br^-(aq)$
(d) $CaF_2(s) + H_2SO_4(l) \longrightarrow CaSO_4(s) + 2HF(g)$
14.129 $HIO < HBrO < HClO < HClO_2$ **14.133** $I_2 < Br_2 < Cl_2$, because Cl_2 is able to oxidize Re to the $+6$ oxidation state, Br_2 only to $+5$, and I_2 only to $+4$.   **14.134** helium; argon
14.136 Only dispersion forces hold atoms of noble gases together.
14.140 $\Delta H_{soln} = -411$ kJ
14.145(a) $MgCl_2 \cdot 6H_2O(s) + 6SOCl_2(l) \longrightarrow$
$$MgCl_2(s) + 6SO_2(g) + 12HCl(g)$$
(b)

14.146(a) Second ionization energies for alkali metals are so high because the electron being removed is from the next lower energy level and these are very tightly held by the nucleus.
(b) $2CsF_2(s) \longrightarrow 2CsF(s) + F_2(g)$; -405 kJ/mol.
14.148(a) hyponitrous acid, $H_2N_2O_2$; nitroxyl, HNO
(b)

(c) In both species the shape is bent about the N atoms.
(d)

14.151 volume $= 2.827 \times 10^{-21}$ cm^3; edge length $= 1.414 \times 10^{-7}$ cm; the low density arises because even though the spheres are closest packed, they are hollow.   **14.154** 13 t   **14.156** In a disproportionation reaction, a substance acts as both a reducing agent and an oxidizing agent because atoms of an element within the substance attain both higher and lower oxidation states in the products. The disproportionation reactions are b, c, d, e, and f.
14.158(a) Group 5A(15)   (b) Group 7A(17)   (c) Group 6A(16)   (d) Group 1A(1)   (e) Group 3A(13)   (f) Group 8A(18)
14.160(a)

(b) three lone pairs on the N atoms   (c) 1.5
14.162 117.2 kJ   **14.165** 25.0 g CS_2. It is not possible to determine how much S_2Cl_2 and SCl_2 formed; the ratio of S_2Cl_2 to SCl_2 is unknown.   **14.170**(a) sp^3 orbitals   (b) tetrahedral   (c) Since the ion is linear, the central Cl atom must be sp hybridized.

(d) The *sp* hybridization means there are no lone pairs on the central Cl atom. Instead, the extra four electrons interact with the empty *d* orbitals on the Al atoms to form double bonds between the chlorine and each aluminum atom.

14.172 $\left[\ddot{\mathrm{O}}{=}\ddot{\mathrm{N}}{-}\ddot{\mathrm{O}}{:} \right]^{-} \longleftrightarrow \left[{:}\ddot{\mathrm{O}}{-}\ddot{\mathrm{N}}{=}\ddot{\mathrm{O}} \right]^{-}$

$\ddot{\mathrm{O}}{=}\dot{\mathrm{N}}{-}\ddot{\mathrm{O}}{:} \longleftrightarrow {:}\ddot{\mathrm{O}}{-}\dot{\mathrm{N}}{=}\ddot{\mathrm{O}}$

$\left[\ddot{\mathrm{O}}{=}\mathrm{N}{=}\ddot{\mathrm{O}} \right]^{+}$

The nitronium ion (NO_2^+) has a linear shape because the central N atom has two surrounding electron groups, which achieve maximum repulsion at 180°. The nitrite ion (NO_2^-) bond angle is more compressed than the nitrogen dioxide (NO_2) bond angle because the lone pair of electrons takes up more space than the lone electron. **14.174** 2.29×10^4 g UF_6 **14.179** O^+, O^-, and O^{2+} **14.180**(a) 39.96 mass % in $CuHAsO_3$; As, 62.42 mass % in $(CH_3)_3As$ (b) 0.35 g $CuHAsO_3$

Chapter 15

15.3(a) Carbon's electronegativity is midway between the most metallic and nonmetallic elements of Period 2. To attain a filled outer shell, carbon forms covalent bonds to other atoms in molecules, network covalent solids, and polyatomic ions. (b) Since carbon has four valence shell electrons, it forms four covalent bonds to attain an octet. (c) To reach the He electron configuration, a carbon atom must lose four electrons, requiring too much energy to form the C^{4+} cation. To reach the Ne electron configuration, the carbon atom must gain four electrons, also requiring too much energy to form the C^{4-} anion. (d) Carbon is able to bond to itself extensively because its small size allows for close approach and great orbital overlap. The extensive orbital overlap results in a strong, stable bond. (e) The C—C bond is short enough to allow sideways overlap of unhybridized *p* orbitals of neighboring C atoms. The sideways overlap of *p* orbitals results in the π bonds that are part of double and triple bonds.
15.4(a) C H, O, N, P, S, and halogens (b) Heteroatoms are atoms of any element other than carbon and hydrogen. (c) More electronegative than C: N, O, F, Cl, and Br; less electronegative than C: H and P. Sulfur and iodine have the same electronegativity as carbon. (d) Since carbon can bond to a wide variety of heteroatoms and to carbon atoms, it can form many different compounds. **15.7** The C—H and C—C bonds are unreactive because electron density is shared equally between the two atoms. The C—I bond is somewhat reactive because it is long and weak. The C=O bond is reactive because oxygen is more electronegative than carbon and the electron-rich π bond makes it attract electron-poor atoms. The C—Li bond is also reactive because the bond polarity results in an electron-rich region around carbon and an electron-poor region around lithium.
15.8(a) An alkane and a cycloalkane are organic compounds that consist of carbon and hydrogen and have only single bonds. A cycloalkane has a ring of carbon atoms. An alkene is a hydrocarbon with at least one double bond. An alkyne is a hydrocarbon with at least one triple bond. (b) alkane = C_nH_{2n+2}, cycloalkane = C_nH_{2n}, alkene = C_nH_{2n}, alkyne = C_nH_{2n-2} (c) Alkanes and cycloalkanes are saturated hydrocarbons.
15.11 a, c, and f

15.14(a)

(b)

(c)

15.16(a)

$CH_2{=}C{-}CH_2{-}CH_2{-}CH_2{-}CH_3$ with CH_3 branch

$CH_2{=}CH{-}CH{-}CH_2{-}CH_2{-}CH_3$ with CH_3 branch

$CH_2{=}CH{-}CH_2{-}CH{-}CH_2{-}CH_3$ with CH_3 branch

$CH_2{=}CH{-}CH_2{-}CH_2{-}CH{-}CH_3$ with CH_3 branch

$CH_3{-}C{=}CH{-}CH_2{-}CH_2{-}CH_3$ with CH_3 branch

$CH_3{-}CH{=}C{-}CH_2{-}CH_2{-}CH_3$ with CH_3 branch

$CH_3{-}CH{=}CH{-}CH{-}CH_2{-}CH_3$ with CH_3 branch

$CH_3{-}CH{=}CH{-}CH_2{-}CH{-}CH_3$ with CH_3 branch

$CH_3{-}CH{-}CH{=}CH{-}CH_2{-}CH_3$ with CH_3 branch

$CH_3{-}CH_2{-}C{=}CH{-}CH_2{-}CH_3$ with CH_3 branch

$CH_3{-}CH_2{-}C{-}CH_2{-}CH_2{-}CH_3$ with CH_2 (=CH_2) branch

(b) $CH_2=C-CH-CH_2-CH_3$ with CH_3 CH_3 substituents

$CH_2=C-CH_2-CH-CH_3$ with CH_3 and CH_3 substituents

$CH_2=CH-\underset{CH_3}{\overset{CH_3}{\underset{|}{\overset{|}{C}}}}-CH_2-CH_3$

$CH_2=CH-CH-CH-CH_3$ with CH_3 CH_3

$CH_2=CH-CH_2-\underset{CH_3}{\overset{CH_3}{\underset{|}{\overset{|}{C}}}}-CH_3$

$CH_2=CH-CH-CH_2-CH_3$ with CH_2-CH_3

$CH_3-C=C-CH_2-CH_3$ with CH_3 CH_3

$CH_3-C=CH-CH-CH_3$ with CH_3 CH_3

$CH_3-CH=C-CH-CH_3$ with CH_3 CH_3

$CH_3-CH=CH-\underset{CH_3}{\overset{CH_3}{\underset{|}{\overset{|}{C}}}}-CH_3$

$CH_3-CH=C-CH_2-CH_3$ with CH_2-CH_3

(c) cyclopentane with $CH-CH_2-CH_3$

cyclopentane with $CH-CH_3$

cyclopentane with $CH-CH_3$ and $H_3C-HC-CH_2$

cyclopentane with $C(CH_3)(CH_3)$

15.18(a) $CH_3-\underset{CH_3}{\overset{CH_3}{\underset{|}{\overset{|}{C}}}}-CH_2-CH_3$

(b) $H_2C=CH-CH_2-CH_3$

(c) $HC\equiv C-CH-CH_3$ with CH_2 / CH_3 (d) Structure is correct.

15.20(a) $CH_3-\underset{CH_3}{\overset{CH_3}{\underset{|}{\overset{|}{CH}}}}-CH-CH_2-CH_2-CH_2-CH_2-CH_3$

(b) cyclohexane with CH_2-CH_3 and CH_3 substituents

(c) 3,4-dimethylheptane   (d) 2,2-dimethylbutane

15.22(a) $CH_3-CH_2-CH_2-\underset{CH_3}{\overset{CH_3}{\underset{|}{\overset{|}{CH}}}}-CH_2-CH_3$

Correct name is 3-methylhexane.

(b) $CH_3-CH_2-CH_2-\underset{CH_2-CH_3}{\overset{|}{CH}}-CH_3$

Correct name is 3-methylhexane.

(c) cyclohexane with CH_3

Correct name is methylcyclohexane.

(d) $CH_3-CH_2-\underset{CH_3\ CH_2-CH_3}{\overset{CH_3}{\underset{|}{\overset{|}{C}}}}-CH-CH_2-CH_2-CH_2-CH_3$

Correct name is 4-ethyl-3,3-dimethyloctane.

15.24(a) cyclopropane ring with H, H and $H-C-C-CH_3$ with Cl, H

(b) $CH_3-CH_2-\overset{H}{\underset{CH_3}{C}}-\overset{CH_3}{\underset{H}{C}}-CH_3$

15.26(a) 3-Bromohexane is optically active.

$CH_3-CH_2-\overset{}{C}H-CH_2-CH_2-CH_3$ with Br

(b) 3-Chloro-3-methylpentane is not optically active.

$CH_3-CH_2-\underset{Cl}{\overset{CH_3}{\underset{|}{\overset{|}{C}}}}-CH_2-CH_3$

(c) 1,2-Dibromo-2-methylbutane is optically active.

$CH_2-\overset{CH_3}{\underset{Br}{C}}-CH_2-CH_3$ with Br

15.28(a)

$\underset{CH_3CH_2}{\overset{H}{C}}=\underset{CH_3}{\overset{H}{C}}$ $\underset{CH_3CH_2}{\overset{H}{C}}=\underset{H}{\overset{CH_3}{C}}$

cis-2-pentene   trans-2-pentene

(b)

cyclohexyl $\overset{H}{C}=\overset{H}{\underset{CH_3}{C}}$ cyclohexyl $\overset{H}{C}=\overset{CH_3}{\underset{H}{C}}$

cis-1-cyclohexylpropene   trans-1-cyclohexylpropene

(c) no geometric isomers

15.30(a) no geometric isomers

(b) $\underset{H}{\overset{CH_3-CH_2}{C}}=\underset{H}{\overset{CH_2-CH_3}{C}}$ $\underset{H}{\overset{CH_3-CH_2}{C}}=\underset{CH_2-CH_3}{\overset{H}{C}}$

cis-3-hexene   trans-3-hexene

(c) no geometric isomers

(d) $\underset{H}{\overset{Cl}{C}}=\underset{H}{\overset{Cl}{C}}$ $\underset{H}{\overset{Cl}{C}}=\underset{Cl}{\overset{H}{C}}$

cis-1,2-dichloroethene   trans-1,2-dichloroethene

15.32

1,2-dichlorobenzene 1,3-dichlorobenzene 1,4-dichlorobenzene
(o-dichlorobenzene)   (m-dichlorobenzene) (p-dichlorobenzene)

15.34

15.35

cis-2-methyl-3-hexene   trans-2-methyl-3-hexene

The compound 2-methyl-2-hexene does not have *cis-trans* isomers.   **15.39** π bond   **15.41**(a) elimination   (b) addition
15.43(a) $CH_3CH_2CH=CHCH_2CH_3 + H_2O \xrightarrow{H^+}$
$$CH_3CH_2CH_2CH(OH)CH_2CH_3$$
(b) $CH_3CHBrCH_3 + CH_3CH_2OK \longrightarrow$
$$CH_3CH=CH_2 + CH_3CH_2OH + KBr$$
(c) $CH_3CH_3 + 2Cl_2 \xrightarrow{hv} CHCl_2CH_3 + 2HCl$
15.45(a) oxidation   (b) reduction   (c) reduction   **15.47**(a) oxidized   (b) oxidized   **15.50**(a) Methylethylamine is more soluble because it has the ability to form H bonds with water molecules. (b) 1-Butanol has a higher melting point because it can form intermolecular H bonds.   (c) Propylamine has a higher boiling point because it contains N—H bonds that allow H bonding, and trimethylamine cannot form H bonds.   **15.53** Both groups react by addition to the π bond. The very polar C=O bond attracts the electron-rich O of water to the partially positive C. There is no such polarity in the alkene, so either C atom can be attacked.

$CH_3-CH_2-\overset{\overset{O}{\|}}{C}-CH_2-CH_3 + H_2O \xrightarrow{H^+}$

$CH_3-CH_2-\overset{\overset{OH}{|}}{\underset{\underset{OH}{|}}{C}}-CH_2-CH_3$

$CH_3-CH_2-CH=CH-CH_3 + H_2O \xrightarrow{H^+}$

$CH_3-CH_2-\overset{\overset{}{|}}{\underset{\underset{OH}{|}}{C}}-CH_2-CH_3 + CH_3-CH_2-CH_2-\overset{\overset{}{|}}{\underset{\underset{OH}{|}}{C}}H-CH_3$

15.56 Esters and acid anhydrides form through dehydration-condensation reactions, and water is the other product.
15.58(a) alkyl halide   (b) nitrile   (c) carboxylic acid   (d) aldehyde
15.60(a)

$CH_3-CH=CH-CH_2-OH$
    alkene   alcohol

(b)

$Cl-CH_2-\langle\text{ring}\rangle-C-OH$
  haloalkane   carboxylic acid

(c)

  amide
alkene

(d) $N\equiv C-CH_2-C-CH_3$
  nitrile   ketone

(e)

ester

15.62 $H_3C-CH_2-CH_2-CH_2-CH_2-OH$
$H_3C-CH_2-CH_2-\overset{\overset{}{|}}{\underset{\underset{OH}{|}}{C}}H-CH_3$

$H_3C-\underset{\underset{CH_3}{|}}{C}H-\underset{\underset{OH}{|}}{C}H-CH_3$
$H_3C-\overset{\overset{OH}{|}}{\underset{\underset{CH_3}{|}}{C}}-CH_2-CH_3$

$H_3C-CH_2-\underset{\underset{OH}{|}}{C}H-CH_2-CH_3$
$H_3C-\underset{\underset{CH_3}{|}}{C}H-CH_2-CH_2-OH$

$H_3C-CH_2-\underset{\underset{CH_3}{|}}{C}H-CH_2-OH$
$H_3C-\overset{\overset{CH_3}{|}}{\underset{\underset{CH_3}{|}}{C}}-CH_2-OH$

15.64 $H_3C-CH_2-CH_2-CH_2-NH_2$
$H_3C-CH_2-\underset{\underset{NH_2}{|}}{C}H-CH_3$

$H_3C-\underset{\underset{CH_3}{|}}{C}H-CH_2-NH_2$
$H_3C-\overset{\overset{CH_3}{|}}{\underset{\underset{CH_3}{|}}{C}}-NH_2$

$H_3C-CH_2-\underset{\underset{H}{|}}{N}-CH_2-CH_3$
$H_3C-CH_2-CH_2-\underset{\underset{H}{|}}{N}-CH_3$

$H_3C-CH_2-\underset{\underset{CH_3}{|}}{N}-CH_3$
$H_3C-\underset{\underset{CH_3}{|}}{C}H-\underset{\underset{H}{|}}{N}-CH_3$

15.66(a) $CH_3-\overset{\overset{O}{\|}}{C}-CH_2-CH_3$ (b) $CH_3-\underset{\underset{CH_3}{|}}{C}H-\overset{\overset{O}{\|}}{C}-OH$

(c)

15.68(a)

$CH_3-\overset{\overset{O}{\|}}{C}-\underset{\underset{H}{|}}{N}-CH_3$

(b)

$CH_3-CH_2-CH_2-\overset{\overset{O}{\|}}{C}-O-\underset{\underset{CH_3}{|}}{C}H-CH_3$

(c)

$H-\overset{\overset{O}{\|}}{C}-O-CH_2-\underset{\underset{CH_3}{|}}{C}H-CH_3$

15.70(a)

$CH_3-(CH_2)_4-\overset{\overset{O}{\|}}{C}-OH$ and $HO-CH_2-CH_3$

(b)

$\langle\text{ring}\rangle-\overset{\overset{O}{\|}}{C}-OH$ and $HO-CH_2-CH_2-CH_3$

(c)

CH_3-CH_2-OH and $HO-\overset{\overset{O}{\|}}{C}-CH_2-CH_2-\langle\text{ring}\rangle$

15.72(a) CH₃—CH₂—OH

(b)

$$CH_3—CH_2—\overset{\overset{O}{\|}}{C}—O—CH_2—CH_3$$

$$CH_3—CH_2—\overset{\overset{C\equiv N}{|}}{CH}—CH_3$$

$$CH_3—CH_2—\overset{\overset{OH}{|}}{\underset{|}{\overset{|}{C}=O}}—CH—CH_3$$

15.74(a) propanoic acid   (b) ethylamine   **15.78** addition reactions and condensation reactions   **15.81** Dispersion forces strongly attract the long, unbranched chains of high-density polyethylene (HDPE). Low-density polyethylene (LDPE) has branching in the chains that prevents packing and weakens the attractions. **15.83** An amine and a carboxylic acid react to form nylon; a carboxylic acid and an alcohol form a polyester.
15.84(a)

$$\left[\begin{array}{c} \overset{H}{|} \ \overset{H}{|} \\ —C—C— \\ \underset{H}{|} \ \underset{Cl}{|} \end{array}\right]_n$$
(b)
$$\left[\begin{array}{c} \overset{H}{|} \ \overset{H}{|} \\ —C—C— \\ \underset{H}{|} \ \underset{CH_3}{|} \end{array}\right]_n$$

15.88(a) condensation   (b) addition   (c) condensation (d) condensation   **15.90** The amino acid sequence in a protein determines its shape and structure, which determine its function. **15.93** The DNA base sequence determines the RNA base sequence, which determines the protein amino acid sequence.
15.94(a)   (b)   (c)

15.96(a)

(b)

15.98(a) AATCGG   (b) TCTGTA   **15.100** ACAATGCCT; this sequence codes for three amino acids.   **15.102**(a) Both R groups are from cysteine, which can form a disulfide bond (covalent bond).   (b) Lysine and aspartic acid give a salt link.   (c) Asparagine and serine will hydrogen bond.   (d) Valine and phenylalanine interact through dispersion forces.

15.104 CH₃—CH=CH—CH₃; decolorization of bromine **15.108**(a) Perform an acid-catalyzed dehydration of the alcohol followed by addition of Br₂.   (b) Oxidize 1 mol of ethanol to acetic acid, then react 1 mol of acetic acid with 1 mol of ethanol to form the ester.
15.110
H₂N—CH₂—CH₂—CH₂—CH₂—CH₂—NH₂
cadaverine
H₂N—CH₂—CH₂—CH₂—CH₂—NH₂
putrescine
Br—CH₂CH₂—Br + 2CN⁻ ⟶
         N≡C—CH₂CH₂—C≡N + 2Br⁻
N≡C—CH₂CH₂—C≡N $\xrightarrow{\text{reduction}}$
         H₂N—CH₂CH₂CH₂CH₂—NH₂

15.111(a)

(b) Carbon 1 is sp^2 hybridized.   Carbon 2 is sp^3 hybridized.
Carbon 3 is sp^3 hybridized.   Carbon 4 is sp^2 hybridized.
Carbon 5 is sp^3 hybridized.   Carbons 6 and 7 are sp^2 hybridized.
(c) Carbons 2, 3, and 5 are chiral centers, as they are each bonded to four different groups.
15.116 The resonance structures show that the bond between carbon and nitrogen will have some double bond character that restricts rotation around the bond.

$$\overset{\overset{:O:}{\|}}{—C—}\overset{H}{\underset{|}{\underset{\cdot\cdot}{N}}—} \longleftrightarrow \overset{\overset{:\ddot{O}:^-}{|}}{—C=}\overset{H}{\underset{|}{\overset{+}{N}}—}$$

15.120(a) Hydrolysis requires the addition of water to break the peptide bonds.   (b) glycine, 4; alanine, 1; valine, 3; proline, 6; serine, 7; arginine, 49   (c) 10,700 g/mol   **15.122** When 2-butanone is reduced, equal amounts of both isomers are produced because the reaction does not favor the production of one over the other.

Chapter 16

16.2 Reaction rate is proportional to concentration. An increase in pressure will increase the concentration, resulting in an increased reaction rate.   **16.3** The addition of water will dilute the concentrations of all dissolved solutes, and the rate of the reaction will decrease.   **16.5** An increase in temperature affects the rate of a reaction by increasing the number of collisions between particles, but more importantly, the energy of collisions increases. Both these factors increase the rate of reaction.
16.8(a) For most reactions, the rate changes as the reaction progresses. The instantaneous rate is the rate at one point, or instant, during the reaction. The average rate is the average of the instantaneous rates over a period of time. On a graph of reactant concentration vs. time of reaction, the instantaneous rate is the slope of the tangent to the curve at any one point. The average rate is the slope of the line connecting two points on the curve. The closer together the two points (the shorter the time interval),

the more closely the average rate agrees with the instantaneous rate. (b) The initial rate is the instantaneous rate at the point on the graph where $t = 0$, that is, when reactants are mixed.

16.10

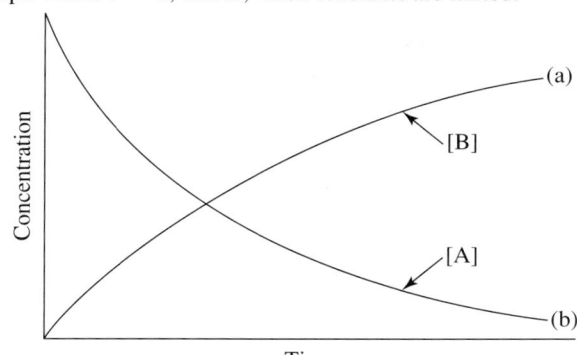

16.12(a) rate $= -\left(\dfrac{1}{2}\right)\dfrac{\Delta AX_2}{\Delta t}$

$= -\left(\dfrac{1}{2}\right)\dfrac{(0.0088\,M - 0.0500\,M)}{(20.0\,s - 0\,s)}$

$= 0.0010\,M/s$

(b)

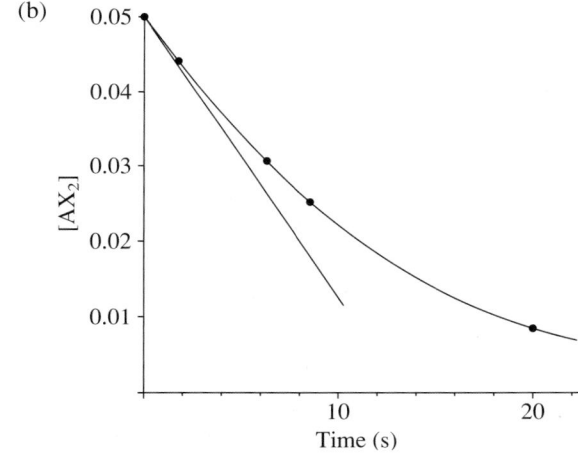

The initial rate is higher than the average rate because the rate will decrease as reactant concentration decreases.

16.14 rate $= -\dfrac{1}{2}\dfrac{\Delta[A]}{\Delta t} = \dfrac{\Delta[B]}{\Delta t} = \dfrac{\Delta[C]}{\Delta t}$; 4 mol/L·s

16.16 rate $= -\dfrac{\Delta[A]}{\Delta t} = -\dfrac{1}{2}\dfrac{\Delta[B]}{\Delta t} = \dfrac{\Delta[C]}{\Delta t}$; 0.2 mol/L·s

16.18 $2N_2O_5(g) \longrightarrow 4NO_2(g) + O_2(g)$

16.21 rate $= -\dfrac{\Delta[N_2]}{\Delta t} = -\dfrac{1}{3}\dfrac{\Delta[H_2]}{\Delta t} = \dfrac{1}{2}\dfrac{\Delta[NH_3]}{\Delta t}$

16.22(a) rate $= -\dfrac{1}{3}\dfrac{\Delta[O_2]}{\Delta t} = \dfrac{1}{2}\dfrac{[\Delta O_3]}{\Delta t}$ (b) 1.45×10^{-5} mol/L·s

16.23(a) k is the rate constant, the proportionality constant in the rate law; it is reaction and temperature specific. (b) m represents the order of the reaction with respect to [A], and n represents the order of the reaction with respect to [B]. The order of a reactant does not necessarily equal its stoichiometric coefficient in the balanced equation. (c) L^2/mol^2·min **16.25**(a) Rate doubles. (b) Rate decreases by a factor of four. (c) Rate increases by a factor of nine. **16.26** first order in BrO_3^-; first order in Br^-; second order in H^+; fourth order overall

16.28(a) Rate doubles. (b) Rate is halved. (c) The rate increases by a factor of 16. **16.30** second order in NO_2; first order in Cl; third order overall **16.32**(a) Rate increases by a factor of 9. (b) Rate increases by a factor of 8. (c) Rate is halved. **16.34**(a) second order in A; first order in B (b) rate $= k[A]^2[B]$ (c) 5.00×10^3 L^2/mol^2·min **16.36**(a) time^{-1} (b) L/mol·time (c) L^2/mol^2·time (d) $L^{3/2}/mol^{3/2}$·time **16.39**(a) first order (b) second order (c) zero order **16.41** 7 s **16.43**(a) $k = 0.0660$ min^{-1} (b) 21.0 min

16.45(a)

x-axis (time, s)	[NH₃]	y-axis (ln [NH₃])
0	4.000 M	1.38629
1.000	3.986 M	1.38279
2.000	3.974 M	1.37977

$k = 3\times10^{-3}$ s^{-1}

(b) $t_{1/2} = 2\times10^2$ s

16.47 Measure the rate constant at a series of temperatures and plot ln k versus $1/T$. The slope of the line equals $-E_a/R$.
16.49 0.033 s^{-1} **16.53** No, other factors that affect the fraction of collisions that lead to reaction are the energy and orientation of the collisions. **16.56** No, reaction is reversible and will eventually reach a state where the forward and reverse rates are equal. When this occurs, there are no concentrations equal to zero. Since some reactants are reformed from EF, the amount of EF will be less than 4×10^{-5} mol. **16.57** At the same temperature, both reaction mixtures have the same average kinetic energy, but not the same velocity. The trimethylamine molecule has greater mass than the ammonia molecule, so trimethylamine molecules will collide less often with HCl. Moreover, the bulky groups bonded to nitrogen in trimethylamine mean that collisions with HCl having the correct orientation occur less frequently. Therefore, the rate of the first reaction is higher. **16.58** 12 unique collisions **16.60** 2.96×10^{-18}
16.62(a)

(b) 2.70×10^2 kJ/mol

(c)

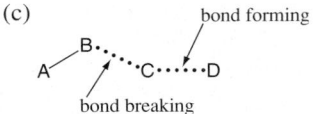

bond forming

bond breaking

16.64(a) Because the enthalpy change is positive, the reaction is endothermic.

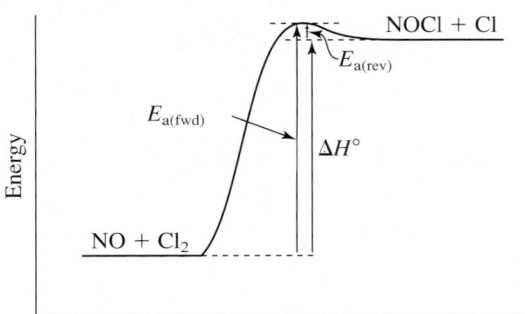

Reaction coordinate

(b) 3 kJ   (c)

:$\ddot{C}l$····$\ddot{C}l$····$\ddot{N}$=$\ddot{O}$:

16.65 The rate of an overall reaction depends on the rate of the slowest step. The rate of the overall reaction will be lower than the average of the individual rates because the average includes higher rates as well.   **16.69** The probability of three particles colliding with one another with the proper energy and orientation is much less than the probability for two particles.   **16.70** No, the overall rate law must contain only reactants (no intermediates), and the overall rate is determined by the slow step.
16.72(a) $A(g) + B(g) + C(g) \longrightarrow D(g)$
(b) X and Y are intermediates.

(c)

Step	Molecularity	Rate Law
$A(g) + B(g) \rightleftharpoons X(g)$	bimolecular	$rate_1 = k_1[A][B]$
$X(g) + C(g) \longrightarrow Y(g)$	bimolecular	$rate_2 = k_2[X][C]$
$Y(g) \longrightarrow D(g)$	unimolecular	$rate_3 = k_3[Y]$

(d) yes   (e) yes   **16.74** The proposed mechanism is valid because the individual steps are chemically reasonable and add to give the overall equation. In addition, the rate law for the mechanism matches the observed rate law.   **16.77** No. A catalyst changes the mechanism of a reaction to one with lower activation energy. Lower activation energy means a faster reaction. An increase in temperature does not influence the activation energy, but increases the fraction of collisions with sufficient energy to equal or exceed the activation energy.   **16.78**(a) No. The spark provides energy that is absorbed by the H_2 and O_2 molecules to achieve the activation energy.   (b) Yes. The powdered metal acts as a heterogeneous catalyst, providing a surface on which the oxygen-hydrogen reaction can proceed with a lower activation energy.
16.82(a) Water does not appear as a reactant in the rate-determining step.
(b) Step 1: $rate_1 = k_1[(CH_3)_3CBr]$
    Step 2: $rate_2 = k_2[(CH_3)_3C^+]$
    Step 3: $rate_3 = k_3[(CH_3)_3COH_2^+]$
(c) $(CH_3)_3C^+$ and $(CH_3)_3COH_2^+$
(d) The rate-determining step is step 1. The rate law for this step agrees with the rate law observed with $k = k_1$.
16.85 4.61×10^4 J/mol   **16.89**(a) Rate increases 2.5 times.
(b) Rate is halved.   (c) Rate decreases by a factor of 0.01.

(d) Rate does not change.   **16.92** second order   **16.94** 7 times faster   **16.96** 57 yr   **16.99**(a) 0.21 h^{-1}; 3.3 h   (b) 6.6 h   (c) If the concentration of sucrose is relatively low, the concentration of water remains nearly constant even with small changes in the amount of water. This gives an apparent zero-order reaction with respect to water. Thus, the reaction is first order overall because the rate does not change with changes in the amount of water.
16.101(a) 0.68 M (b) 0.57   **16.104** 71 kPa   **16.108** 7.3×10^3 J/mol   **16.110**(a) 2.4×10^{-15} M (b) 2.4×10^{-11} mol/L·s
16.113(a) 2.8 days   (b) 7.4 days   (c) 4.5 mol/m^3
16.116(a) $rate_1 = 1.7 \times 10^{-5}$ M s^{-1}; $rate_2 = 3.4 \times 10^{-5}$ M s^{-1}; $rate_3 = 3.4 \times 10^{-5}$ M s^{-1} (b) zero order with respect to $S_2O_8^{2-}$; first order with respect to I^- (c) 4.3×10^{-4} s^{-1} (d) rate = $(4.3 \times 10^{-4}$ s$^{-1})[KI]$ **16.119**(a) 7×10^4 cells/L   (b) 2.0×10^1 min
16.122(a) Use the Monod equation.

(b) 8.2×10^3 cells/m^3 (c) 8.4×10^3 cells/m^3
16.126(a) 8.0×10^1 s   (b) 59 s   (c) 2.6×10^2 s

Chapter 17

17.1 If the change is one of concentrations, it results temporarily in more products and less reactants. After equilibrium is re-established, K_c remains unchanged because the ratio of products and reactants remains the same. If the change is one of temperature, [product] and K_c increase and [reactant] decreases.
17.7 The equilibrium constant expression is $K = [O_2]$. If the temperature remains constant, K remains constant. If the initial amount of Li_2O_2 present is sufficient to reach equilibrium, the amount of O_2 obtained will be constant.

17.8(a) $Q = \dfrac{[HI]^2}{[H_2][I_2]}$

The value of Q increases as a function of time until it reaches the value of K.   (b) no   **17.11** Yes. If Q_1 is for the formation of

1 mol NH_3 from H_2 and N_2, and Q_2 is for the formation of NH_3 from H_2 and 1 mol of N_2, then $Q_2 = Q_1{}^2$.

17.12(a) $4NO(g) + O_2(g) \rightleftharpoons 2N_2O_3(g)$;

$$Q_c = \frac{[N_2O_3]^2}{[NO]^4[O_2]}$$

(b) $SF_6(g) + 2SO_3(g) \rightleftharpoons 3SO_2F_2(g)$;

$$Q_c = \frac{[SO_2F_2]^3}{[SF_6][SO_3]^2}$$

(c) $2SClF_5(g) + H_2(g) \rightleftharpoons S_2F_{10}(g) + 2HCl(g)$;

$$Q_c = \frac{[S_2F_{10}][HCl]^2}{[SClF_5]^2[H_2]}$$

17.14(a) $2NO_2Cl(g) \rightleftharpoons 2NO_2(g) + Cl_2(g)$;

$$Q_c = \frac{[NO_2]^2[Cl_2]}{[NO_2Cl]^2}$$

(b) $2POCl_3(g) \rightleftharpoons 2PCl_3(g) + O_2(g)$;

$$Q_c = \frac{[PCl_3]^2[O_2]}{[POCl_3]^2}$$

(c) $4NH_3(g) + 3O_2(g) \rightleftharpoons 2N_2(g) + 6H_2O(g)$;

$$Q_c = \frac{[N_2]^2[H_2O]^6}{[NH_3]^4[O_2]^3}$$

17.16(a) 7.9 (b) 3.2×10^{-5}

17.18(a) $2Na_2O_2(s) + 2CO_2(g) \rightleftharpoons 2Na_2CO_3(s) + O_2(g)$;

$$Q_c = \frac{[O_2]}{[CO_2]^2}$$

(b) $H_2O(l) \rightleftharpoons H_2O(g)$; $Q_c = [H_2O(g)]$

(c) $NH_4Cl(s) \rightleftharpoons NH_3(g) + HCl(g)$; $Q_c = [NH_3][HCl]$

17.20(a) $2NaHCO_3(s) \rightleftharpoons Na_2CO_3(s) + CO_2(g) + H_2O(g)$;

$Q_c = [CO_2][H_2O]$

(b) $SnO_2(s) + 2H_2(g) \rightleftharpoons Sn(s) + 2H_2O(g)$;

$$Q_c = \frac{[H_2O]^2}{[H_2]^2}$$

(c) $H_2SO_4(l) + SO_3(g) \rightleftharpoons H_2S_2O_7(l)$;

$$Q_c = \frac{1}{[SO_3]}$$

17.23(a) (1) $Cl_2(g) + F_2(g) \rightleftharpoons 2ClF(g)$

(2) $2ClF(g) + 2F_2(g) \rightleftharpoons 2ClF_3(g)$

overall: $Cl_2(g) + 3F_2(g) \rightleftharpoons 2ClF_3(g)$

(b) $Q_{overall} = Q_1Q_2 = \dfrac{[\cancel{ClF}]^2}{[Cl_2][F_2]} \times \dfrac{[ClF_3]^2}{[\cancel{ClF}]^2[F_2]^2}$

$$= \frac{[ClF_3]^2}{[Cl_2][F_2]^3}$$

17.25 K_c and K_p are equal when $\Delta n_{gas} = 0$. **17.26**(a) smaller (b) Assuming that $RT > 1$ ($T > 12.2$ K), $K_p > K_c$ if there are more moles of products than reactants at equilibrium, and $K_p < K_c$ if there are more moles of reactants than products. **17.27**(a) 3 (b) -1 (c) 3 **17.29**(a) 3.2 (b) 28.5 **17.31**(a) 0.15 (b) 3.6×10^{-7} **17.33** The reaction quotient (Q) and equilibrium constant (K) are determined by the ratio [products]/[reactants]. When $Q < K$, the reaction proceeds to the right to form more products. **17.35** no; to the left **17.38** At equilibrium, equal concentrations of $CFCl_3$ and HCl exist, regardless of starting reactant concentrations, because the product coefficients are equal. **17.40**(a) The approximation applies when the change in concentration from initial concentration to equilibrium concentration is so small that it is insignificant; this occurs when K is small and initial concentration is large. (b) This approximation should not

be used when the change in concentration is greater than 5%. This can occur when [reactant]$_{initial}$ is very small or when change in [reactant] is relatively large due to a large K. **17.41** 50.8

17.43

Concentration (M)	$PCl_5(g)$	$\rightleftharpoons$	$PCl_3(g)$ +	$Cl_2(g)$
Initial	0.075		0	0
Change	$-x$		$+x$	$+x$
Equilibrium	$0.075 - x$		x	x

17.45 28 atm **17.47** 0.33 atm **17.49** 3.5×10^{-3} M **17.51** $[I_2]$ = 0.00328 M; [HI] = 0.0152 M **17.53** $[I_2]_{eq}$ = $[Cl_2]_{eq}$ = 0.0200 M; $[ICl]_{eq}$ = 0.060 M **17.55** 6.01×10^{-6} **17.58** Equilibrium position refers to the specific concentrations or pressures of reactants and products that exist at equilibrium, whereas equilibrium constant is the overall ratio of equilibrium concentrations or pressures. Equilibrium position changes as a result of a change in reactant and product concentrations. **17.59**(a) B, because the amount of product increases with temperature (b) A, because the lowest temperature will give the least product **17.63**(a) shifts toward products (b) shifts toward products (c) does not shift (d) shifts toward reactants **17.65**(a) more F and less F_2 (b) more C_2H_2 and H_2 and less CH_4 **17.67**(a) no effect (b) less H_2 and O_2 and more H_2O **17.69**(a) no change (b) increase volume **17.71**(a) amount decreases (b) amounts increase (c) amounts increase (d) amount decreases **17.73** 2.0 **17.76**(a) lower temperature, higher pressure (b) Q decreases; no change in K (c) Reaction rates are lower at lower temperatures, so a catalyst is used to speed up the reaction. **17.79**(a) 2.26×10^{-2} atm (b) 4.46×10^3 μg/L **17.81** 0.204 atm **17.84**(a) 3×10^{-3} atm (b) high pressure; low temperature (c) 2×10^5 (d) No, because water condenses at a higher temperature. **17.89**(a) 0.016 atm (b) $K_c = 5.6 \times 10^2$; $P_{SO_2} = 0.16$ atm **17.91** 12.5 g $CaCO_3$ **17.95** Both concentrations increased by a factor of 2.2. **17.97**(a) 3.0×10^{-14} atm (b) 0.013 pg CO/L **17.100**(a) 98.0% (b) 99.0% (c) 2.60×10^5 J/mol **17.102**(a) $2CH_4(g) + O_2(g) + 2H_2O(g) \rightleftharpoons 2CO_2(g) + 6H_2(g)$ (b) 1.76×10^{29} (c) 3.19×10^{23} (d) 48 atm **17.104**(a) 4.0×10^{-21} atm (b) 5.5×10^{-8} atm (c) 29 N atoms/L; 4.0×10^{14} H atoms/L (d) The more reasonable step is $N_2(g) + H(g) \longrightarrow NH(g) + N(g)$. With only 29 N atoms in 1.0 L, the first reaction would produce virtually no $NH(g)$ molecules. There are orders of magnitude more N_2 molecules than N atoms, so the second reaction is the more reasonable step. **17.105**(a) P_{N_2} = 31 atm; P_{H_2} = 93 atm; P_{total} = 174 atm (b) P_{N_2} = 18 atm; P_{H_2} = 111 atm; P_{total} = 179 atm; not a valid argument **17.108**(a) P_{N_2} = 0.780 atm; P_{O_2} = 0.210 atm; $P_{NO} = 2.67 \times 10^{-16}$ atm (b) 0.990 atm (c) $K_c = K_p = 4.35 \times 10^{-31}$ **17.110**(a) 1.26×10^{-3} (b) 794 (c) -51.8 kJ (d) 1.2×10^4 J/mol **17.113**(a) 5 (b) 8 (c) Once sufficient F was produced, neither branch would proceed. (d) The lower branch would not proceed once sufficient F was made. **17.115**(a) 1.52 (b) 0.9626 atm (c) 0.2000 mol CO (d) 0.01128 M

Chapter 18

18.2 All Arrhenius acids contain hydrogen in their formula and produce hydronium ion (H_3O^+) in aqueous solution. All Arrhenius bases produce hydroxide ion (OH^-) in aqueous solution. Neutralization occurs when each H_3O^+ ion combines with an

OH$^-$ ion to form two molecules of H$_2$O. Chemists found the reaction of any strong base with any strong acid always produced 56 kJ/mol ($\Delta H = -56$ kJ/mol). **18.4** Strong acids and bases dissociate completely into ions when dissolved in water. Weak acids and bases dissociate only partially. The characteristic property of all weak acids is that a significant number of the acid molecules are undissociated. **18.5** (a), (c), and (d) **18.7** (b) and (d)

18.9(a) $K_a = \dfrac{[CN^-][H_3O^+]}{[HCN]}$

(b) $K_a = \dfrac{[CO_3^{2-}][H_3O^+]}{[HCO_3^-]}$

(c) $K_a = \dfrac{[HCOO^-][H_3O^+]}{[HCOOH]}$

18.11(a) $K_a = \dfrac{[NO_2^-][H_3O^+]}{[HNO_2]}$

(b) $K_a = \dfrac{[CH_3COO^-][H_3O^+]}{[CH_3COOH]}$

(c) $K_a = \dfrac{[BrO_2^-][H_3O^+]}{[HBrO_2]}$

18.13 CH$_3$COOH < HF < HIO$_3$ < HI   **18.15**(a) weak acid (b) strong base   (c) weak acid   (d) strong acid   **18.17**(a) strong base   (b) strong acid   (c) weak acid   (d) weak acid **18.22**(a) The acid with the smaller K_a (4×10^{-5}) has the higher pH, because less dissociation yields fewer hydronium ions. (b) The acid with the larger pK_a (3.5) has the higher pH, because a larger pK_a means a smaller K_a.   (c) Lower concentration (0.01 M) gives fewer hydronium ions.   (d) A 0.1 M weak acid solution gives fewer hydronium ions.   (e) The 0.01 M base solution has a lower concentration of hydronium ions.   (f) The solution has pOH = 6.0 because pH = 14.0 − 6.0 = 8.0 **18.23**(a) 12.05; basic   (b) 11.13; acidic   **18.25**(a) 2.212; acidic (b) −0.708; basic
18.27(a) $[H_3O^+] = 1.4 \times 10^{-10}$ M, pOH = 4.15, $[OH^-]$ $= 7.1 \times 10^{-5}$ M

(b) pH = 4.57, $[H_3O^+] = 2.7 \times 10^{-5}$ M, $[OH^-]$ $= 3.7 \times 10^{-10}$ M

18.29(a) $[H_3O^+] = 1.7 \times 10^{-5}$ M, pOH = 9.23, $[OH^-]$ $= 5.9 \times 10^{-10}$ M

(b) pH = 8.35, $[H_3O^+] = 4.5 \times 10^{-9}$ M, $[OH^-]$ $= 2.2 \times 10^{-6}$ M

18.31 4.8×10^{-4} mol OH$^-$/L   **18.33** 1.4×10^{-4} mol OH$^-$ **18.36**(a) Rising temperature increases the value of K_w.   (b) $K_w =$ 2.5×10^{-14}; pOH = 6.80; $[OH^-] = 1.6 \times 10^{-7}$ M **18.37** The Brønsted-Lowry theory defines acids as proton donors and bases as proton acceptors, while the Arrhenius definition looks at acids as containing ionizable hydrogen atoms and at bases as containing hydroxide ions. In both definitions, an acid produces hydronium ions and a base produces hydroxide ions when added to water. Ammonia and carbonate ion are two Brønsted-Lowry bases that are not Arrhenius bases because they do not contain hydroxide ions. Brønsted-Lowry acids must contain an ionizable hydrogen atom in order to be proton donors, so a Brønsted-Lowry acid is also an Arrhenius acid.   **18.40** An amphiprotic substance is one that can lose a proton to act as an acid or gain a proton to act as a base. The dihydrogen phosphate ion, H$_2$PO$_4^-$, is an example. H$_2$PO$_4^-$(aq) + OH$^-$(aq) $\longrightarrow$ H$_2$O(l) + HPO$_4^{2-}$(aq) H$_2$PO$_4^-$(aq) + HCl(aq) $\longrightarrow$ H$_3$PO$_4$(aq) + Cl$^-$(aq)

18.41(a) H$_3$PO$_4$(aq) + H$_2$O(l) $\rightleftharpoons$ H$_2$PO$_4^-$(aq) + H$_3$O$^+$(aq)

$K_a = \dfrac{[H_3O^+][H_2PO_4^-]}{[H_3PO_4]}$

(b) C$_6$H$_5$COOH(aq) + H$_2$O(l) $\rightleftharpoons$

C$_6$H$_5$COO$^-$(aq) + H$_3$O$^+$(aq)

$K_a = \dfrac{[H_3O^+][C_6H_5COO^-]}{[C_6H_5COOH]}$

(c) HSO$_4^-$(aq) + H$_2$O(l) $\rightleftharpoons$ SO$_4^{2-}$(aq) + H$_3$O$^+$(aq)

$K_a = \dfrac{[H_3O^+][SO_4^{2-}]}{[HSO_4^-]}$

18.43(a) Cl$^-$ (b) HCO$_3^-$ (c) OH$^-$ **18.45**(a) NH$_4^+$ (b) NH$_3$ (c) C$_{10}$H$_{14}$N$_2$H$^+$
18.47(a) HCl + H$_2$O $\rightleftharpoons$ Cl$^-$ + H$_3$O$^+$
    acid   base   base   acid
    Conjugate acid-base pairs: HCl/Cl$^-$ and H$_3$O$^+$/H$_2$O
(b) HClO$_4$ + H$_2$SO$_4$ $\rightleftharpoons$ ClO$_4^-$ + H$_3$SO$_4^+$
    acid   base   base   acid
    Conjugate acid-base pairs:
    HClO$_4$/ClO$_4^-$ and H$_3$SO$_4^+$/H$_2$SO$_4$
(c) HPO$_4^{2-}$ + H$_2$SO$_4$ $\rightleftharpoons$ H$_2$PO$_4^-$ + HSO$_4^-$
    base   acid   acid   base
    Conjugate acid-base pairs: H$_2$SO$_4$/HSO$_4^-$ and
    H$_2$PO$_4^-$/HPO$_4^{2-}$
18.49(a) NH$_3$ + H$_3$PO$_4$ $\rightleftharpoons$ NH$_4^+$ + H$_2$PO$_4^-$
    base   acid   acid   base
    Conjugate acid-base pairs:
    H$_3$PO$_4$/H$_2$PO$_4^-$ and NH$_4^+$/NH$_3$
(b) CH$_3$O + NH$_3$ $\rightleftharpoons$ CH$_3$OH + NH$_2^-$
    base   acid   acid   base
    Conjugate acid-base pairs:
    NH$_3$/NH$_2^-$ and CH$_3$OH/CH$_3$O$^-$
(c) HPO$_4^{2-}$ + HSO$_4^-$ $\rightleftharpoons$ H$_2$PO$_4^-$ + SO$_4^{2-}$
    base   acid   acid   base
    Conjugate acid-base pairs:
    HSO$_4^-$/SO$_4^{2-}$ and H$_2$PO$_4^-$/HPO$_4^{2-}$
18.51(a) OH$^-$(aq) + H$_2$PO$_4^-$(aq) $\rightleftharpoons$ H$_2$O(l) + HPO$_4^{2-}$(aq)
    Conjugate acid-base pairs:
    H$_2$PO$_4^-$/HPO$_4^{2-}$ and H$_2$O/OH$^-$
(b) HSO$_4^-$(aq) + CO$_3^{2-}$(aq) $\rightleftharpoons$
SO$_4^{2-}$(aq) + HCO$_3^-$(aq)
    Conjugate acid-base pairs:
    HSO$_4^-$/SO$_4^{2-}$ and HCO$_3^-$/CO$_3^{2-}$
18.53 $K_c > 1$: HS$^-$ + HCl $\rightleftharpoons$ H$_2$S + Cl$^-$
    $K_c < 1$: H$_2$S + Cl$^-$ $\rightleftharpoons$ HS$^-$ + HCl
18.55 $K_c > 1$ for both a and b   **18.57** $K_c < 1$ for both a and b **18.59**(a) A strong acid is 100% dissociated, so the acid concentration will be very different after dissociation.   (b) A weak acid dissociates to a very small extent, so the acid concentration before and after dissociation is nearly the same.   (c) same as b, but with the extent of dissociation greater   (d) same as a **18.60** No. HCl is a strong acid and dissociates to a greater extent than the weak acid CH$_3$COOH. The K_a of the acid, not the concentration of H$_3$O$^+$, determines the strength of the acid. **18.63** 1.5×10^{-5} **18.65** $[H_3O^+] = [NO_2^-] = 2.1 \times 10^{-2}$ M; $[OH^-] = 4.8 \times 10^{-13}$ M **18.67** $[H_3O^+] = [ClCH_2COO^-] =$ 0.041 M; $[ClCH_2COOH] = 1.21$ M; pH = 1.39   **18.69**(a) $[H_3O^+] =$ 6.0×10^{-3} M; pH = 2.22; $[OH^-] = 1.7 \times 10^{-12}$ M; pOH = 11.78   (b) 1.9×10^{-4} **18.71** 2.2×10^{-7} **18.73**(a) 2.37 (b) 11.53   **18.75**(a) 2.290   (b) 12.699   **18.77** 1.1%

18.79 $[H_3O^+] = [HS^-] = 9 \times 10^{-5}$ M; pH = 4.0; $[OH^-] = 1 \times 10^{-10}$ M; pOH = 10.0; $[H_2S] = 0.10$ M; $[S^{2-}] = 1 \times 10^{-17}$ M
18.82 1.5% **18.83** All Brønsted-Lowry bases contain at least one lone pair of electrons, which binds an H^+ and allows the base to act as a proton acceptor.
18.86(a) $C_5H_5N(aq) + H_2O(l) \rightleftharpoons OH^-(aq) + C_5H_5NH^+(aq)$
$$K_b = \frac{[C_5H_5NH^+][OH^-]}{[C_5H_5N]}$$
(b) $CO_3^{2-}(aq) + H_2O(l) \rightleftharpoons OH^-(aq) + HCO_3^-(aq)$
$$K_b = \frac{[HCO_3^-][OH^-]}{[CO_3^{2-}]}$$
18.88(a) $HONH_2(aq) + H_2O(l) \rightleftharpoons OH^-(aq) + HONH_3^+(aq)$
$$K_b = \frac{[HONH_3^+][OH^-]}{[HONH_2]}$$
(b) $HPO_4^{2-}(aq) + H_2O(l) \rightleftharpoons H_2PO_4^-(aq) + OH^-(aq)$
$$K_b = \frac{[H_2PO_4^-][OH^-]}{[HPO_4^{2-}]}$$
18.90 11.79 **18.92** 11.45 **18.94**(a) 5.6×10^{-10} (b) 2.5×10^{-5}
18.96(a) 12.04 (b) 10.77 **18.98**(a) 11.19 (b) 5.56
18.100(a) 8.78 (b) 4.66 **18.102** $[OH^-] = 5.5 \times 10^{-4}$ M; pH = 10.74 **18.104** As a nonmetal becomes more electronegative, the acidity of its binary hydride increases. The electronegative nonmetal attracts the electrons more strongly in the polar bond, shifting the electron density away from H, thus making the H^+ more easily transferred to a water molecule to form H_3O^+.
18.107 Chlorine is more electronegative than iodine, and $HClO_4$ has more oxygen atoms than HIO. **18.108**(a) H_2SeO_4
(b) H_3PO_4 (c) H_2Te **18.110**(a) H_2Se (b) $B(OH)_3$
(c) $HBrO_2$ **18.112**(a) 0.25 M $Al_2(SO_4)_3$ (b) 0.3 M $PbCl_2$
18.114(a) 0.2 M $Ni(NO_3)_2$ (b) 0.35 M $Al(NO_3)_3$ **18.117** NaF contains the anion of the weak acid HF, so F^- acts as a base. NaCl contains the anion of the strong acid HCl.
18.119(a) $KBr(s) \xrightarrow{H_2O} K^+(aq) + Br^-(aq)$; neutral
(b) $NH_4I(s) \xrightarrow{H_2O} NH_4^+(aq) + I^-(aq)$
$NH_4^+(aq) + H_2O(l) \rightleftharpoons$
$NH_3(aq) + H_3O^+(aq)$; acidic
(c) $KCN(s) \xrightarrow{H_2O} K^+(aq) + CN^-(aq)$
$CN^-(aq) + H_2O(l) \rightleftharpoons HCN(aq) + OH^-(aq)$; basic
18.121(a) $Na_2CO_3(s) \xrightarrow{H_2O} 2Na^+(aq) + CO_3^{2-}(aq)$
$CO_3^{2-}(aq) + H_2O(l) \rightleftharpoons$
$HCO_3^-(aq) + OH^-(aq)$; basic
(b) $CaCl_2(s) \xrightarrow{H_2O} Ca^{2+}(aq) + 2Cl^-(aq)$; neutral
(c) $Cu(NO_3)_2(s) \xrightarrow{H_2O} Cu^{2+}(aq) + 2NO_3^-(aq)$
$Cu(H_2O)_6^{2+}(aq) + H_2O(l) \rightleftharpoons$
$Cu(H_2O)_5OH^+(aq) + H_3O^+(aq)$; acidic
18.123(a) $SrBr_2(s) \xrightarrow{H_2O} Sr^{2+}(aq) + 2Br^-(aq)$; neutral
(b) $Ba(CH_3COO)_2(s) \xrightarrow{H_2O} Ba^{2+}(aq) + 2CH_3COO^-(aq)$
$CH_3COO^-(aq) + H_2O(l) \rightleftharpoons$
$CH_3COOH(aq) + OH^-(aq)$; basic
(c) $(CH_3)_2NH_2Br(s) \xrightarrow{H_2O} (CH_3)_2NH_2^+(aq) + Br^-(aq)$
$(CH_3)_2NH_2^+(aq) + H_2O(l) \rightleftharpoons$
$(CH_3)_2NH(aq) + H_3O^+(aq)$; acidic
18.125(a) $NH_4^+(aq) + H_2O(l) \rightleftharpoons NH_3(aq) + H_3O^+(aq)$
$PO_4^{3-}(aq) + H_2O(l) \rightleftharpoons HPO_4^{2-}(aq) + OH^-(aq)$;
$K_b > K_a$; basic
(b) $SO_4^{2-}(aq) + H_2O(l) \rightleftharpoons HSO_4^-(aq) + OH^-(aq)$;
Na^+ gives no reaction; basic

(c) $ClO^-(aq) + H_2O(l) \rightleftharpoons HClO(aq) + OH^-(aq)$;
Li^+ gives no reaction; basic
18.127(a) $Fe(NO_3)_2 < KNO_3 < K_2SO_3 < K_2S$
(b) $NaHSO_4 < NH_4NO_3 < NaHCO_3 < Na_2CO_3$
18.129 Since both bases produce OH^- ions in water, both bases appear equally strong. $CH_3O^-(aq) + H_2O(l) \longrightarrow OH^-(aq) + CH_3OH(aq)$ and $NH_2^-(aq) + H_2O(l) \longrightarrow OH^-(aq) + NH_3(aq)$. **18.131** Ammonia, NH_3, is a more basic solvent than H_2O. In a more basic solvent, weak acids such as HF act like strong acids and are 100% dissociated. **18.133** A Lewis acid is an electron-pair acceptor while a Brønsted-Lowry acid is a proton donor. The proton of a Brønsted-Lowry acid fits the definition of a Lewis acid because it accepts an electron pair when it bonds with a base. All Lewis acids are not Brønsted-Lowry acids. A Lewis base is an electron-pair donor and a Brønsted-Lowry base is a proton acceptor. All Brønsted-Lowry bases can be Lewis bases, and vice versa. **18.134**(a) No, $Zn(H_2O)_6^{2+}(aq) + 6NH_3(aq) \rightleftharpoons Zn(NH_3)_6^{2+} + 6H_2O(l)$; NH_3 is a weak Brønsted-Lowry base, but a strong Lewis base.
(b) cyanide ion and water (c) cyanide ion **18.137**(a) Lewis acid (b) Lewis base (c) Lewis acid (d) Lewis base
18.139(a) Lewis acid (b) Lewis base (c) Lewis base
(d) Lewis acid **18.141**(a) Lewis acid: Na^+; Lewis base: H_2O
(b) Lewis acid: CO_2; Lewis base: H_2O (c) Lewis acid: BF_3; Lewis base: F^- **18.143**(a) Lewis (b) Brønsted-Lowry and Lewis (c) none (d) Lewis **18.147** 3.5×10^{-8} to 4.5×10^{-8} M H_3O^+; 5.2×10^{-7} to 6.6×10^{-7} M OH^-
18.148(a) Acids vary in the extent of dissociation depending on the acid-base character of the solvent. (b) Methanol is a weaker base than water since phenol dissociates less in methanol than in water. (c) $C_6H_5OH(\text{solvated}) + CH_3OH(l) \rightleftharpoons$ $CH_3OH_2^+(\text{solvated}) + C_6H_5O^-(\text{solvated})$
(d) $CH_3OH(l) + CH_3OH(l) \rightleftharpoons CH_3O^-(\text{solvated}) + CH_3OH_2^+(\text{solvated})$; $K = [CH_3O^-][CH_3OH_2^+]$ **18.151**(a) $SnCl_4$ is the Lewis acid; $(CH_3)_3N$ is the Lewis base (b) $5d$
18.152 pH = 5.00, 6.00, 6.79, 6.98, 7.00 **18.155** H_3PO_4
18.159 3×10^{18} **18.161**(a) $[CH_3O^-] = 4 \times 10^{-9}$ M
(b) 4×10^{-16} **18.163** 10.48 **18.165** 2.41 **18.168** amylase, 2×10^{-7} M; pepsin, 1×10^{-2} M; trypsin, 3×10^{-10} M
18.172 1.47×10^{-3} **18.174**(a) Ca^{2+} does not react with water; $CH_3CH_2COO^-(aq) + H_2O(l) \rightleftharpoons CH_3CH_2COOH(aq) + OH^-(aq)$; basic (b) 9.08 **18.180** 4.5×10^{-5} **18.183**(a) The concentration of oxygen is higher in the lungs so the equilibrium shifts to the right. (b) In an oxygen-deficient environment the equilibrium shifts to the left to release oxygen. (c) A decrease in $[H_3O^+]$ shifts the equilibrium to the right. More oxygen is absorbed, but it will be more difficult to remove the O_2. (d) An increase in $[H_3O^+]$ shifts the equilibrium to the left. Less oxygen is bound to Hb, but it will be easier to remove it.
18.186(a) 1.012 M (b) 0.004 M (c) 0.4% **18.187**(a) 10.0
(b) The pK_b for the 3° amine group is much smaller than that for the aromatic ring, thus the K_b is significantly larger (yielding a much greater amount of OH^-). (c) 4.7 (d) 5.1

Chapter 19

19.2 The acid component neutralizes added base and the base component neutralizes added acid, so the pH of the buffer solution remains relatively constant. The components of a buffer do not neutralize one another because they are a conjugate acid-base pair.
19.7 The buffer range, the pH over which the buffer acts

effectively, is greatest when the buffer-component ratio is 1; the range decreases as the component ratio deviates from 1. **19.9**(a) Ratio and pH increase; added OH^- reacts with HA. (b) Ratio and pH decrease; added H^+ reacts with A^-. (c) Ratio and pH increase; added A^- increases $[A^-]$. (d) Ratio and pH decrease; added HA increases [HA]. **19.11** $[H_3O^+]$ = 5.6×10^{-6} M; pH = 5.25 **19.13** $[H_3O^+] = 5.2\times10^{-4}$ M; pH = 3.28 **19.15** 3.89 **19.17** 10.03 **19.19** 9.47 **19.21**(a) K_{a2} (b) 10.55 **19.23** 3.6 **19.25** 0.20 **19.27** 3.37 **19.29** 8.82 **19.31**(a) 4.81 (b) 0.49 g KOH **19.33**(a) $HOOC(CH_2)_4COOH/$ $HOOC(CH_2)_4COO^-$ or $C_6H_5NH_3{}^+/C_6H_5NH_2$ (b) $H_2PO_4{}^-/HPO_4{}^{2-}$ or $H_2AsO_4{}^-/HAsO_4{}^{2-}$ **19.35**(a) $HOCH_2CH(OH)COOH/HOCH_2CH(OH)COO^-$ or $CH_3COOC_6H_4COOH/CH_3COOC_6H_4COO^-$ (b) $C_5H_5N/C_5H_5NH^+$ **19.38** 1.6 **19.40** To see a distinct color in a mixture of two colors, you need one to have about 10 times the intensity of the other. For this to be the case, the concentration ratio $[HIn]/[In^-]$ has to be greater than 10/1 or less than 1/10. This occurs when pH = pK_a − 1 or pH = pK_a + 1, respectively, giving a pH range of about two units. **19.42** The equivalence point in a titration is the point at which the number of moles of base equals the number of moles of acid. The endpoint is the point at which the added indicator changes color. If an appropriate indicator is selected, the endpoint is close to the equivalence point, but they are not usually the same. The endpoint, or color change, may precede or follow the equivalence point, depending on the indicator chosen. **19.44**(a) initial pH: *strong acid*–strong base < *weak acid*–strong base < strong acid–*weak base* (b) pH at equivalence point: strong acid–*weak base* < *strong acid*–strong base < *weak acid*–strong base **19.46** At the center of the buffer region, the concentrations of weak acid and conjugate base are equal, so the pH = pK_a of the acid. **19.48** pH range from 7.5 to 9.5 **19.50**(a) bromthymol blue (b) thymol blue or phenolphthalein **19.52**(a) methyl red (b) bromthymol blue **19.54**(a) 1.0000 (b) 1.6368 (c) 2.898 (d) 3.903 (e) 7.00 (f) 10.10 (g) 12.05 **19.56**(a) 2.91 (b) 4.81 (c) 5.29 (d) 6.09 (e) 7.41 (f) 8.76 (g) 10.10 (h) 12.05 **19.58**(a) 59.0 mL and 8.54 (b) 66.0 mL and 7.13, total 132.1 mL and 9.69 **19.60**(a) 123 mL and 5.17 (b) 194 mL and 5.80 **19.63** Fluoride ion is the conjugate base of a weak acid and reacts with H_2O: $F^-(aq) + H_2O(l) \rightleftharpoons$ $HF(aq) + OH^-(aq)$. As the pH increases, the equilibrium shifts to the left and $[F^-]$ increases. As the pH decreases, the equilibrium shifts to the right and $[F^-]$ decreases. The changes in $[F^-]$ influence the solubility of BaF_2. Chloride ion is the conjugate base of a strong acid, so it does not react with water and its concentration is not influenced by pH. **19.65** The compound precipitates. **19.66**(a) $K_{sp} = [Ag^+]^2[CO_3{}^{2-}]$ (b) $K_{sp} =$ $[Ba^{2+}][F^-]^2$ (c) $K_{sp} = [Cu^{2+}][HS^-][OH^-]$ **19.68**(a) $K_{sp} =$ $[Ca^{2+}][CrO_4{}^{2-}]$ (b) $K_{sp} = [Ag^+][CN^-]$ (c) $K_{sp} =$ $[Ni^{2+}][HS^-][OH^-]$ **19.70** 1.3×10^{-4} **19.72** 2.8×10^{-11} **19.74**(a) 2.3×10^{-5} M (b) 4.2×10^{-9} M **19.76**(a) 1.7×10^{-3} M (b) 2.0×10^{-4} M **19.78**(a) $Mg(OH)_2$ (b) PbS (c) Ag_2SO_4 **19.80**(a) $CaSO_4$ (b) $Mg_3(PO_4)_2$ (c) $PbSO_4$ **19.82**(a) $AgCl(s) \rightleftharpoons Ag^+(aq) + Cl^-(aq)$. The chloride ion is the anion of a strong acid, so it does not react with H_3O^+. No change with pH. (b) $SrCO_3(s) \rightleftharpoons Sr^{2+}(aq) + CO_3{}^{2-}(aq)$. The strontium ion is the cation of a strong base, so pH will not affect its solubility. The carbonate ion acts as a base: $CO_3{}^{2-}(aq) + H_2O(l) \rightleftharpoons HCO_3{}^-(aq) + OH^-(aq)$; also $CO_2(g)$ forms and escapes: $CO_3{}^{2-}(aq) + 2H_3O^+(aq) \longrightarrow$ $CO_2(g) + 2H_2O(l)$. Therefore, the solubility of $SrCO_3$ will

increase with addition of H_3O^+ (decreasing pH). **19.84**(a) $Fe(OH)_2(s) \rightleftharpoons Fe^{2+}(aq) + 2OH^-(aq)$. The OH^- ion reacts with added H_3O^+: $OH^-(aq) + H_3O^+(aq) \longrightarrow$ $2H_2O(l)$. The added H_3O^+ consumes the OH^-, driving the equilibrium toward the right to dissolve more $Fe(OH)_2$. Solubility increases with addition of H_3O^+ (decreasing pH). (b) $CuS(s) + H_2O(l) \rightleftharpoons Cu^{2+}(aq) + HS^-(aq) + OH^-(aq)$. Both HS^- and OH^- are anions of weak acids, so both ions react with added H_3O^+. Solubility increases with addition of H_3O^+ (decreasing pH). **19.86** yes **19.88** yes **19.91**(a) $Fe(OH)_3$ (b) The two metal ions are separated by adding just enough NaOH to precipitate iron(III) hydroxide. (c) 2.0×10^{-7} M **19.94** No, because it indicates that a complex ion forms between the lead ion and hydroxide ions: $Pb^{2+}(aq) + nOH^-(aq) \rightleftharpoons$ $Pb(OH)_n{}^{2-n}(aq)$ **19.95** $Hg(H_2O)_4{}^{2+}(aq) + 4CN^-(aq) \rightleftharpoons$ $Hg(CN)_4{}^{2-}(aq) + 4H_2O(l)$ **19.97** $Ag(H_2O)_2{}^+(aq) +$ $2S_2O_3{}^{2-}(aq) \rightleftharpoons Ag(S_2O_3)_2{}^{3-}(aq) + 2H_2O(l)$ **19.99** 8.1×10^{-15} M **19.101** 0.05 M **19.103** 1.0×10^{-16} M Zn^{2+}; 0.025 M $Zn(CN)_4{}^{2-}$; 0.049 M CN^- **19.105** 0.035 L of 2.00 M NaOH and 0.465 L of 0.200 M HCOOH **19.107**(a) 0.99 (b) Assuming the volumes are additive: 0.468 L of 1.0 M HCOOH and 0.232 L of 1.0 M NaOH (c) 0.34 M **19.110** 1.3×10^{-4} M **19.114**(a) 14 (b) 1 g **19.117**(a) 0.088 (b) 0.14 **19.118** 0.260 M TRIS; pH = 8.53 **19.120** Lower the pH below 6.6. **19.124** 8×10^{-5} **19.126**(a)

V(mL)	pH	ΔpH/ΔV	V_{avg} (mL)
0.00	1.00		
10.00	1.22	0.022	5.00
20.00	1.48	0.026	15.00
30.00	1.85	0.037	25.00
35.00	2.18	0.066	32.50
39.00	2.89	0.18	37.00
39.50	3.20	0.62	39.25
39.75	3.50	1.2	39.63
39.90	3.90	2.7	39.83
39.95	4.20	6	39.93
39.99	4.90	18	39.97
40.00	7.00	200	40.00
40.01	9.40	200	40.01
40.05	9.80	10	40.03
40.10	10.40	10	40.08
40.25	10.50	0.67	40.18
40.50	10.79	1.2	40.38
41.00	11.09	0.60	40.75
45.00	11.76	0.17	43.00
50.00	12.05	0.058	47.50
60.00	12.30	0.025	55.00
70.00	12.43	0.013	65.00
80.00	12.52	0.009	75.00

(b)

Maximum slope (equivalence point) is at $V_{avg} = 40.00$ mL.

19.128 $K_b = \dfrac{[BH^+][OH^-]}{[B]}$

Rearranging to isolate $[OH^-]$:

$[OH^-] = K_b \dfrac{[B]}{[BH^+]}$

Taking the negative log:

$-\log[OH^-] = -\log K_b - \log \dfrac{[B]}{[BH^+]}$

Therefore, $pOH = pK_b + \log \dfrac{[BH^+]}{[B]}$

19.129 4.05 **19.136** H_2CO_3/HCO_3^- and $H_2PO_4^-/HPO_4^{2-}$; $[HPO_4^{2-}]/[H_2PO_4^-] = 5.8$ **19.139** 3.8 **19.141**(a) 58.2 mL (b) 7.4 mL (c) 6.30 **19.143** 170 mL **19.146**(a) 65 mol (b) 6.28 (c) 4.0×10^3 g **19.148** 5.68 **19.150** 3.9×10^{-9} μg $Pb^{2+}/100$ mL blood **19.152** No NaCl will precipitate. **19.153**(a) A and D (b) $pH_A = 4.35$; $pH_B = 8.67$; $pH_C = 2.67$; $pH_D = 4.57$ (c) C, A, D, B (d) B

Chapter 20

20.2 A spontaneous process occurs by itself, whereas a nonspontaneous process requires a continuous input of energy to make it happen. It is possible to cause a nonspontaneous process to occur, but the process stops once the energy source is removed. A reaction that is nonspontaneous under one set of conditions may be spontaneous under a different set of conditions. **20.5** The transition from liquid to gas involves a greater increase in dispersal of energy and freedom of motion than does the transition from solid to liquid. **20.6** In an exothermic reaction, $\Delta S_{surr} > 0$. In an endothermic reaction, $\Delta S_{surr} < 0$. A chemical cold pack for injuries is an example of an application using a spontaneous endothermic process. **20.8**(a), (b), and (c) **20.10**(a) and (b) **20.12**(a) positive (b) negative (c) negative **20.14**(a) positive (b) positive (c) positive **20.16**(a) negative (b) negative (c) positive **20.18**(a) positive (b) negative (c) positive **20.20**(a) positive (b) negative (c) positive **20.22**(a) Butane. The double bond in 2-butene restricts freedom of rotation. (b) $Xe(g)$. It has the greater molar mass. (c) $CH_4(g)$. Gases have greater entropy than liquids. **20.24**(a) $C_2H_5OH(l)$. It is a more complex molecule. (b) $KClO_3(aq)$. Ions in solution have their energy more dispersed than those in a solid. (c) $K(s)$. It has a greater molar mass. **20.26**(a) Diamond < graphite < charcoal. Freedom of motion is least in the network solid; more freedom between graphite sheets; most freedom in amorphous solid. (b) Ice < liquid water < water vapor. Entropy increases as a substance changes from solid to liquid to gas. (c) O atoms < $O_2 < O_3$. Entropy increases with molecular complexity. **20.28**(a) $ClO_4^-(aq) > ClO_3^-(aq) > ClO_2^-(aq)$; decreasing molecular complexity (b) $NO_2(g) > NO(g) > N_2(g)$. N_2 has lower standard molar entropy because it consists of two of the same atoms; the other species have two different types of atoms. NO_2 is more complex than NO. (c) $Fe_3O_4(s) > Fe_2O_3(s) > Al_2O_3(s)$. Fe_3O_4 is more complex and more massive. Fe_2O_3 is more massive than Al_2O_3. **20.31** For a system at equilibrium, $\Delta S_{univ} = \Delta S_{sys} + \Delta S_{surr} = 0$. For a system moving to equilibrium, $\Delta S_{univ} > 0$. **20.32** $S^{\circ}_{Cl_2O(g)} = 2S^{\circ}_{HClO(g)} - S^{\circ}_{H_2O(g)} - \Delta S^{\circ}_{rxn}$ **20.33**(a) negative; $\Delta S^{\circ} = -172.4$ J/K (b) positive; $\Delta S^{\circ} = 141.6$ J/K (c) negative; $\Delta S^{\circ} = -837$ J/K **20.35** $\Delta S^{\circ} =$

93.1 J/K; yes, the positive sign of ΔS is expected because there is a net increase in the number of gas molecules. **20.37** $\Delta S^{\circ} = -311$ J/K; yes, the negative entropy change matches the decrease in moles of gas. **20.39** -75.6 J/K **20.41** -242 J/K **20.44** -97.2 J/K **20.46** A spontaneous process has $\Delta S_{univ} > 0$. Since the absolute temperature is always positive, ΔG_{sys} must be negative ($\Delta G_{sys} < 0$) for a spontaneous process. **20.48** ΔH°_{rxn} is positive and ΔS°_{sys} is positive. Melting is an example. **20.49** The entropy changes little within a phase. As long as the substance does not change phase, the value of ΔS° is relatively unaffected by temperature. **20.50**(a) -1138.0 kJ (b) -1379.4 kJ (c) -224 kJ **20.52**(a) -1138 kJ (b) -1379 kJ (c) -226 kJ **20.54**(a) Entropy decreases (ΔS° is negative) because the number of moles of gas decreases. The combustion of CO releases energy (ΔH° is negative). (b) -257.2 kJ or -257.3 kJ, depending on the method **20.56**(a) -0.409 kJ/mol·K (b) -197 kJ/mol **20.58**(a) $\Delta H^{\circ}_{rxn} = 90.7$ kJ; $\Delta S^{\circ}_{rxn} = 221$ J/K (b) at 28°C, $\Delta G^{\circ} = 24.3$ kJ; at 128°C, $\Delta G^{\circ} = 2.2$ kJ; at 228°C, $\Delta G^{\circ} = -19.9$ kJ (c) For the substances in their standard states, the reaction is nonspontaneous at 28°C, near equilibrium at 128°C, and spontaneous at 228°C. **20.60** $\Delta H^{\circ} = 30910$ J, $\Delta S^{\circ} = 93.15$ J/K, $T = 331.8$ K **20.62**(a) $\Delta H^{\circ}_{rxn} = -241.826$ kJ, $\Delta S^{\circ}_{rxn} = -44.4$ J/K, $\Delta G^{\circ}_{rxn} = -228.60$ kJ (b) Yes. The reaction will become nonspontaneous at higher temperatures. (c) The reaction is spontaneous below 5.45×10^3 K. **20.64**(a) ΔG° is a relatively large positive value. (b) $K >> 1$. Q depends on initial conditions, not equilibrium conditions. **20.67** The standard free energy change, ΔG°, applies when all components of the system are in their standard states; $\Delta G^{\circ} = \Delta G$ when all concentrations equal 1 M and all partial pressures equal 1 atm. **20.68**(a) 1.7×10^6 (b) 3.89×10^{-34} (c) 1.26×10^{48} **20.70**(a) 6.57×10^{173} (b) 4.46×10^{-15} (c) 3.46×10^4 **20.72** 4.89×10^{-51} **20.74** 3.36×10^5 **20.76** 2.7×10^4 J/mol; no **20.78**(a) 2.9×10^4 J/mol (b) The reverse direction, formation of reactants, is spontaneous, so the reaction proceeds to the left. (c) 7.0×10^3 J/mol; the reaction proceeds to the left to reach equilibrium. **20.80**(a) no T (b) 163 kJ (c) 1×10^2 kJ/mol **20.83**(a) spontaneous (b) + (c) + (d) − (e) −, not spontaneous (f) − **20.87**(a) 2.3×10^2 (b) Administer oxygen-rich air to counteract the CO poisoning. **20.90** $-370.$ kJ **20.92**(a) $2N_2O_5(g) + 6F_2(g) \longrightarrow 4NF_3(g) + 5O_2(g)$ (b) $\Delta G^{\circ}_{rxn} = -569$ kJ (c) $\Delta G_{rxn} = -5.60 \times 10^2$ kJ/mol **20.95** $\Delta H^{\circ}_{rxn} = -137.14$ kJ; $\Delta S^{\circ}_{rxn} = -120.3$ J/K; $\Delta G^{\circ}_{rxn} = -101.25$ kJ **20.100**(a) 1.203×10^{23} molecules ATP/g glucose (b) 3.073×10^{23} molecules ATP/g tristearin **20.106**(a) 1.67×10^3 J/mol (b) 7.37×10^3 J/mol (c) -4.04×10^3 J/mol (d) 0.19 **20.110**(a) 465 K (b) 6.59×10^{-4} (c) The reaction rate is higher at the higher temperature. The shorter time required (kinetics) overshadows the lower yield (thermodynamics).

Chapter 21

21.1 Oxidation is the loss of electrons and results in a higher oxidation number; reduction is the gain of electrons and results in a lower oxidation number. **21.3** No, one half-reaction cannot take place independently because there is a transfer of electrons from one substance to another. If one substance loses electrons, another substance must gain them. **21.6** To remove H^+ ions from an equation, add an equal number of OH^- ions to both sides to neutralize the H^+ ions and produce water. **21.8** Spontaneous

reactions, for which $\Delta G_{sys} < 0$, take place in voltaic cells (also called galvanic cells). Nonspontaneous reactions, for which $\Delta G_{sys} > 0$, take place in electrolytic cells.   **21.10**(a) Cl^-
(b) MnO_4^- (c) MnO_4^- (d) Cl^- (e) from Cl^- to MnO_4^-
(f) $8H_2SO_4(aq) + 2KMnO_4(aq) + 10KCl(aq) \longrightarrow$
$$2MnSO_4(aq) + 5Cl_2(g) + 8H_2O(l) + 6K_2SO_4(aq)$$
21.12(a) $ClO_3^-(aq) + 6H^+(aq) + 6I^-(aq) \longrightarrow$
$$Cl^-(aq) + 3H_2O(l) + 3I_2(s)$$
Oxidizing agent is ClO_3^- and reducing agent is I^-.
   (b) $2MnO_4^-(aq) + H_2O(l) + 3SO_3^{2-}(aq) \longrightarrow$
$$2MnO_2(s) + 3SO_4^{2-}(aq) + 2OH^-(aq)$$
Oxidizing agent is MnO_4^- and reducing agent is SO_3^{2-}.
   (c) $2MnO_4^-(aq) + 6H^+(aq) + 5H_2O_2(aq) \longrightarrow$
$$2Mn^{2+}(aq) + 8H_2O(l) + 5O_2(g)$$
Oxidizing agent is MnO_4^- and reducing agent is H_2O_2.
21.14(a) $Cr_2O_7^{2-}(aq) + 14H^+(aq) + 3Zn(s) \longrightarrow$
$$2Cr^{3+}(aq) + 7H_2O(l) + 3Zn^{2+}(aq)$$
Oxidizing agent is $Cr_2O_7^{2-}$ and reducing agent is Zn.
   (b) $MnO_4^-(aq) + 3Fe(OH)_2(s) + 2H_2O(l) \longrightarrow$
$$MnO_2(s) + 3Fe(OH)_3(s) + OH^-(aq)$$
Oxidizing agent is MnO_4^- and reducing agent is $Fe(OH)_2$.
   (c) $2NO_3^-(aq) + 12H^+(aq) + 5Zn(s) \longrightarrow$
$$N_2(g) + 6H_2O(l) + 5Zn^{2+}(aq)$$
Oxidizing agent is NO_3^- and reducing agent is Zn.
21.16(a) $4NO_3^-(aq) + 4H^+(aq) + 4Sb(s) \longrightarrow$
$$4NO(g) + 2H_2O(l) + Sb_4O_6(s)$$
Oxidizing agent is NO_3^- and reducing agent is Sb.
   (b) $5BiO_3^-(aq) + 14H^+(aq) + 2Mn^{2+}(aq) \longrightarrow$
$$5Bi^{3+}(aq) + 7H_2O(l) + 2MnO_4^-(aq)$$
Oxidizing agent is BiO_3^- and reducing agent is Mn^{2+}.
   (c) $Pb(OH)_3^-(aq) + 2Fe(OH)_2(s) \longrightarrow$
$$Pb(s) + 2Fe(OH)_3(s) + OH^-(aq)$$
Oxidizing agent is $Pb(OH)_3^-$ and reducing agent is $Fe(OH)_2$.
21.18(a) $5As_4O_6(s) + 8MnO_4^-(aq) + 18H_2O(l) \longrightarrow$
$$20AsO_4^{3-}(aq) + 8Mn^{2+}(aq) + 36H^+(aq)$$
Oxidizing agent is MnO_4^- and reducing agent is As_4O_6.
   (b) $P_4(s) + 6H_2O(l) \longrightarrow$
$$2HPO_3^{2-}(aq) + 2PH_3(g) + 4H^+(aq)$$
P_4 is both the oxidizing agent and the reducing agent.
   (c) $2MnO_4^-(aq) + 3CN^-(aq) + H_2O(l) \longrightarrow$
$$2MnO_2(s) + 3CNO^-(aq) + 2OH^-(aq)$$
Oxidizing agent is MnO_4^- and reducing agent is CN^-.
21.21(a) $Au(s) + 3NO_3^-(aq) + 4Cl^-(aq) + 6H^+(aq) \longrightarrow$
$AuCl_4^-(aq) + 3NO_2(g) + 3H_2O(l)$ (b) Oxidizing agent is NO_3^- and reducing agent is Au.   (c) HCl provides chloride ions that combine with the gold(III) ion to form the stable $AuCl_4^-$ ion.
21.22(a) A   (b) E   (c) C   (d) A   (e) E   (f) E   **21.25** An active electrode is a reactant or product in the cell reaction. An inactive electrode does not take part in the reaction and is present only to conduct a current. Platinum and graphite are commonly used as inactive electrodes.   **21.26**(a) A   (b) B   (c) A
(d) Hydrogen bubbles will form when metal A is placed in acid. Metal A is a better reducing agent than metal B, so if metal B reduces H^+ in acid, then metal A will also.
21.27(a) Oxidation: $Zn(s) \longrightarrow Zn^{2+}(aq) + 2e^-$
   Reduction: $Sn^{2+}(aq) + 2e^- \longrightarrow Sn(s)$
   Overall: $Zn(s) + Sn^{2+}(aq) \longrightarrow Zn^{2+}(aq) + Sn(s)$

(b)

21.29(a) left to right   (b) left   (c) right   (d) Ni   (e) Fe
(f) Fe   (g) $1\,M$ $NiSO_4$ (h) K^+ and NO_3^- (i) neither   (j) from right to left
(k) Oxidation: $Fe(s) \longrightarrow Fe^{2+}(aq) + 2e^-$
   Reduction: $Ni^{2+}(aq) + 2e^- \longrightarrow Ni(s)$
   Overall: $Fe(s) + Ni^{2+}(aq) \longrightarrow Fe^{2+}(aq) + Ni(s)$
21.31(a) Reduction: $Fe^{2+}(aq) + 2e^- \longrightarrow Fe(s)$
   Oxidation: $Mn(s) \longrightarrow Mn^{2+}(aq) + 2e^-$
   Overall: $Fe^{2+}(aq) + Mn(s) \longrightarrow Fe(s) + Mn^{2+}(aq)$

(b)

21.33(a) $Al(s)\,|\,Al^{3+}(aq)\,\|\,Cr^{3+}(aq)\,|\,Cr(s)$
   (b) $Pt(s)\,|\,SO_2(g)\,|\,SO_4^{2-}(aq), H^+(aq)\,\|\,Cu^{2+}(aq)\,|\,Cu(s)$
21.36 A negative E_{cell}° indicates that the redox reaction is not spontaneous, that is, $\Delta G^\circ > 0$. The reverse reaction is spontaneous with $E_{cell}^\circ > 0$.   **21.37** Similar to other state functions, E° changes sign when a reaction is reversed. Unlike ΔG°, ΔH°, and S°, E° (the ratio of energy to charge) is an intensive property. When the coefficients in a reaction are multiplied by a factor, the values of ΔG°, ΔH°, and S° are multiplied by that factor. However, E° does not change because both the energy and charge are multiplied by the factor and thus their ratio remains unchanged.
21.38(a) Oxidation: $Se^{2-}(aq) \longrightarrow Se(s) + 2e^-$
   Reduction: $2SO_3^{2-}(aq) + 3H_2O(l) + 4e^- \longrightarrow$
$$S_2O_3^{2-}(aq) + 6OH^-(aq)$$
   (b) $E_{anode}^\circ = E_{cathode}^\circ - E_{cell}^\circ = -0.57\,V - 0.35\,V = -0.92\,V$
21.40(a) $Br_2 > Fe^{3+} > Cu^{2+}$ (b) $Ca^{2+} < Ag^+ < Cr_2O_7^{2-}$
21.42(a) $Co(s) + 2H^+(aq) \longrightarrow Co^{2+}(aq) + H_2(g)$
   $E_{cell}^\circ = 0.28\,V$; spontaneous
   (b) $2Mn^{2+}(aq) + 5Br_2(l) + 8H_2O(l) \longrightarrow$
$$2MnO_4^-(aq) + 10Br^-(aq) + 16H^+(aq)$$
   $E_{cell}^\circ = -0.44\,V$; not spontaneous
   (c) $Hg_2^{2+}(aq) \longrightarrow Hg^{2+}(aq) + Hg(l)$
   $E_{cell}^\circ = -0.07\,V$; not spontaneous
21.44(a) $2Ag(s) + Cu^{2+}(aq) \longrightarrow 2Ag^+(aq) + Cu(s)$
   $E_{cell}^\circ = -0.46\,V$; not spontaneous
   (b) $Cr_2O_7^{2-}(aq) + 3Cd(s) + 14H^+(aq) \longrightarrow$
$$2Cr^{3+}(aq) + 3Cd^{2+}(aq) + 7H_2O(l)$$
   $E_{cell}^\circ = 1.73\,V$; spontaneous

(c) $Pb(s) + Ni^{2+}(aq) \longrightarrow Pb^{2+}(aq) + Ni(s)$
$E^{\circ}_{cell} = -0.12$ V; not spontaneous
21.46 $3N_2O_4(g) + 2Al(s) \longrightarrow 6NO_2^-(aq) + 2Al^{3+}(aq)$
$E^{\circ}_{cell} = 0.867$ V $-$ $(-1.66$ V$) = 2.53$ V
$2Al(s) + 3SO_4^{2-}(aq) + 3H_2O(l) \longrightarrow$
$\qquad\qquad\qquad 2Al^{3+}(aq) + 3SO_3^{2-}(aq) + 6OH^-(aq)$
$E^{\circ}_{cell} = 2.59$ V
$SO_4^{2-}(aq) + 2NO_2^-(aq) + H_2O(l) \longrightarrow$
$\qquad\qquad\qquad SO_3^{2-}(aq) + N_2O_4(g) + 2OH^-(aq)$
$E^{\circ}_{cell} = 0.06$ V
Oxidizing agents: $Al^{3+} < N_2O_4 < SO_4^{2-}$
Reducing agents: $SO_3^{2-} < NO_2^- < Al$
21.48 $2HClO(aq) + Pt(s) + 2H^+(aq) \longrightarrow$
$\qquad\qquad\qquad Cl_2(g) + Pt^{2+}(aq) + 2H_2O(l)$
$E^{\circ}_{cell} = 0.43$ V
$2HClO(aq) + Pb(s) + SO_4^{2-}(aq) + 2H^+(aq) \longrightarrow$
$\qquad\qquad\qquad Cl_2(g) + PbSO_4(s) + 2H_2O(l)$
$E^{\circ}_{cell} = 1.94$ V
$Pt^{2+}(aq) + Pb(s) + SO_4^{2-}(aq) \longrightarrow Pt(s) + PbSO_4(s)$
$E^{\circ}_{cell} = 1.51$ V
Oxidizing agent: $PbSO_4 < Pt^{2+} < HClO$
Reducing agent: $Cl_2 < Pt < Pb$
21.50 Yes; $C > A > B$ **21.53** $A(s) + B^+(aq) \longrightarrow A^+(aq) +$
$B(s)$ with $Q = [A^+]/[B^+]$. (a) $[A^+]$ increases and $[B^+]$
decreases. (b) E_{cell} decreases. (c) $E_{cell} = E^{\circ}_{cell} - (RT/nF)$
$\ln ([A^+]/[B^+])$; $E_{cell} = E^{\circ}_{cell}$ when $(RT/nF) \ln ([A^+]/[B^+]) = 0$.
This occurs when $\ln ([A^+]/[B^+]) = 0$, that is, $[A^+]$ equals $[B^+]$.
(d) Yes, when $[A^+] > [B^+]$. **21.55** In a concentration cell, the
overall reaction decreases the concentration of the more concentrated electrolyte because it is reduced in the cathode compartment.
21.56(a) 3×10^{35} (b) 4×10^{-31} **21.58**(a) 1×10^{-67} (b) 6×10^9
21.60(a) -2.03×10^5 J (b) 1.73×10^5 J **21.62**(a) -3.82×10^5 J
(b) -5.6×10^4 J **21.64** $\Delta G^{\circ} = -2.7 \times 10^4$ J; $E^{\circ} = 0.28$ V
21.66 $E^{\circ}_{cell} = 0.054$ V; $\Delta G^{\circ} = -1.0 \times 10^4$ J **21.68** 8.8×10^{-5} M
21.70(a) 0.05 V (b) 0.50 M (c) $[Co^{2+}] = 0.91$ M; $[Ni^{2+}] =$
0.09 M **21.72** A; 0.083 V **21.74** Electrons flow from the anode, where oxidation occurs, to the cathode, where reduction
occurs. The electrons always flow from the anode to the cathode
no matter what type of battery. **21.76** A D-sized alkaline battery
is much larger than an AAA-sized one, so the D-sized battery
contains greater amounts of the cell components. The cell potential is an intensive property and does not depend on the amounts
of the cell components. The total charge, however, depends on
the amount of cell components, so the D-sized battery produces
more charge than the AAA-sized battery. **21.78** The Teflon
spacers keep the two metals separated so that the copper cannot
conduct electrons that would promote the corrosion (rusting) of
the iron skeleton. **21.81** Sacrificial anodes are made of metals
with E° less than that of iron, -0.44 V, so they are more easily
oxidized than iron. Only (b), (f), and (g) will work for iron;
(a) will form an oxide coating that prevents further oxidation;
(c) would react with groundwater quickly; (d) and (e) are less
easily oxidized than iron. **21.83** To reverse the reaction requires
0.34 V with the cell in its standard state. A 1.5 V cell supplies
more than enough potential, so the cadmium metal is oxidized to
Cd^{2+} and chromium plates out. **21.85** The oxidation number
of N in NO_3^- is $+5$, the maximum O.N. for N. In the nitrite ion,
NO_2^-, the O.N. of N is $+3$, so nitrogen can be further oxidized.
21.87(a) Br_2 (b) Na **21.89** I_2 gas forms at the anode; magnesium (liquid) forms at the cathode. **21.91** Bromine gas forms at

the anode; calcium metal forms at the cathode. **21.93** copper
and bromine **21.95** iodine, zinc, and silver
21.97(a) Anode: $2H_2O(l) \longrightarrow O_2(g) + 4H^+(aq) + 4e^-$
$\qquad$ Cathode: $2H_2O(l) + 2e^- \longrightarrow H_2(g) + 2OH^-(aq)$
(b) Anode: $2H_2O(l) \longrightarrow O_2(g) + 4H^+(aq) + 4e^-$
$\qquad$ Cathode: $Sn^{2+}(aq) + 2e^- \longrightarrow Sn(s)$
21.99(a) Anode: $2H_2O(l) \longrightarrow O_2(g) + 4H^+(aq) + 4e^-$
$\qquad$ Cathode: $NO_3^-(aq) + 4H^+(aq) + 3e^- \longrightarrow$
$\qquad\qquad\qquad\qquad\qquad\qquad NO(g) + 2H_2O(l)$
(b) Anode: $2Cl^-(aq) \longrightarrow Cl_2(g) + 2e^-$
$\qquad$ Cathode: $2H_2O(l) + 2e^- \longrightarrow H_2(g) + 2OH^-(aq)$
21.101(a) 3.75 mol e^- (b) 3.62×10^5 C (c) 28.7 A
21.103 0.275 g Ra **21.105** 9.20×10^3 s **21.107**(a) The sodium
and sulfate ions make the water conductive so that current will
flow through the water, facilitating electrolysis. Pure water,
which contains very low (10^{-7} M) concentrations of H^+ and
OH^-, conducts electricity very poorly. (b) The reduction of
H_2O has a more positive half-potential than does the reduction of
Na^+; the oxidation of H_2O is the only reaction possible because
SO_4^{2-} cannot be oxidized. Thus, it is easier to reduce H_2O than
Na^+ and easier to oxidize H_2O than SO_4^{2-}. **21.109** 62.6 g Zn
21.111(a) 3.3×10^{11} C (b) 4.8×10^{11} J (c) 1.2×10^4 kg
21.114 64.3 mass % Cu **21.116**(a) 8 days (b) 32 days
(c) \$250 **21.119** (a) 2.4×10^4 days (b) 2.1 g (c) 3.4×10^{-5}
dollars **21.122**(a) Pb/Pb^{2+}: $E^{\circ}_{cell} = 0.13$ V; Cu/Cu^{2+}: $E^{\circ}_{cell} =$
0.34 V (b) The anode (negative electrode) is Pb. The anode in
the other cell is platinum in the standard hydrogen electrode.
(c) The precipitation of PbS decreases $[Pb^{2+}]$, which increases
the potential. (d) -0.13 V **21.125** The three steps equivalent
to the overall reaction $M^+(aq) + e^- \longrightarrow M(s)$ are
(1) $M^+(aq) \longrightarrow M^+(g) \qquad \Delta H$ is $-\Delta H_{hydration}$
(2) $M^+(g) + e^- \longrightarrow M(g) \qquad \Delta H$ is $-IE$
(3) $M(g) \longrightarrow M(s) \qquad \Delta H$ is $-\Delta H_{atomization}$
The energy for step 3 is similar for all three elements, so the difference in energy for the overall reaction depends on the values
for $\Delta H_{hydration}$ and IE. The Li^+ ion has a much greater hydration
energy than Na^+ and K^+ because it is smaller, with large charge
density that holds the water molecules more tightly. The energy
required to remove the waters surrounding the Li^+ offsets the
lower ionization energy, making the overall energy for the reduction of lithium larger than expected. **21.127** The very high and
very low standard electrode potentials involve extremely reactive
substances, such as F_2 (a powerful oxidizer) and Li (a powerful
reducer). These substances react directly with water because any
aqueous cell with a voltage of more than 1.23 V has the ability to
electrolyze water into hydrogen and oxygen.
21.129(a) 1.073×10^5 s (b) 1.5×10^4 kW·h (c) 6.8¢
21.131 $F < D < E$. If metal E and a salt of metal F are mixed,
the salt is reduced, producing metal F because E has the greatest
reducing strength of the three metals.
21.134(a) Cell I: 4 mol electrons; $\Delta G = -4.75 \times 10^5$ J
$\qquad$ Cell II: 2 mol electrons; $\Delta G = -3.94 \times 10^5$ J
$\qquad$ Cell III: 2 mol electrons; $\Delta G = -4.53 \times 10^5$ J
(b) Cell I: -13.2 kJ/g
$\qquad$ Cell II: -0.613 kJ/g
$\qquad$ Cell III: -2.62 kJ/g
$\qquad$ Cell I has the highest ratio (most energy released per
$\qquad$ gram) because the reactants have very low mass, while
$\qquad$ Cell II has the lowest ratio because the reactants have
$\qquad$ large masses.

21.138(a) 9.7 g Cu   (b) 0.56 M Cu^{2+}
21.140$Sn^{2+}(aq) + 2e^- \longrightarrow Sn(s)$
$\quad\quad Cr^{3+}(aq) + e^- \longrightarrow Cr^{2+}(aq)$
$\quad\quad Fe^{2+}(aq) + 2e^- \longrightarrow Fe(s)$
$\quad\quad U^{4+}(aq) + e^- \longrightarrow U^{3+}(aq)$
21.144(a) 3.6×10^{-9} M (b) 0.3 M
21.146(a) Nonstandard cell:
$$E_{waste} = E°_{cell} - (0.0592\ V/1)\ \log\ [Ag^+]_{waste}$$
Standard cell:
$$E_{standard} = E°_{cell} - (0.0592\ V/1)\ \log\ [Ag^+]_{standard}$$
(b) $[Ag^+]_{waste} = antilog\left(\dfrac{E_{standard} - E_{waste}}{0.0592}\right)[Ag^+]_{standard}$

(c) $C_{Ag^+,\ waste} = antilog\left(\dfrac{E_{standard} - E_{waste}}{0.0592}\right)C_{Ag^+,\ standard}$,

where C is concentration in ng/L
(d) 900 ng/L
(e) $[Ag^+]_{waste} =$

antilog $\left[\dfrac{(E_{standard} - E_{waste})\dfrac{nF}{2.303R} + T_{standard}\ \log\ [Ag^+]_{standard}}{T_{waste}}\right]$

21.148 (a) 1.08×10^3 C   (b) 0.629 g Cd, 1.03 g NiO(OH), 0.202 g
H_2O; total mass of reactants = 1.86 g   (c) 10.1%   **21.150** Li >
Ba > Na > Al > Mn > Zn > Cr > Fe > Ni > Sn > Pb > Cu
> Ag > Hg > Au. Metals with potentials lower than that of wa-
ter (-0.83 V) can displace H_2 from water: Li, Ba, Na, Al, and
Mn. Metals with potentials lower than that of hydrogen (0.00 V)
can displace H_2 from acid: Li, Ba, Na, Al, Mn, Zn, Cr, Fe, Ni,
Sn, and Pb. Metals with potentials greater than that of hydrogen
(0.00 V) cannot displace H_2: Cu, Ag, Hg, and Au.
21.153(a) 5.3×10^{-11} (b) 0.20 V   (c) 0.43 V   (d) 8.2×10^{-4} M
NaOH   **21.156**(a) -1.18×10^5 kJ   (b) 1.20×10^4 L
(c) 2.31×10^6 s   (d) 3.84×10^3 kW·h   (e) 4.15×10^3 cents
21.158 2.94

Chapter 22

22.2 Fe from Fe_2O_3; Ca from $CaCO_3$; Na from NaCl; Zn from
ZnS   **22.3**(a) Differentiation refers to the processes involved in
the formation of Earth into regions (core, mantle, and crust) of
differing composition. Substances separated according to their
densities, with the more dense material in the core and the less
dense in the crust.   (b) O, Si, Al, and Fe   (c) O   **22.7** Plants
produced O_2, slowly increasing the oxygen concentration in the
atmosphere and creating an oxidative environment for metals.
The oxygen-free decay of plant and animal material created large
fossil fuel deposits.   **22.9** Fixation refers to the process of converting a substance in the atmosphere into a form more readily
usable by organisms. Carbon and nitrogen; fixation of carbon
dioxide gas by plants and fixation of nitrogen gas by nitrogen-
fixing bacteria.   **22.12** Atmospheric nitrogen is fixed by three
pathways: atmospheric, industrial, and biological. Atmospheric
fixation requires high-temperature reactions (e.g., initiated by
lightning) to convert N_2 into NO and other oxidized species.
Industrial fixation involves mainly the formation of ammonia,
NH_3, from N_2 and H_2. Biological fixation occurs in nitrogen-
fixing bacteria that live on the roots of legumes. Human activity
is an example of industrial fixation. It contributes about 17% of
the nitrogen fixed.   **22.14**(a) the atmosphere   (b) Plants excrete
acid from their roots to convert PO_4^{3-} ions into more soluble

$H_2PO_4^-$ ions, which the plant can absorb. Through excretion and
decay, organisms return soluble phosphate compounds to the
cycle.   **22.17**(a) 1.1×10^3 L   (b) 4.2×10^2 m^3 **22.18**(a) The
iron ions form an insoluble salt, $Fe_3(PO_4)_2$, that decreases the
yield of phosphorus.   (b) 8.8 t   **22.20**(a) Roasting involves
heating the mineral in air at high temperatures to convert the
mineral to the oxide.   (b) Smelting is the reduction of the metal
oxide to the free metal using heat and a reducing agent such as
coke.   (c) Flotation is a separation process in which the ore is
removed from the gangue by exploiting the difference in density
in the presence of detergent. The gangue sinks to the bottom and
the lighter ore-detergent mix is skimmed off the top.   (d) Refining is the final step in the purification process to yield the pure
metal.   **22.25**(a) Slag is a byproduct of steel-making and contains the impurity SiO_2.   (b) Pig iron is the impure product of iron
metallurgy (containing 3–4% C and other impurities).   (c) Steel
refers to iron alloyed with other elements to attain desirable
properties.   (d) The basic-oxygen process is used to purify pig
iron and obtain carbon steel.   **22.27** Iron and nickel are more
easily oxidized and less easily reduced than copper. They are
separated from copper in the roasting step and converted to slag.
In the electrorefining process, all three metals are in solution, but
only Cu^{2+} ions are reduced at the cathode to form Cu(s).
22.30 Le Châtelier's principle says that the system shifts toward
formation of K as the gaseous metal leaves the cell.
22.31(a) $E°_{half-cell} = -3.05$ V, -2.93 V, and -2.71 V for Li^+,
K^+, and Na^+, respectively. In all of these cases, it is energetically
more favorable to reduce H_2O to H_2 than to reduce M^+ to M.
(b) $2RbX + Ca \longrightarrow CaX_2 + 2Rb$, where $\Delta H = IE_1(Ca) +$
$IE_2(Ca) - 2IE_1(Rb) = 929$ kJ/mol. Based on the IEs and positive ΔH for the forward reaction, it seems more reasonable that
Rb metal will reduce Ca^{2+} than the reverse.   (c) If the reaction
is carried out at a temperature greater than the boiling point of
Rb, the product mixture will contain gaseous Rb, which can be
removed from the reaction vessel; this would cause a shift in
equilibrium to form more Rb as product.   (d) $2CsX + Ca \longrightarrow$
$CaX_2 + 2Cs$, where $\Delta H = IE_1(Ca) + IE_2(Ca) - 2IE_1(Cs) =$
983 kJ/mol. This reaction is more unfavorable than for Rb,
but Cs has a lower boiling point.   **22.32**(a) 4.5×10^4 L
(b) 1.30×10^8 C   (c) 1.69×10^6 s   **22.35**(a) Mg^{2+} is more difficult to reduce than H_2O, so $H_2(g)$ would be produced instead of
Mg metal. $Cl_2(g)$ forms at the anode due to overvoltage.
(b) The $\Delta H°_f$ of $MgCl_2(s)$ is -641.6 kJ/mol. High temperature
favors the reverse (endothermic) reaction, the formation of magnesium metal and chlorine gas.   **22.37**(a) Sulfur dioxide is the
reducing agent and is oxidized to the +6 state (SO_4^{2-}).
(b) $HSO_4^-(aq)$ (c) $H_2SeO_3(aq) + 2SO_2(g) + H_2O(l) \longrightarrow$
$Se(s) + 2HSO_4^-(aq) + 2H^+(aq)$ **22.42**(a) O.N. for Cu: in
Cu_2S, +1; in Cu_2O, +1; in Cu, 0   (b) Cu_2S is the reducing
agent, and Cu_2O is the oxidizing agent.   **22.44**(a) 1.3×10^6 C
(b) 1.2×10^3 A   **22.47** $2ZnS(s) + C(graphite) \longrightarrow 2Zn(s) +$
$CS_2(g)$; $\Delta G°_{rxn} = 463$ kJ. Since $\Delta G°_{rxn}$ is positive, this reaction is
not spontaneous at standard-state conditions. $2ZnO(s) + C(s) \longrightarrow$
$2Zn(s) + CO_2(g)$; $\Delta G°_{rxn} = 242.0$ kJ. This reaction is also not
spontaneous, but is less unfavorable.   **22.48** The formation of
sulfur trioxide is very slow at ordinary temperatures. Increasing
the temperature can speed up the reaction, but because the reaction is exothermic, increasing the temperature decreases the
yield. Adding a catalyst increases the rate of the reaction, so a
lower temperature can be used to enhance the yield.

22.51(a) Cl_2, H_2, and NaOH   (b) The mercury-cell process yields higher purity NaOH, but releases some Hg, which is discharged into the environment.   **22.52**(a) $\Delta G° = -142$ kJ; yes (b) The rate of the reaction is very slow at 25°C.   (c) $\Delta G°_{500} = -53$ kJ, so the reaction is spontaneous.   (d) $K_{25} = 7.8 \times 10^{24} > K_{500} = 3.8 \times 10^3$ (e) 1.05×10^3 K   **22.53** 7×10^2 lb Cl_2
22.57(a) $P_4O_{10}(s) + 6H_2O(l) \longrightarrow 4H_3PO_4(l)$ (b) 1.52
22.59(a) 9.006×10^9 g CO_2 (b) The 4.3×10^{10} g CO_2 produced by automobiles is much greater than that from the blast furnace.
22.61(a) If $[OH^-] > 1.1 \times 10^{-4}$ M (i.e., if pH > 10.04), $Mg(OH)_2$ will precipitate.   (b) 1 (To the correct number of significant figures, all the magnesium has precipitated.)
22.63(a) K_{25}(step 1) $= 1 \times 10^{168}$; K_{25}(side rxn) $= 7 \times 10^{228}$
(b) K_{900}(step 1) $= 4.6 \times 10^{49}$; K_{900}(side rxn) $= 1.4 \times 10^{63}$
(c) 7.2×10^7 dollars   (d) 5.2×10^7 dollars
22.67(1) $2H_2O(l) + 2FeS_2(s) + 7O_2(g) \longrightarrow$
$$2Fe^{2+}(aq) + 4SO_4{}^{2-}(aq) + 4H^+(aq)$$
increases acidity
(2) $4H^+(aq) + 4Fe^{2+}(aq) + O_2(g) \longrightarrow$
$$4Fe^{3+}(aq) + 2H_2O(l)$$
(3) $Fe^{3+}(aq) + 3H_2O(l) \longrightarrow Fe(OH)_3(s) + 3H^+(aq)$
increases acidity
(4) $8H_2O(l) + FeS_2(s) + 14Fe^{3+}(aq) \longrightarrow$
$$15Fe^{2+}(aq) + 2SO_4{}^{2-}(aq) + 16H^+(aq)$$
increases acidity
22.69 density of ferrite: 7.86 g/cm^3; density of austenite: 7.55 g/cm^3
22.71(a) Cathode: $Na^+(l) + e^- \longrightarrow Na(l)$
Anode: $4OH^-(l) \longrightarrow O_2(g) + 2H_2O(g) + 4e^-$
(b) 50%
22.74(a) $nCO_2(g) + nH_2O(l) \longrightarrow (CH_2O)_n(s) + nO_2(g)$
(b) 27 L   (c) 7.6×10^4 L   **22.75** 73 mg/L   **22.77** 891 kg Na_3AlF_6 **22.78**(a) 23.2 min   (b) 14 effusion steps
22.80(a) 1.890 t Al_2O_3 (b) 0.3339 t C   (c) 100%   (d) 74%
(e) 2.813×10^3 m^3 **22.85** Acid rain increases the leaching of phosphate into the groundwater, due to the protonation of $PO_4{}^{3-}$ to form $HPO_4{}^{2-}$ and $H_2PO_4{}^-$. As shown in calculations for parts (a) and (b), solubility of $Ca_3(PO_4)_2$ increases from 6.4×10^{-7} M (in pure water) to 1.1×10^{-2} M (in acidic rainwater).
(a) 6.4×10^{-7} M   (b) 1.1×10^{-2} M   **22.86**(a) 1.00 mol %
(b) 238.9 g/mol   **22.88** density of silver = 10.51 g/cm^3; density of sterling silver = 10.2 g/cm^3

Chapter 23

23.2(a) $1s^2 2s^2 2p^6 3s^2 3p^6 4s^2 3d^{10} 4p^6 5s^2 4d^x$
(b) $1s^2 2s^2 2p^6 3s^2 3p^6 4s^2 3d^{10} 4p^6 5s^2 4d^{10} 5p^6 6s^2 4f^{14} 5d^x$ **23.4** Five; examples are Mn, [Ar] $4s^2 3d^5$, and Fe^{3+}, [Ar] $3d^5$.   **23.6**(a) The elements should increase in size as they increase in mass from Period 5 to Period 6. Because 14 additional elements lie between Periods 5 and 6, the effective nuclear charge increases significantly; so the atomic size decreases, or "contracts." This effect is significant enough that Zr^{4+} and Hf^{4+} are almost the same size but differ greatly in atomic mass.   (b) The atomic size increases from Period 4 to Period 5, but stays fairly constant from Period 5 to Period 6.   (c) Atomic mass increases significantly from Period 5 to Period 6, but atomic radius (and thus volume) increases slightly, so Period 6 elements are very dense.   **23.9**(a) A paramagnetic substance is attracted to a magnetic field, while a diamagnetic substance is slightly repelled by one.   (b) Ions of transition elements often have half-filled d orbitals whose unpaired electrons make the ions paramagnetic. Ions of main-group

elements usually have a noble-gas configuration with no partially filled levels.   (c) Some d orbitals in the transition element ions are empty, which allows an electron from one d orbital to move to a slightly higher energy one. The energy required for this transition is small and falls in the visible wavelength range. All orbitals are filled in ions of main-group elements, so enough energy would have to be added to move an electron to the next principal energy level, not just another orbital within the same energy level. This amount of energy is very large and much greater than the visible range of wavelengths.
23.10(a) $1s^2 2s^2 2p^6 3s^2 3p^6 4s^2 3d^3$
(b) $1s^2 2s^2 2p^6 3s^2 3p^6 4s^2 3d^{10} 4p^6 5s^2 4d^1$ (c) [Xe] $6s^2 4f^{14} 5d^{10}$
23.12(a) [Xe] $6s^2 4f^{14} 5d^6$ (b) [Ar] $4s^2 3d^7$ (c) [Kr] $5s^1 4d^{10}$
23.14(a) [Ar], no unpaired electrons   (b) [Ar] $3d^9$, one unpaired electron   (c) [Ar] $3d^5$, five unpaired electrons   (d) [Kr] $4d^2$, two unpaired electrons   **23.16**(a) +5   (b) +4   (c) +7
23.18 Cr, Mo, and W   **23.20** In CrF_2, because the chromium is in a lower oxidation state.   **23.22** Atomic size increases slightly down a group of transition elements, but nuclear charge increases much more, so the first ionization energy generally increases. The reduction potential for Mo is lower, so it is more difficult to oxidize Mo than Cr. In addition, the ionization energy of Mo is higher than that of Cr, so it is more difficult to remove electrons from Mo.   **23.24** CrO_3, with Cr in a higher oxidation state, yields a more acidic aqueous solution.   **23.28**(a) seven   (b) This corresponds to a half-filled f subshell.   **23.30**(a) [Xe] $6s^2 5d^1$
(b) [Xe] $4f^1$ (c) [Rn] $7s^2 5f^{11}$ (d) [Rn] $5f^2$ **23.32**(a) Eu^{2+}: [Xe] $4f^7$; Eu^{3+}: [Xe] $4f^6$; Eu^{4+}: [Xe] $4f^5$. The stability of the half-filled f subshell makes Eu^{2+} most stable.   (b) Tb^{2+}: [Xe] $4f^9$; Tb^{3+}: [Xe] $4f^8$; Tb^{4+}: [Xe] $4f^7$. Tb should show a +4 oxidation state because that gives the half-filled subshell.
23.34 Gd has the electron configuration [Xe] $6s^2 4f^7 5d^1$ with eight unpaired electrons. Gd^{3+} has seven unpaired electrons: [Xe] $4f^7$.
23.36 Valence-state electronegativity is the apparent electronegativity of an element in a given oxidation state. As Mn is bonded to more O atoms in its different oxides, its O.N. becomes more positive. In solution, an H_2O molecule is attracted to the increasingly positive Mn, and one of its protons becomes easier to lose.
23.40 Mercury exists in the liquid state at room temperature due to relatively few and relatively weak interactions among Hg atoms. Mercury occurs in the +1 oxidation state, unusual for its group, because two mercury 1+ ions form a dimer, $Hg_2{}^{2+}$, by sharing their unpaired 6s electrons in a covalent bond.
23.42 $2Cr^{3+}(aq) + Cr(s) \longrightarrow 3Cr^{2+}$; $E° = +0.50$ V; yes, the reaction is spontaneous.   **23.44** $3Mn^{2+}(aq) + 2MnO_4{}^-(aq) + 2H_2O(l) \longrightarrow 5MnO_2(s) + 4H^+(aq)$; $E° = +0.28$ V; the reaction is spontaneous, so the Mn^{2+} will react with remaining $MnO_4{}^-$ to form MnO_2.   **23.47** The coordination number indicates the number of ligand atoms bonded to the metal ion. The oxidation number represents the number of electrons lost to form the ion. The coordination number is unrelated to the oxidation number.   **23.49** 2, linear; 4, tetrahedral or square planar; 6, octahedral   **23.52** The complex ion has a negative charge.
23.55(a) hexaaquanickel(II) chloride
(b) tris(ethylenediamine) chromium(III) perchlorate
(c) potassium hexacyanomanganate(II)   **23.57**(a) 2+, 6
(b) 3+, 6   (c) 2+, 6   **23.59**(a) potassium dicyanoargentate(I)
(b) sodium tetrachlorocadmate(II)   (c) tetraammineaquabromo-cobalt(III) bromide   **23.61**(a) 1+, 2   (b) 2+, 4   (c) 3+, 6
23.63(a) $[Zn(NH_3)_4]SO_4$ (b) $[Cr(NH_3)_5Cl]Cl_2$

(c) $Na_3[Ag(S_2O_3)_2]$ **23.65**(a) 4, two ions   (b) 6, three ions (c) 2, four ions   **23.67**(a) $[Cr(H_2O)_6]_2(SO_4)_3$ (b) $Ba[FeBr_4]_2$ (c) $[Pt(en)_2]CO_3$ **23.69**(a) 6, five ions   (b) 4, three ions   (c) 4, two ions   **23.71**(a) The nitrite ion forms linkage isomers because it can bind to the metal ion through the lone pair on the N atom or any lone pair on either O atom.

$$[\ddot{\text{O}}-\text{N}=\ddot{\text{O}}\,]^-$$

(b) Sulfur dioxide molecules form linkage isomers because the lone pair on the S atom or any lone pair on either O atom can bind the central metal ion.

$$\ddot{\text{O}}=\ddot{\text{S}}=\ddot{\text{O}}$$

(c) Nitrate ions have an N atom with no lone pair and three O atoms, all with lone pairs that can bond to the metal ion. But all of the O atoms are equivalent, so these ions do not form linkage isomers.

$$\left[\ddot{\text{O}}-\text{N}=\ddot{\text{O}} \atop :\ddot{\text{O}}: \right]^-$$

23.73(a) geometric isomerism

Br Br		Br NH₂CH₃
Pt	and	Pt
CH₃NH₂ NH₂CH₃		CH₃NH₂ Br
cis		*trans*

(b) geometric isomerism

H₃N NH₃		H₃N F
Pt	and	Pt
Cl F		Cl NH₃
cis		*trans*

(c) geometric isomerism

H₂O NH₃		H₂O NH₃		H₂O F
Pt	and	Pt	and	Pt
Cl F		F Cl		Cl NH₃

23.75(a) geometric isomerism

$\left[{\text{Cl} \atop \text{Br}} {\text{Cl} \atop \text{Br}} \right]^{2-}$	and	$\left[{\text{Cl} \atop \text{Br}} {\text{Br} \atop \text{Cl}} \right]^{2-}$
Pt		Pt
cis		*trans*

(b) linkage isomerism

$$\begin{bmatrix} & \text{NH}_3 & \\ \text{H}_3\text{N} & | & \text{NH}_3 \\ & \text{Cr} & \\ \text{H}_3\text{N} & | & \text{NO}_2 \\ & \text{NH}_3 & \end{bmatrix}^{2+} \quad \text{and} \quad \begin{bmatrix} & \text{NH}_3 & \\ \text{H}_3\text{N} & | & \text{NH}_3 \\ & \text{Cr} & \\ \text{H}_3\text{N} & | & \text{ONO} \\ & \text{NH}_3 & \end{bmatrix}^{2+}$$

(c) geometric isomerism

$$\begin{bmatrix} & \text{NH}_3 & \\ \text{H}_3\text{N} & | & \text{I} \\ & \text{Pt} & \\ \text{H}_3\text{N} & | & \text{I} \\ & \text{NH}_3 & \end{bmatrix}^{2+} \quad \text{and} \quad \begin{bmatrix} & \text{NH}_3 & \\ \text{H}_3\text{N} & | & \text{I} \\ & \text{Pt} & \\ \text{I} & | & \text{NH}_3 \\ & \text{NH}_3 & \end{bmatrix}^{2+}$$

cis   *trans*

23.77 The compound with the traditional formula is $CrCl_3\cdot 4NH_3$; the actual formula is $[Cr(NH_3)_4Cl_2]Cl$.   **23.79**(a) $K[Pd(NH_3)Cl_3]$ (b) $[PdCl_2(NH_3)_2]$ (c) $K_2[PdCl_6]$ (d) $[Pd(NH_3)_4Cl_2]Cl_2$

23.81(a) dsp^2 (b) sp^3 **23.84** absorption of orange or yellow light   **23.85**(a) The crystal field splitting energy (Δ) is the energy difference between the two sets of d orbitals that result from the bonding of ligands to a central transition metal atom.   (b) In an octahedral field of ligands, the ligands approach along the x, y, and z axes. The $d_{x^2-y^2}$ and d_{z^2} orbitals are located *along* the x, y, and z axes, so ligand interaction is higher in energy. The other orbital-ligand interactions are lower in energy because the d_{xy}, d_{yz}, and d_{xz} orbitals are located *between* the x, y, and z axes. (c) In a tetrahedral field of ligands, the ligands do not approach along the x, y, and z axes. The ligand interaction is greater for the d_{xy}, d_{yz}, and d_{xz} orbitals and lesser for the $d_{x^2-y^2}$ and d_{z^2} orbitals. Therefore, the crystal field splitting is reversed, and the d_{xy}, d_{yz}, and d_{xz} orbitals are higher in energy than the $d_{x^2-y^2}$ and d_{z^2} orbitals.   **23.88** If Δ is greater than E_{pairing}, electrons will pair their spins in the lower energy d orbitals before adding as unpaired electrons to the higher energy d orbitals. If Δ is less than E_{pairing}, electrons will add as unpaired electrons to the higher energy d orbitals before pairing in the lower energy d orbitals. **23.90**(a) no d electrons   (b) eight d electrons   (c) six d electrons **23.92**(a) five   (b) ten   (c) seven **23.94**

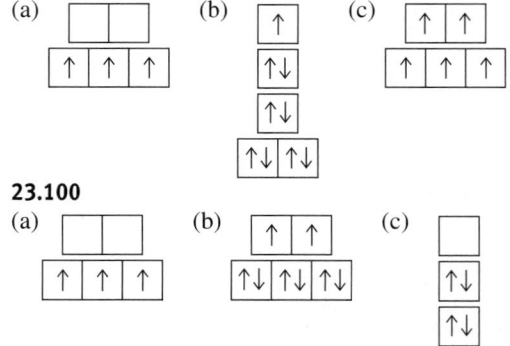

In an octahedral field of ligands, the ligands approach along the x, y, and z axes. The $d_{x^2-y^2}$ orbital is located along the x and y axes, so ligand interaction is greater. The d_{xy} orbital is offset from the x and y axes by 90°, so ligand interaction is less. The greater interaction of the $d_{x^2-y^2}$ orbital results in its higher energy. **23.96** (a) and (d) **23.98**

(a) [☐ ☐]   (b) [↑]   (c) [↑ | ↑]

[↑ | ↑ | ↑]   [↑↓]   [↑ | ↑ | ↑]

[↑↓]

[↑↓ | ↑↓]

23.100

(a) [☐ ☐]   (b) [↑ | ↑]   (c) [☐]

[↑ | ↑ | ↑]   [↑↓ | ↑↓ | ↑↓]   [↑↓]

[↑↓]

[↑↓ | ↑↓]

23.102 $[Cr(H_2O)_6]^{3+} < [Cr(NH_3)_6]^{3+} < [Cr(NO_2)_6]^{3-}$ **23.104** A violet complex absorbs yellow-green light. The light absorbed by a complex with a weaker field ligand would be at a lower energy and higher wavelength. Light of lower energy than yellow-green light is yellow, orange, or red. The color observed would be blue or green.   **23.107** The H_2O ligand is weaker than

the NH_3 ligand. The weaker field ligand results in a lower splitting energy and absorbs visible light of lower energy. The hexaaqua complex appears green because it absorbs red light. The hexaammine complex appears violet because it absorbs yellow light, which has higher energy (shorter λ) than red light. **23.111** Hg^+ is [Xe] $6s^1 4f^{14} 5d^{10}$ and Cu^+ is [Ar] $3d^{10}$. The mercury(I) ion has one electron in the $6s$ orbital that can form a covalent bond with the electron in the $6s$ orbital of another mercury(I) ion. In the copper(I) ion there are no electrons in the s orbital, so these ions cannot bond with one another. **23.113**(a) 6 (b) +3 (c) two (d) 1 mol
23.123 geometric (*cis-trans*) and linkage isomerism

cis-diamminedithiocyanatoplatinum(II)

trans-diamminedithiocyanatoplatinum(II)

cis-diamminediisothiocyanatoplatinum(II)

trans-diamminediisothiocyanatoplatinum(II)

cis-diammineisothiocyanatothiocyanatoplatinum(II)

trans-diammineisothiocyanatothiocyanatoplatinum(II)

23.125 $[Co(NH_3)_4(H_2O)Cl]^{2+}$

tetraammineaquachlorocobalt(III) ion

2 geometric isomers

cis   *trans*

$[Cr(H_2O)_3Br_2Cl]$
triaquadibromochlorochromium(III)

3 geometric isomers

Br's are *trans*   Br's are *cis*   Br's and H_2O's are *cis*
                     H_2O's are *cis* and *trans*

$[Cr(NH_3)_2(H_2O)_2Br_2]^+$
diamminediaquadibromochromium(III) ion

6 isomers (5 geometric)

all ligands are *trans*   only NH_3 is *trans*   only Br is *trans*

only H_2O is *trans*   optical isomers
                                all three ligands are *cis*

23.128(a) in MnO_4^{2-}, +6; in MnO_4^-, +7; in MnO_2, +4
(b) $3MnO_4^{2-}(aq) + 4H^+(aq) \longrightarrow$
$2MnO_4^-(aq) + MnO_2(s) + 2H_2O(l)$
23.135(a) no optical isomers (b) no optical isomers (c) no optical isomers (d) no optical isomers (e) optical isomers
23.136 $Pt[P(C_2H_5)_3]_2Cl_2$

cis-dichlorobis(triethylphosphine)platinum(II)

trans-dichlorobis(triethylphosphine)platinum(II)

23.140(a) The first reaction shows no change in the number of particles. In the second reaction, the number of reactant particles is greater than the number of product particles. A decrease in the number of particles means a decrease in entropy. Based on entropy change only, the first reaction is favored.
(b) The ethylenediamine complex will be more stable with respect to ligand exchange in water because the entropy change for that exchange is unfavorable.

Chapter 24

24.1(a) Chemical reactions are accompanied by relatively small changes in energy; nuclear reactions are accompanied by relatively large changes in energy. (b) Increasing temperature increases the rate of a chemical reaction but has no effect on a nuclear reaction. (c) Both chemical and nuclear reaction rates increase with higher reactant concentrations. (d) If the reactant is limiting in a chemical reaction, then more reactant produces more product and the yield increases. The presence of more radioactive reactant results in more decay product, so a higher reactant concentration increases the yield. **24.2**(a) 95.02% (b) The atomic mass is larger than the isotopic mass of ^{32}S. Sulfur-32 is the lightest isotope. **24.4**(a) Z down by 2, N down by 2 (b) Z up by 1, N down by 1 (c) no change in Z or N (d) Z down by 1, N up by 1 (e) Z down by 1, N up by 1; A different element is produced in all cases except c. **24.6** A neutron-rich nuclide decays by beta decay. A neutron-poor nuclide undergoes

positron decay or electron capture.   **24.8**(a) $^{234}_{92}\text{U} \longrightarrow ^{4}_{2}\alpha +$
$^{230}_{90}\text{Th}$ (b) $^{232}_{93}\text{Np} + ^{0}_{-1}\text{e} \longrightarrow ^{232}_{92}\text{U}$ (c) $^{12}_{7}\text{N} \longrightarrow ^{0}_{1}\beta + ^{12}_{6}\text{C}$
24.10(a) $^{27}_{12}\text{Mg} \longrightarrow ^{0}_{-1}\beta + ^{27}_{13}\text{Al}$ (b) $^{23}_{12}\text{Mg} \longrightarrow ^{0}_{1}\beta + ^{23}_{11}\text{Na}$
(c) $^{103}_{46}\text{Pd} + ^{0}_{-1}\text{e} \longrightarrow ^{103}_{45}\text{Rh}$ **24.12**(a) $^{48}_{23}\text{V} \longrightarrow ^{48}_{22}\text{Ti} + ^{0}_{1}\beta$
(b) $^{107}_{48}\text{Cd} + ^{0}_{-1}\text{e} \longrightarrow ^{107}_{47}\text{Ag}$ (c) $^{210}_{86}\text{Rn} \longrightarrow ^{206}_{84}\text{Po} + ^{4}_{2}\alpha$
24.14(a) $^{186}_{78}\text{Pt} + ^{0}_{-1}\text{e} \longrightarrow ^{186}_{77}\text{Ir}$ (b) $^{225}_{89}\text{Ac} \longrightarrow ^{221}_{87}\text{Fr} + ^{4}_{2}\alpha$
(c) $^{129}_{52}\text{Te} \longrightarrow ^{129}_{53}\text{I} + ^{0}_{-1}\beta$ **24.16**(a) Appears stable because its N
and Z values are both magic numbers, but its N/Z ratio (1.50) is
too high; it is unstable.   (b) Appears unstable because its Z value
is an odd number, but its N/Z ratio (1.19) is in the band of stability, so it is stable.   (c) Unstable because its N/Z ratio is too high.
24.18(a) The N/Z ratio for ^{127}I is 1.4; it is stable.   (b) The N/Z
ratio for ^{106}Sn is 1.1; it is unstable because this ratio is too low.
(c) The N/Z ratio is 1.1 for ^{68}As. The ratio is within the range of
stability, but the nuclide is most likely unstable because there is
an odd number of both protons and neutrons.   **24.20**(a) alpha decay   (b) positron decay or electron capture   (c) positron decay
or electron capture   **24.22**(a) beta decay   (b) positron decay or
electron capture   (c) alpha decay   **24.24** Stability results from a
favorable N/Z ratio, even numbered N and/or Z, and the occurrence of magic numbers. The N/Z ratio of ^{52}Cr is 1.17, which is
within the band of stability. The fact that Z is even does not account for the variation in stability because all isotopes of
chromium have the same Z. However, ^{52}Cr has 28 neutrons, so N
is both an even number and a magic number for this isotope only.
24.28 seven α emissions and four β^- emissions   **24.31** No, it is
not valid to conclude that $t_{1/2}$ equals 1 min because the number of
nuclei is so small. Decay rate is an average rate and is only meaningful when the sample is macroscopic and contains a large
number of nuclei. For the sample containing 6×10^{12} nuclei, the
conclusion is valid.   **24.33** 2.56×10^{-2} Ci/g   **24.35** 1.4×10^8 Bq/g
24.37 1×10^{-12} day^{-1} **24.39** 2.31×10^{-7} yr^{-1} **24.41** 1.49 mg
24.43 2.2×10^9 yr   **24.45** 30 dpm   **24.47** 1.0×10^6 yr
24.50 Both γ radiation and neutron beams have no charge, but
a neutron has a mass approximately equal to that of a proton.
24.52 Protons are repelled from the target nuclei due to interaction with like (positive) charges. Higher energy is required to
overcome the repulsion.   **24.53**(a) $^{10}_{5}\text{B} + ^{4}_{2}\alpha \longrightarrow ^{1}_{0}\text{n} + ^{13}_{7}\text{N}$
(b) $^{28}_{14}\text{Si} + ^{2}_{1}\text{H} \longrightarrow ^{1}_{0}\text{n} + ^{29}_{15}\text{P}$
(c) $^{242}_{96}\text{Cm} + ^{4}_{2}\alpha \longrightarrow 2^{1}_{0}\text{n} + ^{244}_{98}\text{Cf}$ **24.58** Ionizing radiation
is more dangerous to children because their rapidly dividing
cells are more susceptible to radiation than adults' slowly
dividing cells.   **24.60**(a) 5.4×10^{-7} rad   (b) 5.4×10^{-9} Gy
24.62(a) 7.5×10^{-10} Gy   (b) 7.5×10^{-5} mrem   (c) 7.5×10^{-10} Sv
24.65 1.86×10^{-3} rad   **24.67** NAA does not destroy the sample,
while chemical analyses do. Neutrons bombard a nonradioactive
sample, inducing some atoms within the sample to be radioactive. The radioisotopes decay by emitting radiation characteristic

of each isotope.   **24.73** Energy is released when a nuclide forms
from nucleons. The nuclear binding energy is the amount of energy holding the nucleus together. This energy must be absorbed
to break up the nucleus into nucleons and is released when nucle-
ons come together.   **24.75**(a) 1.861×10^4 eV   (b) 2.981×10^{15} J
24.77 7.6×10^{11} J   **24.79**(a) 7.976 MeV/nucleon   (b) 127.6
MeV/atom   (c) 1.231×10^{10} kJ/mol   **24.81**(a) 8.768 MeV/
nucleon   (b) 517.3 MeV/atom   (c) 4.99×10^{10} kJ/mol
24.85 Radioactive decay is a spontaneous process in which
unstable nuclei emit radioactive particles and energy. Fission
occurs as the result of high-energy bombardment of nuclei with
small particles that cause the nuclei to break into smaller
nuclides, radioactive particles, and energy. All fission events are
not the same. The nuclei split in a number of ways to produce
several different products.   **24.88** The water serves to slow the
neutrons so that they are better able to cause a fission reaction.
Heavy water is a better moderator because it does not absorb
neutrons as well as light water does; thus, more neutrons are
available to initiate the fission process. However, D_2O does not
occur naturally in great abundance, so its production adds to the
cost of a heavy-water reactor.   **24.91** In the s-process, a nucleus
captures a neutron, emitting a γ ray. Then the nucleus emits a β
particle to form another element. The stable isotopes of most
heavy elements form by the s-process. The r-process forms less
stable isotopes by multiple neutron captures, followed by multi-
ple β decays.   **24.95**(a) 1.1×10^{-29} kg   (b) 9.8×10^{-13} J
(c) 5.9×10^8 kJ/mol. This is approximately 1 million times larger
than a typical heat of reaction.   **24.97** 8.0×10^3 yr
24.101 1.35×10^{-5} M **24.103** 6.2×10^{-2} **24.106**(a) 5.99 h
(b) 21%   **24.108**(a) 0.999   (b) 0.298   (c) 5.58×10^{-6}
(d) Radiocarbon dating is more reliable for the fraction in part b
because a significant quantity of ^{14}C has decayed and a
significant quantity remains. Therefore, a change in the amount
of ^{14}C will be noticeable. For the fraction in part a, very little
^{14}C has decayed, and for c very little ^{14}C remains. In either case,
it will be more difficult to measure the change, so the error will
be relatively large.   **24.110** 6.58 h   **24.114** 4.904×10^{-9} L/h
24.117 81 yr   **24.119**(a) 0.15 Bq/L   (b) 0.27 Bq/L
(c) 3.3 days; a total of 12.8 days   **24.122** 7.4 s   **24.125** 1926
24.128 6.27×10^5 eV/reaction, 6.05×10^7 kJ/mol reaction
24.131(a) 2.07×10^{-17} J   (b) 1.45×10^7 H atoms
(c) 1.4960×10^{-5} J   (d) 1.4959×10^{-5} J   (e) No, the captain
should continue using the current technology.
24.134(a) 0.043 MeV, 2.9×10^{-11} m   (b) 4.713 MeV
24.138(a) 3.26×10^3 days   (b) 3.2×10^{-3} s   (c) 2.78×10^{11} yr
24.140(a) 1.80×10^{17} J   (b) 6.15×10^{16} kJ   (c) The procedure in
part (b) produces more energy per kilogram of antihydrogen.
24.141 9.316×10^2 MeV   **24.145** 0.357 day

Numbers in parentheses refer to the page(s) on which a term is introduced and/or discussed.

A

absolute scale (also *Kelvin scale*) The preferred temperature scale in scientific work, which has absolute zero (0 K, or $-273.15°C$) as the lowest temperature. (25) [See also *kelvin (K)*.]

absorption spectrum The spectrum produced when atoms absorb specific wavelengths of incoming light and become excited from lower to higher energy levels. (281)

abundance The amount of an element in a particular region of the natural world. (981)

accuracy The closeness of a measurement to the actual value. (31)

acid In common laboratory terms, any species that produces H^+ ions when dissolved in water. (150) [See also *Brønsted-Lowry, classical (Arrhenius),* and *Lewis acid-base definitions*.]

acid anhydride A compound, sometimes formed by a dehydration-condensation reaction of an oxoacid, that yields two molecules of the acid when it reacts with water. (658)

acid-base buffer (also *buffer*) A solution that resists changes in pH when a small amount of either strong acid or strong base is added. (832)

acid-base indicator A species whose color is different in acid and in base, which is used to monitor the equivalence point of a titration or the pH of a solution. (154, 793, 842)

acid-base reaction Any reaction between an acid and a base. (150) (See also *neutralization reaction*.)

acid-base titration curve A plot of the pH of a solution of acid (or base) versus the volume of base (or acid) added to the solution. (841)

acid-dissociation (acid-ionization) constant (K_a) An equilibrium constant for the dissociation of an acid (HA) in H_2O to yield the conjugate base (A^-) and H_3O^+:

$$K_a = \frac{[H_3O^+][A^-]}{[HA]} \qquad (786)$$

actinides The Period 7 elements that constitute the second inner transition series (5f block), which includes thorium (Th; $Z = 90$) through lawrencium (Lr; $Z = 103$). (316, 1030)

activated complex (See *transition state*.)

activation energy (E_a) The minimum energy with which molecules must collide to react. (706)

active site The region of an enzyme formed by specific amino acid side chains at which catalysis occurs. (723)

activity ($\mathscr{A}$) (also *decay rate*) The change in number of nuclei ($\mathscr{N}$) of a radioactive sample divided by the change in time (t). (1074)

activity series of the metals A listing of metals arranged in order of decreasing strength of the metal as a reducing agent in aqueous reactions. (171)

actual yield The amount of product actually obtained in a chemical reaction. (118)

addition polymer (also *chain-reaction,* or *chain-growth, polymer*) A polymer formed when monomers (usually containing $C=C$) combine through an addition reaction. (663)

addition reaction A type of organic reaction in which atoms linked by a multiple bond become bonded to more atoms. (647)

adduct The product of a Lewis acid-base reaction characterized by the formation of a new covalent bond. (817)

adenosine triphosphate (ATP) A high-energy molecule that serves most commonly as a store and source of energy in organisms. (908)

alchemy An occult study of nature that flourished for 1500 years in northern Africa and Europe and resulted in the development of several key laboratory methods. (10)

alcohol An organic compound (ending, *-ol*) that contains a $-\overset{|}{\underset{|}{C}}-\overset{..}{\underset{..}{O}}-H$ functional group. (650)

aldehyde An organic compound (ending, *-al*) that contains the carbonyl functional group ($C=\overset{..}{\underset{..}{O}}$) in which the carbonyl C is also bonded to H. (655)

alkane A hydrocarbon that contains only single bonds (general formula, C_nH_{2n+2}). (635)

alkene A hydrocarbon that contains at least one $C=C$ bond (general formula, C_nH_{2n}). (640)

alkyl group A saturated hydrocarbon chain with one bond available. (646)

alkyl halide (See *haloalkane*.)

alkyne A hydrocarbon that contains at least one $C\equiv C$ bond (general formula, C_nH_{2n-2}). (643)

allotrope One of two or more crystalline or molecular forms of an element. In general, one allotrope is more stable than another at a particular pressure and temperature. (585)

alloy A mixture with metallic properties that consists of solid phases of two or more pure elements, a solid-solid solution, or distinct intermediate phases. (369, 506, 997)

alpha (α) decay A radioactive process in which an alpha particle is emitted from a nucleus. (1069)

alpha particle (α or ^4_2He) A positively charged particle, identical to a helium nucleus, that is one of the common types of radioactive emissions. (1067)

amide An organic compound that contains the $-\overset{\overset{\displaystyle :O:}{\|}}{C}-\overset{|}{\underset{..}{N}}-$ functional group. (657)

amine An organic compound (general formula, $-\overset{|}{\underset{|}{C}}-\overset{..}{\underset{|}{N}}-$) derived structurally by replacing one or more H atoms of ammonia with alkyl groups; a weak organic base. (652, 806)

amino acid An organic compound [general formula, $H_2N-CH(R)-COOH$] with at least one carboxyl and one amine group on the same molecule; the monomer unit of a protein. (507, 667)

amorphous solid A solid that occurs in different shapes because it lacks extensive molecular-level ordering of its particles. (463)

ampere (A)   The SI unit of electric current; 1 ampere of current results when 1 coulomb flows through a conductor in 1 second. (965)

amphoteric   Able to act as either an acid or a base. (327)

amplitude   The height of the crest (or depth of the trough) of a wave; related to the intensity of the energy. (270)

angular momentum quantum number (*l*) (also *orbital-shape quantum number*)   An integer from 0 to $n - 1$ that is related to the shape of an atomic orbital. (289)

anion   A negatively charged ion. (60)

anode   The electrode at which oxidation occurs in an electrochemical cell. Electrons are given up by the reducing agent and leave the cell at the anode. (929)

antibonding MO   A molecular orbital formed when wave functions are subtracted from each other, which decreases electron density between the nuclei and leaves a node. Electrons occupying such an orbital destabilize the molecule. (422)

apatite   A compound of general formula $Ca_5(PO_4)_3X$, where X is generally F, Cl, or OH. (990)

aqueous solution   A solution in which water is the solvent. (75)

aromatic hydrocarbon   A compound of C and H with one or more rings of C atoms (often drawn with alternating C—C and C=C bonds), in which there is extensive delocalization of π electrons. (644)

Arrhenius acid-base definition   [See *classical (Arrhenius) acid-base definition.*]

Arrhenius equation   An equation that expresses the exponential relationship between temperature and the rate constant: $k = Ae^{-E_a/RT}$. (706)

atmosphere   The mixture of gases that extends from a planet's surface and eventually merges with outer space; the gaseous portion of Earth's crust. (218, 983) (For the unit, see *standard atmosphere.*)

atom   The smallest particle of an element that retains the chemical nature of the element. A neutral, spherical entity composed of a positively charged central nucleus surrounded by one or more negatively charged electrons. (47)

atomic mass (also *atomic weight*)   The average of the masses of the naturally occurring isotopes of an element weighted according to their abundances. (54)

atomic mass unit (amu) (also *dalton, Da*)   A mass exactly equal to $\frac{1}{12}$ the mass of a carbon-12 atom. (54)

atomic number (*Z*)   The unique number of protons in the nucleus of each atom of an element (equal to the number of electrons in the neutral atom). An integer that expresses the positive charge of a nucleus in multiples of the electronic charge. (53)

atomic orbital (also *wave function*)   A mathematical expression that describes the motion of the electron's matter-wave in terms of time and position in the region of the nucleus. The term is used qualitatively to mean the region of space in which there is a high probability of finding the electron. (287)

atomic size   A term referring to the atomic radius, one-half the distance between nuclei of identical bonded elements. (317, 555) (See also *covalent radius* and *metallic radius.*)

atomic solid   A solid consisting of individual atoms held together by dispersion forces; the frozen noble gases are the only examples. (470)

atomic symbol (also *element symbol*)   A one- or two-letter abbreviation for the English, Latin, or Greek name of an element. (53)

aufbau principle   The conceptual basis of a process of building up atoms by adding one proton (and one or more neutrons) at a time to the nucleus and one electron around it to obtain the ground-state electron configurations of the elements. (308)

autoionization (also *self-ionization*)   A reaction in which two molecules of a substance react to give ions. The most important example is for water:

$$2H_2O(l) \rightleftharpoons H_3O^+(aq) + OH^-(aq) \qquad (789)$$

average rate   The change in concentration of reactants (or products) divided by a finite time period. (688)

Avogadro's law   The gas law stating that, at fixed temperature and pressure, equal volumes of any ideal gas contain equal numbers of particles, and, therefore, the volume of a gas is directly proportional to its amount (mol): $V \propto n$. (197)

Avogadro's number   A number (6.022×10^{23} to four significant figures) equal to the number of atoms in exactly 12 g of carbon-12; the number of atoms, molecules, or formula units in one mole of an element or compound. (90)

axial group   An atom (or group) that lies above or below the trigonal plane of a trigonal bipyramidal molecule, or a similar structural feature in a molecule. (392)

B

background radiation   Natural ionizing radiation, the most important form of which is cosmic radiation. (1084)

balancing coefficient (also *stoichiometric coefficient*)   A numerical multiplier of all the atoms in the formula immediately following it in a chemical equation. (105)

band of stability   The narrow band of stable nuclides that appears on a plot of number of neutrons versus number of protons for all nuclides. (1071)

band theory   An extension of molecular orbital (MO) theory that explains many properties of metals, in particular, the differences in electrical conductivity of conductors, semiconductors, and insulators. (474)

barometer   A device used to measure atmospheric pressure. Most commonly, a tube open at one end, which is filled with mercury and inverted into a dish of mercury. (189)

base   In common laboratory terms, any species that produces OH^- ions when dissolved in water. (150) [See also *Brønsted-Lowry, classical (Arrhenius),* and *Lewis acid-base definitions.*]

base-dissociation (base-ionization) constant (*K*_b)   An equilibrium constant for the reaction of a base (B) with H_2O to yield the conjugate acid (BH^+) and OH^-:

$$K_b = \frac{[BH^+][OH^-]}{[B]} \qquad (805)$$

base pair   Two complementary bases in mononucleotides that are H bonded to each other; guanine (G) always pairs with cytosine (C), and adenine (A) always pairs with thymine (T) (or uracil, U). (671)

base unit (also *fundamental unit*)   A unit that defines the standard for one of the seven physical quantities in the International System of Units (SI). (18)

basic-oxygen process   The method used to convert pig iron to steel, in which O_2 is blown over and through molten iron to oxidize impurities and decrease the content of carbon. (1002)

battery   A self-contained group of voltaic cells arranged in series. (952)

becquerel (Bq) The SI unit of radioactivity; 1 Bq = 1 d/s (disintegration per second). (1076)

bent shape (also *V shape*) A molecular shape that arises when a central atom is bonded to two other atoms and has one or two lone pairs; occurs as the AX_2E shape class (bond angle < 120°) in the trigonal planar arrangement and as the AX_2E_2 shape class (bond angle < 109.5°) in the tetrahedral arrangement. (390)

beta (β) decay A radioactive process in which a beta particle is emitted from a nucleus. (1069)

beta particle (β, β⁻, or $_{-1}^{0}β$) A negatively charged particle identified as a high-speed electron that is one of the common types of radioactive emissions. (1067)

bimolecular reaction An elementary reaction involving the collision of two reactant species. (715)

binary covalent compound A compound that consists of atoms of two elements in which bonding occurs primarily through electron sharing. (70)

binary ionic compound A compound that consists of the oppositely charged ions of two elements. (60)

biomass conversion The process of applying chemical and biological methods to convert plant matter into fuels. (256)

biosphere The living systems that have inhabited Earth. (983)

blast furnace A tower-shaped furnace made of brick material in which intense heat and blasts of air are used to convert iron ore and coke to iron metal and carbon dioxide. (1000)

body-centered cubic unit cell A unit cell in which a particle lies at each corner and in the center of a cube. (464)

boiling point (bp or T_b) The temperature at which the vapor pressure of a gas equals the external (atmospheric) pressure. (447)

boiling point elevation (ΔT_b) The increase in the boiling point of a solvent caused by the presence of dissolved solute. (529)

bond angle The angle formed by the nuclei of two surrounding atoms with the nucleus of the central atom at the vertex. (389)

bond energy (BE) (also *bond enthalpy* or *bond strength*) The enthalpy change accompanying the breakage of a given bond in a mole of gaseous molecules. (353, 557)

bond length The distance between the nuclei of two bonded atoms. (354, 557)

bond order The number of electron pairs shared by two bonded atoms. (352, 557)

bonding MO A molecular orbital formed when wave functions are added to each other, which increases electron density between the nuclei. Electrons occupying such an orbital stabilize the molecule. (422)

bonding pair (also *shared pair*) An electron pair shared by two nuclei; the mutual attraction between the nuclei and the electron pair forms a covalent bond. (352)

Born-Haber cycle A series of hypothetical steps and their enthalpy changes needed to convert elements to an ionic compound and devised to calculate the lattice energy. (346)

Boyle's law The gas law stating that, at constant temperature and amount of gas, the volume occupied by a gas is inversely proportional to the applied (external) pressure: $V \propto 1/P$. (194)

branch A side chain appended to a polymer backbone. (488)

bridge bond (also *three-center, two-electron bond*) A covalent bond in which three atoms are held together by two electrons. (583)

Brønsted-Lowry acid-base definition A model of acid-base behavior based on proton transfer, in which an acid and a base are defined, respectively, as species that donate and accept a proton. (793)

buffer (See *acid-base buffer.*)

buffer capacity A measure of the ability of a buffer to resist a change in pH; related to the total concentrations and relative proportions of buffer components. (838)

buffer range The pH range over which a buffer acts effectively. (838)

C

calibration The process of correcting for systematic error of a measuring device by comparing it to a known standard. (32)

calorie (cal) A unit of energy defined as exactly 4.184 joules; originally defined as the heat needed to raise the temperature of 1 g of water 1°C (from 14.5°C to 15.5°C). (240)

calorimeter A device used to measure the heat released or absorbed by a physical or chemical process taking place within it. (247)

capillarity (or *capillary action*) A property that results in a liquid rising through a narrow space against the pull of gravity. (457)

carbon steel The steel that is produced by the basic-oxygen process, contains about 1% to 1.5% C and other impurities, and is alloyed with metals that prevent corrosion and increase strength. (1002)

carbonyl group The C=O grouping of atoms. (655)

carboxylic acid An organic compound (ending, *-oic acid*)

$$\text{that contains the} -\overset{\displaystyle :O:}{\overset{\displaystyle \|}{C}}-\ddot{O}H \text{ group. (657)}$$

catalyst A substance that increases the rate of a reaction without being used up in the process. (720)

cathode The electrode at which reduction occurs in an electrochemical cell. Electrons enter the cell and are acquired by the oxidizing agent at the cathode. (929)

cathode ray The ray of light emitted by the cathode (negative electrode) in a gas discharge tube; travels in straight lines, unless deflected by magnetic or electric fields. (49)

cation A positively charged ion. (60)

cell potential (E_{cell}) (also *electromotive force*, or *emf; cell voltage*) The potential difference between the electrodes of an electrochemical cell when no current flows. (934)

Celsius scale (formerly *centigrade scale*) A temperature scale in which the freezing and boiling points of water are defined as 0°C and 100°C, respectively. (25)

ceramic A nonmetallic, nonpolymeric solid that is hardened by heating it to high temperatures and, in most cases, consists of silicate microcrystals suspended in a glassy cementing medium. (481)

chain reaction In nuclear fission, a self-sustaining process in which neutrons released by splitting of one nucleus cause other nuclei to split, which releases more neutrons, and so on. (1094)

change in enthalpy (ΔH) The change in internal energy plus the product of the constant pressure and the change in volume: $\Delta H = \Delta E + P\Delta V$; the heat lost or gained at constant pressure: $\Delta H = q_P$. (243)

charge density The ratio of the charge of an ion to its volume. (516)

Charles's law The gas law stating that at constant pressure, the volume occupied by a fixed amount of gas is directly proportional to its absolute temperature: $V \propto T$. (195)

chelate A complex ion in which the metal ion is bonded to a bidentate or polydentate ligand. (1040)

chemical bond The force that holds two atoms together in a molecule (or formula unit). (60)

chemical change (also *chemical reaction*) A change in which a substance is converted into a substance with different composition and properties. (5)

chemical equation A statement that uses chemical formulas to express the identities and quantities of the substances involved in a chemical or physical change. (104)

chemical formula A notation of atomic symbols and numerical subscripts that shows the type and number of each atom in a molecule or formula unit of a substance. (64)

chemical kinetics The study of the rates and mechanisms of reactions. (685)

chemical property A characteristic of a substance that appears as it interacts with, or transforms into, other substances. (5)

chemical reaction (See *chemical change.*)

chemistry The scientific study of matter and its properties, the changes it undergoes, and the energy associated with those changes. (4)

chiral molecule One that is not superimposable on its mirror image; an optically active molecule. In organic compounds, a chiral molecule typically contains a C atom bonded to four different groups (asymmetric C). (639)

chlor-alkali process An industrial method that electrolyzes concentrated aqueous NaCl and produces Cl_2, H_2, and NaOH. (101)

chromatography A separation technique in which a mixture is dissolved in a fluid (gas or liquid), and the components are separated through differences in adsorption to (or solubility in) a solid surface (or viscous liquid). (77)

cis-trans isomers (See *geometric isomers.*)

classical (Arrhenius) acid-base definition A model of acid-base behavior in which an acid is a substance that has H in its formula and produces H^+ in water, and a base is a substance that has OH in its formula and produces OH^- in water. (784)

Clausius-Clapeyron equation An equation that expresses the relationship between vapor pressure P of a liquid and temperature T:

$$\ln P = \frac{-\Delta H_{vap}}{R}\left(\frac{1}{T}\right) + C, \text{ where } C \text{ is a constant} \quad (446)$$

coal gasification An industrial process for altering the large molecules in coal to sulfur-free gaseous fuels. (256)

colligative property A property of a solution that depends on the number, not the identity, of solute particles. (527) (See also *boiling point elevation, freezing point depression, osmotic pressure,* and *vapor pressure lowering.*)

collision frequency The average number of collisions per second that a particle undergoes. (217)

collision theory A model that explains reaction rate as the result of particles colliding with a certain minimum energy. (708)

colloid A suspension in which a solute-like phase is dispersed throughout a solvent-like phase. (539)

combustion The process of burning in air, often with release of heat and light. (11)

combustion analysis A method for determining the formula of a compound from the amounts of its combustion products; used commonly for organic compounds. (101)

common-ion effect The shift in the position of an ionic equilibrium away from an ion involved in the process that is caused by the addition or presence of that ion. (833)

complex (See *coordination compound.*)

complex ion An ion consisting of a central metal ion bonded covalently to molecules and/or anions called ligands (862, 1037)

composition The types and amounts of simpler substances that make up a sample of matter. (4)

compound A substance composed of two or more elements that are chemically combined in fixed proportions. (42)

concentration A measure of the quantity of solute dissolved in a given quantity of solution. (121)

concentration cell A voltaic cell in which both compartments contain the same components but at different concentrations. (948)

condensation The process of a gas changing into a liquid. (438)

condensation polymer A polymer formed by monomers with two functional groups that are linked together in a dehydration-condensation reaction. (664)

conduction band In band theory, the empty, higher energy portion of the band of molecular orbitals into which electrons move when conducting heat and electricity. (474)

conductor A substance (usually a metal) that conducts an electric current well. (475)

conjugate acid-base pair Two species related to each other through the gain or loss of a proton; the acid has one more proton than its conjugate base. (795)

constitutional isomers (also *structural isomers*) Compounds with the same molecular formula but different arrangements of atoms. (637, 1043)

contact process An industrial process for the manufacture of sulfuric acid based on the catalyzed oxidation of SO_2. (1012)

controlled experiment An experiment that measures the effect of one variable at a time by keeping other variables constant. (13)

conversion factor A ratio of equivalent quantities that is equal to 1 and used to convert the units of a quantity. (14)

coordinate covalent bond A covalent bond formed when one atom donates both electrons to provide the shared pair and, once formed, is identical to any covalent single bond. (1046)

coordination compound (also *complex*) A substance containing at least one complex ion. (1037)

coordination isomers Two or more coordination compounds with the same composition in which the complex ions have different ligand arrangements. (1043)

coordination number In a crystal, the number of nearest neighbors surrounding a particle. (464) In a complex, the number of ligand atoms bonded to the central metal ion. (1038)

copolymer A polymer that consists of two or more types of monomer. (489)

core The dense, innermost region of Earth. (981)

core electrons (See *inner electrons.*)

corrosion The natural redox process that results in unwanted oxidation of a metal. (956)

coulomb (C) The SI unit of electric charge. One coulomb is the charge of 6.242×10^{18} electrons; one electron possesses a charge of 1.602×10^{-19} C. (934)

Coulomb's law A law stating that the electrostatic force associated with two charges A and B is directly proportional to the product of their magnitudes and inversely proportional to the square of the distance between them:

$$\text{electrostatic force} \propto \frac{\text{charge A} \times \text{charge B}}{\text{distance}} \quad (348)$$

counter ion A simple ion associated with a complex ion in a co-ordination compound. (1037)

coupling of reactions The pairing of reactions of which one releases enough free energy for the other to occur. (907)

covalent bond A type of bond in which atoms are bonded through the sharing of two electrons; the mutual attraction of the nuclei and an electron pair that holds atoms together in a molecule. (62, 351)

covalent bonding The idealized bonding type that is based on localized electron-pair sharing between two atoms with little difference in their tendencies to lose or gain electrons (most commonly nonmetals). (342, 556)

covalent compound A compound that consists of atoms bonded together by shared electron pairs. (60)

covalent radius One-half the distance between nuclei of identical covalently bonded atoms. (318)

critical mass The minimum mass needed to achieve a chain reaction. (1095)

critical point The point on a phase diagram above which the vapor cannot be condensed to a liquid; the end of the liquid-gas curve. (448)

crosslink A branch that covalently joins one polymer chain to another. (488)

crust The thin, light, heterogeneous outer layer of Earth. (981)

crystal defect Any of a variety of disruptions in the regularity of a crystal structure. (477)

crystal field splitting energy (Δ) The difference in energy between two sets of metal-ion *d* orbitals that results from electrostatic interactions with the surrounding ligands. (1050)

crystal field theory A model that explains the color and magnetism of coordination compounds based on the effects of ligands on metal-ion *d*-orbital energies. (1048)

crystalline solid Solid with a well-defined shape because of the orderly arrangement of the atoms, molecules, or ions. (463)

crystallization A technique used to separate and purify the components of a mixture through differences in solubility in which a component comes out of solution as crystals. (76)

cubic closest packing A crystal structure based on the face-centered cubic unit cell in which the layers have an *abcabc . . .* pattern. (466)

cubic meter (m³) The SI-derived unit of volume. (19)

curie (Ci) The most common unit of radioactivity, defined as the number of nuclei disintegrating each second in 1 g of radium-226; $1 \text{ Ci} = 3.70 \times 10^{10}$ d/s (disintegrations per second). (1076)

cyclic hydrocarbon A hydrocarbon with one or more rings in its structure. (637)

D

d orbital An atomic orbital with $l = 2$. (294)

dalton (Da) A unit of mass identical to *atomic mass unit*. (54)

Dalton's law of partial pressures A gas law stating that, in a mixture of unreacting gases, the total pressure is the sum of the partial pressures of the individual gases: $P_{total} = P_1 + P_2 + \cdots$. (206)

data Pieces of quantitative information obtained by observation. (13)

de Broglie wavelength The wavelength of a moving particle obtained from the de Broglie equation: $\lambda = h/mu$. (284)

decay constant The rate constant k for radioactive decay. (1076)

decay series (also *disintegration series*) The succession of steps a parent nucleus undergoes as it decays into a stable daughter nucleus. (1074)

degree of polymerization (*n*) The number of monomer repeat units in a polymer chain. (484)

dehydration-condensation reaction A reaction in which H and OH groups on two molecules react to form water as one of the products. (602)

delocalization (See *electron-pair delocalization.*)

density (*d*) An intensive physical property of a substance at a given temperature and pressure, defined as the ratio of the mass to the volume: $d = m/V$. (22)

deposition The process of changing directly from gas to solid. (439)

derived unit Any of various combinations of the seven SI base units. (18)

desalination A process used to remove large amounts of ions from seawater, usually by reverse osmosis. (542)

deuterons Nuclei of the stable hydrogen isotope deuterium, ^{2}H. (1081)

diagonal relationship Physical and chemical similarities between a Period 2 element and one located diagonally down and to the right in Period 3. (575)

diamagnetism The tendency of a species not to be attracted (or to be slightly repelled) by a magnetic field as a result of its electrons being paired. (330, 1029)

differentiation The geochemical process of forming regions in Earth based on differences in composition and density. (981)

diffraction The phenomenon in which a wave striking the edge of an object bends around it. A wave passing through a slit as wide as its wavelength forms a circular wave. (273)

diffusion The movement of one fluid through another. (216)

dimensional analysis (also *factor-label method*) A calculation method in which arithmetic steps are accompanied by the appropriate canceling of units. (16)

dipole-dipole force The intermolecular attraction between oppositely charged poles of nearby polar molecules. (451)

dipole–induced dipole force The intermolecular attraction between a polar molecule and the oppositely charged pole it induces in a nearby molecule. (503)

dipole moment (μ) A measure of molecular polarity; the magnitude of the partial charges on the ends of a molecule (in coulombs) times the distance between them (in meters). (399)

disaccharide An organic compound formed by a dehydration-condensation reaction between two simple sugars (monosaccharides). (666)

disintegration series (See *decay series.*)

dispersion force (also *London force*) The intermolecular attraction between all particles as a result of instantaneous polarizations of their electron clouds; the intermolecular force primarily responsible for the condensed states of nonpolar substances. (454)

disproportionation A reaction in which a given substance is both oxidized and reduced. (600)

distillation A separation technique in which a more volatile component of a mixture vaporizes and condenses separately from the less volatile components. (76)

donor atom An atom that donates a lone pair of electrons to form a covalent bond, usually from ligand to metal ion in a complex. (1039)

doping Adding small amounts of other elements into the crystal structure of a semiconductor to enhance a specific property, usually conductivity. (477)

double bond A covalent bond that consists of two bonding pairs; two atoms sharing four electrons in the form of one σ and one π bond. (352, 419)

double helix The two intertwined polynucleotide strands held together by H bonds that form the structure of DNA (deoxyribonucleic acid). (513, 671)

Downs cell An industrial apparatus that electrolyzes molten NaCl to produce sodium and chlorine. (999)

dynamic equilibrium The condition at which the forward and reverse reactions are taking place at the same rate, so there is no net change in the amounts of reactants or products. (173)

E

e_g orbitals The set of orbitals (composed of $d_{x^2-y^2}$ and d_{z^2}) that results when the energies of the metal-ion d orbitals are split by a ligand field. This set is higher in energy than the other (t_{2g}) set in an octahedral field of ligands and lower in energy in a tetrahedral field. (1050)

effective collision A collision in which the particles meet with sufficient energy and an orientation that allows them to react. (710)

effective nuclear charge (Z_{eff}) The nuclear charge an electron actually experiences as a result of shielding effects due to the presence of other electrons. (306)

effusion The process by which a gas escapes from its container through a tiny hole into an evacuated space. (215)

elastomer A polymeric material that can be stretched and springs back to its original shape when released. (488)

electrochemical cell A system that incorporates a redox reaction to produce or use electrical energy. (923)

electrochemistry The study of the relationship between chemical change and electrical work. (923)

electrode The part of an electrochemical cell that conducts the electricity between the cell and the surroundings. (928)

electrolysis The nonspontaneous lysing (splitting) of a substance, often to its component elements, by supplying electrical energy. (961)

electrolyte A substance that conducts a current when it dissolves in water. (142, 527) A mixture of ions, in which the electrodes of an electrochemical cell are immersed, that conducts a current. (928)

electrolytic cell An electrochemical system that uses electrical energy to drive a nonspontaneous chemical reaction ($\Delta G > 0$). (928)

electromagnetic radiation (also *electromagnetic energy* or *radiant energy*) Oscillating, perpendicular electric and magnetic fields moving simultaneously through space as waves and manifested as visible light, x-rays, microwaves, radio waves, and so on. (269)

electromagnetic spectrum The continuum of wavelengths of radiant energy. (271)

electromotive force (emf) (See *cell potential*.)

electron (e$^-$) A subatomic particle that possesses a unit negative charge (-1.60218×10^{-19} C) and occupies the space around the atomic nucleus. (52)

electron affinity (EA) The energy change (in kJ) accompanying the addition of one mole of electrons to one mole of gaseous atoms or ions. (324)

electron capture A type of radioactive decay in which a nucleus draws in an orbital electron, usually one from the lowest energy level, and releases energy. (1069)

electron cloud An imaginary representation of an electron's rapidly changing position around the nucleus over time. (289)

electron configuration The distribution of electrons within the orbitals of the atoms of an element; also the notation for such a distribution. (303, 554)

electron deficient Referring to a bonded atom, such as Be or B, that has fewer than eight valence electrons. (384)

electron density diagram (also *electron probability density diagram*) The pictorial representation for a given energy sublevel of the quantity ψ^2 (the probability density of the electron lying within a particular tiny volume) as a function of r (distance from the nucleus). (288)

electron-pair delocalization (also *delocalization*) The process by which electron density is spread over several atoms rather than remaining between two. (382)

electron-sea model A qualitative description of metallic bonding proposing that metal atoms pool their valence electrons into a delocalized "sea" of electrons in which the metal cores (metal ions) are submerged in an orderly array. (369)

electron volt (eV) The energy (in joules, J) that an electron acquires when it moves through a potential difference of 1 volt; 1 eV = 1.602×10^{-19} J. (1092)

electronegativity (EN) The relative ability of a bonded atom to attract shared electrons. (363, 555)

electronegativity difference (ΔEN) The difference in electronegativities between the atoms in a bond. (366)

electrorefining An industrial electrolytic process in which a sample of impure metal acts as the anode and a sample of the pure metal acts as the cathode. (997)

element The simplest type of substance with unique physical and chemical properties. An element consists of only one kind of atom, so it cannot be broken down into simpler substances. (41)

elementary reaction (also *elementary step*) A simple reaction that describes a single molecular event in a proposed reaction mechanism. (715)

elimination reaction A type of organic reaction in which C atoms are bonded to fewer atoms in the product than in the reactant, which leads to multiple bonding. (648)

emission spectrum The line spectrum produced when excited atoms return to lower energy levels and emit photons characteristic of the element. (281)

empirical formula A chemical formula that shows the lowest relative number of atoms of each element in a compound. (64)

enantiomers (See *optical isomers*.)

end point The point in a titration at which the indicator changes color. (154, 844)

end-to-end overlap Overlap of s, p, or hybrid atomic orbitals that yields a sigma (σ) bond. (557)

endothermic process A process that occurs with an absorption of heat from the surroundings and therefore an increase in the enthalpy of the system ($\Delta H > 0$). (244)

energy The capacity to do work, that is, to move matter. (8) [See also *kinetic energy* (E_k) and *potential energy* (E_p).]

enthalpy (H) A thermodynamic quantity that is the sum of the internal energy plus the product of the pressure and volume. (243)

enthalpy diagram A graphic depiction of the enthalpy change of a system. (244)

enthalpy of hydration (ΔH_{hydr})   (See *heat of hydration.*)

enthalpy of solution (ΔH_{soln})   (See *heat of solution.*)

entropy (S)   A thermodynamic quantity related to the number of ways the energy of a system can be dispersed through the motions of its particles. (517, 884)

environmental cycle   The physical, chemical, and biological paths through which the atoms of an element move within Earth's crust. (986)

enzyme   A biological macromolecule (usually a protein) that acts as a catalyst. (723)

enzyme-substrate complex (ES)   The intermediate in an enzyme-catalyzed reaction, which consists of enzyme and substrate(s) and whose concentration determines the rate of product formation. (724)

equatorial group   An atom (or group) that lies in the trigonal plane of a trigonal bipyramidal molecule, or a similar structural feature in a molecule. (392)

equilibrium constant (K)   The value obtained when equilibrium concentrations are substituted into the reaction quotient. (740)

equivalence point   The point in a titration when the number of moles of the added species is stoichiometrically equivalent to the original number of moles of the other species. (154, 843)

ester   An organic compound that contains the $-\overset{\overset{\displaystyle :O:}{\|}}{C}-\overset{..}{\underset{..}{O}}-\overset{|}{\underset{|}{C}}-$ group. (657)

exact number   A quantity, usually obtained by counting or based on a unit definition, that has no uncertainty associated with it and, therefore, contains as many significant figures as a calculation requires. (30)

exchange reaction   (See *metathesis reaction.*)

excited state   Any electron configuration of an atom or molecule other than the lowest energy (ground) state. (278)

exclusion principle   A principle developed by Wolfgang Pauli stating that no two electrons in an atom can have the same set of four quantum numbers. The principle arises from the fact that an orbital has a maximum occupancy of two electrons and their spins are paired. (305)

exothermic process   A process that occurs with a release of heat to the surroundings and therefore a decrease in the enthalpy of the system ($\Delta H < 0$). (233)

expanded valence shell   A valence level that can accommodate more than eight electrons by using available d orbitals; occurs only with large central nonmetal atoms from Period 3 or higher. (385)

experiment   A clear set of procedural steps that tests a hypothesis. (13)

extensive property   A property, such as mass, that depends on the quantity of substance present. (22)

extraction   A method of separating the components of a mixture based on differences in their solubility in a nonmiscible solvent. (76)

F

face-centered cubic unit cell   A unit cell in which a particle occurs at each corner and in the center of each face of a cube. (464)

Faraday constant (F)   The physical constant representing the charge of 1 mol of electrons: $F = 96{,}485$ C/mol e$^-$. (943)

fatty acid   A carboxylic acid that has a long hydrocarbon chain and is derived from a natural source. (658)

filtration   A method of separating the components of a mixture on the basis of differences in particle size. (76)

first law of thermodynamics   (See *law of conservation of energy.*)

fission   The process by which a heavier nucleus splits into lighter nuclei with the release of energy. (1090)

fixation   A chemical/biochemical process that converts a gaseous substance in the environment into a condensed form that can be used by organisms. (987)

flame test   A procedure for identifying the presence of metal ions in which a granule of a compound or a drop of its solution is placed in a flame to observe a characteristic color. (281)

flotation   A metallurgical process in which oil and detergent are mixed with pulverized ore in water to create a slurry that separates the mineral from the gangue. (994)

formal charge   The hypothetical charge on an atom in a molecule or ion, equal to the number of valence electrons minus the sum of all the unshared and half the shared valence electrons. (383)

formation constant (K_f)   An equilibrium constant for the formation of a complex ion from the hydrated metal ion and ligands. (865)

formation equation   An equation in which 1 mole of a compound forms from its elements. (253)

formula mass   The sum (in amu) of the atomic masses of a formula unit of an ionic compound. (72)

formula unit   The chemical unit of a compound that contains the relative numbers of the types of atoms or ions expressed in the chemical formula. (66)

fossil fuel   Any fuel, including coal, petroleum, and natural gas, derived from the products of the decay of dead organisms. (256)

fraction by mass (also *mass fraction*)   The portion of a compound's mass contributed by an element; the mass of an element in a compound divided by the mass of the compound. (44)

fractional distillation   A process involving numerous vaporization-condensation steps used to separate two or more volatile components. (536)

free energy (G)   A thermodynamic quantity that is the difference between the enthalpy and the product of the absolute temperature and the entropy: $G = H - TS$. (899)

free radical   A molecular or atomic species with one or more unpaired electrons, which typically make it very reactive. (385, 1084)

freezing   The process of cooling a liquid until it solidifies. (438)

freezing point depression (ΔT_f)   A lowering of the freezing point of a solvent caused by the presence of dissolved solute particles. (531)

frequency (ν)   The number of cycles a wave undergoes per second, expressed in units of 1/second, or s^{-1} [also called *hertz* (Hz)]. (270)

frequency factor (A)   The product of the collision frequency Z and an orientation probability factor p that is specific for a reaction. (710)

fuel cell (also *flow battery*)   A battery that is not self-contained and in which electricity is generated by the controlled oxidation of a fuel. (955)

functional group   A specific combination of atoms, typically containing a carbon-carbon multiple bond and/or carbon-heteroatom bond, that reacts in a characteristic way no matter what molecule it occurs in. (631)

fundamental unit   (See *base unit.*)

fusion   (See *melting.*)

fusion (nuclear) The process by which light nuclei combine to form a heavier nucleus with the release of energy. (1090)

G

galvanic cell (See *voltaic cell*.)

gamma emission The type of radioactive decay in which gamma rays are emitted from an excited nucleus. (1070)

gamma (γ) ray A very high-energy photon. (1067)

gangue In an ore, the debris, such as sand, rock, and clay, attached to the mineral. (993)

gas One of the three states of matter. A gas fills its container regardless of the shape. (6)

Geiger-Müller counter An ionization counter that detects radioactive emissions through their ionization of gas atoms within the instrument. (1075)

genetic code The set of three-base sequences that is translated into specific amino acids during the process of protein synthesis. (671)

geometric isomers (also *cis-trans isomers* or *diastereomers*) Stereoisomers in which the molecules have the same connections between atoms but differ in the spatial arrangements of the atoms. The *cis* isomer has similar groups on the same side of a structural feature; the *trans* isomer has them on opposite sides. (641, 1044)

Graham's law of effusion A gas law stating that the rate of effusion of a gas is inversely proportional to the square root of its density (or molar mass):

$$\text{rate} \propto \frac{1}{\sqrt{\mathcal{M}}} \qquad (215)$$

gray (Gy) The SI unit of absorbed radiation dose; 1 Gy = 1 J/kg tissue. (1083)

green chemistry Field that is focused on developing methods to synthesize compounds efficiently and reduce or prevent the release of harmful products into the environment. (120)

ground state The electron configuration of an atom or ion that is lowest in energy. (278)

group A vertical column in the periodic table. (57)

H

Haber process An industrial process used to form ammonia from its elements. (771)

half-cell A portion of an electrochemical cell in which a half-reaction takes place. (930)

half-life ($t_{1/2}$) In chemical processes, the time required for half the initial reactant concentration to be consumed. (702) In nuclear processes, the time required for half the initial number of nuclei in a sample to decay. (1076)

half-reaction method A method of balancing redox reactions by treating the oxidation and reduction half-reactions separately. (924)

haloalkane (also *alkyl halide*) A hydrocarbon with one or more halogen atoms (X) in place of H; contains a $-\overset{|}{\underset{|}{C}}-\ddot{\underset{\cdot\cdot}{X}}\!:$ group. (652)

hard water Water that contains large amounts of divalent cations, especially Ca^{2+} and Mg^{2+}. (542)

heat (q) (also *thermal energy*) The energy transferred between objects because of differences in their temperatures only. (24, 238)

heat capacity The quantity of heat required to change the temperature of an object by 1 K. (246)

heat of formation (ΔH_f) The enthalpy change occurring when 1 mol of a compound is produced from its elements; designated ΔH_f° at the standard state. (245)

heat of fusion (ΔH_{fus}) The enthalpy change occurring when 1 mol of a solid substance melts; designated ΔH_{fus}° at the standard state. (245)

heat of hydration (ΔH_{hydr}) The enthalpy change occurring when 1 mol of a gaseous species is hydrated. The sum of the enthalpies from separating water molecules and mixing the gaseous species with them; designated ΔH_{hydr}° at the standard state. (516)

heat of reaction (ΔH_{rxn}) The enthalpy change of a reaction; designated ΔH_{rxn}° at the standard state. (244)

heat of solution (ΔH_{soln}) The enthalpy change occurring when a solution forms from solute and solvent. The sum of the enthalpies from separating solute and solvent molecules and mixing them; designated ΔH_{soln}° at the standard state. (515)

heat of sublimation (ΔH_{subl}) The enthalpy change occurring when 1 mol of a solid substance changes directly to a gas. The sum of the heats of fusion and vaporization; designated ΔH_{subl}° at the standard state. (439)

heat of vaporization (ΔH_{vap}) The enthalpy change occurring when 1 mol of a liquid substance vaporizes; designated ΔH_{vap}° at the standard state. (245)

heating-cooling curve A plot of temperature vs. time for a substance when heat is absorbed or released by the system at a constant rate. (440)

Henderson-Hasselbalch equation An equation for calculating the pH of a buffer system:

$$pH = pK_a + \log\left(\frac{[\text{base}]}{[\text{acid}]}\right) \qquad (837)$$

Henry's law A law stating that the solubility of a gas in a liquid is directly proportional to the partial pressure of the gas above the liquid: $S_{gas} = k_H \times P_{gas}$. (522)

Hess's law of heat summation A law stating that the enthalpy change of an overall process is the sum of the enthalpy changes of the individual steps of the process. (251)

heteroatom Any atom in an organic compound other than C or H. (631)

heterogeneous catalyst A catalyst that occurs in a different phase from the reactants, usually a solid interacting with gaseous or liquid reactants. (722)

heterogeneous mixture A mixture that has one or more visible boundaries among its components. (75)

hexagonal closest packing A crystal structure based on the hexagonal unit cell in which the layers have an *abab . . .* pattern. (466)

high-spin complex Complex ion that has the same number of unpaired electrons as in the isolated metal ion; contains weak-field ligands. (1052)

homogeneous catalyst A catalyst (gas, liquid, or soluble solid) that exists in the same phase as the reactants. (721)

homogeneous mixture (also *solution*) A mixture that has no visible boundaries among its components. (75)

homologous series A series of organic compounds in which each member differs from the next by a $-CH_2-$ (methylene) group. (635)

homonuclear diatomic molecule A molecule composed of two identical atoms. (424)

Hund's rule A principle stating that when orbitals of equal energy are available, the electron configuration of lowest energy has the maximum number of unpaired electrons with parallel spins. (309)

hybrid orbital   An atomic orbital postulated to form during bonding by the mathematical mixing of specific combinations of nonequivalent orbitals in a given atom. (412)

hybridization   A postulated process of orbital mixing to form hybrid orbitals. (412)

hydrate   A compound in which a specific number of water molecules are associated with each formula unit. (69)

hydration   Solvation in water. (515)

hydration shell   The oriented cluster of water molecules that surrounds an ion in aqueous solution. (502)

hydrocarbon   An organic compound that contains only H and C atoms. (632)

hydrogen bond (H bond)   A type of dipole-dipole force that arises between molecules that have an H atom bonded to a small, highly electronegative atom with lone pairs, usually N, O, or F. (452, 563)

hydrogenation   The addition of hydrogen to a carbon-carbon multiple bond to form a carbon-carbon single bond. (722)

hydrolysis   Cleaving a molecule by reaction with water, in which one part of the molecule bonds to the water —OH and the other to the water H. (659)

hydronium ion (H_3O^+)   A proton covalently bonded to a water molecule. (784)

hydrosphere   The liquid portion of Earth's crust. (983)

hypothesis   A testable proposal made to explain an observation. If inconsistent with experimental results, a hypothesis is revised or discarded. (13)

I

ideal gas   A hypothetical gas that exhibits linear relationships among volume, pressure, temperature, and amount (mol) at all conditions; approximated by simple gases at ordinary conditions. (193)

ideal gas law (also *ideal gas equation*)   An equation that expresses the relationships among volume, pressure, temperature, and amount (mol) of an ideal gas: $PV = nRT$. (198)

ideal solution   A solution whose vapor pressure equals the mole fraction of the solvent times the vapor pressure of the pure solvent; approximated only by very dilute solutions. (528) (See also *Raoult's law.*)

indicator   (See *acid-base indicator.*)

induced-fit model   A model of enzyme action that pictures the binding of the substrate as inducing the active site to change its shape and become catalytically active. (724)

infrared (IR)   The region of the electromagnetic spectrum between the microwave and visible regions. (271)

infrared (IR) spectroscopy   An instrumental technique for determining the types of bonds in a covalent molecule by measuring the absorption of IR radiation. (357)

initial rate   The instantaneous rate occurring as soon as the reactants are mixed, that is, at $t = 0$. (689)

inner electrons (also *core electrons*)   Electrons that fill all the energy levels of an atom except the valence level; electrons also present in atoms of the previous noble gas and any completed transition series. (315)

inner transition elements   The elements of the periodic table in which *f* orbitals are being filled; the lanthanides and actinides. (316, 1030)

instantaneous rate   The reaction rate at a particular time, given by the slope of a tangent to a plot of reactant concentration vs. time. (689)

insulator   A substance (usually a nonmetal) that does not conduct an electric current. (475)

integrated rate law   A mathematical expression for reactant concentration as a function of time. (699)

intensive property   A property, such as density, that does not depend on the quantity of substance present. (22)

interhalogen compound   A compound consisting entirely of halogens. (614)

intermolecular forces (also *interparticle forces*)   The attractive and repulsive forces among the particles—molecules, atoms, or ions—in a sample of matter. (437)

internal energy (E)   The sum of the kinetic and potential energies of all the particles in a system. (226)

ion   A charged particle that forms from an atom (or covalently bonded group of atoms) when it gains or loses one or more electrons. (60)

ion-dipole force   The intermolecular attractive force between an ion and a polar molecule (dipole). (450)

ion exchange   A process of softening water by exchanging one type of ion (usually Ca^{2+}) for another (usually Na^+) by binding the ions on a specially designed resin. (542)

ion–induced dipole force   The intermolecular attractive force between an ion and the dipole it induces in the electron cloud of a nearby particle. (502)

ion pair   A pair of ions that form a gaseous ionic molecule; sometimes formed when a salt boils. (350)

ion-product constant for water (K_w)   The equilibrium constant for the autoionization of water:

$$K_w = [H_3O^+][OH^-] \qquad (789)$$

ionic atmosphere   A cluster of ions of net opposite charge surrounding a given ion in solution. (537)

ionic bonding   The idealized type of bonding based on the attraction of oppositely charged ions that arise through electron transfer between atoms with large differences in their tendencies to lose or gain electrons (typically metals and nonmetals). (342, 556)

ionic compound   A compound that consists of oppositely charged ions. (60, 563)

ionic radius   The size of an ion as measured by the distance between the centers of adjacent ions in a crystalline ionic compound. (332)

ionic solid   A solid whose unit cell contains cations and anions. (471)

ionization   The process by which a substance absorbs energy from high-energy radioactive particles and loses an electron to become ionized. (1082)

ionization energy (IE)   The energy (in kJ) required to remove completely one mole of electrons from one mole of gaseous atoms or ions. (321, 555)

ionizing radiation   The high-energy radiation that forms ions in a substance by causing electron loss. (1082)

isoelectronic   Having the same number and configuration of electrons as another species. (328)

isomer   One of two or more compounds with the same molecular formula but different properties, often as a result of different arrangements of atoms. (103, 637, 1043)

isotopes   Atoms of a given atomic number (that is, of a specific element) that have different numbers of neutrons and therefore different mass numbers. (53, 1066)

isotopic mass   The mass (in amu) of an isotope relative to the mass of carbon-12. (54)

J

joule (J)   The SI unit of energy; $1\ J = 1\ kg\cdot m^2/s^2$. (240)

K

kelvin (K)   The SI base unit of temperature. The kelvin is the same size as the Celsius degree. (25)

Kelvin scale   (See *absolute scale*.)

ketone   An organic compound (ending, *-one*) that contains a carbonyl group bonded to two other C atoms, $-\overset{|}{\underset{|}{C}}-\overset{:O:}{\underset{}{\overset{||}{C}}}-\overset{|}{\underset{|}{C}}-$. (655)

kilogram (kg)   The SI base unit of mass. (21)

kinetic energy (E_k)   The energy an object has because of its motion. (8)

kinetic-molecular theory   The model that explains gas behavior in terms of particles in random motion whose volumes and interactions are negligible. (210)

L

lanthanide contraction   The additional decrease in atomic and ionic size, beyond the expected trend, caused by the poor shielding of the increasing nuclear charge by *f* electrons in the elements following the lanthanides. (1026)

lanthanides (also *rare earths*)   The Period 6 (4*f*) series of inner transition elements, which includes cerium (Ce; $Z = 58$) through lutetium (Lu; $Z = 71$). (316, 1030)

lattice   The three-dimensional arrangement of points created by choosing each point to be at the same location within each particle of a crystal; thus, the lattice consists of all points with identical surroundings. (464)

lattice energy ($\Delta H^\circ_{lattice}$)   The enthalpy change (always positive) that accompanies a solid ionic compound forming separate gaseous ions. (346)

law   (See *natural law*.)

law of chemical equilibrium (also *law of mass action*)   The law stating that when a system reaches equilibrium at a given temperature, the ratio of quantities that make up the reaction quotient has a constant numerical value. (741)

law of conservation of energy (also *first law of thermodynamics*)   A basic observation that the total energy of the universe is constant: $\Delta E_{universe} = \Delta E_{system} + \Delta E_{surroundings} = 0$. (240)

law of definite (or constant) composition   A mass law stating that, no matter what its source, a particular compound is composed of the same elements in the same parts (fractions) by mass. (44)

law of mass action   (See *law of chemical equilibrium*.)

law of mass conservation   A mass law stating that the total mass of substances does not change during a chemical reaction. (44)

law of multiple proportions   A mass law stating that if elements A and B react to form two compounds, the different masses of B that combine with a fixed mass of A can be expressed as a ratio of small whole numbers. (46)

Le Châtelier's principle   A principle stating that if a system in a state of equilibrium is disturbed, it will undergo a change that shifts its equilibrium position in a direction that reduces the effect of the disturbance. (761)

leaching   A hydrometallurgical process that extracts a metal selectively, usually through formation of a complex ion. (994)

level (also *shell*)   A specific energy state of an atom given by the principal quantum number *n*. (290)

leveling effect   The inability of a solvent to distinguish the strength of an acid (or base) that is stronger than the conjugate acid (or conjugate base) of the solvent. (816)

Lewis acid-base definition   A model of acid-base behavior in which acids and bases are defined, respectively, as species that accept and donate an electron pair. (817)

Lewis electron-dot symbol   A notation in which the element symbol represents the nucleus and inner electrons, and surrounding dots represent the valence electrons. (343)

Lewis structure (also *Lewis formula*)   A structural formula consisting of electron-dot symbols, with lines as bonding pairs and dots as lone pairs. (378)

ligand   A molecule or anion bonded to a central metal ion in a complex ion. (862, 1037)

like-dissolves-like rule   An empirical observation stating that substances having similar kinds of intermolecular forces dissolve in each other. (502)

limiting reactant (also *limiting reagent*)   The reactant that is consumed when a reaction occurs and therefore the one that determines the maximum amount of product that can form. (114)

line spectrum   A series of separated lines of different colors representing photons whose wavelengths are characteristic of an element. (276) (See also *emission spectrum*.)

linear arrangement   The geometric arrangement obtained when two electron groups maximize their separation around a central atom. (389)

linear shape   A molecular shape formed by three atoms lying in a straight line, with a bond angle of $180°$ (shape class AX_2 or AX_2E_3). (389, 393)

linkage isomers   Coordination compounds with the same composition but with different ligand donor atoms linked to the central metal ion. (1044)

lipid   Any of a class of biomolecules, including fats and oils, that are soluble in nonpolar solvents. (658)

lipid bilayer   An extended sheetlike double layer of phospholipid molecules that forms in water and has the charged heads of the molecules on the surfaces of the bilayer and the nonpolar tails within the interior. (511)

liquid   One of the three states of matter. A liquid fills a container to the extent of its own volume and thus forms a surface. (6)

liquid crystal   A substance that flows like a liquid but packs like a crystalline solid at the molecular level. (478)

liter (L)   A non-SI unit of volume equivalent to 1 cubic decimeter ($0.001\ m^3$). (19)

lithosphere   The solid portion of Earth's crust. (983)

lock-and-key model   A model of enzyme function that pictures the enzyme active site and the substrate as rigid shapes that fit together as a lock and key, respectively. (724)

lone pair (also *unshared pair*)   An electron pair that is part of an atom's valence shell but not involved in covalent bonding. (352)

low-spin complex Complex ion that has fewer unpaired electrons than in the free metal ion because of the presence of strong-field ligands. (1052)

M

macromolecule (See *polymer*.)

magnetic quantum number (m_l) (also *orbital-orientation quantum number*) An integer from $-l$ through 0 to $+l$ that specifies the orientation of an atomic orbital in the three-dimensional space about the nucleus. (289)

manometer A device used to measure the pressure of a gas in a laboratory experiment. (191)

mantle A thick homogeneous layer of Earth's internal structure that lies between the core and the crust. (981)

mass The quantity of matter an object contains. Balances are designed to measure mass. (21)

mass-action expression (See *reaction quotient*.)

mass fraction (See *fraction by mass*.)

mass number (A) The total number of protons and neutrons in the nucleus of an atom. (53)

mass percent (also *mass %* or *percent by mass*) The fraction by mass expressed as a percentage. (44) A concentration term [% (w/w)] expressed as the mass in grams of solute dissolved per 100. g of solution. (524)

mass spectrometry An instrumental method for measuring the relative masses of particles in a sample by creating charged particles and separating them according to their mass-charge ratio. (54)

matter Anything that possesses mass and occupies volume. (4)

mean free path The average distance a molecule travels between collisions at a given temperature and pressure. (217)

melting (also *fusion*) The change of a substance from a solid to a liquid. (438)

melting point (mp or T_f) The temperature at which the solid and liquid forms of a substance are at equilibrium. (447)

metabolic pathway A biochemical reaction sequence that flows in one direction and in which each reaction is enzyme catalyzed. (770)

metal A substance or mixture that is relatively shiny and malleable and is a good conductor of heat and electricity. In reactions, metals tend to transfer electrons to nonmetals and form ionic compounds. (58)

metallic bonding An idealized type of bonding based on the attraction between metal ions and their delocalized valence electrons. (343, 556) (See also *electron-sea model*.)

metallic radius One-half the distance between the nuclei of adjacent individual atoms in a crystal of an element. (317)

metallic solid A solid whose individual atoms are held together by metallic bonding. (472)

metalloid (also *semimetal*) An element with properties between those of metals and nonmetals. (58)

metallurgy The branch of materials science concerned with the extraction and utilization of metals. (993)

metathesis reaction (also *double-displacement reaction* or *exchange reaction*) A reaction in which atoms or ions of two compounds exchange bonding partners. (148)

meter (m) The SI base unit of length. The distance light travels in a vacuum in 1/299,792,458 second. (18)

methanogenesis The process of producing methane by anaerobic biodegradation of plant and animal waste. (257)

milliliter (mL) A volume (0.001 L) equivalent to 1 cm^3. (19)

millimeter of mercury (mmHg) A unit of pressure based on the difference in the heights of mercury in a barometer or manometer. Renamed the *torr* in honor of Torricelli. (191)

mineral The portion of an ore that contains the element of interest, a mineral is a naturally occurring, homogeneous, crystalline inorganic solid, with a well-defined composition. (993)

miscible Soluble in any proportion. (502)

mixture A group of two or more elements and/or compounds that are physically intermingled. (40, 489)

MO bond order One-half the difference between the number of electrons in bonding and antibonding MOs. (411)

model (also *theory*) A simplified conceptual picture based on experiment that explains how an aspect of nature occurs. (11)

molality (m) A concentration term expressed as number of moles of solute dissolved in 1000 g (1 kg) of solvent. (511)

molar heat capacity (C) The quantity of heat required to change the temperature of 1 mol of a substance by 1 K. (235)

molar mass ($\mathcal{M}$) (also *gram-molecular weight*) The mass of 1 mol of entities (atoms, molecules, or formula units) of a substance, in units of g/mol. (89)

molar solubility The solubility expressed in terms of amount (mol) of dissolved solute per liter of solution. (834)

molarity (M) A concentration term expressed as the moles of solute dissolved in 1 L of solution. (121, 522)

mole (mol) The SI base unit for amount of a substance. The amount that contains a number of objects equal to the number of atoms in exactly 12 g of carbon-12. (90)

mole fraction (X) A concentration term expressed as the ratio of moles of one component of a mixture to the total moles present. (206, 525)

molecular compounds Compounds that consist of individual molecules and whose physical state depends on intermolecular forces. (563)

molecular equation A chemical equation showing a reaction in solution in which reactants and products appear as intact, undissociated compounds. (145)

molecular formula A formula that shows the actual number of atoms of each element in a molecule. (65, 99)

molecular mass (also *molecular weight*) The sum (in amu) of the atomic masses of a formula unit of a compound. (72)

molecular orbital (MO) An orbital of given energy and shape that extends over a molecule and can be occupied by no more than two electrons. (422)

molecular orbital (MO) diagram A depiction of the relative energy and number of electrons in each MO, as well as the atomic orbitals from which the MOs form. (423)

molecular orbital (MO) theory A model that describes a molecule as a collection of nuclei and electrons in which the electrons occupy orbitals that extend over the entire molecule. (422)

molecular polarity The overall distribution of electronic charge in a molecule, determined by its shape and bond polarities. (399)

molecular shape The three-dimensional structure defined by the relative positions of the atomic nuclei in a molecule. (388)

molecular solid A solid held together by intermolecular forces between individual molecules. (470)

molecularity The number of reactant particles involved in an elementary step. (715)

molecule A structure consisting of two or more atoms that are chemically bound together and behave as an independent unit. (42)

monatomic ion An ion derived from a single atom. (60)

monomer A small molecule, linked covalently to others of the same or similar type to form a polymer, on which the repeat unit of the polymer is based. (484)

mononucleotide A monomer unit of a nucleic acid, consisting of an N-containing base, a sugar, and a phosphate group. (512, 670)

monosaccharide A simple sugar; a polyhydroxy ketone or aldehyde with three to nine C atoms. (513, 666)

N

nanotechnology The science and engineering of nanoscale (1–100 nm) systems. (489)

natural law (also *law*) A summary, often in mathematical form, of a universal observation. (13)

Nernst equation An equation stating that the voltage of an electrochemical cell under any conditions depends on the standard cell voltage and the concentrations of the cell components:

$$E_{cell} = E^{\circ}_{cell} - \frac{RT}{nF} \ln Q \qquad (946)$$

net ionic equation A chemical equation of a reaction in solution in which spectator ions have been eliminated to show the actual chemical change. (146)

network covalent solid A solid in which all the atoms are bonded covalently. (472, 563)

neutralization reaction An acid-base reaction that yields water and a solution of a salt; when a strong acid reacts with a stoichiometrically equivalent amount of a strong base, the solution is neutral. (150, 784)

neutron (n⁰) An uncharged subatomic particle found in the nucleus, with a mass slightly greater than that of a proton. (52)

nitrile An organic compound containing the $-C\equiv N:$ group. (661)

node A region of an orbital where the probability of finding the electron is zero. (293)

nonbonding MO A molecular orbital that is not involved in bonding. (429)

nonelectrolyte A substance whose aqueous solution does not conduct an electric current. (144, 527)

nonmetal An element that lacks metallic properties. In reactions, nonmetals tend to bond with each other to form covalent compounds or accept electrons from metals to form ionic compounds. (58)

nonpolar covalent bond A covalent bond between identical atoms such that the bonding pair is shared equally. (365)

nuclear binding energy The energy required to break 1 mole of nuclei of an element into individual nucleons. (1092)

nuclear magnetic resonance (NMR) spectroscopy An instrumental technique used to determine the molecular environment of a given type of nucleus, most often ¹H, from its absorption of radio waves in a magnetic field. (646)

nuclear transmutation The induced conversion of one nucleus into another by bombardment with a particle. (1080)

nucleic acid An unbranched polymer consisting of mononucleotides that occurs as two types, DNA and RNA (deoxyribonucleic and ribonucleic acids), which differ chemically in the nature of the sugar portion of the mononucleotides. (512, 670)

nucleon A subatomic particle that makes up a nucleus; a proton or neutron. (1066)

nucleus The tiny central region of the atom that contains all the positive charge and essentially all the mass. (51)

nuclide A nuclear species with specified numbers of protons and neutrons. (1066)

O

observation A fact obtained with the senses, often with the aid of instruments. Quantitative observations provide data that can be compared objectively. (13)

occurrence (also *source*) The form(s) in which an element exists in nature. (985)

octahedral arrangement The geometric arrangement obtained when six electron groups maximize their space around a central atom; when all six groups are bonding groups, the molecular shape is octahedral (AX_6; ideal bond angle = 90°). (393)

octet rule The observation that when atoms bond, they often lose, gain, or share electrons to attain a filled outer shell of eight electrons. (344)

optical isomers (also *enantiomers*) A pair of stereoisomers consisting of a molecule and its mirror image that cannot be superimposed on each other. (639, 1044)

optically active Able to rotate the plane of polarized light. (640)

orbital diagram A depiction of electron number and spin in an atom's orbitals by means of arrows in a series of small boxes, lines, or circles. (308)

ore A naturally occurring compound or mixture of compounds from which an element can be profitably extracted. (985)

organic compound A compound in which carbon is nearly always bonded to itself and to hydrogen, and often to other elements. (629)

organometallic compound An organic compound in which carbon is bonded covalently to a metal atom. (656)

osmosis The process by which solvent flows through a semipermeable membrane from a dilute to a concentrated solution. (532)

osmotic pressure (Π) The pressure that results from the inability of solute particles to cross a semipermeable membrane. The pressure required to prevent the net movement of solvent across the membrane. (532)

outer electrons Electrons that occupy the highest energy level (highest n value) and are, on average, farthest from the nucleus. (315)

overall (net) equation A chemical equation that is the sum of two or more balanced sequential equations in which a product of one becomes a reactant for the next. (112)

overvoltage The additional voltage, usually associated with gaseous products, that is required above the standard cell voltage to accomplish electrolysis. (963)

oxidation The loss of electrons by a species, accompanied by an increase in oxidation number. (159)

oxidation number (O.N.) (also *oxidation state*) A number equal to the magnitude of the charge an atom would have if its shared electrons were held completely by the atom that attracts them more strongly. (159)

oxidation number method A method for balancing redox reactions in which the change in oxidation numbers is used to determine balancing coefficients. (163)

oxidation-reduction reaction (also *redox reaction*) A process in which there is a net movement of electrons from one reactant (reducing agent) to another (oxidizing agent). (158)

oxidation state   (See *oxidation number.*)

oxidizing agent   The substance that accepts electrons in a reaction and undergoes a decrease in oxidation number. (159)

oxoanion   An anion in which an element is bonded to one or more oxygen atoms. (68)

P

***p* block**   The main-group elements that lie to the right of the transition elements in the periodic table and whose *p* orbitals are being filled. (554)

***p* orbital**   An atomic orbital with $l = 1$. (293)

packing efficiency   The percentage of the available volume occupied by atoms, ions, or molecules in a unit cell. (466)

paramagnetism   The tendency of a species with unpaired electrons to be attracted by an external magnetic field. (330, 1029)

partial ionic character   An estimate of the actual charge separation in a bond (caused by the electronegativity difference of the bonded atoms) relative to complete separation. (366)

partial pressure   The portion of the total pressure contributed by a gas in a mixture of gases. (205)

particle accelerator   A device used to impart high kinetic energies to nuclear particles. (1081)

pascal (Pa)   The SI unit of pressure; $1 \text{ Pa} = 1 \text{ N/m}^2$. (191)

penetration   The process by which an outer electron moves through the region occupied by the core electrons to spend part of its time closer to the nucleus; penetration increases the average effective nuclear charge for that electron. (307)

percent by mass (mass %)   (See *mass percent.*)

percent yield (% yield)   The actual yield of a reaction expressed as a percentage of the theoretical yield. (118)

period   A horizontal row of the periodic table. (57)

periodic law   A law stating that when the elements are arranged by atomic number, they exhibit a periodic recurrence of properties. (303)

periodic table of the elements   A table in which the elements are arranged by atomic number into columns (groups) and rows (periods). (57)

pH   The negative common logarithm of $[\text{H}_3\text{O}^+]$. (791)

phase   A physically distinct portion of a system. (437)

phase change   A physical change from one phase to another, usually referring to a change in physical state. (437)

phase diagram   A diagram used to describe the stable phases and phase changes of a substance as a function of temperature and pressure. (448)

phlogiston theory   An incorrect theory of combustion proposing that a burning substance releases the undetectable material phlogiston. (11)

photoelectric effect   The observation that when monochromatic light of sufficient frequency shines on a metal, an electric current is produced. (274)

photon   A quantum of electromagnetic radiation. (275)

photovoltaic cell   A device capable of converting light directly into electricity. (259)

physical change   A change in which the physical form (or state) of a substance, but not its composition, is altered. (4)

physical property   A characteristic shown by a substance itself, without interacting with or changing into other substances. (4)

pi (π) bond   A covalent bond formed by sideways overlap of two atomic orbitals that has two regions of electron density, one above and one below the internuclear axis. (419)

pi (π) MO   A molecular orbital formed by combination of two atomic (usually *p*) orbitals whose orientations are perpendicular to the internuclear axis. (425)

Planck's constant (*h*)   A proportionality constant relating the energy and the frequency of a photon, equal to 6.626×10^{-34} J·s. (274)

plastic   A material that, when deformed, retains its new shape. (488)

polar covalent bond   A covalent bond in which the electron pair is shared unequally, so the bond has partially negative and partially positive poles. (365)

polar molecule   A molecule with an unequal distribution of charge as a result of its polar bonds and shape. (141)

polarimeter   A device used to measure the rotation of plane-polarized light by an optically active compound. (640)

polarizability   The ease with which a particle's electron cloud can be distorted. (454)

polyatomic ion   An ion in which two or more atoms are bonded covalently. (63)

polymer (also *macromolecule*)   An extremely large molecule that results from the covalent linking of many simpler molecular units (monomers). (484)

polyprotic acid   An acid with more than one ionizable proton. (803)

polysaccharide   A macromolecule composed of many simple sugars linked covalently. (513, 666)

positron (β⁺)   The antiparticle of an electron. (1069)

positron emission   A type of radioactive decay in which a positron is emitted from a nucleus. (1069)

potential energy (E_p)   The energy an object has as a result of its position relative to other objects or because of its composition. (8)

precipitate   The insoluble product of a precipitation reaction. (146)

precipitation reaction   A reaction in which two soluble ionic compounds form an insoluble product, a precipitate. (146)

precision (also *reproducibility*)   The closeness of a measurement to other measurements of the same phenomenon in a series of experiments. (31)

pressure (*P*)   The force exerted per unit of surface area. (189)

pressure-volume work (*PV* work)   A type of work in which a volume change occurs against an external pressure. (239)

principal quantum number (*n*)   A positive integer that specifies the energy and relative size of an atomic orbital. (289)

probability contour   A shape that defines the volume around an atomic nucleus within which an electron spends a given percentage of its time. (289)

product   A substance formed in a chemical reaction. (105)

property   A characteristic that gives a substance its unique identity. (4)

protein   A natural, linear polymer composed of any of about 20 types of amino acid monomers linked together by peptide bonds. (507, 667)

proton (p⁺)   A subatomic particle found in the nucleus that has a unit positive charge (1.60218×10^{-19} C). (52)

proton acceptor   A substance that accepts an H^+ ion; a Brønsted-Lowry base. (794)

proton donor   A substance that donates an H^+ ion; a Brønsted-Lowry acid. (793)

pseudo–noble gas configuration The $(n-1)d^{10}$ configuration of a p-block metal atom that has emptied its outer energy level. (328)

Q

quantum A packet of energy equal to hv. The smallest quantity of energy that can be emitted or absorbed. (274)

quantum mechanics The branch of physics that examines the wave motion of objects on the atomic scale. (287)

quantum number A number that specifies a property of an orbital or an electron. (274)

R

rad (radiation-absorbed dose) The quantity of radiation that results in 0.01 J of energy being absorbed per kilogram of tissue; 1 rad = 0.01 J/kg tissue = 10^{-2} Gy. (1083)

radial probability distribution plot The graphic depiction of the total probability distribution (sum of ψ^2) of an electron in the region near the nucleus. (289)

radioactivity The emissions resulting from the spontaneous disintegration of an unstable nucleus. (1066)

radioisotope An isotope with an unstable nucleus that decays through radioactive emissions. (1078)

radioisotopic dating A method for determining the age of an object based on the rate of decay of a particular radioactive nuclide. (1078)

radius of gyration (R_g) A measure of the size of a coiled polymer chain as the average distance from the center of mass of the chain to its outside edge. (486)

random coil The shape adopted by most polymer chains and caused by random rotation about the bonds joining the repeat units. (485)

random error Error that occurs in all measurements (with its size depending on the measurer's skill and the instrument's precision) and results in values *both* higher and lower than the actual value. (31)

Raoult's law A law stating that the vapor pressure of a solution is directly proportional to the mole fraction of solvent: $P_{\text{solvent}} = X_{\text{solvent}} \times P°_{\text{solvent}}$. (528)

rare earths (See *lanthanides*.)

rate constant (k) The proportionality constant that relates the reaction rate to reactant (and product) concentrations. (692)

rate-determining step (also *rate-limiting step*) The slowest step in a reaction mechanism and therefore the step that limits the overall rate. (716)

rate law (also *rate equation*) An equation that expresses the rate of a reaction as a function of reactant (and product) concentrations. (691)

reactant A starting substance in a chemical reaction. (105)

reaction energy diagram A graph that shows the potential energy of a reacting system as it progresses from reactants to products. (712)

reaction intermediate A substance that is formed and used up during the overall reaction and therefore does not appear in the overall equation. (715)

reaction mechanism A series of elementary steps that sum to the overall reaction and is consistent with the rate law. (714)

reaction order The exponent of a reactant concentration in a rate law that shows how the rate is affected by changes in that concentration. (692)

reaction quotient (Q) (also *mass-action expression*) A ratio of terms for a given reaction consisting of product concentrations multiplied together and divided by reactant concentrations multiplied together, each raised to the power of their balancing coefficient. The value of Q changes until the system reaches equilibrium, at which point it equals K. (741)

reaction rate The change in the concentrations of reactants (or products) with time. (687)

reactor core The part of a nuclear reactor that contains the fuel rods and generates heat from fission. (1096)

redox reaction (See *oxidation-reduction reaction*.)

reducing agent The substance that donates electrons in a redox reaction and undergoes an increase in oxidation number. (159)

reduction The gain of electrons by a species, accompanied by a decrease in oxidation number. (159)

refraction A phenomenon in which a wave changes its speed and therefore its direction as it passes through a phase boundary. (272)

rem (roentgen equivalent for man) The unit of radiation dosage for a human based on the product of the number of rads and a factor related to the biological tissue; 1 rem = 10^{-2} Sv. (1083)

resonance hybrid The weighted average of the resonance structures of a molecule. (382)

resonance structure (also *resonance form*) One of two or more Lewis structures for a molecule that cannot be adequately depicted by a single structure. Resonance structures differ only in the position of bonding and lone electron pairs. (381)

reverse osmosis A process for preparing drinkable water that uses an applied pressure greater than the osmotic pressure to remove ions from an aqueous solution, typically seawater. (542)

rms (root-mean-square) speed (u_{rms}) The speed of a molecule having the average kinetic energy; very close to the most probable speed. (215)

roasting A pyrometallurgical process in which metal sulfides are converted to oxides. (994)

round off The process of removing digits based on a series of rules to obtain an answer with the proper number of significant figures (or decimal places). (29)

S

s block The main-group elements that lie to the left of the transition elements in the periodic table and whose s orbitals are being filled. (554)

s orbital An atomic orbital with $l = 0$. (292)

salt An ionic compound that results from a classical acid-base reaction. (153)

salt bridge An inverted U tube containing a solution of a nonreacting electrolyte that connects the compartments of a voltaic cell and maintains neutrality by allowing ions to flow between compartments. (930)

saturated hydrocarbon A hydrocarbon in which each C is bonded to four other atoms. (635)

saturated solution A solution that contains the maximum amount of dissolved solute at a given temperature in the presence of undissolved solute. (519)

scanning tunneling microscopy An instrumental technique that uses electrons moving across a minute gap to observe the topography of a surface on the atomic scale. (468)

Schrödinger equation An equation that describes how the electron matter-wave changes in space around the nucleus. Solutions of the equation provide allowable energy levels of the H atom. (287)

scientific method   A process of creative thinking and testing aimed at objective, verifiable discoveries of the causes of natural events. (13)

scintillation counter   A device used to measure radioactivity through its excitation of atoms and their subsequent emission of light. (1075)

screening   (See *shielding*.)

second (s)   The SI base unit of time. (26)

second law of thermodynamics   A law stating that a process occurs spontaneously in the direction that increases the entropy of the universe. (887)

seesaw shape   A molecular shape caused by the presence of one equatorial lone pair in a trigonal bipyramidal arrangement (AX_4E). (392)

selective precipitation   The process of separating ions through differences in the solubility of their compounds with a given precipitating ion. (861)

self-ionization   (See *autoionization*.)

semiconductor   A substance whose electrical conductivity is poor at room temperature but increases significantly with rising temperature. (475)

semimetal   (See *metalloid*.)

semipermeable membrane   A membrane that allows solvent, but not solute, to pass through. (532)

shared pair   (See *bonding pair*.)

shell   (See *level*.)

shielding (also *screening*)   The ability of other electrons, especially those occupying inner orbitals, to lessen the nuclear attraction for an outer electron. (306)

SI unit   A unit composed of one or more of the base units of the Système International d'Unités, a revised metric system. (18)

side reaction   An undesired chemical reaction that consumes some of the reactant and reduces the overall yield of the desired product. (118)

side-to-side overlap   Overlap of p with p (or sometimes with d) atomic orbitals that leads to a pi (π) bond. (557)

sievert (Sv)   The SI unit of human radiation dosage; 1 Sv = 100 rem. (1083)

sigma (σ) bond   A type of covalent bond that arises through end-to-end orbital overlap and has most of its electron density along the bond axis. (419)

sigma (σ) MO   A molecular orbital that is cylindrically symmetrical about an imaginary line that runs through the nuclei of the component atoms. (423)

significant figures   The digits obtained in a measurement. The greater the number of significant figures, the greater the certainty of the measurement. (28)

silicate   A type of compound found throughout rocks and soil and consisting of repeating —Si—O groupings and, in most cases, metal cations. (590)

silicone   A type of synthetic polymer containing —Si—O repeat units, with organic groups and crosslinks. (590)

simple cubic unit cell   A unit cell in which a particle occurs at each corner of a cube. (464)

single bond   A bond that consists of one electron pair. (352)

slag   A molten waste product formed in a blast furnace by the reaction of acidic silica with a basic metal oxide. (1001)

smelting   Heating a mineral with a reducing agent, such as coke, to obtain a metal. (995)

soap   The salt of a fatty acid and a strong base, usually a Group 1A(1) or 2A(2) hydroxide. (510)

solid   One of the three states of matter. A solid has a fixed shape that does not conform to the container shape. (6)

solubility (S)   The maximum amount of solute that dissolves in a fixed quantity of a particular solvent at a specified temperature when excess solute is present. (502)

solubility-product constant (K_{sp})   An equilibrium constant for the dissolving of a slightly soluble ionic compound in water. (852)

solute   The substance that dissolves in the solvent. (121, 502)

solution   (See *homogeneous mixture*.)

solvated   Surrounded closely by solvent molecules. (142)

solvation   The process of surrounding a solute particle with solvent particles. (515)

solvent   The substance in which the solute(s) dissolve. (121, 502)

source   (See *occurrence*.)

sp hybrid orbital   An orbital formed by the mixing of one s and one p orbital of a central atom. (412)

sp^2 hybrid orbital   An orbital formed by the mixing of one s and two p orbitals of a central atom. (414)

sp^3 hybrid orbital   An orbital formed by the mixing of one s and three p orbitals of a central atom. (414)

sp^3d hybrid orbital   An orbital formed by the mixing of one s, three p, and one d orbital of a central atom. (415)

sp^3d^2 hybrid orbital   An orbital formed by the mixing of one s, three p, and two d orbitals of a central atom. (416)

specific heat capacity (c)   The quantity of heat required to change the temperature of 1 gram of a substance by 1 K. (246)

spectator ion   An ion that is present as part of a reactant but is not involved in the chemical change. (146)

spectrochemical series   A ranking of ligands in terms of their ability to split d-orbital energies. (1051)

spectrophotometry   A group of instrumental techniques that create an electromagnetic spectrum to measure the atomic and molecular energy levels of a substance. (281)

speed of light (c)   A fundamental constant giving the speed at which electromagnetic radiation travels in a vacuum: $c = 2.99792458 \times 10^8$ m/s. (270)

spin quantum number (m_s)   A number, either $+\frac{1}{2}$ or $-\frac{1}{2}$, that indicates the direction of electron spin. (305)

spontaneous change   A change that occurs by itself, that is, without an ongoing input of energy. (881)

square planar shape   A molecular shape (AX_4E_2) caused by the presence of two lone pairs at opposite vertices in an octahedral arrangement. (394)

square pyramidal shape   A molecular shape (AX_5E) caused by the presence of one lone pair in an octahedral arrangement. (393)

standard atmosphere (atm)   The average atmospheric pressure measured at sea level and 0°C, defined as 1.01325×10^5 Pa. (191)

standard cell potential ($E°_{cell}$)   The potential of a cell measured with all components in their standard states and no current flowing. (935)

standard electrode potential ($E°_{half-cell}$) (also *standard half-cell potential*)   The standard potential of a half-cell, with the half-reaction written as a reduction. (935)

standard entropy of reaction (ΔS°_{rxn}) The entropy change that occurs when all components are in their standard states. (893)

standard free energy change (ΔG°) The free energy change that occurs when all components are in their standard states. (900)

standard free energy of formation (ΔG°_f) The standard free energy change that occurs when 1 mol of a compound is made from its elements. (901)

standard half-cell potential (See *standard electrode potential*.)

standard heat of formation (ΔH°_f) The enthalpy change that occurs when 1 mol of a compound forms from its elements, with all substances in their standard states. (253)

standard heat of reaction (ΔH°_{rxn}) The enthalpy change that occurs during a reaction, with all substances in their standard states. (253)

standard hydrogen electrode (See *standard reference half-cell*.)

standard molar entropy (S°) The entropy of 1 mol of a substance in its standard state. (888)

standard molar volume The volume of 1 mol of an ideal gas at standard temperature and pressure: 22.4141 L. (197)

standard reference half-cell (also *standard hydrogen electrode*) A specially prepared platinum electrode immersed in 1 M H$^+$(*aq*) through which H$_2$ gas at 1 atm is bubbled. $E^\circ_{half-cell}$ is defined as 0 V. (936)

standard states A set of specifications used to compare thermodynamic data: 1 atm for gases behaving ideally, 1 M for dissolved species, or the pure substance for liquids and solids. (253)

standard temperature and pressure (STP) The reference conditions for a gas:

$$0°C \ (273.15 \ K) \ and \ 1 \ atm \ (760 \ torr) \qquad (197)$$

state function A property of the system determined by its current state, regardless of how it arrived at that state. (241)

state of matter One of the three physical forms of matter: solid, liquid, or gas. (6)

stationary state In the Bohr model, one of the allowable energy levels of the atom in which it does not release or absorb energy. (277)

steel An alloy of iron with small amounts of carbon and usually other metals. (1000)

stellar nucleosynthesis The process by which elements are formed in the stars through nuclear fusion. (1099)

stereoisomers Molecules with the same connections of atoms but different orientations of groups in space. (639, 1044) (See also *geometric isomers* and *optical isomers*.)

stoichiometric coefficient (See *balancing coefficient*.)

stoichiometry The study of the mass-mole-number relationships of chemical formulas and reactions. (90)

strong-field ligand A ligand that causes larger crystal field splitting energy and therefore is part of a low-spin complex. (1050)

strong force An attractive force that exists between all nucleons and is about 100 times stronger than the electrostatic repulsive force. (1072)

structural formula A formula that shows the actual number of atoms, their relative placement, and the bonds between them. (65)

structural isomers (See *constitutional isomers*.)

sublevel (also *subshell*) An energy substate of an atom within a level. Given by the n and l values, the sublevel designates the size and shape of the atomic orbitals. (290)

sublimation The process by which a solid changes directly into a gas. (439)

substance A type of matter, either an element or a compound, that has a fixed composition. (41)

substitution reaction An organic reaction that occurs when an atom (or group) from one reactant substitutes for one in another reactant. (648)

substrate A reactant that binds to the active site in an enzyme-catalyzed reaction. (723)

superconductivity The ability to conduct a current with no loss of energy to resistive heating. (475)

supersaturated solution An unstable solution in which more solute is dissolved than in a saturated solution. (519)

surface tension The energy required to increase the surface area of a liquid by a given amount. (457)

surroundings All parts of the universe other than the system being considered. (236)

suspension A heterogeneous mixture containing particles that are distinct from the surrounding medium. (539)

syngas Synthesis gas; a combustible mixture of CO and H$_2$. (256)

synthetic natural gas (SNG) A gaseous fuel mixture, mostly methane, formed from coal. (256)

system The defined part of the universe under study. (236)

systematic error A type of error producing values that are all either higher or lower than the actual value, often caused by faulty equipment or a consistent fault in technique. (31)

T

t_{2g} orbitals The set of orbitals (composed of d_{xy}, d_{yz}, and d_{xz}) that results when the energies of the metal-ion d orbitals are split by a ligand field. This set is lower in energy than the other (e_g) set in an octahedral field and higher in energy in a tetrahedral field. (1050)

T shape A molecular shape caused by the presence of two equatorial lone pairs in a trigonal bipyramidal arrangement (AX$_3$E$_2$). (393)

temperature (T) A measure of how hot or cold a substance is relative to another substance. (24)

tetrahedral arrangement The geometric arrangement formed when four electron groups maximize their separation around a central atom; when all four groups are bonding groups, the molecular shape is tetrahedral (AX$_4$; ideal bond angle 109.5°). (391)

theoretical yield The amount of product predicted by the stoichiometrically equivalent molar ratio in the balanced equation. (118)

theory (See *model*.)

thermochemical equation A chemical equation that shows the heat of reaction for the amounts of substances specified. (249)

thermochemistry The branch of thermodynamics that focuses on the heat involved in chemical and physical change. (236)

thermodynamics The study of heat (thermal energy) and its interconversions. (236)

thermometer A device for measuring temperature that contains a fluid that expands or contracts within a graduated tube. (24)

third law of thermodynamics A law stating that the entropy of a perfect crystal is zero at 0 K. (888)

titration A method of determining the concentration of a solution by monitoring its reaction with a solution of known concentration. (153)

torr A unit of pressure identical to 1 mmHg. (191)

total ionic equation   A chemical equation for an aqueous reaction that shows all the soluble ionic substances dissociated into ions. (146)

tracer   A radioisotope that signals the presence of the species of interest by emitting nonionizing radiation. (1087)

transition element (also *transition metal*)   An element that occupies the *d* block of the periodic table; one whose *d* orbitals are being filled. (313, 1023)

transition state (also *activated complex*)   An unstable species formed in an effective collision of reactants that exists momentarily when the system is highest in energy and that can either form products or re-form reactants. (711)

transition state theory   A model that explains how the energy of reactant collision is used to form a high-energy transitional species that can change to reactant or product. (711)

transuranium element   An element with atomic number higher than that of uranium ($Z = 92$). (1082)

trigonal bipyramidal arrangement   The geometric arrangement formed when five electron groups maximize their separation around a central atom. When all five groups are bonding groups, the molecular shape is trigonal bipyramidal (AX_5; ideal bond angles, axial-center-equatorial 90° and equatorial-center-equatorial 120°). (392)

trigonal planar arrangement   The geometric arrangement formed when three electron groups maximize their separation around a central atom. When all three groups are bonding groups, the molecular shape is trigonal planar (AX_3; ideal bond angle 120°). (390)

trigonal pyramid   A molecular shape (AX_3E) caused by the presence of one lone pair in a tetrahedral arrangement. (391)

triple bond   A covalent bond that consists of three bonding pairs, two atoms sharing six electrons; one σ and two π bonds. (352)

triple point   The pressure and temperature at which three phases of a substance are in equilibrium. In a phase diagram, the point at which three phase-transition curves meet. (449)

Tyndall effect   The scattering of light by a colloidal suspension. (540)

U

ultraviolet (UV)   Radiation in the region of the electromagnetic spectrum between the visible and the x-ray regions. (271)

uncertainty   A characteristic of every measurement that results from the inexactness of the measuring device and the necessity of estimating when taking a reading. (27)

uncertainty principle   The principle stated by Werner Heisenberg that it is impossible to know simultaneously the exact position and velocity of a particle; the principle becomes important only for particles of very small mass. (286)

unimolecular reaction   An elementary reaction that involves the decomposition or rearrangement of a single particle. (715)

unit cell   The smallest portion of a crystal that, if repeated in all three directions, gives the crystal. (464)

universal gas constant (*R*)   A proportionality constant that relates the energy, amount of substance, and temperature of a system; $R = 0.0820578$ atm·L/mol·K $= 8.31447$ J/mol·K. (198)

unsaturated hydrocarbon   A hydrocarbon with at least one carbon-carbon multiple bond; one in which at least two C atoms are bonded to fewer than four atoms. (640)

unsaturated solution   A solution in which more solute can be dissolved at a given temperature. (519)

unshared pair   (See *lone pair.*)

V

V shape   (See *bent shape.*)

valence band   In band theory, the lower energy portion of the band of molecular orbitals, which is filled with valence electrons. (474)

valence bond (VB) theory   A model that attempts to reconcile the shapes of molecules with those of atomic orbitals through the concepts of orbital overlap and hybridization. (411)

valence electrons   The electrons involved in compound formation; in main-group elements, the electrons in the valence (outer) level. (315)

valence-shell electron-pair repulsion (VSEPR) theory   A model explaining that the shapes of molecules and ions result from minimizing electron-pair repulsions around a central atom. (388)

van der Waals constants   Experimentally determined positive numbers used in the van der Waals equation to account for the intermolecular attractions and molecular volume of real gases. (223)

van der Waals equation   An equation that accounts for the behavior of real gases. (223)

van der Waals radius   One-half of the closest distance between the nuclei of identical nonbonded atoms. (450)

vapor pressure (also *equilibrium vapor pressure*)   The pressure exerted by a vapor at equilibrium with its liquid in a closed system. (445)

vapor pressure lowering (ΔP)   The lowering of the vapor pressure of a solvent caused by the presence of dissolved solute particles. (528)

vaporization   The process of changing from a liquid to a gas. (438)

variable   A quantity that can have more than a single value. (13) (See also *controlled experiment.*)

viscosity   A measure of the resistance of a liquid to flow. (458)

volatility   The tendency of a substance to become a gas. (76)

volt (V)   The SI unit of electric potential: 1 V $= 1$ J/C. (934)

voltage   (See *cell potential.*)

voltaic cell (also *galvanic cell*)   An electrochemical cell that uses a spontaneous reaction to generate electric energy. (928)

volume (*V*)   The space occupied by a sample of matter. (19)

volume percent [% (v/v)]   A concentration term defined as the volume of solute in 100. volumes of solution. (524)

W

wastewater   Used water, usually containing industrial and/or residential waste, that is treated before being returned to the environment. (542)

water softening   The process of removing hard-water ions, such as Ca^{2+} and Mg^{2+}, from water. (542)

wave function   (See *atomic orbital.*)

wave-particle duality   The principle stating that both matter and energy have wavelike and particle-like properties. (286)

wavelength (λ)   The distance between any point on a wave and the corresponding point on the next wave, that is, the distance a wave travels during one cycle. (270)

weak-field ligand   A ligand that causes smaller crystal field splitting energy and therefore is part of a high-spin complex. (1050)

weight   The force exerted by a gravitational field on an object. (21)

work (*w*)   The energy transferred when an object is moved by a force. (238)

X

x-ray diffraction analysis   An instrumental technique used to determine spatial dimensions of a crystal structure by measuring the diffraction patterns caused by x-rays impinging on the crystal. (468)

Z

zone refining   A process used to purify metals and metalloids in which impurities are removed from a bar of the element by concentrating them in a thin molten zone. (533, 997)

Background used for all Chapter Openers: Courtesy of Dr. Robert Sum/Nanosurf AG. About the Author Photo: Courtesy of Ruth Melnick.

Chapter 1

Opener: © McGraw-Hill Higher Education; **Page 1.1A:** © Paul Morrell/Stone/Getty Images; **1.1B, Table 1.1A:** © The McGraw-Hill Companies, Inc./Stephen Frisch Photographer; **Table 1.1B:** © Ruth Melnick; **Table 1.1C–E:** © The McGraw-Hill Companies, Inc./Stephen Frisch Photographer; **Page 6(left):** Gallileo Project, SSI, University of Arizona; **1.4:** © Giraudon/Art Resource, NY; **Page 11:** © Culver Pictures; **1.5A:** © George Haling/Photo Researchers, Inc.; **1.5B:** © Javier Soriano/AFP/Getty Images; **1.5C:** © Phanie Agency/Photo Researchers, Inc.; **1.5D:** © Krista Kennell/ZUMA/Corbis; **1.7:** © Loren Santow/Stone/Getty Images; **1.9A:** © The McGraw-Hill Companies, Inc./Stephen Frisch Photographer; **1.9B:** © BrandTech Scientific, Inc.; **1.13:** © 2005 Geoffrey Wheeler Photography; **1.14B:** © Hart Scientific, Inc.; **1.15A,B:** © The McGraw-Hill Companies, Inc./Stephen Frisch Photographer; **B1.1A:** © AP/Wide World Photos; **B1.1B:** © Bill Bachman/Alamy; **B1.1C:** © Bruce Forester/Stone/Getty Images.

Chapter 2

Opener: © PhotoDisc/Getty Images; **Table 2.1, 2.2:** © The McGraw-Hill Companies, Inc./Stephen Frisch Photographer; **2.3(top):** © The Art Archive/Archaeological Museum Istanbul/Dagli Orti; **2.3(bottom):** © Tim Ridley/DKimages.com; **Page 47:** © Oesper Collection, University of Cincinnati; **Page 49(right):** © Owaki - Kulla/Corbis; **B2.2D:** © James Holmes/Oxford Centre for Molecular Sciences/SPL/Photo Researchers, Inc.; **2.10, 2.11A, 2.11E:** © The McGraw-Hill Companies, Inc./Stephen Frisch Photographer; **2.16:** © Dane S. Johnson/Visuals Unlimited; **2.20, B2.3, B2.4, B2.9:** © The McGraw-Hill Companies, Inc./Stephen Frisch Photographer.

Chapter 3

Opener: © 2007 Richard Megna/Fundamental Photographs, NYC; **3.1, 3.2, Page 98:** © The McGraw-Hill Companies, Inc./Stephen Frisch Photographer; **Page 101:** © Ruth Melnick; **3.8:** © Richard Megna/Fundamental Photographs, NYC; **3.9, 3.13, 3.14:** © The McGraw-Hill Companies, Inc./Stephen Frisch Photographer.

Chapter 4

Opener: © Joe Drivas/Stone/Getty Images; **4.3, 4.4:** © The McGraw-Hill Companies, Inc./Stephen Frisch Photographer; **4.5:** © The McGraw-Hill Companies, Inc./Richard Megna, Photographer; **4.6, 4.8, 4.9** © The McGraw-Hill Companies, Inc./Stephen Frisch Photographer; **4.11:** © Ed Degginger/Color-Pic; **Page 163, 4.15, 4.16, 4.17, 4.18, 4.19, 4.20:** © The McGraw-Hill Companies, Inc./Stephen Frisch Photographer.

Chapter 5

Opener: © Gerolf Kalt/zefa/Corbis; **5.1, 5.2, 5.9:** © The McGraw-Hill Companies, Inc./Stephen Frisch Photographer; **Page 203:** © Jorge Uzon/AFP/Getty Images; **Page 209:** © The McGraw-Hill Companies, Inc./Stephen Frisch Photographer; **Page 220:** © Julian Baum/SPL/Photo Researchers, Inc.

Chapter 6

Opener: © Metcalfe-Thatcher/Photographer's Choice/Getty Images; **Page 236:** © David Malin/David Malin Images; **6.1:** © The McGraw-Hill Companies, Inc./Stephen Frisch Photographer.

Chapter 7

Opener: © Kevin Schafer/The Image Bank/Getty Images; **Page 271:** © Ralph Wetmore/Getty Images; **Page 273:** © Galen Rowell/Mountain Light; **7.5A:** © Richard Megna/Fundamental Photographs, NYC; **7.6A:** © Ed Degginger/Color-Pic; **7.6B:** © Phil Degginger/Color-Pic; **7.6C:** © Paul Dance/Getty Images; **B7.1A:** © The McGraw-Hill Companies, Inc./Stephen Frisch Photographer; **B7.1B:** Washington DC Convention & Visitors Association; **Page 283:** © SPL/Photo Researchers, Inc.; **Page 284:** © Dennis Kunkel/Phototake;

7.14A,B: PSSC Physics @1965, Education Development Center, Inc.; D.C. Heath & Company/Education Development Center, Inc.

Chapter 8

Opener: © Jana Leon/Stone/Getty Images; **Page 303:** © Corbis; **Page 304:** © Oesper Collection, University of Cincinnati; **8.10, Page 326, 8.23:** © The McGraw-Hill Companies, Inc./Stephen Frisch Photographer.

Chapter 9

Opener: Courtesy of Wim Vanroose; **Page 344:** © Bancroft Library, University of California; **9.5, 9.8A, 9.9, 9.14, 9.25, 9.28A:** © The McGraw-Hill Companies, Inc./Stephen Frisch Photographer; **Page 371:** © Diane Hirsch/Fundamental Photographs, NYC.

Chapter 10

Opener: Courtesy of www.puzzlemaster.ca; **Page 385:** © Richard Megna/Fundamental Photographs, NYC; **10.2A, 10.6:** © The McGraw-Hill Companies, Inc./Stephen Frisch Photographer.

Chapter 11

Figure 11.22: © Richard Megna/Fundamental Photographs, NYC.

Chapter 12

Opener: © Raymond Gehman/Corbis; **Page 438:** © Michael Lustbader/Photo Researchers, Inc.; **Page 439:** © Jill Birschbach Photo Services; **12.8:** © Richard Megna/McGraw-Hill Higher Education; **Page 457:** © Yoav Levy/Phototake; **12.20:** © The McGraw-Hill Companies, Inc./Stephen Frisch Photographer; **Page 459(top left):** © Rod Planck/Photo Researchers, Inc.; **Page 459(middle left):** © Chris Sorensen Photography; **Page 459(bottom middle):** © Ruth Melnick; **Page 459(middle right):** © Nuridsany & Perennou / Photo Researchers, Inc.; **Page 459(top right):** © Phil Jude/SPL/Photo Researchers, Inc.; **12.22B:** © Scott Camazine/Photo Researchers, Inc.; **12.23:** © Tom Pantages; **12.25A:** © Jeffrey A. Scovil Photography (Wayne Thompson Minerals); **12.25B:** © Jeffrey A. Scovil Photography (Francis Benjamin Collection); **12.25C:** © Jeffrey A. Scovil Photography; **12.25D:** © The McGraw-Hill Companies, Inc./Doug Sherman, Photographer; **12.25E:** © Mark Schneider/Visuals Unlimited; **Page 466:** © Michael Herron/Woodfin Camp & Assocs.; **B12.3A:** © Philippe Plailly/Eurelios/Phototake; **B12.3B:** Courtesy National Institute of Standards and Technology; **12.39:** © IBM Research/Peter Arnold, Inc.; **Page 478:** © Photofusion Picture Library/Alamy; **12.44A:** © J. R. Factor/Photo Researchers, Inc.; **12.44B:** © James Dennis/Phototake; **Page 488:** © Ingram Publishing/Alamy; **12.50:** Courtesy of Benoit Dubertret; **12.51:** © & by permission from Ferrotec (USA) Corporation. May not be reproduced without written permission; **12.52:** T. Sasaki/Rice University (Image courtesy of Rice University Office of Media Relations and Information).

Chapter 13

Opener: © David Bishop/FoodPix/Jupiter Images; **Table 13.1:** © A. B. Dowsett/SPL/Photo Researchers, Inc.; **13.4A:** © Ruth Melnick; **13.4B:** © Ed Degginger/Color-Pic; **Page 507:** © MacDonald Photography/Root Resources; **13.20:** © The McGraw-Hill Companies, Inc./Stephen Frisch Photographer; **13.22:** © Shiela Terry/SPL/Photo Researchers, Inc.; **Page 522:** © Peter Scoones/Photo Researchers, Inc.; **13.24:** © Darwin Dale/Photo Researchers, Inc.; **13.25:** © The McGraw-Hill Companies, Inc./Stephen Frisch Photographer; **Page 533(top right):** © Traverse City Record-Eagle, John L. Russell/AP/Wide World Photos; **Page 533(top left):** © OSF/Doug Allan/Animals, Animals; **Page 533(middle right):** © Chris Sorensen Photography; **Page 533(bottom left):** © Westinghouse/Visuals Unlimited; **Page 534(top):** © David M. Phillips/Photo Researchers, Inc.; **Page 534(middle right):** © Kenneth W. Fink/Root Resources; **Page 534(center):** © John Lei/Stock Boston; **Page 534(bottom right):** © Osentoski & Zoda/Envision; **13.32:** © Ed Degginger/Color-Pic; **B13.3A:** © Robert Essel/Corbis; **Page 543:** © Earth Satellite Corporation/SPL/Photo Researchers, Inc.

Interchapter

Opener: © Richard Megna/Fundamental Photographs, NYC; **Page 560:** © The McGraw-Hill Companies, Inc./Stephen Frisch Photographer.

Chapter 14

Opener: © mindfrieze/flickr.com; **Table 14.1, Page 570, 572, 573(top):** © The McGraw-Hill Companies, Inc./Stephen Frisch Photographer; **Page 573(bottom):** © Jeff Hunter/The Image Bank/Getty Images; **Page 576, Page 577(top):** © The McGraw-Hill Companies, Inc./Stephen Frisch Photographer; **Page 577(middle left):** © Jeffrey A. Scovil Photography; **Page 577(bottom left):** © Annebicque Bernard/Corbis Sygma; **Page 577(bottom right):** © Nick Green/ Photolibrary.com; **Page 580:** © The McGraw-Hill Companies, Inc./Stephen Frisch Photographer; **Page 581(top):** © Scott Camazine/Photo Researchers, Inc.; **Page 581(bottom left):** © Phil Degginger/Color-Pic; **Page 581(center):** © Ed Degginger/Color-Pic; **Page 581(bottom right):** © AT&T Bell Labs/SPL/Photo Researchers, Inc.; **Page 582:** © The McGraw-Hill Companies, Inc./Stephen Frisch Photographer; **14.13A:** © Gordon Vrdoljak, UC Berkeley Electron Microscope Lab; **14.13B:** © Norbert Speicher/Alamy; **Page 586:** © The McGraw-Hill Companies, Inc./Stephen Frisch Photographer; **Page 587(top right):** © Charles Winters/Photo Researchers, Inc.; **Page 587(center):** © Charles Falco/Photo Researchers, Inc.; **Page 587(bottom left):** © Phil Jude/SPL/Photo Researchers, Inc.; **14.14:** © The McGraw-Hill Companies, Inc./Stephen Frisch Photographer; **Page 592(bottom & middle left):** © Scott Camazine/Photo Researchers, Inc.; **Page 593(top right):** © Barry Runk/Grant Heilman Photography; **Page 593 (bottom right):** © Tom Pantages; **Page 594(top):** © Photri-Microstock; **Page 594(center):** © Mark Richards/PhotoEdit; **Page 594(bottom):** Boston Public Library/Rare Books Department. Courtesy of the Trustees; **Page 596:** © The McGraw-Hill Companies, Inc./Stephen Frisch Photographer; **Page 597(middle right):** © Robert Galbraith/Peter Arnold, Inc.; **Page 597(bottom right):** © The McGraw-Hill Companies, Inc./Pat Watson Photographer; **Page 604:** © The McGraw-Hill Companies, Inc./Stephen Frisch Photographer; **Page 605(left):** © Robert Mathena/Fundamental Photographs, NYC; **Page 605(right):** © Gregory Dimijian/Photo Researchers, Inc.; **Page 608:** © Steve Cole/Getty Images; **14.26:** © The McGraw-Hill Companies, Inc./Stephen Frisch Photographer; **14.27:** © Wards Earth Science Catalog/Wards Natural Science Establishment; **14.30, Page 612:** © The McGraw-Hill Companies, Inc./Stephen Frisch Photographer; **Page 613(left):** © Jeffrey A. Scovil Photography; **Page 613(right):** © Gene Blevins/LA Daily News/Corbis; **Page 616:** © Jeff Hunter/The Image Bank/Getty Images; **14.33:** © Ed Degginger/Color-Pic; **Page 618:** © The McGraw-Hill Companies, Inc./Stephen Frisch Photographer.

Chapter 15

Opener: © Steve Gschmeissner/Photo Researchers, Inc.; **Page 629:** © SPL/Photo Researchers, Inc.; **15.8:** © Martin Bough/Fundamental Photographs, NYC; **B15.5:** © Zephyr/Photo Researchers, Inc.; **15.13:** © Richard Megna/McGraw-Hill Higher Education; **Page 659:** © Fred Hansen; **15.26:** © Ed Degginger/Color-Pic; **Page 666:** © The McGraw-Hill Companies, Inc./Pat Watson Photographer; **15.30A:** © J. Gross/SPL/Photo Researchers, Inc.; **15.30B:** © Ed Degginger/Color-Pic; **B15.9:** © & Courtesy of Orchid Cellmark.

Chapter 16

Opener: © Dave King/Dorling Kindersley/Getty Images; **16.2A:** © Crown Copyright / Health & Safety Lab./Photo Researchers, Inc.; **16.2B:** © Ruth Melnick; **16.2C:** © Paul Silverman/Fundamental Photographs, NYC; **16.2D:** © Ruth Melnick; **16.3:** © The McGraw-Hill Companies, Inc./Stephen Frisch Photographer; **Page 687:** © Roberto Schmidt/AFP/Getty Images; **B16.1:** © Varian Inc.; **B16.2, B16.3:** © The McGraw-Hill Companies, Inc./Stephen Frisch Photographer; **B16.6:** NASA.

Chapter 17

Opener: © Richard Megna/Fundamental Photographs, NYC; **17.1:** © The McGraw-Hill Companies, Inc./Stephen Frisch Photographer; **Page 761:** © Kennan Ward 2004; **B17.3** © Grant Heilman Photography.

Chapter 18

Opener: © Organica/Alamy; **Table 18.1(Acids & Bases), 18.3, 18.7:** © The McGraw-Hill Companies, Inc./Stephen Frisch Photographer.

Chapter 19

Opener: © Ruth Melnick; **19.1, 19.2, 19.6:** © The McGraw-Hill Companies, Inc./Stephen Frisch Photographer; **19.11:** © Jackie Lewin, Royal Free Hospital/Photo Researchers, Inc.; **Page 851:** © The McGraw-Hill Companies, Inc./Stephen Frisch Photographer; **Page 854:** © Jeffrey A. Scovil Photography; **19.12:** © The McGraw-Hill Companies, Inc./Stephen Frisch Photographer; **Page 857** © CNRI/Science Source/Photo Researchers, Inc.; **19.13, Page 858:** © The McGraw-Hill Companies, Inc./Stephen Frisch Photographer; **B19.1:** © Ruth Melnick; **B19.2:** © Kevin Downey Photography; **B19.4:** © Will & Deni McIntyre/Corbis Images; **B19.5(left):** © Hulton Archive/Getty Images; **B19.5(right):** © Richard Megna/Fundamental Photographs, NYC; **Page 867:** © Phil Degginger/Color-Pic; **19.16:** © The McGraw-Hill Companies, Inc./Stephen Frisch Photographer.

Chapter 20

Opener: © Tony Wright/earthscapes/Alamy; **20.1:** © The McGraw-Hill Companies, Inc./Stephen Frisch Photographer; **B20.1:** © Bettmann/Corbis; **B20.2:** © Vanderbilt University-Clinical Nutrition Research; **Page 899:** © SPL/Photo Researchers, Inc.; **Page 901:** © Karl Weatherly/PhotoDisc/Getty Images.

Chapter 21

Opener: © imagebroker/Alamy; **Page 923:** © AP/Wide World Photos; **21.1, 21.2, 21.4, 21.5B:** © The McGraw-Hill Companies, Inc./Stephen Frisch Photographer; **21.7:** © Richard Megna/Fundamental Photographs, NYC; **21.9:** © The McGraw-Hill Companies, Inc./Stephen Frisch Photographer; **Page 946:** © Physikalisch - Technische Bundesanstalt, courtesy AIP Emilio Segre Visual Archives; **Page 950:** © Science VU/ Look Up/Visuals Unlimited; **Page 951:** Photo © F.W. Goro; **21.14:** © Chris Sorensen Photography; **21.15:** © The McGraw-Hill Companies, Inc./Pat Watson Photographer; **21.16:** Courtesy of the Guidant Corporation; **21.18:** © McGraw-Hill Higher Education/Stephen Frisch Photographer; **21.19:** © AP/Wide World Photos; **21.21A:** © Alfred Dias/Photo Researchers, Inc.; **21.22:** © David Weintraub/Photo Researchers, Inc.; **21.26:** © Chris Sorensen Photography; **21.27, Page 964:** © The McGraw-Hill Companies, Inc./Stephen Frisch Photographer; **Page 965:** © The Royal Institution, London, UK/Bridgeman Art Library.

Chapter 22

Opener: © James Holmes, Hays Chemicals/Photo Researchers, Inc.; **22.3A:** © Doug Sherman/Geofile; **22.3B:** © University of Cambridge, Department of Earth Sciences; **Page 984:** © Ruth Melnick; **Page 985:** © Robert Holmes/Corbis; **Page 994:** © Patrick W. Grace/Photo Researchers, Inc.; **22.12:** © Richard Megna/Fundamental Photographs, NYC; **Page 999:** © Christine L. Case/ Visuals Unlimited; **Page 1000:** © Remi Benali; **22.17B:** Courtesy Bethlehem Steel Corporation; **Page 1002:** © Lester Lefkowitz/Tech Photo, Inc.; **22.18B:** © Tom Hollyman/Photo Researchers, Inc.; **Page 1004:** © James L. Amos/Peter Arnold, Inc.; **22.20A:** © Rob Crandall/Rainbow; **22.20C:** © ALCOA/Alcoa Technical Center; **22.20D:** © Ed Degginger/Color-Pic; **Page 1006:** © Hank Morgan/Rainbow; **Page 1008:** © Ken Whitmore/Getty Images; **Page 1010:** © The McGraw-Hill Companies, Inc./Stephen Frisch Photographer; **22.22C:** © Nathan Benn/Woodfin Camp & Associates.

Chapter 23

Opener: © Lawrence Lawry/Photo Researchers, Inc.; **23.2, 23.5, 23.6:** © The McGraw-Hill Companies, Inc./Stephen Frisch Photographer; **23.7B:** © Richard Megna/Fundamental Photographs, NYC; **Table 23.5:** © The McGraw-Hill Companies, Inc./Stephen Frisch Photographer; **Page 1033:** © Ed Degginger/Color-Pic; **Page 1034:** © Peter Ryan/Photo Researchers, Inc.; **23.8(left):** © F.S. Judd Research Laboratories; **23.8(right):** © Kathy Shorr; **Page 1037:** © The Granger Collection; **Page 1044:** © Richard Megna/Fundamental Photographs, NYC; **Page 1049:** © Pete Turner/The Image Bank/Getty Images; **23.20A:** © The McGraw-Hill Companies, Inc./Pat Watson Photographer; **23.21:** © The McGraw-Hill Companies, Inc./Stephen Frisch Photographer.

Chapter 24

Opener: NASA/JPL-Caltech/O. Krause (Steward Observatory); **Page 1067:** © AIP Emilio Segre Visual Archives, W. F. Meggers Gallery of Nobel Laureates; **B24.1:** © Hank Morgan/Rainbow; **B24.2:** © Hewlett-Packard; **Page 1078:** © Getty Images; **Page 1080:** © Hulton Archive/Getty Images; **24.9A:** © Scott Camazine/Photo Researchers, Inc.; **24.9B:** © T. Youssef/Custom Medical Stock Photo; **24.10:** © Dr. Robert Friedland/SPL/Photo Researchers, Inc. **24.11:** © Dr. Dennis Olson/Meat Lab, Iowa State University, Aimes, IA; **Page 1094:** © Emilio Segre Visual Archives; **24.16A:** © Albert Copley/Visuals Unlimited; **24.17:** © Dietmar Krause/Princeton Plasma Physics Lab; **B24.4:** Ben Dilday and the SDSS Collaboration, www.sdss.org.

Page numbers followed by *f* indicate figures; *mn*, margin notes; *n*, footnotes; and *t*, tables. Page numbers preceded by "A-" indicate pages in the *Appendix*. Page numbers preceded by "G-" indicate pages in the *Glossary*.

A

AB block copolymer, 489
Absolute temperature scale, 195, G-1
Absolute zero, 25
Absorption spectrum, 281–282, 281*f*, 282*f*, G-1
Abundance, of elements, 981–986, 982*f*, 984*t*, G-1
Accuracy, 31–32, 32*f*, G-1
Acetaldehyde
  boiling point, 452, 452*f*
  dipole moment, 452, 452*f*
  molecular structure, 651*t*, 656*f*
Acetamide, molecular structure, 651*t*
Acetaminophen, molecular structure, 660*f*
Acetate ion, 68*t*, 833–837, 833*t*
Acetic acid
  acid-dissociation constant (K_a), A-8
  as a base, 817
  boiling point elevation constant, 530*t*
  depiction of molecule, 74
  dissociation constant, 787*t*
  dissociation of, 175, 833–837, 833*t*
  equilibrium constant, 791*t*
  formula, 103*t*
  freezing point depression constant, 530*t*
  molecular structure, 651*t*
  uses, 103*t*, 783*t*
Acetone
  bonds and orbitals, 421
  molecular shape, 397
  molecular structure, 651*t*, 656*f*
  nuclear magnetic resonance spectroscopy, 646, 647*f*
Acetonitrile
  boiling point, 452*f*
  dipole moment, 452*f*
  molecular structure, 651*t*
Acetylene
  molecular shape, 418, 420, 420*f*
  molecular structure, 651*t*
  production of, 207, 574*mn*, 587
Acetylsalicylic acid, acid-dissociation constant (K_a), A-8
Acid(s)
  acidic solutions, balancing redox reactions in, 925–926, 925*f*
  Arrhenius acids, 784, 820
  Brønsted-Lowry acids, 156, 793–798, 816–817, 820
  buffers, 832–841
  concentration, finding from titration, 154–155
  conjugate acid-base pairs, 794–795, 795*t*, 797*f*, 808–809, 833, 834*f*, 837, 839–840
  defined, 150, 156, 783, 784, 793, G-1
  dissociation, 144, 151–152, 785–787, 785*f*, 786*f*, 787*t*, 802, 808–809, 812*t*, 813, 817, 833–837, 833*t*, A-8–A-10
  as electrolytes, 151, 151*f*
  equilibria. *See* Acid-base equilibria
  household uses, 783*t*
  indicators, 154, 154*f*, 793, 793*f*, 842, 842*f*
  leveling effect, 816–817
  Lewis acids, 817–820, 862
  monoprotic, 787, 787*t*
  names, 70
  pH. *See* pH
  polyprotic, 803–805, 803*t*, 849–851, 850*f*
  proton transfer, 155–157, 156*f*
  strength. *See* Acid strength
  strong, 151–152, 151*t*, 156, 784, 785*f*, 786, 786*f*, 788, 843–845, 848–849
  titrations, 841–851
  in water, 784–789
  weak, 151–152, 151*t*, 157, 175, 786, 786*f*, 788, 798–810, 814–815, 845–848

weak-acid equilibrium, 798–805
  weak bases and, 805–810
Acid anhydride, 658, 661, 661*f*, G-1
Acid-base behavior, of element oxides, 326–327, 327*f*, 559
Acid-base buffer systems, 832–841
  buffer capacity, 838, 838*f*
  buffer range, 838
  common-ion effect, 833–837
  conjugate acid-base pair, 833, 834*f*
  defined, 832, G-1
  essential features, 834–835
  Henderson-Hasselbalch equation, 837
  preparing, 839–841
Acid-base catalysis, 724
Acid-base equilibria, 782–821
  acid-base indicators, 793, 842, 842*f*
  acid-dissociation constant (K_a), 785–787, 787*t*, 812*t*, A-8–A-10
    weak-acid equilibrium problem solving, 798–805
    weak bases and weak acids, 808–810
  autoionization of water, 789–793, 790*f*, 808
  base-dissociation constant (K_b), 805–809, 806*t*, A-11–A-12
  Brønsted-Lowry acid-base definition, 793–798, 816–817, 820
  conjugate acid-base pairs, 794–795, 795*t*, 797*f*, 808–809, 833, 834*f*, 837, 839–840
  hydrated metal ions, 811–812, A-12
  hydronium ion, 784, 791–793
  ion-product constant for water (K_w), 789–93, 790*f*, 809
  Lewis acid-base definition, 817–820, 862
  pH, 791–793, 791*f*, 792*f*, 793*f*
  polyprotic acids, 803–805, 803*t*
  salt solutions, 812–816
    amphiprotic anions, 814–815
    of weakly acidic cations and weakly basic anions, 814
    yielding acidic solutions, 813, 815*t*
    yielding basic solutions, 815*t*, 813–814
    yielding neutral solutions, 813, 815*t*
  variation in acid strength, 785–789, 785*f*–786*f*, 787*t*
  weak-acid equilibrium problem solving, 798–805
    acid-dissociation constant (K_a), 798–805, 803*t*
    concentrations, 799–802
    notation system, 799
    polyprotic acids, 803–805, 803*t*
Acid-base indicator
  acid-base titration, 154, 154*f*, 842, 842*f*
  color and pH range of common, 842, 842*f*
  defined, 793, 842, G-1
  pH paper, 793, 793*f*
Acid-base (neutralization) reactions, 150–157
  acid-base indicator, 154, 154*f*, 842, 842*f*
  defined, 150, G-1
  of element oxides, 326–327, 327*f*
  end point, 154
  equivalence point, 154
  gas-forming reactions, 156–157, 157*f*
  key event, 152–153
  proton transfer, 155–157, 156*f*
  relative acid-base strength and the net direction of reaction, 796–798
  salt formation, 152–153
  titrations, 153–155, 154*f*, 841–851
  water formation, 152–153
  writing ionic equations for, 153
Acid-base titration, 153–155, 154*f*
Acid-base titration curves
  acid-base indicators, 842, 842*f*
  defined, 841, G-1
  end point, 844
  equivalence point, 843–845
  polyprotic acids, 849–850, 850*f*
  strong acid–strong base, 843–845, 843*f*
  weak acid–strong base, 845–848, 845*f*
  weak base–strong acid, 848–849, 849*f*
Acid-dissociation constant (K_a)
  defined, 786, G-1
  of hydrated metal ions, A-12

meaning of, 786–787
  relation with K_b in conjugate pair, 808–810
  relation with pK_a, 791*f*
  tables of, A-8–A-10
  values for monoprotic acids, 787*t*
  weak acid equilibrium problem solving, 798–805
Acid-ionization constant. *See* Acid-dissociation constant (K_a)
Acid rain
  carbonates as protection from effects of, 589, 832, 864
  chemistry's role in solving, 33, 33*f*
  Hess's law applied to, 251
  overview of problem, 863–864
  sulfur dioxide, 113, 608*mn*
Acid strength
  acid-dissociation constant and, 785–787, 787*t*
  classifying, 788
  conjugate acid-base pairs, 797*f*
  direction of reaction and, 796–798
  hydrated metal ions, 811–812, 812*f*, 812*t*
  molecular properties and, 810–812
  of nonmetal hydrides, 810, 810*f*
  of oxoacids, 811, 811*f*
  strong acids, 784, 785*f*, 786, 786*f*, 788, 843–845, 848–849
  weak acids, 786, 786*f*, 788, 798–810, 814, 845–848
Acidic solutions, balancing redox reactions in, 925–926, 925*f*
Acrylonitrile
  infrared (IR) spectrum of, 358*f*
  molecular structure, 589*f*
Actin, arrangement of filaments, 480*f*
Actinides
  defined, 316, 1031, G-1
  electron configuration, 314*f*, 316
Actinium, 1030
Activated complex, 711
Activation analysis, 1088
Activation energy (E_a), 706–710, 706*f*, 707*f*, 709*f*, 710*f*, 712*f*, 718*f*, 720–721, 720*f*, G-1
Active metal
  activity series of the metals, 170–172, 171*f*, 941–943, G-1
  displacing hydrogen from water, 170, 170*f*
  reactivity, 942
  reduction with, 996, 996*f*
Active nonmetal, oxidation with, 996
Active site, enzyme, 723–724, 723*f*, 724*f*, 770, 770*f*, G-1
Activity ($\mathcal{A}$), 1074, G-1
Activity series of the halogens, 172
Activity series of the metals, 170–172, 171*f*, 941–943, G-1
Actual yield, 118–119, G-1
Addition, significant figures and, 28
Addition polymers, 663–664, 663*t*, 664*f*, G-1
Addition reaction
  with benzene, 655
  defined, 647, 663–664, G-1
  organic, 647–648
Adduct and Lewis acid-base definition, 817–819, G-1
Adenine, 653*f*, 670
Adenosine triphosphate. *See* ATP
Adhesive forces, 457–458, 459
Adipic acid, acid-dissociation constant (K_a), A-8
Advanced materials, intermolecular forces and, 476–491
Aerosol, 539*f*
Air pollution
  acid rain, 251, 589, 863–864
  chemistry's role in solving, 33, 33*f*
  chlorofluorocarbons (CFCs), 589, 589*mn*, 725
  lead, 587
  smog, 203*mn*, 600, 600*f*
  sulfur dioxide, 608*mn*
  thermal pollution, 521, 521*mn*, 1097
Airplane, de-icers, 533, 533*f*
-al (suffix), 651*t*, 655
Alanine
  optical isomers, 639*f*, 640
  structure of, 667*f*
Alchemy, 10, 10*f*, G-1

Alcohols
　defined, 650, G-1
　functional group, 72, 650, 651t, 652
　grain alcohol, density of, 23t
　solubility, 504–505, 504t, 505f
Aldehydes, 651t, 655–656, 656f, G-1
Alizarin, 842f
Alkali metals
　alkaline earth metals compared, 574–575
　anomalous behavior of lithium, 571
　electron configuration, 312, 328, 570, 572
　halides, 985, 985f
　　lattice energy, 348, 348f
　important compounds, 573
　important reactions, 573
　industrial production, 998–1000
　ionization energy, 321f
　isolation and industrial uses, 999–1000
　melting point, 370, 370f
　periodic table, 59
　properties, 570–572
　reactions with halogens, 209
　reactivity, 312f, 570–571
Alkaline battery, 952, 952f
Alkaline earth metals, 574–577
　alkali metals compared, 574–575
　anomalous behavior of beryllium, 575
　diagonal relationships, 575
　electron configuration, 574–576
　important compounds, 577
　important reactions, 577
　melting point, 370, 370f
　periodic table, 59
　properties, 574–576
Alkanes
　chiral molecules, 639, 639f, 640mn
　cycloalkanes, 637, 637f
　defined, 71, G-1
　depicting with formulas and models, 636,
　　636f, 637f
　general formula, 635
　homologous series, 635
　isomers
　　constitutional, 637–638, 638t
　　optical, 639–640, 639f, 640f
　naming, 71, 71t, 635–636, 635t, 636t, 643
　properties, 637–638, 639f, 640
　saturated hydrocarbons, 635
　straight-chain, 71
Alkenes
　defined, 640, G-1
　functional group, 651t, 655
　general formula, 640
　isomerism, 641, 641t
　naming, 641, 643
　reactivity, 655
Alkoxide ion, 650
Alkyl group, 646, G-1
Alkyl halides, 652, 654, G-8
Alkylmagnesium halides, 577
Alkynes, 643, 651t, G-1
Allotropes
　carbon, 585, 585f
　defined, 585, G-1
　entropy and, 892
　oxygen, 606
　phosphorus, 595, 595f
　selenium, 606
　sulfur, 606, 645
　tin, 585
Alloys
　aluminum, 574, 1005
　binary, 997–998, 998f
　composition and uses of common, 997t
　crystal defects in, 477
　defined, 369, 997, G-1
　interstitial, 506, 507f
　magnesium, 1008
　metallurgy, 997–998
　samarium-cobalt, 1031
　as solid-solid solutions, 506, 507f
　substitutional, 506, 507f
Alpha (α) decay, 1068t, 1069, G-1
Alpha (α) particles
　behavior in electric field, 1067, 1067f
　de Broglie wavelength, 284t
　defined, 1067, G-1
　nuclear reactions, 1080–1081
　penetrating power, 1083, 1083f

radioactive decay, 1068t, 1069
Rutherford's α-scattering experiment, 51,
　51f, 56mn
Rutherford's scattering experiment, 1069
transmutation by, 1080–1081
Aluminosilicates, 592, 863
Aluminum
　acid strength of hydrated ion, 812, 812f, 812t
　alloys, 574
　boiling point, 390t
　bond type, 584t
　density, 23t
　diagonal relationship with beryllium, 584
　diffraction pattern, 284, 285
　electron configuration, 311, 311t, 312f, 1005mn
　fluoro-oxy ions and, 1005
　isolation, 1004–1005, 1005f, 1005mn
　melting point, 390t, 584t
　metallic radius, 317f
　metallurgy, 1004–1006, 1005f, 1005mn
　mineral sources of, 993t
　properties, 327, 578, 580
　recycling, 1006
　specific heat capacity, 246t
　uses, 783t, 1005–1006, 1005f, 1006f
Aluminum chloride, 579, 818
　electron density distribution, 368f
　ionic properties, 368
Aluminum hydroxide, 153, 853t, 868–869, 869f
Aluminum ion, 65t
Aluminum oxide, 581, 1004–1006
Aluminum salts, uses of, 783t
Aluminum sulfate, 581
Amalgams, mercury, 1036
Americium, 1082mn
-amide (suffix), 651t, 659
Amides, 651t, 657, 659–660, 660f, 661, G-1
-amine (suffix), 651t, 653
Amines
　defined, 652, 806, G-1
　functional group, 72, 651t, 652–653, 653f
　as weak bases, 788, 806, 806t
Amino acids. See also Proteins
　defined, 507, 667, G-1
　displacement reactions, 153mn
　optical isomers, 640
　physiological form, 508f
　polarity of side chains, 508
　as polyprotic acids, 850–851
　structures of, 507, 667–668, 667f
　uses, 597
Amino functional group, 72
Ammonia
　base-dissociation constant (Kb), A-11
　as covalent hydride, 566
　dissociation, 175
　fertilizer, 566, 771f
　fixed mass ratio of, 42
　formation, 598, 894
　Haber process for synthesis of, 771, 771f, 988
　hybrid orbitals, 415, 415f
　Lewis structure, 806t
　mole fraction of air at sea level, 219t
　molecular shape, 391–392
　origin and industrial use, 187t
　polarity, 400
　titration of, 848, 848f
　uses, 566, 597, 783t
　as weak base, 788, 806, 806t, 818
Ammonium carbonate, 96
Ammonium cyanide, 814
Ammonium ion, 68t, 391–392
Ammonium nitrate, 516f, 517, 518, 883, 883f
Ammonium perchlorate, 613, 616
Ammonium phosphates, 602mn
Amontons's law, 196
Amorphous solids, 463, 473, 473f, G-1
Amount of gas, Avogadro's law and, 196–197, 211, 213
Amount of substance, 18t
Ampere (A) (unit), 18t, 965, G-2
Amphoteric
　amphoteric hydroxides, complex ions of,
　　868–869, 869f
　anions, 814–815
　defined, 327, G-2
　nature of water, 794
Ampicillin, 104, 104f
Amplitude, wave, 270, 270f, G-2
Analysis. See Chemical analysis

Analytical balance, 30, 30f
Angel Falls, 15–16, 15f
Angstrom (Å) (unit), 270
Angular momentum quantum number (l), 289–291,
　290t, 305t, G-2
Aniline
　base-dissociation constant (Kb), 806t, A-11
　Lewis structure, 806t
Anionic ligands, 1040, 1041t
Anions
　amphiprotic, 814–815
　common monatomic, 65t
　common polyatomic, 68t
　complex, 1041t
　defined, 60, G-2
　ionic radius, 332–333
　names, 70
　salts of weakly acidic cations and weakly basic
　　anions, 814
　size, 558
Anisotropic, liquid crystal phase, 478
Anode
　defined, 929, G-2
　half-reaction at, 929, 929mn
　sacrificial, 958, 958f
Antacid, 125–126, 153
Antarctic ozone hole, 725, 725f
Antibiotics, dual polarity of, 512, 512f
Antibonding MO, 422–424, 423f, G-2
Anticancer drugs, 1044mn
Antifluorite structure, crystals, 472
Antifreeze, 531, 533
Antilogarithms, A-2
Antimony
　bond type, 584t
　electron configuration, 596
　melting point, 584t
　properties, 326–327, 327f, 595–596
Antineutrino, 1069mn
Apatites, 990, G-2
Aqueous equilibria, 831–870
　acid-base buffer systems, 832–841
　acid-base titration curves, 841–851
　acid rain and, 863
　complex ions, 862, 865–869
　slightly soluble ionic compounds, 851–862, 866
Aqueous ionic reactions
　acid-base. See Acid-base (neutralization) reactions
　molecular equations, 145, 145f
　net ionic equation, 145f, 146
　spectator ions, 146
　total ionic equation, 145f, 146
Aqueous solutions
　defined, 75, G-2
　determining moles of ions in aqueous ionic solution,
　　143–144
　electrolysis of, 963–964
　salts, 812–816
　standard state, 253
Arginine, structure of, 667f
Argon
　cubic closest packing, 470f
　electron configuration, 311, 311t, 312f, 328, 328f
　heat of vaporization and fusion, 439f
　mole fraction in air at sea level, 219t
　properties, 617–618
Aristotle, 41
Arithmetic operations, significant figures and, 29
Aromatic carcinogens, 645mn
Aromatic hydrocarbons, 644–645, 644f, G-2
Arrhenius, Svante, 706, 784
Arrhenius acid-base definition, 783, 784–785, 820, G-4
Arrhenius acids, 784, 820
Arrhenius bases, 784, 820
Arrhenius equation, 706, 708, 710, G-2
Arsenic
　bond type, 584t
　electron configuration, 313t, 596
　melting point, 584t
　properties, 326, 327, 327f, 595–596
Arsenic acid
　acid-dissociation constant (Ka), 803t, A-8
　lime reaction with, 574mn
Asbestos, 592
Asbestos diaphragm, 1014–1015, 1015f
Ascorbic acid, 804–805, A-8
Asparagine, structure of, 667f
Aspartic acid, structure of, 667f
Aspirin, 74

Astatine
    electron configuration, 312, 612
    properties, 610, 612
-ate (suffix), 68, 68f, 1041
Atmosphere
    carbon dioxide, 986, 988
    defined, 218, 983, G-2
    Earth
        atmosphere-biosphere redox interconnections, 187mn
        carbon cycle and, 986–988, 987f
        composition, 218f, 219, 219t
        impact of life on, 983, 983f
        nitrogen cycle and, 988–989, 989f
        ozone, stratospheric, 219, 380, 606, 725
        ozone depletion, 219, 380, 725, 725f
        pressure variation, 218, 218f
        primitive atmosphere, 220
        regions of, 219
        temperature, 218–219, 218f, 257–258
    planetary, 220, 220t
    pollution. See Air pollution
Atmosphere (atm) (unit), 191, 192t
Atmospheric pressure
    boiling point and, 447, 447mn
    effect of, 189, 189f
    measurement, 189–191, 190f, 191f
    units of measurement, 191–192, 192t
    variation in, 218, 218f
Atom
    chemical bonding and properties of, 341–344
    defined, 47, 53, G-2
    depiction of, 53f
    energy, 274, 277–280, 279f
    history of concept, 41, 47, 48mn
    mass of, 48, 54, 56
    nucleus. See Nucleus
    structure. See Atomic structure
Atom economy, 120
Atomic bomb, 1095–1096, 1095n, 1096f
Atomic clock, 26, 27f
Atomic mass
    calculating, 56
    defined, 54, 92t, G-2
    fixed, 91
    mole concept and, 91, 91f
    relative, 92n
Atomic mass unit (amu), 54, 91, 92t, G-2
Atomic nucleus. See Nucleus
Atomic number (Z)
    defined, 53, G-2
    periodic table organization and, 57, 303, 304mn
Atomic orbital, 287–295
    aufbau principle, 308, 329
    d orbital, 294, 294f, 1049–1050, 1049f, 1050f
    defined, 287, G-2
    f orbital, 294, 294f
    filling order, 315
    g orbital, 294
    hybrid orbital, 412–418
    molecular orbital. See Molecular orbital (MO)
    overlap. See Orbital overlap
    p orbital, 293, 293f, 425–426, 425f, 426f
    Pauli exclusion principle, 305, 308–309, 411
    quantum numbers, 289–291, 290t
    s orbital, 292–293, 292f
Atomic properties. See also specific properties
    alkali metals, 572
    alkaline earth metals, 576
    atomic size. See Atomic size
    boron family elements, 579f, 580
    carbon family elements, 586
    electron configuration. See Electron configuration
    electronegativity. See Electronegativity (EN)
    halogen family elements, 612
    ionization energy. See Ionization energy (IE)
    key, 554–555
    nitrogen family elements, 596
    noble gases, 618
    oxygen family elements, 604
    Period 2 elements, 567, 568t–569t
    transition elements, 1026–1027, 1026f, 1027f
Atomic radius
    alkali metals, 572
    alkaline earth metals, 576
    boron family elements, 579f, 580
    calculation from bond length, 354
    carbon family elements, 586
    determining from crystal structure, 469

entropy and, 891
halogen family elements, 612
main-group elements, 318, 319f
nitrogen family elements, 596
oxygen family elements, 604
Period 2 elements, 569t
periodic trends, 317–320, 320f, 555
transition elements, 318, 319f
Atomic size. See also Atomic radius
    bond length relationship, 354
    covalent radius, 450, 450f
    defined, 555, G-2
    electronegativity, 364, 365f
    entropy and, 890–892
    ionic size compared, 332–333, 332f
    main-group elements, 318, 319f
    periodic trends, 317–320, 320f, 555
    transition elements, 318, 319f, 1026, 1026f
Atomic solids, 470, 470f, 470t, G-2
Atomic spectra, 276–280
    Bohr model of the hydrogen atom, 277–279, 278f
    energy states of the hydrogen atom, 279–280
    excited state, 278
    ground state, 278
    line spectrum, 276–277, 276f, 277f, 277n
    quantum staircase, 278f
    Rydberg equation, 277, 278f, 279
    stationary state, 277–278
Atomic structure. See also Electron (e⁻); Neutron (n°); Proton (p⁺)
    Bohr model, 277–279
    chemical reactivity and, 325–333
    general features, 52, 52f, 53t
    many-electron atoms, 304–308
    quantum-mechanical model, 287–295, 308–317
Atomic symbol, 53, G-2
Atomic theory
    atomic nucleus, discovery of, 50–51, 51f
    Dalton's theory, 41, 47
    definite composition, law of, 44–45, 45f, 47
    electron, discovery of, 48mn, 49–50
    law of definite (constant) composition, 44–45, 45f, 47, G-10
    law of mass conservation, 44, 47, G-10
    many-electron atoms, 304–308
    Millikan's oil drop experiment, 49–50, 50f
    modern theory, 52–57
    nuclear atom model, 48–51
    Rutherford's α-scattering experiment, 51, 51f
Atomic weight, 54. See also Atomic mass
Atomization, 346
ATP (adenosine triphosphate), 597, 603, 908–909, 908f, 909f, 967–968, 968f, G-1
Aufbau principle, 308, 329, G-2
Aurora, 49mn
Autoionization of water, 789–793, 790f, 808, G-2
Automobiles
    battery, 953–954, 954f
    battery-gasoline hybrid, 923mn
    catalytic converter, 722mn, 864
    motor oil, viscosity of, 459
    tire pressure, 196
Average reaction rate, 687, 688–689, G-2
Avogadro, Amedeo, 90
Avogadro's law
    breathing and, 197mn
    defined, 197, G-2
    Earth's atmosphere and, 219
    ideal gas law, relationship to, 198f
    kinetic-molecular theory, 211, 213, 213f, 214
    molecular view of, 213, 213f
    volume-amount relationship, 196–197, 196f
Avogadro's number, 214–215
    Boltzmann constant and, 884
    as conversion factor, 90, 93–94, 94f, 95f
    defined, 90, G-2
Axel, Richard, 403
Axial group, VSEPR theory, 392, G-2

B

Background radiation, 1084–1085, G-2
Bacterium, approximate composition of, 501t
Balanced equations
    information contained in, 110t
    nuclear equations, 1070
    process of balancing, 104–109
    reaction quotient and, 742–743

redox reaction, 162–164, 924–928
sequence, 112–113
states of matter, 106
stoichiometrically equivalent molar ratios, 109–112
using gas laws to determine, 202
Balances, 21, 30, 30f
Balancing (stoichiometric) coefficient, 105–108, G-2
Ball-and-stick models, 74, 636, 636f
Ballpoint pen, action of, 459
Band of stability, 1071–1072, G-2
Band theory, 474–476, 474f, 475f, G-2
Bar (unit), 191, 192t
Barite, 463f
Barium
    boiling point, 390t
    electron configuration, 315
    ionic compounds, 326
    isotope, 1094
    melting point, 390t
    mineral sources of, 993t
    properties, 326, 574–576
Barium hydroxide, 883, 883f
Barium ion, 65t
Barium sulfate, 857, 857f
Barometer, 189–190, 190f, G-2
Base(s). See also Acid-base equilibria; Acid-base (neutralization) reactions; Acid-base titration curves
    anions of weak acids as, 808
    Arrhenius base, 784, 820
    Brønsted-Lowry base, 793–798, 816–817, 820
    Brønsted-Lowry definition, 156
    buffers, 832–841
    conjugate acid-base pairs, 794–795, 795t, 797f, 808–809, 833, 834f, 837, 839–840
    defined, 150, 156, 783, 784, 794, G-2
    as electrolytes, 151, 151f
    household uses, 783t
    indicators, 154, 154f, 793, 793f, 842, 842f
    leveling effect, 816–817
    Lewis base, 817–820, 862
    strong, 151–152, 151t, 156, 784, 788, 843–848
    titration, 841–851
    in water, 784–789
    weak, 151–152, 151t, 175, 788, 805–810, 814, 848–849
Base-dissociation constant (Kb)
    of ammonia and amines, 806–807, 806t
    defined, 805, G-2
    pH determination from, 807
    relation to Ka in conjugate acid-base pair, 808–809
    table of, A-11–A-12
Base-ionization constant. See Base-dissociation constant (Kb)
Base pairs, DNA, 671, G-2
Base units, SI, 18, 18t, G-2
Basic-oxygen process, 1002, 1002f, G-2
Basic solutions, balancing redox reactions in, 926
Battery
    alkaline, 952, 952f
    aluminum, 1005mn
    button, 953, 953f
    defined, 952, G-2
    dry cell, 952
    flow batteries, 955
    free energy change and, 903
    fuel cells, 955–956, 955f
    lead-acid, 854, 953–954, 954f, 960, 960f
    lithium, 953, 953f, 955, 955f
    mercury, 953, 1036
    nickel-cadmium (nicad), 954
    nickel–metal hydride (Ni-MH), 954, 954f
    nonrechargeable batteries, 952–953
    primary batteries, 952–953
    rechargeable batteries, 953–955
    secondary batteries, 953–955
    silver, 953, 953f
Battery-gasoline hybrid cars, 923mn
Bauxite, 1004, 1004f
Bayer process, 1004
Becquerel, Antoine-Henri, 1066, 1067
Becquerel (Bq) (unit), 1076, G-3
Beeswax, 507mn
Bends, 522mn
Bent shape, VSEPR theory, 390, 390f, 391f, 392, 394f, 415, G-3
Benzaldehyde, molecular structure, 656f

Benzene
addition reaction, 655
boiling point elevation constant, 530*t*
freezing point depression constant, 530*t*
heat of vaporization and fusion, 439*f*
for maleic anhydride production, 120
molecular orbitals, 430, 430*f*
reactivity, 655
resonance forms, 382
structure, 583*f*, 644, 644*f*
Benzo[*a*]pyrene, 645*mn*
Benzoic acid, acid-dissociation constant (*K*ₐ), A-8
Benzoic acid, molecular structure, 658*f*
Beryl, 463*f*, 577, 590, 591*f*
Beryllium
anomalous behavior of, 575, 575*f*
diagonal relationship with aluminum, 584
diatomic, 425, 425*f*
electron configuration, 309, 311*f*, 312*f*
electron-deficient molecules, 384–385
ionization energy, 323*f*, 323*t*
isotopes, decay constants, and half-lives of, 1077*t*
mineral sources of, 993*t*
properties, 326, 384*n*, 568*t*, 574–576
Beryllium chloride
electron deficiency, 575, 575*f*
hybrid orbitals, 412, 413*f*
Lewis structure, 384–385
linear shape, 389
Beta (β) decay, 1068*t*, 1069, 1069*n*, 1071, G-3
Beta (β⁻) particles
behavior in electric field, 1067, 1067*f*
defined, 1067, G-3
penetrating power, 1083*f*, 1084
radioactive decay, 1068*t*, 1069
Bevatron, 1082*mn*
Bhopal (India), industrial accident, 203*mn*
Bicarbonate ion, 68*t*, 542
Bidentate ligands, 1039–1040, 1039*t*
Big Bang theory, 1099
Bile salts, 539*mn*
Bimolecular reaction, 715, 715*t*, G-3
Binary acids, 70
Binary alloys, 997–998, 998*f*
Binary covalent compounds
defined, 70, G-3
formation, 70
names and formulas, 70–71
redox reactions, 168
Binary ionic compounds
defined, 60, G-3
formation, 167
formulas, 66–67
naming, 66
Binding energy. *See* Nuclear binding energy
Binnig, Gerd, 468
Biodiesel, 257
Biological antifreeze, 533
Biological energetics
ATP, 908–909, 908*f*, 909*f*, 967–968, 968*f*
electrochemistry, 967–968, 967*f*–968*f*
Biological fixation
carbon, 987
nitrogen, 989
Biological macromolecules
amino acids, 507–508, 508*f*, 667–668, 667*f*
base pairs, 671
cellulose, 513–514, 514*f*, 666
chromosomes, 671, 674
disaccharides, 666, 666*f*
DNA, 512–513, 512*f*–513*f*, 670–675, 671*f*–675*f*
DNA replication, 672–673, 673*f*
double helix, 671, 671*f*
fibrous proteins, 669, 669*f*
genetic code, 671
globular proteins, 669, 669*f*
glycogen, 667
intermolecular forces and, 507–514
mononucleotides, 670
monosaccharides, 666, 666*f*
mRNA, 672
as natural polymers, 665
nucleic acids, 512–513, 513*f*, 670–675
nucleotides, 512, 670–671, 670*f*, 674–675, 674*f*, 675*f*
peptide bond, 668
phospholipids, 511, 511*f*
polysaccharides, 665–667, 666*mn*
protein synthesis, 671–672, 672*f*

proteins, 507–509, 508*f*, 509*f*, 667–669
ribosome, 672
RNA, 670, 672
starch, 666–667
sugars, 665–666, 666*f*
tRNA, 672
Biological receptor, 402–403, 402*f*
Biological systems
breathing and the gas laws, 197*mn*
diffusion and movement of oxygen from lungs to
blood, 217
Biomass conversion, 256–257, G-3
Biopolymers, 484
nucleic acids, 670–673
polysaccharides, 666–667
proteins, 667–669
Biosensor, 491
Biosphere, 187*mn*, 983, 984, 984*t*, 986, 990, G-3
Bismuth
bond type, 584*t*
electron configuration, 328–329
ionic compounds, 326
melting point, 584*t*
oxides, 327
properties, 326, 327, 328–329, 595–596, 598
Bismuth subsalicylate, 597
Bisulfate ion, 68*t*
Bisulfite ion, 608
Blackbody radiation, 274, 274*f*, 274*n*
Blast furnace, 1000–1002, 1001*f*, G-3
Bleach, 574*mn*, 611, 613
Block copolymer, 489
Body-centered cubic unit cell, 464, 465*f*, 466, 467*f*, G-3
Bohr, Niels, 277, 287*mn*, 304*mn*, 1030*mn*
Bohr model of the hydrogen atom, 277–279, 278*f*
Bohrium, 1082*mn*
Boiling point
alkali metals, 572
alkaline earth metals, 576
alkanes, 639*f*
boron family elements, 580
carbon family elements, 586
defined, 447, G-3
dipole moment and, 452, 452*f*
halogen family elements, 612
hydrogen bonding and, 452, 452*f*
ionic compounds, 350*t*
metals, 370, 370*t*
molar mass and, 455, 455*f*
molecular shape and, 455, 455*f*
nitrogen family elements, 596
noble gases, 618
oxygen family elements, 604
phase changes, 447
solubility and, 506*t*
solution, 529–530, 531
vapor pressure and, 447
water, 25*f*
Boiling point elevation (Δ*T*ᵦ), 529–530, 530*t*, 531, G-3
Boltzmann, Ludwig, 210, 884
Boltzmann constant, 884
Boltzmann equation, 884
Bomb calorimeter, 248, 248*f*
Bond angle, 388–397, 388*f*, 394*f*, G-3
Bond energy (BE)
calculating heat of reaction, 359–361, 359*f*, 360*f*
cellular electrochemistry and, 967–968, 967*f*, 968*f*
chemical change, 356, 359–363
covalent bonding, 346
defined, 353, 557, G-3
halogens, 611*f*
relation to bond order and bond length, 353–354,
353*t*, 354*t*
Bond enthalpy. *See* Bond energy
Bond length, 353–354, 353*t*, 354*f*, 354*t*, 557, 611*f*, G-3
Bond-line formulas, 74
Bond order, 352, 354*t*, 382, 557, G-3
Bond polarity, 141, 365–366, 366*f*
Bond strength, 353–355, 557. *See also* Bond energy
changes in, 356, 359
in fuels and foods, 361–363, 362*f*, 363*t*
Bonding. *See also* Covalent bonds; Ionic bonding;
Metallic bonding
alkali metals, 572
alkaline earth metals, 576
atomic properties, 341–344, 341*f*
bond energy. *See* Bond energy
bond length, 557, 611*f*

bond polarity, 365–366, 366*f*
bond strength, 557
boron family, 580
carbon family, 586
covalent. *See* Covalent bonds
defined, 60, G-4
electron-pair delocalization, 382
electronegativity, 363–369, 364*f*, 365*f*
in Group 4A(14) elements, 588
in homonuclear diatomic molecule, 424–429
intermolecular forces compared, 450, 451*t*
ionic. *See* Ionic bonding
Lewis electron-dot symbols, 343–344, 343*f*
metallic bonding, 451*t*, 474–476, 474*f*, 475*f*, 556
molecular orbital band theory, 474–476, 474*f*, 475*f*
octet rule, 344
oxygen family, 604
physical properties, effect on elements, 584
types, 341–343, 342*f*, 451*t*, 556
Bonding MO, 422–424, 422*f*, 423*f*, G-3
Bonding pair, 352, G-3
Boranes, 583–584, 584*f*, 645
Borates, 582*mn*, 985*f*
Borax, 581, 582
Borazine, 583*f*
Borazon, 483*f*, 583*f*
Boric acid, 581, 582, 591, 783*t*
Born-Haber cycle, 346–347, 347*f*, 570, G-3
Boron
bond type, 584*t*
chemistry, 582–584, 583*f*
diagonal relationship with silicon, 591
diatomic molecule, orbitals of, 426–428, 426*f*, 427*f*
electron configuration, 309, 311*f*, 312*f*
electron-deficient molecules, 384–385
ionization energy, 322, 323, 323*t*
melting point, 584*t*
properties, 568*t*, 578, 580, 582–584
Boron family
important compounds, 581
important reactions, 581
properties, 578–580, 582–584
Boron nitride, 482, 483, 483*f*, 582, 583, 583*f*
Boron oxide, 581
Boron trifluoride
electron pair acceptance, 818
hybrid orbitals, 414, 414*f*
Lewis structure, 384–385
molecular shape, 390
polarity, 400
Boyer, Paul D., 968
Boyle, Robert, 41, 193
Boyle's law
application of, 199
breathing and, 197*mn*
combined gas law, 196
defined, 194, G–3
Earth's atmosphere and, 218
ideal gas law, relationship to, 198*f*
kinetic-molecular theory, 210, 212, 212*f*
molecular view of, 212, 212*f*
volume-pressure relationship, 193–194, 193*f*, 199
Bragg, W. H., 468
Bragg, W. L., 468
Bragg equation, 468
Branch, of polymer chain, 488, G-3
Brand, Hennig, 10*f*
Brass, 507, 997*t*, 998, 998*f*
Breathing, gas laws and, 197*mn*
Breeder reactors, 1000, 1097*mn*
Bridge bond, 583, G-3
Brine, 999, 1008
British thermal unit (Btu), 241
Bromcresol green, 842*f*
Bromide ion, 65*t*
Bromine
chemistry, 614
electron configuration, 313*t*, 328
isolation, 1008
oxoacids, 616*t*
ozone and, 725
properties, 328, 610–612, 611*f*, 612, 614
stratospheric, 725
uses, 1008
Bromine trifluoride, molecular shape, 393
Bromphenol blue, 842*f*
Bromthymol blue, 842, 842*f*
Brønsted, Johannes, 156

Brønsted-Lowry acid-base definition, 156, 783, 793–798, 820, G-3
Brønsted-Lowry acids, 793–798, 816–817, 820
Brønsted-Lowry bases, 793–798, 816–817, 820
Bronze, 997t
Brownian motion, 283, 540
Buck, Linda B., 403
Buckminsterfullerene ("bucky ball"), 398, 398f, 585, 585f
Buffer, G-3. *See also* Acid-base buffer systems
Buffer capacity, 838, 838f, G-3
Buffer range, 838, G-3
Buffered solution, 122
Building-block elements, 64
But- (numerical root), 635t
Butane
    boiling point, 638t, 639f
    constitutional isomers, 638t
    depiction of molecule, 74
    formula, 71t, 638t
    isomers, 103t
    for maleic anhydride production, 120
    model, 71t
    properties, 638t, 639t
Butanoic acid, molecular structure, 658f
Butanol, 72, 72f, 652
    infrared spectrum, 358f
    solubility, 504t
    surface tension, 457t
Butanone, 652, 656f
Butene, 641, 641t
Button battery, 953, 953f
Butylamine, base-dissociation constant (K_b), A-11
Butyric acid, molecular structure, 658f

C

Cadmium
    nickel-cadmium (nicad) battery, 954
    oxidation state, 1028
    toxicity, 1037
Cadmium ion, 65t, 1037
Calcium
    abundance, 984, 984t
    electron configuration, 313, 313t
    mineral sources of, 993t
    properties, 574, 576
    reaction in water, 942
    redox titrations and ion concentration in blood, 165–166
    soaps, 510mn
    use in hydrogen production, 1010
Calcium bromide, 66
Calcium carbide, 207
Calcium carbonate, 174, 192
    acid rain and, 864, 864f
    law of definite composition, 45, 45f
    limestone caves, 859, 859f
    polyatomic ion, 63
    solubility, 857, 859
    uses, 577
Calcium chloride, to melt ice, 533
Calcium fluoride
    crystal structure, 472, 472f
    precipitation of, 147f, 858
Calcium hydroxide, 855, 856–857
Calcium hypochlorite, 613
Calcium ion, 65t, 542, 542f
Calcium oxide, 174, 574mn, 1001
Calcium phosphates, 602mn, 995
Calcium sulfate, 852
Calculations, significant figures in, 29–31
Calculator, significant figures and, 30
Calibration, 32, 32f, G-3
Calomel, 951, 951f
Calorie (cal) (unit), 240, 241, G-3
Calorimeter
    bomb, 248–249, 248f
    coffee-cup, 247, 247f
    defined, 247, G-3
    whole-body, 897, 897f
Calorimetry
    bond energy values and, 360
    constant-pressure, 247–248
    constant-volume, 248–249
    specific heat-capacity and, 246–247, 246t
Calvin, Melvin, 1087
Calx, 11–12
Calx, metal, 11–12, 12mn
Cameroon, 1986 disaster in, 203mn

Cancer
    anticancer drugs, 1044mn
    treatment with ionizing radiation, 1090
Candela (cd), 18t
Capillarity, 457–458, 459, 461, G-3
Carbohydrates
    disaccharides, 666, 666f
    energy from, 362–363
    heat of reaction from combustion, 363t
    monosaccharides, 513, 513f, 666, 666f
    photosynthesis and, 240
    polysaccharides, 513–514, 513f, 665–667, 666mn
Carbon. *See also* Diamond; Graphite; Organic chemistry;
        Organic compounds; Organic molecules;
        Organic reactions
    abundance, 984, 984t
    allotropes, 585, 585f
    bond type, 584t
    chemistry, 588–589, 631
    diatomic molecule, orbitals of, 426–427, 426f, 427f
    electron configuration, 309, 311f, 312f
    granular activated, 541
    ionization energy, 323t
    isotopes, 53
    Lewis symbol, 344
    melting point, 584t
    molecular stability, 630
    nanotubes, 490–491, 567mn
    as network covalent solids, 472–473, 473t
    in periodic table, 630f
    phase diagram, 585, 585f
    properties, 568t, 584, 586, 630
    radioactive decay of ^{14}C, 1076–1077, 1076f
    radiocarbon dating, 1078–1080, 1078n, 1079f
    as reducing agents, 995
    skeletons, 632–633, 633f
    standard state, 253
Carbon compounds, naming, 635t
Carbon cycle, 986–988, 987f
Carbon dioxide
    atmospheric, 986, 988
    as blowing agent, 203
    in carbon cycle, 986–988, 987f
    CO-shift reaction, 256
    dipole moment, 399
    dissolved in water, 460
    dry ice, 447, 449, 885
    formation from methane and oxygen, 8mn
    from gas-forming reactions, 156–157, 157f
    greenhouse effect and global warming, 257–258, 257f
    law of multiple proportions, 46, 47
    from limestone decomposition, 746–747
    linear shape, 389
    mole fraction of air at sea level, 219t
    molecular shape, 399
    phase changes, 439
    phase diagram, 448–449, 448f
    release from limestone, 192
    sequestration, 258
    in soda pop, 522mn
    suffocation from, 203mn
    supercritical, 449mn
    uses, 587
Carbon disulfide
    boiling point elevation constant, 530t
    freezing point depression constant, 530t
Carbon family
    allotropes, 585
    important compounds, 587
    important reactions, 587
    properties, 584–586, 588–591
Carbon fixation, 987
Carbon monoxide
    atmospheric concentration in volume percent, 524
    boiling point, 506t
    depiction of molecule, 74
    formation, 589
    law of multiple proportions, 46, 47
    mole fraction of air at sea level, 219t
    solubility in water, 506t
    toxicity, 1056
    uses, 587, 589
Carbon-skeleton formula, 636, 636f
Carbon steel, 507f, 1002, G-3
Carbon tetrachloride
    boiling point elevation constant, 530t
    freezing point depression constant, 530t
    polarity, 399–400
    specific heat capacity, 246t

Carbonate buffer, 840–841
Carbonate ion, 68t, 382
Carbonates
    conversion to oxides, 994
    in coral reefs, 460
    as protection from effects of acid rain, 589, 832, 864
    solubility-product constant (K_{sp}), A-13
    sources of elements, 985, 985f
    test for presence of, 857, 857f
    uses, 589
Carbonic acid, acid-dissociation constant (K_a), 803t, A-8
Carbonyl functional group, 655–656, 656f, G-3
Carbonyl sulfide, polarity, 400
Carborundum, 587
Carboxyl group, 72
Carboxylate ion, 658
Carboxylic acids
    defined, 657, G-3
    functional group, 72, 651f, 657–658, 658f
    homogeneous catalysis and, 721
    molecular structures, 658f
    odor, 659mn
    properties, 658
    strength, 788
Carcinogens, aromatic, 645mn
Carnauba wax, 507mn
Cars. *See* Automobiles
Cast iron, 1001
Catalysis, 720–724
    acid-base, 724
    in biological systems, 720, 723–724
    heterogeneous, 722, 724, 725
    homogeneous, 721–722, 724, 725
    ozone depletion and, 725
    perpetual motion, 768mn
Catalyst
    biological (enzymes), 720, 723–724, 723f–724f
    defined, 720, G-3
    heterogeneous, 721t, 722, 722mn, 724, 725
    homogeneous, 721–722, 721t, 724, 725
    lack of effect on equilibrium, 768
    in modern processes, 721t
Catalytic converter, automobile, 722mn, 864
Catalytic hydrogenation, 722, 722mn
Catenation, 588, 630, 645
Cathode, 929, 929mn, G-3
Cathode ray, 49, 49f, 49mn, G-3
Cathode ray tube, 49, 49f
Cathodic protection, 958, 958f
Cation-electron pairs, 1082–1083
Cations
    common monatomic, 65t
    common polyatomic, 68t
    defined, 60, G-3
    ionic radius, 332–333
    metal cations as Lewis acids, 819
    salts of weakly acidic cations and weakly basic
            anions, 814
    size, 558
Cave, limestone, 832, 859, 859f
Cell membrane, structure of, 511–512, 511f, 512f
Cell potential (E_{cell}), 934–951
    concentration and, 946–951
    defined, 934, G-3
    standard cell potential ($E°_{cell}$), 935–937
    voltaic cells, 934–943
Cellular electrochemistry, 967–968, 967f, 968f
Cellular metabolism, metabolic pathways,
        770, 770f
Cellulose, 513–514, 514f, 666
Celsius temperature scale, 25–26, 25f, G-3
Cement, specific heat capacity, 246t
Centi- (prefix), 19t
Ceramic materials, 481–483
    defined, 481, G-3
    preparation, 482
    structures, 483, 483f
    superconducting, 350
    uses, 482t, 483
Ceria, 1030mn
Cerium, 1030
Cerium sulfate, 520, 520f
Cesium
    atomic clock, 26, 27f
    atomic radius, 320mn
    electron configuration, 315
    properties, 570–572, 571f
Cesium ion, 65t

Cetyl palmitate, molecular structure, 659f
CFCs (chlorofluorocarbons), 589, 589mn, 725
Chadwick, James, 51, 1080
Chain reaction, nuclear reactions, 1094–1095, 1095f, G-3
Chain-reaction (chain-growth) polymers, 663–664, 663t, 664f
Chain silicates, 592, 592f
Chain silicones, 593, 593f
Chair conformation, of cyclohexane, 637
Chalcocite, 111, 113, 906
Chalcopyrite, 1002, 1002f
Change in enthalpy (Δ H). See Enthalpy change (ΔH)
Charcoal, as amorphous solid, 473
Charge, electron, 49–50, 50f, 52, 52f
Charge density, 516, G-3
Charge-induced dipole forces
　　polarizability and, 454
　　in solutions, 502–503
　　types, 454
Charles, J. A. C., 194
Charles, Jacques, 204mn
Charles's law
　　breathing, 197mn
　　combined gas law, 196
　　defined, 195, G-3
　　Earth's atmosphere and, 219
　　ideal gas law, relationship to, 198f
　　kinetic-molecular theory, 211, 213, 213f
　　molecular view of, 213, 213f
　　volume-temperature relationship, 194–195, 195f
Chelate, 1040, G-3
Chemical analysis
　　minimicroanalysis, 951mn
　　nuclear magnetic resonance (NMR) spectroscopy, 646–647, 647f
　　radioactive emissions, detection, 1075, 1075f
　　scanning tunneling microscopy, 468–469, 469f
　　spectrophotometry, 281–282, 281f, 282f
　　X-ray diffraction analysis, 468, 468f
Chemical bond. See Bonding
Chemical change. See also Chemical reactions
　　bond energy, 356, 359–363
　　chemical equation, 104
　　defined, 5, G-4
　　irreversibility, 6–7
　　physical change compared, 4f, 7
　　universality of, 140f
Chemical elements. See Element(s)
Chemical equations
　　aqueous ionic reactions, 145–146, 146f
　　calculating amounts of reactant and product, 109–112
　　defined, 104, G-4
　　formation, 253–254
　　molecular, 145, 145f, 148, 153
　　net ionic, 145f, 146, 148, 153
　　products. See Products
　　reactants. See Reactants
　　thermochemical equations, 249–250
　　total ionic, 145f, 146, 148, 153
　　writing and balancing, 104–109
Chemical equilibrium. See also Equilibrium
　　dynamic equilibrium, 173, 445, 739
　　law of, 741, G-10
Chemical formulas
　　alkanes, 636, 636f–637f
　　binary covalent compounds, 70–71
　　by combustion analysis, 101–102, 101f
　　coordination compounds, 1040–1042
　　defined, G-4
　　depiction of molecules, 74
　　empirical formula, 64, 98–99
　　ionic compounds, 66–70
　　mass percent, 96–98
　　molecular formula, 65, 99–102
　　molecular masses from, 72–73
　　molecular structures, 102–104
　　structural formula, 65
　　types, 64–65
　　unknown compounds, 98–104
Chemical kinetics, 685–726
　　defined, 685, G-4
　　reaction rates, 685–726
Chemical names. See Nomenclature
Chemical problem solving. See Problem solving
Chemical properties. See also individual chemical elements
　　of copper, 6t
　　defined, 5, G-4

Period 2 elements, 567, 568t–569t
　　transition metals, 1027–1030
Chemical reactions. See also Chemical equations; specific reactions
　　acid-base reactions, 150–157
　　aqueous ionic reactions, 145–146, 146f
　　combination reaction, 167–168, 167f, 172
　　coupling of reactions to drive a nonspontaneous change, 907
　　decomposition reaction, 168–169, 169f, 173
　　defined, 5
　　dehydration-condensation reaction, 602, 603f, 658, 659, 664, 666, 666f, 668, 670
　　displacement reaction, 169–172, 170f, 171f, 173
　　disproportionation reaction, 600, 611, 613
　　elementary steps, 715–716, 715t
　　elimination reactions, 648
　　endothermic. See Endothermic processes
　　equilibrium, 173–175, 174f
　　exothermic. See Exothermic processes
　　heat of reaction, 356, 359–361, 359f, 360f
　　ideal gas law and stoichiometry, 208–210
　　limiting reactants, 114–118, 115f
　　mass conservation, 44, 44f, 47
　　mass-mole-number relationship, 111f
　　metathesis, 148, 153
　　neutralization, 783, 784
　　nuclear. See Nuclear reactions
　　nuclear reactions compared, 1065, 1065t
　　organic reactions
　　　　addition reactions, 647–648, 663–664
　　　　alkyl group, 646
　　　　elimination reactions, 648
　　　　oxidation-reduction reactions, 649
　　　　substitution reactions, 648
　　　　types, 646–648
　　overall (net) equation, 112–113
　　oxidation-reduction reactions, 158–173
　　precipitation reactions, 146–150, 147f
　　radioactive tracer use, 1087–1088
　　rate. See Reaction rate
　　redox. See Redox reactions
　　reversibility, 173–175, 709, 738
　　sequences, 112–113
　　side reactions, 118, 118f
　　spontaneous direction, 882
　　standard entropy of reaction, 893–894
　　stoichiometry. See Stoichiometry
　　stoichiometry in solution, 125–127
　　titration. See Titration
　　unimolecular reaction, 715
　　yield, 118–120, 118f
Chemical reactivity, atomic structure and, 325–333
Chemical shift, 646
Chemical system, 237, 237f
Chemistry
　　analytical. See Chemical analysis
　　central theme, 8
　　defined, 4, G-4
　　in everyday life, 3
　　history, 10–12, 10f, 11mn, 12mn
Chernobyl disaster, 1082, 1097
Chiral molecule, 639, 639f, 640mn, G-4
Chitin, 666f
Chlor-alkali process, 573, 1014–1015, 1015f, G-4
Chlorate ion, 68t
Chloride ion, 65t
Chlorides, 368–369, 368f
Chlorine
　　abundance, 984t
　　covalent radius, 317f
　　diatomic, 317, 326
　　electron configuration, 311, 311t, 312f
　　gas, 209–210
　　industrial production, 1014–1016, 1015f
　　origin and industrial use, 187t
　　oxoacids, 616t
　　ozone depletion and, 725
　　properties, 42, 42t, 327, 610–612
　　standard state, 253
　　water treatment and, 541
Chlorine dioxide, 615, 615f
Chlorine oxides, 615–616, 615f
Chlorine trifluoride, 115–116
Chlorite ion, 68t
Chloroacetic acid, acid-dissociation constant (Kₐ), A-8
Chlorofluorocarbons (CFCs), 589, 589mn, 725
Chloroform
　　boiling point elevation constant, 530t

freezing point depression constant, 530t
　　intermolecular forces, 502f
　　polarity, 399–400
Chloromethane, molecular structure, 651t
Chlorophyll, 282, 282f, 819, 819f, 1048–1049, 1056
Chlorous acid, 787t, 811, A-8
Cholesteric, liquid crystal phase, 480, 480f, 481
Cholesterol, molecular structure, 650f
Chromate ion, 68t, 856, 856f
Chromates, solubility-product constant (Kₛₚ), A-13
Chromatography, 77, 77f, G-4
Chromium
　　abundance, 984t
　　acid strength of hydrated ion, 812t
　　appearance of, 1024f
　　chromium(II) and chromium(III), 67t
　　as dietary trace elements, 1056t
　　electron configuration, 313, 313t, 316, 1024, 1025t
　　oxidation states, 1028, 1028t
　　properties, 1030, 1030t, 1032–1033, 1032f
Chromosomes, 671, 674
Chymotrypsin, 723f, 724
cis-trans isomers, 401, 641–642, 641t, 1044, 1044mn, 1045f
Cisplatin, 1044mn
Citric acid, 783t, A-9
Claus process, 256, 1012
Clausius-Clapeyron equation, 446, G-4
Clay, 984
Clay ceramics, 481
Clock, atomic, 26, 27f
Closed-end manometer, 191, 191f
CO-shift reaction, 256
Coal, 256, 864
Coal gasification, 256, 767, G-4
Cobalt
　　abundance, 984t
　　acid strength of hydrated ion, 812t
　　appearance, 1025f
　　cobalt(II) and cobalt(III), 67t
　　coordination compounds, 1042–1043, 1042t
　　as dietary trace elements, 1056t
　　electron configuration, 313t, 1025t
　　oxidation states, 1028, 1028t
Cobalt(II) carbonate, solubility-product constant, 853t
Cobalt(II) chloride hexahydrate, 1029f
Cocaine, molecular structure, 653f
Coffee-cup calorimeter, 247, 247f
Cohesive forces, 457–458, 459
Coke process, 1000
Cold packs, 517mn
Collagen, 669, 669f
Colligative properties of solutions
　　antifreeze, 531, 533
　　boiling point elevation, 529–530, 530t, 531
　　defined, 527, G-4
　　electrolytes, 527, 537–538, 537f
　　fractional distillation, 536–537, 536f
　　freezing point depression, 530t, 531, 533, 533f
　　nonelectrolytes, 527–532, 527f, 535, 536–537
　　osmotic pressure, 532, 532f, 534, 534f, 535–536
　　solute molar mass and, 535–536
　　underlying theme of, 535
　　uses in industry and biology, 533–534, 533f–534f
　　vapor pressure lowering, 528–529, 528f
Collision energy and reaction rate, 687, 687f
Collision frequency, 217, 217mn, 219, G-4
Collision theory, 708–711, 708f–711f, G-4
Collisions, of gas particles
　　elasticity of, 211
　　gas laws and, 211–213
Colloids
　　Brownian motion in, 540
　　defined, 501, 539, G-4
　　global, 543f
　　soaps, 539mn, 540
　　structure and properties, 539–540
　　Tyndall effect, 540, 540f
　　types, 539, 539t
　　in water purification, 541–542, 541f–542f
Color, 1048–1049, 1048f, 1050t, 1050–1051, 1051f
Column chromatography, 77f
Combination reactions, 167–168, 167f, 172
Combined gas law, 196
Combustion. See also Fuels; Heat of reaction from combustion
　　biodiesel, 257
　　coal, 256
　　defined, 11, 172, G-4
　　of foods, 172

gasoline, 241, 903
hydrogen, 172*mn*, 258
Lavoisier's theory, 11–12
mass conservation, 44
methane, 360–361, 360*f*
octane, 108–109, 242, 242*f*, 903
phlogiston theory, 11
propane, 109–110, 894
spontaneous change, 883
Combustion analysis
  defined, 101, G-4
  molecular formula determination from, 101–102, 101*f*
Combustion reactions, 172
Common-ion effect, 832, 833–837, G-4
Complementary color, 1048
Complex. *See* Coordination compounds
Complex ions
  of amphoteric hydroxides, 868–869, 869*f*
  calculating concentration of, 866
  coordination compounds, 1037–1046, 1038*f*, 1039*t*
  defined, 862, 1037, G-4
  formation, 862, 865, 865*t*
  formation constant (K_f), A-12
  geometry of, 1038–1039, 1039*t*
  isomerism, 1043–1046, 1043*f*, 1045*f*
  names and formulas, 1040–1042, 1041*t*
  precipitates, solubility of, 863–864
  structure, 862, 862*f*
  valence bond theory, 1046–1048, 1047*f*, 1048*f*
    octahedral complexes, 1047, 1047*f*
    square planar complexes, 1047, 1047*f*
    tetrahedral complexes, 1048, 1048*f*
Composition, of matter, 4, G-4
Compounds
  bonding, 60–63
  calculating the molecular mass of, 72–73
  converting moles, 95–96
  covalent. *See* Covalent compounds
  defined, 42, 42*f*, G-4
  definite composition, law of, 44–45, 47
  determining the formula of unknown, 98–104
  fixed ratio, 42, 47
  formation
    covalent, 62–63, 62*f*, 158*f*, 159
    ionic, 60–61, 60*f*, 158–159, 158*f*
    from polyatomic atoms, 63, 63*f*
  formulas and names, 64–73
  ionic. *See* Ionic compounds
  law of multiple proportions, 46, 47
  mass, 72–73
  mass fraction, 44–45
  mass-mole-number relationships, 95*f*
  mass of element in, calculating, 45–46
  mass percent, 44–45
  mass percent of element in, 96–98
  mixture distinguished from, 42–43
  mixtures distinguished from, 75, 75*f*
  molar mass, 93
  physical properties of, 563
  properties, 42
  specific heat capacity, 246*t*
  standard state, 253
Compton, Arthur, 285
Computer chips. *See* Semiconductors
Concentration, 522–527
  cell potential and, 946–951
  defined, 121, 522, 523*t*, G-4
  effect of change on equilibrium, 761–764, 763*f*,
    763*t*, 768*t*
  expressing molarity, 121–122
  finding by a redox titration, 165
  Le Châtelier's principle and, 761–764, 763*f*, 763*t*, 768*t*
  mass percent [%(w/w)], 524
  molality, 523, 523*t*
  molarity, 522–523, 523*t*
  mole fraction, 523*t*, 525
  parts by mass, 523*t*, 524, 525
  parts by volume, 523*t*, 524–525
  preparing and diluting molar solutions, 122–124, 123*f*
  reaction rates and, 686
  solution stoichiometry, 121
  units of, 522
    converting, 526–527
  visualizing changes in, 125
  volume percent [%(w/v)], 524
  weak-acid equilibrium problem solving, 799–802
Concentration cell
  action of, 949, 949*f*
  applications, 950–951, 950*mn*, 951*f*

calculating potential of, 950
  defined, 948, G-4
Condensation, 438, 441, 441*f*, G-4
Condensation polymers, 664–665, 665*f*, G-4
Condensed electron configuration, 311, 311*t*, 312*f*, 313*t*
Condensed formula, 636, 636*f*
Condensed phases, 437
Conduction band, 474, 475, G-4
Conductivity. *See* Electrical conductivity
Conductometric methods, for reaction rate measurement,
  692, 693*f*
Conductors
  defined, G-4
  description, 475, 475*f*
  metals, 371
  superionic, 350
Conjugate acid-base pair
  acid-base buffer system, 833, 834*f*, 837, 839–840
  Brønsted-Lowry acid-base definition, 794–795, 795*t*
  defined, 795, G-4
  identifying, 795
  K_a and K_b relationship, 808–809
  strength, 797*f*
Conservation of energy, law of, 240, 882, G-10
Constant-pressure calorimetry, 247, 247*f*
Constant-volume calorimetry, 248, 248*f*
Constitutional (structural) isomers
  alkanes, 637–638, 638*t*
  coordination compounds, 1043–1044
  defined, G-4
  description, 103–104, 103*t*
Contact-lens rinses, 534
Contact process, 1012, G-4
Contour map, 367*f*
Control rods, 1097
Controlled orientation, 490
Convective mixing, 219
Conversion factors, 14–16, 17
  Avogadro's number as, 90, 93–94, 94*f*, 95*f*
  defined, G-4
  interconverting wavelength and frequency, 270,
    271–272
  molar mass (M) as, 93–96
  units, 14–16, 17
  use in problem solving, 14–16, 17
Coordinate covalent bond, 1046, G-4
Coordination compounds. *See also* Transition metal
    complexes (coordination compounds)
  bidentate ligands, 1039–1040, 1039*t*
  chelate, 1040
  complex ions, 1037–1046, 1038*f*, 1039*t*
  components of, 1038, 1038*f*
  constitutional isomers, 1043–1044
  coordination isomers, 1043
  coordination numbers, 464, 465*f*, 1038, 1043
  coordination theory, 1042–1043
  counter ions, 1037
  defined, 1037, G-4
  donor atoms, 1039
  enantiomers, 1044–1045, 1045*f*
  formulas and names, 1040–1042, 1041*t*
  geometric isomers, 1044, 1044*mn*, 1045*f*
  history of, 1042–1043
  isomerism, 1043–1046, 1043*f*, 1045*f*
  ligands, 1037, 1038–1040, 1039*t*
  linkage isomers, 1044
  monodentate ligands, 1039, 1039*t*
  optical isomers, 1044–1045, 1045*f*
  polydentate ligands, 1039–1040, 1039*t*
  stereoisomers, 1044–1046
  Werner, Alfred, 1042–1043
Coordination isomers, 1043, G-4
Coordination number, 464, 465*f*, 1038, 1043, G-4
Copolymer, 489, 664, G-4
Copper
  abundance, 984*t*, 985
  acid strength of hydrated ion, 812*t*
  appearance, 1025*f*
  boiling point, 390*t*
  in coins, 111
  in concentration cells, 949, 949*f*
  copper(I) and copper(II), 67*t*
  cubic crystal structure, 469
  as dietary trace element, 1056*t*
  electrical conductance, 371
  electron configuration, 313, 313*t*, 314, 316, 331,
    1024, 1025*t*
  flame test, 281, 281*f*
  flotation for recovery of, 994

isolation and electrorefining, 1002–1003, 1003*f*
  melting point, 390*t*
  oxidation states, 1028, 1028*t*
  properties of, 6*t*
  roasting, 111–112, 113
  specific heat capacity, 246*t*
  in voltaic cells, 929–933, 929*f*, 931*f*, 935–939,
    947–949, 959, 959*f*
Copper(I) oxide, 111–112, 113, 906
Copper(II) sulfate pentahydrate, 1029*f*
Copper(I) sulfide, 111–112, 113, 994, 1002–1003
Core, Earth, 981, 982*f*, G-4
Core electrons, 315
Corrosion, 956–958, 957*f*, 958*f*, G-4
Corundum, 1004
Cosmic radiation, 1084
Cosmology, element synthesis, 1099, 1100*f*
Cottrell precipitator, 540, 540*f*
Coulomb (C) (unit), 50, 934, G-4
Coulomb's law, 61, 306, 307, 348, 437, G-4
Counter ions, 1037, G-5
Counters, for detection of radioactive emissions,
  1075, 1075*f*
Coupling of reactions, 907, G-5
Covalent bonds
  bond energy, 557
  bond length, 557
  bond order, 352, 354*t*, 382, 557
  bonding pairs, 352
  characteristics of, 342, 342*f*
  coordinate, 1046
  defined, 342, 351, 556, G-5
  double, 381, 390, 640–641, 655–661, 818–819
  formation, 62–63, 62*f*, 351–352, 351*f*
  intermolecular forces compared, 451*t*
  lone pairs, 352
  model of, 351–356
  nonpolar, 365
  polar, 365, 366–367
  single, 378–380, 418–420, 635–637, 650–654, 657–661
  triple, 420, 643, 661–662
  valence bond theory. *See* Valence bond (VB) theory
Covalent compounds
  defined, 60, G-5
  dissociation, 144, 151
  formation, 62–63, 62*f*, 158*f*, 159
  ionic compounds compared, 63
  ionic compounds distinguished from, 63
  molar mass, 93
  names and formulas, 70–71
  network covalent solids, 355–356, 355*f*
  oxidation number, 159
  properties, 355–356, 355*f*
  redox reactions, 158*f*, 159, 168
  in water, 144
Covalent hydrides, 566–567
Covalent radius, 318, 354*f*, 450, 450*f*, G-5
Critical mass, 1095, 1095*f*, G-5
Critical point, 448, G-5
Critical pressure (P_c), 448–449
Critical temperature (T_c), 448
Crosslinks, in polymers, 488, G-5
Crust, Earth, 981–984, 982*f*, 984*f*, 998, G-5
Crutzen, Paul J., 725
Cryolite, 581, 1004–1005
Crystal defect, in electronic materials, 477, G-5
Crystal field splitting energy, 1050, G-5
Crystal field theory
  chlorophyll, 1048–1049
  color and, 1048–1049, 1048*f*, 1049*t*, 1050–1051,
    1050*f*, 1051*f*
  complementary color, 1048
  crystal field splitting in tetrahedral and square planar
    complexes, 1053–1054, 1053*f*
  defined, 1048, G-5
  e_g orbitals, 1050
  high-spin complexes, 1052–1053, 1052*f*
  ligand field-molecular orbital theory, 1054
  low-spin complexes, 1052–1053, 1052*f*
  magnetic properties, 1052
  octahedral complexes, 1049–1050, 1049*f*, 1050*f*, 1052
  spectrochemical series and, 1051, 1051*f*
  splitting of *d* orbitals in an octahedral field of ligands,
    1049–1050, 1049*f*, 1050*f*
  square planar complexes, 1054, 1054*f*
  strong-field ligands, 1050, 1050*f*, 1052–1053, 1052*f*
  t_{2g} orbitals, 1050
  tetrahedral complexes, 1053, 1053*f*
  weak-field ligands, 1050, 1050*f*, 1052–1053, 1052*f*

Crystal lattice, 464, 464f
Crystal structure
   determination from atomic radius, 469
   fluorite, 472, 472f, 854f
   metals, 472, 472f
   silicon dioxide, 473
   sodium chloride, 471, 471f
   zinc blende, 471, 471f
Crystal violet, 842f
Crystalline solids
   atomic radius determination, 469
   beauty of, 463f
   coordination number, 464, 465f
   crystal lattice, 464, 464f
   defined, G-5
   dissolution and, 890–891, 890f
   entropy, 890–891, 890f
   packing efficiency, 466, 467f
   types
      atomic, 470, 470f
      characteristics, 470t
      ionic, 470t, 471–472, 471f, 472f, 558
      metallic, 470t, 472, 472f
      molecular, 470, 471f
      network covalent, 470t, 472–473, 473t
   unit cell, 464, 465f, 466, 467f, 468
   x-ray diffraction analysis of, 468, 468f, 469f
Crystallinity, polymer, 486, 486f
Crystallization, 76, 76f, G-5
Crystals. See Crystalline solids
Cubanes, 398, 398f
Cubic closest packing, 466, 467f, 470f, G-5
Cubic meter (m³), 19–20, G-5
Curie, Marie Sklodowska, 1066–1067, 1067mn, 1080
Curie, Pierre, 1067, 1080
Curie (Ci) (unit), 1076, G-5
Curium, 1082mn
Current. See also Electrical conductivity
   electric, 18t, 49
   photoelectric effect and, 274–275, 275f
   strong electrolytes and, 527, 527f
Cyanate ion, resonance forms, 384
Cyanide ion, 68t, 589, 814
Cyanides, solubility-product constant (Ksp), A-13
Cyclic hydrocarbons, 637, 637f, G-5
Cyclo-S₈, 606, 606f
Cycloalkanes, 637, 637f
Cyclobutane, 398
   decomposition of, 700–701
   molecular structure, 637f
Cyclohexane, molecular structure, 637f
Cyclone separator, 993–994, 994f
Cyclopentane, 637f, 894
Cyclopropane
   half-life, decomposition of, 704
   molecular structure, 637f
Cyclotron, 1081, 1081f
Cysteine, structure of, 508f, 667f
Cytochromes, 1056
Cytosine, 670

D

d block, 578, 1023, 1023f, 1024. See also Transition
      elements
d orbital
   defined, 294, G-5
   description, 294, 294f
   orbital overlap, 411–412
   splitting of d orbitals in an octahedral field of ligands,
      1049–1050, 1049f, 1050f
Dacron, 665
Dalton, John
   atomic theory, 41, 47
   career highlights, 47mn
   humidity studies, 205
   law of multiple proportions, 46, 47
Dalton (Da), 54, G-5
Daltonism, 47mn
Dalton's law of partial pressures, 536, 753
   applying, 206–208
   defined, 206, G-5
   discovery, 205
   Earth's atmosphere and, 219
   kinetic-molecular theory, 211, 212, 212f
   molecular view of, 212, 212f
Darmstadtium, 1082mn

Data, G-5
Davisson, C., 284
de Broglie, Louis, 283
de Broglie wavelength, 284, 284t, G-5
Debye, Peter, 399
Debye (D) (unit), 399
Dec- (numerical root), 635t
Deca- (prefix), 69t
Decane
   boiling point, 639f
   formula and model, 71t
Decay constant, 1076, 1077t, G-5
Decay rate, 1074–1077, 1076f
Decay series, 1074, 1074f, G-5
Deci- (prefix), 19t
Decimal point, significant figures and, 28
Decimal prefixes, 18, 19t
Decomposition reactions, 168–169, 169f, 173
Decompression sickness, 522mn
Definite composition, law of, 44–45, 45f, 47
Degree of polymerization (n), 484–485, 484t, G-5
Dehydration-condensation reaction, 602, 603f, 658, 659,
      664, 666, 666f, 668, 670, G-5
Deka- (prefix), 19t
Demeton, 992mn
Democritus, 41
Dendrimers, 398, 398f, 488
Denitrification, 989
Density (d)
   alkali metals, 572
   alkaline earth metals, 576
   boron family elements, 580
   calculating from mass and volume, 24
   carbon family elements, 586
   of common substances, 23t
   defined, 22, G-5
   gas, 188, 203–204, 203mn
   halogen family elements, 612
   as intensive property, 22
   nitrogen family elements, 596
   noble gases, 618
   oxygen family elements, 604
   SI units, 23
   transition elements, 1027, 1027f
   of water, 461–462, 461f
Dental amalgam, 997t
Deoxyhemoglobin, 1055
Deoxyribonucleic acid. See DNA (deoxyribonucleic acid)
Deoxyribose, 670
Deposition, 439, G-5
Derived units, 18, G-5
Desalination, 542, G-5
Desulfurization, 256
Detergents, 510, 992
Deuterium, 1008–1009, 1008t, 1011, 1081, 1098
Deuterons, 1081, G-5
Dextrorotatory isomer, 640
Di- (prefix), 69t
Diagonal relationships in the periodic table, 575, 575f,
      584, G-5
Diamagnetism, 330–331, 330f, 426–427, 1024,
      1029, G-5
Diamond
   covalent bonding, 356, 356f
   properties, 461–462, 461f
   standard molar entropy, 892
   structure, 583f
Diaphragm cell, 1015f
Diastereomers, 1044
Diatomic molecules, 92, 357
   beryllium, 425, 425f
   bond polarity, 399
   chlorine, 317, 326
   defined, 42
   halogens, 312
   heteronuclear, 429, 429f
   homonuclear, 424–429, 425f–428f
   lithium, 425, 425f
   orbital overlap, 411f
   oxygen, 42
Diberyllium, 425, 425f
Diborane, 581, 583f
Dichlorine heptaoxide, 615f, 616
Dichlorine monoxide, 615, 615f
Dichloroethylene, 401, 421, 421f
Dichlorofluoromethane, 391, 391f
Dichromate ion, 68t, 925–926, 925f
Dietary trace elements, 1055–1056, 1055f, 1056f, 1056t

Diethyl ether, 439f, 457t
   boiling point elevation constant, 530t
   freezing point depression constant, 530t
   infrared spectrum, 358f
Diethylamine, 653, 806t, A-11
Differentiation, geochemical, 981–982, 983f, G-5
Diffraction
   defined, 273, G-5
   electron, 284, 285f
   pattern of light, 273, 273f
   wave, 272f, 273
   X-ray, 284, 285f, 468, 468f, 469f
Diffusion
   defined, 216, G-5
   environmental flow, 438mn
   gas, 216–217, 216f
   in liquids, 216–217
Digestion of food and displacement reactions, 153mn
Dihydrogen phosphate ion, 68t
Dilithium, 425, 425f
Dimensional analysis, G-5
Dimers, 578, 579f
Dimethoxymethane, 646–647, 647f
Dimethyl ether
   boiling point, 452f
   dipole moment, 452f
   isomers, 103t
Dimethylamine, 806t, 807, A-11
Dimethylbenzene
   infrared spectrum, 358f
   structure, 644, 644f
Dimethylbutane, 456, 456f
Dimethylformamide, molecular structure, 660f
Dimethylmethanamide, molecular structure, 660f
Dinitrogen monoxide, 219t, 599t, 600
Dinitrogen pentaoxide, structures and properties, 599t
Dinitrogen tetraoxide, 117, 599t, 738–742, 739f,
      742f, 742t
Dinitrogen trioxide, structures and properties, 599t
Dinitrophenol, 842f
Dioxin, 524mn
Diphosphate ion, 603, 603f
Diphosphoric acid, 661
Dipole-dipole forces, 451–452, 451f, 451t, 502, 502f, G-5
Dipole-induced dipole forces, 451t, 454, 502f, 503, G-5
Dipole moment (μ), 399–400
   boiling point and, 452, 452f
   defined, 399, G-5
Direction of reaction, 749–751, 750f, 759–760
Disaccharide, 666, 666f, G-5
Disintegration series, 1074, 1074f
Dispersion force, 273, 273mn, 451t, 454–455, 454f, 502f,
      503, 512, 514, 518, G-5
Displacement reactions, 153mn
   double, 148, 153
   redox reactions, 169–172, 170f, 171f, 173
   single, 169
Disproportionation reaction, 600, 611, 613, G-5
Dissociation. See also Acid-dissociation constant (Ka);
      Base-dissociation constant (Kb)
   acids, 144, 151–152, 175, 785–787, 785f, 786f, 787t,
      802, 808–809, 812t, 817, 833–837, 833t,
      A-8–A-10
   bases, 175
   ionic compounds, 142–143, 142f
   photodissociation, 219
Dissolution
   covalent compound, 144
   enthalpy change, 515–517, 515f, 516f, 518f
   entropy change, 517–518, 890–891, 890f–891f
   ionic compound, 142–144, 142f
   solution process, 514–518
   solvent properties of water, 460
Dissolved air flotation, 541
Distillation, 76, 76f
   defined, G-5
   fractional, 536–537, 536f
   metallurgy and, 997
Disulfide bridge, 509
Disulfuric acid, 661, 1013, 1013f
Divers, decompression sickness, 522mn
Division, significant figures and, 28
DNA (deoxyribonucleic acid)
   depiction of molecule, 74
   replication of, 672–673, 673f
   structure of, 512–513, 512f, 513f, 664–671, 671f
   transcription of, 672, 673f
DNA fingerprinting, 674–675

DNA polymerase, 673, 675
DNA sequencing, 674–675, 674f–675f
Döbereiner, Johann, 303
Donor atom, 1039, G-5
Doped semiconductors, 477–478, 477f
Doping, 477, G-6
Double bonds
  alkenes, 640–641
  bond energy, 353t
  bond order, 352
  cis and trans structures and, 421
  color test for, 648, 648f
  defined, G-6
  functional groups with, 655–657
  functional groups with single and double bonds, 657–661
  inorganic compounds with, 657
  Lewis acid, 818–819
  Lewis structure, 381
  molecular shapes, 390
  pi (π) bond, 419
  sigma (σ) bond, 419
Double-displacement reactions, 148, 153
Double helix, DNA, 513, 513f, 671, 671f, G-6
Dow process, 1007
Downs cell, 999, 999f, G-6
Downs process, 573
Dry cell, 952
Dry ice, 447, 449, 885
dsp^2 hybrid orbitals, 1047, 1047f
d^2sp^3 hybrid orbitals, 1047, 1047f
Dual polarity
  of antibiotics, 512, 512f
  of soaps and detergents, 510, 510f
Dubnium, 1082mn
Dumas, J. B. A., 204
Dynamic equilibrium, 173, 444, G-6

E

Earth
  abundance of elements, 981–986, 982f, 984t
  atmosphere. See Atmosphere
  aurora display, 49mn
  biosphere, 187mn, 983, 984, 984t, 986, 990
  core, 981, 982f
  crust, 981–984, 982f, 984t, 988
  differentiation, 981–982, 983f
  element cycling, 986–993
  formation and layering of, 981–982, 983f, 983t
  hydrosphere, 983, 984, 984t
  life's impact on, 983–985
  lithosphere, 983, 984, 984t
  mantle, 981, 982f
EDTA, 1040mn
Effective collision, 710, G-6
Effective nuclear charge (Z_{eff}), 306–307, 307f, 318, 554, G-6
Effusion, 215–216, 215f
e_g orbitals, 1050, G-6
Einstein, Albert
  on the atomic bomb, 1095
  biography of, 283mn
  Brownian motion, 283
  mass change into energy, 44mn
  mass-energy relationship, 1091
  photon theory, 275
  on the uncertainty principle, 287mn
Eka silicon, 303, 304t
Elastomers, 488–489, 489t, 594, G-6
Electric current, 18t, 49
Electrical conductivity
  band theory and, 474–475, 475f
  colligative of solutions, 527, 527f
  ion mobility, 350f
  in ionic solutions, 142, 143f
Electrical potential, 933, 933mn
Electrocatalyst, 955–956
Electrochemical cells, 923, 928–929, G-6
Electrochemical potential, 967–968
Electrochemical redox, 996, 1007
Electrochemistry, 922–969
  batteries, 952–956, 952f–955f
  biological energetics, 967–968
  cellular, 967–968, 967f, 968f
  corrosion, 956–958, 957f, 958f
  defined, 923, G-6

electrochemical cells, 923, 928–929
electrolysis, 961–964, 962f, 996
electrolytic cells. See Electrolytic cells
  free energy and electrical work, 943–951
  galvanic cells. See Voltaic cells
  half reactions, 924–928, 949
  minimicroanalysis, 951mn
  voltaic cells. See Voltaic cells
Electrodes, 928–929, 929mn
  defined, 928, G-6
  ion-specific, 951, 951f
  microelectrode for minimicroanalysis, 951mn
  pH, 793, 793f, 950–951, 951f
Electrolysis
  aluminum production, 1004–1005, 1005f
  of aqueous ionic solutions, 963–964
  chlorine production, 1014–1016, 1015f
  defined, 961, G-6
  described, 169
  magnesium production, 1007, 1007f
  metallurgy and, 996
  of mixed molten salts, 961–962
  overvoltage, 963–964
  of pure molten salts, 961
  stoichiometry, 964–966
  of water, 962–963, 962f, 1009
Electrolytes
  acids and bases as, 151
  defined, 142, 928, G-6
  electrical conductivity of, 527, 527f
  ionic compounds as, 142
  strong, 527, 527f, 537–538, 537f
  weak, 527, 527f
Electrolytic cells
  in aluminum manufacture, 1005f
  aqueous ionic solutions, electrolysis of, 963–964
  construction and operation, 959–960
  defined, 923, 928, G-6
  Faraday, Michael, 965, 965f
  lead-acid battery, 960, 960f
  mixed molten salts, electrolysis of, 961–962
  nonstandard half-cell potentials, 962–963
  overvoltage, 963–964
  pure molten salts, electrolysis of, 961
  stoichiometry of electrolysis, 964–966
  voltaic cells compared, 928–929, 928f, 959–960, 961t
  water, electrolysis of, 962–963, 962f
Electrolytic decomposition, 169
Electrolytic enrichment, deuterium, 1011
Electromagnetic radiation, 269–271, 271mn, G-6
Electromagnetic spectrum, 271, 271f, 277, 277f, G-6
Electrometallurgy, 993
Electromotive force (emf), 934
Electron (e⁻)
  categories, 315
  cathode rays, 49
  charge, 49–50, 50f, 52, 52f, 53t, 158f, 159
  as current, 275, 275f
  de Broglie wavelength, 284, 284t
  defined, 52, G-6
  diffraction pattern, 284, 285f
  discovery of, 48mn, 49–50
  energy, 278
  filling order, 308f, 312, 423
  inner (core), 315
  mass, 50, 53t
  movement in oxidation-reduction reactions, 158–159, 158f
  outer, 315
  probable location, 287–289, 288f
  properties, 53t
  quantum numbers, 289–291, 290t, 304–305, 305t
  valence, 315, 378–379
  wave nature of, 283–286, 283f, 285f
Electron affinity (EA), 349
  defined, 324, G-6
  electronegativity compared, 363n
  trends in, 324–325, 324f, 324n, 325f
Electron capture, 1068t, 1069, G-6
Electron cloud representation, 288, 292, 292f–294f, G-6
Electron configuration
  alkali metals, 572
  alkaline earth metals, 576
  atomic properties and, 317–325
    atomic size, 317–320, 319f, 320f, 325f, 332–333, 332f
    electron affinity, 324–325, 324f, 325f
    ionization energy, 321–324, 321f–323f, 323t, 325f

aufbau principle, 308, 329
  boron family elements, 580
  carbon family elements, 586
  chemical reactivity and, 312f
  condensed, 311, 311t, 312f, 313t
  defined, 303, 554, G-6
  determining, 316–317
  filling order, 308f, 315, 315f
  within groups, 312, 314–315
  halogens, 612
  Hund's rule, 309, 311, 427, 567
  magnetic properties, 330–331
  main-group ions, 328
  many-electron atoms, 304–308
  nitrogen family elements, 596
  noble gases, 618
  notation, 308–310
  orbital diagram, 308–310, 309f
  oxygen family elements, 604
  Period 1 and Period 2 elements, 308–310, 311f, 312f
  Period 3 elements, 310–311, 311t, 312f
  Period 4 elements, 312–314, 313t
  periodic table and, 314f, 315f, 554
  principles, 314–315
  transition elements, 312–314, 315–316, 329–330
  transition metal ions, 329–331
Electron deficient, 384–385, G-6
Electron-deficient atoms, Lewis acids and, 818
Electron-deficient molecules, 384–385
Electron density, 352, 352f, 365, 366f
Electron density contour map, 352, 352f
Electron density diagram, 288, 288f, G-6
Electron density distributions, 366f
Electron-density models, 74
Electron density relief map, 292, 292f, 352, 352f, 368–369, 368f, 556f
Electron-dot symbols. See Lewis structure
Electron-group arrangements, 388–397
Electron microscope, 284mn
Electron-pair delocalization, 382, G-6
Electron pooling, 342–343, 342f
Electron repulsion, 306–307
Electron-sea model, 369–371, G-6
Electron sharing, and covalent bonding, 342, 342f, 351
Electron-spin quantum number, 304–305
Electron transfer, ionic bonding and, 158–159, 342, 342f, 344
Electron transport chain (ETC), 967–968, 968f
Electron volt (eV) (unit), 1092, G-6
Electronegativity (EN)
  alkali metals, 572
  alkaline earth metals, 576
  atomic size, 364, 365f
  bond polarity, 365–366
  boron family elements, 579f, 580
  carbon family elements, 586
  defined, 363, 555, G-6
  electron affinity compared, 363n
  halogens, 612, 616
  nitrogen family elements, 596
  noble gases, 618
  oxidation number, 364–365, 384
  oxygen family elements, 604
  Pauling scale, 363–364, 364f
  Period 2 elements, 569t
  periodic trends, 555
  periodic trends in, 364
  Robert Mulliken and, 364n
  transition elements, 1026–1027, 1026f, 1027f
  valence-state, 1033
Electronegativity difference (ΔEN), 366–369, 366f, 367f, 368f, 555, G-6
Electronic (analytical) balance, 21, 30, 30f
Electronic calculators, significant figures and, 30
Electronic materials
  crystal defects, 477
  doped semiconductors, 477–478, 477f
  doping, 477
  p-n junction manufacture, 478, 478f
  solar cells, 478mn
Electrophoresis, 674–675
Electrorefining, 997, 1002–1003, 1003f, G-6
Electrostatic interactions
  Coulomb's law, 348
  defined, 9
  and energy-level splitting, 306–308
  ionic compound properties and, 349f

Electrostatic interactions—*Cont.*
  lattice energy, 348
  nuclear stability, 1072
Element(s). *See also* Periodic table; *specific elements*
  absorption and emission spectra, 281, 281*f*
  abundance, 981–985, 982*f*, 984*t*
  of ancient Greek philosophy, 187
  atomic mass, 54, 56, 91
  atomic number, 53
  atomic size. *See* Atomic size
  atomic theory, 47
  building-block, 64
  converting moles of, 94–95
  defined, 41, G-6
  differentiation, 981–982, 983*f*
  electron configuration. *See* Electron configuration
  emission and absorption spectra, 281, 281*f*
  environmental cycles, 986–993
    carbon, 986–988, 987*f*
    defined, 986
    nitrogen, 988–989, 989*f*
    phosphorus, 990–992, 991*f*
  geochemical differentiation, 981–982, 983*f*
  ionization energy, 321–324, 321*f*–323*f*, 323*t*, 325*f*
  isolation and uses of, 998–1011
  isotopes, 53–54
  law of multiple proportions and, 46, 47
  of life, 63–64
  line spectra, 276–277, 276*f*, 277*f*
  mass, calculating, 45–46
  mass-mole-number relationships, 94*f*
  mass number *(A)*, 1066
  mass percent, 44–45
  mass percent in compound, 96–98
  metallurgy, 993–998
  mineral sources of, 993*t*
  molar mass, 92
  molecule, 42, 42*f*
  as molecules, 62, 62*f*
  name, 41
  origin in stars, 1099, 1100*f*
  oxidation number, 160
  oxides of, and acid-base behavior, 326–327, 327*f*, 559
  periodic table. *See* Periodic table
  phase changes, 563
  physical states, 562
  properties, 41–42
  in redox reactions, 166–173
  refining and alloying, 997–998, 997*t*
  sources of, 985–986, 985*f*, 993*t*
  specific heat capacity, 246*t*
  symbol, 53
  trace, 64, 1055–1056, 1055*f*, 1056*f*, 1056*t*
  uniqueness, 41
Element symbol, 53
Elementary reactions (steps), 715–716, 715*t*, G-6
Elimination reactions, 648, G-6
Emission spectrum, 280*mn*, 281, 281*f*, G-6
Empirical formula
  defined, 64, G-6
  determination of, 98–99
  different compounds with same formula, 102–103, 103*t*
  molecular formula obtained from, 99–100
Emulsion, 539, 539*t*
Enantiomers, 639, 1044–1045, 1045*f*, G-6. *See also*
  Optical isomers
End point, titration, 154, 844, G-6
End-product feedback inhibition, 770
End-to-end overlap, 419, 557, G-6
Endothermic processes
  bond energy, 353, 359
  collision theory and, 710
  defined, 244, G-6
  enthalpy change, 244–245, 244*f*
  entropy change in the surroundings, 895
  phase changes, 439
  solution process, 515, 515*f*, 517
  spontaneous reactions, 882–883, 898–899, 899*f*
-ene (suffix), 651*t*
Energy
  activation energy (*E*ₐ), 706–710, 706*f*, 707*f*, 709*f*,
    710*f*, 712*f*, 718*f*, 720–721, 720*f*
  atom, 274, 277–280, 279*f*
  blackbody radiation and the quantization of energy,
    274, 274*f*, 274*n*
  bond energy. *See* Bond energy
  common quantities of, 241*f*
  conservation of, 8, 10, 237, 240, 259, 882, 882*n*, G-10

conversion of potential to kinetic, 8–10, 9*f*
crystal-field splitting, 1050–1051
defined, 8, G-6
dispersal, 884–887
due to motion, 8
due to position, 8
electromagnetic, 269–271, 271*f*, 271*mn*
as extensive property, 24
first ionization energy. *See* First ionization energy
flow to and from a system, 237, 237*f*
forms and interconversion, 236–242
free. *See* Free energy
free energy change. *See* Free energy change
heat, 238, 238*f*
interconversion of mass and energy, 1090–1093
internal energy *(E)*, 237, 237*f*, 239*f*, 241–242, 882
ionization energy. *See* Ionization energy (IE)
kinetic energy (*E*ₖ), 236
  in chemical reaction, 356, 359
  defined, 8, G-10
  kinetic-molecular theory and, 211–215
  potential energy converted to, 8–10, 9*f*
  temperature and, 219
  temperature relationship, 213–215
lattice, 345–349, 347*f*, 348*f*, 571, 571*f*
mass change into, 44*mn*
matter distinct from, 272–273
nuclear, 259. *See also* Nuclear fission; Nuclear fusion
nuclear binding energy, 1091–1093, 1093*f*
potential energy (*E*ₚ), 236, 341
  in chemical reaction, 356, 359
  conversion to kinetic energy, 8–10, 9*f*
  in covalent bond, 351, 351*f*
  defined, 8, G-13
quantization of energy, 274
radiant, 269, 271
solar, 259
as state function, 241–242
system energy, 236–237, 237*f*, 240
thermal, 238, 982
units of, 240–241
wave-particle duality, 283–287
work, 238, 239, 239*f*
Energy efficiency, 903*mn*
Energy-level diagram, 709*f*, 710*f*
Energy-level splitting, 306–308
Energy threshold, reaction, 708
Energy transfer (Δ*E*)
  as heat, 238, 238*f*
  sign conventions, 239*t*
  as work, 238, 239, 239*f*
Energy use, future of, 256–259
Enthalpy *(H)*
  bond energy, 353, 359
  defined, 243, G-6
  endothermic processes, 244–245, 244*f*
  exothermic processes, 244–245, 244*f*
  heat of reaction, 359
  Hess's law of heat summation, 251–252
  lattice energy, 346, 347
  as state function, 244
Enthalpy change (Δ*H*)
  defined, 243, G-3
  energy change compared, 243–244
  free energy change and, 903–907, 905*t*
  Hess's law of heat summation, 251–252
  phase changes, 438–439, 439*f*, 440*f*
  solution process, 515–517, 515*f*, 516*f*, 518*f*
  spontaneous change and, 882–883
  standard heat of formation, 253–255, 253*t*
  standard heat of reaction, 253–255
  types, 245
Enthalpy diagram, 244–245, 244*f*, 515*f*, 516*f*, 518*f*, G-6
Entropy *(S)*
  change and the solution process, 517–518
  defined, 884, G-7
  number of microstates and, 884–887
  second law of thermodynamics and, 887–888
  standard molar entropy (*S*°), 888–892
Entropy change (Δ*S*)
  with atomic size or molecular complexity, 891–892
  calculating in reactions, 893–896, 898–899
  defined, 885
  with dissolution of a gas, 891
  with dissolution of solid or liquid, 890–891, 890*f*, 891*f*
  equilibrium state and, 896, 898
  with physical states and phase changes, 890, 890*f*
  predicting relative values, 888–893

quantitative meaning of, 885–887
spontaneous, 885*f*, 895, 896, 898–899
standard entropy of reaction, 893–894
in surroundings, 895–896
with temperature changes, 888, 889*f*
Environmental cycles, 986–993
  carbon, 986–988, 987*f*
  defined, 986, G-7
  nitrogen, 988–989, 989*f*
  phosphorus, 990–992, 991*f*
Environmental issues
  acid rain, 33, 33*f*, 251, 589, 832, 863–864
  air pollution, 33, 33*f*, 203*mn*
    chlorofluorocarbons (CFCs), 589, 589*mn*, 725
    lead, 587
    smog, 600, 600*f*
    sulfur dioxide, 608*mn*
    thermal pollution, 521, 521*mn*, 1097
  chemistry's role in solving, 33, 33*f*
  fuels, 256–259
  global warming, 257–258, 257*f*
  green chemistry, 256, 505*mn*, 663
  greenhouse effect, 257–258, 257*f*
  ozone depletion, 380, 725, 725*f*
  pollution
    in food chain, 652*mn*
    phosphates and, 992
    polychlorinated biphenyls (PCBs), 613
Enzyme-substrate complex (ES), 724, 724*f*, G-7
Enzymes
  as catalysts, 720, 723–724, 723*f*–724*f*
  defined, 723, G-7
  hydrogen bonding in, 452
  metabolic pathways, 770, 770*f*
  molecular shape, 403, 403*f*
Epinephrine, molecular structure, 653*f*
Epsom salt, 575
Equations. *See* Chemical equations
Equatorial group, VSEPR theory, 392, G-7
Equilibrium, 737–772
  acid-base. *See* Acid-base equilibria
  disturbances
    catalyst presence, 768, 768*t*
    concentration changes, 761–764, 763*f*, 763*t*, 768*t*
    pressure changes, 764–766, 765*f*, 768*t*
    temperature changes, 766–768, 768*mn*, 768*t*
  dynamic, 173, 445, 739
  entropy change and, 896, 898
  free energy change and, 909–910
  heterogeneous, 746–747, 747*f*
  homogeneous, 746
  law of chemical equilibrium, 741, G-10
  Le Châtelier's principle, 761–768
  metabolic pathways, 770
  phase changes
    equilibrium nature of, 444–447, 444*f*
    liquid-gas equilibria, 444–445, 445*f*
    solid-gas equilibria, 447
    solid-liquid equilibria, 447
  pressure units, 748–749
  reaction direction, 749–751, 750*f*, 759–760
  reaction quotient, 741–748
  reversible reactions, 173–175
  solubility and, 519–522
  solving problems, 752–760, 760*f*
Equilibrium constant (*K*)
  based on concentration (*K*c), 742
  based on pressure (*K*p), 748–749
  defined, 740, G-6
  free energy change, 943–945, 944*f*
  range of, 740, 740*f*
  reaction direction and, 749–751, 750*f*, 759–760
  reaction quotient and, 741–748, 909–910, 913
  solving equilibrium problems, 752–760, 760*f*
  standard cell potential and, 943–945
  in van't Hoff equation, 767–768
Equilibrium state, 174*f*
Equilibrium vapor pressure, 445
Equivalence point, 154, 843–845, G-7
Erlenmeyer flask, 20, 20*f*
Error
  random, 31–32
  systematic, 31–32
Erythrose
  formula, 103*t*
  uses, 103*t*
Esterification, 659
Esters, 651*t*, 657, 658–659, 659*f*, 659*mn*, 661, 721, G-7

ETC (electron transport chain), 967–968, 968*f*
Eth- (numerical root), 635*t*
Ethanal
    intermolecular forces, 502*f*
    molecular structure, 651*t*, 656*f*
Ethanamine, 72, 72*f*, 651*t*
Ethane
    boiling point, 639*f*
    formula and model, 71*t*
    heat of reaction, 362*t*
    molecular shape, 396, 397*f*, 418–419, 419*f*
    structure, 583*f*
Ethanenitrile, molecular structure, 651*t*
Ethanoic acid, molecular structure, 651*t*
Ethanol
    boiling point elevation constant, 530*t*
    dissolution in water, 891, 891*f*
    freezing point depression constant, 530*t*
    gasohol, 256
    heat of reaction, 362*t*
    heat of vaporization and fusion, 439*f*
    isomers and, 103*t*
    molecular shape, 396, 397*f*
    solubility, 504*t*
    specific heat capacity, 246*t*
    surface tension, 457*t*
Ethanolamine, A-11
Ethene. *See* Ethylene
Ethers, 650
Ethylamine
    base-dissociation constant (K_b), A-11
    molecular structure, 651*t*
Ethylene
    double and single bonds, 352
    free-radical polymerization of, 664, 664*f*
    hydrogenation of, 722, 722*f*
    Lewis structure, 381
    molar mass, 484
    molecular shape, 418, 419, 419*f*
    molecular structure, 651*t*
    orbital overlap in, 557
    origin and industrial use, 187*t*
    ozone, reaction with, 688–690, 688*t*, 689*f*, 690*f*
Ethylene glycol
    airplane de-icers, 533
    molecular structure, 650*f*
    specific heat capacity, 246*t*
Ethylenediamine, A-11
Ethylenediaminetetraacetate (EDTA), 1040*mn*
Ethyne, molecular structure, 651*t*
Europa (Jupiter moon), 832
Eutrophication, 989
Evaporation, of sweat, 439*mn*
Exact numbers, 30–31, G-7
Exchange reactions, 148, 153
Excited state, 278, 289, G-7
Exclusion principle, 305, 308–309, 411, G-7
Exosphere, 218, 218*f*
Exothermic processes
    bond energy, 353, 359
    collision theory and, 709–710, 709*f*, 710*f*
    defined, 244, G-7
    enthalpy change, 244–245, 244*f*
    entropy change in the surroundings, 895
    phase changes, 438
    solution process, 515, 515*f*, 517, 517*mn*
    spontaneous reactions, 882–883, 898, 899*f*
Expanded formula, 636, 636*f*
Expanded valence shells, 385–386, G-7
Expansion of gas, 188*mn*
Experiment
    controlled, 13
    defined, G-7
    in the scientific method, 13–14, 13*f*
Explosives, 398, 1096
Exponential (scientific) notation, 18, 19*t*, A-2–A-3
Extensive properties, 22, G-7
Extraction, 76, 76*f*, G-7

F

f block, 578, 1023, 1023*f*. *See also* Inner transition
        elements
f orbital, 294, 294*f*
Face-centered cubic unit cell, 464, 465*f*, 466, G-7
Fahrenheit temperature scale, 25*f*, 26
Faraday, Michael, 943, 965, 965*mn*

Faraday constant (*F*), 943, G-7
Fats
    emulsification of, 539*mn*
    energy from, 361–363
    ester group, 658–659
    heat of reaction from combustion, 363*t*
    removal by supercritical carbon dioxide, 449*mn*
Fatty acids, 510, 658, G-7
Feedstocks, hydrocarbons as, 632
Feldspars, 593
Femto- (prefix), 19*t*
Fenn, John B., 55
Fermi, Enrico, 1094
Ferrofluid, 490, 490*f*
Fertilizer, 566, 600, 602, 771*f*, 992
Feynman, Richard, 489
Fibrous proteins, 669, 669*f*
Filtration, 76, 76*f*
    defined, G-7
    for water treatment, 541, 541*f*
Fireworks, 281*f*, 616*mn*, 901, 901*f*
First ionization energy, 321–323, 321*f*, 322*f*, 345
    alkali metals, 572
    boron family elements, 580
    carbon family elements, 586
    halogens, 612
    nitrogen family elements, 596
    noble gases, 618
    oxygen family elements, 604
    Period 2 elements, 569*t*
    transition elements, 1026, 1026*f*, 1027, 1027*f*
First law of thermodynamics, 240, 882
First-order reactions
    half-life, 702–704, 703*f*
    integrated rate law, 699–700, 701*f*, 705*t*
    rate constant (*k*) for, 699*t*
    rate law, 694
    zero- and second-order reactions compared, 705*t*
Fischer-Tropsch process, 256
Fission. *See* Nuclear fission
Fixation
    carbon, 987
    defined, 987, G-7
    nitrogen, 597, 771, 988–989, 989*f*
Flame tests, 281, 281*f*, G-7
Flashbulb, chemical reaction in, 105
Floc, 541
Flotation, 994, 994*f*, G-7
Flow, environmental, 438*mn*
Flow batteries, 955
Fluoride ion, 65*t*
Fluorinating agents, 614
Fluorine
    chemistry, 614–615
    diatomic molecule, orbitals of, 411, 411*f*, 426–428,
        426*f*, 427*f*
    electron configuration, 310, 311*f*, 312*f*
    electron density, 365, 366*f*
    ionization energy, 323*t*
    orbital overlap, 411*f*
    oxidation number, 160*t*
    as oxidizing agent, 611
    oxoacids, 616*t*
    production of, 613
    properties, 569*t*, 610–612
Fluorite
    crystal structure, 472, 472*f*, 854*f*
    uses, 613
Fluoro-oxy ions, in aluminum isolation, 1005
Fluorspar (fluorite), 613
Foam, 539, 539*t*
Food
    bond strength, 361–363, 363*t*
    combustion, 172
    digestion and displacement reactions, 153*mn*
    irradiation of, 1090, 1090*f*
    preservation, hypertonic, 534
Food chain, pollution in, 652*mn*
Fool's gold, 609, 609*f*
Formal charge, 383–384, 387, G-7
Formaldehyde
    formula, 103*t*
    molecular shape, 390
    molecular structure, 656*f*
    uses, 103*t*
Formate ion, 840
Formation constant (K_f), 865–866, 865*t*, A-12, G-7
Formation equation, 253–254, G-7

Formic acid, 658*f*, 787*t*, 840, A-9
Formula. *See* Chemical formulas
Formula mass, 72, 73, G-7
Formula unit, 66, G-7
Fossil fuels, 256, 257, G-7
Fraction by mass (mass fraction), 44–45, G-7
Fractional distillation, 536–537, 536*f*, G-7
Framework silicates, 593, 593*f*
Framework silicones, 594, 594*f*
Francium, 570, 570*n*, 571*f*, 572
Frasch process, 1012, 1013*f*
Free energy (*G*)
    in biological systems, 908–909
    defined, 899, G-7
    electrical work and, 943–951
    electrochemistry, 943–951
    Gibbs free energy, 899
Free energy change (ΔG)
    calculating, 900–902
    defined, 899
    equilibrium, 909–910
    extent of reaction, 913, 913*f*
    reaction direction, 909–910, 913
    reaction spontaneity and, 899–900, 904–907, 905*t*
    standard cell potential and, 943–945, 944*f*
    standard free energy change (ΔG°), 900–902
    standard free energy of formation ($ΔG_f°$), 901–902
    temperature and, 904–907, 904*t*
    work and, 902–903
Free-radical polymerization, 664, 664*f*
Free radicals
    biological systems, 385*mn*, 1084
    defined, 385, 1084, G-7
    ionizing radiation and, 1084
    Lewis structure, 385
    nitrogen monoxide, 600*mn*
    ozone depletion and, 725
    photochemical smog, 600, 600*f*
Freezing, 438, 442, G-7
Freezing point depression ($ΔT_f$), 530*t*, 531, 533, G-7
Freons (chlorofluorocarbons), 587
Frequency (*v*), of wave, 270, 270*f*, 271–272, 275, G-7
Frequency factor (*A*), 710–711, G-7
Frisch, Otto, 1094
Fructose, structure, 666*f*
Fuel
    bond strength, 361–363, 362*f*
    combustion, 172*mn*
    nuclear, 215*mn*
    rocket, 117
Fuel cells, 172*mn*, 955–956, 955*f*
    defined, 955, G-7
    electric cars, 923*mn*
    microbial, 257
Fuel rods, 1096
Fuels. *See also* Combustion
    biomass, 256–257
    cleaner, 256–257
    coal, 256
    energy conservation, 259
    fossil, 256, 257
    greenhouse effect and global warming, 257–258, 257*f*
    hydrogen, 258–259
    methanol, 380
    nuclear energy, 259
    solar energy, 259
    wood, 256
Fuller, R. Buckminster, 398, 398*f*, 585
Fullerenes, 398, 398*f*, 585, 585*f*
Functional groups, organic
    alcohol, 650, 651*t*, 652
    aldehyde, 651*t*, 655–656, 656*f*
    alkene, 651*t*, 655
    amide, 651*t*, 658, 659–660, 660*f*
    amine, 651*t*, 652–653, 653*f*
    carbonyl, 655–656, 656*f*
    carboxylic acid, 657, 658, 658*f*
    defined, 631, G-7
    with double bonds, 655–657
    ester, 651*t*, 658–659, 659*f*
    haloalkane, 651*t*, 652
    ketone, 651*t*, 656, 656*f*
    list of important, 651*t*
    names, 72
    nitrile, 651*t*, 661
    with single and double bonds, 657–661
    with single bonds, 650, 652–654
    with triple bonds, 661–662

Fundamental units, SI, 18, 18*t*
Fusion, nuclear. *See* Nuclear fusion
Fusion (melting), 438

G

g orbital, 294
Galileo, 190*mn*
Gallium
 bond type, 584*t*
 doped semiconductor, 477, 477*f*
 electron configuration, 313*t*
 melting point, 584*t*
 properties, 370, 578–580
 quantum dots, 490
Gallium arsenide, 258, 578*mn*, 581
Galvanic cell. *See* Voltaic cells
Galvanizing process, 958
Gamma (γ) emission, 1068*t*, 1070, G-8
Gamma (γ) ray
 behavior in electric field, 1067, 1067*f*
 defined, 1067, G-8
 penetrating power, 1083, 1083*f*
 radioactive decay, 1068*t*, 1070
 wavelength, 271, 271*f*
Gangue, 993–994, 994*f*, G-8
Gas
 behavior at standard conditions, 197–198
 behavior of liquids and solids compared, 187–188
 collecting over water, 207–208
 defined, 6, 189, G-8
 density, 23, 23*t*, 188, 203–204, 203*mn*
 deviation from ideal behavior, 221–224
 diffusion, 216–217
 dissolution and entropy change, 891, 891*f*
 effusion, 215–216
 expansion
  rapid, 188*mn*
  work from, 902
 extreme conditions and, 221–223
 gas laws
  Amontons's, 196
  Avogadro's, 196–197, 211, 213, 213*f*, 219
  Boyle's, 193–194, 210, 212, 212*f*, 218
  Charles's, 194–195, 211, 213, 213*f*, 219
  combined, 196
  Dalton's law of partial pressures, 205–208, 211, 212, 212*f*, 219, 536, 753
  Henry's law, 522
  ideal, 198, 203–210, 208–210, 748
  kinetic-molecular theory, 210–217
  molecular view of, 211–213
  solving problems, 199–202
 ideal, 198, 887
 industrial uses, 187*t*
 kinetic-molecular theory, 210–217, 438
 miscibility, 188
 molar mass, 204–205
 molar volume of common gases, 221*t*
 particle arrangement in, 6, 7*f*
 pressure, 188*mn*, 189–192, 205–208, 210–213
 properties, 438*t*
 solubility
  Henry's law and, 522
  pressure and, 521–522, 521*f*, 522*mn*
  temperature and, 521
 spontaneous expansion, 885–887, 885*f*, 886*f*
 standard state, 253, 253*n*
 universal gas constant, 198
 van der Waals constant for common gases, 223, 223*t*
 van der Waals equation, 223–224
 viscosity, 188
 volume-amount relationship, 196–197, 196*f*, 200–201
 volume-pressure relationship, 187, 193–194, 193*f*, 199, 210, 212
 volume-temperature relationship, 188, 194–195, 195*f*, 200, 211, 213
Gas-forming reactions, 156–157, 157*f*
Gas-gas solutions, 506
Gas laws
 Amontons's, 196
 Avogadro's, 196–197, 211, 213, 213*f*, 219
 Boyle's, 193–194, 210, 212, 212*f*, 218
 Charles's, 194–195, 211, 213, 213*f*, 219
 combined, 196
 Dalton's law of partial pressures, 205–208, 211, 212, 212*f*, 219
 ideal, 198, 203–210, 208–210

kinetic-molecular theory, 210–217
 molecular view of, 211–213
 solving problems, 199–202
Gas-liquid chromatography (GLC), 77, 77*f*
Gas-liquid equilibria, 444–445, 444*f*, 445*f*
Gas-liquid solutions, 506
Gas-solid equilibria, 447
Gas-solid solutions, 506, 567
Gasification, of coal, 256, 767
Gasohol, 256
Gasoline, 107–108, 241
Gay-Lussac, J. L., 194, 204*mn*
Geiger-Müller counter, 1075, 1075*f*, G-8
Genes, 513, 671
Genetic code, 671, G-8
Genome, sequencing of, 674–675
Geology
 environmental flow, 438*mn*
 formation and layering of Earth, 981–982, 983*f*, 994*t*
 limestone caves, 832, 859, 859*f*
 volcanoes, 605*f*, 725
Geometric isomers, 641–642, 641*t*, 1044, 1044*mn*, 1045*f*, G-8
Geraniol, 98*mn*
Germanium
 bond type, 584*t*
 electron configuration, 313*t*
 melting point, 584*t*
 properties, 303, 304*t*, 584, 586
Germer, L., 284
Gibbs, Josiah Willard, 899, 899*mn*
Gibbs equation, 899, 900
Gibbs free energy, 899
Giga- (prefix), 19*t*
Glass
 as amorphous solid, 473
 borosilicate, 582*mn*
 etching with hydrogen fluoride, 613
 polymer, 488
 specific heat capacity, 246*t*
Glass electrode, 950
Glass transition temperature (T_g), 488
Glassware, volumetric, 20–21, 20*f*
Global warming, 257–258, 257*f*, 988
Globular proteins, 669, 669*f*
Glucosamine, 666*mn*
Glucose
 in cellulose, 513–514, 514*f*
 combustion, 172
 cyclic structure in aqueous solutions, 513*f*
 formula, 103*t*
 information contained in the chemical formula, 93*t*
 metabolism of, 44, 113
 optical isomers, 640
 production by photosynthesis, 220
 structure, 666*f*
 uses, 103*t*
Glucose-6-phosphate, 661*f*
Glutamic acid, 508*f*, 667*f*, 851
Glutamine, structure of, 667*f*
Glyceric acid, acid-dissociation constant (K_a), A-9
Glycerol, 533, 658, 659
Glycine, 121, 667*f*, 851
Glycogen, 667
Glyoxylic acid, acid-dissociation constant (K_a), A-9
Gold
 alchemy, 10
 alloy composition of jewelry, 997*t*
 density, 23*t*, 1027
 electron configuration, 316
 leaching, 994
 malleability of, 371*mn*
 molar mass of, 92
 oxidation state, 1028
 panning and fleecing for, 994*mn*
 specific heat capacity, 246*t*
Graham's law of effusion, 215–216, 215*mn*, 1097, G-8
Grain alcohol, density of, 23*t*
Gram, 91
Gram-molecular weight. *See* Molar mass
Gramicidin, 512, 512*f*
Granite, 246*t*, 593
Granular activated carbon, 541
Graphing data, A-4
Graphite
 allotropism, 585
 electrodes, 931–932, 932*f*
 in electrolytic cell for aluminum manufacture, 1005, 1005*f*

properties, 472–473, 473*t*
 specific heat capacity, 246*t*
 standard molar entropy, 892
 standard state, 253
 structure, 583*f*
 in voltaic cells, 931–932, 932*f*
Gravity, atmospheric pressure and, 189
Gray (Gy) (unit), 1083, G-8
Greases, silicones, 593
Green chemistry, 113, 203
 atom economy, 120
 defined, 120, G-8
 future of energy use, 256
 organic solvents, 505*mn*
 synthetic polymers, 663
Green vitriol, 1012
Greenhouse effect, 257–258, 257*f*, 988
Ground state, 278, 289, 308, G-8
Groups. *See also specific groups*
 electron configurations within, 312
 ionization energies trends, 321–322
 metallic behavior, 327*f*
 number, 315
 periodic table, 57, 59
 reactivities within, 312*f*
Guanine, 670
Guldberg, Cato, 741
Gypsum, 256, 575, 859, 859*f*
Gyration, radius of, 485*f*, 486, 486*n*

H

H-bond. *See* Hydrogen bond (H bond)
Haber, Fritz, 771
Haber process, 771, 771*f*, 988, G-8
Hafnium, 1030
Hahn, Otto, 1094, 1094*mn*
Halate, 616
Half-cell, 930, G-8
Half-cell potentials, 935–938, 936*f*, A-14
 nonstandard, 962–963
Half-life ($t_{1/2}$)
 defined, 1076, G-8
 first-order reactions, 702–704, 703*f*
 radioactive decay, 1076–1077, 1077*t*
 of a reaction, 702–704, 703*f*
 second-order reactions, 705
Half-reaction method, for balancing redox reactions, 924–928, G-8
Halic acid, 616, 616*t*
Halides, 654
 formation, 168, 597, 598, 605
 Group 6A elements, 605, 607
 halogen, 614
 hydrogen, 614
 lattice energy of alkali-metal halides, 348, 348*f*
 percent ionic character, 367
 solubility-product constant (K_{sp}), A-13
 sources of elements, 985, 985*f*
Halite, 616, 999, 999*mn*
Hall-Heroult process, 1004–1006
Halo- (prefix), 651*t*
Haloalkanes, 651*t*, 652, G-8
Halogens
 acid strength of hypohalous acids, 811, 811*f*
 activity series, 172
 bond energies and bond lengths, 611*f*
 chemistry, 614–616
 combination reactions, 168
 diatomic molecules, 312
 important compounds, 613
 important reactions, 613
 oxidizing ability, 611, 611*f*
 periodic table, 59
 properties, 610–612
 reaction with alkali metals, 209
 reactivity, 312*f*, 610–611
Halomethanes, 589
Halous acid, 616*t*
Hard water, 510*mn*, 541, G-8
Hassium, 1082*mn*
"Hatter's shakes," 1037*mn*
HCFCs (hydrochlorofluorocarbons), 725
HCl (hydrogen chloride), 613, 614
 acid-base reactions, 151, 155–157
 formation, 158*f*, 159
 production, 614
 stomach release of, 125–126

titration of, 843f, 844
uses, 613, 783t
HDPE (high-density polyethylene), 488
Heat (q)
  defined, 24, G-8
  endothermic processes, 244–245, 244f
  energy transfer, 238, 238f
  exothermic processes, 244–245, 244f
  kinetic energy as, 236
  phase changes and, 440–443
  sign conventions, 239t
Heat capacity
  defined, 246, G-8
  molar heat capacity, 246
  specific heat capacity, 246–247, 246t, 249mn
  of water, 460
Heat of atomization (ΔH_{atom}), 570
Heat of formation (ΔH_f)
  boron family elements, 579f
  defined, 245, G-8
  standard heat of formation, 253–255
Heat of fusion (ΔH_{fus}), 245, 439, 439f, 440f, 442, G-8
Heat of hydration (ΔH_{hydr}), 516–517, 516f, 516t, 571, G-8
Heat of reaction (ΔH_{rxn})
  calculating using bond energies, 359–361, 359f, 360f
  calorimetry, 246–249, 247f, 248f
  defined, 244, G-8
  exothermic and endothermic processes, 244
  from combustion
    calculating, 249
    carbon compounds, 362t
    fats and carbohydrates, 363t
  origin of, 356, 359
  stoichiometry of thermochemical equations, 249–250
Heat of solution (ΔH_{soln}), 515, 515f, 517mn, G-8
Heat of sublimation (ΔH_{subl}), 439, 440f, G-8
Heat of vaporization (ΔH_{vap}), 245, 439, 439f, 440f, 461, G-8
Heat summation, Hess's law of, 251–252
Heating-cooling curve, 440–442, 441f, G-8
Heavy-metal ions, 542, 1037
Hecto- (prefix), 19t
Heisenberg, Werner, 286
Heisenberg uncertainty principle, 286–287, 287mn
Helium
  in blimp, 200
  boiling point, 506t
  burning of in stars, 1099
  dihelium, molecular orbitals of, 424, 424f
  discovery of in stars, 280mn
  electron configuration, 305, 308–309, 311f, 312f
  from hydrogen burning, 1099
  mole fraction of air at sea level, 219t
  properties, 617–618
  in scuba breathing mixture, 522mn
  solubility in water, 506t
Hematite, 983f, 1000t
Heme, 74, 1055–1056, 1055f
Hemimorphite, 590, 591f
Hemoglobin, 851, 1055–1056, 1055f
Henderson-Hasselbalch equation, 837, G-8
Henry's law, 522, G-8
Hept- (numerical root), 635t
Hepta- (prefix), 69t
Heptane
  boiling point, 639f
  formula and model, 71t
Hertz (Hz) (unit), 270
Hess's law of heat summation, 251–252, 346, 347, 360, 894, G-8
Heteroatoms, 631, G-8
Heterogeneous catalysis (catalyst), 721t, 722, 722mn, 724, 725, G-8
Heterogeneous equilibrium, 746–747, 747f
Heterogeneous mixture, 75, 501, 539, G-8
Heteronuclear diatomic molecules, 429, 429f
Heterosphere, 218f, 219
Hex- (numerical root), 635t
Hexa- (prefix), 69t
Hexaamminechromium(III) ion, 1047, 1047f
Hexagonal closest packing, 466, 467f, G-8
Hexagonal cubic unit cell, 466, 467f
Hexane
  boiling point, 639f
  formula and model, 71t
  intermolecular forces, 456, 456f, 502f
  as solvent, 503–505, 504t, 518, 518f
Hexanol, solubility, 504t
HFCs (hydrofluorocarbons), 725
High-density polyethylene (HDPE), 488

High-performance (high-pressure) liquid chromatography (HPLC), 77, 77f
High-spin complexes, 1052–1053, 1052f, G-8
High-surface-area materials, 490–491
Histidine, structure of, 667f
History, of chemistry, 10–12, 10f, 11mn, 12mn
Homogeneous catalysis (catalyst), 721–722, 721t, 724, 725, G-8
Homogeneous equilibrium, 746
Homogeneous mixture, 75, 501, 539, G-8
Homologous series, 635, G-8
Homonuclear diatomic molecules, 424–429
  atomic p-orbital combinations, 425–426, 425f, 426f
  bonding in p-block, 426–428
  bonding in s-block, 425, 425f
  defined, 424, G-8
Homopolymer, 489
Homosphere, 218f, 219
Hot packs, 517mn
Hubble Space Telescope, 280mn
Hund's rule, 309, 311, 412, 427, G-8–G-9
Hybrid orbital, 412–418
  composition and orientation, 416t
  defined, 412, G-9
  molecular formula and, 416, 417f
  dsp^2 hybridization, 1047, 1047f
  d^2sp^3 hybridization, 1047, 1047f
  sp hybridization, 412, 413f, 416t
  sp^2 hybridization, 414, 414f, 416t
  sp^3 hybridization, 414–415, 414f, 415f, 416t, 1048, 1048f
  sp^3d hybridization, 415, 415f, 416t
  sp^3d^2 hybridization, 416, 416f, 416t
Hybridization, 412, G-9
Hydrargyrum (mercury), 1036
Hydrated metal ions
  dissociation constants, table of, A-12
  trends in acid strength, 811–812, 812f, 812t
Hydrated proton, 151f
Hydrates, 69, 69t, G-9
Hydration, 515, G-9
Hydration, heat of (ΔH_{hydr}), 516–517, 516f, 516t, 571
Hydration shells, 502, 503f, G-9
Hydrazine, 117, 597, 598mn
Hydride ion, 65t, 566
Hydrides
  bridge bond, 583
  catenated inorganic, 645
  covalent, 566–567
  Group 5A(15) elements, 598, 598mn
  Group 6A elements, 607, 607mn
  ionic, 566
  metal, 567
  trends in acid strength, 810, 810f
  VSEPR theory and, 418
Hydrocarbons, 632–645
  alkanes, 71, 71t, 635–639, 636f–637f, 636t, 638t, 639t
  alkenes, 640–641, 651t, 655
  alkynes, 643, 651t
  aromatic, 644–645, 644f, 645mn
  cyclic, 637, 637f
  defined, 632, G-9
  double bonds, 640–642
  isomers
    constitutional isomers, 637–638, 638t
    geometric isomers, 641–642, 641t
    optical isomers, 639–640, 639f, 640f, 640mn, 641f
  naming, 635–636, 635t, 636t, 643
  saturated, 635
  structure, 632–635, 633f
  triple bonds, 643
  unsaturated, 640
Hydrochloric acid. See Hydrogen chloride
Hydrochlorofluorocarbons (HCFCs), 725
Hydrocyanic acid, 787t, A-9
Hydrofluoric acid, 787t, A-9
Hydrogen
  abundance, 984t
  chemistry, 566–567
  combustion, 172mn, 258
  covalent bonding, 62, 62f, 351–352, 351f
  density, 23t
  deuterium, 1008–1009, 1008t, 1011
  displacement by metals, 170, 170f
  electron configuration, 308–309, 311f, 312f
  electron density, 365, 366f
  electron distribution, 141f
  formation from decomposition of water, 258–259

as fuel, 258–259, 567mn
  fuel cells, 955–956, 955f
  industrial production, 1009
  isotopes, 1008–1009, 1008t, 1011
  laboratory production, 1010
  mole fraction of air at sea level, 219t
  molecular orbitals, 422–423, 423f
  orbital overlap, 411f
  oxidation number, 160t
  periodic table, position in, 565–566, 565f
  polar nature of water, 141
  reduction with, 995
  tritium, 1008–1009, 1008t
  uses, 1010–1011
Hydrogen atom
  Bohr model, 277–279, 278f
  energy levels of the, 295, 295f
  energy states, 279–280, 279f
  ionization energy, 321
  line spectra, 276–277, 276f, 277f
  quantum number, 304
Hydrogen bisulfate ion, 609
Hydrogen bond (H bond)
  in biological systems, 453, 509, 512, 671, 673
  boiling point and, 453, 453f
  defined, 452, G-9
  intermolecular forces, 451t, 452–454, 453f
  physical properties, effect on, 563
  significance of, 452–453
  in solutions, 502, 502f
  surface tension and, 457, 457t
  water, 459, 461
Hydrogen bromide, 151
Hydrogen carbonate ion, 68t
Hydrogen chloride, 613, 614
  acid-base reactions, 151, 155–157
  formation, 158f, 159
  production, 614
  stomach release of, 125–126
  titration of, 843f, 844
  uses, 613, 783t
Hydrogen fluoride, 104, 105f, 352, 356, 359
  bond polarity, 365, 366f
  electronegativity, 363
  molecular orbitals, 429, 429f
  orbital overlap, 411f
  structure, 614mn
  uses, 613, 614mn
Hydrogen halides, 614
Hydrogen ion, 65t
  in acids, 150, 151
  Arrhenius acid-base definition and, 784–785
  concentration measurement, 950–951, 951f
  molarity determination, 152
Hydrogen peroxide
  as disinfectant, 385mn
  molecular formula, 99
  properties, 607mn
  uses, 605, 607mn
Hydrogen phosphate ion, 68t
Hydrogen sulfate ion, 68t, 609, 791t
Hydrogen sulfide
  bacterial use of, 859
  on Io (Jupiter moon), 6mn
  mole fraction of air at sea level, 219t
  removal from coal and natural gas, 763–764
  uses, 605
  VSEPR theory and, 418
Hydrogen sulfide ion, 852, 861–862
Hydrogen sulfite, 608
Hydrogenation
  catalytic, 722, 722f
  defined, 722, G-9
  of liquid oils, 1010
Hydrohalic acids, 614, 788, 810
Hydrolases, 724
Hydrolysis, 659, G-9
Hydrolysis constant (K_h), 808n
Hydrometallurgy, 993
Hydronium ion (H_3O^+), 68t
  balancing redox reaction using, 926
  calculating concentration in aqueous solution, 789–790, 790f
  defined, 784, G-9
  formation of, 151, 151f
  pH scale and, 791–793, 792f
  water and, 784–787
Hydroquinone, 1035
Hydrosphere, 983, 984, 984t, G-9

Hydrosulfuric acid, acid-dissociation constant (K_a), 803t, A-9
Hydroxide ion, 68t, 650, 852
  in bases, 150, 152
  calculating concentration in aqueous solution, 789–790, 790f
    Arrhenius acid-base definition and, 784–785
    molarity determination, 152
Hydroxides, 868–869, 869f, A-13
Hydroxyl functional group, 72
Hypertonic solution, 534
Hypo- (prefix), 68, 68f, 70
Hypobromous acid
  acid-dissociation constant (K_a), A-9
  equilibrium constant, 791t
Hypochlorite ion, 68t, 541
Hypochlorous acid, 787t, A-9
Hypofluorous acid, 611
Hypohalate, 613, 616
Hypohalous acids, 616t, 811, 811f
Hypoiodous acid, A-9
Hypothesis
  defined, 13, G-9
  in the scientific method, 13–14, 13f
Hypotonic solution, 534

I

-ic (suffix), 67, 70
Ice, 461, 461f
-ide (suffix), 66, 70
Ideal gas, 887
  defined, 193, G-9
  standard molar volume, 197–198, 197f, 198f
Ideal gas law, 203–210, 748
  defined, 198, G-9
  in limiting-reactant problem, 209
  reaction stoichiometry, 208–210, 208f
  relationship to individual gas laws, 198
  van der Waals equation and, 223–224, 223t
Ideal solution, 528, G-9
IGCC (integrated gasification combined cycle), 256
Ilmenite, 1000t
Inch, historic physical standard, 18mn
Indicator
  acid-base, 154, 154f
  in redox titration, 164, 165f
Indium
  bond type, 584t
  melting point, 584t
  properties, 578, 580
Indium oxide, 582
Induced-fit model, of enzyme action, 724, 724f, G-9
Inert-pair effects, 579
Infrared (IR) radiation, 271, 271f, G-9
Infrared spectroscopy, 357–358, 357f, 358f, G-9
Initial reaction rate, 689, 693, G-9
Inner (core) electrons, 315, 318, G-9
Inner transition elements
  actinides, 316, 1031
  defined, 1030, G-9
  electron configurations, 315–317
  lanthanides, 1030–1031, 1031mn
  periodic table, 57, 58f, 1023, 1023f
Inorganic chemistry, 59
Inorganic compounds, double bonds in, 657
Inorganic polymers, 665
Insects, irradiation of, 1090
Insoluble compounds, 142, 148t
Instantaneous reaction rate, 689, G-9
Instrument calibration, 32, 32f
Insulator
  band theory and, 475, 475f
  defined, 475, G-9
Integrated gasification combined cycle (IGCC), 256
Integrated rate law, 699–705
  defined, 699, G-9
  first-order reactions, 699–701, 701f, 705t
  half-life, 702–705, 703f
  reaction order, determining, 701–702, 701f, 702f
  second-order reactions, 700, 701f, 705t
  zero-order reactions, 700, 701f, 705t
Intensive properties, 22, G-9
Interference, 273
Interhalogen compounds, 614–615, 615f, G-9
Intermetallic compounds, 997
Intermolecular attractions, gas pressure and, 222, 222f

Intermolecular forces, 436–491
  advanced materials, 476–491
  biological macromolecules, 507–514
  in covalent compounds, 355
  defined, 437, G-9
  dipole-dipole forces, 450, 451–452, 451f, 451t
  dipole–induced dipole forces, 451t, 454, 459
  dispersion (London) forces, 451t, 454–455, 454t
  enzyme-substrate binding, 723
  hydrogen bonds, 451t, 452–454, 453f
  ion-dipole forces, 450–451, 451t, 460
  ion–induced dipole forces, 451t, 454
  liquids, 457–462
  membrane structure, 511f
  phase changes, 437–449
  polarizability, 454
  solids, 463–476
  in solution, 502–503, 502f
  summary diagram for analyzing a sample, 455f
  types compared, 451t
  van der Waals radius, 450, 450f
  vapor pressure and, 445–446, 446f
  water, 457, 457t, 460–462, 461f–462f
Internal energy (E), 237, 237f, 239f, 241–242, 882, G-9
International System of Units, 18. See also SI units
International Union of Pure and Applied Chemistry (IUPAC), 1082mn
Interstitial alloys, 506, 507f
Interstitial hydrides, 567
Intramolecular forces, 437, 450. See also Bonding
Io (Jupiter moon), 6mn, 220t
Iodates, solubility-product constant (K_{sp}), A-13
Iodide ion, 65t, 925–926, 925f
Iodine
  oxoacids, 616t
  properties, 610–612
  radioactive tracers, 1088–1089, 1089f
  sublimation, 447, 447f
Iodine pentafluoride, 393
Ion(s). See also Anions; Cations
  common monatomic, 65t
  common polyatomic, 68t
  defined, 60, G-9
  depicting formation of, 345
  heats of hydration, 516t
  names and formulas, 65, 65f, 65t
  relationship to nearest noble gas, 61, 61f
  selective precipitation, 861–862
  size, 558
  solvated, 142, 146
  spectator, 146
Ion-dipole forces, 450–451, 451t, 460
  defined, G-9
  in solutions, 502, 502f, 503f
Ion exchange, 542, 542f, G-9
Ion–induced dipole force, 451t, 454
  defined, 502, G-9
  in solutions, 502–503, 502f
Ion pairs, 350, 509, G-9
Ion-product constant for water (K_w), 789–793, 790f, 809, G-9
Ion-product expression (Q_{sp}), 852
Ion-selective (ion-specific) electrode, 951, 951t
Ionic atmosphere, 537, 537f, G-9
Ionic bonding, 451t
  Born-Haber cycle, 346–347, 347f
  characteristics of, 342, 342f
  charge, 61
  Coulomb's law, 348
  defined, 342, 556, G-9
  formation, 60–61
  lattice energy, 345–349, 348f
  model of, 344–350, 344f
  octet rule, 344
  strength, 61, 61f
Ionic charge, 159–160, 349
Ionic compounds. See also Ionic bonding
  binary, 60
  covalent compounds compared, 63
  crystal structure, 470t, 471–472, 471f, 472f
  defined, 60, G-9
  electrical conductivity, 142, 143f
  entropy change with dissolution, 890–891, 890f
  equilibria of slightly soluble ionic compounds, 851–862, 866
  formation, 60–61, 60f, 158–159, 158f
  formula mass, 72, 73
  hydrated, 69, 69t

  molar mass, 93
  names and formulas of, 66–70
  properties, 349–350, 349f, 350f, 350t, 563
  redox reactions, 158–159, 167
  solubility and temperature, 520, 520f
  solubility-product constant, A-13
  solubility rules, 148t
  spontaneous change, 883
  in water, 142–144, 142f, 148t
Ionic equations. See also Net ionic equations; Total ionic equations
  for acid-base reactions, 153, 157, 157f
  for aqueous ionic reactions, 145f, 146
Ionic equilibria, 831–870
  acid-base buffer systems, 832–841
  acid-base titration curves, 841–851
  complex ions, 862, 865–869
  slightly soluble ionic compounds, 851–862, 866
Ionic hydrides, 566, 577
Ionic radius, 332–333, 332f
  alkali metals, 571, 572
  boron family elements, 580
  carbon family elements, 586
  defined, 332, G-9
  nitrogen family elements, 596
  oxygen family elements, 604
Ionic size, 348
Ionic solids, 470t, 471–472, 471f, 472f, G-9. See also Ionic compounds
Ionic solutions
  corrosion of iron and, 957, 957f
  electrical conductivity, 142, 143f
Ionization
  defined, G-9
  photoionization, 219
  radioactive emissions, 1082
Ionization counter, 1075, 1075f
Ionization energy (IE)
  alkali metals, 571, 572
  alkaline earth metals, 574, 576
  boron family elements, 579f, 580
  carbon family elements, 586
  defined, 321, 555, G-9
  electron configuration and, 321–324, 321f–323f
  halogens, 612
  of the hydrogen atom, 280
  of metals, 326
  nitrogen family elements, 596
  noble gases, 618
  oxygen family elements, 604
  periodic trends, 321–324, 555
  successive, 323–324, 323f
  transition elements, 1026, 1026f, 1027, 1027f
Ionizing radiation
  applications, 1082–1083, 1090
  background radiation, 1084–1085
  defined, 1082, G-9
  free radical formation, 1084
  penetrating power of, 1083, 1083f
  risk from, 1085–1086, 1085mn, 1086f, 1086t
  sources, 1084–1085, 1084t
  units of radiation dose, 1083
Iridium, density of, 1027
Iron
  abundance, 984t, 985
  acid strength of hydrated ion, 812t
  alloy, 997
  appearance, 1025f
  corrosion of, 956–958, 957f, 958f
  as dietary trace element, 1055, 1056t
  electron configuration, 313t, 331, 1025t
  hematite, 983f
  iron(II) and iron(III), 67t
  metallurgy, 1000–1002, 1001f, 1002f
  minerals, 993t, 1000t
  oxidation reactions, 883
  oxidation states, 1028, 1028n, 1028t
  radioactive, 1092
  smelting, 995
  specific heat capacity, 246t
Iron phase, 982
Iron(II) hydroxide, solubility-product constant, 853t, 868–869
Irradiation of food, 1090, 1090f
Isobutane
  constitutional isomers, 638t
  formula, 638t
  properties, 638t

Isoelectronic, 328, G-9
Isoleucine
  biosynthesis of, 770, 770f
  structure of, 667f
Isomers
  chirality, 639, 639f, 640mn
  cis-trans isomerism, 401, 641–642, 641t
  constitutional (structural), 103–104, 103t, 637–638,
      638t, 1043–1044
  in coordination compounds, 1043–1046, 1043f, 1045f
  coordination isomers, 1043
  defined, 103, 637, G-10
  enantiomers, 639, 1044–1045, 1045f
  geometric, 641–642, 641t, 1044, 1044mn, 1045f
  linkage, 1044
  molecular polarity, 401
  optical, 639–640, 639f, 640f, 640mn, 641f,
      1044–1045, 1045f
  stereoisomers, 639–640, 1044–1046
Isopentane
  constitutional isomers, 638t
  formula, 638t
  properties, 638t
Isopropylamine, base-dissociation constant (K_b), A-12
Isotonic solution, 534
Isotopes. *See also* Radioisotopes
  defined, 53, 1066, G-10
  hydrogen, 1008–1009, 1008t, 1011
  mass spectrometry, 54, 55
  uranium, 215mn
Isotopic mass, 54, 92t, 1087n, G-10
Isotropic, liquid crystal phase, 478
-ite (suffix), 68, 68f
-ium (suffix), 66

J

Janssen, Pierre, 280mn
Joliot-Curie, Frederic, 1080
Joliot-Curie, Irene, 1080
Joule (J) (unit), 240, G-10
Jupiter (planet), 6mn, 220t, 832

K

K. See Equilibrium constant *(K)*
K_a. *See* Acid-dissociation constant (K_a)
Kaolinite, 481
K_b. *See* Base-dissociation constant (K_b)
Kelvin (K), 18t, 25, G-10
Kelvin (absolute) temperature scale, 25–26, 25f, 195
Ketobutyrate, 770
Ketones, 651t, 655–656, G-10
Kevlar, 480
K_f. *See* Formation constant (K_f)
Kilo- (prefix), 19t
Kilocalorie (kcal) (unit), 241
Kilogram (kg)
  defined, 21, G-10
  as SI base unit, 18t
Kilojoule (kJ) (unit), 240
Kinetic energy (E_k), 236
  in chemical reaction, 356, 359
  defined, 8, G-10
  kinetic-molecular theory and, 211–215
  potential energy converted to, 8–10, 9f
  temperature relationship, 213–215, 219
Kinetic isotope effect, 1009, 1087n
Kinetic-molecular theory
  collision frequency, 217, 217mn
  defined, 210, G-10
  diffusion, 216–217, 216f
  effusion, 215–216
  gas laws and, 210–215
  mean free path, 217
  molecular speed, 211, 211f, 213, 214–215,
      214f, 217
  molecular view of the gas laws, 211–213, 212f, 213f
  origin of pressure, 210, 211
  phase changes, 437–438, 438t, 440–442
  postulates of, 211
  rms speed, 215
  temperature and kinetic energy, 213–215
Kinetics. *See* Chemical kinetics
Kolbe, Adolf, 48mn
Krypton
  electron configuration, 313t

mole fraction of air at sea level, 219t
properties, 617–618
K_{sp}. *See* Solubility-product constant (K_{sp})
K_w. *See* Ion-product constant for water (K_w)
Kyoto Protocol (1997), 258

L

Laboratory glassware, 20–21, 20f
Lactic acid
  acid-dissociation constant (K_a), A-9
  formula, 103t
  uses, 103t
Lanthanide contraction, 1026, G-10
Lanthanides
  defined, 1030, G-10
  discovery, 1030mn
  electron configuration, 314f, 316
  properties, 1030–1031
  uses, 1031
Lanthanum, 315, 1030
Lattice, 464, 464f, G-10
Lattice energy ($\Delta H^\circ_{lattice}$)
  alkali metal chlorides, 571, 571f
  defined, 346, G-10
  importance of, 345–347
  periodic trends in, 345–347, 348f
  solution process and, 516–517, 516f
Lauterbur, Paul C., 647
Lavoisier, Antoine
  combustion theory, 11–12, 44
  element list, 57
  respiration experiments, 897, 897f
  scientific contributions, 11mn
  thermal decomposition, 168
Law of chemical equilibrium, 741, G-10
Law of conservation of energy, 240, 882, G-10
Law of definite (constant) composition, 44–45, 45f,
      47, G-10
Law of heat summation, Hess's, 251–252
Law of mass action, 741, G-10
Law of mass conservation, 44, 47, G-10
Law of multiple proportions, 46, 47, G-10
Laws of thermodynamics
  first, 240, 882
  living things and, 897
  second, 887–888
  third, 888
Layer silicates, 592, 592f
LDPE (low-density polyethylene), 488
Le Châtelier's principle, 761–768, 833, 856, 857,
      999–1000, 1012
  catalyst effect, 768, 768t
  concentration changes, 761–764, 763f, 763t, 768t
  defined, 761, G-10
  Haber process for ammonia synthesis, 771, 771f
  pressure changes, 764–766, 765f, 768t
  temperature changes, 766–768, 768t
  universality of, 761mn
Leaching, 988, 990, 994, G-10
Lead
  acid strength of hydrated ion, 812t
  bond type, 584t
  density, 23t
  electron configuration, 317, 328–329
  lead(II) and lead(IV), 67t
  melting point, 584t
  mineral sources of, 993t
  properties, 584, 586, 588, 588f
  tetraethyl, 587
Lead-acid battery, 953–954, 954f, 960, 960f
Lead(II) chloride, 851–852, 851f
Lead(II) chromate, 856, 856f
Lead(II) fluoride, 853t, 854
Lead(II) iodide, 147, 147f
Lead(II) nitrate, 147, 147f, 852
Lead(II) sulfate, 852, 853t, 854, 855t
Lecithin, 511f, 659, 659f
Leclanché cell, 952
Length
  common examples, 23f
  converting units of, 17
  English equivalents, 19t
  SI units, 18–19, 18t, 19t

Leucine, structure of, 508f, 667f

Level, 290, G-10
Leveling effect, of solvent, 816–817, G-10
Levitating power, of a superconducting oxide, 476f
Levorotatory isomer, 640
Lewis, Gilbert Newton, 343, 344mn
Lewis acid-base definition, 817–820
  adduct, 817–819
  complex ions and, 862
  defined, 817, G-10
Lewis acids, 817–820
  complex ions, 862
  with electron-deficient atoms, 818
  metal cations as, 819, 819f
  molecules as, 818–819
  with polar multiple bonds, 818–819
Lewis bases, 817–820, 862
Lewis electron-dot symbols, 343–344, 343f, G-10
Lewis structure (formula), 378–387
  defined, 378, 378n, G-10
  electron-deficient molecules, 384–385
  electron-pair delocalization, 382
  expanded valence shells, 385–387
  formal charge, 383–384, 387
  free radicals, 385
  molecules with multiple bonds, 380–381
  molecules with single bonds, 378–380
  octet rule, 378–381
      exception to, 384–387
  odd-electron molecules, 385
  resonance, 381–383
  valence-shell electron-pair repulsion theory, 388
Libby, Willard F., 1078
Life, elements of, 63
Ligand field-molecular orbital theory, 1054
Ligands
  bidentate, 1039–1040, 1039t, 1050
  common types of, 1039–1040, 1039t
  coordination compounds and, 1037–1040, 1039t
  defined, 862, 1037, G-10
  monodentate, 1039, 1039t
  polydentate, 1039–1040, 1039t
  spectrochemical series, 1051f
  splitting *d* orbitals in an octahedral field of ligands,
      1049–1050, 1049f, 1050f
  strong-field ligands, 1050, 1050f, 1052–1053, 1052f
  weak-field ligands, 1050, 1050f, 1052–1053, 1052f
Light
  as electromagnetic radiation, 269–271, 271mn
  particle nature of, 274–276
  photon theory, 274–275
  rainbows, 273mn
  speed of, 270
  from stars, 280mn
  wave nature of, 270–273
Light-water reactor, 1096f, 1097
Lightning, 988
Like-dissolves-like rule, 502, 503, 505f, G-10
Lime
  from limestone, 746–747
  slaked, 1007
  uses, 574mn, 855
Limestone, 192
  acid rain and, 589, 864
  caves, 832, 859, 859f
  decomposition, 746–747
Liming, lakes and rivers, 864
Limiting reactant
  chemical reactions involving, 114–118
  defined, 114, G-10
  in everyday life situations, 115mn
  ice cream sundae analogy, 114–115, 115f
Limiting-reactant problem
  ideal gas law use in, 209
  reactant and product amounts, calculating, 117–118
  reaction tables in, 117, 126
  solving in solution, 126–127
  solving with molecular depictions, 115–117
Line spectrum, 276–277, 276f, 277f, 277n, 306f, G-10
Linear accelerator, 1081, 1081f
Linear arrangement
  defined, 389, G-10
  hybrid orbitals, 416t
  VSEPR, 389, 389f, 394f
Linear shape, 389, 389f, 393, 394f
  complex ion, 1038, 1039t
  defined, 389, G-10
  hybrid orbitals, 412, 416t
Linkage isomers, 1044, G-10

Lipid bilayer, 511, 511*f*, G-10
Lipids, 658–659, G-10
Liquid
　defined, 6, G-10
　density, 23, 23*t*
　entropy change with dissolution, 890–891, 891*f*
　intermolecular forces, 457–462
　kinetic-molecular view, 438
　particle arrangement in, 6, 7*f*
　properties, 438*t*, 459
　　capillarity, 457–458, 459, 461
　　surface tension, 457, 457*f*, 457*t*, 459, 461
　supercooled, 473
　viscosity, 458–459, 458*t*
Liquid crystal display (LCD), 481, 481*f*
Liquid crystals
　applications of, 481, 481*f*
　defined, 478, G-10
　properties, preparation, and types, 478–480,
　　479*f*, 480*f*
Liquid-gas equilibria, 444–445, 445*f*
Liquid-liquid solutions, 503–505
Liquid-solid equilibria, 447
Liquid solutions, molecular polarity and, 503–506
Liter (L) (unit), 19–20, G-10
Lithium
　anomalous behavior, 571, 571*f*
　boiling point, 390*t*
　diagonal relationship with magnesium, 575
　diatomic, 425, 425*f*
　electron configuration, 306, 309, 309*f*, 311*f*, 312*f*
　ionic bonding, 344–347
　ionization energy, 323*t*
　melting point, 390*t*
　molecular orbitals, 474, 474*f*
　nuclear fusion and, 1098
　organometallic compounds, 656–657
　properties, 568*t*, 570–572, 570*f*, 571*f*
　soaps, 510*mn*
Lithium battery, 953, 953*f*, 955, 955*f*
Lithium bromide, 573
Lithium carbonate, 573
Lithium chloride, 573
Lithium fluoride
　Born-Haber cycle, 346–347, 347*f*
　ionic bonding, 344, 344*f*
　lattice energy, 345–349, 348*f*
　percent ionic character, 367*f*
Lithium ion, 65*t*
Lithosphere, 983, 984, 984*t*, G-10
Lock-and-key model, of enzyme action, 724,
　　724*f*, G-10
Logarithms, A-1–A-2
London, Fritz, 454
London forces, 451*t*, 454. *See also* Dispersion force
Lone pairs, 352, 390, 392, 394, G-10
Low-density polyethylene (LDPE), 488
Low-spin complexes, 1052–1053, 1052*f*, G-11
Lowry, Thomas, 156
LSD (lysergic acid diethylamine), molecular structure, 660*f*
Luminous intensity, 18*t*
Lutetium, 1030
Lyotropic, liquid crystal phase, 479
Lysergic acid diethylamine (LSD), molecular structure,
　　660*f*
Lysine, 508*f*, 589*f*, 653*f*, 667*f*
Lysozyme, 724

M

Mach, Ernst, 48*mn*
Macromolecules. *See also* Biological macromolecules;
　　Polymers
　defined, 484
　synthetic, 662–665
Macronutrients, 64
Macroscopic properties, 8, 41
"Mad hatter," 1037*mn*
Magnesium
　abundance, 984*t*
　alloy, 574*mn*
　diagonal relationship with lithium, 575
　electron configuration, 311, 311*t*, 312*f*
　in flashbulbs, 105–106, 106*f*
　isolation, 1007, 1007*f*
　metallurgy, 1007–1008
　properties, 326, 327, 574–576
　uses, 1008

Magnesium chloride
　electron density distribution, 368*f*
　ionic properties, 368
Magnesium hydroxide, 125–126
Magnesium ion, 65*t*, 819, 819*f*
Magnesium oxide, 105–107, 349
　in flashbulbs, 105–106, 106*f*
　formation of, 158–159
　uses, 577
Magnetic properties
　crystal field theory, 1052
　diamagnetism, 330–331, 330*f*, 426–427, 1024, 1029
　paramagnetism, 330–331, 330*f*, 427, 428, 428*f*,
　　1024, 1029
　of transition elements, 1029
　of transition metal complexes, 1052
　of transition metal ions, 330–331
Magnetic quantum number (*m*$_l$), 289–291, 290*t*, 305*t*, G-11
Magnetite, 1000*t*
Main-group elements. *See also specific elements*
　electron affinities, 324*f*
　electron configuration, 314, 314*f*
　ionic versus atomic radii, 332*f*
　ionization energies, 322*f*
　Lewis electron-dot symbols, 343
　monatomic ion names and formulas, 65*f*
　oxidation number, 160, 160*f*, 561
　periodic patterns, 564–619
　periodic table, 57, 58*f*
Main-group ions, electron configurations of, 328–329, 328*f*
Malathion, 603*mn*
Maleic acid, acid-dissociation constant (*K*$_a$), A-9
Maleic anhydride, 120
Malleability, 371*mn*, 475
Malonic acid, acid-dissociation constant (*K*$_a$), A-10
Manganese
　abundance, 984*t*, 985
　appearance of, 1024*f*
　in batteries, 952, 952*f*
　as dietary trace elements, 1056*t*
　electron configuration, 313*t*, 1025*t*
　ocean source, 1034*mn*
　oxidation states, 1028, 1028*t*, 1033*t*
　properties, 1033–1034
Manganese(II) chloride tetrahydrate, 1029*f*
Manganese(II) sulfide, 852–853
Manometer, 191, 191*f*, G-11
Manometric methods, for reaction rate measurement,
　　693, 693*f*
Mansfield, Peter, 647
Mantle, Earth, 981, 982*f*, G-11
Many-electron atoms, 304–308
Mars (planet), atmosphere, 220*t*
Mass
　common examples, 23*f*
　conservation, 44, 44*f*, 47
　converting units of, 22
　defined, 21, G-11
　English units, 19*t*
　as extensive property, 22
　interconversion of mass and energy, 44*mn*, 1090–1093
　molality and, 523
　molar mass of gas, 204–205, 205*f*
　mole and, 90–96
　SI units, 18*t*, 19*t*, 21
　solute in solution, 122
　terminology, 92*t*
　weight distinguished from, 19
Mass action, law of, 741, G-10
Mass-action expression. *See* Reaction quotient *(Q)*
Mass/charge ratio *(m/e)*, 54
Mass-energy equivalence, 882*n*
Mass fraction, 44–45, 46
Mass laws
　definite composition, 44–45, 47
　explained by atomic theory, 47
　mass conservation, 44, 44*f*, 47
　multiple proportions, 46, 47
　visualization of, 48
Mass-mole-number relationships, 93–96, 94*f*, 95*f*, 111*f*
　chemical reaction, 111*f*
　compounds, 95*f*
　elements, 94*f*
　overview of key, 128*f*
Mass number (*A*), 53, 54, 1066, G-11
Mass percent (mass %), 44–45, 96–98, 524, G-11
Mass ratio, fixed, 42
Mass spectrometry, 54, 55, 55*f*, G-11
Mass standard, 48, 54

Matches, 603*mn*, 616
Material flow, tracers and, 1088
Mathematical operations, A-1–A-4
Matter
　ancient Greek view of, 41
　atomic theory. *See* Atomic theory
　classification, 78*f*
　composition of, 4
　conservation, 44
　de Broglie wavelength, 284, 284*t*
　defined, 4, G-11
　effects of nuclear radiation on, 1082–1086
　element. *See* Element
　energy distinct from, 272–273
　properties of, 4–6
　states of, 6–7, 7*f*, 106, 187–188, 188*f*
　types, 41–43
　wave-particle duality, 283–287
Maxwell, James Clerk, 210
Mean free path, 217, G-11
Measurement
　accuracy, 31–32, 32*f*
　density, 22–24
　length, 18–19
　mass, 21–22
　precision, 31–32, 32*f*
　SI units, 18–27
　significant figures, 28–31
　standards, 18, 18*mn*
　temperature, 24–26
　time, 26, 27*f*
　uncertainty, 27–32
　volume, 19–21
Mechanical balance, 21
Medical diagnosis, radioactive tracer use, 1088–1089,
　　1089*f*, 1089*t*
Medical tradition, in chemistry, 11
Mega- (prefix), 19*t*
Meitner, Lise, 1094, 1094*mn*
Meitnerium, 1082*mn*, 1094*mn*
Melting, 438, 440*f*, G-11
Melting point
　alkali metals, 572
　alkaline earth metals, 576
　boron family elements, 580
　carbon family elements, 586
　defined, 447, G-11
　effects on, 447
　halogens, 612
　ionic compounds, 350*t*
　metals, 370, 370*f*, 370*t*
　nitrogen family elements, 596
　noble gases, 618
　oxygen family elements, 604
Membrane-cell process, of chlorine production,
　　1015–1016
Mendeleev, Dmitri, 57, 303, 303*mn*
Mendelevium, 1031
Meniscus, 458, 458*f*
Mercury
　barometer, 189–190, 190*f*
　in batteries, 953, 1036
　Boyle's experiments with, 193, 193*f*
　calx, 11, 12*mn*
　heat of vaporization and fusion, 439*f*
　Lavoisier's experiments with, 11
　line spectra, 276*f*
　manometer, 191, 191*f*
　meniscus shape in glass, 458, 458*f*
　mercury(I) and mercury(II) ions, 67*t*
　mineral sources of, 993*t*
　oxidation state, 1028
　properties, 1036–1037
　surface tension, 457*t*, 458
　toxicity, 1036, 1037, 1037*mn*
　uses, 126
Mercury (planet), atmosphere, 220*t*
Mercury-cell process, 1015
Mercury ion, 1037
Mercury(I) iodide, solubility-product constant, 853*t*
Mercury(II) nitrate, 126–127
Mercury(II) oxide, 168, 169*f*
Mercury(II) sulfide, 126–127
Mesosphere, 218, 218*f*
Messenger RNA (mRNA), 672
Metabolic pathways, 113, 769, 769*f*, G-11
Metal(s)
　acid strength of hydrated metal ions, 811–812, 812*f*, 812*t*
　activity series, 170–172, 171*f*, 941–943

alloys, 369, 997–998
boiling point, 370, 370t
calx, 11–12, 12mn
conductivity, 371, 475, 475f
crystal structure, 471, 471f
defined, G-11
deformation of, 370, 370f
displacement reactions, 170–172, 170f, 171f
electron affinity, 325
electron configuration, 329–330
electronegativity, 364, 367
examples, 59f
forming more than one ion, 67–68, 67t
ionic bonding, 167, 342, 344
ionization energy, 325
Lewis electron-dot symbols, 343
magnetic properties, 330–331
melting point, 370, 370f, 370t
metal cations as Lewis acids, 819
metallic bonding, 342–343
names of ionic compounds, 67–68
nonmetals compared to, 341f
periodic table and, 58, 58f, 558, 562
properties, 325–327, 325f, 327f, 370–371, 370f, 370t, 558
    acid-base behavior of element oxides, 326–327, 327f
    tendency to lose electrons, 326
relative reactivities, 941–943
Metal sulfides, 609, 609f
Metallic bonding
    alkali metals, 572
    alkaline earth metals, 576
    band theory, 474–476, 474t, 475f
    breaking, 346
    characteristics of, 342–343, 342f, 556
    defined, 343, 556, G-11
    electron-sea model, 369–371
    intermolecular forces, 451t
    in mercury, 458, 458f
Metallic (interstitial) hydrides, 567
Metallic radius, 317, 317f, G-11
Metallic solids, 470t, 472, 472f, G-11
Metalloid(s)
    defined, G-11
    examples, 59f
    oxides, 327
    periodic table and, 58, 58f, 558, 562
    properties, 325
    as semiconductors, 475, 475f
    zone refining of, 997
Metallurgy, 993–998
    alloying, 997–998, 997t
    aluminum, 1004–1006, 1005, 1005mn
    bromine, 1008
    converting mineral to element, 994–996, 996f
    copper, 1002–1003, 1003f
    defined, 993, G-11
    iron and steel, 1000–1002, 1001f, 1002f
    magnesium, 1007–1008
    ore pretreatment, 993–994
    potassium, 999–1000
    refining (purifying) the element, 997
    sodium, 999
    steps in, 993f
Metathesis reactions, 148, 153, G-11
Meter (m) (unit)
    defined, 18–19, G-11
    English equivalents, 19t
    history, 18mn
    as SI base unit, 18t
    wavelength expressed in, 270
Meth- (numerical root), 635t
Methanal, molecular structure, 656f
Methane
    boiling point, 639f
    combustion of, 360–361, 360f
    cubic closest packing, 470f
    formula and model, 71t
    heat of reaction, 362t
    heat of vaporization and fusion, 439f
    hybrid orbitals, 412, 414, 414f
    for hydrogen production, 1009, 1010
    mole fraction of air at sea level, 219t
    molecular shape, 391
    origin and industrial use, 187t
    syngas, 256
    uses, 587
Methanogenesis, 257, G-11

Methanoic acid, molecular structure, 658f
Methanol
    heat of reaction, 362t
    hybrid orbitals, 417
    intermolecular forces, 502f
    Lewis structure, 380
    molecular structure, 650f, 651t
    production, 1011
    solubility, 504–505, 504f, 505f
Methionine, structure of, 667f
Methyl acetate, molecular structure, 651t
Methyl chloride
    boiling point, 452, 452f
    dipole moment, 452, 452f
    molecular structure, 651t
Methyl cyanide, molecular structure, 651t
Methyl ethanoate, molecular structure, 651t
Methyl red, 842f, 843f, 844, 845f, 849, 849f
Methylamine, 175, 806, 806f, 806t, A-11
Methylbenzene, 644, 644f
Methylbutane, 71, 71f
Methylhexane, 639f
Methylisocyanate, 203mn
Methylpropane, 103t
Metric system, 18, 18mn
Mexico City, smog of, 203mn
Meyer, Julius Lothar, 303
Mica, 592
Micelles, 540
Micro- (prefix), 19t
Microbial fuel cell, 257
Micronutrients, 64
Microstate, 884–887
Milli- (prefix), 19, 19t
Millikan, Robert, 49–50
Millikan's oil-drop experiment, 49–50, 50f
Milliliter (mL) (unit), 19, G-11
Millimeter of mercury (mmHg), (unit), 191, 192t, G-11
Minerals
    converting to elements, 994–996, 996f
    defined, 993, G-11
    as element source, 993, 993t
    iron, 1000t
    major (macronutrients), 64
Minimicroanalysis, 951mn
Misch metal, 1031
Miscibility, 188, G-11
Miscible substances, 502
Mitochondrion, 967, 967f
Mixing, convective, 219
Mixtures, 500–543. See also Colloids; Solution
    classification and separation, 75–77
    compounds distinguished from, 75, 75f
    defined, 42–43, 501, G-11
    heterogeneous, 75, 501, 539
    homogeneous, 75, 501, 539
    partial pressure of gas in gas mixture, 205–208
    racemic, 640, 640mn
    reaction direction and, 758–759
MO. See Molecular orbital
MO bond order, 423, G-11
Mobile phase, chromatography, 77, 77f
Model (theory), in the scientific method, 13–14, 13f, G-11
Molal boiling point elevation constant, 530, 530t
Molal freezing point depression constant, 530t, 531
Molality (M)
    calculating, 523
    defined, 523, 523t, G-11
Molar heat capacity (C), 246, G-11
Molar mass (M)
    boiling point and, 455, 455f
    of compounds, 93
    as conversion factor, 93–96
    defined, 92, 92t, G-11
    determining for unknown volatile liquid, 205–206, 205f
    dispersion (London) forces and, 455, 455f
    of elements, 92
    entropy and, 891
    gas, 204–205, 205f
    Graham's law of effusion, 215–216
    in molecular formula determination, 99–100
    molecular speed relationship, 214, 214f
    polymer, 484–485, 484t
    of solutes, from colligative properties, 535–536
Molar ratio, stoichiometrically equivalent, 109–112
Molar solubility, 853–855, 855t, G-11
Molar solutions, preparing and diluting, 122–124, 123f

Molar volume, of common gases, 221t
Molarity (M)
    defined, 121, 522, G-11
    expressing concentration as, 522–523
Mole (mol)
    Avogadro's number and, 90, 93–94, 94f, 95f
    of common substances, 92f
    definition, 90–92, 90f, G-11
    ions in aqueous ionic solutions, 143–144
    molar mass, 92–93
    mole-mass-number relationships, 93–96, 94f, 95f, 111f
    mole-mass-number relationships in solution, 122, 122f
    as SI base unit, 18t
    standard molar volume, 197–198, 197f
Mole fraction (X), 206, 523t, 525, G-11
Molecular compounds, 563, G-11
Molecular equation
    for acid-base reactions, 153, 157, 157f
    for aqueous ionic reactions, 145, 145f, 148
    defined, 145, G-11
Molecular formula
    converting to Lewis structure, 378–381, 378f
    defined, 65, G-11
    determining from combustion analysis, 101–102
    determining from empirical formula, 99–100
    different compounds with same formula, 103–104, 103t
    hybrid orbitals and, 416, 417f
Molecular hydrides, 566–567
Molecular (formula) mass
    calculating, 72, 73
    defined, 92t, G-11
Molecular orbital (MO)
    antibonding, 422
    bond order, 423
    bonding, 422
    defined, 422, G-11
    energy and shape of, 423, 423f
    formation, 422–423
    of hydrogen, 422–423, 423f
    of lithium, 474, 474f
    nonbonding, 429
Molecular orbital (MO) band theory, 474–476, 474f, 475f
Molecular orbital (MO) diagram, 423, 429f, G-11
Molecular orbital theory, 422–430
    antibonding MO, 422
    atomic 2p-orbital combinations, 425–426, 425f, 426f
    benzene, 430
    bonding MO, 422
    central themes, 422–424
    defined, 422, G-11
    formation of molecular orbitals, 422–423
    heteronuclear diatomic molecules
        bonding in hydrogen fluoride, 429, 429f
        bonding in nitrogen oxide, 429, 429f
    homonuclear diatomic molecules, 424–429
        atomic p-orbital combinations, 425–426, 425f, 426f
        bonding in p-block, 426–428
        bonding in s-block, 425, 425f
    ligand field-molecular orbital theory, 1054
    MO bond order, 423
    nonbonding, 429
    ozone, 430
    pi (π) MOs, 425–426, 425f
    sigma (σ) MOs, 423, 425f
Molecular polarity
    amino acid side chains, 508
    bond angle, 399
    bond polarity, 399
    defined, 399, G-11
    dipole moment, 399–400
    effect on behavior, 401
    liquid solutions and, 503–506
    molecular shape and, 399–401
    pi (π) bonds and, 421
Molecular shape, 377–403
    beautiful and practical shapes, 398, 398f
    biological receptors and odor molecules, 402–403, 402f–403f
    cubanes, 398, 398f
    defined, 388, G-11
    dendrimers, 398, 398f
    dispersion forces and, 455, 455f
    fullerenes, 398, 398f
    Lewis structure (formula), 378–387
        defined, 378, 378f
        electron-deficient molecules, 384–385
        expanded valence levels (shells), 385–386

Molecular shape—*Cont.*
  formal charge, 383–384, 387
  molecules with multiple bonds, 380–381
  molecules with single bonds, 378–380
  odd-electron molecules, 385
  resonance, 381–383
molecular polarity and, 399–401, 399*f*
nanotubes, 398, 398*f*
valence-shell electron-pair repulsion (VSEPR) theory, 388–398, 557
  bent shape, 390, 390*f*, 391*f*, 392, 394*f*, 415
  defined, 388
  electron-group arrangements, 388–389, 389*f*
  linear arrangement, 389, 389*f*, 394*f*
  molecules with multiple central atoms, 396–397, 397*f*
  octahedral arrangement, 389*f*, 393–394, 393*f*, 394*f*
  seesaw shape, 392–393, 394*f*
  square planar shape, 393*f*, 394, 394*f*
  square pyramidal shape, 393, 393*f*, 394*f*
  T shape, 392*f*, 393, 394*f*
  tetrahedral arrangement, 389*f*, 391–392, 391*f*, 394*f*
  trigonal bipyramidal arrangement, 389*f*, 392–393, 392*f*, 394*f*
  trigonal planar arrangement, 390, 390*f*, 394*f*
  trigonal pyramidal shape, 389*f*, 391, 391*f*, 394*f*
  using VSEPR to determine molecular shape, 395–396
  V shape, 390, 390*f*, 391*f*, 392, 394*f*, 415
viscosity, effect on, 458
Molecular solids, 470, 470*f*, G-11
Molecular speeds, 211, 213, 214–215, 217
Molecular volume, gas pressure and, 222, 223*f*
Molecular weight. *See* Molecular mass
Molecularity, elementary reactions and, 715–716, 715*t*, G-11
Molecules
  bonds. *See* Bonding
  chemical formula and structure of, 102–104
  defined, 42, G-11
  depictions of, 74
  elements, 42, 42*f*, 62, 62*f*, 92
  Lewis structures, 378–387
  orbitals. *See* Molecular orbital (MO)
  organic. *See* Organic molecules
  polarity. *See* Molecular polarity
  reaction rates and structure, 710–711, 711*f*
  shape. *See* Molecular shape
Molina, Mario J., 725
Molybdenum, 316, 1030, 1030*t*
Monatomic element, molar mass of, 92
Monatomic ions
  compounds formed from, 66
  defined, G-12
  formation, 60–62
  metals, 67, 67*t*
  names and formulas, 65, 65*f*, 65*t*
  properties of, 328–333
  size trends, 558
Mono- (prefix), 69*t*
Monochromatic light, 271
Monodentate ligands, 1039, 1039*t*
Monomers, 484, G-12
Mononucleotides, 512, 670, G-12
Monoprotic acids, 787, 787*t*
Monosaccharides, 513, 513*f*, 666, 666*f*, G-12
Moon, atmosphere of, 220*t*
Moon rocks, radiodating, 1079*mn*
Moseley, Henry, 304, 304*mn*
Moseley's law, 304*mn*
Motor oil, viscosity of, 459
Mulliken, Robert S., 364*n*
Multiple proportions, law of, 46, 47, G-10
Multiplication, significant figures and, 28
Muriatic acid, 783*t*
Muscovite mica, 592, 592*f*
Mylar, 665
Myosin, arrangement of filaments, 480*f*

N

n-type semiconductor, 476, 477–478, 478*f*
N/Z (neutron-to-proton) ratio, 1070–1071, 1071*f*
NAA (neutron activation analysis), 1088
NAD (nicotinamide adenine dinucleotide), 967
Naming. *See* Nomenclature
Nano- (prefix), 19*t*
Nanocomposites, 490
Nanomachines, 491, 491*f*
Nanometer (nm) (unit), 270

Nanostructuring, 490
Nanotechnology, 489–491
  defined, 489, G-12
  high-surface-area materials, 490–491
  nanomachines, 491
  nanostructured materials, 490
  optical materials, 490
  quantum dots, 490, 490*f*
  self-assembly, 490
Nanotubes, 398, 398*f*, 585*f*
Nanotubes, carbon, 490–491
Naphthalene, 645*mn*
Naproxen, 640*mn*
Natta, Giulio, 664
Natural law, G-12
Nematic, liquid crystal phase, 480, 480*f*
Neon
  atomic size, 320*mn*
  boiling point, 506*t*
  diatomic molecule, orbitals of, 426–428, 426*f*, 427*f*
  electron configuration, 310, 311*f*, 312*f*
  ionization energy, 323*t*
  mass spectrometry of, 55, 55*f*
  molar mass of, 92
  mole fraction of air at sea level, 219*t*
  properties, 569*t*, 617–618
  sign, 49*mn*
  solubility in water, 506*t*
Neopentane
  constitutional isomers, 638*t*
  formula, 638*t*
  molecular shape, 455, 455*f*
  properties, 638*t*
Neoprene, 489*t*
Neptune (planet), atmosphere, 220*t*
Nernst, Walther Hermann, 946, 946*mn*
Nernst equation, 946–947, 950, G-12
Nerve cells, as concentration cells, 950*mn*
Nerve poison, phosphorus, 992*mn*
Net ionic equations
  for acid-base reactions, 153, 157, 157*f*
  for aqueous ionic reactions, 145*f*, 146
  defined, 146, G-12
Network covalent compounds, 563
Network covalent solids, 470*t*, 472–473, 473*t*, G-12
Neutralization, 783, 784, 841. *See also* Acid-base (neutralization) reactions
Neutralization reaction. *See also* Acid-base reactions
  defined, 150, G-12
  salt formation, 152–153
Neutrino, 1069*mn*
Neutron (n^0)
  defined, 52, G-12
  discovery, 51, 1080–1081
  mass, 53*t*
  nuclide and, 1066
  number of, 53
  properties, 53*t*
Neutron activation analysis (NAA), 1088
Neutron-capture processes, 1099
Neutron star, 1099, 1100*f*
Neutron-to-proton (*N/Z*) ratio, 1070–1073, 1071*f*
Newlands, John, 303
Nickel
  abundance, 984*t*
  acid strength of hydrated ion, 812*t*
  appearance, 1025*f*
  electron configuration, 313, 313*t*, 1025*t*
  oxidation states, 1028, 1028*t*
Nickel-cadmium (nicad) battery, 954
Nickel–metal hydride (Ni-MH) battery, 954, 954*f*
Nickel(II) nitrate hexahydrate, 123, 1029*f*
Nile River, 543*mn*
Niobium, 567*mn*, 1030
Nitrate ion, 68*t*
  molecular shape, 390
  resonance structure, 383
Nitric acid, 152, 600–601, 601*f*, 863
Nitric oxide
  atmospheric concentration, 600*f*
  biochemical uses, 600*mn*
  boiling point, 506*t*
  formation, 600
  molecular orbitals, 429, 429*f*
  solubility in water, 506*t*
  structures and properties, 599*t*
  uses, 597
Nitride ion, 65*t*
-nitrile (suffix), 651*t*

Nitriles, 651*t*, 661, G-12
Nitrite ion, 68*t*
Nitrogen
  abundance, 984, 984*t*
  boiling point, 506*t*
  bond type, 584*t*
  chemistry, 599–601
  critical temperature, 449
  diatomic molecule, orbitals of, 426–428, 426*f*, 427*f*
  electron configuration, 309, 311*f*, 312*f*
  fixation, 597, 771, 988–989, 989*f*
  gas, the bends and, 522*f*
  ionization energy, 323*t*
  irradiation of, 1098
  Lewis structure, 381
  melting point, 584*t*
  mole fraction in air at sea level, 219*t*
  properties, 326, 569*t*, 595–596
  solubility in water, 506*t*
  triple bond, 352
Nitrogen cycle, 988–989, 989*f*
Nitrogen dioxide
  atmospheric concentration, 600*f*
  depiction of molecule, 74
  equilibrium with dinitrogen tetraoxide, 738–742, 739*f*, 742*f*, 742*t*
  formation, 600
  Lewis structure, 385
  mole fraction of air at sea level, 219*t*
  structures and properties, 599*t*
Nitrogen family
  chemical behavior, 598–599
  important compounds, 597
  important reactions, 597
  oxygen family compared, 603, 606–608
  physical behavior, 595–596
  properties, 595–596, 598–603
Nitrogen fixation, 597, 771
Nitrogen monoxide, 219*t*, 600*mn*. *See also* Nitric oxide
Nitrogen oxides, 599–600, 599*t*, 863, 864, 892, 892*f*
Nitrogen trifluoride, 378–379
Nitrous acid, 600–601, 601*f*, 787*t*, 791*t*, A-10
NMR (nuclear magnetic resonance) spectroscopy, 646–647, 646*f*, 647*f*, G-12
Nobelium, 1031
Noble gases
  as atomic solids, 470
  electron affinity, 325
  electron configuration, 312, 328, 328*f*
  ionization energy, 321*f*, 325
  ions formed and nearest noble gas, 61, 61*f*
  periodic table, 59
  properties, 617–618
  reactivity, 617
Node, 293
Nomenclature
  acids, 70
  alkanes, 635–636, 635*t*, 636*t*
  binary covalent compounds, 70–71
  binary ionic compounds, 66
  coordination compounds, 1040–1042
  hydrates, 69, 69*t*
  ionic compounds, 66–70
  ions, 65, 65*t*, 68*t*
  metals that form more than one ion, 67–68, 67*t*
  numerical roots, 635, 635*t*
  organic compounds, 71–72, 71*t*
  oxoanions, 68, 68*f*
  Roman numerals, 67
  transuranium elements, 1082*mn*
Non- (numerical root), 635*t*
Nona- (prefix), 69*t*
Nonane
  boiling point, 639*f*
  formula and model, 71*t*
Nonbonding MO, 429, G-12
Nonelectrolytes
  colligative properties
    nonvolatile nonelectrolyte solutions, 528–532, 535
    volatile nonelectrolyte solutions, 536–537
  defined, 144, G-12
  nonconductivity of, 527, 527*f*
Nonmetal(s)
  covalent bonding, 168, 342
  defined, G-12
  electron affinity, 325
  electronegativity, 364, 367
  examples, 59*f*

as insulators, 475, 475*f*
ionic bonding, 167, 342, 344
ionization energy, 325
Lewis electron-dot symbols, 343
metals compared to, 341*f*
oxide, 326
periodic table and, 58, 58*f*, 558, 562
properties, 325, 558
Nonmetal hydrides, trends in acid strength, 810, 810*f*
Nonpolar covalent bond, 365, G-12
Nonrechargeable batteries, 952–953
Nonstandard half-cell potentials, 962–963
Nonvolatile nonelectrolyte solutions, colligative
    properties of, 528–532, 535
Nuclear atom model, 48–51
Nuclear binding energy, 1091–1093, 1093*f*, G-12
Nuclear charge *(Z)*, effect on orbital energy,
    306, 306*f*
Nuclear energy reactors, 1096–1098, 1096*f*, 1097*mn*
Nuclear equations. *See* Nuclear reactions
Nuclear fission
    atomic bomb, 1095–1096, 1095*n*, 1096*f*
    binding energy per nucleon and, 1092, 1093*f*
    chain reaction, 1094–1095, 1095*f*
    defined, 1090, G-7
    energy from, 259
    nuclear energy reactors, 1096–1098, 1096*f*, 1097*mn*
Nuclear fuel, 215*mn*
Nuclear fusion
    binding energy per nucleon and, 1093, 1093*f*
    defined, 1090, G-8
    as energy source, 259
    promise of, 1098
Nuclear magnetic resonance (NMR) spectroscopy,
    646–647, 646*f*, 647*f*, G-12
Nuclear reactions, 1064–1101
    atomic bomb, 1095–1096, 1095*n*, 1096*f*
    balancing, 1070
    chain reaction, 1094–1095, 1095*f*
    chemical reactions compared, 1065, 1065*t*
    fission, 1094–1098
    fusion, 1090, 1098
    nuclear binding energy, 1091–1093
    nuclear energy reactors, 1096–1098, 1096*f*, 1097*mn*
    nuclear radiation, effect on matter, 1082–1086
    nuclear stability, 1070–1072 1071*f*, 1072*t*
    radioactive decay
        decay series, 1074, 1074*f*
        half-life, 1076–1077, 1076*f*, 1077*t*
        kinetics, 1074–1080
        nuclear equations, 1070
        nuclear stability and, 1070–1072, 1071*f*, 1072*t*
        predicting mode of, 1073
        radioisotopic dating, 1078–1080, 1079*f*
        rate of, 1074–1077, 1076*f*
        types of, 1067–1070, 1068*t*
    radioisotope. *See* Radioisotopes
    transmutation. *See* Nuclear transmutation
    transuranium elements, 1082, 1082*mn*, 1082*t*, 1094
Nuclear stability, and the mode of decay, 1070–1072,
    1071*f*, 1072*t*
Nuclear transmutation, 1080–1082
    bevatron accelerator, 1082*mn*
    cyclotron accelerator, 1081, 1081*f*
    defined, 1080, G-12
    early transmutation experiments, 1080–1081
    linear accelerators, 1081, 1081*f*
    particle accelerators, 1081–1082, 1081*f*, 1082*mn*
    synchrotron accelerators, 1082, 1082*mn*
    transuranium elements, 1082, 1082*mn*, 1082*t*, 1094
Nuclear waste disposal, 1097–1098
Nucleic acids, 512, 670–675, G-12
Nucleons
    binding energy per, 1092–1093, 1093*f*
    defined, 1066, G-12
Nucleoside triphosphates, 670, 670*f*, 674–675, 674*f*
Nucleotides, 670, 674
Nucleus
    components of, 52, 52*f*, 53*t*, 1066
    defined, 49, 51, G-12
    discovery, 50–51, 51*f*
    notation, 1066
    size, 52, 1066*mn*
    strong force and, 1072
Nuclide
    defined, 1066, G-12
    notation, 1066
    parent and daughter, 1067
    stability, 1070–1074, 1071*f*, 1072*t*

Nylon-66, molecular model, 74
Nylons, 665, 665*f*

O

-oate (suffix), 651*t*, 658
Observations, in the scientific method, 13–14, 13*f*, G-12
Occurrence, element, 985, G-12
Oceans
    abundance of elements, 984
    carbon cycle and, 987, 987*f*
    desalination of seawater, 542
    environmental flow, 438*mn*
    as transition metal source, 1034*mn*
Oct- (numerical root), 635*t*
Octa- (prefix), 69*t*
Octadecanoic acid, molecular structure, 658*f*
Octahedral arrangement
    defined, 393, G-12
    hybrid orbitals, 415, 416, 416*f*, 416*t*
    VSEPR, 389*f*, 393–394, 393*f*, 394*f*
Octahedral complexes
    coordination compounds, 1044, 1045*f*
    crystal field theory and, 1049–1050, 1049*f*, 1050*f*
    hemoglobin, 1055, 1055*f*
    valence bond theory and, 1047, 1047*f*
Octahedral shape, 393, 393*f*, 394*f*
    complex ion, 1039, 1039*t*
    hybrid orbitals, 416*t*
Octane
    boiling point, 639*f*
    combustion, 242, 242*f*, 903
    combustion of, 107–108
    dissolving in hexane, 518, 518*f*
    formula and model, 71*t*
    intermolecular forces, 502*f*
Octanitrocubane, 398
Octet rule, 344
    defined, 378, G-12
    exceptions to, 384–387
    Lewis structures, 378–381
Odd-electron molecules, 385
Odor
    molecular shape and, 402–403, 402*f*–403*f*
    of organic compounds, 659*mn*
-oic acid (suffix), 651*t*, 658
Oil of vitriol, 1012
Oils, silicones, 593
-ol (suffix), 650, 651*t*
Olfaction, 402–403, 402*f*
-one (suffix), 651*t*, 656
Open-end manometer, 191, 191*f*
Optical isomers, 639–640, 639*f*, 640*f*, 640*mn*, 641*f*,
    1044–1045, 1045*f*, G-12
Optically active substance, 640, 640*f*, G-12
Orbital diagram, 308–310, 309*f*, G-12
Orbital energy
    electron repulsion and, 306–307
    nuclear charge and, 306
    orbital shape and, 307–308
    penetration, 307–308, 307*f*
    shielding, 306–307, 307*f*
Orbital overlap, 557
    description, 411–412, 411*f*
    molecular rotation and, 421, 421*f*
    in single and multiple bonds, 418–421, 419*f*–420*f*
Ores
    defined, 985, G-12
    elements in, 985–986, 985*f*
    gangue, 993–994, 994*f*
    mineral, 993
    pretreating, 993–994
Organic chemistry
    carbon chemistry, 588–589, 631
    defined, 59
    history, 629, 629*mn*
Organic compounds, 628–675
    alcohols, 650, 651*t*, 652
    aldehydes, 651*t*, 655–657, 656*f*
    alkanes, 635–639, 636*f*–637*f*, 636*t*, 638*t*, 639*t*
    alkenes, 640–641, 651*t*, 655
    alkynes, 643, 651*t*
    amides, 651*t*, 658, 659–660, 660*f*
    amines, 651*t*, 652–653, 653*f*
    aromatic compounds, 644–645, 644*f*, 645*mn*
    biological macromolecules. *See* Biological
        macromolecules
    carbonyl group, 655–657, 656*f*

carboxylic acid, 651*t*, 657, 658, 658*f*
chemical diversity, factors in, 631, 631*f*
defined, 629, G-12
esters, 651*t*, 658–659, 659*f*, 659*mn*
haloalkanes, 651*t*, 652
hydrocarbons. *See* Hydrocarbons
ketones, 651*t*, 655–656, 656*f*
naming, 70–72, 71*t*, 635–636, 635*t*, 636*t*
nitriles, 651*t*, 661
odor, 659*mn*
organic molecules. *See* Organic molecules
organic synthesis, 657
organometallic compounds, 656–657
reactions. *See* Organic reactions
vitalism, 629
Wöhler, Friedrich, 629, 629*mn*
Organic molecules
    bond properties, 630
    chemical diversity, factors in, 631, 631*t*
    electron configuration, 630
    functional groups, 631
    heteroatoms, 631
    molecular stability, 630
    structural complexity, 630
Organic reactions
    addition reactions, 647–648, 663
    alkyl group, 646
    elimination reactions, 648
    oxidation-reduction reactions, 649
    substitution reactions, 648
    types, 646–648
Organic solvents, 505*mn*
Organic synthesis, 657
Organisms, reaction sequences in, 113
Organometallic compounds, 656–657, G-12
Organotin compounds, 587
Orientation probability factor *(p)*, 710–711
Orthosilicate, 590
Osmium, density of, 1027
Osmoregulator, 534
Osmosis, 532, G-12
    reverse, 542, 542*f*
Osmotic pressure, 532, 532*f*, 534, 534*f*, 535–536, G-12
Ostwald, Wilhelm, 48*mn*
Ostwald process, 600, 988
-ous (suffix), 67, 70
Outer electrons, 315, 318, G-12
Outgassing, 982, 984
Overall (net) equation, 112–113, G-12
Overvoltage, 963–964, G-12
Oxalates, solubility-product constant (K_{sp}), A-13
Oxalic acid, acid-dissociation constant (K_a), 803*t*, 805,
    A-10
Oxidation, 923, 924*f*
    defined, 159, G-12
    in early Earth atmosphere, 983
    as increase in oxidation number, 161
Oxidation number (O.N.)
    defined, 159, 560, G-12
    determining, 159–160
    electronegativity and, 364–365
    formal charge compared to, 384
    of main-group elements, 160*f*
    rules for assigning, 160, 160*t*
    transition elements, 1028, 1028*t*
Oxidation number method for balancing redox equations,
    163–164, G-12
Oxidation-reduction reaction. *See* Redox reactions
Oxidation state. *See also* Oxidation number (O.N.)
    alkali metals, 572
    alkaline earth metals, 576
    boron family elements, 579, 580
    carbon family elements, 586
    coordination compound, 1041
    defined, 560
    halogens, 612, 614–615
    main-group elements, 561
    nitrogen family elements, 596
    noble gases, 618
    oxygen family elements, 604
    primary valence, 1043
    transition elements, 1028, 1028*t*
Oxide ion, 65*t*
Oxide superconductors, 456*f*, 476
Oxides, 985*f*, 986
    acid-base behavior, 559
    boron family elements, 582
    carbonates converted to, 994
    combination reactions, 168

Index

Oxides—*Cont.*
    halogen, 610–611, 613, 615–616, 615*f*
    nitrogen, 599–600, 599*t*
    phosphorus, 602, 602*f*
    properties, 608
Oxidizing agents, 560
    defined, 159, 923, G-13
    halogens, 610–611
    in metal displacement reactions, 170–171
    recognizing, 161–162
    in redox titration, 164
    relative strengths, 937–943
Oxoacids, 654, 661
    formation, 597
    halogen, 614, 616, 616*t*
    names, 70
    nitrogen, 600–601, 600*f*
    phosphorus, 602
    strength of, 788
    trends in acid strength, 811, 811*f*
Oxoanions, 1028, 1028*f*
    defined, G-13
    halogen, 616
    naming, 68, 68*f*
    nitrogen, 600–601, 600*f*
Oxygen
    abundance, 982*f*, 983–984, 984*t*
    allotropes, 606
    atmospheric, 983, 983*f*
    boiling point, 506*t*
    chemistry, 608
    combustion reactions, 172
    critical temperature, 449
    density, 23*t*
    as diatomic molecule, 42
    diffusion movement from lungs to blood, 217
    in Earth's atmosphere, 220
    electron configuration, 309–310, 311*f*, 312*f*
    ionization energy, 323*t*
    molar mass of, 92
    mole fraction in air at sea level, 219*t*
    naming by Lavoisier, 12
    orbitals of diatomic molecule, 426–428, 426*f*, 427*f*
    oxidation number, 160*t*
    paramagnetic properties of diatomic, 428, 428*f*
    polar nature of water, 141
    properties, 569*t*, 603–604, 606–607
    solubility in water, 506*t*
    standard state, 253
    thermal decomposition of potassium chlorate, 168
Oxygen family
    allotropism, 606
    important compounds, 605
    important reactions, 605
    nitrogen family compared, 603, 606–608
    periodic table, 59
    properties, 603–604, 606–610
Oxyhemoglobin, 1055
Ozone
    decomposition, 606
    depiction of molecule, 74
    depletion of, 219, 380, 725, 725*f*
    as disinfectant, 541
    formation, 219, 600
    hybrid orbitals, 414
    Lewis structure, 381–382, 383
    mole fraction of air at sea level, 219*t*
    molecular orbitals, 430, 430*f*
    nitrogen fixation and, 988
    reaction with ethylene, 688–690, 688*t*, 689*f*, 690*f*
    stratospheric, 219, 380, 606, 725

P

p block, 425, 426–428, 554, 578, G-13
p-n junction, 478, 478*f*
p orbital, 293, 293*f*, 425–426, 425*f*, 426*f*, G-13
    orbital overlap, 411–412
p-scale, 791
p-type semiconductor, 476
Packing efficiency, unit cells and, 466, 467*f*, G-13
Palladium, 506, 567*mn*
PANs (peroxyacylnitrates), 600, 600*f*
Paracelsus, 11
Paramagnetism, 330–331, 330*f*, 427, 428, 428*f*, 1024, 1029, G-13
Partial ionic character, 366, G-13

Partial pressure
    Dalton's law of, 536, 753
        applying, 206–208
        defined, 206, G-5
        discovery, 205
        Earth's atmosphere and, 219
        kinetic-molecular theory, 211, 212, 212*f*
        molecular view of, 212, 212*f*
    defined, G-13
    of gas in gas mixture, 205–208
Particle accelerator, 1081–1082, 1081*f*, 1082*mn*, G-13
Particles
    dispersal of particle energy, 883–884
    freedom of particle motion, 883–884
    nature of light/photons, 274–276, 283–286
    uncertainty principle, 286–287
    wave behavior compared to, 272–273, 272*f*
Parts by mass, 523*t*, 524, 525
Parts by volume, 523*t*, 524–525
Pascal (Pa) (unit), 191, 192*t*, G-13
Path independence of energy change, 241–242
Pauli, Wolfgang, 305
Pauli exclusion principle, 305, 308–309, 411
Pauling, Linus, 363–364, 412
Pauling electronegativity (EN) scale, 363–364, 364*f*
PCBs (polychlorinated biphenyls), 589*f*, 613, 652*mn*
PEM (proton exchange membrane), 955
Penetration, orbital energy and, 307–308, 307*f*, G-13
Pent- (numerical root), 635*t*
Penta- (prefix), 69*t*
Pentane
    boiling point, 638*t*, 639*f*
    constitutional isomers, 638*t*
    covalent bonding forces, 355, 355*f*
    formula, 71*t*, 638*t*
    intermolecular forces, 355, 355*f*
    model, 71*t*
    molecular shape, 455, 455*f*
    properties, 637–638, 638*t*, 639*t*
Pentanoic acid, 72, 72*f*
Pentanol, solubility, 504*t*
Peptide bond, 508, 508*f*, 660, 668
Per- (prefix), 68, 68*f*, 70
Percent by mass (mass percent, mass %), 44–45
Percent ionic character, 367, 367*f*
Percent yield (% yield), 118–119, G-13
Perchlorate ion, 68*t*
Perchlorates, 616, 616*mn*
Perhalates, 616
Perhalic acid, 616*t*
Period 2 elements, atomic properties of, 567, 568*t*–569*t*
Period numbers, 315
Periodic law, 303, G-13
Periodic table of the elements. *See also specific elements; specific groups*
    atomic number, 303, 304*mn*
    atomic properties, 554–555
    biological, 63*f*
    defined, G-13
    development of, 303–304
    electron configuration, 314*f*, 554
        atomic properties and, 317–325, 319*f*–325*f*, 332–333, 332*f*
        chemical reactivity and, 312*f*
        filling order, 315, 315*f*
        within groups, 312, 314–315
        Period 1 and Period 2, 308–310, 311*f*, 312*f*
        Period 3, 310–311, 311*t*, 312*f*
        Period 4, 312–314, 313*t*
        principles, 314–315
        transition elements, 312–314, 315–316, 329–331
    group number, 315
    groups, 312, 312*f*, 314, 315
    history, 57, 303–304
    inner transition elements, 1023, 1023*f*
    lattice energy, 348–349, 348*f*
    mass value in, 92*n*
    metal and nonmetal comparison, 341–342, 341*f*
    molar mass, 92
    organization of, 57–59
    oxidation numbers and, 561
    Pauling electronegativity scale, 364*f*
    physical states of elements, 562
    quantum-mechanical model, 308–317
    transition elements, 1023, 1023*f*
    trends across, 567, 568*t*–569*t*
    trends in covalent and van der Waals radii, 450*f*
Periods
    continuum of bonding across, 342–343

    defined, 57, G-13
    ionization energies trends, 321–322
    metallic behavior, 327*f*
Permalloy, 997*t*
Permanganate ion, 68*t*, 164, 165*f*, 927, 1034
Peroxide ion, 68*t*
Peroxyacylnitrates (PANs), 600, 600*f*
Perpetual motion, 768*mn*
Perspective drawing, 391
Pesticide, organophosphorus, 603*mn*
PET (positron-emission tomography), 1070, 1089, 1089*f*
pH
    buffer capacity and pH change, 838, 838*f*
    calculation, 844–845, 846–848
    common aqueous solutions, 791*f*
    defined, 791, G-13
    measurement, 793, 793*f*, 950–951, 951*f*
    monitoring with acid-base indicators, 842, 842*f*
    pH meter, 793, 793*f*, 950–951, 951*f*
    pH scale, 791, 791*f*
    relation with pOH and pK_w, 792
    solubility and, 857–858
pH meter, 793, 793*f*, 950–951, 951*f*
pH paper, 793, 793*f*
Phase, 437, G-13
Phase change
    boiling point, 447
    condensation, 438, 441, 441*f*
    cooling from, 439*mn*
    critical point, 448, 448*f*
    defined, 437, G-13
    deposition, 439
    of elements, 563
    entropy change, 885
    equilibrium nature of, 444–447, 444*f*
    freedom of particle motion and dispersal of particle energy, 883–884
    freezing, 442
    fusion, 438
    heat involved in, 440–443
    kinetic-molecular view, 437–438, 438*t*, 440–442
    liquid-gas equilibria, 444–445, 444*f*, 445*f*
    melting, 438
    periodic trends in, 563
    quantitative aspects, 440–449
    solid-gas equilibria, 447
    solid-liquid equilibria, 447
    standard molar entropy and, 890, 890*f*
    sublimation, 447, 447*f*
    triple point, 448*f*, 449
    types, 438–440
    vapor pressure, 445–447, 445*f*, 446*f*
    vaporization, 438–439
Phase diagram
    of carbon, 585, 585*f*
    of carbon dioxide, 448–449, 448*f*
    defined, 448, G-13
    of solvent and solution, 530*f*
    of water, 448*f*, 449
Phenol, 791*t*, A-10
Phenol red, 842*f*
Phenolphthalein, 154*f*, 842*f*, 843*f*, 844, 845*f*, 849, 849*f*
Phenylacetic acid, 800, 800*f*, A-10
Phenylalanine, structure of, 667*f*
Pheromones, 524, 524*f*
Phlogiston theory, 11, 12*mn*, 14, G-13
Phosgene, 758
Phosgene gas, 203*mn*
Phosphate esters, 661
Phosphate fertilizers, 992
Phosphate ion, 68*t*, 597, 602, 858
Phosphates
    element cycling, 990–991
    formation, 602–603
    polyphosphates, 602–603, 603*f*
    solubility-product constant (K_{sp}), A-13
    sources of elements, 985
    uses, 602*mn*
Phosphite ion, 602
Phosphodiester bonds, 670, 670*f*, 674–675
Phospholipids, 511, 511*f*
Phosphor, 1075, 1075*f*
Phosphoric acid, 602
    acid-dissociation constant (K_a), 803, 803*t*, A-10
    uses, 783*t*
Phosphorous acid, 602
Phosphorus
    abundance, 984*t*
    allotropes, 595, 595*f*

bond type, 584*t*
chemistry, 602–603
doped semiconductor, 447, 447*f*
electron configuration, 311, 311*t*, 312*f*
industrial uses of, 992*f*
melting point, 584*t*
nerve poisons, 992*mn*
occurrence in outer space, 990*mn*
properties, 326, 327, 595–596
radioisotope, 1080–1081
standard state, 253*n*
sulfides and nitrides, 603*mn*
Phosphorus cycle, 990–992, 991*f*
Phosphorus pentachloride, 386, 392, 415, 415*f*
Phosphorus trichloride, 597
Photochemical smog, 600, 600*f*
Photodissociation, 219
Photoelectric effect, 274–275, 275*f*, 1075, G-13
Photography, chemistry of black-and-white, 1034–1036, 1035*f*
Photoionization, 219
Photomultiplier tube, 1075
Photon theory of light, 274–275
Photons
    defined, 275, G-13
    particle nature of, 283–286
    ping, analogy in photoelectric pong effect, 275*mn*
Photosynthesis, 220, 240, 987
Photovoltaic cell, 259, G-13
Physical change
    chemical change compared, 4–5, 4*f*, 7
    defined, 4, G-13
    on Io (Jupiter moon), 6*mn*
    reversal, 6–7
Physical properties
    alkali metals, 572
    alkaline earth metals, 576
    bonding effect on, 584
    carbon family elements, 584, 586
    of copper, 6*t*
    defined, 4, G-13
    halogens, 610, 612
    nitrogen family elements, 595–596
    noble gases, 618
    oxygen family elements, 603–604, 606
    Period 2 elements, 567, 568*t*–569*t*
    physical change and, 4–5
    transition elements, 1026–1027, 1026*f*, 1027*f*
Physical states, 6–7, 7*f*, 187–188, 188*f*
    of elements, 562
    gas. *See* Gas
    kinetic-molecular view, 437–438, 438*t*
    liquid. *See* Liquid
    reaction rate, influence on, 686
    solid. *See* Solid
Pi (π) bond, 419–421, 420*f*, G-13
Pi (π) MO, 425–426, 425*f*, G-13
Pico- (prefix), 19*t*
Picometer (pm) (unit), 29, 270
Pig iron, 1001–1002
Piperidine, base-dissociation constant (K_b), A-11
Pipets, automatic, 20*f*, 21
Pitchblende, 45–46, 1067
Planck, Max, 274
Planck's constant *(h)*, 274, G-13
Plasma, fusion, 1098
Plastics, 488, 591, G-13
Platinum
    cancer therapy and, 1044*mn*
    density of, 1027
    electrodes, 931
Pleated sheet structure, 669, 669*f*
Polar arrow, 141
Polar covalent bond, 365, 366–367, G-13
Polar molecules, 141
    defined, G-13
    dipole-dipole forces, 451, 451*f*
Polarimeter, 640, 640*f*, G-13
Polarity, molecular. *See* Molecular polarity
Polarizability, 454, G-13
Pollution
    in food chain, 652*mn*
    phosphates and, 992
    thermal pollution, 521, 521*mn*, 1097
Polonium
    discovery of, 1067, 1067*mn*
    properties, 603–604, 606–607
    radioactive poison, 1084*mn*
Poly (dimethyl siloxane), 489*t*

Poly- (prefix), 662
Polyamides, 664–665, 665*f*
Polyatomic ions
    compounds formed from, 66, 66*f*
    defined, 63, G-13
    Lewis structure, 382
    molecular shape, 390
    names and formulas, 68, 68*f*, 69
    oxoanions, 68, 68*f*
Polybutadiene, 489*t*
Polychlorinated biphenyls (PCBs), 589*f*, 613, 652*mn*
Polychloroprene, 489*t*
Polychromatic light, 271
Polydentate ligands, 1039–1040, 1039*t*
Polyesters, 664–665
Polyethylene, 209, 484–485, 485*f*, 488, 663*t*, 664, 664*f*
Polyisoprene, 489*t*
Polymer glass, 488
Polymeric viscosity index improvers, 459
Polymers, 484–489. *See also* Macromolecules
    addition polymers, 663–664, 663*t*, 664*f*
    biopolymers, 484
    branches, 488
    chain dimensions, 484–487
    condensation polymers, 664–665, 665*f*
    copolymer, 489
    crosslinks, 488
    crystallinity, 486, 486*f*
    defined, 484, G-13
    degree of polymerization, 484–485, 484*t*
    dendrimers, 488
    elastomers, 488–489, 489*t*
    flow behavior, 487–488
    glass transition temperature, 488
    homopolymer, 489
    inorganic, 665
    liquid crystal, 481
    mass, 484–485, 484*t*
    molecular architecture, 488–489
    monomers, 484
    naming, 662
    nucleic acids, 512, 670–673
    plastic, 488
    polysaccharides, 513, 666–667
    proteins, 507, 667–669
    radius of gyration, 485*f*, 486, 486*n*
    random coil, 485, 485*f*
    silicone, 590, 593–594
    synthetic, 662–665
    thermoplastic elastomers, 489
    thermoplastic polymer, 488
    thermoset polymer, 488
    viscosity, 487–488
Polyphosphates, 602–603, 603*f*
Polyphosphazene, 603*mn*, 665
Polypropylenes, 663*t*, 664
Polyprotic acids
    acid-base titration curves, 849–850, 850*f*
    acid-dissociation constant (K_a), 802–804, 802*t*, A-8–A-10
    amino acids, 850–851
    defined, 803, G-13
Polysaccharides, 513–514, 513*f*, G-13
Polystyrene containers, 203
Polysulfanes, 645
Porphin, 1055–1056, 1055*f*
Positron (β⁺)
    defined, 1069, G-13
    penetrating power, 1084
Positron decay, 1068*t*, 1069
Positron emission, 1068*t*, 1069, 1084, G-13
Positron-emission tomography (PET), 1070, 1089, 1089*f*
Potassium
    abundance, 984, 984*t*
    distribution of, 984
    electron configuration, 313, 313*t*
    industrial production, 999–1000
    properties, 570–572, 571*f*
    soaps, 510*mn*
    uses of, 1000
Potassium bromide, electrical conductivity and, 142–143, 143*f*
Potassium chlorate, 168, 616, 901, 901*f*
Potassium chloride, formation of, 167
Potassium ferricyanide, 1029*f*
Potassium iodide, 613
Potassium ion, 65*t*
    in nerve cells, 950*mn*
Potassium nitrate, 147, 573

Potassium sulfide, 93
Potassium superoxide, 1000, 1000*f*
Potential energy (E_p), 236, 341
    in chemical reaction, 356, 359
    conversion to kinetic energy, 8–10, 9*f*
    in covalent bond, 351, 351*f*
    defined, 8, G-13
Pounds per square inch (lb/in² or psi) (unit), 192*t*
Precipitate
    complex ion effect on solubility, 866–867
    defined, 146, G-13
    predicting formation of, 858
Precipitation, selective, 861–862, G-15
Precipitation reactions
    aqueous ionic reactions, 146–150, 147*f*
    defined, 146, G-13
    key event, 146
    metathesis reactions, 148
    predicting, 147–150, 148*t*
Precipitator, Cottrell, 540, 540*f*
Precision, in measurement, 31–32, 32*f*, G-13
Predictions, in the scientific method, 13–14, 13*f*
Prefixes. *See also* Nomenclature
    alkanes, 71
    coordination compounds and, 1040–1041
    decimal, 18, 19*t*
    numerical, 69*t*, 70
    organic compounds, 635, 636*t*
    oxoanions, 68
Pressure *(P)*
    Amontons's law, 196
    atmospheric, 189–190, 189*f*, 218, 218*f*
    Boyle's law, 193–194, 193*f*, 199
    Dalton's law of partial pressures, 536, 753
    defined, 189, G-13
    effect of change on equilibrium, 764–766, 765*f*, 768*t*
    equilibrium constant and, 748–749
    extreme, 221–222, 221*f*, 222*f*, 223*f*
    gas, 187, 188*mn*, 189–192, 193–194, 196–202, 205–208, 210–213
    Le Châtelier's principle, 764–766
    origin of, 210, 211
    solubility and, 521–522, 521*f*, 522*mn*
    units of, 191–192, 192*t*
Pressure-volume work (*PV* work), 239, G-13
Priestley, 168
Priestley, Joseph, 12*mn*
Primary batteries, 952–953
Primary structure, of protein, 668*f*, 669
Principal quantum number *(n)*, 289–291, 290*t*, 305*t*, G-13
Probability contour, 288*f*, 289, 293*f*, 294*f*, G-13
Probability density, 288–289, 288*f*, 292*f*
Problem solving
    conversion factors, 14–16, 17
    gas law problems, 199–202
    mass-mole-number relationships, 94
    in real world, 33, 33*f*
    systematic approach, 16–17
    tips for, 17
    units, 14–16, 17
Product-favored reaction, 913, 913*f*
Products
    calculating amount of, 109–112
    calculating amounts in limiting-reactant problem, 117–118
    definition, 105, G-13
    determining heat of reaction from heat of formation values, 254–255, 254*f*
    gaseous, 208–210
    using gas variable to find amounts of, 209
Proline, structure of, 667*f*
Prop- (numerical root), 635*t*
Propane
    boiling point, 452*f*, 639*f*
    combustion, 109–110, 894
    dipole moment, 452*f*
    formula and model, 71*f*
Propanoic acid, 787*t*, 801
    acid-dissociation constant (K_a), A-10
    titration of, 845–847, 845*f*
Propanol, 504*t*, 649
Propanone, 649, 651*t*, 656*f*
Properties. *See also specific properties*
    chemical, 5, 6*t*
    defined, 4, G-13
    of matter, 4–6
    physical, 4–5, 6*t*
Propylamine, base-dissociation constant (K_b), A-11
Propylenediamine, base-dissociation constant (K_b), A-12

Protactinium, 1094*mn*
Protein synthesis, 671–672, 672*f*
Proteins
  amino acids, 667–668, 667*f*
  composition, 669
  defined, 667, G-13
  digestion and displacement reactions, 153*mn*
  fibrous proteins, 669, 669*f*
  globular proteins, 669
  intermolecular forces and protein shape, 509, 509*f*
  mass spectrometry studies, 55
  polarity of amino acid side chains, 508
  size, 668
  structure, 507–509, 508*f*, 509*f*, 668–669, 668*f*
  synthesis, 671–672, 672*f*
Proton (p$^+$)
  atomic number, 53
  charge, 52, 52*f*, 53*t*
  defined, 52, G-13
  discovery, 51
  hydrated, 151*f*
  mass, 53*t*
  in NMR spectroscopy, 646–647, 646*f*, 647*f*
  nuclide and, 1066
  properties, 53*t*
  transfer and acid-base reactions, 155–157, 156*f*
Proton acceptor, 794, G-13
Proton donor, 793, G-13
Proton exchange membrane (PEM), 955
Proton transfer, in Brønsted-Lowry acid-base definition,
  793–798
Pseudo–noble gas configuration, 328, G-14
Pure substance, in standard state, 253
Purines, 670, 671
Pyridine
  base-dissociation constant (K_b), 806*t*, A-12
  Lewis structure, 806*t*
Pyrimidines, 670, 671
Pyrite, 463*f*, 609, 609*f*, 1000*t*
Pyrometallurgy, 993
Pyrrhotite, 1000*t*
Pyruvic acid, acid-dissociation constant (K_a), A-10

Q

Quadratic formula, use in problem solving, 756–757, A-3
Quantitative theories, 12
Quantization of energy, 274
Quantum, 274, G-14
Quantum dots, 490, 490*f*
Quantum mechanics
  atomic orbital (wave function), 287–295
    quantum numbers, 289–291, 290*t*
    shapes, 292–294, 292*f*, 293*f*, 294*f*
  band theory, 474
  defined, 287, G-14
  electron cloud representation, 288, 292, 292*f*–294*f*
  electron density diagram, 288, 288*f*
  energy levels in the hydrogen atom, 295, 295*f*
  level (shell), 290
  molecular orbital (MO) model, 422, 423
  nodes, 293
  periodic table and, 308–317
  probability contour, 288*f*, 289, 293*f*, 294*f*
  radial probability distribution plot, 288*f*, 289, 289*mn*,
    292–293
  Schrödinger equation, 287
  sublevel (subshell), 290
Quantum number, 274, 289–291, 290*t*, 304–305, 305*mn*,
  305*t*, 310, G-14
Quantum theory, 283, 285*f*
Quartz
  covalent bonding, 356, 356*f*
  crystal structure, 463*f*, 473, 473*f*, 593
Quaternary structure, of protein, 668*f*, 669
Quinone, 1035

R

R. See Universal gas constant *(R)*
R group, amino acid, 667, 667*f*, 669
r-process, 1099
Racemic mixture, 640, 640*mn*
Rad (radiation-absorbed dose) (unit), 1083, G-14
Radial probability distribution plot, 288*f*, 289, 289*mn*,
  292–293, 292*f*–294*f*, 307, 307*f*, 312, G-14

Radiant energy, 269, 271
Radiation
  background, 1084–1085
  blackbody, 274, 274*f*, 274*n*
  cosmic, 1084
  counters for detection of, 1075, 1075*f*
  electromagnetic, 269–271, 271*mn*
  infrared (IR), 271, 271*f*
  intensity, 270
  ionizing
    applications, 1082–1083, 1090
    background radiation, 1084–1085
    defined, 1082
    free radical formation, 1084
    penetrating power of, 1083, 1083*f*
    risk from, 1085–1086, 1085*mn*, 1086*f*, 1086*t*
    sources, 1084–1085, 1084*t*
    units of radiation dose, 1083
  nuclear radiation, effect on matter, 1082–1086
  sources of, 271*mn*
  ultraviolet (UV), 271, 271*f*, 381, 725
Radiation-absorbed dose (rad) (unit), 1083, G-14
Radioactive decay
  decay series, 1074, 1074*f*
  as first-order process, 703
  half-life, 1076–1077, 1076*f*, 1077
  kinetics, 1074–1080
  nuclear equations, balancing, 1070
  nuclear stability and, 1070–1073, 1071*f*, 1072*t*
  predicting mode of, 1073
  radioisotopic dating, 1078–1080, 1079*f*
  rate of, 1074–1077, 1076*f*
  types of, 1067–1070, 1068*t*
Radioactive emissions
  alpha (α) particles, 56*mn*, 1067, 1068*t*, 1069,
    1080–1081, 1083, 1083*f*
  behavior in electric field, 1067, 1067*f*
  beta (β) particles, 1067, 1068*t*, 1069, 1083*f*, 1084
  detection, 1075, 1075*f*
  gamma rays, 1067, 1068*t*, 1070, 1083, 1083*f*
  ionization, 1082
  penetrating power of, 1083, 1083*f*
  positron emission, 1068*t*, 1069, 1069*n*, 1084
  x-ray photon, 1068*t*, 1069
Radioactive tracers, 1087–1089, 1089*f*, 1089*t*
Radioactivity. *See also* Nuclear reactions
  of actinides, 1031
  defined, 1066, G-14
  discovery, 50, 1066–1067
Radiocarbon dating, 1078–1080, 1078*n*, 1079*f*
Radioisotope heater unit (RHU), 1090
Radioisotopes
  applications, 1087–1090
  artificial, 1080–1081
  dating, 1078–1080, 1079*f*
  defined, 1078, G-14
  ionizing radiation, 1090
  tracers, 1087–1089, 1089*f*, 1089*t*
Radioisotopic dating, 1078–1080, 1078*mn*, 1078*n*,
  1079*f*, G-14
Radium
  discovery of, 1067, 1067*mn*
  Ernest Rutherford and, 56*mn*
  paint, 1083*mn*
  properties, 574, 576
  radioactive decay, 1069
Radius of gyration (R_g), 485*f*, 486, 486*n*, G-14
Radon
  Ernest Rutherford and, 56*mn*
  properties, 617–618
  risk from, 1085, 1085*mn*
Rainbows, diffraction of light and, 273*mn*
Ramsey, William, 280*mn*
Random coil, polymer, 485, 485*f*, G-14
Random error, 31–32, G-14
Raoult's law, 528–529, 536, G-14
Rare earth elements. *See also* Lanthanides
  electron configuration, 314*f*, 316
Rate constant *(k)*
  defined, 692, G-14
  determining, 699
  temperature and, 705–706, 705*f*
  units, 699, 699*t*
Rate-determining step, 716–717, G-14
Rate law (rate equation)
  collision theory, 708–711, 708*f*, 711*f*
  defined, 691, G-14
  elementary steps, 715*t*

  first order, 699–700, 701*f*, 705*t*
  initial rate, 689, 693
  integrated, 699–705, 701*f*
  rate constant, 692, 699
  rate-determining step and, 716–717
  reaction mechanism and, 716–719
  reaction order terminology, 694–695
  reaction orders, 692, 695–698
  second order, 700, 701*f*, 705*t*
  zero order, 700, 701*f*, 705*t*
Rate-limiting step. *See* Rate-determining step
Rate of reaction. *See* Reaction rate
Reactant-favored reaction, 913, 913*f*
Reactants
  calculating amount of, 109–112
  calculating amounts in limiting-reactant problem,
    117–118
  definition, 105, G-14
  determining heat of reaction from heat of formation
    values, 254–255, 254*f*
  gaseous, 208–210
  limiting, 114–118, 115*f*, 126–127, 209–210
  using gas variable to find amounts of, 209
Reaction direction
  equilibrium, 749–751, 750*f*, 759–760
  reaction quotient and, 749–751, 750*f*, 759–760
Reaction energy diagram, 712–714, 712*f*–714*f*, 718*f*,
  720*f*, G-14
Reaction intermediate, 715, G-14
Reaction mechanism
  defined, 714, G-14
  elementary reaction, 715–716, 715*t*
  molecularity, 715–716, 715*t*
  rate-determining step, 716–717
  rate law and, 716–719
Reaction order
  defined, 692, G-14
  determining experimentally, 695–697, 696*t*
  determining from the integrated rate law, 699–701,
    701*f*, 702*f*
  first-order reactions, 699–700, 701*f*, 705*t*
  rate constant *(k)* and, 699, 699*t*
  rate law and, 692, 695–698
  second-order reactions, 700, 701*f*, 705*t*
  terminology, 694–695
  third-order reactions, 699, 699*t*
  zero-order reactions, 700, 701*f*, 705*t*
Reaction quotient *(Q)*, 909–910
  defined, 741, G-14
  law of mass action, 741
  reaction direction, 749–751, 750*f*, 759–760
  relation to equilibrium constant, 741–748
  variations in form, 743–747
  writing, 742–743, 744
Reaction rate
  average, 687, 688–689
  catalysis, 720–724, 720*f*–724*f*, 721*t*, 722*mn*
  concentration and, 686
  defined, 685, 685*f*, 687, G-14
  expressing, 687–691
  initial, 689, 693
  instantaneous, 689
  measurement of, 692–693
    conductometric methods, 692, 693*f*
    direct chemical methods, 693
    manometric methods, 693, 693*f*
    spectrometric methods, 692, 692*f*
  molecular structure and, 710–711, 711*f*
  physical state and, 686
  range of, 685–686, 685*f*
  surface area and, 686, 686*f*
  temperature and, 687, 687*f*, 705–710, 705*f*,
    709*f*–709*t*
  terminology, 694–695
Reaction table, 117, 126, 752–754
Reactions. *See* Chemical reactions; Nuclear reactions;
  Organic reactions
Reactor core, nuclear plant, 1096, G-14
Real gases
  extreme conditions, 221–223, 221*f*, 222*f*, 223*f*
  intermolecular attractions, 222, 222*f*
  molecular volume, 222, 223*f*
  van der Waals constants, 223, 223*t*
  van der Waals equation, 223–224
Receptor, biological, 402–403, 402*f*
Receptor site, 402–403, 402*f*
Rechargeable batteries, 953–955
Rectifier, 478

Recycling, of aluminum, 1006
Red blood cells, 851, 851*f*
Redox behavior, of the elements, 560
Redox couple, 967
Redox reactions
 atmosphere-biosphere interconnections, 187*mn*
 balancing, 162–164, 924–928
  in acidic solutions, 925–926, 925*f*
  in basic solutions, 926
  half-reaction method, 924–928
  half-reaction method, hydronium ion in, 926
  oxidation-number method, 162–164
 combination reaction, 167–168, 167*f*, 172
 combustion reactions, 172
 decomposition reaction, 168–169, 169*f*, 173
 defined, 158, G-12
 displacement reaction, 169–172, 170*f*, 171*f*, 173
 disproportionation, 600
 electrochemistry
  acidic solutions, 925–926, 925*f*
  basic solutions, 926
 electrolytic cells, 928, 928*f*
 elements in, 166–173
 identifying, 161
 identifying the type of, 172–173
 key event, 158–159, 158*f*
 in metallurgy, 995–996, 996*f*
 movement of electrons between reactants, 158–159, 158*f*
 organic reactions, 649
 oxidation, 159, 161
 oxidation numbers. *See* Oxidation number
 oxidizing agent, 159, 161–162, 923, 937–943
 redox titrations, 164–166, 165*f*
 reducing agent, 159, 161–162, 560, 570, 573, 574, 577, 923, 937–943
 reduction, 159, 161
 terminology, 159, 162*f*, 924*f*
 voltaic cells, 928, 928*f*, 929–943
  relative reactivities of metals, 941–943
  spontaneous redox reactions, 929–934, 929*f*
Redox titrations, 164–166, 165*f*
Reducing agent, 560
 alkali metals, 570, 573
 alkaline earth agents, 574, 577
 defined, 159, 923, G-14
 metals as, 170–171
 recognizing, 161–162
 relative strengths, 937–943
Reduction, 923, 924*f*, 935
 as decrease in oxidation number, 161
 defined, 159, G-14
Refining (purifying), elements, 997
Refraction, 272–273, 272*f*, G-14
Relative atomic mass, 92*n*
Relativity, theory of, 283
Relief map, 367*f*
Rem (roentgen equivalent for man) (unit), 1083, G-14
Replication, of DNA (deoxyribonucleic acid), 672–673, 673*f*
Reproducibility, 31
Resins, 594
Resonance hybrid, 382–384, 382*mn*, 386, 390, G-14
Resonance structures (resonance forms)
 defined, 381, G-14
 formal charge and, 383–384
 Lewis structures, 381–384
Respiration, 172, 897, 897*f*, 987
Retinal, 642
Reverse osmosis, 542, 542*f*, G-14
Rhenium, density of, 1027
Rhodopsin, 642, 642*f*
RHU (radioisotope heater unit), 1090
Ribonucleic acid (RNA), 670, 672
Ribose
 formula, 103*t*
 in RNA, 670
 uses, 103*t*
Ribosome, 672
rms speed (*u*rms), 215, G-14
RNA (ribonucleic acid), 670, 672
Roasting
 defined, 994, G-14
 of sulfides, 994
Rock salt, 349
Rocket fuel, 117
Roentgen equivalent for man (rem) (unit), 1083
Roentgenium, 1082*mn*

Rohrer, Heinrich, 468
Roman numeral, in compound names, 67
Root-mean-square speed (*u*rms), 215
Roots, in organic compound naming, 635, 635*t*, 636*t*
Rose, 98*mn*
Rosenberg, Barnett, 1044*mn*
Rounding off, 29, 30, 31, G-14
Rowland, F. Sherwood, 725
Rubber, 473
Rubidium, 570–572, 571*f*
Rusting of iron, 956–958, 957*f*, 958*f*
Rutherford, Ernest, 50–51, 56*mn*, 269, 287*mn*, 1067, 1080
Rutherfordium, 1030, 1082*mn*
Rutherford's α-scattering experiment, 51, 51*f*
Rydberg equation, 277, 278*f*, 279, 280

S

s block, 425, 425*f*, 554, G-14
s orbital, 292–293, 292*f*, 411–412, G-14
s-process, 1099
Sacrificial anodes, for prevention of corrosion, 958, 958*f*
Salt(s)
 corrosion and, 957, 957*f*, 958
 defined, 153, G-14
 dissolution and entropy change, 890–891, 890*f*
 electrolysis
  of mixed molten salts, 961–962
  of pure molten salts, 961
 of weakly acidic cations and weakly basic anions, 814
 yielding acidic solutions, 813, 815*t*
 yielding basic solutions, 813–814, 815*t*
 yielding neutral solutions, 813, 815*t*
Salt bridge, 930–931, 931*f*, G-14
Salt link, 509
Saltwater, 43
Samarium, 1031
Sanger chain-termination method, of DNA sequencing, 674–675, 675*f*
Saponification, 659
Saturated calomel electrode, 951
Saturated hydrocarbons, 635, G-14
Saturated solution, 519, 519*f*, 519*mn*, G-14
Saturn (planet), atmosphere, 220*t*
Scale, 542
Scandium
 appearance of, 1024*f*
 electron configuration, 313, 313*t*, 1024, 1025*t*
 oxidation states, 1028*t*
Scandium oxide, 1029*f*
Scanning tunneling microscopy, 468–469, 469*f*, G-14
Scatter, 32
Schrödinger, Erwin, 287
Schrödinger equation, 287, 304, G-14–G-15
Scientific method, 13–14, 13*f*, 14*mn*, G-15
Scientific notation, A-2–A-3. *See also* Exponential notation
Scintillation cocktail, 1075, 1075*f*
Scintillation counter, 1075, G-15
Scrubbers, and clean-coal technology, 256, 864
Scuba diving, decompression sickness, 522*mn*
Seaborgium, 1082*mn*
Second (s)
 defined, 26, G-15
 as SI base unit, 18*t*
Second law of thermodynamics, 887–888, G-15
Second-order reactions
 half-life, 705
 integrated rate law, 700, 701*f*, 705*t*
 rate constant (*k*) for, 699*t*
 rate law, 694
 zero- and first-order reactions compared, 705*t*
Secondary batteries, 953–955
Secondary structure, of protein, 668*f*, 669
Seesaw shape, VSEPR theory, 392–393, 394*f*, G-15
Selective precipitation, 861–862, G-15
Selenium
 allotropes, 606
 electron configuration, 313*t*
 properties, 603–604, 606–607
 in xerography, 606*mn*
Self-assembly, 490
Self-ionization. *See* Autoionization
Semiconductors
 band theory and, 475, 475*f*
 defined, 475, G-15
 doped, 477–478, 477*f*

 gallium arsenide, 578*mn*
 n-type, 477–478, 478*f*
 p-n junction, 478, 478*f*
 p-type, 477–478, 478*f*
Semicrystallinity, of polymers, 486
Semimetals, 58. *See also* Metalloid(s)
Semipermeable membrane, 532, 542, G-15
Sense of smell (olfaction), 402–403, 402*f*, 403*f*
Separation techniques, 76–77, 76*f*
Sequestration, of carbon dioxide, 258
Serine
 molecular structure, 650*f*
 structure of, 508*f*, 650*f*, 667*f*
Sertraline, 119, 119*f*
Shape. *See* Molecular shape
Shared electron pair, 352, 359
Sheet silicates, 592, 592*f*
Sheet silicones, 594, 594*f*
Shell. *See* Level
Shielding, 306–307, 307*f*, 318, G-15
Short tandem repeat (STR) analysis, 675
Shroud of Turin, radiocarbon dating of, 1078*mn*
SI units
 base (fundamental), 18, 18*t*
 decimal prefixes, 18, 19*t*
 defined, 18, G-15
 density, 22–24
 derived, 18
 English equivalents, 19*t*
 length, 18–19
 mass, 21–22
 temperature, 24–26
 time, 26, 27*f*
 volume, 19–21
Sickle cell anemia, 851, 851*f*
Side reaction, 118, 118*f*, G-15
Side-to-side overlap, 419, 421, 557, G-15
Siderite, 1000*t*
Sievert (Sv) (unit), 1083, G-15
Sigma (σ) bond, 419–421, 420*f*, G-15
Sigma (σ) MO, 423, 425*f*, G-15
Significant figures
 in arithmetic operations, 29
 calculator use, 30
 defined, 28, G-15
 exact numbers, 30–31
 measuring device, dependence on, 28*f*, 30, 30*f*
 number of, 28–29
 rounding of, 29, 30, 31
Silanes, 645
Silicate phase, 982
Silicates, 473, 590, 591*f*, 592–593, 985, 985*f*, G-15
Silicic acid, 591
Silicon
 atomic mass, 54
 bond type, 584*t*
 chemistry, 590–591
 diagonal relationship with boron, 591
 electron configuration, 311, 311*t*, 312*f*
 isotopes, 54
 melting point, 584*t*
 properties, 326, 584, 586
 in semiconductors, 477, 477*f*
 zone refining, 533, 533*f*
Silicon carbide, 119, 482, 483, 483*f*, 587
Silicon dioxide
 amorphous structure, 473, 473*f*
 crystalline structure, 473, 473*f*
 glass, 457
 uses, 587
Silicon nitride, 482–483
Silicone polymers, 590, 593–594
Silicones, 590, 593–594, 665, G-15
Silk fibroin, 669, 669*f*
Silver
 in batteries, 953, 953*f*
 boiling point, 390*t*
 in concentration cells, 950, 951, 951*f*
 electron configuration, 316, 330
 isotopes, 56
 melting point, 390*t*
 oxidation state, 1028, 1034
 photography, 1034–1036, 1035*f*
 properties, 1034
Silver chloride, 857, 891
Silver chromate, 145–146, 145*f*
Silver iodide, 350
Silver ion, 65*t*

Silver nitrate, 145–146, 145f
Silver sulfide, solubility-product constant (K_{sp}), 853t
Simple cubic unit cell, 464, 465f, 466, 467f, G-15
Single bonds
  alkanes, 635–637
  bond energy, 353t
  bond length, 354
  covalent bonding, 352
  defined, G-15
  functional groups with, 650, 652–654
  functional groups with single and double bonds, 657–661
  Lewis structures for molecules with, 378–380
  orbital overlap, 418–420
  organic compounds, 650–654, 657–661
Single-displacement reactions, 169
Skou, Jens C., 950mn
Slag, 1001, 1001mn, G-15
Slightly soluble ionic compounds
  common ion effect on solubility, 856–857
  equilibria, 851–862
  ion-product expression, 852
  molar solubility, 853–855, 855t
  pH effect on solubility, 857–858
  precipitate, 858, 860–862
  solubility-product constant, 852–855, 853t, 855t, A-13
Smectic, liquid crystal phase, 480, 480f
Smell, sense of, 402–403, 402f, 403f
Smelting, 995, G-15
Smog, 203mn, 385, 505mn, 600, 600f
SNG (synthetic natural gas), 256
Snowflakes, crystal structure of ice and, 461f
Snowshoes, weight distribution and, 189mn
Soaps
  defined, 510, G-15
  dual polarity, 510, 510f
  hardness and softness, 510mn
  saponification, 659
  structure and function, 510f
Soda water, 12mn
Soddy, Frederick, 1067
Sodium
  abundance, 984t
  electron configuration, 311, 311t, 312f, 328
  industrial production, 999
  ionization energy, 323t
  mineral sources of, 993t
  properties, 42, 42t, 326, 327, 570–572, 571f
  soaps, 510mn
  standard state, 253
  uses of, 1000
Sodium acetate, 809
Sodium atoms, emission and absorption spectra, 281f
Sodium bromide, 345f
Sodium carbonate, 573, 783t
Sodium chloride
  crystal structure, 469–470, 470f, 471, 471f
  dissolving, 516–517, 516f, 518f
  electrolysis, 964–965
  electron density distribution, 368f
  formation, 60–61, 60f
  ionic properties, 368
  isotonic, 534
  to melt ice, 533
  ocean supply of, 999mn
  preparing a dilute solution from a concentrated solution, 124
  properties, 42, 42t
  solubility, 502, 505
  uses, 533, 573
Sodium chromate, 145–146, 145f, 1029f
Sodium fluoride, 541
Sodium formate, 840
Sodium hydrogen carbonate, 573, 783t
Sodium hydroxide
  from chlor-alkali process, 1014–1015
  dissolving, 516f, 517
  proton transfer, 155
  uses, 573, 783t
Sodium hypochlorite, 611, 613
Sodium iodide, 147, 147f
Sodium ions, 65t, 534, 542, 542f, 950mn
Sodium phosphate, 783t
Sodium stearate, 510
Sodium thiosulfate pentahydrate, 609
Sodium tripolyphosphate, 597
Solar cells, 478mn
Solar emission spectrum, 280mn
Solar energy, 258, 259

Solar system, age of, 1079mn
Solder, 997t, 1037
Solid
  advanced materials
    ceramics, 481–483, 482t, 483f
    electronic materials, 477–478, 477f, 478f
    liquid crystals, 478–481, 479f–481f
    nanotechnology and, 489–491, 490f, 491f
    polymeric materials, 484–489, 484t, 485f–487f, 489t
    semiconductors, 477–478
  amorphous, 463, 473, 473f
  bonding in, molecular orbital band theory, 457f, 474–476, 474f
  crystalline, 463–473
  defined, 6, G-15
  density, 23, 23t
  entropy change with dissolution, 890–891, 890f
  kinetic-molecular view, 438
  particle arrangement in, 6, 7f
  phase changes, 447, 447f, 530f, 585f
  properties, 438t
  solubility and temperature, 520, 520f
  structural features, 463–469
Solid-gas equilibria, 447
Solid-liquid equilibria, 447
Solid-liquid solutions, 503
Solid-solid solutions, 506–507
Sols, 539, 539t
Solubility
  boiling point and, 506t
  chromatography, 77
  common-ion effect, 856–857
  comparing, 855
  complex ion effect on, 866–867
  defined, 502, G-15
  equilibria of slightly soluble ionic compounds, 851–862
  equilibrium, 519–522
  extraction, 76
  gas-gas solution, 506
  gas-liquid solution, 506
  gas-solid solution, 506
  of ionic compounds, 142–144, 142f, 148t
  liquid-liquid solution, 503–505
  pH effect on, 857–858
  of precipitates, 858, 860–862
  prediction of, 505
  pressure and, 521–522, 521f, 522mn
  solid-liquid solutions, 503
  solid-solid solutions, 506
  temperature and, 520–521, 520f
Solubility-product constant (K_{sp}), 852–855, 853t, 855t, A-13, G-15
Solubility rules for ionic compounds in water, 148t
Soluble compounds, 142, 148t
Solute
  defined, 121, 502, G-15
  entropy and, 890–891, 890f
  molar mass determination from colligative properties, 535–536
  parts by parts of solution, 524–525
  vapor pressure and, 528–529, 528f
Solution. See also Solubility; Solute; Solution process; Solvent
  acid-base properties of salt solutions, 812–816
  aqueous, 75
  balancing redox reactions in
    acidic solutions, 925–926, 925f
    basic solutions, 926
  buffered, 122
  calculating the molarity of, 121
  colligative properties, 527–539
    in biology, 533–534, 533f–534f
    boiling point elevation, 529–530, 530t, 531
    electrolyte solutions, 527, 537–538, 537f
    freezing point depression, 530t, 531, 533, 533f
    in industry, 533–534, 533f–534f
    nonelectrolytes, 527, 527f, 528–532, 535–537
    osmotic pressure, 532, 532f, 534, 534f, 535–536
    solute molar mass from, 535–536
    vapor pressure lowering, 528–529, 528t
  concentration, 522–527
  defined, 75, 501
  gas-gas, 506
  gas-liquid, 506
  gas-solid, 506, 567
  ideal, 528
  intermolecular forces in, 502–503, 502f
  liquid-liquid, 503–505

mixtures, 75
  mole-mass-number relationships, 122
  molecular polarity and, 503–506
  preparing and diluting molar, 122–124, 123f
  saturated, 529, 529f, 529mn
  solid-liquid, 503
  solid-solid, 506–507, 507f
  solution process, 514–518
  stock, 123
  stoichiometry. See Solution stoichiometry
  supersaturated, 519, 519f
  tonicity, 534
  unsaturated, 519
  water purification and, 541–542
Solution, heat of (ΔH_{soln}), 515, 515f, 517mn. See also Solubility; Solution
Solution process, 514–518. See also Solubility; Solution
  enthalpy change, 515–517, 515f–518f
  entropy change, 517–518
  heat of hydration, 515–517, 516f, 516t
  heat of solution, 514–515, 515f
Solution stoichiometry
  chemical reactions in solution, 125–127
  concentration, 121
  molar solutions, preparing and diluting, 122–124, 123f
  mole-mass-number conversions, 122
Solvated
  defined, G-15
  ions, 142, 146
  proton, 151
Solvation, 515, G-15
Solvent
  defined, 121, 502, G-15
  hexane as, 503–505, 504t, 518, 518f
  volatile organic, 505mn
  water as, 141–144, 460, 504, 504t, 505f
Source, element. See Occurrence, element
sp hybridization, 412, 413f, 416t, G-15
sp^2 hybridization, 414, 414f, 416t, G-15
sp^3 hybridization, 414–415, 414f, 415f, 416t, 1048, 1048f, G-15
sp^3d hybridization, 415, 415f, 416t, G-15
sp^3d^2 hybridization, 416, 416f, 416t, G-15
Space-filling models, 636, 636f
  depiction of molecule, 74
Space shuttle, 172mn, 613, 616
Specific activity, 1076
Specific heat, 246n
Specific heat capacity (c), 246–247, 246t, 249mn, 460, G-15
Specific rotation, isomer, 640
Spectator ions, 146, G-15
Spectra
  absorption, 281–282, 281f, 282f
  atomic, 276–280
  emission, 280mn, 281, 281f
  line, 276–277, 276f, 277f, 277n
Spectrochemical series, 1051, 1051f, G-15
Spectrometry, measuring reaction rates by, 692, 692f
Spectrophotometry, 281–282, 281f
Speed of light (c), 270, G-15
Spin quantum number (m_s), 305, 305t, G-15
Spontaneous change
  defined, 881, G-15
  enthalpy change and, 882–883
  exothermic and endothermic reactions, 882–883, 883f, 898–899, 899f
  free energy change and, 899–900
  redox reactions, 929–934, 929f
  second law of thermodynamics and, 887–888
  temperature and, 904–907, 905t, 907f
  voltaic cells, 929–934, 938–941
Square planar complex, 1038, 1039t
  crystal field splitting, 1054, 1054f
  valence bond theory, 1047, 1047f
Square planar shape, 393f, 394, 394f, 416, G-15
Square pyramidal shape, 393, 393f, 394f, 416, G-15
Stalactite, 859
Stalagmite, 859
Standard atmosphere (atm) (unit), 191, 192t, G-15
Standard cell potential ($E°_{cell}$), 935–937, 943–951, G-15
Standard electrode potential ($E°_{half-cell}$), 935–937, 936f, 938, 938t, 1029t, A-14, G-15
Standard entropy of reaction ($\Delta S°_{rxn}$), 893–894, G-16
Standard free energy change ($\Delta G°$), 900–902, G-16
Standard free energy of formation ($\Delta G°_f$), 901–902, G-16
Standard half-cell potential ($E°_{half-cell}$), 935–937, 936f, 938, 938t, 1029t, A-14
Standard heat of formation ($\Delta H°_f$), 253–255, 253t, G-16
Standard heat of reaction ($\Delta H°_{rxn}$), 253–255, G-16
Standard hydrogen electrode, 935–937, 936f

Standard molar entropy (S°), 888–892, G-16
Standard molar volume, 197–198, 197f, G-16
Standard reference half-cell, 936–937, 936f, G-16
Standard states, 253, 253n, G-16
Standard temperature and pressure (STP), 197, G-16
Starch, 666–667
Stars, element synthesis in, 1099, 1100f
State function, 241–242, 242mn, G-16
States of matter, 6–7, 7f, 106, 187–188, 188f
    changes in state. See Phase change
    defined, 6, G-16
    gas. See Gas
    kinetic-molecular view, 437–438, 438t
    liquid. See Liquid
    solid. See Solid
Stationary phase, chromatography, 77, 77f
Stationary states, 277–278, 278f, G-16
Steam-carbon reaction, 256
Steam-reforming reaction, hydrogen production,
    1009, 1010
Stearic acid, molecular structure, 658f
Steel
    alloy composition, 997, 997t
    defined, 1000, G-16
    metallurgy, 1000–1002, 1001f, 1002f
    specific heat capacity, 246t
Stellar nucleosynthesis, 1099, G-16
Stereochemical theory of odor, 402
Stereoisomers, 639–640, G-16
Stereoselective catalysts, 664
Sterling silver, 997t
Stock solution, 123
Stoichiometrically equivalent molar ratios, 109–112
Stoichiometry
    balancing (stoichiometric) coefficient, 105–108
    defined, 90, G-16
    electrolysis, 964–966
    ideal gas law and, 208–210, 208f
    molar ratios and balanced equations, 109–112
    redox titration, 164
    solution, 121–127
        chemical reactions in solution, 125–127
        concentration, 121
        molar solutions, preparing and diluting,
            122–124, 123f
        mole-mass-number conversions, 122
    thermochemical equations, 249–250
Strassmann, F., 1094
Stratosphere, 218, 218f, 219
Strong acids, 151–152, 151t, 784, 785f, 786, 786f, 788
    Brønsted-Lowry concept, 156
    dissociation extent, 785, 785f
    strong acid–strong base titration curves, 843–845, 843f
    weak base–strong acid titration curves, 848–849, 849f
Strong bases, 151–152, 151t, 784, 788
    Brønsted-Lowry concept, 156
    strong acid–strong base titration curves, 843–845, 843f
    weak acid–strong base titration curves, 845–848, 845f
Strong electrolyte solutions
    colligative properties of, 537–538
    nonideal behavior of, 537f
Strong-field ligand, 1050, 1050f, 1052–1053,
    1052f, G-16
Strong force, 1072, G-16
Strontium
    flame test, 281, 281f
    line spectra, 276f
    properties, 574, 576
    radioactivity, 1077
Strontium ion, 65t
Structural formula, 65, G-16
Structural isomers, 103–104, 103t. See also
    Constitutional isomers
Subcritical mass, 1095, 1096, 1096f
Sublevel, 290, 307–308, 308–310, 308f, 315f, G-16
Sublimation, 439, 447, 447f, G-16
Sublimation, heat of (ΔH_subl), 439, 440f
Submicroscopic properties, 8
Substance
    defined, 4, 41, G-16
    mixture distinguished from, 43
    standard state, 253
Substitution reactions, 648, G-16
Substitutional alloys, 506, 507f
Substrate, enzyme, 723–724, 724f, G-16
Subtraction, significant figures and, 28
Succinic acid, acid-dissociation constant (K_a), A-10
Sucrose, 666f
Suction pump, 190mn

Suffix, organic compound, 635, 636t
Suffixes. See also Nomenclature
    alkanes, 71
    functional group, 72
    oxoanions, 68
Sulfanilamide, 661f
Sulfate ion, 68t, 386, 609, 852
Sulfates, solubility-product constant (K_sp), A-13
Sulfide ion, 65t, 326, 852–853
Sulfide phase, 982
Sulfides, 985f, 986, A-13
Sulfite ion, 68t, 608
Sulfonamides, 661, 661f
Sulfur
    abundance, 984, 984t
    acid-base behavior, 327, 559
    allotropes, 606, 645
    catenation, 645
    chemistry, 608–609
    in coal, 256
    electron configuration, 311, 311t, 312f
    production, 605, 1012, 1013f
    properties, 327, 603–604, 606–607
    standard state, 253
Sulfur dioxide, 93, 112
    acid rain and, 113, 608mn, 863, 864
    on Io (Jupiter moon), 6mn
    Lewis structure, 387
    mole fraction of air at sea level, 219t
    power plant emission of, 33, 33f
    release during roasting, 994
    roasting copper and, 111–112, 113
    uses, 605
Sulfur hexafluoride, 605, 607, 608f
    hybrid orbitals, 416, 416f
    Lewis structure, 386
    molecular shape, 393
Sulfur tetrafluoride, 417, 607–608, 608f
Sulfur trioxide, 387, 605, 608–609
Sulfuric acid, 657
    acid-dissociation constant (K_a), A-10
    acid rain, 608mn, 863
    applications, 1014f
    cave formation and, 859
    chemical manufacturing, 1012–1014
    formation, 608–609
    Lewis structure, 386
    uses, 605, 609, 609f
Sulfurous acid, 608, 849–850, 850f
    acid-dissociation constant (K_a), 803t, A-10
    acid rain, 863
    Lewis acid and, 819
Superconducting oxides, 476, 476f
Superconductivity, 475–476, 476f, G-16
Supercooled liquids, 473
Supercritical fluid (SCF), 449, 449mn
Superionic conductors, 350
Supernova, 1099, 1100f
Supersaturated solution, 519, 519f, G-16
Surface area, reaction rate and, 686, 686f
Surface tension, 457, 457f, 457t, 459, 461, G-16
Surfactants, 457
Surroundings, 236–237, 237f
    defined, 236, G-16
    entropy change in, 895–896
Suspension, 539, G-16
Sylvite, 999
Synchrotron, 1082, 1082mn
Synthesis gas (syngas), 256, 589, 767, G-16
Synthetic macromolecules, 662–665
Synthetic natural gas (SNG), 256, G-16
System
    defined, 236–237, 236mn, G-16
    energy flow to and from, 237, 237f
System energy, 236–237, 237f, 240
Systematic error, 31–32, G-16

T

T shape, VSEPR theory, 392f, 393, 394f, G-16
t_2g orbitals, 1050, G-16
Table salt, density of, 23t
Tanaka, Koichi, 55
Tantalum, 1030
TCDD (tetrachlorodibenzodioxin), 524mn
Technetium-99, 1088, 1089f
Technological tradition, in chemistry, 11
Tellurium, properties of, 603–604, 606–607

Temperature (T). See also Endothermic processes;
    Exothermic processes
    absolute scale, 195
    Amontons's law, 196
    atmospheric, 218–219, 218f
    Celsius scale, 25–26, 25f
    Charles's law, 194–195, 195f
    converting units of, 26
    defined, 24, G-16
    Earth's atmosphere, 257–258
    effect of change on equilibrium, 766–768, 768mn,
        768t
    of example objects, 25f
    Fahrenheit scale, 25f, 26
    gas, 188, 194–202
    heat distinguished from, 24
    as intensive property, 24
    Kelvin scale, 25–26, 25f
    kinetic energy relationship, 213–215
    molecular speed and, 211
    reaction rate and, 687, 687f, 705–710, 705f, 709f–709t
    reaction spontaneity, effect on, 904–907, 905t, 907f
    SI units, 18t, 25
    solubility and
        gas solubility, 521
        solid solubility, 520, 520f
    standard molar entropy and, 888, 889f
    vapor pressure, effect on, 445–446, 445f, 446f
Temperature inversion, 219
Tera- (prefix), 19t
Termolecular, 715, 715t
Tertiary structure, of protein, 668f, 669
Tetra- (prefix), 69t
Tetraethyl lead, 587
Tetrahedral arrangement
    defined, 391, G-16
    hybrid orbitals, 414–415, 416t
    VESPR, 389f, 391–392, 391f, 394f
Tetrahedral complexes
    carbonic anhydrase, 1056, 1056f
    crystal field splitting, 1053, 1053f
    valence bond theory, 1048, 1048f
Tetrahedral shape, 391, 391f, 394f
    complex ion, 1039, 1039t
    hybrid orbitals, 416t
Tetramethylsilane, 646
Tetranitrocubane, 398, 398f
Tetraphosphorus decaoxide, 326, 326f, 327, 597
Thalidomide, 640mn
Thallium
    bond type, 584t
    electron configuration, 328–329
    melting point, 584t
    properties, 578–580
    radioactive poison, 1084mn
Theoretical yield, 118–119, G-16
Theory, in the scientific method, 13–14, 13f
Thermal decomposition, 168, 169f
Thermal energy, 238, 982, 983t
Thermal pollution, 521, 521mn, 1097
Thermals, 219
Thermite, 112
Thermite reaction, 996, 996f
Thermochemical equation, 249–250, G-16
Thermochemistry, 235–259
    calorimetry, 246–249
    defined, 236, G-16
    energy forms and interconversion, 236–242
    energy use, future of, 256–259
    enthalpy, 243–245
    Hess's law of heat summation, 251–252
    standard heats of reaction, 253–255
    stoichiometry of thermochemical equations, 249–250
Thermodynamics
    defined, 236, G-16
    in the kitchen, 238mn
    laws
        first, 240, 882
        living things and, 897
        second, 887–888
        third, 888
    spontaneous change and, 881–893
    standard values for common substances, A-5–A-7
Thermometer, 24, 28f, G-16
Thermoplastic elastomers, 489
Thermoplastic polymer, 488
Thermoset polymer, 488
Thermosphere, 218, 218f
Thermotropic, liquid crystal phase, 479, 481

Thiosulfate ion, 605, 609
Thiosulfuric acid, 609
Third law of thermodynamics, 888, G-16
Third-order reactions, 699, 699t
Thomson, J. J., 49, 50
Thomson, William (Lord Kelvin), 25, 27mn, 195
Thorium, 56mn, 1030, 1031, 1066
Three Mile Island, 1097
Threonine, structure of, 667f
Threshold frequency, 275
Thymine, 670
Thymol blue, 842f
Time, 18t, 26, 27f
Tin
　acid strength of hydrated ion, 812t
　allotropes, 585
　boiling point, 390t
　bond type, 584t
　in electrolytic cells, 959, 959f
　melting point, 390t, 584t
　mineral sources of, 993t
　organotin compounds, 587
　properties, 584, 586, 588, 588f
　solder, 997t
　tin(II) and tin(IV), 67t
Tin(II) chloride, molecular shape, 390
Tin(IV) oxide, 995
Titan (Saturn moon) atmosphere, 220t
Titanic, liner, 1001mn
Titanium
　appearance of, 1024f
　electron configuration, 313t, 330–331, 1024, 1025t
　oxidation states, 1028t, 1029
Titanium(IV) chloride, 254–255
Titanium(IV) oxide, 1029f
Titration. See also Acid-base titration curves
　acid-base, 153–155, 154f
　defined, 153, G-16
　end point, 844
　equivalence point, 843–845
　redox, 164–166, 165f
TNT (trinitrotoluene), 645, 645f, 1087, 1096
Tobacco mosaic virus, 480f
Tokamak design, 1100, 1100f
Toluene, 644, 818
Tonicity, 534
Torr, 191, 192t, G-16
Torricelli, Evangelista, 189
Total ionic equations
　for acid-base reactions, 153, 157, 157f
　for aqueous ionic reactions, 145f, 146
　defined, 146, G-17
Trace elements, 64, 1055–1056, 1055f, 1056f, 1056t
Tracers, radioactive, 1087–1089, 1089f, 1089t, G-17
trans isomer, 401. See also cis-trans isomers
Transfer RNA (tRNA), 672
Transistors, 478
Transition elements
　actinides, 316, 1031
　complexes. See Transition metal complexes
　　(coordination compounds)
　coordination compounds, 1037–1046
　defined, 313, 1023, G-17
　as dietary trace elements, 1055–1056, 1055f,
　　1056f, 1056t
　electron configuration, 1024–1025, 1025t
　influence on Group 3A(13) properties, 578, 579f
　inner. See Inner transition elements
　lanthanide, 1030–1031
　lanthanide contraction, 1026
　metallic hydrides, 567
　ocean sources of, 1034mn
　periodic table, 57, 58f, 1023, 1023f
　properties, 1026–1030
　　chemical, 1027–1030
　　color and magnetism, 1029, 1029f
　　metallic behavior and reducing strength, 1029,
　　　1029f
　　oxidation states, 1028, 1028f, 1028t
　　physical and atomic, 1026–1027, 1026f, 1027f
Transition metal complexes (coordination compounds)
　color, 1050–1051, 1050f, 1051f
　crystal field theory, 1048–1054
　as dietary trace elements, 1055–1056, 1055f,
　　1056f, 1056t
　formulas and names, 1040–1042, 1041t
　isomerism, 1043–1046, 1043f, 1045f
　magnetic properties, 1052

structure, 1038–1040, 1039t
valence bond theory, 1046–1048
Transition metal ions
　electron configurations of, 329–330
　magnetic properties, 330–331
Transition metals, abundance of, 984–985, 984t
Transition state, 711, 713, 713f, 724, G-17
Transition state theory, 708, 711–714, G-17
Transmutation
　nuclear. See Nuclear transmutation
　Rutherford experiments in, 56mn
Transuranium element, 1082, 1082mn, 1084t, 1094,
　G-17
Trees, hypotonic watering of, 534
Tri- (prefix), 69t
Trial and error, 12
Triatomic molecule, 357
Trichlorofluoromethane (Freon-11), 587
Triethylamine, base-dissociation constant (K_b), A-12
Triglycerides, 658, 659
Trigonal bipyramidal arrangement
　defined, 392, G-17
　hybrid orbitals, 416t
　VSEPR, 389f, 392–393, 392f, 394f
Trigonal bipyramidal shape, 392, 392f, 394f, 416t,
　712, 712f
Trigonal planar arrangement
　defined, 390, G-17
　hybrid orbitals, 414, 416t
　VSEPR, 390, 390f, 394f
Trigonal planar shape, 390, 390f, 394f
　alkenes, 641
　defined, 390
　hybrid orbitals, 416t
Trigonal pyramidal shape, 391, 391f, 394f, 415, 653f, G-17
Triiodide ion, 393
Trimethylamine, 590, 590f, 653, A-12
Trinitrotoluene (TNT), 645, 645f, 1096
Triple bonds
　alkynes, 643, 651t
　bond energy, 353t
　bond order, 352
　defined, G-17
　functional groups with, 661–662
　molecular shape and, 420
　pi (π) bond, 420
　sigma (σ) bond, 420
Triple point, 448f, 449, G-17
Trisilylamine, 590, 590f
Tristearin, 659f
Tritium, 1008–1009, 1008t, 1098
Troposphere, 218, 218f
Tryptophan, structure of, 667f
Tungsten
　density, 1027
　electron configuration, 316
　properties, 1030, 1030t
Twisted nematic phase, liquid crystal phase, 480, 481
Tyndall effect, 540, 540f, G-17
Tyrosine, structure of, 667f

U

Ultraviolet light, effect on DNA, 541
Ultraviolet (UV) radiation, 271, 271f, 381, 725
　absorption by stratospheric ozone, 219
　defined, G-17
Uncertainty, in measurement, 27–32, G-17
Uncertainty principle, 286–287, 287mn, G-17
Unimolecular reaction, 715, 715t, G-17
Unit cell
　body-centered cubic, 464, 465f, 466, 467f
　defined, 464, G-17
　edge length, 468
　face-centered cubic, 464, 465f, 466
　hexagonal, 466, 467f
　packing efficiency, 466, 467f
　simple cubic, 464, 465f, 466, 467f
Units. See also SI units
　arithmetic operations, 14
　conversion factors, 14–16, 17
Universal gas constant (R), 198, 706, 706mn, G-17
Unnilquadium, 1082mn
Unsaturated hydrocarbons, 640, G-17
Unsaturated solution, 519, G-17
Unshared pair, 352
Uracil, 670

Uranium
　boiling point, 390t
　decay series, 1074, 1074f
　enrichment, 215mn
　induced fission, 1094, 1094f, 1095f
　isotopes, 215mn
　mass fraction, calculating, 45–46
　melting point, 390t
　nuclear energy reactors and, 1096, 1097, 1097mn
　oxidation states, 1031
　pitchblende, 1067
　radioactivity, 1066–1067, 1079mn
Uranus (planet), atmosphere, 220t
Urea, 629, 989
Urease, 724

V

V agent, 992mn
V shape, VSEPR theory, 390, 390f, 391f, 392, 394f
Vacancies, crystal defects, 477
Valence band, 475, G-17
Valence bond (VB) theory
　central themes, 411–412
　complexes, 1046–1048, 1047f, 1048f
　　hybrid orbitals in, 1046–1048, 1047f, 1048f
　　octahedral complexes, 1047, 1047f
　　square planar complexes, 1047, 1047f
　　tetrahedral complexes, 1048, 1048f
　defined, 411, G-17
　hybrid orbitals, 412–418, 413f–417f, 416t, 1046–1048
　molecular rotation, 421, 421f
　orbital overlap, 411–412, 411f, 418–421, 419f–421f
　single and multiple bonds, 418–421
Valence electrons, 315, 352, 378–379
　defined, G-17
　in metallic bonding, 369
Valence shell, expanded, 385–386
Valence-shell electron-pair repulsion (VSEPR) theory,
　388–398
　bent shape, 390, 390f, 391f, 392, 394f, 415
　defined, 388, G-17
　electron-group arrangements, 388–389, 389f
　hydrides and, 418
　linear arrangement, 389, 389f, 394f
　molecular shape and, 557
　molecules with multiple central atoms, 396–397,
　　397f
　octahedral arrangement, 389f, 393–394, 393f, 394f
　seesaw shape, 392–393, 394f
　square planar shape, 393f, 394, 394f
　square pyramidal shape, 393, 393f, 394f
　T shape, 392f, 393, 394f
　tetrahedral arrangement, 389f, 391–392, 391f, 394f
　trigonal bipyramidal arrangement, 389f, 392–393,
　　392f, 394f
　trigonal planar arrangement, 390, 390f, 394f
　trigonal pyramidal shape, 389f, 391, 391f, 394f
　using VSEPR to determine molecular shape, 395–396
　V shape, 390, 390f, 391f, 392, 394f, 415
Valence-state electronegativity, 1033
Valine, structure of, 667f
van der Waals, Johannes, 223, 449
van der Waals constants, 223, 223t, G-17
van der Waals equation, 223–224, G-17
van der Waals (intermolecular) forces, 450
van der Waals radius, 450, 450f, G-17
Vanadium
　abundance, 984t
　appearance of, 1024f
　as dietary trace elements, 1056t
　electron configuration, 313, 313t, 1025t
　oxidation states, 1028, 1028t
Vanadium carbide, 998, 998f
Vanadyl sulfate dihydrate, 1029f
van't Hoff, Jacobus, 537
van't Hoff equation, 767–768
van't Hoff factor, 537–538
Vapor pressure
　boiling point and, 447
　defined, 445, G-17
　equilibrium, 445
　solute effect on vapor pressure of solution,
　　528–529, 528f
　temperature and intermolecular forces effect on,
　　445–446, 445f, 446f
　water, 207, 207t

Vapor pressure lowering (ΔP), 528–529, G-17
Vaporization, 438–439
  defined, 438, G-17
  heat of vaporization (ΔH_{vap}), 245, 439, 439f, 440f, 461
Vaporizing an ionic compound, 350, 350f
Variables, 13, G-17
Venus (planet), atmosphere, 220t
Vibrational energy, infrared spectroscopy and, 357, 357f
Vibrational motion, entropy and, 892
Villard, P., 1067
Vinegar, 783
Viscosity
  defined, 458, G-17
  gas, 188
  intermolecular forces and, 458
  liquid, 458
  liquid crystal phase, 479
  motor oil, 459
  polymer, 487–488, 487f
  water, 458t
Vision, chemistry of, 642, 642f
Vitalism, 629
Vitamin A, 642
Vitamin B$_{12}$, 819
Vitamin C, 101–102, 534, 804–805
Vitamin E, 385mn
Volatile nonelectrolyte solutions, colligative properties of, 536–537
Volatility
  defined, G-17
  distillation, 76
Volcanoes, 605f, 725
Volt (V) (unit), 934, G-17
Voltage, 934. See also Cell potential (E_{cell})
Voltaic cells. See also Battery
  action of, 933, 933f
  cell potential, 934–951
  construction and operation, 930–932, 931f, 932f
  defined, 928, G-17
  dental, 943mn
  diagramming, 932–933
  electrodes, 928–929, 929mn
  electrolytic cells compared, 928–929, 928f, 959–960, 961t
  galvanic cell, 928
  half-cell, 930
  notation, 932–933
  oxidizing and reducing agents
    relative reactivities of metals, 941–943
    relative strengths, 937–943
    spontaneous redox reactions, 929–934, 929f
  spontaneous redox reactions, 929–934, 938–941
  standard cell potential, 935–937, 943–951
  standard electrode potentials, 935–937, 936f, 938, 938t
  standard hydrogen electrode, 935–937, 936f
  voltages, 934t
Volume (V)
  Avogadro's law, 196–197, 196f, 200–201
  Boyle's law, 193–194, 193f, 199
  Charles's law, 194–195, 195f
  common examples, 23f
  converting units of, 21
  defined, 19, G-17
  effect of change on equilibrium, 764–766, 765f, 768t
  English units, 19t
  as extensive property, 22
  gas, 187–188, 193–202, 210–213
  SI units, 19–20, 19t, 20f
  standard molar, 197–198, 197f
Volume percent [% (v/v)], 524, G-17
Volumetric glassware, 20–21, 20f
von Laue, Max, 946mn
VSEPR theory. See Valence-shell electron-pair repulsion (VSEPR) theory

W

Waage, Peter, 741
Walker, John E., 968
Wastewater, 542, G-17
Water
  amphiprotic nature of, 794
  aqueous ionic reactions, 145–146, 146f
  autoionization, 789–793, 790f, 808
  boiling point, 25f
  chemical change, 5

  collecting a gas over, 207–208, 207f
  cooling curve, 441f
  covalent compounds in, 144
  decomposition, 258–259
  density, 23, 23t
  dipole moment, 399
  Earth's temperature and, 249mn
  electrolysis, 169, 962–963, 962f, 1009
  formation in acid-base reactions, 152–153
  freezing point, 25f
  hybrid orbitals, 415, 415f
  hydrogen bonds, 460–461, 460f, 461f
  hydrogen displacement from by metal, 170, 170f
  intermolecular forces, 502f
  ionic compounds in, 142–144, 142f, 148t
  meniscus shape in glass, 458, 458f
  molecular mass, 72
  molecular shape, 392, 399
  phase changes, 439mn, 440–443, 441f
  phase diagram for, 448f, 449
  physical change, 5
  polar nature, 141, 141f
  polar nature of, 399
  properties
    atomic, 462, 462f
    boiling point elevation constant, 530t
    capillarity, 459
    density, 461–462, 462f, 463f
    freezing point depression constant, 530t
    gas solubility in, 521
    heat of vaporization, 461
    heats of vaporization and fusion, 439f
    specific heat capacity, 246t, 249mn, 459
    surface, 461
    surface tension, 457, 457t, 461
    thermal, 460–461
    triple point, 449
    viscosity, 458t
  as solvent, 141–144, 460, 504, 504t, 505f, 890–891, 890f
  states of, 6
  supercritical, 449mn
  uniqueness of, 460–462
  vapor pressure, 207, 207t
Water-gas reaction, 256
Water-gas shift reaction, 1009
Water purification, 541–542, 541f, 542f
Water softening
  defined, G-17
  via ion exchange, 542, 542f
Water-splitting, for hydrogen production, 1010
Water treatment plants, 541, 541f
Water vapor, 461
Wave
  nature of electrons, 283–286, 283f, 286f
  nature of light, 270–273
  particle behavior compared to, 272–273, 272f
Wave function, 287
Wave-particle duality, 283–287, G-17
Wavelength (λ), 270–272, 270f, 271f, G-17
Waxes, 506–507, 507mn
Weak-acid equilibrium problem solving, 798–805
  acid-dissociation constant (K_a), 798–805, 803t
  concentrations, 799–802
  notation system, 799
  polyprotic acids, 803–805, 803t
Weak acids, 151–152, 151t, 786, 786f, 788, 798–810, 814
  anions as weak bases, 808
  dissociation, 175, 786, 786f
  K_a values, 787, 798t
  proton transfer, 157
  weak-acid equilibrium, 798–805
  weak acid–strong base titration curves, 845–848, 845f
  weak bases and, 805–810
Weak bases, 151–152, 151t, 788, 814
  dissociation, 175
  weak acids and, 805–810
  weak base–strong acid titration curves, 848–849, 849f
Weak-field ligand, 1050, 1050f, 1052–1053, 1052f, G-18
Weathering, phosphorus cycle and, 990
Weight, 524–525
  defined, 21, G-18
  mass distinguished from, 21
Werner, Alfred, 1042–1043
Whiskers, silicon carbide, 483
Wöhler, Friedrich, 629, 629mn
Wood
  cellulose, 666

  fuel, 256
  hydrogen bonding in, 514
  specific heat capacity, 246t
Work (w)
  defined, 236, 238, G-18
  energy defined as, 8
  energy transfer as, 238, 239, 239f
  free energy and electrical work, 943–951
  free energy change and, 902–903
  pressure-volume work, 239, 243, 243f
  sign conventions, 239t
Wulfanite, 463f
Wüthrich, Kurt, 647

X

X-ray diffraction analysis, 468, 469f, G-18
X-ray photon, 1068t, 1069
X-rays
  diffraction pattern, 284, 285f
  opacity of barium sulfate suspension, 857, 857f
  radiation dose, 1085, 1085t
  wavelength, 271–272, 271f
Xanthophylls, 1048–1049
Xenon
  intermolecular forces, 502f
  mole fraction of air at sea level, 219t
Xenon tetrafluoride, 394, 617, 617f
Xerography, 606mn
Xylenes, 644

Y

Yield
  actual, 118–119
  green chemistry perspective on, 120
  in multistep syntheses, 119
  overall, 119
  percent, 118–119
  side reaction effect on, 118, 118f
  theoretical, 118–119
-yne (suffix), 651t
Yttrium, 1030mn

Z

Zeolites, 593, 1009, 1031
Zero, absolute, 25
Zero-order reactions
  enzyme-catalyzed reactions, 724
  first- and second-order reactions compared, 705t
  integrated rate law, 700, 701f, 705t
  rate constant (k) for, 699t
  rate law, 694
Zewail, Ahmed H., 711
Ziegler, Karl, 664
Ziegler-Natta catalysts, 664
Zinc
  abundance, 984t, 985
  acid strength of hydrated ion, 812t
  appearance, 1025f
  in batteries, 952–953, 952f, 953f
  as dietary trace elements, 1056r
  electron configuration, 313t, 314, 331, 1024, 1025f
  in enzyme active site, 1056, 1056f
  in galvanizing process, 958
  mineral sources of, 993t
  oxidation of, 924f
  oxidation states, 1028, 1028t
  reaction with a strong and a weak acid, 786, 786f
  in voltaic cells, 929–933, 929f, 931f, 935–939, 936f, 947–948
Zinc blende, crystal structure, 471, 471f
Zinc formate, 815–816
Zinc hydroxide, 868
Zinc iodate, solubility-product constant (K_{sp}), 853t
Zinc ion, 65t, 1037, 1056, 1056f
Zinc oxide, 481–482, 994, 1037
Zinc sulfate heptahydrate, 1029f
Zircon, 590, 591f
Zirconia, 483
Zirconium, 1023, 1030
Zone refining, 533, 533f, 587, 997, G-18
Zwitterion, 851

Fundamental Physical Constants (six significant figures)

Avogadro's number	N_A	$= 6.02214 \times 10^{23}$/mol
atomic mass unit	amu	$= 1.66054 \times 10^{-27}$ kg
charge of the electron (or proton)	e	$= 1.60218 \times 10^{-19}$ C
Faraday constant	F	$= 9.64853 \times 10^4$ C/mol
mass of the electron	m_e	$= 9.10939 \times 10^{-31}$ kg
mass of the neutron	m_n	$= 1.67493 \times 10^{-27}$ kg
mass of the proton	m_p	$= 1.67262 \times 10^{-27}$ kg
Planck's constant	h	$= 6.62607 \times 10^{-34}$ J·s
speed of light in a vacuum	c	$= 2.99792 \times 10^8$ m/s
standard acceleration of gravity	g	$= 9.80665$ m/s^2
universal gas constant	R	$= 8.31447$ J/(mol·K)
		$= 8.20578 \times 10^{-2}$ (atm·L)/(mol·K)

SI Unit Prefixes

p	n	μ	m	c	d	k	M	G
pico-	nano-	micro-	milli-	centi-	deci-	kilo-	mega-	giga-
10^{-12}	10^{-9}	10^{-6}	10^{-3}	10^{-2}	10^{-1}	10^3	10^6	10^9

Conversions and Relationships

Length
SI unit: meter, m

1 km	$= 1000$ m
	$= 0.62$ mile (mi)
1 inch (in)	$= 2.54$ cm
1 m	$= 1.094$ yards (yd)
1 pm	$= 10^{-12}$ m $= 0.01$ Å

Volume
SI unit: cubic meter, m^3

1 dm^3	$= 10^{-3}$ m^3
	$= 1$ liter (L)
	$= 1.057$ quarts (qt)
1 cm^3	$= 1$ mL
1 m^3	$= 35.3$ ft^3

Pressure
SI unit: pascal, Pa

1 Pa	$= 1$ N/m^2
	$= 1$ kg/m·s^2
1 atm	$= 1.01325 \times 10^5$ Pa
	$= 760$ torr
1 bar	$= 1 \times 10^5$ Pa

Mass
SI unit: kilogram, kg

1 kg	$= 10^3$ g
	$= 2.205$ lb
1 metric ton (t)	$= 10^3$ kg

Energy
SI unit: joule, J

1 J	$= 1$ kg·m^2/s^2
	$= 1$ coulomb·volt (1 C·V)
1 cal	$= 4.184$ J
1 eV	$= 1.602 \times 10^{-19}$ J

Math relationships

$\pi = 3.1416$

volume of sphere $= \frac{4}{3}\pi r^3$

volume of cylinder $= \pi r^2 h$

Temperature
SI unit: kelvin, K

0 K	$= -273.15$°C
mp of H_2O	$= 0$°C (273.15 K)
bp of H_2O	$= 100$°C (373.15 K)
T (K)	$= T$ (°C) $+ 273.15$
T (°C)	$= [T$ (°F) $- 32]\frac{5}{9}$
T (°F)	$= \frac{9}{5}T$ (°C) $+ 32$

Useful Data and Information

Atomic and Molecular Properties

Atomic radii	Figure 8.15, p. 319
Bond energies	Table 9.2, p. 353
Bond lengths	Table 9.2, p. 353
Ground-state electron configurations	Figure 8.11, p. 314
Electronegativity values	Figure 9.20, p. 364
Ionic radii	Figure 8.29, p. 332
First ionization energies	Figure 8.18, p. 322
Molecular shapes	Figure 10.9, p. 394
Oxidation states	Interchapter, p. 561

Equilibrium Constants and Thermodynamic Data

K_a of hydrated metal ions	Table 18.7, p. 812, and Appendix C, p. A-12
K_a of selected acids	Appendix C, pp. A-8 to A-10
Strengths of conjugate acid-base pairs	Figure 18.9, p. 797
K_b of amine bases	Appendix C, p. A-11
K_f of complex ions	Table 19.4, p. 865, and Appendix C, p. A-12
K_{sp} of slightly soluble ionic compounds	Appendix C, p. A-13
Standard electrode potentials, $E^\circ_{\text{half-cell}}$	Appendix D, p. A-14
Standard free energies of formation, ΔG°_f	Appendix B, pp. A-5 to A-7
Standard heats of formation, ΔH°_f	Appendix B, pp. A-5 to A-7
Standard molar entropies, S°	Appendix B, pp. A-5 to A-7

Names and Formulas

Ligands	Table 23.8, p. 1041
Metals in complex anions	Table 23.9, p. 1041
Metals with more than one monatomic ion	Table 2.4, p. 67
Monatomic ions	Table 2.3, p. 65
Organic functional groups	Table 15.5, p. 651
Polyatomic ions	Table 2.5, p. 68

Properties of the Elements

Group 1A(1): Alkali metals	pp. 572–573
Group 2A(2): Alkaline earth metals	pp. 576–577
Group 3A(13): Boron family	pp. 580–581
Group 4A(14): Carbon family	pp. 586–587
Group 5A(15): Nitrogen family	pp. 596–597
Group 6A(16): Oxygen family	pp. 604–605
Group 7A(17): Halogens	pp. 612–613
Group 8A(18): Noble gases	p. 618
Period 4 transition metals, atomic properties	Figure 23.3, p. 1026
Period 4 transition metals, oxidation states	Table 23.2, p. 1028

Miscellaneous

Rules for assigning an oxidation number	Table 4.3, p. 160
SI-English equivalent quantities	Table 1.4, p. 19
Solubility rules for ionic compounds in water	Table 4.1, p. 148
Vapor pressure of water	Table 5.3, p. 207